BRICK AND BLOCK MASONRY - FROM HISTORICAL TO SUSTAINABLE MASONRY

PROCEEDINGS OF THE 17TH INTERNATIONAL BRICK AND BLOCK MASONRY
CONFERENCE (17TH IB^{2}MaC 2020), KRAKÓW, POLAND, 5–8 JULY 2020

Brick and Block Masonry - From Historical to Sustainable Masonry

Editors

Jan Kubica
Silesian University of Technology, Gliwice, Poland

Arkadiusz Kwiecień
Cracow University of Technology, Cracow, Poland

Łukasz Bednarz
Wroclaw University of Science and Technology, Wroclaw, Poland

CRC Press
Taylor & Francis Group
Boca Raton London New York

CRC Press is an imprint of the
Taylor & Francis Group, an **informa** business

A BALKEMA BOOK

CRC Press/Balkema is an imprint of the Taylor & Francis Group, an informa business

© 2020 Taylor & Francis Group, London, UK

Typeset by Integra Software Services Pvt. Ltd., Pondicherry, India

Library of Congress Cataloging-in-Publication Data

Applied for

Published by: CRC Press/Balkema
                Schipholweg 107C, 2316XC Leiden, The Netherlands
                e-mail: Pub.NL@taylorandfrancis.com
                www.crcpress.com – www.taylorandfrancis.com

ISBN: 978-0-367-56586-2 (Hbk)
ISBN: 978-1-003-09850-8 (eBook)
DOI: 10.1201/9781003004042
DOI: https://doi.org/10.1201/9781003004042

Brick and Block Masonry - From Historical to Sustainable Masonry –
Kubica, Kwiecień & Bednarz (eds)
© 2020 Taylor & Francis Group, London, ISBN 978-0-367-56586-2

Table of contents

Sustainability & innovation of masonry

Brick and Block Masonry - From Historical to Sustainable Masonry –
Kubica, Kwiecień & Bednarz (eds)
© 2020 Taylor & Francis Group, London, ISBN 978-0-367-56586-2

Preface

Masonry is one of the oldest type of structures with above 6000 years of history. However, it is still one of the most popular and traditional building material, demonstrates new and more attractive properties and using possibilities. Modern masonry, based on the new and modified traditional materials and solutions, offer the higher quality of living, energy saving and more sustainable development. Masonry became more environmental friendly building structures.

Over the past five decades, the successful IB^{2}MaC series of conferences has been a very good platform for people from around the world to share knowledge, exchange experiences, and learn about new materials and technologies related to masonry structures. These conferences allowed people of science to meet with representatives of industry and building management, as well as conservators of monuments.

The first edition of IB^{2}MaC Conference was held in 1967 in Austin, Texas, USA. Since then, it has been fruitfully organised every three or four years in the: Stoke-on-Trent, UK (1970); Essen, Germany (1973); Bruges, Belgium (1976); Washington, USA (1979); Rome, Italy (1982); Melbourne, Australia (1985); Dublin, Ireland (1988); Berlin, Germany (1991); Calgary, Canada (1994); Shanghai, China (1997); Madrid, Spain (2000); Amsterdam, The Netherlands (2004); Sydney, Australia (2008); Florianopolis, Brazil (2012) and Padua, Italy (2016).

This 17th edition of the Conference is held in Kraków, Poland (July 5-8, 2020). It is first time, where this conference is organised in Central-East Europe. Three Technical Universities located at the south part of Poland, i.e. Cracow University of Technology, Silesian University of Technology and Wroclaw University of Science and Technology, which became the main and the most active research centres in masonry construction in Poland, decided to be organisers of this Conference. Unfortunately, by the reason of the Covid-19 became not possible to organise this Conference with physically participation in one of the beautiful and historical city in Poland. At the request of many participants, we decided not to postpone the date but to organise an online conference. We hope this experiment was successful.

The book contains the full texts of lectures and papers accepted to publication and on-line presented at the 17th IB^{2}MaC. They include 4 keynote lectures, 7 semi-keynote lectures and more than 130 technical papers received from over 30 countries from six continents, demonstrating the worldwide interest of the scientific community on the proposed topics. During the conference, some special sessions have been organised on important problems of sustainable TRM composites, organised with cooperation with the RILEM TC IMC and problems concerning repair and strengthening of masonry infill walls working under in-plain and out-of-plain forces, organised with cooperation with partners involved in the SERA Project INMASPOL.

The submitted papers represent an up-to-date overview and significant novel contributions to the analysis of masonry structures, material and structure testing, advanced analytical and numerical analysis, codes and standards including new drafts of Eurocode 6, design and construction problems of masonry structures. The topics cover major aspects of construction practice; technology and earthen structures; composite materials in masonry; earthquake resistance and retrofitting; seismic and limit design; problems concerning monitoring; inspection, repair and strengthening; conservation of historic buildings; numerical modelling of materials and structures; masonry repair and strengthening; energy, moisture and thermal performance; codes and standards; reinforced and confined masonry; masonry arches and bridges; masonry partitions and infill walls; sustainability and innovation of masonry and case studies.

The editors would like to thank all authors, keynote and semi-keynote speakers, organizers of special sessions and participants for their contributions, members of the International Scientific Committee and Review Panel for their hard work and the Conference Secretariat for their dedicated teamwork, especially during this exceptional pandemic time. We would like cordially thank to all the sponsors of 17th IB[2]MaC 2020, that they supported the organization of the conference also in this difficult time for companies and institutions.

Jan Kubica
Arkadiusz Kwiecień
Łukasz Bednarz

Brick and Block Masonry - From Historical to Sustainable Masonry –
Kubica, Kwiecień & Bednarz (eds)
© 2020 Taylor & Francis Group, London, ISBN 978-0-367-56586-2

Organising committee

Jan Kubica, *Silesian University of Technology, Gliwice, Poland*
Arkadiusz Kwiecień, *Cracow University of Technology, Kraków, Poland*
Łukasz Bednarz, *Wroclaw University of Science and Technology, Wrocław, Poland*

International Scientific Committee

Daniel Abrams (USA)
Josep M. Adell (Spain)
Gilberto Artioli (Italy)
Katrin Beyer (Switzerland)
David Biggs (USA)
Vlatko Zvonimir Bosiljkov (Slovenia)
Gianmarco de Felice (Italy)
Manicka Dhanasekar (Australia)
Miloš Drdácký (Czech Republic)
Geoffrey Edgell (UK)
Fernando Fonseca (USA)
Carl-Alexander Graubner (Germany)
Michael Craig Griffith (Australia)
Barry Haseltine (UK)
Mehrdad Hejazi (Iran)
Jason Ingham (New Zeland)
Wolfram Jäger (Germany)
Jerzy Wojciech Jasieńko (Poland)
Stanisław Jemioło (Poland)
Andrzej Kadłuczka (Poland)
Paulo Barbosa Lourenço (Portugal)
Guido Magenes (Italy)
Dirk R. W. Martens (The Netherlands)
Mark John Masia (Australia)
Arun Menon (India)
Gabriele Milani (Italy)
Claudio Modena (Italy)
Nebojsa Mojsilović (Switzerland)
Daniel V. Oliveira (Portugal)
Adrian Wiliam Page (Australia)
Guilherme Aris Parsekian (Brazil)

Francesca da Porto (Italy)
John Roberts (UK)
Humberto Roman (Brazil)
Phillip Samblanet (USA)
Michael P. Schuller (USA)
Nigel G. Shrive (Canada)
Hipólito Sousa (Portugal)
Tadeusz Tatara (Poland)
Miha Tomazević (Slovenia)
Yuri Totoev (Australia)
Thanasis C. Triantafillou (Greece)
Maria Rosa Valluzzi (Italy)
Elizabeth Vintzileou (Greece)
Rob van der Pluijm (The Netherlands)
Maria Wesołowska (Poland)

International Reviewer Panel

Rafael Aguilar (Peru)
Maria Antonietta Aiello (Italy)
Dariusz Alterman (Australia)
Marco Andreini (Switzerland)
Łukasz Bednarz (Poland)
Andrea Benedetti (Italy)
Roberto Capozucca (Italy)
Catherina Papanicolaou (Greece)
Sergey Churilov (Macedonia)
Camilla Colla (Italy)
Marco Corradi (Italy)
Stefano De Santis (Italy)
Dmytro Dizhur (New Zealand)
Łukasz Drobiec (Poland)
Robert Drysdale (Australia)
Elena Dumova-Jovanowska (Macedonia)
António Sousa Gago (Portugal)
Matija Gams (Slovenia)
Natalino Gattesco (Italy)
Cristina Gentilini (Italy)
Bahman Ghiassi (United Kingdom)
Łukasz Hojdys (Poland)
Alper Ilki (Turkey)
Marta Kałuża (Poland)
Jaromir Klouda (Czech Republic)
Piotr Krajewski (Poland)
Gian Piero Lignola (Italy)
Sergio Logomarsino (Italy)

Cristián Sandoval Mandujano (Chile)
Fabio Matta (USA)
Piotr Matysek (Poland)
Claudio Mazzotti (Italy)
Udo Meyer (Germany)
Francesco Micelli (Italy)
Paolo Morandi (Italy)
John Morton (United Kingdom)
Corina Papanicolaou (Greece)
Fernando Peña (Mexico)
Jan Rots (The Netherlands)
Theodoros Rousakis (Greece)
Cristián Sandoval (Chile)
Yaacov Schafer (Israel)
Marek Skłodowski (Poland)
Leyla Tanaçan (Turkey)
Yuri Totoev (Australia)
Graça Vasconcelos (Portugal)
Ad T.Vermeltfoort (The Netherlands)
Alberto Viskovic (Italy)
Radek Zigler (Czech Republic)

Organizing Committee

Krzysztof Chudyba (Cracow University of Technology)
Cristina Gentilini (University of Bologna)
Łukasz Hojdys (Cracow University of Technology)
Andrzej Kadłuczka (Cracow University of Technology)
Marta Kałuża (Silesian University of Technology)
Bernard Kotala (Silesian University of Technology)
Piotr Krajewski (Cracow University of Technology)
Małgorzata Krystek (Silesian University of Technology)
Piotr Kuboń (Cracow University of Technology)
Piotr Matysek (Cracow University of Technology)
Witold Misztal (Wrocław University of Science and Technology)
Izabela Murzyn (Cracow University of Technology)
Tomasz Nowak (Wrocław University of Science and Technology)
Marek Skłodowski (Institute of Fundamental Technological Research PAS)
Krzysztof Raszczuk (Wrocław University of Science and Technology)
Teresa Stryszewska (Cracow University of Technology)
Marcin Tekieli (Cracow University of Technology)
Bogusław Zając (Cracow University of Technology)

Keynotes

Design of masonry structures (General rules): Highlights of the new European masonry code

P.B. Lourenço & R. Marques
ISISE, Institute of Science and Innovation for Bio-Sustainability (IB-S), University of Minho, Guimarães, Portugal

ABSTRACT: Despite the wide use of masonry buildings in Europe, existing rules for its structural design remains very scattered and lacking in clarity and understandability. Furthermore, since the last version of Eurocode 6 – Part 1-1 (2005), many research on the structural behavior of masonry has been developed, in way that the specifications in the code need to be updated. Following the European Commission Mandate M/515, different subtasks were identified to be addressed in the new version of Eurocode 6 – Part 1-1, aimed to improve general aspects amongst the Eurocodes (e.g. reduction in NDPs and enhanced ease of use) and specific issues of masonry design (e.g. material properties and rules for reinforced masonry). There is still a long way to put the European codes in line with more recent design philosophies, like risk- and resilience-based design. Nevertheless, a significant evolution of Eurocode 6 – Part 1-1 has been achieved. At the current stage, this work is intended to give an overview of international masonry codes and present the main developments in the new version of the European masonry code. A discussion is also made on needs and ways for further improvement of the standard, in terms of code philosophy, research lines and technical aspects.

1 INTRODUCTION

Masonry has been historically used as an easy and affordable solution for buildings. Today, masonry construction is more complex, because it needs to satisfy modern requirements of structural safety and sustainability. Unreinforced masonry (URM) is largely disseminated in low seismicity countries like UK, Germany and Brazil. In other regions, building construction with structural masonry is abandoned, mainly because of threats related to earthquakes. In Europe, even in low seismicity countries masonry has been replaced by different solutions, particularly reinforced concrete (RC), see Table 1. However, considering that solutions for reinforcement of masonry can be used, i.e. reinforced masonry (RM) and confined masonry (CM), it has large potential for resumption in Europe. To this end, the current European masonry code (EN 1996-1-1:2005) needs to be updated to a new context of development and research on masonry, construction practices and materials, as well as trends in building design.

The development of comprehensive standards for structural design of buildings is nowadays recognized as fundamental to ensure an adequate performance both in terms of safety and serviceability requirements. According to Anwar et al. (2016), there are different design approaches which have been considered in the evolution stairs of building standards, i.e. intuitive design, code-based design, performance-based design, consequences and risk-based design, and resilience-based design. The

Eurocodes can be considered as a design approach in between code- and performance-based design. In the particular case of EN 1996-1-1, the definition of minimum dimensions based on calculation, and certain practical limits, e.g. slenderness of loadbearing walls, can be seen as instruments for a performance-based approach (van der Pluijm 2009).

A major challenge in developing design codes is the conversion of research results to practical rules, according to a given design philosophy. Moreover, the subject of standardization needs to receive more attention and even be taken as a research topic. Such a lack may also raise an issue of understandability, because many of the design rules that are put into the codes are provided to the practitioner without a clear background. The new version of the European masonry code was developed as a compromise between the complexity of research results, the pragmatism of practical experience and the practitioner's capabilities. The new code should also contribute to make masonry more competitive, because only in this way it has the force to move trends and contribute for the regulation of the building sector.

In the case of EN 1996-1-1, given aspects were identified as a priority to deal with in masonry design, in particular the prescription of masonry properties, resistance criteria, and design rules for specific cases. The harmonization and extension of rules according to other European standards, especially EN 1992-1-1:2004 for concrete, was also a claim. For example, RM is widely disseminated, but its design rules are not so developed when

Table 1. Listing of main building typologies used in European countries with different seismicity.

Countries	Seismicity	Main structural typologies
Set	Low	URM
Malta	Low	URM, RC frames
Spain	Low	RC frames
Luxembourg	Low	RC and steel frames
Sweden	Low	Prefabricated wooden panels
Norway	Low-to-medium	Wooden apartment buildings
France, Portugal	Low-to-medium	RC frames
Switzerland	Low-to-medium	RC and steel frames
Slovakia	Low-to-medium	URM, RC frames
Hungary	Medium	RC frames
Romania	Medium	Lightly RC shear walls
Slovenia	Medium	URM, CM, RM, RC, mixed
Cyprus, Greece	Medium-to-high	RC frames
Iceland	Medium-to-high	RC frames, lightly RC walls
Italy	Medium-to-high	URM, RM, RC, mixed

Set: AT, BE, CZ, DK, EE, FI, DE, IE, LV*, LT, NL, PL, UK
* Precast and large concrete panels are also largely used.

compared with the ones for RC. New materials and solutions for masonry construction have been developed which need also to the framed in the code. In the following, an overview of international masonry codes, the main developments in the new version of the European masonry code, as well as a discussion for further improvement of the standard are presented.

2 OVERVIEW OF MASONRY CODES

The design of masonry structures has followed different trends worldwide, according to dissimilar building techniques and materials that are locally used. Different hazards and functional requirements for buildings, e.g. earthquakes and insolation, in each region are also a reason for dissimilar masonry solutions. The codes and guidelines for masonry structures developed worldwide are different in what concerns technical aspects, but also in terms of design philosophy and approaches. Most existing codes are based on the individual design of the structural members, i.e. verification is made that the resistance of each structural element is higher than the load to which it is subjected. More recent design approaches are differently based on the overall

resistance of the structure, i.e. the safety check is made by comparing the global capacity of the building against its demand, which is known as performance-based design.

At international level, the development of masonry codes has been particularly promoted across the Americas and in Europe, and design normative from these two regions have been adopted and/or used as benchmark for the building codes of many countries around the world. In the following, an overview of European and American codes for design of masonry structures is presented, as well as of their influential aspects on new practices and developments for masonry design.

2.1 European codes

The modern design of masonry in the UK and Europe was particularly stimulated after the publication of the British standard BS 5628-1 in 1978, called as code of practice for use of masonry. The promotion of the limit state approach, testing in compression of full sized walls instead of 'prisms', design of walls for lateral loads (primarily wind), new requirements for accidental design situations (e.g. explosions), where the main highlights of this standard (Haseltine 2012). Later, new parts of the standard were developed for use of reinforcement and prestressing in masonry, and to deal with ancillary components for masonry walls and with workmanship. The BS 5628-1 was largely the basis of the first version of the European masonry code.

Masonry is also widely disseminated for modern construction in countries like Germany and Italy, where design codes were developed following similar approaches to the ones in the British standard, i.e. DIN 1053-1 (1996) in Germany and DM 1987-11-20 (1987) in Italy. However, in the British standard, the design of walls for lateral loads was not applicable to seismic loading. In Italy, a country with high seismic hazard, in the sequence of some destructive earthquakes (e.g. Molise 2002), a modern code for seismic assessment of masonry buildings was introduced, the OPCM 3274 (2004). Since then, the advances in both topics of masonry structures and seismic design, as well as in other building typologies, like RC and steel frames, allowed the establishment of a comprehensive building code, the Italian building code (NTC 2018).

Within the European Union (EU), an attempt to adopt common rules in design of masonry structures has been made with the introduction, since 1995, of Eurocode 6. Indeed, the structural design is integrated through the consideration of the EN 199x series, specifically the EN 1996-1-1 (2005) for the general design rules of masonry structures, and the EN 1998-1 (2004) to what concerns its seismic design. The allowance of given numbers or methods to be left open for national choice is made through the definition of Nationally Determined Parameters (NDPs). Then, each country

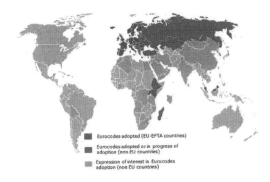

Figure 1. Status of worldwide interest in the Eurocodes [https://eurocodes.jrc.ec.europa.eu].

must publish, unaltered, the Eurocodes, but a National Annex may be further added that gives the chosen values or methods listed as NPDs for that country. The Eurocodes have also been adopted in non EU countries (e.g. the Balkan region, Russia, Georgia and Egypt), as mapped in Figure 1.

In EU countries, although most of the Eurocodes Parts are published in the national language, the application of EN 1996-1-1 is usually non-mandatory, once the current national building codes can still be used. The issue is that in most EU countries no code exists allowing for design of masonry structures, so the application of masonry is difficult against solutions like RC and steel frames. Contrarily, in Italy, where a comprehensive building code is available, modern masonry is a common practice. For instance, in the Italian region of Emilia, low-rise masonry buildings built after the introduction of OPCM 3274 (2004), so considering seismic design and proper detailing, resulted in most cases almost undamaged (Penna et al. 2014). This also demonstrates the need for harmonization of the masonry and seismic codes, to allow an efficient design in earthquake prone regions, beyond the general rules as per the masonry code.

2.2 American codes

Masonry construction is widely used in the Americas, both in non-seismic countries, like Brazil, and in regions with high seismic hazard, particularly North America. In the USA, the first integrated masonry code was published by the Masonry Standards Joint Committee, the MSJC Code in 1988. This former standard was based on allowable-stress and empirical design, and seismic design requirements were also included into an annex to be locally applied. Just later, based on the results of the TCCMAR Program, aimed to develop analytical and experimental research to improve masonry technology and to establish the basis for modern masonry design, strength-based provisions were added to the MSJC Code in 1999 (Klingner 2012). In the subsequent versions of this standard, new masonry applications were incorporated, e.g. autoclaved aerated masonry and infill walls, particularly

harmonized with the strength design of reinforced concrete, up to the current MSJC Code (2013).

The USA code has similarities in terms of content with the EN 1996-1-1 (2005), although the strength formulas are in general different and the MSJC Code (2013) includes provisions for seismic design. The research and development of masonry materials and building codes in the USA was also a benchmark for the sustainable development of masonry in Brazil (Parsekian et al. 2012). In this country, masonry construction has been mostly based on use of concrete blocks and clay bricks, respectively according to standards NBR 15961-1 (2011) and NBR 15812-1 (2010). These standards adopt a design philosophy based on the limit states, with the walls designed to primarily resist vertical loading and combined vertical and wind loading, and further considering criteria for serviceability limit states. In Brazil, the universities are very close to the building market and joint research projects for masonry development have been promoted, as well as exists a tradition of strong education in masonry design, resulting in a large sustainability for masonry application (Corrêa 2012).

In the other Americas, the use of masonry has also been influenced by the USA regulations, particularly in what concerns the application of reinforced masonry and its seismic design. Some Latin American countries, like Chile and Peru, adopted USA practice to their own conditions, which was not always successful as many buildings presented earthquake-related issues; contrarily, when RM was integrally built according to USA standards, its behavior was exceptional, which stresses the need of suitable design standards (Casabonne 2000). In these countries there is a strong tradition of construction with confined masonry. Indeed, the masonry codes in Latin American countries are mostly intended for design of CM buildings. The Mexican masonry code, NTC-M (2017), is probably the most comprehensive standard for application of masonry in this region, based on limit state design. This code includes rules for infill masonry, CM and RM, which design is supported by rational formulas and illustrative rules for detailing.

From Latin America, reference is also made to the Peruvian masonry standard NTE-E.070 (2006). This code, beyond the general specifications and design rules for both CM and RM, emphasizes in assuring an adequate seismic behavior of the buildings. For CM, this is made through further application of the Peruvian seismic standard, NTE-E.030 (2003), by establishing a procedure in which the walls are designed to fail in shear, and maintaining their lateral load capacity. Then, the design procedure is based on strength and the seismic performance is verified for two levels of earthquakes: moderate and severe (San Bartolomé et al. 2006). In the USA,

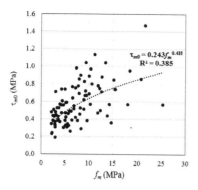

Figure 2. Scatterplot of the initial diagonal shear strength vs. compressive strength of masonry (Marques & Lourenço 2019).

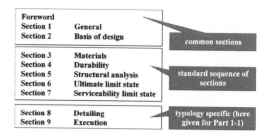

Figure 3. Structure of the EN-1996-1-1 (2005).

the seismic design of masonry buildings can essentially be performed by considering the standard ASCE/SEI 7-16 (2017). Also in this case, as in the European seismic code EN 1998-1 (2004), the seismic design requirements are given in general terms for building structures, with only a few specifications given for masonry structures.

The purpose here is to give an overview of international masonry codes, but just to exemplify the different assumptions in European and American codes, reference is made to the way how the masonry shear strength is taken in the formula for the shear resistance of a masonry wall. While in Europe the shear strength of masonry is either related to its pure shear strength or initial diagonal shear strength, in design codes from the Americas, the square root of the masonry compressive strength f_m is taken as an estimate of the initial diagonal shear strength of masonry, τ_{m0}. However, when considering a dataset collected by Marques & Lourenço (2019), a large scatter and poor correlation between f_m and τ_{m0} are observed (Figure 2). So, it will result in different estimates of the shear resistance of a masonry wall.

3 EUROCODE 6: GENERAL REMARKS

The design of masonry structures in EU countries is supposed to be made according to Eurocode 6, once the last version is from 15 years ago. This code includes the parts of general rules for design, EN-1996-1-1 (2005); fire design, EN-1996-1-2 (2006); materials and workmanship, EN-1996-2 (2006b); and simplified calculation methods, EN-1996-3 (2006c). Eurocode 6 is however still unused in many EU countries, despite the large promotion for its use. In any case, the design methods in Eurocode 6 have been used for reference and even incorporated in national codes of non-EU countries (Athanasopoulou et al. 2019). The EN-1996-1-1 (2005) has nine sections (Figure 3); the ones common to all Eurocodes:

general, basis of design; the ones establishing the standard sequence: materials, durability, structural analysis, ultimate limit state, serviceability limit state; and the ones which are typology specific: detailing, execution. It also includes ten informative annexes, to support and complement the content of the code.

The approach in Eurocode 6 is based on limit state design, in conjunction with a partial factor method to calculate the design values for loads and material strengths from the characteristic ones. An ultimate limit state is considered concerned to events of structural collapse, loss of equilibrium, buckling and loss of stability of masonry members, while the serviceability limit state is related to deflection and cracking of masonry. Then, the design of a structural member is based on verifying that the design value of the load applied to the element (E_d) is less than or equal to the design value of its resistance (R_d). In a perspective of optimal design the value of R_d for each member of a given structure should be a little higher than the value of E_d, but it is not always easy, because individual checks are basically made to each structural member. In some cases, redistribution of internal forces is allowed, e.g. for masonry beams reinforced with steel bars, the linear elastic distribution of internal forces may be modified, assuming equilibrium, if the members have sufficient ductility.

Although the inclusion of NDPs is a general issue in the EN 1996x-series standards, Eurocode 6, despite masonry is a widely variable material, is one of these standards that has fewer number of NDPs. Considering the trend of free market and internationalization of construction companies within the EU, a reduction in number of NDPs is expected to occur both in codes and industry, as well as increased interest and application of masonry buildings. The lack of harmonization of some topics among the Eurocodes, in particular between EN 1996-1-1 and EN 1992-1-1 for concrete structures and between EN 1996-1-1 and EN 1998-1 for seismic design, and even within the content of EN 1996-1-1, is another issue that has hindered application of masonry. A claim which is also made is that the design rules in EN 1996-1-1 are not so easy to understand and use. To this concern, reference is made to the website at https://www.eurocode6.org (Roberts 2020), which presents documents and examples for guidance on application of the code.

Since the 2005 version of EN 1996-1-1, the industry of masonry is changed and many research on the structural behavior of masonry has been developed. Furthermore, there is a new feedback concerning the application of the code, so an update of EN 1996-1-1 is needed. Under the European Commission Mandate M/515 for development of the 2nd generation of Eurocodes (Fardis 2015), task SC6.T1 was established for revision of EN 1996-1-1. In the following, the main developments in the revised EN 1996-1-1 are presented.

4 DEVELOPMENTS IN 1996-1-1

Since the publication of the Eurocodes (finished in 2007), the context for their application is significantly changed, as regards the building sector, research results and design approaches. In the case of EN 1996-1-1 (2005) for masonry, new materials have been introduced and different methods and requirements for design of masonry structures have been disputed. As part of Mandate M/515 for development of the next generation of Eurocodes, ten subtasks were identified as having priority to improve the EN 1996-1-1:

1. Reduction in number of National choices (NDPs): The objective was to review the standard and the contents of all Countries' National Annexes, in order to develop proposals to reduce the number of NDPs and/or enable better consensus on values adopted by Countries.
2. Enhanced ease of use: It was aimed to enhance ease of use by improving clarity, simplifying routes through the standard, avoiding or removing rules of little practical use in design and avoiding additional and/or empirical rules for particular structures or structural-element types.
3. Masonry material properties: The objective was, considering new available experimental data, to establish a more reliable way for calculation of the masonry compressive strength, based on the properties of units and mortar used, and by simplifying the grouping of units.
4. Structural analysis - general: It was aimed, given the similarities between the behavior of unreinforced and reinforced concrete with respectively unreinforced and reinforced masonry, to harmonize rules to calculate the second-order effects, and for shear and braced walls.
5. Structural analysis - complex shapes: The objective was, since complex shapes are common in masonry buildings, to propose new rules for complex shaped members, allowing these to contribute to the stability of masonry structures, and so to an increased resistance with less material.
6. Confined masonry: It was intended to extend and improve the general rules and provisions for design of confined masonry structures, and also to include rules for what concerns seismic-induced forces, i.e. shear and bending.

7. Limit state design: The objective was, based on recent research, to propose more clear rules for walls subjected to mainly vertical or combined vertical and lateral loading, to improve strength models for walls under concentrated loads, and the enhancement of rules for reinforced and confined masonry in shear and bending.
8. Annex for out-of-plane eccentricity of loading: It was intended, because of new knowledge, results of research and experiences of the practice, to revise the provisions dealing with floors supported over a part of the thickness of a wall and their spans.
9. Robustness: The objective was, considering the revision of robustness rules in EN 1990 and EN 1991-1-7, to work out specific masonry related rules, probably related to consequence classes.
10. Sustainability: It was aimed, due to increasing demands on energy efficiency, to extend rules for given cases, e.g. floors partly supported by walls due to included insulation, inclusion of filling materials to avoid thermal bridges, and wider cavity walls with increasing difference in stiffness of the outer and inner leaves.

Nine out of these ten subtasks were addressed in the work of the Project Team (PT) for revision of EN 1996-1-1, once the subtask Robustness was leave for a specific task. The most relevant advances among the several subtasks are presented next with reference to specific aspects.

4.1 Nationally Determined Parameters

The reduction of NDPs is a flag for the revision of the Eurocodes, although it may be a difficult task, because of local threats and construction practices in certain countries. A critical analysis and comparison of the NDPs of various National Annexes (NAs) of Eurocode 6 is made by Graubner and Koob (2015). According to these authors, most NDPs in Eurocode 6 have reduced chances of harmonization, particularly the strength properties, because they are very different in each country. Recently, a systematic report of NDPs in the NAs of EN 1996-1-1 was carried out by Briceño et al. (2019), to assess their influence on the specification of masonry mortar.

Table 2, taken from Briceño et al. (2019), reports the values of parameters used in different countries to calculate the characteristic compressive strength of masonry, f_k, according to the indicated formula. A significant variance of values is observed, whose adoption results in different values of f_k, as shown in Figure 4 for clay brick masonry with different mortar classes. The recommended values for these parameters in EN 1996-1-1 (REF in Figure 4) provide values of f_k which are in general higher than the ones obtained by adopting the parameters in NAs, with the exception of Estonia for mortar class M2.5. So, the countries are in general requesting for more conservative values of f_k.

Table 2. Values of K, α, and β for use with general purpose mortar set in NAs (Briceño et al. 2019).

Parameters	Countries	B/L	AT	FI	NL	PL	LT	UK	EE	DE
K Clay	G1	0.5	0.6	0.6	0.6	0.45	0.5	0.5	0.55	
	G2	*	0.55	0.5	0.5	0.4	0.45	0.4	0.45	0.54-
	G3	**	0.5	0.4	-	0.3	0.35	-	0.35	-0.79
	G4	-	-	0.35	-	0.3	0.3	-	0.35	
K Concrete	G1	0.6	0.6	0.65	0.6	0.4	0.5	0.75	0.55	
	G2	0.5	0.55	0.55	0.5	0.35	0.45	0.7	0.45	0.74-
	G3	0.45	0.5	0.5	-	0.3	0.3	-	0.4	-0.95
	G4	-	-	0.45	-	0.25	0.3	-	0.35	
α		0.65	0.65	0.65	0.65	0.7	0.7	0.7	0.85	0.585- -0.630
β		0.25	0.25	0.25	0.25	0.3	0.3	0.3	0	0.100- -0.162

B/L: Belgium/Luxembourg * $0.5\delta^{-0.65}$ ** $0.4\delta^{-0.65}$
† $f_k = K f_b^{\alpha} f_m^{\beta}$, where f_b is the normalized mean compressive strength of units and f_m is the mortar compressive strength, in N/mm^2.
§ δ is the form factor as defined in the standard EN 772-1:2011.

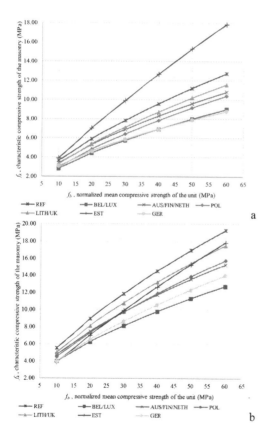

Figure 4. Values of f_k for clay brick masonry with general purpose mortar of classes (a) M2.5 and (c) M10. REF corresponds to recommended values in EN 1996-1-1 (Briceño et al. 2019).

Indeed, the masonry materials used in each country are very different, i.e. masonry units with very different geometry and compressive strength, even for units made of similar material, and mortars with mix proportions and compressive strength very varied. This is even more relevant for the calculation of the shear resistance, because the value of the initial shear strength of masonry and the limiting value of the shear strength are specified in EN 1996-1-1 depending on the material of the unit and mortar compressive strength, with some fixed values in correspondence with a very wide range of unit-mortar possible combinations. To deal with this problem, some countries consider additional factors to define the shear strength, like the type of shear failure mechanism and the tensile strength of the units (Briceño et al. 2019).

In the revised version of EN 1996-1-1 worked out by the PT, two NDPs have been removed:

1. Clause 2.4.4(1): The value to be ascribed to the partial factor γ_M, for all material properties for serviceability limit states is now set according to the recommended value of 1.0.
2. Clause 6.1.2.2(2): The slenderness ratio below which creep may be ignored, following a decision of SC6/WG1 to remove creep from the calculations in EN 1996-1-1, is now deleted as NDP.

For most NDPs, local practices and materials for design and construction in different countries makes the harmonization difficult, particularly in what concerns the specification of masonry mortars and use of different masonry units, whose parameters are either based on test results and/or calculation rules.

Despite the little reduction of NDPs in the revised EN 1996-1-1, its small number and the improvement in the way they are specified allow for a better guidance to support decision in design, without largely restraining for specific requirements needed in any country.

4.2 Shear resistance

The shear resistance of a masonry wall is one of the topics that has raised more interest and discussion for the design of masonry structures. The consideration in EN 1996-1-1 of a strength formulation based on different failure mechanisms, considering background of experimental research since the 1980's, has been claimed by some countries. More recently, the ESECMaSE project (http://www.esecmase.org, 2004-2008) was particularly devoted to the experimental and theoretical investigation of the shear resistance of masonry walls, in way to develop design procedures to include in standards and simplified rules for application of modern masonry. The growing use of unusual materials, e.g. shear walls with glued joints and damp proof courses, and adopting particular details, e.g. weaker mortar joint at the wall bottom or top, introduces an extra complication in design because different wall sections need to be verified in shear.

The formulation of the out-of-plane shear resistance of masonry walls is another aspect to consider, although its interpretation is still unclear, since the out-of-plane response is normally controlled by bending. Indeed, the out-of-plane behavior does not contribute much to the load path in masonry buildings, but the failure of a masonry wall for out of plane can lead to a partial or even global collapse of the structure, if it involves a load bearing wall. The failure due to out-of-plane shear is more likely to occur in the case of flanged walls and/or as the result of a combined in-plane and out-of-plane shear mechanism. However, there is also the need to keep the code simple and easy to use. In the following, the main considerations for revision of the rules for shear in EN 1996-1-1 are presented.

4.2.1 In-plane shear

The formula in EN 1996-1-1 to calculate the shear strength of masonry is basically based on the Mohr-Coulomb failure criterion, in which the characteristic shear strength f_{vk} is calculated by adding the characteristic initial shear strength f_{vk0} (taken as cohesion) to the frictional strength. For this last portion, a friction coefficient μ_f equal to 0.4 is assumed, i.e. it has a value of $0.4\sigma_d$, where σ_d is the design compressive stress perpendicular to the shear. Indeed, the μ_f of 0.4 is taken after the application of a reduction factor corresponding to the Mann-Müller (1982) theory, in way that the frictional strength of the wall depends on the friction of the joint and geometry of the units, in a multi-criteria failure surface (Jäger & Schöpes 2008). Although different shear failure

Table 3. Mean values of shear strength parameters.

Reference	pdc membrane	f_{vm0} (MPa)	tanϕ
Mojsilović (2012)	none	0.30	0.87
	elastomer-based	0.04	0.71
	polyester-based	0.11	0.75
	bitumen-based	0.17	0.06
Martens & Bertram (2008)	none	0.51	0.63
	polyethylene	0.10	0.38

mechanisms have been theoretically formulated, i.e. diagonal shear cracking through units and mortar joints and shear-stepped diagonal cracking along the mortar bed joints, there is some argumentation that the current formula is able to provide a conservative estimation of the shear strength according to different mechanisms, e.g. Marques & Lourenço (2019).

Another aspect that has raised discussion is the influence of damp proof courses in the shear resistance of masonry walls. Indeed, the use of a damp proof course (dpc) membrane is, in given countries, required to prevent groundwater rising in masonry. Experimental tests have been performed in order to evaluate the shear strength parameters according to the Mohr-Coulomb criterion, e.g. Mojsilović (2012) and Martens & Bertram (2008). Mojsilović (2012) tested 10 series of precompressed masonry triplets with a dpc membrane made of different materials. A low initial shear strength (0.04 MPa) is reported for specimens with an elastomer dpc when compared to the ones with other dpc membranes and particularly with plain masonry, see Table 3. Contrarily, a low friction coefficient is reported when using a bitumen pdc membrane. This stresses the need to set the values of f_{vk0} and μ_f to be used in the verification of the shear resistance, for different courses of a masonry wall, according to the current approach for shear strength in EN 1996-1-1.

The shear strength of reinforced masonry is also a topic that has been debated, particularly in what concerns the effect of the vertical reinforcement. There are three main points which have been considered in the revision of EN 1996-1-1 in order to provide a more consistent approach for design of RM walls in shear:

1. The formulation of the URM contribution for the shear resistance should include the effective length of the wall instead of its full length. By considering the experimental works carried out for RM, it is concluded that the double simplification of taking the shear strength over the effective depth of the section and considering a 45° (fixed) inclination of the crack yields to a good estimate of the value of the experimental shear strength (Tomaževič 1999). An even more precise approach could consist in taking the compressed part of the masonry wall, as it is considered in URM. However, this would yield to overcomplicated design.

9

2. The definition of the total area A_{sw} of the horizontal shear reinforcement over the part of the wall being considered needs clarification. If a shear resistance plane acting at a 45° angle across the effective wall depth (d) is assumed, the rate of reinforcement should be expressed by dividing d by the spacing of reinforcement, i.e. d/s. Besides, the effectiveness ratio of shear reinforcement (considering that the reinforcement is not fully yielding) should, according to comprehensive experimental works (e.g. Tomaževič 1999, da Porto et al. 2011), be limited to a value in the range 0.5-0.7, so a value of 0.6 has been adopted.

3. The current limitation of the shear strength of RM walls is inconsistent and strongly overestimated. Indeed, the shear strength of RM must be related with the masonry compressive strength, since the presence of reinforcement allows a redistribution of stresses that involves the entire masonry wall, and not only the single unit. So, assuming a truss mechanism (presence of horizontal and vertical reinforcements), the limitation should be in terms of the maximum value of the shear strength inducing failure of the compressed strut of the truss. The limit value of the shear resistance for RM walls is now defined as $0.3f_d td$, where f_d is the design compressive strength of masonry or concrete infill, whichever is the lesser, t is the wall thickness and d is wall effective depth. This formula is similar to the current approach in EN 1996-1-1 for RM beams.

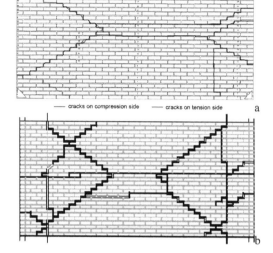

Figure 5. Crack patterns in URM: (a) single wall (van der Plujm 1999) and (b) wall with flanges (Griffith et al. 2007).

4.2.2 *Out-of-plane shear*

The out-of-plane behaviour of URM walls is mostly determined by flexure, so the definition of an out-of-plane shear strength is discussible. In out-of-plane bending, the support conditions have a deep influence on the orientation of internal principal stresses and the resulting crack pattern (Griffith et al. 2007). The occurrence of out-of-plane shear may also be largely dependent on the boundary conditions. Many experimental tests have been conducted to investigate the out-of-plane behaviour of masonry walls (e.g. Griffith et al. 2007, van der Pluijm 1999), namely to assess the influence of return walls (flanges) in the shear response and crack pattern of the loaded panels. Indeed, the flanges increase the out-of-plane flexural strength of the wall and amplify the effect of out-of-plane sliding when compared to the case of a single panel (Figure 5). Griffith et al. (2007) report that, during out-of-plane cyclic tests of masonry walls with flanges, sliding between consecutive courses of adjacent panels was observed, causing a reduction in the overlapping cross-section of the wall. In Figure 5b it is observed that a wall with flanges presents a larger extension of sliding cracks, beyond the typical yield lines due to out-of-plane bending in a single panel as in Figure 5a.

In EN 1996-1-1 no differentiation is made between in-plane and out-of-plane shear. Such a distinction is only made in the German NA (DIN 2012), where the out-of-plane shear resistance due to friction failure is calculated by taking the compressed thickness of the wall after considering the eccentricity of the axial load, similarly to what is made in the verification of buckling. In the revised EN 1996-1-1, a verification rule of the out-of-plane shear resistance is included, similar to the one for the in-plane shear resistance, but with inverted roles of the thickness and length of the wall. So, the design value of the out-of-plane shear resistance of a masonry wall should be calculated based on the characteristic initial shear strength, f_{vk0}, and characteristic friction coefficient of the bed joint or of the damp proof course, μ_f.

4.3 *Second-order effects*

Architectural trends are progressively demanding for smaller cross-sections and larger spans in masonry structures, in way that second-order effects are an increasingly important aspect in design. In EN 1996-1-1 a specific section is included to deal with second-order effects, which presents a procedure based on the hypothesis that the structure under design has their parts braced together, so it is applicable to a global structure to check if sway may be disregarded. Indeed, instability of individual walls related to second-order effects is a more disputed topic in masonry design, both in the case of URM and RM structures. These effects are particularly important concerning the out-of-plane behavior of slender walls. Despite its importance, most building codes do not present a consistent approach for calculating second-order effects in masonry structures. According to Donà et al. (2018), in the European codes (EN 1996-1-1 and EN 1998-1) these effects are not explicitly considered, while limit values of slenderness for pin support conditions at both ends are fixed, without providing rational procedures.

4.3.1 URM walls

Buckling of a wall is related to a reduction of the axial load capacity due to eccentric vertical loading. The approach for buckling in EN 1996-1-1 has been studied by Bakeer & Jäger (2017). These authors claim that such approach, which is based on an empirical exponential formula for the capacity reduction factor Φ, is not suitable for softer masonry used in some countries, e.g. Denmark. This is because that formula was defined after considering a constant ratio of elastic modulus to characteristic compressive strength, E/f_k, equal to 1000. So, when calculating the axial load capacity N_{Rd} as a function of f_k, there is a point of the N_{Rd}–f_k relation after which the calculated values are inconsistent, once N_{Rd} decreases as f_k increases. This is debatable because the formulation was proposed for use with a constant E/f_k ratio, but the claim was raised by several countries.

Bakeer (2016), based on numerical studies, proposed an empirical formula for the N_{Rd} of URM walls under buckling, which is derived for perfectly plastic material and makes a distinction between material failure and stability failure. In the revised EN 1996-1-1, based on some simplifications to that approach and best-fitting of experimental data, a new formulation has been included. The values of the capacity reduction factor at the mid-height of a wall, Φ_m, plotted versus the slenderness (effective height to thickness h_{ef}/t) ratio and calculated using the formulations in the current (EC6) and revised (PT SC6.T1) EN 1996-1-1, are presented in Figure 6a for the case with $E/f_k = 1000$,

where it is observed that the new proposal allows larger N_{Rd} for small eccentricities, and contrarily it is more conservative for large eccentricities.

The Φ_m reduction factor is for walls subjected to mainly vertical loading. If the seismic action is to be taken into account, combined vertical and lateral loading needs to be considered. Different approaches have been developed for the out-of-plane stability of URM walls subjected to seismic excitation, e.g. Griffith et al. (2004). In this case, the post-cracking behavior of the out-of-plane vertical bending mechanism is dominated by large displacements, where second-order effects play an important role. Morandi et al. (2008) proposed, based on a parametric study varying the mechanical and geometric characteristics of URM walls in a very wide range, a capacity reduction coefficient Φ_M to reduce the first-order resisting moment.

This last approach is adopted in the revised EN 1996-1-1, where graphs have been introduced in which Φ_M is plotted as a function of wall slenderness (h_{ef}/t) for a range of the dimensionless axial load ratio v (= $N_{Rd}/(A\ f_d)$), where A is the horizontal gross cross-sectional area of the wall). The graph for the case with $E/f_k = 1000$ is presented in Figure 6b. It can be observed that the second-order effects are particularly evident for a slenderness ratio greater than 12, while the capacity reduction can be up to 70% in correspondence with the limiting value of h_{ef}/t for URM walls in EN 1996-1-1 equal to 27. So, when the acting bending moment M_{Ed} is calculated with a first-order analysis, the second-order effects are taken into account by defining the reduced moment of resistance of the wall, $\Phi_M M_{Rd}$.

4.3.2 RM walls

The use of RM has allowed the design of structures with higher slenderness ratios. Some researchers have claimed for the inadequacy of the current method in EN 1996-1-1 to consider second-order effects in design, e.g. in the scope of DISWall project (http://diswall.dic.unipd.it, 2006-2008). Indeed, according to the code, RM members with a slenderness ratio greater than 12 may be designed according to the principles and application rules for URM. In this case, neither their higher resistance nor their capacity in controlling the failure mechanism, allowed by the presence of reinforcements, can be exploited (Donà et al. 2018). When the slenderness is greater than 12, an additional design moment M_{ad} can be calculated to take into account second-order effects. In EN 1996-1-1 a direct formulation for M_{ad} is provided, based on the nominal curvature method, in which a fixed value of curvature ($1/r$) is considered corresponding to a balanced failure of the section, i.e. the masonry reaches its ultimate compressive strain when the steel reaches its yielding strain. So, the formula for calculation of the design bending moment applied is as follows.

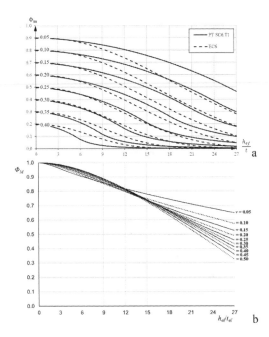

Figure 6. Φ versus h_{ef}/t for $E = 1000\ f_k$, from current (EC6) and revised (PT SC6.T1) EN 1996-1-1: (a) mainly vertical loading and (b) combined vertical and lateral loading.

$$M_{Ed} = M_{1Ed} + M_{ad} = M_{1Ed} + N_{Ed}e_2 \qquad (1)$$

where M_{1Ed} is the first-order design value of the moment applied, N_{Ed} is the design value of the vertical load, and e_2 is the second-order eccentricity of the vertical load.

Although Equation 1 is suitable, the way how e_2 is calculated has been revised in EN 1996-1-1, as following a section analysis for RM sections the values for $(1/r)$ are not fixed as given in the code, and they are differentiated for type of masonry according to the values of strain in masonry at crushing (Donà et al. 2018). This approach is harmonized with the one based on a nominal curvature given by EN 1992-1-1 for RC sections. Furthermore, the evaluation of RM sections with a slenderness ratio greater that 12 as if they are unreinforced, as prescribed in the current EN 1996-1-1, is very conservative when considering experimental and numerical investigations (e.g. Ferracuti et al. 2016, da Porto et al. 2011).

In the revised EN 1996-1-1 it is also left the possibility to obtain $(1/r)$ from a more general moment-curvature $(M-1/r)$ analysis of the section, so allowing for a more precise evaluation of the actual section behavior under the second-order effects. In this case, if M_{Rd} is the resisting moment of the critical (base) section of the wall in the $(M-1/r)$ diagram, calculated for the design axial load N_{Ed}, the first-order moment M_{1Ed} for that section, available to absorb the design actions, is the maximum difference between the ordinate M_{Rd} of the $(M-1/r)$ curve and the ordinate of an assumed linear equation representing the second-order effect (Donà et al. 2016), see Figure 7.

4.4 Confined masonry

Although CM presents advantages against usual RC framed structures, e.g. all material is exploited and RC cross-sections can be smaller, it has little usage in Europe. The lack and/or dispersion of application rules in design codes has contributed to this. In CM, tie-columns and tie-beams are used around the boundaries of the URM panels to impart confinement and additional tensile strength, thus delaying formation of diagonal tension and hence contributing to the lateral load resisted by the walling system. Despite CM is prescribed in EN 1996-1-1, the specifications for design of CM members are too general and/or refer to the ordinary clauses in the code for design of URM and/or RM.

A lot of research results and guidelines to support design of CM structures is available, particularly from Latin American countries, but a large dispersion of rules is observed, because of very dissimilar building contexts (Meli et al. 2011, Marques & Lourenço 2019). In the following, the main rules for CM as proposed in the revised EN 1996-1-1 are presented, namely in terms of resistance to vertical loading, shear forces and in-plane bending.

4.4.1 Vertical resistance

The vertical load capacity of a CM wall may be calculated as the sum of contributions from the masonry panel, concrete cross-section of tie-columns, and longitudinal steel reinforcement in tie-columns, still considering a reduction factor allowing for the effects of slenderness and eccentricity. The contribution of each material depends, however, on the relative stiffness and strength of the materials, because of the dissimilar deformation and stress to strength ratio in the different materials. Despite, in EN 1996-1-1:2005 the vertical resistance was specified to be computed based only on the compressive strength of the masonry, with the steel reinforcement in compression being ignored, because a perfectly composite behavior of the masonry and tie-elements may not occur. This approach is however very conservative, as verified by Marques & Lourenço (2019) when looking at experimental results by León et al. (2004). Therefore, a new formula has been included in the revised EN 1996-1-1, as described below.

In calculating the design vertical resistance of a CM wall, N_{Rd}, the full cross-section of the wall is assumed to have the masonry compressive strength, and the contribution of steel bars in compression is considered, as given by Equation (2). This is also the approach in the Mexican masonry code (NTC-M 2017), which according to León et al. (2004) provides a sufficiently conservative estimation. For a set of benchmark CM walls tested by these authors, the tie-columns take most of vertical load (at failure, about 50% the concrete and 20% the steel).

$$N_{Rd} = \Phi \left(A_w f_d + A_{st} f_{yd} \right) \qquad (2)$$

where Φ is the capacity reduction factor allowing for the effects of slenderness and eccentricity of loading; A_w is the gross cross-sectional area of the wall, including the tie-columns; f_d is the design compressive strength of masonry; A_{st} is the total area of longitudinal steel reinforcement in tie-columns; and f_{yd} is the design yield strength of reinforcement.

Indeed, as regards the axial loading, failure due to vertical shear may be a more critical situation in CM

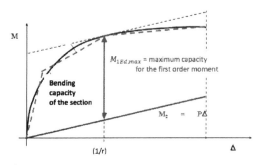

Figure 7. Moment-curvature $(M-1/r)$ diagram illustrating the consideration of second-order effects (Donà et al. 2016).

walls, so it may be needed to verify the maximum compressive stress at the wall edges against the masonry compressive strength, to avoid damage due to crushing and vertical shear, because the stress to strength ratio is significantly greater for masonry than for concrete (Marques & Lourenço 2019).

4.4.2 Shear resistance

The shear failure mechanism of a CM wall is mostly characterized by composite masonry–tie-columns action and diagonal cracking, with the cracks propagating into the columns, so its shear resistance is the sum of contributions from the masonry panel and adjacent RC tie-columns (San Bartolomé et al. 2010, Meli et al. 2011). However, the maximum shear resistance of tie-columns can only be reached after severe cracking of the masonry panel. Thus, only part of the resistance of tie-columns should be considered for the shear resistance of the wall. A conservative approach is to assume that the tie-columns are integrated with the masonry panel, thus calculating a cross-sectional area by taking into account the total wall length (Meli et al. 2011). For this condition, the assumption of a Coulomb-based approach for calculation of the wall shear strength, like in Equation (3), provides an enough good estimation for design purposes (Meli et al. 2011, Marques & Lourenço 2013, Pérez-Gavilán et al. 2015).

$$f_{vk} = f_{vk0} + 0.4\sigma_d \qquad (3)$$

where f_{vk0} is the characteristic initial shear strength of masonry, under zero compressive stress; and σ_d is the design compressive stress perpendicular to the shear in the member at the level under consideration.

The calculated shear strength using Equation (3) (same formula as in EN 1996-1-1 for URM) versus the experimental shear strength of CM walls, for a dataset collected by Marques & Lourenço (2013), is plotted in Figure 8; $R^2 = 0.83$ denotes its accuracy. So, the proposal in the revised EN 1996-1-1 is to calculate the shear strength of a CM wall as for URM, but considering the total wall length for the definition of the cross-section.

4.4.3 Flexural resistance

The flexural failure of CM walls is not a common mechanism, so only a few studies concerning the flexural behavior of CM walls have been developed; e.g. Bustos et al. (2000), Varela-Rivera et al. (2019). Indeed, for regular quality masonry materials, the flexural failure should only be an issue for relatively slender walls with tie-elements spaced at relatively short intervals (Tomaževič 1999).

For the revised EN 1996-1-1, a formulation for the bending resistance M_{Rd} is considered based on flexure theory, adapted from Tomaževič (1999). Assuming the full wall section made of a same material, and a rectangular stress distribution based on the design compressive strength of masonry or concrete, whichever is the lesser, the equilibrium of loads is as shown in Figure 9 and formulated in Equations (4) and (5). Here, yielding of steel in tension is assumed and reinforcement in compression is ignored.

$$N_{Ed} = F_c - F_s \Leftrightarrow \frac{N_{Ed} + A_s f_{yd}}{0,8\eta_f f_d\, t} \qquad (4)$$

$$M_{Rd,CM} = A_s f_{yd}\,(d - 0,4x) + N_{Ed}\left(\frac{L}{2} - 0,4x\right) \qquad (5)$$

where N_{Ed} is the design value of the vertical load; F_c is the resultant of compressive stresses on the wall; F_s is the tensile force of tensioned reinforcement at yielding; x is the depth to the neutral axis of the wall section; A_s is the area of vertical reinforcement, symmetrically placed at both ends; f_{yd} is the design yield strength of reinforcement; f_d is the design compressive strength of masonry or concrete, whichever

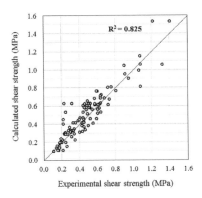

Figure 8. Calculated versus experimental shear strength of CM walls (Marques & Lourenço 2019).

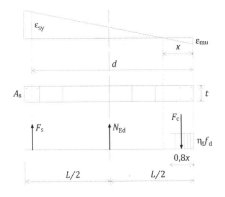

Figure 9. Stress and strain distributions assumed for of a CM wall in bending.

13

is the lesser; t is the thickness of the wall; η_f is the factor defining the equivalent rectangular stress block, assumed equal to 0.85; d is the effective depth of the wall cross-section; L is the length of the wall; and σ_d is the design compressive stress on the wall.

The formulation above is similar to the one for RM, see da Porto et al. (2011b). However, based on the results of a set of selected tests from the literature, it is observed that Equation (5) avoids a significant overestimation of the in-plane moment of resistance obtained with the formula for RM, against the experimental resistance. So, since Equation (5) provides in general an estimate of the flexural resistance on the safety side, it has been adopted in the revised EN 1996-1-1. The proposed formulations, as well as requirements for detailing of CM, are presented in detail by Marques & Lourenço (2019).

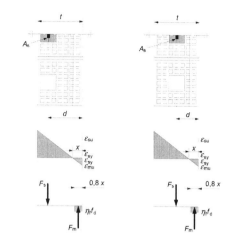

Figure 10. Stress and strain distributions for reinforced walls with single reinforcement.

5 FUTURE DEVELOPMENTS

Most of the changes and new approaches in the revised EN 1996-1-1 were based on existing experimental and numerical studies in the literature, as well as in research and practical work carried out by the participants in the project team for revision of the standard. Each identified subtask for enhancement of EN 1996-1-1 was addressed, although further improvements are still possible, particularly for ease of use and harmonization with other standards.

There are procedures in the code which are still too time-consuming and complex, e.g. the check of vertical resistance for a wall, when compared against more influential issues for structural safety. For sure that many aspects for masonry design can be improved in EN 1996-1-1. Indeed, many proposals were received from different countries to implement in the code, which need however to be further validated against experimental and practice evidence. Despite the number of NDPs in EN 1996-1-1 is reduced, very different figures for design are included in the NAs of CEN country members, so ways for further harmonization should be promoted.

The approach for verification of serviceability limit states (SLS) is one of the aspects that can be improved in EN 1996-1-1, since the current procedure is based on very general requirements. Although prescriptions are given in the code related to deflection and cracking, a deep analysis of the avoidance of cracking in SLS is not possible (Sousa et al. 2015). EN 1996-1-1 includes design criteria for strength/deflection control and movement joints, but rules for deformation limits, tensile limits and ancillary reinforcement are not included. An interesting benchmark for improvement of the code in this topic are the Brazilian standards, i.e. NBR 15961-1 (2011) and NBR 15812-1 (2010) for concrete block and clay brick masonry, respectively. The guide by CIB Commission W023 - Wall Structures (Sousa et al. 2015) is also a relevant tool to consider complementarily to the revised EN 1996-1-1, and to take as reference in future revisions of EN 1996-1-1.

The harmonization of rules in the code with provisions given in EN 1998-1 for earthquake resistance needs to be further improved, towards a comprehensive approach for modern performance-based design. Some guidance to the practitioner in the application of design rules is also needed, for which the inclusion in the code of illustrative pictures and examples can be made, like Figure 10 showing the stress and strain distributions to consider in the calculation of the out-of-plane bending resistance of reinforced walls. A more comprehensive view of sustainability related issues, namely the promotion of given structural typologies and constructive solutions for buildings, should also be considered. The feedback from stakeholders will be also an important input for future revisions of the code.

6 CONCLUDING REMARKS

Relevant aspects in the drafting of the new European masonry code have been presented and discussed above, including the main changes in the revised EN 1996-1-1. There are however a set of improvements to the code which should be further highlighted, in each of the worked subtasks, as described below:

- NDPs: The National choices in each country have been reviewed. Although only two NDPs are removed from the code, the way how these parameters are now specified allows for a better guidance on how to deal with NDPs in design.
- Enhanced ease of use: The articulation of Section 'Ultimate limit states' is improved in clarifying the loading types and safety verification rules, and also to harmonize with EN 1998-1, for beyond the improvement of writing and cross-referencing.
- Masonry material properties: A new architecture of the clauses to deal with the grouping of

masonry units is implemented, particularly to consider units not fitting in the current groups.
- Structural analysis - general: Design rules for RM are improved as far as possible harmonized with EN 1992-1-1, e.g. second-order effects and verification of walls subjected to in-plane horizontal loads. The specifications for lap and anchorage lengths of steel bars are also revised.
- Structural analysis - complex shapes: A new method for design of complex shaped members has been developed and is now included as an annex to the code.
- Confined masonry: Specific rules and clauses for design of CM walls subjected to vertical, shear and axial-bending loadings are now included in the code. Terminology and provisions for detailing of CM are also improved.
- Limit state design: The rules for concentrated loads and design of RM members subjected to shear and axial-bending loadings are improved. The annex for out-of-plane eccentricity of loading is extended to consider the cases with floors supported over a part of the wall, and further considering the node stiffness, application to timber floors and wind loading in the formulation. A new annex is added to calculate the capacity reduction factor when taking the influence of the flexural strength into account.
- Annex for buckling: The formulation has been revised to provide a suitable value of the capacity reduction factor for the full range of masonry stiffness and E/f_k ratios. The annex is also extended to the case of masonry walls subjected to combined vertical and lateral loading.
- Sustainability: Design rules are now included in the code to allow using new materials and constructive details to deal with energy efficiency issues, as well as to consider the possible positive effect of the boundary conditions and eccentricity of loading.

The revised EN 1996-1-1 includes the agreed viewpoint of the contributors in the work of the PT for revision of EN 1996-1-1, after a large consolidation period after several stages for solving the comments and suggestions by CEN members to improve the code. Despite the potential for improvement of the EN 1996-1-1, the new version of the European masonry code is considered to be an evolution of the previous standard. It is to believe that application of masonry structures is now more possible in Europe.

ACKNOWLEDGMENTS

The overview here is based on the work for task SC6.T1 'Masonry - Revised version of EN 1996-1-1', in the scope of EC Mandate 'M/515 phase 1 tasks for the development of the 2nd generation of EN Eurocodes' coordinated by NEN. The PT members in the work for SC6.T1 are gratefully acknowledged: Denis Camilleri, Francesca da Porto, John Roberts, Poul Dupont Christiansen, Wolfram Jäger. Special thanks are given to the Chairman of TC250/SC6, Rob van der Pluijm, and to the Convener of TC250/SC6/WG1, Barry Haseltine.

The comments and suggestions by the individuals of the CEN country members and mirror committees for EN 1996-1-1 are also acknowledged, as they were the foundation for most of the changes and improvements to the standard.

This work was financed by FEDER funds through the Competitivity Factors Operational Programme – COMPETE and by national funds through FCT – Portuguese Foundation for Science and Technology, within ISISE (project UID/ECI/04029).

REFERENCES

ABNT 2010. NBR 15812-1:2010, Alvenaria estrutural – Blocos cerâmicos, Parte 1: Projeto. Associação Brasileira de Normas Técnicas, São Paulo.

ABNT 2011. NBR 15961-1:2011, Alvenaria estrutural – Blocos de concreto, Parte 1: Projeto. Associação Brasileira de Normas Técnicas, São Paulo.

Anwar, N., Htut-Aung, T. & Najam, F. 2016. From prescription to resilience: Innovations in seismic design philosophy. *Technology*, December 2016: 9-13.

ASCE/SEI 7-16 2017. Minimum design loads and associated criteria for buildings and other structures. American Society of Civil Engineers, Reston VA.

Athanasopoulou, A., Formichi, P., Spehl, P., Dabizheva, I., Gacesa-Moric, V. et al. 2019. *The implementation of the Eurocodes in the National Regulatory Framework*. Luxembourg: Publications Office of the European Union.

Bakeer, T. 2016. Empirical estimation of the load bearing capacity of masonry walls under buckling - Critical remarks and a new proposal for the Eurocode 6. *Construction and Building Materials* 113: 376–394.

Bakeer, T. & Jäger, W. 2017. Buckling of reinforced and unreinforced masonry walls - A unified solution for Eurocode 6. *Proceedings of the 13th Canadian Masonry Symposium*, Halifax.

Briceño, C., Azenha, M. & Lourenço, P.B. 2019. *Systematic Report: Current situation of the influence of masonry mortar in the Eurocode 6 Part 1-1*. Report 2019-DEC/E-15, University of Minho, Guimarães.

BSI 2005. BS 5628-1:2005. Code of practice for use of masonry – Part 1: Structural use of masonry. The British Standards Institution, London.

Bustos, J.L., Zabala, F., Masanet, A.R., Santalucía, J.R., Estudio del comportamiento dinámico de un modelo de mampostería encadenada mediante un ensayo en mesa vibratoria. *Anales de las XXIX Jornadas Sudamericanas de Ingenieria Estructural*, Punta del Este, Uruguay.

Casabonne, C. 2000. Masonry in the seismic areas of the Americas, recent investigation and developments. *Progress in Structural Engineering and Materials* 2(3): 319–327.

CEN 2004. EN 1992-1-1:2004, Eurocode 2: Design of concrete structures – Part 1-1: General rules and rules for buildings. European Committee for Standardization, Brussels.

CEN 2004b. EN 1998-1:2004, Eurocode 8: Design of structures for earthquake resistance – Part 1: General rules, seismic actions and rules for buildings. European Committee for Standardization, Brussels.

CEN 2005. EN 1996-1-1:2005, Eurocode 6: Design of masonry structures – Part 1-1: General rules for reinforced and unreinforced masonry structures. European Committee for Standardization, Brussels.

CEN 2006. EN 1996-1-2:2006, Eurocode 6: Design of masonry structures – Part 1-2: General rules - Structural fire design. European Committee for Standardization, Brussels.

CEN 2006b. EN 1996-2:2006, Eurocode 6: Design of masonry structures – Part 2: Design considerations, selection of materials and execution of masonry. European Committee for Standardization, Brussels.

CEN 2006c. EN 1996-3:2006, Eurocode 6: Design of masonry structures – Part 3: Simplified calculation methods for unreinforced masonry structures. European Committee for Standardization, Brussels.

CEN 2011. EN 772-1:2011, Methods of test for masonry units – Part 1: Determination of compressive strength. European Committee for Standardization, Brussels.

Corrêa, M.R.S. 2012. The evolution of the design and construction of masonry buildings in Brazil. Gestão & Tecnologia de Projetos 7(2): 3–11.

DIN 1996. DIN 1053-1:1996, Masonry – Part 1: Design and construction. German Institute for Standardisation, Berlin.

DIN 2012. EN 1996-1-1/NA, National Annex – Nationally determined parameters – Eurocode 6: Design of masonry structures – Part 1-1: General rules for reinforced and unreinforced masonry structures. German Institute for Standardisation, Berlin.

Donà, M., Tecchio, G. & da Porto, F. 2016. Report on II order effects in reinforced masonry walls. Università degli Studi di Padova, Padua.

Donà, M., Tecchio, G. & da Porto, F. 2018. Verification of second-order effects in slender reinforced masonry walls. Materials and Structures 51: 69.

Fardis, M. 2015. Towards a second generation of European Standards on Eurocodes. Pers. comm. in 4th EU Standardization Summit 'How standardization can support the transition to a cleaner and smarter economy', 4 June 2015, Riga.

Ferracuti, B., Bacci, L. & Savoia, M. 2016. Out-of-plane behavior of slender reinforced masonry walls for tall single storey buildings: Design procedure. Proc. of the 16th International Brick and Block Masonry Conference, Padua.

Graubner, C.A. & Koob, B. 2015. Analysis and comparison of the NDPs of various national annexes of Eurocode 6. Mauerwerk 19(6): 427–440.

Griffith, M.C., Lam, N.T.K., Wilson, J.L. & Doherty, K. 2004. Experimental investigation of URM walls in flexure. Journal of Structural Engineering 130(3): 423–432.

Griffith, M.C., Vaculik, J., Lam, N.T.K., Wilson, J. & Lumantarna, E. 2007. Cyclic testing of unreinforced masonry walls in two-way bending. Earthquake Engineering & Structural Dynamics 36(6): 801–821.

Haseltine, B. 2012. The evolution of the design and construction of masonry buildings in the UK. Gestão & Tecnologia de Projetos 7(2): 20–26.

Jäger, W. & Schöpes, P. 2008. D9.1 Proposals for an advanced design model of masonry under lateral loads for the implementation in Eurocode 6-1-1. ESECMaSE project, Deliverable 9.1, Technical University of Dresden.

Klingner, R. 2012. The evolution of the design and construction of masonry buildings in the United States. Gestão & Tecnologia de Projetos 7(2): 12–19.

León, I.I., Flores, L.E. & Reyes, C. 2004. Estudio experimental de muros de mampostería de barro sometidos a compresión pura. Anales del XIV Congreso Mexicano de Ingenieria Estructural, Acapulco.

Mann, W. & Muller, H. 1982. Failure of shear-stressed masonry - An enlarged theory, tests and application to shear walls. Proceedings of the British Ceramic Society 30: 223–235.

Marques, R. & Lourenço, P.B. 2013. A model for pushover analysis of confined masonry structures: implementation and validation. Bulletin of Earthquake Engineering 11(6): 2133–2150.

Marques, R. & Lourenço, P.B. 2019. Structural behaviour and design rules of confined masonry walls: Review and proposals. Construction and Building Materials 217:137–155.

Martens, D. & Bertram, G. 2008. Shear strength of clay brick masonry including damp proof course. Proc. of the 14th International Brick and Block Masonry Conference, Sydney.

Meli, R., Brzev, S., Astroza, M., Boen, T., Crisafulli, F., Dai, J. et al. 2011. Seismic design guide for low-rise confined masonry buildings. Earthquake Engineering Research Institute (EERI), Oakland CA.

MLP 1987. D.M. 1987-11-20.Norme tecniche per la progettazione, esecuzione e collaudo degli edifici in muratura e il per il loro consolidamento. Ministero dei Lavori Pubblici, Rome.

Mojsilović, N. 2012. Masonry elements with damp-proof course membrane: Assessment of shear strength parameters. Construction and Building Materials 35: 1002–1012.

Morandi, P., Magenes, G. & Griffith, M. 2008. Second order effects in out-of-plane strength of unreinforced masonry walls subjected to bending and compression. Australian Journal of Structural Engineering 8(2): 133–144.

MSJC 2013. ACI 530/530.1-13: Building code requirements and specification for masonry structures and companion commentaries. ACI/ASCE/TMS Masonry Standards Joint Committee, Farmington Hills MI.

NTC 2018. Norme tecniche per le costruzioni. Ministero delle Infrastrutture e Trasporti, Rome.

NTC-M 2017. Normas técnicas complementarias para el diseño y construcción de estructuras de mampostería. Gobierno del Distrito Federal, Ciudad de México.

NTE-E.030 2003. Reglamento Nacional de Edificaciones, Norma Técnica E.030: Diseño Sismorresistente (Peruvian seismic code). Ministerio de Vivienda, Construcción y Saneamiento, Lima.

NTE-E.070 2006. Reglamento Nacional de Edificaciones, Norma Técnica E.070: Albañilería (Peruvian masonry code). Ministerio de Vivienda, Construcción y Saneamiento, Lima.

Parsekian, G.A., Hamid, A.A. & Drysdale, R.G. 2012. Behavior and design of structural masonry. São Paulo: EdUFS-Car - Editora da Universidade Federal de São Carlos.

PCM 2004. O.P.C.M. 3274. Norme tecniche per il progetto, la valutazione e l'adeguamento sismico degli edifici. Presidenza del Consiglio dei Ministri, Rome.

Penna, A., Morandi, P., Rota, M., Manzini, C. F., Da Porto, F., & Magenes, G. (2014). Performance of masonry buildings during the Emilia 2012 earthquake. Bulletin of Earthquake Engineering 12(5), 2255–2273.

Pérez-Gavilán, J.J., Flores, L.E. & Alcocer, S.M. 2015. An experimental study of confined masonry walls with varying aspect ratios. *Earthquake Spectra* 31(2): 945–968.

van der Pluijm, R. 1999. *Out-of-plane bending of masonry: behaviour and strength*. PhD thesis, Eindhoven University of Technology, Eindhoven.

van der Pluijm, R. 2009. Eurocode 6, Design of masonry structures. Personal communication in workshop 'Dissemination of information for training', 2-3 April 2009, Brussels.

da Porto, F., Mosele, F. & Modena, C. 2011. Cyclic out-of-plane behaviour of tall reinforced masonry walls under P-Δ effects. *Engineering Structures* 33(2): 287–297.

da Porto, F., Mosele, F. & Modena, C. 2011b. In-plane cyclic behaviour of a new reinforced masonry system: Experimental results, *Engineering Structures* 33(9): 2584–2596.

Roberts, J. 2020. eurocode6.org, Web site for Eurocode 6: Design of Masonry Structures. Prof. John Roberts, London.

San Bartolomé, A., Quiun, D., Casabonne, C. & Torrealva, D. et al. 2006. New Peruvian desing masonry code. *Proceedings of the 14th International Brick and Block Masonry Conference*, Sydney.

San Bartolomé, A., Bernardo, J., Peña, M. 2010. The effect of column depth on seismic behavior of confined masonry walls. *Proceedings of the Chilean Conference on Seismology and Earthquake Engineering*, Valdivia-Santiago.

Sousa, H., Thomaz, E., Roman, H., Morton, J., Silva, J.M. et al. 2015. *Defects in masonry walls, Guidance on cracking: Identification, prevention and repair*. CIB Commission W023 - Wall Structures, Rotterdam.

Tomaževič, M. 1999. *Earthquake-resistant design of masonry buildings*. Series on Innovation in Structures and Construction. London: Imperial College Press.

Varela-Rivera, J., Fernandez-Baqueiro, L., Gamboa-Villegas, J., Prieto-Coyoc, A., Moreno-Herrera, J.M. 2019. Flexural behavior of confined masonry walls subjected to in-plane lateral loads. *Earthquake Spectra* 35(1): 405–422.

Brick and Block Masonry - From Historical to Sustainable Masonry –
Kubica, Kwiecień & Bednarz (eds)
© 2020 Taylor & Francis Group, London, ISBN 978-0-367-56586-2

Persian adobe heritage: Construction technology, characterisation and protection

M. Hejazi

Department of Civil Engineering, Faculty of Civil Engineering and Transportation, University of Isfahan, Isfahan, Iran

S. Hejazi

Department of Architecture, Faculty of Architecture and Urban Design, Art University of Isfahan, Isfahan, Iran

ABSTRACT: Adobe construction has a history of thousands of years in Iran. A large part of residential buildings in rural areas of Iran in addition to a large number of historical building in the country are earthen architecture and made of adobe. For this reason, it is necessary to get a deep insight into the construction technology, characterisation and protection of adobe. In this paper, a broad overview on Persian adobe heritage is carried out. The method of adobe making in Iran, the construction technology of traditional adobe structures including foundation, wall and lintel, floor, wooden roofs and different types of adobe roofs, characteristics of Persian traditional and new adobe bricks, and damage to adobe and protection are described.

1 INTRODUCTION

The advantage of earthen structures is that their construction material is sustainable and has little impact on the environment. Building with earth expends little or none of earth's finite resources, such as fossil fuels. Their embodied costs are low such as cost of creating, storing, distributing, using, and disposing. In hot-dry climates the high thermal mass of earth houses can render them substantially more energy-efficient than stick-built ones. However, earthen buildings have some disadvantages. They do not span open spaces or window and door openings very well, so they tend to crack near windows and doors that have inadequate wooden lintels. In the case of any failure in the roof, moisture can seep in and quickly erode the walls.

The availability of clay in Persia promoted the use of amorphous mud and later sun-dried moulded mud in the sixth millennium B.C. Archaeological surveys show the usage of earth and mud brick in houses from the fourth millennium B.C. Adobe buildings are still used in different parts of Iran, in particular in hot-arid areas. Adobe buildings include houses, bazaars, caravanserais, mosques, cisterns, icehouses, etc. The number of existing adobe buildings is more than that of brick, stone and wooden ones in the country. Hence, Persian adobe heritage is an important part of Persian architecture. The most important features of Persian adobe buildings including the method of adobe making, construction technology, characteristics of Persian adobe and causes of deterioration are explained as follows.

2 LITERATURE REVIEW

There are a number of studies on the architecture of Persian adobe heritage performed by a number of researchers such as Pope & Ackerman (1938), Hejazi (1997), Hejazi et al. (2014a, b, c, d, 2015), and Zomarshidi (1995). A number of studies have also been done to improve the properties of adobe. After the Bam earthquake in 2003, studies were carried out in the Bam citadel to design adobe bricks with improved characteristics for using in the restoration of the citadel (Esrafili 2006). Ismaili & Ghalehnowi (2012) investigated the effect of adding palm fibres and lime on the mechanical properties of adobe prepared from Zahedan soil. Hejazi et al. (2015) studied effect of additives including sand, gypsum, lime, brick powder and straw with different amounts and combinations on mechanical properties of adobe made from Isfahan soil. Eskandari (2020) characterised Persian historical and new adobe bricks and compared them. In his work, mechanical properties, stress-strain relationship, and soil properties were determined.

3 ADOBE MAKING

Adobe, or mud brick, is simply a mixture of clayey soil and water dried under the sun. The percentage of the components depends on the type of soil. In general, the clayey soil should contain about 30% to 40% clay and about 60% to 70% soil. In some cases, straw, equal to 0.5% of mass of dry soil, may be added to the mixture. Figure 1 shows the steps of adobe making. After water is added to it is left for a couple of days in order to let water to penetrate into soil and activate the clay. Then it is mixed thoroughly by a shovel and then by feet and hands before moulding. Wooden moulds are used for forming the mixture. After one or two days the adobe bricks are brought out the moulds and placed vertically in shadow to dry completely.

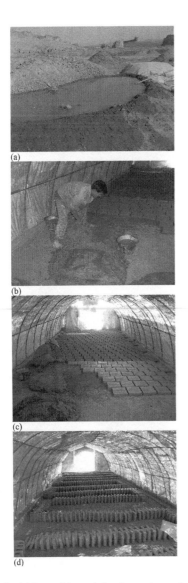

Figure 1. Adobe making: a) leaving water and soil for a couple of days, b) mixing, c) moulding by hands, d) placing vertically for drying, Bam citadel, 2007.

4 CONSTRUCTION TECHNOLOGY

4.1 *Foundation*

The foundation is generally made of lime concrete containing about 200 to 225 kg/m^3 lime wash, 25% clay, 50% to 60% sand, gravel and cursed stone and water. The connection between the foundation and the base (footing) of the wall above is provided by narrow stones put vertically (Figure 2a). Footing is made by using stone set into sand-lime mortar (Figure 2b). A stone dado is then constructed on the footing, with a height of about 1 m. The dado supports the adobe wall and prevents the wetting of the adobe wall.

4.2 *Wall and lintel*

The adobe wall is constructed on the stone dado. Special connectors, such as wooden vertical or horizontal elements, are inserted inside the wall during the construction in order to provide enough connection between adjoining walls, and between the wall and doors and windows (Figure 3). Usually arched adobe lintels are constructed over the doors. For building the lintel, a gypsum form, which is placed on brick or stone supports, is used (Figure 4).

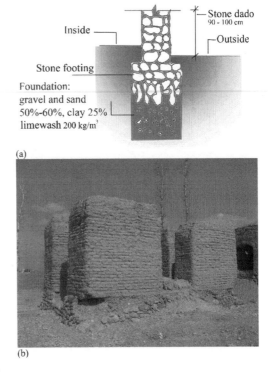

Figure 2. Foundation, footing and dado: a) details, b) adobe wall on the dado, Kerman, 2011.

(a)

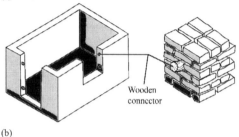

(b)

Figure 3. A) Wall construction, Bam citadel, 2007, b) wooden connectors inside the wall.

(a)

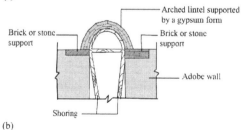

(b)

Figure 4. A) An arched lintel, Bam citadel, 2004, b) method of lintel construction.

4.3 *Floor*

The floor may be constructed in two ways. In the first method, the floor can is built directly on the soil. Stone set in lime mortar is placed on the ground and it is plastered by lime paste. Then it is covered with mud-straw plaster (Figure 5a). In the second method, the floor is made of wooden beams and laths with a distance above the ground to prevent moisture. The laths are nailed to the beams and covered with mat, reed and mud-straw plaster (Figure 5b).

4.4 *Roof*

There are two types of roofs for adobe buildings. The first type is a wooden roof that is flat. The second type is an adobe roof that has an arched shape.

4.4.1 *Wooden roof*
A wooden roof is not heavy and acts as a rigid diaphragm that causes the wall have equal lateral displacement, which is preferable in the earthquakes. Wooden posts are placed inside the walls to be connected to the roof. Wooden ring beams are placed on the top of the

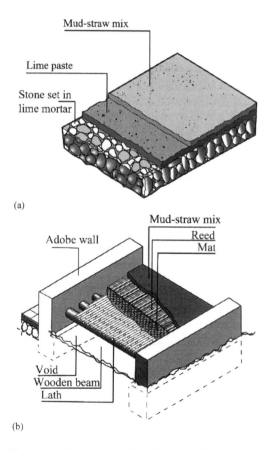

(a)

(b)

Figure 5. A) Lime concrete floor, b) wooden floor.

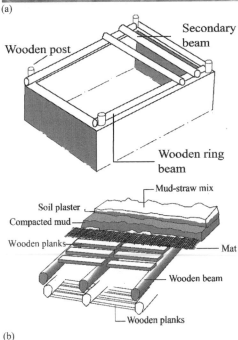

Figure 6. Wooden roof: a) Najafi house, Isfahan, 2004, b) details.

walls and nailed to the wooden posts. Then secondary wooden beams with a 50 cm to 60 cm distance are put perpendicular to the ring beams and nailed to them. Laths, or alternatively wooden planks, are connected to the bottom and top of secondary beams. The top lathing/plank is covered with mat, compacted mud, muddy soil, and mud-straw mix. The bottom lathing/plank is covered with mud-straw mix covered with gypsum. The double-layer wooden is already a rigid diaphragm causing the same movement of the top of walls in an earthquake and it is also a good heat insulation layer during hot summers and cold winters (Figure 6).

4.4.2 Adobe roof

If the roof is made of adobe it should be built in the form of a vault in order to reduce tensile stresses in adobe. There are different types of vaults used as the roof, a few of them is explained as follows.

4.4.3 Spherical vault

When the plan of the building is square it is possible to use a spherical vault to cover it. The main task is to transit from a square plan to a circular one. For this transition, four inclined supports are constructed at four corners. Then courses of adobe bricks are placed on the top of each other until a quarter of a circle is built at each corner, i.e. a complete circle of adobe bricks is available. Then adobe brick courses are laid layer by layer to complete the vault. A shoring system is used during the construction (Figure 7).

4.4.4 Quadripartite vault

This type of vault does not need shoring and must be built on a square plan. The vault consists of four

(a)

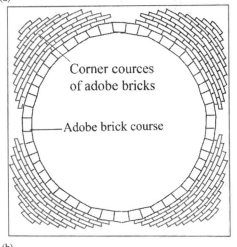

(b)

Figure 7. Spherical vault on a square plan: a) under construction, b) details.

(a)

(a)

(b)

Figure 8. Quadripartite vault: a) construction, Bam citadel, 2009, b) details.

parts at four corners. A quarter of the vaults is constructed at each corner, and the vault is complete when the four parts intersect (Figure 8).

4.4.5 *Groined quadripartite vault*

This type of vault can be built on a square or a rectangular plan and does not need shoring. Adobe brick courses are placed parallel to the walls (Figure 9).

4.4.6 *Barrel vault*

Use of barrel vaults is extensive in Persian adobe buildings. In a barrel vault the shape of the cross-section is constant along the vault and it is built over two longitudinal lead bearing walls. At one end a supporting wall and at the other end and arch is used. The construction of the barrel vault starts from the end wall, on which inclined courses of are laid that makes it unnecessary to use shoring. The construction continues until reaching the other end. When two or more adjacent vaults are constructed the space between the vaults is filled with soft soil (Figure 10).

4.4.7 *Ribbed vault*

A ribbed vault consists a number of ribs with vaults upon them. To construct a rib, gypsum forms, which are made on the ground, are used. Gypsum forms are made of gypsum and reed inside to resist tensile stresses. Two gypsum forms are placed on the load bearing walls and fixed with wooden supports and holders. Adobe bricks are then placed in the space between the

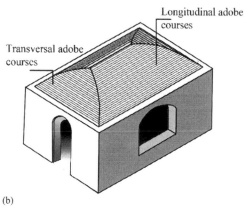

(b)

Figure 9. Groined quadripartite vault: a) construction, Bam citadel, 2009, b) details.

(a)

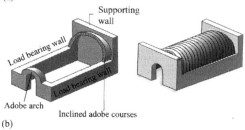

(b)

Figure 10. Barrel vault: a) construction, Bam citadel, 2007, b) details.

Figure 11. Ribbed vault: a) construction, Bam citadel, 2005, b) and c) gypsum form, Bam citadel, 2007 and 2009, d) details of the rib.

two gypsum forms. After the ribs are built, vaults are constructed on them. The role of ribs is to transfer the load from the vaults to the ground (Figure 11).

5 CHARACTERISTICS OF PERSIAN ADOBE

5.1 Papoli house in Yazd

There are few reports about various characteristics of Persian adobe. The results of a recent series of experimental tests on old adobe bricks taken from a historical house and new adobe bricks for its restoration show interesting findings. The Papoli adobe house (probably mid-eighteenth century A. D.) is in the city of Yazd, Central Iran (Figure 12). The house needs restoration. To select the suitable new adobe to be compatible with old adobe, a number of experimental tests were carried out. Four original types of adobe bricks and three types

Figure 12. Papoli adobe house, Yazd, probably mid-eighteenth century A.D.

Table 1. Codes of tested adobe bricks in Papoli restoration project.

Code	Type of adobe
A	Old (original): basement wall at dry part
B	Old (original): body of southern wind tower
E	Old (original): basement wall at previously wet and currently dry part
G	Old (original): roof of the basement
I	New: with 70% clay
J	New: with 80% clay
K	New: with 90% clay

of new adobe bricks were selected as indicated in Table 1. A number of mechanical and non-mechanical tests were conducted (Figure 13), and results for compressive, flexural, and split tests are presented in Figure 14.

Figure 14a shows the compressive strength for dry and wet tests, for two different test methods, i.e. a single adobe brick or two adobe bricks on top of each other. Two types of old adobe bricks (A, B, E and G) were decomposed in water and it was not possible to conduct tests on them (Figure 13i). Adobe type E had become previously wet in the basement due to water drainage and had been re-dried. For one brick test method, the compressive strength of adobe types A and B was about 0.8 MPa, that of adobe type G was about 1.5 MPa, whereas that of adobe type E was about 3.9 MPa. For new adobe bricks (I, J and K) the dry and wet compressive strengths were respectively about 6 MPa and 2 MPa.

According to Figure 14b the same trend holds for the modulus of elasticity (E). The values of E for dry old adobe types A, B and G are respectively 44 MPa, 31 MPa and 190 MPa. These values for new adobe types J and K are respectively 274 MPa and 128 MPa. For wet samples, these values reduce by 55% to 81%.

Figure 14c compares flexural strength and split (Brazilian) strength. For flexural strength, the value for old adobe bricks ranges between 0.35 MPa to 0.47 MPa. It ranges from 1.54 MPa to 2.18 MPa for new adobe bricks. For split strength, the value for old adobe bricks ranges from 0.11 MPa to 0.18 MPa.

(a)

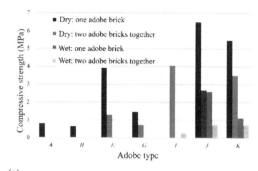

(a)

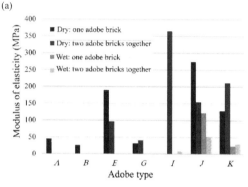

(b)

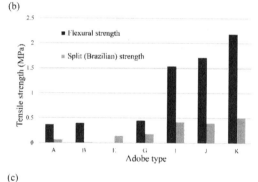

(c)

Figure 13. Experimental tests on adobe bricks of Papoli house in Yazd: a) adobe type A before cutting, b) cutting adobe brick, c) adobe type A to be prepared for flexural test, d) adobe type E before cutting, e) adobe type G before cutting, f) compressive test samples, g) test set-up, h) compressive test on adobe type B, i) adobe bricks in water after 10 minutes, one row of old adobe bottom, three rows of new adobe top, j) adobe bricks in water after 2 hours (23 April 2019).

For new adobe bricks it ranges between 0.4 MPa and 0.51 MPa.

The average flexural strength to split ratio is 2.8 for old adobe bricks and 4 for new adobe bricks. The average compressive strength to flexural strength ratio is 4.3 for old adobe bricks and 3.2 for new adobe bricks. The average compressive strength to split strength ratio is 12 for old adobe bricks and 12.8 for new adobe bricks.

Figure 14. Results for adobe bricks of Papoli house in Yazd: a) compressive strength, b) modulus of elasticity, c) tensile strength, d) stress-strain diagram for adobe type A, e) stress-strain diagram for adobe type E, f) stress-strain diagram for adobe type I.

Stress-strain diagrams for old adobe bricks types A and B and new adobe brick type I are shown in Figures 14d-f.

5.2 Adobe bricks in the Bam citadel

The Bam citadel is on the largest adobe complexes in the world that was severely damage during the Bam earthquake in 2003. A huge amount of adobe bricks was needed for its restoration. For this reason, studies were carried out on the geology of the Bam region to select a soil quarry for producing adobe to be used in restoration of the Bam citadel (Esrafili 2006). The soil of Nizam-Abad, region near Bam, was selected as soil quarry. The grading of Nizam-Abad soil is shown in Figure 15a. Mechanical properties of new adobe bricks produced from Nizam-Abad soil are shown in Table 1.

In order to improve the properties of adobe bricks made of Nizam-Abad soil different components such as sand, quicksand, brick powder, gypsum powder, crushed ceramics, palm fibres, straw and light weight aggregates with various percentages were added to the soil and experimentally tested (Figure 15b). On the basis of results from laboratory tests, the compounds Nizam-Abad soil + 30% quicksand, and Nizam-

Table 2. Mechanical properties of different compounds of new adobe bricks made of Nizam-Aabad soil (N = Nizam-Abad soil, SS = quicksand, ST = straw, after Esrafili 2006).

Property	N	N+30%SS	N+40%SS +1.5%ST
Water absorption (%)	1.25	3.74	2.69
Water erosion (%)	-	9.4	6.4
Abrasion resistance (MPa)	-	4.78×10^{-4}	7.59×10^{-4}
Longitudinal shrinkage (%)	7.5	5	2.33
Bulk density (kg/m³)	1,890	1,770	1,700
Compressive strength (MPa)	3.4	4.19	3.4
Flexural tensile strength (MPa)	1.13	1.22	1.37

Abad soil + 40% quicksand + 1.5% straw had the most satisfactory properties (Table 2). The latter compound was not selected due to the existence of straw that might be damaged by termites. The former compound, whose grading is shown in Figure 15c, was selected for producing new adobe bricks. Five compounds of mortar were selected according to the best surface condition, i.e. with least cracks. The properties of suitable designed mortars are presented in Table 3.

6 DAMAGE TO ADOBE AND PROTECTION

6.1 *Damage*

Damage to adobe can be due to several reasons such as structural issues, water-related problems, wind erosion, vegetation, insects, vermin, and material incompatibilities. Structural damage and water-related damage are two important causes of deterioration of adobe.

Structural damage usually originate from improper design or construction, insufficient foundations, weak or inadequate materials, and the effects of external forces such as wind, water, snow and earthquakes.

Water-related problems are mostly because of moisture due to either disproportionate rainwater or ground water. Deep cracks and uneven surfaces may be produced by erosive action of rainwater. Coving may be produced by accumulation of rainwater at foundation level and rain splash (Figure 16a). High underground water table, lack of drain or plant watering may cause water to rise through capillary action into the wall and cause adobe to erode, bulge and. Cove is also caused by spalling through the freeze-thaw cycles. Dissolved minerals or salts deposited on

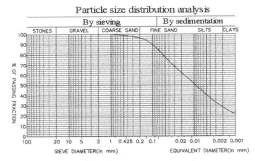

(a)

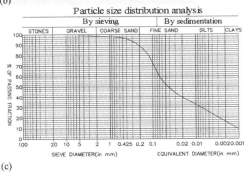

(b)

(c)

Figure 15. A) Grading of Nizam-Abad soil (after Esrafili 2006), b) adobe bricks made of Nizam-Abad soil with different components in the laboratory, Bam citadel, 2006, c) grading of Nizam-Abad soil + 30% quicksand (after Esrafili 2006).

Table 3. Properties of suitable new mortars with respect to surface condition (N = Nizam-Abad soil, SS = quicksand, NS = new sand, after Esrafili 2006).

Property	N+75%NS (NS/N=3/1)	N+66.66%NS (NS/N=2/1)	N+60%NS (NS/N=3/2)	N+66.66%SS (SS/N=2/1)	N+75%SS (SS/N=3/1)
Surface condition	excellent (without crack)	very good (very fine cracks)	good (fine cracks)	good (fine cracks)	good (fine cracks)
Moisture content (%)	1.25	1.83	1.29	1.42	1.53
Water absorption (%)	7.63	12.13	8.5	5.4	10
Water erosion (%)	11.26	9.3	9.89	17.37	-
Abrasion resistance (MPa)	4.22×10^{-5}	4.71×10^{-5}	3.83×10^{-5}	3.04×10^{-5}	4.22×10^{-5}
Longitudinal shrinkage (%)	1.66	2.83	2.16	2.66	2.33
Bulk density (kg/m^3)	1,740	1,920	1,810	1,660	1,700
Compressive strength (MPa)	1.75	2.41	16.7	1.77	1.69
Flexural tensile strength (MPa)	0.59	0.63	0.91	0.92	0.83

the surface of the wall due to moisture evaporation can deteriorate the adobe. In adobe with high amount of clay, when the adobe dries out shrinkage cracks occur. The remedies for water-related problems include use of a water-tight roof with proper drainage, maintenance of adobe wall and roof surfaces with traditional plaster, regrading the ground and foundations levels, and installation of drains. The construction of canals along the wall (Figure 16b), and construction of canals under the floor is a very popular and effective solution for moisture in traditional buildings.

6.2 Protection

Repair and replacement of damaged parts should be carried out with traditional or original materials. These include: a) brick: patching and repairing adobe brick with a texture and colour similar to the original fabric, b) mortar: patching and replacing mortar similar to the original material, colour and texture; adobe mud mortar should never be replaced by lime mortar or Portland cement mortar, c) wooden elements: repairing and replacing wooden members by wood, d) coating: patching and replacing surface coatings by same material, e) roof: flat adobe roofs should be repaired and maintained with their original form and materials, f) floors, windows, doors, etc.: floors, doors, windows and other original details of the older adobe building should be preserved whenever possible.

Mud-straw is the most commonly used plater for protection of adobe surfaces. The mud-straw plaster is between 2 cm to 5 cm thick based on weather conditions. Mud-straw is a mixture of clayey soil, straw and water, left for a couple of days before mixing (Figure 17).

(a)

(b)

Figure 16. A) Coving in adobe walls, Soukias adobe house, seventeenth century A.D., Isfahan, b) canals along the wall to resolve moisture problems.

7 CONCLUSION

The most important features of Persian adobe buildings were described. The method of adobe making, construction technology, characteristics of Persian adobe and causes of deterioration where explained. It is important to keep the traditional methods of construction and to use traditional materials for the maintenance of adobe buildings.

REFERENCES

Eskandari, A. (2020). *Characterisation of Persian historical and new adobe bricks* (in Farsi). Master Thesis, Faculty of Civil Engineering and Transportation, University of Isfahan, Isfahan.

Esrafili, A. 2006. *Introducing the activities in the soil mechanics laboratory of the Bam citadel from 2004 to 2006.* Bam: Bam Cultural Heritage Rescue Project.

Hejazi, M. 1997. *Historical buildings of Iran: their architecture and structure.* Southampton and Boston: Computational Mechanics Publications (WIT Press).

Hejazi, M., Hashemi, M., Jamalinia, E. & Batavani, M. 2015. Effect of additives on mechanical strengths of adobe made of Isfahan soil (in Farsi). *Journal of Housing and Rural Environment* 34 (151): 67–80.

Hejazi, M., Hejazi, B. & Hejazi, S. 2015. Evolution of Persian traditional architecture through the history. *Journal of Architecture and Urbanism* 39(3): 188–206.

Hejazi, M. & Mehdizadeh Saradj, F. 2014a. *Persian architectural heritage: architecture.* Southampton and Boston: WIT Press.

Hejazi, M. & Mehdizadeh Saradj, F. 2014b. *Persian architectural heritage: architecture, structure and conservation.* Southampton and Boston: WIT Press.

Hejazi, M. & Mehdizadeh Saradj, F. 2014c. *Persian architectural heritage: conservation.* Southampton and Boston: WIT Press.

Hejazi, M. & Mehdizadeh Saradj, F. 2014d. *Persian architectural heritage: structure.* Southampton and Boston: WIT Press.

Ismaili, A. & Ghalehnowi, M. 2012. The effect of palm fibres and lime on mechanical properties of adobe (in Farsi). *Journal of Housing and Rural Environment* 31 (138): 53–62.

Pope, A.U. & Ackerman, P. 1938. (eds). *A survey of Persian art: from prehistoric times to the present*, 6 Vols. London and New York: Oxford University Press (and 3rd edn, 1965).

Zomarshidi, H. 1995. *Iranian architecture: building with traditional materials* (in Farsi). Tehran: Khajeh Press.

(a)

(b)

(c)

(d)

Figure 17. Mud-straw: a) mixing clayey soil, straw and water, b) close-up view, c) deteriorated mud-straw plaster, d) plastered new mud-straw mix.

Brick and Block Masonry - From Historical to Sustainable Masonry –
Kubica, Kwiecień & Bednarz (eds)
© 2020 Taylor & Francis Group, London, ISBN 978-0-367-56586-2

Masonry façades in Australia and challenges for engineering research and design

M.J. Masia
Centre for Infrastructure Performance and Reliability, The University of Newcastle, Australia

ABSTRACT: The paper describes challenges for engineering research and design relevant to Australian masonry façade construction. Research aimed at addressing such challenges, being conducted by the author and his colleagues at The University of Newcastle is discussed. The structural safety of new and older existing masonry cavity and veneer wall systems is considered. In particular, the influence on structural reliability, of the spatial and temporal variability of material properties, including aspects such as corrosion of steel wall ties, is addressed. The paper also presents examples of innovations in the design of new masonry façade systems. In recent years architects have lead a revival in the use of masonry in building façade systems in Australia. They have incorporated into their designs textured masonry, stack bonded masonry, and hit and miss (lattice) masonry; they have curved walls in plan and elevation and they have used these various forms in both loadbearing and non-loadbearing applications. This has resulted in challenges for structural engineering design, as many of these forms of masonry construction are not addressed in national design standards. The paper discusses examples and overviews research being conducted to help inform engineering solutions needed to bring architectural visions to life.

1 PERFORMANCE OF EXISTING MASONRY FAÇADE SYSTEMS

1.1 Introduction

Structural masonry, and unreinforced masonry (URM) walls in particular, are highly vulnerable to damage during seismic and extreme wind events. In past earthquakes and storms in Australia and elsewhere, masonry façade collapses have resulted in human injury and death and significant damage to property (Moon et al. 2014). In unreinforced masonry exterior walling (façade) systems, non-loadbearing veneers and outer leaves in cavity wall construction are supported by an internal structural framing system or parallel structural wall via a series of closely spaced ties of small gauge wire or flat steel. The structural stability and safety of such veneer and cavity wall façades when subjected to out-of-plane loads due to wind and earthquakes is reliant on the structural integrity of the ties. Where ties have inadequate strength, are excessively spaced or deteriorated due to corrosion, the veneer or outer masonry leaf may collapse with potentially disastrous consequences (Figure 1). Such collapses represent a significant life safety risk due to falling debris, which usually lands on footpaths and streets immediately adjacent to the exterior of the building.

Ongoing research at The University of Newcastle aims to develop an improved understanding of the structural behaviour and structural reliability of masonry veneer and cavity wall façade systems, considering: spatial variability of masonry strength, and strength and stiffness of the ties, veneers with stiff or flexible structural supports, and time dependent strength deterioration of the ties due to corrosion.

1.2 Background

Unreinforced masonry construction is used extensively in Australia for residential dwellings such as houses and low rise apartments and townhouses, and as facades in low-rise and multi-storey framed buildings. To achieve weather proofing in exterior walls, the two most commonly used construction systems are the masonry veneer wall system (Figure 2a) and the masonry cavity wall system (Figure 2b). In both systems, the air gap formed between the exterior and interior layers provides a barrier to moisture movement across the cavity.

For a masonry veneer wall system (Figure 2a), the stud framing (timber or steel) provides lateral support to the external leaf of masonry (veneer) which is supported on a footing beam or shelf angle at its base and is laterally supported by wall ties attached to the stud frame. The veneer does not support any vertical loads other than its own self weight and it is laterally supported only by the wall ties. The ties themselves are typically fabricated from galvanised

(a)   (b)   (c)

(d)   (e)

Figure 1. (a) Collapsed outer leaf of a masonry cavity wall after the newcastle Earthquake, 1989 (photo: A.W. Page), (b) Severe tie corrosion in outer leaf of the wall shown in (a) (photo: A.W. Page), (c) Collapsed masonry veneer following high winds in newcastle 2015 (note that similar tie corrosion to that shown in (b) was observed also in the rubble of (c)) (photo: G. Simundic), (d) Tie with insufficient strength and stiffness (photo: A.W. Page), and (e) Ties bent down during construction and then not engaged (photo: A.W. Page).

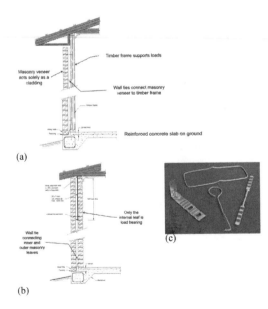

(a)

(b)

(c)

Figure 2. (a) Masonry veneer wall construction, (b) Masonry cavity wall construction, (c) Typical veneer and cavity wall ties (Think Brick Australia).

or stainless steel (Figure 2c). The ties are attached to the masonry (typically spaced at up to 600 mm centres horizontally and vertically) by being built into the mortar bed joints as the masonry is constructed and are screw or nail fixed to the stud framing. The function of the ties is to transfer compressive and tensile axial forces across the cavity from the masonry veneer to the internal stud frame when the wall is subjected to wind, seismic or other out-of-plane lateral loading. The masonry cavity wall system functions in a similar fashion to the masonry veneer system described above, with the role of the stud framing being replaced by a second (interior) leaf of masonry (Figure 2b). In the cavity wall system, the interior masonry leaf provides the loadbearing role and the exterior masonry leaf acts as a veneer.

When these wall systems are subjected to lateral out-of-plane loads, even though the exterior masonry leaf (veneer) is intended only as a weather enclosure to the building, it does contribute to the resistance of wind and seismic load by virtue of its stiffness and to the complex load sharing interaction which is established between the masonry veneer and its supporting backup. The way in which the load is shared within the wall system depends on the relative stiffnesses of the veneer and its support, and the stiffness and distribution of the wall ties. At the ultimate limit state, the behaviour is further complicated by non-linear behaviour, and ultimately progressive failure, of the various materials. The Australian Masonry Structures Standard AS3700 (Standards Australia 2018) requires that the wall ties transfer 100% of the out-of-plane lateral load on the veneer to the structural backup. However, the current design provisions do not consider the influence on the system performance and structural reliability, of the spatial and temporal variability of the strengths and stiffnesses of the masonry components, the ties and the stud framing. This lack of a rigorous stochastic and structural reliability assessment means that the actual level of safety of masonry veneer systems is not known. The problem is compounded by the fact that the strength properties of masonry are highly variable. High unit-to-unit spatial variability is also observed, due to variations in the quality of workmanship, the weather during construction, and materials, all within the one structure.

Lawrence (2007) reviewed Australian design practices and demonstrated that the assumption of elastic brittle behaviour in masonry and ties and a "weakest link" approach to AS3700 design through the use of lower 5th percentile characteristic strengths was overly conservative. Lawrence also acknowledged that further research is required (i) in the rational assessment of appropriate capacity reduction factors (ϕ) for the design of wall ties and unreinforced masonry in bending, and (ii) in the *"detailed examination, by analysis and test, of the behaviour of wall ties connecting two masonry leaves (cavity wall system) and the interaction between a single masonry leaf and its supporting frame in a masonry veneer wall, leading to an improved understanding of the appropriate forces to be used to design the ties."* Significant recent progress has been made in understanding the influence of spatial variability of material properties on the bending strength of unreinforced masonry panels (Heffler et al. 2008, Li et al. 2014, 2016a, b). This research has demonstrated that when cracking initiates at weaker than average joints in a masonry wall, the presence of stronger joints nearby allows stress redistribution resulting in peak wall strengths higher than those predicted by a deterministic model based on a lower 5% characteristic strength. Furthermore, the coefficient of variation (COV) of peak wall strength is lower than the COV of material strengths. However, the research gaps (i) and (ii) as defined by Lawrence are yet to be addressed.

It is well known that many older existing buildings do not meet current design standards. Masonry wall systems may deteriorate over time due to corrosion of wall ties. It was observed following the 1989 Newcastle Earthquake that in some walls, galvanised steel ties had corroded through their complete thickness resulting in zero tie strengths (Figure 1b). Also observed were numerous examples of poor workmanship such as misalignment of ties, excessive spacing of ties and ties which were not engaged into the masonry during construction, rendering them completely ineffective Figure 1d, e) (Page 1992). Such observations from the Newcastle Earthquake prompted subsequent design code revisions aimed at improving construction practices and durability in new construction (Beattie & Thurston 2011), but many existing masonry buildings in Australian cities predate these post 1989 standards. It is almost certain that numerous existing masonry buildings across Australia have the same types of deficiencies, making them highly vulnerable to earthquakes and severe storms. Further highlighting this concern, just last year the outer leaf of a masonry cavity wall in an early 1900s constructed building in Perth collapsed without any unusual loading event (Carmody 2019). Inspection of debris revealed heavily corroded wall ties.

The following sections overview a program of experimental testing and stochastic numerical modelling currently underway at The University of Newcastle which aims to develop an improved understanding of the behaviour of masonry veneer and cavity wall systems when subjected to out-of-plane wind and earthquake loading. The study considers spatial and temporal variability of material properties to quantify the structural reliability and fragility of such systems.

1.3 Experimental characterization of masonry strength, wall tie strength and stiffness and stud frame stiffness

Stochastic nonlinear Finite Element Analysis (FEA) of masonry veneer and cavity wall systems requires probabilistic material models for the masonry, the wall ties and the stud framing. For the experimental program reported herein, only veneer walls were tested, using a single masonry type (brick and mortar combination), a single type of wall tie and a single timber stud size and stress grade. The computational framework (stochastic FEA) developed can then be used to consider other material types as well as the behaviour for masonry cavity wall systems.

To characterize the masonry, the author and his colleagues constructed masonry walls and then used the bond wrench test to remove each brick, one by one, from the walls. The resulting masonry flexural tensile strength data enabled characterization of the unit-to-unit spatial variability and correlation of masonry material properties (Heffler et al. 2008).

For the wall ties, an experimental testing program was conducted (Muhit et al. 2020) to develop probabilistic material models of wall tie strengths and stiffnesses in axial tension and compression, and to develop suitable nonlinear failure models for wall ties for use in the stochastic FEA of veneer wall systems.

AS/NZS 2699.1 (Standards Australia/Standards New Zealand 2000) provides a test method to determine the mean and characteristic axial tensile and compressive strengths for wall ties for use in Australia and New Zealand. This test method was adapted to characterize the behaviour for the chosen ties. A total of 50 veneer wall tie specimens (25 in compression and 25 in tension) were tested (Figure 3). The complete load versus displacement response for

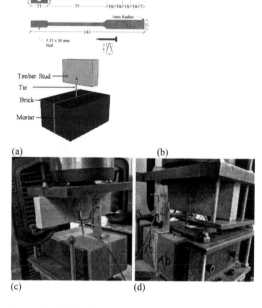

Figure 3. (a) Side fixed veneer wall tie, (b) Test specimen for wall tie characterization tests, (c) Test setup for application of compression force, and (d) Tension force (Muhit et al. 2020).

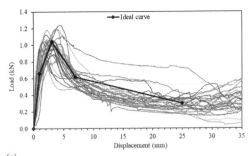

(a)

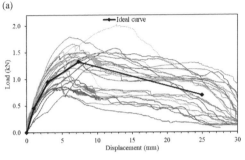

(b)

Figure 4. Axial load versus displacement response, together with the average of all responses ("Ideal curve") for wall ties loaded in (a) Compression, and (b) Tension (Muhit et al. 2020).

characterise the load versus displacement behaviour and to infer probabilistic distributions of collapse loads for masonry veneer wall systems. In total, twenty full scale single storey masonry veneer wall specimens are being constructed using new materials and then subjected to lateral out-of-plane distributed loading using an airbag system (Figure 5). Ten specimens will be subjected to "inward" loading on the masonry veneer (wall ties subjected to net compressive load) and ten specimens will be subjected to "outward" loading (wall ties in net tension). At the time of writing, eight of the ten inward loaded specimens have been tested. All of the veneer specimens will be theoretically identical (that is, the same geometry, type of masonry, ties, stress grade of timber framing, etc) to allow the determination of probabilistic data for the wall system collapse load for each loading direction.

The construction dimensions were selected to represent typical Australian construction practice and are compliant with current Australian standards. Due to the vertical orientation of wall

each specimen was recorded allowing the determination of elastic stiffness, peak strength and displacement capacity and the mean, COV and probability distribution associated with each property across all the specimens tested (Figure 4).

Under compression loading, almost all specimens failed via tie buckling, with just two cases of combined tie buckling and nail pull-out from the timber. Under tension loading, all 25 specimens failed via nail pull-out from the timber.

The axial load versus displacement response (Figure 4) is non-linear with significant post peak displacement capacity under both compression and tension loading. This indicates that a wall system subjected to out-of-plane lateral loading incorporating many such ties should possess a significant ability for load redistribution once the axial strength capacity of the most heavily loaded and/or lowest strength ties is reached.

Finally, a probabilistic material model for timber stud flexural stiffness will be developed from bending tests to be conducted on every timber stud used in the masonry veneer wall system tests described in Section 1.4 below.

1.4 Experimental testing of masonry veneer wall systems

An experimental testing program including multiple repeat specimens is currently being used to

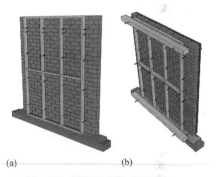

(a)   (b)

(c)

Figure 5. Testing of masonry veneer wall assemblies: (a) Veneer wall specimen (2400 mm x 2400 mm), (b) Test setup – edge of airbag is just visible on far side of veneer, (c) Photo of test setup.

studs in the framing systems of veneer walls, they are primarily one way vertically spanning structural systems, with theoretically equal tie forces along the length of the system at any given height. If the behaviour is considered deterministically, a specimen length of only 1200 mm including two studs and two vertical lines of ties spaced at the AS3700 maximum spacing of 600 mm would suffice to represent walls of any length. However, in the current testing program, a four stud system (total length 2400 mm) is being used to experimentally capture the influence of spatial variability of material properties across the length and height of the system, and hence the progressive failure and load redistribution which is expected to take place.

The data recorded for each test includes the total distributed load resisted by the wall system, displacements at various points on the masonry veneer and support stud framing, as well as displacements across the cavity at each wall tie location to deduce, from the wall tie constitutive law, the forces in the ties.

In the testing conducted to date, the failure sequence has included masonry veneer cracking along a single bed joint close to wall mid-height (Figure 6a, b), followed by progressive buckling of the ties (Figure 6c). The load versus out-of-plane displacement responses showed an initial peak defined by the cracking of the masonry, after which

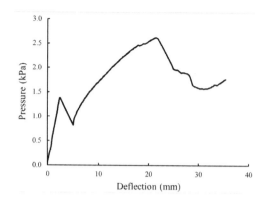

Figure 7. Experimentally observed applied pressure versus displacement at the mid-height of the masonry veneer for one of the eight tested veneer walls.

the applied pressure increased again before reaching a maximum load defined by progressive buckling of the ties. A typical response is shown in Figure 7. In just one specimen, which was found to have higher than average masonry bond strength, tie buckling preceded veneer cracking.

1.5 Stochastic nonlinear finite element analysis (FEA) of masonry veneer and cavity wall systems

Stochastic nonlinear FEA models of masonry veneer and cavity wall systems are being developed (Muhit et al. 2019). The modelling framework considers random and spatial variability in material strengths and stiffnesses for the masonry, the ties and the supporting frame. The models consider nonlinear material behaviour to allow simulation of the complete load versus displacement response of the wall systems to capture peak load, post peak softening and collapse. Model error will be characterised via comparison of FEA strength predictions with the experimental wall capacities obtained from the testing described above in Section 1.4.

The commercially available FEA software DIANA is being used to implement the stochastic non-linear finite element modelling strategy including the following key features: (i) the masonry is modelled using a simplified micro model in which masonry units are modelled as linear elastic continuum elements connected by non-linear joint interface elements representing the potential for flexural or out-of-plane shear failure at the mortar joints in the masonry, (ii) the wall ties are modelled as tension/compression elements with non-linear constitutive laws established from the observed test data, (ii) the timber wall studs are modelled as elastic beam elements as testing to date shows that they remain elastic during the wall system tests. The material properties for use in the FEA will be established

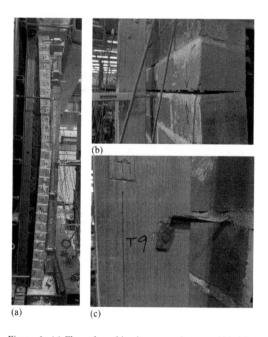

Figure 6. (a) Flexural cracking in veneer close to mid-height, (b) close up of veneer cracking, and (c) tie buckling.

from the experimental testing program described above in Section 1.3.

As was discussed in Section 1.4, a masonry veneer wall system with vertically spanning timber frame members is essentially a one way vertically spanning system and could be modelled deterministically using a single timber stud with the appropriate tributary length of the masonry wall. However, a wall length of 2400 mm, including four stud framing members, is being used for the stochastic FEA to allow for the potentially higher chance of a weak joint or tie in the wall system at which cracking/buckling could initiate, compared to a wall with a shorter length.

At the time of writing, a deterministic FEA model of the tested masonry veneer walls has been developed (Figure 8). The next step is to commence the stochastic modelling. The following key material properties will be treated as spatially varying random variables: masonry tensile bond strength; masonry tensile fracture energy (correlated to tensile bond strength); masonry elastic modulus; timber flexural stiffness; wall tie axial strength; and wall tie axial stiffness. This will allow the numerical generation of probability distributions of collapse loads for the wall systems. The model error statistics will be characterised by comparison with the experimentally derived distributions from the wall system testing described above in Section 1.4.

1.6 Corrosion of mortar embedded wall ties in masonry veneer and cavity wall construction

In parallel with the program of experimental testing and numerical modelling of masonry façade systems described above, research is currently underway at The University of Newcastle to better understand the mechanisms for corrosion of metal wall ties in masonry veneer and cavity walls.

Observations of damage following wind and earthquake events (Page et al. 1990, Ingham and Griffith 2011), as well as in-situ field studies of cavity ties (Jardim do Nascimento et al. 2017) show that galvanized steel wall ties are highly susceptible to corrosion. Corrosion can be severe at the mortar/cavity interface and within the mortar bed joints of the outer masonry leaf (Figure 1b), particularly in South facing walls, which in the Southern hemisphere can experience prolonged periods of dampness from the prevailing weather conditions. This has led to the specification of stainless steel ties in near coastal zones in Australia since the early 1990s, however, light galvanized ties still exist in older masonry buildings and masonry façades.

Jardim do Nascimento et al. (2018, 2019) describe an experimental program in which identical masonry couplets with mortar embedded ties (Figure 9) were subjected to a highly aggressive (near coastal) natural environment and an accelerated artificial salt spray environment to study the mechanisms of corrosion. The aim of the study is to predict time-dependent corrosion loss, which in turn will affect strength and stiffness loss of wall ties. The study considers three mortar types with increasing cement content and seven types of wall ties of varying corrosion resistance ratings achieved via galvanizing with varying weights of zinc coating or by using stainless steel. Six of the seven tie types tested comply with AS/NZS 2699.1 (Standards Australia/Standards New Zealand 2000). However, the tie type with the least weight of zinc coating is not commercially available as a masonry wall tie; instead these ties were cut from galvanized wire and included in the study to represent ties typical of older existing (pre 1990s) construction in Australia.

The testing program is designed to run for 2.5 years in the natural environment and 24 weeks in

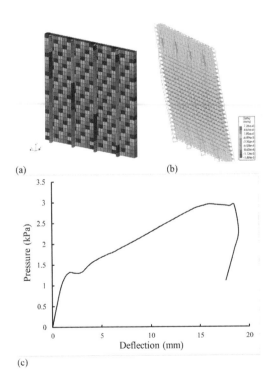

(a)   (b)

(c)

Figure 8. FEA of masonry veneer wall (a) FE mesh, (b) Displaced shape, and (c) Pressure versus out-of-plane displacement.

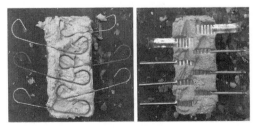

Figure 9. Wall tie corrosion specimens showing mortar embedded ties immediately before placing upper brick of masonry couplet.

the accelerated salt-spray chamber tests. At the time of writing the first year of specimen recoveries have been made for the natural environment along with the full 24 weeks from the accelerated salt-spray chamber tests. The results to date indicate that corrosion rates are highly dependent on the tie material and to a lesser extent on the cement content of the mortar. Significant differences were observed in the corrosion behaviour between the accelerated artificial (salt-spray chamber) and natural environments, meaning that accelerated testing could result in misleading corrosion loss estimates. However, it is noted that the accelerated testing returned higher mass losses than the natural environment and hence is likely to be conservative. Consistent with field observations, corrosion damage was observed to be localized at the mortar/atmosphere interface (Figure 10). Measurements of mortar pH through the thickness of the bed joints indicate that there is a small region of low pH at the surface of the mortar and this is involved in initiating the local corrosion process.

1.7 Structural reliability and fragility of new and existing masonry veneer and cavity wall systems

Modern codes of practice for structural engineering design are almost exclusively presented in a "Limit States" format. *Structural reliability* is the measure of level of safety for these limit state design codes.

Figure 10. Close-up on the mortar/atmosphere interface for a 950 g/m^2 galvanized tie exposed for 1 year in the natural environment.

Such approaches use probability distributions of loads and resistances to calculate probabilities of failure.

The stochastic FEA models developed will be used to calculate the structural reliability and fragility (probability of failure conditional on a specific wind speed, peak ground acceleration or other hazard) of masonry veneer and cavity wall systems for both new construction and scenarios considering various levels of deterioration during the service life of a structure.

The probability of failure is $p_f = \Pr[G(\mathbf{X}) \leq 0]$ where $G(\mathbf{X}) = ME \times R - S$.

$G(\mathbf{X})$ represents the limit state function for failure, $\mathbf{X}$ represents the uncertainty and variability of parameters that influence the structural performance for each limit state, ME is the model error, R is the resistance obtained from the spatially variable stochastic FEA, and S is the 50-year peak load effect (seismic or wind loading). $G(\mathbf{X}) \leq 0$ denotes failure of the structure. Statistical parameters for dead load and 50-year peak wind and seismic loads for Australian conditions can be inferred from the available literature. Fragility $\Pr(D|H)$ is the probability of failure (or damage D) for a specific hazard H (wind speed or peak ground acceleration).

The assessment of older existing masonry façade systems will be achieved by considering several scenarios believed to be common amongst older existing construction, such as corroded wall ties, loss of anchorage of ties due to deterioration of mortar joints, and/or inadequate detailing of ties at the time of construction. These scenarios can be incorporated into the stochastic FEA by replacing the probability distributions for new materials with distributions better reflecting the material performance of deteriorated materials (corroded ties for example). Added to this, the modelling framework could allow consideration of the effects of climate change on wall tie corrosion rates and wind speeds, the latter affecting the loading in the structural reliability calculation.

1.8 Summary

The research described in Sections 1.1 to 1.7 above aims to provide a rigorous theoretical basis for assessing the strength and reliability of new and existing masonry wall systems subjected to seismic and wind loads. It is hoped that the framework developed will provide tools for decision makers to help identify at risk building stock and hence inform inspection regimes and retrofit programs.

2 ENGINEERING CHALLENGES IN CONTEMPORARY MASONRY DESIGN

2.1 Introduction

In recent decades, Australian architecture has revelled in an environment of increased emphasis on

34

building façade systems. This has translated into increased budgets for façade design and construction, not only in terms of façade energy and acoustic performance but also in terms of aesthetic appeal. This has created competition between building owners leading to rapid growth in the number of showcase façade systems in Australian buildings. The trend has influenced structures of all sizes across the residential and commercial building landscape.

Masonry has played a major role in this trend, encouraged strongly via the Think Brick Australia architectural awards for innovation and excellence in masonry architecture. The number of projects nominated has grown rapidly since 2005 when the Think Brick Awards were first introduced. Architects have incorporated into their designs textured masonry, stack bonded masonry, and hit and miss (lattice) masonry. They have curved walls in plan and elevation and they have used these various forms in both loadbearing and non-loadbearing applications. Examples of exemplary Australian masonry architecture are shown in Figure 11, including some past winners in the annual Think Brick Awards.

The often irregular shapes, non-conventional bonding patterns and the presence of openings in the case of hit and miss masonry has resulted in challenges for structural engineering design, as many of these forms of masonry construction are not

addressed in national design standards. Façade systems such as the Gehry Partners designed Dr Chau Chak Wing Building (Figure 11e, f) clearly require first principles design and bespoke detailing solutions, often supported by prototype testing and advanced numerical modelling (Turley et al. 2014). However, smaller commercial and residential projects must be designed on budgets which preclude such approaches. Researchers at The University of Newcastle have been approached via Think Brick Australia on numerous occasions by structural designers seeking guidance on the design of non-codified applications of masonry. The following sections overview research being conducted at The University of Newcastle to help inform structural engineering solutions required to bring architectural visions to life.

2.2 Stack bonded masonry

Stack bonded masonry is a form of construction in which the masonry units in adjacent courses are aligned vertically above one another (Figure 12). This bonding pattern leads to continuous vertical joints running the full height of the wall resulting in a weak form of construction, particularly under horizontal out-of-plane bending. For this reason, the use of stack bonded masonry in Australia has not been encouraged, and although examples of stack bonding can be found among older Australian buildings, until

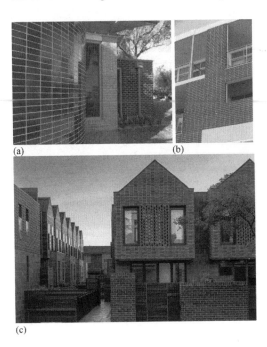

Figure 11. Contemporary masonry architecture in Australia (Photos: Think Brick Australia, A.W. Page): (a) Alternative bonding and (b) Texture in residential construction, (c) Hit and miss masonry, (d) Glass brick façade (e), (f) Dr Chau Chak Wing Building, University of Technology, Sydney.

Figure 12. (a), (b) Examples of stack bonded masonry (photos: Cathy Inglis), (c) Residential façade showing stack bonded masonry (above and below window) and hit and miss masonry (between windows) (photo: Billy Kavellaris, Kavellaris Urban Design).

the most recent (2018) edition of the Australian masonry structures code AS3700 (Standards Australia 2018), the design of stack bonded masonry was not specifically addressed. However, increasing interest from architects in the use of stack bonded masonry has driven demand by engineers for specific guidance for structural design. This led the author and his colleagues to conduct research into the behaviour of stack bonded masonry subjected to out-of-plane pressure loading which resulted in new provisions being introduced into AS3700 in 2018 (Standards Australia 2018).

The research, which is detailed in Masia et al. (2014, 2015, 2016) involved experimental testing of masonry panels to assess the performance of typical Australian stack bonded clay brick masonry subjected to lateral out-of-plane pressure loading using an airbag apparatus. The study considered unreinforced running bond (control) specimens and stack bonded specimens with and without varying levels of stainless steel bed joint reinforcement tested under different spanning configurations (Figure 13).

The experimental program included testing of 1200 mm long x 1200 mm high x 110 mm thick clay brick masonry panels as follows:

1. One way vertical bending (x 3 specimens), one way horizontal bending (x 3) and two way bending (x 6) of traditional half unit overlap (running) bonded unreinforced panels. These were tested as control specimens.
2. One way vertical bending of stack bonded panels (unreinforced) (x 3),
3. One way horizontal bending of stack bonded panels, unreinforced (x 3) and with bed joint reinforcement (3 reinforcement ratios x 3 specimens each = 9 specimens),
4. Two way bending of stack bonded panels, unreinforced (x 7) and with bed joint reinforcement (x 3).

Alongside the experimental program, the AS3700 code equations for unreinforced and reinforced masonry were applied, adapted where necessary to consider stack bonding, to assess their ability to predict the panel strengths (using mean measured material strengths and a capacity reduction factor (ϕ) = 1) and to provide safe designs (using code default values for characteristic strengths and ϕ factors). From this, recommendations for code provisions for stack bonded masonry were proposed.

The study concluded that under one way vertical bending the flexural behaviour and capacity is similar for running and stack bond. This allowed the existing code equation for strength prediction in vertical bending to be recommended also for stack bond.

Under one way horizontal bending, the strength of stack bond is much smaller than running bond, owing to the continuous vertical joints (planes of weakness) in the stack bonding pattern. The stack bonded specimens subjected to horizontal bending displayed a strength similar to the vertical bending strength. However, the horizontal bending strength is highly variable in stack bond, probably because it relies directly on the correct filling of the perpend joints, which is difficult to control. Therefore, the study recommended not to rely on flexural strength of unreinforced perpends in stack bond consistent with the existing AS3700 provisions.

For stack bonded panels subjected to one way horizontal bending, the inclusion of bed joint reinforcement significantly increased the peak pressures resisted by the stack bonded panels, but more notably, also allowed the panels to continue to resist close to their peak loads well into the post peak range, with all reinforced panels displaying a very ductile and predictable load versus displacement response. The peak pressure resisted increased with increases in the area of reinforcement provided. For one way horizontal bending, the flexural strength provisions in AS3700 were shown to provide accurate and consistent strength prediction for the reinforced stack bonded specimens.

The unreinforced stack bonded panels subjected to two way bending resisted peak pressures only

(a)

(b)

(c)

Figure 13. Experimental testing of running bond (control) and stack bonded masonry panels under out-of-plane pressure loading: (a) One way vertical bending, (b) One way horizontal bending (after first rotating panels through 90°), and (c) Two way bending.

slightly less than the running bonded panels subjected to two way bending and hence were much stronger than the stack bonded panels in one way horizontal bending. The use of bed joint reinforcement improved the behaviour of the two way bending stack bonded panels, however, the improvement was not as marked as for one way horizontal bending.

The virtual work method for strength prediction of unreinforced masonry in two way bending is not appropriate for use with stack bonded masonry because the failure patterns assumed for the virtual work method do not arise for stack bonding.

Based on this research, as well as comparison with international design codes (Canadian Standards Association 2004, Masonry Standards Joint Committee 2013), AS3700 (Standards Australia 2018) now includes clauses specifically addressing stack bonded masonry. The code defines stack bonded masonry as *"Masonry in which the overlap of masonry units in successive courses is less than ¼ of the unit length or 50 mm, whichever is greater"* and requires that it be reinforced using anchored bed joint reinforcement, even if used as a veneer. The reinforcement must meet specific durability requirements, a specified minimum reinforcement ratio, and limitations on the maximum spacing and the location of reinforcement in the vicinity of openings, wall edges and lateral supports. The new code provisions specify that stack bonded masonry be designed as unreinforced for compression, shear and vertical bending, and designed as reinforced for one way horizontal bending.

Future research proposes to study the dispersion of concentrated compression loads through stack bonded masonry to determine whether the 45° dispersion angle assumed for running bond is appropriate, as well as determining the strength enhancement, due to confinement by the surrounding masonry, immediately beneath concentrated loads and how this compares to that assumed for running bond.

2.3 Hit and Miss (lattice) masonry

Hit and miss masonry (also referred to herein as lattice masonry) is a form of construction in which the mortar perpend joints are left unfilled and the masonry units are spaced along the courses to leave gaps between adjacent units. This creates a wall which has a lattice type appearance, allowing the passage of light and air through the wall, while still providing a level of privacy. A range of examples are shown in Figures 11c, 12c and 14 and online searches using keywords such as "hit and miss masonry", "perforated masonry" or "masonry screens" reveal that the use of such masonry is widespread internationally with many spectacular examples of varying geometries in a range of applications.

Figure 14. Examples of lattice masonry construction (photos: (a, b) Think Brick Australia and (c) A.W. Page).

Regardless of whether this masonry is used in a loadbearing or nonloadbearing application, it must still be capable of spanning between supports to meet minimum robustness requirements and to resist out-of-plane lateral loading due to wind and/or earthquake actions, and hence it requires structural design. Australian Standard AS3700: Masonry Structures (Standards Australia 2018) provisions for one way vertical bending of unreinforced masonry can be applied in the case of hit and miss masonry by using a section modulus based on the net bedded area. However, the AS3700 provisions for horizontal bending require that the masonry be constructed with all perpends completely filled and therefore hit and miss masonry falls outside the scope of AS3700 for horizontal bending. By extension, in their present form, the AS3700 provisions for two way bending action, which rely on the horizontal bending capacity, also cannot be applied for hit and miss masonry. As a result of this apparent lack of guidance for structural design, Think Brick Australia receives numerous enquiries each year from architects and structural engineers wishing to specify and design hit and miss masonry.

A review of overseas codes and literature also reveals limited existing information on the design of this form of construction, thus further justifying the need for research in this area. Trimble (2013) and Ortlepp and Schmidt (2014) both reported a lack of guidance for structural design of hit and miss masonry. Industry fact sheets such as Brick Development Association (2019) and Think Brick Australia (2019), while providing guidance on hit and miss brickwork for architects and builders, defer the reader to their structural engineer to ensure that walls are robust and structurally sound. Think Brick Australia (2019) even goes as far as noting that hit and miss masonry falls outside the scope of the AS 3700 (Standards Australia 2018) provisions for one way horizontal and two way bending. The result is that at this point in time, structural engineers must resort to first principles for the design of hit and miss masonry walls. Personal communications between the author and practicing structural engineers (Collis 2020, Curren 2020, Walkowicz 2018) as well as case study projects (Mullins 2020) indicate that where hit and miss masonry has been designed, elaborate close spaced support systems and/or reinforcement placed in bed joints or hidden in vertical cores has been required to ensure that the masonry is able to resist design out of plane loading scenarios. For commercial projects of significant scale such as the façade system shown in Figure 14c, or for highly unusual hit and miss patterns, this is justified and necessary. However, for more regular hit and miss patterns (Figure 12c, Figure 14a, b) the development of relatively simple code based strength design equations seems like a realistic, and hopefully achievable, research goal.

Ortlepp and Schmidt (2014) performed a pilot numerical study to assess the loadbearing capability for hit and miss masonry walls. They concluded that the absence of contact along part of the unit length results in stress concentrations which influence the loadbearing behaviour meaning that it is not possible to transfer the principles for compression design of solid masonry to hit and miss masonry walls simply by reducing the bedded area used in design. This conclusion was supported by Edgell (2020) who found via testing that the compressive capacity for hit and miss walls was lower than predicted by reducing the capacity of solid walls based on the reduction in bedded area. Under in-plane shear loading, Ortlepp and Schmidt (2014) demonstrated that the individual units are subjected to flexural stresses, again rendering inappropriate, the use of existing design approaches for solid masonry.

Although hit and miss masonry has been used in loadbearing walls in some instances, it is most commonly used in non-loadbearing applications. In such applications, it is the ability of the masonry to resist out-of-plane lateral loading which is most likely to limit its structural capacity. Although Edgell (2020) was able to show via pilot testing that a reduction in wind pressure due to the holes in the masonry may be justified, pressure does still exist and robustness and earthquake loading require that hit and miss walls must resist out-of-plane pressures.

This prompted research by the author and his colleagues at The University of Newcastle which is designed to better understand the behaviour of unreinforced hit and miss masonry walls when subjected to lateral out-of-plane pressure loading.

The research to date which is detailed in Masia et al. (2017, 2018 and 2020) involved an experimental study to assess the behaviour of unreinforced hit and miss masonry walls subjected to out-of-plane pressure loading using an airbag system. Twenty one single leaf hit and miss masonry walls, of varying aspect (height: length) ratios, were constructed using extruded clay bricks (230 mm long x 110 mm wide x 76 mm high) and 1:1:6 (cement: lime: sand) mortar. Three walls were tested in one way vertical bending. Twelve walls, with varying overlap between units in adjacent courses, were tested in one way horizontal bending. Six walls, with two different aspect ratios (height:length), were tested in two way bending. The study considered the load versus deformation behaviour and the observed failure modes. The AS3700 provisions for solid masonry were used to predict the panel strengths and an assessment of the suitability of the provisions for the design of hit and miss masonry was made.

The study found that panels subjected to one way vertical bending failed in a non-ductile mode via bed joint cracking which occurred suddenly at the peak load along a single course close to panel mid-height (Figure 15a).

Panels subjected to one way horizontal bending also displayed non-ductile failure modes with no observable damage prior to a sudden failure surface developing at peak load. The failure surface varied between specimens from stepped failure exclusively through the mortar joints to line failure (fracture of the units) and combinations of stepped and line failure (Figure 15b). As the overlap between units in adjacent courses was increased, the panel capacities increased.

For the aspect ratio 1.0 panels subjected to two way bending, non-ductile failure modes were observed with no detectable damage prior to a sudden failure surface developing at peak load. The failure patterns included stepped diagonal cracking extending from the panel corners, joined by a vertical stepped crack through the middle of the panel (Figure 15c). For the aspect ratio 0.5 panels subjected to two way bending, horizontal or vertical cracking initiated at loads lower than peak, prior to the full failure surface, which included diagonal cracking, developing suddenly at peak pressure (Figure 15d).

The specimens supported in horizontal and two way bending configurations resisted much larger pressures than those subjected to one way vertical bending. This difference in strengths (also observed in fully bonded masonry) highlights the benefit in providing vertical lines of support to help engage horizontal and diagonal bending action.

The failure modes observed for each of the support configurations are similar to those observed in fully bonded unreinforced masonry. This should allow existing design methodology for one and two way bending of fully bonded masonry to be adapted to reflect the behaviour observed in hit and miss masonry walls.

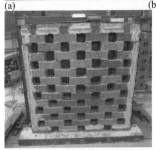

(a) (b)

(c)

(d)

Figure 15. Observed failure modes for hit and miss masonry panel tests (a) One way vertical bending (b) One way horizontal bending (after first rotating panels through 90°), (c) Two-way bending (aspect ratio 1.0), and (d) Two-way bending (aspect ratio 0.5).

In Masia et al. (2017, 2018) the author compares strength predictions made using AS3700 (Standards Australia 2018) code equations for solid masonry with the experimental panel strengths, despite hit and miss masonry falling outside the scope of the code equations for one way horizontal and two way bending. It was found for the masonry tested in the study, that the code provisions for one vertical and one way horizontal bending return safe designs. However, for two way bending the strength predictions made by applying directly the AS3700 provisions (intended for solid masonry) were non-conservative indicating that it is not appropriate to apply the provisions, in their present form, to hit and miss masonry. In particular, the AS3700 expression for torsional section modulus of the diagonal crack lines requires modification to account for the gaps along the mortar bed joints in hit and miss masonry.

Reliable methods to predict the horizontal and diagonal bending capacity of hit and miss masonry walls are needed in order for designers to best capitalize on the considerably higher strengths when the masonry is supported in this way compared to simple vertical bending.

The work is ongoing and includes a parametric study using finite element analyses to assess the influence on failure behaviour and load capacity of the presence of the gaps in hit and miss masonry.

3 CONCLUSION

Masonry holds a special place in Australian architecture and construction, being used extensively in many building forms since the very beginning of European settlement 230 years ago. Although its use in loadbearing applications has reduced in recent decades, giving way to alternative structural systems, its application in building façades remains strong. In recent years architects have pushed the limits of what is possible in façade design and masonry has played a major role, encouraged strongly via the annual Think Brick Awards.

In older existing masonry cavity and veneer wall systems, spatial and temporal variability of material properties, including aspects such as corrosion of steel wall ties, has threatened the structural reliability of façades and collapses have occurred, sometimes under loading scenarios far below normal design level events.

Ensuring the structural safety of new and innovative masonry façade systems, as well as those in older existing buildings has created challenges for engineering research and design. Research aimed at addressing such challenges, being conducted by the author and his colleagues at The University of Newcastle has been described in this paper. In the view of this author structural engineers are in an exciting game of catchup and many challenges remain.

ACKNOWLEDGEMENTS

The author gratefully acknowledges the financial support provided by the Australian Research Council via Discovery Project DP180102334 and Linkage Project LP0669538, as well as the cash and in-kind support of Think Brick Australia, and Brickworks Ltd.

The research was performed in the Civil Engineering laboratories of The School of Engineering at The University of Newcastle. The assistance of the laboratory staff is gratefully acknowledged.

Finally, the research summarized in this paper is the result of the collaboration over several years of a large team of staff and research students, the contributions of whom are gratefully acknowledged.

REFERENCES

Beattie, G. & Thurston, S. 2011. Twenty years of improvement in the seismic performance of masonry veneer construction, *Australian Journal of Structural Engineering*, 11(3): 225–229.

Brick Development Association 2019. Hit and miss brickwork, BDA Design Note 02, February, 2019.

Canadian Standards Association. 2004. *CSA A371-04 2004: Masonry construction for buildings*. Canadian Standards Association.

Collis, K. 2020. Personal communication between author and K. Collis, Northrop Consulting Engineers, regarding design of hit and miss façade system, Newcastle, Australia.

Curren, B. 2020. Personal communication between author and B. Curren, MPC Consulting Engineers, regarding design of hit and miss façade system, Maitland, Australia.

Edgell, G.J. 2020. The design of brickwork facades. *Proc. 17th International Brick and Block Masonry Conference*, Kraków, Poland, 5-8 July, 2020.

Heffler, L.M., Stewart, M.G., Masia, M.J. & Correa, M.R.S. 2008. Statistical analysis and spatial correlation of flexural bond strength for masonry walls, *Masonry International*, 21(2): 59–70.

Ingham, J. & Griffith, M. 2011. Performance of unreinforced masonry buildings during the 2010 Darfield (Christchurch, NZ) earthquake, *Australian Journal of Structural Engineering*, 11(3): 1–18.

Jardim do Nascimento, B., Chaves, I.A., Masia, M.J. & Melchers, R.E. 2017. Masonry brick veneer and cavity brick wall tie corrosion. *Proc. ACA Conference: Corrosion and Prevention 2017*, Sydney, Australia – Australasian Corrosion Association Inc., 2017.

Jardim do Nascimento, B., Chaves, I.A., Masia, M.J. & Melchers, R.E. 2018. Long term atmospheric corrosion of mortar embedded wall ties. *Proc. ACA Conference: Corrosion & Prevention 2018*, Adelaide, Australia – Australasian Corrosion Association Inc., 2018.

Jardim do Nascimento, B., Chaves, I.A., Masia, M.J. & Melchers, R.E. 2019. Mortar embedded wall tie corrosion in natural and artificial environments. *Proc. ACA Conference: Corrosion & Prevention 2019*, Melbourne, Australia – Australasian Corrosion Association Inc., 24-27 November, 2019.

Lawrence, S.J. 2007. The use of characteristic strengths in masonry design. *Australian Journal of Structural Engineering*. 7(3): 225–237.

Li, J., Masia, M.J., Stewart, M.G. & Lawrence, S.J. 2014. Spatial variability and stochastic strength prediction of unreinforced masonry walls in vertical bending, *Engineering Structures*, 59(1): 787–797.

Li, J., Masia, M.J., Stewart, M.G. & Lawrence, S.J. 2016a. Spatial correlation of material properties and structural strength of masonry in horizontal bending, *ASCE Journal of Structural Engineering*, 142(11).

Li, J., Masia, M.J. & Stewart. M.G. 2016b. Stochastic spatial modelling of material properties and structural strength of unreinforced masonry in two-way bending, *Structure and Infrastructure Engineering*, 13(6): 683–695.

Masia, M.J. Simundic, G. & Page, A.W. 2014. The flexural strength of stack bonded masonry: Stage 1 – A preliminary study. *Proc. 9th International Masonry Conference*, Guimaraes, Portugal, 2014.

Masia, M.J. Simundic, G. & Page, A.W. 2015. The flexural strength of stack bonded masonry – an ongoing study. *Proc. 12th North American Masonry Conference*, Denver, Colorado, 17-20 May 2015.

Masia, M.J. Simundic, G. & Page, A.W. 2016. Flexural strength of stack bonded masonry in one way horizontal bending: influence of bed joint reinforcement. *Proc. 16th International Brick and Block Masonry Conference*, Padova, Italy, 26-30 June 2016.

Masia, M.J. Simundic, G. & Page, A.W. 2017. Flexural strength of unreinforced lattice masonry walls subjected to lateral out-of-plane loading. *Proc. 13th Canadian Masonry Symposium*, Halifax, Canada, 4-7 June 2017.

Masia, M.J. Simundic, G. & Page, A.W. 2018. Behaviour of unreinforced lattice masonry walls under one way horizontal and two way out-of-plane bending. In (eds.) G. Milani, A. Taliercio & S. Garrity, *Proc. 10th International Masonry Conference*, Milan, Italy, July 9-11, 2018.

Masia, M.J. Simundic, G. & Page, A.W. 2020. Experimental testing of unreinforced lattice masonry walls subjected to out-of-plane pressure loading. *Proc. 17th International Brick and Block Masonry Conference*, Kraków, Poland, 5-8 July, 2020.

Masonry Standards Joint Committee. 2013. *TMS-0402-11 2013: Building code requirements and specification for masonry structures*. Masonry Standards Joint Committee (MSJC – TMS, ACI & SEI).

Moon, L., Dizhur, D., Senaldi, I., Derakhshan, H., Griffith, M., Magenes, G. & Ingham, J. 2014. The demise of the URM building stock in Christchurch during the 2010/2011 Canterbury earthquake sequence. *Earthquake Spectra*, 30(1): 253–276.

Muhit, I.B., Masia, M.J. & Stewart, M.G. 2019. Nonlinear finite element analysis of unreinforced masonry veneer wall systems under out-of-plane loading. *Proc. 13th North American Masonry Conference (NAMC 2019)*, Salt Lake City, USA, 16-19 June 2019.

Muhit, I.B., Stewart, M.G. & Masia, M.J. 2020. Experimental evaluation and probabilistic analysis of the masonry veneer wall tie characteristics. *Proc. 17th International Brick and Block Masonry Conference (IB2MaC 2020)*, Krakow, Poland, 5-8 July, 2020.

Mullins, P. 2020. *Infrastructure Thought Leaders Series: Façade Design – Engineering the Impossible*, Engineers Australia in association with Brickworks Building Products, webinar, 17 March 2020.

Ortlepp, S. & Schmidt, F. 2014. Perforated masonry – light weight construction. *Proc. 9th International Masonry Conference*, Guimarães, Portugal, 7-9 July, 2014.

Page, A.W. 1992. The design, detailing and construction of masonry – The lessons from the Newcastle Earthquake. *Civil Engineering Transactions*, The Institution of Engineers, Australia, Vol. CE 34, No. 4, December 1992, pp. 343–353.

Page, A.W., Kleeman, P.W., Stewart, M.G. & Melchers, R.E. 1990. Structural aspects of the Newcastle Earthquake. *Proc. Second National Structural Engineering Conference 1990*, Institution of Engineers, Australia, Adelaide, 1990.

Standards Australia 2018. *AS 3700: 2018: Masonry structures*. Standards Association of Australia, Sydney, 2018.

Standards Australia/Standards New Zealand 2000. *AS/NZS 2699.1 (2000), Built-in Components in Masonry Construction, Part I: Wall Ties*, Standards Australia, Sydney.

Think Brick Australia 2019. Hit and miss brick screen, Fact Sheet No. 1, 2019.

Trimble, B.E. 2013. Design of unique landscape walls and their use in building facades. *Proc. 12th Canadian Masonry Symposium*, Vancouver, Canada, 2-5 June, 2013.

Turley, J., Er, M. & Morkaya, K. 2014. Project case study: Structural analysis, design and laboratory testing of a complex masonry façade. *Proc. Australasian Structural Engineering Conference*, Auckland, New Zealand, 2014.

Walkowicz, S.W. 2018. Personal communication between author and S.W. Walkowicz, Walkowicz Consulting Engineers, regarding design of hit and miss façade system, USA.

Brick and Block Masonry - From Historical to Sustainable Masonry –
Kubica, Kwiecień & Bednarz (eds)
© 2020 Taylor & Francis Group, London, ISBN 978-0-367-56586-2

Recent advances of Italian recommendations and standards on structural safety of existing masonry structures

C. Modena

Emeritus Professors, Department of Environmental, Civil and Architectural Engineering, Padova, Italy

ABSTRACT: The continued use of the huge stock of existing buildings, that makes the everyday life possible in a country like Italy, requires that not only continuum research efforts are made to set up reliable and specific methodologies and technologies to conduct controls and investigations regarding their structural efficiency and to execute adequate maintenance plans and, when necessary, repair and strengthening interventions, but also indispensable adjustments of the principles of the structural safety and of the ways how it is quantified. Clear traces of what above can be found in the recent updating of the Italian national structural codes where particular importance is assumed by some innovations regarding structural safety issues. The very critical issue is that the general recognized context of limited resources really available to make systematic interventions aimed to improve the structural safety of existing structures obliges to find a right balance between "structural safety" and "sustainability", as clearly stated in the very important international standard ISO 13822 Bases for design of Structures – Assessment of existing structures. In this respect, major results are obtained on one side limiting as much as possible the effects of "overdesign" of structural interventions that can be connected to the use of consolidated, necessarily "simplified"; procedures in use for verifying the structural safety of new structures and, on the other side, maximizing the effects, measured in terms of "risks reductions", connected to the use of the effectively available resources to execute repair and strengthening interventions. From the designer's operative point of view a particular importance is assumed in this context by the so called "knowledge phase", i.e. a deep, reliable knowledge of which are the real capacities of an existing structure to resist any type of possible action.

1 INTRODUCTION

The most recent updates of the Italian national structural codes and standards mainly deals with the management of structural safety of existing constructions thus progressively introducing into the everyday praxis of all those having any possible role in such field – e.g. public authorities, professionals, owners and users – necessary, even profound changes in the way how they are required to operate compared to the way how structural safety is normally dealt with when new constructions are to be built.

The scope of such changes, and the reasons why they are made "necessary", are more and more explicitly evidenced in official documents, also at international level, and is very well synthetized in a fundamental standard, produced by the ISO TC 98 (Bases for design of structures), i.e. the *ISO 13822 –*

Bases for the design of structures, evaluation of existing structures, where just in the introduction is stated that: "*The continued use of existing structures is of great importance because the built environment is a huge economic and political asset, growing larger every year. The assessment of existing structures is now a major engineering task. The structural engineer is increasingly called upon to devise ways for*

extending the life of structures whilst observing tight cost constraints.

The establishment of principles for the assessment of existing structures is needed because it is based on an approach that is substantially different from design of new structures, and requires knowledge beyond the scope of design codes.

*This document is intended not only as a standard of principals and procedures for the assessment of existing structures but also as a guide for use by structural engineers and clients. **Engineers may apply specific methods for assessment in order to save structures and to reduce a client's expenditure.** The ultimate goal is to limit construction intervention to a strict minimum, a goal that is clearly in agreement with the principles of sustainable development.*"

*The basis for the reliability assessment is contained in the performance requirements for safety and serviceability of ISO2394. **Economic, social, and sustainability considerations, however, result in a greater differentiation in structural reliability for the assessment of existing structures than for the design of new structures.***

Such considerations find evident justification in what the common practice tells us, i.e. that structural verifications of existing constructions conducted in accordance with modern structural codes most frequently (if not always) demonstrate that their safety level results in being lower than that would be required for new ones (if built in the same place with the same function), and that the costs of the interventions that would be required to completely eliminate this gap are "too high", e.g. "socially unsustainable".

An afterthought, and a concern, on this very critical issue – the "costs of structural safety" – is justified considering that it is not clear, and even ambiguous, the way how minimum safety levels have been prescribed in modern codes without any apparent consideration of their cost, being indeed evident that higher costs are connected to higher safety levels. Such an ambiguity is strictly connected to the origin itself of the concept of structural safety, born in a "deterministic world", that didn't allow to evaluate the "consequences" of the choice of a prescribed "safety level". It however persisted even after appropriate "probabilistic" approaches to structural safety were introduced, when it became clear that the scope of fixing prescribed safety levels was not to avoid structural damages-collapses, but to limit the probability of their occurrence, and that such probability could be reduced as much as one wants provided that the corresponding increasing costs are "socially sustainable".

The building codes dealing with structural design of new buildings have evolved over the last century, prescribing from time to time "minimum safety levels" which, although defined in an increasingly "sophisticated" manner, have in fact been "quantified" certainly taking into account the evolution of scientific knowledge and of the construction technologies. An example of that is the progressive evolution (actually reduction) of the "safety coefficients" used first in a deterministic way and then in a semi-probabilistic one.

Those codes however by no doubt evolved also, in a decisive way, based on very empirical considerations, namely the fact that their practical, continued use always demonstrated that the costs connected to them have been considered "socially acceptable" compared to the obtained results in terms of protection of economical values and of human lives .

What compels us to "do the math" again with the "costs of safety" when defining the rules for the design of structural interventions on existing constructions are the following considerations, strongly supported by data coming from practical experiences in our country, namely that:

- on one hand, the extremely high costs of interventions needed to increase the safety level of an existing construction, getting moreover higher and higher the higher is the increase to be attained, compared to the costs of similar interventions on a new construction;
- on the other hand, the resources available to carry out structural interventions on existing buildings, both privately and publicly, are limited. This because, also considering the high levels of vulnerability of existing buildings against seismic actions, the economic resources are not enough, at least those that can be made available in a reasonable time, to ensure to the enormous built heritage stock the same level of safety "formalistically" quantified in the same way for new buildings.
- It is then necessary to re-consider the entire process that leads to the definition of a structural safety verification method, first of all in the way how the real enormously complex behavior of the many different typologies of structures in all their possible operating conditions is "necessarily" simplified, carefully considering which are the consequences of all the choices are possible in this field on all the aspects of structural safety itself, i.e. on the capacity to furnish acceptable levels of protection of properties and of human lives and on the corresponding costs.

But also the objectives themselves of the structural design are to be re-defined an explicit and conceptually correct way, referring to the different possible "risks" to which the existing structures are subject with the aim of using the "limited resources", actually available, in order to maximize obtainable results measured in terms of risk mitigation for the whole community, first of all, related to the human losses.

The above mentioned changes introduced in the Italian standards are actually the result of a very long and complex process, particularly critical for a country like Italy characterized by very vulnerable historical centers – due to their original construction characteristics, to the damages suffered in the past and to not always adequate interventions executed to repair them, and finally to unavoidable deterioration processes, frequently emphasized by the lack of appropriate maintenance - that are at the same time subjected to high (especially seismic) hazards.

Such process actually started in Italy since 1986, when for the first time the national structural codes system made it possible, and even recommended for historical constructions, to execute repair/strengthening interventions on an existing construction (when needed according to the above mentioned procedure) such that its structural safety level is "improved", without necessarily attaining its "complete retrofitting", i.e., as above said, the same safety level that "would be required for a new one built in the same place with the same function".

The "break" that was in such way introduced in the principles themselves of the structural safety as they was formalistically developed for new constructions in the structural codes starting from the beginning of the 20th century, progressively lead to a more and more refined definition of the above mentioned specific approaches to structural safety in the field of existing constructions.

Such approaches found more and more "theoretical" support in the advancements of specifically financed, by national agencies, scientific researches, especially those allowing for the reduction of the uncertainties that characterize all the phases of the structural analyses, i.e. criteria and methods for the mechanical characterization and for the structural modelling of various possible types of existing materials and construction typologies.

The main innovations introduced with the aforementioned updates of the Italian technical regulations concerning the structural safety of existing buildings are clearly inspired by these considerations, especially in its more general aspects but also in specific aspects relating to individual structural typologies, among which historical buildings and bridges are particularly meaningful.

2 THE ITALIAN FRAMEWORK OF STRUCTURAL STANDARDS

Before analyzing and discussing about the novelties introduced into the Italian standards, and in order to clarify for the readers the references made in the text to the structural codes, an overview of the general framework of such codes is herein reported. The name of codes, all in Italian as the content, are literally translated here just to improve the comprehension.

In 2008, the Ministry of Infrastructures and Transport, which is responsible for this field, issued the first version of structural standard for constructions that made it compulsory the use of the limit state approach of Eurocodes:

[1] N.T.C. 2008 –*Technical standard for constructions* D.M. 14/01/2008 (Ministry of Infrastructures and Transport, 2008).

After one year, the Ministry approved a document to extend and better explain the concepts of the codes:

[2] Circolare n. 617 02/02/2019. *Instructions for the applications of "Technical standard for constructions"* Ministry of Infrastructures and Transport (2009).

The same approach has been adopted in the update of the 2018 of the code (N.T.C.) and of the explanatory document:

[3] N.T.C. 2018 – *New Technical standard for constructions* D.M. 17/01/2018 (Ministry of Infrastructures and Transport, 2018);
[4] Circolare n. 7/2009 *"Instructions for the application of the "New Technical standard for constructions* (High Council of Public Works, 2019).

Other important documents of the Ministry of Infrastructures and Transport and its High Consultive Council are the laws called "Sismabonus"

providing measures to boost the voluntary interventions for obtaining seismic risks mitigations:

[5] Decree of Ministry 58/2017 *SismaBonus - Guidelines for the classification of the seismic risk of constructions and modality of attestation by practitioners of the interventions effectiveness* (Ministry of Infrastructures and Transport, 2017a);
[6] Decree of Ministry 65/2017 *SismaBonus - Guidelines for the classification of the seismic risk of constructions and relative annexes. Modification of article 3 of DM58/2017* (Ministry of Infrastructures and Transport, 2017b);

Finally a guideline will be shortly available and released by the High Council of Public Works:

[7] *Guidelines on the risk classification, verification and monitoring of existing bridges* (High Council of Public Works, 2020)

In the following to improve clarity, each laws will be defined by the listed number.

3 STANDARD UPDATES OF GENERAL ASPECTS OF STRUCTURAL SAFETY

What essentially characterizes the general approach to structural safety in the current codes and standards is that:

- their pursued and declared aim can be expressed only in probabilistic terms, whatever is the conventional way it is formalized, i.e. the positive outcomes of the verifications indicate that the "probability" of "unsatisfactory" structural responses is not greater then fixed values, typically the "worst" possible response, i.e. the collapse, should have a probability of occurrence in the order of $1/1.000.000$ in a fixed period of time, conventionally named "nominal life" of the structure under consideration, normally assumed equal to 50 years, a very short period indeed when considering the real duration in use of any type of structures (not to mention the historical constructions);
- all the numerical parameters used to achieve this aim, from the "nominal life" of the structure to the "return period" of the considered variable actions (or, what is the same, the probability of exceeding of an action in that return period), does not have any peculiar, individual and autonomous meaning other than theoretically defining as a whole and in the general framework of the safety verification procedure that is proposed in the code itself (including for example the type of control of the construction methods and materials, the type of analyses and models to be used, etc..) the fixed values of probability of occurrence of defined "unsatisfactory" structural responses (the so called "limit states").

Acceptable in the field of new structures, this approach becomes clearly not usable in the case of existing structures.

On one side, in fact, becomes almost impossible, even in the most favorable conditions where extensive on-site test campaigns are executed, to have enough data on the mechanical properties of the employed materials allowing for statistically meaningful evaluations of basic design parameters.

On the other side it is well know how inadequate can be even the most sophisticated structural models normally in use, to reliably represent the real behavior of existing constructions, especially when dealing with their "ultimate limit states", i.e. when the structural performances to which the safeguard of human lives is entrusted. It is then in fact that, particularly in the case of masonry made constructions, the contribution of some not reliably identifiable and quantifiable resistance mechanisms that, for such reasons, is normally not taken into account, may instead effectively play decisive roles.

What happens in such conditions is that on one side the values of fundamental mechanical parameters are not the result of precise calculations but they are instead more or less reliable "estimates", largely bases on "qualitative" judgments, and on the other side that the structural models become too simplified representations of very complex and difficultly identifiable combinations and interactions of the various possible resistance mechanisms.

It becomes crucial at this point the way how uncertainties and simplifications are dealt with in the current approaches to structural safety, i.e. by making conservative hypotheses, both when defining the "design" values of mechanical parameters and when constructing the structural models. Such hypotheses are furthermore, and unavoidably in such context, made increasingly conservative the greater are the uncertainties and the simplifications, what actually happens in the case of existing, especially masonry made constructions compared to the case of new constructions made of reinforced concrete and steel.

The not surprising, really unavoidable consequence of all what above is that the most frequent result of structural verifications of existing constructions – really almost the only one obtained in the case of historical masonry made constructions – is that the formal verifications proposed in current structural codes are not satisfied.

There are then enough reasons to understand and to justify why in the above mentioned international standard ISO 13822 the concept of "Plausibility check" has been introduced (§ 7.4), thus in fact defined:

"The conclusion from the assessment shall withstand a plausibility check. In particular, discrepancies between the results of structural analysis (e.g. insufficient safety) and the real structural condition (e.g. no sign of distress or failure, satisfactory structural performance) shall be explained.

NOTE Many engineering models are conservative and cannot always be used directly to explain an actual situation. See also clause 8."

In conclusion, the over-designed structural solutions that are the implicit and normally accepted consequences of the approaches to structural safety currently in use when designing new constructions are no more acceptable when designing structural interventions on existing constructions, considering the above mentioned general context of limited available resources to do that, as in fact:

- besides being, as previously stated, the costs of interventions required to obtain a given increase of the safety level of an existing construction much higher than the increase of costs that would be connected to the same increase of safety of a new construction;
- the effects in terms of "over design" of the above-mentioned uncertainties and simplifications are furthermore much higher in the case of existing constructions compared to the case of new constructions, being uncertainties and simplifications much higher in the first case then in the second one.

No matter connected to all what above said are some of the most significant innovations recently introduced into the national structural codes, considering in particular that, based on them:

- it is explicitly accepted and allowed the use of existing constructions whose structural safety level is lower than that prescribed for new constructions §8.2 [4], in case adopting, under the responsibility of the owners and users, limitations to the use, including for example the change of use;
- when the "complete seismic retrofitting" of an existing construction, i.e. the same "seismic safety conditions" of a new one, is required, such goal is to be considered fully achieved, provided that structural interventions are designed respecting some prescribed conditions, when a resistance is attained against seismic actions equal to 80% of the resistance would be required of a new construction (§ 8.4.3 [3]);

- verifications are not required regarding "serviceability limit states" (for all structures except strategic ones §8.3 [3]) i.e. only the protection of human lives is to be mandatorily taken into consideration when designing structural interventions on existing constructions (clearly in light of optimization criteria of limited resources);
- besides explicitly recognizing the limits of currently used structural models in representing the real structural behavior and real structural performances in the various possible service conditions of existing constructions, the role of the so called "knowledge phase" (based on a variety of activities, starting form archive-historical

investigations to accurate and extended survey and on-site investigations, besides on-site and laboratory tests) is strongly emphasized in the verification procedure, being evidently its scope that of allowing for systematic and continuous "plausibility checks". What it is said in fact about this is that such phase really characterizes and differentiates the structural verification of an existing construction compared to a new construction §8.1.1 [4]. *The principal difference between new and existing constructions is represented, in terms of design procedure, front the peculiarity and specific problems connected to their knowledge*) and at the same time that it is essential and integrated part of the verification process, not a simple execution according standard procedure of tests on mechanical properties of the employed materials to be introduced into structural models § 8.5.3 [4]. *The adequate knowledge of the construction is fundamental premise and essential phase for the comprehension of any possible single critical local structural performance and of the global structural behavior; the reliability of the resets of the structural analyses is then strictly connected to the level of knowledge. It is to be emphasized that the phases of the knowledge and of the structural analysis are not sequential but strictly connected"*).

A particularly important impact is expected to be produced on the structural design procedures of existing constructions by its, quite newly defined (§1.1 [4] the distinction of this design phase between new and existing constructions "*..was not clearly evidenced in the previous versions on the structural codes...*"), "knowledge phase", aimed to fill what past experiences have demonstrated to be the major shortcoming of the current design procedure in ensuring the expected safety levels of existing constructions, e.g. the attainment of information and data allowing to simulate their real structural behavior (*..§1.1 [4] "...it is precisely the lack of knowledge of the behavior of existing constructions the major cause of frequently unsatisfactory results produced by the executed interventions"*).

The big issue is the substantial differences between the types of information and data required by current codes to design new constructions and those that are needed firstly to understand and then to reliably reproduce via any possible theoretical model the real structural performances of an existing constructions in all the possible operating conditions.

Very significant are in this sense some sentences written in the codes under consideration, in particular the following ones that can be found in §1.1 of [4].

When referring to new constructions it is in fact said that:

- "*…. It is easy to quantify the number and the distribution of the samples of the employed materials to be subjected to laboratory tests..*";
- the required knowledge is connected to "*…the mechanical properties of the materials really employed paying little attention to the way how different components are really interacting among themselves as the way how interact is prescribed by the project and ensured by the construction procedures, unless gross errors are made both in the design and in the construction phases*";
- "*..the intrinsic uncertainties of the structural model are merged with the uncertainties of the actions…*" and in this way such uncertainties "*…disappear from the awareness (of the designer)*".

When referring to existing constructions it is on the other way said that:

- It is "*..extremely difficult to define…*" design parameters "*..connected to the real structural behavior of the construction. Even more complex is to identify the typologies of structural components capable to condition, in a positive or negative way, such behavior, especially considering that they can combine and interact in a variety of ways. It is not possible in fact to directly compare the various possible structural components of an existing construction to the technical-construction characteristics that structural components of a new construction must by law fulfil in order to be automatically accepted for use*";
- the knowledge has to pay "*…little attention to the mechanical characteristics of the employed materials……and a particular attention to the way how the different structural components .. interact with each other*";
- in the case of new constructions the model actually doesn't control the uncertainties "*..while on the contrary, especially in the case of exiting constructions, it (*the model*) is their main container…*" and consequently when operating according to the current design approaches, i.e. overcoming the uncertainties by simply reducing, via a "safety coefficient", the strength characteristics of the employed materials "*…the importance is underestimated of the investigations regarding the construction structural details, the connections between the various structural components, the way how they interact and fail…*" while such aspects actually play a "*...fundamental role in order to identify possible structural deficiencies…*" and are "*..indispensable in order to identify the structural models capable to represent both the global response and any possible local failure mechanism*".

Having in such a way dealt with the "uncertainties" issue, further specific indications are introduced on how the design objectives are to be re-defined in the general terms of optimizing the use of "limited resources" in order to obtain the maximum possible "risks mitigation" as mentioned above.

In such context a particular attention is paid to seismic risk, being it very high in Italy but at the same time connected to a "rare" action .

Taking this into account, the above mentioned optimization criteria clearly are at the basis of the different decisions are to be taken when structural verifications are not satisfied against normal service, non-seismic, loading conditions and against seismic actions (§ 8.3 [4]).

In the first case in fact "...it is necessary to adopt appropriate measures, for example limiting the maximum allowed service loads, limiting the uses and/or executing interventions capable to increase the structural safety, thus allowing the use of the construction under the safety conditions prescribed by the code. The interventions needed to eliminate the most important vulnerabilities can be partial and/or temporary, while waiting for their completion during subsequent, more extensive interventions, capable to increase/adequate the entire construction and/or parts of it."

While in the second case "considering the aleatory characteristics of the actions, ... the conditions for using, the necessity and consequent planning of interventions on the constructions are to be established taking into consideration a variety of factors, such as: how serious are the structural deficiencies, their consequences also in terms of public safety, the available resources, etc.".

The same origin has a seemingly mere formal novelty introduced into the most recent version of the structural code when presenting the classification of the type of structural interventions that can executed on existing constructions (§ 8.4 [3] and §8.4 [4]).

Such classification was for the first time introduced in the version of the national code published in 2008, and was presented in the following order:

- first, interventions capable to make an existing construction as safe as a new one that would be constructed in the same place for the same use, thus requiring extensive and global seismic verifications;
- second, interventions capable to "improve", without any more specification, the original structural safety of an existing construction,;
- third, "repairing" and "local" interventions, i.e. interventions that don't necessarily increase the "global" safety level of an existing construction – and for such reason not requiring the extensive calculations that are required in the first two cases- being asked not to decrease it while solving local deficiencies without changing the global structural behavior of the construction itself.

In the 2018 version of the codes ([3] and [4]) the presentation was exactly in the opposite order, and was in such way justified: "The importance that local structural deficiencies take on in the case of existing construction in terms of damages caused to people and to properties has led, among other things, to pay greater attention to local interventions and to interventions aimed to improve the structural safety. Such greater attention is demonstrated also by the different order of presentation..." of the types of intervention.

Even more explicitly is then said (§ 8.1.1 [4]) that: "The different way how (structural safety) is dealt with in the case of new and existing constructions.... is justified by the will of obtaining, in conditions of "limited resources" available, the maximum reduction of the average seismic risk. In the proposed way in fact it is possible to intervene, by using the same resources, on a greater number of exiting constructions, greater and greater of the number in which would be possible to intervene by attaining the safety level of new constructions. The advantages that in such way derive for the entire community in terms of reduction of casualties, injured and damages are evident. In particular it is possible to eliminate mostly local structural deficiencies capable to activate failure mechanisms having even relevant effects by intervening locally with modest costs. Then the programmed elimination of even modest structural deficiencies can be a reasonable and economically sustainable intervention strategy in order to obtain a diffused risk reduction."

The fact that better results are considered to be obtained in order to "mitigate "the seismic risk to the huge stock of exiting constructions by executing, through the use of a given amount of available resources, "modest/limited" interventions but at large scale rather than "exhaustive" ones on few constructions, actually represents what has been learnt by studying the effects, frequently destructive in Italian historical centers, of the past earthquakes.

Further steps have been recently done in the same sense by the Italian government by approving a law [5] and [6] (Ministry of Infrastructures and Transport 2017a & 2017b) that boosts voluntary initiatives of the owners of existing buildings to execute on them interventions capable of obtaining prefixed, conventionally defined, reductions of the levels of "seismic risk" to which they are subjected.

The incentive are very significant for people owning a typical apartment of a residential building, as it consists of tax reductions ranging from 70% to 80% of the cost of the interventions.

What makes such government initiative perfectly in line with the above mentioned approach to the reduction of seismic risk at a general, national level, is that the incentive is obtained even when interventions are made on buildings having a very low seismic safety level and very low is also the seismic safety level attained after the interventions have been executed: e.g. an incentive equal to 70% of the cost of interventions is ensured even when they are executed on a building characterized by a seismic safety level equal to 15% of the safety level required for a new building, being furthermore considered enough the attainment of an increase of such level to 25%.

4 PARTICULAR ASPECTS REGARDING SPECIFIC STRUCTURAL TYPOLOGIES

4.1 Historical buildings

Much has been done and is still being done by the scientific community and by the official standardization bodies on the very critical issue of the seismic safety of the architectural heritage. The ways how it is evaluated and improved continue in fact to be the subject of intense debates under the pressure of the effects of the frequent earthquakes that occur on the national territory: every time in fact the results until then obtained are to be compared, in the post-earthquake emergency experiences, with almost always destructive effects on the historical centers.

One fundamental and unquestionable conclusion has been derived from the above experiences, namely that even considering the problem of conservation of architectural heritage from the sole structural safety point of view, i.e. even ignoring its (discussed and controversial) implications on the conservation of historical and artistic values, every idea and possibility is now to be considered gone of obtaining durable and reliable results by means of technological solutions aimed to substantially modify original mechanical properties of materials and structural behaviors of historical constructions.

This is because a historical construction, whatever the way it has been "strengthened", exhibits local mechanical behaviors and global structural responses that are in any case determined by in series resistance mechanism: the result is that it behaves like a "chain", whose resistance actually is that of its weakest ring, and this will be in any one of the "original rings", being evidently impossible to intervene on all them, whatever is the employed technology, being indeed even possible to damage them by improperly executed interventions.

Therefore, the maximum achievable goal when designing and executing interventions aimed to increase the structural safety of a historical construction is to do such that the "weakest ring" that inevitably remains after the interventions - i.e. the original resistance mechanism on which it was not possible to execute improvement interventions and on which then the possibility of structural collapses depends - will be the most efficient possible among the "original" resistance mechanism (the "rings"), being its "capacity" actually and in any case an insurmountable limit to the real possibility of increasing the safety of the considered building.

From all what above it comes out that:

- the approach to structural safety of historical constructions based on the concept of "improvement" is to be intended as first of all the identification of all the possible (local and global) failure mechanisms and then as the execution of interventions capable to "respect" and "valorize" their original construction characteristics, i.e. making them as much as possible efficient by using, where possible and strictly necessary, minimum interventions of minimal impact, then necessarily "local" and very well "targeted";
- such a normative provision is not simply and "expedient" to prevent too invasive interventions, such as to compromise the fulfillment of fundamental conservation criteria, it really is the most appropriate way to operate from the "structural mechanics" point of view.

Very important are then some specifications given on this very delicate issue where it is said § 1.1 [4] that:

"The substantial unitarity of the design process, provided that its knowledge phase is that actually necessary as previously highlighted, is not compromised by the historical, artistic or environmental constraints that often characterize the existing constructions.

The historical buildings, came to us through centuries as the result of long and complex processes of transformation, adaptation, damage and re-pair /reconstruction (also after having been stricken by earthquakes whose intensity was not lower than that prescribed by modern codes for the design of new construction, i.e. corresponding to a given probability of occurrence during the conventional "service life" of a new building); every time interventions followed that were executed according to the available constructive traditions at that time and location (not necessarily analytical, but not less effective and decisive for this). In this context, the conditions have gradually matured due to which any attempt to improve the capacity/ demand ratio by changing the original structural behavior of existing instructions have proved to produce not always satisfactory results.

Design approaches based on the identification - through an adequate and rigorous knowledge processes, of all the possible structural vulnerability factors of a historical construction and on interventions aimed at reducing them, if not entirely eliminating them, modifying as less as possible the original structural behavior of the exiting construction - indeed are not only more respectful of the preservation criteria of historical/artistic values, but also more reliable and effective from the point of view of structural safety, as evidenced also by the experiences gained in occasion of the most recent earthquakes."

4.2 Bridges

Statistics regarding road accidents consequences in terms of casualties, injuries and damages, undoubtedly demonstrate that the higher risks for the public safety come from the "functional" features of bridges (e.g. lanes width, presence of dedicated lanes for bicycles or pedestrians, radius of curvature, road junctions, edge barriers) rather than their seismic resistance.

This result in an absolute priority of interventions aimed at improving the functionality of existing bridges rather than of intervention aimed at reducing their seismic vulnerability. In the repeatedly emphasized framework of "limited available resources", better results are attained in terms of risks reduction by means of a spread and systematic interventions on functional deficiencies rather than a complete elimination of both functional and structural (static and seismic) ones on a limited number of bridges.

How much preferred is the first approach by the most recent Italian structural codes is clearly demonstrated by very innovative paragraph where the "repairing or local interventions" are specifically defined for bridges.

At § 8.8.7 [4] "Repairing or local interventions" is in fact written that:

"Provided that the number of physical lanes has not increased, and conditions don't exist that make it compulsory to carry out retrofitting or "improvement" interventions, the following interventions may be considered belonging to this category:

- *Replacement of bearings, provided that this does not involve a variation in the stiffness of the "piers-constraints" system greater than 10%;*
- *widening the platform dedicated to the main function of the infrastructure (in order to increase the driveway surface, or the area available for cycle-pedestrian transit). In this case, the deck verifications must be carried out, in general, considering the live loads required by the new code [3]. It is also possible, with adequate motivation and by adopting specific limitations of use, to consider the live loads originally adopted in the deck design.*
- *complete replacement of bridge decks. The verifications s of the newly built deck must be, in general, carried out by considering the traffic loads required by the new code [3]. In the case of railway bridges, considering that all existing railway lines are classified according to the maximum loads actually allowed to circulate, the verifications of the newly built deck can be carried out by adopting the traffic loads required in the original deck design, or by using the loads corresponding to the category of line to which the infrastructure belongs.*
- *any combination of the interventions listed in previous points, provided that the conditions are not in such way created that make it compulsory the design of "retrofitting" or "improvement" interventions.*

5 CASE STUDIES AND EXAMPLES

5.1 *Historical buildings*

In the following, a series of examples are presented to display a practical application of the concepts before mentioned and introduced in the new update of the Italian building code.

5.1.1 *Basilica of Saint Benedict in Norcia*
The Saint Benedict basilica in Norcia is sadly famous as emblem of earthquakes occurred in the center of Italy on 2016.

After the main shock for Norcia, occurred on the 30th of October (M_W 6.5) with epicenter few hundreds of meters from the church, the basilica completely collapsed leaving stand only the façade (Figure 1).

The ongoing research on the remains of the church is helping to think by a retrospective way to the possible causes of the collapse of such type of structures and at which could be the correctives to reduce the vulnerabilities before the destructive event. To date this case-study stresses the importance of:

- the knowledge of the construction phases and the possible effects on the overall behavior;
- the knowledge of constructive details and the connection among structural elements;
- the design of compatible interventions that aim at improving the weakest mechanism of the structure.

What seems clear looking at the church remains after the earthquake is that the belfry of the bell tower collapsed into the church producing the failure of the entire church and breaking though the underground crypt. The questions that arise are related to

Figure 1. Saint Benedict church after the 30th of October earthquake.

Figure 2. Collapse mechanism of the bell tower in the nave.

the analysis of vulnerabilities and to the effects of the collapse. The historical analysis reveals that the stratification of many transformations provided some intrinsic vulnerabilities not easy to be overcome by interventions.

In the case of Saint Benedict the most crucial point was related to the bell tower. Before 1730, the church had a bell tower about 60 meters high that, due to the severe earthquake of that year, it collapsed into the church with similar effects of today. During the reconstruction, the belfry has rebuilt more wide and on a wall inserted between the tower and the church without any connection among these three parts (Figure 3, left). Therefore, the belfry has the two columns on the transept side supported by another wall separated by the rest of the tower. Moreover, the historical analysis pointed out that on the nave side the first meters at the base were made of a very weak wall probably dating back to the

Romanic phase of the church. Hence, also the nave side has a strong vulnerability at the base.

During the earthquake, these vulnerabilities demonstrated their relevance to the overall stability of the tower with the collapse of the involved walls on the transept and nave side that provided the loss of supports of the belfry and part of the tower that collapsed into the church.

Figure 4 reports the two walls described, in solid green the one between the transept and the tower; and in solid blue the height of the wall made in the Romanic phase.

Figure 4 also shows the interventions carried out before the 2000 jubilee, after the 1997 earthquake that slightly damaged the basilica. The most relevant are three: the connection by injected rebar of all voussoirs; the realization of a reinforced concrete caps with a ring beam on the roof; the introduction in the four belfry piers of three injected bars that stopped at a certain height.

It is still ongoing the analysis of possible worsening in actions by these interventions but it is clear that the collapsed part overturned exactly at the height in which the vertical rebar ends (Figure 4 bottoms right) and only a few centimeters remained free. In addition, this intervention also helps to hold together the entire top part of the bell tower that fallen down as a whole with consequent massive effects on the rest of the church.

About the injected rebar of voussoirs (Figure 4 tops right) they demonstrated to be ineffective with a slip out often seen that pointed out the difficulties to realize them in a proper way since no much injection remains on rebar after the slip out.

The analysis of ruins also highlighted that the intervention of reinforcement carried out in the 18th century was weakly connected with the rest of the original structure. With a thickness of 30-40

Figure 3. The bell tower before (l) the earthquake and after (r).

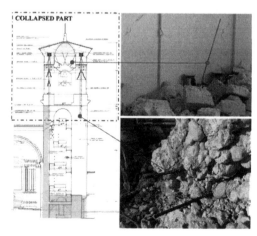

Figure 4. Intervention carried out on the Saint Benedict bell tower (left) injected rebars uneffective (right).

centimeters, these reinforcements should improve the overall thickness of the wall with a consequent better behavior of the structure (Figure 5). In the realization, only few metallic connectors were used and some new vaults and arches were only supported on these elements. This, instead of improving the structural behavior, as in the original goal, it creates some slender elements that in the earthquake completely collapse detaching from the original structure.

In conclusion, this case-study exemplify some aspects previously discussed:

• the characterization of the real role of each portion of the structure is a key aspect to understand the actual behavior of the structure and to point out eventual local vulnerabilities;

• some weaknesses, such as the absence of connection inside the cross-section of the wall, are almost impossible to be observed during the assessment phase and they provide an example of those weak rings that are difficult to be recognized and that limit the overall capacity of the structure;

• the interventions that has the goal to change significantly the equilibrium of the existing structure modifying stiffnesses, masses or bearing mechanism may results in more severe damages than the unstrengthened condition.

5.1.2 Castelsantangelo sul Nera (MC)

An interesting case to present was observed during the last earthquake swarm in Center of Italy in 2016.

Figure 6 shows the emblematic case of two churches, San Martino dei Gualdesi (r) and Santo Stefano (l) which are very similar and are only two hundred meters far but they presented a very different seismic behavior.

Both churches where stroked by a first earthquake on the 24th of August (Amatrice, M_w 6.0) but the

Figure 6. S. Martino bel tower (l) and S. Stefano bell tower (r).

most severe one was on the 26th of October (M_w 5.9) with Castelsantangelo sul Nera (MC) as epicenter that resulted in a Peak Ground Acceleration of 5.31 m/s². This event caused a damage on both bell towers with diagonal cracks and the mechanism activation of the belfry. In general, damages were more severe in Santo Stefano. A following stroke on the 30th of October (Norcia, M_w 6.5) with a smaller PGA of 4.35 m/s² produced the effects reported in Figure 6 with the global collapse of the bell tower and the consequent collapse of the entire church below in Santo Stefano whereas in San Martino only a small damage extension has been observed.

Especially because occurred on two geometrically equivalent cases and very close between them, this case is interesting from many points of views: the combination effects of many earthquakes, the effects that some mechanisms have on the entire stability as reported also for the case of Saint Benedict, and the presence of interventions.

In this case the latter is the stressed one because, on one side it demonstrates that the detrimental difference that turn the behavior is not the masonry resistance but the presence of the hidden hooping system embedded into the masonry. This remark the point that, although important, the material tests are not the only task of the knowledge phase, but also the analysis and study of the structural details and element connections is crucial (see §1.1 [4]) (Binda, 1999). In fact, the capacity by means of non-destructive tests or guided by the historical research on interventions, the awareness of the presence of the hooping system might change the overall outlook on the structure.

On the other hand the intervention, which is a traditional one and that only aim at reducing the weakest ring of the structural "chain", such as the

Figure 5. Constructive details of two different construction phases.

resistance in traction and the prevention of the macro element mechanisms activation though the confinement, proved to fulfil the concept of those local interventions supported by the Italian code. It is here clear the reduction in vulnerability and the impact of the local intervention on the global behavior.

Even if very common in the scientific community, especially in Italy, in which there was the possibility to observe the effects of earthquakes on already strengthened structures, an example of the effects of an intervention that try to change the structural behavior of the historical construction is herein reported.

In the municipality of Castelsantangelo sul Nera (MC), the Saint Martin's church in Gualdo presents some effective interventions carried out on the church resulting in an overall small damage index recorded after all strokes. Conversely, the bell tower has been strengthened by a reinforced concrete caps that increase significantly the masses on the top of the tower and hence the seismic action and instead of improve the overall behavior of the structure highlights the weakness of the masonry which has crumbled on the ground (Figure7).

5.1.3 The masonry chimney in San Bonifiacio (Verona)

In San Bonifacio (Verona, Italy) there was an ancient and abandoned sugar factory that required in the production process the presence of a chimney (Figure 8).

The factory, made of clay brick masonry, dates back to the 20th century and is 26 m wide and 83 long with a chimney 44.2 m height form the ground with a squared base of 5x5m. The circular flue has

Figure 8. The masonry chimney before the interventions.

a tapered geometry that goes from 4.35 m at the base to 2.80 m on the top.

The entire factory was part of a refurbishment project of an investor to make a mall and the requirement was to retrofit the chimney in order to make it safe against winds and seismic actions (Cescatti et al. 2019).

This case study is here reported to make an example of the importance of the knowledge phase considered in parallel with the design phase, and sometimes also with the execution, as remarked in §2.1 and implemented in the Italian code (§8.1.1 [3]).

This reasonable concept points out that if you design a knowledge phase is probably because you do not know a lot of the structure and for this reason you could not design a testing campaign satisfactorily only by a first guess even if you prescribe a massive number of tests. Therefore, in order to optimize efforts and resources is better to manage and change the type or the extensions of tests once you have first results.

This distributed knowledge approach provides, in the case of the chimney, a resource save in tests and satisfactorily results. The first attempt aim at characterizing the masonry mechanical properties and at describing the chemical and physical effects of high temperatures on mortar inside the flue.

During this campaign appears the need to understand the constructive details of the structures next to the chimney and a series of inspections have made. Thanks to them, and by means of some historical pictures, it was possible to figure out that the basement was not structurally connected to the next parts allowing the isolation of the chimney without structural consequences.

Figure 7. The bell tower of Saint Martin's church in Gualdo after the 2016 earthquake swarm.

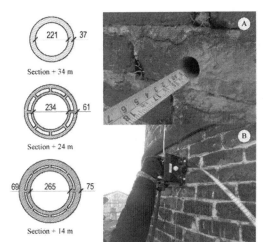

Figure 9. Cross-sections (left), borescopy (a) and GPR test (b).

In a second phase, in order to understand the thickness of the flue, a series of endoscopies (Figure 9 a) at different height have performed discovering an internal chamber that separate an external and an internal shell. At this point, to describe precisely the internal geometry a Ground Penetration Radar (GPR) test has required (Figure 9 b). The tests describe the location and the thickness of eight internal walls that connect the two shells.

This further information helps in defining a series of point in which was necessary to study the bricks arrangement to evaluate the connection between the radial walls and the two shells. This additional investigation pointed out that the radial walls were only juxtaposed to the external shell.

Together with the local repairing of cracks and broken bricks, the designed intervention was the installation inside the flue of FRCM to improve the flexural behavior of the chimney. Without the possibility to extend the test by the information given by the previous one this information was completely hidden reducing the effectiveness of the intervention itself.

Finally, in this case, due to the key role played by the dynamic behavior of the structure, a series of three dynamic identification has carried out. The first was used to fix the structural model by the modal updating, the second one after the masonry restoration with 'scuci-cuci' and local repointing to assess their impact on the dynamic behavior and the last one after all interventions to estimate the impact on the dynamic behavior of interventions.

This approach of testing during the entire restoration phase has already used on historical constructions (Gaudini, 2008) with important results. The approach allows to evaluate the changes in the structural behavior to validate the design expectation of changes in terms of stiffnesses (frequencies) and mode-shapes.

5.2 Bridges

5.2.1 The Alpini's Bridge in Bassano

The timber bridge of Bassano, also called "Ponte degli Alpini" (Alpini's Bridge – Figure 10), is one of the most famous monumental bridge in Italy because built up by Palladio in the second half of the 16th century and because of the Alpini contributions, an Italian army corps, to reconstruct a part of the bridge destroyed after the world war two. Moreover, the entire timber structure, very rare in Italy and in the resto of the world gives it even more attention.

In this case, the effects of the approaches transposed in the Italian codes, are related to the analysis of the actual behavior of the structure.

After an in-depth knowledge phase, based on the research of the historical transformations, analysis of the constructive details, timber testing and the installation of a structural health monitoring (SHM), a more clear interpretation of the structural behavior has been achieved. Therefore, also this case helps to figure out the code approach on real cases.

The first problem of the bridge was about the settlements in normal conditions, with a considerable magnitude (up to 40 cm), and hence related to the structural safety on vertical and static loads. The analysis of the decay and timber degradation reveals a correlation between settlements and decay pointing out the independence among piers about the vertical load distribution.

Thanks to the decay analysis, the interpretation of the deformed structure (see Figure 11) and by the historical analysis of interventions a changing in the structural scheme has recognized.

During the '90s a consolidation intervention has performed and based on the addition of external piles and timber elements to bear the existing beam under the columns. The beam which worked with a distribute support, also by hydraulic impact abrasion, in years became a beam working on few supports with resulting bending failures. Moreover, the refurbishment performed on columns through the addition of "innovative" epoxy resins displayed bad compatibility with timber also causing a degradation speed up. The second problem concerns the structural behavior against horizontal action like the flood. The bridge has

Figure 10. View of the Ponte degli Alpini Bridge.

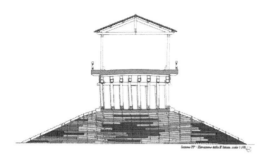

Figure 11. Cross-section of the bridge with the decay analys.

been overwhelmed by a severe flood in 1966 that by abrasion destroyed the lateral triangular elements, here called "rostri" (Figure 12 & 13).

The study of this event pointed out that the bridge does not rely only on the "rostri" to bear the horizontal action because, once they fail, even with large displacements (1 m estimated for the '66 event), the entire deck bring the horizontal force directly on abutments. In this light the strengthening project also looked to improve this original mechanism of

Figure 12. The bridge after the flood in 1966.

Figure 13. Deformation of the bridge deck after the 1966 flood.

the structure with the addition of a lattice beam to redistribute among piers the horizontal action.

In the analysis of the transversal capacity against floods, a deep analysis of the role of the ballast has been performed. In past decades, when the traffic was limited, it was a driveway bridge with a ballast 40 cm thick. Usually, this mass helps in facing the horizontal load by the equilibrium of piers but for solid piers solely. In this case (see Figure 11) the eight columns could not work together and behaving like a monolithic element. For this reason, the ballast does not have a positive contribution against floods and rather it may lead to P-Δ effects in the case of a column tilting. These evaluations provide an example of the importance in getting the real structural behavior to avoid trivial interventions with negative results due to the misinterpretation of the mechanisms.

This case study, because of the severe extent of damage was also interesting to stress the role of SHM and the incremental approach, already diffused in cultural heritage applications (Gaudini et al. 2008), also applied here to interventions on the ballast. Indeed, the serious extent of the decay and settlement suggest to the public owner to install a SHM system to evelute continuosly the bridge behavior with the achivement of two goals:

- evaluate the stability of the bridge and eventual damage trends to design and intervene to counterbalance these effects;
- minimize the counteract measures on the structure which will be object of restoration.

With this approach on one side is possible to limit interventions before the overall refurbishment project with resource savings and on the other providing immediately when necessary the counteract measures to keep the structure safe. In this case, due to the emergency given by the severe conditions, this operative approach is quite evident in terms of benefits but it can be enlarged to common conditions of buildings with more conservation and resource-saving for the same safety conditions.

An example of this approach was the removal of the ballast which provides an essential stop to the vertical settlements that have speed up significantly. The SHM firstly, plotting the settlement speed shows the problem, and after the ballast removal, it evaluated the intervention effectiveness.

5.2.2 The Gresal Bridge in Belluno

The Gresal Bridge is a stone masonry arch bridge with three spans and an overall length of 67.4 m and a road width of 6.1 m. The arches have a semicircular shape with a thickness of 50 cm, a span of about 15 m and a rise of 7.4 m. Spans are supported by two central piers with a maximum height of 12.8 m with a rectangular shape of 3.5x7.0 m (Figure 14).

First of all, an on-site campaign has carried out to characterize the internal composition of the bridge also updating the structural geometry such as the thickness

Figure 14. The Gresal Bridge, view from the stream riverbed.

of arches, walls, filling material, etc. The corings figured out the presence of a backfill made of loose coarse limestone that has a not negligible resistance in compression.

The on-site campaign was also composed of a dynamic identification that was performed in order to execute a modal updating with the dynamic characterization of the structure obtaining a good agreement between the model and the actual bridge (Islami et al. 2011). The setup used for the estimation is shown in (Figure 15, top)

Table 1 reports the first ten modes frequencies and the comparison with those obtained after the manual calibration in the software Strauss. As is possible to notice, the percentage errors in frequencies are very limited as well as mode shapes are properly

Table 1. Results of the dynamic identification and comparison with the finite element model.

Mode	FDD	Strauss7 model		
	f	f	Δ	MAC
	[Hz]	[Hz]	[%]	
1	4.93	4.92	0.17	0.98
2	8.35	7.92	5.39	0.92
3	9.62	8.20	17.28	0.81
4	10.50	10.99	- 4.48	0.83
5	11.08	11.19	-1.01	0.96
6	11.43	12.73	-10.22	-
7	16.65	15.99	4.12	0.91
8	18.60	16.42	13.31	0.67
9	19.24	17.17	12.03	-
10	19.53	18.88	3.45	-

estimated by the model with a MAC (Allemang, 1983) values higher than 0.9 in the first two modes.

The use of such types of non-destructive tests is remarkable in existing structures because it allows nearing significantly the model with the real structure with a consequent correct estimation of stiffnesses and boundary conditions that lead to more reliable evaluations.

The analyses made with the calibrated model pointed out an insufficient seismic capacity in both the transversal and longitudinal directions. This was due to the activation of a mechanism in both situations as reported in Figure 15 center and bottom respectively.

Therefore, from these analyses it appears clearly that the main vulnerability concerns the mechanisms activation and for this reason the interventions design aim at restraining this activation with a local and limited intervention. The design goal was to involve all the intrinsic capacity of the original structures based on the original bearing mechanisms although upgraded by the intervention. The restrain of mechanisms also allows possible increases of vertical loads that might be the case of a future enlargement of the road for improve the bridge functionality.

Figure 16 (above) displays the interventions, very simple, that meanly substitute the road thickness with a new slab maintaining as much as possible the original backfill of vaults and connecting the slab by micro piles and ties to the abutments.

In the transversal direction, the slab has the following roles:

- to restrain the overturn out of plane of the spandrel walls through its connection at the slab;
- to provide a new lateral boundary condition on the top very effective, although deformable, that restrains the transversal mechanism of piers (Figure 15, center), otherwise acting as a cantilever with only the hinge at the base.

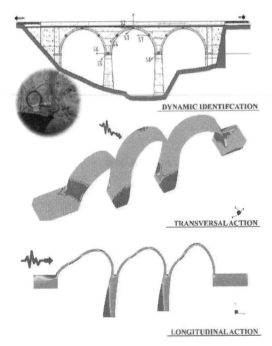

Figure 15. Dynamic identification setup (top), cracked non linear model under trasversal (center) and longitudinal (bottom) actions.

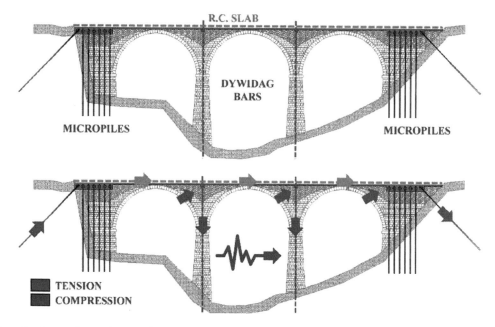

Figure 16. Interventions and structural concept of the retrofit carried out on the Gresal Bridge in Belluno.

The latter point is the same concept implicated by tie-rods in the out of plane mechanism of walls.

In the longitudinal direction, the slab acts together with a series of vertical tie-rods inserted in each pier and on a trestle made of micropiles on abutments. Figure 15 (bottom) shows the bearing mechanisms for the longitudinal actions where the slab on one hand, it transfers the horizontal actions form piers to abutments in both compression and traction and on the other hand, together with the vertical tie-rods in piers, it holds the vertical compressive strut generated in the backfill by the mechanisms.

It should be stressed that the latter bearing mechanism with vertical tie-rods, restraining the in-plane mechanism of the arch also results in an increased capacity also in case of static load.

REFERENCES

Allemang R.J.,Brown D.L.. (1983) *A Correlation Coefficient for Modal Vector Analysis*. 1ˢᵗ Int. Modal Analysis Conference, 247–261.

Binda L., Gambarotta L., Lagomarsino S., Modena C., (1999); *A multilevel approach to the damage assessment and seismic improvement of masonry buildings in Italy*. Seismic damage to masonry buildings, Balkema, Rotterdam, Netherlands.

Cescatti E., Lorenzoni F., Sgaravato M. and Modena C. (2019) *Analisi dinamica e sismica di una ciminiera alta in muratura* Proceedings of the XVIII A.N.I.D.I.S conference in Ascoli Piceno 15–19 September (in Italian).

European Committee for Standardization (2002) *EN 1990: Eurocode - Basis of structural design* 305/2011, Directive 98/34/EC, Directive 2004/18/EC.

High Council of Public Works (2019) Circ. 7/2019 – *Circolare n.7 del 21 gennaio 2019* Istruzioni per l'applicazione dell'«Aggiornamento delle "Norme tecniche per le costruzioni"» di cui al decreto ministeriale 17 gennaio 2018.

High Council of Public Works (2020) – Linee Guida sulla classificazione del rischio, verifica e monitoraggio dei ponti esistenti (in Italian).

Islami, K., Tecchio, G., & Modena, C. (2011). Seismic Intervention and Dynamic Testing of an Arch Bridge. Applied Mechanics and Materials, 105–107, 1159–1164. https://doi.org/10.4028/www.scientific.net/amm.105-107.1159.

International Organization for Standardization ISO 13822 (2001) *Bases for design of structures-Assessment of existing structures* – ANNEX I (Informative) Historic structures.

International Organization for Standardization ISO CD 13822 –ANNEX I (Informative) Historic Structures.

Gaudini G., Modena C., Casarin F., Bettio C., Lucchin F. (2008) *Monitoring and strengthening interventions on the stone tomb of Cansignorio della Scala, Verona, Italy*. In proceeding of the SAHC conference 02–04 July, Bath, United Kingdom.

International Organization for Standardization ISO 2394 (2015) General principles on reliability for structures Techincl Committee: ISO/TC 98/SC 2 Reliability of structures.

Ministry of Infrastructures and Transport (2009) Circolare esplicativa 02/02/2009 n. 617. *Istruzioni per l'applicazione delle "Nuove norme tecniche per le costruzioni"*. D.M. 14/02/2008 (In Italian).

Ministry of Infrastructures and Transport (2008) N.T.C. 2008 – *Norme tecniche per le costruzioni. D.M. 14/01/2008* [In Italian].

Ministry of Infrastructures and Transport (2017a) D.M. 58/2017 – *Sisma Bonus - Linee guida per la classificazione* del rischio sismico delle costruzioni nonché le modalità per l'attestazione, da parte di professionisti abilitati, dell'efficacia degli interventi effettuati. (In Italian).

Ministry of Infrastructures and Transport (2017b) D.M. 65/2017 *Sisma Bonus - Linee guida per la classificazione del rischio sismico delle costruzioni e i relativi allegati Modifiche all'articolo 3 del Decreto Ministeriale numero 58 del 28/02/2017* (In Italian).

Ministry of Infrastructures and Transport (2018) N.T.C. 2018 – *Nuove norme tecniche per le costruzioni. D.M. 17/01/2018* (In Italian).

Modena C., Defina A., Cescatti E., Viero D., Russo D. (2015) *Relazione illustrativa* – Convenzione per lo svolgimento di attività di ricerca e sperimentazione sullo stato di conservazione e sulle condizioni di sicurezza, sotto azioni statiche e sismiche, e sui possibili interventi per la riparazione, il rinforzo e il restauro del "Ponte degli Alpini" di Bassano del Grappa (in Italian).

Modena C., da Porto F., Valluzzi M.R., Carapezza Guttuso F., Iannelli P., Rubino C. (2016) Sustainable approaches to the assessment and mitigation of seismic risk and of the effects of earthquake induced damages to historical urban centers. Keynote in Structural Analysis of Historical Constructions –Anamnesis, diagnosis, therapy, controls – Van Balen & Verstrynge (Eds) © 2016 Taylor & Francis Group, London, ISBN 978-1-138-02951–4.

Semi Key notes

Validation of design tools for the prediction of mechanical behaviour of masonry arches strengthened with inorganic matrix-based composite systems

M.R. Valluzzi & L. Sbrogiò
Department of Cultural Heritage, piazza Capitaniato, Padova, Italy

E. Cescatti
Department of Environmental, Civil and Architectural Engineering, Padova, Italy

ABSTRACT: Strengthening of masonry arches and vaults with composite materials has become a quite common strategy of intervention, especially in seismic area. In the last decades, fibre reinforced polymers (FRP) have been conveniently replaced with inorganic matrix-based composites (mostly known as FRCM, fibre reinforced cementitious matrix). Nowadays, several experimental works provide a significant dataset to characterize and interpret the mechanical behaviour under various strengthening configurations on different types of arches and vaults. From one hand, these studies contributed to the definition of design and assessment approaches, which have been only recently agreed in the scientific community for FRCM (e.g., ACI-Rilem recommendations, to be issued). From the other hand, design tools currently available for the design and assessment of curved structures in strengthened conditions need to be validated and/or upgraded according to the current scientific state-of-the-art. In this paper, based on the experimental results obtained in twenty-six literature cases, a comparative study among different conditions of masonry arches strengthened with FRCM systems is proposed. The main parameters affecting the structural behaviour of components in both plain and strengthened conditions are identified, and their influence in analytical procedures implemented in common software able to predict failure modes and bearing capacity (either limit state or rigid block analyses-based) are discussed. The pro and cons, as well as the strategies for representing at best the experimental outcomes are also presented. This paper is intended to support the choices required to professionals approaching the design and assessment issues of arches and vaults strengthened with the new generation of composites, in absence of recommendations and standards, and by using the simplified methods implemented in the available engineering tools.

1 INTRODUCTION

Since the '80s, fibre reinforced polymers (FRP) including various type of fibres (e.g. carbon, glass, basalt) have been used on masonry elements to increase the load-carrying capacity and improve their structural behaviour, thus avoiding the most critical failure modes. Such composite materials are particularly effective on arches and vaults, as they provide a tensile contribution and allow for developing a pseudo-ductile behaviour, thus preventing the brittle collapse of the curved structures. In the last decades, fibre textiles embedded into inorganic matrices (known as FRCM - fibre reinforced cementitious matrix, TRM - textile reinforced mortar, IMG - inorganic matrix grid, etc.; FRCM will be used in the following to identify all these systems) have been increasingly applied to existing masonry buildings as alternative to FRPs. This is mainly due to the higher compatibility (including air-permeability) of inorganic matrix than epoxy with respect to the existing materials (Papanicolaou et al. 2007, Valluzzi et al. 2014).

This work analyses the behaviour of a series of masonry arches and vaults strengthened with FRCM systems as reported in twenty-six contributions currently available in literature. They refer to the decade 2007-2017 and include eighteen experimental campaigns, which differ among them in terms of masonry materials (types of units and mortars), geometry (e.g. catenary, barrel, segmental), test type (monotonic, cyclic, on shaking table), and strengthening material (e.g. basalt, carbon, glass, or steel fibres). Among these, only ten were analysed in this study, as experimental conditions and specimens features were comparable.

The experimental configurations of the above-mentioned campaigns were implemented in common design tools, as 'Limit state RING' (Gilbert 2005) and 'Arco' (Gelfi & Metelli 2006), as well as limit analysis approach. Results obtained by the three assessment methods were compared with the experimental outcomes, thus providing several points to discuss about their reliability.

The eighteen experimental campaigns available in literature aimed at characterizing the structural behaviour of both unreinforced and reinforced masonry arches/vaults and at studying the effect of reinforcing systems applied with various type of fibres and matrices on the load capacity and the failure mechanism of the structures. Polymer (PBO - poliparafenilenbenzobisoxazole), carbon, steel, basalt and glass FRCM were applied to the masonry curved specimens.

The reinforcing systems were installed either at the intrados or at the extrados of the arch/vault, and the structural elements were mainly tested under vertical loads, which were commonly applied at the quarter or at the middle span and were distributed over the whole surface of the voussoirs. In few cases, fibres were applied to both intrados and extrados, and horizontal loads were applied to the specimens, in either monotonic or cyclic (either inverted, loading-unloading or alternate) modes. Tests were mainly performed in laboratory on full-scale mockups. In few cases, reduced scale models were tested (Alecci et al. 2016, 2017a, b, Briccoli Bati et al. 2007, Pantò et al. 2017) or, in other cases, tests were performed onsite on existing structures (Cescatti et al. 2017, De Santis & De Felice 2015, 2016, Incerti et al. 2017). In Giamundo et al. (2015, 2016) and Ramaglia et al. (2015, 2017a, b) shaking table tests were carried out to simulate earthquake actions.

The majority of arches were made of solid clay bricks, except for the campaigns by Garmendia et al. (2011, 2014) and Borri et al. (2007, 2009), where sandstone and concrete bricks were used, respectively. Mortar types were mostly reproduced as representative of existing materials (hydraulic lime-based), thus having preferably low strengths.

As of the geometric features, segmental round arches with uniform thickness were mostly tested, except for Borri et al. (2007, 2009) which refer to a catenary arch, De Santis & De Felice (2015, 2016) to a three centred arch, whereas Cescatti et al. (2017) and Incerti et al. (2017) used existing arches with different thicknesses between the haunches and the crown.

Arches/vaults were generally bare, except for Bednarz et al. (2011), Bertolesi et al. (2017), Briccoli Bati et al. (2007) and De Santis et al. (2017) experimental campaigns, in which spandrel walls were present. In De Santis et al. (2017) and Hoydis et al. (2013, 2015, 2016a, b) a backfilling was also included.

Boundary conditions of the arches (e.g. vertical walls, skewbacks, abutments) vary among tests. Mechanical anchors (e.g. spikes, metal devices) were used to connect the textile onto the substrate in Garmendia et al. (2011, 2014, 2015), Girardello et al. (2013), Hojdys et al. (2013, 2015, 2016a, b), De Santis et al. (2017), Borri et al. (2007) and Cescatti et al. (2017).

Campaigns by Girardello et al. (2013a, b) and Garmendia et al. (2015) shared identical geometry for the specimens but different materials. Pantò et al. (2017) focussed on an analytical approach applied to the results of the Alecci et al.'s tests.

Appendix 1 lists the main characteristics of the eighteen experimental campaigns. Data concerning reinforcing systems different from FRCM were omitted (this apply throughout the paper).

However, according to the main objective of this study, the comparison of the experimental results was possible for ten experimental campaigns, as described in the following. They refer to the application of quasi-static vertical point loads and comparable geometric configurations. The features of the tested arches with their registered failure modes are included in Figure 5, whereas Appendix 2 and Appendix 3 report, among other data computed in the analysis, the main mechanical properties of the specimens and the results of the related experimental campaigns, respectively. When not available, significant parameters of basic materials were assumed or computed by the Authors (criteria are described where needed).

2 ANALYSIS OF EXPERIMENTAL RESULTS

Results of the ten selected experimental campaigns were compared, to illustrate possible trends according to the variables under investigation.

Figure 1 shows the comparison among the main geometrical characteristics of the specimens. Except for Garmendia et al. (2011), where a squat arch was tested, all the others presented a comparable ratio between thickness (t) and span (s) of about 1/20 (see App. 2 for values). On the contrary, the ratio between rise (r) and span varied significantly; by excluding the results from Bertolesi et al. (2017) and Incerti et al. (2017), which related to depressed arches, the average value of all the others was about 0.30, which corresponds to an angle of about 30° on the haunches for a circular shape. The width of arches also varied in the range of one (Alecci et al. 2017) to eight times (Hojdys et al. 2013) of thicknesses.

As the failure load F_0 measured from the experimental tests in unreinforced conditions depends on the geometry of the arches, to make results comparable, a normalization was carried out. It is based on the bending moment M^*_0 acting along the arch and computed according to Equation 1 (x is the load position):

$$M^*_0 = x(s - x) \cdot F_0/s \qquad (1)$$

Therefore, the normalized compressive force N^*_0 is evaluated at the crown by diving that moment by the rise (Equation 2) (see App. 2 for values):

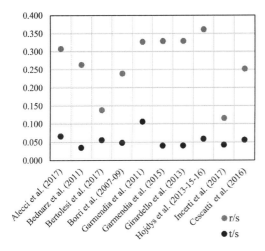

Figure 1. Plot of main geometrical ratios of vaults (*r* rise, *t* thickness, *s* span).

$$N_0^* = \frac{M_0^*}{r} \qquad (2)$$

Figure 2 shows the comparison between the cross section area (A) of the arches and the maximum axial force obtained from the experimental tests. In the chart, the two cases with different thicknesses along the arch, i.e. Cescatti et al. (2017) and Incerti et al. (2017) were neglected. Although Bednarz et al. (2011), Girardello et al. (2013) and Garmendia et al. (2015) referred to the same cross section, results exhibited a quite different resisting force. In addition, the latter two experimental campaigns corresponded to identical geometry of specimens, but the resulting normalized loads differed of more than three times. This stresses how experimental results may vary and that no sound relationships of correlations can be found. However, a reasonable increasing trend of the resisting axial force according to higher cross sectional area was recognizable.

The effect of strengthening on the maximum load (F_s) was computed through the ratio F_s/F_0 (see App. 3 for values). To evaluate the quantity of reinforcement used, the fraction ω was defined as the ratio between the resisting cross sections of FRCM (E_f Young's modulus, A_f section area, ε_f strain of reinforcement) and of masonry (f_m compressive strength, w width of arch), according to Equation 3 (see App. 3 for values):

$$\omega = \frac{E_f \cdot A_f \cdot \varepsilon_f}{f_m \cdot t \cdot w} \qquad (3)$$

Figure 3 correlates the increase of strength F_s/F_0 with ω according to the type of fibers. To improve readability, due to the high amount of reinforcement, the result of IN+EX.01 test (see App. 3) of Borri et al. (2007) was not included in the graph.

In case of carbon, polymer and especially steel fibres (particularly when applied locally as strips), a trend of increasing strength with increasing reinforcement was recognizable. Nevertheless, only in few cases, the crushing of masonry was reported in the literature; therefore, it was not possible to see the limitations due to the failure of masonry, which typically relates to an excessive application of reinforcement.

The use of glass and, especially, of basalt fibres showed a very low sensitiveness to the amount of fibres. In case of basalt fibres, the load increased up to 20 times with respect to the same amount of reinforcement. Indeed, the application of grids width a large amount of inorganic matrix (e.g. if compared with strip installations) may provide a higher strength due to the matrix tensile resistance, as also reported in Garmendia et al. (2014). Moreover, the use of cyclic tests is probably detrimental to evaluate the reinforcement influence on the overall behaviour of the vaults.

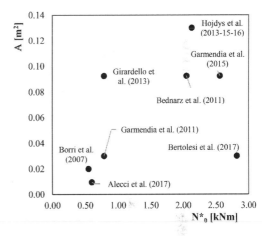

Figure 2. Comparison between maximum experimental axial force and cross-section area.

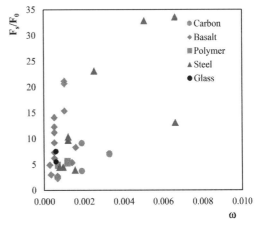

Figure 3. Strenght increase versus quantity of reinforcement according to fibre types.

3 ANALYTICAL APPROACHES IMPLEMENTED IN AUTOMATED PROCEDURES

To compare the various experimental conditions and results available in literature with respect to analytical evaluations for the assessment of the arches, some common software, i.e. 'Arco' (Gelfi & Metelli 2006) and 'Limit state RING' (Gilbert 2005), were taken into consideration (ARCO and RING will be used in the following to mention these tools).

The unreinforced arches were modelled with both software, whereas reinforced arches were modelled with RING only, because ARCO does not allow reinforcing system to be implemented. As further comparison, the collapse conditions of both unreinforced/reinforced arches were determined through the limit analysis (Focacci 2008).

ARCO analyses masonry arches and vaults based on the 'safe theorem' of the plastic theory (Heyman 1966, 1982). By means of an iterative method, it identifies the thrust line of the arch that minimise the eccentricity, thus providing a minimal geometrical safety factor. Required input data are: the geometry of the arch (span, thickness, width, rise, number of voussoirs), the unit weight of masonry and backfill, the protrusion of the upper wall, the depth of solid backfill, and the load. As output, the software provides the thrust line, the resultant of the distributed load on the arch, and the diagrams of the maximum stresses. If the stresses are lower than the resisting ones the arch is verified and the resulting load is the collapse one.

RING is a software for the analysis of masonry arch bridges that implement the rigid block analysis to rapidly check the capacity factor of an arch. It is able to model multiple load cases (particularly for rail and highway loading), backfill, mortar loss, support movements, and reinforcement. RING also defines properties of all materials and partial factors on load and material, and identifies critical failure modes of bridges (hinges development or masonry failure), the equivalent thrust line, and the force diagrams.

Both tools are able to simulate arches with other shapes from segments of circle (e.g. catenary, three centred) and with different thicknesses between haunches and crown. However, they cannot simulate interlocking masonry textures: therefore, in this study, arches from Alecci et al. (2016, 2017a, b) and Bertolesi et al. (2017) were simplified as a row of voussoirs.

3.1 Unreinforced arches

ARCO provides the thrust line and the stress distribution on the arch, but, since it assumes an infinite compressive strength, these stresses need to be checked with respect to the compressive strength of masonry. Conversely, RING allows to model the compressive strength of materials and the reinforcement system. Although the infinite compressive

strength is a valid hypothesis, to make results more reliable a limited compressive strength was assumed here (see App. 2). Some authors (Garmendia et al. 2014) (Hojdys et al. 2016b) suggest using an additional reinforcement to model the tensile resistance of masonry. Nevertheless, according to the design rules of masonry, and to make results among experimental campaigns comparable, no tensile contribution was assumed in the models (according to experimental evidence, the tensile strength of mortar in masonry is generally lost just after the first cycle of loading).

Table 1 and Figure 4 show the comparison between experimental and calculated values with the three methods for the selected experimental campaigns (see data in App. 1-3).

Table 1. Results of experimental tests and of models (ARCO, RING an limit analysis) for unstrengthened arches.

Reference	F_0			
	Exp	Arco	Ring	Lim. An.
	kN	kN	kN	kN
Alecci et al.	0.98	0.20	0.19	0.17
Bednarz et al.	2.88	1.85	1.78	1.79
Bertolesi et al.	2.08	1.98	1.82	1.68
Borri et al.	0.70	0.55	0.53	0.54
Cescatti et al.	19.04	22.30	18.80	23.50
Garmendia et al. (11)	1.38	1.60	1.57	1.08
Garmendia et al. (15)	4.50	1.23	1.46	1.11
Girardello et al.	1.38	1.31	1.21	1.27
Hojdys et al.	4.10	2.50	2.44	2.81
Incerti et al.	17.80	5.20	4.79	3.38

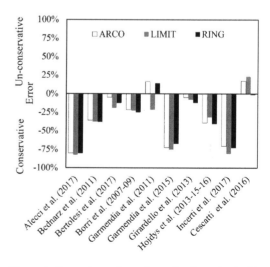

Figure 4. Comparison of experimental collapse load with values computed by ARCO, RING and limit analysis.

Although the methods refer to different approaches, the results were very similar. Evaluations were generally conservative with an average error of -27 % for RING, -30 % for ARCO and -36 % for the limit analysis with similar standard deviation of 0.30, 0.36, and 0.35 respectively for each approach.

The results obtained by limit analysis were generally more un-conservative compared to the other methods. In fact, in the limit analysis the hinges position of the other approaches were implemented and no material strength was considered. To overcome this problem a hinge regression considering the material strength could improve the solution.

Figure 5 shows the localization of plastic hinges resulting from the experimental tests (where available) and the ones computed by ARCO and RING. The position is satisfactorily correct for both approaches. In case of Bertolesi et al. (2017) and Garmendia et al. (2015) some differences between experimental and computed layouts of hinges were detected. A different output between ARCO and RING was observed for Incerti et al. (2017) (no layout of hinges were available), probably due to the changing in thickness. For the other onsite test campaign (Cescatti et al. 2017), compared to the experimental evidence, RING seems to performs better than ARCO.

In some cases (Alecci et al. 2017) (Garmendia et al. 2015), tests were performed monotonically; this did not allow during tests to assess the influence of tensile strength of the matrix that may lead to an overstrength of the arch until the onset of cracking.

It is also worth noting that the experimental campaigns presented by Girardello et al. (2013) and Garmendia et al. (2015), referred to specimens of identical geometric features, but the experimental collapse loads resulted in a ratio 1:3. Such a difference was probably due to the different strength of the embedding mortar in both tension and compression (they are in a ratio of 1:2), which affected the masonry properties (cfr. App. 2). Moreover, the last hinge opened in Garmendia et al. (2015) (Figure 5) was in a different position than the one derived from calculation; hence, the arch exhibited a flexural resistance. However, the analytical values were always close to the lowest value obtained by Girardello et al. (2013).

A more precise estimation of this approach should be performed based on additional experimental tests considering cyclic loading.

3.2 Reinforced arches

The reinforced arches were calculated by RING and by the limit analysis. In the first approach, the block analysis requires a tensile force of the reinforcement (T_f), not related to the strain properties of the system, according to Equation 4:

$$T_f = E_f \cdot A_f \cdot \varepsilon_f \qquad (4)$$

where the value of ε_f should be defined.

This can be evaluated by the characterisation of the reinforcing system, i.e. either from the ultimate strain of the fibre or the debonding load. Due to the lack of information about the latter, only the ultimate strain of the system characterisation was assumed here.

As for the limit analysis, the issue in the strength modelling depends on the correct evaluation of the virtual work of the fibres. According to Cescatti (2016) the work should be defined as follow:

$$\Delta E_i = \frac{1}{2} \cdot \frac{E_f \cdot A_f}{L} \cdot \Delta l_i^2 \qquad (5)$$

where L is length of the deformed fibre and l is the virtual deformation related to the hinge opening. Due to still open issues on the actual bond behaviour, the effective length of bond models, and the distance between two points of null strain, a simplified formula was used (Equation 6):

$$\Delta E_i = T_f \cdot \Delta l_i \qquad (6)$$

where T_f is the same of above.

Figure 6 plots the error in the estimation of the two approaches. It is possible to notice that all estimations are conservative, except for few cases.

The selection of this type of strain is not accurate in case of carbon fibres, as detected for Bednarz et al. (2011), which corresponds to errors of 421% and 271% related to the overestimation of the fibres contribution. In the case of Girardello et al. (2013) the overestimation (171%) is due to the localisation of hinges along the arch that is not accurately assessed by the models (the test setup applied alternate forces in two positions, thus leading hinges to occur under the jacks).

Likewise the cases of unreinforced arches, the two approaches provided analogous results even with a high variability of cases.

Figure 7 and Table 2 highlight the average error in the prediction of the ultimate load according to the fibre materials. It is worth noting that for carbon FRCM the ultimate resistance should take into account the entire reinforcing system, which provide lower performances than the sole fibre (no other data were available for these cases in literature).

As for steel fibres, both approaches provided satisfactorily results with a low average error. Glass, polymer and basalt fibers showed a higher underestimation, probably due to the matrix contribution (tensile strength can reach up to 2.5 MPa).

It is clear that by evaluating a debonding resistance (that it is certainly lower than the experimental ultimate

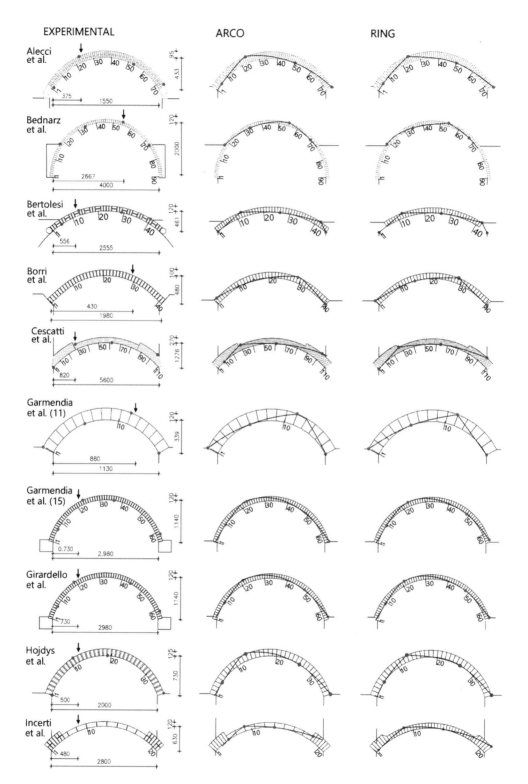

Figure 5. Dimension, plastic hinges, and thrust lines as determined by experimental tests, ARCO and RING (unreinforced cases). Note that spans are reduced to same dimension; measures are in mm.

strain of the FRCM system) the estimated maximum load of all approaches will be more conservative, thus underestimating the experimental load. This could be

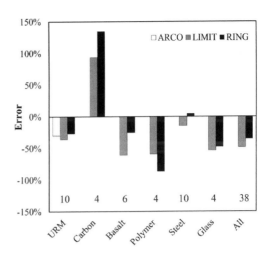

Figure 7. Average error in estimation of ultimate load of strengthened vaults according to reinforcement type (numbers accounts for cases implemented; *URM* unreinforced masonry).

Table 2. Average error and standard deviation (σ) compared among models.

	Arco		Ring		Limit Analysis	
	Error	/σ	Error	/σ	Error	/σ
URM	-30% (36%)		-27%	(30%)	-36%	(35%)
Carbon	-		135%	(218%)	93%	(122%)
Basalt	-		26%	(56%)	61%	(32%)
Polymer	-		87%	(0%)	59%	(5%)
Steel	-		5%	(50%)	5%	(37%)
Glass	-		-48%	(41%)	-53%	(8%)
ALL	-		36%	(57%)	49%	(36%)

the case of the extrados reinforcements, where an additional contribution of the curvature is given to the strength by the reinforcement at the hinge positions. In fact, the fibres activate there a stabilizing vertical component due to the inclination occurring over the hinge (Cescatti 2016), but to solve this issue more experimental results are needed.

At last, it is worth pointing out an important issue concerning the static admissibility of the solutions obtained by the approaches here applied. RING does not consider the masonry crushing in a strengthened cross-section; therefore, the check of the achieved level of compression on the most stressed cross sections is always advisable. The same works for the limit analysis, taking into account the hinge regression, in either plain or reinforced arches, as the regression increases due to the internal bending moments. Moreover, the limit analysis does not consider the abovementioned resisting contribution of the fibres due to their inclination when cracks open and hinges develop.

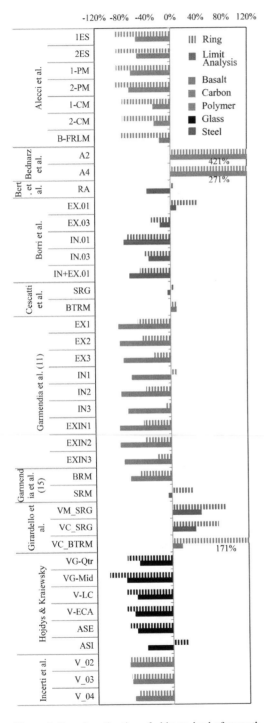

Figure 6. Error in estimation of ultimate load of strengthened vaults with limit analysis and RING.

4 CONCLUSIONS

The validation of design tools commonly used among professionals for the evaluation of the mechanical behavior of masonry arches and vaults in plain and strengthened conditions by using inorganic matrix-based composite systems has been proposed. In particular, experimental data derived from ten complete experimental campaigns referred to the application of vertical quasi-static loads, whose results were compared with the ones obtained from RING and ARCO tools, as well as limit analysis.

Results showed an overall good agreement among the different approaches, either for hinges layout or ultimate load, and this make them reliable for practical use; in addition, they resulted mostly conservative and therefore suitable for design purposes. However, it is important to remind that the check of compression stresses in the masonry must be carried out to satisfy the static admissibility of the solution. Moreover, both ARCO and RING consider only vertical loads and this is a severe limitation in case of seismic assessment of arches, either in plain or reinforced conditions.

An underestimation of 30% in the resisting load (cfr. Figure 4), probably due to additional contributions which are not considered in the models, was detected for unreinforced arches. Therefore, models currently used also for strengthened solutions, where a higher variability of results is expected, does not seem unreliable, although further improvements are desirable.

A fundamental still open issue concerns the definition of the working tensions (and related resisting force) of the whole FRCM system rather than the high-performing fibers. As an example, in the case of carbon strips results are clearly overestimated, but in case of grid installations, e.g. for basalt fibers, the results mostly depends on the matrix characteristics.

More data from complete (i.e. including all data on materials, testing, and observed behavior) experimental campaigns would increase this validation, especially in case of cyclic tests, where the contribution of either the embedding mortar or the inorganic matrix can be estimated.

ACKNOWLEDGEMENTS

Authors wish to acknowledge A. Rossin for her contribution in the preliminary data processing phase.

REFERENCES

Alecci V. et al. 2016. Extrados strengthening of brick masonry arches with PBO-FRCM composites: Experimental and analytical investigations. *Composites Structures* 149: 184–196.

Alecci V. et al. 2017a. Strengthening Masonry Arches with Lime-Based Mortar Composite. *Buildings* 7(2).

Alecci V. et al. 2017b. Intrados strengthening of brick masonry arches with different FRCM composites: Experimental and analytical investigations. *Composites Structures* 176: 898–909.

Bednarz L. et al. 2011. Simulations and analysis of arched brick structures. *Automation in construction* 20(7): 741–754.

Bertolesi E. et al. 2017. In Situ- Tests and Advanced Numerical Modelling for Masonry Arches Retrofitted With Steel Reinforced Grout. *Key Engineering Materials* 747: 242–249.

Borri A. et al. 2007. Rinforzo di archi in muratura con materiali compositi innovativi. *Atti XII Convegno Nazionale L'Ingegneria Sismica in Italia (ANIDIS), Pisa 10-14 June 2007*.

Borri A. et al. 2009. Strengthening of brick masonry arches with externally bonded steel reinforced composites. *Journal of Composites for Construction* 13(6): 468–475.

Briccoli Bati S. et al. 2007. Strengthening Masonry Arches with Composite Materials. *Journal of Composites for Construction* 11(1): 33–41.

Cescatti E. 2016. Combined experimental and numerical approaches to the assessment of historical masonry structures. PhD Thesis. University of Trento.

Cescatti E. et al. 2017. In-Situ Destructive Testing of Ancient Strengthened Masonry Vaults. *International Journal of Architectural Heritage* 12(3): 350–361.

De Santis S. et al. 2015. Prove di distacco in situ su rinforzi in SRG applicati sulla superficie estradossale di volte in muratura. *Atti XVI Convegno Nazionale L'Ingegneria Sismica in Italia, L'Aquila 13-17 September 2015*.

De Santis S. & De Felice G. 2016. Bond behaviour of steel reinforced grout strengthening systems applied to the extrados of masonry vaults. *Proc. Int. Conf. Structural Analysis of Historical Constructions 13-15 September 2016, Leuven*.

De Santis S. et al. 2017. Retrofitting Masonry Vaults with Basalt Textile Reinforced Mortar. *Key Engineering Materials* 747: 250–257.

EN 1996- 1-1. 2005. Eurocode 6: Design of masonry structures. Part 1-1: General rules for reinforced and unreinforced masonry structures.

Garmendia L. et al. 2011. Rehabilitation of masonry arches with compatible advanced composite material. *Construction and Building Materials* 25: 4374–4385.

Garmendia L. et al. 2014. Strengthening of masonry arches with Textile-Reinforced Mortar: experimental behaviour and analytical approaches. *Material and Structures* 47: 2067–2080.

Garmendia L. et al. 2015. Strengthening masonry vaults with organic and inorganic composites: an experimental approach. *Materials and Design* 85: 102–114.

Gattesco N. et al. 2017. Experimental Behavior of Masonry Vaults Strenthened with Thin Extradoxal or Intradoxal Layer of Fiber Reinforced Lime Mortar. *Key Engineering Materials* 747: 274–281.

Gelfi, P. & Metelli, G. 2006. *Arco, Analysis of masonry arches and vaults. Help Manual*. University of Brescia, Brescia.

Giamundo V. et al. 2015. Experimental investigation of the seismic performances of IMG reinforcement on curved masonry elements. *Composites: Part B* 70: 53–63.

Giamundo V. et al. 2016. Shaking table tests on a full-scale unreinforced and IMG-retrofitted clay brick masonry barrel vault. *Bull. Earthquake Eng.* 14: 1663–1693.

Gilbert, M. 2005. *RING Theory and modelling guide. Computational limit analysis and design unit. University of Sheffield*, Sheffield.

Gilbert, M. 2007. Limit analysis applied to masonry arch bridges: state-of-art and recent developments. *5th International conference on Arch Bridges*. 12-14 september, Madeira.

Girardello P. et al. 2013a. Comportamento sperimentale di volte in muratura rinforzate con materiali compositi a matrice inorganica. *Atti XV Convegno L'Ingegneria Sismica in Italia, 30 June- 4July, Padova*.

Girardello P. et al. 2013b. Experimental testing and numerical modelling of masonry vaults. *Proc. Int. Conf. Rehabilitation and Restoration of Structures, 13-16 February, Chennai*.

Heyman J. 1966. The stone skeleton. *International Journal of Solid Structures* 2:274–279.

Heyman J. 1982. *The masonry arch*. New York: Ellis Horwood.

Hojdys L. & Krajewski P. 2013. Behaviour of masonry arches strengthened with TRM. *Proc. 7th Int. Conf. on Arch Bridges*.

Hojdys L. & Krajewski P. 2015. Laboratory Tests on Masonry Vaults with Backfill Strengthened at the Extrados. *Key Engineering Materials* 624: 510–517.

Hojdys L. & Krajewski P. 2016a. Experimental tests on masonry arches strengthened with a mortar-based strengthening system. *Proc. 8th Int. Conf. Arch Bridges*.

Hojdys L. & Krajewski P. 2016b. Glass fiber grids embedded in a cement-based matrix as strengthening of masonry structures. *Proc. Int. Conf. Structural Analysis of Historical Constructions 13-15 September 2016, Leuven*.

Incerti A. et al. 2017. Destructive In Situ Tests on Masonry Arches Strengthened with FRCM Composite Materials. *Key Engineering Materials* 747: 567–573.

MIT 2019: Ministry of Infrastructures and Transportations. Regulation n. 7 2019: Istruzioni per l'applicazione dell'aggiornamento delle "Norme tecniche per le costruzioni" di cui al decreto ministeriale 17 gennaio 2018. (In Italian).

Pantò B. et al. 2017. Non-Linear Modelling of Masonry Arches Strengthened with FRCM. *Key Engineering Materials* 747: 93–100.

Papanicolaou C.G. et al. 2007. Textile-reinforced mortar (TRM) versus FRP as strengthening material of URM walls: in-plane cyclic loading. *Material and Structures* 40: 1081–1097.

Ramaglia G. et al. 2015. A simplified approach to evaluate retrofit effects on curved masonry structures. *Proc. 3rd Conf. Smart Monitoring Assessment and Rehabilitation of Civil Structures, 7-9 September 2015, Antalaya*.

Ramaglia G. et al. 2017a. Seismic strengthening of Masonry Vaults with Abutments Using Textile-Reinforced Mortar. *Composite Construction* 21 (2).

Ramaglia G. et al. 2017b. Numerical Modelling of Mansory Barrel Vaults Reinforced with Textile Reinforced Mortars. *Key Engineering Materials* 747: 11–19.

Valluzzi M.R. et al. 2014. Current practice and open issues in strengthening historical buildings with composites. *Material and Structures* 47: 1971–1985.

Appendix 1. Geometry data of masonry arches and properties of composite materials for FRCM systems of whole set of experimental campaigns available in literature.

Reference	Shape	Boundary condition	t mm	w mm	s mm	r' mm	Fibre	Setting	t_f mm	b_f mm	σ_f MPa	Test and load type (position)	Notes
Alecci et al. 2016, 2017a, b,	S	B	95	95	1500	432.5	Polymer	E, I	0.014	95	3328	Mono	Scale 1:2
		B	95	95	1500	432.5	Polymer		0.014	95	5700	VP (s/4)	
		B	95	95	1500	432.5	Carbon		0.047	95	4800		
		B	95	95	1500	432.5	Basalt		0.032	95	3000		
Bednarz et al. 2011	R	B, PSW	120	770	4000	2000	Carbon	E	0.047	770	4800	Mono VP (s/3)	
Bertolesi et al. 2017	S	B, PSW	120	250	2555	450	Steel	E	0.118	95	1570	Cyclic VP (s/4)	
Borri et al. 2007, 2009	C	B	100	200	1980	490	Steel	E, I, B	0.383	150	1658	Mono VP (s/4)	
		B	100	200	1980	490	Steel		0.098	150	2469		
		B	100	200	1980	490	Steel		0.098	300	2469		
Briccoli Bati et al. 2007	S	SW	100	100	1500	435	Glass	I	-	80	-	Mono VD	Scale 1:2
Cescatti et al. 2017	S	B, DT	270	800	5600	1100	Steel	E	0.084	250	-	Cyclic VP (s/4)	On site
		B, DT	270	800	5600	1100	Basalt	E	0.032	800	-		
De Santis & De Felice 2015, 2016	3C	B, W	-	-	-	-	Steel	E	0.254	150	3070	Mono	On site
De Santis et al. 2017	S	SW, BF	55	500	2790	650	Basalt	E	-	500	880	Cyclic VP (s/3)	
Garmendia et al. 2011, 2014	S	B	120	250	1130	339	Basalt	E, I, B	0.042	500	415.8	Mono VP (s/4)	
Garmendia et al. 2015	S	B	120	770	2980	1140	Basalt	E	0.053	770	505	Mono VP (s/4)	
		B	120	770	2980	1140	Steel	E	0.750	240	3165		
Gattesco et al. 2017	S	B	120	770	4000	1500	Glass	E	0.005	770	1735	Cyclic H	
Giamundo et al. 2015	S	B	120	2200	2980	1140	Basalt	E	0.045	2200	3000	ST	
Girardello et al. 2013	S	B	120	770	298	1140	Steel	E	0.086	228	2820	Mono/ cyclic	
		B	120	770	02980	1140	Basalt	E	0.053	770	1735	cyclic VP (s/4)	
Hojdys et al. 2013, 2015, 2016a, b	S	B, BF, W	125	1040	2000	730	Glass	E	0.035	900	1285.7	Mono VP (s/4)	
		B, BF, W	125	1040	2000	730	Carbon		-	1000	-		
Incerti et al. 2017	S, DT	B, W	120	800	2800	630	Basalt	E, I	0.032	200	1700	Cyclic VP (s/4)	On site
Pantò et al. 2017	S	-	95	95	1500	432.5	Steel	E, I	0.084	95	3060	Mono VP (s/3)	Scale 1:2
Ramaglia et al. 2015	S	B	120	2200	2980	1140	Basalt	E	-	-	-	ST	
Ramaglia et al. 2017	R	B, W	120	1160	2980	1140	Basalt	E	0.039	1160	60	ST	

Notations: t arch thickness, w arch width, s arch span and r arch rise (measured from the base of the arch as in Figure 5); t_f equivalent thickness of fibre provided by manufacturer, b_f fibre width, σ_f fibre maximum tensile strength. As for boundary conditions: B bare (i.e. with skewbacks or horizontal imposts); SW spandrel walls; PSW partial spandrel walls; BF backfilling; W vertical walls at abutments. As for arch shape: S segmental, C catenary, $3C$ three centred, R round, DT different thickness between haunches and crown. As for installation of fibres: E extrados, I intrados, B both extrados and intrados. As for test type: $Mono$ quasi-static monotonic; $Cyclic$ quasi-static cyclic; VP vertical point load; VD vertical distributed load;

Appendix 2.   Mechanical properties of materials and geometric features of selected experimental campaigns.

Reference	Masonry type	f_{cb} MPa	f_{tm} [1] MPa	f_{cm} MPa	f_m [2] MPa	ρ kg/m³	r [6] mm	r/s -	t/s -	N_o* kN
Alecci et al. 2016, 2017a, b	Clay bricks / cement-lime mortar	24.08	1.49	3.22	7.24	1800 [3]	462	0.31	0.07	0.60
Bednarz et al. 2011	Clay bricks / lime mortar	10.00 [4]	-	1-5 [4]		1800 [3]	893	0.26	0.04	2.43
Bertolesi et al. 2017	Clay bricks / lime mortar	10.70	0.12	2.50	3.80	1800 [3]	297	0.14	0.06	2.82
Borri et al. 2007, 2009	Concrete bricks / cement mortar	43.30	-	0.80	7.19	2000	490	0.24	0.05	0.55
Cescatti et al. 2017	Clay bricks / lime mortar	19.47	-	3.00	3.00	1800 [3]	1411	0.25	0.06	14.17
Garmendia et al. 2011, 2014	Sandstone / lime mortar	21.3	0.98	2.03	5.79	2000	370	0.33	0.11	0.79
Garmendia et al. 2015	Clay bricks / NHL mortar	19.80	1.90	7.30	8.07	1800 [3]	969	0.33	0.04	2.56
Girardello et al. 2013	Clay bricks / NHL mortar	17.68	0.74	1.75	4.86	1800 [3]	969	0.33	0.04	0.79
Hojdys et al. 2013, 2015, 2016a, b	Clay bricks / NHL mortar	24.40	-	1.10	5.30	1800 [3]	762	0.36	0.06	2.13
Incerti et al. 2017	Clay bricks / lime mortar	-	-	-	3.45 [5]	1800 [3]	324	0.12	0.04	28.84

Notations: f_{cb} compressive strength of brick, f_{tm} tensile strength of mortar, f_{cm} compressive strength of mortar, f_{cm} compressive strength of masonry, ρ density of masonry, N_o* normalized compressive force according to Equation 2; NHL natural hydraulic lime

[1] from flexural test
[2] calculated by Authors as $0.55 f_{cb}^{0.7} \cdot f_{cm}^{0.3}$ according to EN 1996-1-1 (2005).
[3] assumed by Authors from full clay bricks and lime mortar masonry type according to MIT (2019)
[4] from technical sheet of manufacturer
[5] assumed by Authors as average value of full clay bricks and lime mortar masonry type according to MIT (2019)
[6] rise of arch measured up to upper hinge position (neglecting buttresses)

Appendix 3. Results of experimental campaigns (monotonic and cyclic tests) and of models according to RING and limit analysis.

Reference	Reinforcing system	Id	F_0 kN	F_s kN	F_s/t kN/mm	Failure mode	ω	F Ring	F Limit Analysis
Alecci et al. 2016, 2017a, b	-	1US	0.91	-	-	M	-	0.19	0.17
	-	2US	1.07	-	-	M	-	0.19	0.17
	PBO-FRCM	1ES	-	4.97	354.9	D	0.0007	0.64	2.24
	PBO-FRCM	2ES	-	4.81	343.8	S+D	0.0007	0.64	2.24
	PBO-FRCM	1-PM	-	5.28	377.1	S	0.0012	0.74	1.94
	PBO-FRCM	2-PM	-	5.67	405.1	S	0.0012	0.74	1.94
	C-FRCM	1-CM	-	7.14	151.9	S	0.0033	1.67	5.18
	C-FRCM	2-CM	-	6.99	148.7	S	0.0033	1.67	5.18
	B-FRLM	B-FRLM	-	5.37	167.7	M+S	0.0014	1.09	4.44
Bednarz et al. 2011	-	A1	2.88	-	-	-	-	1.78	1.79
	CFRCM	A2	-	10.89	231.6	D	0.0019	56.70	37.78
	CFRCM	A4	-	26.37	561.0	-	0.0019	97.90	73.77
Bertolesi et al. 2017	-	US	2.08	-	-	M	-	1.82	1.68
	SRG	RA	-	8.29	70.2	D	0.0015	8.74	5.20
Borri et al. 2007, 2009	-	UN.01	0.70	-	11.4	-	-	0.53	0.54
	SRG	EX.01	-	9.20	29.0	S	0.0066	13.10	10.06
	SRG	EX.03	-	23.5	20.0	M	0.0066	16.30	19.59
	SRG	IN.01	-	16.2	28.4	D	0.0025	4.83	4.17
	SRG	IN.03	-	23.00	40.5	S+D	0.0050	13.20	15.05
	SRG	IN+EX.01	-	32.8	14.9	S	0.0163	16.90	11.43
Cescatti et al. 2017	-	01_URM	19.04	-	-	M	-	18.80	23.49
	SRG	04_SRG		86.64	-	C	0.0009		83.22
	BTRM	05_BTRM	-		-	M	0.0003	90.60	63.06
			-	57.75				63.50	
Garmendia et al. 2011, 2014	-	A1	0.98	-	-	M	-	1.57	1.08
	-	A2	1.45	-	-	M	-	1.57	1.08
	BTRM	EX.1	-	19.30	214.4	L	0.0005	9.22	3.20
	BTRM	EX.2	-	16.83	187.0	D	0.0005	9.22	3.20
	BTRM	EX.3	-	12.65	140.6	C	0.0005	9.22	3.20
	BTRM	IN1	-	8.52	94.7	S	0.0005	9.29	3.20
	BTRM	IN2	-	15.33	170.3	S	0.0005	9.29	3.20
	BTRM	IN3	-	10.07	111.9	D	0.0005	9.29	3.20
	BTRM	EXIN1	-	28.96	332.9	D	0.0005	16.30	5.31
	BTRM	EXIN2	-	28.30	314.4	C	0.0005	16.30	5.31
	BTRM	EXIN3	-	21.05	233.9	S	0.0005	16.30	5.31
Garmendia et al. 2015	-	R1	4.50	-	416.9	M	-	1.46	1.11
	BRP	BRM	-	22.10	300.0	M	0.0003	11.00	7.65
	SRM	SRM	-	20.40	22.9	M	0.0008	26.90	19.20
Girardello et al. 2013	-	VM	1.38	-	-	M	-	1.21	1.27
	SRG	VM SRG	-	13.47	156.6	S	0.0012	24.70	19.61
	SRG	VC_SRG	-	14.33	166.6	S	0.0012	24.70	19.61
	BTRM	VC_BTR	-	11.49	216.8	M	0.0016	31.10	13.29
Hojdys & Krajewski 2013, 2015, 2016a, b	-	AU	4.10	-	-	M	-	2.44	2.81
	GTRM	ASE	-	31.00	885.7	L	0.0006	9.99	13.51
	GTRM	ASI	-	22.80	651.4	L	0.0006	28.20	13.51
	GTRM	V_LC	24.70	40.10	-	L	0.0006	7.99	13.51
		V-ECA	24.70	80.00	-	S	0.0006	8.99	13.51
	-	V-Qtr	4.50	-	-	M	-	-	-

(Continued)

Appendix 3.   (Cont.)

Reference	Reinforcing system	Id	F_0 kN	F_s kN	F_s/t kN/ mm	Failure mode	ω	F Ring	F Limit Analysis
	GTRM	VG-Qtr	-	28.20	805.7	D	0.0060	7.99	13.51
	C-FRCM	VC	-	37.90	-	L	0.0060	-	-
	C-FRCM	VC_AN	-	62.50	-	-	0.0060	-	-
Incerti et al. 2017	-	V 01	17.80	-	-	M	-	4.79	3.38
	B-FRCM	V_02	-	45.77	1430.3	M	0.0007	14.40	8.60
	B-FRCM	V_03	-	40.60	1268.8	M	0.0007	13.60	8.60
	B-FRCM	V_04	-	49.51	1547.2	M	0.0007	22.80	13.82

Notations: F_0 collapse load of unreinforced arches; Fs collapse load of reinforced arches; Fs/t ratio between collapse load and arch thickness; ω ratio between resisting cross sections of FRCM and masonry; *F Ring* collapse load calculated by Ring; *F Limit Analysis* collapse load calculated through limit analysis. As for failure mode: *M* kinematic mechanism, *D* debonding of reinforcement, *S* sliding at hinges, *C* masonry crushing, *L* failure of strengthening layer.

Brick and Block Masonry - From Historical to Sustainable Masonry –
Kubica, Kwiecień & Bednarz (eds)
© 2020 Taylor & Francis Group, London, ISBN 978-0-367-56586-2

Unconventional measurement techniques in experiments on masonry

S. De Santis

Department of Engineering, Roma Tre University, Rome, Italy

ABSTRACT: A number of advanced measurement techniques are available for experimental mechanics, which are named as *unconventional* in the sense that do not make use of *conventional* displacement and strain transducers. They generally need careful setup preparation, good knowledge of fundamental operation principles and critical analysis of results. But, if so, they can provide a large number of data with relatively simple and cost-efficient implementation. A conscientious experimenter might question: (i) what useful information can be derived by unconventional techniques that would be unavailable otherwise? (ii) To what extent, or under what conditions, detected measurements are reliable? (iii) How can these measurements be used to validate or integrate those recorded by traditional devices? The paper discusses the use of Digital Image Correlation (DIC), passive 3D motion capture systems (3DVision), and acoustic emission (AE) monitoring, applied to laboratory and field investigations on masonry structures to establish the knowledge gained so far and identify some future research needs.

1 INTRODUCTION

Unconventional measurement techniques include the extremely wide range of methods used in experimental mechanics that do not make use of traditional sensors, such as potentiometers, LVDTs, laser transducers, extensometers, strain gauges, etc. These non-standard methods generally provide a large number of data and are relatively easy to use, even when they are developed using complex and advanced numerical strategies. Some of them also require limited implementation efforts in the laboratory and, all things considered, are cost efficient. Their application, however, is only apparently that simple. The reliability of results depends on careful setup preparation, good knowledge of fundamental operation principles and appropriate analysis of output data. It is worth asking what useful information can be derived by such techniques that would be unavailable otherwise, and to what extent this information can validate or integrate measurements recorded by traditional devices.

This paper discusses these issues through an overview on some unconventional measurement techniques applied to experimental tests on masonry, namely Digital Image Correlation (DIC), passive 3D motion capture systems (3DVision), and acoustic emission (AE) monitoring. Digital Image Correlation (DIC) was used for the mechanical characterization of composite materials for structural rehabilitation, providing information on displacement and strains, strain distribution and effective bond length, and crack pattern. DIC was also applied to laboratory tests on full-scale vault specimens to measure deflections and investigate arch-fill interaction. The

passive 3D motion capture system named 3DVision was used to record the displacements of full-scale masonry specimens tested on the shake table, to integrate measurements provided by displacement transducers and accelerometers, derive information of the development and opening of cracks, and run experimental modal analyses. The acoustic emission (AE) monitoring technique was used to investigate the fatigue deterioration process of small-scale samples of brick masonry in the laboratory and for the condition assessment of a masonry arch bridge, before and after a retrofitting work, under traffic loads.

2 DIGITAL IMAGE CORRELATION

2.1 *Principles*

Digital Image Correlation (DIC) is a full-field optical technique developed in the early 1980s to measure surface displacements (Peters & Ranson 1982). It has been used in experimental mechanical engineering and material science (Sebastiani et al. 2011, Sadowski & Kneć 2013), structural health monitoring (Malesa et al. 2010) and biomechanics (Sztefek et al. 2010). The DIC method is based on the correlation of digital images taken before and after deformation during a test. Many correlation criteria exist to calculate strain and displacement fields, using pre-defined shape functions, which interpolate the data measured in some points of the specimen surface. Data are obtained in pixels and converted into millimeters, so resolution and accuracy depend on the sensor of the camera and on the distance between camera and specimen, that is, on how big is a pixel. Sub-pixel interpolation algorithms can further improve resolution to 1/100th of the pixel (Pan et al. 2009).

2.2 Applications

DIC was applied to experimental tests on composite materials for structural rehabilitation, including fibre reinforced polymer/polyurethane (FRP/FRPU) and fabric reinforced cementitious matrix (FRCM) systems. The studies devoted to the use of DIC in the laboratory were carried out in Roma Tre University within research projects and international cooperation initiatives, including those of the RILEM Technical Committee 250-CSM Composites for the Sustainable Strengthening of Masonry, devoted to the development of composite materials structural rehabilitation. Externally bonded to structural members, these systems can enhance ultimate strength with negligible mass increase. Their effectiveness has fostered the scientific research, the industrial development and the spreading of field applications. Laboratory tests are carried out to derive tensile behaviour (Figure 1) and substrate-to-composite load transfer capacity (Figure 2). The reliability of the material properties obtained from the tests is crucial for both the acceptance and the design of retrofitting systems. Main data include ultimate strength and corresponding strain, stiffness, and crack spacing under tension, as well as reinforcement-to-substrate strength and relative displacement (slip). Nevertheless, cracking, fabric slippage within the matrix, sliding in the gripping areas, local damage/strain concentrations and uneven load distributions may complicate the measurement of displacements and strains accurately with traditional sensors and affect the reliability of test results. DIC has been used to overcome some of these drawbacks in experiments devoted to FRP-to-concrete (Carloni & Subramaniam 2013, Napoli et al. 2016), FRP-to-masonry (Ghiassi et al. 2013, Caggegi et al. 2016) and FRPU-to-masonry (De Santis et al. 2018b) bond, and in the characterization of FRCM tensile (Tekieli et al. 2017) and bond (De Santis 2017) behaviour.

In FRCM composites, the development of cracks in the matrix poses specific challenges to the measurement of displacements in tensile tests. The reliability of strain and stiffness data relies on the number of cracks that develop within the selected measurement length. When displacement transducers are used, such length is fixed, so, if cracking or rupture/sliding occur out of the measurement length, the portion of the specimen monitored by the instrument may be not representative of the actual overall behaviour. On the other hand, strain measurements provided by strain gauges may be too local and not particularly useful, and the damage of the textile or of the matrix may disrupt their reading, making data unavailable or unreliable. In this framework, DIC offers the advantage that the measurement points can be selected after the test, based on the crack pattern (Figure 1b). DIC was used in direct tensile tests and the obtained measurements were compared with those provided by traditional devices, namely, the LVDT integrated in the testing machine, linear potentiometers, extensometers (Figure 1a). A speckle pattern made of randomly distributed black dots on a white background was realized by means of spray painting. Photographs were taken at 10 s time interval, with a digital camera positioned

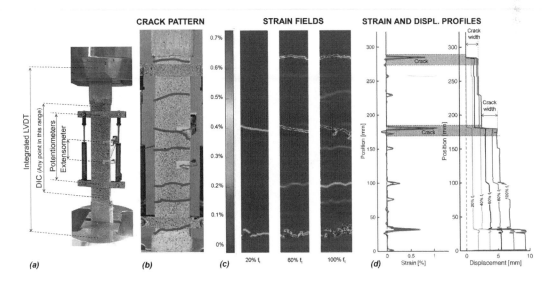

Figure 1. Direct tensile tests on FRCM coupons: experimental setup (a), crack pattern at failure (b), strain field (c) and strain and displacement profiles (d) calculated by Digital Image Correlation (DIC) during test execution (f_t denotes the FRCM tensile strength).

75

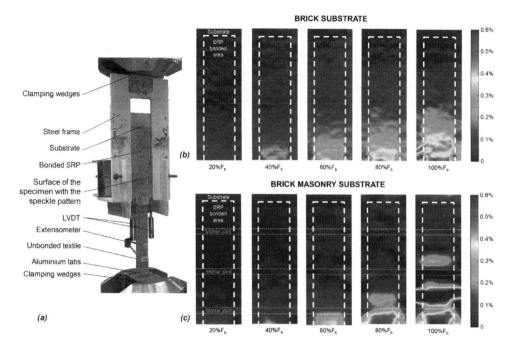

Figure 2. Steel reinforced polymer (SRP)-to-substrate shear bond: experimental single-lap setup (a) and strain fields on clay brick (b) and masonry (c) substrates calculated by DIC during text execution (*Fb* denotes the SRP-to-substrate bond capacity).

on a stiff frame at 1.20 m from the specimen, taking care of ensuring correct alignment to minimize image distortions. Four LED spotlights were used to keep stable and even illumination. Pictures had 6016 × 4016 pixels, which corresponds to a pixel size of 0.11 mm. DIC data proved reliable when the points are selected in the middle between two cracks, where, due to symmetry, the slip between fabric and matrix is negligible and, therefore, the displacement measured on the outer surface corresponds to that of the textile embedded within the matrix. Additionally, by selecting more couples of points, slippage in the gripping areas and stress concentrations were recorded. Finally, by further refining the analyses, location and width of cracks were identified before they became visible to the naked eye (Figure 1c and d).

DIC was used in shear bond tests on FRP reinforcements consisting of a steel textile applied to clay bricks and to masonry prisms by epoxy resin. The substrate-to-FRP relative displacement (slip) was calculated as the relative displacement between two points, one on the substrate and one on the unbonded portion of the steel textile. It was compared with that measured by two LVDTs, and a good agreement was found (the error was lower than 2%). Additionally, the strain fields on the surface of the specimen provided information on effective bond length and stress distributions. The strain fields (Figure 2b and c) show that the alternation of bricks

and mortar joints induces an uneven strain distribution with a periodic variation, in which stresses are concentrated at the brick surfaces and strongly reduce at the bed joints, due to their lower stiffness (Napoli et al. 2016). This full-field measurement would have been unavailable even if an extremely large number of strain gauges was glued to the cords before installation, which, on the other hand, on would have been expensive and very difficult to implement.

Figure 3 shows the use of DIC in shear bond tests on an Steel Reinforced Grout (SRG) system comprising a steel textile and a lime mortar applied to curved masonry substrates (De Santis 2017). A double-lap/double-prism setup was used. The crack pattern and the portions of reinforcement detached from the substrate changed from test to test, so the displacements measured by LVDTs and potentiometers resulted unsuitable to describe the SRG-to-substrate bond behaviour at a sufficiently local scale. Taking advantage of the possibility offered by DIC to select the measurement points in the post-processing phase, the relative displacements between the reinforcement and the upper and lower portions of the substrate were calculated accurately (Figure 3c). In order to prevent the results from being affected by possible matrix-to-textile sliding, the measurement points on the SRG strip were chosen at a distance of at least 20 mm from the cracks, which is a reasonable distance beyond which the displacement of the outer layer of mortar is expected to match that of the steel cords. Moreover, to improve the reliability of the correlation, DIC did not correlate just two points (i.e. two pixels), but two circular subsets of pixels, centered at each point and having, in this case, 30 pixel radius. Errors related to radial lens distortion (caused by the optical distortion of the image in the zones at a larger distance from its centre) were corrected to improve accuracy. Nevertheless, it has to be mentioned that 2D DIC was used to measure the relative displacements of points that

are in the middle of the specimen and close to each other, so the reliability of slip data was not affected by aberration due to surface curvature.

In addition to applications to small scale tests, DIC was used in laboratory tests also in an experimental investigation on full-scale brick vaults, provided with buttress walls and fill to reproduce the characteristic of interstorey vaults carrying a floor on top (Figure 4a). A Plexiglas panel was placed on the front side of the specimen to contain the fill soil, leaving the arch barrel, the buttresses and the gravel

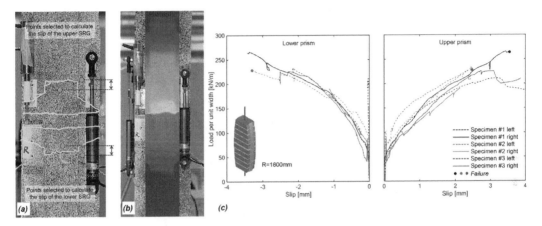

Figure 3. Steel reinforced grout (SRG)-to-curved masonry substrate shear bond test (double-lap/double-prism setup): crack pattern and identification of the measurement points (a), displacement field (b) and load versus slip response curves (c) calculated by DIC.

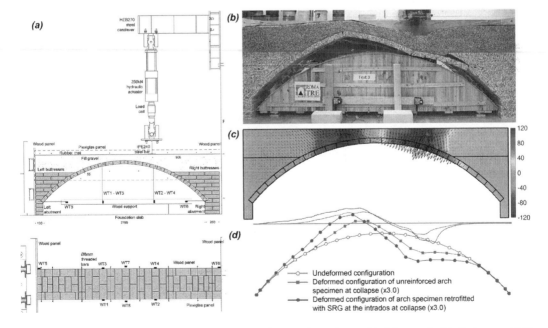

Figure 4. Laboratory tests on real scale brick masonry vaults: experimental setup (a), failure mode (b), displacement field on vault, spandrel walls and fill soil at collapse (c) and deformed configuration of the arch barrel at collapse (d) measured by DIC.

visible (De Santis et al. 2018a, 2019b). DIC was used, in addition to wire transducers, to measure the displacements of the front side of the specimen. A speckle pattern was realized on the side of the arch barrel and of the buttresses, and the backfill was made of white and black grains (Figure 4b). DIC provided the vertical and the horizontal displacements of the entire surface of the arch and of the buttresses, which would have been unfeasible unless an extremely large number of transducers was used, with unsustainable efforts in terms of cost of equipment and test preparation (Figure 4c). The deformed configuration of the arch was detected at each time of the test automatically, providing information on stiffness and deflection concentration (Figure 4d), useful for comparing the behaviour of the unreinforced arch with that of the specimens retrofitted with steel (De Santis et al. 2018a) and basalt (De Santis et al. 2019b) FRCM systems. DIC also detected crack occurrence (even before it was visible to the naked eye) and provided information on displacements and strains in the backfill (Figure 4c and d), which were associated with the spreading of the load from the top of the fill gravel to the extrados of the vault and with the volume of backfill involved in the structural response. It was found that, with respect to the unreinforced specimen, all FRCM systems limited the opening of cracks on their side (either extrados or intrados) and the onset of the four-hinge failure mechanism, avoided displacement concentrations and entailed a larger volume of mobilized backfill. Additionally, whilst the initial stiffness was not significantly modified, the displacement capacity was improved in terms of both peak and ultimate displacements.

2.3 *Advantages and limitations*

In experiments on composite materials for structural retrofitting and masonry structures reinforced with composites, DIC proved useful for detecting information on displacements, crack location and width, and composite-to-substrate load transfer mechanism, which would have been unavailable with traditional devices within reasonable costs and efforts for setup implementation. The possibility of selecting several measurement points after the test allowed overcoming the drawbacks of displacement transducers when the crack pattern influenced the response of the material and the reliability of the instruments whose measurement length is fixed a prori. A huge number of virtual markers can be selected, with negligible increase of time and no additional cost. On the other hand, the use of DIC needs particular care in the preparation of experimental setup. A speckle pattern, consisting of black dots randomly distributed over a white surface, needs to be realized on the specimen surface, by means of spray or airbrush painting. A high-resolution camera is needed, to be placed on a stiff frame, ensuring the parallelism between the surface of the specimen and the sensor

in order to minimize errors. Stable lighting conditions are also needed. Measurement points need to be selected carefully, considering that only the outer surface of the specimen is monitored, so possible internal slippage is not detected, and that the strain value provided within the crack is unreliable. DIC generally provides data with lower resolution, higher noise, and at lower sampling frequency (in the order of 1Hz or less with ordinary digital cameras) than traditional devices. Finally, with 2D-DIC, in order to avoid errors associated to lens distortion, the surface of the specimen should be plane and only the central portion of the image should be processed. 3D-DIC, instead, offers the possibility of monitoring spatial displacements of more complex specimens (Di Benedetti et al. 2015), at higher costs for equipment and software tools for image processing.

3 PASSIVE 3D MOTION CAPTURE SYSTEM

3.1 *Principles*

3DVision is a 3D motion capture system for the measurement of displacements in dynamic tests. It was developed by De Canio and coworkers for applications to shake table tests (De Canio et al. 2013), to overcome some drawbacks of conventional instrumentation, such as encumbrance, limitations in the number of sensors, range restrictions, and risk of damage of the devices. Other computer vision and image processing techniques had already tried to reach the same goal specifically for shake table tests (Beraldin et al. 2004, Lunghi et al. 2012). 3DVision makes use of spherical retro-reflecting markers installed on the specimen and of near infrared (NIR) digital cameras. NIR light emitted by LEDs illuminates the specimen and the radiation reflected by the markers is acquired by the cameras and analyzed. Setup is designed in order for each marker to be seen by at least three cameras, such that its position is determined by triangulation. Spatial displacement, velocity and acceleration of each marker at each time instant of the test (sampling frequency can reach 2000 Hz) is then calculated through a displacement data processing procedure, which includes a smoothing filter for reducing the noise (De Canio et al. 2016). Processed measurements are used for the analysis of the response of the specimen tested on the shake table, including the determination of fundamental frequencies and modal shapes, determined through experimental modal analysis (EMA).

3.2 *Applications*

3DVision was applied to a shake table test carried out on full-scale masonry walls (Figure 5a), a two-leaves irregular stone masonry wall and a single-leaf wall built with squared tuff blocks. The tests were performed in cooperation with ENEA at Casaccia Research Centre, near Rome. Natural accelerograms (labelled with the codes of the record stations BGI, MRN, AMT, NCR, AQV in Figure 6) were applied at the base of the specimens. The amplitude of input signals was scaled and progressively increased to collapse (De Santis et al. 2019a). The experimental setup was designed to induce out-of-plane vertical bending under earthquake base motion (Figure 5b). The walls were tested unreinforced, retrofitted with mortar-based composites (basalt meshes on the stone masonry wall and steel reinforced grout strips on the tuff wall) and tested again. A total of 50 wireless retro-reflecting spherical markers were glued to the specimens (Figure 5c) to measure displacements and derive information on acceleration and velocity time histories and on dynamic identification parameters. For validation of 3DVision measures, some markers were placed in the same position of accelerometers, both on the shake table and on the specimens.

Moreover, the fundamental frequencies calculated on the basis of 3DVision data were compared with those provided by velocimeters placed on the specimens under environmental noise. Both the comparisons showed a good agreement and were used to refine the calibration of the filtering parameters during displacement data processing.

The measurements taken on the tuff wall are shown in Figure 6. Thanks to the possibility of monitoring a large number of markers, 3DVision provided information on the out-of-plane displacements (δ) at different heights (z) in each time instant. The deflection profiles were built and compared, to develop a deeper understanding of the effect of the retrofitting systems on the seismic response of the walls. The response of the unreinforced wall was characterized by the development of a crack in the middle, behaving as a hinge. Another crack developed at the

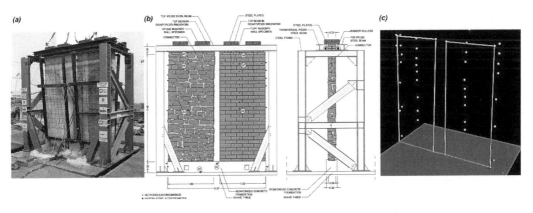

Figure 5. Shake table tests on masonry walls: experimental setup (a), instrumentation (b) and 3DVision system (c).

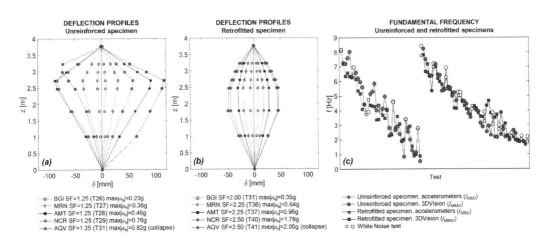

Figure 6. Shake table tests on a tuff masonry wall: deflection profiles of unreinforced (a) and SRG-retrofitted (b) specimen measured by 3DVision and fundamental frequency calculated through Multi/Input/Multi-Output (MIMO) experimental modal analyses starting from 3DVision data for all the tests of the investigation, before and after retrofitting (c).

base, whereas rotations were free on top. Therefore, the dynamic behaviour was similar to that of a two nearly-rigid block system (Figure 6a). The profiles of the retrofitted wall show smaller and more distributed displacements both under analogous seismic input and at the end of the test series (Figure 6b), when much higher intensities of earthquake base motion were attained. This indicates that displacements and damage were reduced thanks to the tensile strength of the reinforcements which made the response similar to that of a simply supported beam (Figure 6b).

3DVision was used also for a Multi-Input/Multi-Output (MIMO) experimental modal analyses. The horizontal displacements of the markers were derived two times to obtain acceleration time-histories. The fundamental frequency of the wall (f) was calculated through the transfer function between the acceleration time-histories recorded by the sensors on the wall (output signals) and on the foundation (input signals). The output signal was calculated as the average of the accelerations of the markers on the upper half of the wall, whereas the input signal was calculated as the average of the accelerations of the markers on the shake table. The use of a large number of input and output points makes the results of a MIMO approach more stable (less noisy) than those of a Single-Input/Single-Output (SISO) one, in which one sensor is taken for the input and one for the output signals. MIMO and SISO analyses (obtained from 3DVision and accelerometers, respectively) are compared for validation in Figure 6c. The plot also shows that the presence of the reinforcements did not significantly modify the initial frequency but entailed a slower damage accumulation with respect to the unreinforced specimen.

3.3 *Advantages and limitations*

3D Vision proved able to record the spatial displacements of a large amount of points during shake table tests on full-scale structural specimens. The accuracy of displacement measurements relies on the careful preparation of the setup and on the calibration of the system. The reliability of velocity and acceleration time histories requires that stable filter algorithms are implemented within the post-processing of displacements. If so, the methods allows, not only displacements and acceleration, but also the dynamic properties of the tested structure to be identified. Setup implementation is not limited by the encumbrance of traditional devices (displacement transducers, accelerometers, velocimeters) and is made easier by the use of wireless markers, so no stiff structures are required for the installation of transducers. Finally, since the markers are very cheap, there is no risk of damage to expensive instrumentation. On the other hand, an important initial investment is required for purchasing the equipment and the software and for installing and calibrating it in the laboratory. Finally, the large output files

generated for each shake table tests need to be processed carefully by expert scientists.

4 ACOUSTIC EMISSION MONITORING

4.1 *Principles*

When cracking occurs in a material, elastic waves are emitted and propagate. Piezoelectric sensors on the surface can detect them in the form of Acoustic Emission (AE) signals. These are pre-amplified, pass-band filtered, amplified, sent to a data logger, and processed. AE events are considered as the portion of the acoustic wave above a minimum amplitude threshold associated with background noise. An AE event is characterized by its amplitude (peak of the signal, detected in µV and converted in dB_{AE}), count (number of times the signal rises and crosses the threshold), energy (area under the envelope of the AE event), duration, and waveform. The AE technique has already been widely used for the condition assessment and structural health monitoring of reinforced concrete and steel structures (Holford 2009), whereas its application to masonry is very limited due to the difficulties in detecting AE signals, which rapidly attenuate as a consequence of the heterogeneity of masonry (Royles 1991, Melbourne & Tomor 2006, Carpinteri & Lacidogna 2007).

4.2 *Applications*

The studies on the use of the AE monitoring were carried out at the University of the West of England at Bristol, UK, within a research project devoted to the Fatigue behaviour and remaining service life of masonry arch bridges. First, the AE monitoring technique was applied in the laboratory to experimental tests on masonry prisms subject to monotonic compression and high-cycle fatigue loading (Figure 7a), in order to detect the development of internal cracks and monitor progressive accumulation of damage (Tomor et al. 2013). Two resonant sensors were attached to the specimens by hot-melt glue (Figure 7b). Figure 7c-f show an AE recording during a quasi-static test in terms of signal amplitude versus applied load (Figure 7c and e) and AE energy versus load (Figure 7d and f) (note that the load is expressed in terms of percentage of its maximum value, F/F_{max}). Under monotonic compression, the amplitude and the energy of detected signals increased with the increase of the load, as a consequence of the development of cracks (Figure 7c) and three crack development stages were identified. In the first stage (I), compaction and elastic behaviour of the mortar occurred, with constant low-level emission, up to 40% F_{max}. In the second stage (II), vertical micro-cracks developed in the mortar joints and in the bricks, with a rapid increase followed by a slight reduction of released energy, up to 95% F_{max}. Finally, in the last

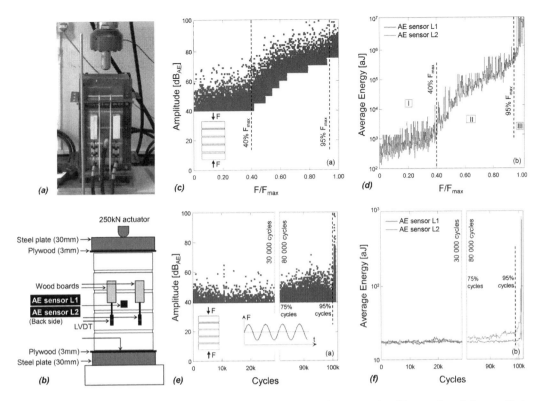

Figure 7. Compression tests on masonry prisms: experimental setup (a), instrumentation (b), acoustic emission amplitude (c,e) and energy (d,f) recorded during compression tests under monotonic (c,d) and high-cycle fatigue (e,f) loading.

stage (III), macro-cracking and bridging took place with a very high energy release associated with the rupture of the bricks (Figure 7d). Fatigue tests were carried out under cyclic compression, varying with a sinusoidal law at 2 Hz frequency between 10% and 70% of the static strength. AE monitoring showed that, as for monotonic tests, three stages of the deterioration process can be distinguished, such as (I) relatively low and constant emission, up to 75% cycles, (II) small increase in emission, up to 95% cycles, and (III) rapid increase in emission and sudden failure. The increase in acoustic emission activity, in terms of both amplitude (Figure 7e) and energy (Figure 7f) prior to failure can warn of collapse.

After laboratory investigations, the AE monitoring technique was tested in the field. A brick masonry arch bridge was monitored under both controlled load (a lorry driving at fixed speeds) and normal road traffic conditions (8h continuous monitoring). Eight permanent resonant sensors were installed on the bridge using a two-component structural adhesive near existing cracks. Data were recorded before and after retrofitting to investigate the differences of crack activity and contribute assess the effectiveness of the work undertaken (De Santis & Tomor 2013). The results of AE monitoring under controlled load tests are shown in Figure 8. Before strengthening, the maximum amplitude was proportional to the vehicle speed. AE records suggested that the bridge was hinging under the moving load (the masonry was crushing during crack opening/closing) and that there was some grinding at a ring separation (sign of sliding in the cracked joint between arch rings). High activity was also detected in the spandrel wall below the crack on the parapet (Figure 8).

The bridge was repaired and strengthened by installing radial pins to prevent ring separation and by inserting flexible joints into the cracks to avoid further damage development while keeping flexibility. The tests under controlled load after retrofitting showed a significant decrease of acoustic emissions for all sensors, indicating that the flexible joints reduced grinding/crushing in the cracks and that radial connectors effectively reinstated the barrel vault.

4.3 Advantages and limitations

The Acoustic Emission monitoring technique makes it possible to identify damage occurrence and crack activity, providing information on the deterioration of masonry even in the inner part of a structure. The method is suitable for condition assessment under

81

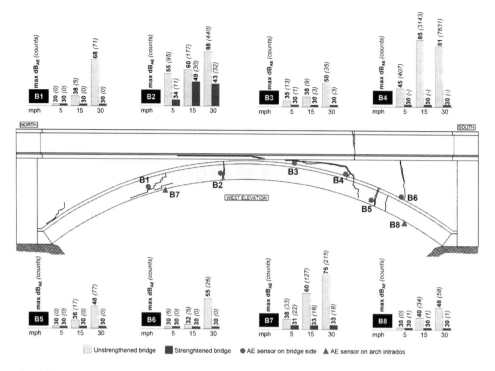

Figure 8. Field monitoring of a masonry arch bridge under controlled load before and after retrofitting: maximum amplitude and count of AE signals detected by the AE sensors installed on the bridge side and on the intrados of the barrel vault.

live loads with relatively low cost, reduced time efforts for the installation of the equipment and without requiring the structure to be closed (e.g. the traffic on a bridge does not need to be stopped). On the other hand, at present stage, it is very difficult to quantify signal characteristics for different types of damage or failure (crushing, sliding, grinding) and identify where damage is developing or what material is experiencing crack activity (bricks/stone units, mortar joints). Finally, a major obstacle to a more confident use of AE monitoring in masonry is represented by signal attenuation in masonry. Sensors need to be placed near existing cracks, so prior knowledge on damage state is essential for an appropriate design of sensors layout.

On the other hand, this information is rarely self-sufficient and data from other instruments are useful or necessary, at least for validation, calibration or for tuning the filtering parameters used in data processing. Some obstacles still remain for the application of unconventional measurement techniques. First, specific expertise is needed for their implementation in the laboratory (setup preparation) and for the analysis of acquired data. Some methods require expensive equipment, which means that an initial important economical investment is needed. Finally, their application in certification protocols or standardized experimental procedures may be limited by the difficulty of calibrating the instruments and getting a calibration certificate.

5 CONCLUSIVE REMARKS

A number of experiences exist, which show the fruitful application of unconventional measurement techniques in several fields of experimental research on masonry. In this paper, the use of three methods, such as Digital Image Correlation, 3D motion capture systems, and Acoustic Emission monitoring, to tests on masonry specimens, is discussed, as the sake of an example, but many other techniques are also available. Most of them are able to provide information that would be unavailable with traditional measurement devices, or only with very high costs and complications.

ACKNOWLEDGEMENTS

The studies presented in this paper were carried out within the research projects "Fatigue behaviour and remaining service life of masonry arch bridges" funded by EPSRC, UK (2008-2011), "Composites with inorganic matrix for sustainable strengthening of architectural heritage" funded by the Italian Ministry for Foreign Affairs (2016-2018, Grant N. PGR00234), "SICURA, Sustainable technologies for the seismic protection of the cultural heritage" funded by Regione Lazio (2018-2020, N. 15136), and "SISMI, Technologies for the improvement of the seismic safety and the restoration of historical centres" funded by Regione Lazio (2018-2019, N. 22535). Funding is acknowledged also from the Italian Ministry of Education, University and Research (MIUR), in the frame of the Departments of Excellence Initiative 2018-2022, attributed to the Department of Engineering of Roma Tre University.

REFERENCES

Beraldin, J.A. et al. 2004. Applications of photo-grammetric and computer vision techniques in shake table testing. *Proceedings of the 13th World Conference on Earthquake Engineering (13WCEE), Vancouver, BC, Canada.*

Caggegi, C. et al. 2016. Strain and shear stress fields analysis by means of Digital Image Correlation on CFRP to brick bonded joints fastened by fiber anchors. *Construction and Building Materials* 106: 78–88.

Carloni, C. & Subramaniam, K.V. 2013. Investigation of sub-critical fatigue crack growth in FRP/concrete cohesive interface using digital image analysis. *Composites Part B: Engineering* 51: 35–43.

Carpinteri, A. & Lacidogna, G. 2007. Damage evaluation of three masonry towers by acoustic emission. *Engineering Structures* 29(7): 1569–1579.

De Canio, G. et al. 2016. Passive 3D motion optical data in shaking table tests of a SRG-reinforced masonry wall. *Earthquakes and Structures* 10(1): 53–71.

De Canio, G. et al. 2013. 3D motion capture application to seismic tests at ENEA Casaccia research center. 3DVision system and DySCo virtual lab. *WIT Transactions on The Built Environment* 134: 803–814.

De Santis, S. 2017. Bond behaviour of Steel Reinforced Grout for the extrados strengthening of masonry vaults. *Construction and Building Materials* 150: 367–382.

De Santis, S. et al. 2019a. Out-of-plane seismic retrofitting of masonry walls with Textile Reinforced Mortar composites. *Bulletin of Earthquake Engineering* 17(11): 6265–6300.

De Santis, S. et al. 2018a. Full-scale tests on masonry vaults strengthened with Steel Reinforced Grout. *Composites Part B: Engineering* 141: 20–36.

De Santis, S. et al. 2019b. Retrofitting of masonry vaults by basalt-textile reinforced mortar overlays. *International Journal of Architectural Heritage* 13(7): 1061–1077.

De Santis, S. et al. 2018b. Durability of Steel Reinforced Polyurethane to substrate bond. *Composites Part B: Engineering* 153: 194–204.

De Santis, S. & Tomor, A.K. 2013. Laboratory and field studies on the use of acoustic emission for masonry bridges. *NDT & E International* 55: 64–74.

Di Benedetti, M. et al. 2015. 3D-DIC for strain measurement in small scale GFRP RC specimens. In: *Proceedings of SMAR 2015, 3rd Conference on Smart Monitoring, Antalya, Turkey.*

Ghiassi, B. et al. 2013. Application of Digital Image Correlation in investigating the bond between FRP and masonry. *Composite Structures* 106: 340–349.

Holford, K.M. 2009. Acoustic emission in structural health monitoring. *Key Engineering Materials* 413-414: 15–28.

Lunghi, F. et al. 2012. Computer vision system for monitoring in dynamic structural testing. *Geotechnical, Geological and Earthquake Engineering* 22: 159–176.

Malesa, M. et al. 2010. Monitoring of civil engineering structures using Digital Image Correlation technique. *The European Physical Journal Conferences* 6: 310–314.

Melbourne, C. & Tomor, A.K. 2006. Application of acoustic emission for masonry arch bridges. *Strain* 42(3): 165–172.

Napoli, A. et al. 2016. Bond behaviour of Steel Reinforced Polymer strengthening systems. *Composite Structures* 152: 499–515.

Pan, B. et al. 2009. Two-dimensional digital image correlation for in-plane displacement and strain measurement: a review. *Measurement Science and Technology* 20(6): 062001.

Peters, W.H. & Ranson, W.F. 1982. Digital imaging techniques in experimental stress analysis. *Optical Engineering* 21(3): 427–431.

Royles, R. 1991. Acoustic emission monitoring of masonry arch bridges. *British Journal of Non-Destructive Testing* 33(7): 339–343.

Sadowski, T. & Kneć, M. 2013. Application of DIC technique for monitoring of deformation process of SPR hybrid joints. *Archives of Metallurgy and Materials* 1 (58): 119–125.

Sebastiani, M. et al. 2011. Depth-resolved residual stress analysis of thin coatings by a new FIB–DIC method. *Materials Science and Engineering: A* 528(27): 7901–7908.

Sztefek, P. et al. 2010. Using digital image correlation to determine bone surface strains during loading and after adaptation of the mouse tibia. *Journal of biomechanics* 43(4): 599–605.

Tekieli, M. et al. 2017. Application of Digital Image Correlation to composite reinforcements testing. *Composite Structures* 160: 670–688.

Tomor, A.K. et al. 2013. Fatigue deterioration process of brick masonry. *Masonry International* 26(2): 41–48.

Brick and Block Masonry - From Historical to Sustainable Masonry –
Kubica, Kwiecień & Bednarz (eds)
© 2020 Taylor & Francis Group, London, ISBN 978-0-367-56586-2

Bond behaviour in lime-based textile reinforced mortars

B. Ghiassi

Centre for Structural Engineering and Informatics, Faculty of Engineering, University of Nottingham, Nottingham, UK

A. Dalalbashi

Department of Civil Engineering, ISISE, University of Minho, Azurém, Guimarães, Portugal

D.V. Oliveira

Department of Civil Engineering, ISISEIB-S, University of Minho, Azurém, Guimarães, Portugal

ABSTRACT: Application of textile-reinforced mortar (TRM) composites have for strengthening of existing structures or for production of new thin structural elements has attracted a growing recent attention. TRMs are made of continuous fibres (in the form of fabric or mesh) embedded in an inorganic matrix forming a composite material. The large variety of available fabric (glass, steel, basalt, PBO, etc.) and mortar types (cement-based, lime-based, etc.) leads to a wide range of mechanical properties making these composites suitable for fit-for-purpose design applications. Due to mechanical and hygrothermal compatibility issues, lime-based TRMs are the preferred choice for application to existing masonry and historical structures. Meanwhile, cement-based TRMs are usually employed for application to existing concrete or new masonry structures. The main characteristic behaviour of these composites is the tension stiffening response and distributed cracking under tensile loads which are highly influenced by the fabric-to-mortar bond behaviour. Fundamental understanding of this mechanism (the fabric-to-mortar bond behaviour) and parameters affecting that are therefore of critical importance of designing TRM composites with desired properties. This paper presents and overview of the recent studies we performed during the lat years for better udnertanding this mechanism.

1 INTRODUCTION

The special feature of textile reinforced mortar (TRM) composites and their advantages with respect to fibre-reinforced polymers (FRPs) have made them an interesting choice for externally bonded reinforcement of masonry and reinforced concrete structures (Triantafillou & Papanicolaou 2005, Papanicolaou et al. 2007, Carozzi & Poggi 2015, Mazzuca et al. 2019).

Textile reinforced mortar (TRM) and textile reinforced concrete (TRC), composed of continuous fibres embedded in mortar or concrete, are novel composite materials recently received an extensive attention for repair of existing masonry and concrete structures and for construction of new structural and non-structural components such as facades, or light structures. These composites show a considerable tensile strength with a pseudo-ductile response, owing to their relatively large post-cracking deformation capacity and multiple cracking response before failure. The large variety of available fabrics and mortar types allow development of TRM composites with a large range of mechanical properties (Caggegi et al. 2017, De Santis et al. 2017, Leone et al. 2017). However, in order to fully utilize the potential of

these materials, it is necessary to understand the mechanisms responsible for structural behaviour of these composites and in particular the role of fabric-to-mortar bond behaviour on their response.

TRMs used for strengthening of masonry structures are usually composed of glass, steel, basalt, PBO or other natural/synthetic fabrics embedded in cementitious or lime-based matrices. While cementitious matrices are preferred for strengthening of new (high quality) masonry structures, lime-based mortars are the preferred choice for strengthening of historical or weak structures due to their hygrothermal and mechanical compatibility.

While several studies have been devoted to investigation and development of TRC composites, there are still several open issues regarding the mechanics and durability of these composites. Meanwhile, investigations on TRMs for strengthening of masonry and concrete structures are very recent. The existing experimental and numerical modelling studies in this field, and consequently understanding of the mechanisms affecting their performance, are therefore limited.

This paper presents an overview of the recent experimental and analytical investigations on the fabric-to-mortar bond behaviour in TRM composites commonly used and recommended for strengthening of masonry structures. (Ghiassi et al. 2016, Dalalbashi et al. 2018a, b, Dalalbashi et al. 2019). The role of a range of critical parameters, such as test setup, fabric type, embedded length and mortar age, on the bond behaviour is investigated and discussed.

2 TRM SYSTEMS USED IN THIS STUDY

A range of fabric and mortar types are available in the market for development of TRMs. TRM composites commonly used for strengthening of masonry structures are usually made of glass, steel, basalt or PBO embedded in a cementitious or a lime-based matrix. Lime-based mortars are the preferred choice when strengthening of existing weak or historical structures are of concern.

The experimental results presented in this paper are performed on two commercially available lime-based TRMs: a steel-based and a glass-based TRM.

The steel-based TRM was composed of a commercially available pure natural NHL 3.5 lime and mineral geo-binder with a 28-day compressive strength of 9.53 MPa (CoV = 11.1%) and a flexural strength of 2.54 MPa (CoV = 9.6%) as the matrix and a unidirectional ultra-high tensile steel sheet (GeoSteel G600, density of 670 g/m^2 and effective area of one cord of 0.538 mm^2) as the reinforcing material. Each steel fibre is made by twisting five individual wires together, three straight filaments wrapped by two filaments at a high twist angle, forming a uniform cord and a non-smooth surface that ensures a good mechanical bond with the matrix. The tensile strength, elastic modulus and ultimate strain of the steel cords were experimentally obtained as 2972 MPa, 189 GPa and 1.88%, respectively.

The glass-based TRM was made of a commercially available hydraulic lime-based mortar (Planitop HDM Restauro) with a 28-day compressive strength of 7.1 MPa (CoV = 10.5%) and a flexural strength of 4.71 MPa (CoV = 7.8%) as the matrix and a woven biaxial fabric mesh made of an alkali-resistance fiberglass (Mapegride G220) with a mesh size and area per unit of width area equal to 25 × 25 mm^2 and 35.27 mm^2/m as the reinforcing material. The tensile strength, elastic modulus and ultimate strain of the glass yarns were experimentally obtained as 875 MPa, 65.94 GPa and 1.77%, respectively.

3 BOND CHARACTERIZATION TEST SETUP

Pull-out tests are the most common testing methodology for characterization of the interface properties in composite materials made of brittle matrices (such as TRC or TRMs). Different pull-out testing configurations have been proposed and adapted by different researchers for performing these tests. These can be generally categorized into pull-pull and pull-push

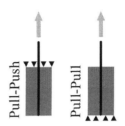

Figure 1. Schematic presentation of the pull-pull and pull-push testing configurations.

testing configurations (Boundary conditions and load application method schematically shown in Figure 1). Some discussions can be found in the literature on the advantages and disadvantages of each of these testing configurations. While it is difficult to draw conclusions on the most suitable test setup, it seems that pull-push tests can be performed with more accuracy and control, but pull-pull tests introduce a more realistic representation of the stress state at the crack surface.

To evaluate the effect of test setup on the experimental force-slip curves, experimental tests were performed on steel-based TRM samples under both testing configurations as presented in Figure 2 (Dalalbashi et al. 2018a). In both cases the yarn free length was embedded in an epoxy resin block to facilitate the gripping and slip measurements during the tests. This strategy also helps in a significant reduction of variations in the experimental results compared to previous tests in which the free yarn length was not embedded as in Ghiassi et al. (2016).

The details of the experimental results and discussions can be found in Dalalbashi et al. (2018a). The results showed specimens tested under a pull-pull configuration present a higher peak debonding load but a lower initial stiffness compared to specimens tested under pull-push test setup. These differences are, in fact, due to the differences in the boundary condition and stress distributions in the samples tested under these two configurations. Clearly this should be

Figure 2. Pull-pull and pull-push test setups.

considered when pull-out results obtaeind from different test setups are compared. Dalalbashi et al. (2018a) showed that when the role of boundary conditions are accurately considered in the numerical or analytical simulations, the extracted bond-slip laws, however, will only show a slight difference. We also observed that the experimental results obtained from the pull-push testing methodology are more repeatable and easier to control and for this reason we used this test setup in our further studies.

4 YARN EMBEDDED LENGTH

The role of yarn embedded length on the pull-out results was investigated by performing pull-put tests on samples with yarn (or cord) embedded lengths of 50 mm, 100 mm, 150 mm and 200 mm for the steel-based TRM and samples with embedded lengths of 50 mm, 75 mm and 100 mm for the glass-based TRM. The tests were performed under a pull-push testing configuration and under displacement controlled condition by pulling the yarns (the epoxy block) with a displacement rate of 1.0 mm/min.

In the steel-based TRM samples (Figure 3), the failure of the samples was yarn slippage from the mortar in all considered embedded lengths. The pull-out curves follow the typical pull-out response of fibres embedded in brittle matrices and can be divided into an initial linear stage, then a nonlinear stage until reaching the peak load, then a drop in the load followed by a slip hardening effect and finally softening of the force until the end of the tests.

It can be observed that peak loads increase until an embedded length of 150 mm and thereafter remain constant. Also, the slope of the slip hardening region increases with increment of the embedded length. Obviously, the toughness (defined as the are under the pull-out curves is also increased with embedded length.

In the glass-based TRM samples, on the other hand, the failure of the samples was yarn slippage at

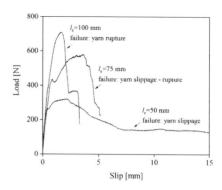

Figure 4. Pull-out response of the glass-based TRM with different embedded lengths.

50 mm embedded length which gradually changed to yarn tensile rupture until the embedded length of 100 mm. This change of failure mode is also obvious in the experimental force-slip curves (Figure 4). Again, where slippage of the yarn has occurred (i.e. in embedded length of 50 mm and 75 mm), a slip hardening is observed in the results.

5 MORTAR AGE

It is clear that the mortar age (or hydration degree) can have a significant influence on the bond performance. With increasing the mortar hydration degree, the bond performance is enhanced as stronger chemical bonds are formed between the fabric and the mortar. Meanwhile the mechanical properties of mortar is improved which affects the mechanical bond as well. Most of the existing literature on characterization of interfaces between fibres and brittle matrices is devoted to cementitious matrices (mortars and concert). Consequently, in most of these studies the pull-out tests are performed after 28 days of curing that is a standard curing time for cementitious matrices for achieving a high cement curing degree.

This curing age, however, might not be suitable for lime-based matrices, especially due to the fact that these matrices have a slower hydration rate. To evaluate this hypothesis, pull-out tests were performed on both steel-based and glass-based TRM systems at different curing ages ranging from 15 to 180 days (Figure 5 and 6).

In the case of steel-based TRM, the results does not show a specific trend with increasing the mortar age. That means that a considerable enhancement of the bond performance is observed from 15 to 30 days of curing, and thereafter, setbacks and enhancements are observed with no correlation to the mortar age. The range of variation of the average pull-out curves, however, seems to be in the same range as the common variation of the experimental results

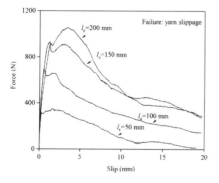

Figure 3. Pull-out response of the steel-based TRM with different embedded lengths.

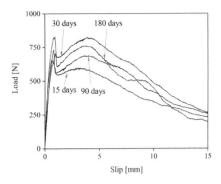

Figure 5. Pull-out response of the steel-based TRM tested at different mortar ages.

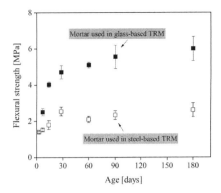

Figure 7. Average flexural strength of the mortars at different ages.

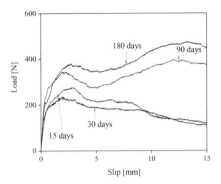

Figure 6. Pull-out response of the glass-based TRM tested at different mortar ages.

observed in pull-out tests performed on identical samples and therefore concluding that in this TRM system, a 30 days aging is a suitable test age seems reasonable.

In the case of glass-based TRM, a clear enhancement of the bond performance is observed until 180 days of mortar curing. Interestingly, the samples tested at 90 and 180 days, show two consecutive slip hardening effects in the pull-put curves. In this case, it is clear that a 30-day curing age is not sufficient for obtaining a response that is representative of the long-term pull-out behaviour of this TRM system.

The results clearly show the importance of mortar age in evaluation of the bond performance in TRM systems. The importance of this difference on the mechanical properties of TRM components (such as tensile strength or tension stiffening) is not clear and require further investigations. The variability of the available mortar types makes definition of a standard testing age complicated and also inaccurate. The results also show that comparison of the bond performance between TRMs made of different mortar should be made with special care and after consideration of the mortar hydration degree.

In the case of TRM systems studied here, there seems to be a clear correlation between the bond performance and flexural strength of the mortar in each TRM system. Figure 7 shows that the flexural strength of the mortar used in the steel-based TRM reaches its peak value at the age of 30 days and after that it fluctuates (it is increased and decreased) around this peak flexural strength. This is consistent with the observed bond performance (shown in Figure 5). In the case of glass based TRM, the mortar flexural strength is increased until 180 days that is also consistent with the observed enhancement of the bond performance in this system.

6 YARN VS FABRIC

To address the effect of fabric type on the bond performance, the differences between the pull-out response of single yarn (or cord) and fabric are investigated in this section for both of the studied systems.

In the glass-based TRM, as a bidirectional glass fabric was used, the pull-out tests were replicated on samples with a 50 mm embedded length made of a single yarn, a single yarn + transverse yarns and two yarns + transverse yarns.

The average experimental force slip curves (presented in Figure 8 in terms of the load/yarn) shows the significant effect of the transverse yarn on the bond behaviour (mainly post-peak response and toughness) in this particular system. Obviously, the effect of transverse yarns is dependent on the type of junction used for connecting the longitudinal and transverse yarns and can be different depending on the type of fabric used. These results, however, show in bidirectional fabrics the role of transverse yarns can be significant in the bond behaviour and need to be considered in the design or in extraction of the bond-slip laws.

In the steel-based TRM, as a unidirectional fabric was used, the pull-out tests were replicated on samples with a 150 mm embedded length made of a single cord, two cords four cords.

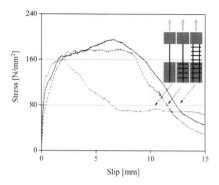

Figure 8. Fabric vs yarn pull-out response in glass-based TRM.

Interestingly, it can be observed that with increasing the number of cords, the pull-out curves show a smaller bond strength and toughness. This observation can be attributed to the group effect and the overlapping of the effect bond area around each cord on each other.

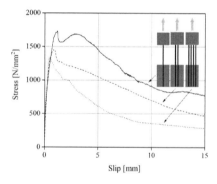

Figure 9. Fabric vs yarn pull-out response in steel-based TRM.

7 CONCLUSIONS

This paper presented an overview of recent studies on the bond performance in lime-based TRM composites. Pull-out tests were used for investigating the role of a number of critical parameters on the bond performance in these composites.

The focus was particularly on the role of test setup, embedded length, TRM type, mortar age and the differences between the yarn and fabric bond behaviour. The results although still limited and preliminary showed the important of these

factors on the bond behaviour in these systems and the need for performing further comprehensive studies for fundamentally understanding the role of each factor.

ACKNOWLEDGEMENTS

The second author wish to thank the Portuguese Scientific Foundation (FCT) for the financial support through grant SFRH/BD/131282/2017.

REFERENCES

Carozzi, F.G. & Poggi, C. 2015. Mechanical properties and debonding strength of Fabric Reinforced Cementitious Matrix (FRCM) systems for masonry strengthening. *Composites Part B: Engineering* 70: 215–230.

Caggegi, C. et al. 2017. Experimental analysis on tensile and bond properties of PBO and aramid fabric reinforced cementitious matrix for strengthening masonry structures. *Composites Part B: Engineering* 127.

De Santis, S. et al. 2017. Round Robin Test on tensile and bond behaviour of Steel Reinforced Grout systems. *Composites Part B: Engineering* 127.

Dalalbashi, A. et al. 2018a. Effect of test setup on the fiber-to-mortar pullout response in TRM composites: experimental and analytical modeling. *Composites Part B: Engineering* 143: 250–268.

Dalalbashi, A. et al. 2018b. Fiber-to-mortar bond behavior in TRM composites: effect of embedded length and fiber configuration. *Composites Part B: Engineering* 152: 43–57.

Dalalbashi, A. et al. 2019. Textile-to-mortar bond behaviour in lime-based textile reinforced mortars. *Construction and Building Materials* 227: 116682.

Ghiassi, B. et al. 2016. Multi-level characterization of steel reinforced mortars for strengthening of masonry structures. *Materials and Design* 110: 903–913.

Leone, M. et al. 2017. Glass fabric reinforced cementitious matrix: Tensile properties and bond performance on masonry substrate. *Composites Part B: Engineering* 127.

Mazzuca, S. et al. 2019. Mechanical Characterization of Steel-Reinforced Grout for Strengthening of Existing Masonry and Concrete Structures. *Journal of Materials in Civil Engineering* 31(5): 04019037.

Papanicolaou, C.G. et al. 2007. Textile reinforced mortar (TRM) versus FRP as strengthening material of URM walls: out-of-plane cyclic loading. *Materials and Structures* 41(1): 143–157.

Razavizadeh, A. et al. 2014. Bond behavior of SRG-strengthened masonry units: Testing and numerical modeling. *Construction and Building Materials* 64: 387–397.

Triantafillou, T.C. & Papanicolaou, C.G. 2005. Textile reinforced mortars (TRM) versus fiber reinforced polymers (FRP) as strengthening materials of concrete structures. *Spec. Publ.* 230: 99–118.

Brick and Block Masonry - From Historical to Sustainable Masonry –
Kubica, Kwiecień & Bednarz (eds)
© 2020 Taylor & Francis Group, London, ISBN 978-0-367-56586-2

A new method for out-of-plane reinforcement of masonry walls using high strength steel strand

M. Corradi
Northumbria University, Newcastle upon Tyne, UK
University of Perugia, Perugia, Italy

E. Speranzini & A. Borri
University of Perugia, Perugia, Italy

G. Bisciotti & S. Agnetti
Self-employed engineers, Perugia, Italy

ABSTRACT: The paper analyses the problem of assessing the seismic capacity of walls subjected to seismic loading, with particular emphasis to out-of-plane loads, before and after reinforcement. A new retrofitting method, consisting in the use of high strength steel strands embedded in the mortar bed joints was studied. The proposed method allows to preserve the fair-faced aspect of the masonry, while improving the out-of-plane capacity of the building's wall panels perpendicular to the direction of the seismic action (face-loaded walls). An experimental investigation using full-scale masonry specimens was carried out in an attempt to assess the walls' structural response when these are subject to out-of-plane loads. Test results demonstrated that it is possible to increase the out-of-plane capacity of the walls with the proposed method. However, it is shown that the out-of-plane capacity of face-loaded walls is sensitive to several parameters, which are usually affected by strong uncertainty when assessing an existing building, namely the boundary conditions, the tensile strength of the mortar and the wall-to-wall connections.

1 INTRODUCTION

Although in the 20th century masonry was largely displaced by steel and concrete in construction, the study of the structural response of old masonry structures, especially when struck by seismic loads, remains of great importance for load bearing walls in historic buildings (D'Ayala & Speranza 2003, Furukawa & Ohta 2009).

Earthquakes are a serious problem not only for the safety of people, but also for their economic cost. The Italian State Aid for reconstructing or retrofitting damaged buildings in Emilia and Abruzzo, after the 2012 and 2009 earthquake, is estimated to be 26 billion euros (Italian Senate Report 2017), and over 115 billion euros have been used between 1987 and 2012 in Italy to fund the repair of public and private building stock and, in many cases, masonry constructions were the object of these repair and retrofit interventions.

It is also important to stress that old masonry constructions often have a heritage value, with architectural and artistic assets. This complicates the work of structural engineers: besides the general requirement of improving the structural safety of the buildings to be repaired or retrofitted, engineers need to consider the important aspects of conservation. They are bound

by the fundamental principles of minimum intervention, compatibility of new materials with old masonry materials, reversibility and like-for-like repair, as defined by ICOMOS in its charters (1964, 2003).

In general, there are numerous reasons for these retrofits but one of the main deficiencies is the low capacity of the buildings' wall panels against out-of-plane actions (Pennazzi et al. 2001, Cangi 2012, Shawa et al. 2012) (Figure 1). Rocking of face-loaded walls is a main reason for fatalities and serious damage to the architectural heritage during earthquakes (Giuffrè 1995, Alexandris et al. 2004). To prevent this, increased requirements were introduced worldwide in new Building Codes.

Borri & Corradi (2018) note that in Italy previous design codes did not specify clear criteria or limits for the out-of-plane behavior of face-loaded walls. These were introduced in Italy after the 1976 Friuli earthquake. As such there is also a need in many old masonry buildings to enhance the out-of-plane capacity of single structural members or the whole building.

The current focus of most out-of-plane retrofitting research is the use of Fibre Reinforced Polymers (FRPs) (Velazquez & Ehsani 2000, Tumialan et al. 2003, Korany & Drysdale 2006) (Figure 2). Many researchers have used FRP sheets bonded to the wall panels with epoxy resin (Gilstrap & Dolan 1998,

Figure 1. Out-of-plane rocking of the façade of a masonry building due to the seismic action.

Figure 3. Traditional wrapping of a bell tower using steel bars.

Figure 4. Detail of the corner protection of a stone masonry bell tower wrapped using a steel wire.

Figure 2. Example of an "innovative" retrofitting solution using FRP strips.

Ghobarah & El Mandooh Galal 2004, Hamed & Rabinovitch 2007). Significant capacity increases have been achieved using this technique, however there are two main drawbacks. First of all, in most applications it is the debonding of the FRP sheet from the wall panel itself that governs the capacity enhancement. Regardless of the number of sheets of FRP used, if the force cannot be transferred between the sheets and the masonry, the enhancement can only reach a maximum dictated by the available bond length.

The use of metal fasteners or profiles to prevent out-of-plane rocking of wall panels is not new (Figure 3). Steel ties have been used from the 19th century for this purpose in many areas of Europe. Domes' wrapping at the base (abutments) using steel bars is another traditional intervention used by structural engineers and architects to absorb the thrust. Previous research used these metal elements to retrofit small buildings, assembled in the laboratory.

Steel cables (Figure 4) and, more recently, FRP strips have been widely used to stabilize historic masonry towers (Vinci 2014). The goal of the stabilization intervention is to hold the top of the tower in place using cables or strips wrapped around the tower. This stabilization procedure is typically employed for provisional interventions (for example to secure post-earthquake structural safety) or when the masonry buttresses need to be removed (Corradi et al. 2007, Jurina 2015, Preciado et al. 2016).

Secondly, the installation of FRP sheets requires a very careful approach. The epoxy must be spread evenly and the sheets laid as flat as possible or else localized flaws could lead to global failure of the retrofit. This is particularly challenging for applications on masonry structures: old masonry walls are often irregular and rugged. Finally, it should be considered that the long term behavior of FRPs is often unsatisfactory, with high reductions of strength with ageing (Tedeschi et al. 2014, Micelli et al. 2017).

To prevent the out of plane collapse of wall panels, traditional retrofitting methods consist in the addition of ring beams at eaves level of historic buildings (Sisti et al. 2016). More recently to overcome the problems associated with ageing and non-reversibility of bonded FRP strips, the use of TRM (Textile Reinforced Mortar) and ECC (Engineered Cementitious Composite) shotcrete were successfully proposed (Papanicolaou et al. 2008, Gattesco & Boem 2017). TRMs are typically composed of composite textiles (glass, carbon, basalt fibres) embedded into a layer of plaster, made of a lime or cement mortar (Figure 5). However, with this method it is not possible to preserve the fair-faced aspect of the masonry.

A technique for out-of-plane retrofitting has been developed that addresses some of these concerns. Early work by Borri at al. (2009) and Corradi et al. (2016) suggested that by using high-strength steel cords or wires embedded in the mortar joints using a new non-organic matrix (a cement or lime mortar), a reinforced joint repointing could be formed. Test results demonstrated that it is possible to increase

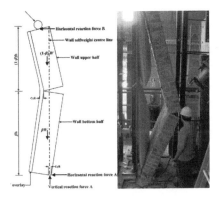

Figure 5. Out-of-plane reinforcement using ECCs and TRMs: these methods do not allow to preserve the fair-faced aspect of the masonry (Lin et al. 2016; Ismail & Ingham 2016).

the shear capacity of wall panels using this retrofitting method.

Using the same a similar method it could be possible to wrap a building or to connect two wall panels to form a strap. The strand can be tensioned with specific clamps, to realize an active retrofitting. This is aim of this research. An experimental campaign was carried out at the Structures laboratory of the University of Perugia, Italy. High-strength steel wires were used in this experiment to wrap brickwork specimens: to preserve the fair-faced aspect of the masonry, strands were embedded into the bed joints and anchored to the return walls, thus preventing the rocking of the wall panel perpendicular to the load.

2 EXPERIMENTAL WORK

Conventional methods of masonry wall construction remained almost unchanged until quite recently, especially when solid bricks or rubble stone are used.

To investigate the effects of the wraps, two full-scale C-shaped full-scale brickwork specimens were constructed and tested at the laboratory. Local skilled labour was employed for masonry construction, using the conventional methods of construction. The C-shaped specimens consisted in three weakly-connected wall panels.

2.1 Specimen geometry

Solid clay bricks (240 × 120 × 55 mm) were used to construct two double-whyte brickwork specimens. The Flemish bond pattern was chosen for construction (Lynch 1993). Each course of brick consists of alternate stretchers and headers, with the headers in alternate courses centered over the stretchers in the intervening courses. Both C-shaped specimens were

tested four times for a total of eight tests: twice without the steel-strand reinforcement (URM: unreinforced and uncracked configuration, and RMA: unreinforced and cracked), and twice, after reinforcement, according to two different loading conditions (see Section 2.2). Geometries of the masonry specimens and wrap applications are shown in Figure 6.

Each masonry specimen consisted of three wall panels to form a C-shaped specimen: two wall panels were connected to the strong RC (Reinforced Concrete) wall of the laboratory. To enhance the understanding of the out-of-plane behavior of actual walls in buildings, it's been tried to construct these walls as height as real existing structures. Thickness of the walls was 240 mm, height 2660 mm. The total length of the wall subjected to the out-of-plane load was 2430 mm.

It is worth noting that there is no brick interlocking between the wall panels at the edges (wall intersection): the gap between two wall was filled with fresh mortar, thus creating a vertical continuous joint (Figure 7). This was made to simulate the weak level of connection between wall panels in historic constructions.

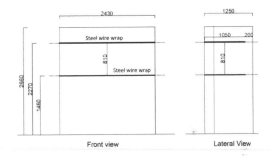

Figure 6. Layout of the strand position (bold lines = high-strength strands). Strands were embedded into the bed joints (dimensions in mm).

Figure 7. The two C-shaped brickwork specimens. In the two boxes, it can be noted the vertical mortar joint used to connect the return and face-loaded walls.

On opposite, the two lateral wall panels (return walls) parallel to the direction of the horizontal load were effectively connected to the RC strong wall using 20 steel anchors (10 anchors/panel) uniformly distributed along the panel's height. One end of the steel anchors was embedded in the bed joints of the brickwork panel and the other end was inserted in holes drilled in the RC wall and filled with cementitious grout.

2.2 Test set up

One single acting hydraulic actuator was placed inside the C-shaped brickwork structure (Figure 8). The actuator was used to apply the out-of-plane horizontal load and to simulate the seismic action [i.e. a triangular-distributed horizontal load]. Two loading configurations have been used: the actuator was initially placed at 2/3 (1880 mm) [i.e. point of application of the equivalent point load of the triangular distributed load] of the wall panel height (2660 mm) (Loading Configuration No. 1, test series RFC_M), and it was subsequently elevated up to the level of the upper steel-strand wrap at a vertical distance of 2270 mm from the panel's base (Loading Configuration No. 2, test series RFC_S).

The actuator acted, on a side, on the RC strong wall and, on the other side, on the C-shaped masonry specimen. Furthermore, a 1200 mm long spreader timber beam (cross section 200 × 200 mm) was placed between the actuator and the specimen to distribute the horizontal load, avoiding the development of high stresses near the point of application of the load. In conclusion, the horizontal load was symmetrically applied over a surface of 200 × 1200 mm.

Electromechanical transducers were used to record the horizontal movements. These were Linear Variable Differential Transformers (LVDTs). Figure 9 shows where they were mounted. LVDT1 to LVDT5 measured absolute horizontal movements of the wall panel subjected to the out-of-plane load, while LVDT6 recorded the relative movement between two intersecting wall panels.

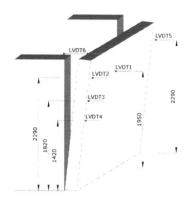

Figure 9. Location of the LVDTs' position (dimensions in mm).

2.3 Material properties

Five steel strands were cut out of the coil for mechanical characterization (Figure 10). These strands were tested in tension and their strains recorded with a 50 mm gauge length mechanical extensometer. Unfortunately, the strain gauges malfunctioned (most probably because of strand surface resulting in defective bonding of the gauge to the strand surface) and no useful results were obtained from these coupon tests for the strain response. The average of the five coupon test values is given in Table 1 and is consistent with the manufacturer's data sheet.

It was also decided to attempt to measure the mechanical properties of the two mortars (the one used for wall construction, and the other used to repoint the joints where the strands were installed) using a three-point simply supported beam bending test arrangement similar to that used in EN 1015-11 (1999). Four three-gang moulds for mortar prisms

Table 1. Properties of high-strength strands (ASTM D2256).

Diameter (mm)	6
Yielding Load (kN)	21.42
Failure Load (kN)	24.58
Yielding Strength (MPa)	882.5

Figure 8. The hydraulic cylinder used for loading. This was placed into the C-shaped brickwork specimen and it acted, on one side, on the RC strong wall, and, on the other side, on the wooden spreader beam.

Table 2. Properties of used construction materials.

Brick	Compressive Strength (MPa)	19.01
	Bending Strength (MPa)	3.61
	Weigth Density (kg/m³)	1623
Mortar No.1	Compressive Strength (MPa)	3.58
(for wall	Weigth Density [hardened] (kg/m³)	1727
construction)	Bending Strenth (MPa)	1.09
Mortar No. 2	Compressive Strength (MPa)	17.27
(for joint	Weigth Density [hardened] (kg/m³)	1917
repointing)	Bending Strenth (MPa)	4.16

Figure 10. Detail of the 6 mm-diameter high-strength steel strand.

Figure 12. A mastic gun was used before and after the application of the steel strand for joint repointing.

$40 \times 40 \times 160$ mm were used to test the two types of mortars in bending and in compression (Table 2).

With regard to the used bricks, these were clay, durable, high strength, solid engineering bricks, specially manufactured by Toppetti (Masserano, Italy) to cope with exposure to aggressive conditions (product denomination *Mattone Pieno*). Bricks were tested in accordance with EN 771-1 Standard (2001). Test results are also reported in Table 2.

2.4 *Retrofitting method*

The proposed retrofitted method consisted in the application of two steel strands to be embedded into the bed joints of the C-shaped brickwork specimen. Each strand was 5.3 m in length. The basic idea was to prevent the rocking of the face-loaded wall (simulating the seismic action) by connecting it to the return walls. The strands were invisible (hidden from view) and fully embedded into the horizontal joints, thus allowing the preservation of the fair-faced aspect of the brickwork masonry.

Two horizontal cuts were made using an electric disc cutter to allocate the steel strands into the mortar joints of the outdoor surface of the brickwork specimen (Figure 11). Attention was focused on cleaning the joints and removing dust and inconsistent material. Figure 6 shows the wall elevation and strand's position with the actual dimensions. These were inserted into two horizontal bed joints located a vertical elevation from the lab's strong floor of 1460 and 2270 mm, respectively. A mastic gun and a trowel were used to apply the new mortar to the prepared cuts, after the application of the strands (Figure 12).

In order to prevent local crushing in the masonry near the edges of the walls, small curved metal strips were interposed between the masonry and the steel strand. This served to distribute the load transmitted from the strand to the masonry, during loading (Figure 13).

Strands were anchored to the return walls using a steel plate and a pre-tensioning device. From the bed joint on the outdoor wall surface, the strand was driven into a hole in the wall and anchored to a plate on the indoor surface of the C-shaped brickwork specimen (Figure 14). The steel plate was perforated in the centre: the strand was inserted in the hole and

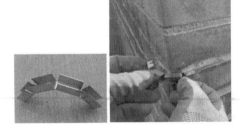

Figure 13. Detail of a curved metal strip and its application: this was applied at wall edges (in the bed joint) and interposed between the masonry and the steel strand.

Figure 14. Detail of the application of the steel strand inside the bed joint and the used anchoring system at the return walls (parallel to the horizontal load). The anchoring system is applied inside (indoor) the C-shape brickwork specimen.

Figure 11. The use of an electric disc cutter to prepare the slots for the installation of the steel strands.

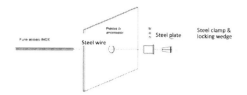

Figure 15. Anchoring & tensioning device. The axial force on the strand was applied by a hollow ram jack, commonly used for pull-out tests in concrete structures.

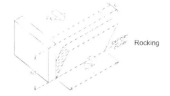

Figure 16. Collapse mechanism of URM and RMA specimens: out-of-plane rocking of the wall panel perpendicular to the horizontal force.

pre-tensioned, using a hollow ram jack, commonly used for pull-out tests in concrete structures, up to stress of 150 MPa (about 4 kN, 20% of the strap capacity), using a tensioning device (Figure 15).

3 TEST RESULTS

3.1 *Unreinforced specimens (URM and RMA tests)*

Both unreinforced specimens have been tested twice. A total of 4 tests were carried out: two tests were conducted on the virgin, undamaged, specimens (URM tests) and, to measure the out-of-plane capacity after the relative separation of face-loaded wall from the return walls, two further tests were carried out (RMA tests).

Experimental capacity values of URM specimens can be seen in Table 3 to be greater than values predicted by the condition of equilibrium of moments of the face-loaded wall (Figure 16). The collapse mechanism consists in the development of a cylindrical hinge at the base of the wall panel: the corresponding rocking load is 2.25 kN. While the limit state (out-of-plane rocking of the wall panel perpendicular to the direction of the horizontal force), used for the analytical calculation, was effectively recorded in the experiment, the corresponding load drastically under-predicts the experimental out-

of-plane capacity by 82% and 102%, with the largest degree of error occurring for the specimen No. 2.

The equilibrium equation under-predicts the experimental results because it does not account for the bonding contribution of the mortar in connecting the wall panels (wall-to-wall connection).

As expected, both unreinforced brickwork specimens (URM-series) failed due to formation of vertical crack in the mortar joint at the connection between face loaded and return walls, and subsequent out-of-plane rocking of the face-loaded wall. The vertical crack spanned from the base of the brickwork specimen to the top. Figures 17 and 18 shows the structural response of the rocking wall panel under horizontal loading (red line): before the development of the vertical crack, the stiffness of the panel (i.e. slope of the curve) was very high, but both out-of-plane load and stiffness dropped down when the mortar connection cracked. When the vertical crack was complete, the stiffness turned to zero and the out-of-plane load reduced to 4.04 kN (residual average load) (Figure 19). From this point, the wall panel initiated to rigidly rotate about the hinge at the base (apparent perfectly plastic behavior).

Figure 20 Shows the horizontal movements vs. time recorded with LVTD2, LVDT3 and LVDT4 located along the panel's height (Figure 9). By considering the ratio between the value of the horizontal

Table 3. Test results.

	Horizontal Load kN	Corresponding deflection mm	Residual load kN
URM_1	11.63	0.669	3.85
URM_2	10.17	0.403	4.24
(mean)	(10.90)		(4.04)
RMA_1	4.03	2.85	4.01
RMA_2	4.38	1.98	4.31
(mean)	(4.21)		(4.16)
RFC_M_1	13.77	8.07	13.61
RFC_M_2	17.13	9.32	15.73
(mean)	(15.45)		(14.47)
RFC_S_1	11.67	10.48	10.40
RFC_S_2	13.27	3.63	7.73
(mean)	(12.47)		(9.05)

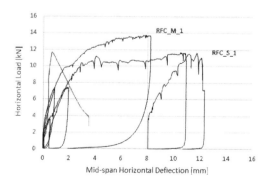

Figure 17. Specimen No.1: horizontal load vs. mid-span deflection.

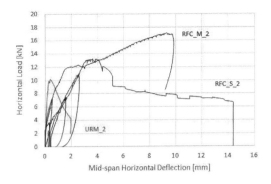

Figure 18. Specimen No.2: horizontal load vs. mid-span deflection (LVDT1).

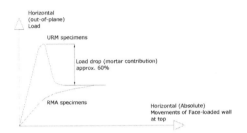

Figure 19. Simplified representation of the structural response of URM and RMA, unreinforced specimens.

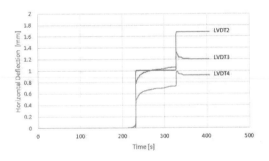

Figure 20. URM2 test: horizontal movements of the three LVTDs located at 2290 mm (LVDT2), 1820 mm (LVDT3) and 1420 mm (LVDT4) above the base of the brickwork specimen. Horizontal movements decrease moving from the panel's top to the bottom (rocking mechanism).

movement and the distance from the panel base (1420, 1820 and 2290 mm for LVDT4, LVDT3 and LVDT2, respectively), it can be noted that this value is almost constant, denoting the out-of-plane rocking of the wall panel perpendicular to the direction of the horizontal force.

Unreinforced brickwork specimens were subsequently re-tested (RMA-series) to assess their structure response without the contribution of the mortar by comparison with previous results (URM-series).

The average horizontal load, producing the activation of rocking mechanism of the face-loaded wall, drastically reduced (from 10.9 to 4.21 kN. When the maximum load was reached, the load remained almost constant with increasing horizontal movements (zero-stiffness). It is worth noting that no load drop was recorded after the maximum load capacity was reached. Furthermore, the calculated rocking load (2.25 kN) is now more similar to the one recorded in the experiments (4.21 kN). The difference can be caused by the unaccounted friction and interlocking resistance along the cracked surface.

The panel's bending deformation can be studied by considering the relative horizontal movement between the LVDTs located along a horizontal plane. It can be noted that the relative horizontal displacements (LVTD1-LVTD2 and LVTD1-LVTD5) (Figures 9 and 21) were negligible (0.62 and 0.35 mm at maximum load for URM_1 and URM_2 respectively). The distance was 1215 mm from each other LVDT, and the resulting curvature on the horizontal plane was very small. This indicated that a flexural failure (or arc-mechanism) was impossible and was part of the reason that such a large horizontal displacement was measured for LVDT1. Although a flexural failure was not produced both the capacity and displacement validated that the method of analysis using masonry macro-elements can be effective in the assessment of the seismic capacity of masonry buildings. The use of equilibrium equations can provide a minimum value of the out-of-plane capacity, and it is able to capture the specimen actual behavior (Figure 22). This conclusion agrees with the outcomes of several another studies and post-earthquake buildings' survey by the authors involving walls subjected to out-of-plane actions, in which it was suggested that the importance of wall-to-wall connections.

3.2 Reinforced specimens (RFC tests)

The two unreinforced, damaged, specimens have been re-tested using two different loading

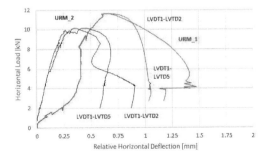

Figure 21. Relative horizontal movements at 2/3 of the panel height (URM-series). The small magnitude of the relative movements demonstrates the limited bending deformation of the wall panel perpendicular to the direction of the horizontal load.

Figure 22. Rocking mechanism of URM specimen.

configurations, as previously described in Section 2.4. A total of four further out-of-plane tests (RFC_M and RFC_S series) have been conducted on the two reinforced brickwork C-shaped structures.

In general, it can be noted that the reinforced specimens shown in Figure 6 and 7 performed well under the cyclic lateral load. The two masonry structures resisted approximately 267% (RFC_M series) and 196% (RFC_S series) more horizontal load compared to the same structures in unreinforced cracked configuration (RMA tests). The average maximum loads were 15.45 and 12.47 kN for tests from RFC_M and RFC_S series, respectively. However, by comparing the out-of-plane capacity with the one of undamaged specimens (URM tests), the load increment is significantly lower, and further analysis is required. It is evident that the mortar and the boundary conditions in general have a critical effect on the capacity of the masonry assemblage.

The mid-span horizontal displacement (LVTD1) accompanying that load level was 8.69 mm (RFC_M series) and 7.05 (RFC_S series). Surprisingly, the stiffness of the structure did not significantly degrade until the lateral load reached a level of 14.47 kN (RFC_M series) and 9.05 kN (RFC_S series).

Apart from considerations about the load bearing capacity of reinforced specimens, it is worth studying the observed failure modes and the specimen deformations under loading. A tension failure of the steel strands was never recorded for all tests on reinforced specimens. Failure was always due to masonry cracking, as described in detail below.

It was difficult to apply an analytical model to reinforced specimens because various aspects that would affect the out-of-plane capacity are not reflected properly in the model e.g. the influence of the concentrated horizontal load, the development of local collapse mechanisms. Nevertheless, several analytical models were used to give an initial prediction of the experimental results of the two reinforced brickwork specimen, employing a number of limiting assumptions, and considering the failure mode recorded in the experiment.

The reinforced walls exhibited three different modes of failure: (1) flexural (simple-supported-beam-type); (2) shear; and (3) flexural (cantilever-beam-type).

Flexural (simple-supported-beam-type). This was the most frequent mode of failure for RFC_M series and it consisted in the formation, at failure, of three horizontal cylindrical hinges (Figure 23). Initial flexural horizontal cracks were only located at mortar bed joints, between two brick courses. Two wall macroelements formed as a consequence of the load application. A cracking noise during the test revealed a progressive cracking of the embedding mortar and the detachment of the face-loaded wall from the two return walls. Since the tensile stresses at the mortar joints were being taken by the strand reinforcement, a redistribution of stresses occurred. Phenomena of deboning of the steel strand from the embedding paste were not recorded.

The corresponding load capacity of the reinforced walls was highly enhanced. By comparing the test results for same wall before and after the application of the strand reinforcement, we can observe that the out-of-plane capacity increased from 4.21 kN (cracked unreinforced specimens) to 15.45 kN (RFC_M series). This failure mode is clearly governed by the mode of application of the load (this is basically a concentrated load, and it can be described as a three-point bending test). Since seismic loads are inertial and thus distributed, it can be said the used mode of application of the load was not able to properly simulate the seismic action. However, it is also evident that this mode of application of the horizontal load was also particularly severe for the out-of-plane stability of the wall itself. The recorded value of out-of-plane capacity of the wall can be regarded as a lower bound value of its capacity under a seismic action.

Figure 24 shows the horizontal movements recorded at 2290 mm (LVDT2), 1820 mm (LVDT3) and 1420 mm (LVTD4) above the base of the brickwork specimen (FRC_M1 test). Contrary to previous tests on unreinforced specimens, after approximately 330 s from the beginning of the test, LVTD3 started measuring very high horizontal movements, denoting the activation of the new failure mode.

To remove the effect of the particular loading configuration, specimens were re-tested using

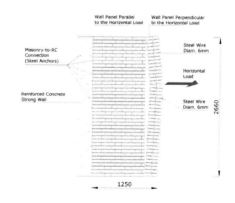

Figure 23. Flexural failure (simple-supported-beam-type). Failure consists in the opening of three plastic hinges (dimensions in mm).

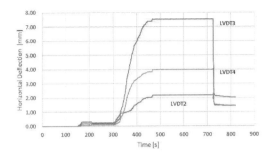

Figure 24. RFC_M1 test: horizontal movements of the three LVTDs located at 2290 mm (LVDT2), 1820 mm (LVDT3) and 1420 mm (LVDT4) above the base of the brickwork specimen. After 330 s from the beginning of the test, horizontal movements recorded with LVDT3 become predominant.

the second configuration (RFC_S series). It is worth noting that the brickwork specimens were partially cracked due to the formation of the hinges (Figure 23) during previous tests. However, while the new loading configuration did not largely affect the overall load capacity of the specimens (load capacity decreased 20% from 15.45 kN to 12.47 kN), new failure modes activated.

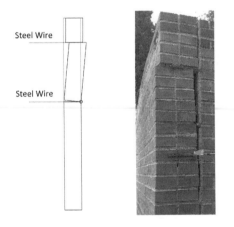

Figure 25. Rocking mechanism of the reinforced specimen (Test No. RFC_S_1).

Figure 23 shows the failure mode of Test No. RFC_S_1. A hinge developed in correspondence of the lower steel strand wrap, and a new macroelement formed. This behaved like a vertical cantilever beam, with sliding phenomena in correspondence of the upper steel strand wrap.

A third failure mode was recorded for Test No. RFC_S_2. This was a shear failure along the mortar head and bed joints (Figures 26 and 27). The overall load capacity did not significantly change (17.13 and 13.27 kN for RFC_M_2 and RFC_S_2, respectively).

Figure 26. Shear failure of reinforced specimen (Test No. RFC_S_2).

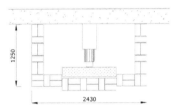

Figure 27. Plan view: shear failure (dimensions in mm).

It is worth noting that the different failure modes observed in reinforced specimens are the consequence of the action of the steel wraps in preventing rocking of the whole panel perpendicular to the direction of the horizontal load. In this respect, it can be concluded that the wraps were effective and functional. It is not easy to identify *a priori* the failure mode of reinforced specimens: this depends on many factors, like the loading configuration, the geometry of the brickwork specimens, the mechanical properties of the masonry material. Although different failure modes were noted for reinforced specimens, a significant increase of the load capacity was always recorded.

4 CONCLUSIONS

This paper has presented an experimental program to examine the out-of-plane behavior of C-shaped wall panel-assemblies consisting of three brickwork panels. The brickwork specimens were intentionally damaged by applying a horizontal, out-of-plane load and repaired with high-strength steel strands, hidden from view and embedded into the bed joints.

Based on the test results of the current experiment, it is possible to install high-strength steel strands inside the mortar bed joints, without affecting the fair-face aspect of the brickwork masonry. The out-of-plane capacity increases from such an installation technique can be quite significant.

Given accurate information on material property evolution and walls' geometry, two brickwork assemblages have been tested in the laboratory under controlled conditions.

The steel strands configuration used in this study showed a 267% (RFC_M series) and 196% (RFC_S series) increases in horizontal load capacity over the control cracked specimen. However, several factors need to be considered when developing a fully-embedded strand configuration.

In addition, we require a better understanding of the boundary conditions. As we continue this part of the program we wish to develop further examples and isolate those details that are critical to making accurate predictions. Within those tests we wish to learn more about the failure modes and observed deformations of the out-of-plane response of both unreinforced and retrofitted brickwork panels.

The analytical models used to predict the out-of-plane capacity of reinforced brickwork specimens were not very accurate. It was also difficult to take into account the various local modes of failure, level of wall-to-wall connections, influence of the boundary conditions in each model. Further tests are currently being carried out at the structures laboratory of the University of Perugia (Italy) to ensure that the available test results can also be also be obtained when the horizontal, out-of-plane load is more distributed, thus validating the results under more realistic conditions.

ACKNOWLEDGEMENTS

This research was funded by the Reluis Consortium through grant line "Linea Murature". Our thanks are also extended to Kimia (Perugia, Italy - supplier of the steel strand reinforcement and special mortars used for joint repointing).

REFERENCES

Alexandris, A. et al. 2004. Collapse mechanisms of masonry buildings derived by the distinct element method. *Proc. 13th World Conference on Earthquake Engineering, 1-6 August 2004, Vancouver, Canada.*

ASTM D2256 2002. Standard test method for tensile properties of yarns by the single-strand method, American Society for Testing and Materials.

Borri, A. & Corradi, M. 2019. Architectural Heritage: A Discussion on Conservation and Safety. *Heritage* 2 (1): 631–647.

Borri, A. et al. 2009. Reinforcement of historic masonry: the "Reticolatus" technique. *Proc. 13th Conference - Remo 2009 Repair, Conservation and strengthening of traditionally erected buildings and historic buildings, 2–4 December 2009, Wroclaw, Poland.*

Cangi, G. 2012. Manuale del recupero strutturale e antisismico. Rome: DEI (in Italian).

Corradi, M. et al. 2016. The Reticulatus method for shear strengthening of fair-faced masonry. *Bulletin of Earthquake Engineering* 14(12): 3547–3571.

Corradi, M. et al. 2007. Confinement of brick masonry columns with CFRP materials. *Composites Science and Technology* 67(9): 1772–1783.

D'Ayala, D. & Speranza, E. 2003. Definition of collapse mechanisms and seismic vulnerability of historic masonry buildings. *Earthquake Spectra* 19(3): 479–509.

EN 771–1: 2011. Specification for masonry units. Clay masonry units, CEN (European Committee for Standardization).

EN 1015–11:1999. Methods of test for mortar for masonry. Determination of flexural and compressive strength of hardened mortar, CEN (European Committee for Standardization).

Furukawa, A. & Ohta, Y. 2009. Failure process of masonry buildings during earthquake and associated casualty risk evaluation. *Natural hazards* 49(1): 25–51.

Gattesco, N. & Boem, I. 2017. Out-of-plane behavior of reinforced masonry walls: Experimental and numerical study. *Composites Part B: Engineering* 128: 39–52.

Ghobarah, A. & El Mandooh Galal, K. 2004. Out-of-plane strengthening of unreinforced masonry walls with openings. *Journal of Composites in Construction* 8(4): 298–305.

Gilstrap, J.M. & Dolan, C.W. 1998. Out-of-plane bending of FRP-reinforced masonry walls. *Composites Science and Technology* 58(8): 1277–84.

Giuffrè, A. 1995. Seismic damage in historic town centres and attenuation criteria. *Annals of Geophysics* 38(5–6).

Hamed, E. & Rabinovitch, O. 2007. Out-of-plane behavior of unreinforced masonry walls strengthened with FRP strips *Composites Science and Technology* 67: 489–500.

ICOMOS Charter, 1964. International Charter for the Conservation and Restoration of Monuments and Sites. 2nd General Assembly of International Council on Monuments and Sites, Venice, Italy.

ICOMOS Charter. 2003. Principles for the analysis, conservation and structural restoration of architectural heritage. 14th General Assembly of International Council on Monuments and Sites, Victoria Falls, Zambia.

Ismail, N. & Ingham, J.M. 2016. In-plane and out-of-plane testing of unreinforced masonry walls strengthened using polymer textile reinforced mortar. *Engineering Structures* 118: 167–177.

Italian Senate. Report No. 01077470, 2017. Earthquakes. Central Italy 2016, Emilia 2012, L'Aquila 2009: reconstruction resources and laws, Rome, Italy.

Jurina, L. 2015. Cerchiatura di strutture murarie: tecniche tradizionali ed innovative. *Ingenio* 596: 1–35.

Korany, Y. & Drysdale, R. 2006. Rehabilitation of masonry walls using unobtrusive FRP techniques for enhanced out-of-plane seismic resistance. *Journal of Composites for Construction* 10(3): 213–222.

Lin, Y. et al. 2016. Out-of-plane testing of unreinforced masonry walls strengthened using ECC shotcrete. *Structures* 7: 33–42.

Lynch, G. 1993. Historic brickwork: part 1. *Structural Survey* 11(4): 388–395.

Micelli, F. et al. 2017. Properties of Aged GFRP Reinforcement Grids Related to Fatigue Life and Alkaline Environment. *Applied Sciences* 7(9): 897.

Papanicolaou, C.G. et al. 2008. Textile reinforced mortar (TRM) versus FRP as strengthening material of URM walls: out-of-plane cyclic loading. *Materials and structures* 41(1): 143–157.

Penazzi, D. et al. 2001. Repair and strengthening of historic masonry buildings in seismic areas. *Proc. Int. Congress, more than two thousand years in the history of architecture safeguarding the structure of our architectural heritage, Bethlehem, Palestine*: 1–6.

Preciado, A. et al. 2016. Seismic vulnerability enhancement of medieval and masonry bell towers externally prestressed with unbonded smart tendons. *Engineering Structures* 122: 50–61.

Shawa, O.A. et al. 2012. Out-of-plane seismic behaviour of rocking masonry walls. *Earthquake Engineering & Structural Dynamics* 41(5): 949–968.

Sisti, R. et al. 2016. An experimental study on the influence of composite materials used to reinforce masonry ring beams. *Construction and Building Materials* 122: 231–241.

Tumialan, J.G. et al. 2003. Fiber-reinforced polymer strengthening of unreinforced masonry walls subjected to out-of-plane loads. *ACI Structural Journal* 100(3): 321–9.

Velazquez-Dimas, J.I. & Ehsani, M.R. 2000. Modeling out-of-plane behavior of URM walls retrofitted with fiber composites. *Journal of Composites in Construction* 4(4): 172–81.

Vinci, M. 2014. I tiranti in acciaio nel calcolo delle costruzioni in muratura. Palermo: Dario Flaccovio Ed.

Brick and Block Masonry - From Historical to Sustainable Masonry –
Kubica, Kwiecień & Bednarz (eds)
© *2020 Taylor & Francis Group, London, ISBN 978-0-367-56586-2*

Performance of TRM-to-masonry joints after exposure to elevated temperatures

C.G. Papanicolaou
Civil Engineering Department, University of Patras, Patras, Greece

ABSTRACT: During the past fifteen years and for a number of well-documented reasons Textile Reinforced Mortar (TRM) systems have been gaining ground as a means of strengthening deteriorated, damaged or seis- mically deficient masonry structures. As for any new material, the bulk of experimental investigations have focused on the mechanical response of TRMs (both as standalone materials and in combination with different types of substrates) under normal service conditions. With the relevant learning curve reaching a plateau the interest of the academia is turning to durability-related aspects and extreme exposure scenarios. Although publications in the field are growing in number, the behavior of these materials under elevated/high temperat- ures and fire conditions is far from being exhaustively investigated and understood. This paper aims at sys- temizing the existing knowledge on the mechanical performance of TRM-to-masonry residual bond characteristics as a function of the exposure temperature of the joints.

1 INTRODUCTION

When first introduced to the market, TRMs – being inorganic matrix composites – were rather hastily and falsely perceived by some practitioners as fire-resistant materials. Compared to more "traditional" externally bonded composites used for strengthening interventions, such as Fiber Reinforced Polymers (FRP), TRMs do exhibit superior performance under elevated/high temperatures. Properties such as non-flammability and incombustibility comprise obvious advantages in regard to composites with organic matrices. Nevertheless, there is still much to learn about the actual behavior of TRM systems under or post their exposure to elevated/high temperatures and fire conditions, especially when these systems are combined with other (substrate) materials.

There exists a rather moderate number of publica- tions discussing the mechanical performance of TRM systems under or post elevated temperatures and fire conditions. These include: (i) material char- acterization tests on tensile and flexure coupons for the derivation of critical mechanical properties under both transient and steady-state thermal conditions (Ehlig et al. 2010, Colombo et al. 2011, de Andrade Silva et al. 2014, Rambo et al. 2015, 2016, 2017, Nguyen et al. 2016, Homoro et al. 2018, Tlaiji et al. 2018, Messori et al. 2019, Truong et al. 2019, Tran et al. 2019, Antons et al. 2012, Xu et al. 2014, 2016, Li et al. 2017, Kapsalis et al. 2018, 2019, Maroudas & Papanicolaou 2017, Donnini et al. 2017); (2) TRM-to-substrate shear bond tests (Ombres 2015,

Maroudas & Papanicolaou 2017, Raoof & Bournas 2017, Donnini et al. 2017, Ombres et al. 2018, Askouni et al. 2019); and (3) tests on TRM-strengthened structural elements (Bisby et al. 2010, 2013, Michels et al. 2014, Raoof & Bournas 2017b, Triantafillou et al. 2017, Tetta & Bournas 2016, Trapko 2013, Papanicolaou et al. 2016 (see Figure1), Al-Salloum et al. 2016 and Ombres 2017).

Most of the afore-mentioned experimental cam- paigns (~55% of total) concern material character- ization tests on tensile and flexure coupons. For the case of TRMs meant to work in conjunction with other materials, these tests provide data of a rather qualitative nature since the thermal inertia of the assemblage (TRM + substrate on which the TRM product is externally bonded) is quite different from that of a standalone coupon (see also Kapsalis et al. 2019). Furthermore, in TRM-furnished structural elements moisture paths and moisture boundary con- ditions are distinctively different than in structural elements comprising textile reinforced concrete. In fact, in the former case, the substrate through phy- sico-chemical interaction with the overlying com- posite alters its properties over time.

Based on the above, TRM-to-substrate bond tests (and, provisionally, single-yarn pull-out ones) during or after exposure to elevated/high temperatures or fire can provide useful information for the design of structural strengthening interventions with TRMs under these conditions. The problem lies on the lack of commonly accepted relevant testing protocols. There is a multitude of factors to be taken into

Figure 1. Plain concrete cylinders furnished with AR glass TRM jackets immediately after removal from the fire resistance furnace: (a) exposure at 400°C and (b) exposure at 1000°C (specimens were left to freely cool down after exposure with the furnace shutter into place; Papanicolaou et al. 2016).

consideration when designing such tests or when conducting comparative analyses between existing sets of experimental data. These factors relate to the testing regime (loading-heating sequence and rates, mechanical boundary conditions, initial moisture content and moisture exchange conditions for all materials in contact, geometry and size of specimens) and to the thermal modes employed [testing prior to heating, for reference; testing under transient temperatures during heating (or re-heating) and cooling; testing under steady-state temperatures (elevated/high or ambient with testing taking place rather shortly after heating-cooling or after a longer post-cooling period of time)].

In the following, tests aiming to assess the post-heating TRM-to-masonry residual bond characteristics are presented.

2 TRM-TO-MASONRY RESIDUAL BOND AFTER EXPOSURE TO ELEVATED TEMPERATURES

2.1 Experimental campaigns

In Maroudas & Papanicolaou (2017), a TRM system consisting of a dry AR glass fiber textile embedded in a polypropylene fiber-reinforced cement-based matrix was applied on masonry prisms made of solid clay fair-faced bricks. Single-lap/single-prism shear bond tests were conducted after specimens' exposure at 100°C, 200°C, and 300°C for 1 hour (heating rate from ambient to target air temperature being equal to 75°C/min; as a comparison, the mean heating rate of the ISO 834 heating curve equals to approx. 295°C/

min). Tests were carried out by varying both bond lengths and exposure temperatures. All free masonry surfaces were covered with the same mortar used for the TRM strip in order to simulate real-life conditions during the heating process and to prevent any unwanted relative thermal expansion which could result to the premature cracking of the prisms. After the completion of the heating process, the cover mortar was removed. The specimens were furnished with heat-shielding covers over areas where the bare textile was exposed (Figure 2a).

In an attempt to equalize moisture content of all specimens prior to subjecting them to the heating regime each specimen was dried in an electrical oven for a duration of 30 min at 40°C. Until testing, all specimens were stored in a way that ensured zero moisture exchange with the atmosphere. After completion of the heating process specimens remained in the switched-off oven (with a small side aperture open to prevent condensation of any moisture left in it); oven's door remained closed until 50°C were reached. Specimens were tested in ambient (lab) temperature and relative humidity conditions. The authors noticed a temperature-dependent shift in failure modes from fibers' slippage at 20°C and 100°C to strip debonding and fibers' rupture at 200°C and 300°C. It was concluded that for adequate bond length the residual shear bond load transferred by this type of TRM/masonry interfaces after exposure at 100°C, 200°C, and 300°C amounts, respectively, to 65%, 60%, and 50% of the respective load reached in ambient conditions. Thermal damage of the bare textile (when reaching 200°C at points where heat shielding failed) is a critical factor for AR glass fibers' rupture.

In the work of Askouni et al. (2019), lightweight aggregate matrices were introduced for the

Figure 2. (a) Heat shielding comprising 30 mm-thick pieces of aluminum-lined mineral wool. (b) Snapshot of the end fixture of a TRM tensile coupon after exposure to elevated temperature; the secondary (stabilizing) polypropylene net melted leaving charring marks on the mortar.

production of TRMs as an alternative to conventional normal-weight ones with the aim to provide a more favorable heat shielding effect to the fibrous grid and, hence, to result in an improved post-heating/cooling residual bond strength of TRM/masonry interfaces. To this purpose, masonry prisms consisting of ridge-faced perforated fired clay bricks and a cement/lime-based mortar were unilaterally furnished with TRM strips comprising two systems; both shared the same type of textile (a dry AR glass fiber textile either in a single-layer or in a double-layer configuration) and different matrices: one normal-weight (TRNM) and another lightweight (TRLM, with pumice sand) of equal compressive strengths. Uniaxial tensile tests on both types of TRMs furnished with two layers of textile revealed an almost identical strain-hardening behavior (Figure 3).

Prior to subjecting the specimens to the prescribed heating regime and again with the aim of equalizing the moisture content in all of them each specimen was dried in an electrical oven for a duration of 24 hours at a constant temperature of 40°C. Until testing, all specimens were stored in a way that ensured zero moisture exchange with the atmosphere. Single-lap/single-prism shear bond tests were conducted after specimens' exposure at 120°C and 200°C for 1 hour (heating rate from ambient to target temperature being equal to 75°C/min). Specimen preparation (covering with extra mortar, heat shielding) was identical to that followed by Maroudas & Papanicolaou (2017). Failure in this case was common for all exposure temperatures and for both types of matrices (fibers' slippage from within the mortar). It was shown that at control conditions (non-heated specimens) and after exposure at a nominal air temperature of 120°C, both single-layer TRM systems (TRNM and TRLM) exhibit similar bond capacities. The same applies for both (TRNM and TRLM) unexposed double-layer TRMs. Nevertheless, after exposure at a nominal air temperature of 200°C, both single-layer and double-layer TRNM overlays outperformed the respective TRLM ones (by 15%-20%, in terms of maximum axial textile stress). This was mainly attributed to larger quantities of evaporable moisture in the lightweight matrix as opposed to the normal weight one buffered in the pumice sand which was used in a saturated state during mixing. Moisture evaporation by heating caused a stress buildup in the dense cement paste (denser for TRLM than for TRNM in order to compensate for the lower crushing strength of the pumice sand) due to high vapor pressure. This stress buildup was considered to be detrimental for the textile-to-mortar bond. The results show that identical uniaxial tensile behavior of different TRMs is a precarious marker for comparable shear bond behavior as the latter is strongly depended upon the response of each TRM system to (hygro-)thermal events (with different TRMs responding in a different way to the same hygro-thermal history).

In the work of Ombres et al. (2018) various TRM systems were applied on solid clay brick masonry prisms: one made of a PBO fiber textile and a cement-based matrix while two other were made of dry basalt fiber textiles (of different areal weight and grid spacing: a 'heavy' and a 'light' one) and a lime-based mortar modified with geo-polymeric binders. (The system comprising a unidirectional ultra-high strength galvanized steel textile and a cement-based mortar is not included in this review). Single-lap/single-prism tests were conducted after specimens' exposure at 100°C, 150°C, and 200°C for 3 hours (heating rate from ambient to target temperature being equal to 2°C/min). Heat shielding measures for the protection of the bare textile were taken. After heating, specimens were left to cool down freely to ambient temperature. Results indicated that the bond response of heat-exposed specimens depended on the fibers' type and textiles' density. The residual shear bond load decreased by up to 50% and 90% in the case of PBO-TRM and 'heavy' basalt-TRM, respectively, while it remained almost unaffected in the case of the 'light' basalt-TRM ('light' having half the areal weight of 'heavy'). Excluding the case of PBO fiber TRMs, exposure to elevated temperatures did not have an effect on the failure modes observed (different for different configurations).

Donnini et al. (2017) proposed a numerical model for the simulation of TRMs' shear bond behavior under different temperatures. For the calibration of the model, double-lap/single-prism shear bond tests were conducted (prisms, in this case, consisting of single clay bricks). The TRM systems comprised textiles with either uncoated (dry) or epoxy-impregnated and sanded carbon fiber yarns embedded in a cement-based mortar. Specimens were tested under two different regimes: (i) while, at 120°C, being conditioned to the same temperature for a duration of 100 min, and (ii) after exposure at 120°C for 60 minutes. It was shown that – when tested under the former heating regime – TRMs with impregnated yarns lost more than half of their bond capacity while TRMs with dry yarns remained almost unaffected. However, TRMs

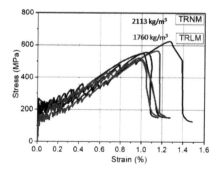

Figure 3. Axial tensile stress versus axial tensile strain curves of TRNM (textile reinforced normal weight mortar) and (b) TRLM (textile reinforced lightweight mortar) coupons with two layers of AR glass textile.

with impregnated yarns retained almost all their initial bond capacity when tested under the second heating regime. Hence, there seems to exist a beneficial thermal effect to the post-heating/cooling mechanical response of polymer-coated carbon fiber TRCs after exposure up to relatively moderate temperatures (~ up to 150°C). This effect was also documented by de Andrade Silva et al. (2014) and was attributed to an interlocking mechanism that develops between the matrix and the molten/re-stiffened polymer, which results in improved bond between the filaments and the matrix.

2.2 Residual bond strength as a function of exposure temperature

A scatter plot of all normalized residual bond strengths (RBS) of TRM overlays referenced in literature after heating/cooling regimes versus the target air temperature at which the TRM/masonry assemblages were exposed is given in Figure 4. RBS values are normalized by dividing the post-exposure bond strength of each system by the one reached at ambient temperatures. Different combinations of TRM/substrate are included from the works of Maroudas et al. (2017), Ombres et al. (2018) and Askouni et al. (2019); failure types of the majority of specimens involved fibers' slippage from within the matrix.

In Figure 4, an indicative lower bound prediction of RBS is proposed as a function of the exposure temperature, following a decay law. However, use of this relationship should be made with caution as it is based on a very small amount of data. each data

point in Figure 4 is the result from either a single specimen or represents the mean value of a very limited number of specimens (2 specimens). It is imperative that the data base is enlarged and expanded to transient thermal conditions present during the shear bond tests. In order to mitigate the inherent variability of shear bond tests and produce statistically reliable results, a larger number (5-7) of identical specimens has to be employed for each set of parameters investigated.

3 CONCLUSIONS

This paper briefly presents the existing knowledge on the mechanical performance of TRM-to-masonry residual bond characteristics (after a heating/cooling regime) as a function of the exposure air temperature of the joints. Before more data is produced, all interested parties must achieve a consensus on the appropriate testing protocols to be followed for TRM-to-masonry joints both under transient and steady-state heating/cooling scenarios. Material-specific experimental campaigns must be run under controlled moisture contents for all porous media comprising the joint. By adopting a material-blind lower bound of the residual bond strength of heated TRM/masonry joints might prove to be a considerably conservative approach.

REFERENCES

Al-Salloum, Y.A. et al. 2016. Effect of elevated temperature environments on the residual axial capacity of RC columns strengthened with different techniques. *Construction and Building Materials* 115: 345–361.

Antons, U. et al. 2012. High-temperature tests on concrete specimens reinforced with alkali-resistant glass rovings under bending loads. In: *Proceedings of the 6th International Conference on FRP Composites in Civil Engineering, Rome, Italy*: 13–15.

Askouni, P.D. et al. 2019. The Effect of Elevated Temperatures on the TRM-to-Masonry Bond: Comparison of Normal Weight and Lightweight Matrices. *Applied Sciences* 9(10): 2156.

Bisby, L. et al. 2010. Comparative performance of fibre reinforced polymer and fibre reinforced cementitious mortar strengthening systems in elevated temperature service environments. *Structural Faults and Repair*.

Bisby, L. et al. 2013. Fire performance of well-anchored TRM, FRCM and FRP flexural strengthening systems. *Advanced Composites in Construction*.

Colombo, I. et al. 2011. Textile reinforced mortar at high temperatures. *Applied mechanics and materials* 82: 202–207.

de Andrade Silva, F. et al. 2014. Effects of elevated temperatures on the interface properties of carbon textile-reinforced concrete. *Cement and Concrete Composites* 48: 26–34.

Donnini, J. et al. 2017. Fabric-reinforced cementitious matrix behavior at high-temperature: Experimental and numerical results. *Composites Part B: Engineering* 108: 108–121.

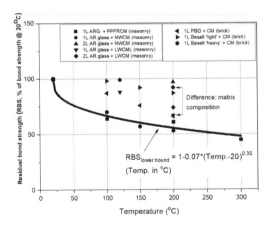

Figure 4. Normalized residual bond strength (% of bond strength at ambient temperature) of TRM/masonry joints after heating/cooling regimes and lower bound prediction as a function of temperature. Abbreviations: L: textile layers; ARG: AR glass fiber textile; CM: cement-based matrix; PPFRCM: PP short fiber-reinforced CM with low content of polymers; NWCM: CM with normal-weight aggregates; LWCM: CM with lightweight aggregates.

Ehlig, D. et al. 2010. High temperature tests on textile reinforced concrete (TRC) strain specimens. In: *International RILEM Conference on Material Science.* RILEM Publications SARL: 141–151.

Homoro, O. et al. 2018. Experimental and analytical study of the thermo-mechanical behaviour of textile-reinforced concrete (TRC) at elevated temperatures: Role of discontinuous short glass fibres. *Construction and Building Materials* 190: 645–663.

Kapsalis, P. et al. 2018. Preliminary High-Temperature Tests of Textile Reinforced Concrete (TRC). *Multidisciplinary Digital Publishing Institute Proceedings* 2(8): 522.

Kapsalis, P. et al. 2019. Thermomechanical Behavior of Textile Reinforced Cementitious Composites Subjected to Fire. *Applied Sciences* 9(4): 747.

Li, T. et al. 2017. Flexural behavior and microstructure of hybrid basalt textile and steel fiber reinforced alkali-activated slag panels exposed to elevated temperatures. *Construction and Building Materials* 152: 651–660.

Maroudas, S.R. & Papanicolaou, C.C.G. 2017. Effect of high temperatures on the TRM-to-masonry bond. *Key Engineering Materials* 747: 533–541).

Messori, M. et al. 2019. Effect of high temperature exposure on epoxy-coated glass textile reinforced mortar (GTRM) composites. *Construction and Building Materials* 212: 765–774.

Michels, J. et al. 2014. Structural strengthening of concrete with fiber reinforced cementitious matrix (FRCM) at ambient and elevated temperature - Recent investigations in Switzerland. *Advances in Structural Engineering* 17(12): 1785–1799.

Nguyen, T.H. et al. 2016. Experimental study of the effect of simultaneous mechanical and high-temperature loadings on the behaviour of textile-reinforced concrete (TRC). *Construction and Building Materials* 125: 253.

Ombres, L. 2015. Analysis of the bond between fabric reinforced cementitious mortar (FRCM) strengthening systems and concrete. *Composites Part B: Engineering* 69: 418–426.

Ombres, L. 2017. Structural performances of thermally conditioned PBO FRCM confined concrete cylinders. *Composite Structures* 176: 1096–1106.

Ombres, L. et al. 2018. Bond analysis of thermally conditioned FRCM-masonry joints. *Measurement* 125: 509–515.

Papanicolaou, C. et al. 2016. Deliverable WP10, 'Prescient' Project: "Paradigm for Resilient Concrete Infrastructures to Extreme Natural and Man-made Threats", National Strategic Reference Framework (NSRF) 2007-2013 (in Greek).

Rambo, D.A.S. et al. 2015. Effect of elevated temperatures on the mechanical behavior of basalt textile reinforced refractory concrete. *Materials & Design* 65: 24–33.

Rambo, D.A.S. et al. 2016. Tensile strength of a calcium-aluminate cementitious composite reinforced with basalt textile in a high-temperature environment. *Cement and Concrete Composites* 70: 183–193.

Rambo, D.A.S. et al. 2017. Experimental investigation and modelling of the temperature effects on the tensile behavior of textile reinforced refractory concretes. *Cement and Concrete Composites* 75: 51–61.

Raoof, S.M. & Bournas, D.A. 2017. Bond between TRM versus FRP composites and concrete at high temperatures. *Composites Part B: Engineering* 127: 150–165.

Raoof, S.M. & Bournas, D.A. 2017. TRM versus FRP in flexural strengthening of RC beams: Behaviour at high temperatures. *Construction and Building Materials* 154: 424–437.

Tetta, Z.C. & Bournas, D.A. 2016. TRM vs FRP jacketing in shear strengthening of concrete members subjected to high temperatures. *Composites Part B: Engineering* 106: 190–205.

Tlaiji, T. et al. 2018. Thermomechanical behaviour and residual properties of textile reinforced concrete (TRC) subjected to elevated and high temperature loading: Experimental and comparative study. *Composites Part B: Engineering* 144: 99–110.

Tran, M.T. et al. 2019. Mesoscale experimental investigation of thermomechanical behaviour of the carbon textile reinforced refractory concrete under simultaneous mechanical loading and elevated temperature. *Construction and Building Materials* 217: 156–171.

Trapko, T. 2013. The effect of high temperature on the performance of CFRP and FRCM confined concrete elements. *Composites Part B: Engineering* 54: 138–145.

Triantafillou, T.C. et al. 2017. An innovative structural and energy retrofitting system for URM walls using textile reinforced mortars combined with thermal insulation: Mechanical and fire behavior. *Construction and Building Materials* 133:1–13.

Truong, G.T. et al. 2019. Tensile Behaviors of Lap-Spliced Carbon Fiber-Textile Reinforced Mortar Composites Exposed to High Temperature. *Materials* 12(9): 1512.

Xu, S.L. et al. 2014. High temperature mechanical performance and micro interfacial adhesive failure of textile reinforced concrete thin-plate. *Journal of Zhejiang University SCIENCE A* 15(1): 31–38.

Xu, S. et al. 2016. The high-temperature resistance performance of TRC thin-plates with different cementitious materials: Experimental study. *Construction and Building Materials* 115: 506–519.

Brick and Block Masonry - From Historical to Sustainable Masonry –
Kubica, Kwiecień & Bednarz (eds)
© 2020 Taylor & Francis Group, London, ISBN 978-0-367-56586-2

Brick walls interventions with FRPU or PUFJ and of RC columns with FR in brick-infilled RC structures with the use of pushover beam-column element analysis and pseudo-dynamic 3d finite element analysis

T. Rousakis
Democritus University of Thrace, Xanthi, Greece

ABSTRACT: This analytical study presents the effects of the strength and position of typical brick wall infills on the behavior of typical low rise deficient RC structure. The RC columns were retrofitted with different kinds of FRPs or Fiber Rope (FR) confinement. Pushover inelastic analyses were performed with SeismoStruct utilizing advanced beam-column elements and inelastic brick infill elements retrofitted with innovative Fiber Reinforced Polyurethane (FRPU). Past and recent pseudo-static and dynamic experiments have validated the increased displacement ductility brick walls retrofitted with FRPU and further strengthen the reliability of the performed analyses. The conclusions of this analytical study suggest that infill wall retrofit with FRPU may enable designers for alternative, more efficient interventions to meet contemporary code requirements. Suitable dynamic 3-dimensional Finite Element Modeling of RC and infills are presented to help advance innovative retrofit design for desirable in- and out-of-plane performance of RC frame – brick infill systems.

1 INTRODUCTION

Advanced resilience of critical buildings and infrastructure urges for further developed and integrated framework. In many regions around the world, resilience related planning and activities were rarely reported in the past while there are some very important ongoing initiatives in Europe through DRMKC (https://drmkc.jrc.ec.europa.eu/). There is great need for further integration of resilience metrics, approaches and concepts to prioritize proactive actions in structures to satisfy resilient society concept (Bocchini et al. 2014, Rousakis 2018). Recent innovative structural retrofits involve advanced materials and techniques (Karantzikis et al. 2005, Triantafillou et al. 2006, Wu et al. 2008, Ilki et al. 2008, Janke et al. 2009, Dai et al. 2011, Kwon et al. 2016, Rousakis et al. 2017, Rouka et al. 2017, Kwiecień et al. 2017, Padanattil et al. 2017, Ispir et al. 2018, Triantafyllou et al. 2019, Qin et al. 2019, Rousakis et al. 2020a,b among else). Rousakis (2018) presents several inherent resilience features of square RC columns retrofitted with composite ropes, under seismic excitations. Rousakis and Tourtouras (2014) applied external rope confinement of deficient RC columns under axial load to achieve enormous upgrade of axial strain ductility of concrete. An extended damage redistribution was observed on the columns while maintaining the axial load capacity. Similar was the case for deficient square RC columns having FRP jacket combined with nonbonded composite rope confinement (Rousakis 2017, 2019). The

interesting redistribution effects were generalized in a design concept of Elastic Redistributable Uniform Confinement (ERUC). This design concept provides upgraded axial load carrying capacity of the member to avoid detrimental collapse. ERUC concept succeeds damage redistribution effects even if the structural component system involves core material with geometric non-uniformities and a second subcomponent that may accumulate damage (i.e slender steel bars or yielded stirrups or fractured FRP jacket) or all of these together (see Rousakis and Tourtouras 2014, Rousakis et al. 2019).

This paper presents the analytical investigation of the performance of typical low-rise reinforced concrete (RC) buildings which are designed with previous generation codes and are retrofitted according to the requirements of the current Eurocodes. The use of novel confinement with fibre rope (FR) reinforcements is investigated. Available experimental results in concrete cylinders and RC columns under axial compression have shown remarkable upgrade of the axial strain ductility of FR confined concrete. The analytical study is focused on the use of vinylon and polypropylene FRs as external unbonded, non-impregnated confinement reinforcement (VFR, PPFR) of RC columns while common FRP sheet confining materials (GFRP, CFRP) and steel straps are also included. It concerns three-dimensional pushover analyses of the buildings with the advanced SeismoStruct software, taking into account the effects of brick wall infill as well. The FR confinement may ensure a further ductile response of the retrofitted structure since FR deformability is higher than the one of FRPs. The FRs could result in comparable displacement ductility levels of the structure, even for very low confinement lateral rigidity that usually leads to inadequate FRP confinement of concrete. Therefore, FR may display a higher upgrading potential than FRPs in cases of column sections with low confinement effectiveness. Further, Different scenarios of brick wall infill retrofits with advanced fiber sheets impregnated with high deformability polymers (FRPU) are examined.

2 INELASTIC STATIC (PUSHOVER ANALYSES) OF AS-BUILT AND FR RETROFITTED BRICK WALL INFILLED RC STRUCTURES

The investigated typical RC structure is displayed in Figure 1. The average concrete compressive strength was $f_{cm}=20$ MPa and the average steel yielding stress was $f_{ym}=253$ MPa (typical values for buildings constructed before 1970 in southern Europe). The inelastic behaviour of brick infill walls and their interaction with the RC frames is modelled with advanced software SeismoStruct (v5.2.2 build 10), according to the approach proposed by Chrisafulli (1997), Figure 2. Cases of buildings with interior and exterior walls were considered with vertical compressive strength $f_{wc}=3.10$ MPa and Young's Modulus $E_w=3000$ MPa (see also Rousakis et al. 2014).

2.1 Characteristics and assessment of the existing structures

Non-linear inelastic analyses in both xx' and yy' directions were carried out for the bare structures and the ones with infill (Figure 3). Target displacements were estimated according to the requirements of Eurocode 8 for seismic hazard zones I and III. The equivalent ground acceleration was 0.16g and 0.36g, respectively.

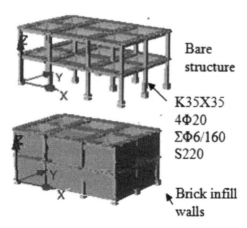

Bare structure

K35X35
4Φ20
ΣΦ6/160
S220

Brick infill walls

Figure 1. Low-rise RC structure.

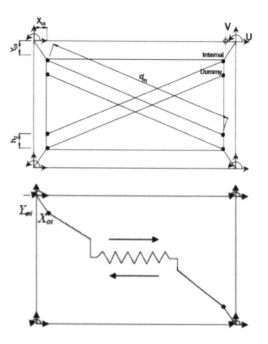

Figure 2. Brick wall modelling of the 4 diagonal compressive struts and of the shear capacity according to Crisafulli (1997).

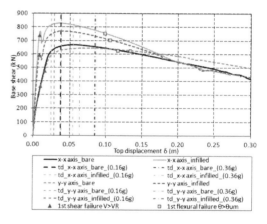

Figure 3. Inelastic behaviour of existing RC structure.

It can be seen from Figure 2 that the effect of the infill walls is significant. As it was expected, the walls increased the stiffness and the overall base shear resistance of the structure (25% higher). A temporary reduction of base shear was evidenced during successive infill wall failures and up to redistribution of loading in the structure. Then a maximum base shear was achieved, followed by degrading behaviour with higher rate than in bare structure. Shear force capacity was exceeded in all ground floor columns and in most of the top floor columns and beams. The curvature at the base of all columns had overdrawn the yielding curvature and the chord rotation capacity has been exhausted for top displacement higher than 0.1 m.

2.2 Characteristics and assessment of the retrofitted structures

Early shear related failures required the strengthening of the existing structure. Different types of confinement of columns with vinylon FR (VFR) or polypropylene FR (PPFR) or carbon FRP (CFRP) or glass FRP (GFRP) and finally steel jackets 0.66 mm thick were investigated. The steel jacket thickness was selected in order to achieve shear strengthening equal to CFRP jacketing. For the rest of the materials the thickness of the confinement was designed in order to provide concrete with equivalent stress-strain curves under compression. This approach is supported by the experimental results in FR wrapped concrete cylinders, presented in Rousakis (2014).

The stress-strain behaviour of confined concrete under compression, in retrofitted structures, differs only by the ultimate stress and strain values that depend on the strain failure of the confining material. CFRP confinement provides 0.012 failure concrete strain while the value for GFRP is 0.015 and for the steel jacket is 0.008. VFRs and PPFR may provide failure strains higher than 10%. The strain failure of concrete confined by extra low quantity of PPFR (3.6 times lower than the above mentioned PPFR) is considered around 0.06. Figure 4 shows the stress-strain curves of confined concrete for different fibre materials. The resilience of the high deformability structural ropes is remarkable.

Figure 5 demonstrates the inelastic behaviour of the retrofitted bare and infill structure. The elastic stiffness and the base shear-top displacement behaviour of both the bare and infill structures remained the same. However, the performance of the retrofitted buildings differed by their ultimate displacement, having suppressed the brittle shear failures (even for the case of reduced PPFR quantity). The lower critical displacements at first section failure were found in the yy loading direction. The critical displacement of 0.13 m corresponds to longitudinal steel bar fractures and occurred in all retrofitted bare structures (except for steel plate confinement). In wall infill

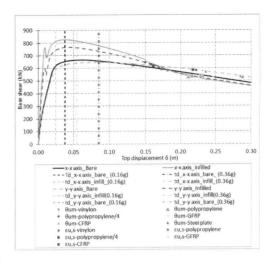

Figure 5. Inelastic behaviour of strengthened structure.

structures, the critical displacement was controlled by θum instead. VFR and PPFR gave similar critical displacements for the structure as well as the conventional materials of GFRP and CFRP (0.2 m to 0.23 m). For steel plates, the structure was less ductile (0.16 m). Similar ductility was observed and in the case that reduced quantity of PPFR was taken into account. In all cases, the external strengthening provided enough structural upgrade to satisfy code requirements for seismic hazard zone III. The ultimate chord end rotation and shear strength capacities of the FR retrofitted columns were calculated according to EC8-3 and the Greek Retrofit Code (GRECO) that is adjusted to the Eurocodes. The ultimate chord end rotations according to GRECO and the shear strengths according to EC8-3 for the columns retrofitted especially with the extra high deformability PPFRs yielded unreliable results.

3 INELASTIC STATIC (PUSHOVER ANALYSES) OF PRESTRESSED FR RETROFITTED RC STRUCTURES CONSIDERING THE EFFECTS OF INFILL POSITION AND STRENGTH

Accurate assessment of the effects of infill walls in reinforced concrete (RC) frame structures is an open issue. Especially in the case of existing structures in need of assessment and retrofit there are different approaches that yield very different results as depicted in section 2. Numerous three-dimensional pushover analyses are performed, investigating different scenarios with respect to the infill walls' strength and position within the structure following a lower bound envelope approach to account for potential infill wall openings or glass partitions. Therefore, the structure investigated already in section 2 is further assessed considering all critical

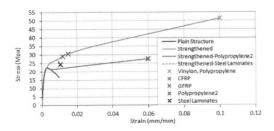

Figure 4. Stress-strain curves of confined concrete.

cases of existence or absence of infill walls and two different infill strengths (Rousakis et al. 2016). The inelastic behaviour of brick infill walls and their interaction with the RC frames is modelled in advanced software SeismoStruct, according to the approach proposed by Chrisafulli (1997). Similarly to section 2, all the infill walls have a vertical compressive strength of 3.10MPa with a Young's Modulus equal to 3000MPa. Their diagonal compressive strength is calculated by multiplying the vertical compressive strength with a reduction factor that takes into account the slenderness of the wall. The physical dimensions (length, height and thickness) are necessary to determine this factor. The diagonal strength of typical plain external infill walls of low strength is taken equal to 0.97MPa and 0.82MPa with the internal infill walls having a strength equal to 0.34MPa. In the cases the strong infill walls are taken into account, the reduction factors are eliminated and a single diagonal compressive strength of 1.40MPa is allocated. In what follows, the study concerns cases of structures retrofitted with prestressed fibre ropes (FR) as developed in Rousakis & Tourtouras 2014. This novel technique can provide columns with shear capacity, ultimate chord rotation capacity and confined concrete axial strain ductility far beyond the required ones for the cases of structures under investigation. Thus, global failure of structures is defined for a decrease of 15% in base shear.

3.1 *Wall strength influence*

Non-linear pushover analyses are conducted in the xx' and the yy' directions for all the investigated cases. As it is mentioned above, normal or strong infill walls are considered as depicted in Figure 6.

As it was expected, the strong infill walls result to an increased stiffness in the structure and also a higher base shear force (xx direction 781kN/1036kN-increased by 32.6%, yy direction 617kN/907kN – increased by 47% in Figure 7). After the peak values, a sudden drop in the base shear force takes place due to the infill walls' failures. This sudden load drop is

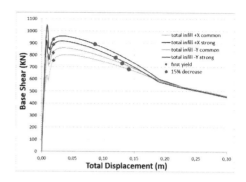

Figure 7. Inelastic behavior of structures with common and strong infill walls in all frames.

higher in the case where strong infill walls are considered (20% drop in xx and 17.9% in yy'). It is worth mentioning that the Greek Retrofit Code (G.RE.CO) recommends that a load drop higher than 15% can be considered as a state of irrecoverable damages in the structure. In the case of structure with normal infill walls, the load drop is limited to only 10.7% in the xx' and 8.2% in the yy' direction. After this temporary load drop, the analyses continues with a redistribution of the load among reinforced concrete elements leading to the recovery of increased base shear force. Figure 2 demonstrates also that the recovery is better in the case of structure with normal infill walls.

3.2 *Infill wall position influence*

Several analyses have been conducted in order to investigate the influence of the position of the infill walls (or their absence) within the structure. These analyses are very crucial as in most of the buildings there are large openings in some of the infill walls or even column to column glass partitions. Besides the cases of bare RC structure or with full infill walls, 15 additional scenarios are considered. Figure 8 presents two characteristic cases out of 15. Only the analyses in x direction are discussed in this study for common infill walls and are depicted in Figure 9. Case 4a (Figure 8) of structure with infill walls exhibits the lowest base shear after the top displacement of 9cm. On the other hand, after this point, Case 9 (Figure 8) has the greatest shear resistance amongst all the considered cases. Finally, the less ductile response is observed for the retrofitted structure with the full infill walls (Figure 6). There, the

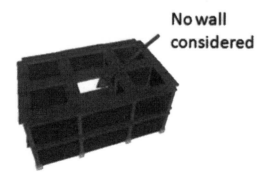

Figure 6. Typical RC framed structure with infill walls.

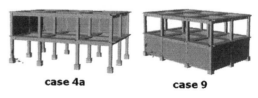

case 4a   case 9

Figure 8. Critical investigated cases (x direction).

108

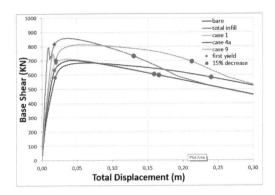

Figure 9. Inelastic behavior of structures with different scenarios for the infill walls' position.

abrupt decrease of the base shear is considerable taking place with the failure of several infill walls. The structure with full infill walls developed the highest base shear with increased stiffness. However, the ultimate top displacement is 13.04 cm (ductility index 7.23). Low ultimate top displacement is also found at the cases where the infill walls are considered only in the first floor with the ground floor having no walls (pilotis). Even though the base shear in those cases is higher than the bare structure, their top displacement was limited to around 15.9 cm compared to the 24.02 cm of the case of bare RC structure (ductility index 7.96 and 12.65 respectively).

4 INELASTIC STATIC (PUSHOVER ANALYSES) OF PRESTRESSED FR RETROFITTED RC STRUCTURES CONSIDERING THE EFFECTS OF INFILL POSITION, STRENGTH AND FRPU RETROFIT

The investigations in sections 2 and 3 suggest the effects of brick wall infill in reinforced concrete (RC) frame structures is a disputable issue because they may have positive effects on structural response or they may trigger undesirable heavy damages or collapse. Especially in the case of existing structures in need of assessment and retrofit there are different approaches that yield very different results on the favourable or unfavourable effects of brick wall infill. This section examines the typical existing reinforced concrete structure used in sections 2 and 3 while taking into account the influence of infill walls when strengthened with fibre sheets impregnated with highly deformable polymer FRPU. Numerous three-dimensional pushover analyses are performed, investigating different scenarios with respect to the infill walls' strength and position within the structure extending the work by Rousakis et al. 2014, Rousakis et al. 2016.

The inelastic behaviour of brick infill walls and their interaction with the RC frames is similarly

modelled with advanced software SeismoStruct (v5.2.2 build 10), according to the approach proposed by Chrisafulli (1997). All the low-strength (typical) infill walls have a vertical compressive strength of 3.10 MPa with a Young's Modulus equal to 3000 MPa. Their diagonal compressive strength is calculated by multiplying the vertical compressive strength with a reduction factor that takes into account the slenderness of the wall. The physical dimensions (length, height and thickness) are necessary to determine this factor. The diagonal compressive strength of low strength plain external infill walls is taken equal to 0.97 MPa and 0.82 MPa with the internal infill walls having strength equal to 0.34 MPa. In the cases that the strong infill walls are taken into account, the diagonal compressive strength of plain external infill walls is taken equal to 2.17 MPa and 1.85 MPa with the internal infill walls having strength equal to 0.76 MPa. In what follows, the cases of typical and of strong infill walls are discussed, retrofitted or not retrofitted with fibre sheets impregnated with highly deformable polymer (FRPU) similarly to the detailing followed in the experimental study by Gams et al. (2014). Therefore, retrofitted low-strength infill walls equal to 2.48 MPa and 2.10 MPa and internal infill wall's strength 0.87 MPa. In the cases of retrofitted strong infill walls, the diagonal compressive strength of plain external infill walls strength of plain external infill walls is 6.27 MPa and 4.53 MPa and the internal infill wall's strength equals 1.87 MPa. The typical force-displacement curves of the infills considered in SeismoStruct for Inelastic Static Analysis of the frames (Static Pushover) are depicted in Figures 10.

The retrofitted structures further receive RC column strengthening with pre-stressed fibre ropes (FR) (Rousakis and Tourtouras 2014, Rousakis 2016) to suppress global shear failure.

4.1 Infill wall position influence

Several analyses have been conducted in order to investigate the influence of the position of the infill walls (or their absence) and the influence of infill walls strengthened with FRPU in different positions within the structure. These analyses are very crucial as in most of the buildings there are large openings in some of the infill walls or even column to column glass partitions. Figure 11a presents six characteristic cases assuming low strength (typical) infill walls. Figure 11b presents six characteristic cases assuming strong infill walls. Case 1+Y and case 7+Y are identical for low strength and strong infills and direct comparison is favored.

4.2 Cases of RC structure with strong brick infills

As far as strong infills are concerned (Figure 11b), the first three cases are retrofitted with fibre sheets in specific positions of infill walls (blue marks) while in the rest of the cases, all infill walls are retrofitted. The analyses of these cases in x and y direction are depicted in

109

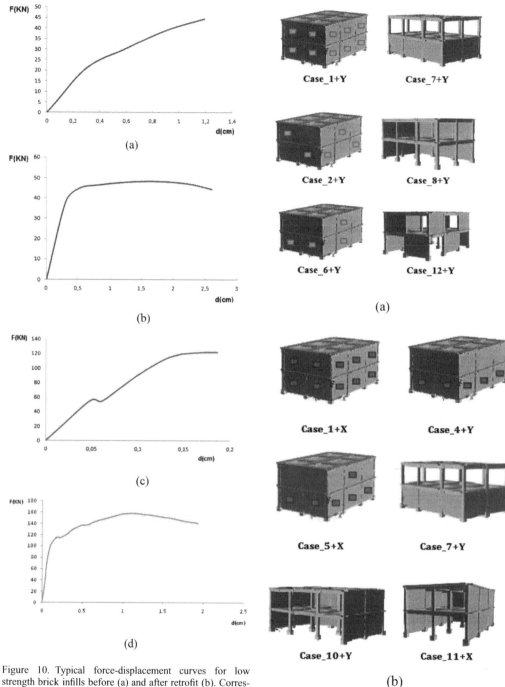

Figure 10. Typical force-displacement curves for low strength brick infills before (a) and after retrofit (b). Corresponding curves for strong infill walls before (c, similar to BM-R-1-QR specimen in Gams et al. 2014) and after retrofit (d, similar to BM-P-2 in Gams et al. 2014).

Figure 11. Critical investigated cases for low strength brick infills (a) and strong brick infills (b).

Figure 12. Non-linear pushover analyses are conducted in the xx' and the yy' directions for all the investigated cases and, for comparison, for the bare RC structure and for the not retrofitted strong infill in all the frame bays. As it was expected, the retrofitted strong infill walls result to an increased stiffness in the structure and a higher base shear force (for case 1 in xx' direction 3210 kN – yy' direction 1700kN, Figure 12).

After the peak values, a sudden drop in the base shear force takes place due to the infill walls'

110

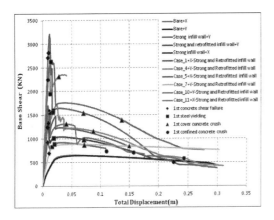

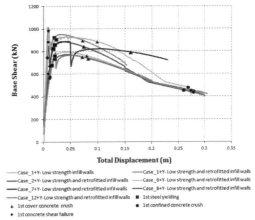

Figure 12. Inelastic behaviour of structures with different scenarios for the strong infill walls' retrofitting position.

Figure 13. Inelastic behaviour of structures with different scenarios for the infill walls' retrofitting position.

failures. This sudden load drop is higher in the case where strong infill walls are considered. The highest drop is happened in the 1st case (50% drop for xx' direction -Figure 12). It is worth mentioning that the Greek Retrofit Code (G.RE.CO) recommends that a load drop higher than 15% can be considered as a state of irrecoverable damages in the structure. Similar results are observed in case 11 in x direction, which has the next highest base shear equal to 3100.45 KN. The difference in this case is the abrupt decrease of the base shear due to the several infill walls' failures and global structure instability issues.

On the other hand, case 5 in x direction, presents high peak shear resistance and the highest residual shear capacity after the first drop. In that case the ductility of the structure is higher than in cases 1 and 11. Further, shear failure of base columns is suppressed.

As regards the y direction, the case 4 with retrofitting of the infill walls of the ground floor only, develops the highest stiffness (higher even than the case with the full retrofitted infill walls). However, cases 4 and 10 present a fluctuation of resistance. Finally, case 7 has the lowest shear resistance but presents the best ductile response. Nonetheless, the base shear in those cases is higher than the bare structure both in x and in y direction.

4.3 Cases of RC structure with low strength brick infills

The cases 1+Y, 2+Y and 6+Y are retrofitted with fiber sheets in specific positions of infill walls (green marks) while in the rest cases all infill walls are retrofitted. Figure 13 presents the analyses results of these specific cases, only in y direction and with low-strength infill walls. Case 1, has the highest shear capacity among all the considered cases, equal to 1015 kN. Nevertheless, after the peak values, a sudden drop in the base shear force takes place due to the infill walls' failures. This sudden base shear force drop is higher in the cases of structure with full infill walls (cases 1, 2, 6 with similar

force drop around 17% - Figure 13). However, the highest force drop occurs in the 7th case (around 24% - Figure 13). Thus, the base shear force drop of the above cases is higher than 15% and denoted a state of irrecoverable damages in the structure according to the Greek Retrofit Code (G.RE.CO).

Furthermore, case 6 presents the next higher peak shear resistance and the highest residual base shear. In the mentioned cases, their maximum base shear appeared after the total displacement of 1cm and after that an abrupt decrease of the base shear took place due to the several infill walls' failures. On the other hand, case 7 exhibits a gradual increase of shear resistance up to 5cm and after this point an abrupt decrease of the base shear occurs. Finally the cases 8 and 12, retrofitted with fibre sheets, exhibit lower shear resistance than the structure with infill walls without retrofitting (case 1). Nonetheless, these specific cases illustrate the maximum ductile response.

Herein, it should be noted that recent experimental results from cyclic pseudo static pushover tests of brick wall infilled RC frames using Orthoblock K100 (KEBE SA, Greece) infills and PUFJs at the RC frame – infill interface or FRPUs at the exterior brick infill surfaces (Akyildiz et al. 2020 and Triller et al. 2020) suggest that in RC frames designed according to Eurocodes, suitable infill protection may lead to horizontal displacement drifts of the brick infilled RC frames as high as 4.4%. Further, recent seismic table tests of RC structure with Orthoblock K100 brick infill, suitably protected with the same PUFJ and FRPU technique suggest that similar horizontal displacement drifts of the brick infilled RC frames up to 3.7% may be achieved at the maximum table acceleration (i.e. maximum base shear capacity) level (Rousakis et al. 2020a,b). Therefore, the temporary load drops presented in sections 2, 3 and 4 may be avoided in some cases or with suitable selective design of the infills.

5 RESILIENT PSEUDO-DYNAMIC FE MODELING

To fully utilize the potential of FR, PUFJ, FRPU and of other techniques that engage strain and damage redistribution phenomena towards advanced structural resilience, suitable mathematical formulations at material, section, member and structure level and their interrelations are necessary.

The pilot study by Fanaradelli et al. (2018) provides a suitable pseudo-dynamic inelastic 3-dimensional FE analysis of composite rope confined concrete (FR technique) reproducing most of the abovementioned effects. Most importantly, it allows for the strain redistribution of the elastic confinement provided by the composite rope throughout loading. Therefore, it could be considered an analytical mechanical analogue to explore the potential of the ERUC design concept. The advanced nonlinear 3 dimensional analytical FE models were developed and analysed with Ansys Workbench R15, EXPLICIT DYNAMICS (ANSYS 15.0) finite element software. Concrete was modelled with eight-node solid element utilizing the RHT advanced plasticity concrete model for brittle materials (Riedel et al. 2009) and suitable for dynamic analyses. The composite rope was modelled as an equivalent hexahedral solid element with "frictionless" interaction with the concrete core. In this study, a pseudo-dynamic imposed axial deformation history on the concrete column was performed.

The study by Rousakis et al. (2018) presents the pilot pseudo-dynamic analytical results for FRP confined columns under combination of axial load and imposed cyclic horizontal displacements. Concrete was modelled with RHT as already presented while the steel reinforcement was modelled using a two-node Beam (Line) element considering a material that yields and hardens.

The study by Thomoglou et al. (2018a, 2018b) presents static structural 3d FE analyses of brick walls with external FRCM or FRP double-sided strengthening. 3D modelling of the bricks and of the mortar joint and of their interactions was followed as well as detailed and advanced modelling of the composite strengthening materials in ANSYS.

5.1 *FE modeling for seismic table experiments of of RC structure with retrofitted brick infills*

The previous sections (see also Rousakis et al 2017 and Rouka et al. 2017) have investigated the effects of the strengthening of the brick infills with composites bonded externally with highly deformable polymers. Such strengthening in masonry walls has been proven very efficient in transforming the masonry walls into sufficiently ductile structural members (Gams et al. 2014).

They highlighted that suitably designed strengthening of brick infills may provide RC structure with higher initial stiffness and base shear as the infills are engaged in a more ductile manner, thus maintaining

(a)

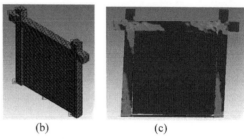

(b)   (c)

Figure 14. Shake table tests within INMASPOL project SERA-TA framework (a). Three-dimensional pseudo-dynamic finite element model (b) and pilot analysis (c) of infilled frame with polymer joint.

sufficient residual base shear despite damage accumulation. As mentioned above, recent experiments examined the in-plane and out of plane performance of orthoblock infills protected with highly deformable polymer joints at the RC frame – brick infill interface or repaired with fibre grids bonded with highly deformable polymer joints (Akyildiz et al. 2020 and Rousakis et al. 2020a,b). The latter experiments were dynamic shake table tests on real scale RC structure (Figure 14a adapted from Rousakis et al. 2020b). All tests revealed remarkable inelastic behaviour of the brick infilled structures, similar to the one considered in section 4. As these advanced retrofits involve strain redistribution phenomena, it is very demanding but promising the pilot analysis of brick infilled RC frames with 3-dimensional FE (Figure 14b and 14c adapted from Rousakis 2020c).

6 CONCLUSIONS

Innovative retrofit materials and techniques may be utilized to upgrade the resilience of critical RC structures. In this analytical study a typical low rise RC structure has been considered and the effects of VFR and PPFR through column confinement were investigated through static inelastic analyses, adopting advanced beam-column elements in SeismoStruct software. Infill walls were also taken into account in the analyses. The inelastic analyses suggest that retrofitted structures in all cases satisfied the performance based requirements of EC8-3 and GRECO for both seismic hazard zones I and III. Even the case of reduced quantity of ultra-high deformability PPFR provided the structures with enough deformation capacity. Conventional strengthening materials (CFRP and GFRP) provided similar results. In bare structure, the critical displacement corresponds to fracture of the longitudinal bars (except for steel plate confinement).

The effects of the strength and position of the infill walls in a structure strengthened with FR reinforcements are of high importance. The higher the strength of the infill wall, the higher the initial stiffness of the structure and the higher the maximum base shear. However, the temporary load drop during failure of the infill walls is higher as the activation of the RC frame at similar top displacement levels is lower. The distribution of the infill walls within the structure may have a significant effect on the maximum base shear, the temporary load drop and the ultimate top displacement of the structure. In several cases the performance of the structure with infill walls is inferior to that of bare one.

FRPU retrofit of infills may provide viable alternative retrofit strategy to designers. The present analytical study investigates also the effects of the strength and position of the strong infill walls in a structure strengthened with FRPU. The higher the strength of the infill wall, the higher the initial stiffness of the structure and the higher the maximum base shear. However, the temporary load drop during successive failure of different infill walls is higher when the strengthening of infill walls is global. The distribution of the infill walls within the structure may have a significant effect on the maximum base shear, the temporary load drop and the ultimate top displacement of the structure. Structure of case 5 x may provide a remarkable base shear capacity increase to resist strong earthquakes. It performs at immediate occupancy level while maintaining a substantial residual strength in cases of overloads against early collapse.

The effects of low-strength infill walls in a structure retrofitted with fibre sheets having highly deformable polymer (FRPU) are similarly interesting. Overall, the distribution of the low-strength infill walls within the structure may affect significantly the maximum base shear, the temporary load drop and the ultimate top displacement of the structure. Especially, case 6 in y direction, may yield a considerable shear resistance so as to resist strong earthquakes while maintaining a substantial residual strength. As compared with the case 5 in x direction with strong infill walls, it is concluded that strong infill walls develop higher initial stiffness of the structure and higher maximum base shear than low-strength infill walls in y direction (case 7, 3210 kN - case 6, 998

kN). However, the temporary load drop during successive failure of different strong infill walls is higher against the load drop of low-strength infill walls: 35% for case 5 of strong infills versus 17% for case 6 of low strength infills. Herein, it should be mentioned these analytical studies were based on the experimental behaviour of retrofitted masonry walls.

As expected, recent experimental results from cyclic pseudo static pushover tests of brick wall infilled RC frames using Orthoblock K100 infills and FRPU strengthening at the exterior brick infill surfaces revealed similarly ductile inelastic behaviour. Further, recent experimental results from cyclic pseudo static pushover tests of brick wall infilled RC frames using Orthoblock K100 infills and PUFJs at the RC frame – infill interface further extend the usable deformability of suitably protected infilled RC frames. Finally, recent seismic table tests of RC structure with Orthoblock K100 brick infill, suitably protected with the same PUFJ and FRPU technique validate similar favourable effects under dynamic excitations. Therefore, the temporary load drops presented in sections 2, 3 and 4 may be avoided in some cases or with suitable selective design of the infills.

To fully utilize the potential of such innovative retrofit techniques that involve strain redistribution phenomena, dynamic 3-dimensional Finite Element Modelling, presented in section 5, could be taken under consideration and further developed.

REFERENCES

Akyildiz, A.T., Kwiecień, A., Zając, B., Triller, P., Bohinc, U., Rousakis, T. & Viskovic, A. 2020 Preliminary in-plane shear test of infills protected by PUFJ interfaces. *17th International Brick and Block Masonry Conference (IB2MaC 2020)*, Cracow, Poland.

ANSYS Release 15.0. Users' manual SAS IP, Inc.

Bocchini, P., Frangopol, D. M., Ummenhofer, T., & Zinke, T. 2014. Resilience and Sustainability of Civil Infrastructure : Toward a Unified Approach. *Journal of Infrastructure Systems*, *20*(2), 1–16. https://doi.org/10.1061/(ASCE)IS.1943-555X.0000177.

CEN Eurocode 8. 2005. Design of structures for earthquake resistance, part 3: Assessment and retrofitting of buildings. European Committee for Standardization, Brussels, Belgium.

Crisafull, F.J. 1997. Seismic Behaviour of Reinforced Concrete Structures with Masonry Infills', PhD Thesis, University of Canterbury, New Zealand.

Dai, J.G, Bai, Y.L. & Teng, JG. 2011. Behavior and modeling of concrete confined with FRP composites of large deformability. *J Compos Construct*. 15(6): 963–73.

Fanaradelli T., Rousakis T. & Pavlou D. 2018. 3d Finite element analyses of axially loaded columns externally strengthened with transverse composite ropes. *26th International Conference on Composites/Nano Engineering Paris, France, 15-21 July, 2018*.

Gams, M., Kwiecien, A., Zajac, B. & Tomazevic, M. (2014) Seismic Strengthening of Brick Masonry Walls With Flexible Polymer Coating. Guimaraes: 9th International Masonry Conference, 2014.

GRECO. Greek retrofitting code. 2017. Ggreek organization for seismic planning and protection. Greece: Greek Ministry for Environmental Planning and Public Works.

Ilki, A., Peker, O., Karamuk, E., Demir, C., Kumbasar, N. 2008. FRP retrofit of low and medium strength circular and rectangular reinforced concrete columns. *Journal of Materials in Civil Engineering* 20(2): 169–188.

Ispir, M., Dalgic, K.D., Ilki, A. 2018. Hybrid confinement of concrete through use of low and high rupture strain FRP. *Compos B Eng* 153: 243–55.ISSN 1359-8368.

Janke, L., Czaderski, C., Ruth, J., Motavalli, M. 2009. Experiments on the residual load-bearing capacity of prestressed confined concrete columns. *Eng Struct* 31(10): 2247–56.

Karantzikis, M., Papanicolaou, C.G., Antonopoulos, C.P., Triantafillou, T.C. 2005. Experimental investigation of nonconventional confinement for concrete using FRP. *J Compos Construct* 9(6): 480–7.

Kwiecień, A., Matija, G., Rousakis, T., Viskovic, A. & Korelc, J. 2017. Validation of a New Hyperviscoelastic Model for Deformable Polymers Used for Joints between RC Frames and Masonry Infills. *Engineering Transactions of Polish Academy of Sciences* 65(1): 113–121.

Kwon, M., Seo, H. & Kim, J. 2016. Seismic performance of RC-column wrapped with velcro. Structural engineering and mechanics. *An International Journal* 58(2) https://doi.org/10.12989/sem.2016.58.2.379

Padanattil, A. & Karingamanna, J., Mini, K.M. 2017. Novel hybrid composites based on glass and sisal fiber for retrofitting of reinforced concrete structures. *Construct Build Mater* 133: 146–53.

Poljansek, K., Marin Ferrer, M., De Groeve, T. & Clark, I. (2017) Science for Disaster Risk Management 2017: Knowing Better and losing less. EU Science Hub, Disaster Risk Management Center. https://ec.europa.eu/jrc/en/publication/science-disaster-risk-management-2017-knowing-better-and-losing-less

Qin, R., Hao, H., Rousakis, T. & Lau D. 2019. Effect of shrinkage reducing admixture on new-to-old concrete interfaces. *Composites Part B: Engineering* 163: 96–106.

Riedel, W., Kawai, N. & Kondo, K.I. 2009. Numerical assessment for impact strength measurements in concrete materials, *International Journal of Impact Engineering* 36: 283–293.

Rouka, D., Kaloudaki, A., Rousakis, T., Fanaradelli, T., Anagnostou, E., Kwiecień, A., Gams, M., Viskovic, A. & Zając B. 2017. Response of RC buildings with Low-strength Infill Walls Retrofitted with FRP sheets with Highly Deformable Polymer – Effects of Infill Wall Strength. *25th International Conference on Composites/Nano Engineering, Rome, Italy* July 16- 22. Editor David Hui.

Rousakis, T., Georgiadis, N. & Anagnostou, E. 2018. 3d Finite element pseudoseismic analysis of CFRP strengthened RC columns. *26th International Conference on Composites/Nano Engineering, Paris, France, 15-21* July.

Rousakis, T., Rouka, D., Kaloudaki, A., Kwiecień, A., Gams, M., Viskovic, A. & Zając B. 2017. Fast Retrofitting of Strong Wall Infill of RC buildings with Fiber Sheets Impregnated with Highly Deformable Polymer. *25th International Conference on Composites/Nano Engineering, Rome, Italy July 16- 22*. Editor David Hui.

Rousakis, T.C. 2016. Reusable and recyclable nonbonded composite tapes and ropes for concrete columns confinement. *Composites Part B: Engineering* 103: 15–22.

Rousakis, T.C. 2017. New Design Concept for Advancing Seismic Structural Resilience of RC Columns Through Hybrid Confinement. *25th International Conference on Composites/Nano Engineering, Rome, Italy July 16-22.* Editor David Hui.

Rousakis, T.C., Kardala, M.K., Moumtzis, I. & Stylianou, M. 2014. Confinement With High Deformability Fiber Ropes In Existing Reinforced Concrete Structure. Proceedings of The *Twenty-second Annual International Conference on COMPOSITES/NANO ENGINEERING, Malta, 13-19 July, 2014.* Editor David Hui.

Rousakis, T., Tsaridis, C. & Moumtzis, I. 2016. Composite Rope Strengthening of Existing Reinforced Concrete Structures - Effects of Infill Wall Position and Strength. *24th International Conference on Composites/Nano Engineering Haikou, Hainan Island, China, July 17- 23, 2016.*

Rousakis, T.C., Panagiotakis, G.D., Archontaki, E.E. & Kostopoulos, A.K. 2019. Prismatic RC columns externally confined with FRP sheets and pretensioned basalt fiber ropes under cyclic axial load. *Composites Part B: Engineering* 163: 96–106.

Rousakis, TC & Tourtouras, IS. 2014. RC columns of square section – passive and active confinement with composite ropes. *J Compos Part B Eng* 58: 573–81.

Rousakis, T. 2013. Hybrid Confinement of Concrete by FRP Sheets and Fiber Ropes Under Cyclic Axial Compressive Loading. *ASCE Journal of Composites for Construction* 17(5): 732–743.

Rousakis, T. 2014 Elastic Fiber Ropes of Ultrahigh-Extension Capacity in Strengthening of Concrete Through Confinement. *ASCE Journal of Materials in Civil Engineering* 26(1): 34–44.

Rousakis, T. C. 2018. Inherent seismic resilience of RC columns externally con fined with nonbonded composite ropes. *Composites Part B* 135: 142–148. https://doi.org/10.1016/j.compositesb.2017.10.023

Rousakis, T., Ilki, A., Kwiecien, A., Viskovic, A., Tiller, P., Ghiassi, B., Benedetti, A., Gams, M., Rakicevic, Z., Halici, O.F., Zając, B., Hojdys, Ł, Krajewski, P., Rizzo F., Colla, C., Sapalidis, A., Papadouli, E., Vanian, V. & Bogdanovic, A. 2020. Quick Reparation of RC Infilled Frames after Seismic Damages – Experimental Tests on Shaking Table. *10th International Conference on FRP Composites in Civil Engineering, Istanbul 1-3 July 2020.*

Rousakis, T., Papadouli, E., Sapalidis, A., Vanian, V., Ilki, A., Halici, O.F., Kwiecień, A., B. Zając, Ł. Hojdys, P. Krajewski, M. Tekieli, T. Akyildiz, A. Viskovic, F. Rizzo, M. Gams, P. Triller, B. Ghiassi, A. Benedetti, C. Colla, Z. Rakicevic, A. Bogdanovic, F. Manojlovski & Soklarovski A. 2020. Flexible Joints between RC frames and masonry infill for improved seismic performance – shake table tests. *17th International Brick and Block Masonry Conference Cracow, Poland.*

Seible, F., Priestley, M.J.N. & Hegemier, G.A. 1997. Innamorato D. (1997) Seismic retrofit of RC columns with continuous carbon fiber jackets. *J Compos Construct* 1(2):52–62. https://seismosoft.com/product/seismostruct/.

Thomoglou, A., Rousakis, T. & Karabinis, A. 2018. Numerical modeling of shear behavior of URM strengthened with FRCM or FRP subjected to seismic loading. *16th European Conference on Earthquake Engineering, Thessaloniki, Greece, 18-21 June 2018.*

Thomoglou, A., Rousakis, T. & Karabinis, A. 2018. Three-dimensional Finite Element Analysis of In-plane Shear Behavior of Masonry Strengthened with TRM/FRCM. *18th Concrete Conference, Athens*, March 29-31, 2018 (in Greek).

Triantafillou, T.C., Papanicolaou, C.G., Zissimopoulos, & P., Laourdekis, T 2006 Concrete confinement with textile-reinforced mortar jackets. *ACI Struct J* 103(1): 28–37.

Triantafyllou, G., Rousakis, T., Karabinis, A. 2019. Corroded RC Beams at Service Load before and after Patch Repair and Strengthening with NSM CFRP Strips. *Buildings* 9(3), 67. https://doi.org/10.3390/buildings9030067.

Triller, P., Kwiecień, A., Bohinc, U., Zając, B., Rousakis, T. & Viskovic, A. 2020. Preliminary in-plain shear test of damaged infill strengthened by FRPU. *10th International Conference on FRP Composites in Civil Engineering, Istanbul 1-3 July 2020*.

Wu, G., Wu, Z.S., Lu, Z.T. & Ando, Y.B. 2008 Structural performance of concrete confined with hybrid FRP composites. *J Reinforc Plast Compos* 27(12): 1323–48.

Brick and Block Masonry - From Historical to Sustainable Masonry –
Kubica, Kwiecień & Bednarz (eds)
© 2020 Taylor & Francis Group, London, ISBN 978-0-367-56586-2

Mining activity in Poland: Challenges

T. Tatara

Cracow University of Technology, Kraków, Poland

ABSTRACT: The study is the review of some results in the field of induced seismicity in Poland due to rock-bursts. Mining shocks are the most intense phenomena generated by human activity and are called paraseismic sources. These vibrations can cause not only significant damage to surface structures but also they can have a negative influence on people occupied buildings. In Poland, rockbursts can usually only be observed in mining regions. Rockbursts are not subject to human control, and they are random events with respect to their time, place, and magnitude. There are many differences between earthquakes and mining tremors. The major differences are e.g. magnitude of energy, intensive phase duration, peak ground accelerations and velocity (*PGA, PGV*), the content of predominant frequencies, frequency of occurrence, depths of hypocentre. Mining-related vibrations stand out by having the greatest intensity of all forms of paraseismic vibrations.

1 INTRODUCTION

Free-field motions can be induced not only by earthquakes but also by so-called paraseismic sources. They accompany the mining of copper ore, hard and brown coal, gold, diamonds, non-ferrous metals and also originating from pile driving, driving sealed walls and car, train, tram traffic and subway traffic. Although these tremors are strictly connected with human activity and can usually be observed only in mining regions, they differ considerably from other paraseismic vibrations. The random nature of mining rockbursts occurrence cause difficulties to measure the surface vibrations. Many countries record mining-induced seismicity in mining areas associated with the underground exploitation of coal, copper, and gold. Active seismicity and rockbursts are substantial problems in mining regions worldwide, for instance, in China, Australia, South Africa, Canada, India, France, Czech Republic, Russia, and Poland (Maciąg et al. 2016). We can observe an increase in the mining-induced seismicity hazard although the energy of mining tremors is smaller than that of natural earthquakes, because of substantial mining activity. Mining-induced ground motion is a hazard to surface structures, as ground motions can result in their damage.

Poland belongs to regions with low and shallow seismic hazard, but from time to time, weak earthquakes occur such as an event in 2004 in the Podhale region (Zembaty 2007). Poland, especially the south of Poland, is exposed to random mining tremors occurring as earthquakes (Zembaty 2004); these events are beyond human control. The maximum magnitudes of underground mining tremors are about 4.0 - 4.6. Such a value of magnitude can be comparable with weak earthquakes. There are, however, many

differences between earthquakes and mining tremors (Tatara et al. 2017). The significant differences are:

(i) Duration of the intensive phase of vibrations – mining tremors last less than earthquakes. The duration of the intensive phase can even be ten times less (Pachla et al. 2019).

(ii) Peak ground accelerations – mining tremors and earthquakes differ concerning the order of magnitude. *PGA* for mining tremors can be about 0.3 g (g – acceleration of gravity) while the *PGA* for earthquakes maybe 1-2 g (Furumura et al. 2011).

(iii) Content of predominant frequencies – generally in the case of earthquakes, the predominant frequencies are low; the predominant frequencies of ground acceleration vibrations in the case of rock bursts are in the bands up to 6.5 Hz, from 6.5 to 12 Hz and higher than 12 Hz (sometimes even up to 40 Hz) (Maciag et al. 2016).

(iv) Frequency of occurrence – strong earthquakes that cause structural damages occurs in the same area with a frequency of the tens to hundreds of years. By contrast, mining tremors with a high value of *PGA* caused by mining activity and underground explosions occur much more often, within a few years of each other.

(v) Focal depths – mining tremors occur at much shallower depths than earthquakes; the depth of the later could be as much as 180 km (Kufner et al. 2017).

(vi) The size of the area affected and the damage in surface structures and underground infrastructure are smaller than for earthquakes (Tatara 2012).

The most critical differences between rockbursts and earthquakes are, e.g. values of peak ground accelerations (*PGA*), peak ground velocity (*PGV*), composite parameter PGV^2/PGA (Furumura et al. 2011, Pineda-Porras & Ordaz 2012), time of significant phase of the free-field vibrations (Pachla et al. 2019), values of dominant frequencies (Maciag et al. 2016), geophysical parameters like, e.g. values of energy magnitude, focal depth, frequency of occurring (Kufner et al. 2017). These two primary sources of free- field vibrations affect surface structures and result in damage them (Deck et al. 2003; Khosravikia et al. 2018; Singh & Roy 2010).

The main subjects focused on the study deal with the essential problems resulting from the impact of mining-related induced seismicity on the environment, structures and people as follows:

i. Analysis of experimentally obtained results in terms of the differences between the mine-induced vibrations described by the response spectra from the free-field near a given building and the simultaneously recorded vibrations in the building foundations.
ii. The impact of the type of building on the transmission of response spectra from the free-field vibrations to the building foundations.
iii. The usefulness of the approximate models of soil-structure interaction (SSI). The contribution presents original, simple, empirical models for the evaluation of the differences in the response spectra originating from free-field and building foundation vibrations in the mining region.
iv. Assessment of the impact of mining-related vibrations on public buildings, industrial and underground structures, and evaluation of their dynamic resistance using results of computer analyses.
v. Assessment of the impact of mining-related vibrations on buildings according to polish standard (PN-B-02170:2016-12, 2016) and the new version of empirical-measurement scale GSIS-2017 (Kuzniar 2018a), and on people staying in dwelling masonry buildings using: a) the RMS method according to British (BS 6472:1992, 1992) and ISO (ISO 2631-2, 2003) and Polish standards (PN-B-02171:2017-06, 2017), b) new version of empirical-measurement scale GSIS-2017 (Kuzniar et al. 2018a).
vi. Damage to surface structures and their causes.

The basis for the studies is the results of in-situ measurements of free-field and building foundations vibrations and vibrations of whole structures.

2 RESPONSE SPECTRA FROM MINING-RELATED VIBRATIONS

Response spectra we use to evaluate the dominant frequencies in the records of vibrations and the difference between free-field vibrations and simultaneously recorded building foundations vibrations. These

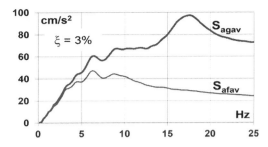

Figure 1. Average response spectra S_{ag} and S_{af} determined based on all registered accelerations at the free-field and foundations of the buildings (Maciag et al. 2016).

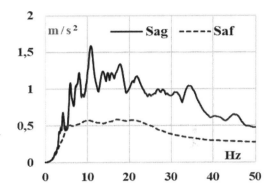

Figure 2. Average response spectra S_{ag} and S_{af} determined based on all registered accelerations at the free-field and foundations of the low-rise building in the USCD region.

records are also useful to assess the transition of vibrations from the ground to the structure. Figure 1 and Figure 2 present example response spectra calculated based on hundreds of horizontal components of free-field and building foundation vibrations in LGCD (for low-rise, medium height and high-rise buildings) and USCD (for low-rise masonry building).

The response spectrum from Figure 1 calculated based on building foundation records substantially differs from the spectrum from free-field vibrations within the almost total range of the investigated frequencies. These spectra are only the same in the frequency range from 2.9 to 3.3 Hz, which corresponds to fundamental natural frequencies of the medium height buildings in the lateral x and y directions. We performed similar calculations for the records from USBD, and Figure 2 shows calculated average response spectra. Analysis of spectral curves shows that the building plays the role of a low-pass filter. The band of dominant ground vibration frequencies is full.

3 KINEMATIC SOIL-STRUCTURE INTERACTION (SSI)

The study used thousands of pairs depicting simultaneously recorded ground and building foundations vibrations. Buildings of various structures were tested. They were single-family masonry buildings and prefabricated buildings of medium height (5 floors) and high buildings up to 12 floors. The multiple interactions (i.e. structure to structure through the soil interaction, so-called "Site-City interaction") also do not appear because of the distance of the other neighbouring buildings (several dozen meters). The accelerograms recorded on the free-field consider the effect of motion amplification.

The test focused on the impact of energy magnitude, epicentre distance and the maximum peak ground acceleration on SSI phenomena. Figure 3 and Figure 4 present examples of the ground and building foundation records, corresponding FFT and coherence functions. We measured vibrations in 5 storey building.

The results of SSI investigation are useful to establish the proper kinematic loading. The works carried out so far at the Chair of Structural and Materials Mechanics at the Cracow University of Technology, and our own experience shows that the adoption of free-field vibration records in most cases leads to over-stated results.

As an example, the articles (Kuzniar & Tatara 2017, Kuzniar et al. 2018c) consist the results of measurements of mining-related surface vibrations in the LGCD, and USCD assessed the impact of mining tremor epicentre distances as well as mining tremors energies on a curve relationship (ratio) of response spectra (RRS) from simultaneously measured free-field vibrations and building foundations vibrations. In LGCD, the sources of considered vibrations were

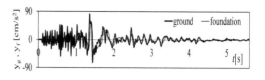

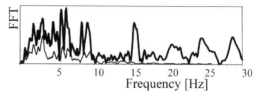

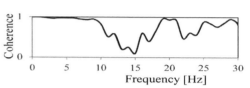

Figure 4. Example of the simultaneously measured horizontal y component of ground and foundation building vibrations and corresponding FFT and coherence function (Maciag et al. 2016).

mine-induced rockbursts resulting from underground exploitation of copper ore. We have analysed only records induced by rockbursts with energy higher than 10^6 J and horizontal components of peak ground accelerations larger than 10cm/s². Epicentre distances of considered mining shocks are in the range re = 270 – 5839 m. The energy magnitudes are in the range En = $1 \cdot 10^6 - 2 \cdot 10^9$ J. The results of analyses indicate that the epicentre distance of mining rockbursts as well as the energy magnitude of mining tremor, can have a significant impact on the transmission of response spectra from the free-field to the apartment medium-rise building and high-rise building foundations.

In the USCD region acceleration records of the free-field vibrations next to the analysed office building and on the building foundations were registered simultaneously using the so-called "an armed partition" gauges. Full-scale tests were performed many times in 8 years (long-term experimental monitoring). Analysis of the results of measurements of mine-induced vibrations in respect to the changes of the maximal values of the horizontal components of acceleration during their transmission from the free-field to the foundation of the low-rise building leads to the conclusion that the influence of many variables: maximum value of ground vibrations, epicentre distance, mining rockbursts energy, the direction of wave propagation, vibration direction (parallel to the longitudinal or transverse axis of the building) on the numerical quantity of these changes (reduction or increase) is visible. But it should be stated that it is complicated to precisely evaluate the way of the ground vibrations transmission to the building foundation. We can formulate only some

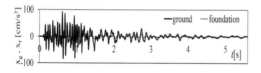

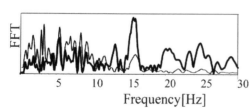

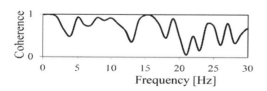

Figure 3. Example of the simultaneously measured horizontal x component of ground and foundation building vibrations and corresponding FFT and coherence function (Maciag et al. 2016).

trends using the experimental data obtained from the full-scale tests. The reason for this conclusion is the scatter of the ratios of the maximum acceleration of the building foundation to ground acceleration, and ambiguous way of the vibration transmission from free-field to the building.

4 APPROXIMATE EVALUATION OF HARMULNESS OF THE MINING-RELATED VIBRATIONS

In Poland approximate evaluation of vibrations harmfulness in case of rockbursts is performed by polish standard (PN-B-02170:2016-12, 2016), scale GSI-2004/11 (Dibinski et al. 2011) and new scale GSIS-2017 (Kuzniar et al. 2018a) combined the intensity of free-field vibrations parameters with damages in buildings.

4.1 Harmfulness for buildings

The GSIS-2017 scale uses the results of registration and observation of several the most substantial shocks occurring in the USCD area in the years 2015-2016, with seismic energies exceeding 10^8 J and the maximum values of vibration velocity $PGV > 0.05$ m/s, and acceleration with $PGA > 1.0$ m/s^2.

The GSIS-2017 scale consists of two versions. The basic is a velocity version (GSIS-2017–V). Acceleration version of the scale (GSIS-2017-A) is an auxiliary. The GSIS-2017 scale associates values of the ground parameters with macroseismic effects in structures and their perceptibility by people. The base for establishing impact evaluation of surface mining-related vibrations is the maximum value of the resultant of ground velocity PGV_H – (velocity version of the scale). The maximum value of the resultant ground acceleration PGA_{H10} calculated using the filtered components of the horizontal vibration in the band to 10Hz (PGA_{H10}) is the base for evaluating the intensity of the rockbursts. The second parameter used in impact evaluation is the duration of the intensive phase of vibrations t_{Hv} and t_{Ha} for the velocity and acceleration records, respectively (Kuzniar et al. 2018a, Kuzniar et al. 2018b). These scales also include the previous measurements of mining-related ground vibrations for weaker rockbursts. In the scale, the technical condition and type of the building structure for its dynamic resistance were taken into account. The GSIS-2017 scale includes the following types of buildings with traditional and traditional-improved construction, i.e. buildings made of brick or other small-size elements, having wall bearing systems; concrete or reinforced concrete bearing structure; frame buildings of reinforced concrete or steel construction. Besides, the scale includes: effects of vibrations on buildings in poor technical condition and subject to the

influence of continuous deformations characteristic of the III–V category of mining area (Popiolek 2009); effects of vibrations on sensitive historic buildings of historical significance; criteria for empirical evaluation of dynamic resilience of buildings subject to seismic effects of mining, developed using the GSIS-2017 scale.

On the GSIS-2017 scale, the impact of the shock is also expressed by measuring the level of seismic intensity. These levels are classified based on the recorded or predicted free-field vibration parameters. These parameters are correlated with macro-seismic observations in buildings described by levels of vibration harmfulness.

The GSIS-2017 scale distinguishes 7 levels of seismic intensity (from 0 to VI), which were assigned the effects of free-field vibrations on buildings and linear underground infrastructure and human percebility on vibrations and the inconvenience of using buildings, in correlation with parameters of free-field vibration (Table 1).

The GSIS-2017 scale gives the level of vibration intensity (IV-VI) at which structural damage to buildings may occur together with a description of the damage to buildings and the level of harmfulness of vibrations. It should be emphasized that the VI level has not been verified by measurement.

Figures 5 and Figure 6 show example results of evaluations of the surface vibrations intensity using velocity and acceleration versions of the GSIS-2017 scale. We used in study free- field and simultaneously recorded building foundations. The significant differences are observed, proving that free-field vibrations give overestimated results.

The Polish standard (PN-B-02170:2016-12, 2016) allows using scales of dynamic influences (SWD) for the approximate assessment of the influence of horizontal components of ground vibrations only for buildings considered as a whole structure. The basis for the development of SWD scales were the results of dynamic analysis of buildings considered as standard. The scales refer to masonry buildings and the buildings made of large blocks and large panels

Table 1. Level of vibration intensity I_{GSIS} vs. observed damages in buildings according to GSIS-2017 scale.

I_{GSIS}	Damages observed in buildings
0	Weakly felt; no damage
I	Noticeable, no damage
II	Intensification of existing damage
III	First new damages of non-structural and architectural elements
IV	Slight damage to structural elements
V	Structural damage that may reduce the dynamic resistance of the building
VI	Substantial structural damage that could lead to the loss of building stability

119

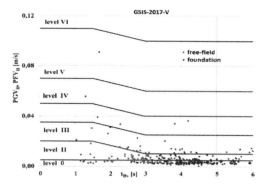

Figure 5. Evaluations of the surface vibrations intensity using velocity version of the scale GSIS-2017-V.

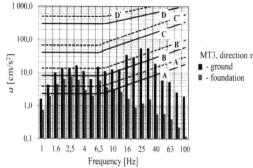

Figure 7. Evaluation of the harmfulness of the horizontal ground and foundation vibrations induced by the MT3 mining tremor using SWD-II scale – LGCD region.

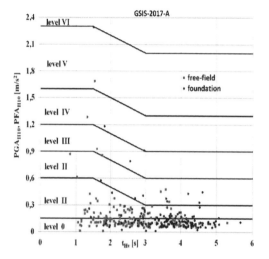

Figure 6. Evaluations of the surface vibrations intensity using velocity version of the scale GSIS-2017-A.

(PN-B-02170:2016-12, 2016). There are two scales: SWD-I and SWD-II. The SWD-I is applied to compact buildings with small dimensions of the horizontal projection (not exceeding 15 m), and one- or two-story height which does not exceed any of the dimensions of the horizontal projection. Indications of the SWD-I scale refer to masonry buildings made of bricks, air bricks, slag concrete blocks or similar. The SWD-II scale is designed for masonry or mixed buildings not higher than 5 storey which height is less than twice the minimum width of the building and for low-rise buildings (up to two storeys), but do not fulfil conditions specified for the scale SWD-I. The scales SWD distinguish five harmfulness zones. Detailed descriptions of the criteria of zoning referring to the vibration harmfulness are presented in the standard (PN-B-02170:2016-12, 2016). The zones boundaries are shown in the two variants referring to

the technical building condition, substrate type and the mode of vibrations. The vibrations, which are eligible for the first three harmfulness zones, are the most common vibrations occurring in engineering practice.

Figure 7 presents example result of the assessment of the harmfulness of the horizontal ground and foundation vibrations induced by one of the most intense rockbursts (denoted as MT3) in the LGCD using SWD-II scale.

The results presented in Figure 7 confirm the essential differences in harmfulness assessment of vibrations on buildings if we use free-field vibrations, not building foundation records. This conclusion has practical value and is the base for applied kinematic loads ion FEM analysis using models of structures.

4.2 Vibrations harmfulness for people

The GSIS-2017 scale is also related to the level of vibration intensity I_{GSIS} with human percebility on vibrations and the inconvenience of using buildings, in correlation with parameters of free-field vibration (Table 2). The scale uses the same parameters as in the evaluation of the intensity of vibration concerning structural damages in buildings.

The scale GSIS-2017 assesses both the intensity of mining-related vibrations versus damages may occur in buildings and perception of vibrations by occupants in buildings.

In the LGCD region, we use a similar scale as the GSIS-2017 scale to assess the surface vibrations intensity on buildings and human. The GSI-2004/11 (Dubinski et al. 2011) scale associate macro-seismic effects of vibrations with the values of the parameters of surface vibrations, selected for the best description of the actual intensity of the vibrations. For buildings, the GSI-2004/11 scale allows an approximate assessment of the impact of mining vibrations from harmless vibrations through causing non-structural damages, up to the limit above which structural damages may occur. The basic parameters

Table 2. Level of vibration intensity I_{GSIS} vs human perception according to GSIS-2017 scale.

I_{GSIS}	Human perception
0	Imperceptible by people or weakly felt
I	Perceptible by people inside buildings, poorly felt by people outside. Hanging objects swinging
II	Inside buildings, people, felt the vibrations very much
III	Strongly felt by most people outside and inside buildings
IV	Very strongly felt by all people. People are terrified. Many scared people run outside. Some lose their balance, especially on the upper floors.
V	Powerful and the nagging sense of vibration. People have problems maintaining their balance in the upper threshold of vibration intensity
VI	General fear and panic

necessary in assessing the impact of surface vibrations caused by rockbursts through the use of the scale GSI-2004/11 are the same as in GSIS-2017 scale. In (Tatara 2012) we can find a detailed description of the effects of the surface mining-related vibrations classified to the degrees (levels) of intensity (four degrees of intensity – from grade 0 to grade III) presented in the seismic scale GSI-2004/11.

In 2017, a new version of the Polish standard appeared, which included the problems of evaluation of the impact of building vibrations on people passively receiving vibrations (PN-B-02171:2017-06, 2017). The standard specifies formal criteria and evaluation methods. These vibrations may cause the reduction of people's quality of life and the effectiveness of their work. The weighting functions referring to people's reaction to building vibrations contained in this standard are taken from the ISO standards (ISO 2631-1, 1997, ISO 2631-2, 2003).

The evaluation of the impact of mechanical vibrations on people staying in buildings uses:

i. Measurement of the corrected acceleration (or velocity) of vibration in the entire frequency band.
ii. Measuring the spectral acceleration (or velocity) of vibration in 1/3 octave bands.

The vibrations excited by mining tremors, do not entirely fall into the vibration classification given in the standard (PN-B-02171:2017-06, 2017). The duration of such vibrations is a few seconds, and they can happen once every few days or less frequently, although they can occur exceptionally in the same building and vibrate from two shocks during the day. The duration time of mining-related vibrations differs even two powers in comparison to duration the duration defined in the standard as sporadic vibrations. The vibration distribution is given in the British standard (BS 6472, 1992) and is more useful in

the case of vibrations caused by mining shocks. The standard adopts the following assumptions:

i. Continuous vibrations are lasting 16 hours a day and 8 hours a night.
ii. "explosive" vibrations, rapidly increasing to the "peak" value and then diminishing due to damping and which may contain several vibration cycles (or not); they may also consist of suddenly operating several cycles of approximately the same amplitude, provided that their duration is short, less than 2 seconds. Also, intermittent vibrations are determined. The evaluation of annoyance of intermittent vibrations is based on the so-called dose vibration values (with continuous vibrations occurring during 16 hours a day and 8 hours at night).

According to polish standard (PN-B-02171:2017-06, 2017), the vibration measurement is carried out at the point where human percept vibration. Unfortunately, practically no vibrations from mining shocks are carried out in buildings, which can currently be used to assess their annoyance on people. Standard (BS 6472, 1992) allows making vibration measurements outside the structure or on the surface at points that are not vibration perception points for human. In such cases, the so-called transition functions, between the measuring point and the vibration perception point by a human, should be used.

In the case of "Explosive" vibrations in the standard (PN-B-02171:2017-06, 2017) is proposed to treat the vertical components of vibration as the dominant one. In single-family houses, the standard assumed the vibrations measured at the building could be treated as vibrations on the ground floor of the building. It can be assumed that vertical components of vibrations (except buildings with wooden floors) in low and high buildings are similar to those at the bottom of the building. Such an approach is a significant simplification which may lead to incorrect evaluation results.

The evaluation of the impact of vibrations on humans given in the standard (PN-B-02171:2017-06, 2017) is based on the root mean square (RMS) method. The RMS method has a physical interpretation that allows the evaluation of vibration energy. RMS is also an essential method of evaluation according to ISO standard (Standard ISO 2631-2, 2003).

The study gives the base for comparing the assessment of the impact of horizontal mining vibration components on humans using the RMS method and the new Mining Intensity Scales GSI-2017. The GSIS-2017 scale is an empirical scale. The questionnaire of the feelings of mining shocks by people was the basis for determining the levels of sensibility. The RMS method is based on experiments made on people in the XX century from early 30tees till 80tees. The results of these experiments were included in the regulations of the second edition of ISO standard (ISO 2631-2, 1989).

The analyses showed that the most intense parameters of vibration remain at the fourth level of intensity, which means that such vibrations are strongly felt by people and causing fear. The results obtained using the GSIS-2017 scale and the RMS methods are comparable. The RMS method is more precise than assumed in GSIS-2017 scale due to allowing determination the threshold of perceptibility and comfort limits for people in buildings in the 1/3 octave bands.

The results of detailed analyses provide the basis for undertaking work to reduce the level of vibrations and their perceptibility by people. Such works should be taken already at the stage of building design. In existing buildings, you can also try to limit the perception of vibrations on people by using vibroisolation in a building.

5 DAMAGES OBSERVED IN SURFACE STRUCTURES IN MINING AREAS IN POLAND

Determining the causes of damage in surface development covered by mining operations is difficult and often requires detailed analysis. Photographic documentation is beneficial in these analyses. Unfortunately, there is often no such documentation before the occurrence of damage, so we do not have a reference system (reference state). In the event of damage resulting from mining tremors, the technical condition of buildings was reviewed after the shock, without knowing the previous state of damage to the building. In such cases, only the deductive method is possible, consisting of linking the possible effects in the building with the shock. The determination of the cause-effect relationship is based on macroscopic examination. Such visual inspection should be carried out by teams involving civil engineers with experience in the field of structural dynamics. Practice shows that in many of the buildings inspected, previously found damages are not removed on an ongoing basis as a result of the impact of previous mining surface deformations. Table 3 presents the synthesis of damages may occur in building due to mining activity.

Determining the causes of damage in surface development covered by mining operations is difficult and often requires detailed analysis. Photographic documentation is beneficial in these analyzes. Unfortunately, there is often no such documentation before the occurrence of damage, so we do not have a reference system (reference state). We review the technical condition of buildings after the shock, but it means we have not known the previous state of damage to the building. In such cases, only the deductive method is possible, consisting of linking the possible effects in the building with the shock. The determination of the cause-effect relationship is based on macroscopic examination by

Table 3. Examples of the mining influence in buildings.

Type of structure	Possible mining damage
Residential and Public buildings	Tilt from the vertical, cracks and shear deformations of the walls, nuisance of use
Industrial halls	Irregular subsidence, loss of proper position by devices, crane track deformation, production stoppages.
Linear technical infrastructure facilities (e.g. gas and water pipelines)	Unsealing
Sewage networks	Unsealing, oversize slopes
Roads, railroad tracks	Non-normative slopes, defective drainage, road damage
Bridges	Construction overload, tightening expansion joints, excessive displacement of sliding bearings
Tower buildings	Vertical deflection
Earth constructions	Risk of loss of stability, gaps, steps
Water tanks	Slopes, faulty overflows, leaks

civil engineers. Practice shows that in many of the buildings inspected, previously found damages are not removed on an ongoing basis as a result of the impact of previous mining surface deformations.

In buildings located in mining areas, damage may also be caused by other reasons - mainly technical or technological - structural. Causes of damage can be very diverse. The most common damage to buildings is damage as a result of processes occurring in the ground, as a result of natural or artificial processes occurring in the structure, including those resulting from design and execution errors, as a result of the impact of the natural environment on the object.

In analyses of damage causes in structures, we may consider taking into consideration many factors, e.g. the nature of the damage and its morphology, location of damage in the building, the used construction solution, possible deviations from the project, general technical condition of buildings and the type of materials used for construction and the condition of maintenance and renovation of buildings.

Speaking of damage to building structures in areas affected by mining exploitation, including mining shocks, we are dealing with many problems not only of a technical nature. Social aspects and mining economics problems cannot be ignored due to surface protection. In many cases, it is unacceptable, irresponsible and risky to indicate mining shocks as the only cause of damage in building facilities in areas affected by mining operations. Very often, the impact of ground deformation due to mining operations can have a significant effect on surface structures. That is the case, for example, in the USCD region (Figure 8).

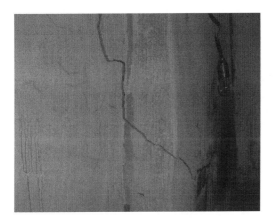

Figure 8. Damages found at basement level in 3-storey masonry multi-family building in USCD.

In this case, the reasons for the occurrence failures in these buildings should be associated with the mine traffic plan adopted by the excessive resistance of buildings to surface deformations. That state concerned an incorrect assessment of the resistance of buildings to deformations before starting underground operation and in particular the lack of assessment on the impact of horizontal compressive deformations and vertical concave curvature. The combined occurrence of the factors mentioned above caused a catastrophe with a substantial social problem, which involved the need to resettle dozens of families from endangered facilities, as well as the closure of the kindergarten, in the building of which the damage threatened the safe presence of children in it (Figure 9).

In the areas of mining plants, own surface development is very diverse. These objects have several damages to the finishing and construction elements. These buildings during their many years of work are exposed to different loads depending on their function. In addition to mining loads (terrain deformations, mining shocks), the impact of operational (technological) loads should also be taken into account. In objects of significant heights, the effect of wind load should also be taken into account. Wind load may affect the technical condition of the load-bearing structure of the building, an example of which is damage, e.g. in the nearly 100-meter skip tower structure in the USCD area (Figure 10). These damages occur in virtually every available tower level.

Damage occurring in building structures in mining areas also includes non-technical factors, e.g. operating time, frequency of repairs and ongoing repairs. Damages with such causes can be found in both LGCD and USCD. That applies above all too old buildings erected using traditional methods with the use of brick and stone, in which, in addition to the cases mentioned above of damage, you can also provide improper joining of fragments of masonry walls made of different materials (Tatara 2012).

Figure 9. Damages found in the outer bearing wall the kindergarten building in USCD.

Figure 10. The vertical crack of skip tower the pylon at ground level.

The great caution and necessity of more in-depth analysis are supported by the fact that often claims related to compensation for damage arising in the opinion of the owners as a result of mining shocks find settlement only in ordinary courts. Often, building owners are aware that, according to mining and geological law, it is the entrepreneur extracting the mineral that is responsible for removing any damage related to the exploitation they are using, abusing these provisions to the requirements of removing damage resulting from natural consumption (Tatara 2012).

Figure 11 and Figure 12 show the special damages that occurred in the church building structure and the wall around the church as a result of intense mining shocks in USCD.

Figure 11. Example damage to the church building due to mining shocks in USCD.

Figure 12. Example damage in the vicinity as a result of mining shocks in USCD.

REFERENCES

BS 6472:1992, 1992. Guide to evaluation of human exposure to vibration in buildings (1 Hz to 80 Hz).

Deck, O., Heib, M.A. & Homand, F. 2003. Taking the soil-structure interaction into account in assessing the loading of a structure in a mining subsidence area. *Eng Struct* 25:435–448.

Dubinski, J., Jaskiewicz K., Lurka A. & Mutke, G. 2011. Mining Seismic Intensity Scale GSI-2004/11 for mining tremors in LGC (in Polish). [In]: Verification of scale GSI-2004, Wrocław: KGHM CUPRUM – CBR.

Furumura, T., Takemura, S., Noguchi, S., Takemoto, T., Maeda, T., Iwai, K. & Padhy, S. 2011. Strong ground motions from the 2011 off-the Pacific-Coast-of-Tohoku, Japan (Mw=9.0) earthquake obtained from a dense nationwide seismic network. *Landslides* 8: 333–338.

ISO 2631-2, 1989. Guide to the evaluation of human exposure to whole body vibration. Part 2- Vibration in buildings.

ISO 2631-1, 1997. Mechanical vibration and shock – Evaluation of human exposure to whole-body vibration – Part 1: General requirements.

ISO 2631-2, 2003. Guide to the evaluation of human exposure to whole body vibration. Part 2- Vibration in buildings.

Khosravikia, F., Mahsuli, M. & Ghannad, M.A. 2018. The effect of soil-structure interaction on the seismic risk to buildings. *Bull Earthq Eng* 16:3653–3673.

Kufner, S.K., Schurr, B., Haberland, Ch., Zhang, Y., Saul, J., Ischuk, A. & Oimahmadov, I. 2017. Zooming into the Hindu Kush slab break-off: a rare glimpse on the terminal stage of subduction. *Earth Planet Sci Lett* 461:127–140.

Kuzniar, K. & Tatara, T. 2017. Impact assessment of epicentral distances and energies of mining shocks on the transmission of free-field vibrations to the building foundations. Proc. COMPDYN 2017 6th ECCOMAS Thematic Conference on Computational Methods in Structural Dynamics and Earthquake Engineering M. Papadrakakis, M. Fragiadakis (eds.), Rhodes Island, Greece, 15–17 June 2017.

Kuzniar, K., Stec, K. & Tatara, T. 2018a. Comparison of approximate evaluations of the harmfulness of mining shocks using ground and building foundation vibrations. *Journal of Measurements in Engineering* 6: 218–225.

Kuzniar, K., Stec, K., & Tatara, T. 2018b Approximate classification of mining tremors harmfulness based on free-field and building foundation vibrations. *E3S Web of Conferences*, 36: 1–8.

Kuzniar, K., Stec, K. & Tatara, T. 2018c. The impact of some parameters of mine-induced rockbursts on the transmission of free-field vibrations to the office building foundation. Proc 16th European Conference on Earthquake Engineering, Thessaloniki, 18-21 June 2018.

Maciag, E., Kuzniar, K. & Tatara, T. 2016. Response Spectra of the Ground Motion and Building Foundation Vibrations Excited by Rockbursts in the LGC Region. *Earthq Spectra* 32:1769–1791.

Pachla, F., Kowalska-Koczwara, A., Tatara, T. & Stypula, K. 2019. The influence of vibration duration on the structure of irregular RC buildings. *Bull Earthquake Eng* 17, 3119–3138.

Pineda-Porras, O.A. & Ordaz, M. 2012. Seismic damage estimation in buried pipelines due to future earthquakes: the case of the Mexico City Water System. *Earthquake resistant structures: design, assessment and rehabilitation* -chapter 5:131–150.

PN-B-02170:2016-12, 2016. Evaluation of the harmfulness of building vibrations due to ground motion (in Polish).

PN-B-02171:2017-06, 2017. Evaluation of the impact of vibrations on people in buildings (in Polish).

Popiołek, E. 2009. *Protection of mining areas.* AGH, Kraków (in Polish).

Singh, P.K. & Roy, M.P. 2010. Damage to surface structures due to blast vibration. *Int J Rock Mech Min Sci* 47:949–961.

Tatara, T. 2012. *Dynamic resistance of buildings under mining tremors* (in Polish). The CUT Press, Krakow. 2012.

Tatara, T., Pachla, F. & Kubon P. 2017. Experimental and numerical analysis of an industrial RC tower. *Bull Earthquake Eng* 15: 2149–2171.

Zembaty, Z. 2004. Rockburst induced ground motion – a comparative study. *Soil Dynamics and Earthquake Engineering* 24: 11–23.

Zembaty, Z., Jankowski, R., Cholewicki, A. & Szulc, J. 2007. Earthquakes in Poland in 2004. *Technical Transactions* 2-B: 115–126.

Analysis of masonry structures

Brick and Block Masonry - From Historical to Sustainable Masonry –
Kubica, Kwiecień & Bednarz (eds)
© 2020 Taylor & Francis Group, London, ISBN 978-0-367-56586-2

Determining the compressive strength of existing brickwork

B. Gigla
Technische Hochschule Lübeck, Germany

ABSTRACT: The compressive strength of brickwork depends on the compressive strength of the bricks, the compressive strength of the hardened mortar and on the bond strength between mortar and units. In the case of existing masonry, the compressive strength of the mortar and the bond strength of the joints are difficult to obtain. At historically significant buildings the amount of samples is limited, following the concept of minimum intervention. Non Destructive Test methods (NDT) like flat jack measurements, sonic testing or Schmidt hammer may not be considered adequate for the assessment of the safety level of structures in service. Approaches for the characterization of existing brickwork are discussed since the 1980s. In the field of applicable Moderately Destructive Tests (MDT) different testing procedures have been introduced, based on tensile splitting or compressive tests on drilled cylinders and compressive tests on smaller brickwork samples. To evaluate their level of significance, experimental tests series have been performed on minimized sandwich cores considering mortar and brick properties and on larger diameter cylinders. The results are compared with the compressive strength tested on a brickwork specimen. Mechanical properties of the mor-tar and the bricks have been assessed to evaluate the compressive strength of the brickwork following EC 6. The results are introduced and discussed.

1 INTRODUCTION

1.1 *Condition assessment of cultural heritage*

The compressive strength of masonry structures is a basic parameter required for the rehabilitation of historic monuments. With the upcoming philosophy of minimum intervention in the 1980ies, concepts for reducing size and number of samples became part of scientific research. At the same time, the developing core drilling technology (Heinz 1985) enabled the sampling of cylindrical brick and masonry cores. Those were considered to be more convenient and monument friendly, than traditional chiseling.

Following EN 1996-1-1, the characteristic compressive strength f_k of new masonry shall be determined from results of tests on masonry specimens according to EN 1052-1. In the case of cultural heritage, samples of this size would not be acceptable. Furthermore, they are not representative, due to alterations from cutting, removal and transportation.

Assessment procedures, applied at cultural heritage, are outlined in Figure 1 with their level of intervention versus the assumed significance of the results. Starting with theoretical structure evaluation and zero-intervention, the level of significance is increasing with laboratory testing of samples that are growingly representative in structure, size and bond.

1.2 *Tests on cored cylinders including joints*

Main concepts of tests on cored cylinders considering brickwork bond and structure are:

- testing the tensile splitting strength of a cylinder with one centered bed joint compared to the properties of solid unit cores (Berger 1987)
- laterally testing the compressive strength of larger cylinders considering the dimensions of the bricks (UIC 2011, Heidel 1989, Brencich et al. 2004, Bilello et al. 2007)

Both procedures were discussed in comparison with constitutive models for masonry compressive strength and have been validated with numerical approaches and experimental results.

With the Berger-method, the sample size is smaller and standard tests are applied. However, the parallel obtained mechanical parameters require due interpretation. With the UIC 778-3 test, the sample size is larger and lateral loading on a cylinder is not a standard test to obtain the compressive strength. In practice, the required cement mortar casting might influence the results. The same applies to laboratory-dependent adaptions of testing machines, e.g. proper leading of the loading plates or performance of concave shaped mouldings. A subsequent approach for lateral loading of smaller cylinders was introduced by (Sassoni & Mazzotti 2014), compare Figure 1.

This paper is focusing on the results of an evaluation of the masonry compressive strength with

Level of intervention in determining the compressive strength of existing brickwork

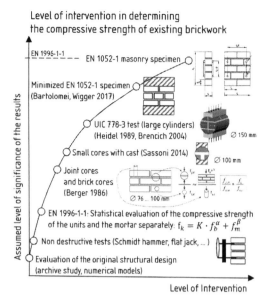

Figure 1. Condition assessment of cultural heritage: Level of intervention in determining the compressive strength of existing brickwork vs. assumed level of significance of the results.

the Berger-method based on own laboratory results. The UIC-procedure was considered comparatively.

2 BERGER-METHOD

2.1 Approach

Berger (1987) assumes the unit compressive strength as the theoretical upper limit of the masonry compressive strength. Based on this limit value, the masonry compressive strength is determined considering the ratio of masonry compressive strength to unit compressive strength. This ratio is referred as 'degree of utilization of the masonry K_m'. The approach is illustrated in Figure 2.

In the case of existing masonry, mortar samples are not seen to be representative to define the required ratio. With the extraction of specimens from bed joints the mortar structure is affected, especially at historical buildings. Therefore, mortar compressive strength is rated imprecise determinable. Additional uncertainties would be the form factor for small mortar samples and the influence of the irregular aggregates size distribution at historical mortars.

To minimize the required sampling, Berger proposes to assess tensile splitting strengths, based on 'joint cores'. A joint core is horizontally separated in two symmetrical brick circle sections through the bed joint. Its diameter is proposed to be about 10 cm, adapting to the geometry of cultural heritage walls. Before drilling, the load direction must be marked. The ratio between joint thickness and joint core diameter should average between 0.10 and 0.15.

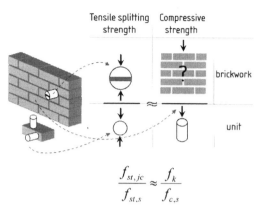

$$\frac{f_{st,jc}}{f_{st,s}} \approx \frac{f_k}{f_{c,s}}$$

Figure 2. Approach of the Berger-method: To assess the masonry compressive strength f_k, the ratio of the tensile splitting strengths of the joint core $f_{st,jc}$ and the units $f_{st,s}$ is required. This ratio is referred as 'degree of utilization of the joint core K_{jc}' and evaluated with the compressive strength of the units $f_{c,s}$ to assess the 'degree of utilization of the masonry K_m.

The tensile splitting strength and the compressive strength of the units may be obtained from an extracted brick. The required total number of samples depends on the particular masonry structure. The ration of the tensile strengths of the joint core and the unit gives the 'degree of utilization of the joint core K_{jc}'. The degree of utilization of the masonry K_m is evaluated with K_{jc} and the compressive strength of the units.

2.2 Required mechanical parameters

The required mechanical parameters are described in Figure 2. For testing the tensile splitting strengths, Berger (1987) proposed diameters of 52 to 68 mm at unit cores and 100 mm at joint cores. Both types of cores are assumed to show a matching type of fracture. Basic types of crack patterns at tensile strength testing are depicted in Figure 3. Berger is not referring to specific testing standards, but to fundamental publications, summarized by Bonzel (1964). At current investigations, it can be recommended to test the tensile splitting strengths in accordance with EN 12390-6: The ratio of length to diameter is $ 1 and the required bearing strips are defined in size and material to ensure tensile fracture (Figure 3). For the joint

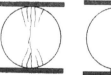

Figure 3. Crack patterns at tensile splitting testing. Types of fracture depending on load application, resp. bearing. Left: compressive fracture, center: tensile fracture, right: shear fracture (in: Bonzel 1964).

cores, a detailed measurement should be recorded for evaluation, due to practical variations like eccentricities and different thicknesses of the bed joints.

The compressive strength of the units was tested by Berger (1987) with diameters of 32 to 52 mm and a ratio of height to diameter = 2. Results of horizontally drilled cores were converted to the strength in load direction. Currently, it can be recommended to test the compressive strength of the units in accordance with EN 12504-1.

2.3 Consideration of brickwork

The compressive strength of masonry depends on the ratio between the thicknesses of the mortar joints and units. The joint core only represents a section of the masonry structure and does not cover the unit heights. Therefore, the influence of the mortar joints has to be approached by a geometric factor. The compression failure stresses in relation to properties of the bed materials have been described by Hilsdorf (1969) and Francis or Khoo (1972). The influences were recently discussed by Como (2013): The larger the ratio between the thicknesses of the mortar joints and bricks is, that is, the thicker the mortar beds are, the lower the masonry compression strength will be. In Jäger & Marzahn (2010) this influence is further examined by a diagram from E. Berndt, covering two types of mortar (Figure 4).

In the case of the joint cores, the stress distribution in the bed joint due to tensile splitting strength testing must also be taken into account. For the Berger-method, Egermann recommends Equation 1 below to calculate the geometric factor, based on Egermann (1995), as described in Wenzel et al. (2000). The equation compares the ratio of mortar to unit thickness at the joint core to the ratio in the masonry (Figure 5).

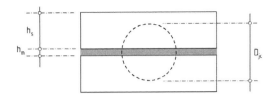

Figure 5. For joint cores, the ratio of mortar thickness h_m to unit thickness h_s is geometrically lower, than in the masonry, because the unit thickness results in: $D_{jc} - h_m$.

$$k = \frac{1 + 3,24 \sqrt{\frac{h_m}{D_{jc} - h_m}}}{1 + 3,24 \sqrt{\frac{h_m}{h_s}}} \qquad (1)$$

k = geometric factor; h_m = mortar thickness
D_{jc} = diameter of joint core; h_s = unit thickness.

The geometric factor k proposed by Egermann is evaluated in Figure 6, depending on the unit thickness and considering the joint core diameter D_{jc}, for a constant mortar thickness of 12 mm. Factor k becomes 1.0 at a joint core diameter of 76 mm and a unit thickness of 65 mm, since then the unit height in the joint core corresponds to the unit height in the masonry. In the case of joint cores with a diameter of 100 mm, the tensile splitting strength test is based on a lower joint proportion compared to the masonry, so that a reduction is required. With diameters of 48 mm the influence is reverted.

2.4 Evaluation of the masonry compressive strength

On basis of the Berger-method (Berger 1987), considering Egermann's suggestions (in: Wenzel et al. 2000), the compressive strength of masonry $f_{c,m}$ is determined according to Equation 2 below:

$$f_{c,m} = c \cdot K_{jc} \cdot k \cdot f_{c,s} \qquad (2)$$

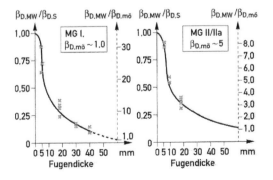

Figure 4. Ratio of masonry compressive strength $\beta_{D,MW}$ to unit compressive strength $\beta_{D,S}$ depending on the bed joint thickness 'Fugendicke', considering the mortar compressive strength $\beta_{D,MW}$ according to E. Berndt. $\beta_{D,S}$ between 38 and 45 N/mm².

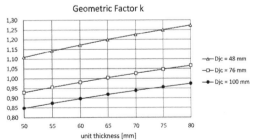

Figure 6. Evaluation of the geometric factor k, recommended by Egermann (1) at constant mortar thickness h_m = 12 mm.

c = reduction factor for the description of structural load-bearing influences of the investigated masonry walls; K_{jc} = degree of utilization of the joint core; k = geometric factor; $f_{c,s}$ = compressive strength of units, tested at cores with a ratio of height to diameter = 2

The reduction factor c has to be defined depending on the local masonry structure and requires due scientific experience. The determination of the reduction factor was one of the goals of the test programs described in the following section. Considering Eurocode 6: EN 1996-1-1 (2012 and 2019) and based on Equation 2 the characteristic compressive strength of masonry f_k according to EN 1996-1-1 might be calculated by Equation 3:

$$f_k = 0.8 \cdot f_{c,m} \qquad (3)$$

$f_{c,m}$ = compressive strength of masonry, determined using the Berger-method.

The factor 0.8 globally considers the reduction from average values to 5% quantile values. effects of platen restraint or eccentricity of loading are omitted and set to 1.0. Assuming that berger's results refer to 5-unit-specimens with a slenderness ratio of 3.5, the factor for the wall slenderness is set to 1.0 on the safe side, instead of 1.1 for the conversion to the theoretical slenderness ratio of 0. To consider the long-term effect, an additional reduction factor of 0.85 has to be taken into account for creep. This factor is usually applied when determining the design compressive strength of masonry.

3 PERFORMED TEST PROGRAMMES

3.1 Investigations into the geometric factor k and eccentricities e of the joint core

Winkelmann (2002) investigated the influence of the geometric factor k and eccentricities e of the joint core. Such eccentricities might occur in practice due to deviations between drilling axis and bed joints. Basis of the investigations were a brickwork test wall made with trass lime mortar and identical three stone specimens. Joint cores with diameters of 100, 74 and 48 mm and eccentricities of 0, 10 and 15 mm were extracted from the wall and examined according to the Berger method. Figure 7 shows the test wall after coring and Table 1 summarizes main properties of the units and the mortar.

The mortar was made by volume with 1 part trass lime (hydraulic lime) and 3.5 parts sand 0/2. The influence of joint core eccentricities was investigated with e = 10 and 15 mm at the joint cores with a diameter of 74 mm. Figures 8, 9 and 10 are showing all 74 mm-joint cores with crack patterns after tensile splitting testing. A total of 75 joint cores and 15 3-unit-specimens were examined.

Winkelmann (2002) determined the brick compressive strength on core samples with a slenderness

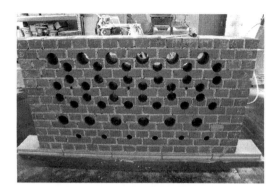

Figure 7. Brickwork test wall with extracted joint cores (Winkelmann 2002). Unit size is (L/W/H) 240/115/71 mm and mortar thickness 12 mm.

Table 1. Main properties of the units and the mortar.

Mortar (EN 998-2 class M1)	
Average flexural strength of hardened mortar (EN 1015-11)	0.04 N/mm²
Average Compressive strength of hardened mortar (EN 1015-11)	0.88 N/mm²

Units: bricks (DIN 105 type VMz 8/1,8/NF)	
Gross dry density (EN 772-13)	1.89 kg/m³
Average compressive strength (units, DIN 105)	13.8 N/mm²
Average compressive strength (cores, d = 30 mm, h = 50 mm)	16.1 N/mm²
Average tensile splitting strength (EN 12390-6, cores, d = 50 mm, l = 50 mm)	2.31 N/mm²
Water absorption coefficient at 1 hour (EN ISO 15148) W_{1h}	9.8 kg/(m²·h^{0.5})

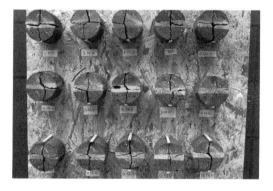

Figure 8. Joint cores with 74 mm diameter and without eccentricity with crack patterns after tensile splitting testing (Winkelmann 2002).

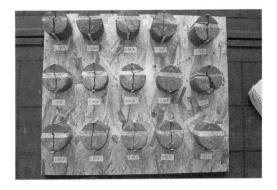

Figure 9. Joint cores with 74 mm diameter and 10 mm eccentricity with crack patterns after tensile splitting testing (Winkelmann 2002).

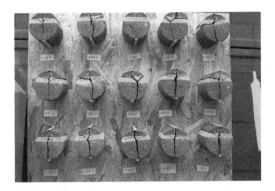

Figure 10. Joint cores with 74 mm diameter and 15 mm eccentricity with crack patterns after tensile splitting testing (Winkelmann 2002).

Table 2. Main results of Winkelmann (2002):Average values of 15 tests each.

Joint cores, tensile splitting strength	$f_{st,jc}$
Diameter D_{jc} = 100 mm, e = 0 mm	1.25 N/mm²
Diameter D_{jc} = 74 mm, e = 0 mm	1.14 N/mm²
Diameter D_{jc} = 74 mm, e = 10 mm	1.04 N/mm²
Diameter D_{jc} = 74 mm, e = 15 mm	0.98 N/mm²
Diameter D_{jc} = 48 mm, e = 0 mm	1.20 N/mm²
3-unit-specimens, compressive strength	
$f_{c,m,3u}$	12.1 N/mm²

of 1.7. According to the form factors for brick cores given by Egermann (1995), a conversion to the slenderness 2.0 applied by Berger (1987) is not required. Based on Mann & Betzler (1994), the coefficient for the conversion of the compressive strength of 3-unit-specimens to the compressive strength of 5-unit-specimens is set to $c_{3u,5u} = 1.05$.

Mann & Betzler (1994) also give a coefficient of $c_{5u,Rilem} = 1.1$ for the conversion between 5-stone-units and DIN EN 1052-1 ('RILEM') specimens. According to own results the coefficient may be up to 1.3, depending on strength and brickwork. Remaining on the safe side, a conversion is not applied.

The results of the investigations of Winkelmann (2002) are summarized in Table 2 and evaluated according to the Berger-method in Table 3 and Figure 11. All tests were performed after 28 days.

Table 3 shows that the Berger-method determines results that remain on the safe side for the low-strength brick and mortar combination investigated. No structural reduction is required. The reduction factor c in Table 3 was therefore set to 1.0 in

accordance with the high quality of the test wall (Figure 7), which was achieved by Winkelmann, a professional bricklayer. Reductions are required in the case of inadequate brickwork, cracks, cavities etc.

Figure 11 Below indicates that the geometric factor k from Equation 1 resp. Figure 7 results in a too high value for the masonry compressive strength at the 48 mm joint core. For smaller joint core diameters, a revision of Equation 1 or an application limit is required.

With the eccentricities and geometry examined, it was found that an average eccentricity of 10 mm leads to a reduction of approx. 8% and an average eccentricity of 15 mm leads to a reduction of approx. 14% of the tensile splitting strength of the joint core (compare Table 2). This conclusion relates to the tensile splitting strength without eccentricity and can be taken into account using a correction factor, if required.

Basically it is shown, that joint cores with eccentricities can also be tested applying the Berger-method. The tests of Winkelmann (2002) resulted in a similar crack pattern to the centered joint core, compare Figures 8 to 10.

3.2 Investigations into the ratio of mortar to unit thickness

A test series based on 3-unit-specimen and 30 joint cores focused on the ratio of mortar to unit thickness. Investigated mortar thicknesses were 5, 10 and 20 mm. The unit size was (L/W/H) 238/114/69 mm. Figure 12 shows the specimen investigated and Figure 13 the process of joint coring in laboratory. Table 4 summarizes main properties of the units and the mortar. All tests were performed after 28 days.

The unit compressive strength was determined at cores with a ratio of height to diameter of 1.0. According to Egermann (1992), a factor of 1.1 is applied for the conversion to the slenderness of 2.0, resulting in an average unit compressive strength of 29.3 N/mm². The diameter of the joint cores was 80 mm. Joint cores detached by the drilling process also were included in the test.

Table 3. Evaluation of the results of Winkelmann (2002) according to the Berger-method.

Tested compressive strength of masonry (5-unit-specimen)	
$f_{c,m,5u} = 12.1/1.05 =$	11.6 N/mm²

Joint cores $D_{jc} = 100$ mm, e = 0 mm	
Degree of utilization of the joint core K_{jc}	0.54
Geometric factor k (Equation 1)	0.94
Compressive strength of masonry $f_{c,m}$	8.2 N/mm²
Structural reduction factor c (Equation 2)	1.0
Characteristic compressive strength of masonry (Berger 1987) f_k	6.6 N/mm²

Joint cores $D_{jc} = 74$ mm, e = 0 mm	
Degree of utilization of the joint core K_{jc}	0.49
Geometric factor k (Equation 1)	1.04
Compressive strength of masonry $f_{c,m}$	8.3 N/mm²
Structural reduction factor c (Equation 2)	1.0
Characteristic compressive strength of masonry (Berger 1987)f_k	6.7 N/mm²

Joint cores $D_{jc} = 48$ mm, e = 0 mm	
Degree of utilization of the joint core K_{jc}	0.52
Geometric factor k (Equation 1)	1,23
Compressive strength of masonry $f_{c,m}$	10.4 N/mm²
Structural reduction factor c (Equation 2)	1.0
Characteristic compressive strength of masonry (Berger 1987)f_k	8.3 N/mm²

Joint cores $D_{jc} = 74$ mm, e = 10 mm	
Degree of utilization of the joint core K_{jc}	0.45
Geometric factor k (Equation 1)	1.04
Compressive strength of masonry $f_{c,m}$	7,6 N/mm²
Structural reduction factor c (Equation 2)	1.0
Characteristic compressive strength of masonry (Berger 1987)f_k	6.1 N/mm²

Joint cores $D_{jc} = 48$ mm, e = 15 mm	
Degree of utilization of the joint core K_{jc}	0.42
Geometric factor k (Equation 1)	1.04
Compressive strength of masonry $f_{c,m}$	7,2 N/mm²
Structural reduction factor c (Equation 2)	1.0
Characteristic compressive strength of masonry (Berger 1987) f_k	5.7 N/mm²

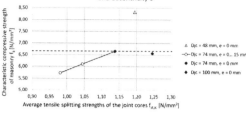

Evaluation of the geometric factor k and eccentricity e

Figure 11. Evaluation of the geometric factor k and eccentricity e. The geometric factor k results in a too high value for the masonry compressive strength in the case of small joint cores. The dashed line indicates the expected result of the characteristic compressive strength of masonry based on the tensile splitting strength of the joint cores.

Figure 12. 3-unit-specimen with mortar thicknesses of 5, 10 and 20 mm and unit heights of 69 mm.

Figure 13. Joint coring in laboratory at the investigated 3-unit-specimen.

The results of the test series are evaluated in Table 5. The coefficient of 1.05 according to Mann & Betzler (1994) was used to convert the compressive strength of the 3-stone units tested and the 5-stone units to be applied according to (Berger 1986).

The results in the Table 5 show that a structural reduction factor of 0.7 is required for the brick and mortar combination investigated. The reduction factor covers inadequate masonry due to the tested 3-unit-specimens, which were not manufactured by a professional bricklayer, as well as reductions in the quality of the joint cores, which here were obtained from the 3-unit-specimens (Figure 13) and not from a masonry wall.

Basically it is shown, that joint cores with mortar thicknesses between 5 and 20 mm giving a joint to

Table 4. Main properties of the units and the mortar.

Mortar (EN 998-2 class M5)

Consistence, flow table (EN 1015-3)	145 mm
Average flexural strength of hardened mortar (EN 1015-11)	3.0 N/mm²
Average Compressive strength of hardened mortar (EN 1015-11)	9.0 N/mm²

Units: bricks (L/W/H) 238/114/69 mm

Gross dry density (EN 772-13)	1.80 kg/m³
Average compressive strength (units, EN 772-1)	42.7 N/mm²
Average compressive strength (cores, d = 50 mm, h = 50 mm)	32.3 N/mm²
Average tensile splitting strength (EN 12390-6, cores, d = 50 mm, l = 50 mm)	3.21 N/mm²
Water absorption coefficient at 1 hour (EN ISO 15148) W_{1h}	0.9 kg/(m²·h^{0.5})

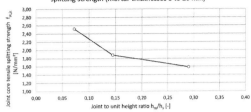

Influence of the joint to unit height ratio versus joint core tensile splitting strength (mortar thicknesses 5 to 20 mm)

Figure 14. Influence of the investigated joint to unit height ratio versus joint core tensile splitting strength (mortar thicknesses 5 to 20 mm at a unit height of 69 mm).

unit ratio between 0.07 and 0.30 can be tested according to the Berger-method. The influence of the joint to unit height ratio versus joint core tensile splitting strength is shown in Figure 14.

3.3 Investigations into high strength bricks and comparing the Berger-method with UIC 778-3

A test series based on 4-unit-specimen and 60 joint cores focused on the influence of high strength bricks. The results are compared with UIC 778-3 (2011) tests. Investigated mortar thicknesses were 10 and 15 mm. The unit size was (L/W/H) 241/118/53 mm. Table 6 summarizes main properties of the units and the mortar. To vary the influence of mortar compressive strength, 10 of the 20 4-stone specimens and 30 of the 60 joint cores were tested 7 days after manufacturing and the remaining samples 28 days after manufacturing.

The results of the test series are evaluated in Table 7. According to the limited size of this publication, the evaluation covers the results for a mortar thickness of 10 mm tested 28 days after manufacturing only.

Table 5. Evaluation of the results of the test series with mortar thicknesses h_m = 5, 10 and 20 mm according to the Berger-method.

Joint cores D_{jc} = 80 mm, h_m = 5 mm, average of 10

Degree of utilization of the joint core K_{jc}	0.78
Geometric factor k (Equation 1)	0.99
Compressive strength of masonry $f_{c,m}$	22,7 N/mm²
Structural reduction factor c (Equation 2)	0.7
Characteristic compressive strength of masonry (Berger 1987) f_k	12.7 N/mm²
Tested compressive strength of masonry (3-unit-specimen)	13,6 N/mm²
Compressive strength of masonry (5-unit-specimen)	12,9 N/mm²

Joint cores D_{jc} = 80 mm, h_m = 10 mm, average of 10

Degree of utilization of the joint core K_{jc}	0.59
Geometric factor k (Equation 1)	1.00
Compressive strength of masonry $f_{c,m}$	17.3 N/mm²
Structural reduction factor c (Equation 2)	0.7
Characteristic compressive strength of masonry (Berger 1987) f_k	9,7 N/mm²
Tested compressive strength of masonry (3-unit-specimen)	10,8 N/mm²
Compressive strength of masonry (5-unit-specimen)	10,3 N/mm²

Joint cores D_{jc} = 80 mm, h_m = 20 mm, average of 10

Degree of utilization of the joint core K_{jc}	0.49
Geometric factor k (Equation 1)	1,04
Compressive strength of masonry $f_{c,m}$	15.1 N/mm²
Structural reduction factor c (Equation 2)	0.7
Characteristic compressive strength of masonry (Berger 1987) f_k	8,4 N/mm²
Tested compressive strength of masonry (3-unit-specimen)	9,0 N/mm²
Compressive strength of masonry (5-unit-specimen)	8,6 N/mm²

Table 6. Main properties of the units and the mortar.

Mortar (EN 998-2 class M5)

Consistence, flow table (EN 1015-3)	155 mm
Average flexural strength of hardened mortar (EN 1015-11), 28 days	2.6 N/mm²
Average Compressive strength of hardened mortar (EN 1015-11), 28 days	8.1 N/mm²
Units: bricks (L/W/H) 241/118/53 mm	
Gross dry density (EN 772-13)	2.03 kg/m³
Average compressive strength (units, EN 772-1)	94.8 N/mm²
Average compressive strength (cores, d = 50 mm, h = 50 mm)	102.6 N/mm²
Average tensile splitting strength (EN 12390-6, cores, d = 50 mm, l = 50 mm)	5.16 N/mm²
Water absorption coefficient at 1 hour (EN ISO 15148) W_1	6.0 kg/(m²·h^{0.5})

Table 7. Evaluation of the results of the test series with mortar thicknesses h_m = 10 mm tested 28 days after manufacturing according to the Berger-method.

Joint cores D_{jc} = 80 mm, h_m = 10 mm, average of 15

Degree of utilization of the joint core K_{jc}	0.46
Geometric factor k (Equation 1)	0.93
Compressive strength of masonry $f_{c,m}$	40.2 N/mm²
Structural reduction factor c (Equation 2)	0.5
Characteristic compressive strength of masonry (Berger 1987) f_k	16.1 N/mm²
Tested compressive strength of masonry (4-unit-specimen)	16.5 N/mm²
Tested compressive strength UIC 778-3	6.7 N/mm²

Figure 15. UIC 778-3 compressive strength testing. The applied concave loading plate geometrically effects cracking.

According to Egermann (1992), a factor of 1.1 is applied for the conversion to the slenderness of 2.0, resulting in an average unit compressive strength of 93.3 N/mm². It was not considered necessary to convert the compressive strength of the 4-unit-specimens to 5-unit-specimens.

The results in the Table 7 show that a structural reduction factor of 0.5 is required for the brick and mortar combination investigated. The reduction factor covers inadequate masonry due to the manufacturing of the tested 4-unit-specimens, analogous to section 3.1, and the brittle fracturing of the high strength bricks. The results of the tested compressive strength in accordance with UIC 778-3 (Figure 15) are underestimating the masonry compressive strength, based on 3 samples. Further research projects are being worked on to describe the structural reduction factor and UIC 778-3-results in an objective manner.

4 CONCLUSIONS

The paper reports about the determination of compressive strength of existing masonry on cultural heritage monuments. The results of three test series based on testing the tensile splitting strength of a cylinder with one centered bed joint compared to the properties of solid unit cores ('Berger-method') are presented and discussed. The method is introduced in detail. Main influences of the properties of joints and units have been investigated in comparison to the compressive strength of 3-unit to 5-unit-specimens.

It is basically shown, that joint cores with eccentricities and joint to unit ratios between 0.07 and 0.30 are giving satisfactory results within the introduced approach. A reduction factor required in the case of inadequate brickwork, cracks, cavities or brittle fracture is evaluated, based on the test results. The factor currently has to be defined on the basis of empirical values. Further research projects are being worked on to describe the structural reduction factor in an objective manner.

It was found, that the compressive strength tested in accordance with UIC 778-3 is underestimating the masonry compressive strength. Further research projects are being worked on in this context.

The results are a contribution to the current discussion in Germany on the question of the most suitable and representative method for determining the compressive strength of existing masonry in the context of minimum intervention.

REFERENCES

Beer, I. & Schubert, P. 2004. Determination of Shape Factors for Masonry Units. *Proc., 13th IBMaC, Amsterdam:* 83–88.

Berger, F. 1987. Zur nachträglichen Bestimmung der Tragfähigkeit von zentrisch gedrücktem Ziegelmauerwerk. *Erhalten historisch Bedeutsamer Bauwerke Jahrbuch 1986*, Verlag Ernst & Sohn, Berlin, Germany.

Bilello, C.; Brencich, A.; Corradi, C.; Di Paola, M.; Sterpi, E. 2007. Experimental Tests and Theoretical Issues for the Identification of Existing Brickwork. *Proc. 10th North American Masonry Conf.*, St. Louis, USA: pp 946–956.

Bonzel, J. 1964. Über die Spaltzugfestigkeit des Betons. *Beton Herstellung & Verwendung Heft 3-1964: 108-114 and Heft 4-1964: 150-156.* Beton-Verlag, Düsseldorf, Germany.

Brencich, A., Corradi, C., Sterpi, E. 2004. Experimental Approaches to the Compressive Response of Solid Clay Brick Masonry. *Proc., 13th IBMaC, Amsterdam:* pp 55–64.

Como, M. 2013. Statics of Historic Masonry Constructions. Berlin, Germany: Springer-Verlag.

DIN 105-1. 1989. Clay bricks; solid bricks and vertically perforated bricks. Standard. DIN, Berlin, Germany.

Egermann, R. 1992. Zur nachträglichen Bestimmung der Mechanischen Eigenschaften von Mauerziegeln. *Erhalten historisch Bedeutsamer Bauwerke Jahrbuch 1990*, Berlin: Verlag Ernst & Sohn, Germany.

Egermann, R. 1995. Tragverhalten mehrschaliger Mauerwerkkonstruktionen. Diss. *Aus Forschung und Lehre 29*, Institut für Tragkonstruktionen, Universität Karlsruhe (TH) Karlsruhe, Germany.

EN 772-1. 2016. Methods of test for masonry units – Part 1: Determination of compressive strength. CEN, Brussels, Belgium.

EN 772-13. 2000. Methods of test for masonry units - Part 13: Determination of net and gross dry density of masonry units (except for natural stone). CEN, Brussels, Belgium.

EN 998-2. 2016. Specification for mortar for masonry – Part 2: Masonry mortar. CEN, Brussels, Belgium.

EN 1052-1. 1998. Methods of test for masonry – Part1: Determination of compressive strength. CEN, Brussels, Belgium.

EN 1015-3. 2007. Methods of test for mortar for masonry - Part 3: Determination of consistence of fresh mortar (by flow table). CEN, Brussels, Belgium.

EN 1015-11. 2020. Determination of flexural and compressive strength of hardened mortar. CEN, Brussels, Belgium.

EN 1996- 1-1. 2012. Eurocode 6: Design of masonry structures – Part 1-1: General rules for reinforced and unreinforced masonry structures. CEN, Brussels, Belgium. With Draft prEN 1996-1-1, Sept. 2019.

EN 12390-6. 2010. Testing hardened concrete – Part 6: Tensile splitting strength of test specimens. CEN, Brussels, Belgium.

EN 12504-1. 2019. Testing concrete in structures – Part 1: Cored specimens – Taking, examining and testing in compression. CEN, Brussels, Belgium.

EN ISO 15148. 2016. Hygrothermal performance of building materials and products – Determination of water absorption coefficient by partial immersion. CEN, Brussels, Belgium.

Heidel, R. 1989. Ermittlung der Materialkennwerte von Mauerwerk als Grundlage zur Beurteilung der Tragfähigkeit von Mauerwerkskonstruktionen. Diss. Technische Hochschule Leipzig, 1989, Leipzig, GDR.

Heinz, W.F. 1985. Diamond Drilling Handbook. Halfway House 1685, RSA.

Hilsdorf, H.K. 1969. Investigation into the Failure Mechanism of Brick Masonry Loaded in Axial Compression. In: *Designing, Engineering and Construction with Masonry Products*, Houston, USA, Gulf Publishing, pp. 34–41.

Jäger W., Marzahn G. 2010. Mauerwerk – Bemessung nach DIN 1053-100. Berlin, Germany, Ernst & Sohn Verlag.

Khoo, C.-L. 1972. A failure criterion for brickwork in axial compression. PhD-thesis, University of Edinburgh, Department of Civil Engineering & Building Science.

Mann, W., Betzler, M. 1994. Investigations on the effect of different forms of Test Samples to test the Compressive Strength of Masonry. *Proc. 10th IBMaC*, Calgary, Canada:1305–1313.

Sassoni, E., Mazzotti, C. 2014. Experimental and numerical study on the determination of masonry compressive strength by means of cores. *Proc. 9th International Masonry Conference July 7- 9, 2014 in Guimarães, Portugal*: ID1440.

UIC 2011. UIC code 778-3R, Recommendations for the inspection, assessment and maintenance of masonry arch bridges, International Union of Railways (UIC), Paris, France.

Wenzel, F.; Gigla, B., Kahle, M.; Stiesch, G. 2000. Empfehlungen für die Praxis, Historisches Mauerwerk: Untersuchen, Bewerten und Instandsetzen. Erhalten historisch Bedeutsamer Bauwerke, Sonderforschungsbereich 315, Universität Karlsruhe, Karlsruhe, Germany.

Winkelmann, J. 2002. Untersuchungen zum Einfluss der Probengeometrie auf die Spaltzugfestigkeit von Fugenbohrkernen. Diploma thesis, University of Applied Sciences, Lübeck, Germany.

Brick and Block Masonry - From Historical to Sustainable Masonry –
Kubica, Kwiecień & Bednarz (eds)
© 2020 Taylor & Francis Group, London, ISBN 978-0-367-56586-2

Seismic fragility and risk of Italian residential masonry heritage

M. Donà* & L. Xu*
Earthquake Engineering Research & Test Center (EERTC), Guangzhou University, China

P. Carpanese, V. Follador & F. da Porto*
Department of Geosciences, University of Padova, Italy

L. Sbrogiò
Department of Cultural Heritage, University of Padova, Italy

ABSTRACT: Seismic risk mitigation at national scale requires the vulnerability assessment of the built stock, along with seismic hazard and exposure evaluation. In this work, the fragilities of more than 500 Italian unreinforced masonry (URM) residential buildings, related to an intermediate damage state DS (between severe and moderate), were calculated based on a mechanical approach (through the Vulnus_4.0 software). The fragility results were then processed for calibrating a macro-seismic vulnerability model available in the literature, in order to propose a fragility model based on the five DSs of the EMS98 scale and on some URM building macro-typologies. The model was then validated using the damage observations of the 2009 L'Aquila earthquake, contained in DaDO (Database of Observed Damage). Lastly, the fragility model was used to derive seismic damage maps and to carry out risk analyses (economic losses) on a national scale.

1 INTRODUCTION

Italian built stock is for the most part historical. This feature, combined with the high seismicity of the country, can lead to catastrophic consequences, which have been seen in the aftermath of the earthquakes that occurred in Italy in the last decades. As a matter of fact, the number of earthquakes casualties in the last two centuries has reached a figure as high as 160,000. Large economic losses are also implied: the cost of Italian earthquakes is about 3.6 billion euro per year excluding the damage to historical and artistic sites and the consequences on cultural heritage and productive sectors (Italian Civil Protection Department - DPC 2018).

For these reasons, Italian scientific community and other institutions (in particular the Department of Civil Protection - DPC) are working in synergy to find effective seismic risk mitigation strategies (Dolce et al. 2019a). The work presented in this paper fits into this context, as it is part of a project coordinated by DPC, to which EUCENTRE (European Centre for Training and Research in Earthquake Engineering) and ReLUIS (Network of University Laboratories for Earthquake Engineering) have joined. The collection of and the mediation between the results of each unit entered in a complete update of the seismic risk maps of the Italian residential heritage, summarized in National Risk Assessment 2018 document (DPC 2018).

Firstly, a fragility model for residential masonry buildings has been elaborated, considering different macro-typologies according to number of storeys and age of construction. Then, implementing this fragility model within a platform developed by EUCENTRE under the direction of the DPC called IRMA (Borzi et al. 2018), it was possible to produce seismic risk maps of the entire national territory.

2 METHOD OF ANALYSIS

2.1 *Mechanical approach for vulnerability assessment*

Many methods to define fragility models can be found in literature. Usually, lognormal fragility curves are used, as they allow to better correlate the probability of exceedance of a given damage state (DS) with parameters of seismic intensity (often Peak Ground Acceleration, PGA).

Fragility curves can be obtained by observational, mechanical or hybrid approaches (Calvi et al 2006). The first type requires a large number of post-earthquake survey data, on the other hand the second one requires detailed and time-consuming modelling.

Hybrid methods use different types of analysis picked from both the previous methods.

In this work a mechanical method was adopted: a number of samples adequate to ensure the representativeness of the Italian building stock was implemented through a simplified mechanical analysis.

The tool chosen for the analysis is Vulnus software (Bernardini et al. 1990), originally developed by A. Bernardini, R. Gori and C. Modena (University of Padova) and later updated to the current version Vulnus vb 4.0 (Munari 2009, Valluzzi 2009). Vulnus analyzes load-bearing URM buildings through limited pieces of information, such as geometry, material properties, construction details and other qualitative information. Through the calculation of three indices and the use of Fuzzy sets theory, Vulnus elaborates three fragility curves: a White curve of central probability and two extreme probability curves of Upper- and Lower-Bounds. The three curves represent the probability of exceeding a moderate-severe damage state, associated to a DS2-3 with reference to the damage scale EMS98 (Grünthal 1998), as a function of PGA.

For a detailed explanation of how Vulnus works, please refer to the user manual (Valluzzi 2009) and to Donà et al. (2019, 2020).

2.2 Definition of macro-typologies

In order to optimize the amount of data to be processed and to select appropriate case studies, the typological distribution of Italian residential masonry buildings was analyzed.

First of all, the whole stock has been subdivided into macro-typologies basing both on expert judgments and classifications already made by census authorities (such as ISTAT) or by survey forms (i.e. AeDES and I level GNDT form). The 10 macro-typologies identified, based on construction age and number of storeys, are shown in Table 1.

It can be observed that the factors that most influence the seismic vulnerability of a building vary considerably from one macro-typology to another. These factors are the following:

– geometric factors, necessary to build a mechanical model and establish regularity in plan and elevation (in particular number of storeys, related to the total inertial seismic forces). In this work the sample was divided into low buildings (1-2 storeys) and medium rise buildings (3-5 storeys); buildings with more than 5 floors were

not taken into consideration because of their small number in the Italian building stock;
– type of resistant system, masonry quality and mechanical properties of materials;
– typological-constructive factors concerning floors and roof structures, in particular with regard to their stiffness or the presence of thrusting roof;
– quality and technology of the construction details, in particular concerning the effectiveness of connections between masonry walls and horizontal diaphragms (tie rods or ring-beams).

2.3 Sampling of buildings: Case studies

In order to obtain representative results, the distribution of each macro-typology in Italy has been firstly evaluated (Italian National Institute of Statistics – ISTAT – 2001) and a sample that could return this distribution was selected (Figure 1). The database is composed of a total of 525 buildings, located in different geographical areas, in order to take into

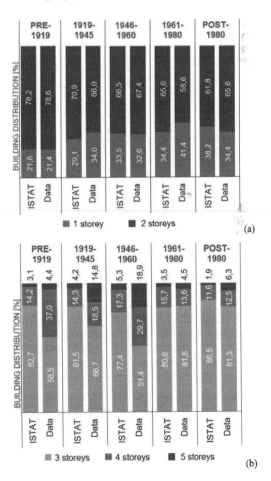

Figure 1. Distribution of buildings by macro-typology and number of storeys (n): comparison between ISTAT data and database of sampled buildings. (a) n ≤ 2; (b) n ≥ 3.

Table 1. Macro-typologies of Italian residential URM buildings.

Construction age	Pre-1919	1919-1945	1946-1960	1961-1980	Post-1980
No. of storeys	≤ 2 ≥ 3	≤ 2 ≥ 3	≤ 2 ≥ 3	≤ 2 ≥ 3	≤ 2 ≥ 3

137

Figure 2. Examples of Italian URM buildings, by construction age and number of storeys.

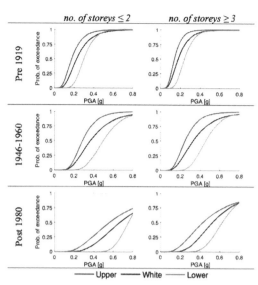

Figure 3. White, Upper and Lower Bounds fragility curves for some building macro-typologies, associated with a DS2-3.

account the possible variations due to territorial factors.

The information required to create the models with *Vulnus* was collected as follows. The geometric information comes from surveys and projects processed by: a) the University of Padova: b) the Municipality of Pordenone; c) engineering firms; d) examples found in historical manuals. Concerning typological and constructive information, reference was made to the national legislations in force and to the technical manuals used by professionals in the different periods, as well as to recent studies on the subject (Donà et al. 2019, 2020).

Figure 2 shows some selected case studies, divided for construction age and number of storeys.

3 DERIVATION OF MECHANICAL-HEURISTIC FRAGILITY MODEL

3.1 Mechanical fragility curves

After collecting all the significant information for the different typologies, the total sample of buildings was implemented in the software *Vulnus 4.0*. The curves obtained in *Vulnus* are discretely defined, so they had to be associated to cumulative lognormal probability density functions that one may to have a continuous representation of fragility.

In order to define a fragility model for all building macro-typologies (considering the DS2-3), the individual curves were averaged according to number of storeys and construction age, considering the weights retrieved by ISTAT data (Figure 1), so as to restore the real distribution of buildings. In doing so, the curves of buildings belonging to the same municipality were first grouped together in order to maintain a geographical representativeness. This

procedure has been implemented separately for the White, Upper- and Lower-Bounds curves (Donà et al. 2019, 2020).

Figure 3 shows the mechanical fragility model obtained for some macro-typologies examined.

These results show that the fragility of the macro-typologies is fairly distributed and it increases as number of storeys and age of buildings increase. This is due to the improvements in technologies, material performances, and construction techniques that occurred in the last century. Furthermore, the dispersion range, defined by the boundary curves, appears to be wider for the less vulnerable macro-typologies. This can be justified considering the better behavior of the most recent masonry buildings, especially to lateral loads. As this dispersion can provide more detailed information when implementing risk assessments, it will be discussed in the next paragraphs.

3.2 Proposed seismic fragility model

In order to represent the seismic fragility in a distributed way, i.e. on five damage states (DS), the fragility model by Lagomarsino & Cattari (2014) was taken as a reference. This macroseismic model was calibrated on the mechanical fragility model obtained for DS2-3, thus deriving a new heuristic fragility model on a mechanical basis (Donà et al. 2019, 2020).

The final sets (from DS1 to DS5) for White, Upper- and Lower-Bounds allow to carry out seismic vulnerability assessments, as they make it possible to obtain probabilistic predictions on average, maximum and minimum damage scenarios.

It is important to note that the dispersion obtained from a mechanical approach depends only on the uncertainties of construction and material characteristics of a building, therefore it usually appears to be much lower than the dispersion of fragility curves obtained with empirical approaches. The latter, indeed, can take into account a larger number of buildings, as well as uncertainties related to damage surveys and ground acceleration values.

With the aim of developing a single fragility set for each building macro-typology that could still take into account the information provided by all three of the fragility sets, the average fragility given by the White curves and the maximum dispersion given by the range of fragility between Upper- and Lower-Bounds curves were considered (Donà et al.2020).

More specifically, for a White probability between 2.5% and 50%, a linear combination of Upper-Bound fragility (100% to 0%) and White fragility (0% to 100%) was computed, while for a White probability between 50% and 97.5% a linear combination of White fragility (100% to 0%) and Lower-Bound fragility (0% to 100%) was calculated. With regard to the external tails, for a White probability lower than 2.5%, the fragility was assumed equal to that of Upper-Bound, while for a White probability greater than 97.5%, the fragility was assumed equal to that of Lower-Bound. The last operation is justified by the need to 'cut the tails' of the lognormal curves.

While maintaining the average fragility defined by the White curves (assumed as the most likely fragility), these curves are characterized by higher dispersion values than the initial mechanical ones, thus being more suitable to represent the vulnerability of the building stock at a large scale.

The final sets of all the 10 macro-typologies are shown in Figure 4.

4 VALIDATION OF THE PROPOSED FRAGILITY MODEL

The resulting fragility model enables large-scale damage forecasts to be made. In particular, the platform IRMA (Italian Risk MAps) was used.

IRMA combines fragility models (vulnerability) with the ISTAT data related to the Italian building stock (exposure) and the Italian seismic hazard, in order to obtain damage and risk scenarios. It allows also to evaluate losses in a simplified way, applying cost and casualties parameters to each DS.

In order to validate the model hitherto presented, the data of the damage surveys collected after the 2009 L'Aquila earthquake have been studied. Reference has been made mainly to the AeDES forms contained in DaDO (Dolce et al. 2019b), an online platform that collects damage surveys related to the seismic events that occurred in Italy in the last decades.

In order to compare the observed data with those obtained through IRMA, only the surveys concerning residential masonry buildings were considered.

However, since AeDES forms aim at safety assessment, they do not provide an overall damage level of the building; therefore, it is necessary to adopt a method for extracting the global damage.

In literature there are different methodologies for the definition of global damage, mainly divided into two categories: those that define an average global damage level and those that define a maximum global damage level. In this study, both types of methods were taken into consideration, in particular the procedures developed by Lagomarsino et al. (2015) and by Rota et al. (2008). The first method evaluates damage by multiplying the damage state of the main

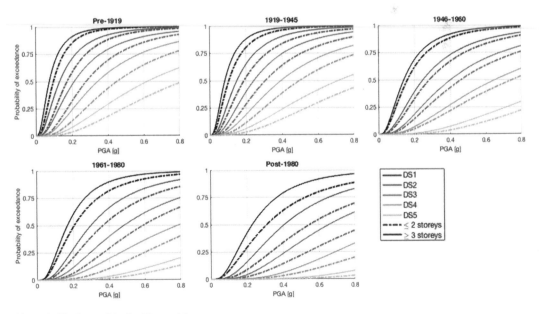

Figure 4. Final sets of the fragility model.

components of a building by a coefficient related to its extension. Subsequently, a weight is assigned to each component, in relation to the accuracy of the survey. On the other hand, the second methodology considers the maximum damage level taking into account every single element. In particular, the damage levels of vertical structures, floors and roof are calculated and the overall damage level of the building is then assessed by taking the maximum of the three.

An example of comparison between the results in terms of damage percentage obtained by the different methods (IRMA and two observational methods) is reported in Figure 5.

Some differences between the two observational methods can be seen. The procedure proposed by Rota is more conservative, since it is developed to capture the damage peaks. On the other hand, the method proposed by Lagomarsino considers an average damage, so it allows to take into account the damage occurred in all structural elements of the building without seizing the levels of severe damage.

Despite the limitations of the different procedures, the results offered by IRMA generally reflect the trends of the two models previously described.

5 SEISMIC DAMAGE MAPS AND RISK ASSESSMENT AT NATIONAL SCALE

Some examples of damage maps developed for Italy at national level are now shown.

The maps in Figure 6 show the conditional probability of damage calculated for a return period (Tr) of 475 years (design return time for ordinary buildings). These maps are generated in the IRMA platform by combining reference PGAs for each Italian municipality and the fragility curves obtained in the last chapters.

From now on, these types of maps will be referred to as "conditional damage maps". They also consider a type A subsoil (bedrock) and are expressed in terms of percentage of damaged masonry buildings.

Then, the results previously obtained for each DS (DS1-DS5) were processed in order to produce maps of average conditional damage (Figure 7): all the damage levels are summed together and multiplied by the probability of them to occur. It must be borne in mind that the maps shown take into account masonry buildings only, so they are not to be considered as explanatory maps of real damage scenarios in Italy.

One can see that the percentage of damaged buildings follows for the most part the trend of the Italian seismic hazard (Figure 8). Whether differences

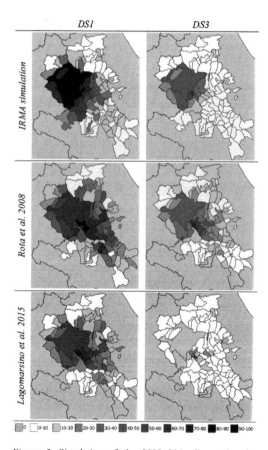

Figure 5. Simulation of the 2009 L'Aquila earthquake through IRMA and comparison with the observed data (% of buildings per municipality that reaches or exceeds DS1 and DS3).

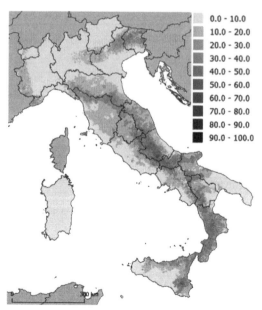

Figure 6. Conditional damage maps, Tr 475 years, type A subsoil, for DS3.

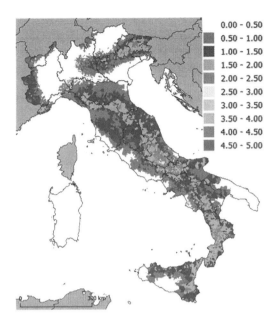

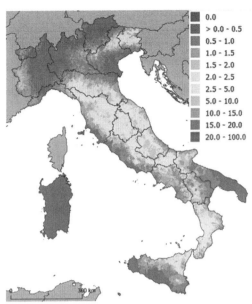

Figure 7. Average conditional damage map, Tr 475 years, type A subsoil.

Figure 9. Unconditional damage maps, Te = 1 year, type A subsoil (bedrock), for DS1 (a) and DS3 (b).

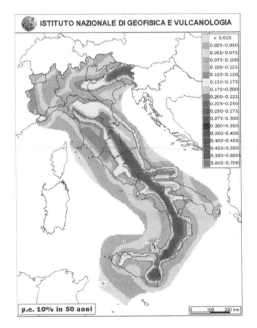

Figure 8. Seismic hazard maps for Italy, Tr = 475 years (OPCM 2006).

between nearby municipalities are found, this may be due to possible non-uniform distributions of buildings by construction period, which can generate maps with uneven damage gradations.

IRMA also produces maps for unconditional probability of damage, as a result of the vulnerability and the seismic hazard convolution, i.e. the whole range of possible PGAs are associated to their own probabilities of occurrence in a certain time window (Te). From now on, these types of maps will be referred as "unconditional damage maps".

The great potential of these maps is the evaluation of the expected average loss per year, which is widely used for financial estimates, or for the lifetime of ordinary buildings.

The results here presented refer to an exposure time of 1 year. Figure 9 also shows that the damage distribution generally follows the seismic hazard, with higher damage percentages in the Central Apennines, Calabria and parts of Sicily and Friuli.

The IRMA platform can produce not only damage maps but also risk maps, which represent economic losses, casualties and impact losses, the latter linked to the usability of buildings. These maps are calculated on the basis of the related damage map. The loss estimates have been calculated in a simplified way, using coefficients based on the observations of past earthquakes. In particular, the economic losses have been evaluated on the basis of expert judgements.

A value of 1350 €/m^2 was assumed as the cost of total replacement for residential buildings; this amount was obtained by comparing the cost of reconstruction faced after past earthquakes, taking into account any additional costs (e.g. costs incurred for disposal of debris). From this value, two sets of coefficients were used to evaluate economic losses, in order to represent a range of possible values for the various damage states (Table 2).

Figure 10 shows the results obtained in terms of economic losses for each Italian region, for both sets

Table 2. Sets of coefficients applied for risk analyses.

Economic losses [% of 1350 €/m²]		
	min	*max*
DS1	2%	5%
DS2	10%	20%
DS3	30%	45%
DS4	60%	80%
DS5	100%	100%

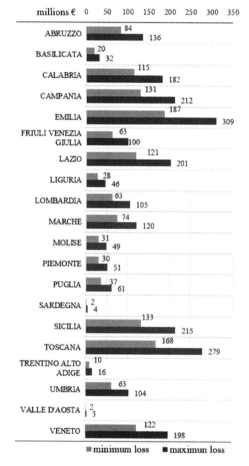

Figure 10. Results of economic losses referred to an unconditional risk map, Te = 1 year, type A subsoil.

of coefficients. These results refer to the unconditional risk map considering a time window of 1 year. Higher economic losses in regions with the highest seismic hazard would be expected, but as one can see this is not always the case. In fact, when analyzing seismic risk, all its components come into play, especially exposure.

As an example, Emilia Romagna and Lazio regions show very high losses despite their low seismic hazard, as they include a significant building stock. Another example is given by Lombardia and Marche regions, whose losses can be very similar despite a deeply different seismic hazard. This stresses the importance of exposure in the assessment of seismic risk: even though seismic hazard in Lombardia is characterized by low PGAs, these may still produce relevant losses as the territory is highly exposed from an economic point of view.

6 CONCLUSIONS

A seismic fragility model was proposed for the Italian residential masonry heritage, based on the five DSs of the EMS98 scale and on ten building macro-typologies. Subsequently, this model was implemented in the IRMA platform for seismic damage and risk simulations. Firstly, the proposed fragility model was validated by comparing the simulated damage scenario (by IRMA) with that observed after the 2009 L'Aquila earthquake (data from DaDO). Then, the model was used to derive seismic damage maps and risk (economic losses) estimates at national scale, related to the Italian residential URM buildings.

The awareness of the complexity of the study and of the uncertainties that characterise the risk analysis, starting from the definition of the fragility model, suggest to use this model and the results obtained with due attention, and highlight the great potential for improvement of the model itself. However, the results obtained seem encouraging.

Future developments could include the expansion of the database of buildings and the refinement of the different macro-typologies. Also, the presence or absence of retrofit interventions should be taken into account, differentiating the vulnerability assessment between the cases "as-built" and "retrofitted", so as to include the possible benefits of the interventions in estimating the vulnerability on a national scale.

All this is aimed at defining and calibrating increasingly reliable seismic risk assessment tools, necessary for the management of this risk at local and national level.

REFERENCES

Bernardini, A., Gori, M., Modena, C. 1990. Application of coupled analytical models and experimental kowledge to seismic vulnerability analyses of masonry buildings, in A. Koridze ed. Earthquake damage evaluation and vulnerability analysis of Building structures, *INEEC, Omega Scientific, Ozon*, 161–180.

Borzi, B., Faravelli, M., Onida, M., Polli, D., Quaroni, D., Pagano, M., Di Meo, A., 2018. Piattaforma IRMA (Italian Risk MAps). Atti del 37° Convegno GNGTS. 19–21 Novembre, Bologna IT. (in Italian).

Calvi, G.M., Pinho, R., Magenes, R., Bommer, J.J., Restrepo-Veléz, L.F., Crowley, H. 2006. The development of seismic vulnerability assessment methodologies for variable geographical scales over the past 30 years. *ISET J Earthq Technol 43(3)*:75–104.

Dolce, M., Borzi, B., Da Porto, F., Lagomarsino, S., Moroni, C., Penna, A., Prota, A., Speranza, E., Zuccaro, G., Verderame, G. M. 2019a. Mappe di rischio sismico per il territorio italiano. XVIII Convegno ANIDIS L'ingegneria Sismica in Italia. Settembre 15–19, Ascoli Piceno, Italia. (in Italian).

Dolce, M., Speranza, E., Giordano, F., Borzi, B., Bocchi, F., Conte, C., Di Meo, A., Faravelli, M., Pascale, V. 2019b. Observed damage database of past Italian earthquakes: the Da.D.O. *WebGIS. Bollettino di Geofisica Teorica ed Applicata, 60(2)*:141–164.

Donà, M., Carpanese, P., Follador, V., da Porto, F. 2019. Derivation of mechanical fragility curves for macro-typologies of Italian masonry buildings. *7th COMPDYN, Crete, Greece*, 24-26 June 2019.

Donà, M., Carpanese, P., Follador, V., Sbrogiò, L., da Porto, F. 2020. Mechanics-based fragility curves for Italian residential URM buildings. *Bulletin of Earthquake Engineering*. (in submission).

Grünthal, G. 1998. European Macroseismic Scale. Chaiers du Centre Européen de Géodynamique et de Séismologie, vol. 15 Luxembourg.

Italian Civil Protection Department (DPC) of the Presidency of the Council of Ministers (PCM) 2018. National risk assessment - Overview of the potential major disasters in Italy: seismic, volcanic, tsunami, hydro-geological/hydraulic and extreme weather, droughts and forest fire risks.

Italian National Institute of Statistics, ISTAT 2001. *https://www.istat.it/it/censimenti-permanenti/censi menti-precedenti/popolazione-e abitazioni/popola zione-2001.*

Lagomarsino, S. & Cattari, S. 2014. Fragility functions of masonry buildings, Chapter 5 in K. Pitilakis, H. Crowley, A.M. Kaynia eds. SYNER-G: *Typology Definition and Fragility Functions for Physical Elements at Seismic Risk, Vol. 27*, 111–156, Springer.

Lagomarsino, S., Cattari, S., Ottonelli, D. 2015. Derivazione di curve di fragilità empiriche per classi tipologiche rappresentative del costruito Aquilano sulla base dei dati del danno dell'evento sismico del 2009. Research Project DPC-ReLUIS, PR1 (Strutture in muratura), WP6 (Vulnerabilità delle costruzioni in muratura a scala territoriale) – Task 6.3. (in Italian).

Munari, M. 2009. Sviluppo di procedure per valutazioni sistematiche di vulnerabilità sismica di edifici esistenti in muratura, Tesi di Dottorato, Università degli Studi di Padova, Scuola di Dottorato in "Studio e Conservazione dei Beni Archeologici e Architettonici" (XXII). (in Italian).

Ordinanza del Presidente del Consiglio dei Ministri (OPCM) n. 3519, 28/04/2006. *Criteri generali per l'individuazione delle zone sismiche e per la formazione e l'aggiornamento degli elenchi delle stesse zone.* (in Italian).

Rota, M., Penna, A., Strobbia, C. 2008. Processing Italian damage data to derive typological fragility curves. *Soil Dyn Earthq Eng 28(10)*:933–947.

Valluzzi, M.R. 2009. User Manual of Vulnus_4.0, original program by Bernardini A, Gori R, Modena C, Vb version edited by Valluzzi MR, with contributions by Benincà G, Barbetta E, Munari M (in Italian).

Brick and Block Masonry - From Historical to Sustainable Masonry –
Kubica, Kwiecień & Bednarz (eds)
© 2020 Taylor & Francis Group, London, ISBN 978-0-367-56586-2

Stochastic finite element model error for unreinforced masonry walls subjected to one way vertical bending under out-of-plane loading

A.C. Isfeld, M.G. Stewart & M.J. Masia
University of Newcastle, Newcastle, Australia

ABSTRACT: The strength of unreinforced masonry (URM) walls subjected to one way bending under out-of-plane loading (no pre-compression) is known to be affected by the masonry unit tensile bond strength. Factors such as batching, workmanship, and environmental exposure alter the strength of this bond, resulting in spatial variability for any URM assembly. This paper focuses on a preliminary stochastic assessment of clay brick URM walls with spatially variable tensile bond strength subjected to uniformly distributed out-of-plane loads in one-way vertical bending. Stochastic computational modelling combining Finite Element Analysis (FEA) and Monte Carlo Simulation (MCS) is used to account for bond strength variability when estimating the walls ultimate failure loads. Previously, this approach has been used to study the effect of material variability on cracking loads, failure loads, and failure modes of URM walls in one and two-way bending. For this preliminary assessment FEA MCS has been applied to a subset of existing test data for walls with consistent properties constructed by ten different masons. For these walls a 3D non-linear FEA model was developed, followed by a stochastic analysis for which the unit tensile bond strength is spatially varied according to the measured flexural tensile bond strength and its coefficient of variation. For a set of simulations the peak load and load-displacement data was extracted and analysed, showing good agreement with the results of wall test data.

1 INTRODUCTION

Masonry is inherently strong in compression and comparatively weak in tension, meaning that failure of masonry assemblies is often initiated by the development of tensile stresses. The tensile strength of the unit mortar interface is a plane of weakness in masonry walls and frequently dictates the points at which failure initiates. Masonry walls are typically constructed in place, and subsequently impacted by variable environmental conditions, workmanship, and batching, resulting in spatial variability of mechanical properties within any wall.

The degree of spatial variability in material properties is known to be higher in masonry than in other common structural materials. This variation increases the complexity when determining the overall strength of a structural element. In limit states design safety factors are used along with deterministic analysis techniques to account for uncertainty and assess the reliability of a structure. The Australian Standard for Masonry Structures (AS 3700, (2018)) follows a limit states approach, using the lower five-percentile values of characteristic strengths with capacity reduction factors. The direct outcome of random and spatial variability cannot be directly quantified with this approach.

When a wall is subjected to vertical flexure the bending moment varies over the height of the wall. As bond strength varies along both the wall height and length, failure will initiate where the ratio of applied bending moment to moment resistance first becomes critical. The importance of quantifying the variability in joint strength, and subsequent variability in wall panel strength has been established by Baker and Franken (1976), Baker (1981), and Lawrence (1983). Lawrence and Stewart (2002) developed a structural reliability model accounting for unit-to-unit variability in bond strength. Heffler et al. (2008), examined the unit-to-unit spatial correlation of flexural tensile bond strength within masonry walls. Analysis of bond wrench testing data gave an average correlation coefficient of $\rho = 0.4$ within courses of masonry, with each unit's flexural bond strength being represented by a truncated Normal distribution.

Several FEA approaches are available to simulate the behaviour of masonry. Models can be developed utilizing macro of micro approaches (Lourenço, 1996, Lourenço et al., 1995) according to the available input and required output, along with size and scope of the modelling effort. Macro modelling approaches are frequently applied when the behaviour of a complex masonry structure is the subject of the modelling effort (Bayraktar et al., 2010, Pande et al., 1989), while simplified and detailed micro modelling have been successfully employed to improve the understanding of local failure mechanisms (Hamid and Chukwunenye, 1986, Bolhassani et al., 2015), or individual structural elements (Lofti and Shing, 1994, Ali and Page, 1988). FEA is most often approached using averaged or characteristic values of material properties with idealized geometry and simplified loads, the FEA approach has also been paired with additional computational tools to account for random variations in structural systems. This combined approached is referred to as Stochastic Finite Element Method, the Random Finite Element Method, or the Probabilistic Finite Element Method (Aggregui-Mena et al., 2016). Randomness can be introduced at the level of material properties, geometry, or loads, and the effect is measured on the response of the system. This randomness can be included globally or vary in space and time.

The Random Field Finite Element Method was developed to include spatial variability in linear and non-linear simulations by Liu et al. (1986). A semi random version of this method has been developed by Moradabadi et al. (2015) and implemented for analysis of eccentrically loaded masonry piers, showing agreement with physical testing. This approach was applied by Isfeld et al. (2016) for the simulation of grout injection in historic stone masonry walls, examining the influence of different material input through sensitivity analysis, and to preform uncertainty analysis whereby a threshold of required material strength was established. Li et al. (2016) utilized the random field process within stochastic FEA and MCS to model the behaviour of brick beams. Spatial variability of the mortar joint properties was considered at different levels of correlation. Statistical independence was considered between joints, with $\rho = 0$, and a realistic correlation of $\rho = 0.4$ (Heffler et al., 2008) was implemented. The results were compared to non-spatial stochastic FEA results, as well as the results of 310 experimental specimens, the peak loads showed good agreement between spatial simulations with $\rho = 0.4$ and the test (Li et al., 2016).

Li et al. (2014) modelled brick masonry walls under one-way bending due to uniform pressure loads as deterministic, non-spatial stochastic, and with unit-to-unit stochastic spatial variability. It was shown that the point of first cracking, mid-height cracking, and peak loads differ between methods. Nonspatial analysis was found to overestimate the probability of wall failure compared to spatial analysis. Thus, demonstrating the importance of considering the unit-to-unit spatial variability of flexural bond strength. This is consistent with the conclusions of Li et al. (2017) when applying the same modelling approach to walls subjected to two-way bending.

In the current study, stochastic FEA have been developed based on walls previously tested by Baker (1981) in order to assess the numerical accuracy of such models. This study focuses on a preliminary stochastic assessment of the tested clay brick URM

walls with spatially variable tensile bond strength subjected to uniformly distributed out-of-plane loads in one-way vertical bending with no vertical pre-compression. The statistical parameters of model error will be quantified in order to assess these techniques for application in further reliability analysis.

2 EXPERIMENTAL PROGRAM

Baker (1981) completed testing of 2/3 scale brick masonry wall panels subjected to one way bending under uniform out-of-plane pressure loads. This testing program is unique in its representation of actual construction conditions, with comprehensive testing being completed on wall panels and bond strength specimens constructed by ten different masons. Through this test program it is possible to establish both the variability of construction for a single mason as well as that seen between masons, and subsequently in masonry structures as they are constructed in practice.

Consistent materials were used for each mason; the small scale bricks had the dimensions $150 \times 67 \times 50$ mm with three 20 mm diameter perforations no other properties were reported. Mortar was a standard 1:1:6 mix by volume of ordinary Portland cement, hydrated lime, and sand. Water was added to achieve consistent flow of 95% between mortar batches. Ten wall panels were constructed (one by each mason) 10 units wide and 24 courses high. In addition to the full size panels, each mason also constructed two 4, 2, and 1 unit wide panels, all 24 courses high. All panels were tested under uniform out-of-plane pressure with the remaining joints of the single unit wide panels being used to determine the flexural bond strength and its variability for each mason. All panels were cured for 28 days prior to testing. During testing the wall panels were supported vertically at the base with a roller along the length of the walls located at mid-thickness. Out-of-plane supports were provided with pipe supports plastered against the wall at a distance of 1.243 m, providing simply supported conditions. The pressure load was applied through an airbag supported with a reaction frame along the full wall height as shown in Figure 1.

Twenty three flexural bond strength tests were completed from the single unit wide panels constructed by each mason, for a total of 230 individual joints tested, using a method designed by Baker (1981) shown in Figure 2. This test is completed by fixing a loading frame to a single brick, a second frame provides a fixed reaction and is secured to the adjacent brick. A moment is applied as equal and opposite jacking forces to the frame until failure occurs.

The average flexural tensile bond strength, f_{mt}, and coefficient of variation (COV) for each mason (1-10), as well as the modulus of rupture, and ultimate load are reported in Table 1. The average bond strengths varied between 0.632 MPa and 0.996 MPa for individual masons with an overall average of 0.754 MPa and average COV for all masons of 0.509, whereas

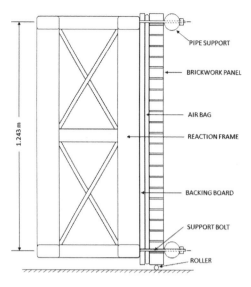

Figure 1. Test set-up for 2/3 scale panels. Adapted from Baker (1981).

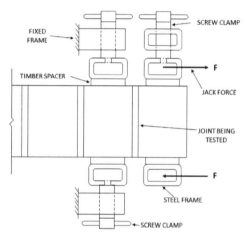

Figure 2. Flexural tensile specimen test. Adapted from Baker (1981).

the COV between masons was 0.177. The ultimate pressure load, w, was calculated from the reported modulus of rupture according to the equation

$$w = \frac{4 f_{mt}}{3(H/t)^2} \qquad (1)$$

where the modulus of rupture was used to represent the average bond strength of the failed course, f_{mt}, the height, H, is 1.243 m and the thickness, t, is 67 mm. The ultimate load varied from 0.89 kPa to 2.32 kPa with a mean of 1.77 kPa having a COV of 0.257 for the set of ten walls.

Table 1. Test results (Baker, 1981).

| | Joint | | | | | Panel | |
Mason	mean f_{mt} MPa	mean f_t MPa	COV -	f_{mtc} MPa	f_{tc} MPa	Mod. Rupt. MPa	Ult. Load kPa
1	0.654	0.436	0.440	0.181	0.120	0.399	1.55
2	0.542	0.361	0.512	0.086	0.057	0.436	1.69
3	0.701	0.467	0.393	0.248	0.165	0.481	1.86
4	0.632	0.421	0.555	0.055	0.037	0.332	1.29
5	0.753	0.502	0.439	0.209	0.139	0.563	2.18
6	0.996	0.664	0.395	0.349	0.233	0.598	2.32
7	0.755	0.503	0.569	0.048	0.032	0.504	1.95
8	0.888	0.592	0.585	0.033	0.022	0.596	2.31
9	0.759	0.506	0.582	0.032	0.022	0.229	0.89
10	0.862	0.575	0.615	0.000	0.000	0.439	1.70
μ	0.754	0.503	0.509	0.123	0.082	0.458	1.77
σ	-	-	-	-	-	0.118	0.456
COV	-	-	-	-	-	0.257	0.257

3 FINITE ELEMENT MODELLING

3.1 Overview

For each wall series a deterministic 3D non-linear finite element model was developed to represent the behaviour of the test walls, considering the full length 1.6 m and loaded height 1.234 m with a thickness of 67 mm. The bricks are modelled as solid expanded units, thus accounting for the mortar joint thickness using the simplified micro modelling approach (Lourenço et al., 1995, Lourenço, 1996). The walls are non-loadbearing, and simply supported at the top and bottom, with quasi-static uniform out-of-plane loading. The DIANA (2017) finite element software is utilized here for its ability to simulate a range of masonry behaviour. The individual clay brick units are modelled as two halves, with linear elastic material properties and non-linear behaviour allocated to the mortar joint and unit crack interface elements. The combined cracking-shearing-crushing model is used to model the mortar joints (Lourenço et al., 1995). This model captures tensile, shear, and crushing fracture, as well as frictional slip. Discrete cracking using the linear tensile softening model is included using interface elements to simulate mid length unit cracking. Thus, mortar joint failure in tension and shear can be simulated, along with unit tensile failure, and combined mortar/unit shear and compressive failure.

3.2 Loading and boundary conditions

The wall height between out-of-plane supports (22 courses) was modelled with loading applied in a quasi-static manner to simulate the testing methodology. Out-of-plane restraints were provided along the edges of the wall, in-plane restraints were provided at a single node on each of these edges, and vertical support was provided at the centre line of the wall base. The boundary conditions and loading are shown in Figure 3.

3.3 Mesh refinement study

A mesh refinement study was completed first on a 2D model then on the full 3D model. When applying the micro modelling approach to masonry walls the mesh density is in part governed by the number of units the wall is comprised of. Each discrete part and interface must contain, at minimum, a single element. The study was subsequently conducted by controlling the number of elements on a single brick, or half brick part and meshing the full wall such that continuity is maintained along all interface elements.

The 2D mesh refinement study used eight-node quadrilateral isoparametric plane strain elements (QU8 CQ16E) for the masonry units and six-node quadratic interface elements (IL33 CL12I) for the mortar joints. The coarsest mesh had one element assigned to a brick unit and a single interface element between each pair of bricks, and the densest mesh had a 10 × 10 (Thickness × Height) mesh or 100 element per brick unit, and 10 elements along an interface, with intermediate asymmetrical meshes also considered. This study showed sufficient accuracy could be achieved with one element along an element height.

In the full wall mesh refinement study two parameters were varied; the number of elements through the wall thickness, and the number of elements along a half brick length. The four combinations are described in Table 2, from the coarsest mesh M1 to the densest mesh M4. Through this study it was found that four elements through the wall thickness, and two

Table 2. 3D mesh refinement study.

Material	Element	M1	M2	M3	M4
Half Brick Unit	HE20 CHX60	2×4×1	4×4×1	2×8×1	4×8×1
Mortar Joint	IS88 CQ48I	2×4×1	4×4×1	2×8×1	4×8×1
Crack Interface	IS88 CQ48I	1×4×1	1×4×1	1×8×1	1×8×1

elements along the length of one half brick unit provided sufficient mesh density while minimizing model run time. The final mesh is shown in Figure 3.

3.4 Analysis procedure

A structural non-linear analysis procedure is required to simulate the material softening and unloading of the walls. Automatic step sizing was used to capture the range of results possible in the stochastic analysis. Arc length controls were used with the spherical path method and a single node at the centre of the tension face (H/2 and b/2) was selected as the control set. The Newton-Raphson modified linear iterative convergence method was applied with the energy convergence norm. Using these controls, the incremental displacement is constrained, and the applied force is increased proportional to the change in the load factor, which can vary in every iteration. The maximum number of iterations was limited to ensure the loading progressed past the peak load, and terminated in the unloading curve to minimize the model run time.

3.5 Material properties

In Diana the combined cracking-shearing-crushing model is used to model the mortar joints for simplified micro-modelling of masonry (Lourenço et al., 1995), with direct tensile strength required as input. Material data provided by Baker (1981) was focused on flexural tensile testing and used to determine the direct tensile strength of the mortar joint. As all bond strengths were measured by means of flexural tensile testing and were thus converted to a direct tensile strength by dividing by the factor 1.5 (Petersen et al., 2012). The direct tensile strength values are given as f_t in Table 1 and the mean value reported for all masons used in a deterministic model was 0.503 MPa, as indicated in Table 1 and included below in Table 3. The tensile fracture energy, G_f^I, was related to the direct tensile strength using Equation (2) established by Heffler (2009) for 1:1:6 mortar.

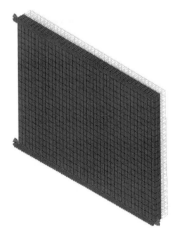

Figure 3. FEA model.

$$G_f^I = 0.01571 f_c + 0.0004882 \qquad (2)$$

The interface cohesion, c, is related to the tensile bond strength according to Equation (3) outlined by (Milani and Lourenco, 2013).

$$c = 1.4f_t \qquad (3)$$

The residual cap must fall outside the tension cut off for all direct tensile strength values. To enforce this the compressive strength is increased to 25 MPa for $f_t > 1.4$ MPa and to 30 MPa for $f_t > 1.7$ MPa. The remaining average deterministic material properties were based values established by Heffler (2009) and refined by Li (2015) shown in Table 3.

3.6 Deterministic FEA model

Deterministic FEA models were completed for the mean direct tensile strength (all values indicated in Table 3), as well as the mean characteristic strength calculated for all masons, and the AS 3700 (2018) prescribed value. The resulting failure mode for each deterministic simulation was a single mid-height crack along the full wall length. The mean direct tensile strength of 0.503 MPa (flexural tensile strength 0.754 MPa) resulted in a FEA model failure load of 3.01 kPa. As expected, this load is significantly higher than the wall test mean failure load of 1.77 kPa. Instead, these results are consistent with the failure load calculated using Equation 1, whereby the mean flexural tensile strength (0.754 MPa) gives a failure pressure of 2.92 kPa. Alternatively, when the deterministic model is updated to use the mean characteristic direct tensile strength calculated for (i) all masons of 0.082 MPa or (ii) that prescribed by AS 3700 (2018) of 0.13 MPa (flexural tensile strength of 0.2 MPa divided by 1.5), the FEA models fail at pressure loads of 0.669 kPa and 1.016 kPa, respectively. Both characteristic strength models conservatively estimate the failure load when compared to the mean test results. With the minimum wall test failure load calculated to be 0.89 kPa, based on the reported modulus of rupture. The actual characteristic strength is conservative for all test cases.

3.7 Stochastic FEA

Stochastic computational modelling combining the FEA and MCS is used to account for bond strength variability when estimating the walls ultimate failure loads. The spatial analysis is generated using the 3D non-linear FEA full wall model and MCS. In a non-spatial stochastic analysis, the parameters of interest would be varied for each simulation, but consistent for each location in the wall – i.e., fully correlated random variables $\rho = 1.0$). Whereas spatial stochastic analysis varies the parameter of interest either randomly at each occurrence (i.e., statistically independent $\rho = 0.0$), or according to a defined relationship with position (spatially correlated $0 < \rho < 1.0$).

Table 3. FEA material data.

Material	Property Name	Units	Model Input
Brick	Young's Modulus	N/mm^2	20 000
	Poisson's Ratio	-	0.15
	Density	kg/mm^3	1800
Brick Crack Fictitious Values	Linear normal stiffness modulus	N/mm^3	1000
	Linear tangential stiffness modulus	N/mm^3	1000
	Direct tensile strength	N/mm^2	2
	Tensile fracture energy	Nmm/mm^2	0.5
Mortar Joints	Linear normal stiffness modulus	N/mm^3	353
	Linear tangential stiffness modulus	N/mm^3	146
	Direct tensile strength	N/mm^2	0.503*
	Tensile fracture energy	Nmm/mm^2	0.00839*
	Cohesion	N/mm^2	0.704*
	Tangent friction angle	-	0.75
	Tangent dilatancy angle	-	0.6
	Tangent residual friction angle	-	0.75
	Confining normal stress	N/mm^2	-1.2
	Exponential degradation coefficient	-	5
	Capped critical compressive strength	N/mm^2	20*
	Shear traction control factor	-	9
	Compressive fracture energy	Nmm/mm^2	15
	Equivalent plastic relative displacement	-	0.012
	Shear fracture energy factor	-	0.15

*Variable properties – values reported here are mean.

For a wall with known dimensions under out-of-plane loading alone, the tensile strength of the mortar-unit interface is the most critical material parameter for determining the load resistance (Stewart and Lawrence, 2002). As such only this value will be randomized, with cohesion and fracture energy calculated as a function of the tensile strength according to equations (2) and (3). When required, for accurate definition of the compression cap, the compressive strength of the mortar will be increased in a step wise manner as indicated in Section 3.5. The remaining material parameters are representative average values as outlined by Heffler (2009) and modified by Li (2015) given in Table 3.

The testing program utilized specialized 2/3 scale bricks (Baker, 1981). Standard bricks range in properties with Young's Modulus values reported between 3.5 and 21 GPa and compressive strength between 5 and 30 MPa (AS, 2018). For scaled bricks values in these ranges would be reasonable, while it would be possible for the bricks to fall outside of standard values. In lieu of more detailed information properties established in previous modelling of clay brick walls were used (Li, 2015, Heffler, 2009). A sensitivity analysis showed that doubling or halving the Young's Modulus resulted in an increase or decrease of the failure load by approximately 5%. This or other properties could be varied and are potential sources of error in the model.

The stochastic FEA models are developed that considers the workmanship and variability of each mason. Hence, next, the individual mean direct tensile strength and COV (Table 1) for each mason was used to define a set of ten stochastic FEA, referred to as 1 through 10. Forty MC simulations were run for each mason, leading to a total of 400 simulation runs.

For the spatial stochastic FEA, each mortar joint is assigned a unique tensile bond strength. Spatial correlation is assumed to occur within courses of masonry on a unit-to-unit basis. A program is written in MATLAB (MathWorks Inc, 2018) in order to automate the procedure of writing DIANA input in the form of python script for batching the FEA MCS. For each simulation the first step is to assign a unique tensile bond strength from a normal distribution truncated at zero (Li, 2015, Heffler, 2009), to the mortar joint below the first unit in each course of masonry. Next, the adjacent joint within a course are assigned tensile bond strengths that are correlated to the first joint. This pairwise correlation is completed for the remaining mortar joints in each course using the correlation coefficient of 0.4 established by Heffler (2009) through bond wrench testing, and validated through FEA of brick beams by Li et al. (2016). The method adopted by Li (2015) was for two correlated random variables X [x1, x2, … xn] and Y [y1, y2, … yn] with equal mean $\mu_x = \mu_y$ and standard deviation $\sigma_x = \sigma_y$. A random number, x, represents the first joint strength in the correlated pair and is used to calculate the updated mean, μ_{y*}, and standard deviation, σ_{y*}, for generation of a correlated distribution as outlined by Ang and Tang, (2007).

All perpend joints are assumed statistically independent from the truncated normal distribution. All parameters that depend on the direct tensile strength are calculated for that interface based on the assigned tensile strength – i.e., they are dependent variables.

3.8 Stochastic FEA results

The MCS results are reported as load displacement plots in Figure 4 and summarized in Table 4 along with the model error (model uncertainty) and statistical analysis of the model error. The detailed results of the modelling for each mason are given in Table 5. The model error, ME, is calculated as

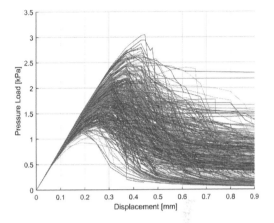

Figure 4. Load displacement curves for stochastic FEA.

Table 4. Results.

| Set | Results | | | | Model Error | | |
| | n | μ | σ | COV | ME | σ_{ME} | V_{ME} |
	-	kPa	kPa	-	-	-	-
FEA	400	1.858	0.379	0.204	0.955	0.134	0.140
Test	10	1.773	0.456	0.248		-	

Table 5. Model results and error.

| | FEA Results | | | |
| | | | | mean ME |
Mason	μ kPa	σ kPa	V -	-
1	1.699	0.253	0.149	0.910
2	1.372	0.190	0.139	1.231
3	1.901	0.209	0.110	0.980
4	1.566	0.227	0.145	0.822
5	1.891	0.265	0.140	1.153
6	2.472	0.287	0.116	0.937
7	1.850	0.246	0.133	1.055
8	2.017	0.245	0.122	1.145
9	1.785	0.306	0.171	0.497
10	2.024	0.295	0.146	0.773

$$ME = \frac{\mu_E}{\mu_M} \qquad (4)$$

where μ_E and μ_M are the mean pressure loads at failure taken from the experimental results and model predictions respectively. The COV for the model error, V_{ME}, can be calculated using the values for the experimental test program, V_E, and stochastic FEA model results, V_M, as

$$V_{ME} = \sqrt{V_E^2 - V_M^2} \qquad (5)$$

In any testing program the variability of the results is attributed primarily to the variability of the material and in part to the variability of the testing approach, V_{test}, and specimen, V_{spec}. The COV of the test accounts for the accuracy of the test measurements and ambiguity regarding the definition of failure. The COV of the test specimen accounts for any deviation between the actual and measured properties. The contribution of each will vary depending on the complexity of the tests. Subsequently, the COV of the experimental test program is corrected according to the Equation 6.

$$V_E = \sqrt{V_{overall}^2 - V_{test}^2 - V_{spec}^2} \qquad (6)$$

The overall COV of the experimental program, $V_{overall}$, was reported as 0.257 (Baker, 1981). Correction values have been established through testing of reinforced concrete beams and columns by Ellingwood et al. (1980) of 0.02 to 0.04 for V_{test} and approximately 0.04 for V_{spec}. Lawrence and Stewart (2009) noted that the variability of masonry testing is likely higher than that of concrete, using $V_{test} = V_{spec} = 0.05$ to quantify the model error for masonry walls subjected to one way bending. Of particular concern is the difficulty in conducting lateral load tests as application of a uniform pressure load through airbags, and assumptions regarding idealized support conditions may introduce error. When the experimental program COV is corrected using the values $V_{test} = V_{spec} = 0.05$, prescribed by Lawrence and Stewart (2009) the final corrected COV value is 0.248 as shown in Table 4.

Hypothesis testing was conducted to determine if the test results could be considered part of the same samples as the stochastic FEA. The null hypothesis was not rejected at a 90% confidence interval using the Student T test for equal variances. The hypotheses that the variances are equal were tested using an F test for equality of variances. The hypothesis that the variances are equal was not rejected. Based on this, the stochastic

FEA can be considered as most likely coming from the same population as the wall test results.

4 CONCLUSIONS

Stochastic FEA was used to calculate the peak pressure load for 2/3 scale URM walls, 10 units wide and 24 courses tall with a 22 course (1.243 m) loaded span. Walls were subjected to uniform out-of-plane pressure loading with no vertical pre-compression. The MCS considered the workmanship and variability of each individual mason, and grouped to assess the model error statistics. A total of 400 simulation runs (40 per mason) were completed. The stochastic FEA utilized probabilistic information in the form of unit-to-unit spatial variability of the mortar joints mechanical properties ($\rho = 0.4$). Results of the simulations were compared with experimental data from ten 2/3 scale walls. The mean model error ME = 0.497 – 1.231 for each mason, with overall mean of 0.955 and COV of 0.14. This quite accurately represents the wall test results and this approach could be applied to additional data sets to conduct a structural reliability analysis.

ACKNOWLEDGEMENTS

The support of the Australian Research Council grant DP180102334 is gratefully acknowledged.

REFERENCES

Aggregui-Mena, J. D., Margetts, L. & Mummery, P. M. 2016. Practical applications of the stochastic finite element method. *Archives of Computational Methods in Engineering*, 23, 171–190.

Ali, S. S. & Page, A. W. 1988. Finite element model for masonry subjected to concentrated loads. *Journal of Structural Engineering*, 114, 1761–1784.

Baker, L. R. 1981. The flexural action of masonry structures under lateral load. *School of Engineering and Architecture*. Melbourne, Australia, Deakin University.

Baker, L. R. & Franken, G. L. 1976. Variability aspects of the flexural strength of brickwork. *4th International Brick and Block Masonry Conference* Bruges, Belgium.

Bayraktar, A., Şahin, A., Özcan, D. M. & Yildirim, F. 2010. Numerical damage assessment of Haghia Sophia bell tower by nonlinear FE modeling. *Applied Mathematical Modelling*, 34, 92–121.

Bolhassani, M., Hamid, A. A., Lau, A. C. W. & Moon, F. 2015. Simplified micro modeling of partially grouted masonry assemblages. *Construction and Building Materials*, 83, 159–173.

DIANA FEA BV (2017) DIANA 10.2 Finite Element Analysis User's Manual. Ed Jonna Manie and Gerd-Jan Schreppers (Ed. Delft, The Netherlands.

Ellingwood, B. R., Galambos, T. V., Macgregor, J. G. & Cornell, C. A. 1980. Development of probability based load criterion for American National Standard A58. Washington, DC., U.S. Government Printing Office.

Hamid, A. A. & Chukwunenye, A. O. 1986. Compression behavior of concrete masonry prisms *Journal of Structural Engineering*, 112, 605–613.

Heffler, L. 2009. Variability of unit flexural bond strength and its effect on strength in clay brick unreinforced masonry walls subjected to vertical bending. *Faculty of Engineering and Built Environment*. Newcastle, Australia, University of Newcastle.

Heffler, L. M., Stewart, M. G., Masia, M. J. & Correa, M. R. S. 2008. Statistical Analysis and Spatial Correlation of Flexural Bond Strength for Masonry Walls. *Masonry International*, 21, 59–70.

Isfeld, A. C., Moradabadi, E., Laefer, D. F. & Shrive, N. G. 2016. Uncertainty analysis of the effect of grout injection on the deformation of multi-wythe stone masonry walls. *Construction and Building Materials*, 126, 661–672.

Lawrence, S. J. 1983. Behaviour of brick masonry walls under lateral loading *School of Civil Engineering*. Sydney, Australia, University of New South Wales.

Lawrence, S. J. & Stewart, M. G. 2009. Structural reliability and partial safety factors for unreinforced masonry in vertical bending. Newcastle, Australia, Centre for Infrastructure Performance and Reliability, University of Newcastle.

Li, J. 2015. Spatial variability and stochastic strength predication of unreinforced masonry walls subjected to out-of-plane bending. *The Faculty of Engineering and the Built Environment*. Newcastle, Australia, University of Newcastle.

Li, J., Masia, M. J. & Stewart, M. G. 2017. Stochastic spatial modelling of material properties and structural strength of unreinforced masonry in two-way bending. *Structure and Infrastructure Engineering*, 13, 683–695.

Li, J., Masia, M. J., Stewart, M. G. & Lawrence, S. J. 2014. Spatial variability and stochastic strength prediction of unreinforced masonry walls in vertical bending. *Engineering Structures*, 59, 787–797.

Li, J., Stewart, M. G., Masia, M. J. & Lawrence, S. J. 2016. Spatial correlation of material properties and structural strength of masonry in horizontal bending. *Journal of Structural Engineering*, 142, 1–11.

Liu, W. K., Belytschko, T. & Mani, A. 1986. Random field finite elements. *International Journal for Numerical Methods in Engineering*, 23, 1831–1845.

Lofti, H. R. & Shing, P. B. 1994. Interface model applied to fracture of masonry structures. *Journal of Structural Engineering*, 120, 63–80.

Lourenço, P. B. 1996. Computational strategies for masonry structures. *Delft Technical University* Delft Technical University.

Lourenço, P. B., Rots, J. G. & Blaauwendraad, J. 1995. Two approaches for the analysis of masonry structures: micro and macro-modelling. *HERON*, 40, 313–340.

MATHWORKS INC (2018) MATLAB. IN MATHWORKS (Ed.

Milani, G. & Lourenco, P. B. 2013. Simple Homogenized Model for the Nonlinear Analysis of FRP-Strengthened Masonry Structures. II: Structural Applications. *Journal of Engineering Mechanics-Asce*, 139, 77–93.

Moradabadi, E., Laefer, D. F., Clarke, J. A. & Lourenco, P. B. 2015. A semi-random field finite element method to predict the maximum eccentric compressive load for masonry prisms. *Construction and Building Materials*, 77, 489–500.

Pande, G. N., Liang, J. X. & Middleton, J. 1989. Equivalent elastic moduli for brick masonry. *Computers and Geotechnics* 8, 243–265.

Petersen, R. B., Ismail, N., Masia, M. J. & Ingham, J. M. 2012. Finite element modelling of unreinforced masonry shear wallettes strengthened using twisted steel bars. *Construction and Building Materials*, 33, 14–24.

STANDARDS AUSTRALIA 2018. AS 3700: Masonry Structures. Sydney, Australia.

Stewart, M. G. & Lawrence, S. J. 2002. Structural reliability of masonry walls in flexure. *Masonry International*, 15, 48–52.

Brick and Block Masonry - From Historical to Sustainable Masonry –
Kubica, Kwiecień & Bednarz (eds)
© 2020 Taylor & Francis Group, London, ISBN 978-0-367-56586-2

Formulation and experimental validation of distributed plasticity macro-element for unreinforced masonry walls

F. Parisi & E. Acconcia

Department of Structures for Engineering and Architecture, University of Naples Federico II, Naples, Italy

ABSTRACT: The equivalent frame modelling (EFM) approach allows the assessment of unreinforced masonry (URM) buildings with a trade-off between realistic damage simulation and computational efficiency. EFM is based on the use of macro-elements by which flexural and shear behavioural modes are simulated through nonlinear force-deformation relationships combined with axial load-shear interaction domains. In this study, a novel macro-element is presented for nonlinear static analysis of URM buildings. The macro-element relies upon force-based fibre modelling approach accounting for geometric and mechanical nonlinearities, the latter through nonlinear constitutive models for masonry in tension and compression. Shear-flexure interaction is taken into account and strain softening is considered according to a smeared crack approach. Numerical results are compared with experimental data, highlighting a good accuracy of the model and its capability of reproducing both shear and flexural behavioural modes under varying aspect ratio of the URM wall and axial load level.

1 INTRODUCTION

Nonlinear incremental static (pushover) analysis is extensively used for seismic assessment of existing unreinforced masonry (URM) buildings or even design optimisation of new URM buildings, because of a number of nonlinearities in their response to earthquake actions. Several computational strategies have been proposed in last decades, including those based on the finite element method in the form of micro- or macro-modelling approaches (Lourenço 1996, Caporale et al. 2014, Zucchini & Lourenço 2009, Lotfi & Shing 1991, Parisi et al. 2016, 2019), discrete element method (Psycharis et al. 2003, DeJong & Dimitrakopoulos 2014) and equivalent frame method (EFM). This latter method is particularly efficient for URM buildings with proper structural detailing, which allow preventing local collapse mechanisms and developing a global box-type response to earthquake loading (Augenti & Parisi 2019). In that case, each URM wall is mainly loaded in its own plane.

EFM is based on the discretisation of walls with openings into the following types of macro-elements modelled as beam elements: pier panels, spandrel panels and spandrel-pier joint panels. Pier and spandrel panels are assumed to be flexible elements, whereas joints panels are supposed to be rigid because they do not usually suffer damage under seismic actions. The identification of such macro-elements becomes rather complex when modelling irregular walls (see e.g. Parisi & Augenti 2013). Nonetheless, numerical-experimental studies have shown that EFM provides good seismic performance predictions for URM buildings having walls with

a quite regular layout of openings. In that context, the use of fibre-based modelling approaches appear an attractive mechanics-based solution to analytically simulate the nonlinear response of URM walls under in-plane lateral loading (e.g. Raka et al. 2015). This study attempts to follow this research line with consideration of axial-bending-shear coupling and rigid-body rocking behaviour, from the initial uncracked stage to the final cracked-crushed stage of the wall.

2 METHODOLOGY

This study is focused on the nonlinear response of cantilevered URM walls to monotonically increasing in-plane horizontal displacement under constant axial load. A force-based fibre capacity model is presented and relies upon the Timoshenko beam theory, where plane cross sections remain plane after deformation, but – differently from the Euler-Bernoulli beam model – they do not keep orthogonality to the beam axis.

This study deals with prismatic walls with rectangular cross section, height H, length L and thickness T. The cross section was discretised in 30×30 fibres, which was found to be an optimal number through a mesh sensitivity analysis (Parisi & Acconcia 2018). Figure 1 shows the wall geometry and discretisation. Young's moduli in tension and compression were assumed to be the same. The self-weight of masonry was neglected, resulting in a constant axial load along the height. This hypothesis allows assuming the same moment–curvature relationship for each section.

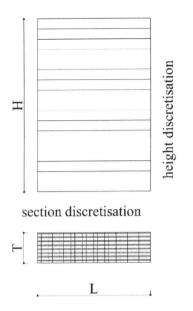

section discretisation

Figure 1. Fibre mesh of URM wall.

at base and subjected to a shear force V and an axial load N on top. The displacement on top of the wall is assessed by integrating curvatures and shear strains along the height of wall. A further displacement due to a pseudo-rigid rotation caused by progressive toe crushing, namely rocking. According to a proposal by Parisi (2010), the total relative displacement of the wall can be assessed through Equation 1:

$$\delta = \delta_f + \delta_s + \delta_r \qquad (1)$$

where δ_f, δ_s and δ_r are the flexural, shear and rocking contributions to the displacement that are calculated as follows:

$$\delta_f = \int_0^H \phi(z)z\,dz \qquad (2)$$

$$\delta_s = \int_0^H \gamma(z)\,dz \qquad (3)$$

$$\delta_r = \theta_r H \qquad (4)$$

where $\phi(z)$ is the curvature at any height z of the wall, $\gamma(z)$ is the shear strain, and θ_r is the rocking rotation that is lumped at the base section.

The procedure for the derivation of shear-drift capacity diagrams consists of the following steps:

– Derivation of moment–curvature diagram corresponding to the base (fixed) section of the wall, according to Parisi et al. (2016).
– Application of bending curvature to base section.
– Calculation of the bending resisting moment of base section through the moment–curvature relationship.
– Evaluation of the bending curvature profile along the height of the wall, according to a linear moment profile.
– Calculation of shear profile and corresponding shear strain profile, accounting for both geometric and mechanical nonlinearities.
– Computation of top displacement via numerical integration of bending curvatures and shear strains along the height of the wall.
– Calculation of rocking rotation angle in order to consider a pseudo-rigid rotation due to toe crushing.
– Evolution of total top displacement associated with flexure, rocking and shear and transformation into element drift.
– Replication of steps above to plot the in-plane capacity curve of the URM wall.

3 NOVEL FIBRE MACRO-ELEMENT

3.1 Analytical procedure

An iterative incremental procedure for nonlinear pushover analysis was developed in MATLAB® to perform numerical integration of the moment–curvature behaviour and evaluation of shear deformations and rocking-induced displacements. Let the wall be fixed

3.2 Material model

According to the fibre modelling approach, a uniaxial material model is assigned to each fibre. The compressive behaviour is simulated through stress–strain equations proposed by Augenti & Parisi (2010) in the direction perpendicular to bed joints. The constitutive model is implemented as a data set.

The tensile behaviour of masonry is modelled as linear elastic up to tensile strength, assuming an exponential softening rule up to zero stress as per a given fracture energy.

Shear sliding and diagonal shear strengths are considered to define the peak resisting force according to failure rules reported below.

The selfweight of masonry is assumed to be negligible compared to the given axial load. This assumption can be also removed (for instance in case of very long walls), resulting in higher computational cost.

3.3 Flexural behaviour

The proposed procedure allows a direct evaluation of flexural behaviour.

The procedure implemented in MATLAB® follows an incremental iterative method that allows evaluating the relative horizontal displacement of the top section and the axial stresses within the wall, once the curvature of the base section and axial load are known.

Given the axial load, the moment–curvature relationship of the base section can be computed and stored for its use during force–drift analysis. Based on the boundary conditions of the wall, the shear force profile can be derived at each curvature step.

As a force-based fibre modelling is adopted, the step-by-step convergence of the iterative procedure is ruled by a linear profile of the bending moment along the height of the wall, according to the fixed-free boundary conditions. This allows the derivation of bending curvature at any other height of the wall. Thus, Equation 2 allows the computation of the top displacement.

When the curvature of the base section exceeds the value corresponding to the maximum resisting moment, the base moment cannot longer increase. Thus, as the curvature further increases, the bending moment gradually drops down according to the post-peak softening branch of the moment–curvature diagram. As per the linear moment profile, the bending moments in upper sections reduce as well. A difference in curvature is then created between the base section and the consecutive upper section. A comparison with finite element analysis highlighted that the curvature profile along the height of the wall concentrates within a zone close to the base section. Thus, in the proposed model, the authors assume a uniform curvature between the two sections nearest the base of the wall. This implies that the contribution of curvature near the base to the lateral displacement of the wall is strongly influenced by the discretisation along the height. In the post-peak phase, the greatest contribution to the lateral displacement of the wall is provided by the curvature of the base section. Figure 2 shows the moment and curvature profiles along the height of the wall, as the lateral drift θ increases. It is observed that the

bending moment grows up until $\theta = 0.40\%$, then reducing as θ progressively attaines the ultimate value corresponding to the ultimate strain of masonry in compression and, equivalently, the ultimate curvature of cross sections (toe crushing failure). The curvature profile is linear only when small drift levels are reached, i.e. until $\theta = 0.09\%$ is attained. As the drift increases, the curvature in the sections closest to the base grows faster than in the other sections and the contribution to the top displacement is concentrated around the base section.

The iterative calculation of bending moments and curvatures is repeated as the curvature of the base section is monotonically increased from zero to the ultimate curvature. This allows a step-by-step computation of the shear force profile (which is uniform according to the boundary conditions), hence deriving the capacity curve of the wall as shear force versus drift. Figure 3 shows the capacity curve of a flexure-dominated URM wall.

3.4 Rocking behaviour

After that the peak compressive strain (ε_p) is reached over the base section, any further increase in curvature is assumed to induce a rigid-body rotation of the wall that is evaluated according to Equation 5:

$$\theta_r = (\phi_i - \phi(\epsilon_P))\Delta z \qquad (5)$$

where: ϕ_i is the curvature of the base section in the i-th analysis step, $\phi(\varepsilon_p)$ is the curvature corresponding to the peak compressive strain of masonry and Δz is the distance between the base section and the consecutive upper section. The value of rocking rotation θ_r is

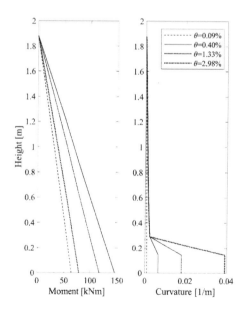

Figure 2. Moment and curvature profiles under increasing drift.

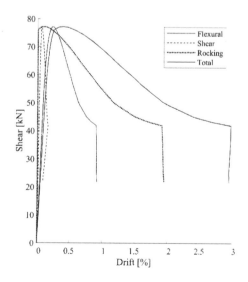

Figure 3. Shear–drift curve of flexure-dominated URM wall with decomposition of total response into its flexural, shear and rocking components.

significantly influenced by the discretisation along the height of the wall. The top displacement contribution due to rocking is assessed by means of Equation 4.

Figure 3 Highlights the notable impact of rocking behaviour on the post-peak branch of the capacity curve, resulting in very large displacement capacity.

3.5 Shear behaviour

As nonlinear analysis proceeds, the secant elastic modulus of masonry reduces according to the adopted stress–strain equations. According to a linear curvature diagram in each cross section, the axial strains in masonry change from a fibre to another, resulting into a non-uniform variation in the secant elastic modulus. Accordingly, an average value of secant elastic modulus is computed as follows:

$$E_{avg} = \sum_A \frac{\sigma_{i,j}}{\epsilon_{i,j}} / n \qquad (6)$$

where $\sigma_{i,j}$ is the normal stress of the fibre (i,j), $\epsilon_{i,j}$ is the corresponding axial strain, n is the number of fibres and A is the gross area of cross section.

The proposed capacity model considers that, in each analysis step, the ratio between E_{avg} and 'average' shear modulus is constant according to Equation 7:

$$G_{avg} = 0.3E_{avg} \qquad (7)$$

that is in agreement with values of secant moduli recommended by the Commentary to the Italian Building Code (IMIT 2019).

Moreover, as the lateral deformation of the wall increases, the cross sections closest to the base experience increasing cracking so the effective length (L_{eff}) of those cross sections decreases. As a result, the shear strain at any height z is calculated through Equation 8:

$$\gamma(z) = \chi \frac{V}{G_{avg} L_{eff} T} \qquad (8)$$

where $\chi = 1.2$ is the shear factor and T is the wall thickness.

The shear contribution to the top displacement of the wall can be assessed by integrating the shear strain along the height of the wall by means of Equation 3.

If the shear resisting force corresponding to either sliding shear or diagonal tension cracking is lower than that associated with the peak resisting moment at the base, the wall suffers a shear failure and the post-peak behaviour is different from that dominated by flexure (Parisi & Augenti 2013). In the proposed method, a horizontal plateau is considered after the attainment of shear strength.

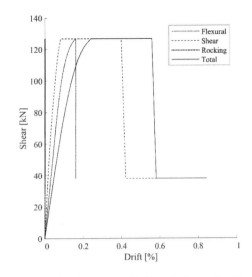

Figure 4. Drift Shear curve with flexural, shear and rocking contribution for a wall dominated by flexural behaviour.

Denoting by δ_e the horizontal displacement associated with the attainment of shear strength, the shear resisting force of the wall is kept constant until the displacement δ_p provided by Equation 9:

$$\delta_p = \mu_{Vmax} \delta_e \qquad (9)$$

where μ_{Vmax} is a displacement ductility factor related to the peak shear strength. Then, a residual shear strength is assumed with a second plateau ending at an ultimate displacement defined by Equation 10:

$$\delta_u = \mu_{Vr} \delta_e \qquad (10)$$

where μ_{Vr} is a displacement ductility factor related to the residual shear strength.

The shear strength of the wall is defined by means of local failure criteria. Namely, diagonal tension cracking is assumed to occur when the principal tensile stress at the centroid of the panel reaches tensile strength of masonry, according to Turnšek & Čačovič (1971). The dimensionless lateral strength corresponding to diagonal tension cracking is defined as follows:

$$\bar{V}_t = \beta \sqrt{1 + \frac{\bar{N}}{p\beta}} \qquad (11)$$

where: β is the ratio between diagonal tensile strength at zero axial load (τ_{c0}) and uniaxial compressive strength of masonry (f_m); and p is a shear stress distribution factor defined as the ratio between the maximum and the average shear stresses (τ_{max} and $\tau_n = V/A$, respectively). The factor p is typically set as follows: $p = 1$ in the case of walls with aspect ratio $H/L \leq 1$; $p = 1.5$ if $H/L \geq 1.5$; and $p = H/L$ if $1 < H/L < 1.5$ (Parisi & Augenti 2013).

Sliding shear strength is predicted through the Mohr-Coulomb failure model, assuming that relative sliding occurs when the maximum shear stress at the mid cross-section of the panel reaches the sliding shear strength. The dimensionless lateral strength corresponding to sliding shear is defined as follows:

$$\bar{V}_a = \frac{1}{p}(\gamma + \mu_a \bar{N}) \qquad (12)$$

where: γ is the ratio between sliding shear strength at zero axial load (f_{v0}) and uniaxial compressive strength of masonry (f_m); and μ_a is the friction coefficient. In the case of bed-joint sliding, both cohesive and frictional strengths have 'local' meaning, as they are related to shear response of mortar bed joints. The friction coefficient reduces as the average normal stress increases, as follows:

$$\mu_a = \frac{0.17}{\sqrt[3]{(\sigma_n/f_m)^2}} \qquad (13)$$

In the case of stair-stepped diagonal sliding, both masonry units and mortar joints are involved so the friction coefficient is regarded as a property of the whole masonry. For this failure mode, past experimental tests have shown that friction coefficient falls in the range [0.3,0.8]. Both Eurocode 6 (CEN 2005) and the Italian Building Code (IMIT, 2018) provide $\mu_a = 0.4$.

4 EXPERIMENTAL VALIDATION

Numerical results are here compared to those of an experimental campaign. Four wall specimens made of tuff stone masonry were tested under constant axial load and cyclically increasing in-plane lateral displacement.

Figure 5 Show the test setup adopted for each experimental test.

Table 1 shows the dimensions, aspect ratio and axial load ratio N/N_u for each specimen.

The direct output of those tests consisted in cyclic shear–displacement curves, allowing the derivation of envelope diagrams in both orientations of lateral loading. For example, Figure 6 shows the cyclic and envelope curves of specimen S1, which was significantly dominated by bending deformations and rocking behaviour. In order to validate the analytical capacity model, each experimental test was numerically simulated. The constitutive model adopted for

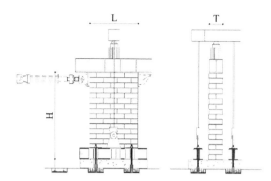

Figure 5. Setup of in-plane shear-compression tests.

Table 1. Specimens dimensions and axial load levels.

Specimen	L [m]	H [m]	T [m]	H/L	N/N_u
S1	1.08	1.88	0.31	1.74	0.1
S2	1.08	1.88	0.31	1.74	0.3
S3	1.5	1.88	0.31	1.25	0.1
S4	1.5	1.88	0.31	1.25	0.3

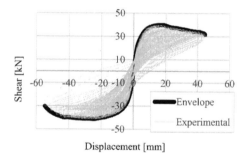

Figure 6. Cyclic and envelope shear–displacement curves of specimen S1.

masonry in compression was that proposed by Augenti & Parisi (2010), which was discretised in 60 points. The peak compressive strength and the corresponding strain were set to 3.96 MPa and $2.44 \cdot 10^{-3}$, respectively. The ratio between peak tensile strength and the peak compressive strength was assumed to be $f_t/f_m = 0.05$, whereas the tensile fracture energy was set to $G_{ft} = 0.015$ N/mm. Both the diagonal tension shear strength at zero axial stress f_{v0} and the sliding shear strength at zero axial stress τ_{c0} were set to $0.0225f_m$. The displacement ductility factors μ_{Vmax} and μ_{Vr} were set to 2.25 and 3.38, respectively. The residual shear force was supposed to be $V_r = 0.3V_{max}$.

Figure 7 shows both numerical and experimental shear–drift diagrams for the four tested specimens, which behaved differently from each other.

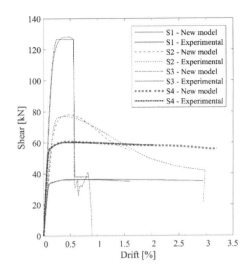

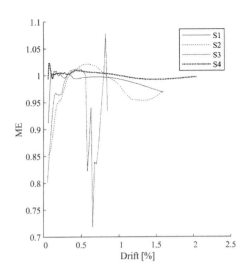

Figure 7. Comparison between analytical and experimental shear–drift curves.

Figure 8. Model error versus drift for each specimen.

Specimens S1and S2 had an aspect ratio greater than 1.5, so they can be considered slender walls. The only difference between those specimens is the applied axial load on top, which was $N/N_u = 0.1$ in the case of specimen S1 and $N/N_u = 0.3$ in the case of specimen S2. The lowest level of axial load produced the minimum force capacity and the largest drift capacity. A 20% increase in axial load caused a higher force capacity, but a significant post-peak strength degradation with a significant reduction in displacement pseudo-ductility of the wall. Nevertheless, both specimens S1 and S2 were flexure-dominated walls.

Specimens S3 and S4 had an aspect ratio lower than 1.5, so they can be considered squat walls. Again, the only difference between those specimens was the axial load ratio. Specimen S3, which was subjected to the minimum level of axial load, was governed by shear failure, as evidenced by the stair-stepped post-peak behaviour.

5 EVALUATION OF MODEL ERROR

In order to use the proposed capacity model in probabilistic procedures for seismic safety assessment of URM buildings, the model (statistical) error was estimated. For each calculated drift level of the capacity curve, the model error was computed as the ratio between the experimental force capacity and that estimated by the proposed method, according to Equation 14:

$$ME = \frac{V_{exp}}{V_{num}} \qquad (14)$$

Figure 8 Shows the model error for each of the four tested walls. In all cases, the capacity model produces an overestimation of the shear force at very

Table 2. Model error for each specimen and over all specimens.

Specimen(s)	Failure mode	μ	σ	CoV
S1	flexural	0.991	0.018	2%
S2	flexural	0.965	0.039	4%
S3	shear	0.948	0.069	7%
S4	flexural	1.001	0.006	1%
All	–	0.988	0.036	4%

small drifts. Nonetheless, after the elastic phase, the model error oscillates around unity, except for specimen S3 that experienced shear failure with significant variations in the residual strength stage. In all simulations,the model error fell in the interval [0.7, 1.1].

To account for uncertainty associated with the capacity model in future probabilistic simulations, the authors calculated the mean value (μ), standard deviation (σ), and coefficient of variation (CoV) of the model error, both for each specimen and over all specimens. The statistics are thus outlined in Table 2, where the failure mode of each wall is also indicated. If a single probability distribution is used for the model error regardless of the failure mode, the statistics show that the proposed capacity model leads to a mean overestimation of the in-plane force capacity equal to 1.2% with a dispersion of 4%.

6 CONCLUSIONS

In this paper, a novel distributed plasticity macro-element has been proposed for its use in incremental nonlinear static (pushover) analysis of URM buildings. The analytical model allows estimating the in-plane lateral load–displacement response of URM walls. The proposed model follows a force-based fibre approach that is integrated with the Timoshenko beam theory for consideration of shear deformations. The model also takes into account the geometric and mechanical nonlinearities, the latter associated with both compressive and tensile behaviour of masonry. The drift capacity is estimated taking into account flexural, rocking and shear contribution to the displacement response of the wall.

Flexural behaviour is simulated through numerical integration of moment–curvature diagrams along the height of the wall. Rocking contribution is considered by adding a further displacement due to a pseudo-rigid rotation of the wall, which is caused by progressive toe crushing. The lateral resisting force corresponding to shear strength is predicted using local strength criteria. The shear stiffness is updated in every analysis step to account for flexural cracking of cross sections and gradual reduction in secant shear modulus as secant Young's modulus reduces under increasing inelastic strains.

The capacity model was experimentally validated, considering the envelope curves of four cantilevered wall specimens that were cyclically tested with displacement control. A robust assessment was performed because those specimens were subjected to two different levels of axial load and two different aspect ratios. The proposed method was able to simulate the experimental load–drift curves. The numerical-experimental comparison allowed the authors to perform a statistical characterisation of the model error, which was found to be very low and suitable for future probabilistic simulations.

Further developments of this model may concern a more complete formulation, which accounts for cyclic loading and explicit integration of inelastic shear strains in the post-peak stage.

The proposed model will be incorporated in equivalent frame models and analytical procedures for pushover analysis of masonry buildings.

ACKNOWLEDGEMENTS

This study was carried out in the framework of DPC-ReLUIS 2019 research project, which was funded by the Italian Civil Protection Department.

REFERENCES

Augenti, N. & Parisi, F. 2019. *Teoria e Tecnica delle Strutture in Muratura.* Milan: Hoepli.

Augenti, N. & Parisi, F. 2010. Constitutive models for tuff masonry under uniaxial compression. *ASCE Journal of Materials in Civil Engineering* 22(11): 1102–1111. DOI: 10.1061/(ASCE)MT.1943-5533.0000119.

Caporale, A., Parisi, F., Asprone, D., Luciano, R. & Prota, A. 2014. Critical surfaces for adobe masonry: micromechanical approach. *Composites Part B: Engineering* 56: 790–796. DOI: 10.1016/j.compositesb.2013.08.087.

Comité Européen de Normalisation (CEN) 2005. *Eurocode 6: Design of masonry structures - Part 1-1:General rules for reinforced and unreinforced masonry structures, EN 1996-1-1.* Brussels.

DeJong, M.J. & Dimitrakopoulos, E.G. 2014. Dynamically equivalent rocking structures. *Earthquake Engineering and Structural Dynamics* 43(10): 1543–1563. DOI: 10.1002/eqe.2410.

Italian Ministry of Infrastructures and Transportation (IMIT) 2018. *D.M. 17. 01.2018: Norme Tecniche per le Costruzioni.* Rome.

Italian Ministry of Infrastructures and Transportation (IMIT) 2019. Circolare 21 gennaio 2019, n. 7 C.S.LL. PP.: Istruzioni per l'applicazione dell'«Aggiornamento delle "Norme tecniche per le costruzioni"» di cui al decreto ministeriale 17 gennaio 2018. Rome.

Lotfi, H.R. & Shing, P.B. 1991. An appraisal of smeared crack models for masonry shear wall analysis. *Computers and Structures* 41(3): 413–425. DOI: 10.1016/0045-7949(91)90134-8.

Lourenço, P.B. 1996. Computational strategies for masonry structures. PhD thesis, Delft: Delft University Press.

Parisi, F. 2010. *Non-linear seismic analysis of masonry buildings.* PhD Thesis, Naples: University of Naples Federico II.

Parisi, F. & Augenti, N. 2013. Seismic capacity of irregular unreinforced masonry walls with openings. *Earthquake Engineering and Structural Dynamics* 42(1): 101–121. DOI: 10.1002/eqe.2195.

Parisi, F., Balestrieri, C. & Asprone, D. 2016. Nonlinear micromechanical model for tuff stone masonry: Experimental validation and performance limit states. *Construction and Building Materials* 105: 165–175. DOI: 10.1016/j.conbuildmat.2015.12.078.

Parisi, F., Balestrieri, C. & Varum, H. 2019. Nonlinear finite element model for traditional adobe masonry. *Construction and Building Materials* 223: 450–462. DOI: 10.1016/j.conbuildmat.2019.07.001.

Parisi, F., Sabella, G. & Augenti, N. 2016. Constitutive model selection for URM cross sections based on best-fit analytical moment–curvature diagrams. *Engineering Structures* 111: 451–466. DOI: 10.1016/j.engstruct.2015.12.036.

Psycharis, I.N., Lemos, J.V., Papastamatiou, D.Y. & Zambas, C. 2003. Numerical study of the seismic behaviour of a part of the Parthenon Pronaos. *Earthquake Engineering and Structural Dynamics* 32: 2063–2084. DOI: 10.1002/eqe.315.

Turnsek, V. & Čačovič, F. 1970. Some experimental results on the strength of brick masonry walls. In: *Proc. 2nd International Brick & Block Masonry Conference.* Stoke-on-Trent.

Zucchini, A. & Lourenço, P.B. 2009. A micro-mechanical homogenisation model for masonry: Application to shear walls. *International Journal of Solids and Structures* 46(3-4): 871–886. DOI: 10.1016/j.ijsolstr.2008.09.034.

Brick and Block Masonry - From Historical to Sustainable Masonry –
Kubica, Kwiecień & Bednarz (eds)
© 2020 Taylor & Francis Group, London, ISBN 978-0-367-56586-2

Characterization of factors determining the durability of brick masonry

T. Stryszewska & S. Kańka
Cracow University of Technology, Cracow, Poland

ABSTRACT: The paper provides characteristic of the factors affecting the durability of brick masonry, such as changing the texture of mineral material, frost destruction, salinity and the development of microorganisms. Their destructive effect occurs as a result of physical, chemical and biological processes. In actual conditions, these impacts usually have a synergistic effect. The range of influence of the factors can be described as superficial, limited to deterioration of the aesthetic appearance of the building, as well as volumetric, more dangerous as it concerns processes resulting in change to entire volume of the material, and thus also affecting the mechanical properties. The condition of all processes related to the reduced durability of materials is the presence of water. Water is a factor that is a threat in itself, because under its influence the texture of the material changes, which may result in deterioration of mechanical properties. In addition, its presence is a necessary condition in the process of chemical corrosion, resulting in efflorescence and subflorescence salting, resulting in a deterioration in the aesthetics and strength of the material, respectively. Water is also a priority for biological corrosion. The development of both organisms and microorganisms is only possible in a humid environment.

1 INTRODUCTION

The durability of construction objects does not mean that its period of use is unlimited. In the longer time perspective, processes related to changes on the surface of materials and development of defects in the structure of masonry take place. When analysing the technical condition of masonry buildings as well as performing restoration and renovation works, it is significantly important to recognize these processes and assess their effects.

The state of preservation or durability of construction objects is influenced by many factors, such as climate, weather conditions and environmental pollution, type and quality of materials, and the construction method. Preventive measures (or failure to take them) are important as well. Basically, these factors which affect the condition of objects during operation can be divided into three categories: physical, chemical and biological factors. Presence of water is crucial to initiate all of the above mentioned corrosive processes, as this factor poses a threat to the material itself. Under the influence of water, the material texture changes, which may result in the deterioration of mechanical properties. In addition, the presence of water is necessary in the process of chemical corrosion, which leads to efflorescence and sub-florescence and thus, deterioration in the aesthetics and durability of the material, respectively. Water plays as well a key role in biological corrosion. The development of both organisms and micro-organisms is only possible in a humid environment.

Figure 1 presents the factors influencing the durability of walls and the effects of their impact.

Under real conditions, the nature of external effects is usually synergistic. The mechanisms that lead to the destruction of objects intertwine and intensify. Sometimes the effects of one impact initiate the development of another, e.g. higher pH in a substrate (bricks and mortar), as a result of chemical corrosion, initiates and promotes biodeterioration and subsequently, its development. Another example consists in the freezing of groundwater, as a result of which shallow foundations start to move, which leads to a number of scratches and cracks, and the walls' deformation. Therefore, the assessment of the durability of a particular construction object or a group of objects requires a comprehensive approach taking account of all parameters affecting the condition and the state of preservation.

Frequently, the only result of the external environment, i.e. fungi, mould, salt, moisture, frost, is the deterioration of the object aesthetics. However, if the intensity of a given type of impact intensifies over a long period of exposure, this environment may lead to significant degradation of the object, and in extreme cases to a failure or even a construction disaster. The assessment of the state of preservation of the construction, which does not take account of all factors, may be subject to error and lead to wrong assumptions adopted for renovation and restoration works. Therefore, the assessment of construction objects, including historic buildings, should be comprehensive and multidimensional.

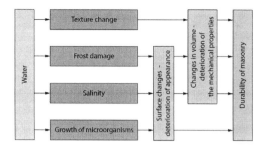

Figure 1. Factors affecting the durability of walls (Stryszewska 2017).

2 THE INFLUENCE OF WATER

The destructive impact of water on masonry is multidimensional. As already mentioned, water is an essential factor in all destructive mechanisms. It might be even stated that it is the greatest "aggressor". Its impact can be either direct or indirect. A direct impact consists in changing the texture of the material under the influence of water, whereas an indirect impact involves water, i.e. a factor that allows and initiates certain corrosive processes, such as frost destruction, chemical and biological corrosion. The source of moisture in real objects includes mainly ground water, rainfall and snowfall, and in the coastal area also the sea breeze. According to the observations and literature reports, masonry buildings, especially historic buildings, due to lack of waterproofing, often get damp over many years of use and in exceptional cases even get reach full saturation point. As a rule, this phenomenon is accompanied by decreased strength. The research on the impact of moisture in bricks, mortars and walls on their mechanical properties presented in the primary sources focus mainly on the relationship between the degree of moisture content in masonry materials, and their strength and deformability when compressed. The first literature reports related thereto come from the beginning of the 20th century and refer to the bricks manufactured in the USA at the turn of the 19th and 20th centuries. Baker (Baker 1909) on the basis of the study of the bricks taken from 10 masonry buildings determined that the strength of wet bricks is about 0.85 of the strength of dry bricks. The results of these studies have been confirmed by other authors (Amade et al. 2004, Foraboschi & Vanin 2014, Matysek & Witkowski 2013, Witzany et al. 2010). Also Matysek & Witkowski (2013) confirmed decreased compressive strength due to water saturation – the reported decreases range from a few to over 20% for ceramic bricks taken from various historical periods. The research conducted recently by the authors of the papers (Matysek et al. 2016, Stryszewska 2014a, Stryszewska & Kańka 2016) made it possible to specify

the reasons for the decreased strength of bricks under the influence of water. The research was conducted for two types of ceramic bricks with very different physical and mechanical properties, i.e. historical and modern bricks. Due to different technological processes and raw materials, those bricks were significantly different in terms of their phase composition and porosity structure. With short-term seasoning in water (1 month), decreased strength of both types of bricks of about 10% was recorded. Long-term seasoning of bricks in water led to a significant decrease in strength only in modern bricks. It was shown that the reason for the strength decrease includes changes in the material texture caused by the impact of water, which include increased porosity as the body elements, such as raw material impurities or salts crystallized during the exposure were washed out (Stryszewska 2014a, Stryszewska & Kańka 2016). The degradation of the silicon-oxygen bridges, which resulted from the fact that the silicon-oxygen bridges (Si-O-) broke due to the prolonged exposure to water, has a significant impact on the strength reduction (Matysek et al. 2016). The breaking of Si-O-Si bridges leads to decreased strength (Mozgawa et al. 1999). As it was observed, this process mainly concerns the amorphous phase, which is the dominant component of bricks (Matysek et al. 2016). Bricks with significant amounts of fly ash may be characterized by a greater susceptibility of the silicon-oxygen bridges to degradation (Stryszewska & Kańka 2016). Also, the aluminosilicate hydrolysis process favors losing brick strength (Awgustynik 1980). Another reason for the decreased strength involves an increase in deformations which accompany the exposure to water (Stryszewska 2014a, Hall & Hoff 2011). As a result of these processes, the structural bonds are loosened and, consequently, also the mechanical properties of bricks are deteriorated. The authors of the paper (Matysek et al. 2016), on the basis of the research conducted to assess the impact of water on the mechanical properties of bricks, introduced a compressive strength reduction factor (η_f) determined as the ratio of strength value in the dry state to the value determined after short or long-term exposure in water. As far as the long-term exposure in water is concerned, it amounted to 0.9 for historical bricks and 0.8 for modern bricks, respectively.

3 FROST DESTRUCTION

Another reason for brick destruction if is the freezing of water in the pores of the material. Taking account of the mechanism of brick destruction as a result of the freezing of water, material factors determining frost resistance can be divided into two groups. The first group includes mechanical properties such as tensile strength and deformation abilities (Maage 1984, Perrin et al. 2011). These parameters determine the material resistance to ice expansion. Greater strength favors increased resistance to stresses that are generated in the process of the freezing of water. The mineral composition of the raw materials applied for the production of bricks as well as the phase composition of the burnt brick seem to be important as well. Products including clay minerals from the kaolinite and illite groups in their original composition are characterized by better frost resistance in comparison to materials the composition of which is dominated by clay minerals from the montmoryllonite group (Bergman & Kurzer 1958). However, according to Elert et al. (2003), the durability of bricks is closely related to the degree of texture vitrification. However, due to the fact that it is difficult to estimate the content of this phase, this feature should be regarded only as a guide and in connection with other parameters such as strength, porosity and the properties of water itself. The second group of factors affecting frost resistance includes physical properties such as permeability and porosity. Permeability allows and conditions the free circulation of water in porous material without damaging the texture. Many researchers create a link between frost resistance of bricks and their porosity, mainly their porosity structure (Maage 1984, Hansen & Kung 1988, Herget et al. 1992, Nieminen & Romu 1988, Wardeh & Perrin 2008). Generally, small and medium pores deteriorate frost resistance, while larger pores improve it (Hansen & Kung 1988). The authors of the paper (Herget et al. 1992) introduced a division of open porosity into pores and spare capillaries with a diameter of 200 μm, the so called safe pores, and pores and capillaries with a diameter of 0.1 to 200 μm, the so called dangerous pores in which the water freezes at 0 to -25°C. The presence of pores with a diameter below 0.1 μm, due to a very low freezing temperature, does not pose a significant threat and has no major impact on the frost resistance of bricks. The form of frost damage of bricks depends on the structure of porosity, mineral composition and strength (Stryszewska & Kańka 2019) as well. Figure 2 shows

the correlation of material features affecting the form of frost damage of the studied bricks. As a parameter describing the porosity structure, the destruction form factor (FFD) was introduced. It is the ratio of the percentage of pores with a diameter of 3-10 μm to the total number of pores with a diameter of 0-10 μm. According to (Maage 1984, Perrin et al. 2011) the total number of pores with a diameter of 0-10 μm, taking account of the number of pores with a diameter of 1-3 μm and 3-10 μm, seems to reflect the impact of the porosity structure on the form of frost damage of bricks in the best way.

Therefore, it seems that the porosity structure is crucial and the most important parameter affecting the form of damage. A clear correlation of the damage form was observed depending on the number of pores with a diameter of 3-10 μm. A parameter that describes the form of frost damage resulting from the porosity structure is the FFD factor (Figure 2), which for bricks with no signs of frost damage falls within the standard range 75-90. However, as far as flaking bricks are concerned, the FFD factor is 4-10, for cracking bricks 2-7, and for powdering bricks 45-50. A similar value of the FFD factor for cracking and flaking bricks with the same amount of binder and similar strength indicates that the reasons for different forms of damage should be observed in the spatial distribution of texture components.

4 SALINITY

Salt sources in masonry buildings can be divided into two main groups. The first includes internal factors resulting from the properties of raw materials and the production process. Harmful substances that cause damage to bricks in their lifetime include gypsum, pyrite, marl introduced into the mass in the form of natural raw material and a weakening industrial additive. Also, the fuel type used in the firing process, due to the potential sulphur content, can be the source of harmful substances, i.e. sulphur compounds. With regard to mortars, especially including cement, the reaction products between calcium oxide and hydroxide, and carbon dioxide, the product of which is white calcium carbonate are the potential source of salt. The second group consists of external factors, such as salts from the external environment which migrate with water deep into the material, and salts introduced into the wall structure with the use of modern repair materials while performing renovation or restoration works.

Determining the type of salt present in the wall is an important element of wall diagnostics as it makes it possible to determine the sources of moisture in the construction object and significantly facilitates this process. Chlorides usually take the form of halite, sylvin, salmiac and bischofite, nitrates in the form of nitronatrite, nitrocalcite, nitrommagnezite, sulphates in the form of mirabilite, acarnite, epsomite, kizerite,

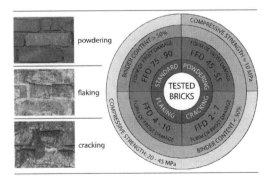

Figure 2. A macroscopic view of damage of the surface of the bricks as a result of powdering, flaking and cracking, and material characteristics having impact on the form of frost damage (Stryszewska & Kańka 2019).

tenardite and gypsum, whereas carbonates often take the form of micrite, thermonatrite and natron (Szostak & Trochonowich 2015). Depending on the structure or degree of hydration, crystals can take various mineralogical forms, e.g. calcium carbonate take the form of calcite, aragonite or micrite.

Figure 3. SE image for chloride and sulphate salts on the surface of bricks taken from existing masonry buildings (from the top).

The presence of salt significantly changes the evaporation process and thus, the drying process. Depending on its type, the drying efficiency, and subsequently, the saturation efficiency, varies significantly. The drying curves and curves showing saturation of bricks with 10% salt solutions are shown in Figure 4.

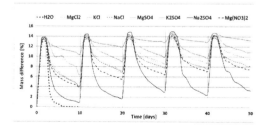

Figure 4. Saturation and drying curves over time.

The research, the results of which were presented above, included 5 ten-day cycles consisting of 2 days of saturation (capillary action) and 8 days of drying under laboratory conditions. For comparison, a cycle of saturation with distilled water was performed analogically. Bricks saturated with distilled water in the process of drying lose their total water content, and their humidity at the end of each cycle is close to zero. It was observed that the saturation and drying process is subject to significant change in the presence of potassium sulphate solution, and to insignificant change in the presence of sodium sulphate. The remaining chloride salts and magnesium sulphate have a comparable impact on the brick drying process, whereas the impact of chloride salts is almost identical, and therefore, does not depend on the cation included in the solution (Stryszewska 2017).

The place of salt crystallization depends on the saturation and drying rate. Under conditions with high moisture, when the evaporation rate is low, the evaporation zone may be close to the surface or even on the surface of the material. The salting on the surface takes the form of a white powder and is defined as efflorescence. Its presence is rather harmless for the material durability, however, it is significant in terms of their aesthetics. Most of salt efflorescence is white, but depending on its composition, conditions of formation and pollution of the external environment, it might turn green, orange, brown and black. However, when the evaporation rate is much higher, the evaporation zone remains inside and the salts crystallize in the pores of the material. This phenomenon is referred to as sub-florescence. The pressure exerted by rapidly crystallizing salts is very high and is sufficient to destroy even very strong masonry materials (Wellman & Wilson 1965). As a result of the crystal growth, the entire texture volume, and not only its surface layers, gets loosened which is dangerous as it destroys the materials not only on the surface, but also inside. The effect of sub-florescence involves crushing, delamination and cracks in the materials and in extreme cases it can clearly destroy the materials (Foraboschi & Vanin 2014, Stryszewska 2014b). Salt crystallization and the pressure related thereto cause internal stresses which eventually lead to the material destruction.

The value of crystallization pressure depends, i.e. on the degree of solution supersaturation, i.e. its concentration, and on the shape and the size of pores (Steigner 2005a, b). The destructive effect of crystallizing salts gets intensified by changes in the volume of salt crystals which crystallize with a varying degree of hydration, e.g. as a result of full hydration of sodium sulphate to $Na_2SO_4 \cdot 10H_2O$, the crystal volume increases by 400% (Domasłowski et al. 2011). A multiple of this process as a result of drying and saturating the material enhances the destructive influence of salt. However, the effects of salt crystallization depend to a great extent on the porosity structure of the material as well as its mechanical properties (Cultrone et al. 2004).

5 DEVELOPMENT OF MICRO-ORGANISMS

One of the reasons of the destruction of construction objects, and thus, the reduction of their durability is the impact of biological factors, the so-called biological corrosion. According to (Perego et al. 1999), it is deterioration, i.e. decomposition of materials caused by the effect of living organisms and their metabolites. Biodeterioration processes include various mechanisms of material destruction, including mechanical destruction, enzymatic degradation, corrosion induced by micro-organisms, and growth of living organisms on the material surface, which is accompanied by the formation of biofilms (Gutarowska 2013). Biological corrosion can be caused by organisms and micro-organisms. The first group includes bryophytes, lichens and algae, while the second group is comprised of fungi and bacteria (Figure 5). Biological corrosion applies to all types of building materials, both organic and inorganic ones such as: wood, plastics, paints, varnishes, bricks, mortar, concrete, glass and metals (Pastuszka et al. 2000). Metabolism-related processes that lead to biodeterioration can be divided into groups, depending on the mechanism of material destruction (Allsopp & Seal 1984, Morton & Surman 1994, Gaylarde & Morton 1999):

☐ mechanical, the material is damaged due to direct effect of organisms,
☐ chemical assimilation, biodeterioration, the material is damaged due to its decomposition for nutritional purposes (applies to organic materials),
☐ chemical dissimilation, biodeterioration, the material is damaged due to the metabolites of microorganisms (applies to organic and inorganic materials),
☐ growth of living organisms on the material surface, formation of biological membranes,
☐ macroscopic symptoms e.g. discolouration, development of organisms on the surface, pitting, cavities, fragility, deformation, structure decomposition,
☐ symptoms of changes in physical properties, e.g. increased water absorption, deterioration of strength and changes in chemical properties, etc.

Bacteria also develop on both types of substrate (Adamiak et al. 2017). It is a group of organisms characterized by a huge variety of metabolic activity, i.e. the ability to live in various, even extreme conditions. The growth of bacteria depends on the presence of water dissolving nutrients which provide them with energy and synthesize the materials included in the cell (Schlegel 2003).

6 CONCLUSION

Basic external factors that lead to the destruction of walls include physical effects resulting from the freezing of water, chemical effects associated with the presence of salt, and biological effects. Each of these destruction mechanisms requires the presence of water, which also deteriorates the characteristics of masonry walls. Freezing water causes the increase in volume which in turn leads to the destruction of the material. In addition, water can damage the internal texture of ceramic materials, reducing their mechanical properties. The presence of water is also necessary in the chemical corrosion associated with the crystallization of salt. Migrating water with the dissolved salts moves in the material and, under favorable conditions, evaporates. This process leads to crystallization of the salt. These salts increase their volume and this causes destruction of the material. The least hazardous are biological corrosion processes, which primarily cause deterioration of aesthetics parameters, less often deterioration of mechanical propertiesThe durability of construction objects is significantly affected by operating conditions, which include climatic factors, possibility of moisture or freezing, exposure to cyclic freeze-thaw actions, and the presence of chemical substances.

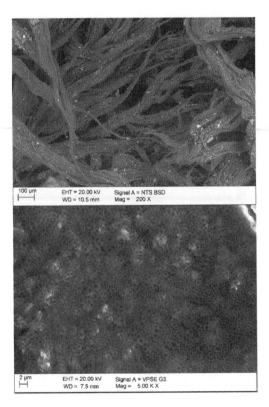

Figure 5. SEM image of moss (mag. 200x) and bacteria (mag. 5000x) on the ceramic substrate (from the top).

REFERENCES

Adamiak J. et al. 2017. Untargeted metabolomics approach in halophiles: understanding the biodeterioration process of building materials. *Frontiers in Microbiology* 11 December 2017.

Allsopp D. & Seal K.J. 1986. *Introduction to Biodeterioration. Edward Arnold, London.*

Amade A.M. et al. 2004. The effect of moisture on compressive strength and modulus of brick masonry. *Proc. 13th International Brick and Block Masonry Conference, Amsterdam 2004.*

Awgustinik A.J. 1980. *Ceramika*. Warszawa: Arkady.

Baker IO. 1909. *A treatise on masonry construction.* New York: John Wiley & Sons Inc.

Bergman K. Kurzer. 1958. Uberblick uber den derzeitingen stand der frostforschung. *Ziegelindustrie* 4/1958.

Cultrone G. et al. 2004. Influence of mineralogy and firing temperature on the porosity of bricks. *Journal of the European Ceramic Society* 24: 547–564.

Elert K. et al. 2003. Durability of bricks used in the conservation of historic buildings—influence of composition and microstructure. *Journal of Cultural Heritage* 4: 91–99.

Foraboschi P. & Vanin A. 2014. Experimental investigation on bricks from historical Venetian buildings subjected to moisture and salt crystallization. *Eng Fail Anal.* 45: 185–203.

Gaylarde C.C. & Morton L.H.G. 1999. Deteriogenic biofilms on buildings and their control: a review. *Biofouling* 14: 59–74.

Gutarowska B. 2013. Niszczenie materiałów technicznych przez drobnoustroje. *LAB Laboratoria, Aparatura, Badania* 18(2):10–14.

Hall Ch. & Hoff W.D. 2011. *Water transport in brick, stone and concrete. 2nd edition.* Spon Press.

Hansen W. & Kung JH. 1988. Pore structure and frost durability of clay bricks. *Matér Constr/Mater Struct.* 21 (126): 443–447.

Herget F.A. et al. 1992. Variability within single projects of physical properties of face brick as related to potential durability. *Proceedings of 6th Canadian Masonry Conference, vol.2 Canada S7N 0W0*: 417–428.

Maage M. 1984. Frost resistance and pore size distribution in bricks. *Matér Constr/Mater Struct* 17(101): 345–350.

Matysek P. et al. 2016. The influence of water saturation on mechanical properties of ceramic bricks – tests on 19th-century and contemporary bricks. *Materiales de Construcción* 66(323).

Matysek P. & Witkowski M. 2013. Ocena wpływu wilgotności na wytrzymałość murów ceglanych. *Materiały Budowlane* 5: 10–12.

Morton L.H.G. & Surman S.B. 1994. Biofilms in Biodeterioration – a review. *International Biodeterioration 35. and Biodegradation* 34 3/4: 203–221.

Mozgawa W. et al. 1999. Spectroscopic studies of different aluminosilicate structures. *Journal of Molecular Structure* 251: 511–512.

Nieminen P. & Romu P. 1988. Porosity and frost resistance of clay bricks. In de Cordy J.W. (ed.): *Brick and block masonry. Vol.1.* London: Elsevier Applied Science: 103–109.

Pastuszka J. S. et al. 2000. Bacterial and fungal aerosol in indor environmental in Upper Silesia. Poland. *Atmospheric Environmental* 34 (22): 3833–3842.

Perego P. & Fabiano B. 1999. Corrosion, microbial. In: Flickinger M. C. & Drew S. W. (eds): *Encyclopedia of Bioprocess Technology: Fermentation, Biocatalysis and Bioseparation.* New York: John Wiley & Sons, Inc.

Perrin B. et al. 2011. Mechanical behaviour of fired clay materials subjected to freeze–thaw cycles. *Construction and Building Materials* 25: 1056–1064.

Schlegel H.G. 2003. *Mikrobiologia ogólna.* Warszawa: Wydawnictwo Naukowe PWN.

Steiger M. 2005a. Crystal growth in porous materials—I: The crystallization pressure of large crystals. *J. Cryst. Growth.* 282: 455–469.

Steiger M. 2005b. Crystal growth in porous materials. II: Influence of crystal size on the crystallization pressure. *J. Cryst. Growth.* 282: 470–481.

Stryszewska T. 2014a. Wpływ pełnego nasycenia wodą na wybrane właściwości fizyko-mechaniczne cegły ceramicznej. *Materiały Ceramiczne* 1: 81–87.

Stryszewska T. 2014b. The change in selected properties of ceramic materials obtained from ceramic brick treated by the sulphate and chloride ions. *Construction and Building Materials* 66: 268–274.

Stryszewska T. 2017. Czynniki determinujące trwałość murów ceglanych. Cracow University of Technology.

Stryszewska T. & Kańka S. 2016. Influence of exposed in water on reduce compressive strength of brick with addition fly ash. *Ochrona przed korozją* 6: 206–209.

Stryszewska T. & Kańka S. 2019. Forms of damage of bricks subjected to cyclic freezing and thawing in actual conditions. *Materials* 12.

Szostak B. & Trochonowicz M. 2015. Analysis of construction salt decomposition within walls of ceramic brick in the midtown tenement houses. *Civil and environmental engineering reports* 4(19): 111–120.

Wardeh G. & Perrin B. 2008. Freezing–thawing phenomena in fired clay materials and consequences on their durability. *Construction and Building Materials* 22: 820–828.

Wellman H.W. & Wilson A.T. 1965. Salt weathering, a neglected geological erosive agent in coastal and arid environmental. *Nature* 205: 1097–1098.

Witzany J. et al. 2010. The effect of moisture on significant mechanical characteristics of masonry. *Eng Struct Technol.* 2: 79–85.

Domasłowski W. (ed.) 2011. *Zabytki kamienne i metalowe, ich niszczenie i konserwacja profilaktyczna.* Toruń.

Brick and Block Masonry - From Historical to Sustainable Masonry –
Kubica, Kwiecień & Bednarz (eds)
© 2020 Taylor & Francis Group, London, ISBN 978-0-367-56586-2

Peridynamic modelling of masonry structures

N. Sau, A. Borbón-Almada, A. López-Higuera & J. Medina-Mendoza
University of Sonora, Hermosillo, Mexico

ABSTRACT: In many cases, masonry structures are subjected to loads that form cracks. Elasticity and damage in quasibrittle structures such as plain and reinforced concrete structures can be modeled with the peridynamic theory, proposed by Stewart Silling in 2000 and 2007. To model these structures, lattice hybrid models with brittle beam elements where used to model concrete, where cracks were expected to appear. One of the problems with lattice models is that they fail to adequately describe compressive behavior of quasibrittle materials. Another shortcoming of lattice and particle models is that they are highly demanding of computational power. Molecular dynamics may be, in some cases an appropriate tool for analyzing microcracks in quasibrittle materials in compression, but molecular dynamics becomes infeasible at scales larger than a few million atoms. In most cases, cracks form in the brick mortar joints, and concrete blocks or bricks can be assumed to have a uniform displacement field. This allows us to use the peridynamic finite element model, which is an improvement over discrete lattice models. This model assumes a continuous displacement field within each finite element, with displacement discontinuities allowed to develop between finite elements. The objective of this work is to model cracks in masonry walls with the peridynamic model. The peridynamic finite element model is shown to be much more computer time- and memory-efficient than the similar discrete particle-based models. Results show that this implementation appears to be more computationally efficient than particle or lattice models.

1   INTRODUCTION

For some cases, masonry structures are subjected to loads that form cracks, and highly nonlinear behavior, where models based on continuum mechanics are no longer applicable. To represent cracks in the material, models based upon Fracture Mechanics concepts were formulated, where the main disadvantage of these models is that they are impractical, since multiple cracks need to be represented in the material. The use of smeared crack models using continuum damage theories appeared to be a more suitable technique to model damage in concrete structures. Nevertheless, spurious mesh sensitivity was one of the many problems associated with the smeared crack approach (ACI 1997). Elasticity and damage in quasibrittle structures such as plain and reinforced concrete structures can be modeled with the peridynamic theory, proposed by Stewart Silling in 2000 and 2007 where lattice hybrid models with brittle beam elements where used to model concrete, where cracks were expected to appear (Gerstle et al. 2007). One of the problems with lattice models is that they fail to adequately describe compressive behavior of quasibrittle materials. Another shortcoming of these models is that they are highly demanding of computational power. Molecular dynamics may be, in some cases an appropriate tool for analyzing microcracks in quasibrittle materials in

compression, but molecular dynamics becomes infeasible at scales larger than a few million atoms.

One of the advantages of the peridynamic model is that governing equations in the peridynamic model do not assume spatial differentiability of the displacement field and permit discontinuities to arise as part of the solution. In addition, the peridynamic theory is non-local, where locality is recovered as a special case (Silling 2000). The micropolar peridynamic model, which is a bond-based model, improves the peridynamic model proposed by Silling in 2000, by allowing moment densities in addition to force densities to interact among particles inside the material horizon. The micropolar peridynamic model is able to model materials with Poisson's ratios different from 1/4 in three-dimensions and 1/3 for two-dimensional plane stress problems; in addition, particles are capable of having rotational as well as translational degrees of freedom (Sau 2008). The peridynamic model starts with the assumption that Newton's second law is true on every infinitesimally small particle within the domain of analysis. A specified force density function, called the pairwise force function (with units of force per unit volume per unit volume) between each pair of particles is postulated to act if the particles are closer together than some specified finite distance, called the material horizon . Within this material horizonδ, the pairwise force function may be assumed a function of the relative position and the relative displacement between the particles. Particles further apart than the material horizon do not interact with each other. The simplest peridynamic damage model, it is assumed that for two particles closer together than the material horizon the pairwise force, f, increases linearly with respect to the stretch, s, but at some stretch, s_o, the pairwise force function is reduced to zero.

In most cases, cracks form in the brick mortar joints, and concrete blocks or bricks can be assumed to have a uniform displacement field. This allows us to use the peridynamic finite element model, which is an improvement over discrete lattice models, where implicit solution schemes and single-processor can be used in order to limit the number of equations that need to be simultaneously solved. By using the displacement-based finite element method, in which a kinematic assumption regarding the displacement field is assumed within each finite element. Use of finite elements also provides the advantage of technology for applying boundary conditions such as work equivalent nodal loads. The use of displacement-based finite elements does not prohibit the use of zero-dimensional elements or rigid in those portions of the problem where high gradients and discontinuities in the displacement field due to damage and crack development. In concrete blocks or masonry bricks, As long as a portion of the domain remains within the strain-hardening regime of peridynamic material behavior, the displacement field within that portion of the domain will be continuous. In those portions of the domain where strain-softening damage is occurring, especially at the brick mortar joints, zero-dimensional finite elements or rigid particles can be used to avoid making an incorrect assumption of continuity of the displacement field. Also, one-dimensional elements can be used in parts of the domain. These one-dimensional elements work as one dimensional bars or beams, in order to simulate reinforced concrete masonry.

2 PERIDYNAMIC MODEL

The peridynamic model postulates a vector function in units of force per unit volume squared fi j that represents the interaction between particles i and j inside a material horizon δ (Silling 2000). The sum of all internal forces per unit volume acting on a particle i is expressed as:

$$\int_{V_J} f_{ij} dV_j + b_i = \rho_i \frac{\partial^2 u_i}{\partial t^2}, \qquad (1)$$

where b_i is the body force acting on particle i; $\frac{\partial^2 u_i}{\partial t^2}$ and ρ_i are the acceleration and density of particle i respectively.

This integral is performed on all particles j within δ. In the micropolar peridynamic model a moment equation must be added (Sau 2008):

$$\int_{V_J} m_{ij} dV_j + m_i = I_i \frac{\partial^2 \theta_i}{\partial t^2}, \qquad (2)$$

where the moment per unit volume squared is m_{ij} and m_i is a moment density function (Gerstle 2007).

These moments and forces can be functions of the relative positions ξ_{ij}, the relative displacements η_{ij}, and the rotations of particles θ_i and θ_j, where forces f_{ij} and moments m_{ij} between particles can be expressed using the following matrix (Equation 3) at the linear microelastic level as:

$$
\begin{bmatrix} \hat{f}_x^i \\ \hat{f}_y^i \\ \hat{m}_z^i \\ \hat{f}_x^j \\ \hat{f}_y^j \\ \hat{m}_z^j \end{bmatrix} =
\begin{bmatrix}
c/L & 0 & 0 & -c/L & 0 & 0 \\
0 & 12d/L^3 & 6d/L^2 & 0 & -12d/L^3 & 6d/L^2 \\
0 & 6d/L^2 & 4d/L & 0 & -6d/L^2 & 2d/L \\
-c/L & 0 & 0 & c/L & 0 & 0 \\
0 & -12d/L^3 & -6d/L^2 & 0 & 12d/L^3 & -6d/L^2 \\
0 & 6d/L^2 & 2d/L & 0 & -6d/L^2 & 4d/L
\end{bmatrix}
\begin{bmatrix} \hat{u}_i \\ \hat{v}_i \\ \hat{\theta}_i \\ \hat{u}_j \\ \hat{v}_j \\ \hat{\theta}_j \end{bmatrix}
$$
$$(3)$$

The parameters c and d are given by $c = E'A$ and $d = E'I'$ respectively, where the cross-sectional area of the link between particles is A'. The variable E' is the modulus of elasticity, I' is the moment of inertia and L is its length (Figure 1). The relative displacements in the x and the y direction are $\eta_{ij}^x = \hat{u}^i - \hat{u}^j$ and $\eta_{ij}^y = \hat{v}^i - \hat{v}^j$ respectively. Similarly, links in a tree-dimensional model have a circular cross-sectional area with a moment of inertia I', polar moment of inertia $2I'$ and an area A', all these links have a finite length with an infinite number of particles j, attached to each particle i (Sau 2008).

Relationships between peridynamic constants (c, d) and conventional linear elastic constants (E, ν) can be obtained by calculating the strain energy density of all the peridynamic links attached a particle inside a material horizon δ. These relationships are given by:

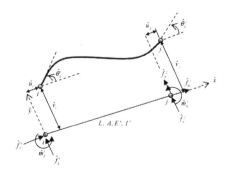

Figure 1. Peridynamic link in the 2-D linear microelastic model in local coordinates.

$$c = \frac{6E}{\pi\delta^3 t(1-\nu)}, \quad d = \frac{E(1-3\nu)}{6\pi\delta t(1-\nu^2)}, \quad (4)$$

for two-dimensional plane stress problems and

$$c = \frac{6E}{\pi\delta^4(1-2\nu)}, \quad d = \frac{E(1-4\nu)}{4\pi\delta^2(1-2\nu)(1+\nu)}, \quad (5)$$

for three-dimensional models (Sau 2008).

Using an energy approach, expressions for the finite element model are developed. This numerical implementation of the micropolar peridynamic model is limited to 2D problems and in-plane loads. Triangular, quadrilateral, one-dimensional, and zero-dimensional elements are included in the model.

The linear portion of the microelastic model in written as:

$$\{df_{ij}\} = [k_{ij}]\{d_{ij}\}dV_i\,dV_j, \quad (6)$$

and the differential strain energy between particles i and j is given by:

$$dU_{ij} = \frac{1}{2}[d_{ij}]\{f_{ij}\} = \frac{1}{2}[d_{ij}][k_{ij}]\{d_{ij}\}dV_idV_j, \quad (7)$$

The total accumulated strain energy within the domain is given by:

$$U = \frac{1}{2}\int_R\int_R\frac{1}{2}[d_{ij}][k_{ij}]\{d_{ij}\}dV_idV_j \quad (8)$$

and partitioning Equation 8 into *num_el* finite elements within the domain, the following expression is obtained:

$$U = \frac{1}{4}\sum_{elemi=1}^{num\_el}\left(\int_{elemi}\left(\sum_{elemj=1}^{num\_el}\left(\int_{elemj}[d_{ij}][k_{ij}]\{d_{ij}\}dV_j\right)\right)dV_i\right) \quad (9)$$

Equation (9) is partitioned as a summation of the strain energy stored in peridynamic links connecting pairs of finite elements i and j, and the strain energy stored in peridynamic links contained within single elements, thus:

$$U = \sum_{eli=1}^{n}\left(\sum_{elj=1}^{eli-1}U_{ij}\right) + \sum_{eli=1}^{n}(U_{ii}). \quad (10)$$

The displacement vector at any point within elementi, $\{d_i\}$, is interpolated from the element nodal displacements $\{D_i\}$ using interpolation functions:

$$\{d_i\} = [N^i]\{D_i\} \quad (11)$$

Similarly, the displacement vector at any point within element j, $\{d_j\}$ is interpolated from element j's element nodal displacements $\{D_j\}$.

$$\{d_j\} = [N^j]\{D_j\}, \quad (12)$$

and for the pair of elements i and j:

$$\{d_{ij}\} = \left\{\begin{array}{c}\{d_i\}\\\{d_j\}\end{array}\right\} = \left\{\begin{array}{c}[N^i]\{D_i\}\\{[N^j]\{D_j\}}\end{array}\right\}$$
$$= \left[\begin{array}{cc}[N^i] & [0]\\{[0]} & [N^j]\end{array}\right]\left\{\begin{array}{c}\{D_i\}\\\{D_j\}\end{array}\right\} = [N^{ij}]\{D_{ij}\} \quad (13)$$

Therefore, Equation 9 can be rewritten as:

$$U = [D_{ij}]\left[\frac{1}{4}\sum_{elemi=1}^{num\_el}\left(\int_{elemi}\left(\sum_{elemj=1}^{num\_el}\right.\right.\right.$$
$$\left(\int_R[N^{ij}]^T[k_{ij}][N^{ij}]dV_j\right)\right)dV_i)]\{D_{ij}\} \quad (14)$$

The strain energy stored within the entire domain R is given by:

$$U = \frac{1}{2}[D][K]\{D\}, \quad (15)$$

where $\{D\}$ is the vector of all nodal displacements, and the matrix $[K]$ is the global stiffness matrix. Comparing Equations (14) and (15), and recognizing that the equality must hold for all possible nodal displacement vectors $\{D\}$, the matrix $[K]$ becomes:

$$[K] = \frac{1}{2}\sum_{elemi=1}^{num\_el}\left(\int_{elemi}\left(\sum_{elemj=1}^{num\_el}\left(\int_R[N^{ij}]^T[k_{ij}][N^{ij}]dV_j\right)\right)dV_i\right) \quad (16)$$

167

The global stiffness matrix $[K]$ is assembled from the stiffness contributions $[K_{ij}]$ from peridynamic links between elements i and j, and from stiffness contributions $[K_{ii}]$ resulting from peridynamic links within elements i:

$$[K] = \sum_{eli=1}^{num\_el} \left(\sum_{elj=1}^{eli-1} [K_{ij}] \right) + \sum_{eli=1}^{n} [K_{ii}], \quad (17)$$

where:

$$[K_{ij}] = \int_{elemi} \left(\int_{elemj} [N^{ij}]^T [k_{ij}] [N^{ij}] dV_j \right) dV_i, \quad (18)$$

and:

$$[K_{ii}] = \frac{1}{2} \int_{elemi} \left(\int_{elemi} [N^{ii}]^T [k_{ij}] [N^{ii}] dV_j \right) dV_i, \quad (19)$$

Equations (18) and (19) are approximately numerically integrated by partitioning finite element i into m_i sub-elements of volume ΔV_i and partitioning finite element j into m_j subelements of volume ΔV_j, thus:

$$[K_{ij}] = \sum_{i=1}^{m_i} \left(\sum_{j=1}^{m_j} \left([N^{ij}]^T [k_{ij}] [N^{ij}] \Delta V_j \right) \Delta V_i \right), \quad (20)$$

$$[K_{ii}] = \sum_{i=1}^{m_i} \left(\sum_{j=1}^{i-1} \left([N^{ii}]^T [k_{ij}] [N^{ii}] \Delta V_j \right) \Delta V_i \right), \quad (21)$$

where $[N_{ii}]$ is given by:

$$\{d_{ij}\} = \left\{ \begin{array}{c} \{d_i\} \\ \{d_j\} \end{array} \right\} = \left\{ \begin{array}{c} [N^i(\xi_i)]\{D_i\} \\ [N^i(\xi_j)]\{D_i\} \end{array} \right\}$$
$$= \left[\begin{array}{c} [N^i(\xi_i)] \\ [N^i(\xi_j)] \end{array} \right] \{D_i\} = [N^{ii}]\{D_i\} \quad (22)$$

The factor of ½ has been removed in Equation (20), after $[K_{ij}]$ is assembled into the global stiffness matrix, since it is not correct to add in $[K_{ji}]$. Also, in Equation (21) the factor of ½ has been removed, with the upper limit on the inner summation being now $i - 1$.

Additional surface and point forces may be contemplated in the boundary terms of the peridynamic formulation. The total potential energy of the system can be expressed as:

$$\pi = U + \pi_1 + \pi_2 + \pi_3, \quad (23)$$

where:

$$U = [D_{ij}] \left[\frac{1}{4} \sum_{elemi=1}^{num\_el} \left(\int_{elemi} \left(\sum_{elemj=1}^{num\_el} \right. \right. \right.$$
$$\left. \left. \left. \left(\int_R [N^{ij}]^T [k_{ij}] [N^{ij}] dV_j \right) \right) dV_i \right) \right] \{D_{ij}\} \quad (24)$$

$$\pi_1 = -[D_{ij}] \left[\sum_{elemi=1}^{num\_el} \left(\int_{elemi} [N^{ij}]^T [b_i] dV_i \right) \right], \quad (25)$$

$$\pi_2 = -[D_{ij}] \left[\sum_{elemi=1}^{num\_el} \left(\int_{elemi} [N^{ij}]^T [t_i] dS_i \right) \right], \quad (26)$$

$$\pi_3 = -[D_{ij}] \left[\sum_{elemi=1}^{num\_el} \{P_i\} \right]. \quad (27)$$

$[b_i]$ and $[t_i]$ are the vectors of body and surface forces respectively and $\{P_i\}$ are the applied external forces.

The following expression is obtained by minimizing the potential energy in Equation (23):

$$\frac{\partial \pi}{\partial \{D_{ij}\}} = \left[\frac{1}{2} \sum_{elemi=1}^{num\_el} \left(\int_{elemi} \left(\sum_{elemj=1}^{num\_el} \right. \right. \right.$$
$$\left. \left. \left. \left(\int_R [N^{ij}]^T [k_{ij}] [N^{ij}] dV_j \right) \right) dV_i \right) \right] \{D_{ij}\}$$
$$- \left[\sum_{elemi=1}^{num\_el} \left(\int_{elemi} [N^{ij}]^T [b_i] dV_i \right) \right]$$
$$\left[\sum_{elemi=1}^{num\_el} \left(\int_{elemi} [N^{ij}]^T [t_i] dS_i \right) \right]$$
$$- \left[\sum_{elemi=1}^{num\_el} \{P_i\} \right] = 0 \quad (28)$$

Assuming zero surface and body forces Equation 5.23 is expressed as:

$$\left[\frac{1}{2}\sum_{elemi=1}^{num\_el}\left(\int_{elemi}\left(\sum_{elemj=1}^{num\_el}\right.\right.\right.$$

$$\left.\left.\left.\left(\int_{R}[N^{ij}]^{T}[k_{ij}]\,[N^{ij}]dV_{j}\right)\right)dV_{i}\right)\right]\{D_{ij}\}$$

$$=\left[\sum_{elemi=1}^{num\_el}\{P_i\}\right], \qquad (29)$$

or also as:

$$[K]\{D\}=[P]. \qquad (30)$$

By inverting matrix [K] with a suitable technique, displacements [D] are obtained, and pair-wise forces, stretches and strains are obtained. Using a maximum principal tensile strain criterion, two dimensional elements are automatically converted into zero-dimensional elements. With the application of a simplest microelastic damage model, peridynamic links are sequentially eliminated to simulate damage. For simplicity and efficiency, it is assumed that only links in between zero-dimensional elements are removed. In addition, nodes on adjacent membrane elements can be disconnected in order to allow crack propagation between two-dimensional elements.

3 APPLICATION TO MASONRY STRUCTURES

An example of a plain concrete masonry wall is analyzed using the peridynamic finite element model described in the previous section. In this case, the analysis is a two-dimensional plane stress model, and the wall is subjected to a uniform compressive load. Material properties for the concrete blocks are: Modulus of elasticity E of 3605 ksi, Poisson ratio v of 0.2, a material horizon of 3" was chosen. By using equations (4) the value of c is equal to 318.8 kips/in^6 and the value of d is 26.56 kips/in^4. The compressive strength of the concrete was 4.3 ksi and a tensile strength of 0.43 ksi was assumed. The maximum compressive stretch is 0.00119 and the maximum tensile stretch is 0.000119. The average block dimension is 15" by 8" and since this is a two-dimensional model a uniform thickness of 1" was selected. The same values of the Modulus of Elasticity of 3605 ksi and the material horizon of 3" were chosen for the mortar joints. Thus, the values of c and d remain the same. Although, the compressive strength of 1.37 ksi and a tensile strength of 0.14 were used. The maximum compressive stretch of the mortar is 0.0038 and the maximum tensile strength is 0.00038. The average thickness of the joints was 0.5". The overall dimension of the masonry wall was

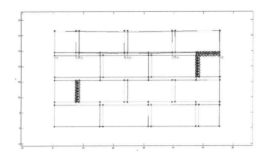

Figure 2. Unreinforced masonry wall subjected to compression.

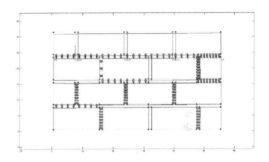

Figure 3. Failure sequence of an unreinforced concrete masonry wall.

56" by 32". Figures 2 and 3 show the model and the failure sequence.

4 CONCLUSIONS

In this work, the peridynamic finite element was explained and applied to masonry structures. It was assumed that the displacement field is continuous inside each finite element with displacement discontinuities allowed to develop between finite elements. The peridynamic finite element model shown is much more computer time- and memory-efficient than the similar discrete particle-based models, due to the fact that the number of degrees of freedom are reduced.

Using the peridynamic finite element model, an unreinforced masonry wall was simulated with different material properties for the concrete blocks and the mortar joints. Results show that the cracks propagate trough the mortar joints. This model shows that the selection of high quality materials is important. Also, it can be used to have better models for reinforced concrete masonry.

Although, in order to obtain better results, mesh convergence studies are required, where more degrees of freedom are included. In addition, three-dimensional computer simulation studies must be performed. In addition, unreinforced and reinforced masonry wall tests should be implemented.

REFERENCES

ACI. American Concrete Institute. 1997 Finite Element Analysis of Fracture in Concrete Structures. Report ACI 446.3 R–97.

Silling S. A. 2000. Reformulation of Elasticity Theory and Long-Range Forces. *Journal of the Mechanics and Physics of Solids* 48: 175–209.

Gerstle, W., Sau N., Silling S. 2007. Peridynamic Modeling of Concrete Structures. *Nuclear Engineering and Design* 237: 1250–1258.

Sau, N. 2008. Peridynamic Modeling of Quasibrittle Structures. Doctoral Dissertation. University of New Mexico.

Brick and Block Masonry - From Historical to Sustainable Masonry –
Kubica, Kwiecień & Bednarz (eds)
© 2020 Taylor & Francis Group, London, ISBN 978-0-367-56586-2

The influence of effective flexural stiffness on slender masonry wall capacity

M. Bogoslavov & N.G. Shrive

University of Calgary, Calgary, Alberta, Canada

ABSTRACT: Procedures regarding slender masonry wall strength capacity design in the Canadian Standard for Design of Masonry Structures (2014) are overly conservative. In the determination of secondary moment the calculated effective stiffness term, EI_{eff}, is believed to be a significant source of error. The lateral deflections of concrete block walls tested by Hatzinikolas et al. (1978a) are compared to deflections calculated following the Standard to predict the error of EI_{eff}. It is found that EI_{eff} is quite inaccurate at low vertical axial load eccentricities. Using data provided by Hatzinikolas et al. (1978a), the relationship between EI_{eff} and axial load (P/P_{cr}) is plotted, with some plots suggesting several slender walls with low load eccentricities are failing in compression rather than by buckling. It is proposed that the vertical axial load eccentricity be considered in addition to the slenderness ratio (kh/t) in defining walls for which buckling must be considered in design.

1 PURPOSE

The objective is to assess the degree to which the effective stiffness, as outlined in the Canadian Standard for the Design of Masonry Structures (2014), contributes to the error in the estimation of the strength of slender masonry walls as they experience vertical axial loading. The results from an extensive testing program conducted by Hatzinikolas et al. (1978a) are used to compare experimental results and calculations for the same parameters following the methods in the design standard.

2 INTRODUCTION

Conservative design procedures with respect to slender masonry walls have led to materially wasteful and economically uncompetitive slender masonry wall designs. Müller et al. (2017) demonstrated quantitatively that the design procedure for slender masonry walls in the Canadian Standard for the Design of Masonry Structures, CSA S304-14, (2014) can significantly underestimate the strength of slender masonry walls, leading to an overly conservative prescription. The underestimation of capacity is particularly high when axial loads are applied at low eccentricities, causing a relatively low initial primary moment. The design procedure has resulted in slender masonry walls generally becoming less economical to build in Canada compared to other materials used in similar situations and therefore a more effective design procedure is required.

In the Standard, a wall is considered slender if its effective height (accounting for fixity conditions) to

thickness ratio exceeds a value of $[10-3.5(e_1/e_2)$, with e_1 and e_2 representing the eccentricity of the applied load at the top and the bottom of the wall, respectively]. In such a case the secondary moment (Pδ effect) is determined using either the moment magnifier or the load displacement method (as outlined below in Equations 2 & 5 respectively). Both these methods account for the reduced stiffness of the wall due to cracking. The cracking is caused by the primary and secondary moments. The primary moment develops from the eccentricity of the axial load as applied to the wall, while the secondary moment develops from the axial load acting over the lateral displacement of the wall as it deforms. It is postulated that the effective stiffness of a wall depends on the applied axial load and its eccentricity, as well as the wall's height, Young's Modulus, critical axial load and cross-sectional radius of gyration or thickness. A 90% axial load reduction factor must be applied if the slenderness ratio (kh/t, where k is the effective length factor based on the top and bottom wall fixity, h is the wall height and t is the wall thickness) exceeds 30.

Slenderness is defined in order to distinguish walls which will experience significant second order moment effects due to axial load being applied eccentrically and therefore the slender walls are expected to experience out-of-plane failure. In their review and analysis of the capacity of slender concrete masonry walls, Müller et al. (2017) chronicle many known testing programs in which slender masonry walls were tested. They demonstrated through calculations that slender masonry wall capacities are significantly underestimated by the Canadian Standard for the Design of Masonry Structures

(2014) and that this was especially the case when axial loads were applied at low eccentricities: the strengths of slender walls as per the Standard were more accurate when walls experienced axial load applied at higher eccentricities.

The reason given by Müller et al. (2017) for the overly conservative slender wall design in the Standard is the large reduction of load-bearing capacity due to the magnification of the acting moment to account for the secondary moment caused by the lateral deflection of the wall. As the magnification of the acting moment is dependent on the applied axial load as well as the critical axial load, itself a function of both a wall's effective stiffness, EI_{eff} (the Young's Modulus multiplied with the reduced second moment of area, I_{eff}, thus accounting for wall cracking under flexure), and effective height (kh), Müller et al. (2017) cite error in the calculation of effective stiffness and consideration of effective height as the two main sources of error in assessing the magnification of the acting moment. The load displacement method can also be used to calculate the secondary moment acting on the wall, although in this case the calculation ultimately depends on the same two variables.

3 CANADIAN STANDARD FOR DESIGN OF MASONRY STRUCTURES CLAUSES FOR CALCULATING SECONDARY MOMENT USING EFFECTIVE STIFFNESS

The clauses in the Canadian Standard for Design of Masonry Structures (CSA S304-14 (2014)) which apply to the definition of slender masonry walls as well as the calculation of the secondary moment and the lateral deflection of the slender masonry walls tested by Hatzinikolas et al. (1978a) are as follows.

The Standard states that the effect of slenderness is a necessary consideration when

$$kh/t \geq \left(10 - 3.5\left(\frac{e_1}{e_2}\right)\right) \quad (1)$$

For the purpose of this study, all the walls assessed are slender according to Equation 1 and therefore CSA S304-14 (2014) Sections 10.7.4.2 (Pδ: load displacement method) & 10.7.4.3 (moment magnifier method) apply. As per the design standard, when the slenderness ratio is above 30 a 90% reduction factor must be applied in addition to the general slenderness considerations which require the calculation of secondary moment, although this is not the case with any of the walls tested by Hatzinikolas et al. (1978a).

Both the load displacement and moment magnifier methods are viable to be used in calculating the secondary moment of a slender masonry wall. In this study however, the data presented have been calculated only using the moment magnifier method.

In CSA S304-14 10.7.4.3, the total design moment calculated by the moment magnifier method is:

$$M_{ftot} = M_{fp} \frac{C_m}{1 - \frac{P_f}{P_{cr}}} \quad (2)$$

where M_{fp} is the primary moment calculated from the applied axial load and eccentricity, P_f is the applied axial load, C_m is the moment diagram factor (below), and P_{cr} is the critical axial compressive load:

$$C_m = 0.6 + 0.4M_1/M_2 \leq 0.4 \text{ where } M2 \geq M1 \quad (3)$$

$$P_{cr} = \pi^2(EI)_{eff} /[(1 + 0.5\beta)(kh)^2] \quad (4)$$

$\beta d = 1$ and $k = 1$ in the case of the Hatzinikolas et al. tests (1978a). k is the effective height factor and βd is the ratio of the total factored dead load moment to total factored moment.

For the load displacement (Pδ) method, CSA S304-14 10.7.4.2 specifies that the total moment shall be determined as:

$$M_{ftot} = M_{fp} + P_f\delta_f \quad (5)$$

where δ_f is the lateral deflection of the wall at the critical section under lateral and axial loads, including secondary moments.

To calculate EI_{eff}, the cracked second moment of area, I_{cr}, is calculated (below) by taking the transformed second moment of area of the cracked section ignoring the effects of axial load, E_m is the modulus of elasticity of masonry, I_o is the second moment of area of the uncracked effective cross-sectional area of a section about its centroidal axis and e_k is the kern eccentricity value:

$$(EI)_{eff} = E_m\left[0.25I_o - (0.25I_o - I_{cr})\left(\frac{e - e_k}{2e_k}\right)\right]$$

$$\leq 0.25E_mI_o \text{ but } \geq E_mI_{cr} \text{ for reinforced walls} \quad (6)$$

$$(EI)_{eff} = 0.4E_mI_o, \text{ for unreinforced walls} \quad (7)$$

In the lateral deflection calculation, the following stiffness is used:

$$EI = (EI)_{eff}/(1 + 0.5\beta) \quad (8)$$

The lateral deflection is calculated using the following equation which is derived using the moment area theorem (from CSA S304-14 Commentary Section 7.5.2):

$$\Delta_f = \frac{M_{ftot}(kh)^2}{8EI} \qquad (9)$$

4 ROLE OF EFFECTIVE STIFFNESS IN SLENDER WALL CAPACITY CALCULATION ERROR

In order to assess the degree to which error in calculating the slender wall capacity is caused by error in the calculation of EI_{eff}, the results from the extensive Hatzinikolas et al. testing program (1978b) were compared to calculations using CSA S304-14 (2014). Specifically, lateral deflections of the slender concrete block walls tested by Hatzinikolas et al. (1978b)

The block thickness is 194 mm, such blocks are not commonly used in Canada today were compared to the lateral deflections of those walls calculated using the effective stiffness as defined in the standard to give an estimate of the inaccuracy caused by the current EI_{eff} calculation. As shown in Equations 8 & 9, the maximum deflection of the wall is a function of the effective stiffness.

Hatzinikolas et al. (1978a) describe an extensive testing program in which 68 slender concrete block walls of varying height, slenderness, reinforcement and load eccentricities were incrementally loaded to failure. In particular, multiple sets of walls of the same height and vertical reinforcement were loaded to failure, with each specimen in every set varying only in the eccentricity of the applied load. As axial load eccentricity, height and reinforcement are individually varied across a large number of tested walls, the relationship between the effective stiffness and these parameters can be examined with some credibility. The study provides detailed results showing strain, vertical and lateral deflections as a function of applied load, allowing also for an investigation on the effect of applied load on the effective stiffness.

The walls shown in Table 1 all experience symmetric single curvature (the load eccentricity at the top and the bottom being the same). Short walls and walls experiencing double curvature were left out of the current analysis. Double curvature tests were omitted in order to simplify the analysis as the margin of safety is known to increase for walls in double curvature as these walls have been found generally to resist greater loads. All walls also had a pinned connection at each end.

As can be observed in Table 1, the calculated lateral deflection, using CSA S304-14 (2014), generally significantly exceeds the lateral deflection at the critical section of the wall (typically around midspan) measured as part of the testing. In each case, the ratio of the calculated over the test deflection is indicative of the degree of the overestimation of the effective stiffness. In particular, the ratio of the calculated to the measured deflection is high as the axial load eccentricity is lower, which is in accordance with the

same trend observed by Müller et al. (2017). At low load eccentricities the deflection determined using the equations in the Standard is typically substantially greater than the measured test deflection. Sometimes the ratio of the displacements is absurdly high (in particular, this is typically the case when the axial load is applied concentrically, but for the purposes of the Canadian Standard a minimum eccentricity of one tenth of the wall thickness, 19.4 mm, must be used). As can be seen in the results shown in Table 1, the eccentricity of the applied load is significantly more relevant than the slenderness of the wall for predicting the accuracy of the effective stiffness calculation. As the axial load eccentricity is greater, the deflection calculated using the Standard becomes distinctly more accurate. This is particularly the case for wall series A, C and D, where in the cases of wall series C and D the ratio of calculated to measured test deflection approaches 1 for the specimens tested with an axial load eccentricity of 76.2 mm or 88.9 mm. In the case of wall series A (unreinforced walls), the standard overestimates the lateral deflection at a higher load eccentricity. This means that the effective stiffness is itself overestimated and thus leads to overly conservative design. The reason for these underestimations of the effective stiffness was explored further. In the following section, the relationship between the applied load and the effective stiffness is plotted.

5 THE VARIATION IN EFFECTIVE STIFFNESS WITH INCREMENTAL LOAD

As the analysis of the Hatzinikolas et al. (1978b) test results indicated that the effective stiffness term is being significantly underestimated in CSA S304-14 (2014), in particular when load is applied at a lower eccentricity, it was of interest to analyze changes in the actual effective stiffness as load is incrementally applied. As the Hatzinikolas et al. (1978b) results provide plots of the changing lateral deflection with increased axial load, the relationship between the eccentrically applied load (and ensuing increasing secondary moment) and the lateral deflection could be used to predict the actual effective stiffness of each wall as it was being tested to failure. In an effort to determine how the effective stiffness varies with applied load and its eccentricity, the detailed results provided by Hatzinikolas et al. (1978b) were used to plot the effective stiffness against the increasing applied axial load. As the bending caused by the initial eccentricity of the axial load and ensuing secondary moment increases, the cross-section of each block wall cracks and the second moment of area decreases. Thus the effective stiffness decreases as the cracking develops with increasing moment. In an effort to predict most accurately EI_{eff} with each load increment, the lateral deflections resulting from gradually applied load were used to back-calculate a realistic effective stiffness to match the load-deflection relationship for

Table 1. Comparison of the Hatzinikolas et al. (1978b) Test Lateral Deflection & the Calculated Lateral Deflection as per CSA S304-14 (2014).

Wall	Slend. Ratio (m/m)	Reinf. (Imp.)	Eccen-tricity (mm)	Failure Load (kN)	Test Defl. (mm)	Calc. Defl. (mm)	Defl. Ratio
A1	13.8	Plain	19.4*	1246	1.3	25.4	19.5
A2	13.8	Plain	32.3	708	7.0	16.5	2.4
A3	13.8	Plain	64.5	357	10.0	13.8	1.4
A4	13.8	Plain	19.4*	1068	1.0	18.9	18.9
A5	13.8	Plain	76.2	116	8.0	4.8	0.6
B1	13.8	3#9	19.4*	1868	4.0	655.7	163.9
B2	13.8	3#9	32.3	1423	10.2	110.5	10.9
B3	13.8	3#9	64.5	622	26.7	37.7	1.4
B4	13.8	3#9	76.2	689	26.7	52.0	2.0
B5	13.8	3#9	88.9	511	19.1	39.3	2.1
C1	15.9	3#9	19.4*	890	1.0	37.3	37.3
C2	15.9	3#9	19.4*	1601	1.0	26.2	26.2
C3	15.9	3#9	32.3	1110	16.5	127.1	7.7
C4	15.9	3#9	64.5	556	25.6	49.1	1.9
C5	15.9	3#9	76.2	545	42.0	56.1	1.3
C6	15.9	3#9	88.9	400	40.6	41.5	1.0
D1	18.0	3#9	19.4*	1068	2.5	364.7	145.9
D2	18.0	3#9	19.4*	1868	3.3	39.1	11.9
D3	18.0	3#9	32.3	890	24.0	143.0	6.0
D4	18.0	3#9	64.5	484	40.6	59.0	1.5
D5	18.0	3#9	76.2	420	48.3	55.1	1.1
D6	18.0	3#9	88.9	369	53.3	52.7	1.0
H1	18.0	3#6	19.4*	1868	0.1	54.7	1215.9
H2	18.0	3#6	32.3	1154	15.0	56.5	3.8
H3	18.0	3#6	64.5	384	34.3	157.5	4.6
H4	18.0	3#6	76.2	290	43.2	134.1	3.1
H5	18.0	3#6	88.9	249	57.0	181.2	3.2
I1	18.0	3#3	19.4*	1423	0.1	41.7	1042.2
I2	18.0	3#3	32.3	965	16.5	47.2	2.9
I3	18.0	3#3	64.5	240	3.0	82.2	27.4
I4	18.0	3#3	76.2	146	4.0	67.8	16.9
I5	18.0	3#3	88.9	108	9.0	110.7	12.3
L1	24.3	3#9	19.4*	1868	15.0	71.3	4.8
L2	24.3	3#9	32.3	667	18.0	42.6	2.4
L3	24.3	3#9	64.5	400	76.2	142.7	1.9
L4	24.3	3#9	76.2	356	79.0	124.9	1.6
L5	24.3	3#9	88.9	326	96.0	120.2	1.3
M1	24.3	Plain	19.4*	623	5.0	95.8	19.2
M2	24.3	Plain	32.3	534	25.0	87.1	3.5

* Load was applied concentrically but the minimum eccentricity of t/10 (19.4 mm) is used in the calculation, as per CSA S304-14 (2014).

each load step. From this procedure, the following graphs were produced. As similar trends are produced across most walls, the following graphs should contribute to the development of a potentially more accurate equation for EI_{eff} which depends on the axial load, eccentricity, wall material properties and the critical Euler axial pin-pin buckling load.

Figures 1-6 each show the relationship between the applied axial load (normalized over the critical

axial buckling load) and the 'actual' effective stiffness as the load is incrementally applied. A general trend can be observed in that the effective stiffness of all the walls first experiences a rapid drop with increasing load, and then drops more gently until failure is reached. Not enough data are available to be able to comment on the difference in behaviour between reinforced and unreinforced walls (wall series 'A' being the only unreinforced set of walls

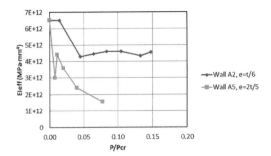

Figure 1. Wall Series 'A' (Unreinforced), kh/t = 13.8.

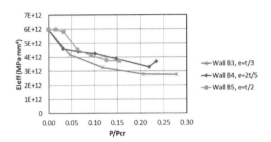

Figure 2. Wall Series 'B' (3#9 Bars), kh/t = 13.8.

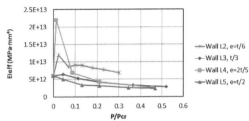

Figure 3. Wall Series 'C' (3#9 Bars), kh/t = 15.9.

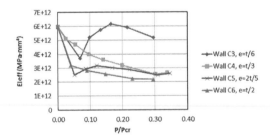

Figure 4. Wall Series 'D' (3#9 Bars), kh/t = 18.0.

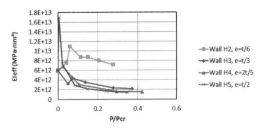

Figure 5. Wall Series 'H' (3#6 Reinf.), kh/t = 18.0.

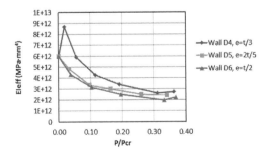

Figure 6. Wall Series 'L' (3#9 Reinf.), kh/t = 24.3.

which were studied), though the reinforced walls evidently resist much higher load as a fraction of their critical axial load.

The responses of walls C3, H2 and L2 are anomalous compared to the rest of the plots. According to the plots for walls C3, H2 and L2, the effective stiffness increases at some point as the walls are loaded and deform. The reason for this apparent error in the calculation of effective stiffness is that these three walls are loaded at low eccentricities and deflect very little laterally for the load which is applied. Any experimental variability in the very small displacements measured is thus highly accentuated. As significant lateral deflection does not occur with increasing load, the calculated EI_{eff} value is reflective of a wall which does not crack significantly. An increasing effective stiffness is physically impossible, as the stiffness is already at the maximum value at the beginning of the test based on the initial second moment of area and Young's modulus ($EI_{eff} = EI$). In fact, these walls do not fail in buckling (as would be expected from a slender wall), but instead fail in compression, with no significant opening of the mortar joints of the concrete block wall – indeed no flexural cracking was reported for these walls. As these walls all have low applied load eccentricities (eccentricity of 32 mm [t/6]), it is speculated that the reason for the anomalies in the graphs are that the walls did not fail with buckling as the dominant mode of failure.

Although not explicitly stated by Hatzinikolas et al. (1978a), the authors do mention that the slender walls loaded concentrically typically failed as a result of vertical cracks forming in the ungrouted cores and that the failure loads of these walls were similar to the failure loads of the short wall equivalents (shorter walls designed to resist the same compressive strength). As such it can be reasonably speculated that walls C3, H2 and L2 did not experience the typical failure of slender walls.

Isfeld et al. (2019) first observed that the failure modes for slender concrete block walls depended not only on the slenderness of the wall but also on the eccentricity of the applied axial load. These findings were based on extensive finite element modelling which explored various height to thickness ratios ranging from h/t = 5 to h/t = 60, as well as eccentricities of the applied axial loads ranging from 0 to t/2. Based on the results of the finite element models a table (solely reporting on the results, not itself prescriptive) was produced which categorized each wall by failure mode, distinguishing between failures during which all bed joints were fully closed, failures with bed joints still in partial contact and failures with some fully opened bed joints. In doing so, Isfeld et al. (2019) distinguished between compressive material failure and buckling failure, aiming to identify at which point the transition between the two modes of failure occur for each wall and eccentricity.

Although the walls modelled by Isfeld et al. (2019) are not analogous to the walls tested by Hatzinikolas et al. (1978a) (varying reinforcement, end conditions, material properties), it is still noteworthy that the three walls mentioned above (C3, H2 and L2) exhibiting an unusual response in the P/P_{cr} vs. EI_{eff} plot would each fall under the category of walls which fail with bed joints still fully closed in the Isfeld et al. study. For reference, Isfeld et al. (2019) found walls with an axial load eccentricity of up to t/6 (as is the case with the eccentricity of applied load on walls C3, H2 and L2) and a height to thickness ratio of 40 and below to fail in compressive material failure. On the other hand, with the exception of unreinforced walls A2 and A5, all other walls plotted in Figures 1 to 6 above would experience buckling failure, according to the Isfeld et al (2019) table.

The Isfeld et al. (2019) results, combined with the walls C3, H2 and L2 not failing in buckling, suggests that categorization of a slender wall should depend on both the slenderness ratio and the eccentricity of an axial load. In such a case, many masonry walls are treated as slender in CSA S304-14 (2014) while in fact they would not behave as walls for which slenderness needs to be considered. Many of the walls with the largest overestimates shown in Table 1 above fall into the category of walls for which the CSA S304-14 (2014) mandates slenderness to be considered but are in fact failing in compression. This is a key factor in the over-design of many slender masonry walls, as the consideration of slenderness requiring the calculation of a secondary moment acting on the wall section leads to a significant underestimation of a wall's capacity to resist an eccentric axial load. As discussed in

Donà et al. (2020), significant lateral loads (i.e. design for wind or a seismic event) may nonetheless require for a slender wall to be designed to resist buckling. In such a case, any lateral deflection caused by lateral loading on the slender masonry wall would also need to be considered in defining the slenderness of the wall, if in fact "slenderness" was taking into account both the height to thickness ratio of the wall and the eccentricity of the axial applied load.

Sandoval & Roca (2013) suggested that a design method which considers the interaction between buckling and compression failure must also account for the tensile strength of the masonry unit-mortar interface. In their paper "Empirical equations for the assessment of the load-bearing capacity of brick masonry walls", Sandoval & Roca (2013) propose equations to calculate brick wall capacity reduction factors to account for wall slenderness and eccentricity of applied axial load. The tensile strength of the brick unit-mortar interface is found not to be negligible and is ultimately accounted for in the proposed equations for the axial capacity reduction factor. While the Canadian Standard for the Design of Masonry Structures (2014) does not consider the tensile strength of the masonry in assessing the masonry design strength, it should be noted that the inaccuracies currently present in the standard significantly outweigh any inaccuracy in design present due to the neglect of tensile strength of the unit-mortar interface.

6 CONCLUSION

While there are several causes for the overly conservative design procedure in the Canadian Standard for the Design of Masonry Structures (2014), the degree to which the conservatism is caused by the inaccuracy of the calculation of the effective stiffness (EIeff) has been examined. Although this parameter is known not to be the only source of error in slender masonry wall design, as per Müller et al. (2017), it does contribute some error to the estimation of capacity, in particular at lower axial load eccentricities. There are two main causes of the inaccuracy in the calculation of effective stiffness: the effect of the eccentricity of the applied axial load on EIeff is not given adequate consideration in the calculation, leading to a much more conservative design when axial load is applied to the wall at a lower eccentricities, and walls which do not fail as slender walls (in buckling) at the lowest load eccentricities are required to be designed accounting for secondary moment effects. While the effective stiffness term does gradually lose accuracy as lower load eccentricities are considered, the largest inaccuracy, causing eventual severe over-design, occurs in walls which do not experience a significant reduction in section stiffness (represented by effective stiffness) because they do not fail by buckling. For this reason, while the effective stiffness term does account for the loss in section stiffness with some error, the most significant error occurs due to the incorrect categorization of slender walls in the Canadian Standard for Design of Masonry Structures (2014).

The most productive course of action for future research, following the findings of this study, is to focus primarily on redefining the concept of slenderness in the Standard to account for both the effective height to thickness ratio (or height to radius of gyration) and the eccentricity of the applied axial load. In particular, the equation for the calculation of effective stiffness should in the future only be modified if it can accurately account for variation in axial load eccentricity, as the current design procedure does not require modifications when dealing with high load eccentricities.

REFERENCES

Canadian Standards Association. 2014. *S304.1-14 Design of Masonry Structures*. Mississauga, Ontario, Canada: CSA Group.

Donà et al. 2020. Second-order effects in URM walls subjected to compression and out-of-plane bending: From numerical evaluation to proposal of design procedures. *Engineering Structures, Elsevier* 209: 110130.

Hatzinikolas, M., Longworth, J., Warwaruk, J. 1978a. Concrete Masonry Walls. *Structures Report No. 70, University of Alberta*, Edmonton.

Hatzinikolas, M., Longworth, J., Warwaruk, J. 1978b. Experimental Data for Concrete Masonry Walls. *Structural Engineering Report No.71, University of Alberta, Edmonton*.

Isfeld, A.C., Müller, A.L., Hagel, M., Shrive, N.G. 2019. Analysis of Safety of Slender Concrete Masonry Walls in Relation to CSA S304-14, *Can. J. Civil Engineering* 46(5): 424–438.

Müller, A.L., Isfeld, A.C., Hagel, M., Shrive, N.G. 2017. Review and Analysis of Capacity of Slender Concrete Masonry Walls. *Proc. 13th Canadian Masonry Symp., Halifax*, Paper 43.

Sandoval, C. & Roca, P. 2013. Empirical equations for the assessment of the load-bearing capacity of brick masonry walls. *Construction and Building Materials* 44: 427–439.

Brick and Block Masonry - From Historical to Sustainable Masonry –
Kubica, Kwiecień & Bednarz (eds)
© 2020 Taylor & Francis Group, London, ISBN 978-0-367-56586-2

Monte Carlo simulation of masonry walls in compression considering spatially variable material properties

L. Bujotzek, D. Müller & C.-A. Graubner
Institute of Concrete and Masonry Structures, Technische Universität Darmstadt, Germany

ABSTRACT: The safety concept and the partial factors for the design of masonry walls in EN 1996 and in particular the National Annexes mostly base on tradition. This is seen as an opportunity to rethink the concept and validate the historical values with studies, taking reliability theoretical approaches into account. The spatial variability of material properties can have a significant influence on the reliability of load-bearing capacity of masonry walls. This influence is highly dependent on the geometry and the material behaviour of the wall. The conducted investigations are based on a finite element model, which is set up following the simplified micro modelling approach. The spatial variability is implemented as a unit-to-unit variability for the compressive strength and the Young's modulus. Those two material properties determine the stress-strain behaviour of the different masonry types. Furthermore, other material properties like the fracture energy must be taken into account in order to generate numerically solid results. Within each run of the conducted Monte Carlo simulation random values for the material properties are generated. The wall is then loaded until failure and a load capacity factor can be calculated. Within the study, the influence of the spatial variability on the load-bearing capacity is investigated by modifying several parameters like the wall slenderness or length. Based on an appropriate distribution function, it is possible to calculate partial safety factors in order to interpret the influence on reliability of structural resistance.

1 INTRODUCTION

1.1 *Motivation*

As masonry construction is one of the oldest disciplines in structural engineering, the dependence of many calculation methods on traditional values is high. In order to improve the performance in design, mechanical models are continuously developed. In EN 1996, the partial safety factor for the resistance side is considered to cover both stress and buckling failure. In order to verify the safety factor for masonry according to EN 1996, a study on the reliability is required.

1.2 *Spatial variability as approach to increase calculated reliability*

Amongst other things, spatially variable material properties are considered to be worth investigated. It can be said for sure that within a wall the material properties like compressive strength and Young's modulus vary, cf. Müller et al. (2017). An impact on the load-bearing capacity can be assumed for both buckling and stress failure for various reasons. This influence is the result of mechanical as well as statistical reasons.

1.3 *Idea of the investigations*

The studies on which this paper is based, aim on the investigation of the influence of spatially variable material properties on the load-bearing capacity of masonry walls. Therefore, a finite element model is set up and provided with random material properties that are generated in the course of a Monte Carlo simulation. The material properties vary from element to element. Within one calculation, the standardized load capacity of the wall is determined by increasing the load step by step. After conducting a range of calculations, results can be statistically evaluated and partial safety factors can be determined.

2 FINITE ELEMENT MODELLING

2.1 *Modelling of the wall*

For creating the numerical model of the wall, the finite element software DIANA FEA 10.2 (2017) is used. The finite element models consist of shell elements that are connected by interface elements. At top and bottom of the wall, the elements are connected to a rigid bar, which is attached to the supports. The system is loaded path-controlled in small

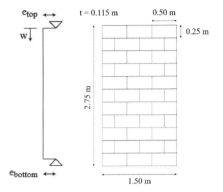

Figure 1. Presentation of the static system and the standard wall.

steps in order to achieve a high accuracy when reading out the results. The rigid bars allow the load to be induced eccentrically, cf. Figure 1.

Figure 1 additionally shows the geometry of the standard wall used to set up most part of the calculation settings.

2.2 Setting the material properties

For the following investigations, walls of calcium silicate and perforated clay brick masonry are examined in order to identify the influence of spatial variability depending on different material behaviour. The most important material properties within the scope of these studies are the compressive strength and the Young's modulus. It can be assumed that the compressive strength is the decisive value concerning cross-sectional stress failure. The Young's modulus is mostly important for buckling failure as it determines the stiffness of the wall. In addition to these two properties, the tensile strength in the bed joints is considered to be an important parameter and therefore worth being part of the variation.

The following investigations are independent on the value of the compressive strength since the examinations are being conducted with standardized stresses. The other material properties are meant to be either constant or within a defined ratio to the compressive strength as Table 1 shows.

where f_{my} = masonry compressive strength perpendicular to the bed joints; E = Young's modulus

Table 1. Material properties used in the investigations.

	calcium silicate unit	clay brick unit
E/f_{my}	850	1100
$G_{fcy}/(f_{my} \cdot h_e)$	0.0034	0.00142
f_{mx}/f_{my}	0.50	0.50
f_{bt}/f_{my}	0.16	0.16

of the element; G_{fcy} = fracture energy in compression perpendicular to the bed joints; h_e = element height; f_{mx} = masonry compressive strength parallel to the bed joints; f_{bt} = tensile strength in the bed joints.

The model follows the simplified micro-modelling approach, which means that it consists of continuous elements displaying the masonry behaviour in compression and interfaces modelling the discrete cracking in the joints. The shell elements contain a complex orthotropic material model, whereas the interface elements display discrete cracking behaviour in the bed joint as soon as the tensile strength is exceeded.

The material model for the shell elements is based on a model by Rankine and Hill. The yield-surface combines compressive failure borders by Hill and tensile failure according to the model of Rankine. The model is set up by Lourenço (1996) and shown in the figure below.

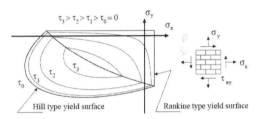

Figure 2. Material-model Rankine-Hill from Lourenço (1996).

The interaction between unit and mortar is implemented implicitly by using the masonry compressive strength for the compressive strength of the shell elements.

The post-cracking failure in tension is described by an approach according to Hordijk (1991). However much more important, in this study is the post-cracking behaviour in compression, as it has a strong effect on the load redistribution behaviour of the wall. With the aim of providing the model with comprehensible values, the fracture energy in compression G_{fc} is calibrated in order to match the stress-strain behaviour of EN 1992-1-1, see Figure 3.

The strain at compressive strength ε_f and standardized Young's modulus k_0 are chosen according to Meyer und Schubert (1992) and displayed in Figure 3. The analytic stress-strain relation according to EN 1992-1-1 is shown below.

$$\frac{\sigma}{f_{cm}} = \frac{k_0 \cdot \eta - \eta^2}{1 + (k_0 - 2) \cdot \eta} \tag{1}$$

where η = strain in relation to strain at compressive strength.

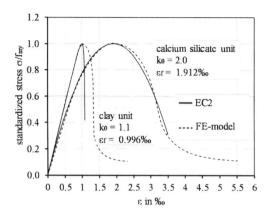

Figure 3. Stress-strain behaviour of masonry in compression.

Figure 3. As a result, the load-bearing capacity according to the FE calculation is slightly smaller.

During path-controlled loading, the vertical reaction forces at the foot of the wall are automatically recorded in each load step. The maximum applied load is determined using an algorithm to identify the maximum value. Finally, the load factor is calculated as follows.

$$\Phi = \frac{F_{Rz,max}}{f_{my} \cdot A} \qquad (2)$$

2.3 *Validation of the numerical model regarding the system load-bearing capacity*

The numerical model is validated as a whole with deterministic material parameters. The following figure shows the load-bearing capacity for calcium silicate masonry at different stages of slenderness.

As Figure 4 shows, the values calculated as a result of the finite element analysis fit well with the analytical approach by Glock (2004). The calculation and material settings therefore constitute a well-founded basis for the investigations with random data in the Monte Carlo simulation. The deviation between the load-bearing capacity from the FE model and the analytical approach in the area of compact walls results from the different stress-strain behaviour. While Glock assumes rigid-plastic behavior after reaching the compressive strength, the modelled material behaviour shows a more realistic post-cracking behaviour and is therefore less ductile, see

3 RELIABILITY THEORETICAL BACKGROUND OF THE INVESTIGATIONS

3.1 *Monte Carlo simulation*

Within the simulation, random material properties are generated in a first step. The random values are then prepared in order to reduce the number of required simulations, cf. 3.2. Files are needed to give the finite element program instructions to build and calculate the walls and then put out the required results. For a better performance of the computing-intensive process, an algorithm is developed, which allows to execute the process fully automatically for a defined number of simulations.

As already mentioned, the random data consists of the masonry compressive strength, the Young's modulus and the tensile strength in the bed joints. The parameters scatter around the mean value with the following coefficients of variation (COV) each following a log-normal distribution function.

3.2 *Reduction of the required number of simulations using Latin Hypercube Sampling*

As the simulation represents a very computational-intensive process, the aim is to reduce the required number of calculations as far as possible without losing accuracy. One possible method to reduce this number in the conducted investigations is called Latin Hypercube Sampling (LHS). Therefore, the assumed probability distribution function is divided into n strips, where n is the number of simulations performed. Each strip contains one point that represents one simulation, cf. Figure 5.

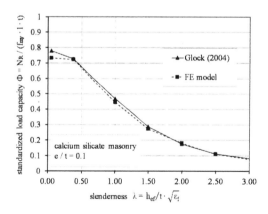

Figure 4. Load-bearing capacity of calcium silicate masonry walls with varying slenderness.

Table 2. Coefficients of variation for the variable input values according to Brehm (2011).

	COV in %
Masonry compressive strength f_{my}	20
Young's modulus E	25
Tensile strength in bed joints f_{bt}	30

180

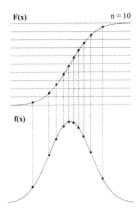

Figure 5. Sampling random data with LHS.

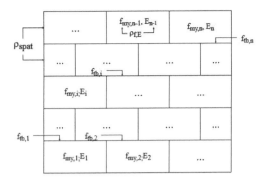

Figure 6. Stochastic model of the wall.

This sampling method sorts the randomly generated data in a way that fits the distribution function, see Fure 5. In this way, it is possible to avoid random accumulations of data points.

As a compromise between accuracy and computing time the number of simulations is set to $n = 100$. After conducting the simulation, the accuracy is proven by making tests with confidence intervals in order to evaluate the accuracy achieved.

3.3 Consideration of correlation

Within the generation of the random values two types of correlation must be considered. The spatial correlation of elements within one wall is described by the coefficient of spatial correlation ρ_{spat} and represents a key value in the course of this examinations. As the actual correlation is not yet known, the spatial correlation coefficient is chosen as study parameter ranging from 0 % over 50 % up to 100 % within most part of the studies. At $\rho_{\text{spat}} = 100$ % the wall is homogenous, which means, that the whole wall consists of elements with the same strength and stiffness. On the other hand, $\rho_{\text{spat}} = 0$ % means, that the spatial scattering within the wall is at its maximum value i.e. there is no correlation. This case enables the stresses inside the wall to be redistributed between the elements. That is one of the reasons why there is a presumption that the difference between the load-bearing capacities of walls within one sample get closer to each other and scattering becomes less.

In addition to spatial correlation this study contains examinations about the correlation between compressive strength and Young's modulus. The correlation of these two material properties is set to $\rho_{\text{f,E}} = 70$ %, according to Schueremans (2001).

Figure 6 shows an extract from the investigated wall as stochastic model with the variable basic values f_{my} = masonry compressive strength, E = Young's modulus and f_{tb} = tensile strength in the bed joints.

3.4 Determining the design value from result data

The final goal of the investigations is to calibrate and evaluate the partial safety factor for resistance γ_M, which is applied in EN 1996. It is therefore necessary to determine the characteristic and the design values of the load factor. For the calculation of the characteristic value, the finite element calculation is carried out by using deterministic material properties. The characteristic strength is chosen as input parameter for the strength. It is defined as 5 % fractile of the corresponding probability distribution function. Eq. (4) shows the calculation of the characteristic value for the compressive strength f_k in relation to the mean value f_{my}.

$$N_{Rk} = N_R(f_k; E_m) \tag{3}$$

$$f_k/f_{my} = e^{-COV_{f_{my}} \cdot 1.645} \tag{4}$$

For the Young's modulus, the mean value is chosen in order to determine the characteristic load-bearing capacity. In contrast, the design value of the load bearing capacity can only be determined after the simulation results for one wall are available. The fractile, that corresponds to the design value can be calculated if a certain probability distribution is assumed. In the case of this study, the probability distribution is expected to be log-normal. With the aid of the sensitivity coefficient $\alpha = 0.8$ and the reliability index $\beta = 3.8$ for a reference period of 50 years, which are given in EN 1990, the design value of the load bearing capacity can be calculated as follows.

$$N_{Rd} = F_{N_R}^{-1}[\Phi(-0.8 \cdot 3.8)] \tag{5}$$

The partial safety factor can subsequently be determined as the ratio of the characteristic and the design value of the load-bearing capacity.

$$\gamma_M = \gamma_m \cdot \gamma_{Rd} = \frac{N_{Rk}}{N_{Rd}} \cdot \gamma_{Rd} \qquad (6)$$

The model uncertainties are considered by multiplying this factor with $\gamma_{Rd} = 1.10$, according to fib Model Code 2010 (2010).

3.5 *Validation of the distribution type*

As described above the design values are calculated based on the assumption that the load capacity is lognormally distributed. In order to verify the type of distribution function a Kolmogoroff-Smirnoff-test is conducted in Bujotzek (2018). Within the test, the results are tested on a normal, log-normal and a Weibull-distribution. Based on the test, none of the distribution types could be rejected. However, the log-normal distribution showed the smallest deviation on the overall results. For consistency reasons an approximation by the log-normal distribution is chosen for all walls.

3.6 *Series and parallel systems*

Two stochastic basic systems are introduced at this point in order to be able to perform a proper interpretation of the examination results. Many of the expected effects due to spatial variability can be explained by the characteristics of these systems. Both of those ideal systems are displayed in Figure 7. Series and parallel systems are used in previous publications on this field, cf. Melchers und Beck (2018).

The main feature of the series system is, that the load capacity of the whole system is determined by the weakest element within the structure. An example for this case is a masonry pillar. Postcracking behaviour of the used material in this case is insignificant as there is no opportunity to redistribute any forces.

On the other hand, the system performance of the parallel system is highly dependent on the post-cracking behaviour. For the case of ideal plastic behaviour after reaching the compressive strength, the load-bearing capacity of the system is the sum of the capacity of all elements inside the structure. It can be assumed that systems that fail due to buckling, are close to a parallel system, as all the elements inside the system determine the decisive stiffness together.

Systems in reality can often be described by a combination of parallel and series systems. The influence of the spatial variability of the material properties on the scatter of the load capacity is interpreted in the subsequent sections on the basis of these simplified subsystems and the simulation results.

4 EXAMINATION RESULTS

4.1 *Wall slenderness*

In order to examine the influence of spatially variable material properties on the load-bearing capacity of the wall for varying wall slenderness, seven walls of different height are investigated considering full spatial correlation ($\rho_{spat} = 100 \%$) and no correlation of the basic variables ($\rho_{spat} = 0 \%$). For convenience, only the results for calcium-silicate walls are shown in the figure below. Similar to the results shown in Figure 4 the load capacity decreases with the slenderness of the construction. The dots connected by the solid lines represent the mean values of the sample of finite element calculations. The dashed lines show the design values, the size of which depends not only on the average load capacity but also on the standard deviations of the result. The distance between the mean and the design values of the standardized load-bearing capacity can be interpreted as a measure of reliability. In order to illustrate the influence on reliability as unaltered as possible, the load-bearing force is related to the mean value of the masonry compressive strength for the mean and the design value.

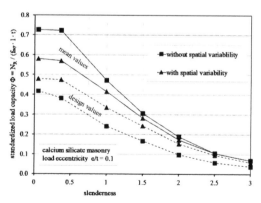

Figure 8. Influence of spatial variability as a function of wall slenderness.

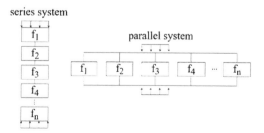

Figure 7. Series and parallel system.

Regarding the mean values of the load capacity one can see that in the range of low walls ($\lambda = 0.1 - 1.0$) the spatial variability reduces the average capacity. The reason therefore is that these walls fail due to stress failure, caused by their compact geometry. Stress failure always occurs within the weakest layer, cf. paragraph 3.5. If spatial variability is applied, the weakest layer is most likely less strong than the average strength of the assumed material.

Regarding the design values, however, something else can be seen. Due to stress redistribution which is made possible by spatially variable material properties, the scattering of the load capacity within the examined sample decreases strongly. As a result of this, the decrease of the mean values can be compensated by the lower deviation, which leads to higher design values.

Considering the slender walls, which fail due to buckling, the effect of spatially variable material properties gets even more positive. According to Glock (2004), the transition between stress and buckling failure is marked by the position of the inflection point in the load curve. Figure 8 shows the first result in the range of buckling failure at a slenderness of $\lambda = 1.50$. It can be seen that the negative influence of spatial scattering material properties on the mean values of the load-bearing capacity is negligible for walls of this or greater slenderness. In the range of very high walls the points lie on top of each other. This sufficiently justifies the statement that a system that fails due to stability represents an almost perfect parallel system. Since the load-bearing capacity is only determined by the average stiffness of the wall, there is no longer any dependence on the weakest layer. For the average load-bearing capacity it does not matter whether a scattering within the wall is applied or not. The smooth transition between stress and stability failure is due to continuously increasing second order effects and can also be observed in terms of reliability as the positive impact of spatial variability increases with the slenderness.

Since the capacity is determined by the average stiffness in this failure case the spatial variation has a positive effect on the scattering of the load-bearing capacity, the scatter of the load-bearing capacity gets close to zero for very slender walls. This looks quite different regarding the walls with full spatial correlation. As the walls are a homogenous system and scatter as a whole, the deviation from the mean value of each wall is significantly greater than with spatial variability.

For clay brick masonry, the influence is rather similar. In qualitative terms, the only significant difference is that in the range of low walls the design values of walls with full spatial variability ($\rho_{spat} = 0$ %) are lower than the homogenous ones ($\rho_{spat} = 100$ %). The explanation for this lies in the lower ability to redistribute stresses because of the brittle post-cracking behaviour, cf. Figure 3.

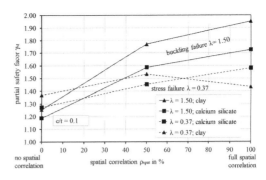

Figure 9. Influence of spatial variability on the partial safety factor γ_M.

The observations can be confirmed when the partial safety factors are evaluated. In Figure 9, the partial safety factor γ_M for resistance is plotted over the spatial correlation coefficient ρ_{spat} for two representative values of slenderness $\lambda = 0.37$ (stress failure) and $\lambda = 1.50$ (buckling failure). The difference in slenderness is shown by the pulled through ($\lambda = 1.50$) and the dotted line ($\lambda = 0.37$).

On the whole it becomes apparent that the required safety factor becomes smaller when spatial variability is assumed. For the walls that fail due to buckling the effect is completely positive for both materials. For the calcium silicate walls, the factor can be reduced from $\gamma_M = 1.72$ for the homogenous wall down to $\gamma_M = 1.19$ for maximum spatial variability. For the calcium silicate walls the impact is also exclusively positive with the compact walls that fail due to stress. This effect is caused by the more ductile material behaviour of the calcium silicate units. For the walls of perforated clay brick masonry the level of reliability decreases between full and 50 % spatial correlation for the compact wall. At $\rho_{spat} = 50$ % the clay units are not capable to redistribute forces sufficiently. As a result of this, the scattering within the sample cannot be reduced adequately to compensate the decreasing mean value. For maximum spatial variability though even the rather brittle clay units are capable to redistribute stresses in order to reduce scattering of the load capacity. Hence, a lower safety factor is required compared to full and 50 % spatial correlation.

4.2 Length of the wall

As the high influence of load redistribution on reliability can be observed in the examination results so far, it is evident to investigate the effect of spatial variability on the load-bearing capacity with regard to the length of the wall or more precisely, the number of elements in one layer. This parameter is crucial for the opportunity that the forces have to redistribute from one element to another. In addition to that the special case masonry pillar can be further illuminated. With only one element per layer the pillar has no possibility to redistribute any load.

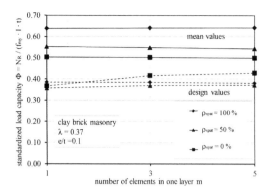

Figure 10. Load-bearing capacity over wall length at different levels of spatial correlation.

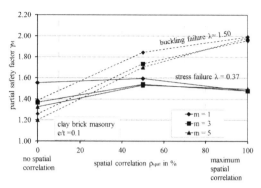

Figure 11. Partial safety factors of clay brick masonry at different wall lengths.

Figure 10 shows the load-bearing capacity for walls of clay brick masonry at different stages of spatial correlation for walls of different length (m = 1;3;5, with m = number of elements in one layer). For convenience, this investigation can only be shown for the compact wall (λ = 0.37) which fails due to stress. The results in Figure 10 point out that the mean values are almost independent on the number of elements in one layer. A closer look at the values reveals that the capacity of the masonry pillar is slightly lower than for the walls with more than 3 elements in one layer. The results also show that the mean values converge fast. From m = 3 to m = 5 there is no further increase apparent. The positive influence of load redistribution can be seen more clearly when considering the results for the calcium silicate walls.

In addition to that it can be objected that the spatial variability has a negative influence on the mean value of the load-bearing capacity which is due to the fact that the system is dominated by its series part and therefore determined by the weakest link inside the structure.

Regarding the design values it can be seen that the positive influence of the increase of the number of units in one layer becomes clearer the more spatial variability is assumed. From no dependence from the number of stones with the homogenous wall (ρ_{spat} = 100 %) the influence can already be guessed at ρ_{spat} = 50 % and becomes clearly apparent with maximum spatial variability (ρ_{spat} = 0 %). It is interesting to note that the design value at ρ_{spat} = 0 % increases clearly stronger from m = 1 to m = 3 than from m = 3 to m = 5. It can therefore be assumed that the range of possible load transfer is quite limited and a further increase of the wall length entails no added value regarding reliability.

Taking a closer look at the resulting partial safety factors, cf. Figure 11, it is apparent why the masonry

pillar is given a special status in EN 1996. Therein the compressive strength of the used material is supposed to be reduced by a factor, which depends on the cross-sectional area A of the pillar, cf. Eq. (7).

$$0.7 + 3 \cdot A \leq 1 \qquad (7)$$

The equation shows that the load capacity masonry pillars with a cross-section less or equal 0.1 m² has to be reduced.

The results in Figure 11 contain both the safety factors for compact and those for slender walls at different numbers of elements in one layer and plotted over the spatial correlation coefficient.

Basically it can be seen that almost without any exception higher safety factors for buckling failure result from the higher COV that is assumed for Young's modulus as well as the fact that the mean Young's modulus is used for the determination auf the characteristic load-bearing capacity. Regarding the required safety factors of the walls which fail due to stress the same observations that were made regarding Figure 10 can be recognized. The level of reliability at first decreases from the homogenous wall to average spatial correlation. Already within that area it becomes obvious that the masonry pillar is significantly less reliable than the walls with several elements per layer. The difference between m = 3 and m = 5 is not recognizable. If one follows the course until maximum spatial variability, the difference between the walls with multiple elements per layer and the masonry pillar gets unambiguously visible. The difference between the two walls with three and five units per layer is negligible if one considers that the simulation results are affected by a certain statistical scattering. Taking maximum spatial variability into account the partial safety factors for the walls become lower than those of the homogenous walls, the masonry pillar being excepted from this observation. It should be mentioned at this point that the results of the calcium silicate masonry show a positive effect of the spatial variability at an average level (ρ_{spat} = 50 %) already.

Regarding the results for the slender walls one can exclusively observe positive effects on the level of reliability. The simulation results show no dependence on the number of elements within one system for full spatial correlation as the dots cover each other. The reason for the visibly higher influence of the spatial variability on the walls of three or more units in one layer is that the absolute number of elements within a parallel system affects the coefficient of variation (*COV*) of the global system as follows, if a parallel system with perfect ductility is assumed.

$$COV_{R,sys} = \frac{COV_{R,i}}{\sqrt{n}} \cdot \sqrt{1 + (n-1) \cdot \rho_{spat}} \qquad (8)$$

where $COV_{R,sys}$ = COV of the system; $COV_{R,i}$ = COV of one element within the system, n = number of elements in the system.

Additionally, the relationship displayed in Eq. (8) depends on the spatial correlation as it is shown in Melchers and Beck (2018). The nonlinear relation between the coefficient of variation of the results and the number of the elements within one investigated system according to Eq. (8) can be guessed when observing the results plotted in Figure 11.

The difference in the required safety factors between m = 1 and m ≥ 3 demonstrate the need of a reduction factor for masonry walls of a small cross-sectional area, cf. Eq. (7). In addition to that there is a visible dependence on the spatial reliability, which, however, remains to be determined. In the most unfavourable case for the masonry pillar (ρ_{spat} = 0 %), the ratio of the partial safety factors of the masonry pillar (*m* = 1) to the wall (*m* = 3) is 1.38/1.58 = 0.87. According to equation 7, this factor (0.875) is approximately calculated for the selected geometry. The normative reduction factor thus covers well the reduced reliability of masonry pillars compared to masonry walls.

5 CONCLUSION

Regarding the results presented in the scope of this paper, one can see that the spatial variability of material properties has a relatively high influence on the reliability. The influence depends a lot on the properties of the investigated wall, such as slenderness, material behaviour and length.

Regarding the slenderness, the main focus must be on the failure mode. Walls that fail due to a stability problem show a clearly higher sensitivity due to spatially variable material properties than walls that fail due to cross-sectional stress failure.

This is because of the fact that the slender wall's load capacity is determined by the stiffness of all its elements and therefore represents an almost perfect parallel system. The scattering in the load-bearing capacity of slender walls becomes almost zero thanks to this effect.

For compact walls, the influence of spatially variable material properties is not as high as for slender walls. Due to the fact that the main reason for reduced scatter is caused by the possible stress redistribution, the material behaviour and especially the post-cracking behaviour have an enormous influence. The differences between perforated clay brick and calcium silicate masonry that are indicated in this paper must be illuminated in further studies.

For further investigations on this topic it is considered to increase the number of simulations within one sample. This measurement makes it possible to reduce the deviation in the results due to statistic scattering as some of the result plots show. In addition to that, if the sample is big enough, it will be possible to determine the design value directly from the result sample, which would lead to further improvement of accuracy. The objective of retaining the global safety factor for the resistance side, which covers both stress and stability failure, requires further investigations. The stress reduction factor that already exists for masonry pillars, according to EN 1996 is a good way to compensate the reduced reliability due to a non-existent load redistribution capability, although the reliability rather correlates to the number of elements within one layer than to the cross-sectional area of the pillar itself.

REFERENCES

EN 191996-1-1. 2009. Eurocode 6. Design of masonry structures - Part 1: General – Rules for reinforced and unreinforced masonry. Brussels: CEN European committee for standardization.

Müller, D., Förster, V., Graubner, C.-A. 2017. Influence of material spatial variability on required safety factors for masonry walls in compression: *Mauerwerk - European Journal of Masonry* 21(4): 209–222.

TNO Diana FEA: Internetdokumentation, DIANA FEA. 2018.

Lourenço, P.B. 1996. Computational strategies for masonry structures: Dissertation, TU Delft.

Glock, C. 2004. Traglast unbewehrter Beton- und Mauerwerkswände Nichtlineares Berechnungsmodell und konsistentes Bemessungskonzept für schlanke Wände unter Druckbeanspruchung. Darmstadt: Dissertation, Technische Universität Darmstadt.

Brehm, E. 2011. Reliability of unreinforced Masonry Bracing Walls – Probabilistic Approach and Optimized Target Values: Dissertation, TU Darmstadt.

Hordijk, D. A. 1991. Local approach to fatigue of concrete. Delft, Delft University.

Meyer, U.; Schubert, P. 1992. Spannungs-Dehnungs-Linien von Mauerwerk: *Mauerwerk-Kalender* 17: 615–622.

Schueremans, L. 2001. Probabilistic evaluation of structural unreinforced masonry. Heverlee (Belgium): KU Leuven, KU Leuven.

DIN EN. 1990. Eurocode 0: Basis of structural design. Brussels: CEN European committee for standardization.

fib Model Code. 2010. Model Code for Concrete Structures: Ernst & Sohn Verlag.

Bujotzek, L. 2018. Monte Carlo-Simulation unbewehrter Mauerwerkswände unter Berücksichtigung räumlich streuender Materialeigenschaften. Technische Universität Darmstadt. Master´s thesis.

Melchers, R.E.; Beck, A.T. 108. *Structural reliability analysis and prediction*. Hoboken NJ: Wiley.

EN 191992-1-1 .2010. Eurocode 2: Design of concrete structures - Part 1-1: General rules and rules for buildings. Brussels: CEN European committee for standardization.

Brick and Block Masonry - From Historical to Sustainable Masonry –
Kubica, Kwiecień & Bednarz (eds)
© *2020 Taylor & Francis Group, London, ISBN 978-0-367-56586-2*

Second-order effects in URM walls subjected to combined vertical and lateral loading

M. Donà
Earthquake Engineering Research & Test Center (EERTC), Guangzhou University, China

P. Morandi & C.F. Manzini
European Centre for Training and Research in Earthquake Engineering (EUCENTRE), Italy

M. Minotto
Department of Civil, Architectural and Environmental Engineering (ICEA), University of Padova, Italy

F. da Porto
Department of Geosciences, University of Padova, Italy

G. Magenes
Civil Engineering and Architecture Department (DICAr), University of Pavia, Italy

ABSTRACT: The ongoing revision of Eurocode 6 is an occasion to reconsider the criteria for verifying the second-order effects in URM walls subjected to combined vertical and out-of-plane loading. The current method proposed in Eurocode 6 is based on an axial load capacity reduction factor (ϕ_m), the values of which are estimated with an approximate model, conservative for a significant wall stiffness range. Moreover, the fact that the safety verification of walls subjected to significant lateral loading (e.g. seismic action) should be carried out in terms of lateral flexural capacity, also accounting for second-order effects, is still not fully recognised. For this verification, a moment capacity reduction factor (ϕ_M) can be defined, similarly to ϕ_m. Therefore, this paper provides: a refined numerical evaluation of the ϕ factors; the demonstration of their one-to-one correspondence; new models for their prediction and the recalibration of a recently proposed model; comparisons with results from literature.

1 INTRODUCTION

Although the load-bearing capacity of unreinforced masonry (URM) walls has received attention from researchers since the 1950s, both analytically (e.g., Kukulski & Lugez 1966, Sahlin 1971, Sandoval & Roca 2013, Bakeer & Jäger 2016) and experimentally (e.g., Hasan & Hendry 1976, Drysdale & Hamid 1982, Cavaleri et al. 2005, Sandoval et al. 2011), the correct methodology to evaluate second-order effects in URM walls subjected to combined vertical and out-of-plane loading is still an important subject of scientific debate.

The axial load capacity of URM walls subjected to mainly vertical loading is affected by the slenderness ratio λ (i.e., the ratio between effective height h_0 and thickness t of the wall), the eccentricity of the applied vertical loads (generally due to the type of slab-to-wall support), the stiffness of masonry, the boundary conditions and the tensile strength of the masonry. In particular, the λ values can vary greatly from country to country, according to the construction traditions. In Northern Europe, for example, walls with λ values of 20 or above are employed even in medium-to-high rise buildings; in this situation, the effects of geometric non-linearity become very important and require appropriate evaluations.

Similarly, the lateral flexural capacity of URM walls subjected to combined vertical and relevant lateral loading (e.g. seismic loading) strongly depends on the slenderness ratio λ, the normalised vertical load v (i.e., $N/(f_k \cdot t \cdot l)$, where N is the total vertical load, l the wall length and f_k the characteristic compressive strength of masonry) and the stiffness of the masonry. In this situation, the out-of-plane (OOP) collapse is mainly due to an excess of OOP displacement rather than the achievement of the OOP strength of the wall, as discussed in previous scientific studies (e.g., Griffith et al. 2003, Morandi et al. 2008, da Porto et al. 2010 and 2011, Graziotti et al. 2016, Donà et al. 2018 and 2019).

Therefore, non-linear geometric effects are significant for walls subjected to mainly lateral actions as are the buckling effects for eccentrically loaded walls, and these effects are correlated.

The current design procedure proposed in EN 1996-1-1 (hereafter called EC6) to evaluate the second-order effects in URM walls is based on an axial load capacity reduction factor ϕ_m, which allows to evaluate the influence of wall slenderness, load eccentricity and boundary conditions on the buckling behaviour of the URM walls. However, the proposed ϕ_m model is outdated and based on a Gaussian bell-shaped approximation to represent both the material and stability failures, which is function of a single set of parameters that includes material strength. According to recent studies (e.g., Sandoval & Roca 2013), this approximation seems too conservative for a wall stiffness range significant for design purposes (i.e., for $E/f_k < 700$, having been calibrated for $E/f_k = 1000$, with E the elasticity modulus of the masonry). Also, as shown in Bakeer & Christiansen (2017) for the first time, this approximation leads to an inconsistency, i.e. the axial capacity of the wall calculated using ϕ_m can decrease as f_k increases while keeping all the other parameters constant (see Sec. 5).

Moreover, the fact that the safety verification of walls subjected to combined vertical and significant lateral loading should be rationally carried out in terms of lateral flexural capacity, has not yet been fully acknowledged and the current version of the EC6 is silent in this regard. To this end, a moment capacity reduction factor ϕ_M (similar to ϕ_m), which takes into account the second-order effects reducing the first-order resisting moment of walls, was initially proposed by Morandi et al. (2008); however, to date it has not yet been implemented in any standard.

The ongoing revision of EC6 is thus an occasion to reconsider the current criteria for verifying the second-order effects in URM walls subjected to combined vertical and out-of-plane loading.

Therefore, this paper offers a refinement of the numerical evaluations of ϕ_m and ϕ_M (removing almost all simplifying hypotheses compared to previous proposals), proves the one-to-one correspondence between them, provides new prediction models and, finally, compares the results obtained with the experimental and numerical ones in the literature.

2 VERIFICATION OF SECOND-ORDER EFFECTS IN URM WALLS

2.1 Axial load capacity reduction factor ϕ_m

In most of the codes for structural design and assessment of URM buildings, as in EC6, the safety verifications at ultimate limit state on walls subjected to vertical compression and lateral (out-of-plane) non-seismic loading, e.g. wind, are performed in terms of vertical strength, comparing the vertically applied design axial load N_{Ed} with the wall section compressive strength reduced by a factor ϕ:

$$N_{Ed} \leq N_{Rd} = \phi \cdot t \cdot l \cdot f_d \qquad (1)$$

where f_d is the design compressive strength of the masonry, equal to the characteristic strength f_k divided by a material partial safety factor γ_M.

According to EC6, the safety checks need to be performed in terms of vertical load per unit length, at the top and at the bottom of the walls applying a coefficient ϕ_i, and at mid-height using a coefficient ϕ_m. The coefficient ϕ_i only depends on the transversal eccentricity e_i at the ends of the walls ($\phi_i = 1 - 2e_i/t$), whereas the coefficient ϕ_m, which implicitly takes into account the second-order effects, depends on the transversal eccentricity at mid-height e_m, on the wall slenderness λ (h_0/t), on the boundary conditions and on the ratio E/f_k. In EC6, the plots of ϕ_m against slenderness λ as a function of e_m/t for different values of E/f_k (1000 and 700) are reported.

2.2 Moment capacity reduction factor ϕ_M

Regarding safety checks on URM walls subjected to vertical loads and lateral out-of-plane seismic actions, Morandi et al. (2008) developed a simplified procedure to properly consider second-order geometric effects through the introduction of a capacity reduction coefficient ϕ_M, applied to reduce the out-of-plane first-order resisting moment M_{Rd}. The values of ϕ_M were evaluated at mid-height of the wall under the hypotheses of non-linear (parabolic) stress-strain relationship for masonry and of curvature distribution proportional to the uncracked one. For each combination of slenderness λ and dimensionless vertical load ν, the reduction coefficient ϕ_M was computed as the ratio between the peak of the displacement-moment curve and the ultimate resisting moment of the first order M_{Rd}. In the proposed methodology, the safety checks can be performed comparing the acting out-of-plane moment M_{Ed} with the first-order resisting moment M_{Rd} reduced by ϕ_M:

$$M_{Ed} \leq M_{Rd,II} = \phi_M \cdot M_{Rd} \qquad (2)$$

The moment M_{Ed} can be computed, under the hypothesis of centred load on hinged wall, as in Equation 3, where q is the equivalent pressure due to the seismic action, whereas, M_{Rd} is provided by Equation 4, where η is the factor defining the equivalent rectangular stress block.

$$M_{Ed} = qh^2/8 \qquad (3)$$

$$M_{Rd} = \frac{N_{Ed}t}{2}\left[1 - \frac{N_{Ed}}{\eta \cdot f_d \cdot t \cdot l}\right] \qquad (4)$$

The ratio ϕ_M was plotted in graphs as a function of slenderness λ for a range of values of normalised axial load $\nu=N_{Ed}/(f_k \cdot t \cdot l)$; therefore, from λ and ν, it is possible to obtain ϕ_M.

EN 1998-1 does not provide any indication on the out-of-plane verification of load-bearing walls, whereas other codes, like the Italian NTC2018, require an out-of-plane bending verification though without specific indications to account for second-order effects.

3 NUMERICAL EVALUATION OF SECOND-ORDER EFFECTS IN URM WALLS

3.1 Definition of the parametric case studies

The parametric evaluation of the capacity reduction factors ϕ_m and ϕ_M was based on the static schemes of simply supported beam of Figure 1, referred to with the ratio between the load eccentricity at the top (e_1) and that at the mid-height (e_m) of the wall.

In particular, the static scheme with $e_1/e_m=1$ (with a constant load eccentricity) is generally the reference one for the verification of the ultimate axial capacity of the wall, although a linear eccentricity profile (i.e. $e_1/e_m=2$) could be more representative in some specific applications (see Donà et al. 2020 for the evaluation of ϕ factors in the latter case). Instead, the static scheme with $e_1/e_m=0$ (with a uniformly distributed lateral load q, well representative of the seismic action) is generally taken as a reference to verify the ultimate OOP bending capacity of the wall. As the two safety verifications refer to different load conditions, different ϕ factors (ϕ_m and ϕ_M, respectively) should be defined to rationally assess the second-order effects in URM walls.

The case studies analysed in this study derive from the combination of the following parameter values (which cover a sufficiently wide range of practical applications): E/f_k [-] = 500, 700, 1000; f_k [MPa] = 1, and from 2.5 to 15, with steps of 2.5; λ [-] = 5 to 30, with steps of 2.5; t [m] = 0.1 to 0.45, with steps of 0.05; l [m] = 1; w [kN/ m³] = 0, 10 (with w the self-weight per unit of volume).

Then, the evaluation of ϕ_m was performed for values of normalised load eccentricity e_m/t from 0.05 to 0.4, with steps of 0.05. Instead, ϕ_M was evaluated for values of normalised vertical load ν_m (at mid-height of the wall) of 0.01, and from 0.05 to 0.5 with steps of 0.05.

The whole study is defined on the basis of the characteristic strength (f_k), so as to obtain code-compliant results that can be directly implemented in standards and applied in practice. Nevertheless, the parametric analysis is carried out with the mean strength (f_m) and the corresponding E/f_m ratios, to realistically evaluate the second-order effects. A conversion factor $f_k/f_m=0.8$ was conservatively assumed based on the indications of EN 1052-01, obtaining E/f_m values of 400, 560 and 800.

The stress–strain (σ–ε) non-linear relationship of masonry was effectively described with a parabolic law (Brencich & de Felice, 2009), which is permitted by EC6 and, moreover, is simple, requiring only two parameters for its definition, E/f_m and f_m. A main assumption was that to interrupt the σ–ε laws at the maximum strain $\varepsilon_{mu}=0.35\%$. An example of σ–ε laws is shown in Figure 2 for three f_m values. In particular, the strain ε_{m1}, related to f_m, is 0.25% for $E/f_k=1000$ and 0.35% for $E/f_k \leq 700$, being limited by ε_{mu}.

Therefore, the values of ε_{m1} and ε_{mu} analysed in this study are slightly larger than those provided by EC6 for design purposes, which are respectively 0.2% and 0.35% for masonry units belonging to Group 1, and 0.175% and 0.2% for the other types of units. However, these values are confirmed by previous experimental research (in particular, the increase of ε_{m1} when E/f_k decreases, see Donà et al. 2020) and, for the purpose of estimating the second-order effects, they are more appropriate or at most conservative (especially for hollow blocks, see Morandi et al. 2018 and 2019).

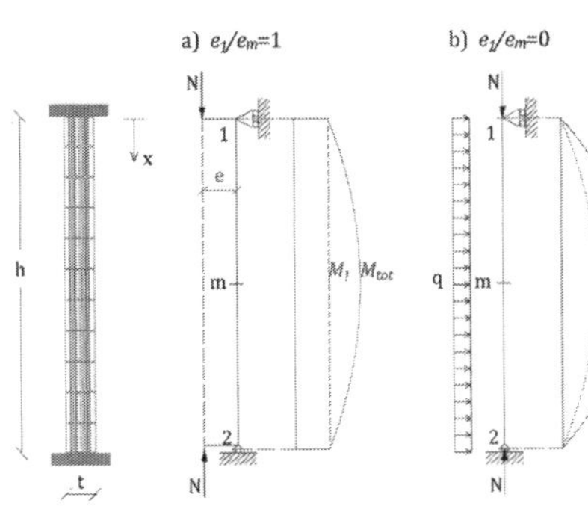

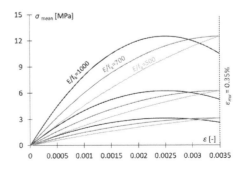

Figure 1. Static schemes assumed to assess second-order effects.

Figure 2. Examples of σ–ε curves evaluated in the analyses.

189

3.2 Calculation of the capacity reduction factors

The reduction factors ϕ_m and ϕ_M were evaluated for both static schemes of Figure 1 and for all the case studies (wall models discretised in 150 mm high elements) through the incremental static procedure summarised in Table 1, whose final output is the ultimate balanced deformed configuration of the wall under second-order effects.

For the evaluation of ϕ_m, the fixed parameter of the analysis is the eccentricity ratio e_m/t and the incremental variable is the normalised vertical load v_m. Instead, for ϕ_M, the fixed parameter is v_m and the incremental variable is e_m/t, which for convenience can be expressed as normalised eccentricity $e'_m \equiv e_m/e_{m,max}$, with $e_{m,max}$ the maximum eccentricity available at the wall mid-height without considering second-order effects (i.e., $e_{m,max} \equiv M_{m,max}/N_m$, with $M_{m,max}$ the maximum resisting moment and N_m the axial load in the relevant section).

The procedure requires the calculation of the moment-curvature (M–χ) diagram for each element of the wall, considering its specific value of v (variable when considering the self-weight of the wall w). For ϕ_m, the M–χ diagrams are variable during the analysis

(as $M_{m,max}$) due to the incremental variable v_m, whereas they are constant in the evaluation ϕ_M.

Then, the first-order moments (M_I) at the centre of each element and the first-order curvature (χ_I) and lateral displacement (Δ_I) profiles along the wall height can be calculated. At this point, the second-order and total moments ($M_{tot}=M_I+M_{II}$, see Table 1) and the actual curvature (χ_{tot}) and displacement (Δ_{tot}) profiles, which include the second-order effects, can be evaluated on the basis of Δ_I and a convergence criterion on the value of maximum displacement. To verify that the external total moment (M_{tot}) does not exceed the maximum bending capacity (M_{max}, from the M–χ diagram), a resistance check is carried out after each update of the displacement profile (Δ_{tot}).

When convergence is attained, and $M_{tot} \leq M_{max}$, the incremental variable is increased and the whole analysis is repeated from the beginning. This is iterated until reaching the last balanced value of the incremental variable, i.e., $v_{m,ult}$ for ϕ_m and $e'_{m,ult}$ for ϕ_M.

Finally, the reduction factors can be evaluated as:

$$\phi_m = N_{m,ult}/N_{max} = \left(v_{m,ult}f_k lt\right)/\left(f_m lt\right) = v_{m,ult}f_k/f_m \tag{5}$$

$$\phi_M = M_{m,ult}/M_{m,max} = e_{m,ult}/e_{m,max} = e'_{m,ult} \tag{6}$$

where $N_{m,ult}$ and $M_{m,ult}$ are the ultimate balanced axial force and moment at mid-height of the wall, and N_{max} is the axial capacity of the wall section without considering the second-order effects. Although $v_{m,ult}$ and ϕ_m are conceptually similar, they differ because v is defined on the characteristic strength (f_k).

Examples of application of this procedure (both for ϕ_m and ϕ_M) are reported in Donà et al. 2020.

3.3 Results of the numerical analyses

Figure 3 shows the values of ϕ_m and ϕ_M for both static schemes analysed and for E/f_k=700.

As for ϕ_m, these considerations can be drawn:

(i) ϕ_m strongly depends on the slenderness λ and, over the entire range of λ, also on the load eccentricity e_m/t.

(ii) The absolute reduction of ϕ_m with respect to λ is greater for low values of e_m/t, because the second-order effects in this case are mainly due to the high value of load ($v_{m,ult}$), which must be considerably reduced to ensure equilibrium as λ increases.

(iii) The ϕ_m curves have different concavity in the range of small λ values, specifically from convex (less sensitive to second-order effects) to concave (more sensitive) as e_m/t increases. This is related to the flexural strength of the section that, being also a function of the vertical load, is greater for lower e_m/t values, which are associated with higher $v_{m,ult}$ values.

Table 1   Main procedural steps to evaluate ϕ_m and ϕ_M.

Common Part	
First-order moment M_I	a) e_I/e_m=1: $M_I = N_{top}e_m$ b) e_I/e_m=0: $M_I(x) = qh\frac{x}{2} - q\frac{x^2}{2}$
Second-order moment M_{II}	$M_{II}(x) = N^{top} \cdot \Delta(x) +$ $+ \sum_{i=1}^{ele(x)-1} w_{ele,i}\left(\Delta(x) - \Delta\left(x_{ele,i}\right)\right)$
Procedure to evaluate ϕ_m	
Analysis input	eccentricity ratio e_m/t: 0.05, 0.1, 0.15, 0.2, 0.25, 0.3, 0.35, 0.4 (for e_I/e_m=0: load q s.t. $M_m(q)/N_m$=e_m)
Incremental variable	normalised vertical load v_m
M–χ diagram	variable during the analysis (as $M_{m,max}$)
Analysis output	ultimate normalised vertical load $v_{m,ult}$
ϕ_m	$\phi_m = N_{m,ult}/N_{max} = v_{m,ult}f_k/f_m$
Procedure to evaluate ϕ_M	
Analysis input	normalised vertical load v_m: 0.01, 0.025, 0.05, 0.075, 0.1, 0.15, 0.2, 0.25, 0.3, 0.35, 0.4, 0.45 0.5
M–χ diagram	constant during the analysis (as $M_{m,max}$) $e_{m,max}$=$M_{m,max}/N_m \rightarrow e'_m \equiv e_m/e_{m,max}$
Incremental variable	normalised eccentricity e'_m (for e_I/e_m=0: e'_m=q/q_{max})
Analysis output	ultimate normalised eccentricity $e'_{m,ult}$
ϕ_M	$\phi_M = M_{m,ult}/M_{m,max} = e_{m,ult}/e_{m,max} = e'_{m,ult}$

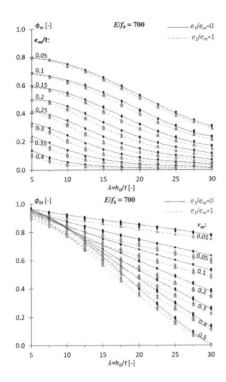

Figure 3. Φ_m and ϕ_M versus slenderness ratio λ, E/f_k=700.

$$\phi_m = c_1 + c_2 \frac{e_m}{t} + c_3 \frac{E}{f_k} + c_4 \frac{e_m}{t} \lambda + c_5 \frac{e_m}{t} \frac{E}{f_k} +$$
$$+ c_6 \lambda^2 + c_7 \frac{e_m}{t} \lambda^2 + c_8 \lambda \left(\frac{e_m}{t}\right)^2 + c_9 \lambda^3 \quad (7)$$

$$\phi_M = c_1 + c_2 \lambda \nu_m + c_3 \frac{E}{f_k} + c_4 \lambda \nu_m \frac{E}{f_k} + c_5 \lambda^2 +$$
$$+ c_6 \nu_m{}^2 + c_7 \lambda \nu_m{}^2 + c_8 \frac{E}{f_k} \nu_m{}^2 + c_9 \nu_m{}^3 + c_{10} \lambda^3$$

$$(8)$$

The comparisons shown in Figure 4 between the numerical results obtained in Section 3 and the ϕ values from Equations 7 and 8 demonstrate the effectiveness of the proposed models, which predict fairly well and conservatively the numerical results. The comparisons are shown only for the models ϕ_m (e_1/e_m=1) and ϕ_M (e_1/e_m=0), and for E/f_k=700; however, for all calibrated response models, the coefficient of determination R^2 is about 99%.

It is worth noting that the factors ϕ were calculated using the tangent elastic modulus E (see Figure 2), whereas EC6 refers to the secant one E_s, at 1/3 of the resistance; the proposed ϕ models can however be used with E_s, obtaining slightly conservative ϕ values.

Having evaluated both ϕ factors for each static scheme, it is possible to numerically demonstrate their one-to-one correspondence for the same static scheme. To this end, each of the Equations 7 and 8 can be expressed as a function of the parameters of the other equation using Equations 9 and 10.

$$\phi_m = v_{m,ult} f_k / f_m \equiv v_m \cdot f_k / f_m \quad (9)$$

$$\phi_M = \frac{e_{m,ult}/t}{e_{m,max}/t} = \frac{e_{m,ult}/t}{0.5 - (\delta_G/\psi) \cdot v_m} \equiv \frac{e_m/t}{0.5 - (\delta_G/\psi) \cdot v_m} \quad (10)$$

As for ϕ_M, analogous considerations can be drawn:

(i) ϕ_M is strongly influenced by λ and the normalised vertical load v_m (especially for high λ values).

(ii) The reduction of ϕ_M with increasing λ is greater for high v_m values, as second-order effects increase as v_m increases.

(iii) For λ<10, a trend reversal of ϕ_M with respect to v_m can be observed (i.e., for v_m>0.2 ϕ_M tends to grow rather than decrease); the reason is that discussed in point iii) of ϕ_m.

In addition, both ϕ_m and ϕ_M decrease as E/f_k decreases, especially for high values of slenderness, and low eccentricities (ϕ_m) or high vertical loads (ϕ_M), i.e. when the second-order effects are more important.

Finally, the static scheme e_1/e_m=1, with constant eccentricity, provides lower values of both ϕ_m and ϕ_M.

4 PREDICTION MODELS OF Φ FACTORS AND ONE-TO-ONE CORRESPONDENCE

Prediction models of the capacity reduction factors were defined as a function of the principal parameters investigated. In particular, a polynomial-type model was chosen for each ϕ factor (see Equations 7 and 8) and was then conservatively calibrated, for both static schemes, on the minimum values obtained for each combination of the parameters analysed. The calibration coefficients are reported in Table 2.

Table 2   Coefficients of the prediction models.

| | ϕ_m | | ϕ_M | |
	e_1/e_m=0	e_1/e_m=1	e_1/e_m=0	e_1/e_m=1
c_1	0.70559	0.7201	0.93365	0.9142
c_2	-1.33954	-1.4541	-0.12775	-0.13159
c_3	3.18E-04	3.09E-04	3.81E-06	4.57E-06
c_4	-0.06724	-0.07107	4.64E-05	3.94E-05
c_5	-7.03E-04	-7.09E-04	-6.26E-04	-7.99E-04
c_6	-9.88E-04	-1.11E-03	3.44555	3.17347
c_7	2.07E-03	2.00E-03	0.05895	0.08202
c_8	0.08996	0.11697	-7.20E-04	-6.30E-04
c_9	1.31E-05	1.70E-05	-4.09565	-3.93825
c_{10}	-	-	1.21E-05	1.77E-05

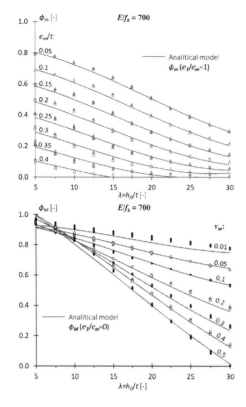

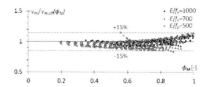

Figure 5. Ratios $v_m/v_{m,ult}$ versus ϕ_M, for the case $e_1/e_m=0$.

ϕ_M model divided output value of the ϕ_m model) is equal to 1 in the case of a one-to-one correspondence between ϕ_M and ϕ_m. Figure 5 shows these ratios versus ϕ_M, for the case $e_1/e_m=0$ and the entire range of the parameter values analysed. Apart from a numerical interpolation error, these ratios tend to 1, proving the one-to-one correspondence between the two factors.

5 VALIDATION OF Φ MODELS WITH AVAILABLE TESTS AND LITERATURE

The comparisons between the ϕ_m values predicted by the model ϕ_m ($e_1/e_m=1$) of Equation 7 and those obtained from tests (Figure 6) and analytical models (Figure 7) available in the literature are shown below. The values of ϕ_m are plotted versus the normalised slenderness $\bar{\lambda}$, defined as $\lambda/(E/f_m)^{0.5}$, as in EC6.

Figure 6 confirms the goodness of the proposed ϕ_m model, which predicts well and conservatively the test results of four different experimentations. The model also succeeds in capturing the different trends of ϕ_m on $\bar{\lambda}$, depending on the material type; this proves the appropriate modelling of the material behaviour (σ–ε laws), which generally does not scale linearly with E/f_k due to the limit strain ε_{mu}. Small differences concern the cases with: very low values of E/f_m, not covered by this study and by EC6; very

Figure 4. Comparisons between numerical and predicted values of ϕ_m ($e_1/e_m=1$) and ϕ_M ($e_1/e_m=0$), for $E/f_k=700$.

In Equation 10, $e_{m,max}/t$ is expressed through the dimensionless parameters δ_G and ψ of the σ–ε diagram: δ_G is the distance of the centroid of the diagram from ε_{mu}, normalised to ε_{mu}; ψ is the ratio between the area of the diagram and the area of the rectangular stress block ($f_k \cdot \varepsilon_{mu}$). For E/f_k respectively of 500, 700 and 1000, δ_G is 0.353, 0.374 and 0.406, and ψ is 0.708, 0.825 and 0.933.

Using the relationships above, the model of ϕ_m (Equation 7) can be expressed as:

$$\frac{v_{m,ult} f_k}{f_m} = c_1 + c_2\left(\frac{\phi_M e_{m,max}}{t}\right) + c_3\frac{E}{f_k} + c_4\left(\frac{\phi_M e_{m,max}}{t}\right)$$
$$\lambda + +c_5\left(\frac{\phi_M e_{m,max}}{t}\right)\frac{E}{f_k} + c_6\lambda^2 + c_7\left(\frac{\phi_M e_{m,max}}{t}\right)$$
$$\lambda^2 + c_8\lambda\left(\frac{\phi_M e_{m,max}}{t}\right)^2 + c_9\lambda^3 T_{ke}\left[1 + (\mu)^{0.5}\right]/2$$

(11)

Therefore, the ϕ_M values predicted by Equation 8 (as a function of E/f_k, λ and v_m) can be used in the model of ϕ_m, rewritten as in Equation 11 (as a function of E/f_k, λ and ϕ_M), to compute the related $v_{m,ult}$ values. Theoretically, the ratio $v_m/v_{m,ult}$ (i.e., input value of the

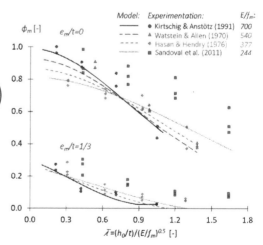

Figure 6. Experimental and predicted (Eq. 7, $e_1/e_m=1$) ϕ_m values.

high values of $\bar{\lambda}$(>1.2, i.e. λ>27 for E/f_k=500) and large eccentricities (e_m/t=1/3), due to the conservative assumption of null tensile strength of masonry (compliant with the EC6 design approach).

Figure 7 compares the proposed ϕ_m model (e_1/e_m=1) with those from EC6 and recent studies. In particular, the model proposed by Sandoval & Roca (2013) is based on a Finite-Element (FE) parametric analysis, with a FE model calibrated on the test results of URM walls with medium-low E/f_m values (as those of Figure 6). The model of Bakeer & Christiansen (2017) is instead based on two analytical laws (i.e. parabolic and hyperbolic, to represent the material and stability failure respectively) and depends on a ζ factor, accounting for the material non-linearity, which was calibrated on experimental ϕ_m values.

From Figure 7, these considerations can be drawn:

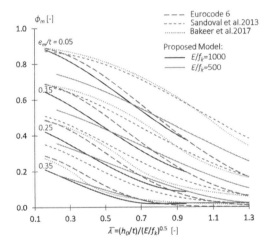

Figure 7. Φ_m values from Eq. 7 (e_1/e_m=1), EC6 and literature.

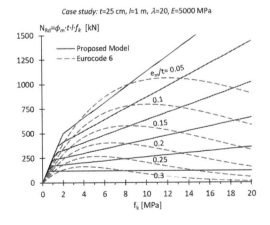

Figure 8. Reduced axial capacity of the wall, calculated by using the proposed ϕ_m model (Eq. 7, e_1/e_m=1) and that of EC6.

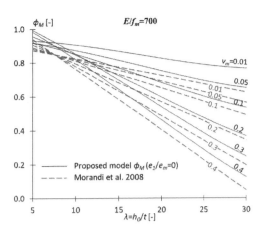

Figure 9. Φ_m values from Eq. 8 (e_1/e_m=0) and literature.

(i)   For high $\bar{\lambda}$ values, EC6 becomes too conservative, as already discussed also by the above authors.

(ii) For low values of $\bar{\lambda}$ and e_m/t (for which the EC6 model was mainly calibrated), the proposed model provides results similar to those of EC6, whereas for high e_m/t values it is more conservative than EC6.

(iii) For e_m/t>0.3, also the model of Bakeer & Christiansen (2017) becomes more conservative than EC6, unlike the Sandoval & Roca (2017) model (being calibrated on low E/f_m values, for which the effects of tensile strength become appreciable).

(iv) The proposed model, calibrated on the minimum results obtained, is generally more conservative than those of the other authors, calibrated on the average of experimental ϕ_m values (with high dispersion).

As mentioned in the Introduction, the EC6 model leads to an inconsistency, which is shown in Figure 8: the reduced axial capacity of the wall, calculated using ϕ_m, can decrease as f_k increases while keeping all the other parameters constant. As can also be seen, the proposed model overcomes this inconsistency.

Lastly, Figure 9 compares the proposed ϕ_M model (e_1/e_m=0) with that of Morandi et al. (2008). As the latter is based on the mean strength f_m, the comparison is shown in terms of E/f_m. The model of Morandi et al. (2008) is more conservative, due to the simplified assumption on curvature discussed in Section 2.1.

6   CALIBRATION OF A MODEL PROPOSED FOR THE REVISION OF EC6

The Bakeer & Christiansen (2017) model discussed above and presented in Equation 12, has recently been proposed for inclusion in the next version of EC6.

$$\phi_m = \begin{cases} A_1 - \dfrac{\bar{\lambda}^2}{3.15\zeta A_1} & \text{if } \bar{\lambda} < A_1\sqrt{1.575\zeta} \\ 0.79\zeta\dfrac{A_1^3}{\bar{\lambda}^2} & \text{if } \bar{\lambda} \geq A_1\sqrt{1.575\zeta} \end{cases} ; A_1 = 1 - 2\dfrac{e_m}{t}$$

$$(12)$$

As discussed before, this model depends on a correction factor ζ that was calibrated on the average of a few experimental ϕ_m values, resulting in $\zeta=1$.

Given the one-to-one correspondence between the ϕ factors, Equation 12 was reversed (using Equations 9 and 10) to determine the associated ϕ_M model:

$$\phi_M = \frac{e_m/t}{(e_m/t)_{max}} = \frac{(1-A_1)}{(1-v_m)} \qquad \text{where:}$$

$$\begin{cases} A_1 = \dfrac{3.15\zeta v_m + \sqrt{(3.15\zeta v_m)^2 + 12.6\zeta\bar{\lambda}^2}}{6.3\zeta} & \text{if } \bar{\lambda} < A1\sqrt{1.575\zeta} \\ A_1 = \left[\dfrac{\bar{\lambda}^2 v_m}{0.79\zeta}\right]^{1/3} & \text{if } \bar{\lambda} \geq A1\sqrt{1.575\zeta} \end{cases}$$

$$(13)$$

Then, the ϕ_m model of Bakeer & Christiansen (2017) and the associated one of ϕ_M were used to recalibrate the ζ factor on the extensive numerical results (minimum values) obtained in this study. In particular, a multi-objective optimisation procedure was applied to simultaneously maximise the R^2 values of both ϕ models, thus obtaining the optimal ζ value for ϕ_m ($\zeta=0.55$), for ϕ_M ($\zeta=0.9$) and for the best compromise between ϕ_m and ϕ_M ($\zeta=0.74$). The comparisons between the different calibrations and the minimum numerical values obtained in this study, for both ϕ factors, are shown in Figure 10.

As for ϕ_m, these considerations can be drawn:

(i) The model with $\zeta=0.55$ (best fit of ϕ_m) has a greater gradient versus λ and turns out to be non-conservative for low to medium values of e_m/t and λ, and conservative for medium to high values.

(ii) The model with $\zeta=0.74$ (best compromise) has a trend more similar to the numerical one, although the results are generally non-conservative.

As for ϕ_M, these considerations can be drawn:

(i) The model with $\zeta=0.9$ (best fit of ϕ_M) provides a fairly good prediction of ϕ_M values, although slightly non-conservative for higher v_m values.

(ii) The model with $\zeta=0.74$ (best compromise) provides a prediction of ϕ_M values globally less good, but conservative on the whole range of v_m and λ.

Therefore, $\zeta=0.74$ seems on average appropriate for both ϕ models, and in any case better than $\zeta=1$, currently proposed to evaluate ϕ_m. Alternatively, it is possible to use $\zeta=0.55$ for ϕ_m and $\zeta=0.9$ for ϕ_M.

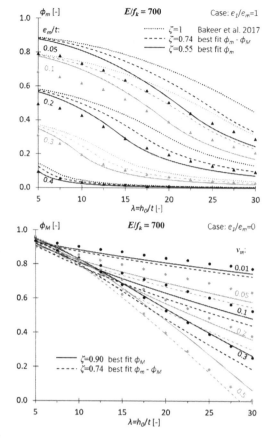

Figure 10. Φ_m values: numerical (dots); Bakeer & Christiansen ($\zeta=1$); calibration with $\zeta=0.74$ and $\zeta=0.55$. ϕ_M values: numerical (dots); calibration with $\zeta=0.74$ and $\zeta=0.90$.

For further details on the proposed models, on the comparisons with the results/models available in the literature (also in other standards) and on the recalibration of the model of Bakeer & Christiansen (2017), please refer to Donà et al. 2020.

7 CONCLUSIONS

A general procedure was presented to numerically evaluate both the axial load capacity reduction factor ϕ_m (used for safety checks of URM walls subjected to gravitational and wind loads) and the moment capacity reduction factor ϕ_M (for walls subjected to seismic actions). Almost all the simplified hypotheses used in previous studies were removed.

For design purposes, polynomial-type prediction models were defined and calibrated for both ϕ factors, based on the results obtained. These models were useful to demonstrate the one-to-one correspondence between ϕ_m and ϕ_M and were validated with the experimental and numerical results available in the literature. Also, a recent ϕ_m model proposed for the revision of EC6 (based on an empirical factor) was recalibrated on the extensive results obtained.

Finally, the lack of an exact theoretical solution for ϕ_m and ϕ_M (therefore the need to adopt approximate models) and the fact that these factors refer to different load conditions, justify the application of both ϕ models for the related safety checks.

REFERENCES

Bakeer, T. & Chrisiansen, P.D. 2017. Buckling of masonry walls A new proposal for the Eurocode 6. *Mauerwerk* 21(2):82–89.

Bakeer, T. & Jäger, W. 2016. Determination the capacity reduction factor of masonry walls under buckling - a numerical procedure based on the transfer-matrix method. *Proc. of the 16th IB²MAC, Padova, Italy*, June 2016.

Brencich, A. & de Felice, G. 2009. Brickwork under eccentric compression: Experimental results and macroscopic models. *Construction and Building Materials* 23:1935–1946.

Cavaleri, L., Failla, A., La Mendola, L. & Papia, M. 2005. Experimental and analytical response of masonry elements under eccentric vertical loads. *Eng Struct* 27(8):1175–84.

da Porto, F., Mosele, F. & Modena, C. 2010. Experimental testing of tall reinforced masonry walls under out-of-plane actions. *Construction and Building Materials* 24:2559–2571.

da Porto, F., Mosele, F. & Modena, C. 2011. Cyclic out-of-plane behaviour of tall reinforced masonry walls under P–Δ effects. *Engineering Structures* 33(2): 287–97.

Donà, M., Minotto, M., Verlato, N. & da Porto, F. 2019. Proposal of simplified design methods to evaluate second-order effects in tall reinforced masonry walls. *Proc. of 13th North American Masonry Conference, Salt Lake*, June 2019.

Donà, M., Morandi, P., Minotto, M., Manzini, C.F., da Porto, F. & Magenes, M. 2020. Second-order effects in URM walls subjected to compression and out-of-plane bending: from numerical evaluation to proposal of design procedures. *Engineering Structures 209 (2020)* 110130.

Donà, M., Tecchio, G. & da Porto, F. 2018. Verification of second-order effects in slender reinforced masonry walls. *Materials and Structures* 51:69.

Drysdale, R. & Hamid, H. 1982. Effect of eccentricity on the compressive strength of brickwork. *Journal of the British Ceramic Society* 30: 140–149.

EN 1052-1. Methods of test for masonry, Part 1: Determination of compressive strength. *European Committee for Standardisation*, Brussels, Belgium; 1998.

EN 1996-1-1.Eurocode 6: Design of masonry structures, Part 1-1: Common rules for reinforced and unreinforced masonry structures. *European Committee for Standardisation*, Brussels, Belgium; 2004.

EN 1998-1. Eurocode 8: Design of structures for earthquake resistance, Part 1: General rules, seismic actions and rules for building. *European Committee for Standardisation*, Brussels, Belgium; 2005.

Graziotti, F., Tomassetti, U., Penna, A. & Magenes, G. 2016. Out-of-plane shaking table tests on URM single leaf and cavity walls. *Engineering Structures* 125:455–470.

Griffith, M.C., Magenes, G., Melis, G. & Picchi, L. 2003. Evaluation of Out-of-Plane Stability of Unreinforced Masonry Walls Subjected to Seismic Excitation. *Journal of Earthquake Engineering* 7 (SI1): 141–169.

Hasan, S.S. & Hendry, A.W. 1976. Effect of slenderness and eccentricity on the compressive strength of walls. *Proc. of the 4th international brick masonry conference, Brugge.*

Kukulski, W. & Lugez, J. 1966. Résistance des murs en béton non-armé soumis a des charges verticals, Théorie - Expérimentation. *Cahiers du Centre Scientifique et Technique du Batiment*, N°79, Cahier 681.

Morandi, P., Albanesi, L. & Magenes, G. 2019. In-plane cyclic response of new URM systems with thin web and shell clay units. *Journal of Earthquake Engineering.*

Morandi, P., Albanesi, L., Graziotti, F., Li Piani, T., Penna, A. & Magenes, G. 2018. Development of a dataset on the in-plane experimental response of URM piers with bricks and blocks. *Construction and Building Materials* 190:593–611.

Morandi, P., Magenes, G. & Griffith. M.C. 2008. Second-order effects in out-of-plane strength of unreinforced masonry walls subjected to bending and compression. *Australian Journal of Structural Engineering* 8:133–144.

NTC2018. Decreto Ministeriale 17/01/2018: Aggiornamento delle "Norme tecniche per le costruzioni", *Ministero delle Infrastrutture e dei Trasporti*, Rome, Italy [in Italian]; 2018.

Sahlin S.1971. Structural masonry. Englewood Cliffs, N.J.: Prentice-Hall.

Sandoval, C. & Roca, P. 2013. Empirical equations for the assessment of the load-bearing capacity of brick masonry walls. *Construction and Building Materials* 44:427–439.

Sandoval, C., Roca, P., Bernat, E. & Gil, Ll. 2011. Testing and numerical modelling of buckling failure of masonry walls. *Construction and Building Materials* 25 (12):4394–402.

Case studies

Brick and Block Masonry - From Historical to Sustainable Masonry –
Kubica, Kwiecień & Bednarz (eds)
© 2020 Taylor & Francis Group, London, ISBN 978-0-367-56586-2

An interactive tool for historic masonry buildings stability studies

P. Nougayrède
ENSA Paris Malaquais, Laboratoire GSA, Paris, France

T. Ciblac
ENSA Paris Malaquais, Laboratoire GSA, Paris, France

F. Guéna
ENSA Paris la Villette, MAP-Maacc/CNRS-MCC UMR, Paris, France

ABSTRACT: These last decades, studies interest about historic masonry buildings has highly increased. The use of limit analysis and more particularly, Jacques Heyman's works in the second half of the twentieth century widely contributed to this development. Considering three hypotheses based on most failures observations (infinite compression resistance, no traction resistance and no sliding between the blocks), the static approach gives a simple and easy criterion of potential stability. The set of all the application points of the forces on each joint, as known as the line of thrust, must be contained within the geometric limits of each joint in order to insure the structure potential stability. The aim of this research is to develop an interactive tool able to consider 2D geometries extracted from a 3D model (surveys for instance) and able to explore various hypotheses. Using the multi-agent system principles of programming and a genetic algorithm, a computing tool which can apply this static theorem for any kind of two-dimensional geometry structure composed of blocks has been developed as a plugin for the 3D modeling software Rhinoceros. The user can set different kind of hypotheses, calculus settings and loadings: fixing chosen points of the thrust line, add forces and seismic solicitations. The tool interactivity allows the user to move the joints force characteristics (point of application, direction and intensity) and see the graphic result immediately with the line of thrust and the force diagram. In addition, the tool provides a security friction coefficient for each joint and the possibility to find the extreme thrusts for any geometrical topologic equivalent of an arch. The tool possibilities are evaluated with the application case of the Sainte-Marie-Madeleine's Basilic in Vézelay.

1 INTRODUCTION

Thanks to the numerical calculation techniques advances, the stability studies of historic masonry buildings has been widely improved. The relevance for using yield design theory and limit analysis in order to define the allowed stability range of these structures has already been proved in different studies in which the observed stones failures correspond most of the time to the failures highlighted by this theoretical approach. Moreover, the difficulties involved in the complexity of stone numerical modelling are overcome because yield design and limit analysis theories are based on easier hypotheses to describe (these theories only require knowing the geometry of the structure and the resistance criterion of the material). We can cite the works of Delbecq (1983) and Smars (2000) who developed numerical tools based on yield design and limit analysis (respectively VOUTE and CALIPOUS) specifically restricted to the study of masonry bridges and Brabant's gothic vaults. The force network method developed by O'Dwyer (1999) has also been used for studies of bidimensional structures by Oikonomopolou (2009). An extension for the force network method has

been implemented by Fantin (2017), with the MANA-COH tool, and has been applied to several case studies of 2D and 3D masonry structures, but this method requires some theoretical constraints (the loads applied to the structure must be parallel and final results are more conservative). For these reasons, and in order to maintain enough flexibility for the different functionalities allowed by the tool, we will prefer the more general use of limit analysis applied to masonry as part of the bidimensional structures studies.

In addition to the possibility of an educational interest, the development of a simple, intuitive and precise interactive tool would make accessibility easier for different kind of users (architects, engineers or historians). Thus, the main objective is to propose a tool which can be used for case studies with various geometries and complexity, while allowing a range of functionalities that supports the structural analysis as much as possible. As part of the MONUMENTUM research, funded by the ANR, whose objective is to establish a diagnosis process of historic masonry buildings, Thierry Ciblac and François Guéna (2019) demonstrated the feasibility of such a tool using an object-oriented programming logic which allows the implementation of a multi-agent system, using a genetic algorithm. From the prototype realized in the Grasshopper working environment (Rhinoceros software's parametric modelling plugin), the interactive tool development and its different functionalities in the form of a Rhinoceros plugin, programmed in Python language, was the subject of Paul Nougayrede's master thesis with the Polytechnic Faculty of Mons.

2 THEORETICAL FRAMEWORK: YIELD DESIGN, LIMIT ANALYSIS AND ALGORITHMIC IMPLEMENTATION

2.1 Yield design theory and limit analysis for masonry stability's study

As formalized by Salençon (1983), yield design theory makes the determination of the structure potential stability range possible by only knowing the geometry of the structure and the resistance criterion of the material, for a defined loading mode. In particular, limit analysis makes it possible to use yield design theory approaches by considering that the material has a perfectly elastic plastic behavior: this hypothesis allows to determine the strict stability domain of the structure (and not only the potential stability domain as in the more general framework of yield design theory), for a defined loading mode. Thus, Heyman (1995) has laid the foundations of limit analysis for masonry structures: assuming three hypotheses of stones structures behavior (resistance of blocks to infinite compression, no tensile strength at joints and no sliding between two blocks) the static approach, with the static theorem, allows us to use a necessary and sufficient geometric criterion for the structure stability, particularly effective and easy to visualize.

2.2 The line of thrust and the static theorem (also known as lower bound theorem)

By evaluating the resistance criterion of the material on the planes of the masonry structure rectilinear joints, the line of thrust is defined as the set of all application points of the resulting forces at the joints, for a forces combination at the structure joints which ensures the static equilibrium of each blocks (examples of pressure lines in Figure 1).

The line of thrust should not be confused with the funicular polygon, which is differently designed. The funicular polygon is used in the more conservative theoretical framework of O'Dwyer force network construction, for masonry structures loaded with parallel forces (often their own weights). Its final form is different from the line of thrust's drawing, as shown by Fantin and Ciblac (2016).

The geometric criterion resulting from the static theorem states that if at least one line of thrust can be found in the boundaries of the masonry joints then the structure stability can be assumed. The knowledge of the actual state of the structure, and

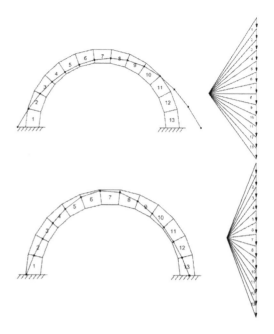

Figure 1. Examples of two different lines of thrust for the same geometry, with their force polygons. The second one is fully contained within the geometry of the stone arch.

therefore of the real structure line of thrust, is not necessary in the theoretical context of limit analysis to conclude on the stability of the structure.

2.3 Algorithmic implementation

The LITHFI tool (LIne of THrust FInder) was developed in Python language as a plugin coupled to a 3D modeler, Rhinoceros. The tool makes it possible to apply the static theorem of limit analysis to any structure drawn in two dimensions in the software's graphic interface, for any kind of loading charges.

2.3.1 How to deal with geometry measurements approximations

The studied section of the building must be drawn blocks after blocks in two dimensions in Rhinoceros' graphic interface. Each block is defined by a closed polyline. The joints are detected at the intersections of two different blocks. When acquiring the geometry of the studied building, the drawing of each block is sometimes approximate. The algorithm automatically manages these kinds of approximations about the detection of joints: if two polylines closes enough are detected, then a joint is automatically defined (Figure 2).

2.3.2 The multi-agent system

The entire algorithm is based on object-oriented programming logic. This allows the establishment of a multi-agent system in which each block represents

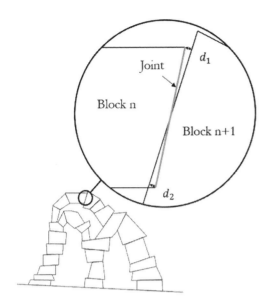

Figure 2. If the algorithm detects two edges from two different blocks relatively close to each other (d1 < ε and d2 < ε, with the ε value chosen by the user) then a joint is defined.

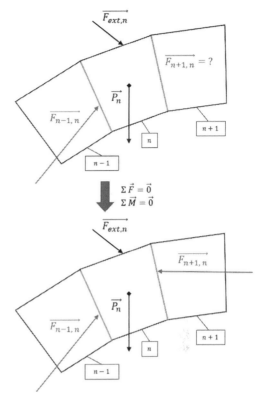

Figure 3. The block-agent n automatically computes the force acting by block-agent n+1 by using the three scalar equations of the equilibrium for the block n in two dimensions.

an agent reacting differently according to its near environment. In other words, each block will check regularly if it is possible to compute any resultant force at one of its joints by using equilibrium equations of forces and momenta applied to the block.

If exactly one joint resultant force applied to a block is not yet calculated in one joint among all the block's joints, then the computation of this last resultant force will be done automatically, using the static equilibrium equations of block (Figure 3).

2.3.3 Choice of thrust line's parameters

Most of masonry structures are statically indeterminate. The structure's degree of indeterminacy directly depends on the structure topology. For example, the degree of indeterminacy of a simple arc on two supports is equal to 3. In order to define any line of thrust for such a structure, it is necessary to firstly define the positions of three distinct points of this line of thrust on the finite lines of corresponding joints. Then, an algorithm automatically computes a resultant force in a joint thanks to these three points positions. Finally, the multi-agent system starts and computes all the resulting forces at the remaining joints, thus defining the entire line of thrust.

For any structure having a more complex topology, knowing the exact number of parameters which could define a thrust line is necessary to determine the exact number of points to be fixed in order to finally deduce the rest of the line of thrust. This number is equal to the degree of indeterminacy of the structure which can be easily determined since it only depends on the number of links between each block (the joints) and the

number of blocks that constitute the structure (Figure 4).

Knowing the exact number of setting points to fully define a line of thrust, an algorithm is tasked to determine on which structure's joints it is necessary and strategically more efficient to set the positions of line of thrust's setting points. This process will avoid static incompatibilities and in addition reduce the size of the search space that will be scanned by the genetic algorithm to find a line of thrust contained within the masonry (Figure 4).

2.3.4 The genetic algorithm

By only changing the setting points positions, which completely define the line of thrust, the line of thrust is entirely modified. The genetic algorithm is programmed to change these setting points positions until a line of thrust contained within the masonry is found.

As defined by Goldberg (1989), a genetic algorithm is a probabilistic optimization process based on the genetic reproduction phenomenon for individuals in a defined environment. First, a set of individuals defined by a fixed number of parameters (the genes) is randomly generated. Then, these individuals are put in

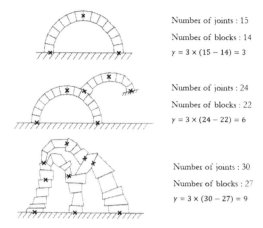

Number of joints : 15
Number of blocks : 14
$\gamma = 3 \times (15 - 14) = 3$

Number of joints : 24
Number of blocks : 22
$\gamma = 3 \times (24 - 22) = 6$

Number of joints : 30
Number of blocks : 27
$\gamma = 3 \times (30 - 27) = 9$

Figure 4. Line of thrust's setting points for three topologies.

competition with a function to be optimized: the highest ranked individuals share their genes by generating new individuals. Finally, a sequence of random mutations makes it possible to further explore a wider range of solutions. Then, these operations are repeated to obtain successive generations of individuals always more efficient regarding the fitness function to optimize.

In our case, when the principles of genetic algorithm are applied to the search of a line of thrust within the masonry, the individuals are the lines of thrust, the variable genes are the setting points positions defining the line of thrust and the fitness function is the sum of the distances between each of the line of thrust's points and their respective joints. The first generation of different lines of thrust are thus randomly generated (examples in Figure 5.4). Each of these individuals are evaluated and classified according to the fitness function (distances between lines of thrust points and their joints edges). The lines of thrust closest to being contained within the boundaries of masonry (in other words with the best grades regarding the fitness function) are saved for the next step. The average values of the selected lines of thrust's setting points positions give new setting points values in order to generate new lines of thrust. After having modified a chosen proportion of random lines of thrust by randomly modifying the setting points positions values which defines them (mutation step), a new complete generation of individuals is obtained. This new generation is then applied again to all genetic algorithm steps: evaluation, classification, selection, crossover and mutation again until one of the stopping conditions of the algorithm is satisfied: either a line of thrust contained entirely within the masonry is found (Figure 5.5) or the maximum number of generations is reached (this parameter is chosen by the user).

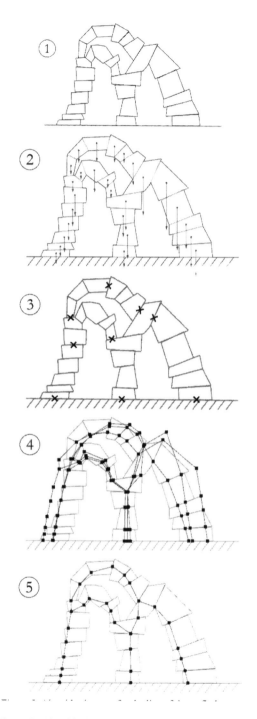

Figure 5. Algorithmic steps for the line of thrust finder: 1. Drawing of a section view. 2. Blocks and joints' automatic detection. 3. Line of thrust's setting points' choice. 4. Lines of thrust generated with genetic algorithm. 5. Final line of thrust fully contained within the masonry found.

202

3 THE LITHFI TOOL FUNCTIONALITIES

The search for a line of thrust fully contained within any bidimensional masonry structure is the main feature of the LITHFI tool. The tool also returns the drawing of the force diagram thus illustrating forces equilibrium for each block and group of blocks in the structure, according to the conventions of graphic statics. Drawing this closed polygonal figure has the benefit of showing in a condensed and intuitive way the various forces' characteristics for a defined line of thrust. For example, the most stressed interface in compression can be immediately identified thanks to the associated force diagram. The user can also modify the previous found line of thrust by changing in an interactive way values of forces characteristics: magnitude, direction and application point for any resultant force at any structure's joint. Both line of thrust drawing and force diagram drawing are consequently dynamically modified in real time.

The flexibility of the implemented algorithmic processes (using multi-agent system and genetic algorithm in order to search for a line of thrust within the masonry) allowed the development of other additional functionalities that will help for the stability studies of masonry structure.

3.1 Multiple loading charges

A feature integrated into the tool makes it possible to add one or more external forces applying to structure's blocks' edges. It is also possible to add a horizontal component to the vertical component of the blocks self-weight in order to model a quasi-static seismic load (Figure 6).

This function can be used simultaneously with the search tool for a line of thrust inside the masonry in order to determine the maximum allowable load of the structure, for example. Proceeding iteratively, the structure is subjected to a force whose intensity is successively more and more high. From the moment when no line of thrust can be found within the masonry boundaries, the force with the greatest intensity that allows at least one line of thrust to be contained inside boundaries of masonry is saved. The intensity of this force is lower or equal to the maximum load corresponding to this loading mode, in the theoretical framework of limit analysis.

Such a procedure is used for the simple example of an arc subjected to seismic loading with a quasi-static model. By successively modifying the Ks value and trying to find a line of thrust fully contained within the masonry, a maximum admissible loading coefficient can be found (Figure 7). This procedure also works in order to solve limit analysis of other kinds of load cases and more complex structures.

3.2 Hinges in a masonry structure: Positions values choices for line of thrust known points

The wish to adapt the tool to the precise goals of a masonry structure historical conservation study has led to identify other kinds of functionalities. The theoretical framework of Heyman's limit analysis for masonry structures points out that the only failure mode consists on relative blocks rotations around specific joints edges (called hinges). In situ observation of the cracks locations on the masonry structure makes it possible to identify these hinges through which the

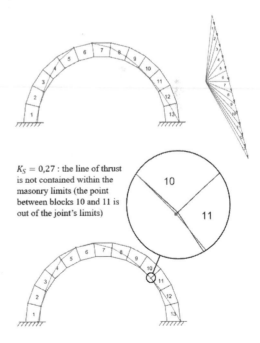

$K_S = 0,27$: the line of thrust is not contained within the masonry limits (the point between blocks 10 and 11 is out of the joint's limits)

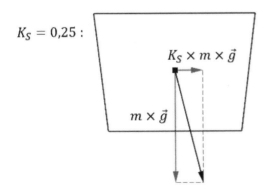

$K_S = 0,25$:

$$K_S \times m \times \vec{g}$$

$$m \times \vec{g}$$

Figure 6. Ks definition for a block loaded with quasi-static seismic charges.

Figure 7. The maximum admissible seismic loading coefficient is Ks = 0.26. For any value above this limit, it becomes impossible to find any line of thrust inside the masonry.

line of thrust must pass. The possibility of considering this specific constraint has been added as another tool's feature.

Thus, a tool command can fix a certain number of the line of thrust's points positions, directly on the structure drawing, in an interactive way. This number of points must be lower or equal to the degree of indeterminacy of the structure. After defining the known points positions for the line of thrust, the genetic algorithm works in the same way as usual by generating as a result a line of thrust crossing exactly through the chosen points and completely contained within the masonry boundaries, if it is possible.

Then, using the visualization of the line of thrust drawing and the force diagram, the user can precisely estimate the magnitudes values of forces involved in the structure's joints when several hinges are identified and evaluate the level of safety of the structure stability, for example.

3.3 Minimum and maximum thrusts

When a line of thrust is contained within the masonry of a complex structure, this structure can be considered as stable in the context of the limit analysis theoretical framework. Then, it becomes possible to estimate the range of permissible horizontal thrusts for this structure, in other words the limit values (minimum and maximum) of horizontal thrusts at the supports of the structure beyond which the structure becomes unstable, in the limit analysis theoretical framework.

For example, let's consider a masonry structure with a degree of indeterminacy equal to 3 (semi-circular arch on two supports for example). The two lines of thrust corresponding to the minimum and maximum thrusts are used in order to find out the more probable kinematic mechanisms if the structure would be subjected to too much or too little horizontal thrust at its supports (Figure 8). These mechanisms are necessarily rotations of sets of blocks around the hinges (the line of thrust's points which are on the joints edges) since it is assumed that sliding between two blocks is impossible.

An algorithm that tests all possible combinations of kinematic mechanisms by successively setting all possible combinations of hinges for the structure has also been developed and integrated into the LIFTHI tool. By automatically identifying the lines of thrust contained within the masonry that have the most extreme thrusts, the two lines of thrust corresponding to the minimum and maximum thrust are determined.

Thus, a LITHFI tool functionality makes it possible to isolate structure's portion having a degree of indeterminacy equal to 3 in order to automatically determine the two lines of thrust corresponding to the minimum and maximum thrusts in this arch portion chosen by the user.

By using this command and each of the features previously described, as part of a user reasoned approach, it becomes possible to determine the more

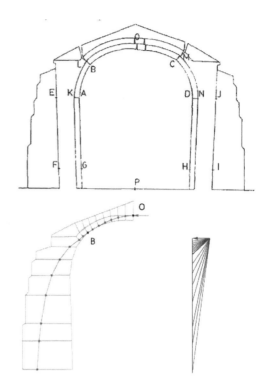

Figure 8. Example of Guimarei church's study by Huerta and Lopez (1997): the line of thrust has to cross points B and O because cracks were identified on the opposite sides of these points.

probable kinematic mechanisms likely to appear and the thrust domains for more complex structures with a degree of indeterminacy greater than 3, as illustrated by the example of the study of the St. Marie-Madeleine's Basilic in Vézelay.

3.4 A concrete example: St. Marie-Madeleine's Basilic in Vézelay

A practical application has been made for the stability study of a geometry inspired by a nave section of the St. Marie-Madeleine's Basilic in Vézelay, in order to verify, using the limit analysis theoretical approach, the disorders descriptions (mentioned especially by Viollet-le-Duc) that would have appeared during its construction when, at first, the nave flying buttresses had not yet been built.

At first, the basilica's section (shown in Figure 9.1) is drawn from different surveys. In order to correctly use limit analysis method for masonry structure, the choice for the joints placement of the studied section is based as much as possible on the actual cutting of the basilica's voussoirs. The resulting structure degree of indeterminacy is equal to 15.

After having defined and integrated the various external forces that rely on the structure (roof load for example), the search tool for a line of thrust contained

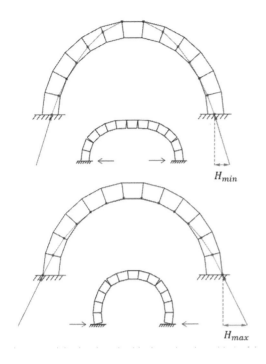

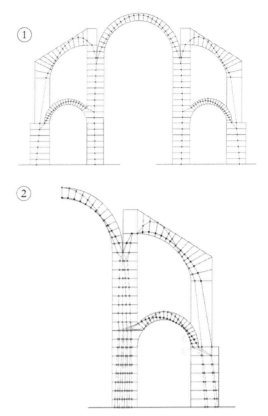

Figure 9. Minimal and maximal horizontal pushes with the following kinematic mechanism.

Figure 10. 1. A line of thrust fully contained within the geometry of Vezelay's basilic. 2. All combinations (8 cases) of extrem pushes in the three chosen structure archs.

within the masonry is used in order to ensure that the structure can be considered stable thanks to the limit analysis theoretical framework and its static theorem (Figure 9.1).

Then, the minimum and maximum thrusts are determined in specific portions of the structure (main vault, flying buttress and aisle vault) in order to determine the kinematic mechanisms most likely to appear in these elements. This also makes it possible to determine the horizontal thrusts solicited in the pillars and abutments (Figure 9.2).

These results are carried out thanks to the simultaneous use of the various features available in the LIFTHI tool: the minimum and maximum thrust search command for arches portions, the command for setting positions of specific line of thrust chosen points and the search command for a line of thrust within the masonry boundaries.

In this specific study, the same process was repeated with the Basilica's geometry before the addition of flying buttresses. Thus, it has been highlighted, as expected, that the range of horizontal thrusts likely to be taken up by the inner pillar is different from the range in the situation of the flying buttress presence. The pillar needs to take more thrust than for a geometry with the flying buttress, since it allows to rebalance the overall structure by providing some thrust on the vault.

This type of method using limit analysis as a theoretical framework is not usually a substitute for a more in-depth analysis of the historic masonry stability studies. For example, the structure should be modelled using finite element methods or discrete elements. However, the advantage lies in the speed of execution and the results rendering of this process thanks to the use of the different functionalities of the LITHFI tool.

4 CONCLUSION

The advantages of the LITHFI tool rely on the interactive study of bidimensional masonry structures for all kinds of geometries, topologies and loading charges. A complex case study must be solved with an iterative process involving the basic features. That's why this tool is not intended to be used as a "black box" in order to automatically solve any given problem of masonry structure in its entirety but rather as a specific tool mobilized in a global approach, structured and built by the user thanks to the use of the basic functionalities.

In this sense, the tool dissemination is intended to allow the exploration of its uses and potentialities to a wide audience. The intuitive approach and the precision of calculations allow both a simplified use for a historian as for an engineer who will look for results precision. The pedagogical use framework could also be considered in order to teach the principles of graphic statics, for example.

A next step will consist to improve and maybe change the tool features according to the identification of new uses. The search for Heyman geometric safety factor for a defined geometric section and the extension of the field of study allowed by the tool to 3D structures cases could be explored, for example.

REFERENCES

Ciblac, T. et Guena, F. 2019. Un outil interactif pour l'étude de la stabilité des maçonneries patrimoniales, Construire ! Entre Antiquité et époque contemporaine, *Actes Sud*, 707–715.

Salençon, J. 2013. *Yield Design*, London, UK; Hoboken, NJ, ISTE – Wiley.

Heyman J. 1995. *The stone skeleton: Structural engineering of masonry architecture*, Cambridge, Cambridge University Press.

Oikonomopoulou, A. 2009. *Approches numériques pour l'étude du comportement des structures maçonnées anciennes. Un outil basé sur le calcul à la rupture et la visualisation graphique*, Ph D thesis, Université Paris-Est.

O'Dwyer, D., 1999. Funicular analysis of masonry vaults, *Computers & structures*, 73 (1-5), 187–197.

Fantin, M. 2017. *Étude des rapports entre stéréotomie et résistance des voûtes clavées*, Ph D thesis, Université Paris-Est.

Delbecq, J.-M. 1983. *Analyse de la stabilité des ponts en maçonnerie par la théorie du calcul à la rupture*, thèse de doctorat, Ecole nationale des ponts et chaussées, Paris.

Smars P. 2000. *Études sur la stabilité des arcs et voûtes - confrontation des méthodes de l'analyse limite aux voûtes gothiques en Brabant*, Ph D thesis, Université Catholique de Louvain, Leuven.

Viollet-Le-Duc, E. 1856. *Dictionnaire raisonné de l'architecture française du XIe au XVIe siècle*, Paris.

GOLDBERG, D. E. 1989. *Genetic algorithms in search, optimization and machine learning*, Addison-Wesley.

Wilensky, U. Rand, W. 2015. *An introduction to agent-based modeling*. The MIT Press.

Fantin, M., Ciblac, T. 2016. Extension of thrust network analysis with joints consideration and new equilibrium states. *International Journal of Space Structures* 31 (2-4), 190–202.

Huerta Fernández S., López Manzanares G., 1997. Stability and consolidation of an ashlar barrel vault with great deformations: the church of Guimarei, *in Structural studies, repairs and maintenance of historical buildings. Computational Mechanics Publications*, Southampton, Reino Unido, 587–596.

Brick and Block Masonry - From Historical to Sustainable Masonry –
Kubica, Kwiecień & Bednarz (eds)
© 2020 Taylor & Francis Group, London, ISBN 978-0-367-56586-2

Best practices for matching replacement brick to historic fired-clay brick

J.C. Dick
Simpson Gumpertz & Heger Inc., Waltham, Massachusetts, USA

ABSTRACT: Brick masonry has been used across the world as a popular building material for thousands of years. This popular and universal building material characterizes historic cities within New England, such as Boston, as well as many other cities and countries across the world. Historic masonry buildings often require localized replacement of historic brick that are damaged. To maintain the historic character and appearance of the existing brick masonry, finding a blend of brick that matches aesthetically is key. However, this is not always easy, as historic brick were manufactured using techniques that are far less common or nonexistent today. This paper discusses brick manufacturing and considerations for replacement brick and draws on several case studies to discuss strategies for matching the technical and aesthetic properties of localized new replacement brick to existing historic brick in repair and restoration projects for historic buildings.

1 BACKGROUND

Brick is one of man's oldest building materials, and specifically, fired-clay brick can trace its history back to the fourth millennium B.C. (Campbell, 2003). The first step in creating this product is to obtain the raw material, clay, which is later used to form the brick. The clay is then tempered, formed into a shape, and fired in a kiln at high temperatures to vitrify the clay, inciting a chemical reaction that creates a changed, durable material. Until it is fired, water can be added to clay to return it to a malleable state, but once it is fired, that is no longer possible.

Ancient Roman brick were made by producing a slab of clay and cutting it into rectangles, which explains the large, flat, Roman brick in comparison with those more typically seen throughout the latter half of the second millennium and into the twenty-first century. Ancient Roman brick often included a maker's stamp (Figure 1), a practice which was later revived in the end of the second millennium.

The concept of utilizing a reusable mold, referred to as a "stock mold," was established by the seventeenth century (Historic England, 2015). This type of mold allowed for faster, more consistent production of brick and was one of the first important steps in the industrialization of the brickmaking process. Various methods of preventing the newly formed brick from sticking to the inside of the mold, such as the use of sand (e.g. sand-molded brick) or of water (e.g. waterstruck brick), contributed to the texture of the brick. The most obvious tell-tale sign of a stock-molded brick is the presence of a frog, or a marking, on one side of the brick (Figure 2). Although the exact or primary purpose of the frog is uncertain (Historic England, 2015), it is certain that by the nineteenth century the frog typically included a maker's stamp (manufacturer's name). This custom continues today in the production of molded brick.

Sizes of brick remained surprisingly consistent throughout the centuries, with the exception of the flat Roman brick and another later variation in the Middle Ages. The approximate ratio of 2:1 of the length of a brick to its width is necessitated by the typical bond patterns, in which headers are utilized to tooth in the various wythes of a brick wall. Since headers are brick oriented in a different direction, this 2:1 ratio is essential to the construction of a brick wall. While many different bond patterns were used historically, the concept of the 2:1 length: width ratio is common throughout all bonds since the Middle Ages, and even in the large Roman brick. One interesting change to brick sizes is that in response to the implementation of the Brick Tax in Great Britain in 1784, which taxed the use of each individual brick, larger brick were manufactured (Historic England, 2015). The 2:1 ratio was retained.

The Industrial Revolution of the nineteenth century brought advances in mechanizing the manufacturing process of brick, including how the clay was harvested, how the raw material was prepared for use, and how the brick themselves were formed. Various types of molds were used, which created different textures of the brick. One specific way to finish the brick and provide texture is to use water to release the brick from the mold, which results in a brick with a waterstruck finish. This technique was developed in the 1600s and is particular to New England, a region in the northeast of the United States (US). Since this paper is primarily based on the author's professional experience in this region, this technique will be further discussed and referenced.

Figure 1. Maker's stamp in roman brick in the pantheon.

Figure 2. Molded brick with frogs consisting of a maker's stamp, indicating that the manufacturer is the stiles and hart brick company (S&H).

The process of pressing brick, in which clay is pressed with a lot of force into steel molds, also evolved in the nineteenth century. The result of this process is a very dense brick with a very consistent shape and low tolerances.

Another fundamental change to brick as a result of the Industrial Revolution resulted from the overall improvement of kilns and consistency of firing temperatures. Development of more efficient kilns in this era meant that the brick product was more uniform and of overall higher durability. Conversely, the frequency of very soft, low-durability, high-absorption "salmons" (pale-salmon colored brick) was greatly reduced as a result of this innovation, as these brick were not produced intentionally, but rather as a byproduct of inconsistent and insufficient firing. As a result, in the US, the frequency of "salmons" is much higher in buildings of the first half of the nineteenth century or earlier than in the latter half of the nineteenth century and later.

Perhaps the single biggest advancement came with the development of the machine to extrude brick, which had a significant impact on the final brick product in the US and other countries. The implementation of this new process often tends to delineate between "historic" or "traditional" hand-molded and "modern" extruded brick (Figure 3). However, while the extruded brick produced in the twenty-first century certainly would not be considered historic, some extruded brick, specifically combed extruded brick (Figure 4) of the early twentieth century in New England, are considered historic.

Figure 3. Twenty-first century non-historic extruded brick.

Figure 4. Combed extruded brick is considered historic in New England.

2 MATCHING PROCESS

When specifying replacement brick for an historic building, it may be inappropriate to use an extruded brick to replace a molded one, a smooth brick to replace a rough one, or a uniform blend of brick colors to replace an area of polychromatic brick. All of these factors, as well as others, must be considered to provide an appropriate brick for repairs to historic brick.

2.1 Reasons and goals for replacement

Many different scenarios exist in which the localized repair or replacement of existing historic brick masonry may be necessary or desirable. Localized damage or deterioration could be due to many different factors, ranging from impact damage, prior alterations, the presence of isolated underfired, less durable brick, or severe exposure conditions.

The aesthetic goals of replacement are not always the same. In some cases, the desire for the repairs to be highly legible, and even the use of different materials for replacement, may be the objective. For the purpose of the rest of this paper, however, the assumed goal is to match the new replacement brick to the existing brick as closely as possible.

2.2 Aesthetic considerations

When matching fired-clay brick, three major aesthetic considerations exist: geometry, texture, and color.

The geometry of a brick is a direct function of the type and shape of the mold. As this can be measured, it is therefore, straightforward to match. However, the importance of matching the geometry of the replacement brick to that of the existing brick should not be underestimated, because without it, the entire bond and construction will be compromised. Using brick of incorrect geometry (e.g. size, tolerance, etc.) means that all the following cannot possibly be matched at the same time:

- Coursing
- Mortar joint width
- Bond pattern

The mortar joints may need to be widened or tightened, or the coursing may gradually become more and more misaligned to accommodate the difference in size. Regardless of the manner in which the difference in geometry is adjusted for, failing to match all of these metrics (coursing, mortar joint width, and bond pattern) precisely will typically result in repairs that are detrimental to the historic character of a building. While matching the geometry of the brick *enables* all these critical characteristics to be maintained, it does not *guarantee* that this will happen. Mockups (e.g. trial repairs), attention to detail, and good craftsmanship during construction are still critical to carry out good replacement efforts.

The texture of a brick also influences its overall appearance. Two brick manufactured of the exact same clay and fired in exactly the same manner may appear very different if their textures are different. The texture or finish of a brick affects the surface area and gloss/matte of the brick and therefore how the brick is perceived. Therefore, accurately identifying and matching the texture is very important. The most common textures of molded brick in New England are sand molded and waterstruck.

Finally, color is a very important consideration. Not only is matching the color imperative, but matching the specific blend or range of colors can greatly affect the overall appearance of a wall (Figure 5).

The color of a brick is dependent on many factors, including the specific type of clay used and its mineral properties (note that metallic oxides contribute greatly to the red color of the brick) (The Brick Industry Association, 2006), the temperature and position of the brick within the kiln, and the length of firing time. The more a brick is fired, whether at a higher temperature, for a longer duration, or because it is in a position closer to the heat source in the kiln, the darker it gets. In general, the relative physical properties (e.g. compressive strength or absorption) can be correlated based on the appearance of the brick, as demonstrated in the set of brick pictured in Figure 6.

The color of a brick may also change over time as it becomes stained and dirty, which in turn affects how its color is perceived. Selecting new brick to aesthetically match dirty brick will never result in a good long-term match, as the new brick will eventually become dirty over time. Thus, matching clean brick to clean brick is crucial to achieve a good long-term match, much as it is when matching other materials such as concrete and mortar (Currie and

Figure 5. Left of pilaster: monochromatic replacement brick is a poor match. Pilaster: polychromatic original brick. Right of pilaster: polychromatic replacement brick reflects a good effort to match the original brick and range of colors.

Figure 6. Compressive strength of the brick pictured is directly correlated to the color of the brick. From left to right, compressive strength is 4.2 MPa (610 psi), 26.1 MPa (3,792 psi), 36.5 MPa (5,292 psi), 40.1 MPa (5,925 psi), 58.7 MPa (8,508 psi), 59.1 MPa (8,577 psi), 81.1 MPa (11,765 psi), and 95.8 MPa (13,894 psi).

Figure 7. Slicked mortar joints at the left, with a lot of paste float, read differently than weathered mortar joints at the right.

Bronski, 2019). Matching replacement brick to cleaned historic brick may result in a temporary mismatch with other, uncleaned areas of historic brick; if this is aesthetically unpalatable, the application of mud or dirt that will not negatively affect the overall long-term performance of the brick could be considered to make the new replacement brick a better initial aesthetic match.

2.3 Other considerations: Mortar joints

The mortar joints can also affect how brick is perceived; specifically, the color, profile, and texture of mortar can all affect how the brick appears. Brick with a dark grey mortar may appear completely different than the same brick with a white mortar – an optical illusion of sorts. The profile of the mortar, whether the mortar joints are recessed, whether mortar is feathered onto the brick, or a difference in a flat vs beaded profile can result in the brick wall being "read" differently. The presence or absence of a recess will increase or decrease shadows at the surface, while feathering mortar onto chipped edges or arrises of surrounding brick will greatly affect the perceived ratio of surface area of mortar to brick. Lastly, the texture of the mortar can affect the overall appearance. For example, a slicked finish appears different in color than a weathered finish (Figure 7).

2.4 Physical properties

The technical performance of brick masonry is affected by several different physical properties, including water absorption, compressive strength, density, and porosity, among others. While matching each of these physical properties may be critical for certain building materials or repairs (e.g. specifying replacement stone for dutchman repairs in sedimentary stone), the case is slightly different with brick. While inserting a much harder material immediately adjacent to a softer material (such as could possibly happen with dutchman repairs in sedimentary stone) may damage the original material, the new replacement brick will never be in contact with the existing brick because of the mortar joints. Furthermore, the mortar joints should always be the weakest link in

brick construction and will act as the sacrificial material. Thus, the concern of using a slightly harder brick or one with slightly different properties is not so critical when introducing replacement brick into existing construction as it is with other materials. The primary objective of matching the physical properties of brick is to replace local individual brick with brick that are of commensurate durability to the typical existing brick. However, oftentimes the brick that must be replaced are of poorer quality. In these instances, in particular, the underperforming, poor-quality brick that are being replaced should not be the standard to which the replacement brick is held.

2.5 Sourcing the replacement brick

Today, approximately 90% of the brick produced in the US is extruded brick (The Brick Industry Association, 2006). Although a relatively small number of manufacturers produce molded brick, quality molded brick is available. Some of these manufacturers are able to make custom shapes and small batch orders and/or match L*a*b (a system which precisely defines the color based on white to black, red to green, and blue to yellow scales) colors to existing brick. Identifying manufacturers with the ability to produce brick of similar color and texture is critical to specifying and selecting replacement brick.

3 CASE STUDIES

3.1 Historic brick walls in Historic Park, Massachusetts, USA

3.1.1 Description of existing brick, scope of work, specification

Brick masonry site walls capped with limestone coping stones flank the sides of an historic park. The existing brick masonry comprises a blend of red waterstruck brick, with darker-colored brick ("clinkers") as headers, and is constructed in Flemish Bond.

Prior to the rehabilitation of these walls, sections of the walls were displaced and localized brick were damaged, spalled, and deteriorated. The walls contained dark staining, graffiti, and biological staining. The primary objective of the rehabilitation project was to restore the walls to appear uniform, to remove staining, and to repair localized areas of damage. The scope of work included performing localized brick masonry spall repairs and rebuilding, 100% repointing, cleaning all surfaces of the wall, and adding sheet metal flashing beneath the coping stones.

Molded fired-clay brick was specified to replace localized damaged brick, and five different US manufacturers of molded clay brick which specifically offer blends of waterstruck brick were suggested as starting points. The brick were tested in a laboratory in accordance with ASTM C67 to determine their ASTM C216 grade. Grade SW brick was specified in accordance with the test results.

3.1.2 *Process of selecting the replacement brick*
In several iterations of submittals, four different options were submitted for the replacement red brick. For each proposed brick option, a range of brick were submitted (either by submitting ten individual brick or a cardboard sample board) to demonstrate the full range of the brick blends. A portable mockup was submitted to promote evaluation of a section of the new wall adjacent to the existing wall in various locations. Four different options were submitted for proposed use to match the dark-colored headers, which were inserted into the wall and evaluated.

3.1.3 *Final blend*
Stretchers: A single blend of waterstruck brick (Figure 8).

Headers: A 50%/50% blend of two different waterstruck brick blends.

Figure 8. Portable mockup of selected brick is an excellent match to existing brick.

Figure 9. Left pillar constructed of original brick, right pillar constructed of replacement brick.

3.1.4 *Final results*
Figure 9 shows two pillars side by side, the left one repointed and the right rebuilt using the new brick blends.

3.2 Historic dormitory building, Massachusetts, USA

3.2.1 *Description of existing brick, scope of work, specification*
A large dormitory building, constructed of brick mass masonry, was recently renovated, and the exterior masonry was rehabilitated. The brick is waterstruck, with red brick stretchers and a blend of red and darker-colored headers. The scope of work for the rehabilitation of the brick masonry included cleaning, repointing, replacing localized single and double wythes of brick, rebuilding approximately 50% of the chimneys, replacing individual cracked or spalled brick, and removing and replacing localized brickwork to allow for the installation of sheet metal flashing. Where salvaged brick could not be reused, molded brick of Grade SW, to match the color and texture of the existing brick, was specified, and specific manufacturers were recommended.

3.2.2 *Process of selecting the replacement brick*
The range of the red brick varies significantly, and a custom blend was desired to match the existing blend. A representative area containing approximately 100 brick was selected and marked. The brick were categorized into one of the following five categories: dark-colored, medium-colored, and light-colored stretchers, and dark- and light-colored headers. The percentage of each category was tabulated in a manner similar to a point count for determining the proportions of a concrete or mortar mixture.

Meanwhile, various replacement brick blends were provided, first in cardboard samples and then in the form of ten full brick to demonstrate the full range of each proposed blend. The best match was selected for each of the five categories within the

Figure 10. Three brick sample panel mockups using different blends of brick.

Figure 11. Original brick masonry left of ruler; new brick masonry blend right of ruler.

existing brick blend described above, and these selected matches were used to create three different brick sample panel mockups of varying percentages (Figure 10). The mockups were evaluated in representative, well-lit areas of the brick masonry, and the best overall match was selected.

The dark-colored headers were particularly difficult to match. For this reason, as many salvaged headers as possible were used, and their use was prioritized at the ground floor and other easily visible areas of the building. Also, due to the lack of availability of darker waterstruck brick, some sand molded brick were selected for use as the darkest brick within the overall blend.

3.2.3 Final blend

Stretchers: Three different blends of molded brick, produced by two different manufacturers, all with precisely specified percentages

Headers: The three different blends used for the stretchers, in addition to one additional darker brick and salvaged existing headers, all with precisely specified percentages

3.2.4 Final results

Figures 11 and 12 show that the selected brick blend is an excellent match. The original brick masonry is to the left of the ruler in Figure 11, while the new brick masonry blend is to the right of the ruler. Figure 12 shows various chimneys and whether they were rebuilt (R) or repointed (P). The difference is indistinguishable.

4 RECOMMENDED BEST PRACTICES

The following best practices are recommended when matching repairs to historic fired-clay brick:

1 Clearly define project objectives with all involved parties.
2 Perform physical testing to identify key physical properties of the brick and determine the grade of the brick; specify brick of equal overall quality.

Figure 12. Chimneys marked with an R were rebuilt. The chimney marked with a P was repointed, and the original brick masonry remains.

3 Survey the existing brick to be matched to determine the original manufacturing process (e.g. sand molded, waterstruck, machine-pressed).
4 Match new replacement brick to cleaned brick, not to dirty brick.
5 Visually evaluate the range of colors of the existing brick and the various percentages of each. Take note of differences between colors in headers and stretchers.
6 Find manufacturers that still use the particular manufacturing processes to be replicated.
7 Obtain a wide range of all available physical samples of each individual color in the existing blend or range.
8 Choose the best match for each individual color.
9 Assemble a dry-laid mockup of the proposed blend, adjacent to an area of existing brick with a good range of lighting conditions during the day (e.g. direct sun, raking sun, shade) to evaluate the brick match.
10 Construct a final aesthetic mockup using the approved mortar, also in an area with a good range of lighting conditions during the day.
11 Ensure that blends are being properly blended in the field during construction.

REFERENCES

Campbell, James W. P. & Pryce, Will. 2003. *Brick: A World History*. London: Thames & Hudson Ltd.

Currie, H.M. & Bronski, M.B. 2019. Concrete Matching for Architecturally Significant Mid-Century Modernist Concrete Facades in the United States. *Journal of the International Masonry Society* Volume 32 (No. 2): 43–54.

Historic England. 2015. *Practical Building Conservation: Earth, Brick & Terracotta Part B*. Dorchester: Henry King Limited.

The Brick Industry Association. 2006. *Technical Note 9 on Brick Construction: Manufacturing of Brick*. Reston, VA: Brick Industry Association.

Brick and Block Masonry - From Historical to Sustainable Masonry –
Kubica, Kwiecień & Bednarz (eds)
© *2020 Taylor & Francis Group, London, ISBN 978-0-367-56586-2*

Assessment of the stability conditions of the tower of Seddülbahir fortress in Turkey

H. Sesigur
Istanbul Technical University, Istanbul, Turkey

M. Alaboz
ARKE Engineering, Istanbul, Turkey

ABSTRACT: In this study, assessment of the two towers of Seddülbahir Fortress are realized by means of linear structural analysis. Dynamic identification is performed for modal updating of structural system which helps to develop intervention proposals. Seddülbahir Fortress is a unique defensive structure of Canakkale Bosphorus which was constructed in the era of Sultan the IV Murat by his mother Hatice Turhan Sultan in 1658. The structure was damaged during Canakkale War and exposed to weathering effects by now. Deteriorations and conscious human harm by replacing the stones are observed. Therefore, the structural integrity of the towers is lost due to the collapse of the domes of the towers and the body walls as well by half. In order to assess the earthquake risk, finite element models were conducted by using 2D and 3D laser scanning survey data. In FEM models, homogeneous and elastic material properties are taken in to account. Dynamic properties of the structures are evaluated by using acceleration data under ambient vibration. Natural frequency and modal shapes obtained by dynamic tests are used to check the consistency of the FEM models constructed with estimations and required modifications are done to match the behavior. In addition to linear elastic analysis results, possible damage mechanisms are studied to estimate damage activation acceleration values which can cause loss of equilibrium. Consequently, structural interventions are proposed by considering different assessment approaches.

1 INTRODUCTION

The damage assessment of historical masonry buildings is often a complex task. It is crucial to distinguish between stable damage patterns and damage evolution leading to a catastrophic structural collapse. Some damage patterns can be subsequently activated by unpredictable events like earthquakes, improper restorations, or even by bombardments during wars. In the following, a survey of several numerical simulations is presented, that have been carried out on two medieval towers in Turkey. The analyses range from linear static to kinematic limit analysis of masonry cracking and crushing. The evaluation of eigenfrequencies and modal shapes, preliminary to further dynamic analyses, has also been accounted for.

2 STRUCTURAL DEFINITION

2.1 Seddülbahir fortress

The Ottoman fortress of Seddülbahir on the European shores of the Dardanelles was built in 1658 by Hatice Turhan Sultan, the queen mother of Sultan Mehmed IV. Since the mid 17th century Seddülbahir has protected the Ottoman, and later Turkish lands, against

threats to the Dardanelles, the strategic waterway which leads to the capital of Istanbul on the Bosphorus (Thys-Şenocak et al. 2009). The Ottoman fortress of Seddülbahir was used in the Ottoman defense during the Gallipoli campaign of World War One and was severely damaged by Allied bombardments during this campaign. After 1923 Seddülbahir fortress served as a Turkish naval outpost until 1997 when the first research team from Koç University and Istanbul Technical University began the preliminary survey of the site after it was demilitarized in that year. The detailed analyses of the materials (wood, stones, mortars. Etc.) were conducted in the project and the conservation and restoration decisions were made accordingly. A precise geodetic and architectural survey of the fortress and the site was conducted from 1997-2005 including a comprehensive 3D laser scanning, the first of its kind in Turkey.

2.2 Architectural and structural definition of towers

Seddülbahir Fortress has a rectangular plan shape and consists of walls and of four towers at the corners of the fortress (Figure 1). North and East towers have slight damages while South and West towers have partial collapses. It was observed that almost half of the

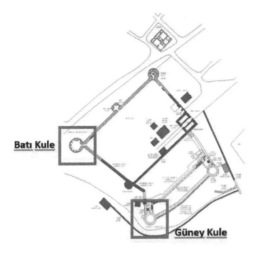

Figure 1. Layout of the fortress.

Figure 3. The West Tower.

Figure 4. The South Tower.

south tower while two third of the west tower were collapsed.

The location of the west tower is 10m high from the sea level while the South Tower is in touch with the sea. The towers have polygonal layout in plan with 3.00m thick three layered masonry walls with an overall height of about 10m (Figure 2). The bearing walls are consist of stone faces enclosing a mixture of rubble bonded with lime mortar. The roof system of the towers consist of dome structures made of stone. Top of the towers are flat as the gaps around walls and dome is filled and covered with stone. Both towers were damaged during the war and the existing damages were progressed more due to weather conditions and deteriorations. Additional losses of materials due to human harm for the purpose of using materials in rural constructions therefore most of the remaining walls were lost. Today, a part of the dome of West tower and lower part of the collapsed walls are still standing.

The South Tower stands more solid (Figure 4) than the West Tower (Figure 3) as the half of the dome exist. However, the shore in front of the tower that protects the tower from waves and sea springs is eroded by time. Today sea has reached to the base of the tower. It might be possible that the effects of the waves and springs could accelerate the deterioration process.

3 ASSESSMENT OF STRUCTURAL SYSTEM

3.1 Damage mapping

Both towers have similar kind of material losses and damages but the confinement of the towers by adjacent walls and the structural equilibriums exhibits some differences. The West Tower has almost two walls fully standing, supporting the dome and form the structure. The fortress wall on north merges with the tower but insufficiently contributes to its structural safety. The south wall connects with the tower at same point was lost. Timber tie beams are visible on the sections in both towers and a grid of timber

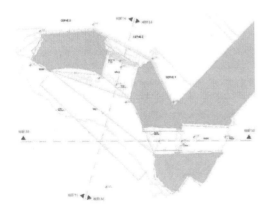

Figure 2. Plan of the West Tower.

joists are observed just underneath the top finishing stone layer of the towers during restoration works. The South Tower is supported by a thick wall of fortress and connected to the south courtyard wall that is in lower level. Accesses between the two perpendicular walls are provided by an arched corridor that is crossing the filling layer of the tower. These corridors seem vulnerable as the arch thickness is low causing by deterioration of the material. Some vertical displacement of arch blocks is also observed.

3.2 Survey

Survey of the structure is made in 2005. As the restoration campaign started, a recent and detailed survey is conducted by laser scanning in order to register any conditional changes up to date. The laser survey data was used both architectural design and structural analysis, where the geometry of towers was difficult to define because of irregular shapes of the towers as a result of their partial collapse.

3.3 Structural analysis

In order to assess the existing risk situation of the structure and to give possible intervention proposals, 3d structural model is prepared by using plan sections of different levels which idealize the geometry and an accurate model by using laser scanning data by using the commercial FE code ABAQUS. Different approaches could be used to numerically model the mechanical behavior of the masonry: (a) by modelling each stone, mortar face and interface separately (micro modelling); (b) by using a smeared model with homogenized properties (macro-modelling); (c) by utilizing a macro-elements modelling (Roca et al.). Each method has its own advantages and drawbacks. Micro-modelling requires a deeper knowledge of the actual geometry of the whole masonry and it is, obviously, inapplicable for large structures. It could be very effective for limited portions of masonry buildings (i.e. a single Wall). In this study, as the aim of numerical modelling is to identify the seismic behavior of the tower only, the macro-modelling approach was used, and the masonry was modelled as an isotropic continuum. Masonry walls and domes were modelled by means tetrahedral elements paying attention to reproduce the main geometric irregularities in the Wall thickness.

With the aim of reproducing the experimental dynamic behavior of the tower, a series of parametric analyses were performed. Variable material properties for the investigated elements were assumed. In particular, due to scattering on the results obtained for the elastic modulus of the internal masonry face, a value of $1000 N/mm^2$ was assumed. Parametric analyses were focused on estimating an effective value for the elastic modulus of the internal unknown filling material. This uncertain parameter was assumed

as an updating parameter and was iteratively modified to minimize the differences in the natural frequencies between numerical and experimental behavior. The experimental ratio between corresponding frequencies is not reached, so that additional analyses were performed by also varying the value of the elastic modulus of the internal filling. Results confirm that no variability is on the value of the ratio due to the mechanical properties of the filling material because of the linear dependence of the frequencies on the elastic properties of materials introduced in the model. The differences on the numerical and the experimental results are probably due to the experimental estimation of the higher frequencies that might be subject to measure errors. To face this drawback the numerical model was identified in order to reproduce the first four flexural main frequencies that are, anyway, the most significant for the dynamic behavior of the towers (Figure 5, 6).

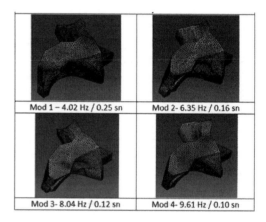

| Mod 1 – 4.02 Hz / 0.25 sn | Mod 2- 6.35 Hz / 0.16 sn |
| Mod 3- 8.04 Hz / 0.12 sn | Mod 4- 9.61 Hz / 0.10 sn |

Figure 5. Mode shapes of The West Tower.

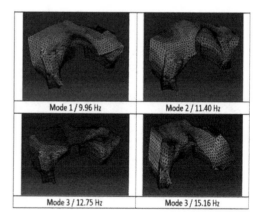

| Mode 1 / 9.96 Hz | Mode 2 / 11.40 Hz |
| Mode 3 / 12.75 Hz | Mode 3 / 15.16 Hz |

Figure 6. Mode shapes of The South Tower.

4 SYSTEM IDENTIFICATION

4.1 Measuring process

System identification test was carried on under ambient vibration on towers to check if the integrity of the structure is satisfied as defined in FEM model as a homogenous modeling technique. In the process, 7 high accuracies, low noise, 3 axial accelerometers were used. In 100 Hz, 10 – 20 min records were collected in setups by placing the sensors at 8 to 10 locations at top and one in the ground (Figure 8, 9).

Figure 7. Layout of accelerometers (left: The West Tower, right: The South Tower).

Figure 8. Layout of accelerometers (left: The West Tower, right: The South Tower).

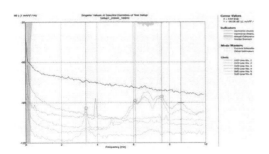

Figure 9. Spectral density diagram.

Table 1. Comparison of the frequencies.

OMA	FEM	Difference %
4.980	5.620	12.85
7.568	7.100	6.18

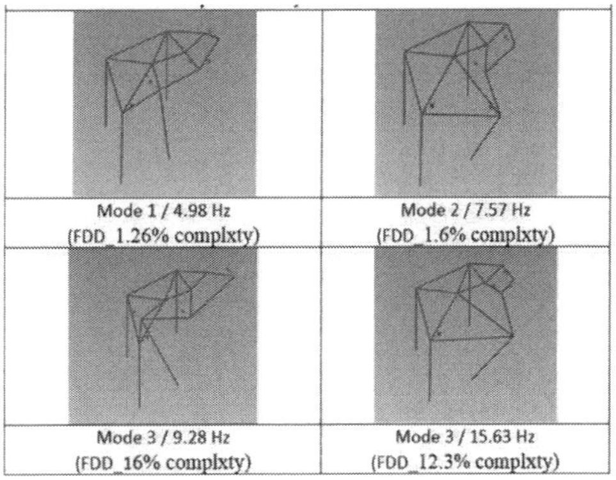

Figure 10. Mode shapes of the South Tower.

4.2 Comparison of FEM/OMA

Obtaining of natural frequency and mode shapes of the tower, the consistency of the models was checked. Considering the frequency values, lower frequency in system identification test might be the result of a lower elastic modulus. But the differences are in an acceptable range to rely on FEM model and assumptions to carry on further analysis (Table 1). When the model shapes are compared, the amplitudes of modal displacements obtained by system identification exhibits higher values then FEM model. Peaks seen in different channels in spectral density graph might be referred to some local discontinuities or irregularities at the inner sections (Figure 9).

5 LIMIT ANALYSIS

5.1 Defining possible mechanisms

The towers are massive structures and the existing situation didn't allow to collect all unknown data except system identification that provide overall outcome parameters of a structure and visual inspection. The damage pattern seen on the structure doesn't follow a specific mechanism. Local cracks seen on the surface of the stones judged as the compressive cracks due to overloading at the skew level of the dome. Loosing of timber tie beams could be visible and the quality of filling could be observed only at the end sections and gave no idea about inner discontinuities and possible loose of material, voids or any separations. At this point, modal analysis results and system identification parameters guided us to make trivial mechanisms by limiting the infinite possibilities. Modal shapes became the major indicators of which part of the structure is vulnerable to move inwards and in tend of behaving freely of the whole structure.

Possible activation of some local mechanisms of the towers has been determined through the linear kinematic analysis. In particular, with reference to the global overturning mechanism, in the case of either an isolated element or a structure portion fixed to the ground, the seismic check is satisfied if the spectral acceleration (a_0^*) activating this mechanism fulfills the following relationship:

$$a_0^* = \frac{a_g(P_{VR})S}{q}$$

Where a_g is the reference acceleration depending on local soil conditions, S is the spectrum coefficient depending on the probability of the seismic event and q is the behavior factor, which can be assumed equal to 2.0 for masonry structures.

Two possibilities of movement; either the whole part lose its equilibrium state where the inner section is solid, either façade layer of the tower stands together where the filling and inward dome part lose its stability were studied. This process was applied to two crossing parts as these edges were tend to move both in FEM and system identification results.

Possible kinematic mechanisms are shown in (Figure 11) for the west tower while (Figure 12) for the south tower. PGA values of minimum 0.16g and maximum 0.4g is calculated for the selected mechanisms where the expected PGA level is 0.313 g for the %10 probability of exceedance in 50 years earthquake according to new Turkish Earthquake code.

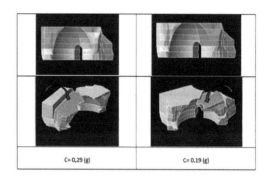

Figure 12. PGA values of selected mechanisms of the South Tower.

6 INTERVENTION PROPOSALS

As in previous sections explained, the main problem of the structure is being in a critical static equilibrium that can be triggered either by earthquake or by losing integrity at inner section. Many different proposals were studied to satisfy equilibrium in place of lost part of the dome such as space truss systems, steel ties that can be fixed to a foundation at the counter side of the collapsed dome in order to limit the displacement at the top of tower. Considering the architectural conservation attitudes, the planned function of the tower, visitor allowance to the tower and surrounding; it was decided to reconstruct the surrounding walls partially in a way to reflect the original geometry and support the risky masses that are critical to move.

In addition to satisfy the integrity of the filling, façade walls and the stone dome, the timber tie beams were suggested to place in their original locations where its possible to access (Figure 13). In order to improve seismic behavior of the towers, repointing and injection were proposed to provide homogeneity of the three layered walls. The effect of adding partial walls/buttresses at sides with its original material and techniques were compared with its efficiency. It

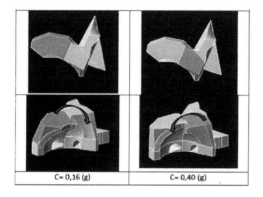

Figure 11. PGA values of selected mechanisms of the West Tower.

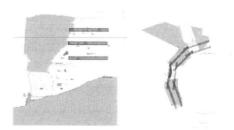

Figure 13. Timber lintel places shown on plan.

Figure 14. Reconstructed walls/buttresses.

is observed that the masses that were studied in mechanism's patterns are changed by this intervention and structure became more solid and supported through the weak direction. Figure 14 shows the reconstruction of the buttresses of the west tower.

7 CONCLUSION

In this study improving the behavior of masonry towers are investigated. As a case study Seddülbahir Fortress is taken into consideration. The structure was damaged during First World War and left uncared resulting with a more than half of the dome is lost. Therefore the structure became vulnerable to any low excitation to lose its static equilibrium. the case was a massive structure with hewn stone façade walls, dome and rubble stone filling with lime mortar. In three levels of the structure, collateral timber tie beams were used to embrace the whole structure. Investigation abilities of such structure is limited with visual inspection, mechanical tests which are not efficient to determine the inner cracks, discontinuities and poorly bind areas at inner section. In this case system identification tests were carried out to check the relevancy of structural model and sense the possible loose areas to study collapse mechanisms. Different proposals were studied to reduce the collapse risk of incomplete pieces. Injection and installment of timber tie beams were proposed to satisfy the integrity of the structure. In addition, the reconstruction of side walls/buttresses increase the lateral load capacity of the tower and also prevent to collapse of the structure.

ACKNOWLEDGEMENTS

This paper has been prepared with the collaboration of the different teams that took part in the Seddülbahir Fortress Restoration Project. The academic team who initiated the project in 1997 and continuing as the scientific advisory board and ÇATAB (Turkish Ministry of Culture and Tourism, the Çanakkale Wars and Gallipoli Historical Area Directorate) are gratefully acknowledged.

REFERENCES

Thys-Şenocak, L. Celik, R.N., Tanyeli, G., Özsavaşçı, A. 2009. Understanding Archeology and Architecture through Archival Records: The Restoration Project of the Ottoman Fortress of Seddülbahir on the Gallipoli Peninsula of Turkey, *The Frontiers of the Ottoman World Fortifications, Trade, Pilgrimage and Slavery*, ed. Andrew Peacock, *The British Academy*, 189–209.

Roca, P., Cervera, M., Gariup, G., Pela, L. 2010. Structural Analysis of Masonry Historical Constructions. Classical and Advamced Approaches. *Arch Comput Methods Eng*, 17: 299–325, Springer.

TBDY2018, Turkish Earthquake Code.

Brick and Block Masonry - From Historical to Sustainable Masonry –
Kubica, Kwiecień & Bednarz (eds)
© 2020 Taylor & Francis Group, London, ISBN 978-0-367-56586-2

Structural analysis of a masonry church with variable cross-section dome

A. Cascardi

ITC - Construction Technologies Institute, CNR - Italian National Research, Bari, Italy

F. Micelli & M.A. Aiello

Department of Innovation Engineering, University of Salento, Lecce, Italy

M. Funari

Department of Civil Engineering, University of Calabria, Rende, Italy

ABSTRACT: The structural capacity of masonry churches is generally hard to be investigated due to the complex geometry that characterize this relevant *Heritage*. Vaults, series of arches, flying buttresses, spiers, apse, dome, are often found in the field. Significant time efforts and costs are commonly expended in the geometrical survey. Traditional measuring results poorly accurate (randomly made within the construction), especially in case of huge buildings. On the other hand, modern technology, such as laser scanner, may result time consuming because of the data acquisition and elaboration. Moreover, the variation of thickness within the masonry members is often averaged and regularized since large endoscopy survey is forbitten for cultural *Heritage* restrictions. For this reason, in those cases in which the structure is affected, or even dominated, by stability issues, the accuracy of the analysis can be compromised without an adequate survey. In this scenario, the present study provides a case study concerning the analysis of a masonry church in which a main dome (> 15 m diameter) is placed on top. Both the base masonry walls and the dome have variable cross-sections. In order to achieve an accurate geometrical model, a drone was used for photogrammetry recognizing both outside and within the church. Then, the digital photos were computer elaborated according to their relative GPS-metadata. As a result, an accurate 3D-solid object was obtained. Finally, FEM analysis was performed in order to quantify the displacement capacity by pushover method.

1 INTRODUCTION

The large variety of masonry types does not allow to make a unified analysis of the mechanical behavior, (ISCARSAH 2013). In other words, each case must be considered unique and strengthening have to met the requirements of compatibility and reversibility as much possible; e.g. (Cascardi et al. 2019, Ombres et al. 2019). Although the masonry is a composite material (mortar and stone), it is generally considered as a homogeneous material, with average properties, in mechanical modeling. Furthermore, the geometry of masonry building is often complicated by modifications, demolitions and reconstructions over time. Finally, the state of degradation is not negligible. The set of these complex aspects represents the degree of uncertainty which affect common structural analysis methods.

The scientific research, firstly focused on resolving the mechanical-linked uncertainties by developing a series of destructive and non-destructive investigations, aimed at estimating mechanical strengths and elastic modulus. Nowadays these investigations demonstrated to converge with good approximation to the same results (Binda et al. 2013).

The geometric uncertainties are still poorly studied, e.g. (Micelli & Cascardi 2020). As a matter of fact, the historic masonry structures are often characterized by complex geometries and thus, simplified approaches may lead to either invalid or inaccurate evaluations. Concerning the full geometry description, thanks to *Digital Photogrammetry* (DP) and *Laser Scanner* (LS) techniques it is possible to define an accurate survey of the geometry. For instance, (Fortunato et al. 2017), proposed a new strategy to obtain a *Finite Elements* (FE) model starting from very dense outlines, which were generated by the point cloud slices. This methodology was verified on a very irregular structural geometry such as the *Baptistery of San Giovanni* in *Tumba* (Italy). However, some limits of this approach appeared in the stage after the geometry generation in which some defects of the *Non-Uniform Rational Basis Spline* (NURBS) model, such as holes, self-intersection etc., were detected. Moreover, (Castellazzi et al. 2017) developed an innovative refined strategy for the survey of historical masonry structures based on Laser Scanner or digital photogrammetry, in which the point cloud is transformed into a 3D-FE model through a semiautomatic procedure.

In many cases, the structural behavior is mostly influenced from problems related to geometry rather than strength: curved elements or variable cross-section members (Heyman 1966, 1982). In order to fill this gap, the present case study aims to apply and validate a structural analysis methodology based on drone survey and numerical simulation in FE (*Finite Elements*) environment.

2 DESCRIPTION OF THE MASONRY CHURCH

The *Church of St. Mary at the Gate*, shown in Figure 1, is a masonry building placed in Lecce (Italy) next to on of the gates of the *old town*. It has a *Pantheon* like structure: a circular inner cross-section is confined by thick walls on which a hemispherical dome is posed. The indoor height of the church is equal to the diameter of the dome (~15 m).

The masonry is characterized by regular limestone blocks and thin lime-based mortar joints (> 10 mm). The dome is externally covered by green and white ceramic finishing. The dome is completed on the top with a *roof lantern*, for a total outdoor height about 22 m. Semi-domes are visible from the indoor. The openings are limited to four semicircular windows placed in front of each other at the base of the main dome. The trampling surface is ~300 m².

The Church was built in 1548 and has undergone partial reconstructions between 1855 and 1858 when the neoclassical style was defined according to the current aspect. Through a visual inspection it was possible to observe the general good actual state, except for few humidity formations at the base of the walls and water infiltrations in the dome's intrados.

3 THE DRONE SURVEY

The relief of the Church was carried out by means of a drone (model *Phantom 3*), from which were taken of the photographs according to the criteria given in (Micelli & Cascardi 2020) both inside and outside building. The flight plans have been programmed in compliance with the *rules of the air* in accordance with the *Italian Civil Aviation Authority* provisions.

The photographic data were collected computed by means of *Agisoft Metashape* software. The computation was implemented throughout the following steps:

1. input of photographic acquisitions within the software for the alignment according to the GPS (*Global Positioning System*) metadata (Figure 2).
2. generation of the sparse point cloud - 146005 points (Figure 3).
3. generation of the dense point cloud - 84.94.50 points (Figure 4).
4. creating the mesh - 2.075.610 faces and 1044046 vertices (Figure 5).
5. creating the textured model (Figure 6).

The next step was to identify matching points between the relief of the internal surfaces and that of the external surfaces: the windows. Due to this method it was possible to derive the real volume of the building as distances between the inner surface and the external as it is possible to appreciate in Figure 7.

Figure 2. Alignment of the drone photos.

Figure 3. Sparse cloud.

Figure 4. Dense cloud.

Figure 1. Bird's eye view of the building.

221

Figure 5. Meshing.

Figure 6. Textured model.

Figure 7. Alignment of the inner and outer surfaces.

4 FROM GEOMETRICAL SURVEY TO FEM

Nowadays, thanks to the available technology used in the field of architectural surveying, Laser Scanning and digital photogrammetry, it is possible to define the almost exact geometry of any case study. The main aim of this section is to highlight the relationship between the geometrical survey, performed using digital photogrammetry, and the generation of a *Finite Element Model* (FEM).

As mentioned above, the photogrammetry's work flow produces a mesh representation of the structures that arises from the interpolation of a dense point cloud. However, this type of three-dimensional model cannot be used for structural numerical analysis, as it is formed by a representation that cannot be directly adopted for FE-analyses. Hence, in order to use the geometric data derived by digital photogrammetry, it is necessary to perform some operations for transforming the mesh arising from photogrammetry's workflow in *Initial Graphics Exchange Specification*

(IGES) format. All the CAD-like (*Computer Aided Design*) software and the FEM pre-processors allow the user to import and to export geometrical information in the form of *text* files. Indeed, the geometry have to save in the IGES format, made by the *American National Standard* (ANS), which is the most used format for exchanging data between CAD, CAM (*Computer Aided Manufacturing*) and CAE (*Computer-aided engineering*).

It is worth of nothing that it is not possible to define a standard procedure that allows the user to create an importable file, correctly. However, a procedure which can be generalized for many situations where monumental masonry structures have to be analyzed, was implemented.

After data acquisition, developed by using the digital photogrammetry technique, the proposed procedure seeks to generate a CAD model starting from the assemblage of the mesh faces generated by the triangulation of the point cloud. For sake of clarity the methodologies expected the following computational steps:

- mesh decimation;
- from mesh faces to assemblage of surfaces;
- generating a closed poly-surface which represent the real geometry of the structure;
- refining the geometry: introduction of the holed that represent doors and/or windows.

The work-flow is all integrated into a computational tool implemented in the visual programming environment offered by *Grasshopper* 3D plug-in for *Rhino* (McNeel 2010). The Figure 8 represents the component assembling which allows to the visual program the generation of the solid model.

The first stage of the proposed procedure consists of the importing process of the mesh generated from the photogrammetric software in Rhinoceros environment. At this stage the intrados surface is defined into *Grasshopper* 3D by adopting a proper mesh parameter. At this stage the work flow needs of a "Mesh Reduce" component. is able to properly decimate face number of the mesh. In order to save the degree of the geometry definition, the fundamental task of this step is the setting of the accuracy parameters. The faces of the mesh are deconstructed into faces and vertices and reconverted into and assemblage of a series of planar surfaces. The edges of this surfaces will be used to construct the computational mesh in the Midas Environment.

It should be noted that visual scripting allows the user to obtained this result immediately without performs others manual operations. Finally, the Surfaces are manipulated into Rhinoceros environment in order to create a closed poly-surface which has to be imported into (Midas-FEA 1989). Then, in Midas-FEA environment the CAD model can be discretized by using 3D element typology, which are based on a tetrahedral or hexahedral geometry with linear interpolation functions as visible in Figure 9.

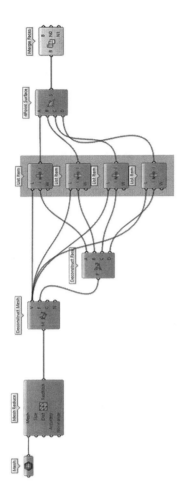

Figure 8. Visual algorithm able to generate a poly-surface starting from mesh arising from digital photogrammetry.

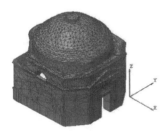

Figure 9. Finite elements model.

5 STRUCTURAL ANALYSIS

Once the model was imported in *Midas FEA Code* (Midas-FEA 1989) a three-dimensional mesh was generated for the analysis (i.e. 273497 tetrahedral elements with 300 mm length). The initial input data are reported in Table 1. The constitutive model was

Table 1. Mechanical properties of the masonry.

Variable	Symbol	Unit	Value
Young's Modulus	E	MPa	2500
Poisson Ratio	v	-	0.2
Weight density	γ	N/m^3	16000
Compressive strength	f_c	MPa	6
Compressive fracture energy	G_c	N/mm	0.1
Tensile strength	f_t	MPa	0.3
Tensile fracture energy	G_f	N/mm	0.012

based on the total strain method with smeared approach for the fracture energy. A concrete damage-plasticity material model for the masonry, already available in the standard software package was adopted. It allowed to reproduce a reliable non-linear behavior with distinct damage parameters in tension (Cornelissen & Hordijk 1986) and compression (Thorenfeldt & Tomaszewicz 1987).

5.1 Modal

In order to obtain a numerical assessment of the dynamic behavior of the *Church of Santa Maria of the Gate*, an eigen-frequency analysis was performed. The ability of such a conventional linear dynamic analysis to represent the actual state of the structure is limited by the behavior of the masonry structures, which may become strongly non-linear after the first seismic excitations. Furthermore, the eigen-frequency analysis is a fundamental prerequisite for the non-linear static procedure adopted in this study. As reported in *Italian Guidelines* for the building *Heritage* (P.C.M. 2011), and confirmed by the recent Italian Technical Code (2018), when the participating mass of the fundamental mode is not less than 60%, the Pushover analysis can be applied to masonry buildings. In the present study, the mathematical formulation utilized to calculate the natural frequencies of the model is consistent with the method reported in (Lanczos 1950), which solves the following eigenvalue problems:

$$(K - \omega_i^2 M)\phi_i = 0 \qquad (1)$$

where K is the stiffness matrix, ω_i^2 is the i-th eigenvalue, M is the mass matrix and ϕ_i is the i-th eigenvector. Tables 2 and 3 show the results in terms of modal period and of mass participation factors of the first main modes whereas Figure 10 represents deformed shapes of the first main vibrating modes.

5.2 Pushover

The Pushover analysis is frequently used to assess the seismic damage expected in case of earthquake.

223

Table 2 . Modal frequencies and periods.

Mode #	Frequency Hz	Period sec
1	5.57	0.180
2	5.62	0.178
3	8.27	0.121
4	8.32	0.120
5	9.01	0.111
6	12.55	0.080
7	12.79	0.078
8	13.71	0.073
9	15.22	0.066
10	15.45	0.065

Table 3 . Modal participating masses.

Mode #	Participating mass X-dir %	Participating mass Y-dir %
1	0.01	80.33
2	78.29	0.01
3	0	0.07
4	0.26	0
5	0	0.05
6	0	0
7	0	0
8	0.02	0
9	3.26	0
10	0	3.41
sum	81.84	83.87

The seismic assessment of the structure is represented by a capacity curve in which the vertical axis defines the base shear, whereas on the horizontal axis the displacement of the control point (typically the node at the top of the structure) is reported. At first, vertical forces including gravity loads are considered. Subsequently the seismic forces are computed in function of the vertical forces. The horizontal forces act separately along the geometric axes of the structure, namely X and Y, in both directions. In the present work, the evaluation of the load carrying capacity is performed by a mass proportional force distribution.

The obtained pushover curves are reported in Figures 11 and 12, representing the capacity of the building and its intermediate behaviour during the increasing force. As expected, the analyses show a stiff uncracked behavior of the structures because its geometry features. In both cases, for the loading distributions along X and Y, the loading displacement curves reaches a maximum load after a pseudo-linear branch. Once that the maximum load is reached, the progressive damage generates a quick loss of stiffness which brings the structures to a hardening branch, subsequently. The structural behavior is quite similar along the two main direction (X and Y). In particular, along the Y axis the loading displacement curves are perfectly overlapped because of the geometry symmetry of the structure.

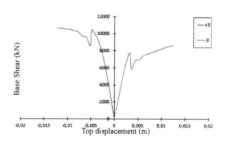

Figure 11. Capacity curves in X-direction.

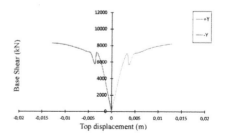

Figure 12. Capacity curves in Y-direction.

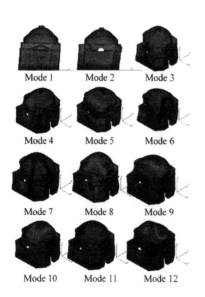

Figure 10. Deformed shapes of the first main modes.

224

In order to numerically evaluate the seismic vulnerability of the case of study, the curve reported in Figures 11 and 12 need to be processed with the aim of converting the *multi-degree of freedom system* (MDOF) into an equivalent system with *single degree of freedom* (SDOF) by means of the modal participation factor defined by the following expression:

$$\Gamma = \frac{\sum m_i \phi_i}{\sum m_i \phi_i^2} \qquad (2)$$

where ϕ_i is the *i*-th component of the eigenvector ϕ and m_i is the mass of the node *i*.

6 DISCUSSION OF THE RESULTS

The *Midas FEA Code* allowed to observe the theoretical crack pattern (delivered from strains pattern) by means of color maps in which the red points mean full crack opening; while blue and violet points individuate the partially opened crack. As expected, the crack developing is affected by the presence of the opening (windows and main door for the Y and X directions, respectively) as illustrated in Figures 13 and 14. In fact, the masonry above the opening is less loaded (in term of compression), since the trust-lines tends to compress the masonry from the top dome to the base-building (Heyman 1966). In this sense, the crack pattern can be considered a sort of dissipation capacity of the building, since the failure mechanisms need further damage along the building to be activated.

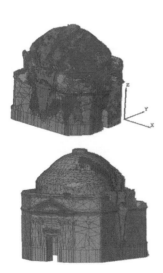

Figure 13. Crack patterns: -Y (up) and +Y (bottom).

Figure 14. Crack patterns: -X (up) and +X (bottom).

7 CONCLUSIONS

In this study, the seismic response of a monumental masonry church has been studied by using numerical analyses. The study started by analyzing the geometrical uncertainties, by using a new drone-assisted survey technique.

The first phase of the research involved a photogrammetric survey conducted by means of remote piloting of a drone equipped with a camera. The photos were then implemented and processed using a computer graphics software in order to get a point cloud. Through a series of processing operations, it was possible reproduce the geometry of both the indoor and outdoor. These surfaces have been made photorealistic through the application of texture resulting from the drone photos. Finally, the two surfaces were positioned with respect to the same reference marker e aligned so as to obtain the correspondence of the surfaces. A further operation was necessary using 3D modeling software to process a solid-type object, ready to be modeled in FEM code.

Structural analysis was so computed by means of linear dynamics and non linear static (pushover) investigations. The capacity curves in the main directions (X and Y) allowed to assess that the performance points of the building were found in the safety side.

The proposed procedure brings the main advantage of a drastically reduction of the pre-processing time. The fast acquisition of the input geometry was possible without losing accuracy. At the end each vulnerability analysis took an estimated time of less than 4 hours by using a common personal computer.

Final remark is about the safety verification by comparing the result in term of seismic displacement demand and check it against displacements limits calculated according to technical code; e.g. in (P.C.M. 2011).

REFERENCES

Binda, L. et al. 2013. Diagnosis of Historic Masonry Structures Using Non-Destructive Techniques: 1089–1102.

Cascardi, A. et al. 2019. Reversible techniques for FRP-confinement of masonry columns. *Construction and Building Materials* 225: 415–428.

Castellazzi, G. et al. 2017. An innovative numerical modeling strategy for the structural analysis of historical monumental buildings. *Engineering Structures* 132: 229–248.

Thorenfeldt, E. et al. 1987. Mechanical properties of high-strength concrete and applications in design. *Proc. Symp. Utilization of High- Strength Concrete*. Trondheim: Stavanger, Norway.

Fortunato, G. et al. 2017. Survey and seismic vulnerability assessment of the Baptistery of San Giovanni in Tumba (Italy). *Journal of Cultural Heritage* 26: 64–78.

Cornelissen, H.A.W. et al. 1986. Experimental determination of crack softening characteristics of normalweight and lightweight concrete. *Heron* 2(31): 45–56.

Heyman, J. 1966. The stone skeleton. *International Journal of solids and structures* 2(2): 249–279.

Heyman, J. 1982. *The masonry arch*. Chichester : Horwood.

ISCARSAH. 2013. Recommendations for the Analysis, Conservation and Structural Restoration of Architectural Heritage.

Lanczos, C. 1950. An iteration method for the solution of the eigenvalue problem of linear differential and integral operators. *J. Res. Nat'l Bur. STD*. 45(4): 255–282.

McNeel, R. & A. 2010. Rhinoceros Version 4.0, Service Release 8.

Micelli, F. & Cascardi, A. 2020. Structural assessment and seismic analysis of a 14 th century masonry tower. *Engineering Failure Analysis* 107: 104198.

Midas-FEA. 1989. Advanced Nonlinear and Detail Program, Analysis and Algorithm.

Ombres, L. et al. 2019. Flexural Strengthening of RC Beams with Steel-Reinforced Grout : Experimental and Numerical Investigation. *J. Compos. Constr*. 23(5): 1–13.

P.C.M. 2011. Linee Guida per la valutazione e riduzione del rischio sismico del patrimonio culturale allineate alle nuove Norme Tecniche per le Costruzioni (No. 47). DM 14/1/2008. Direttiva del Primo Ministro, 9/02/2011. GU. (d.m. 14 gennaio 2008).

Codes & standards

Brick and Block Masonry - From Historical to Sustainable Masonry –
Kubica, Kwiecień & Bednarz (eds)
© 2020 Taylor & Francis Group, London, ISBN 978-0-367-56586-2

Numerical modelling of two-way out-of-plane bending tests on URM walls: The influence of lateral boundary conditions

L. Chang, J.G. Rots & R. Esposito

Faculty of Civil Engineering and Geosciences, Delft University of Technology, Delft, The Netherlands

ABSTRACT: Research has shown that lateral boundary conditions can have a large influence on the force capacity of two-way spanning unreinforced masonry (URM) walls subjected to out-of-plane (OOP) loading. Differently than laterally free one-way spanning walls, they show a higher force capacity, which however is underestimated by current analytical formulations. By means of nonlinear finite element analyses adopting a detailed 3D brick-to-brick model, the influence of lateral boundary conditions on two-way out-of-plane failure of a single wythe masonry wall is studied. Results indicate that the cracking pattern varies as lateral boundaries become stiffer, accordingly the force capacity increases. Numerical results are compared with analytical formulation proposed in the Australian Standard AS3700. These preliminary results will serve to evaluate how to consider the lateral boundary conditions, provided by the wall-to-wall connection, for two-way spanning walls in existing buildings.

1 INTRODUCTION

Out-of-plane (OOP) failure of two-way spanning unreinforced masonry (URM) walls can be possibly the most dangerous failure mechanism for masonry structures during strong earthquakes (Sorrentino et al., 2016). Though this type of failure is attracting more interest in academia, some crucial factors having major influence on the force capacity of walls remain to be further studied, for example, lateral boundary conditions.

Recently, researchers have put effort in evaluating the force capacity of two-way spanning walls through both experiments and analytical formulations. Nevertheless, focuses on boundary conditions are quite limited. Griffith et al. (2007) carried out tests on two-way spanning walls subjected to cyclic loads with lateral boundaries restrained by return walls. This is close to practice however the flexibility of return wall was not estimated. A similar testing campaign was conducted at Delft University of Technology (Damiola et al., 2018) in which the lateral edges of walls were constrained with steel tubes and were assumed as hinged. In fact, the lateral boundaries of walls in reality are mostly partially clamped considering they are restrained by return walls which are able to transfer part of bending moments. This situation leads to a difficulty in accurately predicting the force capacity of walls. In Eurocode 6 (EN1996-1-1, 2012) where yield line method is applied, boundary conditions of two-way spanning walls are calculated as either hinged or clamped. This obviously misevaluate the boundary conditions. Australian Standard – Masonry Structures (AS3700-2011, 2011) which is based on virtual work

method quantifies lateral boundaries with coefficient R_f. The lateral boundaries are hinged with $R_f = 0$ or clamped with $R_f = 1$. Still, the code notifies that walls constrained within intermediate situations (e.g. with return walls) should be evaluated by users and an intermediate value R_f has not been recommended. Willis (2004) proposed an update for virtual work method where torsional strength of masonry was introduced to evaluate diagonal bending capacity. Griffith and Vaculik (2007) found that when R_f is assumed as 0.5 the prediction of the updated method could be the most close to testing results for two-way spanning walls with return walls. This was also confirmed by parametric study by Damiola et al. (2018). However, according to testing results by Graziotti et al. (2019), the value of masonry torsional strength can influence the accuracy of evaluation on the wall capacity. Based on aforementioned discussion, a research gap can be identified that experiments focusing on lateral boundary conditions of two-way spanning URM walls are limited and evaluation of those by analytical formulations in current codes need to be improved.

The aim of this paper is to study the influence of lateral boundary conditions on the force capacity of two-way spanning URM walls and provide basis for improving related part in current codes. With this goal, numerical simulation was firstly applied to model two-way OOP bending test for URM wall performed at Delft University of Technology (Damiola et al., 2018). Subsequently, a parametric study was carried out considering various flexibility of lateral boundaries based on previous model. Finally, numerical results were compared with Australian Standard and revised formulations using experimentally derived torsional shear strength provided by Graziotti et al. (2019).

2 TWO-WAY OOP BENDING TESTS

The two-way OOP bending tests were carried out at Delft University of Technology (Damiola et al., 2018). Testing sample TUD_COMP-11 was selected as reference for numerical models in this study. This sample was made of calcium silicate bricks and cement based mortar. The dimension of bricks was 214 × 72 × 102 mm³ while the thickness of mortar joint was 10 mm. The dimension of the wall was 3874 × 2765 × 102 mm³. The test set-up of the wall is shown in Figure 1.

The wall was constrained on four sides. Along top and bottom boundaries, the wall was glued to strengthened steel beams (Figure 2(a)). At the top the wall was allowed to move vertically. Along lateral sides, steel tubes were applied and wooden edges were inserted between steel tubes and the wall to prevent local damage (Figure 2(b)).

The wall was firstly loaded with a pre-compression of 0.05 N/mm² at the top. Afterwards, horizontal cyclic load was uniformly applied to the wall by means of

Table 1. Material properties from material tests.

Material properties	Symbol	Units	Average	C.o.V	Testing standard
Density of masonry	ρ	kg/m³	1910	-	-
Elastic modulus of brick units	E_b	MPa	8990	0.36	EN 772-1
Poisson's ratio	v	-	0.16	-	EN 772-1
Flexural bond strength	f_w	MPa	0.27	0.43	EN 1052-5
Cohesion	c	MPa	0.14	-	EN 1052-3
Friction angle	ϕ	rad	0.406	-	EN 1052-3
Compressive strength	f_c	MPa	5.93	0.09	EN 1052-1
Compressive fracture energy	$G_f^{\,c}$	N/mm	31.5	0.16	EN 1052-1

airbags placed against both front and back sides of the wall. Initially airbags on both sides were pumped to certain pressure. Then pressure of airbag on back side of the wall was kept unchanged while pressure within the airbag on the front side was varied in a cyclic manner. The cyclic load was displacement-controlled with regard to the very central point of the wall. Additional tests were carried out to determine increment caused by friction between the wall and the airbags. More details about boundary conditions and loading protocol were discussed in Ravenshorst and Messali (2016) and Damiola et al. (2018). Table 1 lists material properties from material tests using the same batch of bricks and mortar.

3 NUMERICAL MODELS

3.1 Modelling technique

Nonlinear finite element analyses were carried out by adopting a 3D brick-to-brick model including geometrical and physical nonlinearity. The simplified modelling technique proposed by Lourenco and Rots (1997) was adopted in which bricks are extended with regard to their original dimensions while mortar joints are modelled with zero-thickness interface elements. 20-node solid elements were used to model the bricks, while 8-node interface elements were adopted for the mortar joints. Symmetric model was adopted (Figure 3). Clamped restraints were imposed at the top and bottom side of the wall, but allowing the vertical displacement of the top side of the wall. In view of the parametric study on the lateral boundary conditions, presented in Section 5, the laterally hinged connections were modelled by means of boundary interface elements with an elastic normal stiffness equal to $k_{n,lateral} = 0.0001$ N/mm³. Self-weight and pre-compression (0.05 N/mm2) were applied to the model. Differently than in the experiment, a monotonic analysis was performed

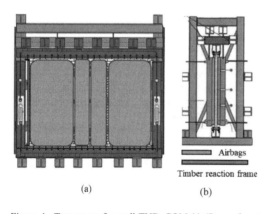

(a)   (b)

Figure 1. Test set-up for wall TUD_COM-11 (Ravenshorst and Messali, 2016).

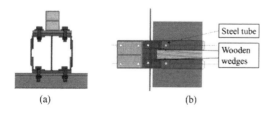

(a)   (b)

Figure 2. Boundary conditions for wall TUD_COMP-11: (a) bottom boundary condition; (b) lateral boundary condition (Ravenshorst and Messali, 2016).

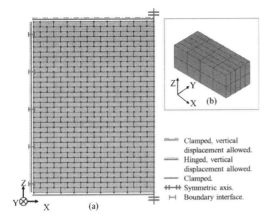

Figure 3. Numerical model of TUD_COMP-11: (a) boundary conditions; (b) meshing for a single block.

applying a uniform pressure on the face of the wall and adopting arc-length control. Quasi-Newton solution method was adopted. The analyses were carried out with FEA software DIANA version 10.3 (DIANA.BV., 2019).

To describe the nonlinear response of the masonry, the Multi-surface Interface Model (Lourenco and Rots, 1997, Rots, 1997) was used, while bricks were modelled as linear elastic. The multi-surface plasticity model comprises a Coulomb friction model, a tension cut-off and an elliptical compression cap (Figure 4). Softening acts in all three modes. In the elastic stage, the normal stiffness and shear stiffness of interface elements were calculated based on elastic modulus of bricks, elastic modulus of mortar, mortar joint thickness and Poisson ratio (Lourenco and Rots, 1997):

$$k_n = \frac{E_{unit}E_{joint}}{h\left(E_{unit} - E_{joint}\right)} \quad (1)$$

$$k_s = \frac{G_{unit}G_{joint}}{h\left(G_{unit} - G_{joint}\right)} \quad (2)$$

$$G_{unit} = \frac{E_{unit}}{2(1 + \nu_{unit})} \quad (3)$$

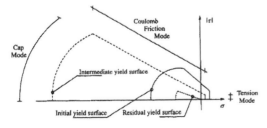

Figure 4. Multi-surface interface model for interface elements (Lourenco and Rots, 1997).

$$G_{joint} = \frac{E_{joint}}{2\left(1 + \nu_{joint}\right)} \quad (4)$$

where E_{unit}, G_{unit} and ν_{unit} the elastic modulus, shear modulus and Poisson's ratio of brick unit respectively; E_{joint}, G_{joint} and ν_{joint} elastic modulus, shear modulus and Poisson's ratio of mortar joint respectively; k_n and k_s normal stiffness and shear stiffness of interface elements respectively; h the thickness of mortar joints.

3.2 Calibration of input parameters

Material properties used as input for numerical models were directly retrieved from or calibrated based on small-scale material tests performed in the same testing campaign (Esposito et al., 2016, Jafari et al., 2019).

Calcium silicate brick units are modelled as linear elastic solid elements, elastic modulus and Poisson's ratio were retrieved from flexural strength test of masonry unit. Since the elastic modulus of mortar joints was not experimentally determined, the normal stiffness of bed joint, $k_{n,bed}$, was calibrated to match the initial stiffness of numerical model with that of experiment. The calibrated value of $k_{n,bed}$ was then used to calculate the elastic modulus of the mortar joint E_{joint} through equation (1). According to construction practice, mortar in bed joints can be generally considered stronger than that one in head joints. In this sense, the normal stiffness of head joints $k_{n,head}$ was defined two times smaller than the normal stiffness of bed joints $k_{n,bed}$. The shear stiffness of bed joints, $k_{s,bed}$, and shear stiffness of head joints, $k_{s,head}$, were calculated through equation (2) - (4).

With regard to material properties related to the nonlinear behaviour, tensile strength f_t and Mode-I fracture energy G_f^I were calibrated according to bond wrench tests and OOP masonry wallet tests respectively; cohesion c, friction angle ϕ and Mode-II fracture energy G_f^{II} were retrieved or calibrated from masonry shear triplet tests; compressive strength f_c and compressive fracture energy G_{fc} were retrieved from masonry compression tests. Note that the same reduction factor α was also applied between $f_{t,bed}$ and $f_{t,head}$. Since shapes of softening and compression cap of bed joints and head joints are assumed to be the same, fracture energy of head joints is approximately ¼ of that of bed joints. The adopted input parameters are given in Table 2 and Table 3.

Table 2. Input parameters for bricks.

Input parameters	Symbol	Units	Value
Young's modulus	E_b	N/mm^2	8990
Poisson's ratio	ν_b	-	0.16
Mass density	ρ	kg/m^3	1910

Table 3. Input parameters for interface elements.

Input parameters	Symbol	Units	Bed joints	Head joints
Normal stiffness	k_n	N/mm^3	28.9	14.22
Shear stiffness	k_s	N/mm^3	12.46	6.13
Tensile strength (2/3 of f_w)	f_t	MPa	0.18	0.09
Mode-I fracture energy	G_f^I	N/mm	0.016	0.004
Cohesion	f_{v0}	MPa	0.14	0.07
Friction angle	ϕ	rad	0.406	0.406
Residual friction angle	ϕ_{res}	rad	0.406	0.406
Mode-II fracture energy	G_f^{II}	N/mm	0.01	0.0025
Compressive strength	f_c	MPa	5.93	2.97
Compressive fracture energy	G_f^c	N/mm	31.5	7.88

4 NUMERICAL RESULTS

In this section, numerical results are presented and compared with experimental results in terms of lateral force vs. mid-span displacement curves (Figure 5) and crack pattern (Figure 6). According to previous study, the revised experimental envelope curve was considered for comparison to exclude the fictitious force increment caused by the friction between the wall and airbags (Damiola et al., 2018). The numerical results show a linear behaviour up to a displacement of approximately 9 mm (point A in Figure 5(b)) followed by a sharp reduction in capacity after the peak (points B and C) and a subsequent hardening behaviour until a displacement of approximately 30 mm (point D). After this point, no convergent results could be achieved.

Compared with experimental curve, the initial stiffness and peak force of numerical model matched well with those of testing results (13.96 kN/mm vs. 12.00 kN/mm and 29.38 kN vs. 28.9 kN). This suggests that calibration of material properties is appropriate and values retrieved directly from material tests are suitable for modelling large-scale walls.

The lateral force vs. mid-span displacement curve was in general in agreement with testing results, except that the peak force in numerical model appeared much earlier than in tests. Besides, a sharp drop in force after peak force was not experimentally observed in testing curve. Additional parametric study with respect to f_t, G_f^I, c and G_f^{II} has shown that as these values varied, there were no obvious changes in this drop of force. This difference can be caused by that in test the lateral load was cyclic while it was monotonic in numerical model. Since the largest displacement for each loading cycle was gradually increased and there was unloading during each cycle, both horizontal cracks and diagonal

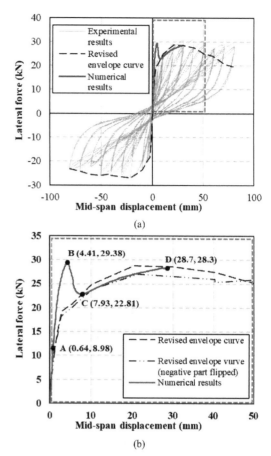

Figure 5. Lateral force vs. mid-span displacement curves of experimental and numerical results: (a) full graph; (b) Close-up graph.

cracks could develop more evenly. Besides, moment resistance capacity along developing cracks could be redistributed over time. This can lead to a continuously increase of force capacity. In contrast, in a monotonic loading test, there was a sequence for cracking at different locations, increase of force capacity is possible after a period of decrease. A similar trend with a sharp drop in force after the attainment of the maximum lateral force was experimentally observed by Griffith et al. (2007) during monotonic tests on full-scale walls.

The evolution of cracking is shown in Figure 6 where crack opening of interface elements is shown. Here four sub-figures corresponding to the four critical points A-D marked in Figure 5(b) are considered. At Point A, two long horizontal cracks firstly develop along bed joints adjacent to top and bottom sides of the wall. This marks the onset of nonlinear behaviour. When horizontal cracks are fully developed along the length of the wall (Point B), the wall reached its peak force (29.38 kN). As displacement increases, the force capacity of the wall starts to decrease till diagonal cracks fully

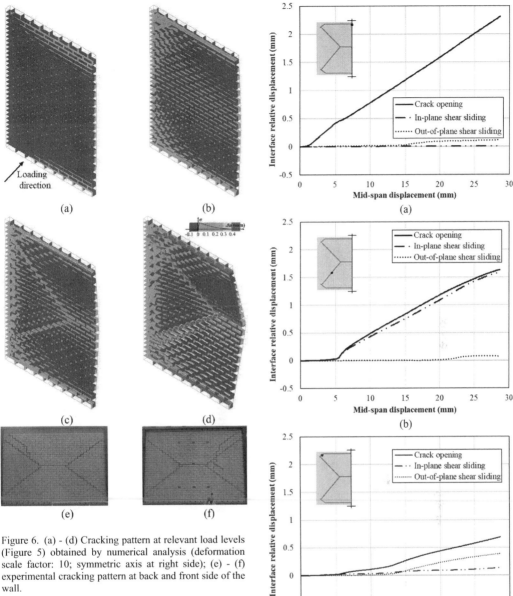

Figure 6. (a) - (d) Cracking pattern at relevant load levels (Figure 5) obtained by numerical analysis (deformation scale factor: 10; symmetric axis at right side); (e) - (f) experimental cracking pattern at back and front side of the wall.

Figure 7. Crack opening and shear sliding at: (a) horizontal crack, (b) diagonal crack and (c) corner crack.

develops (Point C, 22.81 kN). Then the force capacity starts to increase again which can be due to an arching effect. The final crack pattern at point D is composed of horizontal cracks at the top, bottom and central mortar bed joint, diagonal cracks starting few courses away from the corner of the wall, and corner cracks; this crack pattern is in agreement with experimental observations (Figure 6(e, f)).

To evaluate the influence of different local failure mechanisms, a detailed analysis was performed by analysing the local failure mechanisms in three main cracks (Figure 7). The horizontal crack at the top of the wall (front side), the diagonal crack (back side)

and the crack at the corner (front side) were considered. Absolute values at integration points are presented. For the horizontal crack, crack opening (normal relative displacement of the interface) and limited in-plane shear and out-of-plane sliding are observed, suggesting as expected, that tensile failure is the main failure mechanism for horizontal cracks

233

(Figure 7(a)). For the diagonal crack (Figure 7(b)), both cracking opening and in-plane shear sliding play an important role. At corner crack (Figure 7(c)), all three failure mechanisms are observed, but their deformation are lower with respect to the other two cracks analysed.

5 PARAMETRIC STUDY ON THE STIFFNESS OF LATERAL BOUNDARY CONDITIONS

The main aim of this paper is to study the influence of lateral boundary conditions on mechanical behaviour of two-way spanning URM walls subjected to OOP loading. For this purpose, the normal stiffness of the lateral boundary interface element, $k_{n,lateral}$, was varied to model lateral boundaries with different rotational stiffness. As $k_{n,lateral}$ increases from 0.0001 N/mm^3 (hinged case already discussed in Section 4) to 10000 N/mm^3, the lateral boundary can be deemed as changing from hinged (free rotation along Z-axis) to clamped (rotation constrained along Z-axis). Five cases are considered, among which three different cases to represent the partially clamped case ($k_{n,lateral}$ = 1 N/mm^3; $k_{n,lateral}$ = 5 N/mm^3; $k_{n,lateral}$ = 10 N/mm^3). Numerical results are shown in Figure 8, together with previously discussed experimental results for the hinged case. Please note that in this section the partially clamped case 2 is further selected for comparison with hinged and clamped cases.

By increasing the stiffness of rotational restraints at the lateral boundaries (from hinged to clamped case) both initial stiffness and force capacity of two-way OOP bending URM wall increased as expected (Table 4 and Figure 8). Furthermore, the sharp drop

Table 4. Comparison of initial stiffness and peak force.

Cases	$K_{n,lateral}$ (N/mm^3)	Initial stiffness (kN/mm)	Peak force (kN)
Hinged	0.0001	13.96	29.38
Partially clamped 1	1	14.25	41.88
Partially clamped 2	5	15.09	54.09
Partially clamped 3	10	15.72	57.71
Clamped	10000	17.95	64.61
Experiment	-	12.00	28.90

in force observed by for the hinged case at the onset of diagonal cracking is gradually reduced for partially clamped cases and it is not observed for the clamped case.

By increasing the stiffness of rotational restraints at the lateral boundaries, besides the formation of horizontal, diagonal and corner cracks as observed for the hinged case, vertical cracks at the lateral sides become predominant. To compare this difference, the crack evolution sequence for the three cases can be summarised as follows (refer to Figure 8):

- Hinged case: horizontal cracks (B)→ corner cracks (not obvious) (C)→ diagonal cracks (D)
- Partially clamped case 2: horizontal cracks (E)→ corner cracks → diagonal cracks (F)→ vertical cracks (G)
- Clamped case: horizontal cracks (H)→ vertical cracks (J)→ corner cracks → diagonal cracks (K)

The presence of vertical cracks for the partially clamped and for the clamped case can be reasonable ascribed to the increase in bending moments with respect to the hinged case. Larger bending moments lead to larger tensile stresses along all joints located near the lateral side of the wall, which form the so-called vertical cracks. Consequently, for an increase in later rotational stiffness, the formation of vertical cracks will occur for a lower mid-height displacement.

It is worthy to note that in the paper by Griffith et al. (2007) a two-way spanning URM wall sample laterally restrained by return walls with similar pre-test set-up as wall TUD_COMP-11 showed a same cracking process as for partially clamped case 2 in this paper. This may suggest that a two-way spanning URM wall in reality is neither clamped nor hinged along lateral boundaries, but at a partially clamped state. Nevertheless, it should be pointed out that in the presented analyses, the formation of the vertical cracks can be influence by the shear stiffness of the lateral boundary interface. Since a relative high value of shear stiffness ($k_{s,lateral}$ =10000 N/mm^3) was adopted, the opening of the bed joint located near the lateral side of the wall may be prevented, thus influencing the crack pattern and the estimation of force capacity. However, it is reasonable to assume that this will not influence the sequence of cracking. Consequently, further studies are needed and these

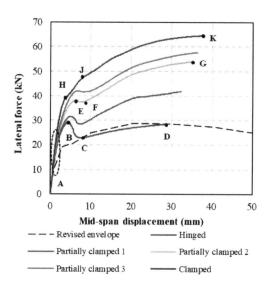

Figure 8. Lateral force vs. mid-span displacement curves of various lateral boundary conditions.

preliminary results can be adopted for a qualitative estimation of the trends.

Five locations, I-V in Figure 9 are selected to check the local failure mechanisms for the three cases. Figure 9(a) shows that as rotational stiffness of lateral

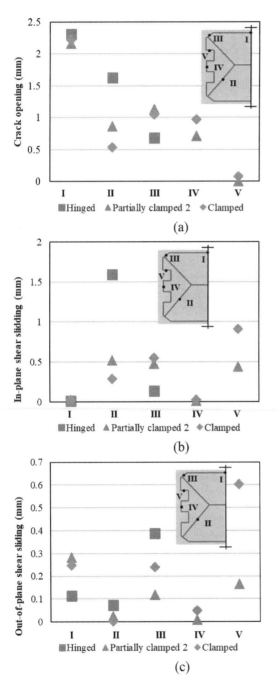

Figure 9. Comparison of cracking mechanisms: (a) crack opening; (b) in-plane shear sliding; (c) out-of-plane shear sliding.

boundaries increases, crack opening decreases at location II while it increase at location IV and V. In Figure 9(b, c) it is shown that as lateral boundaries become stiffer, both in-plane and out-of-plane shear sliding decrease at location II. At location III, in-plane shear sliding increases while out-of-plane shear sliding decrease as lateral boundaries become stiffer. At location IV and V, both in-plane and out-of-plane shear increase as the rotational stiffness of lateral boundaries increase. This is because stiffer lateral boundaries on one hand constrain developing of diagonal cracks, on the other hand they lead to larger bending moment along vertical edges that causes larger crack opening and shear sliding.

6 COMPARISON WITH ANALYTICAL FORMULATIONS

Australian standard AS3700 (AS3700-2011, 2011) currently provides the most advanced analytical formulations on predicting the force capacity of two-way spanning URM walls. This is because it applies virtual word method and comprehensively evaluates various crucial factors such as boundary conditions, aspect ratio and openings that have major influence on wall behaviour. However, in some cases, it still largely underestimates the wall capacity. According to Damiola et al. (2018), AS3700 underestimated the capacity of single wythe clay wall up to -56% when lateral boundaries were considered as hinged. The large error was caused by an incorrect prediction of cracking pattern by AS3700, which considered the formation of a vertical crack at the centre of the wall instead of the experimentally observed horizontal crack. This raised the doubt on the accuracy of the analytical formulation, in particularly questioning the definition of the restrain coefficient R_f and the definition of the torsional strength.

To further study the influence of lateral boundary conditions, force capacity of the wall was calculated based on Australian Standard AS 3700 (AS3700-2011, 2011) and revised formulas proposed by Willis (2004). The torsional shear strength in later formulas were replaced with experimental results provided by Graziotti et al. (2019). From Figure 10 it can be seen that if lateral boundaries are considered as hinged ($R_f = 0$) or clamped ($R_f = 1$), AS3700 either underestimates (-22%) or overestimates (+48%) the wall capacity. If lateral boundaries are considered as partially clamped ($R_f = 0.5$), the standard provides a more accurate prediction with an error of +13%. By now an intermediate value for R_f has not been proposed by AS3700 (AS3700-2011, 2011). Griffith and Vaculik (2007) and (Graziotti et al., 2019) suggested that R_f can be 0.5, but this requires further verification.

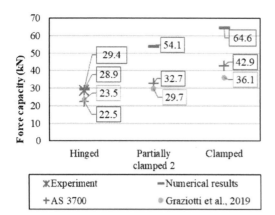

Figure 10. Comparison of force capacity with analytical formulations.

7 CONCLUSIONS

This paper focuses on the influence of lateral boundary conditions on response of two-way spanning URM walls subjected to out-of-plane (OOP) loading. For this purpose, nonlinear finite element analyses adopting a detailed 3D brick-to-brick model were carried out to evaluate the influence of lateral boundary conditions on the force capacity of two-way spanning URM walls. A parametric study was conducted to evaluate the influence of lateral boundary conditions and results were compared with analytical formulations proposed by Australian Standard AS3700 and Willis (2004) respectively. Here torsional stiffness in formulations by Willis (2004) were replaced with experimental values by Graziotti et al. (2019). The following observations and conclusions can be drawn:

- Using material properties directly retrieved from small-scale material tests combined with minor calibrations, a good agreement between numerical and experimental results in terms of initial stiffness, force capacity and cracking pattern was obtained. This suggests that material properties measured in small-scale material tests are suitable for modelling large-scale two-way OOP spanning URM wall components.
- Differently than observed in the experiment, the numerical results for the case with hinged lateral boundary conditions shows an initial drop in force before the formation of the diagonal cracks. This can be caused by different loading protocol between numerical model and experiment. In experiment the loading was cyclic while in numerical model it was monotonic.
- The parametric study shows that as the rotational stiffness of lateral boundaries increased, both initial stiffness and force capacity of two-way spanning walls increased. Also, the evolution of cracks and

the final cracking pattern varied as lateral boundary conditions changed. Vertical cracks were not observed for the hinged case ($k_{n,lateral}$ = 0.0001 N/mm^3), while they were observed both for the partially clamped case 2 ($k_{n,lateral}$ = 5 N/mm^3) and the clamped case ($k_{n,lateral}$ = 10000 N/mm^3). Furthermore, as the lateral boundaries become rotationally stiffer, vertical cracks occurred for a lower midspan displacement. This causes an internal redistribution of the moments leading to an increase initial stiffness and force capacity of the wall.

- A quantitative analysis of local failure mechanisms occurring in the main cracks was carried out and comparison between the three cases was made. A general tendency is that as rotational stiffness of lateral boundaries increases, crack opening, in-plane shear sliding and out-of-sliding increase at vertical cracks along lateral boundaries accordingly while they decrease at diagonal cracks.
- Either considering lateral boundaries as hinged or clamped when applying AS 3700 can be inaccurate. In contrast, assuming lateral boundary as partially clamped provides the most accurate prediction. A determined R_f value needs to be quantified for walls in practice. Meanwhile, the accuracy in terms of predicting cracking pattern by AS3700 should also be improved.
- The results presented in this paper for the partially clamped and the clamped case may be influenced by the shear stiffness of the boundary interface element used to simulate different rotational restraints. Consequently, results should be treated as preliminary and further studies are needed to evaluate the influence of such a parameter.

Based on aforementioned conclusions, it implies that assuming lateral boundaries of two-way spanning URM walls as either hinged or clamed is inaccurate when applying analytical formulations. More research is suggested to quantitatively evaluate the stiffness of lateral boundary conditions and further improve the accuracy of analytical formulations.

REFERENCES

AS3700-2011 (2011) Australian standard of masonry structures. *AS 3700.*

Damiola, M., Esposito, R., Messali, F. & Rots, J. G. (2018) Quasi-static cyclic two-way out-of-plane bending tests and analytical models comparison for URM walls. *10th International Masonry Conference.* Milan, Italy.

Diana. BV. (2019) *DIANA user's manual - Release 10.3,* Delft, The Netherlands.

EN1996-1-1 (2012) Eurocode 6: Design of masonry structures—Part 1-1: General rules for reinforced and unreinforced masonry structures. *Comité Européen de Normalisation: Brussels, Belgium.*

Esposito, R., Messali, F. & Rots, J. G. (2016) Tests for the characterization of replicated masonry and wall ties. Delft, The Netherlands.

Graziotti, F., Tomassetti, U., Sharma, S., Grottoli, L. & Magenes, G. (2019) Experimental response of URM single leaf and cavity walls in out-of-plane two-way bending generated by seismic excitation. *Construction and Building Materials*, 195, 650–670.

Griffith, M. C. & Vaculik, J. (2007) Out-of-plane flexural strength of unreinforced clay brick masonry walls. *TMS Journal*, 25, 53–68.

Griffith, M. C., Vaculik, J., Lam, N. T. K., Wilson, J. & Lumantarna, E. (2007) Cyclic testing of unreinforced masonry walls in two-way bending. *Earthquake Engineering & Structural Dynamics*, 36, 801–821.

Jafari, S., Esposito, R. & Rots, J. G. (2019) From Brick to Element: Investigating the Mechanical Properties of Calcium Silicate Masonry. Cham, Springer International Publishing.

Lourenco, P. B. & Rots, J. G. (1997) Multisurface interface model for the analysis of masonry structures. *Journal of Structural Engineering-Asce*, 123, 660–668.

Ravenshorst, G. J. P. & Messali, F. (2016) Out-of-plane tests on replicated masonry walls. Delft University of Technology.

Rots, J. G. (1997) *Structural Masonry*, Netherlands, A.A. Balkema.

Sorrentino, L., D'ayala, D., De Felice, G., Griffith, M. C., Lagomarsino, S. & Magenes, G. (2016) Review of Out-of-Plane Seismic Assessment Techniques Applied To Existing Masonry Buildings. *International Journal of Architectural Heritage*, 1–20.

Willis, C. (2004) Design of Unreinforced Masonry Walls for Out-of-plane Loading. Adelaide, Australia, The University of Adelaide.

Building acoustics – calculation of sound insulation in buildings with single leaf clay unit masonry

K. Naumann

Arbeitsgemeinschaft Mauerziegel im Bundesverband der Deutschen Ziegelindustrie e. V., Berlin, Germany

ABSTRACT: In Germany a high percentage of residential buildings are erected using clay unit masonry for internal and external walls. The external walls are made of monolithic, vertically perforated clay units. The prognosis of the sound reduction index of the separating components between adjacent apartments in Germany is quantified using DIN 4109 'Sound insulation in buildings'. According to the editions of DIN 4109 until 2016 there was no normative calculation method for buildings with outside walls made of hollowed clay brick masonry. In 2000 EN 12354-1 'Estimation of acoustic performance of buildings' has been published. Based on this document the German clay brick industry developed in 2010 a technical approval that included an official calculation procedure for buildings with flanking components of monolithic, vertically perforated clay units. In 2016 a completely revised issue of DIN 4109 has been published that includes now several calculation rules of EN 12354-1. At the same time the Association of the German Brick and Tile Industry launched the software 'Modul Schall 4.0' that enables calculations in building acoustics with a high ease of use. This article provides results about the certainty of prognosis of sound reduction indices of separating components in residential buildings with outside walls of monolithic, vertically perforated clay units in comparison with results of field measurements.

1 INTRODUCTION

In Germany the calculation of sound reduction indices is regulated in DIN 4109 'Sound insulation in buildings'. In the past no calculation method exists for buildings with outside walls made of masonry with monolithic, vertically perforated clay units in this standard. Because of the inhomogeneous design of this type of clay units the acoustic performance is not only depending on the mass per area of the wall component.

In 2010 the German brick industry developed the technical approval Z-23.22-1787, that is based on the rules of EN 12354-1:2000 [4]. This standard opened up the opportunity to implement a calculation method to forecast the airborne sound insulation of separating elements with flanking components made of masonry with monolithic, vertically perforated clay units.

In 2016 completely revised versions of series of standards of DIN 4109 have been published. The rules of EN 12354-1:2000 [4] resp. EN ISO 12354-1:2017 were included now. It is possible to forecast the sound reduction index for apartments in buildings with outside walls of masonry with monolithic, vertically perforated clay units. To applicate this calculation method the knowledge of individual acoustic properties like sound reduction index, loss factor and the vibration reduction index of the clay brick wall is required.

To simplify the design and calculation of building acoustic performance of ceilings and separating walls in massive buildings the members of German clay brick association provide the software 'Modul Schall 4.0' that includes a large, permanently growing database of products of monolithic, vertically perforated clay units as well as other constructions of massive components or lightweight elements. All relevant acoustic properties of each product are available in the software.

The values of the prognosis according to DIN 4109:2016 are compared with several results of measurements in residential or office buildings. The assessment of calculated value vs. measured result will be shown. It could be shown, that the reliability of the calculation method for buildings with outside walls of monolithic, vertically perforated clay units is very high.

2 CONSTRUCTION METHOD OF MULTI-STOREY BUILDINGS WITH MONOLITHIC CLAY UNIT MASONRY

The outside walls of single-family houses as well as multi-storey residential and office buildings were build using masonry with monolithic, vertically perforated clay units. The benefit of this construction type is, that the outside walls are load bearing in vertical direction as well as against shear forces and fulfil also thermal requirements. That means, that no additional thermal insulation is needed and the masonry can be erected easily and economically. But there are other aspects that are important for safety and comfort for the users of the buildings, e.g. non-combustibility as well as regulation of air humidity and room climate.

monolithic, vertically perforated clay units are usually available in a range of characteristic compressive strength f_k from 1.5 N/mm² to 6.0 N/mm². Typically these clay units are produced in density classes from 0.65 to 0.90. The range of thermal conductivity λ is from 0.07 W/(m · K) to 0.12 W/(m · K). The fire resistance classification is up to REI-M 90. Single leaf outside walls made of monolithic, vertically perforated clay units is used in a thickness from 300 mm up to 490 mm. To protect the masonry against rainfall the outside surface is plastered. Inside the building an interior plaster is applicated on the masonry.

In the last decade the members of the German Brick association carried out a lot of research to optimize the different parameters of structural design and building physics of monolithic, vertically perforated clay units. Several internal research departments of each producers and external institutes as well as universities supported this work. It was shown, that a high synergy effect for the interaction of the parameters mentioned above is created, when the vertical holes of the clay units were filled with thermal insulation material.

3 PROGNOSIS OF SOUND INSULATION FOR BUILDINGS WITH MONOLITHIC HIGHLY INSULATED CLAY UNITS

In Germany requirements and calculation of sound reduction indices are regulated in the series of standards of DIN 4109 'Sound insulation in buildings'. Due to the fact, that monolithic, vertically perforated clay units are inhomogeneous structured and the different producers provide a broad spectrum of variations of the design of clay web and holes, in the past it was not possible, to calculate the sound reduction index for these walls in accordance with a normative procedure. This was described explicitly in Supplement 1 to DIN 4109:1989 [2] with the explanation, that bricks in the density class $\leq$ 0,80 kg/dm³ with an – according to building acoustic parameters – inconvenient shape of hollows are excluded. The calculation method of Supplement 1 to DIN 4109:1989 [2] was very generalized and was based on tables respectively simple equations.

The publication of the European standard EN 12354-1:2000 [4] revolutionises the approach of calculation of building acoustic properties of separating elements in massive buildings generally. The calculation method of EN 12354-1:2000 [4] considers distinctively each path of flanking sound transmission.

Decisive parts of the assessment procedure according to EN 12354-1:2000 respectively later the technical approval Z-23.22-1787 [5] and DIN 4109-2:2016/2018 [1] is the consideration of the following parameters:

- Sound reduction index of separating element R_S/dB,
- Sound reduction index for (flanking) element in source room $R_{i,w}$/dB
- Sound reduction index for (flanking) element in receiving room $R_{j,w}$/dB
- Vibration reduction index for each transmission path K_{ij}/dB
- Geometry

For calculation of each flanking sound transmission path the variables $R_{i,w}$, $R_{j,w}$ and K_{ij} are obligatory to be used in DIN 4109-2:2018 [1], equation (10), shown here in equation (1).

$$R_{ij,w} = \frac{R_{i,W}}{2} + \frac{R_{j,W}}{2} + \Delta R_{ij,W} + K_{ij} + 10 \cdot \log \frac{S_s}{l_0 \cdot l_f} \text{ [dB]}$$

$$(1)$$

Additional layers, that are fixed on the massive flanking components, may influence the sound reduction index of the massive component positively or negatively. This effect has to be considered by the variable $\Delta R_{ij,w}$. Finally the geometry influences the result. Thus the separating area S_S/m² between the two adjacent rooms and the length of the joint between each flanking component and the separating component l_f/m are part of the equation.

For homogeneous massive building materials DIN 4109-32:2016 [3] provides equations to perform the calculations, that are only dependent from the mass per area of the components. Because of a wide range of test results of weighted sound reduction indices R_w and vibration reduction indices K_{ij} of masonry with monolithic, vertically perforated clay units, the knowledge of these individual parameters for each clay unit product is necessary, regardless the mass per area or other characteristics. This requires, that these values have to be determined by doing measurements.

In the early stage of development of new draft of DIN 4109 threatens danger, that there will be in a renewed version again no regulation for monolithic, vertically perforated clay units. For this reason, German producers of this building material decided to develop at least for an interim period a separate technical approval that covers the gap of calculation procedure. Numerous studies have been conducted together with University of Applied Sciences (HFT Stuttgart, Germany). Finally the technical approval Z-23.22-1787 was published in 2010. Proofs according to this document have to be accepted by developers, building owners or public authorities.

To apply the calculation method of technical approval Z-23.22-1787 it is required to evaluate test results of the weighted sound reduction indices $R_{w,\square,ref}$ for each product, determined in an acoustic laboratory. This means, that the tested value of weighted sound reduction index R_w includes the correction of total loss factor η_{tot} related to the average of total loss factor on site. Further on the vibration reduction indices K_{ij} according to series of standards of ISO 10848 of the joints of hollowed clay units and separating elements like ceilings or partition walls are mandatory.

The producers of monolithic, vertically perforated clay units, that are organised in the German Brick association, carried out a huge number of measurements of sound reduction indices $R_{w,\eta,ref}$ and vibration reduction indices K_{ij} in the past and still ongoing. For each new product test specimens have to be erected to realise these investigations. The measurements were carried out by independent acoustic engineers.

In case there are no test results of weighted sound reduction index R_w respectively $R_{w,\eta,ref}$ or vibration reduction indices K_{ij} it is also possible to calculate the weighted sound reduction index $R'_{w,cal}$. But then the final result of $R'_{w,cal}$ will decrease in most cases extremely, because according to DIN 4109-2:2018 there is a rule implemented in the calculation method, that effects that the vibration reduction index K_{ij} has to be reduced. That means, the final calculation result $R'_{w,cal}$ is clearly conservatively and most of the examples are extremely on the safe side.

Clearly it is the fact, that the calculated weighted sound reduction index $R'_{w,cal}$ matches better the measured value of $R'_{w,meas}$ in buildings, if the calculation is done, using the measured sound reduction indices $R_{w,\eta,ref}$ in the laboratory test stand.

4   ACCURACY OF FORECAST OF CALCULATION METHOD ACCORDING TO TECHNICAL APPROVAL AND RENEWED DIN 4109

While the technical approval Z-23.22-1787 has been used since 2010 the standardization work went on to finalise the new series of standards of DIN 4109. In this time several proofs to calculate the weighted sound reduction index on site R'_w were done.

It was a big advantage to exercise the new calculation method in this early stadium of standardization development, because many experiences could be gathered. Thereby it was also possible to compare the calculated values of weighted sound reduction indices $R'_{w,cal}$ with proof results $R'_{w,meas}$ measured on site. In study [6] several assessments have been carried out for buildings with outside masonry with monolithic, vertically perforated clay units. Figure 1 shows the prognosticated values $R'_{w,cal}$ on the x-axis. The tested value on site $R'_{w,meas}$ is faced on the y-axis. It should be recalled, that the weighted sound reduction indices were calculated using the individual sound insulating parameters R_w and K_{ij} of masonry with monolithic, vertically perforated clay units measured in a laboratory.

The bisecting line (black) means, that the calculated value is equal to the proof result. Dots above the black line achieved on site higher values compared

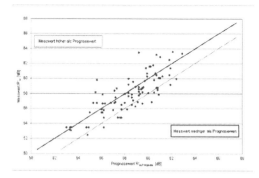

Figure 1. Comparison of calculated values of weighted sound reduction indices $R'_{w,cal}$ and proof results $R'_{w,meas}$ for ceilings measured on site.

with the calculated value and dots beneath the bisecting line are on site worse than the calculated value.

According to DIN 4109-2:2018 [1] it is required to consider the uncertainty of the calculation method. Therefore the calculated weighted sound reduction index $R'_{w,cal}$ has finally to be reduced by the global uncertainty u_{prog} = 2.0 dB. It is the aim of German standardization committee, that 85 % of all projected separating components between third party inhabitants fulfil the legal requirement of airborne sound insulation. The red line in Figure 1 is shifted parallel to the bisecting line about 2 dB and includes therefore the uncertainty. This effects, that approximately 85 % of the pairs of values are at least as high as the calculated value $R'_{w,cal}$.

In the study [6] also the tested separating walls are included in the assessment.

5   BUILDING ACOUSTICS SOFTWARE 'MODUL SCHALL 4.0'

As mentioned above it could be shown, that it is possible to forecast the sound reduction index for apartments in buildings with outside walls of masonry with monolithic, vertically perforated clay units now. To applicate this calculation method the knowledge of individual acoustic properties like sound reduction index $R_{w,\eta,ref}$, total loss factor η_{tot} and the vibration reduction index K_{ij} of the clay brick wall is required.

It was also described, that the new calculation method according to EN 12354-1 respectively DIN 4109-2:2016/2018 [1] is very detailed, because a huge equation system has to be solved.

To simplify the design and calculation of building acoustic performance of ceilings and separating walls in massive buildings the members of German clay brick association provide the software 'Modul Schall 4.0' that uses the calculation method of DIN 4109-2:2016/2018 [1].

The following forecasts can be made using the software:

- Weighted sound reduction index R'_w of single leaf separating walls and ceilings inside single-family houses and multi-storey houses
- Weighted sound reduction index R'_w against outside noise
- Weighted sound reduction index R'_w of double leaf separating walls of row- and twin-houses
- Weighted impact sound pressure level $L'_{n,w}$.

Each result is also displayed as weighted standardized level difference $D_{nT,w}$ respectively weighted standardized impact sound pressure level $L_{nT,w}$.

The proof of each project can be printed out with a detailed overview about all components and chosen constructions.

Figure 2 shows the compact user interface of the software. All relevant parameters are displayed.

The biggest asset of the software is the large, permanently growing database of products of monolithic, vertically perforated clay units as well as other constructions of massive components or lightweight elements. All relevant acoustic properties of each product are easily available in the software. The measured sound reduction indices $R_{w,\eta_{ref}}$ were assigned automatically, if clay unit has been chosen. The user gets a selection about possible details of joints between the separating element and the flanking component made of monolithic, vertically perforated clay units. The corresponding values of vibration reduction indices K_{ij} are also automatically assigned.

Figure 3 shows in general a example of a horizontal section of a T-joint of a monolithic, vertically perforated clay unit outside wall as a flanking component and a filling brick as a separating component.

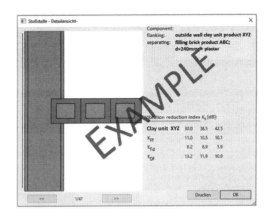

Figure 3. Example of database of vibration reduction indices K_{ij} included in the database of software 'Modul Schall 4.0'.

This selection is provided for each manufacturer of clay units of the German clay brick association. Planners can easily access this selection.

It is also possible to make the prognosis of sound reduction indices for buildings with other massive building materials. Further on users can extend the database with own building materials and constructions.

The application 'Modul Schall 4.0' is freeware and available on the internet sites www.schallrechner.de or www.ziegelrechner.de.

6 EXAMPLE OF THE PROGNOSIS OF THE SOUND REDUCTION OF A CEILING IN A MULTI-STOREY HOUSE

The viewed project is a multi-storey residential building. The ground plan is shown in Figure 4.

The outside masonry walls are made of monolithic, vertically perforated clay units with the relevant parameters shown in Table 1.

The inside wall between the considered room and the corridor is made of clay unit masonry with a thickness of 115 mm and a density class 1.20. The other inside wall adjacent to the bath room is a dry wall construction. The ceiling is made of 220 mm concrete with a floating screed.

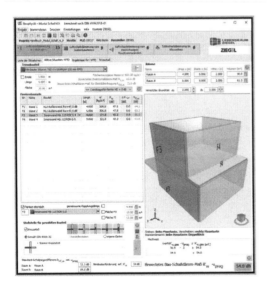

Figure 2. User interface of building acoustics software 'Modul Schall 4.0'; example of calculation of a ceiling.

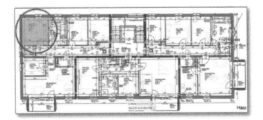

Figure 4. Ground plan of a multi-storey residential building.

241

Table 1. Parameters of outside clay unit masonry used in the regarded building.

Parameter	Physical Value	Value	Unit
Thickness	t	365	mm
Density class	-	0.70	-
Thermal conductivity	λ	0.09	W/(m · K)
Sound reduction index	$R_{w,\eta,ref}$	50.0	dB

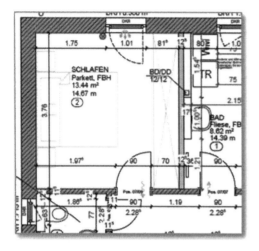

Figure 5. Ground plan of considered room.

Figure 5 shows detailed one of the two rooms under consideration, that are positioned symmetrically on top of each other. The separating area is about $S_S = 13.4$ m².

The calculated sound reduction index of the ceiling above the ground floor is determined with the software 'Modul Schall 4.0' and is about $R'_{w,cal} = 58.8$ dB. After finalisation of the building acoustic measurements have been carried out. The proof value of the sound reduction index is $R'_{w,meas} = 61$ dB. This result is 2.2 dB better, than the prognosis. The calculated value of weighted impact noise level is about $L'_{n,w,cal} = 47.3$ dB. The measured value is about $L'_{n,w,meas} = 35$ dB. This is 12.3 dB better than the prognosis.

Numerous other measurements were carried out in this building. It could be shown, that the airborne sound insulation and the impact noise level in other parts of the building are also as good as shown in the example in the paragraph above.

7 SUMMARY

The launch of EN 12354-1:2000 [4] enables the implementation of a calculation method to forecast the airborne sound insulation of separating building elements in multi-storey facilities for residential or office use with flanking components made of masonry with monolithic, vertically perforated clay units.

Since 2010 the technical approval Z-23.22-1787 [5] is available and could be used to realise official calculated proofs of airborne sound insulation for building projects with monolithic masonry made of monolithic, vertically perforated clay units.

Especially using the acoustic optimized monolithic, vertically perforated clay units in the masonry of the outside walls, it is possible to reach airborne sound insulation between adjacent foreign living areas in buildings with a high acoustic comfort for the inhabitants.

It is verified, that there is a high and satisfying conformance comparing computed values of airborne weighted sound reduction indices $R'_{w,cal}$ and test results $R'_{w,meas}$.

To increase the ease of use of calculated acoustic performance of buildings the German Brick Association provides the software 'Modul Schall 4.0'. In the database of this application a huge database is available, that includes the relevant acoustic parameters of monolithic, vertically perforated clay units. Further on also proofs for constructions with other massive building materials are possible.

The freeware 'Modul Schall 4.0' is available on the internet sites www.schallrechner.de or www.ziegelrechner.de.

REFERENCES

DIN 4109-2:2016/2018 'Sound insulation in buildings – Part 2: Verification of compliance with the requirements by calculation' (Remark: issue 2018 substitutes issue 2016).
Supplement 1 to DIN 4109:1989 'Sound insulation in buildings; Construction examples and calculation method' (Remark: Document withdrawn in 2016).
DIN 4109-32:2016 'Sound insulation in buildings – Part 32: Data for verification of sound insulation (component catalogue) – Solid construction'.
DIN EN 12354-1:2000 'Building acoustics – Estimation of acoustic performance of buildings from the performance of elements – Part 1: Airborne sound insulation between rooms; German version EN 12354-1:2000' (Remark: Document withdrawn and substituted by DIN EN ISO 12354-1:2017).
Z-23.22-1787 Technical Approval 'Mauerwerk aus Hochlochziegeln nach DIN V 105-100 oder DIN EN 771-1 in Verbindung mit DIN V 20000-401; Nachweis der Luftschalldämmung von Mauerwerk aus Hochlochziegeln in Wohngebäuden in Massivbauart'.
Fischer, Gierga, Schneider. 2016. Airborne sound insulation in multi-storey buildings with monolithic, highly insulated clay units – Prognosis in accordance with DIN 4109: 2016and comparison with proof values on site. *Bauphysik 38* (2016), 183–192. https://doi.org/10.1002/bapi.201610024.

Brick and Block Masonry - From Historical to Sustainable Masonry –
Kubica, Kwiecień & Bednarz (eds)
© 2020 Taylor & Francis Group, London, ISBN 978-0-367-56586-2

Critical review of normative framework for seismic assessment of existing masonry buildings

S. Krishnachandran & M. Arun
Department of Civil Engineering, IIT Madras, Chennai, India

ABSTRACT: The multiple elements that receive and transfer the lateral loads and their interconnectivity form the lateral load path of a structure. When a structure is subjected to seismic loads, every element in the load path is tested. Load path determines the order of component capacities and the hierarchy of failure in a structure. Hence, identification of load path forms the fundamental step in the assessment of any structure. However, for unreinforced masonry (URM) buildings, the lack of a well defined and continuous load path makes it difficult to set an assessment protocol. The fundamental load paths and the role of participating elements are not always properly understood and therefore, structural assessment of existing unreinforced masonry (URM) buildings has been an area of challenge. Further, the variation in construction typology and non-uniformity in material properties across the world created difficulties in bringing uniformity to these assessment guidelines. Although the global scheme of assessment is similar to other structural types, the thin line of separation that exists between the onset of different mechanisms and the multiple interactions that happen between these mechanisms and associated elements alter the load paths in an existing structure. However, from time to time, observations made during earthquakes have necessitated revisiting the load paths and mechanisms in URM structures and modifying the assessment guidelines. The fundamentals of seismic assessment procedures across the world show similarities with each other but the differences observed are also substantial. Hence, a critical review of the normative assessment procedures is carried out to understand the similarities and limitations in code-based assessment procedures and to identify the scope for refinement in existing methodologies.

1 INTRODUCTION

Strength hierarchy of different components of a structure follows the load path (Murty et al. 2012) and hence, identification of load path forms the fundamental step in the assessment of any structure. The seismic load path of an unreinforced masonry building is explained by Priestley (1985) as shown in Figure 1. The ground acceleration initiates an in-plane wall response depending on the stiffness of the in-plane lateral load resisting elements. These serve as the inputs for the diaphragm response which filters the input signal and feeds it to the face loaded URM walls (Menon & Magenes, 2011). Thus, the out-of-plane(OOP) walls are subjected to inertial excitation and further, these inertial loads are taken by the diaphragm and transferred back to the in-plane walls if the wall-diaphragm connections are capable of transferring the load. In other words, as per the fundamental seismic energy path for URM buildings, the seismic input for the OOP walls is defined by the diaphragm; given proper diaphragm-wall connectivity exists. Hence, the deformation of OOP walls can be considered as a combination of induced diaphragm displacements and the inertial displacements of the wall.

Traditionally, structural assessment of URM structures assumes them as a global system with strength capacity derived from the IP response of piers, connected together by the box-action provided by spandrel beams and a rigid diaphragm; where the assumptions of rigid diaphragm action, stability of OOP walls and adequate coupling of spandrels were implicit. However, seismic assessments post earthquakes highlighted the premature failure of URM buildings due to local failure mechanisms (Bruneau,1994) and necessitated the need of research in this area.

There has been considerable uniformity across codes achieved till date in defining and formulating the component level OOP response (local mechanism), which forms the first step of the seismic load path and the component level in-plane response which completes the loop and forms the last step of the lateral load path. However, the accuracy of these estimates is subject to the validity of the aforementioned implicit assumptions. If these assumptions are not confirmed on a quantitative basis, it could result in interaction of mechanisms and these could limit the structure attaining its estimated peak capacity. Therefore, recent assessment frameworks including the Italian guidelines (2007) and New Zealand(NZ)

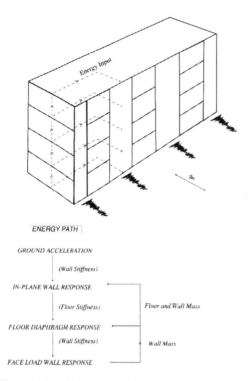

ENERGY PATH

GROUND ACCELERATION

(Wall Stiffness)

IN-PLANE WALL RESPONSE

(Floor Stiffness)   Floor and Wall Mass

FLOOR DIAPHRAGM RESPONSE

(Wall Stiffness)   Wall Mass

FACE LOAD WALL RESPONSE

Figure 1. Seismic Energy Path in a URM Building (After Priestlet, 1985).

guidelines (2017) have considered URM as an assemblage of local sub systems identifying the multiple interactions occurring between these component level mechanisms at different points of the load path. This includes the effects of the orthotropic nature of diaphragms and diaphragm flexibility in developing interactions, the role of OOP effects in restricting the IP capacity and the effect of spandrels in providing adequate coupling for the piers.

2 FROM PRESCRIPTIVE TO PREDICTIVE ASSESSMENT SCHEMES FOR URM BUILDINGS: CRITICAL FACTORS

2.1 *Role of seismic input*

The first seismic code in Japan (1924) had used a seismic coefficient of 0.1, US (1927) had 0.075 to 0.1 and Italy had 0.075 to 0.13 (1909) (Fajfar, 2018). By 1950's, the dependence of seismic loads on time period of a structure was established by the California state and the modal response spectrum method for analysis was introduced in 1957 (Fajfar, 2018). Currently, most seismic assessment guidelines are using probabilistic hazard framework or "beyond the code" resilience-based approach (FEMA P-58, 2017) where scenario-based assessment procedures with an ensemble of ground motions matching the site-specific

spectrum of a particular region defines the seismic demand. Also, current research is moving in the direction of using the term 'seismic risk' for carrying out the assessment of a structure.

Risk is a function of the input motion (hazard) and the building characteristics (vulnerability). Resilience based seismic risk assessment results when both hazard and vulnerability are expressed in probabilistic framework. Vulnerability is defined using the creation of fragility curves for each component and assessment of the structure using probabilistic hazard framework for loading conditions corresponding to multiple damage limit states. However, vulnerability characterization using a probabilistic framework is still not well accepted by many codes (Fajfar, 2018) and hence seismic assessment guidelines are still prescriptive in nature. A failure probability not exceeding 1% in 50 years is defined as an acceptable risk for the design of new structures. For defining seismic hazard, the acceptable limits for exceeding a seismic event is expressed as 10% in 50 years for design level earthquake and 2% in 50 years for maximum considered earthquake in majority of the codes. However, for existing structures, resorting to these limits could lead to unreasonable strengthening interventions. Hence, the acceptable probability limits for a seismic event (return period) should be different for new structures and existing structures. In this regard, the American, Italian and the NZ guidelines have methods for scaling the seismic input with respect to return periods specified for different limit states. Depending on the category of use and category of importance, for a span of 50 years, the Italian code has probabilities of exceedance ranging from 6.5% to 40% with the corresponding scaling factors from 1.2 to 0.5. Similarly, the NZ code uses return period scaling factors (McVerry, 2003). When global design codes are in a direction towards risk-based assessment methodologies, quantifying the risk for existing buildings and benchmarking it with new constructions should be of prime importance.

2.2 *Towards methodologies considering component level interaction*

Seismic assessment frameworks across the world follow initial steps involving gathering information including construction documentation and as built drawings followed by visual inspection and on-site investigations to identify potential structural weaknesses. While providing assessment guidelines, most of the assessment frameworks are prescriptive in nature. All three frameworks (ASCE 41, Italian, EC8) do not explicitly consider the URM structure to be an assemblage of local sub-systems, although they state the presence of local collapse mechanisms and the importance to address those. Italian code addresses the effect of these local mechanisms in simplified mechanical models and has clearly articulated that in cases where a well-defined system-level response cannot be ensured (especially, URM with flexible diaphragms, multiple local mechanisms, etc), the global-level assessment is conducted by collection of local mechanisms, provided that the transfer of forces between these sub-systems are adequately captured (Italian Guidelines, 2007). However, the interaction of these local mechanisms with the system-level response is not properly accounted for. An assessment methodology based on *"weakest link analogy"*, in close alignment with the aforementioned objective is defined by the NZ guidelines. The NZ guidelines recommend deciding on the level of assessment as per the building complexity. A less sophisticated analysis is recommended for initial assessment and a detailed analysis scheme is warranted only if the complexity of the structure requires and reflects the analysis effort (especially in cases where stiffness compatibility issues are there if the building is subjected to previous retrofits). However, the most important step is with respect to the strategy incorporated for seismic assessment. It redefines the structural assessment protocol from the flow of loads, observed failure modes and strength hierarchy carrying out assessment as per weakest link analogy. Hence, the assessment follows a hierarchy in the order of increasing vulnerability where the weakest element in the load path is assessed by the capacity of the weakest element (Figure 2).

A weak link in the chain limits the maximum capacity attainment of a component which in turn reduces the global capacity estimate of the structure. Thus, this methodology of addressing elements in the increasing order of vulnerability brings in significant improvement in the existing assessment schemes to arrive at better capacity estimates for the structure. This is justified in the code stating that there is little point in spending effort on refining the existing capacities since it will be influenced by the capacity of a more vulnerable item which needs to be addressed. Particularly with respect to existing URM buildings categorized by limited redundant load paths, there is much meaning in studying component-level interactions rather than redefining the global capacity estimates.

2.3 *Effect of diaphragm behavior*

Diaphragms are considered to be primary elements in the overall load path of a structure. The role of diaphragm is defined as to provide support to the OOP walls and to distribute the inertial loads to the IP lateral load resisting elements. The failure modes of URM buildings after past earthquakes have proved that there is a significant effect of diaphragms on their seismic performance and hence, diaphragm-related failure modes are considered as one of the common failure mode for URM structures (FEMA P-74, 2009). Partial or complete OOP wall failures due to diaphragm induced displacements were

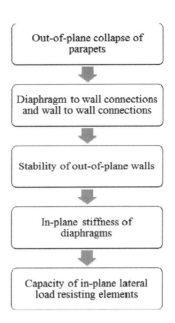

Figure 2. Strength hierarchy-based assessment scheme (NZ Guidelines).

observed in URM failures during many earthquakes. (Ingham & Griffith, 2010, 2011).

The definition of diaphragm rigidity is based on the relative IP stiffness of the diaphragm with respect to the IP lateral stiffness of the lateral load resisting elements. The dominant mode of response in buildings with flexible diaphragms is the response of diaphragms with inertial forces from the diaphragms and the forces transferred to the face loaded walls responding out-of-plane. The diaphragm deformation limits should be within the allowable limits so that the OOP walls are laterally stable and adequately supported. Further, they should possess adequate capacity to transfer the shears to the IP walls. The effect of diaphragms in the OOP stability of walls and their overall role in load transfer was identified by (Kariotis et al. 1984) in a series of experimental studies and analytical formulations. Later, the seismic performance of URM structures in many earthquakes necessitated studying the role of diaphragm flexibility in the response of URM structures. Simsir et al. (2003), Penner et al. (2004) an Gallonelli et al. (2007) have conducted extensive experiments including dynamic tests to investigate the effect of floor diaphragm flexibility on the OOP response of URM walls and the results indicated that diaphragm flexibility significantly increases the OOP displacement of walls. Similarly, large scale experiments with full building tests also were conducted to study the response of URM buildings with flexible diaphragms (Paquette et al. 2004). Penner (2004) recommended an assessment procedure with the classification of diaphragms as rigid or flexible being the first step. In the present scenario, many research works are done in characterizing diaphragm flexibility, orthotropic nature of diaphragms and have highlighted the importance of these in defining component level interactions (Aleman & Whittakker, 2014, Wilson et al. 2014, Giongo et al. 2016). The test results of Giongo et al. (2016) are incorporated in the NZ Assessment Guidelines for the calculation of shear stiffness of straight sheathed timber diaphragms and suitable multipliers are used for other sheathing types from ASCE 41. Simplified numerical models (Derakshan etal. 2016) have also been developed, which account for the effect of diaphragm flexibility on the OOP response of URM walls. Further, Nakmura et al. (2016) has investigated the effect of non-rigid diaphragm behavior on the seismic demand of IP wall by classifying diaphragms as rigid, stiff, semi-flexible and flexible. The recent version of ASCE 41-17 has incorporated the assessment of OOP walls for life safety based on the works of Penner & Elwood (2016) considering the dynamic stability by including diaphragm flexibility as well.

The limit for the deformation capacity is specified in NZ guidelines as the minimum of half the thickness of the supported OOP wall from deformation considerations and the deformation associated with the attainment of full strength capacity subject to the maximum building drift of 2.5%, which should ideally include diaphragm deformations as well. Hence, from the perspective of considering URM as an assemblage of sub-systems and accounting for component-level interactions, considerable research is currently underway to characterize diaphragms and their connections.

2.4 Role of spandrels

In URM buildings, piers are interconnected by horizontal elements called spandrels. The role of spandrels in the seismic assessment of existing URM is a link which is subject to extensively research, however not considered by the assessment guidelines explicitly. Cattari & Beyer (2015) have pointed out the issue of the role of spandrel modeling and the lack of progress in codes, existing guidelines and literature with regard to spandrel models despite the extensive experimental investigations (Beyer & Dazio, 2012, Parisi & Augenti., 2010, 2016, Grazziotti etal. 2011) showing their importance. Further, Foraboschi (2009) had identified the need for addressing the degree of coupling provided by masonry spandrels in defining the behavior factor for URM structures. Depending upon the strength and stiffness of spandrels, the coupling offered by them could be categorized as weak (negligible force transfer and equal displacements imposed on the piers), intermediate (limited force transfer between piers) or strong (significant force transfer developing framing action) (Petry & Beyer, 2014). Hence, the behaviour of spandrels

affect the end restraints for the piers thus affecting their overall response. Also, there is less uniformity across assessment guidelines for spandrel failure modes (Beyer & Mangalathu, 2013).

One drawback of the assessment schemes based on weak link analogy is that spandrels are considered as part of the IP capacity estimation and not considered as a separate element in the load path. The failure mechanism of a pier can be altered and its capacity can be significantly affected by the deformability of a spandrel. A weak spandrel results in increased rotations at the ends of piers and hence reduced moments and axial loads on them. Thus, deformability of spandrel acts as another weak link which can restrict the full IP capacity attainment of piers and this aspect needs to be considered explicitly in the assessment process. Currently, in the assessment schemes based on strength hierarchy, the weak spandrel-strong pier effect is considered in the assessment scheme by the use of a force reduction factor of 1 as the lateral load resistance is restricted by the spandrel capacity. However, since restoration follows the strength hierarchy, it is essential to make sure that the spandrel performs adequately and utilizes its full strength capacity to ensure that the capacity estimates of the IP walls are reliable. Hence, in future assessment frameworks, spandrel elements need to be considered as an important link in the seismic load path.

2.5 Stiffness of URM walls

Ambiguity exists in different codes on the calculation of effective stiffness of masonry walls (Lu et al., 2013). According to Petry and Beyer (2015), displacement capacity and effective stiffness are two parameters which require considerably improved estimates. Once the focus has shifted from force-based assessment method to displacement-based methods, there has been significant amount of research in quantifying the effective stiffness. Petry & Beyer (2015), Wilding & Beyer (2016, 2018) have developed analytical models to estimate the effective stiffness of URM piers stating the inadequacy of current approaches in predicting the effective stiffness of URM piers. Force-based assessment procedures in codes recommend using elastic stiffness of the wall subjected to IP forces, during linear analysis for calculating the shear distribution. For rocking and sliding mechanisms, the fact that force reduction factors are applied in the capacity estimates inherently indicates that non-linear range is entered. However, during the analysis, stiffness reduction of masonry walls is not carried out to reflect this effect. EC8 recommends the use of a 50% reduction in the stiffness of masonry walls. Experimental results by Wilding & Beyer (2018) observed the increase in effective stiffness for URM walls with increase in axial stress and have recommended the effective stiffness in URM walls to be 75% of the initial stiffness. In both these analysis schemes, at different performance levels (Life Safety level in NZ guidelines & operational, life safety and collapse prevention), the force capacity of the structure is compared with the demand level seismic forces. However, in EC 8, except at the first level, deformation demands are compared with the deformation capacity of the structure. Hence, if there is an overestimation of the stiffness of the structure, the deformation demand will be underestimated. For a force-based assessment, if every wall is cracked, there would not be a change in the relative force distribution for the walls and hence the demand calculation. However, masonry is an anisotropic material in which the extent of cracking depends extensively on the mechanisms. The reduction in IP stiffness of URM walls under cyclic loading was observed in experimental studies (Salmanpour et al. 2013). However, for existing URM buildings, the extent of cracking observed after preliminary inspections can have an influence on the relative force distribution along IP walls. In reality, there are significant chances of masonry walls undergoing cracking due to the effect of OOP displacements. The extent of cracking can depend on multiple factors like the eccentricity of loads, diaphragm flexibility, etc., and the extent of cracking varies across walls depending on the causative mechanisms. For example, a wall with increased OOP eccentricities and low levels of axial precompression can undergo cracking at very low levels of input displacements and hence should have reduced stiffness compared to a wall with very low eccentricities and high levels of axial compression. This could alter the load paths and mechanisms induced and provide more precision on the structural behavior. A proper quantification of URM stiffness becomes particularly important assessment approaches utilizes dynamic identification techniques like operation modal analysis and effective stiffness of masonry needs to be calibrated based on damage based degrading functions (Zimmermann et al. 2013). Hence, guidelines should be developed defining the stiffness modification parameters to be employed depending on the wall conditions as per visual inspections and condition assessment.

2.6 IP strength behavior of URM walls: With and without the effects of interaction

Another important aspect which is not considered in the current assessment framework is the bidirectional interaction effects which are difficult to quantify reliably (ASCE 41, 2013). IP shear damage can increase the OOP instability of walls and OOP displacements can reduce the IP lateral load resistance of the walls (Dolatshahi etal. 2014). The sources of this interaction effects include diaphragm flexibility, diaphragm deformation modes and structural eccentricities.

In URM walls, interaction response can be generated at two levels in the seismic load path; At the intermediate level, where the diaphragm transfers seismic input to the OOP wall which excites and transfers the forces back to the diaphragm if the connections are adequate. Hence, the OOP wall stability is affected by the diaphragm which is properly addressed in the codes. Similarly, allowable diaphragm displacements are also limited by the maximum usable deflections of the OOP walls. At the final level of load path hierarchy, the diaphragm distributes the inertial forces to IP walls. The response of flexible diaphragms can induce OOP displacements at the top of the lateral load resisting IP walls which can cause premature cracks on the wall and reduce its stiffness and load carrying capacity. The IP capacity of the wall can be compromised in the presence of OOP displacements and this effect is not considered in the seismic assessment guidelines. In addition, the inertial response of OOP walls and multidirectional nature of the ground motions could result in the interaction between IP and OOP response in URM walls. ASCE 41 explicitly states that the sequence of IP actions, combinations of actions and bidirectional effects are difficult to be considered and quantified reliably in the assessment scheme. However, few works (Dolatshahi etal. 2014, 2015, Najafgholipour etal. 2014, Kollerathu & Menon, 2017, Noor-E-Khuda & Dhanasekar, 2017) have investigated into defining the multi-directional interaction response for URM walls.

2.7 Ductility and force reduction factors

Characterization of mechanisms are based on their relative ductility and force reduction factors are used to account for ductility, inelastic displacements, damping effects and period elongation. The value of behaviour factor/response reduction factor/force reduction factor to be used in scaling the URM response has been an area of uncertainty. The fundamental difference between the US guidelines and others is in the categorization of mechanisms into force-controlled and deformation-controlled. Force reduction factor as per the NZ guidelines comes from the increased damping and not from ductility considerations as it is negligible for URM buildings and hence recommends that the reduction factors shall be applied only to the forces and not the deformations. However, ductility derived from energy dissipation characteristics could be high, particularly in the case of rocking walls IP and OOP walls which can sustain much high accelerations than predicted by elastic analysis.

ASCE 41 defines deformation-controlled actions as the ones which have an associated deformation limit which is not allowed to exceed and force-controlled actions are the ones which are not allowed to exceed its own limiting values. The categorization of mechanisms into deformation or and force-controlled action factors help to account for the mechanism ductility. IP shear is considered as a deformation-controlled action if the capacity of wall in toe crushing or diagonal cracking using lower bound material properties exceeds the capacity of the wall using flexural rocking or sliding mechanism using expected material strengths. Due to the inherent ductility associated with these mechanisms, the capacities for deformation-controlled mechanisms are scaled by multiplying with modification factors ('m' factors) that account for the element behavior, ductility and knowledge factor which accounts for the reliability of material and element characterizations. Hence, more ductile mechanisms, namely rocking and sliding have increased scaling factors of the structural component due to their inherent ductile response. Similarly, the dependence on the deformability associated with different mechanisms is considered through the use of a force reduction factor (K_r) in the NZ assessment guidelines. For deformation-controlled mechanisms like rocking and sliding, demand calculated from lateral load analysis is scaled down by a factor of 3. Similarly, toe crushing mechanism has a reduction factor of 1.5 and diagonal cracking mechanism does not have a reduction factor. Rocking mechanism exhibits high levels of post yield deformation, but low levels of hysteretic damping. URM buildings can have a damping as high as 10% for diagonal cracking mode, 15% for rocking mode and 20% for sliding mode (NZ Guidelines, 2017). The use of force-reduction factor is restricted to 1.0 when there is a strong pier-weak spandrel mechanism, as the mechanism then will be governed by the spandrel capacity. The behavior factors provided by the EC8 for existing masonry buildings is considered to be conservative (Allen et al., 2013). Also, for URM structures, behavior factor is a function of the geometry and layout of the building (Manojlovic et al. 2018). In addition, the effect of spandrels on the behavior factor is also an area currently being investigated. Hence, existing research in the direction of characterizing spandrel behavior and stiffness of URM could provide better results for the quantification of behavior factors.

3 CONCLUSIONS & WAY FORWARD: FROM PRESCRIPTIVE GUIDELINES TO PREDICTIVE ASSESSMENT SCHEMES

Over the years, review of existing assessment guidelines (Morandi & Magenes,2008, Magenes & Penna, 2009, 2011) has pointed out different aspects requiring refinement for existing seismic guidelines for the assessment of URM. These resulted in improving analysis methods, material characterization and component behavior Considerable research has been done in standardizing preventing local mechanisms. At the global level, the limitations for code based prescriptive assessment methodologies were mainly coming from the reliability of available data for existing structures. However, in the current scenario, sophisticated techniques like operational modal analysis are being used in most of the cases for structural identification, documentation and condition assessment for quantifying the existing condition of a URM structure. Similarly, more reliable models for

defining seismic demand and characterizing losses are also the procedures for estimating the global capacity and available (FEMA P-58, 2017). In future, the advent of more robust data analysis and processing tools could simplify most of the work related to system identification and documentation. Hence, future assessment frameworks should be capable of utilizing the advantages provided by these toolsets in providing better capacity estimates by addressing the component level interactions taking place at each step of load-path hierarchy. A potential URM assessment framework based on the existing code-based assessment schemes and current research focus is provided as an attempt for redefining the existing guidelines (Figure 3) involving the following additional factors:

1. Checking the diaphragm stiffness, strength and stability to distribute the loads to IP load resisting walls.
2. Checking the deformability and strength capacity of spandrels and classify it as weak, intermediate or strong;
3. Estimating the capacity of IP walls by considering the effect of OOP demands induced due to the bidirectional interaction resulting from multiple sources. There are potential areas of refinements, especially at the connecting links between different steps in the load path and interactions occurring within the structure. Majority of the current research in URM assessment focuses on fixing these weak links and improving the predictions for the structural behavior.

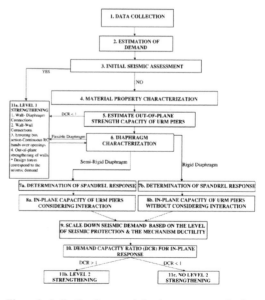

Figure 3. Indicative framework for the assessment of existing URM.

REFERENCES

Abrams, D.P., 1992. Strength and behavior of unreinforced masonry elements. *Proceedings of the 10th World Conf. Earthq. Eng. Balkema, Rotterdam*, 3475–3480.

Adham SA. 1985. Static and dynamic out of plane response of brick masonry walls. Proceedings of the 7th International Brick Masonry Conference, Melbourne, Australia, 1218–122.

Allen, C., Masia, M.J., Derakhshan, H., Griffith, M.C. 2013. What ductility value should be used when assessing unreinforced masonry buildings, Proceedings of NZSEE Conference 2013, Paper no:41.

ASCE 41-13. 2013 *Seismic Evaluation and Retrofit of Existing Buildings*, American Society of Civil Engineers, Reston, VA, USA.

ASCE 41- 17. 2017 *Seismic Evaluation and Retrofit of Existing Buildings*, American Society of Civil Engineers, Reston, VA, USA, 2017.

Augenti, N., Parisi, F. 2010. Ultimate flexural strength of unreinforced masonry spandrel panels, *8th Int. Mason. Conf. Dresden, Germany*. 1653–1662.

Beyer, K., Dazio, A. 2012. Quasi-static cyclic tests on masonry spandrels. *Earthq. Spectra* 28, 907–929.

Beyer, K., Mangalathu, S. 2013. Review of strength models for masonry spandrels. *Bull. Earthq. Eng.* 11, 521–542.

Bruneau, M. 1994, State of the Art Report on Seismic Performance of Unreinforced Masonry Buildings, *J. Struct. Engg*, 120(1), 230–251.

Bruneau, M. 1995, Performance of masonry structures during the 1994 Northridge(Los-Angeles) earthquake, *Can.J. Civ.Engg*, 22, 378–402.

Cattari, S., Beyer, K. 2015. Influence of spandrel modelling on the seismic assessment of existing masonry buildings, *Proc. Tenth Pacific Conf. Earthq. Eng. Build. an Earthquake-Resilient Pacific* 88–95.

Derakhshan, H, Griffith, M.C, Ingham, J.M. 2016, Out of plane seismic response of vertically spanning URM walls connected to flexible diaphragms. *Earthquake Engg Struct. Dyn*, 45:563–580.

Dolatshahi, K. M., Aref, A. J., Whittaker, A.S. 2015, Interaction curves for IP and OOP behaviors of unreinforced masonry walls. *J Earthq Eng*, 19(1):60–84.

Dolatshahi, K.M., Yekrangnia, M., Mahdizadeh, A. 2014, On the influence of IP damages on the OOP behavior of unreinforced masonry structures. *NCEE 2014-10th U.S. Natl. Conf. Earthq. Eng. Front. Earthq. Eng.*

Eurocode 8 (2005). 2005, Assessment and retrofit of buildings-Part 3: General rules, seismic actions and rules for buildings. *European Committee for Standardization*.

Fajfar, P. 2018, Analysis in Seismic Provisions for Buildings: Past, Present and Future, *The Fifth Prof. Nicholas Ambraseys Lecture*, Springer Netherlands).

FEMA P-58(2017). 2017, Seismic Performance Assessment of Buildings Volume 1 – Methodology, *Federal Emergency Management Agency.*

FEMA P-774(2009). 2009, Unreinforced Masonry Buildings and Earthquakes. *Federal Emergency Management Agency.*

Foraboschi, P. 2009, Coupling effect between masonry spandrels and piers, *Mater. Struct. Constr.* 42, 279–300.

Giongo, I., Wilson, A., Dizhur, D.Y., Derakhshan, H., Tomasi, R., Griffith, M.C., Quenneville, P., Ingham, J. M. 2014, Detailed seismic assessment and improvement procedure for vintage flexible timber diaphragms. *Bull. New Zeal. Soc. Earthq. Eng.* 47, 97–118.

Graziotti, F., Magenes, G., Penna, A., Fontana, D. 2011, Experimental cyclic behaviour of stone masonry spandrels. *15th World Conf. Earthq. Eng.* Lisbon, Portugal.

Ingham, J., Griffith, M. 2010. Performance of unreinforced masonry buildings during the 2010 Darfield (Christchurch, NZ) earthquake, *Aus. J. Struct. Eng.*, 11(3), 207–224.

Ingham, J., Griffith, M. 2011. The performance of earthquake strengthened URM buildings in the Christchurch CBD in the 22 February 2011 earthquake, Addendum Report to the Royal Commission of Inquiry.

Kollerathu, J.A., Menon, A. 2017. Interaction of IP and OOP responses in unreinforced masonry walls under seismic loads. *J. Struct. Eng.*, 44, 422–441.

Lu. S, Beyer. K, Bosiljkov. V, Butenweg. C, D'Ayala. D, Degee. H, Gams. M, Klouda. J, Lagomarsino. S, Penna. A, Mojsilovic. N, F. da Porto, Sorrentino. L, Vintzileou. E. 2016. Next generation of Eurocode 8 Masonry chapter., *Proc. 16th Int. Brick Block Mason. Conf., Padova, Italy, 695-700.*

Magenes, G. and G.M. Calvi, 1997. In plane seismic response of brick masonry walls, *Earthq Eng Struct Dynam*, 26, 1091–1112.

Magenes, G., Calvi, G. M. 1992, Cyclic behavior of brick masonry walls, *Proceedings of the 10th World Conf. Earthq. Eng. Balkema, Rotterdam*, 3517–3522.

Magenes, G., Penna, A. 2009. Existing masonry buildings: general code issues and methods of analysis and assessment, *Eurocode 8 Perspectives from the Italian Standpoint Workshop, Naples, Italy*, 185–198.

Magenes, G., Penna, A. 2011. Seismic Design and Assessment of Masonry Buildings in Europe: Recent Research and Code Development Issues, 9th *Australas. Mason. Conf.* 583–603.

Manojlovic D, Radujkovic, A., Kocetov, T. 2018. Evaluation of the behaviour factor of unreinforced masonry buildings, *Sixth International Conference Seismic Engineering and Engineering Seismology*, 267–274.

McVerry, G.H. 2003. From hazard maps to code spectra for New Zealand. *Proc. Pacific Conf. Earthq. Eng.*, 1–9.

Menon, A., Magenes, G. 2011. Definition of seismic input for OOP response of masonry walls: I. parametric study. *J. Earthq. Eng.* 15, 165–194.

Ministry for cultural heritage and activities.2007, *Guidelines for Evaluation and Mitigation of Seismic Risk to Cultural Heritage*, Gangemi.

Morandi, P., Magenes, G. 2008. Seismic design of masonry buildings: current procedures and new perspectives. *14th World Conf. Earthquake Eng.*

Murty, C.V.R., Goswami, R., Vijayanarayanan, A.R., Mehta, V. 2012 *Earthquake Behavior of Buildings*. Gujarat State Disaster Management Authority, Gandhinagar.

Najafgholipour, M, Maheri, M.R.and Lourenço, P. 2014, Definition of interaction curves for the IP and OOP capacity in brick masonry walls, *Construction and Building Materials* 55, 168–18.

Nakamura, Y, Derakhshan, H, Griffith, M.C, Magenes, G. 2017, Influence of Diaphragm Flexibility on Seismic Response of Unreinforced Masonry Buildings, *Journal of Earthquake Engineering*, 21:935–960.

Noor-E-Khuda, S., Dhanasekar, M. 2018. Masonry Walls under Combined IP and OOP Loadings, *J. Struct. Eng. (United States)*, 144, 1–10.

Parisi, F., Augenti, N. 2011, The role of spandrels within masonry walls with openings: an experimental investigation., *Proc. Ninth Pacific Conf. Earthquake*

Eng. *Building an Earthquake-Resilient Society*, Auckland, New Zealand, Paper no.232.

Penner, O. 2014. OOP Dynamic Stability of Unreinforced Masonry Walls Connected to Flexible Diaphragms., PhD Thesis, *University of British Columbia*.

Penner, O., Elwood, K.J., 2016, OOP dynamic stability of unreinforced masonry walls in one-way bending: Parametric study and assessment guidelines, *Earthq. Spectra*, 32, 1699–1723.

Petry, S., Beyer, K. 2014. Influence of boundary conditions and size effect on the drift capacity of URM walls, *Eng. Struct.* 65, 76–88.

Petry, S., Beyer, K. 2015. Force-displacement response of unreinforced masonry walls to seismic design, *Mason. Int.* 28, 19–36.

Priestley, M.J.N. 1985, Seismic Behaviour of Unreinforced Masonry Walls. *Bulletin of New Zealand National Society for Earthquake Engineering*,18, No. 2, 191–205.

Salmanpour, A.H. 2017. Displacement capacity of structural masonry., PhD thesis, *ETH Zurich*.

Salmanpour. B. A., Mojsilovic N., Schwartz, J., 2013. Experimental Study of the Deformation Capacity of Structural Masonry, *12th Can. Mason. Symp.*

Simsir, Can, C., Aschheim M, and Daniel Abrams D. 2004. Out of Plane Dynamic Response of Unreinforced Masonry Bearing Walls Attached to Flexible Diaphragms, *Proceedings of World Conference on Earthquake Engineering*, Canada, Paper No. 2045.

Technical Guidelines for Engineering Assessments, New Zealand. 2017. The Seismic Assessment of Existing Buildings: Part A, Part B, Part C. *NZSEE*.

Wilding, B. V., Beyer, K. 2016. Effective stiffness of unreinforced brick masonry walls, *Proc. 16th Int. Brick Block Mason. Conf., Padova, Italy*, 1993–2002.

Wilson, A., Quenneville, P.J.H., Ingham, J.M. 2014. In plane orthotropic behavior of timber floor diaphragms in unreinforced masonry buildings. *J. Struct. Eng.*,140.

Zhang, S., Petry, S., Beyer, K. 2014. Investigating the in plane Mechanical Behavior of URM Piers Via DSFM, *Second Int. Conf. Earthq. Eng. Seismol*, 1–12.

Zimmermann, T, Strauss, A & Bergmeister, K. 2012. Energy Dissipation and Stiffness Identification of Unreinforced Masonry, *Proc. 15th Int. Brick Block Mason. Conf, Brazil*.

Brick and Block Masonry - From Historical to Sustainable Masonry –
Kubica, Kwiecień & Bednarz (eds)
© 2020 Taylor & Francis Group, London, ISBN 978-0-367-56586-2

Experimental testing of unreinforced lattice masonry walls subjected to out-of-plane pressure loading

M.J. Masia, G. Simundic & A.W. Page
Centre for Infrastructure Performance and Reliability, The University of Newcastle, Australia

ABSTRACT: In lattice masonry (also known as hit and miss brickwork) the mortar perpend joints are left unfilled and the masonry units are spaced along the courses to leave a gap between adjacent units, allowing the passage of light and air through the wall. Regardless of whether this masonry is used in a loadbearing or nonloadbearing application, it must still be capable of spanning between supports to resist out-of-plane lateral loading to satisfy robustness requirements and to resist wind and/or earthquake actions and hence requires structural design. The Australian Standard AS3700: Masonry Structures (Standards Australia 2018) provisions for one way vertical bending can be applied in the case of lattice masonry by using a section modulus based on the net bedded area. However, the provisions for horizontal bending and two way bending require that the masonry be constructed with all perpends completely filled and therefore lattice masonry falls outside the scope of AS3700 for horizontal and two way bending. Internationally, there also exists a lack of guidance for the structural design of this form of masonry.

The paper describes an experimental study to assess the behaviour of unreinforced lattice masonry walls subjected to lateral out-of-plane pressure loading. Twenty one single leaf lattice masonry walls, of varying aspect (height: length) ratios, were constructed using extruded clay bricks (230 mm long x 110 mm wide x 76 mm high) and 1:1:6 (cement: lime: sand) mortar. Three walls were tested in one way vertical bending. Twelve walls, with varying overlap between units in adjacent courses, were tested in one way horizontal bending. Six walls, with two different aspect ratios, were tested in two way bending. The load versus deformation behaviour and the observed failure modes are reported.

1 INTRODUCTION

Lattice masonry (also known as hit and miss brickwork) is a form of construction in which the mortar perpend joints are left unfilled and the masonry units are spaced along the courses to leave gaps between adjacent units. This creates a wall which has a lattice type appearance, allowing the passage of light and air through the wall, while still providing a level of privacy. Two examples are shown in Figure 1 and online searches using keywords such as "hit and miss masonry", "perforated masonry" or "masonry screens" reveal that the use of such masonry is widespread internationally with many spectacular examples of varying geometries in a range of applications.

Regardless of whether this masonry is used in a loadbearing or nonloadbearing application, it must still be capable of spanning between supports to meet minimum robustness requirements and to resist out-of-plane lateral loading due to wind and/or earthquake actions, and hence it requires structural design. Australian Standard AS3700: Masonry Structures (Standards Australia 2018) provisions for one way vertical bending of unreinforced masonry can be applied in the case of lattice masonry by using a section modulus based on the net bedded

area. However, the AS3700 provisions for horizontal bending require that the masonry be constructed with all perpends completely filled and therefore lattice masonry falls outside the scope of AS3700 for horizontal bending. By extension, in their present form, the AS3700 provisions for two way bending action, which rely on the horizontal bending capacity, also cannot be applied for lattice masonry. As a result of this apparent lack of guidance for structural design, Think Brick Australia receives numerous enquiries each year from architects and structural engineers wishing to specify and design lattice masonry. This prompted the current study which is designed to better understand the behaviour of unreinforced lattice masonry walls when subjected to lateral out-of-plane pressure loading.

A review of overseas codes and literature also reveals limited existing information on the design of this form of construction, thus further justifying the need for research in this area. Trimble (2013) and Ortlepp and Schmidt (2014) both reported a lack of guidance for structural design of lattice masonry. Ortlepp and Schmidt (2014) performed a pilot numerical study to assess the loadbearing capability for lattice masonry walls. They concluded that the absence of contact along part of the unit length

Figure 1. Examples of lattice masonry construction (photographs supplied by Think Brick Australia).

of a further three horizontal bending tests using lattice masonry panels with reduced overlap between the units in adjacent courses, and a further three two way bending tests using panels with a reduced aspect ratio.

2 EXPERIMENTAL PROGRAM

Twenty one unreinforced clay brick lattice masonry panels were constructed by an experienced mason using one type of vertically cored extruded clay brick (230 mm long x 110 mm thick x 76 mm high) with 10 mm thick bed joints using a 1:1:6 cement:lime: sand (by volume) mortar. The dimensions chosen for the panels are shown in Figure 2. All panels were single leaf (110 mm thick). Fifteen of the panels were 1190 mm long x 1194 mm high (aspect ratio 1). Of these, nine panels were constructed with a unit overlap of 70 mm (gap between units of 90 mm), three

results in stress concentrations which influence the loadbearing behaviour meaning that it is not possible to transfer the principles for compression design of solid masonry to lattice masonry walls simply by reducing the bedded area used in design. This conclusion was supported by Edgell (2020) who found via testing that the compressive capacity for lattice walls was lower than predicted by reducing the capacity of solid walls based on the reduction in bedded area. Under in-plane shear loading, Ortlepp and Schmidt (2014) demonstrated that the individual units are subjected to flexural stresses, again rendering inappropriate, the use of existing design approaches for solid masonry.

Although lattice masonry has been used in loadbearing walls in some instances, it is most commonly used in non-loadbearing applications. In such applications, it is the ability of the masonry to resist out-of-plane lateral loading which is mostly likely to limit its structural capacity. Although Edgell (2020) was able to show via pilot testing that a reduction in wind pressure due to the holes in lattice masonry may be justified, pressure does still exist and robustness and earthquake loading require that lattice walls must resist out-of-plane pressures. The authors conducted previous studies (Masia et al. 2017, Masia et al. 2018) in which lattice masonry panels were tested in one and two way out-of-plane bending. The current paper summarizes the experimental results from the two previous studies and reports the results

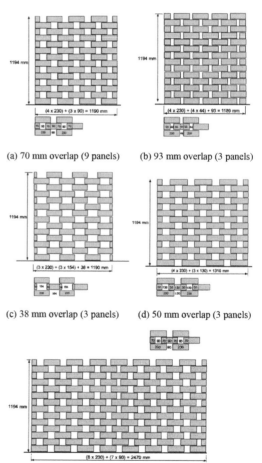

(a) 70 mm overlap (9 panels)

(b) 93 mm overlap (3 panels)

(c) 38 mm overlap (3 panels)

(d) 50 mm overlap (3 panels)

(e) 70 mm overlap for aspect ratio 0.5 (3 panels)

Figure 2. Lattice masonry panel dimensions.

were constructed with an overlap of 93 mm (gap of 44 mm) and three were constructed with an overlap of 38 mm (gap of 154 mm). A further three panels were 1310 mm long x 1194 mm high, with overlap of 50 mm (gap of 130 mm). The final three panels were 2470 mm long x 1194 mm high (aspect ratio 0.5), with overlap of 70 mm (gap of 90 mm).

The lateral modulus of rupture of the masonry units was tested in accordance with AS/NZS4456.15 (Standards Australia 2003) and found to be 2.90 MPa (mean) with a COV of 0.16. For each batch of mortar mixed during the panel construction, two x 6 unit high (10 mortar joints) stack bonded piers were constructed and the flexural tensile strength of the masonry was determined by bond wrench testing in accordance with AS3700 (Standards Australia 2018). The bond wrench tests were performed on the same date of testing of the corresponding panel constructed using that mortar batch. The panels and bond wrench piers were cured under ambient conditions in the laboratory for a minimum of 28 days prior to testing.

2.1 One way bending tests

Three of the panels with 70 mm overlap (Figure 2a) were supported in a one way vertical bending configuration (bending stresses acting normal to the mortar bed joints) as shown in Figure 3a, b. Twelve of the panels (three of each of the 93mm (Figure 2b), 70 mm (Figure 2a), 50 mm (Figure 2d) and 38 mm (Figure 2c) unit overlaps) were designated for testing under one way horizontal bending. Unfortunately, the three panels with just 38 mm overlap between units failed during handling and could not be tested. The remaining nine panels were supported in a one way horizontal bending configuration (bending stresses acting parallel to the mortar bed joints) as shown

in Figure 3c. The panels were simply supported and subjected to out-of-plane pressure loading using an airbag system as shown in Figure 3.

All twelve one way spanning panels (vertical and horizontal bending) were tested in a vertical spanning orientation (simply supported along the top and bottom edges with centre to centre span of 1104 mm) as shown in Figure 3a. Therefore, to induce horizontal bending, this required first rotating the horizontal bending specimens through 90° as shown in Figure 3c. This allowed the use of the same apparatus for all twelve specimens and avoided the problems associated with providing a frictionless base support. The panels were supported on stiff water filled rubber tubes to absorb any specimen irregularities (Figure 3a). The air bag pressure was applied monotonically at a constant rate until failure was observed. The displacements of each panel at mid-height and at the supports were continuously recorded using potentiometers throughout each test.

2.2 Two way bending tests

The remaining three panels with 70 mm unit overlap and aspect ratio 1 (Figure 2a) and three panels with 70 mm unit overlap and aspect ratio 0.5 (Figure 2e) were tested in a two way bending configuration. The apparatus used for the one way bending tests was modified by fabricating support frames (one frame for each aspect ratio) which provided simple support along all four edges of the panels (Figures 4a and 4c). The centre to centre span was 1104 mm in each

(a) Aspect ratio 1.0   (b) Plaster bedding

(c) Aspect ratio 0.5

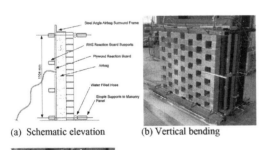

(a) Schematic elevation   (b) Vertical bending

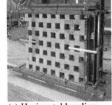

(c) Horizontal bending

Figure 3. Test setup for one way bending tests.

Figure 4. Test setup for two way bending tests.

direction for the aspect ratio 1.0 specimens. For aspect ratio 0.5, the vertical span was 1050 mm and the horizontal span was 2370 mm. To ensure full contact was achieved between the supports and the masonry panels in both spanning directions, the support was set against the panel using plaster, in place of the water filled tubes (Figure 4b). The plaster was set using a flat surface, which was then greased so that the panels could slide in-plane relative to the steel tube supports, which were welded to the support frame. This was necessary to avoid arching action being induced in the panels. As for the one way bending tests, the air bag pressure was applied monotonically at a constant rate until failure was observed. The displacements of each panel at the panel centre and at the mid-length of supports were continuously recorded using potentiometers throughout each test.

For later use in the presentation and discussion of results, the specimen naming convention is as follows: the first letter denotes the bonding pattern, L for lattice bond; the second letter (or letters) denotes the bending direction, V for one way vertical bending, H for one way horizontal bending and HV for two way bending; the first number 93, 70, 50 or 38 denotes the overlap (mm) of units in adjacent courses. For the two way bending specimens, the aspect ratio "1.0" or "0.5" is included in the naming convention. Finally, three replicates of each specimen type were constructed leading to the inclusion of a number (1, 2 or 3), or lower case letter (a, b, c) in the naming convention.

3 RESULTS AND DISCUSSION

The experimental results are summarized in Tables 1 and 2 and the recorded plots of load versus mid-height displacement for all panels are shown in Figures 5 and 6, for the one way and two way bending tests, respectively. The displacements reported in Figures 5 and 6 are the net displacements after subtracting any movement occurring at the supports. The inset in Figure 5 shows an expanded view of the small displacement region of the observed behaviour.

3.1 One way bending tests

The specimens subjected to one way vertical bending (LV70) displayed linear elastic response with no damage observed prior to a crack developing suddenly at peak pressure. The failure line in all three panels extended completely along a single bed joint close to the panel mid-height (region of maximum bending moment under the imposed loading). The same behaviour would be expected of fully bonded masonry panels when subjected to one way vertical bending. In the case of Specimens LV70_1 and LV70_3, the crack occurred in the bed joint one course below the panel mid-height (Figure 7a). For Specimen

Table 1. Results for one way bending tests.

Test ID	f_{mt}* (COV) (MPa)	w_{exp}** (kPa)	Observed failure mode
LV70_1	0.82 (0.24)	1.94	Flexural failure at bed joint near mid-height
LV70_2	0.82 (0.24)	2.69	Flexural failure at bed joint near mid-height
LV70_3	0.49 (0.14)	3.32	Flexural failure at bed joint near mid-height
Mean (COV)		2.65 (0.26)	
LH38	0.45 (0.33)	-	Panels failed during handling – not tested
LH50_1	0.73 (0.15)	5.88	Step/line failure – mostly line failure
LH50_2	0.73 (0.15)	7.68	Sudden collapse – line failure through units
LH50_3	0.73 (0.15)	7.47	Combined step/line failure
Mean (COV)		7.0 (0.14)	
LH70_1	0.75 (0.34)	13.32	Sudden collapse – step/line failure
LH70_2	0.49 (0.14)	10.50	Stepped failure through joints
LH70_3	0.75 (0.34)	17.25	Sudden collapse - step/line failure
Mean (COV)		13.7 (0.25)	
LH93_1	0.38 (0.27)	17.93	Sudden collapse – step/line failure
LH93_2	0.93 (0.45)	16.41	Sudden collapse – line failure through units
LH93_3	0.38 (0.27)	17.38	Sudden collapse – step failure through joints
Mean (COV)		17.2 (0.04)	

*Masonry flexural tensile strength determined by bond wrench in accordance with Standards Australia (2018)
**Experimentally observed wall panel capacity (peak pressure resisted)

LV70_2 the crack occurred in the bed joint two courses below mid-height. Post cracking, the pressure resisted by the panels dropped suddenly but the panels did not collapse. Rather, they were able to continue to resist small post peak pressures (Figure 5) by virtue of the restoring moment generated due to the self weight of the two "halves" of the panel acting as rigid blocks rotating about the contacting edges.

The three LH38 specimens failed during handling and could not be tested. The other nine panels subjected to one way horizontal bending (LH50, LH70 and LH93 specimens) displayed linear elastic response with no damage observed prior to a sudden failure at peak pressure (Figure 5). Six of

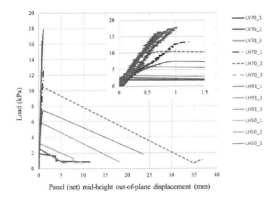

Figure 5. Load versus displacement for one way bending tests.

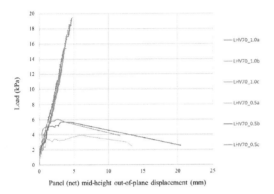

Figure 6. Load versus displacement for two way bending tests.

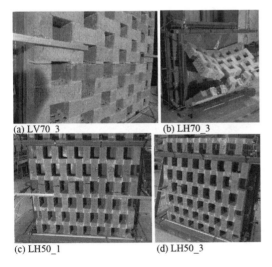

(a) LV70_3

(b) LH70_3

(c) LH50_1

(d) LH50_3

Figure 7. Observed failure modes for one way bending (a) LV70_3, (b) LH70_3, and (c) LH50_1, and (d) LH50_3.

the nine specimens (the exceptions being LH50_1, LH50_3, and LH70_2) collapsed completely at peak pressure making it difficult to observe the failure surface. Inspection of video footage and panel debris revealed that the failure surfaces varied between exclusively stepped failure through the mortar joints, to complete line failure via rupturing of the units, and combinations of stepped and line failure (Figure 7b). Likewise, for the three panels which did not collapse, the same various possible failure modes were represented (Figure 7c, d). The failure mode for each specimen is noted in the last column in Table 1. It is noted that horizontal bending in fully bonded masonry also results in stepped, line or combined failure modes.

As shown in Table 1, the LH specimens were able to sustain considerably higher pressures prior to failure (average peak pressures between 7.0 kPa and 17.2 kPa) than the LV specimens (average peak pressure of 2.65 kPa). Failure of the LH specimens engages the flexural strength of the units and/or the torsional shearing strength of the bed joints as a result of the unit overlap in the bonding pattern, whereas failure of the LV specimens is limited only by the relatively low flexural strength of the mortar bed joints. This difference in strengths (also observed in fully bonded masonry) highlights the need for reliable methods to predict the horizontal bending capacity of lattice masonry walls in order for designers to best capitalise on the considerably higher strength when spanning in the horizontal direction.

Increasing the overlap between units in adjacent courses resulted in increasing peak pressures (averages of 7.0 kPa, 13.7 kPa and 17.2 kPa for overlaps of 50 mm, 70 mm and 93 mm, respectively (Table 1)). This result is to be expected given that the failure of seven of the nine panels tested involved at least some stepped cracking through the mortar bed joints and increased overlap provides increased torsional shearing area on the bed joints. Prior to testing it was hypothesized that increasing the overlap might also lead to a greater number of line failures via unit rupture as the torsional shearing area of the bed joints was increased. However, the effect of overlap on failure mode was inconclusive, with a similar spread between the possible failure modes being recorded at all overlap values (Table 1). The absence of a clear relationship between overlap and failure mode can perhaps be attributed to the large variations in masonry flexural tensile strength (f_{mt} in Table 1) observed between specimens. Comparing the observed failure modes with the measured f_{mt} for each specimen shows that stepped failures through the joints coincided with lower values of bond strength and line failures through masonry units coincided with high bond strengths. This outcome is to be expected. The bond strength (f_{mt}) will influence the failure mode in horizontal bending and variations in f_{mt} between specimens make it difficult to isolate the influence of unit overlap.

3.2 Two way bending tests

The three aspect ratio 1.0 panels subjected to two way bending (LHV70_1.0 specimens) also displayed non-ductile behaviour. The load versus panel mid-height displacement (Figure 6) was essentially linear, and no visible cracking was observed, prior to the full failure pattern developing suddenly at peak pressure. Once the full cracking pattern developed, the pressure resisted by the panels dropped suddenly and testing was discontinued. The failure mode shown in Figure 8 for Specimen LHV70_1.0a is typical of all three LHV70_1.0 specimens; that is, diagonal cracking extending from the panel corners to meet a short section of stepped vertical cracking in the middle of the panel. Specimens LHV70_1.0a and LHV70_1.0b displayed exclusively joint failures. Specimen LHV70_1.0c displayed mostly joint failures, with just two of the units being ruptured. The failure patterns observed are also typical of those reported for fully bonded unreinforced masonry subjected to two way bending.

As expected, the peak pressures resisted by the two way bending specimens were higher, on average, than those resisted by the one way horizontal bending specimens of the same aspect ratio and unit overlap (17.8 kPa for LHV70_1.0 specimens compared to 13.7 kPa for LH70 specimens (Tables 1 and 2)).

The three aspect ratio 0.5 panels subjected to two way bending (LHV70_0.5 specimens) displayed initial cracking prior to reaching peak pressure as evidenced by the nonlinear pre-peak load versus displacement responses shown in Figure 6. Specimen LHV70_0.5a developed a horizontal crack along the complete length of the bed joint at panel mid-height at a lateral pressure of approximately 3 kPa. The pressure resisted by the panel was then able to increase further until diagonal and vertical cracking extending

Figure 9. Observed failure mode for Specimen LHV70_0.5a.

to the corners of the panel developed suddenly at peak pressure as shown in Figure 9. The failure mode shown in Figure 9 is similar to what would be expected for a fully bonded masonry panel of the same aspect ratio.

Specimens LHV70_0.5b and LHV70_0.5c displayed slightly different behaviour to that described above. The first cracking to be observed was vertical cracking in some of the masonry units. This occurred prior to reaching peak pressure. At peak pressure, the full crack patterns developed in each case and included a horizontal bed joint crack at panel mid-height, combined with diagonal and further vertical cracking.

It was considered that the vertical cracking which occurred prior to reaching peak load could be due to excessive deflection of the support lines along the top and bottom edges of the panels, which would allow horizontal bending stresses, and hence possibly vertical cracking, to develop. To assess this concern, a numerical analysis was conducted to predict the deflection of the top and bottom edge supports when a pressure of 6 kPa (equal to largest peak pressure observed) was applied to the panel. The analysis returned a mid-span deflection of 3.7 mm. AS3700 (Standards Australia 2018) specifies that lateral supports for unreinforced masonry should not deflect by more than span/500, or 4.7 mm for the 2370 mm horizontal span used in the current testing. Therefore, the stiffness of the supports used were confirmed to meet AS3700 requirements for lateral support. Despite meeting the code requirement it is still considered that the support flexibility may be responsible for the unexpected vertical cracking prior to reaching peak pressure in two of the three specimens. However, the peak pressures achieved are considered to correctly reflect the two way bending behaviour of the tested panels.

The average peak pressure resisted by the aspect ratio 0.5 specimens was just 5.2 kPa, compared to 17.8 kPa for the aspect ratio 1.0 panels (Table 2). This is expected given the much longer horizontal span. However, the LHV70_0.5 specimens were able to resist, on average, twice the pressure recorded for

Figure 8. Observed failure mode for Specimen LHV70_1.0a.

Table 2. Results for two way bending tests.

Test ID	f_{mt}* (COV) (MPa)	w_{exp}** (kPa)	Observed failure mode
LHV70_1.0a	0.74 (0.46)	15.41	sudden cracking at peak load, failure confined to joints
LHV70_1.0b	0.93 (0.45)	19.47	sudden cracking at peak load, failure confined to joints
LHV70_1.0c	0.74 (0.46)	18.57	sudden cracking at peak load, mostly in joints
Mean (COV)		17.8 (0.12)	
LHV70_0.5a	0.37 (0.20)	3.92	horizontal cracking first, then diagonal at peak load
LHV70_0.5b	1.54 (0.19)	5.69	vertical cracking first, then horizontal & diagonal at peak load
LHV70_0.5c	1.54 (0.19)	6.07	vertical cracking first, then horizontal & vertical at peak load
Mean (COV)		5.2 (0.22)	

*Masonry flexural tensile strength determined by bond wrench in accordance with Standards Australia (2018)
**Experimentally observed wall panel capacity (peak pressure resisted)

the one way vertical bending specimens (LV70) of almost identical vertical span. This clearly illustrates the benefit achieved by providing vertical lines of support to help engage horizontal and diagonal bending action in the lattice masonry.

4 CONCLUSION

An experimental study was conducted to assess the behaviour of unreinforced lattice masonry walls subjected to lateral out-of-plane pressure loading. Twenty one single leaf lattice masonry walls, of varying aspect ratios, were constructed. Three walls were tested in one way vertical bending. Nine walls (three failed during handling), with varying overlap between units in adjacent courses, were tested in one way horizontal bending. Six walls, with two different aspect ratios, were tested in two way bending.

Panels subjected to one way vertical bending failed in a non-ductile mode via bed joint cracking which occurred suddenly at the peak load along a single course close to panel mid-height.

Panels subjected to one way horizontal bending also displayed non-ductile failure modes with no observable damage prior to a sudden failure surface developing at peak load. The failure surface varied between specimens from stepped failure exclusively

through the mortar joints to line failure (fracture of the units) and combinations of stepped and line failure. As the overlap between units in adjacent courses was increased, the panel capacities increased.

For the aspect ratio 1.0 panels subjected to two way bending, non-ductile failure modes were observed with no detectable damage prior to a sudden failure surface developing at peak load. The failure patterns included stepped diagonal cracking extending from the panel corners, joined by a vertical stepped crack through the middle of the panel. For the aspect ratio 0.5 panels subjected to two way bending, horizontal or vertical cracking initiated at loads lower than peak, prior to the full failure surface, which included diagonal cracking, developing suddenly at peak pressure.

The specimens supported in horizontal and two way bending configurations resisted much larger pressures than those subjected to one way vertical bending. This difference in strengths (also observed in fully bonded masonry) highlights the benefit in providing vertical lines of support to help engage horizontal and diagonal bending action. Reliable methods to predict the horizontal and diagonal bending capacity of lattice masonry walls, which do not currently exist, are needed in order for designers to best capitalize on the considerably higher strengths when the masonry is supported in this way compared to simple vertical bending.

The failure modes observed for each of the support configurations are similar to those observed in fully bonded unreinforced masonry. This should allow existing design methodology for one and two way bending of fully bonded masonry to be adapted to reflect the behaviour observed in lattice masonry walls. This is the subject of ongoing research by the authors.

ACKNOWLEDGEMENTS

This research has been performed in the Civil Engineering laboratories of The School of Engineering at The University of Newcastle. The assistance of the laboratory staff is gratefully acknowledged, as is the support of Think Brick Australia, and Brickworks Ltd.

REFERENCES

Edgell, G.J. 2020. The design of brickwork facades. *Proc. 17th International Brick and Block Masonry Conference, Kraków, Poland, 5-8 July, 2020.*

Masia, M.J. Simundic, G. & Page, A.W. 2017. Flexural strength of unreinforced lattice masonry walls subjected to lateral out-of-plane loading. *Proc. 13th Canadian Masonry Symposium, Halifax, Canada, 4-7 June 2017.*

Masia, M.J. Simundic, G. & Page, A.W. 2018. Behaviour of unreinforced lattice masonry walls under one way horizontal and two way out-of-plane bending. In (eds.) G. Milani, A. Taliercio & S. Garrity, *Proc. 10th*

International Masonry Conference, Milan, Italy, July 9- 11,2018.

Ortlep p, S. & Schmidt, F. 2014. Perforated masonry – light weight construction. *Proc. 9th International Masonry Conference, Guimarães, Portugal, 7-9 July, 2014.*

Standards Australia 2003. *AS/NZS 4456.15: 2003: Masonry units, segmental pavers and flags – Methods of test, Method 15: Determining lateral modulus of rupture*, Standards Association of Australia, Sydney, 2003.

Standards Australia 2018. *AS 3700: 2018: Masonry structures*. Standards Association of Australia, Sydney, 2018.

Trimble, B.E. 2013. Design of unique landscape walls and their use in building facades. *Proc. 12th Canadian Masonry Symposium, Vancouver, Canada, 2-5 June, 2013.*

Brick and Block Masonry - From Historical to Sustainable Masonry –
Kubica, Kwiecień & Bednarz (eds)
© 2020 Taylor & Francis Group, London, ISBN 978-0-367-56586-2

Design of masonry panels subjected to fire in Europe: An overview on the new draft of EN 1996-1-2

U. Meyer
Arbeitsgemeinschaft Mauerziegel im BV der Deutschen Ziegelindustrie e. V., Berlin, Germany

R. van der Pluijm
Wienerberger AG, Zaltbommel, The Netherlands

M. Andreini
Studio Masiello Strutture, Pisa, Italy

G. Pettit
Concrete Block Association, Leicester, UK

L. Miccoli
Xella Technologie- und Forschungsgesellschaft, Kloster Lehnin, Germany

ABSTRACT: This paper gives an overview on the new draft of EN1996-1-2 (Eurocode 6 part 1-2) being prepared by Project Team SC6.T2 of CEN/TC250. The prescriptive and performance-based approach for the design of masonry panels subjected to fire is described. An improvement to the tabulated data for the determination of the fire rating of non-loadbearing masonry panels is described. The problems associated with the calculation methods in the current version of EN1996-1-2 for the assessment of loadbearing walls are explained. A proposal is made for the improvement of the input parameters for mechanical models based on experimental tests. Finally, the results of an experimental programme carried out on some common masonry materials (units and mortar) are presented.

1 INTRODUCTION

The general objectives of the Eurocodes for fire protection are to limit risks to the individual and society, neighboring property, and where required, the directly exposed property, in the case of fire.

In the same safety domain, the European Construction Product Regulation (CPR) 305/2011 (European Parliament and Council, 2011) mentions, as an essential requirement for the limitation of fire risks, that the construction works must be designed and built in such a way that, in the event of an outbreak of fire:

- the load-bearing resistance of the construction can be assumed for a specified period of time
- the generation and spread of fire and smoke within the works are limited
- the spread of fire to neighboring construction works is limited
- the occupants can leave the works or can be rescued by other means
- the safety of rescue teams is taken into consideration.

This essential requirement may be observed by following various possibilities for fire safety strategies like conventional fire scenarios (nominal fires), characterizing a prescriptive design approach, or 'natural' (parametric) fire scenarios, including passive and/or active fire protection measures, characterizing the performance-based design methodologies.

The European Standard EN 1996-1-2. Design of masonry structures – General rules – Structural fire design (CEN 2005b, 2010b) deals with specific aspects of passive fire protection in terms of designing masonry structures and parts thereof for adequate loadbearing resistance that could be needed for the safe evacuation of occupants and fire rescue operations and for limiting fire spread as relevant. The possible options for the structural fire design allowed in EN 1996-1-2 (CEN 2005b, 2010b) are illustrated in Figure 1. Although the current version of this standard offers the possibility to adopt simple or advanced calculation methods, only the tabulated data approach is considered as a reliable design method. This has made the use of EN 1996-1-2 (CEN 2005b, 2010b) much more limited than Parts 1-2 of the other Eurocodes. This limitation has been

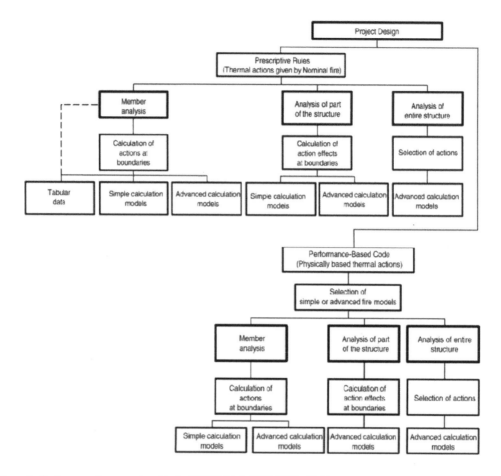

Figure 1. Design procedures mentioned in the EN 1996-1-2 (CEN 2005b, 2010b).

considered in the context of the on-going revision of the structural Eurocodes.

The revision of EN 1996-1-2 (CEN 2005b, 2010b) started with the appointment of the responsible Project Team (PT) in July 2018. A first draft was delivered in April 2019, the second draft is due April 2020 and the final PT document will be delivered in October 2021. This paper gives an overview of the current work in the Project Team SC6.T2 dedicated to the revision of EN 1996-1-2 (CEN 2005b, 2010b).

2 REVISION OF THE MAIN TEXT

2.1 General

The work of the PT is focused on editorial improvements and clarifications arising from the use of the current version, an improvement of the background information concerning the tabulated values of fire resistance, (see chapter 3 of this paper) and an improvement of the information on relevant material properties as a basis for a future calculation method. The current Annex A will be removed.

2.2 Nationally Determined Parameters (NDPs)

Two NDPs from the previous version have been changed into a prescriptive value or deleted. The emissivity of the masonry surface is now normative with the value ($\varepsilon_m = 0.8$) and as there is no need to give information about the nationally determined safety parameter γ_{Global}, the requirement has been deleted.

2.3 Technical clarifications concerning surface finishes

As there is no available test evidence justifying the previously existing exclusion of General Purpose Mortar according to EN 998-1 (CEN 2016a) as a suitable finish for fire protection, this type of mortar is now included.

It has now been clarified that a second leaf of masonry (i.e. the external leaf of a cavity wall or the second leaf of a double leaf wall) provides at least the equivalent fire protection as an applied suitable plaster or render. In these cases, the tabulated values for plastered walls can be applied. It is also clarified that an appropriately applied non-combustible thermal insulation system is a suitable fire-resistant finish.

3 TABULATED DATA METHOD FOR NON-LOADBEARING WALLS

3.1 General

Current Annex B gives tabulated data for the verification of non-loadbearing and loadbearing walls and provides design solutions for standard fire exposures of up to 240 minutes. In the tables, minimum values of thickness are given for walls with and without plaster for different types of masonry (clay, calcium silicate, dense and lightweight concrete, autoclaved aerated concrete and manufactured stone). Other parameters taken into account are gross density, percentage of voids, types of masonry mortar and plaster. For loadbearing walls, the fire resistance can depend on the level of the applied vertical load. Using tabulated data, a designer can quickly verify whether the wall thickness for a normal temperature design, is acceptable under fire conditions.

3.1.1 Current status

The tables currently reported have been developed based on a significant number of experimental results made in accordance with former national test standards. These results, were confirmed by experience and theoretical evaluations of tests. According to Meyer (2013), the experimental results considered in the development of the current tabulated values, were mainly provided by Belgium, Germany and the United Kingdom. In the current version of EN 1996-1-2 (CEN 2005b, 2010b), no agreement could be reached about minimum wall thicknesses at the European level, so the tables in current Annex B contain ranges of values, covering experience from the countries involved in the standardization process. More details are reported by Schlundt & Meyer (2014).

3.2 Revision process

As mentioned, one of the goals of the revision of EN 1996-1-2 (CEN 2005b, 2010b) is to reduce the number of NDPs. For this reason, Project Team SC6.T2 will attempt to find agreement on minimum wall thicknesses at the European level. To reduce the range of values, the revision of tabulated data will focus only on non-loadbearing walls, since for this type of wall no global safety factors are reported. In this framework, Member States have been asked to provide explanations/background documents on the tabulated data currently reported in their National Annexes. For the time being, only results from tests carried out according to EN 1363-1 (CEN 2012b) and EN 1364-1 (CEN 2015) will be taken into account as reference data. The revised tabulated data will be presented in a new Annex A.

4 NEW ANNEX B ON THE INPUT PARAMETERS FOR DESIGN BY CALCULATION

4.1 Motivations

The current version of EN1996-1-2 (CEN 2005b, 2010b) offers only tabulated data as a reliable method for structural fire assessment. The current Annex C presents an assessment method of masonry piers based on the definition of parts of the cross-section with reduced and zero resistance due to the rise in temperature. The current Annex D presents some functions of temperature for the thermal and mechanical properties to be used in advanced calculation methods that are not defined. The content of these two Annexes is generally considered as not being reliable for design because of the absence of validation by experimental tests.

There is a need to replace the existing calculation method Annexes with a new Annex B to maintain consistency with the main text of EN1996-1-2 (CEN 2005b, 2010b) and parts 1-2 of the other Eurocodes. In particular, the possibility to follow a performance-based design approach needs to be preserved. On the other hand, although technical literature provides contributions to the prediction of the mechanical behavior of masonry panels subject to fire (Andreini & Sassu 2011), it has been recognized that currently there are no calculation methods which have been calibrated against full scale wall samples. Thus, the new Annex B is intended to provide an experimental procedure for the determination of the input parameters to be adopted in future advanced calculation models, which subsequently would require to be validated by experimental tests on wall panels in accordance with EN 1364 (CEN 2015) and EN 1365 (CEN 1999, 2013a). After such validation, the calculation models can be adopted to predict the thermo-mechanical behavior of more complex elements (e.g. walls with openings), parts or the entire building structure.

As a further useful tool, the new Annex B will provide pre-set stress-strain curves for some specific materials.

4.2 Thermo-mechanical response of masonry materials

As stated in the current version of the Annex D, advanced calculation methods for thermal and mechanical response should be based on the acknowledged principles and assumptions of the theory of heat transfer and structural mechanics, considering the changes of mechanical properties with temperature.

The thermal response model should include consideration of the relevant thermal actions specified in EN 1991-1-2 (CEN 2002b, 2013b) and the temperature dependent thermal properties of the materials.

The influence of moisture content and of migration of moisture within masonry may conservatively be neglected. The effect of non-uniform thermal exposure and of heat transfer to adjacent building components may be included where appropriate. The pre-set curves of thermal conductivity, specific heat and density as a function of temperature provided in the current Annex D for some given materials are expected to be reported in the new Annex B.

The values of a generic mechanical property X at a certain material temperature θ, is indicated with X_θ and it is given by the product $k_\theta X$, where k_θ is the reduction factor expressed as a function of θ.

In particular, referring to EN 1990 (CEN 2002a, 2005a, 2010a), the design values of the material strength, $X_{d,fi}$, is defined as follows:

$$X_{d,fi} = \frac{k_\theta X_k}{\gamma_{M,fi}} \tag{1}$$

where X_k is the characteristic value of the strength of the material for normal temperature design to EN 1996-1-1 (CEN 2012c) and $\gamma_{M,fi}$ is the partial safety factor for the relevant material property, for the fire situation. The effects of thermally induced strains and stresses both due to temperature rise and due to temperature differentials, should be considered.

According with Andreini et al. (2014, 2015), X_θ can be expressed as a second order polynomial function:

$$X_\theta = A_0 + A_1\theta + A_2\theta^2 \tag{2}$$

where A_0, A_1 and A_2, are parameters to be defined by a regression on the results of experimental tests carried out in accordance with the section 4.3 and 4.4.

The reduction factors k_θ can be, therefore, defined as:

$$k_\theta = 1 + \frac{A_1}{A_0}\theta + \frac{A_2}{A_0}\theta^2 \tag{3}$$

Still referring to Andreini et al. (2014), the stress-strain relationship in function of temperature is given by the following equation:

$$\sigma_d(\varepsilon, \theta, n_{mat}) = \frac{k_{fmat,k}(\theta)f_{mat,k}}{\gamma_{M,fi}} \times \frac{\varepsilon}{k_{\varepsilon mat,u}(\theta)\varepsilon_{mat,u}} \times \frac{n_{mat}}{n_{mat} - 1 + \left(\frac{\varepsilon}{k_{\varepsilon mat,u}(\theta)\varepsilon_{mat,u}}\right)^{n_{mat}}} \tag{4}$$

where $k_{fmat,k}(\theta)$ and $k_{\varepsilon mat,u}(\theta)$ are respectively the reduction factors k_θ for the characteristic compressive strength $f_{mat,k}$ and the ultimate strain $\varepsilon_{mat,u}$.

The parameter n_{mat} characterizes each type of material, regardless of the temperature, and can be determined by regression analysis of experimental test results. Pre-set values of the parameters n_{mat}, A_0, A_1 and A_2 for some given materials are reported in section 4.5.

As mentioned, the stress-strain relationship in Equation 4 can be used in calculation models of masonry members and the safety verification is carried out by checking that

$$E_{fi,d}(t) \leq R_{fi,t,d} \tag{5}$$

in which $E_{fi,d}$ is the design effect of actions for the fire situation, determined in accordance with EN 1991-1-2 (CEN 2002b, 2013b), including effects of thermal expansion and deformation, $R_{fi,t,d}$ is the corresponding design resistance in the fire situation and t is the designed duration of fire exposure.

4.3 Experimental procedure to determine the material stress-strain relationship as a function of temperature

A brief description of the experimental determination of the mechanical property functions of temperature is described in this paragraph. A more extensive description is reported by Andreini et al. (2015). The experimental tests are carried out on cylindrical samples having a diameter of 100 mm and a height of 200 mm, in accordance with EN 12390-1 (CEN 2012a). There have to be at least 3 samples for i-th value of the temperature θ_i, for which the mechanical property value X_θ has to be investigated.

Each sample is heated in accordance to the regime indicated in Table 1. Once the test temperature θ_i is achieved, the free thermal strain ε_{th} of the sample is measured and a compressive test having the same accuracy and load rates of EN 12390-3 (CEN 2019b) is performed, recording displacement and applied load, in a period of time short enough to limit the heat dispersion: the temperature decay on the lateral surface of the sample has not to exceed 5% of θ_i. From the applied load and the displacement readings, the related stress and strain values are evaluated.

4.4 Data treatment to determine the mechanical properties variation as a function of temperature

The determination of the parameters A_0, A_1 and A_2, and n_{mat} requires that the experimental tests described in paragraph 4.3 be executed at least at 2 different temperatures θ_i in addition to the room conditions (e.g. $\theta_1 = 20°C$, $\theta_2 = 300°C$ and $\theta_3 = 600°C$). From

Table 1. Heating regime for the experimental tests.

Temperature	Period of time hours	Temperature rate °C/min
From 20 to 100°C	0.5	2.67
100°C	2.0	0.00
From 100°C to the test temperature θ_i	1.5	Test temperature $\theta_i/90$ min
Test temperature θ_i	2.5	0.00

the stress-strain curve recorded for the j-th sample tested at the temperature θ_i, the maximum stress is identified as the compressive strength $f_{mat,\theta i,j}$ and the corresponding strain as the ultimate strain $\varepsilon_{mat,u,\theta i,j}$. The mean and the characteristic compressive strength and the ultimate strain at θ_i are given by the following formulae:

$$f_{mat,\theta i,m} = \frac{\sum_{j=1}^{n_s} f_{mat,\theta i,j}}{n_s}$$

$$f_{mat,\theta i,k} = f_{mat,\theta i,m} \exp\left(-\frac{1.64 SD_{\theta i}}{f_{mat,\theta i,m}}\right) \quad (6)$$

$$\varepsilon_{mat,u,\theta i,m} = \frac{\sum_{j=1}^{n_s} \varepsilon_{mat,u,\theta i,j}}{n_s}$$

where $n_s \geq 3$ is the number of samples tested at the temperature θ_i and $SD_{\theta i}$ is the standard deviation of the compressive strength readings $f_{mat,\theta i,j}$ given by:

$$SD_{\theta i} = \sqrt{\frac{\sum_{j=1}^{n_s}\left(f_{mat,\theta i,j} - f_{mat,\theta i,m}\right)^2}{n_s - 1}} \quad (7)$$

Taking the characteristic compressive strength as an example, the coefficients A_0, A_1 and A_2 can be determined by means of the least squares method for a quadratic regression, which is based on minimizing the minimum of the sum $S(A_0, A_1, A_2)$ given by:

$$S(A_0, A_1, A_2) = \sum_{i=1}^{n_\theta}\left[f_{mat,k,\theta i} - \left(A_0 + A_1\theta_i + A_2\theta_i^2\right)\right]^2 \quad (8)$$

In the same way, the coefficients A_0, A_1 and A_2 for the mean compressive strength and the ultimate strain can be determined. It is worth mentioning that the polynomial regression model shall have the intercept close to the room temperature value.

Analogously, the coefficient n_{mat} can be determined minimizing the sum $S(n_{mat})$ given by:

$$S(n_{mat}) = \sum_{i=1}^{n_\theta}\sum_{j=1}^{n_s}\left[\frac{\sigma_j}{f_{mat,\theta i,j}} - \frac{n_{mat}\frac{\varepsilon_j}{\varepsilon_{mat,u,\theta i,j}}}{n_{mat} - 1 + \left(\frac{\varepsilon_j}{\varepsilon_{mat,u,\theta i,j}}\right)^{n_{mat}}}\right]^2 \quad (9)$$

where n_θ is the number of temperatures investigated in the tests, σ_j and ε_j are respectively the stress and the strain from the readings of the j-th tested sample.

Table 2. Ratios A_1/A_0 and A_2/A_0 of Equation 3 for some masonry unit materials, tested by Andreini et al. (2014, 2015).

Mechanical Property	A_1/A_0	A_2/A_0
AAC with a density of 530 kg/m³		
$f_{mat,m}$	9.34E-04	-1.67E-06
$f_{mat,k}$	7.13E-04	-1.42E-06
$\varepsilon_{mat,u}$	1.12E-03	0.00E+00
LWC with a density of 1600 kg/m³		
$f_{mat,m}$	9.65E-05	-1.08E-06
$f_{mat,k}$	2.16E-04	-1.11E-06
$\varepsilon_{mat,u}$	6.02E-04	2.08E-06
LWC-LAP with a density of 1600 kg/m³		
$f_{mat,m}$	-5.58E-04	0.00E+00
$f_{mat,k}$	-6.91E-04	0.00E+00
$\varepsilon_{mat,u}$	2.27E-03	-1.19E-06
LWC-FV with a density of 1600 kg/m³		
$f_{mat,m}$	-2.34E-04	0.00E+00
$f_{mat,k}$	-8.79E-05	0.00E+00
$\varepsilon_{mat,u}$	1.23E-03	4.60E-07

4.5 Pre-set stress-strain curves for some given materials

The new Annex B is foreseen as providing some pre-set stress-strain curves for some specificmaterials. Such curves come from the experimental campaign carried out by Andreini et al. (2014, 2015) on about 200 samples, following the procedure described above in paragraphs 4.3 and 4.4. The materials covered are:

- Aerated Autoclaved Concrete (AAC) with a density of 530 kg/m³;
- Lightweight aggregate concrete (LWC) with a density of 1600 kg/m³;

Table 3. Ratios A_1/A_0 and A_2/A_0 of Equation 3 for some mortars, tested by Andreini et al. (2014, 2015).

Mechanical Property	A_1/A_0	A_2/A_0
Mortar M5		
$f_{mat,m}$	1.72E-03	-3.89E-06
$f_{mat,k}$	2.14E-03	-4.92E-06
$\varepsilon_{mat,u}$	6.11E-04	2.24E-06
Mortar M10		
$f_{mat,m}$	9.01E-04	-1.92E-06
$f_{mat,k}$	1.43E-03	-2.97E-06
$\varepsilon_{mat,u}$	2.50E-03	8.12E-07

Table 4. Shape parameter n_{mat} for masonry unit materials and mortars, tested by Andreini et al. (2014; 2015).

Material	n_{mat}
AAC	7.9
LWC	8.2
LWC-LAP	10.8
LWC-FV	9.1
Mortar M5	3.4
Mortar M10	5.6

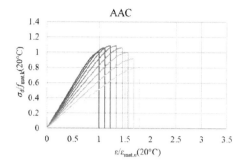

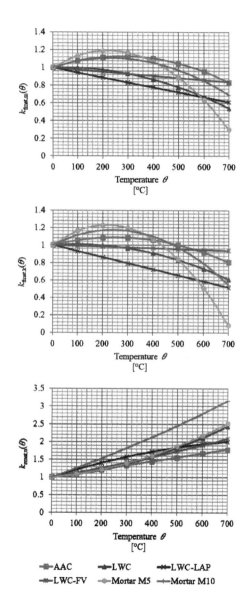

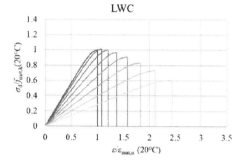

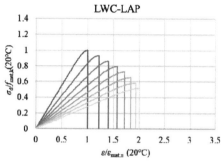

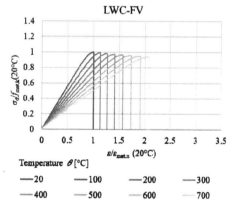

Figure 3. Stress-strain curves in function of temperature for some block materials, tested by Andreini et al. (2014, 2015).

Figure 2. Reduction factors $k_{fmat,m}(\theta)$, $k_{fmat,k}(\theta)$ and $k_{\varepsilon mat,u}(\theta)$ for some given materials, tested by Andreini et al. (2014, 2015).

– Lightweight aggregate concrete with Lapillus (LWC-LAP) with a density of 1832 kg/m³;
– Lightweight aggregate concrete (LWC-FV) for façades with a density of 2012 kg/m³;

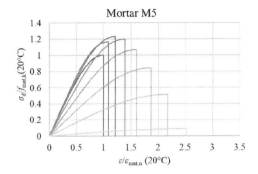

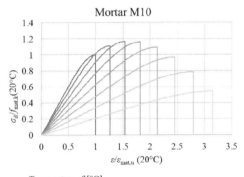

Temperature θ [°C]

——20   ——100   ——200   ——300
——400   ——500   ——600   ——700

Figure 4. Stress-strain curves in function of temperature for some mortars, tested by Andreini et al. (2014, 2015).

– Mortar M5 in conformity with EN 998-2 and EN 1015-11 (CEN 2016b, 2019a);
– Mortar M10 in conformity with EN 998-2 and EN 1015-11 (CEN 2016b, 2019a).

The values obtained by Andreini et al. (2014, 2015) for the coefficients A_1/A_0 and A_2/A_0 in Equation 3 for the mean compressive strength $f_{mat,m}$, the characteristic compressive strength $f_{mat,k}$ and the ultimate strain $\varepsilon_{mat,u}$ for the mentioned materials are indicated in Tables 2 and 3, and the graphs of the reduction factors as a function of temperature are shown in Figure 2. The values of n_{mat} for the same materials are given in Table 4 and the related stress-strain functions of temperature from Equation 4 are illustrated in Figures 3 and 4.

5 CONCLUSIONS

This paper presents the work of PT SC6.T2 of CEN/TC250 for the revision of the EN 1996-1-2 (CEN 2005b, 2010b). The potential reduction of some NDPs is described. The improvement of the background information concerning the tabulated values of fire resistance is also presented. Finally, the content of a new Annex B for the improvement

of the information on relevant material properties as a basis for a future calculation method is described.

ACKNOWLEDGEMENTS

The Authors kindly acknowledge the European Commission (EC), which mandated the European Committee for Standardization (CEN) for amending the existing Eurocodes and extending the scope of the structural Eurocodes (mandate M/515).

REFERENCES

Andreini, M. et al. 2015. Mechanical behavior of masonry materials at high temperatures. *Fire and Materials 39*(1): 41–57.

Andreini, M. et al. 2014. Stress-strain curves for masonry materials exposed to fire action. *Fire Safety Journal* 69.

Andreini, M. & Sassu, M. 2011. Mechanical behaviour of full unit masonry panels under fire action. *Fire Safety Journal*.

CEN. 1999. EN 1365-4:1999 - Fire resistance tests for loadbearing elements - Part 4: Columns.

CEN. 2002a. EN 1990:2002 - Eurocode - Basis of structural design.

CEN. 2002b. EN 1991-1-2:2002 - Eurocode 1: Actions on structures - Part 1-2: General actions - Actions on structures exposed to fire.

CEN. 2005a. EN 1990:2002/A1:2005 - Eurocode - Basis of structural design.

CEN. 2005b. EN 1996-1-2:2005 - Eurocode 6 - Design of masonry structures - Part 1-2: General rules - Structural fire design.

CEN. 2010a. EN 1990:2002/A1:2005/AC:2010 - Eurocode - Basis of structural design.

CEN. 2010b. EN 1996-1-2;2005/AC:2010 - Eurocode 6 - Design of masonry structures - Part 1-2: General rules - Structural fire design.

CEN. 2012a. EN 12390-1:2012 - Testing hardened concrete - Part 1: Shape, dimensions and other requirements for specimens and moulds.

CEN. 2012b. EN 1363-1:2012 - Fire resistance tests - Part 1: General Requirements.

CEN. 2012c. EN 1996-1-1:2005+A1:2012 - Eurocode 6 - Design of masonry structures - Part 1-1: General rules for reinforced and unreinforced masonry structures.

CEN. 2013a. EN 1365-1:2012/AC - Fire resistence tests for loadbearing elements - Part 1: Walls.

CEN. 2013b. EN 1991-1-2:2002/AC:2013 - Eurocode 1: Actions on structures - Part 1-2: General actions - Actions on structures exposed to fire.

CEN. 2015. EN 1364-1:2015 - Fire resistance tests for non-loadbearing elements - Part 1: Walls.

CEN. 2016a. EN 998-1:2016 - Specification for mortar for masonry - Part 1: Rendering and plastering mortar.

CEN. 2016b. EN 998-2:2016 - Specification for mortar for masonry - Part 2: Masonry mortar.

CEN. 2019a. EN 1015-11:2019 - Methods of test for mortar for masonry - Part 11: Determination of flexural and compressive strength of hardened mortar.

CEN. 2019b. EN 12390-3:2019 - Testing hardened concrete - Part 3: Compressive strength of test specimens.

Meyer, U. 2013. Fire resistance assessment of masonry structures - Structural fire design of masonry buildings according to the Eurocodes. *Mauerwerk 17*(3): 143–148.

Regulation (EU) No 305/2011 of the European Parliament and of the Council of 9 March 2011 laying down harmonised conditions for the marketing of construction products and repealing Council Directive 89/106/EEC. (2011). Official Journal of the European Union.

Schlundt, A. & Meyer, U. 2014. Structural design of masonry for fire resistance according to Eurocode 6/ Brandschutzbemessung von Mauerwerk nach Eurocode 6. *Mauerwerk 18*(3–4): 258–266.

Brick and Block Masonry - From Historical to Sustainable Masonry –
Kubica, Kwiecień & Bednarz (eds)
© 2020 Taylor & Francis Group, London, ISBN 978-0-367-56586-2

Pultruded-FRP for retrofitting purposes: Mechanical characterization

O. Tamborrino & M.A. Aiello
Department of Engineering for Innovation, University of Salento, Lecce, Italy

C.R. Passerino
Fibre Net S.P.A., Pavia di Udine, Italy

ABSTRACT: Unreinforced masonry structures have proven to be vulnerable during seismic events. Therefore, provisional or permanent interventions are often needed in post-seismic scenarios. In response to this request, a large interest has been given to fiber-reinforced polymers (FRPs) in the recent past. Among all the use of pultruded FRPs is becoming an attractive solution for realizing structural profiles for civil engineering applications. In fact, they guarantee structural reliability and efficiency under different loading conditions. As a consequence, the issue related to the mechanical characterization of pultruded FRPs is a key-aspect for their proper design. The present paper aims to report the testing procedure for the determination of the main properties of glass pultruded specimens: inter-laminar shear, flexural, pin-bearing and tensile strength, as well as, tensile elastic modulus. This paper discusses the influence of some parameters on the obtained results and the relevance of the investigated properties in relation to the design requirements and the available standards and codes.

1 INTRODUCTION

Polymer composites were first developed during the 1940s for military and aerospace applications. Afterward, the global production speedily increased, reaching the current development in the late 1960s.

Interest in the use of FRP materials for structures has been increasing steadily since the 1990s. This increase in usage is chiefly attributed to the numerous advantages that FRPs have over conventional materials such as reinforced concrete and structural steelwork. In particular, FRP pultruded profiles (Figure 1) present several advantages, such as high strength-to-weight ratio, lightness, electromagnetic neutrality, ease of installation and improved durability under aggressive conditions (Correia et al. 2009). However, there are drawbacks preventing FRP composites to be fully embraced for structural applications, notably the relatively high production and material costs, the behavior of the material under fire conditions and most crucially the lack of consensual or "official" design codes and guidelines that are easily accessible and readily applicable by engineers. Some studies and technical solutions have been assessed also in the field of the retrofitting of existing masonry structures through FRP pultruded profiles. Particularly in seismic areas, ancillary FRP structures represent an effective solution, not yet sufficiently explored, which improve the structural performance with a very small mass addition. These frames are adjacent to the masonry structure and connected to it with mechanical fasteners. The numerical results evidenced increased in-plane strength and stiffness, as well as, a more dissipative failure mode

(Casalegno et al. 2017, Cecchi et al. 2017). A well-known tested structure of this type was built in order to temporarily protect the church of Santa Maria Paganica in L'Aquila (Figure 2) that was badly damaged during the 2009 earthquake (Boscato et al. 2015). This structure was widely studied from the point of view of seismic behavior.

In this paper a wide experimental campaign aiming to investigate the influence of relevant parameters (profile geometry, profile stratigraphy, resin properties) on the mechanical properties of pultruded profiles, is reported and discussed. Direct tension, three-point flexural, pin-bearing and inter-laminar shear tests were performed according to recommendations provided by UNI EN 13706-3:2003. In addition, some considerations regarding the implications of the mechanical properties evaluation and the design requirements are reported referring to the provisions of Italian Guideline (CNR-DT 205/2007). Therefore, the purpose of the paper is to evaluate the influence of different characteristic material parameters on its mechanical properties and to explore the possible implications in structural design.

2 MATERIALS AND METHODS

An extended experimental campaign was conducted on numerous samples extracted from web parts of different GFRP pultruded materials, having a typical stratigraphy of the pultruded section as illustrated in Figure 3. The reinforcement in the longitudinal/

Figure 1. FRP pultruded profiles.

Figure 2. FRP structure in Santa Maria Paganica (L'Aquila).

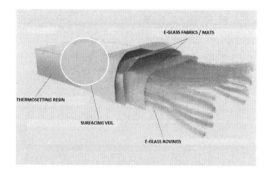

Figure 3. Pultruded section stratigraphy.

Table 1. Components properties.

	Tensile elastic modulus GPa	Tensile strength MPa	Tensile failure strain %	Density
Rovings	81	2600	4.9	9600 g/km
Fabrics	81	2600	4.9	800 g/m²
Mats	81	2600	4.9	450 g/m²
Resin	3.9	75	2.0	

different process parameters: (i) profile geometry; (ii) profile stratigraphy; (iii) resin properties.

In particular:

(i)   the samples were extracted from three different profiles, with three different web thicknesses (5, 10 and 15 mm);

(ii) each material is made from a variable number of rovings, fabrics and mats, in order to vary the fiber contents in the various directions;

(iii) the employed resin is always the same but the content of its two kinds of fillers was changed.

These parameters were varied to obtain 18 different materials in order to analyze their influence on the mechanical properties. Another scope was to define a stratigraphy which could guarantee a good mechanical behavior.

In Table 2 a synthesis of all tested materials is reported, indicating with:

- Id, the univocal denomination given to each material;
- th [mm], the web thickness;

Table 2. Pultruded materials parameters.

Id	th	A_0	A_{90}	$A_{\pm45}$	A_{ran}	C_1	C_2
M1	15	36.4	0.1	2.1	5.4	12.2	0.0
M21	10	24.4	2.2	2.3	6.1	9.8	9.8
M22	5	36.7	0.0	0.0	5.0	7.3	7.3
M25	10	33.3	0.2	3.4	8.7	4.9	4.9
M26	10	43.9	0.0	0.0	4.6	9.8	9.8
M27	10	34.3	2.2	2.3	6.1	9.8	9.8
M32	5	33.6	0.0	0.0	5.0	4.9	4.9
M33	5	33.2	1.0	1.1	5.0	4.9	4.9
M34	5	29.7	2.1	2.1	5.0	4.9	4.9
M36	15	34.8	0.0	0.0	3.8	12.2	12.2
M37	15	34.8	0.0	0.0	5.8	12.2	12.2
M38	15	36.3	1.4	1.4	5.8	12.2	12.2
M39	15	36.3	1.4	1.4	5.8	12.2	12.2
M40	15	36.3	1.4	1.4	3.8	12.2	12.2
M43	10	30.3	2.2	2.3	6.1	4.9	4.9
M44	10	30.3	2.2	2.3	6.1	4.9	4.9
M47	10	24.4	2.2	2.3	6.1	4.9	4.9
M48	10	20.8	2.2	2.3	6.1	4.9	4.9

pultrusion direction is given by the rovings; the reinforcement in the other directions is given by fabrics and mats; while the surfacing veil protects the section from chemical attacks, UV rays, etc.

The tested materials, the apparatus and test methods adopted are described in the following subsections.

2.1   Materials

The pultruded GFRP C-profiles used in the experimental program, supplied by Fibre Net S.P.A., are made of an isophtalic polyester matrix reinforced with rovings, fabrics and mats made of E-glass fibers. Physical and mechanical properties of these components are reported in Table 1. Numerous types of GFRP material have been produced varying three

- A_0 [%], fiber volume content in longitudinal/pultrusion direction (given from rovings);
- A_{90} [%], fiber volume content in transversal direction (given from fabrics);
- $A_{\pm45}$ [%], fiber volume content in 45° direction (given from fabrics);
- A_{ran} [%], fiber volume content in random direction (given from mats);
- C_1 [%], inert weight content of the resin;
- C_2 [%], fire retardant weight content of the resin.

2.2 Experimental program

The mechanical characterization of an orthotropic material involves the determination of its elastic constants and strength when subjected to different actions and in different directions. Experimental tests were carried out on small-scale specimens in order to determine the most relevant mechanical properties of the GFRP material in longitudinal/pultrusion and in transversal direction. This experimental study comprised inter-laminar shear, flexural, pin-bearing and tensile tests on GFRP coupons with rovings orientation on the longitudinal direction (to determine the longitudinal properties) and with rovings orientation on the transversal direction (to determine the transversal properties); whose dimensions are reported in Table 3.

For each similar test, a minimum number of 5 specimens was utilized.

2.3 Material characterization tests

Transversal tensile tests on 5 and 10 mm, all flexural tests and all inter-laminar shear tests were executed

Table 3. Specimens dimensions.

Determined property	Specimens dimensions (l × b)		
	th: 5 mm	th: 10 mm	th: 15 mm
Tensile modulus-axial	250 × 25 mm	250 × 25 mm	500 × 50 mm
Tensile modulus-transversal	170 × 25 mm	170 × 25 mm	250 × 50 mm
Tensile strength-axial	250 × 25 mm	250 × 25 mm	500 × 50 mm
Tensile strength-transversal	170 × 25 mm	170 × 25 mm	250 × 50 mm
Pin-bearing strength-axial	220 × 45 mm	450 × 90 mm	670 × 135 mm
Pin-bearing strength-transversal	170 × 45 mm	170 × 90 mm	250 × 135 mm
Flexural strength-axial	150 × 15 mm	300 × 15 mm	450 × 30 mm
Flexural strength-transversal	150 × 15 mm	170 × 15 mm	250 × 30 mm
Inter-laminar shear strength-axial	50 × 25 mm	100 × 50 mm	150 × 75 mm

Table 4. Test speeds adopted.

Test	Test speed [mm/min]		
	th = 5 mm	th = 10mm	th = 15mm
Tensile	2.0	2.0	2.0
Flexural	2.0	5.0	10.0
Int. shear	1.0	1.0	1.0
Pin-bearing	1.0	1.0	1.0

by a universal machine with displacement control and capacity of 25 kN. The other tensile tests (transversal on 15 mm and longitudinal) and all pin-bearing tests were executed by another machine with displacement control and capacity of 600 kN. Furthermore, other tensile tests, in order to compare some obtained results, were made by a universal machine with displacement control and capacity of 100 kN. For tensile tests, the deformations were measured by a contact extensometer. The test speeds (displacement ratio) adopted for each test are reported in Table 4.

Tensile tests
Tensile tests, performed according to UNI EN ISO 527-4:1999, allowed determining the tensile strength and the tensile elastic modulus in longitudinal ($\sigma_{tu,x}$; $E_{t,x}$) and in transversal direction ($\sigma_{tu,y}$; $E_{t,y}$).
Tensile strength was determined as:

$$\sigma_{tu} = F_u/bh \ [MPa] \tag{1}$$

with:
F_u [N] ultimate load;
b [mm] specimen width;
h [mm] specimen thickness.
Tensile elastic modulus should be determined as secant modulus between two values of deformation respectively equal to 0.0005 and 0.0025. However, these values generally fall in the very initial test stage, where the influence of the setup stabilization may affect the results. For this reason, the calculation was made between two higher values of deformation, respectively corresponding to 10% and 40% of ultimate tension. To assess the reliability of this method, further specific tensile tests were carried out using another universal machine, which is not affected by the problems of initial stabilization and which allows the elastic modulus to be calculated according to the standards. The results are almost close to those obtained using the modified method, which has therefore been deemed suitable for the purpose.

Flexural tests
Three-point bending tests, performed according to UNI EN ISO 14125:2011, allowed determining the flexural strength in longitudinal ($\sigma_{fu,x}$) and in

269

transversal direction ($\sigma_{fu,y}$). The sample has a length such as to avoid failures due to inter-laminar shear. It is placed on two supports (radius 5 mm), of adequate distance ($L \geq 20h$), and loaded in the center by means of a loading pin (radius 5 mm) until failure.

Flexural strength was determined as:

$$\sigma_{fu} = 3F_uL/2bh^2 \text{ [MPa]} \qquad (2)$$

with:

F_u [N] ultimate load;
$L = 20h$ [mm] span length;
b [mm] specimen width;
h [mm] specimen thickness.

Inter-laminar shear tests

Inter-laminar shear tests, performed according to UNI EN ISO 14130:1999, are three-point bending tests which allowed determining the inter-laminar shear strength in longitudinal direction (τ). This property constitutes a measure of the shear strength of the matrix layer, located between the reinforcing fibers.

The sample has a length such as to allow failures due to simple or multiple inter-laminar shear. It is placed on two supports (radius 2 mm), of adequate distance ($L \leq 5h$), and loaded in the center by means of a loading pin (radius 5 mm) up to the failure.

Inter-laminar shear strength was determined as:

$$\tau = 3F_u/4bh \text{ [MPa]} \qquad (3)$$

with:

F_u [N] ultimate load;
b [mm] specimen width;
h [mm] specimen thickness.

Pin-bearing tests

Pin-bearing tests, performed according to Annex D – UNI EN 13706-2:2003, allowed determining the pin-bearing strength in longitudinal ($\sigma_{p,x}$) and in transversal direction ($\sigma_{p,y}$). All specimens have a circular hole, positioned as indicated in Figure 4a, where a metallic pin is inserted. The specimen is inserted between two metallic sheets and is subjected to tensile test (Figure 4b).

Pin-bearing strength was determined as:

$$\sigma_p = F_u/hd \text{ [MPa]} \qquad (4)$$

with:

F_u [N] ultimate load;
d [mm] pin diameter;
h [mm] specimen thickness.

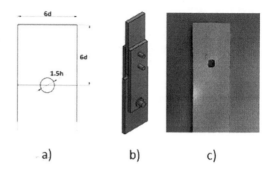

a)   b)   c)

Figure 4. Pin-bearing test: a) specimen dimensions; b) specimen set-up; c) failure M39.

3 RESULTS AND DISCUSSION

This section presents and discusses the results of the experimental investigation.

3.1 *Mechanical behavior and failure modes*

Tensile tests

Figure 5 plots an example of stress-strain curves, which were obtained based on the measurements of the applied load and the axial deformation. All specimens exhibited linear-elastic behavior up to failure, with a visible and well defined ultimate stress. In all tests, tensile failure was brittle and observed in a region of the specimen sufficiently far from the grips, therefore being considered valid (Figure 6).

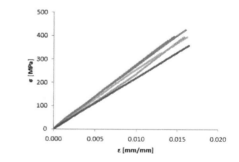

Figure 5. Longitudinal tensile stress-strain curves: M32.

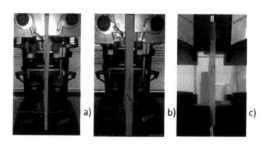

Figure 6. Transversal tensile failures: a) M33; b) M27; c) M40.

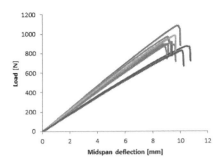

Figure 7. Longitudinal flexural load-deflection curves: M32.

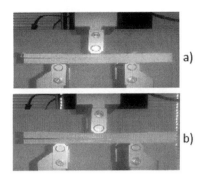

Figure 9. Simple shear failures: a) M21; b) M25.

Flexural tests

Figure 7 plots an example of load-deflection curves, which were obtained based on the measurements of the applied load and the crossbar displacement. All specimens exhibited linear-elastic behavior up to failure, which is a typical characteristic of the GFRP material, when subjected to this type of mechanical loading. Failures occurred in three different valid modes: tensile fracture at outermost layer (Figure 8a); tensile fracture including inter-laminar shear (Figure 8b); compressive fracture (Figure 8c).

Inter-laminar shear tests

In all tests, failures occurred for simple shear (Figure 9).

Pin-bearing tests

Failures occurred for typical pin-bearing rupture or for tensile rupture. Only the first was considered as valid (Figure 4c).

3.2 Mechanical properties

Tables 5 and 6 present a summary of the mechanical properties obtained in the tests, namely: tensile elastic modulus longitudinal ($E_{t,x}$) and transversal ($E_{t,y}$);

Table 5. Summary of results (with CoV): longitudinal properties [MPa].

	$E_{t,x}$	$\sigma_{tu,x}$	$\sigma_{fu,x}$	τ	$\sigma_{p,x}$
M1				22.6 (7.4%)	
M21	21,200 (7.7%)		299.6 (11.4%)	16.4 (5.2%)	210.9 (9.3%)
M25	26,900 (9.3%)	370.7 (4.3%)	419.5 (4.8%)	18.2 (6.5%)	249.7 (4.2%)
M27	28,700 (10.2%)		387.9 (4.5%)	15.4 (6.4%)	255.3 (7.8%)
M32	30,300 (11.3%)	357.8 (7.5%)	407.1 (7.6%)	35.5 (5.4%)	267.3 (8.7%)
M34	28,800 (8.8%)	334.3 (9.1%)	257.7 (14.4%)	27.9 (5.4%)	233.4 (12.6%)
M37		396.6 (2.3%)			281.3 (5.4%)
M39		384.4 (3.6%)			297.2 (10.5%)

tensile strength longitudinal ($\sigma_{tu,x}$) and transversal ($\sigma_{tu,y}$); flexural strength longitudinal ($\sigma_{fu,x}$) and transversal ($\sigma_{fu,y}$); inter-laminar shear strength longitudinal (τ); pin-bearing strength longitudinal ($\sigma_{p,x}$) and transversal ($\sigma_{p,y}$). In particular, the average values of modulus and the minimum values of strengths are reported.

On the basis of the reported results the following considerations can be made:

-   transversal flexural strength $\sigma_{fu,y}$ is always higher than transversal tensile strength $\sigma_{tu,y}$ (Figure 10). Such differences, which are typical in specimens derived from GFRP pultruded profiles, but are not observed in homogeneous and isotropic materials (such as steel), are due to the layered-wise architecture of the GFRP material, with different types of reinforcements (rovings, fabrics and mats), positioned in successive layers;
-   the low inter-laminar shear strength of some materials (with thickness 10 or 15 mm) is

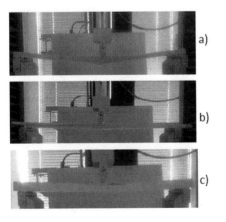

Figure 8. Flexural failures: a) tension (M25); b) tension+ shear (M21); c) compression (M27).

271

Table 6. Summary of results (with CoV): transversal properties [MPa].

	$E_{t,y}$	$\sigma_{tu,y}$	$\sigma_{fu,y}$	$\sigma_{p,y}$
M21	4700 (14.9%)	59.7 (5.8%)		156.7 (6.5%)
M22	3400 (22.9%)	34.0 (6.1%)		
M25	4300 (12.4%)	38.0 (9.4%)	97.9 (7.7%)	166.6 (5.9%)
M26	2300 (22.0%)	22.8 (9.4%)		
M27	5700 (11.6%)	52.3 (6.3%)	88.2 (6.5%)	162.0 (3.9%)
M32	3400 (8.6%)	32.8 (15.3%)	103.2 (12.5%)	
M33	3700 (10.3%)	55.5 (4.3%)	163.9 (5.7%)	
M34	5900 (3.9%)	72.5 (7.6%)	130.8 (9.1%)	
M36		23.7 (3.8%)	46.9 (13.4%)	
M37		31.9 (5.2%)	77.3 (5.9%)	
M38		46.8 (4.4%)		
M39		48.7 (4.7%)	87.3 (8.4%)	
M40		40.8 (6.0%)	56.0 (7.3%)	
M43	4800 (15.1%)	74.9 (3.0%)		
M44	5200 (27.1%)	67.2 (5.1%)		
M47	5200 (8.6%)	62.2 (8.5%)		
M48	5500 (27.3%)	73.7 (13.8%)		

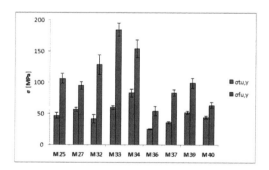

Figure 10. Comparison between $\sigma_{tu,y}$ and $\sigma_{fu,y}$ (average values).

probably due to the presence of an intermediate layer composed of mats and fabrics, that becomes the weakest part of the sample. This problem underlines the need to optimize the

material configuration, for instance by arranging mats and fabrics only in external layers;
- there is a clear difference in the distribution of the results obtained between the longitudinal and transversal properties. In fact, for the latter, coefficients of variation in some cases higher than 15% have been obtained, especially for tensile characteristics. This trend is typical of those properties more dependent on the matrix, such as the transversal one with respect to the longitudinal ones.

3.3 Mechanical properties: Influence of analyzed parameters

In the previous subsection, the influence of the loading type conditions on the mechanical properties of GFRP pultruded profiles was analyzed. But, it's also necessary to study what are the parameters (Tab. 2) which significantly affect these properties.

For each property, except $\sigma_{p,y}$ (of which only 3 results were obtained) a correlation was made with the analyzed parameters. In particular:

- longitudinal mechanical properties have been correlated to A_0 and transversal ones to A_{90};
- the parameter C, as the sum of C_1 and C_2 has been introduced indicating the total quantity of fillers into the resin.

In Figure 11 the linear correlation coefficients R^2 obtained are reported. All properties have been correlated to one parameter (A_0 or A_{90}: A_0/A_{90}, th and C).

Analyzing these graphs, it's possible to observe that:

- tensile and pin-bearing properties have a good correlation ($R^2 > 75\%$) to the fiber content in the specific direction (A_0 or A_{90}), while flexural and shear properties have a scarce correlation to these parameters, in fact, as mentioned in the previous subsection, these properties are largely influenced by the material configuration;
- the specimen thickness (th) does not affect the mechanical properties analyzed, given that, for

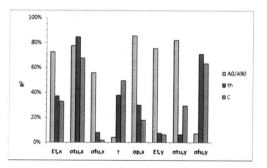

Figure 11. Correlation mechanical properties vs A_0/A_{90}, th and C.

the thicknesses analyzed, there is no great variation in the quantity of fiber as the thickness changes;
- any property presents a good correlation to resin fillers content C. The fillers are generally used to reduce the quantity of resin saving money. Within the range of filler content utilized it seems that the effect may be neglected.

3.4 Design requirements (CNR DT-205/2007)

The growing demand for structural solutions with elements of pultruded composite material led, in 2003, to the national implementation of the regulations European EN 13706-1, EN 13706-2 and EN 13706-3, which define the minimum requirements necessary to classify pultruded profiles as "structural". It is evident how the drafting of an instruction document for the design, execution and control of structures built with FRP profiles could no longer be postponed. The Italian Guideline (CNR DT-205/ 2007) is intended to satisfy this requirement limited to the case of glass fibers (GFRP), for which the scientific bibliography is able to provide consolidated and widely shared answers.

The design approach is in accordance with semi-probabilistic limit state method.

As an example, flexural resistance and deformability verifications for a C-shaped GFRP profile (with web and flange thicknesses equal to 15 mm) are reported. It's highlighted how the outcome of these verifications varies as parameter A_0 varies, which, as seen in the previous subsection, is the most influent parameter on the tensile longitudinal strength $\sigma_{tu,x}$ and tensile longitudinal elastic modulus $E_{t,x}$.

Flexural resistance

A simply supported beam ($L = 6.5$ m) with a distributed load ($q = 6.5$ kN/m) is considered. The flexural verification is satisfied if $M_{rd}/M_{sd} \geq 1$, with:

M_{sd} design bending moment;

$M_{rd} = \min (W^*f_{t,d}; W^*f_{c,d})$ resistance bending moment.

W is the resistance modulus of the section, while $f_{t,d}$ and $f_{c,d}$ are the design tensile and compressive strengths, calculated dividing the characteristics strengths for the partial safety factors.

In particular, compressive strength was assumed equal to the tensile one.

In Figure 12 the progress of the flexural check is reported as a function of A_0 (in a range 25-35%). For the verification to be satisfied, A_0 must be at least 30%.

Deformability

A simply supported beam ($L = 3.0$ m) with a distributed load ($q = 5.5$ kN/m) is considered. The deformability verification is satisfied if $\delta_{sd}/\delta_{max} \leq 1$,

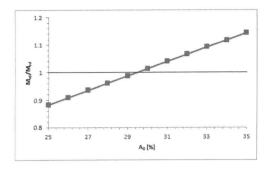

Figure 12. Flexural verification progress as a function of A_0.

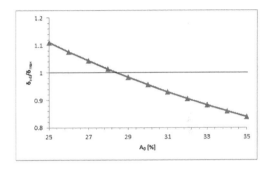

Figure 13. Deformability verification progress as a function of A_0.

with:

δ_{sd} design displacement, calculated as the sum of flexural and shear displacement, considering the reducing viscosity effect ($t = 50$ years) on tensile elastic modulus $E_{t,x}$ and shear modulus G; $\delta_{max} = L/500$, maximum displacement.

According to CNR Guideline, G is assumed equal to $0.12*E_{t,x}$.

In Figure 13 the progress of the deformability check is reported as a function of A_0 (in a range 25-35%). For the verification to be satisfied, A_0 must be at least 28.5%.

3.5 Technical design: Connections

The experimental analyses of the beam-column connections highlight the greater dissipation capacity of the bolted joint compared with the adhesive joint. In particular, the design of steel-bolted joints with GFRP pultruded plates is very frequent.

Experimental and numerical studies demonstrated that the plates have considerable influence on the moment-rotation response of joints so that the joint's failure mode involves mainly the bolted plate. For this reason, ultimate limit states design is controlled by local failure of the GFRP pultruded plate. So, the design of beam-column joints with a higher rotational stiffness can be considered as an effective way to improve the global dynamic behavior (Boscato et al. 2015).

Even if at the design phase, it's necessary to consider all the possible crisis including structural elements and their connections, in case of higher dissipation requirements the joints failure is expected to occur firstly.

Since the requirements regarding the ductility a pin-bearing crisis mode is preferred for the joint; therefore at the design phase the pin-bearing strength of the pultruded material constituting the plate becomes very important. In particular, as seen in subsection 3.2, there is a significant difference between the longitudinal and transversal pin-bearing strength, hence the effect of the orientation of the fibers in the pultruded plate (Feroldi et al. 2016).

4 CONCLUSIONS

This paper reported the available experimental results on the mechanical properties of GFRP structural elements. The specimens used in this work were extracted from web parts of different C-shaped GFRP profiles, made by a polymeric resin reinforced with E-glass (rovings, fabrics and mats). All materials were obtained varying the following three parameters: (i) profile geometry; (ii) profile stratigraphy; (iii) resin properties. Therefore, experimental tests were carried out on small-scale specimens in order to determine the mechanical properties in longitudinal/pultrusion and in transversal direction and, the effect of the mentioned parameters has been investigated. In addition to the influence of the studied parameters on the mechanical properties of the pultruded profiles, their influence on some common checks required during the design phase of GFRP structures was also analyzed.

REFERENCES

Correia, J. R. et al. 2009. Material characterization and structural behaviour of GFRP pultruded profiles. *IABSE Symposium Report* 96 (17): 34–41.
Casalegno, C. et al. 2017. Preliminary numerical analysis of a masonry panel reinforced with pultruded GFRP profiles. *Materials Science Forum* 902: 20–25.
Cecchi, A. et al. 2017. Preliminary investigation on FRP profiles for the structural retrofit of masonry structures. *Key Engineering Materials* 747: 77–84.
Boscato, G. & Russo, S. 2015. Dynamic parameters of pultruded GFRP structures for seismic protection of historical building heritage. *Key Engineering Materials* 624: 461–469.
UNI EN 13706-3:2003. Reinforced plastics composites – Specifications for pultruded profiles – Part 3: Specific requirements.
CNR-DT 205/2007. Istruzioni per la Progettazione, l'Esecuzione ed il Controllo di Strutture realizzate con Profili Pultrusi di Materiale Composito Fibrorinforzato (FRP).
UNI EN ISO 527–4:1999. Plastics – Determination of tensile properties – Part 4: Test conditions for isotropic and orthotropic fibre-reinforced plastic composites.
UNI EN ISO 14125:2011. Fibre-reinforced plastic composites – Determination of flexural properties.
UNI EN ISO 14130:1999. Fibre-reinforced plastic composites – Determination of apparent interlaminar shear strength by short-beam method.
UNI EN 13706-2:2003. Reinforced plastics composites – Specifications for pultruded profiles – Part 2: Methods of test and general requirements.
Feroldi, F. & Russo, S. 2016. Structural behavior of all-FRP beam-column plate-bolted joints. *Journal of Composites for Construction* 20(4): 04016004.

Brick and Block Masonry - From Historical to Sustainable Masonry –
Kubica, Kwiecień & Bednarz (eds)
© 2020 Taylor & Francis Group, London, ISBN 978-0-367-56586-2

The revision of EN 1996-2

J.J. Roberts

Emeritus Professor of Civil Engineering, Kingston University and John Roberts Consultancy, UK

ABSTRACT: This paper looks at the revision of EN 1996-2:2006 (Eurocode 6: Design of masonry structures – Part 2: Design considerations, selection of materials and execution of masonry) by Project Team 3 of CEN TC 250/SC6. Some of the key areas identified in the review of the document by National Standards Bodies in 2014 are noted. The Project Team has reviewed twenty-four National Annexes to EN 1996-2 as part of its work and used the results to inform the proposed revisions to the code. The current draft reduces the number of Nationally Determined Parameters from two to one and recommends the retention of the three annexes as Informative.

1 INTRODUCTION

Under a contract issued by NEN through CEN TC250 on behalf of the European commission a project team has been established to review the current version of EN 1996-2:2006 (January 2006 + AC September 2009 Eurocode 6: Design of masonry structures – Part 2: Design considerations, selection of materials and execution of masonry), respond to comments raised during the 2014 review of the Eurocode and update the document to reflect current practice. Technically this is Project Team T3 (PT3) of CEN TC250/SC6/WG4. It is important to recognize that Part 2 of Eurocode 6 addresses only those requirements that the designer needs to take into account when carrying out design to EN 1996-1-1 and it is not intended to be a comprehensive guide to the construction of masonry.

2 PROGRESS TO DATE

The requirements for the contract included the following:

a) Review the contents of all Countries' National Annexes and key supporting documents provided to the Project Team. Following guidance provided by CEN/TC 250, agree NDPs to consider for detailed review with the relevant SC/WG/HG. Develop proposals to reduce the number of NDPs and/or enable better consensus on values adopted by Countries to be achieved. Incorporate those proposals agreed with the relevant SC/WG/HG into task deliverables.
b) Enhance ease of use by improving clarity, simplifying routes through the Eurocode, avoiding or removing rules of little practical use in design and avoiding additional and/or empirical rules for particular structure or structural-element types, all to the extent that it can be technically justified whilst safeguarding the core of essential technical requirements.
c) Since finalising the current EC 6 an enormous development with a lot of innovations has happened and it is the time to check if new provisions for execution are needed and the current ones should be reviewed. Harmonisation with part 1-1, 2 and 3.

The over-riding timetable for the revision is set by the contract document. The timetable for the Project Team is shown in Table 1.

To date PT3 has reviewed the document as required and produced a new version which was considered by CEN C250/WG4 on the 1st July 2019. PT3 met to discuss the points raised by WG4 and has incorporated them, where appropriate, into a revised draft of Part 2. This latest draft incorporates the changes required to comply with the new format for Eurocodes which was agreed by CEN TC 250 and described in document CEN/TC 250 N1250.

3 REVIEW OF THE ANNEXES

EN 1996-2 contains three annexes each of which are Informative Annexes and their use in a particular state is therefore determined by each state. The three Annexes are:

a) Annex A
   Classification of micro conditions of exposure of completed masonry.
b) Annex B
   Acceptable specifications of masonry units and mortar for durable masonry in various exposure conditions.
c) Annex C
   Selection of material and corrosion protection specifications for ancillary components according to exposure class.

Table 1. Timetable for the project team.

Date	Activity
30/10/2018	Kick off meeting London
11/12/2018	Videoconference 1
31/01/2019	Videoconference 2
28/02/2019	Meeting 2 Brussels
08/04/2019	Videoconference 3
30/04/2019	Deliver First Draft to NEN
01/07/2019	Receive review comments from SC6/WG4
17/10/2019	Meeting 3 Brussels
18/12/2019	Videoconference 4
31/02/2020	Complete Second Draft taking into account WG4 comments
30/04/2020	Deliver Second Draft to NEN
31/06/2020	Receive review comments from SC6/WG4
30/09/2020	Complete Final Draft taking into account WG4 comments
31/10/2020	Deliver Final Draft to NEN
31/01/2021	End of Formal Enquiry
31/03/2021	Preparation of final documents taking into account Enquiry
30/04/2021	Delivery of Final Standard and Background docs to NEN
31/12/2021	Deal with enquiries etc
28/02/2022	PT stands down

Table 2. National use of the annexes.

Country	A	B	C
UK	P	No	No
Germany	P	No	No
France	P	P	P
Italy	P	P	P
Finland	P	No	No
Denmark	P	P	No
Netherlands	P	No	No
Belgium	P	P	P
Austria	P	P	P
Poland	P	P	P
Ireland	P	No	No
Iceland	P	P	P
Greece	P	P	P
Bulgaria	P	P	P
Sweden	P	P	P
Czech	P	P	P
Cyprus	P	P	P
Switzerland	P	P	P
Luxembourg	P	P	P
Estonia	P	P	P
Croatia	P	P	P
Lithuania	P	P	P
Romania	P	P	P
Slovenia	P	P	P

where: P means the use of the Annex is permitted
No means the use of the Annex is not pemitted.

National Annexes from 24 member states were made available to PT3 and these were reviewed in detail to identify the acceptability and use of the Annexes. The result of the review is shown in Table 2. It was found that a number of countries had not included a definitive statement regarding the applicability of each National Annex and it is assumed that the use of the three Annexes was therefore accepted in that country.

In the case of Annex A, dealing with micro conditions of exposure of completed masonry, all the National Annexes accepted the use of Annex A. Only one country, the Netherlands, indicated that Annex A should be read as Normative.

In response to comments the definition of exposure condition MX4 has been extended from:

MX4 - Exposed to saturated salt air or seawater; to

MX4 - Exposed to saturated salt air seawater or deicing salts.

The situation with Annex B was more diverse with five countries not accepting the use of the Annex. These countries are the UK, Germany, Finland, Netherlands and Ireland. A closer examination of the requirements for these five countries showed wide disparity in the way that acceptable combinations of masonry unit and mortar for a given exposure condition were specified. This is not surprising given the wide variations in the types of unit, construction practice and weather conditions in different member states. At the moment the masonry unit standards rely on experience in the place of use to establish the suitability of a masonry unit in a given location but a freeze thaw durability test has now been published for clay masonry units and other test methods are being reviewed for most of the remaining unit types. No comparable test has been established for mortar. Given the current availability of test methods PT3 has decided that it is not yet possible to further develop Annex B in a way which would make it more widely acceptable. Annex B has therefore been updated and made a little easier to use.

In the case of Annex C six countries do not accept the use of the Annex, namely, UK, Germany, Finland, Denmark, Netherlands and Ireland. Again, it comes down to differences in construction practice and exposure conditions and the PT felt that the best it could achieve was to update the current requirements. The Tables in this Annex have therefore been updated to reflect current standards and best practice.

4 PERMISSIBLE DEVIATIONS

Table 3 and Figure 3.1 of EN 1996-2 show permissible deviations in masonry. Some of the comments received on the Code questioned whether these requirements are consistent with the requirements in 7.3 of EN 1996-1-1. It was noted during the review process that six National Annexes have permissible deviations

Table 3. Permissible deviations for masonry walls.

Position	
Verticality	Maximum deviation
in any one storey	± 20 mm
in total height of building of three storeys or more [*]	± 50 mm or ± h_{tot}/250 whichever is the lesser
vertical alignment	± 20 mm or ± 10% of the wall thickness of the storey below whichever is the lesser
Straightness/flatness [**] in any one metre	± 10 mm
in 10 metres	± 40 mm
Thickness of wall leaf [***]	± 5 mm or ± 5 % of the leaf thickness whichever is the greater
of overall cavity wall	± 10 mm.

[*] h_{tot} is the total height of the building.
[**] Deviation from straightness/flatness is measured from a straight reference line between any two points.
[***] Excluding leaves of single masonry unit width or length, where the dimensional tolerances of the masonry units govern the leaf thickness. The permissible tolerances for dimensions of masonry units are specified in EN 771.

lower than those recommended in the EN 1996-2. The current limitation of 15mm on the overhang of the first course of masonry from the edge of a floor or roof also required clarification. Progress has been made by comparing the intention of EN 1996-1-1 with the requirements in EN 1996-2 and adding more information regarding the height of masonry. The current proposals are shown in Table 3. WG4 did express the view that the PT3 proposals might be too difficult to achieve in practice and they are being further considered.

5 POINTING

The requirements for pointing in 3.5.3.1 were closely reviewed in the National Annexes because it is one of only two NDPs in EN 1996-2. It was found that there were a number of National variations and an attempt has been made to eliminate the need for this to be an NDP. This has been achieved by introducing more accurate recommendations and relating the depth of pointing to the thickness of the wall and whether the wall is loadbearing or non-loadbearing. Loadbearing columns are not currently covered in EN 1996-2 and requirements have been added to the revised draft.

The revised text currently is:

a) Where joints are to be pointed, the unhardened mortar joints should be raked out so as to have clean sides to a depth of at least d_p, depending on the application.
b) Before pointing loose material should be brushed out. The whole area should be cleaned and if

necessary wetted to give the best practicable adhesion.
c) For non-loadbearing walls, the maximum value of d_p, is 15 mm but the remaining width of the joint should be at least 70% of the wall thickness.
d) For loadbearing walls, the maximum value of d_p, is 15 mm but no more than 15% of the wall thickness, measured from the finished surface of the joint.
e) For loadbearing columns, the maximum value of d_p, is 15 mm but no more than 10% of the column thickness, measured from the finished surface of the joint.
f) Higher maximum values of d_p, may be used when the capacity of the masonry is checked by calculation taking into account the reduced cross-section of the masonry when the joint is raked.

6 MOVEMENT

The guidance provided in EN 1996-2 regarding the movement of masonry requires clarify and attention was drawn to this through a number of comments. The current Table is contained in a note to 2.3.4.(2) and shown here as Table 4 and is the second NDP in the code. It is meant to give the maximum spacing of movement joints in non-loadbearing walls. The term "non-loadbearing" has led to different interpretations with some countries considering the outer leaf of a cavity wall as "non-loadbearing" and others considering the outer leaf as part of a loadbearing wall.

The sections detailing the requirements for design for movement have been significantly revised and clarified. The proposed revised wording for movements is as follows:

a) The horizontal spacing of vertical movement joints in masonry walls should take into account the type of wall, masonry units, mortar and the specific construction details.
b) The positioning of movement joints should take into account the need to maintain structural integrity of load bearing internal walls.
c) In unreinforced loadbearing masonry walls vertical movement joints should be considered at:
   - strong discontinuities in the geometry of the wall;
   - strong discontinuities in the loads of the walls.
d) The distance between the movement joints should take into account characteristics of the mortar: e.g. the adhesion of the mortar to the masonry unit and the compressive strength of the mortar.
e) The distance between movement joints in external non-loadbearing walls also depends on the type of wall ties, allowing or not independent in plane movement between masonry leaves or between the external masonry leave and other structures to which the external masonry is attached.

Table 4. Maximum recommended horizontal distance, l_m, between vertical movement joints for unreinforced, non-loadbearing walls.

Type of masonry	l_m (m)
Clay masonry	12
Calcium silicate masonry	8
Aggregate concrete and manufactured stone masonry	6
Autoclaved aerated concrete masonry	6
Natural stone masonry	12

Table 6. Recommended horizontal distance, l_m, between vertical movement joints for unreinforced, non-loadbearing external walls.

Type of masonry	l_m (m)
Clay masonry	10 to18
Calcium silicate masonry	5 to 9
Dense aggregate concrete and manufactured stone masonry	5 to 9
Lightweight aggregate concrete	4 to 8
Autoclaved aerated concrete masonry	4 to 8
Natural stone masonry	10 to 20

f) The need for vertical movement joints in loadbearing unreinforced walls should be considered.

g) Unless determined by calculation the maximum horizontal spacing between vertical movement joints in loadbearing unreinforced walls should not exceed those given in Table 5.

NOTE: The vertical movement joints in loadbearing unreinforced walls depend on local building traditions, type of foundation and floors used, masonry typology and other construction details, humidity and temperature variation.

h) Unless determined by calculation the horizontal distance between vertical movement joints in external non-loadbearing unreinforced masonry walls should not exceed lm.

NOTE 1 The values for l_m are given in Table 5, unless the National Annex of a country gives different values. For each type of masonry, a Country can select a single value or a range of values.

NOTE 2: The table is applicable for unreinforced non-loadbearing external single-leaf walls, for unreinforced non-loadbearing leaf of cavity walls, and for unreinforced veneer walls, etc.

It will be noted that in response to comments the current draft includes a range of values for movement as shown in Table 5 but the requirements of Table 6 remain an NDP.

Table 5. Maximum horizontal distance, between vertical movement joints for unreinforced, loadbearing walls.

Type of masonry	Cavity wall Internal load-bearing leaf	Single leaf wall Filled head-joints	Single leaf wall Unfilled head-joints
Clay masonry	40	30	25
Other types of masonry	30	25	20

7 EASE OF USE

A key aspect of the review has been to improve the ease with which EN 1996-2 can be used by designers. This is being done by removing unnecessary notes, checking the clarity of each clause and making sure all tables and figures are clear and easily understood.

The wording in the document has been reviewed and the use of shall, should, may and can now conform with the TC250 recommendations expressed in N1250, namely:

"**shall**" expresses a requirement strictly to be followed and from which no deviation is permitted in order to comply with the Eurocodes.

"**should**" expresses a highly recommended choice or course of action. Subject to national regulation and/or any relevant contractual provisions, alternative approaches could be used/adopted where technically justified.

"**may**" expresses a course of action permissible within the limits of the Eurocodes.

"**can**" expresses possibility and capability; it is used for statements

"**is**" means a statement of certainty, a fact. It should be used sparingly, if at all of fact and clarification of concepts.

8 CONCLUSION

PT3 has made significant progress in reviewing the content of EN 1996-2 and responding to the comments received. The draft has now been prepared in the revised format in which it will be made available for further review by April 2020 following which the final draft leading to formal review will be produced.

In carrying out the updating of EN 1996-2 the opportunity has been taken to review the developments that have taken place in masonry construction since the document was originally published in 2006. Whist there have been a number of interesting developments in the design of masonry units and some innovative wall constructions, none of these developments required changes to be incorporated into the draft revision of EN 1996-1-1.

All the comments received to date have been carefully considered and replied to in response documents. It is apparent that many detailed comments are very valid for local construction practice but more difficult to apply generically through the Eurocode. As always it is very helpful to have a detailed proposal for a proposed revision when comments are submitted.

ACKNOWLEDGEMENT

This work has been undertaken under a contract addressing Mandate M515 Eurocodes Phase 3 and the financial support provided is gratefully acknowledged. As project team leader I recognize the contributions made by the whole team would like to acknowledge the help and support given by the other members of CEN TC/250/SC6 PT3 who are:

Dieter van Rossem

Paolo Morandi

Jan Kubica

My thanks are also due to the Chairman of CEN TC 250/SC6 Rob van der Pluijm for his ongoing contribution to the work of the PT.

REFERENCES

CEN/TC 250 N1250 version 8 Policy Guidelines and Procedures.

EN 1996-2. 2006. Eurocode 6: Design of masonry structures Part 2: Design considerations, selection of materials and execution of masonry.

EN 1996- 1-1. 2012. Eurocode 6: Design of masonry structures Part 1-1: General rules for reinforced and unreinforced masonry structures. British Standards Institution. London. 2005 +A1: 2012.

Brick and Block Masonry - From Historical to Sustainable Masonry –
Kubica, Kwiecień & Bednarz (eds)
© 2020 Taylor & Francis Group, London, ISBN 978-0-367-56586-2

Three story CM buildings with joint reinforcement: Shaking table tests

J.J. Perez Gavilan E.
Universidad Nacional Autónoma de México (UNAM), Mexico City, México

L.E. Flores
Centro Nacional de Prevención de Desastres (CENAPRED), Mexico City, México

ABSTRACT: Two three story confined masonry buildings scaled 1:2 were tested in the shaking table. The structures represent one axis of a typical building for housing. The walls of the structures were built with concrete multi-perforated units confined with tie-columns and tie-beams. The scaled units were obtained by sawing the ones of the prototype; although they did not have the same perforation pattern they preserved the net area. The floor system was a solid concrete slab. The first building tested did not include joint reinforcement while the second included a high strength wire every three courses. Live load was included with lead ingots and the walls were posttensioned to achieve the required axial stress. The structures were subjected to a several motions based on a synthetic accelerogram with amplitudes progressively scaled up until failure. Preliminary results in terms of displacement capacity and strength are given.

1 INTRODUCTION

Confined masonry (CM) is a construction system based on masonry walls that are confined by concrete ties, vertical tie-columns and horizontal tie-beams. These confining elements provide additional shear strength and considerable displacement capacity to the walls. Unlike unreinforced masonry (UM) walls that abruptly reduce their shear strength capacity after diagonal cracks appear, CM walls may resist additional lateral force after cracking. In addition, after peak strength degradation is slower with additional lateral displacement (Meli 1975). This behavior allows the system to dissipate a considerable amount of energy before failure. These features make CM a suitable earthquake resistant system, a fact confirmed by after quake inspection reports (Astroza et al. 2012)

Including joint reinforcement (JR) to CM walls very much enhances their performance, increase their shear strength and their displacement capacity (Curz et al. 2019). How much the JR may increase the shear strength of a wall depend on the masonry's compressive strength and the masonry units' net area (Rubio 2017). Height to length aspect ratio (H/L) is a parameter that may affect the efficiency of the reinforcement, however this possibility is under investigation. In order for the JR to contribute to the shear strength of a wall requires that inclined cracks appear and cross the reinforcement. When that happens, the JR deform around the crack and consequently the tension force generated tends to close it. In doing so, the

internal forces in the wall may find new paths to the supports causing new cracks away from the main diagonal. This type of cracking appears due to tension perpendicular to a compressive stress field. It is the predominant damage mechanisms of CM walls when subjected to lateral forces. However, cracks in long walls (H/L<1) and in walls that have a large amount of reinforcement in the bed joints, may appear parallel to a bed joint causing sliding. In that case, the JR cannot contribute the shear strength of the wall, but it can still make a difference regarding its displacement capacity. This fact is attributed to the JR keeping the confinement of the wall. After cracking, tie-columns tend to bend out in the plane of the wall, separating from each other at their mid height. This phenomenon allow the cracks in the wall to widen; however, the JR, which is anchored to the tie-columns, prevents this by preserving the distance between the tie columns, and the confinement of the wall panel.

Until recently, the effect of JR in the performance of the walls was investigated based on pseudo-static tests of full scale masonry walls subjected to cyclic loads. The first study of a confined masonry structure with JR tested in a shaking table was presented by Flores (Flores et al. 2017).

This study is part of an investigation to understand the contribution of the JR to the performance of the walls in terms of strength and displacement capacity. The investigation includes pseudo-static and dynamic tests of which the study of Flores was the first dynamic experimental program. It included two half scale structures, with three stories, one with and the other without joint reinforcement, built with traditionally crafted solid clay units and solid concrete slabs.

It is very difficult to simulate the actual behavior beyond the elastic limit of a building through a reduced scale sample. However, important qualitative information may be obtained through the models.

The experimental program presented in this study include two structures that were very similar to the ones tested by Flores. The main difference was the type of the masonry units that, in this study, were multi-perforate concrete units. The details of the experimental program and the main results are presented next. The displacements of the structures are compared to the allowable drift limits prescribed by the seismic Mexican code, released late in 2017 (NTCM, 2017).

1 EXPERIMENTAL PROGRAM

Two half scale structures, with three stories, one with (MPC-CR) and the other without JR reinforcement (MPC-SR) were tested in UNAM Institute of Engineering's shaking table. The structures represent one axis of a confined building for housing.

1.1 Prototype

Because the models represent only one axis of a structure, the virtual prototype was selected to have the usual spans between walls, without having to describe all the structure details. Its geometry was useful to figure out the prototype loads and the normal stresses on the walls included in the model. A schematic plan view of the prototype is shown in Figure 1.

The direction of analysis was parallel to axis B.

The central wall (axis 2) was not represented in the model. It was considered that, because of its position, it would have only a small participation to the resisting overturning moment and no participation to resist shear force. However, it was taken into account for estimating the load carried by the walls in axis B; it reduces their axial stress.

1.2 Scaling similitude laws

The models were scaled according to simple similitude law. Accordingly, the materials density and failure stresses were the same in the model and the prototype as it was the gravity acceleration. Those assumptions lead to the following scaling factors, given in Table 1, that relate different quantities of the prototype to the model.

It can be verified that the shear stress in the prototype and the model are the same. This is a basic requirement for the analysis of the results. However, the axial stress in the model is just half of that of the prototype. To overcome this difference, post-stressing of the walls was used, as other authors did (Alcocer, et al., 2004).

Acceleration for the model should be two times the ground acceleration of the prototype. A fact that has to be taken into account to ensure that the shaking table is able to produce the desired accelerations. This is also dependent on the total weight of the specimen.

1.3 Models geometry

The model geometry is shown in Figure 2. Story height was 125 cm. Tie-column dimensions were 6×7.5 cm (12×15 cm in prototype) which are the minimum dimensions allowed by the Mexican code. Tie-beams were 6×12.3 cm that are the scaled dimensions of the typically used (t×25 cm) for those elements. An 8 cm thick concrete slab was used. Although thicker than it should to represent the typical 12 cm prototype slab thickness, it was decided that it was convenient for the construction of the models. Using an elastic analysis of the model, it was verified that it had minimal impact in the lateral stiffness of the model, it lowered the fundamental period less than 4%. The extra weight was considered part of the live load.

Transverse walls were 160 cm long, considerably shorter than those used by Flores (2017). They were shorter to reduce the overall weight of the model.

Foundation beams were 16×22 cm, which were built on top of a steel frame, provided with clips made of 4"×1/4" channel sections to anchor the foundation to the frame.

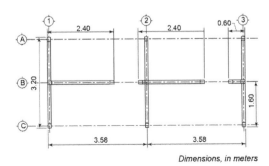

Dimensions, in meters

Figure 1. Prototype configuration and geometry.

Table 1. Similitude scaling factors.

Quantity	Factor
Length	1/2
Area	1/4
Volume	1/8
Axial Stress	1/2
Period (time)	1/2
Lateral Stiffness	1/2
Acceleration	2

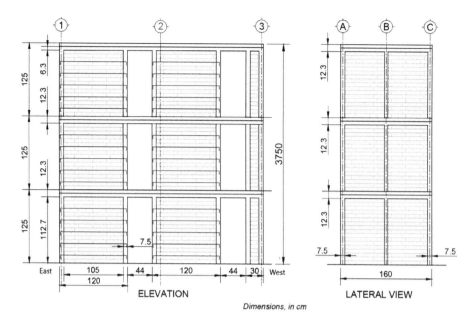

ELEVATION   LATERAL VIEW

Dimensions, in cm

Figure 2. Model geometry. Position of the horizontal reinforcement in specimen MC-CR. Lead Ingots position.

1.4 Reinforcement

Tie-columns were reinforced with 4, 6 mm dimeter bars having fy=412 MPa yielding stress. The commercial 6.3 mm bars in Mexico have a yielding stress 248 MPa. In order for them to have the required 412 MPa yielding stress, the 6.3 mm diameter bars were cold drawn up to the hardening region of the stress strain curve of the steel up to the desired stress. After this process, a sample of the bars were tested in tension. It was verified that the yielding stress was the one required and it was observed that the bars could still deform more than 6% before failure. However, the bars had no ribs, they were smooth. From a preliminary pseudo-static test of a wall with this type of reinforcement, and careful inspection after the tests in the shaking table, it was verified that the smoothness of the bars did not have any effect on the performance of the reinforcement and there were no signs of sliding.

Stirrups were provided with 3.25 mm diameter wires that had a yielding stress of 210 MPa, which is similar to the standard 248 MPa nominal yielding stress of typical 6.3 mm diameter stirrups.

Stirrups spacing was 90 mm, which agree with the 1.5 t maximum spacing of the code, where t is the thickness of the wall. The spacing was reduced to half at both ends in every tie-column and tie-beam (Figure 3)

The horizontal reinforcement consisted of ϕ5/32" wires, placed every three courses, starting in the second course from the bottom (Figure 2); Its nominal yielding stress was f_{yh}=600 MPa. Altogether, the amount of reinforcement was $p_h f_{yh}$

=0.6 MPa, which is in the Mexican code's allowable range [0.3,1.25]. The steel percent p_h=A_s/(s×t), where A_s=12.4 mm^2 is the JR area, s=199 mm its spacing and t=60 mm is the wall thickness. The allowable prototype joint thickness should be in the range [6,15] mm, with a typical value of 10 mm. For the model, the joint without reinforcement was 6 mm and the one with JR was 7 mm. Also the diameter of the bar meet the code requirement in the sense that it should be less than 3/4 of the joint thickness (NTCM, 2017).

1.5 External instruments

Accelerometers were installed in each floor including the base, just on top of the foundation beam near the center of the structure. Three accelerometers were installed in each floor, in the south-west and north-east corners and one near the center of the structure over axis B (Figure 3), so that eventual torsion could be detected and measured.

Lateral displacements were measured at the base and at each floor level with a transducer at the este side. Redundant instruments were installed in the west side. In both cases on axis B.

Deformation of the diagonal in both directions were measured in most walls. Vertical deformation was also measured in all the walls of the first story and one of the second story (Figure 3).

1.6 Internal instruments

Stain gages were installed at the end of tie-columns of the first story and some tie-beam ends.

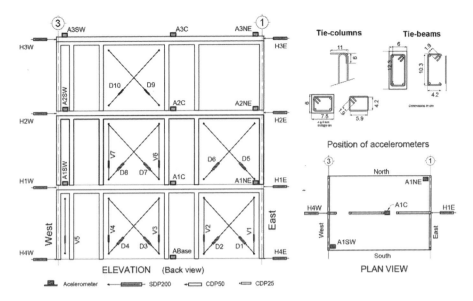

Figure 3. Reinforcement in tie-columns and tie-beams. External instruments and location of accelerograms.

Extensometers were also installed in the JR. The location of these strain gages is not shown as no results from those instruments are presented in this report.

1.7 Materials

The masonry units were multi-perforated concrete units. Although the manufacturer of the units tried to produce a scaled version of their units, they were unsuccessful for many technical reasons. It was then decided to saw the full scale units to the scaled sized. After some analysis, a sawing procedure was devised so that the net area was almost preserved. This fact guaranties that the density of the masonry is the same in the model than in the prototype. It took six cuts to produce each scaled unit. The sawing of the units had to be constantly readjusted as the wearing out of the sawing disc produced, eventually, two or three milli-meter deviations from the required dimension. That variation was considered too large as it imply a 6 mm variation in the prototype. The largest allowable vari-ation was 1 mm. The full scale and sawed units are shown in Figure 4.

During the constructions of the specimens the materials were sampled so that at least 3 piles and wallets were built, for each story, using the mortar that was being used for the walls of the specimen. The wallets and piles were tested using the Mexican standard (ONNCCEE 464, 2010) to obtain the diag-onal compression strength, v_m, and shear modulus, G_m, and the masonry compressive strength, f_m and modulus of elasticity E_m. Similarly the mortar was sampled using 5 cm side cubes from which the mortar compressive strength, f_j, was obtained.

Figure 4. Prototype and model masonry units.

Concrete cylinders from the tie-columns and slab were tested in compression to obtain the concrete compressive strength f_c. The average properties for each specimen are given in Tables 2 and 3 for the model SPC-SR and SPC-CR respectively.

1.8 Vertical loads

Two objectives were sought. That the shear stress produced by the scaled base acceleration would be the same as the shear stresses in the prototype for the ground motion. And that the axial stress in the walls due to the scaled vertical loads in the model be equal to the axial stress in the walls due to self-weight and live loads considered in the prototype.

283

Table 2. Model without joint reinforcement (MPC-SR), material average properties.

Story	Slab f_c [MPa]	Tie-columns f_c [MPa]	E_c [MPa]	Masonry f_j [MPa]	f_m [MPa]	v_m [MPa]	E_m [MPa]	G_m [MPa]
1	28.2	17.56	6050.75	1.92	10.17	0.84	5649.95	1884.11
2	31.13	23.5	6137.6	2.08	9.71	0.96	3420.56	-
3	33.17	19.14	5809.9	1.86	11.99	1.07	6105.75	2067.99

Table 3. Model with joint reinforcement (MPC-CR), material average properties.

Story	Slab f_c [MPa]	Tie-columns f_c [MPa]	E_c [MPa]	Masonry f_j [MPa]	f_m [MPa]	v_m [MPa]	E_m [MPa]	G_m [MPa]
1	30.94	15.64	8302.41	1.57	10.35	0.59	7737.16	2234.61
2	30.38	32.21	12365.20	1.97	11.13	0.81	7965.29	2477.26
3	25.62	23.96	23478.33	1.92	10.24	0.95	6401.94	2037.48

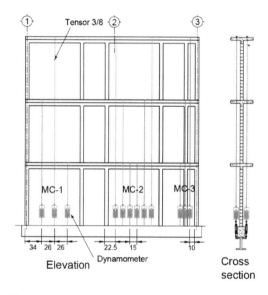

Figure 5. Post-stressing of model walls.

Table 4. Number of post-stressing cables and their force.

Wall	Stress difference with prototype (MPa)	No of cables	force (N)
MC-1	0.044	6	529.7
MC-2	0.099	12	588.6
MC-3	0.155	6	461.1

To achieve the desired mass for the dynamic analysis 11 lead ingots, 50 kg each, were installed in

each floor. Their position was decided so that the vertical stress in the walls produced by his weight was as uniform as possible. An elastic analysis of the model was used for that purpose. The transverse walls along axis 1 and 3 tend to absorb a considerable amount of the vertical load on the slab.

The axial stress of the prototype walls was calculated considering the transverse wall along axis 2. This has the effect to reduce the axial stress on the walls in axis B.

As mentioned before, the scaling of the loads that produce the correct shear stress in the model, will produce a reduced axial stress in the walls. To achieve the desired axial stress, the walls of the model in axis B were post-stressed using cables and dynamometers. The number of post-stressing cables and the load in each one is shown in Figure 5 and Table 4.

1.9 Ground motion

A synthetic accelerogram was developed to ensure that the frequency content of the signal will damage and subsequently fail the structures. The accelerogram is shown in Figure 6. The target response spectrum and the spectrum obtained with the synthetic accelerogram are shown in Figure 7. With an elastic analysis using finite elements and the average material properties shown in Tables 2-3 the fundamental period was 0.068 s.

2 RESULTS

The mass multiplied by the absolute acceleration in each floor give the lateral forces and adding them the base shear. Plots of the base shear and the displacements of the first floor and the roof, relative to the

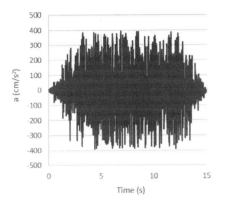

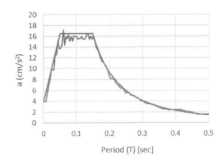

Figure 7. Target response spectrum and the actual response spectrum calculated with the synthetic signal.

Figure 6. Synthetic accelerogram.

base, are presented in Figure 8, for the case of the structure without joint reinforcement, MPC-SR, and the one with horizontal reinforcement, MPC-CR. Crack patterns of the two main walls in the first story of both models are presented in Figure 9. On top of those curves, the envelope can be appreciated.

2.1 Strength

It can be observed that the use of horizontal reinforcement had no impact in the shear strength of the structure (Table 5). This is attributed to the failure sliding mechanism. The later can be observed in Figure 9, MPC-CR East and Central walls. A sliding plane developed on the 5th bed joint counting from the bottom. The joint included JR.

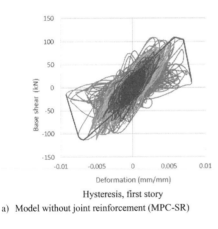

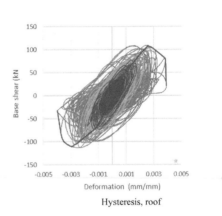

Hysteresis, first story

Hysteresis, roof

a) Model without joint reinforcement (MPC-SR)

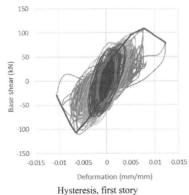

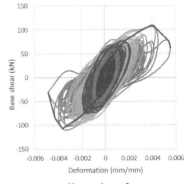

Hysteresis, first story

Hysteresis, roof

b) Model with joint reinforcement (MPC-CR)

Figure 8. Hysteresis curves using first story and roof lateral displacements relative to the base, for the specimens a) without joint reinforcement (MPC-SR) and b) with joint reinforcement (MPC-CR). a) Model without joint reinforcement (MPC-SR), Model with joint reinforcement (MPC-CR).

285

| East wall | Central wall |
| Model without joint reinforcement (MPC-SR) | |

| East wall | Central wall |
| Model with horizontal reinforcement (MPC-CR) | |

Figure 9. Crack patterns near peak strength, a) Model with and b) Model without joint reinforcement.

Table 5. Positive and negative peak strength.

Model	V_R+ kN	VR- kN
MPC-SR	108.8	-108.9
MPC-CR	112.8	-107.2

This result does not agree with full scale pseudo-static tests of confined masonry walls subjected to cyclic loads (Curz O., et al., 2019). In that study, the same type of multi-perforated units, as in this study, were used. Six square walls were tested with increasing amount of JR, starting with no JR at all. The JR was installed every two courses. The contribution of the JR to the shear strength of the walls grew proportionally with its amount up to a point, after which, no additional increase in strength was observed, although a larger amount of JR was installed. The interpretation

of that behavior was that the forces developed in the JR were no longer able to be transmitted by the masonry, so that a limit of reinforcement was found based on the compressive strength of the masonry. The exhaustion of the masonry's capacity was observed on the density of the cracking and damage of the walls as the amount of JR increased. In that study, sliding did occur; however, only for the walls that had a large amount of JR ($p_{hf_{yh}} \geq 1.21$ MPa). So that, in those cases, sliding limited the participation of the JR. Both phenomena, crushing of the masonry and sliding may interact to limit the increase in strength for large amounts of JR. However, in this study sliding appeared for a much less amount of JR.

One of the differences between the pseudo-static test and the dynamic ones is the boundary conditions of the walls. In the referred pseudo-static tests, the walls were tested in cantilever, i.e., they bend in simple curvature while in the dynamic tests they bend in double curvature. However is not clear how this difference could have produce sliding.

Table 6. Global and first story positive deformation with and without JR.

Model	First story mm/mm	Roof mm/mm
MPC-SR	0.00588	0.00259
MPC-CR	0.00790	0.00416

2.2 Displacement capacity

The comparison of the lateral displacements divided by the corresponding height, measured from the base, of the first story and the roof show that the global displacement demand is less than that of the first story.

The deformations at peak strength in the first story and roof with and without JR are given in Table 6.

The first floor deformation can be compared to the inter-story drift limits of the Mexican code, that for solid units without JR is 0.005 and with JR is 0.01. The masonry units used in this study may be considered solid, according to the Mexican code, as they have a net area larger than 75%. The drift observed in MPC-CR fell 20% short compared to the allowable one.

3 CONCLUSIONS

The preliminary results show that the horizontal reinforcement had no effect with regard to the strength of the structure. The results disagree with past pseudo-static experiments that have shown otherwise. It is clear, however, that the contribution of the reinforcement to the strength of the model was limited by the sliding mechanism of failure. A failure mode that was observed in pseudo-static tests only when the walls included large amounts of JR. It was pointed out a boundary condition difference of the walls in the dynamic and pseudo-static experiments. However, it is not clear how that difference could have produced sliding in the dynamic test.

The displacement capacity was enhanced by the horizontal reinforcement, judging by the first floor drift. However, the drift at peak strength of the model with JR was 20% smaller than the allowable drift limit prescribed by the Mexican code. However, the model without JR reached a drift larger than the allowed one.

ACKNOWLEDGMENTS

This work was funded by the Mexico City Government through grant ISCDF/CEC-04/2017-02. Many thanks to Industrial Bloquera Mexicana its CEO Daniel Vasquez and Manager Felipe Garcia that provided the materials the masons and also the workers that saw the units. Many thanks to Ariel Flores and undergraduate student from the Technological Institute of Tehuacán Puebla that did an internship in the Institute of Engineering assign as an assistant to this project. Roberto Duran gave assistance and operated the shaking table.

REFERENCES

Alcocer, S. M., Arias, J. G. & Vásquez, A. 2004. *Response assesment of Mexican confined masonry structures through shaking table tests*. Vancouver, Canada, s.n., p. paper 2130.

Alcocer, S. M. & Casas, N. 2019. *Shake-Table Testing of a Small-Scale Five-Story Confined Masonry Bulding*. Salt Lake City, Utah, The Masonry Society, pp. 121–131.

Alcocer, S. M. & Zepeda, J. A., 1999. *Behavior of multiperforated clay brick walls under earthquake-type loading*. Austin, Texas, USA, s.n.

Astroza, M., Moroni, O., Brzev, S. & Tanner, J., 2012. Seismic Performance of Engineered Masonry Buildings in the 2010 Maule Earthquake. *Earthquake Spectra*, June, 28 (S1), pp. 385–406.

Curz O., A., Pérez Gavián, J. J. & Flores C., L., 2019. Experimental study of in-plane shear strength of confined concrete masonry walls woth joint reinforcement. *Engineering Structures*, Volume 182, pp. 213–226.

Flores, L. E., Pérez Gavilán, J. J. & Alcocer, S., 2017. *Displacement capacity of confined masonry structures reinforced with horizontal reinforcement: Shaking Table Tests*. Santiago de Chile, s.n.

Marques, R. & Laurenco, P., 2019. Structural behaviour and design rules of confined masonry walls: Review and proposals. *Construction and Building Materials*, Volume 217, pp. 137–155.

Meli, R., 1975. *Comportamiento Sísmico de Muros de Mampostería*, Mexico: Instituto de Ingeniería. Universidad Nacional Autónoma de México.

NTCM, 2017. *Normas Tecnicas Complementarias para el Diseño y Construcción de Estructuras de Mampostería*, Ciudad de México, Mexico: s.n.

ONNCCEE 464, 2010. *Determination of diagonal compressive strength and shear modulus of masonry assemblages, as well as compressive strength and modulus of elasticity of prisms for clay or concrete masonry - testing methods*. Mexico: Organismo Nacional de Normalización y certificación de la construcción y la edificación S. C.

Pérez Gavilán, J. J., 2019. *Ductility of confined masonry walls: results from several experimental campaigns in Mexico*. Salt Lake City, s.n.

Pérez Gavilán, J. J., Flores, L. E. & Alcocer, S. M., 2015. An experimental study of confined masonry walls with varying aspect ratio. *Earthquake Spectra*, pp. 945–968.

Rubio P., L., 2017. *Contribución del refuerzo horizontal a la resistencia a corte de muros confinados de piezas de arcilla extruida*, México: s.n.

Brick and Block Masonry - From Historical to Sustainable Masonry –
Kubica, Kwień & Bednarz (eds)
© 2020 Taylor & Francis Group, London, ISBN 978-0-367-56586-2

Improvements in EN 1996-3 – Explanations and background information

C.-A. Graubner & B. Purkert
Institute of Concrete and Masonry Structures, Technische Universität Darmstadt, Germany

ABSTRACT: In 2017, Phase 2 of mandate M515 has started, in which Project Team 2 (PT2) dealt with the systematic review of EN 1996-3. In this paper, the main improvements in prEN 1996-3 are presented. First, the new equations for determination of the capacity reduction factor are addressed, which now also applies to walls bearing partially supported slabs. Other improved design equations, e.g. for basement walls, for walls subjected to mainly lateral loading, and for concentrated load resistance as well as the improved method for simplified verification of overall stability are explained. In closing, the paper takes a look at the number of national choices.

1 INTRODUCTION

1.1 General background information

In accordance with task SC6.T4 of Mandate M515 Phase 2, Project Team 2 (PT 2) has created the revised version of EN 1996-3 in October 2019. After having received the comments of the National Standardization Bodies (NSBs) in January 2020, the final version is now available for the start of the formal enquiry.

There are five main tasks the PT had to deal with. Besides of the common two topics "reduction of NDPs" and "enhanced ease of use", the third main issue is the inclusion of rules for the design of walls with partially supported slabs. The fourth main concern affects the incorporation of changes in EN 1996-1-1 in order to obtain conservative results when EN 1996-3 is applied. In this context, the last task is to add clauses for robustness in accordance with part 1-1.

1.2 Scope of this paper

This paper presents the most important changes and improvements in the revised version of EN 1996-3 mostly related to the third and fourth task. In addition to the new rules, background information is given and models or calculations, which form the basis of the new rules, are explained.

On the whole, this paper follows the structure of EN 1996-3 starting with the vertical load resistance with special regard to walls with partially supported slabs and the effects on the conditions for application. In addition, the required minimum vertical load on walls subjected to lateral loads is considered. After that, the focus lies on the extended design equation for basement walls and the revised and improved simplified calculation method for shear

walls. The paper concludes with a short view on some minor amendments.

2 DESIGN OF WALLS SUBJECTED TO VERTICAL AND WIND LOADING

2.1 Vertical load resistance

2.1.1 Methodology and boundary conditions

The simplified design of vertically loaded walls is the most important section in EN 1996-3 because this case will occur most frequently in practice. That's why this issue got special attention and took a lot of time during revision. In order to ensure conservative results according to EN 1996-3 compared to part 1-1 a lot of comparative calculations has been carried out. These calculations always follow the same principle: the vertical resistance of a wall with clearly defined boundary conditions is determined both according to EN 1996-1-1 and according to part 3. Then, the values are compared to each other to check whether the resistance according to part 1-1 is exceeded or not. Afterwards, this procedure repeats with modified parameter combinations. The results of this comparison identify those parameter combinations, which require an adjustment of the rules in EN 1996-3.

In this iterative process, walls with partially supported slabs are considered as well during all comparative calculations – the bearing length of the slab on the wall is handled as one additional parameter.

Table 1 shows the value range the comparisons are based on.

self-weight of the masonry is neglected.

Due to the complex interaction of wall-slab-connections in combination with wind loads – especially with regard to partially supported floors – the

Table 1. Parameter and values considered in comparative calculations between part 1-1 and part 3.

Parameter	Symbol	Considered value(s)	Unit
clear wall height	h	variable parameter	m
wall thickness	t	0.15 0,5; $\Delta t = 0.05$	m
bearing length	t_b/t	2/3; 3/4; 4/5; 1	-
clear slab span	l	4; 5; 6; 6.5; 7	m
slab thickness	d		
(simply		0.18; 0.22; 0.27; 0.29	m
supported)		0.16; 0.17; 0.20; 0.22;	
(continuous)		0.24	
masonry compressive			
strength	f_k	1; 1.5; 2; 3; 4; 5	MPa
young's modulus			
masonry	E_M	$700 \cdot f_k$; $1000 \cdot f_k$	MPa
concrete	E_c	30000	MPa
additional			
permanent			
load*	Δg_k	1.7	kPa
live load	q_k	5	kPa
wind load	w_{Ek}	0.9; 1.3	kPa
partial factor for			
materials	γ_M	2.0	-
permanent load	γ_G	sup: 1.35; inf: 1.0	-
live/wind load	γ_Q	sup: 1.5; inf: 0	-
combination coef-			
ficients for			
live load	$\psi_{0,Q}$	0.7	-
wind load	$\psi_{0,w}$	0.6	-

* self-weight of the slab is calculated from $d \cdot 25$ kPa;

investigations mainly focus on exterior walls. In general, the newest draft of part 1-1 prEN 1996-1-1 (2019) applies for comparison using Annex C for determination of the load-eccentricity at the top or bottom of the exterior wall:

$$M_1 = -\frac{w_1 h_1^2}{4(n_1 - 1)} + \frac{\frac{n_1 E_1 I_1}{h_1}}{\frac{n_1 E_1 I_1}{h_1} + \frac{n_2 E_2 I_2}{h_2} + \frac{n_4 E_4 I_4}{l_4}} \cdot \left[\frac{w_1 h_1^2}{4(n_1 - 1)} - \frac{w_2 h_2^2}{4(n_2 - 1)} - \frac{q_4 l_4^2}{4(n_4 - 1)} \right] \quad (1)$$

where n_i = stiffness factor; I_i = second moment of area with $I_i = t_b^3 \, l/12$.

Thereby, different load combinations are considered taking live load or wind load, respectively, as predominant action according to Eq. (2) and Eq. (3).

$$E_{d,1} = 1.35 G_k + 1.5 Q_k + 1.5 \psi_{0,W} W_k \quad (2)$$

$$E_{d,2} = 1.35 G_k + 1.5 \psi_{0,Q} Q_k + 1.5 W_k \quad (3)$$

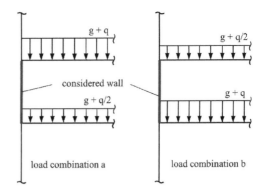

Figure 1. Arrangement of the live load accounting for its unfavourable effects.

The comparison considers both wind pressure as well as wind suction. Additionally, the live load is variably arranged as shown in Figure 1 to account for its unfavourable effects. The results show, that full live load at the top of the wall and half the live load at the bottom (case a)) leads to the most unfavourable design situation in mid-storey height. Similarly, wind suction is governing in general.

Beyond that, both simply supported slabs as well as continuous slabs are considered. Thereby, the thickness of the concrete floor is assumed in accordance with the simplified deflection control for cases where calculations may be omitted according to 7.4.2 of EN 1992-1-1 (2010).

Finally, it is important to mention that the comparison assumes a Young's modulus of $E_M = 700 \cdot f_k$ as well as $E_M = 1000 \cdot f_k$. If buckling is the governing design case (in mid-storey height), the lower Young's modulus is unfavourable while the higher one is governing design of thick walls with low height and low compressive strength.

2.1.2 Resulting capacity reduction factor

Eqs. (4), (5), and (6) show the new rules for determination of the capacity reduction factor. While Eqs. (4) and (5) are adopted from the previous code EN 1996-3 (2009) and only slightly adjusted, Eq. (6) is introduced newly. The vertical resistance results from Eq. (7).

$$\Phi_s = 0.85 \frac{t_b}{t} - 0.0011 \left(\frac{h_{ef}}{t} \right)^2 \quad (4)$$

$$\Phi_s = \left(1.2 - \frac{l_{f,ef}}{l_{ref,c}} \right) \frac{t_b}{t} \geq 0.33 \frac{t_b}{t} \quad (5)$$

$$\Phi_s = \left(1.2 - \frac{l_{f,ef}}{l_{ref,t}} \right) \left(\frac{t_b}{t} \right)^2 + 0.09 \frac{t_b}{t} \frac{h}{t} \geq 0.33 \frac{t_b}{t} \quad (6)$$

$$N_{Rd} = \Phi_s \, t \, l \, f_d \qquad (7)$$

where h_{ef} = effective wall height; $l_{f,ef}$ = effective span; $l_{ref,c}$, $l_{ref,t}$ = reference values according to Table 2.

Compared to EN 1996-3 (2009), Eq. (4), which covers stability failure, now contains the ratio bearing length of the slab to thickness of the wall to account for the systematic eccentricity in mid-storey height which occurs in case of partially supported slabs. This ratio is also integrated in Eq. (5) which considers the effects of slab rotation at the top or bottom of a wall. Here a reduced bearing length has a big influence on the vertical resistance. Further, this equation is adjusted to improve accordance with prEN 1996-1-1 (2019), especially in case of partially supported floors and lower masonry compressive strength as Figure 2 illustrates. In this context, the reference value $l_{ref,c}$ is introduced as stated in Table 2. In addition, a minimum value similar to EN 1996-1-1, Annex C is given.

The design of intermediate walls with similar slab spans on both sides follows Eq. (4). However, if the spans differ significantly ($l_1 - l_2 > 3,0$ m), the capacity reduction factor results of the minimum of Eqs. (4) and (5) because the load-bearing behaviour of the wall more and more equals the behaviour of walls

Table 2. Reference values $l_{ref,c}$ and $l_{ref,t}$ for use in (5) and (6)*.

Compressive Strength	Thickness	$l_{ref,c}$	$l_{ref,t}$
$f_k = 1$ MPa	$t \leq 0.3$ m	2.8 m	8.0 m
$f_k = 1$ MPa	$t = 0.5$ m	3.7 m	8.0 m
$f_k \geq 5$ MPa	$t \leq 0.3$ m	4.4 m	9.5 m
$f_k \geq 5$ MPa	$t = 0.5$ m	8.1 m	9.5 m

* linear interpolation is permitted.

bearing slabs only on one side. Such walls which act as end support to floors, e.g. exterior walls, always are designed using the lesser of Eqs. (4) and (5) for determination of the vertical resistance.

For stocky walls, Eq. (5) is governing design (see Figure 2), whereas Eq. (4) determines the vertical resistance of slender walls as Figure 3 illustrates.

Only in special cases, when both the masonry compressive strength and the slenderness are very low (cf. empirical Eq. (8)), Eq. (6) has to be considered additionally when determining the reduction factor. This is due to the very high bending moment resulting from Eq. (1) when very thick walls ($t \geq 0.30$ m) with low compressive strength ($f_k < 2$ MPa) and low wall heights are designed. Figure 4 and Figure 5 show the effects of Eq. (6). However, these cases are of rather academic nature and quite rare in building practice.

$$h/t < 2\left(l_{f,ef} + 1\right) - f_k \qquad (8)$$

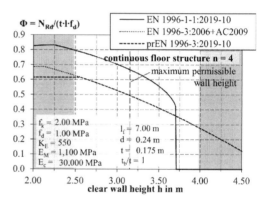

Figure 3. Comparison of load-bearing capacity of slender walls with fully supported slab.

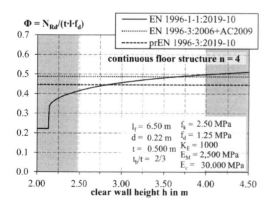

Figure 2. Load-bearing capacity of walls with partially supported slab and lower compressive strength.

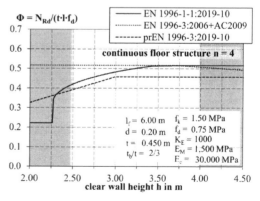

Figure 4. Load-bearing capacity of very thick walls with very low compressive strength and partially supported slab.

2.2 Conditions for application

As shown in Figure 3, the load-bearing capacity decreases with increasing slenderness due to stability failure. In several cases of walls acting as end support to a floor and additionally subjected to wind loads, this leads to maximum permissible wall heights below 4 m, even though the allowed limit in the conditions for applications is not exceeded yet. In Figure 3, this limit is about 3.7 m.

For this reason, a new table containing a limit for the permissible clear wall height depending on the wind load, the compressive strength and the bearing length is introduced as nationally determined parameter (NDP) in section 6.2 of prEN 1996-3 (2019). The values are derived as described in 2.1.1.

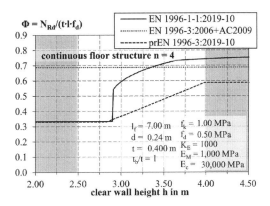

Figure 5. Load-bearing capacity of very thick walls with very low compressive strength and fully supported slab.

Table 3. Limit of the clear wall height of walls acting as end support to a slab.

compressive strength*	wind load	
	$w_{Ek} \leq 0.90$ kN/m²	$w_{Ek} \leq 1.30$ kN/m²
$f_k = 1$ MPa	$h \leq 16\,t_b$	$h \leq 14\,t_b$
$f_k \geq 5$ MPa	$h \leq 24\,t_b$	$h \leq 22\,t_b$

* linear interpolation is permitted for 1 MPa $\leq f_k \leq 5$ MPa.

2.3 Required minimum vertical force

Due to the very low tensile strength of masonry, which is not considered in most normative design models, masonry walls usually need a certain vertical load in order to resist lateral bending moments. In EN 1996-3 (2009), this problem is covered by determination of a required minimum wall thickness. In order to improve the ease of use of this method, in prEN 1996-3 (2019) the design format has changed. Now, according to Eq. (9), it should be verified that the acting minimum normal force in mid-storey height exceeds a required minimum value:

$$N_{Ed,min} \geq \frac{w_{Ed}\, l\, h^2}{7\, t_b} \qquad (9)$$

where l = length of the wind-loaded wall or the width of the wind load to be considered in case of pillars, respectively.

This approach is the result of a comprehensive analysis of several design approaches, e. g. Schmitt et al. (2015), Jäger & Salehi (2016), finally taking the proposal of Jäger & Salehi (2016) as basis. This approach is based on an arch model as shown in Figure 6 and was derived for wind suction, which is the governing design case – especially in combination with a reduced bearing length. Thereby, the

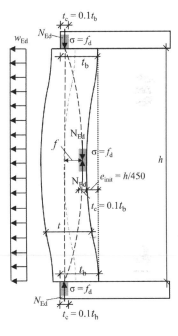

Figure 6. Arch model which forms the basis of the determination of the required minimum vertical load.

rotation of the slab is assumed to be quite low compared to the rotation of the wall due to wind suction. Further, the model assumes an acting normal force of 10 percent of the cross-sectional load-bearing capacity, coming from the self-weight of the slab. Neglecting the self-weight of the wall and considering an initial eccentricity of $e_{init} = h/450$ when determining the arch rise f, the bending resistance can be calculated according to Eq. (10).

$$N_{Ed,min} \geq \frac{w_{Ed}\, l\, h^2}{8f} = \frac{w_{Ed}\, l\, h^2}{8\left(0.9\, t_b - \frac{h}{450}\right)} \qquad (10)$$

For simplification purposes the arch rise in the denominator is approximated on the safe during derivation of Eq. (9).

3 DESIGN OF BASEMENT WALLS

The equations for the design of basement walls subjected to lateral earth pressure are only slightly modified compared to EN 1996-3 (2009), but this modification turns out as big improvement. In the current edition of EN 1996-3 (2009), the given rule applies only if the earth pressure coefficient does not exceed 0.33, which is the fixed value, the equation was calibrated for. Additionally, the clear wall height is limited to 2.6 m. With the new approach for verification of the required minimum normal force according to Eq. (11), clear wall heights up to 3.0 m are possible. This equation now considers the present earth pressure coefficient as variable parameter up to 0.5. Further, the depth of the fill may exceed the wall height up to 1.15 times the clear wall height to enable entry and exit on ground level. Thus, the application range of this design equation has increased significantly and design is more efficient for lower earth pressure coefficients.

$$N_{Ed,min} \geq 0.25 K_e^{1.5} \frac{\rho_e \, b \, h \, h_e^2}{\beta_e \, t} \qquad (11)$$

where K_e = earth pressure coefficient; ρ_e = weight of the soil; h_e = depth of the soil from ground level to bottom of the wall; β_e = factor accounting for uniaxial ($\beta_e = 1$) or biaxial ($\beta_e = 2$) load transfer.

The derivation of Eq. (11) is made quite similar to the approach in 2.3 because stress situations are equal: In order to ensure sufficient bending capacity a certain normal force is required. Taking an analytical arch model as shown in Figure 7 as basis and formulating equilibrium, the equations are solved for

the required normal force. This is done for both bending and shear resistance because shear at the bottom of the wall can be governing for design, too – in contrary to common walls subjected to wind loads. Subsequently, the maximum value of those two failure modes is the required minimum vertical load.

By varying the compressed part of the wall to be considered in the analytical model the required minimum vertical load can be reduced slightly in some cases.

Based on these results for the required minimum vertical load, the analytical model is used for calibration of Eq. (11), which is a semi-empirical formula. Lots of comparative calculations with varying parameter combinations are carried out resulting in a factor 0.25 and a superscript of 1.5 for K_e. Figure 8 and Figure 9 exemplarily illustrate the calibration for different compressive strengths and earth pressure coefficients as indicated (see boundary conditions).

For more detailed background information see Förster & Graubner (2016), Graubner et al. (2017), Mazur et al. (2018).

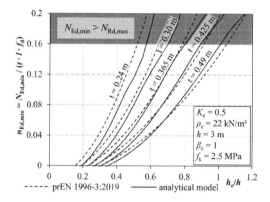

Figure 8. Required minimum vertical load to ensure bending and shear resistance of basement walls with $f_k = 2.5$ MPa and $K_e = 0.5$ based on Graubner et al. (2017).

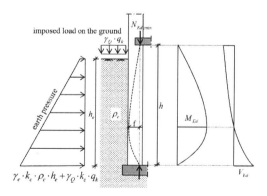

Figure 7. Actions, effects of actions, and assumed arch model for the design of basement walls subjected to lateral earth pressure according to Förster & Graubner (2016).

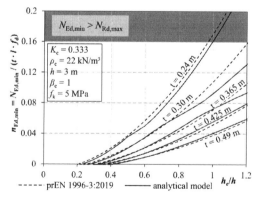

Figure 9. Required minimum vertical load to ensure bending and shear resistance of basement walls with $f_k = 5$ MPa and $K_e = 0.33$ based on Graubner et al. (2017).

4 SIMPLIFIED CALCULATION METHOD FOR SHEAR WALLS

As agreed to by the majority of CEN member states, rules for shear design have been removed from prEN 1996-3 (2019) because they equalled those stated in prEN 1996-1-1 (2019), just written in a different form. However, informative Annex A provides another simplified calculation method for shear walls which can be applied to omit a detailed verification of wind load resistance.

Generally, this method is applicable if the conditions of application of prEN 1996-3 (2019) as well as additional requirements of Annex A are fulfilled. These additional conditions are in particular:

- at least two shear walls in each direction which are mainly located close to the edges;
- shear walls are loadbearing whose vertical load resistance is verified using a reduced masonry compressive strength of $0.8 f_d$.

Then, shear resistance is assumed to be sufficient if the condition in Eq. (12) is fulfilled for both perpendicular directions of the building. Herein, c_t is a dimensionless constant given in Table 4 depending on the actual load level of the wall α_r. This means, only the load ratio of each wall according to Eq. (13) as well as the masonry compressive strength, the total wind load and the geometrical dimensions in each direction must be known for design.

$$w_{Ed}\ l_b\ h_{tot}^2 \leq \sum c_t\ f_d\ t\ l_s^2 \qquad (12)$$

$$\alpha_r = \frac{N_{Gk}}{A\,f_d} \qquad (13)$$

where l_b = length of the building in the considered direction; h_{tot} = total height of the building; c_t = constant according to Table 4; l_s = length of the considered shear wall not to be taken greater than h_{tot}.

The new design method is based on Annex A.3 of EN 1996-3 (2009). However, the current Annex only considers slender walls where the in-plane bending capacity is governing design. The new approach considers both in-plane bending and in-plane shear resistance as derived by Graubner et al. (2017) and therefor applies for stocky walls too, which is why

Table 4. Values of c_t depending on the load level α_r *.

h_{tot}/l_s	load level α_r						
	0	0.05	0.10	0.15	0.20	0.25	0.30
1.0**	0.0	0.011	0.022	0.031	0.038	0.045	0.050
≥ 2.0	0.0	0.023	0.043	0.061	0.077	0.090	0.101

* linear interpolation is permitted
** values lower than 1.0 cannot occur due to the definition of l_s

this approach is more precise and more general. During calibration against the final draft of prEN 1996-1-1 (2019), several assumptions concerning the compressive and shear strengths as well as the partial factor to be used in calibration are necessary.

In general, biaxial bending is considered in calibration assuming a capacity reduction factor in lateral direction of $\Phi_s = 0.6$. Furthermore, imperfections are considered implicitly by reducing the maximum permissible bending moment to 85 % of the actual value according to prEN 1996-1-1 (2019), assuming that 15 % of the total horizontal loading result from imperfections. For comparison, all design equations (both bending and shear) are solved for the maximum permissible characteristic bending moment resulting from wind, which serves as comparative figure. This is done in accordance with the procedure and the relevant combinations of actions described by Graubner & Schmitt (2013).

Another advantage of the new approach is, that the verification format changed to $E_d \leq R_d$ to be consistent with the common format throughout the Eurocodes.

Figure 10 shows an example comparing the load-bearing capacity according to Annex A with that resulting from prEN 1996-1-1 (2019) for those ratios of f_{vk0}/f_k and f_b/f_k and a partial factor for materials of $\gamma_M = 2.0$ which are used for calibration of Annex A. These values are conservative approximations assuming a low initial shear strength and a low compressive strength of the units. Actually, strength values are often higher in practice. This is why Annex A provides slightly conservative results for some practical cases.

Figure 10 uses normalized values of the characteristic permanent load for comparison of shear and bending resistance assuming a common ratio of live load to permanent load $q_k/g_k = 0.5$. Thus, the maximum value of the normalized characteristic value of the permanent load n_{Gk} gives 0.24 as the derivation in Eqs. (14) to (16) demonstrates ($n_{Rk} > 1$ is not

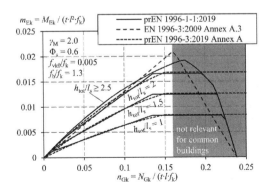

Figure 10. Comparison of load-bearing capacity for strength values used for calibration.

possible). However, due to the fact, that only $0.8f_d$ may be considered in design and the maximum value of the capacity reduction factor for wall is 0.85 according to prEN 1996-3 (2019), the actual maximum value that can occur reduces according to Eq (17). That's why higher values of the normalized characteristic permanent load are highlighted in grey color.

$$\frac{N_{Ed}}{t\,l\,f_k} \leq \frac{N_{Rd}}{t\,l\,f_k} \Leftrightarrow \frac{\gamma_G N_{Gk} + \gamma_Q N_{Qk}}{t\,l\,f_k} \leq \frac{N_{Rk}}{t\,l\,f_k \gamma_M} \quad (14)$$

$$\Leftrightarrow \gamma_G n_{Gk} + \gamma_Q n_{Qk} = \gamma_G n_{Gk} + \gamma_Q \frac{n_{Gk}}{2} \leq \frac{n_{Rk}}{\gamma_M} \quad (15)$$

$$\Leftrightarrow \left(1.35 + \frac{1.5}{2}\right) n_{Gk} \leq \frac{n_{Rk}}{2} \Leftrightarrow n_{Gk} \leq 0.24 n_{Rk} \quad (16)$$

$$n_{Gk} \leq 0.24 \cdot 0.8 \cdot 0.85 = 0.162 \quad (17)$$

More detailed background information can be found in Graubner & Müller (2019).

5 FURTHER MINOR CHANGES

5.1 *Concentrated loads*

The design of wall subjected to concentrated loads is slightly different in prEN 1996-3 (2019) compared to the previous edition. As simplification, only fix increase factors are defined, which only depend on the group of the units and the size of the loaded area. For group 1 units, Eq. (18) applies if the loaded area falls below 1/3 of the effective area of bearing. In all other cases Eq. (19) is to be used.

$$N_{Rdc} = 1.2 f_d A_b \quad (18)$$

$$N_{Rdc} = f_d A_b \quad (19)$$

where A_b = loaded area

This new definition additionally fixes a certain lack of safety where EN 1996-3 (2009) provides higher concentrated load resistance than prEN 1996-1-1 (2019). This is the case for group 1 units as soon as the ratio of loaded area to effective area of bearing exceeds a value of 0.3. Figure 11 illustrates this issue.

5.2 *Reduction of NDP*

As explained in the beginning, one main task during Phase 2 of M515 is the reduction of national choices (NDPs). Due to the fact, that only 7 choices are allowed only in EN 1996-3 (2009), this is quite challenging. Due to the elimination of shear design in

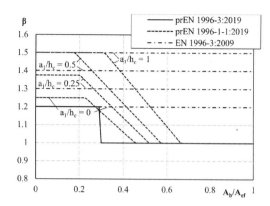

Figure 11. Comparison of the increase factor for concentrated load resistance for different ratios a_1/h_c (distance of the load to the edge of the wall to height wall under level of the concentrated load.

the code, there is no need for values of the initial shear strength, which were previously given in Annex D.2. Additionally, agreement is found on the abandonment of an effective thickness t_{ef}. For simplification purpose – to enhance ease of use – the thickness to be used in the equations is always the full thickness of the single-leaf wall or the full thickness of the inner leaf of a double-leaf or cavity wall. Furthermore, the NDP concerning the choice of the method for verification of the overall stability is omitted because no proposal is given where national choices are possible.

On the contrary, one additional NDP is necessary in order to ensure conservative results compared to part 1-1. Due to the national choices in prEN 1996-1-1 (2019) the maximum permissible wall height of walls subjected to vertical and wind load varies a lot. Therefore, the values given in Table 3 must be determinable nationally.

All other NDPs, which are only very few, remain as they are necessary for national determination of the safety level.

6 CONCLUSIONS

The present paper points out the main improvements and changes in prEN 1996-3 (2019) compared to EN 1996-3 (2009). Beyond a pure comparison of the code, background information on each issue are given and the models and calculations that form the basis of the new rules are explained.

As main change, the new formulation of the capacity reduction factor is to mention, which now also applies to partially supported slabs. Extensive comparative calculations between prEN 1996-1-1 and prEN 1996-3 have been carried out to ensure that the simplified calculation methods always lead to conservative results. This is shown exemplarily for a few cases. In this context, the newly introduced limit of the clear wall height of walls acting as end support to floors or roof is indicated and explained.

In addition to the verification of the maximum vertical load, also the verification of the minimum vertical load has changed. Here, design is based on an arch model and the required minimum value of the normal force can be determined easily.

With regard to basement walls, an arch model forms the basis, too. The new approach is a slight modification of the existing equation considering the earth pressure coefficient directly and accounting for earth pressure coefficients up to 0.5. Additionally, the allowed wall height increases to 3.0 m.

While in the code the existing rule for shear design was skipped, the simplified procedure in Annex A has been modified and now covers both bending and shear resistance. This method turns out as an easily applicable tool for a simplified proof of overall stability of masonry buildings. Naturally, due to calibration with conservative assumptions, there might be cases where the results are very conservative and a more precise design using part 1-1 could be worth it.

Finally, concentrated load resistance is simplified, too, and safety deficits are eliminated at the same time.

On the whole, prEN 1996-3 (2019) proves to be a quite practically orientated standard being easy to apply and providing conservative results compared to part 1-1. In conclusion, ease of use is enhanced significantly and even the number of national choices is reduced from seven to five which is about 25 %.

ACKNOWLEDGEMENTS

The authors want to thank all members of Project Team 2 for their collaborative work and their active contribution to the revised version of EN 1996-3.

REFERENCES

EN 1992- 1-1 (2010). Eurocode 2: Design of concrete structures - Part 1-1: General rules and rules for buildings. Brussels: CEN European committee for standardization.

EN 1996-3 (2009). Eurocode 6: Design of masonry structures - Part 3: Simplified calculation methods for unreinforced masonry structures. Brussels: CEN European committee for standardization.

Förster, V., & Graubner, C.-A. (2016). Design of basement walls under lateral earth pressure. In C. Modena, F. da Porto, & M. R. Valluzzi (Eds.), *Brick and Block Masonry: Proceedings of the 16th International Brick and Block Masonry Conference, Padova, Italy, 26-30 June 2016* (pp. 2225–2230). Boca Raton: CRC Press.

Graubner, C.-A., Faust, T., Purkert, B., Mazur, R., Krieger, L., Müller, D., & Förster, V. (2017). *Erweiterung des Anwendungsgebiets der vereinfachten Berechnungsmethoden nach EN 1996-3/NA: Research report (in German)*: unpublished.

Graubner, C.-A., & Schmitt, M. (2013). Tragfähigkeit unbewehrter Mauerwerksscheiben nach DIN EN 1996-1-1/NA. *Mauerwerk*, *17*(5), 296–306. https://doi.org/10.1002/dama.201300595

Graubner, C.-A., & Müller, D. (2019). Simplified verification method for unreinforced masonry shear walls. *Mauerwerk - European Journal of Masonry*, *23*(5), 300–305. https://doi.org/10.1002/dama.201900011

Jäger, W., & Salehi, H. (2016). *Parameterstudie zur Mindestauflast nach dem vereinfachten Verfahren nach DIN EN 1996-3/NA unter Beachtung der neueren Erkenntnisse zum Bogenmodell: Ausführlicher Endbericht (in German)*: unpublished.

Mazur, R., Purkert, B., Graubner, C.-A., & Förster, V. (2018). Proposal for the simplified design of basement walls under lateral earth pressure. *Mauerwerk - European Journal of Masonry*, *22*(3), 162–174. https://doi.org/10.1002/dama.201800008

PrEN 1996- 1-1 (2019). *Eurocode 6: Design of masonry structures - Part 1-1: General rules for reinforced and unreinforced masonry structures*. Brussels: CEN European committee for standardization.

PrEN 1996-3 (2019). *Eurocode 6 - Design of maonry structures - Part 3: Simplified calculation methods for unreinforced masonry structures*. 3rd draft: CEN/TC 250/SC 6/WG 2/PT 2 - Project Team 2.

Schmitt, M., Graubner, C.-A., & Förster, V. (2015). Minimum vertical load on masonry walls - a realistic view/ Mindestauflast auf Mauerwerkswänden - eine realitätsnahe Betrachtung. *Mauerwerk European Journal of Masonry*, *19*(4), 245–257. https://doi.org/10.1002/dama.201500664.

Brick and Block Masonry - From Historical to Sustainable Masonry –
Kubica, Kwiecień & Bednarz (eds)
© 2020 Taylor & Francis Group, London, ISBN 978-0-367-56586-2

Quantifying surface deterioration: Exemplified on fired clay bricks

C.P. Simonsen & I. Rörig-Dalgaard

Department of Civil Engineering, Technical University of Denmark, Kgs. Lyngby, Denmark

ABSTRACT: For durability examinations of new materials, various resistance tests exists. The present work examines the resistance towards two accelerated salt crystallization methods (European Standard EN 12370 and RILEM test MS-A.2) on four types of fired clay bricks. In line with previous work, different deterioration patterns came into existence with each of these standards followed by an evaluation based on the evaluation criteria´s in the respective standards: weight changes, number of exposure cycles and photo documentation. However, a precise and true evaluation based on these existing evaluation criteria was found challenging to perform. This present work suggest a new method for quantifying initial surface deterioration, by coloring the exposed surface in a contrast color to ease distinction between non-deteriorated and deteriorated areas, followed by quantification with the point count method (best method out of three examined). This simple methodology seems to offer quantification of the surface deterioration.

1 INTRODUCTION

Salt induced deterioration is both of relevance in existing constructions and new materials (Charola & Rörig-Dalgaard, 2019). For new materials, the issue is to test their resistance towards a saline environment originating from costal vicinity, deicing exposure etc. and thereby ensuring production and selection of materials with the needed properties. For this purpose, various standard tests exists such as the European Standard EN 12370 and the RILEM test MS-A.2. These standards basis on different physical phenomena, and consequently results in different degradation patterns (Lubelli et al., 2018). At the same time, the degradation patterns obtained in these tests have shown to differ from degradation patterns seen in praxis. Presently a new RILEM test is in preparation within the RILEM TC 271-ASC, with the aim to develop a test method reflecting natural degradation patterns to a higher extend.

Regardless of the test method, it is essential to evaluate the results obtained from the tests. For EN 12370, the evaluation criteria's are mass change and photographic documentation. In MS-A.2 the evaluation is based on photographic documentation and the number of cycles until initiated deterioration. Further, to unify damage descriptions, two different damage deterioration atlases ICOMOS-ISCS (ICOMOS-ISCS, 2008) and MDCS (MDCS, 2019) are used.

In practice, it is challenging to make conclusions based on the presently proposed evaluation methods: mass change, photographic documentation and number of cycles. In regards to the evaluation of mass changes it has been found challenging to wash out all present salts within the porous material to get a clear picture of actual mass degradation. For the photographic documentation and the number of cycles, it is challenging to make a clear distinction between non-deteriorated materials and deteriorated material and especially to make a precise description of the deterioration degree.

Therefore, a more objective method is lacking. The present work suggests two steps to quantify surface deterioration with the aim to obtain an unambiguous interpretation of the surface deterioration, divided into I) coloring of the exposed surface in a contrast color following by II) quantification. Three different quantification methods are studied.

Similar challenges have been identified for quantifying deterioration in depth, and to overcome this challenge a new method has been proposed, using the state of carbonate (with significantly different pH values) as a tracer to detect surface areas with increased porosity just as cracks starting out from the surface. Thereby, estimation of the condition of the fired clay brick basis on the state of the carbonate just as changes in the silicon content (Rörig-Dalgaard & Charola, 2020).

2 MATERIALS AND METHODS

2.1 *Materials*

Four types of non-calcareous fired bricks figures in the present examinations. Two types were fired at a top temperature of respectively 960°C ±4°C and 1060°C ±4°C at the Danish Technological Institute (DTI, not commercial available). Firing of those two brick types in pilot scale oven proceeded at a constant heating rate of 40°C/hour until the top temperature.

Table 1. Selected material properties, determined in six-fold for each brick type.

Standard	EN 1936 Open porosity [-]	EN 1925 Water absorption coeff.,[kg/m^{2}s$^{1/2}$]	EN 772-1 Compressive strength, [MPa]
Red Neutral*	0.34	0.24	20 [+]
DTI 960	0.33	0.28	-
Black Beauty**	0.32	0.28	20 [+]
DTI 1060	0.28	0.34	-

* Exposure class: salt non-resistant.
** Exposure class: salt resistant.
+ From datasheet.

The top temperature existed for 6 hours. Thereafter, rapid cooling by around 140°C/hour until 600°C. Following, a switch off the oven resulted in natural cooling and in an approximately cooling rate of 20°C/hour around the quartz point. Two commercial available bricks "Red Neutral" (2207A00M) and "Black Beauty" (347A00M) from Egernsund Brickyard with known salt resistant exposure classes were included in the examinations too. The brick specimens had the dimensions: 0.04m x 0.04m x 0.04m, where one of the surfaces were the original exposed surface (here termed exposed test surface). See Table 1 for selected material properties.

For the water vapor permeability measurements "Red Neutral" was cut into cylindrical specimens with a diameter of 0.08m, a height of 0.025 ± 0.001m and one exposed test surface.

Mirabilite (NaSO$_4$·10H$_2$O) from ACROS Organics with a purity > 99 % and Thenardite (NaSO$_4$) from VWR with a purity > 98% was used for the accelerated salt crystallization tests.

Epoxy resin sealed the samples exposed to test method MS-A.2 and the cup method.

STAEDTLER Lumocolor permanent marker 352 with a line width of approx. 2 mm established the contrast color on the test surface. The naming of the brick specimens is according to their name-eventual addition of color on the exposed test surface and specimen number, e.g. DTI 960-C-1 (brick type "DTI 960", colored exposed test surface, specimen number 1).

2.2 Methods

Accelerated salt crystallization tests were performed according to the European Standard EN 12370 and the RILEM test MS-A.2.

A balance, Sartorius ED323S, with an accuracy of ±0.001g was used to follow potential mass changes and report paper masses.

The three methods to quantify the deteriorated surface area from the non-deteriorated surface area: a systematic manual point count, paper indentation and determination using a CAD software.

The systematic point count is according to ASTM E562-19 a method to statistical estimate the volume fraction (expressed in percentage) of a constituent of interest (e.g. deterioration) from any two-dimensional section. Presently mortar analysis make use of this method. In this work, the number of equally spaced points constituted of 100 as a compromise between limitation of labor and reasonably insurance to obtain a reliable result.

The paper weighing method from the European Standard EN 772-2 determines the deterioration fraction by the mass difference of a paper cutout of the deteriorated surface area related to the paper mass of the total surface area. Voids in concrete masonry can be determined with EN 772-2 and the method suggests a determination of specific areas with the use of a planimeter (not performed in the present work).

To use a 2D CAD software, photo documentation is attached and scaled according to original sizes. Following calculation of the deterioration fraction is based on the initial non-deteriorated surface area. The used software was AutoCad in the present work.

Examination of possible change in water vapor permeability due to surface coloring followed the cup method ISO 12572:2016.

Macro pictures were taken with a Nikon D610 camera. The camera was placed on a stand at a distance of 0.12m from the photographed surface. The specimens were illuminated with Kaiser RB 5000 Daylight Copy Light Set. The camera settings were F16; ISO 100.

The thickness of the colered surface layer was studied on a crosssection in a stereomicroscope SMZ25 NIKON with NIS-Elements D software.

3 RESULTS

3.1 Durability assessment based on current methods

The quantitative results from the test methods current evaluation criteria are seen in Table 2 for the four types of test specimens. EN 12370 was performed as six-fold determination, MS-A.2 as a double determination for each brick type. A mass increase was found for three out of the four types of test specimen (after washing). The numbers of cycles until initiated deterioration are subjectively determined by the executor's definition of deterioration.

All test specimens where photo documented. "DTI 960" exposed to both EN 12370 and MS-A.2 appear in Figure 1 and Figure 2 respectively as examples on the photo documentation.

Table 2. Results related to the evaluation criteria's in EN 12370 and MS-A.2 respectively.

Standard	EN 12370 Mass change, [%]	MS-A.2 Number of Cycles
Red Neutral*	0.28± 0.12	8
DTI 960	-0.20± 0.25	8
Black Beauty**	.0.43± 0.22	11
DTI 1060	.0.45± 0.09	11

* Exposure class classified as salt non-resistant.
** Exposure class classified as salt resistant.

Figure 1. Photo documentation for DTI 960-2, exposed to EN 12370.

Figure 2. Photo documentation for DTI 960-3, exposed to MS-A.2.

3.2 A more clear distinction between non-deteriorated and deteriorated areas

Following the approach of the present RILEM TC 271-ASC for development of an improved accelerated salt crystallization method (Flatt et al., 2017), the deterioration is divided into an initial accumulation phase which continues until the first deterioration occurs, followed by a propagation phase proceeding until the end of lifetime of the construction and/or material. To follow the shift between the accumulation phase and propagation phase, it is sufficient to study surface deterioration since this is assumed to be the onset of salt induced deterioration. Further, for application to e.g. accelerated salt crystallization tests, a simple and cheap technique highlighting surface deterioration is needed to ease this distinction. For this purpose coloring of the exposed test surface in a contrast color is proposed. The potential of this proposal is examined with the one brick type resulting in the most extensive visual deterioration in the present work, being "DTI 960". Two brick specimens (red) were colored with a permanent marker (black) on the exposed test surface and subjected to test method EN 12370, just as identical preparations were made for test specimens subjected to test method MS-A.2. See Figure 3 and Figure 4.

Having a significant distinction between non-deteriorated areas (black) and deteriorated areas (red in reality, brighter grey in Figure 3 and Figure 4), makes possible quantification of the deteriorated areas.

3.3 Three methods to quantify the deteriorated surface area

Examination of existing methods for quantification, though presently used for different purposes and their applicability for salt induced surface deterioration follows.

The first method is EN 772-2. A print is made of the test surface as basis for determination of the deteriorated fraction of the surface area according to EN 772-2 and following cut out according to the black lines. See Figure 5. Recording of the paper masses for both the entire test surface and following only the deteriorated area serve as basis for determination of deteriorated area fraction.

See Table 3 for the recorded masses of the deteriorated and non-deteriorated areas along with the deteriorated area in percent.

The second method is use of the software CAD for quantifying non-deteriorated and deteriorated area receptively.

After importing the photo documentation into the CAD software, Figure 6, a division of the photographed exposed test surface takes place into a deteriorated area (hatched) and non-deteriorated surface (black line). The respective areas are

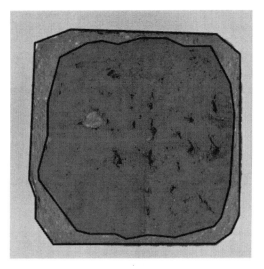

Figure 3. Photo documentation for DTI 960-C-1 with black colored test surface, exposed to EN 12370 (15 cycles).

Figure 5. Division of non-deteriorated areas from deteriorated areas on DTI 960-C-1 (same underlying picture as Figure 3) made more visible with a black line, prior to paper cut out with the indentation method.

Figure 4. Photo documentation for DTI 960-C-3 with black colored test surface, exposed to MS-A.2 (8 cycles).

Table 3. Determination of deteriorated percentage area according to EN 772-2, the paper indentation method, tested on "DTI 960". Both results from the double determination appear, before and after the slash back sign respectively.

Standard	EN 12370	MS-A.2
Total area, [g]	0.25/0.25	0.27/0.27
Deteriorated area, [g]	0.06/0.05	0.19/0.25
Non-deteriorated area, [g]	0.19/0.20	0.07/0.02
Deteriorated area, [%]	23/19	72/92

Manually, the number of points related to deteriorated and non-deteriorated areas adds up and following recalculated to percentage, see Table 5.

For a comparison of the results for percentage-deteriorated area obtained with the three different methods, see Figure 8.

3.4 Water vapor permeability

To examine possible changed surface properties as a consequence of the colored surfaces, which potentially e.g. could influence the result of the accelerated salt crystallization test, water vapor permeability measurements were performed with the brick type "Red Neutral". Following the standard, minimum 5 identical samples were examined. In this case permeability measurements were carried out on 6 reference specimens and 6 specimens colored on the exposed test surface, including both fired fronts and backs just as originating from 3 different bricks to ensure a representative result. The results are shown in Table 6.

following measured in the correct scale [mm] with the CAD software.

Depending on the deterioration pattern, either measurement of the deteriorated area or non-deteriorated area took place. See Table 4.

The third method, the systematic manual point count followed ASTM E562-19 consisting of placing a grid with points in the intersections. For each point it was manually determined whether the point was placed on a deteriorated area (here marked with a black point) or non-deteriorated area (here marked with a bright grey color), see Figure 7.

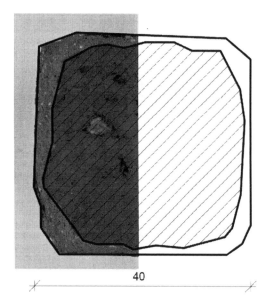

40

Figure 6. Division of non-deteriorated area (hatched) from deteriorated area on "DTI 960-C-1" (same underlying picture as Figure 3). Measurement of the areas are in the correct scale [mm] with 2D CAD software. The left side of the figure shows the original appearance with the picture whereas the right side of the figure only shows the division of areas based on the picture.

Table 4. Results from determination of the deteriorated area with CAD software. Method tested on "DTI 960" as a double determination.

Standard	EN 12370	MS-A.2
Initial total area, [mm^2]	1600/1600	1600/1520
Measured det. area, [mm^2]	-/-	1194/1379
Measured non-det. area, [mm^2]	1179/1257	-/-
Deteriorated area, [%]	26/21	75/91

3.5 Thickness of colored layer

The thickness of the colored layer on the exposed test surface was measured on a crosssection, see Figure 9.

The thickness of the colored layer on the exposed test surface was measured to 228 ± 65 μm, based on 20 measurements on the section shown in Figure 9.

4 DISCUSSION

4.1 Challenges with existing assessment methods

During the accelerated salt crystallization test, each cycle exposure of the sample to a saline solution occurs, resulting in accumulation of salts within the test sample. When following the weight changes in EN 12370, the mass increases until the mass loss resulting

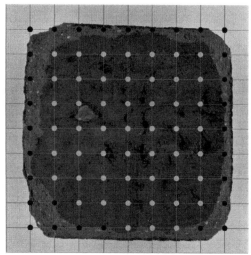

Figure 7. Test grid with 100 points on "DTI 960-C-1" (same underlying picture as Figure 3). Bright grey dots represent non-deteriorated areas, whereas black dots represent deteriorated areas.

Table 5. Determination of percentage deteriorated surface area by point count. Method tested on "DTI 960" as a double determination.

Standard	EN 12370	MS-A.2
Deteriorated area, [%]	30/28	68/86
Non-Deteriorated area, [%]	70/72	32/14

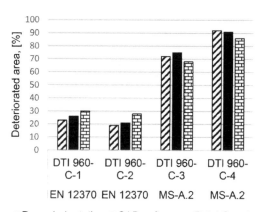

☑ Paper indentation ■ CAD software ⊞ Point Count

Figure 8. Comparison of the results obtained with the three different methods: Paper indentation, CAD software and point count. The percentage-deteriorated areas are a function of sample and accelerated test method.

from salt induced deterioration has a higher impact on the total mass than the accumulation of salt. This is challenging for the evaluation of the test, despite

Table 6. Measured water vapor permeability in six-fold on the brick type "Red Neutral".

	Reference samples	With colored surface
[10⁻¹²kg/ (m·s·Pa)]	Permeability [10⁻¹²kg/ (m·s·Pa)]	Permeability
Red Neutral*	11.4± 0.96	11.3± 0.74

* Exposure class, classified as salt non-resistant.

Figure 9. Cross section for measurement of the colored layer on the exposed test surface. On the picture, at the outer most right side the epoxy resin appears black. Following from the right, the dark grey area visualize the colored layer.

washing of the samples with distilled water before the final weighing. In a previous study (Bartholdy, 2019) the samples were washed thoroughly and dried at 105° C 5 or 6 times respectively before a correlation was found between mass change and visually seen deterioration. However, it must be pointed out, that these washing outs with distilled water and though still partially filled with salt followed by drying at 105°C also affect the result of the test methods (and adds on additional cycles), stressing the challenges with evaluation based on mass changes.

According to MS-A.2, the accelerated salt crystallization test must proceed until initial surface deterioration occurs. During the test, efflorescence comes into existence on the exposed test surface. However, no removal of efflorescence after each cycle took place to mimic real conditions. The drawback was, however, difficulties to follow exactly, after which cycle deterioration initiated. Furthermore, no description on, how initial surface deterioration is defined in the standard and it is therefore up to each executer of the test to evaluate. This complicate the use of number of cycles as a reliable evaluation criterion.

Based on the photographic documentation, the deteriorated areas on the test specimens in Figure 1

and Figure 2 are ambiguous to distinguish from the non-deteriorated. Based on the photographic documentation and using existing damage atlases (ICOMOS-ISCS, 2008) (MDCS), "Red Neutral" and "DTI 960" were found relative less durable in comparison to "Black Beauty" and "DTI 1060" according to both test methods. A comparison with the selected material properties in Table 1, shows relative higher open porosity for "Red Neutral" and "DTI 960" than "Black Beauty" and "DTI 1060". On the contrary "Red Neutral", "DTI 960" and "Black Beauty" possesses relatively lower water absorption coefficients than "DTI 1060". This together indicate that there might be a link between open porosity, water absorption and salt resistance in fired clay bricks.

However, it is challenging to make a precise description of the damages and the extent of these objectively based on present evaluation criteria´s.

4.2 Methodology for clear distinction between non-deteriorated and deteriorated areas

The photographic documentation exemplified in Figure 1 and Figure 2 clearly showed the need for a clear distinction between non-deteriorated and deteriorated areas. The new suggestion in the present work, to add a contrast color (here with a permanent marker) to the surface, clearly offered the missing distinction between non-deteriorated and deteriorated surface, see Figure 3 and Figure 4. This is an important first step to quantify surface deterioration.

4.3 Permeability measurement

It is of outmost importance to ensure coloring of the test surface does not change transport of water vapor through the specimens since this would change the deterioration pattern. Therefore, measurement of the water vapor permeability took place.

It is well-known (Feng et al. 2019), that the reliability of the permeability tests are dependent on the experimental preparations. Therefore, epoxy resin sealing took place twice.

The results from the permeability measurements on 6 reference specimens and 6 specimens colored on the exposed test surface, showed a non-significant difference (see Table 6), indicating that the surface coloring is not affecting the specimens water vapor permeability properties. This result is not surprising since a permanent marker consists of ink and evaporable alcohol leaving only a layer of ink on the exposed test surface after evaporation of the alcohol. The absence of significant differences in water vapor permeability and thereby absence of changed deterioration patterns between with and without colored exposed test surface is further made likely by similar deterioration patterns in Figure 1 (without coloring of the test surface) and Figure 3 (with coloring of the test surface).

4.4 Thickness of the colored layer

When applying a colored layer to the exposed test surface for detection of surface deterioration, obviously deterioration is visualized only when the surface deterioration exceeds the thickness of the colored exposed surface layer. In section 3.5, the thickness of the colored exposed surface layer was found to 228 ± 65 μm, 62 μm as a minimum when limited by aggregates and maximum 312 μm. Surface deterioration beyond 0.3 mm most be expected, and therefore the thickness of the colored exposed test surface layer seems unproblematic in regard to detecting the deteriorated areas.

4.5 Methods of quantifying the deterioration area

The present work examines the applicability of three methods, originally proposed for different purposes, but also for quantify surface deteriorated areas. The examined methods rely on an unbiased approach for a clear distinction between deteriorated and non-deteriorated area and point out the need for the present proposed coloring of the test surface prior to possible use of quantification of surface deterioration. To facilitate comparison between laboratories, narrowing down to one out of the three present examined methods is preferred.

The paper indentation method and point count method are both established standards, well tested for their abilities to determine a percentage area, whereas the method of using a CAD program is a tool and not an acknowledged method. Both the point count and the CAD method can be performed digital after finalized photo documentation, whereas the paper indentation method requires print out, cut out and a balance to record the paper masses. The paper indentation and CAD methods are properly the most precise methods with the present used conditions, since the found values between these two methods are consistent with ±4% points, whereas deviation increased to maximum ±9% when comparing with the point count method. Increased precision of the point count method is likely when increasing the number of points, e.g. a 1000 rather than a 100.

Depending on the demand for precision, the point count method is considered most suitable in regards to simplicity of needed equipment, execution procedure just as ability to obtain consistent results.

4.6 Outlook

When offering a methodology for quantifying surface deterioration, this enables a common understanding of how to define initial surface deterioration not only in relation to test methods but also between producers, advising professionals and consumers. Further, it can be the basis for a common understand of which degrees of surface deterioration is acceptable.

5 CONCLUSION

Within the present work, the suggested evaluation criteria´s in EN 12370 and MS-A.2 was applied on four fired clay brick types. In all cases, it was challenging to offer an objective evaluation with the existing methods. To meet this challenge, a new methodology was proposed consisting of I) coloring the exposed test surface in a contrast color with a permanent marker to enable a clear distinction between non-deteriorated and deteriorated areas II) quantifying the non-deteriorated and deteriorated areas with the point count method described in ASTM E562-19.

The present proposed methodology seems like a significant improvement compared to existing evaluation criteria´s.

ACKNOWLEDGEMENT

DTI is acknowledged for donating the bricks "DTI 960" and "DTI 1060" (non-commercial available) and Torben Moos from Egernsund Brickyard is acknowledged for donating the brick types "Red Neutral" and "Black Beauty".

REFERENCES

ASTM E562-19. Standard Test Method for Determining Volume Fraction by Systematic Manual Point Count.
Bartholdy, J. 2019. Tøsaltes nedbrydende effekt på natursten. En undersøgelse af en række tøsaltsprodukter og deres nedbrydende effekt på natursten brugt i bevaringsværdige bygningsfacader og monumenter. Copenhagen, The Royal Danish Academy of Fine Arts, Schools of Architecture, Design and Conservation.
Charola AE, Rörig-Dalgaard I, 2019. Brick. III. Conservation. Oxford Art online. https://www.oxfordartonline.com/groveart/view/10.1093/gao/9781884446054.001.0001/oao-9781884446054-e-2000000153?rskey=WJSLW9.
EN 772-1. 2015. Methods of test for masonry units – Part 1: Determination of compressive strength.
EN 1925. 1999. Natural stone test methods: Determination of water absorption coefficient by capillarity.
EN 12370. 1999. Natural stone test methods: Determination of resistance to salt crystallization.
Feng C., Guimarães A. S., Ramos, N., Sun L., Gawin D., Konca, P., Konca P, Zhao J, Grunewald J, Hansen K.K, Hall C, Fredriksson M, Pavlík, Z. 2019. A round robin campaign on the hygric properties of porous building materials. Paper presented at 4th Central European Symposium on Building Physics, Prague, Czech Republic.
Flatt R. et al., 2017. Predicting salt damage in practice: A theoretical insight into laboratory tests. RILEM Technical Letters: 108.
ICOMOS-ISCS. 2008. Illustrated glossary on stone deterioration patterns. ISBN:978-2-918086-00-0.
ISO 12572. "Hygrothermal performance of building materials and procucts – Determination of water vapour transmission propertids – Cup method.
Lubelli B, Cnudde V, Diaz-Goncalves T et al. 2018. Towards a more effective and reliable salt crystallization

test for porous building materials: state of the art. *Materials and Structures* 51: 55.

MDCS. 2019. The Monument Diagnosis and Conservatortion system. URL: mdcs. Monumentkennis.nl (visited on 10/ 14/2019).

MS-A.2. 2006. Uni-directional salt crystallization test for masonry units. *Materials and Structures* 31.1: 10–11.

Rörig-Dalgaard I, Charola AE. 2020. A new method for evaluating the condition of fired clay bricks: Preliminary results. *In: Siegesmund, S. & Middendorf, B. (ed.): Monument future: Decay and conservation of stone. – Proceedings of the 14th international congress on the deterioration and conservation of stone – volume i and volume ii. Mitteldeutscher verlag 202*0. *(Submitted).*

Brick and Block Masonry - From Historical to Sustainable Masonry –
Kubica, Kwiecień & Bednarz (eds)
© 2020 Taylor & Francis Group, London, ISBN 978-0-367-56586-2

In-plane shear resistance of thermal insulating monolithic clay unit masonry

U. Meyer
Arbeitsgemeinschaft Mauerziegel im Bundesverband der Deutschen Ziegelindustrie, Germany

D. Schermer & J. Schmalz
Ostbayerische Technische Hochschule Regensburg, Fakultät Bauingenieurwesen, Germany

M. Gams
Faculty of Civil and Geodetic Engineering, University of Ljubljana, Slovenia

M. Lutman & P. Triller
Slovenian National Building and Civil Engineering Institute, Slovenia

ABSTRACT: Thermal insulating single leaf clay unit masonry is a common type of construction in central Europe, a region with very low to moderate seismic exposure. The in-plane shear resistance is the most relevant parameter to describe seismic performance of that type of masonry. A shear test set-up for full-scale storey-high masonry walls was developed within the European-Commission-sponsored research project ESECMaSE. 35 tests with that ESECMaSE-test method were carried out in laboratories in Ljubljana (Slovenia) and Kassel (Germany) to identify the shear capacity for 5 typical perforation patterns of that type of masonry. The tests comprised different wall geometries, different load levels and different types of units, as well as different support lengths of the slab on top of the shear walls. The results are presented and discussed.

1 INTRODUCTION

For exterior walls without additional external insulating layers, hollow clay masonry walls are commonly used in Germany. The clay unit masonry walls have to take vertical and horizontal loads resulting in the structure. For combined loadings the design approaches in the current codes and technical approvals seemed to be very conservative, so 35 shear tests on full scale walls have been performed at the ZAG in Slovenia and Uni Kassel in Germany. The chosen five different types of units as well as the tested wall geometries and load levels represent the most common wall constructions used for housing in Germany.

The aim of the tests was the verification of the shear resistance under realistic boundary conditions and the comparison with the design in the codes (Eurocode 6 with the National Annex) and national technical approvals.

2 MATERIALS

The tests were performed using a total of five types of high precision hollow clay masonry bricks: large-chamber clay masonry units (according to national technical approvals No. 17.1-1006 (ABZ (2014)) Figure 1, left, and 17.1-1084 (ABZ 2018), Figure 1,

right) and clay masonry unit with small coring (according to national technical approvals No. 17.1-1021 (with thermal insulation ABZ (2016)), Figure 2, left, 17.1-946 ABZ 2019, Figure 2, middle and 17.1-889, Figure 2, right). The flatness and plane parallelism of bed faces of the high precision units are adequate for the use of thin layer mortar bed joints with a thickness of 0,5 to 3 mm – the head joints remain generally unfilled. The mechanical properties of the units and thin layer mortar are gi-ven in Tables 1 and 2, respectively.

3 TEST SET-UP

For tests on full scale masonry walls under combined N-M-V-loadings a test-set-up shown in Figure 3 was used.

Bracing walls in usual structures are loaded by axial forces and appropriate horizontal in-plane-loadings. Interaction with the (concrete) floor slabs leads to restraining-effects which lead to counteracting in-plane-bending-moments at top of the wall. As a simplification, this effect can be described through the two limit cases of fixed against rotation at the top of the wall or a cantilever wall without any restraining moments. In the presented tests, fixity at the top of the wall was assumed.

Figure 1. High precision clay masonry units filled with insulating material; large-chamber clay masonry unit (ZAG (2017) and ZAG (2018b)).

Figure 2. High precision clay masonry units with small coring ((ZAG (2018a), and ZAG (2019) Uni Kassel (2010)).

The horizontal load was applied with a cyclic load history, to represent earthquake loadings. The specified loading history was the result of the research project ESECMaSE with the proposal of a unified test procedure (Schermer 2008). The vertical load in the tests was applied corresponding to the specification of moment at the top of the wall and the total amount of vertical load – it was kept constant during each test.

Throughout the tests, on one side of the wall, the displacement field of the entire surface of the walls, including foundation blocks, was measured using optical Digital Image Correlation (DIC) system with measuring accuracy of 2/100 px (pixels) or 0.04 mm. To facilitate the displacement field analysis, the observed side of the wall was painted with a contrasted random speckle pattern.

Table 2. Material properties of used thin layer mortar types "maxit 900 D" and "ZP99".

Brick type Approval No.	Compressive strength of the mortar	flexural strength of the mortar
	[MPa]	[MPa]
17.1.1006	11,4	3,9
17.1.1084 (ZP99)	16,5	6,3
17.1-1021	11,2	4,0
17.1-946	10,5	2,2

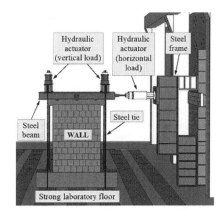

Figure 3. Test set-up for in-plane combined N-M-V-loadings on full scale masonry walls (ZAG (2017)).

4 TEST RESULTS

This paper is dealing with the maximum shear load bearing capacity of the masonry walls (ZAG (2017), ZAG (2018a), ZAG (2018b), ZAG (2019) and UNI K (2009)) under combined N-M-V loadings (N: axial force, M: in-plane bending, V: shear loadings).

Table 1. Material properties of the clay units and strength parameters of the masonry using thin layer mortar.

Brick type Approval No.	Unit Dimensions	Compressive strength of the unit f_b	Density	Volume of all vertical holes (% of the gross volume)	Compressive strength f_k of the masonry according to EN 1052-1	Initial shear strength f_{v0} according to EN 1052-3
	l × t × h [mm]	[MPa]	[kg/m³]	[-]	[MPa]	[MPa]
17.1-1006	248 × 365 × 248	13,0	800	53%	6,6	0,44
17.1-1084	248 × 365 × 248	7,3	800	64%	2,5	0,06
17.1-1021	248 × 300 × 248	12,8	870	43%	7,7	0,52
17.1-946	248 × 365 × 248	9,0	650	56%	2,9	0,18
17.1-889	248 × 300 × 248	10,3	650	57%	nvg	nvg

4.1 Maximum load bearing capacities

In Tables 3 to 6 the maximum measured horizontal loadings can be found.

4.1.1 Clay units with large chambers 17.1-1006

The hysteresis of the cyclic tests on wall 3 and wall 5 are given in Figure 4. In Figure 5 the mean strains at the state of maximum horizontal force, obtained by DIC system, are plotted.

4.1.2 Clay units with small coring

The hysteresis of the cyclic tests on wall 11 and wall 13 are given in Figure 6. In Figure 7 the mean strains at the state of maximum horizontal force, obtained by DIC system, are plotted.

4.1.3 Scatter of the results and determination of characteristic resistance level

Several walls were tested with identical conditions (loadings, boundary conditions). The comparison is given in Table 7. It can be found, that the scatter generally is very small.

For the lowest layer a general purpose mortar was used. The compression strength was 11,6 MPa, 12,4MPa and 12,1 MPa respectively.

For the determination of the characteristic resistance the following equation (1) can be used:

$$V_{Rk,Experiment} = \frac{V_{R,m}}{f_{SsvF}} \qquad (1)$$

Table 3. Maximum horizontal force within the tests on full scale masonry walls (large-chamber units (ABZ (2014))).

Test specimen	W1	W2	W3	W4	W5	W6	W7	W8	W9	W10
Length of the wall [m]	2,00	2,00	2,00	2,00	2,00	1,00	1,00	1,00	1,00	1,00
Overlapping (units) × h_{Unit}	0,4	0,4	0,4	0,5	0,4	0,4	0,4	0,4	0,5	0,4
Vertical load MPa; kN/m	1,27; 471	2,07; 748	0,72; 247	2,07; 748	2,07; 748	2,07; 748	2,07; 748	0,82; 247	1,35; 494	1,35; 501
Maximum horizontal force V_{max} [kN]	339	349	212	370	376	132	123	78	140	122

Table 4. Maximum horizontal force within the tests on full scale masonry walls (large-chamber units (ABZ (2013))).

Test specimen	W1-ZAG	W2-ZAG	W3-ZAG	W4-ZAG	W5-ZAG	W6-ZAG	W1-UNIK
Length of the wall [m]	1,00	1,00	1,50	2,00	2,00	2,00	1,50
Vertical load MPa; kN/m	0,89; 321	0,30; 107	0,32; 113	0,30; 107	0,59; 214	0,89; 321	0,32; 113
Maximum horizontal force V_{max} [kN]	84	39	83	109	182	215	88

Table 5. Maximum horizontal force within the tests on full scale masonry walls (unit with small coring (ABZ (2016))) 17.1-1021.

Test specimen	W11	W12	W13	W14	W15	W16	W17	W18	W19	W20
Length of the wall [m]	2,00	2,00	2,00	2,00	2,00	1,50	1,50	1,00	1,00	1,00
Vertical load MPa; kN/m	0,62; 187	1,25; 374	1,87; 561	0,62; 187	1,87; 561	0,62; 187	1,87; 561	0,62; 187	1,25; 374	1,87; 561
Maximum horizontal force V_{max} [kN]	183	317	368	187	356	129	236	63	113	138

Table 6. Maximum horizontal force within the tests on full scale masonry walls (unit with small coring 17.1-946 and 17.1-889).

Test specimen	W1-946	W2-946	W3-946	W4-946	W5-946	W6-946	W2-889	W6-889
Length of the wall [m]	1,00	1,00	1,50	2,00	2,00	2,00	2,00	1,00
Vertical load MPa; kN/m	0,46; 166	1,37; 498	0,46; 166	0,46; 166	0,91; 332	1,37; 498	0,47; 140	0,20; 60
Maximum horizontal force Vmax [kN]	53	84	102	144	177	223	125	32

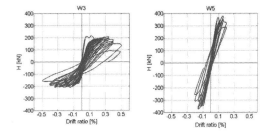

Figure 4. Hysteresis of the tests on walls W3 (L=2 m, vertical load level 0,72 MPa/left) and W5 (L=2m, vertical load level 2,07 MPa/right)/large-chamber clay masonry unit (ZAG (2017)).

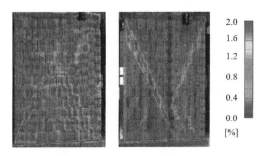

Figure 7. Graphic plot of the mean strains at maximum load level: walls W11 (L=2 m, vertical load level 0,62 MPa/left) and W13 (L=2 m, vertical load level 1,87 MPa/middle)/unit with small coring (ZAG (2018a)); principal strain scale (right).

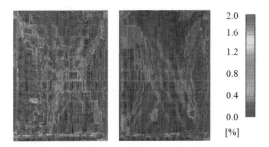

Figure 5. Graphic plot of the mean strains at maximum load level: wall W3 (L=2 m, vertical load level 0,72 MPa, left) and wall W5 (L=2 m, vertical load level 2,07 MPa, middle)/large-chamber clay masonry unit (ZAG (2017)); principal strain scale (right).

Table 7. Comparison of tests with identical conditions.

Comparable Tests	Length of the wall [m]	Vertical load level [MPa]	Difference of the maximum horizontal forces [%]
W2/W5	2,00	2,07	8%
W6/W7	1,00	2,07	7%
W9/W10	1,00	1,36	15%
W11/W14	2,00	0,62	2%.
W13/W15	2,00	1,87	3%

5 COMPARISON WITH THE CODE-DESIGN

A comparison of the test results and the results using the provisions in the design codes is decisive for the quality of the design rules currently used for these products, In the following consideration, the verification is performed with the measured mean values of the material properties (initial respectively adhesion shear strength, brick compression strength, brick tension strength derived from the brick compression strength, masonry compressive strength derived from the brick compression strength).

The characteristic values of shear loadbearing capacity are considered with the reduction factor of f_{SsvF} = 1.2 and the partial safety factor γ_M (resistance) taken to be 1,5.

The determination of the shear resistance is done applying following equation (design-level) according EUROCODE 6 (2005):

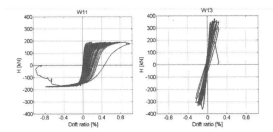

Figure 6. Hysteresis of the tests on walls W11 (L=2 m, vertical load level 0,62 MPa, left) and W13 (L=2m, vertical load level 1,87 MPa, right)/unit with small coring (ZAG (2018a)).

$$V_{Rdlt} = l_{cal} \cdot f_{vd} \cdot \frac{t}{c} \qquad (2)$$

The mean value of the tests is $V_{R,m}$ and the factor f_{SsvF} gives the relation between mean and characteristic value $V_{Rk,Experiment}$ – it can be taken to be equal to f_{SsvF} = 1,2.

where l_{cal} is the design length of the wall, f_{vd} the design shear strength, t the wall thickness and c a factor for the distribution of shear stresses.

Regarding the two design equations for the two different failure modes

- friction failure (equation NA.4 in DIN EN 1996-1-1/NA (EUROCODE 6 (2005))

$$f_{vlt1} = f_{vk0} + 0.4 \cdot \sigma_{Dd} \qquad (3)$$

where f_{vk0} is the characteristic bond strength and σ_{Dd} the design value of vertical stress
and

- tension failure of the bricks (equation NA.4 in DIN EN 1996-1-1/NA (EUROCODE 6 (2005))

$$f_{vlt2} = 0.45 \cdot f_{bt,cal} \cdot \sqrt{1 + \frac{\sigma_{Dd}}{f_{bt,cal}}} \qquad (4)$$

where $f_{bt,cal}$ is the design value of the tensile strength of the unit and σ_{Dd} the design value of vertical stress.

It can be found, that in the regarded cases the tension failure f_{vlt2} was always decisive. This corresponded with the observed failure modes within the tests (Figures 4, 5, 6 and 7).

For the brick failure criteria the tension strength $f_{bt,cal}$ was taken equal to 2,6% of the compression strength of the bricks f_b. As the head joints always were unfilled, this effect had to be taken into account by reducing the initial shear strength f_{vk0} to 50% of the experimentally determined value (Table 1). An additionally reducing of the initial shear strength caused by cyclic loadings and the possible degradation of this strength was not taken into account.

Usually according the German National Annex of Eurocode 6 (EUROCODE 6 (2015)) the distribution of the shear stresses is not assumed to be constant. For design a so called "shear-stress-distribution factor c" has to be taken into account – it depends on the shear slenderness and is between 1,0 and 1,5. For the presented tests the value could be taken to be c=1,5 (wall length 1,0 m), c=1,417 (wall length 1,5 m) and c=1,118 (wall length 2,0 m).

The verification of the bending compression zone was performed using equation (5):

$$N_R = t \cdot 1 \cdot \Phi \cdot f > N_E \qquad (5)$$

Where t and l are the thickness respectively length of the wall, f is the compression strength of the wall and Φ describes the assumption of a rectangular stress block when eccentric loadings are appearing.

$$\Phi = 1 - 2 \cdot M \cdot 1 / N \qquad (6)$$

Within these tests, the equations (5) and (6) were not relevant for design.

Since the check of shear load-bearing capacity has always to be performed under combined axial force and bending, and also that action and resistance depend on each other through (compressed length of a wall depends on the vertical load level but also on the bending moment, which depends on the horizonal load), only an iterative procedure is possible for the determination of the calculated loadbearing capacity.

Both mentioned products are covered by technical approvals (ABZ (2014) respectively ABZ (2016)) for the use in load-bearing walls in Germany. In these documents additional requirements for design and construction are given. For the shear design the approvals reduce the shear strength value to 33% to 50% of the values obtained from the code equations (3) and (4).

5.1 Large chamber clay masonry unit walls

With the large-chamber clay masonry units it was found, that the code equations in DIN EN 1996-1-1/NA (EUROCODE 6 (2015)) (equation (2)) with the assumed mean experimentally determined material parameters describe the results of the tests well and on the safe side. For all, the design equations underestimate the real loadbearing capacity by 20 % (range of values 5 % to 42 %) – no value was overestimated by the calculated prediction, as is shown in Tables 8 and 9.

The specification of the approvals (ABZ (2014), ABZ (2013)), taking the shear strengths only to 50% of the value of the code, is extremely conservative and uneconomic.

Table 8. Comparison of the experimental results with the results of the design code (EUROCODE 6 (2015)).

Test specimen	W1	W2	W3	W4	W5	W6	W7	W8	W9	W10
$V_{R,mean,Experiment}$ [kN]	339	349	212	370	376	132	123	78	140	122
$V_{R,k,Experiment}$ [kN]	282	291	177	308	313	110	103	65	117	102
$V_{R,d,Experiment}$ [kN]	189	194	118	205	209	73	69	43	78	68
$V_{R,dlt,EUROCODE\ 6}$ [kN]	136	166	110	166	166	66	66	40	55	55
$V_{R,d,Experiment}/V_{R,dlt,EUROCODE\ 6}$	139%	117%	107%	123%	126%	111%	105%	108%	142%	124%

Table 9. Comparison of the experimental results with the results of the design code (EUROCODE 6 (2015)).

Test specimen	W1-ZAG	W2-ZAG	W3-ZAG	W4-ZAG	W5-ZAG	W6-ZAG	W1-UNIK
$V_{R,mean,Experiment}$ [kN]	84	39	83	109	182	215	88
$V_{R,k,Experiment}$ [kN]	70	32,5	69,2	90,8	151,7	179,2	73,3
$V_{R,d,Experiment}$ [kN]	46,7	21,7	46,1	60,6	101,1	119,4	48,7
$V_{R,dlt,EUROCODE\ 6}$ [kN]	33	19	34	55	71	83	41
$V_{R,d,Experiment}/V_{R,dlt,EUROCODE\ 6}$	142%	116%	134%	111%	143%	144%	119%

Table 10. Comparison of the experimental results with the results of the design code (EUROCODE 6 (2015)).

Test specimen	W11	W12	W13	W14	W15	W16	W17	W18	W19	W20
$V_{R,mean,Experiment}$ [kN]	182	317	368	187	356	129	236	63	113	138
$V_{R,k,Experiment}$ [kN]	153	264	307	156	297	108	197	53	94	115
$V_{R,d,Experiment}$ [kN]	102	176	205	104	198	72	131	35	63	77
$V_{R,dlt,EUROCODE\ 6}$ [kN]	86	110	130	86	130	53	82	30	44	51
$V_{R,d,Experiment}/V_{R,dlt,EUROCODE\ 6}$	119%	160%	158%	121%	152%	136%	160%	117%	143%	151%

Table 11. Comparison of the experimental results with the results of the design code (EUROCODE 6 (2015)).

Test specimen	W1-946	W2-946	W3-946	W4-946	W5-946	W6-946	W2-889	W6-889
$V_{R,mean,Experiment}$ [kN]	53	84	102	144	177	223	125	32
$V_{R,k,Experiment}$ [kN]	44,2	70	85	120	147,5	185,8	104,2	26,7
$V_{R,d,Experiment}$ [kN]	29,4	46,8	56,7	79,8	98,1	124	69,4	17.8
$V_{R,dlt,EUROCODE\ 6}$ [kN]	26	45	46	74	95	113	66	14
$V_{R,d,Experiment}/V_{R,dlt,EUROCODE\ 6}$	113%	104%	123%	108%	103%	110%	105%	127%

5.2 Small coring clay masonry unit walls

For the clay masonry units with small coring, in Tables 10 and 11 it can be found, that the code equations in DIN EN 1996-1-1/NA (EUROCODE 6 (2005)) with the assumed mean experimentally determined material parameters tend to describe the results of the tests very well, but considerably under-estimate the absolute shear loadbearing capacities. On average, the design equations for vertically cored clay masonry units underestimate the real shear load-bearing capacity by 41 % (range of values 17 % to 60 %) – no value was overestimated by the calcula-ted prediction.

The specification of the approvals (ABZ (2016), ABZ (2014), ABG (2018)), taking the shear strengths only to 50% of the value of the code, is extremely conservative and uneconomic.

6 SUMMARY

In this paper, the results of 35 shear tests on full-storey-height masonry walls of high precision clay masonry units have been presented and discussed.

The objective of the extensive investigations was to evaluate the loadbearing behavior under realistic boundary conditions and to check the quality of the design equations currently given in the code and the requirements in the technical approvals. It was found, the design equations in the current code DIN EN 1996-1-1/NA tend to describe the tests well and on the safe side.

Regarding these test results and the comparison to the current design approaches, it can be concluded, that a reduction of the shear loadbearing capacity of clay masonry units within the technical approvals compared to the values specified in DIN EN 1996-1-1/NA for clay-unit masonry is not necessary.

REFERENCES

ABG. 2018. Allgemeine Bauartgenehmigung des Deutschen Instituts für Bautechnik. Berlin 17.1-889. Mauerwerk aus POROTON-Planhochlochziegeln T10-T11 „Mz 33"im Dünnbettverfahren vom 09.02.2018.

ABZ. 2014. Allgemeine Bauaufsichtliche Zulassung des Deutschen Instituts für Bautechnik. Berlin 17.1-1006:

Mauerwerk aus THERMOPOR Planhochlochziegeln mit integrierter Wärmedämmung im Dünnbettverfahren mit gedeckelter Lagerfuge vom 14.10.2014.

ABZ. 2016. Allgemeine Bauaufsichtliche Zulassung des Deutschen Instituts für Bautechnik. Berlin 17.1-1021: Mauerwerk aus Planhochlochziegeln UNIPOR-WS10 CORISO im Dünnbettverfahren mit gedeckelter Lagerfuge vom 12.10.2016.

ABZ. 2013. Allgemeine Bauaufsichtliche Zulassung des Deutschen Instituts für Bautechnik. Berlin 17.1-1084: Mauerwerk aus Planhochlochziegeln mit integrierter Wärmedämmung bezeichnet als „Thermoplan MZ70"im Dünnbettverfahren vom 13.02.2013.

ABZ. 2014. Allgemeine Bauaufsichtliche Zulassung des Deutschen Instituts für Bautechnik. Berlin 17.1-946: Mauerwerk aus Ott Klimatherm PL Ultra Planhochlochziegeln im Dünnbettverfahren mit gedeckelter Lagerfuge vom 30.06.2014.

EUROCODE 6. 2005: DIN EN 1996-1-1: 2005 EUROCODE 6: Bemessung und Konstruktion von Mauerwerksbauten - Teil 1-1: Allgemeine Regeln für bewehrtes und unbewehrtes Mauerwerk; Deutsche Fassung EN 1996-1-1:2005+A1:2012 + Nationaler Anhang NA: 2012-05.

Schermer, D. 2008. Vorschlag für ein neues Schubprüfverfahren. *Mauerwerkskalender*. Berlin: Ernst & Sohn.

Schermer, D. 2019. Schubtragfähigkeit von zulassungsgeregeltem wärmedämmendem Planziegelmauerwerk. (2019-7015). Burghausen: 27.11.2019.

UNI KASSEL. 2009. Horizontalkraft-Tragfähigkeit von Außenwandkonstruktionen aus Wärmedämmziegeln. Juli.

ZAG. 2017. Slovenian National Building and Civil Engineering Institute. Ljubljana. ZAG - Report P 455/17-610-1 (19.12.2017). Seismic response of masonry walls by experimental testing.

ZAG. 2018a. Slovenian National Building and Civil Engineering Institute. Ljubljana. ZAG - Report P 455/17-610-2 (27.02.2018). Seismic response of masonry walls by experimental testing.

ZAG. 2018b. Slovenian National Building and Civil Engineering Institute. Ljubljana. ZAG - Report P 577/18-610-1 (11.10.2018). Seismic response of masonry walls by experimental testing.

ZAG. 2019. Slovenian National Building and Civil Engineering Institute. Ljubljana. ZAG - Report 160/19-610-1 (23.08.2019). Seismic response of masonry walls by experimental testing.

Composite materials in masonry

Brick and Block Masonry - From Historical to Sustainable Masonry –
Kubica, Kwiecień & Bednarz (eds)
© 2020 Taylor & Francis Group, London, ISBN 978-0-367-56586-2

In-plane behaviour of FRCM-strengthened masonry panels

A. Incerti, A.R. Tilocca & A. Bellini
CIRI Buildings and Construction, University of Bologna, Bologna, Italy

M. Savoia
Department of Civil, Chemical, Environmental and Materials Engineering, University of Bologna, Bologna, Italy

ABSTRACT: The evolution of seismic design regulations together with the degradation of the monumental heritage made necessary to increase the structural safety of a large variety of masonry buildings. A recent strengthening technique consists in the application of a fiber grid/unidirectional sheet with lime or cementitious mortar (FRCM), thanks to several advantages of this retrofitting system in comparison to traditional epoxy-based materials (FRPs), such as better compatibility with the masonry substrate, resistance to high temperatures and reversibility. In this paper, the results of in-plane tests performed on panels strengthened with different types of FRCM systems (that differ in layout, matrix, grid materials and spacing) are presented, after a complete mechanical characterization of FRCM materials, carried out through direct tensile tests and single-lap shear tests. In diagonal compression tests (DCT), failure modes and global behavior of the panels were also analyzed using Digital Image Correlation (DIC) technique, for a better understanding of their in-plane behaviour.

1 INTRODUCTION

In the last years, many studies analyzed Fiber Reinforced Cementitious Matrix (FRCM) systems for retrofitting masonry structures vulnerable to seismic actions, as an alternative to the more common Fiber Reinforced Polymer (FRP) based strengthening techniques. FRCM retrofitting systems are preferred nowadays thanks to their better compatibility with the substrate, in particular when strengthening of ancient buildings, with a particular historical relevance, is required, and the use of epoxy resins is forbidden by the authorities with specific rules for buildings subject to heritage protection.

Several researchers performed experimental campaigns in order to characterize the mechanical properties of FRCMs (De Felice et al. 2014, Bellini et al. 2016, 2019a, b, Carozzi et al. 2017). The in-plane behaviour of masonry element retrofitted with composite materials with different matrices and/or type of fibers is studied in (Papanicolaou et al. 2011, Bellini et al. 2015, Tilocca et al. 2019) and in (Faella et al. 2010, Ferretti et al. 2019, Incerti et al. 2019a, b, Parisi et al. 2013), considering brick and stone masonry panels. Other studies in literature show the comparison of experimental results of campaigns where both FRCMs and FRPs were used for strengthening the masonry panels (Babaeidarabad et al. 2014, Ferretti et al. 2016). Authorities and institutions, especially in Italy in the last year, indicated peculiar aspects and details concerning the application of FRCMs in guidelines and/or codes, in order to regulate the growing diffusion of these new retrofitting systems.

Therefore, mechanical characterization of the properties of each component can be now properly defined, with the aim of qualifying the final product according to a widely shared scientific approach. In fact, any manufacturer must specify all the details concerning the retrofitting system, like fiber/textile material, density, grid spacing, matrix composition and performances, thickness to be adopted during the application on the different substrates, before testing and qualifying the product. In this framework, the present experimental work is devoted to analyzing, through preliminary characterization tests and large-scale shear tests, some different combinations of matrix and fiber, in order to evaluate the behaviour of the strengthened element and the performances of these retrofitting systems when applied on masonry structures realized with poor quality materials.

2 EXPERIMENTAL CAMPAIGN

2.1 *Experimental program, retrofitting systems and layout*

During the experimental activities, seven double-leaf masonry panels ($129 \times 129 \times 25$ cm³) were built in laboratory by using standard clay bricks ($12 \times 25 \times 5.5$ cm³) and hydraulic lime-based mortar (with 10 mm thickness) for the bed joints. After the curing time in controlled environmental conditions ($20 \pm 5°C$, $60 \pm 10\%$ RH) for 28 days, specimens were retrofitted on both sides according to several layouts with the aim of comparing different combinations of fibers and matrices available on the market. In particular, as for the fabrics/textiles used in the experimental campaign, unidirectional steel fiber sheets with a density equal to 650 g/m^2 (S), basalt grids with 170 (B2) or 264 g/m^2 (B1) density and glass fiber grid with 150 g/m^2 density (G) were adopted. The retrofitting layout used for unidirectional sheets can be seen in Figure 1a, where four strips, applied on the masonry substrate with a spacing equal to 580 mm in both (vertical and horizontal) directions, are shown. Basalt and glass fiber grids, instead, were applied adopting a continuous layout (Figure 1b) covering the entire surface of the

Table 1. FRCM strengthening systems adopted in the study.

Sample	Fiber type	Grid spacing [mm]	Equivalent thickness [mm]	Matrix
DCT_URM	-	-	-	-
DCT_S_NHL1	Steel	-	0.083	NHL1
DCT_S_NHL2	Steel	-	0.083	NHL2
DCT_B1_NHL1	Basalt	8.5	0.049	NHL1
DCT_B1_NHL2	Basalt	8.5	0.049	NHL2
DCT_B2_NHL2	Basalt	17.5	0.027	NHL2
DCT_G_CEM1	Glass	16.5	0.028	CEM1

wall, considering an overlapping zone equal to 200 mm centered in the middle of the panel.

As far as the matrices adopted are concerned, two kinds of hydraulic lime-based mortars were used for coupling steel sheets and basalt fiber grids, with a granulometry of fine aggregates < 3 mm and < 1.5 mm, respectively, indicated in the following with the codes NHL1 and NHL2. For glass fiber grids, instead, a cement-based mortar (CEM1) was adopted. After the curing time, reinforced panels were tested according to ASTM E519 for analyzing the shear behavior of strengthened masonry walls through a diagonal compression test set-up, in order to evaluate the performances of the retrofitting systems and the resulting failure modes. In the following, the panels will be named DCT_X (fiber type)_Y (matrix type) for what concerns diagonal compression tests, T_X (fiber type)_Y (matrix type) for direct tensile tests and B_X (fiber type)_Y (matrix type) for bond tests. Table 1 summarizes the characteristics of the strengthening systems under study together with the adopted sample code.

2.2 Masonry texture and mechanical properties of the materials

It is well known, from literature, that a typical masonry element is an heterogeneous and anisotropic medium composed of brick/stone and mortar. So, the monolithicity of the panel, when subject to in-plane actions, is very important in order to prevent local collapse mechanisms. Taking into account these considerations, with the purpose of realizing poor quality masonry panels in order to highlight their possible weaknesses, in the experimental campaign the masonry panels were not built using a classical layout, but the presence of bricks connecting the two leafs was drastically reduced to only four elements (see Figure 2). In addition, a poor-quality mortar was adopted for the bed-joints. The mortar was mechanically characterized through standard laboratory tests conducted on 21 mortar specimens, according to EN 1015-11, obtaining a mean compressive strength equal to 1.51 MPa (CoV = 23.8%) and a mean flexural strength of 0.51 MPa (CoV = 28.8%).

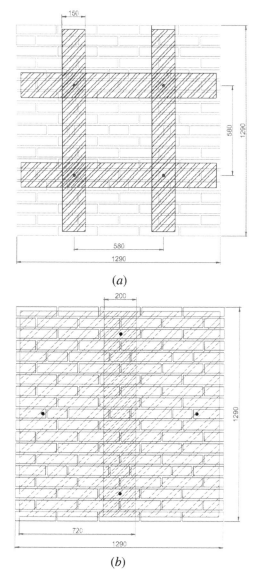

(a)

(b)

Figure 1. Reinforcement layouts adopted in experimental tests on FRCM strengthened masonry panels.

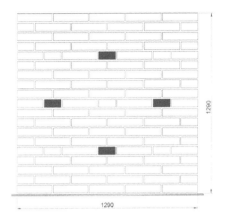

Figure 2. Panel texture: bricks (in red) connecting the two leafs.

Brick specimens were also tested, obtaining results in terms of compressive strength (according to EN 772-1), tensile strength (brazilian splitting test) and flexure capacity of 18.66 MPa (CoV = 5.3%), 3.14 MPa (CoV = 15.3%) and 4.79 MPa (CoV = 6.9%), respectively.

2.3 Mechanical characterization of the FRCM systems

Mechanical characterization of mortars used as matrices within the FRCM strengthening systems was performed according to EN 772-1, EN 1015-11 and EN 12390-13 and provided the results shown in Table 2.

Mechanical characterization of the strengthening systems was carried out through tensile tests on FRCM coupons and bond tests on masonry substrate under controlled laboratory conditions (de Felice et al. 2018, FRCM Italian Guidelines 2018).

Uniaxial tensile tests were performed on samples characterized by the number of strands/yarns indicated in Table 3, with three test repetitions for each FRCM type. Tests were carried out using a MTS servohydraulic testing machine with a maximum capacity of 100 kN, equipped with an onboard class 0.5 load cell. Tests were performed in displacement control at a rate of 0.2 mm/min, recording strain by means of a MTS extensometer, adopting a 200 mm gage

Table 2. Mechanical properties of FRCM matrices.

Matrix	Compressive strength f_c [MPa]	Tensile strength f_t [MPa]	Elastic modulus E_f [GPa]
NHL1	17.21	4.10	13.00
NHL2	20.44	5.75	10.44
CEM1	21.19	6.74	11.10

Table 3. Tensile tests results.

Sample type	Number of strands/yarns	σ_f [MPa]	$\sigma_{f, mean}$ [MPa]	CoV [%]
T_S_NHL1	9	2463	2576	6.60
		2494		
		2772		
T_S_NHL2	9	2472	2470	1.28
		2500		
		2437		
T_B1_NHL1	7	1647	1631	0.95
		1616		
		1630		
T_B1_NHL2	7	1712	1663	2.67
		1650		
		1626		
T_B2_NHL2	4	1720	1649	5.35
		1550		
		1676		
T_G_CEM1	4	1324	1301	1.62
		1283		
		1295		

length. Samples were strengthened at their extremities with composite tabs before the tests. The obtained stress-strain curves are shown in Figures 3, 4, where a typical curve for each

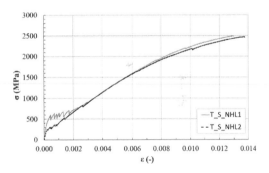

Figure 3. Tensile behavior of steel FRCM samples.

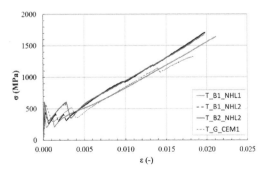

Figure 4. Tensile tests on basalt and glass FRCM coupons.

315

strengthening system type is reported. Experimental results in terms of maximum stresses, calculated taking into account the area of dry fibers only, are indicated in Table 3, together with mean values and Coefficient of Variation (CoV). Tensile tests conducted on steel strengthening systems showed a reduced scattering, in particular if the steel fiber system is used in combination with NHL2 matrix. Stress corresponding to first cracking was found to be lower for this mortar type in comparison to NHL1 matrix (see Figure 3), but NHL2 mortar was able to promote a better stress redistribution, with more distributed cracks along the specimen and a smoother transition between the first and the second branch. Tests carried out on basalt FRCM reinforcements showed similar ultimate stresses, regardless of the type of matrix used or the grid density. The third branch (Figure 4) was found to be rather similar for the three basalt strengthening systems. However, unlike steel, tensile tests highlighted, during the second (cracking) phase, the formation of a reduced number of macro-cracks along the extensometer gage length (between 2 and 4). All basalt samples showed, as typical failure mode, fiber rupture affecting all the longitudinal yarns (Figure 5a). In steel specimens, instead, fiber failure mainly affected the central part of the unidirectional sheet, with the slippage of one or both external strands (Figure 5b). Finally, as for glass FRCM, tensile tests showed a reduced scattering in terms of ultimate stresses, with a mechanical behaviour characterized by the formation of three or four macro-cracks along the gage length of the extensometer and fiber rupture occurring in correspondence of one of the cracks already opened in the central part of the sample.

Bond tests were performed adopting a previously validated single-lap set-up (Mazzotti et al. 2015, Carozzi et al. 2017, Bellini et al. 2019a, b), in which the specimen was placed inside a rigid steel frame designed to fasten the sample avoiding displacements and rotations. Tests were conducted using a MTS servo-hydraulic testing machine with

a maximum capacity of 100 kN and a class 0.5 load cell. Bond tests were conducted in displacement control, adopting a maximum rate of 0.2 mm/min. Slip, defined as the relative displacement between the strengthening grid and the adjacent substrate, evaluated at the beginning of the reinforcement cross-section, was measured by means of two 20 mm displacement transducers (LVDTs). Results of single-lap shear tests, in terms of maximum stresses (referred to the area of dry fibers) and failure modes, are reported in Table 4, whereas some typical stress-slip curves are shown in Figures 6,7 (a single reference curve for each strengthening system).

Steel FRCMs showed two different failure modes: debonding at the matrix-to-textile interface (Failure mode A, see Figure 8a) or

Table 4. Results of single-lap shear tests.

Sample type	Failure mode	σ_f [MPa]	$\sigma_{f,\ mean}$ [MPa]	CoV [%]	η [%]
B_S_NHL1	A	2358	2347	3.46	91.11
	B	2261			
	B	2422			
B_S_NHL2	A	2440	2429	1.63	98.34
	A	2462			
	A	2385			
B_B1_NHL1	C	860	960	9.80	58.86
	D	1047			
	C	972			
B_B1_NHL2	D	1035	1090	6.00	65.54
	D	1074			
	D	1162			
B_B2_NHL2	C	1133	1190	12.85	72.16
	C	1074			
	C	1364			
B_G_CEM1	C	1145	1170	7.66	89.93
	C	1095			
	C	1269			

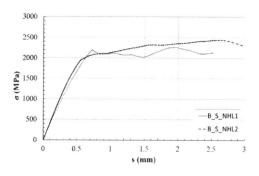

Figure 6. Single-lap shear tests on steel FRCMs: stress-slip curves.

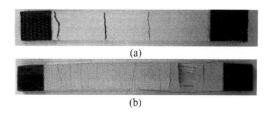

Figure 5. Failure modes identified during tensile tests on FRCM systems: (a) basalt fibers, (b) steel fibers.

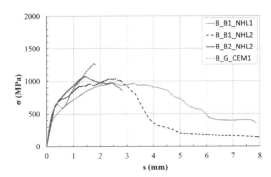

Figure 7. Single-lap shear tests on basalt and glass strengthening systems: stress-slip curves.

debonding at the matrix-to-textile interface with partial debonding at the support interface (Failure mode B, Figure 8b). The typical behavior of this class of samples, in terms of stress-slip curves, is shown in Figure 6 where a first almost linear branch is followed by a second pseudo-horizontal part characterizing the debonding phenomenon.

Basalt strengthening systems highlighted two different failure modes: slippage and final tensile failure of the fiber (Failure mode C, Figure 8c) or simple slippage of the fiber within the matrix

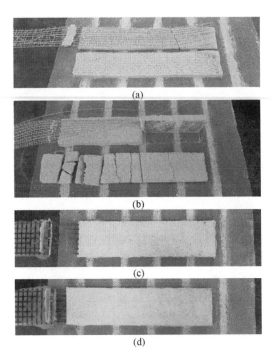

(a)

(b)

(c)

(d)

Figure 8. Bond tests: failure modes of steel and basalt strengthening systems.

(Failure mode D, see Figure 8d). Failure mode C typically occurred also in glass FRCM system (B_G_CEM1) tests. Regardless of the failure mechanism, considering the same strengthening grid, results obtained adopting NHL2 matrix to realize the FRCM system were, on average, higher than those obtained using NHL1 mortar (see Table 4). If the outcomes of bond tests and tensile tests are compared, it is possible to define the efficiency factors η reported in Table 4, defined as the ratio between maximum stresses recorded during bond and tensile tests, respectively.

2.4 Diagonal compression test: Set-up and instrumentations

The main experimental program concerned seven diagonal compression tests (DCTs) on masonry panels, performed, according to ASTM E519. A vertical load (displacement control) was applied along the principal diagonal direction of the square panel, after the installation of two steel constraint shoes (one for each corner), as shown in Figure 9. The instrumentation adopted during the DCT tests includes a servo-hydraulic actuator equipped with a 500kN load cell for the

Figure 9. Diagonal compression test set-up on masonry panels.

application of the vertical load, capable of operating in displacement control (at a constant rate equal to 1.2 mm/min) in order to evaluate the shear behavior of the sample also during the post-peak phase.

Deformations along the two principal directions were recorded by using linear potentiometric transducers (50 mm stroke), using a physical base with 1200 mm gage length, on both surfaces of the specimens. Moreover, the Digital Image Correlation (DIC) optical technique was used on one side of the sample, with the aim of evaluating the full deformation field during the DCT test. Thanks to this optical method, in fact, after a preliminary calibration phase, it is possible to obtain 2D-3D maps of displacement and deformations of the entire surface analyzed, increasing the knowledge in comparison to the physical instruments located along the principal diagonals.

According to the elastic solution (Frocht et al. 1931, Yokel & Fattal 1976, Calderini et al. 2010, Incerti et al. 2016), the masonry diagonal tensile strength f_t, starting from the evaluation of the stress state (σ_f, τ_f) in the center of the masonry square element, can be calculated with the following formulation:

$$f_t = 0.5 \frac{P_u}{A_n} \qquad (1)$$

where P_u is the maximum vertical load applied during the DCT test and A_n is the net cross-section of the panel. Therefore, deformations along the principal diagonals $(\varepsilon_c, \varepsilon_t)$ recorded during the test are summed up to evaluate the shear strain γ, i.e.:

$$\gamma = \varepsilon_c + \varepsilon_t \qquad (2)$$

Experimental results will be reported and discussed in the following using τ-γ diagrams, expressed according to this formulation.

3 EXPERIMENTAL RESULTS

3.1 Diagonal compression tests DCT

The results obtained from the laboratory tests are reported in Table 5, in terms of peak load P_u, shear stress τ_f, tensile strength f_t and shear elastic modulus $G_{10\%-30\%}$ (secant elastic modulus between 10% and 33% of the maximum peak load). The increment of tensile strength f_t provided by the FRCM strengthening with respect to the unreinforced specimen DCT_URM is also reported.

Table 5. Results of diagonal compression tests.

Sample	P_u [kN]	f_t [MPa]	τ_f [MPa]	$G_{10\%-30\%}$ [MPa]	Increment [%]
DCT_URM	123	0.190	0.400	1868	-
DCT_S_NHL1	166	0.258	0.542	3331	+ 36
DCT_S_NHL2	181	0.280	0.587	2863	+ 47
DCT_B1_NHL1	286	0.445	0.934	2798	+ 133
DCT_B1_NHL2	303	0.471	0.989	2021	+ 147
DCT_B2_NHL2	239	0.371	0.780	2427	+ 95
DCT_G_CEM1	311	0.485	1.019	2263	+ 155

As indicated in the Table, panels retrofitted by using unidirectional steel sheets (S) showed performance increments of about 36-47% with respect to unreinforced panels, with a common failure mode that typically characterizes discontinuous strengthening layouts, with one initial cracking pattern in the center of the panel and a subsequent redistribution of stresses occurring when steel sheets start to be effective. After that, a second phase starts, involving the contribution of the retrofitting system up to its detachment from the masonry substrate, without reaching fiber rupture (Figure 10).

For panels reinforced with a continuous layout, instead, the increment in terms of peak load obtained in DCT tests ranges from 133% to 147%. For specimens strengthened with high density basalt fiber (B1), detachments of the external layer of the matrix from the fiber grid occurred during the tests, probably due to the reduced grid spacing that did not permit a proper impregnation of the fibers. The associated failure mode showed debonding at the substrate level close to the loaded corners at high level of imposed vertical displacement, along the compressed diagonal, without fiber tensile failure phenomena (Figure 11).

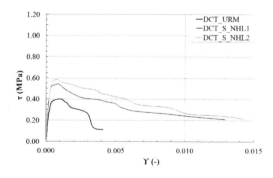

Figure 10. Diagonal compression tests on masonry panels with steel FRCM: shear stress – strain diagrams.

318

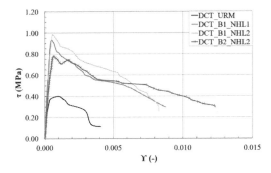

Figure 11. Diagonal compression tests on masonry panels with basalt FRCM: shear stress – strain diagrams.

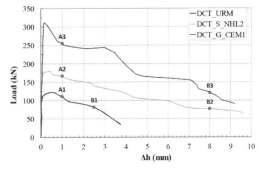

Figure 13. Identification of load levels considered to select results from DIC strain maps.

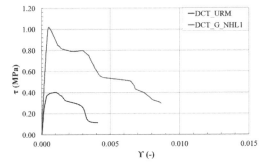

Figure 12. Diagonal compression tests on masonry panels with glass FRCM: shear stress – strain diagrams.

Finally, for specimens retrofitted with low density fiber grids and a continuous layout (B2, G), fiber rupture was detected during the tests and, independently from the fiber type, both panels were able to increase their load bearing capacity (from 95% to 155%) with respect to unreinforced masonry panels.

Thanks to the use of DIC optical technique, it was possible to evaluate 2D strain maps for each masonry panel tested. Figure 14, Figure 15 and Figure 16 show horizontal deformations corresponding to imposed vertical displacements equal to 1 mm, 2.5 and 8 mm, corresponding to the load levels identified with points in Figure 13. The obtained strain maps highlight the higher strain concentration in the central portion of the panel (without reinforcement) occurring for the discontinuous steel layout, with respect to the continuous layouts, thus preventing the achievement of similar peak loads. Moreover, according to results highlighted by

post-peak strain maps at high displacement levels, the use of a continuous layout can allow a better stress redistribution in comparison to a discontinuous layout.

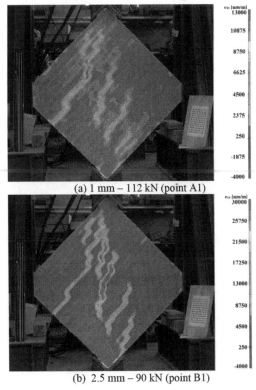

Figure 14. DCT_URM: strain maps for two different displacement levels: (a) 1 mm (A1), (b) 2.5 mm (B1).

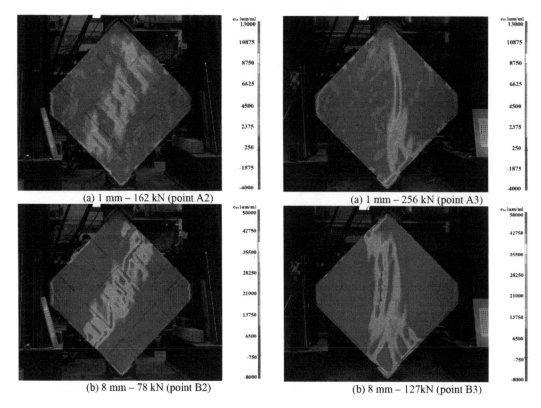

(a) 1 mm – 162 kN (point A2)

(a) 1 mm – 256 kN (point A3)

(b) 8 mm – 78 kN (point B2)

(b) 8 mm – 127kN (point B3)

Figure 15. DCT_S_NHL2: strain maps for the assigned displacement level: (a) 1 mm (A2), (b) 8 mm (B2).

Figure 16. DCT_G_CEM1: strain maps for two different displacement levels: (a) 1 mm (A3), (b) 8 mm (B3).

4 CONCLUSIONS

The experimental study presented here is focused on diagonal compression tests carried out on masonry panels strengthened with FRCMs. A complete characterization of the strengthening systems by means of tensile and single-lap shear tests was performed. All the reinforcement types provided, during DCTs, performance increments ranging between 36% and 155% in comparison to the tensile strength f_t of the unreinforced wall. In more detail, NHL2 matrix, based on lime mortar with a fine particle size, allowed the reinforcement to achieve better performances in comparison to NHL1 matrix, confirming the results obtained in single-lap shear tests. Thanks to the improved adhesion between matrix and fiber within the reinforcement, in fact, no early debonding phenomena at textile-matrix or matrix-substrate interface were observed. CEM1 cementitious matrix, in combination with glass fiber, achieved better performance in terms of maximum capacity and improved also the initial behavior until the peak load, thanks to a more efficient stress redistribution. This phenomenon was not detected on the specimens reinforced with unidirectional fiber sheets, where the first cracking phase appears in the central portion without reinforcement, preventing the achievement of peak loads similar to the panels reinforced with a continuous layout. In the post-peak phase, instead, the presence of the composite sheets reduces crack propagation and determines a behavior similar to that observed on the panels reinforced with continuous strengthening systems.

ACKNOWLEDGEMENT

The financial support of the Italian Department of Civil Protection (ReLUIS 2019-2021 Grant – Innovative Materials) is gratefully acknowledged. The Authors would like to thank Fassa S.r.l. for providing FRCM materials and for the realization of strengthened systems of masonry panels. Those reported here are only the first results of a wide and still in progress experimental campaign. The Authors would like also to thank Mr. Diiterihs Erra and Mr. Michele Esposito (CIRI Building and Construction technicians) for their collaboration in laboratory tests preparation.

REFERENCES

ASTM International. 2015. ASTM E519/E519M-15 Standard Test Method for Diagonal Tension (Shear) in Masonry Assemblages. West Conshohocken, PA; ASTM International. https://doi.org/10.1520/E0519_E0519M-15.

Babaeidarabad, S., De Caso, F., Nanni, A. 2014. URM Walls Strengthened with Fabric-Reinforced Cementitious Matrix Composite Subjected to Diagonal Compression, *Journal of Composites for Construction*, 18(2).

Bellini, A., Ferracuti, B., Mazzotti, C. 2015. Effect of matrix on bond between FRCM and masonry. *Proc. of FRPRCS-12 & APFIS-2015 Joint Conference, Nanjing, China, 14-16 December 2015.*

Bellini, A., Mazzotti, C. 2016. Bond behavior and tensile characterization of FRCM strengthening systems applied on masonry panels. *Proc. of the 10th SAHC Conference, Van Balen & Verstrynge Eds., Taylor & Francis Group, London, 2016.*

Bellini, A., Bovo, M., Mazzotti, C. 2019a. Experimental and numerical evaluation of fiber-matrix interface behaviour of different FRCM systems, *Composites Part B 161:* 411–426.

Bellini, A., Kahangi Shahreza, S., Mazzotti, C. 2019b. Cyclic bond behavior of FRCM composites applied on masonry substrate, *Composites Part B 169:* 189–199.

Borri, A., Castori, G., Corradi, M., Speranzini, E. 2001. Shear Behavior of Unreinforced and Reinforced Masonry Panels Subjected to In Situ Diagonal Compression Tests, *Construction and Building Materials* 25: 4403–4414.

Calderini, C., S. Cattari, and S. Lagomarsino. 2010. The use of the diagonal compression test to identify the shear mechanical parameters of masonry. *Construction and Building Materials* (24): 677–685.

Carozzi, F.G., Bellini, A., D'Antino, T., de Felice, G., Focacci, F., Hojdys, L., Laghi, L., Lanoye, E., Micelli, F., Panizza, M., Poggi, C. 2017. Experimental investigation of tensile and bond properties of Carbon-FRCM composites for strengthening masonry elements. *Composites Part B 128*: 100–119.

de Felice, G., De Santis, S., Garmendia, L., Ghiassi, B. et al. 2014. Mortar-based systems for externally bonded strengthening of masonry, *Materials and Structures* 47(12): 2021–2037.

de Felice, G., Aiello, M.A., Caggegi, C., Ceroni, F., De Santis et al. 2018. Recommendation of RILEM Technical Committee 250-CSM: Test method for Textile Reinforced Mortar to substrate bond characterization, *Materials and Structures* 51: 95.

EN 772 (European Norms). 2011. EN 772-1:2011 +A1:2015 Methods of test for masonry units - Part 1: Determination of compressive strength.

EN 1015 (European Norms). 2006. EN 1015-11:1999/ A1:2006 Methods of test for mortar for masonry - Part 11: Determination of flexural and compressive strength of hardened mortar.

EN 12390 (European Norms). 2013. EN 12390-13:2013 Testing hardened concrete. Determination of secant modulus of elasticity in compression.

Faella, C., Martinelli, E., Nigro, E., Paciello, S. 2010. Shear Capacity of Masonry Walls Externally Strengthened by a Cement-based Composite Material: An Experimental Campaign. *Construction and Building Materials 24*: 84–93.

Ferretti, F., Incerti, A., Ferracuti, B., Mazzotti, C. 2016. Diagonal compression tests on masonry panels strengthened by FRP and FRCM. *Proc. of the 10th SAHC Conference, Van Balen & Verstrynge Eds., Taylor & Francis Group, London, 2016.*

FRCM Italian Guidelines 2018. Linea Guida per la identificazione, la qualificazione ed il controllo di accettazione di compositi fibrorinforzati a matrice inorganica (FRCM) da utilizzarsi per il consolidamento strutturale di costruzioni esistenti, *Consiglio Superiore dei Lavori Pubblici, Servizio Tecnico Centrale*, Rome, Italy. 2018.

Frocht, M. 1931. Recent advances in photoelasticity, *ASME Trans.* (55): 135–153.

Mazzotti, C., Ferracuti, B., Bellini, A. 2015. Experimental bond tests on masonry panels strengthened by FRP, *Composites Part B 80*: 223–237.

Ferretti, F., Incerti, A., Tilocca, A.R., Mazzotti, C. 2019. In-Plane Shear Behaviour of Stone Masonry Panels Strengthend through Grout Injection and Fiber Reinforced Cementitious Matrices, *International Journal of Architectural Heritage* (In press).

Incerti, A., Rinaldini, V., Mazzotti, C. 2016. The evaluation of masonry shear strength by means of different experimental techniques: A comparison between full-scale and laboratory tests. *Proceeding of the 16th international brick and block masonry conference*, Padova, Italy.

Incerti, A., Tilocca, A.R., Ferretti, F., Mazzotti, C. 2019. Influence of Masonry Texture on the Shear Strength of FRCM Reinforced Panels, *RILEM Bookseries* 18: 1623–1631.

Incerti, A., Ferretti, F. Mazzotti, C. 2019. FRCM strengthening systems efficiency on the shear behavior of pre-damaged masonry panels: an experimental study. *J Build Rehabil* 4: 14. https://doi.org/10.1007/s41024-019-0053-9.

Papanicolaou, C., Triantafillou, T., Lekka, M. 2011. Externally bonded grid as strengthening and seismic retrofitting materials of masonry panels, *Construction and Building Materials* 25(2): 504–514.

Parisi, F., Iovinella, I., Balsamo, A., Augenti, N., Prota, A. 2013. In-plane behaviour of tuff masonry strengthened with inorganic matrix-grid composites, *Composites Part B 45*: 1657–1666.

Tilocca, A.R., Incerti, A., Bellini, A., Savoia, M. 2019. Influence of matrix properties on FRCM-CRM strengthening systems, *Key Engineering Materials* 817 KEM: 478–485.

Yokel, F.Y., Fattal, S.G. 1976. A Failure hypothesis for masonry shearwalls. *Journal of Structural Divisions - ASCE* (102): 515–532.

Brick and Block Masonry - From Historical to Sustainable Masonry –
Kubica, Kwiecień & Bednarz (eds)
© 2020 Taylor & Francis Group, London, ISBN 978-0-367-56586-2

Nonlinear flexural capacity model for FRCM-strengthened masonry walls under in-plane loading

F. Parisi, G.P. Lignola & A. Prota
Department of Structures for Engineering and Architecture, University of Naples Federico II, Naples, Italy

ABSTRACT: Recent strong earthquakes produced heavy structural damages to existing unreinforced masonry (URM) buildings, particularly in the case of historic urban centres such as L'Aquila (2009) and Amatrice (2016). Spatially-distributed damages at urban-to-regional scale, as well as the need to speed up the recovery process, have motivated an extensive use of fabric-reinforced cementitious matrix (FRCM) systems for external retrofitting of URM buildings. This has significantly motivated the development of ad-hoc guidelines for a mechanics-based design of FRCM retrofitting systems. In this paper, the authors present a novel flexural capacity model for incremental static analysis of URM walls subjected to in-plane lateral loading. The formulation makes use of a fibre-based approach that explicitly accounts for geometric and mechanical nonlinearity sources at sectional level. Capacity modelling was carried out using data sets available in the 2019 Italian building code commentary and experimental review studies. The output of this study is focused on moment–curvature diagrams remarking a number of performance limit states. This study attempts to move beyond the classical strength-based verification formats implemented in current guidelines for the ultimate limit state. The proposed capacity model allows evaluating the evolution of flexural behaviour as cracking and crushing propagate throughout the URM-FRCM cross section.

1 INTRODUCTION

Fibre reinforced composites are more and more frequently used in structural strengthening interventions. New materials are joining the "classic" Fibre Reinforced Polymer (FRP) made of long fibres immersed in organic matrices (e.g. epoxy resins). The Fabric Reinforced Cementitious Mortar (FRCM) is the result of substituting organic with inorganic matrix based on lime or cement mortars. This provides a variety of combinations that allow designers to tailor the intervention to specific needs, but at the same time, it represents the main difficulty encountered in drafting general rules and guidelines (Papanicolaou et al. 2007).

Mortar matrices make the FRCM more compatible to masonry in comparison to polymeric matrices, as mortar is at the same time a component of masonry. However, the mortar has lower bond performance compared to resins, but the enlargement of the surfaces involved by the intervention recovers global bond capacity even if local bond stress is lower (Kwiecień et al. 2016) moving the bond issue inside the mortar matrix (i.e. telescopic failure at the fibre-mortar interface, de Felice et al. 2018). First attempts to adapt the wide knowledge gained for FRPs revealed these limitations and drawbacks. For this reason, research is expected to provide peculiar requirements for the optimal design of FRCM strengthening interventions.

2 METHODOLOGY

A flexural capacity model for incremental nonlinear static (pushover) analysis of URM walls subjected to in-plane lateral loading was developed. The formulation makes use of a fibre-based approach that explicitly accounts for geometric and mechanical nonlinearity sources at sectional level. Geometric nonlinearities originate from tensile cracking of masonry and ineffectiveness of fibres in compression, hence resulting in an effective cross section that depends on the magnitude of loading. Mechanical nonlinearities are associated with the nonlinear shape of the stress–strain relationships assigned to materials in tension or compression. A rectangular cross section is considered, whereas the FRCM is supposed to be applied over both sides of the URM cross section. Experimental tests have shown that double-side FRCM strengthening produces the best mechanical behaviour of the strengthened masonry (e.g. Parisi et al. 2013). A uniaxial constitutive model is assigned to masonry, whereas different options are alternatively used to model the mechanical behaviour of either the whole FRCM or its constituents (i.e. matrix mortar and fibres). Performance limit states (PLSs) are then assumed to capture progressive damage to the FRCM-strengthened URM cross section under increasing bending curvature. The Euler-Bernoulli hypothesis of plane cross section after flexural deformation is used, according to previous studies (e.g. Parisi & Augenti 2013a, Parisi et al. 2016, Kouris & Triantafillou 2019). A perfect bond at the FRCM-masonry interface is assumed, because fracture or sliding of the fabric is more likely to occur than fibre-matrix debonding within the composite (Bilotta et al. 2017). Therefore, the capacity model consists in the discretisation of the cross section in a number of fibres, in order to account for different materials at each depth.

2.1 Masonry properties

Let masonry be considered as equivalent homogenous material with no tensile strength and linear elastic-perfectly plastic behaviour in compression. In such conditions, the contribution of masonry to the sectional response to axial load and uniaxial bending depends only on compressive strength (f_m), Young's modulus (E_m) and ultimate strain (ε_{um}). According to Table C8.5.I of the 2019 Italian building code commentary (IBCC 2019), Table 1 outlines how the mean values of f_m and E_m change with the masonry type under consideration, e.g. assuming either regular or irregular tuff stone masonry (TSM) and clay brick masonry (CBM). In the calculations, the effective compressive strength was set to $f_d = 0.85\, f_m$, whereas the limit elastic and ultimate compressive strains were assumed to be $\varepsilon_{em} = 0.10\%$ and $\varepsilon_{um} = 0.35\%$, respectively (Augenti & Parisi 2019). In the following, irregular tuff stone masonry is considered.

2.2 FRCM properties

In this research, FRCM is modelled through different stress–strain relationships for the matrix and fibres.

The matrix, when taken into account in the compressed zone of cross section, is assumed linear elastic up to (brittle) failure: the compressive strength is f_M = 16 MPa and Young's modulus is E_M = 8 GPa, resulting in a limit elastic strain ε_{eM} = 0.2%. The fabric of the FRCM system is assumed to consist of glass fibres, the latter having a linear elastic behaviour in tension with strength f_f = 807 MPa and Young's modulus E_f = 72 GPa.

The mechanical behaviour of the FRCM system comes from the interaction between the two

Table 1. Mean compressive strength and Young's modulus of selected masonry types.

Property	Irregular TSM	Regular TSM	CBM
f_m [MPa]	1.80	2.60	3.45
E_m [GPa]	1.53	2.21	2.93

constituents, so it was modelled according to four alternative strategies (Figure 1) as follows.

The first model, here called 'mesomodel' (Parisi et al. 2011), takes into account the geometry of the reinforcement by differentiating the matrix from the fibres. The other three alternative models consider the FRCM system as a homogeneous material, representing its mechanical behaviour through a single trilinear, bilinear or linear stress–strain relationship, the latter in agreement with the recently issued Italian guidelines CNR-DT 215/2018. Tension stiffening after cracking of the matrix provides such

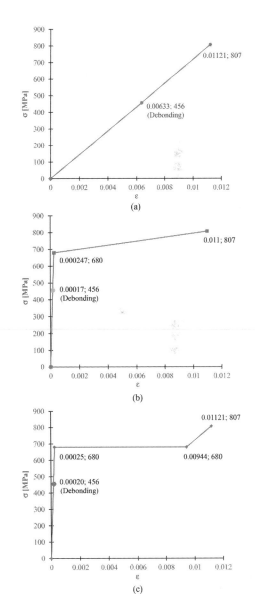

Figure 1. Stress–strain relationships for glass FRCM: a) mesomodel & CNR model; b) bilinear model; c) trilinear model.

differences. However, the CNR model neglects the matrix contribution, reducing the FRCM system to fibres only (hence very similarly to procedures used in case of FRP strengthening). In addition to the failure mechanisms that can arise in individual materials, interaction yields to a further failure mode due to the debonding between the fibres and the matrix.

The mechanical properties were derived from a European Round Robin Test, involving both tensile and single shear (bond) tests. Experimental properties for glass FRCM system are mean values derived from Leone et al. (2017), as reported in Table 2.

2.3 *Geometric properties*

Moment–curvature analysis was run by assuming the cross section shown in Figure 2.

The width and depth of the masonry cross section were respectively set to $t_m = 300$ mm and $h_m = 1000$ mm. The thickness of the FRCM on each side

Table 2. Characteristic points in stress–strain relationships of glass FRCM systems.

Model/Limit point	Stress	Strain
Mesomodel & CNR		
Ultimate point	$f_u = 807$ MPa	$\varepsilon_{uF} = 1.12\%$
Debonding point	$f_D = 456$ MPa	$\varepsilon_D = 0.63\%$
Bilinear		
First cracking point	$f_1 = 680$ MPa	$\varepsilon_{1M} = 0.025\%$
Ultimate point	$f_u = 807$ MPa	$\varepsilon_{uF} = 1.12\%$
Debonding point	$f_D = 456$ MPa	$\varepsilon_D = 0.017\%$
Trilinear		
First cracking point	$f_1 = 680$ MPa	$\varepsilon_{1M} = 0.025\%$
Full cracking point	$f_2 = 680$ MPa	$\varepsilon_{2M} = 0.94\%$
Ultimate point	$f_u = 807$ MPa	$\varepsilon_{uF} = 1.12\%$
Debonding point	$f_D = 456$ MPa	$\varepsilon_D = 0.017\%$

of the cross section and fabric spacing were respectively assumed to be $t_r = 10$ mm and $s_r = 25$ mm. The latter property is explicitly taken into account when using the mesomodel (Parisi et al. 2011).

2.4 *Performance limit states*

If the FRCM-strengthened cross section of a URM wall subjected to axial load and in-plane lateral loading is assumed to fail in flexure, its progressive damage under monotonically increasing curvature demand may be lumped into a number of PLSs.In this study, the following PLSs are considered:

– the elastic limit state associated with the masonry (here denoted as 'm-ELS'), which is defined by the onset of crushing of the masonry in compression and/or cracking of the matrix in tension;
– the elastic limit state associated with the matrix (here denoted as 'M-SLS'), which is related to the crushing of the matrix in compression and/or cracking of the matrix in tension;
– the ultimate limit state (ULS), which is related to the crushing of the masonry in compression and/or rupture of the fibres in tension.

For each of PLS, the failure mode consisting in the loss of fibre-matrix adhesion (debonding) in the FRCM reinforcement was also considered (Bilotta et al. 2017). Table 3 shows, for each considered approach, the maximum and minimum strain limits adopted in the capacity modelling and recorded as nonlinear moment–curvature analysis goes on.

3 RESULTS AND DISCUSSION

The fibre-based capacity modelling of FRCM-strengthened URM cross sections outlined in previous section was applied to derive moment–curvature diagrams at different axial loads, including or not the debonding failure mode. This failure mode has

Table 3. Maximum and minimum strain limits for each performance limit state.

Model	PLS	ε_{max}	ε_{min}
Mesomodel	m-SLE	ε_{em}	0
	M-SLE	ε_{eM}	0
	ULS	ε_{um}	$\varepsilon_{uF}/\varepsilon_D$
CNR	m-SLE	ε_{em}	0
	M-SLE	ε_{eM}	0
	ULS	ε_{um}	$\varepsilon_{uF}/\varepsilon_D$
Bilinear	m-SLE	ε_{em}	ε_{1M}
	M-SLE	ε_{eM}	ε_{1M}
	ULS	ε_{um}	$\varepsilon_{uF}/\varepsilon_D$
Trilinear	m-SLE	ε_{em}	ε_{1M}
	M-SLE	ε_{eM}	ε_{1M}
	ULS	ε_{um}	$\varepsilon_{uF}/\varepsilon_D$

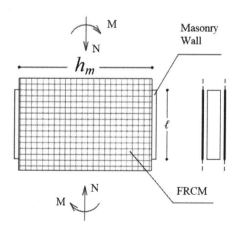

Figure 2. Comparison between moment–curvature diagrams corresponding to different models.

significant impact on the global performance depending on the assumption on FRCM behaviour. Nonlinear moment–curvature analysis was run by keeping constant the axial load, according to the procedure proposed by Parisi et al. (2016). When using the mesomodel, a further nonlinearity was associated with the number of fibres in tension. Different ultimate strain thresholds were assumed, depending on whether the debonding failure was considered or not.

The moment–curvature diagrams presented herein were derived by considering the entire sectional depth. Such analytical models may be used in the following alternative cases: (i) as a basis for the formulation of flexural hinges, to be used in simplified macroelement models of loadbearing masonry walls with openings; and (ii) direct integration throughout the macroelement volume for derivation of force–displacement curves in advanced (distributed plasticity) models (Parisi & Augenti 2013b).

3.1 Moment–curvature diagrams without consideration of debonding

In this set of moment–curvature diagrams, the ultimate strain thresholds correspond to masonry crushing and fibre tensile failure. The impact of the four alternative constitutive models is evident (Figure 3).

In case of pure bending (i.e. when the ratio of the axial load to the ultimate load is $N/N_u = 0$), the mesomodel and CNR model provides a quasi-linear behaviour of the cross section, whereas both the bilinear and trilinear models result in a nonlinear M–ϕ curve with significant pseudo-ductility.

The moment–curvature diagrams corresponding to the mesomodel and CNR model follow roughly the same trend as N/N_u increases up to 0.3. As the axial load ratio further increases up to 0.5, the mesomodel provides a moment–curvature diagram with significantly increasing peak resisting moment and post-peak softening. A similar outcome is observed in the case of bilinear and trilinear models. Furthermore, this is line with other studies where post-peak softening in URM cross sections was investigated (e.g. Parisi & Augenti 2013a, Parisi et al. 2016).

The softening branch of the mesomodel, bilinear model and trilinear model begins when the matrix is cracked and subsequently softens, since the contribution of the matrix begins to weaken. It ends at an ultimate curvature associated to:

- masonry crushing, for axial load ratios $N/N_u \geq 0.2$;
- fibre tensile failure, for $N/N_u < 0.2$.

For the model based on CNR guidelines, the softening branch does not appear. The diagram is characterised by a monotonically increasing curve. This happens because this model neglects the contribution of the matrix. Consequently, it is not possible to define the limit curvature associated to the failure of the matrix.

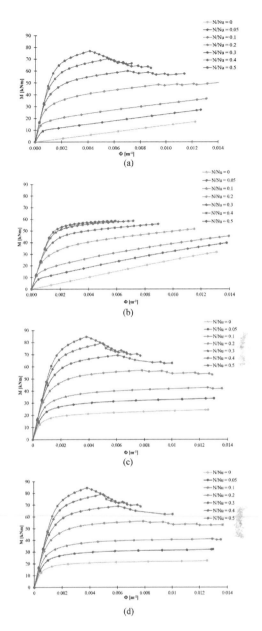

Figure 3. Moment–curvature diagrams without consideration of debonding: a) mesomodel; b) CNR; c) bilinear; d) trilinear.

3.2 Moment–curvature diagrams with consideration of debonding

In case of glass FRCM, the debonding strength is rather lower than the ultimate strength corresponding to tensile rupture of fibres (Figure 1). This causes a major influence of debonding failure on flexural behaviour of cross sections, as highlighted in Figure 4 particularly in the case of bilinear and trilinear homogenized models.

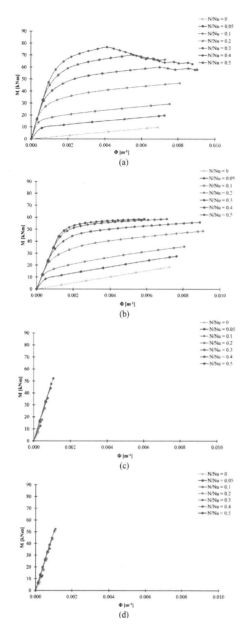

(a)

(b)

(c)

(d)

Figure 4. Moment–curvature diagrams with consideration of debonding: a) mesomodel; b) CNR; c) bilinear; d) trilinear.

In the latter cases, the occurrence of debonding fully neutralises the nonlinearity of moment–curvature diagrams. This is not observed in the case of diagrams corresponding to the mesomodel and CNR model, where most of post-peak strength degradation is associated with gradual failure of the matrix in compression. However, in general, the moment–curvature diagram depends on strain thresholds corresponding to masonry crushing in compression and FRCM debonding in tension.

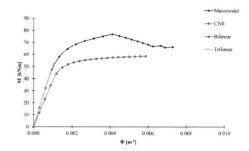

Figure 5. Comparison between moment–curvature diagrams corresponding to different models.

Figure 5 allows a direct comparison between moment–curvature diagrams corresponding to $N/N_u = 0.5$. As the CNR model does not account for the matrix, its implementation in nonlinear moment–curvature analysis leads to a significant underestimation of initial stiffness and peak resisting moment, also resulting in a lower ultimate curvature compared to the mesomodel. The latter appears to produce the most realistic behaviour of the cross section. In this respect, it is noted that the CNR model could be suitable for safety verifications at the ultimate limit state, i.e. the case for which that model was proposed in CNR-DT 215/2018 guidelines. By contrast, the contribution of the matrix should be taken in due consideration when performing a pushover analysis of the retrofitted URM structure under earthquake actions. Indeed, incorrect estimations of stiffness and bending resisting moment may produce errors in the elasto-plastic distribution of seismic demand among masonry macro-elements, as well as a wrong identification of failure modes (e.g. Parisi & Augenti 2013b).

4 CONCLUSIONS

A fibre-based capacity modelling of FRCM-strengthened URM cross sections was implemented in nonlinear moment–curvature analysis. Four alternative options were used to model the FRCM in tension and compression, with or without consideration of debonding failure. A mesomodelling approach was also implemented to distinguish between the contributions from fibres and matrix at different depths of the cross section.

If the occurrence of debonding is neglected, the adoption of the CNR model produces monotonically increasing moment–curvature diagrams up to either tensile rupture of fibres or masonry crushing. By contrast, the other selected models induce nonlinear moment–curvature diagrams with post-peak strength degradation associated with progressive damage to the matrix.

In the case of glass FRCM (which was considered in this study), the impact of the debonding appears to be significant, particularly in bilinear and trilinear constitutive models. The latter produce quasi-linear moment–curvature diagrams. Conversely, the meso-model and CNR model result in nonlinear moment–curvature diagrams, in the former case with post-peak strength degradation associated with gradual failure of the matrix in compression. The definition of three performance limit states allowed the authors to capture the progressive damage to masonry and FRCM constituents, which is the physical phenomenon behind the shape of moment–curvature diagrams.

Analysis results show that the CNR model could be conservative for ULS safety verifications. Conversely, the mesomodelling approach firstly proposed by Parisi et al. (2011) can provide more realistic predictions of bending resistance and deformation capacity, including the beneficial contribution from the matrix. In this respect, the mesomodel is recommended for pushover analysis of retrofitted URM buildings subjected to earthquake actions.

Further outcomes of this study could be the derivation of interaction domains for safety verifications, for instance in terms of bending moment versus axial load. In addition, this study could be extended to other combinations of masonry and FRCM, in order to carry out cross-comparisons between various FRCM-URM combinations and to identify the relative impact of the different failure modes associated with masonry and FRCM.

REFERENCES

Augenti, N. & Parisi, F. 2019. *Teoria e Tecnica delle Strutture in Muratura*. Milan: Hoepli.

Bilotta, A., Ceroni, F., Lignola, G.P. & Prota A. 2017. Use of DIC technique for investigating the behaviour of FRCM materials for strengthening masonry elements. *Composites Part B: Engineering* 129: 251–270. DOI: 10.1016/j.compositesb.2017.05.075.

CNR-DT 215/2018. Guide for the Design and Construction of Fiber Reinforced Inorganic Matrix Systems for Strengthening Existing Structures. Rome: Consiglio Nazionale delle Ricerche (CNR). https://www.cnr.it/it/node/9347 (January 15, 2019).

de Felice, G., Aiello, M.A., Caggegi, C., Ceroni, F., De Santis, S., Garbin, E., Gattesco, N., Hojdys, Ł., Krajewski, P., Kwiecień, A., Leone, M., Lignola, G.P., Mazzotti, C., Oliveira, D.V., Papanicolaou, C.G., Poggi, C., Triantafillou, T.C., Valluzzi, M.R. & Viskovic, A. 2018. Recommendation of RILEM Technical Committee 250-CSM: Test method for Textile Reinforced Mortar to substrate bond characterization.

Materials and Structures 51(4): 1–9. DOI: 10.1617/s11527-018-1216-x.

IBCC 2019. Italian Building Code Commentary: Circolare 21 gennaio 2019, n. 7 Istruzioni per l'applicazione dell'«Aggiornamento delle "Norme tecniche per le costruzioni"» di cui al decreto ministeriale 17 gennaio 2018. Rome: Consiglio Superiore dei Lavori Pubblici.

Kouris, L.A.S. & Triantafillou, T.C. 2019. Design methods for strengthening masonry buildings using textile-reinforced mortar. *ASCE Journal of Composites for Construction* 23(1): 04018070. DOI: 10.1061/(ASCE)CC.1943-5614.0000906.

Kwiecień, A., de Felice, G., Oliveira, D.V., Zając, B., Bellini, A., De Santis, S., Ghiassi, B., Lignola, G.P., Lourenço, P.B., Mazzotti, C. & Prota, A. 2016. Repair of composite-to-masonry bond using flexible matrix. *Materials and Structures* 49: 2563–2580. DOI: 10.1617/s11527-015-0668-5.

Leone, M., Aiello, M.A., Balsamo, A., Carozzi, F.G., Ceroni, F., Corradi, M., Gams, M., Garbin, E., Gattesco, N., Krajewski, P., Mazzotti, C., Oliveira, D., Papanicolaou, C., Ranocchiai, G., Roscini, F. & Saenger, D. 2017. Glass fabric reinforced cementitious matrix: Tensile properties and bond performance on masonry substrate. *Composites Part B: Engineering* 127: 196–214. DOI: 10.1016/j.compositesb.2017.06.028.

Papanicolaou, C.G., Triantafillou, T.C., Karlos, K. & Papathanasiou, M. 2007. Textile-reinforced mortar (TRM) versus FRP as strengthening material of URM walls: in-plane cyclic loading. *Materials and Structures* 40: 1081–1097. DOI: 10.1617/s11527-006-9207-8.

Parisi, F. & Augenti, N. 2013a. Assessment of unreinforced masonry cross sections under eccentric compression accounting for strain softening. *Construction and Building Materials* 41: 654–664. DOI: 10.1016/j.conbuildmat.2012.12.039.

Parisi, F. & Augenti, N. 2013b. Seismic capacity of irregular unreinforced masonry walls with openings. *Earthquake Engineering and Structural Dynamics* 42(1): 101–121. DOI: 10.1002/eqe.2195.

Parisi, F., Iovinella, I., Balsamo, A., Augenti, N. & Prota, A. 2013. In-plane behaviour of tuff masonry strengthened with inorganic matrix-grid composites. *Composites Part B: Engineering* 45(1): 1657–1666. DOI: 10.1016/j.compositesb.2012.09.068.

Parisi, F., Lignola, G.P., Augenti, N., Prota, A. & Manfredi, G. 2011. Nonlinear behavior of a masonry sub-assemblage before and after strengthening with inorganic matrix-grid composites. *ASCE Journal of Composites for Construction* 15(5): 821–832. DOI: 10.1061/(ASCE)CC.1943-5614.0000203.

Parisi, F., Sabella, G. & Augenti, N. 2016. Constitutive model selection for URM cross sections based on best-fit analytical moment–curvature diagrams. *Engineering Structures* 111: 451–466. DOI: 10.1016/j.engstruct.2015.12.036.

Brick and Block Masonry - From Historical to Sustainable Masonry –
Kubica, Kwiecień & Bednarz (eds)
© 2020 Taylor & Francis Group, London, ISBN 978-0-367-56586-2

Study of the effectiveness of anchorages applied to SRG strips bonded to masonry blocks

G. Baietti
Department of Civil, Chemical, Environmental, and Materials Engineering, University of Bologna, Italy

F. Focacci
eCampus University, Novedrate, Como, Italy

C. Gentilini
Department of Architecture, Civil, Chemical, Environmental, and Materials Engineering, University of Bologna, Italy

C. Carloni
Case Western Reserve University, Cleveland, Ohio, USA

ABSTRACT: Steel reinforced grout (SRG) is considered part of the broader family of fiber-reinforced cementitious matrix (FRCM) composites employed to strengthen and retrofit concrete and masonry structures. The main advantages are low cost of the fibers, excellent resistance to high temperatures, and compatibility with different types of support. As premature debonding is the main failure mode of these composites, an anchorage system can be required to improve the bond capacity, i.e. the maximum force transferable between the composite and the substrate. This paper presents the results of single-lap shear tests carried out to study the effectiveness of two different anchorage systems: 1) a separate spike that interacts with the bonded strip; and 2) an extension of the fibers of the SRG strip that are inserted into the masonry block. Test results demonstrate that bond behavior of the SRG strip is positively affected by both types of anchorage systems.

1 INTRODUCTION

During the last two decades, the topic of strengthening masonry and concrete structures has become very popular among civil engineers in Europe. This was mainly due to the fact that some regions of Italy and other European countries had to face several seismic events in recent years. In the late 90s and 2000s, many researchers focused their studies on the use FRP composites, which consists of fiber sheets embedded into an epoxy matrix, for retrofitting and reinforcing structural elements (Pellegrino et al. 2009, Carloni et al. 2012, Carrara et al. 2011, Carrara et al. 2013). FRP composites have been proven to be very effective in terms of flexural and shear capacity enhancement. However, to overcome some drawbacks of FRP composites, such as low vapor permeability, poor behavior at elevated temperatures and inapplicability onto wet surfaces, an alternative type of composite has been proposed in the last 15 years. SRG composite system consists of steel fibers embedded into a cement-based and hydraulic lime-based matrix. Some studies have shown that this system can be successfully used to improve the shear and flexural capacity of reinforced concrete beams, the axial capacity

of masonry and reinforced concrete columns by means of their confinement, and to increase the in-plane and out-of-plane capacity of masonry walls (Sneed et al. 2018, Sneed et al. 2017, Bellini et al. 2017, Incerti et al. 2020, Sneed et al. 2019).

One of the topics that has drawn the attention of researchers is the debonding between the substrate and the SRG composite, which compromise the increase of capacity due to the composite itself. Anchorage systems can be used to improve the bond behavior of the SRG in situations where debonding occurs, or when the lack of effective length (Carloni et al. 2017) at the end of the SRG strip could compromise the response of the strengthened element.

2 MATERIALS AND METHODS

2.1 *Materials*

Brick units used to build the masonry blocks had nominal dimensions of 120 mm (width) × 250 mm (depth) × 55 mm (height). The average compressive strength of bricks, determined according to UNI EN 772-1, resulted equal to 20.3 MPa and the corresponding coefficient of variation (CoV) was 0.170. The mortar used to bind the bricks was a hydraulic lime-based mortar. The average flexural and compressive strength of mortar evaluated according to UNI EN 1015-11, and their corresponding CoV, resulted equal to 1.1 MPa (CoV = 0.183) and 2.6 MPa (CoV = 0.140), respectively.

SRG composite strip bonded to the masonry block was comprised of high-strength steel fibers embedded into a hydraulic lime-based mortar matrix. Mechanical properties of steel fibers were provided by the manufacturer (http://products.kera koll.com). Steel fibers used for SRG system had a density equal to 1200 g/m². The average flexural and compressive strength of mortar evaluated according to UNI EN 1015-11, and the corresponding CoV, resulted equal to 1.8 MPa (CoV = 0.112) and 6.6 MPa (CoV = 0.141), respectively.

The fiber sheet used for the anchor spike system had a density of 670 g/m². The mortar used for the injections was recommended by the SRG manufacturer to realize connections employed for structural reinforcement and seismic improvement (http://products.kerakoll.com). The flexural and compressive strength of the injectable mortar and the corresponding CoV were evaluated according to UNI EN 1015-11, and they resulted equal to 6.4 MPa (CoV = 0.229) and 22.7 MPa (CoV = 0.029), respectively.

2.2 Specimen preparation

Twelve masonry blocks were constructed using solid clay bricks. Each block consisted of six layers of bricks arranged in a stretcher and header bond configuration. Five 10 mm-thick mortar joints were employed. Each masonry block had the same nominal dimensions of 250 mm (width) × 250 mm (depth) × 380 mm (height).

The effectiveness of the anchorages was studied by conducting single-lap shear tests on SRG strips bonded to a masonry block with the presence of an anchorage either at the loaded end of the strip or at the free end. The longitudinal fibers of the SRG strip bonded to the masonry substrate were left bare outside the bonded area.

Two types of anchorage were considered: an extension of the fibers of the SRG strip was inserted into the masonry block (Figure 1a). The fibers were pre-bent and inserted into a drilled through-thickness hole filled with the injectable mortar after the placement of the fibers. The second type of anchorage was a separate spike that interacted with the bonded strip (Figure 1b and c). The fibers were placed into the drilled through-thickness hole, which was then filled with the injectable mortar. The fiber strip of the spike was longer than the hole. The fibers were pre-bent at two locations corresponding to the end of the hole in order to create a double umbrella-like spike that, on the side where the SRG strip was applied, was interweaved with

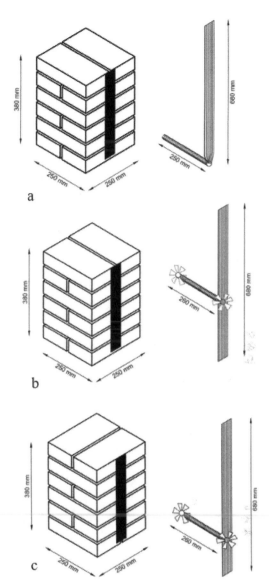

Figure 1. Types of anchorage: a) extension of the fibers of the SRG strip inserted into the masonry block at the free end; b) separate spike at the loaded end (second top layer of bricks); and c) separate spike at the free end (second bottom layer of bricks).

the longitudinal fibers of the SRG, and on the opposite side were covered with mortar. For all masonry blocks, the hole was drilled in the center of the brick, although the location varied based on the anchorage configuration. The hole, whose dimensions were 18 mm (diameter) × 250 mm (length), was performed orthogonally to the height of the block. Specimens were named following the notation DS_A_B where A refers to the type of anchorage system (i.e. G1 = bonded strip with extension of the fibers into the hole, G2 = separate spike in the second top layer of bricks, and G3 = separate spike in the second bottom layer of bricks), and B is the number of the specimen. Dimensions of bonded area of the SRG strip was the same for all specimens and it was equalto 50 mm (width) × 310 mm (length), and it started 35 mm away from the top edge of the masonry block to avoid shear failure, i.e. spalling of the brick.

2.2.1 Bonded strip with extension of the SRG fibers into the masonry block

The first group G1 was comprised of four masonry blocks. For this group the hole was drilled in the bottom layer of bricks. Steel fiber sheets were cut into the desired length, with dimensions of 50 mm (width) × 930 mm (length). Fiber sheet strip was bent at a distance of 250 mm from the end of the fiber strip opposite to the loaded end, and then bundled in order to be inserted into the hole. After the insertion of fibers (Figure 2a), the hole was filled with mortar

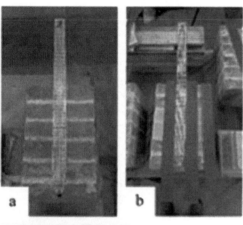

Figure 2. Preparation of G1 specimens: a) Insertion of fibers b) marked bonded area c) application of the second layer of mortar.

and then the masonry blocks were left at room temperature for 24 hours before the application of the mortar to realize the SRG strip. The bonded area was previously marked (Figure 2b) and then the first layer of 4 mm thick mortar was applied. Steel fibers were lifted up as the mortar was applied and then laid over while exerting a small pressure with a trowel. As the fibers strip was embedded into the mortar, the second layer of mortar was applied to fully cover the fibers (Figure 2c). All the G1 SRG-masonry joints were cured at room temperature under wet cloths for 28 days.

2.2.2 Separate spike and bonded strip

For the second and third groups of specimens, G2 and G3, eight masonry blocks were employed, four per each group.

For 4 masonry blocks, the hole was drilled in the second top layer of bricks, while for the remaining 4 blocks, the hole was in the second bottom layer.

A separate fiber strip, with dimensions of 100 mm (width) × 460 mm (length), was used for the spike that was inserted into the hole. After the insertion of the fibers into the hole, the spike strands were opened radially in the back face of the masonry block at the pre-bent location. The fibers were covered with mortar on the back face. The spike strands on the front face were also opened. The hole was then filled with mortar, and all the masonry blocks were left to cure for 24 hours at room temperature before the application of the SRG strip. Steel fiber sheets with a density of 1200 g/m^2 were used for the bond strip, with dimensions of 50 mm (width) × 680 mm (length). Prior to applying the first layer of the mortar of the SRG strip, the spike fibers were interweaved with the longitudinal fibers of the SRG strip (Figure 3a). The fiber of the strip were then lifted to apply the first 4 mm-thick layer of mortar (Figure 3b).

Subsequently, the steel fibers were delicately pressed onto the layer of mortar by means of a trowel and fully covered with the second layer of the mortar matrix (Figure 3c). All the SRG-masonry joints were cured at room temperature under wet cloths for 28 days.

2.3 Test set-up and procedure

SRG-masonry joints were tested using a single-lap shear test set-up (Figure 4). Tests were conducted under displacement control using a closed-loop servo-hydraulic universal testing machine with a capacity of 100 kN.

Prior to testing, 7.5 mm long epoxy tabs were applied at the end of the bare fibers to facilitate the gripping of the fibers themselves inside the jaws. The average of the readings of two linear variable displacement transformers (LVDT) was named

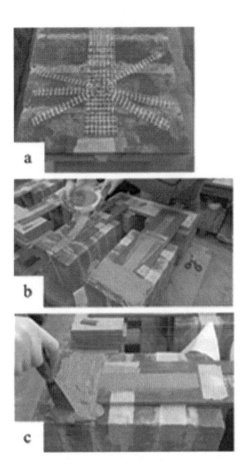

a

b

c

Figure 3. Preparation of G3 specimens: a) spike fibers interweaved with the longitudinal fibers of the SRG strip b) application of first layer of mortar c) application of the second layer of mortar to fully cover the fibers.

global slip *g*, which is the relative displacement at the beginning (at the loaded end) of the bonded area. The LVDT holders were glued on the masonry surface adjacent to the strip. The LVDTs reacted off an aluminum plate mounted on the bare fibers close to the beginning of the bond area. The test was controlled by the displacement *g* that was increased at a constant rate equal to 0.00084 mm/s.

3 RESULTS AND DISCUSSION

3.1 *Failure modes and discussion of results*

Specimens of G1 experienced two types of failure modes: interlaminar failure at the matrix-fiber interface (MF) and a mixed failure mode (Figure5a), which consists of interlaminar failure at the matrix-fiber interface and debonding of part of the SRG strip with a thin layer of brick attached to the matrix (MF/SF). It should be noted that after the fibers in the hole were engaged and the SRG strip was fully

Figure 4. Single lap shear test set-up.

a   b

c

Figure 5. Example of failure modes for a) G1 b) G2 and c) G3 specimens.

debonded, the test was stopped because of the significant backward rotation of the masonry block.

All the specimens of G2 and G3 showed the interlaminar failure with delamination at the matrix-fiber interface (Figure 5b and c).

However, for both groups, it was not possible to see the complete detachment of the SRG strip due to the presence of the anchor spike system. Also for these specimens, the test was stopped after the engagement of the spike because of the magnitude of the rotation of the block.

3.2 Load responses

The applied load versus global slip responses are plotted in Figure 6. All specimens exhibited an initial linear branch. For G1 specimens, except for DS_G1_3, the end of the linear branch is

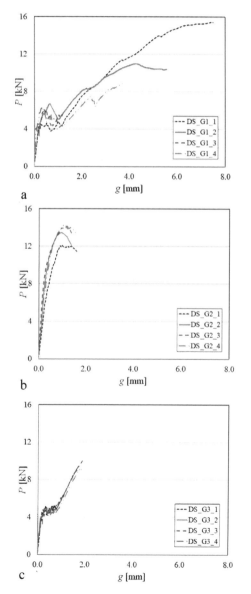

Figure 6. Load responses of a) G1 b) G2 and c) G3 specimens.

characterized by a drop of the load that indicates the beginning of the debonding process of the strip.

If no anchorage was present, the load drop would be followed by a constant branch (or plateau) that is associated with the progressive debonding of the strip towards the free end.

With the addition of an anchorage at the free end, as the stress transfer between the strip and the masonry block reaches the free end, the force is partially transferred within the anchorage, which entails for an increase of the load. A second major load drop is observed when the strip fully debonds. Once the strip is fully debonded, the only stress transfer mechanism occurs within the anchorage and the load increases again.

For DS_G1_3, the test was prematurely stopped because of a control issue in the loading rate. Load responses of G3 specimens are similar to those of specimens of G1. The linear branch is followed by a constant branch (plateau) marked by a load drop.

The load started to increase again when the complete detachment of the composite occurred and the spike was engaged. For G3 specimens, the plateau branch was more evident than G1 specimens, because the engagement of the spike was associated with the splice obtained by interweaving the fibers of the strip and the spike and therefore more gradual.

For G2 specimens, no load drop was observed. A non-linear branch followed directly the linear branch because the spike was engaged from the beginning of the test. The load increases until the peak is reached, which is probably associated with large slips between the strip and the spike. As the load level is too high for the bonded strip, a sudden debonding occurs for it, which is associated with a decrease of the load.

For each specimen g_1 was determined as the global slip that corresponds to the first drop of the load, while g_2 was determined as the global slip at the end of the plateau, i.e. the second load drop for G1 specimens. The critical load P_{crit} is the average of the load values evaluated within the range g_1 and g_2. $P*$ is the absolute peak of the load response. All values for the specimens herein presented are reported in Table 1.

The activation of the anchor system started at different points of the load response for each type of anchorage. For G1 and G3 specimens, the activation of the anchorage can be associated with the value g_2 of the global slip g. The values of g_2 for G1 specimens are slightly larger than the values for G3 specimens because the hole was placed further away from the loaded end in G1 specimens.

For G3 specimens, the activation of the anchorage occurred from the very beginning of the test since the end of the linear response is associated with values of g smaller than the values of g_1 for G1 and G3 specimens.

Table 1. Test results in terms of g_1, g_2, P_{crit}, P^* and failure modes.

Specimen	g_1 [mm]	g_2 [mm]	P_{crit} [kN]	P^* [kN]	Failure mode
DS_G1_1	0.49	1.22	4.29	15.44	SF/MF
DS_G1_2	/	/	/	10.99	MF
DS_G1_3	0.74	0.84	4.40	6.14	SF/MF
DS_G1_4	0.58	1.32	4.83	8.91	SF/MF
DS_G2_1	/	/	/	12.07	MF
DS_G2_2	/	/	/	13.44	MF
DS_G2_3	/	/	/	14.09	MF
DS_G2_4	/	/	/	14.20	MF
DS_G3_1	0.35	0.74	4.80	9.41	MF
DS_G3_2	0.28	0.75	4.62	8.97	MF
DS_G3_3	0.27	0.65	4.63	10.04	MF
DS_G3_4	0.38	0.73	4.53	9.25	MF

4 CONCLUSIONS

In this study, single-lap shear tests of an SRG strip bonded to a masonry block were conducted. Two types of anchorage were considered: 1) an extension of the SRG fibers into the masonry block obtained by bending the fibers and inserting them into a hole filled with a mortar; and 2) a separate spike that interacts with the bonded strip. Test results demonstrate that effectiveness of the SRG strip is positively affected by both types of anchorage systems. The anchorages were able to carry additional load with respect to the load-carrying capacity of the SRG–masonry interface. In general, depending on the use of the anchorage systems, it would be more appropriate to choose one system over the other. One advantage of the first type of anchorage system, i.e. the bonded strip with extension of the SRG fibers into the masonry block, is the ease of realization. In fact, a separate spike requires more work, in particular when spike fibers have to be interweaved with the longitudinal fibers of the SRG. However, in some applications, such as repairing of continuous masonry wall, a separate spike will be required.

Authors of this paper are now working on the mechanical modeling of the anchorage system.

REFERENCES

Bellini, A., A. Incerti, and C. Mazzotti. 2017. Out-of-Plane Strengthening of Masonry Walls with FRCM Composite Materials. *Key Engineering Materials 747*: 158–165.

Carloni, C., and K.V. Subramaniam. 2012. Application of Fracture Mechanics to Debonding of FRP from RC Members. *Special Publication 286*: 1–16.

Carloni, C., M. Santandrea, and I.A. OmarImohamed. 2017. Determination of the interfacial properties of SRP strips bonded to concrete and comparison between single-lap and notched beam tests. *Engineering Fracture Mechanics 186*: 80–104.

Carrara, P., and D. Ferretti. 2013. A finite-difference model with mixed interface laws for shear tests of FRP plates bonded to concrete. *Composites Part B: Engineering 54*: 329–342.

Carrara, P., D. Ferretti, F. Freddi, and G. Rosati. 2011. Shear tests of carbon fiber plates bonded to concrete with control of snap-back. *Engineering fracture mechanics 78, no. 15*: 2663–2678. http://products.kerakoll. com/index.asp. n.d.

Incerti, A., A. Tilocca, A. Bellini, and M. Savoia. 2020. In-plane shear behaviour of FRCM-strengthened masonry panels. *17th International Brick and Block Masonry Conference (IBMAC 2020)*.

Pellegrino, C., e C. Modena. 2009. Influence of FRP Axial Rigidity on FRP-Concrete Bond Behaviour: An Analytical Study. *Advances in Structural Engineering 12, n. 5*.

Sneed, L.H., C. Carloni, G. Baietti, and G. Fraioli. 2017. Confinement of Clay Masonry Columns with SRG. *Key Engineering Materials 747*: 350–357.

Sneed, L.H., C. Carloni, G. Fraioli, e G. Baietti. 2018. Confinement of Brick Masonry Columns with SRG Jackets. *Special Publication 324*.

Sneed, L.H., G. Baietti, G. Fraioli, and C. Carloni. 2019. Compressive Behavior of Brick Masonry Columns Confined with Steel-Reinforced Grout Jackets. *J. Compos. Constr. 23, no. 5*.

UNI EN 1015-11. 2007: *Metodi di prova per malte per opere murarie - Parte 11: Determinazione della resistenza a flessione e a compressione della malta indurita*.

UNI EN 772-1. 2011: *Metodi di prova per elementi per muratura - Parte 1: Determinazione della resistenza a compressione*.

Brick and Block Masonry - From Historical to Sustainable Masonry –
Kubica, Kwiecień & Bednarz (eds)
© *2020 Taylor & Francis Group, London, ISBN 978-0-367-56586-2*

Shear behavior of masonry triplets with damp proof course and thermal insulating layer

M. Vanheukelom, B. Vandoren, D. Dragan & H. Degée
Hasselt University, Hasselt, Belgium

ABSTRACT: Energy efficiency is an important aspect in today's society. In Belgium, the first layer placed directly on the floor is commonly executed using masonry units with higher insulating properties, preventing the occurrence of thermal bridges between the slab and the rest of the wall. Moreover, a damp proof course (DPC) is often placed between the first two layers with two main functions: (i) to direct water outside the wall out of the cavity and (ii) to prevent moisture from rising into the building.

Thermal conditions are improved with a thermal break layer. However, both interventions have an impact on the structural behavior. As of yet, hardly any research has been performed on the shear strength of a masonry wall where both layers are included.

In this contribution, an experimental campaign is presented in which shear tests are performed on masonry triplets with varying combinations of bricks and with or without a DPC.

1 INTRODUCTION

Masonry is one of the oldest construction methods that is still used in today's buildings. Materials are easy to produce and widely available. It is also a cheap and very durable technique. The main purpose of a masonry wall is to support the load of the construction above and withstand wind loads.

Nowadays sustainability and durability has become increasingly important in the construction sector. This makes that a masonry wall has to fulfill multiple tasks. One aspect of durability can be achieved if water/damp can be directed to the outside of the wall, especially in areas potentially sensitive to freezing. This prevents water getting trapped in the bricks and cracking of the bricks from the inside.

A damp proof course (DPC), located as shown in Figure 1, ensures that water exits the cavity of the wall through the buttjoints This DPC layer also prevents moisture from rising up into the wall.

This non-structural measure has structural consequences on the shear behavior. In Eurocode 6 (EN 1996-1-1 2005), a design procedure is given regarding the shear strength of a masonry wall, but this procedure is not suited for a masonry wall with a DPC layer. The influence on the shear strength of a DPC layer should be assessed according to EN1052-4 (EN 1052-4 2000). In this testing standard a small masonry wall panel is constructed with three layers. In every layer at least one vertical joint is present. A DPC layer is placed in the middle of every mortar layer. After precompressing the wall panel, it is tested on shear up to failure. The general setup of this test is shown in Figure 2.

Martens et al. (2008) performed experiments in accordance with EN 1052-4 to evaluate the applicability of masonry walls with DPC layer as structural elements. In these tests, although the poly-ethylene-sheets (PE) had a double sided embossed diamond shape for improved adherence, it was found that the characteristic initial shear strength should be taken as null (i.e. $f_{vk0} = 0$) as a safe lower bound value.

Another commonly used test procedure to evaluate masonry shear strength is EN 1052-3 (2002). Although this procedure is not officially suited for masonry with DPC layers, it can be anyway used for that purpose. Mojsilović & Simundic (2009, 2012) performed such experiments to obtain a better understanding on the shear behavior of a masonry wall with a DPC layer. Triplet tests in accordance with EN 1052-3 and masonry panels with PE-sheets as DPC layer were tested. It was again observed that there was hardly any adherence between the PE-sheet and the mortar.

A second aspect of sustainability is related to energy performances. New buildings in Belgium and other northern countries need to be almost energy neutral. To help in achieving this goal, masonry walls are more and more constructed with a thermal break layer at their base. These units have better insulating properties than a standard brick but their compressive strength is usually lower. Only one layer is needed to prevent a cold bridge, as illustrated in Figure 1. The red continuous line shows the thermal break between the outside and inside.

Hardly any research has been performed on masonry walls with an insulation layer. Deyazada (2019) investigated the compression strength of

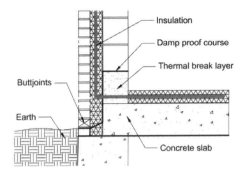

Figure 1. Section of masonry walls with insulating layer and damp proof course.

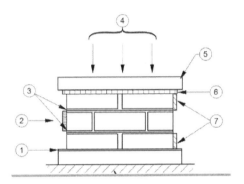

Figure 2. Test setup EN 1052-4 (2000).

masonry walls with a thermal break layer. The compressive strength was tested between homogeneous masonry panels only consisting of aerated autoclaved concrete (AAC) and composite masonry panels consisting of one layer of AAC and the other layers were constructed with standard clay bricks.

Figure 3 illustrates the mean compressive strength for the homogeneous and composite specimens. The bars show the compressive strength of the specimens.

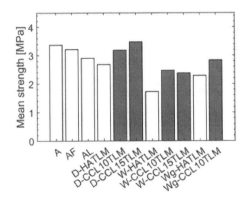

Figure 3. Mean compressive strength of masonry panels (Deyazada 2019).

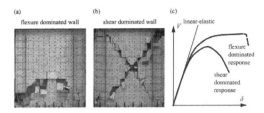

Figure 4. (a) Photo of flexure dominated wall at failure, (b) Photo of shear dominated wall at failure, (c) Shear load-drift curves qualitatively representing typical flexure and shear response of an in-plane loaded URM wall (Wilding & Beyer 2017).

The white bars are homogeneous speci mens (H) and the grey ones consist in a combination of two types of masonry units (C). The figure shows that the compressive strength of a composite wall is higher compared to a masonry wall built with only the weaker material (AAC).

The shear behavior of such a composite wall, however, is not known. The shear behavior of unreinforced masonry walls (URM) is in general a non-linear behavior. The failure mode can be categorized into three groups, namely flexure dominated walls, shear dominated walls and hybrid failure (Wilding & Beyer 2017). Flexure dominated walls exhibit a rocking behavior with toe crushing as failure mode. Shear dominated walls exhibit a typically staircase stepped failure and hybrid failure shows modes of both categories, as shown in Figure 4 (Wilding & Beyer 2017).

2 EXPERIMENTAL CAMPAIGN

The current paper is part of a larger research program focusing on the structural shear behavior of non-standard masonry wall panels where the focus is put on the behavior of walls with a thermal break layer at the bottom and a damp proof course between the first and second layer. The starting point of this global experimental campaign is shown in this paper and consists in 54 triplet tests according to EN 1052-3 carried out to evaluate the initial shear stress and the friction coefficient between the different layers.

According to this procedure, a Mohr-Coulomb failure envelope can be constructed, from which the initial shear strength and the friction angle can be derived as shown in Figure 5. Nine tests were performed for every configuration. Relatively low pre-compression levels needed to be chosen to make sure the experiment fails in sliding and not in pore collapse of the masonry units.

2.1 Experimental program

Figure 6 shows the setup designed according to EN 1052-3 (2002). These triplets are performed to

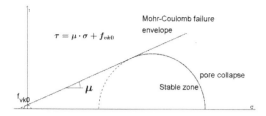

Figure 5. Mohr-Coulomb failure curve.

Figure 7. Diamond embossed shape of damp proof course.

$$\tau = \mu \cdot \sigma + f_{vk0}$$

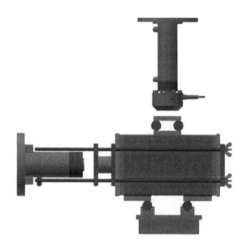

Figure 6. Setup triplet tests.

investigate the shear behavior with and without a DPC layer. Two types of clay bricks are considered: perforated clay bricks assembled with general purpose mortar (GPM) and perforated clay bricks with a thin layer mortar (TLM). The properties of the materials as declared by the manufacturer are listed in Table 1.

Table 1. Specifications of materials.

Specimen	Abbreviation	Dimensions mm	Compressive strength* N/mm²
Clay bricks with GPM	t	287 x 138 x 138	10
Clay bricks with TLM	p	500 x 138 x184	15
AAC block	a	600 x 140 x 150	2.97
General purpose mortar	n	/	8
Thin layer mortar	g	/	15

* declared by the manufacturer

The mortars used in the experiments were ready-made mortars. The DPC layers used in this project are polyethylene sheets with an embossed diamond shape on both sides aiming at improving the adherence between the sheet and the mortar, see Figure 7. The total thickness of the sheet is one millimeter and it is placed in between two layers of mortar.

Six different series are tested and for each series, three different precompression levels are chosen. For each precompression level three tests are performed, which makes a total of 54 experiments as can be seen in Table 2. The targeted failure mode was sliding of the specimens to allow for calibrating the Mohr-Coulomb failure envelope.

The name of each test is based on the abbreviations from Table 1 and has the format $X_1X_2X_3X_4/Y/N$, where X_1 represents the bricks on the outside (t or p), X_2 represents the bricks in the middle (t, p or a), X_3 represents the mortar type (n or g), X_4 indicates the use of DPCs (d), Y refers to the level of precompression and N is the number of the test.

2.2 Test procedure

The specimens made with the bricks for GPM had a length of 287 mm and the bricks for TLM and AAC blocks were cut to 300 mm of length. When AAC blocks were used a glue was added to the mortar mixture for extra adhesion between the mortar and the AAC, as recommended by the manufacturer. After preparation, the specimens cured for at least 28 days and were then placed in the test setup as can be seen in Figure 8.

The specimen is placed between two wooden plates (indicated by number 12 on Figure 8) for

Table 2. Different series for triplet tests.

Series				
Without DPC	With DPC	Precompression level in MPa		
ttn-	/	0.05	0.1	0.2
ppg-	/	0.01/0.04	0.02	0.2
tan-	tand	0.1	0.3	0.5
pan	pand	0.1	0.3	0.5

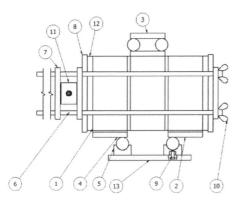

Figure 8. Parts triplet test setup (legend explained in text below).

a better transfer of the precompression load. Steel plates (7-8) are placed behind it to ensure a distributed pre-compression load. This load is measured with a load cell (11) between two steel plates (7). The precompression system is hold together with four rods (6) and nuts (10) at the end. The shear load is applied at the top steel plate (3) and the four steel rollers (5) are placed according to EN 1052-3. Where the steel plates touch the bricks cardboard is placed for a better transfer of the applied forces.

Most of the experiments with a DPC layer lost adhesion during transport, even when it was placed cautiously in the test setup. Only three specimens (out of 18 specimens) could be tested with an intact connection on both DPC layers.

The mortar joints of the specimens with glue mortar had an average thickness of 1mm and for the specimens with a normal purpose mortar was the average joint thickness 14 mm.

During the tests hydraulic jacks were used to create precompression and apply the shear force. LVDTs were placed on top of the outside bricks and on top of the steel plate above the rollers for the

Figure 9. Test setup with LVDTs showing in the blue ellipses.

middle brick. The LVDT in the middle measured the relative displacement between the central masonry unit and the ones on the outside. The LVDTs on the side were used for the purpose of test control as can be seen in Figure 9.

The precompression levels were based on the gross area (length x width) of the perforated clay bricks. The net area is less than half of the gross area (47%) but the calculations for the shear stress remains the same principle.

3 RESULTS

Figures 10 and 11 show the load-displacement diagrams for "tan-" and "tand". They both show a linear elastic behavior until the occurrence of a first crack in the mortar joint. After the first crack the specimens with a DPC layer decreased in stiffness compared to the ones without DPC layer. The first crack for all specimens happens around 15 kN. For all specimens with a DPC layer, there was only slip between the mortar and the DPC layer. The peak in tand-03-1 is probably from the ripping of the DPC which is shown in Figure 12. Also tand-03-2 had a small rip in the DPC layer.

The tan-03 specimens failed mostly in sliding between the AAC and the mortar. Tan-03-1/2 also had a small crack in the AAC before sliding as can be seen in Figure 13.

For each experiment the shear strength f_{v0i} and the average precompression stress f_{pi} are calculated:

$$f_{v0i} = \frac{F_{i,max}}{2 \cdot A_i} \qquad (1)$$

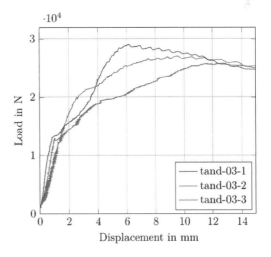

Figure 10. Load - Displacement diagram for specimens with a precompression level of 0.3MPa and with a DPC layer.

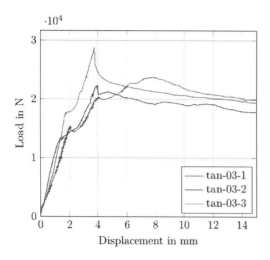

Figure 11. Load - Displacement diagram for specimens with a precompression level of 0.3MPa and without a DPC layer.

Figure 12. Ripping of the DPC layer (tand-03-3).

Figure 13. Crack in AAC in top left corner after failure of tan-03-1.

Figure 14. Brick failure from specimen ppg-02-3.

Table 3. Mean value of precompression stress and shear strength.

Test	f_{pi} N/mm²	f_{voi} N/mm²
ttn-005	0.053	0.241
ttn-01	0.104	0.279
ttn-02	0.160	0.352
tan-01	0.105	0.253
tan-03	0.312	0.324
tan-05	0.507	0.400
tand-01	0.103	0.160
tand-03	0.302	0.344
tand-05	0.517	0.410
ppg-001	0.011	0.190
ppg-002	0.024	0.203
ppg-004	0.04	0.208
ppg-02	0.164	0.156
pan-01	0.083	0.214
pan-03	0.241	0.320
pan-05	0.403	0.289
pand-01	0.111	0.166
pand-03	0.237	0.276
pand-05	0.393	0.345

$$f_{v0i} = \frac{F_{i,\max}}{2 \cdot A_{i,1}} \qquad (2)$$

where $F_{i,\max}$ = the maximum shear force in N, A_i = the area based on the global external dimensions where slip occurs in mm², $F_{pi,av}$ = the average precompression force before sliding occurs in N and $A_{i,1}$ = the area on top of the middle brick in mm².

With the precompression and shear strength for every specimen, a Mohr-Coulomb envelope can be derived and the initial shear strength and friction angle can be determined. Figures 15 until 20 show the results of the experiments.

Figure 18 shows no trend line. To ensure that sliding was the failure mode, very low precompression

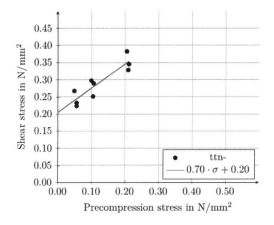

Figure 15. Individual precompression shear tests ttn-.

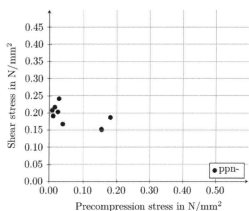

Figure 18. Individual precompression shear tests ppn-.

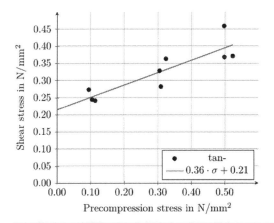

Figure 16. Individual precompression shear tests tan-.

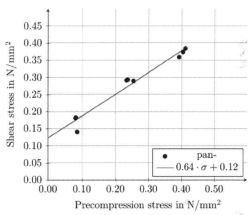

Figure 19. Individual precompression shear tests pan-.

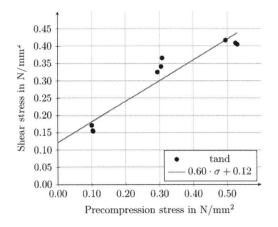

Figure 17. Individual precompression shear tests tand.

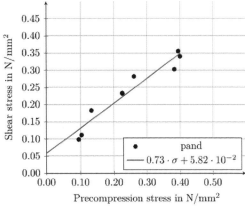

Figure 20. Individual precompression shear tests pand.

Table 4. Initial shear strength and friction angle (mean values).

Test	Initial shear strength MPa	Friction angle rad
ttn-	0.2	0.7
tan-	0.21	0.36
tand	0.12	0.6
ppn-	0.21	/
pan-	0.12	0.64
pand	0.06	0.73

levels were needed for the perforated bricks with TLM. If the precompression level is too high, the specimen will not fail in sliding but pore collapse will be the failure mode as can be seen in Figure 5. The highest precompression level for this series, however, showed cracking in the bricks without sliding as shown in Figure 14. Deriving a trendline for this series is, therefore, not representative. However, averaging the bulk of experimental points close to the vertical axis leads to an initial shear strength of 0.21 MPa.

In Table 3 an overview is given of all results. This clearly shows an increase in friction resistance in the specimens with a DPC layer compared to the ones without. Finally, Table 4 shows the mean values of these results.

4 CONCLUSIONS

In order to evaluate the influence of a thermal insulating layer and a DPC, possibly in combination, on the shear strength of unreinforced masonry walls, this contribution presents the first outcomes of an ongoing research program. 54 triplet tests were carried out to evaluate the initial shear strength and friction coefficient for different combinations of materials and mortar. Various precompression levels were considered. Although the initial shear strength of masonry units with a DPC layer should normally be tested according to EN 1052-4, it is chosen to test all configurations according to EN1052-3 for the sake of a consistent comparison. The behavior of DPC in the presence of head joints, which is the main added value of EN 1052-4, will be further investigated in upcoming tests on panels.

The masonry specimens with a DPC layer show a rather low initial shear strength. This results in a characteristic initial shear strength f_{vk0} of almost zero. This result was also evidenced by Martens (2008) and Mojsilović & Simundic (2009). Their friction coefficient (tan α) is however in average slightly higher than for the specimens without a DPC layer.

According to Eurocode 6 (EN 1996-1-1 2005) a value of 0.2 MPa is recommended for the initial shear strength of masonry with normal purpose mortar. This value appears in accordance with the results obtained in this campaign. For glue mortar, according to the standard, a value of 0.3 MPa is recommended while the current experimental results lead to an f_{vk0} of about 0.2 MPa. It has, however, to be mentioned that, although the precompression level was kept as low as possible to target a pure sliding failure, the observed failure mode of glued specimens was actually combined sliding with pore collapse, making the outcomes of this set of tests less reliable than those with general purpose mortar.

Moreover, the value of the friction coefficient recommended by Eurocode 6 is 0.4. This value was always reached in the reported tests, with the exception of the specimens combining AAC units with clay bricks and GPM. This can possibly be related to the use of an additive to improve the contact between mortar and AAC, but requires further investigation.

REFERENCES

Deyazada, M. 2019. Experimental investigations on the resistance of masonry walls with AAC thermal break layer. *Constr. Build. Mater. 224*: 474–492.

EN 1052-3 2002. EN_1052-3: Methods of test for masonry- Part 3: Determination of initial shear strength. European Union norm on construction.

EN 1052-4 2000. EN_1052-4: Methods of test for masonry - Part 4: Determination of shear strength including damp proof course. European Union norm on construction.

EN 1996-1-1 2005. Eurocode 6: Design of masonry structures - Part 1-1: General rules for reinforced and unreinforced masonry structures. European committee for standardization.

Martens, D.R.W. & Bertram, G. 2008. Shear strength of clay brick masonry including damp proof course 8.

Mojsilović, N. & Simundic, G., 2009. Static-Cyclic Shear Tests on Masonry Wallettes with a Damp-Proof Course Membrane.

Mojsilović, N., 2012. Masonry elements with damp-proof course membrane: Assessment of shear strength parameters. *Constr. Build. Mater. 35*: 1002–1012.

Wilding, B. & Beyer, K. 2017. Drift capacity of URM walls: Are analytical models an alternative to empirical approaches? *Proc. 16th World Conf. Earthq.*

Brick and Block Masonry - From Historical to Sustainable Masonry –
Kubica, Kwiecień & Bednarz (eds)
© 2020 Taylor & Francis Group, London, ISBN 978-0-367-56586-2

Experimental characterization of PBO-FRCM composites for masonry structures retrofit

I.E. Senaldi, G. Guerrini, A. Bruggi & A. Penna
Department of Civil Engineering and Architecture, University of Pavia, Pavia, Italy

M. Quaini
Ruregold S.r.l., Milan, Italy

ABSTRACT: Fabric Reinforced Cementitious Matrices (FRCM), have recently emerged as innovative materials for the retrofit of existing unreinforced masonry structures. The variability of their constituent materials results in a wide range of mechanical properties of these composite materials, and consequently can affect the structural performance of the retrofitted masonry. The Italian Ministry of Public Works guidelines specify standard tensile, bond and durability tests to evaluate a number of mechanical properties for the quali- fication and acceptance of FRCM materials. Within this context, an experimental testing program was per- formed at University of Pavia on FRCM composites incorporating polyparaphenylene benzobisoxazole (PBO) meshes and fibre-reinforced hydraulic-lime mortar with pozzolan, to investigate their tensile behaviour as well as the one of the PBO textile. The stability of the mechanical properties with respect to degradation, induced by environmental actions, was evaluated by means of durability tests performed on conditioned sam- ples. A series of single-lap tests allowed determining the bond properties between FRCM and two different masonry supports.

1 INTRODUCTION

In recent years, innovative strengthening solutions with externally bonded Fibre Reinforced Cementi- tious Matrix (FRCM) materials, also known as TRC (Textile Reinforced Concrete), TRM (Textile Reinforced Mortars), FRM (Fabric Reinforced Mortar) or IMG (Inorganic Matrix-Grid Compos- ites), have been proposed for the retrofit of existing structures (Papanicolau et al. 2008, de Felice et al. 2014, Ascione et al. 2015). The use of inorganic matrices is gradually substituting the epoxy resins adopted in Fibre Reinforced Polymers (FRPs), in particular in heritage masonry buildings. The use of FRPs is currently discouraged for applications to masonry structures since the presence of epoxy matrix may compromise the preservation and dur- ability of the structure (de Santis & de Felice 2015). Mortar-based strengthening contributes to the fulfil- ment of the preservation criteria required to applica- tions to cultural heritage assets, such as the respect of authenticity in terms of materials and structural behaviour, or the design of interventions based on the principle of minimum intervention. Additionally, the use of suitable mortars may provide reversibility and physical/chemical compatibility with the ori- ginal masonry substrate, hence ensuring the durabil- ity of the intervention (Valluzzi et al. 2014b; Ascione et al. 2015; Papanicolau et al. 2008).

Although the development of FRCM composites is still at a relatively early stage, a fair number of experi- mental studies have been performed in the last decade to investigate their mechanical characteristics and their effectiveness as strengthening materials. Fundamental information has been recently provided by research studies on the tensile response (Bertolesi et al. 2014; D'Antino & Papanicolau 2017, de Santis & de Felice 2015), the durability (Nobili & Signorini 2017; Trian- tafillou et al. 2017), and the bond performance on both masonry (d'Ambrisi et al. 2013a; D'Antino et al. 2014; de Felice et al. 2014; Malena & de Felice 2014; Razavizadeh et al. 2014; Carozzi et al. 2017; de Santis et al. 2017) and concrete substrates (d'Ambrisi et al. 2013b).

The influence of FRCM seismic retrofit on the in- plane response of masonry walls has been studied experimentally and numerically by several researches (Prota et al. 2006, Augenti et al 2010; Parisi et al. 2013, Giaretton et al. 2018, Del Zoppo et al. 2019, Garcia-Ramonda et al. 2020). Experimental tests were also devoted to the study of FRCM strengthening of walls subjected to out-of-plane loading, since they can enhance the response of masonry structural elem- ents thanks to the tensile strength provided by the tex- tile. Tests have been performed on medium-scale panels (Papanicolaou et al. 2008; Valluzzi et al. 2014a; Kariou et al. 2018) or on full-scale specimens with different set-up and testing procedures (Bellini et al. 2017; Babaeidarabad et al. 2014a; D'Antino et al. 2018; de Santis et al. 2014; de Santis et al. 2016; de Santis et al. 2019a; Gattesco & Boem 2017; Meriggi et al. 2020; Stempniewski & Urban 2014). Although less numerous, medium- to large-scale investigations addressed also the effectiveness of FRCM systems at strengthening masonry arches and vaults (Borri et al. 2009; Giamundo et al. 2015; Alecci et al. 2016; Gattesco et al. 2018; de Santis et al. 2019b) or confining columns (Carloni et al. 2015). Nevertheless, the large number of textiles and inorganic matrices and the different applications result in a significant variability in terms of mechanical response, tensile and bond strength and expected failure modes that need to be further investigated.

Currently, FRCM systems available in the European market and used in construction practice are undergoing standardized mechanical characterization tests to become formally acceptable for the approval of the structural design (Ascione et al. 2015), as required by design codes. In Italy, the Ministry of Public Works and the National Research Council have recently published two new guidelines, to provide acceptance criteria and design provisions for Fabric-Reinforced Cementitious Matrix (FRCM) systems used for repair and strengthening of existing buildings, in particular masonry and reinforced concrete structures (CNR 2018, CSLLPP 2018). The guidelines regarding the qualification of FRCM systems establish the testing procedures to determine the main mechanical properties that are required for the design of the application.

This paper describes part of an extensive experimental program performed at the University of Pavia to investigate the tensile response of FRCM and the influence of artificial ageing and exposure to aggressive environments on the mechanical properties of FRCM systems with polyparaphenylene benzobisoxazole (PBO) textiles.

2 CHARACTERISTICS OF FRCM COMPOSITES

The FRCM systems are externally bonded composite materials, constituted by an inorganic matrix, a reinforcing mesh, connectors or anchoring devices and, if present, organic additives. The thickness, as defined in the Italian Guidelines (CSLLPP 2018) is below 30 mm, typically between 5 to 20 mm.

The inorganic matrix is constituted by mortar, either cement- or hydraulic lime-based, to which polymeric microfibers are added be dispersed in the matrix to prevent premature cracking due to shrinkage or to improve the response of the mortar. The mesh is an open grid fabric of strands made of fibres, woven as weft or warp yarns. Different materials constitute the continuous fibres, such as high strength steel, aramid, basalt, carbon, glass, PBO, or a combination of them.

The typical stress-strain relationship of FRCM under tensile load can be schematically represented as a tri-linear (Ascione et al. 2018, Carozzi and Poggi 2015, D'Antino and Papanicolaou 2017), where the three branches are associated to the different response stages. The first branch represents the un-cracked state of the specimen (stage A, as defined in CSLLPP 2018), where the response is linear and the initial slope of the stress-strain curve identifies the modulus of elasticity of the mortar matrix. The occurrence of the first crack identifies the transition to the second stage B, which corresponds to a decrease of stiffness and to the progressive development of cracks in the mortar. The response in the first stages is related to the mechanical characteristics of both textile and matrix and to the tensile stress transfer from the mesh to the mortar. Such properties influence the crack pattern, in terms of crack number, width and distribution that in turn may affect the durability once the textile is exposed to aggressive environment (e.g. alkaline, saline environment) (Ascione et al. 2015, Carozzi & Poggi 2015). Once the number of cracks is stabilized, any additional strain increase determines an increase of stiffness, tensile strength and widening of the existing cracks up to failure of the textile (stage C).

The mechanical response of FRCM systems is characterized not only by the tensile strength of textile and matrix, but also by the bond strength with respect to the support to which the reinforcement is applied and by the failure modes that may occur. The Italian Guidelines (CSLLPP 2018) identify the following modes: debonding with cohesive failure in the support (mode A), debonding at the reinforcement-to-substrate interface (B) or at the mesh-to-matrix interface (C), sliding of the textile within the reinforcement thickness (D), tensile rupture of the textile within the mortar matrix (E), tensile rupture of the textile in the un-bonded portion (F).

3 QUALIFICATION PROCESS ACCORDING TO ITALIAN GUIDELINES

3.1 Qualification process

The Italian Guidelines (CSLLPP 2018) specifies the methods and criteria to assess the performance of FRCM systems, allowing the mechanical characterization of FRCM composites and providing the fundamental parameters for the design of the externally bonded reinforcement. The prescriptions are meant to be applicable to possibly most of the available FRCM systems.

The purpose of the testing procedure is to identify the ultimate tensile strength and the shear bond strength of the reinforcement, which are fundamental parameters for the design of the most recurrent interventions with FRCM (for example, the confinement of columns/pillars and nodes, the extrados reinforcement of arches and vaults or the out-of-plane retrofit of masonry structural elements). The guideline recommends performing direct tensile tests on textiles and composite together with shear bond test on three conventional supports (concrete, tuff masonry, solid clay brick masonry).

The mechanical properties of FRCM systems should be sufficiently stable with respect to degradation induced by aggressive environments. The stability of the mechanical characteristics with respect to the degradation induced by environmental actions is evaluated by means of tensile tests performed after durability tests in which the FRCM specimens are subjected to different artificial environments, including freeze-thaw cycles and artificial aging tests. For qualifying the FRCM system, the mean tensile strength of the conditioned specimens should not be lower than a specific percentage of the mean strength of unconditioned specimens.

3.2 Experimental program performed at University of Pavia

The extensive experimental program performed at the University of Pavia aims to investigate the response of FRCM systems and to address the influence of artificial ageing and exposure to aggressive environments on their mechanical properties.

Direct tensile tests were performed on FRCM composite specimens, constituted by hydraulic lime based mortar and bi-directional PBO textiles. The direct tensile tests were performed on prismatic specimens both unconditioned and subjected to durability tests, according to the procedures defined in the Italian Guidelines (CSLLPP 2018). Shear bond tests were performed to assess the evaluate the bond strength of the investigated FRCM system, by applying the reinforcement both on clay bricks and on tuff masonry supports, in a single-lap testing scheme. The single-lap set up allows identifying the peak load at failure, nevertheless requiring particular attention in specimen alignment, to ensure the application of a pure shear stress to the reinforcement hence prevent the result from being affected by parasitic normal stresses on the substrate-to-mortar interface.

Table 1 summarizes the number of tests performed during the experimental program.

4 MATERIAL PROPERTIES AND GEOMETRY OF SPECIMENS

The FRCM composite employed is comprised of an inorganic matrix composed of a fiber-reinforced hydraulic-lime mortar with pozzolan. The mortar is characterized by a compressive mean strength f_m of 35.8 MPa and a means tensile strength of 6.2 MPa, determined from 3-point bending tests and vertical compression tests on mortar samples cured for at least 28 days (EN1015-11:2007).

The dry fiber net is a bi-directional balanced mesh constituted of PBO fibers together with a percentage of glass fibers. Rovings are spaced of approximately

Table 1. Summary of tests performed.

Test	Number of specimens
Mechanical tests:	
Tensile tests on dry mesh	10
Tensile tests on FRCM coupons	14
Debonding tests:	
Clay support	9
Tuff support	9
Durability tests:	
Moist environment	5+5
Saline environment	5+5
Alkaline environment	5+5
Freeze-Thaw cycles	5+5
Dry heat	5

13 mm on center in the two directions. The weight per unit area of PBO fibers is 22.0 g/m² both in warp and weft directions. The nominal resisting area of PBO fibers per unit length is equal to 14.10 mm²/m.

Dry fiber specimens had all nominal size of 500 × 100 mm (excluding the clamping portions).

FRCM coupons for tensile tests were made in flat molds applying a first layer of mortar (5mm thick), the PBO fiber mesh and a second mortar layer (5mm thick). The curing of specimens lasted 28 days. The coupons were all rectangular (nominal size 4000 × 100 x 10 mm, excluding the clamping portions).

The specimens for shear bond tests are characterized by the application of PBO-FRCM reinforcement on one side (single-lap configuration) of clay brick or tuff masonry prisms. The reinforcement is applied at 30 to 35 mm from the upper face of the masonry support and at least at 20 mm from the sides. The bond length is 300 mm while the bond width is equivalent to that of the FRCM coupons (100 mm). The masonry supports are composed by either seven stack-bonded clay bricks (brick compressive strength: 22.8 MPa) or three-stack bonded tuff blocks (tuff compressive strength: 9.0 MPa). Mortar joints with thickness ranging between 5 to 10 mm were used, with compressive strength ranging between 5.5 MPa to 7.7 MPa.

The extremes of both dry mesh and FRCM specimens are clamped in the wedges of the testing machine, avoiding sliding in the griping areas thanks to the applied normal stress. To limit possible concentration of stresses and crushing of the mortar, fiber reinforced tabs (dimensions 130 × 100 mm for dry mesh specimens or 80 × 100 mm for FRCM coupons) were glued to the extremities of the specimens with epoxy resin. Since the clamping method may affect significantly the evaluation of the mechanical properties of the FRCM (De Santis & De Felice, 2015; Carozzi et al, 2017), Carozzi & Poggi (2015) suggest to apply the aforementioned tabs, since they have proven to allow fully exploiting the ultimate tensile stress of the fiber mesh during the third phase of the FRCM stress-strain relationship.

5 EXPERIMENTAL SET-UP AND LOADING PROTOCOLS

Direct tensile tests were performed on a Material Testing Systems (MTS) load frame, with 500 kN loading capacity. The load was applied monotonically under displacement control and recorded by the load cell integrated in the MTS.

The loading velocity depends the type of test performed:

- Tensile tests on dry mesh specimens: test velocity of 0.05 mm/min up to failure.
- Tensile tests on FRCM specimens: constant rate equal to 0.02 mm/min in the non-cracked and crack-formation phase, increased to 0.05 mm/min at the end of the cracked stage up to failure.

Several instruments were used to measure strain/displacements, as shown in Figure 1. The relative displacement between the extremities of the MTS testing machine were acquired by an internal LVDT and, only for PBO mesh specimens, by two linear potentiometers (± 12.5mm range) fixed on the end tabs of the net specimen with steel plates (Figure 1, top). The redundancy of global displacement measures helps identifying possible sliding of the specimen in the gripping areas. The elongation of two single PBO yarns is recorded by two further potentiometers (± 12.5 mm range). In case of FRCM specimens (Figure 1, bottom), two potentiometers (± 12.5 mm range) are fixed to the mortar matrix by steel plates with a base length of 355 mm approximately.

Shear bond tests were conducted in a single-lap configuration, on the same MTS testing machine used for tensile tests (Figure 2). The specimen is located on a supporting steel frame, gripped from below to the testing machine, while the PBO net strip is pulled from above. Tests were performed monotonically under displacement control at 0.02 mm/min constant rate up to failure. Two spring-loaded potentiometer (12.5 mm range) monitor the slip between the substrate and the reinforcement, in the section at the loaded edge, while

Figure 2. Shear bond tests on clay brick (left) or tuff (right) masonry supports.

two linear potentiometers (± 12.5 mm range) measure the strain in the un-bonded textile.

6 TEST RESULTS

6.1 *Evaluation of mechanical properties of PBO-FRCM systems*

The Italian Guidelines qualifies the FRCM composites based on the following predefined mechanical properties (CSLLPP 2018):

- Characteristic tensile strength σ_u and average tensile strength σ_{max} of the FRCM sample;
- Ultimate tensile strain ε_u (mean value) of the FRCM sample;
- Elastic modulus E_1 (mean value) of the FRCM sample in the stage A, if detectable;
- Tensile strength σ_{uf} (characteristic value) of the dry mesh;
- Ultimate tensile strain ε_{uf} of the dry mesh, evaluated as the ratio between σ_{uf} and E_f;
- Elastic modulus E_f (mean value) of the dry mesh, computed between the 10% and the 50% of the peak stress;
- Conventional limit stress (characteristic value), $\sigma_{lim,conv}$ and conventional limit strain, $\varepsilon_{lim,conv}$, calculated as the ratio between $\sigma_{lim,conv}$ and E_f;
- Compressive strength $f_{c,mat}$ (characteristic value) of the matrix.

The stress and strength values refer to the cross-section area of the bare mesh (A_f) present in the sample, without considering the matrix. In the present study, only the nominal resisting area of PBO fibers was considered, neglecting the presence of glass fibers. The characteristic values are computed

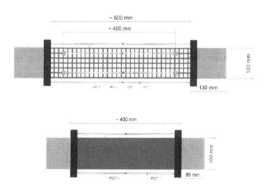

Figure 1. Potentiometer location for PBO net (top) and FRCM specimen (bottom).

as defined in EN1990:2002, Annex D, Table D1, depending on the number of specimens.

6.2 Textile specimens: Monotonic tensile tests

Monotonic tensile tests were performed on ten specimen of dry mesh. The stress-strain curve evaluated for each specimen (Figure 3) depicted well the brittle tensile failure of PBO yarns in the warp direction.

Table 2 presents the results from monotonic tensile tests on dry mesh specimens in terms of average and characteristic tensile strength, ultimate tensile strain, average strain in correspondence of the peak stress $\varepsilon(\sigma_{maxf})$ and the elastic modulus, together with the corresponding values of the coefficient of variation (C.o. V.). The response of the dry mesh is consistent in terms of stiffness and average peak stress, with a characteristic tensile strength close to 2480 MPa, while the strain at failure is on average more dispersed with a C.o.V. of 17.3%.

6.3 PBO-FRCM composites: Tensile tests

Tensile tests on PBO-FRCM specimen exhibited the typical tri-linear response, as shown in Figure 4 where the stress-strain curve is plotted. In the first linear branch the specimen is uncracked and the slope reflect the stiffness of the mortar matrix. The elastic modulus E_1 is computed, when possible, between the 10% and 20% of the peak stress. Once the peak stress is reached a brittle failure usually occurs in the PBO fibres in the warp direction.

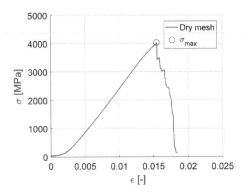

Figure 3. Tensile test on a PBO dry mesh specimen.

Table 2. Tensile tests on PBO textile specimens.

	σ_{maxf} [MPa]	σ_{uf} [MPa]	$\varepsilon(\sigma_{maxf})$	ε_{uf}	E_f [GPa]	Sample Nr.
Value	3208	2479	1.35%	0.88%	282	10
C.o.V.	11.8%	-	17.3%	-	7.5%	

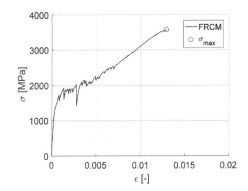

Figure 4. Tensile test on a PBO-FRCM coupon.

Table 3. Tensile tests on FRCM specimens.

	σ_{max} [MPa]	σ_u [MPa]	ε_u	E_1 [GPa]	Sample Nr.
Value	3065	2354	1.21%	3369	14
C.o.V.	11.7%	-	14.1%	35.9%	

Table 3 reports the results from tensile tests in terms of average peak stress and characteristic tensile strength, ultimate tensile strain in correspondence of the peak stress and elastic modulus.

Aggressive environmental conditions are known to affect the performance of construction materials. In particular, the exposure to freeze-thaw cycles is deleterious for cementitious materials, while alkalinity, salt water and moist and are critical for composites (Arboleda et al, 2014). Hence, monotonic tensile tests were also performed on PBO-FRCM specimen after the following artificial conditioning:

- Freezing and thawing (FT): one week of conditioning in a humidity chamber (temperature 38 °C, relative humidity 95%) followed by 20 cycles, with each cycle consisting in 4 h at -18 °C, followed by 12 h in a humidity chamber (temperature 38 °C, relative humidity 95%);
- Ageing in a humidity chamber (temperature 38 °C, relative humidity 95%) for 1000 h (M1000) and 3000 h (M3000);
- Ageing in oceanic salt water, with immersion at 23 °C (salt concentration 4% in weight) for 1000 h (S1000) and 3000 h (S3000);
- Ageing in alkaline solution, with immersion at 23 °C (pH = 9.56) for 1000 h (A1000) and 3000 h (A3000).

For each condition, five specimens were tested under monotonic tensile load applying the same loading protocol used for unconditioned specimens (UC).

Table 4 reports the values of average peak stress and average strain at peak stress recorded after durability tests with the corresponding coefficient of variations, in comparison with those computed for

Table 4. Tensile tests on FRCM specimens after durability tests.

Artificial environment	Duration [hours]	ε_u Value [%]	C.o.V. [%]	σ_{max} Value [MPa]	C.o.V. [%]
Unconditioned	-	*1.21*	*14.1*	*3065*	*11.7*
Moist	1000	1.25	21.1	3314	13.5
	3000	1.02	18	2793	9.9
Saline	1000	1.07	11.9	2908	9.6
	3000	1.07	15.2	2907	11.6
Alkaline	1000	1.07	21.2	2870	12.6
	3000	1.21	5.3	3223	8.7
Freeze-thaw (20 cycles)	528	1.25	8.4	2998	3.3

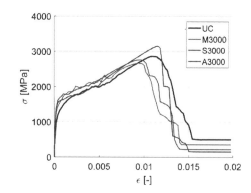

Figure 6. Mean stress-strain curves for conditioned specimens (3000 h artificial conditioning).

unconditioned PBO-FRCM specimens. It is noticeable the decrease in strength on average as effect of the artificial conditioning, with a reduction of maximum 9% less of the mean value, independently of the typology of ageing process. This occurs with the exception few cases (for M1000 and A3000 conditioning) in which the increase in strength, ranging between 5% and 8%, may be associated to the continued hydration of the mortar after 28 day of curing, although only one M1000 test result is outside the range of variability of the UC set of data.

The higher loss of mean strength is recorded, for ageing procedures relatively short duration, for specimens in alkaline environment (A1000) with a reduction of 6% while for long duration conditioning specimens subjected to moist environment (M3000) are characterized by a value of σ_{max} that is 91% of the reference unconditioned specimens.

The response of conditioned samples was studied in terms of average stress-strain curve and compared with the performance of unconditioned specimens for each strain level up to failure. Figure 5 and

Figure 6 show the average curves for short or long duration durability tests, respectively. The negative effect of aggressive environments on the response is confirmed by the reduction in strength, in particular for specimens subjected to alkaline environment for 1000 h and to moist and saline conditioning for 3000 h.

Independently from the typology or duration of the artificial ageing process, first cracking appears at a strain level of 0.02%. Also, the transition between the stage B and the stage C occurs approximately at the same level of elongation, close to 0.5%. The variation of response with respect to unconditioned samples is evident in stage C, were the mesh is not able to develop fully its tensile capacity together with a decrease of the ultimate stiffness.

6.4 PBO-FRCM composites: Bond tests

Shear bond tests were performed on nine specimens for each types of support, namely clay brick and tuff masonry, in a single-lap test set-up. Specimens were tested up to failure. Table 5 reports the mean peak stress σ_{max} and the conventional limit stress and strain. In case of clay masonry support, the values of peak stress are quite dispersed (C.o.V. = 14.2%) with a value of $\sigma_{lim,conv}$ lower more than 30% with respect to tuff masonry.

Independently from the type of masonry support, in most cases (78% for clay and 67% for tuff masonry) the collapse was characterized by

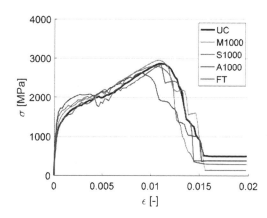

Figure 5. Mean stress-strain curves for conditioned specimens (1000h artificial conditioning, freeze-thaw cycles).

Table 5. Bond tests on clay and tuff masonry supports.

Support	σ_{max} Value [MPa]	C.o.V. [%]	$\sigma_{lim,conv}$ [MPa]	$\varepsilon_{lim,conv}$ [%]
Clay	2322	14.2%	1662	0.59%
Tuff	2833	6.5%	2467	0.87%

a progressive failure of the dry mesh (failure mode F), first in the weft rovings followed by the yarns in the warp direction with no slippage between the mesh and the matrix. The failure mode F is associated to instantaneous load reductions at the tensile failure of the bundles of the textile. Approximately 20% of the specimens exhibited a failure mode characterized by the sliding of the textile within the reinforcement thickness (failure mode D). Such failure is usually related to a soft load decrease due to the progressive loss of friction of the textile, which is sliding within the mortar thickness.

7 CONCLUSIONS

An extensive experimental program was carried out with the purpose to characterize the tensile response of PBO-FRCM materials and to investigate the influence of aggressive environmental condition on their mechanical properties.

The direct tensile tests confirmed the typical trilinear stress-strain response of FRCM materials found in literature. Independently of the durability test performed, the first two branches of the curve, corresponding to the uncracked and crack-development stages, were comparable with those recorded for unconditioned specimens. The main differences were recorded in the third branch associated with the fully cracked response of the composite system: a decrease in stiffness and strength was related to the deterioration of the textile and of its bond with the mortar matrix, due to the exposure to aggressive environments. Nevertheless, the decrease in strength was less than 10% with respect to the average peak strength of unconditioned specimens.

Single-lap bond tests showed good adhesion between the PBO-FRCM composite and both masonry substrates, with failure of the textile in the unbonded portion between the testing machine grip and the FRCM region. Bond tests on clay brick masonry support showed a lower characteristic strength compared to those on tuff block masonry, because of higher variability of the results.

Given the significant variability of FRCM mechanical properties, associated with the large number of textiles and mortar matrices available, a deeper knowledge is needed on the overall behaviour of masonry structures retrofitted with different types of FRCM. Static and dynamic tests on structural elements or building prototypes would indeed provide better insight on the effectiveness of these materials in enhancing the structural performance of heritage buildings.

ACKNOWLEDGEMENTS

The experimental program was performed at the Testing Laboratory of Material of the Department of Civil Engineering and Architecture of the University of Pavia on specimens produced using reinforcing materials and mortars distributed by Ruregold S.r.l. The authors would like to thank Simone Girello and Claudio Pozzi for the technical support during the tests.

REFERENCES

Alecci, V., Misseri, G., Rovero, L., Stipo, G., De Stefano, M., Feo, L., & Luciano, R. 2016. Experimental investigation on masonry arches strengthened with PBO-FRCM composite. *Composites Part B: Engineering, 100*, 228–239.

Ascione L., de Felice g., & De Santis S. 2015. A qualification method for externally bonded Fibre Reinforced Cementitious Matrix (FRCM) strengthening systems, *Composites Part B 78*:497–506.

Ascione, L., Carozzi, F. G., D'Antino, T., & Poggi, C. 2018. New Italian guidelines for design of externally bonded Fabric-Reinforced Cementitious Matrix (FRCM) systems for repair and strengthening of masonry and concrete structures. *Procedia Structural Integrity, 11*, 202–209.

Augenti, N., Parisi, F., Prota, A., & Manfredi, G. (2010). In-plane lateral response of a full-scale masonry subassemblage with and without an inorganic matrix-grid strengthening system. Journal of Composites for Construction, 15(4), 578–590.

Babaeidarabad S, De Caso F, Nanni A 2014a. Out-of-plane behavior of URM walls strengthened with Fabric-Reinforced Cementitious Matrix composite. *J Compos Constr 18(4)*:04013057.

Bellini, A., Incerti, A., Bovo, M., Mazzotti, C., 2017. Effectiveness of FRCM reinforcement applied to masonry walls subjected to axial force and out-of-plane loads evaluated by experimental and numerical studies. *International Journal of Architectural Heritage, 12(3)*, 376–394.

Bertolesi, E., Carozzi, F. G., Milani, G., & Poggi, C. 2014. Numerical modeling of Fabric Reinforce Cementitious Matrix composites (FRCM) in tension. *Construction and Building Materials, 70*, 531–548.

Borri, A., Casadei, P., Castori, G., & Hammond, J. 2009. Strengthening of brick masonry arches with externally bonded steel reinforced composites. *Journal of composites for construction, 13(6)*, 468–475.

Carloni, C., Mazzotti, C., Savoia, M., & Subramaniam, K. V. 2015. Confinement of masonry columns with PBO FRCM composites. *Key Engineering Materials, 624*: 644–651.

Carozzi, F. G., & Poggi, C. 2015. Mechanical properties and debonding strength of Fabric Reinforced Cementitious Matrix (FRCM) systems for masonry strengthening. *Composites Part B: Engineering, 70*, 215–230.

Carozzi, F. G., Bellini, A., D'Antino, T., de Felice, G., Focacci, F., Hojdys, Ł., Laghi, L., Lanoye, E., Micelli, F., Panizza, M. & Poggi, C. 2017. Experimental investigation of tensile and bond properties of Carbon-FRCM composites for strengthening masonry elements. *Composites Part B: Engineering, 128*, 100–119.

CEN, European Committee for Standardization. (2002) EN1990:2002 Eurocode - Basis of structural design.

CEN, European Committee for Standardization. (2019) EN1015-11:2019 Methods of test for mortar for masonry - Part 11: Determination of flexural and compressive strength of hardened mortar.

CNR, Consiglio Nazionale delle Ricerche. CNR DT 215/2018. 2018. Istruzioni per la Progettazione, l'Esecuzione ed il Controllo di Interventi di Consolidamento Statico mediante l'utilizzo di Compositi Fibrorinforzati a Matrice Inorganica (in Italian).

CSLLPP, Consiglio superiore dei Lavori Pubblici. 2018. Linea Guida per la identificazione, la qualificazione ed il controllo di accettazione di compositi fibrorinforzati a matrice inorganica (FRCM) da utilizzarsi per il consolidamento strutturale di costruzioni esistenti (in Italian).

D'Ambrisi, A., Feo, L., & Focacci, F. 2013. Experimental and analytical investigation on bond between Carbon-FRCM materials and masonry. *Composites Part B: Engineering*, 46, 15–20.

D'Ambrisi, A., Feo, L., & Focacci, F. 2013. Experimental analysis on bond between PBO-FRCM strengthening materials and concrete. *Composites Part B: Engineering*, 44(1), 524–532.

D'Antino, T., Carloni, C., Sneed, L. H., & Pellegrino, C. 2014. Matrix–fiber bond behavior in PBO FRCM composites: A fracture mechanics approach. *Engineering Fracture Mechanics*, 117, 94–111.

D'Antino T, Carozzi FG, Colombi P, Poggi C. 2018. Out-of-plane maximum resisting bending moment of masonry walls strengthened with FRCM composites. *Compos Struct* 202:881–896.

D'Antino, T., & Papanicolaou, C. 2017. Mechanical characterization of textile reinforced inorganic-matrix composites. *Composites Part B: Engineering*, 127, 78–91.

De Felice, G., De Santis, S., Garmendia, L., Ghiassi, B., Larrinaga, P., Lourenço, P. B.,... & Papanicolaou, C. G. 2014. Mortar-based systems for externally bonded strengthening of masonry. *Materials and structures*, 47(12), 2021–2037.

Del Zoppo, M., Di Ludovico, M., Balsamo, A., & Prota, A. 2019. Experimental in-plane shear capacity of clay brick masonry panels strengthened with FRCM and FRM composites. *Journal of Composites for Construction*, 23(5), 04019038.

De Santis S, Casadei P, De Canio G, de Felice G, Malena M, Mongelli M, Roselli I 2016a. Seismic performance of masonry walls retrofitted with Steel Reinforced Grout. *Earthq Eng Struct Dyn* 54(2):229–251.

De Santis, S., Ceroni, F., de Felice, G., Fagone, M., Ghiassi, G., Kwiecien, A., Lignola, G.P., Morganti, M., Santantandrea, M., Valluzzi, M.R., Viskovic, A., 2017. Round robin test on tensile and bond behavior of steel reinforced grout systems. *Composites Part B Engineering* 127, 100–120.

De Santis, S., De Canio, G., de Felice, G., Meriggi, P., & Roselli, I. 2019a. Out-of-plane seismic retrofitting of masonry walls with Textile Reinforced Mortar composites. *Bulletin of Earthquake Engineering*, 17(11), 6265–6300.

De Santis, S., & de Felice, G. 2015. Tensile behaviour of mortar-based composites for externally bonded reinforcement systems. *Composites Part B: Engineering*, 68, 401–413.

De Santis S, de Felice G, Casadei P, De Canio G, Mongelli M, Roselli I. 2014. Shake table tests on masonry walls strengthened with tie-bars and new generation mortar-based composite materials. *In: Proceedings of 9th International Conference on Structural Analysis of Historic Constructions SAHC'14*.Mexico City.

De Santis S., Roscini F., de Felice G. 2019a. Strengthening of Masonry Vaults with Textile Reinforced Mortars. In: Aguilar R., Torrealva D., Moreira S., Pando M.A., Ramos L.F. (eds) *Structural Analysis of Historical Constructions. RILEM Bookseries, vol 18*. Springer, Cham.

Garcia-Ramonda, L., Pelà, L., Roca, P., & Camata, G. 2020. In-plane shear behaviour by diagonal compression testing of brick masonry walls strengthened with basalt and steel textile reinforced mortars. *Construction and Building Materials*, 240, 117905.

Gattesco N, Boem I 2017. Out-of-plane behavior of reinforced masonry walls: experimental and numerical study. *Compos B Eng* 128:39–52.

Gattesco, N., Boem, I., & Andretta, V. 2018. Experimental behaviour of non-structural masonry vaults reinforced through fibre-reinforced mortar coating and subjected to cyclic horizontal loads. *Engineering Structures*, 172, 419–431.

Giamundo, V., Lignola, G. P., Maddaloni, G., Balsamo, A., Prota, A., & Manfredi, G. 2015. Experimental investigation of the seismic performances of IMG reinforcement on curved masonry elements. *Composites Part B: Engineering*, 70, 53–63.

Giaretton, M., Dizhur, D., Garbin, E., Ingham, J. M., & da Porto, F. 2018. In-plane strengthening of clay brick and block masonry walls using textile-reinforced mortar. *Journal of Composites for Construction*, 22(5), 04018028.

Kariou FA, Triantafyllou SP, Bournas DA, Koutas LN 2018. Out-of-plane response of masonry walls strengthened using textile-mortar system. *Constr Build Mater* 165:769–781.

Malena, M., & de Felice, G. 2014. Debonding of composites on a curved masonry substrate: Experimental results and analytical formulation. *Composite Structures*, 112, 194–206.

Meriggi, P., de Felice, G., & De Santis, S. 2020. Design of the out-of-plane strengthening of masonry walls with Fabric Reinforced Cementitious Matrix composites. *Construction and Building Materials*, 240, 117946.

Nobili, A., & Signorini, C. 2017. On the effect of curing time and environmental exposure on impregnated Carbon Fabric Reinforced Cementitious Matrix (CFRCM) composite with design considerations. *Composites Part B: Engineering*, 112, 300–313.

Papanicolaou CG, Triantafillou TC, Papathanasiou M, Karlos K. 2008. Textile reinforced mortar (TRM) versus FRP as strengthening material of URM walls: out-of-plane cyclic loading. *Mater Struct* 41(1):143–157.

Parisi, F., Iovinella, I., Balsamo, A., Augenti, N., & Prota, A. 2013. In-plane behaviour of tuff masonry strengthened with inorganic matrix–grid composites. *Composites Part B: Engineering*, 45(1), 1657–1666.

Prota, A., Marcari, G., Fabbrocino, G., Manfredi, G., & Aldea, C. 2006. Experimental in-plane behavior of tuff masonry strengthened with cementitious matrix–grid composites. *Journal of Composites for Construction*, 10(3), 223–233.

Razavizadeh, A., Ghiassi, B., & Oliveira, D. V. 2014. Bond behavior of SRG-strengthened masonry units: Testing and numerical modeling. *Construction and Building Materials*, 64, 387–397.

Stempniewski L., Urban M. 2014. Shaking Table Tests of a Full-Scale Two-Storey Pre-Damaged Natural Stone Building Retrofitted with the Multi-Axial Hybrid Textile System "Eq-Grid". In: Ilki A., Fardis M. (eds) *Seismic Evaluation and Rehabilitation of Structures*.

Geotechnical, Geological and Earthquake Engineering, vol 26. Springer, Cham.

Triantafillou, T. C., Karlos, K., Kefalou, K., & Argyropoulou, E. 2017. An innovative structural and energy retrofitting system for URM walls using textile reinforced mortars combined with thermal insulation: Mechanical and fire behavior. *Construction and Building Materials, 133,* 1–13.

Valluzzi, M.R., da Porto, F., Garbin, E., Panizza, M., 2014. Out-of-plane behavior of infill masonry panels strengthened with composite materials. *Materials and Structures. 47(12):*2131–2145.

Valluzzi, M. R., Modena, C., & de Felice, G. 2014. Current practice and open issues in strengthening historical buildings with composites. *Materials and structures, 47(12),* 1971–1985.

Brick and Block Masonry - From Historical to Sustainable Masonry –
Kubica, Kwiecień & Bednarz (eds)
© 2020 Taylor & Francis Group, London, ISBN 978-0-367-56586-2

Effect of FRCM on sliding shear failure of masonry

C. D'Ambra

Department of Civil, Environmental and Architectural Engineering, University of Padua, DICEA, Padua, Italy

G.P. Lignola, A. Prota & E. Sacco

Department of Structures for Engineering and Architecture, University of Naples "Federico II", Naples, Italy

ABSTRACT: Fiber Reinforced Cementitious Matrix (FRCM) has been already successfully adopted for strengthening masonry elements. In the scientific literature, most of the investigation and studies were focused on strengthening against flexural failure by means of bending tests and against shear, diagonal cracking by means of diagonal compression tests. Results confirmed the great potential of such FRCM systems; however, few results are available in terms of bed-joints sliding shear failure. This lack of knowledge jeopardizes a sound and safe design of strengthening interventions as sliding shear failure could occur as a premature failure compared to the others. In particular, squatter walls (quite common in practice) may run into this type of failure when the vertical compression stress is low. The main focus of this study concerns the experimental determination of the capacity of clay brick masonry triplets under bed-joints sliding shear loads, at different normal stress levels. A Mohr-Coulomb type failure theory has been adopted for evaluating the shear capacity as function of the applied normal load, considering the angle of internal friction and the cohesion detected for unstrengthened masonry. The FRCM consisted of a basalt fiber grid with NHL mortar matrix applied on the sides of the triplet specimens. To perform the tests, the well-known UNI EN 1052-3 has been adapted to strengthened masonry triplets, for a wider range of normal loads to account for the strengthening. Experimental tests have been performed on unstrenghtened and strengthened specimens, comparing the obtained results, demonstrating that FRCM is able to significantly increase the performance of masonry against bed-joints sliding shear failure.

1 INTRODUCTION

The FRCM (Fabric Reinforced Cementitious Mortar) retrofit method, characterized by the application of composite grids into inorganic mortar layers (Nerilli et al. 2020) onto the surface of masonry walls, represents today a valid solution to improve the performance especially of masonry structures. The high ratio between mechanical strength and weight allows to increase the global response, in terms of capacity and ductility, of structural elements without increments of stiffness and mass, in order to leave unchanged the global behavior of the structures. The use of a matrix made of cement or pozzolanic-based mortar (FRCM materials) allow to obtain a good compatibility with the masonry substrate and a better fire resistance than FRP (Kwiecień et al. 2016, de Felice et al. 2018). Theoretical and experimental results have shown the effectiveness of this technique to increase performances of masonry walls (da Porto et al. 2010, 2018, Garofano et al. 2016, Giamundo et al. 2015, 2016, Giaretton et al. 2018, Incerti et al. 2019, Lignola et al. 2016, Parisi et al. 2013, Ramaglia et al. 2016, Younis et al. 2017).

The large use and spread of this innovative techniques, after recent earthquakes too, led to the introduction of Italian Guidelines (CNR-DT 215/2018) that standardized the application and offer the design criteria. The Guidelines, for each failure mode, associate equations to calculate the increment of capacity related to the application of reinforced system. In particular, for the masonry structures, the Italian Guidelines provide a specific formulation to calculate the benefits lead by the FRCM system both to shear, diagonal cracking, and flexure but they do not allow to calculate the contribution of reinforced system to the shear sliding failure. In this paper the shear sliding behavior of clay brick masonry and the efficacy of FRCM system as shear sliding strengthening system have been evaluated performing an experimental program on unreinforced and reinforced tests at the laboratory of Department of Structures for Engineering and Architecture at the University of Naples "Federico II".

2 THE EXPERIMENTAL PROGRAM

2.1 *Experimental tests*

The experimental investigation consisted in shear sliding tests on 33 specimens made of clay brick masonry.

The specimens were built according to the UNI EN 1052-3 (2007), each test is composed of three clay bricks of dimensions $12 \times 12.5 \times 5.5$ cm^3 (half brick) connected by means of a 10 mm thick layer of mortar. Particular attention was used to build the specimens, each half brick was first cleaned before application of mortar, each specimen is composed of three half bricks and two mortar joints and their thickness was checked carefully.

During the hardening period of the mortar (28 days), the specimens were subjected to a compression load of 0.05 kN applied perpendicularly to the mortar joints. The FRCM reinforcement, composed of a double mortar layer of 5 mm with interposed a basalt grid, was applied on the two parallel faces after the hardening period of the mortar.

Figure 1. Specimens: a) Unreinforced; b) FRCM Reinforced.

A total of 33 masonry test specimens were tested, as follows:

- 20 Unreinforced tests;
- 10 FRCM Reinforced tests (mortar matrix and basalt grid on two faces);
- 3 Mortar Reinforced tests (mortar matrix only on two faces);
- According to UNI EN 1052-3 (2007) standard, three steps of compressive load are suggested (see Tab. 1) to characterize the shear strength of the masonry; in the case of a compressive

Table 1. Pre-compressive lateral load.

Characteristic Compressive strength	I Load MPa	II Load MPa	III Load MPa
> 10 N/mm^2	0.2	0.6	1
< 10 N/mm^2	0.1	0.3	0.5

strength of brick higher than 10 MPa the suggested compressive stresses are 0.2 MPa, 0.6 MPa and 1 MPa; in this research the behavior at higher compressive load have been considered, too, considering additional values of 3 MPa and 5 MPa.

For each compressive load, three specimens have been tested assuming the following format for IDs: TN_1-N_2, where the letter T indicates the specimen type, in particular:

- U: Unreinforced tests;
- R: Reinforced tests (cementitious matrix and basalt grid);
- M: Matrix reinforced only tests.

The number N_1 indicates the number of the repetition (i.e. specimen subjected to the same compressive load).

The number N_2 indicates the compressive horizontal stress applied.

2.2 Material properties

Mechanical properties of mortar for joints, and of mortar as matrix of the FRCM were determined by means of experimental tests according to EN 1015-11 (2006) standard. Tensile and compressive strengths were computed by means of bending and compressive tests: three $40 \times 40 \times 160$ mm^3 mortar prisms were tested under flexure with three-point bending and 6 blocks, obtained from failed mortar specimens in flexure, were subjected to compression tests. The 28-day average strength results were (see Tab. 2): Mortar for joints, 1.95 MPa with Coefficient of Variation (CoV) (12.64%) for flexural tests and 10.49 MPa (7.85%) for those in compression. Mortar as matrix 6.84 MPa (11.98%) for flexural tests and 20.9 MPa (8.82%) for compression. The mortar used as matrix is a bi-component premixed pozzolanic based grout made also of hydraulic natural lime, sand, special additives, polymers, and short glass fibers spread in the matrix. Clay brick properties are taken from technical datasheet; the average compressive strength is equal to 30 MPa while the tensile strength is equal to 6 MPa. The basalt grid has a square mesh having 6 mm x 6 mm dimensions made of basalt fibers (equivalent thickness of dry fabric 0.039 mm). The nominal tensile strength of the dry fibers is 1542 MPa and the elastic modulus is 89 GPa.

Table 2. Mechanical properties of materials.

Material	$f_{fl,m}$ MPa	f_{cm} Mpa
Mortar for joints	1.95 (12.6%)	10.5 (7.9%)
Mortar for matrix	6.84 (12.0%)	20.9 (8.8%)

2.3 Experimental setup

The masonry specimens were tested by means of a steel frame set-up consisting of two vertical load contrast steel frames connected by an horizontal beam utilized to apply the shear load, as depicted in Figure 2. To distribute the compressive lateral load on the entire faces of the specimens, two 35 mm thick steel plates have been used. The shear load was applied on three points and it has been increased up to the specimen's failure.

In detail the load setup of the test is composed by:

- Two manual hydraulic pumps to apply compressive and shear loads, once the intended loading pressure was achieve, a valve allowed to keep a constant load;
- Two hydraulic jacks, for the application of the vertical and horizontal load, fixed on the steel frame set-up;

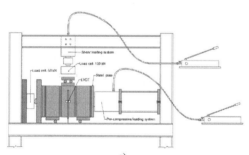

a)

b)

Figure 2. A) Setup scheme; b) Setup of test.

- Two load cells respectively of 50 and 100 kN to measure the level of applied compressive and shear stress;
- Steel plates to distribute the compressive load on the entire faces of the specimens;
- Three steel loading devices, composed by 12 mm thick plates and steel cylinders to apply the shear load only along the mortar joints;
- Two linear displacement transducers (LVDTs) with a gauge length of 200 mm were installed on the two sides of the specimen, on the central brick, to measure the vertical (relative) displacement;
- Only for the reinforced specimens, one strain gauge to record the stress in the reinforcement was applied.

For each test, initially a compressive load was applied on the lateral sides of the specimens by means of a hydraulic jack up to the intended value according to UNI EN 1052-3 (2007). At this point the valve was closed to keep compressive load during the test. The shear load was applied on the central brick, under force control with a rate between 0.1 N/(mm²/min) and 0.4 N/(mm²/min) and it was stopped at the complete sliding of the central brick.

The experimental shear stress τ, the shear strain γ, the shear elastic modulus G and the compressive stress σ were evaluated as follows, according to UNI EN 1052-3 (2007):

$$\tau_{voi} = \frac{F_{i,max}}{2 \cdot A_i} \qquad (1)$$

$$\gamma = \delta_v / s \qquad (2)$$

$$G = \frac{\tau_{voj} - \tau_{voi}}{\gamma_j - \gamma_i} \qquad (3)$$

$$\sigma_i = \frac{F_{ci}}{A_i} \qquad (4)$$

where:
F_i is the vertical shear load;
F_{ci} is the lateral compressive load;
A_i is the cross-sectional area of the specimens parallel to the shear force;
δ_v is the vertical displacement;
s is thickness of the mortar joint.

3 EXPERIMENTAL RESULTS

For each compressive lateral load, the unreinforced tests showed a similar failure mode characterized by a vertical sliding between mortar joint and bricks, according to failure mode A1 reported in UNI EN 1052-3 (2007), Figure 3a. The interface failure determined a premature collapse of the specimens and a sliding of the central brick.

Figure 3. Failure modes: a) Unreinforced; b) FRCM Reinforced.

Table 3. Experimental results.

Test ID	σ_1 N/mm²	γ mm/mm	τ_{voil} N/mm²	G N/mm²
U1-0.2	0.24	1.55	0.70	16.20
U2-0.2	0.20	2.04	0.56	16.16
U3-0.2	0.20	1.64	0.60	37.79
U1-0.6	0.59	1.67	1.23	20.79
U2-0.6	0.58	1.55	0.99	50.91
U1-1	1.00	1.76	1.78	106.81
U2-1	1.42	1.96	1.89	39.99
U3-1	1.07	1.64	1.71	27.62
U4-1	0.97	1.63	1.31	67.25
U5-1	1.13	0.62	1.23	47.26
U6-1	0.91	1.84	1.10	17.58
U7-1	0.95	0.81	1.40	15.92
U1-2	2.24	1.93	2.40	17.49
U2-3	2.87	1.61	2.93	107.67
U3-3	2.83	1.68	2.83	57.61
U4-3	3.19	1.05	3.45	60.53
U1-5	4.81	1.19	4.06	46.20
U2-5	5.04	1.47	4.57	59.02
U3-5	4.86	0.94	4.17	91.45
U4-5	4.67	0.60	3.94	62.23
R1-0.2	0.21	0.88	0.93	17.74
R2-0.2	0.23	1.56	1.32	34.64
R3-0.2	0.20	1.43	1.08	25.13
R1-0.6	0.55	0.83	1.13	24.55
R2-0.6	0.55	0.77	1.66	62.43
R3-0.6	0.66	1.48	1.57	38.27
R1-1	0.97	0.08	2.02	58.31
R2-1	1.02	1.14	1.93	43.67
R3-1	0.78	1.57	2.11	33.01
R1-5	4.82	0.19	4.63	90.13
M1-0.2	0.26	0.98	1.20	25.50
M2-0.2	0.23	1.97	1.12	9.18
M1-1	1.09	1.56	1.51	29.62

The FRCM reinforcement, applied on the two sides of the specimens, allowed to obtain an increase of shear capacity of the specimens, delaying the sliding between mortar joint and brick. The effective contribution of reinforcement is shown in Figure 3b, the failure mode indicates a contribution opposed to the sliding of the central brick, indeed the cracks localized in the central zone of the specimen following the line of mortar joints.

In Table 3 the experimental results of the tests are listed for twenty unreinforced specimens, ten specimens retrofitted with basalt grid and three retrofitted only with matrix mortar: compressive lateral load, shear strain, maximum shear stress - evaluated all according to Equation 1 - and shear elastic modulus for each specimen.

The compressive lateral load of reinforced tests has been calculated as a difference between the value red on the horizontal load cell and strain gauges estimations, assuming uniform stress in the FRCM thickness:

$$\sigma_{cell} - \sigma_{sg} = \sigma_1 \qquad (5)$$

In Figure 4 and Figure 5 the stress strain curves for each compressive lateral load are summarized; the

experimental results showed an increase in terms of shear capacity as normal stress increased, for all the tests, the complete sliding of the brick center determined the end.

All the tests showed a similar behavior in terms of shear stress strain curve, an elastic initial branch has been followed by a second branch characterized by a sliding descending branch. The shear capacity of the masonry is related to the cohesion and friction, while once such strength is exceeded, the behavior was influenced exclusively by the friction.

The application of the reinforcement system on the two faces of the specimens allowed to obtain an increment of shear capacity unchanging the stiffness of the first and second branches; the reinforcement yielded to a cohesion increase between mortar and brick.

In Figure 5c a comparison between the results obtained from all the tests is shown. As expected, the

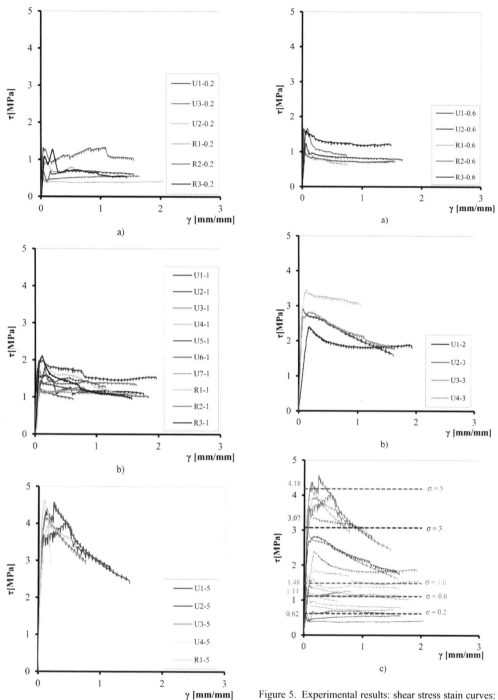

Figure 4. Experimental results: shear stress stain curves: a) Compressive lateral load 0.2 MPa; b) Compressive lateral load 1 MPa; c) Compressive lateral load 5 MPa.

Figure 5. Experimental results: shear stress stain curves: a) Compressive lateral load 0.6 MPa; b) Compressive lateral load 3 MPa; c) Summarized diagram for each lateral load.

compressive load applied on the two sides of the specimens did not influence the initial stiffness of the tests.

4 COMPARISON OF RESULTS

In this section the results obtained from the experimental tests are compared to investigate on the

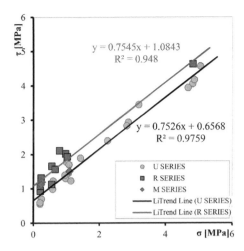

Figure 6. Experimental results in the plane τ-σ.

Inside the figure:
$y = 0.7545x + 1.0843$
$R^2 = 0.948$

$y = 0.7526x + 0.6568$
$R^2 = 0.9759$

○ U SERIES
■ R SERIES
◆ M SERIES
—— LiTrend Line (U SERIES)
—— LiTrend Line (R SERIES)

effects of the FRCM system. The comparison is based on a representation of the τ-σ points to individuate the mechanical parameters to describe observed the physical phenomenon. In Figure 6, in the τ-σ plane, the results are summarized: the orange circles represent the results of unreinforced specimens under different compressive lateral loads (0.2; 0.6; 1; 3; 5 MPa), the gray squares are the results of reinforced specimens (cementitious matrix and basalt grid), while the green rumbles are the specimens reinforced only with cementitious matrix (Matrix tests). Both the unreinforced and reinforced series, showed a linear distribution to different values of examined compressive lateral loads. The trend lines of the series U and R have been determined to describe the trend of shear capacity as a function of compressive lateral load. The two lines showed a very low dispersion of results $R^2 \approx 1$, in black is plotted the best fitting line of unreinforced tests and in red of reinforced tests.

The trend lines, obtained from best fitting of experimental tests, allowed to express, by means a Mohr Coulomb criterion, the sliding behavior of the analyzed masonry; the coefficients in the equation of the lines reported in Figure 6 represent cohesion and friction coefficient of the analyzed material.

$$\tau = c + \sigma f \qquad (6)$$

The parallelism of the two lines indicate an influence of the reinforcement system on the cohesion only of the interface between mortar joints and brick; for the reinforced specimens an increase of cohesion of about 68% has been recorded:

$$\chi = \frac{c_R}{c_U} = 1.687 \qquad (7)$$

5 CONCLUSIONS

The results of the experimental program on the shear sliding capacity characterization of unreinforced and reinforced masonry specimens showed the effectiveness of FRCM systems in term of increments of shear capacity. The comparison of the obtained results, in a τ-σ plane allowed to estimate the effect of the FRCM reinforcement system in terms of increase of shear sliding capacity of clay brick masonry specimens.

The two parallel lines indicate a constant increment of shear capacity independent on the compressive lateral load; in particular, for the analyzed masonry typology, an increment of about 0.45 MPa compared to unreinforced specimens has been recorded: $\tau_R = \tau_U + 0.45$ MPa.

The experimental results showed an increment of the cohesion only at the interface between mortar joints and bricks.

M series tests were not enough to estimate the effective contribution of the grid to the increase of the shear sliding capacity in a FRCM system. To analyze the grid contribution, a larger experimental program is required, based on different typologies of grids. The dependency between the number of reinforced sides and the increase in shear sliding capacity represents another aspect to investigate in the future.

REFERENCES

CNR-DT 215. 2018. Guide for the Design and Construction of Fiber Reinforced Inorganic Matrix Systems for Strengthening Existing Structures. Italian Council of Research (CNR).

da Porto, F. et al. 2010. In-Plane Behavior of Clay Masonry Walls: Experimental Testing and Finite-Element Modeling. *Journal of Structural Engineering* 136(11): 1379–92.

da Porto, F. et al. 2018. Strengthening of Stone and Brick Masonry Buildings. In: Costa A., Arêde A., Varum H. (eds), *Strengthening and Retrofitting of Existing Structures. Building Pathology and Rehabilitation, vol 9*: 59-84. Singapore: Springer.

de Felice, G. et al. 2018. Recommendation of RILEM Technical Committee 250-CSM: Test method for Textile Reinforced Mortar to substrate bond characterization. *Mater Struct* 51, 95.

EN 1015-11. 2006. Methods of Test for Mortar for Masonry. Part 11: Determination of Flexural and Compressive Strength of Hardened Mortar.

Garofano, A. et al. 2016. Model-ling of the In-Plane Behaviour of Masonry Walls Strengthened with Polymeric Grids Embedded in Cementitious Mortar Layers. *Composites Part B: Engineering* 85: 243–58.

Giamundo, V. et al. 2015. Experimental Investigation of the Seismic Performances of IMG Reinforcement on Curved Masonry Elements. *Composites Part B: Engineering* 70: 53–63.

Giamundo, V. et al. 2016. Shaking Table Tests on a Full-Scale Unreinforced and IMG-Retrofitted Clay Brick Masonry Barrel Vault. *Bulletin of Earthquake Engineering* 14(6): 1663–93.

Giaretton, M. et al. 2018. In-Plane Strengthening of Clay Brick and Block Masonry Walls Using Textile-Reinforced Mortar. *Journal of Compo-sites for Construction* 22(5): 04018028.

Kwiecień, A. et al. 2016. Repair of composite-to-masonry bond using flexible matrix. *Mater Struct* 49: 2563–2580.

Incerti, A. et al. 2019. FRCM Strengthening Systems Efficiency on the Shear Behavior of Pre-Damaged Masonry Panels: An Experimental Study. *Journal of Building Pathology and Rehabilitation* 4(1): 14.

Lignola, GP. et al. 2016. Modelling of Tuff Masonry Walls Ret-rofitted with Inorganic Matrix–Grid Composites. In *Brick and Block Masonry*, CRC Press: 2127–35.

Nerilli, F. et al. 2020. Micromechanical modeling of the constitutive response of FRCM composites. *Construction and Building Materials* 236: 1-16, 117539.

Parisi, F. et al. 2013. Rocking response assessment of in-plane laterally-loaded masonry walls with openings. *Engineering Structures* 56: 1234–1248.

Ramaglia, G. et al. 2016 Collapse analysis of slender masonry barrel vaults. *Engineering Structures* 117: 86–100.

UNI EN 1052-3. 2007. Methods of Test for Masonry - Determination of Initial Shear Strength.: *9*.

Younis, A. et al. 2017. Different FRCM Systems for Shear-Strengthening of Reinforced Concrete Beams. *Construction and Building Materials* 153: 514–26.

Brick and Block Masonry - From Historical to Sustainable Masonry –
Kubica, Kwiecień & Bednarz (eds)
© 2020 Taylor & Francis Group, London, ISBN 978-0-367-56586-2

Durability of steel reinforced grout composites

S. De Santis, P. Meriggi & G. de Felice
Department of Engineering, Roma Tre University, Rome, Italy

ABSTRACT: Mortar-based composites have proved effective for the external strengthening of existing structures and have already been widely applied in the field. To date, however, their long-term performances have not been sufficiently investigated, even if they are of the utmost importance for mitigating the risks and the costs associated to damage and repair/substitution. This is crucial for the sustainability of the building stock and of the rehabilitation measures developed for its lasting safeguarding. This paper collects the studies performed on the durability of Steel Reinforced Grout (SRG) systems, comprising ultra-high tensile strength steel textiles and inorganic matrices. Brass and zinc coated steel cords have been tested after accelerated artificial aging in saltwater solutions, salt mist, acid solutions and freeze-thaw cycles. Laboratory investigations include tensile and bond tests. Available results are discussed to make the point on current state-of-knowledge, contribute to the improvement of artificial aging protocols and test procedures, and promote future research and standardization activities.

1 INTRODUCTION

Mortar-based composite materials are an effective solution for the improvement of the ultimate strength of existing structural members. They consist of a textile bonded to the external surface of the structure by means of an inorganic matrix, such as a lime, cement or geopolymer mortar. The available textiles include basalt, carbon, glass, aramid and PBO open meshes or unidirectional textiles of ultra-high tensile strength steel (UHTSS) cords or ropes. Different names and acronyms are used, such as fabric reinforced cementitious matrix (FRCM), textile reinforced mortar (TRM), inorganic matrix composites (IMC) or, when comprising steel textiles, steel reinforced grout (SRG). These reinforcement systems have a great potential for structural rehabilitation thanks to their high strength-to-weight ratio, which provides a significant strength improvement without increase of mass. The use of inorganic matrices allows for installation on irregular and wet surfaces, and entails good behaviour at high temperatures and cost-efficiency.

In the last 10-15 years, many research studies have deeply investigated the tensile and bond behaviour of FRCM composites when applied to concrete (Matana et al. 2005, Ascione et al. 2020) and masonry (Razavizadeh et al. 2014, De Santis et al. 2017) substrates. Lessons learned from experiments led to the development of testing (de Felice et al. 2018) and acceptance (De Santis et al. 2018b) procedures. Investigations on large scale-specimens have proved the effectiveness of mortar-based composites for the structural upgrade of reinforced concrete (Huang et al. 2005, Napoli & Realfonzo 2015, Thermou et al. 2015, Ombres & Verre 2019) and

masonry (Borri et al. 2009, Valluzzi et al. 2014, Sneed et al. 2017, De Santis et al. 2018c) structures and contributed to the development of design guidelines (ACI-RILEM 2020). Finally, many applications have been done in the field, especially for the seismic retrofitting of masonry structures, including architectural heritage.

Nevertheless, an improved knowledge still needs to be gained on the durability of mortar-based composites, which is fundamental for ensuring the effectiveness of externally bonded reinforcements (and therefore the safety level of the retrofitted structures) in the long term. This is of the utmost importance for for mitigating the risks and the costs associated to damage and repair/substitution and, more in general, for the sustainability of the building stock and of the rehabilitation measures for its lasting safeguarding. This paper collects the studies performed on the durability of SRG systems, which have proven to be particularly effective for structural retrofitting in combination with inorganic matrices (de Felice et al. 2020). Steel textiles have a tensile strength up to $3000 \div 3200$ N/mm^2 ($1500 \div 2000$ N/mm^2 for stainless steel textiles), a Young's modulus of $170 \div 190$ kN/mm^2 and a strain at failure of $1.8 \div 2.2\%$. The layout of the cords generally offers good interlocking within mortar matrices, which leads to good bond performances. The bond capacity (axial stress in the textile at detachment) is in the order of $60 \div 80\%$ of the tensile strength and the SRG-to-substrate load transfer capacity ranges between 130 kN/m and 220 kN/m (depending on textile thickness and cord spacing, De Santis et al. 2017, de Felice et al. 2020).

Due to the small diameter (0.2 mm$\div$0.5 mm, De Santis et al. 2016) of the wires used for manufacturing the cords, durability is a particularly crucial issue for SRG systems, since the deterioration of the material may cause a significant reduction of the mechanical properties of the reinforcement. Available experimental outcomes include direct tensile tests on bare textile specimens and on SRG coupons, and SRG-to-substrate shear bond tests. Tested SRG systems include brass and zinc coated steel cords, embedded in lime, cement and geopolymer mortars. The considered aging conditions are immersion in saltwater and acid solutions, exposure to saline mist or outdoor environment, freeze-thaw cycles and partial wetting/drying cycles in water and salt solutions. More studies are needed but, due to the complexity of the issue and the efforts of durability investigations, it is worth, for SRGs and for the other mortar-based composites, making the point on current state-of-knowledge, in order to orient future research and standardization activities.

2 DIRECT TENSILE TESTS

2.1 Durability of steel textiles

The first studies devoted to the durability of SRG systems consisted in direct tensile tests on aged textile specimens. Exposing the bare cords, i.e. not embedded within the mortar matrix, to artificial aging was assumed to be the most critical condition, also in consideration of the fact that, in real applications, the possible cracking of the matrix would have exposed the cords anyway.

The investigation performed by Borri & Castori (2011) was devoted to brass coated and zinc coated (galvanized) steel textiles. Both bare textile specimens and SRG coupons were tested before and after aging and the deterioration was measured in terms of tensile strength reduction. As for the bare textiles, natural and accelerated artificial aging were considered. Natural aging consisted in leaving brass coated steel specimens outdoors for 57 weeks, one set of 4 specimens near a thermoelectric station and one set of 4 specimens near a cement factory. The high concentrations of carbon dioxide and nitrogen monoxides in the atmosphere were considered particularly aggressive (Hulatt et al. 2002, Tavakkolizadeh & Saadatmanesh 2001). Tensile tests revealed a strength reduction of 33% for the specimens near the thermoelectric station and of 27% for those near the cement factory (Figure 1).

Accelerated artificial aging consisted in immersing specimens of brass coated steel textile for 24 h at 110°C in neutral and acid pH aqueous solutions (3 specimens per solution), prepared according to ASTM G36 (1994, current version 2018). The former (neutral) comprised magnesium chloride $MgCl_2$ (45% wt.), whereas the latter (acid) was obtained by adding a small amount (1.5 cm^3) of nitric acid HNO_3 (67% wt.) to the former (the overall amount of water is not specified), which leads to a pH of 1.2÷2.0. Then, the specimens were tested under tension to investigate the effect of acid attack.

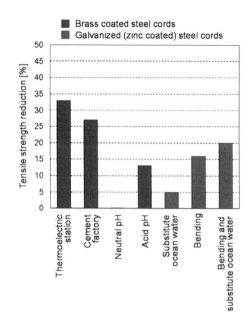

Figure 1. Deterioration of tensile strength exhibited by steel textile specimens after different aging conditions.

The strength reduced by 13% after aging with acid pH solution, whereas it did not change after the treatment in neutral pH solution (Figure 1).

De Santis & de Felice (2015) tested galvanized steel textiles after aging in Substitute Ocean Water, prepared according to ASTM D1141 (2013). It is an aqueous solution comprising a mix of salts, the more concentrated of which are sodium chloride NaCl (ca. 24.5 g/L) and anhydrous sodium sulphate Na_2SO_4 (ca. 41 g/L), with a pH of 8.2. Textile strips were immersed for 15, 30 42, 84 and 126 days in tanks kept at 23°C (3 specimens per aging duration). Note that 42 and 126 days correspond to 1000 h and 3000 h respectively, which are the aging durations required by certification standards (ICC-ES 2016, EOTA 2019, CSLLPP 2019). After such aging periods, the reduction of tensile strength was 2.3%, 3.9%, 4.5%, 6.3% and 6.2%; and that of the Young's modulus was 2.9%, 3.8%, 4.1%, 3.5% and 4.0% (Figure 1 and 2). Additionally, five specimens were bent at 90 degrees and tested under tension, which revealed a decrease of 16% for the strength and of 10% for the Young's modulus. Finally, five specimens were first bent at 90 degrees and then immersed for 1000 h in Substitute Ocean Water and the strength and Young's modulus reductions were 19% and 12%, which approximately correspond to the sum of the effects of bending and aging (Figure 1 and 2). This outcome suggested that bending the cords, which is necessary in many applications (such as the reinforcement of a concrete beam in shear, the confinement of a column or the preparation of a connector) does not significantly damage the zinc coating.

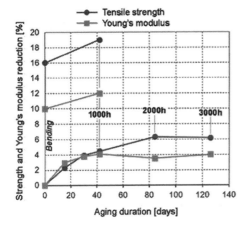

Figure 2. Deterioration of tensile strength and Young's modulus exhibited by galvanized steel textiles after different durations of complete immersion in substitute ocean water.

2.2 Durability of SRG composites

In addition to bare textile specimens, Borri & Castori (2011) tested also SRG coupons manufactured using brass coated steel cords and either a hydraulic lime mortar or a cement mortar. Twelve coupons (6 per mortar type) were left outdoors and exposed to natural aging for 57 weeks (as for the textile strips). Direct tensile tests revealed a strength reduction of 49% for the specimens left near the thermoelectric station and of 36% for those left near the cement factory (Figure 3), which are both higher than those observed on bare textile specimens. Accelerated

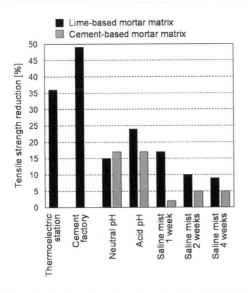

Figure 3. Deterioration of tensile strength exhibited by SRG coupons after different aging conditions.

artificial aging on SRG coupons caused a strength reduction of 17% (after exposure to neutral or acid pH solutions) for the composites manufactured with the cement mortar and of 15% (neutral pH) and 24% (acid pH) for those comprising the lime based mortar. Also in this case, the deterioration was higher than that of bare textiles.

Finally, SRG coupons comprising either brass or zinc coated cords and hydraulic lime mortar were exposed to saline mist in a climatic chamber at 35°C for 1, 2 or 4 weeks (3 specimens for cord type and aging duration, for a total of 18 specimens). According to ASTM B117 (2009, current version 2019), the aqueous solution was prepared using sodium chloride NaCl (5% wt.) and had a pH between 6.5 and 7.2. The strength reduction exhibited by the specimens with the brass coated cords was 9÷17% (higher for the shorter aging periods), whereas that of the composites comprising the galvanized cords was nearly negligible (2÷5%) (Figure 3).

3 SRG-TO-SUBSTRATE SHEAR BOND TESTS

The durability of the SRG-to-masonry bond capacity was investigated by Franzoni and coworkers and published in two papers (Franzoni et al. 2017, 2018). In the first study (2017), twenty-five specimens were manufactured by bonding an SRG strip (with a width of 50mm and a total thickness of 8 mm) to a brickwork prism. The reinforcement was comprised of a textile made of galvanized steel cords with 0.635 mm spacing and of a lime-based mortar matrix. The specimens were partially immersed in an aqueous saline solution, comprising sodium sulfate decahydrate $Na_2SO_4 \cdot 10H_2O$ (8% wt.), for two days (although it was observed that 8 hours were sufficient to attain complete saturation) and then dried in ventilated oven (at 60°C) for two days, in order to reach a dry condition (moisture < 2% wt.). Four wetting/drying cycles were carried out to induce salt crystallization on the surface of the masonry prism where the SRG was applied. The number of cycles was selected to produce the amount of efflorescence typically surveyed in the field. In order to concentrate the salts on the reinforced face of the substrate, the sides of the prisms were covered with duct tape in five specimens. Other five specimens were covered with the tape also on the front near the SRG strip, to investigate whether or not this leads to a more localized concentration of salt crystals inside and on the outer surface of the reinforcement layer. Finally, ten specimens (five wrapped laterally and five fully wrapped) were subject to the same cycles but immersion was in deionized water (rather than in the saline solution), in order to detect possible curing effects associated with the additional water provided in weathering cycles, that is, in order to separate the effects of curing from those of salt crystallization. Finally, five control specimens were not subjected to any conditioning.

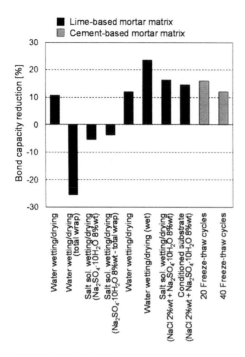

Figure 4. Deterioration of bond capacity exhibited by externally bonded SRG reinforcements after different aging conditions.

The tests showed no reduction of the shear bond capacity. In fact, some sets of specimens exhibited even higher average peak loads (Figure 4) and this was attributed to the additional curing induced by the provision of water in the cycles. The high porosity of the matrix helped the capillary flow of the solution, preventing local salt accumulations by precipitation or sub-efflorescence within the prisms. Salt efflorescence was instead observed on their surface. The crystallization of salts did not cause microcracks and, since failure occurred between textile and mortar in both unaged and aged specimens, it did not compromise the masonry-to-matrix bond capacity.

From a methodological standpoint, no significant difference was found between the specimens wrapped only on the sides and those covered with the tape also on the front surface. As for the number of cycles, four cycles were considered adequate for inducing the desired amount of efflorescence. On the other hand, since this process is sensitive to the porosity of the specimens and the salts of the solutions, it was suggested that the amount of salts is quantitatively measured after each cycle by ion chromatography analyses, at least for some specimens.

The study of the effects of salt crystallization was continued by the same group in another investigation on 19 specimens (Franzoni et al. 2018), on the same SRG system and substrate. In the light of previous

experience, particular attention was paid to ensure high humidity conditions during the 4 weeks of curing to attain a nearly complete hardening of the matrix. Also, all the specimens were wrapped with duct tape on the four sides and not on the front. Apart from three control specimens, four specimens were subjected to 6 cycles, each consisting of 2 days of partial immersion in a saline solution comprising sodium chloride NaCl (2% wt.) and sodium sulfate decahydrate $Na_2SO_4 \cdot 10H_2O$ (8% wt.), followed by 3 days of drying in ventilated oven at 60°C. The observed deterioration of bond capacity was 16.3%. Failure occurred by detachment at the textile-to-matrix interface as in the control specimens. No salt accumulation was detected in the pores of the matrix whose mechanical properties were considered not compromised. On the other hand, rust was detected on the cords and the decrease of bond capacity was attributed to the deterioration of the textile.

Eight specimens underwent the same cycles but in deionized water. Four of them were tested after drying, whereas the other four ones were tested in saturated conditions. The decrease in bond capacity was 12% and 23.5%, respectively, the latter being higher than the former due to the overpressures of water inside the pores of the matrix. Finally, in four specimens the SRG strips were bonded to the prisms after these latter ones underwent the same 6 wetting/drying cycles of the other specimens. The bond capacity decreased by 14.6% and failure occurred also in this case by detachment at the interface between textile and matrix. Despite the surface of the prism was cleaned and salt efflorescence was removed before SRG installation, the strength deterioration and the rust on the cords revealed that the salts in the substrate moved to the composite overlay during curing. This result suggests that the masonry be carefully analyzed and cleaned before externally bonded reinforcements are installed in the field.

Based on ion chromatography and mercury intrusion porosimetry analyses performed on conditioned specimens, number of cycles, concentration of salts in the solution and oven temperature were considered appropriate for reproducing the conditions surveyed in the field on masonry structures.

Meriggi & co-workers (2019) performed twelve SRG-to-concrete shear bond tests. SRG reinforcements were composed of a galvanized steel textile (with cords having 0.635 mm spacing) and a geopolymer mortar matrix. Curing under high humidity conditions lasted 3 months to ensure a complete hardening of the mortar. Ten specimens were subjected to freeze-thaw cycles, 20 cycles for 5 specimens and 40 cycles for other 5 specimens. Each cycle consisted in four hours at -18°C followed by 12 hours at +38°C and 98% relative humidity. The cycles were preceded by one week conditioning at +38°C and 98% relative humidity. Two specimens were not conditioned and tested for reference. All the specimens were tested at the same time, so

Figure 5. Signs of oxidation of the steel cords near the loaded end of the bonded area after 40 freeze-thaw cycles.

curing of unaged specimens continued while the other ones were undergoing freeze-thaw cycles.

The shear bond capacity reduced by 16% after 20 cycles and 12% after 40 cycles (Figure 4). Note that, since the coefficients of variation of test results are in the order of 11÷17%, the deteriorations induced by the two aging durations may be considered analogous. All the specimens (both the unaged and the aged ones) exhibited the same failure mode consisting in the detachment at the textile-to-matrix interface. The strength reduction was therefore attributed to a modification of the internal structure of the matrix, which weakened its bond with the steel cords. On the other hand, the deterioration of the steel cords, which exhibited signs of oxidation near the loaded end of the bonded area (the worst case is shown in Figure 5), did not significantly affect the SRG-to-concrete bond capacity in these tests.

4 CONCLUSIVE REMARKS AND OPEN CHALLENGES

In the wake of the promising outcomes of scientific research and industrial developments, SRG reinforcements are considered an effective, sustainable and cost-efficient solution for the upgrade of existing structures. The knowledge gathered so far, however, is insufficient for a comprehensive understanding of their long-term performances, even though it is acknowledged that this is crucial for their confident use in structural rehabilitation. SRG systems share this knowledge gap with the other mortar-based composites, as witnessed by the increasing scientific interest in this challenging field.

Available experimental evidence shows that the mechanical properties of the steel textiles used in SRG composites exhibit non-negligible decrease after the accelerated artificial aging performed, either in the laboratory or outdoors, within the investigations carried out so far on the tensile behaviour of bare textiles. The first studies revealed that the protection with zinc coating (galvanization) is more effective than brass coating and, indeed, brass coated steel textiles have been replaced by galvanized ones in all the following studies and even in the market (De Santis et al. 2018a). On the other hand, stainless steel textiles have been recently made commercially available for SRG composites, but there are no published studies on their durability. Apparently, despite they are expected to exhibit promising durability performances, the lower strength and Young's modulus (De Santis et al. 2017) and, especially, their higher cost make them less competitive than galvanized steel textiles.

Tensile tests on SRG prismatic specimens (coupons) indicated that the protection offered by the matrix layer may be ineffective, especially when lime-based mortars are used, which are more porous than cement-based ones. As a result, test outcomes were in line with those obtained on bare textiles. On the other hand, if SRG coupons are tested for investigating the textile-to-matrix bond capacity (instead of performing more complicated shear bond tests), pre-cracked specimens could be considered, especially for matrices with low porosity, assuming that externally bonded SRG overlays are expected to crack when cracks develop in the substrate. No experimental evidence is however yet available to state whether or not this corresponds to testing bare textile specimens (which would clearly be simpler). In principle, localized exposure could even lead to more severe consequences than fully immersing textile strips.

The studies performed on the durability of the SRG-to-substrate bond provided some information, but much more research is needed to develop a deeper understanding of this issue and to develop standardized protocols. To this purpose, it is worth noting that most existing guidelines for the certification of mortar-based composites (ICC-ES 2016, CSLLPP 2019) foresee the assessment of durability through direct tensile tests on SRG coupons, whereas bond tests after artificial aging are not needed. The assessment protocol recommended in (EOTA 2019), on the contrary, includes shear bond tests on specimens aged in saline or alkaline solutions or exposed to humidity, but not after freeze-thaw cycles. On the one hand, manufacturing, conditioning and testing specimens for shear bond tests (on the different types of substrates foreseen for structural applications) is expensive, whereas qualification protocols need to be economically sustainable for suppliers. On the other hand, it appears questionable whether direct tensile tests on aged coupons provide reliable information for the assessment of the durability of externally bonded reinforcements, whose effectiveness often relies on the interaction with the substrate, even in terms of long-term performances.

Bond tests on aged specimens showed a deterioration of the SRG-to-substrate load transfer capacity up to 15÷20% and an increase of the scatter of test results with respect to unaged specimens, due to the deterioration of the mortar matrix, which is much less relevant in direct tensile tests on coupons.

Matrices with high porosity (with lime-based binders) exhibited the advantageous feature that the capillary flow of the solutions was not impeded, which prevented the local accumulation of salts. On the other hand, the low porosity of cement-based matrices may be considered preferable when the embedded textile is expected to deteriorate under the attack of aggressive agents coming from the environment or from the substrate by rising damp, and therefore may benefit from the protection offered by the matrix layer.

Although EOTA aging protocol (2019), the only one including tests on bond durability, requires that specimens be completely immersed in the solutions, available experimental outcomes (on salt solutions) suggest that this may be not representative of real exposure conditions. The absence of oxygen may prevent the activation of oxidation processes, whereas the lack of wetting/drying cycles does not allow for salt crystallization.

The implementation of weathering processes (be they cyclic or not) that include water or water-based solutions should carefully account for the continuing curing of the mortar matrices, which may lead to a higher strength of aged specimens with respect to unaged ones (which is not particularly useful for a durability study). Reference specimens should be tested at the same time of aged ones, that is, their curing should continue while the other specimens undergo artificial aging. But this might not be enough and, therefore, longer curing durations of all specimens (both the reference ones and those destined to conditioning) under high humidity could be essential, especially for composites with lime-based matrices.

Future research efforts should be devoted to the identification of (i) aging protocols that reasonably represent the conditions experienced by externally bonded reinforcements in real structural applications and (ii) mechanical tests that investigate (all and only) the main features that affect the long term effectiveness of structural reinforcements. Anything but trivial challenges are posed by the variability of the possible exposure conditions combined with that of the available strengthening systems (textiles, matrices) and substrates. Additionally, the large number of parameters at stake, the long durations of curing and aging processes, and, last but not least, their costs make durability studies particularly demanding. This requires that a rationale planning of wide and well-coordinated experimental investigations is set, that involves many research institutions and industrial partners.

ACKNOWLEDGEMENTS

This work was carried out within the Research Projects "SiCura, Sustainable technologies for the seismic protection of the cultural heritage" (Years 2018-2020, N. 15136) funded by Regione Lazio and "ReLUIS" (2019-2021) funded by the Italian Civil Protection Department. Funding is acknowledged also from the Italian Ministry of Education, University and Research (MIUR), in the frame of the Departments of Excellence Initiative (2018-2022), attributed to the Department of Engineering of Roma Tre University.

REFERENCES

ACI 549 0L – RILEM TC 250-CSM: Guide to Design and Construction of Externally Bonded Fabric-Reinforced Cementitious Matrix (FRCM) and Steel-Reinforced Grout (SRG) Systems for Repair and Strengthening Masonry Structures. 2020.

Ascione, F. et al. 2020. Experimental bond behavior of Steel Reinforced Grout systems for strengthening concrete elements. *Construction and Building Materials* 232: 117105.

ASTM B117. 2019. Standard Practice for Operating Salt Spray (Fog) Apparatus. ASTM International, West Conshohocken, PA, US.

ASTM D1141. 2013. Standard Practice for the Preparation of Substitute Ocean Water. ASTM International, West Conshohocken, PA, US.

ASTM G36. 2018. Standard Practice for Evaluating Stress-Corrosion-Cracking Resistance of Metals and Alloys in a Boiling Magnesium Chloride Solution. ASTM International, West Conshohocken, PA, US.

Borri, A. & Castori, G. 2011. Indagini sperimentali sulla durabilità di materiali compositi in fibra d'acciaio. In: *Proceedings of XIV Convegno ANIDIS, Bari, Italy* (in Italian).

Borri, A. et al. 2009. Strengthening of brick masonry arches with externally bonded steel reinforced composites. *Journal of Composites for Construction* 13(6): 468–475.

CSLLPP (Italian High Council of Public Works). 2018. Guidelines for the identification, the qualification and the acceptance of fibre-reinforced inorganic matrix composites (FRCM) for the structural consolidation of existing constructions (in Italian).

Da Porto, F. et al. 2012. SRG application for structural strengthening of RC beams. ACI SP 286: 119–132.

de Felice, G. et al. 2018. Recommendation of RILEM TC 250-CSM: Test method for Textile Reinforced Mortar to substrate bond characterization. *Materials and Structures* 51(4): 95.

de Felice, G. et al. 2020. Lessons learned on the tensile and bond behaviour of fabric reinforced cementitious matrix (FRCM) composites. *Frontiers in Built Environment* 6: 5.

De Santis S. et al. 2018a. Durability of Steel Reinforced Polyurethane-to-substrate bond. *Composites Part B: Engineering* 153: 194–204.

De Santis, S. et al. 2017. Round Robin Test on tensile and bond behaviour of Steel Reinforced Grout systems. *Composites Part B: Engineering* 127: 100–120.

De Santis, S. & de Felice, G. 2015. Steel reinforced grout systems for the strengthening of masonry structures. *Composite Structures* 134: 533–548.

De Santis, S. et al. 2018b. Acceptance Criteria for Tensile Characterization of Fabric Reinforced Cementitious Matrix (FRCM) Systems for Concrete and Masonry Repair. *Journal of Composites for Construction* 22(6): 04018048.

De Santis, S. et al. 2018c. Full-scale tests on masonry vaults strengthened with Steel Reinforced Grout. *Composites Part B Engineering* 141: 20–36.

EOTA (European Organization for Technical Assessment). 2019. Externally bonded composite systems with inorganic matrix for strengthening of concrete and masonry structures – European Assessment Document (EAD) 17-34-0275-01.04.

Franzoni, E. et al. 2018. Effects of rising damp and salt crystallization cycles in FRCM-masonry interfacial debonding: Towards an accelerated laboratory test method. *Construction and Building Materials* 175: 225–238.

Franzoni, E. et al. 2017. Durability of steel FRCM-masonry joints: effect of water and salt crystallization. *Materials and Structures* 50: 201.

Huang, X. et al. 2005. Properties and potential for application of steel reinforced polymer and steel reinforced grout composites. *Composites Part B Engineering* 36: 73–82.

Hulatt, J. et al. 2002. Preliminary investigations on the environmental effects on new heavyweight fabrics for use in civil engineering. *Composites Part B Engineering* 33: 407–414.

ICC-ES. 2016. Acceptance criteria for masonry and concrete strengthening using Fabric-Reinforced Cementitious Matrix (FRCM) and Steel Reinforced Grout (SRG) composite systems AC434. Whittier, CA, US.

Matana, M. et al. 2005. Bond Performance of steel reinforced polymer and steel reinforced grout. In: *Proceedings of International Symposium on Bond Behaviour of FRP in Structures (BBFS), Hong Kong, China.*

Meriggi, P. et al. 2019. Durability of Steel Reinforced Grout systems subjected to freezing-and-thawing conditioning. In: *Proceedings of Fib symposium, Parma, Italy.*

Napoli, A. & Realfonzo, R. 2015. Reinforced concrete beams strengthened with SRP/SRG systems: Experimental investigation. *Construction and Building Materials* 93: 654–677.

Ombres, L. & Verre, S. 2019. Flexural strengthening of RC beams with steel-reinforced grout: Experimental and numerical investigation. *Journal of Composites for Constructions* 23(5): 04019035.

Razavizadeh, A. et al. 2014. Bond behavior of SRG-strengthened masonry units: Testing and numerical modeling. *Construction and Building Materials* 64: 387–397.

Sneed, L.H. et al. 2017. Confinement of clay masonry columns with SRG. *Key Engineering Materials* 747: 350–357.

Tavakkolizadeh, M. & Saadatmanesh, H. 2001. Galvanic corrosion of carbon and steel in aggressive environments. *Journal of Composites for Constructions* 5(3): 200–210.

Thermou, G.E. et al. 2015. Concrete confinement with steel-reinforced grout jackets. *Materials and Structures* 48(5): 1355–1376.

Valluzzi, M.R. et al. 2014. Out-of-plane behavior of infill masonry panels strengthened with composite materials. *Materials and Structures* 47(12): 2131–2145.

Brick and Block Masonry - From Historical to Sustainable Masonry –
Kubica, Kwiecień & Bednarz (eds)
© 2020 Taylor & Francis Group, London, ISBN 978-0-367-56586-2

Experimental investigation of the long-term behaviour of fabric reinforced matrix systems

A. Bonati, A. Franco & L. Schiavi
Construction Technologies Institute – Italian National Research Council, Milan, Italy

A. Occhiuzzi
Construction Technologies Institute – Italian National Research Council, Milan, Italy
Department of Engineering, University of Naples "Parthenope", Naples, Italy

ABSTRACT: The considerable increment in use of Fabric Reinforced Matrix (FRM) strengthening systems in recent years has led to the need of investigating their durability when exposed to long-term actions. Single lap shear tests were performed on concrete and masonry specimens reinforced with different types of fabrics embedded in an inorganic matrix. A new test setup was developed with the aim of applying a sustained load for up to six months simultaneously to considerable amounts of specimens under the same controlled temperature and relative humidity conditions. The paper will describe the proposed setup and the instrumentation. It will also describe the results in terms of measured displacements for clay and natural stone masonry as well as concrete reinforced with fabrics made of different materials. The experimental campaign is part of an ongoing extensive assessment plan for the analysis of the durability of fabric reinforced inorganic matrix composites.

1 INTRODUCTION

Retrofitting of existing civil structures by externally bonded composite materials is considered to be one of the fastest and easiest strengthening methods compared to other traditional procedures. Such types of systems rely on the bond between the composite and the substrate to transfer stresses through the matrix to the fibre/fabric. To solve the problems associated with compatibility of epoxy resin with masonry structures in the well-known Fibre Reinforced Polymer (FRP) systems, new composite systems made of fabrics/meshes embedded in inorganic matrixes have been developed. In Fabric Reinforced Matrix (FRM) systems, the matrix, usually a cementitious or lime-based mortar, offers compatibility with substrates, lower costs and better performance at high temperatures while enhancing permeability and achieving reversibility. This is a great advantage especially for the preservation and safeguarding of historical structures that constitute an important part of many countries' cultural heritage (Bonati et al. 2019).

Civil structures are exposed to different types of loads during their service life. The majority of these loads are sustained loads, which lead to creep of the structure and consequently of the strengthening system. This, in turn, corresponds to a potential reduction in time of the effectiveness of the strengthening system. Therefore, the need to investigate the long-term behaviour of strengthened members became increasingly more evident. Moreover, for qualification and design concepts, the question of strength losses due to long-term loading and environmental exposure needed to be addressed.

It is well known that intensive research has been conducted in the past years on the bond strength of FRP systems bonded to concrete and on the parameters that influence it (De Lorenzis et al. 2001, Yao et al. 2005, Chen & Teng 2001, Sayed-Ahmed et al. 2009, Franco & Royer-Carfagni 2014), together with various investigations on the mechanisms that govern the interfacial debonding failure, in terms of both experimental analyses and analytical studies. Research on the effect of anchoring devices to prevent or delay debonding of such systems was also conducted (Ceroni & Pecce 2010, Ceroni 2017).

FRM systems recently drew the interest of researchers for their application on masonry structures and retrofitting of historical buildings, both in terms of mechanical behaviour (Arboleda et al. 2016, De Santis & de Felice 2015, Donnini & Corinaldesi 2017), and bond properties with respect to concrete and masonry substrates (Täljsten & Blanksvärd 2008, Triantafillou & Papanicolaou 2006, Basalo et al. 2009, Babaeidarabad et al. 2014, Awani et al. 2015, Arboleda et al. 2016).

Research focus is also on the effectiveness of externally bonded systems under harsh environments (Sen 2015, Myers 2007, Ombres 2015, Arboleda et al. 2014, Ceroni et al. 2018, Nobili & Signorini 2017), in order to determine environmental conversion factors for design purposes.

However, little investigation on the long-term performance of strengthening systems has been found in literature, most of which regards FRP-to-concrete bonded systems (Jeong et al. 2015, Choi et al. 2007, Soliman et al. 2012, Houhou 2014, Meshgin et al. 2009, Mazzotti & Savoia 2009).

Consequently, long-term performance of FRM systems still remains a key concern to be addressed. It has to be recalled that the term "creep" is intended here as the behaviour of the composite system under constant loading. This is identified through the measurement of the displacements mainly associated with the crack opening of the inorganic matrix, rather than viscous deformations, which, it is known, are negligible or totally absent in inorganic materials, especially in the considered testing conditions.

In the framework of an extensive experimental assessment plan for the analysis of FRM materials, the Construction Technologies Institute of the National Research Council of Italy is investigating the long-term durability of different types of fabrics embedded in mortar and bonded to different types of substrates. To fulfil the need of testing large amounts of specimens in particular conditions of temperature and relative humidity (RH), a new test setup has been developed.

Preliminary results are here presented even if further insights are still needed and single-lap shear tests are currently ongoing to determine the residual bond strength after the applied sustained load.

2 EXPERIMENTAL ASSESSMENT

2.1 Material and test specimens

Five different types of fabric bonded on different substrates were prepared. UHTSS (Ultra High Tensile Strength Steel) fabrics characterised by two different weights per unit area (low density, LD and high density, HD), a hybrid fabric made of alkali-resistant glass and aramid fibres, and basalt fabrics (characterised by two different weights per unit area, low and high density) were analysed. A summary of fabric properties is reported in Table 1.

The fabrics were bonded on concrete blocks and clay and tuff masonry specimens. Concrete prisms of 150 mm (width) × 150 mm (depth) × 600 mm (height) were prepared considering an MC (0.40) concrete type according to EN 1766 (EN1766, 2017). The average compressive strength f_{cm} and the average axial surface strength f_{hm} measured from tests are reported in Table 1 as well. Masonry (clay and tuff) test specimens were composed of 6 layers of bricks so that the total length was around 380 mm. Compressive tests

Table 1. Properties of materials.

Fabric	Mass g/m²	Width Mm	Spacing mm	No. cords/ yarns -
UHTSS Steel LD	670	50	-	8
UHTSS Steel HD	1200	50	-	16
AR glass+aramid	250	50	15	4
Basalt LD	200	50	17	3
Basalt HD	400	50	15	3

Substrate	size mm	f_{cm} MPa	f_{hm} MPa
Concrete	150 x 150 x 600	57.5	4.4
Clay masonry	125 x 125 x 380	24.5	1.5
Tuff	125 x 125 x 380	4.3	0.6

Mortar	Strength class -	f_{cm} MPa	E GPa
Cementitious	R4	53	19.5
Lime-based	M15	21	11

and pull-off tests were also conducted on clay and tuff bricks and the respective results are shown in Table 1.

The different fabrics were bonded to the substrates with a bonded length of 300 mm through the wet lay-up procedure. A first layer of mortar (average thickness 3-5 mm) was applied on the surface of the substrate, then the fabric was pressed on it and completely covered by a second layer of mortar (average thickness 2-5 mm), such to obtain a total thickness for the FRM composite strip of 8-10 mm. The width of the applied system b_f was such to include at least two grid openings (minimum 3 threads), i.e. around 55 mm on average.

A cementitious mortar was used to apply fabrics on concrete, while a lime-based mortar was used for the application on masonries. Compressive strengths and Young's moduli of mortars were given by the manufacturer and are reported in Table 1. In the following sections, we will refer to SRG (Steel Reinforced Grout) systems when dealing with steel fabrics and to the more generic FRM systems, when dealing with the other types of fabrics.

The distance between the beginning of the bonded area and the top edge (at loaded end of the composite strip) of the masonry element was 40 mm for concrete specimens and 20 mm for masonry specimens. Such distance was left to avoid stress concentration and cone formation during the tests. Outside the bonded area, fibres were left bare in the overhang length. A schematic representation of the test specimens is shown in Figure 1.

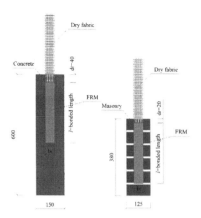

Figure 1. Specimens' dimensions and details of FRM system.

2.2 Short-term tests

To determine the reference load to be applied for the long-term durability test, short-term tests were carried out at room temperature on twelve test series, each consisting of five test specimens. The single-lap shear test was conducted using a push-pull configuration, where the substrate block is restrained while the strip bonded on it is pulled. The specimens were loaded with 0.2 mm/min in displacement control until failure. The peak load and the debonding load were determined. The maximum force P_{max} corresponds to the maximum applied force, while the bond capacity P_{deb} corresponds to the nominally constant load observed after the maximum load drop. Values are reported in Table 2 for all different combinations of fabrics bonded to substrates. The reference load value P_{creep} for the long-term tests has also been reported. This value was set equal to a possible service level, that is the 30% of the averaged maximum load P_{max} obtained from short-term tests. Such percentage was verified to be a common value found in literature in long-term investigations of composite materials.

2.3 Long-term tests

The long-term durability tests were carried out in especially developed test rigs. A total of 36 specimens (12 series of three specimens each) were prepared with the same properties as reported in Section 2.1. The rigs were then positioned in a climatic room with temperature of 23°C and 80% RH for six months (Figure 2). A high RH value was chosen to reproduce the effect of ageing which naturally accompanies long term loading in service conditions instead of controlled conditions typical of a laboratory. This represents a first set of conditions that will be monitored and changed in future investigations.

The test setup is shown in Figure 3 and was designed to meet the following requirements: *i)* limited dimensions, in order to store a large number of

Table 2. Short-term test results.

Substrate	Fabric	P_{max} N Average	CoV	P_{deb} N Average	CoV	P_{creep} N
Concrete	Steel LD	8281	0.02	7780	0.03	2484
	Steel HD	8317	0.06	7509	0.05	2495
Clay	Steel LD	9522	0.15	7989	0.02	2857
	Steel HD	7651	0.02	6794	0.07	2295
	AR glass + aramid	2194	0.08	-*	-*	658
	Basalt LD	1799	0.06	-*	-*	540
	Basalt HD	3004	0.08	-*	-*	901
Tuff	Steel LD	7454	0.02	6520	0.09	2236
	Steel HD	7423	0.02	7130	0.02	2227
	AR glass + aramid	2078	0.04	-*	-*	623
	Basalt LD	1658	0.03	-*	-*	497
	Basalt HD	3601	0.06	-*	-*	1080

* P_{deb} was not determined because failure of fibres occurred outside the bonded length.

Figure 2. Test rigs in the climatic room.

specimens in a climatic room with controlled conditions of temperature and RH, *ii)* compatibility with the specimens' geometry, chosen such to evaluate the residual bond capacity after long-term tests in the same configuration of the short-term tests, *iii)* capacity to maintain a constant load for the established period of time and *iv)* capacity to connect/disconnect the load cell in order to be able to check the load applied on each specimen with a measuring instrument and to

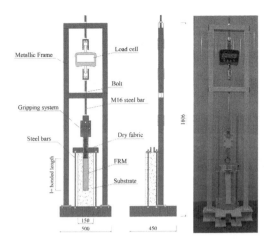

Figure 3. Test set-up.

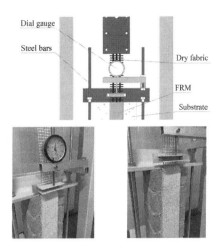

Figure 4. Detail of the dial gauge.

avoid its permanence in not ideal conditions for a long period of time.

The single-lap shear equipment consisted of a metallic frame made of two vertical and two transverse elements with square hollow sections bolted to the bottom side of the rig (I-section beams welded together). The size of the metal frame was designed to accommodate different heights of test specimens and to allow for a correct alignment of the specimen with the loading system, in order to avoid bending of the strip and undesired normal stresses along the bonded joint. The specimen block was restrained against movement by two L-shaped steel elements placed against the cross-sections of the prism. Such elements were connected to the bottom side of the rig through four steel bars bolted to the top and bottom steel elements. Care was taken to position the top elements at a certain distance from the edge of the substrate block to avoid stress concentration in correspondence of the bonded composite system. All the steel elements were designed in order to sustain the loads to be applied for the long-term tests.

The load P_{creep}, evaluated through the short-term tests and reported in Table 2, was applied via clamping devices made of steel plates and calibrated through the tensioning of an M16 steel bar. A load cell with 10 kN capacity was connected in line with the bar to measure the load. Once the target load was reached by tensioning the bar, the bolt immediately above the central transverse beam was tightened so to disconnect the load cell from the system.

A pre-load of 100 N was applied to every specimen in order to align the sample before reaching the target load, which then was maintained and controlled for six months.

Displacements were measured through a dial gauge which was positioned as to react-off of a thin C-shaped plate on the bare fibres at a distance of 2 cm from the end of the support (Figure 4). This positions was chosen to also take into account

possible slips between fabric and matrix. The dial gauge was fixed to the L-shaped contrast plates which block the specimen (substrate).

3 EXPERIMENTAL RESULTS

Scope of the research was to monitor the relative displacement of each composite strip with respect to the substrate during the whole period of sustained loading. In addition, the retained bond strength of the system after six months is the key factor to determine the influence of the long-term loading on the resistance of the bonded joint.

As with regards to the displacements, these were measured and recorded:

- before applying the reference tensile load;
- immediately after applying the reference tensile load;
- after one day, two days and every seven days until the test was completed at six months.

The load was checked during the whole time period, independently from the fixed time displacements' measurements, to verify that no load drop occurred. No specimen failed during the tests.

Recordings are shown in Figure 5 and 6 for SRG and FRM systems, respectively.

From such results, one can infer that no appreciable displacements occurred during the whole examination period and values are in the range of the measurement system resolution.

In particular, Figure 5 shows the average displacements for low and high density steel fabrics bonded to concrete, clay and tuff masonry, respectively. Furthermore, it can be noted that displacements are generally distributed around 0.01 mm and the behaviour is almost stable especially for high density steel fabrics (Steel HD) for all substrates. Very low displacements (nearly zero) can be evidenced on concrete,

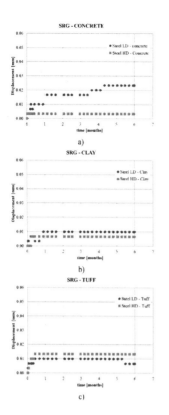

a)

b)

c)

Figure 5. Long term average displacement measured during the reference period (six months) on SRG systems bonded to: a) concrete; b) clay; c) tuff.

except for the Steel LD that instead shows a little higher values of slip (Figure 5a).

Figure 6 represents the behaviour of FRM systems on clay and tuff masonry. A different distribution is shown with respect to SRG systems, with higher values especially in the case of basalt fibres (maximum values around 0.05 mm) bonded on clay masonry. Hybrid fabrics (AR glass+aramid) show a more stable behaviour with average values in the range of 0.01 mm, which is coherent with the higher resistance of aramid fibres. Generally lower displacements are shown instead for FRM systems bonded on tuff substrates.

Such results demonstrate that the composite systems do not show any creep behaviour under sustained loading with the considered reference load, i.e. the 30% of the maximum transmissible load.

Further analysis should be performed by increasing the sustained load to a higher percentage of the short-time peak load and considering different exposure conditions.

The residual bond strength is currently being evaluated for all the specimens and will be object of future discussions, in order to establish the behaviour of different combinations of FRM systems and substrates after sustained loading.

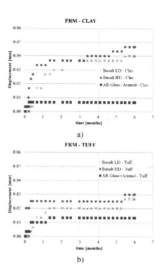

a)

b)

Figure 6. Long term average displacement measured during the reference period (six months) on FRM systems bonded to: a) clay; b) tuff.

4 CONCLUSIONS

Among different retrofitting techniques of civil structures, strengthening through composite materials plays a fundamental role for their ease of application and high performance.

The spread of such innovative technique established the necessity of assessment especially in terms of durability. Indeed, while extensive work has been devoted to developing a sound knowledge of bond properties and debonding mechanisms, limited information is currently available in literature concerning FRM durability.

A major gap has been found especially in terms of long-term behaviour of inorganic matrix systems, where the main investigations are instead on the well-known polymeric based strengthening systems.

In the framework of a larger experimental assessment plan involving FRM systems, the long-term behaviour of different fabrics bonded to different types of substrates was investigated and the results here presented. To be more specific, a new test setup was designed in order to meet a few requirements necessary for qualification purposes of such systems. Tests were performed on specimens made of concrete prisms and clay and tuff masonry over which different FRM composite strips (made of steel, aramid and AR-glass, and basalt fibres) were bonded through two types of mortar (cementitious and lime-based). Short term tests allowed to determine the reference loads for long-term tests to be applied to each combination of substrate and FRM strip in a single-lap shear configuration.

The designed test setup was able to apply the target loads (30% of the short-term peak loads) to considerable amounts of specimens for the prescribed duration of time (six months) in a climatic room with fixed values of temperature and relative humidity.

Recordings of displacements and continuous check of the load have shown the capability of the system to maintain the reference load without appreciable load drops (maintained in the limit of a fixed tolerance).

Results of the long-term tests have shown however that no significant increase of displacement occurred in the reference period with the indicated percentage of maximum transmissible load.

Further investigations will therefore be carried out to monitor the relative displacement under a higher percentage of the ultimate bond strength, to define limits for admissible service load levels, and the effect of different conditions in terms of temperature and relative humidity.

Bond strength tests in single-lap shear configurations are currently being performed to assess the residual capacity of the bonded joints. Results will therefore provide the differences among the various systems and substrates tested and a deeper insight of the effect of the sustained loading on the strengthening system object of analysis.

REFERENCES

Arboleda, D. et al. 2014. Durability of fabric reinforced cementitious matrix (FRCM) composites. *Proceedings 7th international conference on FRP composites in civil engineering, CICE, 2014.*

Arboleda, D. et al. 2016. Testing Procedures for the Uni-axial Tensile Characterization of Fabric-Reinforced Cementitious Matrix Composites. *Journal of Composites for Construction* 20.

Awani, O. et al. 2015. Bond characteristics of carbon fabric-reinforced cementitious matrix in double shear tests. *Construction and Building Materials* 101: 39–49.

Babaeidarad, S. et al. 2014. Flexural strengthening of RC beams with an externally bonded fabric-reinforced cementitious matrix. *Journal of Composites for Construction* 18: 04014009.

Basalo, F. et al. 2009. Fiber reinforced cementitious matrix composites for infrastructure rehabilitation. *Composites & Polycon 2009.*

Bonati, A. et al. 2019. Strengthening of masonry structures: Current national and international approaches for qualification and design. *Key Engineering Materials.*

Ceroni, F. 2017. Bond tests to evaluate the effectiveness of anchoring devices for CFRP sheets epoxy bonded over masonry elements. *Composites Part B-Engineering* 113: 317–330.

Ceroni, F. et al. 2018. Effects of Environmental Conditioning on the Bond Behavior of FRP and FRCM Systems Applied to Concrete Elements. *Journal of Engineering Mechanics* 144: 04017144.

Ceroni, F. & Pecce, M. 2010. Evaluation of Bond Strength in Concrete Elements Externally Reinforced with CFRP Sheets and Anchoring Devices. *Journal of Composites for Construction* 14: 521–530.

Chen, J. F. & Teng, J. G. 2001. Anchorage strength models for FRP and steel plates bonded to concrete. *Journal of Structural Engineering* 127: 784–791.

Choi, K. K. et al. 2007. Shear creep of epoxy at the concrete-FRP interfaces. *Composites Part B-Engineering* 38: 772–780.

De Lorenzis, L. et al. 2001. Bond of FRP laminates to concrete. *ACI Materials Journal* 98: 256–264.

De Santis, S. & De Felice, G. 2015. Tensile behaviour of mortar-based composites for externally bonded reinforcement systems. *Composites Part B: Engineering* 68: 401–413.

Donnini, J. & Corinaldesi, V. 2017. Mechanical characterization of different FRCM systems for structural reinforcement. *Construction and Building Materials* 145: 565–575.

EN1766 2017. Products and systems for the protection and repair of concrete structures - Test methods - Reference concretes for testing. Comite Europeen de Normalisation.

Franco, A. & Royer-Carfagni, G. 2014. Effective bond length of FRP stiffeners. *International Journal of Non-Linear Mechanics* 60: 46–57.

Houhou, N. B. et al. 2014. Durability of concrete/FRP bonded assemblies subjected to hydrothermal coupled creep ageing mechanisms: experimental and numerical investigations. *Transport Research Arena 2014, Paris.*

Jeong, Y. et al. 2015. Modeling and Measurement of Sustained Loading and Temperature-Dependent Deformation of Carbon Fiber-Reinforced Polymer Bonded to Concrete. *Materials* 8: 435–450.

Mazzotti, C. & Savoia, M. 2009. Stress Redistribution Along the Interface Between Concrete and FRP Subject to Long-term Loading. *Advances in Structural Engineering* 12: 651–661.

Meshgin, P. et al. 2009. Experimental and analytical investigations of creep of epoxy adhesive at the concrete-FRP interfaces. *International Journal of Adhesion and Adhesives* 29: 56–66.

Myers, J. J. 2007. Durability of external fiber-reinforced polymer strengthening systems. In: Karbhari, V. M. (ed.) *Durability of Composites for Civil Structural Applications.* Woodhead Publishing.

Nobili, A. & Signoroni, C. 2017. On the effect of curing time and environmental exposure on impregnated Carbon Fabric Reinforced Cementitious Matrix (CFRCM) composite with design considerations. *Composites Part B: Engineering* 112: 300–313.

Ombres, L. 2015. Analysis of the bond between Fabric Reinforced Cementitious Mortar (FRCM) strengthening systems and concrete. *Composites Part B: Engineering* 69: 418–426.

Sayed-Ahmed, E. Y. et al. 2009. Bond Strength of FRP Laminates to Concrete: State-of-the-Art Review. *Electronic Journal of Structural Engineering* 9: 45–61.

Sen, R. 2015. Developments in the durability of FRP-concrete bond. *Construction and Building Materials* 78: 112–125.

Soliman, E. et al. 2012. Limiting shear creep of epoxy adhesive at the FRP-concrete interface using multi-walled carbon nanotubes. *International Journal of Adhesion and Adhesives* 33: 36–44.

Täljsten, B. & Blanksvärd, T. 2008. Strengthening of concrete structures with cement based bonded composites. *Nordic Concrete Research* 2: 133–153.

Triantafillou, T. C. & Papanicolaou, C. G. 2006. Shear strengthening of reinforced concrete members with textile reinforced mortar (TRM) jackets. *Materials and structures* 39: 93–103.

Yao, J. et al. 2005. Experimental study on FRP-to-concrete bonded joints. *Composites Part B: Engineering* 36: 99–113.

Brick and Block Masonry - From Historical to Sustainable Masonry –
Kubica, Kwiecień & Bednarz (eds)
© 2020 Taylor & Francis Group, London, ISBN 978-0-367-56586-2

Environmental durability of FRCM strengthening systems and comparison with dry fabrics

A. Bellini, A.R. Tilocca & I. Frana
CIRI Buildings and Construction, University of Bologna, Bologna, Italy

M. Savoia & C. Mazzotti
DICAM - Department of Civil, Chemical, Environmental and Materials Engineering, University of Bologna, Bologna, Italy

ABSTRACT: Fiber Reinforced Cementitious Matrix (FRCM) composites have been recently introduced for repairing and strengthening masonry structures. Even if they proved to be an effective solution for structural applications, their durability is still an open issue, which is fundamental to guarantee the long-term effectiveness of the strengthening intervention. In fact, FRCMs may be subjected to a combination of different environmental conditions that may affect their performances: humidity, rainfall, freeze/thaw cycles, exposure to saline and alkaline environments are some of the conditions that may promote deterioration over time. In order to investigate FRCMs durability, an extensive experimental campaign has been carried out on different FRCM systems (basalt, glass and steel fibers) through tensile tests on conditioned and unconditioned samples. Tests have been performed both on FRCM coupons and on dry fabrics, with the aim of analyzing the effect of the matrix in terms of mechanical behavior and external protection of the fabric under aggressive environments. The creation of an experimental database on the effects of environmental conditions is a first step to analyze long-term properties of these composite materials and to suggest appropriate strength reduction factors to be taken into account for the design of durable retrofitting interventions.

1 INTRODUCTION

Extensive damages caused to masonry buildings by possible earthquakes, structural modifications or simply by ageing led to the need for strengthening and protecting the existing building heritage and, in particular, the historical ones. The demand for high performance structural materials, which must be also compatible with the masonry substrate, led, at first to the development of composite FRP-based solutions (Ceroni et al. 2014, Mazzotti et al. 2015a, b) and then to the substitution of epoxy resins with inorganic matrices, pushing for the development of FRCM strengthening systems. In particular, many studies focused on the tensile properties and the bond behavior of FRCM retrofitting systems (Papanicolaou et al. 2011, de Felice et al. 2014, Bellini et al. 2015, 2019a, b, Caggegi et al. 2017, Carozzi et al. 2017), analyzing also the in-plane and the out-of-plane behavior of full scale structures, through diagonal compression tests (Del Zoppo et al. 2019, Ferretti et al. 2019) and flexural out-of-plane tests (Bellini et al. 2018, De Santis et al. 2018). This large body of studies have conducted to the drafting of guidelines (FRCM Italian Guidelines 2018) and to the proposal of design formulations (CNR DT 215/2018). After the definition of mechanical properties of the strengthening systems, based on

the coupling of different types of fibers and matrices, it is currently becoming more and more important to study durability of FRCM composites after exposure to different aggressive environmental conditions, such as high relative humidity, freeze/thaw cycles, alkaline and saline solutions. The type of fabric considered or its external coating can lead to different cracking and damage states in relation to the fiber used, as pointed out by several experimental tests on glass (Donnini 2019, Rizzo et al. 2019), carbon (Signorini et al. 2018), basalt (Micelli & Aiello 2019) and steel fibers (Borri et al. 2014, De Santis et al. 2015).

The effect of mortar was also found to be not negligible on the whole behavior of conditioned FRCM samples, as highlighted by different experimental researches (Arboleda et al. 2014, Nobili et al. 2017).

With the aim of evaluating more in detail the durability of FRCM materials and to study the role of the matrix, during the present experimental campaign different glass, basalt and steel systems were analyzed through tensile tests on conditioned and control (unconditioned) samples, taking into account both FRCM coupons and dry fibers only.

After the conditioning period and before the tensile tests, FRCM samples were inspected in order to identify any possible matrix surface alteration such as erosion, desquamation or cracking. The tensile behavior of specimens was fully characterized in terms of failure modes, stress-strain curves and ultimate tensile strength. The comparison between the results obtained on FRCM coupons and on dry fabrics allowed to analyze the effect of matrix in terms of tensile behavior of the samples and possible interaction between fibers and mortar.

2 EXPERIMENTAL CAMPAIGN

2.1 *Experimental program, material properties and environmental conditionings*

The experimental program consists of a total number of 100 tensile tests on dry fabrics and 125 tests on FRCM samples (performing 5 test repetitions for each specimen type). In detail, for both dry fabrics and FRCM coupons, 5 unconditioned (control) samples were tested, for each fiber/matrix type, in order to obtain the reference strength values and those after conditioning. The environmental conditions chosen for durability tests were: freeze-thaw cycles, humidity, alkaline and saline environments.

The effects of exposure to freeze-thaw cycles were evaluated performing a total number of 20 cycles, consisting in 4 hours at a temperature of -18°C followed by 12 hours at 38°C with a relative humidity RH > 90%, after conditioning the samples for 7 days (168 hours) in a humid chamber (temperature 38°C, RH > 90%), according to (FRCM Italian Guidelines 2018). Humidity resistance was instead evaluated after 1000 hours of exposure in a humid chamber characterized by a temperature of 38 ± 2°C and a relative humidity RH > 90% (according to ASTM D2247-11). The effects of exposure to alkaline environments were studied after 1000 hours of immersion of the samples in an alkaline solution, with a pH equal to 9.5 (according to ASTM D7705/D7705M-12), maintaining the immersion chamber in controlled temperature conditions (23 ± 2°C). Finally, the effects of exposure to saline environments were evaluated after an immersion period of 1000 hours in salt water, preparing the solution according to ASTM D1141-98, ASTM C581-03.

The described environmental conditions were adopted for both dry fabrics and FRCM coupons, performing then a comparison with results obtained on unconditioned (control) samples.

Four different types of fibers were chosen during the experimental campaign for evaluating their durability properties: a unidirectional galvanized steel fiber sheet with a density equal to 650 g/m^2 (S650), a bi-directional, Styrene-Butadiene Rubber (SBR) external coated, glass fiber grid with a density of 150 g/m^2 (G150) and two basalt fiber grids with densities of 170 and 274 g/m^2 (B170, B274). These four fabrics were coupled with cementitious (C1, C2) or lime-based (N) mortars, obtaining five different FRCM types (see Table 1). Analyzing more in detail the mechanical properties of the matrices used to realize the FRCM strengthening systems, C1 is a commercial

Table 1. Fabrics and matrices used to realize the FRCM strengthening systems subject to environmental durability tests.

Samples	Fiber type	Density g/m^2	Number of yarns	FRCM Matrix
G150	Glass	150	4	C1
B170	Basalt	170	4	N
B274	Basalt	274	7	N
S650	Steel	650	9	N-C2

cementitious matrix used together with coated glass fibers characterized by a compressive strength of 21.19 MPa, whereas C2 cementitious matrix is a high performance mortar to be coupled with steel fibers with a compressive strength of 51.13 MPa. The natural hydraulic lime (NHL) based mortar used together with basalt and steel fibers was instead characterized by a lower compressive strength (20.44 MPa).

Results of mechanical characterization of dry fabrics and FRCM systems will be reported in the following together with the discussion of the mechanical properties of unconditioned (control) samples.

2.2 *Tensile test: Set-up and instrumentations*

Uniaxial tensile tests were performed on conditioned and unconditioned samples using the set-up shown in Figure 1a, b. The tensile force was monotonically applied by means of a universal MTS servohydraulic testing machine with a maximum capacity of 100 kN, equipped with hydraulic wedge grips and an onboard class 0.5 load cell. The axial strain was measured by means of a MTS extensometer adopting a gage length of 200 mm for both tests on FRCM coupons and dry fabrics. Specimens were strengthened with composite tabs at their extremities before performing tensile tests, to avoid possible local failures caused by grip clamping. All the tests were carried out under displacement control at a rate of 0.2 mm/min (during the FRCM

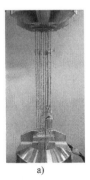

a)   b)

Figure 1. Experimental set-up used for uniaxial tensile tests: a) dry fabrics; b) FRCM coupons.

cracking phase) which was increased up to a maximum of 0.5 mm/min during the FRCM cracked phase and when performing tensile tests on dry fabrics.

The results in terms of maximum stresses and stress-strain curves reported in the following refer to the area of dry fibers only for both types of specimens and are, according to this approach, directly comparable. The adopted experimental set-up was previously used and validated on several campaigns performed on FRCM materials (Bellini & Mazzotti 2016, Bellini et al. 2019a, b).

3 EXPERIMENTAL RESULTS

3.1 Tensile tests on dry fabrics

Tensile tests were carried out both on conditioned and unconditioned dry fabrics characterized by the number of yarns/strands indicated in Table 1 and with a total length (with the tabs included) of 500 mm. The sample code adopted in Tables 2-6 includes the type of fiber (glass G150, basalt B170 and B274, steel S650) followed by the type of conditioning (C = control specimens, FT = freeze/thaw cycles, HU = humidity, AK = alkaline environments, SW = saline environments). Table 2 shows the results of tensile tests on unconditioned (control) specimens in terms of mean ultimate tensile strength $\sigma_{fu,m}$, referred to the area of dry fibers only and mean elastic modulus $E_{f,m}$, together with their Coefficient of Variation (CoV). As shown in the table, all the reference values, used in comparison to conditioned dry fabrics, can be considered reliable since they are affected by a reduced statistical variation. Tables 3-6 show the same types of results, but obtained on conditioned samples. Again, a reduced scattering of results can be observed, and this made possible a reliable comparison between conditioned and unconditioned samples, which will be discussed in detail in the following. Examples of typical stress-strain graphs coming from tensile tests are reported in Figure 2, where, as expected, the two types of basalt fibers, which differ only in density and spacing of the yarns, show similar elastic moduli and tensile strengths.

3.2 Tensile tests on FRCM systems

Unconditioned FRCM coupons were subjected to uniaxial tensile tests (FRCM Italian Guidelines 2018,

Table 2. Dry fabrics: results of tensile tests on unconditioned (control) samples.

Samples	$\sigma_{fu, m}$ MPa	CoV %	$E_{f, m}$ MPa	CoV %
G150_C	1218	2.2	68434	0.8
B170_C	1559	2.6	82048	1.3
B274_C	1524	2.9	80257	2.0
S650_C	2825	2.3	205286	1.4

Table 3. Dry fabrics: conditioned samples – freeze/thaw cycles (FT).

Samples	$\sigma_{fu, m}$ MPa	CoV %	$E_{f, m}$ MPa	CoV %
G150_FT	1234	5.4	68532	1.0
B170_FT	1562	4.9	81205	1.5
B274_FT	1519	4.2	80461	1.1
S650_FT	2807	1.5	208013	1.0

Table 4. Dry fabrics: conditioned samples – humidity (HU).

Samples	$\sigma_{fu, m}$ MPa	CoV %	$E_{f, m}$ MPa	CoV %
G150_HU	1210	4.5	68137	2.8
B170_HU	1527	2.0	82069	0.9
B274_HU	1504	3.9	78584	0.8
S650_HU	2789	1.7	204332	2.0

Table 5. Dry fabrics: conditioned samples – exposure to alkaline environments (AK).

Samples	$\sigma_{fu, m}$ MPa	CoV %	$E_{f, m}$ MPa	CoV %
G150_AK	1233	4.2	67888	0.5
B170_AK	1365	4.7	79019	0.7
B274_AK	1382	2.1	76658	0.8
S650_AK	2610	1.2	201361	3.0

Table 6. Dry fabrics: conditioned samples – exposure to saline environments (SW).

Samples	$\sigma_{fu, m}$ MPa	CoV %	$E_{f, m}$ MPa	CoV %
G150_SW	1226	5.1	68757	1.9
B170_SW	1357	4.7	78068	6.0
B274_SW	1362	1.4	75744	1.2
S650_SW	2387	3.5	196449	1.4

CNR DT 215/2018, de Felice et al. 2018) in order to obtain the reference values, in terms of mean maximum tensile strength $\sigma_{fu,m}$, reported in Table 7.

The sample code used contains the type of fiber, with the same meaning already discussed, and the type of matrix used within the FRCM strengthening system (cementitious C1, lime-based N or cementitious C2). As for tensile tests on dry fabrics, FT, HU, AK, SW indicate the type of conditioning (freeze/thaw cycles, humidity, alkaline and saline environments, respectively).

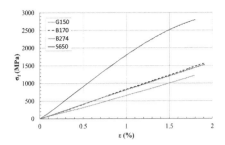

Figure 2. Example of stress-strain graphs obtained from tensile tests on dry fabrics.

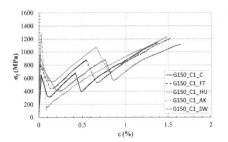

Figure 3. Tensile tests on glass FRCM samples (G150_C1): examples of stress-strain curves.

Table 7. FRCM coupons: results of tensile tests on unconditioned (control) specimens.

Samples	$\sigma_{fu,\,m}$ MPa	CoV %
G150_C1_C	1183	5.1
B170_N_C	1554	3.8
B274_N_C	1358	6.2
S650_N_C	2636	1.4
S650_C2_C	2654	3.4

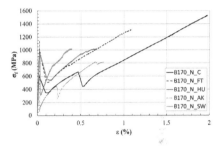

Figure 4. Tensile tests on basalt FRCM samples (B170_N): examples of stress-strain curves.

Table 8 shows the outcomes obtained from tests on conditioned FRCM coupons, highlighting, in general, a reduced statistical variation, even if results tends to be more scattered when dealing with conditioning involving immersion in water (AK and SW environments).

Figures 3-7 show the typical stress-strain graphs obtained during tensile tests on glass, basalt and steel FRCM coupons, respectively, where a single representative curve for each type of conditioning, including the control specimens, is reported. Results show an increment of the stress corresponding to first cracking for conditioned samples, as a consequence of the curing of mortar in humid environments or in water, which was evident in all the samples tested. In particular, the effect of curing

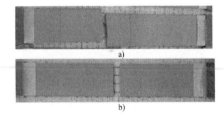

Figure 5. Cracking pattern in glass FRCM specimens: a) control specimen; b) specimen conditioned in alkaline environment.

Table 8. FRCM coupons – conditioned specimens.

Samples	Environmental conditions							
	FT		HU		AK		SW	
	$\sigma_{fu,\,m}$ MPa	CoV%	$\sigma_{fu,\,m}$ MPa	CoV%	$\sigma_{fu,\,m}$ MPa	CoV%	$\sigma_{fu,\,m}$ MPa	CoV%
G150_C1	1224	4.7	1109	2.8	1252	7.8	1211	8.2
B170_N	1323	2.4	1038	2.5	1181	9.0	1031	7.6
B274_N	1182	4.0	955	4.7	988	8.6	983	5.7
S650_N	2622	4.6	2562	2.1	2611	5.5	1859	7.7
S650_C2	2644	2.6	2603	1.5	2708	4.1	1700	9.9

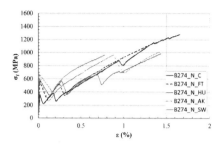

Figure 6. Tensile tests on basalt FRCM samples (B274_N): examples of stress-strain curves.

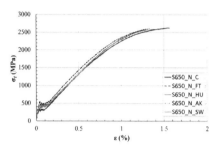

Figure 7. Tensile tests on steel FRCM samples (S650_N): examples of stress-strain curves.

was generally more evident in water (AK or SW environments) than under humidity conditions with RH > 90%, as expected, and influenced the overall behavior of the specimens. In fact, for glass G150 and basalt B170 samples, in particular, the achievement of a very high first peak caused the formation of a single central crack without an efficient stress redistribution after the first FRCM uncracked phase. Fiber tensile failure then occurred, at the end of the test, exactly in correspondence of this single and widely open crack, with a behavior very different from unconditioned samples. This behavior (see Figures 3-5) was detected, in particular, in samples conditioned in alkaline environments.

The effect of mortar curing was evident also in high density fabrics (basalt B274 and steel S650), where the same higher first phase peak was detected and the cracking phase developed at generally higher

stress levels, especially when samples were conditioned under water (see Figures 6 and 7). As for low density fabrics, the achievement of better mechanical properties of the matrix, in the first phase, generally led to a cracked phase with a high stress concentration and less distributed cracks.

The effect of environmental conditions was significant not only on matrix behavior, but also and above all in reducing the ultimate tensile strength of the fibers, in particular in steel specimens when conditioned in saline environments and in basalt samples under all environmental conditions. These results will be discussed in detail in the following.

3.3 Discussion of experimental results

Results of tensile tests on control and conditioned dry fabrics specimens are first analyzed in order to obtain the variation of ultimate tensile strength and elastic modulus after the conditioning period. Table 9 shows the residual percentages of previous quantities (X_{cond}/X_{ref}) for each fiber type after the four different durability exposures.

As shown in the table, SBR coated glass fibers seem to be not affected by aggressive environmental conditions, proving the effectiveness of their external coating, especially in alkaline environments, in preserving both tensile strength and elastic modulus of the fiber.

Steel fibers, instead, seem to be not affected by freeze-thaw cycles and humidity. They are only slightly degraded by alkaline environments and suffer some corrosion after immersion in salt water, with a preserved tensile strength percentage of about 84% after 1000 hours of conditioning period.

Both basalt fibers (B170 and B274) show a similar behavior, with small reduction of tensile strength and, partially, of elastic modulus, under alkaline and saline environments (see Table 9). This final remark will be essential to analyze and understand the results of durability tests on basalt FRCM systems.

The outcomes coming from durability tests on dry fabrics are also presented in graphic form in Figures 8-10 for each type of fabric tested, where the residual percentages of tensile strength are reported together with the mean values of ultimate tensile strengths and a graphical representation of CoV.

Table 9. Dry fabrics: residual percentages after conditioning in different environments.

Samples	$\sigma_{fu,\ m}$ %				$E_{f,\ m}$ %			
	FT	HU	AK	SW	FT	HU	AK	SW
G150	101.4	99.4	101.3	100.7	100.1	99.6	99.2	100.5
B170	100.2	97.9	87.5	87.0	99.0	100.0	96.3	95.2
B274	99.7	98.7	90.7	89.3	100.3	97.9	95.5	94.4
S650	99.3	98.8	92.4	84.5	101.3	99.5	98.1	95.7

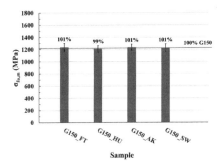

Figure 8. Glass grids: tensile strength percentage preserved after conditioning.

Table 10. FRCM strengthening systems: residual percentage of tensile strength after conditioning.

Samples	$\sigma_{fu,\,m}$ %			
	FT	HU	AK	SW
G150_C1	103.5	93.7	105.8	102.4
B170_N	85.1	66.8	76.0	66.3
B274_N	87.1	70.3	72.8	72.4
S650_N	99.5	97.2	99.0	70.5
S650_C2	99.6	98.1	102.0	64.1

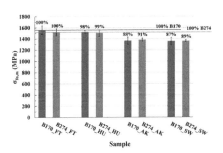

Figure 9. Basalt grids: tensile strength percentage preserved after conditioning.

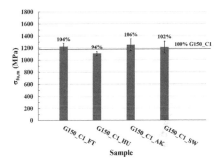

Figure 11. Glass FRCM coupons: residual tensile strength percentage after conditioning.

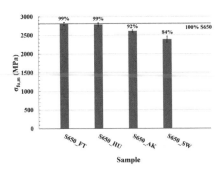

Figure 10. Unidirectional S650 steel sheets: tensile strength percentage preserved after conditioning.

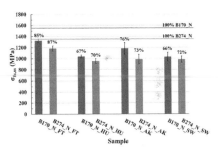

Figure 12. Basalt FRCMs: residual tensile strength percentage after conditioning.

The effects of aggressive environmental conditions on FRCM samples are summarized in Table 10, where the parameter considered for the analysis is the residual percentage of the mean ultimate tensile strength $\sigma_{fu,m}$. Results, presented also in graphic form in Figures 11-13 for a more immediate comparison, show that the performance of glass fiber FRCM is almost not influenced by aggressive exposures, whereas remarkable effects can be observed on basalt FRCMs for different conditioning types and on steel FRCM samples conditioned in saline environments.

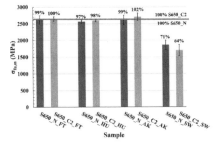

Figure 13. Steel FRCM coupons: residual tensile strength percentage after conditioning.

Analyzing the results in detail, especially when dealing with basalt fibers, the interaction between fibers and matrix seems to considerably affect the overall durability performance of the system. Moreover, the results coming from dry fibers are not conservative.

In fact, as noticed during conditioning of FRCM samples, the interaction between mortar and water, in immersion or exposed to high RH environments, produces, as expected, an alkaline environment with a local pH higher than 9.5, thus increasing the effect of exposure on alkali-sensitive fibers.

This effect, not revealed by durability tests on dry fibers performed under non-alkaline environments (freeze-thaw cycles, humidity, saline environments) because of the absence of the lime matrix, is able to cause a degradation of basalt fibers and of their corresponding mechanical performance.

After the conditioning period and before tensile tests, dry fabrics and FRCM samples were carefully inspected in order to identify any possible alteration and Figures 14 and 15 show the most important effects of conditioning on basalt and steel fibers. Figure 14 highlights the color change of the external part of the basalt grid after exposure to alkaline environments, which resulted, as previously discussed, in a reduced residual tensile strength after conditioning. Figure 15 shows instead the effect of saline environments on galvanized steel fibers, highlighting a diffused surface corrosion after 1000 hours of exposure, both on dry fibers (Figure 15a) and steel FRCM system (Figure 15b).

Further durability tests, still in progress on steel samples in order to reach 3000 hours of exposure, are showing sometimes a diffuse corrosion and sometimes a partial non-uniform spot corrosion,

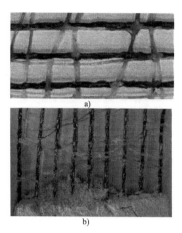

Figure 15. Durability tests on steel fibers: external corrosion on a) dry fibers and b) steel FCRM system after conditioning under saline environments.

pointing out the sensibility of the adopted electrogalvanization process, strongly dependent on the realization quality.

Results of these additional durability tests will be object of future in-depth analyses.

4 CONCLUSIONS

The experimental study presented here is focused on the analysis of durability of FRCM composite materials when exposed to aggressive environmental conditions. To this purpose, during the experimental campaign, different basalt, glass and steel FRCM systems have been considered, performing tensile tests on both conditioned and unconditioned samples and analyzing also the results in comparison to performance of dry fabrics, with the aim of evaluating any possible effect of the matrix under aggressive environments.

The effect of conditioning on dry fibers coupons was found to be negligible on coated glass fibers, but significant on basalt fibers under alkaline and saline environments and on steel fibers when immersed in salt water.

Results of durability tests on FRCM systems show a different tensile behavior of the specimens after conditioning and amplified degradation effects due to the interaction between fibers and matrix. In fact, curing of mortar in water or under high RH environments caused the modification of stress-strain curves with a remarkable increase of the stress at first cracking, due to the higher matrix tensile strength and a different, less distributed, cracking pattern.

If residual percentages of tensile strength after conditioning are analyzed, the matrix alkaline environment, determined by the interaction between lime mortar and water, with pH locally higher than 9.5,

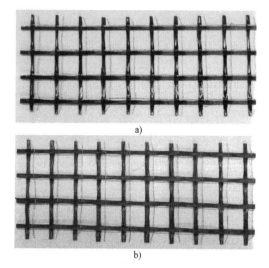

Figure 14. Effects of immersion of basalt fibers in alkaline environments: color change of the external part of the textile: a) control specimen; b) conditioned sample.

was found to increase the effect of wet conditioning on alkali-sensitive fabrics, such as those based on basalt fibers. In order to confirm these preliminary results, further durability tests on basalt fibers with a higher percentage of SBR external coating are currently in progress.

In case of galvanized steel fibers, exposure to saline environments produced some corrosion effects and a correspondent reduced residual tensile strength. Possible unpredictable variability of the quality of electrogalvanization process could have an influence on the diffused or non-uniform spot corrosion; in fact, the scattering of the corresponding results (9.9%) was the largest among all the others types of exposure.

The paper aims at promoting the creation of an experimental database on the effects of environmental conditions on FRCM materials, with the purpose of suggesting reliable strength reduction factors to be taken into account during the design of strengthening interventions. At the same time, it is important to understand the more appropriate test procedure.

The provided analyses, together with the given results and suggestions, can be directly applicable for modifying or improving the actually required materials qualification procedures.

ACKNOWLEDGEMENT

The financial support of the Italian Department of Civil Protection (ReLUIS 2019-2021 Grant – Innovative Materials) is gratefully acknowledged.

REFERENCES

Arboleda, D. et al. 2014. Durability of fabric reinforced cementitious matrix (FRCM) composites. *Proc. of the 7th International Conference on Fibre-Reinforced Polymer (FRP) Composites in Civil Engineering - CICE 2014, Vancouver, Canada, 20-22 August 2014.*

ASTM C581-03. Standard Practice for Determing Chemical Resistance of Thermosetting Resins Used in Glass-Fiber-Reinforced Structures Intended for Liquid Service. American Society for Testing Material, Reapproved 2008.

ASTM D1141-98. Standard Practice for the Preparation of Substitute Ocean Water. American Society for Testing Material, Reapproved 2013.

ASTM D2247-11. Standard Practice for Testing Water Resistance of coatings in 100% Relative Humidity. American Society for Testing Material, 2011.

ASTM D7705/D7705M-12. Standard test method for Alkali Resistance of Fiber Reinforced Polymer (FRP) Matrix Composite Bars used in Concrete Construction. American Society for Testing Material, Reapproved 2019.

Bellini, A. et al. 2015. Effect of matrix on bond between FRCM and masonry. *Proc. of FRPRCS-12 & APFIS-2015 Joint Conference, Nanjing, China, 14-16 December 2015.*

Bellini, A. & Mazzotti, C. 2016. Bond behavior and tensile characterization of FRCM strengthening systems applied on masonry panels. *Proc. of the 10th SAHC Conference, Van Balen & Verstrynge Eds., Taylor & Francis Group, London, 2016.*

Bellini, A. et al. 2018. Effectiveness of FRCM reinforcement applied to masonry walls subject to axial force and out-of-plane loads evaluated by experimental and numerical studies, *International Journal of Architectural Heritage* 12(3): 376–394.

Bellini, A. et al. 2019a. Experimental and numerical evaluation of fiber-matrix interface behaviour of different FRCM systems, *Composites Part B* 161: 411–426.

Bellini, A. et al. 2019b. Cyclic bond behavior of FRCM composites applied on masonry substrate, *Composites Part B* 169: 189–199.

Borri, A. et al. 2014. Durability Analysis for FRP and SRG Composites in Civil Applications. *Key Engineering Materials* 624: 421–428.

Caggegi, C. et al. 2017. Experimental analysis on tensile and bond properties of PBO and aramid fabric reinforced cementitious matrix for strengthening masonry structures, *Composites Part B* 127: 175–195.

Carozzi, F.G. et al. 2017. Experimental investigation of tensile and bond properties of Carbon-FRCM composites for strengthening masonry elements. *Composites Part B* 128: 100–119.

Ceroni, F. et al. 2014. Analytical and numerical modeling of composite-to-brick bond, *Materials and Structures* 47(12): 1987–2003.

CNR-DT 215/2018. Istruzioni per la Progettazione, l'Esecuzione ed il Controllo di Interventi di Consolidamento Statico mediante l'utilizzo di Compositi Fibrorinforzati a Matrice Inorganica, *Consiglio Nazionale delle Ricerche*, Rome, Italy. 2018.

De Felice, G. et al. 2014. Mortar-based systems for externally bonded strengthening of masonry, *Materials and Structures* 47(12): 2021–2037.

De Felice, G. et al. 2018. Recommendation of RILEM Technical Committee 250-CSM: Test method for Textile Reinforced Mortar to substrate bond characterization, *Materials and Structures* 51: 95.

De Santis, S. & de Felice, G. 2015. Steel reinforced grout systems for the strengthening of masonry structures, *Composite Structures* 134: 533–548.

De Santis, S. et al. 2018. Design of the out-of-plane strengthening of masonry walls with textile reinforced mortar composites. *Proc. of the 9th International Conference on Fibre-Reinforced Polymer (FRP) Composites in Civil Engineering - CICE 2018, Paris, France, 17-19 July 2018.*

Del Zoppo, M. et al. 2019. Analysis of FRCM and CRM parameters for the in-plane shear strengthening of different URM types, *Composites Part B* 171: 20–33.

Donnini, J. 2019. Durability of glass FRCM systems: Effects of different environments on mechanical properties, *Composites Part B* 174, 107047.

Ferretti, F. et al. 2019. In-plane shear behavior of stone masonry panels strengthened through grout injection and fiber reinforced cementitious matrices, *International Journal of Architectural Heritage.*

FRCM Italian Guidelines 2018. Linea Guida per la identificazione, la qualificazione ed il controllo di accettazione di compositi fibrorinforzati a matrice inorganica (FRCM) da utilizzarsi per il consolidamento strutturale di costruzioni esistenti, *Consiglio Superiore dei Lavori Pubblici, Servizio Tecnico Centrale, Rome, Italy. 2018.*

Mazzotti, C. et al. 2015a. Experimental bond tests on masonry panels strengthened by FRP, *Composites Part B* 80: 223–237.

Mazzotti, C. & Murgo, F.S. 2015b. Numerical and experimental study of GFRP-masonry interface behavior: Bond evolution and role of the mortar layers, *Composites Part B* 75: 212–225.

Micelli, F. & Aiello, M.A. 2019. Residual tensile strength of dry and impregnated reinforcement fibres after exposure to alkaline environments, *Composites Part B* 159: 490–501.

Nobili, A. & Signorini, C. 2017. On the effect of curing time and environmental exposure on impregnated Carbon Fabric Reinforced Cementitious Matrix (CFRCM) composite with design considerations, *Composites Part B* 112: 300–313.

Papanicolaou, C. et al. 2011. Externally bonded grid as strengthening and seismic retrofitting materials of masonry panels, *Construction and Building Materials* 25(2): 504–514.

Rizzo, V. et al. 2019. Influence of Alkaline Environments on the Mechanical Properties of FRCM/CRM and their Materials. *Key Engineering Materials* 817: 195–201.

Signorini, C. et al. 2018. Mechanical performance and crack pattern analysis of aged Carbon Fabric Cementitious Matrix (CFRCM) composites, *Composite Structures* 202: 1114–1120.

Brick and Block Masonry - From Historical to Sustainable Masonry –
Kubica, Kwiecień & Bednarz (eds)
© 2020 Taylor & Francis Group, London, ISBN 978-0-367-56586-2

Experimental study of comparative efficacy of out-of-plane strengthening of masonry walls using FRP and wire/textile reinforced mortar

P.K.V.R. Padalu & Y. Singh
Indian Institute of Technology Roorkee, Uttarakhand, India

S. Das
University of Windsor, Ontario, Canada

ABSTRACT: To improve the seismic out-of-plane performance of unreinforced masonry (URM) walls, three different types of strengthening techniques using welded wire mesh (WWM), a low cost material prevalent in India, basalt fibre mesh (BFM) and basalt fibre reinforced polymer (BFRP), sustainable composites comprising of natural fibres, are considered in the present study. The experimental study includes mechanical characterization of masonry and retrofit materials using uniaxial compression and tension tests, respectively. One-way out-of-plane flexural behaviour of URM and strengthened masonry is studied by testing a large number of masonry wallettes perpendicular and parallel to bed-joints under two-point (line) loading. The study also includes testing of full-size H-shape masonry walls simulating real boundary conditions, in two-way reversed cyclic bending under airbag loading. The experimental results for flexural strength, deformation, and energy absorption capacity are compared. The present study also identifies appropriate non-dimensional parameters to quantify the relative amount of reinforcement in different specimens for direct comparison of test results.

1 INTRODUCTION

The majority of past experimental and theoretical research was concentrated significantly on the in-plane behaviour of walls; whereas the studies on out-of-plane behaviour are limited. Because of the poor performance of URM buildings in past earthquakes, there is a pressing need for studies on the out-of-plane behaviour of unreinforced and strengthened masonry walls. Such studies, especially for strengthened masonry walls, are limited in the literature. Two-way spanning walls (Griffith et al. 2007) have received limited attention, even though these are most commonly encountered in practice, compared to one-way spanning walls (Derakhshan et al. 2013, Gattesco & Boem 2017). Consideration of geometry and boundary conditions is crucial in a two-way out-of-plane simulation of masonry walls.

On the other hand, there has been a surge of interest in developing retrofit techniques that can be applied to improve the seismic performance of URM walls. In past, several retrofit techniques have been applied to URM for enhancement in strength and deformation capacity. Application of inorganic matrix using wire mesh to make a composite to improve the out-of-plane behaviour of URM walls has been a commonly used method for the strengthening of existing URM buildings in the Indian subcontinent and some other South Asian countries (Kadam et al. 2014,

Shermi et al. 2017). The behaviour of masonry buildings after application of such strengthening system is still not clearly understood and requires extensive experimental studies. Basalt fibre composites is an emerging and green fibre material that has the potential to be used in seismic retrofits, but very few studies (Monni et al. 2015, Marcari et al. 2017) have been conducted in the past to investigate its efficacy.

The main objective of this study is to develop effective and sustainable strengthening techniques for unreinforced masonry (URM) buildings and to study the behaviour of masonry wallette/wall specimens, before and after retrofits, subject to one-way and two-way out-of-plane loading. To achieve this objective, URM specimens with and without retrofits have been tested. These results have been analyzed to evaluate the effectiveness of the selected retrofit techniques. Three innovative, sustainable, easy-to-apply, and cost-effective seismic retrofit techniques have been tested and their efficacy is validated using experiments.

2 CHARACTERIZATION OF MASONRY BASED ON COMPRESSIVE STRENGTH

Estimation of basic material properties of masonry and understanding its behaviour are the initial crucial steps in analyzing the performance of masonry buildings. The compressive strength and elastic modulus of bricks and mortar are the primary factors governing the properties of masonry. In the present study, masonry specimens have been constructed in the laboratory, with representative materials (bricks and 1:4 cement-sand mortar), that are prevalent in Northern part of India and to test them as per relevant standards. Uniaxial monotonic compressive tests have been carried out to understand the stress-strain behaviour of brick masonry and its constituent materials. The average stress-strain curve and the mechanical properties of masonry and its constituent materials obtained in the present study are presented in Figure 1. The observed mechanical properties are found to be in good agreement with the past studies in Indian conditions. The ratio of the observed modulus of elasticity to the observed compressive strength of the masonry is also within the range specified in the available literature (Kaushik et al. 2007).

3 CHARACTERIZATION OF COMPOSITE MATERIALS

In the present study, three types of reinforcements along with the composites have been studied: (a) bidirectional welded wire mesh (WWM) consisting of 2.96 mm diameter wire with two different grid spacings, 25 and 50 mm (identified as 25W and 50W); (b) bi-directional basalt fibre mesh (BFM) made of basalt yarn strands of cross-sectional area of 0.75 mm^2 and 1.0 mm^2 in 'warp' and 'weft' directions, respectively, with two different grid spacings of 25 and 50 mm (identified as 25FP, 50FP, 25FT, and 50FT); and (c) 0.35 mm thick bi-directional basalt fibre reinforced polymer coupons (identified as BFRP). Wire/fibre reinforced cementitious matrix (WRCM/FRCM) composites have been prepared using an inorganic matrix (cement-sand mortar), whereas, the basalt fibre reinforced polymer (BFRP) composites have been prepared using an organic

matrix (epoxy resin). The composite specimens have been identified by suffixing 'C' at the end of the specimen identifier. The length, width, and thickness of the composite specimens ranged from 420 to 437 mm, 78 to 88 mm, and 11 to 12 mm, respectively. Several tensile tests on a single wire (Figure 2a), single basalt fibre roving specimens (Figure 2b), basalt FRP coupons (Figure 2c), and WRCM/FRCM composites (Figure 2d) have been carried out to determine the mechanical properties of the strengthening materials.

From the test results, it has been observed that the strength of wires (585-853 MPa) is significantly higher and the modulus of elasticity (66-123 GPa) is lower as compared to mild steel. The wire specimens of the WWM have shown a non-linear behaviour (Figure 3a), however, with much lower ductility in comparison to structural steel and steel rebar, especially in case of the 25 mm spaced grid mesh, before rupture of the wires (Padalu et al. 2018a, Padalu et al. 2020a). The basalt fibre strands of roving in mesh have shown linear stress-strain curves and brittle failure. The sequential load-drops and partial regains (Figure 3b and c) in the stress-strain curves of the basalt fibre mesh near the failure, have been observed due to progressive breaking of individual fibres or yarns of a roving (Padalu et al. 2018b, Padalu et al. 2020a). The mean tensile strength of basalt fibre yarns (856-1472 MPa) in warp and weft directions are found to be different (Figure 3b and c). The difference in mechanical properties in the warp and weft directions is due to the influence of the weaving process of the grid (Padalu et al. 2020a). The maximum tensile strength of BFRP varies from 311 to 366 MPa. The test results are quite repetitive and the stress-strain curves are linear right up to the failure (Figure 3d). The BFRP specimens have shown a COV of less than 10% in mechanical properties (Padalu et al. 2019a).

The tensile strength, rupture strain, and elastic modulus of WRCM and basalt FRCM composites are invariably lower than the corresponding reinforcement. The difference is particularly large in case of basalt FRCM (nearly 63%, 37%, and 60%, in the tensile strength, breaking strain, and elastic modulus, respectively), because of the slippage of fibres within the mortar matrix. In case of the WRCM coupons, 18% and 27% of reduction in strength and strain, respectively, (no reduction in modulus of elasticity) has been observed, compared to that of wire reinforcement, as the failure mode involves rupture of wires in a sequential pattern (Padalu et al. 2020a).

4 EXPERIMENTAL INVESTIGATION: ONE-WAY BENDING

In the present study, unreinforced and strengthened masonry wallettes have been subjected to four-point bending, i.e. two-point (line) flexural loading, in one-way out-of-plane action. The masonry wallettes are tested in horizontal position without applying

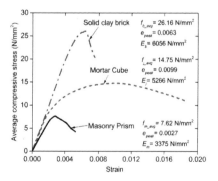

Figure 1. Average stress-strain curves of masonry and its constituent materials.

Figure 2. Test-setup for specimens in direct tension: a) wire in WWM; b) fibre roving in BFM; c) basalt FRP coupon; and d) wire/fibre reinforced cementitious matrix (WRCM or FRCM) composite.
Note: '25' and '50' denote the grid spacing of wire/fibre-mesh; 'W' denotes the wire; 'F' denotes the fibre; 'P' denotes the warp direction; 'T' denotes the weft direction; and 'C' denotes the composite.

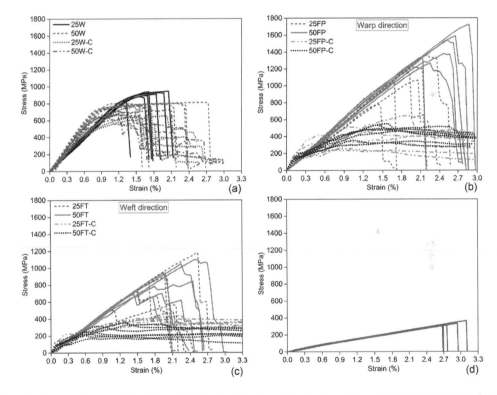

Figure 3. Stress-strain curves of specimens tested in direct tension: a) single wire reinforcement and WRCM composites; b), c) single basalt fibre yarn and basalt FRCM composites – warp and weft direction, respectively; and d) basalt FRP coupons.
Note: The nomenclature in the legend is the same as in caption of Figure 2.

any axial load and hence the flexural capacity represents the contribution of tensile reinforcement offered by the WWM, BFM, and BFRP.

The influence of different parameters, e.g. reinforcement ratio, the effect of a shear span, and loading direction, i.e. bending tension perpendicular to bed-joints (Figure 4a representing the walls spanning in a vertical direction) and bending tension parallel to bed-joints (Figure 4b representing the walls spanning in the horizontal direction) has been investigated. The load-displacement curves and failure modes of wallettes are studied in detail. The experimental results have been analyzed for flexural strength, deformation, and energy absorption capacity.

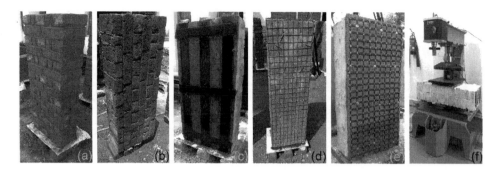

Figure 4. Unreinforced and strengthened masonry wallettes: a) for bending tension perpendicular to bed-joints; b) for bending tension parallel to bed-joints; c) basalt FRP strengthened wallette; d) WWM strengthened wallette; and e) BFM strengthened wallette; and f) tested in two-point (line) loading test setup.

The experimental investigation consists of total 80 specimens, in which 8 number of URM wallettes (control specimens), 28 numbers of URM wallettes strengthened using WWM, 16 numbers of URM wallettes strengthened using BFM, and 28 numbers of URM wallettes strengthened using BFRP, have been tested. All the masonry wallettes have been built by an experienced mason using solid clay bricks and cement-sand mortar of 1:4 proportion with a nominal thickness of the mortar joint as 10 mm. The length of wallettes varied from 1200 mm (in case of short shear span 'L') to 1700 mm (for long shear span 'H'), whereas the width and thickness of masonry are 500 and 230 mm, respectively. English bond, which is the prevalent bond of URM buildings in India, has been used. The specimens have been cured for 28 days before testing.

4.1 Application of strengthening technique

The FRP strengthening of masonry wallettes has been done in four steps: (i) smoothening/grinding of the masonry surface; (ii) application of a layer of epoxy (resin and hardener mixed in the specified proportion) over the masonry substrate; (iii) application of the layer of textile fabric of 75 mm wide; (iv) removal of entrapped air between the fabric strip and masonry by tapping with hand and using a grooved roller. The fabric was saturated with the epoxy by applying another layer at the top of it (Figure 4c) (Padalu et al. 2019a, b).

The strengthening of masonry wallettes using externally bonded grid/mesh has been done in three steps: (i) application of the first layer of mortar (15 mm thick) as a base; (ii) application of fibre/wire mesh, over the layer of mortar; and (iii) application of second layer of mortar (15 mm thick) covering the wire/fibre mesh. To improve the connection of wire-reinforced cementitious matrix (WRCM) composite with the masonry substrate, dowel bars of 4 mm diameter passing through previously drilled holes in masonry have been used to interconnect the wire mesh on the opposite faces (Figure 4d) (Padalu et al. 2018a). In case of

basalt fibre reinforced cementitious matrix (FRCM) composite, the textile mesh has been connected to the URM substrate using mechanical anchors (threaded screws of 6 mm diameter and 50 mm length with washers) embedded into previously drilled holes in the masonry (Figure 4e) (Padalu et al. 2018b).

4.2 Details of the test setup

The wallettes have been tested in a horizontal position under two-point (line) loading setup (Figure 4f). The monotonic loading has been applied using a 300 kN capacity flexural testing machine. The out-of-plane deformation of the wallettes has been measured at the mid-span of the wallette using a 50 mm stroke linear variable differential transducer (LVDT).

4.3 One-way bending behaviour and failure modes

In URM wallettes subjected to bending tension perpendicular to bed-joints, a sudden strength degradation and brittle failure have been observed, once the cracking has been initiated. The cracks have been formed across the width of the wallette in the constant moment zone, due to debonding of the mortar from the brick unit along the bed-joints indicating weak interfacial bond. For bending tension parallel to bed-joints, it has been observed that cracks initiated in the head joints and progressed either in a toothed pattern along with a combination of head and bed-joints or through the units in the alternate courses. Due to interlocking of brick units, the strength degradation in URM wallettes tested with bending tension parallel to bed-joints was a little slower and the load-carrying capacity also increased to about 2.3 times of that in case of bending tension perpendicular to bed-joints.

The masonry wallettes strengthened using 25 mm grid wire mesh, either failed in sliding shear along a bed-joint or due to diagonal shear crack, joining the support and the loading line. In case of the wallettes strengthened with 50 mm grid wire mesh, flexural failure has been observed in two wallettes, sliding shear failure has been observed in one wallette, and diagonal shear failure has been observed in another wallette (Padalu et al. 2018a). The shear failure of the specimens can be mainly attributed to the fact that these are being tested in a horizontal position without any axial force. The short shear span to depth ratio (1.17) in these specimens, is also responsible for shear failure. Therefore, wallettes have also been tested with a longer shear span to depth ratio of 2.24. In this case, the failure mode of wallettes strengthened using 25 and 50 mm grid wire mesh has been changed from shear to flexural-shear or flexure. It is interesting to note that the flexural failure is brittle (refer W25-L-S shown in Figure 5a), whereas the shear failure has some ductility (refer W25-L-D shown in Figure 6a). In the case of shear failure, the gradual sliding has been resisted by the catenary action of the WWM (Figure 7a), resulting in a slightly ductile failure (Padalu et al. 2018a). On the other hand, the brittle flexural failure is accompanied by a rupture of WWM (Figure 7b), can be attributed to very low reinforcement ratio in the strengthened specimens, resulting in a very shallow depth of the neutral axis.

All the strengthened wallettes with 25 as well as 50 mm grid basalt mesh tested in bending tension perpendicular and parallel to bed-joints, failed in flexure. Intermittent load drops and strength gains occurred due to sequential rupture of individual fibre yarns and redistribution of tensile force to the intact yarns. The wallettes ultimately failed, because of

rupture of fibre mesh (Figure 7c). However, the bond failure, i.e. debonding of the fibre mesh from the masonry has not been observed. The mechanical anchors are helpful in the good bond of basalt fibre reinforced cementitious matrix (FRCM) with the masonry. It has been observed that the two grid spacing used in the present study do not have any appreciable difference in the capacity of the strengthened wallettes (Figure 5b and 6b), as the reinforcement ratio (number of strands in unit width is unchanged. However, a smaller grid with an equal number of strands per unit width is preferable for more uniform stress distribution and better bond with the masonry substrate (Padalu et al. 2018b).

All the wallettes strengthened using BFRP tested in bending tension perpendicular and parallel to bed-joints, failed in flexure. Vertical cracks initiated at the bottom face of wallettes in the constant moment zone, resulting in the transfer of tensile stresses to the BFRP. The wallettes ultimately failed due to either rupture (Figure 7d) of BFRP strips which can be attributed to the relatively better bond characteristics, or due to debonding (Figure 7e) of BFRP strips from the masonry substrate followed by slippage of the BFRP strip at one of the ends. The debonding failure can be classified as the 'intermediate crack (IC)' debonding (Vasquez & Seracino 2010), as it started from the flexural cracks in the maximum moment region and then propagated towards the

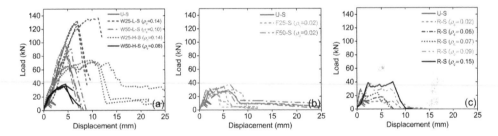

Figure 5. Load-displacement curves of URM and strengthened wallettes tested in bending tension parallel to bed-joints – 'S': a) WWM; b) BFM; and c) BFRP. Note: In the legend, '25' and '50' denote the grid spacing of wire/fibre-mesh; 'U' denotes the URM wallette; 'W' and 'F' denote the WWM and BFM strengthened wallette; 'ρ_c' denotes the reinforcement ratio; and 'R' denotes the BFRP strengthened wallette; 'L' denotes the short shear span – 1.17; and 'H' denotes the long shear span – 2.24.

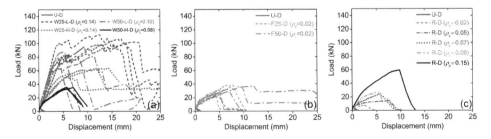

Figure 6. Load-displacement curves of URM and strengthened wallettes tested in bending tension perpendicular to bed-joints – 'D': a) WWM; b) BFM; and c) BFRP.
Note: The nomenclature in the legend is the same as in caption of Figure 5.

Figure 7. Failure modes of reinforcement in strengthened wallettes: a) catenary action of wires; b) rupture of wires; c) rupture of fibre mesh in tension; d) FRP rupture; and e) debonding of FRP strip.

supports. Also, in case of few BFRP strengthened wallettes, both rupture and debonding failure modes have been observed.

4.4 Test results

The experimental study demonstrated that all the strengthening techniques, viz. WWM, BFM, and BFRP are highly effective against out-of-plane loads with considerably enhanced moment capacity. The moment capacity of the strengthened wallettes increased up to about 17 and 10 times compared to URM wallettes, for bending tension perpendicular and parallel to the bed-joints, respectively. These strengthening methods also significantly improved the deformability of the wallettes, resulting in large displacements before failure under out-of-plane action. The maximum mid-span deflection of strengthened wallettes was up to 9 and 61 times higher than that for the corresponding URM wallettes for bending tension parallel and perpendicular to the bed-joints, respectively. The energy absorption capacity of strengthened masonry wallettes has also shown significant enhancement, up to 61 and 1026 times of that of URM wallettes, for bending tension parallel and perpendicular to the bed-joints, respectively (Padalu et al. 2018a, b, 2019a).

The test results for all strengthening techniques have shown a very promising solution for the structural upgrade against out-of-plane loading of masonry wallettes. As a result, the designer has difficulty in selecting the most appropriate technique. Therefore, to quantify the relative amount of reinforcement provided in different strengthening systems, a parameter, termed as reinforcement index ($\omega = \rho_c f_c / f_m$), has been defined. It represents the combined effect of the reinforcement ratio (ρ_c), the compressive strength of masonry (f_m), and tensile strength of composite material (f_c). For comparison of the relative efficacy of different strengthening composites, their strength ratio ($\dot{M} = M_{str}/M_{urm}$), displacement ratio ($\dot{D} = D_{str}/D_{urm}$), and energy ratio ($\dot{E} = E_{str}/E_{urm}$) have been compared for a uniform reinforcement index.

Table 1 shows the comparative efficacy of different strengthening techniques for a uniform reinforcement index of 0.031 in terms of the different non-dimensional parameters. It is interesting to note that the strengthening technique using WWM has demonstrated better performance in terms of average strength ratio (up to 2.14 and 2.76 times of BFM and BFRP),

Table 1. Average non-dimensional parameters for wallettes strengthened using different composites.

Type of composite	Specimen Identifier	ω	$\dot{M}$	$\dot{D}$	$\dot{E}$
WWM	W50-H-D	0.031	10.01	22.90	124.3
	W50-H-S	0.031	5.47	3.95	15.0
BFM	F25-D	0.031	4.79	19.33	86.5
	F25-S	0.031	2.49	4.05	10.1
BFRP	R-D	0.031	3.20	13.46	41.2
	R-S	0.031	2.28	2.00	8.2

Note: 'ω' denotes the reinforcement index; '$\dot{M}$' denotes the strength ratio; '$\dot{D}$' denotes the displacement ratio; and '$\dot{E}$' denotes the energy ratio.

average deformation ratio (up to 1.08 and 1.84 times of BFM and BFRP), and average energy absorption ratio (up to 1.46 and 2.42 times of BFM and BFRP) by making full use of the composite material. However, for BFRP strengthened wallettes, the enhancement has been observed to be lower because of the debonding failure mode which can be avoided using mechanical anchorage.

5 EXPERIMENTAL INVESTIGATION: TWO-WAY BENDING

The experimental investigation consists of total five full-scale H-shaped masonry walls, in which the first one is a control specimen of unreinforced masonry (designated as URM-F), other three walls are the identical URM walls retrofitted using the three chosen types of composites, viz. WWM, BFM, and BFRP. These three walls are designated as WWM-F, BFM-F, and BFRP-F. The H-shape test wall consists of a main wall (length, $L = 3000$ mm, height, $H = 3000$ mm, and $t = 230$ mm) and two cross-walls (length, $l = 1255$ mm, height, $H = 3000$ mm, and $t = 230$ mm) with configuration and geometry as shown in Figure 9a. The main and the cross-walls have been interlocked at the junctions using staggered overlapping bricks (toothed connection).

In case of WWM strengthened technique, the masonry wall has been retrofitted using WWM in the form of a horizontal bandage at lintel height as per the guidelines of IS 13935:2009. The bandage is provided continuously on both faces of the main-and the cross-

384

wall. The dowel bars have been used to connect masonry and WWM to ensure a composite action and transfer of shear forces at the interfaces. The horizontal bandage integrates the orthogonal walls and supports the main wall in the out-of-plane direction. In case of BFM strengthened technique, the BFM has been applied on the full area of both the faces of the wall using a hand layup procedure. The fibre mesh has been placed and attached to the wall using mechanical anchors. In case of BFRP strengthened wall, the FRP has been applied on both sides of the test wall in the form of a horizontal bandage at the lintel level, similar to the WWM. The area of the WWM has been provided according to the guidelines of IS 13935:2009, and that of BFRP has been provided to keep the reinforcement index same, for a direct comparison. However, BFM being applied on the full face the reinforcement index of BFM could not be matched.

All the five test walls have been subjected to identical quasi-static, reversed cyclic out-of-plane uniformly distributed loading using an airbag system on both faces of the main wall, alternatively, along with a pre-compression load. The details of the test set-up are presented in Figure 9a. The lateral load has been applied to the wall faces in two phases: (i) pre-cracking phase, consisting of the first half cycle of load up to significant cracking of the wall (i.e. up to having a well-defined collapse mechanism), and (ii) post-cracking phase, consisting of reversed cyclic load with the pre-defined displacement loading history.

The cracking patterns (Figure 10a-d) of all the retro-fitted walls, except for WWM-F are similar to the URM wall. The failure of wall BFM-F has occurred due to rupture of reinforcement, whereas failure of BFRP-F has been due to debonding. However, for wall WWM-F, propagation of diagonal cracks has been prevented by the bandage at the lintel height and cracking pattern in the wall has been divided into two distinct zones - above and below the bandage. The upper zone has been characterized by predominantly horizontal cracks, whereas the lower zone developed diagonal cracks similar to other walls (Padalu et al. 2020b).

The experimental load-displacement plots of all the tested walls are shown in Figure 11. Application of all the three retrofitting techniques substantially enhanced the lateral out-of-plane performance of URM walls in terms of strength, stiffness, and energy dissipation. The largest enhancement has been observed in the wall, BFM-F retrofitted with basalt fibre mesh applied on the entire face of the test wall. For this retrofitting technique, the flexural strength in pre- and post-cracking phases increased up to 1.65 times and 1.73 times, respectively, and the stiffness increased up to 5.82 and 3.76 times, respectively, in comparison with the corresponding URM wall. Similarly, the energy dissipation capacity increased up to 1.72 times. Among the two-way walls strengthened using WWM and FRP (having same reinforcement index), the WWM strengthened wall performed better with 5% higher strength, 52% higher stiffness, and 20% higher energy absorption capacity.

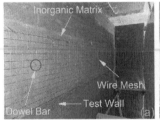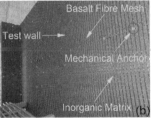

Figure 8. Strengthening of masonry walls using: a) WWM in form of a lintel bandage; b) BFM reinforcement; and c) BFRP bandage at lintel level and vertical splints along the junctions of test wall.

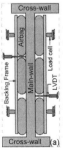

Figure 9. Two-way out-of-plane flexural test setup: a) sketch showing a schematic arrangement in the plan; b) photograph showing test arrangement; c) photograph showing side view; and d) photograph showing front view.

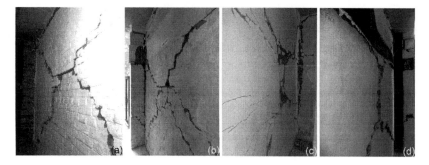

Figure 10. Crack patterns of tested walls: a) URM-F; b) WWM-F; c) BFM-F; and d) BFRP-F.

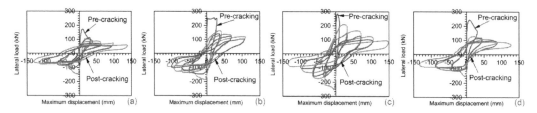

Figure 11. Reversed cyclic load-displacement response of URM and strengthened masonry walls: a) URM-F; b) WWM-F; c) BFM-F; and d) BFRP-F.

6 CONCLUSIONS

The major conclusions of the presented research work are as following:

- The experimental study in one-way and two-way bending has demonstrated that all the strengthening techniques, viz. WWM, BFM, and BFRP are highly effective against out-of-plane loads.
- A reinforcement index along with the other non-dimensional parameters have been defined to compare the efficacy of strengthened walls. WWM strengthening system performed better compared to the BFRP and BFM for the same reinforcement index. Strengthening using WWM resulted in the highest flexural strength. This method also significantly improved the deformability of the wallettes/walls undergoing relatively larger displacements before failure in out-of-plane action. The energy absorption capacity of masonry wallettes/walls has also shown significant enhancement using WWM strengthening system.
- The cracking patterns of all the strengthened walls in two-way bending are similar to the URM wall. The failure of wall BFM-F and WWM-F occurred due to rupture of reinforcement, whereas failure of BFRP-F has occurred due to debonding.
- In the case of two-way bending tests, the cyclic behaviour of all the walls, strengthened as well as unreinforced, has been observed to be asymmetric in the two directions of loading. The strengths in the two directions, at a given displacement, differed by up to 23%.

REFERENCES

ACI 440.7R-10. 2010. Guide for the design and construction of externally bonded fiber-reinforced polymer systems for strengthening unreinforced masonry structures. Farmington Hills, MI, U.S.A.

Derakhshan, H. et al. 2013. Airbag testing of multi-leaf unreinforced masonry subjected to one-way bending. *Engineering Structures* 57: 512–522.

Gattesco, N. & Boem, N. 2017. Out-of-plane behavior of reinforced masonry walls: Experimental and numerical study. *Composites Part B: Engineering* 128: 39–52.

Griffith, M.C. et al. 2007. Cyclic testing of unreinforced masonry walls in two-way bending. *Earthquake Engineering and Structural Dynamics* 36(6): 801–821.

IS 13935. 2009. Seismic evaluation, repair and strengthening of masonry buildings - Guidelines. New Delhi, India.

Kadam, S. et al. 2014. Out-of-plane behaviour of unreinforced masonry strengthened using ferrocement overlay. *Materials and Structures* 48(10): 3187–3203.

Kaushik, H.B. et al. 2007. Stress-strain characteristics of clay brick masonry under uniaxial compression. *Journal of Materials in Civil Engineering* 19(9): 728–739.

Marcari, G. et al. 2017. Experimental investigation of tuff masonry panels reinforced with surface bonded basalt textile-reinforced mortar. *Composites Part-B: Engineering* 108: 131–142.

Monni, F. et al. 2015. Dry masonry strengthening through basalt fibre ropes: experimental results versus out-of-plane actions. *Engineering Materials* 624: 584–594.

Mosallam, A.S. 2007. Out-of-plane flexural behavior of unreinforced red brick walls strengthened with FRP composites. *Composites Part-B: Engineering* 38(5-6): 559–574.

Padalu, P.K.V.R. et al. 2018a. Experimental investigation of out-of-plane behaviour of URM wallettes strengthened using welded wire mesh. *Construction and Building Materials* 190: 1133–1153.

Padalu, P.K.V.R. et al. 2018b. Efficacy of basalt fibre reinforced cement mortar composite for out-of-plane strengthening of unreinforced masonry. *Construction and Building Materials* 191: 1172–1190.

Padalu, P.K.V.R. et al. 2019a. Out-of-plane flexural strengthening of URM wallettes using basalt fibre reinforced polymer composite. *Construction and Building Materials* 216: 272–295.

Padalu, P.K.V.R. et al. 2019b. Out-of-Plane flexural behaviour of masonry wallettes strengthened using FRP composites and externally bonded grids: Comparative study. *Composites Part B: Engineering* 176: 107302 (1–23).

Padalu, P.K.V.R. et al. 2020a. Tensile properties of wire and fibre reinforced cementitious matrix composites for strengthening of masonry structures. *Structures* 23: 164–179.

Padalu, P.K.V.R. et al. 2020b. Cyclic two-way out-of-plane testing of unreinforced masonry walls retrofitted using composite materials. Construction and Building Materials 238: 117784 (1–24).

Shermi, C. & Dubey, R.N. 2018. Performance evaluation of a reinforced masonry model and an unreinforced masonry model using a shake table testing facility. *Journal of Performance of Constructed Facilities* 32(1): 04017121.

Vasquez, D. & Seracino, R. 2010. Assessment of the predictive performance of existing analytical models for debonding of near-surface mounted FRP strips. *Advances in Structural Engineering* 13(2): 299–308.

Brick and Block Masonry - From Historical to Sustainable Masonry –
Kubica, Kwiecień & Bednarz (eds)
© 2020 Taylor & Francis Group, London, ISBN 978-0-367-56586-2

The influence of FRCM system with a basalt mesh on the shear properties of AAC masonry walls

M. Kałuża

Department of Civil Engineering, Silesian University of Technology, Gliwice, Poland

ABSTRACT: Effective damage prevention of structural elements is a basic maintenance activities. In particular, this applies to masonry structures sensitive to various types of ground deformations. The application of FRCM strengthening systems, which are based on the materials compatible with masonry substrate is a very good solution here. The paper presents an analysis of the impact of selected FRCM system on the behavior of walls made of AAC blocks. The strengthening in the form of basalt mesh was applied on the masonry surface - in two configurations - using pozzolanic reaction cementitious mortar. Test results showed a significant increase in shear capacity of the walls strengthened on both sides, as well as a desirable mode of failure with a safety margin after obtaining the load-bearing capacity. One-sided application of the system allowed for only a slight improvement in shear properties in comparison with unstrengthened walls.

1 INTRODUCTION

Excessive damage to masonry walls is a very common problem, especially in areas with unusual ground conditions. One such case is the mining exploitation area. This activity causes severe deformation of the ground of different nature. These are discontinuous deformations in the form of cavity, continuous interactions occurring during the passage of a mining basin or dynamic effects associated with a mining shock. All these ground reactions generate additional loads or non-mechanical influences that lead to premature damage of structures. Shearing of the walls, both horizontally and vertically, is the most adverse effect observed in such situations.

Research on shear strengthening methods for masonry wall have been conducted for a long time. A common method is the application of reinforced plasters on the wall surface. Various types of reinforcement – both metallic (Farooq et al. 2006, Shermi & Dubey 2018, Wang et al. 2018) and non-metallic (Ismail & Ingham 2016, Lin et al. 2016, Parisi et al. 2013) – can be used here. In recent years, due to the fast development of composite materials, their popularity in building applications has increased significantly, especially for strengthening of the structures.

The positive effect of the composite strengthening of brick and tuff walls has been presented many times. The analyzes were conducted in both qualitative and quantitative aspects (Kouris & Trantafillou 2018, Nezhad & Kabir 2017, Sathiparan 2015, Valluzzi et al. 2002).

The use of modern strengthening materials (based on carbon, glass or basalt meshes) for untypical masonry components – such as autoclaved aerated concrete (AAC) blocks – becomes problematic. This article presents selected studies of this issue conducted by the author as a part of her research works.

The analysis of the impact of FRP strengthening system on masonry walls made of AAC blocks was presented in Kubica & Kałuża (2010). The use of a typical plastering system was discussed in Kałuża & Kubica (2019). This time, the use of basalt mesh and mineral strengthening mortar (system FRCM solution) is discussed.

2 RESEARCH ASSUMPTION

2.1 Material specification

Masonry walls were made of typical autoclaved aerated concrete (AAC) blocks with the dimension of $600\times200\times240$ mm. The compressive strength of the AAC block was tested according to the EN 772-1 standard, and amounted to 4.81 MPa. Its density was 600 kg/m^3, according to the manufacturer's data. This type of AAC blocks allows to construction the load-bearing walls in low-rise building as well as stiffening and infill walls in any type of buildings.

All elements tested were 900 mm wide, 815 mm high and 240 mm thick. The wall was erected in the technology of thin bed joints (with a thickness of about 3 mm) and unfilled head joints. The mortar used during erection of the walls was dedicated to thin joints. Its compressive and flexural strength – tested according EN 1015-11 standard – amounted to 18.8 MPa and 5.2 MPa respectively.

The walls were strengthened using FRCM method i.e. FRP mesh is applied using cementitious matrix.

Figure 1. Basalt fiber mesh.

Figure 2. View of specimen with optical measurement prepared to test.

Both materials form a strengthening system. The strengthening material is pre-primed basalt fibre mesh with a mesh size of 6 × 6 mm. The properties of the mesh, given by the manufacturer, are: density – 250 g/m^2, tensile strength – 3000 kN/m, elongation at failure – ca. 2%. The strengthening mortar is two-component, high ductility and fiber reinforced, pozzolanic reaction cementitious mortar. The strength parameters of the mortar – tested according EN 1015-11 standard – amounted to 25.8 MPa in compression and 6.7 MPa in bending.

The basalt fiber mesh was laid on the wall surface on a thin layer of a system mortar, and then covered with another layer of mortar. The thickness of each layer was about 10 mm (as recommended by the manufacturer). The basalt fiber mesh is visible in Figure 1.

2.2 Testing procedure and measurement

Three testing series were analyzed. The first one consisted of unstrengthened walls (Y-US), the other two series included walls superficial strengthened in different configurations. The first type of strengthening was one-sided application of basalt system (Y-BS1), the second one – application of basalt mesh on both sides of the wall (Y-BS2).

All specimens were subjected to diagonal compression according to Rilem Lumb 6 standard (Diagonal tensile strength tests of small wall specimens). All tested elements were loaded in one cycle up to the failure. Load ration for all models amounted to approximated 0.1 kN/s.

During the tests a traditional measurement of vertical and horizontal displacement using inductive gauges located on both sides of the walls was made. Additionally, one specimen from each series had an optical measurement of wall deformation using DIC system. Such a measurement was made only on one side of the wall. On the other side of the walls the measurement was carried out using inductive gauges. In the case of one-sided strengthening the optical system measured the deformation only on the strengthened wall surface.

Figure 2 presents the specimen with optical measurement prepared to test.

3 UNSTRENGTHENED WALLS - RESULTS

3.1 Behavior of the wall

All unstrengthened walls made of AAC blocks behave in an elastic manner up to the cracking. Immediately after the appearance of first crack the walls were completely destroyed. A diagonal crack ran through the wall and divided it into two independence parts. The damage was sudden and indicated a brittle nature of unstrengthened walls. There was no visible cracking of the wall before failure, so the element has no previously signaled the moment of destruction. Such a mode of failure resulted from the test method, in which the lateral displacement of the wall was not limited.

The optical measurement of the wall deformation in the subsequent loading steps made it possible to determine the places, which generates destruction. The concentration of stresses (just before failure) was observed in all unfilled head joints – Figure 3. This indicated potential places of destruction. The presented strain map (Figure 3) reveals a large heterogeneity of the structure. In the observed case, the failure was initiated near the support site and unfilled head joint.

Figure 3. Strain map just before failure recorded using optical system.

Based on the analysis of the damage of all unstrengthened specimens it should be considered that the erecting technique (with thin bed and unfilled head joints) enforced very adverse and premature failure of AAC masonry walls. Both types of joints generated an increase in strains and led to rapid and unexpected damage.

3.2 Shear stresses and strains

The values that describe the shear parameters of diagonal compressed wall are shear stresses and shear strains.

The shear stress is calculated as a quotient of loading force (P_i) and shear area:

$$\tau_i = \frac{P_i}{\sqrt{b^2 + h^2} \cdot t} \tag{1}$$

where b = width of the wall; h = height of the wall; and t = thickness of the wall.

The shear strain was calculated as:

$$\gamma_i = \frac{\Delta V_i + \Delta H_i}{l_m} \tag{2}$$

where ΔV = vertical shortening; ΔH = horizontal extension; and l_m = length of the measuring base that amounted to 900 mm in both directions. The changes in diagonal length were measured during the tests.

As described in point 3.1 there was no visible cracking before failure of the walls, so it was assumed, that stresses at cracking and maximum stresses, taken as load-bearing capacity are identical. The shear strains – in the case of using only traditional measurement (specimens from Y-US-1 and Y-US-2) – were calculated as the average value of measurements made on both sides of the wall. In specimen Y-US-3, with optical measurement on one side of the wall, the shear strain was calculated only on the basis of the reading from inductive gauges (from one side of the wall).

The calculated values are listed in Table 1.

The calculated values of stresses and strain as well as mode of failure characterized the behavior of walls without any strengthening. These values are treated as reference ones and allowed to estimate the impact of strengthening made in two configurations, described below.

4 WALLS STRENGTHNENED ON ONE SIDE

4.1 Cracking

In all walls with one-sided strengthening two clear stages of work – before cracking and after cracking – were observed. The appearance of first crack was characterized be a temporary drop in load as well as the appearance of a diagonal crack – visible on the unstrengthened wall side. Up to the cracking the increase in force was linear, so the structure behaved in an elastic way. After the cracking, the walls showed plastic behavior; much faster increase in deformation was recorded.

Shear stresses corresponding to the cracking forces in all tested walls were determined and listed in Table 2.

The shear strains at cracking were calculated (according to the inductive gauges readings) independently for both sides of the walls due to differences in values, caused by unsymmetrical arrangement of strengthening. The values are presented in Table 2, where *side A* is the strengthened wall surface and *side B* is the surface without strengthening. Also the average strains of entire wall are listed in Table 2.

The shear strains recorded on strengthened and unstrengthened wall surfaces differed almost twice. Of course, higher values were recorded on the wall side without mesh. The basalt mesh applied on relatively strong mortar caused significant reduction in deformation. However, in both cases the values were much smaller than the average strains calculated in walls with no strengthening.

4.2 Load-bearing capacity

Load-bearing capacity of strengthened walls was determined for the maximum recorded compressive force. In Table 3 the shear stresses calculated at this point are presented.

Table 1. Main stresses in unstrengthened walls.

Unstrengthened specimens	Cracking moment		Load-bearing capacity	
	stress [MPa]	strain [mm/m]	stress [MPa]	strain [mm/m]
Y-US-1	0.264	1.073	0.264	1.073
Y-US-2	0.261	1.033	0.261	1.033
Y-US-3	0.270	1.068	0.270	1.068
mean value	0.265	1.058	0.265	1.058
CoV	0.027	0.020	0.027	0.020

Table 2. Shear parameters at cracking.

Specimens strengthened on one side	Cracking moment			
	stresses [MPa]	strains side A [mm/m]	side B [mm/m]	average [mm/m]
Y-BS1-1	0.260	0.406	0.652	0.529
Y-BS1-2	0.251	-	0.642	-
Y-BS1-3	0.262	0.321	0.754	0.537
Mean value	0.258	0.363	0.683	0.533
CoV	0.023	0.164	0.090	0.011

Table 3. Shear parameters at maximum force.

| Specimens strengthened on one side | Load-bearing capacity | | | |
	stresses [MPa]	strains side A [mm/m]	side B [mm/m]	average [mm/m]
Y-BS1-1	0.343	3.858	5.452	4.655
Y-BS1-2	0.354	-	5.397	-
Y-BS1-3	0.334	3.446	4.913	4.179
Mean value	0.344	3.652	5.254	4.417
CoV	0.030	0.080	0.056	0.076

As in the case of cracking, shear strains recorded on both wall surfaces and also average strains, which correspond to the maximum forces, were determined. All values are listed in Table 3. Also here, significant differences between the strains calculated on both walls surfaces were noted.

The average maximum stresses increased more than 30% in relation to average cracking stresses. Deformations recorded at maximum load, compared to cracking stage, were more than 10 times higher on the strengthened surface and almost 8 times higher on the unstrengthened surface. As described earlier, this stage should be considered more plastic in comparison with the elastic stage appearing up to the cracking.

4.3 Failure

The failure of one-sidedly strengthened walls followed shortly after reaching the maximum load. The force decreased rapidly because all walls underwent significant deformations.

The mesh significantly reduced deformation of the walls, but only in the area where it was used. The other side (unstrengthened) deformed much more. The failure pattern of wall surface without strengthening is shown in Figure 4.

It can be seen, that the main diagonal cracks was not running through the unfilled head join, as it was

Figure 4. Crack pattern observed on the unstrengthened wall surface (side B).

observed in the unstrengthened walls. The mesh prevented against complete damage of the walls.

5 WALLS STRENGTHNENED ON BOTH SIDES

5.1 Cracking

The cracking of the symmetrically strengthened walls was clearly visible only on the gauges readings, since the appearance of cracks on the strengthening mortar was not observed. As in the case of walls strengthened on one side, the cracking changed the characteristic of the wall. Up to the cracking, all three walls behaved linear-elastic, and then a more rapid increase in deformation was observed. The cracking stresses and strains were calculated for the forces at which a sudden increase in displacement was noted at a constant force value. The values are listed in Table 4. Shear strains were calculated as the average value of strains on both surfaces, because the component values were almost identical.

Two-sided strengthened walls were cracked later, at higher stresses than both previous described series. The moment of cracking was almost imperceptible here, because the mesh and reinforced strengthening mortar effectively reduced the cracking pattern.

5.2 Load-bearing capacity

After cracking of the walls, a further significant increase in the load was observed. At the same time, a faster increase in deformation was recorded. This behavior continued in all walls until the maximum load was obtained. At the maximum load the stresses and average strains were calculated – Table 5. Values of these stresses were considered the load-bearing capacity of the two-sided strengthened walls.

Load-bearing capacity of the walls analyzed here was significant higher than in both previous described series (unstrengthened and one-sided strengthened walls).

Deformations (shear strains) at the maximum forces of double-sided strengthened walls were comparable

Table 4. Shear properties at cracking.

| Specimens strengthened on both sides | Cracking moment | |
	stresses [MPa]	strains [mm/m]
Y-BS2-1	0.314	0.546
Y-BS2-2	0.295	0.539
Y-BS2-3	0.285	0.481
Mean value	0.298	0.512
CoV	0.050	0.104

Table 5. Shear properties at maximum load.

| Specimens strengthened on both sides | Load-bearing capacity | |
	stresses [MPa]	strains [mm/m]
Y-BS2-1	0.457	4.214
Y-BS2-2	0.436	3.718
Y-BS2-3	0.463	4.429
Mean value	0.452	4.120
CoV	0.032	0.089

Figure 5. Wall surface after failure of two-sided strengthened specimen.

with the average deformations of walls strengthened only on one side. It should be pointed, however, that these values were recorded at much higher forces than in one-sided strengthened specimens.

5.3 Failure

After reaching the maximum load the deformation of the wall increased, what resulted in rapid decrease in load. After load stabilization, another small increase in load was observed until the final failure of the element. This happen in all two-sided strengthened walls. Just before damage of the wall the average shear stresses and strains were calculated and listed in Table 6.

A small decrease in load (ca. 20%) and its fast stabilization after exceeding the load capacity indicated a safety reserve of such strengthened structure. Also the process of wall damage was not sudden and rapid. All two-sided strengthened walls did not shown significant damage. Locally the mesh detached from the wall surface and the external layer of mortar dropped out. This is visible on the wall surface presented in Figure 5.

6 COMPARISON OF TESTED ELEMENTS

6.1 Shear stresses

Comparison of characteristic stress values, i.e. cracking stresses and load-bearing capacities, obtained in

the tests, clearly indicates the positive effect of basalt meshes application. Of course, the quantitative impact of strengthening depends on its arrangement. Figure 6 presents the percentage changes in stresses of strengthened walls, taking as reference value (1.00) the average stress recorded in unstrengthened walls, in which cracking meant damage of the elements.

One-sided strengthened walls are cracked at forces almost identical to cracking forces of unstrengthened walls, but their load-bearing capacity is on average 30% higher. Strengthening applied on both wall surfaces insignificant delays the wall cracking, however increases load capacity by almost 70%.

A two-fold increase in the degree of strengthening (meshes arranged on two wall surfaces) resulted in more than a double increase in load capacity in comparison with the one-sided mesh application. This excludes the influence of the strengthening eccentricity on the wall capacity.

6.2 Shear strain

The deformation analysis of the walls tested in diagonal compression scheme according Rilem Lumb 6 standard is inadequate to the actual behavior of shear walls. In the actual structure, the deformation of the analyzed wall is limited by another piece of wall or transverse wall. Therefore, the shear strains obtained during the test are much higher than in reality. In addition, the possibility of unlimited wall deformation

Table 6. The values at failure.

| Specimens strengthened on both sides | Failure | |
	stresses [MPa]	strains [mm/m]
Y-BS2-1	0.386	7.113
Y-BS2-2	0.301	8.398
Y-BS2-3	0.373	7.889
Mean value	0.354	7.800
CoV	0.129	0.083

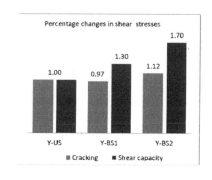

Figure 6. Changes in characteristic stresses.

during loading does not generate internal stresses in the structure, which could cause its earlier failure. However, the standard procedure allows a comparative assessment of changes in wall deformation and this is the only way to interpret the results obtained.

In Figure 7 shows the changes in shear strains of strengthened walls in relation to the average deformation of unstrengthened walls, which was the relevant value (1.00).

Both strengthening arrangement reduced twice wall deformation at the cracking and allowed a significant increase in deformation when exceeding the load capacity of walls. The shear strains at these characteristic moments were almost identical, regardless of the mode of strengthening. However, it should be remembered that the deformations of the walls of each series occur at different loads.

A comparison of average shear strains of all tested series recorded at stresses 0.200 MPa, 0.250 MPa (before cracking of all specimens) and 0.344 MPa (load-bearing capacity of one-sided strengthened walls) are listed in Table 7.

The above values show a huge influence of the strengthening on the wall stiffness. The use of meshes significantly increase the stiffness of the structure. The strains calculated at every load stage are much smaller than those obtained in unstrengthened walls. The difference in strains calculated at the stress amounted to 0.344 MPa indicates much higher stiffness of walls strengthened on both sides.

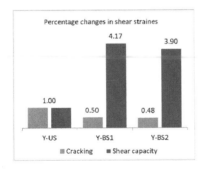

Figure 7. Changes in characteristic strains.

Table 7. Strains calculated at selected stresses.

Specimens	Average strains [mm/m]		
	at 0.200MPa	0.250MPa	0.344MPa
Y-US-i	0.590	0.830	failed
Y-BS1-i	0.389	0.474	4.417
Y-BS2-i	0.276	0.356	1.389

Figure 8. Strain map at maximum force recorded using.

6.3 Failure

As expected, the desirable and relatively safe mode of failure is observed in the walls strengthened on both sides. The visible damage are minor; no significant cracks are noticed. Huge deformations recorded at damage were possible to achieve due to the high tensile strength and elongation of the basalt mesh. The use of bilateral strengthening caused that the structure worked as a homogeneous material, without discontinuities in the form of unfilled head joints, what was observed in unstrengthened walls (see Figure 3). This phenomenon confirms the image of wall surface deformation recorded by Aramis system. Figure 8 shows the strain pattern of two-sided strengthened wall at maximum force. The color scales (from 0.00 to 1.00) were adopted identical for both deformation images (Figure 3 and 8).

optical system

Destruction of a one-sided strengthened walls is rapid, with a significant damage observed on unstrengthened wall surface (see Figure 5). Final failure occurs almost immediately after reaching the load-bearing capacity of the walls. This happened just after reaching the full load-bearing capacity of the walls. The negative impact of unilateral mesh application reveals in the mode of failure of such a strengthened walls.

The unstrengthened walls failed just after appearance of first crack. All tested elements has split into two parts along diagonal. This type of damage was considered the most dangerous without prior warning. The stresses noticed at failure was the lowest of measured during all tests.

7 CONCLUSIONS

The presented laboratory tests show the influence of the surface application of basalt mesh laid on a special cementitious mortar on the changes in shear properties of walls made of AAC blocks.

The results obtained indicate an increase in the load-bearing capacity of strengthened walls, regardless of the mode of application (on one or both sides). Qualitative analysis of the issue allows to formulate the following conclusions:

- unstrengthened walls failed when the first crack appeared and it happened suddenly, without warning;
- the strengthening did not significantly affect the value of the cracking forces, but in both cases it ensured further work of the elements;
- both strengthening configurations almost twice reduced the deformation at cracking in comparison with the unstrengthened walls;
- application of FRCM system increased the load-bearing capacity by 30% in walls strengthened on one side and 70% in walls with bilateral strengthening;
- only two-sided application of strengthening allowed for safe failure (a lot of small cracks) that was prolonged in time.

Based on the presented research, it can be concluded that the use of two-sided wall strengthening with basalt mesh is an effective way to strengthen the AAC walls in mining areas or seismic areas with a low magnitude. However, the one-sided application of strengthening is not recommended here, because of the slight increase in load capacity and the lack of a safety margin after obtaining the load-bearing capacity.

REFERENCES

Farooq, S., Ilyas, M. & Ghaffar, A. 2006. Technique for strengthening of masonry wall panels using steel strips. *Asian Journal of Civil Engineering – Building and Housing*. 7(6): 621–638.

Ismail, N. & Ingham, J. 2016. In-plane and out-of-plane testing of unreinforced masonry walls strengthened using polymer textile reinforced mortar. *Engineering Structures*. 118: 167–177.

Kałuża, M & Kubica, J. 2019. The effects of strengthening the AAC walls using glass mesh arranged in different configurations. *Key Engineering Materials*. 817: 442–449.

Kouris, L. & Triantafillou, T. 2018. State-of-the-art on strengthening of masonry structures with textile reinforced mortar (TRM). *Construction and Building Materials*. 188: 121–1233.

Kubica, J. & Kałuża, M. 2010. Diagonally compressed AAC block's masonry – Effectiveness of strengthening using CFRP and GFRP laminates. *Proceeding of the British Masonry Society*. 11: 419–428.

Lin, Y., Lawley, D., Wotherspoon, L. & Ingham, J. 2016. Out-of-plane testing of unreinforced masonry walls strengthened using ECC shotcrete. *Structures*. 7: 33–42.

Nezhad, R. & Kabir, M. 2017. Experimental investigation on out-of-plane behavior of GFRP retrofitted masonry panels. *Construction and Building Materials*. 131:630–640.

Parisi, F., Iovinella, I., Balsamo, A., Augenti, N. & Prota, A. 2013. In-plane behaviour of tuff masonry strengthened with inorganic matrix-grid composites, *Composites Part B*. 45: 1657–1666.

Sathiparan, N. 2015. Mesh type seismic retrofitting for masonry structures: Critical issues and possible strategies. *European Journal of Environmental and Civil Engineering*. 19: 1136–1154.

Shermi, C. & Dubey, R. 2018. In-plane behavior of unreinforced masonry panels strengthened with welded wire mesh and mortar. *Construction and Building Materials*. 178: 195–203.

Wang, X., Chiu, Ch. & Iu, V. 2018. Experimental investigation of in-plane shear behavior of grey clay brick masonry panels strengthened with SRG. *Engineering Structures*. 162: 84–96.

Valluzzi, M., Tinazzi, D. & Modena, C. 2002. Shear behavior of masonry panels strengthened by FRP laminates. *Construction and Building Materials*. 16: 409–416.

Construction, practice, technology & earthen masonry

Brick and Block Masonry - From Historical to Sustainable Masonry –
Kubica, Kwiecień & Bednarz (eds)
© 2020 Taylor & Francis Group, London, ISBN 978-0-367-56586-2

Mortar-auxetic rendered drystack masonry walls under concentric and eccentric compression

T. Zahra & M. Asad
School of Civil and Environmental Engineering, Queensland University of Technology, Australia

R. Dhanasekar
School of Engineering and Technology, CQ University Australia, Melbourne Campus, Australia

J.A. Thamboo
Department of Civil Engineering, South Eastern University of Sri Lanka, Oluvil, Sri Lanka

ABSTRACT: Despite the existence of many historic buildings containing drystack walls, this walling system is not regarded as a legitimate masonry in many national standards. Unlike the thick historic drystack walls, the thin contemporary drystack walls require rendering to survive unavoidable eccentricities in the vertical compression loading and lateral loads. Commonly used fibre reinforced mortar composite (FRMC) render fails due to delamination and brittle splitting. Authors have developed mortar-auxetic fabric composite (MAFC) render as an alternative rendering that eliminates delamination. To illustrate the difference between the MAFC and FRMC, wallettes with these renders were analysed under compression and the numerical model results were validated through available experimental data on FRMC rendered drystack wallette. The mortar-auxetic composite (MAFC) render has shown to improve the response of the wallette under concentric compression through elimination of delamination. This render was then employed to analyse the mortar-auxetic composite rendered drystack wallettes subjected to eccentric compression. A traditional mortared masonry wallette without any render was also analysed as a benchmark case. Mortar-auxetic fabric rendered wallettes exhibited less sensitivity to eccentricity compared to the unrendered traditional mortared masonry under eccentric compression. It is shown that the drystack masonry walls rendered on both sides can resist much larger eccentricity to compression loading compared to the traditional mortared masonry walls. Considering the economic benefits of dry stack construction and based on the structural benefits illustrated in this paper, it appears sensible to recognise dry stack as a valid masonry system in the design standards.

1 INTRODUCTION

Drystack masonry is shown to exhibit compressive strength close to the strength of units in (Zahra & Dhanasekar, 2016; Thamboo et al. 2020). However, as compression is seldom applied without eccentricity, drystack masonry is vulnerable to early failure due to opening of bed and head joints due to lack of bond in comparison to mortared masonry under eccentric compression. Surface bonding through rendering is one of many strategies to overcome this vulnerability. Research on the behaviour of unrendered and rendered drystack masonry under eccentric compression is very limited. Anand and Ramamurthy (2005), Dhanasekar et al. (2008) and Thanoon et al. (2007) tested unrendered drystack masonry under eccentric loads and reported that the prisms failed due to combination of web splitting and face-shell spalling for lower eccentricity, while web splitting was replaced by tilting of prism for higher load eccentricity. Rendered drystack masonry was tested under out-of-plane loading by Qamar et al. (2018) and they reported improved lateral load resistance for rendered wallettes in comparison to

unrendered specimens. Most commonly, fibre reinforced mortar composite (FRMC) renders are employed for strengthening and increasing the resistance of masonry against various loads. However, these fibre reinforced composites failed by delaminating in a brittle manner with fracturing of fabric/mesh as observed by Dhanasekar et al. (2008), Huang et al. (2017), Alotaibi & Galal (2018) and Wang et al. (2020). To address this issue, authors have developed and characterised an alternative type of rendering composite for strengthening purposes which utilises auxetic foams and fabric in place of fibre reinforced polymers (Zahra & Dhanasekar, 2017; Asad et al. 2019). Auxetic materials exhibit negative Poisson's ratio (NPR) which improves their shear modulus, impact resistance through high energy absorption and enhanced bonding behaviour (Asad et al. 2018). Auxetic fabric was also employed by the authors to mitigate the surface unevenness of drystack masonry by reducing the stress peaks caused by dirt and interstices between the drystack blocks in absence of mortar (Zahra & Dhanasekar, 2018). The output of these researches yielded encouraging results and auxetic fabrics and mortar-auxetic composites outperformed conventional fibre reinforced composites by absorbing more energy with rectification of delamination issues. The aim of this paper is to exploit the benefits of mortar-auxetic composite render in enhancing the compression resistance of drystack masonry walls subject to various eccentricities.

A 3D finite element model was established for the analysis of rendered drystack wallettes subject to concentric and eccentric compression in this paper. The modelling method was first validated using an experimental dataset on a fibre reinforced mortar composite (FRMC) rendered drystack wallette tested by Dhanasekar et al. (2008) under concentric and eccentric compression. The model predicted the stress-strain curves and failure modes close to experimental results. In order to compare the performance mortar-auxetic fabric composite (MAFC) render against fibre reinforced mortar composite (FRMC), the same wallette model was reanalysed with the mortar-auxetic composite layer on both exposed surfaces of the wallette. MARC outperformed FRMC with complete elimination of debonding failure and reasonable load bearing capacity. Drystack wallettes rendered with MAFC render were then analysed subject to various eccentricities: 0 mm, 30 mm, 60 mm and 90 mm. For comparison, conventional mortared masonry wallettes without any render were also analysed for the same eccentricities. Capacity reduction in both types of masonries were observed, however, rendered drystack walls exhibited a higher load resistance (up to 30%) in comparison to traditional mortared masonry. Sensitivity of the rendered drystack masonry to eccentricity was also lower. For the eccentricity of 90mm (eccentricity/thickness ratio, e/t = 0.45), rendered drystack wallettes showed 50% reduction in compressive load capacity, while 75% decline was realised for mortared masonry. These results proved that mortar-auxetic rendered drystack walls can perform better than unrendered mortared masonry walls and should be considered as a legitimate type of structural masonry.

2 MODEL VALIDATION UNDER CONCENTRIC COMPRESSION

A 3D finite element model was developed to analyse a FRMC rendered drystack wallette under concentric compression loads tested by Dhanasekar et al. (2008). The analysed wall consisted of two types of blocks with different geometry and mechanical properties. Quarter of the wallette was modelled as marked in Figure 1 by exploiting the symmetry. The behaviour of the blocks and FRMC render was simulated using the concrete damage plasticity model in ABAQUS. Separate properties were input for half and full blocks based on the experimental results reported in Dhanasekar et al. (2008). The FRMC render used by Dhanasekar et al. (2008) is a composite which consisted of fibre reinforced cement polymer mix (FRCPM) and fibreglass mesh. The properties of the FRMC render were adopted from Dhanasekar et al. (2008). The input values are listed in Tables 1 and 2.

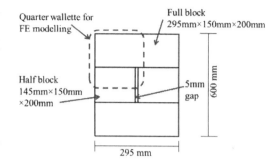

Figure 1. Drystack wallette tested by Dhanasekar et al. (2008) (shown without render).

Table 1. Details of concrete block and render material properties used in the analysis.

Material properties	Half block	Full block	FRMC
Elastic Modulus (MPa)	33000	20000	10000
Poisson's ratio (υ)	0.22	0.2	0.2
Compressive strength (MPa)	35	30	25
Tensile strength (MPa)	3.5	3.0	6.0
Biaxial stress ratio	1.16	1.16	1.16
Dilatation angle (°)	15	15	15
Flow potential eccentricity (ε)	0.1	0.1	0.1
Viscosity parameter	0.01	0.01	0.01

Table 2. Compression and tension failure stress – inelastic strain data.

Compression		Tension	
Stress (MPa)	Strain	Stress (MPa)	Strain
Half block data			
10	0.001	3.0	0
17	0.00125	2.5	0.0005
35	0.002	2.0	0.0010
30	0.0035	1.5	0.0015
10	0.0045	1.0	0.0020
Full block data			
10	0.001	3.0	0
15	0.00125	2.5	0.0005
30	0.002	2.0	0.0010
27	0.0035	1.5	0.0015
10	0.0045	1.0	0.0020
FRMC render data			
9	0	6	0
18	0.0010	5	0.0005
21	0.00125	4	0.0010
25	0.0020	3	0.0015
20	0.0035	2	0.0020
7	0.0045	1	0.0025

Table 3. Interface properties.

Parameter	Magnitude
Normal stiffness (N/mm^3)	28
Shear stiffness (N/mm^3)	32
Friction coefficient	0.6
Maximum tensile stress (MPa)	0.68
Maximum shear stress (MPa)	0.82

The interaction between the render and the block surface and between the vertical and the horizontal joints of the blocks were defined using a constitutive law accounting for the traction-separation of the interface from the ABAQUS library. This model considers initially a linear elastic behaviour of the interface which is followed by the initiation and evolution of interface damage. Mohr-Coulomb failure criterion was used to model the friction behaviour. The selected interface properties used in the FE model are shown in Table 3.

The model was developed using 3D, 8 nodded (C38DR) solid elements. The FE mesh, boundary conditions and loading are shown in Figure 2. Prescribed displacements were applied normal to the top surface of the masonry face-shell as shown in Figure 2. There were 21,223 nodes and 14,236 elements in the mesh.

The vertical stress distribution of the FRMC rendered wallettes is shown in Figure 3(a). The experimental failure mode is also shown in Figure 3(b). It can be observed from Figure 3 that the FRMC render excessively deformed at the bed joints and the edges. This response agrees with the experimental observations as shown in Figure 3(b). The onset of delamination at the edges due to excessive deformation matched well with the experimental failure modes. The average stress-strain response of the analysed rendered wallette with comparison of experimental results is plotted in Figure 4. The FE results of the FRMC rendered wallette are in good agreement with the experimental data (Dhanasekar et al. 2008).

The validated model was then used to study the behaviour of MAFC render by replacing the FRMC render. 5 mm thick MAFC render consisting of 2 mm thick polymer cement mortar (PCM) layers and a 1 mm thick auxetic fabric layer embedded in the middle of render was modelled to replace the FRMC render on the wallette as shown in Figure 5(a). The material properties of the PCM mortar for the mortar-auxetic composite render were taken from Thamboo & Dhanasekar (2015), while auxetic fabric material data was input from Zahra & Dhanasekar (2018). The interaction properties between the render and the wallette were kept unchanged as given in Table 3. The failure mode obtained from FE analysis is shown in Figure 5 (b). It is evident that the failure mode improved without any compressibility at dry joints and no sign of delamination of the render. This outcome proves the effectiveness of mortar-auxetic renders in comparison

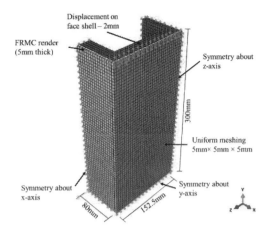

Figure 2. 3D model of Dhanasekar et al. (2008) wallette with FRMC render.

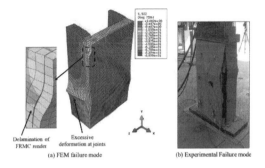

Figure 3. Vertical stress distribution and failure mode of FRMC rendered wallette.

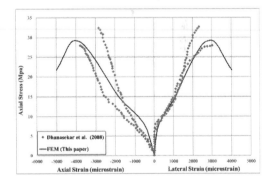

Figure 4. Average stress-strain response of Dhanasekar et al. (2008) wallette.

to fibre reinforced polymer renders with effective protection and eradication of debonding problem.

3 FE ANALYSIS OF RENDERED WALLS UNDER ECCENTRIC COMPRESSION

Rendered drystack wallette of size 600 mm wide × 200 mm thick × 800 mm high were modelled using

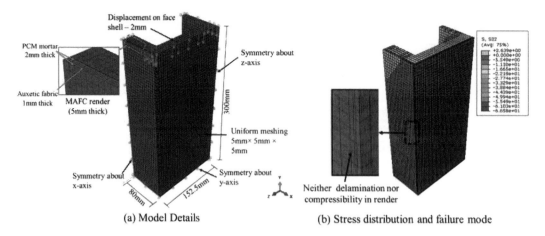

(a) Model Details

(b) Stress distribution and failure mode

Figure 5. Wallette model with MAFC render.

400 mm wide × 190 mm thick × 200 mm high generally available drystack concrete blocks. The 200 mm wall thickness incorporated the 190 mm thick block and 5 mm thick render layers on front and back sides. Due to symmetry about X and Y axis only quarter of the wallette was modelled as shown in Figure 6(a). Wallette configuration without render and loading plate is also shown in Figure 6(b). Conventional masonry wallette of same size without any render was also modelled using the conventional 390 mm wide × 190 mm thick × 190 mm high concrete blocks with 10 mm vertical and horizontal mortar joints as shown in Figure 6(c). The loading plate was modelled with linear elastic properties of high tensile steel to apply the eccentric load. The material properties of the drystack blocks were adopted from Zahra & Dhanasekar (2018). The interactions between the render and the wall and block to block were defined using the properties listed in Table 3. The mortar-auxetic fabric render properties were input as described in the Section 2. The geometry and the boundary conditions are shown in Figure 6(a). An incremental displacement of 4 mm on each of 60 nodes along X-axis on the centreline of the loading plate was applied for the concentric loading (eccentricity, e = 0). For other eccentricities of the applied compressive displacement, the loading line was shifted at:

i   30 mm eccentricity along the Z-axis, left of the centreline (C.L)
ii. 60mm eccentricity along the Z-axis, left of the centreline (C.L)
iii. 90mm eccentricity along the Z-axis, left of the centreline (C.L)

The failure mode, stress distribution and load-deflection characteristics were studied and are presented in this section.

The vertical stress profile of the rendered drystack wallette for 0 mm and 60 mm eccentricities are shown in Figure 7. Due to space limits, stress profile of two eccentricities are shown, however, ultimate load and capacity reduction with increasing eccentricity is shown for all eccentricities later in this section.

For concentric compression at (e = 0 mm), the stresses were unfirmly distributed with high concentration in webs which changed to nonuniform stress for e = 60 mm with high stresses in web and minor tilt of the wallette as observed by Dhanasekar et al. (2008). Similar results were observed for unrendered mortared walls as shown in Figure 8. It can be concluded that mortar-auxetic composite rendered drystack wallette has performed equivalent to an unrendered mortared masonry wallette under eccentric compression.

Variation of the ultimate load of rendered drystack wallette and mortared wallette with the increase in the eccentricity is shown in Figure 9. As expected, the ultimate load reduced with the increase in eccentricity. Drystack wall rendered with mortar-auxetic composite layer consistently exhibited higher load capacity for all levels of eccentricity. Rendered drystack wallette showed 30% higher capacity on an average for all eccentricities in comparison to the mortared wallette. The load capacity of this wallette exhibited linear reduction of 22% for eccentricity of 30 mm, 38% for eccentricity of 60 mm and 50% for eccentricity of 90 mm. While the capacity of unrendered mortared masonry wall reduced by 26% for the eccentricity of 30 mm, 56% for the eccentricity of 60 mm and 74% for the eccentricity of 90 mm. With these results, it can be concluded that rendering with the mortar-auxetic composite layers can protect drystack walls from brittle collapse in case of eccentric compression loads with better performance in comparison to unrendered mortared masonry.

Relationship between the normalised ultimate load (P_{ue}/P_u) of the rendered drystack walls and mortared masonry walls with the increase in the ratio of

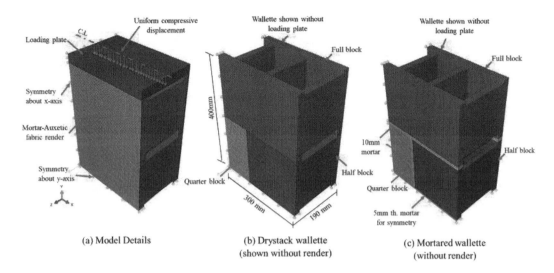

(a) Model Details

(b) Drystack wallette
(shown without render)

(c) Mortared wallette
(without render)

Figure 6. Model details for eccentric compression analysis.

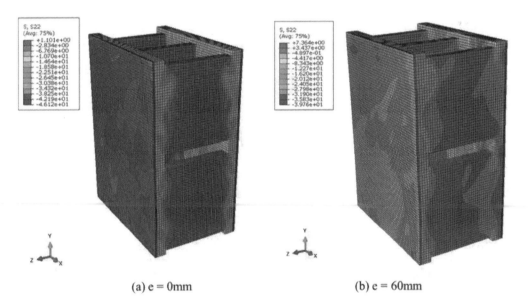

(a) e = 0mm

(b) e = 60mm

Figure 7. Stress profile of rendered drystack wallette at different eccentricities.

eccentricity to the wall thickness (e/t) is shown in Figure 10.

Regression lines to fit the data were plotted as shown in Figure 10; linear relationships given in Equation 1 and Equation 2 were obtained between the normalised load and the eccentricity ratio for the rendered drystack walls and mortared walls, respectively.

$$\frac{P_{ue}}{P_u} = 1.0 - 1.1\left(\frac{e}{t}\right) \tag{1}$$

$$\frac{P_{ue}}{P_u} = 1.0 - 1.7\left(\frac{e}{t}\right) \tag{2}$$

Equation (1) shows the reduction in strength for rendered drystack masonry is lower compared to the reduction of mortared masonry (without render) as given in Equation (2). This expression for mortared masonry wallettes in Equation (2) is close to the theoretical relationship reported in the literature $\frac{P_{ue}}{P_u} = 1.0 - 2.0\left(\frac{e}{t}\right)$ (Brencich & Gambarotta, 2005;

401

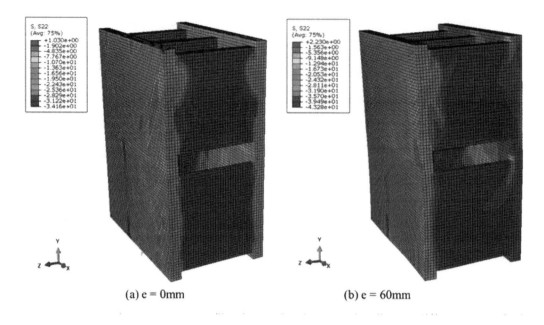

(a) e = 0mm   (b) e = 60mm

Figure 8. Stress profile of unrendered mortared wallette at different eccentricities.

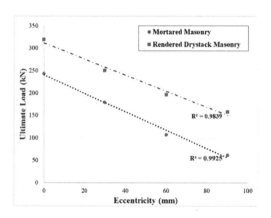

Figure 9. Variation of ultimate load with increase in eccentricity.

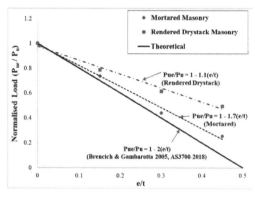

Figure 10. Relationship between normalised load and eccentricity ratio.

Koltsida et al. 2014; Dhanasekar et al. 2017) and Australian Masonry Standards (AS3700-2018) predictions for hollow block mortared masonry. It seems that the drystack masonry walls rendered on both sides can respond to much larger eccentricity in the loading and hence designing rendered dry stack wall as a viable structural member need not be discouraged as currently done in the Australian Masonry Standards (AS3700, 2018).

4   CONCLUSIONS

3D finite element analyses of the rendered drystack wallettes under concentric and eccentric compressive loads have been presented in this paper. The modelling method was first validated with the results reported in Dhanasekar et al. (2008) for FRMC rendered drystack wallette. The results of mortar-auxetic fabric composite (MAFC) rendered wallette have been compared with the FRMC rendered drystack wallettes subjected to concentric load. The mortar-auxetic composite rendered drystack wallettes have also been analysed under eccentric compression and comparisons were made with conventional mortared masonry walls. Following main conclusions have emerged from this investigation.

- The mortar-auxetic composite render improves the failure pattern of the wallette by reducing the delamination potential which is the main drawback of the FRMC rendered drystack wallettes.
- Under eccentric compression, mortar-auxetic fabric rendered wallettes resistance was better than mortared masonry with 30% higher loads. Thus, the drystack wallettes rendered with mortar-auxetic fabric composite appear to resist more effectively under eccentric compression.
- An eccentric load reduction expression of $\frac{P_{ue}}{P_u} = 1.0 - 1.1\left(\frac{e}{t}\right)$ was obtained from the analysis of rendered drystack masonry wallettes subject to eccentric compression. This expression shows the reduction is quite modest compared to the reduction of mortared masonry (without render) reported in this paper $\frac{P_{ue}}{P_u} = 1.0 - 1.7\left(\frac{e}{t}\right)$ and the literature $\frac{P_{ue}}{P_u} = 1.0 - 2.0\left(\frac{e}{t}\right)$. It appears that the drystack masonry walls rendered on both sides can respond to much larger eccentricity in loading and hence designing rendered dry stack wall as a viable structural member could be included in the design standards.

ACKNOWLEDGEMENT

The authors thankfully acknowledge the technical support provided by the high-performance computing facility at Queensland University of Technology, Australia. Testing was carried out at CQ University Australia, Rockhampton.

REFERENCES

Alotaibi, K.H. & Galal, K. 2018. Experimental study of CFRP-confined reinforced concrete masonry columns tested under concentric and eccentric loading. *Composites Part B* 155:257–271.

Anand, K.B. & Ramamurthy, K. 2005. Development and performance evaluation of hollow concrete interlocking block masonry system, *TMS Journal*, 11–20.

Asad, M., Dhanasekar, M., Zahra, T. & Thambiratnam, D. 2018. Mitigating impact failure of masonry boundary walls using auxetic composites. *In Proceedings of 10th Australasian Masonry Conference AMC, 11-14 February 2018*. Sydney: Australia.

Asad, M., Dhanasekar, M., Zahra, T. & Thambiratnam, D. 2019. Characterisation of polymer cement mortar composites containing carbon fibre or auxetic fabric overlays and inserts under flexure. *Construction and Building Materials* 224: 863–879.

Dhanasekar, M., Song, M. & Ring, J. 2017. Response of reinforced concrete masonry walls to eccentric compression. *In proceedings of 13th Canadian Masonry Symposium, 4-7 June 2017*. Halifax: Canada.

Dhanasekar, R., Ferozkhan, M., Dhanasekar, M. & Holt, W. 2008. Behaviour of dry stack concrete masonry blocks under eccentric compression, *In proceedings of the 14th International Brick & Block Masonry Conference, 17-20 February 2008*. Sydney: Australia.

Fortes, E.S., Parsekian, G.A., Fonseca, F.S. & Camacho, J.S. 2018. High-Strength Concrete Masonry Walls under Concentric and Eccentric Loadings. *Journal of Structural Engineering* 144(6): 04018055.

Huang, L., Gao, C., Yan, L., Li, X., Ma, G. & Wang, W. 2017. Experimental and Analytical Modeling of GFRP Strengthened Grouted Mortarless Masonry Prisms. *Fibers* 5(18). DOI:10.3390/fib5020018.

Koltsida, S.I., Tomor, A.K. & Booth, C.A. 2014. Insights of experimental studies on the load carrying capacity of masonry under eccentric loading. *Masonry International* 27(3): 49–54.

Qamar, F., Thomas, T. & Ali, M. 2018. Use of natural fibrous plaster for improving the out of plane lateral resistance of mortarless interlocked masonry walling. *Construction and Building Materials* 174: 320–329.

Thamboo, J.A. & Dhanasekar, M. 2015. Characterisation of thin layer polymer cement mortared concrete masonry bond. *Construction and Building Materials* 82: 71–80.

Thamboo, J.A., Zahra, T. & Dhanasekar, R. 2020. Development of design methodology for mortarless masonry system: Case study – a resettlement housing colony. *Journal of Building Engineering* 27: 100973.

Thanoon, W.A., Jaafar, M.S., Noorzaei, J., Abdulkadir, M. R. & Fares, S. 2007. Structural Behaviour of Mortar-Less Interlocking Masonry System Under Eccentric Compressive Loads, *Advances in Structural Engineering*, 10(1): 11–24.

Zahra, T. and Dhanasekar, M. 2016. A generalised damage model for masonry under compression. *International Journal of Damage Mechanics* 25(5): 629–660.

Zahra, T. and Dhanasekar, M. 2017. Characterisation of cementitious polymer mortar – Auxetic foam composites. *Construction and Building Materials* 147: 143–159.

Zahra, T. and Dhanasekar, M. 2018. Characterisation and strategies for mitigation of the contact surface unevenness in dry-stack masonry. *Construction and Building Materials* 169: 612–628.

Wang, J., Wan, C., Zeng, Q., Shen, L., Malik, M.A. & Yan, D. 2020. Effect of eccentricity on retrofitting efficiency of basalt textile reinforced concrete on partially damaged masonry columns. *Composite Structures* 232:111585.

Brick and Block Masonry - From Historical to Sustainable Masonry –
Kubica, Kwiecień & Bednarz (eds)
© *2020 Taylor & Francis Group, London, ISBN 978-0-367-56586-2*

Load-bearing capacity of slender earth masonry walls under compression

M. Brinkmann & C.-A. Graubner

Institute of Concrete and Masonry Structures, Technische Universität Darmstadt, Germany

ABSTRACT: Previous compression tests have shown that earth masonry has a significantly lower Young's modulus than common masonry made of fired clay, calcium silicate or autoclave aerated concrete units. As this effect favours stability failure, this paper examines whether the design equation for slender masonry walls according to EN 1996-3 can be applied to the design of slender earth masonry walls. For this purpose, a numerical model of earth masonry is created and calibrated on the basis of component tests. Subsequently, its slender-dependent load-bearing capacity under eccentric compression is examined and compared with the load-bearing capacity of common masonry materials and the design capacity according to EN 1996-3. Afterwards, the applicability of the existing normative regulations for the design of slender masonry to the design of earth masonry is evaluated.

1 INTRODUCTION

Earth masonry is a traditional building material, which has been used all over the world for thousands of years. In Europe, it lost importance due to industrialization and has almost completely been replaced by industrially produced masonry materials. However, the use of earth masonry offers several advantages. Regarding the ongoing climate change, using earthen building materials can make a contribution to the protection of the environment, given the fact that their production generates less carbon dioxide and additionally needs less energy than the production of concrete, steel or other kinds of masonry. Furthermore, earth is completely recyclable and can easily be reused for the production of new earth units or earth mortar. Apart of the environmental advantages, earth masonry is particularly suitable for residential construction due to its positive influence on the indoor climate, its appealing architectural appearance and its high availability all over the world.

Unfortunately, the load-bearing capacity of earth masonry has not yet been sufficiently researched for its widespread application in the European construction industry. To replace the outdated design regulations of earth masonry and simplify its use, it would be useful if the existing standard EN 1996-3 was applicable to its design. This paper focuses on the global load-bearing capacity of slender earth masonry walls. The slender-dependent load-bearing capacity of earth masonry is examined numerically and compared with that of conventional masonry made of fired clay, calcium silicate and autoclave aerated concrete units. Subsequently, the numerical results are contrasted with the design equation for slender masonry walls according to EN 1996-3 in order to evaluate its applicability to the design of load-bearing earth masonry.

The cause for the following investigations is the low bending stiffness of earth masonry in comparison to common masonry materials. Compression tests on four different combinations of earth units and earth mortar conducted by (Müller et al. 2017) at the Federal Institute for Materials Research and Testing have shown that the ratio between secant Young's modulus at 33% of compressive strength $E_{0.33}$ and characteristic compression strength f_k of earth masonry is approximately $E_{0.33}/f_k \approx 350 - 500$. In contrast, EN 1996-1-1 generally recommends a significantly higher ratio $E_{0.33}/f_k$ of 1000 for the design of load-bearing masonry walls. The ratios $E_{0.33}/f_k$ of conventional masonry materials, which are provided by the German National Annex of EN 1996-1-1, as well as the experimentally determined values of earth masonry are summarized in Table 1.

Due to their lower bending stiffness, the transverse deformations of slender earth masonry walls under out-of-plane bending loads are larger than that of common masonry. Regarding second order effects, additional bending stresses can occur, which in turn reduce their load-bearing capacity and favour stability failure.

As the design equation for slender masonry walls according to EN 1996-3 (1) provides no possibility to take the Young's modulus into account, it is questionable if this equation is suitable for the design of earth masonry walls.

$$\phi_s = 0.85 - 0.0011 \cdot (h_{ef}/t)^2 \qquad (1)$$

In order to examine whether equation (1) can be applied to the design of slender earth masonry, a finite element model is created and calibrated on the basis of compression test results, taking into

Table 1. $E_{0.33}/f_k$ of masonry materials according to the German National Annex of EN 1996-1-1 and experimental results.

| Material | $E_{0.33}/f_k$ | |
	Range	Calculative value
Aggregate concrete	2050 – 2700	2400
Fired clay	950 – 1250	1100
Calcium silicate	800 – 1250	950
Lightweight concrete	800 – 1100	950
Autoclave aerated concrete	500 – 650	550
Earth masonry*	≈ 350 – 500	

* experimental results

account its nonlinear material behaviour. Afterwards the slender-dependent load-bearing capacity of earth masonry under eccentric compression is examined.

To compare the load-bearing capacity of earth masonry with that of common masonry, slender walls made of fired clay, calcium silicate and autoclave aerated concrete are also modelled numerically. Finally, a comparison between the various numerical results and the design equation for slender masonry walls according to EN 1996-3 (1) is carried out in order to investigate whether the existing regulations are applicable to the design of slender earth masonry walls.

2   NUMERICAL MODELLING

2.1   *General*

The numerical model is intended for the investigation of the global load-bearing capacity of slender masonry walls under compression and out-of-plane bending stresses. As the investigation does not focus on local failure or cracking, the approach of a simplified micro modelling is pursued, which means that every unit is modelled mutually with half of its surrounding mortar joints as one solid block. It is assumed that the earth units have standard format (240 x 115 x 71 mm³) and the bed and head joints have a thickness of 10 mm each. Every block of unit and mortar is assigned the smeared earth masonry properties under compression, whereby the simulation is generally based on mean values.

In order to take the discreteness of masonry into account, interface elements are inserted in the bed joints, which consider the stress transmission between every layer of blocks and determine the flexural strength of the masonry wall. Due to the investigated uniaxial stresses, which are invariable in longitudinally direction, the head joints do not influence the global load-bearing capacity in the following examinations (Brinkmann 2019). For that

reason, there is no need to insert interface elements into the head joints of the numerical model.

Line supports are attached at the top and bottom of the model, which both allow free rotations around the x axis and prevent displacements in y direction. Hence, the buckling length of the model equals its total height. In order to avoid singularities at the load application points, load introduction plates are attached at top and bottom of the model. The material properties of these plates are based on steel, which is why they are assigned linear-elastic material behaviour with a Young's modulus of $E_p = 210000$ N/mm². The plates have a height of 50 mm each. The geometry of the numerical model is displayed in Figure 1.

As shown in Figure 2, each masonry block is discretized by 1000 mesh elements, which in this case ensures a good convergence of the numerical results (Brinkmann 2019). Every block edge is divided equally into ten parts, so the dimensions of the mesh elements are 25 x 11.5 x 8.1 mm³ (x,y,z).

2.2   *Modelling of the earth masonry blocks*

The blocks of the model represent the properties of the composite material consisting of earth units and earth mortar. Each block is therefore modelled by using an eight-node solid element. Given the fact that no horizontal loads are applied and the difference between the Poisson's ratios of the load introduction plate and masonry is neglected, it is not necessary to consider

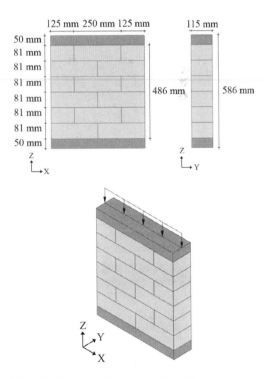

Figure 1. Geometry of the numerical model.

405

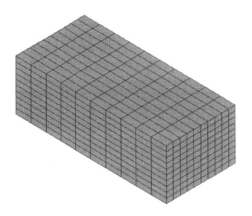

Figure 2. Discretization of a masonry block.

orthotropic material behaviour. On this account, the blocks are assigned a total strain crack model, which is well capable of simulating uniaxial stress states and can also consider a different tensile and compressive behaviour (see Figure 3).

During the numeric simulation, the total strain crack model expects smeared strains and stresses, which is why no local failure or cracking of the blocks but only the global load-bearing capacity can be examined. The compressive behaviour of earth masonry is modelled by two different approaches. The first approach is that of fib Model Code 2010, which is originally intended to represent the non-linear stress-strain relation of concrete (see equation (2)). The same approach is also adopted in EN 1992-1-1.

$$\frac{\sigma}{f} = \begin{cases} \frac{k \cdot \eta - \eta^2}{1 + (k-2) \cdot \eta} & \text{with } \eta \leq \frac{\varepsilon_{ult}}{\varepsilon_f} \\ 0 & \text{with } \eta \leq \frac{\varepsilon_{ult}}{\varepsilon_f} \end{cases} \quad (2)$$

Where η is the ratio between the strain ε and the strain at maximum stress ε_f and k is the ratio between the initial Young's modulus E_0 and the secant Young's modulus at maximum stress E_f. Similar to EN 1992-1-1, the initial Young's modulus is approximated with $E_0 \approx 1.05 \cdot E_{0.33}$. Furthermore, the failure of the material under compression is defined by the ultimate strain ε_{ult}.

According to EN 1996, the cross-sectional load-bearing capacity of masonry walls may generally be determined with the assumption of ideal-plastic compressive behaviour, which is why this is considered as a second approach in order to get comparable numerical results. The ascending part of the stress-strain relation is therefore similar to the first approach according to fib Model Code 2010. Hence, the deformation behaviour and the corresponding second order effects are considered similar. In contrast to the approach of fib Model Code 2010, the stress however remains constant after reaching its peak without any strain limitation (see Figure 4). Due to the missing strain limitation, the load-bearing capacity converges against the ideal-plastic capacity under the condition that high cross-sectional strains are obtained. However, high strains cannot be reached by walls with large slenderness because stability failure occurs before the cross-sectional load-bearing capacity is obtained. In contrast, walls with low slenderness can reach high strains without obtaining large deformations due to their high bending stiffness. This prevents stability failure and results in an almost ideal-plastic cross-sectional load-bearing capacity, given the fact that the stress remains constant after reaching its maximum and the strains are not restricted. At higher slenderness, the bending stiffness of the model decreases which in turn rises its transverse deformations. If the transverse deformations exceed a critical value before the cross-sectional load-bearing capacity is reached, stability failure is governing instead of cross-sectional failure. In this case, only the ascending part of the stress-strain relation is relevant.

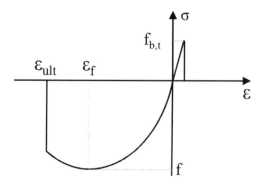

Figure 3. Stress-strain relation of earth masonry blocks with compression behaviour according to fib Model Code 2010 including an ultimate strain.

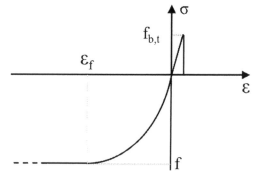

Figure 4. Stress-strain relation of earth masonry blocks with compression behaviour according to fib Model Code 2010 without an ultimate strain.

The input values to determine the stress-strain relation under compression are taken from component tests on earth masonry with standard format earth units and earth mortar M2 (Müller et al. 2017). To account for second order effects adequately, the geometrical and physical nonlinearities are always considered during the numerical simulation.

As the flexural strength of the model is defined by the interface elements in the bed joints, the blocks are assigned the estimated tensile behaviour of the earth units. Therefore, the stress-strain relation is assumed to be linear-elastic with brittle post-cracking characteristics. The tensile strength $f_{b,t}$ of earth units is approximately 10% - 20% of their compressive strength f_b (Schroeder 2019). In the following, $f_{b,t}$ is estimated by equation (3) according to fib Model Code 2010, where f_b is the experimentally determined compressive strength of earth units in standard format (Müller et al. 2017).

$$f_{b,t} = 0.3 \cdot f_b^{2/3} \quad (3)$$

The input values, which are used for defining the material properties of the earth masonry blocks are summarized in Table 2.

2.3 Modelling of the interface elements

The interface elements in the bed joints define the force transmission between each layer of masonry blocks and simulate the opening of the joints. For that reason, the interfaces are assigned the discrete crack model. The estimated relation between the relative vertical displacement of the bed joints and the tensile stress in vertical direction is displayed in Figure 5. Similar to the earth masonry blocks, the tensile behaviour is assumed to be linear-elastic with brittle post-cracking characteristics. The bed joints of masonry fail in general either if the tensile stress exceeds the tensile strength of the mortar or the bond strength between mortar and unit. Since these failure modes usually occur before the tensile strength of

Table 2. Material properties of the earth masonry blocks.

f	f_k	f_b	$f_{b,t}$	ε_f	ε_{ult}
N/mm²	N/mm²	N/mm²	N/mm²	‰	‰
3.46	3.34	4.68	0.84	5.74	6.50

$E_{0.33}$		E_0		E_f		k
N/mm²		N/mm²		N/mm²		-
1367		1435		608		2.36

the units is obtained, they determine the flexural strength of masonry. In comparison to the compressive strength of masonry, its flexural strength is significantly lower. That is why the flexural strength can only increase the global load-bearing capacity of walls with high slenderness, due to the enlarged uncracked cross-sectional area which in turn leads to a higher bending stiffness. In the following, the material-independent flexural strength of masonry $f_{x1} = 0.075$ N/mm² (Schmidt & Schubert 2004) is assumed.

In order to calibrate the stiffness of the interface elements, the linear stiffness moduli in every spatial direction have to be defined. As the vertical deformations of the bed joints under compression are already included in the smeared material behaviour of the masonry blocks, the interface elements are assumed to be almost rigid. For this purpose, the vertical stiffness modulus is set to $k_z = 10^5$ N/mm³, which disables significant deformations. Nevertheless, as soon as the tensile strength of the interface elements is exceeded, they open up and simulate a cracked bed joint.

In the following investigations, the transmission of shearing forces in the bed joints is not relevant as there are no horizontal loads applied to the model. For that reason, the shear stiffness moduli in x and y direction are set as well to a high value ($k_x = k_y = 10^5$ N/mm³) in order to create an almost rigid shear connection between the layers of masonry blocks.

The joint failure under compression does not have to be considered because the interaction between the mortar and the units is already taken into account at the smeared material characteristics of the masonry blocks. The properties to define the material behaviour of the interfaces in the bed joints are summarized in Table 3.

Table 3. Material properties of the interface elements in the bed joints.

f_{x1}	k_x	k_y	k_z
N/mm²	N/mm³	N/mm³	N/mm³
0.075	100000	100000	100000

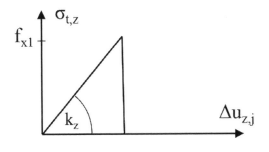

Figure 5. Relation between vertical relative displacement and normal tensile stress of the interface elements in the bed joints.

2.4 Verifying the compression behaviour of the numerical model

To verify the compression behaviour of the numerical model, in Figure 6 the simulated stress-strain relation of fib Model Code 2010 with an ultimate strain is compared with the experimental test results of masonry made of standard format earth units and earth mortar M2 under centric compression, which were conducted by (Müller et al. 2017).

The numerically determined stress-strain relation is completely enveloped by the experimental results in the ascending part of the curve and adequately fits their nonlinearity. After reaching maximum stress, the experimental results scatter widely due to the different specimen. Hence, the simulated stress-strain relation can just approximate the average course of the experimental curves in the area of post-cracking. Nevertheless, the total strain crack model sufficiently fits with the experimental results under centric compression, which is why it is assumed to be suitable for further examinations.

3 LOAD-BEARING CAPACITY OF SLENDER EARTH MASONRY WALLS

3.1 Numerical and analytical analysis of the slender-dependent load-bearing capacity of earth masonry

To investigate the slender-dependent load-bearing capacity of earth masonry, several eccentric compression loads are applied to the numerical model and the related capacity reduction factor Φ_s, which is defined by equation (4), is determined.

$$\phi_s = \frac{N_R}{f \cdot t \cdot l} \quad (4)$$

Where N_R is the vertical resistance, f the mean value of the compressive strength, t the thickness and l the length of the numerical model. The results of the

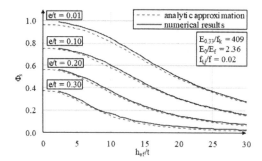

Figure 7. Slender-dependent load-bearing capacity of earth masonry under eccentric compression loads.

examination are displayed in Figure 7. Each presented load-bearing curve consists of 41 linear connected values, which are generated by various numerical models with gradually increasing slenderness. The model with the lowest slenderness consists of two layers of masonry blocks.

The curves presented in Figure 7 show the typical reduction of the global load-bearing capacity of masonry with increasing slenderness. In the beginning, the load-bearing capacity is almost unaffected by second order theory but depends on the ratio between the eccentricity e and the thickness of the model t. The higher the ratio e/t, the smaller the load-bearing capacity due to the increased bending stresses. As the curves proceed, the capacity reduction factor decreases nonlinear, which can be explained by second order effects. To validate the numerical results, the slender-dependent capacity reduction factor is also approximated analytically, using the approach by (Glock 2004) for the determination of the load-bearing capacity of unreinforced masonry walls under compression. The numerical and the analytical curves converge well, which is why the numerical simulation is assumed to approximate the load-bearing capacity of earth masonry sufficiently and can be used for a comparison with the design equation of EN 1996-3 (1).

3.2 Determination of the relevant eccentricity for the comparison with EN 1996-3

The applicability of EN 1996-3 is limited by certain restrictions, as e.g. the wall height, the span of the floors or the imposed load. These restrictions offer the advantage, that the design of EN 1996-3 is simplified and the eccentricities at the design locations do not have to be explicitly calculated but are already included in the provided design equations. Nevertheless, most of the common buildings made of load-bearing masonry can be designed by using EN 1996-3. In order to create comparable numerical results, the eccentricity which is included in equation (1) has to be elaborated analytically. According to EN 1996-1-1 the cross-sectional capacity reduction

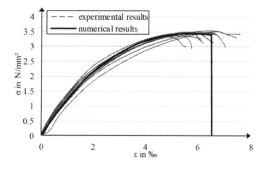

Figure 6. Comparison of the experimental and numerical stress-strain relation of earth masonry under centric compression.

factor is defined by equation (5) with the approach of ideal-plastic material behaviour.

$$\phi = 1 - 2 \cdot \frac{e}{t} \qquad (5)$$

If a slenderness of $h_{ef}/t = 0$ is assumed, the capacity reduction factor according to equation (1) results in $\Phi_s = 0.85$. This factor represents the cross-sectional load-bearing capacity, because it does not consider any slender-dependent reduction of the load-bearing capacity. On this account, $\Phi_s = 0.85$ can be inserted in equation (5) to resolve the corresponding ratio between eccentricity and wall thickness e/t:

$$\frac{e}{t} = \frac{1 - \Phi}{2} = \frac{1 - 0.85}{2} = 0.075 \qquad (6)$$

For further investigations, it is assumed that $e = 0.075 \cdot t$ is the eccentricity on which the design equation for slender masonry according to EN 1996-3 is based. Provided that the normative application conditions are observed, this value globally considers typical bending loads due to slab rotation, wind loads or initial eccentricities. In the following, the compression loads as well as the supports are applied to the model with regard to this eccentricity in order to generate numerical results, which are comparable to the design equation of EN 1996-3.

3.3 Comparison between the load-bearing capacity of earth masonry and EN 1996-3

In Figure 8, the numerically determined slender-dependent load-bearing capacity of earth masonry is compared with the design equation for slender masonry according to EN 1996-3. If a minimum wall height of 200 cm and a maximum wall thickness of 40 cm is assumed, the lowest relevant slenderness of load-bearing masonry walls in residential construction is $h_{ef}/t \approx 5$. Taking into account the application conditions according to the German National Annex of EN 1996-3, the maximum wall height of 275 cm in combination with the minimum wall thickness of 15 cm limit the relevant slenderness to a maximum of $h_{ef}/t \approx 18$. In the following figures, the area which is usually not relevant in practice is highlighted with a grey background.

The numerically generated curve with the approach of fib Model Code 2010 including an ultimate strain starts below the design capacity of EN 1996-3. For $h_{ef}/t < 23$, EN 1996-3 provides a continuously higher load-bearing capacity than the simulation of earth masonry. At $h_{ef}/t \approx 23$, the numerically determined load-bearing capacity equals the design capacity according to EN 1996-3. After this point the capacity reduction factor of earth masonry is bigger than the normative reduction factor.

The approach of fib Model Code 2010 without an ultimate strain starts with a good approximation of the design value according to EN 1996-3. Due to the ideal-plastic cross-sectional load-bearing capacity, this approach slightly extends the load-bearing capacity of earth masonry for $h_{ef}/t < 8$. After this point, the maximum stress cannot be obtained anymore and stability failure is governing. In this case, the load-bearing capacity is determined by the ascending part of the stress-strain relation which is why the reduction factors of both approaches are equal.

The ratio of the numerical results and the design equation according to EN 1996-3 is displayed in Figure 9. The comparison yields, that EN 1996-3 overestimates the load-bearing capacity of earth masonry at $h_{ef}/t < 23$. The numerical results decrease down to $\approx 80\%$ of the normative design values at certain slenderness. Only walls with $h_{ef}/t > 23$ provide higher load-bearing capacities than EN 1996-3. As load-bearing masonry walls, which are used in residential construction usually have a slenderness between $h_{ef}/t \approx 5 - 18$, it seems to be unlikely that the existing equation (1) can be applied to the design of earth masonry without modification.

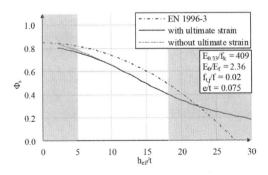

Figure 8. Comparison between the load-bearing capacity of earth masonry and the design capacity according to EN 1996-3.

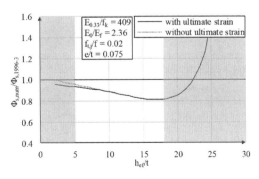

Figure 9. Ratio of the load-bearing capacity of earth masonry and the design capacity according to EN 1996-3.

3.4 Comparison between the load-bearing capacity of earth masonry and other masonry materials

In addition to the comparison with the design equation according to EN 1996-3, in the following the slender-dependent load-bearing capacity of earth masonry is as well compared to that of other masonry made of fired clay, calcium silicate and autoclave aerated concrete units. Since masonry made of autoclave aerated concrete units may be designed by EN 1996-3, although it has a comparatively low Young's modulus, it is investigated whether its global load-bearing capacity differs significantly from that of earth masonry.

During the numerical modelling of the different masonry materials, the boundary conditions outlined above, e.g. the geometry and the tensile behaviour of the model or the calibration of the interfaces in the bed joints, remain unmodified. Only the stress-strain relation under compression is adjusted, which is displayed in Figure 10.

The assumed Young's moduli of masonry made of fired clay, calcium silicate and autoclave aerated concrete units are based on the suggested calculative values in Table 1. According to EN 1052-1, the ratio of the mean value and the characteristic value of compressive strength is approximated with $f/f_k = 1.2$. The ascending part of the stress-strain relation of fired clay and autoclave aerated concrete is estimated to be linear elastic while that of calcium silicate is approximated with a quadratic parabola (Meyer & Schubert 1992). After reaching their compressive strength, the stress-strain relations remain constant without any ultimate strain in order to converge against the normatively permitted ideal-plastic cross-sectional load-bearing capacity at small slenderness.

The numerically determined slender-dependent load-bearing capacities of the several unit materials are displayed in Figure 11. The comparison yields that the global load-bearing capacity of earth masonry descends significantly faster than that of common masonry. Even masonry made of autoclave aerated concrete units provides a higher

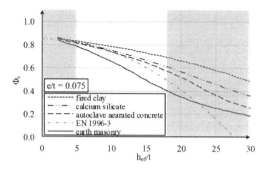

Figure 11. Comparison between the load-bearing capacities of different masonry materials.

capacity reduction factor than earth masonry at any slenderness.

In Figure 12 the ratio of the numerical and the normative load-bearing capacities is presented. In contrast to earth masonry, conventional masonry always provides a ratio $\Phi_{s,num}/\Phi_{s,1996-3} \geq 1.0$, which implies that their load-bearing capacity is continuously higher than the design capacity according to EN 1996-3. This indicates that the current design regulations offer a safe design of conventional masonry.

Although the Young's modulus of autoclave aerated concrete is significantly lower than the recommended value according to EN 1996-1-1, it provides a higher load-bearing capacity than EN 1996-3. As earth masonry shows an up to 30% smaller capacity reduction factor than masonry made of autoclave aerated concrete units at certain slenderness, the lower Young's modulus of earth obviously has a decisive influence on its load-bearing capacity, which cannot be neglected in design.

4 CONCLUSIONS

The presented numerical examinations point out that slender earth masonry provides a lower global load-

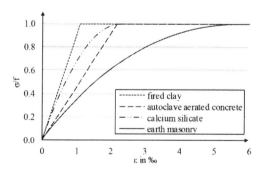

Figure 10. Assumed stress-strain relations under compression of the compared masonry materials.

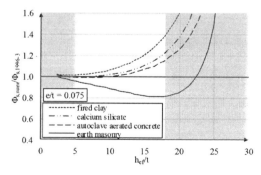

Figure 12. Ratio of the load-bearing capacity of different masonry materials and the design capacity according to EN 1996-3.

bearing capacity than conventional masonry made of fired clay, calcium silicate and autoclave aerated concrete units. Moreover, the design capacity of slender masonry walls according to EN 1996-3 mostly exceeds the load-bearing capacity of earth masonry.

The reason for the lower load-bearing capacity of earth masonry is its low Young's modulus which leads to a reduced bending stiffness, enlarged transverse deformations and, hence, increased capacity reduction due to second order effects. Even autoclave aerated concrete, which has the lowest stiffness of the normative approved masonry materials, provides a continuously higher load-bearing capacity than earth masonry and EN 1996-3. For those reasons, it is questionable whether the existing design regulations according to EN 1996-3 are applicable to the design of slender earth masonry walls without modification.

However, some of the assumptions have to be verified by further research in order to finally assess the applicability of EN 1996-3 to the design of slender earth masonry walls. It must be ensured, whether the assumed stress-strain relation and Young's modulus of earth masonry is a proper approximation for arbitrary combinations of earth units and earth mortar. Furthermore, the numerical results have to be validated by additional experimental compression tests, examining eccentric loads in combination with varying slenderness.

Moreover, the interaction between earth building materials and moisture has not been taken into account yet. Previous research has shown that moisture influences the compressive strength of earth masonry (Heath et al. 2009, Steingass 2002). It is conceivable that moisture affects the stiffness of earth masonry walls, too. In order to investigate the influence of moisture on the material behaviour of earth masonry in depth, further tests will be carried out.

Additionally, the creep behaviour of earth masonry has to be investigated since creep deformations can also decrease the load-bearing capacity of slender walls. To restrict the influence of creep, the application conditions of EN 1996-3 limit the final creep coefficient to $\Phi_\infty \leq 2.0$. Whether earth masonry can meet this restriction is examined in future.

If further investigations confirm that the design equation for slender masonry walls according to EN 1996-3 cannot be applied to the design of load-bearing earth masonry, a new equation has to be developed, taking into account the characteristics of earth masonry sufficiently and guaranteeing a safe design.

ACKNOWLEDGEMENTS

The authors would like to thank the German Federal Environmental Foundation for funding the research project "Development of design regulations for earth masonry on the basis of EN 1996-3 by experimental and numerical investigations" which is conducted in cooperation with the Federal Institute for Materials Research and Testing and ZRS Ingenieure.

REFERENCES

Brinkmann, M. 2019. Tragverhalten unbewehrten Lehmmauerwerks unter Druckbeanspruchung. Master Thesis, Technische Universität Darmstadt.

EN 1052-1:1998-12. Methods of test for masonry – Part 1: Determination of compressive strength. Brussels: European Committee for Standardization (CEN).

EN 1992-1-1:2011-01. Design of concrete structures – Part 1-1: General rules for buildings. Brussels: European Committee for Standardization (CEN).

EN 1996-1-1:2013-02. Design of masonry structures – Part 1: General rules for reinforced and unreinforced masonry structures. Brussels: European Committee for Standardization (CEN).

EN 1996-1-1/NA:2019-12. Nationally determined parameters – Eurocode 6: Design of masonry structures – Part 1: General rules for reinforced and unreinforced masonry structures. Brussels: European Committee for Standardization (CEN).

EN 1996-3:2010-12. Design of masonry structures – Part 3: Simplified calculation methods for unreinforced masonry structures. Brussels: European Committee for Standardization (CEN).

Glock, C. 2005. Traglast unbewehrter Beton- und Mauerwerkswände: Nichtlineares Berechnungsmodell und konsistentes Bemessungskonzept für schlanke Wände unter Druckbeanspruchung. Dissertation, Technische Universität Darmstadt.

Heath, A.; Lawrence, M.; Walker, P. 2009. The compressive strength of modern earth masonry. *11th International Conference on the Study and Conservation of Earthen Architecture Heritage in Lima, Peru.*

International Federation for Structural Concrete 2010. fib Model Code for Concrete Structures 2010. Berlin: Ernst & Sohn.

Meyer, U.; Schubert, P. 1992. Spannungs-Dehnungs-Linien von Mauerwerk. *Mauerwerk-Kalender 1992: 615-622.* Berlin: Ernst & Sohn.

Müller, P.; Miccoli, L.; Fontana, P. & Ziegert, C. 2017. Development of partial safety factors for earth block masonry. *Material and Structures 50.* Dordrecht: Springer Netherlands.

Schmidt, U. & Schubert, P. 2004. Festigkeitseigenschaften von Mauerwerk – Teil 2: Biegezugfestigkeit. *Mauerwerk-Kalender 2004: 31-63.* Berlin: Ernst & Sohn.

Schroeder, H. 2019. Lehmbau – Mit Lehm ökologisch planen und bauen. Wiesbaden: Springer Vieweg.

Steingass, P. 2002. Moderner Lehmbau 2002 – Internationale Beiträge zum modernen Lehmbau. Stuttgart: Fraunhofer IRB Verlag.

Brick and Block Masonry - From Historical to Sustainable Masonry –
Kubica, Kwiecień & Bednarz (eds)
© 2020 Taylor & Francis Group, London, ISBN 978-0-367-56586-2

In-plane shear strength characteristics of masonry walls with varying mortar types and aspect ratios

T. Aoki, K.C. Shrestha, U. Nonaka & H. Aoki

Graduate School of Design and Architecture, Nagoya City University, Nagoya, Japan

ABSTRACT: This paper presents an experimental campaign to understand the in-plane shear characteristics of clay-unit masonry walls subjected to quasi-static horizontal cyclic loading. The test matrix involves a total of 9 full-scale masonry walls with two major parameters studied, first three different mortar types and second three different aspect ratios. Cement mortar used in modern masonry construction to lime mortar used in historical masonry structures are utilized. Further, the specimens cover aspect ratios of 0.63, 0.95 and 1.93. Walls with cement mortar showed flexural rocking failure mode and the ones with lime mortar showed diagonal shear cracking. The observed failure modes were irrespective of the different aspect ratios. The peak horizontal shear loads recorded for modern construction walls were more than 40% of the historical walls. Historical walls with lime mortar with extensive diagonal shear cracking showed higher energy absorption and equivalent damping compared to their modern counterparts.

1 INTRODUCTION

Unreinforced masonry (URM) involve historical constructions as well as modern structures. A common methodology includes clay units arranged to form walls joined together using mortar. The used mortar generally involves cement mortar in modern construction practices and lime mortar with no cement in case of historical constructions. With relatively weak mortar compared to the stones and bricks used in such historical constructions, and further absence of reinforcing bars, these structures pose high risk of damage during earthquake occurrences (Modena et al. 2010). Further, URM walls are also extensively used as partition walls or outer walls within the reinforced concrete columns and beams. These walls are also vulnerable under earthquake excitations (Beyer et al. 2015).

Several researches on URM walls have been extensively done to understand their seismic vulnerabilities under earthquake loading (Magenes & Calvi 1992, Magenes & Calvi 1997, Abrams 1992, Tomazevic 1990, Tomazevic 1999, Shrestha et al. 2011a, Shrestha et al. 2011b, Shrestha et al. 2013, Petry & Beyer 2014, Beyer et al. 2015). Majority of previous works involve study on a single mortar type or solitary aspect ratio for tested walls. Rarely, works have focused on influence of both the mortar type as well as aspect ratio in a single series of tests.

The present study focuses on understanding the strength characteristics and failure modes of masonry walls through full-scale experimentations with the parametric investigation involving walls made of different mortar types and also the walls located at different locations within a building, reflected through different aspect ratios. In addition to understanding the strength characteristics, the observations made in this experimental program will also make one identify the potential weak regions in URM which shall facilitate in development of proper intervention techniques and strengthening solutions.

2 TEST PROGRAM

2.1 Test specimens

Nine URM full-scale brick walls were built with the test matrix involving two different parameters, first the three different mortar types and second three different aspect ratios. The details on the test specimens are summarized in Table 1. The schematic drawing for each specimen type is illustrated in Figure 1. The adopted mortar types comprise: (i) cement mortar (CM) with cement and sand at 1:3, (ii) lime/cement mortar (LCM) with lime, cement and sand at 2:1:5, and (iii) lime mortar (LM) with lime and sand at 1:5. Here, LM (with no cement) and LCM mortars represent the mortar mix proportions of historical masonry structures in Japan constructed until the early 20th century. CM represents the mortar used in modern masonry construction. Further, the specimens cover aspect ratios of 0.63, 0.95 and 1.93.

2.2 Material properties

The specimens were constructed with solid clay bricks of nominal size 210 mm × 100 mm × 60 mm with

Table 1. Test specimens' matrix.

Specimen ID	Mortar type*	Aspect ratio (H/B)*	Vertical load (kN)
CM-TypeA	CM	1.93	164 (σ_v =1.2 MPa)*
LCM-TypeA	LCM	1.93	164 (1.2 MPa)
LM-TypeA	LM	1.93	164 (1.2 MPa)
CM-TypeB	CM	0.95	332 (1.2 MPa)
LCM-TypeB	LCM	0.95	332 (1.2 MPa)
LM-TypeB	LM	0.95	332 (1.2 MPa)
CM-TypeC	CM	0.63	164 (0.4 MPa)
LCM-TypeC	LCM	0.63	164 (0.4 MPa)
LM-TypeC	LM	0.63	164 (0.4 MPa)

* CM is cement mortar, LCM is lime/cement mortar, LM is lime mortar, H is wall's height, B is wall's width, σ_v is vertical stress.

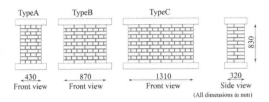

Figure 1. Test specimen details.

average compressive strength of 7.8 MPa, and elastic modulus of 17.9 MPa (JIS R 1250 2011). Eighteen cylindrical mortar samples of 50 mm diameter and 100 mm height were tested (JIS A 1108 2018) with 6 sample number for each mortar type. Table 2 summaries the results of the material characteristic tests for cylindrical mortar samples and brick units. CM mortar understandably has highest compressive strength with LM mortar having very low strength characteristics.

Furthermore, prism tests to understand compressive characteristics of the masonry assemblage were also done. Three samples were tested for each mortar type. Table 3 lists out the test results for the masonry prism

Table 2. Summary of material characterization tests.

Specimen type	f_c (MPa) *		f_t (MPa) *		E_m (GPa) *	
	Mean	Dev.	Mean	Dev.	Mean	Dev.
CM	23.1	3.90	3.70	0.72	21.20	0.60
LCM	11.0	1.50	1.50	0.05	10.80	0.31
LM	0.70	0.10	0.13	0.01	2.90	0.45
Brick	7.80	0.40	-	-	17.97	2.3

* f_c is compressive strength, f_t is tensile strength, E_m is elastic modulus.

Table 3. Summary of masonry prism tests.

Specimen type	ρ_b (kg/m³) *	f_{cp} (MPa)		E_m (GPa)	
		Mean	Dev.	Mean	Dev.
CM-P	1900	8.30	1.30	9.74	2.70
LCM-P	1866	8.10	1.20	10.55	3.41
LM-P	1800	7.30	1.10	0.35	0.05

* ρ_b is bulk modulus of masonry assemblage.

tests. The maximum compressive strength of the masonry assemblages from the prism tests is governed by crushing of bricks, hence little difference was seen among the three specimen types. The elastic modulus for LM specimen showed comparatively lower value.

2.3 Test setup and instrumentation

A series of nine wall tests were performed to investigate the effect of three different mortar types and three different aspect ratios. The walls were tested using the test setup shown in Figure 2. The walls were fixed against rotation at both their bottom and top. Vertical force was applied to the specimens using vertical hydraulic jack (500 kN capacity) in force controlled mechanism. Vertical stress (σ_v) of 1.2 MPa was applied to TypeA and TypeB specimens and 0.4 MPa to TypeC specimens as listed in Table 1. The vertical stress (σ_v) of 1.2 MPa represents the average normal stress acting on an internal wall at the first floor of a typical 4-storey reference masonry building. 0.4 MPa average normal stress would be typical for third floor internal wall of the same reference building. It should be noted that the lower level of applied vertical stresses in TypeC specimens of 0.4 MPa were also governed by the limitation in the capacity of the horizontal hydraulic jack.

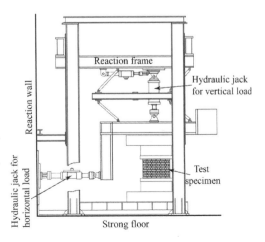

Figure 2. Test setup for in-plane shear loading.

413

After applying the corresponding vertical load, a displacement controlled cyclic lateral load was applied using horizontal hydraulic jack (300 kN capacity). A typical cyclic loading history adopted for the test program is shown in Figure 3. The first 22 loading cycles involved two repeated cycles for drifts of 0.05%, 0.1%, 0.2%, 0.3%, 0.4%, 0.5%, 0.6%, 0.7%, 0.8%, 0.9% and 1.0% respectively. From the 23rd cycle onwards, a single cyclic loading for drifts of 1.1%, 1.2%, 1.4%, 1.6%, 1.8% and 2.0% were applied. It should be noted that for CM and LCM walls of Type B and Type C, the first 22 loading cycles as shown in Figure 3 could not be executed either due to sudden rocking failure or slippage along the joint.

The test instrumentations involved devices including load cells attached to the actuators to measure the forces and a set of laser sensors and horizontal linear variable differential transducers (LVDTs) to measure the displacements as illustrated in Figure 4. The shear and vertical deformations of the wall were measured by 4 LVTDs (Sensors 1 to 4) along the diagonal and vertical direction of specimens respectively. The horizontal deformations in the wall were recorded using 4 laser sensors (Sensors 5 to 8) at the push and pull directions and at the top and bottom of the wall. The drift-force hysteresis plotted in the later sections utilizes inter-storey drift computed as the ratio of difference in measurements in sensors 5–6 and difference in height of the sensors 5 and 6.

3 TEST RESULTS

The test results with the drift-force hysteresis and failure mode pictures are show in Figures 5-7. In general, two different failure mechanisms were distinct, first the flexural mode where rocking response of the wall was observed and second the diagonal shear cracking mechanism dominated by shear cracking.

3.1 *TypeA specimens (Aspect ratio – 1.93)*

As shown in Figure 5, the predominant failure mode observed for CM and LCM walls were flexural rocking with cracks at the bottom joint of the walls. For CM-TypeA a single bed joint at the base of the wall was observed. LCM-TypeA wall showed crack initiation from the bed joint a brick unit above the wall's base which later extended towards the base of the wall as shown in Figure 5. LM-TypeA wall on the other hand showed diagonal shear cracking. The cracks were both along the mortar joint as well as through the brick units as illustrated in Figure 5.

CM and LCM walls, with their typical flexural rocking response, showed moderate/narrow hysteresis with lower energy dissipation. The peak lateral force was 80 kN, same for both CM-TypeA and LCM-TypeA walls. Further, there was negligible strength degradation in subsequent increasing drift cycles. LM-TypeA wall, dominated by shear cracking, showed

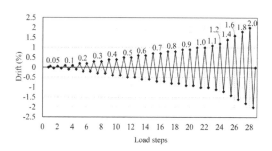

Figure 3. Loading history adopted for the in-plane shear loading.

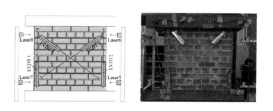

Figure 4. Instrumentations for displacement measurements.

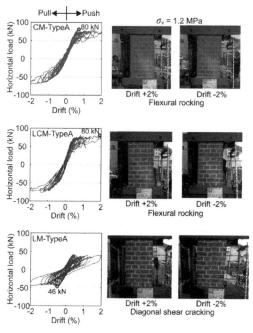

Figure 5. Drift-horizontal force hysteresis and failure modes for CM-TypeA, LCM-TypeA and LM-TypeA specimens.

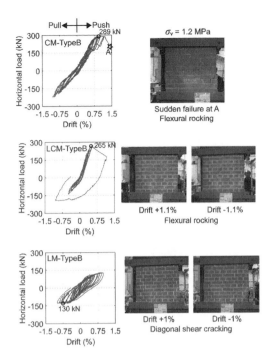

Figure 6. Drift-horizontal force hysteresis and failure modes for CM-TypeB, LCM-TypeB and LM-TypeB specimens.

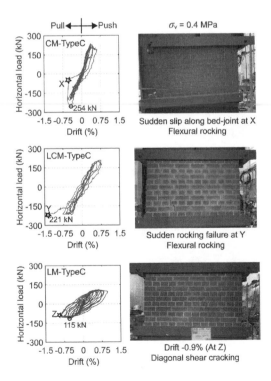

Figure 7. Drift-horizontal force hysteresis and failure modes for CM-TypeC, LCM-TypeC and LM-TypeC specimens.

larger hysteresis and energy dissipation. The first crack for LM-TypeA wall was within the mortar joint observed at 0.4% drift at about 90% of the peak lateral force. Next, at 0.5% drift, the crack extended through the brick units, and the peak lateral force (46 kN) was observed at this load step. The post peak response for LM-TypeA was characterized by gradual degradation of the lateral strength and higher energy dissipation. In contrary, CM-TypeA and LCM-TypeA walls' post peak response did not show degradation of the lateral strength.

3.2 TypeB specimens (Aspect ratio – 0.95)

Figure 6 shows the hysteresis and the failure mechanisms for the TypeB specimens. Both CM-TypeB and LCM-TypeB walls showed rocking response, rocking about the base of the wall. LM-TypeB wall showed extensive diagonal shear cracks with cracks along the mortar as well as within the brick units.

Both the CM and LCM walls showed narrow hysteresis curves due to their rocking response. The peak horizontal force was 289 kN at 0.9% drift cycle for CM-TypeB and 265 kN at 0.5% drift cycle for LCM-TypeB walls. CM-TypeB wall showed first crack at 0.3% drift cycle and a sudden drop in strength beyond the 1% drift cycle, with extensive flexural cracks along the bed joint of the wall. LCM-TypeB wall also showed first crack at 0.3% drift cycle and a sudden drop in strength and extensive crack along the bed joint at 1.1% drift cycle. For LM-TypeB wall, first shear crack along the mortar joint was observed at 0.4% drift cycle, with extension of crack diagonally for subsequent increment in drift. The peak lateral force for LM-TypeB wall was 130 kN at 0.8% drift cycle.

3.3 TypeC specimens (Aspect ratio – 0.63)

For TypeC walls, both CM and LCM walls showed flexural failure modes with rocking about the bed joint at the base of the walls as shown in Figure 7. LM-TypeC wall showed clear diagonal shear cracking confirming a pure shear failure mechanism.

The force-drift hysteresis for CM-TypeC and LCM-TypeC was relatively narrow with low energy dissipation characteristic. The peak lateral force was 254 kN at 0.6% drift cycle for CM-TypeC wall and 221 kN at 1.4% drift for LCM-TypeC wall. CM-TypeC wall showed extensive bed-joint slip along the base of the wall at 0.7% drift cycle. For LCM-TypeC wall, there was a sudden rocking failure with extensive crack extending from 6th brick unit level towards the bottom of the wall. For LM-TypeC wall, the peak lateral force was 115 kN at 0.6% drift cycle. The first shear crack appeared at 0.3% drift and the cracks extended diagonally over the height of the wall with subsequent increment in the drift cycle. Crack both in the mortar as well as within the brik units were observed for LM-TypeC wall.

4 DISCUSSIONS

4.1 Energy absorption, equivalent damping ratio and secant stiffness

Plots for several performance indicators namely energy dissipation (W_a), equivalent damping ratio (ξ_e) and secant stiffness (K_s) of each test specimen are presented in Figures 8-10. The performance of each wall type is evaluated based on these parameters.

The energy dissipation for each hysteresis loop (W_a) is calculated as the area enclosed by a hysteresis loop in one cycle of loading. It is understandable that the energy absorption capacity for all the specimens increased with the subsequent increment in loading cycles. Furthermore, for LM walls, the energy absorption is almost same or more than CM and LCM walls, even with having lower lateral loading capacity. The primary reason is due to wider hysteresis loop for LM walls and narrower in case of CM and LCM walls. It should be noted that there is an apparent increment in energy dissipation in LM walls after diagonal cracking.

The equivalent damping ratio (ξ_e) is calculated as a function of the dissipated energy (W_a) and the elastic energy (W_e) at the peak displacement for the particular loading cycle (Magenes & Calvi 1997), given by $\xi_e = W_a/2\pi(W_e^+ + W_e^-)$, where, $W_e^+ + W_e^-$ represent the elastic energy at the peak displacement for push and pull directions respectively. It is clear that the equivalent damping for CM and LCM walls are relatively lower, in the range below 10%, where failure mode is largely dominated by flexural rocking.

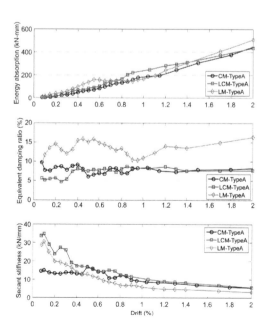

Figure 8. Energy dissipation, equivalent damping and secant stiffness plotted against drift for CM-TypeA, LCM-TypeA and LM-TypeA specimens.

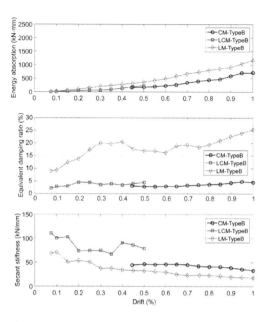

Figure 9. Energy dissipation, equivalent damping and secant stiffness plotted against drift for CM-TypeB, LCM-TypeB and LM-TypeB specimens.

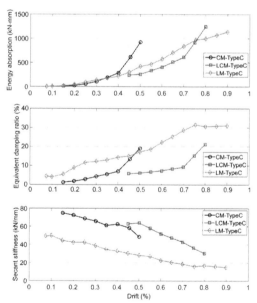

Figure 10. Energy dissipation, equivalent damping and secant stiffness plotted against drift for CM-TypeC, LCM-TypeC and LM-TypeC specimens.

There is also no notable increment in damping ratio with the increment in the drift cycles. LM walls on the other hand showed relatively higher equivalent damping close to more than 20%. Furthermore, with the increment in the drift cycles and subsequent damage to the wall, the equivalent damping ratio also increases.

There is gradual decrement in the secant stiffness (K_s) for all walls tested. The secant stiffness for TypeA walls was the least, for obvious reasons for walls with highest aspect ratio 1.93. The trend of decrement in secant stiffness was higher for LM walls with subsequent reduction of lateral strength at higher drift levels. TypeB walls have higher secant stiffness than TypeC walls primarily due to difference in the applied vertical stress.

4.2 Effect of different mortar types and aspect ratio

CM mortar represents mortar used in modern construction with high strength characteristics and LCM mortar also has moderately high strength characteristics. LM mortar on the other hand signifies historical construction with very low strength characteristics. The lateral load carrying capacity for CM and LCM walls were very similar with about 5% variance in average. For apparent reasons, LM walls showed comparatively lower lateral strength characteristics. The peak horizontal shear load for LM wall was about 56%, 47% and 42% of the CM and LCM walls for TypeA, TypeB and TypeC walls respectively.

The test results on all 9 walls as shown in Figures 5-7 indicated that the failure modes observed were primarily governed by the mortar types used in the walls. Walls with different aspect ratios and same mortar type, on the other hand, showed no change in the failure mode. For CM and LCM walls, the failure mode was flexural rocking and LM walls showed diagonal shear cracking failure. The failure modes observed for each mortar type governed the performance indicators as well. The LM walls with extensive diagonal shear cracking showed higher energy absorption and equivalent damping. CM and LCM wall counterparts, with narrow drift-force hysteresis showed comparatively lower energy absorption and damping.

4.3 Lateral strength comparison

The maximum lateral strengths observed during the tests are compared with the analytical formulations (Magenes & Calvi 1997). Here, two different formulations are utilized, one flexural rocking failure mode and the other for diagonal shear cracking failure mode.

The maximum horizontal shear for flexural rocking failure mode of wall is given by,

$$V_r = \frac{Bt\,\sigma_v}{\alpha_v}\frac{1}{2}\left(1 - \frac{\sigma_v}{Kf_c}\right) \qquad (1)$$

where V_r is the rocking shear resistance in N, B is the wall length in mm, t is the wall thickness in mm,

Table 4. Comparison between experimental and theoretical values for lateral strength.

Specimen ID	Experimental, E (kN)	Theoretical, T (kN)	E/T	Eqn.
CM-TypeA	80	71	1.1	(1)
LCM-TypeA	80	70	1.1	(1)
LM-TypeA	46	69	0.6	(2)
CM-TypeB	289	290	0.9	(1)
LCM-TypeB	265	289	0.9	(1)
LM-TypeB	130	140	0.9	(2)
CM-TypeC	254	250	1.1	(1)
LCM-TypeC	221	249	0.9	(1)
LM-TypeC	115	74	1.4	(2)

parameter $(\alpha_v = H_0/D)$, $(H_0 = \psi' H)$ is the effective height of wall, ψ' is the parameter whose value is 1 for cantilever wall and 0.5 for both ends fix wall, H is the height of the wall in mm, σ_v is the vertical normal stress in N/mm^2, f_{cp} is the masonry compressive strength in N/mm^2, $\kappa = 0.85$ is a coefficient that takes into account the vertical stress distribution at the compressed toe.

The maximum horizontal shear for diagonal shear cracking failure mode of wall is given by,

$$V_d = Dt\ \min\left(\frac{1.5c + \mu\sigma_v}{1 + 3c\frac{\alpha_v}{\sigma_v}}, \frac{c + \mu\sigma_v}{1 + \alpha_v}\right) \qquad (2)$$

where V_d is the diagonal cracking shear resistance in N, c is the mortar joint cohesion in N/mm^2 and μ is the joint coefficient of friction angle.

Table 4 shows the comparison between the experimentally observed and theoretically computed horizontal load carrying capacities for all the wall specimens. For CM and LCM walls, where flexural rocking failure mode was observed, Equation 1 was used to compute the maximum horizontal shear. Equation 2 was utilized to calculate the lateral shear strength for diagonal shear cracking in LM walls. Here, c (0.038 MPa) and μ (0.83) were assumed based on Lumantarna et al. 2012, joint cohesion, $c = 0.055$ f_c, where f_c is the mortar compressive strength in N/mm^2. The adopted formulations predicted the lateral strength reasonably well for the flexural rocking walls, but with some deviance for LM walls, possibly due to higher variability of joint strength characteristics.

5 CONCLUSIONS

In an attempt to understand in-plane shear resistance of clay-brick masonry walls with different mortar types and aspect ratios, a series of tests were performed. Three different mortar types (from modern cement mortar to historical lime mortar) and three different aspect ratios (H/B from 0.63 to 1.93) were investigated. The experimentally observed lateral load strengths were also compared with simple theoretical formulations. Based on the finding, the following conclusions can be drawn.

- CM and LCM walls with higher strength characteristics showed flexural rocking failure mode. LM walls with low strength mortar showed diagonal shear cracking in the wall. Interestingly, the failure modes observed for the walls were irrespective of the different aspect ratios and were completely governed by the mortar types.
- The peak in-plane lateral load carrying capacity were very close for CM and LCM walls (only about 5% variance in average). LM walls showed comparatively lower lateral strength characteristics. The peak horizontal shear load for LM wall was about 56%, 47% and 42% of their respective counterparts of CM and LCM for TypeA, TypeB and TypeC walls.
- A clear trend was observed where, the LM walls with extensive diagonal shear cracking showed higher energy absorption and equivalent damping. CM and LCM wall counterparts, with narrow drift-force hysteresis showed comparatively lower energy absorption and damping. The average damping ratio for LM walls were generally more than 15% after the initiation of cracks, whereas CM and LCM walls showed damping ratio less than 10% in average.
- The theoretical formulations satisfactorily predicted the in-plane lateral load carrying capacities for all the tested walls with an average ratio of experimental over theoretical value, E/T of 0.9.

ACKNOWLEDGEMENTS

This work was supported by JSPS KAKENHI Grand Number 16H06363. We acknowledge the assistance of the Research Equipment Sharing Center at the Nagoya City University.

REFERENCES

Abrams, D.P. 1992. Strength and behavior of unreinforced masonry elements. In *Proc. of the 10th World Conference on Earthquake Engineering, Madrid, 1992.*

Beyer, K. et al. 2015. Dynamic testing of a four-storey building with reinforced concrete and unreinforced masonry walls: prediction, test results and data set. *Bulletin of Earthquake Engineering* 13(10): 3015–3064.

JISA1108. 2018. Method of test for compressive strength of concrete. Japanese Industrial Standard. Tokyo: Japan Standards Association.

JISA1250. 2011. Common bricks and facing bricks. Japan Industrial Standard. Tokyo: Japan Standards Association.

Lumantarna, R. et al. 2012. Compressive, flexural bond, and shear bond strengths of in situ New Zealand unreinforced clay brick masonry constructed using lime mortar between the 1880s and 1940s. *ASCE Journal of Materials in Civil Engineering* 26(4): 559–566.

Magenes, G. & Calvi, M. 1997. In-plane seismic response of brick masonry walls. *Earthquake Engineering and Structural Dynamics* 26: 1091–1112.

Magenes, G. & Calvi, M. 1992. Cyclic behavior of brick masonry walls. In *Proc. of the 10th World Conference on Earthquake Engineering, Madrid.*

Modena, C. et al. 2010. L'Aquila 6th April 2009 earthquake: Emergency and post-emergency activities on cultural heritage buildings. In M. Garevski & A. Ansal (eds), *Earthquake Engineering in Europe*: 495–521. New York: Springer.

Petry, S. & Beyer, K. 2014. Influence of boundary conditions and size effect on the drift capacity of URM walls. *Engineering Structures* 65: 76–88.

Shrestha, K.C. et al. 2011a. Finite element study on pinning retrofitting technique of masonry walls with opening subjected to in-plane shear load. *ACEE Journal* 4(4): 81–96.

Shrestha, K.C. et al. 2011b. Finite element modeling of cyclic out-of-plane response of masonry walls retrofitted by inserting inclined stainless steel bars. *Journal of Disaster Research* 6(1): 36–43.

Shrestha, K.C. et al. 2013. Effectiveness of superelastic bars for seismic rehabilitation of clay-unit masonry walls. *Earthquake Engineering and Structural Dynamics* 42(5): 725–741.

Tomazevic, M. 1990. Masonry structures in seismic areas – a state-of-the-art report. In *Proc. of the 9th European Conference on Earthquake Engineering, Moscow.*

Tomazevic, M. 1999. *Earthquake-resistant design of masonry buildings.* London: Imperial College Press.

Brick and Block Masonry - From Historical to Sustainable Masonry –
Kubica, Kwiecień & Bednarz (eds)
© 2020 Taylor & Francis Group, London, ISBN 978-0-367-56586-2

Shear bond strength of manufactured thin stones considering various adhesion methods

S. Rizaee
Civil Engineering Department, University of Calgary, Calgary, Alberta, Canada

M.D. Hagel
Alberta Masonry Council, Calgary, Alberta, Canada

N.G. Shrive
Civil Engineering Department, University of Calgary, Calgary, Alberta, Canada

ABSTRACT: The strength of the shear bond between manufactured thin stone units and Type S mortar substrates with various surface textures was investigated to assess how and if the type of substrate texture affects the bond strength. Three different types of substrate were considered: mortar block and scratch coat, mortar block with metal lath and scratch coat, and wood block with metal lath and mortar scratch coat. The scratch coat surface was either smoothed, left unfinished, brush scratched, or scratched with a V-Notch trowel. Preblended Type S (1:0.5:4.5) and polymer modified mortar were used as the setting-bed. Analysis of the results demonstrated that the mortar block with encased metal lath and mortar scratch coat achieved the highest strength especially when combined with polymer modified mortar. The substrate surface texture did not affect the strength very much. Finally, the polymer modified mortar achieved higher bond strengths than Type S mortar.

1 INTRODUCTION

In modern Canadian construction, manufactured thin stone (MTS) units are being used as veneers on the exterior of buildings from single-storey residential to multi-storey commercial. Each stone unit is individually adhered to a backup wall through setting-bed mortar. An adhered MTS veneer consists of thin stone units, setting-bed mortar, a substrate, a weather resistant barrier (optional) and a back-up wall (MVMA 2016). As the height of the adhered masonry veneer increases the risk of damage to property or injury if the bond fails increases. Therefore, the reliance on the mortar bond to keep the MTS unit attached to the support structure increases. Since no specific regulations or standards for the design and construction of MTS systems are available, in some cases the mortar has not provided enough bond between the MTS units and the structural backing to prevent the units from detaching and falling off. This is especially true when the wall is subjected to high wind loads.

Types N and S mortars (CSA-A179-14 2014) with average compressive strengths of 3.5 and 8.5 MPa (Hatzinkiolas & Korany 2005) are commonly used in MTS veneer applications. The mix proportions are (1:1:6) for Type N and (1:0.5:4.5) for Type S (cement: lime: sand by volume). These mortars are traditionally designed to transfer compressive loads in the vertical direction and to transfer the lateral loads to the lateral load resisting system. However, when used as setting-bed mortar in MTS applications, they are subject to shear, flexure and tensile loads. Therefore, using these two types of mortar for MTS applications has resulted in some bond failures (Garagliano 2014, Hagel et al. 2017). Some examples of failures observed in practice are shown in Figure 1.

Two types of bond failure can occur. One is cohesive failure which occurs through only one material; here that could be through the stone, the substrate, the scratch coat or the setting-bed mortar. The second is adhesive failure which occurs through an interface; the mortar interface with the stone or with the scratch coat as well as at the metal lath interface if it is used. Various types of failure have been observed both in the field and in the laboratory (Hagel et al. 2017). Whenever investigating bond failure in masonry, the type of failure (adhesive or cohesive) should always be considered (Sugo, 2000).

Failures occur when the applied loads exceed the bond strength. Bond strength can be compromised by factors that affect the formation and durability of the bond. These factors include unit, mortar and substrate properties together with workmanship, curing and exposure conditions. Unit properties more specifically include unit surface texture, water absorption properties, moisture content (Lawrence & Page 1994) and modulus of rupture (Sise 1984). Mortar properties

Figure 1. MTS bond failures in practice.

include mortar type, flow, water retentivity and air content (Lawrence & Page 1994). Curing and exposure conditions include curing at and exposure to extreme hot and cold temperature and exposure to extensive freezing and thawing cycles (Litvan, 1980, Al-Omari et al. 2015).

Shear bond strength in adhered MTS veneer applications has received little attention (e.g.: Garagliano 2014, Hagel et al. 2017). Both studies report that polymer modified mortars showed improved performance in adhered MTS veneer applications. Polymer modified mortar was initially invented for tile and then used for MTS veneer applications. The mortar is not commonly used in adhered MTS veneer because the standards allow the use of traditional types of mortar and Hi-Bond mortars are more expensive. In a previous study (Hagel et al. 2017) Hi-Bond setting-bed mortar was compared with Type N and Type S mortars. The results showed that Hi-Bond mortar consistently exceeded the minimum required bond strength of 0.35 MPa required by ASTM C1780-16 (2016) and strength values were 7-10 times greater than with Type N and S while the latter two types

had difficulty achieving 0.35 MPa. These findings are in agreement with those of Gradliano (2014).

In the work reported here mortar type, and variation in substrate type and surface texture were examined. Unit properties complied with "Standard Specifications for Adhered Manufactured Stone Masonry Veneer Units" C1670/C1670M-15 (ASTM 2016).

2 TEST SAMPLES AND TEST SETUP

Test samples were constructed using three substrates, four scratch coat surface textures, two setting-bed mortars and one type of MTS. The substrates include Type S mortar block and scratch coat (S), Type S mortar block with encased metal lath and scratch coat (SL), and wood block with encased metal lath and Type S mortar scratch coat (WSL). The surface of the scratch coat was textured by smoothing it with a trowel (SMT), providing no finish and leaving it rough (RGH), broom/brush scratching the surface (BRM) or texturing the surface with a V-Notch (V-N), as shown in Figure 2. A pre-blended Type S mortar (S) and polymer modified cement mortar (Hi-Bond), complying with ANSI A108/A118 (ANSI 2015), were used as the setting-bed mortars. All test parameters are summarized in Table 1.

The tests have been conducted at an ISO/IEC 17025:2005 accredited laboratory. Mortar blocks for substrate were cast using Type S mortar in $150 \times 110 \times 50$ mm moulds and cured for a minimum of 7 days in an ambient laboratory condition, at 21 ℃ and 30% relative humidity. Although "Standard Practice for Installation Methods for Adhered Manufactured Stone Masonry Veneer", ASTM C1780-16a (ASTM 2016) recommends curing at 70-77° F (21-25 ℃) and 45-55% relative humidity, the relative humidity in this laboratory is slightly lower. The same size wood blocks were cut, then a piece of metal lath was nailed to the block between the mortar scratch coat

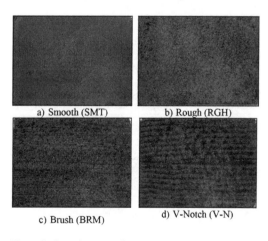

a) Smooth (SMT)   b) Rough (RGH)

c) Brush (BRM)   d) V-Notch (V-N)

Figure 2. Scratch coat surface textures.

Table 1. Test parameters.

Substrate	Setting-bed mortar	Scratch coat texture
Type S mortar block with scratch coat (S)	Type S (S)	Smooth (SMT)
Type S mortar block with encased metal lath and scratch coat (SL)		Rough (RGH)
Wood block with encased metal lath and Type S scratch coat (WSL)	Polymer modified cement mortar (HiB)	Brush (BRM)
		V-Notch (V-N)

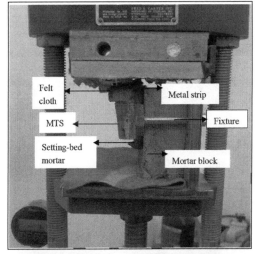

Figure 4. Current Shear bond strength test setup picture and sketch; modified version of ASTM C482-02 (2014).

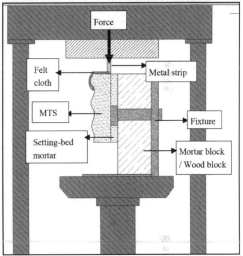

and the wood. Type S mortar was then used to create a scratch coat 24 hours before MTS unit attachment. Stone units were then cut into approximate dimensions of 75 to 100 × 100 to 120 mm and attached to the substrate using Type S or Hi-Bond mortar. Three samples of each combination were cast and tested.

Although a specific standard for shear bond strength testing of adhered MTS units is not available, three are available for shear bond strength testing in ceramic tile applications, Standard Test Method for Bond Strength of Ceramic Tile to Portland Cement in ASTM C482-02 (2014), Figure 3, MR 14 Determination of the bond of renderings by shear tests and MR 20 Determination of the bond strength of renderings by torsion test (RILEM TC 1994). We used a modified version of ASTM C482-02 according to ASTM C1670/C1670M-15 (2015), Figure 4 (picture and sketch). The actual test normally requires a metal plate to be used as the loading fixture, however, in these tests the loading fixture was replaced with a piece of felt cloth, to distribute point loads and a metal strip approximately 25 × 200 × 6 mm. The metal strip was positioned on the edge of the stone closest to the bedding mortar to localize the shear load and minimize the introduction of

flexural moment. The loading rate is recommended as 0.9 +/- 0.09 kN/min, but since load was applied here using a manual hydraulic press, conformance to the required loading rate was difficult to achieve (Hagel et al. 2017).

3 RESULTS

Possible failure surfaces include failure through the scratch coat and the setting-bed mortar (SC/M), the stone and the setting-bed mortar (M/S), through the scratch coat (SC) and at the metal lath (Lath) interface, see Figure 5.

A summary of the failure types of the various samples considering different combinations is shown in Figure 6. Failures through the SC/M interface occurred most frequently followed by failure through the metal

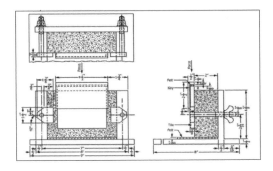

Figure 3. Shear bond strength test setup for ceramic tile to Portland cement paste, ASTM C482-02 (2014).

a) Failure through the scratch coat and setting-bed mortar interface (SC/M)

b) Failure through the stone and setting-bed mortar interface (S/M)

c) Failure in the scratch coat (SC)

d) At the metal lath interface (with wood block)

e) At the metal lath interface (without Mortar block)

Figure 5. Examples of observed failure surfaces.

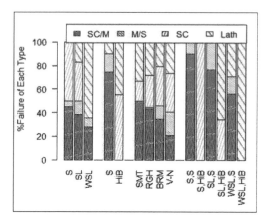

Figure 6. Percentage of failure of each type.

lath, scratch coat (SC) and finally through the S/M interface.

Considering the different substrates, in the case of Type S substrate, more than 80% of failures occurred through the scratch coat (SC) or through the scratch coat and mortar interface (SC/M). When metal lath is used with Type S mortar block and scratch coat (SL), all four possible failure modes occurred, but failure through the SC/M interface was still the most frequent. In the case of WSL substrate more than 60% of the

failures occurred at the metal lath interface showing that the weakest plane is mostly where the metal lath is connected to the wood block (Figure 5d): no failures through the scratch coat were recorded.

Considering the mortar types, Type S mortar resulted mostly in failures through the M/S interface, while when Hi-Bond (HiB) mortar was used failures only occurred in SC or through the metal lath interface, meaning that Type S mortar block/substrate is weaker than the setting-bed mortar and the weak plane is where there is some discontinuity due to the use of metal lath (Figure 5e).

Considering the four different types of scratch coat surfaces, one can say that smooth surface failures mostly occur at the SC/M interface showing that increasing the roughness of the surface can reduce this type of failure. With V-Notch surfaces the bonding between the substrate and the setting-bed mortar is improved and the least number of failures occurred at this interface.

When combinations of substrates and setting-bed mortars are considered, in cases where both are Type S more than 90% of the time the weakest plane was the SC/M interface. When the Hi-Bond mortar was used, 100% of failures occur in SC meaning that the Type S mortar block was the weakest plane. When metal lath is added to the mortar block, with Type S setting-bed mortars, failures at the SC/M interface decrease to approximately 75% and the rest happen at the M/S interface. However, in the case of Hi-Bond mortar failures get distributed between in the lath interface and in the SC, showing that the setting-bed mortar has strong adhesion to both the stone and the substrate. When wood block substrate is used with Type S mortar most of the failures still occur at the SC/M interface, while metal lath and M/S interfaces show weakness as well. However, with Hi-Bond mortar the metal lath to wood block becomes the weakest plane and all the failures occur at this plane (Figure 5d).

As with many other types of masonry applications and tests, there is large variation in the bond strength results. Therefore, comparing the average shear strength values is not the best way to make conclusions. One alternative to compare the results is to use box plots. In a box plot the distribution of the results is shown as medians, quartiles and possible outliers, as shown in Figure 7. To make graphical presentation and comparison easier, a naming system is used in which the substrate types are named from 1 to 3, surface textures are named from 1 to 4 and types of mortar are named 1 and 2. The detail of the naming system is presented in Table 2.

The shear bond strength values achieved using two types of setting-bed mortar, Type S and Polymer modified (Hi-Bond), are presented in Figure 8. According to the box plots, one can see that Hi-Bond mortar improves the bond strength and the values mostly exceed the minimum 0.35 MPa (50 psi) required by ASTM- C1670/C1670M (ASTM 2016).

Considering the types of scratch coat surfaces, one cannot make specific conclusions on which

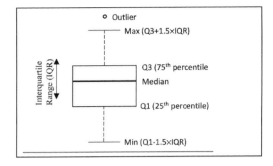

Figure 7. Box plot components and definitions.

Table 2. Various sample combinations.

Substrate (label)	Setting-bed mortar (label)	Surface texture (label)
Type S Mortar (1)	Type S (11)	Smooth (111)
		Rough (112)
		Brush (113)
		V-Notch (114)
	Hi-Bond (12)	Smooth (121)
		Rough (122)
		Brush (123)
		V-Notch (124)
Type S/Metal Lath (2)	Type S (21)	Smooth (211)
		Rough (212)
		Brush (213)
		V-Notch (214)
	Hi-Bond (22)	Smooth (221)
		Rough (222)
		Brush (223)
		V-Notch (224)
Wood Block/Type S Mortar/Metal Lath (3)	Type S (31)	Smooth (311)
		Rough (312)
		Brush (313)
		V-Notch (314)
	Hi-Bond (32)	Smooth (321)
		Rough (322)
		Brush (323)
		V-Notch (324)

texture works best as the values are very close to each other, as shown in Figure 9.

The variation of strength when considering the three different substrate types is shown in Figure 10: Type S mortar block and scratch coat (S), Type S mortar block with metal lath and scratch coat (SL), and wood block with metal lath and Type S scratch coat (WSL). The results show that the SL substrate performs better than the other two. The minimum strength achieved using the SL substrate is less than when using WSL, because in the case of the SL substrate, when the setting-bed mortar was Type S the bond strength values were very low, but when the Hi-Bond mortar was used the highest values were achieved.

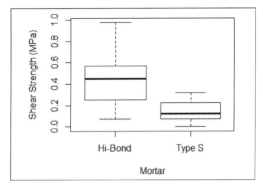

Figure 8. Shear bond strength, considering Type S and Hi-Bond mortar ($n^1 = 36$).

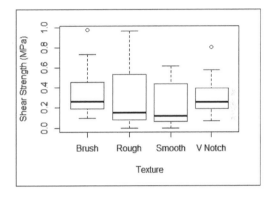

Figure 9. Shear bond strength, considering different scratch coat surfaces (n=18).

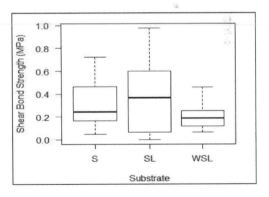

Figure 10. Shear bond strength, considering different substrates (n=24).

The strength values differ when combinations of three different substrates and two different types of mortar are considered, as shown in Figure 11. In samples with the same type of substrate the ones with Hi-

---

[1] n is the number of samples in each group

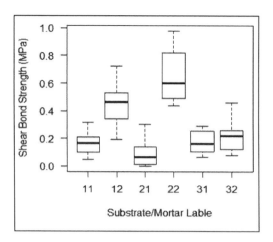

Figure 11. Shear bond strength, considering different substrate/mortar combinations (n =12).

Bond setting-bed mortar achieve higher strengths. The difference is more pronounced when the substrate is the Type S mortar block with metal lath (SL). While the Hi-Bond mortar shows the greatest values compared to the other groups, the Type S mortar shows the lowest strength amongst all. When considering Hi-Bond mortar the second substrate (SL) is recommended to be used. Use of Type S mortar is not recommended at all for the setting bed. Additionally, when the wood backing and metal lath (WSL) is used with Hi-Bond mortar, all failures occur through the metal lath with loads lower than 0.35 MPa. Therefore, regardless of the type of setting bed mortar, wood backing with metal lath encourages failures through

the metal lath and the strengths are less than the minimum required. Use of this type of backing is therefore not recommended. The strength values with Hi-Bond are very close to what is achieved when Type S setting-bed mortar is used. Wood backing is therefore not recommended for Hi-Bond mortar. If Type S mortar has to be used as the setting bed material, Type S mortar with metal lath (SL) as the back-up should be avoided: the other back-up systems (S and WSL) are about equal, but do not meet minimum expectations.

The test results are presented more comprehensively in Figure 12. In all groups Hi-Bond mortar reaches higher strengths than Type S. The highest and lowest values of shear bond strength are achieved by Hi-Bond and Type S mortar respectively when using SL substrate. Looking at the different scratch coat textures, the same trend is only seen in samples 111-114 and 311-314. When Type S mortar is used as the setting-bed the smooth surface gives the lowest and the V-Notch or brushed surface gives the highest strength. With Type S setting-bed mortar, bond strengths are generally very low, and the type of substrate does not make a significant difference.

4 SUMMARY

The shear bond strength of manufactured thin stone to two types of setting-bed mortar (Type S and polymer modified (Hi-Bond)) when attached to three different types of substrate textured in four different forms has been assessed. The three different substrates including Type S mortar block and scratch coat (S), Type S mortar block with encased metal lath and scratch coat (SL) and wood block with

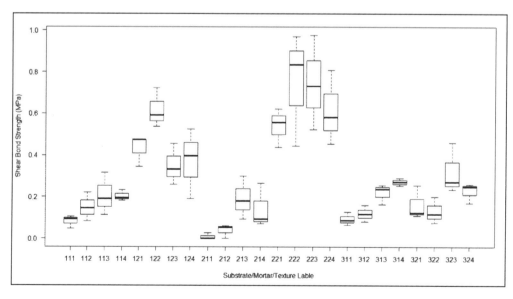

Figure 12. Shear bond strength, considering different substrate, mortar and surface texture (n=3). Notation is in Table 2.

encased metal lath and Type S mortar scratch coat (WSL). The scratch coat surface texture was varied between smooth, unfinished (rough), brushed and V-Notch. Hi-Bond setting-bed mortar and its interfaces had the strongest adhesion while Type S mortar and metal lath had the lowest. Failures tended to occur more frequently in the Type S mortar (either as substrate or scratch coat) or at the metal lath. Considering the four different types of scratch coat surfaces, with smooth surfaces the failures mostly occurred at the SC/M interface showing that increasing the roughness can reduce this type of failure. When the V-Notch surface was used, the bond between the substrate and setting-bed mortar was improved and the least number of failures occurred through this interface. The results show that the SL substrate with Hi-Bond mortar performs the best. However, polymer mortar with the wood block and metal lath (WSL) back up should not be used. With Type S setting-bed mortar the shear bond strength values were below the minimum expectation of 0.35 MPa, so use of this mortar as a setting bed mortar is not recommended.

ACKNOWLEDGEMENT

The authors would like to thank the MITACS, the University of Calgary and the Alberta Masonry Council for their support and Adanac Global Testing for the laboratory tests.

REFERENCES

Al-Omari, A., Beck, K., Brunetaud, X., Török, Á., Al-Mukhtar, M. 2015. Critical Degree of Saturation: A Control Factor of Freeze-Thaw Damage of Porous Limestone at Castle of Chambord, *Engineering Geology*, Volume 185: 71–80.

ANSI 2015. A118.4/A118.15 American National Standard Specifications for the Installation of Tile. *Materials and Installation Standards*. South Carolina: American National Standards Institute.

ASTM 2016. *C1670/C1670-15 Standard Specification for Adhered Manufactured Stone Masonry Veneer Units*, Conshohocken: American Standard Test Method.

ASTM 2016. *C1780-16a Standard Practice for Installation Methods for Adhered Manufactured Stone Masonry Veneer*, Conshohocken: American Standard Test Method.

ASTM 2014. *C482-02 Standard Test Method for Bond Strength of Ceramic Tile to Portland Cement Paste*. Pennsylvania: American Standard Test Method.

CSA 2014. *CSA-A179-14 Mortar and Grout for Unit Masonry*. Missisauga: Canadian Standards Association.

Garagliano, R. 2014. Evaluating Manufacturing and Installation Practices of Adhered Manufactured Stone Veneer to Optimize Product Performance and Appearance. *Masonry 2014*: 182–193.

Hagel, M., Isfeld, A. & Rizaee, S. 2017. Shear and Tensile Bond Strength of Manufactured Stone Veneer Units Individually secured by Mortar Adhesion. *13th Canadian Masonry Conference*, Halifax: Canada Masonry Design Centre.

Hatzinikolas, M. & Korany, Y. 2005. *Masonry Design for Engineers and Architects*. Edmonton: Canadian Masonry Publications.

Lawrence, S. J. & Page, A. W. 1994. *Bond Studies in Masonry*. 10th IB²MaC, Calgary, 909–918.

Litvan, G. 1980. Freeze-Thaw Durability of Porous Building Materials. *ASTM-Durability of Building Materials and Components*, Volume 691: 455–463.

MVMA 2016. *Installation Guide and Detailing Options for Compliance with ASTM C1780 for Adhered Manufactured Stone Veneer*, Herndon: Masonry Veneer Manufacturers Association.

Sise, A. 1984. *Flexural Bond Strength of Masonry*. Calgary: The University of Calgary.

Sugo, H. O. 2000. *Strength and Microstructural Characteristics of Brick/Mortar Bond*. Newcastle: The University of Newcastle, Australia.

RILEM TC 1994. MR 14 Determination of the bond of renderings by shear tests. In: RILEM, ed. RILEM *Recommendations for the Testing and Use of Constructions Materials*: 527–527.

RILEM TC 1994. MR 20 Determination of the bond strength of renderings by torsion tests, 1982. In: RILEM, ed. *RILEM Recommendations for the Testing and Use of Constructions Materials*: 534–534.

Brick and Block Masonry - From Historical to Sustainable Masonry –
Kubica, Kwiecień & Bednarz (eds)
© 2020 Taylor & Francis Group, London, ISBN 978-0-367-56586-2

Intelligence prediction system of semi-interlocking masonry panels behavior

O. Zarrin
The University of Newcastle, Australia

M. Ramezanshirazi
Department of Structural and Geotechnical Engineering, Sapienza University of Rome, Italy

ABSTRACT: A Semi-Interlocking Masonry (SIM) system has been developed by the Masonry Research Group at The University of Newcastle, Australia. This is a system of seismic resistance for framed structures with masonry panels. SIM panels have the capability of dissipating energy during earthquake excitation through the friction on sliding joints between courses of units. The main objective of this study is to investigate experimentally the out-of-plane load-displacement capacity of SIM panels, and predict its capacity using Artificial Neural Network (ANN). Attempting to introduce several ANNs to predict the maximum displacement of the SIM panels lead to find the best network. The results show that ANNs have the ability of accurate prediction the maximum displacement of SIM panels. In addition, the outcomes of feature selection algorithm help to arrange the most effective factors on maximum displacement to optimize time, safety and cost of projects.

1 INTRODUCTION

Dissipating earthquake energy is one of the important issues in structures in a seismic region. The performance of dampers leads the engineers to focus on different type of dampers such as active, semi-active and passive devices. However, issues such as cost, skill of the installer, and maintenance affect its popularity. In order to find an affordable system to dissipate the seismic energy, engineers tried to develop a new method using an inspiration the oldest method, which is the dry-stack masonry system.

Masonry is a popular material used throughout the world because of its special structure and building characteristics, as well as cost effective and availability (Zhiyu Wang, et al. 2015.). However, unreinforced masonry has some limitations when used in seismic regions due to its inherent weakness in tension and shear. The high rigidity and low ductility of masonry panels cause brittle behaviour during an earthquake excitation. Therefore, reinforcing masonry panels with different materials and methods has been used to compensate for the brittle behavior of unreinforced masonry. In this regard, framed structure combines brittle masonry panels with concrete or steel ductile frame. (Kun Lin & Hongjun Liu, 2011). Despite the popularity of framed structure, some concerning problems still remain unsolved. During an earthquake, the main damage in a masonry-framed structure occurs because of the high potential of masonry panels to

absorb the energy. Such a damage could be unsafe and costly to repair. In order to avoid or significantly minimize the damage while improving the energy dissipation, the Semi-Interlocking Masonry (SIM) panel has been developed (Totoev et al. 2011). This new mortarless masonry system consists of specially designed bricks capable of relative in-plane sliding, but restrained against out-of-plane sliding. The initial low rigidity of the panel is the key role for the semi-interlocking brick to reach its design objective (Totoev et al. 2011).

Over the last two decades, rapid urbanization has increased and improved the construction industry and standards of living. At the same time, mathematical modelling of structural systems has increased tremendously. Artificial intelligence, particularly, Artificial Neural Networks (ANNs) have been used successfully to solve several civil engineering problems. The ANN is inspired by biological neurons to facilitate the solution of many practical engineering projects (Figure 1).

The capability of the ANN depends on different parameters, such as input data, output data, topologies of the network, connection weights, and threshold limit (bias factor). The position of each factor is different and related to the duty of the parameter.

Generally, an ANN consists of three main layers: input, output and hidden layer (Qingdong 2014, Hecht-Nielsen 1987). The ability of a trained network to reproduce the results is called generalization capability (Jung-Huai Chou 2001). One of the

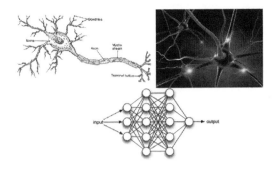

Figure 1. Essence of ANNs.

Figure 2. Topological interlocking SIM bricks (Zarrin et al. 2018).

important features of an ANN is the process of error during its learning phase. Nowadays, artificial intelligence supports all processes of the physical hardware and software to solve large problems.

The total behaviour of the network consists of five senses:

• scrutiny of the database;
• design structure;
• learning process;
• training of the network;
• test and generalization of the network.

Warren McCulloch and Walter Pitts are known as the pioneers of the ANN model. They presented the details of the ANN model and the exact formulation of the network. Donald Hebb and Pavlov followed their work and studied the traditional conditioning and properties of each neuron.

Frank Rosenblatt developed the first application of the ANN models at the end of the 1950s. Recently, many researchers have developed different algorithms (Jung-Huai Chou 2001, Wang Ji-Zong & Jin-Yun 1999, Lai 1997, Kasperkiewicz & Dubrawski 1995, Yeh 1998, Mansour et al. 2004, Ni Hong-Guang 2000, Lee 2002, Bai et al. 2003).

In this paper a full-scale test of a SIM panel with an out-of-plane loading is presented. The out-of-plane loading of the panel was exerted by a hydraulic jack. The load-displacement of the panel was recorded at regular increment. The results of the test show the high out-of-plane displacement capability of the SIM panel. The maximum displacement of the panel was approximately 175cm, which is greater than the panel's thickness. An artificial neural network (ANN) was used and was able to predict the SIM panels' out-of-plane displacement. The first step of the neural network development was its training with the input data, which was the main factors of the tested panel's behaviour. The prediction of panels' displacement shows the use ANN in masonry construction.

2 LABORATORY INVESTIGATION (EXPERIMENTAL TEST SETUP)

The panel dimensions were 1980×2025×110 mm, length × height × thickness, respectively. The panel

was mortarless and made of topological semi-interlocking bricks with dimensions of 220×110×76 mm (Figure 2). The lateral load was applied at the center of the panel by a hydraulic jack. Every 5 mm of applied displacement, the deflection and load on the panel were recorded.

3 NUMERICAL A PILOT STUDY AND ANN MODELS

In order to find the optimal network, the research presented herein assessed the limitation as well as the capability of typology, architecture and critical inner features of several neural networks - from the simplest available in MATLAB to an improved ANN that had an algorithms implemented to increase the precision and reliability of the predicted results.

The ANNs use were:

• Multi-layer Perceptron (MLP-Toolbox);
• Radial Basis Function (RBF-MATLAB code);
• Extreme Learning Machine (ELM-MATLAB code);
• MLP-Delta (MATLAB code);
• Multi-layer Perceptron Levenberg–Marquardt Sequential Forward Feature Selection" (MLP-LM-SFS feed forward MATLAB code).

In the design phase, the effect of different structure and configuration needed a significant consideration. Therefore, Pilot Training Method (PTM) was applied to find the optimal network (Ramezanshirazi et al. 2019, Ramezanshirazi et al. 2018).

The following list presented the details of variation in each network:

• different number of hidden neurons;
• different number of the epoch;
• feature selection (Sequential forward Feature Selection SFS);

To achieve the optimal performance of the network, all the models were trained and tested using the same database. The outcome of the networks were compared to the results of the experimental tests. In this research, the coefficient of determination (R^2) and the Mean Square Error (MSE) presented in the Equation 1 and 2 were used as an index for comparison between the predicted and measured results (Zarin et al. 2018).

$$R^2 = 1 - \left(\frac{\sum_i (Yi - \hat{Y}i)^2}{\sum_i (Y - \hat{Y})^2} \right) \qquad (1)$$

$$MSE = \frac{1}{n} \sum_{i=1}^{n} (\hat{Y}i - Yi)^2 \qquad (2)$$

where:

Ŷi is predicted data;

Yi is measured data.

The important factors of the experiment were summarized in Table 1 and were introduced to the networks as input parameters:

Table 1. Network input parameters for prediction of maximum displacement.

Input parameters	Minimum	Maximum
Height of Brick (mm)	73	76
The thickness of Brick (mm)	110	110
Length of Brick	220	220
Geometry Angle of Brick	25	35
Friction Angle of Brick	37.95	40
Compressive Strength (N/mm^2)	33.35	38.01
Force (KN/m^2)	0	43.67
Length of Panel (mm)	1980	2400
Height of Panel (mm)	2025	2400

4 EXPERIMENTAL RESULTS

Figure 7 illustrates the response of the mid-point of the panel. The loading of the panel continued until 174.5 mm displacement without collapsing the panel. This displacement is significantly greater than the panel thickness. Due to filling the gap between the top of the panel and frame with grout, both vertical and horizontal arching were activated. However, only the thrust force due to horizontal arching caused some brick crushing near the head joints. The vertical thrust induced stresses below the bricks compressive strength, hence no damage was observed due to vertical arching. The tested SIM panel was more flexible than traditional masonry panels tested by other researchers (Griffith 2007, Griffith 2005, Griffith2003). The panel stability at the maximum displacement indicate that the interlocking in the SIM panel could compensate for the lack of mortar and dissipate more energy without collapsing (Zarrin et al. 2018).

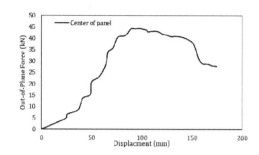

Figure 3. Maximum force-displacement (Zarrin et al. 2018).

5 ANNS RESULTS

Finding the optimum network to predict the out-of-plane displacement of the SIM panel started by using the simplest network in the MATLAB toolbox. This network was the Multi-Layer Perceptron (MLP).

By applying the PTM method that has been introduced by (Ramezanshirazi et.al.2018) for the prediction of Tunnel Boring Machine (TBM) performances, the predicted results of maximum displacement presented in Figure 4 (Ramezanshirazi & Miliziano 2018).

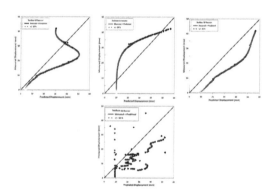

Figure 4. Predicted results of maximum displacement by MLP network-toolbox of MATLAB with different hidden neuron (10, 20, 30 and 40 neuron).

Table 2. Detaille of selected network by using PTM.

optimal Network	
Network	Multi-Layer Perceptron (*MLP*)
Number of Neurons	30
Number of epochs	2000
Type of Learning Rule	MLP-backpropagation delta
Type of design	*MATLAB*-code
Number of input node	9

These figures contain a comparison between measured displacement (experimental tests) and predicted displacement of the ANN. Each figure shows the predicted results with a different number of neurons. The trend of improvement is not completely obvious even after increasing the number of neurons. There is poor agreement and instable results of MLP network. Each dot on the figure represent a couple of data and compare the measured with predicted displacement (based on R^2 and MSE).

The results of using the RBF network is presented in Figures 5. The RBF network has poor capability to predict this type of database due to the high error of prediction of displacement.

The third network has been designed based on the Extreme Learning Machine (ELM) producer. Figures 6 illustrate the predicted results of displacement by ELM. In these figures, the improvement of results is visible. The accuracy of the network as well as the precision have been improved by increasing the number of the neuron.

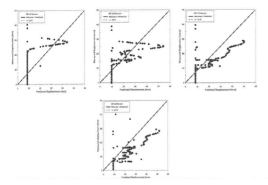

Figure 5. Prediction results of maximum displacement by RBF network with different hidden neuron (10, 20, 30 and 40 neuron).

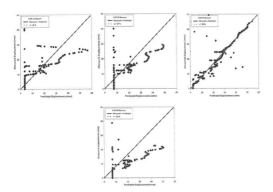

Figure 6. Prediction results of maximum displacement by ELM network with different hidden neuron (10, 20, 30 and 40 neuron).

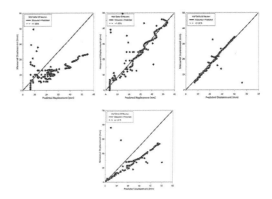

Figure 7. Prediction results of maximum displacement by MLP Delta network with different hidden neuron (10, 20, 30 and 40 neuron).

Figure 7 presents the predicted results by MLP-Delta network. In this model, two important issues are addressed. First, the high capability of back-propagation that have a significant influence to increase the network performances; second, the necessity of using optimal parameters.

The good agreement can be observed between predicted and measured displacement in the network with 30 neurons. However, by increasing the number of neurons to 40 neurons the "overtraining" problem occurred.

6   EFFECT OF CRITICAL INNER ELEMENTS

6.1   Hidden neurons

In order to understand the inner mechanism of ANNs, it is essential to investigate the behavior of all critical parameters in the hidden layers of the network. One of the effective inner parameters on ANNs' performance is a hidden neuron. Since there is not a specific method to calculate the number of neurons, the best way is empirical as shown in Figure 8.

The epoch in the learning part is based on training the iteration. For each epoch, the learning algorithm generates a different set of weights. If the network trains to 2000 epochs, the learning algorithm investigates the performance of the network 2000 times and calculate the error for each iteration. The objective of this process is to find the global minimum instead of local minima. The procedure of finding the optimum number of epochs is similar to the number of neuron's calculation. Finding the optimum network can be based on empirical method. Figure 9 depicts the process of learning and testing in the network with the back-propagation algorithm. The gradient of the error of the network in the back propagation calculated relates to the network's modifiable weights. This gradient is used in a simple stochastic gradient descent algorithm to find the weights to minimize the error.

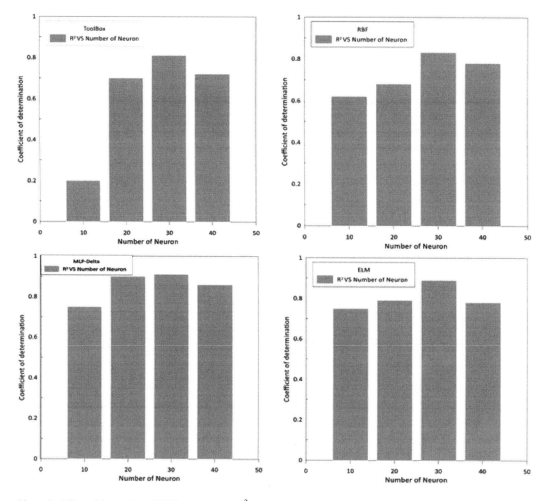

Figure 8. Effect of the number of hidden neurons on R^2.

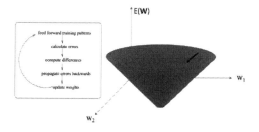

Figure 9. The process of each epoch in one network.

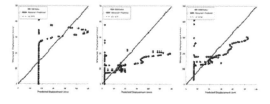

Figure 10. Improvement trends of predicted results due to an increase in the number of epoch (RBF network).

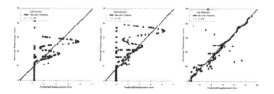

Once the network has learned, it is ready to perform the new examples. Just the MLP-Delta network (due to its ability of minimizing the error) has been tested using different number of epochs. Figures 10-12 show the effect of increasing the epochs' number on the prediction results. Contrary to the number of the neuron that has an optimum number for each network, the epoch turns to a constant error after reach-

Figure 11. Improvement trends of predicted results due to an increase in the number of epoch (ELM network).

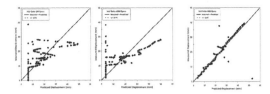

Figure 12. Improvement trends of predicted results due to an increase in the number of epoch (MLP- Delta network).

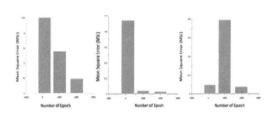

Figure 13. Effect of number of epoch on MSE.

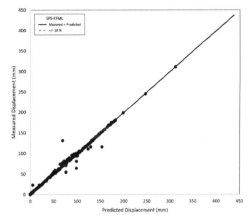

Figure 14. Prediction results of maximum displacement by MLP-LM-SFS network.

ing the maximum accuracy. By increasing the epoch number, the network takes more computing time.

Figures 13 show the relationship between MSE value and number epoch. The degradation of MSE is obvious increasing the number of epoch from 100 to 2000. This behavior leads to improve in the performance of the network to predict an accurate displacement.

7 FINAL CONFIGURATION OF THE MODEL AND FUTURE SELECTION STRATEGY

Each masonry panel has limitations in construction such as maximum load and displacement capacity. Therefore, the sensitivity parameters leads to find the maximum displacement and minimize the collapse threat and cost. The Multi-layer Perceptron with Levenberg–Marquardt and Sequential Forward Feature Selection (MLP-LM-SFS feed forward) model was implemented in MATLAB to find the most influential input parameters of the maximum displacement of the SIM panels. Figure 14 shows the improvement of predicted results and the agreement between measured and predicted displacement.

This network could improve its performance by finding the priority of input parameters and improving the capability of real-time prediction. Figure 14 shows the minimum error in finding the maximum displacement of the SIM panel.

The most influential parameters on the maximum displacement behavior of the SIM panels was summarized in Table 3.

Figure 15 shows the most effective features of the panel's displacement that the network correctly

Table 3. Feature selection results.

	Selected futures by SFS
1	Length of Panel
2	Strength of SIM
3	Height of Panel

selected. This issue can be proved by removing each feature from database. For instance, the lowest network performance was obtained when the length of the panel was eliminated from the input database.

8 MODEL VALIDATION

The validation of models is essential to evaluate the model's performance. There are several methods to validate the network such as Hold-out, K-Fold, and Random-Subsampling (RS).

The RS algorithm has been selected due to the high capacity in number of folds. Figure 16 presents the variation of the prediction performance through the RS validation on the MLP-LM-SFS network. The process is based on the calculation of correlation percentage (representative of model performance) between the measured and predicted results. For each number of test, the RS algorithm trained and tested the network 100 times. The total accuracy of the model was 86.22 % and the maximum and minimum accuracy was reached to 97.3 % and 71.06 %, respectively (Table 3).

9 CONCLUSION

A test of out-of-plane Semi-Interlocking Masonry (SIM) panel was performed on a full-scale infill panel

431

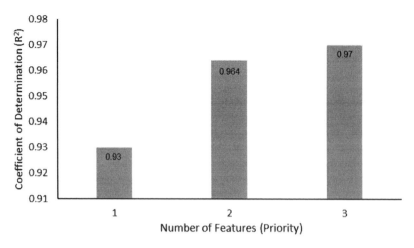

Figure 15. Validation of feature selection.

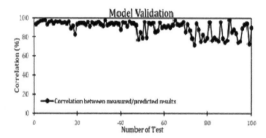

Figure 16. Performance validation of the model by applying the RS algorithm (100 iterations).

Table 4. Results of the RS method.

Total accuracy	Max accuracy	Min accuracy
86.22 (%)	97.3 (%)	71.06 (%)

subjected to the lateral load applied by a hydraulic jack. The jack was operated in displacement control. The SIM panel displayed high flexibility and high capacity for absorbing energy. The interlocking in the panel increased the panel's displacement capacity and decreased the collapse threat compared to traditional mortared masonry construction. The maximum displacement of the panel was more than 170 mm, which is greater than the panel thickness. After unloading, the panel remained steading.

In this research attempted to introduce the application of advanced Artificial Neural Networks (ANNs) to predict SIM panel displacement. The result of the particular network showed a promising capacity of out-of-plane displacement prediction. In addition, the ANNs proved a significant ability to arrange the most influential factors to optimize the experimental time of projects. The most effective factors on the panel's displacement are defined by the SFS algorithm:

- length of the panel;
- strength of SIM;
- height of the panel.

The comparison of the results of the feature selection with typical evidence from the literature showed the capability of the algorithm to select the proper features that strongly affect the relevant target. This mathematical model is a reliable way to optimize the time, cost and safety of masonry construction.

REFERENCES

Zhiyu Wang, et al. 2015. Numerical Simulation of Earthquake Response of Multi-Storey SteelFrame with SIM Infill Panels. *Advances in Structural Engineering and Mechanics (ASEM15)*.

Kun Lin, Y.T., and Hongjun Liu, 2011. In-Plane Cyclic Test on Framed Dry-Stack Masonry Panel. *Advanced Materials Research,. Vols. 163-167*: 167.

Totoev, Y., K. Lin, and A. Page. 2011. Numerical modeling of framed dry stack masonry panels. in *11th North American Masonry Conference*.

Qingdong Wu, B.Y., Chao Zhang, Lu Wang, Guobao Ning, and B. Yu, 2014. Displacement Prediction of Tunnel Surrounding Rock: A Comparison of Support Vector Machine and Artificial Neural Network. *Hindawi Publishing Corporation Mathematical Problems in Engineering Volume, 6*.

Hecht-Nielsen R., 1987. Kolmogorov's mapping neural network existence theorem. *Proceedings of the IEEE 1st International Conference on Neural Networks. 4*.

Jung-Huai Chou, J.G., 2001. Genetic algorithm in structural damage detection. *Computers & Structures, 79*.

Wang Ji-Zong, N.H.-G., He Jin-Yun, 1999.The application of automatic acquisition of knowledge to mix design of concrete. *Cement and Concrete Research, 29*: 6.

Lai, S.S., M, 1997.Concrete strength prediction by means of neural network. *Construction and Building Materials, 11*: 8.

Kasperkiewicz, J.R., Dubrawski A. 1995, HPC strength prediction using artificial neural network. *Journal of computing in civil engineering*. 9.

Yeh, I.-C., 1998. Modeling Of Strength Of High-Performance Concrete Using Artificial Neural Networks. *Cement and Concrete Research, 28*.

Mansour M.Y., Lee J.Y., Zhang J. 2004. Predicting the shear strength of reinforced concrete beams using artificial neural networks. *Engineering Structures. 26*.

Ni Hong-Guang, W.J.-Z. 2000. Prediction of compressive strength of concrete by neural networks. *Cement and Concrete Research, 30*: 6.

Lee, S.-C., 2002.Prediction of concrete strength using artificial neural networks. *Engineering Structures, 25*:849–857.

Bai J., Ware J.A., Sabir B.B. 2003. Using neural networks to predict workability of concrete incorporating metakaolin and fly ash. *Advances in Engineering Software, 34*: 7.

Totoev Y., F.R., Kanjanabootra S. and Alterman, D., 2015. Experimental investigationof thermal insulation properties of Semi Interlocking Masonry (SIM) Walls, *in Proc., 16th International Brick and Block Masonry Conference*. Padova, Italy.

Zarrin, O., Y. Totoev, and M. Masia, 2018. Evaluating Thrust Force Induced in Semi-Interlocking Masonry Panel during Horizontal and Vertical Arching, *in 10th International Masonry Conference International Masonry Conference*. Milan, Italy.

Totoev, Y.a.W., Zhiyu, 2013. In-Plane and Out-of-Plane Tests on Steel Frame with SIM Infill, *in 12th Canadian Masonry Symposium*. Vancouver, British Columbia.

Zarrin, O., Y. Totoev, and M. Masia, 2018. The Out-of-Plane Capacity of Semi-Interlocking Infill Panels: Experimental Investigation, *in 10th Australasian Masonry Conference*.: Sydney, Australia.

Totoev, Y. and Z. Wang. 2013. In-plane and out-of-plane tests on steel frame with SIM infill. *in 12th Canadian Masonry Symposium*.

Hossain, A., et al., 2017. In-Plane Cyclic Behavior of Semi Interlocking Masonry Panel Under Large Drift, *in 13th Canadian Masonry Symposium*.: Halifax, Canada.

C.R.KennedyCompany, *Leica Cyclone Basic User Manual*. 2016.

EDF R&D, T.P., *CloudCompare* (version 2.6.1) [GPL software]. Retrieved from http://www.danielgm.net/cc/. 2017.

Ramezanshirazi, M., D. Sebastiani, and S. Miliziano, 2019. Artificial Intelligence for predictions of Maximum Surface Settlements induced by Mechanized Tunnelling. *CNIRG 2019*.

Ramezanshirazi, M. and S. Miliziano, 2018. Application of Artificial Neural Networks to Predict Excavation Performances of Mechanized Tunnelling Machines, *in 2Department of Structural and Geotechnical Engineering*. Sapienza University of Rome, Rome, Italy.

Zarrin, O., Y. Totoev, and M. Ramezanshirazi, 2018. Introduce Applicability of "Radial Basis Function Network (RBF)" to Predict the Behaviour of Semi-Interlocking Masonry (SIM) *Panel. The International Masonry Society, Vol 31, No 1*: 27–32.

Griffith, M. and J. Vaculik, 2007. Out-of-plane flexural strength of unreinforced clay brick masonry walls. *TMS Journal, 25(1):* 53–68.

Griffith, M. and J. Vaculik. 2005. Flexural strength of unreinforced clay brick masonry walls. *in Proceedings of the 10th Canadian Masonry Symposium*, Banff, Alberta.

Griffith, M.C., et al., 2003. Evaluation of out-of-plane stability of unreinforced masonry walls subjected to seismic excitation. *Journal of Earthquake Engineering, 7(spec01):* 141–169.

Ramezanshirazi, M. and S. Miliziano. 2018. Application of Artificial Neural Networks to Predict Excavation Performances of Mechanized Tunnelling Machines, Sapienza University of Rome: Rome, Italy.

Zarrin, O., Y. Totoev, and M. Masia, 2018. Evaluating Thrust Force Induced in Semi-Interlocking Masonry Panel during Horizontal and Vertical Arching. *in 10th International Masonry Conference International Masonry Conference*, Milan, Italy.

Brick and Block Masonry - From Historical to Sustainable Masonry –
Kubica, Kwiecień & Bednarz (eds)
© 2020 Taylor & Francis Group, London, ISBN 978-0-367-56586-2

Earthen mortar in the walled city of Ahmedabad (India). Analysis of construction techniques and damages of two portion of masonry

A.G. Landi & A. Tognon
Politecnico di Milano, DAStU, Milan, Italy

K. Shah
CEPT University, Ahmedabad, India

ABSTRACT: Carefully constructed brick masonry walls, built using earthen mortar are common in vast areas of India (where they coexisted with the most widespread lime mortar masonry) at least until the middle of the nineteenth century. Studies on earthen mortar masonry, though in an embryonic state, already show significant differences compared to the many well studied raw earth buildings. It is believed that this kind of masonry was spread in India, particularly in the Gujarat region. However, the use of local materials and the technological refinement achieved by craftsmen over the years produced heterogeneous applications, also depending on the time period and external influences. In particular, in Ahmedabad, Gujarat, the mixture of earth, sand and quicklime is commonly observed, and this catalogue has documented a basic knowledge of the different types of construction and damages which is currently unavailable in published research.

1 INTRODUCTION

Founded in the 15th Century, the historic city of Ahmedabad was encircled by heavy multi-leaf brick masonry walls measuring around 10 km in length and enclosing an area of some 5.4 km². As was the case of many cities' world over, towards the end of the 19th Century, new town planning schemes proposed demolition of the city walls of Ahmedabad to enable the decongestion and expansion of the city. As per the Bombay Government records, the city Collector Mr. H.E.M. James suggested the removal of the city walls to permit the expansion of the city towards the railway station as early as 1888 (Gillion 1968, p. 149). In the early 1900s, the proposal of demolition of city walls was seriously considered by the government and became the subject of City Wall Improvements Scheme, 1920 drawn up in accordance with the Bombay Town Planning Act of 1915 (Yagnik & Sheth, 2011, p. 225). A large portion of the fort wall was subsequently demolished and replaced by roads, open spaces and new construction during the 1930s & 1940s. However, a long stretch along the river, which also served as retaining wall and flood protection; most of the city gates; boundary walls of the Gaekwad Haveli citadel (18th Century) and some portions of the Bhadra citadel (15th Century) walls were spared the destruction. Most of the City gates are protected as monuments of national importance by the Archaeological Survey of India (ASI). The remaining portions of the walls and a few gates on the riverside which did not have any

statutory protection during the 20th Century were listed as Grade-1 heritage structures by the Ahmedabad Municipal Corporation (AMC) in 2015 and form an important architectural and urban element of the World Heritage Site of the Historic City of Ahmadabad, inscribed on the UNESCO WHS list in 2017.

The western river side edge of the wall has undergone many maintenance and repair works over the past century and most notably since the first decade of the 21st Century. As seen in Figures 1 & 2, the authors have observed that around the Bhadra Citadel, the wall consisted of a traditional triple leaf masonry with one or two bricks thick outer leaves built either with lime mortar or earthen mortar and lime pointing and the bulk of the core with coursed masonry of bricks and brickbats(broken pieces of bricks) in earthen mortar (Figure 1). Given its location on the river edge the wall also had its lower section constructed in sandstone for additional flood protection. The authors have observed that during the course of recent restorations, the outer leaves have often been rebuilt with rich lime mortar up to a depth of almost 1 m, completely changing the morphology of the wall (Figure 2). Only a few portions of the original type of construction though with layers of earlier modifications remain visible today.

This analysis is an attempt to document the traditional building technology and materials of the city walls of Ahmedabad and record the widespread use of earthen mortar in the region. As seen in Figure 3, two segments of the wall have been chosen for the study. They are: 1) A segment of the wall located

Figure 1. Riverside wall of Bhadra Citadel and a view of the collapsed section before restoration. Given its vulnerable position on the river's edge the wall has been restored many times in the past and some of the portions even showed traces of cement plaster. Nevertheless, when observed carefully as in image 1b, one can observe thick lime mortar used for the outermost brick leaf while the rest was constructed with earthen mortar stabilized with lime (source: Khushi Shah, 2017).

Figure 2. Restoration of the riverside wall of Bhadra citadel showing reconstruction of outer leaf with lime mortar (source: Khushi Shah, January 2018).

around 30m South of Manek Burj the South-West corner of Bhadra citadel, originally constructed in the 15th Century; 2) A segment of the wall of Gaekwad Haveli, the Maratha period citadel from the 18th Century.

These sites have been chosen as they have not undergone any major restoration work in the past 4 to 5 decades, and hence are more representative of the wall as it was in the early 20ᵗʰ Century. Secondly, these fragments represent an identifiable historical period and belong to two different phases of expansion of the city. Therefore, the two fragments studied here allow us to observe and compare the differences in material and construction technology attributable to historical developments. There are very few such

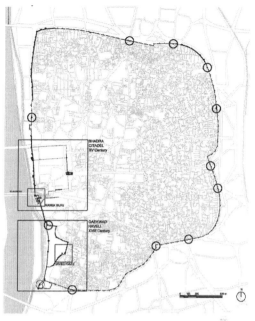

Figure 3. Map of the Historic city of Ahmedabad showing the remains of the fort wall and gates, the Bhadra Citadel & Gaekwad Haveli citadel. The location of both the sites selected for the study are highlighted (source: authors' elaboration based on the Cadastral map of the city available from AMC, 2016).

sites accessible for architectural surveys and this study provided an occasion to document this rapidly disappearing evidence of the city's history.

2 THE WALLS – CONSTRUCTION THROUGH THE TIME

The city of Ahmedabad was established in 1411 AD as the new capital of the Sultanate of Gujarat in western India by Sultan Ahmad Shah who shifted his capital from Patan to *'Ahmad'*abad, literally meaning the city of Ahmad (Commissariat 1938, p. 91). The site chosen on the eastern banks of river Sabarmati provided the ideal conditions for the growth of the city due to its location at the crossroads of important sub-continental trade routes and availability of water. At the beginning, the 'Bhadra' citadel on a more or less square plan was constructed on the riverbanks to house the royal family and nobility. The citadel is believed to have been completed in 1413 AD following which the city walls to protect the various markets and settlements of the population were built and are believed to have been completed around 1487 during the rule of Mahmud Begada, the grandson of Ahmadshah (Commissariat 1938, p. 96-97). As described in Mirat-i-Ahmadi, the wall at the time of its construction had "a circumference of 6 miles, 12 gates, 189 towers and 6,000 battlements" (Commissariat 1938, p. 97). The fort wall followed the natural topography and enclosed the extent of earlier settlement within an irregular shape with its western boundary defined by the river banks. The urban structure was governed by the strong East-West axis established by Sultan Ahmad Shah by building of the Bhadra citadel on the West and Jama masjid to its East with the ceremonial royal route connecting the two. The Gaekwad haveli, the second citadel of the city was established in 1760s as the headquarters of Maratha Gaekwad rulers who shared the revenue of the city with the Peshwas of Pune who had their own haveli in the Bhadra area (Commissariat 1980, p. 741). It uses the original riverside city wall defining its boundary on the west with new walls constructed in the 18th Century on the other three sides enclosing an irregular parcel of land (Commissariat 1980, p. 741). The citadel has four access gates on North, East and South and one postern on the river side (Figure 3). The contour map of the city (UNESCO, World Heritage Nomination Dossier documents, Historic City of Ahmadabad, 2015) shows that the terrain of the city slopes gently from North - North East to South - South West with Bhadra and Maidan-i-Shahi occupying a flat higher ground and the lowest points occurring in the South-West quadrant around Gaekwad Haveli and Raikhad Gate.

Records of repairs to the walls are available in published sources since Mughal period, particularly on the river side which often got affected by seasonal floods of the Sabarmati River. The author of Mirat-i-Ahmadi (18th Century) has recorded the instance of repair to the walls post the devastating floods of 1684 under the rule of Emperor Aurangzeb (Yagnik & Sheth 2011, p. 65-66). The worst floods in the river date to 1973 and 2005 AD. Before the Sabarmati River Front Development project (around 2010) was implemented, the river used to flow along the fort wall to the south of Manek Burj and entire city's stormwater was let into the river north of Raikhad Gate. In 1780 AD, the British under the leadership of General Goddard captured the city by storming in through the Khanjahan gate on the South-West and creating a breach in the wall which remained unrepaired for a long time (Commissariat 1980, p. 792-793). The governance of the city was momentarily restored to the Marathas as part of a treaty with the British East India Company till 1817. British finally took over the city in 1817 AD under the leadership of John Andrew Dunlop who was aghast at the deplorable condition of city walls (Gillion 1968, p. 42). Between 1831 and 1842, the Town Wall Fund Committee, a citizen's initiative funded by a special duty on certain commodities, undertook the task of repairs of the breaches and regular maintenance of the city walls (Gillion 1968, p. 111). This work laid the path of establishment of local self-government in the form of Ahmedabad Municipal Commission in 1856 (Gillion 1968, p. 116). As seen earlier a large part of the wall was demolished during 1930s and 1940s to make way for the modern Ahmedabad which grew many folds outside the walls in less than a Century. The remaining fragments of the city walls are nevertheless recognized as a significant heritage and structural assets for the city and have been regularly maintained by the AMC which has embarked on an ambitious restoration project for the wall since 2008. Before the Sabarmati River Front Development project (around 2010) was implemented, the river used to flow the fort wall to the south of Manek Burj and the entire city's stormwater was let into the river north of Raikhad Gate.

3 ANALYSIS OF CONSTRUCTION TECHNIQUES AND DAMAGES OF THE MANSORY

As highlighted in earlier descriptions, a few portions of the city wall have retained their historic fabric despite various transformations and damages due to natural calamities (floods, earthquake, …) as well as human actions (lack of maintenance, vandalism, repairs and so on). As mentioned in the introduction, two distinct sites built at different times have been chosen for arriving at a scientific understanding of the fort wall. The first site is near the Manek Burj, adjacent to the Hari Vishnu Temple, South of Ellis Bridge. The second is a portion of the Eastern, city side wall of Gaekwad Haveli citadel and located a few meters North of its imposing gateway. While both these sections have undergone repairs and have been added to over a period of time, they have not been significantly altered in recent times and represent the condition of the wall in the early 20th Century, making them relevant to the study.

3.1 Site 01 – Bhadra citadel

The first study site is located a few meters to the South of Bhadra citadel. This fragment near Manek Burj and adjacent to Hari Vishnu Temple [72° 34′ 39.96″ E 23° 1′ 18.34″ N] is a portion of the wall which could have survived the demolition of the city walls due to its significance as a retaining wall. Over the years it got integrated within the new construction that came up around it, notably the temples and hostels on the riverbank. The Victoria (Tilak) Garden and Municipal stores lie on its other side within the city. The temples on the riverbanks date mostly from the Maratha period (18th Century) and later. The section connecting this wall to Manek Burj on its North was destroyed many decades ago for constructing a road leading to the riverbanks. A new road was constructed to its South a few years ago, further isolating the site from the remaining section of the wall to its South. It is possible to trace the extents of the citadel and city walls on the current site based on historical maps, notably the City

Survey map of 1881 prepared by the British Government and available with the AMC. Given the many additions, alterations and damages to the wall due to its changing context, natural disasters and repair/restoration attempts in the past, it is difficult to recognise the original fabric of the wall. Nevertheless, as elaborated in the previous sections of this paper the original artefact can be dated to the end of the 15th Century when the rest of the city walls were built.

As seen in the Figure 4a the external riverside façade of the section under study overlooks a narrow passage between the historic city on the higher ground and the temple, a residence and a new hostel building (20th Century) on the riverbank. The

residence and hostel buildings are attached to the historic fabric of the wall a situation which could have helped the preservation of this section of the wall during the recent restoration drive by the AMC, while also contributing to its deterioration.

The primary construction materials visible are burnt clay bricks with lime mortar, lime pointing as well as earthen mortars. In the section under study, bricks measuring 23/27 x 15/18 x 4,5 cm (LxBxH), with approximately 1,5 cm thick lime and earthen mortar laid out in well-defined regular horizontal courses are observed. This fragment does not exhibit the use of sandstone base as observed in many fragments on the river side including parts of Bhadra citadel wall and the Manek Burj which are just a few meters to the North of the selected site.

The upper portion of the wall is articulated with two distinct layers of battlement patterns with well-defined arched merlons and recessed embrasures (indicating gun hole slits) in two levels, indicating two phases of construction. Based on the site observations, one can deduce that, during a past intervention the height of the wall was increased by approximately 1,20 m by adding the second level of battlement wall. This additional wall is some 50 cm wide and primarily built using burnt clay bricks of the same dimensions as the earlier ones and lime mortar. The collapsed part of the lower portion which would be the earlier construction phase exhibits presences of clay mortar in the core masonry along with lime mortar and pointing on the external face.

The dimensions of the merlons (95 cm wide) and embrasures (16 cm wide) remain the same in both sections. The merlons and embrasures in both levels are vertically staggered with the centre of the merlons of the top level being aligned with the embrasures of the lower level. Surprisingly, the gun hole slits in both cases have been filled up with bricks. At present, it is difficult to say if they were functional as defensive elements in the past and later filled up or were just intended as a visual relief from the beginning. The dimensions of the bricks used in both sections are the same and some patches of lime plaster corresponding to later interventions are observed on the upper levels. A thin lime mortar moulding is observed on the second layer of merlons. The top of the wall is capped and finished with cement plaster. During visual inspection, large patches of cement plaster in different stages of deterioration were observed on both faces of the wall, while small patches of older lime plaster were observed on the upper portions of the wall on the riverside.

On observing the cut section of the wall from the bridge near Manek Burj, it is possible to notice the difference of around 2,10 m in the ground level within and outside the city walls. On the inner part of the wall, a walkway approximately measuring between 35 to 70 cm wide runs along the entire wall. The average thickness of the wall above ground level varies between 83 to 118 cm, while the top

Figure 4. General view of selected study site 1: segment of the wall around 30 m South of Manek Burj. It is important to note that new buildings have been constructed very close to the wall and a portion of the wall has been demolished to make way for a new road. General view of selected study site 2: section of the wall of Gaekwad Haveli, along the Gaekwad Haveli Road, to the north of its eastern gateway. It is important to note that the wall is in poor condition and exhibits a variety of issues including cracks, biological colonization, material loss, graffiti and unsympathetic use and activites (source: Alisia Tognon, 2019).

battlement section of the wall is around 45 to 50 cm wide.

The brick wall is well constructed with staggered vertical joints and with a slightly wider section at the base providing stability to the construction. The stretch of the wall has an orderly brickwork texture with cooked bricks and mortar joints.

Traces of plaster are visible on the battlements and on the overhangs, where the surface is exposed to weathering and subject to degradation. The damages affect limited portions of the walls facing and where the masonry has been broken, it is due to anthropic events. Here, the nucleus of the masonry built with burnt clay bricks and earthen mortar is visible, but made up of bricks laid in a disorderly fashion.

As mentioned earlier the external surface of the wall has been extensively damaged by past floods and repair interventions, the chief being application of cement pointing, plaster and capping. Heavy deterioration of bricks, rising damp and partial collapse of sections of the outer leaf of the wall are observed which can be related to the application of cement plaster. The wall exhibits major loss of mortar (both lime and earth), loss of structural material (bricks), erosion of bricks and mortar and partial mechanical damage due to collapse and local settlement within the structure (Figure 5). The situation is further exacerbated by efflorescence, biological colonization (algae), and growth of vegetation.

The facing presents traces of degradation due to weathering: crumbling bricks, partly powdering or cracked, traces of plaster and lime mortar pointing are still visible. However, the most evident effect is washing away of the joints in earthen mortar, which loose consistency due to the presence of water. The main risks foreseen is the loss of cohesion between the bricks, which precedes the weakening and collapse of

Figure 6. Bhadra citadel, view of the wall adjacent to the temple - the top of the walls have often been rebuilt or integrated with new parts, not always faithful to the existing ones. On the top portions of the walls can be noticed a more evident degradation, due to the weathering: as a consequence this portions have been reconstructed more frequently (source: authors, 2019).

the whole structure. This decay can be dangerous when weathering affect the nucleus of the masonry, where the proportion of earthen mortar and bricks increases considerably.

This section while saved from losing its historical traces during the most recent restoration works is highly vulnerable to ongoing damage and deterioration due to forces of nature (wind & rain) and potential human interventions in the future.

Frequent maintenance for this fragment of the wall façade, affected by weathering and the floods of the Sabarmati River was essential. Due to these reasons, the original state of materials and finishes are challenging to identify. However, the traces of tamperings and maintenance can be clearly noticed.

3.2 Site 02 - Gaekwad haveli

The second site of the wall analysis is located 25 m north of the Eastern Gateway [72° 34′ 50.748″ E - 23° 0′ 59.22″ N]. This portion is part of the Gaekwad Haveli, the Maratha period citadel located in Raikhad ward, in the South-Western quarter of the historic city of Ahmedabad. Prof. M. S. Commissariat in his history of Gujarat, Vol. III, dates the building of Gaekwad Haveli to sometime between 1761 and 1768 (Commissariat 1980, p. 741).

As mentioned earlier, between 1753 and 1819 AD Ahmedabad was jointly ruled by Gaekwad rulers of Baroda and Peshwas of Pune. The haveli was built to house the army and political agents of the Gaekwads in the city. They chose the South Western quarter of the city, along the banks of the river due to availability of open land and its proximity to the Jamalpur Gate which was under their control (Yagnik & Sheth 2011, p. 77). The citadel has an irregular plan (Figure 3) and still occupies a large open ground between the battlements along the river and the main road to its east. The entire length of the

Figure 5. Bhadra citadel - different textures between the external façade and the nucleus can been noticed. The internal wall section has a higher quantity of earthen mortar just to fill the voids and to bind the cooked bricks (source: authors, 2019).

Figure 7. Gaekwad Haveli, inner façade - extensive renovations and maintenance were carried out on the masonry over the past two centuries (source: authors, 2019).

citadel wall still survives today, though in varying states of conservation characterized by encroachment (informal settlements, abandoned detritus, spontaneous and unsympathetic activities, ...) along the portions on the city side. In the cadastral maps of 1881 (AMC), one can see the citadel divided into three distinct parts. According to the Bombay presidency Gazetteer (Campbell, 1879), the central portion resting on the river bank was the original citadel. During the British occupation, the citadel continued to be used as an arsenal and ordnance depot and, since Independence, it has been utilized as headquarters of the city police.

The section between the Raikhad and the Khanjahan gate which also served as the western wall of the citadel was built with stone and brick masonry, interspersed with ramparts, watchtowers and battlements and a series of entrances with sturdy wooden gates. The original city wall on the river side could have been modified by opening up of new posterns and construction of strong bastions with stone during the Maratha period as a lot of re-use of old materials from different buildings or the original wall can be observed in this section of the wall. As mentioned earlier the river flowed right along this edge of the wall. This was one of the most dilapidated sections of the city walls and has been entirely restored by AMC in the recent past (2017-2020). Major interventions and redevelopment works have also been undertaken to its surrounding under the Sabarmati Riverfront Development Project (since 2010).

On the other side, along the Gaekwad Haveli Road within the city, the wall has not been restored in the recent past except for the application of cement plaster and paints in many patches. On this façade, it is possible to see the relatively original condition of the fortification. On the basis of visual observation and measured survey of the rampart, it is possible to highlight four horizontal layers of construction along the outer part of the citadel (Figure 8).

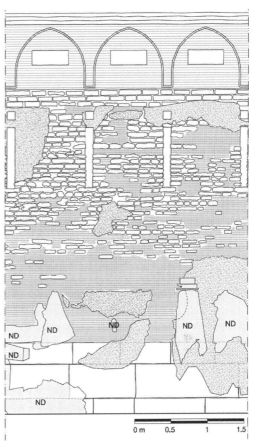

Figure 8. Gaekwad Haveli, survey of the external façade – (source: A. Tognon, 2019).

The base of the wall has been built with stones of variable dimensions (about 40 cm long by 20/25 in height). Some of the stones show decorative details, which attest the reuse of material from older structures (Figure 9). Above the stone base, masonry walls built of burnt brick (size: 20/24 x 16/17 x 4,5) and mortar (thickness 2 cm), which are laid in horizontal and regular gae. The external height of the wall is approximately 5 meters. At a height of about 3 meter the presence of a sequence of embrasures (gun-holes) is observed (Figure 10). Above this, a three-brick thick band runs along the façade, and a series of merlons with horizontal slit gun holes in their center and vertical slits for gun holes between the merlons details the battlements. The internal level of the fortification is higher than the altitude along the Gaekwad Haveli road. A walkway (1,43 m wide) with battlement walls on the exterior runs along the inner rampart and the top of the lower gun-holes open up at its base. The dimension of the wall section is approximately 2 m thick with the top battlement wall being approximately 0,6 m thick.

Figure 9. Gaekwad Haveli, external façade along Gaekwad Havely Rd. - the lower part of the masonry is made of squared stone blocks: probably, these have been reused, as it can be noticed by the numerous molded ashlars and the singular workmanship (source: authors, 2019).

Figure 10. Gaekwad Haveli, inner façade, merlons detail - the wall has a slot for embrasures, placed at regular intervals between the ogival merlons. Over the time, the rainwater has caused the washout of the external facade, perhaps protected by a smooth plaster. Today some lime mortar joints are in good condition as a result of "recent" maintenance (source: authors, 2019).

At present, lack of maintenance and encroachments are main conservation issues for this part of the citadel. Many instances of damage and deterioration like the growth of vegetation, loss of material (mortar as well as bricks), and disintegration of bricks can be observed. Most of the wall surface has been significantly altered using cement mortar pointing and cement plaster, leading to loss of original lime mortar and spalling of bricks. In some parts, scraps of cement plaster with paintings and billboard covers the façade. Encroachment and informal uses diminish the state of conservation: temporary shelters, commercial activities, small temples and infrastructure installations use the historic walls as the backdrop for their activities.

4 CONCLUSION

Considering the continuous tampering and reconstruction of Ahmedabad's urban walls, the findings and analyses carried out on the two masonry samples take on value not only in terms of method - however to be refined and adjusted on the surviving walls - but also in cultural terms: this documentation is aimed at recording valuable information useful for future studies and is also a useful work-base for restoration interventions, more respectful of historical artifacts. The analysis of construction techniques, previous restorations, materials and different forms of degradation have provided the basis for an effective and coherent intervention capable of guiding and designing the restoration projects in all their phases.

The walls, if analyzed analytically, can constitute a very rich source of information on local construction techniques and, with regard to instrumental analyses, they can give rise to further studies and researches with repercussions on such rich Indian heritage; this is all usually built of brick walls and mortar based either on earth or lime: the materials used, the different construction techniques, the mechanical strength, the seismic vulnerability, the correct maintenance and restoration methods are just some technical aspects which show a synthesis and a fair chance of further study on these structures.

Studies and investigations constitute the basis for the appreciation of these artifacts of extraordinary landscape, historical and architectural value: wherever these values have not been well identified and classified yet, the city government has interpreted the artefacts just as constitutive elements of the urban fabric according to merely functional criteria: in fact they are used as fence walls, supports for buildings, new bridges and road axes or as barriers against floods. Wherever urban regeneration projects have affected the city walls, without investing them with new meanings and values and without any preliminary studies - as for example it occurred in correspondence of the RiverFront on the Sabarmati River - they have maintained the role of a backdrop and are subject to a rapid degradation also owing to improper uses and anthropic acts of vandalism. So, a further possible development of the research may also consist in involving the local communities in the project, so that they can appreciate and actively participate in cataloguing and taking care of their precious remains of the past.

ACKNOWLEDGMENT

The research was carried out as part of the project "Territorial fragilities", funded by the Italian Ministry of Education, Universities and Research (MIUR) and by DAStU – Department of Excellence (2019/22) – Politecnico di Milano. CEPT University participated in the research as part of an academic cooperation agreement with Politecnico di Milano.

REFERENCES

Briggs, H. G. 1849. *The Cities of Gujarashtra*. Bombay: The times Press.

Campbell, J. M. 1879. *Gazetteer of The Bombay Presidency Vol IV*. Bombay: Government Central Press.

Commissariat, M. S. 1938. *History of Gujarat Vol. I, From AD 1297-8 to to AD 1573*, Longmans. Green & Co.

Commissariat, M. S. 1957. *A History of Gujarat With a Survey of its Monuments and Inscriptions Vol. II*, The Mughal Period: From 1573 to 1758. Orient Longmans.

Commissariat, M. S. 1980. *History of Gujarat. Vol.3: The Maratha period: 1758A.D. to 1818A.D.* Ahmadabad: Gujarat Vidya Sabha.

Gillion, K. 1968. *Ahmadabad: A study in Indian urban history Book*. London: Uni. of California Press.

Hope, T., Biggs, T., & Fergusson, J. 1866. *Architecture at Ahmadabad, the Capital of Goozerat, photographed by Colonel Biggs, With an historical and descriptive sketch by T.C.H., architectural notes by J. Fergusson.* London.

Yagnik, A., & Sheth, S. 2011. *Ahmedabad, From Royal City to Megacity*. New Delhi: Penguin books India.

Heritage Listing. (n.d.). Retrieved May 26, 2016, from http://www.egovamc.com/heritage/heritage.aspx

Historic city of Ahmadabad. (n.d.). Retrieved March 24, 2018, from https://whc.unesco.org/en/list/1551/documents/

Brick and Block Masonry - From Historical to Sustainable Masonry –
Kubica, Kwiecień & Bednarz (eds)
© 2020 Taylor & Francis Group, London, ISBN 978-0-367-56586-2

Shear-compression experimentation of full-scale wood-cement shuttering block walls

M. Minotto & N. Verlato
Department of Civil, Architectural and Environmental Engineering (ICEA), University of Padova, Italy

M. Donà
Earthquake Engineering Research & Test Center (EERTC), Guangzhou University, China

F. da Porto
Department of Geosciences, University of Padova, Italy

ABSTRACT: The shuttering block construction system is used for the construction of loadbearing walls and consists of special blocks with cavities suitable for hosting vertical and horizontal reinforcements and the cast-in-situ concrete. These walls present a resistant concrete grid made of vertical columns joined together by horizontal squat beams with regular spacing. The present paper discusses the results of the experimental activity conducted at the University of Padua on 6 full-scale wall specimens of diverse aspect ratio (1:1 or 3:4) and openings (i.e. doors and windows). The blocks used in the current research activity are made of petrified wood chips and cement and have the dual function of being a disposable form-work for the cast-in-situ concrete and being a thermal-acoustic surface insulation device of the wall. Cyclic In-Plane displacements of increasing amplitude until failure were imposed (with constant vertical load) to characterize the shear-compression behaviour of the specimens.

1 INTRODUCTION

The shuttering block construction system is used nowadays for the construction for the construction of ecological/eco-sustainable residential, commercial and green buildings. The construction system allows to realize loadbearing walls based on the use of formwork blocks and weakly Reinforced cast-in-situ Concrete (RC) with small percentage of steel reinforcement. Such solution demonstrated a good response against strong earthquake ground motions and, in the last years, a remarkable expansion of this solution was observed especially in Italy and other European countries as Austria and Germany. This is related to the multiple advantages that the wood-chip and cement hollow blocks can offer, as the excellent sound/thermal insulation and the vapor permeability.

From the experimental point of view, few studies were conducted to assess the effectiveness of such construction system and the current paper aims at showing some results of an extensive experimental campaign carried out at the Laboratory of the University of Padova to assess the performance of shuttering block walls. Past experimental tests were carried out at the University of Bologna by Ceccoli et al. 2000, Tullini 2000 and Malavolta 2008. More recent experimental tests were carried out at the University of Padova by Girardini 2015.

The main problem for the construction of structures based on such system concerns the lack of adequate design references. The guidelines for load bearing panel construction systems based on the use of formwork blocks and weakly reinforced cast-in-situ concrete, approved in 2011 by the Superior Council of Public Works (LL.GG. 2011), is the only official code (not mandatory) which provides indications for the experimental and design purposes.

The following paper addresses the experimental results of shear-compression tests carried out on several full-scale wall specimens that differ each other for the wall length to height geometric ratio (1:1 or 4:3) and for the introduction of openings (doors and windows) at the centre of the wall. The samples were tested by firstly applying a constant vertical load to the wall and secondly imposing cyclic In-Plane (IP) displacements of increasing amplitude until failure. The results, especially in terms of capacity curves, will be discussed in detail.

2 DESCRIPTION OF THE CONSTRUCTION SYSTEM

The construction system described in the current paper allows to realize load-bearing walls made of special blocks with cavities suitable for hosting vertical and horizontal reinforcements and the cast-in-situ concrete.

Figure 1. Details of shuttering block construction system.

The blocks are made of petrified wood chips and cement and have the dual function of being a disposable form-work for the concrete and being a thermal-acoustic surface insulation device of the wall. Figure 1 shows the configuration and the arrangement of such blocks, which are placed adjacent to each other on several overlapping dry courses as for ordinary masonry. The geometry of the blocks allows the hosting of vertical and horizontal reinforcement and, if it is required, it is possible to insert additional reinforcing bars at the openings and the ends of the walls in order to increase their resistance under horizontal actions. Particular attention must be paid to ensuring adequate anchorage and lap length of the reinforcing bars, using the same criteria that are used for reinforced concrete structures (see EN1992-1-1: 2004).

The correct construction of the wall takes place by positioning the first course of blocks on two layers of mortar; starting from the corner blocks, proceed with the dry laying of the following courses keeping the blocks well attached to avoid thermal and acoustic bridges. At the reaching of the height corresponding to six courses (around 1.5 m) the concrete is cast into the wall through a well or a pump, being careful not to exert excessive pressure that could shift the alignment of the blocks. The concrete must have an adequate degree of workability in order to ensure a complete filling of all the formwork blocks. According to EN 206-1: 2000, concrete with a consistence S4-S5 (fluid and superfluid) with aggregate size of 12÷15 mm should be.

The proposed construction solution is characterized by multiple advantages. Firstly, it guarantees excellent sound/thermal insulation and vapor permeability. The dry installation of the blocks allows a precise and fast execution. The blocks used in the current research activity are developed in the respect of the green building philosophy and are certified by the manufacturer as fire and burst resistant.

3 EXPERIMENTAL TESTS

3.1 Specimens

For the execution of the experimental tests the authors refereed to the "Guidelines for load bearing panel construction systems based on the use of formwork blocks and weakly reinforced cast-in-situ concrete" approved by the Superior Council of Public Works (LL.GG. 2011). According to the standard prescriptions, the construction system has to be characterized from a structural point of view by means of experimental tests to demonstrate an effective behaviour of the load-bearing elements to withstand to both vertical and horizontal actions, even cyclical ones. The current paper focuses on the results of test type 3 (shear-compression) that represents the In-Plane shear-compression test carried out on full-scale wall specimens. LL.GG. 2011 provides that 8 specimens should be tested at least including 4 walls without openings, 2 with door and 2 with central window. In case of full walls (without openings), two shape ratios (base to height, i.e. b:h) 1:1 and 4:3 are requested. Such test type aims at assessing the IP response in terms of strength and deformability capacities and damping. At this purpose, the tests are conducted applying a constant vertical compressive load and IP horizontal quasi-static cyclic displacements as described in detail in the following paragraphs. Figure 2 shows the geometry of the blocks involved in the experimental campaign whereas Figure 3 the geometry of the specimens.

The specimens were constructed using concrete type C25/30 (f_{ck} = 25 N/mm^2) and steel type B450C (f_{yk} = 450 N/mm^2) for the reinforcing bars. The following Table 1 summarizes the main construction properties of the wall specimens in terms of horizontal and vertical reinforcement and area of each internal concrete column A_{col}. In case of vertical reinforcement, the symbol (1-2) means that for each internal columns one bar was used in the central part of the wall and two bars were placed in the outer pillars or near door/window. It is worth to mention that in case of walls without openings, only two specimens were tested at the University of Padova (one 1:1 and one 4:3) whereas the remaining samples were tested at another laboratory (University of Perugia).

3.2 Set-up

The experimental set-up for shear-compression tests shown in Figure 4 was fixed to the laboratory floor by means of Ø42 tie rods tensioned and fixed to the intrados of the floor itself. The steel reaction frame was made of four columns type HE360 which support an upper steel beam for the application of the preload.

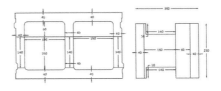

Figure 2. Geometry of wood-cement blocks.

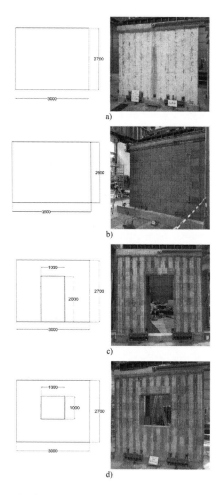

Figure 3. Geometry of wall specimens for test type3:
a) and b) without openings (*b*:*h*=1:1 and 4:3). c) door.
d) window.

Table 1. Main construction properties of wall specimens.

Specimen ID	Horizontal reinforcement	Vertical reinforcement	A_{col} cm^2
1_Door	1Ø10/25	2Ø10/25	288
2_Window	1Ø10/25	2Ø10/25	288
3_Window	1Ø10/25	2Ø10/25	288
4_Door	1Ø10/25	2Ø10/25	288
5_Full (1:1)	1Ø10/25	2Ø10/25	288
6_Full (4:3)	1Ø10/25	2Ø10/25	288

The connection between the beam and the columns
was realized by two flanged beams ("*C*" shape)
800 mm high which were fixed to the columns
through 28 class 8.8 bolts. Two solutions were carried
out to fix wall specimens to the laboratory floor. Such
approach was necessary because in case of 1:1
sample the base RC beam had a limited thickness and

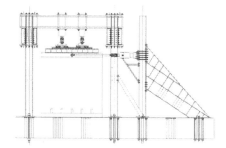

Figure 4. Shear-compression test set-up.

special steel jaws were specifically designed and realized (see Figure 5a). In case of the wall with 4:3
shape ratio, the bottom RC beam was designed allowing the fixing of the specimen by means of tie rods as
shown in Figure 5b. The vertical load was applied in
force control by means of two 600 kN capacity jacks
connected to each other by means of a flanged beam
type HE300 with the interposition of a linear guide to
allow horizontal displacement at the top of the
sample. The horizontal actuator, fixed to a contrast
system with an oblique strut fixed to the laboratory
floor, has a capacity of ±620 kN at pressure of 250
bar. The horizontal actuator is linked to the wall by
steel rods hinged to a 60 mm diameter steel pin into
a jacketed hole made in the centre of the RC upper
beam.

The so conceived setup allows the in-plane rotation of the top beam, simplifying, by geometric symmetry, the typical double interlocking behaviour of
bearing masonry walls. Compared to other shear-
compression test procedures developed in recent
years, such as the ESECMaSE procedure (Fehling
et al. 2008), this approach allows a considerable simplification of the control system since there is no
need for real time adjustments of both the applied
vertical load and the top beam in-plane rotation.

3.3 *Procedure*

The tests were carried out by applying a monotonic
vertical compressive load at the top of the specimen
and a subsequent IP horizontal quasi-static cyclic
displacement history. Each sample was subjected to
a preload of about 220 kN which was kept constant
during the tests. The vertical load value was chosen

Figure 5. A) Steel jaws system. b) Tie rods system.

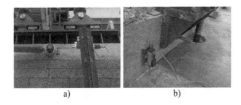

a)   b)

Figure 6. A) Upper MTS (±100 mm). b) Bottom potentiometers (±25 mm).

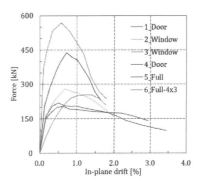

Figure 7. Average envelope curves at the 3rd hysteresis cycle.

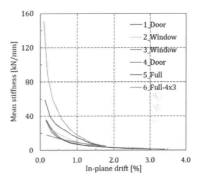

Figure 8. Average stiffness at the 3rd hysteresis cycle.

to simulate the compressive stresses acting on the ground floor walls of an ideal three-storey building. The history of applied In-Plane displacements consist of a sequence of inter-story drifts with an increment of 0.2% up to 2.2% and subsequently with an increment of 0.4% up to 3.8%. The IP test was quasi-static with 3 cycles for each drift level with a maximum stroke speed of less than 0.5 mm/s. The IP cyclic tests were in displacement control thanks to the installation of one magnetostrictive transducer MTS (±100 mm) on the upper RC beam as shown in Figure 6a.

One potentiometer (±25 mm) was installed in the central part of the lower beam to monitor relative sliding movements of the specimen on the laboratory floor. Lastly, other two potentiometers (±25 mm) were placed near the steel jaws to measure the lifting of the lower RC beam from the reaction floor and thus to check the rotation of the specimen (see Figure 6b).

4 EXPERIMENTAL RESULTS

The experimental results will be described in the following paragraphs. For each tested specimen, the main mechanical and hysteretic properties were calculated, i.e. the maximum IP strength and deformation capacity, the IP stiffness degradation due to increasing imposed displacements, the energy dissipation capacity and the equivalent viscous damping (ξ_{eq}).

4.1 Evaluation of in-plane capacity and stiffness

As reported in Table 1, six specimens were tested IP until failure such that the ultimate drift of the samples can vary. Figure 7 contains the IP envelope capacity curves of the walls whereas Figure 8 shows the reduction of the secant stiffness as the imposed displacement increases. Both values were determined, for each drift level, at the third hysteretic cycle (the most stabilized). Table 2 summarizes the main information by tests as the maximum IP capacity of specimens F_{max} and its corresponding drift level θ_{max}, the ultimate force F_{ult}, corresponding to a 20% reduction of the maximum capacity, and its drift θ_{ult}. The last column of the table shows the ductility µ of the systems obtained as the ratio between the IP drifts θ_{max} and θ_{ult}.

Table 2. Summary of IP test results.

Specimen ID	F_{max} kN	θ_{max} %	F_{ult} kN	θ_{ult}%	$\theta_{max}/\theta_{ult}$ -
1_Door	204.7	0.64	163.7	2.48	3.9
2_Window	264.4	0.67	211.5	1.48	2.2
3_Window	253.9	1.20	203.1	1.78	1.5
4_Door	219.5	0.50	175.6	1.85	3.7
5_Full (1:1)	438.0	0.72	350.4	1.23	1.7
6_Full (4:3)	568.3	0.57	433.2	1.17	2.1

Regarding to the specimens of shape factor 1:1, the panel that has reached the highest IP strength is the 5_Full, with a maximum load of around 438 kN due to the absence of openings.

The two specimens with central door reached maximum load values of 204.7 kN (Door_1) and 219.5 kN (Door_2), with drift levels at the maximum load of, respectively, 0.64% and 0.5%. These specimens, compared to the other tested specimens, showed lower maximum strength values and higher ductility, with envelope curves characterised by a post-peak phase with a very low softening slope. Such behaviour is due to the development of a predominant flexural resistant mechanism which characterise the high slenderness bearing elements of this type of specimens.

The two specimens with central window reached maximum load values of 264.4 kN (Window_1) and 253.9 kN (Window_2), with drift levels at the maximum load of, respectively, 0.67% and 1.2%. These specimens, compared to those with central door, presented a slightly greater capacity, with a strength increment of about 22%, and, on the other hand, a considerable ductility reduction. As a matter of fact, once the peak strength is reached, the panels with central window show a steeper softening branch due to the predominant shear resistant mechanism, characterised by a more brittle failure than the flexural one.

It is worth noting that, taking as a reference the 5_Full specimen, the introduction of a central door or a central window resulted in, respectively, a 51% and a 41% decrease in IP capacity.

Figures 9a and b show the typical failure modes of the specimens with window and door respectively. Because of such behaviour, the walls with door opening reached a greater ductility than the samples with the window. The specimen with b:h = 4:3 reached an IP maximum load of 610 kN, with an increase of about 30% compared to the 5_Full sample. This specimen showed also a significantly higher initial stiffness than the other specimens. Such behaviour is related to the different aspect ratio and, partly, to a more efficient fixing system to constrain the specimen to the laboratory floor slab.

In the sample without openings (5_Full) the cracks appeared in the lower part of the panel, where the maximum bending moment was reached, without however manifesting themselves with significant amplitudes (not very visible). The rigid rotation effect associated to the detachment of the panel from the base RC beam was very evident. The specimen reached combined compressive and flexural failure with the slipping of the longitudinal reinforcing bars (see Figure 10a). Lastly, the 6_Full specimen showed a significant concentration of damage at the base, especially at the external columns. During the test, the shearing of the vertical reinforcing bars occurred, a symptom of a combined compressive and flexural failure (see Figure 10b).

Figure 10. A) Failure of specimen 5_Full (1:1). b) Failure of specimen 6_Full (4:3).

4.2 Evaluation of the energy dissipation capacity and of the equivalent viscous damping

Figures 11 and 12 show the average trends of, respectively, the ratio between the dissipated and the input energy, and the equivalent viscous damping (ξ_{eq}). All the reported values were calculated, for each drift level, at the third hysteretic cycle so as to have values as stable as possible, i.e. not susceptible to the fluctuations due to the sudden propagation of the damage pattern. The equivalent viscous damping was calculated following the simplified method originally proposed by Jacobsen (1930) for SDOF (Sigle Degree Of Freedom) structures subjected to sinusoidal loads. This approach allows to evaluate the dynamic behaviour of the specimens starting from hysteretic curves obtained through quasi-static procedures. Considering a single load-displacement cycle, ξ_{eq} can be evaluated as a function of the dissipated energy E_{diss} and of the elastic energy stored at the peak displacement E_{sto} and is calculated as follows:

$$\xi_{eq} = \frac{1}{4\pi} \cdot \frac{E_{diss}}{E_{sto}} \qquad (1)$$

The equivalent viscous damping coefficient assumes values in the range 3-10% and in case of low IP displacement values, i.e. when an elastic-linear

a) b)

Figure 9. A) Shear failure of specimen 2_Window. b) Flexural failure of specimen 4_Door.

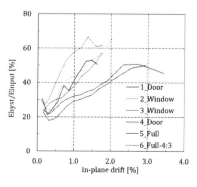

Figure 11. Ratio between hysteresis and input energy at the 3rd cycle.

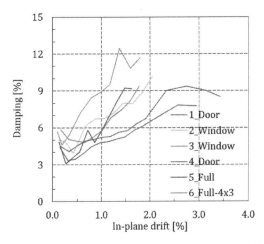

Figure 12. Average equivalent viscous damping at the 3rd hysteresis cycle.

behaviour of the material can be assumed, the damping coefficient ξ assumes lower values of about 5%.

5   CONCLUSIONS

The research activity addresses the shear-compression experimental tests carried out on six full-scale wall specimens that differ each other for the wall height to length geometric ratio (1:1 or 4:3) and for the introduction of openings (doors and windows) at the centre of the wall. The samples were tested by firstly applying a constant vertical load to the wall and secondly imposing cyclic IP displacements of increasing amplitude until failure.

The experimental research proved the effectiveness of the analysed construction system in terms of In-Plane capacity and energy dissipation characteristics. Each specimen was mechanically characterised by means of cyclical shear-compression tests and the elaboration of the results in terms of In-Plane capacity curves. Further elaborations on the obtained In-Plane hysteretic cycles were performed to evaluate the energetic properties of the specimens in terms of dissipated energy to input energy ratio and in terms of equivalent viscous damping.

Furthermore, the research activity investigated how the introduction of a central opening, influencing the slenderness of the resistant elements, entails a substantial alteration both in the IP capacity and in the failure mode of the specimens. In particular, the specimens with central door, characterized by a predominant flexural failure, showed a lower post-peak strength degradation than the window specimens, characterised, instead, by an almost pure shear failure.

Acknowledgements

The research activity was funded by Legnobloc s.r.l. Experimental testing was carried out at the Laboratory of Structural Materials Testing of the University of Padova, Italy.

REFERENCES

Ceccoli, C. et al. 2000. Indagini sperimentali effettuate su pareti realizzate con blocchi cassero in legno-cemento (in Italian). *Atti del XIII Convegno CTE, Pisa*: 327–336.

EN1992-1-1
: 2004. Eurocode 2: Design of Concrete Structures, Part 1-1: General Rules and Rules for Buildings. European Committee for Standardisation, Brussels, Belgium.

EN206-1: 2000. Concrete - Part 1: Specification, performance, production and conformity. European Committee for Standardisation, Brussels, Belgium.

Fehling, E. & Schermer, D. 2008. ESECMASE - Shear test method for masonry walls with realistic boundary conditions. *Proceedings of the 14th Brick and Block Masonry Conference. Sydney, Australia*.

Girardini, D. 2015. Static and seismic performances of R.C. shear walls cast into wood chip and cement formworks - Experimental tests, theoretical interpretation and numerical validations. *Padova*.

Jacobsen, L.S. 1930. An approximate solution of the steady forced vibration of a system of one degree of freedom under the influence of various types of damping. *Bulletin of the Seismological Society of America* 20: 28.

LL.GG. 2001. Guidelines for load bearing panel construction sys-tems based on the use of formwork blocks and weakly reinforced cast-in-situ concrete. *Superior Council of Public Works, Italy*.

Malavolta, D. 2008. Strutture a pareti portanti in C.A. caratterizzate da elevate prestazioni sismiche (in Italian). *Bologna*.

Tullini, N. 2000. Interpretazione delle prove sperimentali eseguite su strutture a pareti portanti realizzate con blocchi cassero in legno mineralizzato (in Italian). *Bologna*.

Earthquake resistance & retrofitting

Brick and Block Masonry - From Historical to Sustainable Masonry –
Kubica, Kwiecień & Bednarz (eds)
© 2020 Taylor & Francis Group, London, ISBN 978-0-367-56586-2

Quantification of damage to masonry structures under seismic conditions

E. Vintzileou, I. Tselios & D. Karagiannaki
National Technical University of Athens, Athens, Greece

ABSTRACT: Solid and perforated brickwork is widely used in construction, both in structural and non-structural elements (e.g. bearing or infill walls, facades, etc.). Masonry elements (especially those unreinforced) are known to be rather brittle and, hence, expected to undergo early damage when subjected to in-plane lateral deformations originated by differential settlements and indeed by seismic actions. In the recent decades, extensive research has been performed on the seismic behaviour of masonry structures, through testing of masonry components. Although valuable relevant information is available, in terms of maximum resistance, ductility properties and hysteretic behaviour, the data on damage and cracking of masonry components is, in most cases, limited to qualitative information. However, relevant quantitative data for various levels of loading (pre- and post-peak) are of importance (a) for the estimation of repair, strengthening, and retrofitting feasibility and cost and (b) for the seismic design of attachments to masonry, such as facade walls, equipment, etc. Thus, the aim of this paper is to collect and assess available published experimental data on the cyclic behaviour of masonry elements, in order to reach, wherever possible, a quantitative assessment of the observed damage (e.g. damage patterns, crack locations and widths, etc.). Limitations of the proposed approach due to limited available data and future research needs are also identified.

1 INTRODUCTION

Masonry structures are known to be quite vulnerable to seismic actions, as they undergo cracking at small displacement values, their ductility is limited, and their cyclic behaviour is rather poor. The seismic vulnerability of masonry structures was repeatedly confirmed during medium and strong earthquakes worldwide. The seismic behaviour of masonry was investigated by numerous researchers through testing of walls, subassemblies or building models. Relevant data on the seismic performance of masonry walls were collected in databases e.g. by Albanesi et al. (2018), Augenti et al. (2012), Gams et al. (2016), Morandi et al. (2018), Vanin et al. (2017-for stone masonry walls). The data provide valuable information on the hysteretic behaviour (in terms of force-response, force-displacement envelopes, stiffness degradation due to cycling and energy dissipation capacity) of masonry walls. On the contrary, limited quantitative information is provided about the cracking pattern and, more importantly, about the evolution of crack widths with loading.

More specifically, in most testing campaigns, crack widths are not systematically monitored, whereas in several cases the crack pattern at the end of testing (representative of advanced post-peak behaviour) is provided in photographs.

However, the cracking pattern and the width of cracks at various levels of the applied displacement width values are of significance for displacement-based design (as they assist the

selection of adequate performance levels, related also to the repairability of structures), as well as for the seismic design of attachments to masonry (including high-cost equipment).

Thus, in this paper the results of experiments on masonry walls, subjected to cyclic in-plane shear, are assessed, in order to draw quantitative information on the location and the evolution of cracks, in function of basic parameters, such as imposed drift, aspect ratio of walls, vertical load, boundary conditions, etc.

2 DESCRIPTION OF THE ASSESSED DATA

The results of one hundred and twenty-nine (129) tests on plain bearing masonry walls were assessed. Those tests cover a large variety of parameters, such as aspect ratio, solid or perforated masonry units (made of clay, calcium silicate, lightweight aerated concrete), mortar (general purpose or thin layer), mechanical properties of masonry units and mortars, type of head joints (filled, unfilled, tongue and groove, unfilled or filled in pocket etc.), boundary conditions of walls (fixed ends, cantilever, intermediate) and value of the imposed normal stress. The number of influencing parameters, as well as the variety of their values, makes the quantitative assessment of the experimental results a difficult task. As for the main purpose of this work, i.e. the identification of crack pattern and the evolution of cracks, further difficulties and uncertainties are encountered due to the limited relevant data included in the respective publications. As explained herein, conclusions of this work remain mostly qualitative.

2.1 Crack width measurements

In the majority of the investigations, global displacements were measured (e.g. horizontal displacement at the top of the wall or relative displacement between the base of the wall and its foundation). In some cases (e.g. Anthoine et al. 1995, Graziotti et al. 2016a, b, Salmanpour et al. 2013, 2015), the shortening/lengthening of the diagonals of the wall were measured. However, those measurements are not included in the respective publications. Nonetheless, selected crack width values are reported by some investigators; those are mostly values measured at an advanced stage of testing.

In general, the data that are available for assessment are: (a) A qualitative description of the failure mode, documented by photographs. An effort was made to estimate crack widths using the photographs. However, the quality of the photos is not adequate, whereas they usually show the wall after the completion of the test, i.e. after the application of large drifts corresponding to the advanced post-peak state of the tested walls. In other cases, drawings depicting the failure mode are provided (with no indication of crack widths though), (b) Presentation of crack width values, corresponding to an advanced stage of loading. In most cases, the crack widths are mentioned with no further indication on whether they refer to a horizontal or to a vertical mortar joint or whether a diagonal/stair stepped crack crosses mortar joints exclusively or masonry units as well (Paparo & Beyer 2014, Graziotti et al. 2016a, b).

In the last decade, several investigators (e.g. Graziotti et al. 2016a, b, Morandi et al. 2013a, Morandi et al. 2013b, Morandi et al. 2014, Paparo & Beyer 2014, Petry & Beyer 2014a, b, Salmanpour et al. 2013, 2015, Rosti et al. 2016) have applied advanced relative displacement measurement procedures like measurement using markers or LEDs. A limited number of these publications (Paparo & Beyer 2014, Petry & Beyer 2014a, b, Salmanpour et al. 2013, 2015) provide detailed continuous displacement measurements. Those experimental results, although limited in number compared to the total number of collected experimental results, are accessible and offer the possibility for quantitative assessment.

2.2 Normal stress level

In bearing walls tested in cyclic in-plane shear, there was a normal compressive stress, uniformly distributed and constant throughout testing. The range of this normal stress is quite large (σ-values between 0.14 and 4 MPa, Figure 1). It is believed

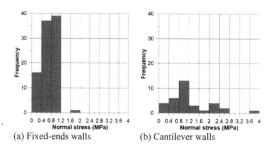

(a) Fixed-ends walls   (b) Cantilever walls

Figure 1. Normal stress values applied to tested bearing walls.

that in the majority of tests (especially in case of cantilever walls) the applied normal stress is unrealistically high. Actually, simple calculations (taking into account the largest allowed distances between walls, according to EC8-Part1 (CEN 2005) show that, for a two-storey unreinforced masonry building, the normal stress at the base of the ground floor does not exceed 0.50MPa. The authors of this paper cannot provide a plausible explanation for the selection of so high vertical stress values that are expected to affect the overall behaviour of the walls, including their failure mode. Indeed, a high normal stress value increases the probability of inclined strut failure, as documented by photographs included in the assessed publications (Figure 2).

2.3 Boundary conditions

With the exception of tests reported by Petry & Beyer (2014a, b), where different levels of fixity are provided at the two ends of the walls (4 out of 129 walls), the assessed experimental results refer either to cantilever (34 out of 129) or to fixed-end walls (91 out of 129). It should be noted that the requirements of current Codes for adequate diaphragm action of the floors and for enhanced connections between the floors and the walls lead to significant degree of fixity.

(a) Normal stress σ= 1.04MPa   (b) Normal stress σ= 1.54MPa

Figure 2. Failure of walls subjected to high normal stress (Petry & Beyer 2014a, b).

3 QUALITATIVE CRACK PATTERN INVESTIGATION

3.1 *Classification of failure modes*

The main failure modes and the associated crack patterns are as follows:

(a) Shear failure: shear diagonal cracks through mortar joints (or through masonry units as well) are formed (Figure 3 (a)). This failure mode is characterized by vertical cracks significantly larger than the horizontal ones (e.g. Petry & Beyer 2014a, b, Salmanpour et al. 2013, 2015). In squat walls (aspect ratio, $\alpha<1.0$), sliding cracks at the base of the walls occurred as well.
(b) Flexural failure: flexural horizontal cracks, at maximum bending moment locations, are dominant (Figure 3 (b)). Flexural cracks (usually combined with toe/corner crushing) reach typically larger width values than cracks in walls failing in shear. At drift levels corresponding to the post-peak behaviour, shear diagonal cracks may also appear (Figure 3 (b)).
(c) Hybrid failure mode: combination of flexural horizontal and shear diagonal cracks. The dominant (shear or flexural) failure mechanism depends on parameters like aspect ratio, normal stress, etc (Figure 3 (c)).

3.2 *Parameters affecting the failure mode*

The main parameters that affect the failure mode and the associated crack pattern are:

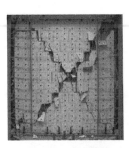 

(a) Shear failure mode   (b) Flexural failure mode

(c) Hybrid failure mode

Figure 3. Typical failure modes (photographs by Petry & Beyer 2014a).

(a) Aspect ratio: Although, expectedly, smaller aspect ratio ($\alpha=h/l$, h being the height and l being the length of the wall) values lead to shear failure mode, whereas a flexural failure is imminent in slender walls, the experimental results show a significant dependence of the failure mode on the value of the normal compressive stress acting on the wall. Indeed,
(b) Normal stress level: As the normal stress level increases, even walls of rather large aspect ratio, expected to fail in bending, may fail in shear. This was observed by several researchers (e.g. Morandi et al. 2013a, b, Morandi et al. 2014, Paparo & Beyer 2014, Petry & Beyer 2014a, Salmanpour et al. 2013, 2015). Some characteristic examples are the walls tested by Graziotti et al. (2016a) (of aspect ratio 2.26), Bosiljkov et al. (2003) (of aspect ratio 1.47): the walls subjected to high compressive stress (larger than 1.20MPa) exhibit a clear shear failure mechanism which involves small width flexural cracks at the two ends of the wall and mainly crushing of the compressed diagonal. On the contrary, when lower normal stresses are imposed (smaller than 1 MPa), flexural failure modes are dominant.
(c) Boundary conditions: The experimental results show that fixed-end walls tend to fail in shear, irrespectively of the aspect ratio and the normal stress values (Figure 4a, 83 walls out of 95 failed in shear). On the contrary, in case of cantilever walls, there is a clear effect of the aspect ratio on the failure mode (Figure 4b), which changes from shear to hybrid and to flexural, as the aspect ratio increases. However, there is a pronounced scatter of the experimental results, due also to the effect of the normal stress on the wall.
(d) Materials and their mechanical properties: This is a parameter that does affect the morphology of the cracks rather than the failure mode. Actually, when perforated units are used, the shear cracks are more likely to cross both units and mortar joints. The same holds true, also for solid bricks, when the mechanical properties of units and mortar are close to one another.

4 QUANTITATIVE CRACK WIDTH INVESTIGATION

4.1 *Introduction*

The data on crack widths correlated to drift values and to damage level being very limited, an effort was made to analyze in detail the measurements provided by Petry & Beyer (2014a, b). The detailed data, made available by the Authors (https://zenodo.org/record/8443?ln=en#.VF9rvslSILM), were obtained by using four LEDs per masonry unit that allow for the monitoring of their coordinates throughout testing. The corresponding photos showing the crack pattern of

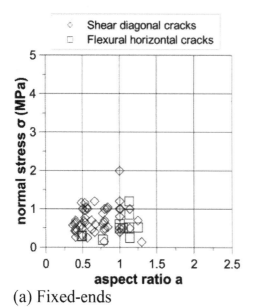

(a) Fixed-ends

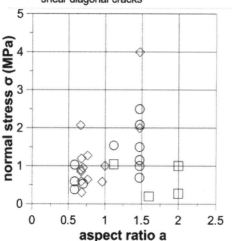

(b) Cantilever

Figure 4. Observed failure modes in function of the normal stress level and the aspect ratio (129 test results).

each specimen at each load step are available as well. Those data were used by the authors of this paper for the estimation of vertical and horizontal crack widths in the relevant locations of each wall, namely, along the diagonals (for shear failure modes) or close to the two ends of the wall (for flexural failure modes) or in both locations (for hybrid failures).

Three specimens were selected for this detailed evaluation of crack widths, one per failure mode, namely PUP1, PUP3 and PUP4. The walls are made of hollow clay bricks and mortar; they are identical in terms of mechanical properties and geometry (aspect ratio= 1.12) They differ in boundary conditions and value of normal compressive stress: (a) boundary conditions: PUP1 is fixed at its ends, while PUP3 and PUP4 are cantilevers and (b) normal stress: PUP1 and PUP3 are subject to a normal stress equal to 1.04MPa and PUP4 to σ=1.54MPa.

4.2 Shear failure mode-PUP1

The crack pattern involves the occurrence of cracks along the main diagonals, followed by secondary ones (parallel to those along the main diagonals, see circles in Figure 5b). At a later stage, after the attainment of the maximum resistance, disintegration of the masonry occurs along the diagonals (Figure 5c) and the wall is unable to sustain its vertical load (termed as vertical failure by the Authors).

The following observations can be made: (a) The crack opening in the vertical portions of the diagonal cracks are significantly larger than the horizontal portions of the cracks (Figure 6 and Table 1).

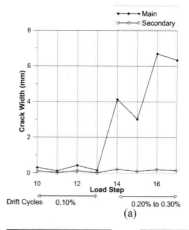

(a)

after failure in shear (d=+0.29%) - In circles, the areas where crack widths were calculated

(b)

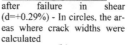

at vertical failure (d=+0.31% / -0.21%)

(c)

Figure 5. PUP1 - (a) Vertical crack widths on the main and secondary diagonals (based on Petry & Beyer 2014a, b), (b) and (c) the wall at failure in shear and at vertical failure (Petry & Beyer 2014a, b).

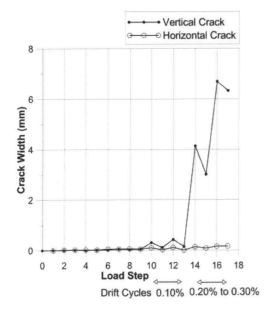

Figure 6. Comparison between vertical and horizontal crack widths along the main diagonals (PUP1, based on Petry & Beyer 2014a, b).

Table 1. Calculated crack width values on the (I) main and (II) secondary diagonals (PUP1, based on Petry & Beyer 2014a, b).

	Vertical cracks (mm)		Horizontal cracks (mm)	
	At max. resistance	Post Peak (at $0.8H_{max}$)	At max. resistance	Post Peak (at $0.8H_{max}$)
(I)	4.00	6.00-9.00	0.20	0.40-0.60
(II)	0.40	0.20-0.40	<0.10	0.10-0.20

This is a typical feature of shear failure, observed (in a qualitative way) in all investigations (e.g. Fehling et al. 2007, Salmanpour et al. 2013, 2015, Magenes et al. 2008, Morandi et al. 2013a, b, Morandi et al. 2015). Furthermore, (b) the attainment of the maximum resistance, as well as the post-peak behaviour is characterized by an abrupt increase of the vertical crack opening alone, (c) Up to the attainment of the maximum resistance, no disintegration of masonry is apparent. Thus, it seems that up to that point, the damage can be considered as repairable. Another interesting observation is that (d) even the vertical portions of the secondary diagonal cracks open at small widths (Figure 5a), thus suggesting that, when attachments to masonry are to be installed, the region along the main diagonals (of an average width of two masonry units length) should preferably be avoided (Figure 5b, c). Finally, (e) the maximum resistance is reached at a drift of

+0.15%/-0.12%, whereas vertical failure occurs at a drift of +0.31%/-0.21% (Petry & Beyer 2014a, b), where "+" and "-" stand for the positive and negative loading direction respectively. It is noted that, according to Petry and Beyer (2014a, b), the horizontal failure defined as the point (on the post-peak branch) corresponding to 20% loss of the peak load, occurs at a drift of +0.29%/-0.17%.

4.3 *Flexural failure mode-PUP3*

As expected, flexural behaviour is characterized by the occurrence of horizontal cracks at one or both end regions of the wall (depending on its boundary conditions). Secondary oblique cracks may occur at an advanced stage of loading (after the attainment of the maximum resistance). The opening of the horizontal cracks at maximum resistance (Figure 7) does not exceed 0.50mm, whereas it abruptly increases to 1.20mm for larger imposed drift.

The occurrence of diagonal cracks was initiated at a drift value of 0.40% (smaller than 0.51% that corresponds to the maximum resistance). However, their opening remains limited (smaller than 1.0mm), until vertical failure occurs at a drift of +0.84%/-0.94% (Figure 7d). Horizontal failure occurs at a drift of +0.72%/-0.93%, at a stage where the wall can be characterized as repairable (Figure 7c).

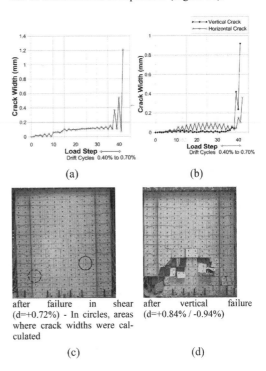

Figure 7. PUP3 - (a) Evolution of the width of flexural crack at the lower region of the wall, (b) vertical vs. horizontal crack width comparison on an oblique crack (based on Petry & Beyer 2014a, b), (c) and (d) the wall at failure in shear and at vertical failure (Petry & Beyer 2014a, b).

It is noted that flexural failures are associated with larger values of drift than shear failures, as observed in other investigations as well (e.g. Albanesi et al. 2018, Morandi et al. 2018).

In general, (a) damage is concentrated at the end regions of the wall and (b) the cracks are of limited opening (and, hence, repairable) until the maximum resistance is reached. However, (c) an abrupt increase of the cracks is observed and local disintegration of masonry occurs, when the imposed drift exceeds that corresponding to the horizontal failure.

4.4 Hybrid failure mode-PUP4

Hybrid failure modes exhibit characteristics of both shear and flexural modes with one of them being dominant depending on the parameters. The wall PUP4, although having the same boundary conditions and aspect ratio as PUP3 which failed due to flexural failure, was subjected to higher normal stress thus exhibiting a hybrid failure mode (Figure 2).

For PUP4, which reached its maximum resistance at a drift of +0.27%/-0.36%, horizontal failure at 0.35% and its vertical failure at a drift of +0.44%/-0.38% (Petry & Beyer 2014a, b), crack widths were calculated at key locations, i.e. at the lower portion of the wall (where flexural cracks occur) and along the cracked diagonals (Table 2). The crack width values and the respective plots (Figure 8) show the same trend as data on PUP1 and PUP3, namely, (a) larger openings in the vertical than in the horizontal portions of diagonal cracks, (b) flexural and diagonal cracks of limited opening (and, hence, repairable) up to the attainment of the maximum resistance and (c) abrupt increase of crack openings (followed by disintegration of masonry) for larger drift values. Nonetheless, (d) it seems (Figure 8) that this mixed failure mode is more unfavourable than the previous ones, in terms of percentage of the area of the wall that is expected to undergo damage.

4.5 Performance levels

According to EC8-Part 3 (CEN 2005) performance levels are set as a function of the failure mode. For example, for shear failure mode the

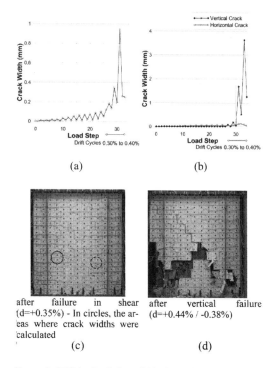

(a)   (b)

after failure in shear (d=+0.35%) - In circles, the areas where crack widths were calculated
(c)

after vertical failure (d=+0.44% / -0.38%)
(d)

Figure 8. PUP4 - Evolution of (a) Flexural crack, (b) vertical vs. horizontal crack width on the diagonal (based on Petry & Beyer 2014a, b), (c) and (d) the wall at failure in shear and at vertical failure (Petry & Beyer 2014a, b).

drift at "Significant Damage-SD" is set at 0.40%, whereas for the "Near Collapse-NC" state, defined at 20% loss of the peak load, a drift value equal to 0.53% is set. For flexural failure, SD-state corresponds to a drift value of $0.80H_0/L$ (H_0 being the distance between the section of max. bending moment and the contraflexure point and L being the length of the wall). NC-state corresponds to a drift value equal $1.07H_0/L$.

For hybrid failure mode, no limits for drift values are set by the Code.

In an effort to compare the Code provisions with the respective experimental values for drift, the authors of this paper have defined, on the basis of experimental load vs. drift curves, the values corresponding to cracking, as well as to SD- and NC-state of walls. As shown in Figure 9, cracking seems to occur at a drift not exceeding 0.20%, independently of any influencing parameter. It seems also that the drift values set by the Code for SD- and NC-states are sensibly conservative. The authors of this paper would like to add that, on the basis of the information provided in the Literature, it seems that the SD-state corresponds actually to a level at which the crack pattern is fully developed, but the walls are still repairable, as no disintegration of masonry has occurred and the openings of the cracks are not excessive.

Table 2. Calculated crack width values at selected locations (I) flexural cracks and secondary diagonals (II) main diagonal (PUP4, based on Petry & Beyer 2014a, b).

	Vertical Cracks (mm)		Horizontal Cracks (mm)	
	At max. resistance	Post Peak (at $0.8H_{max}$)	At max. resistance	Post Peak (at $0.8H_{max}$)
(I)	0.50	2.00	0.30	1.00
(II)	1.70	4.00	<0.10	0.10

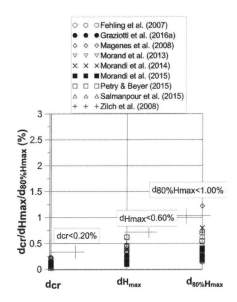

(a) Shear failure mode

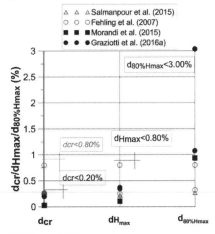

(b) Flexural failure mode

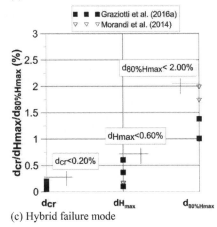

(c) Hybrid failure mode

Figure 9. Drift values at cracking, SD- and NC-states.

5 CONCLUSIONS

The survey of literature reporting results of testing unreinforced masonry walls in in-plane cyclic shear aimed at identifying the crack pattern and the evolution of crack widths in function of the imposed drift. This (definitely, not exhaustive) bibliographical investigation allows for the following conclusions to be drawn:

(1). Although cracks occurring in shear walls depend (in terms of pattern and width) on several geometrical and mechanical parameters (e.g. aspect ratio, type of masonry units and mortar, relative masonry unit to mortar strength, vertical load, etc.), it seems that up to the attainment of the maximum shear resistance of the walls, the opening of the cracks (either shear or flexural) remains limited. Thus, one may assume that up to that stage, the walls are easily repairable.

(2). However, for larger imposed drifts, leading to a limited decrease of the maximum resistance (by 20%), an abrupt increase of crack widths, associated with more or less localized disintegration of masonry is observed.

(3). It is, therefore, sensible to adopt (as EC8-Part 3 does) this stage as a near collapse state, as the related damage may be considered not to be easily repairable.

(4). Nonetheless, depending on the failure mode of the wall, one may distinguish areas that remain slightly damaged or free of damage and, hence, they are adequate for the installation of attachments, such as heavy equipment, plumbing and electrical facilities, etc.

(5). The assessed experimental results allow for design drift values to be adopted for various performance levels, depending on the expected failure mode of the wall. In this respect, it seems that the values adopted by EC8, taking into account the scatter of the experimental drift values, are in agreement with the experimental results.

(6). Finally, in most of the publications, limited quantitative data is provided on the evolution of the crack pattern and the width of cracks with increasing imposed drift. Due to the significance of such information (allowing for provisions for the repairability of walls, as well as for the design of attachments to masonry), it is believed that future researches should preferably include more quantified data on the development and evolution of cracks.

REFERENCES

Albanesi, L., Morandi P., Graziotti F., Li Piani T., Penna A., Magenes G. 2018. Database collecting in-plane test results of urm piers with bricks and blocks.

Proceedings of the 16th European Conference on Earthquake Engineering, Thessaloniki, 18-21 June 2018.

Anthoine, A., Magonette, G., Magenes, G. 1995. Shear-compression testing and analysis of brick masonry walls. Proceedings of the 10th European Conference on Earthquake Engineering, Vienna, 28 August-2 September 1994.

Augenti, N., Parisi, F., & Acconcia, E. 2012. MADA: Online experimental database for mechanical modelling of existing masonry assemblages. Proceedings of the 15th World Conference on Earthquake Engineering, Lisbon, 24-28 September 2012.

Bosiljkov, V., Page, A., Bokan-Bosiljkov, V., Žarnić, R. 2003. Performance based studies of in-plane loaded unreinforced masonry walls. Masonry International, 16(2), 39–50.

CEN (2005). Eurocode 8: design of structures for earthquake resistance - Part 1: General rules, seismic actions and rules for buildings. EN 1998-1:2004. Brussels (Belgium): European Committee for Standardisation; 2004.

CEN (2005). Eurocode 8: Design of structures for earthquake resistance – Part 3: Assessment and retrofitting of buildings. EN 1998-3:2004. Brussels (Belgium): European Committee for Standardisation; 2004.

Fehling, E., Stürz, J. & Emami, A. (2007). Enhanced safety and efficient construction of masonry structures in Europe. Test results on the behaviour of masonry under static (monotonic and cyclic) in plane lateral loads. Report ESECMaSE D7.1a. ESECMaSE-Report, Institute of Structural Concrete.

Gams, M., Triller, P., Lutman, M., & Snoj, J. 2016. Seismic behaviour of urm walls: Analysis of a database. In Modena, da Porto & Valluzzi (eds), Brick and Block Masonry: Trends, Innovations and Challenges - Proceedings of the 16th International Brick and Block Masonry Conference, IBMAC, Padova, 26-30 June 2016: 1593–1600. https://doi.org/10.1201/b21889-198.

Graziotti, F., Tomassetti, U., Rossi, A., Kallioras, S., Mandirola, M., Penna, A., & Magenes, G. 2016a. Experimental campaign on a clay urm full-scale specimen representative of the Groningen building stock. Report EUC128/2016U, EUCENTRE, Pavia, IT.

Graziotti, F., Rossi, A., Mandirola, M., Penna, A., Magenes, G. 2016b. Experimental characterization of calcium-silicate brick masonry for seismic assessment. Proceedings of the 16th International Brick and Block Masonry Conference, IBMAC, Padova, 26-30 June 2016.

Magenes, G., Morandi, P., Penna, A. 2008. Test results on the behaviour of masonry under static cyclic in plane lateral loads. Deliverable D7.1c, ESECMaSE Project, www.esecmase.org.

Morandi, P., Albanesi, L., & Magenes, G. 2013a. In-plane experimental response of masonry walls with thin shell and web clay units. C. Adam, R. Heuer, W. Lenhardt & C. Schranz (eds), Vienna Congress on Recent Advances in Earthquake Engineering and Structural Dynamics, Vienna, 28-30 August 2013.

Morandi, P., Magenes, G., & Albanesi, L. 2013b. Prove sperimentali per la valutazione della risposta sismica nel piano di pareti murarie in blocchi di laterizio a setti sottili. XV Convegno Nazionale ANIDIS - "L'INGEGNERIA SISMICA IN ITALIA," Padova, July.

Morandi, P., Albanesi, L., & Magenes, G. 2014. Urm walls with thin shell/web clay units and unfilled head-joints : Cyclic in-plane tests. Proceedings of the 2nd European Conference On Earthquake Engineering and Seismology, Istanbul, 25-29 August 2012: 1–12.

Morandi, P., Albanesi, L., Graziotti, F., Li Piani, T., Penna, A., & Magenes, G. 2018. Development of a dataset on the in-plane experimental response of URM piers with bricks and blocks. Construction and Building Materials, 190, 593–611. https://doi.org/10.1016/j.conbuildmat.2018.09.070.

Paparo, A. & Beyer, K. 2014. Quasi-static cyclic tests of two mixed reinforced concrete-unreinforced masonry wall structures. Engineering Structures, 71, 201–211. https://doi.org/10.1016/j.engstruct.2014.04.002 https://www.epfl.ch/labs/eesd/data-sets/data_sets/.

Petry, S. & Beyer, K. 2014a. Cyclic test data of six unreinforced masonry walls with different boundary conditions [Data set]. Earthquake Spectra. published online Zenodo. http://doi.org/10.5281/zenodo.8443.

Petry, S. & Beyer, K. 2014b. Influence of boundary conditions and size effect on the drift capacity of URM walls. Engineering structures, 65, 76–88. https://doi.org/10.1016/j.engstruct.2014.01.048 https://www.epfl.ch/labs/eesd/data-sets/data_sets/.

Rosti, A., Penna, A., Rota, M., & Magenes, G. 2016. In-plane cyclic response of low-density AAC URM walls. Materials and Structures/Materiaux et Constructions, 49 (11), 4785–4798. https://doi.org/10.1617/s11527-016-0825-5.

Salmanpour, A., N. Mojsilović, N. & Schwartz J. 2013. Experimental study of the deformation capacity of structural masonry. Proceedings of the 12th Canadian Masonry Symposium, Vancouver, 2-5 June 2013.

Salmanpour, A. H., Mojsilović, N., & Schwartz, J. 2015. Displacement capacity of contemporary unreinforced masonry walls: An experimental study. Engineering Structures, 89, 1–16. https://doi.org/10.1016/j.engstruct.2015.01.052.

Vanin, F., Zaganelli, D., Penna, A., & Beyer, K. 2017. Estimates for the stiffness, strength and drift capacity of stone masonry walls based on 123 quasi-static cyclic tests reported in the literature. Bulletin of Earthquake Engineering, 15(12), 5435–5479. https://doi.org/10.1007/s10518-017-0188-5.

Zilch, K., Finck, W., Grabowski, S., Schermer, D., Scheufler, W., (2008). Test results on the behaviour of masonry under static cyclic in plane lateral loads. Deliverable D7.1.b, ESECMaSE Project, www.esecmase.org.

Brick and Block Masonry - From Historical to Sustainable Masonry –
Kubica, Kwiecień & Bednarz (eds)
© 2020 Taylor & Francis Group, London, ISBN 978-0-367-56586-2

Virtual work method for evaluation of out-of-plane load/displacement capacity of topological semi-interlocking masonry infill panel

Y.Z. Totoev, O. Zarrin & M.J. Masia
The University of Newcastle, Australia

ABSTRACT: The Semi Interlocking Masonry (SIM) infill panels have been developed in Masonry Research Group at the University of Newcastle, Australia. The improving seismic performance is the main purpose of the SIM panels that is achieved by dissipating earthquake energy through in-plane sliding of brick courses on bed joints. The out-of-plane relative sliding of bricks in SIM panels is prevented by topological or mechanical interlocking. The SIM panels have reduced in-plane stiffness and increased displacement ductility and energy dissipation capacity compared to the traditional masonry panels. One of the problems of traditional masonry panels during earthquake excitation is their out-of-plane failure. The out-of-plane behavior of semi-interlocking masonry panel has not been extensively researched yet. The main objective of this paper is to present an analitical method for evaluation of the out-of-plane load/displacement capacity of a SIM panel. The developed method is based on the virtual work approach. Analytical predictions were compared to the experimental results for a topological SIM panel. The results show that the proposed method has good correlation with experimental results for the topological SIM panel and could be used potentially for estimation of the out-of-plane strength.

1 INTRODUCTION

Modern slender URM panels have inherently poor earthquake resistant ability due to their high mass and stiffness combined with low tensile and shear strength (Totoev & Wang. 2013). During the last decades engineers have been working on a solution to minimize the earthquake hazard for masonry infill panels. Masonry panel with some improvement in the units can change the attitude of masonry construction and can be considered as structural element (Elvin & Uzoegbo 2011). By removing the mortar from the joints of panels and changing the geometry of bricks, make dry-stack masonry panel more reliable with high ductility and absorbing energy capability (Dyskin et al. 2012). The main objective of this paper is to formulate an analitical method for evaluation of the out-of-plane load/displacement capacity of a SIM panel. The developed method is based on the virtual work approach. Analytical predictions were compared to the experimental results for a topological SIM panel. The panel was 1980×2025 mm (width x height) respectively, and 110 mm thick with full contact with the frame. A lateral load was applied to the central area of SIM panel. The results show that the proposed method has good correlation with experimental results for the topological SIM panel and could be used potentially for estimation of the out-of-plane strength.

2 SIM PANEL

SIM panels fall into category of segmented plates with topological or mechanical interlocking of units. SIM panels are made of large number of small segments, which are the masonry units identically connected. Therefore, on a macro scale, such panels can be considered quasi homogeneous. Because of interlocking in one direction only (hence the term semi interlocking) they are anisotropic. Several out-of-plane tests were performed on SIM panels (Zarrin et al. 2019), (Zarrin et al. 2018a). Deflections recorded in these tests were greater than the thickness of the panel. Almost none of the fundamental assumptions of the small deflection elastic theory of bending for thin plates is applicable to SIM panels.

3 BOUNDARY CONDITIONS

SIM panels are designed attached to the frame in order to ensure their out-of-plane support. One simple way to achieve this is to fasten to the frame metal angles along the entire perimeter of the panel on both sides. These angles could also serve as guides during the panel construction. Although this connection provides some limited restrain to rotation of panel's edges it would be more reasonable and conservative to assume that the panel is simply supported on the edges.

4 DEFLECTED SHAPE

Sinusoidal OOP deflected shape is assumed in both, vertical and horizontal direction for the panel. Where A is the area of compressed bricks; P is the thrust force; F is the reaction force. Deflection of the centre of each brick with coordinates (x, y) in terms of the maximum deflection in the centre of the panel Δ_{max} is

$$\Delta = \Delta_{max} \cdot \sin \frac{\pi x}{L} \cdot \sin \frac{\pi y}{H}. \tag{1}$$

The deflection of bricks in the row distance d from the bottom of the panel or in the column distance k from the left side of the panel is

$$\Delta_d = \Delta_{max} \cdot \sin \frac{\pi x}{L} \cdot \sin \frac{\pi d}{H};$$
$$\Delta_k = \Delta_{max} \cdot \sin \frac{\pi k}{L} \cdot \sin \frac{\pi y}{H}. \tag{2}$$

The slope of the horizontal tangent to the deflected row is

$$\frac{d\Delta_d}{dx} = \frac{\pi \Delta_{max}}{L} \cdot \cos \frac{\pi x}{L} \cdot \sin \frac{\pi d}{H}. \tag{3}$$

The maximum deflection in each row of bricks distance d from the bottom of the panel is in the middle of the row. The maximum deflection in each column of bricks distance k from the left side of the panel is in the middle of the column.

$$\Delta_{md} = \Delta_{max} \cdot \sin \frac{\pi L}{2L} \cdot \sin \frac{\pi d}{H} = \Delta_{max} \cdot \sin \frac{\pi d}{H};$$
$$\Delta_{mk} = \Delta_{max} \cdot \sin \frac{\pi k}{L} \cdot \sin \frac{\pi H}{2H} = \Delta_{max} \cdot \sin \frac{\pi k}{L}. \tag{4}$$

5 WORK METHOD

Work method is used to get an equation relating the maximum displacement of the panel Δ_{max} to the intensity of external uniformly distributed OOP load q. In this method, the work done by the external load over the OOP displacement Δ

$$W_{ext} = \sum_{\substack{for\ all \\ SIM\ units}} (q \cdot l \cdot h \cdot \Delta) \tag{5}$$

is equated to the work done by internal forces resisting OOP displacement W_{int} over corresponding relative (with respect to the neighbouring units)

displacements and rotations for all units in the panel. In addition to the self-weight, following in-plane membrane forces are induced in the panel when it is bent out-of-plane:

- Horizontal thrust force P_x due to horizontal arching of the panel between supports;
- Vertical thrust force P_y due to vertical arching of the panel between supports.

These forces together with the material properties and geometry of SIM units determine the magnitude of internal forces such as friction and interlocking resisting OOP relative displacements of units.

6 INTERNAL FORCES ON HEAD JOINTS

The horizontal thrust force P_x is constant in the compressed arch. It relates to the compressive stress induced on head joints during arching. The compressive stress in the middle of the arch is highest because the thickness of the arch $(t - \Delta)$ is the lowest there (Figure 1).

Let us adopt the following linear stress model for evolution of the average compressive stress on head joints $\bar{\sigma}$ in the row of bricks at the distance d from the bottom of the panel during arching (Zarrin et al. 2018b). In this model, s is the deflection in the middle of the row of bricks at which all head joints get closed on the loaded side of the panel and horizontal arching sets up, and $\bar{\sigma}_{md}$ is the maximum average compression that can be achieved in the horizontal compressed arch (Figure 2).

$$s = n_{hj} \cdot const_1, \tag{6}$$

$$\bar{\sigma}_{md} = f_{SIM} \cdot const_2, \tag{7}$$

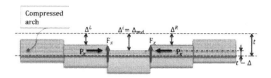

Figure 1. Horizontal arching.

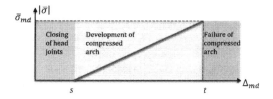

Figure 2. Linear stress model.

Where, n_{hj} is the number of head joints in the row of bricks, f_{SIM} is the compressive strength of SIM units. There are three distinct stages of arching in this linear stress model: $0 < \Delta_{md} \leq s$, no arching of the row because head joints are not closed;

1. $s < \Delta_{md} \leq t$, arching of the row;
2. $t < \Delta_{md}$, no arching because the compressed arch has failed.

The horizontal thrust force is a product of the compressive stress, which is a linear function of Δ_{md}, and the area of the horizontal compressed arch in the middle of the row, which is also a linear function of Δ_{md}. As a result, the horizontal thrust force is a quadratic function of Δ_{md}:

$$\forall(s < \Delta_{md} \leq t) : P_x = \bar{\sigma} \cdot A_{HCA}$$
$$= \left(\bar{\sigma}_{md} \frac{\Delta_{md} - s}{t - s}\right) \cdot h(t - \Delta_{md})$$

(8)

The thrust force is constant in the entire compressed arch for every value of Δ_{md} (Figure 3).

7 WORK OF FRICTION FORCES ON HEAD JOINTS

Friction forces on head joints are calculated from thrust force and are also constant in the compressed arch

$$\forall(s < \Delta_{md} \leq t) : F_x = P_x \cdot \tan \varphi$$
$$= \left(\bar{\sigma}_{md} \frac{\Delta_{md} - s}{t - s}\right) \cdot h(t - \Delta_{md}) \cdot \tan \varphi$$

(9)

Work done by friction forces on head joints over the relative displacement $\widehat{\Delta}^i$ of brick i with respect to its neighbours in the compressed arch is

$$W_x^i = \int F_x \cdot d\widehat{\Delta}^i$$

(10)

It varies in the row depending on the relative displacement of bricks. Let us evaluate the relative displacemen

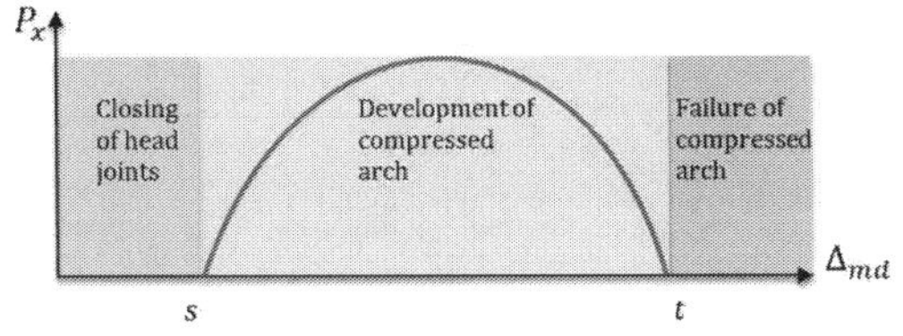

Figure 3. Thrust force model.

$$\Delta^L = \Delta_{md} \cdot \sin \frac{\pi(x - l)}{L} = \Delta_{md} \cdot \sin\left(\frac{\pi x}{L} - \frac{\pi l}{L}\right)$$
$$= \Delta_{md} \cdot \left(\sin \frac{\pi x}{L} \cdot \cos \frac{\pi l}{L} - \sin \frac{\pi l}{L} \cdot \cos \frac{\pi x}{L}\right),$$
$$\Delta^R = \Delta_{md} \cdot \sin \frac{\pi(x + l)}{L}$$
$$= \Delta_{md} \cdot \left(\sin \frac{\pi x}{L} \cdot \cos \frac{\pi l}{L} + \sin \frac{\pi l}{L} \cdot \cos \frac{\pi x}{L}\right),$$
$$\frac{\Delta^L + \Delta^R}{2} = \frac{2 \cdot \Delta_{md} \cdot \sin \frac{\pi x}{L} \cdot \cos \frac{\pi l}{L}}{2}$$

(11)

$$= \Delta_{md} \cdot \sin \frac{\pi x}{L} \cdot \cos \frac{\pi l}{L},$$
$$\widehat{\Delta}^i = \Delta^i - \frac{\Delta^L + \Delta^R}{2} = \Delta_{md} \cdot \sin \frac{\pi x}{L} - \Delta_{md} \cdot$$
$$\sin \frac{\pi x}{L} \cdot \cos \frac{\pi l}{L}$$
$$= \Delta_{md} \cdot \sin \frac{\pi x}{L}\left(1 - \cos \frac{\pi l}{L}\right).$$

Substituting Equations (10) and (12) into Equation (11) we can get equations for the work of friction forces on head joints of a brick:

$$\forall(0 < \Delta_{md} \leq s) : W_x^i = 0;$$
$$\forall(s < \Delta_{md} \leq t) : W_x^i$$
$$= \int_s^{\Delta_{md}} \left(\bar{\sigma}_{md} \frac{\Delta_{md} - s}{t - s}\right) \cdot h(t - \Delta_{md}) \cdot \tan \varphi \cdot$$
$$\sin \frac{\pi x}{L}\left(1 - \cos \frac{\pi l}{L}\right) \cdot d\Delta_{md} =$$
$$= \tan \varphi \cdot \sin \frac{\pi x}{L} \cdot \left(1 - \cos \frac{\pi l}{L}\right) \int_s^{\Delta_{md}} \left(\bar{\sigma}_{md} \frac{\Delta_{md} - s}{t - s}\right) \cdot$$
$$h(t - \Delta_{md}) \cdot d\Delta_{md} =$$
$$= \tan \varphi \cdot \sin \frac{\pi x}{L} \cdot \left(1 - \cos \frac{\pi l}{L}\right) \cdot \frac{\bar{\sigma}_{md} h}{6}$$
$$\cdot \frac{(s - \Delta_{md})^2(s - 3t + 2\Delta_{md})}{s - t};$$
$$\forall(t < \Delta_{md}) : W_x^i$$
$$= \int_s^t \left(\bar{\sigma}_{md} \frac{\Delta_{md} - s}{t - s}\right) \cdot h(t - \Delta_{md}) \cdot \tan \varphi \cdot$$
$$\sin \frac{\pi x}{L}\left(1 - \cos \frac{\pi l}{L}\right) \cdot d\Delta_{md} =$$
$$= \tan \varphi \cdot \sin \frac{\pi x}{L}$$
$$\cdot \left(1 - \cos \frac{\pi l}{L}\right) \int_s^t \left(\bar{\sigma}_{md} \frac{\Delta_{md} - s}{t - s}\right)$$
$$\cdot h(t - \Delta_{md}) \cdot d\Delta_{md} =$$
$$= \tan \varphi \cdot \sin \frac{\pi x}{L} \cdot \left(1 - \cos \frac{\pi l}{L}\right) \cdot \frac{\bar{\sigma}_{md} h}{6} \cdot (t - s)^2$$
$$= const.$$

(12)

461

Now we could calculate the work of friction forces on head joints for all bricks in a row at the fixed distance d from the bottom of the panel:

$$\forall(s<\Delta_{md}\leq t):W_x^d=\sum_{i=1}^{\frac{L}{l}}W_x^i=$$

$$=\tan\varphi\cdot\frac{\bar{\sigma}_{md}h}{6}\cdot\frac{\Delta_{md}^3-3s\Delta_{md}^2-s^3+3ts^2}{t-s}$$

$$\cdot\sum_{i=1}^{\frac{L}{l}}\sin\frac{\pi\left(il-\frac{l}{2}\right)}{L}\cdot\left(1-\cos\frac{\pi l}{L}\right)=$$

$$=\tan\varphi\cdot\frac{\bar{\sigma}_{md}h}{6}\cdot\frac{(s-\Delta_{md})^2(s-3t+2\Delta_{md})}{s-t}\cdot2\sin\frac{\pi l}{2L}=$$

$$=\frac{\bar{\sigma}_{md}h}{3}(s-\Delta_{md})^2\frac{s-3t+2\Delta_{md}}{s-t}\sin\frac{\pi l}{2L}\tan\varphi.$$

$$\forall(t<\Delta_{md}):W_x^d=\sum_{i=1}^{\frac{L}{l}}W_x^i=$$

$$=\tan\varphi\cdot\frac{\bar{\sigma}_{md}h}{6}\cdot(t-s)^2$$

$$\cdot\sum_{i=1}^{\frac{L}{l}}\sin\frac{\pi\left(il-\frac{l}{2}\right)}{L}\cdot\left(1-\cos\frac{\pi l}{L}\right)=$$

$$=\tan\varphi\cdot\frac{\bar{\sigma}_{md}h}{6}\cdot(t-s)^2\cdot2\sin\frac{\pi l}{2L}$$

$$=\frac{\bar{\sigma}_{md}h}{3}(s-t)^2\sin\frac{\pi l}{2L}\tan\varphi.$$

$$(13)$$

The work of friction forces on head joints for all bricks in the panel is

$$W_x^\Sigma=\sum_{j=1}^{\frac{H}{h}}W_x^d=\sum_{j=1}^{\frac{H}{h}}W_x^{\left(jh-\frac{h}{2}\right)}\qquad(14)$$

8   INTERNAL FORCES ON BED JOINTS

There are two main differences between internal forces on head and bed joints:

- Presence of self-weight G, which does not depend on arching effect;
- Extra resistance of bed joints to relative displacement due to topological interlocking, which depends on the maximum inclination angle α of the interlocking surface.

Equations for bed joints corresponding to equations 7 to 15 for head joints are slightly altered because of these two differences.

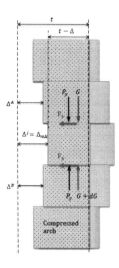

Figure 4. Vertical arching.

The vertical thrust force P_y is constant in the compressed arch. It relates to the compressive stress induced on head joints during arching. In addition to the thrust force, there is also compression due to self-weight (Figure 4).

We will use the same linear stress evolution model for the average compressive stress on bed joints $\bar{\sigma}^b$ in the "column" of bricks at the distance k from the left side of the panel due to arching. In this model, s^b is the deflection in the middle of the column of bricks at which all bed joints get closed on the loaded side of the panel and vertical arching sets up, and $\bar{\sigma}_{mk}$ is the maximum average compression that can be achieved in the vertical compressed arch.

$$s^b=n_{bj}\cdot const_3,\qquad(15)$$

$$\bar{\sigma}_{mk}=f_{SIM}\cdot const_4,\qquad(16)$$

Where, n_{bj} is the number of bed joints in the column of bricks, f_{SIM} is the compressive strength of SIM units, and constants should be determined experimentally. There are three distinct stages of arching:

1. $0<\Delta_{mk}\leq s^b$, no arching of the column because bed joints are not closed;
2. $s^b<\Delta_{mk}\leq t$, arching of the column;
3. $t<\Delta_{mk}$, no arching because the compressed arch has failed.

The vertical thrust force is a product of the compressive stress, which is a linear function of Δ_{mk}, and the area of the vertical compressed arch in the middle of the column, which is also a linear function of Δ_{mk}. As a result, the vertical thrust force is a quadratic function of Δ_{mk}:

$$\forall\left(s^b < \Delta_{mk} \leq t\right): P_y = \bar{\sigma} \cdot A_{VCA}$$

$$= \left(\bar{\sigma}_{mk} \frac{\Delta_{mk} - s^b}{t - s^b}\right) \cdot (l(t - \Delta_{mk})) \tag{17}$$

The thrust force is constant in the entire compressed arch for every value of Δ_{mk}. The self-weights is calculated for the part of the column above the centre of the brick in row at the distance d from the bottom of the panel

$$G_y = l \cdot t \cdot (H - d) \cdot \gamma \cdot g \tag{18}$$

were, l is the length of SIM brick, γ is its density, and g is the acceleration due to gravity. The friction forces on bed joints are calculated from the thrust force and the weight:

$$\forall\left(0 < \Delta_{mk} \leq s^b\right): F_y = G_y \cdot \tan(\varphi + \alpha);$$
$$\forall\left(s^b < \Delta_{mk} \leq t\right): F_y = (P_y + G_y) \cdot \tan(\varphi + \alpha);$$
$$\forall(t < \Delta_{mk}): F_y^{\Sigma} = G_y \cdot \tan(\varphi + \alpha) \tag{19}$$

9 WORK OF FRICTION FORCES ON BED JOINTS

Work done by friction forces on bed joints over the relative displacement $\widehat{\Delta}^j$ of brick j with respect to its neighbours above and below in the vertical compressed arch is

$$W_y^j = \int F_y \cdot d\widehat{\Delta}^j \tag{20}$$

It varies in the brick column depending on the relative displacement of bricks. Let us evaluate the relative displacement

$$\Delta^B = \Delta_{mk} \cdot \sin\frac{\pi(y - h)}{H} = \Delta_{mk} \cdot \sin\left(\frac{\pi y}{H} - \frac{\pi h}{H}\right) = \Delta_{mk} \cdot$$
$$\left(\sin\frac{\pi y}{H} \cdot \cos\frac{\pi h}{H} - \sin\frac{\pi h}{H} \cdot \cos\frac{\pi y}{H}\right),$$

$$\Delta^A = \Delta_{mk} \cdot \sin\frac{\pi(y + h)}{H} = \Delta_{mk} \cdot \left(\sin\frac{\pi y}{H} \cdot \cos\frac{\pi h}{H} + \sin\frac{\pi h}{H} \cdot \cos\frac{\pi y}{H}\right),$$

$$\frac{\Delta^B + \Delta^A}{2} = \frac{2 \cdot \Delta_{mk} \cdot \sin\frac{\pi y}{H} \cdot \cos\frac{\pi h}{H}}{2} = \Delta_{mk} \cdot \sin\frac{\pi y}{H} \cdot \cos\frac{\pi h}{H},$$

$$\widehat{\Delta}^j = \Delta^j - \frac{\Delta^B + \Delta^A}{2} = \Delta_{mk} \cdot \sin\frac{\pi y}{H} - \Delta_{mk} \cdot \sin\frac{\pi y}{H} \cdot \cos\frac{\pi h}{H}$$

$$= \Delta_{mk} \cdot \sin\frac{\pi y}{H}\left(1 - \cos\frac{\pi h}{H}\right) \tag{21}$$

It is convenient to calculate the contribution to the work from self-weight separately from the contribution from the vertical thrust force. Since the self-weight on bed joints does not change during out-of-plane bending of the panel, its contribution to the work of friction forces is a product of Equations (21) and (18):

$$W_{G_y}^j = l \cdot t \cdot (H - d) \cdot \gamma \cdot g \cdot \tan(\varphi + \alpha) \cdot \Delta_{mk} \cdot$$
$$\sin\frac{\pi y}{H}\left(1 - \cos\frac{\pi h}{H}\right) \tag{22}$$

Substituting Equations (20) into Equation (21) we can get equations for the contribution of the vertical thrust force to work of friction forces on bed joints of a SIM brick:

$$\forall\left(0 < \Delta_{mk} \leq s^b\right): W_{P_y}^j = 0;$$
$$\forall\left(s^b < \Delta_{mk} \leq t\right): W_{P_y}^j$$

$$= \int_{s^b}^{\Delta_{mk}} \left(\bar{\sigma}_{mk}\frac{\Delta_{mk} - s^b}{t - s^b}\right) \cdot l(t - \Delta_{mk}) \cdot \tan(\varphi + \alpha) \cdot$$

$$\sin\frac{\pi y}{H}\left(1 - \cos\frac{\pi h}{H}\right) \cdot d\Delta_{mk} =$$

$$= \tan(\varphi + \alpha) \cdot \sin\frac{\pi y}{H} \cdot$$

$$\left(1 - \cos\frac{\pi h}{H}\right) \int_{s^b}^{\Delta_{mk}} \left(\bar{\sigma}_{mk}\frac{\Delta_{mk} - s^b}{t - s^b}\right) \cdot l(t - \Delta_{mk}) \cdot d\Delta_{mk} =$$

$$= \tan(\varphi + \alpha) \cdot \sin\frac{\pi y}{H} \cdot \left(1 - \cos\frac{\pi h}{H}\right) \cdot$$

$$\frac{\bar{\sigma}_{mk}l}{6} \cdot \frac{\Delta_{mk}^3 - 3s^b\Delta_{mk}^2 - \left(s^b\right)^3 + 3t\left(s^b\right)^2}{t - s^b};$$

$$\forall(t < \Delta_{mk}) : W_{P_y}^j = \int_{s^b}^{t} \left(\bar{\sigma}_{mk}\frac{\Delta_{mk} - s^b}{t - s^b}\right) \cdot l(t - \Delta_{mk}) \cdot$$

$$\tan(\varphi + \alpha) \cdot \sin\frac{\pi y}{H}\left(1 - \cos\frac{\pi h}{H}\right) \cdot d\Delta_{mk} =$$

$$= \tan(\varphi + \alpha) \cdot \sin\frac{\pi y}{H} \cdot$$

$$\left(1 - \cos\frac{\pi h}{H}\right) \int_{s^b}^{t} \left(\bar{\sigma}_{mk}\frac{\Delta_{mk} - s^b}{t - s^b}\right) \cdot l(t - \Delta_{mk}) \cdot d\Delta_{mk} =$$

$$= \tan(\varphi + \alpha) \cdot \sin\frac{\pi y}{H} \cdot \left(1 - \cos\frac{\pi h}{H}\right) \cdot \frac{\bar{\sigma}_{mk}l}{6} \cdot \left(t - s^b\right)^2 = const \tag{23}$$

Now we could calculate the work of friction forces on bed joints for all bricks in a column at the fixed distance k from the left side of the panel:

$$\forall\left(s^b < \Delta_{mk} \le t\right) : W_y^k = \sum_{j=1}^{\frac{H}{h}} W_y^j =$$

$$= \tan(\varphi + \alpha)\cdot$$

$$\frac{\bar{\sigma}_{mk} l}{6} \cdot \frac{\Delta_{mk}^3 - 3s^b\Delta_{mk}^2 - s^{b^3} + 3ts^{b^2}}{s^b - t}\cdot$$

$$\sum_{j=1}^{\frac{H}{h}} \sin\frac{\pi\left(jh - \frac{h}{2}\right)}{H}\cdot\left(1 - \cos\frac{\pi h}{H}\right)$$

$$= \tan(\varphi + \alpha)\cdot\frac{\bar{\sigma}_{mk} l}{6}\cdot$$

$$\frac{(6 - \Delta_{mk})^2\left(s^b - 3ts + 2\Delta_{mk}\right)}{s^b - t}\cdot 2\sin\frac{\pi h}{2H}$$

$$= \frac{\bar{\sigma}_{mk} l}{3}\left(s^b - \Delta_{mk}\right)^2\cdot\frac{s^b - 3t + 2\Delta_{mk}}{s^b - t}\cdot\sin\frac{\pi h}{2H}\cdot\tan\varphi$$

$$\forall(t < \Delta_{mk}) : W_y^k = \sum_{j=1}^{\frac{H}{h}} W_y^j =$$

$$= \tan(\varphi + \alpha)\cdot\frac{\bar{\sigma}_{mk} l}{6}\cdot(t - s)^2\cdot$$

$$\sum_{i=1}^{\frac{H}{h}} \sin\frac{\pi\left(jh - \frac{h}{2}\right)}{H}\cdot\left(1 - \cos\frac{\pi h}{H}\right)$$

$$= \tan(\varphi + \alpha)\cdot\frac{\bar{\sigma}_{mk} l}{6}\cdot(t - s)^2\cdot 2\sin\frac{\pi h}{2H}$$

$$= \frac{\bar{\sigma}_{mk} l}{3}\cdot(t - s)^2\cdot\sin\frac{\pi h}{2H}\cdot\tan\varphi \tag{24}$$

The work of friction forces on bed joints for all bricks in the panel is

$$W_y^\Sigma = \sum_{i=1}^{\frac{L}{l}} W_y^k = \sum_{i=1}^{\frac{L}{l}} W_y^{\left(il - \frac{l}{2}\right)} \tag{25}$$

The total work of the SIM panel can be calculated by a combination of each head and bed joints work.

$$W^\Sigma = W_x^\Sigma + W_y^\Sigma = \sum_{j=1}^{\frac{H}{h}} W_x^{\left(jh - \frac{h}{2}\right)} + \sum_{i=1}^{\frac{L}{l}} W_y^{\left(il - \frac{l}{2}\right)} \tag{26}$$

10 COMPARISON TO THE EXPERIMENT

It is well known that predictive capacity of methods based on the virtual work approach (for example, the yield line method) depend strongly on the correct assumption about the crack pattern. In this respect,

one big advantage of a SIM panel is that its "crack pattern" is known exactly and consists of a regular pattern of hed and bead joints. There was only one test performed so far on a topological SIM panel, where all the required data for comparison with the developed analytical method are available. Hence, this section should not be considered as a verification of the presented method, but rather a comparison of results. Full set of experimental results for OOP test on topological panel was presented elsewhere [5]. The tested panel has been laterally loaded in the central area by a hydraulic jack through the steel plate. For comparison to the presented method, the the force needs to be converted from kN to the equivalent uniformly distributed load (kPa). This conversion was based on the equivalent displacement in the center of the panel (Eq. 28) (Timoshenko 1939). Where E is the elastic modulus of units, η is the ratio of length to height of the panel, L is the length of the panel, t it is the thickness of the panel, μ is the Poisson ratio, and q is distributed load.

$$q = \frac{Et^3\Delta}{\eta L^4 12(1 - \mu^2)} \tag{27}$$

Figure 7 Shows the original lateral load/deflection experimental results, the results converted to the equivalent UDL, and the analytical prediction based on the presented method. In the topological SIM panel test which include putty joint filler, all four expected distinct stages were observed, however, experimental results beyond the failure point (accepted at 20% reduction to the load capacity) are not shown in Figure 5. An initial low stiffness stage refers to the squeezing of the joint filler (putty). At this stage (up to about 25 mm deflection in the middle of the panel) the compressive membrane force between the bricks are very small and the panel deflects easily under a low pressure. The second stage (up to about 100 mm deflection) on the graph shows an increase in stiffness of the panel. At this stage the membrane forces were

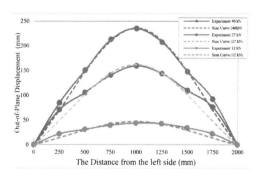

Figure 5. Deflection along the centreline of the pane.

464

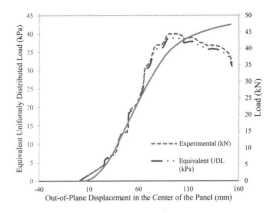

Figure 6. Load-deflection diagrams for topological SIM panel.

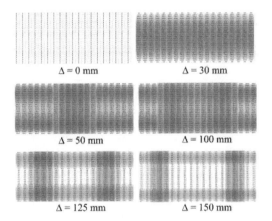

$\Delta = 0$ mm $\Delta = 30$ mm

$\Delta = 50$ mm $\Delta = 100$ mm

$\Delta = 125$ mm $\Delta = 150$ mm

Figure 7. Evolution of horizontal and vertical arching in the SIM panel.

gradually activated in both vertical and horizontal direction. The last stage of the graph shows a gradual degradation of compressive arches. The virtual work method presented in this paper assumes sinusoidal deflected shape of the panel in both directions. To check this assumption, some of the experimental and simulated deflected shapes are compared in Figure 6. The comparison demonstrates that the assumption about the deflected shape is valid. The virtual work method does not model softening of the brick material explicidly. As a result, it is unable to predict the forth – failure stage of the observed behaviour. The virtual work simulations were carried out to the 3% convergence of the load capacity. To calculate the work in horizontal and vertical direction two constants have been determined experimentally [6].

For topological SIM units, $const_1 \cong 0.4$mm, $const_2 \cong 0.31$Pa. And for the vertical direction the constants have been found: $const_3 \cong 1.4$mm, $const_4 \cong 0.17$Pa.

The experimental failure point was assumed at 31.1 kPa corresponding to the 20% drop in the OOP load capacity of the panel. The load capacity of tested panel was 38.7 kPa and the displacement capacity (at failure) was 150 mm. The maximum displacement of virtual work method related to the convergence of 1 percent of ultimate load was also 150 mm. The virtual work method predicted the load capacity of 42.5 kPa. The difference between the experimental and predicted load capacity is 9%, and for displacement capacity is less than 1%. Figure 7 shows the development and evolution of compressive membrane forces in the simulated SIM panel. The green colour represents the horizontal arching forces and the brown colour represent the vertical arching. The darcker the shade the larger the force. At zero displacement there is no arching. At 30 mm displacment just the horizontal arching has been activated; it is graeter in the middle of the panel height. At 50 mm displacement the vertical arching also activated in the middle of the panel. However, the horizontal arches are already deteriorating in the middle height. The maximum horizontal arching effect is gradually moving up and down from the middle. When the deflection in the middle approaches (at 100mm) the panel's thickness, the maximum arching effect moved from center toward the supports in both directions. After the deflection in the middle (at 125mm) exceeds the panel's thickness, arching through the central part of the panel is not possible and we can see only two bands of residual arching near supported adges of the panel.

11   CONCLUSION

An analytical method for evaluating the load/displacement capacity of topological SIM panels was presented. The method is based on the virtual work approach. Its prediction were compared to the experimental results for one SIM panel. The comparison show that the proposed method has good correlation with experimental results and could be used potentially for estimation of the out-of-plane strength of topological SIM panels.

REFERENCES

Totoev, Y. and Z. Wang. 2013. In-plane and out-of-plane tests on steel frame with SIM infill. in *12th Canadian Masonry Symposium*. 2013.

Elvin, A. and H. Uzoegbo. 2011. Response of a full-scale dry-stack masonry structure subject to experimentally applied earthquake loading. *Journal of the South African Institution of Civil Engineering= Joernaal van die Suid-Afrikaanse Instituut van Siviele Ingenieurswese, 53(1)*: 22–32.

Dyskin, A.V., E. Pasternak, and Y. Estrin, 2012. Mortarless structures based on topological interlocking. *Frontiers of Structural and Civil Engineering, 2012*: 1–10.

Zarrin, O., Y. Totoev, and M. Masia. 2019. The out-of-plane behaviour of SIM infill panel with mechanical interlocking: experimental investigation, in *The 13th North American Masonry Conference*: Salt Lake City, UT.

Zarrin, O., Y. Totoev, and M. Masia, 2018. The Out-of-Plane Capacity of Semi-Interlocking Infill Panels: Experimental Investigation, in *10th Australasian Masonry Conference*. Sydney, Australia.

Zarrin, O., Y. Totoev, and M. Masia. 2018. Evaluating Thrust Force Induced in Semi-Interlocking Masonry Panel during Horizontal and Vertical Arching, in *10th International Masonry Conference International Masonry Conference*. Milan, Italy,.

Timoshenko, S.P. 1939. Bending of rectangular plates with clamped edges. in *Proc of 5th Int Congress for Applied Mechanics*.

Brick and Block Masonry - From Historical to Sustainable Masonry –
Kubica, Kwiecień & Bednarz (eds)
© 2020 Taylor & Francis Group, London, ISBN 978-0-367-56586-2

Fully automatic evaluation of local mechanisms in masonry aggregates through a NURBS-based limit analysis procedure

N. Grillanda, M. Valente & G. Milani
Department of Architecture, Built Environment and Construction Engineering (ABC), Politecnico di Milano, Milan, Italy

A. Chiozzi & A. Tralli
Department Engineering, University of Ferrara, Ferrara, Italy

ABSTRACT: Masonry aggregates constitute a large part of the structural typologies in historical centers worldwide. The peculiar characteristics of ancient aggregates (such as structural heterogeneity, uncertainties about walls connections, geometric irregularities and, usually, diffuses damage) suggest the use of analysis methods that investigate local failure mechanisms under seismic actions. In this work, the study of masonry aggregates subjected to horizontal actions through local analyses is presented. An upper bound limit analysis based on a NURBS (Non-Uniform Rational B-Spline) representation of the geometry of the whole construction and a mesh adaptation by Genetic Algorithm (GA) is adopted. The great efficiency of NURBS in the representation of curved geometries allows analyzing aggregates characterized by masonry arches and vaults with low computational effort. Since this procedure is fully automatic, local mechanisms on complex geometries can be assessed in an easy way. The analysis of a complex masonry aggregate is presented as meaningful case study.

1 INTRODUCTION

A high number of historical masonry constructions is constituted by masonry aggregates, i.e. buildings erected in continuity each other during time. The seismic behavior of these structures is quite difficult to predict, because of the several factors which affect the response to horizontal actions: among others, geometrical irregularities, uncertainties on connections between walls, difficulties on the identification of masonry mechanical parameters, and lack of seismic design details.

Despite the complexity characterizing masonry structures, the high number of aggregates of constructions that can be found in historical centers required the development of simplified methods for quick estimations of the seismic vulnerability. In order to satisfy this request, some methods based on the use of appropriate indexes have been recently presented (Lourenço & Roque, 2006; Rota, Penna & Magenes, 2014; Lagomarsino & Cattari, 2015). However, more detailed analyses are generally needed to achieve a thorough knowledge and an accurate evaluation of the seismic behavior and vulnerability of different typologies of masonry constructions, especially in the case of conventional or innovative retrofitting interventions (Valente & Milani, 2019a; Habieb et al., 2019a, 2019b).

Several methods for the seismic assessment of historical masonry constructions can be found in the technical literature (Roca et al., 2010). It is worth mentioning the numerical strategies based on micro- and macro-modeling approaches for the representation of the mechanical behavior of the masonry material (Lourenço, 2002). In particular, the use of the macro-modeling representation of masonry, in some cases enriched with homogenization procedures (Scacco et al., 2020), has allowed to perform advanced global pushover analyses and nonlinear dynamic analyses on global finite element models of masonry aggregates (Grillanda, Valente, et al., 2020; Valente et al., 2019), even in presence of complex geometries (see for example (Scacco et al., 2019)). Among pushover analyses, it is also worth to mention the Equivalent Frame Method (Roca, Molins & Marì, 2005), which has been largely applied on masonry aggregates after the implementation in the software TREMURI (Lagomarsino et al., 2013).

However, taking into account the typical characteristics of masonry aggregates, in the great majority of cases these structures do not show global response when subjected to horizontal loads. Moreover, among masonry structures damaged during recent seismic events in Italy, a high number of local failures have been observed (Valente & Milani, 2018, 2019b). Therefore, one of the most used methods for investigations on the seismic vulnerabilities of historical masonry aggregates is local analysis. The analysis of partial failure mechanisms is straightly recommended by the Italian code (NTC2018, Circolare 21/01/2019) for historical masonry buildings. Here, a kinematic limit analysis is applied to pre-assigned mechanisms and the classical no-tension material is adopted (Heyman, 1966). A series of typical partial failure mechanisms can be quickly evaluated through the well-known application C.I.N.E. (ReLUIS, DPC and CNR-ITC, 2009) developed by ReLUIS (ReLUIS, 2010). Despite the effectiveness and the correspondence with the observed failures during past seismic events, such a method presents some drawbacks, which are the overestimations due to the use of pre-assigned mechanisms, the limitation to a standard set of failures, and finally the use of the no-tension material. Recent research showed that the use of a suited three-dimensional failure surfaces is more suited to some typologies of masonry constructions. A series of limit analysis procedures based on the use of failure surfaces obtained through homogenization techniques (Milani, Lourenço & Tralli, 2006a, 2006b, 2007) have been presented. Moreover, an innovative recently published procedure (Tiberti & Milani, 2019) is aimed to determine the failure surface of a generally irregular masonry texture (which is a typical situation in historical buildings).

On the frontier of limit analysis tools, the Authors have recently proposed a new limit analysis procedure initially applied to masonry vaults. Such a method, the so-called GA-NURBS limit analysis (Chiozzi, Milani & Tralli, 2017), apply an upper bound formulation on a model discretized through NURBS (Non-Uniform Rational B-Spline) surfaces. NURBS surfaces are common in the CAD environment and are particularly suited to represent curved geometries. Since the method uses an upper bound limit analysis, a mesh adaptation through Genetic Algorithm is implemented in order to optimize the load multiplier by finding the real collapse mechanisms. Some applications of this procedure can be found in (Chiozzi, Grillanda et al., 2018; Chiozzi, Milani et al., 2018; Grillanda, Chiozzi, Bondi et al., 2019; Grillanda, Chiozzi, Milani et al., 2019).

A very recent paper of the Authors (Grillanda, Valente et al., 2020) showed the effectiveness of limit analysis, and in particular of the GA-NURBS limit analysis, in the seismic assessment of historical masonry aggregates. Therefore, in this work, the GA-NURBS approach has been implemented in a fully automatic tool that assesses the seismic vulnerability of aggregates through advanced evaluations of partial failure mechanisms. The study of a complex masonry aggregate (belonging to the historical center of Arsita (Indirli et al., 2014)) containing a high number of vaulted elements is presented in order to show the efficiency of the tool.

2 NURBS LIMIT ANALYSIS

This local failure analysis procedure relies on a discretization of the whole masonry aggregates through few NURBS surfaces. A NURBS surface (Non-Uniform Rational B-Spline) is a parametric surface defined by a bidirectional net of control points $\mathbf{B}$ and the NURBS basis functions, whose mathematical formulation is the following:

$$R_{i,p}(\xi) = \frac{N_{i,p}(\xi).W_i}{\sum_{i=1}^n N_{i,p}(\xi).W_i} \qquad (1)$$

where:
$\Xi = \{\xi_1, \xi_2, \ldots \xi_{n+p+1}\}$ is a non-uniform knots vector, $w_i \in$ define a set of weights, and $N_{i,p}$ are the spline basis functions of degree p Therefore, a NURBS surface $\mathbf{S}$ in the parametric coordinates u and v is defined as follows:

$$\mathbf{S}(u,v) = \sum_{i=0}^n \sum_{j=0}^m R_{i,j}(u,v)\mathbf{B}_{i,j} \qquad (2)$$

Authors remand to (Piegl & Tiller, 1995) for details on the mathematical formulation about NURBS objects. The application of the NURBS modeling strategy on masonry aggregates allow to take into account curved elements, such as arches, vaults, and circular openings, in easy way. The NURBS model of an aggregate is realized in Rhinoceros and then imported into MATLAB, where each surface is converted into a 3D macro-element once thickness and offset properties are assigned. Moreover, each initial surface can be subdivided into few sub-surfaces in order to obtain a mesh of NURBS elements in which the initial geometry remains still the same.

By considering these elements as infinitely rigid and assigning a rigid-perfectly plastic behavior along interface only, an upper bound limit analysis can be formulated on this model. Since relative displacements can occur only along interfaces, these assume the meaning of possible fracture lines. Each interface is discretized through points, in which a local reference system $(\mathbf{n}, \mathbf{s}, \mathbf{t})$ (Figure 1) is defined: once a suited failure surface is assigned, on each point the correlation between relative displacements and internal dissipation power is imposed by implementing the plastic flow rule. In this model, a Mohr-Coulomb behavior with tension cut-off and linear cap in compression is considered (Figure 2, (Milani & Taliercio, 2015)).

Once a load configuration composed with dead- and live-load is defined (respectively $\mathbf{F}^0$ and), the live-load multiplier Γ is found by applying the Principle of Virtual Powers. The procedure can be written according to the following linear programming formulation:

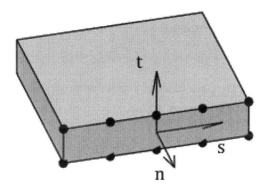

Figure 1. Discretization of interfaces and local reference system.

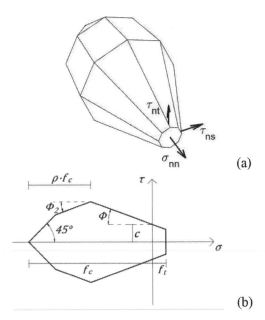

Figure 2. (a) Linearized 3D failure surfaces and (b) 2D section.

$$\min\left\{\Gamma = \sum_i P_{int}^i - P_{F_0}\right\} \text{ such that } \left\{\begin{array}{c} \mathbf{A}_{eq}\, \mathbf{X} = \mathbf{b}_{eq} \\ \dot{\lambda} \geq 0 \end{array}\right.$$

(3)

where:
$\sum_i P_{int}^i$ is the total amount of internal power;
$P_{F_0} + P_\Gamma$ is the power dissipated by the external loads;

$\mathbf{X}$ is the vector of unknowns, which include the velocity components of each element and the non-negative plastic multipliers $\dot{\lambda}$;

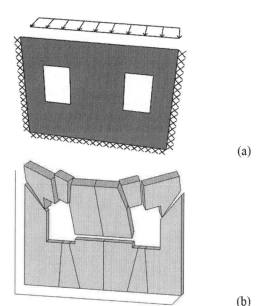

(a)

(b)

Figure 3. Example of an out-of-plane loaded wall constrained in external edges: (a) load condition and (b) collapse mechanism obtained.

$\mathbf{A}_{eq}$, $\mathbf{b}_{eq}$ are respectively the overall equality constraints matrix (containing geometric constraints, plastic compatibility and normalization condition of live loads power) and the corresponding right-hand side vector.

As a result, the linear programming provides a load multiplier and a mechanism which takes place starting from the hypothesized position of fracture lines. According to the kinematic theorem of limit analysis, if the correct collapse mechanism is not reproduced, the associated multiplier is an overestimation of the collapse multiplier. Therefore, the initial mesh has to be adjusted in order to find the real position of fracture lines. The mesh adaptation is performed through a Genetic Algorithm (GA). In this way, a good estimation of the live load multiplier and the collapse mechanism is provided (see Figure 3 for a simple example).

More details about this procedure can be found in (Chiozzi, Milani & Tralli, 2017).

3 APPLICATION ON A HISTORICAL MASONRY AGGREGATE

The masonry aggregate here studied is located in the historical center of Arsita, a small village in Central Italy which was hit by the 2009 earthquake. Widespread damage has been observed on masonry constructions (Formisano et al., 2012) as a consequence of the seismic event. After this event, 17 masonry aggregates have been inspected by a scientific team set up by ENEA (Italian National Agency for New Technologies, Energy and Sustainable Economic Development), resulting in the publication of the Post-Earthquake Reconstruction Plan for the Arsita Municipality (Chiarini & Indirli, 2013; Indirli et al., 2014). One of this building, the masonry aggregate named "Il Corso", has been chosen as meaningful case study for this work (see Figure 4). This construction is the final result of the aggregation of some residential buildings, which constitute the structural units A and B (see Figure 4(d)), and a masonry church, the Church of Santa Vittoria, which is the unit C. The masonry observed is characterized by the presence of irregular stones and bad quality mortar. The whole structure can be inscribed into a rectangle 47 m x 15 m, whereas the maximum height is about 15 m in correspondence of the bell tower. All the structural units are characterized by thick walls: an average thickness of 70 cm has been observed on units A and B, whereas in unit C values vary from a minimum of 90 cm to a maximum of 200 cm (which is the façade). No chains and concrete edgings are present, suggesting the complete absence of box behavior under horizontal load.

The presence of a high number of vaults makes this aggregate a very meaningful example. Several semicircular barrel vaults can be found in the central nave of the church, plus another one located adjacently to the triumphal arch. Moreover, some masonry arches are located in unit B, supported by the perimeter walls.

Failure mechanisms on the masonry aggregate "Il Corso" have been studied through the proposed upper bound limit analysis. According to the prescriptions of the Italian Code (NTC2018, Circolare 21/01/2019), the live load applied consists in a configuration of horizontal load proportional to masses. Therefore, once a group of macro-blocks has been selected and boundary conditions are defined, the analysis provides the horizontal load multiplier α_0 and the collapse mechanisms involving the selected elements. By assigned different boundary conditions, which can consist of geometric constraints or different resistance parameters specifically for the external edges of the selected macro-blocks, different hypotheses on the interlockings can be taken into account. It is clear that this versatility becomes useful when a complete knowledge of design details of the aggregates has been reached, i.e. when the quality of interlocking on each intersection between walls has been properly quantified.

Since no information about connections is available for the presented masonry aggregate, no connection has been supposed for interfaces between different structural units, whereas good interlocking is considered between orthogonal walls belonging to the same unit. Mechanical parameters of masonry are reported in Table 1 (see Figure 2 for the meaning of symbols). Finally, since the perimeter walls are usually the most critical elements of masonry aggregates, as shown in a previous study (Valente et al., 2019), here, for each structural unit, analyses have been focused on these elements.

Results in terms of local failures, horizontal load multipliers and spectral accelerations are depicted in Figure 5. The use of NURBS has been particularly useful in the evaluation of local failures when curved shaped openings (Figure 5(a)), arches (Figure 5(b)) and vaults (Figure 5(d,e)) assume a fundamental role in the collapse.

The most evident cases are the mechanisms obtained for the church. The main façade collapses by overturning with participations of the supported masonry vaults (Figure 5(d)), which present the typical failure of arches subjected to horizontal

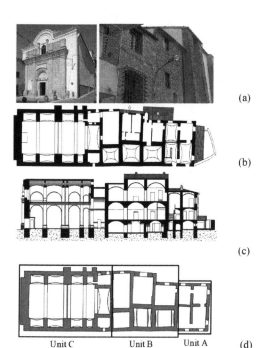

(a)

(b)

(c)

Unit C   Unit B   Unit A   (d)

Figure 4. Aggregate "Il Corso": (a) some photos (pictures taken from (Chiarini & Indirli, 2013), (b) plan view, (c) section along the longitudinal direction and (d) identification of the structural units.

Table 1. Parameters defining the adopted failure surface (referring to Figure 2(b)).

Parameter	Value
Ultimate tensile stress (f_t)	0.05 *walls* - 0.1 *vaults* MPa
Ultimate compression stress (f_c)	2 MPa
Cohesion (c)	0.035 MPa
Friction angle (Φ)	22°
Linear cap in compression (ρ, Φ_2)	0.5, 10°

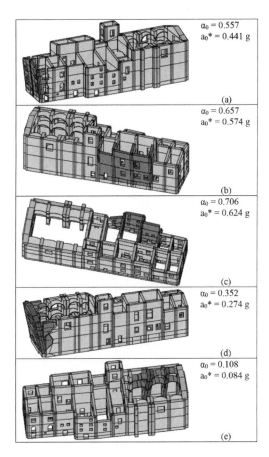

$\alpha_0 = 0.557$
$a_0^* = 0.441$ g

(a)

$\alpha_0 = 0.657$
$a_0^* = 0.574$ g

(b)

$\alpha_0 = 0.706$
$a_0^* = 0.624$ g

(c)

$\alpha_0 = 0.352$
$a_0^* = 0.274$ g

(d)

$\alpha_0 = 0.108$
$a_0^* = 0.084$ g

(e)

Figure 5. Results obtained: collapse of perimeter walls of units (a) A, (b, c) B, and (d) C, and (d) failure of vaults in unit C.

settlements at the base (Tralli et al., 2019). Moreover, the portions of nave walls involved in the failure collapse through a combination of sliding and tensile failure. This is a natural consequence of the low quality of mechanical properties assumed for masonry compared with the high dimensions of the blocks characterizing the church. When the horizontal response of the vaults in the central nave is studied, the collapse occurs because of the overturning of supporting walls, whereas vaults come to failures because of the movement of their supports.

The mechanism presenting the lowest horizontal load multiplier is the horizontal response of the vaults in the central nave, which is characterized by a horizontal load multiplier equal to 0.108 and a corresponding spectral acceleration of 0.084 g. As regards the other results obtained, the high values of horizontal load multiplier are mainly due to the high dimensions of walls and the internal dissipation included in the computations.

It is worth noting that these results cannot be achieved a-priori by using the C.I.N.E. application or following standard hand calculations. The main advantage of this method is the possibility to find automatically the shape of the failure mechanisms, avoiding the overestimation of the seismic capacity due to the use of pre-assigned mechanisms. Moreover, the computational effort remains low even in presence of vaulted structures, which do not require subdivision into a high number of elements thanks to NURBS properties.

4 CONCLUSIONS

A computational procedure for the seismic assessment of local failures in historical masonry aggregates has been presented. The procedure relies on an upper bound limit analysis based on NURBS geometries and a GA mesh adaptation. The NURBS discretization coupled with the mesh adjustment through GA allows performing local analyses in an efficient way even in case of complex geometries and by taking into account different interlocking conditions. Differently from the well-known CINE application or standard hand calculations, which require the use of pre-assigned mechanisms, this method provides failure mechanisms of general shape in a fully automatic way. The presence of vaults does not affect the computational efforts, which remains low thanks to the use of NURBS. Moreover, the procedure is not limited to the classical no-tension material but permits to use one or more Mohr-Coulomb failure surfaces, allowing to take into account also the presence of different masonry textures.

The analysis of a complex historical masonry aggregate is presented as proof of the potentiality of the procedure. Good results in terms of local failures, horizontal load multipliers and spectral accelerations have been found.

Therefore, the local failure analysis through the GA-NURBS upper bound limit analysis has proven to be effective in providing the seismic assessment of historical masonry aggregates.

REFERENCES

Chiarini, R. & Indirli, M. 2013. Piano di ricostruzione del comune di Arsita (Te) a seguito dell'evento sismico del 6/04/2009. (in Italian).

Chiozzi, A., Milani, G., Grillanda, N. & Tralli. A. 2018. A fast and general upper-bound limit analysis approach for out-of-plane loaded masonry walls, *Meccanica*, 53(7):1875–1898. https://doi.org/10.1007/s11012-017-0637-x.

Chiozzi, A., Grillanda, N., Milani, G. & Tralli. A. 2018. UB-ALMANAC: An adaptive limit analysis NURBS-based program for the automatic assessment of partial failure mechanisms in masonry churches, *Engineering Failure Analysis*, 85:201–220. https://doi.org/10.1016/j.engfailanal.2017.11.013.

Chiozzi, A., Milani, G. & Tralli, A. 2017. A Genetic Algorithm NURBS-based new approach for fast kinematic limit analysis of masonry vaults, *Computers & Structures*, 182:187–204. https://doi.org/10.1016/j.compstruc.2016.11.003.

Circolare 21/01/2019, n. 7 C.S.LL.PP. Istruzioni per l'applicazione dell'«Aggiornamento delle 'Norme tecniche per le costruzioni'» di cui al decreto ministeriale 17 gennaio 2018. (in Italian).

Formisano, A., Florio, G., Fonti, R., Candigliota, E., Immordino, F., Marzo, A., Moretti, L., Indirli, M., Abate, D., Furini, G., Pierattini, S., Screpanti, A., Angelini, M., Gambatesa, T. & Massaia, C. 2012. Seismic vulnerability of historic centres: identification of damage mechanisms occurred in Arsita (TE) after L'Aquila earthquake, *15th World Conference of Earthquake Engineering (15th WCEE)*, 24–28.

Grillanda, N., Valente, M., Milani, G., Chiozzi, A. & Tralli, A. 2020. Advanced numerical strategies for seismic assessment of historical masonry aggregates, *Engineering Structures*, 212(110441). https://doi.org/10.1016/j.engstruct.2020.110441.

Grillanda, N., Chiozzi, A., Milani, G. & Tralli, A. 2019. Collapse behavior of masonry domes under seismic loads: an adaptive NURBS kinematic limit analysis approach, *Engineering Structures*, 200(109517). https://doi.org/10.1016/j.engstruct.2019.109517.

Grillanda, N., Chiozzi, A., Bondi, F., Tralli, A., Manconi, F., Stochino, F. & Cazzani, A. 2019. Numerical insights on the structural assessment of historical masonry stellar vaults: the case of Santa Maria del Monte in Cagliari, *Continuum Mechanics and Thermodynamics*. 1–24. https://doi.org/10.1007/s00161-019-00752-8.

Habieb, A.B., Valente, M. & Milani, G. 2019a. Effectiveness of different base isolation systems for seismic protection: numerical insights into an existing masonry bell tower. *Soil Dynamics and Earthquake Engineering*, 125 (105752). https://doi.org/10.1016/j.soildyn.2019.105752.

Habieb, A.B., Valente, M. & Milani, G. 2019b. Base seismic isolation of a historical masonry church using fiber reinforced elastomeric isolators. Soil Dynamics and Earthquake Engineering, 120:127–145. https://doi.org/10.1016/j.soildyn.2019.01.022.

Heyman, J. 1966. The stone skeleton, *International Journal of Solids and Structures*, 2(2):249–256.

Indirli, M., Bruni, S., Geremei, F., Marghella, G., Marzo, A., Moretti, L., Formisano, A., Castaldo, C., Esposito, L., Florio, G., Fonti, R., Spacone, E., Biondi, S., Miccadei, E., Vanzi, I., Tralli, A., Vaccaro, C. & Gambatesa, T. et al. 2014. The reconstruction plan of the town of Arsita after the 2009 Abruzzo (Italy) seismic event, *9th International Conference of Structural analysis of Historical Constructions*, 14–17.

Lagomarsino, S., Penna, A., Galasco, A. & Cattari, S. 2013. TREMURI program: An equivalent frame model for the nonlinear seismic analysis of masonry buildings, *Engineering Structures*, 56:1787–1799. https://doi.org/10.1016/j.engstruct.2013.08.002.

Lagomarsino, S. & Cattari, S. 2015. Seismic Performance of Historical Masonry Structures Through Pushover and Nonlinear Dynamic Analyses, in. *Perspectives on European Earthquake Engineering and Seismology*, 265–292. https://doi.org/10.1007/978-3-319-16964-4_11.

Lourenço, P.B. 2002. Computations on historic masonry structures, *Progress in Structural Engineering and Materials*, 4(3):301–319. https://doi.org/10.1002/pse.120.

Lourenço, P.B. & Roque, J.A. 2006. Simplified indexes for the seismic vulnerability of ancient masonry buildings, *Construction and Building Materials*, 20(4):200–208. https://doi.org/10.1016/j.conbuildmat.2005.08.027.

Milani, G., Lourenço, P.B. & Tralli, A. 2006a. Homogenised limit analysis of masonry walls, Part I: Failure surface, *Computers & Structures*, 84(3–4):166–180. https://doi.org/10.1016/j.compstruc.2005.09.004.

Milani, G., Lourenço, P.B. & Tralli, A. 2006b. Homogenised limit analysis of masonry walls, Part II: Structural examples, *Computers & Structures*, 84(3–4):181–195. https://doi.org/10.1016/j.compstruc.2005.09.004.

Milani, G., Lourenço, P.B. & Tralli, A. 2007. 3D homogenized limit analysis of masonry buildings under horizontal loads, *Engineering Structures*, 29(11):3134–3148. https://doi.org/10.1016/j.engstruct.2007.03.003.

Milani, G. & Taliercio, A. 2015. In-plane failure surfaces for masonry with joints of finite thickness estimated by a Method of Cells-type approach, *Computers & Structures*, 150:34–51. https://doi.org/10.1016/j.compstruc.2014.12.007.

NTC2018. D.M. 17/01/2018 Aggiornamento delle «Norme tecniche per le costruzioni». S.O. alla G.U. n. 42 del 20/02/2018; 2018. (in Italian).

Piegl, L. & Tiller, W. 1995. *The NURBS Book*, Berlin: Springer. https://doi.org/10.1007/978-3-642-59223-2.

ReLUIS. 2010. Linee guida per il rilievo, l'analisi ed il progetto di interventi di riparazione e consolidamento sismico di edifici in muratura in aggregato, 3:105. (in Italian).

ReLUIS, DPC & CNR-ITC. 2009. C.I.N.E. Condizioni d'Instabilità degli Edifci—Applicativo per le verifche sismiche dei meccanismi di collasso fuori dal piano negli edifci esistenti in muratura mediante analisi cinematica lineare. Available at: http://www.reluis.it/doc/emergenza_terremoto_abruzzo/%0ACINE_1.0.4.xls. (in Italian).

Roca, P., Cervera, M., Gariup, G. & Pelà, L. 2010. Structural analysis of masonry historical constructions. Classical and advanced approaches, *Archives of Compututational Methods in Engineering*, 17(3):299–325. https://doi.org/10.1007/s11831-010-9046-1.

Roca, P., Molins, C. & Marí, A.R. 2005. Strength capacity of masonry wall structures by the equivalent frame method, *Journal of Structural Engineering*, 131 (10):1601–1610. https://doi.org/10.1061/(ASCE)0733-445(2005)131:10(1601).

Rota, M., Penna, A. & Magenes, G. 2014. A framework for the seismic assessment of existing masonry buildings accounting for different sources of uncertainty, *Earthquake Engineering and Structural Dynamics*, 43 (7):1045–1066. https://doi.org/10.1002/eqe.2386.

Scacco, J., Salazar, G., Bianchini, N., Mendes, N., Cullimore, C. & Jain, L. 2019. Seismic assessment of the church of Carmo convent, in *Congresso de Métodos Numéricos em Engenharia. 1–3 July 2019 Guimarães, Portugal*.

Scacco, J., Ghiassi, B., Milani, G. & Lourenço, P.B. 2020. A fast modeling approach for numerical analysis of unreinforced and FRCM reinforced masonry walls under out-of-plane loading, *Composites Part B:*

Engineering, 180:107553. https://doi.org/10.1016/j.compositesb.2019.107553.

Tiberti, S. & Milani, G. 2019. 2D pixel homogenized limit analysis of non-periodic masonry walls, *Computers & Structures* 219:16–57. https://doi.org/10.1016/j.compstruc.2019.04.002.

Tralli, A., Chiozzi, A., Grillanda, N. & Milani, G. 2019. Masonry structures in the presence of foundation settlements and unilateral contact problems, *International Journal of Solids and Structures*. https://doi.org/10.1016/j.ijsolstr.2019.12.005.

Valente, M. & Milani, G. 2018. Seismic response and damage patterns of masonry churches: seven case studies in Ferrara, Italy, *Engineering Structures*, 177:809–835. https://doi.org/10.1016/j.engstruct.2018.08.071.

Valente, M. & Milani, G. 2019a. Advanced numerical insights into failure analysis and strengthening of monumental masonry churches under seismic actions, *Engineering Failure Analysis*, 103:410–430. https://doi.org/10.1016/j.engfailanal.2019.05.009.

Valente, M. & Milani, G. 2019b. Earthquake-induced damage assessment and partial failure mechanisms of an Italian Medieval castle, *Engineering Failure Analysis*, 99:292–309. https://doi.org/10.1016/j.engfailanal.2019.02.008.

Valente, M., Milani, G., Grande, E. & Formisano, A. 2019. Historical masonry building aggregates: advanced numerical insight for an effective seismic assessment on two row housing compounds, *Engineering Structures*, 190:360–379. https://doi.org/10.1016/j.engstruct.2019.04.025.

Brick and Block Masonry - From Historical to Sustainable Masonry –
Kubica, Kwiecień & Bednarz (eds)
© 2020 Taylor & Francis Group, London, ISBN 978-0-367-56586-2

Characterization of mortar and stone masonry quality in Amatrice historical buildings hit by the 2016 central Italy earthquake

E. Cescatti, M. Secco, F. da Porto* & G. Artioli*
Inter-Departmental Research Centre for the Study of Cement Materials and Hydraulic Binders, Padova, Italy

C. Modena*
ITC-CNR, Italy

L. Xu*
Earthquake Engineering Research & Test Center – Guangzhou University, Guangzhou, China

ABSTRACT: The 2016 Central Italy seismic swarm, and in particular the first stroke on 24th of August in Amatrice, hit a rural zone on Apennine composed by a typical masonry made of rubble stone, arranged in two leaves with weak mortar. The first weakness of rubble masonry is related to masonry disaggregation, which does not allow a macro-element activation.

The paper presents the research carried mainly in the Amatrice region to characterize the masonry quality and its properties. Firstly, physical and chemical characterization of the constituting mortars was performed analysing 33 mortar specimens. The bulk characterization was performed through X-ray powder diffraction (XRPD) coupled with scanning electron microscopy and X-ray microanalysis (SEM-EDS). Those analyses, compared with the analysis of the masonry quality looking at the arrangement of the entire panel, provided an evaluation of the quality and gave an insight of the main issues of the materials, defining three different mortar groups summarized in two masonry typologies. A case study is presented to focus on the accelerations which lead to the masonry disaggregation compared to those associated with the mechanism activation.

The paper aims at characterizing the masonry pointing out the weaknesses which led to the building collapse, with the goal to propose data and information crucial to design efficient and effective strengthening interventions which can avoid the dramatic effects observed.

1 INTRODUCTION

The central Italy seismic swarm occurred in 2016 highlighted the weaknesses of the typical rural masonry of the area, made of rubble stone.

The main event of the swarm which stroke the Amatrice area was on the 24th of august in Accumuli (M_w 6.0), a village close to the main centre. Other relevant events occurred on the 26th of October (M_w 5.4 and 5.9) with epicentre in Castelsantangelo sul Nera (MC), the 30th of October (M_w 6.5) in Norcia (PG), and on the 18th of January 2017 four strokes with M_w greater than 5 had epicentre between Rieti and L'Aquila provinces.

A clear relationship between damages observed and the masonry vulnerability is evident comparing collapses observed in Amatrice after the 24th of august and the following stroke of the 30th of October with the scenario of Norcia, in which damages were more limited thanks to consolidation interventions previously carried out.

Some studies carried out in similar regions due to past earthquakes (e.g. Binda et al. (1999)) provided some description of this masonry typology. Furthermore, injection interventions are proposed in literature to retrofit these elements (Valluzzi et al. 2004). Considering the typical ranges of Italian masonry typologies defined in the Italian guidelines (Boschi et al. 2016), this could be associated to the weaker class.

The following study focuses on the masonry quality in the Amatrice area, looking at its arrangement and at the mortars used, in order to understand the observed damage and to implement a strengthening approach to avoid fragile collapses of the entire structures.

---

* China-Italy International Research Center for Protection of Historical Architectures and Cultural Relics

2 DESCRIPTION OF THE STUDY

The Amatrice plateau lies between 900 and 1000 m on the sea level and it is located in-between the Laga mountains, reaching heights of 2400 m. The system of villages and construction typologies are strongly influenced by the environment. For instance, the mountains are constituted of marls and sandstones, whereas the other Apennine groups are mainly composed of limestone. Villages are spread over the area and are composed of few isolated structures with few habitants each. The municipality is 174 km^2 wide with a total population of 2461, spread in 97 villages.

During the emergency surveys for building usability after some earthquakes occurred in the region (Abruzzo 1984, Umbria-Marche 1997 and L'Aquila 2009), a dataset of 150'000 buildings was built. In this dataset the 83% of buildings are made of masonry and more than a half have an irregular masonry or a poor masonry quality (Rosti et al. 2017). Moreover, the 2011 ISTAT data on Amatrice reports that 90% of the buildings are made of masonry and 60% of them were constructed before the 1960's. Due to this representativeness of the masonry and due to the observed collapses (Figure 1) the need for the local population to study and to propose interventions is high.

A multi-analytical campaign was conducted for the characterization of the structural mortars constituting the binding agents of the typical masonry of the Amatrice area. 30 mortar samples were taken from historical residential buildings of several villages in the area showing different degrees of damage. Care was taken in choosing sampling areas not characterized by evident degradation of the binders, in order to analyse the pristine properties of the mortars. The samples were taken from deeper masonry portions, removing superficial renders when present, and sampling the materials manually through hammers and chisels. A complete list of the analysed samples is reported in Table 2, together with the village of provenance. The table reports also all specimen labels and locations, with the ID of the village that refers to Figure 2. For each place it is reported also the maximum PGA recorded during the seismic swarm with indication of the data. As is

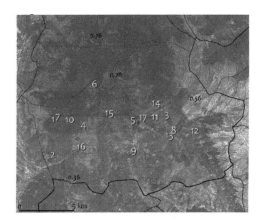

Figure 2. Position of villages in the Amatrice area.

possible to observe, for the entire area the worst event was the one of the 24th of august.

Figure 2 highlights the municipality boundary (dashed) and the region one (continuous line). In red In red is reported a contour of accelerations and Amatrice is the spot in dark blue and spot number 17.

3 ANALYSIS OF BINDERS

3.1 *Materials and methods*

The samples were characterized through a combination of petrographic, mineralogical and microchemical-microstructural analyses, in order to obtain a thorough definition of their compositional and textural properties.

First of all, mineralogical analyses were performed on all the bulk samples through quantitative phase analysis based on X-ray powder diffraction data (XRPD-QPA), clustering the materials according to their composition through multivariate statistical treatments, according to the methodology described by Secco et al. (2019). The materials were subject to micronization by a McCrone mill, using a plastic jar with agate grinding elements and ethanol 99% as micronizing fluid. Furthermore, 20 wt% of ZnO was added to the powders for the quantification of the paracrystalline fraction through the internal standard method. Data were collected using a Bragg-Brentano θ-θ diffractometer (PANalytical X'Pert PRO, Co Kα radiation, 40 kV and 40 mA) equipped with a real-time multiple strip (RTMS) detector (X'Celerator by Panalytical). Diffraction patterns were interpreted using the X'Pert High-Score Plus 3.0 software by PANalytical, qualitatively reconstructing mineral profiles of the compounds by comparison with PDF databases from the International Centre for Diffraction Data (ICDD). Then, QPAs were performed using the Rietveld (1969) method. Refinements were

Figure 1. Damages in Sant'Angelo (l) and Retrosi (r).

accomplished with TOPAS software (version 4.1) by Bruker AXS. The starting structural models for the refinements were taken from the International Crystal Structure Database (ICSD).

Subsequently, a selection of samples pertaining to the defined mineralogical groups were subject to a preliminary petrographic study, performed following the macroscopic and microstratigraphic analytical procedures for the study of mortar-based building materials described in Standard UNI 11176:2006 "Cultural heritage - Petrographic description of a mortar". Mass color was defined through Munsell soil color charts (Munsell 2000). The study was performed both on untreated portions, through macroscopic and stereomicroscope observations, and by transmitted light optical microscopy (TL-OM) on 30 μm thin sections, obtained by vacuum impregnating portions of the materials with epoxy resin and sectioning them transversally.

Finally, the same samples were microstructurally and microchemically characterized by scanning electron microscopy and energy-dispersive microanalysis (SEM-EDS). A CamScan MX2500 scanning electron microscope has been used, equipped with a lantanium hexaboride cathode and a four-quadrant solid state BSE detector for imaging. The analytical conditions were: accelerating voltage 20 kV; filament current 1.80 A; emission current 20 μA; aperture current 300 nA; working distance 20-30 mm. Furthermore, an EDAX-EDS energy dispersive X-rays fluorescence spectrometer was used for chemical microanalysis, mounting a Sapphire Detector composed by a LEAP+ Si(Li) crystal and a Super Ultra-Thin Window. Qualitative interpretation of spectra and semiquantitative chemical analysis were performed through SEM Quant Phizaf software.

3.2 Result of the mineralogical characterisation

The results of the quantitative mineralogical analysis on the selected binding material samples are shown in Table 2.

The mortars all have similar qualitative mineralogical profiles, given by a clear prevalence of silicate phases, with dominant quartz and associated feldspars of the oligoclase and microcline type. Carbonate phases are also present, with calcite clearly predominant over dolomite, as well as substantial amounts of clay minerals of the illite and chlorite type.

Given these similarities in the qualitative mineralogical profiles, which indicate a local supply of raw materials for all the samples analysed (Scarsella et al. 1955), quantitative mineralogical analysis has revealed significant compositional differences between the materials, as well as the constant occurrence of significant amounts of amorphous phase, due to the presence of nanostructured clay fractions dispersed in the binder matrices.

On the whole, the data obtained made it possible to group the mineralogical phases according to the fundamental constituent of relevance in the initial mortar mix design:
- Fraction of aggregate, consisting of a predominantly silicate sand (quartz, oligoclase and microcline) with an associated carbonate fraction (dolomite and part of calcite);
- Calcic binder fraction, consisting of anthropogenic calcite derived from the carbonation process of lime during setting and hardening (Cizer et al. 2010);
- Clayey binder fraction, consisting of illite, chlorite and nano-clays of the smectite type, derived from local soil aliquots deliberately added to the mixture or contained within the aggregate fraction not purified by washing and decantation.

Based on this subdivision and the results obtained, the samples analysed were grouped into three groups characterized by similar compositional profiles:
- Group 1: earthen mortars (see Table 2). These samples are characterized by calcite contents always below 10 wt%, with silicate aggregate fractions in-between 60 and 70 wt%, together with clay fractions in-between 30 and 40 wt%. From the data obtained, it is evident that the use of lime as binding material in this group of mortars is negligible if not null, therefore the cohesive properties of the mixtures are due only to the mild physical interactions between the clay particles (Onnis et al. 2009);
- Group 2: earthen-lime mortars (see Table 2) These samples are characterized by calcite contents in-between 10 and 20 wt%, with lower silicate and clayey aggregate fractions than the previous group. Such experimental evidence suggests for this class of binding materials the combined occurrence in the original mix design of a fraction of aerial lime combined with an aliquot of clayey soil, mixed with a silicate sandy aggregate.

The cohesive properties of the mixtures are for this group guaranteed by the chemical carbonation process of aerial lime (Cizer et al. 2010), together with the physical interactions between the dispersed clay particles;
- Group 3: lime mortars (see Table 2). In these samples calcite is always significantly higher than the other materials analysed, at values tending to exceed 25 wt%, counterbalanced by significantly lower levels of silicate aggregate and clay fraction. The presence of higher rates of dolomite compared to those found in the samples of the other groups makes it possible to hypothesize the occurrence of a more relevant fraction of carbonate aggregate, and therefore to reallocate part of the calcite fraction to the aggregate itself.

Nevertheless, the data obtained clearly indicate that these mortars can be considered as standard aerial binding materials, in which the cohesive properties are guaranteed by the chemical process of carbonation of calcic lime during the setting and hardening phases.

3.3 Petrographic, microstructural and microchemical characterisation

Optical and electronic microscopy analyses confirmed the subdivision into groups obtained by mineralogical analyses, providing further evidence on the textural properties of the materials.

Group 1 mortars can be macroscopically distinguished by their greenish-brown color (5Y 5/6 Munsell), arenaceous-silty dimensional appearance and very poor cohesion. Microscopically (Figure 3a), they are characterized by a binder/aggregate ratio of 1:3. The binder matrix has a microcrystalline texture and a homogeneous structure, consisting exclusively of optically isotropic clayey phases. The occurrence of negligible fractions of carbonated lime is further confirmed by the total absence of lime lumps. The inert fraction is always poorly sorted, with a unimodal grain size distribution centred in the fine sands granulometric range according to the Wentworth (1922) scale (max. size: 800 μm; mean size: 140 μm). Furthermore, a relevant silty fraction is always present. The aggregate is generally highly spherical and sub-rounded/sub-angular, evenly distributed in the mortar and not oriented. From a mineral-petrographic point of view, it is characterized by the exclusive occurrence of silicate minerals, with a prevalence of quartz and associated plagioclase feldspar and potassic feldspar. The general porosity of the mortar is generally 5-7 % of the total volume, with a clear prevalence of irregular vughs with an average size of 1 mm and subordinate spherical voids. The materials are not showing features related to ongoing alteration processes. Electron microscopy and point chemical microanalyses have clearly confirmed the optical microscopy observations.

The materials (Figure 4a, b) are composed of a fine silicate aggregate blandly bound by a microporous matrix with lamellar texture, strongly fractured and largely composed of clayey phases, as confirmed by the EDS microanalyses (Figure 4c, d), which indicate dominant aliquots of silicon and aluminium, associated with alkalis, magnesium and iron, typical compositional profiles for clayey phases (Meunier, 2005). The extremely low calcium values confirm that the calcic binder is present in these materials in absolutely negligible aliquots, if not completely absent.

Group 2 mortars are macroscopically distinguishable by their yellowish color (5Y 7/6 Munsell), arenaceous-silty dimensional appearance and medium-low

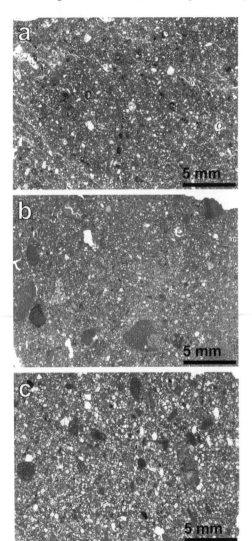

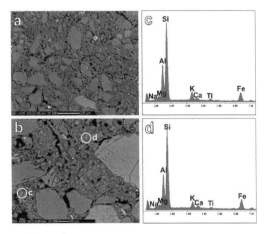

Figure 3. Sample CC_M1 (group 1), transmitted-light optical micrographs (parallel nicols); b) Sample CG_M1 (group 2), transmitted-light optical micrographs (parallel nicols), lime lumps highlighted in red; c) Sample RI_M1_A (group 3), transmitted-light optical micrographs (parallel nicols).

Figure 4. Sample CC-M1 (group 1): a, b) backscattered electron micrographs; c, d) EDS microanalyses of the binder matrix points indicated in b).

cohesion. Microscopically (Figure 3b), they are characterized by a binder/aggregate ratio of 1:3. The binder matrix is characterized by a heterogeneous microcrystalline texture, with portions characterized by typical interference colors of anthropogenic calcite associated to areas clearly showing the occurrence of optically isotropic clay fractions, both dispersed and clustered to form lumps. The two components of the binder fraction present evident mixing problems, as evidenced by the occurrence of numerous millimeter-sized lime lumps (highlighted in red in Figure 3b). The inert fraction is always poorly sorted, with a unimodal grain size distribution centred in the fine sands granulometric range (max. size: 800 μm; mean size: 140 μm) and the presence of a relevant silty fraction. The aggregate is generally highly spherical and sub-rounded/sub-angular, evenly distributed in the mortar and not oriented. From a minero-petrographic point of view, it is characterized by the exclusive occurrence of silicate minerals, with a prevalence of quartz and associated plagioclase and potassic feldspar. The general porosity of the mortar is 5-7 % of the total volume, with a clear prevalence of irregular vughs with an average size of 1 mm and subordinate spherical voids. The materials are not characterized by clear ongoing alteration processes. Electron microscopy and point chemical microanalyses have clearly confirmed the optical microscopy observations, in particular the poor homogeneity of the binding matrices. The materials (Figure 5a, b) are made up of a fine silicate aggregate bound by a strongly heterogeneous matrix, consisting of portions with a dominant calcic composition, strongly cohesive and microporous, composed of aerial lime carbonation products, associated with lamellar-textured microporous matrices, strongly fractured and consisting of an association of clayey phases and anthropogenic carbonates, as confirmed by EDS microanalyses (Figure 5c, d), showing the occurrence of comparable aliquots of silicon and calcium,

associated with aluminium, alkalis, magnesium and iron. The occurrence of numerous lime lumps is also evident through electron microscopy analyses, confirming the poor homogenization of the binding components.

Finally, group 3 mortars can be macroscopically distinguished by their yellow-greyish color (5Y 8/4 Munsell), by their fine conglomerate dimensional appearance and by their medium-high cohesion. Microscopically (Figure 3c), they are characterized by a binder/aggregate ratio of 1:3. The binder matrix has a homogeneous microcrystalline texture, with interference colours typical of calcite, confirming the calcic nature of the binder used. A limited presence of lime lumps is observed, in the order of 5 vol% of the total binder matrix. The inert fraction is always poorly sorted, with a unimodal grain size distribution centred in the medium sands granulometric range (max. size: 4 mm; mean size: 400 μm) and the presence of a limited silty fraction. The aggregate is generally highly spherical and sub-rounded/sub-angular, evenly distributed in the mortar and not oriented. From a minero-petrographic point of view, it is characterized by a prevalence of silicate minerals (quartz and associated plagioclase and potassic feldspar), associated with fragments of sandstones with carbonate cement and microcrystalline carbonate rocks, generally constituting the coarse fraction of the aggregate. The pore network of the mortars generally constitutes 3-5 % of the total volume, with a clear prevalence of spherical voids with an average size of 600 μm, and subordinate irregular vughs. The materials don't show any alterative phenomena. Electron microscopy and point chemical microanalyses have clearly confirmed the optical microscopy observations. The materials (Figure 6a, b) consist of a mixed silicate-carbonate aggregate strongly bound by a microporous matrix consisting of globular agglomerates of anthropogenic calcite, as clearly confirmed by EDS microanalyses

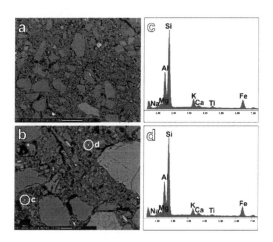
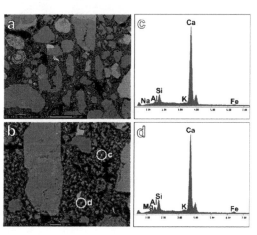

Figure 5. Sample CG_M1 (group 2): a, b) backscattered electron micrographs; c, d) EDS microanalyses of the binder matrix points indicated in b).

Figure 6. Sample RI_M1_A (group 3): a, b) backscattered electron micrographs; c, d) EDS microanalyses of the binder matrix points indicated in b).

(Figure 6c, d), which show a homogeneous and predominantly calcic composition. The presence of secondary aliquots of silicon, aluminium, alkalis, magnesium and iron indicates the occurrence of a scarce clay fraction intimately dispersed in the carbonation products of the calcic binder.

4 MASONRY QUALITY INDEX

The Masonry Quality Index (MQI) is a qualitative evaluation of masonry by means of seven parameters that evaluate the fulfilment of the rules of art (Borri et al. 2015).

The definition of the MQI has been performed on some specimens in order to evaluate a possible relationship with the mortar results. By comparing the data obtained from petrographic, mineralogical, micro-spectroscopic and microchemical-microstructural analyses, with those found in by MQI forms some data emerged.

The index, which is a number, can be linked by ranges in three categories A, B or C where the first is good quality and C the poorest.

Figure 7 shows one example of each group used to evaluate the masonry arrangement. The same application is made also in the cross-section to evaluate if leaves are connected or not.

The evaluation of masonry quality of specimens in Group 1, provides a grade C for all panels, corresponding to an inadequate behaviour of masonry and then prone to crumbling and collapses. Taking into account the very low calcite quantity, less than 10%, the cohesive contribution is due only to the physical interaction of the clay particles, and it is therefore very low. The low mechanical characteristics are also combined with the possible liquefaction effect by saturating fluids inside the masonry.

Group 2 mortars are comparable with the first category, although the amount of calcite is higher. The masonry quality was very poor as in the first case, with all analysed specimens falling in the category C.

Panels analysed in Group 3 shows a better behaviour, with more connections among stones in the frontal view and with a more regular arrangement. Furthermore, stones were less variable, allowing a better positioning of the masonry elements. In this group, all specimens fall within class B that means a behaviour of average quality, better than the other panels analysed.

The cross-section analysis of all panels demonstrated a well-known vulnerability (Giuffré 1993, Figure 8 a,b) of such type of masonry, that is the lack of transversal wall connections among leaves.

Figure 8 (c, d) shows an example of the CV-M1 panels, were is possible to recognise a sort of vertical joint in the middle of the leaves that does not allow any interlocking behaviour. From a mechanical point of view, when subjected to horizontal loads those masonry typologies lead to local failures or to the overturning of one leaf as frequently occurred in Amatrice (Figure 1 b).

With the average indexes obtained, using the correlation provided for the MQI (Borri et al. 2015), an indicative range of compressive strength could be between 1.4 and 2.4 MPa for masonries constituted of Group 1 and 2 mortars (MQI 1.86), and between 2 and 3 MPa for the Group 3 typology.

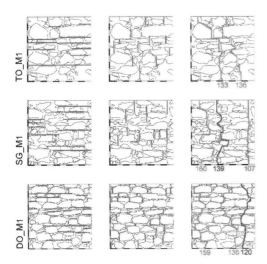

Figure 7. Panels used for the MQI evaluation for each group.

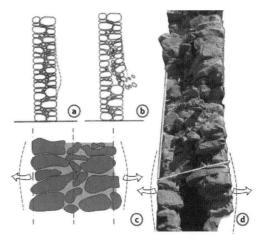

Figure 8. Failure mechanisms of three leaf masonry (a,b) Giuffrè (1993). Drawing (c) and photo (d) of the specimens CV-M1 cross-section.

This comparison shows a good agreement between the mortar typology and the overall quality of the masonry, although the mortar is only one parameter among seven.

5 APPLICATION ON A CASE STUDY

5.1 San Giovanni Battista in Macchie

The church of San Giovanni Battista in Macchie (MC) is an emblematic case of the key role played by the masonry quality and its associated weakness. The church is 9 m wide and 19 m long for an average height of 4.5 m. The cross section of walls is significantly squat due to a tapered thickness from 166 cm to 90 cm at the top, which also improves the stability under lateral loads. In a vulnerability analysis, the first step before the global analysis is the verification of the out-of-plane mechanisms with the hypothesis that those macro-elements behave as rigid blocks (Doglioni et al. 1994). This hypothesis, that is often rapidly assumed by practitioners, should be carefully evaluated in order to avoid macroscopic errors in the estimation of the seismic vulnerability.

The mechanisms considered in the evaluation are the most common in simple churches and concern: the overturning of the façade (M1a-North; M1b-south) or of its higher part (M2), the out-of-plane behaviour of the nave walls (M5a/b), and the mechanisms of the bell tower (M27). The mechanisms are defined according the Italian form for the evaluation of seismic damage of churches (DCP 2001). The seismic capacity is estimated according the kinematic analysis and the Italian National code for structures. Evaluations are made neglecting any possible hold action of the roof.

Table 1 reports the indication of the macro-elements involved and two verification ratios. The first is between the evaluation acceleration and the one required at the Ultimate Limit State for this church (V_R 75 years, soil type B), observing a safety evaluation considering mechanisms. The PGA acceleration required by the code in this case is 1.55 m/s^2

The second column reports the ratio between the acceleration related to the expected failure and the earthquake one (estimated by shake-maps) which was considered as 2.58 m/s^2. In this case it is possible to observe that out-of-plane mechanisms could appear for the tower and the façade, although the non-linear analysis looking at displacements provided all verified mechanisms for both scenarios.

The damage highlights a complete collapse of the tower but not of the façade. However, it is important to underline that with a compact masonry the observed damage after the earthquake, according

Table 1. Mechanisms percentage of verification.

Mechanism	Element	Check	ULS	Quake
M1a	Façade-North	✓	1.73	1.04
M1b	Façade-South	✓	1.40	0.84
M2	Top façade-	✓	1.65	0.99
M5a	Lateral wall	✓	2.12	1.28
M5b	Lateral wall	✓	2.20	1.32
M27	Tower	✓	1.38	0.83

Figure 9. The church before the earthquake (l) and after (r).

with the kinematic analysis, should be slight, whereas Figure 9 (r) shows, after the earthquake, the complete collapse of the entire structure due to dis-aggregation of masonry.

6 CONCLUSIONS

The presented paper, following an already established methodological path, provides a precise characterization of masonry typologies in the central Italy area. This preparatory work aims to provide to the scientific community a data background to design, and hopefully testing, effective interventions on this masonry, and it should be a base for future researches on interventions and tests.

The multi-analytical characterization process performed on the analysed binding materials allowed defining their compositional and textural characteristics, clustering the mortars into groups with homogeneous characteristics and correlating their properties with the performances of masonry typologies.

In the Amatrice region, it was possible to observe two masonry typologies. The first one is very vulnerable, with a chaotic arrangement and constituted of mortars mainly made of clay, as identified for Group 1 and 2 classes of materials.

The second typology is better arranged in the leaves, but still without transversal connections. Concerning mortars, they present first of all a coarser aggregate fraction with an associated carbonate component. Furthermore, the binder is basically made up of aerial lime, with a limited clay

Table 2. Specimens ID, Locations and its ID, resulting group, highest PGA occurred and its date and mineralogic composition.

Spec.	Location	Location ID	Group	PGA (g)	Date	Calcite	Dolomite	Quartz	Oligoclase	Microcline	Illite	Chlorite	Smectite nanoclays
						Mineralogic composition [% wt]							
AL_M1	Aleggia	1	3	0,682	24-aug	26,1	0,9	28,1	14,2	3,1	3,9	3,6	20,1
AL_M2	Aleggia	1	3	0,682	24-aug	38,5	0,6	20,2	9,4	3,7	4,2	2,5	21,0
CS_M1	Casali di Sotto	2	1	0,371	24-aug	8,7	0,3	43,3	12,1	4,7	5,3	2,6	23,1
CC_M1	Collecreta	3	1	0,576	24-aug	0,0	0,2	45,6	14,4	6,2	4,8	3,5	25,3
CC_M2a	Collecreta	3	1	0,576	24-aug	9,8	0,4	49,7	13,7	4,8	3,1	2,7	15,6
CC_M2b	Collecreta	3	1	0,576	24-aug	13,0	1,9	31,2	17,9	3,8	6,4	3,8	22,0
CG_M1	Collegentilesco	4	2	0,576	24-aug	13,1	0,1	39,4	13,0	5,4	4,3	2,8	21,9
CG_M2	Collegentilesco	4	2	0,576	24-aug	28,3	1,5	29,8	11,4	3,4	3,9	3,3	18,4
CG_M3	Collegentilesco	4	3	0,576	24-aug	2,8	0,1	39,8	19,1	6,1	5,8	4,1	22,1
CV_M1	Cornillo Vecchio	5	3	0,721	24-aug	24,0	2,3	36,9	11,4	2,8	2,9	3,1	16,5
DO_M1	Domo	6	3	0,773	24-aug	32,4	1,6	29,2	11,8	4,3	2,9	2,6	15,1
FO_M1	Forcelle	7	3	0,712	24-aug	27,2	2,1	27,6	12,8	4,0	3,3	2,5	20,5
FO_M2	Forcelle	7	3	0,712	24-aug	37,9	4,5	21,4	11,6	3,2	3,2	2,9	15,4
MO_M1	Moletano	8	1	0,555	24-aug	0,5	0,3	41,7	13,3	5,0	5,9	3,4	29,9
MO_M2	Moletano	8	2	0,555	24-aug	15,1	0,4	41,0	12,6	4,8	3,3	2,3	20,3
NO_M1	Nommisci	9	1	0,406	24-aug	6,1	0,1	34,7	10,6	6,5	7,2	2,1	32,6
PA_M1	Pasciano	10	3	0,705	24-aug	35,7	1,9	26,9	11,1	3,1	3,0	2,6	15,8
PA_M2	Pasciano	10	3	0,705	24-aug	24,4	6,0	26,7	11,2	3,2	4,1	3,1	21,3
PR_M1	Prato	11	1	0,650	24-aug	2,5	1,5	41,1	16,6	4,5	7,7	3,6	22,5
PR_M2	Prato	11	2	0,650	24-aug	12,1	0,7	35,2	13,7	4,5	7,2	3,0	23,5
PE_M1	Preta	12	1	0,402	24-aug	4,2	0,3	40,5	17,3	5,4	6,1	4,6	21,6
PE_M2	Preta	12	2	0,402	24-aug	12,5	0,2	48,8	15,0	4,9	1,8	2,3	14,4
RI_M1_a	Retrosi	13	3	0,526	24-aug	28,8	2,9	30,9	10,4	3,5	2,5	2,3	18,7
RI_M1_b	Retrosi	13	3	0,526	24-aug	26,3	3,2	35,0	12,1	3,4	2,4	2,5	15,1
RI_M2	Retrosi	13	1	0,526	24-aug	0,1	0,6	34,9	13,1	5,6	6,4	2,4	36,9
RI_M3	Retrosi	13	3	0,526	24-aug	32,6	1,8	31,1	9,6	3,9	2,3	2,4	16,4
SA_M1	Sant' Angelo	14	1	0,730	24-aug	0,4	0,6	40,6	16,6	5,3	8,6	4,4	23,4
SA_M3	Sant' Angelo	14	1	0,730	24-aug	8,6	1,8	36,8	14,8	4,8	6,6	4,1	22,5
SG_M1	Santa Giusta	15	2	0,738	24-aug	17,6	1,0	33,6	15,5	4,1	4,3	2,8	21,2
TO_M1	Torrita	16	1	0,422	24-aug	7,3	0,5	39,1	15,7	6,5	4,2	3,0	23,7

fraction homogeneously dispersed in the fine matrix. These characteristics indicate the good quality of these materials, which have mechanical and cohesive properties much higher than those of the mortars of the previous groups.

The study helps to show that mortar properties are important to be studied even to understand the overall behaviour of masonry panels and helps also in underlining that mortar is even more relevant in weak masonry where the mechanical interaction among blocks is lower and where a tensile contribution could be provided by the cohesive behaviour.

The analysis of the cross-section and the presented case study, as many other observed in Central Italy after the earthquake, confirmed and stressed that the quality of masonry is the first step that should be accounted in the design of intervention to improve the seismic behaviour. Those interventions should aim at homogenising the entire cross-sections as well as at connecting leaves in the transversal direction. All of them should obviously consider carefully the physical and chemical characteristics of blocks and mortars, herein precisely described.

ACKNOWLEDGMENTS

The authors wish to acknowledge Laura Benetti and Anna Bernardi for their support in the on-site survey and processing phase.

REFERENCES

Binda L., Baronio G., Penazzi D., Palma M, Tiraboschi C. 1999 Characterization of stone masonry walls in seismic areas: data-base on the masonry sections and material investigation. *Proceedings of the 9ᵗʰ National Conference, Turin (in Italian)*.

Borri A., Corradi C., Castori G., De Maria A. 2015. A method for the analysis and classification of historic masonry *Bulletin of Earthquake Engineering* 13(9):2647–2665.

Cizer Ö, Van Balen K., Van Gemert, D. 2010. Competition Between Hydration and Carbonation in Hydraulic Lime and Lime-Pozzolana Mortars, *Advanced Materials Research*, Vol. 133-134, 241–246.

Department of Civil Protection 2001 Survey form for the cultural heritage—damage to churches. G.U. no. 116, 21/05/2001 (in Italian).

Doglioni F., Moretti A., Petrini V, 1994 Le chiese ed il terremoto ed. LINT. Trieste ItalY. (in Italian).

Giuffrè, A. 1993. Sicurezza e conservazione dei centri storici - Il caso Ortigia, *Laterza, Bari, Italy* (in Italian).

Meunier A. 2005. Clays, *Springer*, Berlin.

Munsell A. 2000, Soil colour chart, *GretagMacbeth*, New Windsor.

Onnis S., Dipasquale L., Paglini M. 2009. Building culture of corbelled dome architecture. *Earthen Domes and Habitats*, 323–351.

Rietveld H. 1969 A profile refinement method for nuclear and magnetic structures, *Journal of Applied Crystallography*, 2:65–71.

Rosti A., Rota M., Penna A., Magenes G. 2017. Statistical treatment of empirical damage data collected after the main Italian seismic events (1980-2009). *In proceedings of the 16th World Conference on Earthquake Engineering, Santiago (Chile)*.

Scarsella F., Alberti A., Beneo E., Manfredini M., Morganti E., Miceli C. 1955, Carta Geologica d'Italia alla scala 1:100000, foglio 139 "L'Aquila", *Istituto Geografico Militare*, Roma.

Secco M., Previato C., Addis A., Zago G., Kamsteeg A., Dilaria S., Canovaro C., Artioli G., Bonetto J. 2019 Mineralogical clustering of the structural mortars from the Sarno Baths, Pompeii: A tool to interpret construction techniques and relative chronologies, *Journal of Cultural Heritage* 40:265–273.

Valluzzi M.R., da Porto F., Modena C. 2004 Behavior and modeling of strengthened three-leaf stone masonry walls *Materials and Structures* 37:184–192.

Vignoli A., Boschi S., Modena C., Cescatti E. 2016. In-situ mechanical characterization of existing masonry typologies: A research project in Italy finalized to update the structural codes. *In proceedings of the 16ᵗʰ International Brick and Block Masonry conference. Padova*.

Wentworth C.K., 1922, A scale of grade and class terms for clastic sediments, *Journal of Geology* 30 (5):377–392.

Brick and Block Masonry - From Historical to Sustainable Masonry –
Kubica, Kwiecień & Bednarz (eds)
© 2020 Taylor & Francis Group, London, ISBN 978-0-367-56586-2

An experimental study on slender reinforced masonry shear walls subjected to in-plane reversed cyclic loading

B.R. Robazza, S. Brzev & T.Y. Yang
University of British Columbia, Vancouver, Canada

ABSTRACT: Reinforced masonry shear walls (RMSWs) consisting of hollow concrete-block units reinforced with vertical and horizontal steel bars have demonstrated good seismic performance in recent experimental studies in terms of ductility potential and energy dissipation capability. This paper presents key results of an experimental study on eight full-scale RMSW specimens with flexure-dominant response under reversed cyclic loading. The test specimens had varying slenderness and height-to-length ratios, vertical and horizontal reinforcement layouts, applied axial stress, and presence of boundary elements. The test specimens were designed per the Canadian masonry design code requirements and performed in a ductile manner, but experienced several failure mechanisms during the tests, including i) ductile-flexure, ii) shear-flexure, iii) toe-crushing, iv) sliding shear, and v) out-of-plane instability. The paper discusses the observed behaviour and damage patterns associated with various failure mechanisms. The test results are compared with the capacity predictions from selected international masonry design codes.

1 INTRODUCTION

1.1 *Background*

Ductile slender reinforced masonry shear walls (DSRMSWs) are often used as the seismic-force-resisting system (SFRS) for masonry buildings in regions of high seismic hazard. Several experimental studies related to the seismic response of DSRMSWs have been performed in the last 40 years, and the failure modes have been well documented [1]. Based on these studies, it has been well established that properly designed, and detailed DSRMSWs can adequately dissipate earthquake energy through flexure-dominated behaviour. The behaviour may culminate either in a pure flexure failure mode or in a combined shear and flexure (shear-flexure) mode. A pure flexural failure is characterized by yielding of the vertical reinforcement at the tension end zone and limited toe crushing at the compression end zones (see Figure 1a), while a combined shear and flexure mode (shear-flexure) is characterized by yielding of both the vertical and horizontal reinforcement (see Figure 1b). However, initial flexure-dominant behaviour may precipitate secondary flexure-related failure modes, including sliding, toe-crushing,

bar-buckling, bar-fracture, rocking/bond-slip, or lateral instability (see Figure 2). Since a pure shear failure mode does not provide good energy dissipation it should be avoided by design and it is outside the scope of this paper.

1.2 *Masonry design code provisions*

Seismic design of DSRMSWs has been addressed by many international design codes since the 1980s, but the design provisions have evolved significantly over time. Most masonry design codes contain provisions to predict the capacity of DSRMSWs for the combined effects of gravity and lateral loading. However, these codes may not contain provisions that address all failure modes. Furthermore, for the failure modes that are addressed by code provisions, the degree of complexity and conservatism may vary significantly between the codes.

In this paper, eight full-scale DSRMSW specimens, which were constructed using grouted concrete block masonry and experimentally tested using displacement-controlled cyclic loading protocol, were used to compare the explicit force-based and displacement-based predictions from the design codes of Canada (CSA S304-14) [2], USA (TMS 402/602-16) [3], New Zealand (NZS 4230:2004) [4], and EU (Eurocodes 6 and 8 – BS EN 1996-1-1:2005/BS EN 1998-1:2004) [5], [6]. Six specimens were rectangular-shaped with varying height-to-length (h_u/L_w) and height-to-thickness (h_u/t_w) ratios, and amount and distribution of reinforcement (vertical and horizontal reinforcement ratios ρ_v and ρ_h). Axial stress level has been normalized in the form of ratio ($P/A_nf'_m$), where P denotes the applied gravity load, and $A_nf'_m$ denotes the product of the wall cross-sectional area and masonry compressive strength f'_m. The remaining two specimens were T-shaped and were subjected

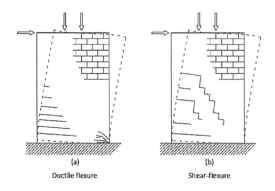

Figure 1. Primary flexural failure modes: (a) ductile flexure and (b) combined shear and flexure (shear-flexure).

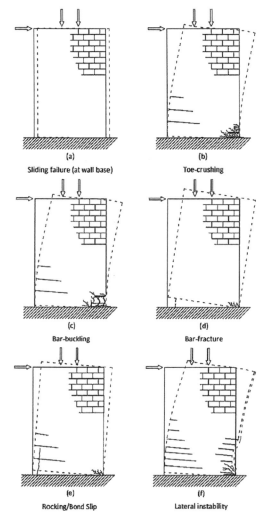

Figure 2. Secondary flexure-related DSRMSW failure modes.

to different loading protocols (symmetrical and asymmetrical cyclic loading). Table 1 presents the specimen test matrix. Detailed commentary and analysis on the test results can be obtained from Robazza et al. (2018) [7] for specimens W1 to W5 and Robazza et al. (2019) [8] for specimens W6 to W8.

2 SPECIMEN BEHAVIOUR AND FAILURE MODES

2.1 *Failure modes and occurrence sequence*

DSRMSWs may experience a wide variety of failure/ behaviour modes, such as: 1) ductile flexural (DF); 2) shear-flexure (SF); 3) sliding (S); 4) toe-crushing (TC); 5) rocking (RO), and 6) lateral instability (LI). In many instances, multiple behaviour modes may be triggered before the failure takes place due to the governing failure mode. The wall design, detailing, and loading influence the behaviour modes which may be encountered during the testing and the failure mode which will likely govern the response.

A graphical summary provided in Figure 3 shows lateral drift ranges corresponding to the failure modes observed for each test. Note that each column chart visually indicates the main failure mode acting over each of the lateral drift ranges shown, i.e. prior behaviour modes typically occurred simultaneously with the subsequent failure modes. For example, specimen W6 experienced six different behaviour types before the testing was terminated at a lateral drift of 4.14% (see Figure 3). In contrast, specimen W1 experienced three behaviour types before the collapse occurred at 0.84% drift.

The top portion of each column in Figure 3 represents the end of the test for a specific specimen, which does not necessarily mean that the lateral capacity of the specimen had been reached since some of the tests had to be terminated before the failure took place due to limitations of the test setup.

Table 1. Specimen test matrix (note: R – Rectangular wall, T – T-shaped wall).

ID	Cross section	h_u [mm]	ρ_v [%]	ρ_h [%]	$\dfrac{h_u}{t_w}$	$\dfrac{h_u}{L_w}$	$\dfrac{P}{A_n f_m}$
W1	R	3800	0.33	0.36	27.1	1.5	0.08
W2	R	3800	0.33	0.36	27.1	1.5	0
W3	R	4000	0.24	0.26	21.1	1.5	0
W4	R	4000	0.15	0.26	21.1	1.5	0
W5	R	4000	0.33	0.36	28.6	1.5	0
W6	T	4000	0.27	0.26	21.1	1.5	0
W7	T	4000	0.27	0.26	21.1	1.5	0
W8	R	4000	0.24	0.26	21.1	1.5	0

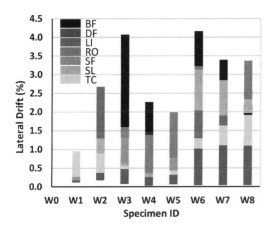

Figure 3. Drift ranges for the test specimen failure modes.

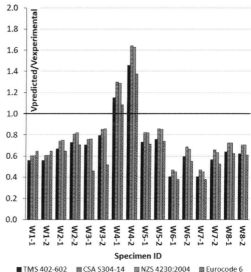

Figure 4. Code predictions for the ductile flexural capacity normalized to the respective experimental results.

3 A COMPARISON OF SEISMIC DESIGN PROVISIONS FOR DSRMSWS IN INTERNATIONAL CODES

3.1 Overview

Most masonry design codes contain provisions for determining the lateral load and displacement capacities of DSRMSWs based on their performance when experiencing the prominent in-plane flexure-dominant failure modes described in the previous section of the paper. These provisions are either explicit (containing quantitative capacity prediction equations) or implicit (based on empirical restrictions). This section provides a detailed comparison of the capacity predictions from the following leading international codes: CSA S304-14 (Canada) [2]; TMS 402/602-16 [3] (USA); NZS 4230:2004 (New Zealand) [4]; and Eurocodes 6 and 8 – EN 1996-1-1:2005/EN 1998-1:2004 (EU) [5], [6]. The results from the testing program on the eight full-scale DSRMSW specimens described earlier in the paper have been used to compare the code predictions with the experimental data. Only unfactored (nominal) capacities from the ULS seismic provisions for DSRMSWs have been considered. Provisions specific to fully-grouted hollow concrete block units have also been used where applicable since the test specimens were fully grouted. Note that the investigated provisions are pertinent to the most ductile DSRMSW class considered by each code. This section provides summary comparisons for both the explicit and implicit code provisions.

3.2 Comparison of explicit force-based code provisions

Figures 4, 5, and 6 show a comparison of code-predicted capacities and experimental values from the eight DSRMSW specimens. Note that the affixes -1 and -2 for the specimen labelling refer to the two

directions of loading for each specimen (W1 to W8) and that unit material factor values were used for the capacity provisions. The horizontal line on the charts shows an exact normalized prediction ($V_{predicted}$/$V_{experimental}$ = 1.0), while normalized predictions greater than 1.0 and less than 1.0 represent overpredictions and underpredictions respectively.

3.2.1 Ductile flexural capacity

Robazza et al. ([7], [8]) found ductile flexure to generally dominate the response of all the specimens during the testing, with the exception of specimen W4. Figure 5 shows that ductile flexural failure generally has the lowest normalized capacity predictions: a median (MED) normalized capacity prediction is 0.70, while a mean (MEAN) prediction is 0.73 with a standard deviation (STD) of 0.16 and a coefficient of variation (COV) of 22.0%. As mentioned earlier, it is not surprising that ductile flexure governed the predicted lateral load capacities considering that the specimens were specifically designed to fail in a ductile flexural failure mode. Some of the codes, e.g. TMS 402/602-16 and Eurocode 6, provide restrictions on the maximum tensile strain permitted in the vertical reinforcement, size of the compression zone, and the tied vertical compression reinforcement in the calculations (see Table 2). However, the general design approach is the same for the considered codes, and it is primarily based on beam theory. The common assumptions are: i) plane sections remain plane, ii) masonry tensile strength is disregarded in the design, iii) steel has a bilinear stress-strain relation, and iv) there is strain compatibility between the masonry, grout and steel. All predictions were relatively close to one another. Ranked

Table 2. Summary of key flexural design requirements for DSRMSWs from different codes.

Code	TMS 402-602	CSA S304-14	NZS 4230: 2004	Eurocode 6
Masonry Material Resistance Factor	0.90	0.60	0.85	0.60
Steel Material Resistance Factor	0.90	0.85	0.85	0.85
Masonry Compression Strain Limit	0.0025	0.0030	0.0030	0.0035
Steel Tension Strain Limit	0.008	None	None	0.0100
Tied Compression Steel in Cal. C.s.	Yes	Yes	No	No

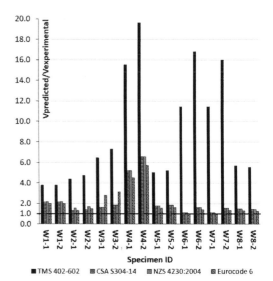

Figure 6. Code predictions for the sliding shear capacity normalized to the respective experimental results.

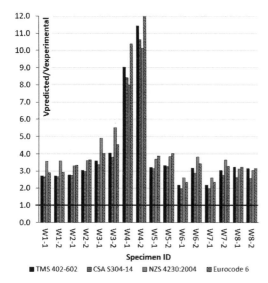

Figure 5. Code predictions for the shear-flexural capacity normalized to the respective experimental results.

in descending order, the predicted values were highest for CSA S304-14, followed by NZS 4230:2004, TMS 402/602-16, and Eurocode 6 (the latter two codes showed consistently lower capacity predictions due to the imposed tensile strain limits).

3.2.2 Shear-flexural capacity

In general, the codes treat shear-flexure and shear failure mechanisms identically and have relatively similar calculation methodologies for determining shear-flexure resistance, e.g. a sum of components attributed to the masonry, effect of horizontal reinforcement, and axial compression stress, where

applicable. In general, the provisions comprise a series of equations to determine each of these components, with the CSA S304-14 and Eurocode 6 equations being the simplest, and the NZS 4230:2004 and TMS 402/602-16 being the most complex. TMS 402/602-16 also has added provisions to place limits on the calculated strength based on the h_u/L_w ratio of a wall. Ranked in descending order, the shear-flexure capacity predictions were typically highest for NZS 4230:2004, followed by Eurocode 6, TMS 402/602-16, and CSA S304-14. The median (MED) normalized capacity prediction for shear-flexure is 3.19, with a mean (MEAN) prediction of 3.21, a standard deviation (STD) of 0.63, and a coefficient of variation (COV) of 19.6%. These statistical values indicate that the code shear-flexure provisions have the best agreement among the capacity predictions. Based on the specimen behaviour, while diagonal cracking did occur in most of the specimens, none of the specimens exhibited a governing shear-flexure failure mode, which may be attributed to the large amount of horizontal reinforcement provided in the specimens.

3.2.3 Sliding capacity

The sliding failure mode has shown somewhat higher normalized capacity predictions than the ductile flexural failure mode: the median (MED) normalized capacity prediction is 1.63, while the mean (MEAN) capacity is 1.59 with a standard deviation (STD) of 0.30 and a coefficient of variation (COV) of 18.6%. These statistical values indicate relatively good agreement among the codes, however high outlier capacity predictions by TMS 402/602-16 for specimens W4 and W6 have been excluded (15.51 and 16.80, respectively). These notably high capacity predictions may be attributed to the TMS 402/602-16 design provisions containing a prescribed allowable stress limit that acts over the area between the neutral axis and maximum compression fibre at the nominal moment capacity for walls with effective h_u/L_w ratios greater than 1.0. These provisions, in conjunction with the tensile strain limits used to determine the size of the compression zone, result in potentially unconservative sliding capacity predictions in some cases. For walls with an effective h_u/L_w ratio less than 1.0, the sliding capacity predictions in all codes are based on Coulomb friction theory, as discussed earlier in the paper. The sliding capacity predictions are typically highest for TMS 402/602-16, followed by NZS 4230:2004, CSA S304-14, and Eurocode 6. It should be noted that the highest code overpredictions were obtained for specimen W4. Despite initially experiencing ductile flexure behaviour during testing, the specimen eventually demonstrated a pronounced sliding failure that governed the response.

The sliding capacity predictions were on average 40 to 150% higher than the corresponding ductile flexural capacities; this is true for all codes, except for TMS 402/602-16 which produced sliding capacity predictions that exceeded ductile flexural capacities by an even larger margin. These results indicate that the sliding failure is less probable than a ductile flexural failure. Nonetheless, significant sliding displacements were recorded in six (out of eight) specimens during the testing, and in some cases contributed with as much as 35% of the total lateral displacement for high-ductility drift cycles. Four out of six specimens (W3, W6, W7, and W8) initially experienced ductile flexural behaviour characterized by toe-crushing before the sliding took place. Once the toe-crushing and face shell spalling was initiated, some of the vertical reinforcing bars became exposed, which effectively eliminated their contribution to the shear-friction resistance and the sliding capacity was reduced by 20%. As a result, an increased propensity for sliding displacements was created. This agrees well with the experimental observations for specimens W3, W6, W7, and W8 (see Figures 4 and 6). The capacity predictions indicated that a ductile flexural failure governs, however the margin between flexural and sliding capacity predictions was rather small.

Based on the test results and specimen behaviour, it appears that the sliding capacity provisions in the considered codes are generally adequate, provided that the specimens have only been subjected to a low number of lateral load cycles. However, after a larger number of cycles and a sustained flexural damage, sliding displacements may become larger than intended by the codes. Only TMS 402/602-16 commentary (Cl.9.3.6.5) addresses the effect of flexural damage on the sliding capacity by prescribing a slight reduction in the design coefficient of friction (from 0.7 to 0.65).

3.3 Comparison of explicit lateral displacement code provisions

As shown in Figure 7, the code provisions for predicting the lateral displacement capacity of DSRMSWs (defined here as the required lateral displacement capacity at the top of the wall required to satisfy the corresponding "ductility check"-type provisions stipulated by each of the codes [11]) vary considerably between the codes, primarily due to the key flexural calculation differences summarized in Table 2. These prediction differences stand out when noting the high standard deviation (STD) of 1.26 and the extreme coefficient of variation (COV) of 97.0% for the examined codes. This is largely because the code displacement capacity predictions (often being linked to "ductility checks" [11]) are generally formulated with the assumption of flexure-dominant behaviour that leads into a toe-crushing failure mode, which as described earlier in the paper, will often govern the ultimate lateral load capacity of flexure-dominant DSRMSWs. On an overall basis, however, the code displacement predictions did agree reasonably well with the experimental test displacements recorded at the end of the tests, as evidenced by the median (MED) normalized capacity prediction of 0.86 and the mean (MEAN) capacity of 1.16.

The code-provided equations for calculating the displacement capacities are similar for CSA S304-14, TMS 402/602-16, and Eurocode 6 in that they require both a relatively small elastic component and a larger inelastic component to the total rotation and displacement. The NZS 4230:2004 code provisions were based on limits on the compression zone length, which appeared to be in agreement with the

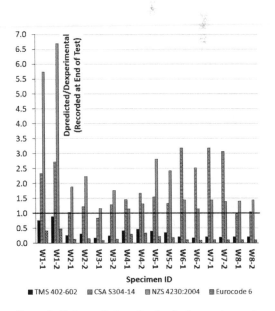

Figure 7. Code predictions for the displacement capacity normalized to the respective experimental results.

experimental data (except for specimen W1, where the predicted capacity was significantly larger than predicted).

The considered codes provide equations for the plastic hinge length/height, which have a significant influence on the lateral displacement capacity. The plastic hinge length/heights, however, varied considerably among the codes and in some cases by a factor of 3. The largest plastic hinge lengths/ heights are derived from the NZS 4230:2004 provisions, the shortest are derived from the Eurocode 6 recommendations, and the CSA S304-14 and TMS 402/602-16 provisions recommend intermediate values. Limits on the masonry compression strain capacity also influenced the predicted displacement capacities. However, the largest contributors to the differences were the permissible tensile strain in the vertical reinforcement, which had resulting effects on the compression zone length limits. This is the primary reason why for a significant difference between the CSA S304-14 and NZS 4230:2004 displacement predictions when compared to the TMS 402/602-16 and Eurocode 6 predictions. The limits on the tensile strains in the vertical reinforcement and compression zone depth had a more pronounced effect on the code prediction differences for the T-shaped specimens (W6 and W7) and the axially-loaded specimen (W1). Since these two code-prescribed limits did not agree well with the test observations, the CSA S304-14 and NZS 4230:2004 displacement predictions were found to be generally closer to the experimental findings, whereas the TMS 402/602-16 and Eurocode 6 provide conservative lateral displacement capacity predictions.

3.4 Comparison of implicit force-based code provisions

Following the explicit code provision comparisons discussed in the previous sections, the implicit code provisions are discussed in this section for the bar-buckling, rocking/bond slip, and lateral instability failure modes. Implicit provisions related to each of these failure modes are comparatively short and simple.

Only TMS 402/602-16 and CSA S304-14 contain provisions for preventing bar buckling, which are in the form of maximum tie spacing around the vertical reinforcement at the wall end zones. The maximum tie spacing is calculated in almost identical manner according to these two codes. If these provisions are satisfied, the designer realizes the benefit of accounting for the compression reinforcement in the determination of flexural capacity. Neither NZS 4230:2004 nor Eurocode 6 contain provisions regarding the bar buckling or tied compression reinforcement.

Rocking/bond-slip provisions are provided in all codes in the form of minimum development/splice lengths. The complexity of the development/splice length calculations varies significantly between the

codes. Both CSA S304-14 and TMS 402/602-16 calculate development/splice length based on several factors, but the CSA S304-14 calculations are more comprehensive. The Eurocode 6 development/splice length calculations are solely based on bar diameter, whereas the NZS 4230:2004 prescribes splice length calculations based on both bar diameter and a product of modification factors. Due to the relative simplicity of the Eurocode 6 and NZS 4230:2004 provisions, these estimates are expected to produce more conservative results than CSA S304-14 and TMS 402/602-16. Moreover, all codes except Eurocode 6 contain limits on the minimum vertical reinforcement ratio or bar size, which can implicitly help designers prevent rocking due to a potential lack of flexural crack redistribution.

Lateral instability is addressed by all codes in an implicit form (through h_u/t_w restrictions). However, TMS 402/602-16 addresses lateral instability by setting the compression stress limits for walls with high h_u/t_w ratios, indicating that these limits are more related to the standard (Euler) buckling caused by uniaxial compression – as opposed to the lateral instability failure mode described earlier in this paper. It is also unclear whether the h_u/t_w limits prescribed by Eurocode 6 are related to Euler buckling or lateral instability since lateral instability is associated with seismic loading and Eurocode 6 is related to the design of non-seismic loading actions. Both CSA S304-14 and NZS 4230:2004 prescribe h_u/t_w limits that are dependent on the expected ductility demand and are likely to be associated with preventing lateral instability in DSRMSWs.

4 CONCLUSIONS

This paper provides an overview of the most prominent in-plane flexure-related failure modes affecting DSRMSWs, including ductile flexural, shear-flexure, sliding, toe-crushing, bar-buckling, rocking/bond-slip, and lateral instability. The seismic code provisions for DSRMSWs from several international masonry design codes were used for a comparison of the nominal code-predicted ULS lateral load capacities and the experimental values obtained from an experimental study on eight full-scale DSRMSW specimens under reversed cyclic loading. The following conclusions can be derived from this study:

1. All DSRMSW specimens exhibited ductile flexural failure mode in the early stages of the tests. However, this failure mode eventually transitioned to toe-crushing, which governed the lateral load capacity for most specimens. Ductile flexural behaviour was initiated with flexural yielding of the vertical reinforcement, which typically occurred at a lateral drift close to 0.20%, whereas toe-crushing typically didn't occur until 0.50 to 1.20% drift.

2. The remaining flexure-dominant failure modes generally occurred only after the ductile-flexural and toe-crushing failure modes were initiated. Aside from specimen W4, sliding occurred at drifts ranging from 0.15 to 2.15%, and rocking occurred at drifts ranging from 0.19 to 3.24%. Shear-flexural behaviour only occurred in specimen W1 at a drift of 0.29%, whereas lateral instability in all affected specimens was initiated at drifts of around 1.6%. Bar fracture occurred at the high drift demands, which ranged from 1.27 to 3.33%.

3. By comparing Figures 4, 5, and 6, it can be seen that CSA S304-14, NZS 4230:2004, and Eurocode 6 generally predict the highest normalized force capacities for shear-flexure; whereas the sliding capacity predictions are the highest for TMS 402/602-16. Ductile flexural capacity predictions were generally the lowest, which is expected considering that this was effectively the intent of the specimen design.

4. Ductile flexural capacity predictions were found to follow basic beam theory for all codes; however, the predicted capacities resulted in a better fit with the test data when tensile strain restrictions on the reinforcement were disregarded. Note that CSA S304-14 and NZS 4230:2004 do not include these strain limits, whereas TMS 402/602-16 and Eurocode 6 do. Tensile strain limits have a beneficial effect of decreasing the chances of brittle secondary flexure-related failure modes, such as bar-buckling and lateral instability, which primarily occur at higher ductility demands.

5. Shear-flexural capacity predictions from the considered codes had a similar methodology of combining the contributions of masonry, horizontal reinforcement, and axial compression to the total shear-flexure resistance mechanism. Although the horizontal reinforcement provided in the test specimens was higher than for a DSRMSW in typical practical applications, the code provisions appeared to have been adequate in preventing a shear-flexural failure mode from occurring in any of the specimens. Overall, shear-flexure capacity predictions also had the best agreement among the codes.

6. Sliding capacity predictions for the codes that use Coulomb friction as their basis, were found to fit better with the test data than codes that incorporate a prescribed allowable stress that acts over the compression zone area of the wall (e.g. TMS 402/602-16). In all cases, the code provisions appeared to provide the most accurate predictions when the number of lateral load cycles was low (less than 5). Sliding displacements were found to increase significantly when a specimen was subjected to higher number of lateral load cycles (more than 8). This may be attributed to a combination of factors, including loss of surface roughness along the sliding interface; increased inelastic tensile strains in the vertical reinforcement, and the effect of the prior occurrence of toe-crushing and flexural damage on reducing the sliding capacity of the wall. Thus, there may be a need to revise code provisions for walls expected to experience high-ductility cyclic loading demands. Some potential simple adjustments could involve a decrease in the coefficient of friction for the Coulomb friction mechanism (in part already covered by TMS 402/602-16), as well as a decrease in the clamping force provided by the vertical reinforcement.

7. Displacement predictions were found to be sensitive to the code prescriptions related to plastic hinge length/height region, as well as restrictions related to the masonry compression strain, tensile strain in the vertical reinforcement, and the length of the compression zone. Considering the significant differences between the codes for these parameters, it is understandable that the code displacement predictions varied considerably. Based on the test specimens examined in this paper, the codes that did not include limits on the vertical reinforcement strains were found to produce displacement predictions that were in better agreement with the experimental findings.

8. Lateral instability due to in-plane lateral loading is intrinsically related to displacement ductility demands, but it is implicitly covered only by NZS 4230:2004 and CSA S304-14 through h_u/t_w limits. Note that the h_u/t_w restrictions are also contained in the TMS 402/602-16 and Eurocode 6, but they appear to be related to buckling due to increasing axial loading.

ACKNOWLEDGEMENTS

The project was generously sponsored with funding from the Natural Sciences and Engineering Research Council of Canada (NSERC) under the Collaborative Research and Development Program, the Canadian Concrete Masonry Producers Association, and the Masonry Institute of British Columbia. The first author acknowledges the support provided by the NSERC Industrial Postgraduate Scholarship program. Wall reinforcement was graciously donated by Harris Rebar Ltd. The invaluable assistance of the UBC and BCIT technicians and students was also critical for the project's success.

REFERENCES

Priestley, M. and Elder, D. 1982. *Seismic behaviour of slender concrete masonry shear walls*. University of Canterbury, Christchurch, New Zealand.

FEMA 306 1999. *Evaluation of Earthquake Damaged Concrete and Masonry Wall Buildings - Basic Procedures Manual*, prepared by the Applied Technology Council (ATC-43), Redwood City, CA, USA.

CSA, S304-14. 2014. *Masonry Design of Buildings*. Canadian Standards Association, Mississauga, ON, Canada.

TMS 402/602-16. 2016. Building Code Requirements and Specification for Masonry Structures. The Masonry Society, Longmont, CO, USA.

NZS 4230:2004. 2004. Design of Reinforced Concrete Masonry Structures. New Zealand Concrete Masonry Association Inc., Wellington, New Zealand.

EN 1996-1-1:2005. 2005. Eurocode 6: Design of Masonry Structures – Part 1-1: Common Rules for Reinforced and Unreinforced Masonry Structures. European Committee for Standardization, Brussels, Belgium.

EN 1998-1:2004. 2004. Eurocode 8: Design of Structures for Earthquake Resistance – Part 1: General Rules, Seismic Actions and Rules for Buildings. European Committee for Standardization, Brussels, Belgium.

Robazza, B.R., Brzev, S., Yang, T.Y., Elwood, K.J., Anderson, D.L, McEwen, W. 2017. Effects of Flanged Boundary Elements on the Response of Slender Reinforced Masonry Shear Walls: An Experimental Study. *Proceedings of the 13th Canadian Masonry Symposium, Halifax*, NS, Canada.

Robazza, B.R., Brzev, S., Yang, T.Y., Elwood, K.J., Anderson, D.L, McEwen, W. 2018. Out-of-Plane Behaviour of Slender Reinforced Masonry Shear Walls under In-Plane Loading: Experimental Investigation. *ASCE Journal of Structural Engineering, 144(3)*, 04018008.

Robazza, B.R., Yang, T.Y., Brzev, S. 2019. Response of Slender Reinforced Masonry Shear Walls with Flanged Boundary Elements under In-Plane Lateral Loading: An Experimental Study. *Engineering Structures 190(2019)*. 389–409.

Brzev, S. and Anderson, D.L. 2018. *Seismic Design Guide for Masonry Buildings*, Second Edition, Canadian Concrete Masonry Producers Association, Toronto, ON, Canada. 354 (www.ccmpa.ca).

Brick and Block Masonry - From Historical to Sustainable Masonry –
Kubica, Kwiecień & Bednarz (eds)
© 2020 Taylor & Francis Group, London, ISBN 978-0-367-56586-2

Mechanical modelling of cavity wall metal ties

O. Arslan
Department Materials, Mechanics, Management & Design, TU Delft, Delft, The Netherlands
Research Centre NoorderRuimte, Groningen, The Netherlands

F. Messali & J.G. Rots
Department Materials, Mechanics, Management & Design, TU Delft, Delft, The Netherlands

E. Smyrou & I.E. Bal
Research Centre NoorderRuimte, Hanze University of Applied Sciences, Groningen, The Netherlands

ABSTRACT: The seismic assessment of unreinforced masonry (URM) buildings with cavity walls is of high relevance in regions such as in Central and Northern Europe, Australia, New Zealand and China because of the characteristics of the masonry building stock. A cavity wall consists of two separate parallel walls usually connected by metal ties. Cavity walls are particularly vulnerable to earthquakes, as the out-of-plane capacity of each individual leaf is significantly smaller than the one of an equivalent solid wall. This paper presents the results of an experimental campaign conducted by the authors on metal wall tie connections and proposes a mechanical model to predict the cyclic behaviour of these connections. The model has been calibrated by using the experimental results in terms of observed failure modes and force-displacement responses. Results are also presented in statistical format.

1 INTRODUCTION

Cavity walls are often used in unreinforced masonry (URM) buildings in many countries, such as in Central and Northern Europe, Australia, New Zealand and China. Double-leaf cavity walls constitute a large portion of the building inventory in the province of Groningen, a large gas field in the north of the Netherlands, where the number of human-induced earthquakes has recently increased. These buildings are subjected to induced earthquakes up to magnitude of 3.6 until now, with the largest recorded horizontal PGA of 0.11g. A cavity wall consists of two separate parallel walls (called leaves), with an inner load-bearing leaf and an outer veneer, that has mostly aesthetic and insulating functions (Figure 1). The inner and outer leaves are interconnected by means of metal ties, as described in NEN-EN 845-1(NEN 2016).

The out-of-plane failure is a common mechanism during an earthquake for this typology of walls, which often stems from poor wall-to-wall, wall-to-floor or wall-to-roof connections, which are unable to provide sufficient restraint and boundary conditions, as well as from the slender geometry of the two parallel leaves.

Lintz & Toubia (2013) proposed a simplified analytical method to determine the amount of load transferred through the ties to the brick veneer and found that placing vertical reinforcement in the outer leaves could allow for an increase of the design strength.

An earlier study by Kobesen (Kobesen et al. 2014) defines the wall tie connection strength based on the pulling out of bars from concrete (Bruggeling et al. 1986, Cement en Beton 2011). In the model of Bruggeling (1986), the reinforcement embedded in concrete and subjected to tension is assumed to have the same strains of concrete. Though it uses slightly different stress and strain profile, Braam and Lagendijk (2011) also provides a similar approach. Kobesen (2014) used the same assumption for metal tie embedded in mortar.

An experimental campaign was performed at the Delft University of Technology (TU Delft) in 2019 (Arslan et al. 2020) to provide a complete characterization of the axial behaviour of metal connections in cavity walls by means of a dataset of 202 couplets. This work discusses the results presented by Arslan et al. (2020) from a statistical point of view with the aim to provide recommendations for design standards and guidelines. Both the mean and the characteristic values of the peak force and the displacement at peak force are computed for each group of tests. In addition, a mechanical model is proposed and calibrated against the load-displacement curves obtained for each group of connections.

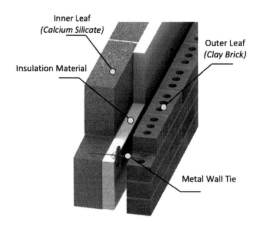

Figure 1. Cavity wall.

2 EXPERIMENTAL CAMPAIGN

An experimental campaign on cavity wall ties was carried out at TU Delft to study the axial behaviour of these type of connections in terms of axial strength, force-displacement curve and dissipated energy (Arslan et al, 2020). A large number of variations was considered in the research in order to provide a complete characterization of the connections: two embedment lengths, four pre-compression levels, two different tie geometries, and five different testing protocols, including monotonic and cyclic loading.

Each specimen (couplet) consisted of two bricks and a mortar joint where a metal wall tie was embedded. The couplet was designed to be representative of a portion of as-built URM cavity walls. L-shaped ties with a diameter of 3.6 mm and a total length of 200 mm were placed inside the mortar bed-joint, as happens in real applications. In practice, the zigzag-end is embedded in the CB masonry, while the L hook-end is embedded in the inner CS walls. 202 couplets were tested in total, consisting of four different typologies: CS70, CB50, CS50 and CS70-15D (Figure 2). A couplet of type CS70 and the test setup is shown in Figure 3.

In Arslan et al. (2020), the findings of the experimental campaign are reported in terms of failure mechanism, average force-displacement curve, peak

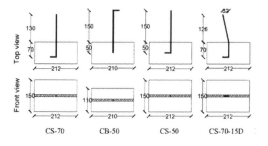

Figure 2. Geometry of tie specimens.

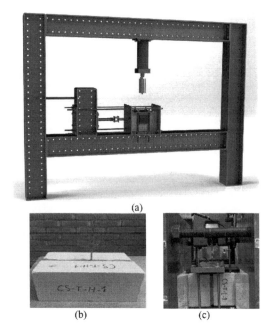

Figure 3. Test setup (a) and couplet (b & c).

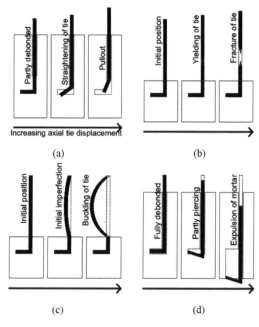

Figure 4. Failure mode sequence: Type A (a), Type B (b), Type C (c) and Type D (d).

force, and displacement at the peak force and at failure, identified as the point of reduction by 20% of the peak force as commonly assumed [Fardis (2009) and Zhang et al. (2015)]. The four different failure modes that were obtained in the experimental campaign are shown in Figure 4. The large majority of the couplets loaded in compression exhibited buckling failure, while in tension the sliding failure prevailed.

A series of companion tests were performed to determine the mechanical properties of the materials used in the experimental campaign, such as the flexural and compressive strength of the mortar, the tensile and compressive capacity of the tie, and the bond strength between masonry units and mortar.

3 STATISTICAL INTERPRETATION OF THE RESULTS

A statistical analysis of the experimental results is performed to compute the median and characteristic values of the peak load and of the displacement at peak load to provide recommendations for guidelines and standards. Besides that, an upper and lower bound for the capacity of the cavity wall tie connection is defined based on the 5th and 95th fractile. The statistical values for each typology (CS70, CB50, CS50 and CS70-15D) are computed according to the procedure proposed in Eurocode 0 (CEN 2005), for both tensile and compression loads, and for monotonic and cyclic loading. When the statistical distribution of a property is not known a priori, the characteristic value X_k can be computed as follows:

$$X_k = exp[m_x - (k_n \times s_y)] \quad (1)$$

where m_x is the mean value of a test series, s_y is the standard deviation and k_n is the factor that can be taken from table D1 in Eurocode 0 (CEN 2005). Table 1 summarizes the obtained experimental results and the characteristic values for each typology.

To have a better understanding of the results, equivalent lognormal distributions are computed for the tensile/compressive strength of each group of connections. A lognormal distribution is selected due to the right skewed nature of the experimental data. The fitting of the original data is conducted by using the method of moments. The obtained distribution is plotted in Figure 5 in terms of cumulative distribution function (CDF) and probability density function (PDF) for the cyclic tests of each examined typology (CS70, CB50, CS70-15D and CB50). The plots show also that the original results exhibit the good correspondence with the lognormal distributions. The CDF curves report the 5th, 50th and 95th fractile values of the dataset distribution.

It was observed that the mean value of the peak force from CB50 is higher than in the case of CS70, CS50 and CS70-15D for tensile loading (Figure 5a), whereas the mean values of the displacement at peak force are all close to each other (Figure 5b). On the contrary, for compressive loading, all the typologies have similar cumulative curves for both the peak force (Figure 5c) and the displacement at peak (Figure 5d).

The PDF of the moments on the experimental results of CS leaf and CB leaf are plotted in the same graph, in order to highlight by which of type of embedment in the two leaves is the overall connection capacity governed (Figure 6). The PDF curves report also the 5th fractile value of the dataset distribution as well as the characteristic value computed according to Equation 1. The two results shown in Figure 6 refer to the ideal connections (Figure 6a) and to imperfect applications in practice, with a bent tie (Figure 6b). The results for the two cases are very similar. It can be concluded that the behaviour of the wall-tie connection is governed by the tie embedment in the CS leaf. The failure mechanism will occur first in CS masonry, followed by CB masonry. The probability of having the failure in CB

Table 1. Summary of the obtained median results and characteristic values for each typology.

| | | Median Results | | | | Characteristic Values | | | |
| | | Tensile | | Comp | | Tensile | | Comp | |
Specimen Type	Loading Protocol	Force (kN)	Disp (mm)	Force (kN)	Disp (mm)	Force (kN)	Disp (mm)	Force (kN)	Disp (mm)
CS70	Mono	2.35	10.63	1.77	3.14	1.99	7.80	1.42	2.30
	Cyclic	1.88	6.88	1.73	2.78	1.31	2.86	1.40	1.85
CB50	Mono	3.54	7.02	1.83	2.13	2.70	2.87	1.50	1.72
	Cyclic	3.59	7.64	1.60	0.39	2.87	4.08	1.49	0.13
CS50	Mono	1.87	8.25	1.80	2.52	1.44	6.47	1.16	2.12
	Cyclic	1.62	4.69	1.90	3.41	1.41	3.07	1.46	1.49
CS70-15D	Mono	2.51	13.07	1.35	4.48	2.25	10.61	1.23	2.97
	Cyclic	2.07	9.67	1.44	3.97	1.68	9.19	1.26	2.72

Note: Disp=Displacement; Comp=Compressive; Mono=Monotonic

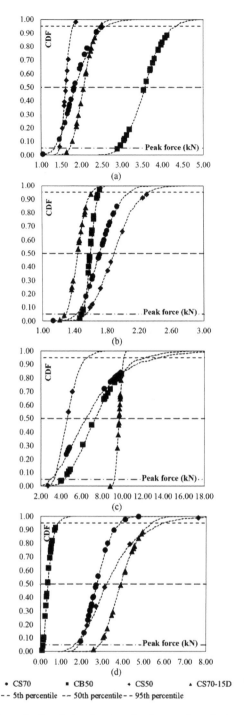

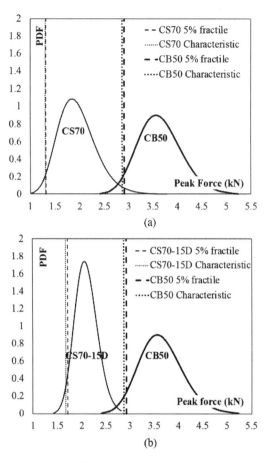

Figure 6. Probability functions for the tested typologies: CS70 and CB50 (a) and CS70-15D and CB50 (b).

Figure 5. Cumulative functions for the tested typologies: Peak force for Tension (a) and Compression (b), Corresponding displacement for Tension (c) and Compression (d). All the curves are defined for the cyclic tests.

masonry before CS masonry is 4.5% for the ideal case (CS70 and CB50), and only 2.4% for the case with bent ties (CS70-15D and CB50).

4 MECHANICAL MODEL

A mechanical model is developed in order to determine the load and the corresponding displacement. The mechanical model must account for CS and CB units in terms of the force-displacement behaviour and the failure mechanism, as explained above. The test results show that sliding failure of ties (Failure mode A) along the tie-mortar interface was the dominant failure mode in tension. While, the main failure mechanism in compression was buckling of the tie (Failure mode C).

Considering all the failure mechanisms and the experimental results, a simplified envelope curve is proposed for each tested typology to fit the results observed from the experimental campaign. The proposed curve has been produced as simplified approximation after averaging data from the results of the experiment so that the elastic, hardening phase and ultimate phase are defined. The force-displacement curve is idealized into trilinear branches in tension whose input parameters are the yielding strength (F_{Te}), yielding displacement (Δ_{Te}), peak strength (F_T), displacement at the peak (Δ_T), ultimate strength (F_{Tu}) and ultimate displacement (Δ_{Tu}). In compression it is approximated by bilinear branches where the input parameters are the peak strength (F_C), displacement at the peak (Δ_C), ultimate strength (F_{Cu}) and corresponding displacement (Δ_{Cu}), as shown in Figure 5. The proposed curve is valid for CS and CB for all failure modes. However, the calibration of these parameters for the envelope curve are different each of the two materials.

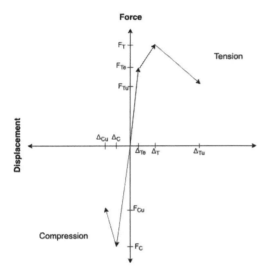

Figure 7. Simplified envelope curve.

4.1 Tensile strength of the connection

The initial stage in tension represents the linear elastic behaviour which is governed by bonding between the tie and mortar. F_{Te} can be computed by using the bond strength equation, adapted from CEB-FIP (1993) as shown below in Equation 2.

$$F_{Te} = T_{b,\max} \times \pi \times \phi \times L_s \qquad (2)$$

$$T_{b,\max} = a \times \sqrt{f_c} \qquad (3)$$

where $T_{b,\max}$ is the maximum bond strength, ϕ is the diameter of the tie, L_s is the embedded length of the tie, fc is the compressive strength of the mortar and a is the modification factor for CS and CB, shown in Table 2. Equation 2 is valid for all typologies. The bond-slip model assumes a different modification factor according to the shape of the bar, i.e. plain (CEB-FIP (1993)) or deformed bar (Verderame et al. (2009)), for the interaction mechanism between tie and mortar. CS couplets can be associated to the

Table 2. a values taken from CEP-FIP (1993).

	CS	Clay
a	0.3	1.5

frictional coefficient between tie and mortar that takes a low value as it represents plain bar condition. In the case of CB, a conservative value has been chosen for the modification factor since the shape of tie where embedded in mortar is zigzag representing thus deformed bar condition.

The peak force in tension (F_T) for CS50 and CS70 can be computed as the summation of the yield strength (F_{Te}) and the hardening force which is the force for straightening of the hooked part computed by the 2nd term in Equation 4. Note that L_H is the hooked length of the tie and σ_S is the tensile strength of the tie.

$$F_T = F_{Te} + \frac{4}{L_H \times} \left(\frac{f_c \times \phi \times L_H^2}{2} + \frac{\sigma y \times \pi \times \phi^3}{32} \right)$$

$$(4)$$

A term accounting for imperfect application in practice is added to Equation 4 yielding in the computation of the peak force for CS70-15D. (Equation 5). The extra term is the deflection due to bending.

$$F_T = F_{Te} + \frac{4}{L_H} \times \left(\frac{f_c \times \phi \times L_H^2}{2} + \frac{\sigma y \times \pi \times \phi^3}{32} \right)$$
$$+ \frac{24 \times E \times I \times \phi}{L_c^2}$$

$$(5)$$

where E is the elastic modulus of the tie, I is the second moment of area of a circle and Lc is the cavity length between two leaves. Regarding CB50, the peak force can be computed using Equation 6, as given below:

$$F_T = F_{Te} + f_c \times \phi \times L_s \qquad (6)$$

The peak strength in tension for CB consists of two terms. The first term, adopted from CEB-FIP (1993), stands for the bond-slip and the last term accounts for the zigzag portion of the tie which crushes the surrounding mortar. The ultimate force in tension (F_{Tu}), identified as the point of reduction by 20% of the peak force, can be computed using Equation 7:

$$F_{Tu} = F_T \times 0.8 \qquad (7)$$

The experimental results for the displacement at peak force for all typologies exhibit a striking variability which makes it difficult to capture by simple mechanical model. For that reason, a fully empirical approach is proposed to fit to the displacement values of the experimental force-displacement curves in tension and compression.

The displacement at elastic force, Δ_{Te}, is equal to 1 mm adapted from CEB-FIP (1993) for all typologies.

The displacement at peak force, Δ_T, is estimated as follows:

$$\Delta_T = \frac{L_H}{4} + 1 \qquad (8)$$

Equation 8 is valid for CS70, CS70-15D and CS50 walls. The displacement at peak force for CB walls, Δ_T, can be computed as follows:

$$\Delta_T = \frac{L_S}{10} + 1 \qquad (9)$$

The ultimate displacement in tension (Δ_{Tu}), identified as the corresponding point of reduction by 20% of the peak force on the proposed curve, can be computed using Equation 10. The ultimate displacement formula was derived by using linear equation between the peak force and zero force. Equation 10 is valid for all typologies.

$$\Delta Tu = 0.8 * \Delta_T + 3 \qquad (10)$$

4.2 Compressive strength of the connection

The typical failure mode in compression was by buckling of the tie. The compression strength, F_C, is estimated in terms of strength at Euler's critical load as follows:

$$F_C = \frac{\pi^2 \times E \times I}{K^2 \times L_C^2} \qquad (11)$$

where K is the column effective length factor. Except for the bent ties, the compression strength can be computed using Equation 11, while regarding the bent ties (CS70-15D), F_C is determined as follows:

$$F_c = \frac{\pi^2 \times E \times I}{K^2 \times L_c^2} - \frac{6 \times E \times I \times \phi}{L_c^2} \qquad (12)$$

The reason that CS70-15D is computed by Equation 12 is due to the geometry of the tie which is bent. K is chosen 0.5 for all typologies due to boundary conditions of the connections (rotation and translation fixed). The ultimate force in compression (F_{Cu}), identified as the point of reduction by 20% of the peak force, can be computed using Equation 13:

$$F_{cu} = F_c \times 0.8 \qquad (13)$$

The displacement at peak force, Δ_C, is equal to 3 mm for CS walls, while it is equal to 1 mm for CB walls adopted from the experimental results. The ultimate displacement in tension (Δ_{Cu}), identified as the corresponding point of reduction by 20% of the peak force on the proposed curve, can be computed using Equation 14, as explained above by using a linear relation.

$$\Delta_{cu} = 0.8 * \Delta_c + 2 \qquad (14)$$

results

A mechanical model is presented based on the experiments conducted. The properties derived from the tests used in the mechanical model are summarized in Table 3.

The estimated values of the mechanical model are compared to the experimental results by grouping

Table 3. Summary of cavity wall tie connection properties.

Material Characteristic	Symbol	CS70	CB50	CS70-15D	CS50
Diameter of the tie (mm)	ϕ	3.6	3.6	3.6	3.6
Embedment length of the tie (mm)	L_S	70	50	70	50
Compressive strength of the mortar (MPA)	fc	5.65	6.47	5.65	5.65
Hooked length of the tie (mm)	L_H	25	-	25	25
Elastic modulus of the tie (MPA)	E	32920	32920	32920	32920
Second moment of area of the tie (mm⁴)	I	8.3	8.3	8.3	8.3
Cavity length (mm)	L_C	80	80	80	80
Effective column factor	K	0.5	0.5	0.5	0.5

Table 4. Obtained results by mechanical model (differences with experimental results between brackets).

Specimen Type	Loading Protocol	Mechanical Model			
		Tensile		Comp	
		Force	Disp.	Force	Disp.
CS-70	Mono	2.23 (-5.13%)	8.56 (-19.51%)	1.68 (-5.25%)	3.54 (12.92%)
	Cyclic	1.89 (0.75%)	7.25 (5.37%)	1.68 (-2.86%)	3.00 (7.94%)
CB50	Mono	3.32 (-6.23%)	6.00 (-14.48%)	1.68 (-8.10%)	1.00 (-53.10%)
	Cyclic	3.32 (-7.51%)	6.00 (-21.51%)	1.68 (5.03%)	1.00 (155.64%)
CS-50	Mono	2.03 (8.42%)	8.56 (3.73%)	1.68 (-6.79%)	3.54 (40.39%)
	Cyclic	1.72 (6.07%)	7.25 (54.49%)	1.68 (-11.77%)	3.00 (-12.05%)
CS70-15D	Mono	2.54 (0.93%)	8.56 (-34.54%)	1.42 (4.87%)	3.54 (-20.99%)
	Cyclic	2.15 (3.94%)	7.25 (-25.05%)	1.42 (-1.72%)	3.00 (-24.45%)

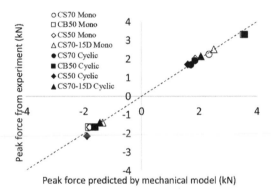

Figure 8. Comparison between experimental results and predicted values by mechanical model for each typology and loading protocol.

them per type of connection and loading protocol, as shown in Figure 8 and in Table 4.

5 CONCLUSIONS

Double-leaf cavity walls are very common in the province of Groningen, an area in the north of the Netherlands subjected to induced earthquakes. The out of plane response of the double-leaf cavity walls is one of the most prominent failure mechanisms for such buildings. The wall-to-wall metallic ties can provide an efficient retain to the out-of-plane collapse of the single leaves, but their strength has not been widely investigated yet.

In this study, cyclic and monotonic tests on CS and CB units are used for proposing the strength and the displacement capacity of the connections according to a statistical interpretation of experimental.

The capacity of the wall-tie connection is mainly governed by the tie embedment in the CS leaf.

Comparing the experimental results with the characteristic values in terms of the peak load, on average the experiments provide 19% larger peak load for tension, and on average 18% larger peak load for compression.

The proposed mechanical model can adequately predict the force-displacement behaviour obtained from the tests.

The Authors believe that the presented model is adequate for structural engineers to model the nonlinear seismic response of such structures.

REFERENCES

Arslan O. & Messali F. & Smyrou E. & Bal I. E. & Rots J. G. 2020. Experimental Characterization of Masonry Wall Metal Tie Connections in Double-Leaf Cavity Walls. Manuscript submitted for publication.

Braam, C.R. ; Lagendijk, ir. P. 2011. *Constructieleer Gewapend beton*; Cement en beton.

Bruggeling, A.S.J. & De Bruijn, W.A. 1986. *Theorie en praktijk van het gewapend beton deel 1*, Den Bosch.

CEB-FIP. Model Code 1990. Final draft, Bulletin d'information No. 213/214; 1993.

CEN (2005) EN 1990:2002. Eurocode 0: Basis of structural design. European Committee for Standardisation, Brussels.

Fardis, M. 2009. Seismic Design, Assessment and Retrofitting of Concrete Buildings Based on EN-Eurocode 8.10 .1007/978-1-4020-9842-0.

Kobesen M.G. & Vermeltfoort A.T. & Mulders S.G.C. 2014 Wall tie research for existing and new structures; Literature Study. *Proceedings of 9th International Masonry Conference, July 7-9, Guimaraes, Portugal.*

Lintz, J.M. & Toubia, E.A. 2013. In-Plane Loading of Brick Veneer over Wood Shear Walls. *Civil and Environmental Engineering and Engineering Mechanics Faculty Publications.* 13.

NEN (2016). NEN-EN 845-1. Specification for ancillary components for masonry - Part 1: Wall ties, tension straps, hangers and brackets. Nederlands Normalisatie-instituut, Delft.

Verderame, G. ; de carlo, G. & Ricci, P. & Fabbrocino, G. 2009. Cyclic bond behaviour of plain bars. Part II: Analytical investigation. *Construction and Building Materials. 23.* 3512–3522.10.1016/j .conbuildmat.2009.07.001.

Zhang, D. & Waas, A.M. & Yen, Chian-Feng. 2015. Progressive Damage and Failure Response of Hybrid 3D Textile Composites Subjected to Flexural Loading, Part II: Mechanics Based Multiscale Computational Modeling of Progressive Damage and Failure. *International Journal of Solids and Structures. 75.* 321–335.10.1016/j .ijsolstr.2015.06.033.

Brick and Block Masonry - From Historical to Sustainable Masonry –
Kubica, Kwiecień & Bednarz (eds)
© 2020 Taylor & Francis Group, London, ISBN 978-0-367-56586-2

Flexible joints between RC frames and masonry infill for improved seismic performance – shake table tests

T. Rousakis, E. Papadouli, A. Sapalidis & V. Vanian
Democritus University of Thrace, Greece

A. Ilki & O.F. Halici
Istanbul Technical University, Turkey

A. Kwiecień, B. Zając, Ł. Hojdys, P. Krajewski, M. Tekieli & T. Akyildiz
Cracow University of Technology, Cracow, Poland

A. Viskovic & F. Rizzo
G. D'Annunzio University of Chieti-Pescara, Italy

M. Gams
Faculty of Civil and Geodetic Engineering, University of Ljubljana, Slovenia

P. Triller
Slovenian National Building and Civil Engineering Institute, Ljubljana, Slovenia

B. Ghiassi
University of Nottingham, UK

A. Benedetti & C. Colla
University of Bologna, Italy

Z. Rakicevic, A. Bogdanovic, F. Manojlovski & A. Soklarovski
ZIIS, Ss. Cyril and Methodius University, Skopje, Republic of North Macedonia

ABSTRACT: The paper reports the first results from seismic tests on shake table on a full scale reinforced concrete (RC) frame building with modified orthoblock brick wall infills, within INMASPOL SERA Horizon 2020 project. The building received innovative protection using polyurethane resin flexible joints (PUFJ) at the frame-infill interface, in different schemes. Further, PUs were used for bonding of glass fiber grids to the weak masonry substrate to form Fiber Reinforced PU (FRPU) as an emergency repair intervention. The test results showed enhancement in the in-plane and out-of-plane infill performance under seismic excitations. The results confirmed remarkable delay of significant orthoblock infill damages at very high RC frame inter-storey drifts as a consequence of the use of PUFJ. Further, the PUFJ protection enabled the repair of the infill even after very high inter-storey drift of the structure up to 3.7%. The applied glass FRPU system efficiently protected the damaged orthoblock infills against collapse under out-of-plane excitation while they restored large part of their in-plane stiffness.

1 INTRODUCTION

A flexible and ductile structure can sustain severe earthquakes exhibiting large displacements. Meanwhile, the brittle components of the structural systems, such as masonry infill walls, may suffer from substantial damages due to excessive drifts (Viskovic et al. 2017, Ricci et al. 2017). An example of in-plane and out-of-plane damages observed after previous earthquakes are shown in Figure 1. To avoid these damages, seismic design documents limit the interstory drifts based on the interface conditions between the infill and surrounding structural frame, and require to check the stability of infills and the safety of their connections to the frames (EN 1998-1 (2004), Turkish Seismic Design Code (2018)).

This paper shows the response of these structural systems can be improved by using a flexible interface between the frame and the infill (Kwiecień et al. 2017, Rouka et al. 2017, Rousakis et al. 2017, Kwiecień 2019, Akyıldız et al. 2019).

Figure 1. In-plane and out-of-plane infill walls damages (Van Earthquake, 2011).

The present study reports the first experimental results of the "INfills and MASonry structures protected by deformable POLyurethanes in seismic areas" (INMASPOL) Project within the SERA, Horizon 2020 framework. INMASPOL investigates the efficiency of innovative PUFJ protection at the frame-infill interface for RC frames with brick infills. Further, PUs are used for bonding of glass fiber grids to the weak masonry substrate to form FRPU as an emergency repair. The methods are applied on full-scale infilled RC building tested on shake table under simulated seismic excitations. The seismic tests validated the improved in-plane and out-of-plane infill performance when modified or repaired with PUFJ and FRPU systems.

2 EXPERIMENTAL SETUP

The tested structure was a fully symmetrical 3D one storey building with 4 RC columns, 4 beams, a slab and 4 infill masonry walls, all designed according to current Eurocodes.

The real scale building had plan dimensions of 3.8x3.8m and a height of 3.3m (foundation and column extensions included, Figure 2).

Figure 2. Picture of the experimental model.

2.1 Materials

Concrete used in construction was produced in a concrete factory. The laboratory testing results for concrete cubes of 15x15x15 cm^3 presented 28 days compression strength of 34.1 MPa for the foundation, 27.1 MPa for the columns and 34.2 MPa for the slab. Average mass density of concrete was 2380 kg/m^3.

Steel for reinforcement was grade B 500B with characteristic yield strength 500 MPa.

Blocks for infill walls were hollow clay units KEBE OrthoBlocks K100 with dimensions of 100x240x250 mm^3 and weight of about 100 kg/m^2 of a wall with vertical holes.

Mortar for construction was OrthoBlocks mounting mortar in the form of a dry ready mix, with nominal strength class of M10. The mortar was laid in thin layers of 3 mm thickness. Both, head and bed joints, were filled with the same mortar thickness.

Polymer for PUFJ (PolyUrethane Flexible Joints) was of type Sika PM. The elastic modulus, strength and ultimate elongation of the polymer were 4 MPa, 1.4 MPa and 110%, respectively obtained from a tensile test.

Polymer for FRPU (Fiber Reinforced PolyUrethane) was of type Sika PS. The elastic modulus, strength and ultimate elongation of the polymer were 16 MPa, 2.5 MPa and 40%, respectively.

GFRP mesh of type Sika Wrap 350G Grid, made of glass fiber reinforced polymer mesh, was used for the repair of the walls. The mesh of a real weight 360 g/m^2 had the elastic modulus, strength and ultimate elongation of 80 GPa, 2.60 GPa and 4%, respectively in a tensile test.

2.2 The structure

The one storey RC frame building with infills had a 1:1 scale and designed according to shaking table capacity. The floor plan dimensions of the frame were 2.7 x 2.7 m, and the 20 x 20 cm columns had a height of 2.5 m (to the top of the slab). The height of infills in the building was 2.3 m (Figure 2). The RC reinforcements were designed according to Eurocode 8. The columns were reinforced with 8 Φ 10 longitudinal rebars and 2 Φ 8/50 mm stirrups. The beam was hidden inside the slab, and reinforced by the same amount of longitudinal reinforcement, and by Φ 8/50 mm stirrups. On top of the columns, an RC slab with a thickness of 20 cm was constructed. The slab was extended beyond the frame beams to serve as additional mass and to attach the additional masses in the form of steel massive elements. In the middle of the slab there was a hole for access to the inside of the model (see Figure 3). The slab was reinforced with Q503 at the top and at the bottom. There was additional reinforcement at the perimeter edges of the slab, and at the edge of the hole. The model structure was constructed on a special foundation, with provided holes for attaching it to the shake table and hooks for lifting and manipulating the structure (Figure 3).

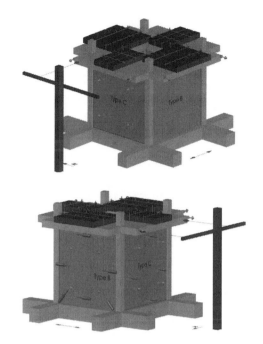

Figure 3. Scheme of instrumentation.

The infills too were designed according to the shaking table capacity and in relation to the RC frame strength and stiffness. They were made of hollow clay blocks (OrthoBlock type) with thickness of 10 cm, typical for internal infills. Two parallel walls were of type B, and the other two walls were of type C. Type B was constructed directly on the foundations, while there was a 2 cm thick in-situ produced PUFJ (by injection) between the infill and the columns, and between the infill and the slab. The joint was thus on three edges (left, right and top) and had a chemical bond due to in-situ application. Type C infill was constructed on a prefabricated 2 cm thick PUFJ joint bonded to the RC foundation beam and the same type of joint was at the sides and top of the infill. The PUFJ was thus bonded on all four edges (top, bottom, left and right) of the frame but through the mortar. On top of the structure, there were 18 steel massive elements with a total mass of 7.2 tonnes. The structure on the shake table is shown in Figure 2.

2.3 Testing facility and equipment

The shake table tests were performed in the laboratory for dynamic testing in the Institute of Earthquake Engineering and Engineering Seismology in Skopje (IZIIS), Republic of North Macedonia. Operational since the year 1980, the shake table is 5.0 m by 5.0 m pre-stressed concrete waffle slab weighing 33.0 tons with payload up to 40 tons. Five degrees of freedom are provided by 2 lateral and 4 vertical MTS hydraulic pistons, controlled by MTS Digital Controller 469D. National instruments PXI

modular system was used as data acquisition for the three different types of transducers: 23 accelerometers, 10 linear variable differential transformers and 2 linear potentiometers (Figure 3).

2.4 Instrumentation

The instrumentation consisted of 23 accelerometers located on points indicated by green arrows in Figure 3. The accelerometers were used to measure in- and out-of-plane accelerations of the infills, and the accelerations of the top slab and of the foundation. Eight (8) LVDTs were used to measure relative displacements between the infills and the RC frame structure on the in-plane and out-of-plane loaded infills. Two additional LVDTs were used to measure diagonal deformation of the infill. In total, there were 10 LVDTs, which are indicated by red markers in Figure 3. Two linear transducers were used to measure the top drift of the structure (blue markers) and two attached to the support, to measure relative displacements of the slab). In addition, optical system was used to measure displacement fields on both in-plane loaded walls, to measure absolute displacements of the frames and the infills.

2.5 Applied testing methodology

For the purpose of the experimental investigations, two types of tests were performed, i.e. dynamic shake table tests and tests for determination of dynamic characteristics. The dynamic shake table tests comprised of gradually increasing input intensity level from 3% to 77% of the adopted earthquake Kefallonia E-W component, simulating a critical excitation, causing out-of-plane failure of the most vulnerable wall infills. The maximum applied load level was dependent on the induced damage of the RC frame and infill walls as well as on the limit state of the shake table (considering the weight of the specimen and the additional load). A total of 23 seismic tests were performed. Only results of in-plane tests related to infills without strengthening and two repaired after significant damage occurrence are presented in this paper.

Resonant frequency tests were examined after selected shake table tests regarding the observed damage. Sine-sweeps and white noise 0.02g intensity tests were performed, confirming the gradual degradation of the bearing elements and infill softening.

2.6 Seismic loading

The structure represents in- and out-of plane loaded walls at the top of 4-storey RC building. In order to generate the excitation for the seismic table, a numerical inelastic model of a typical 4-storey RC frame building was subjected to the actual Kefallonia 2014 earthquake record and the acceleration response (with $a_{max} = 2.12$ g) at the top floor was decided to be the most detrimental for out-of-plane failure of the infills (Figure 4 – top). This top floor excitation was slightly modified to conform to the shake table capacity and get a better reproduction of the input signal. Its dominant frequency range was 2,5-4 Hz. The applied maximum acceleration history (input signal), scaled to 77% of the Kefallonia 2014-based earthquake record (77% KEF-1) is shown in Figure 4 - bottom.

2.7 Experimental program

The structure was always tested by a uniaxial dynamic excitation and after selected earthquake runs, the eigen-frequencies were measured by loading the structure with white-noise (random) excitation.

The original position of the model was such that both type B walls were loaded in-plane, and both type C walls were loaded out-of-plane (Figure 2 and 3). In this position (PHASE 1), the dynamic load was gradually increased until there were significant in-plane damages in infills B (Figure 5), which occurred during 77% KEF-1 – see Table 1. There were also damages in concrete columns (at the top and at the bottom).

Fissures in head and bed mortar joints of infills C were also detected at this stage, caused by additional shear forces during out-of-plane excitation with 77% KEF-1. The fissures pattern is presented in Figure 6.

In PHASE 2, the type B walls were repaired by glass FRP mesh, bonded to the damaged wall using flexible adhesive of type Sika PS (FRPU repair), without any special treatment of the infill face and no crack repair. The mesh was applied in two diagonal strips and at all the edges of the infills, without

Figure 5. Infill type B wall with significant damages.

Table 1. Test phases and list of earthquake runs.

	PHASE 1	PHASE 2 (after emergency repair of damaged walls)	PHASE 3 (after rotating the building 90 degrees)
INTENSITY	3% KEF-1 (0.06g)	6% KEF-2 (0.12g)	6% KEF-3 (0.12g)
	6% KEF-1 (0.12g)	11% KEF-2 (0.24g)	10% KEF-3 (0.22g)
	11% KEF-1 (0.23g)	18% KEF-2 (0.39g)	16% KEF-3 (0.35g)
	25% KEF-1 (0.53g)		
	38% KEF-1 (0.80g)		
	55% KEF-1 (1.16g)		
	72% KEF-1 (1.54g)		
	74% KEF-1 (1.57g)		
	69% KEF-1 (1.47g)		
	77% KEF-1 (1.64g)		

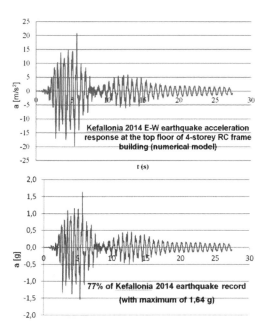

Figure 4. Kefallonia 2014 E-W earthquake acceleration response at the top floor of 4-storey building (reference signal) – top and scaled acceleration loading for 77% of the reference signal (KEF) - bottom.

direct connection to the RC frame (the repaired infills B were connected with the RC frame only through the previously injected PUFJs).

The width of all applied glass FRP strips was 50 cm. The layout of the FRP intervention is shown in Figure 7. The FRP application was made on both sides of damaged infills B, without additional connections between inner and outer sides. The aim of this strengthening was to protect the damaged infills of the model against collapse during out-of-plane excitation, after rotation of the building by 90 degrees. This emergency retrofitting of such damaged infill can be applied in specific situation in real applications, as protection against out-of-plane collapse before aftershocks occurrence.

Once quick emergency repairing was applied, the model was again loaded only 10 hours after FRPU application. First, out-of-plane tests with harmonic resonance frequencies (16 Hz and 32 Hz) of the infills C were carried out up to 40% KEF. No significant damages to the infills C were observed (only widening of the existing fissures in mortar (Figure 6). Next, gradually increasing seismic loading (see Table 1 for details) was applied. It should be noted that the building was not rotated during the second stage of excitations. Therefore, the walls that were subjected to in-plane actions in the first stage, were again subjected to in-plane actions in the second stage. At 18% KEF-2 intensity, the testing was stopped in order to avoid uncontrolled damage to the RC structure (crushing of concrete at the top and the bottom of columns was observed – see Figure 8) that could jeopardize the third phase of the tests.

In the next step, the entire structure was rotated by 90 degrees on the shake table.

In PHASE 3, the strengthened type B walls were loaded in the out-of-plane direction, and the type C walls (with the crack pattern presented in Figure 6) were now loaded in the in-plane direction. The intensity was gradually increased until 16% KEF-3

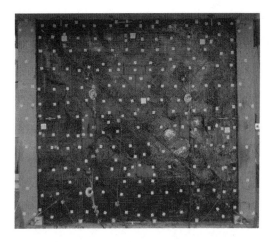

Figure 7. Strengthened type B walls.

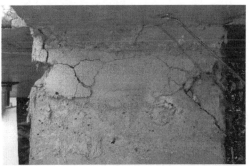

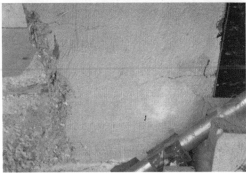

Figure 8. Cracked concrete at the RC column top and bottom.

intensity, when observable damages were noticed (by visual inspection and by changes in the eigen-frequency of the structure). The sequence of intensities is presented in Table 1. It should be noted that the building was rotated during the third stage excitations. Therefore, the walls subjected to in-plane actions in the first and second stages (Infills B), were subjected to out-of-plane actions in the third stage.

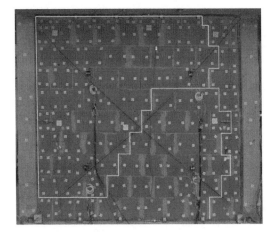

Figure 6. Fissures pattern in the infill C after out-of-plane tests, here in position before in-plane tests (after rotation).

At 16% KEF-3 intensity, the testing on the rotated structure was stopped in order to avoid further damage to the RC structure and the masonry infill. Before stopping the test, an increase of softening and a further increase of fissure widths in head and bed mortar joints (shown in Figure 6) were observed. Another reason of stopping the test was to protect the weakened infills C to be able to test their performance under in-plane action after a new FRPU strengthening (not presented in this paper).

3 MAIN RESULTS

Shake table tests were carried out for various earthquakes load levels (% KEF), described in Table 1. Direction of the shake table excitation was marked as Y and the perpendicular direction as X. Analysis of results was carried out first for horizontal relative displacements of the slab related to the building foundation, measured by two linear transducers attached to the support and the shake table displacement (Figure 3). The relevant drifts of the structure for the in-plane tested infills (presented in %) were calculated. Next, changes in eigen-frequencies were analyzed and the change of the global stiffness of the structure was calculated.

3.1 *Setup with infills B tested in-plane*

In PHASE 1, the infills B were tested in-plane in an initial phase up to 72% KEF-1 intensity (see Table 1), where damage of the infills B was hardly observed. When the structure was again loaded up to 74% KEF-1 intensity, the capacity of the table was exhausted and additional damages in the building occurred due to sudden stop of the shake table, but without visible degradation of the specimen. After checking and correction in the shake table system, the infills B were tested further in the damage phase with 69% KEF-1 and 77% KEF-1 intensity (Table 1). Formation of significant damages (Figure 5) were observed during the last one excitation. The observed crack pattern was different from the classical cross shape (><) going through corners. The corner zones were protected against cracking by the PUFJs and thus damages were localized in the middle of the infill height, forming the horizontally extended cross shape (>−<).

Changes in measured maximum horizontal relative displacement of the slab (Y-direction) were used to calculate the RC frame drift using the height of the columns up to the center of the joint with the beams 240 cm (Figure 9). Maximum drift obtained during 72% KEF-1 and 74% KEF-1 was (2.87/240) 1.20% and (3.20/240) 1.33%, respectively. Maximum drift obtained during 69% KEF-1 and 77% KEF-1 (before and after occurrence of significant damage to the infills B) was (2.52/240) 1.05% and

(8.89/240) 3.70%, respectively based on the average values of the two horizontal displacement meters.

The behavior of the structure was almost linear up to 38% KEF-1 intensity (0.80 g). Softening started once plastic hinges were developed on the top and at the bottom of the columns and cracking occurred in the head and bed mortar joints at 55% KEF-1 intensity (1.16 g), which further developed when excitation continued up to 77% KEF-1 intensity (1.64 g) with meantime repetition of levels 74% KEF-1 and 69% KEF-1.

Observation related to the drift changes are confirmed by the reduction of the main eigenfrequency of the building in the Y-direction from 7.2 Hz to 1.8 Hz, obtained from the random tests (Figure 10). Significant damage to infills B appeared in the 77% KEF-1 intensity. These correspond to reduction of the eigenfrequency below 4 Hz and interfering in resonance with the dominant excitation frequency of the earthquake signal (2,5-4 Hz). Changes of the global stiffness of the building are presented in Figure 11, in the form of a stiffness ratio, defined by the equation: (EI$_{current}$)/(EI$_{initial}$) = (fy$_{current}$)^2/(fy$_{initial}$)^2, where EI$_{current}$ = current stiffness; EI$_{initial}$ = initial stiffness; fy$_{current}$ = current eigen-frequency; fy$_{initial}$ = initial eigenfrequency. Reduction of the global stiffness up to 31% was caused mainly by formation of plastic hinges in the RC columns (Figure 8) and loss of cohesion in head and bed mortar joints.

Drop of stiffness to 6% (Figure 11) and significant damages to the infills B (Figure 5) occurred within 77% KEF-1 intensity excitation, but the structure did not collapse, and no significant permanent deflection was observed. Moreover, the blocks were kept by the PUFJs on the position with only single blocks rushing out. The infills B (in such condition) were able to be repaired and strengthened using GFRPU system (Figure 7). After strengthening the infills B, the structure revealed increased eigen-frequency in the Y direction up to 3.6 Hz (recovery of stiffness to the level of 25%).

After in-plane testing of the strengthened infills B up to 18% KEF-2 (0.39 g) in PHASE 2, no additional damages were observed, but the eigen-frequency dropped to 2.6 Hz (decrease of stiffness to the level of 13%). The strengthened infills B withstood all further out-of-plane earthquake excitations applied to the structure after rotation of the building by 90 degrees. The presented results related to the infills B indicate that PUFJ and glass FRPU systems effectively protect RC frames and infill walls during strong earthquakes.

3.2 *Setup with infills C tested in-plane*

In PHASE 1 (the infills B tested in-plane), the infills C were tested out-of-plane in the initial phase up to

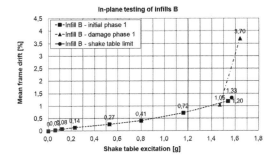

Figure 9. Changes of calculated drift of the slab.

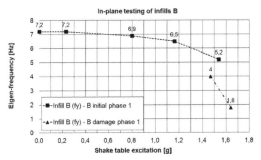

Figure 10. Changes of the main eigen-frequency of the building in the Y-direction.

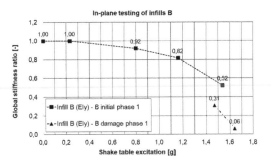

Figure 11. Changes of the global stiffness of the building in the Y-direction.

77% KEF-1 intensity and then in the damage phase with 69% KEF-1 and 74% KEF-1 intensity (Table 1). Practically no serious damages to the infills C (only fissures in mortar – Figure 6) were observed at this stage and also after the harmonic resonance tests with 16 Hz and 32 Hz frequencies. The infill was protected by PUFJs against detachment from the RC frame, which allowed the infill to move out-of-plane like a stiff slab with the eigen-frequency of 16 Hz and like a bending slab with the eigen-frequency of 32 Hz. The crack pattern visible in Figure 6 corresponds to

damage of a simple supported slab. Degradation of RC columns resulted also in changes of eigen-frequencies in the perpendicular X-direction, from 6.4 Hz to 4.0 Hz (Figure 12), close to the resonance frequency range of the KEF earthquake excitation. Excitation of the structure in PHASE 2 did not change the eigen-frequency of the building in X-direction (left 4.0 Hz).

Reduction of stiffness in the X-direction up to 39% after PHASES 1 and 2 was calculated from Figure 13. Influence of the building rotation and of action of additional shear forces were observed in change of eigen-frequencies and stiffness. The building had variable frequencies (and stiffness): in-plane of the infills C - decreased from 4.0 Hz (stiffness of 39%) to 3.5 Hz (stiffness of 30%), whereas in-plane of the infills B - increased from 2.6 Hz (stiffness of 13%) to 3.7 Hz (stiffness of 26%). In PHASE 2, damages of the repaired infills B were hardly observed.

After rotation of the building, the infills C presented slightly non-linear behavior during in-plane tests in PHASE 3. The drift changes calculated based on the maximum relative horizontal displacement of the slab are presented in Figure 14 for different excitation intensities up to 16% KEF-3.

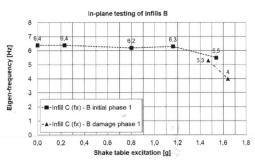

Figure 12. Changes of the main eigen-frequency of the building in the X-direction.

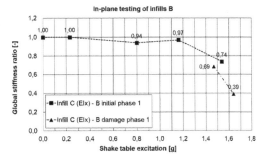

Figure 13. Changes of the global stiffness of the building in the X-direction.

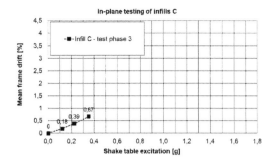

Figure 14. Changes of calculated drift of the slab.

Observation related to the drift changes are confirmed by small reduction of the eigen-frequency - Figure 15 (and stiffness - Figure 16) in the Y-direction: from 3.5 Hz (stiffness of 30%) to 2.9 Hz (stiffness of 21%), even if the structure worked in the resonance frequency range of the KEF earthquake excitation (2.5-4 Hz).

It is worth to notice that the initial building stiffness in the direction of the infills C plane (related to 6.4 Hz frequency – Figure 12) is 21% lower than the initial building stiffness in the direction of the infills

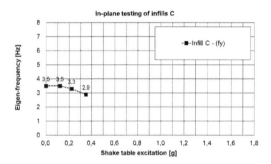

Figure 15. Changes of the main eigen-frequency of the building in the Y-direction.

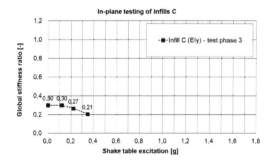

Figure 16. Changes of the global stiffness of the building in the Y-direction.

B plane (related to 7.2 Hz frequency – Figure 10). This is caused by the influence of the PUFJ working at the bottom of the infills C.

4 CONCLUSIONS

The tests of a full-scale RC frame with infills on a shake table showed that infills have a significant influence on the seismic response of the structure. The detrimental effects of strong earthquakes on the stiffness and the eigen-frequencies of the structure – both denoting damage accumulation - are better controlled in case the orthoblock infill-RC frame boundaries are protected with PUFJ joints.

It was observed that due to the flexibility of the 2 cm thick polymer joint, the interaction between the special orthoblock infill and the RC frame can be manipulated so as to achieve the delay of significant infill damages at very high RC frame inter-storey drifts. The present tests suggest that first brick disintegration (that may cause injuries) occurs at 2.5% drift while avoiding undesirable effects on the RC columns, caused by the infills. Further, it is validated, during the same run of excitation, that the PUFJ protection enabled the infill to be repaired even after the frame has been subjected to very high inter-storey drifts up to 3.7%. This drift level is higher than the one corresponding to repairable damages for ordinary infilled RC frames.

Finally, the applied glass FRPU system efficiently protects the damaged infills against collapse under out-of-plane excitation while they restore large part of their in-plane stiffness.

The tests offer a direct comparison between the cases with PUFJs on three sides (left, top, and right) and on all four sides for low excitation levels. Both PUFJ systems protected efficiently the orthoblock brick infills against out-of-plane failure for very high inter-storey drifts and accelerations.

ACKNOWLEDGMENTS

This research activity was within the framework of the project Seismology and Earthquake Engineering Research Infrastructure Alliance for Europe – SERA, INfills and MASonry structures protected by deformable POLyurethanes in seismic areas (INMASPOL). The project leading to this application has received funding from the European Union's Horizon 2020 research and innovation programme under grant agreement No 730900.

REFERENCES

Akyıldız, A.T., Kowalska-Koczwara, A. & Kwiecień, A. 2019. Stress distribution in masonry infills connected with stiff and flexible interface. Journal of Measurements in Engineering Vol. 7, 1, pp. 40–46.

EN 1998-1 2004. Eurocode 8: Design of Structures for Earthquake Resistance – Part I: General Rules, Seismic Actions and Rules for Buildings.

Kwiecień, A. 2019. Reduction of stress concentration by polymer flexible joints in seismic protection of masonry infill walls in RC frames. Materials Science and Engineering Vol. 474, pp. 1–7.

Kwiecień, A., Gams, M., Rousakis, T., Viskovic, A. & Korelc, J. 2017. Validation of a new hyperviscoelastic model for deformable polymers used for joints between RC frames and masonry infills. Engineering Transactions Vol. 65, No 1, pp. 113–121.

Ricci, P., Di Domenico, M. & Verderame, G.M. 2018. Empirical-based out-of-plane URM infill wall model accounting for the interaction with in-plane demand. Earthquake Engineering and Structural Dynamics Vol. 47, pp. 802–827.

Rousakis T., Rouka D., Kaloudaki A., Kwiecień A., Gams M., Viskovic A., Zając B. Fast Retrofitting of Strong Wall Infill of RC buildings with Fiber Sheets Impregnated with Highly Deformable Polymer. 25th International Conference on Composites/Nano Engineering, July 16-22, 2017 Rome.

Rouka D., Kaloudaki A., Rousakis T., Fanaradelli T., Anagnostou E., Kwiecień A., Gams M., Viskovic A., Zając B. (2017): Response of RC buildings with Low-strength Infill Walls Retrofitted with FRP sheets with Highly Deformable Polymer – Effects of Infill Wall Strength. 25th International Conference on Composites/Nano Engineering, July 16-22, 2017 Rome.

Turkish Seismic Design Code 2018. Turkish Ministry of Interior, Disaster and Emergency Management Presidency.

Viskovic, A., Zuccarino, L., Kwiecień, A., Zając, B. & Gams, M. 2017. Quick seismic protection of weak masonry infilling in filled framed structures using flexible joints. Key Engineering Materials 747, pp. 628–637.

Brick and Block Masonry - From Historical to Sustainable Masonry –
Kubica, Kwiecień & Bednarz (eds)
© 2020 Taylor & Francis Group, London, ISBN 978-0-367-56586-2

An innovative timber system for the seismic retrofit of unreinforced brick masonry buildings

N. Damiani & M. Miglietta
UME School, University School for Advanced Studies (IUSS), Pavia, Italy

G. Guerrini & F. Graziotti
Department of Civil Engineering and Architecture (DICAr), University of Pavia, Italy
European Centre for Training and Research in Earthquake Engineering (EUCENTRE), Pavia, Italy

ABSTRACT: A new timber-based seismic retrofit technique was investigated at the EUCENTRE laboratories (Pavia, Italy) within an extensive experimental campaign on the vulnerability of existing Dutch unreinforced masonry (URM) cavity-wall terraced houses. These structures typically consist of a single-wythe, calcium-silicate-brick inner leaf with load-bearing function, and an external clay-brick veneer with no structural purposes, connected to each other by steel ties. The first floor is usually built with precast reinforced concrete slabs, while the second floor and roof often consist of timber joists and planks. The main objective of the retrofit was to improve the seismic capacity of Dutch URM terraced houses with a light, cost-effective and low-invasive intervention. The proposed retrofit system included timber frames fastened to the internal surface of masonry piers and to the building floors, on which oriented-strands boards (OSB) are nailed. The retrofit was designed to increase both in-plane and out-of-plane capacities of masonry piers as well as to improve the overall connections between masonry elements and floor diaphragms. After performing quasistatic in-plane cyclic tests on two calcium-silicate URM piers, two identical full-scale two-storey buildings were tested dynamically on the shake-table, one in bare and the other one in retrofitted configuration. This paper focuses on the experimental performance of these buildings, with emphasis on the improved seismic response of the specimens, and proposes simple design equations.

1 INTRODUCTION

Unreinforced masonry (URM) cavity-wall systems are commonly used as structural solution for residential construction in several parts of the world characterised by a low seismic hazard. During the last decades it has been demonstrated that natural phenomena, (i.e. the slip of an unknown fault; Horton et al. 2012) or human activities (i.e. gas extraction; Bourne et al. 2015, Crowley et al. 2019, Graziotti et al. 2019) can expose a site to unexpectedly intense seismic events. These buildings were often designed with insufficient seismic details, and therefore these occurrences have prompted the interest on possible retrofit techniques to reduce their seismic vulnerability.

Various strengthening techniques for unreinforced masonry walls have been proposed by several researchers, to enhance their in-plane and out-of-plane capacities. The most common solution consists in the application of an additional material layer to the masonry, such as fiber-reinforced composites (Tomaževič et al. 2015), fiber-reinforced cementitious matrices (Babaeidarabad et al. 2014) and externally bonded grids (Giaretton et al. 2018). Further possibilities are represented by the application of steel elements mechanically connected to the masonry (Darbhanzi et al. 2014) and by the use of post-tensioning (Ma et al. 2012). Moreover, the enhancement of connections between intersecting walls (Podestà et al. 2019) and between floor diaphragms and masonry (Senaldi et al. 2019) has proved to be effective on the improvement of URM buildings seismic behaviour.

Overall, the strengthening of masonry buildings in low-seismicity areas needs to be light, cost-effective, sustainable, and possibly reversible. To meet these requirements, timber represents a viable alternative due to its ability to add tensile strength to the masonry and its high strength-to-density ratio. Only recently the application of timber as a retrofit solution for masonry structures has been investigated, to enhance the in- and out-of-plane capacities of masonry piers (Riccadonna et al. 2019, Dizhur et al. 2017, Giaretton et al. 2016).

Starting from these studies, an innovative retrofit system consisting of timber frames and oriented-strand boards (OSB), mechanically connected to the masonry, has been conceived and experimentally investigated at the EUCENTRE laboratories in Pavia, Italy. The retrofit solution was first applied to a masonry pier subjected to in-plane quasi-static cyclic tests (Guerrini et al. 2020), and then to a full-scale two-storey building tested dynamically on the shake-table (Damiani et al. 2019, Miglietta et al. 2019). Unstrengthened counterparts were also tested to appreciate the effectiveness of the retrofit on an individual component and on a building system. This paper describes the basic concepts of the proposed system, and validates simple analytical design equations against the shake-table test results. All processed data and instrumentation schemes are available at http://www.eucentre.it/nam-project.

2 PROPOSED RETROFIT FEATURES

The proposed retrofit system stems from the out-of-plane strengthening solution developed by Giarretton et al. (2016) and by Dizhur et al. (2017), based on the use of vertical timber posts termed strong-backs. This solution has been further extended to improve also the in-plane capacity of piers and the connection between masonry and floor diaphragms. Indeed, the lack of masonry-to-floors connections represents one of the most relevant weaknesses showed by existing URM buildings not designed for seismic loads (Tomassetti et al. 2019, Miglietta et al. 2019).

The innovative retrofit system consists of a timber frame mechanically connected to the masonry internal surface, which includes vertical posts and horizontal nogging elements (Figure 1). Top and bottom sill plates complete the frame allowing its connection to floors and foundations. OSB are nailed to the frame.

The effectiveness of the proposed system depends on the quality of connections between timber elements, masonry walls, and floor diaphragms: C1 indicates the tie-down anchorages between posts and floors (or

foundations) through the sill plates; C2 refers to additional anchorages between sill plates and floors or foundations for sliding-shear strength; and C3 identifies connections between timber frames and masonry walls, while C4 between timber frame components.

The masonry wall out-of-plane response is enhanced by the vertical posts which act as strong-backs in flexure, transferring inertia forces from the masonry panels to the adjacent floor diaphragms. The pier in-plane capacity is increased by both timber frames and OSB, which collaborate with the masonry as they are forced to deform with it by connections C3. In particular, the vertical posts and their tie-down connections C1 provide in-plane flexural strength, while the OSB nailed to the frame contribute to the in-plane shear strength.

Moreover, the connection enhancement between masonry walls and floor diaphragms offered by the retrofit system plays an essential role in the improvement of the seismic performance. In fact, it favours a global box-type response, preventing the occurrence of undesired local mechanisms, which can strongly reduce the building seismic capacity. This aspect, however, may require the assessment of the strength of existing footings or diaphragms.

3 FULL-SCALE RETROFITTED BUILDING SHAKE-TABLE TEST

Following the experimental investigation of the retrofit effects on the lateral response of an individual masonry pier (Guerrini et al. 2020), the same solution was applied to a full-scale two-storey URM building specimen, which represented the end-unit of a typical Dutch terraced house.

The seismic response of the bare URM building was first investigated through a unidirectional incremental dynamic shake-table test up to near collapse conditions (Miglietta et al. 2018, 2019). The experiment evidenced a significant lack of connections between the masonry walls and the timber second-floor diaphragm: this triggered a local sliding mechanism which prevented the development of a global response with full exploitation of pier capacities. A second specimen was then built and the retrofit installed on it. The same dynamic excitation protocol was followed, with additional increments made possible by the effectiveness of the retrofit intervention.

3.1 Bare specimen overview

The prototype was 5.9-m long and 5.6-m wide, with a large amount of openings asymmetrically distributed on the front (West) and back (East) façades (Figure 2). No openings were provided in the South wall, simulating a party wall, while a small window was realized in the North gable. The structure was oriented with the most vulnerable East and West sides parallel to the shaking direction. The URM cavity-walls consisted of an internal single-wythe, 100-mm-

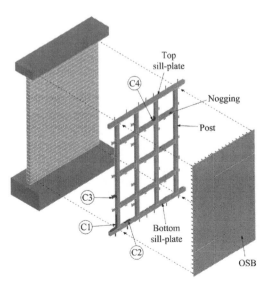

Figure 1. Retrofit components and layout.

509

Figure 2. Bare URM building specimen; arrows indicate the shaking direction.

thick, loadbearing calcium-silicate (CS) leaf and an external single-wythe, 100-mm-thick, clay veneer. 3.1-mm-diameter steel ties provided connection between the two leaves with a density of 1 tie/m^2.

The specimen was built on a composite concrete/ steel foundation firmly bolted to the shake-table. It

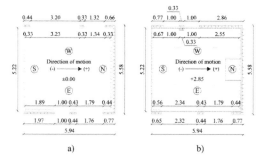

a)   b)

Figure 3. Bare URM specimen plan views: a) first storey; b) second storey. Units of m.

included a rigid 160-mm-thick reinforced concrete (RC) first-floor diaphragm and a flexible timber second-floor diaphragm, with a staircase opening and four holes to accommodate a reference steel frame (Figure 3). A 39° pitched timber roof completed the building. Floors and roof were supported only by the North and South CS transverse walls. L-shape steel anchors, with diameter of 14 mm, connected the second-floor and roof 100x240-mm timber joists with the supporting walls. 185x18-mm tongue-and-groove planks were nailed to the joists by pairs of 2-mm-diameter, 60-mm-long nails.

3.2 Retrofit system details

The retrofit frames consisted of timber elements with a cross-section of 80x60 mm, oriented with the 60-mm side perpendicular to the walls (Figures 4a, c). 18-mm-thick OSB were fastened to all frame elements using 4-mm-diameter, 75-mm-long anker nails at 100-mm-spacing (Figure 4b), following the American specifications for timber shear-walls (AWC 2008). These sizes were selected to combine structural efficiency with minimal invasiveness. Due to the significant in-plane capacity of the transverse North and South masonry walls, nogging elements and OSB were not provided there, to keep the intervention as light as possible and to allow visual inspection of the masonry during the test (Figure 4c).

The second-floor timber diaphragm was stiffened with a 18-mm-thick OSB layer, fastened to existing joists and additional blocking beams by 4-mm-diameter, 75-mm-long anker nails, according to the American provisions for timber diaphragms (AWC 2008). The roof diaphragm was not stiffened initially, but loose safety steel cables were subsequently tied to form cross-braces between the ridge beam and the second-floor, when the gable out-of-plane displacements became excessive.

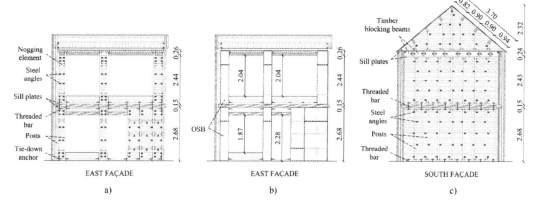

Figure 4. Retrofitted specimen elevations: a) timber frames applied to longitudinal façades; b) OSB nailed to frames on longitudinal façades; c) timber frames applied to transverse façades. Units of m.

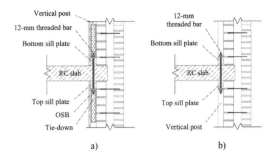

Figure 5. Connection details with the first floor: a) longitudinal walls; b) transverse walls.

Connections C1, C2, C3 and C4 (see section 2) were realized with steel angles (Rothoblaas 2015, 2018), fastened to the timber frames by 5-mm-diameter, 70-mm-long screws and to masonry or RC elements by class 8.8 threaded rods with diameter of 10 mm and 12 mm, respectively. An epoxy adhesive bonded the threaded bars within masonry walls and RC foundation. For connections C1 and C2 with the first-floor diaphragm, threaded bars crossing the RC slab through core-holes were bolted to the first-storey top sill plate and to the second-storey bottom sill plate (Figures 5a, b). Along the transverse walls, vertical posts were connected to sill plates through simple C4 timber-to-timber connections, while C2 connections tied the sill plates to the RC slab (Figure 5b) or to the foundation.

To improve the connection between the second-floor diaphragm and the masonry, the second-storey top sill plates were screwed densely to the inner-leaf spreader beams along the longitudinal walls (Figure 6a). Timber blocking beams inserted between floor-joists along the transverse walls allowed connecting the second-storey top sill-plate and the gable bottom sill plate to the diaphragm (Figure 6b). The vertical posts attached to the gables were connected directly to the roof purlins, since no top sill-plate was provided there (Figure 4c).

To simulate the connection enhancement between inner and outer leaves, the density of steel ties was increased to 5 tie/m^2, and the corresponding spreader beams were fastened together by pairs of diagonally crossing, 8-mm-diameter, 360-mm-long timber screws at 500-mm spacing.

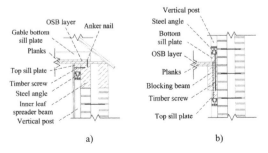

Figure 6. Connection details with the second floor: a) longitudinal walls; b) transverse walls.

3.3 Mechanical properties

Material characterization tests were performed on masonry materials at the DICAr laboratory of the University of Pavia, Italy, on specimens that reached at least 28 days of maturation. The calcium silicate and clay masonry compressive strengths (f_m) and the secant elastic moduli at 33% of their compressive strengths (E_m) were determined according to EN 1052-1 (CEN 1998); the corresponding mortar tensile (f_t) and compressive (f_c) strengths were determined according to EN 1015-11 (CEN 1999). The initial shear strength (f_{v0}) and the friction coefficient (μ) were obtained according to EN 1052-3 (CEN 2007). Table 1 summarizes the results.

The timber of the frames was red solid fir (Picea abies) of class S10/C24 according to EN 14081-1 (CEN 2016) with a density of 517 kg/m^3. The specified characteristic compressive strength parallel to fibers ($f_{c,0}$) was 21 MPa, the characteristic tensile strength parallel to fibers ($f_{t,0}$) was 14 MPa, and the characteristic Young modulus ($E_{0.05}$) was 7400 MPa. The OSB were classified as OSB/3 according to EN 300 (CEN 2006), with a density of 572 kg/m^3.

Connections C1 had characteristic tensile strength of 11.6 kN (Rothoblaas 2015), while connections C4 had characteristic strength of 11 kN considering screw flexural yielding (Rothoblaas 2018).

3.4 Masses

Calcium silicate and clay masonry had average densities of 1862 kg/m^3 and 2072 kg/m^3, respectively. The mass of the inner masonry leaf was 17.7 t while the one of the outer leaf was 13.9 t. First and second floors had masses of 11.2 t and 1.9 t, respectively, including additional weights. The complete roof had mass of 2.8 t. The retrofit system mass was estimated as 1.6 t. The specimen total mass was 49.1 t.

3.5 Testing protocol

The specimen was subjected to an incremental dynamic test, applying a series of motions of

Table 1. Masonry and mortar mechanical properties.

	Calcium silicate	C.o.V.	Clay	C.o.V.
	MPa	-	MPa	-
f_m	10.05	0.11	17.62	0.21
E_m	7319	0.15	5686	0.31
f_t	1.39	0.38	0.86	0.43
f_c	3.97	0.41	3.02	0.38
f_{v0}	0.49	-	0.18	-
μ	0.39*	-	0.71*	-

* Dimensionless property

increasing intensity through the shake-table, to assess damage evolution, failure modes, and ultimate capacity of the building. The testing sequence was initially identical to the one used for the bare building (Miglietta et al. 2019, Miglietta et al. 2018), to allow investigating the retrofit system benefits, then continued with larger acceleration amplitudes.

The input signal consisted of an acceleration time history representing a realistic induced ground motion for the Groningen region of the North-East part of The Netherlands. A single-component earthquake accelerogram with $PGA = 0.31$ g and short significant duration $D_{s,5-75} = 1.82$ s (Figure 7), termed EQ-NPR, was selected upon spectrum-compatibility with the uniform hazard spectrum (UHS) at 2475-years return period for the site of Loppersum (lat. +53.33, long. +6.75; Bommer et al. 2017). The acceleration amplitude of the original input signal was progressively scaled to 33%, 50%, 66%, 85%, 100%, 133%, 166%, 200%, and 266%.

3.6 Test results and retrofit benefits

The retrofit intervention allowed the building prototype to exhibit a global box-type response with increased both in-plane and out-of-plane capacities of masonry walls. The enhancement of connections between masonry and diaphragms prevented the onset of local mechanisms, as observed during the bare specimen shake-table test where the second-floor diaphragm slid above the longitudinal walls (Miglietta et al. 2019, Miglietta et al. 2018) and imposed large out-of-plane displacements on the transverse façades.

The retrofit system did not influence the onset of damage, due to its negligible stiffness compared to the masonry structure. However, up to run EQ-NPR-100% ($PGA = 0.31$ g) the retrofitted specimen accumulated damage only in the roof-gable subsystem, which showed higher flexibility compared to the first and second storey (Graziotti et al. 2017, Tomassetti et al. 2019).

After test EQ-NPR-133% ($PGA = 0.39$ g), the roof-gable subsystem was stiffened by steel cables, to limit gable out-of-plane displacements. More importantly, at this stage diagonal stair-stepped cracks were observed on the first-storey transverse walls, denoting the onset of a global torsional response. This behaviour was strongly desired in the retrofit conception since it indicated exploitation of the in-plane capacity of the transverse walls. Moreover, the bare specimen reached near-collapse conditions under the same shaking intensity.

The following increment of ground motion amplitude accentuated the first-storey damage due to torsional effects and caused a significant accumulation of residual displacements. The test was stopped after run EQ-NPR-266% ($PGA = 0.78$ g), as the specimen was deemed very close to near-collapse conditions with a weak-storey mechanism, given the severity of the first-storey damage. The retrofitted building was able to sustain twice the ground-motion intensity compared to the bare structure.

Figure 8 shows the experimental backbone curves. The base shear was computed ignoring the inertia of the longitudinal clay veneers and of the bottom half of the first-storey walls, which were assumed not to engage the retrofit system. The base-shear coefficient was obtained dividing the shear force by the corresponding weight. The global drift ratio was computed dividing the second-floor average displacement by its height above the foundation. The points corresponding to the maximum and minimum base shear of each run were considered, as well as the ones associated with the maximum and minimum displacement of the final test.

As expected, the application of the retrofit system did not affect the specimen initial stiffness but allowed a significant strength increase. In particular, the retrofitted building exhibited a maximum absolute base shear of 202 kN, that is 132% higher than the bare specimen one of 87 kN. Further details on the retrofitted specimen response can be found in Miglietta et al. (2019) and in Damiani et al. (2019).

4 ANALYTICAL PREDICTION OF THE EXPERIMENTAL RESULTS

In this section, simple analytical equations are proposed to capture the experimental base shear of the

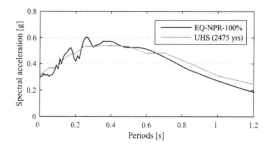

Figure 7. Elastic response spectrum of input signal EQ-NPR for 5% viscous damping ratio and uniform hazard spectrum at 2475-years return period for Loppersum, The Netherlands.

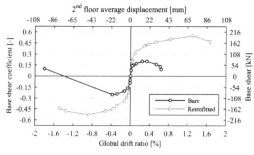

Figure 8. Bare and retrofitted specimen backbone curves.

retrofitted building. Only the contribution of the calcium-silicate inner leaf was considered since the outer-leaf clay veneer had no intended structural function. The obtained analytical results are validated against the experimental backbones curves showed in Figure 8.

4.1 Masonry contribution to pier lateral strength

For masonry piers in double-curvature boundary conditions, the lateral flexural strength can be estimated with the following equation (Magenes et al. 1997):

$$V_{R,mas,f} = \frac{\sigma_v L^2 t}{H_{eff}} \left(1 - \frac{\sigma_v}{\eta f_m}\right) \qquad (1)$$

where σ_v is the vertical compressive stress at mid-height of the considered pier; L the wall length; t the wall thickness; H_{eff} the pier effective height; η the stress-block magnitude parameter assumed equal to 0.85; and f_m the masonry compressive strength.

The lateral shear strength, due to frictional and cohesive contributions, can be calculated as (Magenes et al. 1997, CEN 2005):

$$V_{R,mas,s} = \mu \sigma_v t L + f_{v0} t l_c \qquad (2)$$

with the contact length (l_c) given by:

$$L \frac{\sigma_v}{\eta \lambda f_m} \le \left[l_c = 1.5.L.\left(1 - \alpha_v \frac{3 f_{v0} + 2 \mu \sigma_v}{\sigma_v + 3 f_{v0} \alpha_v}\right)\right] \le L \qquad (3)$$

where μ is the friction coefficient; f_{v0} the masonry initial shear strength; λ the stress-block depth parameter assumed equal to 0.85; and $\alpha_v = H_{eff}/(2L)$ the shear-span ratio.

A recent study on the seismic response of a full-scale one-storey URM terraced house with RC floor (Tomassetti et al. 2019) has demonstrated that the in-plane rocking response of longitudinal corner piers can result in slab uplifting from the supporting transverse walls, with consecutive redistribution of gravity loads to these piers. This effect is more pronounced with flexurally stiffer slabs, while it disappears if the floor system deforms and maintains contact with the transverse bearing walls.

In light of this, because of the rigid first-floor RC slab of the building specimen, the whole overburden load acting on the first-storey walls can be evenly shared between the four CS corner piers (W1, W3, W4, W6) in dynamic conditions, leaving the intermediate longitudinal piers (W2, W5) and the transverse walls to carry only their self-weight, as shown in Figure 9.

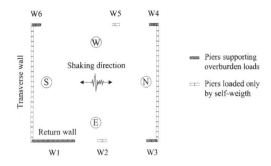

Figure 9. Gravity loads distribution among first-storey piers in dynamic conditions.

4.2 Retrofit contribution to pier lateral strength

The axial forces acting on vertical timber posts provide the flexural strength of the retrofit system. Assuming that no axial load is carried by the timber frame and that one outermost post is in compression, while the remaining posts are stressed to their tensile strength, the lateral flexural strength provided by the timber frame is evaluated as follows for double-curvature conditions (Guerrini et al. 2020):

$$V_{R,tim,f} = \frac{2 \sum_i (T_i b_i)}{H_{eff}} \qquad (4)$$

where T_i is the minimum between the tensile strength of the ith post and the one of the tie-down connection C1, while b_i is the the distance between the ith post and the compressed one.

The retrofit system lateral shear strength contributed by the nailed OSB and is given by:

$$V_{R,tim,s} = v_s L \qquad (5)$$

where v_s is the nominal unit shear capacity of timber shear-walls, typically expressed as a function of OSB thickness and fastener type, size, and spacing.

The rigid slab uplift in dynamic conditions affects also the retrofit system response. In particular, this induces tension in the timber posts along transverse walls, giving an additional flexural contribution to the longitudinal corner piers which engage the transverse walls as flanges (Figure 10). For each corner pier, this additional strength can be quantified as:

$$V_{R,tim,t} = \sum_j T_j \frac{b_t}{H_{eff}} \qquad (6)$$

where T_j is the minimum between the tensile strength of the jth post and the one of the post-to-sill plate

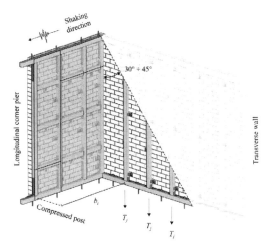

Figure 10. Contribution of the retrofit along transverse walls to the longitudinal base shear.

connection C4 along the transverse wall, while b_t is the distance between the row of posts and the compressed post of the longitudinal corner pier (Figure 10). The number of posts contributing to the additional strength can be identified considering a stress diffusion angle between 30° and 45° (Figure 10).

4.3 Pier lateral strength

Assuming that masonry and retrofit behave as springs in parallel, their individual flexural or shear contributions can be directly summed up, and the lateral strength of each pier (V_R) is given by the minimum between flexural and shear capacities:

$$V_R = \min\left(V_{R,mas,f} + V_{R,tim,f}; +V_{R,mas,s} + V_{R,tim,s}\right) + V_{R,tim,t}$$

(7)

4.4 Retrofitted building base shear

The analytical results obtained for the first-storey piers of the building specimen are reported in Table 2, considering the contribution of three posts to $V_{R,tim,t}$ for each corner pier. Characteristic strengths of steel connectors and fasteners were transformed into expected strengths by a factor of 1.10 (ASCE 2017). Characteristic strengths of timber members and nominal unit shear strengths of nailed OSB (AWC 2008) were converted into expected values by a factor of 1.50 (ASCE 2017).

The predicted base-shear capacity is 204 kN, very close to the maximum absolute experimental base shear of 202 kN. The masonry-only contribution was predicted as 76 kN, which approximates the experimental base shear of the bare masonry building (87 kN) with an underestimation of 13%.

Table 2. Analytical prediction of pier strengths.

Pier ID	$V_{R,mas,f}$ kN	$V_{R,mas,s}$ kN	$V_{R,tim,f}$ kN	$V_{R,tim,s}$ kN	$V_{R,tim,t}$ kN	V_R kN
W1	49	45	33	42	49	131
W2	0	1	5	10	-	5
W3	13	31	4	10	12	29
W4	8	31	2	7	7	17
W5	0	1	3	7	-	3
W6	9	31	3	7	8	19

5 CONCLUSIONS

This paper discussed an innovative timber-based retrofit solution for the seismic retrofit of unreinforced masonry buildings. The proposed system aims at enhancing in-plane and out-plane masonry pier capacities as well as wall-to-diaphragm connections. It consists of timber frames mechanically connected to the masonry, to the foundation, and to the floor systems, with OSB panels nailed to them.

The retrofit intervention was applied to a two-storey full-scale unreinforced masonry building specimen, identical to another specimen previously tested in bare conditions. The prototype represented the end unit of a typical Dutch terraced-house. The specimens were subjected to an incremental dynamic shake-table test sequence up to near-collapse conditions, scaling the same unidirectional ground motion record.

The proposed retrofit induced significant improvements in the building seismic response. It allowed the specimen to exhibit a global box-type behaviour, avoiding the occurrence of local mechanisms which strongly affected the response of the bare building. The retrofitted specimen sustained a maximum base shear 132% higher than the bare specimen did, and did not experience static instability as shown by the unstrengthened building.

Simple analytical equations were proposed to predict the in-plane strength of the retrofitted piers. Considering the retrofit system and the masonry as two springs in parallel, their contributions were evaluated separately and then summed together. The analytical prediction of the specimen base-shear capacity resulted in very good agreement with the experimental value, with a negligible error.

The proposed equations will be further extended for the development of detailed design guidelines, useful for practical applications. All processed data and instrumentation schemes can be downloaded at http://www.eucentre.it/nam-project.

ACKNOWLEDGEMENTS

This work is part of the EUCENTRE project "Study of the vulnerability of masonry buildings in

Groningen", within the research programme framework on hazard and risk of induced seismicity in the Groningen province, sponsored by the Nederlandse Aardolie Maatschappij BV (NAM). The data post-processing was also partially funded by the DPC-ReLUIS within the framework of the Work-Package -5 2019-2021: "Interventi di rapida esecuzione a basso impatto ed integrati". The authors would like to thank all parties involved in this project: the DICAr Laboratory of the University of Pavia and the EUCENTRE Laboratory, which performed the tests, the DPC-ReLUIS, and the partner NAM. The valuable advice of R. Pinho, G. Magenes and A. Penna was essential to the project and is gratefully acknowledged.

REFERENCES

American Society of Civil Engineers (ASCE) 2017. ASCE/SEI 41-17: Seismic evaluation and retrofit of existing buildings. Reston, Virginia, USA.

American Wood Council (AWC) 2008. ANSI/AF&PA SDPWS-2008: Special design provisions for wind and seismic. Washington, DC, USA.

Babaeidarabad, S., Arboleda, D., Loreto, G., & Nanni, A. 2014. Shear strengthening of un-reinforced concrete masonry walls with fabric-reinforced-cementitious-matrix. *Construction and Building Materials*, 65, 243–253.

Bommer, J.J., Dost, B., Edwards, B., Kruiver, P.P., Meijers, P., Ntinalexis, M., Rodriguez-Marek, A., Ruigrok, E., Spetzler, J., & Stafford, P.J., 2017. V4 ground-motion model (GMM) for response spectral accelerations, peak ground velocity, and significant durations in the Groningen field. *Research Report submitted to NAM*.

Bourne, S.J., Oates, S.J., Bommer, J.J., Dost, B., van Elk, J., & Doornhof, D. 2015. A Monte Carlo method for probabilistic hazard assessment of induced seismicity due to conventional natural gas production. *Bulletin of the Seismological Society of America*, 105(3), 1721–1738.

Crowley, H., Pinho, R., van Elk, J., & Uilenreef, J. 2019. Probabilistic damage assessment of buildings due to induced seismicity. *Bulletin of Earthquake Engineering*, 17(8), 4495–4516.

Damiani, N., Miglietta, M., Mazzella, L., Grottoli, L., Guerrini, G., & Graziotti, F. 2019. Full-scale shaking table test on a Dutch URM cavity-wall terraced-house end unit – A retrofit solution with strong-backs and OSB boards – EUC-BUILD-7. *Research report EUC052/2019U*, EUCENTRE, Pavia, Italy.

Darbhanzi, A., Marefat, M. S., & Khanmohammadi, M. 2014. Investigation of in-plane seismic retrofit of unreinforced masonry walls by means of vertical steel ties. *Construction and Building Materials*, 52, 122–129.

Dizhur, D. Y., Giaretton, M., Giongo, I., & Ingham, J.M. 2017. Seismic retrofit of masonry walls using timber strong-backs. *SESOC Journal*, 30(2), 30.

European Committee for Standardization (CEN) 1998. EN 1052–1: Methods of test for masonry. Part 1: Determination of compressive strength. Brussels, Belgium.

European Committee for Standardization (CEN) 1999. EN 1015-11: Methods of test for mortar for masonry - Part 11: Determination of flexural and compressive strength of hardened mortar. Brussels, Belgium.

European Committee for Standardization (CEN) 2005. Eurocode 6 - Design of masonry structures - Part 1.1: General rules for reinforced and unreinforced masonry structures. Brussels, Belgium.

European Committee for Standardization (CEN) 2006. EN 300: Oriented Strand Boards (OSB). Definitions, classification and specifications. Brussels, Belgium.

European Committee for Standardization (CEN) 2007. EN 1052–3: Methods of test for masonry units. Part 3: Determination of initial shear strength. Brussels, Belgium.

European Committee for Standardization (CEN) 2016. EN 14081-1: Timber Structures. Strength graded structural timber with rectangular cross section. Part I: General requirements. Brussels, Belgium.

Giaretton, M., Dizhur, D., & Ingham, J.M. 2016. Shaking table testing of as-built and retrofitted clay brick URM cavity-walls. *Engineering Structures*, 125, 70–79.

Giaretton, M., Dizhur, D., Garbin, E., Ingham, J. M., & da Porto, F. 2018. In-plane strengthening of clay brick and block masonry walls using textile-reinforced mortar. *Journal of Composites for Construction*, 22(5), 04018028.

Graziotti, F., Penna, A., & Magenes, G. 2019. A comprehensive in situ and laboratory testing programme supporting seismic risk analysis of URM buildings subjected to induced earthquakes. *Bulletin of Earthquake Engineering*, 17(8), 4575–4599.

Graziotti, F., Tomassetti, U., Kallioras, S., Penna, A., & Magenes, G. 2017. Shaking table test on a full scale URM cavity wall building. *Bulletin of Earthquake Engineering*, 15(12), 5329–5364.

Guerrini, G., Damiani, N., Miglietta, M., & Graziotti, F. 2020. Cyclic response of masonry piers retrofitted with timber frames and boards. *Structures and Buildings*. DOI: 10.1680/jstbu.19.00134.

Horton Jr., J.W. & Williams, R.A. 2012. The 2011 Virginia earthquake: What are scientists learning? *Eos, Transactions American Geophysical Union*, 93(33), 317–318.

Ma, R., Jiang, L., He, M., Fang, C., & Liang, F. 2012. Experimental investigations on masonry structures using external prestressing techniques for improving seismic performance. *Engineering Structures*, 42, 297–307.

Magenes, G. & Calvi, G.M. 1997. In-plane seismic response of brick masonry walls. *Earthquake engineering & structural dynamics*, 26(11), 1091–1112.

Miglietta, M., Damiani, N., Grottoli, L., Guerrini, G., & Graziotti, F. 2019. Shake-table investigation of a timber retrofit solution for unreinforced masonry cavity-wall buildings. *Proceedings of XVIII Conference of ANIDIS*, 117–127. Ascoli Piceno, Italy.

Miglietta, M., Mazzella, L., Grottoli, L., Guerrini, G., & Graziotti, F. 2018. Full-scale shaking table test on a Dutch URM cavity-wall terraced-house end unit – EUC-BUILD-6. *Research report EUC160/2018U*, EUCENTRE, Pavia, Italy.

Podestà, S., & Scandolo, L. 2019. Earthquakes and Tie-Rods: Assessment, Design, and Ductility Issues. *International Journal of Architectural Heritage*, 13(3), 329–339.

Riccadonna, D., Giongo, I., Schiro, G., Rizzi, E., & Parisi, M. A. 2019. Experimental shear testing of timber-masonry dry connections for the seismic retrofit of unreinforced masonry shear walls. *Construction and Building Materials*, 211, 52–72.

Rothoblaas 2015. Wood connectors and timber plates. https://www.rothoblaas.com/catalogues-rothoblaas.

Rothoblaas 2018. Screws and connectors for wood. https://www.rothoblaas.com/catalogues-rothoblaas.

Senaldi, I. E., Guerrini, G., Comini, P., Graziotti, F., Penna, A., Beyer, K., & Magenes, G. 2019. Experimental seismic performance of a half-scale stone masonry building aggregate. *Bulletin of Earthquake Engineering*, 1–35.

Tomassetti, U., Correia, A.A., Candeias, P.X., Graziotti, F., & Costa, A.C. 2019. Two-way bending out-of-plane collapse of a full-scale URM building tested on a shake table. *Bulletin of Earthquake Engineering*, 17(4), 2165–2198.

Tomaževič, M., Gams, M., & Berset, T. 2015. Strengthening of stone masonry walls with composite reinforced coatings. *Bulletin of earthquake engineering*, 13(7): 2003–2027.

Brick and Block Masonry - From Historical to Sustainable Masonry –
Kubica, Kwiecień & Bednarz (eds)
© 2020 Taylor & Francis Group, London, ISBN 978-0-367-56586-2

Seismic performance assessment of existing stone masonry school building in Croatia using nonlinear static procedure

M. Uroš, J. Atalić & M. Šavor Novak
Faculty of Civil Engineering, Department of Engineering Mechanics, University of Zagreb, Zagreb, Croatia

K. Kuk
Faculty of Science, Department of Geophysics, University of Zagreb, Zagreb, Croatia

ABSTRACT: Seismic performance assessment of a stone masonry school building in Croatia is presented in this paper. The building is located in the seismically active southern part of the country, with the reference peak ground acceleration on the bedrock for the return period of 475 years of 0.214g. Measurements of the soil fundamental frequency were conducted. The lateral load-resisting system of the structure comprises stone masonry walls typical for this region. Seismic analyses with nonlinear static methods were performed including material non-linearities of structural elements. Criteria for evaluation and performance requirements were accounted and relevant engineering demand parameters were compared in relation to design criteria. The seismic performance assessment was made based on the obtained results, including the building collapse mechanisms and the identification of critical structural elements. It is shown that main failure mode of stone masonry walls in the building is shear diagonal cracking.

1 INTRODUCTION

1.1 Description of the building

Pucisca Elementary School on the island of Brac (Croatia) was built at the beginning of the 20th century. It is protected as cultural heritage of the urban landscape in Pucisca (Figure 1). The plan dimensions of the building are 20.90×17.95 m. The building originally comprised ground floor and 1st floor. At the beginning of the 21st century the building was reconstructed and upgraded with another storey and the attic. The height of the building from the ground level to the roof top is about 15 m.

Figures 2 and 3 show the building plan and cross-section from the school reconstruction project (Conex-ST 2000).

The load-bearing system consists of longitudinal and transversal massive stone masonry walls in the ground floor and the first storey, which are relatively regularly distributed in plan. The walls are three-leaf of 48 and 69 cm thickness. They were built with a combination of concrete, rubble stone and limestone (dressed and semi-dressed rock). In the original structure the floor system was wooden and foundation was made as a combination of the crushed stone and weak concrete mortar in a 60 cm thick layer.

As previously mentioned, the school was reconstructed in 2000 (Figure 4). In the scope of the reconstruction project the rehabilitation of the foundation was carried out by constructing new reinforced concrete beam at the elevation of the existing foundation, as well as constructing jet-injected piles in the ground underneath the load-bearing walls and columns up to a depth of 13.5 m.

By field investigations, it was revealed that the soil comprised heterogeneous soil embankment of very low compression modulus and that the ground-water level was very high. Considering the performed soil investigations, the soil type D according to Eurocode 8 (CEN 2005) was selected.

According to the reconstruction project, wooden roof structure, wooden floors, load-bearing walls in the original attic, some parts of the load-bearing walls in the 1st floor and the attic were removed. The new RC floor system of 15 and 18 cm thickness was constructed. Stone masonry walls were additionally linked by RC tie beams of 20 cm depth at the level of the floor system, with built-in longitudinal reinforcement 4Φ14 and stirrups Φ 6/15 cm. The position and area of stone walls in the ground and 1st floor are presented on the Figure 4.

Newly constructed walls of the 2nd floor and the attic are of reinforced concrete and they continue from the existing walls. The inner walls are 20 cm thick, while the outer ones are 40 cm thick. New RC columns in the existing walls were constructed in their intersections and along the opening edges. Minimal reinforcement amounts to 0.6 % of the cross-section area of the column. Concrete grade of all new structural elements is MB30 (C25/30).

Figure 1. Photograph of the school.

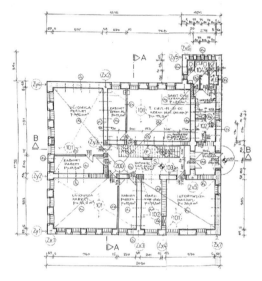

Figure 2. Plan view of the ground floor.

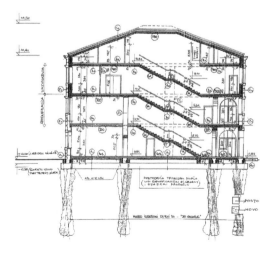

Figure 3. Characteristic building cross-section.

Figure 4. Photograph taken during reconstruction.

area of three-leaf stone masonry walls / floor area

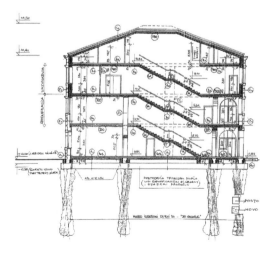

X : 36,83 m^2 (9,40%)   floor area: X : 37,58 m^2 (9,6%)
Y : 38,97 m^2 (9,96%)   A = 391 m^2 Y : 34,34 m^2 (8,8%)

Figure 5. The layout of the stone masonry walls in the ground floor and 1st floor.

This paper describes the part of the study on seismic performance assessment of the school building, made in the scope of the Interreg Readiness project in 2019, in collaboration of the Faculty of Science and Faculty of Civil Engineering in Zagreb.

2 NUMERICAL MODELS AND ANALYSIS METHODS

2.1 *Numerical model*

Numerical model of the building was created based on available documentation (Conex-ST 2000, Galic 2000) visual inspection and on-site testing of limited scope, in the Etabs software (CSI 2011).

The material properties adopted in the analysis of stone masonry were in accordance with Italian standards (MIT 2009): the compression strength f_m=2.0 MPa, shear strength τ_0=0.04 MPa, modulus of elasticity E=1200 MPa, shear modulus G=480 MPa, density γ=20 kN/m^3 and tensile strength f_T=0.06 MPa. The construction details of the stone masonry were assumed according to the rules of the profession from the time of construction and the data obtained by visual inspection. According to (CEN 2005) the confidence factor was set to 1.2. The complete numerical model with the section and material characteristics is presented in the Figure 6. Also, the coordinate axes are shown what will be important in interpretation of the results.

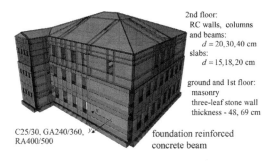

2nd floor:
RC walls, columns
and beams:
$d = 20, 30, 40$ cm
slabs:
$d = 15, 18, 20$ cm

ground and 1st floor:
masonry
three-leaf stone wall
thickness - 48, 69 cm

C25/30, GA240/360,
RA400/500

foundation reinforced
concrete beam

Figure 6. Numerical model of the school building.

For observing the limit state of near collapse, continuous cracking of cross sections during seismic loads was defined by reducing stiffness. Therefore, the beam flexural stiffness was reduced to 30% stiffness of uncracked elements, while the reduced stiffness for columns and walls was 50%. Beams were defined with their effective width. Shear stiffness of all elements was reduced to 40% of the initial stiffness of uncracked elements.

Nonlinear behaviour of structural elements was taken into account by concentrated plastic hinges, accounting for moment and shear failure mechanisms. The load-bearing capacity and ductility of the stone masonry walls and spandrels were determined for all failure modes. Capacity curve of every structural wall was based on the predominant failure mechanism, determined from all observed mechanisms. The backbone curves and acceptance criteria for performance requirements were taken into account using Eurocode 8-3 (CEN 2005) but considering also guidelines provided by (NIST 2017a, 2017b & ASCE 2014). Because of unreliability of the failure mechanisms and the peculiarities of the construction method of stone walls, many scientific papers were consulted (e.g. Corradi & Borri 2018, Kržan et al. 2015, Magenes & Della Fontana 1998, Pasticier et al. 2018, Tomaževič 2011, Vanin et al. 2017 for stone masonary, and Grammatikou et al. 2015, Fardis 2009, Fardis et al. 2015, FIB 2003 for reinforced concrete elements). Load-bearing capacities of concrete reinforced structural elements were calculated based on the concrete strength and as-built reinforcement. Different design regulations (CEN 2005 & ASCE 2014) were considered and critically compared, based on engineering judgement. The primary components of the ground floor and 1st floor are shown in Figure 7.

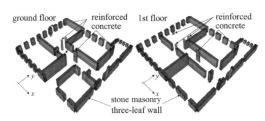

ground floor   reinforced concrete   1st floor   reinforced concrete

stone masonry three-leaf wall

Figure 7. Structural elements in the ground floor and 1st floor.

In numerical models all vertical loads on the structure were defined based on collected data, professional experience and rules of practice. Vertical load comprises self-weight of all elements, additional dead loads from non-structural elements and 20% of live load.

Seismic load was determined according to valid Croatian standards, with reference peak ground acceleration on the bedrock of 0.214g for the return period of 475 years, 0.159g for the return period of 225 years and 0.115g for the return period of 95 years. The importance factor of the building was taken with the value of 1.0 (class II). All load combinations were in accordance with valid standards.

2.2 *Analysis methods*

In the study on seismic performance assessment, analyses were conducted according to CEN 2005, using lateral force method, response spectrum method, nonlinear static method and nonlinear dynamic method using time-histories. However, only the results of the nonlinear static procedure will be presented in this paper, as well as very few results of the response spectrum method (only for comparison). Response spectrum method was conducted with the behaviour factor q=1.5, as the building has no required mechanisms for dissipation of energy during an earthquake.

Important requirement for application of pushover analysis is that the structure mainly responds in its first vibration mode and that the contribution of the first mode to activation of total structural mass is significant. The procedure was conducted under constant gravity loads and monotonically increasing horizontal loads, with a control of target displacement of the elements and structure. It was assumed that horizontal load pattern shape does not change while increasing the loads. The method was performed on a three-dimensional numerical model, separately applying loads in every horizontal direction, using displacement control. Material nonlinearity was accounted for in the analysis, while geometric one was not, because it had been shown that this effect was minor due to the large building stiffness and small displacements at failure.

Procedure for determination of the target displacement using equivalent SDOF model was not performed, because the aim was to show and explain the pushover curve which defined the behaviour of the structure.

Figure 8 presents the acceptance criteria of the relative displacements of the structural walls, defining the corresponding limit states. The different criteria were applied for flexural and shear failure mechanisms for each structural element. Damage state of the structure is defined through three limit states: Near Collapse (BR), Significant Damage (ZO) and Damage Limitation (OO).

It should be mentioned that failure mechanisms of majority of elements according to the initial state of vertical stress, geometrical and material characteristics are brittle and correspond to shear failure.

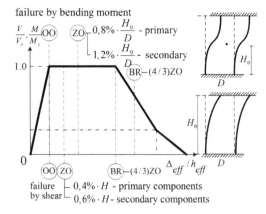

failure by bending moment

$\frac{V}{V_y}, \frac{M}{M_y}$ (OO) (ZO) $-0{,}8\% \cdot \frac{H_0}{D}$ - primary

$-1{,}2\% \cdot \frac{H_0}{D}$ - secondary

(BR)$-(4/3)$ZO

1.0

0

(OO)(ZO)   (BR)$-(4/3)$ZO Δ_{eff}/h_{eff} D

failure $\left[\begin{array}{l} 0{,}4\% \cdot H \text{ - primary components} \\ 0{,}6\% \cdot H \text{ - secondary components} \end{array}\right.$
by shear

Figure 8. Definition of limit states on the backbone curve.

However, large structural redundancy causes significant robustness which is favourable from the viewpoint of internal force distribution.

Structural elements are classified into primary and secondary ones. The primary elements have a load-bearing function for vertical and horizontal loads, while the secondary elements are not essential for load-bearing capacity, although they participate in transfer of loads. In this building, the secondary elements are the stone masonry spandrels and partition walls. Their failure should not cause the partial collapse of the slab, because the horizontal tie-beams of 30/30 cm on each wall were constructed together with a reinforced concrete slab in the scope of reconstruction.

2.3 Experimental investigations

A seismological measurement was conducted to gather instrumental recordings of ambient noise throughout the town of Pucisca, with the highest measurement density in the immediate vicinity of the school. Horizontal to Vertical Spectral Ratio (HVSR) was computed for all free-field measurements (Nakamura 1989). The obtained soil eigen-frequency was 9.19 Hz.

Moreover, experimental measurements of the dynamic properties of the building were performed. Continuous instrumental data had been collected in two locations inside the building (on the ground floor and top floor) using seismographs. The building eigen-frequencies of 7.3 Hz in longitudinal direction and 8.1 Hz in the transversal direction were obtained.

During the time period when the instruments were operational, the building's response to earthquakes that occurred showed the release of energy predominantly between 7 and 10 Hz, with a strong contribution from the soil fundamental frequency at 9 Hz.

Based on the obtained results, the possibility of the occurrence of resonance between the structure and the soil, which can cause significant increases in the amplitude of vibration during an earthquake, was estimated. Due to the close values of the eigen-frequencies of the

soil and the building, a high resonance hazard rating was obtained for both horizontal components. The possibility of resonance occurrences and the importance of the building was the reason for a more detailed and accurate building seismic assessment.

3 ANALYSIS RESULTS

3.1 Dynamic properties

The first model was created in order to additionally check the geometric and material properties of the structure and to confirm the results of ambient vibrations measurements. This model has a limited condition of cracking, because it is only loaded with vertical loads. The building eigen-frequencies and corresponding shapes are displayed in Figure 9.

The second model for calculation of the vibration modes was defined for the near-collapse state of the structure (before systematic failures of structural elements appeared) in order to determine dynamic properties in the moments of the occurrence of the failures. Stiffness of the structure is significantly reduced, what can be observed in lengthened vibration periods (Figure 10).

3.2 Linearly elastic analysis for vertical loads

Structural analysis for vertical loads is essential for further analyses (static or dynamic) and it is necessary to determine internal forces and generally, structural condition before applying earthquake loading. The total vertical load of the building is 26316 kN. The vertical stresses in the walls of the ground floor due to vertical loads are presented in Figure 11.

1st period (uncracked)   2nd period (uncracked)   3rd period (uncracked)
$T_1 = 0.16$ s $T_2 = 0.14$ s $T_3 = 0.11$ s

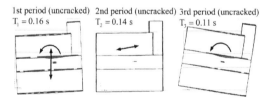

Figure 9. Eigen-periods of the structure (cracked state).

1st period (cracked)   2nd period (cracked)   3rd period (cracked)
$T_1 = 0{,}40$ s $T_2 = 0{,}32$ s $T_3 = 0{,}28$ s

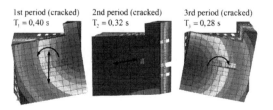

Figure 10. Eigen-periods of the structure (cracked state).

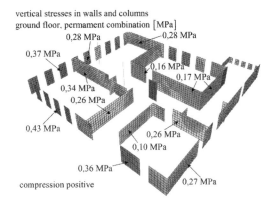

vertical stresses in walls and columns
ground floor, permanent combination [MPa]

0,28 MPa 0,28 MPa
0,37 MPa
0,16 MPa
0,17 MPa
0,34 MPa
0,26 MPa
0,43 MPa
0,26 MPa
0,10 MPa
0,36 MPa
0,27 MPa

compression positive

Figure 11. Wall stresses in the ground floor for vertical loads.

The stresses in the structural walls are in the range of 10% to 20% of the compressive strength of the stone masonry, which in addition to the geometric characteristics of the walls, affects the failure mechanism of the wall. Also, level of axial force significantly affects rotation and shear capacity of columns and walls.

3.3 Modal response spectrum analysis

As an introduction to the nonlinear static procedure, the most important results obtained by response spectrum method are presented, in order to observe the main features of the structure. Figure 12 shows the distribution and level of the shear force in the structure for directions X and Y for 475 years return period of earthquake. Moreover, the absolute and relative storey displacements are presented for 95 years return period of earthquake. A difference in stiffness between the first two floors and an additional reinforced concrete floor may be observed.

The reference force is determined for a return period of 475 years. The importance factor of the building is 1.0 and the behaviour factor is 1.5. The obtained base shear amounts to 10400 kN what is

39% of the building weight in seismic combination (39% BS in the further text) for direction X and 7800 kN (30% BS) for direction Y.

3.4 Nonlinear static procedure

In this section, only the most important and relevant results for determining the load-bearing capacity of the structure will be highlighted. Besides obtaining the pushover curve, the objective was to monitor the gradual forming of the mechanisms and failure of the structure. Since the structure is relatively stiff and non-ductile, the classical procedure with a target displacement has not been fully performed. The pushover curves of the building were determined specifically for each load direction. The adopted generalised coordinate was the displacement of the 2nd storey plan centre.

All specific points on the curve, corresponding to different limit states of the structure, are presented. In addition, the sketch of the ground floor with marked critical elements was given for each pushover curve point, in order to gain insight into the potential structural damage at that force level. Figure 13 presents the pushover curve for direction X.

It may be observed that with the increase in lateral load certain elements reach the condition of limited damage, which is manifested on the curve as a reduction of global stiffness. This happens at the base shear level of 5800 kN, i.e. the base shear coefficient is 22% (BS). The initial opening of cracks and reducing the stiffness of the individual elements occurs. The elements on the western building façade in which the cracking occurs are shown on the scheme below the characteristic curve point in Figure 13. This does not mean that they lose the load-bearing capacity or fail, but they reached the yielding point and possess the ability to further deform. As it will be confirmed later, the elements on the edge of the building have proved the most vulnerable due to the influence of the torsion in the structural response, but also due to the relatively short wall elements on the critical façade.

The beginning of the failure of the structural elements occurs for the base shear level of 6400 kN (25% BS). This refers to local failures of the façade short walls on the western part of the building (the part sticking out of the main rectangular shape in plan),

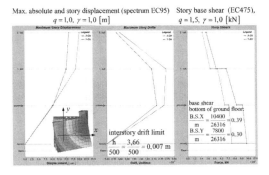

Figure 12. Displacements and shear force in x and y directions.

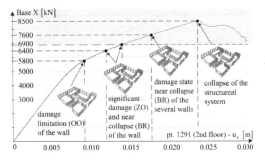

Figure 13. Pushover curve for X direction.

occurring at the base shear level of 6900 kN (27% BS). This failure mode is initiated by very unfavourable torsional effect (Figure 14). Following the wall failure, the local instability of the slab above may occur. This corresponds to near-collapse limit state and it may be classified with the highest degree of damage. However, even in the event of a local collapse the rest of the structure is unlikely to collapse.

The next key point on the pushover curve marks the complete failure of the load-bearing system, causing the collapse of the larger part of the structure. The critical elements are marked. Global failure occurs at the base shear level of 7600 kN (29% BS), initiated by failures of the structural walls in the western façade, when the lateral capacity of the structure is exploited. In order to describe the damage of the critical structural elements in more detail, Figure 14 presents the condition of the structural elements for the near-collapse state.

Figure 15 shows the pushover curve for direction Y, in the same manner as for direction X. Some structural elements reach the state of damage limitation at the base shear level of 5500 kN (21% BS) where-by the initial opening of cracks and the reduction of the stiffness of individual elements occurs. In the scheme, the elements in which cracks occur are displayed below the characteristic point. These are the inner walls and the façade wall. These walls reach the yielding point but have the capacity of further deformation.

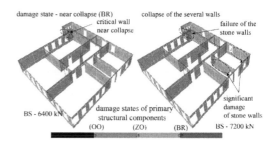

Figure 16. Conditions of the structural system for the near-collapse state (pushover Y).

The structural failure begins at the base shear level of 6400 kN (25% BS). It is a local failure of the façade short wall on the southern part of the building, which was initiated by the torsional response of the building (Figure 16).

The next key point marks the failure of the load-bearing system, causing the collapse of the larger part of the structure. The critical elements are marked. Global failure occurs at the base shear level of 7200 kN (28% BS), initiated by failures of the structural walls in east-west direction. Figure 16 presents the condition of the elements for the near-collapse state.

4 CONCLUSIONS

Based on the performed analyses, conclusions on the state of the structure and assessment of its behaviour for earthquakes may be given.

It should be pointed out that the accuracy of the calculation primarily depends on the accuracy of the input data. It should be taken into account that there are possible discrepancies between the assumed and the real values. All this contributes to the unreliability of the model which cannot be completely eliminated but should be accounted for when interpreting the results and drawing conclusions. The main findings are presented in the following text.

Due to irregularities in plan there is an eccentricity of the centre of stiffness and centre of mass that causes torsional effects. Critical elements are placed on the extension of the main rectangular plan shape of the building in the west. These elements are also the most distant from the centre of the building.

The irregularity of the building in height due to the new reinforced concrete floor and attic causes an uneven distribution of stiffness in height, and thus uneven distribution of the inter-storey drifts, which is unfavourable since most of the displacements of the structure occur on the ground floor and the first floor which are already critical spots.

Floor systems anchored to the walls act as diaphragms and have a favourable effect on the behaviour

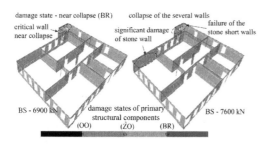

Figure 14. Conditions of the structural system for the near-collapse state (pushover X).

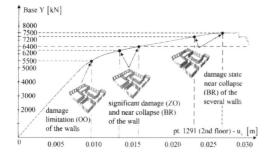

Figure 15. Pushover curve for Y direction.

of the structure. The diaphragms evenly distribute the loads on walls at the floor-level and reduce the unsupported height of the stone masonry wall, which is extremely important when determining the wall failure mechanism. The new RC tie-beams act favourably and connect the walls. Although their bending resistance is relatively small and plastic hinges are formed at the edge, the cross section is still able to transfer the axial force and the part of the shear force without failure.

Due to the high degree of static indeterminacy in the structure, the redistribution of forces to the adjacent elements occurs after cracking of the critical load-bearing walls. In this way, the sudden partial collapse of the structure is avoided. Therefore, it may be concluded that the system possesses limited but not insignificant capacity reserves.

The main failure mode of stone walls is shear diagonal cracking or sliding shear failure. Moreover, the toe compression failure mode initiated by rocking had also been observed.

The calculation using the elastic and design response spectrum was conducted to make a comparison with nonlinear static analysis. The obtained base shear for design spectrum is 10400 kN (39% BS) for direction X and 7800 kN (30% BS) for direction Y. It should be emphasized that such base shear level in elastic analysis causes considerable local exceeding in relation to the estimated properties of the load-bearing capacity of materials and elements.

The more comprehensive analyses were performed using non-linear static procedure. The earthquake load in the direction Y (east-west) was shown to be critical. By monotonically increasing horizontal loads on the structure, some structural elements reach the state of limited damage at the base shear level of 5500 kN (21% BS). Then the initial opening of cracks and reducing the stiffness of individual elements occurs. Local failures of façade short walls on the western part of the building (the building extension in the floor plan) occur at the base shear level of 6400 kN (25% BS). The building reaches a near-collapse state and can be classified with the highest degree of damage. This type of failure is initiated by torsion of the building. As a consequence, a local collapse of the slab above may occur, although it is assumed that the slab with horizontal tie-beams has a sufficient load-bearing capacity and that no collapse would occur even after losing the support on its edge. The failure of the load-bearing system with the collapse of a large part of the structure occurs at the base shear level of 7200 kN (28% BS), and is initiated by failures of the load-bearing walls on the western façade (direction X) and the walls in the east-west direction (direction Y).

On the basis of conducted seismological measurements, it was found out that the soil-eigen frequency (approximately 9 Hz) is close to the building eigenfrequencies (7.3 Hz and 8.1 Hz), what may cause occurrence of resonance. This phenomenon may induce the significant increase in vibration amplitudes during an earthquake. However, the cracking of structural elements will occur, and consequently the eigenfrequencies will decrease to values of 2.5 Hz and 3 Hz.

REFERENCES

ASCE 2014. Seismic Evaluation and Retorofit of Existing Buildings, ASCE/SEI 41-13, American Society of Civil Engineers, Reston, Virginia, USA.

CEN 2005. European Standard EN 1998-3:2005: Eurocode 8: Design of structures for earthquake resistance. Assessment and retrofitting of buildings. CEN, Brussels.

Conex-ST 2000. Glavni projekt rekonstrukcije, nadogradnje i dogradnje Osnovne škole Pučišća – Brač, statički proračun, P-10/00, Split.

Corradi, M. & Borri, A. 2018. A database of the structural behavior of masonry in shear, Bulletin of Earthquake Engineering, 16, pp 3905–3930.

CSI 2011. CSI Analysis Reference Manual For SAP2000, ETABS, SAFE and CSiBridge, Computers and Structures, Berkeley, California, USA.

Fardis, M.N. 2009. Seismic Design, Assessment and Retrofitting of Concrete Buildings Based on EN-Eurocode 8, Geotechnical, geological and earthquake engineering, Vol. 8, Springer.

Fardis, M.N., Carvalho, E.C., Fajfar, P. & Pecker, A. 2015. Seismic Design of Concrete Buildings to Eurocode 8, CRC Press, Taylor & Francis.

fib 2003. Seismic assessment and retrofit of reinforced concrete buildings, State-of-the-art report prepared by Task Group 7.1, fib bulletin 24, fib.

Galić, P. 2000. Rekonstrukcija, nadogradnja i dogradnja Osnovne škole u Pučišćima, glavni arhitektonski projekt, T.D.-2/2000, Split.

Grammatikou, S., Biskinis, D. & Fardis, M.N. 2015. Strength, deformation capacity and failure modes of RC walls under cyclic loading, Bulletin of Earthquake Engineering, 13 (11), 3277–3300.

Kržan, M., Gostič, S., Cattari, S. & Bosiljkov V. 2015. Acquiring reference parameters of masonry for the structural performance analysis of historical buildings, Bulletin of Earthquake Engineering, 13, 203–236.

Magenes, G. & Della Fontana, A. 1998. Simplified Non-linear Seismic Analysis of Masonry Buildings, Proc. of the 5th International Masonry Conference, London.

MIT 2009. Instruzioni per l'applicazione delle nuove norme tecniche di cui al Decreto Ministeriale 14 Gennaio 2008, G.U.S.O.n.27 of 26.6.2009, No. 47, Ministry of Infrastructures and transportation, Circ. C.S.L1.Pp. No. 617, 2009. (in Italian).

Nakamura, Y. 1989. A method for dynamic characteristics estimation of subsurface using microtremor on the ground surface. Quart. Rep. Railway Tech. Res. Inst. (RTRI) 30, 25–33.

NIST 2017a. Guidelines for Nonlinear Structural Analysis and Design of Buildings, Part I – General, NIST GCR 17-917-46v1, prepared by the Applied Technology

Council for the National Institute of Standards and Technology, Gaithersburg, Maryland.

NIST 2017b. Recommended Modeling Parameters and Acceptance Criteria for Nonlinear Analysis in Support of Seismic Evaluation, Retrofit, and Design, NIST GCR 17-917-45, prepared by the Applied Technology Council for the National Institute of Standards and Technology, Gaithersburg, Maryland.

Pasticier, L., Amadio, C. & Fragiacomo, M. 2008. Nonlinear seismic analysis and vulnerability evaluation of a masonry building by means of the SAP2000 V.10 code, Earthquake Engineering and Structural Dynamics, 37, 467–485.

Tomaževič, M. 2011. Seismic resistance of masonry buildings in historic urban and rural nuclei: lessons learned in Slovenia, International Journal of Architectural Heritage, 5(4-5), 436–465.

Vanin, F., Zaganelli, D., Penna, A. & Beyer, K. 2017. Estimates for the stiffness, strength and drift capacity of stone masonry walls based on 123 quasi-static cyclic tests reported in the literature, Bulletin of Earthquake Engineering, 15, 5435–5479.

Brick and Block Masonry - From Historical to Sustainable Masonry –
Kubica, Kwiecień & Bednarz (eds)
© 2020 Taylor & Francis Group, London, ISBN 978-0-367-56586-2

A simple deformation-based damage index for shake table testing of historic masonry prototypes

I. Roselli, V. Fioriti & A. Colucci

Italian National Agency for New Technologies, Energy and the Economic Sustainable Development (ENEA), Rome, Italy

ABSTRACT: The present works illustrates a simple deformation-based damage index that exploits the potentialities of advanced 3D optical systems for markers displacements measurements in shaking table testing of historic masonry prototypes. The application of passive 3D motion capture systems to shaking table testing proved very efficient in recording a large number of measurement points, which provide accurate and complete description of the field of displacements of the studied structure. The proposed index was built as a function of the global residual deformation obtained from the measured distances between each couple of adjacent markers located on the tested structure. The results were validated by correlation with a consolidated stiffness-based index calculated through modal analysis of the structure. The proposed index considers only the residual deformations, while neglecting the effects of energy absorption and of different collapse mechanisms on the assessment of damage. Nonetheless, its simplicity of computation directly from the markers measurements represents a big advantage with respect to other more used damage indices that were mainly developed for numerical analysis and need to estimate structural capacity parameters. Besides, the stiffness-based indices might sometimes provide controversial results in shake table tests, as prototypes dynamic behavior is dependent on the boundary conditions, i.e. the observed stiffness reduction might be caused by fixing loosening and/or changes in the prototype constraints to the table. In order to avoid misleading conclusions, the crack pattern and the deformations arisen in the tested specimen should always be considered of utmost importance for assessing the actual state of structural damage. After application with several historic masonry prototypes, encouraging results were obtained in terms of good correlation ($R^2 > 0.9$) with the consolidated index.

1 INTRODUCTION

In the past, several damage indices (DIs) were developed and described in the literature (Sinha & Shiradhonkar 2012). They were built on the base of different kinds of properties (structural or dynamic), at different scale (local or global), and some of them take into account also fatigue effects through cumulative formulations (Williams & Sexsmith 1995). Such DIs have been mainly developed for numerical simulations and make wide use of theoretical parameters that need to be estimated. This makes them of difficult feasibility for experimental applications.

Moreover, while most of consolidated DIs were developed through extensive theoretical and experimental studies on reinforced concrete buildings, which are the most common building typology worldwide, much less effort was dedicated to historic masonry. In fact, reinforced concrete structures are designed with somewhat simple schemes and have a quite homogeneous and controllable behavior, whereas historic masonry constructions are more heterogeneous and generally present large cracking damage before the hysteretic one, which makes the

definition of proper and effective DIs more complex (Parisi & Augenti 2013).

On the other hand, the damage conditions of structures after earthquakes are commonly assessed with qualitative or semi-quantitative parameters through quite consolidated procedures that are widely accepted for expert inspections. A major example in Europe is the use of the European Macroseismic Scale (EMS-98), which provides damage grades (Grünthal 1998) that are usually considered as reference for calibration and validation of results obtained with DIs developed for numerical simulations.

In EMS-98, as well as in similar scales, the damage observed on the field after earthquakes in masonry buildings is classified in grades defined mainly in terms of cracks distribution and size. Once it is assumed that crack pattern and its geometrical characterization is of primary importance for damage assessment in masonry buildings, a DI can be developed on the basis of the measurements of the residual deformations arising in the masonry after a seismic excitation. Such DI is conceptually simple and does not require any estimation of theoretical parameters, as many other more complex DIs do. Moreover, a simple deformation-based DI would be particularly feasible for dynamic laboratory testing, in which advanced 3D motion measurements systems based on machine-vision technologies are available nowadays (Roselli et al. 2019b). In particular, the use of passive 3D motion capture systems proved to be very efficient in the measurement of displacements of more than a hundred of markers positioned on large-scale prototypes tested on shaking table experiments (De Canio et al. 2016).

In fact, the most widely used DIs in shaking table testing are the based on global parameters that can be easily estimated by experimental measurements, such as the tested structure modal parameters. In particular, the resonance frequency (or period) is commonly monitored as the structures encounter softening (period elongation) when damage increases (DiPasquale & Cakmak 1988). Frequency-based DIs are simple, do not require theoretical parameters, and can be calculated by processing the experimental motion measurements recorded during characterization tests on shaking table (e.g. typically low-intensity white-noise input can be used) with consolidated modal analysis algorithms (De Canio et al. 2011, Roselli et al. 2015). Frequency-based DIs are consolidated and widely accepted indicators of the global state of health of the overall structure on the basis of the observation that damage processes induce a measureable decrease of the system's stiffness. Nonetheless, stiffness-based DIs have a serious inconvenient in the application to laboratory testing: providing the stiffness of the system they are dependent on the structure boundary conditions. The above inconvenient is regularly neglected in the applications for structural health monitoring to on-the-site structures, because boundary conditions of real buildings are assumed as invariant. On the contrary, it is not negligible in laboratory testing, where the fixing between the structure base and the shaking table is not always perfectly efficient (e.g. a bolt loosening may occur during shakes), then the decrease of the system stiffness may indicate a change in the constraints rather than a damage to the structure.

Consequently, in a previous study aimed at exploring the possibility of developing an alternative DI for laboratory testing, a simple deformation-based DI was defined and validated through the measurements of markers distances (MD) in experimental tests of an historic masonry prototype (Roselli et al. 2019a). Such pilot study gave encouraging results, but it was based on very few experimental data, which came from just one experimental campaign conducted on only one prototype. So they could not be considered much representative of the general behavior of historic masonry buildings. For the above reason, in the present work further shaking table experiments on different prototypes of historic masonry typologies typical of the Italian cultural heritage were analyzed and discussed. In particular, 5 different experimentations were considered, which were conducted on prototypes

built with tuff or limestone units and clay bricks, and representing different architectural typologies (2-storey and 1-storey building, and cross-vault). The above experiments were carried out on the 4.00 × 4.00 m shaking table at the ENEA Casaccia laboratory, where a passive 3D motion capture systems, named 3DVision, based on Vicon technology, is installed (De Canio et al. 2013).

2 METHODOLOGY

2.1 Development of DI_m damage index

In all considered shaking table experiments the tested prototypes are instrumented with a number of 3DVision markers comprised between 50 and 80. They were located in order to have a well distributed and comprehensive cover of the external surfaces of the main structural parts of each prototype. In fact, being the 3DVision a machine-vision system, markers must be visible from outside the shaking table area, which implies that no measurement points internal to the structure are detectable.

For each couple of adjacent markers in horizontal direction the distance (MD) was computed through the markers 3D coordinates. Vertical couples were neglected, as gravity tends to close cracks so that vertical deformations generally give little contribution to the global residual deformations. The monitoring of each MD during the seismic test permits automatic detection and location of cracks. Then, a global residual deformation (GRD) was calculated by the following expression:

$$GRD = \sum_{i=1}^{n} \left(\frac{\Delta MD_i}{MD_i} \right) \qquad (1)$$

where n = number of considered couples of adjacent markers. If markers positions cover well the overall volume of the structure, then GRD is representative of the global state of the structure.

As GRD was calculated at the end of each seismic test, it takes account of cumulative deformations. On the other hand, residual deformations at the end of a shake correspond to the crack pattern observed in on-the-field post-seismic inspections for the assessment of the damage state through international regulations and guidelines (e.g. EMS-98, FEMA etc.). A deformation-based index calculated through markers measurements, DI_m, can be developed as an empirical function f of GRD as follows:

$$DI_m = f(GRD, a, b) \qquad (2)$$

where a and b are empirical coefficients that are defined by correlation between GRD and a consolidated DI that is evaluated for the same test. According to previous research (Roselli et al. 2019a) a well fitted model for function f was obtained as follows:

$$DI_m = a.\ln(GRD) + b \qquad (3)$$

2.2 Correlation with DIf damage index

The empirical coefficients a and b for DI_m were obtained by correlation with a consolidated frequency-based global index DI_f defined as follows:

$$DI_f = 1 - \left(\frac{\omega_d}{\omega_o}\right)^2 \qquad (4)$$

where ω_o = initial fundamental frequency of the undamaged structure; and ω_d = fundamental frequency of the damaged structure.

The above frequencies were estimated by Experimental Modal Analysis (EMA) of markers displacements through the computation of the Frequency Response Function (FRF). In particular, markers at the prototype base were taken as input signals and the remaining markers located on the structure were taken as output signal. Consequently, a MI-MO (Multi-Input Multi-Output) FRF procedure was carried out. Even if machine-vision motion measurements are substantially less accurate than conventional accelerometers, the high number of available measurement points (typically from 50 to 150 markers are used on each prototype) permits a quite clean and accurate estimate of modal frequencies.

Most importantly, the efficiency of prototype fixing to the shaking table was verified after each seismic test, in order to exclude that the observed frequency decrease (system softening) was caused by base fixing loosening (changing boundary conditions). This verification was achieved by checking that the relative displacements between markers positioned at the prototype base and the markers located on the table were within measurement error (0.1 mm in terms of RMS error).

3 EXPERIMENTAL

3.1 Shaking table tests

Shaking table experiments were conducted on the basis of seismic inputs taken from natural recordings at real seismic stations. In particular, a selection of recordings of the main earthquakes occurred in the last 50 years was considered. The recorded accelerograms were scaled in intensity and uploaded to the shaking table control system starting from 0.05 g of nominal PGA. The sequence of seismic tests was obtained by increasing gradually the seismic input intensity (e.g. with steps of 0.05 g of PGA) at each successive step of the sequence until the final failure of the masonry prototype. Each step of the sequence was alternated with a characterization test (typically a low-intensity white-noise test) in order to facilitate modal identification of the structure after each shake. The values of GRD and DI_f index were calculated at each seismic step of the sequence in order to for perform the correlation to assess coefficients a and b to obtain DI_m.

3.2 Tested prototypes

In the present sub-section the considered tested masonry prototypes are described. The five prototypes are shown in Figure 1. They are identified as follows:

- CVR: cross-vault in roman bricks;
- 2SHC: 2-storey in hollow-clay blocks;
- 2STR: 2-storey in tuff rubble masonry;
- 2SAS: 2-storey in limestone ashlar masonry;
- 1STA: 1-storey in coursed tuff ashlar.

The CVR prototype was a real-scaled reproduction of one of the 10 cross-vaults at the Mosque of Dey. This was a private mosque of the king of Algiers, built inside the Palace of the Dey, located in the Citadel of Algiers, Tunisia. It was built during several construction phases, probably from the 16th to the 19th centuries. The real vaults are supported by columns that were neglected in the prototype, as their influence on the dynamic response of the vault wanted to be avoided.

The size of the CVR is 3.50 × 3.00 m in plan and the overall height is 2.90 m. The materials were reproduced after on-the-site non-destructive

Figure 1. Tested masonry structures (from left to right): cross-vault in roman bricks (CVR), 2-storey in hollow-clay blocks (2SHC), 2-storey in tuff rubble masonry (2STR), 2-storey in limestone ashlar (2SAS) and 1-storey in tuff coursed ashlar (1STA). Markers appear as bright points on the structures, as they reflect photo camera flash light.

tests and laboratory characterization of samples of bricks and mortar taken from within the citadel, where buildings are made up of typical masonry similar to the one observed in the mosque (Rossi et al. 2020). The number of markers located on CVR was 67.

The 2SHC prototype is a 2-storey 2/3-scaled building made up of regular masonry texture with hollow-clay blocks as units (Pepi et al. 2019). 2SHC dimensions are 3.00 × 3.50 m in plan and 2.20 m for each storey in height. Walls were built using M10 mortar. Bearing hollow-clay units have 45% porous density, 25 × 18 cm size and 16 cm thickness. They were obtained by cutting typical Italian load-bearing thermal blocks. The building presents several openings with asymmetric configuration and a balcony at each storey. The overall number of markers were 70.

The 2STR is geometrically similar to 2SHC, but is made of tuff rubble masonry (Betti et al. 2015). Openings and balconies are geometrically different from 2SHC. The walls were double-leaf with no transversal connections and had a constant thickness of 0.25 m. The horizontal structures were made with flexible timber floors, built using wooden joists (with a section of 0.10 m × 0.18 m) and wooden boards with a thickness of about 20 mm nailed to the wooden beams. The masonry was constructed with calcareous tuff stones and lime-cement mortar with weight ratios of the components of 10.5 (sand), 11 (lime) to 1 (cement). The masonry units had irregular shape and dimensions varying from medium to large. Here the number of used markers was 141 (Roselli et al. 2017).

Another tested 2-storey 2/3-scaled building, named 2SAS, was in limestone ashlar (Valluzzi et al. 2009, Silva et al. 2012). The building had regular floors of 2.40 × 2.80 m and a total height of 3.60 m. Double planking wooden floors were employed to simulate a non-rigid diaphragm. Three-leaf limestone masonry typology with natural hydraulic lime mortar was adopted for walls. The structure was characterized by absence of transversal connection between external layers and incoherence of the internal leaf, constituted by stone fragments and characterized by 12% of voids. 2SAS was monitored with 110 markers.

Finally, a 1-storey prototype, named 1STA, was also tested (De Santis et al. 2019). It was a full-scale U-shaped tuff ashlar structure, provided with an asymmetric plan distribution of openings and with a 12-deg inclined wooden roof. The prototype height was 3.41 m and its plan was 2.73 × 3.39 m.

The walls thickness was 0.25 m and they were obtained with 29 courses of tuff blocks whose size was 37 × 25 × 11 cm. The stone used for the units is the yellow tuff (*tufo giallo*) of Via Tiberina. This kind of tuff was widely used as a building material in ancient Rome from about the fourth century BCE until the Augustan period. The mortar was made up of hydraulic lime and sand. The number of markers located on 1STA was 69.

4 RESULTS

Initially, GRD values obtained from all five shaking table experiments were considered together and correlated with corresponding DI_f values in order to understand whether an overall behavior was detectable. This general behavior is illustrated in Figure 2, where a quite good correlation is shown ($R^2 = 0.8$). Subsequently, the empirical coefficients a and b were determined and used to compute the DI_m.

Also, separated correlations between DI_m and DI_f for each prototype were assessed. In this way, much better correlations could be achieved as shown in Figure 3, where R^2 is higher than 0.9 for all five studied cases.

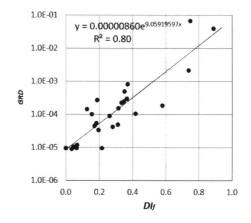

Figure 2. Empirical correlation between GRD and DI_f, according to Equation 3, considering all masonry structures tested on shaking table.

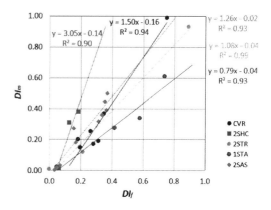

Figure 3. Deviation from the identity equation between DI_m and DI_f considering each tested masonry structure separately.

5 DISCUSSION

The correlation depicted in Figure 2 utilizes all the available experimental results from all prototypes. This helps the consistency of statistics, which depend on the number of correlation points (sample size) and their distribution. However, it is interesting to analyze also the different behaviors of each masonry typology separately.

From this point of view, the slopes and the y-intercepts of the correlation equations illustrated in Figure 3 can be seen as corrections from the identity equation between DI_m and DI_f for the different tested masonry typologies. On the one hand, the results of such partial correlations are more homogeneous because they refer to points regarding the same building material. On the other hand, they are obtained with fewer points and not always well distributed, therefore their statistics are less consistent. Nevertheless, experimental points are mainly concentrated at low DI values, which makes the correlations more relevant where it is more needed. In fact, at DI values higher than 0.4-0.5 the structure is to be considered severely damaged, which is clearly more visible and easily assessable. On the contrary, at lower DI values the damage is characterized by fine cracks that are still hardly visible and collapse mechanisms are still far from being clearly developed. The above considerations make evident that the assessment of DI by quantitative measurements in laboratory tests is particularly relevant to detect and analyze the activation of damage processes at the early stages.

The physical interpretation of such corrections is not simple. As a general indication, a slope higher than 1 means that the increase in residual deformation of the structure overestimates damaging processes with respect to stiffness decay. More in detail, Figure 3 shows that masonry structures with clay units (2SHC and CVR) gave the steepest corrections (3.05 and 1.5 slopes, respectively) and provided higher negative values of y-intercepts (-0.14 and -0.16 intercepts, respectively). This indicates that masonry with clay units tend to keep its initial stiffness, though cracks openings increase. On the other hand, prototypes built with tuff units (1STA and 2STR) provided less steep correlation equations (0.79 and 1.08 slopes, respectively), which means that they are more prone to softening phenomena. The only prototype built with limestone units provided intermediate values of slope correction (1.26) and the smallest intercept (-0.2).

As for the geometrical properties of prototypes, 2-storey buildings provided a steeper slope than 1-storey types built with similar materials.

6 CONCLUSIONS

In the present paper a simple deformation-based damage index, named DI_m, for application to shaking table testing on historic masonry structures was illustrated. It was developed through empirical correlation with a widely used and consolidated global DI based on frequency decay of five different tested structures. The methodology exploits the potentialities of advanced 3D optical systems to measure a large number of markers displacements with high accuracy, since measurement error lower than 0.1 mm can be achieved.

Even if the proposed index considers only the residual deformations, while neglecting the quantification of energy absorption, it provided very encouraging results in its potential use to estimate global damage in masonry prototypes. Moreover, its simplicity of computation based uniquely on 3D motion data by machine-vision systems represented a big advantage in laboratory testing with respect to traditional and more sophisticated DIs, which are mainly intended for numerical analysis and require to estimate theoretical parameters or to carry out more difficult experimental measurements.

It is worth noting that a simple deformation-based index, like DI_m, can not discriminate from cracks openings that arise in different parts of the structure and may signal the arising of different collapse mechanisms with different damage severity. Consequently, a more advanced version of DI_m should be implemented in order to take into account the effect of different collapse mechanisms.

Moreover, the effect of different masonry typologies in terms of materials and geometries were analyzed and discussed. In particular, the presented approach of data analysis made possible to hypothesize corrections to DI_m in order to improve its performance ($R^2 > 0.9$) in the application to specific typologies. However, further data from future experimental cases on different masonry typologies are welcome. They would provide a more comprehensive understanding of the potentiality of the used methodology and help to refine the correlations found here with even more consistent statistics.

REFERENCES

Betti, M. et al. 2015. Time-History Seismic Analysis of Masonry Buildings: A Comparison between Two Non-Linear Modelling Approaches. *Buildings* 5: 597–621.

De Canio, G. et al. 2011. Displacement based approach for a robust operational modal analysis. In *Conference Proceedings of the Society for Experimental Mechanics Series 6; 29th IMAC, a Conference on Structural Dynamics, Jacksonville, FL, USA 31 January - 3 February 2011.*

De Canio, G. et al. 2013. 3D motion capture application to seismic tests at ENEA Casaccia research center: 3Dvision system and DySCo virtual lab. *WIT Transactions on the Built Environment* 134: 803–814.

De Canio, G. et al. 2016. Passive 3D motion optical data in shaking table tests of a SRG-reinforced masonry wall. *Earthquakes and Structures* 40(1): 53–71.

De Santis, S. et al. 2019. Shake table tests on a masonry structure retrofitted with composite reinforced mortar. *Key Engineering Materials* 817: 342–349.

DiPasquale, E. & Cakmak, A.S. 1988. Identification of the serviceability limit state and detection of seismic structural damage. *Technical Report No. NCEER-88-0022.* Buffalo, NY: National Center for Earthquake Engineering Research.

Grünthal, G. 1998. European Macroseismic Scale 1998 (EMS-98). *Cahiers du Centre Européen de Géodynamique et de Séismologie* 15. Luxembourg: Helfent-Betrange.

Parisi, F. & Augenti, N. 2013. Earthquake damages to cultural heritage constructions and simplified assessment of artworks. *Engineering Failure Analysis* 34: 735–760.

Pepi, C. et al. 2019. Performance comparison between unreinforced and confined masonry buildings subjected to shaking table tests. In M. Papadrakakis & M. Fragiadakis (eds.), *7th ECCOMAS thematic conference on Computational Methods in Structural Dynamics and Earthquake Engineering; COMPDYN 2019, Crete, Greece, 24-26 June.*

Roselli, I. et al. 2015. Analysis of 3D motion data from shaking table tests on a scaled model of Hagia Irene, Istanbul. *Key Engineering Materials* 624: 66–73.

Roselli, I. et al. 2017. Processing of 3D optical motion data of shaking table tests: filtering optimization and modal analysis. In *6th ECCOMAS thematic conference on Computational Methods in Structural Dynamics and Earthquake Engineering; COMPDYN 2017, Rhodes, Greece, 15-17 June.*

Roselli, I. et al. 2019. Relative displacements of 3D optical markers for deformations and crack monitoring of a masonry structure under shaking table tests. *International Journal of Computational Methods and Experimental Measurements* 7(4): 350–362.

Roselli, I. et al. 2019. Machine Vision-Based Application to Structural Analysis in Seismic Testing by Shaking Table. In M. Rivas-Lopez, O. Sergiyenko, W. Flores-Fuentes & J. C. Rodríguez-Quiñonez (eds), *Optoelectronics in Machine Vision-Based Theories and Applications:* 269-300. IGI Global ed.

Rossi, M. et al. 2020. Seismic analysis of a masonry cross vault through shaking table tests: the case study of the Dey Mosque in Algiers. *Earthquakes and Structures* (in press).

Silva, B. et al. 2012. Calibration of a material behavior model for the simulation of multi-leaf stone masonry structures: Experimental case study application. In *Proc. 15th World Conference on Earthquake Engineering (15WCEE), Lisbon,* 24-28 September 2012. Lisbon: SPES.

Sinha, R. & Shiradhonkar, R. 2012. Seismic damage index for classification of structural damage–closing the loop. In *Proc. 15th World Conference on Earthquake Engineering (15WCEE), Lisbon,* 24-28 September 2012. Lisbon: SPES.

Valluzzi, M.R. et al. 2009. Effectiveness of injections evaluated by sonic tests on reduced scale multi-leaf masonry building subjected to seismic actions. In *Non-Destructive Testing in Civil Engineering, Nantes; Proc. of NDTCE '09, France,* 30th June –3rd July.

Williams, M.S. & Sexsmith, R.G. 1995. Seismic Damage Indic es for Concrete Structures: A State-of-the-Art Review. *Earthquake Spectra* 11(2): 319–349.

Brick and Block Masonry - From Historical to Sustainable Masonry –
Kubica, Kwiecień & Bednarz (eds)
© *2020 Taylor & Francis Group, London, ISBN 978-0-367-56586-2*

Seismic behaviour of RC frames with uncoupled masonry infills having two storeys or two bays

M. Marinković

Department of engineering mechanics and theory of structures, Faculty of Civil Engineering, University of Belgrade, Belgrade, Serbia

C. Butenweg

Center for Wind and Earthquake Engineering (CWE), RWTH-Aachen University, Aachen, Germany

ABSTRACT: Reinforced concrete (RC) structures with masonry infills are widely used for several types of buildings all over the world. However, it is well known that traditional masonry infills constructed with rigid contact to the surrounding RC frame performed rather poor in past earthquakes. Masonry infills showed severe in-plane damages and failed in many cases under out-of-plane seismic loading. As the undesired interactions between frames and infills changes the load transfer on building level, complete collapses of buildings were observed. A possible solution is uncoupling of masonry infills to the frame to reduce the infill contribution activated by the frame deformation under horizontal loading. The paper presents numerical simulations on RC frames equipped with the innovative decoupling system INODIS. The system was developed within the European project INSYSME and allows an effective uncoupling of frame and infill. The simulations are carried out with a micro-modelling approach, which is able to predict the complex nonlinear behaviour resulting from the different materials and their interaction. Each brick is modelled individually and connected taking into account nonlinearity of a brick mortar interface. The calibration of the model is based on small specimen tests and experimental results for one bay one storey frame are used for the validation. The validated model is further used for parametric studies on two storey and two bay infilled frames. The response and change of the structural stiffness are analysed and compared to the traditionally infilled frame. The results confirm the effectiveness of the INODIS system with less damage and relatively low contribution of the infill at high drift levels. In contrast to the uncoupled system configurations, traditionally infilled frames experienced brittle failure at rather low drift levels.

1 INTRODUCTION

1.1 *Performance of RC frame structures with masonry infill walls*

As shown (Kappos & Ellul 2000, Kose 2009, Ricci et al. 2011, Asteris et al. 2015, Varum et al. 2017) infill walls in reinforced concrete frame buildings cause an 'increase' in lateral stiffness, strength, and energy dissipation capacity. Therefore, infills should be taken into account during the design; however this is not the case. This results in damage and failure of infill walls and sometimes RC elements during the earthquakes (Dazio et al. 2009, Braga et al. 2011, Manfredi et al. 2014). Damage of masonry infills may contribute significantly to economic losses and cause considerable threats to human lives, even in the case of infills in newly constructed buildings (Hermanns et al. 2012). Vicente et al. (2012) and Hermanns et al. (2012) pointed out that there is an urgent need for improvements in the current approach for the verification and detailing of singular points of infills. Whether this behaviour is favourable or not, the infill walls are usually the first

elements to be damaged in seismic events. Villaverde (1997) showed that the cost related to the failure of a non-structural component in a building may easily exceed the replacement cost of a building, due to the loss of inventory, loss of business, repair and reconstruction costs, downtime, injuries and casualties. These facts have pushed research and development of solutions that disable these negative effects resulting from the interaction between the masonry infill walls and concrete structures.

1.2 *Literature review*

The behaviour of masonry infilled frames subjected to in-plane lateral loads was investigated by a number of researchers, both numerically and experimentally (Mehrabi et al. 1996, Al-Chaar et al. 2002, Shing & Mehrabi 2002, Drysdale & Hamid 2005, Stylianidis 2012, Morandi et al. 2014, Hak et al. 2017).

The behaviour of infill walls under out-of-plane loads was examined by McDowell et al. (1956a, b), Dawe & Seah (1989b), Asteris et al. (2017) and Walsh et al. (2017).

Figure 1. Damage of two bay infilled frame during the $M_w = 6.4$ earthquake on 26.11.2019 in Albania.

However, in-plane/out-of-plane interaction was noticed and investigated by a few authors (Hashemi & Mosalam 2007, Di Trapani et al. 2017, Pasca et al. 2017, Butenweg et al. 2019).

Recently, several researchers developed innovative systems for improvement of behaviour of infilled frames (Verlato et al. 2016, Morandi et al. 2018, Preti et al. 2019).

Most of the studies considered one bay and one storey frame, just a few authors investigated frames with more bays and/or more storeys. Liauw & Lo (1988) showed that there is significant increase of ultimate load in a case of two bay frame in comparison to the one bay frame. Al-Chaar (1998) and Al-Chaar et al. (2002) stated that the presence of stiffer masonry infill wall enables the system of carrying more load than the case of the bare frame. Moreover, the stiffness increases with the increase of the number of infilled bays of a multi bay frame.

Figure 1 shows damaged two bay infilled Rc frame. Therefore the aim of this study was to investigate the behaviour of two storey and two bay infilled frames. Additional novelty of this paper is that it shows the findings on uncoupled infilled frames too. The system for uncoupling presented in Marinković & Butenweg (2019a) was numerically investigated using already calibrated and validated model (Marinković & Butenweg 2018). Ridington & Smith (1977) investigated three bay frame and concluded that the behaviour is dependent on the distribution of loading.

2 UNCOUPLED INFILL/FRAME CONNECTION

2.1 Description of uncoupled connection

From 2013 until 2017, the European Project INSYSME (Innovative Systems for Earthquake Resistant Masonry Enclosures in RC Buildings) has been funded within the 7th Framework Programme by the European Commission. The research presented here is part of this project and its aim was developing a constructive measure that solves the above mentioned problems and provides its simple application in practice, thus enabling engineers to apply the system easily and without any complicated numerical models. The objective of the study was the development of a new system for the improvement of seismic safety of infills made of bricks. And this was done, so at the end of the project system INODIS (Innovative Decoupled Infill System), for improving the seismic behaviour of masonry infilled reinforced concrete frames was developed and patented on the European level.

The conceptual idea is to uncouple infill by applying elastomers between RC frame and infill panel such that the brittle behaviour of the infill walls will be avoided. This way activation of infill walls due to RC frame in-plane deformations is postponed to higher drifts, thus disabling high stresses in both RC frame and infill wall. The elastomer bearings are designed to allow the design drift of the RC frame without inducing damages to the infill wall. Moreover, due to the viscoelastic behaviour of the elastomeric joints, overall damping capacity of the building is enhanced. Furthermore, the load transfer mechanism of out-of-plane load is limited by uncoupling, and to overcome this, alternative mechanisms for the out-of-plane load transfer have been provided with the shear key and U-shaped elastomeric profile. More detailed description of the INODIS system and its performances tested experimentally are given in Marinković & Butenweg (2019b) and Marinković & Butenweg (2018)

3 NUMERICAL MODELLING

3.1 Overall approach

A three dimensional finite element model was developed (Figure 2) to investigate the strength, lateral displacement and stress distribution throughout the infill wall. Finite element analyses were conducted using software Abaqus (2013). Some simplifications are introduced in the model, such as not taking into account bond slip effect directly by defining interaction between reinforcement and concrete. Also material model used for bricks does not take into consideration orthotropic behaviour of masonry and mortar joints were modelled using interaction contact between the bricks instead of full size continuum elements.

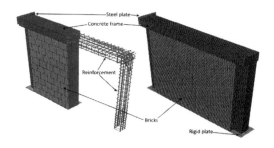

Figure 2. Geometry and finite element mesh of the numerical model of infilled frame (Marinković 2018).

According to recommendations (Abaqus 2013), three-dimensional 8-node hexahedral continuum finite elements, with the reduced integration (C3D8R) are most appropriate for the explicit dynamic analyses and they are used for modelling concrete, bricks and elastomer. For reinforcement, truss elements (T3D2) have been used and they are embedded in solid concrete elements.

In the next sections just a rough description of numerical modelling approach is given, while detailed description can be found in Marinković (2018) and Marinković & Butenweg (2019).

3.2 Material modelling

For material definition, built-in material models in Abaqus (2013) were used to describe the behaviour of concrete, masonry units, reinforcing steel and elastomer. Similar to the approach used by Stavridis & Shing (2010), smeared crack models are enforced on the concrete and brick elements. The constitutive model adopted for concrete is the concrete damage plasticity (CDP) model for quasi-brittle material implemented in Abaqus (2013). The model is a continuum, plasticity-based, damage model suitable for quasi brittle materials.

The elastic response is assumed to be linear and isotropic. It can be assessed through the use of modulus of elasticity and Poisson's ratio. Modulus of elasticity is taken from the experimental tests, while the value of 0.2 is used for Poisson's ratio as suggested in Eurocode 8 (EN 1998-1, 2004). CDP model allows describing separately the nonlinear tensile and compressive behaviour of plain concrete. In compression, the stress-strain curve (Figure 3) is first linear until $0.4f_{cm}$ according to Eurocode 2 (EN 1992-1-1, 2004). Beyond this point, concrete is in the plastic region in which plastic strain is input to define the stress-strain relationship in the finite element model in Abaqus (Abaqus 2013). For this part, expression given in Eurocode 2 (EN 1992-1-1, 2004) is used, which can be regarded as a specialisation of the non-linear stress-strain curve according to Sargin (1971). This curve is defined only up to the nominal ultimate

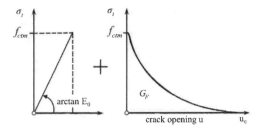

Figure 4. Stress-displacement curves for concrete in tension (Hordijk 1992).

strain. To reach experimental levels of deformation, the curve is extended with respect to the equation proposed by Pavlović (2013). The behaviour in tension is defined using a fracture energy criterion and a stress-displacement curve (Figure 4) instead of a stress-strain curve. The relationship between the crack width and the corresponding tension stress is based on an equation proposed by Hordijk (1992).

The same as for concrete material, the constitutive model for masonry used in this study is concrete damage plasticity (CDP) model. Stavridis & Shing (2010) recommended that the material characteristics of masonry units based on that of masonry prisms should be used, rather than that of individual brick units. This approach has been accepted in this study. Also, the approach according to Stavridis & Shing (2010) is used to generate the required stress curves in both compression and tension.

Elastomer material used for uncoupling infill wall from the RC frame is rubber-based material, with hyperelastic behaviour. Since the material used is highly compressible, it can be characterized as elastomeric foam. Therefore, it was decided to use hyperfoam material available in Abaqus (2013) to model its behaviour.

3.3 Contact definition

All the joints between the bricks, both vertical and horizontal, as well as joints between the frame and infill have been defined using general contact with the specified interaction properties. Three interaction properties have been defined, with the first defined as global property assignment to the all elements that are in contact. For this interaction, "hard" contact normal behaviour is defined together with penalty friction formulation with the friction coefficient being 0.6. This means that just compression stresses and frictional forces are transferred, when the surfaces are in contact, without having any tensile strength. Since head joints did not have mortar applied, global property assignment has been used for them.

Second interaction property was defined to represent bed joint behaviour and it was assigned to the horizontal surfaces of the bricks being in contact. This interaction property, beside "hard" contact and

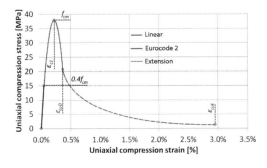

Figure 3. Stress-strain curve for concrete in compression (Marinković 2018).

Table 1. Values used for interaction definition.

Bed joint Interaction	Value	Frame/infill Interaction	Value
k_{nn} [GPa/m]	35.2	k_{nn} [GPa/m]	35.2
k_{ss} [GPa/m]	1.48	k_{ss} [GPa/m]	1.48
k_{tt} [GPa/m]	1.48	k_{tt} [GPa/m]	1.48
t_n [MPa]	0.19	t_n [MPa]	0.06
t_s [MPa]	0.15	t_s [MPa]	0.05
t_t [MPa]	0.15	t_t [MPa]	0.05
G_n [N/m]	20	G_n [N/m]	1
G_s [N/m]	20	G_s [N/m]	10
G_t [N/m]	20	G_t [N/m]	10
η [-]	2	η [-]	2
μ [-]	0.7	μ [-]	0.7

penalty friction assignment, contains surface-based cohesive interaction.

Third interaction property is interaction at the infill/frame connection, for which the same approach is used but with the reduced characteristics (Table 1). For more details check Marinković (2018).

4 NUMERICAL SIMULATIONS

4.1 Two storey frame

Numerical model was used for simulating in-plane behaviour of infilled RC frames. First, the model was calibrated and validated according to the experiments (Butenweg et al. 2019) on full scale bare frame (BF), traditionally infilled frame (TIF) and infilled frame with the INODIS system (IIF). Traditionally infilled frame was constructed in a way that connection between frame and infill is made with the mortar.

The whole process of calibration and validation of the model is described in details in Marinković (2018). This model is used as a base for the simulations shown in this paper. The model is extended to the frames of two stories and two bays.

First a model with the two stories is made, with the height twice as the basic one storey frame. The same size of the beams, columns and bricks is used as described in Marinković (2018). Now, the horizontal displacement load is applied on the beam at the top of

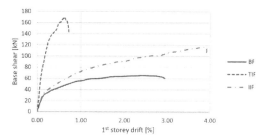

Figure 5. Comparison of force-displacement curves for interstorey drift of the first storey.

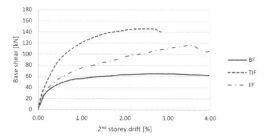

Figure 6. Comparison of force-displacement curves for interstorey drift of the second storey.

the second storey. Vertical forces of 200 kN are applied on each column at the top beam of the second storey.

The effect of infill uncoupling from the frame can be seen (Figure 5) in the case of two storey frame too. Here, it is obvious that the frame with the INODIS system (IIF) is less activated in-plane than traditionally infilled frame (TIF) and therefore reaches higher in-plane drifts before collapse. TIF model is characterized with high stiffness and brittle failure at low in-plane drifts. In contrary, IIF model has stiffness slightly higher than BF. This stiffness increases with the increase of in-plane drift, due to the high compression of the elastomer used for uncoupling. The same trend can be seen in Figure 6, where base shear force is plotted against the interstorey drift of the second storey.

The results of the numerical simulations on the two storey frames are given in Table 2, where again it can

Table 2. Forces and drift levels for two storey model.

	First crack			Max. load			Max. displacement		
	F kN	1st storey drift %	2nd storey drift %	F kN	1st storey drift %	2nd storey drift %	F kN	1st storey drift %	2nd storey drift %
BF	32.6	0.17	0.17	65.4	2.40	2.55	62.4	2.98	3.97
TIF	134.8	0.27	1.37	169.1	0.64	2.21	142.1	0.72	2.71
IIF	86.41	1.68	1.71	126.2	4.65	3.61	104.2	3.95	3.98

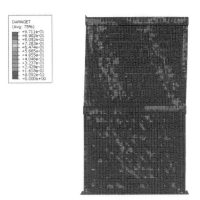

Figure 7. Deformed shape with the tensile damage distribution at the level of maximum load for TIF model.

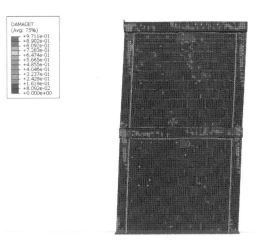

Figure 8. Deformed shape with the tensile damage distribution at the level of maximum load for IIF model.

be seen that IIF and BF have reached high in-plane drifts while TIF is heavily damaged at quite lower in-plane drifts. This is especially pronounced at the first storey.

First visible crack is the moment of appearance of the first crack in infill wall (for TIF and IIF) or in RC frame (for BF). Figures 7 and 8 show that at the level of maximum force resistance of the TIF model is several times more damaged than IIF model.

4.2 *Two bay frame*

For the modelling of the two bay frame, one bay frame model validated according to experimental results (Marinković 2018) is used. The model is just extended in the length by adding one infill wall and one column at the end of it. Everything else was kept the same.

Figure 9 confirms beneficial effect of application of the INODIS system, which can significantly reduce infill/frame interaction, thus allowing reaching high in-plane drifts with low or no damage.

Tensile damage distribution of the simulated specimens is presented in Figure 10. Here it can be seen that infill wall at maximum load of specimen IIF is almost no damaged in contrast to TIF specimen, even

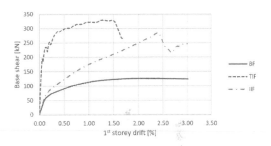

Figure 9. Comparison of force-displacement curves for two bay frames.

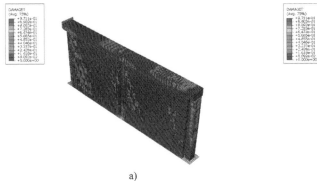

a)

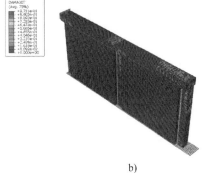

b)

Figure 10. Deformed shape with the tensile damage distribution at the level of maximum load for a) TIF and b) IIF model.

Table 3. Forces and drift levels for two bay model.

	First crack		Max. load		Max. displacement	
	F kN	drift %	F kN	drift %	F kN	drift %
BF	62.0	0.16	126.2	1.89	124.2	3.02
TIF	193.4	0.06	329.2	1.51	265.45	1.69
IIF	278.5	2.40	283.4	2.42	246.81	3.00

high in-plane drifts reached for TIF are much lower than for IIF. This again shows the advantage of application of the INODIS systems, which provides infill/frame uncoupling and thus reduces damage in infills.

Summary of the results is given in Table 3 where it can be seen that traditional system exhibits first crack at the very low drift and reaches quite low maximal drift.

5 CONCLUSION

The paper presents results of numerical simulations on two bay and two storey frames, with the focus on uncoupled masonry infill walls. The aim of the paper was to extend experimental campaign of testing of the INODIS system used for infill/frame uncoupling. The extension was done by investigating the behaviour of two storey and two bay frames. This was done using already developed and validated numerical model.

Results show that INODIS system is effective not on just one bay one storey frame, but also in a case of more bays and stories. It highly reduces infill activation, thus the stiffness of the system is several times lower than traditionally infilled frame. And models with the uncoupling systems always reached higher in-plane drifts.

For further work, numerical studies on the building levels using macro modelling approach, is planned in order to study the effects of the INODIS system on the global level. This will move us a one step closer to the design concept for application of INODIS system in the design of RC frame buildings with masonry infills.

REFERENCES

Abaqus 2013. User Manual. Version 6.13. Providence, RI, USA: DS SIMULIA Corp.

Al-Chaar, G. 1998. Non-Ductile Behaviour of Reinforced Concrete Frames with Masonry Infill Panels Subjected to In-Plane Loading. *PhD thesis, University of Illinois at Chicago, Chicago, USA.*

Al-Chaar, G. et al. 2002. Behavior of masonry-infilled nonductile reinforced concrete frames. *J. Struct. Eng.* 128(8): 1055–1063.

Asteris, P. G. et al. 2015. Parameters affecting the fundamental period of infilled RC frame structures. *Earthquakes and Structures* 9(5): 999–1028.

Asteris, P. G. et al. 2017. Numerical modelling of out-of-plane response of infilled frames: State of the art and future challenges for the equivalent strut macromodels. *Engineering Structures* 132: 110–122.

Braga, F. et al. 2011. Performance of non-structural elements in RC buildings during the L'Aquila, 2009 earthquake. *Bulletin of Earthquake Engineering* 9(1): 307–324.

Butenweg, C. et al. 2019. Experimental results of reinforced concrete frames with masonry infills under combined quasi-static in-plane and out-of-plane seismic loading. *Bulletin of Earthquake Engineering* 17(6): 3397–3422.

Dawe, J. L. & Seah, C. K. 1989b. Out-of-plane resistance of concrete masonry infilled panels. *Journal of the Canadian Society of Civil Engineering* Vol. 16: 854–864.

Dazio, A. et al. 2009. Das Mw=6.3 Erdbeben von L'Aquila am 6. April 2009. *Report of SGEB-Erkundungsmissionvom*: 15–18.

Di Trapani, F. et al. 2017. Macroelement Model for In-Plane and Out-of-Plane Responses of Masonry Infills in Frame Structures. *Journal of Structural Engineering* 144(2): 04017198.

Drysdale, R. G. & Hamid, A. A. 2005. Masonry Structures Behaviour and Design. *Canada Masonry Design Center, Mississauga, Ontario, Canada:* 531–562.

EN 1992-1-1: 2004. Eurocode 2: Design of concrete structures. Part 1-1: General rules and rules for buildings. CEN, Brussels, Belgium.

EN 1998-1: 2004. Eurocode 8: Design of structures for earthquake resistance. Part 1: General rules, seismic actions and rules for buildings. CEN, Brussels, Belgium.

Hak, S. et al. 2017. Prediction of inter-storey drifts for regular RC structures with masonry infills based on bare frame modelling. *Bulletin of Earthquake Engineering*: 1–29.

Hashemi, A. & Mosalam, K.M. 2007. Seismic evaluation of reinforced concrete buildings including effects of masonry infill walls, *Pacific Earthquake Engineering Research Center, University of California, Berkeley, PEER, Report* 2007/100.

Hermanns, L. et al. 2012. Performance of masonry buildings during the 2011 Lorca earthquake. *Proceedings of the 15th World Conference on Earthquake Engineering, Lisbon, Portugal.*

Hordijk, D.A. 1992. Tensile and tensile fatigue behavior of concrete; experiments, modeling and analyses. *Heron* 37(1): 3–79.

Kappos, A. J. & Ellul, F. 2000. Seismic design and performance assessment of masonry infilled RC frames. In: *Proceedings of the 12th world conference on earthquake engineering*, paper No. 989.

Kose, M. M. 2009. Parameters Affecting the Fundamental Period of RC Buildings with Infill Walls. *Engineering Structures* vol. 31: 93–102.

Liauw, T. C. & Lo, C.Q. 1988. On Multubay Infilled frames, *Proceedings of the Institution of Civil Engineers* 85(2): 469–183.

Manfredi, V. & Masi, A. 2014. Combining in-plane and out-of-plane behaviour of masonry infills in the seismic analysis of RC buildings. *Earthquakes and Structures* 6(5): 515–537.

Marinković, M. 2018. Innovative system for seismic resistant masonry infills in reinforced concrete frame structures. *PhD Thesis, University of Belgrade – Faculty of Civil Engineering.*

Marinković, M. & Butenweg, C. 2019a. Experimental and Numerical analysis of RC Frames with decoupled masonry infills. *COMPDYN 2019-7th ECCOMAS Thematic Conference on Computational Methods in Structural Dynamics and Earthquake Engineering, Crete, Greece, 24-26 June, 2019.*

Marinković, M. & Butenweg, C. 2019b. Innovative decoupling system for the seismic protection of masonry infill walls in reinforced concrete frames. *Engineering Structures* 197: 109435.

Marinković, M. & Butenweg, C. Innovative System for Earthquake Resistant Masonry Infill Walls. *16th European Conference on Earthquake Engineering, Thessaloniki, Greece, 18-21 June, 2018.*

Mehrabi, A. B. et al. 1996. Experimental evaluation of masonry-infilled RC frames. *Journal of Structural Engineering* 122(3): 228–237.

McDowell, E. L. et al. 1956a. Arching action theory of masonry walls. *J. Struct. Div.* 82(ST2): 915/1–915/18.

McDowell, E. L. et al. 1956a. Arching action theory of masonry walls. *J. Struct. Div.* 82(ST2): 915/1–915/18.

Morandi, P. et al. 2018. Performance-based interpretation of in-plane cyclic tests on RC frames with strong masonry infills. *Engineering Structures* 156: 503–521.

Morandi, P. et al. 2018. Innovative solution for seismic-resistant masonry infills with sliding joints: in-plane experimental performance. *Engineering Structures* 176: 719–733.

Pavlović, M. 2013. Resistance of bolted shear connectors in prefabricated steel-concrete composite decks. *University of Belgrade.*

Pasca, M. et al. 2017. Reliability of analytical models for the prediction of out-of-plane capacity of masonry infills. *Structural engineering and mechanics* 64(6): 765–781.

Preti, M. et al. 2019. Seismic infill–frame interaction of masonry walls partitioned with horizontal sliding joints: analysis and simplified modeling. *Journal of Earthquake Engineering* 23(10): 1651–1677.

Ricci, P. et al. 2011. Analytical Investigation of Elastic Period of Infilled RC MRF Buildings. *Engineering Structures* vol. 33: 308–319.

Ridington, J.R. & Stafford Smith, B. 1977. Analysis of Infilled Frames Subject to Racking with Design Recommendations. *Structural Engineer* 55 (6): 263–268.

Sargin, M. 1971. Stress-Strain Relationship for Concrete and the Analysis of Structural Concrete Section. *University of Waterloo.*

Shing, P.B. & Mehrabi, A.B. 2002. Behaviour and analysis of masonry-infilled frames. *Prog. Struct. Eng. Mater.* 4: 320–331.

Stavridis, A., & Shing, P. B. 2010. Finite-element modelling of nonlinear behavior of masonry-infilled RC frames. *J. Struct. Eng.* 136(3): 285–296.

Stylianidis, K. C. 2012. Experimental investigation of masonry infilled RC frames. *Open Constr Build Technol J* 6(1): 194–212.

Varum, H. et al. 2017. Seismic performance of the infill masonry walls and ambient vibration tests after the Ghorka 2015, Nepal earthquake. *Bulletin of Earthquake Engineering* 15(3): 1185–1212.

Verlato, N. et al. 2016. Innovative systems for masonry infill walls based on the use of deformable joints: combined in-plane/out-of-plane tests. In: *Brick and Block Masonry.* CRC Press:1359–1366.

Vicente, R. S. et al. 2012. Performance of masonry enclosure walls: lessons learned from recent earthquakes. *Earthquake engineering and engineering vibration* 11(1): 23–34.

Villaverde, R. 1997. Seismic design of secondary structures: state of the art. *J Struct Eng, ASCE* 123(8): 1011–9.

Walsh, K. Q. et al. (2017). Effect of boundary conditions and other factors on URM wall out-of-plane behaviour: Design demands, predicted capacity, and in situ proof test results. *SESOC Journal* 30(1): 57.

Brick and Block Masonry - From Historical to Sustainable Masonry –
Kubica, Kwiecień & Bednarz (eds)
© 2020 Taylor & Francis Group, London, ISBN 978-0-367-56586-2

Older clay masonry can be more earthquake-resistant than calcium-silicate masonry for light damage

P.A. Korswagen & J.G. Rots
University Delft University of Technology, Delft, The Netherlands

ABSTRACT: Seismic events in the northern part of the Netherlands over the past few decades have led to an intense study on the seismic performance of unreinforced masonry structures, ubiquitous in the region. Studies have focused on the safety aspect characterized by the near collapse behavior of the structures, but also on the aesthetic and functional aspect denoted serviceability or light damage state. The latter, triggered by multiple small seismic events until now, has led to economic losses and societal unrest. Experimental studies, detailed in this paper, have looked at the initiation and propagation of visible cracks in clay and calcium-silicate unreinforced masonry specimens. Both materials, replicating samples from existing structures, were surveyed with high-resolution Digital Image Correlation while being subjected to repetitive loading causing horizontal, vertical and diagonal stair-case cracks. Small and full-scale experiments evidenced a difference in behavior between clay and calcium-silicate masonry: where the former only developed cracks along the mortar joints, the latter also exhibited brick-splitting cracks. This mechanism proved extremely brittle and led to a reduced capacity of energy release for the calcium-silicate specimens during cyclic loading, and to their sudden failure. The higher bond-strength between the silicate brick and mortar coupled with the lower strength of the bricks themselves and the overall higher stiffness of the calcium-silicate masonry, in the case of the replicated masonry specific to this study, are suspected of being responsible for the brick-splitting failures. These brittle failures, occurring at low drift levels, set calcium-silicate masonry as more vulnerable than comparable clay-brick masonry in regards to light damage, leading to wider and larger cracks.

1 INTRODUCTION

The north of the Netherlands is exposed to frequent, light earthquakes induced by the extraction of natural gas (NAM 2016). These seismic events lead to vibrations in the order of 2 to 10 mm/s, with 32 mm/s being the maximum peak ground velocity (PGV) recorded to date (den Bezemer & van Elk 2018). The vibrations, in turn, have the potential to cause light damage, or damage corresponding to state 1 (DS1), to the ubiquitous, unreinforced masonry structures in the region. Masonry structures built before 1970 were commonly constructed with baked-clay bricks, with double wythe walls or cavity walls. Post-1970 structures, however, most often present cavity walls with an inner, load-bearing leaf of calcium-silicate bricks, and an outer clay-brick veneer. This trend continued until the 90's when calcium-silicate bricks were replaced with larger blocks or elements.

Consequently, a large number of masonry houses sport a combination of calcium-silicate brick walls with baked-clay brick veneers. Hence, it is necessary to understand the damage behaviour of both walls when characterising or determining the expected light damage of the structures in response to the earthquakes. As a first step, the behaviour of both materials needs to be understood and compared for small, similar values of inter-storey drift independently, that is, assuming that the walls behave together through their connection at the level of the floors, a reasonable compromise given that light damage is expected to occur mostly due to actions in the plane of the walls, while out-of-plane effects are more a concern of higher damage states beyond light damage (Van Staalduinen et al. 2018).

A first comparison is provided herein by the means of experimental tests on full-scale walls and smaller experiments on both clay and calcium-silicate specimens, with a focus on the latter.

2 FULL-SCALE WALL TESTS

2.1 *Setup of the wall tests*

Two nominally-identical full-scale masonry walls of about 3.1 m wide and 2.7 m tall (see Figure 1) were tested in-plane under a small, constant vertical stress of 0.12 MPa. The tests were performed displacement-controlled and considered cyclically increasing lateral drift. The two walls were built of calcium-silicate bricks with a cementitious general mortar in joints of 10 mm. An opening for a window, located asymmetrically towards the centre of the wall and covered

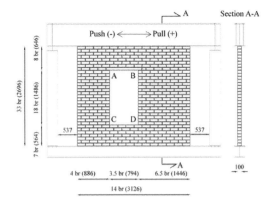

Figure 1. Geometry of the calcium-silicate walls and convention for the direction of testing in-plane.

Figure 2. Photograph of a calcium-silicate wall in the setup and zoomed-in detail of a wall corner showing the DIC pattern.

with a precast concrete lintel, divided the walls in two piers and two spandrels, and would later serve for the initiation of cracks.

The walls were constructed on steel beams, later fixed to the setup, and were finished off with a top steel beam glued to the last row of bricks. The top beam was free to rotate in its plane and was attached to the testing setup which exerted horizontal displacements in the plane of the wall.

The quasi-static tests started at an enforced drift of 0.26‰ which was increased by a value of 0.073‰ after 30 one-way cycles in the positive direction. This means that 30 cycles were performed at a drift of 0.26‰ in step 1. The drift was increased four times for a total of 5 one-way steps, before it was reduced back to 0.26‰ and the step-wise increasing procedure was repeated for two-way cycles (positive and negative in-plane directions), with a total of 7 additional steps, ultimately reaching a drift of 0.7‰ in the last, twelfth step; see Table 1 and later Figure 4.

The specimens were painted white and a black pattern of random dots was applied to monitor displacements on the surface of the wall using Digital Image Correlation. Figure 2 presents a photograph of the wall with a corner zoomed-in to show the speckle pattern. The 2D, in-plane displacements of approximately one million gridpoints, spaced about 2.8 mm from each other, were surveyed simultaneously at an accuracy of 20 μm; this is herein denoted high-

resolution DIC. The setup was configured so as to track the initiation and propagation of cracks in the masonry; the former was considered to occur when cracks reached a width of 100 μm, as lines of this thickness start to become visible to the naked eye of a careful observer. Additionally, LVDTs and laser sensors at the back of the wall verified in-plane and out-of-plane deformations, respectively.

Finally, the walls were compared to five previously- and identically-tested walls of similar geometry, albeit made of baked-clay bricks.

2.2 Results of the wall tests

The lateral force opposed by the walls when applying the prescribed displacements was recorded. From these, the maximum contours in the positive and negative directions of all 360 cycles are drawn in Figure 3, these can be recognised as envelope curves. The graph shows the contours of both calcium-silicate walls shaded, while the envelopes of force-drift for the clay walls are plotted with dash-dotted lines. First, it can be observed that the envelopes of the calcium-silicate walls reach a higher force both in the positive and the negative directions compared to the clay walls; this points to the higher initial strength of the calcium walls. Nonetheless, at higher drift values, the capacity of both clay and calcium-silicate walls is similar, hinting at the starker strength degradation occurring in the case of the calcium walls. Moreover, the significantly larger area enclosed by the envelope of the calcium-silicate

Table 1. Overview of wall tests.

Type	Name	Protocol	Control
Full-scale wall Lateral, in-plane drift	TUD-Component 49 TUD-Component 50	5x30 one-way cyclic followed by 7x30 two-way cyclic	Drift (displacement) controlled

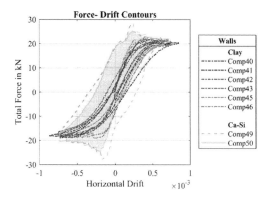

Figure 3. Envelope or contours of the cyclic force versus drift relationships of the experimental walls.

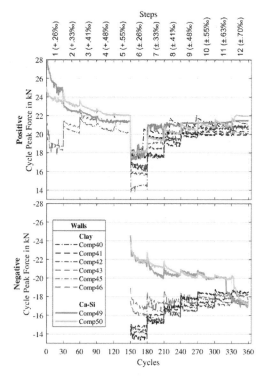

Figure 4. Peak force at each cycle in the positive (top) and negative (bottom) testing directions.

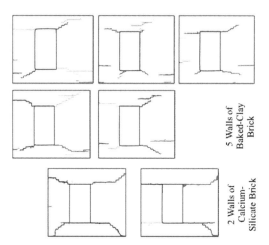

Figure 5. Cumulative crack pattern for the clay walls (top five) and calcium-silicate walls (bottom two).

walls means that a higher amount of energy has been released plastically by the walls, suggesting that these have also sustained more damage.

The strength degradation is visualised with more detail in Figure 4, where the maximum positive and minimum negative force reached in each cycle is plotted for all walls. The higher initial strength of the calcium-silicate walls is again evident, but it quickly degrades away even within the same steps where the amplitude of the enforced drift is kept constant: As the

drift increases, a jump up is seen in the lateral force, but the strength degrades again during the step. A similar behaviour is observed in the case of the clay walls; however, the degradation is not as pronounced as for the calcium-silicate walls. Figure 4 also shows that during the last step, a drop in the capacity in the negative direction of the calcium-silicate walls takes place.

The cumulative crack pattern is illustrated in Figure 5. These are cracks wider than 0.1 mm, occurred throughout each of the tests and gathered onto a single frame. The illustration shows that clay walls developed a larger number of (smaller) cracks, propagating mostly in a diagonal stair-case pattern from out of the window corners; while, the two calcium-silicate walls, present only four clear-defined cracks, spanning out horizontally from the window, and then propagating upwards, or down-wards, at a steeper angle, sometimes cutting verti-cally through the bricks. These four cracks concentrated most of the deformation in the walls and were thus comparatively wider than the cracks in the clay walls.

Accordingly, the width and length of the cracks was measured automatically with DIC for zero, positive, and negative drift positions for every cycle in the tests. Then, the Psi damage parameter (Ψ) was computed for every point with Equation 1. This parameter gathers into a single scalar the number, width, and length of the cracks such that comparisons regarding the inten-sity of the light damage can be made. The par-ameter expresses the presence of visible, yet easy-to-repair cracks in the order of 0.1 mm with a value of $\Psi \approx 1$, represents larger cracks around 1 mm in width with a value of $\Psi \approx 2$, and cracks larger than 4 mm (exceeding light damage) produce a value of $\Psi \approx 3$.

$$\psi = 2 \cdot n_c^{0.15} \cdot \hat{c}_w^{0.3} \ with \ \hat{c}_w = \frac{\sum_{i=1}^{n_c} c_{w,i}^2 \cdot c_{L,i}}{\sum_{i=1}^{n_c} c_{w,i} \cdot c_{L,i}} \quad (1)$$

where:

n_c is the number of cracks in the wall/specimen
$\hat{c}_w$ is the width-weighted and length-averaged crack width (in mm) calculated with:

c_w the maximum crack width along each crack in mm, and

c_L the crack length in mm.

For $n_c = 1$, $\hat{c}_w = c_w$, in Equation 1, the crack width of each crack is measured at their widest point, see Korswagen et al. (2019) for more details.

Then, the Psi damage parameter was plotted in Figure 6 against the drift values of the walls in the experiment. It can be observed that the calcium-silicate walls display higher values of damage for both positive and negative drifts, but also at values close to or at zero drift where the walls remain damaged even when no lateral drift (or force) is applied. Attempting a linear fit of the non-zero data to produce a relationship between drift and damage of clay and calcium-silicate walls, confirms the steeper slope of the latter. Here, if values of Ψ between 1 and 3 including one standard deviation of the linear fit are considered DS1, then the clay walls are expected to be in DS1 between 0.25‰ and 1.1‰, while the calcium-silicate walls incur into light damage earlier at 0.15‰ and exceed it at 0.65‰.

Finally, while visible in Figure 6, Figure 7 presents an example of the brick-splitting crack in a photograph on the back side of the first wall; as opposed to the cracks running along mortar joints, the crack through the brick is easily detectable and its repair would necessitate the replacement of the units.

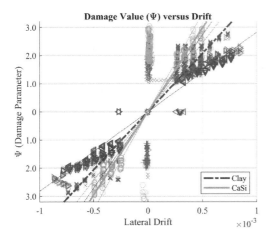

Figure 6. Damage at various values of drift for all the walls tests. Damage is measured in Ψ and registered at zero, positive and negative drift. Thin lines show one standard deviation. Circular points denote Calcium-Silicate tests and angular shapes represent Clay tests.

Figure 7. Photograph of the backside of the first calcium-silicate wall showing a crack splitting a brick (to the left of the dashed highlight).

3 SPANDREL TESTS

3.1 Setup of the spandrel tests

A modified four-point-bending test was used to test the behaviour of the masonry when bending in-plane and producing a vertical crack. Such a crack, not present in the wall tests where horizontally and diagonally propagating cracks were investigated, is characteristic underneath windows when, for example, (differential) settlement actions damage the masonry. The modifications to the standard four-point bending tests are the larger dimensions of the specimen, the reversal of loading such that the crack propagates downwards and not upwards, and the inclusion of a constant force (CW) aligned with the jacks (F) to counter the weight of the specimen. See Figure 8. These modifications permit 1) a controlled loading of the samples, as when these start to break, the counterweights prevent them from fully collapsing; and 2) the possibility of cyclic testing, as the counterweights produce a restitutive force capable of bringing the deformation of the specimens (almost) back to the zero position. These tests are thus also displacement controlled, albeit on the crack mouth opening displacement (CMOD) measured by a sensor spanning the three top head joints, where a vertical crack was expected to begin.

Five calcium-silicate samples, about 560 mm tall, 1330 mm wide, and 100 mm in thickness, were tested cyclically by enforcing a CMOD of 50 μm for 30 cycles, then 100 μm for another 30 repetitions, 150 μm until a total of 90 cycles, and finally by driving the specimens monotonically until failure. See Table 2.

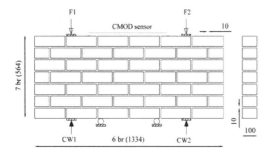

Figure 8. Geometry and testing scheme of the spandrel tests (specimen of calcium-silicate is shown).

Table 2. Overview of calcium-silicate specimens tested as spandrels.

Type	Name	Protocol	Control
Spandrel - Modified 4-point bending test	Sample A Sample B Sample C Sample D Sample E	3x30 one-way cyclic followed by monotonic until failure	CMOD-controlled

As for the walls, DIC was used to monitor crack propagation throughout the tests using about 75′000 measurement points at a precision of 5 μm for displacements in the plane of the specimens. To achieve this, the pattern on the samples consisted of much smaller dots, see Figure 9.

Figure 9. Photograph of a calcium-silicate spandrel with a detail of one of the top head joints where the pattern for DIC is visible.

3.2 Results of the spandrel tests

All the spandrels developed a vertical crack that started at the top of the samples and found its way down, see Figure 10. In the case of the calcium-silicate specimens, the crack bifurcated one or two rows of bricks, before cutting straight down through the specimens. This failure was brittle, so much so that in two of the experiments it was not possible to capture the failure mechanism with DIC, but a photograph of the spandrel reveals the same straight vertical crack cutting bricks, see Figure 11. In comparison, the clay specimens had a smoother failure with cracks

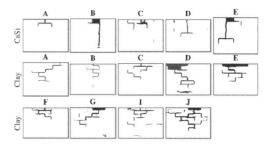

Figure 10. Latest registered crack pattern for each of the spandrel tests.

Figure 11. Photographs of two of the calcium-silicate tests that were too brittle to register the ultimate crack pattern with DIC.

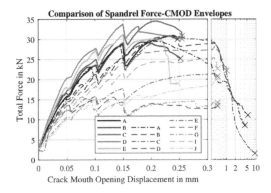

Figure 12. Envelope of contour of the force against crack mouth opening displacement (CMOD) of the spandrel tests. Solid lines represent calcium-silicate tests and interrupted lines mark the clay tests.

zig-zagging down through the specimens and resulting in a toothed failure mechanism. Note that sample H was broken before testing.

Moreover, the total force applied by both jacks is plotted against the crack mouth opening displacement of the specimens in Figure 12. All calcium-silicate samples exhibited a brittle failure around 0.25 mm, while the clay samples reached crack openings twice as high with some even reaching 8 mm before collapsing. Conversely, both materials presented a similar amount of strength degradation due to the cyclic nature of the tests. In this figure, the drops at 50, 100 and 150 μm of the envelope curves, reveal the degradation accumulated over the three steps of 30 cycles each. However, the degradation of the clay tests was associated with an increased displacement of the jacks, suggesting that the toothed failure mechanism accumulated sliding at the bed-joints, while the calcium-silicate tests displayed a degradation in strength at more or less (repetitively) constant jack displacements. This is visible in Figure 13 where the average jack displacement, instead of the CMOD, is plotted against the total vertical force of each experiment independently so as to show the cyclic character of each test clearly. These graphs also allow the computation of the fracture energy of an idealised vertical crack of 500 mm in height and 100 mm in width. This approach is not entirely correct since the tests are cyclic, which allows the release of a greater amount of energy; nevertheless, it is used here as a further comparison between clay and calcium-silicate tests. Accordingly, a significantly higher fracture energy of the clay tests in comparison to the calcium-silicate material is observed. This is attributed to the different failure mechanism of the material, the first with a toothed crack which allows a high energy release via sliding at the bed joints, the second with a brittle crack that also splits bricks. A summary of the average material properties obtained from the spandrel tests is collected in Table 3.

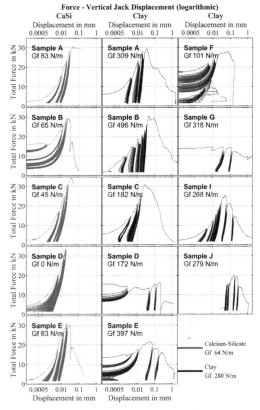

Figure 13. Average vertical displacement of the jacks against the total force applied for each of the spandrel tests. Additionally, the fracture energy ($G_{f\,l}$) is computed for each sample and the average is shown (bottom right).

Table 3. Summary of material properties obtained from the spandrel tests. The coefficient of variation is shown in percent next to the values.

Description	Symbol		Calcium-Silicate		Clay	
Maximum Capacity	F_{max}	kN	31.9	6%	24.2	24%
Flexural strength	f_{x3}	MPa	0.59	7%	0.52	24%
Direct tensile strength	f_t	MPa	0.27	15%	0.24	12%
Stiffness	E	MPa	3705	13%	3199	25%
Mode-I Composite Fracture Energy	$G_{f\,l}$	Nm/m²	64	24%	280	43%

4 DISCUSSION

Two types of experiments have been used to show that specimens made from calcium-silicate brick masonry behave poorly in comparison to similar baked-clay masonry specimens; this appraisal is done in terms of light damage and is thus more concerned with the aesthetic nature of the damage in contrast to other measures of damage more representative of structural safety. Both types of masonry were built so as to be representative of dutch masonry, with experiments used to characterise building samples and replicate both clay and calcium-silicate masonry in the laboratory (Jafari et al. 2017).

While the calcium-silicate masonry was stronger and stiffer than the clay masonry, for light damage, it is the capacity of the materials against displaying visible cracks what is considered paramount. Moreover, for light damage or 'damage state 1', the ease of repair of the damage is also relevant. In this light, cracks in calcium-silicate specimens were consistently wider, and hence more visible, than cracks in clay specimens. Additionally, the cracks in calcium-silicate masonry also compromised bricks, as opposed to only following the mortar joints, where the cracks in clay masonry were focused. This fact made cracks in calcium-silicate masonry more easily detectable and more difficult to repair. These observations were obtained by looking at horizontal, diagonal, and vertical cracks in masonry using tests on walls and spandrels, both for actions in the plane of the specimens. Out-of-plane effects or damage were monitored and appeared to be negligible and were thus not considered in these experiments nor in this study under the assumption that light damage occurs mostly in-plane and out-of-plane damage is more characteristic of damage states beyond DS1.

At identical inter-storey drift, which is expected if the two materials are assumed to correspond to the outer leaf (clay) and inner leaf (calcium-silicate) of a cavity wall, typical in the Netherlands, the wall tests showed greater damage for the calcium-silicate leaf, with wider and longer cracks. Moreover, the calcium-silicate brick masonry walls were seen to exceed light damage at a lower drift than clay walls. The brick-splitting failure, only seen in calcium-silicate walls, occurred well before light-damage was exceeded.

The precise cause for the difference in behaviour is uncertain. It is clear that the relationships between brick and mortar are key in determining whether cracks will focus at the interface or will also go through bricks. The stiffer and stronger calcium-silicate masonry, with a higher bond strength and cohesion between brick and mortar, seems to offer the straight vertical crack as a weaker alternative, while for the clay masonry, the toothed vertical crack appears weaker. It is expected that certain combinations of material properties will favour one failure mechanism over the other; this requires further study.

The strength degradation was also observed in both types of test. Here, displacements were applied repeatedly (or cyclically), and the reduction in force was measured. In both cases, the force seemed to stabilise after approximately 30 repetitions and consisted ultimately of about a 6% reduction for horizontal and diagonal cracks (from the wall tests) and about 20% for vertical cracks (from the spandrel tests). This is explained partly by the vertical cracks activating sliding at the bed joints and by the size of the specimens, where the single crack in the smaller spandrel involves a greater relative amount of material.

The drift limits for exceeding DS1 determined experimentally herein, 0.65‰ for calcium-silicate and 1.1‰ for baked-clay, appear in the same order of magnitude as those defined by other authors for similar masonry. For example, Hak et al. (2017) observed a drift limit for operational damage of masonry infill walls in RC buildings of 2‰; while Kallioras et al. (2018) modelled typical dutch terraced houses and found a 'damage limitation' value of 0.73‰. Moreover, Del Gaudio et al. (2017) compiled information from 60 tests from 28 authors for in-plane damage of hollow brick infill masonry. In their analysis, they defined damage state 1 as cracks in the order of 1 to 2 mm in the panels. From the data, they discovered a minimum value of 0.2‰, a maximum of 3.5‰ and an average of 1.0‰, very much in line with the values from this study. Yet, the focus of these authors is in many cases on the structural performance of the masonry and not on its aesthetic behaviour or its repairability, boundary conditions also differ, and materials are not identical. In this light, comparisons become difficult but give a frame upon which to judge the drift values obtained.

5 CONCLUSIONS

Wall and spandrel tests on calcium-silicate brick masonry were compared to similar specimens of baked-clay brick masonry. Experiments showed that cracks, characteristic of light or aesthetic damage, were wider and more difficult to repair for the former than for the latter. Comparisons were carried out at equal values of inter-storey drift, both for the calcium-silicate masonry representative of dutch buildings built mainly between 1970 and 1990 and for the baked-clay masonry also present in the outer veneer of the cavity walls of the same buildings. The difficulty in repairing the light damage in this calcium-silicate masonry is due to the propagation of cracks through bricks whereas cracks in the representative clay masonry were limited to the mortar joints.

ACKNOWLEDGEMENTS

This research was funded by Nederlandse Aardolie Maatschappij (NAM) under contract number UI67339 'Damage sensitivity of Groningen masonry building structures – Experimental and computational studies', contract holders: Jan van Elk and Jeroen Uilenreef. This cooperation is gratefully acknowledged. The authors also express their gratitude to: Edwin Meulman for his support in designing, conducting and interpreting the laboratory experiments; Lucia Licciardello for performing the spandrel tests; and Michele Longo for his (computational) analyses before and after the experiments.

REFERENCES

den Bezemer T. & van Elk J. 2018. Special Report on the Zeerijp Earthquake - 8th January 2018. NAM.

Del Gaudio C. et al. 2017. Drift-based fragility functions for hollow clay masonry infills in RC buildings under in-plane seismic actions. Anidis, Pistoia.

Hak S. et al. 2017. Verification Of Drift Demands in the Design Of RC Buildings With Masonry Infills. *16th World Conference on Earthquake Engineering, 16WCEE 2017 Santiago Chile, January 9th to 13th 2017.*

Jafari S. et al. 2017. Characterizing the Material Properties of Dutch Unreinforced Masonry. *Procedia engineering* 193: 250–257.

Kallioras S. et al. 2018. Numerical assessment of the dynamic response of a URM terraced house exposed to induced seismicity. *Bull Earthquake Eng.*

Korswagen P.A. et al. 2019. Crack initiation and propagation in unreinforced masonry specimens subjected to repeated in-plane loading during light damage. *Bulletin of Earthquake Engineering.*

NAM. 2016. Production, Subsidence, Induced Earthquakes and Seismic Hazard and Risk Assessment in the Groningen Field. NAM, Technical Addendum to the Winningsplan Groningen 2016.

Van Staalduinen P.C. et al. 2018. Onderzoek naar de oorzaken van bouwkundige schade in Groningen Methodologie en case studies ter duiding van de oorzaken. Delft University of Technology. Report number CM-2018-01, 11 July 2018.

Brick and Block Masonry - From Historical to Sustainable Masonry –
Kubica, Kwiecień & Bednarz (eds)
© 2020 Taylor & Francis Group, London, ISBN 978-0-367-56586-2

Change in stiffness of damaged RC frame with masonry infills connected with stiff and flexible interfaces

A.T. Akyildiz, A. Kowalska-Koczwara, Ł. Hojdys & P. Krajewski
Cracow University of Technology, Cracow, Poland

ABSTRACT: This study investigates the dynamic behaviors of RC frames with masonry infills which are constructed by different connection methods along the frame – wall boundaries. Two different connection types are considered namely, stiff connection to represent the traditional construction technique and flexible connection provided by Polyurethane PM as an innovative method. PolyUrethane Flexible Joints (PUFJ) are intended to use on both old (existing) and new (to-be-built) structures. Therefore, three single-bay and single-story frames are numerically created; Frame A has stiff, Frame B has PUFJ around the three sides of wall except the bottom one and Frame C has entire perimeter of the wall PUFJ connections, respectively. Among the frames of PUFJ implemented ones, Frame B is for representing the usage of PUFJ on existing structures whereas Frame C shows the implementation of method in new buildings. All of the frames were exposed to constant vertical loads affecting on beams. Various horizontal loads caused by seismic excitations were also used to understand the behavior of frames under different lateral load stages. Frame stiffness values were changed by means of arranging artificial column hinges that are located either only on the bottom or on both the bottom and top of columns. Thus, different damage scenarios were able to be investigated. The results are presented in terms of natural mode frequencies and maximum displacement values.

1 INTRODUCTION

Masonry infill walls are largely used in all around the world. They are preferred especially in RC framed buildings for various purposes such as, for creating partitions or as insulation members. In most instances, their direct influence on building dynamic characteristics is neglected. Designers consider their effects as mere vertical loads but ignoring the horizontal load carrying capacities of those walls during the design phase. This is mostly due to the complexity of reflecting the infill behavior on design calculations. In addition, as per the common practice in structural engineering field, infill walls are perceived as extra loads on buildings yet they are standing as auxiliary load carrying members against the lateral forces, e.g. earthquake loads. Thus, contribution of the infill walls is widely acclaimed as a positive impact on seismic resistant structural designing. However, many studies revealed that the masonries might have adverse impact on buildings when they lose their load carrying strengths (Hermanns et al. 2014, Akhoundi et al. 2015, Pasca & Liberatore 2015). Masonry units are mostly chosen from brittle materials such as concrete or clay bricks. These are vulnerable beyond their elastic ranges and therefore sudden damages occur on the infill walls. Consequently, partial or total collapse of the masonries take place due to such incidents which lead to irregularities on building dynamic characteristics. The main reason behind of it is the interaction effects arise around the boundary zones of RC frame members and masonry infill walls. Cyclic horizontal loads cause damage on these parts and reducing the bonding strength between the frames and walls. It also happens very often that masonries and frames get damages, too.

Some past researches focused on preventative solutions against such failures and proposed multiple solutions (Dafnis et al. 2002, Preti et al. 2015, Vailati & Monti 2015, Muthu Kumar & Satyanarayanan, 2018). In this paper, a flexible joint solution replacing the classical type of mortar usage is investigated. PolyUrethane Flexible Joints (PUFJ) method is used for this purpose. Polymer based material Polyurethane PM is already tested previously for various purposes and experiments done under different load types (Kwiecień 2012, Kisiel 2015, Kwiecień et al. 2017a, b). It has high bonding feature that enables to maintain connection of different members, namely RC frames and infill walls. In addition, the material is capable of withstanding large deformations thanks to its hyper-elastic features. Details of the materials are given in this paper.

In order to have an overview on PUFJ influence on buildings, single-bay and single-story RC frames were numerically created. All of them consist of masonry infill walls. The frames were categorized in terms of having different frame-to-masonry connection configuration. The frames were divided into three types and among them Frame A has the stiff connection to represent the traditional construction technique, e.g. mineral mortar. Frame B and Frame C have PUFJ material along their infill boundaries. Frame B has three sides of the wall covered by flexible joints except the bottom one whereas the masonry of Frame C is entirely surrounded by the PUFJ. Frame B is designed to exhibit the implementation of innovative method on old (existing) and Frame C is for the new (to-be-built) buildings, respectively. Schematic views of the frames are given in Figure 1. The dimensions are identical for all of the frames, therefore these values are given only once in the Figure 1. Dynamic behaviors of the frames were firstly investigated in

terms of natural frequencies. After that, frame top displacement values were analyzed under constant vertical loads and cyclically effecting loads from real earthquake records. The results are presented with details in the following sections.

2 DESCRIPTION OF THE FRAMES AND MATERIALS

The frames consist of two columns, one top beam and one strong beam at the bottom. Masonry infill walls were created by solid clay bricks. Both column and top beam cross sectional dimensions are 25×25 cm^2. The wall length in both vertical and horizontal direction is 220 cm, thus it is formed in a square shape. Brick units were arranged in one layer along the vertical direction for proving 12 cm wall thickness. PUFJ thickness used for the Frame B and Frame C was determined as 2 cm.

The material properties of the bricks and PUFJ are taken from previous studies (Kisiel 2018, Gams et al. 2017, Viskovic et al. 2017). RC frames have concrete class of C30/37 and B500C type of reinforcement steel (EC2 2004). Masonry elastic properties were calculated according to Eurocode-6 (EC6 2005). These values are given in Table 1 and Table 2.

3 DESCRIPTION OF THE NUMERICAL MODELS

The models were created by FEM software SAP2000 (Computers & Structures Inc. 2005). Material and geometrical (second order effects) nonlinearities are taken into account during all of the analyses. Beam and columns were simulated by one dimensional bar elements whereas two dimensional shell phenomenon was preferred whilst modeling the walls and PUFJ. Null weight rigid links that solely designated to provide load transfer were used as connectors between those bar and shell elements, see Figure 2.

Masonries were created with macro model technique in a similar approach presented in Cavaleri & Di

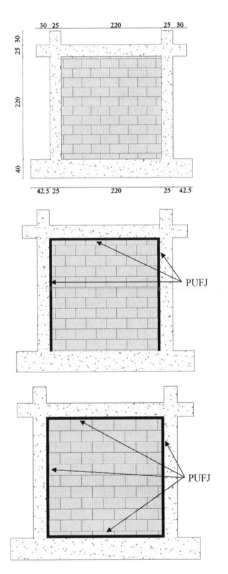

Figure 1. Schematic views of the samples; Frame A (top), Frame B (middle) and Frame C (bottom) [cm].

Table1. Masonry mechanical properties.

f_b [MPa]	E_m [MPa]	G_m [MPa]	ν
15.0	6620	2648	0.25

Table 2. Polyurethane PM mechanical properties.

Material	E_{pm} [MPa]	ν	ε
Polyurethane PM	4.5	0.47	0.5 - 1.5*

* Values dependent on the strain ratio during a tensile test.

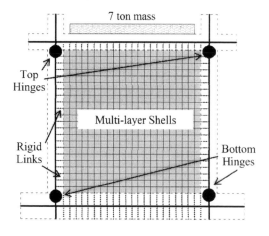

Figure 2. Details of the numerical modeling.

Trapani (2015). The material nonlinearities were provided at the both end of bar elements by plastic hinges which have the length of half cross sectional height. On the other hand, multilayer nonlinear model is used for the shells in order to actualize nonlinearity (Miao et al. 2006). The nonlinear material properties are adopted from the past studies (Cavaleri & Di Trapani 2015, Kisiel 2018) and compression/tension stress-strain curves of them are shown here in Figure 3. The material properties given in Table 1 and Table 2 are reflected on the analysis software in order to calculate the respective shear rigidity of the frame members by using the stress-strain values provided in Figure 3.

Since the global behavior was aimed to be investigated, following simplifications were made; walls were modeled with isotropic mechanical properties,

namely the same stress-strain capacities were assumed in horizontal and vertical directions. Continuum wall model was preferred and therefore different masonry units i.e. brick, mortar and interfaces were modeled as a whole. Nonlinearity of the masonry is reflected by means of using the compressive stress-strain curve. Frictional effects around the boundary zones are not taken into account in the model, since it is previously stated that friction has a marginal effect on overall structural behavior (Fiore et al. 2012).

4 LOADING SCHEME

The frames were exposed to constant vertical static loads during the analyses. Uniformly distributed 7 tons of vertical weight was placed on the top beam, see Figure 2. Seismic loads taken from real earthquake records were used as horizontal loads. For this purpose, two records were chosen with the details given in Table 3. Time-acceleration values of those are also presented in Figure 4.

In addition to the dynamic earthquake loads, quasi-static lateral loads were used for determining the modal frequency shifting of the frames. Ramp loads in terms of "g" gravity forces were used for

Table 3. Details of the seismic records.

Earthquake Name	Date	Depth [km]	Magnitude [M_w]	PGA [g]
El Centro	18.05.1940	16	6.9	0.32
Petrolia	25.04.1992	11	7.2	0.59

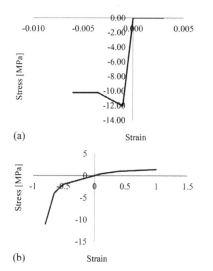

(a)

(b)

Figure 3. Nonlinear material properties: (a) masonry, (b) polyurethane PM.

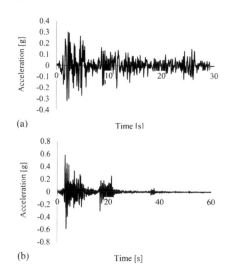

(a)   Time [s]

(b)   Time [s]

Figure 4. Time-acceleration graphs of the earthquake records: (a) El Centro, (b) Petrolia.

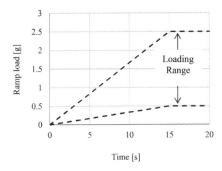

Figure 5. Ramp loads [g].

Table 4. Fundamental mode frequencies.

Frame Type	Initial phase - Fundamental Frequency (In-Plane) - [Hz]		
	No Hinge	Column Bottom Hinge	Column Bottom-Top Hinge
Bare Frame	5.30	N/A	N/A
A	26.04	25.56	25.52
B	12.02	10.94	10.51
C	9.69	8.15	7.69

exciting the frames horizontally. Since the loads were related to inertia forces of the frames, the increments of the ramp loads were executed by means of changing the "g" amplitude. The load range was starting from the 0.5 g and ending at the ultimate value of 2.5 g. Total loading time is 20 seconds for each "g" level, where 15 seconds belong to the incremental ramp period and last 5 seconds shows the steady state part at the maximum force. Loading protocol of the ramp is shown in Figure 5.

5 RESULTS OF THE ANALYSES

5.1 Modal frequency results

Frequency results presented here are taken from the end state of each nonlinear analysis case. Only in-plane modal results are taken into account. Initial phase corresponds to the modal frequency of the frames only under vertical loads. The other modal results are labeled by the name of relevant loadings. According to the results, the frames had the highest frequency values at the initial phases. While the inertial gravity forces gradually increased, the frequencies dropped to lower values due to the nonlinear changes of the structural elements. Among the frames with different connection types, Frame A which represents the stiff connection had the highest frequencies in all different load states. On the other hand bare-frame without the infill wall, which is given in this study as a reference frame, had the lowest frequency results due to the absence of infill walls. For the frame types of B and C, it is seen that PUFJ presence makes the frames more ductile and the modal values are rather closer to the bare-frame in comparison to the stiffly connected Frame A. Although the values are close, Frame B had slightly higher frequency response in comparison to the Frame C. This is expected since the bottom part of Frame B designed as stiffly connected whereas Frame C has the flexible joints across all frame-wall interfaces. Outcomes of the results revealed that artificially placed column hinges also did not make

substantial difference in terms of frequency results for the relevant loadings. This is due to the contribution of the infill walls on structural response. Modal frequency results are given in Table 4.

As mentioned in the previous section, ramp loads were operated on the frames. The loads were steadily increased from the initial phase, which indicates the only vertical load domination and therefore no horizontal forcing. The increment was performed up to 2.5 g. Modal frequency results were recorded at the end of each increment state and given in Figure 6. According to the outcomes, all of the infilled frames experienced slight natural frequency drops. On the other hand, analysis for the bare-frame could not reach to the loads above 1.75 g due to the lack of load carrying capacity. Nonlinear damages were observed beyond that point. Presence of artificial hinges at the end of columns caused lower modal frequencies. For the frame with only bottom column ends hinged (Bot Hinge), the frequency values were slightly higher than for the frame with both ends of columns hinged (Bot-Top Hinge). This difference was relatively more marginal for the Frame A, which has stiff connection.

Using the below well-known Equation 1 for determining the natural frequency of undamped systems, it is seen that frequency is directly proportional to the square root of stiffness.

$$f_n = \frac{1}{2\pi}\sqrt{\frac{k}{m}} \qquad (1)$$

where f_n = natural frequency; k = stiffness; and m = mass.

Deriving from the Equation 1, one can say that flexural stiffness for the structural systems has correlation as shown in Equation 2, where E and I indicate the Young's modulus and moment of inertia, respectively. This equation has good agreement with the majority of civil engineering systems since they have low damping ratios (Bernal et al. 2015) as per the Equation 3, which shows the correlation between f_n: natural frequency and f_d: damped frequency for the under-damped systems. The symbol of ζ corresponds to the damping coefficient.

$$\frac{EI_{current}}{EI_{initial}} = \frac{(f_{current})^2}{(f_{initial})^2} \qquad (2)$$

$$f_d = f_n\sqrt{1 - \zeta^2} \qquad (3)$$

In the light of aforementioned equations, stiffness changes of the frames were calculated by the information taken from Figure 6, and presented in Figure 7. Initial phase was assumed to have 100% of the EI capacity and reduction was made based on the frequency changes due to the ramp loads. According to the results, sudden drop was observed for the Bare Frame and Frame A regardless of the column hinge configuration. The stiffness loss for those frame types were 2.5% and 5% respectively. On the other hand, PUFJ implemented ones could keep the stiffness loss trend more smoothly without such sudden drop until higher loads, namely above 1.5 g. However, larger amount of the stiffness degradation was experienced in both Frame B and Frame C with the relative stiffness values of 35% and 25% respectively as the highest amounts. The models of columns without hinges were the most vulnerable for those frames in terms of stiffness maintaining capacities at the high load stages (> 1.5 g). For lower loads (< 1.5 g), the models with column hinges exhibited higher stiffness. Another fact is, columns with artificial hinges for the Frame C had a slight stiffness increment while increasing the lateral load until 0.5 g. This behavior can be interpreted as a positive influence of the horizontal loading in terms of resisting against to the constant vertical loads that affecting on this type of frames with column hinges. However, for the larger values of ramp loads, the contribution could not be effective and had adverse impact on the frame regardless of the column hinge configuration.

5.2 Displacement results

Displacement results are taken from the maximum values of dynamic earthquake loadings. The tendency of frame structural responses had similar outcomes as of the modal frequency ones in terms of the dynamic behaviors. Stiffly connected Frame A stands as a robust structure and therefore had relatively lower displacement values. Frame B and Frame C reached to the higher displacements because of the PUFJ. Displacements were low for all type of frames since the neither earthquake record could give a nonlinear damage, therefore the frames stayed in elastic range. Even for the bare-frame no inelastic response was observed and the

horizontal drift ratio at the maximum displacement was less than 0.30 %. It should be noted that dynamic time history analyses are directly related to the mass and inertia of the structural systems. Therefore, additional loadings might change the results drastically. For this study, it can be concluded that the combination of total mass and structure withstand safely against the given earthquake loads.

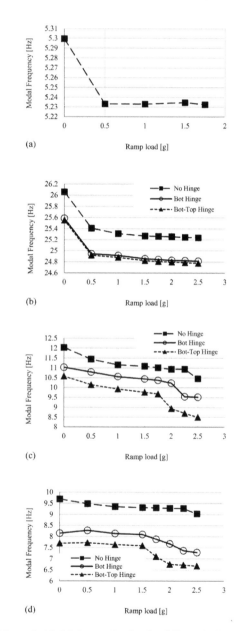

Figure 6. Modal frequencies for the different load states: (a) Bare Frame, (b) Frame A, (c) Frame B and (d) Frame C.

550

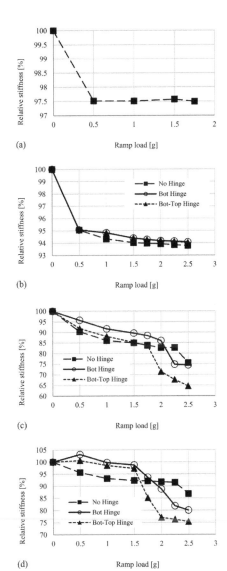

(a)

(b)

(c)

(d)

Figure 7. Stiffness changes for the different load states: (a) Bare Frame, (b) Frame A, (c) Frame B and (d) Frame C.

Despite the slight differences, assumed column hinges also could not increase the displacement values. Frame A had almost zero displacement change dismissing the fact that decimals were unimportantly different, whereas Frame B and Frame C had very low displacement increments while placing the artificial hinges on columns. Infill wall contribution is the main reason for this which played a significant role on dynamic response. Maximum displacements for the reference point given in Figure 8 is presented below, see Table 5.

Table 5. Maximum displacement values.

| Frame Type | Earthquake | Max. Displacement (In-Plane) - [mm] | | |
		No Hinge	Column Bottom Hinge	Column Bottom-Top Hinge
Bare Frame	El Centro	3.07	N/A	N/A
	Petrolia	5.59	N/A	N/A
A	El Centro	0.12	0.12	0.12
	Petrolia	0.24	0.24	0.24
B	El Centro	0.56	0.61	0.69
	Petrolia	1.08	1.20	1.32
C	El Centro	0.92	1.36	1.57
	Petrolia	1.74	2.41	2.76

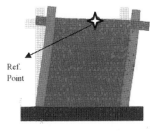

Ref.
Point

Figure 8. Typical fundamental mode shape and reference point.

6 CONCLUSIONS

In this paper, effectiveness of an innovative method for protecting the RC frame buildings with infill walls is investigated. The proposed solution of using PUFJ is compared with the traditional method, namely stiffly connected frame-masonry interfaces. The results are given below:

- Modal frequency results indicate that Frame A with stiff connection has the highest values. This values are approximately the twice and three times as much as of the frames B and C, respectively. PUFJ leads to ductile behavior on structure and therefore both Frame B and Frame C have substantially lower frequencies than the Frame A.
- The frames were exposed to gradually increasing ramp loads that reached up to 2.5 g in order to determine frequency shift while the nonlinear damages occurring. The results are interpreted also in terms of stiffness degradation. Despite the decrease of frequencies, none of the infilled

frames passed to the inelastic stage. On the other hand, bare-frame which is the reference frame here, could not withstand such high lateral disturbance and the stability of the frame was lost above the ramp load of 1.75 g.

- Maximum displacement results were insignificantly low for the frames. Apparently, seismic records were not able to create a destructive effect on the frames. The bare-frame with no infill wall was also stayed in elastic range. Even though it was not a concern of this paper, masonry damage status was also checked roughly by means of stress concentration in a similar way previously studied by Akyildiz et al. (2019). The results indicate that no inelastic damage occurred on the infill walls.

- Column hinges were configured either only at the bottom or both at the bottom and top ends in a hypothetical way. The assumption was made in order to investigate the inconvenient scenarios once the columns lost their load carrying capacities. It is seen there was no major change in the results of frequency and displacement values in comparison to the non-hinged frames. It reveals the contribution of strong infill masonry in terms of horizontal load carrying capacity.

REFERENCES

Akhoundi, F., Vasconcelos, G., Lourenco P.B., Palha C., Martins, A. 2015. Out-of-plane behavior of masonry infill walls. *7th International Conference on Seismology & Earthquake Engineering, Tehran, 18-21 May 2015*.

Akyildiz, A.T, Kowalska-Koczwara A., Kwiecien, A. 2019. Stress distribution in masonry infills connected with stiff and flexible interface, *Journal of Measurements in Engineering* 7(1): 40–46.

Bernal, D., Döhler, M., Kojidi, SM., Kwan, K., Liu, Y. 2015. First mode damping ratios for buildings, *Earthquake Spectra* 31(1): 367–381.

Cavaleri, L., Di Trapani, F. 2015. Prediction of the additional shear action on frame members due to infills, *Bulletin of Earthquake Engineering* 13(5):1425–1454.

Computers & Structures Inc. SAP2000 v15.2.1. 2005. *CSI Analysis Reference Manual*, Berkeley, CA.

Dafnis, A., Kolsch, H., Reimeders, H.G. 2002. Arching in masonry walls subjected to earthquake motions, *Journal of Structural Engineering* 128(2): 153–159.

European Committee for Standardization. 2004. *Eurocode 2: Design of concrete structures - Part 1-1: General rules and rules for buildings*, Brussels.

European Committee for Standardization. 2005. *Eurocode 6: Design of masonry structures - Part 1-1: General rules for reinforced and unreinforced masonry structures*, Brussels.

Fiore, A., Netti, A., Monaco, P. 2012. The influence of masonry infill on the seismic behaviour of RC frame buildings. *Engineering Structures* 44: 133–145.

Gams, M., Kwiecień, A., Korelc, J., Rousakis T., Viskovic, A. 2017. Modelling of deformable polymer to be used for joints between infill masonry walls and R.C. frames. *Procedia Engineering* 193: 455–461.

Hermanns, L., Fraile, A., Alarcon, E., Alvarez, R. 2014. Performance of buildings with masonry infill walls during the 2011 Lorca earthquake. *Bull Earthquake Eng* 12(5): 1977–1997.

Kisiel, P. 2015. The stiffness and bearing capacity of polymer flexible joint under shear load, *Procedia Engineering* 108: 496–503.

Kisiel, P. 2018. *Model approach for polymer flexible joints in precast elements joints of concrete pavements*. Doctoral Dissertation. Cracow: CUT.

Kwiecień, A. 2012. Polymer flexible joints in masonry and concrete structures, *Monography No. 414, Wyd. Politechniki Krakowskiej, Seria Inżynieria Lądowa, Kraków*, (in Polish).

Kwiecień, A., Gams, M., Rousakis, T., Viskovic, A., Korelc, J. 2017a. Validation of a new hyperviscoelastic model for deformable polymers used for joints between rc frames and masonry infills, *Engineering Transactions* 65(1): 113–121.

Kwiecień, A., Gams, M., Viskovic, A., Kisiel, P., Korelc, J., Rousakis, T. 2017b. Use of polymer flexible joint between RC frames and masonry infills for improved seismic performance, Zurich: SMAR.

Miao, Z.W., Lu, X.Z., Jiang, J.J., Ye, L.P. 2006. Nonlinear FE model for RC shear walls based on multi-layer shell element and microplane constitutive model. *In Tsinghua University Press & Springer-Verlag. Computational Methods in Engineering and Science, 21–23 August 2006, Sanya, Hainan, China*.

Muthu Kumar, S. & Satyanarayanan, K.S. 2018. Study the effect of elastic materials as interface medium used in infilled frames, *Materials Today: Proceedings* 5: 8986–8995.

Pasca, M. & Liberatore, L. 2015. Predicting models for the evaluation of out-of-plane ultimate load carrying capacity of masonry infill walls, *Earthquake Resistant Engineering Structures X*. Southampton: WIT Press.

Preti, M., Migliorati, L,. Giuriani, E. 2015. Experimental testing of engineered masonry infill walls for post-earthquake structural damage control. *Bull Earthquake Eng* 13: 2029–2049.

Vailati, M. & Monti, G. Earthquake-Resistant and Thermo-Insulating Infill Panel with Recycled-Plastic Joints. 2016. *In: D'Amico S. (eds) Earthquakes and Their Impact on Society, Springer Natural Hazards*, pp. 417–432.

Viskovic, A., Zuccarino, L., Kwiecień, A., Zając, B., Gams M. 2017. Quick seismic protection of weak masonry infilling in filled framed structures using flexible joints, *Key Engineering Materials* 747: 628–637.

Energy, moisture & thermal performance

Brick and Block Masonry - From Historical to Sustainable Masonry –
Kubica, Kwiecień & Bednarz (eds)
© 2020 Taylor & Francis Group, London, ISBN 978-0-367-56586-2

State of art: Evaluating thermal resistance of masonry walls

M. Ismaiel & Y. Chen
University of Alberta, Edmonton, Canada

ABSTRACT: Increasing the thermal resistance of masonry exterior walls can reduce the buildings' energy consumption. Different challenges have been confronted with implementing the relevant changes to the design of these masonry walls. One of them is the thermal bridging, which occurs typically in places where highly conductive structural components penetrate insulating materials. Another challenge is to estimate the effective thermal resistance with sufficient accuracy. A survey of the literature conducted here identifies the major technical and practical challenges, the corresponding advancements, and the important influencing factors on the thermal performance of masonry walls. It will summarize the state of the thermal performance design and evaluation of masonry walls. The paper will also focus on how overall R-values are obtained using numerical calculations and computer simulation, and experimental measurements.

1 INTRODUCTION

The demand for energy-efficient housing is increasing in Canada. In the recent decade, many regulations have been established to improve the thermal performance of building envelopes. Two of the main aspects that need further consideration when improving the design is; the thermal properties of the materials used and thermal bridging.

Large thermal bridging takes place at the structural floor or slabs and partition penetrations through the insulation plane. A more serious thermal bridge occurs when the design has the floor slab extending beyond the exterior wall to form the balcony slab as reported by the Canadian Concrete Masonry Producers Association (C.C.M.P., 2013). Thermal bridging should be minimized in the design and calculated carefully. This review summarizes some of the challenges as well as techniques used to solve and estimate the thermal resistance of different types of masonry walls.

2 COMPONENTS OF MASONRY WALLS

The main components of concrete masonry walls are the concrete blocks, mortar, grout, reinforcement (vertical, horizontal and joint), insulation boards, shelf angles, and connectors. In general, the masonry concrete cavity walls consist of inner wythe made of concrete block units and the insulation, the type and thickness of which are according to the required thermal resistance. Insulation should be placed in the wall cavity along with an air space. It should completely cover the face of structural elements, like columns and beams, to avoid thermal bridges and to protect the structure from temperature variations.

Furthermore, it consists of an outer wythe of brick veneer to protect the insulation and concrete blocks. Also, the presence of the brick veneer improves the thermal resistance of the wall. This section addresses each component and its effect on the overall thermal performance of masonry walls according to previous studies.

2.1 Concrete blocks and veneer clay bricks

The unit shape, size, aggregate size, moisture content, and concrete density of concrete blocks affect the thermal resistance of the concrete masonry units to a great extent. By optimizing the parameters and the shape of the blocks, the thermal resistance value could be significantly different. Many investigations were performed to optimize the design and obtain an accurate thermal resistance value of concrete blocks. Fogiatto et al. examined different cavity configurations of concrete blocks using CFD simulations. The results of the simulation were compared with the calculations mentioned in ACI (ACI committee, 2002). The study concluded that large cavities provide higher transmittance values and that radiation effect may play an important role in the overall heat transfer through concrete hollow blocks (Fogiatto, et al. 2016).

Many studies discussed the effect of the arrangement of the insulation inside the blocks as well as the shape of the blocks and the size of the webs on the thermal resistance of the whole wall. Szoko et al. studied different shapes for masonry units as shown in Figure 1 and found that the reduced web area provides a smaller cross-sectional area for heat flow through the wall (Szoko, 2010). Hence, some manufacturers have developed concrete masonry units with two cross-webs, instead of three.

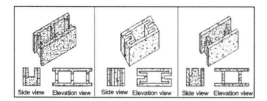

Figure 1. Concrete masonry units specially designed to accommodate insulation.

Detailed R-value simulations performed on several types of commercially available CMU showed that the arrangement of thermal insulation influences the overall thermal resistance of concrete and masonry wall systems (Kosny, 2011). Calculations were done to determine the thermal efficiency of the insulation inserts. The study concluded that although insulation inserts can increase a CMUs thermal resistance, thermal bridging through the solid webbing reduces the effectiveness of the insulation. The method of estimating Thermal efficiency (TE) value is based on the R-value comparison of insulated and uninsulated masonry units — each having the same face area. TE is expressed as shown in equation (1);

$$TE = \left(\frac{Ri - Ru}{Re}\right) \times 100\% \qquad (1)$$

Where: R_i= R-value of insulated CMU; R_u= R-value of uninsulated CMU and R_e=R-value of insulation material alone.

2.2 Grout and mortar

The grout fills the hollow cores in masonry concrete units to bond the vertical and the horizontal steel bars in masonry walls for structural purposes. In an investigation into the effect of the grout on the thermal performance, Kosny et al. observed that the grout effect decreases when the concrete blocks thermal resistivity increases (Kosny, 2011) The study shows that the cut-web units are less sensitive to the grout effect (the grout effect varies from 3% to 7%). In normal-density two-core concrete units, the reduction of the R-value caused by the grout is 10%; in lightweight two-core concrete units, the grout effect was 5%. For uninsulated multicore CMUs, the grout effect remains in the range of 6% to 12%. The R-value of insulated multicore units is very sensitive to the local thermal bridges caused by cores filled with grout. The reduction of the R-value for these units may reach 30% for normal-density concretes and 25% for lightweight concretes.

The mortar occupies only a small proportion of the total wall area in concrete masonry construction (approximately 7 %) (NCMA, 2004), it has a considerable influence on the wall performance. Mortar has many crucial functions: it bonds units together into the structural assembly; seals joints against penetration by air and moisture; and bonds to joint reinforcement, ties and anchors so that all elements perform one an assembly. An investigation into the effect of the mortar on the thermal performance of the masonry concrete walls showed that the R-value reduction can exceed 12% for two-core units and the mortar effect increases when the thermal resistivity of concrete block increases (Kosny, 2011) They recommended using less conductive mortars or decreasing the area of mortar joints to decrease the amount of heat loss through the mortar. Recently, interlocking concrete is replacing side mortar as means to connect adjacent units.

2.3 Ties and shelf angles

The metal ties connect multiple wythes of masonry and hold the veneer in place. The ties have evolved in shapes, sizes, materials, and configurations to fit both structural and thermal purposes. A thermal bridge occurs at the locations where ties breach the insulation layer, which leads to a loss in the thermal performance. Several commercial tie types, made with different materials, have been produced in order to lower the thermal bridging while maintaining structural functions. These include unit ties, adjustable ties, and re-anchoring systems.

The adjustable tie systems permitted the use of face brick whose bed joints did not align vertically with interior masonry wythes. Nowadays, the Slotted Tie, which transfers tensile and compressive lateral loads, is commonly used. The L-shaped plate is fastened to the facing surface of the structural backing as shown in Figure 2. After set-up, the V-Tie provides the veneer wall with a constant connection to the structure without any risk of veneer misalignment. The function of holes through the body of the Slotted L-Plate minimizes thermal conductivity through the tie system. Another tie type is the block shear connector (FERRO, 2015). This type of tie eliminates the need for the fasteners which helps in decreasing the thermal bridging effect. At the same time, it

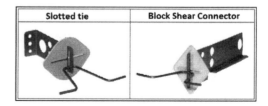

Figure 2. Slotted ties and block shear connector.

provides uniform behavior of the cavity walls. It also helps the structure to sustain more loads, with thinner walls and less reinforcement. So, it has many benefits thermally and structurally.

The tie material and tie design can have a significant impact on the effective R-value of masonry veneer walls. The effective reduction ranges from 3% to 25% depending on the thickness of the insulation and backup wall. Wilson et al. studied two types of ties: the Stainless steel tie with holes, and Galvanized tie without holes (Wilson & Higgins, 2019). They concluded that stainless steel performs better than galvanized steel, with exterior insulation reductions in the order of 3% to 9% for stainless steel over concrete/steel wall backup versus 8% to 25% for galvanized steel.

Another source of thermal bridging in concrete masonry walls is the shelf angles (Mark Lawton, 2014). A thermal bridging problem usually occurs when these shelf angles are fixed to the slabs or beams using the fasteners. Botticelli investigated masonry veneers and observed that the actual R-values observed using a thermal camera are less than the theoretical values by 25-60% (Botticelli, 2015). Also, attachments form substantial thermal bridges that can decrease the overall thermal performance. Other types of shelf angles were introduced to the market to improve thermal performance and decrease the thermal bridging effect. The analysis was performed by using different shelf angels (Wilson & Higgins, 2019) as shown in Figure 3.

The amount of insulation displaced by the knife plates, HSS tubes, and the FAST System is similar, and as such the reduction factor of the exterior insulation R-value by 14% to 16%. Table 1 presents the results of the analysis of the influence of the shelf angle on the effective R-value of the assembly with four types of shelf angles. The study concluded that the effective R-values for wall assemblies with these details are all around or above R-16 (R--16=2.816 m^{2}K/W) (Michael Wilson & James Higgins, 2019).

3 CURRENT STANDARDS FOR EVALUATING THERMAL RESISTANCE

Current Canadian building codes are influenced by energy considerations. Recently, building codes integrate many of the energy and thermal performance requirements from earlier standards. The building industries have adopted and are still developing new building techniques to comply with these continuously changing code requirements. There are many paths which a designer can use to achieve the required energy performance: prescriptive path, trade-off path, and whole-building energy modeling path. Table 2 summarizes the maximum allowed U-value as a function of climate zone for common example energy codes (ASHRAE, 2019) and (NECB, 2017) and for the perspective path (Straube, 2017)

Restrictions to window area are considered in the window-to-wall ratio (WWR) because window areas affect the reduction of lighting energy or heating losses with useful solar gains (Carmody et al. 2004). So, the (WWR) is limited to 40% in the prescriptive compliance method. The NECB specifies a maximum fenestration-and-door-to-wall Ratio (FDWR) equation that relates to the Heating Degree Days (18°C), starting at 40% and dropping to 20% for Zone 8.

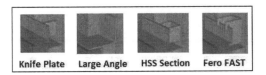

Knife Plate   Large Angle   HSS Section   Fero FAST

Figure 3. Different types of shelf angles, knife plate, Ferro FAST, HSS section, and large angle.

Table 1. Summary of R-values and thermal degradation for different shelf angles.

	HSS section	Ferro FAST	Knife Plate	Large angle
Effective R-value	RSI-2.84	RSI-2.87	RSI-2.89	RSI-1.87
Reduction %	15.7%	14.6%	14%	43%

Table 2. Prescriptive enclosure wall U-value for different energy codes.

U-values (W/m^{2}K)

		ASHRAE 90.1-2019		NECB-2017
Climate Zone	HDD* (18C)	Non-Residential	Residential	All
		mass	mass	any
4	< 3000	0.104	0.09	0.315
5	3000-4000	0.09	0.080	0.278
6	4000-5000	0.08	0.071_	0.247
7/7A	5000-6000	0.071	0.071	0.210
7/7B	6000-7000	0.071	0.071	0.210
8	> 7000	0.048	0.048	0.183

* HDD: Heating Degree Days

4 METHODS USED TO DECREASE THE THERMAL BRIDGING EFFECT

When exterior insulated sheathings are installed, they reduce thermal bridging in assemblies and decrease overall energy consumption. Love et. al. investigated common North American masonry veneer wall systems that are dependent on shelf angles and a grid of tie-backs that structurally stabilize the assembly (Love Andre, 2011). With the infrared images, the researchers observed that the masonry veneers performed at a 25–60% decrease in R-value when compared to theoretical calculations. Removing the shelf angles was suggested as a possible solution to this thermal problem as the masonry veneers could be supported on the foundation, albeit for limited heights. Unfortunately, the continuous shelf angles are required to support heights over two stories, and supporting every story is common to minimize deflection joints. These steel shelf angles accounted for an approximate 35% decrease in the R-value. Love et al. recommends making the entire angle out of stainless steel as the material change could reduce the performance impact of the shelf angle from 35% down to 29%. Metal ties were also discussed in this investigation with typical spacing between 400 and 600 mm on center, horizontally and vertically. The ties can contribute up to a 15% decrease in the thermal performance. Because spacing, material conductance, and type of tie all impact the R-value for masonry walls, three types of ties were studied at both 400 and 600 mm. a screw-on tie, a barrel tie, and a thermally broken tie. The choice of steel or stainless steel proved to have the biggest impact on performance, with an average of 6% improvement in the R-values, whereas the larger spacing of the ties and the choice of tie type both showed an average of a 4% improvement in thermal performance. Stainless steel ties spaced 600 mm on center, which have a minimum diameter of material penetrating the insulation, were shown to have a negligible impact on the thermal performance, decreasing the R-value by 2%.

5 PRACTICAL AND CONSTRUCTION CHALLENGES FACING THERMAL PERFORMANCE OF MASONRY WALLS

5.1 *Technical challenges*

Many technical challenges are considered when improving the thermal performance of masonry. There are also some construction limitations for the structural behavior of the elements as the structural capacity and the strength of the block itself may contradict the thermal performance. To resolve these challenges new elements and different shapes and sizes of concrete blocks and ties, intended to lower

the thermal bridging values and to sustain its structural and thermal performance, were introduced.

5.2 *Practical and construction challenges*

Construction of the cavity wall and the insulation boards is one of the main practical challenges discussed by the European Insulation Manufacturers Association (Eurima, 2008). The study states that the insulation boards should be installed in intimate contact with masonry to avoid air circulation degrading the thermal resistance. The misalignments are mainly due to the excess of mortar between blocks, unclean wall surface and other construction issues. Any misalignment of the insulation boards is a potential source of heat loss. Gaps create air movement that degrades thermal resistance of the insulation. The external insulations bonded to the tapered edge, vapor check plasterboard must be continuously sealed at all perimeter edges to eliminate air infiltration between the panel and the external wall. To avoid and control this problem, there should be high construction quality control on the site and highly qualified experienced technicians in the construction field.

6 ESTIMATION OF THE MASONRY WALLS THERMAL PERFORMANCE

6.1 *The thermal resistance of concrete masonry units*

More than 100 concrete masonry unit walls were experimentally tested with two commonly used calculation techniques to acceptable results (Valore, 1980). These methods are the parallel path method and the series-parallel method. The series-parallel method is also known as isothermal planes. The parallel path is applicable for calculating the thermal resistance of concrete masonry units only if the units are not insulated. For this method, the heat flows are assumed to be transferred through the CMU in straight parallel lines. If the CMU are hollowed, the heat flows depend on the ratio between the web area and the core area. Based on this ratio the parallel method assumes that the heat flow passes through the core and the webs by the same ratio. This method is reasonable and accurate only in case of hollow and not insulated CMU. The series-parallel method or isothermal planes is used to calculate the thermal resistance for both insulated and uninsulated CMU. For this method, the heat flow follows the path of the least resistance. The methods takes into consideration the lateral heat flows in the face shells and heat which passes the area of high thermal resistance as air space or insulated cores. Therefore, the CMU webs are considered to be thermal bridging elements.

An information series from the national authority on concrete masonry technology (NCMA, 2013) discussed the calculation method of the R-value by using the series-parallel/isothermal planes. The calculation method, recognized and discussed in references (ACI committee, 2002; ASHRAE, 2010), considers the thermal bridging that occurs through the webs of concrete masonry units. The series-parallel calculation method is recommended for estimating R-values of concrete masonry walls. This calculation divides the block as a series of thermal layers. The face shells form continuous outer layers, which are in series with the layer containing webs and cores. The total R-value is the sum of the R-values of each layer. The equation is described as follows;

$$R_T = R_i + \frac{Rf \times Rm}{af \times Rm + am \times Rf} + \frac{Rw \times Rc}{ac \times Rw + aw \times Rc} + Ra + Rv + Ro \quad (2)$$

Where: a_c = fractional core area; a_f = fractional face shell area; a_m = fractional mortar joint area; a_w = fractional web area ; R_a = thermal resistance of cavity; R_c = thermal resistance of cores; R_f = thermal resistance of both face shells, r_cx $(2t_{fs})$; Ri = thermal resistance of inside air surface film; R_m = thermal resistance of mortar joint, r_m x $(2t_{fs})$; R_o = thermal resistance of outside air surface film ; R_T = total thermal resistance of wall; Rv = thermal resistance of veneer; Rw = thermal resistance of concrete webs, r_c x t_w; r_c = thermal resistivity of concrete ; r_m = thermal resistivity of mortar; t_{fs} = face shell thickness; t_w = length of webs.

The tables provided in (NCMA, 2013) show R- values for the grouted cores; they consider grouted walls with different materials as Perlite, Vermiculite, Polyurethane foamed-in-place, hollow cores and solid. Also, different concrete block densities were considered. The U-factor of the wall is calculated from the area-weighted average of the U-factors of the grouted area and un-grouted areas as follows:

$$U = \left(a_{gr} \times U_{gr}\right) + \left(a_{ungr} \times U_{ungr}\right) \quad (3)$$

Where: a_{gr} = fractional grouted area of wall; a_{ungr} = fractional un-grouted area of wall; R = total thermal resistance of wall, $(m^2.K/W)$; U = thermal transmittance of a partially grouted wall, $(W/m^2.K)$; U_{gr} = thermal transmittance of fully grouted wall, $(W/m^2.K)$; U_{ungr} = thermal transmittance of un-grouted wall, $(W/m^2.K)$.

Clear field assembly is an assembly with uniformly distributed thermal bridges, which are not practical to consider on an individual basis for U-value calculations, such as walls with brick ties and structural studs. The thermal performance of clear field assemblies can be determined through the calculation provided in (ENISO, 2017) provides an approximate procedure for assessing the effect of mechanical fasteners, which can be used if fasteners are not accounted for by other methods. When an insulation layer is penetrated by mechanical fasteners, such as wall ties between masonry wythes, roof fasteners or fasteners in composite panel systems, the correction to the thermal transmittance is given by

$$\Delta U_f = \alpha \times \frac{\lambda_f \times A_f \times n_f}{d_1} \times \left(\frac{R_1}{R_{tot}}\right)^2 \quad (4)$$

Where the coefficient $\alpha = 0.8$ if the fastener fully penetrates the insulation layer. In these expressions λ_f is the thermal conductivity of the fastener W/m·K; n_f is the number of fasteners per m²; A_f is the cross-sectional area of one fastener in m²; d_1 is the length of the fastener that penetrates the insulation layer in m; R_1 is the thermal resistance of the insulation layer penetrated by the fasteners in m²·K/W; R_{tot} is the total thermal resistance of the component ignoring any thermal bridging.

6.2 Thermal performance of intersection details area-weighted

The area-weighted technique is commonly used to calculate the effective R-values of the details at the intersection of building envelope and structural components. These elements at the intersections disturb the uniformity of a clear field assembly and the additional heat loss should be considered. Generally, this method is done by weighting the heat flow through the materials by the area they occupy. There is an assumption that the heat flow paths through an intersection detail are one-dimensional and parallel. It is calculated by creating an "area-weighted average" R-value. This represents the proportion of area occupied by each element of the wall, multiplied by the element's R-value concerning the total area.

Unfortunately, this is not accurate and we cannot neglect the highly conductive building components which can conduct lateral heat flows to other components in 3D that are not considered for in basic parallel flow assumptions. In conclusion, this method is more reliable and recommended when analyzing structures with low thermal conductive structural members. But it is difficult and not accurate for conductive materials as concrete or steel (B.C. Hydro, 2019).

6.3 Thermal performance of intersections using linear transmittance

Linear thermal bridges typically occur at intersection details that have a continuous uniform connection with the assembly as shelf angles, slab edges, and balconies. Using the linear transmittance approach, the heat flow through the intersection detail assembly is investigated in two cases: with and without the thermal bridge. The difference in heat flow is related to the detail as heat flow per a linear length. The building envelope thermal bridging guide (B.C. Hydro, 2019) provides an example of an exterior insulated steel stud wall with a cantilevered balcony slab. This assembly was first studied as a whole unit with the slab intersection. Then it was studied as clear fields assembled with no interruptions or intersections. The difference in the heat flow between the two assemblies was the effect caused by the presence of the intersection (slab and balcony) which interrupts the thermal insulation. By dividing the difference heat flow value obtained from both assemblies over the assembly width, which represents the linear length of the slab, the linear transmittance of the slab is obtained, which is a heat flow per linear length. Another concept called point transmittance, which is similar to the linear transmittance approach, does not depend on area or length. This includes building components such as beam penetrations and intersections between linear details. By using these two approaches along with the clear field transmittance the overall heat flow for any size of wall or roof can be determined.

6.4 Overall thermal performance

The thermal performance values of each of the envelope elements are involved in computing the overall thermal transmittance for the building envelope. The U-values used in all calculations are categorized into three groups: Clear field transmittance; represented by a U-value denoted as the clear field (U_o), Linear transmittance; represented by psi (Ψ) and Point transmittance; represented by chi (χ) (B.C. Hydro, 2019).

The overall U-value for any building envelope is addition and multiplication process it is simply calculated as follows;

$$U_T = \frac{\sum (\psi \times L) + \sum \chi}{A_{Total}} + U_o \qquad (5)$$

Where: U_T = total effective assembly thermal transmittance (W/m^{2}K); Uo = clear field thermal transmittance (W/m^{2}K); A_{total} = the total opaque wall area (m^2); ψ = heat flow from linear thermal bridge (W/m K); L = length of linear thermal bridge, i.e. slab width (m); χ = heat flow from point thermal bridge (W/K). From the equation above, it is concluded that by finding the heat flows separately in each component, it can be evaluated to find their relative contribution to the overall heat flow. The overall U-value for a building can be determined as long as the thermal performance values for the clear field (U_o), linear (Ψ) and point (χ) transmittances are known as well as the dimensions and the quantities of each element.

7 EXPERIMENTAL ESTIMATION FOR R-VALUE

Accurate As the energy codes became stricter, the accurate predictions of the thermal performance of different elements have become increasingly required in addition to the use of computer modeling. The thermal performance of any building envelope is highly dependent on the amount of insulation installed (Schumacher, Straube, Ober, & Grin, 2013). However, it can also be influenced by many other factors such as air leakage, thermal bridging, moisture content, operating conditions, and installation defects. These factors are not accurately considered when insulation R-value is used as the main measuring term for element thermal performance. Therefore, the experimental testing is highly required to verify the modeling results and to assure than the properties of the assemblies agrees to an acceptable extent with the code requirements.

The Hot box apparatus is the most common experimental test to get the thermal performance of assemblies with larger dimensions. It can simulate the three types of heat transfer: conduction, convection, and radiation and it can be adjusted with the required environmental conditions. The temperature, humidity, and pressure could be controlled to provide a realistic environmental condition for the testing procedures. The minimum requirements for accurate testing of the thermal performance of building assemblies under steady-state conditions are outlined in the specifications (ASTM, 2013). This test can be used for homogenous and non-homogeneous specimens.

Hot boxes are designed to measure the heat transfer through a specimen when the environmental conditions on both sides of the specimen are held constant in a steady-state condition. The equation given to describe the heat balance of the apparatus is given by the ASTMC1363 as follows

$$Qaux + Qmw + Qft = Q = \frac{A \times \Delta t}{R} \qquad (6)$$

Where: Q_{aux} = net heat flow due to the fan, heater, cooling coil, W, = Qc + Qh + Qf, W; Qc = net heat removed by the cooling coil, W; Qh = net heat added by the heaters, W; Qf = net heat added by the fans, W; Q_{mw} = metering box wall loss, W; Q_{fl} = flanking loss, W; Q = heat flow through the specimen, W; R = thermal resistance of the specimen, m^{2}K/W; A = metered area of heat flow, m^2, and Δt = surface temperature difference across the specimen, K. Note: The flanking loss is defined to be the quantity of heat, which flows between the metering and climatic chambers through the surround panel or test frame, which holds the specimen.

Many other regulations discuss the instrumentation used in field measurements as calculations for using heat flux transducers (HFTs) and temperature transducers (TTs) for in-situ dynamic or steady-state thermal behavior of walls (ASTM, 2013). All descriptions of the apparatus and the test procedure have been described. The calculations of heat flux q, is given according to the following equation:

$$q = S(T_i) \times E_i \qquad (7)$$

Where: E_i = the averaged voltage reading, of the ith measurement, and T_i = the corresponding temperature of the ith measurement. The conversion factor (S) is a function of the HFT design and the thermal environment surrounding the HFT.

Many recommendations were stated for the location and the placement of the HFTs; it was suggested that placement of a small HFT over a fastener can help represent the contribution of fasteners to heat transfer. Experience indicates, that the face of a concrete masonry unit distributes heat flux sufficiently that HFT placement is insensitive to the location on the block.

The thermal resistance of a building envelope subsection, from measured in-situ temperature and heat flux data, can be calculated using the mathematical procedures outlined in the standards (ASTM C1046 - 95, 2013). The procedures are the summation technique as shown in equations (8):

$$R_e = \frac{\sum\limits_{k-1}^{M} \Delta T_{sk}}{\sum\limits_{k-1}^{M} q_k} \qquad (8)$$

Where: M = number of values of ΔT and q in the source data; k = counter for summation of time-series data and s = surface. This calculation procedure employs an accumulation of data on heat flux and differences in surface temperatures over time. It requires a significant difference in temperatures and constant temperature on one side for rapid convergence.

8 R-VALUE SIMULATION ESTIMATE

The finite element modeling is used in accurate analysis of thermal behavior of elements (Zienkiewicz, Taylor, & Zhu, 2005). Real tests to monitor the thermal behavior of elements require full size concrete walls and cavity walls and therefore can be expensive and sometimes not impractical. When compared, the three-dimensional results were comparable to the experimental measurements, while the two-dimensional results consistently overestimate the results (Yao & Qu, 1999). As usual, the size effect of the elements has a significant impact on the overall results. There are many thermal analysis finite element programs introduced to simulate thermal problems. The finite element programs have already proved its accuracy and reliability in thermal performance simulation. Del Coz Díaz et al. applied a FEM to the non-linear heat transfer analysis of light concrete hollow brick walls. The conduction and convection were taken into account. According to their study, the

difference between numerical and experimental results has been lower than 2.6% (Del Coz Diaz, J J, Nieto, Rodriguez, Martínez-Luengas, & Biempica, 2006). Huygen et al. used a hot box apparatus to determine steady-state R-value measurements of different wall assemblies; one in particular was the brick veneer cavity wall. In conjunction with 3D finite element simulations in ANSYS, the study was carried out under steady-state conditions to provide numerical verification of the experimental results (Huygen N.C. & John P., 2019). It was concluded that the excellent agreement between model and experiment will allow future simulations of wall systems without the necessity of running length hot box tests.

9 RESEARCH GAP AND FUTURE RECOMMENDATIONS

Many studies investigated the thermal efficiency of concrete masonry unit blocks, introduced insulation patterns and represented the thermal efficiency using different formulas and techniques. There have been a few attempts at studying the thermal performance of the complete wall including all its components: air gaps, ties and shelve angles as well as, the overall thermal behavior of the building envelope. The significance of such work would help provide guidelines to the construction industry. In addition, it would help designers predict the total R-value of masonry walls with different conditions and analyze the effect of the interruptions of the intersections of the slabs or balconies on the overall thermal performance. Providing easy formulas to the designer to compute the R-values will help in having a reliable estimation of energy needs for the buildings and a first step for improving the thermal envelope or calculating the HAVC equipment required without any complicated experimental investigations.

REFERENCES

ACI committee. 2002. *Guide to thermal properties of concrete and masonry systems*. Michigan: American Concrete Institute.

ASHRAE. 2013. *ASHRAE.90.1 Energy standard for buildings except low-rise residential buildings*. New York: ASHRAE.

ASHRAE. 2019. *ASHRAE 90.1- energy standard for buildings except low-rise residential buildings*. New York: ASHRAE.

ASTM Standard C1046 - 95. 2013. *Standard practice for in-situ measurement of heat flux and temperature on building envelope components*. West Conshohocken: ASTM International.

ASTM Standard C1046 - 95. 2013. *Standard practice for in-situ measurement of heat flux and temperature on building envelope components*. West Conshohocken: ASTM International.

Hydro, B.C. 2019. *https://www.bchydro.com/powersmart/business/programs/new-construction.html*.

Botticelli, M. 2015. Thermal bridging research masonry veneerwalls. *https://www.payette.com/researchinnovation/thermal-bridging-research-masonry-veneer-walls/*

C.C.M.P. 2013. *Thermal properties & design details.* Ontario; National Concrete Masonry Association.

Carmody, J., Selkowitz, S., Lee, E., Arasteh, D., & Willmert, T. 2004. *Window system for high-performance buildings.* New York; W. W. Norton & Company.

Del, C., Nieto, P.G., Rodriguez, A.M., Martínez-Luengas, A., & Biempica, C. 2006. Non-linear thermal analysis of light concrete hollow brick walls by the finite element method and experimental validation. *Applied Thermal Engineering* 26: 777–786.

ENISO6946. 2017. *Building components and building elements. Thermal Resistance and Thermal Transmittance. Calculation Methods. (Polish Version PN-EN ISO 6946: 2017).* Ireland: National Standards Authority of Ireland.

Eurima. 2008. *https://Www.eurima.org/energy-efficiency-in-buildings.html.*

FERRO. 2015. *Block shear connector.* Edmonton: Fero Corporation.

Fogiatto, M., Santos, G., & Mendes, N. 2016. Thermal transmittance evaluation of concrete hollow blocks. *12th International Conference on Heat Transfer, Fluid Mechanics and Thermodynamic. Costa del Sol, 11-13 July 2016.* Spain.

Kosny, J. 2011. Thermal performance of concrete masonry unit wall systems. *Thermal Envelopes VI/Heat Transfer in Walls I-Principles: 40(1)*: 139–149.

Love A. 2011. *Material impacts on operational energy usage.* Massachusetts: Massachusetts Institute of Technology.

Mark, L. 2014. *Design of shelf angles for masonry veneer.* Calgary: Structure magazine.

Wilson, M., & Higgins, J. 2019. Fero rap tie and fast systemthermal analysis. *American Journal of Hematology* 94(1).

Huygen, N.C. & John, P. Airflow within a brick veneer cavity wall. *13th North American Masonry Conference at the Salt Lake City, 16–19 June 2019.* USA.

NCMA. 2004. *Mortars for concrete masonry.* Virginia: National Concrete Masonry Association.

NCMA. 2013. *R-values of single-wythe concrete masonry walls.* Virginia: National Concrete Masonry Association.

NECB. 2017. *https://Nrc.canada.ca/en/certifications-evaluations-standards/codes-canada/codes-canada-publications/national-energy-code-canada-buildings-2017.*

Schumacher, C., Straube, J., Ober, D., & Grin, A. 2013. Development of a new hot box apparatus to measure building enclosure thermal performance. *Proceedings of Buildings XII*: 1–19.

Straube, J. 2017. Meeting and exceeding building code thermal performance requirements. *Canadian Precast/Prestressed Concrete Institute.*

Szoko, S.S. 2010. *Insulating masonry walls.* Virginia: National Concrete Masonry Association.

Valore, R. C. 1980. Calculations of U-values of hollow concrete masonry. *Concrete International* 2(2): 40–63.

Yao, Q. & Qu, J. 1999. Three-dimensional versus two-dimensional finite element modeling of flip-chip packages *ASME. J. Electron. Packag* 121(3): 196–201.

Zienkiewicz, O. C., Taylor, R. L., & Zhu, J. Z. 2005. *The finite element method.* Massachusetts: Butterworth-Heinemann.

Brick and Block Masonry - From Historical to Sustainable Masonry –
Kubica, Kwiecień & Bednarz (eds)
© 2020 Taylor & Francis Group, London, ISBN 978-0-367-56586-2

Humidity consequences for the application of the contemporary repairs into historic walls

M. Wesołowska
UTP University of Science and Technology, Bydgoszcz, Poland

ABSTRACT: Brick fences are an indispensable element of historic buildings. They are often the only remnants of the historic buildings of cities. Due to the exposure, they are displayed to extreme environmental conditions: driven rain, splash water, snow cover, ground moisture and frost. As a consequence, intensive biological and chemical corrosion develops. The purpose of the conservation and restoration works is to remove the damage caused during the operation of the fence, stop the corrosive processes and restore aesthetic values. This is connected with the application of modern materials. The decision on the selection of solutions is made on the basis of typical tests: the composition of mortars, the volume density of mortars as well as bricks. As a consequence, solutions that can intensify damage are created. One of the main causes of degradation is the subflorescence of soluble mineral salts. It involves the flow of moisture in the wall, which is disturbed as a result of the application of contemporary materials. The article presents the results of experimental and in situ research on the mechanism of moisture flow in fencing walls. Moisture maps were made for selected objects and subflorescence areas were established. Laboratory tests of historic and contemporary materials were carried out: thermal and moisture properties, microstructure parameters. The obtained results served as the output data for the simulation analysis of heat and mass transport in the WUFI 2D computer application. The simulations covered three basic variants: filling the joints, re-profiling the bricks and introducing new bricks and mortars to the wall. The result of the research is the distribution of moisture in the brick wall. A detailed analsis was made of the masonry area of the wall (range 0 to 50 mm), which is exposed to subflorescence. The obtained results clearly indicate that each of these elements has an impact on the moisture status.

1 INTRODUCTION

Proper functioning of historical walls is ensured by limiting intervention, entering reversible changes which do not interfere with general wall substance. The compatibility of a wall is created mostly by mortar which plays service role to wall elements. Kubica et al (2010) claim that mortar should be a link, flexible base for a wall element. In this case lime fulfills a special role influencing mortar flexibility and its ability to absorb stresses resulting from wall movements (Lourenco, 2010). Basic requirements for conservation mortars (type of binder, additives and additions, type, form and dimensions of aggregates, workability, hardening behavior in wet and dry conditions, available working time, salt content, physical and mechanical parameters (porosity, capillarity, water vapor permeability, strength and deformability, drying behavior, cracking resistance, adherence to the support), durability (resistance to salt crystallization, if relevant, etc.) were defined in work of Domasłowski et al (1998) and Delgado Rodrigues et al. (2007). Their requirement also

needs using traditional materials and technologies (Moropoulou, 2000). The same refers to bricks for supplementing and re-masonry. The selection of materials must include actual parameters of the renovated object (Lourenco, 2010). Because of individual character of historical ceramic goods the researches often concern a specific object (Cardiano et al, 2004), a group of features associated with durability (Baronio et al, 1985) or environmental pollutant effects (Cultrone et al, 2000) There are also considered issues associated with transport of mass (Castellazzi 2014, Franzoni 2014).

Despite of broad knowledge concerning selection of materials for monument buildings, often there are solution which intensify degradation processes. Such examples are defense walls and fences which are burdened with extreme climatic and ground humidity conditions.

Often in direct neighborhood of modern materials there is increase of humidity and efflorescence. In this work there are presented the analyses of influence of most often used supplements on humidity condition of a wall.

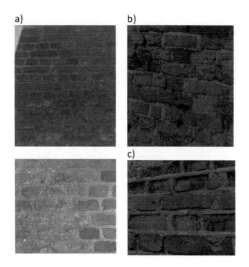

a) b)

c)

Figure 1. Example of wall destruction resulting from using contemporary materials: a=re-masonry of historical wall with contemporary bricks b=replacement of defected elements with contemporary ones, c=joints.

In order to establish the dynamics of these changes the WUFI 2D simulation program was used. The program gives opportunity to model a wall as a system of wall elements and mortar. It lets to charge external wall sides with appropriate climate conditions (Zirkelbach, 2010). It concerns driving rain and capillary rise of humidity from soil (Holm, 2000). In performed simulations the basic system was a wall 120cm made of gothic brick on lime mortar. Material data was supplemented with own laboratory results in area of: heat conduction, microstructure parameters and capillarity. Further modification referred to introduction of present materials for: joints, partial replacement of degraded mortar in a joint, replacement of single bricks with contemporary ones and re-masonry of a wall. Assuming the least favorable simulation period (from October do April) humidity distributions were evaluated in the most sensitive brick layer – at contact place with modern materials.

2 LABORATORY TESTS

2.1 *Test specimens*

The specimens for laboratory tests were halves of historic bricks sized 135x140x85mm. The range of laboratory tests included evaluation of:

- heat conductivity at average reference temperature 1ÔC and conversion to humidity and temperature conditions,
- selected microstructure parameters (bulk density, porosity),
- capillary absorption and initial water absorption (in two ways: through base and face surface of a brick).

2.2 *Thermal conductivity coefficient*

The measurement of heat conduction coefficient was made with TCA300DTX, according to EN 12667 on 12 specimens dried to solid mass. The mean reference temperature was 1ÔC.

90% of quintiles at 90% level of confidence for heat conduction coefficient was calculated with a formula:

$$\lambda_{90/90} = \lambda_{mean} + k \cdot s_\lambda \qquad (1)$$

Where: λ_{mean} = mean heat conductivity coefficient calculated from 12 values, k = coefficient associated with available numbers of test results – assumed as k = 1.65 for 12 results, and s_λ = standard deviation.

For n=12, s_λ was calculated with a formula:

$$S_\lambda = \sqrt{\frac{\sum_{i=1}^{12} (\lambda_i - \lambda_{mean})^2}{11}} \qquad (2)$$

Where: λ_i = heat conduction coefficient for a single specimen.

The value of heat conduction coefficient was rounded up with accuracy of 0.001 W/(m·K).

The conversion factor of heat conduction associated with humidity content on a volume unit was defined as follows:

$$F_m = e^{f_u(U_1 - U_2)} \qquad (3)$$

where: f_u = the moisture conversion coefficient mass by mass for the ceramic (according to ISO 10456 fu = 4.0 kg/kg); U_1 is the moisture content mass by mass of the first set of conditions; and U_2 is the moisture content mass by mass of the second set of conditions.

Humidity content in sorption range (from 0 to 90% RH) was set based on experimental tests (Garbalińska et al, 2007). The analyses indicate that the best match of adsorption isotherm equation to experimental data was obtained in case of equations by Chen Y. and Chen Z. (Garbalińska et al, 2007):

$$u_L = \frac{a \cdot \varphi}{(1 + b \cdot \varphi) \cdot (1 - c \cdot \varphi)} \qquad (4)$$

Where: φ = relative humidity at reference temperature; a, b, c = coefficients depending on reference temperature (assumed for: t = +5̂C –a = 0.0201, b = -0.4748, c = 0.9853; t = +2ÔC –a = 0.041609, b = 1.923164, c = 1.006998; t = +35̂C –a = 0.4447, b = 9.4599, c = -0.5467) (Garbalińska et al, 2007).

Within the range of 90 to 100% relations introduced in WUFI 2D software were applied.

The conversion factor of heat conduction because of temperature was defined from a formula:

$$F_T = e^{f_T \cdot (T_2 - T_1)} \qquad (5)$$

Where: f_T = the temperature conversion coefficient (according to ISO 10456 f_T = 0,001; T_1 = the temperature of the laboratory test (T_1 = 10°C); and T_2 = the real temperature of the brick.

2.3 Selected microstructure parameters

The definition of specific and bulk density and porosity was performed on laboratory specimens created by taking material from six bricks. From the general sample two laboratory specimens were selected. Such prepared specimens were dried to solid mass at +40°C for 120 hours and then put into an exicator. The value of specific density necessary for definition of porosity was measured by method of gas pycnometry using AccuPyc II 1340 series device. After weighing and entering the specimen into the chamber 10 cycles of chamber rinsing, and 10 cycles of measuring were performed. One cycle of measurement included zeroing to atmospheric pressure, filling the chamber with helium to $1.344 \cdot 10^5$ Pa (above atmospheric pressure), stabilization of pressure (at balance level of 34.47 Pa/min) and removing of gas to total lowering of pressure. The result of the test is an arithmetic average from 10 measurements. For measurement of bulk density and porosity the method of mercury porosymetry was used with AutoPore IV porosimeter of 9500 series, equipped with a port of low and high pressure and pressure range enabling measurement of pores from 2nm to 360µm. Before the measurement the calibration and „blank test" were performed – which defined volume, compressibility and thermal effect of the penemometer. Based on control measurements the set time was defined as 30 s.

2.4 Water absorption

The research of capillary absorption was performed according to guidelines of EN 772-11 standard, on brick halves. The specimens dried to solid mass were put in a cuvette with a grill so they were immersed in water at constant depth of 5 +1 mm. After soaking time (1, 4, 9, 16, 25, 36, 49... min respectively) the specimens were taken out, immersed surfaces were wiped, and weighed. By measuring the mass of absorbed water it was defined:

– the initial water absorption in kg/m²min

$$c_{wis} = \frac{m_{so,s} - m_{dry,s}}{A_s \cdot t} \cdot 10^3 \qquad (6)$$

– water absorption caused by capillary rise in kg/m²√min

$$c_{ws} = \frac{m_{so,s} - m_{dry,s}}{A_s \cdot \sqrt{t_{so}}} \cdot 10^3 \qquad (7)$$

Where: $m_{so,s}$ = mass of a specimen after soaking for time t, [g]; $m_{dry,s}$ = mass of a specimen after dying, [g]; t_{so} – time of soaking, [min]; and A_s – total area of a specimen immersed in water

– for a clinker brick by capillary rise through the base A_s = 135x135 = 18 225 mm²,
– for a clinker brick by capillary rise through the face side A_s = 85x135 = 11 475 mm².

3 LABORATORY TESTS RESULTS

3.1 Heat conduction coefficient

The results of heat conduction coefficient for a historic brick were showed in Table 1. For comparison, additional tests were made for contemporary bricks manually formed (Table 2).
Variation coefficient of results:
$v_\lambda = s_\lambda / \lambda_{10, \text{ mean}} = 0.0589/0.3532 = 0.167$
The typical result variability area: X_{typ} = [0.2943 W/(m·K); 0.4121 W/(m·K)]
The results obtained from 12 specimens indicate a high variation of specimens associated with historical technology and inhomogeneity raw material (Figure 2).
For contemporary bricks formed manually the heat conduction coefficient is 25% higher and it

Table 1. Heat conductivity for historic bricks.

Specimen number	Measurement value [W/mK] $\lambda_{10,i}$	Mean value [W/mK] $\lambda_{10, \text{ mean}}$	Standard deviation [W/mK] s_λ	Declared value [W/mK] $\lambda_{10, \text{ mean}}$
1	0.4235			
2	0.4233			
3	0.4324			
4	0.2757			
5	0.2660			
6	0.2703			
7	0.3456	0.3532	0.0589	0.416
8	0.3533			
9	0.3329			
10	0,3766			
11	0,3668			
12	0,3721			

Table 2. Heat conductivity of contemporary brick formed manually.

Specimen number	Measurement value [W/mK] $\lambda_{10,i}$	Mean value [W/mK] $\lambda_{10, mean}$	Standard deviation [W/mK] s_λ	Declared value [W/mK] $\lambda_{10, mean}$
1	0.5285			
2	0.5157			
3	0.5234			
4	0.4394			
5	0.4211			
6	0.4232			
7	0.5102	0.4808	0.0434	0.515
8	0.5237			
9	0.5184			
10	0.4392			
11	0.4505			
12	0.4761			

Figure 2. Example of historic bricks in the wall.

amounts to $\lambda_{10} = 0{,}515$ W/(m·K). The variation coefficient of results is almost 2 times lower and it is $v_\lambda = 0{,}090$.

In order to converse thermal values to temperature and humidity conditions F_U and F_T coefficients were set for temperatures and humidity indicated in chapter 2 gained from (Table 3)

Depending on humidity and temperature, the heat conduction for a historical brick will change from

Table 3. Conversion coefficients depending on temperature and humidity.

	Conversion coefficient				
T,°C	depending on temperature F_T	depending on humidity F_U, at RH=			
		0.1	0.4	0.8	0.9
5	0.99104	1.00009	1.00065	1.00491	1.01122
20	1.01816	1.00016	1.00063	1.00270	1.00587
35	1.04603	1.00087	1.00122	1.00115	1.00113

Table 4. Heat conduction coefficient variation at assumed temperature and humidity conditions.

T,°C	Heat conduction coefficient, at RH =			
	0.1-0.3	0.4-0.7	0.8	0.9
5	0.412	0.413	0.414	0.417
20	0.424	0,424	0.425	0.426
35	0.436	0.436	0.436	0.436

0.412 to 0.436 W/mK (Table 4). Based on results gained the Thermal Conduction Supplement can be defined as 0,002 W/mK.

3.2 Selected microstructure parameters

From direct measurements the following parameters were obtained for historical bricks:
Specific density – 2,482 kg/m³,
Bulk density - 1528 kg/m³,
Open porosity 32%
The obtained results indicate that the brick characterizes itself with significantly higher porosity than contemporary bricks. It is the effect of a formation technology.

3.3 Water absorption

The results of capillary absorption (Figure3) for 3 specimens indicate at significant differences at capillarity through the face surface. It is particularly evident at initial phase when dynamics of moistening increase is high.

The capillary absorption coefficient is higher than values of moisture capillary movement through the brick base. This fact is caused not only by forming

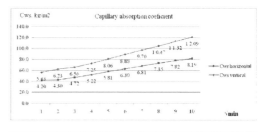

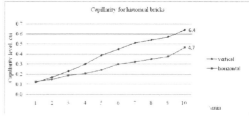

Figure 3. Capillary absorption of historic bricks.

technology of gothic bricks, but also by microstructure changes resulting from long term exposition to environmental factors. For samples absorbing by base surface the range of results is low and dynamics significantly lower.

As consequence, the initial water absorption through the base for historic bricks is about 30% lower than through the face surface. Such tendency stays also for capillarity coefficient and for capillary level.

4 SIMULATION MODEL

Changing climate parameters directly influence the wall humidity condition. In order to define the dynamics of these changes a WUFI 2D simulation program was applied. The program gives opportunity to model the wall as a system of wall elements and mortar. This allows charging the external surfaces with appropriate climate conditions. It takes into account driving rain and capillary humidity rising (Holm et al, 2000). In order to define influence of improve hand-made contemporary brinks into historic wall a wall section was analyzed with height of six bricks and four brick thick, made from wall units sized 280x140x85 mm (Figure 4).

Joint thickness was assumed based on field measurements: horizontal – 20 mm, vertical – 20mm. In the wall the overlapping of environmental humidity (air humidity, rainfall) and humidity from long term building existence for historical bricks and technological humidity (for contemporary bricks and construction mortar) was simulated. It was assumed that capillary humidity from ground was eliminated.

In order to define influence of modern additions for humidity condition of the wall there were assumed 6 variants:

Variant 1 –historic wall of gothic bricks on lime mortar (simulation of natural drying after cutting the capillary humidity)

Variant 2 –historic wall of gothic bricks on lime mortar with refilling of cement-lime mortar

Variant 3 - historic wall of gothic bricks on lime mortar with filling of contemporary bricks manually formed on cement-lime mortar

Variant 4 - historic wall of gothic bricks on lime mortar with filling of contemporary bricks manually formed on cement-lime mortar with removed degraded joints to 7cm in depth and filled with cement-lime mortar

Variant 5 - historic wall of gothic bricks on lime mortar with filling of contemporary bricks manually formed on cement-lime mortar

Variant 6 - historic wall of gothic bricks on lime mortar with filling of contemporary bricks manually formed on cement-lime mortar with refilling of cement-lime mortar

The left and right areas of the wall were exposed to external climate characteristic for central Poland while for upper and lower ones, adiabatic surface was assumed. It corresponds conditions for historical defense walls, gates and fences. The boundary conditions are gathered in Table 6.

For simulation, the researched properties of historical bricks were used as well as data bases in program library. The input data for simulation were gathered in Table 5.

The simulation period was assumed as 6 months: from 15.09.2019 to 15.04.2020. The changes of humidity in wall elements were analyzed as well as humidity distribution in historic bricks and in materials around

Table 5. Hygrothermal parameters of the used materials.

Parameters	Historic brick	Hand-made brick	Lime mortar	Cement-lime mortar
Bulk density, [kg/m3]	1528	1725	1785	1910
Heat capacity, [kJ/(kg·K)]	0.85	0.85	0.85	0.85
Heat conductivity, [W/m·K]	0.416	0.6	0.7	0.8
Porosity, [Vol, %]	32	34	28	25
Free saturation, [Vol, %]	326	236	248	200
Diffusion resistant factor, [-]	10.0	9.5	15.0	45.89
A-value (hor.), [kg/(m2·√s)]	0.677	0.283	0.153	0.085
A-value (vert.), [kg/(m2·√s)]	0.420	0.119	0.102	0.057
Temp-dep. Thermal Cond Supplement [W/mK]	0.004	0.0002		

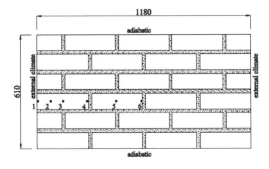

Figure 4. Wall area modelled for simulation with WUFI 2D software.

Table 6. Boundary conditions for masonry.

Properties	Surface		
	Brick	Lime Mortar	Cement Mortar
1 Surface coefficients			
Sd Value	Sd =0[m] (No coating)		
Heat Transfer Coefficient [W/(m²·K)]	25		
Short-Wave Radiation Absorptivity	0.68	0.4	0,6
Long-Wave Radiation Emissivity	0.9	0.9	0.9
2 Climate			
Localization	Central part of Poland		
Temperature	Tmean=+8.1°C Tmax=+30.4°C Tmin=-16.0°C		
Humidity	Hmean=80% Tmax=100% Tmin=29%		
Rain Sum	Normal Rain Sum = 493 [mm/m²] Driving Rain Sum [mm/m²]		

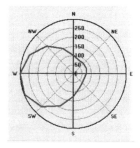

Azimuth	West/East
Driving Rain	R1=0
Coefficients	R2=0,07s/m

them (Figure 4) in two time steps (every 3 months). The results obtained are shown as humidity changes at distances from the face side to stabilization area.

5 SIMULATION RESULTS

Humidity distribution for particular time steps for the whole area of selected variants were shown in Figure 5. Distribution of water content in function of distance from wall face side was shown in Figure 6-8. Case 1 refers to a wall entirely made of historical bricks on lime mortar (Figure 6a). Case 2 shows changes in humidity of a brick in direct contact with

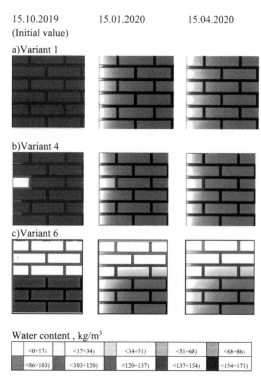

Water content , kg/m³

<0÷17)	<17÷34)	<34÷51)	<51÷68)	<68÷86)
<86÷103)	<103÷120)	<120÷137)	<137÷154)	<154÷171)

Figure 5. Humidity distribution in the analyzed wall area.

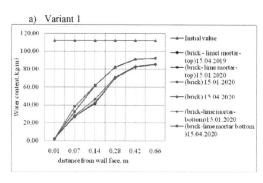

Figure 6. Water content in bricks for a wall built entirely from historical bricks.

a) Variant 3

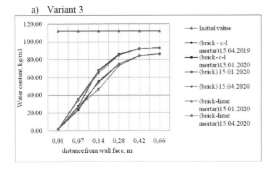

b) Variant 4

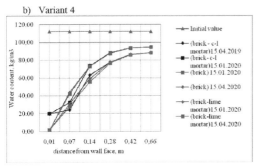

Figure 7. Water content in historical bricks for a wall filled with contemporary bricks.

a) Variant 5

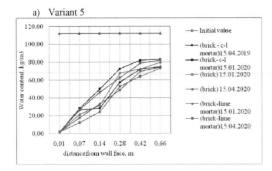

b) Variant 6

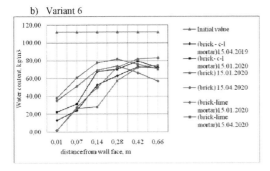

Figure 8. Water content in historical bricks for a wall with interleaved with contemporary bricks.

a contemporary filling of a brick manually formed (Figure 6). Case 3 is an example of historical wall with inserted contemporary bricks at entire width (Figure 8).

6 DISCUSSION OF RESULTS

The above simulations show that mortar type is an essential factor influencing humidity conditions of a historical wall made of gothic bricks. In case of lime mortar (Figure 6a) drying takes place in entire wall section and the process is most intensive within 1.5 brick distance from the face surface. Humidity distribution in layers close to the brick surface contacting with mortar (top and bottom) have similar shape – differences in particular points are small and amount to max. 0.4 kg/m³ at level of 92.25kg/m³. In the middle part of bricks to14cm from the face side, humidity is higher for about 5kg/m³ which confirms capillary activity of intensively drying mortar in area of contact with ceramic. Inside the wall this process is practically unnoticeable. The introduction of cement-lime joint partly closed the gaps (Figure 6b). This is the effect of tightening of the joint which blocks drying through the mortar.

Filling with a single contemporary brick with cement-lime mortar introduces changes of humidity transfer in contact zone. The difference in relation to even values for middle layer and lime mortar layer is: after 3 months - 8 kg/m³, after 6 months – 3.5 kg/m³. The humidity levels reached finally are different about 0.6 kg/m³in relation to variant 1(Figure 7a). Moreover, introduction of joint supplement with cement-lime mortar until 7cm in depth causes changes in brick humidity in direct contact to with supplement (6 kg/m³in relation to other two patterns). When comparing the other two patterns to variant 1 it can be noticed the humidity content increased. Maximum increase is at $W(0.07m) = 9$ kg/m³and then it gradually decreases to $W(0.66m) = 2$ kg/m³. The humidity distribution along the brick height also changed– opposite than in variant 1 (Figure 6a) its value is higher at central part.

Repairing the wall with a contemporary brick manually formed on cement-lime mortar lowers humidity by 10 kg/m³in a layer of historical brick for entire pattern. (Figure 8a). It is the effect of a fact that a vast volume of wall is made of material in dry state. In Figure 5c there is shown redistribution of humidity resulting such a repairing. Introducing stronger mortar 2 cm deep into joint causes total change in humidity distribution: there is high humidity in a layer under brick surface: in contact zone of rebuilding $W(0.01m) = 22$ kg/m³ and central part $W(0.01m) = 38$kg/m³ In subsequent period these values decrease respectively to $W(0.01m) = 13$ kg/m³ and in central part $W(0.01m) = 35$kg/m³. In comparison to variant 5, the range of brick drying after rebuilding decreased from 42cm to 14cm (Figure 8b). Also the humidity profile of the brick changed.

7 SUMMARY AND CONCLUSIONS

Drying of walls is a complex process in which both environmental as built-in material properties and their configuration play roles. The proper approach to evaluation of this phenomenon requires association of laboratory tests with capabilities of calculation tools. The WUFI program lets generate own material and climatic data. It offers a good tool for comparison analysis. In this case a range of laboratory tests were made on gothic bricks coming from objects as specific as defense walls. Their functioning for hundreds of years in severe climate conditions resulted in physical properties of ceramic. It results in drying processes. In this work 5 variants of wall systems four brick thick were analyzed. As comparison basis, the solution of non-repaired wall was assumed, made of gothic bricks on lime mortar. Each of subsequent variants was a modification using contemporary solution: manually formed brick and cement-lime mortar of typical composition. The assumed period of simulation results from author experience in this field tests on walls (Wesołowska, 2016).The analysis covered 3 layer in geometric model which directly contacted with basic treatments of historical walls which include: joints, partial change of degraded joints, replenishment and repairing of wall made with contemporary bricks manually formed. The obtained distributions of water content indicate that each contemporary modification leads to interference into humidity flow. The biggest changes are made with joining and partial replacement of mortar in joints. It is a treatment which introduces blockage in humidity flow and directs the humidity flow through the brick. As consequence the brick is destroyed. Under the new filling, humidity gathers which further damages already weak lime mortar. As a result, the new material delaminates and degradation process enters deeper into the wall interior.

The above presented analysis is the first step toward a detailed research which will include changes of thermal and humidity parameters of historical bricks (because of microstructure changes caused by environment influence), wall humidity profile, capillary rise.

ACKNOWLEDGEMENTS

The work was done using apparatuses within the frame of „Realization of 2nd Stage of Regional Centre of Innovation" project co-financed from means of European Fund for Regional Development within the frame of Regional Operational Program of Kujawsko-Pomorskie Voievodship for years 2007 – 2013.

REFERENCES

Baronio, G. & Binda, L. 1985. Physico-mechanical characteristics and durability of bricks from some monuments in Milan. *Masonry International*, Issue 4: 29–35.

Castellazzi, G., et al. 2014. Modelling of Non-Isothermal Salt Transport and Crystallization in Historic Masonry. *Key Engineering Materials* Issue 624: 222–229.

Cultrone, G. et al. 2000. Behavior of brick samples in aggressive environments. *Water, Air and Soil Pollution*, Issue 119: 191–207.

Cardiano, P. 2004. Study and characterization of the ancient bricks of Monastery of "San Filippo di Fragalà" in Frazzanò (Sicily). *Analytica Chimica Acta*, 519(1): 103–111.

Delgado Rodrigues, J. &. Grossi, A. 2007. Indicators and ratings for compatibility assessment of conservation actions. *Journal of Cultural Heritage* Issue 8: 32–43.

Domasłowski W. et al. 1998. Research on maintenance brick walls. Nicolaus Copernicus University in Toruń (in Polish).

Franzoni, E. 2014. Rising damp removal from historical masonries: A still open challenge, *Construction and Building Materials* Issue 54: 123–136.

Garbalińska, H. & Siwińska, A. 2007. Isotherm of sorption of ceramic brick, silica brick and cellular concrete. *Building Physics in Theory and Practice*. Issue II: 41–46 (in Polish).

Holm, A. & Kunzel, H.M. 2000. Two-Dimensional Transient Heat and Moisture Simulations of Rising Damp with WUFI 2D. Fraunhofer-Institut fur Bauphysik.

Kubica J. & Gąsiorowski S. 2010. Mortar selectin in design practice – description of the problems, solutions and requirements. *Architecture Civil Engineering Environment* No 1: 53–61.

Lourenco, P. et al. 2010. Handmade Clay Bricks: Chemical, Physical and Mechanical Properties International. *Journal of Architectural Heritage* 4(1): 38–58.

Moropoulou, A. 2000. Reverse engineering to discover traditional technologies: a proper approach for compatible restoration mortars". *PACT 58*: 81–107.

Schueremansa, L. et al. 2011. Characterization of repair mortars for the assessment of their compatibility in restoration projects: Research and practice. *Construction and Building Materials*. Volume 25, Issue 12: 4338–4350.

Wesołowska, M. 2016. The analysis of facing wall construction date on its humidity condition. *Proceedings of the Central European Symposium on Building Physics and BauSIM/2016*: 775–780.

Zirkelbach, D. 2010. WUFI® 2D Calculation Example Step by Step. Fraunhofer. Institut Bauphysik.

Brick and Block Masonry - From Historical to Sustainable Masonry –
Kubica, Kwiecień & Bednarz (eds)
© 2020 Taylor & Francis Group, London, ISBN 978-0-367-56586-2

The effect of high temperatures on the mechanical performance of concrete blocks made with gneiss aggregates

W.A. Medeiros & G.A. Parsekian
Federal University of Sao Carlos, São Carlos, Brazil

A.L. Moreno Jr
University of Campinas, Campinas, Brazil

ABSTRACT: In Brazil, one of the most used aggregates in the concrete blocks production is gneiss. There is few information in literature about the behavior of concrete blocks with this type of aggregate at high temperatures. Besides that, the test procedures to determine the concrete stress-strain curve is generally prescribed for cylindrical cast specimens. The concrete's properties in this type samples are different from those produced in concrete block industries that use large vibrating presses with strong compaction energy and low cement consumption. This paper describes the effects of temperature on the mechanical properties of concrete blocks made with gneiss aggregate. For this, the CMUs were heated to 800°C and their residual compression strength and deformability was determined. The test method was adapted from the RILEM TC 200-HTC recommendations. In conclusion, the aggregate was thermally stable, and the residual compressive strength of the concrete block was basically unchanged below 400°C.

1   INTRODUCTION

Concrete masonry is known to have a good fire resistance performance. However, since the 1950s, research such as Saemannt & Washat (1958) reports that, in general, the mechanical properties of concrete deteriorate as the material heats up to high temperatures. Likewise, Abrams (1971) noted that the type of aggregate used in concrete also affects thermal performance and mechanical properties at high temperatures, as each aggregate has characteristics such as thermal expansion, density, conductivity and capacity heat different.

There are variations in mineralogical rock compositions used as aggregates for each country. According La Sena & Rezende (2009), in Brazil the gneiss is one of the most used aggregates in the concrete production, inclusive at manufacture of concrete masonry units (CMU). This aggregate is a metamorphic rock, resulting from the deformation of arkose or granites sediments and is found throughout Brazilian territory, mainly in coastal regions, but also in various inland areas.

Another issue concerns the data found on the concrete's properties at high temperatures. The most information is from fire tests use a concrete cast cylinder. The concrete's properties in a laboratory cast sample are different from a CMU produced in concrete block industries that use large vibrating presses with strong compaction energy and low cement consumption.

Information on the variation of the compressive strength with temperature, for conventionally produced normal-weight concretes, is presented in Figure 1 (for a silicate aggregate concrete). This information is already well consolidated, being adopted by several International Standards and Codes. Unfortunately, there is few information in the literature on the behavior at high temperatures of concrete blocks manufactured with this type of aggregate, requiring further research on this subject. Thus, this paper aims to describe the effects of temperature on the mechanical properties of concrete blocks made with gneiss aggregate, considering its geometry and manufacturing process.

The modulus of elasticity (E) dependent mainly on the water-cement ratio in the mixture, the age of concrete, the method of conditioning, and the amount and nature of the aggregates (Figure 2).

The water/cement ratio, type of aggregate, aggregate/cement ratio, load level, heating rate, evaporation rate, age of the concrete, type of curing must have its importance highlighted in obtaining the modulus of elasticity of the concrete at high temperatures (Schneider 1988). Concrete blocks are manufactured with low cement consumption when

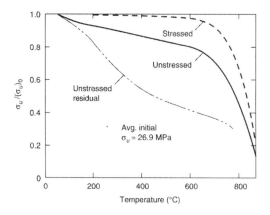

Figure 1. The effect of temperature on the compressive strength of a normal-weight concrete with silicate aggregate (Abrams 1971).

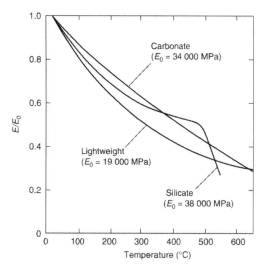

Figure 2. The effect of temperature on the modulus of elasticity of concretes with various aggregate (Kodur & Harmathy 2016).

compared to wet cast concrete. A high aggregate/cement ratio tends to increase the water/cement ratio, causing more porosity in the blocks, as can be seen in Figures 3 and 4.

Figure 3. Face shell cross section.

Figure 4. Enlarged image of the cross section.

2 EXPERIMENTAL DETAILS

2.1 Aggregate properties

To characterize the initial state of aggregates, petrophysical study was performed before any thermal-mechanical test. A macroscopic recognition was made in order to identify the petrographic characteristics. In this analysis it was possible to identify that gneisses are metamorphic rocks, whose structure has foliation characterized by compositional banding (light bands rich in quartz and feldspathic and dark bands rich, for example in biotite and hornblende) with minerals visible to the naked eye. The granoblastic texture presents rock with granular minerals, without orientation and gray color, whitish with bands (Figure 5).

An X-ray Fluorescence spectrometry (XRF) was used in this work to assess the microstructure of the aggregates by characterizing them chemically and ensuring correct identification. XRF is an established technique for elementary analysis, process monitoring and quantitative and qualitative chemical

Figure 5. Gneiss aggregate.

Table 1. Spectrometry XRF - Gneiss.

Element	% m/m	Oxides	% m/m
Si	25.294	SiO_2	44.052
Al	4.592	Al_2O_3	8.065
Fe	4.897	Fe_2O_3	3.904
K	3.465	K_2O	2.716
Ca	2.820	CaO	2.420
Ti	0.451	TiO_2	0.441
Mn	0.107	MnO	0.078

characterization of metallic, ceramic and polymeric materials. The chemical elements found and their respective percentage in the aggregate's composition are detailed in Table 1 ordered according to the predominance in the composition, being possible to observe that the gneiss is mostly a siliceous aggregate.

Thermogravimetric analysis (TGA) is a technique of thermal analysis in which the variation of the sample mass (loss or gain) is determined as a function of temperature and time, while the sample is subjected to a controlled temperature programming. This technique makes it possible to know the changes that heating can cause in the mass of substances, allowing to establish the temperature range in which they start to decompose and to follow the progress of dehydration reactions (moisture loss). The aggregate was subjected to a heating rate of 1°C/min up to an approximate temperature of 1000°C. In Figure 6, the aggregate obtained a small mass loss, less than 2%.

The aggregates used in the CMUs production (sand, stone powder and gravel) were also characterized in terms of granulometry, unit mass, water absorption, specific mass, powdery material and moisture content as prescribed by Brazilian standards (ABNT NBR NM: 2003a, 2003b, 2006, 2009).

2.2 Concrete blocks production

The blocks were produced by an industry installed in the city of Guarulhos, Sao Paulo, Brazil equipped with vibro-press concrete block machines and with an installed capacity of more than 4 million blocks per month. The company has been in the market for 48 years and its products are all certified by the appropriate institutions. All the research blocks were produced on the same day, by the same machine following the proportions of the line already used in the manufacture of the blocks that the company supplies to the market. The factory uses high early strength Portland cement, industrial crushing sand (same supplier of gneiss aggregate) and super plasticizer additive. Figure 7 shows the vibro-press machine that produced the blocks.

It is important to note the dry-cast concrete used in concrete masonry blocks is manufactured using a very dry zero-slump concrete that is placed in steel molds and vibrated in shape (Parsekian et al. 2013). In Brazil, it is still common to find a wet curing process, in which case there is greater concern in curing control and with the potential retraction of the product (Parsekian et al. 2013). Most of the research carried out on the mechanical strength of concrete at elevated temperatures is done on wet cast concrete. Some of the main differences between the two concrete types are the amount of cement, aggregate sizes, water cement ratios and compaction energy. Therefore, it is important to note that there will be some variation in the results due to the different mixtures of concrete and curing methods.

After the curing period and with the desired resistance reaching, the blocks were released by the manufacturer and sent to the university for testing.

Figure 7. Vibro-press machine used in the blocks manufacture.

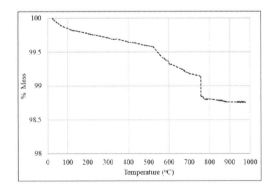

Figure 6. TGA - Gneiss.

The evaluation characteristics of concrete masonry units followed the methods prescribed for ASTM C140/C140M (2017), being carried out: measurement of dimensions, compressive strength, absorption, unit weight (density) and moisture content.

2.3 Test method and parameters

Concrete does not conduct heat quickly through the material; thus, the outer layers of concrete are hotter than the inner layers. When small cylinders are used for testing, they are usually exposed on all sides, so that they can warm up to the desired temperature relatively quickly. After being exposed to the same temperature for an enough period, the temperature of the entire cylinder becomes uniform (Pope & Zalok 2019).

The concrete masonry units' size and geometry makes temperature uniformity in the concrete blocks difficult. If most of the concrete is not affected by the high temperature, the results are not indicative of its fire resistance properties. To ensure there was no lack of temperature uniformity in the blocks, several tests were carried out with different thermocouple positions (Figure 8). Adapting to the values prescribed by RILEM TC 200-HTC (2005), it was established that maximum temperature differences between any two points temperature readings shall not exceed 1°C at 20°C, 5°C at 100°C and 20°C at 750°C. For intermediate values, the maximum temperature differences permitted shall be calculated by linear interpolation between the two points.

The specimen should be at least 90 days old before testing. The blocks had a heating and cooling rate of 1°C/min, from room temperature to 100, 200, 300, 400, 500, 600, 700 and 800°C. After reaching the maximum test temperature as indicated by the

read points temperature, the temperature was fixed for a period of 60 minutes. The tests procedure to determination of compressive strength and modulus of elasticity followed the methods of EN 1052-1 (1999).

3 TEST RESULTS AND DISCUSSION

3.1 Residual compressive strength

As already evidenced by several authors and presented in ACI/TMS 216.1 (2014), the compressive strength is not only a function of temperature but is also affected by the applied load. Thus, there are three test methods available to finding the residual compressive strength of concrete at elevated temperatures: stressed test, unstressed test and unstressed residual strength test. The latter has the lowest strength and, therefore, is the most suitable to obtain the limit values, being the one used in this research. Six blocks were tested for each temperature, the results of residual compressive strength are shown in Figure 9.

An important observation is to compare Figures 1 and 9. We noticed that the dry cast concrete used in the manufacture of CMUs maintains its resistance to higher temperatures, when compared to wet cast concrete.

The test result indicates that there was a small resistance gain at lower temperatures, in addition to maintaining resistance up to 400°C. The temperature have a negligible effect on their compressive strength up to 400°C but suffer a significant loss of strength subsequently, due to the loss of crystal water resulting in the reduction of $Ca(OH)_2$ content, besides the changes in the morphology and the microcrack formation (Handoo et al. 2002). The strength gains up to 300°C may be due to the relief of pressures by drying, which also creates greater van der Waals forces, resulting a closer configuration of capillary pores. In addition, strength gain at this temperature level is greater for water cured mortars than autoclave cured ones (Aydın et al. 2008); which is consistent with the wet curing process carried out by the block manufacturer.

The capillary water is lost completely at 400°C. Up to 300°C, hydration of unhydrated cement grains is improved due to an internal autoclave condition as

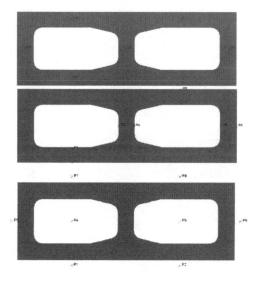

Figure 8. Concrete blocks specimens showing the location of the temperature measuring points.

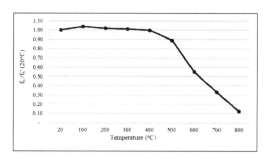

Figure 9. Residual compressive strength vs temperatures.

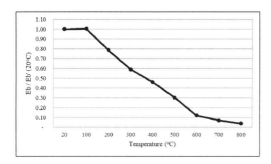

Figure 10. Residual modulus of elasticity *vs* temperatures.

a result of high temperature and water evaporation. This is particularly true, principally, for high strength concrete, as its low permeability resists moisture flow. This can be used to explain the constant compressive strength when the temperature is below 300°C (Ma et al. 2015).

3.2 *Residual modulus of elasticity*

The residual modulus of elasticity *vs* temperature relationships for concrete blocks with gneiss aggregate are presented in Figure 8. The deteriorating effect of elevated temperatures on modulus of elasticity was more severe than compressive strength case. The residual modulus was maintained without minor variation up to 100°C, subsequently an almost quasi-linear reduction was reached over 100°C. At 400°C, although there are almost no significant amounts of compressive strength loss, there is a great reduction in modulus of elasticity. The residual modulus of elasticity is less than 50% at this temperature level.

Comparing Figures 2 and 10, there are also differences between the behavior of the elastic modulus of dry and wet cast concrete. Considering that the gneiss is siliceous aggregate, according to the XRF results presented in Table 1, there are some differences mainly in the first temperatures. The modulus of elasticity of the concrete block did not change until 100°C, differently from what occurs with wet molded concrete. It is believed that this difference is also related to geometry, in addition to porosity and manufacturing process.

4 CONCLUSIONS

The residual compressive strength of concrete is generally not changed below 400°C. Above 400°C, the strength begins to decrease due to a combination of moisture loss, micro cracks, chemical decomposition and thermal stresses. The chosen aggregate and the additional cementitious materials added can help to reduce these negative effects, thus improving the mechanical performance of the concrete masonry unit. The results of residual compressive strength can also be affected by the test procedure, so it is important to take this into account when comparing the results. The low cement consumption and high aggregate/cement ratio have been found to improve the residual compressive strength of concrete, which means that dry cast concrete masonry would have better residual compressive strength and modulus of elasticity. The geometry and porosity of the CMUs can influence the mechanical properties of the blocks.

ACKNOWLEDGMENTS

The authors thank the São Paulo Research Foundation (FAPESP) by the support in the project 2018/19748-9.

They also thank the support of the company 'Glasser Floors and Precast' for the block's donation.

REFERENCES

Abrams, M.S. 1971. Compressive strength of concret at temperatures to 1600F. In: *Temperature and concrete*: 33–58. Detroit: ACI SP-25.

American Concrete Institute and The Masonry Society. 2014. ACI/TMS 216.1: Code requirements for determining fire resistance of concrete and masonry construction assemblies. ACI/TMS. Farmington Hills.

American Society for Testing and Materials. 2017. ASTM C140/C140M: Standard test methods for sampling and testing concrete masonry units and related units. ASTM International.

Associação Brasileira de Normas Técnicas. 2003a. ABNT NBR NM 52: Agregado miúdo - Determinação da aparente. ABNT.

Associação Brasileira de Normas Técnicas. 2003b. ABNT NBR NM 248: Agregados - Determinação da composição granulométrica. ABNT.

Associação Brasileira de Normas Técnicas. 2006. ABNT NBR NM 45: Agregados – Determinação da massa unitária e do volume de vazios. ABNT.

Associação Brasileira de Normas Técnicas. 2009. ABNT NM 53: Agregado graúdo - Determinação de massa específica, massa específica aparente e absorção. ABNT.

Aydın, S. et al. 2008. High temperature resistance of normal strength and autoclaved high strength mortars incorporated polypropylene and steel fibers. *Construction and Building Materials* 22: 504–512.

European Committee for Standardization. 1999. *BS EN 1052-1:1999 Methods of test for masonry. Determination of compressive strength.*

Handoo, S.K. et al. 2002. Physicochemical, mineralogical, and morphological characteristics of concrete exposed to elevated temperatures. *Cement and Concrete Research* 32 (7): 1009–1018.

Kodur, V.K.R. & Harmathy, T.Z. 2016. Properties of building materials. In: M. Hurley et al. (eds): *SFPE Handbook of Fire Protection Engineering, Fifth Edition*: 277–324. New York: Springer.

Ma, Q. et al. 2015. Mechanical properties of concrete at high temperature-A review. *Construction and Building Materials* 93: 371–383.

Parsekian, G.A. et al. 2013. *Comportamento e dimensionamento de alvenaria estrutural*. 2nd ed. São Carlos: EdUFSCar.

Pope, H. & Zalok, E. 2019. The Effect of Fire Temperatures on the Mechanical Performance of Concrete Masonry Materials. In: *13th North American Masonry Conference*: 252–264. Salt Lake City: TMS.

RILEM Technical Committee. 2005. Recommendation of RILEM TC 200-HTC: mechanical concrete properties at high temperatures-Modelling and applications Part 10: Restraint stress. *Materials and Structures* 38 (284): 913–919.

Saemannt, J.C. & Washat, G.W. 1958. Variation of mortar and concrete properties with temperature. *ACI Journal Proceedings* 29 (5): 385–395.

Schneider, U. 1988. Concrete at high temperatures - A general review. *Fire Safety Journal* 13 (1): 55–68.

La Sena, H.A. & Rezende, M.M. 2009. Agregados para a Construção Civil. *Sumário Mineral do Brasil* (11): 602–635.

Brick and Block Masonry - From Historical to Sustainable Masonry –
Kubica, Kwiecień & Bednarz (eds)
© 2020 Taylor & Francis Group, London, ISBN 978-0-367-56586-2

Thermal inertia improvement and acoustic impact of walls with projecting facing bricks

D. Palenzuela, S. Ciukaj & N. Coin
Centre Technique de Matériaux Naturels de Construction (CTMNC), France

ABSTRACT: The study consists in determining the impact of projecting facing brickwork on thermal inertia and Sound Transmission Loss (STL) of walls. Specific 3D thermal and acoustic calculation methods have been developed in order to determine the areal heat capacities and STL of walls with non-coplanar faces. An optimization process was carried out to determine the arrangements of projecting bricks allowing a maximum areal heat capacity to be obtained. This approach led to the definition of a wall, called "S", for which the areal heat capacity is significantly increased compared to the reference wall with coplanar faces, with an almost constant areal mass. The structural modification of the wall is no longer consistent with the acoustic mass law. This change improves significantly the STL at low frequencies, however it degrades at medium and high frequencies compared to the 22 cm thick reference wall.

1 INTRODUCTION

The thermal inertia of a wall depends on the characteristics of the materials that compose it: thermal conductivity, density and specific heat capacity. But an additional characteristic is important: the heat transfer surface of the wall with the surroundings. If this surface is not flat, then the wall will be able to exchange more heat with its environment. By considering a masonry brick wall made of fired clay facing bricks, it is possible to increase the heat transfer surfaces overlooking the indoor thermal environment by working on the brick arrangement to get projecting and recessed bricks. A specific inner wall system made of projecting bricks therefore also has an impact on its thermal inertia, arguments to which architects may be sensitive, particularly in the design of tertiary buildings. The vibro-acoustic behavior of this type of wall is different from a masonry wall with coplanar faces; therefore, it is useful to assess its acoustic impact.

2 CALCULATION OF THERMAL INERTIA

Thermal inertia is an input data used in the French Thermal Regulation (RT2012) to determine the bioclimatic need, energy consumption and assess exposure to thermal discomfort during hot periods of a building or part of it. It is the subject of a specific booklet of the "Th-bat Rules" associated with the French thermal regulations (Commission Th-bat 2017).

Two types of inertia are required to calculate the thermal behavior of a building: daily inertia to characterize the damping of the daily wave of temperature and sunshine in warm season as well as to characterize the recovery rate of heat input in winter (24 hours period); sequential inertia to characterize the damping of the sequential temperature wave. It allows inertia to be considered during summer heat sequences. The outside temperature wave, considered in a conventional manner, has a period of 14 days. Two other thermal inertias can be useful to evaluate: hourly inertia for a one-hour period, and annual inertia for one-year period, i.e. 365 days.

To determine these inertias in detail, it is necessary to know the areal heat capacities of each wall of the building. The areal heat capacity of each wall for a given time period is calculated by applying NF EN ISO 13786 standard (AFNOR 2017). The concept behind this standard is the similarity between an isotropic homogeneous one-dimensional thermal system and an electrical circuit. These two systems can be modeled in transient mode by a quadrupole, the temperature corresponding to the electric potential and the heat flux corresponding to the electric current. The principle has been described by Carslaw & Jaeger (1959) and Degiovanni (1999) has completed the thermal quadrupole method. The calculation method uses complex numbers representing a sinusoidal change in temperature and heat flux to calculate the dynamic thermal properties of a building component.

The procedure described in the NF EN ISO 13786 standard applies to building components made up of homogeneous and isotropic flat layers. However, this standard opens the possibility of calculating the dynamic thermal characteristics of non-planar components and components containing significant thermal bridges. It indicates that this calculation must be carried out by numerical resolution of the heat transfer equation under periodic boundary conditions, but without specification of these boundary conditions. The method to be applied to switch from heat flows and temperatures obtained by numerical calculation to dynamic thermal characteristics is also not specified. The CTMNC developed this numerical method. The validation of the method was carried out positively by comparison with the two examples in annex D (informative) of NF EN ISO 13786 in its version of July 2008; the first one is a single-layer component made of concrete, the second one is a multi-layer component consisting of a concrete wall associated with a layer of thermal insulation and a coating layer.

3 SOUND TRANSMISSION LOSS CALCULATION

The Sound Transmission Loss (STL) calculation is inspired by a decoupling method (Jean & Roland 2001), transposed to the code of finite element method. This approach was simplified and performed for each frequency in the third octave range, from 50 Hz to 2000 Hz. It requires sequentially the determination of:

- the incident sound power on the emission side wall, using an image source method, obtained with a code developed by CTMNC in the CASTEM (CEA) software, (Guyader 2002, Funkhouser et al. 2004, Lehnert 1993, Allen & Berkley 1978, Mechel 2002),
- the response in vibratory velocity of the wall reception face to the incident complex acoustic pressure, calculated with a linear method of vibrational structure response by modal transposition of Finite Elements of Volumes (FEM), obtained with a code developed by CTMNC in CASTEM (CEA) software, (Guyader 2002),
- the transmitted sound power by the reception side wall, using a Boundary Element Method (BEM), obtained with FastBEM Acoustics® 5 R4 (Advanced CAE Research, LLC) software, (Guyader 2002).

The STL of the wall is finally obtained by difference between the incident sound power and the transmitted sound power levels. We call this STL calculation method "FEM/BEM".

4 REFERENCE BRICK WALL

A reference brick wall is defined to allow future comparisons. It has a thickness of 22 cm, with flat and parallel faces. It consists of solid fired clay bricks with dimensions 220 x 105 x 60 mm. This size is the one used for facade bricks by the "opaque

walls" booklet of the guide for the application of the Th-Bat Rules (Bouchié et al. 2018). Horizontal and vertical mortar joints between bricks are 10 mm thick. The thermal and mechanical characteristics of the materials of the reference wall are indicated in Table 1.

The reference wall is available in three configurations or models depending on the type of brick arrangement: stretcher bond, header bond, Flemish bond. These three bricks bonding arrangements have the same horizontal joints, but different arrangements of the vertical joints (Figure 1).

The density, the surface mass, the thermal resistance and the daily areal heat capacity of the wall are calculated for the three models. The heat flux densities used to determine the thermal resistance and the areal heat capacities are calculated numerically in three dimensions (3D) using the TRISCO software published by the Belgian company Physibel. Table 2 presents the results of the calculations carried out on the three reference configurations.

The density (1821.7 kg/m^3) and the surface mass (400.8 kg/m^2) are identical for the 3 models

Table 1. Thermal and mechanical characteristics of materials.

Characteristic	Unit	Fired clay	Joint mortar
Design thermal conductivity	W/(m.K)	0.69	1.30
Specific heat capacity	J/(kg.K)	1000	1000
Density	kg/m^3	1800	1900
Young modulus	GPa	7.5	12
Poisson factor	-	0.25	0.2

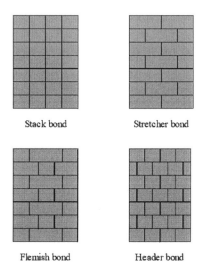

Stack bond   Stretcher bond

Flemish bond   Header bond

Figure 1. Front face of the four reference walls with different brick arrangements.

Table 2. Results of calculations on three configurations of the reference wall.

Wall calculations	Unit	Stretcher bond	Flemish bond	Header bond
Density	kg/m^3	1821.7	1821.7	1821.7
Surface mass	kg/m^2	400.8	400.8	400.8
Thermal resistance	m^2.K/W	0.271	0.269	0.267
Daily areal heat capacity	kJ/(m^2.K)	157.9	158.5	159.6

considered. The relative deviations of thermal resistance and daily areal heat capacity from the average are less than 1%.

The STL obtained by third octave bands for four different bricks arrangements based on the three reference models considered above and an additional one with stack bond (non-conventional brickwork, without brick offset), are showed in Figure 1.

Their STL does not present any significant deviation (less than 1.5 dB in each frequency band analyzed). No matter what the brick arrangement is, for a constant wall thickness of 22 cm, the result indicates a minor deviation of the STL of these walls, as the mass law would indicate. In addition, the STL obtained with the ACOUSYS (FTMM calculation) and ACOUBAT (mean of several measurements) software from CSTB, are showed in Figure 2.

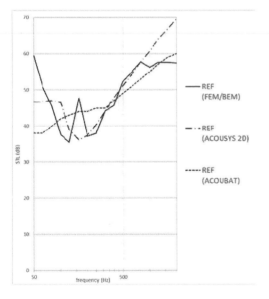

Figure 2. STL of the reference wall for the FEM/BEM calculation, FTMM (2D) calculation and mean of several measurements.

5 IMPACT OF PROJECTING BRICKS ON THE AREAL HEAT CAPACITY

The impact of projecting bricks is evaluated on a simplified two-dimensional (2D) model, which diagram is given in Figure 3. This 2D configuration represents a horizontal section of the brick wall.

The dimensions X and Y vary in the following chosen domain: 0 cm ≤ X ≤ 12 cm and 6 cm ≤ Y ≤ 22 cm. The number of bricks is unchanged whatever the values of X and Y. The values of X and Y for the reference wall are X = 0 cm and Y = 10,5 cm.

The areal heat capacities of the model are calculated by dividing heat capacity (in kJ) by the projected surface of the wall on the Y plane (in m²). We designate this quantity: apparent areal heat capacity. The apparent areal heat capacities are calculated in the domain (X, Y) in steps of 2 cm for three time periods: one day (Figure 4), one hour (Figure 5), fourteen days (Figure 6). Apparent areal heat capacity contour lines are plotted for each time period.

In the considered (X, Y) domain, the highest daily apparent areal heat capacity is obtained for X = 12 cm and Y ≈ 10 cm, which is the optimal solution. The highest hourly apparent areal heat capacity is

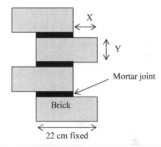

Figure 3. Diagram of the 2D configuration of projecting brickwork.

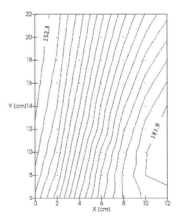

Figure 4. Apparent areal heat capacity contour lines – one-day period.

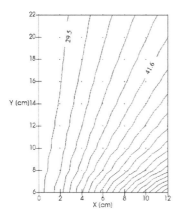

Figure 5. Apparent areal heat capacity contour lines -one-hour period.

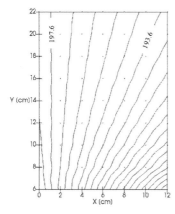

Figure 6. Apparent areal heat capacity contour lines - two-weeks period.

obtained for X = 12 cm and Y = 6 cm. The sequential apparent areal heat capacity is the highest for X = 0 cm, which correspond to a wall with flat faces.

6   PROPOSAL FOR A 3D MASONRY WALL WITH PROJECTING BRICKS

Based on the calculations made in 2D, a reflection on a 3D wall model leads to the model presented in Figure 7.

We call this wall, wall "S". The letter S refers to the "S" shape of the brick arrangement. It is designed to get an apparent areal heat capacity optimized for a one-day period.

The proposed wall is based on 220 x 105 x 60 mm bricks, 105 x 105 x 60 mm bricks (half bricks) and 1 cm thick mortar joints. This wall has X dimension equal to 11.5 cm and Y dimension equal to 10.5 cm. Its surface mass is 382 kg/m².

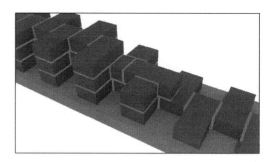

Figure 7. 3D view of the "S" masonry.

The apparent areal heat capacities of the "S" wall are calculated for different periods of time. They are presented with those associated with the reference wall in Table 3.

Table 4 presents the results in the form of relative difference.

Tables 3 and 4 show that "S" wall has an apparent areal heat capacity significantly higher than that of the reference wall for the hourly (+95%) and daily (+17%) periods. The slight decrease (-5%) observed for the sequential period and the annual period is directly linked to the reduction in the apparent surface mass: 382 kg/m² for the "S" wall, 400.8 kg/m² for the reference wall, i.e. a 5% reduction in mass.

In comparison with the reference wall, the acoustic improvement of STL (ΔR) obtained by third

Table 3.   Results of thermal calculations on the "S" wall and on the reference wall.

Period	"S" wall apparent areal heat capacity kJ/(m².K)	Reference wall (Flemish bond) areal heat capacity kJ/(m².K)
1 hour	56.1	28.8
24 hours	185.8	158.5
14 days	191.0	200.0
365 days	191.7	201.1

Table 4.   Relative difference between areal heat capacities of "S" wall and reference wall.

Period	"S" wall/Reference wall areal heat capacity relative difference %
1 hour	+95
24 hours	+17
14 days	-5
365 days	-5

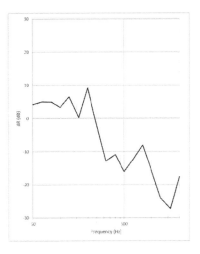

Figure 8. Acoustic improvement ΔR of the « S » wall in comparison with the reference wall.

octave bands on the wall with projecting bricks show significant deviations of several decibels (Figure 8). STL values are calculated using FEM/BEM method both for reference wall and "S" wall.

Gap of STL in low frequency range (frequency below 300 Hz) is significantly improved from +5 dB to +10 dB. However, the acoustic improvement decreases till -26 dB for frequency above 300 Hz.

7 CONCLUSION

The study shows the significant impact of the heat transfer surface on the thermal inertia of exposed brick walls. Increasing this surface using a device of projecting bricks significantly increases the apparent areal heat capacity for periods of time of one hour and one day.

From the acoustic point of view, at practically constant surface mass (i.e. without acoustic impact on a homogeneous planar wall), the mass law is no longer applicable when calculating STL of the "S" wall. The STL is greatly degraded in mid and high frequency ranges from 300 Hz, and this wall type should be the subject of a specific acoustic study.

A continuation of the study will consist in assessing the sound diffusion index, in evaluating the impact on intelligibility (i.e. acoustic correction) in a room, and in evaluating the impact of this type of wall on the thermal behavior of a room.

REFERENCES

AFNOR 2017. Performance thermique des composants de bâtiment – Caractéristiques thermiques dynamiques – Méthodes de calcul. NF EN ISO 13786. Saint-Denis La Plaine: AFNOR.

Allen, J. B. & Berkley, D. 1978. A. Image method for efficiently simulating small-room acoustics. Journal of the Acoustical Society of America. 65(4): 943–950.

Bouchié, R. et al. 2018. RT: valeurs et coefficients pour l'application des règles Th-Bat. Marne-la-Vallée: CSTB Éditions.

Carslaw, H.S., Jaeger, J.C. 1959. Conduction of Heat in solids. New York: Oxford University Press.

Commission Th-bat 2017. Règles Th-bat. Fascicule Inertie. Downloadable from the website http://www.rt-batiment. fr/.

Degiovanni, A. 1999. Transmission de l'énergie thermique. Conduction. BE8200. Saint-Denis: Techniques de l'Ingénieur.

Funkhouser, T. et al. 2004. A Beam Tracing Method for Interactive Architectural. Acoustics. Journal of the Acoustical Society of America. 115(2): 739–756.

Guyader, JL. 2002. Vibrations des milieux continus. Cachan: Hermes - Lavoisier.

Jean, P. & Roland, J. 2001. Application of the Green Ray Integral Method GRIM to sound transmission problems. Building Acoustics. 8(2): 139–156.

Lehnert, H. 1993. Systematic Errors of the Ray-Tracing Algorithm. Applied Acoustics. 38 (2-4): 207–221.

Mechel, F.P. 2002. Improved Mirror Source Method in Room Acoustics. Journal of Sound and Vibration. 256(5): 873–940.

Brick and Block Masonry - From Historical to Sustainable Masonry –
Kubica, Kwiecień & Bednarz (eds)
© 2020 Taylor & Francis Group, London, ISBN 978-0-367-56586-2

Critical overview of problems with existing buildings made in the "big block" technology with insulation from AAC blocks. Monitoring and repair possibilities

L. Bednarz
Wroclaw University of Science and Technology, Wroclaw, Poland

D. Bajno
UTP University of Science and Technology, Bydgoszcz, Poland

ABSTRACT: The masonry structures have been evaluated in many directions during several thousand years of their use. They have changed in terms of quality, material, strength and heat. From traditional typical (historical) wall structures, pillars, arches, ceilings, domes and vaults to, sometimes very sophisticated, but also impractical, solutions used in the 20th century. The article concerns untypical masonry constructions such as fillings of residential buildings made in the "big block" technology, as well as external insulation layers from AAC wall blocks. Structures of this type were used in residential buildings for many years. Nowadays, these technologies are being abandoned but the existing buildings remain. The article presents the experience gained during the research and analysis of residential buildings made in the above described technology. It also proposes ways of monitoring technical condition and technologies of repair of this type of buildings.

1 INTRODUCTION

The system of constructing multi-family buildings mentioned in the abstract was burdened with many technological defects, including low thermal insulation of external partitions and numerous thermal bridges (Marciniak 2018, Rybka 2005). At present, the technical efficiency of this type of external partitions in terms of their heat resistance far exceeds the requirements of current regulations. Moreover, at present, the fixing of thermal insulation (autoclaved aerated concrete - AAC blocks) to prefabricated wall blocks has become very controversial.

The analyzed buildings were built in the first half of the 1970s (Biliński et al. 1975). Their foundation is a reinforced concrete foundation slab permanently connected to reinforced concrete basement walls, which was to create a box foundation. The ceilings above the basement storey and partially above the ground floor are made as monolithic - reinforced concrete, while on the higher storeys and at the level of the roof there are multi-channel (prefabricated) ceiling slabs. The structure of external walls, which are the subject of the article, consists of prefabricated, externally insulated multi-channel wall blocks with a thickness of 24 cm (constructional part) together with a 14 cm layer of external facade. The lining consists of: 12 cm thick cellular concrete (autoclaved aerated concrete - AAC blocks) and a 2 cm thick graft layer of cement mortar. From the outside, the facades are covered with cement

plaster. The total thickness of the external supporting wall is 38 cm. The external 24 cm thick curtain walls and internal partition walls are made of AAC blocks and finished with cement plaster on both sides.

2 MASONRY STRUCTURES AS HEAT-PROTECTIVE BUILDING ENVELOPES

2.1 *Technologies of thermal insulation of vertical wall panels, used in the second half of the 20th century*

Among the many used as the first one, the method of insulating with insulation panels covered with traditional plaster, on metal mesh should be mentioned. This method has been developed as a response to technological defects occurring in large panel construction, i.e. the need to protect against freezing and leakage, and therefore it was widely used in the 1970s. In the technology of industrialised large-panel construction, both at the design and construction stage, the main concern was to ensure structural safety, ignoring the importance of physical processes taking place inside the walls and their impact on the safety and durability of facilities. This method consisted in "wrapping" the surface of the walls (using cement mortar with a polyvinyl acetate additive) with at least two centimeters of polystyrene foam, which was protected by a three-layer cement and lime plaster, laid on a pre-wall steel mesh of rebar, fastened with steel pins. The idea of such wall insulation is shown in Figure 1.

The POSS/70 method (Figure 2), used in the 1980s, can also be regarded as the heavy wet method, in which the insulating material was composite two-layer panels consisting of polystyrene foam and chipboard (in the construction practice different panel thicknesses were used).

In the 1980s, the so-called dry method was commonly used, in which the external finish (façade) was a cladding made of flat asbestos-cement (AC) panels, attached to a wooden frame. Alternatively, the cladding made of trapezoidal steel sheets fixed to a galvanized steel grate was used here. Both methods were used mainly on facades of large panel buildings up to a height of 35 m, with so-called technological defects. An example of the method of fastening the grate and cladding in the dry method is shown in Figure 3a, with AC plate cladding, and Figure 3b with coated trapezoidal sheet cladding.

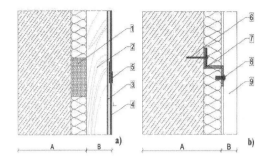

Figure 3. Scheme of the light-dry method used for renovation of large panel buildings;A- existing wall, B- protective layer, 1-waterproof plywood backing, 2-plywood grate (or profile of galvanized metal tee)/mineral wool, 3-crubber seal, 4- asbestos-cement plate, 5- steel connector, 6- anchoring pin, 7- profile made of galvanized metal tee, 8- single-sided rivet, 9- fold coated sheet metal T-55 or T-30 (Arendarski 1988).

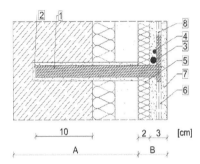

Figure 1. Scheme of the heavy wet method used for renovation of large panel buildings; A- existing wall, B- protective layer, 1-anchoring pin Ø 20 mm, 2-cement mortar, 3-bottom bar of the lower construction mesh Ø 10 mm, 4-bottom bar of the upper construction mesh Ø 8 mm, 5-styrofoam, 6-Rabitza wire mesh, 7-plaster 3 cm thick, 8-vertical bar of the construction mesh (Arendarski 1988).

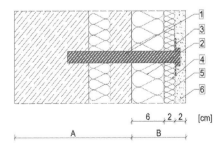

Figure 2. Scheme of the heavy wet method used for the renovation of large panel buildings; A- existing wall, B-protective layer, 1-segment expansion joint, 2-plaster, 3-plaster of galvanized metal tee, 4- chipboard, 5- Rabitza wire mesh, 6-plaster 2 cm thick (Arendarski 1988).

One of the problems that became the main reason for withdrawing this technology (with AC panel cladding) from use in the construction industry was the harmfulness of asbestos, which contained façade panels.

2.2 Insulation technologies for vertical wall panels, currently used

Currently, there are jointless systems of external thermal insulation in several dozen different varieties. Insulation consists in attaching the thermal insulation material (Styrofoam, mineral wool) to the walls and then covering it with an external layer, which usually includes a layer reinforced with glass cloth and a thin layer of plaster or ceramic lining. In the construction practice, it has become a common phenomenon to mix different (often incompatible with each other) elements of insulation systems dictated by the price criterion, i.e. using the cheapest solutions and materials (mesh, glue, plaster, other accessories), which has become a determinant of many damages and, most of all, has affected the durability of the facade. The result of such actions were numerous damages to thermal insulation systems. Research works led to the creation of ETICS (External Thermal Insulation Composite System), i.e. an external complex thermal insulation system, significantly different from the traditional one. Publications describing this method, e.g. (Barreira et al. 2013, Amaro et al. 2014.) include important aspects in terms of design, including, among others: how to determine the thickness of insulation taking into account the influence of mechanical fasteners (point bridges) and taking into account the influence of linear bridges, the principle of taking into account the durability of the system under different operating conditions and the level of fire safety.

3 USE OF MASONRY PRODUCTS FOR NON-BEARING THERMAL INSULATION LAYERS

3.1 Design guidelines for the construction of layered walls using masonry elements

Originally, wall blocks were made as uninsulated concrete elements in steel, demountable forms. Then, in the lying position, cellular concrete blocks were laid on the cement mortar, additionally flooding from the top with cement slurry. After the prefabricated elements had been assembled on the storey, they were finished in the final phase with plasterwork on both sides.

A newer solution was the production of prefabricated concrete elements including insulation. They were manufactured in moulds. External wall blocks were produced in demountable forms, in a horizontal position, where blocks of cellular concrete were placed on their expected surface, then rods of ribbed steel #6 ÷ #8 were driven into them, in the line of inter-channel ribs. Next, two longitudinal reinforcing bars were laid and steel cores were introduced to profile the channels. The prepared form was filled with concrete. An example of such a large block with a layer of insulation from AAC blocks is shown in Figure 4.

Figure 5 shows a model of the pre-fabricated section described above with the fastening rods visible. These bars were the only suspension structure for AAC blocks with a volume-weight of 700 kg/m³ to 900 kg/m³.

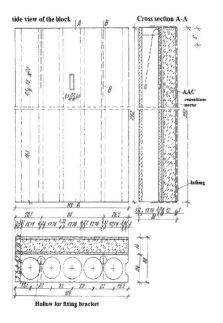

Figure 4. External wall block with mortar-reinforced insulation.

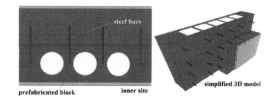

Figure 5. A scheme of a prefabricated wall with bars fixing the insulation from AAC blocks.

Table 1. Mandatory heat transfer coefficients for walls in individual years.

| | Heat transfer coefficient U_{max} |
Year	$[W/(m^2 \cdot K)]$
1957	1.16 – 1.42
1964	1.16 – 1.42
1974	1.16 – 1.42
1982	0.75
1991	0.55 – 0.70
2002	0.30 – 0.50
2008	0.30
2014	0.25
2017	0.23
2021*	0.20

* Estimated value.

3.2 Real conditions of manufacture of prefabricated elements

However, the implementation reality was much more than just the design assumptions. The low quality of workmanship led to a situation where a large part of the elements did not meet the dimensional standards and the elements were damaged or missing. Despite many improvements, there were numerous problems with joints and insulations. The wall thermal insulation standards that have been in force for years (see Table 1) were many times lower than those required today.

4 TECHNICAL EFFICIENCY OF THE PARTITIONS, METHODS OF THEIR MONITORING AND REPAIR

4.1 Technical efficiency of "large block" partitioning with thermal insulation from AAC blocks

From the static side, the structure of these already over 50 years old buildings made in "big block" technology is usually not damaged. However, due to the full covering of the external load-bearing walls with a curtain layer, it is not possible to assess the

584

technical condition of the external load-bearing walls of buildings during testing. This may only be possible after the existing cladding has been removed and concrete surfaces have been cleaned of mortar and plaster. However, numerous defects of the existing curtain layer of external walls can be observed, indicating a much lower than currently required effectiveness in limiting heat losses through external partitions. There can be no question of obtaining minimum thermal parameters of partitions. External walls with the facade made of AAC blocks, finished with cement plaster, should be considered as too "modest" insulation in relation to the requirements of currently binding regulations (Table 1). This is due to the low efficiency of the applied insulation.

This type of buildings has many thermal bridges, located in the areas of balcony slab fixings, in the vicinity of lintels, in the places where windows are embedded and in the corners of buildings. In addition, there may be bands of walls with clearly scratching fields with significantly higher heat losses in relation to other places, which may indicate local dampness of the walls and a significant decrease in their heat protection parameters. The technical condition of buildings will not improve without improving the thermal insulation parameters of these surfaces and at the same time maintaining full continuity of thermal insulation with the required lowest possible heat transfer coefficient and required thickness.

A very important element of correct and safe operation is also ensuring efficient ventilation. During many air flow measurements in this type of facilities, many inoperative ventilation openings with even zero draught were found, as well as many sealed ventilation grilles. Such a situation not only does not allow for the required replacement of used air, but also does not allow the partitions to dry out. The execution of insulation (made of foamed polystyrene) from the inside on external walls may periodically improve its thermal parameters, but due to the movement of the low temperature zone towards the insulation (from the outside) it will cause periodic freezing and condensation of water vapour, causing permanent dampness at the contact between the wall and the insulation. Such a situation in the absence of the possibility of effective moisture release to the interior of the room (high diffusion resistance of polystyrene foam) and in the absence of efficient ventilation is the main cause of the appearance and development of mould fungi in rooms.

4.2 Procedures for monitoring technical condition and repair possibilities

The technical condition of external wall claddings (insulation) is bad in most buildings with structural tees. They are heavily cracked and damp, which means that they have already lost their thermal continuity and at the same time threaten to break away

Figure 6. UAV with thermal imaging camera used for facade inspection.

from the facade. The current technical possibilities allow to use for this purpose also flying drones (UAV), also equipped with thermovision cameras (Figure 6).

In order to locate faults, mainly technological, in the objects, e.g. in the scope of e.g. heat resistance of their external partitions, together with the indication of surface, linear and point thermal bridges, their external and internal visual inspection under visible light and thermovision should be performed (Figure 7 - 9).

The purpose of such inspections is not only to show and determine the size of thermo-humidity parameters of weak points of buildings, but above all to indicate the heterogeneity of their external structures, which protect buildings from excessive heat loss. Performing such observations (measurements) on a continuous basis, if the procedures of proceeding in case of exceeding the defined limit values are also implemented, can be called monitoring. Monitoring of this type should be complemented by humidity

Figure 7. External wall insulated with AAC blocks with visible thermal bridges (in thermovision and visible light).

Figure 8. Thermal bridges between the insulation with AAC.

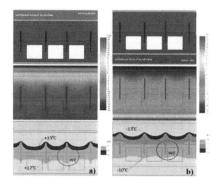

Figure 11. The distribution of isotherms of temperature in the wall: a) insulated from outside, b) insulated from inside.

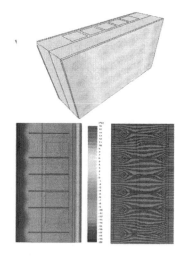

Figure 9. Thermal bridges between the insulation with AAC.

and temperature measurements carried out with non-destructive methods outside and inside the walls (Bajno 2020) and by simulation calculation models in order to confirm the diagnosis made. The results of calculations, made in software for thermal analysis of building components (also in graphic form) should be included in the report for owners and managers.

Figure 10 - 11 shows the temperature distribution in sections of prefabricated elements in their existing state and after thermomodernization. Due to the

Figure 12. Isothermal distribution and adiabate of heat streams concentrated in rods locations.

doubts about the connections of the outer insulation layer, the possibility of insulation from the inside has also been taken into account here, which, as the last diagram of Figure 11b shows, will not be beneficial for the existing fixings. The bars may be in a very cold zone during the winter, which may lead to frost damage and the fall of façade fragments. Figure 12 illustrates the adiabat streams of compacted heat in the locations of the bars (pins) fixing the existing insulation.

Depending on the condition in which the façade is standing, the correct direction for further repair must be taken. Some suggestions are set out below.

4.2.1 Proposal no. 1

It is proposed to remove the 12 cm thick AAC layer together with a layer of mortar (approx. 2 cm) and replace it with a layer of approx. 12-16 cm of mineral wool. It will be an optimal solution connected with significant improvement of heat transfer

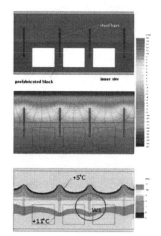

Figure 10. Isothermal distribution in cross-sections of pre-fabricated elements in the existing state.

coefficient and correct (safe) migration of moisture inside partitions. It is suggested to anchor to the mandrels of reinforced concrete external walls of the building and inter-channel ribs. It is necessary to remember about the necessity to perform additional static calculations of the building related to significant uneven load.

4.2.2 *Proposal no. 2*

It is proposed to dismantle the 12 cm thick AAC layer together with a layer of mortar (approx. 2 cm) and replace them with a layer of approx. 12 cm of mineral insulating panels made of a very lightweight variety of AAC (115 kg/m^3) with heat-protective parameters comparable to mineral wool or polystyrene. From the point of view of building physics, it is also an interesting and effective solution involving a significant improvement in the heat transfer coefficient, and thus reducing heat loss. It is necessary to remember about the necessity of additional panel anchoring to walls. It is suggested to anchor the reinforced concrete pins of the external walls of the building and the inter-channel ribs.

4.2.3 *Proposal no. 3*

It is proposed to strengthen the 12 cm thick AAC layer with local injections of resin composition, strengthening of the whole existing facade made by its covering with C-FRCM. In the next stage, insulation of the reinforced facade with mineral wool, approx. 5-10 cm thick. From the point of view of building physics, this solution is also connected with the improvement of heat transfer coefficient and reduction of moisture condensation in the partition. It is necessary to remember about the necessity of anchoring composite mesh to walls. It is suggested to anchor external walls to the concrete pillars. This solution does not require any additional static calculations of the building related to its load.

5 CONCLUSION

Conducting a thermomodernization procedure is an absolutely necessary activity to ensure proper thermal management of buildings with such construction. At present it is not possible to speak about external building partitions, having identical thermal insulation properties over their entire surface. They have a number of thermal bridges, occurring at the joints of "large block" panels, in the joints of curtain layers made of AAC blocks, as well as on stiffening elements, such as pins or reinforced concrete transoms. As thermal imaging tests have shown, the density of linear and point thermal bridges on external walls is so high that it would be impossible to provide even the minimum required thermal insulation parameters.

The chosen technology for the future thermomodernisation procedure is primarily intended to ensure the continuity of thermal insulation of the external partitions and to reduce the dense mesh of thermal bridges on the facade. Replacement of the existing AAC layer with local hollow brick, lightweight AAC panels and mineral wool panels with a low thermal conductivity coefficient must be carried out in accordance with the required minimum values for thermal transmittance coefficients for external partitions. However, it is recommended, thinking ahead, to make the insulation with the highest possible parameters, bearing in mind that such treatments are performed on average once every 30 years.

The adopted solution of replacing of the external wall curtain layers with a much lighter material, i.e. mineral wool, requires a full and detailed static analysis of the structure in terms of reducing the weight of the structure, transferred to the foundation. It is necessary to see it in the construction and executive design and to make static-durability calculations, taking into account the foundation conditions. In addition, a plan for the organisation of the construction work must be drawn up, which will indicate the exact stage of the thermal insulation work. This will prevent situations in which, e.g. after the removal of protective layers from one side of the building, the load transferred to the foundation will be uneven.

REFERENCES

Amaro, B., Saraiva, D., de Brito, J., Flores-Colen, I. 2014. Statistical survey of the pathology, diagnosis and rehabilitation of ETICS in walls. Journal of Civil Engineering and Management, 20(4): 511–526.

Arendarski, J. 1988. *Poprawa izolacyjności cieplnej budynków mieszkalnych*. Arkady, Warszawa.

Bajno, D., Bednarz, L., Matkowski, Z., Raszczuk, K. 2020. Monitoring of Thermal and Moisture Processes in Various Types of External Historical Walls. Materials 13: 505.

Barreira, E., de Freitas, V.P. 2013. Experimental study of the hygrothermal behaviour of External Thermal Insulation Composite Systems(ETICS). Building and environment 63.

Biliński, T., Gaczek, W., Klorek, E. 1975. *Systemy prefabrykowanego budownictwa ogólnego*. Politechnika Poznańska. Poznań.

Marciniak, P. 2018. Prefabricated elements and typification in communist Poland. In *6th International Congress on Construction History (6ICCH 2018), Brussels, Belgium July 9-13, 2018.*

Rybka, A. 2005. Using MBT in transformation of multi-family prefabricated buildings. In *Improvement of Buildings' Structural Quality by New Technologies: Proceedings of the Final Conference of COST Action C12, 20-22 January 2005*, Austria, Innsbruck: CRC Press.

Existing masonry - monitoring, inspection, repair & strengthening

Brick and Block Masonry - From Historical to Sustainable Masonry –
Kubica, Kwiecień & Bednarz (eds)
© 2020 Taylor & Francis Group, London, ISBN 978-0-367-56586-2

Impact of type of mortar on shear bond strength of brick masonry

M. Ramesh, C. Briceño, M. Azenha & P.B. Lourenço

Department of Civil Engineering, ISISE, Universidade do Minho, Portugal

ABSTRACT: When masonry is subjected to different loading conditions, quite often its failure occurs due to failure in bond between the unit and mortar. Among other factors such as surface roughness and porosity of the brick, the type of mortar used plays a significant role in influencing the shear bond strength of masonry. This work is aimed at understanding the impact of the type of mortar used on the shear strength of masonry triplets, tested according to EN 1052-3:2002. Clay bricks (solid and frogged) were used with three different mortars - 1:5 (Cement: Sand), 1:1:6 and 1:2:9 (Cement: Lime: Sand), by volume, all designed to have a flow value of 175±10 mm. Bond strength has been discussed as a function of presence of lime in the binder as well as the strength of the mortar, at 90 days of age.

1 INTRODUCTION

The structural performance of masonry relies heavily on the bond between brick and mortar, which often dictates its mode of failure as well (Resketi *et al.* 2019). Amongst the different types of failures that are known to occur in masonry joints, shear failure is recognized as one of the most common types (Rahman & Ueda 2014). It is, therefore, possible to find works from the past that have studied the shear bond strength between unit and mortar using either couplet or triplet specimens with different materials (Rahman & Ueda 2014, Resketi *et al.* 2019, Lourenço *et al.* 2004). These studies exhibit the linear dependency of shear stress (τ) on normal stress (σ) and use the Mohr-coulomb criterion to estimate values of parameters like cohesion (c) and angle of internal friction (ϕ), expressed by the equation $\tau = c + \tan \Phi . \sigma$, where *tan ϕ* may also be denoted as (μ). This principle, however, may be adopted only for low to moderate values of normal stresses, because in higher values, the relationship becomes non-linear (Lourenço *et al.* 2004). The general range of values for the coefficient of friction (denoted as μ or *tan ϕ*) in literature is found to vary from 0.5 to 1.2 (Lourenço *et al.* 2004, Vasconcelos & Lourenço 2009, Bei & Papayianni 2004, Pluijm 1999). Zhang *et al.* (Zhang *et al.* 2018) studied different test setups for measuring shear strength and discovered that the general accuracy in estimation of the coefficient of friction (μ) is higher than that of cohesion (c). Abdou *et al.* (Abdou *et al.* 2006) studied the shear behavior of joints using a commercial mortar with two different types of clay bricks (solid and hollow) and found that the type of brick did not influence the angle of internal friction in the mortar joint. Based on this information, it may then be worthwhile to test if the type of mortar employed, has an impact on

the coefficient of friction. Such studies are not easily found in the literature. Alecci *et al.* (Alecci *et al.* 2013) did study the shear strength of triplet specimens with different mortars but measured only the initial shear strength (or cohesion, denoted by c), in the absence of any normal stresses. Qualitatively, it is possible to observe from their study, that a stronger mortar (~8 MPa, compressive strength) compared to a weaker (~3 MPa, compressive strength) leads to higher values of initial shear strength in the masonry joint. However, the absence of information on the frictional coefficient leaves room to study the impact of the type of mortars on different shear parameters. This is what this research aims to contribute to. The choice of binder composition in this work has been guided by two of the most commonly used components in conventional mortars, namely lime, and cement. It is known that the presence of lime tends to make masonry more deformable in compression and improves flexural bond strength (Haach *et al.* 2010, Pavia & Hanley 2010), but information on shear bond strength is not easily available.

Therefore, this research has chosen two mortars with similar compressive strength, 1:5 (cement: aggregate) and 1:1:6 (cement: lime: aggregate) to understand the influence of the presence of lime in the mortar on shear strength of masonry. Furthermore, a mortar 1:2:9 (cement: lime: aggregate) has also been studied to compare two mortars with similar binder compositions but different compressive strengths. Triplets were formed with these mortars and solid clay bricks, to perform shear tests according to EN 1052-3, with three specimens tested in each case for three different levels of pre-compression: 0.2 MPa, 0.6 MPa, and 1 MPa. Parallel mechanical characterization of the mortars for strength and density have also been presented.

2 RESEARCH PROGRAM

2.1 Raw materials for mortars

For the binders, air lime (CL 90-S, bulk density 0.36 g/cm^3) and Portland cement (CEM I-42.5 R, bulk density 0.93 g/cm^3) were chosen. The choice of CEM I was guided by repeatability and consistency in the results obtained from the experimental campaign. Because even though CEM II is more often used in real applications, its chemical constituency has greater variation than that of CEM I, which would lead to more complicated chemical reactions and possibly greater difficulty in interpretation of resulting mechanisms, as well. The aggregate chosen was well-graded and siliceous with a particle size up to 0/4 mm.

2.2 Mortar compositions and masonry construction

Three mortars were chosen for this experimental campaign, the compositions of which were 1:0:5 or 1:5, 1:1:6 and 1:2:9 (Cement: Lime: Aggregate) expressed by volume. The notations 1C5S, 1C1L6S and 1C2L9S respectively have been used to denote each of the mortar mixes. All measurements were done by weight to increase accuracy in the mix designs i.e. volume batched by weight (Table 1). The water-binder ratios were chosen such that all mortars had a flow table value of 175±10 mm, to facilitate the use of the mortar by masons in real applications. The mixing process was performed according to EN 196-1 (2005) and was carried out by a professional mason in the laboratory.

All mortars were used within one hour of mixing. For each mortar that was made, a small quantity was used to cast prisms for characterization of its compressive strength, flexural strength, and hardened density. For the sake of consistency, these prismatic specimens were cured in the same conditions as those of the masonry specimens; in the basement of the laboratory, with relatively stable temperature and humidity conditions, around ~21±1°C and 70±5% RH. The brick used was a solid, frogged clay brick with dimensions 215×102×65 mm and supplied by Wienerberger, DoP number 152110-B1W1210 (Table 2). All bricks were well dusted and soaked in water for ~30 minutes before construction. The size of the mortar joints was kept at 10 mm in all the specimens which were constructed as triplets.

Table 1. Composition of mortars used in this campaign.

Nomenclature of mixes (Ratio by volume)	Cement (kg)	Lime (kg)	Water/binder (By weight)
1C2L9S (1:2:9)	128.4	97.9	1.3
1C1L6S (1:1:6)	192.6	73.4	1.09
1C5S (1:0:5)	233.5	0	1.2

Table 2. Characteristics of the brick used in this campaign, as specified in the datasheet (EN 771-1:2011+A1:2015).

Property	Value
Compressive strength	12 N/mm^2
Water absorption	~15%
Initial rate of absorption	1-5 kg/(m^2.min)
Net density	1800 kg/m^3
Gross density	1650 kg/m^3

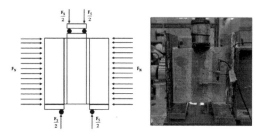

Figure 1. Illustration of shear bond test and image of specimen.

2.3 Details of the experimental setup

The mortar specimens were tested according to EN 1015-11 (2007), in flexure and compression. A previous article by this research team, focuses on mortar level studies in greater detail, quantifying and correlating different mechanical properties (Ramesh et al. 2019).

For masonry, 27 triplet specimens were tested according to EN 1052-3 (2007), with 9 specimens for each type of mortar. Thereafter, each type of mortar was subjected to three levels of pre-compression: 0.2 MPa, 0.6 MPa, and 1 MPa, three specimens being tested in each case. A manual hydraulic pump was used to apply the pre-compression and the load was kept constant through the duration of the test. The vertical load applied was meant to shear the specimens and a different actuator was employed for such purpose with displacement control at a rate of 3 microns/second, whereas acquisition frequency was of 4 Hz. Two LVDTs were used on either side of the specimen to measure the relative slip in the bricks (Figure 1). All tests were carried out at 90 days of age.

3 RESULTS

3.1 Mortar characterization

Prismatic specimens of the mortars were tested at 90 days of age for compressive strength and hardened density (Table 3). It may be observed that though the mix 1C5S has the highest compressive strength, 12.1

Table 3. Composition of mortars used in this campaign.

Mortar mixes	Compressive strength (MPa) (CoV %)	Flexural strength (MPa) (CoV %)	Density (kg/m³) (CoV %)
1C2L9S	5.3 (5.2%)	1.9 (2.1%)	2.06 (.5%)
1C1L6S	10.1 (8.5%)	3.5 (7.8%)	2.12 (.1%)
1C5S	12.1 (6.0%)	3.8 (11.3%)	2.07 (.2%)

MPa, the strength of the mix 1C1L6S is also in the same range, 10.1 MPa. The mix 1C2L9S has a strength of 5.3 MPa which is almost half the value of the strength of the other two mixes. Flexural strength of the mixes also seems to follow the same pattern as that of compressive strength, whereas density is observed to be similar for the different mortars.

Based on the values strength and density obtained, it is possible to compare the performance of two mortars 1C5S and 1C1L6S with the same brick in shear bond strength, from a mechanical point of view. A major difference in the two mortars lies in their composition since 1C5S has only cement in the binder, while 1C1L6S has an equal amount of cement and lime in the binder by volume. Furthermore, it may also be interesting to compare the performance of the mixes 1C1L6S and 1C2L9S since both have a binder-aggregate ratio of 1:3 and a binder with lime and cement in them, but with different values of compressive strength. Since the mix 1C2L9S has a greater quantity of lime in its binder, by volume, 67%, compared to 1C1L6S which has 50%, it is expected to have a lower value of strength at the mortar level (Arandigoyen *et al.* 2007, Ramesh *et al.* 2019).

3.2 *Experimental results of masonry specimens*

Results of shear stress versus relative slip in the bricks have been graphically shown for each of the three mortars, with results chosen from one sample specimen for each pre-compression level (Figure 2, 3, 4).

The two curves in each case of pre-compression, exhibit results from the two LVDTs that were used. It may be observed that both joints did not always shear simultaneously . It is also easy to notice, that in all mortars, as the level of pre-compression was increased, the maximum shear capacity of the specimen also increased.

From the graphs, it may also be observed that while specimens with mortar 1C2L9S exhibit lower values of shear stress, the specimens with mortars 1C5S and 1C1L6S exhibit relatively similar values of shear stress for each level of precompression. The numeric values of average shear stress have been presented in Table 4. Shear stress was calculated by averaging values obtained from three different specimens, for each level of vertical precompression.

To obtain values of parameters like the coefficient of friction and cohesion, shear stress for each mortar

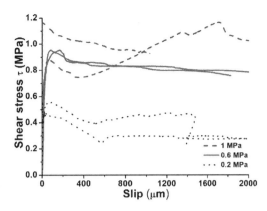

Figure 2. Shear stress versus slip for mortar 1C5S (1:0:5).

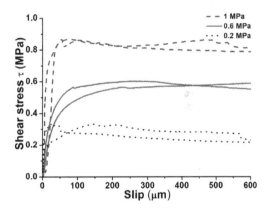

Figure 3. Shear stress versus slip for mortar 1C1L6S (1:1:6).

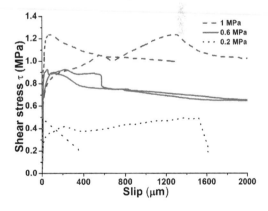

Figure 4. Shear stress versus slip for mortar 1C2L9S (1:2:9).

type was plotted versus normal stress (or vertical precompression) (Figure 5).

Additionally, to ensure that the Mohr-coulomb criterion could be adopted, the linearity of the

Table 4. Joint mortar characteristics in masonry specimens.

Type of mortar/Shear stress (MPa)	Vertical precompression/ Normal stress		
	0.2 MPa	0.6 MPa	1 MPa
1C2L9S	0.37	0.73	0.83
1C1L6S	0.43	0.87	1.12
1C5S	0.54	0.89	1.11

Table 5. Joint mortar characteristics in masonry specimens.

Type of mortar	Cohesion (c)	Coefficient of friction (tan φ)	Angle of friction (φ)
1C2L9S	0.298	0.58	29.99
1C1L6S	0.286	0.87	40.98
1C5S	0.414	0.72	35.81

The second observation is that despite two of the mortar mixes having similar compressive strength 1C5S and 1C1L6S (Table 3), and similar values of shear strength for different levels of precompression (Table 4), the mix with lime, 1C1L6S seems to result in a slightly higher value of the coefficient of friction. The difference in the values of the coefficient of friction does not appear to be very large for the two mixes at this age of 90 days. However, since air lime is known to carbonate with time and gain strength (Arandigoyen et al. 2007), it may be worth testing the same values after perhaps, 180 days or 365 days to explore the change in shear strength parameters with time.

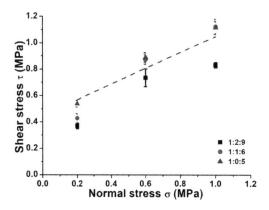

Figure 5. Shear stress versus normal stress for different mortars.

relationship between shear stress and normal stress had to be confirmed, for each mortar type. Linear regression was performed and R^2 values were found to be 0.90, 0.97 and 0.98 for the mortars 1C2L9S, 1C1L6S, and 1C5S respectively, thus making it acceptable to use the Mohr-coulomb equation $\tau = c + \tan\Phi.\sigma$, where tan φ or (μ) refer to coefficient of friction, c refers to cohesion or initial shear stress, (τ) denotes shear stress and (σ) normal stress or vertical precompression. The values of the obtained parameters for each of the mortar types have been displayed in Table 5. The mix 1C5S exhibits the highest value of cohesion or initial shear strength, while the mix 1C1L6S results in the highest value of the coefficient of friction or angle of friction.

The first observation is that for the two mortars 1C1L6S and 1C2L9S, with similar binder composition, compressive strength (Table 3) of the mortar may be an indication of the shear strength of the joint between mortar and brick. The mix 1C1L6S consistently exhibits higher values of shear strength than the mix 1C2L9S for all levels of vertical precompression (Table 4). It also results in a higher value of the coefficient of friction, albeit a similar value of cohesion (Table 5).

4 CONCLUSIONS

This paper examines shear strength and joint mortar characteristics for three different mortars. Triplet masonry specimens were built using the same clay brick in each case and the specimens were subjected to different levels of vertical precompression (0.2, 0.6 and 1MPa). Globally, it was found that values of shear strength varied between 0.4 to 1.1 MPa, cohesion was found to vary between 0.3 to 0.4 MPa and the coefficient of friction was found to vary from 0.58 to 0.87. Based on the results obtained, the following conclusions may be made:

1) For mortars with similar binder composition and binder-aggregate ratio, greater compressive strength appears to have a direct impact on shear strength, increasing the value of bond strength between brick and mortar, as well as increasing the value of internal friction.

2) For two mortars with similar compressive strength, the presence of lime in the binder appears to increase the value of the coefficient of friction, compared to a cement only mortar, when tested at 90 days of age.

It must be mentioned that this study is preliminary in nature, to understand the impact of the type of mortar on the mortar joint characteristics in shear. More experiments need to be conducted before a generalization of any nature. Nevertheless, initial results show that this subject merits further investigation, with different types of bricks and mortars.

ACKNOWLEDGEMENTS

Funding provided by the Portuguese Foundation for Science and Technology (FCT) to the scholarship SFRH/BD/137358/2018, the research project PTDC/ECM-EST/1056/2014 (POCI-01-0145-FEDER-016841), and to the Research Unit ISISE (POCI-01-0145-FEDER-007633) is acknowledged. The authors are also grateful to the European Lime Association for funding this project.

REFERENCES

Resketi, N.A. & Toufigh, V. 2019. Enhancement of brick-mortar shear bond strength using environmental friendly mortars. *Construction and Building Materials* 195: 28–40.

Rahman, A. & Ueda, T. 2014. Experimental investigation and numerical modeling of peak shear stress of brick masonry mortar joint under compression. *Journal of Materials in Civil Engineering* 26(9).

Lourenço, P.B., Barros, J.O. & Oliveira, J.T. 2004. Shear testing of stack bonded masonry. *Construction and Building Materials* 18: 125–132.

Vasconcelos, G. & Lourenço, P.B. 2009. Experimental characterization of stone masonry in shear and compression. *Construction and Building Materials* 23 (11):3337–3345.

Bei, G. & Papayianni, I. 2004. Experimental study of shear bond strength of traditional masonry. *Proc. 13th International Brick and Block Masonry Conference, Amsterdam, 4-7 July 2004.*

Pluijm, R. van der. 1999. Out-of-plane bending of masonry: behaviour and strength. *Eindhoven: Technische Universiteit Eindhoven.*

Zhang, S., Richart, N. & Beyer, K. 2018. Numerical evaluation of test setups for determining the shear strength of masonry. *Materials and Structures* 51:110.

Abdou, L., Saada, R.A., Meftah, F. & Mebarki, A. 2006. Experimental investigations of the joint-mortar behaviour. *Mechanics Research Communication* 33(3):370–384.

Alecci, V., Fagone, M., Rotunno, T. & Stefano, M.D. 2013. Shear strength of brick masonry walls assembled with different types of mortar. *Construction and Building Materials* 40:1038–1045.

Haach, V.G., Vasconcelos, G., Lourenço, P.B., Mohamad, G. 2010. Influence of the mortar on the compressive behavior of concrete masonry prisms. *Mecânica Experimental* 18:79–84.

Pavia, S. & Hanley, R. 2010. Flexural bond strength of natural hydraulic lime mortar and clay brick. *Materials and Structures* 43:913–922.

CEN, EN 196-1:2005. *Methods of testing cement-Part 1: Determination of strength.*

CEN, EN 1015-11:(1999). 2007. *Methods of Test for Mortar for Masonry, Part 11: Determination of Flexural and Compressive Strength of Hardened Mortar.*

Ramesh, M., Azenha, M. & Lourenço, P.B. 2019. Quantification of impact of lime on mechanical behaviour of lime cement blended mortars for bedding joints in masonry systems. *Construction and Building Materials* 229:116884.

CEN, EN 1052-3(2002). 2007 *Methods of test for masonry - Part 3: Determination of initial shear strength.*

Arandigoyen, M. & Alvarez, J.I. 2007. Pore structure and mechanical properties of cement–lime mortars. *Cement and Concrete Research* 37(5):767–775.

Brick and Block Masonry - From Historical to Sustainable Masonry –
Kubica, Kwiecień & Bednarz (eds)
© 2020 Taylor & Francis Group, London, ISBN 978-0-367-56586-2

Assessment of brick masonry strength using tests on core samples cut from the structures

P. Matysek & S. Seręga
Cracow University of Technology, Cracow, Poland

ABSTRACT: A common diagnostic task for masonry structures is to assess compressive strength of existing brickwork. This article presents a method of assessing masonry compressive strength using tests on core samples extracted from masonry walls and pillars. This method can be classified as minor - destructive one (MDT) and is especially useful for historical buildings. In the research, core samples with diameters of 150 mm and 100 mm were tested. The core samples were cut from brick masonry made at the laboratory and from historical structures erected in the 19th and 20th century. The component materials were ceramic solid bricks and lime or cement - lime mortars. The size effect on compressive behaviour of core samples was analysed for different types of masonry. A similar ratio between the compressive strength measured on two geometries of samples was found. Based on test results and numerical simulations practical recommendations have been proposed.

1 INTRODUCTION

The masonry walls and pillars subjected to compression are in many cases the most important elements of historical buildings. The masonry walls and pillars were erected according to the traditional rules passed down and developed by successive generations of builders. Various materials and technologies were used for their construction. In the analysis of historical structures, determination of mechanical material characteristics is an important issue. One of the most reliable method to determine the masonry strength is testing masonry samples cut from a structure. However, it is rarely possible to take masonry prismatic samples of sufficiently large dimensions which are representative for massive walls or pillars. Extraction of such samples causes significant damage to the masonry structure and decreases its bearing capacity. The extraction of large samples is also limited for conservation reasons. UIC 778 - 3R (2008) recommends a method for determination of masonry strength using tests on core samples with a diameter of 150mm. Tests on masonry core samples can be qualified as a minor destructive method. UIC 778 - 3 recommends cylindrical samples consisting of two bed joints and one centrally located vertical joint. The core samples extracted from the structure are compressed in the laboratory perpendicularly to the generatrix using regularization caps. The load direction in the laboratory test is identical to that in the masonry structure.

In the tests carried out by Billelo et al. and by Brencich and Sterpi (2006) steel loading plates were used.

These tests were conducted on the masonry samples made in the laboratory from the contemporary bricks and mortars. Identical steel caps were use also by Matysek (2016) but tested samples were extracted from a masonry building erected at the end of the 19th century. Many tests on masonry core samples with a diameter of 150mm were done in Spain by Pela et al. (2015, 2019). In this research regularization caps made of high strength mortar were also used. The width of the high strength mortar regularization caps was greater than the width of steel caps used by Billelo et. al (2006), Brencich and Sterpi (2006) and Matysek (2016). Therefore, direct comparison of the results of tests on core samples with a diameter of 150 mm is not an easy task and has not yet been carried out.

For many historical buildings, due to their cultural heritage value, collecting samples with diameter of 150 mm is not acceptable. For this reason, attempts are made to test smaller core samples. In a study performed by Matysek et al. (2016) 100 mm core samples were tested. The damage fields that arise after cutting out cylindrical samples with diameters of 100 mm are more than twice smaller than damage fields after cutting out samples with a diameter of 150mm (Figure 1).

Core samples with a diameter of 90 mm were used by Segura et al (2019). The results of these tests appear very promising but require extension and additional analyses.

In this paper compression tests on 150 mm and 100 mm core samples are presented. The core samples were cut from brick masonry made at the laboratory and from historical structures erected in the 19th and 20th century. Additionally, the results of FEM analyses for core samples subjected to

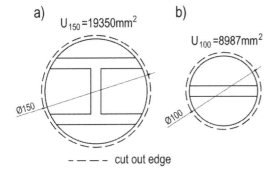

a) $U_{150} = 19350 \text{mm}^2$

b) $U_{100} = 8987 \text{mm}^2$

Ø150 Ø100

– – – – cut out edge

Figure 1. Damage fields for 150mm (U_{150}) and 100mm (U_{100}) core samples.

compression were also presented. The experiments on 150 mm samples as well as 100 mm samples were simulated. The effect of size on behaviour of core samples subjected to compression was analysed.

2 TESTS ON BRICKS, MORTARS AND CORE SAMPLES

The results of strength tests carried out on bricks, mortars and core samples with a diameter of 150 mm are shown in Table 1 and Table 2. Bricks, mortars and core samples were collected from four types of masonry (marked MAS1, MAS2, MAS3 and MAS4). The masonry MAS1 were erected at the end of the 19th century while masonry MAS2 in the middle of the 20th century. The masonry MAS3 was made in the laboratory. The original bricks from the first half of the 20th century and typical lime-cement mortar were used. The bricks were taken during demolition works of an existing building. The masonry MAS4 was made in laboratory from contemporary materials.

The compressive strength tests for bricks were obtained in compliance with the requirements of EN 772-1 (2011), while mortars were tested using the DPT test according to the guidelines of DIN 18555-9 DIN (1999).

The 150 mm core samples were drilled using water-cooled tools with a strongly limited amount of water. Figure 2 shows the masonry after cutting out the core samples. In the laboratory, core samples were cut to the assumed lengths. The shapes of core samples were in accordance with UIC 778 - 3 recommendations (2008). The core samples were seasoned for 3 months at the temperature of 18 - 22°C and relative humidity below 65%. The core samples were compressed using steel loading plates. The width of the contact zone of the steel loading plates in contact with the core sample (s) was 75 mm (Figure 3a).

The displacement control loading procedure was applied in the experiments. The vertical displacement was applied monotonically at a rate of 0.2 mm/min. A minimum of 5 core samples were tested for each masonry. The compressive strength $f_{cyl1,150}$ was

Table 1. Results of strength tests for bricks (f_b) and mortars (f_j) collected from the masonry structures.

Type of masonry	f_b MPa	f_j MPa
MAS1	20.7 (0.21)	2.3 (0.28)
MAS2	21.1 (0.19)	17.8 (0.30)
MAS3	11.4 (0.26)	5.8 (0.23)
MAS4	16.6 (0.17)	14.5 (0.14)

(…) - COV
f_b – normalized compressive strength of bricks

Table 2. Test results for cylindrical samples and masonry columns – mean values and ranges of values.

Type of masonry	$f_{cyl1,150}$ MPa	$f_{cyl2,150}$ MPa	f MPa
MAS1	$1.9^{1/2c}$ (1.2-2.8) 1.7^{1c} (1.2-2.5)	3.8 3.4	3.3
MAS2	$4.2^{1/2c}$ (3.3-5.1)	8.4	6.9
MAS3	$3.5^{1/2c}$ (2.6-4.2)	7.0	(-)
MAS4	$5.1^{3/4c}$ (4.1-5.9)	10.2	9.8

(…) – ranges of values.
$^{(1/2c)}$ $^{(3/4c)}$, $^{(1c)}$ - the lengths of core samples, c is length of brick
$f_{cyl1,150} = P_{ult}/dl$; $f_{cyl2,150} = P_{ult}/sl$
d, l – diameter and length of samples, s – width of steel loading plates in contact with samples

Figure 2. Masonry MAS3 after sampling.

calculated as the ratio of the maximum compressive load to the cross-sectional area in the middle of the cylindrical samples. The compressive strength $f_{cyl2,150}$ was obtained as the ratio of the maximum compressive load to the cross-section in which the load is transferred to the sample. The failure mode of the 150 mm sample is presented in Figure 4a.

For the sake of comparison in Table 2 results of strength tests on masonry columns (English bond columns) are also shown. A minimum of 3 columns were tested for determining the mean values of compressive strength (f). The height of the columns was 5 masonry layers.

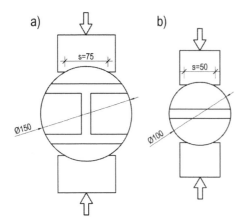

Figure 3. The cylindrical masonry samples subjected to compression a) samples with a diameter of 150 mm, b) samples with diameter of 100 mm.

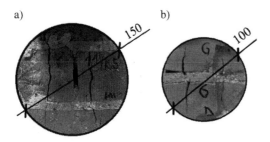

Figure 4. The failure modes of cylindrical samples.

Table 3. Test results for masonry samples with a diameter of 100mm.

Type of masonry	$f_{cyl1,100}$ MPa	$f_{cyl2,100}$ MPa
MAS3	$4.6^{1/2c}$	9.2
MAS4	$7.2^{1/2c}$	14.4

From masonry MAS2 and MAS3 core samples with a diameter of 100 mm were also extracted. Each core sample contained one bed joint and two pieces of bricks (Figure 3b). The 100 mm samples were tested according to the same procedure as 150 mm cylinders. The compression load was transferred on 100 mm masonry samples by steel loading plates. The contact width between masonry samples and steel plates was equal 50 mm. The results of tests for 100 mm core samples are presented in Table 3.

A minimum of 9 core samples with a diameter of 100mm were tested for each masonry. The compressive strength $f_{cyl1,100}$ and $f_{cyl2,100}$ was calculated in the same manner as compressive strength $f_{cyl1,150}$ and $f_{cyl2,150}$

3 DISCUSSION

3.1 Experimental research

Figure 5 compressive strength obtained core samples ($f_{cyl2,150}$) vs. compressive strength determined from masonry column (f) is presented. The test results performed by Bilello et al. (2006) and Pela et al. (2015), (2019) are also shown in this figure.

Tests presented by Pela et al. (2015), (2019) were performed on a stack-bonded prism. To calculate compressive strength (f) from these tests a correction factor was used. Research on stack-bonded prism and English bond columns published by Gumaste et al. (2007) and Nanjunda Rao and Pavan (2014) showed that the correction factor is equal to 1.02 (the mean value from experiments on 7 types of brick masonry). The regression analysis of data presented in Figure 5 provides the relationship:

$$f = 0.914 f_{cyl2.150} \qquad (1)$$

This relationship has an R2 value of 0.974. The coefficient of determinations R2 equal to 0.974 is very satisfactory.

The test results on core samples with different diameters are shown in Figure 6. In addition to our own research on cylindrical samples with diameters of 150 mm and 100 mm (see Table 2, Table 3), Figure 6 also shows the results of research performed by Segura et al. (2019) on 150 mm and 90 mm samples.

The linear regression for the test results showed in Figure 5 is described by the formula (2):

$$f_{cyl2.150} = 0.74 f_{cyl2.100} \qquad (2)$$

with a very satisfactory coefficient of determination R2 equal to 0.969

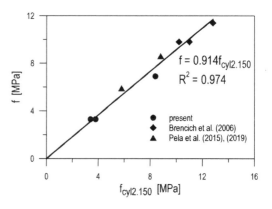

Figure 5. Compressive strength obtained on 150mm cylindrical samples ($f_{cyl2,150}$) vs. compressive strength of masonry determined on masonry columns (f).

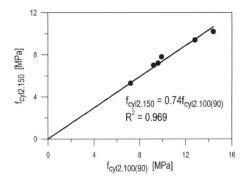

Figure 6. Compressive strength obtained on 100 mm and 90 mm samples ($f_{cyl2,100(90)}$) vs. compressive strength determined on 150 mm samples ($f_{cyl2,150}$).

The failure modes of 150 mm and 100 mm cylindrical samples were similar (see Figure 4). The failure of the cylindrical samples was initiated by the appearance of cracks at the ends of the steel loading plates. These cracks developed during the increase of the compression load. After the test, the remaining part of the samples was sandglass shaped. Very similar failure modes were observed also in the tests carried out by Pela et al. (2015), (2019) and by Segura et al. (2019) in which the regularization caps were made of high strength mortar. In the tests with regularization caps made of high strength mortar generally lower CoV values of strength f_{cyl2} were obtained. This is probably due to a reduction in the effects of uneven application of compressive load compared to testing with steel loading plates. In the future comparative studies using different types of regularization caps on samples cut from the same masonry are necessary to explain this problem in detail.

3.2 Numerical simulations

Numerical simulations were carried out for the cylindrical samples with a diameter of 150 mm and 100 mm compressed perpendicularly to the generatrix using steel loading plates. The geometry of the samples and steel plates was the same as in the experimental studies – Figure 3. In the simulations, a thickness of joints in masonry samples equal to 15 mm was assumed. In Figure 7 finite element models of analysed cylindrical samples are presented.

The simulations were performed for two types of mortars in masonry joints. Masonry with lime mortar and masonry with cement-lime mortar were considered. It was assumed that the compressive strength of bricks is 20 MPa, while the compressive strengths of the lime mortars are equal to 2.0 MPa and 5.0 MPa (for the lime mortar and cement-lime mortar, respectively). The moduli of elasticity for bricks (E_b) and mortar (E_j) were determined on the basis of relationships given by Kaushik et al. (2007): $E_b=300f_b$, $E_j=200f_j$. The Poisson's ratios for bricks $v_b=0.18$ and for mortars $v_j=0.20$ were assumed.

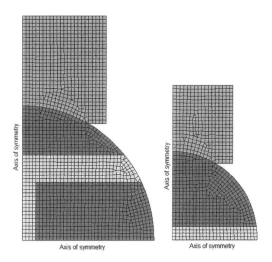

Figure 7. Finite element models of analysed samples.

The loading was realized by concentrated force of 22.5 kN and 15 kN for the samples with diameters of 150 mm and 100 mm, respectively. The loadings referred to the horizontal dimensions of the steel plates provide the same vertical stress level of about 2.5 MPa.

Results of numerical simulations are presented in Figures 8 to 10.

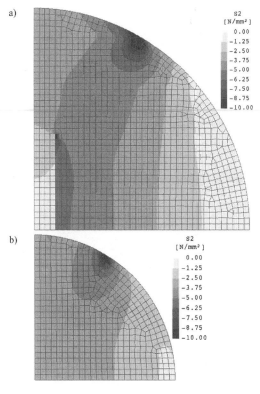

Figure 8. Principal compressive stress for material set 1 (f_b=20MPa, f_j=5MPa): a) d=150mm, b) d=100mm.

In Figure 8 and Figure 9 the concentration of stresses in the cylindrical samples at the contact points with steel plates are clearly visible both for 150 mm and 100 mm samples. The concentration of stresses initiating cracks were observed in experimental study.

The stress distribution in the cross-section located in the middle of the samples is clearly more uneven for samples with diameter of 150 mm (Figure 10). The vertical joint located in the centre of sample significantly influences the stress distribution. The parts of the 150 mm cylindrical samples outside the area defined by the width of steel loading plates have limited contributions to the transfer of compressive load. The type of mortar in joints of 150 mm masonry samples is irrelevant for distribution of vertical stresses.

For the samples with a diameter of 100 mm the distribution of vertical stresses in cross-section in the middle of samples is more uniformly distributed than for samples with a diameter of 150 mm. In 100 mm samples the compressive loads are transferred on to parts located outside the area defined by the width of loading plates to a much greater extent.

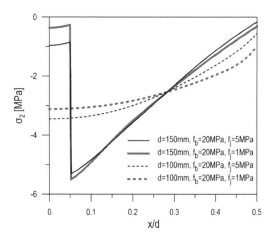

Figure 10. Horizontal distributions of principal compressive stress along the diameter of the core samples (origin of the x-axis is located at the centre of the sample).

4 CONCLUSIONS

In the paper experimental and numerical studies on compressive strength of masonry in existing buildings were presented. The experimental research was carried out on core samples drilled from masonry walls. This method can be classified as a minor-destructive one (MDT) and is especially useful for historical buildings. In the research, samples with diameters of 150 mm and 100 mm cut from brick masonry were tested. The core samples were drilled from brick masonry made at the laboratory and from historical structures erected in the nineteenth and twentieth centuries. In laboratory tests compressive load was applied by steel regularization caps.

Based on experimental research a formula for determining the compressive strength of masonry using tests on 150 mm core samples was proposed – see formula 1. This formula is valid for brick masonry with lime mortars or cement-lime mortars.

In this paper it has also been shown that tests on cylindrical samples with a diameter of 100 mm are very useful for determination of masonry compressive strength.

Smaller core samples are recommended for historical buildings. The failure modes of both the 150 mm and the 100 mm were very similar. The 100 mm diameter core samples provided higher values of compressive strength. The sample geometry strongly influences this effect. Based on numerical simulations, it was shown that in samples with a diameter of 100 mm without a vertical joint, the stress distribution is more uniformly distributed. In 100 mm samples the compressive load is evenly transferred over a significant part of the sample diameter.

For different types of masonry similar ratios between the compressive strength obtained with the

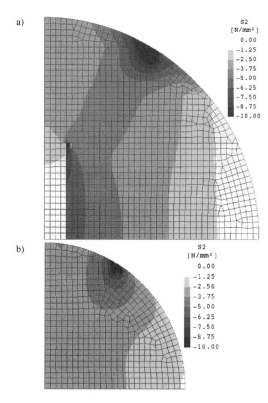

Figure 9. Principal compressive stress for material set 2 (f_b=20MPa, f_j=2MPa): a) d=150mm, b) d=100mm.

two geometries of samples were found. Based on linear regression analysis a 0.74 ratio was established (with the coefficient of determination R2 equalled 0.969) – see formula (2).

REFERENCES

UIC – International Union of Railways. 2008. UIC Code. Recommendations for the inspection, assessment and maintenance of masonry arch bridges. Final draft, Paris.

Billelo C., Brencich A., Di Paola M., Sterpi E. 2006. Compressive strength of solid clay brickwork: Calibration of experimental tests, *In Proceedings of 7th International Masonry Conference*, London.

Brencich A, Sterpi E. 2006. Compressive strength of solid clay brick masonry: calibration of experimental tests and theoretical issues. *In Proceedings of the 5th International Conference on Structural Analysis of Historical Constructions, ed. P.B. Lourenco. P. Roca, C. Modena, S. Agrawal*, 757–765. Macmillan India, New Delhi.

CEN - European Committee for Standardization. 2011. EN 772-1:2000 - Methods of tests for masonry units - determination of compressive strength. CEN, Brussels.

Deutsches Institut für Normung. 1999. DIN 18555-9:1999 - Prüfung von Mörteln mit mineralischen Bindemitteln – Teil 9, Festmörtel: Bestimmung der Fugendruckfestigkeit. Berlin.

Pela L., Roca P., Benedetti A. 2015. Mechanical characterization of historical masonry by core drilling and testing of cylindrical samples, *International Journal of Architectural Heritage, 10 (2-3)*.

Matysek P., Stryszewska T., Kańka S. 2016. Experimental research of masonry compressive strength in the Auschwitz II – Birkenau former death camp buildings, *Engineering Failure Analysis, 68*, 263–274.

Matysek P. 2016. Compressive strength of brick masonry in existing buildings – research on samples cut from the structures, *Proceedings of the 16th International Brick Block Masonry Conference, Padova*.

Segura J., Pela L., Roca P., Cabane A. 2019. Experimental analysis of the size effect on the compressive behaviour of cylindrical samples core-drilled from existing brick masonry, *Construction and Building Materials, 228* (116759).

Pela L., Saloustros S., Roca P. 2019 Cylindrical samples of brick masonry with aerial lime mortar under compressive: Experimental and numerical study, *Construction and Building Materials, 227* (116782).

Gumaste K.S., Nanjunda Rao K.S., Venkatarama Reddy B.V., Jagadish K.S. 2007. Strength and elasticity of brick masonry prisms and wallettes under compression. *Materials and Structures*, 241–253.

Nanjunda Rao K.S., Pavan G.S. 2014. FRP – confined clay brick masonry assemblages under axial compression: Experimental and analytical investigations, *Journal of Composites for Construction 19 (4)*.

Kaushik H.B., Rai D.C. 2007. Stress – strain characteristics of clay brick masonry under uniaxial compression, *Journal of Materials in Civil Engineering, ASCE, 09/2007*, 728–739.

Brick and Block Masonry - From Historical to Sustainable Masonry –
Kubica, Kwiecień & Bednarz (eds)
© 2020 Taylor & Francis Group, London, ISBN 978-0-367-56586-2

In-plane capacity of beam and block floor systems: An in-field experimental study

E. Casprini, C. Passoni, A. Marini & A. Belleri
University of Bergamo, Bergamo, Italy

E. Giuriani
University of Brescia, Brescia, Italy

ABSTRACT: The evaluation of the in-plane capacity of existing floors is of fundamental importance both in the seismic vulnerability assessment and in the retrofit design, especially when retrofit interventions are carried out from the outside of the buildings. Such interventions are particularly appreciated nowadays as they overcome the major problem of the relocation of the inhabitants, but they may be hindered by the need to strengthen the floors. In this paper, the outcomes of an in-field experimental study are discussed, which is aimed at investigating the in-plane capacity of a floor, consisting in RC beam and clay block floor systems without additional extrados RC slab. The conceptual design of the testing bench and the main results of the experimental study are discussed. Based on the experimental results, a simplified model is proposed to estimate the ultimate capacity of existing floors.

1 INTRODUCTION

The existing building heritage requires deep renovation actions to ensure structural resilience and avoid future human and economic losses. In Europe, 50% of the actual building stock is made of reinforced concrete structures built after the World War II (BPIE 2011, Marini 2014). Those buildings are usually conceived as one-way frames, designed only for gravity loads, with low ductility structural details, which make them highly vulnerable against the seismic actions. Besides structural issues, those buildings present poor architectural features and low energy efficiency. Despite deep renovation of such building stock is an urgent issue, the renovation rate is extremely low.

One of the major barriers to the renovation of these buildings, used as residential multi-storey buildings or public facilities such as schools, hospitals or offices, is the need to relocate their inhabitants (La Greca and Margani 2018). For this reason, recent researches proposed retrofit interventions carried out from the outside of the building, introducing the so-called exoskeletons (Takeuchi et al. 2009; Marini et al. 2017). Under a structural point of view, such exoskeletons may be organized as wall or shell structures. In the former, additional shear walls are located parallel or perpendicular to the façade of the building; in the latter, the whole extension of the façade is exploited as to have a seismic resistant box enveloping the building, with high stiffness but low thickness.

In the seismic retrofit of buildings, the presence of floor diaphragms represents a crucial aspect that governs the seismic behaviour of existing and retrofitted structures. In fact, the in-plane diaphragm action is needed to transfer the floor inertial loads to the lateral force resisting system.

In the past, the diaphragm flexibility and capacity, and the floor load distribution have been studied (Kolston and Buchanan 1980; Paulay and Priestley 1992; Naeim and Boppana 2001; Rodriguez et al. 2002; Bull 2004; Blandon and Rodriguez 2005; Rodriguez et al. 2007; Lee et al. 2007; Schoettler et al. 2009; Mohele et al. 2010; Tena-Colunga et al. 2015). Recently, an ongoing experimental and numerical research carried out by the authors on beam and block floor systems has shown that the behaviour of such diaphragms relies on the onset of an in-plane arch bridging the span between the vertical seismic resistant elements (Figure 1, top). The floor system behaviour may be considered as rigid-brittle as long as the openings are not too large and their layout does not impair the correct in-plane behaviour of the floor (Marini et al. 2020).

In order to both evaluate the capacity of existing structures and to select and design the most suitable intervention strategy, the floor layout, the in-plane load transfer mechanism and the resulting in-plane capacity are acknowledged as crucial parameters. While in the building as-is condition the floor in-plane capacity may not be a major issue, it may become critical after a retrofit intervention,

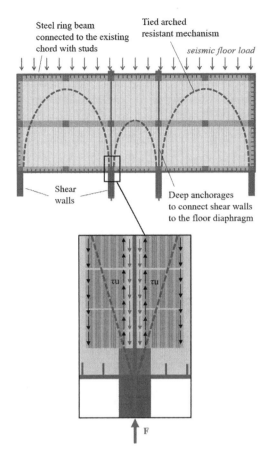

Steel ring beam connected to the existing chord with studs

Tied arched resistant mechanism

seismic floor load

Shear walls

Deep anchorages to connect shear walls to the floor diaphragm

τu τu

F

Figure 1. Earthquake resistant exoskeleton (top) and detail of the diffusion of a lumped force into the floor system (e.g. in correspondence of exterior shear wall) (bottom).

particularly when shear walls or shell exoskeletons are implemented; these strengthening interventions entail a major increase of both the total stiffness of the retrofitted building and the resisting arch span, which varies from the existing single-bay span to the distance between the new and stiffer lateral resisting structural elements. The evaluation of the floor capacity and the in-plane distribution of loads may thus be critical in determining whether the existing floors need to be strengthened, especially after the retrofit intervention, and which is the best layout of the retrofit intervention. For example, in the design of the retrofit, the in-plane floor capacity may define the maximum distance between the new shear walls to avoid the strengthening of the floors.

Within this framework, the present paper addresses an in-field experimental campaign on a portion of an existing beam and block floor system. The aim of the research is to evaluate the actual capacity of the floor and its capability to redistribute an in-plane load, such as the seismic reaction force, in correspondence of a new shear wall. A brief description of test set-up is presented, and the main

results are summarized, together with a simple analytical model, which enables estimating the floor system capacity.

2 IN-PLANE CAPACITY OF BEAM AND BLOCK FLOOR SYSTEMS

Two main floor typologies can be found in the European RC Post-World War II buildings: RC slabs and one-way composite beam and block floor systems with or without additional extrados RC overlay. The latter system, made of parallel RC joists and rows of hollow clay blocks, is widespread in the existing building heritage. The floor in-plane capacity, both in the building as-is or retrofitted condition, is governed by a tied-arch mechanism likely to develop within the thickness of the floor (Bull 2004).

In the case of cast-in-place structural overlay, the mechanism mainly develops within the concrete overlay due to its higher stiffness compared to the clay blocks. Anytime the RC overlay is lacking, the blocks need to transfer in-plane actions, and the maximum strength of the system must be assessed.

Failure of the system can be associated with failure of the tie, or failure of the blocks subjected to high compressive stresses in the direction orthogonal to the holes, or for shear failure by the supports. The latter mechanism is often the most critical, and it is the only one that cannot be inhibited through specific retrofit measures (Marini et al. 2020).

As a preliminary evaluation of the in-plane capacity of the beam and block floor system lacking the structural concrete overlay, previous laboratory local tests on specimens composed of a single hollow block and two lateral joists were carried out (Feroldi 2014, Marini et al. 2020). The ultimate shear strength of the block was evaluated for three different lateral confining loads. Average ultimate shear stresses $\tau_u = 1.6 \div 1.9$ MPa were obtained, always associated with the shear failure of the clay block, independently of the strut inclination.

Starting from the mentioned result, the capacity of a beam and block floor system lacking the concrete overlay in the presence of a concentrated force can be evaluated referring to a simplified shear stress distribution (see Figure 1, bottom):

$$F = 2 \, \tau_u t_{eq} L \tag{1}$$

where τ_u = ultimate shear resistance of the block; L = length of the joist; t_{eq} = equivalent thickness of the block, considered as the net thickness of the single horizontal webs: β = shear stress distribution coefficient, depending on the ratio between the axial stiffness of the RC joist and the shear stiffness of the clay blocks. β may be considered = 1 if the shear stress is uniformly distributed along the beam-to-floor interface, whereas $\beta = 0.5$ if the shear stress follows a triangular distribution. The multiplier 2 in Equation 1 indicates a shear stress distribution on both the side of the joist.

3 IN-FIELD EXPERIMENTAL STUDY

A first promising experimental test was carried out on the floor of an abandoned building built in 1960 in Brescia, Italy, which is going to be demolished soon. The structure is composed by a reinforced concrete frame with three levels above the ground, and the floors are composed by a beam and block floor system, which is one of the most common floor typologies in the existing reinforced concrete buildings (Figure 2).

The experimental campaign was aimed at determining the in-plane capacity of the floor system and its load diffusion capacity. The test was carried out on a sample composed by a central concrete beam and 4 blocks on each side, as shown in Figure . By considering the geometry of the sample and by applying Equation 1, the maximum in-plane shear capacity of the system is expected to range between 128 kN and 152 kN. In the previous evaluation the average ultimate shear stress obtained by Feroldi 2014 (ranging between 1.6 MPa and 1.9 MPa) is adopted. The net thickness of the clay webs is equal to 40 mm (considering the central portion of the block, which is the thinnest). A uniform shear stresses distribution (β=1) is assumed due to the reduced length of the joist (L=1m).

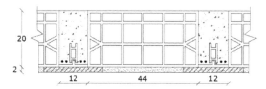

Figure 2. Beam and block floor system (measures in cm). Horizontal web thickness is equal to 1cm.

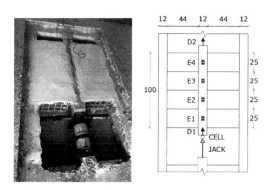

Figure 3. Test set-up (left) and instrument layout (right). E: strain gauges; D: Linear variable displacement transducer (LVDT) (measures in cm).

3.1 Analytical prediction of the experimental results

A preliminary modelling of the experimental test was carried out for the prediction of the possible behavior of the floor specimen. A simplified analytical model accounting for both the axial stiffness of the concrete joist and the shear stiffness of the clay blocks was proposed, in order to obtain a more reliable prediction of the tested sample maximum capacity with respect to Equation 1. The system was modelled as illustrated in Figure 4. All the components are modelled as springs; in detail, K_{beam} is the axial stiffness of the concrete beam segment, K_{block} is the shear stiffness of each block. Furthermore, in the presence of a slab overlay or of a plaster layer below the blocks, K_{block} can also be calibrated as to include the stiffness of these additional resisting layers.

The model allows to calculate the displacement of n defined nodes along the beam length, assumed in correspondence of each row of blocks (1 to 4 in Figure 4). In each node (i), the stiffness of the system may be modelled as three springs in parallel: two springs represent the blocks of the row (i) ($2K_{block,i}$) and the third spring represents the portion of the specimen beyond the node (i). The latter is calculated as the stiffness of the beam between nodes (i) and (i+1) ($K_{beam,i}$) and the stiffness of the system beyond (i+1) (K_{i+1}) acting in series. As a consequence, the stiffness in correspondence of the last node (n) is simply equal to $2K_{block,n}$, and the stiffness in correspondence of the generic node (i) may be obtained with Equation 2. $K_{beam,i}$ and $K_{block,i}$ are obtained with Equation 3-4. In in the presence of an overlaying RC thin slab, a layer of plaster, or other resistant layers contributing to the stiffness of the blocks, the m layers are assumed to act in parallel.

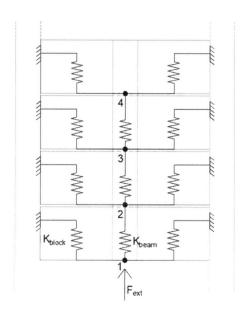

Figure 4. Scheme of the proposed analytical model.

$$\begin{cases} K_i = 2K_{\text{block},i} + \dfrac{1}{\frac{1}{K_{beam,i}} + \frac{1}{K_{i+1}}} & \text{for } i = 1 : n-1 \\ K_n = 2K_{\text{block},n} & \text{for } i = n \end{cases} \quad (2)$$

$$K_{beam,i} = E_c \cdot A_c / L_i \quad (3)$$

$$K_{block,i} = \sum_{j=1}^{m} \left(G \cdot \frac{A_v}{B} \right)_j \quad (4)$$

where E_c = concrete elasticity modulus; A_c = beam sectional area; G = shear modulus of the block layer; A_v = shear area of the block layer; B = width of the block; L_i = length of the portion of the beam associated with each row of blocks. The model allows to estimate both the displacement η_i of each node (i) and the share of the shear action $V_{block,i}$ transferred to each block starting from the node 1, as in Equation 5-7:

$$\eta_i = F_i / K_i \quad (5)$$

$$V_{block,i} = \eta_i \cdot K_{block,i} \quad (6)$$

where

$$\begin{cases} F_1 = F_{ext} & \text{for } i = 1 \\ F_i = F_{i-1} - 2V_{block,i-1} & \text{for } i = 2 : n \end{cases} \quad (7)$$

The ultimate capacity of the system can be finally estimated by considering the mechanical properties of the m layers in a single block. The force acting on the block $V_{block,i}$ is divided between the layers proportionally to their stiffness; the shear force in each layer (j) $V_{layer,j}$ can be obtained as in Equation 8, and the equivalent average shear stress in each layer (j) is obtained with Equation 9:

$$V_{layer,j} = V_{block,i} \frac{\left(G \cdot \frac{A_v}{B} \right)_j}{\sum_{k=1}^{m} \left(G \cdot \frac{A_v}{B} \right)_k} \quad (8)$$

$$\tau_{layer,j} = V_{layer,j} / (t_{eq} L_i)_j \quad (9)$$

Thanks to this simplified model, a preliminary estimate of the in-plane stiffness of the system, of the forces acting in each floor component, and of the shear stresses in each layer of the block can be obtained. Similarly, the ultimate force to be applied to the specimen ($F_{ext,u}$) can be calculated by considering the ultimate shear resistance of each layer of the block. Equations 10-13, used to estimate the maximum capacity of the tested

system are obtained by reversing equations 9-8-5-6 from the proposed model.

$$V_{layer,j,u} = \tau_{layer,j,u} (t_{eq} L_i)_j \quad (10)$$

$$V_{block,layer,j,u} = \frac{V_{layer,j,u}}{\dfrac{\left(G \cdot \frac{A_v}{B} \right)_j}{\sum_{k=1}^{m} \left(G \cdot \frac{A_v}{B} \right)_k}} \quad (11)$$

$$V_{block,i,u} = \min \left(V_{block,layer,j,u} \right) \quad (12)$$

$$F_{ext,u} = V_{block,1,u} \frac{K_1}{K_{block,1}} \quad (13)$$

3.2 On-site experimental test

Some in field preliminary operations were needed to prepare the specimen to be tested. Once the specimen boundaries were defined, the extrados nonstructural layers were removed. A portion of a joist, together with the adjacent blocks, were isolated from the floor by means of two cuts at the beginning and at the end of the specimen (see Figure 3). The plaster layer was removed by the joist-to-block interface, in order to weaken the shear action transfer through it (see shaded area in Figure 2). A propping system was placed to support the system, and a strip of Teflon© was introduced to minimize friction at the joist-to-propping support interface.

A hydraulic jack with 326 kN capacity and a 1000 kN load cell were used to apply and control an in-plane load to the central beam. The load cycles were defined as shown in Figure . A thin layer of gypsum was applied to the specimen extrados to enable detecting crack onset and propagation.

In order to assess the effective distribution of the applied in-plane load through the sample components, four strain gauges were glued in the center of each concrete beam segment, delimited by each block row (i) (see Figure 3). Strain measurements are reported in Figure 6. The displacements at the beginning (D1) and

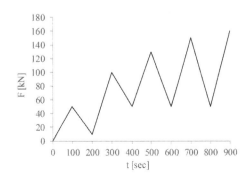

Figure 5. Load – time plot.

the end of the beam (D2) were also recorded with respect to the external reference system, as illustrated in Figure 7.

A very brittle shear failure of the internal part of the clay blocks (having an equivalent net thickness of 40 mm) occurred for an external load of 160 kN, as shown in Figure 8. At collapse, displacements

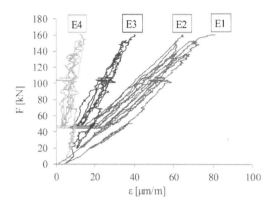

Figure 6. Total load-axial strain curves at the four sections of the concrete beam.

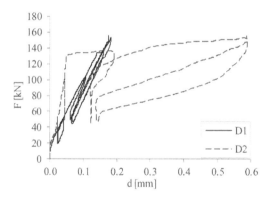

Figure 7. Displacement measured at the beginning (D1) and the end (D2) of the beam.

Figure 8. Shear failure of the clay blocks.

lower than 1mm were recorded, which may be considered as negligible, and no visible cracks were detected before the collapse of the sample.

The evaluation of the axial strains along the joist length, registered by means of strain gauges, allowed to investigate the ability of the floor system to distribute the external in-plane load among its components. Although the quantitative interpretation of such measures is not straightforward, particularly for the variation of the conventional area along the specimen length, following the load diffusion, and for the many uncertainties encountered in the physical sample (as for example the variable nature of the contact between adjacent clay blocks, the uneven distribution of cracks in the hollow block webs, and the variable thickness of the concrete slab), it may be clearly observed that the joist is able to distribute the load among all the row of blocks. An almost regular spacing can be observed between the four strain gauge measurements (Figure 6), which entails an almost uniform distribution of the external in-plane load from the joist to the hollow blocks. The maximum average shear force in the blocks at failure can thus be simply estimated by dividing the external load by the number of the blocks (160kN/8=20kN). This force should then be divided between the two layers composing the block, consisting in the clay hollow block and the plaster, proportionally to the shear stiffness of each layer, as in Equation 8. Accordingly, a force of 16.2 kN is obtained in the clay hollow block, which corresponds to an average shear stress equal to 1.62 MPa. The material parameters adopted in the model are reported in Table 1; as for the clay blocks, an equivalent net thickness of 50 mm is adopted to calculate the shear stiffness (to take into account the beneficial presence of a fifth horizontal web at the joist-to-block interface, Figure 2), while a thickness of 40 mm referred to the internal part is used to estimate the maximum strength (where also the plaster layer is present).

3.3 Comparison of analytical and experimental results

After the test, the experimental and analytical results were compared. The parameters used to calibrate the model are reported in Table 1. Some laboratory tests were carried out in order to better estimate the mechanical properties of the system components.

Table 1. Material parameters.

Properties	Concrete	Clay	Plaster
E [MPa]	36000	15000	8000
v [-]	0.2	0.1	0.24
G [MPa]	15000	6818	3226
t [mm]	200	40-50	20
σ_{lim} [MPa]	35	18	7
τ_{lim} [MPa]	3.5	1.6-1.9	0.7

A concrete sample was core drilled from the beam to define the properties of the concrete. The elasticity modulus was found to be 36000 MPa and the concrete cubic compressive strength 35 MPa (Figure 9). The properties of the other components of the system were assumed as mean common reference values. In this specific case, the plaster layer is considered to contribute in terms of stiffness of the block because of its noticeable thickness and good quality. A test of absorption by total immersion was carried out on the plaster material, resulting in 10% of water absorption with respect to its weight and showing further evidence of the good quality of the plaster.

The results obtained from the model in terms of node displacements for an external load Fext=160 kN (the maximum force recorded during the test) are reported in Figure 10 (top) and compared with the actual displacements obtained in the experimental campaign. The displacement of the node 1 corresponds to the one measured at the beginning of the beam (D1), where the load is applied. The displacements of the other nodes are computed from that of the first one, by adding the beam shortening measured by the strain gauges in the beam sections.

As for the shear force distribution (Figure 10, bottom), in the analytical model, shear actions in the blocks are computed as the product of the relative node displacement by the block stiffness (Equation 6); while, in the experimental campaign, they are calculated as the difference of the conventional axial force in two consecutive beam portions at collapse divided by two blocks. The conventional axial force in the four monitored segments of the beam can be estimated by assuming a uniform distribution of stresses inside each portion and by multiplying the measured axial deformation ε_i of each segment by the equivalent conventional area of the segment itself (assumed

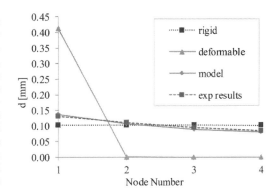

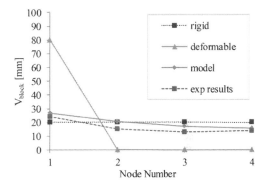

Figure 10. Comparison between analytical model outputs and experimental test results in terms of displacements (top) and estimated shear force in the single block (bottom).

as the area of the concrete joist, 120 mm x 200 mm). For the reasons explained before, only strain measures obtained from the first stage of the test can be considered quantitatively correct since, at the final stages of the test, internal and non-visible damages may have occurred. In order to estimate the shear forces in the blocks at collapse, the same axial force distribution calculated at the first load peak (Fext=50.7 kN) is thus considered as constant up to collapse (Fext=160 kN) (Table 2). This way, when the maximum capacity is reached (Fext=160 kN), the distribution of the conventional axial force in the joist is calculated adopting the

Figure 9. Laboratory test on the beam concrete core.

Table 2. Forces distribution.

Row	Ax. force 1st peak	Ax. force	Ax. force at collapse	V_{block}
	[kN]	[%]	[kN]	[kN]
F_{ext}	50.7	100	160	
1	35.4	70	112.0	24.0
2	26.1	51	81.6	15.2
3	17.9	35	56.0	12.8
4	9.3	18	28.8	13.6

607

same rates, and the shear forces transferred to the single blocks (V_{block}) are computed as the difference between the axial forces in two consecutive segments of the beam, divided by two blocks.

Adopting such a procedure, a maximum average shear force in a block is estimated as 24 kN. This value should be then divided between the two layers composing the block, consisting in the clay hollow block and the plaster, proportionally to the shear stiffness of the layer, as in Equation 8. A force of 19.4 kN is obtained in the clay hollow block, which corresponds to a maximum average shear stress equal to 1.94 MPa. Comparison between analytical and experimental results are reported in Figure 10 (bottom).

In order to better understand the behavior of the system, experimental results are also compared with two limit behaviors: a system with an infinitely rigid beam with respect to the other components (i.e. with a uniform load distribution, referred in figure as 'rigid'), and a quite deformable beam (referred in figure as 'deformable'). The upper-bound and lower-bound behaviors are obtained by amplifying or reducing the beam axial stiffness by three order of magnitude, respectively. As expected, both the experimental results and analytical model show that the behavior of the tested specimen is very close to the rigid one, and the forces are thus quite well distributed among the beam portions, and consequently among the blocks. Such behavior is mainly due to the limited length of the specimen, i.e. due to the high axial stiffness of the beam compared to the shear stiffness of the blocks in the absence of a concrete slab. When accounting for the actual strain gauge measurements, a small increase of shear action in the first row of blocks is observed with respect to the rigid configuration, which leads to an increase in the maximum average shear stress estimated in the clay block.

As mentioned in paragraph 3.1, Equation 10-13 of the model can also be applied to obtain the maximum capacity of the system, by assuming that the shear action acting on the block is divided between the block layers proportionally to their stiffness. In this case, the estimated maximum capacity is 118 kN with 1.6 MPa and 141 kN with 1.9 MPa.

It is worth noting that the model accounts for a system in which blocks are not in direct contact with each other; however, in the specimen, this happened just for some blocks. For this reason, the system maximum capacity was also estimated in the case of presence of perfect contact between blocks, which can be modelled as 2 single blocks of 1 m length (one for each side of the joist). In this case, the estimated maximum capacity is 152 kN with 1.6 MPa and 180 kN with 1.9 MPa. The actual behavior of the tested portion of the floor lies between these two models.

4 CONCLUDING REMARKS

This paper deals with the assessment of the actual in-plane capacity of beam and block floor systems subjected to seismic actions. Floor in-plane capacity is relevant when dealing with existing structures, both in the building as-is condition and in relation to a possible retrofit intervention. For example, based on in-plane capacity of the floor, feasibility of a retrofit carried out from outside can be evaluated, or spacing of possible shear walls can be determined. Floor in-plane behaviour can be modelled as a tied-arch developing between adjacent seismic resistant elements. A floor may fail for overcoming tensile resistance of the tie (corresponding to the reinforcement of possible side beams), for overcoming compressive resistance of the blocks at the key section, or for shear failure by the supports.

An experimental campaign aimed at estimating the in-plane capacity of a portion of a beam and block floor system was carried out on an existing structure. Among the possible failure modes, focus was made on the possible shear failure by the supports with reference to a floor lacking the RC overlay (Marini et al. 2020), which is the most critical. The test set up together with the preliminary results were presented and discussed. As a major result, it was observed that the behaviour of the tested floor specimen was almost rigid up to a brittle collapse.

A simple analytical model, accounting for the ratio between the beam axial stiffness and the floor shear stiffness, was developed to model the experimental test, particularly to describe the diffusion of the point load applied to the RC beam to the adjoining clay blocks and to calculate the shear action inside the clay blocks. Analytical predictions matched fairly well with the experimental results. The maximum average shear stress in the clay blocks was derived and it ranges between 1.6 and 1.9 MPa. This result complies with the outcomes of a previous experimental campaign characterizing the behaviour of the beam-to-block interface through local tests. In existing floors, hollow clay blocks may be locally weakened by cracks or damages due to the casting operations or to fabrication defects. Moreover, the presence or absence of mortar between adjacent blocks strongly influence the behavior of the system. These features, which have been noted in the analyzed floor portion, are very common in the existing structures, and they cannot be exactly controlled to predict the behavior.

Finally, the adoption of a fairly large safety factor is suggested in the prediction of the in-plane capacity of existing composite floors, to avoid brittle in-plane failures and to account for possible uncertainties and inherent defects characterizing existing floor composite systems.

The presented preliminary experimental campaign is part of a broader research aimed at investigating the role of existing beam and block floor system in the seismic behaviour of existing and retrofitted RC buildings. Some future experimental tests are planned on other composite floor structural typologies, both with or without additional extrados RC slab. Experimental results will be adopted to calibrate both analytical and numerical models.

ACKNOWLEDGEMENTS

This research was carried out in the framework of RelUIS-DPC 2019–2021 project. The authors gratefully acknowledge the municipality of Brescia, particularly arch. M. Azzini, and eng. C. Lazzaroni and A. Caporali, who supported the research project and were proactive in the experimental campaign. A special thank goes to C. A. Arcuri and to the technical staff of the Structural Testing Laboratory of the University of Bergamo for their contribution and assistance during the test.

REFERENCES

Blandón, J.J. & Rodriguez, M.E. 2005. Behavior of Connections and Floor Diaphragms in Seismic Resisting Precast Concrete Buildings. *PCI Journal* 50 (2): 56–75.

BPIE (Building Performance Institute Europe) 2011. *Europe's buildings under the microscope: A country-by-country review of the energy performance of the buildings*. Brussel.

Bull, D.K. 2004. Understanding the complexities of designing diaphragms in buildings for earthquakes. *Bulletin of the New Zealand Society for Earthquake Engineering* 37 (2): 70–88.

Feroldi, F. 2014. Riqualificazione sostenibile del patrimonio edilizio del secondo dopo guerra mediante doppia pelle ingegnerizzata per il rinforzo strutturale, l'efficientamento energetico e la riqualificazione architettonica e urbana. PhD thesis, University of Brescia (in Italian).

Kolston, D. & Buchanan, B.W. 1980. Diaphragms in Seismic Resistant Buildings. *Bulletin of the New Zealand Society for Earthquake Engineering* 12 (2): 162–170.

La Greca, P. & Margani, G. 2018. Seismic and Energy Renovation Measures for Sustainable Cities: A Critical Analysis of the Italian Scenario. *Sustainability* 10 (1): 254.

Lee, H.J., Kuchma, D. & Aschheim, M.A. 2007. Strength-based design of flexible diaphragms in low-rise structures subjected to earthquake loading. *Engineering Structures* 29 (7): 1277–1295.

Marini, A., Passoni, C., Riva, P., Negro, P., Romano, E. & Taucer, F. 2014. Technology options for earthquake resistant, eco-efficient buildings in Europe: Research needs. Report EUR 26497 EN. JRC87425. ISBN 978-92-79-35424-3. doi:10.2788/68902. Publications Office of the European Union.

Marini, A., Passoni, C., Belleri, A., Feroldi, F., Preti, M., Metelli, G., Giuriani, E., Riva, P. & Plizzari, G. 2017. Combining seismic retrofit with energy refurbishment for the sustainable renovation of RC buildings: a proof of concept. *European Journal of Environmental and Civil Engineering*, DOI:10.1080/19648189.2017.1363665

Marini, A., Belleri, A., Passoni, C., Feroldi, F. & Giuriani, E. 2020. In-plane capacity of existing Post-WWII beam and block floor systems. Submitted to *Engineering Structures*.

Moehle, J.P., Hooper, J.D., Kelly, D.J. & Meyer, T.R. 2010. Seismic design of cast-in-place concrete diaphragms, chords, and collectors: a guide for practicing engineers. NEHRP Seismic Design Technical Brief No. 3.

Naeim, F. & Boppana, R.R. 2001. Seismic design of floor diaphragms. 2001. *The seismic design handbook*. Kluwer Academic Publishers. ISBN 978-0-7923-7301-8. Chapter 8: 373–407.

Paulay, T. & Priestley, M.J.N. 1992. *Seismic Design of Reinforced Concrete and Masonry Buildings*. New York, N.Y., Wiley.

Rodriguez, M.E., Restrepo, J.I. & Carr, A.J. 2002. Earthquake Induced Floor Horizontal Accelerations in Buildings. *Earthquake Engineering and Structural Dynamics* 31 (3): 693–718.

Rodriguez, M.E., Restrepo, J.I. & Blandón, J.J. 2007. Seismic design forces for rigid floor diaphragms in precast concrete building structures. *Journal of Structural Engineering*, ASCE, 133 (11): 1604–1615.

Schoettler, M.J., Belleri, A., Zhang, D., Restrepo, J.I. & Fleishman, R.B. 2009. Preliminary results of the shake-table testing for the development of a diaphragm seismic design methodology. *PCI Journal* 54 (1): 100–124.

Takeuchi, T., Yasuda, K. & Iwata, M. 2009. Seismic Retrofitting using Energy Dissipation Façades. *Proceedings of ATC & SEI Conference on Improving the Seismic Performance of Existing Buildings and Other Structures*, 1000–1009.

Tena-Colunga, A., Chinchilla-Portillo, K.L. & Gelacio Juárez-Luna, G. 2015. Assessment of the diaphragm condition for floor systems used in urban buildings. *Engineering Structures* 93: 70–84.

Brick and Block Masonry - From Historical to Sustainable Masonry –
Kubica, Kwiecień & Bednarz (eds)
© 2020 Taylor & Francis Group, London, ISBN 978-0-367-56586-2

Investigation of the structural response through simple numerical models derived automatically from LiDAR scanning: The case of masonry vaults in historical buildings

G. Angjeliu & G. Cardani

Department of Civil and Environmental Engineering Politecnico di Milano, Milan, Italy

ABSTRACT: Structural response of masonry vaults is closely connected with their geometric shape. LiDAR scanning has created a new possibility for a fast surveying of their complex geometry. The aim of this paper is to investigate the use of an approximated surface created from a registered point cloud to develop a structural model for the fast analysis of masonry vaults. The study combines scanning with LiDAR sensor, automatic surface reconstruction methods and structural analysis with the finite element method. The results highlight that the simplified models created in a short time, although preliminary, allow to give important results on the structural response of masonry vaults as the magnitude of forces in the system, the force flow or damage predictions. The elements of a future methodology are developed alongside the case of the church of St. Marta, located in the town of Arona, in the shores of Lake Maggiore in Italy.

1 INTRODUCTION

LiDAR (Light Detection And Ranging) measurements have created a new perspective for the fast survey of historic buildings. It is a remote sensing method that uses light in the form of laser pulses to measure point coordinates with respect to the sensor origin. The geometry is of significant interest for masonry vaults since the structural response is closely connected with the possibility to consider the real vault shape for structural analysis.

The acquired point cloud cannot be used in a straightforward way to obtain a geometrical model. In general, the geometric model can be developed over the point cloud in three ways: a) manual modelling (Costa-Jover et al. 2017, Bednarz et al. 2014), b) parametric modelling based on user defined inputs (Angjeliu et al. 2019b, Angjeliu et al. 2020, Barazzetti 2016), c) automatic reconstruction methods, mostly based on point cloud segmentation and surface fitting methods over a set of point clouds (Thomson & Boehm 2015, Macher et al. 2017, Li et al. 2020). Manual modelling allows to create very detailed models (including the thickness of the sectioned structural elements which is not possible to be surveyed with LiDAR scanning), but it is very time consuming. Whereas, automatic reconstruction methods, are very fast, provide high geometric accuracy on the exterior surface, while remain poor on the reconstruction of the elements thickness.

In this study we investigate the possibility to use an approximated surface over the recorded point cloud as a geometrical model suitable for a fast-structural analysis.

An original procedure for the analysis of the safety of masonry vaults and the use of point cloud data derived from scanning was applied in the church of St. Jacobs in Belgium (Schueremans & Van Genechten 2009). The procedure featured a low level of automation, since the point cloud was fitted with a surface, which was later sliced manually in arches to perform limit analysis. A similar approach based on point cloud slicing and limit analysis was considered also by Acikgoz et al. (2017). A more advanced procedure was considered by Argiolas et al. (2019) which involved the use of a vault geometry created through parametrized surfaces and NURBS (Non-Uniform Rational Basis Spline) approximation in FE modelling.

In the present study an approach, implemented in Matlab and Abaqus software, featuring a higher level of automation is considered. It involves point cloud registration through LiDAR scanning, point cloud approximating through fitting functions, surface meshing and finite element simulations.

The results demonstrate that the proposed simplified models can give important results on the structural response and safety of the masonry vaults. Their future use directly on site could improve significantly the quality of the structural survey, as they can give hints during structural inspections, as well as improve the decision-making process.

2 THE CASE STUDY: THE CHURCH OF ST. MARTA IN ARONA

2.1 Overview

The small church of St. Maria di Loreto, also known as St. Marta (Figure 1), built in 1592, is located in Arona, on a square in the historic center overlooking the old harbour on Lake Maggiore. The importance of this church is given by a second construction that is enclosed within its walls: a full-scale reproduction of the Holy House of Loreto (Figure 2). The church was built according to the requests and wishes of Cardinal Federico Borromeo and his mother. They employed important designers and artists from the construction site of Milan Cathedral, which at the time was still under construction, such as: Martino Bassi, Tolomeo Rinaldi, Marco Antonio Prestinari and Francesco Maria Richino (Di Bella 2018, Casati 2018).

2.2 Structural evolution

The original design of Bassi was modified after his death and when the construction had already started, but not completed, when a lowered barrel vault with lunettes was added in 1626 (Casati 2018). This addition led to several changes in the building, such as the closing of some windows on the longitudinal walls, still visible. The irregular shape of the vault is divided into 6 bays by twin arches and intrados iron tie-rods. This shape seems to demonstrate that the wooden truss roof was already present when the masonry vault was built, and the masons had to adapt it to the available space. (Figure 3).

a)

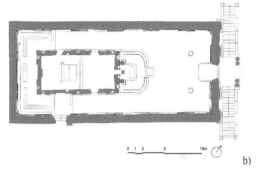

b)

Figure 2. A) Facade of the Holy House, inside the church, with the high altar leaning against it and b) plan of the church.

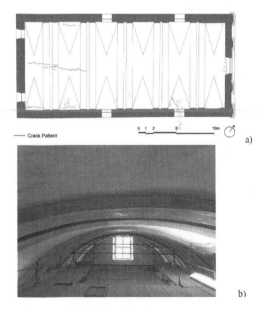

a)

b)

Figure 3. A) Plan of the barrel vault and b) Irregular masonry barrel vault observed from the scaffold at a height of about 9 meters. A longitudinal crack is visible in the middle.

Figure 1. Façade of the St. Maria di Loreto church in Arona.

The bell tower, as originally designed, was completed in 1708. A restoration in 1820 due to high moisture problems, modified the interior decorations and in 1838 the new marble altars of the church and of the Holy House were built (Bertolazzi 2006).

The small rectangular church (24x10x13 m) has been in the last years in a condition that requires restoration. The works started in 2013 on the façade, including also the restoration of the interior surfaces and the installation of a structural health monitoring system of the cracks. On the ceiling there is a non-passing through crack that runs along the key of the vault (Figure 3). Cracks are also present at the lower level of the church (Westward) along the longitudinal walls, highlighting the excessive vault thrust, not balanced by the iron tie-rods which appear to be poorly tensioned.

3 LIDAR SCANNING AND POST-PROCESSING

3.1 *Data acquisition*

The geometric survey in the church of St. Marta was carried out using a high-resolution LiDAR sensor (OS1-64). During the survey with the LiDAR sensor it is possible to view the scan in real time, which allows to minimize problems with missing parts due to obstacles in the post-processing phase. The configuration during the acquisition phase was set out to a horizontal resolution of 2048 points and a vertical resolution of 64 point at a rotation rate of 10Hz.

The survey was focused on the first 3 bays of the church from the façade (Figure 2). Only 2 scans were extracted and used to create the final point cloud (Figure 4). The scan alignment was carried out by using the Iterative Closed Point algorithm (Glira et al. 2015). The final point cloud includes approximately 60′000 points and an alignment error of 0.015m.

3.2 *Identification of geometric shapes and deformation*

The procedure is based on a mathematical fitting of the equation of a circular or an elliptic arch in a subset of

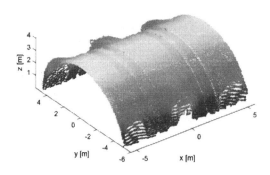

Figure 4. Point cloud measurements of the analysed vaults.

the registered point cloud. The identification of the geometric shapes was discussed previously in detail (Angjeliu et al. 2019a, Angjeliu et al. 2019b). The data can be used for manual geometric modelling of the vault or for understanding the structural deformations.

Four sections were extracted from the point cloud, in order to understand the geometric shape of the twin arches (Figure 5). It was possible to calculate the radius of the twin arches accordingly: 5.197m, 5.222m, 5.303m, 5.266m. It is noted a difference by about 3 cm between the first and second arch, 8 cm between the second and third arch and 2 cm between the third and fourth arch. These small differences must be further investigated, but can be probably related to construction irregularities, due to the subsequent addition of the vault, or to structural deformations.

The barrel vault thickness is 20 cm on average. It does not have a constant thickness. In the intrados it is present a groove above the twin arches, which is 170 cm wide. It creates space for the timber trusses over the barrel vault, that would otherwise lie directly on the vault.

A longitudinal section of the point cloud allows to see a 3 - 4 cm deflection extended between the 1st and 2nd bay (Figure 6).

4 GEOMETRIC MODELLING

In this section it is studied the possibility to use an approximated surface over the recorded point clouds as a geometrical model suitable for a fast-structural FE analysis.

4.1 *Surface fitting procedure for automatic modelling*

The point cloud recorded be means of LiDAR scanning cannot be used directly to create a surface since they are not in a gridded form z (x,y). We chose to apply the surface fitting procedure over a point cloud as developed by D'Errico (2005), and later on adopted by Angjeliu et al. (2019b), (Angjeliu et al. 2019c) for the approximation of masonry vaults geometry starting from a point cloud.

The problem of surface interpolation is posed as a linear combination of the values at the grid nodes x with the value of the local measured points y:

$$A \mathbf{x} = \mathbf{y} \qquad (1)$$

$\mathbf{x}$ - vector of length nx · ny
A - has n rows and nx · ny columns
nx – number. of nodes in the x– direction
ny - number of nodes in the y– direction
n - is corresponding to each data point supplied by user.

The final output would be a gridded surface of the form z (x,y). It is a straightforward matter to obtain,

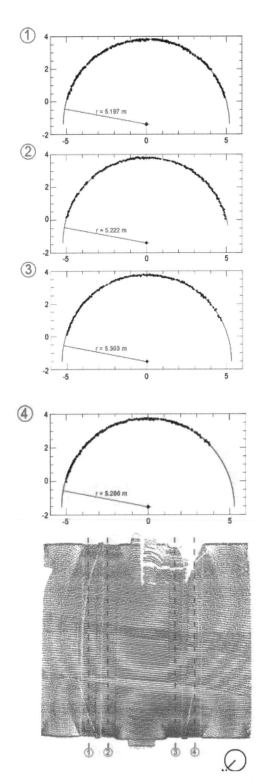

Figure 5. Identification of geometric shape of the twin arches by least square minimization procedure.

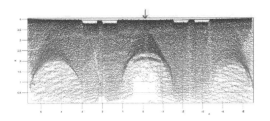

Figure 6. Side view of the point cloud with deformation indication.

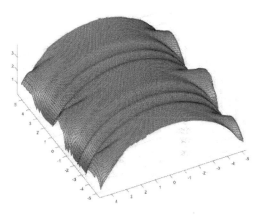

Figure 7. Re-meshing of the automatically created geometry in Abaqus software.

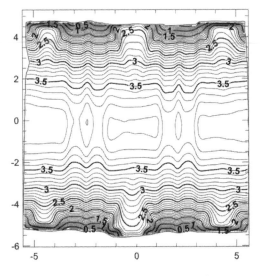

Figure 8. Contour plot of the approximated surface.

meshing, the contour plot and the rendering of the obtained gridded surface using the basic functions available in Matlab (Figure 7, 8).

613

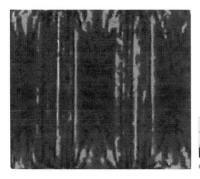

0 m   0.044 m

Figure 9. Geometric modelling accuracy (Hausdorff distance).

4.2 Geometric modelling quality

In the following it is investigated the geometric modelling accuracy, by computing the Hausdorff distance between the approximated surface and the point cloud measurements (Cignoni et al. 1998).

The computations show that the mean difference is 0.0046m (Figure 9). The greater difference is noted on sharp borders, where the lunettes intersect the barrel vault, with a maximum difference of 0.044m (Figure 9).

It is worth commenting that during tests carried out by the authors for the geometric difference between point cloud and manually created models, the estimated error was around 0.25 m. Therefore, the achieved mean error of 0.0046m is extremely good.

5 STRUCTURAL ANALYSIS

5.1 Structural modelling

The automatically created geometrical model (see section 4) is used here directly to create FE model in the Abaqus software. The geometry is re-meshed with 3500 quadrilateral linear elements, since the previous mesh is not suitable for FE simulations. The vault is considered with a constant thickness of 20 cm (Figure 10). On site measurement with an electromagnetic impulse device measuring the strength of the magnetic field between two units (Hilti PX 10 Transpointer) showed variation between 18 - 22 cm.

The simplified model features only the masonry web and does not include the rubble fill, since it cannot be created automatically in the present state of the project. The brick masonry is assumed to have an elastic modulus, 2000 MPa, compressive strength, 5 MPa and fracture energy in compression 1.2 Nmm/mm^2, while the tensile branch with a strength 0.25 MPa and a fracture energy 0.012 Nmm/mm^2.

5.2 Interpretation of the structural response

Under self-weight condition the horizontal thrust is 18.9 kN/m, equal to 189 kN for the considered

Figure 10. Re-meshing of the automatically created geometry in Abaqus software.

length of 10 m in the present structural model. Therefore, the horizontal thrust is 63 kN/support.

Under this condition the masonry vault is stable. The principal compressive stresses are 320 kPa, while the principal tensile stresses are 75 kPa (Figure 11a). The principal tensile stresses are very close to the masonry strength in tension. Furthermore, the displacements in the central zone of the vaults are 10 times

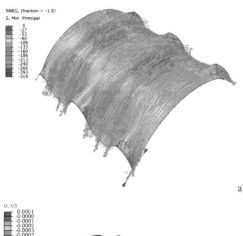

a)

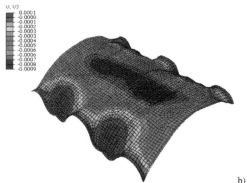

b)

Figure 11. Self-weight configuration: a) Principal compressive stresses unders self-weight [Units: kPa], b) Vertical displacements, [Units: m].

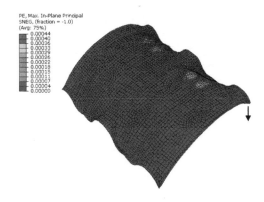

PE, Max. In-Plane Principal
SNEG, (fraction = -1.0)
(Avg: 75%)

Figure 12. Plastic strain due to a 2 cm settlement.

higher compared to the rest of the parts (Figure 11b). As expected, on site observations show that in this zone the vault is subject to cracking (Figure 3).

A further possible kinematics is connected with a soil settlement in the second span of the vault, which is further corroborated by a crack in the north wall.

The results of the support settlements are shown in Figure 12. Plastic strain localization is observed between the central line and the settled support, which is different from the crack surveyed on site. This response is also different from the self-weight simulation where a crack following the central line could be expected. The fundamental reason is the fact that the rubble-fill is not included in the numerical model, which could have provided buttressing, and the possibility to have strain localization almost in the vault centreline.

6 DISCUSSION

As highlighted in our previous studies, accurate structural models of masonry vaults should include all the components (arch, rib, web, infill, tas-de-charge) as well as need to consider their hierarchy when organizing the structural model (Angjeliu et al. 2019b, Angjeliu et al. 2019a). Clearly, a model based on the surface approximation cannot have this level of detail, hence cannot be used for detailed simulations.

However, results of this study show that it can be used to get fast and preliminary information on the structural response right after the in-situ survey is completed.

7 CONCLUSION

The elements of a future methodology for a fast-structural assessment of masonry vault are discussed. The study concluded on the necessity to combine LiDAR scanning and the structural analysis methods to study the structural response of historic masonry

vaults. It can be summarized in three main stages: a) registration with LiDAR scanning, b) point cloud approximated surface, c) use of this surface as geometry for a FE structural model for a fast analysis of masonry vaults.

The procedure is tested in the barrel vault with lunette of the church of St. Marta, located in in the shores of Lake Maggiore in Italy. The results highlight that the simplified model created in a short time, allow to give important results on the horizontal thrust of the vault, the compressive force flow, level of stress, or part of the structure with higher probability of cracking.

In terms of predictions of observed damage, we had positive matching under self-weight conditions, while further improvement needs to be implemented when considering soil settlements. In particular, the rubble-fill needs to be considered.

Finally, the results improve significantly the quality of the structural survey, as they can guide further inspections directly after the first visit, as well as improve the decision-making process.

ACKNOWLEDGMENTS

The authors are grateful to Ouster Inc. for donating the sensor used during our survey in the church of St. Marta in Arona. We also thank Arch. M. Ziggiotto in charge of the restoration, the restorer Giorgio Perino and Ester Mondini for her contribution in the survey and drawings.

REFERENCES

Acikgoz, S., Soga, K., & Woodhams, J. 2017. Evaluation of the response of a vaulted masonry structure to differential settlements using point cloud data and limit analyses. *Construction and Building Materials*, *150*, 916–931. doi:10.1016/j.conbuildmat.2017.05.075

Angjeliu, G., Cardani, G., & Coronelli, D. 2019a. Digital Modelling and Analysis of Masonry Vaults. *ISPRS - International Archives of the Photogrammetry, Remote Sensing and Spatial Information Sciences*, *XLII-2/W11*, 83–89. doi:https://doi.org/10.5194/isprs-archives-XLII-2-W11-83-2019.

Angjeliu, G., Cardani, G., & Coronelli, D. 2019b. A parametric model for ribbed masonry vaults. *Automation in Construction*, *105*, 102785. doi:https://doi.org/10.1016/j.autcon.2019.03.006.

Angjeliu, G., Coronelli, D., & Cardani, G. 2019c. *Challenges in Modelling Complex Geometry in Historical Buildings for Numerical Simulations*. Paper presented at the The 18th International Conference on Geometry and Graphics, Milan, Italy.

Angjeliu, G., Coronelli, D., Cardani, G., & Boothby, T. 2020. Structural assessment of iron tie rods based on numerical modelling and experimental observations in Milan Cathedral. *Engineering Structures*, *206*, 109690. doi:https://doi.org/10.1016/j.engstruct.2019.109690

Argiolas, R., Cazzani, A., Reccia, E., & Bagnolo, V. 2019. From Lidar Data Towards Hbim for Structural Evaluation. *ISPRS - International Archives of the*

Photogrammetry, Remote Sensing and Spatial Information Sciences, XLII-2/W15, 125–132. doi:10.5194/isprs-archives-XLII-2-W15-125-2019.

Barazzetti, L. 2016. Parametric as-built model generation of complex shapes from point clouds. *Advanced Engineering Informatics, 30*(3), 298–311. doi:10.1016/j.aei.2016.03.005

Bednarz, Ł. J., Jasieńko, J., Rutkowski, M., & Nowak, T. P. 2014. Strengthening and long-term monitoring of the structure of an historical church presbytery. *Engineering Structures, 81*, 62–75. doi:10.1016/j.engstruct.2014.09.028

Bertolazzi, D. 2006. *La chiesa di Santa Maria di Loreto di Arona: problemi storici e di conservazione*. (Ms. Thesis in Architecture), Politecnico di Milano.

Casati, M. L. 2018. *Da Martino Bassi e Tolomeo Rinaldi a Francesco Maria Richini: Santa Maria di Loreto, un santuario per il borgo di Arona e per il lago Maggiore*. Paper presented at the La chiesa di Santa Maria di Loreto e la confraterni-ta di Santa Marta di Arona dai Borromeo a oggi. Storia, restauro e valorizzazione, Novara: Interlinea.

Cignoni, P., Rocchini, C., & Scopigno, R. 1998. *Metro: measuring error on simplified surfaces*. Paper presented at the Computer Graphics Forum.

Costa-Jover, A., Ginovart, J., Coll-Pla, S., López Piquer, M., Samper-Sosa, A., Moreno García, D., & Solís Lorenzo, A. 2017. 3D surveying and geometric assessment of a gothic nave vaulting from point clouds. *ISPRS-International Archives of the Photogrammetry, Remote Sensing and Spatial Information Sciences*, 203-208.

D'Errico, J. 2005. Surface fitting using gridfit. *MATLAB central file exchange, 643.*

Di Bella, G. 2018. *Indagine storico-urbanistica sull'antica piazza del porto di Arona tra XIV e XIX secolo*. Paper presented at the La chiesa di Santa Maria di Loreto e la confraternita di Santa Marta di Arona dai Borromeo a oggi. Storia, restauro e valorizzazione, Novara: Interlinea.

Glira, P., Pfeifer, N., Briese, C., & Ressl, C. 2015. A Correspondence Framework for ALS Strip Adjustments based on Variants of the ICP Algorithm. *Photogrammetrie - Fernerkundung - Geoinformation, 2015*(4), 275–289. doi:10.1127/pfg/2015/0270.

Li, Y., Li, W., Tang, S., Darwish, W., Hu, Y., & Chen, W. 2020. Automatic Indoor as-Built Building Information Models Generation by Using Low-Cost RGB-D Sensors. *Sensors (Basel), 20*(1), 293. doi:10.3390/s20010293.

Macher, H., Landes, T., & Grussenmeyer, P. 2017. From Point Clouds to Building Information Models: 3D Semi-Automatic Reconstruction of Indoors of Existing Buildings. *Applied Sciences, 7*(10), 1030. doi:10.3390/app7101030.

Schueremans, L., & Van Genechten, B. 2009. The use of 3D-laser scanning in assessing the safety of masonry vaults—A case study on the church of Saint-Jacobs. *Optics and Lasers in Engineering, 47*(3-4), 329–335. doi:10.1016/j.optlaseng.2008.06.009.

Thomson, C., & Boehm, J. 2015. Automatic Geometry Generation from Point Clouds for BIM. *Remote Sensing, 7*(9), 11753–11775. doi:10.3390/rs70911753.

Brick and Block Masonry - From Historical to Sustainable Masonry –
Kubica, Kwiecień & Bednarz (eds)
© 2020 Taylor & Francis Group, London, ISBN 978-0-367-56586-2

Making decision on repointing of clay brick facades on the basis of moisture content and water absorption tests results – a review of assessment methods

S.K. Shahreza, M. Molnár, J. Niklewski & I. Björnsson
Department of Building and Environmental Technology, Division of Structural Engineering, Lund University, Lund, Sweden

T. Gustavsson
Tomas Gustavsson konstruktioner AB, Lund, Sweden

ABSTRACT: Use of clay brick masonry in façades is often motivated by its aesthetic values and durability. Yet, mortar joints exposed to climate agents erode over time, expected to cause elevated moisture content and water absorption. Thus, it is often recommended that 40- to 50-year-old facades should be repointed – a measure which is intrusive and costly. Decision is in many cases taken without a clear evidence that repointing will diminish water absorption and moisture content in the renovated walls. This paper presents the results of a state-of-the-art study on field and laboratory methods to measure moisture content and water absorption in clay brick masonry. For common buildings, use of low cost and time efficient measurement methods is feasible. However, prior to measurements, analysis of technical and climate data combined with a visual inspection might give a rational basis for decision on repointing or other alternative maintenance measures.

1 INTRODUCTION

Clay brick masonry is one of the most common building materials in the facades of residential buildings in the Nordic countries. The ubiquitous use of clay brick masonry as façade material is due to its aesthetic values, good durability and low maintenance needs. Although the expected technical life time of a clay brick façade is more than hundred years, maintenance can still be needed due to inevitable deterioration caused by climate and ambiance actions. Important climate actions in a Nordic climate include wind-driven rain (WDR) and freeze-thaw-cycles – actions that individually or in conjunction can cause spalling, delamination or cracking of bricks and erosion and cracking of mortar joints.

The focus in this paper is on the repointing of mortar joints, since it is an intrusive and costly maintenance measure. A common argument for repointing is that the erosion of mortar joints facilitates water up-take in facades exposed to WDR (Fried et al., 2014). Further, erosion of mortar joints is, at least in the Nordic countries, regarded as detrimental from an aesthetic point of view, since it creates, seen superficially, the impression of poor technical condition of the building. Understanding that aging of clay brick facades can be perceived as an aesthetic value, e.g. through exposure of fossil shells in the surface of the mortar joints, is generally poor (Tägil et al., 2011).

According to the present practice in the Nordic countries, repointing shall be carried out as part of a regular maintenance scheme, after 40-50 years from erection or when limited façade parties with more or less eroded mortar joints are observed (Tindall, 1987, Brief, 2017). No further investigations, e.g. concerning factual water up-take, are usually carried out. Nor are alternative measures, such as partial repointing of eroded façade parties, considered - full repointing is regularly carried out without a more in-depth analysis of the possible technical, economic or aesthetic implications of this measure. In the light of the presented practices it can be objected that decision concerning repointing of clay brick facades usually is not based on rational grounds.

In the present paper the results of a state-of-the-art study concerning field and laboratory methods to assess water content and water up-take caused by WDR are presented. Using information on water content and water up-take to rationally analyse whether repointing can improve the technical condition of clay brick facades in relation to WDR action is discussed and research and development needs are identified.

2 RESEARCH APPROACH

A literature review concerning repointing of clay brick façades has been carried out with keywords including repointing, masonry, clay brick façade, masonry façade, brick masonry, mortar joints, mortar, environmental factors, wind-driven rain, impingement, water up-take, water penetration, moisture content, durability, erosion and deterioration. Other search terms include destructive and non-destructive tests, study of WDR, field, laboratory, etc. Main literature sources and databases include Lund University Library, National Library of Sweden including the libraries of all Swedish universities, ASCE Library, Engineering Village (Elsevier), Wiley Online Library. In addition to library searches, meetings with Swedish and German researchers and industry representatives provided additional sources of information.

3 RESULTS

3.1 *Tests measuring moisture content*

A number of non-destructive and destructive experimental procedures for assessing moisture conditions of brick facades have been reviewed based on previous research studies (Emerisda, 2014, Bison et al., 2011, Litti et al., 2015, Hola, 2017, Larsen, 2012). Experimental procedures can be categorized into three groups based on their destructiveness and their type of output. The groups are described as follows: *group A* - destructive tests (DT) measuring moisture content quantitatively; *group B* non-destructive tests (NDT) measuring moisture content quantitatively; and *group C* - NDT indicating moisture content qualitatively. The potential for making on-site measurements using each experimental procedure is also assessed.

Tests belonging to *group A* include, among others, gravimetric tests, the calcium carbide test and the chemical method (Karl Fisher). Although destructive tests are generally seldom carried out on historical buildings, there is no consensus whether or not they are appropriate for residential buildings.

Gravimetric testing involves sampling by core drilling, after which the samples are dried in an oven at a specified temperature. Finally, the actual moisture content (MC) is generally derived from the difference in weight of the sample before and after drying (Camuffo and Bertolin, 2012, EN, 1993). Gravimetric testing is considered being a precise and reliable method to measure moisture content in masonry walls; however, sampling by core drilling is perceived as a drawback.

The calcium carbide test involves grinding a sample from the masonry wall and mixing the powder with a certain amount of calcium carbide (Blystone et al., 1962, ASTM, 2011, Camuffo and Bertolin, 2012). Subsequently, the moisture content can be derived from the pressure of the gas released during the reaction between calcium carbide and water, by the use of a calibration curve (Binda et al.,

1996). While the test is appropriate for on-site measurements, it is less reliable than the gravimetric test.

The chemical method, invented by Karl Fisher, is based on the reaction between iodine and water, producing a non-conductive chemical substance. It is possible to carry out titration on site to determine trace amounts of water in a sample. The capabilities of this test are measuring accurately small amounts of moisture and determining the water content level from low values till saturation (Schöffski, 2006, Bruttel and Schlink, 2003). However, it is stated that the chemical method would be helpful for small samples and not reliable for masonry walls (Hola et al., 2012).

Among tests belonging to *group B*, nuclear magnetic resonance (NMR) and neutron radiography are of special interest. *Nuclear magnetic resonance* (NMR) is a non-contacting, fast, accurate, and reliable technology to measure the water content in masonry walls and to evaluate the distribution of moisture content along the wall surface (Pel et al., 1996, Wolter and Krus, 2005, Litti et al., 2015). *Neutron radiography* records the radiation passing through an object by a position sensitive detector. Although both methods quantitatively and non-destructively measure the water content in walls, their high cost and limited availability (Hola, 2017) make their applicability to common buildings rather limited. They could, however, be employed under certain circumstances, such as for buildings with great cultural or economic values.

Tests in *group C* include, among others, the paper indicator method, infrared thermography (IRT), holographic radar and the dielectric and microwave methods. The *paper indicator* method is a simple and inexpensive method to qualitatively evaluate the moisture content in a masonry wall. In this test, contact between chemical papers and the moist surface of a facade provides indications of the moisture content based on the subsequent change in colour of the paper; similar to the litmus paper test for evaluating acidity (Hola, 2017).

IRT uses thermal imagery to map the location of damp areas and the existence of voids (Griffin, 2013), yet without the possibility of quantitative evaluation of the moisture content. This non-destructive test is carried out in-situ with at a relatively low cost. The time when the test can be performed is critical and limits its use, since it is strongly affected by environmental conditions [high relative humidity and low temperature] (Emerisda, 2014, Bison et al., 2011).

Holographic radar has the capability to detect moisture in the range of 50 to 200 mm beneath the surface as a function of continuous wave frequency (Litti et al., 2015, Bison et al., 2011). Also, detection of voids is possible. In contrast with IRT, the holographic radar technique is not influenced by relative humidity or air temperature (Litti et al., 2015).

The *dielectric* method works on the principle of variation of the dielectric constant of a material in the presence of water. The dielectric constant increases with increasing moisture content, making differences in moisture content detectable. This test is commonly employed by surveyors to determine the moisture distribution along the height of masonry walls. The method is, however, limited to depths of 50 to 100 mm (Hola, 2017). The *microwave* method is another non-destructive method which works on the principle of reduction of the radiation intensity as microwaves pass through a damp material (Hola, 2017). Thus, the more water the specimen contains, the bigger the energy loss. This method is procedurally similar to the dielectric one, however the microwave method can be used to depths up to 300 mm (Hola, 2017). The advantages of these two techniques are low cost of the equipment and ease of use (Emerisda, 2014, Hola, 2017).

The main features of the test methods described in this section are, together with methods to be described in section 3.2, summarized in Table 1.

3.2 Tests measuring water absorption

There are several quantitative methods available for measuring the amount of water being absorbed through a brick masonry wall. They are divided into two groups - *group D* comprising low-intrusive methods, while group E including NDT methods. Again, the potential for making on-site measurements using each experimental procedure is also assessed.

A newly developed, *group D* technique to measure water absorption is named Wasseraufnahme Messgerät – WAM (*Instrument for measurement of water up-take* – the author's translation), which measures water absorption in the absence of wind pressure

(Möller and Stelzmann, 2013, Stelzmann et al., 2015). The apparatus includes a scale, a storage tank and a pipe with nozzles. The apparatus is attached to a section of the wall with the edges and sealed in order to create a closed system (Figure 1a). The apparatus then projects a water film on the entire section. Runoff water that is not absorbed is collected. The rate of absorption can be calculated and monitored in real time by continuously weighing the amount of moisture in the closed system. The apparatus is portable and can easily be used in-situ. If attached to the lower end of a wall, the apparatus can rest on the ground or alternatively on a small support. In order to test the upper parts of a wall, the apparatus needs to be attached by 8 screws, making it semi-intrusive (*group D*).

Among NDT methods (*group E*), the RILEM tube test, the Franke plate and the Stockbridge method are reviewed here. The *RILEM tube test* is widely used to quantify water absorption during a specified period of time through an up-take tube, with possible application in laboratory and on site (RILEM, 1978, Crissinger, 2005). The uptake tube is first sealed to the substrate with a putty and then filled with water. The amount of absorbed water is recorded during specified time intervals. This simple NDT test is helpful for assessing the water absorption rate before and after repointing. However, it only provides results for a small area of masonry wall and results are not precise when the tube is applied on mortar joints. In order to speed up the procedure, it is suggested to use several tubes in different locations (Figure 1.b).

The Franke-Platte method consists of a plate (25 cm × 8.1 cm absorption area) and a tube like the RILEM tube (Figure 1.c). The procedure is similar to the RILEM tube test with the exception that rather than attaching a tube to a small area, a plate is attached to an area including bricks and mortar

Table 1. Summary of test method features.

Test	Group	Destructiveness[1]	Output[2]	Application	On site
Gravimetric	A	De	QN	moisture content and its distribution	-
Calcium carbide	A	De	QN	moisture content	✓
Chemical method (Karl Fisher)	A	De	QN	moisture content indirectly	✓
Nuclear magnetic resonance (NMR)	B	N	QN	surface (flat) moisture content	✓
Neutron radiography	B	N	QN	moisture content	✓
Paper indicator method	C	N	QL	moisture content level	✓
Infrared thermography (IRT)	C	N	QL	surface moisture content	✓
Holographic radar	C	N	QL	moisture content (flat surface)	✓
Dielectric method	C	N	QL	moisture content	✓
Microwave method	C	N	QL	moisture content	✓
Wasseraufnahme Messgerät (WAM)	D	L	QN	water penetration	✓
RILEM tube test	E	N	QN	water penetration	✓
Franke-Platte	E	N	QN	water penetration	✓
Stockbridge (Stockbridge, 1989)	E	N	QN	water penetration	✓

1. De – destructive, N – non-destructive, and L – low intrusive
2. QL – qualitative and QN – quantitative

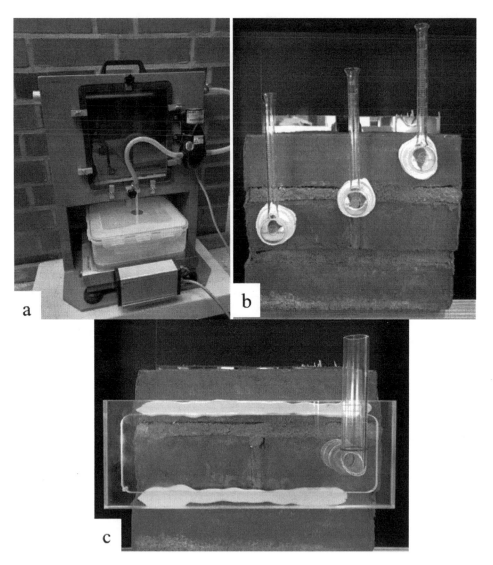

Figure 1. Apparatus and set-up of test methods measuring water absorption; (a) WAM Device, (b) RILEM tube test, (c) Franke Platte.

joints (Franke and Bentrup, 1991, Neumann et al., 2014, Stelzmann et al., 2015).

Stockbridge developed a watertight frame (91 × 122 cm area of the wall) to be attached to a masonry façade while measuring the water absorption (Stockbridge, 1989, ASTM, 2014), without refilling the absorbed water. It is recommended that if the rate of absorption is less than one litre per hour, no repointing is needed. Further, it is stated that if the absorption rate is larger than five litres per hour, repointing will result in a substantial decrease in water absorption. Unfortunately, it is not shown how these criteria have been established.

4   DISCUSSION AND STRATEGIES

The results presented in the previous section show that there are a couple of methods to assess moisture content and water absorption in clay brick facades. Each method is associated with costs related to investment in equipment and operation. Qualitative or quantitative information possible to obtain must be valued in relation to its usefulness. In the following sections a brief analysis and discussion are presented concerning circumstances when decision on repointing can be more rational by using information obtained by the presented methods.

4.1 Preliminary investigations

Prior to carrying out potentially costly and time-consuming experimental studies, either in laboratory or on-site, gathering basic information concerning the building, local climate and weather history can provide useful information such as:

- Age of the building; previous façade maintenance measures; type and/or brand of the bricks and of the mortar;
- Occupants' or building owners' reports on problems with dampness of external walls; dampness or discoloration of facades;
- Local climate data indicating temperature, precipitation, wind intensity and direction; current, reliable weather records.

Based on the above information, previous experience and knowledge of the performance of similar facades, a competent inspector might conclude whether increased dampness of facades and external walls depends on moisture and water absorption characteristics of the bricks and mortar, seasonal increase in WDR or recent heavy driving rain events. Further investigations might not be needed nor repointing.

A visual inspection can further shed light on the general condition of the façade, including moisture and moisture related aspects, by registering incidence of:

- Eroded mortar joints with respect to erosion depth and cracks;
- Damaged bricks with respect to spalling and cracks;
- Efflorescence, discoloration and microbiolo-gical growth.

Presence of eroded mortar joints indicate that climate actions have a tangible impact on the facade, a hypothesis that can be further underpinned if the incidence and degree of erosion is correlated with the exposure of the façade to WDR. Efflorescence, discolorations and microbiological growth concentrated to façade parties with eroded mortar joints might indicate that the erosion of the mortar joints constitute the root cause of these phenomena, making repointing, especially of the eroded parties, justifiable. Yet, there is a lack of knowledge concerning to what extent water absorption from WDR can be diminished by repointing.

4.2 On-site and laboratory testing

When preliminary examinations are considered inconclusive, on-site and laboratory testing of moisture content and water absorption might be justifiable. By taking into consideration benefits and drawbacks, one or more suitable test methods among those presented in *section 3* might be chosen.

Despite their high accuracy, the usability of Group A tests is somewhat limited, since they damage the examined buildings. Similarly, although the nuclear magnetic resonance (NMR) method can provide accurate results concerning moisture content, its high cost will limit its usage when it comes to common residential buildings. Thus, in most residential projects, NDT methods determining moisture content qualitatively (Group C) are recommended, though their accuracy is lower than that of other methods'.

Although high moisture content is not necessarily an indicator that repointing is needed, alteration of moisture content over time might indicate erosion of both mortar joints and of bricks. Thus, recurring measurements or continuous monitoring over time of moisture content of brick facades of high cultural or economic value might, in spite of high costs, be justifiable.

Methods measuring water absorption, named group D and E in this paper, can indicate the degree of erosion of different façade parties, since both larger, protrusive cracks and surfaces with micro-cracks are expected to absorb more water. Clay brick facades with eroded and recessed joints are further believed to absorb more water, yet, to the knowledge of the authors, no quantitative models have been established.

4.3 Criteria for decision on repointing

Qualitative and quantitative criteria concerning the need for repointing have been proposed by e.g. (Griffin, 2013, Tindall, 1987, Holland, 2012, Brief, 2005, Stockbridge, 1989), recommending repointing when a) the surface of the mortar joints contain hairline cracks, b) eroded mortar joints to a certain depth [a quarter of an inch, i.e. 6.4 mm] have been observed, c) high suction/retention mortar has been used, d) crack widths larger than 2 mm have been measured, e) the rate of water absorption is more than 4.5 litre/hour/m^2, or f) presence of voids has been detected, e.g. by means of the IRT test.

Considering the suitability of the mentioned criteria and that during repointing joints are generally raked out to approximately 25 mm or 2.5 times of the mortar joint thickness (Maurenbrecher et al., 2008, Young, 2015), it should be investigated to what extent high moisture content and water absorption are related to the condition of the outer part of the mortar joints and whether a repointing can make a difference. In this context, the relation between the depth of erosion of the mortar joints and the possible increase in water absorption from WDR should be quantified.

Furthermore, the rationality of some of the proposed criteria can be questioned, e.g. concerning acceptable crack width, since it has been shown that water ingress in cementitious materials increases exponentially when the crack width exceeds 0.2 mm (Wang et al., 1997, Aldea et al., 1999).

Eventually, possible benefits and drawbacks of other maintenance techniques rather than repointing to restore the appearance and technical condition of a facade should be considered as well.

4.4 *Alternative maintenance techniques*

Cleaning and plant removal techniques can be used to postpone the need of costlier maintenance actions. Furthermore, their implementation may uncover potential hidden defects or problems.

Cleaning techniques can be categorized into three different groups: abrasive cleaning, chemical cleaning and water cleaning. Cleaning dirt, soil, stains and paints is not only a way to restore aesthetics of a facade; it is also a method to maintain the structure and postpone repointing (Mack and Grimmer, 2000). However, if inappropriate cleaning techniques are adopted, damage to the masonry facade may result. Generally, washing gently with low pressure water is a lenient cleaning technique. Application of mechanical or chemical cleaning is not recommended, particularly not in the case of historic façades, since it might damage the masonry surface.

Plant removal can even be considered as an alternative technique to repointing. The presence of biological growths like ivy, lichens, and mosses affect water penetration, water evaporation and drying process. As such, removing these growths will result in a reduced moisture content and potentially eliminate the need for repointing.

Superficial hairline cracks in mortar joints can be repaired by surface grouting. Texture, colour, and properties of the repair grout must be chosen to match the existing mortar. Bricks with larger cracks can be replaced.

High water content in combination with freeze-thaw cycles over the service life of the facade may cause spalling (with the brick face flaking and crumbling) due to the volume increase of water when it is freezing. Damaged bricks should be with new units with similar properties. However, to limit future damages, the root cause of high-water content has to be identified and dealt with appropriately, if possible.

Water-repellent (WR) coating has been considered as a technique to reduce water penetration (Brown, 1982, Coney and Stockbridge, 1988), although there is a debate about its efficiency. In some cases it has been argued that applying water repellents cannot protect the brick-mortar interfacial zone from water ingress (Slapø and AL, 2017).

5 CONCLUSIONS

To reach a rational decision concerning repointing, different methods to measure moisture content and water absorption in clay brick façades were discussed. A systematic review of the available techniques, as presented in this paper, might contribute to improve current recommendations with respect to maintenance of clay brick masonry facades. To sum up, a rational strategy including following steps can reveal the real need for repointing:

- Preliminary studies prior to conducting costly and time-consuming measurements might clarify whether repointing is needed.
- Measurements of moisture content and water absorption can deliver data for deeper analyses. In selecting the most appropriate measuring technique, the stakeholder should consider the purpose of the measurements and the value of the data.
- Non-destructive, qualitative and inexpensive measurement techniques such as the RILEM tube or the dielectric method may therefore be favourable over more complex ones.
- Criteria available in the literature, can be used, with due engineering judgement, to reach a rational decision on repointing.
- Other maintenance techniques such as removal of microbiological growth or cleaning by water, have the potential to reduce moisture content and water absorption into brick façades, and thus to postpone the need of more fundamental maintenance measures such as repointing.

ACKNOWLEDGMENTS

The authors gratefully acknowledge financial support from SBUF - The Development Fond of the Swedish Construction Trade (grant 13576) and TMPB - The Masonry and Render Construction Association.

REFERENCES

Aldea, C.-M., Shah, S. P. & Karr, A. 1999. Permeability of cracked concrete. *Materials and structures*, 32, 370–376.

ASTM 2011. Standard Test Method for Field Determination of Water (Moisture) Content of Soil by the Calcium Carbide Gas Pressure Tester, D4944-18. West Conshohocken, Pennsylvania: American Society for Testing and Materials.

ASTM 2014. Standard test method for field determination of water penetration of masonry wall surfaces, C1601-14a. West Conshohocken, PA: ASTM International.

Binda, L., Squarcina, T. & Van Hees, R. 1996. Determination of moisture content in masonry materials. Calibration of some direct methods.

Bison, P., Cadelano, G., Capineri, L., Capitani, D., Casellato, U., Faroldi, P., Grinzato, E., Ludwig, N., Olmi, R., Priori, S., Proietti, N., Rosina, E., Ruggeri, R., Sansonetti, A., Soroldoni, L. & Valentini, M. 2011. Limits and Advantages of Different Techniques for Testing Moisture Content in Masonry. *Materials Evaluation*, 69, 111–116.

Blystone, J., Pelzner, A. & Steffens, G. 1962. Moisture content determination by the calcium carbide gas pressure method. *Highway Research Board Bulletin*.

Brief, B. 2005. Repointing (Tuckpointing) Brick Masonry. *Brick Industry Association*.

Brief, B. 2017. Maintenance of Brick Masonry. *Brick Industry Association, Technical Notes on Brick Construction*, 46, 1–11.

Brown, R. H. 1982. Initial effects of clear coatings on water permeance of masonry. *Masonry: Materials, Properties, and Performance*. ASTM International.

Bruttel, P. & Schlink, R. 2003. Water determination by Karl Fischer titration. *Metrohm monograph*, 8, 50003.

Camuffo, D. & Bertolin, C. 2012. Towards standardisation of moisture content measurement in cultural heritage materials. *E-Preserv. Sci*, 9, 23–35.

Coney, W. B. & Stockbridge, J. G. 1988. The effectiveness of waterproofing coatings, surface grouting, and tuckpointing on a specific project. *Masonry: Materials, Design, Construction, and Maintenance*. ASTM International.

Crissinger, J. 2005. Measuring moisture resistance to wind-driven rain using a RILEM tube. Tech. Rep.

EMERISDA 2014. Summary report on existing techniques, procedures and criteria for assessment of effectiveness of interventions. TU Delft: Emerisda.

EN 1993. Wood-based panels – Determination of moisture content, 322. Brussels: European Committee for Standardisation (CEN TC 346).

Franke, L. & Bentrup, H. 1991. Einfluss von Rissen auf die Schlagregensicherheit von hydrophobiertem Mauerwerk und Prüfung der Hydrophobierbarkeit, Teil 2. *Bautenschutz Bausanierung*, 14, (117–121).

Fried, A., Tovey, A. & Roberts, J. 2014. *Concrete masonry designer's handbook*, CRC Press.

Griffin, I. M. 2013. *Deterioration mechanisms of historic cement renders and concrete*. Doctoral dissertation, University of Edinburgh.

Hola, A. Measuring of the moisture content in brick walls of historical buildings–the overview of methods. IOP Conference Series: Materials Science and Engineering, 2017. IOP Publishing, 012067.

Hola, J., Matkowski, Z., Schabowicz, K., Sikora, J., Nita, K. & Wójtowicz, S. 2012. Identification of moisture content in brick walls by means of impedance tomography. *COMPEL: Int J for Computation and Maths. in Electrical and Electronic Eng.*, 31.

Holland, M. 2012. *Practical Guide to Diagnosing Structural Movement in Buildings*, John Wiley & Sons.

Larsen, P. K. 2012. Determination of Water Content in Brick Masonry Walls using a Dielectric Probe. *Journal of Architectural Conservation*, 18, 47–62.

Litti, G., Khoshdel, S., Audenaert, A. & Braet, J. 2015. Hygrothermal performance evaluation of traditional brick masonry in historic buildings. *Energy and Buildings*, 105, 393–411.

Mack, R. C. & Grimmer, A. E. 2000. Assessing cleaning and water-repellent treatments for historic masonry buildings. *Preservation briefs*.

Maurenbrecher, A. H. P., Trischuk, K., Rousseau, M. Z. & Subercaseaux, M. I. 2008. Repointing mortars for older masonry buildings: design considerations. *Construction Technology Update; no. 67*.

Möller, U. & Stelzmann, M. 2013. Neue Messmethode zur Bewertung der kapillaren Wasseraufnahme von Fassaden. *wksb*, 69, 62–65.

Neumann, H.-H., Niermann, M. & Steiger, M. 2014. *Methodenentwicklung zur zerstörungsfreien Prüfung des Wassertransportes für die Planung und zum Bautenschutz in historischem Ziegelmauerwerk bei dem Einsatz von Innenraumdämmungen: Abschlussbericht zu dem DBU-geförderten Vorhaben, Förderkennzeichen: 28751-45*, Universität Hamburg, Fachbereich Chemie, Anorganische und Angewandte Chemie.

Pel, L., Kopinga, K. & Brocken, H. 1996. Moisture transport in porous building materials. *Heron*, 41, 95–105.

RILEM, D. Protection of Stone Monuments. Experimental Methods. Test No. II. 4. Water absorption under low pressure (pipe method). International Symposium UNESCO-RILEM Paris, 1978.

Schöffski, K. S., D. 2006. Karl Fischer Moisture Determination. *Encyclopedia of Analytical Chemistry*.

Slapø, F. & Al, E. 2017. Masonry's Resistance to Driving Rain: Mortar Water Content and Impregnation. *Buildings*, 7, 70.

Stelzmann, M., Möller, U. & Plagge, R. 2015. Waterabsorption-measurement instrument for masonry façades. *ETNDT6, Emerging Technologies in Non-Destructive Testing*, 6, 27–29.

Stockbridge, J. G. 1989. Repointing masonry walls. *APT bulletin*, 21, 10–12.

Tägil, T., Gustavsson, T., Bergkvist, K. & Staaf, B. M. 2011. *Modernismens tegelfasader (The clay brick facades of the Modernism) in Swedish*, Arkus Publication.

Tindall, S. M. 1987. Repointing Masonry—Why Repoint? *Old-House Journal*, 24–31.

Wang, K., Jansen, D. C., Shah, S. P. & Karr, A. F. 1997. Permeability study of cracked concrete. *Cement and concrete research*, 27, 381–393.

Wolter, B. & Krus, M. 2005. Moisture Measuring with Nuclear Magnetic Resonance (NMR). *In:* KUPFER, K. (ed.) *Electromagnetic Aquametry: Electromagnetic Wave Interaction with Water and Moist Substances*. Berlin, Heidelberg: Springer Berlin Heidelberg.

Young, D. Repointing mortar joints: some important points. Australia ICOMOS Conference, 5-8 November 2015 Adelaide Australia.

Brick and Block Masonry - From Historical to Sustainable Masonry –
Kubica, Kwiecień & Bednarz (eds)
© 2020 Taylor & Francis Group, London, ISBN 978-0-367-56586-2

Modelling of historic brickwork walls strengthened with GFRP strips

R. Capozucca & E. Magagnini
DICEA, Polytechnic University of Marche, Ancona, Italy

ABSTRACT: In recent years, the strengthening of masonry structures with external bonded (EB) fiber reinforced polymers (FRPs) has been increased to improve the strength and ductility of walls under seismic actions. The presence of FRP elements strongly affects the structural response through complex interaction mechanisms between masonry and strengthening elements. Many aspects of this strengthening method are not yet completely known; in particular, the debonding mechanisms of FRP strips need to be analyzed through more investigation. In this paper, an iterative, incremental non linear three dimensional Finite Element (FE) model is developed to analyze the response of historic masonry wall, with and without strengthening by Glass-FRP strips, under in-plane cyclic loading, addressing particular regard to the delamination phenomenon. A numerical macroscopic model obtained from the combination of constitutive laws is adopted to simulate the behaviour of the unreinforced and reinforced masonry. Both progressive local failure and nonlinear stress-strain relationship of masonry are taken into account. For the GFRP strips, a model that considers the reinforcement perfectly glued to the masonry support has been used. Finally, the developed analytical model is calibrated and validated by comparison with experimental results and interesting aspects are remarked.

1 INTRODUCTION

In the last decades numerous researchers have focused on the study of load-bearing walls under seismic action in order to verify strengthening solutions designed to prevent collapse and/or severe damages of masonry structures. Knowledge regarding the behaviour of masonry walls under horizontal motion of soil is, without doubt, the basis for preserving both monumental historic masonry buildings and/or minor masonry buildings.

Recently, the scientific community has demonstrated great interest in the development of sophisticated numerical tools as an opposition to the tradition of rules-of-thumb or empirical formulae adopted to evaluate the safety of masonry buildings. In particular, nonlinear models implemented in suitable Finite Elements (FE) formulations currently represent the most common advanced strategy to simulate the structural behaviour of masonry structures. The main problem in the development of accurate stress analysis for masonry structures is the definition and the use of suitable material constitutive laws. In the last twenty years several authors (Berto et al. 2002, Gambarotta & Lagomarsino 1997, Giambanco et al. 2001, Lourenço & Rots 1997, Rots 1991, Capozucca 2011) have proposed different modelling strategies to predict the structural response of masonry structures and, consequently, to assess the safety level of existing buildings.

Recently, the strengthening techniques with composite materials able to improve the strength and ductility to masonry structures under seismic actions have increased (Ehsani & Saadatmanesh 1999, Shrive

2002). In particular, external bonded (EB) fiber reinforced polymers (FRPs) have seen remarkable development (Capozucca 2011). The above mentioned complexity of masonry structural response increases when the presence of these additional elements is taken in to account. Starting from this point, the use of EB FRP strips raises opening new venues for theoretical and numerical analyses (Fedele & Milani 2010, Ghiassi et al. 2012, Grande et al. 2013).

In this paper, an iterative, incremental non linear three dimensional FE model is developed to analyse historic masonry wall under in-plane cyclic loading. The nonlinear code ANSYS has been used; a macromodel obtained from the combination of constitutive laws is performed to simulate the behaviour of the unreinforced and strengthened masonry. Finally, the developed analytical model is calibrated and validated by comparison with experimental results.

2 EXPERIMENTAL TESTS

2.1 *Brickwork models and set-up*

A set of experimental tests were carried out on masonry walls with full clay bricks in scale 1/3rd (Bejamin & Williams 1958) obtained from full scale handmade bricks recovered from the restoration of a historic 18th century building from the south of Marche Region in Italy, invested by earthquake in 2016. Static tests were performed on two models identified as W1 and W2. The masonry walls were tested under combined compression and shear force as described below (Capozucca & Magagnini 2019).

Shear tests involved only one model without strengthening, W1, while W2 model was strengthened by GFRP strips just before cyclic shear tests; diagonal GFRP strips were glued on only one web surface. W1 after damage due to diagonal cracking has been also strengthened with GFRP strips in the same way of W2.

A cement: lime: sand (1:1:5) mortar by volume was used for the construction of specimens. The sand grain size was three times smaller in order to reproduce the friction at the interface of brick and mortar in the 1:3rd scale models. Nine brick specimens were taken from the original historic clay bricks and they were tested under compression till failure. The test results show a remarkable uncertainty due to the type of clay. The average compressive strength of the clay bricks was $f_b = 27.8$ N/mm^2. Three prismatic specimens of dimensions $40 \times 40 \times 160$ mm were made in order to establish the strength of mortar. Flexural and compressive tests were performed. The mortar was characterized by the following values of strength: average compressive strength $f_m = 3.56$ N/mm^2 and bending tensile strength $f_{mt} = 1.80$ N/mm^2. A series of preliminary tests were done on brickwork wallettes of 200 mm x 200 mm in plane and thickness of 50 mm, were tested under compression until failure; the obtained average compressive strength was $f_w \approx 13.5$ N/mm^2 with Young's modulus of elasticity, $E_w = 8550$ N/mm^2 and Poisson ratio $v = 0.20$.

The dimensions of the tested model are shown in Figure 1. The choice of using double T shape sections was connected to the need to avoid bending cracking at the base of the wall model so that failure is initiated only due to shear; furthermore, flanges reduce the maximum value of shear stress distribution in the whole section, making its trend almost constant.

The unreinforced wall (U-W1) damaged and reinforced (DR-W1) and wall W2 reinforced (R-W2) have been subjected to the same pre-compression normal stress equal to $\sigma_v = 1.2$ N/mm^2, kept constant during the cyclic shear loading. Vertical load was applied to the wall through a system of load distribution with a steel plate placed on the top of the model (Figure 1).

The steel plate, positioned on the top of the wall, distributed over the entire panel both vertical load and horizontal cyclic load applied by hydraulic jacks. The horizontal load, F, was applied using double phase jack. Bases to measure vertical strains were located at five positions along the length, labelled A, B, C on the web. Measuring bases were also used to evaluate principal strains in the centre of the wall; for this reason, Rosetta (D-E-F) was positioned in order to monitor the evolution of the wall's strains completely every cyclic load. The measurement of lateral deflection under horizontal cyclic load was achieved using three inductive linear displacement transducers (LVDTs) (Figure 1). LVDTs no. 1 and 2 were applied on each flange in order to measure maximum displacement from top of the wall equal about 25 mm.

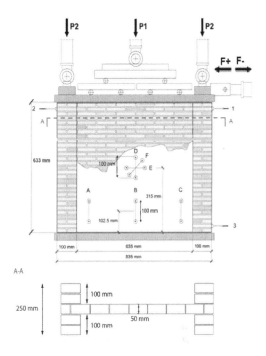

Figure 1. Experimental brickwork wall model with double T shape section and load distribution system with hydraulic jacks.

2.2 Experimental tests on unreinforced wall U-W1

Unreinforced wall model (U-W1) was tested in a first phase under combined vertical load and cyclic horizontal shear force, F, up to cracking damage controlling deflection at stages till damage; deflection and strains were also measured at various steps. Six complete loading cycles with an increase of about ±5 kN for each cycle have been performed up to a lateral force equal to 55 kN. After the sixth cycle, wall U-W1 was subjected to other two load steps with increasing load only in the positive direction. The first crack appeared when wall was subjected to a horizontal load value equal to $F+ = 30.66$ kN.

The cracks occurred at the mortar-brick interface and showed a prevalently diagonal trend with an angle of about $\pi/4$ (Figure 2). The test was carried out up to horizontal forces value equal to about $F = 62.1$ kN and $\delta = 3.39$ mm, with an average shear stress value, referred for the resistant area of the web, equal to $\tau_u \cong 1.90$ N/mm^2. In Figure 3 the experimental diagrams of cyclic load, F, vs deflection, δ, at the top of the model wall (LVDT no. 2) is shown.

2.3 Experimental tests on strengthened walls

The wall model W1 damaged and, after, strengthened was subjected to same loading path. A glass fiber reinforced polymer (type FIDGLASS UNI 300 HT73)

Figure 2. Diagonal cracking distribution on unreinforced wall U-W1.

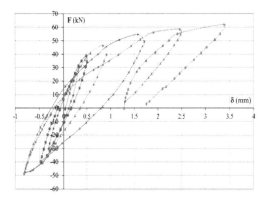

Figure 3. Exp. diagrams cyclic force, F, vs lateral deflection, δ, at LVDT no. 2 (U-W1).

was used and it was characterized by the following mechanical parameters: tensile strength f_{fi} = 200 N/mm² and Young's modulus E_f = 73000 N/mm², with a value of width equal to 50 mm. An epoxy resin (type Kimitech EP-IN) was adopted as the matrix; the epoxy resin presented an average tensile strength equal to f_{resin} = 30 N/mm² and Young's modulus E_{resin} = 1760 N/mm². The walls subjected to shear tests have been strengthened adopting six diagonal GFRP strips with inclination of $\pm\pi/4$ glued only one surface of web (Figure 4)

In addition to all instruments used for the test on the un-reinforced walls, strain gauges were used in order to value axial strains along the reinforced strip. For these measures, strain gauges of type LY11-10/120 with a nominal resistance of 120 Ω were adopted (Figure 4).

The test was carried out on DR-W1with the same procedure adopted for the test on the unreinforced wall: horizontal shear force was applied with increasing intensity cycles with an increase of ±5 kN for each cycle. Fifteen complete loading cycles were applied to DR-W1 model. During the 16th cycle, wall DR-W1

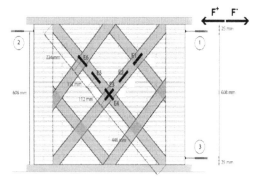

Figure 4. DR-W1 with strain gauges E1,…,E6 on the GFRP strips of main diagonals.

was subjected to increasing load until failure. During the first loading cycles, panel DR-W1 initially presented linear elastic behaviour and no new cracks, not even small ones, appeared. Load cracking was reached during the 9th cycle equal to F = 40 kN. The wall reached failure for the delamination of the GFRP fibre with debonding of the diagonal strips due to tension at the 16th cycle at a horizontal load value of approximately F = 92.2 kN. In Figure 5 experimental diagrams force, F, vs lateral deflection, δ, for all cycles are shown.

Shear test on the W2 model undamaged and strengthened (R-W2) with the GFRP strips on one side was carried out using the same methodology and with the same loading path as the DR-W1 model. The same instruments used in DR-W1 to measure deflection and vertical strain were adopted here, with six strain gauges E1,…,E6 placed on the main diagonals of the GFRP strips.

Eleven complete loading cycles were applied to R-W2 model. The wall was without cracking until the 10th cycle and the first crack appeared for a load equal to $F+ \cong 37.0$ kN. While the crack pattern became widespread and diffuse at the maximum value of the shear load about $F- = 55$ kN. The test was carried out up to horizontal force value equal to about $F = 45$ kN and $\delta= 8$ mm, Collapse of R-W2 wall

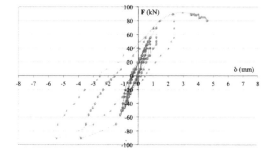

Figure 5. Exp. diagrams cyclic force, F, vs lateral deflection, δ, at LVDT no. 2 (DR-W1).

Figure 6. Experimental failure of R-W2.

model has been assumed at value of horizontal load equal about $F- \cong 55$ kN with complete detachment of GFRP strip due to compression and displacement of flange with damage of masonry (Figure 6). In this case, after the diagonal cracking, when the tensile capacity of masonry has been increased, detachment of GFRP strips happened close to the bottom of wall with cracks on the mortar joints of flange.

Further the results obtained by tests on strengthened walls, allow observing that the presence of GFRP strips changes the mechanical response and the failure mechanism of walls. In particular, the strengthening conferred, above all, a greater capacity to undergo ample horizontal deflection and, hence, to dissipate energy through the progression of ample and widespread cracking.

3 ANALYSIS BY FE MODELLING

3.1 Modelling of unreinforced wall model

Masonry is a material that exhibits an anisotropy behaviour due to the presence of mortar joints (Milani 2011, Cecchi & Milani 2006, Bertolesi et al. 2016). In general, the approach towards its numerical representation can focus on the micro- modelling of the individual components of unit and mortar, or on the macro-modelling of masonry as a composite material (Lourenço 2002). For the objectives of this work, the theoretical FE analysis of brickwork walls was developed by a macro-modelling approach, considering the masonry as a homogeneous and isotropic material.

The behaviour of unreinforced U-W1 wall was analyzed by a non linear procedure implemented in the finite element code ANSYS (2019) using brick solid finite elements. Specifically, 8-nodes isoparametric finite elements having three degrees of freedom at each node, Solid65, were employed to model the historic masonry panel's assemblage (Figure 7).

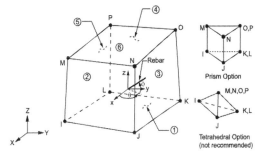

Figure 7. Geometry, node locations and coordinate system for the Solid65 element.

With the aim of simulating the mechanical behaviour of the historic brickwork masonry, a 3D model obtained through the combination of constitutive laws have been adopted. Taking into account the available material laws in the ANSYS code, the Drucker-Prager plasticity material model has been assumed for masonry. This is typically employed for pressure-dependent inelastic materials such as soils, rocks and concretes, and it is a modification of the Von Mises yield criterion that accounts for the hydrostatic stress component (confinement pressure). In addition, a smeared crack model has been introduced through the use of the William-Warnke failure surface (Ansys 2019). According to this criterion the element is capable of cracking in tension and crushing in compression. Under uniaxial tension the stress-strain response follows a linear elastic relationship until the value of the failure stress, is reached. The failure stress corresponds to the onset of micro-cracking in the material. Beyond the failure stress the formation of micro-cracks is represented macroscopically with a softening stress-strain response, which induces strain localization in the material structure. In the non elastic regime the response is typically characterized by stress hardening followed by strain softening beyond the ultimate stress. The failure surface shows an elliptic trace on the deviatoric sections in each sextant, and a parabolic trace in the meridian sections (Figure 8). This representation, although somewhat

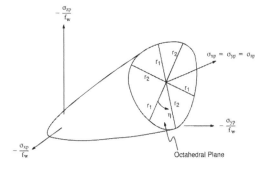

Figure 8. William-Warnke failure surface in the principal stress state.

simplified and initially developed for concrete structures, can well capture even the main features of the masonry response.

A total of five input strength parameters are needed to define the Willam-Warnke failure surface as well as an ambient hydrostatic stress state: the uniaxial compressive strength, f_w, the uniaxial tensile strength, f_{wt}, the biaxial compressive strength f_{wb} and, two additional parameters f_1 and f_2, respectively, the ultimate compression strengths for states of biaxial and uniaxial compression superimposed on hydrostatic stress state. In most practical cases when the hydrostatic stress is limited by $\sqrt{3}f_w$ the definition of the failure surface can be specified by means of only two constants, f_w and f_{wt}, since the three other constants can be assumed as follows:

$$f_{wb} = 1.2f_w \quad (1)$$

$$f_1 = 1.45f_w \quad (2)$$

$$f_2 = 1.725f_w \quad (3)$$

The model allows for the introduction of two additional coefficients, denoted as β_t and β_c, that account for a shear strength reduction of the stress producing sliding across the crack face for open (β_t) or reclosed cracks (β_c).

In this way, the material behaves as an isotropic medium with plastic deformation, cracking and crushing capabilities. The proper combination of these parameters allows reproducing an elastic-brittle behaviour in case of biaxial tensile stresses or biaxial tensile-compressive stresses with low compression level. On the contrary, the material is elastoplastic in case of biaxial compressive stresses or biaxial tensile-compressive stresses with high compression level. The mechanical parameters adopted for the FE modeling of masonry are shown in Table 1.

Non-linear push-over analysis was performed, by considering a progressive increase of the horizontal shear force, maintaining constant the precompression vertical load.

Table 1. Mechanical parameters adopted for the FE modelling of masonry.

Young's modulus	E_w	8550 N/mm²
Poisson's coefficient	V	0.19
Shear modulus	G	3700
Cohesion	c	0.31 N/mm²
Friction angle	φ	30°
Dilatancy angle	δ	10°
Uniaxial compressive strength	f_w	13.42 N/mm²
Uniaxial tensile strength	f_{wt}	0.4602 N/mm²
Open shear transfer coefficient	β_t	0.25
Closed shear transfer coefficient	β_c	0.75

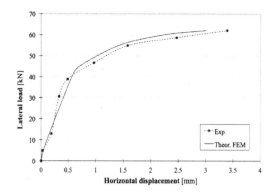

Figure 9. Comparison between envelope diagrams F versus displacements for U-W1 model.

The accuracy of the numerical results obtained by means of macro-modelling approach is assessed through a comparison with experimental results (Figure 9); in particular, the evolution of crack pattern has been recorded and analysed. The presence of a crack at an integration point is represented through modification of the stress-strain relations by introducing a plane of weakness in a direction normal to the crack face. If cracking occurs at an integration point, the cracking is modeled through an adjustment of material properties, which effectively treats the cracking as a "smeared band" of cracks, rather than discrete cracks. In particular, the crack propagation has been tracked graphically by the calculation of the principal stresses at the integration point. When a principal tensile stress exceeds the ultimate tensile strength of masonry, cracking signs appear represented by a circle in the plane of the crack. Crushing, instead, occurs when all the principal stresses are of compression and exceed the limit value. This type of cracking is represented with an octahedron outline.

The cracking load predicted by the FE analysis is equal to $F = 31.18$ kN which is reasonable compared with the experimental results equal to $F = 30.66$ kN. The cracking orientation development in theoretical FE model is presented in Figure 10 and it is possible to notice that it is consistent with the experimental observations. As the load level increases and therefore as the state of damage increases, there is an increase in the crack pattern. It can be noticed that the final crack distribution lies in a diagonal band connecting the toe and the horizontally loaded point of the wall. Diagonal cracking is predicted and no crushing failure occurred, in agreement with the experimental evidences.

3.2 Modelling of GFRP strengthened wall model

Experimental tests described above allow focusing attention on the great influence that the property of materials and the cracking behaviour of brickwork have on the mechanisms of adhesion between GFRP strips and masonry support. The decohesion of the

628

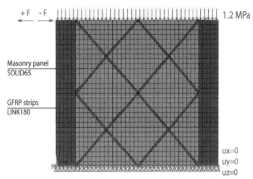

Figure 11. FE macromodelling of R-W2 wall: typical mesh for FE analysis and load's assignment.

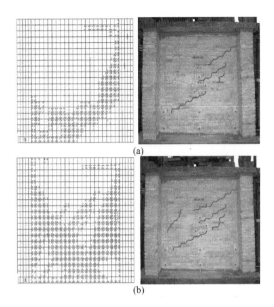

(a)

(b)

Figure 10. Cracking orientation development in theoretical FE model and in the experimental prototype U-W1: (a) $P = 40$ kN and (b) $P = 60$ kN.

strengthening system is a very complex phenomenon that involves different aspects to consider, such as the quality of masonry, FRP strengthening and adhesive layer, the development of local cracking mechanisms, the presence of local roughness, etc.

Different approaches can be developed in order to reproduce with FEM the presence of the FRP strengthening and its possible detachment from the masonry support. One of these involves the use of "special" interface elements which connect the nodes of the mesh of masonry to the nodes of the FRP elements (Sacco & Toti 2010). In this case, the forces acting on the interface are related to the relative displacement between masonry and FRP. The use of this approach is quite difficult because it implies a rigorous assessment of the parameters that characterized the interface behaviour and it involves the development of sophisticated models able to consider the complex interaction between FRP and masonry. The delamination mechanism can be also taken into account, in a simplified manner, considering a perfect adhesion between the two layers and adopting special constitutive laws for the GFRP-strengthening elements (Grande et al. 2008).

Theoretical analysis with modeling by FE has been developed to analyze the strengthened wall model R-W2 under shear tests and to compare experimental results of strain measured on GFRP strips at E1,…,E6 with numerical data. Also in this case, a macro-modelling approach and a non-linear static analysis have been developed (Figure 11).

The historic masonry wall's assemblage has been modeled with the employment of the same type of solid brick element used for the unstrengthened wall, U-W1. For what concerns the strengthening system,

the GFRP strips are modeled using truss elements (Link 180) directly connected to the nodes of the mesh of the panels, imposing a perfect adhesion between the nodes of trusses and the corresponding nodes of the mesh of the panels (Bertolesi et al. 2016). Also for the strengthened model, masonry has been modeled through the combination of constitutive laws, that is the Drucker-Prager plasticity material model together with William-Warnke failure surface.

With the aim of simulating the mechanical behaviour of the GFRP strengthened masonry and the inter-action between masonry/FRP elements, for what concerns the mechanical parameters adopted for GFRP strips, a simplified approach is proposed adopting a special constitutive law for truss elements and considering the reinforcement perfectly glued to the masonry support. An elastic-brittle behaviour has been considered by assuming an elastic stiffness value equal to Young's modulus of the GFRP strengthening and different values for the strength of the FRP elements.

In particular, the maximum strength of FRP elements, linked to the bond length, l_b, of each element, has been evaluated according to the relations proposed by Italian Code of CNR (2004):

$$f_{fdd} = \frac{1}{\gamma_{fd}\sqrt{\gamma_m}}\sqrt{\frac{2E_{FRP}\Gamma_{FK}}{t_{FRP}}} \text{ if } l_b \geq l_e \qquad (4)$$

$$f_{fdd,rid} = f_{fdd}\frac{l_b}{l_e}\left(2 - \frac{l_b}{l_e}\right) \text{ if } l_b < l_e \qquad (5)$$

$$l_e = \sqrt{\frac{E_{FRP} \cdot t_{FRP}}{2 \cdot f_{wt}}} \qquad (6)$$

where γ_{fd} is a factor that takes into account the modality of the application of the reinforcement system; γ_m is the partial safety factor for masonry material (Italian Code for Construction 2018), assumed to be equal to 1.0 for obtaining the characteristic value of bond strength; l_e is the optimal bond length of FRP corresponding to the minimal bond length able to carry the maximum anchorage force and l_b is the effective bond lengh. In CNR-DT 200 (2004), the term Γ_{Fk} is also introduced and it represents the characteristic value of the specific fracture energy of the FRP strengthened masonry. In particular, when the debonding involves the first masonry layers, the following relation is proposed:

$$\Gamma_{Fk} = c_1 \sqrt{f_w} \cdot f_{wt} \qquad (7)$$

where c_1 is a constant depending on the geometrical factors. In the light of these considerations, for each truss element (node-to-node of the mesh of the panel) the bond length l_b has been assumed equal to the distance between the middle section of the truss element and the closest edge of the FRP. The fracture energy value has been derived from the results of pull–push shear tests on bonded GFRP-to-historic brickwork masonry wallets (Capozucca & Ricci 2016) and it was assumed equal to 0.375 N/mm.

The identification of f_{fdd} is crucial, as it represents the parameter strictly related to the efficiency of the connection between reinforcement and masonry.

In Figure 12, the envelope of the results of FEM and experimentation in terms of external force–displacement curves of the point of application of the load is reported. It is interesting to notice that the results obtained exhibit a curve that is reasonable compared with experimental data.

The FE analysis has been controlled also by comparison of experimental and theoretical strain values at the GFRP strips. Figure 13 contains the maximum

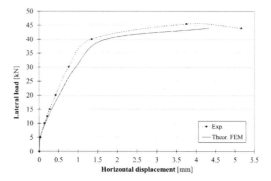

Figure 12. Comparison between envelope diagrams F versus displacements for U-W1 model.

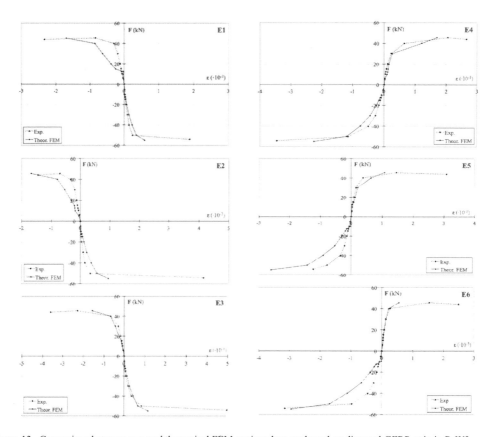

Figure 13. Comparison between exp. and theoretical FEM strain values evaluated on diagonal GFRP strip in R-W2.

strain recorded for each load cycle for R-W2 wall model in function of the horizontal load, F, at each position of strain gauges compared with the theoretical axial strains evaluated by FEM. It is possible to observe that the proposed simplified approach assuming an elastic-brittle behavior of the GFRP strips, allows obtaining a good estimate of the axial deformations, in agreement with what emerged from the experimental campaign.

4 CONCLUSIONS

This paper presents a numerical analysis on the structural response of historical brickwork masonry wall with and without GFRP strengthening. Macromodelling by FEM have been adopted considering elastoplastic materials focusing attention on the cracking development in masonry and on the bond behavior of strengthening. A simplified model able to take into account the delamination failure of GFRP strips from the masonry support has been proposed and data have been compared with experimental results. The macromodelling approach based on a smeared crack model for masonry and on an elastic-brittle behaviour for FRP strengthening, reproduces adequately the experimental behaviour well both for elastic and plastic range.

REFERENCES

ANSYS Inc. 2019. *ANSYS manual*. USA: Southpoint.
Benjamin, R. & Williams, H.A. 1958. The behaviour of one-story brick shear walls. *Jour. of Struct. Division ASCE* 84(1723): 1–30.
Bertolesi, E. et al. 2016. Simple holonomic homogenization model for the non-linear static analysis of in-plane loaded masonry walls strengthened with FRCM composites. *Composite Structures* 158: 291–307.
Berto, L. et al. 2002. An orthotropic damage model for masonry structures. *Int. Journal for Numerical Methods in Engineering* 55: 127–57.
Capozucca, R. 2011. Shear behavior of historic masonry made of clay bricks. *Open Construction and Building Technology Journal* 5, Special Issue 1: 89–96.
Capozucca, R. 2011. Experimental analysis of historic masonry walls reinforced by CFRP under in-plane cyclic loading. *Composite Structures* 94: 277–289.
Capozucca, R. & Magagnini, E. 2019. Experimental response of masonry walls in-plane loading strengthened with GFRP strips.
Capozucca, R. & Ricci, V. 2016. Bond of GFRP strips on modern and historic brickwork masonry. *Comp Struct.* 140: 540–555.
Cecchi, A. & Milani, G. 2008. A kinematic FE limit analysis model for thick English bond masonry walls. *Int. Journal of Solids and Structures* 45(5): 1302–1331.
CNR-DT 200/2004. Guide for the design and construction of externally bonded FRP systems for strengthening existing structures. (in Italian).
Ehsani, M. et al. 1999. Behavior of retrofitted URM walls under simulated earthquake loading. *J. Compos. Constr. ASCE* 3(3): 134–42.
Fedele, R. & Milani, G. 2010. A numerical insight into the response of masonry reinforced by FRP strips. The case of perfect adhesion. *Compos. Struct.* 92: 2345–57.
Gambarotta, L. & Lagomarsino, S. 1997. Damage Models for the Seismic Response of Brick Masonry Shear Walls, Part I: The Mortar Joint Model and its Applications. *Earthquake Engineering and Structural Dynamics* 26: 423–439.
Ghiassi, B. et al. 2012. Numerical analysis of bond behavior between masonry bricks and composite materials. *Eng. Struct.* 43: 210–20.
Giambanco, G. et al. 2001. Numerical analysis of masonry structures via interface models. *Comp. Mech. Appl. Mech. Eng*, 190: 6493–6511.
Grande, E. et al. 2013. Modeling and numerical analysis of the bond behavior of masonry elements strengthened with SRP/SRG. *Composites: Part B* 55: 128–138.
Grande, E. et al. 2008. Modelling and analysis of FRP-strengthened masonry panels. *Eng. Struct.* 30: 1842–1860.
Italian Code for Constructions. DM 17/01/2018. Norme Tecniche per le Costruzioni. (in Italian).
Lourenço, P. 2002. Computations of historical masonry constructions. *Progr in Struct Eng and Mat* 4(3): 301–319.
Lourenço, P.B. & Rots, J.G. 1997. Multisurface interface model for analysis of masonry structures. *J Eng. Mech. (ASCE)* 123: 660–668.
Milani, G. 2011. Simple lower bound limit analysis homogenization model for in- and out-of-plane loaded masonry walls. *Con. Build. Mat.* 25: 4426–4443.
Rots, J.G. 1991. Numerical simulation of cracking in structural masonry. *Heron* 36(2): 49–63.
Shrive, N.G. 2006. The use of fibre reinforced polymers to improve seismic resistance of masonry. *Con. Build. Mat.* 20: 269–277.
Sacco, E. & Toti, J. 2010. Interface Elements for the Analysis of Masonry Structures. *Int. Jour. for Computational Methods in Engineering Science and Mechanics* 11(6): 354–373.

Brick and Block Masonry - From Historical to Sustainable Masonry –
Kubica, Kwiecień & Bednarz (eds)
© 2020 Taylor & Francis Group, London, ISBN 978-0-367-56586-2

Evaluation of the out-of-plane performances of masonry walls strengthened with composite reinforced mortar

N. Gattesco & I. Boem

Department of Engineering and Architecture, University of Trieste, Italy

ABSTRACT: In historical masonry structures subjected to earthquakes frequently the out-of-plane partial or global collapse of some walls occurs, due to their limited flexural resistance. The application of Composite Reinforced Mortars (mortar coating with composite meshes embedded) is an innovative, effective strengthening solution for increasing the out-of-plane capacities of masonry walls and also fit with the compatibility requirements for interventions on the construction heritage. A numerical study is presented in the paper, aimed at the prediction, through nonlinear static analysis, of the bending response of entire walls before and after the strengthening intervention. The influence of different parameters is investigated and the results, in terms of capacity curves, are compared. An evaluation in terms of resisting peak ground acceleration, based on the Capacity Spectrum Method with reference to the Floor Spectrum, is also presented, with application to a case study, so to allow the comparison with the seismic demand for practice design purposes.

1 INTRODUCTION

In historical masonry structures subjected to earthquake excitation frequently occurs the out-of-plane overturning or bending collapse of some walls, due to their very limited flexural resistance (Fiorentino et al. 2018). Effective connections between the walls and between the walls and the floors can considerably reduce these failures. However, the out-of-plane shortcomings remain relevant in case of buildings with large interstorey (> 4 m) and large distances between orthogonal walls, especially in upper storeys, where the axial load is lower and the cross section thinner.

The use of composite materials for the structural rehabilitation of historical masonry buildings is gradually spreading. Experimental tests demonstrated a considerable increase in the brickwork flexural resistance and displacement capacities (Papanicolaou et al. 2008, Willis et al. 2009, Anil et al. 2012, Dizhur et al. 2014, Ismail & Ingham 2016, Nezhad et al. 2016, Bellini et al. 2017, Sorrentino et al. 2017, De Santis et al. 2019). Typically, the benefits are related to the ability of the composite material in contrasting the opening of masonry cracks, due to the high tensile strength. Different retrofitting methods have been developed: the reinforcement can be externally bonded (EB) to the masonry surface, through either epoxy resin or inorganic matrix, or near surface mounted (NSM) in the masonry mortar joints (reinforced repointing) or in thin slots cut into the surface and bonded with resins or grouts.

In particular, the application of Composite Reinforced Mortars (CRM, mortar coating with composite meshes embedded), fit with the effectiveness, durability and compatibility requirements for interventions on the building heritage. The authors already performed experimental tests to investigate the out-of-plane performances of a CRM strengthening strategy based on fiberglass meshes (Gattesco & Boem 2017a), proving its effectiveness. It clearly emerged that the reinforcement is able to provide a significant increase to the out-of-plane resistance and the displacement capacity of the wall.

However, experimental results solely do not permit directly to reproduce accurately the actual effect of CRM, as based on simplified test arrangements and load patterns as well as frequently on reduced scale samples and isolated elements, due to the complexity in recreate in the laboratory the actual stress state of the structures. The numerical approach is thus fundamental, as permits to extend the experimental evidences to more numerous and realistic configurations, allowing also to investigate on the optimization of the strengthening intervention. The authors already developed a detailed, numerical 2D finite element model for CRM strengthened masonry (Gattesco & Boem 2017a), whose reliability was proved by comparison with the simplified laboratory bending tests. In this paper, the extension of the detailed 2D model to a simplified 3D one is presented, with the aim to investigate the behavior of entire walls and more articulated configurations subjected to the common actual conditions. The influence of different parameters, such as the geometry, the boundary conditions, the presence of openings, the axial load and the material properties, is investigated. The results are compared in terms of capacity curves, representing the trend of the horizontal seismic load at the varying of the wall out-of-plane deflection. Moreover, to facilitate the practice design, an evaluation in terms of resisting peak ground acceleration, based on the Capacity Spectrum Method with reference to the Floor Spectrum, is also carried out for CRM strengthened walls, so to allow also the comparison with the seismic demand.

2 THE CRM STRENGTHENING TECHNIQUE

The CRM strengthening technique (Figure 1) consists in the application, on both faces of the masonry, of a 30 mm thick mortar coating with GFRP (Glass Fiber Reinforced Polymer) meshes embedded and connected to the masonry through L-shaped, GFRP connectors ($6/m^2$). The mechanical characteristics of the considered GFRP elements are summarized in Table 1. The grid dimension is 66×66 mm^2.

The tensile performances of CRM layers (Figure 2) were evaluated experimentally (Gattesco & Boem 2017b). A lime and cement mortar was utilized for the plaster (compressive strength $f_{c,c} = 6.3$ MPa, Young modulus $E_c = 14.4$ GPa, tensile strength $f_{t,c} = 1.1$ MPa). The results showed an approximately trilinear behavior, with a first elastic branch, as the mortar was uncracked. Sub-horizontal cracks gradually appeared in the mortar, located, for the most, in correspondence of the transversal wires of the mesh. The presence of the intact mortar between the cracks induced a stiffening of the behavior of the CRM layer, in respect to that of the wire only, similarly to concrete sections with a steel reinforcement embedded (CEN 2014). The mortar "equivalent" softening law (accounting also for the tension stiffening effect) can be derived by subtracting, from the stress-strain curve of the CRM coupon, that of the GFRP wire alone.

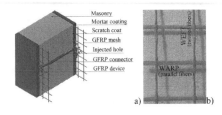

Figure 1. CRM strengthening technique: a) main application characteristics and b) detail of the GFRP mesh.

Table 1.   Characteristics of GFRP wires and connectors: A_{tot} equivalent cross section area, A_{fib} fiber area, T tensile resistance and E Young modulus.

	A_{tot}	A_{fib}	T	E
	[mm^2]	[mm^2]	[kN]	[GPa]
Parallel fiber wires	9.41	3.80	5.62	39.9
Twisted fiber wires	7.29	3.80	4.49	36.2
Connectors	96.0	57.6	36.01	-

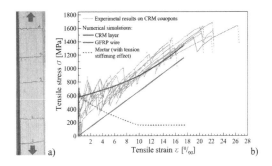

Figure 2. CRM layer coupons subjected to tensile tests: a) typical crack pattern at the end of the tests and b) experimental stress-strain curves (the stresses were evaluated in respect to the cross section of dry fibers, equal to 3.8 mm^2 per wire). The numerical performances of the CRM layer, of the GFRP wire and of the "equivalent" mortar are also reported.

To compare the out-of-plane bending performances of masonry walls before and after the application of CRM, four point bending tests on full-scale samples ($0.25 \times 1 \times 3$ m^3), made of solid brick masonry (compressive strength $f_{c,m} = 9.3$ MPa, Young modulus $E_c = 5.1$ GPa), were performed (Figure 3). In the unreinforced masonry wall, the opening of a single horizontal crack occurred abruptly (Figure 4a and c), at the reaching of the

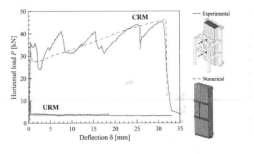

Figure 3. Four point bending tests: experimental capacity curves and comparison with the numerical results.

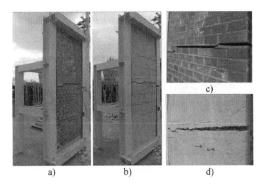

Figure 4. Experimental four point bending tests: crack pattern of the a), c) URM and b), d) CRM strengthened samples.

masonry flexural tensile strength ($f_{f,m}$ = 0.32 MPa); then, the load dropped down to a residual value corresponding to the rocking of rigid blocks, and gradually decreased till the overturning. Differently, in the CRM strengthened masonry wall, several horizontal cracks sequentially formed at the increasing of the deflection (Figure 4b) and the collapse occurred when the GFRP wires crossing a crack reached the ultimate strain (Figure 4d). The absence of slides or detachments between the masonry and the CRM layer was noted: this was basically due to the good chemical bond between materials, rather than to the action of the connectors. In fact, the resistance mechanism related to the chemical bond is stiffer than that of the shear connectors and, as long as it resists, makes negligible the connectors contribution. However, the connectors proved to be necessary as allow the masonry-CRM collaboration in case of lack of chemical bond (e.g. the bond resistance is reduced, for the low roughness of the masonry surface, or unexpected, consistent thermal variations induce tangential stresses at the interface which compromise the bond). Moreover, the connectors constitute a fundamental confining device in case of multiple-leafs masonry, so to avoid leafs separation (Gattesco et al. 2015).

An analytical study on CRM strengthened walls (Gattesco & Boem 2017a) showed that the cracking and the ultimate bending moment can be easily predicted using the well-known relationships used in the design of reinforced concrete beams subjected to combined axial and bending action for uncracked and cracked sections, respectively.

3 NUMERICAL SIMULATIONS

3.1 Modelling method

In the developed simplified 3D model, 8-nodes solid elements were adopted for representing both the masonry and the mortar coating (mesh dimensions about 66 × 60 × 30 mm³). Perfect adhesion was assumed (coincident nodes), as no evident masonry-mortar slip was monitored in the experimental tests. The GFRP wires were modelled by means of truss elements having a length equal to that of the solid elements and were connected to the nodes of the solid element representing the coating; thus, a perfect bond was considered also between the GFRP wires and the mortar. Truss elements were adopted also for the GFRP connectors, but the end nodes only were connected to the solid elements nodes of the coating. A smeared crack model, with total strain crack criteria (rotating crack model), was considered for all the materials; the main parameters, in accordance with those adopted in the detailed 2D numerical model (Gattesco & Boem 2017a), are summarized in Figure 5. Considering the wider mesh size, in respect to the previous models, the mortar softening behavior, which accounts also for the tension stiffening effect

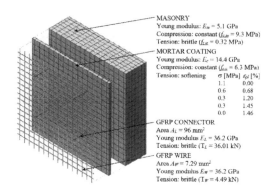

Figure 5. Main material properties adopted in the numerical model.

of the mortar between cracks, was re-calibrated. For this purpose, the simulation of the tensile tests on CRM layer coupons was performed and the mortar softening law was calibrated so to fit the experimental results, as shown in Figure 2b (the numerical model characteristics were those described in these section).

To reproduce the experimental bending test configuration, the vertical displacement at the base was constrained and the out-of-plane horizontal displacements in correspondence of the contrasts provided by the apparatus were avoided. The horizontal restraint due to steel-to-steel friction at the base was also accounted (coefficient of static friction μ = 0.74). Non–linear static analyses were performed by applying, at first, the vertical loads (self-weight of 18 kN/m³ for the masonry, 20 kN/m³ for the mortar coating) and incrementing then step-by-step the imposed horizontal displacement. The Newton-Rapshon iterative method was considered (energy convergence criteria with a tolerance of 0.001).

The comparison with the experimental capacity curves, shown in Figure 3, evidences the good reliability of the simplified numerical model.

3.2 Case study

The basic case study (Case A), consisted in a wall with dimensions 6.9 × 4.7 × 0.25 m³ (widthx height x thickness), without openings, subjected to the vertical loads induced by the self-weight and those transmitted at the top, equal to 45.8 kN/m (upper wall g_1 = 15 kN/m, upper floor $G_1 + G_2$ = 6.4 kN/m², $\psi_{2,j}\cdot Q_k$ = 2.4 kN/m², floor span 7 m). A horizontal load proportional to the wall mass was considered; the out-of-plane horizontal displacement was restrained along the whole perimeter, assuming an effective connection with the orthogonal walls and with the floors - Figure 6.

The performances of the URM and CRM strengthened masonry walls are compared in Figure 7 in terms of α-δ capacity curves, representing the ratio between the horizontal load F and the wall self-weight W at the varying of the displacement δ monitored in the middle

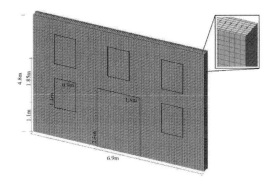

Figure 6. Main characteristics of the wall model.

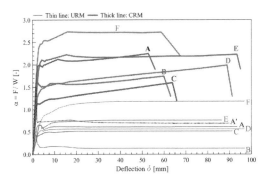

Figure 7. Capacity curves related to the analyzed configurations.

of the wall. The resistance increased significantly with the reinforcement (about 3.7 times). Moreover, the role of the GFRP mesh in maintaining the resistance till considerable values of deflections is evident when comparing Case A with A' (no mesh, mortar with brittle tensile behavior), in which the load rapidly dropped down to the residual value when the first cracking of the mortar coating was attained.

Different, varied configurations were analyzed: only self-weight considered (B), absence of lateral restraint at wall horizontal ends (C), doubled length wall (D), wall with a 1.8 × 2.4 m² central door (E) and with door and five 0.9 × 1.2 m² windows (F – Figure 6). In Figure 8, the ultimate displacements and the tensile stresses acting on the GFRP mesh on the wall face, just before the rupture, are illustrated.

The ultimate values $\alpha_{u,URM}$ and $\alpha_{u,CRM}$ and the ratios $\alpha_{u,CRM}/\alpha_{u,URM}$ are indicated in Table 2 (for $\alpha_{u,URM}$, the residual load corresponding to the kinematic mechanism is considered). The wall deflection δ_u, in correspondence of α_u, was also reported for the CRM strengthened configurations.

In both Cases A and B, $\alpha_{u,CRM}$ was reached just before the occurrence of the failure of the vertical wires in the central wall strip (about 1.4 m width), just above the mid-height of the wall, in correspondence of a deflection of about 1/90 of the wall's

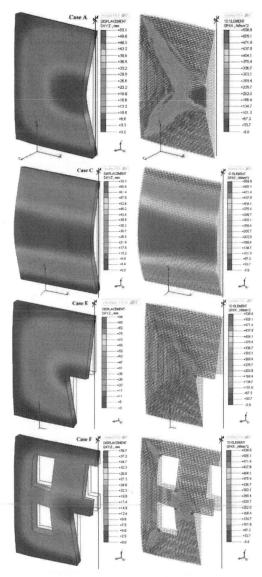

Figure 8. Out-of-plane behavior of CRM strengthened walls: displacements (left) and stresses on the GFRP meshes on the tensed wall face (right).

Table 2. Numerical results: ultimate values α_u and correspondent ultimate displacement δ_u.

Case	A	A'	B	C	D	E	F
$\alpha_{u,URM}$	0.61	-	0.14	0.52	0.56	0.76	1.18
$\alpha_{u,CRM}$	2.25	1.52	1.74	1.60	1.99	2.23	2.73
$\alpha_{u,CRM}/\alpha_{u,URM}$	3.69	2.49	12.4	3.08	3.55	2.93	2.31
$\delta_{u,CRM}$	52.9	-	59.9	63.8	88.8	93.4	58.3

height. Then, the consecutively, rapid tensile failure of other wires, along the panel diagonals, determined a progressive load drop and the wall failure. α_u resulted lower in Case B, both in URM and CRM, due to the lower axial load, but the effectiveness of the intervention ($\alpha_{u,CRM}/\alpha_{u,URM}$) resulted significantly greater. In respect to Case C, where the wall was subjected to vertical bending due to the failure of the vertical wires involved simultaneously the entire wall width, the plate effect of Case A led to a higher resistance and a greater effectiveness. Intermediate values were attained in Case D, as the plate effect is reduced by the wider wall width. In presence of openings (Cases E and F) the ultimate deflection resulted in general greater than Case A, due to the higher wall deformability, but the ratios $\alpha_{u,CRM}/\alpha_{u,URM}$ were reduced. A significant stress concentration in the GFRP wires resulted close to the openings' corners.

4 RESISTING PGA ESTIMATION

4.1 Method

Differently from URM, the lateral performances of CRM reinforced masonry walls are governed by the bending strength of the reinforced masonry, rather than the rocking mechanism. Therefore, the kinematic analysis, typically adopted for URM, is not appropriate and the F-δ capacity curve before the mechanism activation has to be considered.

Among the different procedures available in the literature for the evaluation of the seismic capacity, the modified Capacity Spectrum Method, based on equivalent viscous damping was adopted (Building Seismic Safety Council 1997; Freeman 2004). This procedure was preferred to the N2-method suggested in Eurocode 8 (CEN 2004), as the hysteretic behaviour of masonry structures is far away from that of reinforced concrete structures, from which the N2-method was derived.

The F-δ capacity curve of the multi degree of freedom (MDOF) system has firstly to be transformed to the acceleration-displacement a_g-δ^* capacity curve of an equivalent single degree of freedom (SDOF) system:

$$a_g = \frac{F^*}{M^*} = \frac{F}{\Gamma}\frac{1}{M^*} \ and \ \delta^* = \frac{\delta s}{\Gamma} \quad (1)$$

where the equivalent mass of the SDOF system M^* and the transformation factor Γ were evaluated by applying Equation 2, which depend on the mass mi and the normalised displacement ϕi of the i-th masonry wall portion:

$$M^* = \sum_{i=1}^{n} m_i\phi_i \ and \ \Gamma = \frac{M^*}{\sum_{i=1}^{n} m_i\phi_i^{2}} \quad (2)$$

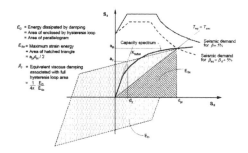

Figure 9. Schematization of the Capacity Spectrum Method and evaluation of the equivalent viscous damping (FEMA 274).

In general, the method, schematized in Figure 9 in a pseudo-acceleration versus spectral displacement diagram, consists in finding the damped Acceleration Displacement Response Spectrum (ADRS) that intersects the a_g-δ^* capacity curve at the target displacement.

The damped spectrum accounts for the dissipative capacity of the analysed element through the equivalent damping β, which can be evaluated accounting for the inherent viscous damping β_{el} and the hysteretic damping β_0:

$$\beta = \beta_{el} + \beta_0 \quad (3)$$

It is observed that, for the considered case, the spectrum to consider for the application of the modified Capacity Spectrum Method is not that referred to the ground: in fact, it is necessary to take into account the variation of the acceleration due to the position of the wall in the building and, thus, to refer to a floor spectrum. In the Commentary of the Italian National Building Code (MIT 2019) the shape of the floor spectrum reported in Eq.7 is provided, considering the peak floor acceleration PFA_Z and the maximum floor spectral acceleration $S_{a,Z}(T_k,\beta)$ evaluated in correspondence of the natural period T_k of the building. In, particular, in case of regular buildings with rigid floors and without torsional problems, it is possible to consider the following simplified formulations:

$$PFA_Z = S_a(T_k)\ \eta(\beta_k)\ \frac{Z}{H}\ \frac{3N}{2N+1}\sqrt{1+4\beta_K^2} \quad (4)$$

$$S_{a,Z}(T_k,\beta) = f_k\ \eta(\beta)\ PFA_Z \quad (5)$$

being $S_a(T_k)$ the ground spectral acceleration in correspondence of T_k, H the building height, N the number of storeys and z the level of the secondary element (average level of the wall restraint lines).

For the evaluation of T_k, it can be approximately considered $T_k = \kappa \cdot C \cdot H^{3/4}$, assuming $C = 0.05$ for masonry buildings; the coefficient κ allows to account for the building's period degradation in correspondence of the reached seismic intensity level.

For the amplification factor, f_k, it can be considered $1.1\beta_k^{-0.5}$. The effective damping correction factor of the building $\eta(\beta_k)$, depends on its equivalent damping β_k, according to Equation 6:

$$\eta(\beta_k) = \sqrt{\frac{0.1}{0.05 + \beta_k}} \tag{6}$$

A similar equation correlates the effective damping correction factor of the secondary element $\eta(\beta)$ to its equivalent damping β.

The damped floor response spectrum is described through the following analytical function (Figure 10), where the coefficients a and b can be assumed equal to 0.8 and 1.1, respectively:

$$(T,\beta) = \begin{cases} \dfrac{f_k\,\eta(\beta)\,PFA_z}{1+\left[f_k\,\eta(\beta)-1\left(1-\frac{T}{aT_k}\right)\right]^{1.6}} & T \angle T_k \\[2ex] f_k\,\eta(\beta)\,PFA & aT_k <; T_k \\[1ex] \dfrac{f_k\,\eta(\beta)\,PFA_z}{1+\left[f_k\,\eta(\beta)-1\left(\frac{T}{bT_k}-1\right)\right]^{1.2}} & T \geq bT_k \end{cases} \tag{7}$$

To evaluate the resisting peak ground acceleration associated to the collapse of the wall subjected to out-of-plane action, at a generic level z, the effective period referred to the wall collapse, T_{eff}, has to be calculated in function of the acceleration and of the equivalent displacement associated to the collapse point of the SDOF capacity curve ($a_{g,u}$, δ^*_u):

$$T_{eff} = 2\pi\sqrt{\frac{\delta^*_u}{a_{g,u}}} \tag{8}$$

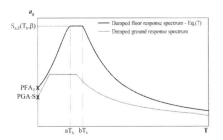

Figure 10. Comparison between ground and floor damped response spectra.

The ground spectral acceleration is evaluated by substituting Equation 4 in Equation 7 and solving for $S_a(T_k)$, assuming $T=T_{eff}$ and $S_{a,Z}(T,\beta) = \delta^*_u/(T_{eff}/2\pi)^2$.

The resisting peak ground acceleration, PGA, associated to the out-of-plane wall collapse is derived accounting for the ground response spectrum shape (e.g. §3.2.2.2, CEN 2004).

4.2 Application to a case study

The four storey, regular masonry building illustrated in Figure 11a is considered (H = 13.4 m). A doubled-height, perforated wall is present at the second floor (z = +10.1 m) - Figure 11b. The wall is made of rubble stone (550 mm thick) in the lower portion and of solid brick (400 mm thick) in the upper one. The roof, the floor and orthogonal walls inhibit the out-of-plane displacements along the perimeter. The wall carries its own weight and the load transmitted by the roof (equal to 12.41 kN/m).

The capacity curves (Figure 12) are obtained through numerical simulations, according to the modelling method described in §3.1. For the URM wall, the ultimate deflection, δ_u, is assumed equal to 40% that associated to the overturning (F = 0). For the CRM strengthened wall, the collapse is attained at the GFRP mesh rupture, which involves mainly the vertical wires at the lower corners of the windows of the third floor, in the wall central portion. This failure rapidly spreads horizontally along the entire masonry strip between the window and the adjacent doors. The ultimate displacements and the tensile stresses acting on the GFRP mesh on the wall face, just before the rupture, are illustrated in Figure 13.

Note that the models are based on the assumption of monolithic behavior of the wall portions; however, the premature rubble stone masonry disaggregation or leaf separation could occur, anticipating significantly the collapse of the URM wall. The

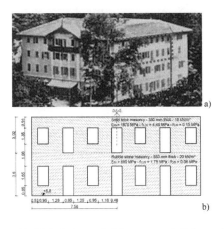

Figure 11. Main features of a) the masonry building and b) details of the wall considered for the case study.

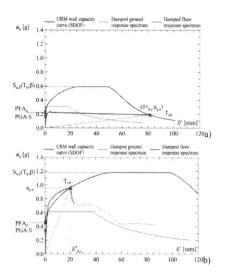

Figure 12. Evaluation of the resisting peak ground acceleration of the a) URM and b) CRM masonry wall subjected to out-of-plane actions, through the Capacity Spectrum Method.

Figure 13. Out-of-plane behavior of CRM strengthened wall: displacements (left) and stresses on the GFRP meshes on the tensed wall face (right).

presence of effective passing-through connector or diatones can inhibit this failure mode.

The evaluations of the resisting PGA are performed through the method described in §4.1. In particular, an equivalent viscous damping β of 7% is considered for the wall, according to the range

5-10% found by Graziotti et al. (2016) for single leaf URM walls subjected to dynamic excitation. Also NZSEE (2006) evidences a limited damping for out-of-plane bending (about 5%), as experiments show that the mating surfaces at hinge lines tend to simply fold onto each other rather than impact. Actually, no calibrated values are available for CRM strengthened walls, so it has been prudentially assumed $\beta = 7\%$, although greater values are reasonably expected, due to the crack diffusion in the reinforced mortar coating.

For the building, it was considered $\beta_k = 15\%$, $\kappa = 1.5$; the soil factor is $S = 1.2$.

The main parameters are summarized in Table 3: the mass (M), the resistance (F_u) and the correspondent deflection (δ_u) of the MDOF system as well as the equivalent mass (M^*), the transformation factor (Γ), the collapse acceleration ($a_{g,u}$) and the displacement (δ^*_u) of the SDOF system. The main results are reported in Table 3 in terms of effective period referred to the wall collapse T_{eff}, peak floor acceleration PFA_Z, ground spectral acceleration $S_a(T_k)$, resisting peak ground acceleration PGA and ratio PGA_{CRM}/PGA_{URM} between PGA of reinforced and unreinforced wall.

The collapse of the plain wall is expected for $PGA_{URM} = 0.145$g, that is a value which is exceeded in areas affected by a moderate seismicity (> 0.15g). The reinforced wall shows a significant improvement in respect to the unreinforced configuration, reaching a value of $PGA_{CRM} = 0.290$g, that affects only very high seismicity areas.

It is worth noting that the estimation of the wall resisting PGA strictly depends on the values of equivalent viscous damping (β_k) and effective period's degradation rate (κ) assumed for the whole building, which are affected by its actual degradation level. Reasonably, for higher PGA the building would be more damaged, so would greater attenuate the acceleration transmitted from the ground to the wall level, as also highlighted by Degli Abbati et al. (2017). However, the values of β_k and κ assumed in the example are consistent with a prominent damaging of both the URM (NZSEE 2006) and the CRM strengthened structure (Gattesco & Boem 2017c).

Table 3. Main parameters and results of the URM and CRM wall configurations.

	M	F_u	δ_u	M^*	Γ	$a_{g,u}$	δ^*_u
	kg	kN	mm	kg	-	g	mm
URM	66·10³	89.9	104.0	39·10³	1.3	0.188	81.9
CRM	74·10³	520.9	26.1	44·10³	1.3	0.726	23.6
	T_{eff}	PFA_Z	PGA	PGA_{CRM}/PGA_{URM}			
	s	g	g	-			
URM	1.326	0.228	0.145	-			
CRM	0.361	0.456	0.290	2.0			

5 REMARKS AND CONCLUSIONS

A numerical study to investigate on the out-of-plane behavior of masonry walls strengthened with CRM was carried out. The simplified numerical model, based on solid elements and smear crack criteria, was validated by comparison with experimental, four-point bending tests; though less refined in respect to a detailed model previously implemented by the authors, it resulted able to reproduce the wall global behavior with good accuracy and reduced time.

An existing masonry wall, $6.9 \times 4.8 \times 0.25$ m^3, with out-of-plane displacements restrained along the perimeter, was analyzed, comparing the performances before and after the application of the CRM strengthening technique. Some variations of this configuration were investigated. Considerable resistance increases were attained; in particular, the reinforcement effectiveness resulted higher for lower axial load and when the bi-dimensional bending effect was more pronounced; differently, in case of openings, the effectiveness tended to reduce. Moreover, the presence of the GFRP mesh permits to maintain the resistance till significant out-of-plane deflection values of the wall (about 1/70 of the wall's height).

The Capacity Spectrum Method based on equivalent viscous damping was applied to estimate the resisting peak ground acceleration associated to the wall out-of-plane collapse in a building. The method accounts of the dissipative capacities of both the wall and the building (by means of the equivalent viscous damping) as well as of the level of the wall in the building (through the floor response spectrum). A case study was analysed, considering the capacity curve obtained from numerical analysis. The results evidenced a significant improvement provided by the CRM strengthening technique in terms of *PGA*, which resulted almost doubled in respect to the URM configuration.

Next steps of the research are aimed at investigating more deeply on the dissipative properties of CRM strengthened walls subjected to out-of-plane bending, so to estimate correctly the actual damping capacities at ultimate state, permitting a more accurate evaluation of their seismic performances.

ACKNOWLEDGMENTS

The financial support of the Italian Department of Civil Protection ("Reluis 2019-2021, WP5") is gratefully acknowledged.

REFERENCES

Anil, O. et al. 2012. Out-of-plane behavior of unreinforced masonry brick walls strengthened with CFRP strips. *Constr. Build. Mater.* 35:614–624.

Bellini, A. et al. 2017. Out-of-plane strengthening of masonry walls with FRCM composite materials. *Key Engineering Materials* 747:158–165.

Building Seismic Safety Council. 1997. FEMA274: NEHRP Commentary on the Guidelines for the Seismic Rehabilitation of Buildings. Washington, DC (US).

CEN. 2014. EN 1992-1-1:2004/A1. Eurocode 2: design of concrete structures - Part 1-1: general rules and rules for buildings. Brussels (BE).

CEN. 2004. EN 1998-1:2004: Eurocode 8: Design of Structures for Earthquake Resistance – Part 1: General Rules, Seismic Actions and Rules for Buildings. Brussels (BE).

De Santis, S. et al. 2019. Out-of-plane seismic retrofitting of masonry walls with Textile Reinforced Mortar composites. *Bulletin of Earthquake Engineering* 17:6265–6300.

Degli Abbati, S. et al. 2017. Proposal of floor spectra for the verification of non-structural elements and local mechanisms in masonry buildings. *Proceedings of the 17th National Conference "ANIDIS, Seismic engineering in Italy"*, 17-21 September, *Pistoia (I)*.

Dizhur, D. et al. 2014. Out-of-plane strengthening of unreinforced masonry walls using near surface mounted fibre reinforced polymer strips. *Engineering Structures* 59: 330–343.

Fiorentino, G. et al., 2018. Damage patterns in the town of Amatrice after August 24th 2016. *Bulletin of Earthquake Engineering* 16(3): 1399–1423.

Freeman, S.A. 2004. Review of the development of the capacity spectrum method. *ISET Journal of Earthquake Technology* 41(1): 1–13.

Gattesco N. et al. 2015. Experimental and numerical study on the shear behavior of stone masonry walls strengthened with GFRP reinforced mortar coating and steel-cord reinforced repointing. *Engineering Structures* 90:143–157.

Gattesco, N. & Boem, I., 2017a. Out-of-plane behavior of reinforced masonry walls: Experimental and numerical study. *Composites Part B* 128: 39–52.

Gattesco, N. & Boem, I., 2017b. Characterization tests of GFRM coating as a strengthening technique for masonry buildings. *Composite Structures* 165: 209–222.

Gattesco, N. & Boem, I. 2017c. Assessment of the seismic capacity increase of masonry buildings strengthened through the application of GFRM coatings on the walls. *IJMRI* 2(4): 300–320.

Graziotti, F. et al. 2016. Out-of-plane shaking table tests on URM single leaf and cavity walls. *Eng. Struct.* 125: 455–470.

Ismail, N. & Ingham, J.M. 2016. In-plane and out-of-plane testing of unreinforced masonry walls strengthened using polymer textile reinforced mortar. *Engineering Structures* 118: 167–177.

MIT. 2019. Circular 21 January 2019, N. 7 "Instructions for the application of the updating of Technical norms for constructions" DM 17 January 2018.

Nezhad, R.S. et al. 2016. Shaking table test of fibre reinforced masonry walls under out-of-plane loading. *Constr. Build. Mater.*120: 89–103.

NZSEE. 2006. Assessment and Improvement of the Structural Performance of Buildings in Earthquakes - Recommendations of a NZSEE Study Group on Earthquake Risk Buildings.

Papanicolaou, C.G. et al. 2008. Textile reinforced mortar (TRM) versus FRP as strengthening material of URM walls: out-of-plane cyclic loading. *Mater. Struct.* 41: 143–157.

Sorrentino, L. et al. 2017. Review of Out-of-Plane Seismic Assessment Techniques Applied To Existing Masonry Buildings. *Int. J. Archit. Heritage* 11(1): 2–21.

Willis, C.R. et al. 2009. Damaged masonry walls in two-way bending retrofitted with vertical FRP strips. *Constr. Build. Mater.* 23: 1591–1604.

Brick and Block Masonry - From Historical to Sustainable Masonry –
Kubica, Kwiecień & Bednarz (eds)
© 2020 Taylor & Francis Group, London, ISBN 978-0-367-56586-2

Pull-off characterization of FRCM composites applied to stones and bricks in on-site conditions

E. Franzoni

DICAM – Department of Civil, Chemical, Environmental and Materials Engineering, University of Bologna, Italy

C. Gentilini

DA – Department of Architecture, University of Bologna, Italy

ABSTRACT: Fiber reinforced cementitious matrix (FRCM) composites represent an alternative for the reinforcement of masonry structures with respect to fiber reinforced polymer (FRP) composites, due to their better compatibility with the substrate and resistance to ageing. Over last years, the behavior of FRCM composites applied over masonry has been studied in detail, but the problem of the adhesion of these composites to masonry substrate in conditions that can be found on-site has not been fully investigated. In fact, real historic masonry can be affected by several problems, such as rising damp and salt deterioration. In this paper, several pull-off tests were conducted, to assess the pull-off behaviour of FRCM composites on different brick and stone substrates, affected by the presence of salts. Composites made of different fibers such as glass and basalt embedded in lime-based or cement-based matrices were tested. Results show that the presence of salts in the substrate influences the bond capacity of the reinforcement.

1 INTRODUCTION

Fiber-reinforced polymer (FRP) composites spread some decades ago as a barely invasive strategy for the reinforcement of historic masonry structures, their main advantages being the low weight, the high mechanical performance, and the flexible design able to address the specific building weaknesses, (Babatunde 2017). However, the adhesion between FRP composite and masonry is a delicate issue, as it is extremely sensitive to the condition of the substrate, and moreover the resin matrices suffer from the ageing processes typical of polymers (Vaculik et al. 2018). As a consequence, fiber-reinforced cementitious-matrix (FRCM) composites have been introduced few years ago (Zampieri et al. 2019). Due to the inorganic nature of the matrices (cement or hydraulic lime), FRCM composites are expected to be more compatible and durable than FRP, besides being easier to remove in case of troubles and/or deterioration (reversibility requirement).

The adhesion and performance of FRCM when applied to masonry walls has been extensively studied through laboratory tests, where small or even real-scale masonry specimens were used (Bellini & Mazzotti 2020). However, in these studies, commercial bricks and stones were commonly employed, which were not necessarily fully representative of real substrates. In fact, real historic masonry walls may exhibit specific features, such as the heterogeneity of bricks or stone blocks, the deterioration of materials owing to salts and/or frost, the presence of

moisture from capillary rise, condensation or infiltration (Maljaee et al 2016). To take into account the adhesion of FRCM composites when applied to historic masonry materials or to specimens in realistic conditions is very important to achieve a better insight on the performance of such composites on-site (Franzoni et al. 2018, Franzoni et al. 2019).

In this paper, different kinds of FRCM composites were tested, encompassing different matrices (cement and hydraulic lime) and fibers (glass and basalt). The composites were applied to two kinds of bricks and two natural stones, testing the adhesion by pull-off test (Ebead & Adel 2019). In order to reproduce the on-site conditions of real masonry, the four substrate materials were subjected to accelerated salt crystallization cycles before the composite application, in order to compare the adhesion with that obtained on unweathered substrates. Pull-off test was used as it is one of the few tests that can be easily carried out on-site, so a better understanding of its capability to assess the adhesion of the composite on weathered substrates of different nature is considered useful.

2 MATERIALS AND METHODS

2.1 *Bricks and stones*

Two different kinds of commercial fired-clay bricks and two natural stones were used as substrates for the tests:

- Solid red fired-clay brick (San Marco, Italy), having size 250 × 120 × 55 mm³ and labelled as RB;
- Solid yellow fired-clay brick (Pica, Italy), having size 250 × 120 × 55 mm³ and labelled as YB. YB bricks belong to the class of the so-called "hand-made bricks", containing rough inclusions in the original clay mix and hence resembling the bricks of many historic buildings (Sandrolini et al. 2007);
- Lecce stone, a porous bioclastic limestone quarried near the city of Lecce, in southern Italy, and composed of calcite, with traces of quartz and phosphates. This stone was cut to the same size of the bricks and labelled LS;
- Yellow Neapolitan tuff, a macroporous volcanic rock containing different amount of pumice, zeolites, analcime and feldspar (Colella et al. 2017). This stone was cut to the same size of the bricks and labelled NT.

2.2 Accelerated salt weathering procedure

The samples of the four different substrates were divided into two halves: one half was left unweathered for comparison (REF samples), while the other half was subjected to an accelerated laboratory procedure aimed at reproducing the salt crystallization damage occurring on-site. The procedure consisted in 5 cycles, each of which composed of:

- A capillary absorption phase, in which the samples were placed horizontally in a saline solution with 8 wt% sodium sulfate decahydrate and 1 wt% sodium chloride in deionized water (solution head = 1 cm) for 24 hours;
- A drying phase in a ventilated oven at 100°C for 21 hours, followed by 3 hours cooling at room temperature.

The cycles were aimed at causing a damage in the top face of the samples, i.e. the face to be reinforced with the composite, similar to that found in historic buildings (Franzoni et al. 2018). The presence of efflorescence and surface deterioration was clearly visible at the end of the cycles, as shown in Figure 1. These samples were labelled as POST, as they were reinforced after the deterioration.

Figure 1. Examples of salt damage on the top surface of RB samples (left) and NT samples (right).

2.3 FRCM composites and application procedure

FRCM composites are usually available in the market as complete reinforcement systems, i.e. as a combination of a specific matrix and fibers, whose compatibility and joint performance has been previously tested by the manufacturer and certified in the technical datasheet. In this study, commercial FRCM composites were used for the tests, namely:

- Basalt fiber net with hydraulic lime-based mortar 1, compression strength Class CS II according to (EN 998-1 2016), having a total thickness equal to 8 mm and labelled BAS-HL1;
- Glass fiber net with hydraulic lime-based mortar 2, compression strength Class M15 according to (EN 772-1 2015), having a total thickness equal to 18 mm and labelled GLASS-HL2;
- Glass fiber net with cement-based mortar, compression strength Class R2 according to (EN 12190 1999), having a total thickness equal to 18 mm and labelled GLASS-CEM.

The composites were applied over the major face of the samples, which was unweathered for REF samples and deteriorated for POST samples. The application was carried out trying to reproduce a typical on-site procedure. Firstly, the salt deteriorated face was cleaned, removing the efflorescence, detaching parts and powder by steel brushing followed by compressed air blowing. Then, the samples were totally immersed in water for 3 minutes. Although all the technical datasheets of FRCM products recommend to completely saturate the masonry before applying the composite, this was not considered realistic and 3 min immersion was chosen, as in previous tests this was found to produce a water absorption equal to about 30% with respect to complete saturation of the samples. In fact, the selected materials were porous and exhibited a fast water absorption rate. Finally, the composite was applied: a first layer of mortar was cast in a wooden frame to ensure the desired thickness and pressed, then the fiber net was laid on the mortar and gently pressed to favor its wetting, and the second layer of mortar was applied and manually pressed, again using a wooden frame for the thickness control (Figure 2).

Figure 2. Application of the first layer of mortar and the fiber net on a major face of the RB brick (left) and application of the final layer of mortar, using the wooden slats for levelling the composite at the desired thickness (right). The two bricks aside the one which is reinforced are aimed at sustaining the slats only.

Reinforced samples were cured in laboratory conditions for two months (under a plastic sheet for the first week, wetting them two times). This curing procedure was regarded as representative of current on-site practice.

The reinforced specimens were labelled following this order: substrate (RB, YB, LS, NT), composite (BAS-HL1, GLASS-HL2, GLASS-CEM), condition (REF, POST) and number of the specimen. At least two replicate samples were used per each group, corresponding to four pull-off tests (see Section 2.4).

2.4 Pull-off test

After curing, the reinforced samples were drilled with a 50 mm diameter core drilling machine, such that the depth in the substrate was equal to 10 mm according to (EN 1542 1999). In each specimen, two circular cores were realized. Steel cylinders were attached to the external surface of the cores using epoxy adhesive and left to cure for 48 hours. Afterwards, the specimens were tested employing the pull-off test set-up shown in Figure 3: specimens were positioned on a steel frame where a steel plate had the function to react against the tensile force applied by the pull-off testing device. A load cell was inserted between the steel cylinder and the tester to record continuously the applied load.

It should be noticed that when the samples reinforced with GLASS-CEM composite were prepared and core drilled, an unexpected detachment of the composite layer occurred, both in REF and POST specimens, notwithstanding the type of substrate. This was attributed to the fact that the cement-based mortar (originally commercialized for concrete structures) has a low compatibility with

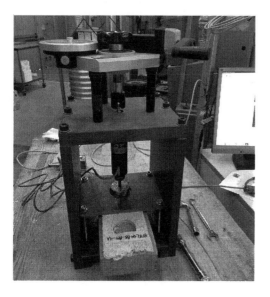

Figure 3. Set-up and execution of the pull-off test (two testing points per brick or stone sample).

porous substrates (due to its high stiffness) and/or to the fact that the substrate absorbed the mixing water, being only partially water saturated. This latter aspect must be carefully considered in on-site applications, where a full saturation of the masonry can be hardly achieved. For this reason, the specimens reinforced with CEM mortar matrix were dismissed.

2.5 Other mechanical tests and determination of salts distribution

During the reinforcement of the specimens, three prisms (nominal dimensions 40 x 40 x 160 mm^3) for each type of mortar were manufactured and left to cure in the same conditions as the strengthened specimens. After curing, prisms were tested for determining compression strength and Young's modulus. In order to determine compression strength and Young's modulus of the substrates according to (EN 14580 2005), 4 cylinders were core drilled from each type of substrate and tested. Each cylinder was subjected to 3 loading/unloading cycles between 2% and 33% of the expected maximum load. At the end of the last cycle, the specimen is loaded till failure. Elastic modulus is calculated as the slope of the last loading/unloading branch.

After the pull-off tests, the amount of salt in the POST specimens was determined. Fragments were taken by chisel from: i) the substrate just under the composite (first cm); ii) the composite matrix (first layer of mortar, in contact with the substrate). Samples were powdered, put in boiling deionized water for 10 minutes and filtered by blue ribbon filter paper. The solution obtained was analyzed by ion chromatography (Dionex, ICS-1000). The salt amount in the original materials, i.e. bricks, stones and mortars, was determined as well, for comparison, collecting the samples from the REF specimens.

3 RESULTS AND DISCUSSION

3.1 Mechanical properties of materials

The results of the mechanical tests carried out on the substrates and mortar matrices are reported in Table 1. The data confirm that the four selected materials (bricks and stones) exhibit a range of very different properties, representative of a variety of historic substrates. YB exhibits much worse mechanical properties than RB, owing to its "handmade" manufacturing technology, which involves a more heterogeneous clay mixture and the shaping by pressing rather than by extrusion. The mechanical properties of LS are the highest, even if this is a porous and soft stone. As expected, NT exhibits the lowest compressive strength due to its extremely porous microstructure.

The two mortars investigated, based on hydraulic lime binders, exhibit a high compressive strength values (10.3 and 15.3 N/mm^2, respectively), but low elastic modulus, this latter being a positive feature for compatibility with historic masonry walls.

Table 1. Characteristics of materials in compression test.

Material strength	Compressive		Young's modulus	
	Average (MPa)	CV (%)	Average (GPa)	CV (%)
RB (red brick)	18.7	2.0	6.2	11.5
YB (yellow brick)	7.6	21.7	3.0	13.5
LS (Lecce stone)	30.8	2.8	13.7	9.8
NT (Neap. tuff)	3.9	19.2	5.3	25.5
Mortar HL1	10.3	4.2	9.3	2.5
Mortar HL2	15.3	3.1	10.1	1.8

3.2 Pull-off tests

The results of pull-off tests obtained with BAS-HL1 and GLASS-HL2 composite reinforcements are reported in Table 2 and Table 3, respectively. Pictures of the corresponding failure modes for representative samples are shown in Figure 4.

Based on the results obtained, the following comments can be made:

- Pull-off loads of REF specimens are on average lower compared to results available in literature on similar tests, see (Carozzi & Poggi 2015). However, in (Carozzi & Poggi 2015) pull-off tests were conducted on FRCM composites made of glass fibers embedded with a cement-based mortar applied on bricks with a very high compressive strength (68.8 MPa). Resulting pull-off loads were comprised between 2.6-2.9 kN, thus closer to LS results, being LS the substrate with the highest compressive strength in the present study;
- All REF substrates reinforced with BAS-HL1 composite exhibited very similar pull-off loads, notwithstanding their different strength, although a higher mean load was found for the strongest substrate (LS). They also exhibited cohesive failure modes, evidencing an adequate bond, except for NT, which was characterized by an adhesive failure;
- The salt deterioration of the substrates in specimens reinforced with BAS-HL1 composite (POST series) did not cause a worsening of the bond, but rather an improvement. This was probably due to the fact that the salt crystallization cycles and the subsequent removal of the decayed layer by steel brush produced a rough surface, with a beneficial effect for adhesion;
- REF substrates reinforced with GLASS-HL2 composite generally exhibited pull-off loads smaller than 1 kN, which were also much lower than those exhibited by BAS-HL1 composite reinforcement.

Table 2. Results of pull-off tests for BAS-HL1 composite (F.M.= failure mode, Adh.= adhesive failure mode, Co.= cohesive failure mode, Mix.= mixed failure mode, i.e. partially adhesive and partially cohesive).

Substrate		Load (kN)	F.M.	Average load (kN)	CV (%)
RB	REF	1.68	Adh.	1.88	10.5
		2.13	Co.		
		1.78	Co.		
		2.05	Co.		
		1.77	Co.		
	POST	3.24	Co.	3.18	5.0
		2.99	Co.		
		3.12	Co.		
		3.36	Co.		
YB	REF	1.48	Adh.	1.35	15.8
		1.58	Adh.		
		1.10	Adh.		
		1.26	Adh.		
	POST	1.65	Adh.	1.21	34.8
		1.18	Adh.		
		0.81	Adh.		
LS	REF	2.74	Co.	2.32	-
		1.89	Co.		
	POST	3.20	Co.	3.17	7.3
		2.98	Co.		
		3.01	Co.		
		3.49	Co.		
NT	REF	1.11	Co.	1.12	4.7
		1.18	Co.		
		1.08	Co.		
	POST	1.39	Co.	1.20	20.2
		1.30	Co.		
		0.93	Adh.		

The failure mode was mostly adhesive, with the exception of NT. For several samples (especially RB and YB), a premature detachment of the FRCM composite occurred, so it was not possible to gain useful data to comment on, but it is interesting to notice that a low adhesion was observed;

- The pull-off loads of the POST specimens reinforced with GLASS-HL2 composite were basically comparable with those obtained for the corresponding REF specimens;
- Notwithstanding the fact that substrate made of single materials were used, the pull-off test gave scattered results for many samples. This suggests that this test, although simple, may be affected by several factors (heterogeneity of the substrate, state of the surface, uneven positioning of the fiber net inside the mortar layer, etc.).

Table 3. Results of pull-off tests for GLASS-HL2 composite (F.M.= failure mode, Adh.= adhesive failure mode, Co.= cohesive failure mode, Mix.= mixed failure mode, i.e. partially adhesive and partially cohesive).

Substrate		Load (kN)	F.M.	Average load (kN)	CV (%)
RB	REF	0.36	Ahd.	-	-
	POST	-	-	-	-
YB	REF	0.76	Adh.	-	-
	POST	0.22	Adh.	-	-
LS	REF	0.63	Adh.		
		0.21	Adh.	0.54	39.1
		0.45	Adh.		
	POST	0.88	Adh.	0.50	-
		0.12	Adh.		
NT	REF	0.62	Mix.		
		0.19	Co.	0.75	62.9
		1.31	Co.		
		0.86	Co.		
	POST	0.60	Mix.		
		0.94	Co.		
		0.73	Co.	0.74	19.9
		0.68	Co.		

Table 4. Amounts of chloride and sulfate ions in the original materials.

Sample	Cl^- (wt%)	$SO_4^=$ (wt%)
YB	0.14	0.58
NT	0.04	0.13
LS	0.04	0.14
RB	0.04	0.18
HL1	0.04	0.97
HL2	0.06	0.46

Table 5. Amounts of chloride and sulfate ions in POST samples after pull-off tests.

Substr.	Composite	Sample	Type	Cl^- (wt%)	SO_4^- (wt%)
RB	BAS-HL1	Substrate (RB)	POST	1.19	2.61
		Mortar (HL1)		2.20	4.41
YB	BAS-HL1	Substrate (YB)	POST	1.42	4.97
		Mortar (HL1)		0.73	1.40
LS	BAS-HL1	Substrate (LS)	POST	0.82	2.24
		Mortar (HL1)		1.03	1.77
NT	BAS-HL1	Substrate (NT)	POST	1.32	5.38
		Mortar (HL1)		0.94	1.21
YB	GLASS-HL2	Substrate (YB)	POST	1.68	5.24
		Mortar (HL2)		0.24	0.64
LS	GLASS-HL2	Substrate (LS)	POST	1.70	4.38
		Mortar (HL2)		0.88	1.08
NT	GLASS-HL2	Substrate (NT)	POST	2.48	6.87
		Mortar (HL2)		1.42	0.38

Figure 4. Representative cylinders of GLASS-HL2 composite detached in the pull-off tests (on the left, a REF NT substrate; in the center, a POST LS substrate, on the right a POST YB substrate).

3.3 Salts distribution

The amounts of ions in the materials used for the tests, i.e. substrates and composite matrices, are reported in Table 4, while the values obtained for the materials after pull-off test are reported in Table 5.

Some ions are present in the original materials, Table 4, especially in terms of sulfate, the highest amounts being in YB and in the two mortars. These amounts are in line with what is currently found in commercial hydraulic lime-based mortars and bricks. The amounts of chloride and sulfate ions in the POST substrates, Table 5, just below the FRCM layer are high, as expected, due to the salt accumulation produced after the accelerated weathering

procedure. Interestingly, the anion amounts are very high also in the mortar matrices of the composites. This suggests that the salts are absorbed by the mortars during their curing and drying, and this aspect must be carefully considered as the salts accumulated in the mortar pores might lead to some subsequent deterioration, upon crystallization cycles. Notably, HL1 mortar exhibited a higher amount of salts with respect to HL2 mortar, probably due to its higher porosity (evidenced by its lower mechanical properties in Table 1).

4 CONCLUSIONS

The present paper presents an experimental campaign carried out on different natural and artificial materials reinforced with two different FRCM composites having hydraulic lime-based matrices. The results allow to make the following remarks:

- The pull-off test allowed a quick and easy investigation of different substrates reinforced with FRCM composites. However, the results obtained by this test are scattered and a high number of testing points seems necessary to achieve sound and conclusive remarks;

- The FRCM having a cement-based mortar detached from all the investigated materials (both in salt-contaminated and unweathered conditions), highlighting the incompatibility of this kind of mortar with historic porous substrates;
- The presence of salt deterioration in the substrates before the FRCM composites application did not cause the worsening of the bond (probably due to the higher surface roughness after deterioration), provided that all the deteriorated and powdering parts on the surface are preventively cleaned;
- The mortars used for the FRCM composites absorbed a very high amount of salts from the substrates, simply by drying. This aspect must be carefully considered for the durability of the mortars, which could be subjected to salt crystallization cycles with time.

ACKNOWLEDGEMENTS

The technicians of DICAM Department (Alma Mater Studiorum-University of Bologna) Paolo Carta and Mario Marcolongo, are gratefully acknowledged for their support. Thanks to Mrs. Isotta Montella for helping with the tests. Kerakoll spa (Sassuolo, Italy) and Biemme srl (Lucrezia di Cartoceto, Italy) are gratefully acknowledged for providing the composite materials.

REFERENCES

Babatunde, S.A. 2017. Review of strengthening techniques for masonry using fiber reinforced polymers. *Composite Structures* 161: 246–255.

Bellini, A., Mazzotti, C. 2020. Bond Behavior of FRCM Composites Applied on Concrete and Masonry. *Lecture Notes in Civil Engineering* 42: 347–359.

Carozzi, F. G., Poggi, C. 2015. Mechanical properties and debonding strength of Fabric Reinforced Cementitious Matrix (FRCM) systems for masonry strengthening. *Composites Part B: Engineering* 70: 215–230.

Colella A., Di Benedetto C., Calcaterra D., Cappelletti P., D'Amore M., Di Martire D., Graziano S.F., Papa L., de Gennaro M., Langella A. 2017. The Neapolitan yellow tuff: An outstanding example of heterogeneity. *Construction and Building Materials* 136: 361–373.

Ebead, U., Adel Y. 2019. Pull-off characterization of FRCM/Concrete interface. *Composites Part B: Engineering* 165: 545–553.

EN 772-1 (2015). Methods of test for masonry units - Part 1: Determination of compressive strength.

EN 998-2 (2016). Specification for mortar for masonry - Part 2: Masonry mortar.

EN 1015-11 (1999). Methods of test for mortar for masonry. Determination of flexural and compressive strength of hardened mortar.

EN 1542 (1999). Products and systems for the protection and repair of concrete structures — Test methods — Measurement of bond strength by pull-off.

EN 12190 (1999). Products and systems for the protection and repair of concrete structures. Test methods. Determination of compressive strength of repair mortar.

EN 14580 (2005). Natural stone test methods. Determination of static elastic modulus.

Franzoni, E. 2014. Rising damp removal from historical masonries: A still open challenge. *Construction and Building Materials* 54: 123–136.

Franzoni, E., Gentilini, C., Santandrea, M., Carloni, C. 2018. Effects of rising damp and salt crystallization cycles in FRCM-masonry interfacial debonding: Towards an accelerated laboratory test method. *Construction and Building Materials* 175: 225–238.

Franzoni, E., Santandrea, M., Gentilini, C., Fregni, A., Carloni, C. 2019. The role of mortar matrix in the bond behavior and salt crystallization resistance of FRCM applied to masonry. *Construction and Building Materials* 209: 592–605.

Maljaee, H., Ghiassi, B., Lourenço, P., Oliveira, D. 2016. FRP–brick masonry bond degradation under hygrothermal conditions. *Composite Structures* 147: 143–154.

Sandrolini, F., Franzoni, E., Cuppini, G., Caggiati, L. 2007. Materials decay and environmental attack in the Pio Palace at Carpi: A holistic approach for historical architectural surfaces conservation. *Building and Environment* 42(5): 1966–1974.

Vaculik, J., Visintin, P., Burton, N.G., Griffith, M.C., Seracino, R. 2018. State-of-the-art review and future research directions for FRP-to-masonry bond research: Test methods and techniques for extraction of bond-slip behavior. *Construction and Building Materials* 183: 325–345.

Zampieri, P., Gonzalez-Libreros, J., Simoncello, N., Pellegrino, C. 2019. Strengthening of masonry arches with frcm composites: A review. *Key Engineering Materials* 817 KEM: 251–258.

Brick and Block Masonry - From Historical to Sustainable Masonry –
Kubica, Kwiecień & Bednarz (eds)
© 2020 Taylor & Francis Group, London, ISBN 978-0-367-56586-2

Masonry elements reinforced with FRCM: Bond behaviour

M. Leone, F. Micelli & M.A. Aiello
Department of Engineering for Innovation - University of Salento, Italy

A. Cascardi
ITC-CNR, Construction Technologies Institute - Italian National Research Council, Bari, Italy

ABSTRACT: Externally bonded reinforcements, made of fibrous meshes, embedded in a cementitious/ hydraulic lime mortar, are now getting a great deal of attention, mostly for strengthening, retrofitting and repair of existing structures. When applied on masonry structures, these innovative materials have shown better behaviour than FRP (Fiber Reinforced Polymer), in terms of compatibility with the substrate, sustainability and reversibility. Moreover, these features are indispensable criteria that need to be fulfilled when the intervention is intended in the field of cultural Heritage, which largely consisting of masonry structures. As well known, the effectiveness of the strengthening is strongly related to the bond behaviour between the reinforcement system and the substrate. With this aim, the present work reports the results of experimental study on GFRCM (Glass Fabric Reinforcement Cementitious Matrix); i.e. a glass open-grid embedded in inorganic matrix, applied on two different masonry substrates. First, the mechanical properties of GFRCM reinforcements were obtained through tensile tests; then, the experimental investigation on bond behaviour was carried out by direct shear bond test. The test results were collected and processed to evaluate both the bond strength and the failure modes.

1 INTRODUCTION

Existing masonry constructions, that represent a relevant part of historical building Heritage, exhibited severe damages due to seismic events in the past decades. Traditional and innovative strengthening systems could be applied to repair or upgrading the structural behaviour of these constructions. In this context a new generation of composite materials with inorganic matrix has appeared, namely FRCM (Fabric Reinforced Cementitious Matrix). In particular, in case of ancient masonry constructions, the new mortar-based reinforcements are considered as a valid alternative to FRP (Fiber Reinforced Polymer), because of the mechanical and the chemical compatibility requirements. In fact, due to the absence of polymeric adhesive, the breathability of the masonry is guaranteed. In addition, the performance at elevated temperature, the vapor permeability and the bonding to irregular substrate is enhanced (Triantafillou & Papanicolaou, 2005).

The fibrous reinforcements used in these materials are often bi-directional grids realized with several types of fiber: carbon, glass, basalt, PBO (poliparafenilenbenzobisoxazolo) and steel. The inorganic matrices are usually lime based or cementitious or a blend of them.

For these types of systems, two important aspects of the behavior are: how the stress is transferred between the reinforcements and the support that

need to be strengthened, and the durability against the external environments. Some studies can be found about the effectiveness of FRCM applied on masonry elements (Caggegi et al., 2017; Carozzi et al., 2017; de Felice et al., 2014; De Santis, Ceroni, et al., 2017; Del Zoppo et al., 2019; Garcia-Ramonda et al., 2020; Gattesco & Boem, 2015; Leone et al., 2017; Lignola et al., 2017).

In particular, the bond between the composite and the substrate plays a relevant role, (Bencardino et al., 2020; De Santis, Ceroni, et al., 2017; Del Zoppo et al., 2019; Gattesco & Boem, 2015; Leone et al., 2017; Triantafillou & Papanicolaou, 2005). The bond depends on many parameters, e.g. mechanical properties of the FRCM and the substrate, environmental conditions, test setup, etc. It is essential to focus the experimental investigations on the role of these parameters and deepen the knowledge in this field. In fact, in (Leone et al., 2017) was observed that the bond test results showed a satisficing level of repeatability in terms of axial strength and failure modes when referring to the same composite, while the data furnished a high level of scatter when two different test set-up were compared, i.e. double and single lap shear test. The authors underlined the need of defining a standard procedure for studying the bond behaviour between GFRCM and masonry substrate. In the same work was also observed that in the performed bond tests a cohesive failure of the reinforcement from the substrate never occurred. It is enough clear that the stress transfer mechanism between GFRCM and a brittle substrate is different from that experienced for FRP materials: the GFRCM are not able to achieve the maximum cohesive bond failure because the failure of the fibers may occur first, due to the lower fiber content or even the slippage between the fabric and the mortar. Even if in (Leone et al., 2017) a total of 130 FRCM/masonry specimens were tested under shear and the results compared, the high dispersion of the test results makes difficult to perform all necessary comparisons and to calibrate design formulas, confirming the difficulties to identify the influence of the several parameters involved and their combined effect.

The present work aims to provide a contribution to the understanding of the general physical phenomena that characterize the mechanical behaviour of GFRCM materials and their bond with masonry structures. To this scope, both tensile and bond test were performed on FRCM realized with dry glass mesh and lime-based mortar varying the type of masonry substrate in the case of shear test. The results were analyzed and discussed. Tensile tests show that the stress-strain curve of the analyzed FCRM is almost bi-linear while the results of the bond test underlined the independency of the substrate by the shear bond strength of the system.

2 MATERIALS

The mechanical capacity of both FRCM composites and FRCM-strengthened structures, is strongly affected by the brittle nature of the inorganic matrix and the mechanical behaviour of the mesh as well as its geometrical scheme. In fact, the manufacturing and curing of the mortar, the accuracy of the FRCM-installation and local defects of the substrate may significantly affect crack development, failure mode, ductility and strength in FRCM-retrofitted members. Generally, direct tensile tests and shear bond tests on each FRCM-system are commonly accomplished in order to understand the retrofitting potentiality.

The herein considered FRCM-system consists of AR-glass (*Alkali Resistant*) mesh and lime-based matrix. The mesh's spacing is 12x12 mm (= 60 mm^2/m) with a 108 GPa, 1929 MPa and 1.8% of *Young's modulus*, tensile strength and maximum elongation respectively, according to (EN-ISO-13934_1, 2013). While the mortar-matrix has a compressive strength of 9.1 MPa, after 28 days curing, according to (UNI1015-11, 2007).

Two types of substrates were considered. *Tuff* blocks and clay bricks, assembled with the same mortar joint. The *Tuff* had an average compressive strength equal to 4.52 ± 2.09 MPa while for the clay brick it was 22.86 ± 3.67 MPa according to a series of four compression tests (40x40x40 mm cubes). The horizontal joints were manufactured 15 mm and 10 mm for the *Tuff* and the clay brick substrates, respectively with an average compressive strength of

the mortar equal to 4.38 ± 0.26 MPa obtained according to (UNI1015-11, 2007).

3 TENSILE TESTS: SET UP AND RESULTS

Tensile tests were carried out on prismatic FRCM-specimens throughout a universal testing machine under displacement control (i.e. 0.2 mm/min) and gripping throughout the bonding of steel tabs (80x60x2 mm) (Consiglio dei Lavori Pubblici, 2015). The load was measured by a load cell and the axial stress was then conventionally calculated by referring to the cross-sectional area of the dry fabric. While, the strains were measured by displacement transducers (see Figure 1 *a* and *b*).

Nine FRCM coupons with dimensions 600x60x 5 mm were used for tensile tests.

The experimental results are reported in Table 1 in terms of tensile strength and ultimate strain. The mean tensile strength was found equal to 890.75 MPa with a relative tensile strain of 1%; the *Coefficient of*

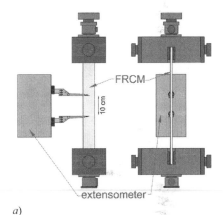

a)

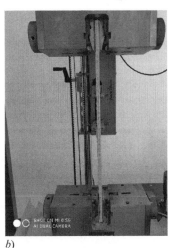

b)

Figure 1. Tensile test set-up: *a*) sketch and *b*) specimen into the machine.

Table 1. Experimental tensile test results.

| # | Tensile strength | Ultimate strain | λ |
	MPa	-	-
1	917.36	0.91	0.48
2	727.18	0.86	0.38
3	946.91	0.95	0.49
4	747.87	0.74	0.39
5	1143.00	1.35	0.59
6	790.25	1.06	0.41
7	1055.30	1.09	0.55
8	836.56	0.78	0.43
9	852.32	0.99	0.44
Mean	890.75	0.97	0.46
CoV	15.7%	19%	16%

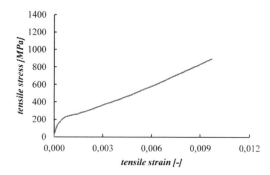

Figure 3. Tensile stress-strain law of the FRCM.

Variation (CoV) is also included in Table 1. The first stage of damage was characterized by the matrix multi-cracking in the transversal direction corresponding to the yard locations (see Figure 2). After that, the combined slippage of the fabric within the matrix and the fiber rupture was manifested.

The average result with the relative dispersion area (yellow) in term of stress strain curves (red) is illustrated in Figure 3. According to literature (CNR, 2018; De Santis, Carozzi, et al., 2017), the tensile stress-strain behaviour is generally characterized by three stages. In the first stage, the mortar is un-cracked and furnishes significant contributions to both the stiffness and the load bearing capacity of the material. In the second stage (i.e. transition phase), the crack pattern progressively develops, while the third stage is characterized by no further cracks development and widening of existing ones. However, in the present work, the results show an almost bi-linear

behavior because the mortar cracks developed almost all at the same time, so the transition phase was not-significant.

The first slope of the stress-strain curves was about 514 GPa; while the second one was much lower due to the mortar cracks evolution. In other words, the second slope (77.45 GPa) was enough close to the dry fabric elastic modulus.

The strength of the FRCM coupons was significantly lower than that of the dry fabric, since the inorganic matrix is not able to completely impregnate the fibers Thus a full composite action is not developed, as found for FRPs. This aspect is quantified in Table 1 by means of the λ coefficient calculated as the ratio between the FRCM tensile strength and the dry fabric tensile strength. In fact, the FRCM composites exploited, in average, the 46% of the tensile capacity of the fabric.

4 SHEAR BOND TESTS: SET UP AND RESULTS

The single face shear tests were performed by a proper designed test set-up as shown in Figure 4 a and b. In fact, a contrast frame, realized with two rigid steel plates (30 mm thick) was built in order to apply the load. A modest eccentric compression force (if compared with the masonry strength) was manually imposed. No cracks within the substrate were observed. The aim was to limit the rotation of the sample. The substrate was made up of 7 bricks and 6 mortar horizontal joints. The bricks were 250x125x55 mm and 250x150x55 mm for the clay and Tuff respectively. Moreover, the thickness of the joints was varied by considering 10 mm and 15 mm for the clay and Tuff masonry respectively. The FRCM-specimen was bonded to the substrate per a length of 300 mm and a width of 60 mm; while a dry fabric was over the substrate per a length over 400 mm; (Consiglio Superiore dei Lavori Pubblici, 2015).

In particular, the FRCM was positioned on the large side in case of clay masonry and on the small

Figure 2. Tensile failure of the FRCM.

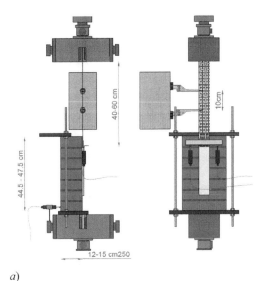

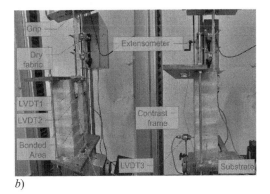

a)

b)

Figure 4. Shear test set-up: *a*) sketch and *b*) specimen into the machine.

side for the Tuff one. Furthermore, the shear bond tests were carried out by means of the same machine used for FRCM tensile tests (under displacement control with a load speed of 0.2 mm/min) by referring to the two types of masonry prisms made off 55mmm thick Tuff and clay brick.

The free end of the FRCM was gripped with the available tool of the machine. In addition, steel tabs (60x100x3 mm) were glued in order to make the gripping action more effective. A 100kN load cell was used for load recordings. The slip between the fabric and the substrate was measured by two Linear Variable Displacement Transducer (LVDT1 and LVDT2 in Figure 4).

A further LVDT3 was adopted in order to control possible rotation of the sample during the loading phase. Finally, the machine was equipped with an extensometer for measuring the dry fabric elongation with respect to an initial length of 100 mm. A total of 10 specimens was manufactured per each substrate and a nomenclature **XY** was used; where

X indicates the type of masonry (C=clay and T=*Tuff*) and **Y** means the number of tests (i.e. from 1 to 10).

The failure mode, the maximum force (Fmax), the slip (s), tensile strength (σf) and the $\sigma f/\sigma u$ (where σu is the tensile strength of the FRCM experimentally evaluated and reported in the previous section) ratio were reported in Table 2. The specimens T4 and T5 where not considered in the results discussion because the outcomes were far from the theoretical expectation, may be due to possible micro-damages of the specimens during the laboratory operations. Even if the CoV related to the bond strength demonstrated the modest scatter of the recordings, the dispersion of data referring to the slips is more important. This dispersion is commonly found in different available studies (Bencardino et al., 2020) basically due to the difficulty related to the specific measure and a proper accurate control. Moreover, an in depth comparison between the recorded failure modes and the ultimate slip seems to suggest a dependence between them. In particular, the matrix cracking and the failure of a dry cross-section affected the larger slip; while inter-phases loss produced more fragile behavior.

Referring to the present work the modes of failure occurred at the reinforcement to-substrate interface (Figure 5 *a*), between the textile and the matrix (Figure 5 *b*), or by textile slippage within the matrix.

Table 2. Experimental bond test results.

Label	Failure mode	F_{max} kN	s mm	$\sigma_f = F_{max}/A_f$ MPa	σ_f/σ_u -
T1	D	1.96	3.97	889.35	1.00
T2	D	2.21	3.76	1006.77	1.13
T3	F	1.32	0.36	602.18	0.68
T6	E	1.89	1.48	859.39	0.97
T7	E	1.96	0.86	890.10	1.00
T8	B + E + A	2.09	0.54	949.29	1.07
T9	E	1.54	0.83	699.84	0.79
T10	B + E + A	1.51	0.52	685.21	0.77
Mean		1.77	2.31	802.48	0.90
CoV		23%	95%	23%	23%
C1	E	1.72	1.33	781.90	0.88
C2	B + E + A	1.71	1.20	778.37	0.87
C3	D	1.96	3.97	889.35	1.00
C4	B + E	2.03	4.18	923.29	1.04
C5	B + E	2.04	8.49	929.02	1.04
C6	E	2.00	0.44	910.39	1.02
C7	D	1.68	0.90	765.79	0.86
C8	E + F	1.79	7.16	813.76	0.91
C9	D	1.53	1.21	694.24	0.78
C10	D	2.00	2.36	906.85	1.02
Mean		1.85	3.12	839.30	0.94
CoV		10%	89%	10%	10%

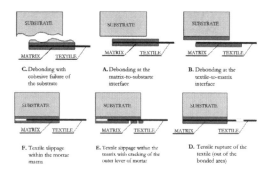

a)

b)

c)

d)

Figure 5. Details of the main failure modes: a) substrate/FRCM, b) fabric/matrix, c) matrix multi-crack and d) fabric rupture.

The failure it often manifests with micro-cracks within the matrix at the interface between the fiber

and the matrix or the matrix and the substrate (Figure 5 c). finally, the rupture of the fabric in the free zone was also observed (see Figure 5 d).

The main failure modes are illustrated in Figure 6, according to (CNR, 2018) and listed in the Table 2. The σf/σu reported in Table 2 demonstrated that the tensile strength of the FRCM was almost reached in all the tests independently from the failure mode.

The main results in term of tensile stress (σf) and slip (s) are illustrated in Figures 7 and 8 for the T and C series, respectively. Thanks to the extensometer measurements (see Figure 4 b), the deformation of the fabric was deducted from the slip recordings. Moreover, the mean and the scatter area of the tensile strength of the FRCM (σu) were both reported with a dash red line and a transparent red rectangle. The experimental curves are characterized by a pre-peak trend, which is pseudo linear, and a post-peak trend varying between softening, plateau and hardening type (see Figure 7). In particular, according to the obtained results, a softening post-peak behavior was evident in case of crisis mostly on the matrix side (A and B); a plateau was observed when the crisis was dominated at the fabric/matrix interface (E and F) and a hardening trend was obtained when the fabric rupture was attained if tensile stress state was increased within the fabric up to failure. In synthesis,

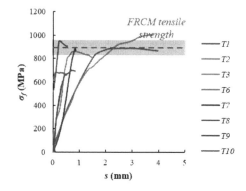

Figure 6. Sketch of the possible failure modes.

Figure 7. Stress-slip law for the *Tuff* series.

650

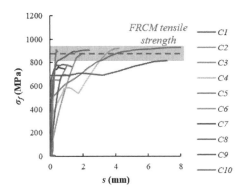

Figure 8. Stress-slip law for the clay series.

there are several cases in which the slip between the masonry and the strengthening material is rather small, and other cases where the maximum resistance is followed by a pseudo-plateau branch depending on the dominant failure mode and its evolution.

Anomaly bond strength was recoded just for T2; which was modestly higher than the scatter range of the tensile strength of the FRCM. This specimen manifested the rupture of the fabric in a dry cross-section.

The data dispersion was significantly less important in terms of bond stress for the C-series as shown in Table 2. The larger slip was evident in C8 and C5 that manifested the larger crack opening in the matrix.

The average bond strength was 802.48 MPa and 839.30 MPa for the T and C series respectively corresponding to the 90% and 94% of the FRCM tensile capacity. For this reason, the investigated bond can be considered independent from the type of substrate.

According to the (dei Lavori Pubblici, 2015) the conventional value of the FRCM strength to be used for design purpose is assumed equal to the strength reached in the bond test. Moreover, this assumption is realistic when the bond strength is found substrate-dependent, as reported in (dei Lavori Pubblici, 2015). In this sense, the present experimental investigation demonstrates the total independency of the bond strength from the type of masonry substrate (see Table 2); so the design strength should be accomplished equal to the tensile strength measured in the direct tensile tests.

5 CONCLUSIONS

In the present paper an experimental program, carried out on dry grids made by glass fibers and coupled with an inorganic mortar was reported and discussed. The goal was to contribute to enlarge the knowledge on the mechanical behaviour of this innovative strengthening system. For this purpose, a GFRCM material was considered and tested under tensile and shear. In the case of bond test the type of masonry substrate was considered as variable parameter.

The following first conclusions can be drawn:

• the stress strain behaviour is almost bi-linear. It is possible to note the presence of two main phases governed by the mortar and the dry fabric respectively.
• the bond strength is very close to the tensile strength of the FRCM itself and it is independent on the different substrates used in this study: tuff and clay.

A more detailed analysis of the experimental results is currently underway with the aim of comparing the obtained results with both the data available in literature and the national and international guidelines. The Results of this study are useful to address the future experimental and theoretical research, for the update of design provisions of FRCM materials.

ACKNOWLEDGMENTS

This study was supported by the ReLUIS-program 2018, PR 5 – "Materiali innovativi per applicazioni su costruzioni esistenti". The authors would acknowledge also *FibreNet s.p.a.* for materials supply.

REFERENCES

Bencardino, F., Nisticò, M., & Verre, S. 2020. Experimental Investigation and Numerical Analysis of Bond Behavior in SRG - Strengthened Masonry Prisms Using UHTSS and Stainless - Steel Fibers. *Fibers 8*(8): 1–16. https://doi.org/10.3390/fib8020008.

Caggegi, C., Carozzi, F. G., De Santis, S., Fabbrocino, F., Focacci, F., Hojdys, Ł., Lanoye, E., & Zuccarino, L. 2017. Experimental analysis on tensile and bond properties of PBO and aramid fabric reinforced cementitious matrix for strengthening masonry structures. *Composites Part B: Engineering 127*: 175–195. https://doi.org/10.1016/j.compositesb.2017.05.048.

Carozzi, F., Bellini, A., D'Antino, T., De Felice, G., Focacci, F., Hojdys, Ł., Laghi, L., La-noye, E., Micelli, F., Panizza, M., & Poggi, C. 2017. Experimental investigation of tensile and bond properties of Carbon-FRCM compo-sites for strengthening masonry elements. *Compos Part B 128*: 100–11.

CNR. 2018. CNR-DT 215/2018 Istruzioni per la Progettazione, l'Esecuzione ed il Controllo di Interventi di Consolidamento Statico mediante l'utilizzo di Compositi Fibrorinforzati a Matrice Inorganica. *CNR: Consiglio Nazionale Delle Ricerche.*

de Felice, G., De Santis, S., Garmendia, L., Ghiassi, B., Larrinaga, P., Lourenço, P. B., Oliveira, D. V., Paolacci, F., & Papanicolaou, C. G. 2014. Mortar-based systems for externally bonded strengthening of masonry. *Materials and Structures/Materiaux et Construction 47*(12): 2021–2037. https://doi.org/10.1617/s11527-014-0360-1.

De Santis, S., Carozzi, F. G., de Felice, G., & Poggi, C. 2017. Test methods for Textile Reinforced Mortar systems. *Composites Part B: Engineering 127*: 121–132. https://doi.org/10.1016/j.compositesb.2017.03.016.

De Santis, S., Ceroni, F., de Felice, G., Fagone, M., Ghiassi, B., Kwiecień, A., Lignola, G. P., Morganti, M., Santandrea, M., Valluzzi, M. R., & Viskovic, A. 2017. Round Robin Test on tensile and bond behaviour of Steel Reinforced Grout systems. *Composites Part B: Engineering 127*: 100–120. https://doi.org/10.1016/j.compositesb.2017.03.052.

dei Lavori Pubblici, C. S. 2015. *Linea Guida per la identificazione, la qualificazione ed il controllo di accettazione di compositi fibrorinforzati a matrice inorganica (FRCM) da utilizzarsi per il consolidamento strutturale di costruzioni esistenti [In Italian]*.

Del Zoppo, M., Di Ludovico, M., Balsamo, A., & Prota, A. 2019. In-plane shear capacity of tuff masonry walls with traditional and innovative Composite Reinforced Mortars (CRM). *Construction and Building Materials 210*: 289–300. https://doi.org/10.1016/j.conbuildmat.2019.03.133.

EN-ISO-13934_1. 2013. *Textiles — Tensile properties of fabrics — Part 1: Determination of maximum force and elongation at maximum force using the strip method*.

Garcia-Ramonda, L., Pelá, L., Roca, P., & Camata, G. 2020. In-plane shear behaviour by diagonal compression testing of brick masonry walls strengthened with basalt and steel textile reinforced mortars. *Construction and Building Materials, 240* 117905. https://doi.org/10.1016/j.conbuildmat.2019.117905.

Gattesco, N., & Boem, I. 2015. Experimental and analytical study to evaluate the effectiveness of an in-plane reinforcement for masonry walls using GFRP meshes. *Construction and Building Materials 88*: 94–104. https://doi.org/10.1016/j.conbuildmat.2015.04.014.

Leone, M., Aiello, M. A., Balsamo, A., Carozzi, F. G., Ceroni, F., Corradi, M., Gams, M., Garbin, E., Gattesco, N., Krajewski, P., Mazzotti, C., Oliveira, D., Papanicolaou, C., Ranocchiai, G., Roscini, F., & Saenger, D. 2017. Glass fabric reinforced cementitious matrix: Tensile properties and bond performance on masonry substrate. *Composites Part B: Engineering 127*: 196–214. https://doi.org/10.1016/j.compositesb.2017.06.028.

Lignola, G. P., Caggegi, C., Ceroni, F., De Santis, S., Krajewski, P., Lourenço, P. B., Morganti, M., Papanicolaou, C. (Corina), Pellegrino, C., Prota, A., & Zuccarino, L. 2017. Performance assessment of basalt FRCM for retrofit applications on masonry. *Composites Part B: Engineering 128*: 1–18. https://doi.org/10.1016/j.compositesb.2017.05.003.

Triantafillou, T. C., & Papanicolaou, C. G. 2005. Textile reinforced mortars (TRM) versus fiber reinforced polymers (FRP) as strengthening materials of concrete structures. *Spec. Publ. 230*: 99–118.

UNI1015-11. 2007. *Metodi di prova per malte per opere murarie - Determinazione della resistenza a flessione e a compressione della malta indurita [In Italian]*.

History & preservation

Brick and Block Masonry - From Historical to Sustainable Masonry –
Kubica, Kwiecień & Bednarz (eds)
© 2020 Taylor & Francis Group, London, ISBN 978-0-367-56586-2

Numerical insights on the structural assessment of typical historical masonry vaults of Cagliari

A. Cazzani, V. Pintus & E. Reccia
Department of Civil and Environmental Engineering and Architecture, University of Cagliari, Cagliari, Italy

N. Grillanda & G. Milani
Department of Architecture, Built Environment and Construction, Politecnico di Milano, Milano, Italy

ABSTRACT: Collapse behavior of masonry vaults is investigated with the aim to evaluate the role of brick patterns in their mechanical behavior. Attention is focused on lowered sail vaults typically built during the nineteenth-century in the city of Cagliari, Italy. A series of rigorous laser scanner surveys has been performed on some of these vaults in order to obtain the effective geometry both at macro-level – i.e. the vault shape – and at micro-level – i.e. brick patterns. Analysis is performed through a NURBS-based upper bound limit analysis approach. The representation of a complex geometry through NURBS (Non-Uniform Rational B-Spline) surfaces is particularly suited to treat curved geometries, such as vaults. A mesh of NURBS element is defined: each element is idealized as a rigid body with dissipation allowed only along interfaces. In order to find the minimum live load multiplier, a procedure of mesh adaptation by a genetic algorithm is applied.

1 INTRODUCTION

The structural behavior of masonry vaults has been the subject of scientific debate for a long time (Boothby 2001, Lucchesi et al. 2007), however, since the lack of specific analysis methods, their mechanical behavior is still not fully understood. The mechanical characterization of masonry materials is of paramount relevance for structural modelling and analysis: inhomogeneity and anisotropy of such kind of material make rather complex a structural assessment, increasing the uncertainty degree in the knowledge of mechanical behavior. Geometry complexity and three-dimensional character of the problem add further difficulties to the study.

Different methods and models may be found in literature and are commonly used in the analysis of masonry vaults up to the collapse. Limit Analysis has proved to be a very effective method for a fast and reliable evaluation of the load bearing capacity of vaulted masonry structures (D'Ayala & Casapulla, 2001, Milani et al. 2008, Pavlovic et al., 2016): classic lower and upper bound theorems recall respectively the concepts of equilibrium and occurrence of failure mechanisms with rigid elements. The so-called Thrust Network Method (O'Dwyer, 1999, Block & Ochserdorf, 2007, Angelillo et al., 2013) moves its steps from lower bound theorems, whereas FE limit analysis approaches with infinitely resistant elements and dissipation on interfaces take inspiration from the upper bound point of view. An alternative to Limit Analysis is represented by traditional FEM combined with either elastic-plastic or damaging models with softening, commonly used for other

materials but recently adapted also to masonry (Roca et al., 2010, Milani & Tralli, 2012). They are able to provide a large set of output numerical information but further studies are still needed to ensure their proper application. An exhaustive and general overview of the State of the Art of the approaches to structural analysis of historic masonry constructions has been provided by (Tralli et al. 2014).

In case of arches and vaults, where equilibrium can be verified provided that the thrust surface is strictly enclosed within the outline of the structure, the exact representation of these geometries is essential in the evaluation of their load-bearing capacity (Cazzani et al. 2016, Chiozzi et al. 2016). The adoption of geometric reconstructions obtained from point cloud data derived by photogrammetry and/or terrestrial laser scanner is crucial (Schueremans & Van Genechten, 2009). Both instrumentation and survey techniques not only provide information on metrics but also give information on the health status, highlighting cracks patterns or deviation from vertical or horizontal directions (Castellazzi et al. 2017, Napolitano & Glisic 2019). Even deviation from ideal shape may be assessed through parametric (Argiolas et al. 2019) or non-parametric functional models which need to be rigorously estimated (Armesto et al. 2010).

In the work, attention is focused on lowered sail vaults typically built during the nineteenth-century in the city of Cagliari, Italy. A series of rigorous laser scanner surveys has been performed on some of these vaults. A NURBS-based upper bound limit analysis approach is adopted to study their collapse behavior.

A preliminary example of the proposed procedure is given by Grillanda et al. (2019a).

2 METHOD: THE GENETIC-ALGORITHM (GA) NURBS-BASED LIMIT ANALYSIS

The procedure starts with the representation of a masonry vault by means of one or more NURBS surfaces. A NURBS surface (Non-Uniform Rational B-Spline) is a parametric surface defined as follows:

$$\mathbf{S}(u,v) = \sum_{i=0}^{n} \sum_{j=0}^{m} R_{i,j}(u,v)\mathbf{B}_{i,j} \qquad (1)$$

Where (u,v) are coordinates in the standard parametric domain; $\mathbf{B}_{i,j}$ is a bidirectional net of control points; n,m are the number of basis functions respectively along the u- and v-direction; $\mathbf{R}_{i,j}$ are the NURBS basis functions where, given a non-uniform knots-vector $\Xi=\{\xi_1,\xi_2,\ldots,\xi_{(n+p+1)}\}$, a set of weights $w_i \in \mathbb{R}$, and the generic i-th B-Spline basis function $N_{i,p}$ of degree p, the generic NURBS basis function can be written as:

$$R_{i,p}(\xi) = \frac{N_{i,p}(\xi).w_i}{\sum_{i=1}^{n} N_{i,p}(\xi).w_i} \qquad (2)$$

A detailed theoretical dissertation on NURBS is reported in (Piegl & Tiller 1995).

NURBS surfaces are particularly suited for historical masonry constructions, in which curved geometries are typically found, and especially for masonry vaults, where equilibrium can be verified provided that the thrust surface is strictly enclosed within the exact outline of the structure. Therefore, the exact representation of vaults' geometry is fundamental in the optic of the evaluation of their load-bearing capacity.

The NURBS representation of a certain masonry vault is performed in the modeling software Rhinoceros. The obtained model is imported into the MATLAB environment. Here, once thickness values are assigned to the surfaces, an assembly of three-dimensional macro-blocks is obtained (Figure 1(a),(b)).

Both regular and irregular mesh of few NURBS elements can be defined directly in MATLAB (Figure 1(c),(d)). NURBS properties allow to discretize an initial surface into elements without affecting its geometry: each element is still a NURBS surface.

An upper bound limit analysis is performed on this model. Each element is idealized as rigid, so its kinematic is completely defined by the six degrees of freedom of its center of mass. Internal dissipation, which is hypothesized to occur only along interfaces, is evaluated assuming an associated flow rule and adopting a Mohr-Coulomb failure domain with tension cut-off and linear cap in compression (Milani & Taliercio 2015). In order to properly evaluate the total amount of the internal

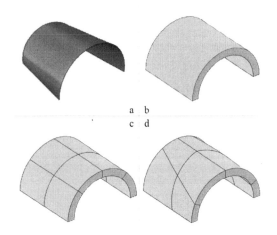

Figure 1. NURBS model generation: (a) NURBS surface in Rhinoceros, (b) NURBS model in MATLAB and its discretization by means of (c) a regular and (d) an irregular mesh.

dissipation, each interface is discretized through a series of points, in which a local reference system (n, s, t) is defined and the compatibility constraint, i.e. the associated plastic flow rule, is imposed (see Figure 2).

Supposing that a configuration of dead (permanent) $\mathbf{F}_0$ and live load Γ (depending on the unknown multiplier Γ) is acting on the vault, in agreement with the Principle of Virtual Works, the total amount of internal dissipation is equal to the power dissipated by the applied external loads $P_{\mathbf{F}0} + P_{\Gamma}$.

Once properly constraints are defined, the upper bound problem can be written through the following linear programming formulation:

$$min\left\{\Gamma = \sum_i P_{int}^i - P_{\mathbf{F}_0}\right\} \text{ such that } \begin{cases} \mathbf{A}_{eq}\mathbf{x} = \mathbf{b}_{eq} \\ \boldsymbol{\lambda} \geq 0 \end{cases}$$

$$(3)$$

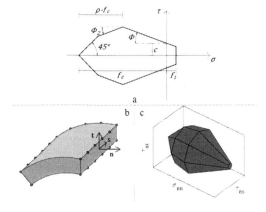

Figure 2. Representation of masonry behavior: masonry-masonry interface and corresponding local reference system (b), parameters of the linearized failure surface (a) and linearized 3D representation of the failure domain (c).

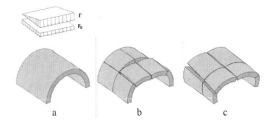

Figure 3. Mesh adaptation procedure: (a) example of load configuration, (b) mechanism associated with the regular mesh, (c) collapse mechanism obtained through GA optimization.

where **x** is the vector of unknowns, which include the velocity components of each element and the non-negative plastic multipliers λ, whereas A_{eq} and b_{eq} represent respectively the overall equality constraints matrix (containing geometric constraints, plastic compatibility and normalization condition of live loads power) and the corresponding right hand side vector. The solution of the linear programming is found through the MATLAB toolbox for optimization over symmetric cones SeDuMi (Sturm 1999).

The obtained load multiplier and the associated solution vector are respectively an upper bound of the real load-bearing capacity of the masonry vault and a mechanism associated with the adopted mesh. In order to find the collapse multiplier and the collapse mechanism, the mesh has to be adjusted until interfaces coincide with the real fracture lines. Therefore, an optimization procedure is needed. A meta-heuristic mesh adaptation scheme based on a Genetic Algorithm (GA) is applied. After a few iterations, the GA optimization automatically provides the mesh in which the correct position of fracture lines is represented and with which the collapse multiplier is associated (see Figure 3).

It is worth mentioning that his method is the natural evolution of the previously published open-source code ArchNURBS (Chiozzi et al. 2016). Authors refer to (Chiozzi et al. 2017) for more details on the GA-NURBS limit analysis and to (Chiozzi et al. 2018a, Chiozzi et al. 2018b, Grillanda et al. 2019b) for application. Moreover, a study of the historical stellar vault of the church of Santa Maria del Monte in Cagliari through this approach has been recently published (Grillanda et al. 2019a).

3 CASE STUDY

Attention is focused on a typology of masonry vaults built around the middle of the XIX cent.: lowered sail vaults – obtained from the intersection of a spherical cap with a prism having a square base – and built with different brick patterns (Figure 4(a)). These vaults have a great diffusion in South Sardinia, and particularly in Cagliari area, but no specific studies about them have been carried out, up to now.

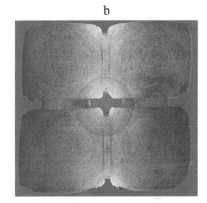

a

b

Figure 4. A typical lowered sail vault in Cagliari (a), result of the laser scanner survey (b).

The structural system involves four lowered sail vaults, obtained from the intersection of a spherical cap with a prism with a square base carved on its base, i.e. with four vertical planes. Each vault appears supported on its whole perimeter: supports are due to masonry arches on two edges, connecting the vaults each other, and walls on the external edges. One of the main characteristics of these vaults is the presence of different orientations in masonry texture. Indeed, even if there are no discontinuities in the curvature of the vaults along the whole surfaces, some triangular portions can be distinguished, each one characterized by a different orientation of the masonry texture. The vaults are composed by a single brick layer, resulting in a thickness value of about 12 cm along the whole surfaces. This is not the case of arches, which are described by a thickness value at least equal to 25 cm. However, by observing the intrados of the structural system, it is not clear if the arch thickness is constant along their longitudinal direction.

A series of rigorous laser scanner surveys have been performed on these vaults, in order to obtain the effective geometry both at macro-level, the vault shape, and at micro-level, the brick patterns. A picture of the obtained geometrical outline is reported in Figure 4(b).

In this work, the analysis of a single vault is reported. Starting from the final result of the laser scanner survey, a series of horizontal and vertical

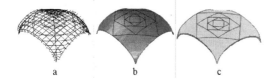

a   b   c

Figure 5. Construction of the NURBS model of the single vault: (a) section lines, (b) assembly of NURBS surfaces in Rhinoceros and (c) obtained NURBS model in MATLAB.

sections of a single vault have been extracted and imported in the modeling software Rhinoceros. Here, an assembly of NURBS surfaces has been created, which describes the surface of the vault with good representation. The sections and the NURBS reconstruction are reported in Figure 5 (a),(b). Let us observe that a different NURBS surface has been adopted for each triangular portion of vault characterized by a defined orientation of bricks. The aim of this choice is the introduction of facilitations in the successive application of the limit analysis in MATLAB. First of all, different properties will be assigned to each surface (in order to take into account the specific orientation of the masonry texture). Secondly, the edge of surfaces will represent the bed joints between the different triangular portions of the vaults, resulting in possible fracture lines under vertical loads. The assemble of NURBS surface has been imported into the MATLAB environment. The obtained NURBS model, after the application of the thickness properties, is depicted in Figure 5(c).

4 NUMERICAL ANALYSES

In all the following analyses the vault is supposed constrained along the external edges and at the base and subjected only to vertical load. However, constraints in these edges have been defined by means of compatibility equations, instead of applying standard geometric constraints: it means that external edges have the same role of interfaces between elements, i.e. displacements form the initial position are evaluated according to the adopted associate plastic flow-rule and the assigned failure surface.

Two configurations of vertical load have been studied. In the first analysis, which is here named LC1 (Load Case 1), the applied live-load consists of the self-weight only. In this way, if the final load multiplier is higher than one, so the vault is definitely safe under self-weight (as it is expected). In the second analysis, LC2, a distributed uniform vertical load, initially equal to 1 kN/m², has been applied at the extrados of the vault as live load (self-weight is applied as dead load, so it is not affected by the load multiplier). For each analysis, a live load multiplier Γ is found. A summary of the analyses here performed is reported in Table 1.

Moreover, these analyses have been repeated by considering different conditions on the NURBS model.

In the first set of analyses, the vault has been studied without introducing additional fracture lines. Therefore, the mechanism is forced to take place starting by failure at the interface between the triangular portions of the vault (i.e. the initial NURBS surfaces on the model defined in Rhinoceros) only. It is clear that in this case the procedure of mesh adaptation through GA is not needed. Moreover, even the different orientations of bricks in each part do not condition the ultimate load-bearing capacity of the whole structure. Indeed, since each interface between the initial surfaces consists in a bed joint, a unique set of material properties, in terms of ultimate stress values, has been assigned. Material properties adopted in these analyses, reported in Table 2 (see Figure 2(a) for a clearer meaning of the symbols), are referred to current Italian technical codes and regulation. The specific value of linear cap in compression is referred to the work (Milani and Taliercio, 2015). It is worth noting that a not null value of ultimate tensile stress has been adopted, cohesion has been chosen as equal to the tensile stress value. Finally, a standard friction angle of 30° has been applied.

Results for both the analyses, in terms of live load multiplier and collapse mechanism, are reported in Figure 6. Live load multipliers are equal to 38.1 and 83.2 respectively for LC1 and LC2. It can be observed that there are no substantial differences between the two collapse mechanisms.

It is clear that the hypothesis of failure only along the changes of bricks orientation is too simplistic. Therefore, a second set of analyses has been performed. Here, each triangular portion has been

Table 1. Presentation of performed analyses.

Load case	Analysis under	Dead load	Live load
LC1	self-weight	-	Γ · self-weight
LC2	uniform vertical load	self-weight	Γ · 1 kN/m²

Table 2. Material properties assigned to interfaces between initial NURBS surfaces.

Property	Symbol	Value
Specific weight [kN/m³]	γ	18
Ultimate tensile strength [MPa]	f_t	0.2
Ultimate compression strength [MPa]	f_c	2.6
Cohesion [MPa]	c	f_t
Friction angle [°]	Φ	30
Linear cap in compression [-,°]	ρ	0.5
	Φ_2	10

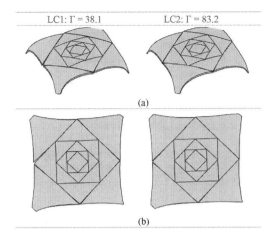

LC1: Γ = 38.1   LC2: Γ = 83.2

(a)

(b)

Figure 6. Analyses without additional fracture lines: collapse mechanisms in (a) axonometric view and (b) top view.

subdivided into elements by adopting a set of lines. The typology of subdivision lines, in terms of number and directions, has been chosen taking into account the shape of each surface and the applied load. A representation of the typology of mesh applied is depicted in Figure 7(a). Each subdivision line starts from a node on one edge and ends on another node belonging to the opposite edge, in which the position of each node on the edge is determined by a numerical variable $t_i \in [0, 1]$ (i is referred to the i-th node). Basically, if $t_i = 0.5$ the i-th node is positioned at the medium point of the edge. A regular mesh in which each value of t has been set equal to 0.5 is shown in Figure 7(b).

Since fracture lines within the triangular portions of vaults have been introduced, appropriate values of ultimate stress which take into account bricks orientation have to be applied.

A simplified orthotropic model for masonry is here adopted. In reference to Figure 8, two main directions have been defined for each NURBS surfaces: direction 1, which is parallel to bed joints, and direction 2, which is orthogonal to the first one. A different set of masonry properties (obtained starting from the standard values reported in Table 2) has been assigned to each

.... Initial NURBS surface
— — Possible mesh

Figure 7. Definition of a possible mesh: (a) schematization of the mesh for the single NURBS surfaces composing the vaults, (b) obtained regular mesh.

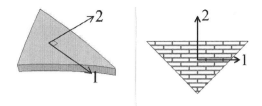

Figure 8. Simplified orthotropic model: main directions on the NURBS surfaces related to the masonry texture.

Table 3. Material properties adopted for masonry in the simplified orthotropic model.

Property	Symbol	Value
Specific weight [kN/m³]	γ	18
Ultimate tensile strength along direction 1[MPa]	f_{t1}	0.4
Ultimate tensile strength along direction 2 [MPa]	f_{t2}	0.2
Ultimate compression strength along direction 1[MPa]	f_{c1}	5.2
Ultimate compression strength along direction 2 [MPa]	f_{c2}	2.6
Cohesion [MPa]	c	f_t
Friction angle [°]	Φ	30
Linear cap in compression [-,°]	ρ	0.5
	Φ_2	10

direction and have been reported in Table 3. Given a certain fracture line, or better an interface, if the orthogonal normal vector to this interface is parallel to one of the two main directions, the assigned ultimate stress values will be that corresponding one (according to Table 3). However, if the orthogonal normal vector is directed along an intermediate direction between 1 and 2, the ultimate stress values are evaluated according to the two main components of the normal vector.

Before applying the mesh adaptation, the analyses LC1 and LC2 have been performed on the model characterized by the regular mesh and the simplified orthotropic model. Results are reported in Figure 9.

In comparison with the previous case, a significant decrease of live load multiplier is observed: for LC1 the live load multiplier decreases from 38.1 to 33.4, whereas for LC2 it changes from 83.2 to 72. The two collapse mechanisms are still quite similar to each other. Finally, in the last set of analyses, the mesh adaptation through GA has been applied. The regular mesh is adaptively modified by changing the positions of the node identifying the additional fracture lines, i.e. the variables t_i, in order to find the minimum value of the live load multiplier. Since the model derives directly from the laser scanner survey, and so

LC1: Γ = 33.4	LC2: Γ = 72.0

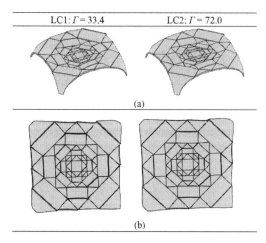

(a)

(b)

Figure 9. Analyses with regular mesh: collapse mechanisms in (a) axonometric view, (b) perspective view and (c) top view.

LC1: Γ = 32.6	LC2: Γ = 70.2

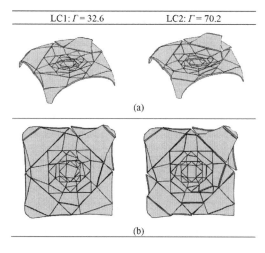

(a)

(b)

Figure 10. Analyses with mesh adaptation: collapse mechanisms in (a) axonometric view, (b) perspective view and (c) top view.

all imperfections and differences from a "perfect shape vault" have been taken into account, no symmetry conditions have been applied in the position of fracture lines. Therefore, a total of 44 variables has been determined through GA. Results are reported in Figure 10. The optimized mesh is quite different from the initial one. However, the optimized live load multipliers are just a little lower than in the previous case: indeed, the final values are 32.6 and 70.2 respectively for LC1 and LC2. Even in this case, the collapse mechanism is substantially the same for the two load configurations applied.

It is worth noting that, because of the lowered shape of the vault, crushing assumes a fundamental role in comparison with tensile and shear failures.

5 CONCLUSIONS

In the paper, the collapse behavior of masonry vaults has been investigated. In particular, attention has been focused on a typology of masonry sail vaults, built with different brick patterns, during the XIX century, that have a great diffusion in Cagliari area (Italy). The geometry of the vaults has been accurately measured through laser scanning techniques and then reconstructed with a three-dimensional NURBS-based model. A limit analysis approach based on an adaptive NURBS modeling coupled with a genetic algorithm, particularly suited for historical masonry vaults, has been performed. This allowed to evaluate the ultimate load of the structure and the associated collapse mechanisms for two different load conditions. The procedure provides fast and reliable assessment of the collapse behavior of these structures.

Due to the lowered shape of the vault and to the boundary conditions adopted, crushing assumes a fundamental role in comparison with tensile and shear failures. Other factors, such as material imperfections, heterogeneity of mortar and bricks or the presence of damages and cracks, may affect the load bearing capacity of the vaults: future advancement of the proposed procedure will take into account these aspects.

Further developments of this research are devoted to evaluate the role played by brick patterns in the construction of such vaults and to investigate sensitivity to bricks arrangement in the structural behavior, through a comparison between several cases of study.

REFERENCES

Angelillo, M., Babilio, E. & Fortunato, A. 2013. Singular stress fields for masonry-like vaults. *Continuum Mechanics Thermod.*, 25: 423–441.

Argiolas, R., Cazzani, A., Reccia, E. & Bagnolo, V. 2019. From LIDAR data towards HBIM for structural evaluation, *ISPRS Archives*, 42: 125–132.

Armesto, J., Roca-Pardiñas, J., Lorenzo, H. & Arias, P. 2010. Modelling masonry arches shape using terrestrial laser scanning data and nonparametric methods, *Eng. Struct.*, 32(2): 607–615.

Block, P. & Ochserdorf, J.A. 2007. Thrust network analysis: a new methodology for three-dimensional equilibrium, *J. IASS*, 48(3): 167–173.

Boothby, T.E. 2001. Analysis of masonry arches and vaults. *Progress in Structural Engineering and materials*, 3(3): 246–256.

Cazzani, A., Malagù, M. & Turco, E. 2016. Isogeometric analysis: a powerful numerical tool for the elastic analysis of historical masonry arches, *Continuum Mech. Thermodyn*, 28: 139–156.

Castellazzi, G., D'Altri, A.M., de Miranda, S. & Ubertini, F. 2017. An innovative numerical modeling strategy for the structural analysis of historical monumental buildings, *Eng. Struct.*, 132: 229–248.

Chiozzi, A., Malagù, M., Tralli, A. & Cazzani A. 2016. ArchNURBS: NURBS-Based Tool for the Structural Safety Assessment of Masonry Arches in MATLAB, *J. Comput. Civ. Eng.*, 30 (2): # 04015010-1-11.

Chiozzi, A., Milani, G. & Tralli, A. 2017. A Genetic Algorithm NURBS-based new approach for fast kinematic limit analysis of masonry vaults. *Computers and Structures*, 182: 187–204.

Chiozzi, A., Milani, G., Grillanda, N. & Tralli, A. 2018a. A fast and general upper-bound limit analysis approach for out-of-plane loaded masonry walls, *Meccanica*, 53(7): 1875–1898.

Chiozzi, A., Grillanda, N., Milani, G. & Tralli, A. 2018b. UB-ALMANAC: An adaptive limit analysis NURBS-based program for the automatic assessment of partial failure mechanisms in masonry churches, *Engineering Failure Analysis*, 85: 201–220.

D'Ayala, D. & Casapulla, C. 2001. Limit state analysis of hemispherical domes with finite friction. In *Structural Analysis of Historical Constructions*, Guimarães (Portugal), 2001.

Grillanda, N., Chiozzi, A., Bondi, F., Tralli, A., Manconi, F., Stochino, F. & Cazzani, A. 2019a. Numerical insights on the structural assessment of historical masonry stellar vaults: The case of Santa Maria del Monte in Cagliari. *Continuum Mech. Thermodyn*, doi:10.1007/s00161-019-00752-8.

Grillanda, N, Chiozzi, A., Milani, G. & Tralli, A. 2019b. Collapse behavior of masonry domes under seismic loads: an adaptive NURBS kinematic limit analysis approach, *Eng. Struct*, 200: 109517.

Lucchesi, M., Padovani, C., Pasquinelli, G. & Zani N. 2007. Static analysis of masonry vaults, constitutive model and numerical analysis. *Journal of Mechanics of Materials and Structures*, 2(2): 221–244.

Milani, E., Milani, G. & Tralli, A. 2008. Limit analysis of masonry vaults by means of curved shell finite elements and homogenization, *Int. J. Solids Struct.*, 45: 5258–5288.

Milani G. & and Tralli, A. 2012. A simple meso-macro model based on SQP for the non-linear analysis of masonry double curvature structures, *Int. J. Solids Struc.*, 49(5): 808–834.

Milani, G., & Taliercio, A. 2015. In-plane failure surfaces for masonry with joints of finite thickness estimated by a method of cells-type approach. *Comp. Struct., 150*: 34–51.

Napolitano, R. & Glisic, B. 2019) Methodology for diagnosing crack patterns in masonry structures using photogrammetry and distinct element modeling, *Eng. Struct.*, 181: 519–528.

O'Dwyer, D. 1999. Funicular analysis of masonry vaults, *Comput. Struct.*, 73(1-5): pp. 187–197.

Pavlovic, M., Reccia, E. & Cecchi, A. 2016. A procedure to investigate the collapse behavior of masonry domes: some meaningful cases. *International Journal of Architectural Heritage*, 10(1): 67–83.

Piegl, L. & Tiller, W. 1995. *The NURBS Book*. Berlin: Springer.

Roca, P., Cervera, M., Gariup, G. & Pelà, L. 2010. Structural analysis of masonry historical constructions. classical and advanced approaches, *Arch. Comput. Method. Eng.*, 17: 299–325.

Schueremans, L. & Van Genechten, B. 2009. The use of 3D-laser scanning in assessing the safety of masonry vaults. A case study on the church of Saint-Jacobs. *Optics Lasers Eng.*, 47: 329–335.

Sturm, J. F. 1999. Using SeDuMi 1.02, a MATLAB toolbox for optimization over symmetric cones. *Optimization Methods and Software*, 11: 625–653.

Tralli, A., Alessandri, C. & Milani, G. 2014. Computational Methods for Masonry Vaults: A Review of Recent Results. *The Open Civil Engineering Journal*, 8: 272–287.

Brick and Block Masonry - From Historical to Sustainable Masonry –
Kubica, Kwiecień & Bednarz (eds)
© 2020 Taylor & Francis Group, London, ISBN 978-0-367-56586-2

The seismic performance of heritage URM buildings: A parametric study

S. Dadras, M.J. Masia & Y.Z. Totoev
The University of Newcastle, Australia

ABSTRACT: Research is currently underway at The University of Newcastle, Australia regarding assessment of the seismic performance of the culturally significant unreinforced masonry (URM) buildings. The first part of the project has been contemplated by performing a series of finite element analyses followed by the equivalent experimental tests on two leaves thick perforated URM walls which vary in the spandrel depths and the level of pre-compression stresses. The shape of the walls and their openings are representative of heritage masonry facades used in prevalent Australian URM structures. The current paper presents a parametric study on small scale one storey URM buildings using the characteristics of the mentioned URM walls to determine the performance of the hypothetical small heritage structure under seismic actions. The various parameters used in the current study are the stiffness of floor diaphragms, the opening aspect ratio which results in shallow or deep spandrels, considering one or two-sided openings in the structure which helps study the torsion effect and the pushover loading direction. The results are provided through comparison of the related structural factors and the equivalent capacity curves.

1   INTRODUCTION

The performance of the buildings should be verified for seismic forces in earthquake-prone areas even though earthquakes are random actions, which may not even occur during the structures' lifetime depending on the seismicity of the region. Unreinforced masonry (URM) buildings are generally rigid structures with little flexibility when subjected to the high accelerations of earthquakes. Many URM buildings are treasured as heritage structures, some of which continue to serve the purposes for which they were constructed and often hold greater importance to societies than their pure functions. The structural elements of these buildings are frequently poorly interconnected and detach from each other, resulting in catastrophic collapse under seismic forces. If they fail under higher or even moderate earthquake accelerations, URM buildings are quite dangerous considering their heavy materials that can inflict serious damage. Aside from retrofit, there are two alternative options for this sort of buildings: ignore their inherent structural weaknesses as well as the probability of earthquake occurrence, or demolish them before they fail in catastrophic events. Neither of these choices addresses the main goal, which is to preserve the mentioned culturally significant structures. Therefore, it is vital to assess the safety of such structures and, if necessary, strengthen them seismically. Moreover, the lack of knowledge in the above fields requires experimental research to measure material characteristics and structural subassembly performance combined with numerical modelling of building system performance (Commission, 2012b, Commission, 2012a, Dizhur et al., 2011, Dizhur et al.,

2010, Ingham and Griffith, 2010). There are several research works in the literature which address the safety issues of heritage structures such as churches and provide a finite element approach to predict the structural damage (Milani and Venturini, 2011, Milani et al., 2018, Milani, 2013). Note that URM in the current paper defines the unreinforced masonry construction, which does not contain any reinforcing materials prior to any retrofitting.

The current study includes a recent joint project performed in The University of Newcastle, Australia on the seismic performance of the heritage URM walls and buildings. In the first part of the project, a series of nonlinear finite element analyses were performed, followed by the equivalent experimental tests on two leaves thick perforated URM walls, which vary in the spandrel depths and the level of pre-compression stresses (Milon K. Howlader, 2019). The shape of the walls and their openings are representative of heritage masonry facades used in prevalent Australian URM structures. The mentioned study covers a comprehensive two-dimensional investigation on the behaviour of the specimens such as the structural ductility and performance criteria and formation of the cracks. However, there exists a need for a detailed study on the seismic behaviour of the actual heritage buildings. The next step of the above joint project intends to extrapolate the study of unreinforced masonry walls to the relevant three-dimensional buildings, which will eventually cover the nonlinear finite element analysis of several culturally significant URM structures located in Australia and New Zealand.

This paper presents a parametric study on small-scale one-storey URM buildings using the characteristics of the mentioned URM walls to determine the performance of the hypothetical small heritage structure under seismic actions. The various parameters used in the current study are the stiffness of floor diaphragms, the opening aspect ratio resulting in shallow or deep spandrels, considering one or two-sided openings as well as adjacent openings in the structure to study the torsion effect and the pushover loading direction. The buildings are simulated numerically using finite element macro models available in the software DIANA 10.2 (Diana, 2005). Each model is simulated considering the above variables to determine their direct impact on the performance of the small-scale masonry buildings. The results from the pushover analysis of the structure include force-displacement graphs and failure modes of the structure. The structural factors are calculated for the different models and the results are compared to each other and to the available values in the literature. The factors to be studied are energy dissipation, structural ductility factor, structural performance factor and displacement capacity.

2 STRUCTURAL DUCTILITY AND PERFORMANCE FACTORS

The seismic design codes such as FEMA-440 and Eurocode 8 accept nonlinear static (Pushover) analysis as a feasible tool to analyse masonry structures due to its simplicity and efficiency (Jiang et al., 2010, FEMA, 2005, Ademovic et al., 2013). However, some standards such as NZS 1170.5 (2004) and AS 1170.4 (2007) do not explicitly reference the details of pushover analysis. An alternative method is to consider the inelastic behaviour implicitly by using a reduction factor through linear analyses to estimate in an approximate way, the results obtained from nonlinear analyses (Davidson et al., 2010, Standard, 2004, Wilson and Lam, 2007). The general equation to calculate the seismic base shear force (V) is as follows:

$$V = C_d \times W_t \ where: \ C_d = C_e / R \qquad (1)$$

The inelastic response factor C_d is calculated through the above equation. The elastic response factor C_e depends on the natural period of the structure, importance level of the structure, seismic hazard factor and the site sub-soil class. W_t is the seismic weight of the structure which equals the full dead load plus a proportion of the live load and R is the force reduction factor (Allen et al., 2013). In some building codes such as OPCM 3274 (2003) and Eurocode 8 (2004), the force reduction factor is called as behaviour factor (q) (Ordinanza, 2003, Eurocode, 1998). The seismic response curves of the actual and ideally elastic structures are displayed in Figure 1. The stiffness of the ideally elastic structure is equal to the initial stiffness of the bi-linearized

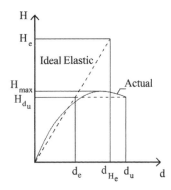

Figure 1. Structural behaviour factor (Tomaževič, 1999).

form of the actual envelope. The dissipated energy of the structure, due to its actual response, is quite high so that it is more reasonable for the structure to be designed for the ultimate design load H_{du} rather than ultimate elastic seismic load H_e (Tomaževič, 1999). According to equation 2, the behaviour factor is equal to the ultimate elastic seismic load divided by the ultimate design load extracted from the actual response envelope.

$$q = H_e / H_{du} \qquad (2)$$

The general equation for the structural force reduction factor is as follows:

$$R = R_\mu \times OSF = R_\mu / S_p \qquad (3)$$

R_μ is a component of the force reduction factor which includes the inherent global ductility factor (μ) accounting for the structural energy dissipation capacity. Ductility factor is determined as ultimate displacement (d_u) of the structure divided by the yield displacement (d_e) of the structure (Allen et al., 2013). R_μ can be calculated from the relationship between force reduction factor and ductility factor through equivalent energy principle (Paulay and Priestley, 1992).

$$R_\mu = (2\mu - 1)^{1/2} \qquad (4)$$

Overstrength factor (OSF) is a factor that represents the amount of reserve strength not considered in structural design. OSF is the maximum base shear of the entire structure divided by the equivalent strength at the first substantial crack which occurs in any component member in the structure. Seismic design codes such as NZS 1170.5 (2004), NZSEE (2006) and AS 1170.4 (2007) use $1/S_p$ rather than OSF while evaluating the force reduction factor (R). S_p is defined as the performance factor in the above standards (Standard, 2004, NZSEE, 2006, Wilson and Lam, 2007, Allen et al., 2013, Taieb and Sofiane, 2014).

3 NUMERICAL MODELLING OF THE STRUCTURE

The numerical macro model of the structure was completed in the commercially available finite element package DIANA 10.2 using the geometrical and material data obtained from the experimentally tested wall specimens mentioned above as the first part of the joint project. The hypothetical one-storey building was simulated using the same properties as for the tested walls in all four sides of the building. The reason for modelling a complete structure is to determine the seismic behaviour of the heritage masonry walls in a three-dimensional form as well as to investigate other factors such as torsion effects during a parametric study. The masonry walls have a thickness of 230 mm. The simulations were performed considering rigid and flexible diaphragms for the roof. The thickness of the diaphragm was considered as 20 mm in flexible diaphragms modelling timber structure. However, the thickness of the rigid diaphragms were equal to 200 mm modelling a single rigid concrete plate. Regarding the materials, linear elastic isotropic models were used for the concrete of the footings and for the roof beams and linear elastic orthotropic models were used for the roof plates. Regular Curved Shell elements were used to model the concrete of the footing and the masonry of the walls. Element Class-II Beams 3D were used to model the timber beams of the roof. The roof plates are made out of either timber or concrete so that Regular Curved Shell elements were used to model the plates. The soil under the building was not modelled, but a Coulomb friction boundary interface model was used at the bottom surface of the footing to represent the contact between the soil and the footing. For modelling the masonry walls, the new orthotropic Engineering Masonry model developed in DIANA 10.2 was used. This is a form of nonlinear homogenized model of the masonry composite. The diagonal stair-case cracks option was chosen as joint failure type. The material properties for the masonry of the walls are displayed in Table 1.

Even though in real buildings the diaphragms are typically lighter in the case of flexible type diaphragms, the dead load for the flexible diaphragm was taken as 5 kN/m^2 which equals the weight of the 200 mm concrete rigid diaphragm. This was done to allow direct comparison between models for flexible and rigid diaphragms. The imposed action considered for both types of diaphragm is equal to 0.6 kN/m^2 according to Australian standards. In the model, Total Lagrange type of nonlinearity was chosen for geometrical settings. The analyses were conducted by

Table 1. Parameters for the masonry walls (Engineering Masonry Model).

Property	Value	Unit
Young modulus in bed joint direction (Ex)	3.0E+09	N/m2
Young modulus orthogonal to bed joint direction (Ey)	6.0E+09	N/m2
Shear modulus (Gxy)	1.875E+09	N/m2
Mass density	1.85E+03	kg/m3
Head joint failure type - Diagonal stair-case cracks		
Bed joint tensile strength (ft)	1.0E+05	N/m2
Fracture energy in tension (GF1)	25	N/m
Compressive strength (fc)	7.0E+06	N/m2
Fracture energy in Compression (Gc)	1.12E+04	N/m
Angle between diagonal stepped crack and bed joint (alpha)	29.68	deg
Factor to strain at compressive strength (E c,fac)	4	
Friction angle (fi)	36.5	deg
Cohesion (c)	1.50E+05	N/m2
Fracture energy shear (Gsh)	20	N/m
Unloading factor (fac,unload)	1	

displacement control method using Secant (Quasi-Newton) method.

The ultimate state was defined when a post peak resistance load drop of 20% of the maximum load was reached, where the load being considered is the total base shear resistance of the building. However, the above load drop did not occur in all of the scenarios. For these cases, a maximum drift limit of 2% (48 mm in-plane lateral displacement for 2.4 m high walls) was imposed for all walls in the finite element model. The 2% limit was based on the experimental wall testing results, in which the test for any wall specimen which did not display a 20% post peak load drop was discontinued at 2% drift. Note that the displacement used in the capacity curves for each building model is equal to the average displacement of the four nodes at the top corners of the building. As an example, Figure 2(a) and (b) demonstrate, respectively, the meshed form for the one-storey building considering a flexible diaphragm, a shallow spandrel and openings in adjacent sides of the building (FX-Sh-A-PX) and for the one-storey masonry building considering a rigid diaphragm, a deep spandrel and a single opening in one of the walls (R-D-O-PX) (See Section 5).

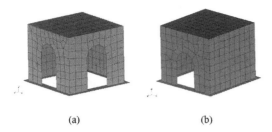

(a)   (b)

Figure 2. (a) Meshed form of scenario FX-Sh-A-PX and (b) scenario R-D-O-PX.

4 IDEALISATION METHODS USED IN THE CURRENT STUDY

To calculate the structural factors for a particular building model, it is expedient to represent the actual lateral load versus displacement response of the structure (capacity curve) by an equivalent bi-linearised response. The idealisation methods offered by Tomazevic (1999), Magenes and Calvi (Magenes and Calvi, 1997), Park (Park, 1989), and ASTM standard (Standard) are all based on the equivalent energy principle. This means that the area under the actual load versus displacement response of the masonry structure should be equal to that of the bi-linearized envelope.

Tomazevic (1999) defines the three limit states of the experimental curve; a) Crack limit (d_{cr}, H_{cr}) b) Maximum resistance of the wall and the corresponding displacement (H_{max}, d_{Hmax}) and c) Ultimate state (d_{max}, H_{dmax}). Figure 3(a) displays the limit states as well as the actual and bi-linearised responses of the wall. As shown, the ultimate displacement of the idealised response envelope (d_u) is a displacement corresponding to a resistance of $0.8H_{max}$ in the experimental response curve. The displacement at the idealised elastic limit state (d_e) is defined as the displacement corresponding to the intersection point of the two lines of the bi-linearised envelope (Toma-ževič 1999).

The displacement ductility factor is the ultimate displacement of the idealised response envelope divided by the displacement at the idealised elastic limit state as represented in Equation 5. Equation 6 calculates the overstrength factor that also provides performance factor.

$$\mu = d_u / d_e \qquad (5)$$

$$OSF = 1 / S_p = H_u / H_{cr} \qquad (6)$$

In the current study, four different methods of idealisation were employed prior to calculation of the structural factors. The first procedure (IM1) refers to Tomazevic method. In this method, the

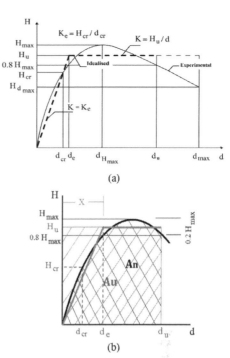

(a)

(b)

Figure 3. (a) Actual and Idealised Response Envelope (b) Details of the idealization method IM4.

ultimate line will form the second part of the bi-linearised envelope. The initial slope of the idealised envelope is determined due to equal dissipated energies. The crack limit is then considered as the intersection point of the actual and idealised responses (Figure 3(a)). The second method (IM2) is the bi-linearisation offered by Magenes and Calvi. They recommended a crack limit resistance of 70 percent of the maximum lateral load. This assumption then defines the stiffness (slope) of the first part of the bi-linearised envelope. The second part of the idealised envelope is calculated through equivalent energy principle. The third method (IM3) is the idealisation offered by ASTM standard. ASTM suggests a crack limit resistance of 40 percent of the maximum lateral load to provide the stiffness (slope) of the first part of the idealised response. The fourth procedure is a visual method (IM4) which is based on the equivalent energy principle. The ultimate resistance of the idealised envelope (H_u) is assumed equal to $0.9H_{max}$. The ultimate line is extended until a displacement corresponding to $0.8H_{max}$. The initial slope of the idealised response is then determined using Equation 7 to ensure equal areas under the idealised and pushover curves. Figure 3(b) displays the details of the IM4 idealization method. In the figure, A_n is the area under the finite element pushover curve and A_u is the area under the ultimate line.

$$A_n = A_u - H_u(d_e/\ 2)$$
$$X = d_e = 2\ (\ d_u - A_n/\ H_u) \tag{7}$$

In the visual method, the crack limit is the point of formation of the first considerable cracks in the structure. It can be also defined as a point at which the slope of the envelope changes. Therefore, the crack limits are determined visually, which will affect the value of the performance factors. The obtained structural factors will be compared and discussed in the next stage.

5 PARAMETRIC STUDY

The nonlinear static (pushover) analyses for the current study of the hypothetical one-storey URM buildings have been designed to consider the influence of variations of several parameters. The various parameters used in the current study are the stiffness of floor diaphragms, the opening aspect ratio which results in shallow or deep spandrels, considering one or two-sided openings in the structure which helps study the torsion effect and the pushover loading direction. As mentioned, the diaphragm of the masonry building is either rigid (R), flexible with a span direction parallel to the pushover load (FX) or flexible with a span direction perpendicular to the applied pushover load (FY). Another variable in the parametric study is the walls' spandrel depth. There are two options of deep (D) and shallow (Sh) spandrels used in the current paper, which will affect the pier and spandrel aspect ratios. Another parameter investigated is the torsion effect, by considering the wall openings in various directions of the building. There are three options in the parametric study: single sided opening (O), Openings in both sides of the same direction (BS) and openings in adjacent sides of the building (A). The last parameter considered in the study is the pushover loading direction which can be sorted as X direction (PX) and Y direction (PY). Considering the above variables, there will be 36 different scenarios in the current parametric study, which will provide the impact of the mentioned variables on the performance of the masonry structure. Table 2 displays the order of the different scenarios used in the current paper.

Figure 4 demonstrates the geometry of the finite element models for some of the above scenarios as 3D and plan views. As shown, the flexible diaphragms with both X and Y span directions, rigid diaphragms, different types of openings and the variety of spandrel depth are quite clear in the figures.

Table 2. Scenarios used in the current study.

FX-Sh-O-PX	FX-Sh-O-PY	FX-Sh-BS-PX
FX-Sh-BS-PY	FX-D-O-PX	FX-D-O-PY
FX-D-BS-PX	FX-D-BS-PY	FX-Sh-A-PX
FX-Sh-A-PY	FX-D-A-PX	FX-D-A-PY
R-Sh-O-PX	R-Sh-O-PY	R-Sh-BS-PX
R-Sh-BS-PY	R-D-O-PX	R-D-O-PY
R-D-BS-PX	R-D-BS-PY	R-Sh-A-PX
R-Sh-A-PY	R-D-A-PX	R-D-A-PY
FY-Sh-O-PX	FY-Sh-O-PY	FY-Sh-BS-PX
FY-Sh-BS-PY	FY-D-O-PX	FY-D-O-PY
FY-D-BS-PX	FY-D-BS-PY	FY-Sh-A-PX
FY-Sh-A-PY	FY-D-A-PX	FY-D-A-PY

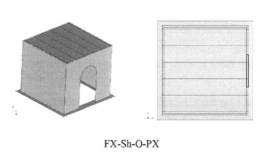

FX-Sh-O-PX

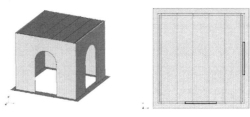

FX-Sh-A-PX

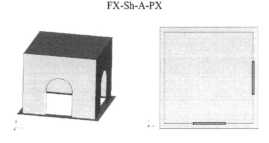

R-D-A-PX

Figure 4. 3D and plan views of simulated scenarios.

666

6 RESULTS AND DISCUSSION

The results from the pushover analyses including the displacement capacity, maximum load resistance, crack width contours, displacement contours, force-displacement graphs and failure modes of the structure were extracted some of which are reported in Table 3. The obtained capacity curves were then idealised to calculate the structural ductility and performance factors as described above. Figure 5(a) displays the meshed form of the case R-Sh-A-PY. The displacement contour of the last step of analysis is shown in Figure 5(b), which provides the maximum displacement of the structure during the analysis. The maximum crack width is obtained through

the related contour displayed in Figure 5(c). Figure 5(d) demonstrates the capacity curve and the bi-linearised response of the building through the idealisation method of Tomazevic (IM1).

Table 3 displays a list of the modelled scenarios including the sway mode of the buildings, displacement capacity, maximum load resistance, maximum crack width and the equivalent failure modes observed in the buildings during analyses. The displacement of the structures are limited to 48 mm to satisfy the maximum drift limit of 2% for any individual wall. The minimum reported displacement capacity is equal to 29 mm for the case scenario FX-Sh-A-PX. That can occur due to the torsion effect and the flexible diaphragm in this scenario. The maximum and minimum load resistances are equal to

Table 3. Comparison of the results of load, displacement and crack width.

Scenarios	Sway Mode	Disp. Capacity (m)	Max Load Resistance (kN)	Max Crack Width (mm)	Failure Mode
FX-Sh-O-PX	2	0.037	225.12	8.05	Mixed
FX-Sh-O-PY	7	0.039	223.98	5.71	Mixed
FX-Sh-BS-PX	2	0.033	218.48	10.80	Toe Crushing
FX-Sh-BS-PY	1	0.048	111.78	8.27	Toe Crushing
R-Sh-O-PX	2	0.048	223.59	11.10	No Failure
R-Sh-O-PY	1	0.048	215.28	16.40	Toe Crushing
R-Sh-BS-PX	2	0.048	227.54	24.00	No Failure
R-Sh-BS-PY	1	0.048	160.72	12.50	Toe Crushing
FY-Sh-O-PX	8	0.048	222.89	4.67	No Failure
FY-Sh-O-PY	2	0.032	178.01	10.90	Shear Failure
FY-Sh-BS-PX	8	0.048	226.85	4.55	Flexural
FY-Sh-BS-PY	2	0.048	156.47	19.40	Mixed
FX-D-O-PX	2	0.034	222.11	7.19	Toe Crushing
FX-D-O-PY	7	0.048	222.05	6.31	Shear Cracks
FX-D-BS-PX	2	0.034	220.32	10.50	Flexural
FX-D-BS-PY	7	0.048	225.34	6.40	Shear Failure
R-D-O-PX	2	0.048	222.82	10.20	Mixed
R-D-O-PY	1	0.048	220.30	21.50	Miexd
R-D-BS-PX	2	0.048	225.91	21.20	Flexural
R-D-BS-PY	1	0.048	186.30	13.30	Mixed
FY-D-O-PX	8	0.048	222.13	7.83	Flexural
FY-D-O-PY	2	0.030	193.16	10.60	Shear Cracks
FY-D-BS-PX	8	0.048	225.29	4.27	Flexural
FY-D-BS-PY	2	0.048	185.18	19.70	Mixed
FX-Sh-A-PX	2	0.030	155.08	10.90	Mixed
FX-Sh-A-PY	2	0.032	185.78	16.20	Toe Crushing
FX-D-A-PX	2	0.034	224.75	9.26	Toe Crushing
FX-D-A-PY	2	0.030	167.52	9.17	Toe Crushing
R-Sh-A-PX	1	0.048	215.13	22.10	Mixed
R-Sh-A-PY	2	0.048	202.19	30.50	No Failure
R-D-A-PX	1	0.048	222.46	16.20	Flexural
R-D-A-PY	2	0.048	213.46	34.10	Flexural
FY-Sh-A-PX	1	0.048	130.04	7.56	Flexural
FY-Sh-A-PY	1	0.048	119.22	9.86	Flexural
FY-D-A-PX	1	0.048	143.50	5.56	Flexural
FY-D-A-PY	1	0.048	125.38	6.18	Flexural

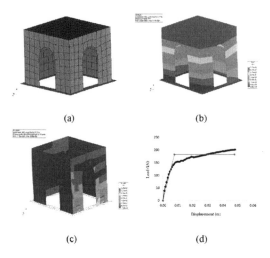

(a)   (b)

(c)   (d)

Figure 5. (a) Meshed form and (b) Displacement contour of the last step of analysis for scenario R-Sh-A-PY (c) Crack width contour of the building (d) Capacity curve and the related idealised response.

227.54 kN and 111.78 kN for the scenarios of R-Sh-BS-PX and FX-Sh-BS-PY, respectively. As seen, the diaphragm flexibility has the largest impact on the load resistance based on the results of the parametric study. The maximum and minimum values reported for the crack width in the last step of the analyses are equal to 34.10 mm and 4.27 mm for the scenarios of R-D-A-PY and FY-D-BS-PX, respectively. The maximum crack width occurs in a scenario with rigid diaphragm and adjacent openings (torsion effect). The minimum crack width occurs in the scenario with flexible diaphragm and the pushover load perpendicular to the span direction. The last column of the table displays the related failure modes. In several scenarios, mixed failure was observed which can be a combination of toe crushing, shear failure and flexural cracks occurred in the masonry units. In some cases, no failure except couple of minor diagonal step-stair cracks were observed on the walls. Shear failure occurred in the buildings with flexible diaphragms. Note that the failure modes can affect indirectly on the structural ductility and performance factors. The maximum values of the ductility factor happen in the case scenarios with no significant

failure reported in the building and the flexural failure mode. The minimum values of the ductility factor occurs in the scenarios with shear failure mode reported in the model. The maximum and minimum values of the performance factor occur in the case scenarios with the failure mode of shear cracks, flexural failure and no significant failure reported in the masonry walls, respectively.

The structural ductility and performance factors were calculated using each of the four idealisation methods. Table 4 merely covers the maximum, minimum and average values of the above structural factors for each idealisation method.

AS 1170.4 (2007) suggests the values of 1.25 and 0.77 for the ductility and performance factors of unreinforced masonry, respectively. The equivalent average values obtained via the Magenes idealisation method (IM2) are the closest to above recommendations, but all idealization methods result in R (μ/S_p) values exceeding the AS 1170.4 values. As the visual method of idealisation is based upon determining the first significant crack which occurred in the modelled structure, it can challenge the value of the performance factor recommended by Australian and NZ codes. The ductility and performance factors obtained through ASTM standard are in a good agreement with the equivalent values through visual method.

After the maximum and minimum values of the structural factors for each idealisation method are determined, the equivalent load-displacement curves are selected for comparison. Figure 6(a) compares the capacity curves belonging to the extremum values of ductility factors determined in Table 4. Figure 6(b) compares the capacity curves belonging to the extremum values of performance factors in Table 4. The scenario IDs are displayed in the figure.

As shown in Table 4, the largest reported ductility factor belonging the scenario R-Sh-O-PX is calculated through IM1 and is equal to 32. The smallest reported ductility factor belonging the scenario FX-D-BS-PY is calculated through IM2 and is equal to 1.25. The largest calculated performance factor belonging the scenario FY-D-O-PX is obtained through IM4 and is equal to 0.99. The smallest performance factor belonging the scenario FY-Sh-O-PX is calculated through IM1 and is equal to 0.14. The one-storey buildings considering

Table 4 . Structural factors calculated due to different idealisation methods.

Idealisation Method	IM1		Tomazevic	IM2		Magenes	IM3		ASTM	IM4		Visual
Structural Factors	μ	S_p		μ	S_p		μ	S_p		μ	S_p	
Max	32.00	0.95		8.98	0.78		11.11	0.55		9.72	0.99	
Min	1.54	0.14		1.25	0.57		2.41	0.40		1.49	0.17	
Average	7.56	0.56		3.42	0.70		6.18	0.47		5.85	0.48	

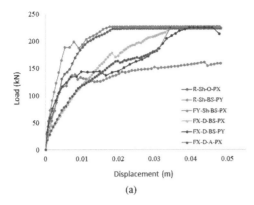

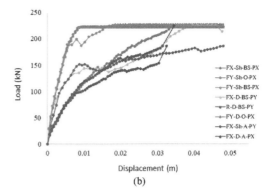

(a)   (b)

Figure 6. Capacity curves belonging to the extremum (a) ductility factors and (b) performance factors.

rigid diaphragm and shallow spandrel resulted in the largest values of ductility factor. The minimum ductility factors in all of the idealisation methods occurred in the buildings with flexible diaphragm and deep spandrel. The extremum values of performance factors through all four idealisation methods occurred in the buildings with flexible diaphragms. According to Figure 6(a), the max and min energy dissipation capacity are for the scenarios FY-Sh-BS-PX and FX-D-A-PX, respectively. Regarding Figure 6(b), the max and min energy dissipation capacity are for the case scenarios FY-D-O-PX and FX-Sh-A-PY, respectively. It is observed that most of the buildings with flexible diaphragm and shallow spandrel dissipate larger amounts of energy. In addition, the buildings including adjacent openings may dissipate smaller amounts of energy as a result of the torsion effect occurring due to asymmetry of wall stiffness. More investigations such as parametric study of a multi-storey structure are needed to accurately determine the relationship of failure modes and equivalent structural factors of heritage URM buildings.

7   CONCLUSION

This paper presents the results of a parametric study performed on small-scale one-storey URM buildings using the characteristics of experimentally tested URM walls to determine the performance of a hypothetical small heritage structure under seismic forces. The various parameters used in the current study are the stiffness of floor diaphragms, the opening aspect ratio resulting in shallow or deep spandrels, considering one or two-sided openings as well as adjacent openings in the structure to study the torsion effect and the loading direction. Nonlinear static (pushover) analyses were performed on the various case scenarios. After performing the pushover analyses and prior to the calculation of the structural factors, four different methods of idealisation were introduced and employed. The concluding remarks are as follows:

• The smallest displacement capacity is equal to 29 mm for the scenario FX-Sh-A-PX. That can occur due to the torsion effect and the flexible diaphragm in this scenario.
• The diaphragm flexibility has the largest impact on the maximum load resistance according to the results of the parametric study.
• The maximum crack width relates to a scenario with rigid diaphragm and adjacent openings. The minimum crack width occurs in the scenario with flexible diaphragm and the pushover load perpendicular to the diaphragm span direction.
• The one-storey buildings considering rigid diaphragm and shallow spandrel resulted in the largest values of ductility factor.
• The minimum ductility factors in all of the idealisation methods occurred in the buildings with flexible diaphragm and deep spandrel.
• The extremum values of performance factors through all four idealisation methods occurred in the buildings with flexible diaphragms.
• In most of the cases, the buildings with flexible diaphragm and shallow spandrel dissipate larger amounts of energy. In addition, the buildings including adjacent openings may dissipate smaller amounts of energy due to the torsion effect occurring in the masonry walls.
• AS 1170.4 (2007) recommends the values of 1.25 and 0.77 for the ductility and performance factors of URM, respectively. The equivalent average values through Magenes idealisation method (IM2) are the closest to above references, but all idealization methods result in R (μ/S_p) exceeding the AS 1170.4 values.
• The ductility and performance factors obtained through ASTM standard are in a good agreement with the equivalent values through visual method.
• The failure modes affect indirectly on the structural ductility and performance factors. The minimum of the ductility factor and the maximum of the performance factor occur in the scenarios with shear failure mode reported in their structural model.

ACKNOWLEDGEMENTS

The authors gratefully acknowledge the financial support provided by the Australian Research Council via Discovery Project DP160102070.

REFERENCES

Ademovic, N., Hrasnica, M. & Oliveira, D. V. 2013. Push-over analysis and failure pattern of a typical masonry residential building in Bosnia and Herzegovina. *Engineering Structures*, 50, 13–29.

Allen, C., Masia, M., Derakhshan, H., Griffith, M., Dizhur, D. & Ingham, J. What ductility value should be used when assessing unreinforced masonry buildings? NZSEE Conference, 2013.

Commission, C. E. R. 2012a. Final report–earthquake-prone buildings. *Canterbury Earthquakes Royal Commission, Wellington, available at:* http://canterbury. royalcommission. govt. nz/Interim-Report (accessed 1 January 2014).

Commission, C. E. R. 2012b. Final report: Volume 2: The performance of Christchurch CBD buildings. *Wellington, New Zealand.*

Davidson, B., Shepherd, R., Twigden, K., Li, X. H., Ali, M., Oyarzo-Vera, C. & Chouw, N. A Nonlinear Static (Pushover) Procedure Consistent with New Zealand Standards. 2010 NZSEE Annual Technical Conference, 2010.

DIANA, T. 2005. DIANA finite element analysis. *The Netherlands.*

Dizhur, D., Ingham, J., Moon, L., Griffith, M., Schultz, A., Senaldi, I., Magenes, G., Dickie, J., Lissel, S. & Centeno, J. 2011. Performance of masonry buildings and churches in the 22 February 2011 Christchurch earthquake. *Bulletin of the New Zealand Society for Earthquake Engineering*, 44, 279–296.

Dizhur, D., Lumantarna, R., Ismail, N., Ingham, J. & Knox, C. 2010. Performance of unreinforced and retrofitted masonry buildings during the 2010 Darfield earthquake. *Bulletin of the New Zealand Society for Earthquake Engineering*, 43, 321.

EUROCODE, C. 1998. 8 (2004). *European Committee for Standardization. Design of Structures for Earthquake Resistance-Part*, 5.

Fema, A. 2005. 440, Improvement of nonlinear static seismic analysis procedures. *FEMA-440, Redwood City.*

Ingham, J. & Griffith, M. 2010. Performance of unreinforced masonry buildings during the 2010 Darfield (Christchurch, NZ) earthquake. *Australian Journal of Structural Engineering*, 11, 207–224.

Jiang, Y., Li, G. & Yang, D. 2010. A modified approach of energy balance concept based multimode pushover analysis to estimate seismic demands for buildings. *Engineering Structures*, 32, 1272–1283.

Magenes, G. & Calvi, G. M. 1997. In-plane seismic response of brick masonry walls. *Earthquake engineering & structural dynamics*, 26, 1091–1112.

Milani, G. 2013. Lesson learned after the Emilia-Romagna, Italy, 20–29 May 2012 earthquakes: a limit analysis insight on three masonry churches. *Engineering Failure Analysis*, 34, 761–778.

Milani, G., Valente, M. & Alessandri, C. 2018. The narthex of the Church of the Nativity in Bethlehem: a non-linear finite element approach to predict the structural damage. *Computers & Structures*, 207, 3–18.

Milani, G. & Venturini, G. 2011. Automatic fragility curve evaluation of masonry churches accounting for partial collapses by means of 3D FE homogenized limit analysis. *Computers & structures*, 89, 1628–1648.

Milon K. Howlader, M. J. M., And Michael C. Griffith. Cyclic in-plane testing of simulated Australian historical perforated URM walls. 13th North American Masonry Conference, 2019. 1615–1627.

NZSEE 2006. Assessment and improvement of the structural performance of buildings in earthquakes: Recommendations of a NZSEE study group on earthquake risk buildings. New Zealand Society for Earthquake Engineering New Zealand.

Ordinanza, P. 2003. 3274. Primi elementi in materia di criteri generali per la classificazione sismica del territorio nazionale e di normative tecniche per le costruzioni in zona sismica. GU.

Park, R. 1989. Evaluation of ductility of structures and structural assemblages from laboratory testing. *Bulletin of the New Zealand national society for earthquake engineering*, 22, 155–166.

Paulay, T. & Priestley, M. N. 1992. Seismic design of reinforced concrete and masonry buildings.

STANDARD, A. E2126–11. *Standard Test Methods for Cyclic (Reversed) Load Test for Shear Resistance of Vertical Elements of the Lateral Force Resisting Systems for Buildings.*

STANDARD, N. Z. 2004. NZS 1170.5: 2004 Structural Design Actions Part 5: Earthquake actions–New Zealand. *Wellington, New Zealand: Standards New Zealand.*

Taieb, B. & Sofiane, B. 2014. Accounting for ductility and overstrength in seismic design of reinforced concrete structures. *Proc of 9th IntConf on Structural Dynamics, EURODYN*, 311–314.

Tomaževič, M. 1999. *Earthquake-resistant design of masonry buildings*, World Scientific.

Wilson, J. & Lam, N. 2007. AS 1170.4 Suppl-2007 Commentary to Structural Design Actions Part 4: Earthquake Actions in Australia. Australian Earthquake Engineering Society, McKinnon, VIC.

Brick and Block Masonry - From Historical to Sustainable Masonry –
Kubica, Kwiecień & Bednarz (eds)
© 2020 Taylor & Francis Group, London, ISBN 978-0-367-56586-2

Documentation and analysis of deformations in a historic church a few years after its conservation and reinforcement

I. Wilczyńska
Wroclaw University of Environmental and Life Sciences, Wroclaw, Poland

A. Brzozowska-Jawornicka & B. Ćmielewski
Wroclaw University of Science and Technology, Wroclaw, Poland

M. Michiewicz
Retired employee of the Building Research Institute, Warsaw, Poland

ABSTRACT: The history of the church dedicated to the Exaltation of the Holy Cross in Ząbkowice Śląskie dates back to the 13th century. Over the centuries, the building was expanded, destroyed, and experienced many construction disaster caused by human or natural factors, which resulted in its numerous reconstructions and not homogeneous structure. The passage of time and neglect of the church generated growing danger of its destruction e.g. cracks in the external walls and vaults. For this reason, in 2011-2014 the church structure was strengthened with interior renovation. Despite the performed protections, the deformation process of the walls and vaults progressed and some new cracks have been appearing. For this purpose, authors used geodetic techniques and the obtained results were compared with the former geotechnical documentation. The research also permits to determine to a certain extent the causes of deformations and pre-indicate the methods of further rescue operations.

1 INTRODUCTION

The aim of the paper is to investigate the church dedicated to the Exaltation of the Holy Cross in Ząbkowice Śląskie. We are especially interested in its recent renovations and reconstructions, in order to establish the causes of recurring damages of its structure. New geodetic measurement techniques used to control the geometry of the church, re-evaluation of its present state of preservation and re-examination of the technical documentation of the previous reinforcement works carried out in the building constitute the subject of our research.

2 CHURCH DEDICATED TO THE EXALTATION OF THE HOLY CROSS

2.1 History

The church dedicated to the Exaltation of the Holy Cross was built as a part of the Dominican Monastery in Ząbkowice Śląskie, a small town in the southern part of Lower Silesia. According to a local legend the erection of the church started in 1221 (Werner 1926), although the first document confirming its foundation is dated to 1302 (Eysymontt & Gliński 2015).

In 1428 the church as well as the monastery were heavily destroyed during the Hussite Wars (Werner 1926) which resulted in their reconstruction conducted in the third quarter of the 15th century. The church burned again in 1474 (Eysymontt & Gliński 2015) and was rebuilt at the turn of the 15th and 16th centuries (Werner 1926).

The dissolution of the Dominican order in Ząbkowice Śląskie in the mid-16th century resulted in acquisition of the convent by the city council. The church was given to the evangelical community (Eysymontt & Gliński 2015).

The Dominicans recouped the convent in 1629 (Kopietz 1885). Soon afterwards a huge fire destroyed the town including the monastery (Werner, 1926) and its consecutive reconstruction lasted until 1669 (Eysymontt & Gliński 2015). During the restoration works in 1688 the foundations slipped causing a collapse of the church western part. Another reconstruction started and was continued until 1714 (Kopietz 1885) including building of a new tower (Eysymontt & Gliński 2015). The latter one and the vaults of the nave were soon (1737) struck by an earthquake only to be followed by a fire caused be a lightning (1757). A new reconstruction started almost immediately (1738), although the tower was rebuilt in two phases, the second one was finished almost a century later (1832) raising the tower and crowning it with a new spire (Werner 1926). The last significant reconstructions before the II World War took place in 1903 (Eysymontt & Gliński 2015) and

1926, the latter one resulted from the catastrophic technical conditions of the vaults (Kramnik 2010).

After the war (1946) the convent was given to the Order of Poor Clares of Perpetual Adoration (PCPA). During next 70 years some repair and maintenance works were undertaken several times, every time caused by problems with cracked walls and vaults. The last extensive reinforcement of the church structure took place in 2011-2014 (Kramnik 2010, 2011), described in more detail below in chapter 3.

2.2 Location

The church dedicated to the Exaltation of the Holy Cross and the convent of the Poor Clares of Perpetual Adoration are located in Ząbkowice Śląskie in the southern part of the Dolnośląskie Voivodship (Poland), on the Sudeten Foothills (Faryna-Paszkiewicz et al. 2001).

The monastery lies in the centre of Ząbkowice, northeast of the Market Square, next to the medieval city walls. The church is located within the convent enclosure, which consists of several buildings arranged around a courtyard. The church constitutes its southern wing, the monastery buildings are the eastern and northern wings, and the eastern side is closed by a wall with a gate leading from the street to the convent.

The area where the monastery complex is built falls eastwards: the street level on the western side of the buildings lies much higher than the monastery gardens located on the eastern side of the church.

Several geological boreholes (Figure 1) were made to establish the soil structure.

The geological cross-section showing the structure of the ground in the area of the convent and the church is presented in Figure 2.

The geological foundation of the convent consists of five layers: 1) anthropogenic embankments; 2) brown silty clay; 3) sandy, gray-brown clays with interbedding of clayey sands with gravel and stones; 4) sandy bottom clay brown, gray, dark gray and 5) medium

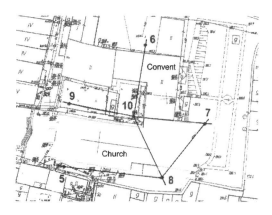

Figure 1. Spatial distribution of geological boreholes (after Sandecki 2009).

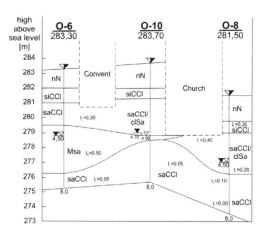

Figure 2. Geological cross-section between boreholes 6–10–8 (after Sandecki 2009).

sands, moderately concentrated with an admixture of gravel fraction. The last layer was found only next to the north-eastern corner of the church. The water table lies 3.8 to 6.5m below the ground level and may vary around +0.2m to -0.8mdue to seasonal changes (Sandecki 2009).

2.3 Description

The building constitutes a galleried hall church. It consists of three clearly distinguished parts (Figures 3, 4). A steep gable roof covers the main body of the church which comprises a central nave with two galleried aisles at each side. A tall tower surmounted by a spire rises on the west side of the nave in the central part of the main façade. On the eastern extension of the nave there is a presbytery topped with a hip roof.

The nave, the side aisles and the galleries consist of five spans, each one covered with a groin vault and separated from each other with a transverse arch. The vaults rest on outer walls and inner massive pillars. The latter are decorated with double Ionic pilasters surmounted by sections of classical entablature. The presbytery is designed in the same manner consisting of three spans with the extreme eastern one separated by a richly decorated altar to create a closed Oratorio of Perpetual Adoration.

The decoration of the outer faces of the church external walls is classical and rather modest with vertical composition consisting of multiplied pilasters supporting the cornice and eaves. Between pilasters there are windows of two heights vaulted with semicircular arches. The architectural decoration is underlined by plain white plaster contrasting with large sections of gray-pink textured plaster.

The church is built over stone and brick foundations which lie on the second geological layer mentioned above, i.e. brown silty clay. Only the north-eastern corner of the church rests on the deeper fifth layer consisting of medium sands, moderately concentrated

Figure 3. The church dedicated to the Exaltation of the Holy Cross in Ząbkowice Śląskie (phot. A. Brzozowska-Jawornicka).

Figure 4. The interior of the church with the steel tie-rods in the nave (phot. A. Brzozowska-Jawornicka).

with an admixture of gravel fraction. The foundations reach from 1.7m to 2.1m below the ground level – high above the water table (Miller 2009, Sandecki 2009).

The walls, pillars and vaults are made of brick with lime mortar. The timber roof truss is covered with metal roofing.

3 FORMER RECONSTRUCTIONS

3.1 *Evaluations of the church technical conditions*

As mentioned above the church dedicated to the Exaltation of the Holy Cross experienced many construction disasters, some made intentionally by men (sieges, battles), others resulting from faulty design (foundation slipping) or natural causes (fire, earthquake, weak ground). Such circumstances have brought about many reconstructions of the building during his long history. The last extensive works strengthening the church started in the beginning of the 21st century and lasted until 2014.

At first (2002) the original roof tiles were replaced with metal roofing. During these works large amounts of debris were removed from the spandrels of the vaults. After few years traces of destruction appeared in several areas of the church, among other the increasing number of new cracks and enlarging of the existing ones. Several research works were performed to establish the condition of the church and the causes of its destruction.

In 2009 three technical expertises were prepared. Firstly, a geological research revealed the characteristic of the ground where the monastery is located (mentioned above: point 2.2 Location, (Sandecki 2009). Secondly, a set of 16 new benchmarks was installed on the external walls (8) and inside pillars (8) of the church. Precise levelling measurements were conducted with a control measure of the "stable" benchmark No. RP2019 fixed on the church façade (Halarewicz 2009). The measurements were repeated several times in 2009 and their results proved that there were no changes in the height of the benchmarks (unpublished documents from the convent archives). Thirdly, a precise stock-taking and an evaluation of the church state of preservation was elaborated (Miller 2009), including the installation of 37 crack meters fixed in the sensitive places in order to monitor the biggest cracks observed on the walls and vaults of the church.

Those three studies constituted the basis for the fourth technical expertise combined with an extensive construction and conservation project of the strengthening of the church structure, especially its vaults. The latter was divided into two phases and took place in 2011-2014 described below (Kramnik 2010, 2011).

All three evaluations of the church condition (Miller 2009, Kramnik 2010, 2011) agreeably indicated the basic damages observed in the building:

1) vertical cracking of the transverse and longitudinal external walls, semi-circular arches above windows and under the window sills. The cracks were characterized by different widths, the biggest ones were observed in the northern and eastern walls of the presbytery. All of them were visible in the church interior and on the attic, not on the outer surfaces of the walls.
2) horizontal cracking of the vaults and transverse arches which were observed in several lines: in the axis of the vault of the nave and the presbytery; and parallel to them symmetrically on both sides of the spans the so-called Sabouret cracks at a distance of about 1/6 of the width of the vaults from the outer edges. The horizontal cracks were visible from the inside of the church and on the ridges of the vaults in the attic. Observation of the latter ones, which are not plastered unlike the interior of church, permitted to establish that the cracks run along the joints and the bricks of the vaults were not fractured. It proved that the mortar joining the vaults had completely lost its binding properties and was only filling the space between the bricks (Kramnik 2010).

The expertise prepared in 2010 included repeated surveying measurements of the church interior. They revealed severe deformations of the church structure. The north-eastern corner of the presbytery deflected north by 11.5cm in relation to the equivalent point on the floor. All the transverse arches of the nave vaults had decreased with increasing deflection towards the presbytery: the extreme western one by 11cm, the extreme eastern one by 18cm (Kramnik 2010).

The technical expertise included also presumed causes of the church damages. The first one (Miller, 2009) pointed to the lack of proper consolidation between the transverse and longitudinal external walls and the lack of appropriate anchoring of the roof truss in the perimeter walls.

The authors of the second and third expertise (Kramnik 2010, 2011) did not agree with the previous ones (Miller 2009) about the bad impact of the roof truss on the church structure. They indicated the foundation of the north-eastern corner of the church on a slope with a large angle of inclination and weak ground (sands) flowing over a deeper layers of clay as the major factor of the destruction. The poor quality and weak condition of the mortar and the unfavourable curvature of the transverse arches (namely they are too low in relation to their span) constituted other suggested reasons of the damages.

All the authors (Kramnik 2010, 2011; Miller 2009) agreed that removing debris from the vaults relieved their spandrels and in consequence violated the statics of the whole structure.

3.2 Reinforcement and reconstruction of the church in 2011-2014

On the basis of the expertise described above a complex construction and conservation project of

Figure 5. The attic with the steel tie-rods installed crosswise over the vaults (phot. A. Brzozowska-Jawornicka).

strengthening the church structure was prepared and executed in two phases in 2011-2014.

The reinforcement of the church vaults constituted the major task of the first stage. The protective work consisted of:

1) the implementation of steel tie-rods and anchors fixed crosswise over the vaults and in the longitudinal and transverse axes of the transverse arches of the nave, galleries and the presbytery (Figures 4, 5);
2) the vault and walls repairs by crack stitching with helical stainless steel reinforcing bars bonded into grooves cut across the cracks;
3) the interior of the church was plastered and painted after the reinforcement of the vaults and walls.

The stiffening of the perimeter walls of the church was the second phase of the project. Two rims were installed in the outer walls of the whole church from the outside on two heights: the reinforced concrete one above and the steel one below.

4 PRESENT STATE AND THE RESULTS OF THE NEWEST RESEARCH

4.1 Current state of church

Unfortunately despite the performed protections the deformation process of the walls and vaults of the church dedicated to the Exaltation of the Holy Cross progressed resulting in new cracks, clearly visible in the freshly renovated interior. Such circumstances made the Poor Clares of Perpetual Adoration look for further help to protect the church.

A new research conducted by the authors of the paper was initiated in 2017 in order to firstly establish the pace and causes of the damages of the building, and secondly indicate possible methods of further rescue operations to stop destruction, preserve the church and enable its safe use. Repeated evaluation of the church condition revealed the presence of new cracks in places of the previously fixed ones described in details in point 3.1, primarily on

Figure 6. The loosen steel tie-rod installed in a transverse arch of the northern gallery (phot. A. Brzozowska-Jawornicka).

the vaults. Moreover a few anchors fixing the steel tie-rods in the transverse arches of the northern gallery have loosened (Figure 6). It seems that they were torn out of the pillar.

4.2 Geodetic survey

According to the standards, geodetic deformation measurements may be conducted using the following methods (Bryś & Przewłocki 1998): polar, orthogonal, angular, linear and angular-linear resection, satellite (GNSS), geometric levelling, trigonometric levelling, satellite levelling. Nowadays also another modern spatial technique becomes popular – the terrestrial laser scanning. For deformation analysis we can also use non-geodetic methods. Such relative observations can be done using: feeler gauges, inclinometers, clinometers.

As it was mentioned above a network of control points was fixed on the study object in May 2009 (Figure 7) by Mr. Halarewicz (Halarewicz 2009). The next measurements epochs were performed in the same year relevant only to the benchmark on the tower (RP2019), which constitutes a substantive and methodical error. Since 2017 authors try to conduct regular analyzes of the church deformation. In 2018 we also fix on the object inclinometers used to relative survey method.

4.2.1 Precise levelling

The geometrical levelling is the most common method for building deformation analysis – in particular precise levelling. This method permits to observe deformation on the sub-millimetre level. Significant deformation values can start on level of 0,2 mm depending on used equipment.

The conducted surveys are referenced to two benchmarks – the base class located on the tower of the city hall and the other on old post office building. Both points are in range of 250 m.

The results of the survey deformation analysis are summarized in Table 1 with the average error of calculated deformations on a level of 0,2 – 0,3 mm.

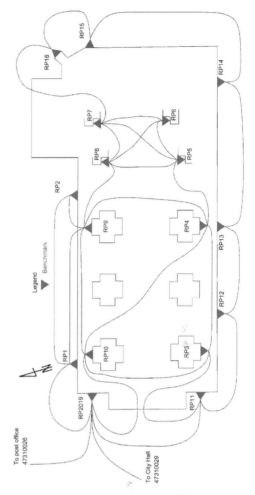

Figure 7. Network of benchmarks fixed on the church.

Table 1. Results of deformation analysis based on precise levelling [mm].

	2017.10	2018.03	2018.07	2018.08
Rp2019	-6,00	-5,65	-5,50	-4,99
Rp1	-1,70	-1,75	-1,55	-0,60
Rp2	-4,80	-3,64	-4,30	-3,41
Rp3	-4,90	-4,51	-3,68	-3,61
Rp4	-5,79	-5,39	-4,32	-4,45
Rp5	-4,56	-4,50	-3,80	-3,96
Rp6	-5,66	-5,27	-3,97	-3,74
Rp7	-4,68	-5,24	-4,00	-3,92
Rp8	-4,92	-4,61	-3,58	-3,70
Rp9	-4,11	-4,02	-3,05	-2,90
Rp10	-4,60	-3,84	-3,30	-3,07
Rp11	-6,70	-6,36	-5,95	-5,54
Rp12	-4,20	-4,99	-3,40	-2,49
Rp13	-5,90	-5,59	-4,50	-3,63
Rp14	-7,75	-6,71	-7,65	
Rp15	-5,11	-8,13		
Rp16	-10,56	-7,84	-6,10	-5,25

If we analyse the deformation results in the period 2017–2018 average vertical deformation is 1,5 mm in uplift. This situation can be explained by the process of cleaning the vault pockets when the church was reinforced.

4.2.2 Inclinometers

The inclinometer was another survey method used in the church. In period of one month we installed four high precise inclinometers (with capability of registration moves of 1 mm at 100 m). The localization of the equipment with the deformation observer in two epochs (beginning and end) of the inclinometer survey presents Figure 8.

Such observations also provide some useful information about the structure of the object. The displacement of the one month interval is presented on Figures 9-12.

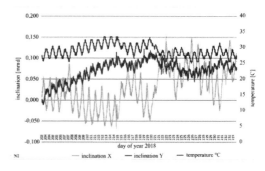

Figure 9. Plot of deformation observation from inclinometer N2.

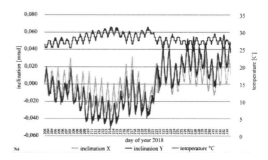

Figure 10. Plot of deformation observation from inclinometer N4.

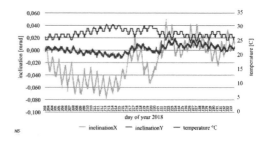

Figure 11. Plot of deformation observation from inclinometer N5.

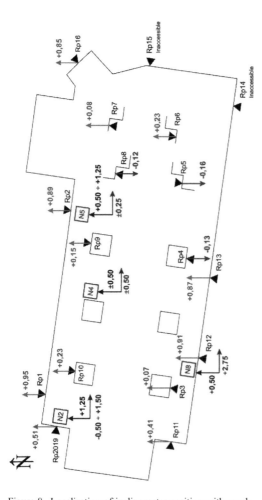

Figure 8. Localization of inclinometer position with graphical interpretation of vertical displacements in the period of 23/07/2018 and 23/08/2018.

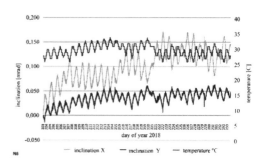

Figure 12. Plot of deformation observation from inclinometer N8.

Plots from Figures 8-11 presents day/night work of object but if we calculate the trend it also indicates movement corelated with results from precise levelling.

5 CONCLUSIONS

It turned out that the extensive reinforcement of the church dedicated to the Exaltation of the Holy Cross in Ząbkowice Śląskie, including its walls and vaults, did not bring the expected results as afterwards new signs of destruction appeared. It seems that without an accurate diagnosis of the complex state of the church it is impossible to stop its progressive deformation as evidenced by constantly reappearing cracks. Such circumstances make the situation of the church very difficult, as the lack of proper actions can lead to its closure caused by the danger, in the best case, or in the worst instance to a construction disaster.

In our opinion the sloping area where the church and monastery are located and weak geological foundations, especially the sands under the north-eastern corner of the presbytery constitute the basic and at the same time the most difficult problem to solve. Without strengthening the ground under the building and its foundations all other methods are ineffective, as proved by the reappearance of the cracks. From the technical point of view the location of the monastery on the slope makes any attempts of reinforcement very difficult.

ACKNOWLEDGMENTS

We are also thankful to Poor Clares of Perpetual Adoration for their support, providing the facility for research and sharing the collected materials, papers, expert opinions, and photos.

REFERENCES

Bryś H., Przewłocki S., 1998, Geodezyjne metody pomiarów przemieszczeń budowli, Wydawnictwo Naukowe PWN, Warszawa 1998.

Eysymontt, R. & Gliński, R. 2015. Badania architektoniczne przy zabytku: podominikański zespół klasztorny, ob. Mniszek Klarysek od Wieczystej Adoracji w Ząbkowicach Śląskich wpisanego do rejestru zabytków decyzją z dnia 05. 09.1960r. pod numerem A/1725/730, access: archives of the Order of Poor Clares of Perpetual Adoration.

Faryna-Paszkiewicz, H., Omilanowska M. & Pasieczny R. 2001. Atlas zabytków architektury w Polsce, PWN Warszawa.

Halarewicz, J. 2009. Protokół założenia reperów i określenia ich wysokości oraz poziomej pracy obiektu, maj 2009r., access: archives of the Order of Poor Clares of Perpetual Adoration.

Kopietz, J.A. 1885, Kirchengeschichte des Fürstentums Münsterberg und des Weichbildes Frankenstein, Frankenstein.

Kramnik, J. 2010. Ekspertyza techniczna wraz z projektem budowlano-konserwatorskim remontu i zabezpieczenia konstrukcji sklepień ceglanych i ścian kościoła pw. Podwyższenia Krzyża Św. w Ząbkowicach Śląskich, Wałbrzych, wrzesień 2010r., access: archives of the Order of Poor Clares of Perpetual Adoration.

Kramnik, J. 2011. Wzmocnienie konstrukcji budynków klasztoru Mniszek Klarysek od Wieczystej Adoracji ze stabilizacją wieży i prezbiterium kościoła przyklasztornego pw. Podwyższenia Krzyża Św. w Ząbkowicach Śląskich, Wałbrzych, grudzień 2011r., access: archives of the Order of Poor Clares of Perpetual Adoration.

Miller, J. 2009. Ekspertyza Techniczna wraz z inwentaryzacją okreslająca stan techniczny budynku kościoła pw. Podwyższenia Krzyża Św. w Ząbkowicach Śląskich, Wrocław, listopad 209r., access: archives of the Order of Poor Clares of Perpetual Adoration.

Sandecki, J. 2009. Dokumentacja Geotechniczna dla budynku Ratusza Miejskiego, Kosciola Rzymskokatolickiego pw. Podwyższenia Krzyża Św. I Klasztoru Sióstr Klarysek opracowana w ramach zadania pn. "Remont Ratusza Miejskiego – etap I" w Ząbkowicach Śląskich, Wrocław, lipiec 209r., access: archives of the Order of Poor Clares of Perpetual Adoration.

Werner, W. 1926. Das Dominikaner-Kloster zum Heiligen Kreuz in Frankenstein Schl.: ein Beitrag zur Geschichte der Evangelischen Stadtpfarrkirche, Frankenstein i. Schles.

Brick and Block Masonry - From Historical to Sustainable Masonry –
Kubica, Kwiecień & Bednarz (eds)
© 2020 Taylor & Francis Group, London, ISBN 978-0-367-56586-2

Benefits of casein additives in historic mortar mixes

K. Falkjar, J. Erochko & M. Santana
Carleton University, Ottawa, Ontario, Canada

D. Lacroix
University of Waterloo, Waterloo, Ontario, Canada

ABSTRACT: Casein, an organic milk protein, was used extensively in masonry mortars centuries ago. It was understood to improve workability, however, little information is available on its effects on material strength. Casein is non-toxic, has minimal environmental impacts, and has no reported health hazards.

Mortar prisms were cast for compressive, tensile, flexural and shear tests. Brick-to-mortar bond strength tests were conducted in flexure and in shear. It was found that adding 0.5% casein by mass yielded a flowable mortar, however, a 75% reduction in strength resulted. Contrarily, brick-to-mortar bond strength substantially improved.

In mortar repair, it has been deemed critically important to appropriately specify mortar with a material strength compatible with both the brick and the original mortar. Casein protein acted as a superplasticizer if 0.5% by mass was added. The application of casein mortar is plausible for repointing work. It may also be injected between wythes to stabilise a wall.

1 BACKGROUND

Casein is a protein naturally occurring in bovine milk; it is divided into three protein subgroups: α-casein, β-casein and κ-casein. Further subgroups of α-casein exist (Atamer, et al., 2017). Bovine casein typically contains 48% α-casein, 34% β-casein and 15% κ-casein (Atamer, et al., 2017). The chemical structure, comprising of a series of amino acids, is the same except for the number of phosphate groups present (Bian & Plank, 2013). Casein is typically extracted from milk by means of acid precipitation: casein precipitates at a pH less than 4.6 (Bian & Plank, 2013), is partially insoluble near a neutral pH—κ-casein is soluble while the other types of casein are insoluble (Zittle & Custer, 1963)—and dissolves completely at a pH greater than 10.0 (Post, Arnold, Weiss, & Hinrichs, 2012). Lime mortar, for which the main constituent is calcium hydroxide, has a pH of 12.6, equivalent to that of a saturated calcium hydroxide solution, therefore, casein dissolves in the mortar and has been shown to increase the workability of the mortar and to improve adhesive properties (Asselin-Boulanger, 2018). In particular, the anionic phosphate component causes the protein to absorb calcium existent in masonry mortar, and thus behave as a superplasticiser additive (Zittle & Custer, 1963). It has been used to increase the workability of mortars since medieval times in many parts of Europe (Ince, 2012) (ASTM International, 2018). It was also used in glues and adhesives in the 19th century (ASTM International,

2013), and is still used to the present day as an additive in certain paints.

Past experience has shown that in many cases, modern codes are inadequate to be applied as written on heritage masonry structures as they are intended for new construction (Canadian Standards Association, 2004) (ASTM International, 2016). Casein mortar had recently been proposed on a mortar conservation project in Canada for masonry wall repair, however, it was not used due to a lack of knowledge of the behavior of material and structural properties (Trischuk, 2018). The casein protein is a plausible candidate for improving the workability in structural rehabilitation applications without adversely affecting the properties of mortar. Furthermore, the cost of casein protein is minimal as it is used in only small quantities (less than 1% by mass) (Trischuk, 2018).

The overarching aim of the current research is to investigate the effects of incorporating casein in mortar for conservation projects and its effects on structural properties. Whereas casein has shown to be a plausible alternative for conservation projects related to masonry wall repair, the lack of research on the mechanical properties beyond the improved workability are scarce. The effects of casein content on the mechanical properties such as compressive, tensile, shear, and flexure of in mortar specimens investigated. Furthermore, the effect of casein on the brick-to-mortar bond performance was investigated through flexure and shear bond tests using brick units. Preliminary results and trends are presented.

2 EXPERIMENTAL PROGRAM

2.1 Overview of experimental program

The experimental program investigating the effect of casein concentration on the structural properties of mortar and bond strength of brick units was divided into two distinct phases, namely: 1) casein additive with a constant water content, 2) casein additive with varying water content. A total of over 500 and 20 mortar specimens and brick stack assemblies were tested, respectively.

The structural properties were evaluated by conducting compression, tension, shear, and bending tests on mortar specimens while flexural and shear bond tests were conducted on brick units.

2.2 Description of materials

The casein sample used for the experiments was a pure Micellar Casein, produced by NKD Nutrition. It was produced by means of milk extract, with no further chemicals added. The mortar used for all mortar samples in the first stage of experimentation was a pre-mixed natural hydraulic lime mortar named HLM-500 (Asselin-Boulanger, 2018), produced by King Masonry Products. The product is a 3:5 ratio by volume of Lime to Masonry Sand (Asselin-Boulanger, 2018). The published compressive strengths from the manufacturer are 1.1 MPa after 7 days, 2.2 MPa after 28 days, 4.3 MPa after 90 days, and 5.5 MPa after 365 days (Asselin-Boulanger, 2018).

The bricks used were clay bricks intended for repair of masonry walls on heritage buildings, manufactured by the Watsontown Brick Company. These bricks are intended to be used in conjunction with lime mortars, as was used in testing. The structural properties of the brick unit in itself were outside of the scope of the research as the focus is on the mortar strength due to the added casein.

2.3 Description of test matrix

The mortar specimens and bonds used in the first phase, as shown in Table 1, used a constant 5.5 L water to 30 kg lime-mortar while varying the casein content. The numbers in the table represent the number of tests conducted for each type of test and casein content. Due to the poor quality, influenced by the mix resulting from the casein, it was not possible to test a consistent number of specimens throughout.

During the second phase, a constant casein content of 0.5% was used with the intent of varying the water to lime-ratio to see the effect it has on the structural properties of the mortar and brick units.

2.4 Preparation of specimens and mortar mixes

All specimens were kept in a high-humidity environment, either by using a humidification cabinet or by completely covering the mortar specimens with wet

Table 1. Summary matrix of all tests conducted.

Test Type	Time (Days)	Casein to mortar weight ratio (%)				
		0	0.1	0.25	1	1.5
Compression	7	3	3	5	3	3
	28	5	4	4	-	5
	56	4	4	4	-	3
	112	3	3	3	-	
	365	8	7	9		
Tension	28	4	4	4	-	5
	56	4	5	5	-	3
	112	3	3	3	-	-
	365	4	5	5		
Shear	28	8	6	5	-	12
	56	8	9	8	-	6
	112	4	5	6	-	-
	365	6	6	8	-	
Bending	28	4	4	5	-	-
	56	3	4	3	-	2
	112	1	3	5	-	-
	365	4	4	4		
Shear bond	56	2	2	3	-	
Flexural bond	56	5	4	4	-	

* No tests were conducted for cells containing hyphens

Table 2. Summary matrix of all tests conducted. The number in the cell represents the number of repeats for each test.

Test Type	Time (Days)	Water used in mix for 0.5% casein per weight ratio (L/30 kg)				
		5.5	5.0	4.5	4.0	3.5
Compression	7		3	3	3	
	28	6	3	3	3	3
	56	4	3	3	3	3
	112	3		3		
	365	6				
Tension	28	5	3	3	3	3
	56	4	3	3	3	3
	112	4		3		
	365	5				
Shear	28	12	6	6	6	6
	56	10	6	6	6	6
	112	3	6	6	8	4
	365	7	5	4		
Bending	28	4	3	3	3	3
	56	3	3	3	3	3
	112	2	3	3	4	3
	365	4	3	4		
Shear bond	56	3				
Flexural bond	56	7		2	1	

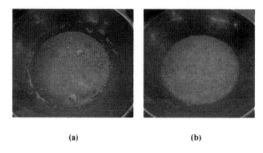

(a)　　　　　　　　　　(b)

Figure 1. Adding casein to water: (a) Before mixing, (b) After mixing. Photo by author.

towels. Inside the high-humidity chamber, the mortar cubes were kept in a saturated lime water solution, and the brick and mortar stacks were covered in moisture-proof bags.

The amount of casein was measured using a centigram balance, and the water was measured using either a graduated cylinder or a balance, based on a density of 1000 kilograms per cubic metre.

The casein protein was added to the water first as shown in Figure 1 and mixed thoroughly by hand as shown in Figure 1.(a), such that the final casein/water mix was as shown in Figure 1.(b). This casein-and-water mix was then added to the mortar, which was measured before the addition of the water and casein mix in a bucket using a decigram balance.

A paddle mixer was used for those mixes involving more than 5 kilograms of mortar. For preliminary test mixes where only a small quantity of mortar, less than 2 kilograms, was needed, mixing was conducted by hand. Mixing was continued until the mortar was visually a near-homogeneous consistency throughout. If it was discovered during pouring that some mortar was not mixed, all poured mortar was returned to the mixing container and re-mixed until a visually homogeneous consistency was obtained.

2.5　Description of test setups

Compressive Strength testing was conducted in accordance with ASTM Standard C109, as shown in Figure 2. This Standard prescribed cubes of 50 millimetres in each Cartesian coordinate direction to be tested by means of a direct compression test (ASTM International, 2012). A minimum of three cubes were to be tested in compression for each time duration and for each different casein content (ASTM International, 2012).

Tensile tests were conducted using the Double-K method outlined by Ince (Ince, 2012). As with the compression test, mortar cubes 50 mm dimensions in each Cartesian coordinate direction were cast. While no dimensions were specified, 50 millimetres was chosen to maintain the same casting procedure and formwork as used for the compression samples. A minimum of three cubes were to be tested in tension for each curing duration and for each different casein content (ASTM International, 2012). The difference between this and the compression tests was the construction of an angled support bracket, applying pressure no more than 10 millimetres from the corner of the mortar cube along opposite edges, initiating a horizontal transverse tensile stress across the centre of the mortar cube. The support system typically uses two brackets comprising of a plate and small angle welded to each other, resembling the letter K, hence the name "Double-K Method". Due to the construction tools available, it was easier to construct an equivalent support bracket out of a solid steel block, maintaining the 10 mm angle dimensions and requiring no welding. See Figure 3 for the experimental test as conducted.

Flexural mortar tests were conducted using the method outlined in ASTM Standard C348 (ASTM International, 2018), as shown in Figure 3.(a). A 1:1:4 prism ratio was used to ensure that bearing stresses through the specimen were not a concern. The dimensions were 160 millimetres in length, 40 millimetres in width and 40 millimetres in height. The span distance was 130 millimetres between supports, with the load exerted 65 millimetres from each support, as shown in Figure 3.(a). The load was applied by the testing machine to the top steel block, which placed a concentrated point load on the top surface at midspan. The supports created counteracting point loads at the edges of the specimen. This caused a bending failure propagating from the underside of the specimen at midspan.

(a)　　　　　　　　　　(b)

Figure 2. Cube tests: (a) Compression, (b) Tension. Photo by author.

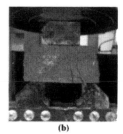

(a)　　　　　　　　　　(b)

Figure 3. Prism tests: (a) Flexure, (b) Shear. Photo by author.

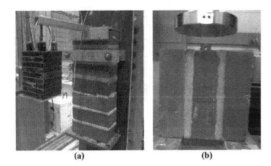

(a)   (b)

Figure 4. Brick unit tests: (a) Flexure, (b) Shear. Photo by author.

Shear mortar prism tests were conducted by means of a direct shear test as shown in Figure 4. The supports were flat surfaces, with a minimal distance—less than 5 millimetres—between facing supports. For the purposes of calculations, exact dimensions in the horizontal plane are not critical. The load was applied by the testing machine to the top steel block, which placed a uniformly distributed load across the mid-surface. The supports also placed a distributed load at the edges of the specimen. This caused a shear failure along the diagonal cracks shown forming in Figure 4.

Flexure bond testing was conducted by means of a mechanical bond wrench tool in accordance with ASTM Standard C1072 (ASTM International, 2013). The primary variation was that instead of a free-standing apparatus, an apparatus attached to a fixed steel column was used, given the conditions available in the structural testing laboratory. The lever arm acted as a clamp supported by the topmost brick. The load from the applied brick is transferred to the top brick. The clamping action induces a downward force on the front of the brick column while simultaneously inducing an upward force on the rear of the brick column.

Shear bond testing was conducted by means of a direct shear test. While the ASTM standards do not cover shear strength as it is not a test used in modern design codes (Canadian Standards Association, 2004), pertinent specimen preparation methods outlined in the ASTM flexural test were followed as closely as possible (ASTM International, 2013). A stack of three bricks was used, with mortar joints being made flush with the face of the bricks as much as possible. The ASTM C1531 shear test of masonry in situ (ASTM International, 2016) uses this method to determine shear bond strength. The load was applied to a small point atop the middle brick in the stack as shown in Figure 4.(b), while the two outer bricks were supported from below.

3 EXPERIMENTAL RESULTS

3.1 *Phase 1*

A 7-day test was conducted in compression. Tests were conducted for all four modes at 28 days, 56 days, 112 days and 365 days. The strength development

curves are shown below in Figure 5 through Figure 8. The compressive strengths were found to be similar to the manufacturer's published values (Asselin-Boulanger, 2018).

3.2 *Phase 1A: Variation in casein content*

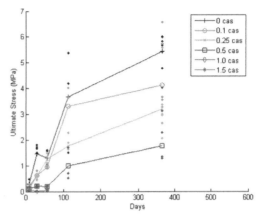

Figure 5. Compressive strength development.

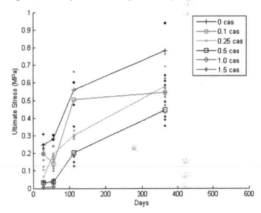

Figure 6. Tensile strength development.

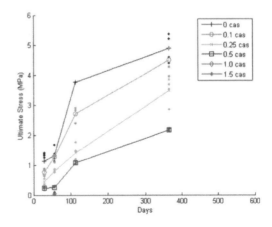

Figure 7. Flexural strength development.

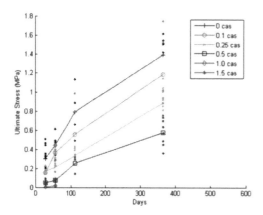

Figure 8. Shear Strength Development.

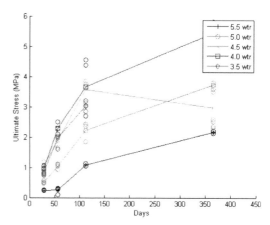

Figure 11. Flexural strength development.

3.3 *Phase 1B: Variation of water content*

Note that 'wtr' in Figure 9 through Figure 12 represents the number of litres of water per 30 kilograms of mortar. A limited number of tests were conducted after 365 days.

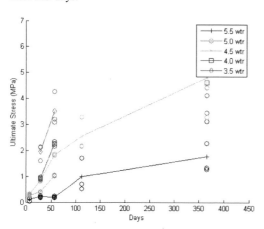

Figure 9. Compressive strength development.

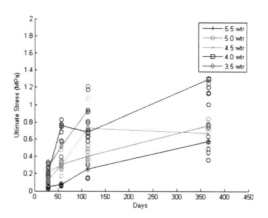

Figure 12. Shear strength development.

3.4 *Phase 2: Brick-to-mortar bond test results*

Flexure brick-to-mortar bond tests were conducted, the results of which are shown in Figure 13 and Figure 14.

3.5 *Flow table test observations*

Qualitative observations of the flowability and diameter increases as outlined in ASTM C109 are shown in Table 3.

4 CHALLENGES ENCOUNTERED

Casein samples would often overrun the flow table. As per the ASTM C109 Standard, there was no means to accommodate this possibility (ASTM International, 2012). Therefore, it was not possible to produce a meaningful quantitative comparison of flow rates; it was only possible to qualitatively indicate that a given specimen was either 'flowable' or 'not flowable'.

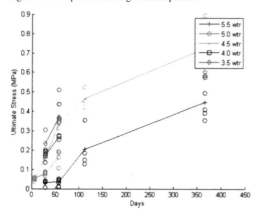

Figure 10. Tensile strength development.

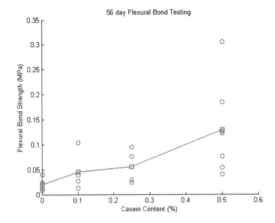

Figure 13. Flexure bond strength development.

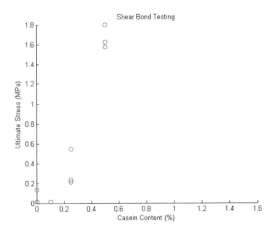

Figure 14. Shear bond strength development.

Table 3. Flow table comparisons.

Mortar Mix	Diameter Increase	Classification (By observation)
No Casein	58%	Not Flowable
0.1% Casein	11%	Not Flowable
0.25% Casein	31%	Not Flowable
0.5% Casein	110%	Flowable
1.0% Casein	67%	Flowable
1.5% Casein	67%	Flowable

5 CONCLUSION

Based upon the results of the experimental tests, it was deduced that casein may serve as a useful historic lime mortar additive in certain circumstances. While compressive strength reduction may seem prohibitive, bond strength typically governs in an unreinforced masonry wall. Casein mortar would likely be best suited to a retrofit application where the existing lime mortar has partially deteriorated

and consolidation of a masonry wall is required. In particular, improvement in flow properties and brick-to-mortar bond strength lead to the use of a casein mortar as a favourable retrofit application. Consolidation in between masonry wythes would be a possible use.

The point at which the tensile strength is equal to the flexural bond strength establishes the optimal casein content. At a casein content less than the effective maximum amount, the bond strength governs, and decreases as the casein content decreases. At a casein content exceeding the effective maximum amount, the mortar strength governs, and decreases as the casein content increases. It is therefore necessary to know both the flexural bond strength and the flexural mortar strength for the casein content being used, the lower of the two strengths would need to be used in design calculations.

The effective maximum casein content was found to vary based on the amount of water in the mortar; the flexural strength was greater in the reduced water specimens, while the flexural bond strength, while not definitively conclusive, changed very little by a reduction in water content. It was found that the best mix involved an 18% reduction in the water content.

A casein content in excess of 0.5% by mass would result in the mortar not hardening beyond an initial set, in which circumstance no strength gain will be observed. Compressive tests at casein contents of 1.0% and 1.5% were attempted, however, the 7-day strength and 56-day strength were nearly identical. It was possible to break these specimens by hand; some specimens collapsed under their own weight during handling.

If casein is desired for its bond properties, it would be advisable to use a higher-strength mortar type than what would normally be specified in order to counteract the strength reduction that casein causes.

6 LIMITATIONS

The scope of this research was concerned with material strength only. Measurement of porosity was not possible with the laboratory equipment available. In an historic building application, porosity is necessary to ensure that water and air movement can occur within the brick wall.

REFERENCES

Asselin-Boulanger, J. B. 2018. NHL Mortar – Information, Boisbriand, Quebec, Canada: King Packaged Materials. – 6.
ASTM International. 2012. C109: Standard Test Method for Tensile Strength of Chemical Resistant Mortar, Grouts, and Monolithic Surfacings. ASTM International, West Conshohocken, Pennsylvania, United States.

ASTM International. 2013. C1072: Standard Test Methods for Measurement of Masonry Flexural Bond Strength. ASTM International, West Conshohocken, Pennsylvania, United States.

ASTM International. 2016. C1531: Standard Test Methods for In Situ Measurement of Masonry Mortar Joint Shear Strength Index. ASTM International, West Conshohocken, Pennsylvania, United States.

ASTM International. 2018. C348: Standard test method for flexural strength of hydrauliccement mortars. ASTM International, West Conshohocken, Pennsylvania, United States.

Atamer, Z., Post, A.E., Schubert, T., Holder, A., Boom R. M & Hinrichs, J. 2017. Bovine betacasein: Isolation, properties and functionality, A review. *International Dairy Journal* no. 66: 115–125.

Bian H. & Plank, J. 2013. Effect of Heat Treatment on the Dispersion Performance of Casein Superplasticizer Used in Dry-Mix Mortar. *Cement and Concrete Research* 51: 1–5.

Canadian Standards Association. 2004. CSA S304.104: Masonry Design, Toronto, Ontario, Canada: Canadian Standards Association.

Ince, R. 2012. Determination of the fracture parameters of the Double-K model using weight functions of split-tension specimens. *Engineering Fracture Mechanics* (96): 416–432.

Post, A. E., Arnold, B., Weiss J. & Hinrichs, J. 2012. Effect of temperature and pH on the solubility of caseins: Environmental influences on the dissociation of alpha- and betacasei. *Journal of Dairy Science* 95(4): 1603–1616.

Post, A. E., Arnold, B., Weiss J. & Hinrichs, J. 2012. Effect of temperature and pH on the solubility of caseins: Environmental influences on the dissociation of alpha- and betacasei. *Journal of Dairy Science* 95(4): 1603–1616.

Trischuk, K. 2018. Interviewee, RE: Questions on mortar testing. Personal e-mail (9 January 2018).

Zittle, C. A. & Custer, J. H. 1963. Purification and some of the properties of alpha-casein and kappa-casein. *Journal of Dairy Science* 46(11): 1183–1188.

Material modelling

Brick and Block Masonry - From Historical to Sustainable Masonry –
Kubica, Kwiecień & Bednarz (eds)
© 2020 Taylor & Francis Group, London, ISBN 978-0-367-56586-2

Masonry walls with multi-layer bed joints subjected to cyclic shear: Analytical modelling

N. Mojsilović & B. Stojadinović
Institute of Structural Engineering, ETH Zurich, Switzerland

M. Petrović
Caprez Ingeniuere, Zurich, Switzerland

ABSTRACT: An analytical rheological model capable of describing the loading speed dependent in-plane shear behavior of the unreinforced masonry (URM) walls with a multi-layer bed joint subjected to static-cyclic shear loading is presented. Such joints consist of a core soft layer protected by two thin extruded elasto-mer membranes, which in turn are placed in a bed mortar joint. The extruded elastomer membranes are employed to prevent and/or limit the deterioration of the core soft layer during the cyclic action. Joint behav-ior is assumed to be linear elastic-perfectly viscoplastic and has been captured by a uniaxial model consisting of three elements: an elastic spring connected in series with the frictional slider and a dashpot (viscous damper). Assuming a linear elastic behavior of masonry, the model was able to describe the in-plane horizon-tal force-displacement behavior of URM walls with multi-layer bottom bed joints and was validated against the experimental results.

1 INTRODUCTION

In Swiss construction practice, different types of deformable layers, i.e. soft layers, are placed at the bottom of unreinforced masonry (URM) walls. The materials used for soft layers are bitumen, cork, polyvinylchloride and different types of rubber (usu-ally extruded elastomer and rubber granulate). The main purpose of such layers is to act as damp-proof course (DPC) and/or as sound insulation. Further-more, soft layers are used to adjust short- or long-term differential movements between the walls and the floors above and beneath them. Thus, these soft layers are not intended for seismic loading. How-ever, these layers are capable of considerably modi-fying the seismic response of masonry walls and structures.

Results from preliminary research on masonry elements with rubber granulate and elastomer soft layers, (Mojsilović et al. 2013, Vögeli et al. 2015) indicated that the presence of such layers in the mortar bed joint can significantly alter the mechan-ical characteristics of URM walls by creating a sliding plane, which, in turn, could influence the seismic response of the entire structure. Rubber granulate soft layers were in some cases heavily damaged during the sliding, whereas the elastomer layers were significantly more durable. As a part of the presented research project, several series of monotonic and static-cyclic, displacement-controlled tests were performed on masonry triplets with

a multi-layer bed joint (Mojsilović et al. 2015). The so-called multi-layer bed joint consisted of a core soft layer (rubber granulate, cork, cork-rubber granu-late, bitumen and PVC-based membranes were used) which, in order to reduce the damage caused by cyclic loading, was protected by two layers of elasto-mer, and placed in the middle of the mortar bed joint. Results indicated that multi-layer bed joints with adequate material properties could change the typical brittle shear response of masonry to a more desirable quasi-ductile one. Further, based on the observed hysteretic behavior, considerable energy dissipation can be achieved in masonry structures with multi-layer bed joints.

Figure 1 shows the structure of the multi-layer bed joint, comprising the top and bottom mortar layers, the 2.2 mm thick protective layers made of extruded elastomer and the core soft layer. The thickness of the multi-layer core-elastomer sand-wich, t_{ml}, was between 6.4 and 7.9 mm depending on the type of core soft layer while the total thick-nesses of a multi-layer bed joints (including the two mortar layers) ranged from 15 to 18 mm.

1.1 Rheological model for multi-layer bed joint

The shear load-deformation behavior of a multi-layer bed joint was assumed linear elastic-perfectly viscoplastic, see Mojsilović et al. (2019) for detailed description of the model. Here, only excerpt needed for the current presentation will be reproduced. Such

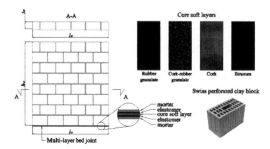

Figure 1. Specimen layout and materials.

behavior can be captured by a uniaxial model consisting of three elements: an elastic spring connected in series with a frictional slider and a dashpot (viscous damper), see Figure 2. Thereby the elastic spring is characterized by the shear modulus of the multi-layer bed joint, G_{ml}, the dashpot by the loading speed sensitive viscosity parameter with the dimension time, ζ, and the frictional slider by the elastic shear stress limit, i.e. sliding resistance, τ_y.

Since the multi-layer bed joint represents a localized zone of intense shearing with constant thickness t_{ml}, the shear strain γ and shear strain rate $\dot{\gamma}$ can be related to the slip in the multi-layer bed joint d and the slip rate $\dot{d}$, respectively, see Oberender & Puzrin (2016). Thus, in addition to the three previously defined parameters of the rheological model, the contraction of the thickness of multi-layer bed joint due to pre-compression, Δt_{ml}, has to be considered, too.

Further, in order to account for the elastic (initial) stiffness degradation, a coefficient ψ is introduced. This coefficient depends on the number of loading cycles performed, n, and will be used as a multiplier of the shear modulus G_{ml}. The data from the static-cyclic shear tests indicate that the evolution of the (relative) degradation of the elastic stiffness, measured at the beginning of each first pushing semi-cycle applied, is independent of the level of pre-compression and that it can be described using a rational function. Thus, the elastic (initial) stiffness degradation dependent on the number of loading cycles can be accounted for by multiplying the shear modulus by a coefficient, namely $\psi = 5.7/(n+5.7)$.

Now applying the coefficient ψ, one obtains, with $H = \tau A_b$, the following relationship between the shear force H and slip d:

$$H = H_y \cdot \left[1 + \frac{\dot{d}}{t_{ml} - \Delta t_{ml}} \cdot \zeta \cdot \left(1 - e^{-\frac{\psi \cdot G_{ml} \cdot A_b}{H_y \cdot \zeta} \cdot \frac{d - d_y}{d}} \right) \right]$$

(1)

where d_y and H_y are horizontal displacement and shear force at the beginning of the sliding, respectively.

The next step in this research is modelling of full-scale unreinforced masonry walls with a multi-layer bed joint. The multi-layer bed joint model was extended to describe the in-plane horizontal force-displacement behavior of URM walls with the multi-layer bed joint at the bottom of the wall. The results from our own tests on URM walls with a multi-layer bottom bed joint (Petrović et al., 2017) have been used for the model validation, as discussed next.

2 TEST PROGRAM

This section presents and discusses the findings obtained by performing a series of static-cyclic shear tests on full-scale URM walls with a multi-layer bed joint. A total of nine walls were tested in two phases. The first (preliminary) testing phase allowed the most suitable core soft layer type to be chosen from the four types investigated. Results from the second (main) phase enabled an analysis of the influence of the pre-compression level, aspect ratio and wall size on the seismic behavior of URM walls with a multi-layer bed joint. The testing program and results have been presented and discussed in detail in Petrović et al. (2017). Here, only excerpt needed for the current presentation will be given.

Table 1 summarizes the test program, where l_w, h_w and t_w are the length, height and thickness of the specimens (see Figure 1), t_{csl} is the core soft layer thickness, f_x is the mean compressive strength of the masonry perpendicular to the bed joints, and σ_{pc} is the pre-compression stress computed with reference to the nominal wall cross section area $A_w = l_w \cdot t_w$. All

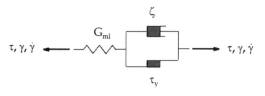

Figure 2. Rheological model for multi-layer bed joint.

Table 1. Test program.

Specimen	t_{csl} [mm]	Dimensions $l_w \times h_w \times t_w$ [mm]	f_x	σ_{pc}/f_x
WG	3	1500×1600×150	4.6	0.10
WGK	3.2	1500×1600×150	5.3	0.10
WK	3.5	1500×1600×150	5.0	0.10
WB	2	1500×1600×150	5.1	0.10
Z1	3	2700×2600×150	4.6	0.10
Z2	3	2700×2600×150	4.6	0.05
Z3	3	2700×2600×150	4.6	0.20
Z5	3	1800×2600×150	4.6	0.10
Z6	3	3600×2600×150	4.6	0.10

walls were tested under fixed-end boundary conditions.

The objectives of the preliminary phase (W-Specimens) were to determine the most promising type of core soft layer for the main testing phase (Z-Specimens) and verify the applied vertical pre-compression levels and measurement system. Four static-cyclic tests on 1600 mm high and 1500 mm long, full-scale URM walls were performed.

The main phase comprised another five tests on large, story-high walls. A rubber granulate soft layer was chosen as a core layer based on the overall behavior and performance exhibited during the preliminary testing phase. Comparison of the tests results with the results of the test on Specimen Z1, chosen as the reference specimen, enabled the investigation of the influence of the pre-compression level (Specimens Z2 and Z3) and the aspect ratio (Specimens Z5 and Z6) on the seismic behavior of URM walls with a multi-layer bed joint. Specimen designation Z4 has been purposely omitted.

Typical Swiss perforated clay blocks, with nominal dimensions of 290 x 150 x 190 mm and void area of 42%, were used to build the specimens for both testing phases, cf. Figure 1. The compressive strength of the blocks was determined according to EN 772-1 (2011) on a sample of 10 blocks, and was equal to 31.2 MPa. The compressive strength of the standard cement mortar for each phase was determined by testing the mortar prisms according to EN 1015-11 (1999) after a curing period of at least 28 days. The mean compressive strengths of 6.9 MPa and 7.9 MPa were determined for the mortar used to prepare the specimens for the first and the second testing phase, respectively.

The masonry compressive strength perpendicular to the bed joint direction, f_x, was determined in accordance with the provisions of EN 1052-1 (1998); the mean values are given in Table 1. For each implemented core soft layer, three specimens with nominal dimensions of 1000 x 600 x 150 mm were constructed at the same time as the walls for the first testing phase. In addition, three such specimens without layers were constructed and served as a reference; the mean value of f_x for these specimens was 5.2 MPa. Note that an 15% reduction of the masonry compressive strength has been observed for the specimens with rubber granulate core layer.

All wall specimens were tested under fixed-end boundary conditions, i.e. the rotation at the top and the bottom of the specimen was restricted. The fixed-end boundary conditions were ensured by a mixed force-displacement control of the vertical actuators which, besides the role of keeping the pre-compression level constant during the tests, were used to keep the loading beam horizontal. Detailed information about the test controlling system can be found in Salmanpour et al. (2015). Each test was stopped either due to critical damage conditions of the specimen, or when the response of the specimen was stable (pure sliding) and no difference in response quantities between several consecutive loading history steps was observed.

Apart from the applied vertical and horizontal loads and the displacements of the servo-hydraulic actuators, measurements comprised vertical, horizontal and diagonal deformations of the specimens. Several potentiometers (POTs) and linear variable differential transformers (LVDTs) were installed on one surface of the specimen to record the specimen's deformation. Besides this conventional hard-wired measuring system, a 2D Digital Image Correlation (DIC) measurement system was used to obtain the deformation field on the opposite surface of a specimen during a test. More details on the implemented DIC system can be found in Mojsilović and Salmanpour (2016). The computer used for data acquisition triggered the DIC cameras at the predefined displacement levels in each cycle performed.

3 MODELLING

Besides the shear deformation of the multi-layer bed joint (sliding motion during the sliding regime) and masonry deformation, another contribution to the total URM wall deformation emerges from the vertical deformability of the multi-layer bed joint, which allows the wall to rotate as a rigid body, i.e. to rock.

Figure 3 shows the deformation contribution factors quantified from the DIC measured vertical and horizontal displacements of the bottom and the top wall cross-sections, i.e. the wall top horizontal displacement components stemming from sliding and, sliding plus rocking motions normalized with respect to the measured wall top horizontal displacement. Vertical displacements of the wall bottom cross-section points can be transformed into the wall base rotation and further into the wall top displacement, until the assumption that the wall bottom cross-section remains plane can be considered as valid. After the appearance of significant (vertical) tensile cracks at the bottom brick course, which change the displacement distribution at the bottom cross-section, a precise quantification of the rocking deformation is not possible anymore. The data in Figure 3 correspond to the first pushing semi-cycle applied, i.e. to the positive horizontal force and displacement. It can be seen that, for each wall,

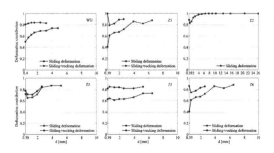

Figure 3. Measured deformation contribution factors.

689

deformation of masonry does not exceed 30% of the total wall deformation, and that this percentage is becoming smaller with the increase of the target wall top displacement.

As presented above, the source of the viscoplastic behavior is multi-layer bed joint. Before the visco-plastic behavior is triggered, a URM wall with a multi-layer bed joint, exhibits elastic behavior, which, in the horizontal force-horizontal displacement space, can be characterized by the initial wall stiffness, K_0. Therefore, starting from a decomposition of the total displacement into elastic (el) and viscoplastic (vp) portions, one obtains:

$$d = d^{el} + d^{vp} \text{ and } \dot{d} = \dot{d}^{el} + \dot{d}^{vp}$$
$$\text{where } d^{el} = H/K_0 \text{ and with} \tag{2}$$

$$\dot{d}^{vp} = (t_{ml} - \Delta t_{ml}) \cdot \dot{\gamma}^{vp} = \frac{(t_{ml} - \Delta t_{ml})}{\zeta} \cdot \left(\frac{H}{H_y} - 1 \right) \tag{3}$$

the model differential equation of equilibrium can be determined as:

$$\dot{H} + \frac{K_0 \cdot (t_{ml} - \Delta t_{ml})}{\zeta \cdot H_y} \cdot H = \frac{K_0 \cdot (t_{ml} - \Delta t_{ml})}{\zeta} + K_0 \cdot \dot{d} \tag{4}$$

Applying standard methods for solving first-order ordinary linear differential equations, while assuming the loading speed as constant, the Equation 4 can be solved and the loading speed-dependent horizontal force-displacement relationship for a URM wall with a multi-layer bed joint can be obtained. Starting from the general differential equation solution

$$H(t) = C_1 \cdot e^{-\frac{K_0 \cdot (t_{ml} - \Delta t_{ml})}{H_y \cdot \zeta} t} + H_y \cdot \left(1 + \frac{\zeta \cdot \dot{d}}{t_{ml} - \Delta t_{ml}} \right) \tag{5}$$

and by substituting $t = d/\dot{d}$, one obtains

$$H(d, \dot{d}) = C_1 \cdot e^{-\frac{K_0 \cdot (t_{ml} - \Delta t_{ml})}{H_y \cdot \zeta} \frac{d}{\dot{d}}} + H_y \cdot \left(1 + \frac{\zeta \cdot \dot{d}}{t_{ml} - \Delta t_{ml}} \right) \tag{6}$$

Further, with known force boundary condition, $H(d = d_y) = H_y$, the constant C_1 can be determined

$$C_1 = -H_y \cdot \frac{\zeta \cdot \dot{d}}{t_{ml} - \Delta t_{ml}} \cdot e^{\frac{K_0 \cdot (t_{ml} - \Delta t_{ml})}{H_y \cdot \zeta} \frac{d_y}{\dot{d}}} \tag{7}$$

And, finally, the loading speed-dependent horizontal force-displacement relationship is

$$H(d, \dot{d})$$
$$= H_y \cdot \left[1 + \frac{\zeta \cdot \dot{d}}{t_{ml} - \Delta t_{ml}} \cdot \left(1 - e^{-\frac{K_0 \cdot (t_{ml} - \Delta t_{ml})}{H_y \cdot \zeta} \frac{d - d_y}{\dot{d}}} \right) \right] \tag{8}$$

The initial stiffness of the URM wall with a rubber granulate core soft layer in the bottom multi-layer bed joint can be approximated well using the elastic stiffness, K_{el}, calculated using Equation 9.

$$K_{el} = \frac{1}{\kappa \cdot \frac{h_w}{G_m \cdot A_w} + \frac{h_w^3 \cdot (\alpha^3 + (1-\alpha)^3)}{3 \cdot E_m \cdot I_w} + \frac{(t_{ml} - \Delta t_{ml})}{G_{ml} \cdot A_w} + \alpha \cdot \frac{h_w^2 \cdot \Delta t_{ml}}{\sigma_{pc} \cdot I_w}} \tag{9}$$

where A_w and I_w denote the area and moment of inertia of the wall horizontal cross-section, respectively; G_m and E_m correspond to the shear and elastic moduli of masonry. Coefficient κ is the shear coefficient and equals to 1.2 for the rectangular cross-section. Coefficient α represents the position of zero bending moment with respect to the bottom cross-section. The ratio between K_0 and K_{el} equals 1.4, see also Petrović (2018). Note that the Equation 9 is written in such manner to distinguish between the components which contribute to the total wall deformation. Looking from left to right, the following deformation components are given: masonry shear deformation, masonry flexural deformation, multi-layer bed joint shear deformation and rocking deformation, respectively.

Table 2 shows the deformation contribution factors calculated for each deformation component in the Equation 9. Keeping the assumption that the values of d_y are the same as those estimated from

Table 2. Calculated deformation contribution factors.

Specimen	Masonry shear	Masonry flexure	Shear of MLBJ	Rocking
WG	14.3	11.3	56.5	17.9
Z1	21.3	13.7	51.7	13.3
Z2	21.3	16.2	52.5	10.0
Z3	21.0	9.6	49.6	19.8
Z5	15.9	23.0	38.6	22.5
Z6	24.2	8.7	58.6	8.5

the monotonic test series on masonry triplets (Mojsilovic et al., 2015), and using the values of initial stiffness estimated using Equation 9, while considering the K_0-K_{el} correlation factor of 1.40, corresponding values of H_y can be calculated for each wall, see Table 3.

Since the major part of the total wall deformation is in fact sliding, see Figure 3, degradation of the initial stiffness, i.e. of the stiffness within the elastic range, with the number of loading cycles can be accounted for through the degradation of the shear modulus of the multi-layer bed joint with the number of loading cycles. The Equation 9 can be modified for that purpose as

$$K_0(n) = 1.4 \cdot \cfrac{1}{\kappa \cdot \cfrac{h_w}{G_m \cdot A_w} + \cfrac{h_w^3 \cdot (a^3 + (1-a)^3)}{3 \cdot E_m \cdot I_w} + \cfrac{t_{ml} - \Delta t_{ml}}{\psi_1(n) \cdot G_{ml} \cdot A_w} + \alpha \cdot \cfrac{h_w^2 \cdot \Delta t_{ml}}{\sigma_{pc} \cdot I_w}}$$

(10)

One can now calibrate the Equation 10 against the parameter ψ_1, to capture the measured data on the degradation of the elastic response range stiffness of the tested URM walls with the number of the loading cycles.

Figure 4 shows the results obtained for walls WG and Z1-Z6 (data on the initial stiffness from each first applied pushing semi-cycle are considered). The results indicate that the evolution of the parameter ψ_1 with the number of loading cycles is independent of the level of pre-compression and the wall aspect ratio, and can be approximated well by using a rational function shown in Figure 4.

With the defined initial stiffness degradation function and known values of H_y, which are kept constant, meaning that the values of d_y change as the initial stiffness degrades with the increase of the number of loading cycles, and previously determined ζ coefficients (Mojsilovic et al., 2015), one can calculate the force resistance of URM walls with a multi-layer bed joint for an arbitrary loading speed, $\dot{d}$, and horizontal displacement, d, using the Equation 8.

As can be seen from Figure 5, the calculated values of horizontal force resistance are in very good agreement with the corresponding measured values (in case of walls Z3 and Z5 to a somewhat lower

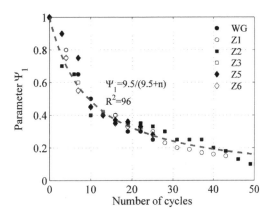

Figure 4. Degradation of the initial wall stiffness with the number of performed loading cycles.

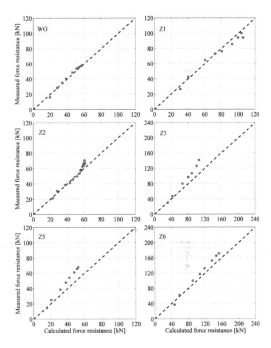

Figure 5. Horizontal force resistance of URM masonry walls with a rubber granulate core soft layer in the multi-layer bed joint: measured vs. calculated values.

extent). Note that only the values of the horizontal force resistance measured before the appearance of significant (vertical) tensile cracks at the bottom brick course or the appearance of shear cracks are considered in case of walls WG, Z3, Z5 and Z6.

Figure 6 shows the model ability to describe the force-displacement relationship of the tested URM walls with multi-layer bed joints. Note that the model developed is capable to describe only the loading hysteresis branches. The unloading branches (going back to the force value of zero) are in fact

Table 3. Calculated values of Hy.

Test	σ_{pc} [MPa]	d_y [mm]	K_0 [kN/mm]	H_y [kN]
WG	0.52	0.27	53.40	14.42
Z1	0.52	0.27	86.11	23.25
Z2	0.26	0.17	85.83	14.59
Z3	1.04	0.35	83.65	29.28
Z5	0.52	0.27	44.91	12.13
Z6	0.52	0.27	127.20	34.34

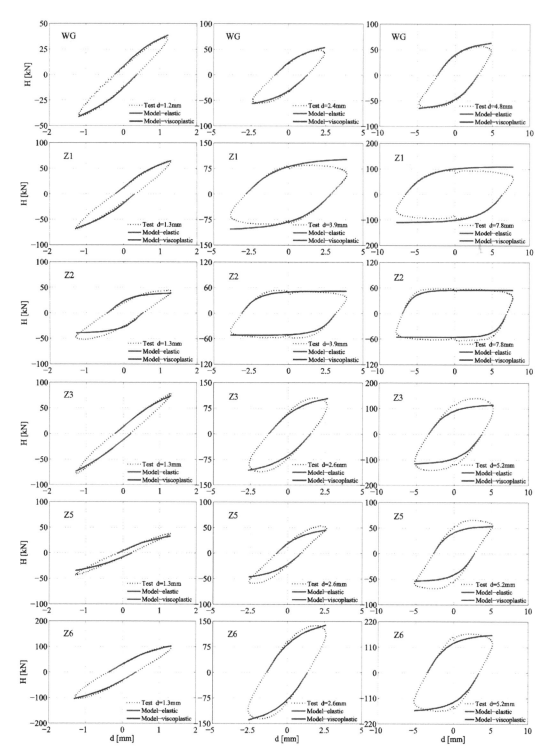

Figure 6. Horizontal force-displacement behavior of URM masonry walls with a rubber granulate core soft layer in the multi-layer bed joint: measured vs. calculated.

parallel to the initial elastic branches. The unloading parts of the measured hysteresis curves are dependent on the (un)loading speed, which can allow for certain horizontal force relaxation. In case of a sinusoidal loading pattern, a drop of the horizontal force appears in the region of peak horizontal displacement, i.e. in the region of almost zero loading speed.

4 CONCLUSIONS

An analytical model of the loading speed-dependent in-plane shear force-slip behavior of the masonry multi-layer bed joint is presented and further calibrated for the multi-layer bed joint with a rubber granulate core soft layer. The influence of the loading speed is considered through the assumption of elastic-perfectly viscoplastic behavior of the multi-layer bed joint. Assuming a linear elastic behavior of masonry, the model is further extended to describe the in-plane horizontal force-displacement behavior of URM walls with such multi-layer bottom bed joints and validated against the experimental results. However, possible separation of the wall into more parts, initiated by the tensile cracks in head joints at the bottom block course, which can lead to significant strength and stiffness degradation of the wall, and/or exceeding of the limit of the elastic masonry deformation, limits the model applicability.

It is shown that, by using the proposed model, for the given loading speed, $\dot{d}$, and horizontal displacement, d, the force resistance of URM walls with a rubber granulate core soft layer in the multi-layer bed joint can be predicted well. Moreover, using the proposed model, the in-plane horizontal force-displacement behavior of URM walls with a rubber granulate core soft layer in the multi-layer bed joint can be described with satisfactory accuracy. Somewhat conservative results are obtained in case of walls Z3 and Z5. It is noteworthy that, although the model is developed and calibrated for URM with a rubber granulate core soft layer in the multi-layer bed joint, it can be extended to the other types of core soft layers once the appropriate test data become available.

REFERENCES

EN 772-1:2011. Methods of test for masonry units, Part 1: Determination of compressive strength. European Committee for Standardization (CEN), Brussels.

EN 1015-11:1999. Methods of test for mortar for masonry, Part 11: Determination of flexural and compressive strength of hardened mortar. European Committee for Standardization (CEN), Brussels.

EN 1052-1:1998. Methods of test for masonry, Part 1: Determination of compressive strength. European Committee for Standardization (CEN), Brussels.

Mojsilović, N., Stojadinović, B., Barandun, A., and Vögeli, C. (2013). Seismic behavior of masonry walls with soft-layer wall bearings. *Proceedings, 5th International Conference on Structural Engineering, Mechanics and Computation*, Cape Town, 1865–1870.

Mojsilović, N., Petrović, M., and Anglada, X. R. 2015. Masonry elements with multi-layer bed joints: Behavior under monotonic and static-cyclic shear, *Construction and Building Materials*, 100, 149–162.

Mojsilović, N., Petrović, M., and Stojadinović, B. 2019. Multi-layer masonry bed joint subjected to shear: Analytical modelling, *Construction and Building Materials*, 205C, 602–610.

Mojsilović, N., and Salmanpour, A. H. 2016. Masonry walls subjected to in-plane cyclic loading: Application of Digital Image Correlation for deformation field measurement, *International Journal of Masonry Research and Innovation*, 1(2),165–187.

Oberender, P.W., & Puzrin, A.M. 2016. Observation-guided constitutive modelling for creeping landslides, *Géotechnique*, 66, 232–247.

Petrović, M., 2018. Use of soft layers for seismic response modification of structural masonry walls, PhD Thesis, ETH Zurich, 210.

Petrović, M., Mojsilović, N., and Stojadinović, B. 2017. Masonry walls with a multi-layer bed joint subjected to in-plane cyclic loading: An experimental investigation, *Engineering Structures*, 143C, 189–203.

Salmanpour, A. H., Mojsilović, N., and Schwartz, J. 2015. Displacement capacity of contemporary unreinforced masonry walls: An experimental study, *Engineering Structures*, 89, 1–16.

Vögeli, C., Mojsilović, N., and Stojadinović, B. 2015. Masonry wallettes with a soft layer bed joint: Behavior under static-cyclic loading, *Engineering Structures*, 86, 16–32.

Brick and Block Masonry - From Historical to Sustainable Masonry –
Kubica, Kwiecień & Bednarz (eds)
© 2020 Taylor & Francis Group, London, ISBN 978-0-367-56586-2

Constitutive model for the nonlinear cyclic behavior of brick masonry structures

M. Sousamli, F. Messali & J.G. Rots

Applied Mechanics, 3Md, Civil Engineering & Geosciences, TU Delft, Delft, The Netherlands

ABSTRACT: The present work focuses on the development of a continuum constitutive model that describes the structural behavior of brick masonry. The model, following a phenomenological approach, is based on a Total Strain Rotating Crack concept and it incorporates the orthotropic behavior of the material. This is achieved by varying the elastic and inelastic parameters (Young's modulus, tensile and compressive strengths and fracture energies) with respect to the angle θ between the principal stresses and the mortar joints. In this way, a different mechanical behavior of the principal stresses is obtained based on the directionality of the mortar joints and the expected failure mechanism along them, i.e. low dissipative behavior for bed-joint's opening and high dissipative behavior for diagonal shear sliding. The material model is implemented in a non-linear finite element program and is validated against quasi-static, cyclic, in-plane tests performed on masonry walls at TU Delft.

1 INTRODUCTION

Masonry is one of the oldest building materials and due to its aesthetics, the low cost and the simplicity of its construction, its use is still widespread around the world. Notwithstanding its popularity, the geometrical arrangement and the different material properties of its constituents make masonry an anisotropic, non-linear material, rendering its modelling a complex task.

Nowadays a wide range of modelling techniques for the structural analysis of masonry structures are available, each one of a different complexity level. The main available strategies for the structural analysis of masonry structures include the limit analysis (Lourenço, 2002, Roca et al., 2010), Macro-Element based methods, like the lumped mass approach and the Equivalent Frame Method (Lagomarsino et al., 2013, Siano et al., 2017), Discrete Element Methods (DEM) (Cundall and Hart, 1992, Lemos, 2007, Malomo et al., 2018) and finally the Finite Element Methods (FEM). For the last case there can be two approaches: micro-modelling (detailed or simplified) and macro-modelling. In micro-modelling, the bricks (and the mortar joints in the case of detailed micro-modelling) are simulated as separate continuous units that are connected with each other through discontinuous interface elements. On the other hand, in macro-modelling an equivalent homogenized material is considered and the properties of the brick units, the mortar and their in-between interface are smeared out in the continuum (smeared-crack models). Macro-models are unable to precisely describe the local effects of the masonry, but they are able to give reliable results with respect to the global behaviour on the basis of a phenomenological approach, the load-bearing capacity and the damage of masonry structures, whereas the computational effort is reduced in comparison to the micro-models (Lourenço 1996). Therefore, they are often preferred for the modelling of large structures, where the computational time is of importance and a detailed description between the interaction of the mortar with the bricks might not be necessary (Roca et al., 2010).

The most well-known, continuous orthotropic constitutive models for masonry based on a phenomenological approach are the models of Lourenco et al (Lourenço et al., 1997), Papa and Nappi (Papa and Nappi, 1997), Berto et al. (Berto et al., 2002) and Pela et al. (Pelà et al., 2013). In the Netherlands, the two most commonly used material models for modelling of masonry are the Total-Strain-Rotating-Crack Model (TSRCM) and the Engineering Masonry Model (EMM) (Rots et al., 2016, Schreppers et al., 2016). However, they both come with certain disadvantages. The TSRCM is able to predict the correct failure mechanism of the structure but it often overestimates the capacity and underestimates the dissipated energy (Schreppers et al., 2016). Moreover, it's an isotropic model, disregarding the orthotropy of masonry. On the other hand, the EMM is able to predict with relatively good accuracy the force capacity of a structure, but it shows very poor crack localization.

The idea behind the constitutive model presented in this paper, is to combine the strengths of the EMM and TSRCM, while tackling their weaknesses.

2 THE CONSTITUTIVE MODEL

The constitutive model is based on a total-strain-rotating-crack (TSRC) concept and the anisotropy of the material is introduced through the dependency of the elastic and inelastic properties on the principal direction. Most TSRC models are isotropic and the direction of the principal strains always coincides with the direction of the principal stresses. However, this is not the case for anisotropic models, where, depending on the elastic (Young's and Shear modulus) and inelastic properties (tensile strength, etc.), the principal direction of the stresses can be different from that of the strains. For this reason, in this model, the principal direction is defined as the one parallel to the direction of the principal strains, by defining an angle α_i, which corresponds to the angle formed by the line parallel to the bed joints and the line parallel to the principal strain ε_i.

2.1 Variation of material properties

According to the existing literature, experiments have been performed for the identification of the material parameters and the description of the behavior of masonry in tension, compression and shear. Most authors (Jafari et al., 2017, Naraine and Sinha, 1989, Page, 1980) identify the compressive and tensile strengths and their corresponding fracture energies, in the directions parallel or perpendicular to the bed joints. Additionally, from shear tests, the cohesion and the friction angle can be defined for loading parallel to the bed joints (Atkinson et al., 1989). Nevertheless, most experiments are limited to testing parallel to the mortar joints and there is a lack of mechanical data for loading under different directions. Given the lack of specific experiments, it is assumed that the mechanical properties (with the exception of the tensile strength) vary linearly with respect to the principal angles α_i, where

$$\alpha_1 = \begin{cases} \theta & \text{,if } \theta > 0 \\ \theta + \pi/2 & \text{,else} \end{cases} \text{ and } \alpha_2 = \alpha_1 - \pi/2 \quad (1)$$

where

$$\theta = 0.5 \text{ atan} \left(\frac{\gamma_{xy}}{\varepsilon_{xx} - \varepsilon_{yy}} \right) \quad (2)$$

For the complete definition of the material model, 13 independent material properties are required as input parameters. Those are the Young's moduli parallel to the bed and head joints (E_{0x}, E_{0y} respectively), the shear modulus G_{xy}, the tensile and compressive strengths alongside the bed and head joints (f_{tx} and f_{ty} for tension and f_{cx} and f_{cy} for compression); the fracture energies in tension and

compression (G_{ftx}, G_{fty}, G_{fcx} and G_{fcy}) and finally the initial cohesion c_0 and friction angle (φ) due to shear.

The Young's moduli $E_{p,i}$, compressive strengths $f_{cp,i}$ and fracture energies in compression $G_{fcp,i}$ in the principal directions are given by Equations (3)-(5). The relationship for the definition of the tensile strength, presented in Figure 1, is derived by fitting the experimental results obtained from (Page, 1980) for uniaxial tensile loading, and is given by Equation (6).

$$E_{p,i} = E_{0,x} + \left(E_{0,y} - E_{0,x} \right) \frac{|\alpha_i|}{\pi/2} \quad (3)$$

$$f_{cp,i} = f_{cp,x} + \left(f_{cp,y} - f_{cp,x} \right) \frac{|\alpha_i|}{\pi/2} \quad (4)$$

$$G_{fcp,i} = G_{fcp,x} + \left(G_{fcp,y} - G_{fcp,x} \right) \frac{|\alpha_i|}{\pi/2} \quad (5)$$

$$f_{tp,i} = f_{tx} - \delta f_t \frac{|\alpha_i|}{\pi/2} + \left(f_{max} - f_{av} \right) \sin(4|\alpha_i|) \quad (6)$$

Where $f_{tp,i}$ is the tensile strength in the direction parallel to the principal angle α_i. Moreover, $\delta ft = f_{tx} - f_{ty}$, $f_{max} = \left(f_{tx}^2 - f_{ty}^2 \right)^{1/2}$ and $f_{av} = 0.5(f_{tx} - f_{ty})$.

2.2 Normalization of strains and stresses

To simplify the definition of the constitutive model in a FORTRAN code, the principal strains are normalized and they are calculated via normalized stress-strain relationships for compression and tension. Eventually, the real principal stresses are calculated by multiplying the normalized stress with the corresponding strength of the direction α_i. The normalized principal strains are given by Equation (7):

$$\bar{\varepsilon}_{p,i} = \begin{cases} \varepsilon_{p,i}/\varepsilon_{cr,i} & \text{if } \varepsilon_{p,i} \geq 0 \\ -\varepsilon_{p,i}/\varepsilon_{pc,i} & \text{if } \varepsilon_{p,i} < 0 \end{cases} \quad (7)$$

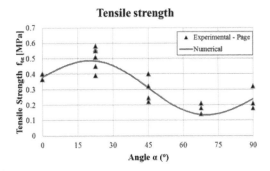

Figure 1. Tensile strength with respect to angle α.

where $\varepsilon_{cr,i}$ is the strain at the onset of tensile cracking, given by $\varepsilon_{cr,i} = f_{tp,i} / E_{p,i}$ and similarly $\varepsilon_{pc,i}$ is the compressive strain that corresponds to the peak compressive strength, given by $\varepsilon_{pc,i} = 5f_{cp,i} / (3E_{p,i})$.

2.3 Tensile behavior

For the description of the tensile behavior, a bilinear curve with linear softening is adopted. However, different softening is defined based on the direction of the principal strains, in order to capture the expected failure mechanism. When the bed-joint opening is expected (for α_i around 90°), the material shows a quasi-brittle behavior. Conversely, for angles associated to cracking due to shear sliding mechanism, the post peak behavior of the material is more ductile. The softening branch is defined through the slope β_i and the normalized ultimate strain $\bar{\varepsilon}_{ult,i}$ (Equations (8) and (9) and Figure 2).

$$\beta_i = \begin{cases} 0 & \text{if } 0 \le |\alpha_i| \le \theta_{cr} \\ \beta_y \sin(\lambda|\alpha_i - \theta_{cr}|) & \text{if } |\alpha_i| > \theta_{cr} \end{cases} \quad (8)$$

$$\bar{\varepsilon}_{ult,i} = \begin{cases} 1000 \cdot \bar{\varepsilon}_{ult,y} & \text{if } \beta_i = 0 \\ \min\left(\frac{1}{\beta_i} + 1; 100\bar{\varepsilon}_{ult,y}\right) & \text{else} \end{cases} \quad (9)$$

Where $\bar{\varepsilon}_{ult,y} = 2G_{ft,y} / (f_{ty} h \varepsilon_{cr,y})$, $\lambda = (\pi/2) / (\pi/4 - \theta_{fr})$ and $\theta_{cr} = \pi/4 + \theta_{fr}$ with $\theta_{fr} = 5\pi/36$, an arbitrarily chosen angle. The crack bandwidth h is used, so that the results are objective with respect to the mesh refinement, given according to (Rots, 1988) as $h = \sqrt(A)$, with A being the full area of the element. The normalized curve for the tensile behavior is then given by:

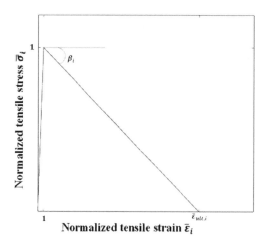

Normalized tensile strain $\bar{\varepsilon}_i$

Figure 2. Normalized principal stress-strain relationship in tension.

$$\bar{\sigma}_{p,i} = \begin{cases} \bar{\varepsilon}_{p,i} & \text{if } \bar{\varepsilon}_{p,i} \le 1 \\ 1 - \frac{\bar{\varepsilon}_{p,i} - 1}{\bar{\varepsilon}_{ult,i} - 1} & \text{if } 1 < \bar{\varepsilon}_{p,i} < \bar{\varepsilon}_{ult,i} \\ 0 & \text{if } \bar{\varepsilon}_{p,i} \ge \bar{\varepsilon}_{ult,i} \end{cases} \quad (10)$$

2.4 Compressive behavior

2.4.1 Envelope curve
For the compressive behavior of masonry the parabolic relationship of the TSRCM as implemented in the finite element software DIANA 10.3 is used (DIANA, 2019). This is expressed in normalized strains as:

$$\bar{\sigma}_{p,i} = \begin{cases} \frac{5}{3}\bar{\varepsilon}_{p,i} & \text{if } \bar{\varepsilon}_{p,i} \ge -0.2 \\ \frac{5}{3}\bar{\varepsilon}_{p,i} + \frac{25}{24}\left(\bar{\varepsilon}_{p,i} + 0.2\right)^2 & \text{if } -0.2 > \bar{\varepsilon}_{p,i} \ge -1 \\ -1 + \left(\frac{\bar{\varepsilon}_{p,i} + 1}{\bar{\varepsilon}_{ulc,i} + 1}\right)^2 & \text{if } -1 > \bar{\varepsilon}_{pi} \ge \bar{\varepsilon}_{ulc,i} \\ -0.1 & \text{if } \bar{\varepsilon}_{p,i} < \bar{\varepsilon}_{ulc,i} \end{cases} \quad (11)$$

Where $\bar{\varepsilon}_{ulc,i} = \varepsilon_{ulc,i} / \varepsilon_{pc,i}$ with $\varepsilon_{ulc,i}$ being the strain corresponding to the full consumption of the fracture energy in compression, given by Equation (12). Finally the compressive strength in direction α_i is then given by $\sigma_{p,i} = \bar{\sigma}_{p,i} f_{pc,i}$.

$$\varepsilon_{ulc,i} = \min\left(\varepsilon_{pc,i} + 1.5\frac{G_{fcp,i}}{h \cdot f_{cp,i}}; \ 2.5\varepsilon_{pc,i}\right) \quad (12)$$

2.4.2 Compressive behavior with lateral cracking
Similar to the models that describe the behavior of cracked concrete, the compressive strength of masonry is reduced due to lateral cracking. This means, that the existence of large tensile strains perpendicular to the principal compressive direction, reduces the masonry compressive strength. The reduction model proposed by Vecchio & Collins (Vecchio and Collins, 1993) for concrete is used. The new compressive strength once cracking has occurred is given by Equation (13).

$$f_{cp,i} = \min\left(\frac{1}{1 + K_{c,i}}f_{cp,i}; 0.1f_{cp,i}\right) \quad (13)$$

with

$$K_{c,i} = \min\left(0.27\left(\frac{\varepsilon_{p,lat}}{|\varepsilon_{pc,i}|} - 0.37\right); 1\right) \quad (14)$$

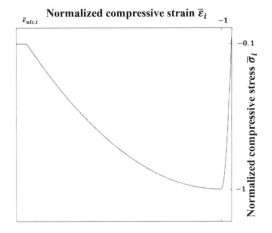

Figure 3. Normalized principal stress-strain relationship in compression.

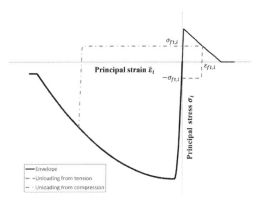

Figure 4. Unloading/reloading from tension to compression and opposite.

Where $\varepsilon_{p,lat}$ is the maximum tensile strain ever reached in the direction perpendicular to the principal compressive.

2.5 Unloading/Reloading

Although most TSC and Damage Models follow a secant unloading and reloading branch, in this model elastic unloading and reloading is adopted. It was observed that by using the TSRCM the energy dissipation during cyclic loading was small and not matching the experimental results. On the other hand, the elastic unloading/reloading of the shear in the EMM increased the amount of dissipated energy significantly. Therefore, including this type of unloading will serve into increasing the total dissipated energy of the structure.

Naturally, there is an upper and lower limit for the stresses during unloading and reloading. This limit is defined as the stress σ_{fl} that corresponds to the maximum tensile strain ever reached ε_{fl}, before unloading occurred. This is presented in Figure 4.

2.6 Shear limit

Once the principal stresses are calculated, they are transformed back to the local element axes xy through Equation (15a). One more limit is applied, concerning the shear stresses τ_{xy}. It was observed that including a limit in the local shear stresses leads to a better estimation of the base shear capacity. It is therefore chosen to apply a limit on the shear stresses based on Coulomb friction. The maximum allowed shear stress is defined as τ_{max} and is given by Equation (16).

$$\begin{Bmatrix} \sigma_{xx} \\ \sigma_{yy} \\ \tau_{xy} \end{Bmatrix} = [T_\sigma]^{-1} \begin{Bmatrix} \sigma_1 \\ \sigma_2 \\ 0 \end{Bmatrix} \quad (15a)$$

with

$$[T_\sigma]^{-1} = \begin{bmatrix} \cos^2\alpha_1 & \sin^2\alpha_1 & -2\cos\alpha_1\sin\alpha_1 \\ \sin^2\alpha_1 & \cos^2\alpha_1 & 2\cos\alpha_1\sin\alpha_1 \\ \cos\alpha_1\sin\alpha_1 & -\cos\alpha_1\sin\alpha_1 & \cos^2\alpha_1 - \sin^2\alpha_1 \end{bmatrix}$$
$$(15b)$$

$$\tau_{max} = \max\left(c_0 - \sigma_{yy}\tan\phi; c_0\right) \quad (16)$$

In case that the absolute value of τ_{xy} exceeds the maximum allowable shear it is limited to $\tau_{xy}' = \tau_{max}$ if τ_{xy} was positive or $\tau_{xy}' = -\tau_{max}$ if it was negative. Naturally, in order to maintain the principal direction in the same angle α_1, it is required to modify also the pair of stresses $(\sigma_{xx}, \sigma_{yy})$ accordingly, so that the principal angle is maintained. Since τ_{max} is dependent of σ_{yy}, it is chosen to recalculate σ_{xx}, so that:

$$\tan 2\alpha_1 = \frac{2\left|\tau_{xy}'\right|}{\sigma_{xx}' - \sigma_{yy}'} \quad (17)$$

Once the local stresses are calculated, the principal stresses in the direction α_i should be reevaluated by rotating the local stresses by the angle α_1. If the newly rotated principal stresses (σ_1', σ_2') are lower than the principal stresses (σ_1, σ_2) calculated in previous step, the stress evaluation is complete and the code proceeds to the update of the stiffness matrix. Otherwise, if one or both of the rotated principal stresses (σ_1', σ_2') exceed their corresponding limit, then they are reduced again to this limit and the local stresses $(\sigma_{xx}, \sigma_{yy}, \tau_{xy})$ need to be reevaluated. This is followed by a new check of the shear stresses. This is repeated until all the conditions are met or until a maximum number of iterations is reached. A maximum number of iterations equal to 500 is also considered, in order to avoid endless loops. For a graphical representation see the flowchart in Figure 5.

697

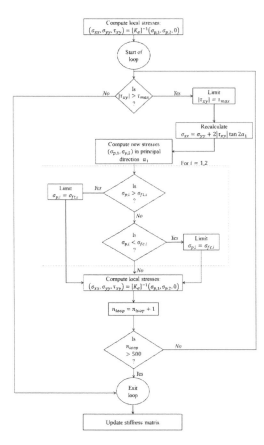

Figure 5. Flowchart of internal limitation of shear stress τ_{xy}.

3   APPLICATION EXAMPLES

In order to assess the validity and applicability of the developed constitutive model, the FORTRAN code is implemented via a user subroutine in the finite element software DIANA 10.3. Three walls having different dimensions, boundary conditions and vertical precompression, that were built and tested at Delft University of Technology in 2015 (Messali et al., 2017), are modelled. The geometric properties of these walls are presented in Table 1.

The walls, consisting of Calcium Silicate (CS) bricks, were tested under quasi-static in-plane loading until they reached the near collapse state or until the experiment could no longer be continued. However, the numerical analyses of the walls are restricted to monotonic in-plane analyses and the numerical results will be compared against the envelope curve of the experimental results.

The material properties chosen as input parameters are the ones obtained from the companion material tests, with the exception of the friction

Table 1.   Summary of geometrical properties of specimens.

Specimen name	Dimensions	Vertical pre-compression	Boundary Condition
	$l_w \cdot h_w \cdot t_w \cdot$ [m]	[MPa]	
TUD-COMP-3	$1.1 \cdot 2.76 \cdot 0.102$	0.40	Double clamped
TUD-COMP-4	$4.0 \cdot 2.76 \cdot 0.102$	0.50	Double clamped
TUD-COMP-6	$4.0 \cdot 2.76 \cdot 0.102$	0.50	Cantilever

angle, that was calibrated to better match the experimental results (Table 2). More information about the experimental in-plane tests can be found in (Messali et al., 2017).

In the following sections the experimental versus the numerical results are presented in terms of shear force versus displacement graph for the three walls, and a comparison is made between the current model and the commonly used TSRCM and EMM. Additionally, special focus is given on the comparison of the crack pattern between the presented model and the EMM, due to the latter's inability to localize damage. For the modelling of the walls the properties that are presented in Table 2 are used for the EMM and the current model, whereas for the TSRCM the properties parallel to the head joints (y direction) are used.

3.1   Slender wall TUD-COMP-3

Specimen TUD-COMP-3 is a slender (aspect ratio of 2.5), double-clamped wall with a precompressive load of 0.4 MPa applied on its top face. In Figure 7

Table 2.   Material properties of tested calcium silicate brick masonry.

Material Properties used in numerical analyses, parallel to			bed joint $i = x$	head joint $i = y$
Modulus of elasticity	E_i	[MPa]	3583	5091
Shear Modulus	G	[MPa]	1571	
Tensile strength	f_{ti}	[MPa]	0.21	0.14
Compressive strength	f_{ci}	[MPa]	7.55	5.93
Fracture energy tension	G_{fti}	[N/ mm]	0.02	0.012
Fracture energy in compression	G_{fci}	[N/ mm]	43.4	31.3
Cohesion	c_0	[MPa]	0.14	
Friction angle	$\tan \phi$	[-]	0.43 (test)	0.63 (num)

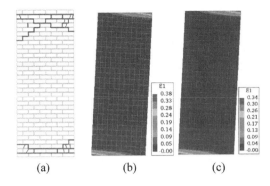

(a)   (b)   (c)

Figure 6. Experimental crack patter (a) and principal strains at maximum horizontal displacement of the EMM (b) and current model (c) for specimen TUD-COMP-3.

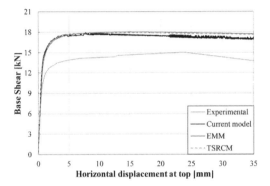

Figure 7. Experimental and numerical shear force-displacement diagram of slender wall TUD-COMP-3.

it can be seen that there is an overestimation of the maximum base shear (17.9 kN) with respect to the experimental results (15 kN). However, similar overestimation is found when modelling the wall by applying both the TSRCM and the EMM. Nevertheless, the behavior of the slender wall and its flexural failure are well captured for all three constitutive models, with the damage localizing at the top and bottom layers of the wall.

Horizontal cracks are formed and the rocking behavior of the EMM and the current model are clearly visible in Figure 6b and c.

3.2 *Shear wall TUD-COMP-4*

Specimen TUD-COMP-4 is a double-clamped squat shear wall (aspect ratio of 0.7) with a vertical precompression of 0.5MPa. The experimental test showed a clear shear failure with diagonal shear cracks forming at the center of the wall and later expanding towards the corners. The maximum experimental shear force is 119.9kN, whereas the predicted numerical force is 136.6kN. There is an overestimation of the shear force for the current

model, reached at a smaller horizontal displacement. The peak numerical force, followed by a small plateau, eventually softens until it reaches a new plateau at a much lower force (75kN). Moreover, the softening rate is slightly faster than in the case of the experimental wall. Nonetheless, the localization of the damage is very good, showing clearly a diagonal crack and the shear sliding along this diagonal. This can be observed not only by the strain tensors, but also from the deformed shape of the wall in Figure 8c that shows an offset of the horizontal displacements of the upper triangle.

By comparing the numerical results of the current model with the ones of the TSRCM and the EMM it is apparent that the prediction of the base shear force with respect to the TSRCM is improved. Moreover, by comparing Figure 8b and c it is evident the sharper and better localization of the current model in comparison to the EMM, where many steep cracks are formed, parallel to each other.

3.3 *Shear wall TUD-COMP-6*

Wall TUD_COMP_6 is a cantilever squat wall (aspect ratio 0.7, same as TUD-COMP-4), with a precompression load of 0.5MPa. The numerical results of the current constitutive model are slightly overestimating the experimental ones, with a predicted peak base shear force of 120kN versus 109.8kN of the experimental value. From the numerical analysis, it can be observed a different type of localization. A diagonal crack is formed at a distance of approximately $l_w/3$ from the bottom right corner. The crack spreads towards the bottom left corner, reaching eventually the base of the wall at a distance $h_w/3$ from the top left corner. This is a different crack pattern than it was observed in the experiment, however it still corresponds to a very localized shear failure.

Similar to wall TUD-COMP-4, the prediction of the base shear force is on one hand improved with respect to the TSRCM (Figure 10), while on the other hand it presents a better localization of damage with respect to the EMM (Figure 11b, c).

4 DISCUSSION AND CONCLUSIONS

A total-strain-crack model, taking into account the anisotropic behavior of masonry, was developed and implemented as user subroutine in DIANA 10.3. The inclusion of orthotropy, as well as the limitation of the shear stresses τ_{xy} alongside the bed joints with a Coulomb friction criterion, give a satisfactory estimation of the base shear force of the walls, as well as their post-peak behavior for walls TUD-COMP-3, 4 and 6. The failure mechanisms match the ones of the experiment and more importantly a very sharp localization of the damage is observed for all the numerical examples. Overall, it can be said that the

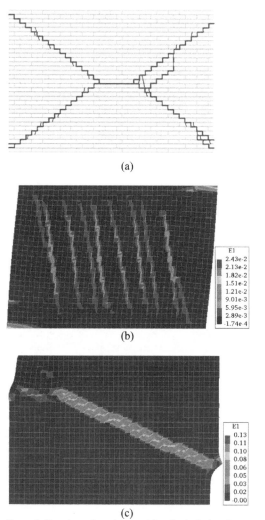

(a)

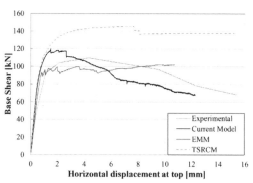

Figure 10. Experimental and numerical shear force-displacement diagram of shear wall TUD-COMP-6.

Figure 8. Experimental crack patter (a) and principal strains at maximum horizontal displacement of the EMM (b) and current model (c) for specimen TUD-COMP-4.

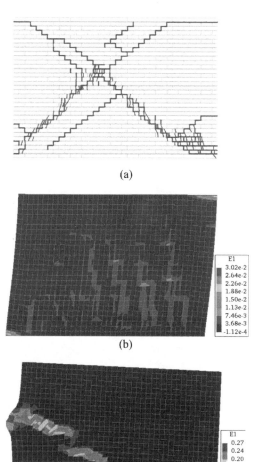

(a)

(b)

(c)

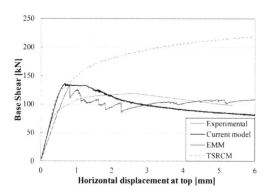

Figure 9. Experimental and numerical shear force-displacement diagram of shear wall TUD-COMP-4.

Figure 11. Experimental crack patter (a) and principal strains at maximum horizontal displacement of the EMM (b) and current model (c) for specimen TUD-COMP-6.

numerical results, at least for monotonic analyses, are promising and that this new approach helps to overcome the limitations and disadvantages of the TSRCM in terms of force estimation and of the EMM in terms of damage localization.

This is only a first step and further research is required. Future steps in the development and refinement of this model will include: the application of the model in cyclic loading; the dependency of the unloading stiffness based on the principal angle α_i so that less energy is dissipated for bed-joint opening and flexural cracks; improvements on the numerical stability and numerical effort of the subroutine.

ACKNOWLEDGEMENTS

This research was funded by Nederlandse Aardolie Maatschappij (NAM) BV under contract number UI46268 "Physical testing and modelling – Masonry structures Groningen", which is gratefully acknowledged.

REFERENCES

Atkinson, R. H., Amadei, B. P., Saeb, S. & Sture, S. 1989. Response of Masonry Bed Joints in Direct Shear. *Journal of Structural Engineering-Asce*, 115, 2276–2296.

Berto, L., Saetta, A., Scotta, R. & Vitaliani, R. 2002. An orthotropic damage model for masonry structures. *International Journal for Numerical Methods in Engineering*, 55, 127–157.

Cundall, P. A. & Hart, R. D. 1992. Numerical Modelling of Discontinua. *Engineering Computations*, 9, 101–113.

Diana, F. (2019) User's Manual Documentation - Release 10.3.

Jafari, S., Rots, J. G., Esposito, R. & Messali, F. 2017. Characterizing the Material Properties of Dutch Unreinforced Masonry. *Procedia Engineering*, 193, 250–257.

Lagomarsino, S., Penna, A., Galasco, A. & Cattari, S. 2013. TREMURI program: An equivalent frame model for the nonlinear seismic analysis of masonry buildings. *Engineering Structures*, 56, 1787–1799.

Lemos, J. V. 2007. Discrete Element Modeling of Masonry Structures. *International Journal of Architectural Heritage*, 1, 190–213.

Lourenço, P. B. 1996. Computational strategies for masonry structures. *Civil Engineering and Geosciences*. Delft, Delft University of Technology.

Lourenço, P. B. 2002. Computations on historic masonry structures. *Progress in Structural Engineering and Materials*, 4, 301–319.

Lourenço, P. B., De Borst, R. & Rots, J. G. 1997. A plane stress softening plasticity model for orthotropic materials. *International Journal for Numerical Methods in Engineering*, 40, 4033–4057.

Malomo, D., Pinho, R. & Penna, A. 2018. Using the applied element method for modelling calcium silicate brick masonry subjected to in-plane cyclic loading. *Earthquake Engineering & Structural Dynamics*, 47, 1610–1630.

Messali, F., Ravenshorst, G., Esposito, R. & Rots, J. G. 2017. Large-Scale Testing Program for the Seismic Characterization of Dutch Masonry Walls. *16th World Conference on Earthquake Engineering*. Santiago Chile.

Naraine, K. & Sinha, S. 1989. Behavior of Brick Masonry under Cyclic Compressive Loading. *Journal of Structural Engineering-Asce*, 115, 1432–1445.

Page, A. 1980. An Experimental Investigation of the Biaxial Strength of Brick Masonry.

Papa, E. & Nappi, A. 1997. Numerical modelling of masonry: A material model accounting for damage effects and plastic strains. *Applied Mathematical Modelling*, 21, 319–335.

Pelà, L., Cervera, M. & Roca, P. 2013. An orthotropic damage model for the analysis of masonry structures. *Construction and Building Materials*, 41, 957–967.

Roca, P., Cervera, M., Gariup, G. & Pela', L. 2010. Structural Analysis of Masonry Historical Constructions. Classical and Advanced Approaches. *Archives of Computational Methods in Engineering*, 17, 299–325.

Rots, J. G. 1988. Computational modeling of concrete fracture. *Civil Engineering and Geosciences*. TU Delft.

Rots, J. G., Messali, F., Esposito, R., Jafari, S. & Mariani, V. 2016. Computational modelling of masonry with a view to Groningen induced seismicity. *Structural Analysis of Historical Constructions: Anamnesis, Diagnosis, Therapy, Controls*, 227–238.

Schreppers, G. M. A., Garofano, A., Messali, F. & Rots, J. G. 2016. DIANA validation report for masonry modelling.

Siano, R., Sepe, V., Camata, G., Spacone, E., Roca, P. & Pelà, L. 2017. Analysis of the performance in the linear field of Equivalent-Frame Models for regular and irregular masonry walls. *Engineering Structures*, 145, 190–210.

Vecchio, F. J. & Collins, M. P. 1993. Compression Response of Cracked Reinforced-Concrete. *Journal of Structural Engineering-Asce*, 119, 3590–3610.

Brick and Block Masonry - From Historical to Sustainable Masonry –
Kubica, Kwiecień & Bednarz (eds)
© *2020 Taylor & Francis Group, London, ISBN 978-0-367-56586-2*

The compressive behaviour of mortar under varying stress confinement

A. Drougkas, E. Verstrynge & K.Van Balen
Building Materials and Building Technology Division, KU Leuven, Leuven, Belgium

ABSTRACT: The confinement of mortar in masonry under compression is one of the key processes influencing the compressive strength of the composite material. It is triggered by the mismatch of elastic properties between units and mortar, coupled with deformation conformity between the two material phases. In cases where the mortar is particularly deformable compared to the units, this confinement results in a peak stress many times the uniaxial compressive strength of the mortar. Therefore, a careful examination of this effect is critical in understanding the failure mechanisms of masonry in compression.

Mortar under compression can be modelled in a damage mechanics context, following the establishment of a) a constitutive stress-strain relation, b) a model for the increase of the compressive failure stress under lateral confinement and c) a model for the development (increase) of the Poisson's ratio of mortar under different stress levels. The first aspect is approached using established hardening-softening curves used for quasi-brittle materials, such as concrete. The second aspect is dealt with through the adoption of a suitable and sufficiently flexible failure criterion. The third aspect is addressed through fitting against experimental data.

The above aspects are expressed in a damage mechanics context, resulting in fast calculations of the compressive stress-strain curves for confined mortar. This approach allows the quantification of the development of damage in compression, the development of the apparent compressive strength and the relation between orthogonal strains in the mortar, leading to a full characterization of the stress, deformation and damage of the material. The analysis results are compared to experimental findings on different mortar types and are used for their interpretation and evaluation. The complexity of the behaviour of confined mortar is demonstrated, motivating the use of advanced numerical models for its accurate simulation and assessment.

1 INTRODUCTION

1.1 *State of the art*

The compressive behaviour and strength of masonry is influenced by numerous geometric and material parameters. Chief among them is the behaviour of the mortar in the bed joints under the confinement effect imposed on it by the units. This effect is a function of the stresses acting in directions perpendicular to the direction of the compressive load. This confinement effect can lead to a substantial increase in the apparent compressive strength of the mortar in the joint, and therefore of the masonry composite.

When studying the behaviour of masonry in compression, the interaction of the units and the mortar is paramount in understanding and quantifying the confinement effect. From a numerical standpoint, the Poisson's ratio of the mortar is the chief property affecting the lateral expansion of the material. Since high compressive stress results in practically zero slip between the units and the mortar, displacement conformity can be assumed between the two phases of the masonry composite. This assumption forms the basis of many analytical models of unit/mortar interaction in masonry (Haller, 1958; McNary & Abrams, 1985).

This phenomenon is demonstrable using numerical means for the simulation of the masonry composite in compression, which highlight the role of the assumed Poisson's ratio in determining the ultimate stress and the failure mode (Drougkas et al., 2019). Nevertheless, measuring the Poisson's ratio in brittle cementitious materials, such as concrete and masonry mortars, is a difficult task (Mohamad et al., 2007; Ottosen, 1979). This problem is compounded by the difficulty in distinguishing between the elastic component of the lateral deformation of the mortar under compression and the component resulting from plastic deformation. It is therefore necessary to also distinguish between the actual Poisson's ratio as a material parameter and the apparent Poisson's ratio as a measured quantity. The potentially substantial difference between the actual and apparent Poisson's ratios of mortar in masonry under compression (Drougkas et al., 2019) needs to be considered when determining the elastic properties of mortar using, for example, electromechanical or optical measurements.

This distinction may be considered by adopting a damage mechanics approach in the behaviour of mortar in compression. By assigning a constitutive law for the development of the actual Poisson's ratio and calculating the apparent Poisson's ratio it is possible to study this complex effect in depth.

1.2 Objectives

In this paper a damage mechanics approach is developed in a numerical context and applied to available experimental data, aiming at simulating the behaviour of mortar in masonry joints. The main focus of the model is the simulation of the lateral expansion of the mortar under vertical loading, governed by a non-constant Poisson's ratio. A model for the development of the Poisson's ratio under different confinement is proposed.

The proposed model is tested against experimental data on cement mortars from the literature and against data from a recent testing campaign.

2 EXPERIMENTAL DATA

2.1 Overview

Two sets of experimental data are used for the validation of the proposed model: a) the experiments by McNary and Abrams (McNary & Abrams, 1985) and b) the experiments by Hayen et al (Hayen et al., 2009), previously reported by Drougkas et al (Drougkas et al., 2019).

The first set, designated as 'Series 1', involves masonry mortars of the standard types M, S, N, O (ASTM, 2019). Information on the lateral expansion of these mortars under different levels of confinement is provided for the M and O varieties, designated here as mortars M_1 and O_1.

The second set, designated 'Series 2', involves a similar series of masonry mortars, with variations on the type of lime and aggregate. Further experimental investigation is ongoing, with complete data on lateral expansion being currently available for one type of mortar under uniaxial compression, designated here as mortar O_2.

The mixtures of the mortars investigated in this paper are presented in Table 1. While not entirely conforming to the mix specifications of typical O mortars, the O_2 mortar had a similar uniaxial compressive strength and lateral expansion behaviour as O_1.

The experimentally obtained uniaxial compressive strength f_c and the Young's modulus E of these mortars are given in Table 2.

Overall, the two data sets do not provide full result data for lateral deformation of mortar under confinement. They have been performed with different materials, different procedures and the results are

Table 1. Summary of mortar mixtures per weight: cement (C), lime (L), sand (S) and water-to-cement ratio (W:C).

Mortar	C	L	S	W:C
M_1	1.00	0.25	3.00	0.55
O_1	1.00	2.00	9.00	1.96
O_2	1.53	1.00	11.86	1.85

Table 2. Uniaxial compressive strength and Young's modulus of mortars.

	f_c (N/mm^2)	E (N/mm^2)
M_1	31.1	11600
O_1	6.2	1750
O_2	5.0	713

presented in different ways. However, it was deemed desirable to attempt the application of the proposed model to as large a number of cases as possible.

The stresses applied on the samples may be normalized using two different approaches. The confining stresses $\sigma_x = \sigma_z = \sigma_c$ can be normalized through division by the vertically applied stress σ_y. The applied vertical stress itself may be normalized by division by the confined compressive strength f_{cc} registered for each level of stress confinement. These normalizations are given by the expressions:

$$\beta_1 = \left| \frac{\sigma_c}{\sigma_y} \right|$$
$$\beta_2 = \left| \frac{\sigma_y}{f_{cc}} \right|$$

(1)

Both normalized parameters assume values between 0.0 and 1.0 and are meaningful for any level of confinement. Absolute values are considered in the normalization since compressive stresses are negative while tensile forces and material properties, such as f_{cc}, are positive.

The initial and final apparent Poisson's ratio, defined at the beginning of loading and at peak stress respectively, are designated v_i and v_f. While the behaviour of masonry in compression is heavily influenced by the Poisson's ratio throughout the loading process, v_f is the parameter that has the greatest influence in the peak load.

2.2 Series 1 mortar

For the study of the Poisson's ratio of masonry mortar in compression, the results of the experiments by McNary are investigated (McNary & Abrams, 1985). This set of results includes four mortars with different elastic and strength properties under different levels of lateral stress confinement. Details on the measured apparent Poisson's ratio are given for two of these mortars.

In addition to uniaxial testing, the mortars were subjected to compression under different levels of lateral stress confinement. This confinement was provided by testing cylindrical samples in a Hoek cell, which provides uniform lateral stress confinement. Deformation measurements were acquired, providing information on the samples' apparent Poisson's ratio.

The apparent Poisson's ratio $\hat{\nu}$ is plotted against the normalized applied vertical stress β_2 for different levels of lateral compression σ_c in Figure 1. No information on the scatter is provided. For all cases, the initial Poisson's ratio is very low, roughly around 0.10, and increases for higher values of vertical load. This increase leads the Poisson's ratio to assume values higher than what normally admissible in elasticity. Additionally, it is immediately apparent that for the same lateral confining stress, the M_1 mortar presents a generally higher apparent final Poisson's ratio than the O_1 mortar. The reason for this is not clear, but could potentially be linked to the higher porosity of the O_1 mortar, which results in lower lateral expansion due to pore collapse in compression.

2.3 Series 2 mortar

Data on the O_2 mortar is limited as the lateral expansion behaviour of the material has only been tested in uniaxial loading conditions in cylindrical samples. However, the data acquired from the test is valuable in this investigation.

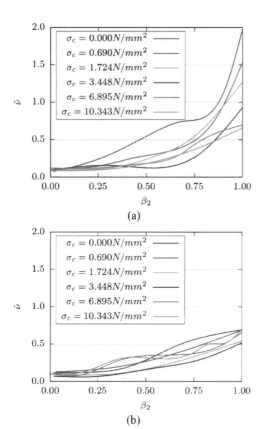

Figure 1. Experimentally determined apparent Poisson's ratio of a) mortar M_1 and b) mortar O_1 (adapted from McNary and Abrams 1985).

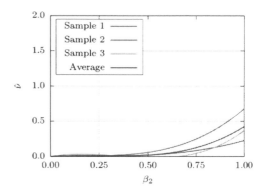

Figure 2. Apparent Poisson's ratio of O_2 mortar.

Table 3. Triaxial compressive test results for O_2 mortar.

β_1 (-)	f_{cc} (N/mm^2)	E (N/mm^2)	ε_{cc} (‰)
0.00	5.25	712.7	12.66
0.05	8.68	937.9	23.74
0.10	11.40	1296.7	32.34
0.15	12.96	1168.0	34.02
0.25	15.94	1134.5	43.21

Three samples of the same material were tested in uniaxial compression, and the resulting apparent Poisson's ratio is shown in Figure 2. The behaviour of this mortar is distinctly different from that of the Series 1 mortars. The initial Poisson's ratio is negligible and the lateral expansion under compression is only mobilized at around 50% of the peak load. However, the final Poisson's ratio is similar to the one obtained for the O_1 mortar, which has similar compressive strength and Young's modulus.

Additional triaxial compression tests were conducted on the O_2 mortar. The obtained confined compressive strength f_{cc} and peak strain ε_{cc} are reported in Table 3.

3 NUMERICAL MODEL

3.1 Overview

The mortar samples are modelled using a simple model for the stress and strain state of the material. This model simulates the stress state in the mortar under uniform triaxial compression, as it arises in a Hoek cell. While stresses through the volume of the mortar in a masonry joint may not be constant, especially near the free edges of the joint, they assume relatively uniform values for a substantial proportion of their volume around the centre of the joint (Drougkas et al., 2015). Therefore, the approach of uniform confinement adopted here is considered adequate for the simulation of mortar in masonry joints.

3.2 Basic expressions

The three-dimensional Hooke's law is used for relating the effective stress tensor σ with the strain tensor ε. Disregarding shear stresses, its matrix form reads:

$$
\begin{bmatrix} \varepsilon_x \\ \varepsilon_y \\ \varepsilon_z \end{bmatrix} = \frac{1}{E} \begin{bmatrix} 1 & -v & -v \\ -v & 1 & -v \\ -v & -v & 1 \end{bmatrix} \begin{bmatrix} \sigma_x \\ \sigma_y \\ \sigma_z \end{bmatrix} \tag{2}
$$

where E is the Young's modulus and v is the actual Poisson's ratio. The term effective stress signifies the stress that is proportional to the strain and is different from the actual stress. Positive values for stress and strain indicate tension.

The actual stress in compression is calculated through the use of a parabolic stress-strain curve based on fracture energy proposed for concrete (Feenstra & Borst, 1996). The increase of the compressive strength due to confinement is taken into account through the use of the Hsieh-Ting-Chen failure criterion (Hsieh et al., 1982). This criterion, based on four numerical parameters derived from different experimental tests, affords substantial flexibility in the definition of the desired failure criterion and can be degenerated to other failure curves, such as the Drucker-Prager criterion. Finally, the variation of the Poisson's ratio of mortar as a function of applied load is modelled using a semi-empirical expression (Drougkas et al., 2019).

Adopting a damage mechanics approach for compression, the actual, or damaged, stress is related to the effective stress from eq.(1) through the use of an integrity variable. According to the constitutive law chosen for compression (negative stress), this variable is calculated according to the piecewise parabolic equation (Drougkas et al., 2019):

$$
C(\varepsilon) = \begin{cases} -\dfrac{f_c}{\sigma} \left(\dfrac{1}{3} + \dfrac{4}{3} \dfrac{\varepsilon - \varepsilon_l}{\varepsilon_c - \varepsilon_l} - \dfrac{1}{2} \right) & \text{if } 0 \le \varepsilon \le \varepsilon_l \\ \dfrac{2}{3} \left(\dfrac{\varepsilon - \varepsilon_l}{\varepsilon_c - \varepsilon_l} \right)^2 & \text{if } \varepsilon_l \le \varepsilon \le \varepsilon_c \\ & \text{if } \varepsilon_c \le \varepsilon \le \varepsilon_u \\ -\dfrac{f_c}{\sigma} \left(1 - \left(\dfrac{\varepsilon - \varepsilon_c}{\varepsilon_u - \varepsilon_c} \right)^2 \right) & \text{if } \varepsilon_u \le \varepsilon_u \\ 0 \end{cases} \tag{3}
$$

where f_c is the compressive strength, ε is the strain and σ is the effective stress in the direction being evaluated. The actual stress tensor σ_d is derived from σ through:

$$
\sigma_d = \sigma C(\varepsilon) \tag{4}
$$

Damage is isotropic, meaning that reduction of stiffness due to damage in one direction leads to the same reduction in stiffness in all loading directions. The strain ε_l is the limit of proportionality in compression:

$$
\varepsilon_l = -\frac{1}{3} \frac{f_c}{E} \tag{5}
$$

ε_c is the peak strain:

$$
\varepsilon_c = -\frac{5}{3} \frac{f_c}{E} \tag{6}
$$

and ε_u is the ultimate strain:

$$
\varepsilon_u = \varepsilon_c - \frac{3}{2} \frac{G_f^c}{f_{cl}} \tag{7}
$$

where G_f^c is the compressive fracture energy and l is the characteristic length, here equal to the sample height, which only affects the post-peak part of the stress-strain curve.

3.3 Triaxial confinement of mortar

The failure criterion, expressed in principal stress terms, reads:

$$
f = A \frac{J_2}{f_c^2} + B \frac{\sqrt{J_2}}{f_c} + C \frac{\sigma_1}{f_c} + D \frac{I_1}{f_c} - 1 \tag{8}
$$

where I_1 and J_2 are the first stress and second deviatoric stress invariants respectively, expressed as:

$$
I_1 = \sigma_1 + \sigma_2 + \sigma_3
$$

$$
J_2 = \frac{1}{6} \left((\sigma_1 - \sigma_2)^2 + (\sigma_2 - \sigma_3)^2 + (\sigma_3 - \sigma_1)^2 \right) \tag{9}
$$

and σ_1 is the maximum of the principal stresses σ_1, σ_2 and σ_3. The numerical parameters A, B, C, D are calculated by solving a linear system of equations derived from eq.(8) for 4 different types of loading: uniaxial compression, uniaxial tension, biaxial compression and triaxial compression. The confined compressive strength f_{cc} is calculated by

705

solving eq.(8) for the given laterally applied confine-ment stresses.

The increase in the peak stress due to confinement also results in an increase in the peak strain. This increase is calculated according to the Eurocode 2 expression for concrete (CEN, 2004):

$$\varepsilon_{cc} = \varepsilon_c \left(\frac{f_{cc}}{f_c}\right)^2 \qquad (10)$$

The equation for the ultimate strain ε_u remains unchanged. This results in a more brittle post-peak response for confined mortar, as the energy expended between peak and ultimate strain does not change.

3.4 Poisson's ratio

For the actual Poisson's ratio v, the following equation, a function of the applied vertical stress, is here proposed:

$$v(\beta_2) = (v_f - v_i)\beta_2^3 + v_i \qquad (11)$$

where v_i and v_f are the initial and final Poisson's ratios, the latter registered at peak stress. For the Series 1 mortars examined in this paper, a value of 0.10 is taken for v_i, while 0.00 is considered for the Series 2 mortar. The value of v_f is, in turn, expressed as:

$$v_f(y) = y(t)\frac{f_{cc}}{f_c} \qquad (12)$$

where y is the vertical abscissa of a quadratic Bézier curve and t is its time parameter, tracing the curve from beginning to end through variation from 0 to 1. The horizontal abscissa x is:

$$x = \varepsilon_{cc} \qquad (13)$$

while for $y(t)$ the piecewise equation is proposed:

$$y(t) = \begin{cases} y_0 & if \quad 0 \le x \le x_0 \\ (1-t)((1-t)y_0 + y_1t) \\ +t((1-t)y_1 + y_2t) & if \quad x_0 \le x \le x_2 \\ y_2 & if \quad x_2 \le x \end{cases} \qquad (14)$$

Inversely to the way in which Bézier curve equations are normally expressed, the time parameter t is expressed as a function of the horizontal abscissa x as:

Table 4. Coordinates of control points for actual Poisson's ratio of mortar v model.

x_0	0.005	y_0	2.000
x_1	0.005	y_1	0.200
x_2	0.025	y_2	0.200

$$t(x) = \frac{\sqrt{(x - x_0) + x_1^2 + x(x_0 - 2x_1)} - x_1 + x_0}{x_2 - 2x_1 + x_0} \qquad (15)$$

The value of $t(x)$ can be plugged into eq.(4) in order to obtain the value of the vertical abscissa.

In its quadratic form, a Bézier curve is characterized by three control points: a starting point $P_0(x_0, y_0)$, an intermediate point $P_1(x_1, y_1)$ and an end point $P_2(x_2, y_2)$. The curve connects the end points P_0 and P_2, while P_1 serves as a control point, which does not necessarily lie on the curve. Bézier curves allow for enhanced flexibility and mathematical convenience in interpolating experimental data compared to polynomial curves. Further, the numerical parameters of the curve, namely the abscissae of the control points, can be directly related to the physical parameters being modelled and evaluated. Hence the adoption of this approach in this paper.

The coordinates for the Bézier curve control points for fitting the experimental data are presented in Table 4.

3.5 Analysis procedure

For a linear elastic analysis the relevant material properties and applied stresses or strains can be simply plugged in eq.(2) and solving the system of linear equations. In the case of a non-linear analysis, the system of equations is solved in steps of applied strain in the vertical y direction. The system of equations $F(X_n)$ is solved iteratively for each load step using a Newton algorithm:

$$X_{n+1} = X_n - J_F(X_n)^{-1}F(X_n) \qquad (16)$$

where n is the iteration number and $J_F(Xn)$ is the Jacobian matrix. The variables assembled in the X_n tensor are the stresses, strains, integrity variables in compression and the actual Poisson's ratio of the material as previously defined in the description of the analytical model.

Due to the analytical formulation of the problem using a small system of simple linear equations, the non-linear analysis can be executed with very low computational cost.

In the first iteration of every load step, the integrity variable and Poisson's ratio of the previous step is used. At every subsequent iteration executed, trial values are calculated for the integrity variable C and the Poisson's ratio v according to the stress and strain increment. These trial values are compared with the initial values for which the iteration is executed. The convergence criterion is considered satisfied when the trial and actual values differ by less than 0.1%. For applied strain load steps equal to 0.0001, convergence is typically reached within 15 iterations.

From eq.(11) it is possible to determine the actual Poisson's ratio v of the material. The apparent Poisson's ratio is calculated in the analysis according to the standard definition of the negative ratio of lateral over vertical deformation:

$$\widehat{v} = -\frac{\varepsilon_x}{\varepsilon_y} \tag{17}$$

In the absence of relevant data, the compressive fracture energy is given a nominal value, calculated as (Drougkas et al., 2015):

$$G_f^c = f_c d \tag{18}$$

where d is a ductility index equal to 1 mm.

4 NUMERICAL ANALYSIS RESULTS

4.1 Overview

The predictive capacity of the proposed model for the final apparent Poisson's ratio is checked. The results of the analytical models compared to the experimental data are presented in Figure 3. Good overall coincidence is obtained between the piecewise expression and the experimental data.

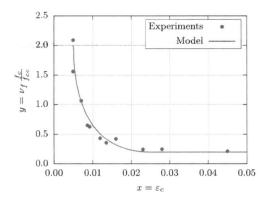

Figure 3. Comparison of analytical model for v_f with experimental data for M_1, O_1 and O_2 mortars.

4.2 Series 1 mortars

Initially, the assumption for the increase in the peak strain due to confinement according to eq.(10) is checked. The experimentally determined stress-strain graphs of the M_1 and O_1 mortars and their comparison with the model results are presented in Figure 4.

While the experimentally determined curves for the O_1 mortar are not completely reported, a good match is obtained for the O_1 mortar up to the peak. The very short post-peak response reported in the O_1 mortar tests indicates that a lower value for the compressive fracture energy should be used. However, conclusive remarks cannot be made in this aspect due to the absence of a complete post-peak curve from the tests as reported. Therefore, no adjustment was deemed necessary at this point.

The comparison of the experimentally determined and numerical derived apparent Poisson's ratio is presented in Figure 5. The model is able to calculate the initial and final apparent Poisson's ratio of the mortar with good accuracy for most cases, while intermediate values are captured well in the cases where the

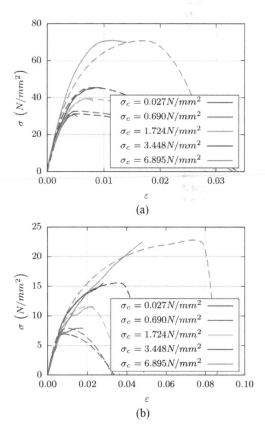

Figure 4. Comparison of experimentally determined (solid lines) and numerically derived (dashed lines) stress-strain curves: a) mortar M_1 and b) mortar O_1 (experimental data adapted from McNary and Abrams 1985).

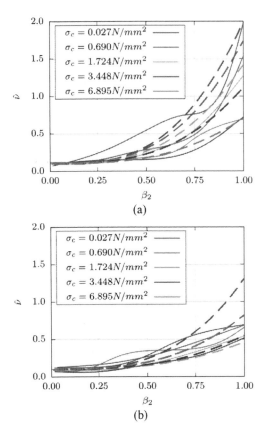

(a)

(b)

Figure 5. Comparison of experimentally determined (solid lines) and numerically derived (dashed lines) apparent Poisson's ratio: a) mortar M_1 and b) mortar O_1.

experimental behaviour presents a regular pattern. It successfully captures the overall difference in the resulting values between the two types of mortar in Series 1.

4.3 Series 2 mortar

The stress-strain curves obtained experimentally and numerically for the O_2 mortar are compared in Figure 6. The compressive fracture energy used in the model was calculated as for the Series 1 mortars, with a very good fit being obtained in this case with the experimental data. More importantly, the peak strain has been calculated with good accuracy and a complete post-peak curve has been registered.

The proposed model is additionally evaluated in terms of predicted Poisson's ratio compared to the experimental average. This comparison is illustrated in Figure 7. Assuming a value of 0.00 for v_i, a good prediction of the behaviour of the mortar is obtained, mainly in terms of v_f at peak load. According to the proposed model, the value elected for v_i does not affect the v_f parameter, and therefore has no real influence in the peak load. It may, nevertheless, affect the mode of damage initiation in the masonry composite.

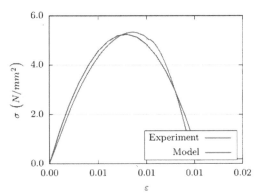

Figure 6. Comparison of experimental and numerically derived stress strains curves for O_2 mortar.

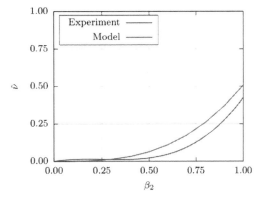

Figure 7. Comparison of analytical model for v_f with experimental data for O_2 mortar.

Finally, the numerically calculated confined compressive strength f_{cc} and peak strain ε_{cc} are presented in Table 5. An excellent approximation of both values is achieved for nearly the entire range of β_1 investigated in the experiments (compare with Table 3). The only notable discrepancy is obtained for a value $\beta_1 = 0.25$, where a lower than anticipated peak strain was obtained in the experiments.

Table 5. Numerical results for O_2 mortar: compressive strength and peak strain under confinement.

β_1 (-)	f_{cc} (N/mm^2)	ε_{cc} (‰)
0.00	5.25	11.88
0.05	6.66	25.50
0.10	8.26	32.34
0.15	10.30	30.14
0.25	16.10	68.00

5 CONCLUSIONS

A model for the development of the Poisson's ratio of mortar under compression is proposed. It is implemented in a numerical damage mechanics context and compared to relevant experimental results. The model accounts for the shift in compressive behaviour and lateral expansion of mortar under different levels of stress confinement. A variety of mortar types are investigated, the main characteristics of their response being successfully captured by the model.

The distinction between the Poisson's ratio of mortar as a material property and as an apparent value is important. While differences in the two parameters in stand-alone samples may not be substantial, the differences in masonry joints need to be considered.

Further work focusing on lime-based mortars is required. Confinement effects are more pronounced in masonry made with deformable mortars with low or zero cement content due to the increased relative deformability of the mortar compared to the units. The initial Poisson's ratio may also be substantially different from the values encountered in cement mortars due to differences in porosity.

REFERENCES

ASTM. (2019). *ASTM C 270 - Standard Specification for Mortar for Unit Masonry.*

CEN. (2004). *EN 1992- 1-1- Eurocode 2: Design of concrete structures - Part 1-1: General rules and rules for buildings.*

Drougkas, A., Roca, P., & Molins, C. 2015. Numerical prediction of the behavior, strength and elasticity of masonry in compression. *Engineering Structures, 90,* 15–28.

Drougkas, A., Verstrynge, E., Hayen, R., & Van Balen, K. 2019. The confinement of mortar in masonry under compression: Experimental data and micro-mechanical analysis. *International Journal of Solids and Structures, 162*(May), 105–120.

Feenstra, P. H., & Borst, R. De. 1996. A composite plasticity model for concrete. *International Journal of Solids and Structures, 33*(5), 707–730.

Haller, P. 1958. Hochhausbau in Backstein: die technischen Eigenschaften von Backstein-Mauerwerk für Hochhäuser. *Schweizerische Bauzeitung, 76*(28), 411–419.

Hayen, R., Van Balen, K., & Van Gemert, D. 2009. Triaxial interaction of natural stone, brick and mortar in masonry constructions. *Building Materials and Building Technology to Preserve the Built Heritage, WTA Schriftenreihe,* 333–352.

Hsieh, S. S., Ting, E. C., & Chen, W. F. 1982. A plastic-fracture model for concrete. *International Journal of Solids and Structures, 18*(3), 181–197.

McNary, W. S., & Abrams, D. P. 1985. Mechanics of Masonry in Compression. *Journal of Structural Engineering, 111*(4), 857–870.

Mohamad, G., Lourenço, P., & Roman, H. R. 2007. Mechanics of hollow concrete block masonry prisms under compression: Review and prospects. *Cement and Concrete Composites, 29*(3), 181–192.

Ottosen, N. S. 1979. Constitutive Model for Short-Time Loading of Concrete. *Journal of the Engineering Mechanics Division, 105*(1), 127–141.

Brick and Block Masonry - From Historical to Sustainable Masonry –
Kubica, Kwiecień & Bednarz (eds)
© 2020 Taylor & Francis Group, London, ISBN 978-0-367-56586-2

3D micro-modelling of brick masonry under eccentric axial load using frictional and cohesive brick-mortar interaction behaviour

G. Iskander & N.G. Shrive
University of Calgary, Canada

ABSTRACT: Masonry is one of the oldest known building materials. Despite the complexity of designing and understanding the behaviour of masonry under many load combinations, masonry still represents a competitive choice to other building materials in many areas of the world because of masonry's low cost, flexibility in creating good classical shapes, fire resistance and thermal properties. The complexity of predicting masonry behaviour results from the use of different materials, units and mortar, as well as the variation in the unit and mortar properties. The interaction between the units and the mortar plays an important role in defining masonry strength. A micro-modeling technique has been developed to provide a non-linear, numerical material model for brick masonry. The model incorporates the non-linear behaviour of the masonry, initial geometric imperfections and non-linear contact properties to represent the normal, tangential and cohesive behaviour of the interaction. 12 different walls with different slenderness ratios and load eccentricities have been modelled using the proposed technique and verified against the experimental test results of Sandoval et al (2011). The model results are found to be in good agreement with the experimental results.

1 INTRODUCTION

Masonry is one of the oldest known building materials. Despite the complexity of understanding the behaviour of masonry under many load combinations, and thus designing efficiently with the material, masonry remains a competitive choice to other building materials in many areas of the world.

Numerical modelling of masonry still represents a challenge in the field of masonry design. Difficulties arise from modelling the non-linear behaviour of the material as well as the interaction between units and mortar. The existence of mortar introduces a cohesive behaviour between separate parts in contact. This cohesive behaviour plays a very important role in determining the strength of masonry, especially in tall walls under high eccentric load.

Lourenco and Rots (2002) proposed three techniques to model masonry walls and discussed the advantages and disadvantages of each technique. In (1997) they applied a macro-modelling technique to model the shear walls tested by Raijmakers and Vermeltfoort (1992) and Vermeltfoort et al. (1993): the results were found to be in good agreement with the experimental results.

The aim here was to develop a numerical finite element technique which can predict the behaviour and the strength of brick masonry walls of different slenderness ratios under axial load with different eccentricities. The model includes material non-linearities, cohesive contact behaviour and initial geometric imperfections. Abaqus 2019 software is

used as the platform to solve the numerical equations and for post processing the results.

The numerical models developed were used to simulate the tests performed by Sandoval et al. (2011) and the results compared to the experimental data: good agreement was observed.

The effects of introducing initial imperfections were studied to find the cases in which considering these imperfections are critical.

2 NUMERICAL MODEL DESCRIPTION

2.1 *Modelling technique*

The micro-modelling approach introduced by Lourenco and Rots (2002) was followed to model the walls. This technique suggests modelling the bricks separately with interaction properties between them. The material properties of the brick should represent the masonry material behaviour while the interaction properties represent the brick-mortar interface.

2.2 *Tested walls description*

Sandoval et al. (2011) tested 36 scaled walls with thickness of 35 mm. The heights of the walls were 237.5, 447.5, 657.5 and 897.5 mm giving slenderness ratios of 6.8, 12.6, 18.7 and 25.6. For each slenderness ratio, the load was applied with 0, t/6 and t/3 eccentricities, where t is the thickness of the wall. For each of the previous 12 combinations, three tests were performed and the mean value was calculated. Walls were tested in single curvature.

2.3 Model assembly and geometry

The bricks used to build the walls had dimensions of 72.5×35×12.5 mm, while the mortar joints were 2.5 mm thick. Since the developed model will only assemble individual units and account for the mortar behaviour by an interaction property described in section 2.6, the bricks used in the model were 75×35×15 mm to account for the mortar joint thickness. Figure 1 shows assemblies of walls with 6.8 and 25.7 slenderness ratios.

2.4 Material model

Sandoval et al. tested the properties of the bricks and the mortar as well as the interaction between them. The compressive strengths of bricks and mortar were 32.5 and 7.3 MPa respectively, while the average prism strength was 14.2 MPa. The direct tensile strength of the mortar was found to be 0.55 MPa.

The brick material was modelled to represent the behaviour of the masonry in both the elastic and the plastic stages. Young's modulus was taken as 8000 MPa and Poisson's ratio as 0.15. Concrete damage plasticity was used to model the plastic behaviour with a maximum of 14.2 MPa compressive strength as found in the experimental tests.

2.5 Mesh and element type used

Eight-node 3D "brick" elements with reduced integration (C3D8R) were used to model the bricks. Mesh sensitivity analysis was performed for different mesh sizes and a 5×5 mm mesh was chosen for meshing the wall as this mesh was found to achieve mesh independency with reasonable model solving time. Figure 2 shows a meshed brick.

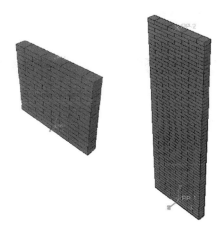

Figure 1. Wall assemblies.

Figure 2. Meshed brick.

2.6 Interaction property

The contact interaction property available in Abaqus was used to model the normal and tangential behaviours of both the head and bed joints. A "Hard contact" normal property was used to model the normal behaviour along with "frictional" tangential behaviour with a coefficient of friction of 0.812 as recommended by the experimental test. Cohesive strength values of 0.55 and 0.6 MPa were taken to represent the cohesive normal and tangential damage initiation stresses since the tensile bond strength was found to be 0.55 MPa for the head joints in the experimental test.

It is important to realize that the model is sensitive to the normal tensile cohesive strength as this parameter controls the buckling failure mode of the wall while the model is relatively insensitive to the shear cohesive strength. In addition, different cohesive elastic stiffnesses were used and the model was found also to be insensitive to the value of this parameter. This finding explains the huge variability of the values recommended by different researchers currently available in literature (Watstein D, Allen MH. (1970), Kirtschig and Anstötz (1991) and Hasan, S. and Hendry A. (1976).

2.7 Boundary conditions

The bottom surface of the wall was connected to a reference point using a coupling constraint. The lower pinned boundary condition was applied to the reference point by restraining movement in the X, Y and Z directions and rotation around the vertical axis. Similarly, the top surfaces of the top bricks were also connected to a reference point using a coupling constraint. The top pinned boundary condition was applied to the reference point by restraining movement in the X and Z directions and rotation around the vertical axis. The displacement in the vertical direction (Y) was not restrained in order to apply load on the top surface of the wall.

2.8 Load application

A vertical displacement was applied to the top reference point in a static step. The load required to apply the assigned displacement was calculated by the software by means of inverting the stiffness matrix. The top and bottom reference points were given different eccentricities of 0, t/6 and t/3 for each wall to model the experimental load eccentricities. Figure 3 shows the load application to the wall.

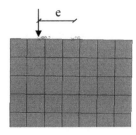

Figure 3. Load application over the width of the wall.

Figure 4. Wall first buckling mode.

2.9 Introducing initial imperfections

Sandoval et al. (2011) constructed their scaled walls using a technique where they glued the bricks to a timber sheet spaced by the mortar thickness value. Then they poured the pre-mixed mortar to fill the spaces and used a vibrating table to compact the mortar. Although the technique allowed them to minimize the effect of workmanship, initial geometric imperfections still existed but with lower values than one might expect from site conditions. To model these imperfections, an elastic buckling analysis was first performed to find the possible buckling modes. The first mode (global buckling) was taken as the assumed initial imperfection with a maximum bow value of brick-thickness/300 (Figure 4). Buckling modes with 3 and 5 half waves were also considered to determine how sensitive the walls are with respect to the initial imperfection profile. As expected, the first mode resulted in slightly lower strength in all walls.

3   RESULTS

3.1 Nomenclature

Walls are defined to show their slenderness ratio and load eccentricity in the following format; W

slenderness ratio-load eccentricity. For example, W25.6-t/3 is a wall with a slenderness ratio of 25.6 subject to a load placed eccentrically at t/3.

Table 1. Numerical model results.

Wall name	Avg. Experimental failure stress	Numerical model failure stress	$\%\frac{Numerical}{Experimental}$
W6.8-0	12.7	12.82	100.94%
W6.8-t/6	8.5	7.9	92.94%
W6.8-t/3	3.7	3.2	86.5%
W12.6-0	11.6	12.8	110.34%
W12.6-t/6	7.6	6.2	81.58%
W12.6-t/3	2.2	2.15	92.73%
W18.7-0	10.4	9.43	90.67%
W18.7-t/6	4.5	4.2	93.33%
W18.7-t/3	1.4	1.68	120%
W25.6-0	7.6	7.1	93.42%
W25.6-t/6	2.8	3	107.14%
W25.6-t/3	1.1	1.22	109.91%

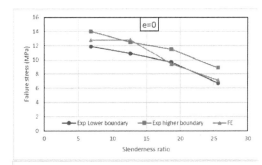

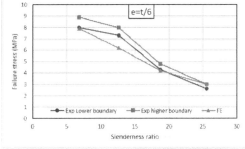

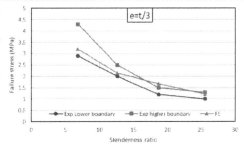

Figure 5. Stress Vs. slenderness ratio.

3.2 *Numerical model results*

A summary of the results of the numerical models and the experimental tests is provided in Table 1. The failure stresses are plotted against the slenderness ratio in Figure 5 for the numerical model results together with the lowest and highest stresses found experimentally.

Walls W6.8-0 and W6.8-t/6 failed experimentally by crushing of the masonry while all other walls failed by out-of-plane buckling. The experimental failure modes for two of the walls are shown against the failure modes predicted by the numerical models in Figures 6 and 7.

The lateral deformations at the mid-height of walls W18.7-t/6 and W18.7-t/3 are plotted against load in Figure 8. From Table 1 and Figures 5 to 8, it is clear that the numerical model results are in good agreement with the experimental results.

3.3 *Effect of introducing the initial imperfections*

The previous models were solved once again with no initial imperfections to determine when the

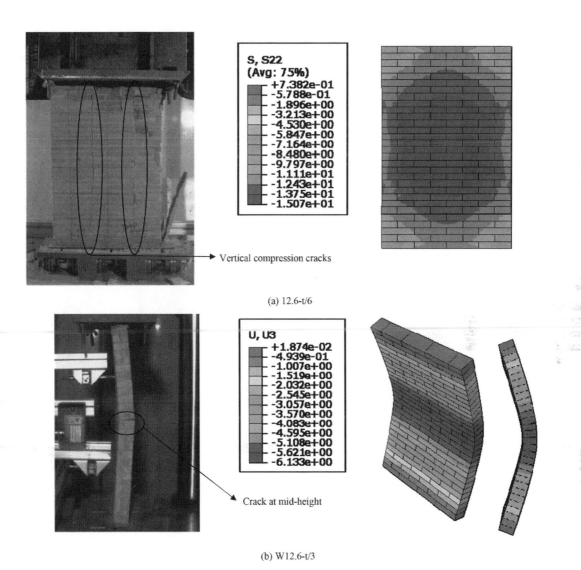

(a) 12.6-t/6

(b) W12.6-t/3

Figure 6. Failure modes of the experimental test as shown in Sandoval et al. (2011) and numerical model's prediction for W12.6-t/6 and W12.6-t/3.

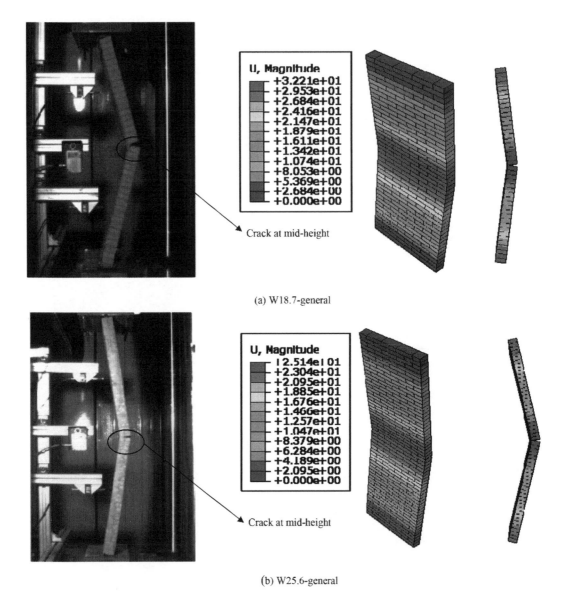

(a) W18.7-general

(b) W25.6-general

Figure 7. Failure modes of the experimental test as shown in Sandoval et al. (2011) and numerical model's prediction for W18.7-general and W25.6-genral.

effect of the imperfections becomes critical. The failure stresses for walls with loads applied at different eccentricities are plotted against slenderness for the numerical models with and without imperfections in Figure 9, together with the experimental test results. It is clear that the initial imperfection is very critical in the case of walls with concentric load failing by buckling as the

imperfections introduce instability to the wall. The imperfections are not critical in the case of walls failing by crushing as the strength of these walls depends mainly on the material properties. Walls subject to axial load (zero eccentricity) are more significantly affected by the introduction of the imperfection compared to the walls with loads at eccentricities of t/6 and t/3.

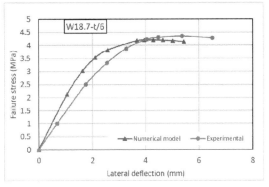

Figure 8. Stress Vs. Lateral deformation at mid-height.

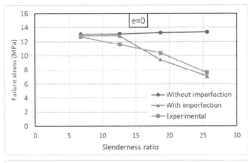

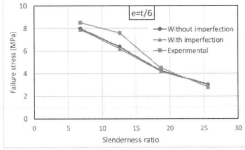

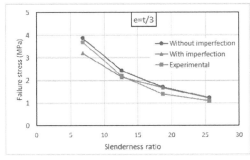

Figure 9. Failure stress vs. slenderness ratio.

4 CONCLUSIONS

- A numerical model was developed which can predict the failure mode and behaviour of solid brick masonry walls, as well as the failure stress, with a mean average of 98% of the experimental results and a Coefficient of Variation of 11.5 %.
- The tensile cohesive behaviour plays an important role for slender walls (h/t=18.7 and 25.6) which fail by out-of-plane buckling while short walls with low eccentricities (W6.8-0 and W6.8-t/3) are affected more by the material strength.
- The Young's modulus of elasticity also highly affects the behaviour of slender walls (h/t=18.7 and 25.6), especially slender walls with zero eccentricity. Young's modulus has a smaller effect on short walls (h/t=6.8 and 12.6) or walls with high eccentricity (W25.6-t/3 and W18.7-t/3).
- The presence of initial geometric imperfections highly affects walls failing by buckling especially walls with low eccentricities (W18.7-0 and W25.6-0).

ACKNOWLEDGEMENT

The authors gratefully acknowledge the financial support of the Natural Sciences and Engineering Research Council of Canada and the University of Calgary for this work.

REFERENCES

Hasan, S. and Hendry A. 1976. Effect of slenderness and eccentricity on the compressive strength of walls. *Proceedings of the Fourth International Brick Masonry Conference*, Brugge, Paper 4.d.3.

Kirtschig, K. and Anstotz, W. 1991 Kinckuntersuchu-ngen an mauerwerksproben. *Proceedings of 9th International brick/block masonry Conference*, p. 202–209.

Lourenco, P. and Rots, J. 1997. Multisurface Interface Modelfor Analysis of Masonry Structures. *Journal of Engineering Mechanics 123*(7):660-668.

Lourenco, P. 2002. Computations on historic masonry structures. *Prog. Struct. Engng Mater. 2002; 4*:301–319.

Raijmakers, T. and Vermeltfoort, A. 1992. Deformation controlled meso shear test on masonry piers. Rep. B-92-1156, TNOBOUW/TW Eidhoven, *Build. And Constr. Res.*, *Eidhoven*, The Netherlands.

Sandoval, C., Roca, P., Bernat, E., Gil, L. 2011. Testing and numerical modelling of buckling failure of masonry walls. *Construction and Building Materials 25 (12)*: 4394–4402.

Vermeltfoort, et al. 1993. Shear test on masonry walls. Proc.,6th North American Masonry conference, Philadelphia, Pa., 1183–1193.

Watstein D, Allen MH. (1970). Structural performance of clay masonry assemblages built with high-bond organic-modified mortars. *Second International Brick Masonry Conference*, 99–112.

Brick and Block Masonry - From Historical to Sustainable Masonry –
Kubica, Kwiecień & Bednarz (eds)
© 2020 Taylor & Francis Group, London, ISBN 978-0-367-56586-2

Efficiency and accuracy of a multiscale domain activation approach for modeling masonry failure

C. Driesen, H. Degée & B. Vandoren
Hasselt University, Hasselt, Belgium

ABSTRACT: Masonry is a composite consisting of two very different materials, which results in complex structural behavior. There exist accurate models in which both constituents are modeled explicitly, such as microscale and mesoscale models, but they require a great amount of computational resources. A faster, alternative, strategy to model masonry structures is the use of macroscale models, where homogenized elements are used which condense constituent behavior into one composite material. This approach also removes the upper bound on finite element sizes, which can lead to a reduction in element numbers. However, using one single material type makes it hard to capture the inherent nonlinear behavior when simulating masonry failure. In order to find a compromise between accuracy and computational efficiency one can use a scale embedding multiscale model where both macro- and microscale elements come into play, combining the advantages both have to offer. In this work a finite element based framework that formulates an adaptive scale embedding multiscale technique is presented, with the goal of both accurately and efficiently simulating large masonry structures. This theory is tested and compared to show its accuracy to rival a fully microscale model, while at the same time comparatively having a higher computational efficiency. In this work, the developed multiscale model is compared to its underlying microscale model using a couple example structures, ranging from small to large scale 2D unreinforced masonry walls with openings, including an application related to failure due to soil settlement, showing a potential for the application of this type of modeling.

1 INTRODUCTION

Despite a long history of working with masonry, the models and construction rules used for this material are less efficient and less clear compared to other building materials like steel or concrete. This is due to the inherent complexity of masonry as a building material, and the accompanying lack of models which accurately portray this material's complex behavior. Most of the methods used even now are based on generalized empirical results instead of verified scientific models. As a result, the use of masonry has declined somewhat compared to more solved materials such as steel and concrete, holding back the development of the material even further. The development of more accurate models could lead to the design of safer and more cost efficient masonry constructions.

The complexity of masonry lies in its constituents: brick units and mortar joints, which are organized on a relatively small scale. These materials form a complex arrangement which proves itself difficult to model due to the large amount of units that make up even a small scale construction. The second major difficulty with masonry is the complex structural response, which can cause computational problems. So, the existence of many elements means that to accurately model a large structure will take considerable resources.

The model discussed in this contribution involves using newer scale embedding multiscale techniques (Heyens, Vandoren, & Schueremans. 2011; Weinan. 2011), a multi-resolution solution (Ghosh. 2001) where the geometry is divided into parts which are modeled using macroscale techniques and other parts which use microscale techniques, as seen in Figure 1. In this figure, the macroscale is presented in green and the microscale in blue. Depending on the technique used, these parts can be explicitly connected through an extended stiffness matrix and solved simultaneously, or they can be implicitly connected via Lagrange multipliers using domain decomposition techniques leading to multiple smaller stiffness matrices (Formica. 2004; Kerfriden, Allix, & Gosselet. 2009; Lloberas-Valls, Rixen, Simone, & Sluys. 2012; Mobasher Amini, Dureisseix, & Cartraud. 2009). Both types have their advantages and disadvantages and are useful for different types of modeling. The macroscale homogenization is formulated via a concurrent type scale transition model. Using scale embedding techniques goes beyond simpler adaptive mesh refinement methods (Zhu & Zienkiewicz. 1988) since not just the mesh resolution changes but also the modelling method itself. An ex-ample is given in Figure 2, where a large wall con-sists partially of macroscale and microscale elements.

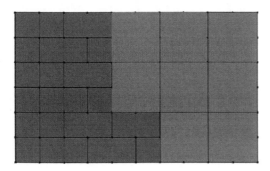

Figure 1. An example of the scale embedding multiscale approach, combining a blue microscale and green macroscale method.

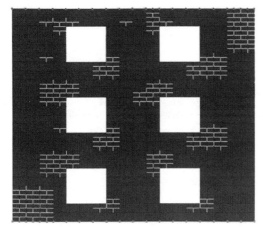

Figure 2. An example of refinement happening in a large wall.

Previous studies on this subject were performed by Greco et al. (Greco, Leonetti, Luciano, & Nevone Blasi. 2016; Greco, Leonetti, Luciano, & Trovalusci. 2017) using a model written in COMSOL. These followed earlier work by Heyens et al. (Heyens et al. 2011). The main novelty in the current contribution is the focus on computational efficiency on top of accuracy, showing that it is possible for such a multiscale model to reach a higher computational effi-ciency than a traditional microscale model.

2 MODELING FRAMEWORK

The model used in this work is based on Greco et al. and is composed of two stages: the initial macroscale and resulting microscale mesh. Both meshes are connected through an adaptive domain activation procedure. The macroscale model is chosen to be linear elastic and analytically homogenized while all the non-linear effects are simulated in the microscale

mesh. Activation happens based on a critical strain surface, calculated as the surface where damage first occurs in the Representative Volume Element (RVE). The microscale model is loosely based on previous work by Vandoren et al. (Vandoren, De Proft, Simone, & Sluys. 2013).

The modeling strategy considers two dimensional in-plane unreinforced masonry, consisting in general of a repeating microstructure, where any force or displacement is applied quasi-statically. Due to a relatively small thickness of the masonry wall, and the negligence of out-of-plane behavior, this work assumes plane stress conditions.

2.1 Microscale damage model

The damage model used in this work is an isotropic damage model in which the stiffness of the interface is scaled with a factor $1 - \omega$ (Mazars & Pijaudier-Cabot, 1989),

$$t = (1 - \omega)D\Delta u, \tag{1}$$

in which t are the tractions and Δu the relative displacements along the interface, D is the elastic constitutive matrix, and $\omega \in [0, 1]$ is called the damage factor. We choose the damage scaling to follow an exponential law using the equivalent displacement jump d_{eq}. This equivalent displacement jump is defined according to a degenerated Drucker-Prager model (Vandoren et al. 2013), with the parameters chosen such that there is a softening response in tension, and a compressive cap.

2.2 Domain activation method

A viable solution to the problem of computationally intensive micro- and mesoscale models, or of inaccurate macroscale models, is the use of scale embedded multiscale modeling techniques (Weinan, 2011). In the global multiscale modeling methods one attempts to achieve the advantages of both scales by combining their working methods, while trying to avoid their drawbacks. In this contribution, the concurrent approach is favored. The three major aspects in this concurrent model are the inter-element connections, the refinement criteria, and the homogenization techniques.

2.2.1 Inter-element connection
Connecting a macroscale element to its neighboring microscale elements is not so simple, due to the non-matching nodes: the macroscale element will usually have a smaller amount of them making straight connections impossible. One possible solution is to use a direct connection method, whereby one applies a hard connection between the outer two corresponding nodes of the microscale and the multiscale, as seen in Figure 3 where only the white nodes are

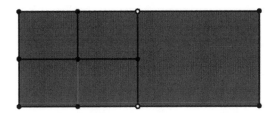

Figure 3. The connections between a microscale and a macroscale region.

connected. In this case, instead of creating new smaller stiffness matrices connected via Lagrange multipliers (Macorini & Izzuddin. 2013), the main stiffness matrix is simply expanded to include the new degrees of freedom. The main advantages are its accuracy (due to added robustness and a possibly tighter connection) and a high stability, but performance might lower in comparison to a Lagrange multiplier approach due to the added equations that need to be solved. Due to its stability and straightforwardness, the direct connection method is used in this work.

2.2.2 Refinement criteria

A crucial part of using a scale embedding multiscale model comes from the right choice of refinement criteria (Ghosh, Lee, & Raghavan. 2001; Lloberas-Valls et al. 2012). Refining the mesh too early means most of the performance gain is lost, but refining too late leads to the use of simple elastic macroscale elements when damage effects should be included. One possibility, which is used in the present research, is the use of a strain surface, where one defines the maximum strains a macro-element should reach before inducing refinement. Finally, these results will be interpolated between different loading conditions leading to the wanted strain surface. Once a macroscale element reaches this threshold it is refined into its accompanying microscale geometry.

There are a few possibilities to calculate this strain surface, and the classical method involves the use of experimental (Dhanasekar, Page, & Kleeman. 1985; Page. 1981, 1983) studies. An alternative is the use of a computational method where one applies a multitude of different loading conditions on the chosen microscale RVE and records for each case the strain present in the model at a chosen damage criterion. This damage criterion can be chosen to be as flexible or restrictive as necessary, and represents the aforementioned balance between performance and accuracy. In this case one chooses the criterion to be a nonzero damage value ω for any interface element in the RVE.

2.2.3 Homogenization techniques

For higher-scale structures it becomes helpful to incorporate macroscopic elements. The use of these elements, however, means that one needs to find a way to condense the complex linear behavior of the undamaged underlying RVE into one continuum element. This is done through homogenization techniques, which can be experimental, computational, or analytical (De Bellis, Ciampi, Oller, & Addessi. 2010; Nguyen. 2011; Weinan. 2011). In this example, a mathematical homogenization technique is used, which combines the geometric and material parameters into an element stiffness matrix (Wang, Li, Nguyen, Sitar, & Asce. 2007).

3 NUMERICAL APPLICATION

The efficiency of the developed model is demonstrated here using two numerical examples. In the first example, shear failure of a large wall, shown in Figure 4, is investigated. This wall is fixed on the bottom and top, with a horizontal displacement to the right applied to the top bricks. In the figure, the fixed boundary conditions are represented in blue, and the applied displacements in green. The model starts from an undamaged rest state, from which displacement increments will be applied until failure of the wall occurs.

In order to define the model an example RVE is chosen. This is represented in Figure 5; it consists of one full brick at the bottom and two half brick on the top. All half bricks are represented by continuous quad elements, while the possible damage pattern is represented by interface elements in the mortar joints and in the middle of the full brick. For the example RVE, the geometrical properties in Table 1 are chosen, with the material properties as in Table 2, in which σ_t is the tensile strength, σ_c denotes the uniaxial compressive strength, σ_b is the biaxial compressive strength, and G_{fI} is the mode I fracture toughness. The results are shown in Figures 6 and 7, which contain the force-displacement and total runtimes of the studied wall, respectively. The results show that

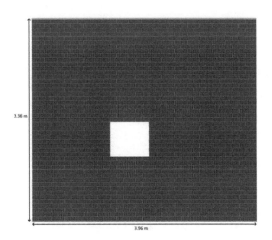

Figure 4. Setup of the first example.

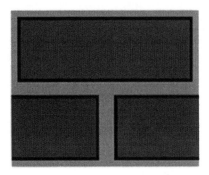

Figure 5. The chosen RVE used in both examples.

Table 1. The geometric properties of the RVE.

	Width (m)	Height (m)	Depth (m)	Mortar width (m)
RVE	0.22	0.14	0.1	0.01

Table 2. The material properties of brick and mortar used in the examples.

Material	Brick	Mortar
Young modulus (GPa)	16	8
Poisson's ratio	0.15	0.15
Gf_I (N/m)	1000	18
σ_t (kN/m^2)	1000	250
σ_c (kN/m^2)	100,000	11,000
σ_b (kN/m^2)	120,000	12,500

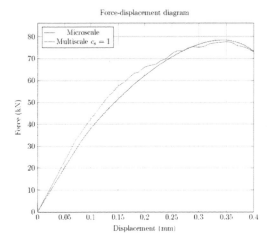

Figure 6. The force-displacement diagrams of the first example using both methods.

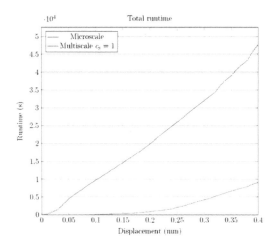

Figure 7. The computational runtimes of the first example using both methods.

a similar peak load can be found for both the multi-scale and the reference microscale model, however, the multiscale model only takes a fraction of the computational time.

Next, a larger real world example is studied, namely a simple masonry residential building situated in Flanders, Belgium, where the foundations have shifted downwards due to both a collapsed cellar and subsidence caused by water drainage. The damaged facade can be seen in Figure 8 with a close-up shown in Figure 9. The simulation uses a quasi-static model with a chosen subsidence function regulating the floor boundaries of the system. The only forces applied to the elements are their respective self-weights. The sides of the geometry are fixed, representing the depth of the side walls.

Figure 8. The facade of the building undergoing soil settlement.

Figure 9. A close-up of the facade.

Figure 10. The final results of the settlement example.

The same material properties are used as the ones from the previous example.

The results of the developed multiscale framework are shown in Figure 10. Here the comparison with a reference microscale model could not be made since the processor running the model could not handle the size of the full stiffness matrix. This is due to the amount of degrees of freedom which, while relatively small in the multiscale model, is up to 24 times larger in the microscale model. This causes a significantly larger stiffness matrix and requires significantly more processing power.

In the result the damage of the interface is shown in its color, from light grey meaning no damage to black meaning very high damage. When there is a completely open crack, meaning a damage value equal to unity, the color pink is used to easily spot the regions of high damage. This result is comparable to the real-life damage occurring in the structure, showing the versatility of the presented multiscale model.

4 CONCLUSIONS

A numerical model is presented, combining the micro- and macroscale approaches in masonry modeling into a single multiscale model. The model is applied to a complicated example of soil settlement under a real-life and very large masonry facade and is seen to show very comparable results. The reference microscale model is unable to obtain any results on the same processor build, due to its large amount of degrees of freedom. This again shows the power and versatility of the presented multiscale method.

REFERENCES

De Bellis, M., Ciampi, V., Oller, S., & Addessi, D. 2010. *Multi-Scale Techniques for Masonry Structures.* CIMNE.

Dhanasekar, M., Page, A. W., & Kleeman, P. W. 1985. Failure of brick masonry under biaxial stresses. *Proceedings of the Institution of Civil Engineers (London), 79*(pt 2), 295–313. https://doi.org/10.1680/iicep.1985.992.

Formica, G. 2004. *Multilevel analysis of masonry buildings.* University of Calabria.

Ghosh, S., Lee, K., & Raghavan, P. 2001. *A multi-level computational model for multi-scale damage analysis in composite and porous materials. International Journal of Solids and Structures* (Vol. 38). https://doi.org/10.1016/S0020-7683(00)00167-0.

Greco, F., Leonetti, L., Luciano, R., & Nevone Blasi, P. 2016. An adaptive multiscale strategy for the damage analysis of masonry modeled as a composite material. *Composite Structures, 153,* 972–988. https://doi.org/10.1016/j.compstruct.2016.06.066.

Greco, F., Leonetti, L., Luciano, R., & Trovalusci, P. 2017. Multiscale failure analysis of periodic masonry structures with traditional and fiber-reinforced mortar joints. *Composites Part B: Engineering, 118*(March), 75–95. https://doi.org/10.1016/j.compositesb.2017.03.004.

Heyens, K., Vandoren, B., & Schueremans, L. 2011. Multiscale Modeling of Masonry Structures Using Domain Decomposition Techniques. *XI International Conference on Computational Plasticity,* 422–431. Retrieved from http://www.scopus.com/inward/record.url?eid=2-s2.0-84858986997&partnerID=tZOtx3y1.

Kerfriden, P., Allix, O., & Gosselet, P. 2009. A three-scale domain decomposition method for the 3D analysis of debonding in laminates. *Computational Mechanics, 44*(3), 343–362. https://doi.org/10.1007/s00466-009-0378-3.

Lloberas-Valls, O., Rixen, D. J., Simone, A., & Sluys, L. J. 2012. *Multiscale domain decomposition analysis of quasi-brittle heterogeneous materials. International Journal for Numerical Methods in Engineering.* TU Delft. https://doi.org/10.1002/nme.3286

Macorini, L., & Izzuddin, B. A. 2013. Nonlinear analysis of masonry structures using mesoscale partitioned modelling. *Advances in Engineering Software, 60–61,* 58–69. https://doi.org/10.1016/j.advengsoft.2012.11.008.

Mazars, J., & Pijaudier-Cabot, G. 1989. Continuum damage theory - application to concrete. *Journal of Engineering Mechanics, 115*(2), 345–365.

Mobasher Amini, A., Dureisseix, D., & Cartraud, P. 2009. Multi-scale domain decomposition method for large-scale structural analysis with a zooming technique: Application to plate assembly. *International Journal for Numerical Methods in Engineering, 79*(4), 417–443. https://doi.org/10.1002/nme.2565.

Nguyen, V. P. 2011. *Multiscale failure modelling of quasi-brittle materials.*

Page, A. 1981. The biaxial compressive strength of brick masonry. *Proc. Institution of Civil Engineers 2*, 2(April), 893–906. https://doi.org/10.1680/iicep.1981.1825.

Page, A. 1983. The strength of brick masonry under biaxial compression-tension. *International Journal of Masonry Construction*, 3(1), 26–31.

Vandoren, B., De Proft, K., Simone, A., & Sluys, L. J. 2013. Mesoscopic modelling of masonry using weak and strong discontinuities. *Computer Methods in Applied Mechanics and Engineering*, (255), 167–182. https://doi.org/10.1016/j.cma.2012.11.005.

Wang, G., Li, S., Nguyen, H., Sitar, N., & Asce, M. 2007. Effective Elastic Stiffness for Periodic Masonry Structures. *Journal of Materials in Civil Engineering*, 19(March), 269–277.

Weinan, E. 2011. Principles of Multiscale Modeling. *Physics Today*, 65(6), 56. https://doi.org/10.1063/PT.3.1609.

Zhu, J. Z., & Zienkiewicz, O. C. 1988. Adaptive Techniques in the Finite Element Method. *Communications in Applied Numerical Methods*, 4, 197–204. https://doi.org/10.1007/978-981-10-8135-4_5.

Brick and Block Masonry - From Historical to Sustainable Masonry –
Kubica, Kwiecień & Bednarz (eds)
© 2020 Taylor & Francis Group, London, ISBN 978-0-367-56586-2

On the effect of tortuosity on the spalling prediction in masonry through a multiphase numerical model

G. Castellazzi, A.M. D'Altri, S. de Miranda, L. Molari & F. Ubertini
DICAM - Department of Civil, Chemical, Environmental, and Materials Engineering, ALMA MATER STUDIORUM - University of Bologna, Italy

H. Emami
ALMA MATER STUDIORUM - University of Bologna, Italy

ABSTRACT: Historical masonry structures are often subjected to degradation processes due to the damp rising, salt transport, and salt crystallization. Pores of building materials, such as natural stones or fired bricks, are filled with gaseous and liquid phases. The liquid phase could then consist of water and dissolved salts, that is then forced to migrate within the pores when external temperature and relative humidity vary. These changes could cause the precipitation of solid phase so called salt crystallization on the masonry surface (efflorescence) or within the material (subflorescence). The first is purely an aesthetic issue, the second can lead to structural damage called spalling: the flaking or peeling of superficial material. In this framework, the present paper aims at highlighting the role of the porous material properties, such as porosity and tortuosity on the whole phenomenon by a multiphase numerical model. Numerical results on benchmark simulations are presented and discussed.

1 INTRODUCTION

The majority of building materials for instance concrete, brick, stone, mortar, etc., have porous structure. During their life-cycle, these structures undergo detrimental processes which reduce their durability and strength. In particular, focusing on historical masonry structures, several are the environmental conditions that can plague these constructions during lifetime, such as rain, ground water absorption, atmospheric pollutant, metabolic activities of algae and microbes and numerous others degradation mechanisms (Franzoni 2014, Franzoni 2018, de Miranda 2019).

When a porous media is exposed to a fluid (sea water, moist air, etc.) the pores are filled by gaseous and liquid phases. These fluids usually contain several salts dissolved within the water which migrate through the pores with temperature and humidity variations.

In some specific circumstances, these salts crystallize generating pressure in the pores.

One of the first formulas regarding pressure due to salt crystallization is the well-known relation that has been introduced by Correns (1949):

$$p_s = \frac{\nu RT}{V_s} \left(ln \frac{\omega}{\omega_{sat}} + ln \frac{\gamma}{\gamma_{sat}} \right) \tag{1}$$

in which ν is the total number of ions released due to complete dissociation of salt, R is the ideal gas constant, T is the temperature, V_s is the molar volume of precipitated salt, ω is the mass of dissolved salt per unit mass of liquid phase ω_{sat} is the mass of dissolved salt per unit mass of liquid phase at saturation, γ is the mean activity coefficient of the dissolved salt and γ_{sat} is the mean activity coefficient of the dissolved salt at saturation. Scherer (2004), mentioned that the trigger for crystallization pressure is the super-saturation ratio in addition to the film between the crystals and pore wall. Steiger (2005), also developed the above-mentioned formula with respect to chemical potential of crystal surfaces accounting that the crystallization pressure and the consequent effects depend on the nature of the salt.

Numerous studies on salt transport and crystallization have been proposed by researchers. Hassanizadeh (1988), presented a salt and moisture transport problem without considering salt phase change.

Nicolai (2008), proposed a coupled heat, air, moisture and salt transport model with phase transition phenomena. Koniorczyk (2012), established a chemo-hydro-thermo-mechanical mathematical model and assessed the moisture content in the masonry.

After that Castellazzi et al. (2013), relates the crystallization pressure to the tensile stress within the pores. Further developments of this model are also discussed by Castellazzi et al. (2015), and

Grementieri et al. (2017a, b) addressing the coupled hygro-mechanical and the hygro-thermal problem respectively.

Recently, similar mathematical model has been implemented by Koniorczyk (2018) assessing in-pore salt crystallization taking into account also the salt damage and its evolution.

2 GOVERNING EQUATIONS

The set of equations used by Castellazzi et al., (2013), to model salt transport and crystallization has been taken into account. The mass balance equations can be written with respect to the moisture and salt mass conservation as follows:

$$\frac{\partial c_w}{\partial t} + \nabla \cdot j_w = -\mu_w^{ls} \tag{2}$$

$$\frac{\partial c_s^l}{\partial t} + \nabla \cdot j_s^l \frac{\partial c_s^s}{\partial t} = 0 \tag{3}$$

where, j_s^l is the flux of salt in liquid phase, $j_w = j_w^g + j_w^l$, being j_w^g the water vapor flux and j_w^l the water liquid flux, c_s^s is the concentration of salt in liquid phase, c_s^s is the concentration of salt in solid phase and c_w is the concentration of moisture (mass of moisture per unit volume of porous material), see Figure 1.

Also μ_w^{ls} is the rate of liquid water trapped in hydrated salt crystals.

The flux of liquid water and dissolved salt are related to their mass fractions and the following expression can be written:

$$j_w^l = (1 - \omega)j_{ws}^l - j_{s,diff}^l \tag{4}$$

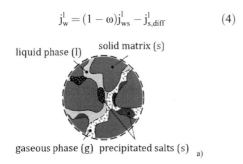

liquid phase (l)   solid matrix (s)

gaseous phase (g)   precipitated salts (s)   a)

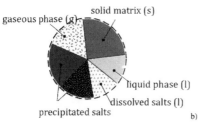

solid matrix (s)

gaseous phase (g)

liquid phase (l)

dissolved salts (l)

precipitated salts   b)

Figure 1. Representative Elementary Volume (REV).

$$j_s^l = \omega j_{ws}^l + j_{s,diff}^l \tag{5}$$

Where ω, as already anticipated in the previous section, is the mass of dissolved salt per mass of liquid phase which is defined in terms of concentration as:

$$\omega = \frac{c_s^l}{c_w^l + c_s^l} \tag{6}$$

2.1 Crystallization and dissolution equations

The salt crystallization or dissolution depends on the supersaturation ratio which is defined by ω over ω_{sat}, which are current mass concentration of dissolved salt and mass concentration at saturation. In particular:

$$\begin{cases} \dfrac{\omega}{\omega_{sat}} > \alpha_0 \Rightarrow \text{Crystallization} \\ \\ \dfrac{\omega}{\omega_{sat}} < 1 \Rightarrow \text{Dissoluation} \end{cases} \tag{7}$$

Where α_0 is the crystallization threshold ($\alpha_0 > 1$ for first cycle, $\alpha_0 = 1$ for further cycles). In general, the threshold of supersaturation ratio for primary crystallization relies on the properties of the porous material and on the type of salt.

The evolution equation which describes the salt precipitation/dissolution kinetics, i.e. it quantifies the amount of salt which precipitates, can be written as:

$$\frac{\partial c_s^s}{\partial t} = \pi r_p^2 \rho_s^s \frac{n}{V_{tot}} K_c \left| \frac{\omega}{\omega_{sat}} - 1 \right|^P \tag{8}$$

where a constant amount of salt nuclei n in the solution, as well as an isotropic distribution of cylindrical pores and cylindrical nuclei of the same radius of the pores (r_p), is assumed. In (12), ρ_s^s is the density of the crystallized salt, K_c is the growth rate coefficient, V_{tot} the pore volume and P is the crystallization process order that relies on the properties of the porous material and on the type of salt.

2.2 Constitutive equations

In order to relate the vapor and capillary pressure to the gas and liquid flow, constitutive equations can be written as below.

$$\mathbf{j}_w^g = -K_g \nabla p_v \tag{9}$$

$$\mathbf{j}_{ws}^l = -K_l \nabla p_c \tag{10}$$

$$\mathbf{j}^l_{s,diff} = -\rho^l_{ws} K_s \nabla \omega \qquad (11)$$

Where K_g is the vapor permeability, K_l the liquid permeability of the salt solution, K_s the salt diffusion coefficient, p_v the vapor pressure, p_c the capillary pressure, and ρ_{ws}^l the mass density of the liquid phase. The vapor permeability K_g can be written as:

$$K_g = \frac{D_v}{R_v T} \qquad (12)$$

where R_v is the gas constant of water vapor, K_l is liquid conductivity of the salt solution and D_v is vapor permeability. Vapor permeability can be defined by vapor permeability of dry air D_v^{air}, vapor resistant factor τ_v which can be obtained by dry-cup and wet-cup diffusion experiment and scale factor f_v depending on S_w^g see Castellazzi et al. (2013). The liquid conductivity of salt solution can be defined as follows

$$K_l = g_\omega(\omega) D_l f_l(S_w^l) \qquad (13)$$

$$f_l = (S_w^l)^{n_l} \qquad (14)$$

where f_l is the correction factor, the exponent n_l ranges from 1 to 6 as suggested by Koniorczyk & Gawin (2011) and D_l is liquid conductivity of pure water proposed by Sykora (2012) and defined as follows:

$$D_l = \left[3.8 \left(\frac{A}{\phi_0 \rho_w^l} \right)^2 10^{3(S_w^l - 1)} \right] \frac{\partial c_w}{\partial h} \qquad (15)$$

being A the water adsorption coefficient. The expression of g_ω is assumed, in agreement with Castellazzi et al. (2015a), as:

$$g_\omega = \frac{\rho_{ws}^l}{\rho_w^l} (1 - 0.03m) \qquad (16)$$

where m is the molality of the solution. When the precipitated salt starts to appear, therefore the effective porosity tends to decrease due to the occupied voids by precipitated salt. This phenomenon can be taken into account in the model by modifying the vapor permeability and liquid conductivity of pure water as reported by Castellazzi et al. (2013):

$$D_v \leftarrow g_v(\phi_{eff}) D_v \qquad (17)$$

$$D_l \leftarrow g_l(\phi_{eff}) D_l \qquad (18)$$

where ϕ_{eff} is the effective porosity defined by total porosity ϕ_0 and solid salt saturation degree S_s^s, as follows:

$$\phi_{eff} = \phi_0 (1 - S_s^s) \qquad (19)$$

moreover, g_v and g_l are correction functions.

$$g_v = (1 - S_s^s)^{n_v^s} \qquad (20)$$

$$g_l = (1 - S_s^s)^{n_l^s} \qquad (21)$$

Note that n_v^s and n_l^s are material dependent parameters is assumed equal to 1.
K_s is assumed to be dependent on the saturation degree of the solution S_{ws}^l:

$$K_s = \frac{D_s}{\tau} f_s(S_w^l) \qquad (22)$$

where D_s is the diffusion coefficient for pure water, τ is tortuosity and f_s is the correction function here defined as:

$$f_s = (S_w^l)^{n_s} \qquad (23)$$

being n_s the saturation exponent.
Tortuosity is a physical property of a porous media defined as ratio of effective path length to the direct length L, see Lu (1992) and Epstein (1989), and its definition can be related to the porosity evolution due to salt crystallization as detailed in the following sections.
The vapor pressure and capillary pressure in Equations 12 ans 13 can be written in terms of relative humidity h as follows

$$p_v = p_{v,sat} h \qquad (24)$$

$$p_c = \rho_w^l R_v T ln(h) \qquad (25)$$

where R_v is the gas constant of water vapor and ρ_w^l is the mass density of liquid phase.

2.3 Model equations and boundary conditions

The main governing equations of the model are moisture balance equation Equation 2, salt mass balance equation Equation 3. By applying the constitutive equations explained in subsection (2.1) and performing algebraic simplifications, the salt mass and moisture balance equations are written as follows:

$$\varphi_h \frac{\partial h}{\partial t} + \nabla \cdot [-C_{hh}\nabla h - C_{h\omega}\nabla\omega] + \varphi_{h\omega}\frac{\partial\omega}{\partial t} +$$
$$\varphi_{hs}\frac{\partial c_s^s}{\partial t} = 0$$

(26)

$$\varphi_\omega \frac{\partial\omega}{\partial t} + \nabla \cdot [-C_{\omega\omega}\nabla\omega - C_{\omega h}\nabla h] + \varphi_{\omega h}\frac{\partial h}{\partial t} +$$
$$\varphi_s\frac{\partial c_s^s}{\partial t} = 0$$

(27)

The formulation of the model is completed by the initial conditions and boundary conditions which can be of Dirichlet type:

$$h = \bar{h}, \quad \omega = \bar{\omega} \qquad (28)$$

or of Neumann type:

$$\mathbf{j}_w \cdot \mathbf{n} = q_w + \gamma_w(A_w h - h_a) \qquad (29)$$

$$\mathbf{j}_s^l \cdot \mathbf{n} = q_\omega \qquad (30)$$

where $\bar{h}$, $\bar{\omega}$ are imposed humidity, salt concentration, q_w, q_ω are prescribed normal fluxes of moisture, salt and $\mathbf{n}$ is the outward unit normal to the boundary, A_w the water activity and γ_w is the convective coefficient.

The differential equations adopted in the model are fully coupled and highly nonlinear. This model has been implemented in COMSOL Multiphysics (2016) using finite elements with quadratic shape functions as detailed in the following sections.

2.4 Updating the tortuosity parameter

A changing of the tortuosity due to the evolution of the porosity can be a proper tool for considering the inter-connection of the voids when altered by salt crystallization. Figure 2 sketches the physical meaning of tortuosity and the possible evolution of the involved parameters.

The minimum value of tortuosity is one and it can increase due to salt crystallization since the effective path increases. In fact, with respect to the pore evolution described in Figure 2 we note that the effective length increase, $L_{e2} > L_{e1}$ updating the tortuosity values $\tau_2 > \tau_1$.

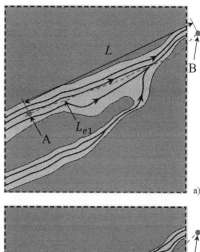

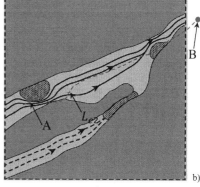

Figure 2. Description of the pore evolution due to salt crystallization: initial configuration of the interconnected pores with $\phi_{eff,1}$, τ_1 and L_{e1} (a); updated configuration of the interconnected pores after solid salt phase formation with $\phi_{eff,2}$, τ_2 and L_{e2} (b).

Several relations have been introduced to relate the tortuosity and porosity in porous media. Here we consider some simple relations proposed for porous media or simplified porous media as presented in Table 1 and plotted in Figure 3.

Table 1. Relation between tortuosity and porosity in porous media.

Tortuosity definition	proposed for	Reference
$1 - p \ln(\phi_{eff})$	random porous media	(Comiti 1989)
$1 - 0.49 \ln(\phi_{eff})$	beds of spheres	(Weissberg 1963)
$0.5 (1 + \phi_{eff}^{-1})$		(Mauret 1997) (Barrande 2007)
$0.8 (1 - \phi_{eff}) + 1$	beds of particles	(Koponen 1996)
$0.5 (1 + \phi_{eff}^{-1})$	porous medium	(Ghafoori 1995)
$\phi_{eff}^{-1.2}$	sedimentary rock	(Luck 2006)

p is the correlation coefficient

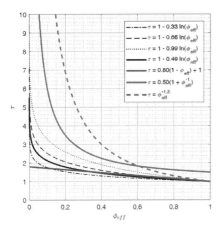

Figure 3. Tortuosity evolution with respect to the effective porosity.

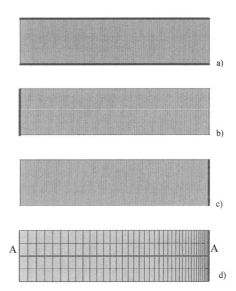

Figure 4. Boundary conditions: no flux (a); prescribed humidity $h = 0.99$ and prescribed mass concentration of dissolved salt $\omega = 0.08$ (b); allowed evaporation with prescribed drop of humidity from 0.9 to 0.5(c). Mesh considered for the simulations: 4×40 elements with 0.1 element ratio biasing along y direction, the cutting line A-A (from 0 to 50 cm) is highlighted in red color (d).

2.5 Spalling prediction

As anticipated in the introduction, the structural damage called spalling is the flaking or peeling of superficial material due to salt crystallization. This phenomena then occurs when salt crystallization focuses just below the evaporating surface and the crystallization pressure exceed the elastic material limit. As previously mentioned, we consider here the Correns law for the pressure definition, accounting for the supersaturation ratio as triggering factor for crystallization. In particular, the proposed model implements the expression proposed by Steiger (2005) for the crystallization pressure, while the tensile stress definition at macroscopic scale is provided considering the amount of precipitated salt, as proposed by Castellazzi et al. (2013):

$$\sigma_C = p_s S_s^\sigma \qquad (31)$$

3 NUMERICAL TESTS

The multiphase model is applied to evaluate the effects of the varying tortuosity on the salt transport and crystallization processes in a fired clay masonry benchmark panel of 20×50 cm exposed to specific boundary conditions as illustrated in Figure. 4. Initial values and boundary conditions are set in order to simulate a portion in time of an ideal weathering real cycle usually composed by wetting and drying phases. Material properties are the same as in de Miranda et al. (2019).

Sodium chloride salt (NaCl) is chosen with mass fraction at saturation $\omega_{sat} = 0.264$ kg$_{salt}$/kg$_{solution}$.

In order to implement the model, the equations are discretized in space using the finite element method. The primary variables h, ω, c_s^s are interpolated based on standard Lagrangian shape functions adopting standard 9-node elements. Mesh biasing is set as described in Figure 4d. The time discretization is carried out by means of the backward finite difference method.

A standard iterative strategy based on the Newton–Raphson method is applied to solve the non-linear system of equations.

The benchmark here proposed is specifically designed to illustrate the relationship between porosity and tortuosity evolution and is accounting for a wetting phase with a drop of humidity on the surface exposed to the evaporation phenomena that is supposed to activate the salt crystallization phenomena just below the evaporating surface.

We consider 30 hours of time to simulate and the following two specific benchmarks in order to highlight the role of the effective tortuosity:

Benchmark #1 that uses constant value of tortuosity $\tau = 1$ and effective porosity varying as illustrated by Equation 19;

Benchmark #2 that uses effective tortuosity τ as described in Table 1. Benchmark #1 – Constant Tortuosity.

The super-saturation ratio is plotted in Figure 5 considering the values computed along the cutting line A-A illustrated in Figure 4d and plotting only the values included within the limit [45 cm - 50 cm] for clarity purpose.

By inspecting Figure 5 we register an initial increase of super-saturation ratio that soon after the 5th hour drops down for the remaining 25 hours of simulation.

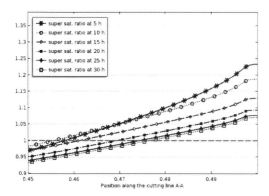

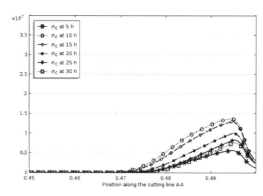

Figure 5. *Benchmark #1* – super-saturation ratio ω/ω_{sat} plotted along the cutting line A-A: the dashed red line is defining the triggering condition for salt dissolution.

Figure 7. *Benchmark #1* – evolution in time of the stress σ_c [Pa] computed along the cutting line A-A.

Figure 6 Illustrates the porosity variation in time due by the variation of the salt saturation degree, Equation 21: a reduction of the porosity is prominent near the evaporating surface with a progressive reduction that slows down after the 15th hour and tends to stop around the 25th hour.

Figure 7 Illustrates the evolution in time of the stress computed along the cutting line A-A using Equation 38. We record an initial increase of stress that reach a maximum peak around the 10th hour of simulation with $\sigma_{c.max}$ = 14 MPa then decreases to 8 MPa at the end of the simulation.

3.1 *Benchmark #2 – Effective tortuosity*

This second benchmark uses an updated definition of the tortuosity parameter using the equations provided in Table 1. For the sake of brevity here we compare the results obtained in terms of porosity and super-saturation ratio for Benchmark #1 only with those

obtained employing the following tortuosity definition provided by Weissberg (1963), Mauret (1997) and Barrande (2007):

$$\tau = 1 - 0.49 \ln(\phi) \qquad (32)$$

given for beds of spheres.

The updated computed solution provides a super-saturation ratio with a trend that is similar to the one obtained in the previous case but with higher ratio at the beginning of the simulation (from 0 to 10 hours) see Figure 8.

The porosity variation in time is now more pronounced: around the 5th hour the porosity drops from 21% to 16% passing from constant to effective tortuosity definition respectively.

The reduction of the porosity is still prominent near the evaporating surface with again a progressive reduction that slows down after 15th hour and now tends to stop around the 20th hour, see Figure 9.

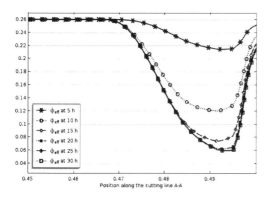

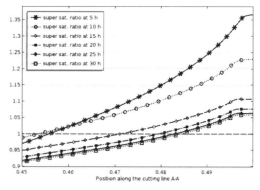

Figure 6. *Benchmark #1* – evolution in time of the porosity ϕ_{eff} [%] computed along the cutting line A-A.

Figure 8. *Benchmark #2* – super-saturation ratio ω/ω_{sat} plotted along the cutting line A-A: the dashed red line is defining the triggering condition for salt dissolution.

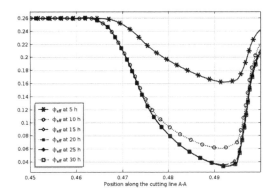

Figure 9. *Benchmark #2* –evolution in time of the porosity ϕ_{eff} [%] computed along the cutting line A-A.

Figure 10 Illustrates the evolution in time of the tortuosity parameter computed using Equation 33.

The evolution in time of the stress is illustrated in Figure 11a.

Moreover, for comparison, in Figures 11b and c the stress evolution for different definitions of the tortuosity parameter is also given. In particular Figures 11b and c refer to bed of particles and porous medium respectively. for the first two definitions, referring to bed of spheres (Koponen 1996) and bed of particles (Ghafoori 1995) respectively, we still record an initial increase of stress that reach a maximum peak around the 10th hour of simulation with and increased value of stress included between $22 \div 25$ MPa. For the third definition, that refers to the generic porous medium, we record a sharp increase of stress that reach the maximum value rapidly soon after the 5th hours with $\sigma_{c.max} = 37$ MPa and then again decreases around 8 MPa at the end of the simulation.

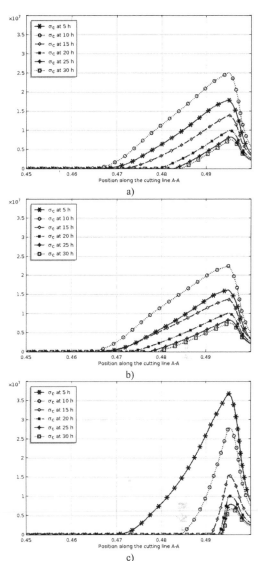

Figure 11. *Benchmark #2* – evolution in time of the stress σ_c [Pa] computed along the cutting line A-A for three different definition of the tortuosity parameter: bed of spheres (a), bed of particles (b) and porous medium (c).

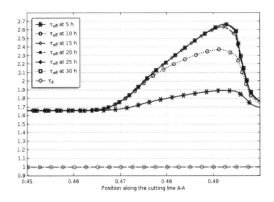

Figure 10. *Benchmark #2* – evolution in time of the tortuosity τ_{eff} [-] computed along the cutting line A-A.

The comparison between the resulting stress obtained using the three different definitions is illustrated in Figure 12 where the computed stress is compared at the 10th, 20th and 30th hours of simulation time. The comparison highlights that the greater is the increase of tortuosity when porosity reduces, the sharper is the increase of stress. Moreover we acknowledge a narrowing of the region affected by the increase of stress due by salt crystallization.

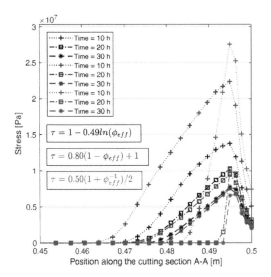

$\tau = 1 - 0.49 ln(\phi_{eff})$

$\tau = 0.80(1 - \phi_{eff}) + 1$

$\tau = 0.50(1 + \phi_{eff}^{-1})/2$

Figure 12. *Benchmark #2* – evolution in time of the stress σ_c [Pa] computed along the cutting line A-A for three different definition of the tortuosity parameter.

4 CONCLUSIONS

The simulation of the spalling damage problem has been carried out through a coupled multiphase model accounting for salt transport and crystallization. The prediction of the salt pressure due to salt crystallization has been computed for several definition of the tortuosity parameter as a function of the actual porosity. As expected it appears that the tortuosity definition affect the resulting stress distribution along the thickness sample. The updated model describes the important variation of the peak stress distribution during the evolution of the multi-phase phenomena. The obtained results suggest that it is important to accurately track the stress distribution evolution in time accounting also for the effective definition of the tortuosity parameter.

REFERENCES

Barrande, M. et al. 2007. Tortuosity of porous particles. *Analytical chemistry* 79 (23): 9115–9121.

Castellazzi, G. et al. 2013 A coupled multiphase model for hygrothermal analysis of masonry structures and prediction of stress induced by salt crystallization. *Construction and Building Materials* (41): 717–731.

Castellazzi, G. et al. 2015. Multiphase model for hygrothermal analysis of porous media with salt crystallization and hydration. *Materials and Structures* 49(3): 1039–1063.

Castellazzi, G. et al. 2015 Coupled hygro-mechanical multiscale analysis of masonry walls. *Engineering Structures* (84): 266–278.

Comiti, J. & Renaud, M. 1989. A new model for determining mean structure parameters of fixed beds from pressure drop measurements: application to beds packed with parallelepipedal particles. *Chemical Engineering Science* 44 (7): 1539–1545.

COMSOL Multiphysics modeling guide, 2016.

Correns, C.W. 1949. Growth and dissolution of crystals under linear pressure. *Discussions of the Faraday society* (5): 267–271.

de Miranda S. et al. 2019. Modeling environmental ageing in masonry strengthened with composites. *Engineering Structures* (201): 109773.

Epstein, N. 1989. On tortuosity and the tortuosity factor in ow and diffusion through porous media. *Chemical engineering science* 44 (3): 777–779.

Koniorczyk, M. & Gawin, D. 2012. Modelling of salt crystallization in building materials with microstructure-poromechanical approach. *Construction and Building Materials* 36: 860–873.

Franzoni, E. et al. 2014. Towards the assessment of the shear behaviour of masonry in on-site conditions: a study on dry and salt/water conditioned brick masonry triplets. *Construction and Building Materials* (65): 405–416.

Franzoni, E. et al. 2018. Effects of rising damp and salt crystallization cycles in FRCM-masonry interfacial debonding: Towards an accelerated laboratory test method. *Construction and Building Materials* (175): 225–238.

Ghafoori, N. & Dutta, S. 1995. Laboratory investigation of compacted no-fines concrete for paving materials. *Journal of materials in civil engineering* 7 (3): 183–191.

Grementieri, L. & Daghia, F. & Molari, L. & Castellazzi, G. & Derluyn, H. & Cnudde V. & de Miranda, S. 2017. A multi-scale approach for the analysis of the mechanical effects of salt crystallisation in porous media, *International Journal of Solids and Structures*, (126): 225–239.

Grementieri, L. et al. 2017. Numerical simulation of salt transport and crystallization in drying Prague sandstone using an experimentally consistent multiphase model. *Building and Environment* (123): 289–298.

Hassanizadeh, S. M. & Leijnse, T. 1988. On the modeling of brine transport in porous media. *Water Resources Research* 24(3): 321–330.

Koniorczyk, M. et al. 2018. Multiphysics model for spalling prediction of brick due to in-pore salt crystallization. *Computers & Structures* (196): 233–245.

Koponen, A. et al. (1996). Tortuous flow in porous media. *Physical Review E* 54 (1): 406.

Lu, B. & Torquato, S. 1992. Lineal-path function for random heterogeneous materials. *Physical Review A* 45 (2): 922.

Luck, J. D. et al. (2006). *Hydrologic properties of pervious concrete*. In 2006 ASAE annual meeting.

Mauret, E. & Renaud, M. 1997. Transport phenomena in multi-particle systems: limits of applicability of capillary model in high voidage beds-application to fixed beds of fibers and fluidized beds of spheres. *Chemical Engineering Science* 52(11): 1807–1817.

Nicolai, A. 2008. *Modeling and numerical simulation of salt transport and phase transitions in unsaturated porous building materials*. Syracuse University.

Scherer, G.W. 2004. Stress from crystallization of salt. *Cement and concrete research* 34(9): 1613–1624.

Skora J. et al. 2012. Computational homogenization of non-stationary transport processes in masonry structures. *Journal of Computational and Applied Mathematics* 236(18): 4745–4755.

Weissberg, H. L. 1963. Effective diffusion coefficient in porous media. *Journal of Applied Physics* 34 (9): 2636–2639.

Brick and Block Masonry - From Historical to Sustainable Masonry –
Kubica, Kwiecień & Bednarz (eds)
© 2020 Taylor & Francis Group, London, ISBN 978-0-367-56586-2

Flemish bond brickwork: Macroscopic elastic properties and nonlinear behaviour

A. Taliercio & F. Midali

Department of Civil and Environmental Engineering, Politecnico di Milano, Milan, Italy

ABSTRACT: Assuming flemish bond brickwork to be periodic, a finite element model of a Representative Volume Element (RVE) is developed to predict its macroscopic behavior using a homogenization approach. In linear elasticity, the numerical results are used to assess the reliability of recently proposed closed-form expressions for the macroscopic elastic properties (Taliercio, 2018). Assuming that both mortar and units experience plastic strains and damage effects, the macroscopic strength domain under in-plane principal stresses parallel to the joints is identified and compared with that predicted by Drougkas et al. (2016). Eventually, the model is applied to predict the homogenized strength of Flemish bond brickwork under elementary macroscopic in-plane stresses and transverse shear. The effect of the collar joint on the macroscopic response is pointed out by comparing the numerical results with those obtained on header bond brickwork. This effect is shown to be particularly significant under horizontal, transverse shear.

1 INTRODUCTION

Flemish bond is a quite common type of brickwork with one stretcher between headers, with the headers centered over the stretchers in the course below (see Figure 1). Headers act as diatons connecting the two wythes of stretchers: accordingly, at the same thickness, a wall built with Flemish bond brickwork (FBB) is much more stable than a wall consisting of two wythes built in running bond pattern. Despite the wide diffusion of FBB, so far very little attention has been devoted to this type of brickwork.

At the authors' knowledge, one of the few experimental test campaign on FBB was carried out by Binda et al. (1988), who investigated the effects of the mortar type on the vertical compressive strength of masonry prisms. Actually, the authors do not explicitly state that their work deals with FBB, but this is what can be inferred from the pictures of the tested specimens.

As far as modelling is concerned, most existing papers consider single-leaf periodic brickwork consisting only of stretchers or headers (see e.g. Cecchi & Sab, 2002; Milani, 2011; Mistler et al., 2007; Pande et al., 1989; Pietruszczak & Niu, 1992; Zucchini & Lourenço, 2002, just to quote a few), and aim at deriving the macroscopic (or homogenized) mechanical properties of masonry according to those of the units and the mortar joints, both within and beyond the elastic limit.

Drougkas et al. (2015, 2016) are among the few researchers that have dealt with the mechanical behavior of FBB: they tried to predict the macroscopic elastic and strength properties of this type of masonry through the analysis of a single unit cell.

Recently, Taliercio (2018) proposed a theoretical approach to derive closed-form expressions for the macroscopic elastic constants of FBB, by using a simplified approach similar to the so-called "Method of Cells" (MoC), proposed by Aboudi (1991) for fiber-reinforced composites and successfully applied to predict the macroscopic elastic (Taliercio, 2014, 2016) and strength properties (Milani & Taliercio, 2015, 2016) of running or header bond brickwork.

This paper aims at giving a contribution towards the prediction of the homogenized response of FBB both under service and ultimate load conditions. After briefly recalling the fundamentals of homogenization theory for periodic media (Sec. 2), a numerical model is developed in Sec. 3 to discretize a Representative Volume Element (RVE) of FBB. Plastic strains and damage effects are allowed both in mortar joints and units. In Sec. 4.1 the model is used to estimate the in- and out-of-plane macroscopic elastic constants of FBB: these estimates are used as benchmark to assess the reliability of the closed-form expressions of the homogenized elastic and shear moduli and the Poisson's ratios of FBB derived by Taliercio (2018). In Sec. 4.2 the macroscopic strength domain of FBB under in-plane loads acting parallel to the mortar joints is predicted and qualitatively compared with that obtained by Drougkas et al. (2016) for running bond brickwork using a theoretical approach. Having assessed the reliability of the model also in the nonlinear field, the response of FBB under increasing elementary macroscopic strains is analyzed in Sec. 4.3. Emphasis is put on the importance of the collar joints by comparing the numerical results with those obtained for header bond brickwork. Also, the failure modes are discussed by examining the damage

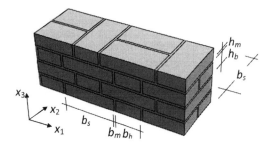

Figure 1. Typical Flemish bond brickwork element.

distribution. Finally, in Sec. 5 the main results of the research are summarized, and possible future perspectives of the research are outlined.

2 HOMOGENIZATION THEORY FOR PERIODIC MEDIA: AN OUTLINE

Referring to Figure 1, FBB can be assumed to be a heterogeneous, periodic medium in the plane (x_1,x_3). From a macroscopic point of view, any Flemish bond masonry wall subjected to uniform boundary conditions behaves like an orthotropic material element. The macroscopic mechanical properties of the equivalent, homogenized medium can be derived from the analysis of a single RVE, also called 'unit cell' for periodic media. Additional details on the fundamentals of homogenization are given e.g. by Nemat-Nasser & Hori (1993). A possible choice for the RVE of FBB is shown in Figure 2.

The homogenized constitutive law relates macroscopic stresses (Σ) and macroscopic strains (E), which are defined as the volume averages of the corresponding microscopic fields over the volume, V, of the RVE:

$$\Sigma = \frac{1}{|V|}\int_V \sigma(x)dV,\ E = \frac{1}{|V|}\int_V \varepsilon(x)dV, \qquad (1)$$

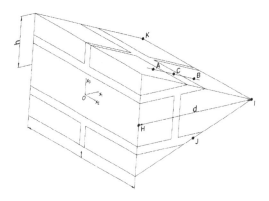

Figure 2. Possible RVE (or unit cell) for Flemish bond brickwork.

where x is any point in the RVE.

Under any given macroscopic stress, the (infinitesimal) microscopic displacement field in the RVE of a periodic, heterogeneous body is of the form (see e.g. Mistler et al., 2007):

$$u(x) = u_0 + \Omega \wedge x + \tilde{E}x + u^P(x), \qquad (2)$$

where u_0 is a rigid displacement, Ω is a rigid rotation, $\tilde{E}$ is defined below, and u^P is the periodic part of the displacement field. As the body is assumed to be periodic only parallel to the x_1,x_3-plane, u^P must fulfil periodicity only with respect to x_1 and x_3. A field of the form (2) is also said to be "strain-periodic" over the RVE.

$\tilde{E}$ is the extensive variable conjugate to the macroscopic stress in the so-called Hill's macro-homogeneity:

$$\Sigma : \tilde{E} = \frac{1}{|V|}\int_V \sigma(x) : \varepsilon(x)dV \qquad (3)$$

and differs from the macroscopic strain E unlike the case of fully periodic media (Nemat-Nasser & Hori, 1993). Eq. (3) applies provided that the microscopic stress field is antiperiodic in (x_1,x_3) over the boundary of the RVE, except for the faces perpendicular to x_2, which are traction-free.

In the linear elastic field, the macroscopic constitutive law can be expressed as

$$\begin{Bmatrix} E_{11} \\ E_{22} \\ E_{33} \\ 2E_{12} \\ 2E_{13} \\ 2E_{23} \end{Bmatrix} = \begin{bmatrix} 1/E_1 & -\nu_{12}/E_2 & -\nu_{13}/E_3 & 0 & 0 & 0 \\ -\nu_{21}/E_1 & 1/E_2 & -\nu_{23}/E_3 & 0 & 0 & 0 \\ -\nu_{31}/E_1 & -\nu_{32}/E_2 & 1/E_3 & 0 & 0 & 0 \\ 0 & 0 & 0 & 1/G_{12} & 0 & 0 \\ 0 & 0 & 0 & 0 & 1/G_{13} & 0 \\ 0 & 0 & 0 & 0 & 0 & 1/G_{23} \end{bmatrix} \begin{Bmatrix} \Sigma_{11} \\ \Sigma_{22} \\ \Sigma_{33} \\ \Sigma_{12} \\ \Sigma_{13} \\ \Sigma_{23} \end{Bmatrix}$$

$$(4)$$

where E_1, E_2 and E_3 are macroscopic Young's moduli, ν_{12}, ν_{13}, ν_{32} are macroscopic Poisson's ratios, and G_{12}, G_{13} and G_{23} are macroscopic shear moduli. Hyperelasticity requires $\nu_{12}E_1 = \nu_{21}E_2$, $\nu_{13}E_1 = \nu_{31}E_3$, and $\nu_{23}E_2 = \nu_{32}E_3$.

Taliercio (2018) recently proposed approximate expressions for the in-plane macroscopic Young's moduli (E_1, E_3), Poisson ratio (ν_{13}) and shear modulus (G_{13}), together with the out-of-plane shear moduli (G_{12}, G_{23}), of Flemish bond brickwork. These constants will be numerically estimated in Sec. 4.1 and compared with the closed-form expressions derived by Taliercio (2018). The remaining independent elastic constants (E_2, ν_{12}, ν_{23}) are of little practical interest, as masonry walls are either in a state of plane stress (if acted upon by in-plane loads), or subjected to transverse shear (if acted upon by lateral loads). Transverse normal stresses are usually disregarded.

Beyond the elastic limit, the homogenized law can be numerically obtained by prescribing the microscopic displacement field to be of the form (2), and submitting the RVE to any macroscopic stress history. This will be done in Sec. 4.2 and 4.3.

3 NUMERICAL MODEL

The macroscopic behavior of FBB will be predicted in Sec. 4 using the FE model shown in Figure 3. It consists of 1638 brick elements and 252 wedge elements with linear modelling of the displacement field, for a total of 6750 degrees of freedom. The model is analyzed using the commercial software Abaqus®.

Referring to Figure 1, the dimensions of units and joints (in mm) are $b_h = 100$, $b_s = 210$, $h_b = 52$, $b_m = h_m = 10$. Units and mortar are supposed to be isotropic. The elastic properties of the materials used in the applications are listed in Table 1.

Both materials were supposed to experience plastic strains and damage effects. These can be incorporated in the FE model by using the so-called Concrete Damaged Plasticity (CDP) model, which was successfully employed by several authors to analyze historic masonry buildings under static and dynamic loads (Milani & Valente, 2015; Valente & Milani, 2016; Condoleo et al., 2018). Details on the theoretical background of the model can be found in the Abaqus Theory guide (vers. 6.13). Here it suffices to recall that the yield surface of the CDP is a modified Drucker-

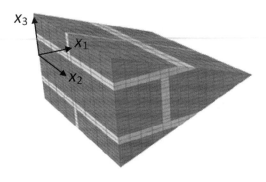

Figure 3. FE mesh of the RVE used in numerical applications. Dark grey: units; light grey: joints.

Table 1. Elastic properties of brick and mortar used in the numerical applications.

Material	Young's modulus, E N/mm^2	Poisson's ratio, v
Brick	20000	0.15
Mortar	10000	0.15

Table 2. Parameters defining the Concrete Damaged Plasticity model used in the numerical applications.

Material	σ_{c0} N/mm^2	σ_{t0} N/mm^2	ψ deg	ε	σ_{b0}/σ_{c0}	Kc	μ
Brick	50	1	10	0.1	1.16	2/3	0.0001
Mortar	6	0.35	10	0.1	1.16	2/3	0.0001

Prager surface with a smoothed tip and a non-circular cross-section in the space of the principal stresses. A non-associated plastic flow rule defined by a dilation angle ψ is employed.

The values of the parameters that define the CDP model are those proposed by Milani and coworkers, and are summarized in Table 2. σ_{b0} denotes the compressive strength in equi-biaxial compression, μ is a viscosity parameter that regularizes the numerical solution, K defines the shape of the cross-section of the yield surface, and ε is related to the flow potential (see Abaqus Theory guide). The remaining parameters were defined previously.

The tensile behavior of masonry is assumed to be linearly elastic up to a peak stress, σ_{t0}. Then micro cracks start developing, and the material behavior is strain softening. To attenuate the mesh-dependency of the results, the strain softening behavior is actually defined by a stress-displacement law. The material is fully cracked when the displacement attains the ultimate value, u_0. The area under the stress-displacement curve is the fracture energy of the material. Displacements are converted into strains by a suitable characteristic length, which is related to the size of the FEs. A damage variable d_t, ranging from 0 (uncracked material) to 1 (fully cracked material), defines the decrease in elasticity modulus after the peak stress. Similar remarks apply to the behavior in compression, which is linear elastic up to a peak stress, σ_{c0}.

Figure 4 Shows the stress-inelastic strain diagrams used in the numerical applications and the assumed evolution of damage in tension (d_t) and compression (d_c) as inelastic strains increase.

In order to enforce strain-periodicity of the microscopic displacement field (2), conditions of the type

$$\frac{u_A + u_B}{2} = u_C, \tag{5}$$

are enforced, where A and B is any pair of points symmetric with respect to C, located on any side of the RVE and sharing the same coordinate x_2 as C (see Figure 2). Rigid translations u_0 (see eq. (1)) are avoided by fixing any point of the RVE. By following a procedure similar to that used by Mistler et al. (2007), rigid rotations Ω of the RVE are avoided by prescribing

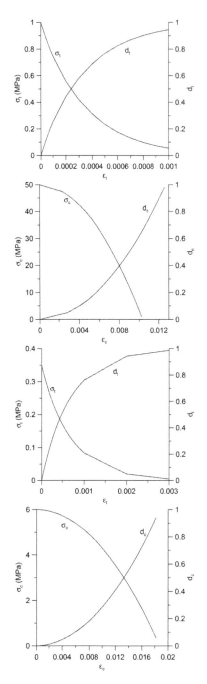

Figure 4. Stress vs. inelastic strain and damage vs. inelastic strain diagrams in tension (a,c) and compression (b,d) used in the numerical applications: (a,b) units; (c,d) mortar.

$$u_{1J} + u_{3I} \frac{h}{2d} = \frac{u_{1I}}{2}, \ u_{1K} = u_{1I} + u_{2I} \frac{t}{d}, \quad \text{(6a,b)}$$

$$u_{3K} + \frac{2u_{2J} - u_{2I}}{h}t = u_{3I}, \quad \text{(6c)}$$

where the points I, J, K are shown in Figure 2, together with the characteristic dimensions d, h, t ($\equiv b_s$).

4 NUMERICAL RESULTS

4.1 Macroscopic elastic properties

The FE model of the RVE was subjected to elementary macroscopic stresses Σ_{11}, Σ_{33}, Σ_{13}, Σ_{12} and Σ_{23}, in order to numerically estimate the 6 homogenized moduli referred to in Sec. 2. For the sake of illustration, Figure 5 shows the deformed model under in-plane shear, Σ_{13}, and horizontal transverse shear, Σ_{12}.

Table 3 summarizes the six homogenized elastic constants estimated both numerically and through the micromechanical model proposed by Taliercio (2018). The agreement is definitely good, the maximum discrepancy being of the order of 1%, except for the in-plane Poisson's ratio.

4.2 Macroscopic strength domain

The FE model was also subjected to macroscopic in-plane normal strains increasing in time following radial paths in the plane (E_{11}, E_{33}). Accordingly, the RVE undergoes principal normal stresses parallel to the bed and head joints ($\Sigma_I = \Sigma_{11}$, $\Sigma_{II} = \Sigma_{33}$). The macroscopic stress paths are radial only when the

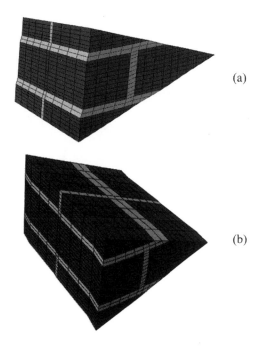

Figure 5. Deformed RVE under (a) macroscopic in-plane shear Σ_{13}; (b) macroscopic horizontal transverse shear Σ_{12}.

734

Table 3. Macroscopic elastic constants for FBB predicted through the proposed numerical model and the theoretical model proposed by Taliercio (2018). The percent difference is also reported (in brackets).

	Present model	Taliercio (2018)	(% diff.)
E_1 (N/mm^2)	17257.1	17198.5	(-0.34)
E_3 (N/mm^2)	16639.1	16588.0	(-0.31)
ν_{13}	0.149	0.1415	(-5.32)
G_{13} (N/mm^2)	7124.5	7060.9	(-0.90)
G_{23} (N/mm^2)	7181.8	7202.5	(+0.29)
G_{12} (N/mm^2)	7453.7	7382.6	(-0.96)

entire RVE is in the linear elastic field, but not beyond. In this way, it is possible to identify both the elastic limit and the boundary of the macroscopic strength domain, which is the envelope of the stress paths (see Figure 6). It is interesting to note that the boundary of the macroscopic elastic domain is not a simple homothetic contraction of the boundary of the strength domain.

Figure 6 also shows the macroscopic strength domain derived by Drougkas et al. (2016) for running bond brickwork (RBB) using a semi-analytical approach. The experimental scatter in terms of biaxial strength for RBB is also shown. Only a qualitative comparison between the domains obtained for FBB and RBB is possible, because of the presence of an interface in the latter case and the different

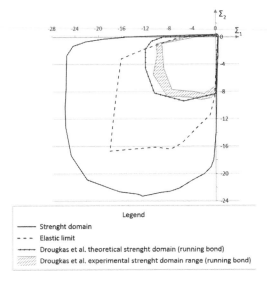

Legend
— Strenght domain
- - - Elastic limit
+—+—+ Drougkas et al. theoretical strenght domain (running bond)
/////// Drougkas et al. experimental strenght domain range (running bond)

Figure 6. Macroscopic strength domains of FBB and RBB under stresses acting along the bed joints and the head joints (dashed area: experimental failure envelope for RBB). The elastic strength domain for FBB is also shown. Stresses are in N/mm^2.

assumptions made in the two theoretical models regarding the evolution of damage in units and joints. Note that both approaches capture the material anisotropy in terms of strength, because of the different failure mechanisms that occur as the ratio of the principal stresses varies (see e.g. Dhanasekar et al., 1985).

4.3 Macroscopic stress-strain behavior

Finally, the numerical model was used to predict the macroscopic response of FBB to elementary stresses, namely, uniaxial horizontal tension ($\Sigma_{11} > 0$), uniaxial vertical compression ($\Sigma_{33} < 0$), in-plane shear (Σ_{13}) or transverse (Σ_{12}, Σ_{23}) shear. The results are presented in terms of macroscopic stress-strain diagrams and contours of the (tensile) damage after the peak stress, when a failure mechanism of the RVE has clearly developed. In order to highlight the effect of the collar joint on the macroscopic response, the stress-strain diagrams and the strength values obtained for header bond brickwork (HBB), with units and joints of the same geometry as FBB, are also reported.

HBB and FBB are found to behave similarly from the macroscopic point of view, except under horizontal transverse shear, Σ_{12}. This can be easily understood by inspection of the contours of the tensile damage, shown in Figure 8. The stress condition in which the collar joint is mostly affected by damage is horizontal transverse shear (Figure 8d): after propagating across the headers, damage basically splits the wall vertically, and the two wythes behave separately. In the other cases, there is virtually no difference in mechanical behavior between FBB and HBB. This remark indirectly confirms the effectiveness of Flemish bond, in which headers give the wall a monolithic behavior typical of single-wythe walls under in-plane loads.

Tensile damage also occurs in the collar joint under vertical compression, but the global strength of the two resulting wythes is comparable to that of a single-wythe wall of equal thickness (provided that buckling phenomena are disregarded). The collapse mechanism is due to activation of tensile damage in the units (see Figure 8a), similarly to the so-called Hilsdorf's mechanisms typical of single-wythe walls.

Table 4 summarizes the values of the macroscopic peak stress numerically estimated for FBB and HBB. Again, the only significant discrepancy in terms of strength is obtained under horizontal transverse shear, Σ_{12}.

5 CONCLUDING REMARKS

A numerical model capable of predicting the macroscopic response of FBB under increasing strains has been developed, within the framework of 3D homogenization theory for 2D-periodic media. The

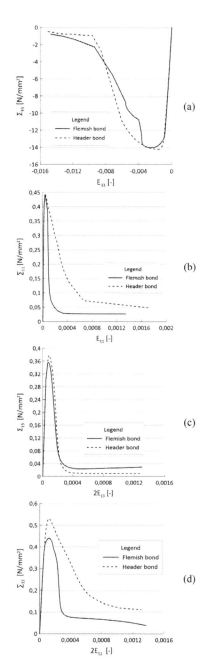

Figure 7. Macroscopic stress-strain diagrams for FBB (solid lines) and header bond brickwork (dashed lines) subjected to (a) uniaxial vertical compression$_3$; (b) uniaxial horizontal tension; (c) in-plane shear; (d) horizontal transverse shear.

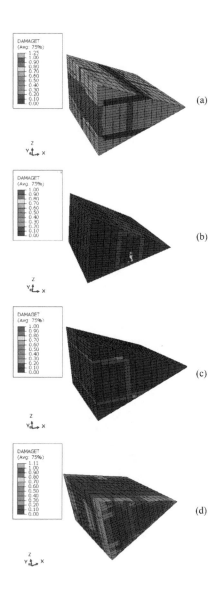

Figure 8. Contours of the tension damage variable at the last increment of the analysis of a RVE under (a) uniaxial vertical compression; (b) uniaxial horizontal tension; (c) in-plane shear; (d) horizontal transverse shear.

reliability of the model was checked through comparisons with available theoretical results in the elastic field and, only qualitatively, with theoretical and experimental results carried out on RBB in terms of macroscopic strength. The lack of tests on FBB is

Table 4. Estimated strength under elementary macroscopic stress (in MPa).

	Flemish bond	Header bond
Σ_{33} (< 0)	14.027	14.247
Σ_{11} (> 0)	0.443	0.445
Σ_{13}	0.357	0.378
Σ_{12}	0.441	0.530
Σ_{23}	0.389	0.398

a serious drawback in the validation of any theoretical or numerical model.

The macroscopic strength of FBB was numerically found to be comparable to that of HBB of the same thickness. This is because in most cases the activation of damage in the collar joint is precluded by the presence of headers, and the behavior of FBB is that of single-wythe masonry. The only exception is the strength under horizontal transverse shear, which is much lower than that of HBB as damage basically splits the wall into two independent wythes.

In the continuation of the research, it would be interesting to investigate the effect of changes in the mechanical properties of units and joints on the macroscopic behavior. Indeed, in the analyses carried out so far the compressive strength of the units was assumed to be much higher than that of the joints (see Table 2), so that units are affected by compression damage to a negligible extent. This may not be true in real brickwork, in which brick crushing is one of the possible contributions to the macroscopic failure mechanism. Also, in the present work the tensile strength of the units was supposed to be of the same order of magnitude of that of mortar: accordingly, in most instances units were found to be affected by tensile damage at failure (see e.g. Figure 8b,d). Further investigation is required to analyze situations in which units are much stronger than mortar, and most failure mechanisms of FBB affect only the mortar joints.

REFERENCES

Aboudi, J. 1991. *Mechanics of composite materials - A unified micromechanical approach.* In: Studies in Applied Mechanics, Vol. 29. Amsterdam: Elsevier.

Binda, L., Fontana, A. & Frigerio, G. 1988. Mechanical behaviour of brick masonries derived from unit and mortar characteristics. *Proc. 8th Int. Brick and Block Masonry Conference, Vol. 1*, Dublin, Eire, pp. 205–216.

Cecchi, A. & Sab, K. 2002. A multi-parameter homogenization study for modeling elastic masonry. *Eur. J. Mech. A/Solids* 21: 249–268.

Condoleo, P., Gobbo, A. & Taliercio, A. 2018. A hybrid masonry and steel mirror-type vault with lunettes: Survey and structural analysis. *Int. J. Arch. Herit.*, doi: 10.1080/15583058.2018.1552997.

Dhanasekar M., Page A.W. & Kleeman P.W. 1985.The failure of brick masonry under biaxial stresses. *Proc. Instn. Civ. Engrs. Part 2* 79: 295–313.

Drougkas, A., Roca, P. & Molins, C. 2015. Analytical micro-modeling of masonry periodic unit cells - Elastic properties. *Int. J. Solids Struct.* 69-70: 169–188.

Drougkas, A., Roca, P. & Molins, C. 2016. Nonlinear micro-mechanical analysis of masonry periodic unit cells. *Int. J. Solids Struct.* 80: 193–211.

Milani, G. 2011. Simple homogenization model for the non-linear analysis of in-plane loaded masonry walls. *Comput. Struct.* 89: 1586–1601.

Milani, G. & Taliercio, A. 2015. In-plane failure surfaces for masonry with joints of finite thick-ness estimated by a Method of Cells-type approach. *Comput. Struct.* 150: 34–51.

Milani, G. & Taliercio, A. 2016. Limit analysis of transversally loaded masonry walls using an innovative macroscopic strength criterion. *Int. J. Solids Struct.* 81: 274–293.

Milani, G. & Valente, M. 2015. Failure analysis of seven masonry churches severely damaged during the 2012 Emilia-Romagna (Italy) earthquake: Non-linear dynamic analyses vs conventional static approaches. *Eng. Fail. Analysis* 54: 13–56.

Mistler M., Anthoine, A. & Butenweg, C. 2007. In-plane and out-of-plane homogenisation of masonry. *Comput. Struct.* 85: 1321–1330.

Nemat-Nasser, S. & Hori, M. 1993. *Micromechanics: Overall Properties of Heterogeneous Materials.* London: North-Holland.

Pande, G.N., Liang, J.X. & Middleton, J. 1989. Equivalent elastic moduli for brick masonry. *Comput. Geotech*, 8: 243–265.

Pietruszczak, S. & Niu, X. 1992. A mathematical description of macroscopic behaviour of brick masonry. *Int. J. Solids Struct.* 29(5): 531–546.

Taliercio, A. 2014. Closed-form expressions for the macroscopic in-plane elastic and creep coefficients of brick masonry. *Int. J. Solids Struct.* 51(17): 2949–2963.

Taliercio, A. 2016. Closed-form expressions for the macroscopic flexural rigidity coefficients of periodic brickwork. *Mech. Res. Comm.* 72: 24–32.

Taliercio, A. 2018. Closed-form expressions for the macroscopic elastic constants of Flemish bond masonry walls. *Proc. 10th Int. Masonry Conf.*, Milan, Italy, July 9-11, 2018.

Valente, M. & Milani, G. 2016. Seismic assessment of historical masonry towers by means of simplified approaches and standard FEM. *Const. Build. Mat.* 108: 74–104.

Zucchini, A. & Lourenço, P.B. 2002. A micro-mechanical model for the homogenization of masonry. *Int. J. Solids Struct.* 39(12): 3233–3255.

Materials & manufacturing

Brick and Block Masonry - From Historical to Sustainable Masonry –
Kubica, Kwiecień & Bednarz (eds)
© 2020 Taylor & Francis Group, London, ISBN 978-0-367-56586-2

Experimental evaluation and probabilistic analysis of the masonry veneer wall tie characteristics

I.B. Muhit, M.G. Stewart & M.J. Masia
Centre for Infrastructure Performance and Reliability, The University of Newcastle, Callaghan, Australia

ABSTRACT: Wall tie strengths and stiffnesses are not constant for all ties in a masonry veneer wall system. This paper uses an Australian standard tie calibration experimental approach to delve into wall-tie probabilistic characterisation by estimating the mean, variance, and characteristic axial tensile and compressive strengths and how they influence failure behaviour. A total of 50 veneer brick-tie-timber subassemblies are tested using an Instron testing machine, 25 samples in compression and 25 samples in tension. Both cross head displacement and displacement across the cavity is recorded along with the complete load versus displacement response, which allows determination of elastic stiffness, peak strength and displacement capacity. Using the maximum likelihood method, a range of probability distributions are fitted to tie strength and corresponding displacement histogram data sets, and a best-fitted probability distribution is selected for each case. A Cumulative Distribution Function plot was also used along with the Anderson-Darling test to infer a goodness-of-fit for the probabilistic models.

1 INTRODUCTION

Brick masonry walls are used as external cladding in both residential and commercial construction in many countries including Australia. Masonry veneer walls comprise of exterior masonry cladding and a flexible structural backing partitioned by an air cavity. The structural backing system varies according to the construction practice being mostly timber, and light steel stud walls or structural masonry in the United States, Australia and New Zealand (Paton-Cole et al. 2012, Reneckis et al. 2004) and reinforced concrete masonry infilled frames in Europe. In Australia, the internal layer of the masonry veneer wall system is composed of timber framing mostly, and provides lateral support by wall ties attached to the external leaf of masonry.

The ties are galvanised or stainless steel depending on the geo-environmental requirement, and typically have axial stiffness and strength in tension and compression, but negligible shear capacity. The out-of-plane mechanisms represent the primary cause of structural failure in unreinforced masonry (URM) buildings under seismic and wind loading, explicitly caused by poor tie connections and strengths. This is because, the ties are responsible for transferring the lateral loads from exterior wall to the backup and allowing in-plane movement to accommodate differential movements, therefore, the properties of the ties have an important role to play in the structural performance of veneer wall system.

Recognizing that tie connections play a crucial role for in-plane and out-of-plane performance of the masonry veneer system under seismic actions, some researchers carried out investigations aimed at assessing the behaviour of the tie connections under shear, tension and compression (Choi & LaFave 2004, Mertens et al. 2014, Page et al. 2009, Reneckis 2009, Ribeiro et al. 2014, Zisi & Bennett 2011). Choi & LaFave (2004) experimented with brick-tie-timber subassemblies for varying tie thickness, initial offset displacement, method of attachment of ties to timber studs, and type of loading (including cyclic), subjected to lateral loads in the in-plane and out-of-plane directions. Reneckis (2009) also conducted subassembly tests akin to Choi & LaFave (2004) to explore tie connection behavior further, primarily when loaded in tension, for various code compliant and non-compliant tie installation methods. As a part of the large-scale testing program carried out at Delft University of Technology to characterise the behaviour of the terraced house typology, characterisation of wall tie connection in cavity walls was reported by Skroumpelou et al. (2018). When the wall system is subjected to a lateral load, the distribution of forces in the ties will be influenced by the deflection of the backup, and maximum forces are experienced by the top and the bottom rows of the ties (Muhit et al. 2019).

Wall tie strengths and stiffnesses are not constant for all ties in a masonry veneer wall system. While an assumption of deterministic material strength properties may be considered in the majority of the masonry design, there is growing realisation that material variability needs to be considered when assessing structural safety. Therefore, it is pivotal to develop probabilistic material models of wall tie strengths and stiffnesses from an ample number of brick-tie-timber subassembly tests. This paper describes an extensive experimental study of the brick-tie-timber subassembly under axial compression and tension loading, considering one of the leaves (brick) is fixed and the relative motion of the free leaf (timber) occur in the perpendicular direction. A range of probability distributions are fitted to tie strength histogram data sets, and a best-fitted probability distribution is selected. The study reported herein is a part of the broader project which is in progress at The University of Newcastle, Australia. The outcome of this paper serves as the basis for performance evaluations of brick masonry veneer wall systems subjected to wind and seismic hazards considering the spatial and random variability of the constituent material properties.

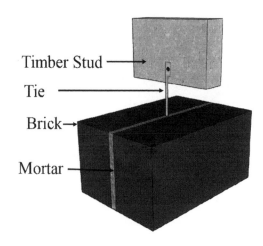

Figure 2. Brick-tie-timber subassembly specimen.

2 EXPERIMENTAL EVALUATION OF TIE CHARACTERISTICS

2.1 Material selection and test specimens

Wall ties in Australia are available in a number of sizes and shapes and are usually made from galvanised mild steel or stainless steel in areas of high corrosion risk. Both face-fixed ties and side-fixed ties are commonly used for veneer wall construction. These veneer ties are often strip ties nailed or screwed to the timber or steel back-up frame. However, side-fixing corrugated sheet metal tie with timber as backup is customary Australian practice and is selected accordingly for this study. The testing of brick-tie-timber subassemblies (alluded to as 'the couplet' onwards) is more realistic and rational than just testing the ties in isolation to characterise the local behaviour of a wall system. Hence, the couplets were constructed with two perforated clay bricks (230 mm long × 110 mm wide × 76 mm high), one machine graded pine (MGP10 grade) timber stud (150 mm in length and 90 mm × 35 mm in cross-section), one corrugated Type-A light-duty side-fixing stainless steel R4 tie (tie dimensions are given in mm in Figure 1), and general purpose M3 mortar

(a)

(b)

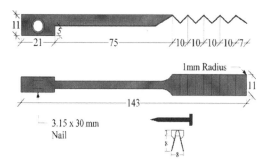

Figure 1. Side-fixed veneer tie details (dimensions are in mm).

Figure 3. Couplet testing setup for (a) compression and (b) tension loading.

(1: 1: 6) (cement: lime: sand by volume). Wall ties were embedded at least 50 mm into the mortar as per AS 4773.2 (Standards Australia, 2015). These ties are side fixed to the timber stud with the supplied nail and a strictly maintained 50 mm cavity width. The complete couplet assemblage is shown in Figure 2. A total of 50 couplet specimens were prepared and left undisturbed for 7 days. Then specimens were randomly divided into two groups (25 specimens each) to be tested for compression and tension, respectively.

2.2 Testing setup and procedure

All tests were performed using a displacement control Instron electromechanical testing system for compression and tension loading at least 7 days after specimen construction in accordance with the test method suggested in AS/NZS 2699.1 (Standards Australia/Standards New Zealand, 2000). The specimen was rotated into a vertical position and clamped in the machine as illustrated in Figure 3. A monotonic compressive load was then induced through a constant displacement of the machine cross head. The load cell in the testing frame was connected to the controller computer to measure and control the load and actuator displacement. In addition to this cross-head displacement, one displacement transducer was attached to measure the displacement of the brick-timber cavity. The actuator displacement was controlled at a rate of 1mm/min for both compression and tension loading. Figures 3(a) and 3(b) shows the setup for compression and tension testing, respectively.

2.3 Test results

2.3.1 Compression tests

A total of 25 specimens were tested in compression, and failure modes slightly varied. Almost all (23 specimens) specimens failed by axial buckling of the tie, and only 2 specimens failed by the combination of tie buckling and pull-out of nail from timber. Each tie started to bend at a 90 degree angle at 20 mm to 30 mm from the nail-tie-timber connection, and scratched the timber surface along its bending path (see Figure 4). After buckling, load was decreased significantly up to 7mm displacement, followed by load fluctuations to a lesser extent. Mean buckling load and corresponding displacement was 1.04 kN and 3.08 mm, respectively. All load-displacement curves, along with an average multi-linear ideal curve for compression specimens are shown in Figure 5. This ideal curve was generated based on the average of all actual load-displacement relationships. This idealisation of

Figure 4. Tie buckling and failure for compression loading.

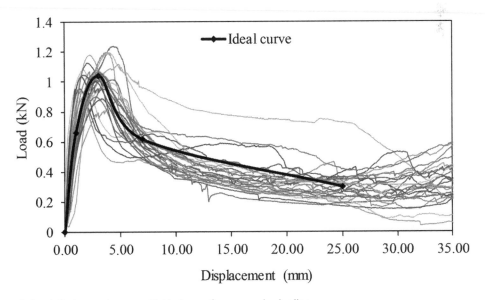

Figure 5. Load-displacement curves with ideal curve for compression loading.

the curve facilitates to input average stiffness data into numerical models of overall brick veneer wall system behaviour. This curve comprises four zones, (a) a line from origin to the inflection point (at 1 mm displacement), (b) an intermediate stage from inflection point to the mean buckling load, (c) a decreasing stage from mean buckling load to 7 mm displacement, and (d) slightly decreasing stage from 7 mm displacement to 25 mm displacement (which represents the half of the air cavity distance). The elastic stiffness was calculated as 0.66 kN/mm from the ideal curve.

2.3.2 Tension tests

Similar to the compression tests, 25 specimens were tested in tension. The failure mode was identical (ductile nail pull-out from the timber stud) for all of the specimens as shown in Figure 6. No pull-out of tie from mortar joint or tie hole yielding were observed. However, in some cases timber was cracked at the nail joint at peak load. The load decreases as the nail started to be pulled out from the timber stud, which represents a ductile failure mode. The variation of peak load and post peak behaviours are notably higher compared to the compression behaviour. Mean peak load and associated displacement was 1.32 kN and 7.36 mm, respectively. All load-displacement curves along with an ideal multilinear curve (akin to compression tests) for tension tests are presented in Figure 7. To capture most of the elastic and peak load behaviour, two intermediate points (at 1 mm and 3 mm)

Figure 6. Nail pull-out from timber in tension specimens.

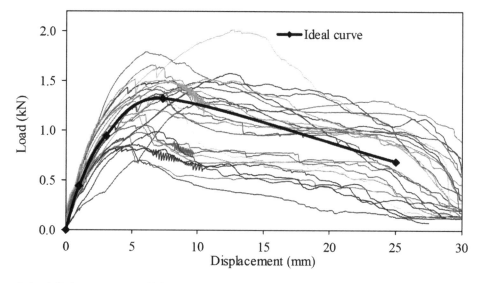

Figure 7. Load-displacement curves with ideal curve for tension loading.

were selected in between the origin and peak load for the ideal curve. Moreover, decreasing stage (from peak load to mean load at 25 mm displacement) represents the post peak behaviour. The elastic stiffness was calculated as 0.45 kN/mm from the ideal curve.

3 PROBABILISTIC MODEL OF WALL TIE CONNECTION PERFORMANCE

3.1 Wall tie connection strength

In a finite element model of a masonry veneer wall system, these behaviours of ties under compression and tension loading would be included with mean and coefficient of variation to represent the stochastic nature of the ties. Using the maximum likelihood method, a range of probability distributions were fitted to tie strength (peak load) data sets under compression and tension loading. The tie strength histograms and fitted probability distributions (normal, lognormal, Weibull, gamma and Gumbel distributions) for compression and tension shown in Figure 8 and Figure 9, respectively. The Anderson-Darling (A-D) test at the 5% significance level was performed to check the goodness-of-fit as the lower tail of the distribution has more importance to wall failure progression compared to the whole distribution.

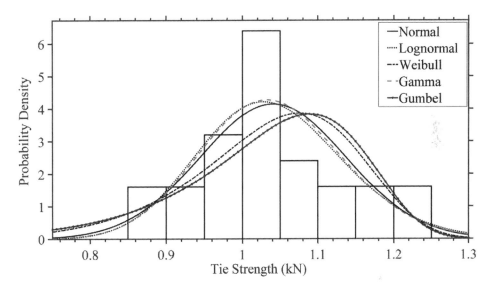

Figure 8. Probability distribution fits of tie connection strength under compression loading.

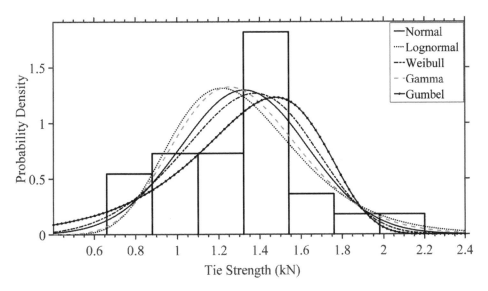

Figure 9. Probability distribution fits of tie connection strength under tension loading.

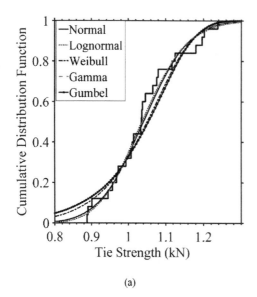

(a)

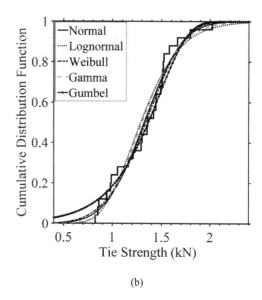

(b)

Figure 10. Comparison of the CDFs and derived data for tie load capacity for (a) compression loading and (b) tension loading.

For compression loaded specimens 'lognormal' ranked highest according to A-D test, whereas 'normal' distribution shows the better fit for tension specimens. Moreover, a visual comparison of CDFs (cumulative distribution functions) with derived data for wall tie connection load capacity (both compression and tension cases) are shown in Figure 10 to infer a goodness-of- fit for the probabilistic models. Statistical parameters for both compression and tension specimens are summarised in Table 1.

Table 1. Statistical parameters for tie strength (peak load).

Sample type	Sample size	Distribution	Mean tie strength	Coefficient of variation (COV)
Compression	25	Lognormal	1.04 KN	0.09
Tension	25	Normal	1.32 KN	0.23

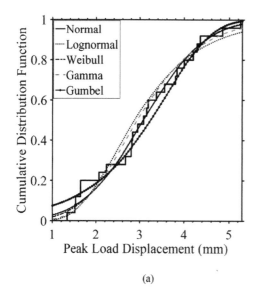

(a)

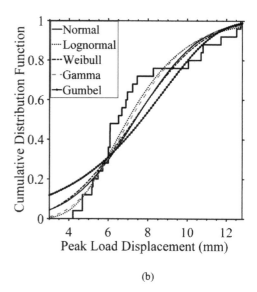

(b)

Figure 11. Comparison of the CDFs and derived data for tie displacement at peak load for (a) compression loading and (b) tension loading.

Table 2. Statistical parameters for tie displacement (at peak load).

Sample type	Sample size	Distribution	Mean tie displacement at peak load	Coefficient of variation (COV)
Compression	25	Normal	3.08 mm	0.35
Compression	25	Lognormal	7.36 mm	0.33

3.2 Wall tie connection displacement at peak load

Lastly, probability distribution parameters were estimated for tie connection displacement at peak load (connection capacity) using the maximum likelihood method. Figure 11 illustrates the resulting CDFs to infer a goodness-of-fit for the distribution models.

According to A-D test, normal and lognormal distribution best fits the displacement histograms for compression and tension loading, respectively. Table 2 summarises the statistical parameters for tie connection displacement at peak load.

It is evident from the statistical parameters that the variation of the tie properties under tension loading is significantly higher compared to compression. This may happen due to the failure pattern of the tie under tension, which is nail pull-out from the timber. Timber is an extremely variable material, and nail pull-out patterns were different from each other, i.e., in some cases, nails were pulled out by creating a wide hole in the timber, while in other cases timbers were cracked, etc. These variable features of failure pattern govern the COV in a significant manner.

4 CONCLUSIONS AND FUTURE WORK

This paper has described part of an on-going investigation into the stochastic behaviour and design of veneer wall systems under out of plane loading. Results and observations are reported in this paper for brick-tie-timber subassemblies under compression and tension loading. Axial buckling of the tie and ductile nail pull-out from the timber stud are the governing failure mode for compression and tension specimens, respectively. For both cases average multi-linear ideal curves were generated to input different parameters for a related finite element model. However, in order to include the variability of the tie strength and stiffness under compression and tension loading a range of probability distributions were fitted to tie strength (peak load) and associated displacement data sets using maximum likelihood method. The study found that, (a) for compression, the tie connection capacity (strength) best fits the lognormal probability distribution whereas corresponding displacement fits normal distribution; (b) for tension, the tie strength best fits the normal probability distribution and corresponding displacement fits lognormal distribution. The mean and COV was calculated for each case and will be included in numerical finite element modelling work along with post peak softening behaviour. Future work will focus on correlation between each of the stages of the load-displacement response. Moreover, using this probabilistic model damage fragility curves will be developed for the tie connections by implementing them in brick veneer walls that are characteristic of buildings located in Australia.

ACKNOWLEDGEMENTS

The authors wish to recognize the financial support provided by the Australian Research Council under Discovery Project DP180102334. The assistance of Mr. Goran Simundic of the University of Newcastle in sample preparations and testing is gratefully acknowledged.

REFERENCES

Choi, Y. H. & LaFave, J. M. 2004. Performance of corrugated metal ties for brick veneer wall systems. *Journal of Materials in Civil Engineering* 16(3): 202–211.

Mertens, S., Smits, A., & Grégoire, Y. 2014. Experimental parametric study on the performance of wall ties. *Proceedings of the 9th International Masonry Conference, Guimarães*, 7-9 July 2014.

Muhit, I. B., Masia, M. J. & Stewart, M. G. 2019. Nonlinear finite element analysis of unreinforced masonry veneer wall systems under out-of-plane loading. In P. B. Dillon & F.S. Fonseca (ed.), *Proceedings of the 13th North American Masonry Conference, Salt Lake City, UT*, 16-19 June 2019: 1769–1781. Longmont, CO: The Masonry Society.

Page, A. W., Simundic, G. & Masia, M. 2009. A study of wall tie force distribution in veneer wall systems (Stage 1). *Proceedings of the 11th Canadian Masonry Symposium, Toronto, Ontario*. May 31- June 3 2009.

Paton-Cole, V. P., Gad, E. F., Clifton, C., Lam, N. T. K., Davies, C. & Hicks, S. 2012. Out-of-plane performance of a brick veneer steel-framed house subjected to seismic loads. *Construction and Building Materials* 28(1): 779–790.

Reneckis, D. 2009. *Seismic performance of anchored brick veneer*. PhD Thesis, Department of Civil and Environmental Engineering, University of Illinois at Urbana-Champaign.

Reneckis, D., LaFave, J. M. & Clarke, W. M. 2004. Out-of-plane performance of brick veneer walls on wood frame construction. *Engineering Structures* 26(8): 1027–1042.

Ribeiro, S., Vicente, R., Varum, H., Graça, J., Lobo, B. & Ferreira, T. 2014. Development of retrofitting solutions: remedial wall ties for masonry enclosure brick walls.

Proceedings of the 9th International Masonry Conference, Guimarães, 7-9 July 2014.

Skroumpelou, G., Messali, F., Esposito, R. & Rots, J. 2018. Mechanical characterization of wall tie connection in cavity walls. *Proceedings of the 10th Australasian Masonry Conference, Sydney*, 11-14 February 2018.

Standards Australia. 2015. Masonry in small buildings - Construction (AS 4773.2:2015). Standards Australia Limited, Australia.

Standards Australia/Standards New Zealand. 2000. Built-in components for masonry construction; Part 1: Wall ties (AS/NZS 2699.1:2000). Jointly published by Standards Australia International Ltd and Standards New Zealand.

Zisi, N. V. & Bennett, R. M. 2011. Shear behavior of corrugated tie connections in anchored brick veneer–wood frame wall systems. *Journal of Materials in Civil Engineering* 23(2): 120–130.

Brick and Block Masonry - From Historical to Sustainable Masonry –
Kubica, Kwiecień & Bednarz (eds)
© *2020 Taylor & Francis Group, London, ISBN 978-0-367-56586-2*

Durability of hemp cords exposed to alkaline environment of lime mortar

M. Zaydan, C. Caggegi, M. Michel & L. Curtil
Laboratory of composite Materials for Construction (LMC2), University Claude Bernard Lyon 1, Villeurbanne, France

ABSTRACT: The use of Textile Reinforced Mortar composites (TRM) for the reinforcement of structural elements is one of the most innovative and effective techniques in the framework of historical masonry rehabilitation. In order to propose an eco-sustainable rehabilitation, this paper focuses on a TRM made of hemp fibers imbedded in a lime mortar. The aim of this study is to identify the effectiveness and the durability of hemp fibers used in TRM characterized by lime mortar to reinforce masonry structures. Tensile tests on hemp fibers were carried out. The accelerated aging effects of the lime mortar on hemp fiber were studied by tensile tests after 7, 14 and 28 days. The aging of fibers was performed by immersing them in a solution that reproduces the same pH of lime mortar, at 23°C. The experimental results show that, in the aging condition analyzed, the hemp fibers don't lose their strength.

1 INTRODUCTION

Historical masonry structures are subject to numerous pathologies, caused in particular by the application of excessive vertical loads, by accidental horizontal actions such as seismic and wind loads, by the movement of the ground, etc. Since these structures present cultural, scientific, historical, artistic and touristic values, they are frequently repaired carefully so that they do not lose any of their values.

During the last two decades, the use of composite materials for the rehabilitation and the reinforcement of historical masonry structures has aroused a great interest. Textile Reinforced Mortar (TRM) is one of these innovative materials, it is a technical solution with many advantages such as light weight, compatibility with masonry material, reversibility, and breathability (Papanicolaou et al. 2007). It consists of a textile grid imbedded in an inorganic matrix. The textiles used in TRM are often made of glass, carbon, polyparaphenylene benzobisoxazole (PBO), aramid or basalt fibers (D'antino et al. 2014). Their role is to ensure good mechanical strength of the composite. While the mineral matrix is frequently manufactured with cement-based mortar, lime, or with a geopolymer, its role is to ensure the cohesion of the composite with the structure and the transfer of forces from the substrate to the textile (Caggegi et al. 2017). Only a few studies address the durability of textile in a mineral matrix (Wei et al. 2014, Micelli et al. 2019, Boulos et al. 2019).

Currently, the major environmental concerns and the awareness of industrial pollution have encouraged the construction and manufacturing industries to look for innovative, reliable and sustainable materials that can replace conventional synthetic fibers (glass, carbon, etc.) as reinforcements for composite materials.

Considerable efforts have been dedicated to the use of natural fibers as reinforcement for inorganic matrix composites. The main reasons for selecting natural fibers are that these fibers have low cost, low density, and are recyclable and biodegradable, unlike synthetic fibers that are expensive to manufacture, have high density, are not renewable and cannot be thermally recycled by incineration (Faruk et al. 2014). According to "RILEM TRC report 36" (Bramshuber 2006), there are three categories of natural fibers: mineral fibers such as basalt and Asbestos, vegetable fibers such as flax, hemp, jute and sisal, and animal fibers such as mohair and silk. As for the mechanical properties, some natural fibers such as basalt, flax, hemp are able to present tensile strengths comparable to those presented by synthetic fibers (Kalia et al. 2009, Ku et al. 2011). However, despite the advantages of natural fibers, their use in mineral matrix composite materials is limited by the relatively low resistance to degradation in alkaline environments (Ardanuy et al. 2015). So, durability studies of mineral matrix and natural fiber composite materials are needed to characterize their long-term behavior before using them in the rehabilitation of structures.

xIn this paper, a research on the durability of hemp cords in an alkaline environment is presented. Single cords were aged by immersion in an alkaline solution. In order to determine the effect of the accelerated aging on the mechanical behavior of hemp fibers, single cords were subjected to tensile tests before and after 7, 14, and 28 days of aging. This study is a part of a durability campaign, at present in progress, which take into account different kinds of natural fibers and different natural and accelerated aging conditions. Tests on single cords and TRM composites will be carried out for an aging duration varying from seven days to two years.

2 EXPERIMENTAL PROGRAM

2.1 Materials

The study proposes an analysis of durability of TRM system characterized by hemp textile and lime mortar matrix.

The hemp cords used in this work were manufactured and supplied in form of coils (Figure 1a) by Stabilimento Militare Produzione Cordami (Agenzia Industrie Difesa, Italy). A single cord consists of three single yarns, having a linear density of 400 tex each, twisted together, resulting in a final cord with a linear density of 1200 tex (Figure 1b).

The average cross-sectional area 'A_f' of the cord has been determined by X-ray tomography, it is equal to 0.751 mm^2 (CoV 0.24 %).

In the durability tests, the alkaline environment produced by the lime mortar around the hemp fibers was reproduced by a NaOH solution characterized by a PH of 12.5.

2.2 Specimens preparation

To assess their mechanical properties, hemp cords were subjected to direct tensile tests. Four series were tested; Table 1 lists the name and the aging related to each series of specimens.

Table 1.   Tensile test series.

Series	Aging duration	pH	Temperature
	days		°C
H-0d	0	12.5	23.3
H-7d	7		
H-14d	14		
H-28d	28		

Six cords with a gauge length of 250 mm were cut from the coil and tested in tension until rupture according to ISO 2062:2009. Three cords of 9 m length were cut from the coil and immersed in a NaOH solution having a pH of 12.5 at a temperature of 23.3 °C (Figure 2). The cords were removed from the solutions at 7, 14, and 28 days, respectively. Then, they were weaved between two nails on a wooden bar (Figure 3), and left to dry. After drying, six specimens with a gauge length of 250 mm were cut from each cord and they were subjected to direct tensile tests.

a)   b)

Figure 1. Hemp coil (a) and hemp cord (b).

Figure 2. Aging of hemp cords in NaOH solution (pH 12.5, T = 23.3 °C).

Figure 3. Drying of the aged cords.

2.3 Experimental setup

According to ISO 2062:2009, tensile tests on hemp cords were carried out by using a universal testing machine Zwick of 50 kN capacity in displacement control.

In order to avoid the concentration of stress which causes the failure at the clamp of the vegetal cord tested in tension, a new grip system has been developed (Figure 4a and b). It consists of a metallic device able to lock the free ends of the cord and reduce the stress concentration by winding the cord two times in the middle of a horizontal cylinder. This latter has the shape of a sandglass to facilitate the positioning of the cord in its middle, which will be superimposed with the axis of the rod where the tensile force will be applied (Figure 4b).

To ensure alignment of the fibers and to have them straight enough, preloads of 6 N were applied to the cords. The gauge length was 25 cm and the loading rate was 250 mm/min (ISO 2062:2009).

3 RESULTS AND DISCUSSION

The results obtained by the tensile test are reported in terms of stress-strain curve (σ - ε), maximum tensile load F_{max}, maximum tensile strength σ_{max}, maximum displacement U_{max}, and elastic modulus E for the cords tested. The stress σ refers to the ratio between the tensile load and the cross-sectional area of the yarn. The strain ε is the ratio between the global displacement of the cross head and the gauge length of the yarn

(250 mm). The elastic modulus refers to the slope of the stress-strain curve. Table 2 shows the average maximum tensile load F_{max}, the average maximum tensile strength σ_{max}, the average maximum displacement U_{max}, and the average elastic modulus E for the different series of hemp cords tested.

Figures 5-8 show the stress-strain curves for the different series of hemp cords tested. As we can see, all the curves are constituted from two branches: the first branch is nonlinear, it is due to the realignment of the single filaments which constitutes each yarn of the cord. This branch is not taken into account for the determination of the Young modulus. The second branch is almost linear up to the rupture of the cord.

Figures 9 and 10 show the evolution of the tensile strength and Young's modulus of hemp cords with aging time, respectively. One can notice that, the tensile strength of series H-7d, H-14d and H-28d is about 10% higher than that of H-0d (Figure 9). It seems that in a first step of aging the alkaline environment of lime mortar has a beneficial effect on hemp cord; the high standard deviation recorded during the tests demands an increment of specimen's number to confirm this evidence. While for the modulus of Young, we can notice an increase of 9.5 times between 0 and 7 days, 11.5 times between 0 and 14 days, and 12.1 times between 0 and 28 days (Figure 10). It can be concluded that the immersion of the hemp cord in a solution of NaOH at pH 12,5 and temperature of 23°C for 7, 14, and 28 days has the effect of increasing the rigidity of the latter while maintaining, nearly, the same tensile strength.

Figure 4. Tensile test setup for hemp cords (a) and details of metallic grip device (b).

Table 2. Tensile test results of all the series tested.

	Series	F_{max}	σ_{max}	U_{max}	ε	E
		N	MPa	mm	%	MPa
Average	H-0d	202.3	269.3	18.2	7.3	4.2
St. dev		32.8	43.6	2.7	1.1	0.4
CoV (%)		16.2	16.2	15.0	15.0	9.2
Average	H-7d	226.4	301.5	23.0	9.2	40.3
St. dev		14.6	19.5	1.2	0.5	1.7
CoV (%)		6.5	6.5	5.3	5.2	4.3
Average	H-14d	255.1	339.6	26.2	10.6	48.6
St. dev		20.2	26.9	2.2	0.8	1.6
CoV (%)		7.9	7.9	8.3	7.6	3.3
Average	H-28d	258.8	344.6	26.0	10.8	51.5
St. dev		17.0	22.6	2.0	0.3	2.8
CoV (%)		6.6	6.6	7.7	3.1	5.5

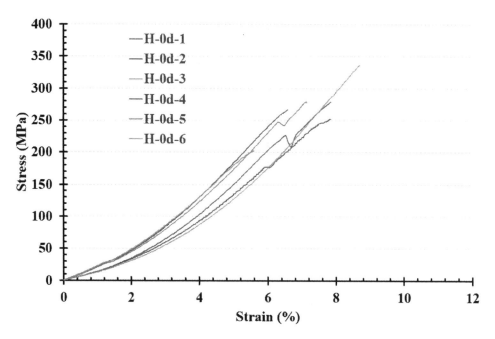

Figure 5. Stress-strain curves of hemp cords tested without any aging.

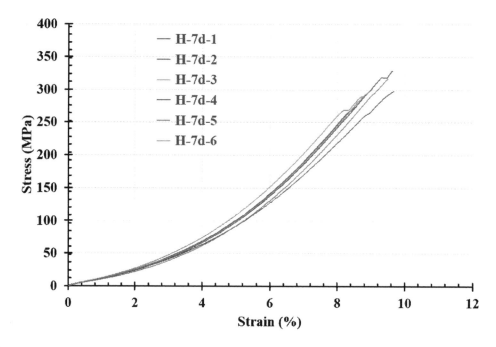

Figure 6. Stress-strain curves of hemp cords aged in NaOH solution for 7 days.

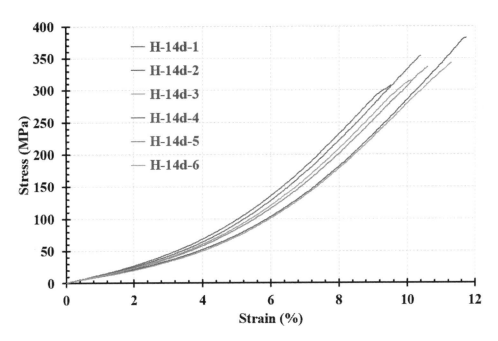

Figure 7. Stress-strain curves of hemp cords aged in NaOH solution for 14 days.

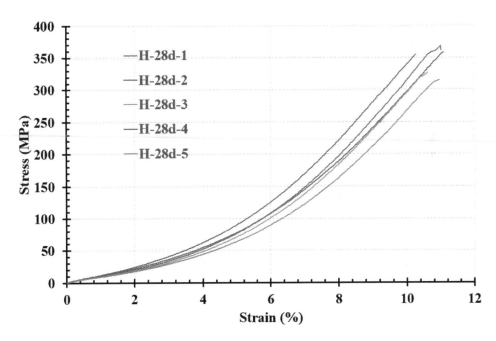

Figure 8. Stress-strain curves of hemp cords aged in NaOH solution for 28 days.

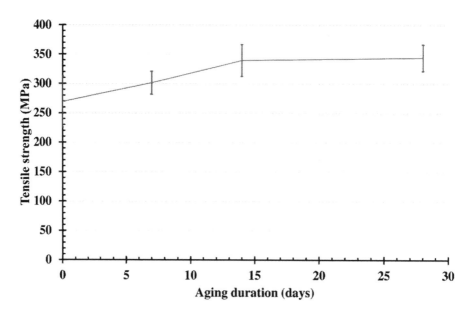

Figure 9. Evolution with time of tensile strength of hemp cords after exposure to alkaline environment.

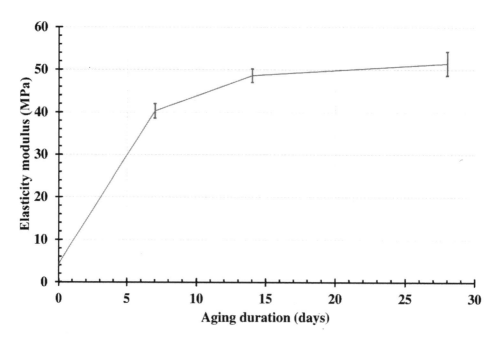

Figure 10. Evolution with time of elasticity modulus of hemp cords after exposure to alkaline environment.

4 CONCLUSIONS

The durability of hemp cords in alkaline environment has been investigated. These cords are used in the fabrication of TRM used for the rehabilitation of the masonry structures. The mechanical performance of as received and aged hemp cords has been determined by the tensile tests. The evolution of the tensile strength and modulus of elasticity along 7, 14, and 28 days has been quantified and discussed. Based on the experimental results, the following preliminary conclusions can be drawn:

- The series tested show a similar behavior in tension. The tensile strength of series H-7d, H-14d and H-28d is about 10% higher than that of H-0d. It seems that in a first step of aging the alkaline environment of lime mortar has a beneficial effect on hemp cord; the high standard deviation recorded during the tests demands an increment of specimen's number to confirm this evidence.
- The modulus of elasticity, increases 9.5 times between 0 and 7 days of immersion, 11.5 times between 0 and 14 days of immersion, and 12.1 times between 0 and 28 days of immersion. Therefore, the immersion in alkaline solution increases the rigidity of the hemp cords.

In the future, results related to the durability test in progress at 3 months, 6 months, 1 year and 2 years will complete the results presented in this manuscript.

ACKNOWLEDGMENT

The authors would like to acknowledge Stabilimento Militare Produzione Cordami (Agenzia Industrie Difesa) – Castellammare di Stabia for supplying the hemp coils.

REFERENCES

Ardanuy, M et al. 2015. Cellulosic fiber reinforced cement based compsosites: A review of recent research. *Construction and Building Materials* 79: 115–128.

Boulos, L. et al. 2019. The effect of a zirconium dioxide sol-gel treatment on the durability of flax reinforcements in cementitious composites. *Cement and Concrete Research* 115: 105–115.

Caggegi, C. et al. L. 2017. Experimental analysis on tensile and bond properties of PBO and aramid fabric reinforced cementitious matrix for strengthening masonry structures. *Composites Part B* 127: 175–195.

Bramshuber, W. 2006. Textile Reinforced Concrete-State-of-the-art report of RILEM TC 201-TRC: RILEM Publications.

D'antino, T. et al. 2014. Matrix-fiber bond behavior in PBO FRCM composites: A fracture mechanics approach. *Engineering Fracture Mechanics* 117: 94–111.

Faruk, O. et al. 2014. Progress report on natural fiber reinforced composites. *Macromolecular Materials and Engineering* 299(1): 9–26.

ISO 2062: 2009. Textiles-yarn packages-determination of single-end breaking force and elongation at break using constant rate of extension (CRE) tester. International Organization for Standardization.

Kalia, S. et al. 2009. Pretreatments of natural fibers and their application as reinforcing material in polymer composites-a review. *Polymer Engineering and Science* 49: 131–135.

Ku, H. et al. 2011. A review on the tensile properties of natural fiber reinforced polymer composites. *Composites Part B: Engineering* 42(4): 856–873.

Micelli, F. & Aiello, M. A. 2019. Residual tensile strength of dry and impregnated reinforcement fibres after exposure to alkaline environments. *Composites Part B* 159: 490–501.

Papanicolaou, C.G. et al. 2007. Textile reinforced mortar (TRM) versus FRP as strengthening material of URM walls: in-plane cyclic loading. *Materials and Structures* 40(10): 1081–1097.

Wei, J. & Meyer, C. 2014. Degradation rate of natural fiber in cement composites exposed to various accelerated aging environment conditions. *Corrosion Science* 88: 118–132.

Brick and Block Masonry - From Historical to Sustainable Masonry –
Kubica, Kwiecień & Bednarz (eds)
© 2020 Taylor & Francis Group, London, ISBN 978-0-367-56586-2

Evolution in the protection of clay-based mortars

A. Karozou & M. Stefanidou

Laboratory of building Materials, Civil engineering Department, Aristotle University of Thessaloniki, Greece

ABSTRACT: Earthen structures have been part of human history and civilization ever since the first formation of agricultural human societies. During medieval ages, earth was used all over Central Europe, while
nowadays, one third of the world's population, dwells in earthen structures. Due to the beneficial role of
earthen materials in social, financial and environmental terms, clay has been reconsidered as a building material in modern structures. In the present paper the physical and mechanical characteristics of old clay mortars
have been analyzed through different research works. The aim was to understand their behavior as to produce
new compatible repair mortars. Moving on a step further, the vulnerability of clay mortars to humidity was
approached by using new materials based on nanotechnology. The behavior of the new produced mortars was
tested through different durability tests. The results are encouraging for the improvement of genuine weakness
of clay mortars.

1 INTRODUCTION

Following human civilization and history, the use of
clay in construction, never ceased. From the ancient
times, in the Mediterranean and China, till today,
earthen structures offered shelter in humanity
throughout the years (Niroumand et al. 2013). Clay
has been the oldest, most used building material,
both as a binder and as the basic constituent of
adobes and bricks. Besides being a part of vernacular
architecture, requiring restoration, earthen building
has returned to the attention of the industry as
a sustainable solution to modern construction (Niroumand et al. 2017; Sameh 2014).

Analyzing numerous clay-based mortars from historic structures in Greece, important technological
data have been recorded. Clay-based mortars were
used for either bonding units such as stones or
adobes (Figure 1a) or as renders/plasters (Figure 1b)
in old structures. They usually contain siliceous
origin aggregates up to 8mm in the case of bonding
mortars or up to 4mm for renders. In some cases,
crushed bricks were also added as aggregates due to
their compatibility and harmonization to the structure. The combination of clay with lime was often
met in order to increase plasticity and strength (Stefanidou et al. 2007; Papayianni et al. 2007). Renders
often contain natural fibers (straw, animal hair or dry
ground vegetable fibers (Figure 1b) in order to
increase mass stability and restrict cracking. For
these mortars, cracking due to shrinkage is often met
in a random distribution and is the main pathology
form. Additionally, biological deterioration, salt
presence and material loss can be recorded. The

compressive strength rarely exceeds 1MPa, while
open porosity is high (25-33%) and the specific gravity ranges between 1.55-1.65, depending also on the
aggregate content. Clay-based mortars are usually
rich in binder and the B:A ratio is 1:1.5 up to 1:2 in
most cases. The color depends on the clay composition, but usually "earthen" colors prevail. They are
well compacted in order to increase the cohesion of
the aggregates to the "soft" clay binder. The technology of clay mortars didn't alter through centuries.

The problematic and vulnerable nature of clay,
towards weathering factors and water, hinder their
use in restoration and modern construction strategies.
They were often substituted by lime or lime-
pozzolana and clay was added mainly for color harmonization. Protective treatments usually applied,
were based on waxes for water proofing, while direct
exposure to water was avoided. To overcome this susceptibility many ways to enhance their structure were
developed. Numerous materials have been also used
as additives and waterproof agents to improve physical and mechanical properties. Mineral and synthetic
additives, as well as biopolymers and animal additives are some of the agents used for reinforcement of
earthen structures (Bahobail 2012; Eires et al. 2017).
Nowadays, the technology offered other possibilities
with new materials being tested for the preservation
of historic materials. Also, the new principles of construction based on recycling and on the exploitation
of cheap, easy to find, local materials gave boost to
the reviving of clay-based technologies.

In this paper, nano clay particles were used as additives in clay mortars and physical, mechanical and
microstructure changes have been recorded in time.

Figure 1. (a) Bonding mortar with brick aggregates and cracks; (b) Render reinforced with straw.

The aim was to study materials of chemical affinity with clays. The idea was based on previous studies showing that different nanoparticles can offer significant advantages to similar in nature materials due to their mechanism of action (Tsardaka et al. 2019). Specifically, nanoclays consist of layered mineral silicate nanoparticles and are categorized by morphology and composition in montmorillonite, kaolinite, halloysite etc. Nanoclays are considered advanced materials that can improve physical and mechanical properties, while reducing permeability and offer protection against environmental factors (Tabarsa et al. 2018; Bastidas-Arteaga et al. 2015).

2 DATA COLLECTED FROM PREVIOUS STUDIES

Analyzing numerous clay-based mortars from historic structures in Greece, important technological data have been recorded. In Table 1 the data collected from analyzing old mortars of the Byzantine Period are presented (Papayianni et al. 2001). The binders of the monuments presented here, are lime with reactive soil of weak pozzolanic properties, with open porosity among 20-24% and compressive strength 0.98-2.45 MPa. The results indicate the high porous nature of these binders and the weak structure with low mechanical strength (Table 1).

Moreover, data from different research studies were collected (Table 2). The archeological site of Logos near the modern town of Edessa is a part of the ancient city located in central Macedonia, Greece. The history of the ancient city starts from the Hellenistic period and continues until the Byzantine era. In this case study it is stated that two types of mortars were used for construction. A hydraulic one containing lime and pozzolanic material and a second type that was a mortar based on clay (Stefanidou et al. 2007). The characteristics of these two types of mortars are presented at Table 2.

Furthermore, the characteristics of old mortars in different historic periods were analyzed in the study of Papayianni and Stefanidou (Papayianni et al. 2007). In this study it is explained that during Roman and Byzantine periods, lime and pozzolana were mostly used in construction combined with brick dust, while in Ottoman period soil was also added and used as a binder. In Table 2 the compressive strength and open porosity of the binders during the Ottoman period are listed.

Results from another case study of the archaeological site of the roman city of Palatiano (4th century B.C -3rd century A.D), located north of the town of Kilkis in north Greece, are reported. The analysis of the old mortars indicated the existence of lime and clay in the binding system (Papayianni et al. 2008). The results of the analysis in terms of compressive strength and open porosity are shown below (Table 2).

From both tables, it is noted that the compressive strength remains low with values ranging from 0.5 to 2.5 MPa, while the porosity values range high between 18-38% (Table 1, Table 2). In all cases, the mortars were exposed to weathering factors and were highly susceptible to water penetration.

Table 1. Results of the old mortars of the Byzantine period analyzed (Papayianni et al. 2001).

Monument	Binder type	Open porosity (%)	Compressive strength (MPa)
Church Acheropoiitos, phase 5th cent. AD	Lime+ reactive soil*	24	0.98-1.96
Church Hagia Aikaterini. 13th cent. AD	Lime+ reactive soil*	20-21	1.47-2.45
Church Hagios Panteleimon, 14th cent.AD	Lime+ reactive soil*	20-21	0.98-1.47

*soil with weak pozzolanic properties

Table 2. Data collected from previous studies on different mortar types (Stefanidou et al. 2007; Papayianni et al. 2007; Papayianni et al. 2008).

Study	Binder type	Compressive strength (MPa)	Open porosity (%)
Logos-Edessa	Lime: soil (1:0.5)	1.2-1.5	18-22
	Lime: soil (1:1)	<1	23-25
Ottoman historic period	Lime, Lime+ pozzolana, Lime+ brick dust, Lime+ soil	1.36-2.50	18.0-27.5
Palatiano	Lime: Clay: Sand (1:0.25:5)	0.5-2.0	25-38

3 NEW MATERIALS AND TECHNIQUES

3.1 *Preparing the specimens*

In the present study, to enhance the properties of clay-based mortars, nanoclay was used as an additive at 5% w.w. of the clay binder. The nanoclay used was provided by Sigma Aldrich (surface modified 0.5-5wt% aminopropyltriethoxysilane, 15-35% octadecyl- amine) and is referred to as NC. Moreover, a reference mortar was manufactured for comparison reasons with no use of additives and is recorded as A.

In all cases, the manufactured specimens were of 40×40×160 mm dimensions for each series of mortars. The clay used was extracted from the island of Crete and has been characterized using XRD analysis, particle size distribution and chemical analysis. The specific gravity of the dry material is 1.96 g/cm^3 (ASTM-C188-95) while the color determination using Munsell charts is 5Y,7/1, light grey. The XRD analysis indicated that the soil particles consisted of quartz, calcite, calcium aluminum hydroxide, cancrinite and a small percentage of muscovite, while a high content of calcium oxide of 25% was reported by the results of the chemical analysis (Table 3). More information on the analysis of the binder, as well as the XRD charts can be found on a previous research paper of the authors (Karozou et al. 2019).

The clay was sieved to have a grain diameter of less than 0.5 mm. The particle size distribution (Malvern 2000, Mastersizer) depicted that the particles of 2-500 μm size prevailed in the binder's mass (Figure 2). The characteristic values of size diameters are d(0.1)= 3.377 μm, d(0.5) = 34.148 μm and d (0.9) = 247.314 μm. The sand used for mortar manufacture was washed river sand of siliceous origin with grain size between 0-4mm and a similar color with that of the binder.

The mixture proportions of the mortars were 1:2.5 (clay:aggregate ratio) by weight, with the same amount of clay in both mixtures. The desired workability was of 15±1cm as tested by flow table (EN 1015-

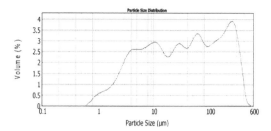

Figure 2. Particle size of cretan clay used.

3:1999), while the curing conditions of the specimens were at ambient conditions of 60%RH and 20±2°C. The workability, the water to binder ratio (w:b), the binder to aggregate ratio and the additives used can be seen in Table 4.

3.2 *Tests performed*

The purpose of this study was to enhance the mechanical and physical properties of the mortars using nanoclay and to compare the results with the old mortars that were analyzed. The tests conducted concerning the behavior of the specimens against water penetration were capillary absorption (UNI EN15801:2010), Karsten tube penetration (EN 16302:2013) and open porosity (RILEM CPC11.3) tests. It should be noted, that the porosity test was conducted using heptane instead of water, due to the great vulnerability of the specimens to water. The duration of the Karsten tube test was ten minutes for each specimen and the water intake per minutes per surface unit was calculated (ml/min/cm^2).

After the conduction of the capillary test the drying test was conducted as to test the breathability of the materials (EVS EN 16322-13:2013). Both the capillary coefficient, indicated as c and the drying index, indicated as ID, were calculated.

The mechanical properties of the specimens were also tested at the age of 28 days. Compressive and flexural strength tests (EN1015-11:1999) were conducted, while the dynamic modulus of elasticity (E_d) was calculated indirectly using ultrasounds (ASTM C597-16). Moreover, as to evaluate volume stability, the linear and volume shrinkage of the specimens were measured (DIN 18947:2013-08). The specimens were preserved at a chamber of constant conditions of temperature and humidity (20±2 °C, 65±5 % RH) and the daily alteration of dimensions was recorded.

Table 3. Chemical analysis of cretan clay used.

Clay	% w.t.
Na$_2$O	0.94
K$_2$O	2.09
CaO	25.33
MgO	7.48
Fe$_2$O$_3$	6.52
Al$_2$O$_3$	13.03
SiO$_2$	30.87
LoI	13.74
Cl$^-$	0.01
NO$_3^-$	0.01
SO$_4^{2-}$	0.21

Table 4. Fresh properties of the mortars.

Mixture	b:a	w.w.	w:b	Workability [cm]
A	1:2.5	-	0.65	15.5
NC	1:2.5	5%	0.71	15.0

Moreover, stereoscopic observation through LEICA WILD M10 microscope was conducted, as to detect cracking, roughness and surface modifications. Microscopic examination by SEM (JEOL840A JSM) equipped with EDS device was also carried out to determine the cohesion of the inner structure. In all cases, optical observation of color change using Munsell charts was determined.

To determine the hydrophobic nature of the specimens, two more tests were conducted. The contact angle measurement test (Kruss Drop Shape Analyzer – DSA100S Surface Tension/Contact Angle meter instrument) and the heuristic test of water drop absorption time.

4 EXPERIMENTAL RESULTS

The results of the physical and mechanical properties of the specimens are presented. In Table 5 the results of the physical properties of the specimens concerning the capillary absorption test, the Karsten tube penetration test and the drying test are presented.

4.1 Capillary absorption - drying test

The capillary absorption test results at 28 days are presented (Figure 3). The absorption trend of the specimens through capillarity with time is presented below. It is noted that the specimens treated with nanoclay, have a lower capillary coefficient value by 73,01% compared to the reference specimen A, showing a low capillary absorption tension (Table 5).

As was mentioned before, after the completion of the capillary test the drying test was immediately carried out as a reverse capillary test. The test was completed after taking 40 daily measurements, that is considered as the necessary time for the specimens to dry and reach equilibrium with the atmosphere. The drying index indicates the resistance of the material to drying, so it can be claimed that a low value of drying index (ID) reflects an overall easier drying behavior (EVS EN 16322-13 2013). The ID values in Table 5 therefore indicate the lower resistance to drying of the NC specimens by 5.34% compared to the reference. It should be noted that despite the lower absorption of the NC specimen, the time for both specimens to dry was approximately the same.

4.1.1 Karsten tube penetration test

Karsten tube penetration test was performed in order to observe the horizontal water absorption. Karsten tubes were applied onto the surface of the specimens and the tubes were filled with 2 ml of water. The test lasted for ten (10) minutes for each specimen while every 15 seconds the water absorbed was recorded. The water absorption through time is presented for both specimens at Figure 4 below, while the surfaces of the specimens after the conduction of the test are also presented (Figure 5).

The high absorption rate of the reference specimen A is proven both by the total value of water penetration (Table 5) and by the water penetration through time (Figure 4). There seems to be a consistency between the results of capillary and Karsten tube tests, since NC specimens show the lowest water intake in both tests conducted. The decrease of water absorption for NC specimens was of 73.75% compared with the reference one.

Table 5. Physical properties of the mortars.

Mixture	Capillary coefficient, c (g/cm^2 *min$^{1/2}$)	Drying Index, ID	Water penetration (ml/ min*cm^2)
A	0.27	0.082	0.053
NC	0.07	0.077	0.014

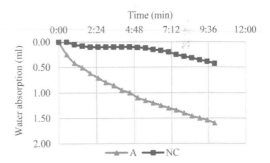

Figure 4. Water absorption through Karsten tube test.

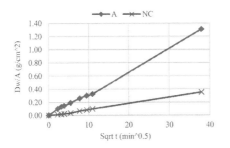

Figure 3. Capillary absorption of the specimens at 28 days.

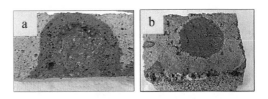

Figure 5. Specimens after Karsten tube test (a) reference specimen A and (b) NC specimen.

759

Table 6. Mechanical properties of the mortars.

	Compressive strength, σ (MPa)		Flexural strength, f_{stm} (MPa)		Dynamic modulus, E_d (GPa)		Open porosity (%)	
	28d	90d	28d	90d	28d	90d	28d	90d
A	0.8	1.6	0.7	0.5	3.3	4.9	16.5	17.6
NC	2.2	1.7	1.2	0.8	3.6	3.5	20.4	21.4

Moreover, it is noted that the surface of the untreated specimen A was damaged during the test, while the one of the NC specimens remained intact (Figure 5).

4.1.2 Mechanical characteristics-open porosity

The compressive and flexural strength were calculated at the age of 28 and 90 days. The results are presented in Table 6 below, while the open porosity of the specimens was also recorded at both ages. The results indicate the positive effect of the addition of nanoclay in terms of early strength development. Specifically, the increase in compressive strength by 165.13% and in flexural strength by 62.14% compared to the reference at the age of 28 days, indicate this positive effect on enhancing the mechanical characteristics.

At the age of 90 days both the compressive and flexural strength for the NC specimen were reduced by 20.89% and 31.5% respectively, while the reference specimen A showed an increase in compressive strength by 95.22% and a decrease in flexural strength by 25.68%. Despite the impressive increase of compressive strength of the reference, the NC specimens showed improved abilities with values greater by 7.44% and 49.45% for compressive and flexural strength respectively.

The values of elasticity modulus for NC specimens, remain practically the same over time, indicating the low values of ultra sonic velocity and implying the existence of voids inside the clay mass. Fact that is justified by the high values of open porosity in both ages, with an increase through time by 4.8%. Moreover, the increase in the dynamic modulus by 46.3% through time for specimen A, complies with the high value of compressive strength at the age of 90 days, implying a more compact structure (Table 6). The open porosity values for mortar A showed a slight increase through by 7.15%.

4.1.3 Shrinkage measurements and hydrophobic properties

Besides the evaluation of the mechanical properties the shrinkage of the specimens was also measured. The values of the linear and volume shrinkage were recorded daily. The results are presented in Table 7, with the total values being calculated at the completion of 90 days. According to DIN 18947:2013-08 the linear shrinkage should be lower than 2%, fact that stands true for both specimens (Table 7). The

volumetric alteration of the specimens shows a great volume loss of the reference, while the NC specimen recorded a 67% reduction compared to the reference.

As far as the hydrophobic properties of the specimens are concerned, the values of contact angle and the water drop absorption time are recorded (Table 7). The measurements of the contact angle show the hydrophobic nature of the NC specimens due to the high value recorded (mean value of three different measurements), that is 75.3% higher than that of the reference. Images of the test are presented below (Figure 6), while it is noted that due to the fast absorption time of the specimen A the contact angle values were changing rapidly.

The results of the heuristic test of the absorption of water droplet indicate the protective role of NC

Table 7. Shrinkage and hydrophobic properties of the mortars.

Mixture	Shrinkage		Contact angle (°)	T(sec)
	Linear, dl/l (%)	Volume shrinkage, dV/V (%)		
A	-1.33	-7.76	63.7	19
NC	-0.98	-2.56	111.7	4380

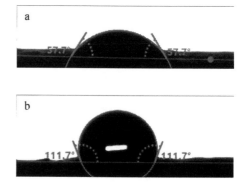

Figure 6. Water drops for (a) A and (b) NC specimens.

against water penetration with 229.52% increase compared to the reference.

4.1.4 *Stereoscopic and microscopic examination*

As to detect alterations on the surface, a stereoscopic observation of the specimens was held. Through stereoscopic observation, color alterations, surface cracks and some evidence concerning roughness can be collected. Images of both specimens after the completion of 90 days are presented in Figure 7. Shrinkage cracks are spotted in both specimens, while some white agglomerations are observed on the surface of NC specimen. A higher microroughness is detected for NC specimen, justifying the higher values of contact angle recorded according to theory (Shirtcliffe et al. 2019). Moreover, the reference specimen presented a rough surface with larger pores on the surface.

The microscopic examination by SEM was held also at the age of 90 days for both specimens. A loose crystal structure was observed for the non-modified mortar (A) that also presented a coarse surface (Figure 8a). The formation of rod like and spherical crystals was detected in the case of NC specimen (Figure 8b). The analysis by the EDS device showed that these crystals were of aluminosilicate nature (Table 8). A good dispersion of nanoclay was observed in the clay mass, presenting, however, small amounts of agglomeration of its nano sized particles.

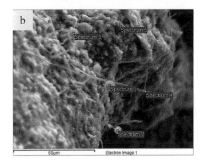

Figure 8. Microscopic observation of the mortars through SEM at the age of 90 days (a) specimen A and (b) specimen NC (scale 60μm).

Table 8. EDS spectrum analysis of NC mortar (all results in atomic %).

Spectrum	Al	Si	S	K	Ca	Fe
Spectrum 1	14.15	37.80	5.59	5.77	20.11	0.00
Spectrum 2	16.10	45.04	3.58	4.07	16.42	0.00
Spectrum 3	13.88	35.54	0.00	3.97	19.34	0.00
Spectrum 4	17.90	45.65	0.00	4.86	13.30	10.36
Spectrum 5	32.99	43.12	0.00	12.20	0.00	2.26

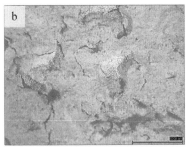

Figure 7. Stereoscopic observation of the mortars at the age of 90 days (a) specimen A and (b) specimen NC (scale 1000μm).

5 CONCLUSIONS

In total, the values of the mechanical properties and open porosity of the reference specimen A and specimen NC are proven to be similar with all the other mortars previously analyzed. As proven by the tests conducted, the addition of nanoclay in the clay mass even at low percentage (5%), affected positively the clay structure especially in terms of enhancing the physical properties. Despite the high porosity values, the mechanical characteristics were enhanced at early age and sustained at similar values as of that of the reference. The low values of capillary coefficient and water absorption, as well as the high values of contact angle and water drop absorption time claim the creation of a hydrophobic structure. Another notable fact was that the surface of the specimens NC remained intact after the conduction of the Karsten tube test.

Overall, the use of nanoclay had an advantageous effect as an additive in the clay mass of the mortar.

ACKNOWLEDGEMENTS

Author Karozou A. would like to thank the General Secretariat for Research and Technology (GSRT) and the Hellenic Foundation for Research and Innovation (HFRI) for funding the research through the scholarship funding program for PhD candidates.

REFERENCES

Bahobail, M.A. 2012. The Mud Additives and their Effect on Thermal Conductivity of Adobe Bricks. *Journal of Engineering Sciences, Assiut University* 40(1): 21–34. Available at: http://www.aun.edu.eg/journal_files/85_J_8907.pdf.

Bastidas-Arteaga, E. et al. 2015. The Future of Civil Engineering with the Influence and Impact of Nanotechnology on Properties of Materials. *Procedia Materials Science* 51(Cnt 2014): 259–266. Available at: http://linkinghub.elsevier.com/retrieve/pii/S2211812815002709%5Cnhttp://dx.doi.org/10.1016/j.engstruct.2013.01.006.

Eires, R., Camões, A., & Jalali, S. 2017. Enhancing water resistance of earthen buildings with quicklime and oil. *Journal of Cleaner Production* 142: 3281–3292.

EVS EN 16322-13. 2013. Conservation of Cultural Heritage -Test Methods- Determination of drying properties. *European standard*.

Karozou, A., Konopisi, S., Paulidou, E., & Stefanidou, M. 2019. Alkali activated clay mortars with different activators. *Construction and Building Materials* 212: 85–91. Available at: https://doi.org/10.1016/j.conbuildmat.2019.03.244.

Niroumand, H., Barcelo, J.A., Kibert, C.J., & Saaly, M. 2017. Evaluation of Earth Building Tools in Construction (EBTC) in earth architecture and earth buildings.

Renewable and Sustainable Energy Reviews 70 (June 2016): 861–866. Available at: http://dx.doi.org/10.1016/j.rser.2016.11.267.

Niroumand, H., Zain, M.F.M., Jamil, M., & Niroumand, S. 2013. Earth Architecture from Ancient until Today. *Procedia - Social and Behavioral Sciences* 89: 222–225. Available at: http://linkinghub.elsevier.com/retrieve/pii/S1877042813029698.

Papayianni, I., & Stefanidou, M. 2001. Microstructural Analysis of Old Mortars of Byzantine Period. In E. C. A. Brebbia (ed) *International Conference STREMAH 28-30 May, Bologna, Italy*, 45–52.

Papayianni, I., & Stefanidou, M. 2007. The influence of mixture design parameters on the long term strength of lime-based mortars. 355-362.

Papayianni, I., & Stefanidou, M. 2008. Pathology symptoms of lime-based repair mortars. *1st Historical Mortars Conference*.

Sameh, S.H. 2014. Promoting earth architecture as a sustainable construction technique in Egypt. *Journal of Cleaner Production* 65: 362–373. Available at: http://dx.doi.org/10.1016/j.jclepro.2013.08.046.

Shirtcliffe, N., Comanns, P., Hamlett, C., Roach, P., & Atherton, S. 2019. 5.11 - The Effect of Roughness Geometry on Superhydrophobicity and Related Phenomena. *Comprehensive Nanoscience and Nanotechnology (Second Edition)* 5: 291–308.

Stefanidou, M.A., & Papayianni, I. 2007. Analysis of the old mortars and proposals for the conservation of the archeological site of Logos-Edessa. *WIT Transactions on the Built Environment* 95: 261–65.

Tabarsa, A., Latifi, N., Meehan, C.L., & Manahiloh, K.N. 2018. Laboratory investigation and field evaluation of loess improvement using nanoclay – A sustainable material for construction. *Construction and Building Materials* 158: 454–463. Available at: http://linkinghub.elsevier.com/retrieve/pii/S0950061817319189.

Tsardaka, E., & Stefanidou, M. 2019. Study of the role of different nanoparticles in lime pastes. In *HMC2019*, Pamplona, Spain.

Brick and Block Masonry - From Historical to Sustainable Masonry –
Kubica, Kwiecień & Bednarz (eds)
© 2020 Taylor & Francis Group, London, ISBN 978-0-367-56586-2

Comparative creep characteristics of lime mortars and brickwork

S. Macharia
Ora Labora Ltd, UK

P. Walker
BRE Centre for Innovative Construction Materials, University of Bath, UK

U. Peter
Lhoist Recherche et Développement S.A., Belgium

ABSTRACT: Lime mortars can be based on air-lime, pure calcium hydroxide, or hydraulic lime being a mixture of air-lime and hydraulic components. Air-lime has traditionally been used both as a binder and as an additive to improve the workability and water retention properties of masonry mortars. The ability to accommodate movements without significant cracking, requiring fewer movement joints, is a commonly cited benefit of mortars containing lime. This paper presents findings from an experimental study into the long-term movement characteristics of air-lime mortared masonry. The creep of cement-lime based ("blended") mortars, containing varying quantities of air-lime, have been compared in an experimental study. Axial movements on mortar specimens and masonry wall panels have been recorded at regular intervals for up to 6 months. The long-term mortar movements are related to the strength and measured creep movement of masonry wall tests.

1 GENERAL INSTRUCTIONS

Until the twentieth century lime was the primary binder used in mortars in masonry construction. Cement brought faster setting times, higher mortar strengths, and more consistent quality compared to traditional small-scale lime production. However, lime still retains many benefits as a binder for masonry mortars. The lower strength and stiffness of lime mortars facilitates masonry unit recycling and is widely considered to be better at accommodating long-term movements, requiring fewer movement joints. Compared to cement mortars, the use of lime also improves mortar workability, mortar water retention properties, and the durability of masonry.

Air lime is also used as a secondary binder together with cement in cement:lime:sand mixes, although in recent years the use of chemical additives, plasticisers, instead has also become commonplace. Over the past 50 or so years research on masonry construction has primarily focussed on cement mortars, with relatively little attention given to developing the wider use of lime in modern construction.

The aim of the work presented in this paper has been to improve understanding of the influence of increasing air lime (calcium hydroxide) content in cement mortars on mechanical properties, including long-term movements, of mortars and brickwork masonry. This aim has been achieved by fulfilling the following objectives:

- Studying mechanical strength development of cement: lime mortars prepared with varying air lime contents.
- Conduct comparative creep tests on small mortar cylinder specimens prepared with varying air lime contents.
- Conduct longer-term comparative creep tests on masonry brickwork wall specimens prepared using mortars of varying air lime contents.

2 SUMMARY OF PREVIOUS WORK

Time dependent deformations are attributed to creep, which is dependent on load, and shrinkage, which is independent of load (Ball and Allen, 2010). Creep is a time dependent deformation occurring on structures under sustained stresses. It is thought to be made up of delayed elastic strain, which is recoverable after removal of the applied stress, and viscous strain, which is permanent (Lenczner, 1981). There are many factors thought to influence the creep of masonry including: unit water absorption characteristics (Forth and Brooks, 2008), (Forth et al., 2000); type of mortar (Brooks and Bakar, 2004), (Brooks, 2001), (Ball et al., 2011).

Time dependent deformations in masonry are not very well understood, with the majority of the data used to estimate this phenomenon in masonry coming directly from concrete research. Similar to concrete, and other viscoelastic materials, masonry exhibits three stages of creep: primary, secondary and tertiary creep. Initially the creep rate of the materials seems to decrease; in the secondary stage, which is also known as the constant creep, the creep rate seems to level out with very small increases in strain with time. The tertiary stage is characterised by an increase in creep rate leading to creep rupture. In masonry structures, it can take hundreds of years for creep rupture to occur depending on the loading rate. When in the primary and secondary creep stages, masonry is considered to be a linear visco-elastic material and permanent strains can be neglected.

Lime mortars are believed to accommodate movement more readily than cement mortars. This is thought to be due to the fact that they remain in a plastic phase for extended periods of time as carbonation slowly progresses. However in stress cycling experiments undertaken by Ball et al. (2007) in hydraulic lime mortars the presence of micro cracks observed in the mortar after loading and unloading of the specimens several times, suggests that hydraulic lime mortars in fact behave in a brittle manner. By absorbing energy through the micro-cracks, those mortars deform more readily without displaying cracking on the macro-scale.

Creep tests using 1:1:6 and 1:¼:3 cement:air-lime:sand mortars, creep was observed to diminish after a longer period of time in the former mortar (Lenczner, 1970). This was however thought to be due to differences in mortar strength. In later experiments, a linear relationship was recorded between calcium hydroxide content of a mortar and the amount of creep deformation observed (Ball, 2009). Deformation was also observed to be a function of carbonation (Ball et al., 2011) confirming a link between mortar type and creep independent of mortar strength.

Current design guidance Eurocode 6 (BSI, 2005) provides a creep coefficient for use in creep predictions. Current creep design prediction models take the form of a creep coefficient derived directly from concrete research BS EN 1992-1-1: 2004 (BSI) which has been shown to acceptably predict masonry creep of cement mortar structures but under predict that of cement:air-lime and cement:plasticiser structures (Kioy et al., 2012). Creep is a function of relative humidity, strength, age at loading factor, notional member size factor, and creep development (time) coefficient.

3 EXPERIMENTAL MATERIALS AND METHODOLOGY

This research has focused on producing formulated lime mortars using a mixture of Portland cement

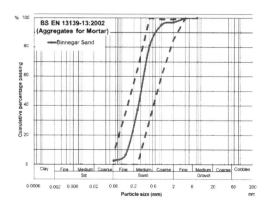

Figure 1. Mortar sand grading.

(CEM I) and air lime (CL90-S), in varying proportions, to investigate the effects of high calcium hydroxide contents of mortar on long term movements of mortars and masonry. The scope of experimental work included mechanical properties and creep performance together with an experimental study of masonry creep.

3.1 Materials

Hydrated Lime (CL90-S, bulk density 550 kg/m³) and cement (CEMI 52.5N, bulk density 1350 kg/m³) were chosen for this study. A silica based mortar sand (bulk density 1600 kg/m³) sourced from a quarry in Dorset, England was used for the study. The sand used was a well graded aggregate widely used for mortars (Figure 1). Wire cut vertically perforated fired clay bricks, nominally 215 mm long, 100 mm wide and 65 mm high with an average normalised compressive strength of 40.0 N/mm² were used. The unit initial rate of water absorption, total water absorption and apparent open porosity of the bricks were measured as 0.49 kg/m²/min, 9.0% and 20.7% respectively.

3.2 Experimental methodology

The experimental study included measuring mechanical property development with age of cement: air-lime mortars, prepared with varying mix compositions, using both prisms and small cylinders. Creep tests on small mortar cylinders were conducted. Following mortar tests long-term creep tests were completed on masonry brickwork wall specimens prepared with the fired clay bricks. Four different mortar mixes were tested as shown in Table 1.

Water content was set using the flow table test method BS EN 1015-3:1999 (BSI, 1999). The different water contents used to achieve a spread of between 160 and 180 mm are outlined in Table 1. Mortar specimens were cast into standard 160 x 40 x 40 mm prisms, in accordance with BS EN 196-1:2005 (BSI, 2005), and also 18 mm diameter x 36 mm high cylindrical specimens (Ball et al (2009)). The mortar specimens were cured in

Table 1. Mortar mix proportions.

Mortar	Mix proportions (by volume)			Water: binder (by mass)	Flow table (mm)	Lime content in binder (by mass)
	CE-MI	CL90-S	Sand			
1:3	1	-	3	0.74	169	0%
1:1:6	1	1	6	1.12	177	30%
1:2:9	1	2	9	1.21	171	45%
1:4:15	1	4	15	1.41	170	60%

Table 2. Lime mortar compressive strengths (N/mm^2).

	Mortar mixes			
Available lime content	1:3 CEM1 0%	1:1:6 30%	1:2:9 45%	1:4:15 60%
25 days				
Average	31.38	7.27	3.85	1.09
Maximum	33.64	7.92	3.98	1.29
Minimum	29.40	5.97	3.67	0.86
No.	6	6	6	6
CoV	4.6%	8.9%	2.5%	14.9%
90 days				
Average	31.33	7.62	5.01	1.19
Maximum	34.57	8.47	4.33	1.45
Minimum	28.31	6.65	3.84	1.05
No.	6	6	6	6
CoV	7.2%	7.6%	3.5%	12.2%
180 days				
Average	33.64	7.51	4.16	1.12
Maximum	35.40	7.93	4.39	1.17
Minimum	32.65	6.63	3.63	1.04
No.	6	6	6	6
CoV	2.8%	6.2%	6.3%	4.4%

a controlled environment in accordance with BS EN 1015-11:1999 (BSI, 1999). They were initially kept sealed in plastic bags for seven days. After the initial seven days, specimens were removed from the plastic wrapping and stored in a room where the environmental conditions were kept constant at temperatures of 20 (± 2) ˚C and relative humidity of 65 (± 5)% until testing. The prisms were tested to establish mortar strengths are various ages, whereas the cylinders were used primarily to establish creep performance.

Compressive strength and creep tests were also carried out on small masonry walls measuring 333 mm (wide) x 750 mm (high) x 102.5 mm (thick). Four separate walls, each with a different mortar mix as shown in Table 1, were constructed and tested for compressive strength and creep performance. The applied stress during the creep tests was maintained at a stress of 2.0 N/mm^2. The walls were cured and tested in a room with environmental conditions kept constant at 20 (± 2) ˚C and relative humidity 65 (± 5)%.

4 MORTAR TEST RESULTS

Compressive strength tests derived from the prisms, carried out at 25, 90 and 180 days, are presented in Table 2; for brevity the accompanying flexural strength tests have not been reported here. Compressive strengths increased with time, initially from hydration and longer term due to carbonation. Mortar compressive strengths were inversely proportional to mortar air lime content by mass, Figure 2. A decrease in strength of 70% is observed with addition of the first 30% air lime content with a further decrease of 84% observed upon addition of the next 30%. A total reduction in strength of 97% was observed when air lime content of the mortar was increased to 60% at all ages tested. The Coefficients of Variations (CoV) are generally below 10%, showing consistent performance for the mortar compressive strength tests. No particular material significance has been given to the higher CoVs for 25 and 98 day 1:4:15 specimens; by 180 days the mix CoV is similar to the other mix types.

The results from the mortar cylinder creep tests are presented in Figure 3. The specimens were loaded at 28 days age. The applied stress levels varied with the relative strengths of the mortar as

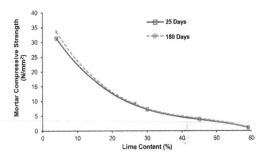

Figure 2. Mortar compressive strength vs air lime content (by mass).

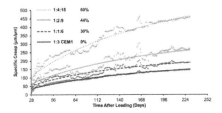

Figure 3. Effects of air lime content on specific creep.

follows: 2.5 N/mm^2 (10% of the average 25 day mortar strength for the 1:3 CEMI mix);1.0 N/mm^2 (18% of the average 25 day mortar strength for the 1:1:6 mix); 0.6 N/mm^2 (17% of the average 25 day

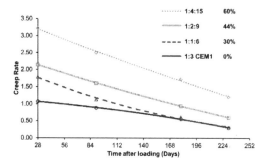

Figure 4. Effect of air lime content on creep rate.

mortar strength for the 1:2:9 mix); and, 0.4 N/mm² (22% of the average 25 day mortar strength for the 1:4:15 mix).

The measured creep strains increased with increasing air lime content in each mortar mix. Three days after application of load, the specific creep strains were 30 for the 1:3 CEM1 mortar, 50 for the 1:1:6 mortar, 60 for the 1:2:9 mortar and 90 for the 1:4:15 mortar corresponding to 67%, 100% and 200% increases in creep strains respectively. By 180 days age, about 5 months after loading, creep strains were 135, 175, 245 and 415 for 1:3 CEM1, 1:1:6, 1:2:9 and 1:4:15 respectively, corresponding to increases of 30%, 80% and 200% with addition of air lime content to the mortar. The results indicated that the addition of 30% and 45% air lime to a cement mortar caused an initial jump in creep strains immediately after loading but this settled and the difference in movement was seen to decrease within 5 months of loading. When the lime content increased to 60%, the movement was consistently 200% higher than the cement mortar throughout the duration of the testing.

The effect of air lime content on creep rate is shown more clearly in Figure 4. An increase in air lime content to 30% resulted in an initial high creep rate which decreased and was similar to the cement mortar creep rate at 180 days. Increasing air lime content to 45% also increased initial creep rate; a change in slope was visible by 180 days, however, the rate did not decrease enough to be similar to the cement mortar creep rate during the duration of the testing. At 60% air lime content the greatest increase in initial creep rate was observed. The high air lime content mortar sustained a steady creep rate for the duration of the testing period.

5 MASONRY TESTS

Table 3 presents the compressive strength results from twenty five small walls tested at 25 days, 90 days and 180 days after construction. The same fired clay brick was used in all walls with the different mortar designations, designed to

Table 3. Wall compressive strength results (N/mm²).

	Mortar mixes			
Available lime content	1:3 CEM1 0%	1:1:6 30%	1:2:9 45%	1:4:15 60%
25 days				
Mortar ave.	31.4	7.3	3.9	1.1
Wall ave.	25.0	15.7	10.9	9.5
Maximum	-	20.6	11.3	9.6
Minimum	-	10.6	10.4	9.3
No. tests	1	3	2	2
CoV	-	25.9%	-	-
90 days				
Mortar ave.	31.3	7.6	5.0	1.2
Wall ave.	41.9	21.2	15.1	11.8
Maximum	-	24.4	16.5	12.4
Minimum	-	18.4	13.1	11.2
No. tests	1	3	3	2
CoV	-	11.5%	9.5%	-
180 days				
Mortar ave.	33.6	7.5	4.2	1.1
Wall ave.	40.9	20.6	15.2	11.0
Maximum	-	30.2	15.2	12.5
Minimum	-	13.4	15.2	9.6
No. tests	1	3	2	2
CoV	-	34.4%	-	-

investigate the effect of varying quantities of air lime on cement mortars.

Initial masonry compressive strength decreased with increasing air lime content. The increase in air lime content of the mortar corresponded with a decrease in cement content, leading to reduced masonry strengths. Brickwork compressive strengths, irrespective of mortar air lime content, increased with time. The 1:3 CEMI walls achieved the greatest compressive strengths at all ages. At 25 days the increase in air lime content resulted in a reduction of wall compressive strength of 37%, 57% and 62% for the 1:1:6, 1:2:9 and 1:4:15 respectively. The reductions in masonry strength are in keeping with the power relationship for f_m used in BS EN 1996-1-1:2005 (BSI). The CoV reported (1:1:6 series only) are higher than the mortar performance (Table 2). This increased variance in performance can be expected for masonry construction, in part due to influence of workmanship and the complex interaction between two materials.

Results presented in Figure 5 show the creep per unit stress (specific creep) recorded on the four specimen walls constructed using mortars with varying air lime content of 0% to 60%, under a constant stress of 2 N/mm². The magnitude of creep strains increased with air lime content of the mortar. After one day from load application the 1:3 CEM 1 wall exhibited 56 µε movement, while the 1:1:6, 1:2:9

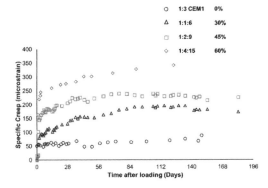

○	1:3 CEM1	0%
△	1:1:6	30%
□	1:2:9	45%
◇	1:4:15	60%

Figure 5. Wall specific creep vs time.

and 1:4:15 walls had strains of 56 με, 102 με and 163 με respectively. By 28 days after stress application the recorded strains were 60 με, 142 με, 224 με and 282 με for the 1:3, 1:1:6, 1:2:9 and 1:4:15 walls respectively. At 90 days after loading the strains were 77 με, 192 με, 239 με and 328 με, and at 140 days after loading they were 88 με, 196 με, 254 με and 342 με for the same walls respectively.

The largest recorded change in gradient was observed between loading (0 days) and 28 days. Some of the initial movement could be attributed to residual elastic movements and bedding-in of the specimen as well as creep. After the initial stage the cement:lime walls continued to experience creep strains, while the 1:3 CEM I wall recorded very little further movement.

Air lime content in the mortar had a directly proportional influence on early creep rates. After just 1 day from loading, creep rates of the cement:air lime mortared walls were two, three and five times greater than those of the 1:3 CEM I wall for the 1:1:6, 1:2:9 and 1:4:15 mortars respectively. On the second day, creep rate reduced by 25% for all mortars. By the fourth day after loading the 1:3 CEM I, 1:1:6 and 1:2:9 had a further reduction in creep rate of 12%, while the 1:4:15 mortar reduced by a further 7%.

6 CONCLUSIONS

The following conclusions have been taken from the experimental work described above:

- An increasing air lime content in the mortar mixes reduces the compressive strength of the masonry prisms and in-situ wall mortars.
- Mortar air lime content of 30% increases initial creep rate of mortar but by 180 days the creep rate is similar to that of cement only mortars. The creep rate of higher air lime content mortars (45% and 60%) remains elevated for beyond 180 days.

- Air lime content in the mortar had a directly proportional influence on early creep rates of the masonry.

ACKNOWLEDGEMENTS

The authors wish to acknowledge the financial support from Lhoist and the University of Bath. The contribution of various colleagues at University of Bath and Lhoist are also gratefully acknowledged.

REFERENCES

Ball, R. J. & Allen, G. C. 2009. Load-dependent deformation and shrinkage in hydraulic lime mortars. *International Journal of Sustainable Engineering* 3: 40–46.

Ball, R. J. & Allen, G. C. 2010. Load-dependent deformation and shrinkage in hydraulic lime mortars. *International Journal of Sustainable Engineering* 3: 40–46.

Ball, R. J., Allen, W. J., Allen, G. C. & El-Turki, A. 2007. The stress cycling of hydraulic lime mortars. *Proceedings of the ICE Construction Materials* 160: 57–63.

Ball, R. J., El-Turki, A. & Allen, G. C. 2011. Influence of carbonation on the load dependent deformation of hydraulic lime mortars. *Materials Science and Engineering: A* 528: 3193–3199.

Brooks, J. J. 2001. Effect of metakaolin on creep and shrinkage of concrete. *Cement & concrete composites* 23: 495–502.

Brooks, J. J. & Bakar, B. H. A. 2004a. Shrinkage and creep of masonry mortar. *Materials and Structures* 37: 177–183.

BS EN 1015-3:1999 Tests for masonry mortar - flow table test.

BS EN 1015-11:1999 Test for masonry mortar -flexural & compressive strength.

BS EN 196-1:2005 - Methods of testing cement - strength.

BS EN 1992-1:2005 Eurocode 2 - Design of concrete structures.

BS EN 1996-1-1:2005+A1:2012 Eurocode 6. Design of masonry structures. General rules for reinforced and unreinforced masonry structures.

Forth, J. P. & Brooks, J. J. 2008. Creep of clay masonry exhibiting cryptoflorescence. *Materials and Structures* 41: 909–920.

Forth, J. P., Brooks, J. J. & Tapsir, S. H. 2000. The effect of unit water absorption on long-term movements of masonry. *Cement & Concrete Comp.* 22: 273–280.

Kioy, S. M., Walker, P., Ball, R. J. & Fodde, E. 2012. Effects of dewatering on long-term movement characteristics of lime mortars. *15th International Brick and Block Masonry Conference*. Brazil.

Lenczner, D. Creep in brickwork. 2nd International conference on brick masonry, 1970 Stoke-on-Trent, UK. British Ceramic Society.

Lenczner, D. 1981. *Movement in buildings*, Pergamon Press; 2nd Revised edition (Jan 1981).

767

Masonry arches & bridges

Brick and Block Masonry - From Historical to Sustainable Masonry –
Kubica, Kwiecień & Bednarz (eds)
© 2020 Taylor & Francis Group, London, ISBN 978-0-367-56586-2

BIM representation and classification of masonry pathologies using semi-automatic procedure

R.A. Bernardello, P. Borin, F. Panarotto & A. Giordano
Department of Civil, Architectural and Environmental Engineering, University of Padova, Italy

M.R. Valluzzi
Department of Cultural Heritage University of Padova, Italy

ABSTRACT: Both the development of national and European directives and the widespread use of BIM (Building Information Modeling) generated the demand to study specific modelling practices for the analysis and conservation of masonry buildings. In this sense, masonry buildings differ from the practices of contemporary production, starting from the typical equivalence between architectural and structural objects. It becomes so important to establish on one hand the geometric rules for the passage of information between the BIM model and the FEM analytical models, on the other, a valid methodology to describe building pathologies for masonry surfaces. Starting from an analysis of analogic and digital methods to classify these conditions, the study shows two different alternatives for integrating information, usually embedded within two-dimensional drawings, in a BIM environment. Both processes allowed to create three-dimensional semantic objects related to specific information of type of building conditions, that could help the conservation and restoration of construction elements in masonry.

1 INTRODUCTION

The digital representation of historical building is mainly aimed at two objectives: the first concerns the construction industry toward the conservation/ restoration design process, the second the heritage question (dissemination of knowledge and enhancement of touristic appeal). Past researches demonstrated how both these two approaches could benefit from the use of BIM (Building Information Modelling) (Giordano et al. 2015): computable and internally coherent information help the restoration process in terms of awareness about technological and structural interventions.

Moreover, the development of national and European directives and the widespread use of BIM generated the demand to study specific modelling practices for the analysis and conservation of masonry buildings. Masonry buildings show how the different subsystems (the architectural vs. the structural, stairs, past mechanical appliances, such as chimneys) are strictly related to each other for the definition of the building itself.

Therefore, BIM constitutes the perfect tool to become a common knowledge management system for these purposes. The integration of BIM with automation systems would positively support the quality control during diagnosis, design and work execution as well as the labour savings (Bruno et al. 2018).

This study focuses on the scientific and technical relations among HBIM (Historic Building Information Modelling) models, diagnosis and performance assessment features.

First, the literature review identifies specific activities and relative tools and methods for knowledge acquisition, semantic enrichment and the application of semi-automatic procedures for decay mapping and defect detection carried out for succeeding conservation and refurbishment operations.

Then, the second part of the article shows two different approaches to facilitate the description of decays in masonry buildings: the first integrates it during the survey process, the second considers the actual process, in which an expert specifies, mainly graphically, the condition survey. Preliminary results and problems are discussed, and conclusion achievements and future studies are proposed.

2 BIM RELATED DOCUMENTS AND DECAY SEMI-AUTOMATIC MAPPING

HBIM process and its strictly related Scan-to-BIM have been the subject of numerous studies in the last decade (Pocobelli et al. 2018).

This multifaceted process is explored by the accessible researches, and it can be organized in two macro-parts: one about the procedure of remote data capturing and processing, which is the base of the following activities, and the other about the object recognition and libraries design of the BIM model (Murphy et al. 2009)(Murphy et al. 2013) (Volk et al. 2014).

The quality of the acquired data has been addressed in various ways, by defining arbitrary levels of quality parameters, by evaluating the density and accuracy of relevant area. Defining than the modelling tolerance, brings to set the level of accuracy of a model related to the data from architectural survey. The references are provided by regulations and international standards and when these are insufficient, by adopted best practises. (Rebolj et al. 2017).

In existing building the accuracy of the HBIM depends on requirements such as the identification of the Level of Development (LOD) and Level of Accuracy (LOA), as defined by USIBD (U.S. Institute of Building Documentation 2016). The guide defines the tolerance limits for survey accuracy, 'measured accuracy', and the accuracy of the BIM model, 'represented accuracy'. (Ali et al. 2018) (Santagati et al. 2018) (Bonduel et al. 2017).

To manage a proper 3D reconstruction from point clouds, BIM experts typically use 3D applications developed for the management of new buildings, which make it difficult to generate complex objects such as historic vaults and irregular and damaged walls.(Banfi 2019) In fact, the application of the process in case studies highlighted some gaps in existing technologies, specially the lack of universal software for the creation of as-built BIM-models and complex structures. (Badenko et al. 2019) The design of the complex geometry libraries are often old construction like masonry vaults. (Angjeliu et al. 2019).

The state-of-the-art in the creation of as-built BIM is fundamentally a manual process. The time-consuming and subjective nature of this process motivates the need for automated, or at least semi-automated, tools for as-built BIM creation. (Tang et al. 2010) Additionally, the possibility of representing and integrating in the same model knowledge related to different domains and to different features of the heritage artefact also enhances the crucial action critical approach, interpretation and assessment.(Simeone et al. 2019).

The document organisation derived from the documentation analysis has some inconveniences: objects composing the building were treated as isolated objects not related to its linked objects. The BIM model can operate as a single archive for the documentation and conservation of architectural heritage, as well as for consultation and query of such data stored by customised databases or automated task programming. (Fai et al. 2011) (Rodrí-guez-Moreno et al. 2018) The amount of information – documents and survey data – is enormous, (Koolen et al. 2009) and problems of data management in an extremely sized file can be solved with the BIM repository, in which building model is used as coordinating centre, synchronised to share data among different applications and databases. (Bruno et al. 2018) Using of HBIM for managing knowledge is about constructive evolution and building conditions, in order to diagnose the real causes of decay and settlements in historic buildings.(Bruno et al. 2019) In fact, many research investigate the application of BIM, from the digital design of the construction to the diagnosis modelling, refurbishment and conservation interventions to the management itself.

As in the common procedure, the first step for restoration and maintenance based on a BIM model is the mapping of the state of conservation and converting the linked information into parametric components. The first issue is that the decays have no specific building categories. However, the traditional orthogonal representation characterized by 2D patterns cannot be semantically enriched by any parameter. This is the reason why the possibility of mapping in 3D the decay analysis was examined in detail in some articles through the use of adaptive component placed manually. (Chiabrando, et al. 2017) (Malinverni et al. 2019) (Turco, et al. 2017) Another approach uses 2D different types filled regions, based on decay classification, which particularly fits with surface deterioration. (Santagati et al. 2019).

Based on the point cloud and on the resulting 3D mesh, some studies were carried out in order to support numerical structural analysis. (Antón et al. 2019) (Yaagoubi et al. 2018).

Other works develop data collection template in a BIM environment for defect management in order to reduce data search time and improve the accuracy of search results as well. (Lee et al. 2016) (Park et al. 2013).

To automatize the process of defect detection in the point cloud, research focused both on concrete infrastructure and masonry walls, thus improving machine learning, according the decay data segmentation to the BIM method, on the basis of a hierarchical organization of the information.(Borin et al. 2019) (Valero et al. 2019).

3 INTEGRATION IN BIM MODELS

One of the main objectives of information modelling is to manage the complexity of the building process. The need to operate, in some cases, with tools that do not have BIM interfaces (in terms of input and output), drives the necessity to address the problem of data management outside the BIM information systems. The information produced outside the BIM database must be traced back through data integration processes. These are processes to be implemented on data coming from different sources to provide the user with a unified vision. This data can be exported in formats such as txt, csv or xml formats.

Afterwards, they can be translated into conversion algorithms through rules based on the conformation of the files. Thanks to these algorithms, the data can be combined according to new structures, making them operable within new platforms. Re-elaboration and transfer of data to other platforms occurs following computational approaches, logics and tools. In this way, data and information coming from a multitude of tools can be integrated, as well as the manual input coming from the designer's expertise, which is often the only way in which the final product of the design can really be qualified.

An example of this is offered by the possibility of implementing the post-processing survey data – such as 3D mesh – or translating CAD databases of a complex building into a BIM database. Figure 1.

The first application thanks to an identification of decay phenomena, directly in 3D reconstruction, can reduce time of drawing in 2D views, keep real appearance, 3D geometry and real extension of decay. This means a higher quality reconstruction in that geometrically complex structures generated by rounded surfaces, or aggregate construction. The mesh-to-HBIM process could be applied in two different methods. On one hand, it segments the surface mesh, thickens the triangle mesh to 3D volume, and transfers the primitives to the BIM model. On the other hand, it uses the 3D mesh as an imported object in the BIM model to describe the state of conservation of each object and the mesh itself can be parametrized and described.

In both cases, the obtained HBIM is then converted to the ontology model, to improve the heterogeneous knowledge. (Yang et al. 2019) In the second scenario the result is based on the ability to read the CAD model of existing buildings, understand the graphic language, translate such symbolism into an algorithm and transcribe this algorithm into code. The development of these algorithms represents a success factor to guarantee the possibility of automating the transformation of CAD floor plan into BIM databases. (Yin et al.2009) (Lim et al. 2018) This result can be achieved by using tools that employ the VPL. VPL is

a language that provides programming through the graphic manipulation of elements. These elements are "functional blocks" representing the codes that, assembled by the programmer in graphic structures, realize the source code. (Burnett 1999).

The study shows the application of these two different alternatives for integrating information, usually embedded within two-dimensional drawings, in a BIM environment. Starting from an analysis of analogic and digital methods to classify these conditions, rules are established for the passage of information between the BIM the geometric model and a valid methodology is explained to describe building pathologies for masonry surfaces.

The first process is tested in a hydraulic system sited in Padova, which main component is a vaulted bridge, it is called Porte Contarine. Starting from photomodelling survey techniques, the obtained mesh surfaces are used to reliably describe and parametrize predominantly superficial conditions. Thanks to machine vision procedures the decays are organized and integrated in the BIM environment. This example demonstrates how it is possible to link the geometry with the associated BIMobject, reducing the error in positioning the elements as in the manual process could happen. The second method is explored in the western part – called 'Barchessa' – of Villa Angaran San Giuseppe, in Bassano del Grappa. Based on ICOMOS condition analysis in a 2Denviroment, a script allows to import polylines into the BIM model. The derived geometric shapes are projected, therefore linked to the architectural elements.

Both approaches are following described, at first focusing on the theory method, on the advantages and peculiarities, then explaining step by step the practical application of the process to each case study. At the end the obtained results are discussed, most explaining the complications and suggesting subsequent applications and improvements.

4 CASE STUDIES

4.1 Decay mapping on vaulted systems through mesh geometry

The bridge object of study is part of a hydraulic system in Padua, that was for long one of the most important. Initially it was an in-built part of the walls of the sixteenth century, where the Naviglio river flowed into the Piovego one. After the internal drainage of the Naviglio the initial context has now been lost. The importance of the bridge of the Porte Contarine is not limited to the construction itself, but it refers to the whole plumbing system, that had a fundamental role in many aspects of the city life with its ancient and complete functioning, such as allowing the boats to come along the two rivers and producing electrical energy. Figure The purpose of this method is to implement a procedure that can be standardized,

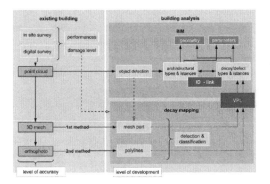

Figure 1. Schema of the two applied processes.

as much automated as possible, and useful for conservation and monitoring process of buildings, especially the historical ones. The two focus of the work were setting up in the smooth workflow of the scan-to-BIM process another succeeding operation without adding new element and keeping inside the model the same reference coordinates between point cloud, BIM model and 3Dmesh. Above all, between the single object and the referred decay-mesh. This brief recap of functions and main role of the Porte Contarine highlights the geometric, architectural, and structural masonry complexity of this historical system, that make it a relevant case of interest for the application of decay diagnosis. Furthermore, the presence of water and humidity and the impactful historic facts influence the state of conservation of the construction material and of the masonry system itself.

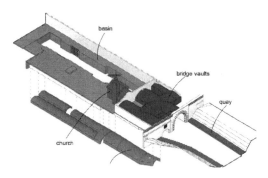

Figure 2. Porte Contarine hydraulic system. BIM model. (credit by Bernardello, R.A.).

For these reasons the conservation analysis procedure was implemented trying to find a solution to those still unresolved problems as a correct representation of the decay state of complex geometric elements (i.e. vaults, cylindrical walls). The issues are: i) inserting the objects indicating the deterioration analysis in an automated way; ii) reducing the positioning error of the damage information and the association with their reference component; iii) and simplifying the decisional process of intervention analysis.

4.1.1 *Decay mesh: Test in a BIM environment*
Following the scan-to-BIM process, the first steps was the digital survey by photogrammetry and topographical survey. The use of the photogrammetry assures to solve some impediments: the presence of high trees which cover a large portion of the system and the strict accessibility permits in some areas. Thanks to the compactness of the instrumentations, the survey was performed on a small boat under the bridge. Due to the different part that form the whole system, a preliminary photo plan was established. The photographic shots were taken on three different

floors: the first one on the decking to the south on the portion delimited by the embankments, the second on the floor of the quay, the third is the water level. Figure Once the photos have been taken, a postprocessing phase were completed to obtain the point cloud, and the texturized 3D mesh. Using the topographical survey, the entire system was aligned. Few problems come up because of the reflectiveness of the water and because of the obstacles of other objects all around. As a result, based on the point cloud, the different objects were visual identify and the libraries for the BIM model were designed. Figure Parallel to the phase of decay diagnosis in the 3D space, the traditional method was applied, in order to compare and verify the results at the end of the process. Using the 2D orthophotos obtained from the texturized mesh of the elevations and the reflected plan for the vaults, the different types of decay were detected and classified. Figure In addition, the diagnosis on the 3D mesh starts in the postprocessing software and ends in the BIM software though two fundamental steps, the first is the identification by BIM component of the decay mesh under examination in the photogrammetric model. The second is the use of a VPL script, aimed at importing the mesh in the BIM software and at assign the parametric information to the deteriorated object. The whole model was divided in the main components of the hydraulic system to obtain better performances. This operation, in fact, reduces considerably the data processing times, as the work files were lighter; moreover, a first decomposition based on the model components was performed. To allow a worthy diagnosis of the decay, and its standard organization, to maximize the automation of the process and to reduce the work-time, different operation to extract the single part of the object were applied choosing practical methods. Some types of decay characterized by a scattered presence of a dominant colour such as the presence of vegetation and biological colonization. Figure For those elements it was possible to operate a selection of polygons through the green colour, with the possibility of indicating a colour range according to certain parameters of saturation, hue and brightness. In other cases, it was necessary selecting manually the portions of the mesh because their recognition depends on the material of the object, e.g., alveolation or sub-

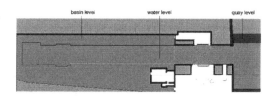

Figure 3. Level of photos (credit by Bernardello, R.A.).

florescence on stone elements. Extracted by component then, for each decay, the object or the modality with which each decay and defect can be better imported and represented in BIM environment, was indicated. The degradations of the case study were mainly superficial, whose corresponding parametric object was identified in a mesh. Once the mesh was isolated, it was saved and exported in obj format, keeping the reference coordinates and imported in the BIM software using a VPL application.

Then, two operations were carried out in succession:

a) the import of the mesh in the BIM software as a generic category, specifying the first two parameters, i.e., the size of the area and the decay type. These two parameters were indicated in an automated way. The parameter of the area was calculated by the software adding the surface areas of each triangle that make up the mesh. Moreover, the import in the correct position of the mesh was exact and automated, since the origin of the 3D mesh is the same as the BIM model (both are created from a common point cloud).

b) the association of the decay-instance with the reference object through the object ID (which is unique).

At the end of these two operations it was possible to query the model by selecting the objects and the related information appeared on the properties table. Moreover, the information were available on spreadsheet-based format or via IFC (Industry Foundation Classes), exported as sub-classes of Ifc Building Element Proxy.

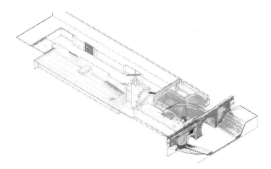

Figure 4. North-east axonometric view. Point cloud and BIM model. Overlapping. (credit by Bernardello R.A.).

4.1.2 Mesh decimation: Results and discussions

A first aim of this work was to develop a working method to analyse the state of conservation of an existing construction. Moreover, the study has the purpose of storing and disseminating the information acquired. Although the degradation analysis has concerned the entire state of conservation of the system of Porte Contarine, the information representation has proved effective especially with surface decays, so there is a strong

correspondence between their perception in the real environment and the representation by mesh in the model. This suggests that more and more progress must be made towards a differentiation of the representation of the different decay types, linked to the various nature of each real object alteration. Good results were obtained importing meshes of complex objects such as vaults, the integration of tools for surveying the real object, such as the laser scanner or photogrammetry, both of which allowed an accurate construction of the details of the historic building. It was possible to return multiple information in an automated way, such as the position of the degraded instance, its relationship between the referenced object, the decay classification through its name, its material and its the quantity take-off (area, perimeter). Many limitations are still represented by the technology tools. The amount of data to be managed is always very high, in this regard the processing time could be very dilated, and the results are not always adequate to expectations. Secondly, the graphic output of the mesh is not satisfactory in popular terms, as it is still very simplified. Figure.

4.2 Decay mapping integration through VPL in Villa Angaran

Based on the above considerations, a study has been undertaken with the objective to use the VPL language for the transferring of information from a decay map, represented in a CAD environment, towards a BIM platform. The research has been performed focusing on a case study, which concerns "Villa Angaran San Giuseppe". Figure 8 This section of the text briefly describes the steps that allowed the development of the proposed process. Firstly, the decay map was treated in the CAD environment (represented in 2D), by

Figure 5. Elevation north. Orthophoto and material decay map. (credit by Bernardello, R.A.).

Figure 6. Elevation south. Biological colonization, decay selection by color 3Dmesh. (credit by Bernardello, R.A.).

using the typical process of redrawing degraded elements (an image is imported into the file and appropriately scaled). The image used in this case is an orthophoto created using a photogrammetric method, performed in Agisoft Metashape software.

Once the various types of degradation were identified, the relative layers were created. Then, with the use of the Polyline command, an open or closed area was created. In the image (processed within the CAD) an origin must be identified. Such origin will be useful to import the decay correctly within the BIM model. Once the redesign of all the degradation elements present in the treated surfaces has been completed, the file is saved in dwg format. Subsequently this file can be read by the script created and reported within the BIM environment. Once this is done the process continues with the script procedure. As mentioned above, the script allows to transfer the data related to the degradation realized in 2D within the BIM model, with the aim of creating BIM objects. In particular, the objects are divided into: open polylines representing cracks and closed polylines, which generate surfaces, representing surface-based decay, such as detachment.

Thus, families and instances in the 3D model are created by the script. These families consist of solid elements represented by a closed surface with a given thickness or by a generic cylindrical surface along a polyline. The script, furthermore, allows the creation

Figure 7. Output. 3D decay-mesh on vaulted system. (credit by Bernardello, R.A.).

of parameters associated with each instance created in the model. The parameters that the script associates to each instance depend on the type of decay: each instance of type 1 (crack) will have a name, which consists of the name of the degradation (name of the layer), an identification number, a length parameter, a thickness parameter, a material and the ID of the referenced wall; each instance belonging to type 2 (surface degradation) will have a unique name, which consists of the name of the degradation (name of the layer), identification number, its area (degradation surface), a material and the ID of the wall to which it is associated. In particular, the "material" parameter assigns a graphic aspect to each degradation based on the type. These materials belong to a library specifically created and pre-loaded in the template. This library is composed of materials with patterns coherent with indications by national standard. Figure 9.

The initial processes of the script solve some problems related to the orientation of the wall in 3D space. These operations allow to guarantee the correct orientation of the wall, obtaining the normal to the surface with the correct direction. This is beneficial to evaluate the external or internal surface of the wall. The intermediate passages of the script realize the projection of the decay (previously roto-translated) on the surface. This ensures that any overhanging or underhanging elements of the wall are taken into consideration.

4.2.1 CAD vs VPL results

The proposed process does not allow to avoid the initial redraw procedures - typical of a decay mapping performed in a CAD environment - but allows to assign to these objects some semantic attributes, to create a BIM information database related to the damage of a building. This database can be consulted to obtain clear and unambiguous computational data. Using this procedure, possible restoration interventions to be performed on the building can be organized. As previously stated, one of the fundamental aspects of the proposed method is that it integrates data from other platforms into a BIM model. This transferring functionality is guaranteed by the use of intermediate tools that use VPL. The degree of precision obtained is the same as in

Figure 8. 3D view of the building modeled in Revit (credit by Callegaro, C.; Marini, M.).

Figure 9. West elevation of the "Barchessa" with the decay mapping. (credit by Panarotto, F.).

the traditional CAD survey. This was expected, because precision is not determined by the process implemented, but by the quality of the modelling. The semi-automated procedure allows for a rapid creation of 3D objects in the BIM model. These 3D objects are complemented by already compiled libraries of materials which ensures a correct reading of the data. From the model it is possible to obtain data related to the decay being present on each wall, with the relative pieces of information regarding the severity of the damage being present. Figure In fact, it is possible to evaluate whether there are more or less serious cracks or the quantity (surface area) of the surface decay present. In this way it is possible to organize the interventions to be performed by assigning priorities based on the seriousness of the observed decay.

5 GENERAL RESULTS AND CONCLUSIONS

This article shows two different approaches for the integration of the condition survey in a BIM model for historic masonry buildings. Both studies examine the process of survey, representation and design of historic buildings: the actual process presents a strong division among stakeholders, achieved by specialized actors, often operating within different companies. Thus, two apparently dissimilar approaches are exposed although they present the same strategy starting from different supports. They also share the same aims, the decay description in several terms: morphological (surface extension), semantic (nomenclature belonging to the ICOMOS standard) and ontological (linking the specific decay to the deteriorated object and vice versa). This last represents a key point to allow the protection of the building: the main objective the model has to describe the performance level of each element, not only the level of deterioration. Only highlighting the relationship between the two elements leads to the automatic definition of the condition and the performance assigned to the specific building element (wall, vault, stone corner). This is the approach also followed by BuildingSmart International defining the decay not as a class but as a property of the technical element.

The first case study deals with a complex hydraulic infrastructure in Padova, a system that acts as a bridge, and as a sluice, with a particular morphological design (brick cross vaults, stone walls). It provided an initial difficulty in integrating geometric degradation information on complex surfaces. Further complexity is added by the extension of the numerous decays in a humid environment. In this first approach, it has been tested the analysis of the mesh, as an outcome of the digital photogrammetric survey process. Thus, it concerns the progressive refinement/integration of the survey techniques and the automatic conversion of 3D surfaces into parametric models, because at the present state models of complex masonry vaults and decorative

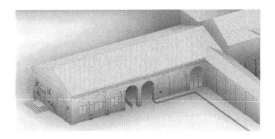

Figure 10. Axonometric view of the building with the decay mapping. (credit by Panarotto, F.).

elements still feature simplifications and inaccuracies or time-consuming procedures.

The second case study applies the process to an Italian villa. The main process is decomposed among those actors which produce the decay map and those that produce the design. After that, it is possible to integrate the information. Thanks to the joint use of procedures in VPL environments and techniques of projective geometry (central and parallel projections) it is possible to automatically create the geometry of the degradations, both in the case of two-dimensional objects (surfaces) and one-dimensional objects (cracks). The result represents a preliminary software for the automatic integration of decays in the BIM environment.

Future developments concern the extension of the same methods of representation starting from images (projection information) and the improvement of the semantic description of decay. A further development of the research is related to the description of the decrease in performance of building elements starting from the number and weight of the single decay, described in the BIM model.

REFERENCES

Ali, M., Ismail, K. M., Hashim, K. S. H. Y., Suhaimi, S., & Mustafa, M.H. 2018.Heritage building preservation through building information modelling: *Reviving cultural values through level of development exploration,16/2:*62–72.

Angjeliu, G., Cardani, G., & Coronelli, D. 2019. A parametric model for ribbed masonry vaults', *Automation in Construction*, 105: 102785. Elsevier.

Antón, D., Pineda, P., Medjdoub, B., & Iranzo, A. 2019. *As-built 3D heritage city modelling to support numerical structural analysis: Application to the assessment of an archaeological remain. Remote Sensing*, Vol. 11.

Badenko, V., Fedotov, A., Zotov, D., Lytkin, S., Volgin, D., Garg, R. D., & Min, L. 2019. Scan-to-bim methodology adapted for different application, *International Archives of the Photogrammetry, Remote Sensing and Spatial Information Sciences-ISPRS Archive*, 42/5/W2:1–7.

Banfi, F. 2019. 'Hbim Generation: Extending Geometric Primitives and Bim Modelling Tools for Heritage Structures and Complex Vaulted Systems, *ISPRS - International Archives of the Photogrammetry, Remote Sensing and Spatial Information Sciences*, XLII-2/W15/September: 139–48.

Bonduel, M., Bassier, M., Vergauwen, M., Pauwels, P., & Klein, R. 2017. Scan-to-bim output validation: Towards a standardized geometric quality assessment of building information models based on point clouds, *International Archives of the Photogrammetry, Remote Sensing and Spatial Information Sciences - ISPRS Arch*, 42/2W8: 45–52.

Borin, P., & Cavazzini, F. 2019. Condition Assessment of Rc Bridges. Integrating Machine Learning, Photogrammetry and Bim, *ISPRS - International Archives of the Photogrammetry, Remote Sensing and Spatial Information Sciences*, XLII-2/W15/September: 201–8.

Bruno, S., De Fino, M., & Fatiguso, F. 2018. Historic Building Information Modelling: performance assessment for diagnosis-aided information modelling and management, *Automation in Construction*, 86/November 2017: 256–76. Elsevier.

Bruno, S., Musicco, A., Fatiguso, F., & Dell'Osso, G. R. 2019. The Role of 4D Historic Building Information Modelling and Management in the Analysis of Constructive Evolution and Decay Condition within the Refurbishment Process, *International Journal of Architectural Heritage*, 00/00: 1–17. Taylor & Francis.

Burnett, M. 1999. Visual programming, *Wiley Encyclopedia of Electrical and Electronics Engineering*, 32/1–3: 275–83.

Chiabrando, F., Lo Turco, M., & Rinaudo, F. 2017. Modeling the decay in an hbim starting from 3d point clouds. A followed approach for cultural heritage knowledge, *International Archives of the Photogrammetry, Remote Sensing and Spatial Information Sciences - ISPRS Archives*, 42/2W5: 605–12.

Fai, S., Graham, K., Duckworth, T., Wood, N., & Attar, R. 2011. *Building Information Modeling and Heritage Documentation. VR kiosk View project Buildin*. Retrieved from http://www.210king.com/.

Giordano, A., Borin, P., & Cundari, M. R. 2015. Which survey for which digital model: critical analysis and interconnections. *XIII International Forum. Le vie dei Mercanti*, pp. 1051–1058. La scuola di Pitagora editrice: Aversa - Capri.

Koolen, M., Kamps, J., & De Keijzer, V. 2009. Information retrieval in cultural heritage, *Interdis. Science Reviews*, 34/2–3: 268–84.

Lee, D.-Y., Chi, H., Wang, J., Wang, X., & Park, C.-S. 2016. A linked data system framework for sharing construction defect information using ontologies and BIM environments, *Automation in Construction*, 68: 102–13. Elsevier.

Lim, J., Janssen, P., & Stouffs, R. 2018. Automated generation of BIM models from 2D CAD drawings, *CAADRIA 2018-23rd International Conference on Computer-Aided Architectural Design Research in Asia: Learning, Prototyping and Adapting*, 2/May: 61–70.

Malinverni, E. S., Mariano, F., Di Stefano, F., Petetta, L., & Onori, F. 2019. Modelling in hbim to document materials decay by a thematic mapping to manage the cultural heritage: The case of "chiesa della pietà" in fermo, *ISPRS Annals of the Photogrammetry, Remote Sensing and Spatial Information Sciences*, 42/2/W11: 777–84.

Murphy, M., Mcgovern, E., & Pavia, S. 2009. Historic building information modelling (HBIM), *Structural Survey*, 27/4: 311–27.

Murphy, M., McGovern, E., & Pavia, S. 2013. Historic Building Information Modelling - Adding intelligence to laser and image based surveys of European classical architecture, *ISPRS Journal of Photogrammetry and Remote Sens.*,76:89–102.

Park, C.-S., Lee, D.-Y., & Kwon, O.-S. 2013. A framework for proactive construction defect management using BIM, augmented reality and ontology-based data collection template, *Automation in Construction*, 33: 61–71. Elsevier.

Pocobelli, D. P., Boehm, J., Bryan, P., Still, J., & Grau-Bové, J. 2018. BIM for heritage science: a review, *Heritage Science*, 6/1: 23–6. Springer International Publishing.

Rebolj, D., Pučko, Z., Babič, N. Č., Bizjak, M., & Mongus, D. 2017. Point cloud quality requirements for Scan-vs-BIM based automated construction progress monitoring, *Automation in Construction*, 84/May: 323–34.

Rodríguez-Moreno, C., Reinoso-Gordo, J. F., Rivas-Lpez, E., Gmez-Blanco, A., Ariza-Lpez, F. J., & Ariza-Lpez, I. 2018. From point cloud to BIM: an integrated workflow for documentation, research and modelling of architectural heritage, *Survey Review*, 50/360: 212–31.

Santagati, C., Laurini, C. R., Sanfilippo, G., Bakirtzis, N., Papacharalambous, & Hermon, S. 2019. Hbim for the surveying, analysis and restoration of the saint john the theologian cathedral in Nicosia (Cyprus), *ISPRS Annals of the Photogrammetry, Remote Sensing and Spatial Information Sciences*, 42/2/W11: 1039–46.

Santagati, C., Lo Turco, M., & Garozzo, R. 2018. Reverse information modeling for historic artefacts: Towards the definition of a level of accuracy for ruined heritage, *International Archives of the Photogrammetry, Remote Sensing and Spatial Information Sciences - ISPRS Archives*, 42/2: 1007–14.

Simeone, D., Cursi, S., & Acierno, M. 2019. BIM semantic-enrichment for built heritage representation, *Automation in Construction*, 97/November 2018: 122–37. Elsevier.

Tang, P., Huber, D., Akinci, B., Lipman, R., & Lytle, A. 2010. Automatic reconstruction of as-built building information models from laser-scanned point clouds: A review of related techniques, *Automation in Construction*, 19/7: 829–43. Elsevier B.V.

Turco, M. Lo, Mattone, M., & Rinaudo, F. 2017. Metric survey and bim technologies to record decay conditions, *International Archives of the Photogrammetry, Remote Sensing and Spatial Information Sciences -ISPRS Archives*,42/5W:261–8.

U.S. Institute of Building Documentation. 2016. USIBD Level of Accuracy (LOA). *Specification Guide 30*.

Valero, E., Forster, A., Bosché, F., Hyslop, E., Wilson, L., & Turmel, A. 2019. Automated defect detection and classification in ashlar masonry walls using machine learning, *Automation in Construction*, 106: 102846. Elsevier.

Volk, R., Stengel, J., & Schultmann, F. 2014. Building Information Modeling (BIM) for existing buildings - Literature review and future needs. *Automation in Construction*.

Yaagoubi, R., & Miky, Y. 2018. Developing a combined Light Detecting and Ranging (LiDAR) and Building Information Modeling (BIM) approach for documentation and deformation assessment of Historical Buildings, *MATEC Web of Conferences*, 149.

Yang, X., Lu, Y. C., Murtiyoso, A., Koehl, M., & Grussenmeyer, P. 2019. HBIM modeling from the surface mesh and its extended capability of knowledge representation, *ISPRS International Journal of Geo-Information*, 8/7.

Yin, X., Wonka, P., & Razdan, A. 2009. Generating 3D Building Models from Architectural Drawings: A Survey, *IEEE*, 29/1: 20–30.

Brick and Block Masonry - From Historical to Sustainable Masonry –
Kubica, Kwiecień & Bednarz (eds)
© 2020 Taylor & Francis Group, London, ISBN 978-0-367-56586-2

Tests on a flat arch concrete block retaining wall

M.C. Kurukulasuriya & N.G. Shrive
University of Calgary, Canada

ABSTRACT: Retaining walls are used in many situations, but most are not usually constructed of masonry. In this study, the potential for concrete blockwork to be used as a low-rise retaining wall was investigated. The wall is a shallow arch in plan view, held between two rigid supports for the abutments. The concept is that the blockwork wall will resist the lateral pressure of the earth through arch action, avoiding "snap-through" of the wall. The response of the wall was measured as the wall was backfilled with soil and the soil compacted. One half of the wall was fully grouted, while the other half was left hollow to compare the behaviour of the grouted and un-grouted masonry. The effects of backfilling are reported via strain and displacement profiles in the wall. The maximum deflection was recorded at the top edge of the wall and the maximum stresses did not exceed the capacity of masonry.

1 INTRODUCTION

Arches have been used in structures for more than 4000 years and many ancient, medieval and modern arches still exist with inherent structural integrity. Although arches have been used in numerous types of structures such as bridges, aqueducts, dams, roofs and wall openings, utilizing an arch shape in a retaining wall is not common. Similar to an arch dam, an arch retaining wall will resist the lateral pressure of soil through arch action, avoiding "snap-through" of the wall and minimizing tensile stresses. Retaining walls are typically constructed using concrete, steel or timber. However, in this instance it is desirable to use concrete blocks as the building material due to ease of construction, cost effectiveness and aesthetic appearance. The shape of the arch is more convenient to be constructed with blockwork as the need for formwork is alleviated. Furthermore, the arch shape provides the wall with ample robustness so that reinforcement is not required. This makes such a wall ideal for low-rise retaining walls in cities, opening a new market in which masonry can compete.

An experimental study was implemented to assess the feasibility of the masonry arch retaining wall system, upon which analytical and numerical methods could be developed for analysis and design of the system. Therefore, a low-rise segmental flat arch (an arch less than a semi-circle) wall was built spanning between two rigid supporting walls. The wall is 2.4 m high with a "rise" of 1.075 m on a - 7.6 m span, as shown in the plan view in Figure 1. The arc length of the centerline is 8 m. The

displacement and strains at selected locations in the wall were monitored as the wall was backfilled with soil. The static response of the wall to the soil pressure and any dynamic effects from the soil compaction are of interest.

2 LITERATURE REVIEW

Earth retaining structures are used to maintain a difference in ground elevation and retain the ground at an angle which it cannot sustain on its own. They are normally classified as gravity walls or embedded walls. Gravity walls resist overturning and sliding by utilizing their self-weight and friction on the bottom face, whereas embedded walls penetrate into the ground and gain lateral support mainly by the passive pressure of the ground in front of the embedded part of the wall (Burland et al. 2012). The idea of an arch retaining wall is novel and such a wall falls into neither of these categories, as its resistance is provided through arch action, where the lateral pressure from the soil applied perpendicularly onto the wall is transferred as axial thrust into the abutments. The abutments must be designed to resist the resultant thrust. Theoretically, the arch wall does not require an enlarged base as with gravity walls or an embedment length, as long as it is held rigidly between properly designed abutments. However, the limitations on the maximum height and the minimum "rise" for which this is applicable should be explored.

The concept of retaining soil using an arch wall is similar to an arch gravity dam retaining

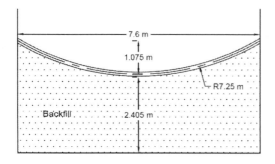

Figure 1. Plan view and dimensions of the retaining wall.

a body of water. However, arch dams are typically made of reinforced concrete: examples of masonry arch dams are almost nonexistent. The Theodore Roosevelt dam in Arizona USA was one such gravity dam; a cyclopean, rubble masonry, thick arch structure which was built in 1911. It was 85 m high and 220 m long until it was modified into a concrete gravity arch dam. Therefore, it can be deemed that analysis and design guidelines for masonry arch walls are not well established.

Zhang & Zhang (2000) discussed the possibility of utilizing arched retaining structures in foundation pits. The proposed design consisted of three rows of soil-cement columns arranged in an in-plane arch on bored piles. This configuration can be categorized as an embedded diaphragm retaining wall. The authors pointed out that the lack of understanding of analysis methods has resulted in these types of structures to be uncommon in engineering and they introduced a new analysis method combining the arch-beam method with the elastic reaction method. The arch-beam method is used to analyze arch dams, where the structure is analyzed by dividing it into a series of horizontal arches and vertical cantilever beams. The elastic reaction method is commonly used in the design of embedded retaining walls. The study explored the influence of subgrade soil characteristics, embedded depth, type of arch, rise to span ratio and buttress stiffness on the behaviour of the cantilever arched retaining wall using the proposed analytical model. It is interesting to note that the influence of the embedment length was found to be insignificant and it is only needed to satisfy the stability of the considered structural system. Also, the shape of the arch (arc of circle, parabolic, catenary, elliptic, hyperbolic, and logarithmic spiral) only had a small influence when the rise to span ratio remained the same. When the rise to span ratio increased the deformations and bending moments in the wall decreased.

Gilmore & Fuentes (2017) analyzed the retaining wall system proposed by Zhang & Zhang (2000) using an analytical method called the 'trial load method', which is extensively used in designing arch dams. The deflections at different locations of the retaining

structure were predicted and compared with numerical results.

Therefore, it is evident that the proposed masonry arched retaining structural system is novel, and an experimental program is ideal to test its efficacy and structural integrity.

3 EXPERIMENTAL PROGRAM

The experimental program was designed to test the stability of the arch masonry retaining wall. The wall was constructed using concrete blocks (190 mm x 190 mm x 390 mm - nominal strength 15MPa & 20MPa) and Type S mortar (CSA A179 -14). The wall spans 7.6 m between two rigid walls in the laboratory and it is 2.4 m (12 courses) high. The wall is a shallow arch in plan - i.e. a flat arch, with a "rise" of 1.075 m, an arc length of 8 m (20 blocks) and a thickness of 0.19 m (1 block). The arch dimensions were calculated based on the available span and the number of blocks which would fit in a circular arc. The thickness of the mortar joints is 10 mm. The wall was designed to retain the compacted soil placed in the space enclosed by the retaining wall and walls of the laboratory (Figure 2).

The profile of the arch was first laid out on the laboratory floor and a skilled mason constructed the wall. The floor was wetted initially, and the first course was bonded to the floor with a mortar layer. The mortar was prepared by mixing Type S premixed mortar (containing 2 parts Portland cement, 1part hydrated lime and 9 parts sand) with water as necessary to achieve the desired consistency; approximately 6 l of water per 80 lb (36.4 kg) bag. First, 2/3rd of the water was mixed with the cement powder for 5 minutes, then the mix was allowed to slake for another 5 minutes before mixing again with the remaining water. An electric hand drill with a mixing paddle was used to prepare the mortar. One half of the wall was fully grouted. The top course was laid with bond beam blocks, grouted and reinforced with one 15M steel bar bent to fit the arch profile along the whole length of the wall. The reinforcement and grout were

Figure 2. The concrete block arch retaining wall after backfilling and compaction was completed.

expected to provide the rigidity and strength required to withstand any accidental impacts applied to the top course during testing. The core closest to each end of the wall was also grouted up the total height to provide better bond with the supporting walls at the ends. The grout mix was prepared using SPEC MIX core fill masonry grout (course) containing Portland cement, pozzolans and dried aggregates, by mixing with water in a batch mixer. A slump of 200 - 275 mm is recommended by the manufacturer, however the consistency required to fill a height of 6 courses (1.2 m) at a time was decided by the mason from his experience. The construction was completed in 5 days. Curing was done daily for 14 days by wetting the wall.

The wall was instrumented with 36 strain gauges (gauge length of 1 inch (25.4 mm)) and 18 linear displacement sensors (strain based) located as shown in Figures 3 and 4 (the front face of the wall, i.e. the intrados is shown). The strain gauges numbered 33-36 (not shown in Figures) are located on the back face of the wall (backfilled side, i.e. the extrados) at the same height and orientation as gauges 29-32, where 33 & 34 are placed in the grouted side and 35 & 36 are placed at the ungrouted side.

The strain gauges were positioned in the lower courses as the maximum stress was expected at the bottom of the wall. The maximum displacement was expected at the free top edge of the wall. Data were obtained from the linear displacement sensors at the top edge (numbered 1-8) and the strain gauges at the two ends of the wall (numbered 1-12) at a frequency of 2000 Hz, to detect any dynamic effects that might result from backfilling or compacting the soil. The remaining gauges were sampled at 1 Hz. The backfilling and compaction of the soil were completed in layers over six days. High frequency data were acquired during the periods of compaction, and the low frequency data were recorded throughout the day. The total height of the backfilled soil layer is 2.3 m.

4 RESULTS

4.1 Material properties

During the construction of the wall, six mortar cube samples (50 mm x 50 mm x 50 mm) were taken from 2 batches (morning and evening) of the mortar mixes and six grout cylinder samples (150 mm x 300 mm) were taken per batch daily. Three course high masonry prisms (hollow and fully grouted) were made at different stages of construction of the wall. The samples were tested in compression for both 28-day strength and the strength at the time of backfilling (200 days). The results are presented in Table 1.

The Young's moduli of the concrete blocks in the directions of perpendicular to bed joints and parallel

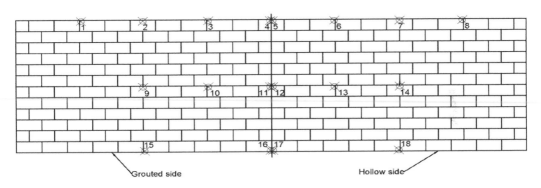

Figure 3. The linear displacement sensor locations.

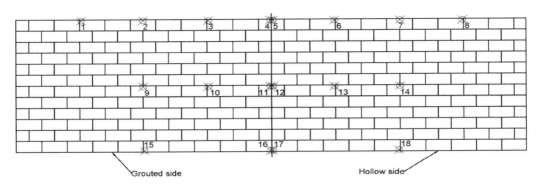

Figure 4. The strain gauge locations.

Table 1. Material properties.

Material	Batch	28 day Strength MPa	S.D.*	200 day Strength MPa	S.D.
Grout	1	9.0	0.45	19.3	0.84
	2	9.7	2.37	20.7	0.76
	3	12.2	2.60	22.8	1.54
	4	12.9	0.54	23.4	0.89
Mortar	1	10.3	0.68	17.8	2.28
	2	12.1	0.59	18.8	1.09
	3	9.9	0.8	17.5	0.93
	4	11.3	0.28	16.8	1.39
	5	9.6	0.11	20.3	0.7
	6	12.2	1.14	22.3	0.24
	7	9.5	0.65	19.8	2.05
	8	12.5	1.32	20.6	1.25
Hollow	1	18.2	1.87	-	-
prisms	2	23.0	2.51	-	-
Grouted prisms	1	13.3	0.27	-	-

* Standard Deviation

to bed joints were calculated using experimental data. The Young's modulus perpendicular to bed joints was 13800 MPa; three 30 mm x 35 mm x 190 mm samples were cut out of concrete blocks, strain gauged (gauge length of 1 inch) and tested in compression to failure. Similarly, three 35 mm x 100 mm x 390 mm specimens were tested to determine the elastic modulus parallel to bed joints, which was 8800 MPa. The average of two samples for each direction is reported due to delamination of strain gauges in some specimens. The sample sizes were chosen to avoid platen effects where the gauges were placed.

4.2 Soil properties

A sieve analysis on a sample of backfill soil revealed that the soil type is SM-SC, i.e. silty-clayey sand (ASTM D2487-11). The plastic index was 5.1 and the liquid limit was 22. To determine the density and the moisture content of the compacted soil, in-situ tests were carried out using a nuclear densometer. The minimum required compaction level was 95% of the maximum dry density. From these tests, the average moisture content was 15% and the average wet density was around 2040 kg/m^3.

4.3 Displacements

Displacements were measured at 18 locations. Out of the transducers, Gauges 6 and 9 were defective, hence they were not considered in the analysis. The data obtained at a frequency of 2000 Hz were smoothed with a 2000 sample running average. Displacements recorded at mid height did not exceed 1 mm and were close to zero at the base. As

expected, the maximum displacement was observed at the top of the wall, although it is negligible in comparison with the scale of the wall. Interestingly, the final displacement after backfilling was completed, was slightly higher in the grouted (stiffer) half than the hollow half of the wall except at the wall center. Generally, the displacement gradually increased as the soil depth behind the wall increased. The variation of the displacements at the wall top with soil depth is illustrated in Figure 5 (Outward deflection is

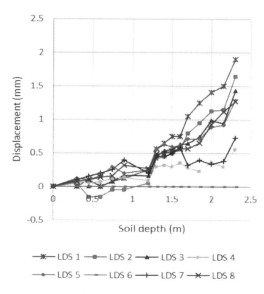

Figure 5. The variation of the displacement at the top edge of the wall with soil depth.

positive). At the wall center, the hollow half deflected slightly more than the grouted half (Figure 6).

The dynamic effect was most prominent when the soil was poured behind the wall from a higher elevation. The variation of the displacement at Gauge 1 (for approximately 2 hours) with the sample number is presented in Figure 7 (here outward deflection is negative). The first and third peaks correspond to the times when the soil was poured behind the wall. The other peaks were recorded during compaction. While at Gauges 1 & 2 the wall displaced outwards when the soil was dumped, the direction was inconsistent at other locations. An instant where the wall displaced inwards is at Gauge 3 when the soil was being poured in as presented in Figure 8.

4.4 Strains

Strains were recorded at 36 locations. Data were obtained from the high frequency gauges (1-12) in

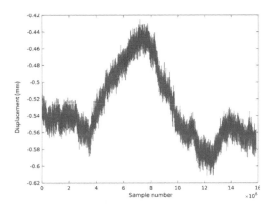

Figure 8. Variation of displacement at Gauge 3 with time.

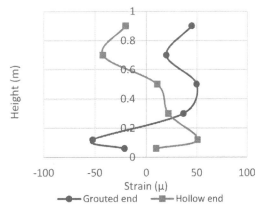

Figure 9. Final strain of gauges placed at wall ends with gauge location (height in the wall).

periods of backfilling and from the low frequency gauges (13-36) throughout the day at each day starting from zero strain. The final strain in the wall was calculated by adding up the final strain values at the end of each day for the low frequency gauges, and by reading the final value of the final day for the high frequency gauges. Final strains near the abutments for the hollow and grouted masonry are presented in Figure 9 (negative strains are compressive). In Figure 10, the strains at the wall center are presented with gauge height. The four curves represent the gauges oriented in the horizontal direction and vertical direction at the grouted and hollow sides. Gauges 25 and 36 were damaged during the experiment and hence data from these gauges are not presented. The final values for gauges at the back face are presented in Figure 11.

It should be noted that during the backfilling process, the strains at all gauge locations also showed significant variations when the soil was poured and compacted; and exhibited higher values than the residual values. The variation of strain at Gauge 17

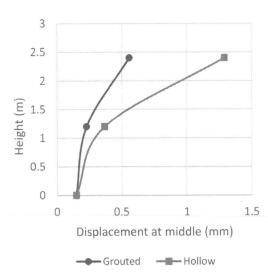

Figure 6. Wall deflection profile at wall center.

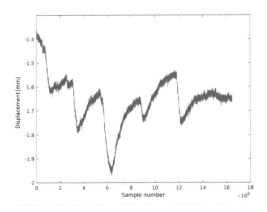

Figure 7. Variation of displacement at Gauge 1 with time.

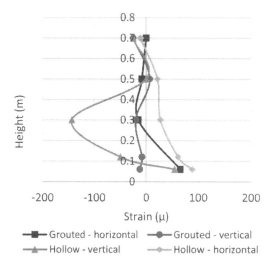

Figure 10. Final strain of guages placed at wall center with gauge loacation.

Figure 11. Final strain recorded at the back face.

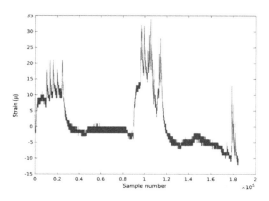

Figure 12. Variation of strain at Gauge 17 with time.

is shown in Figure 12 (recorded over the last two days of backfilling; i.e. soil depth of 1.85 m – 2.3 m, starting from zero strain) where the strain reaches 30 με, whereas the final value adds up to -8 με.

Table 2. Elastic moduli of concrete blocks.

	Experimental MPa	CSA MPa
Perpendicular to bed joints	13800	17000
Parallel to bed joints	8800	11050

4.5 Stress

The stresses in the face of the wall were calculated using the elastic moduli presented in section 4.1. These values were applied to both grouted and ungrouted masonry, considering the fact that the strain gauges were attached to the face shell and also, the strength of the grout at the time of loading is similar to that of the concrete blocks.

To validate the experimental values, the elastic modulus was also calculated according to the equation given in CSA S304-14 (Equation 1).

$$E_m = 850f'_m \qquad (1)$$

Where f'_m is the specified compressive strength of masonry.

f'_m in the perpendicular direction was taken as 20 MPa, the mean compressive strength of the concrete blocks tested. For the parallel direction the mean compressive strength of the tested samples; 13 MPa was considered as f'_m. A comparison of these values is presented in Table 2.

The stresses were calculated by multiplying the strain with the respective experimental elastic modulus. The stress distributions follow trends similar to those of the strains shown in Figures 9 & 10. The maximum values of compressive and tensile stresses calculated are presented in Table 3.

5 DISCUSSION

The overall behaviour of the wall was stable during and after backfilling. There were no observable cracks in the visible region of the wall, except for shrinkage cracks in the grout in the top course. The maximum displacement recorded was 2 mm which is only 1% of the smallest block dimension. The slightly higher values in the stiffer (grouted) end could be due to slight imperfections of the wall; patently, the wall will not be geometrically perfect, and properties are not uniform or homogeneous. The final maximum horizontal tensile strain was 90 με at Gauge 32 (hollow side center, 0.06 m up from the base) and the maximum horizontal compressive strain was -52 με at Gauge 5 (grouted side end 0.12 m up from the base). The maximum tensile vertical strain was 55 με at Gauge 31 (hollow side center 0.06 m up from the base) and the maximum compressive vertical strain was -144 με at Gauge 23 (hollow side center 0.12 m up from the

Table 3. Maximum stresses in the wall faces.

Location	Direction	Maximum compressive stress		Maximum tensile stress	
		Stress MPa	Height m	Stress MPa	Height m
Grouted wall end	H	0.46	0.12	0.44	0.50
Hollow wall end	H	0.37	0.70	0.45	0.12
Front face	H	0.13	0.30	0.58	0.06
center	V	0.35	0.70	0.10	0.50
Grouted half					
Front face	H	0.10	0.70	0.78	0.06
center	V	1.99	0.30	0.76	0.06
Hollow half					
Back face	H	-	-	0.41	0.06
center	V	-	-	-	-
Hollow half					
Back face	H	0.26	0.06	-	-
center	V	-	-	0.47	0.06
Grouted half					

* H= Horizontal, V= Vertical

base). These strains are well below the maximum limit for strains in masonry.

Considering the vertical strains at the center of the wall (Figure 10), at the grouted side the magnitude of the strains are smaller than that in the ungrouted side, and the strains are mostly in compression on both sides. Furthermore, the vertical strains at the backfilled side of the wall (Gauges 34 and 35) near the base are tensile, see Figure 11. This implies that the wall is deflecting outwards as a cantilever, which results in the back face to be in tension and the front face to be in compression in the vertical direction. This phenomenon is also evident in the deflected shape of the arch at the center (Figure 6).

Observing the axial strain at the center, on both grouted and hollow sides the strain becomes tensile closer to the base. This profile could be explained by the thrust line moving out of the kern towards the extrados at the center as the pressure (consequently the thrust) is increased and causing tensile stresses in the horizontal direction in the intrados. The shape of the thrust line could be obtained by creating the funicular polygon for the applied load. In this case, the thrust line will coincide with the funicular polygon for the applied force because the self-weight is not acting in the direction of the soil pressure. The soil pressure is applied perpendicular to the arch extrados; therefore, the vertical component is a cosine distribution, and the horizontal component is cancelled out. The differential equation of the funicular polygon for a vertical load (Equation 2) could be derived by the graphical method of constructing rays for the applied loads and vertical reactions.

$$-p(x)/H = d^2y/dx^2 \qquad (2)$$

Where p(x) is the load function, x is the horizontal distance from the crown of the arch and H is the horizontal thrust.

The vertical load function can be expressed in terms of the magnitude of the load at a specific height (P), the span (s) and horizontal distance along the arch x measured from the crown as shown in Equation 3.

$$p(x) = P\cos(\pi x/s) \qquad (3)$$

The resulting equation for the funicular polygon; hence, the shape of the thrust line will be a cosine function (Equation 4), where P increases with the increasing depth of soil. This means that at the center, the thrust line could initially reside in the kern, and when the soil pressure increases, P increases, therefore at the center the thrust line moves towards the extrados.

$$y = \frac{Ps^2}{\pi^2 H}\cos(\pi x/s) \qquad (4)$$

In the wall ends near the supports, the hollow and grouted sides have opposite effects, where the strain in the grouted side becomes compressive from tensile as it moves towards the base, and the strain in the hollow side becomes tensile from compressive (Figure 9). These changes imply that in the hollow side the thrust line moves towards the intrados whereas in the grouted side it moves in the opposite direction with increasing pressure. This could be a result of the difference in stiffness at the two sides. Also, even though the last two cores at the ends were fully grouted to maintain contact with the supports, the grout might not have filled the cores uniformly, causing the wall not to touch the supports fully at some places (see Figure 13). This might have contributed to the inconsistent strain distribution along the height as well.

Since there were no observable cracks in the arch (the masonry can withstand the low tensile strains without cracking), the thrust line remains within the arch cross section and no hinges were formed throughout the backfilling process. Thus, even though the dynamic effects from soil movement and compaction resulted in sudden variations in the strains at all locations monitored, there was nothing significant enough to form a hinge in the arch.

Another approach to predict the stresses within the arch would be a thick wall cylinder analysis (Benham & Warnock, 1973). The axial stress (hoop stress) varies along the thickness and is inversely proportional to the square of the radial distance from the center of the cylinder. However, in this arch case the segment of the cylinder is restrained at the two ends (the abutments) inducing bending moments. The thrust line would therefore vary in location within the section because of these moments and therefore such an analysis could produce an approximate prediction of its location.

The compressive stresses in the arch are well within the elastic limit; the maximum compressive stress perpendicular to bed joints was 2 MPa and maximum stress parallel to bed joints was 0.5 MPa. If the tensile capacity of the masonry is taken as 10% of the specified compressive strength, the maximum tensile stresses perpendicular to the bed joints (0.8 MPa in the hollow and 0.5 MPa in the grouted masonry) do not exceed the capacities. However, masonry is weaker in the direction parallel to bed joints. CSA S304-14 specifies the capacity of masonry parallel to bed joints as 50% for zones where grout is not continuous. Even though the maximum compressive stresses are below this limit, the tensile capacity parallel to bed joints might have been exceeded at some parts of the wall. The hollow half was subjected to higher stresses than the grouted half. In order to investigate the potential of adopting fully hollow masonry for the arch wall, further experimentation with surcharge loading will be conducted.

Furthermore, with the use of masonry, one should account for the possibility of water saturation, which would lead to freeze thaw damage. To avoid moisture movement into the wall, copings, drainage, damp-proof courses and waterproofing methods should be adapted. Also, in this experiment, the foundation design for the wall was not considered as the arch wall was constructed on a rigid concrete floor. In practice the wall will rest on soil and be supported by two abutments. Therefore, appropriate foundation and abutment design should be developed.

Further studies should be carried out to investigate the influence of the soil type and the soil structure interaction. The effect of boundary conditions and geometry of the wall should be studied by means of analytical or numerical methods.

6 CONCLUSIONS

1. The proposed concrete block arch wall is suitable to be adopted as a low-rise retaining wall along with proper designs for foundations and abutments.
2. The residual deflections and stresses in the wall after backfilling was completed were within acceptable limits.
3. The dynamic effects when the soil was deposited behind the wall from a higher elevation caused measurable strains. However, there were no cracks visible in the intrados of the arch.
4. Further experiments are to be conducted with surcharge loading on the backfilled soil, to assess the feasibility of implementing the use of hollow concrete blocks for the entire arch wall.

ACKNOWLEDGEMENTS

The authors would like to express their gratitude to the Alberta Masonry Council (AMC) and the Masonry Contractor's Association of Alberta (MCAA) (South), for their kind donations of

Figure 13. An instance where the wall is not fully in contact with the support.

resources and skilled labour. We also extend our deepest thanks to Dr. Doug Phillips in Research Computing Services, Information Technologies, University of Calgary, for his technical support on processing the large amount of data. The authors are further grateful for the valuable support from Dr. Ron Wong and Yadong Zhang from the Department of Civil Engineering, Dr. Fengman Jia and Dr. Derek Lichti from the Department of Geomatics Engineering at the University of Calgary.

REFERENCES

ASTM D2487-11. 2011. *Standard practice for classification of soils for engineering purposes (unified soil classification system).* West Conshohocken, PA, USA, ASTM International.

Benham, P. P. & Warnock, F. V. 1973. *Mechanics of solids and structures.* London: PITMAN publishing.

Burland, J., Chapman, T., Skinner, H., & Brown, M. 2012. *ICE manual of geotechnical engineering: Volume II.* London: ICE publishing.

CAN/CSA-A179-14. 2019. *Mortar and grout for unit masonry.* Mississauga, Ontario: Canadian Standards Association Group.

CAN/CSA-S304-14. 2014. *Design of masonry structures.* Mississauga, Ontario: Canadian Standards Association Group.

Gilmore, D. & Fuentes, R. 2000. Predicting the behaviour of non-circular, curved-in-plan retaining walls using the trial load method. *Proceedings of the 19th International Conference on Soil Mechanics and Geotechnical Engineering, Seoul.* 1987-1990.

Zhang, Y. & Zhang, T. 2000. Behaviour Analysis of Cantilever Arched Retaining Structures in Foundation Pits. *Journal of Zhejiang University.* 2(3), 309–312.

Brick and Block Masonry - From Historical to Sustainable Masonry –
Kubica, Kwiecień & Bednarz (eds)
© 2020 Taylor & Francis Group, London, ISBN 978-0-367-56586-2

Analysis framework for seismic assessment of single-span masonry arch bridges

G. Jofin & M. Arun
Department of Civil Engineering, Indian Institute of Technology Madras, Chennai, India

ABSTRACT: Around 20,000 masonry arch bridges are currently operational with the Indian Railways of which 16,000 bridges are at least 100 years old. There is a constant need to assess and if necessary, retrofit and upgrade these bridges to meet the current demands of new seismic performance requirements, increased axle speeds and loads, traffic volume, etc. The geometric dependence of failure mechanisms has made limit analysis an obvious choice for the assessment of masonry arch bridges. In this context, a simple, yet robust, first-level quantitative assessment procedure which can be applied to a large stock of masonry arch bridges with minimum input parameters is proposed. The collapse factor is calculated based on equilibrium of internal and external work, resulting from the virtual displacement of the rigid block. For a given single span masonry arch bridge under seismic action, the governing collapse factor is identified as the minimum of all possible failure mechanisms.

1 INTRODUCTION

Masonry arch bridges continue to be an integral part of the road and railway network in India. Despite their age and deterioration due to external agents, a broad spectrum of these bridges continues to be in service condition with increased traffic volume and axle loads and speeds, but a significant number of these bridges are located in earthquake-prone regions. As it is mandatory to design all public infrastructure to be seismic-resistant, existing infrastructure should be checked for seismic adequacy. The current assessment methodology adopted by most of the bridge owners is based solely on visual inspection. The visual inspection could give the actual geometric data, existing structural distresses and material deterioration, which could be used for simplified assessment procedures.

Limit analysis based on kinematic theorem has been identified as a robust tool in the assessment of masonry arch bridges. The use of funicular polygon concept based on elastic theory gave a stable configuration of the arch under applied loading conditions. Adding voussoir thickness to the arch geometry ensured the stability of the structure (Baker,1909). Unlike elastic analysis, Heyman (1966,1969) used plastic theory for the arch analysis which gave a unique solution of the orientation of the thrust line. The safety was ensured by forcing the thrust line within the middle third of the arch

thickness (Middle third rule). With this background of rigid block analysis, various research methodologies are formulated to ensure the safety and stability of masonry arch bridges under various loading conditions. The most popular and commonly adopted tool for the analysis of masonry arch bridges is limit analysis. The method has proved robust and efficient in estimating the collapse conditions of masonry arches under the assumption of rigid block behavior of masonry voussoirs. The failure of masonry arches was found to be reliant on collapse mechanisms rather than exceedance of material strength at subsequent sections of the arch.

The overall objective is to develop a framework of assessment that considers both gravity and seismic loads and other critical aspects such as support settlement for a given masonry arch bridge. This component of the research focuses on the assessment of masonry arch bridges under seismic loads, while rest is out of the scope of the present paper. A simplified procedure for the seismic assessment of single-span masonry arch bridges is proposed, extendable to multi-span bridges along with further modifications/simplifications without compromising the accuracy of the estimation scheme. The dependency of the collapse mechanism on geometry is established using a parametric study. Further adequacy of the rigid block assumption is verified by comparing it with a finite strength material model.

2 LIMIT ANALYSIS BASED FRAMEWORK

2.1 *Theoretical background*

The proposed assessment framework is based on the fundamental assumption that the failure of the arch is dependent on its geometric parameters rather than material properties (Clemente et al., 1995). This assumption was verified by Melbourne and Gilbert (1995) by static gravity testing of a full-scale model of a single span masonry arch bridge. Hence the limit behavior of arch bridges can be considered as a rigid block equilibrium problem. The kinematic method based on limit analysis states that *the arch is on the verge of collapse if an equilibrium solution can be found such that the thrust line is contained inside the masonry arch and sufficient number of non-dissipative hinges form, turning the structure into a mechanism* (De Luca et al., 2004). The uniqueness theorem ensures that this solution exists and is unique (Clemente et al., 1995). When subjected to ground acceleration, the arch carries gravity loads as well as horizontal inertial forces. Under constant vertical loads, if the ground acceleration and correspondingly inertial forces are increased up to the collapse, the value of acceleration at collapse is unique. Hence if the line of thrust changes, and at collapse, if at least four hinges are formed, the thrust line should then pass through these hinges and should remain in the arch barrel. The vertical component of the ground acceleration is neglected as it is typically uncorrelated to the horizontal forces (Lagomarsino and Resemini (2009), Lagomarsino (2015)). The equation of equilibrium under this condition is written as:

$$P_v + \alpha P_h = 0 \qquad (1)$$

Limit analysis is based on the following fundamental assumptions:

- *No tensile strength of masonry*: This assumption holds true if the arch is made up of weak mortar. Hence there cannot be a transfer of tensile load from one block to another despite the block itself having some amount of tensile strength.
- *Absence of sliding in masonry:* As the frictional resistance developed between blocks is significantly higher, the assumption is reasonable in practical cases.
- *Infinite compressive strength of masonry*: The assumption implies that the crushing failure of masonry cannot occur. Usually, the stress level in the arch is very low relative to compressive strength, which makes this assumption valid until the arch reaches its collapse state.

In limit analysis based seismic assessment, the capacity of the bridge can be stated in terms of a collapse multiplier from the equilibrium of internal and external work. Here the lateral resistance is assumed proportional to inertial forces under seismic action. The collapse multiplier (α_o) is obtained by applying the virtual work principles to the initial configuration. Solving the virtual work equation, a unique solution that satisfies the equilibrium, mechanism and yield condition is obtained as (NTC-08)

$$\alpha_o \left(\sum_{i=1}^{n} V_i \delta_{x,i} + \sum_{j=n+1}^{n+m} S_j \delta_{x,j} \right) - \sum_{i=1}^{n} V_i \delta_{y,i} - \sum_{b=1}^{o} F_b \delta_b = L_{fi}$$

$$(2)$$

For a masonry arch bridge under seismic action, different failure mechanisms are possible depending on geometric properties. The failure is considered to be local if the arch alone is participating in the collapse mechanism. If the possible mechanism includes the arch as well as the abutments, a global mechanism is said to be formed. Single-span bridges with high abutments were identified vulnerable to medium-high seismic action in the longitudinal direction. The overall arch-abutment longitudinal mechanism was found to be critical for slender single-span bridges (Modena et al., 2016). For single-span bridges with squat abutment, the arch was identified as the most vulnerable element to seismic action in the longitudinal direction. Hence an arch mechanism in the longitudinal direction (AL) was proposed (Clemente, 1998). For transverse loading, the spandrel wall transverse mechanism (SWT) was found to dominate (De Luca et al., 2004; da Porto et al., 2007).

For possible longitudinal mechanism, simplifying assumptions are proposed increasing the applicability of the proposed procedure to a wide spectrum of bridges. The spandrel wall, as well as the infill material, can be considered as a homogenous applied mass acting on the arch segment (da Porto et al., 2015). Neglecting the stabilizing effect of the infill material (Cavicchi, 2006), the analysis should give an upper bound solution, hence safe. In the case of a single span masonry arch bridge with slender abutments, lateral earth pressure develops when the abutments are displaced (Zampieri et al., 2016). Hence the effect of active and passive earth pressure due to instability of the soil needs to be considered in the analysis. For a single-span masonry arch bridge, the position of two hinges can be assumed to occur at support, either at the arch springing or at the abutment. The other two hinges can occur at any point on the arch barrel, which needs to be identified. Hinges are typically located by the calculation of thrust line/graphical methods followed by the application of limit analysis. Here an iterative scheme based on limit analysis is proposed for locating the actual hinge location, which is illustrated in the flow chart in Figure 1. Keeping the number of voussoirs

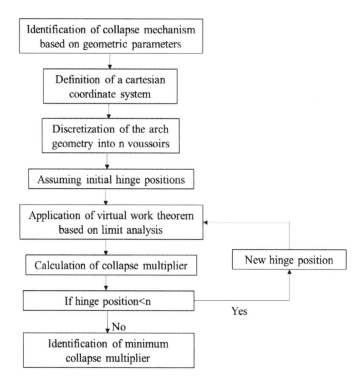

Figure 1. Iterative scheme to identify governing collapse factor.

in the arch discretization equal to the actual number of voussoirs in the arch bridge, a better estimate of hinge locations for a governing collapse mechanism can be obtained. For an assumed hinge location, equilibrium equations based on virtual work are derived and the seismic collapse factor is calculated for varying hinge locations. Governing value is identified as the minimum of all collapse factors for a given configuration of hinges.

2.2 Formulation of possible collapse mechanisms

Depending on the direction of seismic action, the possible collapse mechanisms can be broadly classified as longitudinal or transverse mechanism. For an assessment scheme, the interest lies in identifying the most critical mode of failure, independent of the directionality of seismic forces. Hence a generalized methodology that reports the criticality of each collapse mechanism is important based on which the prioritization of further retrofit strategies shall be based on.

2.2.1 Arch in Longitudinal Mechanism (AL)
Figure 2 Shows a discretized arch geometry with the weight of arch voussoir and spandrel acting at their center of gravity. The entire span of the arch is discretized into a finite number of elements. The spandrel wall is also divided into an equal number of elements and is considered as a superimposed load

acting on each arch element. The arch is then divided into three rigid blocks, as shown in Figure 2. According to the theorem of kinematic chains, for a group of three rigid bodies, their relative centers of rotations (C_1, C_2 and C_3), as well as the relative center of rotations (C12, C23) for an arch bridge in the mechanism state, is shown. The virtual displacements of the entire arch can be characterized based on the rotation of one of the rigid blocks. Once the

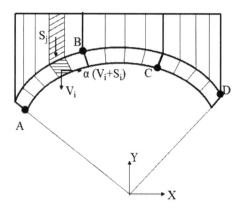

Figure 2. Arch discretization.

790

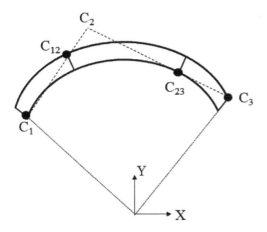

Figure 3. Centre of rotation of rigid blocks.

virtual displacements δx and δy are formulated, the collapse multiplier can be obtained as

$$\alpha_o = \frac{\sum S_j \delta_{y,j} + \sum V_i \delta_{y,i}}{\sum (V_i + S_j) \delta_{x,i}} \quad (3)$$

2.2.2 Arch Abutment Longitudinal mechanism (AAL)

In Arch-Abutment-Longitudinal (AAL) mechanism, the abutments also take part in the mechanism along with the arch. The failure occurs by the formation of three rigid segments rotating about their center of rotation. This results in the formation of four hinges, with the location of hinges A and D in abutments as shown in Figure 4. The location of hinges B and C are initially unknown, which can be identified using the iterative scheme proposed in Figure 1. Since the soil is in contact with the abutments, significant lateral earth pressure is expected to develop when the arch bridge is subjected to lateral load due

to the displaced abutments. For lateral force acting towards the left, active earth pressure develops in the right abutment, and passive earth pressure develops in left abutment (Figure 4) and vice versa. Considering the equilibrium of external and internal virtual work, the seismic collapse factor is obtained as,

$$\alpha_0 = \frac{\sum_{i=1}^{n} V_i \delta y_i + \sum_{j=1}^{n} S_j \delta y_j + \sum_{i=1}^{2} A_i \delta y_{Ai} + P_p \delta y_p - P_a \delta y_a}{\sum_{i=1}^{n} (V_i + S_j) \delta x_i + A_i \delta x_{Ai}}$$

$$(4)$$

2.2.3 Spandrel wall transverse mechanism (SWT)

For transverse seismic action, the spandrel wall is considered as the most vulnerable element. The out of plane collapse occurs when hinge forms at the base of the wall and the spandrel wall rotate as shown in Figure 5. Before the formation of the collapse mechanism, at-rest earth pressure acts on the spandrel wall resulting in uniform distribution of earth pressure on the wall. At the limit state, the infill material wedges behind the spandrel wall generating passive and active earth pressure distribution on the spandrel wall (da Porto et al., 2007). A simplified analysis is formulated considering a typical element of spandrel wall of height S_H and thickness, t as shown in Figure 5. Considering the moment equilibrium of the spandrel section under seismic action and assuming rigid body rotation, the infill mass exerts earth pressure in the direction of seismic action. Then the moment equilibrium equation can be written as,

$$\alpha_o = \frac{S\frac{t}{2} - E_p \frac{S_H}{3}}{S\frac{S_H}{2}} \quad (5)$$

As the height of the spandrel wall varies along the arch barrel, considering the section with maximum

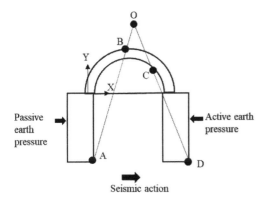

Figure 4. Arch abutment longitudinal mechanism.

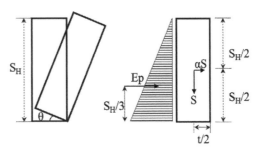

Figure 5. Representation of spandrel wall transverse mechanism.

spandrel wall height, a conservative estimate of collapse multiplier is obtained.

The material and geometric data used for further study are shown in Table 1. A single-span semicircular masonry arch bridge is considered in the analysis with 20 voussoirs in the arch barrel for studying the effect of material strength.

2.3 Influence of material strength

The location where hinges are formed in an arch barrel is independent of stress at that section but dependent on arch geometry. Figure 6 shows the effect of material strength on an arch section. Across the section, with material strength included, a hinge is now assumed to be distributed over a region of depth, s/2-e (plastic region), instead of a single point in the intrados/extrados of the arch barrel. For a typical section XX' of the arch barrel (Figure 6), adopting a rigid plastic model for the cross-section, the axial thrust on the arch section due to finite material strength, N_u is obtained as,

$$N_u = \sigma_c B\left(\frac{s}{2} - e\right) \qquad (6)$$

The virtual work equation for AL mechanism is modified incorporating the material strength.

Table 1. Bridge parameters used.

Parameter	Adopted value
Span length (L)	5 m
Arch thickness (s)	0.5 m
Abutment width (A_B)	1.5 m
Abutment height (A_H)	3.9 m
Unit weight of masonry	19.2 kN/m^3
Unit weight of infill	17 kN/m^3
Width of the bridge	10 m

$$\alpha_o = \frac{\sum P_j \delta_{y,j} + \sum P_i \delta_{y,i}}{\sum (P_i + P_j)\delta_{x,i} - \left(\sum\limits_{j=1}^{r} M_j \theta_j\right)} \qquad (7)$$

Where M_j is the moment acting at the hinge section, j due to eccentricity, e and θ_j is the rotation at the corresponding section, and 'r' is the number of hinges considered. With material strength included, the hinge locations are again identified using the iterative procedure mentioned in Figure 1. The effect of compressive strength on the collapse mechanism is shown in Figure 7 for a single span semicircular masonry arch bridge for AL mechanism. With the increase in compressive strength, the finite strength model converges to the rigid block model.

2.4 Simplification of spandrel contribution

The spandrel wall is typically considered as a superimposed weight acting on the arch barrel in limit analysis. Simplifying this assumption further, the possibility of considering the spandrel wall as a lumped mass acting on arch voussoirs is verified. This simplifies equation 3 as

$$\alpha_o = \frac{\sum\limits_{i=1}^{n} (P_i + P_j)\delta_{y,i}}{\sum\limits_{i=1}^{n} (P_i + P_j)\delta_{x,i}} \qquad (8)$$

The collapse factor obtained with both these assumptions is plotted in Figure 8 against the increasing number of voussoirs in the arch for a semicircular arch bridge for AL mechanism. There is minimal change in the collapse factor by considering spandrel and infill weight as a lumped mass in the arch. For an

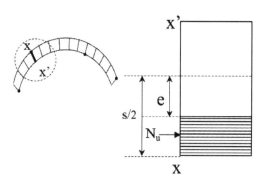

Figure 6. Adopted rigid plastic model.

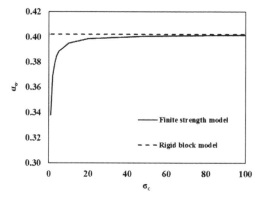

Figure 7. Effect of compressive strength on collapse mechanism.

792

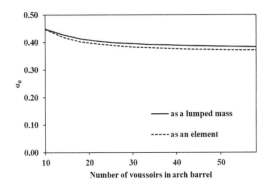

Figure 8. Effect of spandrel wall and infill considered as a lumped mass on arch.

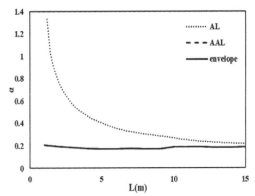

Figure 9. Dependence of collapse mechanism on Span length (L).

arch with 20 voussoirs, the collapse factor estimated using both assumptions differ by 2.5%, which is conservative. This is because the virtual displacement δx_j of the spandrel has a minimum contribution to the total external work done, and hence spandrel and infill can be safely assumed as a lumped mass system acting on the arch.

3 PARAMETRIC STUDY

As already reviewed, the failure mechanism of an arch bridge is well dependent on its geometry than its material parameters (Clemente et al., 1995). Hence the effect of various geometric parameters on arch collapse is discussed. The adopted geometric, as well as material properties in the parametric study, are shown in Table 2. The dependency of failure mechanism on s, L and s/L ratios are considered in the parametric study for a single span semicircular arch. With the increase in span length, L, AAL remains the governing mechanism for spans up to 15m (Figure 9). The collapse factor for AAL is independent of span length, arch thickness, or s/L ratios but is dependent on A_B/A_H ratios of the abutment. With the increasing thickness and decreasing length, the collapse factor tends to increase for the AL

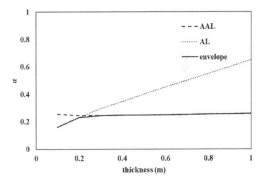

Figure 10. Dependence of collapse mechanism on arch thickness (s).

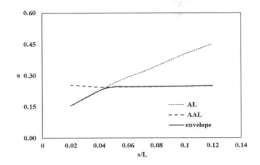

Figure 11. Dependence of collapse mechanism on s/L ratio.

mechanism (Figure 9, Figure 10). To consider this opposing effect, using s/L as a single parameter, the effect on the collapse factor is illustrated in Figure 11. For single-span masonry arch bridges with s/L <0.045, AL mechanism was found to dominate, but when s/L ratio exceeds 0.045, the hinges can form at abutment base and hence AAL mechanism

Table 2. Arch bridge parameters adopted.

Parameter	Adopted value
Span length (L)	1-15 m
Arch thickness (s)	0.1-1 m
Abutment width (A_B)	1.5 m
Abutment height (A_H)	0.5-6 m
Unit weight of masonry	19.2 kN/m^3
Unit weight of infill	17 kN/m^3
Width of the bridge	10 m
Number of voussoirs (n)	20

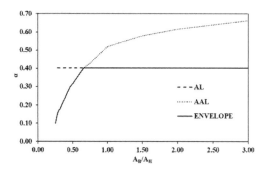

Figure 12. Shift in collapse mechanism with A_B/A_H ratio.

dominates. For the same geometry, with the change in A_B/A_H, the governing collapse mechanism shifts from AAL to AL at A_B/A_H =0.67. Hence for slender abutments (A_B/A_H <0.67), the AAL mechanism can be considered to be the governing mechanism. For a single-span masonry arch bridge, the failure envelope for increasing A_B/A_H values is shown in Figure 12. With the increase in A_B/A_H ratios, the abutments are more vulnerable for lateral seismic action than the arch supports and hence a shift in governing mechanism is observed.

3 CONCLUSION

The dependency of structural response on geometry has made limit analysis an obvious choice for the assessment of masonry arch bridges. An iterative scheme was formulated based on limit analysis which can calculate collapse factor depending on the level of information available. For a given single-span masonry arch bridge, the proposed algorithm is capable of identifying the governing collapse mechanism, from all possible collapse mechanisms. If the material compressive strength is available from field test data, a more accurate estimate of the collapse factor can be obtained using a finite strength model. But for a first-hand estimation, the collapse factor based on rigid block assumptions are sufficient as variation is minimal. Further, assuming the spandrel wall as a lumped mass on the arch barrel has simplified the algorithm with reasonable accuracy. The critical geometric parameters were identified based on the parametric study. The AAL mechanism was affected by the change in A_B/A_H ratios of the abutment while AL mechanism was governed by the change in arch thickness as well as the span lengths. Finally, an envelope curve with governing collapse factor is presented which identifies the governing collapse factor for the given geometric proportions for a single span semicircular masonry arch bridge within the considered range of geometric as well as material parameters.

4 NOTATIONS

A_B and A_H	Width and height of the abutment respectively
A_i	Weight of the abutment
e	Eccentricity at which the axial force is acting
E_p	The equivalent force caused by Earth pressure from infill material
F_b	External force applied
L	Span length
L_{fi}	Work done by internal forces.
M_j	The moment acting at a section due to eccentricity, e
m	Number of weight forces not directly acting on the blocks but generate horizontal forces on the elements of the kinematic chain
n	Number of dead loads applied on the kinematic chain
N_u	Axial thrust on the arch section
o	Number of external forces applied
P_p and P_a	Passive and active earth pressure and δ_{yp} and δ_{ya} are their corresponding virtual displacements
P_v and P_h	Work done by vertical and horizontal forces respectively
S	Weight of the spandrel per unit span length
S_j	Member not directly acting on blocks but generating horizontal forces under seismic action
s	Thickness of the arch barrel
S_H	Equivalent height of the spandrel wall at the considered cross-section
t	Thickness of the spandrel wall
V_i	Weight force involved directly in the mechanism
δ_{Ai}	Virtual displacement corresponding to Ai
α_0	Seismic collapse factor
$\delta_{x,i}$ and $\delta_{y,i}$	Virtual displacement of the point of application of weight $P_{i,j}$ in x and y directions
σ_c	Compressive strength of masonry
θ_j	Rotation at a section corresponding to M_j
δ_b	Virtual displacement corresponding to F_b

REFERENCES

Baker I.O. (1909). *A treatise on masonry construction.*

Clemente, P., Occhiuzzi, A., and Raithel, A. 1995. Limit Behavior of Stone Arch Bridges. *Journal of Structural Engineering,* 121(7), 1045–1050.

Clemente, P. 1998. Introduction to dynamics of stone arches. *Earthquake Engineering and Structural Dynamics,* 27(5), 513–522.

Cavicchi, A., and Gambarotta, L. 2006. Two-dimensional finite element upper bound limit analysis of masonry bridges. *Computers and Structures,* 84(31–32), 2316–2328.

Da Porto, F., Franchetti, P., Grendene, M., Ranzato, L., Valluzzi, M., and Modena, C. 2007. Structural capacity of masonry arch bridges to horizontal loads. *5th International Conference on Arch Bridges (ARCH'07).*

Da Porto, F., Tecchio, G., Zampieri, P., Modena, C., and Prota, A. 2016. Simplified seismic assessment of railway masonry arch bridges by limit analysis. *Structure and Infrastructure Engineering*, Taylor & Francis, 12(5), 567–591.

De Luca, A., Giordano, A., and Mele, E. 2004. A simplified procedure for assessing the seismic capacity of masonry arches. *Engineering Structures*, 26(13), 1915–1929.

Heyman, J. 1966. The stone skeleton. *International Journal of Solids and Structures*, 2(2), 249–279.

Heyman, J. 1969. The safety of Masonry Arches. 11 (November 1968), 363–385.

Lagomarsino, S., and Resemini, S. 2009. The assessment of damage limitation state in the seismic analysis of monumental buildings. *Earthquake Spectra*, 25(2), 323–346.

Lagomarsino, S. 2015. Seismic assessment of rocking masonry structures. *Bulletin of Earthquake Engineering*, 13(1), 97–128.

Melbourne, C., and Gilbert, M. 1995. Behaviour of multiring brickwork arch bridges. *Structural engineer London*, 73(3), 39–47.

Modena, C., Tecchio, G., and da Porto, F. 2016. Masonry arch bridges in the Italian railway engineering experience: state of condition, static reliability, seismic verification and retrofit strategies. *Maintenance, Monitoring, Safety, Risk and Resilience of Bridges and Bridge Networks - Proceedings of the 8th International Conference on Bridge Maintenance, Safety and Management, IABMAS 2016*.

NTC-08. Italian National Technical Code, 2008.

Zampieri, P., Zanini, M. A., and Faleschini, F. 2016. Derivation of analytical seismic fragility functions for common masonry bridge types: methodology and application to real cases. *Engineering Failure Analysis*, Elsevier Inc., 68, 275–291.

Brick and Block Masonry - From Historical to Sustainable Masonry –
Kubica, Kwiecień & Bednarz (eds)
© 2020 Taylor & Francis Group, London, ISBN 978-0-367-56586-2

Structural performance and seismic response for Chinese ancient stone arch bridge – a case study of the Putang bridge

Qing Chun, Hui Jin & Shiqi Zhang
School of Architecture, Southeast University, Nanjing, China.

ABSTRACT: In order to study the structural performance and seismic response of Chinese ancient stone arch bridge, a typical Chinese ancient stone arch bridge, the Putang Bridge, was taken as a study case in this paper. Firstly, the finite element model of this bridge was built up by the software of ANSYS according to the on-site survey and test, the structural performance of this bridge under the vertical static loads was studied, resulting in the weak parts of the whole structure under the static loads. Then, the seismic response of this bridge under different seismic wave excitations were studied, and the weak parts of this bridge under the seismic waves were obtained. The study results can provide basis and reference for scientific conservation for the similar types of stone arch bridges.

1 INTRODUCTION

In China, a lot of ancient stone arch bridges have been used for hundreds of years till now. Chinese ancient stone arch bridges generally have single, three, five, seven and nine arches, etc. The architectural forms, structures and construction techniques of them are different from those of ancient stone bridges in Western countries. Most of the Western stone arch bridges are in the forms of three-center-circular arch, semi-circular arch and Gothic arch, but Chinese traditional stone arch bridges are mainly semi-circular arches and shallow arches. In addition, the construction methods used in Chinese ancient stone arch bridges are various, these methods can be generally divided into three types, the first type of the arches is lack of lateral linkage, the second type of arches improve lateral linkage by laying voussoirs in a staggered way, the third type of arches improve lateral linkage by inserting long voussoirs (Figure 1).

The Putang Bridge is a Chinese typical stone arch bridge which was built in Ming Dynasty (1386-1644), and it is located in Nanjing, China. The materials of the bridge are tuff sandstones with the color of light purple-red. The bridge is one of the largest ancient stone arch bridges around China, which has been listed as a major historical and cultural site protected at the national level since 2013. The length of the whole bridge is 91.7m and the width is 6.68m. The Putang Bridge has nine arches, and the middle arch spans 9.85 m and rises 5.1 m, and the thickness of the arch is 0.34m. The arches on both sides gradually shrink. The sections of the upstream cutwater stones are triangular, and the purpose is to reduce the water flow impact pressure. The bridge has been used for over 500 years since 1512. It is necessary to evaluate the structural performance and seismic response of the bridge before the following conservation. The Putang Bridge is shown in Figure 2.

Many studies on the structural performance of the stone arch bridges have been published around the world. Yang (1980) analyzed the structural mechanical properties of the stone arch bridge with a theoretical derivation method. Qian (1987) took Zhaozhou Bridge as a case study, and calculated the minimum thickness of the arch and the change of the safety of the arch thickness under concentrated live loads. Yao (1996) proposed an assessment method for the safety inspection of stone arch bridges. Fanning and Boothby (2001) verified the applicability of the solid element for stone and the Drucker–Prager material for fill through some three-dimensional nonlinear finite element models. Liu and Chen (2003) used the finite spring method to numerically simulate a multi-span stone arch bridge and studied its structural performance under vertical live load and horizontal seismic load. Du et al. (2005) studied the seismic performance of a stone arch bridge with the mode-superposition response spectrum method. DeJong *et al.* (2008) imposed five different earthquake time histories on model arches, resulting in a suite of failure curves which can be used to determine the rocking stability of masonry arches under a primary base acceleration impulse which has been extracted from an expected earthquake motion. According to some common standards, Luca Pelà et al. (2009) analyzed the seismic performance of a stone arch bridge based on inelastic pushover analysis and response spectrum approaches. Shen (2010) took the multi-span catenary open-belt arch bridge as the research object, and compared the seismic response analysis of multi-arch and single-arch with the finite element method. Bayraktar et al.

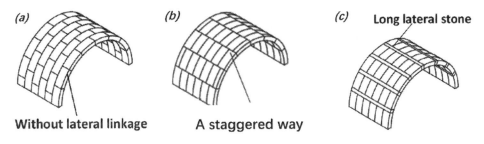

Figure 1. The different construction methods of Chinese ancient stone arch bridges.

Figure 2. The photo of the Putang bridge.

(2015) carried out in-situ investigations for the eight stone arch bridges, and determined the frequencies, damping ratios and mode shapes of the bridges by the Enhanced Frequency Domain Decomposition (EFDD) and Stochastic Subspace Identification (SSI) Techniques. Conde et al. (2017) proposed a multidisciplinary approach for the structural assessment of masonry arch bridges including the nondestructive testing techniques (such as integrates laser scanning, ground penetration radar, sonic tests and ambient vibration testing) and analysis of finite element model, and an optimization algorithm was also presented to minimizes the discrepancies with respect to the experimentally obtained modal properties.

Overall, although some studies on the structural performance of the stone arch bridge had been published, the stone arch bridges studied by foreign researchers were different from the Chinese ancient stone arch bridges, such as bridge forms, bridge structures, and bridge construction techniques; while the studies on the structural performance and seismic response of the ancient stone arch bridges from domestic researchers were in a small quantity. In this study, a typical Chinese ancient stone arch bridge was taken as a case study. The structural performance and seismic response were studied with the finite element method. The results can provide a basis for the scientific conservation of this type of stone arch bridge.

2 THE FINITE ELEMENT MODEL OF THE PUTANG BRIDGE

The main structure of the Putang Bridge consists of the masonry and the earth fill, and both of them were modeled by Solid 45 Element of ANSYS. This type of element is hexahedral element with 8 nodes, which is suitable for 3D solid structure with an acceptable calculation speed. Both of these two materials can be simplified as continuously homogeneous isotropic material for finite element model (Chun et al., 2016). The material strengths of the stone and mortar were tested with rebound method, including 12 arch stones and the nearby mortars, and the tested compressive strengths were 20MPa and 2MPa respectively. For finite element model, the stone block and mortar were thought to work together as the masonry structure, thus the material of the bridge model was isotropic. The *Code for design of masonry structures (GB50003-2011)* provides the calculation method for the equivalent material properties of masonry according to the strengths of stone and mortar. The equivalent compressive strength of the masonry was conservatively taken as 1.75 MPa, the tensile strength was 0.07 MPa, and the elastic modulus was 1524MPa. For the earth fill, some samples were collected for X-ray diffraction test, and the results provided the fill was trinity mixture fill (consisting of lime, clay and fine sand). The elastic modulus of this kind of fill was taken as 40MPa (Chun et al., 2016), which was adopted in this study. The values of the density were 2000kg/m³ and 1800kg/m³ for the masonry and the earth fill respectively. The mesh was resulted by sweep mode, which is shown in Figure 3. The restrictions of the foundations were considered as the fixed constraints. The contact between the masonry and the earth fill was set as displacement coupling.

3 THE STRUCTURAL PERFORMANCE OF THE PUTANG BRIDGE UNDER STATIC LOAD

The main dead load of the Putang Bridge is the weight of the slabs on the bridge deck, and the value was taken as 5.0 kN/m² with 1.2 load factor for load combination. The bridge is used only for pedestrians, and the value of live load was taken as 3.5 kN/m² with the 1.4 load factor. Four typical load conditions were analyzed in this study, and the load conditions were as followings:

Condition 1: only dead load;

Condition 2: dead load + uniform live load for the whole bridge;

Condition 3: dead load + live load on half width of the bridge;

Condition 4: dead load + live load on half length of the bridge.

The results of the stress analysis are listed in Table 1.

It can be found that the Condition 2 (dead load + uniform live load) is the critical load condition. The stress contours for Condition 2 are presented in Figure 4 and Figure 5.

According to the results from Figure 3, it can be found that the tensile stresses on the area of the spandrel walls between two arches, and the arch-back and the arch-waist of each arch are larger than the tensile strength of the masonry. According to the results from Figure 5, it can be found that the maximum compressive stress occurs at the middle-arch's feet, and the value is larger than the compressive strength of the masonry. These weak parts need to be strengthened in the following conservation.

4 THE SEISMIC RESPONSE OF THE PUTANG BRIDGE

In order to evaluate the earthquake performance of the Putang Bridge, the time history analysis method was used to analyze the seismic response of the bridge. The design earthquake (including two natural earthquake waves and one artificial earthquake wave) were

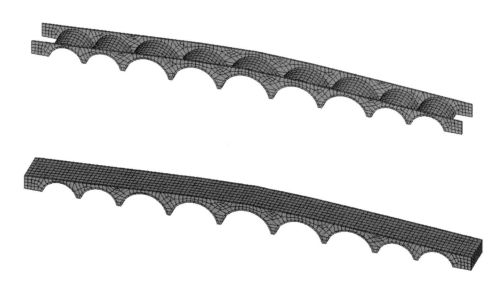

Figure 3. A&b The finite element model of the Putang bridge.

Table 1. Analysis results of different load conditions.

Load condition	Condition 1	Condition 2	Condition 3	Condition 4
Maximum principal tensile stress	0.40MPa	0.45MPa	0.40MPa	0.45MPa
Maximum principal compressive stress	1.78MPa	1.93MPa	1.79MPa	1.88MPa

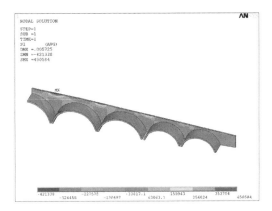

Figure 4. The principal tensile stress contour for condition 2.

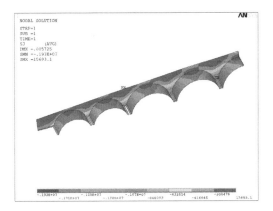

Figure 5. The principal compressive stress contour for condition 2.

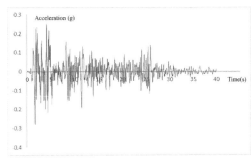

Figure 6. The acceleration-time curve of the natural earthquake wave I (IMPVALL.I_I-ELC180) (source: http://nga west2.berkeley.edu/).

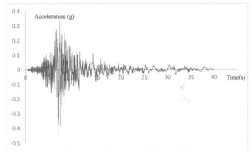

Figure 7. The acceleration-time curve of the natural earthquake wave II (NIIGATA_NIGH11NS) (source: http://nga west2.berkeley.edu/).

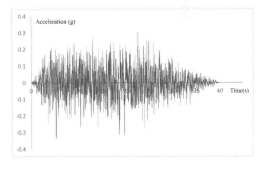

Figure 8. The acceleration-time curve of the artificial earthquake wave.

determined by *Code for seismic design of buildings (GB50011-2010)*. The peak values of the seismic acceleration time history of 7-degree frequent earthquake, design earthquake, and rare earthquake are 0.35 m/s^2, 0.98 m/s^2, and 2.20 m/s^2, respectively, and the design acceleration response spectrum characteristic period is 0.45s. According to the requests above, two natural earthquake waves were chosen online and one artificial earthquake wave was created through a software. The three acceleration-time curves are presented in Figure 6, Figure 7 and Figure 8.

In this study, two seismic conditions were analyzed, and they were as followings:

Seismic condition 1: 1.2 * bridge weight + 1.3 * horizontal seismic excitation on longitudinal direction + 0.5 * vertical seismic excitation;

Seismic condition 2: 1.2 * bridge weight + 1.3 * horizontal seismic excitation on transversal direction + 0.5 * vertical seismic excitation.

The vertical seismic excitation was taken as 0.65 time of horizontal seismic excitation. For the

Seismic condition 1, the arch-waist of the middle-arch (AW1) was taken as the control part for deformation, while the arch-feet of the middle-arch (AF1) were taken as the control part for tensile stress. For the Seismic condition 2, the spandrel wall above the middle-arch (SW2) was taken as the control part for

Table 2. The comparisons of displacements and stresses for the Putang bridge under different seismic excitations.

Categories	Seismic condition 1		Seismic condition 2	
	Maximum deformation of AW1 (mm)	Maximum tensile stress on AF1 (MPa)	Maximum deformation of SW2 (mm)	Maximum tensile stress on AF2 (MPa)
Natural Earth-quake Wave I	3.16	0.63	2.72	0.56
Natural Earth-quake Wave II	3.65	0.68	3.34	0.68
Artificial Earth-quake Wave	3.24	0.58	4.70	1.00

deformation, while the arch-feet of the middle-arch were taken as the control part for tensile stress (AF2). The results under the 7-degree rare earthquake are listed in Table 2.

The results under Seismic condition 1 and Seismic condition 2 are in little difference for the two natural waves, while for the artificial earthquake wave, the results under Seismic condition 2 are obviously larger than the results under Seismic Condition 1. The phenomenon is related to the character of the artificial earthquake wave and the character of the stone arch bridge. The artificial earthquake wave can more easily excite the low-order modes of the bridge than the two natural waves, the bridge stiffness along the transversal direction is relatively weaker than that along the longitudinal direction. Thus, the integrity of the side walls of this bridge needs to be strengthened in the following conservation.

5 CONCLUSIONS

According to the analysis of the structural performance and seismic response of the Putang Bridge, some conclusions can be drawn as followings:

1. For the static load condition, the critical load condition is "dead load + uniform live load on the whole bridge". The maximum compressive stress occurs at the middle-arch's feet. The spandrel walls between two arches, and the arch-back and the arch-waist of each arch are in the risk of tensile failure. All these parts mentioned above are recommended to be strengthened in the conservation work.
2. The arch-feet are the most sensitive parts of the whole bridge under the seismic excitations, and the arch-feet of the second to the eighth arches would be damaged under the 7-degree rare earthquake. These parts are also recommended to be strengthened in the conservation work.
3. The bridge stiffness along the transversal direction is relatively weaker than that along the longitudinal direction. Thus, the integrity of the side walls of this bridge needs to be strengthened in the conservation work.

ACKNOWLEDGEMENTS

This study was supported by the National Key Research and Invention Program of the 13th Five-year Plan (Grant No. 2019YFD1100904).

REFERENCES

Bayraktar, A., Türker, T. & Altunişik, A. C. (2015) Experimental frequencies and damping ratios for historical masonry arch bridges. *Construction and Building Materials*, 75, 234–241.
Chun, Q., Tang, Y., Pan, J. & Dong, Y. (2016) Reasearch on conservation of Jin Hu Bridge, a stone atch bridge built in the Ming dynasty. *Sciences of Conservation and Archaeology*, 28, 65–72.
Conde, B., Ramos, L. F., Oliveira, D. V., Riveiro, B. & Solla, M. (2017) Structural assessment of masonry arch bridges by combination of non-destructive testing techniques and three-dimensional numerical modelling: Application to Vilanova bridge. *Engineering Structures*, 148, 621–638.
Du, Y., Liu, J., Nie, J. & Fan, J. (2005) Seismic analysis of Jin-shui stone arch bridge. *Building Structure*, 08, 43-45+51.
DeJong, M. J. et al. (2008) 'Rocking Stability of Masonry Arches in Seismic Regions', *Earthquake Spectra*, 24(4), pp. 847–865.
Fanning, P.J. and Boothby, T.E., 2001. Three-dimensional modelling and full-scale testing of stone arch bridges. *Computers & Structures*, 79(29-30), pp.2645–2662.
Liu, Y. & Chen, A. (2003) Analysis for finite spring method of stone arched bridge and its safety evaluation. *China Civil Engineering Journal*, 08, 69–73.
Ministry of Housing and Urban-Rural Development of the People's Republic of China. (2011) Code for design of masonry structures, China Architecture Publishing &Media Co., Ltd., Beijing, China.
Pelà, L., Aprile, A. & Benedetti, A. (2009) Seismic assessment of masonry arch bridges. *Engineering Structures*, 31, 1777–1788.
Qian, L. (1987) The carrying capacity of Zhaozhou stone arch bridge. *China Civil Engineering Journal*, 04, 39–48.
Shen, K. (2010) Seismic response analysis on long-span continous arch bridge. *Chang'an University*.
Yang, D. (1980) Preliminary study on structural calculation method of curved arch bridge. *Journal of Chang'an University (Natural Science Edition)*, 01, 1–11.
Yao, Q. (1996) Limited behavior of stone arch bridge. *Southwest Highway*, 02, 26–37.

Brick and Block Masonry - From Historical to Sustainable Masonry –
Kubica, Kwiecień & Bednarz (eds)
© 2020 Taylor & Francis Group, London, ISBN 978-0-367-56586-2

Performance of Persian open spandrel brick masonry barrel vaults under uniform and linear loads

M. Hejazi & Y. Soltani
Faculty of Civil Engineering and Transportation, Department of Civil Engineering, University of Isfahan, Isfahan, Iran

ABSTRACT: In this paper, the structural behaviour of Persian brick masonry barrel vaults with open spandrels under uniformly distributed load over the entire structure, linear load at the centre of the middle pier, and linear load at the apex of one of the vaults is studied using non-linear finite element analysis. Seven different shapes have been considered for the cross-section of the vaults. Results indicate that the best position for the spandrel vaults to achieve the maximum load carrying capacity is the same level as the main vaults. The semi-circular vault has the best performance under uniform load and the pointed vault has the best behaviour under linear load.

1 INTRODUCTION

The construction of adjacent barrel vaults has been a common practice in Persian architecture for many centuries. In order to neutralise the lateral thrust of barrel vaults two methods may be used to interconnect them. In the first method, the space between them is filled, which makes the construction heavy and vulnerable against the earthquake. In the second method, smaller vaults are constructed on the top of one another making hollow or open spandrels, called *konu*, in the space between the main barrel vaults and parallel to them (Figure 1). Barrel vaults with open spandrels are lighter than those with spandrel fills. In this paper, the structural performance of Persian open spandrel brick masonry barrel vaults under distributed and linear loads is studied.

2 LITERATURE REVIEW

Calderoni et al. (2011) suggested a simplified theoretical model for determination of structural behaviour of masonry spandrels subjected to shear and formulated expressions for different failure mechanisms. Amodio et al. (2019) used the discontinuity layout optimisation numerical limit analysis approach to study masonry bridges with internal and external spandrel walls. Forgács et al. (2019) used the district element method to study the effect of spandrel walls on load carrying capacity and structural stiffness of a masonry arch bridge and the failure modes of spandrels due to excessive lateral displacement of the backfill. Hokelekli & Yilmaz

(2019) used non-linear finite element analysis to determine the shear and opening conditions at the spandrels-fill and arch-fill interfaces of a masonry bridge under the earthquake. Bayraktar & Hökelekli (2020) studied transverse performance of strengthened spandrel walls of masonry arch bridges by using tapered section instead of prismatic section, using transverse tie bars and covering with fabric reinforced cementitious matrix composite. No investigation on structural behaviour of Persian open spandrel brick masonry barrel vaults have been done so far.

3 STUDIED VAULTS, METHOD, LOADING AND MATERIALS

The geometry of vaults studied is shown in Figure 1b. The structure consists of two barrel vaults, three piers and three spandrel vaults. Barrel vaults have seven difference shapes of semi-circular, drop, ordinary and raised pointed, and drop, ordinary and raised four-centred. Spandrel vaults have a semi-circular cross-section. Studied parameters are the vault span (D) of 4.5 m, three pier width to vault span ratios (b/D) of 0.4, 0.6 and 0.8, three vault width to vault span ratios (W/D) of 0.75, 1 and 1.25, three spandrel vault rise to vault span ratios (Z/D) of -0.15, 0 and 0.15, three pier height to vault span ratios (h_1/D) of 0.75, 1, 1.25, three thickness over the vault to vault span ratios (h_3/D) of 0.25, 0.33 and 0.5, five thickness of the main vault to vault half span ratios (t_0/R) of 0.07, 0.1, 0.13, 0.15 and 0.18, and five thickness of the spandrel vault to vault half span ratios (t_1/R) of 0.07, 0.1, 0.13, 0.15 and 0.18.

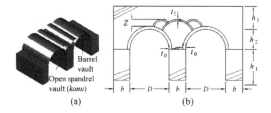

Figure 1. (a) Barrel vault with open (hollow) spandrels, b) studied parameters.

Three-dimensional non-linear finite element analysis was performed by the ANSYS (2017) software. The four-node tetrahedral and eight-node hexahedral SOLID 65 elements, with three translational degrees of freedom at each node, which incorporate the Willam-Warnke failure criterion (Willam & Warnke 1957) capable of modelling brittle materials such as brick masonry, were used. The boundary conditions were such that the displacements of the bottom of the piers were constrained. A previous study (Soltani 2018) proved that if the lateral displacements of external faces of side piers from half to three-fourth of their height are constrained it creates the maximum load carrying capacity. Therefore, this was also used as additional boundary conditions.

Three loading cases, i.e. uniformly distributed load over the entire structure, linear load at the centre of the middle pier (Figure 2a), and linear load at the apex of one of the vaults (Figure 2b) are studied.

The vaults are made of traditional clay brick and clay-gypsum mortar. The mechanical properties and stress-strain diagram (Figure 2c) of gypsum mortar brick masonry previously determined by Hejazi et al. (2014, 2016) were used in modelling. The compressive strength, tensile strength, modulus of elasticity, Poisson's ratio and bulk density were respectively f_c=2.73 MPa, f_t=0.27 MPa, E=2730 MPa, ν=0.17, and ρ=1530 kg/m³. Willam-Warnke failure criterion parameters corresponding to obtained mechanical materials to be used in the software were calculated by relevant equations (Chen & Han 2007). Shear transfer coefficient across the open crack, shear transfer coefficient across the closed crack, uniaxial tensile strength,

uniaxial compressive strength, biaxial compressive strength, hydrostatic pressure, biaxial compressive strength for the case of hydrostatic pressure, uniaxial compressive strength for the case of hydrostatic pressure, and stiffness multiplier for cracked tensile condition were calculated as $\beta_t = 0.15$, $\beta_c = 0.75$, $f_t = 0.27$ MPa, $f_c = 4.71$ MPa, $f_{cb} = 1.2f_c = 3.276$ MPa, $|\sigma_h^a| \leq \sqrt{3}f_c = 4.73$ MPa, $f_1 = 1.45f_c = 3.96$ MPa, $f_2 = 1.725f_c = 10$ MPa, and TCF =0.6, respectively.

4 SEMI-CIRCULAR VAULT

4.1 Results for t_0/R and t_1/R

4.1.1 Position of the spandrel vaults at Z/D=0.15
The values of the thickness of the main vaults (t_0) and spandrel vaults (t_1) for the maximum load carrying capacity under uniformly distributed load over the entire structure are obtained in this section. The half span R=2.25 m is constant and the thickness is constant from the bottom to the apex. The ratio t_0/R has been taken 0.07, 0.1, 0.13, 0.15 and 0.18, respectively, while keeping t_1/R=0.07 constant. Then, t_1/R has been taken 0.07, 0.1, 0.13, 0.15 and 0.18, while keeping t_0/R=0.07 constant. Finally, both t_0/R and t_1/R have been simultaneously increased from 0.07 to 0.18 to reach the maximum load carrying capacity. Figure 3 shows the failure uniform load versus thickness to half span ratio. The most appropriate values for the ratio is according to Equation (1).

$$0.13 \leq t/R \leq 0.16, \quad t = t_0 \text{ or } t_1 \qquad (1)$$

Figure 4 shows the structure at failure. The white spots shown in Figure 4b are related to the cracks in the structure. Cracks have grown in the area where the spandrel vaults exist and at side supports. The maximum tensile and compressive stresses are respectively 63% and 96.34% of relevant strengths. The maximum tensile stress occurs at the spandrel

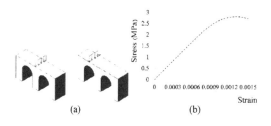

Figure 2. A) Linear load at the centre of the middle pier, b) linear load at the apex of a barrel vault, c) stress-strain diagram of gypsum mortar brick masonry.

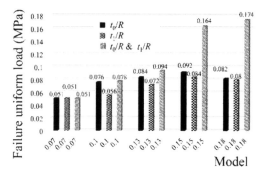

Figure 3. Results for Z/D=0.15 under uniformly distributed load for the semi-circular vault.

802

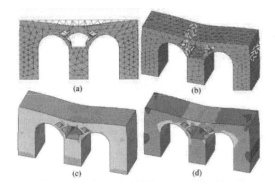

Figure 4. Results for Z/D=0.15 under uniformly distributed load for the semi-circular vault: a) vertical displacement, b) expansion of crack, c) distribution of the first principal stress, d) distribution of the third principal stress.

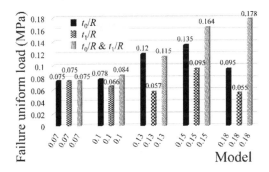

Figure 5. Results for Z/D=0 under uniformly distributed load for the semi-circular vault.

vaults and the maximum compressive stress occurs at the area where the spandrel vaults sit on the vaults.

4.1.2 Position of the spandrel vaults at Z/D=0

When the top level of the spandrel vaults is the same as the top level of main vaults, i.e. Z/D=0, the largest load bearing capacity is obtained for the thickness to half span ratio of 0.18 (Figure 5). The suggested range to obtain the largest values of failure uniformly distributed load when the two ratios change independently is as Equation (2).

$$0.14 \leq t/R \leq 0.16, t = t_0 \text{ or } t_1 \qquad (2)$$

4.1.3 Position of the spandrel vaults at Z/D=-0.15

Figure 6 Illustrates the variation of failure uniformly distributed load with the thickness of spandrel vault/ main vault to vault half span for the case Z/D=-0.15, where spandrel vaults are constructed lower than the main vaults. The best ratio is 0.13. The suitable

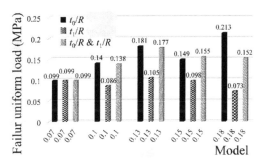

Figure 6. Results for Z/D=-0.15 under uniformly distributed load for the semi-circular vault.

range for independent variation of the two ratios for this case is as Equation (3).

$$0.12 \leq t/R \leq 0.14, t = t_0 \text{ or } t_1 \qquad (3)$$

4.2 Results for Z/D and b/D

Obtained results for the semi-circular vault for three ratios Z/D and b/D under three loading cases are as follows. Models are described in Table 1.

4.2.1 Uniformly distributed load over the entire structure

According to Figure 7, when the spandrel vaults are constructed above the main vault (Z/D=0.15) or below the main vault (Z/D=-0.15), the pier width to vault span ratio of b/D=0.6 gives the maximum load bearing capacity. When the spandrel vaults are built at the same level as the main vault (Z/D=0), the best ratio is b/D=0.4.

4.2.2 Linear load at the centre of the middle pier

In this case, the maximum value of b/D ratio, i.e. b/D=0.8, always gives the maximum load capacity.

Table 1. The codes of models for different values of Z/D.

Model	Positioning of spandrel vaults and loading
US	Z/D=0.15, Uniformly distributed load
NS	Z/D=0, Uniformly distributed load
DS	Z/D=-0.15, Uniformly distributed load
USL	Z/D=0.15, Linear load at the centre of the middle pier
NSL	Z/D=0, Linear load at the centre of the middle pier
DSL	Z/D=-0.15, Linear load at the centre of the middle pier
UNSL	Z/D=0.15, Linear load at the apex of one of the vaults
NNSL	Z/D=0, Linear load at the apex of one of the vaults
DNSL	Z/D=-0.15, Linear load at the apex of one of the vaults

803

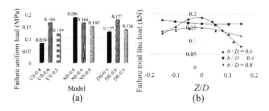

Figure 7. Failure load for the three ratios of b/D=0.4,0.6 and 0.8 under uniformly distributed for the semi-circular vault based on: a) the model, b) Z/D ratio.

In other words, for linear load at the centre of the middle pier, a wider pier provides mode load bearing capacity (Figure 8).

4.2.3 Linear load at the apex of one of the vaults

In this case, for spandrel vaults above, at the same level, and below the main vault b/D ratios of 0.8, 0.4 and 0.6 create largest load bearing capacities, respectively (Figure 9).

4.3 Results for W/D, h_1/D and h_3/D

Figure 10a shows the effect of vault width to vault span ratio (W/D) on load carrying capacity for three load cases. For uniformly distributed load on the entire structure and linear load at the centre of the middle pier, a wider vault creates a more load capacity. But when the linear load is applied to the apex of one the vaults it does not significantly increase the load carrying capacity of the structure. In other words, if the location of linear load is far from the

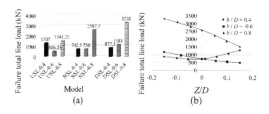

Figure 8. Failure load for the three ratios of b/D=0.4,0.6 and 0.8 under linear load at the centre of the middle pier for the semi-circular vault based on: a) the model, b) Z/D ratio.

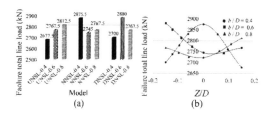

Figure 9. Failure load for the three ratios of b/D=0.4,0.6 and 0.8 under linear load at the apex of one of the vaults for the semi-circular vault based on: a) the model, b) Z/D ratio.

spandrel vaults, increasing the vault width does not have a significant effect on the load carrying capacity. According to Figure 10b the effect of the change of pier height to vault span ratio (h_1/D) on load carrying capacity does not follow a particular trend. For the thickness over the main vault to vault span ratio (h_3/D), in the case of applying linear load at the centre of the middle pier, the more this ratio is, the higher load carrying capacity of the structure is. But when the linear load is applied to the apex of one of the vaults, the change of this ratio does not significantly increase the load carrying capacity of the structure. In the case of applying uniformly distributed load over the entire structure, the ratio of h_3/D=0.33 is the most appropriate one. In other words, when the linear load is applied to the part of structure that is the location of spandrel vaults, the higher this ratio, the greater load carrying capacity of the structure. But when the linear load is applied to the part of structure that is not the location of spandrel vaults, changing the ratio does not affect remarkably the load carrying capacity of the structure (Figure 10c).

5 POINTED VAULT

5.1 Ordinary pointed vault

Obtained results for the ordinary pointed vault for three ratios of the spandrel vault rise to main vault span (Z/D=-0.15, 0 and 0.15) and pier width to main vault span (b/D=0.4, 0.6 and 0.8) under three loading cases are as follows.

5.1.1 Uniformly distributed load over the entire structure

It can be seen from Figure 11 that for all three ratios of b/D, when spandrel vaults are at the same level as the main vaults of the structure Z/D=0, the load

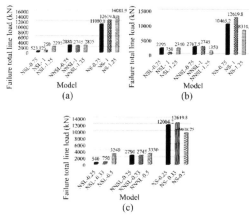

Figure 10. Failure load for the semi-circular vault for the three ratios of: a) W/D=0.75, 1 and 1.25, b) $h1/D$=0.75, 1 and 1.25, c) $h3/D$ =0.25, 0.33 and 0.5.

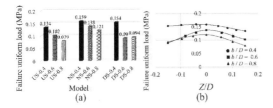

Figure 11. Failure load for the three ratios of b/D=0.4,0.6 and 0.8 under uniformly distributed load for the ordinary pointed vault based on: a) the model, b) Z/D ratio.

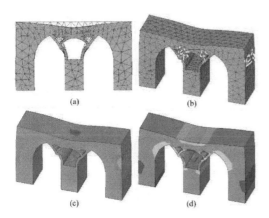

Figure 12. Results for Z/D=0 under uniformly distributed load for the ordinary pointed vault: a) vertical displacement, b) expansion of crack, c) distribution of the first principal stress, d) distribution of the third principal stress.

carrying capacity of the structure is more than the other two cases. On the other hand, for the three ratios of b/D, it can be said that the lower this ratio, the greater the load carrying capacity of the structure. The best ratio is then b/D=0.4.

The most appropriate range of Z/D is according to Equation (4):

$$-0.05 \leq Z/D \leq 0.05 \qquad (4)$$

Figure 12 illustrate the vertical displacement, crack distribution, and principal stress distribution in the structure at failure. The maximum tensile and compressive stresses are respectively 67% and 71% of tensile and compressive strengths. The maximum tensile stress occurs in the spandrel vaults and the maximum compressive stress occurs in the area where the spandrel vaults sit on the vaults.

5.1.2 Linear load at the centre of the middle pier
In this case of loading, spandrel vaults at the same level as vaults of structure, is the most suitable arrangement in comparison with other two conditions and also it has the highest load carrying

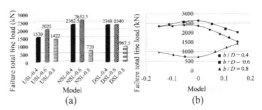

Figure 13. Failure load for the three ratios of b/D=0.4,0.6 and 0.8 under linear load at the centre of the middle pier for the ordinary pointed vault based on: a) the model, b) Z/D ratio.

capacity. In addition, according to Figure 13, it is seen that the ratio of b/D=0.6 is the best ratio. The most appropriate range of Z/D is as Equation (5):

$$-0.05 \leq Z/D \leq 0.05 \qquad (5)$$

5.1.3 Linear load at the apex of one of the vaults
Analysis results in this section show that in this case of linear loading similar to the other two cases of loading, spandrel vaults at the same level as vaults of structure and the ratio of b/D=0.6 are the most suitable conditions and have the most load carrying capacity (Figure 14). The suggested range for this loading condition is according to Equation (6):

$$-0.05 \leq Z/D \leq 0.05 \qquad (6)$$

5.2 Drop pointed vault
In two loading cases of uniformly distributed load over the entire structure and linear load at the centre of the middle pier, the positioning of spandrel vaults as low as 0.15D below the main vaults of the structure, in comparison with the other two positions is the most appropriate one (Figure 15). For linear load at the apex of one of the vaults, the spandrel vaults position does not greatly affect the load carrying capacity of the structure.

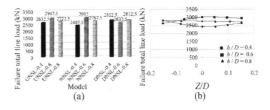

Figure 14. Failure load for the three ratios of b/D =0.4, 0.6 and 0.8 under linear load at the apex of one of the vaults for the ordinary pointed vault based on: a) the model, b) Z/D ratio.

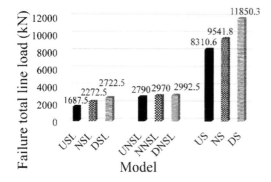

Figure 15. Failure load for the three ratios of Z/D under three cases of loading for the drop pointed vault.

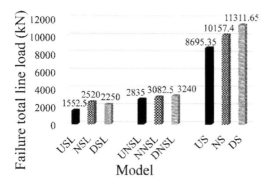

Figure 16. Failure load for the three ratios of Z/D under three cases of loading for the raised pointed vault.

5.3 Raised pointed vault

Figure 16 indicates that for the case of linear load at the centre of the middle pier, the positioning of spandrel vaults at the same level as structure main vaults and in the case of applying uniformly distributed load, the positioning of spandrel vaults 0.15D below the main vaults, are the best suited to position the spandrel vaults. For linear load applied at the apex of one of the vaults, the effect of the position of spandrel vaults on the load carrying capacity is not important.

6 FOUR-CENTRED VAULTS

6.1 Ordinary four-centred vault

6.1.1 Uniformly distributed load over the entire structure

As shown in Figure 17, for the locations of spandrel vaults 0.15D above and 0.15D below the main vaults, b/D=0.6 is the best ratio for maximum load carrying capacity. For the location of spandrel vaults at the same level as the main vaults, b/D=0.8 is the

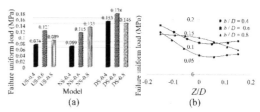

Figure 17. Failure load for the three ratios of b/D=0.4, 0.6 and 0.8 under uniformly distributed load for the ordinary four-centred vault based on: a) the model, b) Z/D ratio.

best ratio. The most appropriate range of Z/D for all three ratios of b/D is according to Equation (7):

$$-0.15 \leq Z/D \leq -0.05 \qquad (7)$$

Figure 18 shows the displacements and principal stresses in the structure at failure. The maximum tensile stresses and compressive stresses are 98% and 96.71% of corresponding strengths, respectively. The maximum tensile stress occurs under the vaults and the maximum compressive stress occurs in the area where the spandrel vaults sit on the vaults and at the place of applying load.

6.1.2 Linear load at the centre of the middle pier

For spandrel vaults location 0.15D above the main vaults the ratio of b/D=0.8, and for spandrel vaults location 0.15D below and at the same level as the main vaults the ratio of b/D=0.6 are the best ratios (Figure 19). The best range of Z/D is according to Equation (8):

$$0.05 \leq Z/D \leq 0.15 \qquad (8)$$

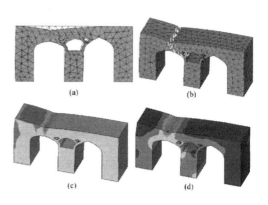

Figure 18. Results for Z/D=0 under uniformly distributed load for the ordinary four-centred vault: a) vertical displacement, b) occurred cracks, c) first principal stress, d) third principal stress.

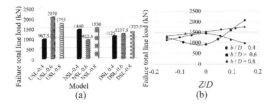

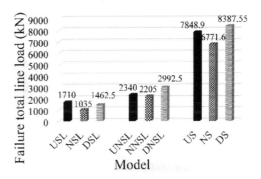

Figure 19. Failure load for the three ratios of b/D=0.4, 0.6 and 0.8 under linear load at the centre of the middle pier for the ordinary four-centred vault based on: a) the model, b) Z/D ratio.

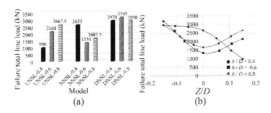

Figure 21. Failure load for the three ratios of Z/D under three cases of loading for the drop four-centred vault.

Figure 20. Failure load for the three ratios of b/D=0.4, 0.6 and 0.8 under linear load at the apex of one of the vaults for the ordinary four-centred vault based on: a) the model, b) Z/D ratio.

ratio of b/D=0.8, and for spandrel vaults 0.15D below the main vaults the ratio of b/D=0.6 are the most suitable ratios.

6.1.3 Linear load at the apex of one of the vaults

Figure 20 illustrates that for the location of spandrel vaults at the same level as the main vaults the ratio of b/D=0.4, for the location of spandrel vaults 0.15D above the main vaults the ratio of b/D=0.8, and for the location of spandrel vaults 0.15D below the main vaults the ratio of b/D=0.6 are the most suitable ratios. The most appropriate range of Z/D is according to Equation (9):

$$-0.15 \leq Z/D \leq -0.05 \qquad (9)$$

In the ordinary four-centred vault, for uniformly distributed load over the entire structure and linear load at the apex of one of the vaults, the positioning of spandrel vaults 0.15D below the main vaults of the structure, and for linear load at the centre of the middle pier, the positioning of spandrel vaults 0.15D above the main vaults are the most appropriate ones. For positioning the spandrel vaults below and above the main vaults, the ratio of b/D=0.6, and for the positioning the spandrel vaults at the same level as the main vaults the ratio of b/D=0.8 are the most suitable ratios. For linear load at the centre of the middle pier for spandrel vaults 0.15D above the main vaults the ratio of b/D=0.6, for spandrel vaults at the same level as the main vaults the ratio of b/D=0.4, and for spandrel vaults 0.15D below the main vaults the ratio of b/D=0.6; for linear load at the apex of one of the vaults, spandrel vaults at the same level as the main vaults the ratio of b/D=0.4, for spandrel vaults 0.15D above the main vaults the

6.2 Drop four-centred vault

For uniformly distributed load over the entire structure and linear load at the apex of one of the vaults the positioning of spandrel vaults 0.15D below the main vaults and for linear load at the centre of the middle pier the positioning of spandrel vaults 0.15D above the main vaults gives maximum load carrying capacity (Figure 21).

6.3 Raised four-centred vault

In the raised four-centred vault, for linear load at the centre of the middle pier and uniformly distributed load the positioning of spandrel vaults at the same level as the main vaults, for linear load at the apex of one of the vaults, the positioning of spandrel vaults 0.15D beyond the vaults are the best positions for load carrying capacity (Figure 22).

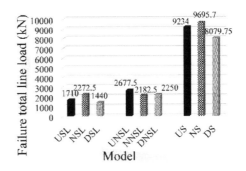

Figure 22. Failure load for the three ratios of Z/D under three cases of loading for the raised four-centred vault.

7 COMPARISION OF SEVEN TYPES OF VAULTS

Compared results for seven types of vaults for three Z/D ratios under three loading cases are presented in Table 2 and Figures 23-25. The following results are obtained.

Table 2. Comparison of seven types of vaults for load carrying capacity.

Vault	Z/D	Uniformly distributed loaded (MPa)	Linear load at the centre of the middle pier (kN)	Linear load at the apex of one of vaults (kN)
Semi-circular	0.15	0.164	506	2767
	0	0.164	750	2745
	-0.15	0.177	1181	2880
Drop pointed	0.15	0.108	1687	2790
	0	0.124	2272	2970
	-0.15	0.154	2722	2992
Ordinary pointed	0.15	0.102	2025	2945
	0	0.138	2632	3015
	-0.15	0.09	2340	2632
Raised pointed	0.15	0.113	1552	2835
	0	0.132	2520	3082
	-0.15	0.147	2250	3240
Drop four-centred vault	0.15	0.102	1710	2340
	0	0.088	1035	2205
	-0.15	0.109	1462	2992
Ordinary four-centred vault	0.15	0.121	2070	2160
	0	0.115	922	1350
	-0.15	0.178	1237	3195
Raised four-centred vault	0.15	0.12	1710	2677
	0	0.126	2272	2182
	-0.15	0.105	1440	2250

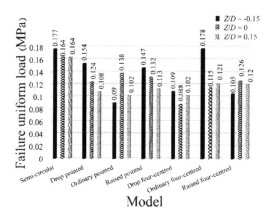

Figure 23. Failure load for seven types of vaults under uniformly distributed load for different Z/D values.

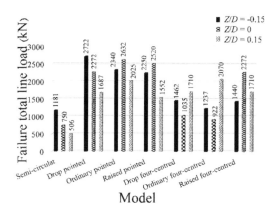

Figure 24. Failure load for seven types of vaults under linear load at the centre of the middle pier for different Z/D values.

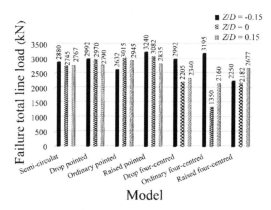

Figure 25. Failure load for seven types of vaults under linear load at the apex of one of the vaults for different Z/D values.

1. In the case of uniformly distributed load over the entire structure, the semi-circular vault shows more load carrying capacity than other vaults.
2. In the case of linear load at the centre of the middle pier, for two locations of spandrel vaults at the same level as the main vaults and 0.15D below the main vaults the pointed vaults, and vaults of the structure) the pointed vaults, and for the location of spandrel vaults 0.15D above the main vaults the four-centred vaults more load carrying capacity.
3. In the case of linear load at the apex of one of the vaults, the pointed vaults have a better load carrying capacity.

8 CONCLUSIONS

Persian brick masonry barrel vaults with hollow spandrels with seven different cross-sectional shapes of semi-circular, drop, ordinary and raised pointed, and drop, ordinary and raised four-centred were studied using the finite element method. They were studied subject to uniformly distributed load over the entire structure, linear load at the centre of the middle pier, and linear load at the apex of one of the vaults. Obtained results indicated that placing the spandrel vaults at the same level of the main results in the maximum load carrying capacity in most cases. The maximum load carrying capacity under uniform load belongs to the semi-circular vault, that of linear load at the centre of the middle pier of linear load at the apex of one the vaults belongs to the pointed vault. Cracks occur near the spandrel vaults. For spandrel vaults $0.15D$ above, $0.15D$ below and at the same level of the barrel vaults, the best thickness to half span (t/R) ratios are 0.15, 0.13 and 0.15, respectively. For the semi-circular barrel vault, the best pier width to barrel vault span ratio is 0.6, which is the dominant ratio in most real structures. For uniformly distributed load over the entire structure and for linear load at the centre of the middle pier, by increasing the barrel vault width to barrel vault span (W/D) ratio the load carrying capacity increases, but for linear load at the apex of one of the vaults changing the ratio does not change load carrying capacity significantly. The pier height to the barrel vault (h_1/D) ratio has a little effect on the load carrying capacity. For uniform load, the best thickness over the barrel vault to barrel vault span (h_3/D) ratio is 0.33. By increasing this ratio, the load carrying capacity increases when the load is near spandrel vaults, whereas changing this ratio does not have not a significant effect on load carrying capacity when the load is far from spandrel vaults.

REFERENCES

Amodio, S., Gilbert, M., & Smith, C. (2019, October). Modelling masonry arch bridges containing internal spandrel walls. In International Conference on Arch Bridges (pp. 315–322). Springer, Cham.

Basic analysis guide for ANSYS 18. (2017). New York: SAS IP Inc.

Bayraktar, A., & Hökelekli, E. (2020). Seismic performances of different spandrel wall strengthening techniques in masonry arch bridges. International Journal of Architectural Heritage 1–19.

Calderoni, B., Cordasco, E. A., Lenza, P., & Pacella, G. (2011). A simplified theoretical model for the evaluation of structural behaviour of masonry spandrels. International Journal of Materials and Structural Integrity 5 (2-3): 192–214.

Chen, W. F., & Han, D. J. (2007). Plasticity for structural engineers. New York: J. Ross Publishing.

Forgács, T., Rendes, S., Ádány, S., & Sarhosis, V. (2019, October). Mechanical role of spandrel walls on the capacity of masonry arch bridges. In International Conference on Arch Bridges (pp. 221–229). Springer, Cham.

Hejazi, M., & Mehdizadeh Saradj, F. (2014). Persian architectural heritage: structure. Southampton and Boston: WIT Press.

Hejazi, M., Moayedian, S. M., & Daei, M. (2016). Structural analysis of Persian historical brick masonry minarets. Journal of Performance of Constructed Facilities 30(2): 04015009.

Hokelekli, E., & Yilmaz, B. N. (2019). Effect of cohesive contact of backfill with arch and spandrel walls of a historical masonry arch bridge on seismic response. Periodica Polytechnica Civil Engineering 63(3): 926–937.

Soltani, Y. (2018). Parametric study of the effect of spandrel fill (kunal) and hollow spandrel (konu) on structural behaviour of Persian brick masonry barrel vaults. M.Sc. Thesis, Department of Civil Engineering, University of Isfahan, Isfahan.

Willam, K.J. & Warnke, E.P. (1957). Constitutive models behaviour of concrete. In International Association for Bridge and Structural Engineering 19:1–30.

Masonry testing

Brick and Block Masonry - From Historical to Sustainable Masonry –
Kubica, Kwiecień & Bednarz (eds)
© 2020 Taylor & Francis Group, London, ISBN 978-0-367-56586-2

Shear response and failure mode of masonry triplets subjected to monotonic and cyclic loading

S. Barattucci
Department of Civil, Chemical, Environmental and Materials Engineering, University of Bologna, Italy

V. Sarhosis
School of Civil Engineering, Leeds University, UK

A.W. Bruno
School of Engineering, Newcastle University, UK

A.M. D'Altri, S. de Miranda & G. Castellazzi
Department of Civil, Chemical, Environmental and Materials Engineering, University of Bologna, Italy

G. Milani
Department of Architecture, Built Environment and Construction Engineering (A.B.C.), Politecnico di Milano, Italy

ABSTRACT: This paper investigates the shear response and failure mode of masonry triplets subjected to monotonic and cyclic shear loadings. In this experimental campaign, different masonry triplets using different mortar compositions have been constructed and subjected to different levels of pre-compression. The cohesion and internal friction angle were derived assuming a Mohr-Coulomb criterion and using a linear regression equation. The influence of the strength of the mortar under monotonic and cyclic conditions was investigated. From the results analysis, it was shown that the mortar composition and pre-compression load strongly affect the shear strength of the masonry triplets. Also, triplets subjected to cyclic shear loading sustained a lower peak strength when compared to triplets subjected to monotonic loading.

1 INTRODUCTION

Masonry is one of the oldest materials which is still widely spread given its ease of construction, low cost and durability (Hendry 1998). Masonry is composed of masonry units (e.g. fired earth bricks, concrete blocks, stones, etc.) that are either bonded together by means of mortar joints or dry assembled without any mortar. Considering this peculiar composition, masonry is a heterogenous material with a strongly non-linear and anisotropic mechanical behaviour. Masonry constructions exhibit an excellent mechanical behaviour under compression while only a limited performance is generally achieved under shearing stress states (Adami et al., 2008). As consequence of this mechanical behaviour, masonry walls are particularly sensitive to horizontal actions (e.g. earthquakes) that can lead to formation of cracks under serviceability conditions, or eventually failure of the structural elements under ultimate limit states.

Shear failure of masonry occurs at: a) the mortar; b) the masonry unit to mortar interface; and/or c) both above (Alecci et al., 2013; Abdou et al. 2006).

Hence, the mortar composition, the properties of the interface brick-mortar as well as the applied precompression stress strongly affect the shear resistance of masonry elements (Rahman et al., 2013).

Past experimental studies have investigated the shear behaviour of masonry elements by means of the triplet shear tests. According to the method, masonry triplet specimens (i.e. a specimen made of three bricks bonded by two interposed mortar layers) are subjected to a certain precompression stress applied on the two external blocks and to a perpendicular shear load simultaneously applied on the intermediate block. Most of these studies solely focused on the monotonic shear behaviour of masonry specimens, i.e. by statically applying shear loading along a unique direction as indicated by the European Standard EN1052-3 (2002).

To fill this gap in literature, this paper presents results from shear tests conducted on masonry triplets subjected to both monotonic and cyclic shear loadings. Masonry triplets were built by using three different mortar compositions with a varying cement to sand ratio were used for the construction of the triplets. Results show that the cyclic loading reduces the peak shear strength of masonry specimens.

2 MATERIALS AND METHODS

2.1 Manufacturing of masonry triplets

Masonry triplets were built by bonding together individual solid bricks with cement-based mortars mixed at the three cement-sand proportions of 1:3, 1:6 and 1:9. These three mortars mixes are respectively named hereafter as 1:3 MC, 1:6 MC and 1:9 MC, i.e. by referring to the sole cement to sand ratio. To investigate the mechanical behaviour of the different mortar mixes, three specimens were casted for each mortar composition in standard moulds of 40 x 40 x 160 mm³. Mortar specimens were left for curing at a constant temperature of 25 °C for 28 days and then tested under three-point bending test. The two parts of the specimen were subsequently collected and tested under unconfined compression tests.

Results from these tests show that both the peak compressive strength and the flexural strength increase with increasing the cement to sand ratio (Figures 1 and 2). As expected, both compressive and flexural strength occurred at higher strain levels as the cement content increased. After the peak strength was attained, all specimens exhibited a brittle failure.

The three mortar compositions 1:3 MC, 1:6 MC and 1:9 MC were used to bond together solid fired

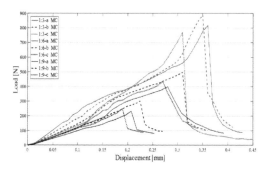

Figure 1. Results from three point-bending tests on specimens of the three mortars 1:3 MC, 1:6 MC and 1:9 MC.

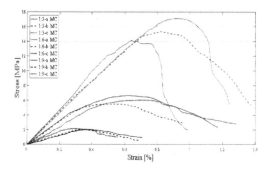

Figure 2. Results from compressive strength tests on specimens of the three mortars 1:3 MC, 1:6 MC and 1:9 MC.

bricks with dimensions 215 x 102.5 x 65 mm³ (commercial name: "Birtley Old English Bricks" produced by J T Dove). A mechanical characterization of the individual bricks was considered outside the scope of the present study as it is expected that, during the shear tests, failure will only occur within the mortar layers.

Before the manufacturing of triplets, individual bricks were submerged in water for a minimum time of 24 hours to ensure a good bonding with the mortar layers and avoid a quick drying during the curing period with a consequent formation of cracks. After submersion in water, bricks were wiped with a dry cloth before the laying the mortar. For each triplet speciment, three solid bricks were bonded together by two interposing mortar layers with a thickness of 10 mm each. Special care was taken to manufacture masonry triplets as straight as possible in order to avoid eccentricities that could compromise the good quality of the experimental results. A total of thirty-three masonry triplets were built with eleven specimens for each mortar composition.

After manufacturing, all masonry triplets were left to cure for a minimum time of 28 days at a room temperature of 25 °C (Figure 3). After curing, masonry triplets were subjected to shear tests performed under both monotonic and cyclic conditions. The testing set-up specifically designed to carry out the shear tests and results from this experimental campaign are presented in the following sections.

2.2 Testing set-up

The standard testing device defined by the European Standard EN 1052-3 (2002) to perform shear tests on masonry specimens has been modified in the present study to apply shear loadings under both monotonic and cyclic conditions.

The bespoke testing set-up designed for the present experimental campaign consists of two independent systems of threaded rods and plates that hold in position the two external bricks of the triplet specimens. In parallel, a similar system was mounted on the intermediate brick and then connected to the upper piston of the hydraulic press. During testing, this latter system applies the shearing loading by imposing

Figure 3. Masonry triplets manufactured with solid bricks bonded by the three mortars 1:3 MC, 1:6 MC and 1:9 MC.

vertical displacement on the intermediate brick. Only downward displacements were imposed under monotonic conditions while upward and downward displacement were imposed during cyclic tests.

A separate frame composed of two rods and two tubular profiles was then used to support a hydraulic pressure jack. This latter system is used to apply the three horizontal precompression pressures of 0.2, 0.6 and 1.0 MPa. This precompression was applied before the shear loading and it was maintained constant during the whole duration of the test. Figure 4 shows the schematic of the testing set-up used for the experimental campaign together with a list of all components.

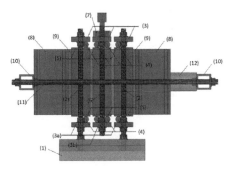

(1) *aluminium base*
(2) *threaded rods*
(3) *thick steel plates*
(4) *cylindrical rods*
(5) *thin plates*
(6) *threaded rods*
(7) *centering device*
(8) *IPE profiles*
(9) *plywood boards*
(10) *steel tubular profiles*
(11) *threaded rods*
(12) *pressure jack*

(a)

(b)

Figure 4. Equipment used to perform monotonic and cyclic tests on masonry triplets: schematic (a) and assembled (b) set-up.

The testing equipment described above was used to perform both monotonic and cyclic shear tests on masonry triplets manufactured with the three mortars 1:3 MC, 1:6 MC and 1:9 MC. Three different levels of precompression pressure (i.e. 0.2, 0.6 and 1.0 MPa) were applied for each set of masonry triplets built with the three mortar mixes. Results from these tests are presented and discussed in the following section.

3 RESULTS

3.1 *Monotonic shear tests*

The monotonic tests were performed by imposing a vertical downward displacement to the intermediate brick at a constant rate of 1 mm/minute. Monotonic shear tests were performed on masonry triplets made of solid bricks bonded by the mortars 1:3 MC, 1:6 MC and 1:9 MC, see Figures 5, 6 and 7, respectively. For each mortar composition, three different specimens were tested at the three precompression levels of 0.2, 0.6 and 1.0 MPa (i.e. one triplet for each precompression level). Inspection of Figures 5, 6 and 7 suggests that the peak strength increases as the precompression stress increases. Additionally, the masonry triplets built with the two mortars 1:3 MC and 1:6 MC exhibit similar level of both strength and stiffness while the triplets bonded with the mortar 1:9 MC (Figure 8) consistently exhibited the weakest mechanical performance. Also, many specimens showed a pre-peak drop of resistance corresponding to the formation of the first crack within one of the two mortar layers. After this first drop of resistance, the applied load increased until the ultimate peak strength was achieved. This behaviour suggests that the masonry elements are capable of redistributing the applied stress during shear loading. This capacity is however more evident in bricks built with the mortars 1:3 MC and 1:6 MC rather than in the triplets bonded with the 1:9 MC mortar. Interestingly, the peak strength occurred at progressively larger displacement as the cement to sand ratio increased, which is in agreement with results of both three-point bending test and unconfined compression tests performed on mortar specimens.

Figures 5, 6 and 7 also show that all masonry triplets exhibit a significant post-peak residual strength. It can be noted that the residual strength is quite similar for the three types of mortar tested in this work. This suggests that the residual strength of masonry elements essentially depends on the applied precompression stress rather than on the specific composition of the mortar joints.

The peak shear strength of all masonry triplets is now calculated as the ratio between the peak shear load and the area of the two vertical cross-sections passing through the mortar joints. The peak shear

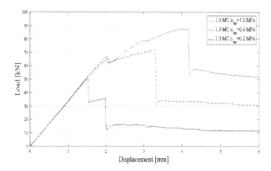

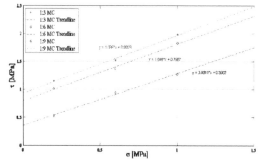

Figure 5. Results from monotonic shear tests on masonry triplets bonded with 1:3 MC mortar.

Figure 8. Relationship between shear strength and precompression pressure of masonry triplets built with 1:3 MC, 1:6 MC and 1:9 MC mortars.

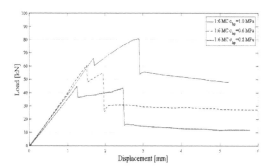

Figure 6. Results from monotonic shear tests on masonry triplets bonded with 1:6 MC mortar.

Table 1. Peak and residual values of cohesion and friction angle for all masonry triplets.

Mortar	c' [MPa]	φ' [°]	c'_{res} [MPa]	φ'_{res} [°]
1:3	0.932	46.1	0.021	50.3
1:6	0.796	45.5	0.038	46.3
1:9	0.356	43.0	0.086	39.4

mortar joints at the low precompression pressure of 0.2 MPa. At higher precompression stress, failure occurred as consequence of cracks formed within the mortar joints.

The next section presents the effect of cyclic loading on the shear behaviour of masonry triplets together with a comparison between the monotonic and cyclic shear behaviour of the tested masonry specimens.

3.2 Cyclic shear tests

Cyclic shear tests were performed by imposing alternatively downward and upward displacements to the intermediate brick over ten subsequent cycles. Displacements were imposed at a constant rate of 1 mm/minute and increased at each cycle by 10% of the displacement that corresponded to the peak strength measured during the monotonic tests. The final cycle was then extended until failure occurred. This procedure was arbitrarily selected to perform a few initial cycles on the range of low stress with an elastic response and then induce progressive damages in the mortar joints over the final cycles.

For the sake of brevity, results from the cyclic shear tests performed on masonry specimens built with the mortar 1:6 MC (i.e. intermediate value of cement content) at 0.2, 0.6 and 1.0 MPa precompression levels are presented in Figures 9a, 9b and 9c, respectively. The conclusions drawn from these tests can also be extended to the specimens constructed with the other two mortars (i.e. 1:3 MC and 1:9 MC). For ease of comparison, Figures 9a, 9b and 9c show results from the monotonic tests by

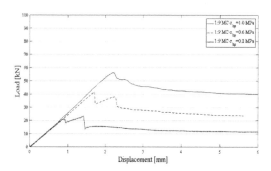

Figure 7. Results from monotonic shear tests on masonry triplets bonded with 1:9 MC mortar.

strength is then plotted against the precompression stress and results are interpolated by the Mohr-Coulomb criterion to determine both the peak cohesion c' and friction angle φ' of the mortar joints.

A similar procedure has also been applied to the residual shear strength in order to evaluate residual cohesion and friction angle. Table 1 summarises both the peak and residual values of cohesion and friction angle.

Finally, all masonry triplets exhibited shear failure within the mortar layers. More specifically, failure occurred at the interface between the bricks and the assuming a symmetric shear behaviour under vertical displacement imposed in both downward and upward directions. Note that cyclic tests could be extended to the post-peak range only if the peak strength was achieved over the last cycle. Conversely, the cyclic residual strength could not be measured if the peak strength was achieved during an intermediate cycle due to experimental limitations.

Inspection of Figures 9a, 9b and 9c indicates that masonry triplets exhibited a similar stiffness under both monotonic and cyclic conditions. On the other hand, the peak strength measured under cyclic conditions is lower than that observed during monotonic tests and this is due to the damaging effect of cyclic shear loading applied before failure.

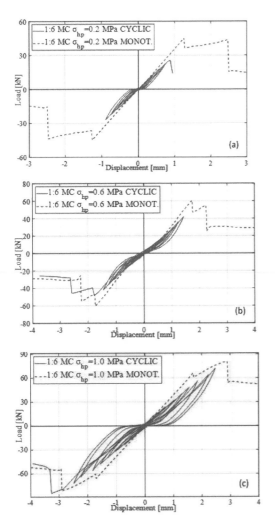

Figure 9. Results from cyclic shear tests on masonry triplets bonded with 1:6 MC mortar and at the three precompression levels of 0.2 MPa (a), 0.6 MPa (b) and 1.0 MPa (c).

Moreover, masonry triplets exhibited a smoother force-displacement relationship compared with the monotonic case without any pre-peak drop of resistance. This can be related to the fact that the cyclic shear loading has produced micro-cracks with a more uniform dissipation of mechanical energy.

The failure mechanisms observed during cyclic testing were similar to those produced by the monotonic shear loading, i.e. slippage at the interface brick-mortar at the low precompression pressure of 0.2 MPa while shear failure of the mortar joints at higher precompression stresses.

Finally, the mechanical energy dissipated during cycles can be estimated from the area underlying the force-displacement curve. Figures 9a, 9b and 9c show that the dissipated energy increased with the increase of the precompression pressures. Results obtained on the other two types of mortar (i.e. 1:3 MC and 1:9 MC) also indicates that the dissipated energy increases with the cement content and this result is again in agreement with the mechanical behaviour observed on mortar specimens (Figures 1 and 2).

4 CONCLUSIONS

This paper presented results from mechanical tests conducted on masonry triplets subjected to both monotonic and cyclic shear loadings. Masonry triplets were bonded by three different cement-based mortars composed by the three cement to sand ratios of 1:3, 1:6 and 1:9. Prior to the application of the shear loading, masonry triplets were compressed at the three precompression pressures of 0.2, 0.6 and 1.0 MPa. Results from this experimental campaign show that:

- The peak shear strength obtained from both monotonic and cyclic shear tests increased with increasing levels of cement content and precompression tests.
- The post-peak residual shear strength depended on the applied precompression stress rather than the specific mortar compositions.
- Masonry triplets exhibited a similar stiffness during both monotonic and cyclic shear tests.
- The failure mechanisms depended on the applied precompression stress: slippage at the interface brick-mortar at the precompression stress of 0.2 MPa and shear failure of the mortar joints at higher precompression stresses.
- The cyclic shear loading induced a certain degree of damages within the mortar joints and this has generally reduced the peak shear strength compared to that measured during monotonic tests.
- The dissipated energy, intended as the area underlying the force-displacement curves, increased as the cement to sand ratio and the precompression stress increased.

Finally, results obtained from this experimental campaign can be used to validate numerical models with the aim of gaining a deeper insight into the shearing behaviour of masonry buildings.

REFERENCES

European Standard EN1052-3. 2002. Methods of test for masonry-Part 3: Determination of initial shear strength EN 1052-3: 2002. European Committee for Standardization: Brussels, Belgium.

Abdou, L., Ami, S.R., Meftah, F. and Mebarki, A. 2006. Experimental investigations of the joint-mortar behaviour. *Mechanics Research Communications*, 33(3), 370–384.

Adami, C.E. & Vintzileou, E. 2008. Investigations of the bond mechanism between stones or bricks and grouts. *Materials and Structures*, 41(2), 255–267.

Alecci, V., Fagone, M., Rotunno, T., & De Stefano, M. 2013. Shear strength of brick masonry walls assembled with different types of mortar. *Construction and Building Materials*, 40, 1038–1045.

Hendry, A.W. 1998. *Structural masonry.* 2nd Edition. Palgrave Macmillan, London, UK.

Rahman, A., & Ueda, T. 2013. Experimental investigation and numerical modeling of peak shear stress of brick masonry mortar joint under compression. *Journal of Materials in Civil Engineering, 26(9)*, 04014061.

Brick and Block Masonry - From Historical to Sustainable Masonry –
Kubica, Kwiecień & Bednarz (eds)
© 2020 Taylor & Francis Group, London, ISBN 978-0-367-56586-2

Development of a flatjack for testing high strength masonry

G.D. Ogden
Atkinson-Noland & Associates, New York, USA

M.P. Schuller & D.B. Woodham
Atkinson-Noland & Associates, Boulder, USA

ABSTRACT: Flatjack methods are commonly used for in situ evaluation of masonry compression behavior, compressive stress within masonry assemblages, and for loading units during tests to evaluate mortar bed joint shear strength. The original flatjack methods considered applications for evaluating historic masonry construction with typical compressive strength in the range of 2.1 to 6.9 MPa (300 to 1000 psi). Some stone masonry construction and most modern masonry have compressive strengths greater than 6.9 MPa (1000 psi). In these cases, masonry compressive strength is estimated based on relationships between compression modulus and compressive strength. A new flatjack design has been developed for evaluating high strength solid-unit masonry with the ability to directly capture nonlinear response and determine peak compressive strength. Laboratory calibration and proof testing shows the new flatjack design can apply stresses of 31 MPa (4400 psi) or greater and has been used in the field for compression and shear testing with good results.

1 INTRODUCTION

Flatjacks are often used as loading devices for in situ measurement of three masonry properties: state of compressive stress, compression behavior, and bed joint shear strength. Using flatjacks is a viable, less destructive alternative to removing wall prisms for laboratory testing. Flatjack tests are often considered as semi-destructive due to the need for the removal of mortar joints to conduct the test. Additionally, there is the possibility of the need to reset or replace units within the test area. No other semi-destructive test method is able to provide results from a direct measurement of masonry properties without an empirical correlation (Noland, Atkinson & Schuller 1990).

Flatjacks used for testing masonry are thin bladders typically made from steel that expand when pressurized with hydraulic fluid. These bladders are inserted into the wall and pressurized to impose loads within the localized test area. The use of flatjacks follows ASTM standards and associated RILEM standards: ASTM C1196 (RILEM LUM. D.2), ASTM C1197 (RILEM LUM.D.3), and ASTM C1531 for the stress, deformability, and shear tests, respectively. The International Existing Building Code (IEBC) also includes procedures for the shear test. Flatjacks are calibrated to convert the gauge pressure to the actual pressure imposed on the masonry using a dimensionless factor, km, specific to each flatjack.

The stress test involves a single flatjack used to determine the state of compressive stress acting normal to the flatjack within a masonry assembly. The test is conducted by the release of stress through the removal of a mortar bed joint and subsequent pressurizing of the flatjack to restore the original state of masonry. Gauge points are placed on the wall above and below the test joint to determine the distance between points before and after cutting the joint and during the pressurization of the flatjacks (Figure 1). The stress in the wall is determined by returning the gauge points to their original distance before the joint was removed through the loading of the flatjack. The pressure in the flatjack at this state, modified by two correction factors, is the compressive stress in the wall.

The deformability test is used to determine the compressive response and compression modulus of the masonry. Two flatjacks are inserted into the wall parallel to each other and separated by four or five masonry courses creating an in situ prism (Figure 2). As the flatjacks are pressurized the corresponding deformations are measured by linear variable differential transformers (LVDTs) or mechanical gauges adhered to the surface of the wall creating a stress-strain curve for the test.

The shear test determines the mortar bed joint shear strength of a brick unit. The test requires the head joints on either side of the test unit to be removed. A small shear flatjack is inserted into the cleared head joint on one side and pressurized (Figure 3). The bed joint shear strength is determined from the pressure needed to create the first visible movement. The use of a shear flatjack is an alternative to a hydraulic ram and eliminates the

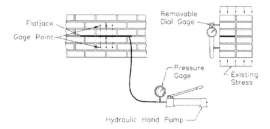

Figure 1. Stress test schematic.

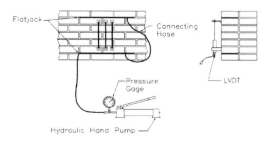

Figure 2. Deformability test schematic.

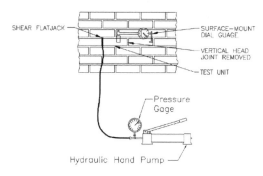

Figure 3. Mortar bed shear strength test schematic.

need for removal of an entire brick for the test (Hamid & Schuller 2019).

The original flatjacks were developed for evaluating historic masonry construction with typical compressive strengths in the range of 2.1 to 6.9 MPa (300 to 1000 psi). The pressure limit of these flatjacks reflected the strength of historic masonry and were not designed to surpass 6.9 MPa (1000 psi). Some stone masonry and most modern masonry have compressive strength in excess of 6.9 MPa (1000 psi). While this does not typically present an issue with stress tests, as the actual compressive stresses in masonry are relatively low, it can impose a significant impact on the deformability and shear tests. If the original flatjack design could not exceed the masonry strength or reach nonlinear behavior

during a deformability test, masonry properties are estimated based on relationships between compression modulus and compressive strength. Previous work shows the relationship between masonry compression modulus and strength to be highly variable (Cargill, Wo & Schuller 2019). For shear tests, the original low-pressure shear flatjacks are often unable to impose great enough load to fail the test specimen, necessitating the partial cut of the tested bed joints, reducing the sensitivity of the results. High bed joint strength is often due to the overbearing pressure, not bed joint shear strength, that warrants large applied loads to create the first movement of the brick. If it is still unable to displace the unit, the test cannot be completed through the flatjack method. Therefore, the hydraulic ram must then be used, which is a more invasive method.

Stone masonry construction and modern masonry have been found to have compressive strengths well in excess of 6.9 MPa (1000 psi), often with strengths of 30 MPa (4350 psi) or greater (Schuller 2001). Even some historic masonry structures constructed with natural hydraulic cement, such as the Brooklyn Bridge, have also been found to have strengths exceeding 6.9 MPa (1000 psi). Hence the design of a flatjack that can produce pressures that exceed masonry strengths of 6.9 MPa (1000 psi) would be beneficial. This equipment would be capable of applying high stress to capture nonlinear response and determine peak compressive strength directly, without having the need to rely on less precise relationships or inability to complete the test altogether.

2 BACKGROUND AND MOTIVATION

Flatjacks in common usage have been designed for historic masonry construction with a maximum capacity of 6.9 MPa (1000 psi) which is not large enough to develop nonlinear behavior that occurs near failure in high strength masonry. Strength is instead estimated based on a relationship between compression modulus (Em) and compressive strength ($f'm$). This relationship varies based on the type of masonry and code provisions followed. In TMS 402 (2016), for clay masonry, the relationship between compression modulus and compressive strength is given as $Em = 700f'm$. For historic masonry, $Em = 550f'm$ using Table 7-1, Default Lower-Bound Masonry Properties, in ASCE/SEI 41-06 (2006). The lower value compared to that of TMS 402 accounts for the lower compressive strengths of historical masonry compared to the strength of modern masonry. Following Chapter 3 of Eurocode 6 (1996), the relationship for modern masonry is $Em = 1000f'm$ (Wo & Schuller 2011).

Using the code relationships can result in errors in the predicated compressive strength compared to the actual masonry strength. In some cases, the code relationship overestimates masonry strength (Figure 4), and, in other cases, masonry strength is underestimated

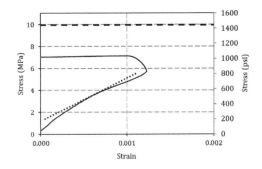

Figure 4. Stress-strain curve from a deformability test with high-pressure flatjacks (solid) on historic masonry, comparing the predicted strength (dashed) based on the calculated compression modulus (dotted) and equation Em = 550f'm. The relationship predicted a strength of approximately 2.9 MPa (420 psi) greater than the actual measured strength.

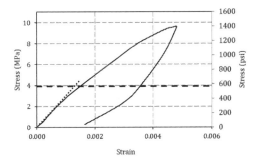

Figure 5. Stress-strain curve from a deformability test with high-pressure flatjacks (solid) on historic masonry, comparing the predicted strength (dashed) based on the calculated compression modulus (dotted) and equation Em = 550f'm. The relationship predicted a strength of approximately 5.9 MPa (850 psi) less than the actual measured strength.

(Figure 5). In some instances, the difference between the estimated value and the actual measured value is greater than 5.5 MPa (800 psi). Therefore, there is a need for developing a more accurate relationship.

Due to the high variability of masonry construction, determining the relationship between compression modulus and compressive strength with a high confidence level has been difficult. Through a statistical study completed by Cargill et al. (2019), the relationships of the test data were highly variable with large standard deviation and low R2 correlation coefficient. From the 250 tests included in their database, 19 of which were from high-pressure flatjacks, only 160 tests could be used in a statistical analysis to determine a relationship between Em and f'm due to the inability of the test result to develop nonlinear response or failure. This represents 39% of tests using original flatjacks unable to reach either nonlinear behavior or failure. Consequently, the data analyzed were heavily weighted toward strengths below 6.2 MPa (900 psi).

Therefore, while the results help better predict the relationship for historical or lower strength masonry, it may not be as beneficial for use with higher strength masonry. More results from higher strength masonry construction are needed to further develop an accurate Em-fm relationship, reinforcing the need for the development of a new flatjack for high strength masonry.

Many factors have been observed to affect masonry strength including unit and joint size and quality, existing deterioration such as weathering and cracking, as well as mortar type and degree of joint filling. This creates high variability between construction eras, location, and mortar hardness (Cargill et al. 2019). Using a range of Em is recommended to estimate f'm for the specific use of the information (Wo & Schuller 2011). The ranges recommended in the analysis vary based on the above properties and are often dissimilar to the existing code recommendations.

Due to the significant variability of construction, code guidelines, and the reliance on an imprecise relationship, there is an impact on the accuracy and utility of flatjack testing results for design considerations. Obtaining a direct measurement of compressive strength provides more accurate results, which can be achieved on significantly stronger constructions with a high-pressure flatjack. This is critical since a small number of tests is typically used to characterize the construction of an entire building.

3 DESIGN

The design and development of the high-pressure flatjack follows complementary research conducted at the Department of Civil Engineering, Federal University of São Carlos, Brazil, where researchers have developed flatjacks for evaluating high strength hollow-unit masonry construction (Soriani 2016). In that research project, tested concrete block walls were found to have individual unit strengths of 20 MPa (2900 psi) and wall strengths of 8 MPa (1160 psi) or greater. These high strengths obtained created a need for design that not only was specific for hollow unit masonry but also capable of generating higher pressures. The flatjack design utilized thin metal sheets with a reinforcing steel bar around the edge of the flatjack to overcome the failure of the welds, which is often observed in the original solid unit masonry flatjack during high pressures (Figure 6). During calibration conditions, the new flatjack was able to withstand 60 MPa (8700 psi) internal pressure before the flatjack ruptured. Results from laboratory testing on high strength hollow-unit masonry walls provided favorable results, which correlated well with compressive strength and compression modulus from prism tests. The in situ tests were able to load the walls to failure with a maximum pressure of 11 MPa (1600 psi)

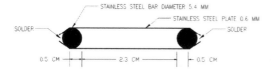

Figure 6. High strength hollow-unit masonry prototype flat-jack cross-section. Modified from Soriani (2016).

being applied to the wall (Soriani 2016). The success of the design prompted the existing flatjack design for solid unit masonry to be adapted to be able to successfully test compressive strengths greater than 6.9 MPa (1000 psi).

The overarching design concept and features of the high-pressure flatjack are similar to the original design of the flatjack for solid masonry construction. Several structural differences have been implemented to achieve greater pressures within the bladders. The shape of the high-pressure flatjacks remains the same: a rectangular envelope with radiused back corners for use with a bed joint slot cut with a circular masonry saw. Currently, no rectangular version has been developed for use with a plunge masonry saw. These types of saws are often rendered inefficient or ineffective on harder mortars characteristic with modern masonry. High-pressure shear jacks have the same rectangular shape as the original flatjacks.

The high-pressure flatjacks are made from two Type 304 stainless steel plates that are milled in the middle to a reduced thickness, leaving an edge with the original thickness around the perimeter. The milled center portion creates the space for the hydraulic fluid or water to pressurize the flatjacks. The two plates are welded together along their edges with the insertion of pressurization and bleed ports (Figures 7-8). The increase of thickness from 3.81 to 6.35mm (0.15" to 0.25") is the major dimensional change of the high-pressure design from the original design.

Figure 8. Final flatjack comprised of two blanks welded together.

The edge design is the critical feature that allows the increase in pressure and durability of the high-pressure flatjacks. The original flatjack design consisting of two thin plates welded at their edges creates a weak point. When pressurized, the flatjacks inflate creating high plastic strains at these hinge-like edge welds. After several large displacement cycles or high pressures, these welds tend to fail due to fatigue or over-pressurization. Unlike the traditional flatjacks, the new design subjects the welds to only direct tension and there is no plastic straining during tests. This is attributed to the thicker, flat edge at which the two halves were welded. The inflation of the flatjacks does not take place directly at the welds (Figure 9). A benefit to the increased thickness of the flatjacks allowed for the attachment of more robust insertion and removal tabs, and hydraulic port connections (Figure 10).

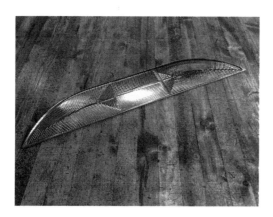

Figure 7. Machined blank flatjack half.

Figure 9. Comparison of inflations of high-pressure (left) and original (right) flatjacks. The inflation of the high-pressure flatjack occurs away from the welded edge while the inflation of the original flatjack occurs at the welded edge.

Figure 10. Comparison of tabs and port connections of high-pressure (left) and original (right) flatjacks.

4 TEST DATA

High-pressure flatjacks have been pressurized to repeated loads of 13.8 MPa (2000 psi) for calibration and re-calibration in a laboratory setting with consistent and adequate behavior. A high-pressure shear jack was loaded to over 31 MPa (4400 psi) to test the ultimate strength of the flatjack without failure.

The high-pressure flatjack has been used in the field with good results. Test results prove that the high-pressure flatjack can provide compressive strength greater than 6.9 MPa (1000 psi) and masonry achieves significant nonlinear deformation and/or failure (Figures 11-12).

Used on over ten projects for a combined 40 tests, the high-pressure flatjack has provided reliable results. All tests were conducted on solid brick masonry with hard mortar and were predominantly deformability tests, with some stress tests also conducted. On average, the in situ tests have been conducted on masonry with a compressive strength greater than 7.9 MPa (1160 psi), with 59% of the

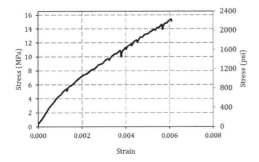

Figure 11. Stress-strain curve from a deformability test with high-pressure flatjacks on high strength historic brick with natural Rosendale cement mortar. The test was halted at approximately 15.2 MPa (2200 psi) due to cracking and brick face spalling.

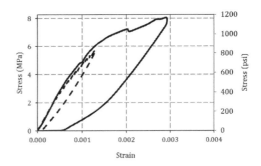

Figure 12. Stress-strain curve directly comparing stress-strain curves on historic brick with hard mortar on the same project. One test was performed with the high-pressure flatjacks (solid line) and the other with the original flatjacks (dashed line). The test with the original flatjacks was unable to cause nonlinear behavior or fail the masonry as the high-pressure flatjack test. Minor variance in stiffness can be attributed to local differences in masonry conditions and construction at each location.

tests conducted on masonry with strengths greater than 6.9 MPa (1000 psi); the compressive strengths ranged from 2.1 to 15.2 MPa (300 to 2200 psi). These tests have ranged from construction dating from the 1870s to the 1930s from across the United States. The use of the flatjack so far indicates that there is an existing need for high-pressure flatjacks even with early 20th century construction. Most tests exceeded the range of the original flatjack design and proved their effectiveness with higher strengths. Their use on masonry constructed on later dates is expected to be more critical, as the compressive strength will be out of range for the capacity of the original flatjack design for a higher percentage of tests.

Forty-one percent of in situ tests were within the range of the original flatjack design with compressive strengths, between 2.1 to 6.9 MPa (300 to 1000 psi). While these tests could have been performed with the original flatjack, they indicate that the high-pressure flatjacks can also perform adequately for lower strength masonry. This can be beneficial on projects where the expected masonry construction strength is unknown or expected to be close to 6.9 MPa (1000 psi). The use of the high-pressure flatjack for all tests ensures the ability to conduct the deformability tests into the nonlinear range and/or failure, without having to rely on the relationship between compression modulus and compressive strength.

In situ tests with the high-pressure shear jack have also provided good results on two projects and over 20 tests on historic brick with high strength mortar. On average, the in situ tests have imposed pressures of 18.6 MPa (2700 psi) on the masonry unit to overcome the bed joint shear strength. The imposed pressures ranged from 1.4 to 28.9 MPa (200 to 4200 psi), indicating again that the high-pressure jack can be used on both low and high strength masonry, and are able to apply high loads necessary for mortar shear failure. The pressures reported above are not the calculated bed joint shear strengths, as these values must be converted into applied pressure removing the friction from the overbearing loads, and divided by the bed joint areas to achieve the final result.

As with the original flatjack, special care must be taken when completing tests to mitigate damage to the surrounding masonry. There must be increased care when using high-pressure flatjacks as it is readily possible to exceed the masonry's compressive strength. Loading should be monitored in real-time to ensure a controlled loading rate and maximum pressure, especially when nearing the expected strength of the masonry. Masonry strain should also be monitored in real-time during load application and testing should be halted at the onset of significant nonlinear deformation (Figure 12). Strain reversal is also an indication of impending failure, as it can indicate that internal damage is occurring within the loaded masonry (Figure 4). Visual and audible indications, such as cracking, especially at the edges of the flatjacks (Figure 13), is an indication of masonry failure and when to stop the test. Modern masonry with hard mortar has been seen to have increased brittle behavior compared to historic construction with softer mortars (Cargill et al. 2019). Therefore, even with the indicators to halt testing described previously, sudden brittle failures can occur in masonry construction with high values of compression modulus (Figure 14). Safety should always be of the utmost importance when performing flatjack tests.

In situ tests have also provided evidence of the increased durability of the high-pressure flatjack compared to that of the original design. For the regular flatjacks, failure, whether it is repairable or not, occurs after several pressurization cycles, typically only five to ten tests. The most common

Figure 14. Example of damage resulting from a sudden brittle failure during testing with high-pressure flatjacks. Nonlinear deformation was not reached, and no signs of distress were observed before failure.

failure mode is due to the fatigue of the welds during testing from repeated loading. Properly shimming and avoiding over inflation can significantly improve the life span of the flatjack. With the increased robustness of the high-pressure flatjack, some have been used for over 30 tests and are still in adequate condition for continued use. Insertion, removal, and general wear also cause damage to the flatjack. High pressure flatjacks have stronger tabs for insertion and removal, and hydraulic port connections that are more resistant to damage aiding in the prolonged serviceability of the flatjack.

Of the six high pressure flatjacks built to date, only one has been removed from service. This flatjack was damaged during the insertion processes rather than during pressurization.

5 DISCUSSION AND CONCLUSIONS

High-pressure flatjacks have been used in the field with good results on masonry strengths greater than 6.9 MPa (1000 psi) to a maximum of 15.2 MPa (2200 psi). Flatjacks have been pressurized up to 31 MPa (4400 psi) during laboratory tests without failure. In situ tests have also shown good results on lower strength masonry from 2.1 to 6.9 MPa (300 to 1000 psi). Overall, the design has proven to be reliable and to work on a wide range of masonry strengths. If the construction permits, the high-pressure flatjack can be utilized on most projects regardless of expected masonry strength. High-pressure flatjacks overlap the pressure range or the original design and can be used exclusively in order to avoid any reliance on the relationship between compression modulus and compressive strength. The direct measurement of compressive strength is achievable for historic, modern and stone masonry.

Figure 13. Example of visual masonry failure with crack formation at the edge of the flatjack slot.

Figure 15. Example of thin bed joints where the high-pressure flatjacks could not be used.

The thickness increase from 3.81 to 6.35 mm (0.15" to 0.25"), however, can restrict the use of high-pressure flatjacks on masonry construction with thinner bed joints (Figure 15).

The new design significantly increased the price of the flatjacks by a factor 3.75 compared to the original design. This high price point can limit their penetration into the testing market and lead to lower use. Although, with their increased durability, the price per test is approximately equivalent to that of the original flatjack design.

Currently, only two designs for high-pressure flatjacks have been developed including the design described in this paper and for high strength hollow-unit masonry construction by researchers at the Department of Civil Engineering, Federal University of São Carlos, Brazil. The original flatjack has been developed for a variety of uses and resulted in multiple layouts, materials, and designs for specific masonry types and applications (Figure 16). The authors encourage the continued design iterations for the high-pressure flatjacks in various layouts and for specific applications and/or materials.

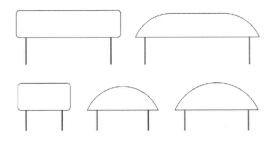

Figure 16. Various flatjack layouts.

The current database of in situ deformability tests using high-pressure flatjacks is limited to approximately 40 tests, all conducted on solid brick masonry, between 1870 and 1930. The continued use and acceptance of this new design is needed in order to create a larger database on a wider variety of construction eras and materials such a stone masonry, as well as for stronger masonry constructions with compressive strengths exceeding 3.8 MPa (2000 psi). These test results can also be used to provide additional data on higher strength masonry for further development of the statistical analysis previously completed by Cargill et al. (2019).

To permit the new high-pressure design, the relevant ASTM standards and corresponding RILEM standards must be updated. Currently, these standards are for the use of the original flatjack and describe their design and dimensions accordingly. The update of these standards with the new design requirements would allow for formal acceptance and use of high-pressure flatjacks. This would aid in developing the high-pressure flatjack further.

REFERENCES

ASCE/SEI 41–06. 2006. *Seismic Rehabilitation of Existing Buildings.* Reston: American Society of Civil Engineers.
BS EN 1996–3. 1996. *Eurocode 6: Design of Masonry Structures – Part 3: Simplified calculations for unreinforced masonry structures.* London: British Standards Institution.
Cargill, N., Wo, S., Schuller, M. 2019. Relationship between compressive strength and modulus for existing masonry construction. *Proceedings of the 13th North American Masonry Conference, Salt Lake City, 16–19 June 2019.*
Hamid, A., & Schuller, M. 2019. *Assessment and retrofit of masonry structures.* Longmont: The Masonry Society.
Noland, J., Atkinson, R., & Schuller, M. 1990. A review of the flatjack method for nondestructive evaluation. *Proceedings of the Nondestructive Evaluation of Civil Structures and Materials, Boulder, October 1990.*
Schuller, M. 2001. Flatjack methods for diagnosis of modern masonry. *Presented at the RILEM International Workshop on Site Control and Nondestructive Evaluation of Masonry – Structures and Materials, Mantova, 12–14 November 2001.*
Soriani, M. 2016. Técnica de macacos planos na avaliação de estruturas de alenaria com blocos vazados de concreto. Doctorate Dissertation, Department of Civil Engineering, Federal University of São Carlos, Brazil.
TMS 402/602. 2016. *Building Code Requirements and Specifications for Masonry Structures.* Longmont: The Masonry Society.
Wo, S. & Schuller, M. 2011. Estimating compressive strength of historical unreinforced masonry using flatjack deformability tests. *Proceedings of the 11th North American Masonry Conference, Minneapolis, June 2011.*

Brick and Block Masonry - From Historical to Sustainable Masonry –
Kubica, Kwiecień & Bednarz (eds)
© 2020 Taylor & Francis Group, London, ISBN 978-0-367-56586-2

Experimental testing of tile vaults

D. López López
Department of Civil and Environmental Engineering, Universitat Politècnica de Catalunya, SNSF Fellow

E. Bernat-Masó
Department of Strength of Materials and Engineering Structures, Universitat Politècnica de Catalunya, Serra Húnter Fellow

L. Gil
Department of Strength of Materials and Engineering Structures, Universitat Politècnica de Catalunya

P. Roca
Department of Civil and Environmental Engineering, Universitat Politècnica de Catalunya

ABSTRACT: Tile vaults (sometimes also referred to as thin-tile, timbrel, Catalan or Guastavino vaults) are masonry structures made with thin bricks (tiles), mortar and fast-setting cement or gypsum. This traditional construction technique has produced a large quantity of built heritage examples with an important historical value. Moreover, in the last few years, its combination with new tools for the design and analysis of masonry structures has led to a rediscovery of the technique resulting in interesting architectural pieces with a new formal language. This paper presents experimental research on tile vaults aiming at the understanding of their structural behaviour for the analysis of both built examples and new architecture. Four full-scale specimens are load-tested until failure at the laboratory, namely, two barrel vaults and two sail domes. The vaults were tested under vertical loading up to failure. Samples of the materials composing the tile vault are also tested for their characterization. The experiments supplied valuable data about the stiffness, peak loads, post-peak behaviour, damage and collapse mechanism of the tested structures.

1 INTRODUCTION

Tile vaults are masonry structures made with thin bricks (tiles), mortar and fast-setting cement or gypsum. The bricks are placed flat, building up to two, three or more courses. Traditionally, tiles are used because of their light weight, which is a necessary condition to build the first course "in space" without supporting falsework (Figure 1). The first course is achieved using the quick adhesion of fast-setting cement or gypsum. The bricks bind within seconds to the edge walls, or the already finished arches/stable sections, eliminating the need for centering. Using this first layer as a stay-in-place formwork, the second and subsequent courses, which build up the necessary structural depth, can be set with lime or Portland cement mortar. The ability of the courses to be built without support and in stable sections, is one of the most relevant characteristics of this technique and what makes it inherently economic. Tile vaults are very efficient, since they have a large load-bearing capacity with high slenderness ratios (Huerta, 2001).

"Such shells [tile vaults] are cheap to make and may be adapted to any desired shape, even to
continuously changing curvatures, [...]. If constructed by bricklayers skilled in the technique, they can be made without the aid of formwork. The only thing necessary is a set of guides to indicate to the craftsman the contours to be followed. These guides are spaced more or less closely, depending on the degree of the curvature." - Torroja, 1958.

The technique reached its peak in terms of expressiveness, versatility and importance in Catalan Modernism with the works of, among others, Antoni Gaudí, Lluís Domènech i Montaner and Lluís Muncunill (González, 2003). No less important in the history of these vaults is the episode of this technique's exportation to the United States by Rafael Guastavino Moreno (1842-1908). Guastavino emigrated to the U.S. with his son in 1881, where they achieved enormous success (Ochsendorf, 2010).

This traditional construction technique has been "rediscovered" in the last few years. The development of new interactive equilibrium methods for the design and analysis of masonry structures have rekindled interest in the versatile tile vaults, resulting in a proliferation of projects worldwide featuring this technique and showing novelty in their shapes and innovation in the fields of construction and materials

Figure 1. Construction of a tile vault "in space" (Truñó, ca. 1951).

Figure 2. Compression tests on tiles in the orthogonal direction.

(Block *et al.*, 2010; López López *et al.*, 2014, 2016, 2019a, 2019b). These recent innovations in construction and design methods for tile vaults have allowed the construction of "free-form" compression-only tile vaults.

Despite the large amount of tile vaults within the built heritage, very little scientific experimental research has been carried out to fully understand the structural behaviour of this construction technique (Bergós, 1965; López López & Domènech, 2017).

This paper presents a set of experiments carried out at the Laboratory for Technological Innovation in Structures and Materials (LITEM) at the Polytechnic University of Catalonia (UPC). The experiments consisted of material characterisation involving compression and bending tests on material samples and load tests on full-scale prototypes, namely, two tile barrel vaults and two sail domes.

2 MATERIAL CHARACTERISATION

The masonry units used for the construction of the tile vaults had sizes of 277 x 134 x 13 mm and a weight of 800 g. The binder for the first layer was a fast-setting cement, whereas that for the second one and for the joint between the two tile layers was dry (pre-mixed) Portland cement mortar (M7.5).

Compression and bending tests were carried out on the materials composing the tile vault, namely tiles, fast-setting cement and mortar. The extruded industrial tiles were grooved in one direction, thus presenting an orthotropic behaviour. Besides, in the built prototypes, as traditionally done, tiles were placed in two different directions. For the barrel vaults, the direction of the extrusion corresponded to the span direction for the first layer and the orthogonal direction for the second. Therefore, four compression tests in each of these two directions were performed (Figure 2). Results of the compressive strength were 111 N/mm^2 for the tiles in the extrusion direction and 87 N/mm^2 for the orthogonal ones, with coefficients of variation of 8.93 % and 2.70 %, respectively.

The natural fast-setting cement and the Portland cement mortar were tested in bending and compression. The average flexural strength of the fast-setting cement was 0.9 N/mm^2, whereas that of the mortar was 2.5 N/mm^2, with coefficients of variation of 1.48 % and 8.47 %, respectively. The average results and coefficient of variation for the compression tests of the binders were, respectively, 4.47 N/mm^2 and 10.16 % for the fast-setting cement and 6.98 N/mm^2 and 14.72 % for the Portland cement mortar. The specific weight of the cement was 1373 kg/m^3 and that of the mortar was 1940 kg/m^3.

3 LOAD TESTS ON TILE BARREL VAULTS

3.1 *Vault's geometry and test configuration*

Two cylindrical, 36-mm-thick, tile barrel vaults with a span of 2.80 m, a rise of 0.26 m and a width of 1 m were tested at the laboratory (Figures 3 and 4). During the tests the supports were pinned (able to rotate), with translation partially constrained by

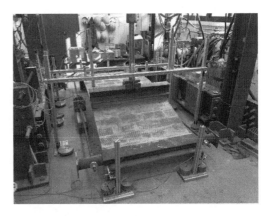

Figure 3. Load test's setup of one of the tile barrel vaults.

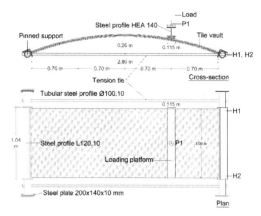

Figure 4. Setup of the monitoring and load test for the tile barrel vaults. Potentiometer indicated as P1. LVDTs indicated as H1 and H2. Up) cross section, down) plan.

means of tension ties, whose stiffness would allow small deformations during the load test. Only one of the two supports was therefore stiffly anchored to the fixed, loading steel frame and the horizontal displacement of the opposite one was measured by means of two LVDTs.

The load was applied at ¼ of the span on a concrete surface, rectangular in plan, 115 mm long in the direction of the span and occupying the entire width of the vault. A single actuator applied a punctual load that was distributed on the entire surface of the loading platform through a HEA 140 steel profile. The load was displacement-controlled at a constant speed of 0.1 mm/min until failure.

The vault's geometry, monitoring and general test setup are shown in Figure 4. The two vaults (5.7 m² in total) were built in one working day by two expert masons and one helper.

3.2 Results

The two vaults had a fragile behaviour with ultimate loads of 4.32 kN and 4.69 kN for vault 1 and 2 respectively (Figure 5). This result is considerably high for a 36-mm-thick masonry structure and evinces the efficiency of the technique. Both vaults developed the same failure mechanism with the formation of two hinges (note that the supports were pinned). The first hinge was in both cases clearly visible as a growing crack at the intrados under the load line in vault 1 and slightly closer to the nearest support in vault 2 (Figure 6). This crack followed mostly a straight path through the width of the vault cutting therefore some masonry pieces. However, in some cases it deviated to follow the joints and describing the tiles' pattern. A crack at the extrados of the opposite side of the load line revealed the second hinge (Figure 7).

The chart in Figure 5 shows ultimate loads corresponding to a displacement at the load point of 2.45 and 3.22 mm for the first and second vault respectively. After the peak load and coinciding with the formation of the first hinge, both vaults suffered a sudden loss of strength revealing their fragile behaviour. The curves show a second drop coinciding with the formation of the second hinge around 5 mm displacement at P1 (Figure 4). After this, the vaults became a mechanism and lost their stiffness. In the case of vault 1, the first drop is much smaller than the second one, being the opposite in the case of vault 2.

Figure 6. Formation of the first hinge. Tile barrel vault 2.

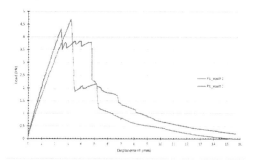

Figure 5. Ultimate load test. Load-displacement curves of the two tile barrel vaults at the loading point (P1, Figure 4).

Figure 7. Crack on the extrados evidencing the development of the second hinge at the tile barrel vault 1.

Additionally, the two LVDTs at H1 and H2 (Figure 4) offered the data about horizontal displacements, with average values of 0,45 mm and 0,54 mm at the peak load for vault 1 and 2 respectively.

It is worth pointing out the relatively high tensile strength of tile vaults considering their nature as masonry structures. The fragile behaviour showed in Figure 5 through the two drops of the load-displacement curves coinciding with the formation of the two hinges already evinced the existence of tensile capacity in these peculiar structures.

Figure 8 Illustrates the limit analysis of the tested tile barrel vaults, showing the uniqueness theorem and a collapse load of 0.38 kN, which is more than 11 times smaller than the lowest peak load of the two tested vaults (4.32 kN). Heyman's limit analysis for masonry structures considers, through one of his three well-known assumptions, that masonry has no tensile strength (Heyman, 1966). This method, although easy, fast and intuitive, is therefore conservative considering the analysis of tile vaults, whose tensile capacity is achieved by its multi-layered condition and the staggered joints.

4 LOAD TESTS ON TILE SAIL DOMES

4.1 Dome's geometry and test configuration

The spherical sail domes had an intrados radius of 2.03 m, were squared in plan, supported on their four corners and with a side of 1.76 m and a rise of 0.40 m. The domes were built on a steel frame, to allow transportation and to partially prevent the displacement of the supports (Figure 9).

Figure 8. Limit analysis of the 36-mm-thick, tile, barrel vault. Uniqueness theorem showing a collapse load of 0.38 kN.

Figure 9. Load test's setup of one of the tile sail domes.

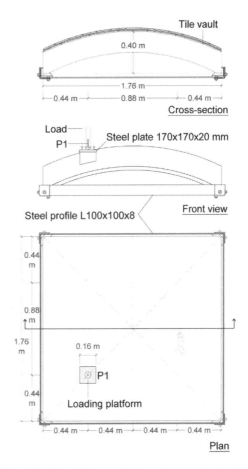

Figure 10. Setup of the monitoring and load test for the tile sail domes. Potentiometer indicated as P1. Up) cross section, middle) front view, and down) plan.

The load was applied at ¼ of the span of both orthogonal directions on a concrete surface, square in plan with an edge of 160 mm. The load was displacement-controlled at a constant speed of 0.2 mm/min until failure.

The dome's geometry, monitoring and general test setup are shown in Figure 10. The two domes (roofing a surface of 6.6 m² in total) were built in two working days by two expert masons and one helper.

4.2 Results

The tested sail domes had peak loads of 19.5 kN and 21.0 kN for the domes 1 and 2, respectively (Figure 11). Ultimate loads were greatly higher than those required by Eurocode 1 (CEN, 2002).

The cracks and crack patterns identified during the load tests (Figure 12), together with the registered displacement data, allow the identification of the failure mechanism, which was the same for both domes. After the load test, the prototypes showed a partial

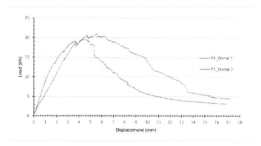

Figure 11. Ultimate load test. Load-displacement curves of the two sail domes at the loading point (P1, Figure 10).

Figure 12. Load test of the first sail dome.

collapse under the applied load, but a mechanism entailing the whole structure (Figure 13).

A crack under the actuator running perpendicular to the dome's sides (Figure 13, a), developed from the loading point to the closest edges, evidenced the appearance of a hinge under the load. Cracks at the corners also manifested the hinges at the supports. A group of cracks following the brick pattern of the dome's extrados revealed the creation of a last

Figure 13. Collapsed sail dome after the load test.

Figure 14. Detachment and fall of some bricks during the load test.

complex hinge. This hinge linked the two dome's sides closest to the actuator (Figure 13, b) and had two additional branches heading to the remaining sides (Figure 13, c), activating the whole structure.

During the unloading response, detachment and fall of some bricks of the first layer was reported (Figure 14).

5 CONCLUSIONS

The construction and load testing of the four full-scale prototypes demonstrated successful structural performance with ultimate loads far higher than required by the codes.

The data registered during the load tests of the prototypes together with the presented further experimental research on the materials' properties provide a valuable benchmark and allow the calibration and validation of structural models.

The experimental research on the tile barrel vaults evinced that the timbrel vault masonry can develop a meaningful tensile strength, which provides these structures with an increased load capacity. The limit analysis carried out on the tested specimens revealed itself as remarkably conservative, predicting a collapse load more than 11 times smaller than those resulting from the load tests.

ACKNOWLEDGEMENTS

This work was supported by the Swiss National Science Foundation (SNSF) [grant number P2EZP2_181591]. It was also sponsored by the construction company URCOTEX, from which Pep Brazo, Antonio Haro and Albert Martí are especially acknowledged. The construction of the prototypes was carried out together with Jordi Domènech, to whom the authors are very thankful. Second author is a Serra Húnter Fellow.

REFERENCES

Bergós, J. (1965). *Tabicados huecos*. Barcelona: Col•legi d'Arquitectes de Catalunya i Balears.

Block, P., DeJong, M., Davis, L., & Ochsendorf, J. (2010). Tile vaulted systems for low-cost construction in Africa. *ATDF Journal (African Technology Development Forum)* 7(1/2), 4–13.

CEN. (2002). *Eurocode 1: Actions on structures - Part 1-1: General actions - Densities, self-weight, imposed loads for buildings. EN 1991-1-1: 2002*. Brussels, Belgium: European Committee for Standardization, CEN.

González, J.L. (2003). *La bóveda catalana en el final del siglo XIX: Guastavino, Domènech i Gaudí*. L'Alguer, Sardenya: Seminari Di Studi Archittetura e Restauro nei Paesi della Corona d'Aragona.

Heyman, J. (1966). The stone skeleton. *International Journal of Solids and Structures 1966, 2*, 249–79.

Huerta, S. (2001). La mecánica de las bóvedas tabicadas en su contexto histórico, con particular atención a la contribución de los Guastavino. *Las bóvedas de Guastavino en América*, 87-112. Madrid: CEHOPU, Instituto Juan Herrera.

López López, D., Domènech, M. & Palumbo, M. (2014). "Brick-topia", the thin-tile vaulted pavilion. *Case Studies in Structural Engineering, 2*, 33–40.

López López, D., Van Mele, T., & Block, P. (2016). Tile vaulting in the 21st century. *Informes de la Construcción, 68*(544), e162. DOI: http://dx.doi.org/10.3989/ic.15.169.m15.

López López, D., & Domènech, M. (2017). Tile vaults. Structural analysis and experimentation. *2nd Guastavino Biennial*. Barcelona: Diputació de Barcelona.

López López, D., Van Mele, T., & Block, P. (2019a). The combination of tile vaults with reinforcement and concrete. *International Journal of Architectural Heritage*, 13:6, 782–798. DOI: 10.1080/15583058.2018.1476606.

López López, D., Roca, P., Liew, A., Van Mele, T., & Block, P. (2019b). Tile vaults as integrated formwork for reinforced concrete: Construction, experimental testing and a method for the design and analysis of two-dimensional structures. *Engineering Structures, 188*, 233–248.

Ochsendorf, J. (2010). *Guastavino Vaulting. The Art of Structural Tile*. Princeton Architectural Press.

Torroja, E. (1958). *The structures of Eduardo Torroja. An autobiography of engineering accomplishment*. Madrid: Centro de Estudios Históricos de Obras Públicas y Urbanismo (CEHOPU), 2000.

Truñó, A. (ca. 1951). *Construcción de Bóvedas Tabicadas*. Barcelona: Library of the Col•legi d'Arquitectes de Catalunya.

Brick and Block Masonry - From Historical to Sustainable Masonry –
Kubica, Kwiecień & Bednarz (eds)
© 2020 Taylor & Francis Group, London, ISBN 978-0-367-56586-2

In-plane flexural behavior of autoclave aerated concrete confined masonry walls

J.A. Moreno-Herrera, J.L. Varela-Rivera & L.E. Fernandez Baqueiro
Universidad Autonoma de Yucatan, Yucatan, Mexico

ABSTRACT: Three full-scale autoclave aerated concrete confined masonry walls were tested under constant axial compressive stress and incremental reverse cyclic loads until failure. The variable studied was the axial compressive stress. The behavior of walls was characterized by horizontal flexural cracks, followed by yielding of the longitudinal steel reinforcement of vertical confining elements. Then flexure-shear and diagonal cracks were observed. As expected, the wall flexural strength increased as the axial compressive stress increased. By the contrary, the displacements ductility increased as the axial compressive stress decreased. Finally, flexural strength of walls was well predicted using flexural theory (kinematics, constitutive models and equilibrium). The ratio between analytical and experimental strengths of walls varied from 0.95 to 1.05.

1 INTRODUCTION

Autoclaved aerated concrete (AAC) is a lightweight cellular material commonly made of Portland cement, lime, sand, water and an expansive agent such as aluminum powder. AAC was first produced commercially in Sweden in 1923. Since that time, its production and use have spread to many countries on all continents. The use of AAC in the United States of America (USA) started in 1990; its use in Mexico, in 1994. Precast AAC products include unreinforced masonry-type units and reinforced panels, among others. Physical requirements for AAC in the US are specified in ASTM C1693 (ASTM 2011).

Confined masonry walls are widely used in some countries of Latin America (Mexico, Chile, Peru, Colombia, Argentina and Venezuela), Asia (Iran, China and Indonesia), Europe (Italy, Slovenia, Romania and Serbia) and Africa (Algeria and Morocco). Confined masonry is a structural system composed of flexible reinforced concrete confining elements cast around an unreinforced masonry wall panel. In this type of construction, the unreinforced wall panel is constructed first and later the confining elements. The use of confined walls in Latin America is very common because of their low construction cost and ease of construction. Confined walls are typically constructed using concrete and clay units. The use of confined walls made of AAC units has spread to some parts of South and Central America (including Mexico) in recent years.

There are some research studies carried out on the behavior of AAC masonry walls subjected to in plane lateral loads (Cancino 2003, Tanner et al. 2005a, b, Varela et al. 2006). These studies focused on the flexure and shear behavior of unreinforced and reinforced AAC walls. Variables considered were the unit type, the wall aspect ratio (height over length) and the wall axial load. These studies formed the basis for the development of design requirements for unreinforced and reinforced AAC walls in the USA (ACI 2013). There are other studies related to the behavior of AAC masonry walls. These studies focused mainly on the in-plane behavior of unreinforced, reinforced and infill AAC walls (Penna et al. 2008, Tomazevic & Gams 2012, Ravichandran & Klingner 2012, Mandirola et al. 2012, Yu et al. 2013, Bose & Rai 2014). Only one study focused on AAC confined walls (Tomazevic & Gams 2012), but wall specimens were constructed at 1:4 scale. Main variables studied by those authors were the type of reinforcement, wall aspect ratio and wall axial load.

On the other contrary, the behavior of confined walls subjected to in-plane lateral load has been widely studied. There are several studies carried out by different authors. The main variables studied were the unit type (Meli 1979, San Bartolome & Quiun 2010), combinations of clay and concrete units (Tena-Colunga et al. 2008), types and quantities of steel reinforcements in confining elements (Treviño et al. 2004, Quiroz et al. 2014), wall axial loads (Urzua et al. 2001), wall aspect ratio (San Bartolome et al. 1992, Perez et al. 2015), wall openings and type of reinforcement around openings (Flores et al. 2004). In general, these studies considered the shear behavior of confined walls with aspect ratios close to one. Few experimental studies have been carried out for walls with different aspect ratios despite the significance of this parameter on the wall shear strength. In addition, walls considered in those studies were constructed using only clay or concrete units.

In Mexico, there are design requirements for unreinforced, reinforced and confined masonry walls (NTCM 2004). These requirements are for walls made of concrete and clay units. The design of unreinforced and reinforced AAC walls in Mexico is carried out using USA requirements (ACI 2013). The design of AAC confined walls is carried out using Mexican design equations (NTCM 2014) developed for walls constructed with clay and concrete units. This situation is similar for those countries of Latin America where AAC confined walls are also used. This could be an inappropriate practice because the behavior of confined walls constructed with clay or concrete units is different than that with AAC masonry-type units; for example, in the first case cracks are formed mainly at joints but in the second case on the AAC units (Tanner et al. 2005a, Perez et al. 2015). Experimental information is still needed on this subject to develop requirements for the structural design of AAC confined walls.

The objective of this paper is to evaluate the flexural in-plane behavior of AAC confined walls subjected to lateral loads. Three full-scale autoclave aerated concrete confined masonry walls were tested in the laboratory. Experimental cracking patterns and lateral load–drift ratio curves are presented. Flexural strength of walls was validated using flexural theory. A discussion on drift ratios and displacement ductilities is presented.

2 EXPERIMENTAL PROGRAM

2.1 Wall specimens and material properties

Three full-scale autoclave aerated concrete masonry walls were tested under constant axial compressive stress and incremental reverse cyclic loads until failure. Walls were designed to induce flexural behavior. Details of each AAC confined wall are presented in Table 1, where H, L and t are the wall height, length and thickness, respectively. The wall height was measured up to the point of load application (Figure 1). The study variable was the wall axial compressive stress (σ). Axial compressive stresses for walls are related to 1, 3 and 5-story AAC structures, respectively. It was assumed during the gravity load analysis that structures are built using AAC walls and AAC floor panels. Class AAC-4 units were considered (ASTM 2011). Nominal dimensions of AAC units were 0.15 × 0.20 × 0.61 m (width × height × length). Cross section (CS) dimensions and steel reinforcement details of confining elements (CE) are presented in Table 2.

Table 1. Details of AAC confined walls.

AAC wall	H (m)	L (m)	t (mm)	H/L	σ (MPa)
M1	2.8	1.24	150	2.26	0.24
M2	2.8	1.24	150	2.26	0.47
M3	2.8	1.24	150	2.26	0.71

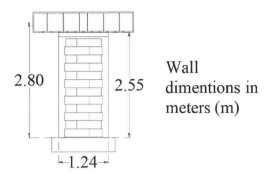

Figure 1. Layout of AAC confined masonry walls.

Table 2. Details of confining elements of walls.

Vertical CE	LR 1#3	
Top CE	LR 6#3 TR #2@200 mm TR 1#4 @200 mm	
Bottom CE	LR 4#4 TR #2@120 mm	

Longitudinal (vertical) reinforcement (LR) consisted of one deformed steel bar with a cross section area (As) of 71 mm². The nominal yield strength (fy) of bars was 412 MPa. Specified axial compressive strength of the concrete of the vertical confining elements was 17.16 MPa.

Compressive strength and splitting tensile strength of AAC were determined in accordance with ASTM C1693 (ASTM 2011) and ASTM C1006 (ASTM 2007), respectively. Compressive strength of mortar and thin bed mortar used in wall construction were determined using ASTM C109 (ASTM 2016b). Compressive strength of the concrete of the vertical confining elements was determined using ASTM C39 (ASTM 2016a). Tensile strength of the steel longitudinal reinforcement of the vertical confining elements was determined using ASTM A370 (ASTM 2015).

2.2 Wall construction

AAC confined walls were constructed in half running bond by a qualified worker. The first block course was laid using both mortar in proportions by volume, 1:3 (Portland cement: sand) and thin bed mortar. The first one was used on the leveling bed joint and the second on the head joints. Subsequent block courses were laid using only thin bed mortar. Properties of thin bed mortar are specified in ASTM C1660 (ASTM 2009).

Construction of walls was as follows, for the case of wall M1, first the six bottom block courses were laid and later the corresponding part of the vertical confining elements was concrete cast. Second, the last six block courses were laid, and the corresponding remaining part of the vertical confining elements was cast. Finally, the top confining element was cast. A 25 mm tooth was used in end blocks of alternating courses. This procedure was repeated for wall M2 to M3, but the first seven block courses were laid first and then the remaining five. Because of this procedure, construction cold joints were formed in the vertical confining elements at heights of 1.10 m for wall M1 and at 1.30 m for walls M2 and M3. The wall layout of AAC confined wall is presented in Figure 1.

2.3 Walls test setup

Each AAC confined wall was tested with constant axial load and reverse monotonic cyclic lateral loads until failure. Axial load for each wall was calculated using the corresponding axial compressive stress (σ), wall length (L) and wall thickness (t) (Table 1). Axial load was applied using a swivel beam, a spreader beam, two threaded rods and a hydraulic actuator (Figure 2). Pressure in the actuator was maintained constant during the test by using a mechanical load maintainer (Edison 1994). Axial load was measured using two donut type load cells. This load was verified using a pressure transducer. Lateral loads were applied using a steel frame, a loading steel beam, and a two-way hydraulic actuator (Figure 3). Lateral load was measured using a tension-compression pin load cell. This load was verified using two pressure transducers. Wall specimens were attached to the lab reaction floor.

2.4 Wall instrumentation and loading history

Horizontal and vertical wall displacements and shortening or lengthening of the wall diagonals were measured using linear string potentiometers (SP). Relative displacements between the loading beam and the

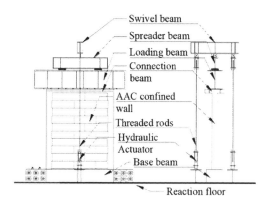

Figure 2. Axial load test setup.

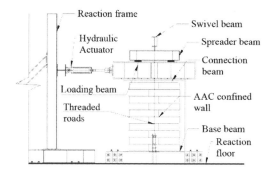

Figure 3. Lateral load test setup.

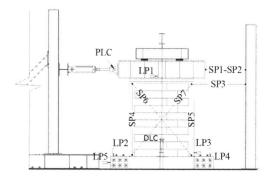

Figure 4. Typical view of wall instrumentation.

wall, the wall and the wall foundation, and the wall foundation and the reaction floor were measured using linear potentiometers (LP). Strain gages were attached to the steel longitudinal reinforcement of both vertical confining elements. Two strain gages were located at the bottom of each bar. A typical view of wall instrumentation is presented in Figure 4. In this figure, PLC and DLC refers to the pin load cell and the donut type load cell, respectively.

The loading history used to test the walls was based on the protocol established in the Mexico City Masonry Technical Norm (NTCM 2004). This loading history has six initial reverse cycles controlled by load and subsequent cycles controlled by drift ratios. The maximum target load was associated with yielding of the longitudinal steel reinforcement of the vertical confining elements.

3 EXPERIMENTAL RESULTS AND DISCUSSION

3.1 Material properties

The average axial compressive strength of the AAC (f_{AAC}) was 4.33 MPa with a coefficient of variation (CV) of 0.11. The average splitting tensile strength of AAC (f_{tAAC}) was 0.38 MPa with a CV of 0.08.

The average compressive strength of mortar was 15.85 MPa with a CV of 0.08. The average compressive strength of thin bed mortar was 15.92 MPa with a CV of 0.05. The average axial compressive strength of the concrete (fc) of walls M1 to M3 was equal to 16.87, 16.57 and 18.34 MPa, respectively. Corresponding CV was equal to 0.03, 0.05 and 0.03, respectively. Mortar and concrete specimens were tested after 28 days. The average yielding tensile strength of longitudinal steel reinforcement was equal to 446 MPa with a CV of 0.05.

3.2 Behavior of walls

The behavior of walls was, in general, similar. First a horizontal flexural crack was observed at the joint between the first block course and the concrete foundation. After this, yielding of the longitudinal steel reinforcement at the bottom end of both vertical confining elements was reached. As the drift ratio increased, flexure-shear cracks formed at both wall ends. Diagonal shear cracks were observed on the wall panels. Failure of walls was associated with the propagation of diagonal cracks into the top and bottom ends of vertical confining elements. The final cracking patterns of walls are presented in Figure 5. The lateral load – drift ratio curves of walls are presented in Figure 6. Diagonal cracks observed on walls were formed on the AAC units and not at the joints; that is, stair-step type inclined cracks were not observed. This means that the walls behaved as monolithic walls. Stair-step type cracks are commonly observed in confined walls constructed with clay or concrete units where mortar is weaker than units.

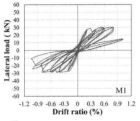
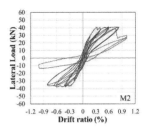
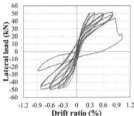

Figure 6. Lateral load–drift ratio curves.

3.3 Flexural strength of AAC confined walls

The flexural strength (M_f) of walls is presented in Table 3. These strengths are related to the average of the two observed maximum lateral loads (V_f). As expected, the flexural strength of walls increases as the wall axial load increases. The ratio of the flexural strength of walls M2 and M3 to that of wall M1 was equal to 1.32 and 1.70, respectively. Analytical flexural strengths of the walls were calculated using flexural theory (kinematics, constitutive models and equilibrium). A rectangular block was used for the compressive stresses of concrete (ACI 2014) (Figure 7). This type of block was used because the neutral axis (c)

Table 3. Analytical and experimental flexural strengths.

AAC wall	P (kN)	V_f (kN)	M_f (kN-m)	M_{fl} (kN-m)	M_{fl}/M_f
M1	46.70	29.77	83.36	87.31.	1.05
M2	85.00	39.43	110.39	107.77	0.98
M3	135.53	50.49	141.36	134.05	0.95

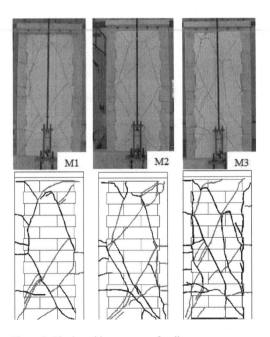

Figure 5. Final cracking patterns of walls.

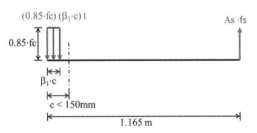

Figure 7. Flexural theory considerations.

835

Table 4. Drift ratios and ductilities for walls.

AAC wall	P (kN)	Δ_y (mm)	Δ_m (mm)	μ	δ (%)
M1	46.7	4.1	22.5	5.49	0.8
M2	85	5.	23.6	4.54	0.84
M3	135.53	5.9	22.3	3.78	0.8

was located within the width of the vertical confining elements. The height of the rectangular block was defined by 0.85·fc and the depth by β_1·c, where β_1 was equal to 0.85. fs is the stress in the steel reinforcement bar. A stress-strain relationship of the steel including strain hardening was considered (Rodriguez & Botero 1997). The analytical flexural strengths of the walls (M_{fl}) are presented in Table 3. This table shows that flexural strength of walls can be well predicted using flexural theory.

3.4 Deformation capacity on confined masonry walls

Deformation capacity of walls was defined by displacement ductilities and drift ratios at flexural strength. Table 4 shows the yielding and maximum displacements of walls. Yielding displacements (Δ_y) were based on the reading of the strain gages. Maximum displacements (Δ_m) were associated with flexural strengths. Corresponding displacement ductilities ($\mu = \Delta_m/\Delta_y$) and drift ratios (δ) of walls are presented in Table 4. As expected, the displacement ductility increased as the wall axial load decreased. Displacement ductilities varied from 3.78 to 5.49. The minimum value of 3.78 observed for AAC confined walls is a little greater than the value of 3.5 reported as ductility capacity for reinforced AAC walls with flexural behavior (Varela et al. 2006). In contrast, the maximum drift ratios of the walls were, in general, similar. Drift ratios varied from 0.8 to 0.84%. This was related to the shear behavior observed after yielding of the steel reinforcement. This shear behavior limited the deformation capacity of the AAC confined walls studied. The minimum value of 0.8% observed for AAC confined walls is smaller than the value of 1% reported as drift ratio capacity for AAC reinforced masonry walls with flexural behavior (Varela et al. 2006). Deformation capacity of confined AAC walls can be improved if shear behavior is avoided.

4 CONCLUSIONS

Based on the results of this study on the flexural behavior of AAC confined walls, the following conclusions are presented:

1. The behavior of walls was characterized by yielding of the longitudinal steel reinforcement followed by flexure-shear and diagonal cracks.

Failure of walls was associated with propagation of diagonal cracks into the vertical confining elements. This shear behavior limited the drift ratios and the displacement ductilities of walls.
2. Flexural strength of walls increases as the axial load increases. Flexural strength of walls was well predicted using flexural theory. The ratio between analytical and experimental strengths of walls varied from 0.95 to 1.05
3. Displacement ductilities of walls increase as axial load decreases. Corresponding drift ratios were similar because of the presence of flexure-shear and diagonal cracks after yielding of the steel longitudinal reinforcement. Flexural behavior of walls can be improved if the shear behavior is avoided.

REFERENCES

ACI Committee 530. 2013. *Building Code Requirements and Specification for Masonry Structures (TMS 402-13/ ACI 530-13/ASCE 5-13; TMS 602-13/ACI 530.1-13/ ASCE 6-13)*. Farmington Hills, MI, USA: American Concrete Institute.

ACI Committee 318. 2014. *Building Code Requirements for Structural Concrete Code and Commentary*. Farmington Hills, MI, USA: American Concrete Institute.

ASTM. 2007. *ASTM C1006/C1006-07 Standard Test Method for Splitting Tensile Strength of Masonry Units*. West Conshohocken, PA, USA: ASTM International.

ASTM. 2009. *ASTM C1660/C1660-09 Standard Specification for Thin-Bed Mortar for Autoclaved Aerated Concrete (AAC) Masonry*. West Conshohocken, PA, USA: ASTM International.

ASTM. 2011. *ASTM C1693/C1693-11 Standard Specification for Autoclaved Aerated Concrete (AAC)*. West Conshohocken, PA, USA: ASTM International.

ASTM. 2015. *ASTM A370/A370-15 Standard Test Methods and Definitions for Mechanical Testing of Steel Products*. West Conshohocken, PA, USA: ASTM International, 2015.

ASTM. 2016a. *ASTM C39/C39-16 Standard Test Method for Compressive Strength of Cylindrical Concrete Specimens*. West Conshohocken, PA, USA: ASTM International.

ASTM. 2016b. *ASTM C109/C109M-16 Standard Test Method for Compressive Strength of Hydraulic Cement Mortars*. West Conshohocken, PA, USA: ASTM International.

Bose, S., & Rai, D. C. 2014. Behavior of AAC Infilled RC Frame Under Lateral Loading. *In Proceedings of the 10th National Conference in Earthquake Engineering*, Anchorage, Alaska, USA, July 2014.

Cancino, U.M. 2003. Behavior of autoclaved aerated concrete shear walls with low-strength AAC. *MS Thesis*, University of Texas at Austin, Austin, Texas, USA.

Edison Hydraulic Load Maintainers. 1994. *Operation and Maintenance Manual*. Edison Hydraulic Load Maintainers, 5657 Sawmill Rd. Paradise, California.

Flores, L. E., Mendoza J. A., & Reyes C. 2004. Ensaye de Muros de Mampostería con y sin Refuerzo Alrededor de la Abertura (Testing of Masonry Walls with and without Reinforcement on Openings). *In Proceedings of the XIV National Congress of Structural Engineering*, Acapulco, Mexico, Oct 2004. (in Spanish).

Mandirola, M., Penna, A., Rota, M., & Magenes, G. 2012. Experimental Assessment of the Shear Response of Autoclaved Aerated Concrete (AAC) Masonry with Flat Truss Bed-Joint Reinforcement. *In Proceedings of the 15th International Brick and Block Masonry Conference*, Florianopolis, Brazil, June 2012.

Meli, R. 1979. *Comportamiento Sísmico de Muros de Mampostería (Seismic Behavior of Masonry walls)*. Serie Azul: Instituto de Ingeniería, No. 352, UNAM, Mexico City, Mexico, 141 pp. (in Spanish).

NTCM. 2004. "Normas Técnicas Complementarias para el Diseño y Construcción de Estructuras de Mampostería (Technical Norms for the Construction and Design of Masonry Structures)," Gaceta Oficial del Distrito Federal, Mexico City, Mexico, 2004, pp. 6–87. (in Spanish).

Penna, A., Magenes, G., Calvi, G. M., & Costa, A. A. 2008. Seismic Performance of AAC Infill and Bearing Walls with Different Reinforcement Solutions. *In Proceedings of the 14th International Brick and Block Masonry Conference*, Sydney, Australia, February 2008.

Perez J., Flores L., & Alcocer S. 2015. An Experimental Study of Confined Masonry Walls with Varying Aspect Ratio. *Earthquake Spectra*, 31(2): 945–968.

Quiroz, L., Maruyama, Y., & Zavala, C. 2014. Cyclic Behavior of Peruvian Confined Masonry Walls and Calibration of Numerical Model Using Genetic Algorithms. *Journal of Engineering Structures*, 75: 561–576.

Ravichandran, S. S., & Klingner, R. E. 2012. Behavior of Steel Moment Frames with Autoclaved Aerated Concrete Infills. *ACI Structural Journal*, 109(1): 83–90.

Rodriguez M., and Botero, J. C. 1997. Evaluación del Comportamiento de Barras de Acero de Refuerzo Sometidas a Cargas Monotónicas y Cíclicas Reversibles Incluyendo Pandeo (Behavior of Steel Reinforcing Bars Subjected to Monotonic and Reversed Cyclic loads Including Buckling). *Revista de Ingeniería Sísmica*, Mexico, 56: 9–27. (in Spanish)

Tanner, J. E., Varela, J. L., Klinger, R. E., & Brightman, M. J. 2005a. Seismic Testing of Autoclaved Aerated Concrete Shear-Walls: A Comprehensive Review. *ACI Structural Journal*, 102(3): 374 382.

Tanner, J. E., Varela, J. L., & Klinger, R. E. 2005b. Design and Seismic Testing of Two-Story, Full Scale Autoclaved Aerated Concrete Assemblage Specimen. *ACI Structural Journal*, 102(1): 114 119.

Tena-Colunga, A, Juarez-Angeles, A. & Salinas-Vallejo, V.H. 2008. Cyclic Behavior of Combined and Confined Masonry Walls. *Journal of Engineering Structures*, 31(1): 240–259.

Tomazevic, M., & Gams, M. 2012, Shaking Table Study and Modelling of Seismic Behavior of Confined AAC Masonry Buildings. *Bulletin of Earthquake Engineering*, Springer Science+Business Media, Berlin, Germany, 10: 863–893.

Treviño, E. L., Alcocer, S. M., Flores, L. E., Larrua, R., Zarate, J. M., & Gallegos L. 2004. Investigation Experimental del Comportamiento de Muros de Mampostería Confinada de Bloques de Concreto Sometidos a Cargas Laterales Cíclicas Reversibles Reforzados con Acero de Grados 60 y 42 (Experimental Study on the Behavior of Confined Walls Constructed with Concrete Blocks Subjected to Reverse Cyclic Loads). *In Proceedings of the XIV National Congress of Structural Engineering*, Acapulco, Mexico, Oct 2004. (in Spanish).

San Bartolome, A., Quiun, D., & Torrealva, D. 1992. Seismic Behaviour of a Three-Story Scale Confined Masonry Structure. *In Proceedings of the 10th World Conference on Earthquake Engineering*, Madrid, Spain, 1992.

San Bartolome, A., & Quiun, D. 2010. Diseño Sísmico de Edificaciones de Albañilería Confinada (Seismic Design of Confined Masonry Structures)." *Revista Ciencia*, Mexico City, Mexico. 13(2): 161–185. (in Spanish).

Urzua, D. A., Padilla, R., & Loza, R. 2001. Influencia de la Carga Vertical en la Resistencia Sísmica de Muros de Mampostería Confinada con Materiales Típicos de Guadalajara (Influence of Axial Load on the Strength of Confined Masonry Walls). *In Proceedings of the XIII National Congress of Structural Engineering*. Guadalajara, Mexico, Oct 2001. (in Spanish)

Varela, J. L., Tanner, J. E., & Klingner, R. E. 2006. Development of Seismic Force Reduction and Displacement Amplification Factors for Autoclaved Aerated Concrete Structures. *Earthquake Spectra*, 22(1): 267 286.

Yu, J., Cao, J., & Fei, T. 2013. Experimental Study on Improving Seismic Behavior of Load-Bearing Masonry Walls Made of Autoclaved Aerated Concrete. *Transaction of Tianjin University*, Tianjin University, Tianjin, China, 19(6): 419–424.

Brick and Block Masonry - From Historical to Sustainable Masonry –
Kubica, Kwiecień & Bednarz (eds)
© 2020 Taylor & Francis Group, London, ISBN 978-0-367-56586-2

Experimental testing of seismic response of brick masonry walls under different boundary conditions

P. Triller & M. Tomaževič

Slovenian National Building and Civil Engineering Institute (ZAG), Ljubljana, Slovenia

M. Gams

Faculty of Civil and Geodetic Engineering, University of Ljubljana, Ljubljana, Slovenia

ABSTRACT: In order to reproduce the seismic response of masonry piers in laboratory cyclic shear tests, the boundary conditions need to be similar to those in actual buildings. If the similarity is not adequate, tests may predict unrealistic behaviour, and in the worst case, even unrealistic failure mechanisms. To study the effect of different boundary conditions, eight walls built with hollow clay units and general purpose mortar were tested under four different boundary conditions. The results show that different boundary conditions have a strong influence on force and displacement capacity of the walls.

1 INTRODUCTION

Boundary conditions in cyclic shear laboratory tests attempt to simulate the actual conditions in a structure during an earthquake. Therefore, the walls are loaded with (usually constant) vertical load to simulate the weight of the structure above the tested wall and horizontal load acting in the plane of the wall to simulate the seismic action. The horizontal forces are usually cyclic and gradually increase until the collapse of the wall. Due to the similarity of damage and failure mechanisms of walls in laboratory tests with observations of buildings after earthquakes, experimental observations and measurements form the basis for the theory of behavior of masonry structures. According to the latter, there are three basic failure mechanisms of masonry walls: sliding shear, shear with diagonal cracks and flexure (Figure 1).

In addition to the dominant mechanism, which is usually shear with diagonal cracks or flexure with compressive failure at corners (also called toe-crushing), other phenomena, such as rocking of the wall are often observed when testing single walls, especially if walls are built from modern units and mortar. Such mechanisms are not observed in earthquake-damaged buildings and in in-situ tests.

The rocking of the wall is characterized by low energy dissipation and the formation of horizontal cracks at the contact with horizontal connecting elements. Although the current standards (Eurocode 8-1, 2005; FEMA 356, 2000) do not contain provisions that would include the contribution of rotation of the wall as rigid body yet, the phenomenon of rocking is more and more often mentioned among experts (Calderini et al., 2009; Parisi et al., 2013).

Despite the experimental evidence, rocking of the wall has never been observed as mechanism, defining the seismic behaviour of masonry buildings.

There are several reasons for rocking in recent tests. One of them are the changes in modern masonry, i.e. in materials and in construction technology. The modern blocks have ground bed surfaces and hole patterns are different. The brick material is also more porous and the mortars are changing. They have higher proportion of cement, are applied in thin(er) layer, and their strength is much higher than in the past. Adhesives based on polyurethane can be used instead of traditional mortar.

Other factors contributing to rocking are the geometry of the wall, the vertical load and, crucially, the boundary conditions in the laboratory test (Tomaževič, 2016). The latter are highly idealized in laboratory tests for several reasons and this idealization may have a large effect on seismic response.

Two types of idealization are the most common in tests: the cantilever type or symmetrically fixed type. In modern masonry buildings, where floor structures are RC slabs with extra reinforcement above the walls, it is reasonable to assume symmetrically fixed boundary conditions. But, not only are these conditions difficult to simulate properly, there are also different ways to achieve them. One is to ensure that the inflection point (zero moment point) is at mid-height of the tested wall during the entire test. In this case, however, the rotations at the top of the wall cannot be constrained. The second approach is based on constraining rotations at the upper edge (Salmanpour and Mojsilović, 2015). In both cases the top of the wall is free to move vertically, because the vertical load remains constant throughout the test.

Figure 1. Typical in-plane masonry failure mechanisms: sliding shear (left), diagonal shear (middle) and flexural mode (right).

Several studies listed below show that the afore-mentioned idealized types of tests fail to represent the real conditions in a structure. The study of bound-ary conditions in experimental tests of masonry walls by Petry and Beyer (2014) is one of few that uses boundary conditions in the entire interval between fixed and free upper edge. Their study shows that the effect of different boundary conditions strongly reflects in the drift capacity of the URM walls.

Wilding and Beyer (2018) studied the influence of different types of boundary conditions numerically. They considered four different types of boundary con-ditions, and showed that their numerical models, covering different boundary conditions, fit well with test results.

The effect of rigidity of the top beam on the seis-mic response of the walls was investigated in the work of Tomaževič and Gams (2009). The results show that the effect is substantial. Same authors later analyzed different types of boundary conditions experimentally and considered three different types of boundary conditions on the seismic response of modern masonry (Gams and Tomaževič, 2012). The tests were performed on a relatively small walls, but show substantial differences in response nevertheless.

A way of eliminating the idealized boundary condi-tions and at the same time taking into account the inter-action between piers, parapets, lintels, and floor structures is to test models of entire structures. Unlike cyclic shear tests of single walls, such tests are rare in the research community due to considerably larger financial costs as well as other difficulties. Each of a few studies of such large specimens which can be found in literature (Leiva, 1991; Magenes et al., 2005; Yi et al., 2016; Triller et al., 2018) is dedicated to spe-cific issue and does not provide an answer to our problem.

In this paper four types of boundary conditions are applied in order to study their effect on seismic response (force and displacement capacity, type of fail-ure) of masonry walls in laboratory cyclic shear tests.

2 MATERIALS

2.1 Units and mortar

Wall specimens were built using hollow clay units with nominal dimensions (length/width/height = 290/190/ 190 mm). The proportion of voids according to the

gross volume of the unit, determined according to EN 772-9:1999 on the sample of 10 blocks, was 55 %. The units are classified as group 2 units according to Euro-code 6 (2005). Declared value of compressive strength of bricks, provided by the manufacturer ($f_{c,b,m}$), was 15.0 MPa. The characteristic compressive strength of units ($f_{c,b,t}$), tested according to EN 772-1 (2011), was 25.1 MPa. The shape of the unit is shown in Figure 2.

Mortar used for laying the units was a typical gen-eral purpose mortar, consisting of cement, lime and sand in volumetric proportion of 0.5:1:8, prepared on site. The flexural and compressive strengths of the mortar have been determined on prisms (40/40/ 160 mm) according to EN 1015-11 (1999) at the age of 28 days. Additionally, the compressive strength of mortar on cubes (70/70/70 mm) was also measured. Average compressive strength on cubes ($f_{m,c,28}$) and prisms ($f_{m,p,28}$) at the age of 28 days was 1.8 MPa and 2.0 MPa, respectively. Flexural strength at the same age ($f_{x,28}$) amounted to 0.7 MPa. Results are presented in Table 1.

 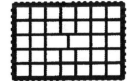

Figure 2. Masonry unit MB 29-19.

Table 1. Mechanical characteristics of materials, used for the construction of test specimens.

Material	Property	Symbol	Unit	Value
Blocks	Compressive strength perpendicular to cells (manufacturer)	$f_{c,b,m}$	(MPa)	15.0
	Compressive strength perpendicular to cells (test)	$f_{c,b,t}$	(MPa)	25.1
Mortar	Compressive strength on cubes at the age of 28 days	$f_{m,c,28}$	(MPa)	1.8
	Compressive strength on prisms at the age of 28 days	$f_{m,p,28}$	(MPa)	2.0
	Flexural strength at the age of 28 days	$f_{x,28}$	(MPa)	0.7
Masonry	Compressive strength perpendicular to bed joints	$f_{c,m}$	(MPa)	3.8
	Modulus of elasticity perpendicular to bed joints	E_m	(GPa)	4.3
Concrete	Compressive strength on cubes at the age of 28 days	$f_{c,c}$	(MPa)	40.8

2.2 Masonry

Compressive strength of masonry ($f_{c,m}$) perpendicular to bed joints was determined on three specimens according to EN 1052-1 (2001). Average compressive strength was 3.8 MPa, while characteristic value amounted to 3.1 MPa. The secant modulus of elasticity of masonry (E_m) perpendicular to bed joints has been evaluated on the basis of stress-strain relationship curve. Average value of E_m at 1/3 of the maximum stress was 4.3 GPa.

2.3 Concrete and steel

Concrete and steel reinforcement were used to construct the top beams. C30/37 concrete was mixed in a concrete plant and transported to the laboratory. To determine compressive strength according to EN 12390-3 (2009), samples of concrete were taken during concreting. The average value of compressive strength of concrete ($f_{c,c}$), measured on cubes (150/150/150 mm) was 40.8 MPa. Steel reinforcement consisted of Φ14 longitudinal bars and Φ8 stirrups.

2.4 Summary of material characteristics

The characteristics of the materials used in the tests are summarized in Table 1.

3 GEOMETRY

Wall specimens for cyclic tests represent a pier in the ground floor of a five-storey unreinforced masonry residential building, built during the reconstruction period after World War II and before 1964, when the first seismic codes were adopted for Slovenia and wider region. In order to investigate the appropriate boundary conditions, which best simulate the behavior of an individual wall in such a building, four types of boundary conditions were chosen and two walls per each type were tested. In total, eight cyclic shear tests of walls were performed.

Wall specimens for cyclic shear tests with dimensions of 1.19/0.79/0.19 m (height/length/thickness) have been built on RC foundation blocks to facilitate the transport. Horizontal and vertical joints were filled with 1.0-1.5 cm thick layer of mortar. Unit overlapping was 33 % of the unit length. On the top of the wall, an RC bond beam was constructed for the purpose of uniform distribution of vertical and in-plane lateral load onto the walls. Schematic presentation of the wall specimen is shown in Figure 3. The first row of bricks was placed on the layer of cementitious mortar, while the next five rows were constructed using general purpose mortar.

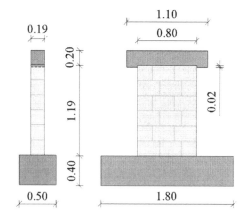

Figure 3. Geometry of the wall; dimensions in meters.

4 TEST DESCRIPTION

4.1 Test setup

The RC foundation block of the wall was fixed to the laboratory strong floor using prestressed bolts. On top of the masonry wall were the RC bond beam and a strong steel beam. The vertical and horizontal loads were applied to the wall via the steel beam. The vertical load was applied using two hydraulic servo controlled actuators at each end of the steel beam, and connected to the strong floor by a set of steel tie rods (see Figure 4). The horizontal load was applied by a hydraulic servo controlled actuator fixed to the steel frame, which served as lateral support (instead of reaction wall). Schematic presentation of the test setup is shown in Figure 4.

4.2 Instrumentation

During the tests the horizontal and vertical displacements at characteristic points of the specimens were measured using LVDTs, while the displacement field

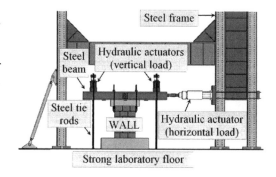

Figure 4. Test setup for cyclic shear testing the wall.

over the entire surface of the walls was measured with an optical digital image correlation (DIC) system with measuring accuracy of 2/100 px or 0.04 mm. Two 5 megapixel resolution camera was used on one side of the wall specimen. In order to ensure optimal performance of the optical system, the monitored surfaces were painted with a random black speckle pattern, while the surface on the opposite side of the walls was painted white to facilitate observation of cracks throughout the tests. Forces and displacements were also measured at all hydraulic actuators. The rotation on the top of the wall was measured using a precise inclinometer. Instrumentation of individual walls is schematically presented in Figure 5.

4.3 Testing procedure

4.3.1 Preload

At the beginning of the test, a vertical load was imposed by hydraulic actuators in order to simulate a stress state in the walls of the bottom storey of a typical five-storey unreinforced masonry residential building. Average compressive stresses amounted to $\sigma_0 = 0.90$-1.0 MPa, which is equivalent to 24-26 % of the average compressive strength of the masonry ($f_{c,m} = 3.8$ MPa), i.e. precompression ratio of $\sigma_0/f_c = 0.24$-0.26.

Depending on the type of boundary conditions, in some cyclic shear tests the vertical load remained constant until the end of the test, while in some cases it varied in accordance with the predetermined boundary conditions.

4.3.2 Seismic load

Seismic action effects have been simulated by imposing horizontal displacements in the form of a series of cyclic reversals. Each displacement amplitude was imposed three times alternately in positive and negative direction. The procedure was repeated with increased displacement amplitudes up to the (near) collapse of the specimen. Imposed drift ratios (Φ) being defined as the ratio between the horizontal displacement (d) and height of the wall (h), are presented in Figure 6.

4.4 Boundary conditions

In cyclic shear tests, the masonry pier is usually fully fixed at the bottom, while at the upper edge there can be different combinations of horizontal and vertical displacements, rotations and vertical (axial) forces. The desired boundary conditions are introduced via regulation of servo controlled hydraulic actuators.

The most commonly used boundary conditions in cyclic shear tests are of cantilever (I) (free vertical displacement and rotations at the top and constant vertical load) and symmetrically fixed (II) (zero rotation at the top and constant vertical load) type as mentioned in the introduction. In our investigation, it was possible to maintain practically zero rotation ($\pm$ 0.01°) at the top of the wall by a set of two computer regulated servo-actuators. When preventing rotations the actuators adjusted the forces in the steel bars so that the total vertical load remained constant throughout the test. The rotation was controlled using a precise inclinometer.

In case of the third type of boundary conditions (III), the rotations as well as the vertical displacements (at a time when the horizontal force H is different from 0 kN) on the top are constrained.

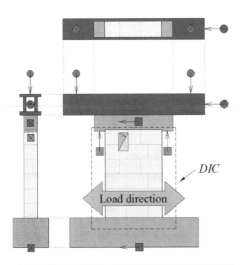

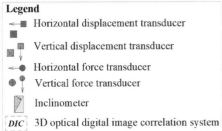

Figure 5. Instrumentation of the wall specimen.

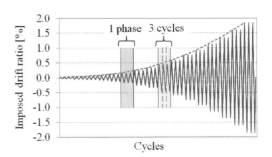

Figure 6. Loading history imposed to the wall specimens.

This causes an increase in the applied vertical load when the imposed seismic load reaches the extreme value (i.e. when the amplitude of a single cycle is reached) and reduction of the vertical load when the damaged wall returns back to zero position. The total vertical force in the wall during the test is therefore not constant. Each time the wall crosses the zero position, the total vertical force is increased to the initially prescribed value, which causes shrinkage of the wall. The fourth type of boundary conditions (IV) represents a combination of the first and the second type: rotations at the top are free and the total vertical load is constant, but the vertical loads at both sides of the wall change so that the zero moment (inflection point) is exactly at the mid height of the wall during the entire test. Schematic representation of the boundary condition types and their description are shown in Figure 7 and in Table 2.

5 TEST RESULTS AND OBSERVED STRUCTURAL BEHAVIOUR

5.1 Damage propagation

The seismic behavior of the walls during cyclic shear tests is described for each type of boundary conditions individually. As there were two walls per each type of boundary conditions, only

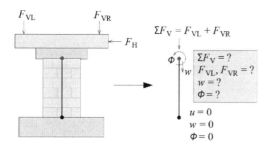

Figure 7. Schematic illustration of the considered boundary condition types.

Table 2. Boundary conditions (b.c.).

Type of b.c.	I	II	III	IV
Symbol		⊤⌐		
ΣF_V	= const.	= const.	≠ const.	= const.
$F_{VL} = F_{VR}$	=	≠	≠	≠
$F_{VL} \neq F_{VR}$				
w	≠ 0	≠ 0	= 0	≠ 0
Φ	≠ 0	= 0	= 0	≠ 0

a general description for each type is given. Reference to specific wall is provided only if necessary. The descriptions of the seismic behaviour of the walls are based on the visual observations made during the tests.

Type I. In the initial loading cycles, the walls W1 and W2, which were tested as cantilevers, behaved as rigid bodies after horizontal cracks at the bottom of the wall appeared due to uplifting of the wall at the contact with the foundation at drift ratio of 0.17 % (horizontal displacement 2.0 mm). The cracks opened and closed according to the load direction. At the rotation of 0.60 % the crushing of both lower corners was observed. As the displacement amplitude increased, the cracks widened and parts of the blocks started to fall off. The damage was mostly of the flexural type. The tests were completed at drift ratio of 1.61 % (W1) and 1.72 % (W2), when the flexural failure occurred.

Type II. A typical shear mechanism was observed in the case walls W3 and W4. The first visible diagonal cracks occurred at drift ratio of 0.04 %. At the same drift ratio, some minor horizontal cracks at the contacts with the RC foundation and RC beam have also been observed. By increasing the lateral load (displacement), the existing diagonal cracks spread and new ones appeared. Most of the cracks were in the mortar joints in a stepwise pattern, but there were also thin cracks in the units. By the end of tests the shells of the hollow units started to crush and to fall off. The walls W3 and W4 failed in shear at wall rotation of 0.34 % and 0.36 %, respectively.

Type III. Walls W5 and W6 first cracked in the central part of the walls at the rotation of 0.06 % and 0.07 %, respectively. The direction of cracks was diagonally across the masonry wall. In subsequent loading cycles cracks spread over the entire height of the wall. In the last loading cycles of the tests, the growth of cracks increased significantly. The failure mechanism in both cases was of shear type and occurred at drift ratio of 0.34 %.

Type IV. Walls W8 and W9 were tested with moment inflection point at the mid-height of the. The first visible cracks occurred at 0.13 % rotation. The extent of the cracks increased proportionally with the imposed seismic load. The damage was concentrated mainly in the central part of the wall W8, while the distribution of cracks along the entire height of the wall W9 was almost uniform. During the test of the wall W8, compressive cracks at the bottom corner appeared although this did not affect the type of the failure mechanism, which was in both cases of shear type. It occurred at drift ratio of 0.26 % and 0.34 %, respectively.

In cases II, III and IV the shear failure mechanism with diagonal cracks, caused by exceeding of tensile strength of masonry prevailed despite the fact that there were some minor horizontal cracks at the bottom and top of the walls and there was some compressive crushing of the corner units. On the other hand, both walls of cantilever type failed in flexure.

5.2 Hysteretic behaviour

Experimentally obtained hysteretic curves, which describe the relationships between the resisting force H and drift ratio Φ, are shown in Figure 8 (note that the scale varies between the walls).

5.3 Limit states and DIC measurements

To compare the seismic behaviour of the tested walls, seismic resistance, displacement and rotation capacity have been evaluated at three characteristic limit states (LS), namely:

- *damage limit state*, defined as the point on the hysteretic envelope, where damage occurs to the walls, leading to noticeable stiffness degradation;
- *maximum resistance limit state*, where the wall reaches maximum resistance to horizontal load;
- *ultimate limit state*, which defines a near collapse limit state at the point where the maximum horizontal displacement of the test is attained.

The values of resisting force (H) and drift ratio (Φ) at limit states, evaluated as the average values obtained at loading in positive and negative direction, are given in Table 3.

Table 4 summarizes the damage propagation and the measurements of optical digital image correlation system, showing the locations where cracks because of stress concentrations appear. Due to the similarity of the damage of the walls, tested under the same boundary conditions, only one figure per each boundary condition type is shown.

Comparison of the photos of damage propagation taken during the cyclic shear tests and of DIC images shows compliance of damage locations and of their extent. DIC measurements as well as the

Table 3. The values of shear force H and drift ratio Φ for tested walls at characteristic limit states (LS).

	Damage LS		Maximum resistance LS		Ultimate LS	
	H_e	Φ_e	H_{max}	Φ_{Hmax}	H_{du}	Φ_u
Wall	kN	%	kN	%	kN	%
W1	32.7	0.17	38.7	0.74	19.1	1.61
W2	34.0	0.17	38.5	0.37	30.3	1.72
W3	57.6	0.04	70.2	0.13	33.2	0.34
W4	65.4	0.04	73.3	0.06	40.8	0.36
W5	70.4	0.04	99.7	0.21	46.5	0.34
W6	61.1	0.04	92.6	0.17	29.9	0.34
W8	46.7	0.07	63.5	0.17	44.8	0.26
W9	56.7	0.06	70.5	0.13	45.9	0.34

observations made during the tests disproved rocking or flexure mechanism as a significant response mechanism whenever the boundary conditions resembled the fixed-fixed type (types II, III and IV).

Results of wall specimens W1 and W2, which were tested under cantilever boundary conditions (type I) were different and responded primarily with rocking and flexure.

In some cases, the DIC system detects the damage even before it is detected visually. The first damage, detected by the DIC system is consistent with the loading stages at which the relationship between the horizontal force H and the drift ratio Φ changes significantly for the first time.

5.4 Reference tensile strength of masonry

The seismic resistance of the wall when shear failure mechanism occurs, can be determined by the maximum value of the principal tensile stress that develops in the wall at the maximum shear force H_{max}, assuming that the wall behaves as an elastic, homogenous and isotropic panel up until the attainment of its maximum resistance. Turnšek and Čačovič (1971) derived the value of the referential principal tensile strength of masonry wall based on the elementary theory of elasticity.

The equation to determine it is:

$$f_t = \sigma_t = \sqrt{\left(\frac{\sigma_0}{2}\right)^2 + (b\tau_{max})^2} - \frac{\sigma_0}{2} \quad (1)$$

where f_t is tensile strength of masonry, evaluated on the basis of the diagonal tension model (Turnšek and Čačovič, 1971), $\sigma_0 = N/A_w$ = the average compression stress due to vertical load N, A_w is the horizontal cross-section area of the wall, b is the shear stress distribution factor which depends on the geometry of the

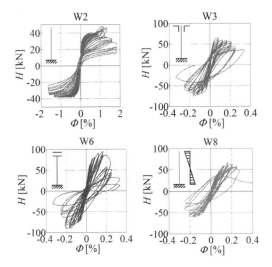

Figure 8. Hysteretic storey shear force H – drift ratio Φ curves for all types of boundary conditions considered.

Table 4. Cracks propagation and major strain distribution measurements at ultimate limit state for each type of boundary conditions (b.c.).

Type of b.c.	Crack propagation	DIC measurements
I		
II		
III		
IV		
Legend	0.0 0.4 0.8 1.2 1.6 2.0 [%]	

Table 5. Calculation of the tensile strength of masonry f_t.

Type of b.c.	II		III		IV	
Wall	W3	W4	W5	W6	W8	W9
σ_0 (MPa)	1.03	1.03	1.85	1.79	1.02	1.03
H_{max} (kN)	70.2	73.3	99.7	92.6	63.5	70.5
f_t (MPa)	0.36	0.38	0.44	0.39	0.30	0.36
$f_{t\_b.c.}$ (MPa)	0.37		0.41		0.33	

2009), the tensile strength of masonry thus defined is still used today as a basic parameter in engineering practice, especially for existing masonry structures.

The values of tensile strength of masonry f_t for the tested walls, which failed in diagonal shear as well as the average f_t values depending on type of boundary condition ($f_{t\_b.c.}$), are shown in Table 5.

When calculating the tensile strength the compressive stresses in the walls at the time of maximum horizontal force were taken into account. Average values for both loading directions are calculated. The average tensile strength of masonry, measured on the W3-W9 wall specimens, is 0.37 MPa. Results in Table 5 indicate that boundary condition have an effect on tensile strength of masonry.

6 CONCLUSIONS

In order to determine the most appropriate boundary conditions for testing masonry walls in cyclic shear, tests of single walls were performed at four different types of boundary conditions. In all cases the walls were fully fixed to the floor and different boundary conditions applied at the top by a controlled regulation of three degrees of freedom at the top of the wall: horizontal and vertical displacements (or forces) and rotations (or moments). Four types were considered: type I (cantilever type), type II (fixed-fixed rotations with constant axial force), type III (fixed-fixed rotations and fixed vertical displacements with variable axial force) and type IV (fixed position of the moment inflection point at mid-height of the wall).

Results of cyclic shear tests have shown that the cantilever boundary conditions (type I) produced flexural failure.

In contrast, all other boundary conditions produced shear type of response and failure mechanism. Rocking as a main mechanism of response was not observed, although there were horizontal cracks at the contact with the foundation and top beam. Rocking might be more relevant for different geometry, and especially lower compressive stress, but for the present conditions, which are deemed representative, this was not the case.

The boundary conditions have an observable effect on the limit states and on the tensile strength evaluated from the T-Č model, which indicates that the choice of the test is not completely arbitrary.

wall. For slender walls with ratio height/length $h/l \geq 1.5$, $b = 1.5$; for squat walls with $h/l < 1.5$; $b = \max\{1.1; h/l\}$. $\tau_{Hmax} = H_{max}/A_w$ = the average shear stress in the wall at the attained maximum resistance H_{max}.

The tensile strength is a material characteristic and should therefore be independent of external loads. Experiments, however, indicate that its value partly depends on the level of compression of the wall. Unfortunately, the impact of compressive stress on the so defined tensile strength of the wall has not been systematically investigated yet. Since the reliability of the predicted shear resistance of masonry buildings using the model of Turnšek and Čačovič has been validated by many researchers (see e.g. Tomaževič,

Type IV tests were the easiest to perform and produced the lowest tensile strength values and ultimate displacement values, which would indicate they are on the safe side.

Finally, it is important to emphasize that the results presented in the paper, relate to the considered geometry, vertical load, material characteristics and construction technology. Other types of masonry and conditions could lead to different conclusions.

ACKNOWLEDGEMENTS

The research, presented in this paper, was carried out within the research projects J2-6749 and ARRS-MR-496, financed by Slovenian Research Agency.

REFERENCES

Calderini, C., Cattari, S. & Lagomarsino, S. 2009. In-plane strength of unreinforced masonry piers. *Earthquake Engineering and Structural Dynamics* 38, 2: 243–267.

CEN. 1998. Methods of test for masonry units. Determination of volume and percentage of voids and net volume of clay and calcium silicate masonry units by sand filling, EN 772-9. Bruxelles.

CEN. 1999. Methods of test for masonry. Determination of compressive strength, EN 1052-1. Bruxelles CEN (2001) Methods of test for mortar for masonry - Part 11: Determination of flexural and compressive strentgth of hardened mortar, EN 1015-11. Bruxelles.

CEN. 2005. Eurocode 6: design of masonry structures - part 1-1: general rules for buildings—rules for reinforced and unreinforced masonry, EN 1996- 1-1. Bruxelles.

CEN. 2005. Eurocode 8: Design of structures for earthquake resistance – part 1: general rules, seismic actions and rules for buildings, EN 1998-1. Bruxelles.

CEN. 2009. Testing hardened concrete. Compressive strength of test specimens, EN 12390-3. Bruxelles.

CEN. 2011. Methods of test for masonry units Part 1: Determination of compressive strength, EN 772-1. Bruxelless.

FEMA 356. 2000. Prestandard and commentary for the seismic rehabilitation of buildings. Washington DC: Federal Emergency Management Agency.

Gams, M. & Tomaževič, M. 2012. Experimental simulation of seismic response of masonry walls. *15th World Conference on Earthquake Engineering, Lisbon, Portugal, September* 24-28, 2012. Lisbon, Sociedade Portuguesa de Engenharia Sismica: p. 1–8.

Leiva, G. H. 1991. *Seismic Resistance of Two Story Masonry Walls with Openings.* Doctoral dissertation. Austin, University of Texas.

Magenes, G., Calvi, G.M. & Kingsley, R. 1995. Seismic testing of a full-scale, two-story masonry building: Test procedure and measured experimental response. In: Experimental and numerical investigation on a brick masonry building prototype – numerical prediction of the experiment – Report 3.0, G.N.D.T. Pavia, Italy.

Parisi, F., Lignola, G.P., Augenti, N., Prota, A. & Manfredi, G. 2013. Rocking response assessment of in-plane laterally-loaded masonry walls with openings. *Engineering Structures,* 56: 1234–1248.

Petry, S. & Beyer, K. 2014. Influence of boundary conditions and size effect on the drift capacity of URM walls. *Engineering Structures,* 65: 76–88.

Salmanpour, A.H. & Mojsilović, N. 2015. Simulation of boundary conditions for testing of masonry shear walls. *Proceedings of the 11th International Conference of Computational Methods in Sciences and Engineering,* Athens, *Greece,* March 20-23, 2015. New York, AIP Publishing: paper 1702.

Tomaževič, M. & Gams, M. 2009. Shear resistance of unreinforced masonry walls. *Ingegneria sismica* 26, 3: 5–18.

Tomaževič, M. 2016. Some considerations on testing and experimental simulation of seismic behaviour of masonry walls and buildings. V: Modena, C. (ed.), Da Porto, F. (ed.), Valluzzi, M. R. (ed.). Brick and Block Masonry: Trends, Innovations and Challenges, *Proceedings of* the 16th International Brick and Block Masonry Conference, Padova, Italy, June 26- 30, 2016. Boca Raton, CRC Press: p. 37–53.

Triller, P., Tomaževič, M. & Gams, M. 2018. Seismic behaviour of masonry buildings built of low compressive strength units. *Bulletin of Earthquake Engineering* 16, 12: 6191–6219.

Turnšek, V. & Čačovič, F. 1971. Some experimental results on the strength of brick masonry walls. In: *Proceedings 2nd international brick-masonry conference.* British Ceramic Society, Stoke-on-Trent, pp 149–156.

Wilding, B.V. & Beyer, K. 2018. Shear-compression tests of URM walls: Various setups and their influence on experimental results. *Engineering structures,* 156: 472–479.

Yi, T., Moon, F.L., Leon, R.T. & Kahn, L.F. 2006. Lateral load tests on a two-story unreinforced masonry building. *Journal of Structural Engineering. Reston,* ASCE: p. 643–652.

Experimental analysis of brick masonry veneer walls under out-of-plane loading

A. Martins & G. Vasconcelos
ISISE, University of Minho, Portugal

A. Campos-Costa
National Laboratory of Civil Engineering (LNEC), Lisbon, Portugal

ABSTRACT: The evaluation of seismic vulnerability of existing buildings with masonry veneer systems has been recognized as a major problem because of the large number of buildings constructed before the development of rational seismic codes. This resulted in the construction of masonry veneers without reference to the design to seismic action and adequate constructive detailing. In order to contribute to increasing of knowledge about seismic behaviour of brick veneer walls, an experimental campaign was developed on testing quasi-statically full-scale systems under in-plane and out-of-plane loading. This paper describes in detail the out-of-plane performance of a constructive system characteristic of Portugal and South of Europe, constituted of brick masonry veneer leaf connected to an infill wall inserted in a reinforced concrete (rc) frame. A description of the test setup for the out-of-plane tests is provided and the main results, including the damage patterns and force displacement diagrams, are presented and discussed.

1 INTRODUCTION

Brick veneer masonry walls are frequently used as a façade finishing in residential construction in several countries in different parts of the world, namely North America, Australia, England and other European countries due to its aesthetic appearance, durability and its thermal performance. In general, brick veneer walls are separated from an air cavity in relation to a backing system to which it is attached. The backing system can be light wood or steel frames, structural masonry or masonry walls enclosed in rc frames. The backup system is considered as the primary lateral load-resisting system and the brick veneer is considered to be non-structural. The brick veneer walls are attached to the backing system through distinct types of ties, most commonly in steel and can have different shapes and geometry, much dependent on the backing system.

Although the veneer walls are regarded as non-structural elements and are not part of the resisting system of a building, they are subjected to different types of loadings, including self-weight, wind or earthquakes in case of seismic hazard regions. With respect to seismic action, the veneer can be considered as an added mass, neither contributing for the stiffness nor for the resistance.

The performance of veneer walls to loads during seismic events is influenced by the interaction of the veneer with the backup through wall ties, their thickness, height, length, and height to width ratio

(Memari et al. 2002). Recent earthquakes occurring in different European countries brought to light fragilities of masonry veneer walls. After many of these events, it was possible to observe common failure mechanisms associated to in-plane diagonal cracking and often the detachment and complete disintegration of the masonry veneer walls. This deficient behaviour should be attributed eventually to the inefficient connections and absence of suitable design rules that consider the effect of the seismic actions on the masonry veneer walls systems (Borchelt 2004).

Usually, masonry walls present particular vulnerability if pushed horizontally in a direction perpendicular to its plane (out-of-plane loading), but offers higher resistance if pushed along its length (in-plane loading). This is not only valid for loadbearing walls but also for non-structural walls that are forced to behave in a structural way in case of seismic actions. Among the non-structural walls, masonry infills and masonry veneers are well known to be used in more modern construction, where reinforced concrete frames as a structural system predominate.

The distribution of the load between the backing support and the brick veneer depends on the type of loading, the stiffness of each element, and the stiffness of the connecting ties. Under wind loads, any in-plane or out-of-lane load in the veneer will to be transferred from the veneers to the backing through the ties. Inertial forces from earthquakes will load both the frame and the veneer. In both cases, the stiffness of the connecting ties should play a key role in the load distribution (Desai & McGinley 2013).

It is considered that a detailed investigation on the seismic behaviour of masonry veneer walls becomes necessary, especially regarding the connection of the masonry veneers to the backing infill masonry walls. The primary gap identified through literature review was the lack of experimental research that addressed the response of masonry veneer walls, whose backing is composed by masonry infill wall inserted in a rc frame (Martins et al. 2017). This represented the major motivation for conducting this research based on experimental characterization of the mechanical behaviour of brick veneer walls attached to brick masonry infills.

The main results of the experimental campaign intended to achieve are: (1) hysteretic force-displacement diagrams under out-of-plane loading; (2) deformation features of the walls and (3) damage patterns and failure mechanism of the masonry veneers and connections under out-of-plane loading.

2 MATERIALS

The system under evaluation is composed of a rc frame with brick masonry infills attached to a brick masonry veneer by steel ties. The brick masonry veneer wall is constituted by ceramic bricks with vertical holes with approximately 237 mm × 115 mm × 70 mm (length × thickness × height) (Figure 1a). The brick masonry infill walls were built with ceramic brick units perforated horizontally with approximately 300 mm × 150 mm × 200 mm (length × thickness × height) (Figure 1b). Notice that, even if the rc frame is built at reduced scale, it was decided to build the brick infill and brick veneer walls with full scale brick units to have better representativeness. The brick veneer walls assemblage was carried out by using a pre-mixed water-repellent cement mortar, usually recommended by the brick unit producer. For the backup, a pre-mixed M5 general purpose mortar was used, following what was used in a previous experimental work carried out on brick infill walls (Akhoundi et al. 2018). The thickness adopted for the mortar bed joints was 15 mm to enable the perfect levelling of the tie.

After a research in the market of steel ties, it was observed that ties with different geometry and shapes are used to attach veneer walls to different backing

systems, see Figure 2. Tie wall T6 is composed by basalt fibre and the other ties are made of stainless steel according to technical notes. Apart from the T5 wall tie, the ties are placed on mortar bed joints in infill and veneer leaves, with suitable embedment length.

For the out-of-plane test on the brick veneer wall tested in this work, it was decided to use the steel tie T2. It has a length of 225 mm, a thickness of 5.5 mm and a cross section area of 23 mm^2.

3 EXPERIMENTAL CAMPAIGN ON MASONRY MATERIALS

3.1 Brick units

Six specimens were tested for each brick typology (brick veneer and brick infill) under compression uniaxial loading according to European standard EN 772-1 (2000). The bricks were tested under three different directions (Figure 3), namely: (a) direction a - parallel to the perforations; (b) direction b – perpendicular to the length of the brick; (c) direction c – perpendicular to the thickness of the bricks.

Regarding the preparation of surfaces, in veneer units, the bed faces of the specimen were cleaned and sanded and any loose grit was removed. For both types of bricks, the surfaces were regularized with levelled thin mortar to achieve an even application of loading. The test machine was a load frame for compression tests with limit capacity of 2500 kN in closed-loop control. The bearing surfaces of testing machine were wiped, and the specimen was aligned carefully in the centre of the ball-seated platen working as an uniform seating. The uniaxial compression testses were caried out in force control with a rate of 2 kN/s. Linear Voltage Displacement Transducers (LDVT) were used in order to record the vertical deformations of specimens during the compressive test.

Figure 2. Wall tie typologies.

Figure 3. Loading direction in uniaxial compression of brick units.

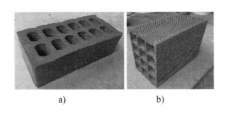

a)   b)

Figure 1. Brick units; a) veneer walls; b) brick infill walls.

3.2 Brick veneer masonry

The uniaxial compression tests carried out to characterize the compressive behaviour of brick veneer masonry was based on EN 1052-1(1999). The top of the specimens was levelled in order to have a uniform vertical load. The test was performed in displacement control at a rate of 5 μm/s. Adequate instrumentation was used so that the stress-strain diagrams and the related mechanical parameters, namely the modulus of elasticity, compressive stress and Poisson coefficient could be obtained.

The flexural tests were based on European standard EN 1052-2 (1999). The flexural strength in pure bending is obtained under four-point loading. There are two typologies of test in order to obtain: (1) the flexural strength with failure parallel to the bed joints (f_{xk1}) and (2) flexural strength with failure perpendicular to the bed joints (f_{xk2}). The specimens support lines were levelled in order to have a uniform load application. The test was performed in displacement control at a rate of 10 μm/s. Five LVDTs were used, two at centre of samples (one in front and other at the back), two at loading application points and a LVDT to control the actuator displacement.

4 OUT-OF-PLANE TESTS ON BRICK VENEER WALL

4.1 Design of masonry specimen

The experimental model of masonry veneer walls was designed taking into account real features of typical brick masonry veneer walls and laboratory conditions. It was defined based on the constructive system composed of a reinforced concrete (rc) frame with brick masonry infills having attached brick veneer walls. This constructive system is not only very common in Portugal but also in south European countries.

The reinforced concrete frames used in the experimental campaign had been previously used in other experimental campaign on the analysis of the out-of-plane behaviour of masonry infill walls (Akhoundi et al. 2020) (Figure 4). The rc frame could be re-used because the damage previously induced was minor given that the out-of-plane loading was directly applied in the brick masonry infill walls. In addition, fixed bottom and upper beams were considered as boundary condition, resulting in the low damage observed.

The experimental program on brick veneer walls was defined in order to get the maximum information about out-of-plane performance (O) of these constructive elements. The attachment of the brick masonry veneer was carried out by beans of steel ties T2 (Figure 2) with a spacing of 2.5 ties per meter square embedded in mortar joints. The air cavity distance is 100 mm. The interface at the base of the brick veneer was defined by using a flashing system (specimen T2_O_100_2.5).

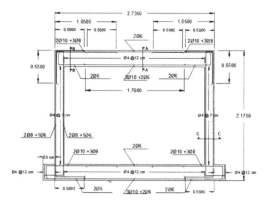

Figure 4. Reinforced concrete frame.

4.2 Construction of masonry walls

The construction of the masonry walls systems (brick masonry infill and brick masonry veneer) is a complex task because it needed be made by phases. In a first phase, the brick infill enclosed in the rc frame was built. In this phase, the positioning of the ties is of major importance to ensure adequate alignment between brick masonry infill and veneer walls (Figure 5a and b).

After this, a shelf angle is bolted to the bottom rc beam just above the foundation, and a flashing is placed on the shelf angle, as shown in Figure 5c and d. This was made to evaluate its role in the friction level developed at the base between the shelf angle

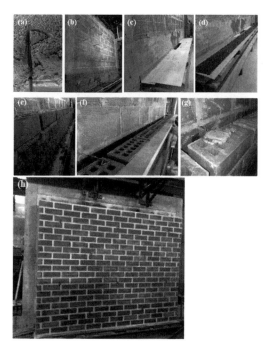

Figure 5. Construction phases of masonry specimens.

and the brick veneer. Finally, the brick veneer walls were built parallel to the masonry infill with similar dimensions of the concrete frame (2.32 length × 1.80 height), see Figure 5e and f.

4.3 Testing setup

For the out-of-plane cyclic test, a complex solution was designed in order to promote the ideal boundary conditions for the brick veneer walls. The out-of-plane loading/reaction system consisted in three parts (Figure 6): (1) a braced loading steel frame; (2) a structure to simulate distributed loading and (3) a steel braced reaction frame. An external steel frame was also placed above the specimens to ensure the restriction of out-of-plane movements at the top beam of rc frame. The restraint was carried out by using four steel rods M40 attached to a steel triangular structure, connected to two HEB 240 steel profiles that were fixed to the lateral reinforced concrete reaction wall. The out-of-plane loading was applied by a structure composed by a welded stiff L-shape profile with a horizontal HEB220 steel profile, an inclined HEB160 steel profile, two perpendicular HEB140 steel profiles and finally a set of tubular elements UNP50.

Four rollers were added at the base of the steel frame to enable its free movement along the horizontal direction without developing friction and thus to prevent additional forces recorded by the horizontal actuator. This framed structure distributes the load from hydraulic actuator into 30 load points (5 rows and 6 columns). Each load point covers an area of about 0.14 m². The framed structure is connected to the veneer wall trough of threaded rods HIT – V 5.8 anchored to the clay masonry veneer using a Hilti HIT – HY 270 adhesive anchoring system in each load point. As mentioned before, the framed structure is also attached to a horizontal actuator, which in turn is coupled to the braced loading frame anchored to the reaction rc concrete reaction slab. This structure is a rigid HEB360 steel profile fixed adequately to reaction floor to completely prevent its uplifting and sliding during the test.

The instrumentation composed of 31 LVDTs applied both at the infill and veneer brick walls was designed to measure the main deformations, see Figure 7. The out-of-plane deformation of the brick infill was monitored in the back side through 11 LVDTs (Figure 7a). LVDTs L1 to L4 were applied to measure the relative displacement between masonry infill from the surrounding rc frame. LVDTs L5 to L11 measured the out-of-plane deformation of the infill panel during loading. Two additional LVDTs were placed to measure de out-of-plane movement of the boundaries, namely at the bottom and top rc beams (L0 and L12). In the brick veneer walls, 12 LVDTs were placed according to the layout

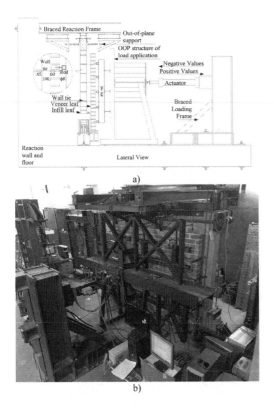

a)

b)

Figure 6. Test setup; a) lateral view; b) frontal view.

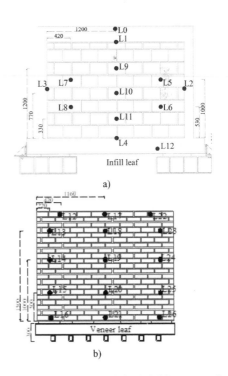

a)

b)

Figure 7. Instrumentation; a) infill; b) brick veneer walls.

presented in Figure 7b to measure the main deformations. An additional LVDT was placed on the connection between actuator and structure of load application to compare the internal displacement of the actuator and the real displacement that is imposed to the veneer wall.

4.4 Loading protocol

The loading protocol was based on FEMA 419 (1997): the displacement amplitude a_{i+1} in step $i+1$ is 1.2 times the amplitude a_i in step i. All levels are repeated twice, with exception of the first cycle that is repeated six times. The measured displacement law applied at the middle of the remaining brick veneer walls is presented in Figure 8. This law was defined in order to apply increasingly displacements during the out-of-plane test.

5 RESULTS

5.1 Mechanical behavior of masonry materials

As expected, the brick units present very different values of compressive strength according to the different directions of loading. The normalized compressive strength obtained for masonry veneer bricks was 21.07 MPa, 15.97 MPa and 9.04 MPa in the loading directions a, b and c respectively. The normalized compressive strength obtained for bricks used in the masonry infill walls was 5.60 MPa, 1.45 MPa and 1.47 MPa in the loading directions a, b and c respectively.

The stress-strain diagrams obtained in brick masonry veneer wallets under uniaxial compression loading can be seen in Figure 9. Some variation was found in specimen comp_02 sample in terms of maximum resistance and especially in terms of modulus of elasticity. The average value of the compressive strength is about 3.95 MPa and the mean elastic modulus is about 8.3 GPa. The load-displacement diagrams for each type of flexural test (loading in parallel and perpendicular direction to the bed joints) are presented in Figure 10.

The stiffness evolution during the cyclic test was assessed in the tests carried out with flexural load according to the perpendicular direction (Figure 10b). The stiffness is increasing until the maximum

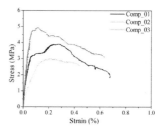

Figure 9. Stress vs strain diagrams.

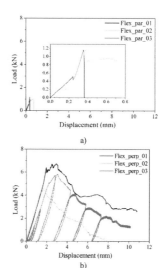

a)

b)

Figure 10. Force-displacement diagrams of masonry under flexure; a) parallel direction to bed joints; b) perpendicular direction to the bed joints.

resistance is achieved. In the post peak regime, the stiffness is decreasing as expected taking into account the degradation of material and decreasing of resistance capacity. As expected, the flexural strength whose failure is parallel to the bed joints is much lower than the flexural strength in specimens where the failure is perpendicular to the bed joints. The experimental results are comparable to values provided in EC6 (2005) of about 0.1 MPa (f_{xk1}) and 0.4 MPa (f_{xk2}).

5.2 Out-of-plane behaviour of the brick veneer wall

The cyclic force-displacement diagrams obtained in the out-of-plane tests for masonry veneer and brick infill walls are presented in Figure 11. For the masonry veneer wall, two force-displacement diagrams are provided, namely considering the out-of-plane displacement measured at the top (L17) and the out-of-plane displacement measured at mid height of the wall (L19). Together to these diagrams, it was also decided to add the force-displacement

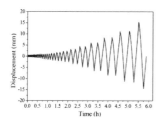

Figure 8. Time displacement history.

850

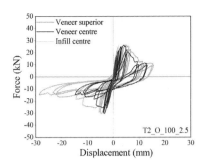

Figure 11. Force-displacement diagrams for infill and veneer walls under out-of-plane loading.

diagrams of the masonry infill wall (backing wall) considering the displacement measured at the centre of the walls. The idea of representing these different diagrams was to: (1) enable the easy comparison of the deformations between the masonry veneer and masonry infill walls; (2) make the comparison between the displacement at the top and centre of the masonry veneer possible. It should be mentioned that the positive and negative values of force induce tension and compression stresses on ties respectively. Due to these different types of loading, the nonlinear hysteretic response was not completely symmetric because the wall tie has no completely symmetric behaviour under compression and tension loading.

For a better assessment of the performance among the different masonry veneer wall, a comparison between the force obtained experimentally and the force calculated by multiplying the obtained tensile/compression maximum force obtained in single tie-masonry prims assemblages (Martins et al., 2016) by the number of connections considered in the specimen testes under out-of-plane loading. Based on this estimation, it is possible to understand in which extent the resistance of the individual tie-brick connection can be reproduced in the masonry veneer wall.

For the wall under analysis, the experimental maximum load was 29.12 kN in the first cycle and 30.6 kN in the second cycle in compression. In tension, the maximum force was 26.19 kN in the first cycle and 23.18 kN in the second cycle. The comparison of these values to the estimated ones enables to conclude that under compression the experimental force was slightly higher than the estimated value, while in tension the experimental and calculated forces are practically the same. The discrepancy between experimental and estimated value can be explained by the workmanship in the application of the wall ties (misalignment in wall ties), differences on boundary conditions of single connections and walls, load application mode and a combination of them. As far is strength concerned, it is observed that a degradation between maximum resistance of first and second cycles.

A considerable difference between the response at middle and top areas of veneer is noticeable, which is related to the different displacement measured at mid height and at the top. The veneer wall was simply supported at base and anchored through wall ties in its perpendicular direction, being the other three sides free to move in out-of-plane direction, meaning that there is trend for the out-of-plane rotation of the wall, particularly in case of the steel ties are compressed. In both cases the wall rotates, being the base of the veneer working as an "hinge". Therefore, the wall presents the highest out-of-plane displacement at the top of the wall and the lowest at the base. However, this difference is much more significant when the ties are compressed. This can be explained by the different behaviour of the ties under tension and compression. It was observed that when the veneer wall is pulled and the steel ties are submitted to tension, the veneer wall presents an initial sliding perpendicular to the masonry infill wall and the steel ties are pulled out across mortar joints of veneer and/or infill leaves. When the brick veneer wall is pushed towards the backing system, the steel ties are submitted to compression and due to constrains caused by compression buckling resistance of wall ties, the veneer wall present an evident rotation around the veneer base, being the maximum rotation observed at the top of wall.

As far as force-displacement diagrams of infill walls are concerned, it is noticed that there is a significant difference with respect to veneer wall. The deformation of infill wall is dependent on the capacity that the steel ties have to transfer the out-of-plane loading to the backing system, taking into account that the load is applied directly in the masonry veneer wall. This is a very important aspect to take into account regarding the seismic behaviour because it shows the interaction between both masonry leaves and can provide some indications for a suitable structural design for resisting the loading.

5.3 *Deformation profiles*

The deformation of the brick veneer and masonry infill walls was also analysed. The lateral deformation profile measured at the centre of the walls is provided in order to understand the interaction between the brick veneer walls and brick masonry infill. The sequential deformations of the walls following the cyclic loading are presented in Figure 12. The deformation profiles show the displacements of masonry infill and veneer walls under tension (OOP positive displacement) and compression (OOP negative displacement). Each deformation profile corresponds to the average displacements recorded in the first and second cycle imposed at each displacement level. It is seen that the central profiles of the infill and veneer wall leaves show higher lateral deformation. It should be mentioned that it is common that the displacements of the veneer walls measured by

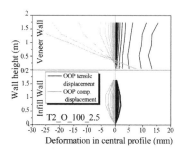

Figure 12. Displacement profiles of infill and veneer walls.

the LVDTs L12-L16 and LVDTS L22 to L26, measure different displacements, meaning that the veneer walls experiments rotation around the central vertical axis.

For the veneer wall, it was decided to put a strain gauge glued at a steel tie of each row to assess the evolution of strains during the out-of-plane loading. The strains recorded in the steel ties along the central vertical line are provided in Figure 13. In general, it is observed that the strain gauges did not record very high deformations, being usually lower than 2.0‰ but close to the yielding strain. Another important aspect that appears to be relevant is the difference in strains at top and bottom rows of the wall.

In all specimens, the values of strain are gradually increasing along the height of wall, being more evident during compression loading. This is mainly related to the higher out-of-plane displacements experienced by the veneer wall, which are mainly controlled by the different boundary conditions at top and bottom borders of the wall.

5.4 Typical damage pattern

The distinct deformational features of the walls discussed previously resulted from the different behaviour of the steel ties, namely as regarding the damage patterns both under compression and tension. As mentioned previously, when the steel ties are under compression (veneer walls is pushed), the veneer walls exhibit a deformation mostly characterized by the rotation along a horizontal axis close to

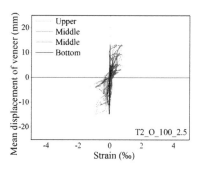

Figure 13. Strains in steel ties along the height of the wall.

the base (rocking), see Figure 12. When the steel ties are working in tension (veneer walls are pulled), the veneer walls also rotate but at much lower grade and mostly slides along the base. When a veneer wall rocks or slides, it can achieve significant displacements without a visible damage. Therefore, the damage is mainly concentrated at the steel ties and at the connection between steel ties and masonry infill and veneer walls. The damages in the steel ties observed consisted of: (1) pull-out of wall tie from embedment bed joint and in more demanding cases the (2) wall tie rupture when veneer wall is subjected to out-of-plane loading under tensile loading.

This justifies the importance these elements in structural safety of buildings with brick veneer walls. The damages on the masonry infill walls are also very reduced, resulting from the low displacements induced by the out-of-plane loads. Notice that the experimental campaign carried out enables to analyse only the out-of-plane loading transfer between brick veneer wall and, thus, the additional out-of-plane deformation induced by an earthquake in case of it has a brick veneer attached.

6 CONCLUSIONS

This paper presents and discusses the experimental results obtained on a quasi-static out-of-plane tests carried out on system composed on an rc frame with brick masonry infill to which a brick veneer walls is attached, The attachment of the brick veneer to the brick masonry infill was carried out by steel ties. The adoption of the rc frame with the masonry infill as the backing system of the brick veneer walls derived from the common use of this structural system in residential buildings in Portugal and in other south European countries.

From the out-of-plane test carried on the brick veneer wall attached to the brick infill wall, it was possible to observed that: (1) nonlinear hysteretic behaviour begins for very early stages of deformation. The hysteretic response is not completely symmetric because the wall ties play a central role on the out-of-plane performance of the system. As steel ties exhibit different behaviour under compression and tension loading, they influence also in the same way the out-of-plane behaviour when tensile and compression loading is induced in the veneer wall; (2) the infill wall develops low deformation levels than the brick veneer and I should be related to the ability of the ties to transfer the out-of-plane loading from the brick veneer walls to the masonry infill wall; (3) the wall ties experienced damages but they were enough to guarantee an adequate post-peak resistance of the veneer wall. The damages observed consisted of: (1) pull-out of wall tie from embedment bed joint and in more demanding cases the (2) wall tie rupture when veneer wall is subjected to out-of-plane loading under tensile loading.

ACKNOWLEDGEMENTS

The authors acknowledge the support of the Portuguese Foundation for Science and Technology (FCT), through the financing of the research project SEVen - Development of Sustainable Ceramic Brick Masonry Veneer Walls for Building Envelops (PTDC/ECI-CON/30876/2017).

REFERENCES

Akhoundi, F. et al. 2018. In-plane behaviour of cavity masonry infills and their strengthening with textile reinforced mortar, *Engineering Structures* 156: 145–156.

Akhoundi, F. et al. 2020. Experimental Out-of-Plane Behaviour of Brick Masonry Infilled Frames, *International Journal of Architectural Heritage* 14(2): 221–237.

Borchelt, J.G. 2004. Building code requirements for brick veneer in seismic areas. in Sísmica 2004 – *6th National Congress in Seismology and Earthquake Engineering*, University of Minho, Guimarães, Portugal.

Desai, N. & McGinley, W. M. 2013. A study of the out-of-plane performance of brick veneer wall systems in medium rise buildings under seismic loads. *Engineering Structures* 48: 683–694.

Eurocode 6 - EN 1996-1-1 Design of masonry structures - Part 1-1: General rules for reinforced and unreinforced masonry structures., Brussels, 2005.

EN 772-1, Method of test for masonry units, Part 1: Determination of compressive strength, 2000.

EN 1052-1, in Methods of test of masonry - Part 1: Determination of compressive strength 1999.

EN 1052-2, in Methods of test of masonry - Part 2: Determination of flexural strength. 1999.

FEMA_273, in NEHRP Guidelines for the seismic rehabilitation of buildings, F.E.M. Agency, Editor. 1997: Washington D.C.

Martins, A. et al. 2016. Experimental assessment of the mechanical behaviour of wall tie connections on anchored brick veneer to masonry infill walls, *Construction and Building Materials* 156: 515–531.

Martins, A. et al. 2017. Brick veneer walls: A review. *Journal of Building Engineering* 9: 29–41.

Memari, A.M. et al. 2002. Seismic response of a new type of masonry tie used in brick veneer walls. *Construction and Building Materials* 16(7): 397–407.

Morandi, P. et al. Simplified Out-of-plane Resistance Verification for Slender Clay Masonry Infills in RC Frames. Dipartimento di Ingegneria Civile ed Architettura, Università degli Studi di Pavia ed EUCENTRE.

Brick and Block Masonry - From Historical to Sustainable Masonry –
Kubica, Kwiecień & Bednarz (eds)
© 2020 Taylor & Francis Group, London, ISBN 978-0-367-56586-2

Response of brickwork wallettes of various bonding patterns under monotonic and cyclic compression

J.A. Thamboo
Department of Civil Engineering, South Eastern University of Sri Lanka, Sri Lanka

M. Dhanasekar
Faculty of Science and Engineering, Queensland University of Technology, Australia

ABSTRACT: Brick masonry walls of varying architectural patterns are commonly seen in many colonial period structures around the world. Subsequently with a view to understanding the monotonic and cyclic compressive characteristics of such brickwork, an experimental programme was carried out in this research. Fourteen brickwork wallettes with the geometry depicting the colonial buildings were built with two types of clay bricks and a lime mortar to simulate the strength characteristics close to the masonry found in those buildings. The failure modes, strengths and deformation characteristics of the masonry are described. It is shown that the monotonic compressive strength of masonry is consistently higher than the cyclic compressive strength. Further the monotonic and cyclic compressive stress-strain characteristics of the brickwork wallettes of varying bond patterns are discussed and their similarities and differences between single-leaf brickwork reported in the literature are examined. An analytical expression for the cyclic compressive stress-strain characteristics of the brickwork with lime mortared wallettes is also presented.

1 INTRODUCTION

Many of the modern and historical building stocks around the world consist of loadbearing masonry walls. Therefore in the past, plenty of research studies were carried out in the focus of assessing and reinstating the loadbearing capacity of different masonries (Binda et al. 1988, Bennett et al. 1997, Sarangapani et al. 2005, Thamboo & Dhanasekar 2016, Zahra & Dhanasekar 2016, Noor-E-Khuda & Albermani 2019). Primarily the strength and deformation characteristics of the constitutive materials (unit and mortar) used to assessable the masonry were shown to significantly affect the compressive strength of masonry.

Masonry walls are designed and built with various forms of architectural bonds depending on the design requirements – especially those found in historical masonry structures. Although the loadbearing walls are constructed with different forms of brickworks, the compressive strength of masonry is largely derived from the testing of single leaf masonry prisms and wallettes (e.g. CSA S304.1-04 2010, AS 3700 2018). As brick masonry response is affected by the presence of mortar joints, the effect of presence of mortar joints in the thickness

direction – especially in some architectural bonds is of interest for the analysis of historic masonry structures (Gumeste & Venkatarama Reddy 2007, Thamboo & Dhanasekar 2019).

Furthermore most of the stone masonry walls in historical structures constitute different bond patterns (Valluzzi et al. 2001, Binda et al. 2005). Such stone masonry walls consist quite complex morphology, where inner and outer leafs are made of different masonry types – and are termed as multi-leaf (also multi-wythe) masonry walls. Compressive characteristics of brickwork built with same brick and mortar but of different architectural bonds are not well explored in the literature. Since such brickwork walls are common elements in many of the colonial period buildings in many parts of the world, including Sri Lanka, their compressive strength characteristics is explored in this paper.

Also masonry structures in historic buildings would have been subjected to significant load cycles due to high wind and earth movements even if earthquakes are rare in their locations. Therefore, assessment of cyclic compressive loading characteristics of brickwork typical of colonial constructions is essential to evaluate the realistic compression capacity of historical masonry structures. There have been few research studies that investigated the cyclic compressive strength characteristics of different types of masonries in the past (Subramaniam & Sinha 1995; Dhanasekar & Shrive 2002, Ispir & llki 2013; Facconi et al. 2018). Only limited studies on brickwork with varied bonding patterns and thickness under monotonic and cyclic compression are explored; this paper bridges the gap by providing experimental data.

In summary, examination of monotonic and cyclic compressive strength characteristics of brickwork of varied bonding pattern and thickness is essential, especially for the lime-mortared masonry as they are found in many of the colonial masonry structures that still serve the community and governments in many nations. Hence this paper provides a new set of experimental data on the monotonic and cyclic behaviour of bonded brickwork. Monotonic and cyclic compression tests were conducted on fourteen masonry wallettes made of two types of clay bricks and a lime mortar. The failure modes, strengths and deformation characteristics of those masonry wallette combinations are presented and discussed in this paper. Further the available predictive analytical models to define the cyclic envelop compressive strength and deformation behaviour of bonded brickwork are assessed with the experimental results obtained in this research.

2 EXPERIMENTAL DETAILS

2.1 Materials

Brickwork wallettes in this experimental programme were constructed using two types of clay bricks denoted as B1 and B2. The dimensions of the B1 and B2 bricks were 200 × 90 × 65 mm and 210 × 100 × 60 mm (length×width×height) respectively. Compression tests were carried on randomly selected six bricks from each type as per BS EN 772-1 (2015) and their average compressive strengths were found to be 5.1 MPa for B1 bricks and 15.5 MPa for B2 bricks. Locally sourced natural hydraulic lime was used as the binder in the mortar. The lime mortar was prepared at a binder to filler ratio of 1:3 by volume. The flow table test consistency of the prepared mortar was maintained as 140 mm. As per BS EN 1015-11 (2019), lime mortar prisms were casted along with the construction of masonry wallettes and then the prism halves were tested under compression after 28 days of casting. The average compressive strength of the prepared lime mortar was found to be 2.4 MPa with the coefficient of variation of 5.4 %.

2.2 Construction and testing

Fourteen bonded brickwork wallettes were constructed and tested to assess the monotonic and cyclic compressive strength and deformation characteristics in this research. For each unit-mortar-masonry combination, seven wallettes were constructed. The overall dimension of the B1 and B2 series wallettes were 410 × 300 × 740 mm and 430 × 320 × 760 mm (length×width×height). Out of those seven wallettes, four were tested under monotonic compression and the rest were used to test under cyclic compression. All wallettes were constructed by a same experienced mason and 10 mm mortar joints were maintained. The wallettes were constructed with English bond pattern as they are the common pattern found in historical masonry structures. The constructed masonry wallettes were air cured for 28 days prior to testing.

Both monotonic and cyclic compression testing of the constructed wallettes were carried out using a 1000 kN capacity servo-controlled universal testing machine (UTM). The wallettes were placed carefully between the platens of the UTM to avoid any accidental eccentricity in loading. Further 5 mm plywood capping was inserted between the contact steel plates and masonry to reduce platen restraint during the compression loading. The testing arrangement is shown in Figure 1. The monotonic testing was conducted under displacement-controlled mode with the rate of 0.25 mm/min. Displacement transducers were attached on both faces of the wallettes to measure the axial and lateral deformation during the testing. The loads and displacements were measured and recorded using a synchronised data acquisition system. The cyclic loading protocol was assigned from the load-displacement responses from the monotonic testing. Each cyclic step was repeated twice to stabilise the readings. However, the rate of loading was assigned as 0.25 min/min for the reloading and unloading stages in the cyclic protocol.

Figure 1. Compression testing of bonded brickwork wallettes.

3 EXPERIMENTAL RESULTS

3.1 Failure mode

Figure 2 shows the typical failure modes observed in the tested wallettes. The failure modes of the wallettes were largely characterised with parallel vertical cracks originating at brick to mortar interface and propagated through the middle of the bricks. This phenomenon is well understood, the incompatible elastic properties between brick and mortar induce vertical tensile cracks in the bricks under compressive stress in masonry. In terms of crack pattern, no distinction could be made between monotonic and cyclic loaded wallettes. Commonly the initial cracks were started to appear at about 70% to 80% of the peak load for the monotonically loaded wallettes, whereas the cyclic loaded wallettes developed initial cracks at around 50-60% of the peak load.

3.2 Compressive strengths

The average monotonic and cyclic compressive strengths determined are presented in Table 1. Apparently B2 series wallettes have shown higher compressive strength than the B1 series wallettes. Further the ratio between unit and masonry compressive strengths is normally defined as efficacy ratio of a particular masonry type. The efficacy ratio of tested bonded brickwork wallettes varies between 0.4 to 0.43 for the tested wallettes. This is slightly lower than the conventional masonry, where the efficacy ratio is generally around 0.5. Previous studies on lime mortared masonry also indicate similar lower compressive resistance than conventional masonry (Verstrynge et al 2011; Drougkas et al 2015).

Obviously the monotonic loaded bonded brickwork wallettes have shown higher compressive strengths than cyclic loaded wallettes. The reduction of cyclic compressive strength varied between 12.5% to 14.4 % for the tested wallettes. The compressive strength reductions observed in cyclic loaded wallettes were mainly attributed to gradual accumulation of lateral tractions and inelastic axial strain in the masonry that lead to lower strength compared to monotonically loaded wallettes

3.3 Stress-strain responses

The average compressive stress-strain curves of the B1 and B2 series monotonically loaded wallettes are presented in Figure 3(a) and (b) respectively. The recorded axial displacements were divided by the gauge length (across one half of the wallette height) to compute the axial strain values and the corresponding stress values from the load measurement were matched to plot the stress-strain curves of the wallettes. All the wallettes under monotonic loading displayed approximately linear stress-strain behaviour up to 60 to 70% of the peak strength and afterward nonlinear behaviour can be observed up to the failure. Moreover it can be seen that the stress-strain curves were obtained up to nearly 30 to 40% drop of

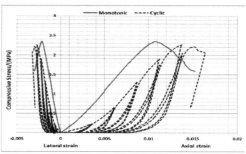

(a) B1 series

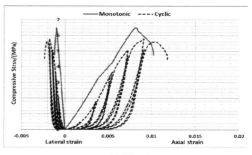

(b) B2 series

Figure 3. Monotonic and cyclic compressive stress-strain curves (a) B1 series and (b) B2 series.

| (a) | (b) |

Figure 2. Typical failure modes of testing wallettes (a) Vertical cracks and (b) Splitting cracks.

Table 1. Compressive strengths of tested wallettes.

Specimen notation	Compressive Strength (MPa)	Variation (%)
B1-M	2.33 (9.6)	-12.5
B1-C	2.04 (6.2)	
B2-M	6.68 (4.3)	-14.4
B2-C	5.72 (10.4)	

peak stress in the post peak region. Obviously the low strength B1 brick series masonry wallettes have shown more deformability than the relatively high strength B2 series wallettes as the axial deformation capacity of the B2 brick is less than B1 brick. Therefore, it can be said that the brick strength and its deformation properties have greatly influenced the overall stress-strain behaviour of masonry.

The stress-strain curves of the B1 and B2 series cyclic loaded wallettes are also presented in Figure 3(a) and (b) respectively for comparison. Even though three wallettes were tested for each combination, only the typical (i.e. closed to average strength) cyclic stress-strain curve of each combination is presented as displaying all curves would be cumbersome for comparison. The initial concave section of the stress-strain curves were corrected to get the proper elastic region as outlined in Costigan et al (2015). Therefore, initial similarities of the monotonic and cyclic curves are not represented in the stress-strain curves.

It can be clearly seen that the non-linear behaviour under cyclic loaded wallettes started nearly around 50% of the peak compressive stress, which associated with the initiation of cracks in the wallettes. The accumulation of non-reversible axial strains in the masonry wallettes on each cycle; especially after nearly 50% peak stress can be clearly observed. Also the strength and stiffness degradation at each step and cycle can be observed in the stress-strain curves. This indicates that gradual damage has occurred in each cycle in the masonry wallettes. Therefore the pattern of cyclic stress-strain curves of the wallettes justifies the reason for the strength reduction compared to monotonic strength. As observed in the monotonic testing results, the B1 series wallettes have shown slightly larger axial deformity under cyclic loading than the corresponding B2 series wallettes.

3.4 Listing and numbering

The average deformation characteristics of the masonry wallettes (1) elastic modulus, (2) Poisson's ratio (3) elastic strain (4) peak strain and (5) ultimate strain were determined from the

stress-strain curves and presented in Table 2. Their COVs of the parameters are provided in parentheses. For the cyclic stress strain responses these parameters were determined from the envelop curves. The elastic moduli of the masonry wallettes were determined at the one-third of the peak stresses and corresponding elastic strains were matched in the stress-strain curve. The Poisson's ratios were computed from the elastic axial strain values and the corresponding lateral strain values. The peak strain was determined conforming to the peak stress of the tested wallettes.

The ultimate strain was acquired corresponding to the 80% of the post-peak stress. The elastic moduli of the B1 and B2 series wallettes vary between 131 MPa to 224 MPa and 431 MPa to 889 MPa respectively. The low elastic moduli are mainly related to the deformation characteristics of lime mortar used. Furthermore, the Poisson's ratio calculated from the stress-strain curves ranged between 0.15 to 0.2. It can be clearly noted from that the elastic, peak and ultimate strains under cyclic conditions are marginally higher than the corresponding strains obtained through monotonic testing.

3.5 Analytical expression

Previously several researchers have developed expressions for envelop curves of masonry under cyclic compression (Alshebani & Sinha 1999; Dhanasekar &Shrive 2002; Facconi et al. 2018). Subsequently those expressions are verified against the experimental data developed in this research. For better comparison, stresses and strains data of the envelop points were normalised. The proposed expression by Alshebani & Sinha (1999) is given in Eq (1). The α and β are constants. Dhanasekar & Shrive (2002) recommended an expression as in Eq (2), where where the u_0 and u_1 are the constants. Further Facconi et al. (2018) suggested two expressions for pre-peak and post-peak portions for the envelop curve and they are given in Eq. 3(a) and (b) respectively, where n is the ratio of initial elastic modulus and the scant modulus at yield point of masonry.

Table 2. Parameters derived from monotonic stress-strain curves of the masonry wallettes.

Specimen notation	Elastic Modulus/(MPa)	Poisson's Ratio	Elastic strain	Peak strain	Ultimate strain
B1-M	219 (10.4)	0.19 (9.8)	0.004 (3.5)	0.010 (11.8)	0.014 (13.6)
B1-C	132 (13.7)	0.15 (11.7)	0.006 (10.1)	0.015 (17.6)	0.018 (10.3)
B2-M	776 (6.2)	0.14 (16.4)	0.003 (5.0)	0.007 (7.6)	0.009 (5.2)
B2-C	446 (18.3)	0.17 (8.6)	0.004 (13.9)	0.010 (11.0)	0.011 (10.1)

$$\sigma = \varepsilon^{\beta} e^{(1-\varepsilon/a)} \qquad (1)$$

$$\sigma = \sigma_{max}\left(\frac{(1+u_0(1+u_1)\varepsilon}{u_0(1+u_1\varepsilon)+\varepsilon^{u_0+1}}\right) \qquad (2)$$

$$\sigma = \varepsilon_p\left[1 - \frac{1}{n}\varepsilon^{n-1}\right] \qquad (3a)$$

$$\sigma = \sigma_{max}\left[1 - \left(\frac{1-\varepsilon}{0.5}\right)^2\right] \qquad (3b)$$

The normalised experimental envelop stresses and strain data are shown in Figure 4.

Also the previously proposed analytical envelop curves are plotted in the same Figure 4 for comparison. It can be noted that the expressions proposed by Alshebani & Sinha (1999) and Dhanasekar and Shrive (2002) to predict the envelop curves are noticeably deviate from the experimental data. The obtained coefficient of correlation of those curves and the data are 0.45 and 0.47. The possible reason for the deviation could be that the unit strengths used in their studies (23.4 MPa and 27 MPa) were relatively higher than the unit strength used in this research. Therefore lower initial stiffness obtained in the data due to softer clay bricks could have led to significant variation between the data and proposed expressions. Nevertheless the two set expressions suggested by Facconi et al. (2018) show good agreement with the pre-peak and post-peak data with the determined coefficient of correlations of 0.88 and 0.82 respectively. Therefore it can be said that the expression proposed by Facconi et al. (2018) can be simply used to define the envelop curve of three-leaf lime mortared masonry.

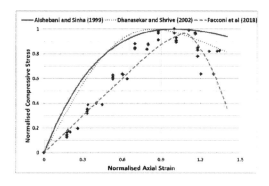

Figure 4. Cyclic envelop point data and variation of analytical expression to predict the envelop curves.

4 CONCLUSIONS

This paper presents the experimental investigation of bonded brickwork under monotonic and cyclic compression. In total, fourteen wallettes were constructed and tested to assess the monotonic and cyclic compression characteristics. The following conclusions can be drawn from the experimental data gathered in this research:

1. The failure modes of the brickwork subject to monotonic and cyclic load were similar.
2. Monotonic compressive strength of the lime mortared masonry was approximately 12-15 % higher than the cyclic compressive strengths for the tested combinations.
3. The axial and lateral deformation capacities of the bonded brickwork under cyclic loading were slightly higher than the monotonic loading condition.
4. Among the expressions proposed to predict the cyclic envelop curve of the masonry, the two set expressions suggested by Facconi et al., (2018) show good argument with the experimental data.

ACKNOWLEDGEMENT

The Authors thank the South Eastern University of Sri Lanka for the financial support to this project (under the research grant of SEU/ASA/RG/2019/02) and provided technical support. The technical assistance provided by Mr. Jiffry, Mr. Farhan and Mr. Imthyas are greatly appreciated.

REFERENCES

AlShebani, M.M. & Sinha, S.N. 1999. Stress-strain characteristics of brick masonry under uniaxial cyclic loading. *Journal of Structural Engineering* 125: 600–604.

AS 3700, 2018. Australian Standards for Masonry Structures, Standards Australia, Sydney.

Bennett, R.M. Boyd K.A, & Flanagan, R.D. 1997. Compressive properties of structural clay tile prisms. *Journal of Structural Engineering* 123(7): 920–926.

Binda, L. Pina-Henriques, J. Anzani, A. Fontana, A., Lourenco, P.B. 2006. A contribution for the understanding of load-transfer mechanisms in multileaf masonry walls: Testing and modelling. *Engineering Structures* 28: 1132–47.

Binda, L., Fontana, A. & Frigerio, G. 1988. Mechanical behaviour of brick masonries derived from unit and mortar characteristics. *Proceeding of 8th International Brick and Block Masonry Conference, Dublin, Ireland.* 205–216.

BS EN 1015-11:1999, Methods of test for mortar for masonry. Determination of flexural and compressive strength of hardened mortar, BSI.

BS EN 772-1:2015, Methods of test for masonry units. Determination of compressive strength, BSI.

Costigan, A. Pavía, S. & Kinnane, O. 2015. An experimental evaluation of prediction models for the mechanical behavior of unreinforced, lime-mortar masonry under compression, *Journal of Building Engineering*. 4: 283–294.

CSA S304.1-04, 2010. Design of masonry structures, Mississauga, Canada.

Dhanasekar, M. & Shrive. N.G. 2002. Strength and deformation of confined and unconfined grouted concrete masonry. *ACI Structural Journal* 99: 819–826.

Drougkas, A. Roca, P. & Molins, C. 2016. Compressive strength and elasticity of pure lime mortar masonry. *Materials and Structures* 49: 983–999.

Facconi, L. Minelli, F. & Vecchio, F.J. 2018. Predicting uniaxial cyclic compressive behavior of brick masonry: new analytical model, *Journal of Structural Engineering* 144: 4017213.

Gumeste, K.S. & Venkatarama Reddy, B.V. 2007. Strength and elasticity of brick masonry prisms and wallettes under compression. *Materials and Structures* 29: 241–253.

Ispir, M. & Ilki, A. 2013. Behavior of historical unreinforced brick masonry walls under monotonic and cyclic compression. *Arabian Journal for Science and Engineering* 38: 1993–2007.

Noor-E-Khuda, S & Albermani, F. 2019. Mechanical properties of clay masonry units: destructive and ultrasonic testing. *Construction and Building Materials* 219: 11–120.

Sarangapani, G. Venkatarama Reddy, B. V. & Jagadish, K. S. 2005. Brick-mortar bond and masonry compressive strength. *Journal of Materials in Civil Engineering* (17)2: 229–237.

Subramaniam, K. & Sinha, S. 1995. Analytical model for cyclic compressive behavior of brick masonry. *ACI Structural Journal* 92(3): 288–294.

Thamboo, J.A. & Dhanasekar, M. 2016. Behaviour of thin layer mortared concrete masonry under combined shear and compression. *Australian Journal of Structural Engineering* 17(1): 39–52.

Thamboo, J.A. & Dhansakar, M. 2019. Correlation between the performance of solid masonry prisms and wallettes under compression. *Journal of Building Engineering* 22: 429–438.

Valluzzi, M.R. Da Porto, F. & Modena, C. 2004. Behavior and modeling of strengthened three-leaf stone masonry walls. *Materials and Structures* 37(3): 184–192.

Verstrynge, E. Schueremans, L. Van Gemert, D. 2010. Time-dependent mechanical behavior of lime-mortar masonry. *Materials and Structures* 44(1): 29–42.

Zahra, T. & Dhanasekar, M. 2016. Prediction of masonry compressive behaviour using a damage mechanics inspired modelling method. *Construction and Building Materials* 109: 128–138.

Brick and Block Masonry - From Historical to Sustainable Masonry –
Kubica, Kwiecień & Bednarz (eds)
© 2020 Taylor & Francis Group, London, ISBN 978-0-367-56586-2

Determination of Ritter constant for hollow clay prisms under compression

G. Mohamad
Federal University of Santa Maria, Santa Maria, Brazil

F.S. Fonseca
Brigham Young University, USA

H.R. Roman
Federal University of Santa Catarina, Florianópolis, Brazil

T. Ottoni & A. Lubeck
Federal University of Santa Maria, Santa Maria, Brazil

ABSTRACT: The main goal of this work was to determine the Ritter constant to predict the modulus of elasticity of different hollow clay block masonry prisms under compression. For this reason, an experimental program with prisms was conducted to evaluate the Ritter constant. Two types of hollow clay blocks with net and gross area ratios of 0.39 and 0.49 and f_{mortar}/f_{block} ratios of 0.3 and 0.7 were used. Strain measurements were obtained on the front and back of the prisms. To estimate the Ritter constant for the masonry, Knutson (1993) equation for stress/strength level below 0.75 was adopted using a normalized stress/strength (σ/f_{prism}) level of 0.3. The conclusion from this work is that the modulus of elasticity of masonry can be estimated using the Ritter constant with reasonable accuracy, but the value depends on the block type used and mortar compressive strength.

1 INTRODUCTION

1.1 *Actual context of masonry modulus of elasticity*

Due to wide range of block typology manufactured in Brazil and the current revision of the Brazilian masonry standard, a clear understanding of the behavior of masonry is important. Thus, the aim of the work presented in this article is to evaluate the stress and strain behavior of masonry prisms and determine the Ritter constant for different types of clay blocks and mortars.

The compressive strength of masonry prisms is the predominant characteristic of masonry design, and it depends on factors such as mortar and block strengths as well as the relationship between the properties of these materials, including the block type as mentioned by several researchers (Mohamad 2005, Sarhat and Sherwood 2014, Costigan et al. 2015). However, a clear description of the mechanism involved in the deformation of masonry prisms has not been given yet. Some masonry codes, like the Brazilian standard NBR 15812-1 (2010), establishes an approximation between characteristic compressive strength of the prism and the masonry modulus of elasticity by a constant that depends on the block type. For instance, for concrete blocks the constant varies from 700 to 800, depending of the characteristic compressive strength of block. For hollow clay masonry, the constant is 600 and neither

depend on the block geometry (solid or perforated) nor on its mechanical strength. On the other hand, the Eurocode 6 relates the masonry elastic modulus (E_m) to the compressive strength of masonry (f_m) and establishes that in the absence of a value determined by tests, the short-term secant masonry modulus of elasticity may be calculated using Equation (1); the recommended value for k is 1000.

$$E_m = k.f_m \qquad (1)$$

CSA 304 (2014) establishes the modulus of elasticity of masonry (E_m) constructed using clay or concrete units as 850 times the compressive strength of the masonry (f_m), but the calculated value cannot be greater than 20 GPa.

The New Zealand masonry code, NZS 4320 (2004), simply states that for concrete masonry the modulus of elasticity (E_m) is 15 GPa for all masonry structures. It also limits the ultimate compressive strain (ε_u) for unconfined masonry to 0.003.

Kaushik et al. (2007) tested clay brick masonry under compression and proposed a simple analytical model to obtain the stress-strain curves for masonry. The authors further explained that it is possible to find in the literature a considerable wide range of values for these constants—from 250 to 1,100 times the compressive strength of masonry.

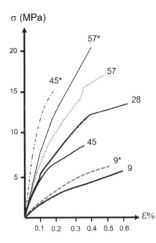

σ (MPa)

Figure 1. Stress-strain diagram of masonry - Knutson (1993).

Figure 2. Mortar test specimen (50mm x 100mm).

1.2 Knutson model for stress-strain diagram

Knutson (1993) evaluated stress-strain diagrams for masonry and showed that they could be formulated using a stress/strength ratio and strain for different unit types and mortars. The author assessed the masonry strain-stress diagram for different combinations of mortar and brick (three solid and one hollow), as shown in Figure 1, and concluded that the stress-strain relationship could be approximated using Eq. (2) and Eq. (3). The author concluded that the interaction between units and mortars is an important characteristic that affect the deformation response and failure mode of masonry structures.

$$\varepsilon = -\frac{f_{cmas}}{E_o} \cdot \ln\left(1 - \frac{\sigma}{f_{cmas}}\right), if \frac{\sigma}{f_{cmas}} \leq 0.75 \quad (2)$$

$$\varepsilon = -4\frac{f_{cmas}}{E_o} \cdot \ln\left(0.403 - \frac{\sigma}{f_{cmas}}\right), if \frac{\sigma}{f_{cmas}} > 0.75 \quad (3)$$

Equation (2) links logarithmically two non-dimensional values, namely the normalized strain $K_r.\varepsilon$ and the normalized stress σ/f_{cmas}. Where K_r is called Ritter constant, which is determined by dividing the masonry modulus of elasticity by the masonry compressive strength; i.e., $K_r = E_o/f_{cmas}$.

2 MATERIALS TESTS AND RESULTS

2.1 Mortar

Two types of mortar specified by the European masonry code were made in the laboratory for the experimental program. The water retention, air content, ratio of aggregate to cementitious material of the mortar and grading of the aggregate were

Table 1. Mortar compressive strength.

Mortar group	Mean compressive strength (MPa) f_{mortar}/f_{block} 0.3	Mean compressive strength (MPa) f_{mortar}/f_{block} 0.7
I – HFS	7.5	17.5
II - SFS	14.5	33.5

determined in accordance with Brazilian codes NBR 13277 (2005), NBR 13278 (2005), 13279 (2005) and 13280 (2005). The mortars were chosen to achieve values of f_{mortar}/f_{block} of approximated 0.3 and 0.7 for prisms with hollow face shell (HFS) and solid face shell (SFS), respectively. The compressive strength of the block, calculated using the net area, was used to calculate f_{mortar}/f_{block} ratios. The mortars were volume-proportioned using cement, lime and sand. Three mortar cylindrical specimens of 50 mm in diameter by 100 mm in height, as shown in Figure 2, were used to determine the mortar shortening until failure. Three strain gauges were spaced 120 degrees around the cylinders to detect any accidental rotation.

Table 1 presents the average compressive strength of the mortars (f_{mortar}), determined from six sample cubes with dimensions of 4 x 4 x 4 cm, according to NBR 13276 (2005).

The results of the tests clearly show the nonlinear stress/strain behavior of the mortars, as it is possible to observe in Figure 3. The stress-strain response for the cylindrical specimens show that the nonlinear behavior began when the load reaches 0.40 and 0.80 of the ultimate value for mortar group II.

2.2 Blocks

Two clay block types, herein namely HFS (hollow face shell) and SFS (solid face shell) were used. The block dimensions were 14 x 19 x 29 cm (thickness, height and length), respectively. Figure 4 shows the

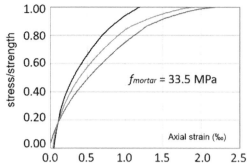

$f_{mortar} = 33.5$ MPa

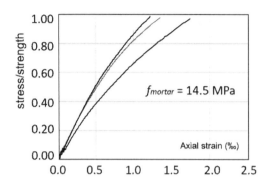

$f_{mortar} = 14.5$ MPa

Figure 3. Mortar stress-strain behavior.

Figure 4. Clay blocks.

blocks with their distinct hollow and solid face shell and cross webs. The net and gross area ratios for the hollow face shell and solid face shell were 0.39 and 0.49, respectively. Eight samples for each block type were tested under compression to calculate the average and characteristic compressive strength. Table 2 presents the compressive strength of the blocks calculated using net and gross areas. For these tests, the standard deviation was reasonably small.

The tests were conducted to evaluate the stress/strength and strain behavior of the blocks under

Table 2. Block compressive strength.

Block type	Mean compressive strength (MPa)		Characteristic compressive strength (MPa)	
	net area	gross area	net area	gross area
HFS	25.2	9.8	23.6	9.2
SFS	48.7	23.9	48.6	23.8

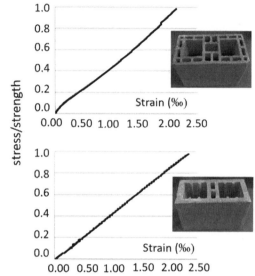

Figure 5. Stress/strength versus axial strain for hollow and solid face shell, respectively (‰).

compression until failure. For measurements, ten strain-gauges on both surfaces were used. For each block type, five samples were prepared and tested. The results indicate that the ultimate axial strain at failure varies from 2 to 2.5‰, as presented in Figure 5. The most interesting fact was the linear behavior for both hollow and solid face shell blocks. Also, a very abrupt and brittle failure was observed for blocks SFS. A linear trend line was fitted to the experimental data to determine the correlation between stress/strength and strain measurement of the blocks until failure.

2.3 Prisms

Experimental tests in stack-bonded masonry prisms under compression were carried out at the Federal University of Santa Maria/Brazil. Tests were done in order to determine the Ritter constant for the hollow (HFS) and solid (SFS) clay prisms and the two different combinations of mortar. Five stack-bonded masonry prisms with two blocks high and mortar

Figure 6. HFS and SFS prisms.

joint of 10 mm were constructed for each mortar-block combination, as shown in Figure 6. The prisms were capped with a thin layer of cement paste at the top and bottom surfaces. The tests were conducted in a 3000 kN servo-controlled universal testing machine. The loading rate used for the test was 0.05 N/mm²/s. Displacement measurements on both side of the prisms were used to evaluate any incidental load plate rotation and the masonry behavior under compression. The strains were calculated from the average measured displacements. Compressive strength results calculated using both net and gross area as well as f_{mortar}/f_{block} ratios are presented in Table 3.

The prism-block strength ratios for each mortar-block strength ratio are presented in Table 4. The compressive strength of prisms built with f_{mortar}/f_{block} of 0.3 was, approximately, 60% of the block compressive strength. When the f_{mortar}/f_{block} ratio increased to 0.7, the f_{prism}/f_{block} relationship reached 1.0 for hollow face shell prisms. The results show that a significant increase in mortar compressive strength affected the f_{prism}/f_{block} relationship. This shows that strength results or strength ratios alone cannot be used to indicate the masonry strength or its failure mode.

3 DISCUSION OF RESULTS

3.1 Prism with HFS

Prisms built with hollow face shell blocks and the two range of mortar strengths are discussed herein. The experimental stress results were normalized and plotted using stress/strength ratio and strain so that the Ritter constant, using Eq. (2), could be determined. For both f_{mortar}/f_{block} ratios, a linear correlation was found between stress/strength and strain for the prisms until failure. Table 5 presents the calculated Ritter constant for stress/strength levels of 0.3. The coefficient of variation for the HFS prism results is considered relatively low for both f_{mortar}/f_{block} ratios. The Brazilian standard NBR 15812-1 (2010) mentions that the elastic properties of masonry can be estimated by multiplying the characteristic strength of the prisms by a factor k of 600. This estimation for the tangent elastic modulus is independent of the block geometry and mortar type. Figure 7 presents the stress/strength vs. strain for the prisms with f_{mortar}/f_{block} ratios of 0.7 and 0.3. The dotted line represents the mean response of stress/strength and strain used to calculate the Ritter constant. To calculate the Ritter constant from experimental test of masonry prisms under compression using Equation (2), it is necessary to transform the masonry compressive strength to prism compressive strength. This is possible using the efficiency ratio of 0.70 (NBR 15812-1:2010) to correlate the characteristic compressive strength of masonry with the compressive strength of prism (f_{cmas}/f_{prism}). The results clearly show that the Ritter constant depends of the compressive strength of mortar. This seems to be very clear, but the Brazilian code does not recognize this important fact. There is a decrease in stiffness for the prisms built with f_{mortar}/f_{block} equal to 0.3 when compared to that of f_{mortar}/f_{block} equal to 0.7.

Table 3. Prism compressive strength.

		Prism type		
		HFS		SFS
f_{prism} (MPa)		net area	gross area	Net area	gross area
f_{mortar}	0.3	14.6	5.7	28.7	14.1
$/f_{block}$	0.7	25.3	9.9	32.9	16.1

Table 4. F_{prism}/f_{block} ratios.

		Block type	
f_{prism}/f_{block}		HFS	SFS
f_{mortar}/f_{block}	0.3	0.58	0.59
	0.7	1.00	0.68

Table 5. Ritter constant for HFS prisms.

		Ritter constant	
Sample	Stress/strength level	f_{mortar}/f_{block} 0.7	f_{mortar}/f_{block} 0.3
1	0.3	441	360
2	0.3	518	476
3	0.3	513	360
4	0.3	459	282
5	0.3	462	314
Mean		479	358
s.d. (MPa)		24.6	40.3
c.v.(%)		5.1	11.2

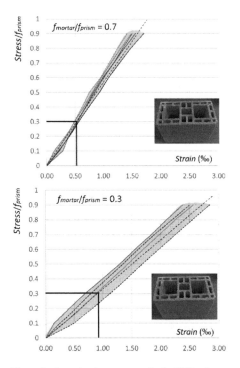

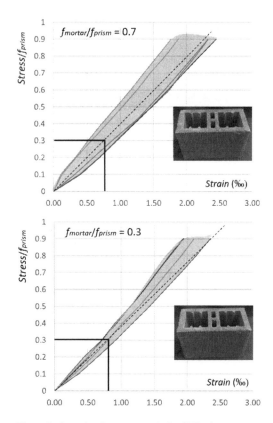

Figure 7. Stress/strain versus strain for HFS prisms.

Figure 8. Stress/strain versus strain for SFS prism.

3.2 Prism with SFS

Prisms built with solid face shell blocks and the two range of mortar strengths are discussed herein. The experimental stress results were normalized and plotted using stress/strength ratio and strain so the Ritter constant, using Eq. (2), could be determined. For the prisms with f_{mortar}/f_{block} equal to 0.7, it was quite difficult to measure the strain until failure due to a sudden crack propagation. Displacement measurements were stopped just before the ultimate load was reached. A linear correlation was found between stress/strength and strain for the prisms until failure. Table 6 presents the calculated

Table 6. Ritter constant for SFS prisms.

Sample	Stress/strength level	Ritter constant f_{mortar}/f_{block} 0.7	f_{mortar}/f_{block} 0.3
1	0.3	687	502
2	0.3	625	492
3	0.3	378	600
4	0.3	403	428
5	0.3	373	442
Mean		493	493
s.d. (MPa)		108.5	38.8
c.v. (%)		22	7.9

Ritter constant for stress/strength levels of 0.3 and 0.7. The coefficient of variation for the SFS prism results is considered very high for prisms with f_{mortar}/f_{block} equal to 0.7. For prisms with blocks with solid face shell, there was no significant difference between Ritter constants for both f_{mortar}/f_{block} ratios. This may be due to the greater thickness of the face shell and cross webs of the SFS block, which may have increased the confinement of the bedding mortar. Hence, this fact changes the mechanical proprieties of masonry under compression. Figure 8 presents the stress/strength vs strain for the prisms with f_{mortar}/f_{block} ratios of 0.7 and 0.3. The behavior may be well represented by a linear relationship. For prisms with SFS blocks, strong and weak mortars typically caused a different failure of mode, brittle when strong mortar was used and ductile when weak mortar was used.

4 CONCLUSIONS

The main conclusions from the work presented herein are:

- The estimation of the masonry modulus of elasticity by the masonry Brazilian code provisions are higher than that obtained from the tests using both types of block;

- Block type and mortar combination affect the Ritter constant calculated using the tests results;
- From this experimental program, the ultimate axial strain of clay blocks at failure were determined to vary from 2 to 2.5‰;
- It was observed a decrease in stiffness for hollow HFS prims with f_{mortar}/f_{block} of 0.3 when compared to those with f_{mortar}/f_{block} of 0.7;
- For SFS prisms, there was no significant difference between Ritter constants for both f_{mortar}/f_{block} ratios.

REFERENCES

CSA S304. 2014. Design of masonry structures. *Canadian Standards Association*. Mississauga, ON, Canada.

Costigan, A.; Pavía, S. and Kinnane, O. 2015. An experimental evaluation of prediction model for the mechanical behavior of unreinforced, lime-mortar masonry under compression. *Journal of Building Engineering*, 4: 283–294.

EN 1996-1. 2005. Eurocode 6-Design of masonry structures. Part 1-1: -General rules for reinforced and unreinforced masonry structures. *European Committee for Standardization CEN*, Brussels.

Knutson, H. H. 1993. The stress-strain relationship for Masonry. *Masonry International- Journal of the British Masonry Society*. Vol. 7(1): 31–33.

Kaushik, H. B.; Rai, D. C. and Jain, S. K. 2007. Stress-strain characteristics of clay brick masonry under uniaxial compression. *Journal of Material Civil Engineering*, 19(9): 728–739.

Mohamad, G.; Lourenço, P. B. and Roman, H. R. 2005. Mechanical behaviour assessment of concrete block masonry prisms under compression. *INCOS - International Conference on Concrete for Structures*, Coimbra: Portugal. 261–268.

NBR 15812-1: 2010. Alvenaria estrutural - Blocos cerâmicos – Parte 1: Projetos. *Associação Brasileira de Normas Técnicas*. Rio de Janeiro.

NBR 13276. 2005. Argamassa para assentamento e revestimento de paredes e tetos – Preparo da mistura e determinação do índice de consistência. *Associação Brasileira de Normas Técnicas*. Rio de Janeiro.

NBR 13277. 2005. Argamassa para assentamento e revestimento de paredes e tetos – Determinação da retenção de água. *Associação Brasileira de Normas Técnicas*. Rio de Janeiro.

NBR 13278. 2005. Argamassa para assentamento e revestimento de paredes e tetos – Determinação de densidade de massa e teor de ar incorporado. *Associação Brasileira de Normas Técnicas*. Rio de Janeiro.

NBR 13279. 2005. Argamassa para assentamento e revestimento de paredes e tetos – Determinação da resistência à tração na flexão e à compressão. *Associação Brasileira de Normas Técnicas*. Rio de Janeiro.

NBR 13280.2005. Argamassa para assentamento e revestimento de paredes e tetos – Determinação da densidade de massa aparente no estado endurecido. *Associação Brasileira de Normas Técnicas*. Rio de Janeiro.

NZS 4320. 2004, Design of reinforced concrete masonry structures. *New Zealand Building Code*. New Zealand.

Sarhat, S.R.; Sherwood, E.G. 2014. The prediction of compressive strength of ungrouted hollow concrete block masonry. *Construction and Building Material Journal* 58: 111–121.

Experimental tests of wall joints

I. Galman & R. Jasiński
Silesian University of Technology, Gliwice, Poland

ABSTRACT: The paper presents the results of experimental tests of joints of walls made of autoclaved aerated concrete (AAC). Different joint types were used to connect two walls: trusses, standards profiles and profiles with widened cross-section at the joint as well as polyurethane glue (foam) commonly used for bricklaying of masonry walls. 18 models were tested altogether in five series. The tests were performed on an original test stand in which shearing of wall joints can be realized with minimized bending. Morphology of cracking and mode of failure were analyzed; the load–deflection relationships of tested models and reference models with traditional masonry bonds were also compared. The tests showed differences in cracking and failure mechanisms as well as in load-bearing capacity of different joint types.

1 INTRODUCTION

The issue of joints in crossing walls was neglected when the joints were realized as traditional masonry bonds. However, since new types of masonry units and connectors have been introduced to the market, such an approach is no longer valid. Numerous damages within the walls joint have been reported ever since due to exceedance of shear stresses.

EC-6 demands that the designer ensures co-operation between the crossing walls but provides only constructional requirements. There are no guidelines or empirical procedures for determination of load-bearing capacity of wall joints both in case of non-uniform vertical loading of the walls (walls loaded mainly vertically) and shear (stiffening walls). The situation is further complicated by the fact that the producers of the connectors provide only the load-bearing capacity of the connectors, not the joints with the connectors. The minimum capacity of the connector, let alone the whole joint is neither provided, thus this solution cannot be practically used.

Therefore, the design of such neuralgic structural elements is often just intuitional, not backed up by any analytical calculations.

Research material on wall joints is also very scarce. Among the very few, the experimental laboratory works of Paganoni S &D'Ayala D. 2014 and Corrêa M.R.S. et al. 2009 can be mentioned. Nevertheless, diversity of the test stands used and a lack of unified testing procedures makes it practically impossible to compare the results.

Therefore, for almost two years now the tests have been carried out at the Laboratory of the Faculty of Civil Engineering, Silesian University of Technology on the joints of walls made of AAC joined with the use of the most popular connectors at the market, as well as with different types of meshes, mats and other elements which allow for fast and easy connection of the walls [Galman&Jasiński 2018a, 2019, Drobiec 2019]. This complex research program aims at investigation of the mechanism of cracking and failure of joints of walls made of AAC as well as at comparison of the load-bearing capacity of wall joints realized as traditional masonry bonds and with the use of different connectors. Understanding of the joint's behavior will allow to design joints based on a concrete knowledge, not intuitionally. Such knowledge is necessary for practical use of different types of connectors.

2 EXPERIMENTAL PROGRAM

The experimental program comprised of five series of models (18 models altogether) of identical shape and dimensions. The models were monosymmetric with a T shape in which the web and the flange were ~89 cm long. A vertical joint was constructed between the loaded and non-loaded wall which differed between the model series. In the series of test models denoted as P a traditional masonry bond was constructed between the flange and the web (Figure 1a). These were the reference elements whose mechanical parameters and behavior during loading and failure was compared to the results of other tests.

In the next three series the joint between the flange and the web was made with the use of steel connectors (geometry of the wall according to Figure 1b). In B10 (Figure 1c) and BP10 (Figure 1d) series flat profiles were used. The solution with a flat profile with widened mid part was proposed by the authors based on their own experimental research of punched hole connectors [Galman&Jasiński 2018b]. The widening serves to increase flexural capacity and stiffness of the

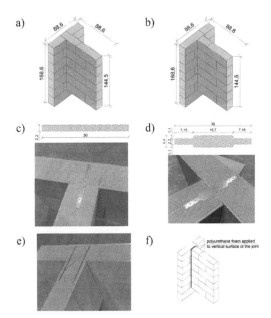

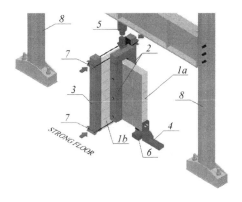

Figure 2. Scheme and view of test stand: 1a - longitudinal wall, 1b - transverse wall, 2 - reinforced concrete column transferring shear load, 3 - reinforced concrete pillars limiting horizontal deformation, 4 - horizontal support, 5 - system of hydraulic cylinder and force gauge used to induce shear stress, 6 - force gauge, vertical reaction, 7 - horizontal tie, 8 - steel frame.

Figure 1. Geometry and details of test models: a) traditional masonry bond, P series, b) masonry walls with steel connectors (B10, BP10 and T series), c) method of connection with punched hole flat profile (B10 series), d) method of connection with widened punched hole flat profile (BP10 series) [dimensions in cm], e) shape and dimensions of flat profile used in T series, f) view of S series test model.

connector. This shape is copyright protected by Polish patent application [Polish Patent Office]. Steel connectors were made of 1-mm thick galvanized steel punched hole flat profile.

In the third series of reinforced masonry walls, denoted as T, steel trusses were used (Figure 1e). Reinforcement was composed of flat profiles with cross-sectional dimensions of 1.5×8 mm spaced by 140 mm, connected with a 1.5 mm wire. Strengthening of the truss was made of normal steel and then galvanized to provide protection against corrosion. Tensile strength of both trusses and punched flat profiles was determined experimentally.

In the last series (S) the joints were realized with the use of polyurethane glue (Figure 1f).

The tests were performed on a specially designed test stand (Figure 2). Models *1a* and *1b*, equipped with confining elements *3* and elements transferring load *2* were mounted on a strong floor (*1b*) and on a force meter *6*, which together with a retaining element *4* formed a pinned support. The models were located under the steel frame *8* to which a hydraulic force meter (with a 1000 kN range) was mounted. The force meter was generating shear at constant increase of displacement of 1 mm/min. Structural response was registered with an inductive force transducer with a 250 kN range and accuracy of ±2.5 kN. To simulate significant length of one of the walls, in a *1b* element initial compression of 0.1

MPa was induced using reinforced concrete elements *3* and steel tendons *7*. The models were loaded in one cycle up to failure. Vertical load generating shear was transferred linearly along the whole height of the wall by elements *2* which allowed to obtain uniform shear stresses in the joint. Values of loads and displacements of the loaded wall with respect to the non-loaded one were registered continuously during the test. Two independent systems were used for this registration. One of the sides of the tested elements was monitored with an optical displacement measurement recorded ARAMIS. The other side was monitored with the use of three PJX-10 LDVTs with a range of 10 mm and accuracy of ±0,002 mm.

The tests were performed on the models made of AAC masonry units and system mortar for thin joints with unfilled head joints. Compressive strength of the masonry wall, determined according to PN-EN 1052-1:2000 and presented in Jasiński & Drobiec (2016) was equal to $f_c = 2.97$ N/mm², modulus of elasticity was equal to $E_m = 2040$ N/mm², while initial shear strength determined according to PN-EN 1052-3:2004 and presented in Jasiński & Drobiec (2015) was equal to $f_{vo} = 0.31$ N/mm². Tests of steel connectors were also performed (according to PN-EN 10002-1:2004). Experimental program is presented in Table 1.

3 MECHANISM OF CRACKING AND FAILURE

The character and morphology of cracking was closely connected with the method of joint formation.

All the tested unreinforced models behaved in a similar manner. In the first phase of loading no

Table 1. Experimental program.

Series name	Joint formation method	No. of models
P	Traditional masonry bond Flat profile 300×22×1	6
B10	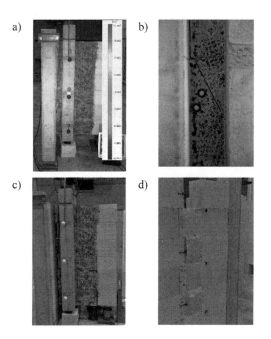	3
BP10	Widened profile 300×44×1	3
T	32-cm long truss	3
S	Polyurethane glue	3

crashes were heard and no spalling was visible on the side surfaces of the elements; plane deformations of part of the wall were dominating. This phase lasted until first slanted cracks appeared in direct vicinity of the joint (Figure 3a and 3b) – at around 70% of the value of the force at failure.

An increase of load induced significant development of cracks in the joint and their propagation towards the reinforced concrete column transferring the load (Figure 3c). In this phase the maximum force was recorded. Further loading caused significant increase of relative displacement and rotation of joined walls. After failure the joint was dismantled (Figure 3d) – almost vertical shear of joined elements was found. No visible damages were observed on the remaining elements.

In the models reinforced with steel connectors (B10, BP10 and T series) cracks characteristic for unreinforced models were not observed in the whole range of loading. In the initial phase of loading relative displacements of joined walls were unnoticeable. At some point a sudden, visible increase of displacement occurred. The T series models were not able to transfer any further loading afterwards. Displacement of the loaded wall with respect to the non-loaded one was of about 10 mm. In case of the walls joined with the use of steel connectors, however, further loading was possible despite the displacements. Later failure had a sudden character and was caused by shearing of the joint and visible further vertical displacement (by about 17 mm) of the web wall. Failure of all models reinforced with steel connectors occurred due to yielding of the steel elements within the joint (Figure 4a, 4b and 4c). Masonry units exhibited local spalling below the connectors. The connectors were not pulled out. Thanks to the holes in the profile connectors did not slip in bed joints. The mortar penetrating the holes was not sheared but acted as a dowel eliminating shift.

A completely different failure model was observed in case of glued walls. Failure of as much as two out of three models in the series was caused by spalling of the column transferring the load. In the third model

Figure 3. Failure of P series models a) first crack in P_2 reference model b) first crack in P_5 reference model c) view of joint in P_3 model after failure c) view of joint in P_1 model after failure.

Figure 4. Failure of reinforced models: a) typical deformation of steel flat profile within the joint (B10_1), b) typical deformation of steel flat profile within the joint (BP10_2), c) typical deformation of steel truss within the joint (T_2).

868

a) 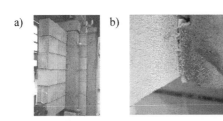 b)

Figure 5. View of of S series models after failure: a) model S_2, b) shift of the column transferring load with respect to the tested model (S_1).

failure was caused by stiff displacement of the web with respect to the flange. The view of damaged model and the detail showing a shift of the test stand column with respect to the tested model is presented in Figure 5.

4 TEST RESULTS

4.1 Models with traditional masonry bond

Up to cracking of the joint, which occurred at the load level of $N_{cr} = 27.3$–54.1 kN, relative displacements u were increasing almost proportionally to the load. After cracking stiffness reduction was observed, but the joints continued to transfer the load. This phase ended under the maximum forces at the level of $N_u = 38.6$–59.8 kN. Further loading attempts caused significant decrease of values of the forces registered by the force meter accompanied with an increase of relative displacements. The forces did not drop to zero and the joints were still able to transfer some load. The values of forces in this phase of joint's work were called the aggregate interlocking forces $N_{ag} = 14.1$–31.1 kN. Further increase of displacements in the joint caused slight increase of load – hardening. The ultimate registered forces, referred to as residual forces, preceded failure which was connected with entire spalling of joined elements and their relative rotation, and had a value at the level of $N_r = 8.4$–42.9 kN. The values of forces and corresponding displacements are collectively presented in Tables 2 and 3, and graphical

Table 2. Results for unreinforced models – forces.

Model	Cracking force $N_{cr,i}$ [kN]	Maximum force $N_{u,i}$ [kN]	Aggregate interlocking force N_{ag} [kN]	Residual force $N_{r,i}$ [kN]
P_1	27.3	56.3	31.1	20.7
P_2	42.6	50.0	14.7	10.2
P_3	31.2	38.6	25.5	13.8
P_4	54.1	59.8	–	8.36
P_5	35.1	48.1	–	–
P_6	45.1	51.6	28.3	27.9

Table 3. Results for unreinforced models – displacements.

Model	Displacement at cracking $u_{cr,i}$ [mm]	Displacement right before failure $u_{u,i}$ [mm]	Displacement at aggregate interlocking force $u_{ag,i}$ [mm]	Residual displacement $u_{r,i}$ [mm]
P_1	0.07	0.31	2.43	6.36
P_2	0.12	0.25	1.95	6.97
P_3	0.12	0.16	2.22	5.64
P_4	0.07	0.17	–	6.72
P_5	0.06	0.10	–	–
P_6	0.08	0.36	1.71	2.22

illustration of the load–displacement relationship is shown in Figure 6.

Moreover, in Table 4 joints' stiffness was given defined as a quotient of the load per joint and corresponding displacement. Knowing the stiffness it is possible to determine relative displacements of the connected walls for known loads as well as the value of the load for known relative displacements.

The mean value of joint's stiffness for all six reference models was equal to 488 MN/m.

4.2 Reinforced and glued models

As in the case of the models with traditional bond, it is also best to analyze the behavior of reinforced

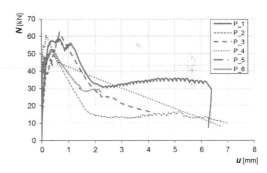

Figure 6. Relationship between total force and mean displacement of P series models.

Table 4. Stiffness of joints with traditional masonry bond.

Model	Stiffness $K_t = N_{cr}/u_{cr}$	
	K_t [MN/m]	$K_{t,mv}$ [MN/m]
P_1	390	
P_2	355	
P_3	260	
P_4	773	488
P_5	585	
P_6	564	

joints on the diagram of the relationship between the load and relative (mutual) displacement of joined walls (Figure 7).

In the models with steel truss (T series) a linear load–displacement relationship was observed. When the force exceeded the value causing yielding of steel reinforcing elements (at the level of 21.5–26.5 kN) a sudden increase of displacements was observed and the joint ceased to perform its function.

A completely different behavior of the joint was observed in the models with steel flat profiles. Up to the moment of cracking of the joint, which occurred at the maximum load of $N_{cr} = N_u = 7.3$–12.3 kN for B10 series models and 12.5–16.5 kN for BP10 series models, displacements u were increasing almost proportionally. After cracking displacements increased and forces decreased to the value of $N_d = 3.4$–5.0 kN for B10 series models and 8.9–10.5 kN for BP10 series models. Nevertheless, the joint was still able to transfer additional load, a slight hardening appeared. Failure induced by excess displacements occurred under the load of $N_r = 2.3$–9.2 kN for B10 models and 10.6–14.8 kN for BP10 models. Therefore, a decrease of the residual force with respect to the maximum force was observed by about 35% for B10 models and only 15% for BP10.

The S series models with polyurethane glue behaved in a relatively brittle manner. The obtained relationships of load N – relative displacement u (Figure 7b) showed almost linear character of the joint's work. After reaching the value of $N_{cr} = $ 50.5–68.7 kN inclination of the line changed (except for the S_3 model) but elastic behavior was still observed. The joint in S_3 model exhibited some load-

Table 5. Results of reinforced and glued joints (forces).

Model	Cracking force $N_{cr,i}$ [kN]	Force at failure $N_{u,i}$ [kN]	Dowel force N_d [kN]	Residual force $N_{r,i}$ [kN]
T_1	21.5	13.1	-	-
T_2	26.5	9.0	-	-
T_3	23.2	16.8	-	-
B10_1	12.3	12.3	5.01	6.68
B10_2	8.41	8.41	5.02	9.20
B10_3	7.27	7.27	3.39	2.32
BP10_1	15.9	15.9	8.86	14.8
BP10_2	16.5	16.5	10.5	12.6
BP10_3	12.4	12.4	9.36	10.6
S_1	64.3	68.3	-	-
S_2	68.7	71.4	-	-
S_3	50.5	57.2	51.6	-

Table 6. Results of reinforced and glued joints (displacements and stiffness).

Model	Displacement at cracking $u_{cr,i}$ [mm]	Displacement right before failure $u_{u,i}$ [mm]	Displacement at aggregate interlocking force $u_{ag,i}$ [mm]	Residual displacement $u_{r,i}$ [mm]	Stiffness N_{cr}/u_{cr} K_t [MN/m]
T_1	0.09	3.89	-	-	239
T_2	0.03	2.52	-	-	883
T_3	0.02	1.84	-	-	1160
B10_1	0.07	0.07	1.83	11.50	176
B10_2	0.08	0.08	0.68	10.33	105
B10_3	0.41	0.41	3.32	12.52	18
B-P10_1	0.04	0.04	0.45	4.15	398
B-P10_2	0.05	0.05	1.61	7.04	330
B-P10_3	0.49	0.49	1.94	6.22	25
S_1	0.14	0.21	-	-	459
S_2	0.09	0.11	-	-	763
S_3	0.08	0.16	0.38	-	631

bearing capacity reserve represented by a diagram branch declining from the value of load of $N_u = 57.2$ kN to $N_r = 51.6$ kN.

5 SUMMARY AND FUTURE RESEARCH

The presented laboratory tests are a part of an experimental program currently realized at the Laboratory of the Faculty of Civil Engineering (Silesian University of technology) on the joints of walls made of AAC. This paper presented five types of joints: with traditional masonry bond, with the use of

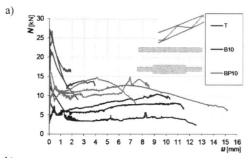

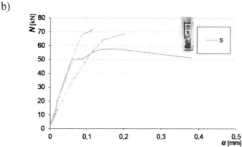

Figure 7. Relationship between total force and mean displacement of a) reinforced models, b) glued models.

two types of punched hole flat profiles, with the use of steel truss and a glued joint.

The process of damage and development of cracking in the walls with traditional bond went on relatively smoothly an in phases. Visible cracks appeared withing the joint before failure. In addition to significantly reduced load-bearing capacity, also the process of cracking and failure of the models with steel connectors was completely different. Failure was not preceded by development of cracks but it was characterized with an instant increase of displacements accompanied with a drop of loads. An undoubtful advantage of the joints made with the use of polyurethane glue was an important increase in the load-bearing capacity and stiffness with respect to the traditional masonry bond. The disadvantage, however, was brittle behavior of the joints which made it impossible to diagnose potential structural safety risks. Having in mind the properties of materials made with polyurethane resins, future tests will focus on the change of joints' properties in time and influence of cyclic loading. Because the analyzed series comprised of only 3 elements no quantitative conclusions for design could have been drawn yet. Therefore, future works should also include analyses of additional models for statistical determination of empirical parameters of the models. Furthermore, currently the works are carried out on the FEM-based model of the joints to determine realistic behavior of the connectors.

REFERENCES

Corrêa M.R.S., Moreira E.M.S., Ramalho M.A. (2009) Experimental small-scale analysis of the connections between structural clay block work masonry walls submitted to vertical loads. *11th Canadian Masonry Symposium*. Toronto (CD-ROM).

Galman I. & Jasiński R. 2018a. Joints in masonry walls. *6th International Conference on Autoclaved Aerated Concrete., Postam, 4 – 3 September 2018.*

Galman, I. & Jasiński, R. 2018b. Tests of joints in AAC masonry walls in: *Architecture Civil Engineering Environment* 11(4): 79.

Galman, I. & Jasiński, R. 2019. Attempt to Describe the Mechanism of Work of Masonry Joints. *IOP Conf. Series: Materials Science and Engineering* 471.

Jasiński R. & Drobiec Ł. 2016. Comparison Research of Bed Joints Construction and Bed Joints Reinforcement on Shear Parameters of AAC Masonry Walls. *Journal of Civil Engineering and Architecture* 10(12): 1329–1343. DOI: 10.17265/1934-7359/2016.12.004.

Drobiec Ł. & Jasiński R. 2015. Influence of the kind of mortar on mechanical parameters of AAC masonry subjected to shear – the basic strength parameters. *Materiały Budowlane* 5: 106–109, 2015. DOI: 10.15199/33.2015.05.44 (in Polish).

Paganoni S & D'Ayala D. 2014. Testing and design procedure for corner connections of masonry heritage buildings strengthened by metallic grouted anchors. *Engineering Structures* 70: 278–293.

Polish Patent Office, ul. Niepodległości 188/192, 00-950 Warsaw Polska. Application from 2019/04/01No. W.128153. Wall joint connector.

PN-EN 1052-1:2000 Methods of tests for masonry. Part 1: Determination of Compression Strenght. (In Polish).

PN-EN 1052-3:2004 Methods of tests for masonry. Part 3: Determination of Initial Shear Strenght. (In Polish).

PN-EN 1996-1-1:2010+A1:2013-05P, Eurocode 6: Design of Masonry Structures. Part 1-1: General rules for reinforced and unreinforced masonry structures. (In Polish).

PN-EN 10002-1:2004 Metallic materials - tensile testing. Part 1. Method of test at ambient temperature.

Numerical modelling & analysis

Brick and Block Masonry - From Historical to Sustainable Masonry –
Kubica, Kwiecień & Bednarz (eds)
© 2020 Taylor & Francis Group, London, ISBN 978-0-367-56586-2

A simple homogenization approach for masonry structures: A discrete approach extension from walls to curved structures

J. Scacco & G. Milani
Department of Architecture Built Environment and Construction Engineering ABC, Politecnico di Milano, Milan, Italy

P.B. Lourenço
Department of Civil Engineering, ISISE, University of Minho, Guimarães, Portugal

ABSTRACT: Curved masonry structures as arches, domes and vaults are the most fascinating features of historical buildings. However, numerical analysis of these structural elements is still a challenge. In the last years the researchers have been proposing several approaches including non- linear analyses by means of micro-modeling or macro-modeling and kinematic limit analyses. Above all, these approaches may result too demanding or not enough accurate. A new discretized homogenized approach is here proposed with the aim to combine easy implementation, accuracy and speed. The model is conceived as an assembly of elastic units joint by non-linear interfaces. These latter are modeled as bricks elements and Concrete Damage Plasticity is used for modeling non-linear mechanical properties obtained from consolidated. homogenization procedures. A validation on brick walls loaded out-of-plane and on a dome is presented in order to show the capability of such an approach.

1 INTRODUCTION

1.1 *Overview of numerical analysis*

In the context of historical heritage vaults and domes can be considered the most fascinating and charming features of masonry building. Since such structural elements began to spread worldwide several centuries ago it is possible to identify a huge variety of materials. Masonry is in fact a heterogeneous material, made of units assembled with or without mortar. Both of these components can have different mechanical, chemical and physical properties, so the definition of reliable mechanical properties is less straightforward in comparison with other construction materials (Huerta, 2001). Moreover, bricks and stones can be arranged in several configurations along with all the different possible shapes peculiar to masonry vaults. For these reasons, numerical simulations of curved masonry structures is still a challenging task in the current research, where a comprehensive numerical strategy seems still lacking (Tralli *et al.*, 2014).

As well know, the design at the base of masonry vaults and arches was based on geometrical principles, which results effective when only vertical loads are accounted for. However, the vulnerability of such elements against horizontal loads has been certified from the experience of previous seismic events, leading to the current issue to preserve structures with cultural value (Formisano and Marzo, 2017).

In order to explore the performances of curved masonry structures, several numerical approaches have been developed in literature starting from the least complex ones based on graphical methods (Heyman, 1966), to advance non-linear simulations (Bianchini *et al.*, 2019). These latter allow to reproduce properly the behavior of brittle materials as masonry over the elastic limit and they are commonly divided into two groups. The first one is identified as macro-modeling, where the different components are substituted by a homogeneous fictitious material, reducing by far the complexity of the problem from the geometrical and numerical point of view. On the other hand, such an approach presents several limitations related to the impossibility to catch the orthotropic behavior peculiar to masonry. As a consequence, the results in terms of damage pattern may result inaccurate. In order to overcome this lack, some orthotropic macro-models were provided along with an elasto-plastic behavior with softening (Lourenço *et al.*, 1997). However, extensive experimental campaigns are required in order to define the mechanical properties of the materials.

The alternative might be a detailed micro-modeling, where both mortar and units are modeled separately (Bove *et al.*, 2019). This approach allows addressing different mechanical properties to each constituent, leading to reliable simulations. On the other hand, when applied to 3D models the computational effort might result huge.

A straightforward alternative to non-linear analyses is the combination of kinematic limit analysis with FE (Chiozzi *et al.*, 2017) or with NURBS - Non Uniform Rational Basis-Splines- (Grillanda *et al.*, 2019). This method provides an upper bound value of the collapse load, along with the collapse mechanism configuration in a computational time significantly lower than the one needed running non-linear analysis. It must be considered that is such a way, no information about the displacement and crack history is known. This may result a strong limitation when the efficacy of strengthening has to be numerically validated.

1.2 *Aims of the present paper*

The aim of the present paper is the proposal and validation of an innovative method for the numerical analysis of masonry curved structures, enable to take advantage of all successful features of the approaches discussed briefly above.

First of all, in order to be a reliable and predictive method, it should be able to account for the heterogeneity of masonry, preserving its orthotropic behavior. The simulation of non-linear capacities of the structure is a key aspect, even for complying with the requirements of international codes, which suggest pushover analyses in order to assess seismic vulnerability (NTC, 2018). On the other hand, with the aim to be easily applied to practical cases, the number of variables involved has to be limited in comparison with micro-modeling. Last but not least, the influence of membrane actions, coming from the vertical load, on the out-of-plane behavior of domes and vaults should be properly reproduced (Tralli *et al.*, 2014).

A macro and micro-modeling do not fulfill these points, making necessary the implementation of an alternative approach. Homogenization procedures proved to be particularly successful in the applications to masonry-like materials with a macroscopically composite nature (Massart and Geers, 2007). Such an approach is based on the definition of a unit cell, the representative volume element (RVE) which represents the whole arrangement of the structure. Several procedures have been implemented in order to catch the elastic and non-linear response of the RVE. Some of them are based on FE methods that provide reliable results (Silva *et al.*, 2017), requiring besides a high computational burden. Alternatively, a straightforward approach is offered by the semi-analytical procedure described in (Bertolesi *et al.*, 2016), where the derivation of homogenized properties results fast due to the displacement formulation and the rough discretization used. Afterward, such properties can be transferred at a structural level as an averaged isotropic model. A valuable alternative is the use of discrete methods where rigid or elastic units are joint by interfaces where the homogenized non-linearities are lumped. One of the most remarkable method to be mentioned is the rigid-spring approach "RBSM", consisting of modeling the interfaces as non-linear springs (Bertolesi *et al.*, 2018). However the implementation may result cumbersome and there is the necessity to keep the in-and-out-plane effects uncoupled, framing its application to walls.

The discrete methods along with previous homogenization steps, seems to be the right choice for modeling curved masonry elements, enabling to satisfy all the key points discussed before. In the proposed approach the elastic properties, coming from homogenization, are applied directly to the units ensuring better stability when the non-linear behavior is explored. The non-linear interfaces are intended to be modeled as smaller and flat brick elements, leading to an easier implementation. These solid elements will be modeled by means of the constitutive model Concrete Damage Plasticity, already available in FEM software Abaqus (Figure 1). The 3D approach allows taking into account automatically of the changing along the cross-section of the normal stress, addressing one of the most important key point regarding domes and vaults. In fact, in such a scenario, the influence on the static behavior of the vaulted structure of the distribution of normal stress inside the section will be automatically taken into account.

Experimental and numerical references are eventually used in order to show the convenience of the proposed approach, going into its capabilities (Figure 2-3).

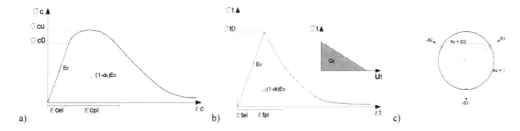

Figure 1. Mono-axial constitutive behavior in compression (a) and in tension (b); CDP yield surface (c).

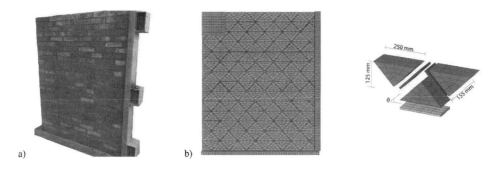

Figure 2. UR wall experimentally tested (a) and mesh discretization of the homogenized panel (b) (Scacco *et al.*, 2020).

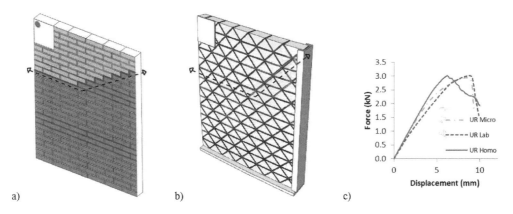

Figure 3. Comparison of experimental crack pattern (a), homogenized crack pattern (b), load-displacement curves (c) (Scacco *et al.*, 2020).

2 METHODOLOGY

2.1 *Preliminary results on a masonry wall (Unreinforced and FRCM-Reinforced)*

In this section the discrete homogenized method is firstly applied in the case of a wall subjected to out-of-plane loading, providing experimentally and numerically comparison to validate the method before its extension to masonry curved structures. Walls loaded out-of-plane is a trend topic, much more studied than curved structures, both from experimental (Murano *et al.*, 2019) and numerical point of view (Casolo and Milani, 2010; Chiozzi *et al.*, 2018), providing an useful starting point for the validation of the method.

The CDP that here is used in order to address non-linearities at the interfaces has been already successful applied for modeling masonry structures (Tiberti and Milani, 2017; Valente and Milani, 2017) Concrete damage plasticity is an isotropic elasto-plastic constitutive model with damage. Its yield surface follows the Drucker-Prager function, which shape is opportunely modified in order to approximate a Mohr-Coulomb surface (Figure 1). Other parameters are input with the scope to smooth the yield surface and tackle computational issues. The degradation of the elastic stiffness is determined by two variables of damage dt and dc, functions of the plastic strains. Such variables vary between a value of 0 for the undamaged material and 1 for the material that has completely lost any stiffness.

The orthotropic behavior is preserved as different mechanical properties are addressed according to the orientation of the interfaces. CDP allows the definition of a distinct behavior under tension and compression, respectively following an exponential softening and a parabolic softening. The use of CDP allows an easy-implementation of the homogenized properties, obtained by means of consolidate procedures, at the structural level resulting straightforward and user-friendly. Moreover, the method is able to provide accurate results with a discretization by far less detailed than the one needed for a micro-modeling (2300 elements against 140000 elements). The interfaces are discretized by 4 elements along the depth and the length, with one element in the thickness. Elastic units are divided by 2 elements in the depth and 4 elements along the edges (Figure 2).

Along with the micro-modeling comparison, even the experimental reference is used for validation. This latter is deeply discussed in (D'Ambra *et al.*, 2018), where a simulation of an infill wall under horizontal seismic loads was carried out. Two configurations of brick walls (Unreinforced-UR and TRM-Reinforced-RE) were tested by the application of a horizontal out-of-plane action at the top left of the specimen. The experimental damage pattern is highly related to the arrangement of the bricks, along with their exceptionally good mechanical properties, driving the damage along the interfaces mortar-bricks with a zig-zag pattern.

With the aim to reproduce the experimental crack-path, a wedge-shape has been assigned to the elastic cells of the homogenized model, concentrating the non-linearities in the diagonal and horizontal interfaces.

The results are shown in terms of load-displacement curves and crack patterns. From the first it is possible to appreciate how the global response is well detected, with a peak carrying capacity around 3 kN and a total displacement of 10 mm, matching the experimental and advanced modeling outcomes (Figure 3). Even the damage pattern is successfully reproduced, with a direction that well approximates the experimental stepped crack. Numerical results of the homogenized model of the reinforced wall are available in (Scacco *et al.*, 2020).

The proposed homogenized method turns out to be highly reliable and it results as a valid alternative to detailed micro-modeling. The numerical performance is even more valued taking into account the reduced computational burden found approximately ten times lower than the corresponding heterogeneous approaches.

2.2   *Extension of the method to curved masonry structures*

Once established the validity of the method, the next stage is its application to curved masonry structures.

The procedure follows the same steps of the analysis concerning the wall. However, in the case of vaults or domes, the preparation of the geometrical model may results problematic. In fact, there is the necessity to leave an empty space at the intersection of the interfaces for ensuring a pure orthotropic behavior. As a consequence, from the geometrical point of view, the use of flat 8-nodes bricks for modeling non-linear interfaces is for sure more demanding in comparison to homogenous models. With the aim to tackle this issue, the authors implemented a MATLAB script capable to interact directly with ABAQUS and to provide a discretized mesh for the homogenized model. In this paper, a dome experimentally tested by Foraboschi in 2006, and extensively numerically reproduced in literature, is used for the validation of the proposed method.

The preparation of the geometrical model starts with a coarse mesh of a shell surface which approximates the intrados of the dome (Figure 4). Each 4-nodes element of the initial mesh represents what will be an elastic cell in the final discrete model.

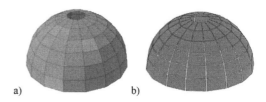

a)   b)

Figure 4. Initial mesh of the inner shell surface of the dome: un-scaled (a), scaled (b).

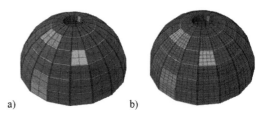

a)   b)

Figure 5. 3D Homogenized mesh configuration: Mesh "2x2" (a), Mesh "4x4" (b).

Each element, described by the 3D coordinates of four vertices is scaled by a fixed value in order to accommodate the creation of the interfaces. The script is able to link automatically the adjacent elements, by means of additional 4-nodes thin elements (which will correspond to the non-linear interface in the final model). At this stage, it is possible to define the wished discretization for each element of the shell-discrete-model. In this work, a coarse mesh of "2x2" elements for each cell, and a more refined one "4x4" are provided (Figure 5).

Once defined this new set of nodes, an extrusion is operated in order to model the thickness of the vault. Each node can be extruded of a different quantity so that vaults with changing in the thickness are easily replicable. Afterward, the discretization of the solid elements along the thickness must be chosen. According to the experience of authors, a number of four elements is sufficient in order to capture properly the flexural behavior of curved structures.

The differential mechanical properties coming from homogenization procedures are directly associated with each interface according to their orientation. This is a future key-point of such an approach as semi-analytical homogenization procedures, see (Milani and Bertolesi, 2017), can be implemented directly in the script leading to a comprehensive homogenization procedure, from the geometrical and material point of view. The authors refer to future research the interested readers.

With the aim to provide even numerically comparison, the work carried out by (Milani and Tralli, 2012) is used as a reference (Figure 6).

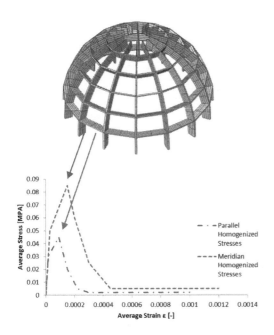

Figure 6. Stress-strain constitutive model of non-linear interfaces under tension.

The above paper provided homogenized numerical simulations at the cell level of the elementary cell of the dome tested by Foraboschi. Here the tensile stress-strain curves are re-proposed for both parallel and meridian stresses. In Abaqus, the post-elastic phase can be described by means of stress-strain or fracture energy input. For the sake of comparison, the first option has been used for the numerical analyses here presented.

3 RESULTS

The dome presents an internal diameter of 2.2 m and a thickness of 0.12 m. It was made of bricks with dimensions equal to 0.12x0.25x0.55 m³. The test consisted of applying an incremental vertical load at the top part of the structures (along the border of the hole). The experimental outcome is described by a load-displacement curve with a peak capacity load close to 48 kN and a total displacement around 4.5 mm. The dome has been already numerically reproduced by means of different software and approaches. In (Milani and Tralli, 2012) a synthesis of the results is provided and re-proposed in Figure 8 as references.

In this section the first results of the proposed method are provided (Figure 7-8). The discretized model is divided into approximately 100 elastic cells with the purpose to balance precision and speed. As mentioned before, two models have been prepared in order to show the reliability of the method with different discretization. The global behavior in terms of load-displacement curves is compared with other methods to gain awareness on the capabilities of the model. As it is evident from Figure 8, even though the homogenized model provides a less reduction in the stiffness during the first stages, the global response is successfully replicated. Moreover, it is noticeable that the discrete model with a rough mesh provides a high-quality result, with a slight difference in the peak-load with the more refined model. The slight difference in the stiffness may be related to the nature of the discrete method. Indeed, as reported in (Milani and Tralli, 2012) sliding occurs under the load-application point in the standard FE non-linear analyses. The proposed method does not allow the simulation of such a marked and local failure leading to a smaller vertical displacement in the first stages of the analysis. The limited softening branch is due to the first occurrence of damage along meridians, which determines a change in the static behavior of the dome.

Even the evolution of damage (see Figure 7) is very close to the one provided by the analysis of Milani, 2012. As expected the symmetry of the problem leads to an equal symmetry in the spread of cracks. The damage starts among the meridians at approximately 2/3 of the height to spread, almost reaching the base, where fixed supports are imposed. At the same time, near the top even the parallels results damaged, identifying a circular flexural hinge.

4 CONCLUSIONS

A simple 3D homogenization procedure for masonry curved structures has been presented. After a brief overview of the numerical methods available in the literature, the needed features for a comprehensive approach have been defined. The proposed methodology relies on a discrete homogenized model where elastic cells are joint by non-linear interfaces. In such a way the orthotropic behavior peculiar to masonry is preserved, and differential non-linear mechanical properties are addressed for the interfaces. The use of solid 8-nodes elements for modeling the non-linear interfaces satisfies the necessity to keep coupled the membrane and flexural actions, a key aspect for curved structures. Moreover, the method took advantage of the constitutive model CDP (concrete damage plasticity), already available in FEM software Abaqus, leading to easy implementation of the homogenized parameters.

At first, the validation of the method is provided for the case of a brick-wall under out-of-plane loading. Experimental and numerical comparisons prove a great advantage in terms of computational time, without loss of accuracy in the results. Afterward, the approach is extended to the case of curved masonry structures. A dome tested by Foraboschi in 2006 and numerically simulated by several authors

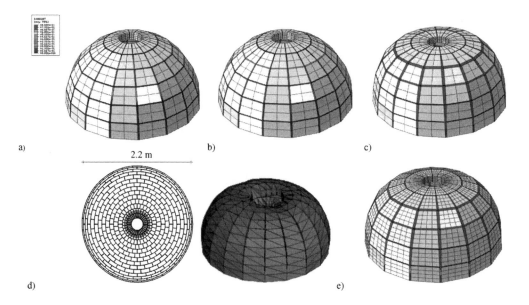

a)　　　　　　　　　b)　　　　　　　　c)

2.2 m

d)　　　　　　　　　　　　　　　e)

Figure 7. Evolution of the damage until the occurrence of a collapse mechanism (a-b-c); collapse mechanism by Milani, 2012 (d); comparison in terms of damage with a more refined mesh of the homogenized model (e).

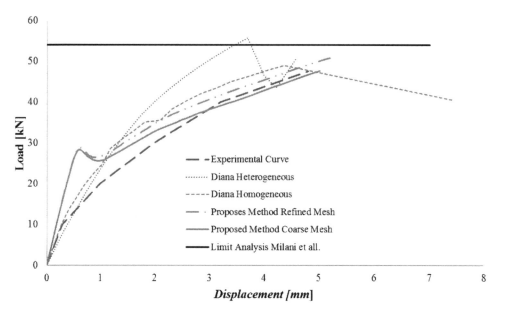

Figure 8. Comparison in terms of Load-displacement curves of the results obtained experimentally, with previous numerical references, and the proposed homogenized models.

has been chosen for the validation. As first step, an automatic procedure for obtaining a discrete mesh of the model is briefly explained, highlighting the possibility to select any kind of discretization. A discrete configuration of 100 elastic cells is chosen, but with two different mesh configuration. The coarse one is made of 2945 elements, whereas the more refined mesh reaches 8960 elements.

Following the experimental tests, the dome has been pushed down until the occurrence of a collapse mechanism. This latter is very close to the one provided in (Milani and Tralli, 2012), and at the same time a good match with the experimental and previous numerical load-displacement curves is obtained. An interesting aspect is the subtle difference in the accuracy provided by the two mesh configuration.

This result opens up to the possibility of straightforward and fast application for practical purposes. Furthermore, the reduced computational effort could allow the application of strengthening materials in a second step (Bertolesi *et al.*, 2016), without leading to unpractical heavy analyses. In this regard, such an extension will be the topic of future works by the Authors.

REFERENCES

Bertolesi, E., Milani, G., Casolo, S., 2018. Homogenization towards a mechanistic Rigid Body and Spring Model (HRBSM) for the non-linear dynamic analysis of 3D masonry structures. *Meccanica 53*, 1819–1855.

Bertolesi, E., Milani, G., Lourenço, P.B., 2016a. Implementation and validation of a total displacement non-linear homogenization approach for in-plane loaded masonry. *Comput. Struct. 176*, 13–33.

Bertolesi, E., Milani, G., Poggi, C., 2016b. Simple holonomic homogenization model for the non-linear static analysis of in-plane loaded masonry walls strengthened with FRCM composites. *Compos. Struct. 158*, 291–307.

Bianchini, N., Mendes, N., Lourenco, P.B., 2019. Seismic assessment of masonry cross vaults through numerical nonlinear static and dynamic analysis, in: *COMPDYN 2019 7th International Conference on Computational Methods in Structural Dynamics and Earthquake Engineering*. Crete, Greece.

Bove, M., Castellano, A., Fraddosio, A., Scacco, J., Milani, G., Piccioni, M.D., 2019. Experimental and Numerical Analysis of FRCM Strengthened Parabolic Tuff Barrel Vault, in: *Key Engineering Materials*. 213–220.

Casolo, S., Milani, G., 2010. A simplified homogenization-discrete element model for the non-linear static analysis of masonry walls out-of-plane loaded. *Eng. Struct. 32*, 2352–2366.

Chiozzi, A., Milani, G., Grillanda, N., Tralli, A., 2018. A fast and general upper-bound limit analysis approach for out-of-plane loaded masonry walls. *Meccanica 53*, 1875–1898. https://doi.org/10.1007/s11012-017-0637-x

Chiozzi, A., Milani, G., Tralli, A., 2017. A Genetic Algorithm NURBS-based new approach for fast kinematic limit analysis of masonry vaults. Comput. Struct. 182, 187–204.

D'Ambra, C., Lignola, G.P., Prota, A., Sacco, E., Fabbrocino, F., 2018. Experimental performance of FRCM retrofit on out-of-plane behaviour of clay brick walls. *Compos. Part B Eng. 148*, 198–206.

Formisano, A., Marzo, A., 2017. Simplified and refined methods for seismic vulnerability assessment and retrofitting of an Italian cultural heritage masonry building. *Comput. Struct. 180*, 13–26.

Grillanda, N., Chiozzi, A., Milani, G., Tralli, A., 2019. Collapse behavior of masonry domes under seismic loads : An adaptive NURBS kinematic limit analysis approach. *Eng. Struct. 200*, 109517.

Heyman, J., 1995. The stone skeleton. Cambridge University Press (UK).

Huerta, S., 2001. Mechanics of masonry vaults: the equilibrium approach. Proc. 1st *Int. Congr. Struct. Anal. Hist. Constr. Guimaraes* 47–70.

Lourenço, P.B., De Borst, R., Rots, J.G., 1997. A plane stress softening plasticity model for orthotropic materials. Int. J. *Numer. Methods Eng. 40*, 4033–4057.

Massart, T.J., Geers, M.G.D., 2007. International Journal of.

Milani, G., Bertolesi, E., 2017. Quasi-analytical homogenization approach for the non-linear analysis of in-plane loaded masonry panels. *Constr. Build. Mater. 146*, 723–743.

Milani, G., Tralli, A., 2012. A simple meso-macro model based on SQP for the non-linear analysis of masonry double curvature structures. *Int. J. Solids Struct. 49*, 808–834.

Murano, A., Ortega, J., Vasconcelos, G., Rodrigues, H., 2019. Influence of traditional earthquake-resistant techniques on the out-of-plane behaviour of stone masonry walls : Experimental and numerical assessment. *Eng. Struct. 201*, 109815.

Scacco, J., Ghiassi, B., Milani, G., Lourenço, P.B., 2020. A fast modeling approach for numerical analysis of unreinforced and FRCM reinforced masonry walls under out-of-plane loading. *Compos. Part B Eng. 180*, 107553.

Silva, L.C., Lourenço, P.B., Milani, G., 2017. Nonlinear discrete homogenized model for out-of-plane loaded masonry walls. *J. Struct. Eng. 143*, 04017099.

Tiberti, S., Milani, G., 2017. Historic city centers after destructive seismic events, the case of Finale Emilia during the 2012 Emilia-Romagna earthquake : advanced numerical modelling on four case studies abstract: 1059–1078.

Tralli, A., Alessandri, C., Milani, G., 2014. Computational methods for masonry vaults: a review of recent results. *Open Civ. Eng. J. 8*, 272–287.

Valente, M., Milani, G., 2017. Effects of geometrical features on the seismic response of historical masonry towers effects of geometrical features on the seismic response of historical masonry towers. *J. Earthq. Eng. 00*, 1–33.

Brick and Block Masonry - From Historical to Sustainable Masonry –
Kubica, Kwiecień & Bednarz (eds)
© 2020 Taylor & Francis Group, London, ISBN 978-0-367-56586-2

Experimental and numerical studies on the shear-sliding behavior of clay brick masonries

F. Ferretti & C. Mazzotti

Department of Civil, Chemical, Environmental and Materials Engineering, University of Bologna, Bologna, Italy

A. Incerti

CIRI Buildings and Construction, University of Bologna, Bologna, Italy

ABSTRACT: The identification of the shear strength parameters of masonry can be carried out through different experimental techniques, aimed at reproducing typical failure modes of masonry structural elements. The objective of this work is to characterize the shear-sliding behavior of brick masonry through the execution of triplet tests on three different masonry typologies. Series of triplet tests (EN 1052-3) were performed in displacement control on standard samples, constituted of clay bricks and lime-based mortar, arranged in a stacked bond. The experimental results were analyzed and compared among the different masonry typologies trying to correlate the shear strength parameters to the mechanical properties of the materials. Numerical simulations of the experimental tests were also carried out, adopting a simplified micro-modeling approach. The nonlinear numerical analyses allowed to properly interpret the experimental outcomes and they were compared with them in terms of failure load, post-peak behavior and specimen deformability, obtaining a good agreement.

1 INTRODUCTION

Brick masonry is characterized by the presence of horizontal and vertical mortar joints, which influence the shear behavior of the structural elements since they represent surfaces of weakness along which the failure can occur (Lourenço 1996, Rots 1997). The shear-sliding failure mode of masonry panels can be described by the Coulomb friction criterion. In this case, the local properties of the mortar-brick interface, such as the initial shear strength and the friction coefficient, are the most important parameters to be defined.

Experimental laboratory tests may be performed to study the shear-sliding behavior of masonry and to evaluate the mortar joint shear capacity. Different test methods were proposed in past researches, varying the geometry of the specimen, the loading arrangement and the boundary conditions (Drysdale et al. 1979, Stöckl et al. 1990, Riddington & Jukes 1994, Van der Pluijm 1999). Indeed, several aspects should be considered in the design of a shear test setup to ensure the reliability of the results (Riddington et al. 1997). First of all, normal and shear stress distributions should be uniform along the sliding mortar joint. Then, the failure should initiate far from the joint edges and should propagate quickly on the entire joint length. Finally, the presence of tensile stresses should be checked and possibly avoided.

Finite element analyses were also performed with the aim of identifying the most reliable test method by investigating peculiar aspects of the different methodologies and highlighting their advantages and disadvantages (Jukes & Riddington 2000, Popal & Lissel 2010).

A particular focus is here devoted to the triplet test, as proposed by the standard EN 1052-3. During the test, a constant axial compression is applied to the specimen, while a shear load is imposed to the central brick to produce its sliding. This test was found to be adequate to provide reliable results (Jukes & Riddington 2001, Vermeltfoort 2010). However, the main issues of this test are: (i) stress concentrations close to the loading points; (ii) presence of bending moment along the joint, which causes non-uniform stress distributions and could lead to undesired failures.

The objective of the present research is to study the shear-sliding behavior of three different masonry typologies through the execution of triplet tests. The tests were conducted in displacement control with the aim of characterizing both the Mohr-Coulomb failure criterion describing the masonry peak shear strength, and the post-peak behavior. Moreover, the transverse expansion upon shearing along the mortar joint, also known as dilatancy, was investigated. Indeed, it was found in previous researches (Van der Pluijm et al. 2000, Van Zijl 2004, Ferretti et al. 2018) that the dilatant behavior of masonry could affect the test results, especially if it is restrained to some extent during the sliding process, e.g. due to the boundary conditions of the setup.

Numerical simulations of the experimental triplet tests were also carried out to better investigate the shear-sliding failure, through the analysis of the stress distributions along the mortar joint and the development of the cracking process.

2 MATERIALS AND METHODS

In the experimental campaign, series of triplet tests (EN 1052-3) were performed on standard samples, built with clay bricks and lime-based mortar, arranged in a stacked bond. Mortar joints were characterized by a thickness approximately equal to 10 mm. In particular, three masonry typologies were studied, differing in the mortar composition from one another. The main characteristics of the triplet samples are presented in Table 1, while the mechanical characterization of the constituent materials is described in the following Sections.

2.1 Bricks

The brick elements chosen for the construction of the triplet samples were handmade fired clay bricks, whose dimensions are reported in Table 2. Cylindrical specimens, characterized by a height-to-diameter ratio h/d equal to 1, were cored from the bricks and tested in compression for their mechanical characterization. The mean brick compressive strength $f_{b,c}$, together with the brick density γ_b, is reported in Table 2.

2.2 Mortars

In the experimental campaign, three different natural hydraulic lime-based mortars (NHL) were adopted. Starting from a pre-mixed mortar available on the

Table 1. Characteristics of the triplet samples.

Triplet Series	Brick Type	Mortar Type	Nr. Specimens	Dimensions (mm³)
A-M1	IBL_BN	M1	4	250×170×120
B-M2	IBL_BN	M2	4	250×170×120
C-M3	IBL_BN	M3	4	250×170×120

Table 2. Mechanical properties of the bricks.

Brick	l_b (mm)	w_b (mm)	h_b (mm)	γ_b (kg/m³)	$f_{b,c}$ (MPa)	CoV (%)
IBL_BN	250	120	50	1640	22.4	15.8

Table 3. Mortar mix designs.

Mortar Type	Mix Design	w/b	$w/(b+s)$	s/b	w/tot
M1	1:1.4:4.7	1.4	0.25	4.7	0.20
M2	1:1.4:6.1	1.4	0.20	6.1	0.17
M3	1:1.7:7.6	1.7	0.20	7.6	0.17

Table 4. Mechanical properties of mortars.

Mortar	$f_{m,c}$ (MPa)	CoV (%)	$f_{m,fl}$ (MPa)	CoV (%)
M1	4.10	4.4	1.30	5.1
M2	3.52	2.7	1.26	0.5
M3	2.67	2.2	0.88	7.5

market, three different mix designs were prepared to obtain low strength mortars but with an adequate workability. In more detail, different growing water/binder ratios w/b were considered to reduce the mortar strength. In parallel, given the significant water content, the amount of fine sand (s), was increased to ensure the mix workability. Characteristics of mortar mixes are presented in Table 3, where proportions (b:w:s) are referred to weights.

Standard laboratory tests, such as monotonic uniaxial compression tests and three-point bending tests (EN 1015-11), were conducted to determine the mortar compressive strength $f_{m,c}$ and the mortar flexural strength $f_{m,fl}$. Results are reported in Table 4 and refer to mortars cured in laboratory conditions (T=22°±1°, RH=50%±5%) for 60 days.

2.3 Triplet Test

In the triplet test, according to the EN1052-3 Standard, a prescribed level of axial compression can be applied to the masonry samples, orthogonally to the mortar joints. Then, a monotonically increasing shear load is imposed to the central brick to produce its sliding. The possible failure modes can be distinguished as: (i) Type A, failure at the mortar-to-brick interface; (ii) Type B, failure within the mortar joints; (iii) Type C, failure within the brick element, and (iv) Type D, diagonal failure of the triplet specimen.

In the experimental campaign, a test setup (Figure 1) was purposely designed for the execution of the triplet tests, allowing for the application of the shear load F_i in displacement control, by using a servo-hydraulic machine, with a maximum capacity of 100 kN (load cell class 0.5). Before the application of the shear displacement, a uniform lateral compression force F_{pi} was applied to the samples, orthogonal to the bed joints, through a hydraulic jack with a maximum capacity of 750 kN. The jack was controlled by an

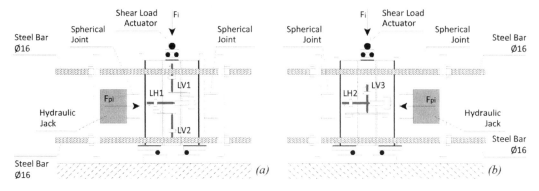

Figure 1. Triplet test setup and instrumentation on (a) side *A* and (b) side *B*.

electric pump, allowing to maintain a constant pre-compression level throughout the test.

The system adopted for the application of the axial compression force F_{pi} was a closed system constituted by four threaded steel bars ($\varnothing16$) and four steel plates. In particular, the steel plates in contact with the lateral surface of the bricks were characterized by a thickness of 15 mm, while the external steel plates were 20 mm thick. Spherical joints were inserted between the hydraulic jack and the internal steel plate, on one side, and between the two steel plates, on the other side, to allow the specimen to rotate, avoiding the occurrence of parasitic stresses which could influence the uniformity of the compressive stress distribution along the mortar joints. Neoprene pads were interposed between the bricks and the steel plates.

For each masonry typology, the following axial compression levels were considered: 0.10, 0.20, 0.60, 1.00 MPa. The shear displacement rate was equal to 1.20 mm/min. In some cases, after the onset and development of the sliding failure, the precompression level was increased and the sliding was produced again, as will be explained in the following.

Linear Variable Differential Transformers (LVDT) were used, on both sides of the triplet samples, to monitor the displacements during the tests. In particular, the relative vertical displacements between the central brick and the lateral ones (shear displacements v) were measured by means of three vertical LVDTs, while the expansion upon shearing due to dilatancy (normal displacement u) was measured by means of two horizontal LVDTs positioned in the center of the specimens (Figure 1).

3 EXPERIMENTAL RESULTS OF THE TRIPLET TESTS

The results of the triplet tests can be interpreted by considering the Mohr-Coulomb failure criterion, which is suitable to describe a shear-sliding failure mode. The criterion is considered here adequate since most of the samples failed with a *Type A* failure

mode, characterized by the sliding along the mortar-to-brick interface (Figure 2).

Typical shear stress τ *vs* shear displacement v_{mean} curves are plotted in Figure 3*a* for the specimen series A-M1. Four curves are reported, one for each axial compression level considered. From the graphs it is possible to notice that the initial branch is characterized by a significant stiffness and it can be considered linear up to the peak load. Only in one case (A-M1-4) a stiffness degradation is visible approaching the peak load, probably due to the formation of some micro-cracks, corresponding to a shear stress equal to 0.59 MPa. After the peak load, the quasi-brittle behavior of the shear sliding failure can be recognized with a softening branch and a residual horizontal plateau, corresponding to a dry friction condition (Van der Plujim et al. 2000). It is worth noticing that the failure brittleness decreases as the axial compression level increases. In general, the expected trend, i.e. higher shear strengths in correspondence with higher axial compressive stresses, was confirmed by the experimental results. The presence of two peak loads, such as for the triplet specimens A-M1-2 and A-M1-3, indicates the non-simultaneous failure of the two mortar joints, thus the non-symmetric behavior of the

Figure 2. Typical failure modes of the triplet tests (*Type A*): (a) pure sliding along the brick-to-mortar interface; (b) sliding with mortar crushing at the joint extremities.

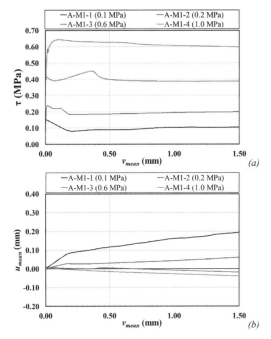

Figure 3. Triplet test results for the specimen series A-M1: (a) shear stress τ vs shear displacement v_{mean} diagram; (b) normal displacement u_{mean} vs shear displacement v_{mean} diagram.

Table 5. Triplet test results: stress state at the peak load and in the residual phase.

Specimen Code	σ (MPa)	τ_u (MPa)	τ_{res} (MPa)
A-M1-1	0.10	0.15	0.09
A-M1-1*	0.38	-	0.27
A-M1-2	0.20	0.24	0.20
A-M1-3	0.62	0.45	0.39
A-M1-3*	1.03	-	0.62
A-M1-4	1.02	0.64	0.58
B-M2-1	0.10	0.14	0.08
B-M2-2	0.21	0.20	0.14
B-M2-3	0.61	0.42	0.39
B-M2-3*	1.02	-	0.60
B-M2-4	1.02	0.64	0.63
C-M3-1	0.11	0.14	0.08
C-M3-1*	0.61	-	0.40
C-M3-1**	1.02	-	0.63
C-M3-2	0.20	0.18	0.14
C-M3-2*	0.65	-	0.39
C-M3-2**	1.02	-	0.51
C-M3-3	0.62	0.38	0.34
C-M3-3*	1.02	-	0.55
C-M3-4	1.02	0.59	-

*, ** further reloading phases

samples, due to intrinsic differences within the mortar or to the construction process of the samples.

Focusing on the dilatant behavior of the samples, in Figure 3b the normal displacement u_{mean} vs shear displacement v_{mean} diagrams are reported for the specimen series A-M1. Positive normal displacements indicate a lateral expansion of the sample. It is clearly noticeable that the expansion upon shearing, due to dilatancy, is the highest for the specimen subject to an axial compressive stress equal to 0.1 MPa, while it is lower, but still positive, for the triplet sample tested with an axial compression level of 0.2 MPa. The normal displacement tends to zero or even becomes negative for higher compressive stresses. In this latter case, a lateral contraction of the specimens is registered during the sliding failure, which can be related to the crumbling or crushing of the mortar in compression. This phenomenon was indeed prevalent for specimens built with the weakest mortars.

The trends here described for the specimen series A-M1 were also observed for the series B-M2 and C-M3, here not fully reported for sake of brevity.

As a result of each triplet test, the stress state (σ; τ) in correspondence of the peak load and in the post peak phase (v_{mean}=1.50 mm) has been measured and reported in Table 5. In particular, for those samples subject to the axial stress increase, after the development of the first sliding, more than one residual stress state was determined. Starting from these

results, it was possible to calibrate a Mohr-Coulomb failure criterion for each masonry typology, according to Equation 1:

$$\tau = f_{v0} + \sigma \tan \varphi, \qquad (1)$$

where f_{v0} is the shear strength at zero compressive stress, i.e. cohesion or initial shear strength, and φ is the friction angle. In a similar way, a residual failure criterion (Van der Pluijm et al. 2000), characterized by a residual shear strength $f_{v0,res}$ and by a residual friction angle φ_{res}, can be determined by plotting the residual stress τ_{res} vs the corresponding compressive stress σ and performing a linear interpolation.

The shear strength parameters obtained from the elaboration of the experimental results are reported in Table 6, for both the peak and the residual Mohr-Coulomb failure criteria, which are plotted in Figures 4a-b, respectively. The values of the cohesion f_{v0} can

Table 6. Triplet test results: shear strength parameters.

Specimen Code	f_{v0} (MPa)	$\tan\varphi$ (-)	$f_{v0,res}$ (MPa)	$\tan\varphi_{res}$ (-)
A-M1	0.12	0.52	0.06	0.53
B-M2	0.09	0.54	0.02	0.59
C-M3	0.08	0.49	0.04	0.52

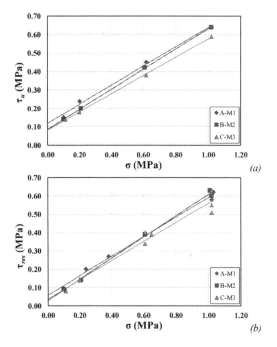

Figure 4. Determination of Mohr-Coulomb failure criteria: (a) peak shear stress τ_u vs compressive stress σ diagram; (b) residual shear stress τ_{res} vs compressive stress σ diagram.

be directly correlated to the mortar properties: the lower the compressive or flexural strength of the mortar, the lower the cohesion. Considering the friction coefficient (tanφ), a clear correlation cannot be found with the mortar strength. Nevertheless, it can be observed that for the specimens built with mortars characterized by the same water binder ratio (M1 and M2), a higher value of the friction coefficient was obtained for the series having the greater amount of fine sand (M2). The friction coefficient can be indeed more related to the mortar composition and to the brick surface characteristics, rather than to the mortar mechanical properties, given the failure mode observed in the present experimental tests. The values of the residual shear strength $f_{v0,res}$, theoretically equal to zero, are quite low and significantly lower than the corresponding values of f_{v0}. A slight friction hardening in the post-peak phase is noticeable for all the specimen series.

4 NUMERICAL ANALYSES

4.1 Numerical model

Modeling of masonry structures can be performed according to different approaches, i.e. macro-modeling or micro-modeling approaches, depending on the desired level of accuracy (Lourenço et al. 1995, Rots 1997). It is, indeed, possible to model

masonry as a composite or to singularly model its constituents. In the numerical analyses here presented, a micro-modeling approach was chosen, given the level of accuracy needed to study the sliding failure along the mortar joint and considering the limited dimensions of the triplet samples. In more detail, a simplified micro-modeling strategy was adopted, which implies to model the mortar joints as zero thickness interface elements and the bricks as continuum element characterized by an expanded geometry to maintain the overall dimensions of the sample unchanged.

A 2D model was adopted for the numerical simulations of the triplet test and, due to the symmetry of the test setup, only half of the specimen was considered. The details of the mesh used in the numerical models are shown in Figure 5. Bricks were modeled using quadratic 8-noded plane stress elements, while line 3-noded interface elements were adopted to model the zero-thickness mortar joint. The steel plates in contact with the sample were also modeled with quadratic 8-noded plane stress elements and interface elements were inserted between the steel elements and the bricks. A linear elastic behavior was considered for bricks and steel, while nonlinearities were only assigned to the interface elements.

To describe the behavior of interface elements in the linear elastic range, it is necessary to define the normal and shear stiffness parameters, k_n and k_t. In particular, for the interface elements adopted to model the mortar joint, the approach proposed by Rots (1997) was considered and the stiffness parameters were determined through Equations 2 and 3:

$$k_n = \frac{E_b E_m}{t_m(E_b - E_m)} \qquad (2)$$

Figure 5. Finite element model.

886

$$k_t = \frac{G_b G_m}{t_m (G_b - G_m)}, \qquad (3)$$

where E_b and G_b are the elastic and shear modulus of the bricks, E_m and G_m are the elastic and shear modulus of the mortar, and t_m is the mortar joint thickness. For the interface elements adopted to model the brick-to-plate contact surfaces, the normal and shear stiffness coefficients were calibrated to allow the transfer of compressive stresses and to avoid the lateral confinement of the bricks, thus a dummy value was given to the normal stiffness coefficient k_n and a very low value was assigned to the shear stiffness coefficient k_t.

The nonlinear behavior along the mortar sliding interface was described with a composite interface model, including a Coulomb friction criterion, a tension cut-off and a compressive cap (Lourenço 1996). Exponential softening for tension and shear failure modes were included, while a hardening/softening behavior was considered in compression. The interface elements between the steel plates and the bricks were considered as a no-tension material.

With reference to the Coulomb friction model, the yielding function is expressed as:

$$f(\sigma, \kappa_2) = |\tau| + \sigma \tan \varphi(\kappa_2) - f_{v0}(\kappa_2), \qquad (4)$$

where κ_2 is a scalar indicating the amount of softening, assumed equal to the plastic shear displacement. In Equation 4, cohesion and friction softening are introduced: an exponential softening for the cohesion, depending on the mode-II fracture energy G_f^{II}, and a friction softening, considered proportional to the cohesion softening, depending on the values of the initial and residual friction angles.

A non-associated plastic potential with a variable dilatancy angle ψ is considered, according to Equations 5 and 6 (Van Zijl 2004):

$$g_2 = |\tau| + \sigma \tan \psi - f_{v0}, \qquad (5)$$

$$\tan \psi = \tan \psi_0 \left\langle 1 - \frac{\sigma}{\sigma_u} \right\rangle e^{-\delta v_p}, \qquad (6)$$

where v_p is the plastic shear displacement along the mortar joints, ψ_0 is the dilatancy angle at zero confining stress and shear slip, σ_u is the pre-compression level at which the dilatancy goes to zero, and δ is the dilatancy shear-slip degradation coefficient. It was demonstrated that the dilatancy is arrested under conditions of high axial compressive stress level and large plastic shear displacement (Van der Pluijm et al. 2000, Van Zijl 2004). These parameters can be calibrated if the expansion upon shearing is measured during the triplet tests, as done in the present experimental campaign, where the specimens were free to expand laterally.

Numerical simulations were performed with the finite element software DIANA FEA (Release 10.3) for the triplet series B-M2, at 4 different axial compression levels (0.10 – 0.20 – 0.60 – 1.00 MPa), to reproduce the loading conditions of the experimental tests. After the application of the lateral compressive distributed load, an increasing vertical displacement was imposed to the top plate. Regular Newton-Raphson method was adopted to solve the nonlinear problem.

The input parameters used in the numerical simulations are reported in Table 7. In particular, the parameters for the Coulomb friction model were calibrated from the experimental results of the triplet tests. The tensile strength and the mode-I fracture energy were determined as a fraction of the initial shear strength and the mode-II fracture energy, respectively (Rots 1997). The compressive strength f_c was evaluated using the formula provided by Eurocode 6, starting from the mechanical properties of the components. The compressive fracture energy was calibrated based on indications found in the literature and on previous experimental results on similar masonry typologies (Lourenço 1996, Rots 1997).

4.2 Numerical results

In this section, the results of the numerical simulations are presented and compared with the experimental results for the triplet series B-M2. In more detail, the comparisons are reported in Figure 6 in terms of shear stress τ vs shear displacement v_{mean} diagram (Figure 6a) and normal displacement

Table 7. Input mechanical parameters.

Parameter	Symbol	Units	Value
Elastic modulus of brick	E_b	(MPa)	5800
Poisson's ratio of brick	v_b	(-)	0.20
Elastic modulus of mortar	E_m	(MPa)	3500
Poisson's ratio of mortar	v_m	(-)	0.20
Elastic modulus of steel	E_s	(MPa)	210000
Poisson's ratio of steel	v_s	(-)	0.30
Interface normal stiffness	k_n	(N/mm³)	882.6
Interface shear stiffness	k_t	(N/mm³)	367.8
Tensile strength	f_t	(MPa)	0.06
Mode-I fracture energy	G_f^I	(N/mm)	0.01
Cohesion	c_0	(MPa)	0.09
Friction angle	ϕ_0	(rad)	0.483
Residual friction angle	ϕ_{res}	(rad)	0.535
Dilatancy angle	ψ_0	(rad)	0.381
Confining normal stress	σ_u	(MPa)	0.50
Exp. degradation coeff.	δ	(-)	1.5
Mode-II fracture energy	G_f^{II}	(N/mm)	0.02
Compressive strength	f_c	(MPa)	7.07
Compr. fracture energy	G_f^c	(N/mm)	20

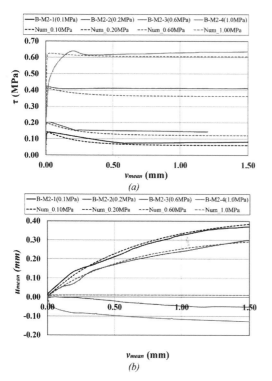

Figure 6. Experimental vs numerical results for the specimen series B-M2: (a) shear stress τ vs shear displacement v_{mean} diagram; (b) normal displacement u_{mean} vs shear displacement v_{mean} diagram.

u_{mean} vs shear displacement v_{mean} diagram (Figure 6b), for each axial compression level chosen.

With reference to Figure 6a, it can be noticed that the numerical models can properly capture the experimental behavior, especially in terms of peak shear stress. A good agreement is also noticed in the initial phase, up to the peak load, for triplets subject to axial compression stresses equal to 0.10, 0.20 and 0.60 MPa, while for the highest axial compression level (1.00 MPa) it can be observed that the numerical model cannot adequately describe the stiffness degradation approaching the peak load. In terms of residual shear stress, a slight underestimation is visible for all the axial compression levels. This can be explained by considering that the residual Mohr-Coulomb failure criterion was calibrated by using also the results obtained in the reloading phases, after the development of the first sliding failure.

Figure 6b shows a very good agreement between numerical and experimental results for axial compression levels equal to 0.10 and 0.20 MPa, while for axial compression levels of 0.60 and 1.00 MPa, the experimental curves show a compression orthogonal to the bed joint which is not well captured by the numerical results. This can be explained by considering some aspects related to the test setup – e.g. specimens not completely free to expand laterally at high axial

compression stress levels – or by considering that the failure mode observed in these experimental tests, especially for the axial compression stress of 1.00 MPa, was not a pure sliding failure along the brick-mortar interface but involved the mortar also, as visible in Figure 2b. Indeed, cracking in the mortar occurred, especially closed to the joint edges. In order to capture the very low – even negative – values of normal displacements u_{mean}, a detailed micro-modeling strategy, not reported in this work, should be adopted.

Considering the results of the nonlinear analyses, it is interesting to investigate the development of the stress distributions along the joint and the propagation of the failure, given that they could influence the reliability of the results.

In Figure 7, the evolution of the normal and tangential stresses along the sliding interface are

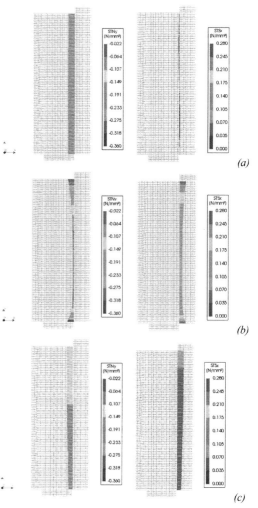

Figure 7. Normal and tangential stress distributions along the sliding mortar joint: (a) application of the axial compressive stress; (b) peak load; (c) shear displacement equal to 1.00 mm – residual phase.

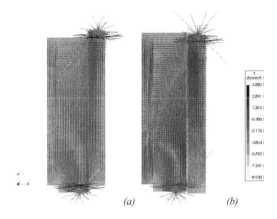

Figure 8. Principal stress distribution: (a) peak load; (b) shear displacement equal to 1.00 mm – residual phase.

shown, for an initial axial compression level equal to 0.10 MPa. It can be noticed that: (i) at the beginning of the test, the normal stress is uniform along the interface, while the tangential stress is negligible; (ii) at the peak, the stress distributions are not uniform along the joint, with concentrations of stresses occurring at the joint edges; (iii) in the residual pure frictional phase, the stress distributions tend to be more uniform along the joint, slightly higher in the upper portion of the specimen. The principal stress distributions for an axial compression level equal to 0.1 MPa are reported in Figure 8, where stress concentrations close to the loading plates are clearly visible, together with the presence of a compressed strut developing within the triplet specimen across the sliding mortar joint. Worth mentioning that the failure started from the joint extremities but then propagated quickly along the entire interface.

5 CONCLUSIONS

The objective of the research was to analyze the shear-sliding behavior of three brick masonry typologies, built using different mortar mixes, through the execution of triplet tests. Numerical simulations of the tests were also carried out considering the mechanical parameters obtained or calibrated from the experimental results.

The experimental tests were performed in displacement control, allowing to evaluate the peak shear strength of masonry and to analyze the post-peak nonlinear behavior. Mohr-Coulomb failure criteria were calibrated for the investigated masonry typologies, both at the peak and in the residual phase. The obtained mechanical properties were compared with the characteristics of the mortar used for the construction of the specimens and a correlation was found between the initial shear

strength and the mortar compressive and flexural strength, i.e. the lower the mechanical properties of the mortar, the lower the initial shear strength. Therefore, it is worth highlighting that the mortar composition can affect the sliding-shear strength of masonry and the quality of the bond along the brick-mortar interface. In the experimental tests, a particular focus was also devoted to the investigation of the dilatant behavior of masonry through the measurements of the lateral expansion of the triplet sample upon shearing. The triplet specimen was, indeed, free to expand laterally during the test.

The numerical analyses allowed to better understand the onset and development of the sliding failure together with the evolution of the normal and tangential stress distributions along the mortar joint. Stress concentrations were noticed at locations corresponding to the joint edges, close to the loading plates. This observation can explain why mortar crushing was noticed at the joint extremities for few triplet samples. In general, it can be concluded that a good agreement was obtained between numerical and experimental results, confirming the suitability of the chosen modeling approach.

ACKNOWLEDGMENTS

This paper was supported by the PRIN 2017 research programme of the Italian Ministry of Education, University and Research, project DETECT-AGING, grant N. 201747Y73L.

REFERENCES

Drysdale, R.G., Vanderkeyl, R. & Hamid, A.A. 1979. Shear strength of brick masonry joints. *Proc. 5th Int. Brick and Block Mas. Conf., Washington D.C., 5-10 October 1979.*

Ferretti, F., Esposito, R., Rots, J.G. & Mazzotti, C. 2018. Shear-sliding behavior of masonry: numerical micro-modeling of triplet tests. *Proc. EURO-C Conf., Austria, 26 February – 01 March 2018.*

Jukes, P. & Riddington, J.R. 2000. Finite element prediction of block triplet shear strength. *Proc. 12th Int. Brick and Block Mas. Conf., Madrid, 25-28 June 2000.*

Jukes, P. & Riddington, J.R. 2001. The failure of brick triplet specimens. *Masonry International*, 15(1).

Lourenço, P.B., Rots, J.G. & Blaauwendraad, J. 1995. Two approaches for the analysis of masonry structures: micro and macro-modeling. *Heron*, 40(4):313–340.

Lourenço, P.B. 1996. Computational strategies for masonry structures, PhD thesis, Delft University of Technology.

Popal, R. & Lissel, S.L. 2010. Numerical evaluation of existing mortar joint shear tests and a new test method. *Proc. 8th International Masonry Conference, Dresden, 2010.*

Riddington, J.R., Fong, K.H. & Jukes, P. 1997. Numerical study of failure initiation in different joint shear tests. *Masonry International*, 11(2).

Riddington, J.R. & Jukes, P. 1994. A masonry joint shear strength test method. *Proc. Instn Civ. Engrs Structs & Bldgs*, 104:267–274.

Rots, J.G. 1997. *Structural Masonry – An experimental/numerical basis for practical design rules*, Rotterdam: Balkema.

Stöckl, S., Hofmann, P. & Mainz, J. 1990. A comparative finite element evaluation of mortar joint shear tests. *Masonry International*, 3(3).

Van der Pluijm, R. 1999 Out of plane bending of masonry. *Ph.D. Thesis. Eindhoven University of Technology, The Netherlands.*

Van der Pluijm, R., Rutten, H. & Ceelen, M. 2000. Shear behaviour of bed joints. *Proc. 12th Int. Brick and Block Mas. Conf., Madrid, 25-28 June 2000.*

Van Zijl, G. 2004. Modeling masonry shear-compression: role of dilatancy highlighted. *Journal of Engineering Mechanics*, 130(11):1289–1296.

Vermeltfoort, A.T. Variation in shear properties of masonry. *Proc. 8th International Masonry Conference, Dresden, 2010.*

Brick and Block Masonry - From Historical to Sustainable Masonry –
Kubica, Kwiecień & Bednarz (eds)
© 2020 Taylor & Francis Group, London, ISBN 978-0-367-56586-2

A macro-element modelling technique to account for IP and OOP interactions in URM infilled RC or steel building

Bharat Pradhan, Maria Zizzo, Vincenzo Sucato & Liborio Cavaleri
University of Palermo, Palermo, Italy

Davorin Penava & Filip Anić
Josip Juraj Strossmayer University of Osijek, Osijek, Croatia

Vasilis Sarhosis
University of Leeds, Leeds, UK

ABSTRACT: Buildings with unreinforced masonry (URM) infills is a common construction practice world-wide. To correctly assess the seismic performance of such structures, the analysis of the masonry infills under in-plane (IP) and out-of-plane (OOP) loading, as well as their interaction, is essential. The aim of this paper is to present a new macro-element modelling technique able to simulate the IP and OOP response of URM infilled frames subjected to earthquakes. The model consists of a horizontal, a vertical and two diagonal struts represented by fiber section beam-column elements. The compressive behavior of the struts are defined by empirical strength and strain parameters. The paper also discusses on derivation of the empirical parameters based on the actual properties of infill wall. The proposed macro-element model has been validated with experimental results available in the literature. From the numerical analysis, a good agreement between the experimental and numerical results was obtained.

1 INTRODUCTION

Unreinforced masonry (URM) infilled frame struc-ture is a common construction practice worldwide. Recent earthquakes have shown that structural per-formance of such buildings is strongly associated with the behavior of the URM infill walls. The damage types in infill walls are either in-plane (IP) or out-of-plane (OOP), although the damage mech-anisms is usually a combination of both IP and OOP damage. Experimental studies were carried out in order to characterize the OOP performance of the URM infill walls considering previous IP damage (Angel 1994, Flanagan & Bennett 1999, Calvi & Bolognini 2001, Ricci et al., 2018, De Risi et al., 2019). Contrariwise, experiments have also been conducted to understand the influence of prior OOP damage in the IP behaviour of the infill walls (Hen-derson et al. 1993, Palieraki et al. 2018). These stud-ies have demonstrated that the interaction between the IP and OOP is important to be considered for evaluating the true capacity of the URM infills and hence should be incorporated in the numerical modelling.

Several analytical and numerical models have been developed to account for the influence of infill walls on the overall response of infilled frame structure. The model of single equivalent diagonal strut has been the most popular and used first in the literature to characterize IP response (Polyakov 1960, Holmes 1961, Smith & Carter 1959, Main-stone 1974, Liauw 1972, Paulay & Priestley 1992, Durrani & Luo 1994, Hendry 1998). Later, models with multiple diagonal struts were proposed to addresses interaction effect between the frame and the infill (Thiruvengadam 1985, Chrysostomou et al. 2002, El Dekhakhni et al. 2003, Crisafulli & Carr 2007). The concept of stiffening effects of infill walls was introduced in the strut based model (Papia et al. 2003, Asteris et al. 2015). Moreover, the non-linear behaviour of the infilled frame were concept-ualized under the action of monotonic loading (Panagiotakos & Fardis 1996, Žarni´c & Gostič 1997, Dolšek & Fajfar 2005) and cyclic loading (Madan et al. 1997, Cavaleri et al. 2005, Rodrigues et al. 2010, Cavaleri & Di Trapani 2014).

The first macro-model that could consider IP as well as OOP resistance of the infill wall under the action of combined IP and OOP load was given by Hashemi & Mosalam (2007). The model consist of two diagonal struts (one in each direction) and each of the struts is modeled by four pin-connected com-pression beam elements jointed by a tension tie. Later, Kadysiewski & Mosalam (2008) and Mosalam and

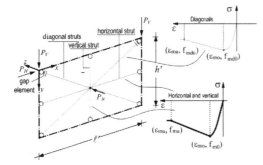

Figure 1. Four strut macro model by Di Trapani et al. (2018).

Günay (2015) proposed a single-strut model consisting of two fiber-section beam-column elements connected at the midpoint node with an assigned lumped mass for the consideration of the response of the infill wall in the OOP direction. Furtado et al. (2015) also proposed macro-model to simulate the IP and OOP behaviour of infill walls. However, in these models, calculation required is rigorous. Di Trapani et al. (2018) proposed a four-strut macro-element model that can take in to account the IP and OOP behaviour of infill wall as well as their interaction (Figure 1). The accuracy of the model in predicting IP response is good but the model in some cases does not consider OOP response effectively. The other drawback of this model is the identification of the mechanical parameters to define the compressive behavior of the struts that is not based on a univocal procedure.

In this paper, the modified version of macro-element model developed by Di Trapani et al. (2018) is presented to simulate the IP and OOP response of URM infill walls. The proposed model has been validated with the experimental results from the literature. Moreover, empirical equations to derive the empirical strength strain parameters which are required to define compressive behaviour of struts used in this macro-element model are also proposed in this paper.

2 NUMERICAL MODELLING: MODIFIED FOUR STRUT MACRO-ELEMENT MODEL

The proposed model is a modification of the macro-element model proposed by Di Trapani et al. (2018). The modified model retains its earlier 4 struts configuration i.e. there are two diagonal struts, one horizontal and one vertical strut. Each strut is represented by two fiber-section beam-column elements connected by a node at the mid-span (Figure 2). These four nodes can move independently in the IP

direction while they are constrained to move together in the OOP direction. Some major changes made in the proposed modified macro –element model are as follows:

a. All struts i.e. diagonals, vertical and horizontal struts, are transformed (rotated) so that all struts contribute in the OOP resistance, compared to transformation of only diagonal struts in the model by Di Trapani et al. (2018).
b. All struts are restrained against rotation at their connections with the frame rather than pin jointed, to enable the struts to take into account the arch mechanism more effectively.
c. A single value of effective compressive strength is used for all struts rather than using the "effect-ive" compressive strength for the diagonal struts and actual compressive strength of infill wall for the horizontal and vertical struts.

In the proposed model, the width of the diagonal struts 'w_d' is taken as one-third of the diagonal length 'd' while the width of the horizontal strut 'w_h' and vertical struts 'w_v' are calculated as a function of w_d, according to Equations 1, 2 and 3.

$$w_d = \frac{d}{3}; d = \sqrt{l'^2 + h'^2} \qquad (1)$$

$$w_h = h - \frac{w_d}{\cos \theta} \qquad (2)$$

$$w_V = 1 - \frac{w_d}{\sin \theta} \qquad (3)$$

Where l and h are the clear length and height of the infill wall, respectively. l' is the centre to centre distance between the columns while h' is the

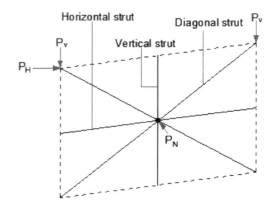

Figure 2. Proposed modified four strut macro model.

distance from top of the lower beam to the centre of the top beam. The dimensioning of the struts are shown in Figure 3.

If 'w_d' is the width of the diagonal strut, '$w_d/\cos\theta$' is the total contact length between the diagonal struts and the columns and '$w_d/\sin\theta$' is the total contact length between the diagonal struts and the beams. The thickness of all the struts are equal to the thickness of the infill wall. For simplicity, the thickness of the diagonal, vertical and horizontal struts is annotated as 't'.

The compressive behavior of masonry is represented using an appropriate constitutive model for the strut elements. The Kent–Park model (Kent & Park 1971) for concrete with tensile strength equal to zero has been chosen for this purpose (Figure 1). To represent the IP resistance, actual thickness of the infill can be used as it yields good numerical results. But, the use of the actual infill thickness as a strut thickness yields comparatively lower OOP strength. Hence, to accurately represent both the IP and OOP resistances of the infill wall, the width and thickness of the diagonal, vertical and horizontal struts were replaced by surrogate values that maintain the same cross-sectional area.

To determine the surrogate width '$\tilde{w}$' and surrogate thickness '$\tilde{t}$' of the struts, it was assumed that the OOP resistance of a strut 'q' is proportional to the compressive strength of the masonry 'f_m' and the strut width 'w' and is inversely proportional to the square of the slenderness ratio 'l_s/t' of the strut (this procedure is also discussed in detail in Di Trapani et al. 2018).

$$q \propto \frac{f_m w}{\left(\frac{l_s}{t}\right)^2} \qquad (4)$$

In Equation 4, 'l_s' is the length of the strut. If f_m is to be replaced by the effective compressive strength (f_{mo}) and the OOP resistance is to remain the same, the strut width 'w' and thickness 't' have to be replaced by '$\tilde{w}$' and '$\tilde{t}$', so that

$$\frac{f_m w}{\left(\frac{l_s}{t}\right)^2} = \frac{f_{mo}\tilde{w}}{\left(\frac{l_s}{\tilde{t}}\right)^2} \qquad (5)$$

To maintain the correct IP resistance, the cross-sectional area of the diagonal strut has to remain unchanged, i.e. the area of the surrogate strut should be equal to the area of original strut.

$$wt = \tilde{w}\tilde{t} \qquad (6)$$

Substituting the expression for w from Equation 6 into Equation 5, one gets

$$\frac{f_m w}{\left(\frac{l_s}{t}\right)^2} = \frac{f_{mo}wt}{\tilde{t}\left(\frac{l_s}{\tilde{t}}\right)^2} \qquad (7)$$

$$\tilde{w} = \frac{f_{mo}}{f_m}w \qquad (8)$$

$$\tilde{t} = \frac{f_m}{f_{mo}}t \qquad (9)$$

The surrogate width of diagonal, vertical and horizontal struts are represented by '$\tilde{w}_d$', '$\tilde{w}_v$' and '$\tilde{w}_h$' respectively while the surrogate thickness for all struts is represented by $\tilde{t}$. The consideration of geometric nonlinearity in the model (in case of slender strut) was done by the use of beam-column elements formulated with the Corotational coordinate transformation. For infill wall with perforated units, the properties like compressive strength 'f_m' and elastic modulus 'E_m' are usually not same in the direction of holes and perpendicular to the holes. The equivalent properties were derived by following the Equations 10 & 11 where f_{m1} and f_{m2} are the compressive strength and E_{m1} and E_{m2} are the elastic modulus of infill masonry in directions parallel and perpendicular to the holes in the masonry unit. For infill

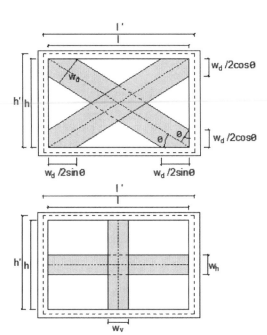

Figure 3. Dimensioning of the diagonal, vertical and horizontal struts.

with solid masonry unit, the properties are considered equal in both vertical and horizontal directions.

$$f_m = \sqrt{f_{m1} \cdot f_{m2}} \qquad (10)$$

$$E_m = \sqrt{E_{m1} \cdot E_{m2}} \qquad (11)$$

3 VALIDATION OF THE PROPOSED MACRO-ELEMENT MODEL

The proposed model has been validated with the experimental data available in the literature. Angel (1994) tested seven full-scale, single-story, single-bay infilled RC frames in both IP and OOP loads. The specimens were first subjected to IP cyclic lateral displacements until the infills reached double the drift of first cracking. Then the masonry infills were applied with monotonically increasing pressure in the OOP direction. From the tests, specimens 2 and 3 subjected to IP drift of 0.34% and 0.22% respectively before applying OOP pressure were taken for numerical simulation. Numerical simulations were carried out in the software platform Open-Sees. Frame elements as well as strut elements were modelled by fiber-section beam-column elements with distributed plasticity. Concrete 02 material model was used to simulate the behavior of the concrete and infill wall while reinforcement in the frames were simulated using Steel02 material model. The loading is applied in a similar pattern as used in the tests. The IP cyclic lateral displacement was applied at the upper nodes of the frame model and the OOP load was applied at the centre of the struts representing the infill centre.

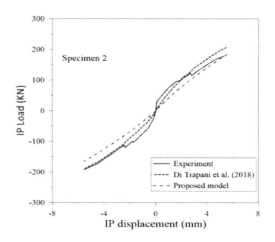

Figure 4. Comparison of IP responses for specimen 2 of Angel (1994).

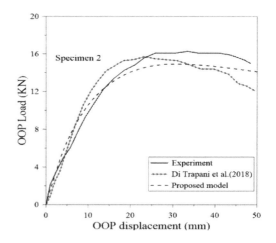

Figure 5. Comparison of OOP responses for specimen 2 of Angel (1994).

Table 1. Geometrical properties of diagonal, vertical and horizontal struts calculated for Angel (1994).

Specimen	$\tilde{w}_d$ mm	$\tilde{w}_v$ mm	$\tilde{w}_h$ mm	$\tilde{t}$ mm
S2	261.3	113.3	75.5	198.6
S3	387.5	168.0	112.0	133.9

Table 2. Mechanical properties of diagonal, vertical and horizontal struts calculated for Angel (1994).

Specimen	f_{mo} N/mm^2	f_{mu} N/mm^2	ε_{mo}	ε_{mu}
S2	2.6	1.56	0.0038	0.018
S3	3.6	2.16	0.0030	0.015

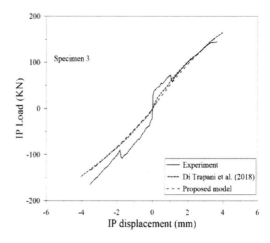

Figure 6. Comparison of IP responses for specimen 3 of Angel (1994).

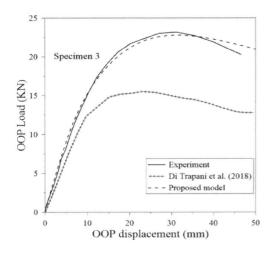

Figure 7. Comparison of OOP responses for specimen 3 of Angel (1994).

The transformed geometrical properties of the struts are given in Table 1. The compressive characteristic of the strut fibers was defined using 4 parameters f_{mo}, f_{mu}, ε_{mo} and ε_{mu} and these parameters are provided in Table 2. The ultimate strength 'f_{mu}' was assigned as 60% of the peak strength 'f_{mo}'. Numerical simulations were performed to match numerical responses with the experimental responses. Numerical results for the IP and OOP responses are shown in Figures 4-7. In spite of the simplicity of the model, the correlation between the experimental and the numerical results is reasonably good. The proposed model's results were also compared with the results from the earlier model given in Di Trapani et al. (2018). It is observed that both models yield similar IP response but the OOP response of the proposed modified model is better captured. This verifies that the proposed modifications enable the model to account for both IP and OOP responses of URM infill wall in a reliable way.

The proposed model was further tested with other sets of experiments (Calvi & Bolognini 2001, Ricci et al. (2018), De Risi et al. 2019), Da Porto et al. 2013). In Calvi & Bolognini (2001), infilled frame was subjected to two different IP drifts; 0.4 % (specimen 6) and 1.2 % (Specimen 2) and was followed by OOP load applied monotonically on four load points. Specimen 6 (#6) from the experiment was used for the numerical simulation. Da Porto et al. (2013) subjected the URM infilled RC frame to IP quasi static cyclic horizontal displacement to a drift of 1.2 % and then applied monotonic out-of-plane load at four load points in the infill wall until collapse. Ricci et al. (2018) experimented two set of infill walls of thickness 80 mm (specimen #1) and 120 mm (specimen #2). Both specimens were tested for pure OOP load and also in IP load followed by OOP load. Specimens were subjected to three different level of IP drift before loading in the OOP direction. IP quasi-static

cycling load was applied at one end of the upper beam while OOP load was applied at four points in the infill wall. For the numerical simulations, specimen #1 and specimen #2 subjected to IP drift of 0.21% and 0.16% prior to the application of OOP load were selected. De Risi et al. (2019) tested infill wall in a similar way of Ricci et al. (2018); a pure OOP test and OOP tests after each of three different IP drifts. For the numerical simulation, OOP response after 0.15% IP drift was taken.

Table 3. Geometrical Properties for the Diagonal, Vertical and Horizontal Struts.

Experiment	$\tilde{w}_d$ mm	$\tilde{w}_v$ mm	$\tilde{w}_h$ mm	$\tilde{t}$ mm
Calvi and Bolognini (2001) #6	1259	581	381	197
Da Porto et al. (2013)	845	436	279	620
Ricci et al. (2018) #1	440	231	180	198
#2	500	263	204	262
De Risi et al. (2019)	400	198	198	191

Table 4. Mechanical Properties for the Diagonal, Vertical and Horizontal Struts.

Experiment	f_{mo} N/mm^2	f_{mu} N/mm^2	ε_{mo}	ε_{mu}
Calvi & Bolognini (2001) #6	0.76	0.45	0.00045	0.0053
Da Porto et al. (2013)	1.29	0.78	0.00069	0.0075
Ricci et al. (2018) #1	0.85	0.51	0.00049	0.0057
#2	0.86	0.51	0.00049	0.0057
De Risi et al. (2019)	1.38	0.83	0.00073	0.0079

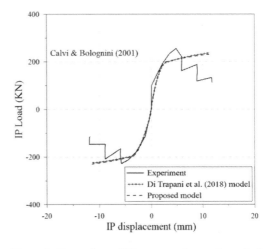

Figure 8. Comparison of IP responses for specimen 6 of Calvi & Bolognini (2001).

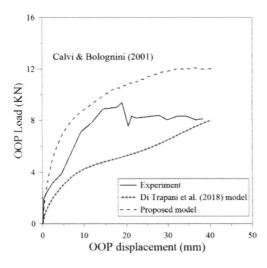

Figure 9. Comparison of OOP responses for specimen 6 of Calvi & Bolognini (2001).

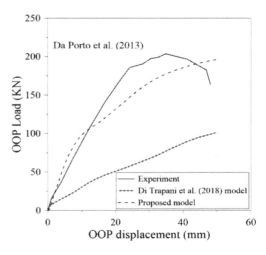

Figure 11. Comparison of OOP responses for Da Porto et al. (2013).

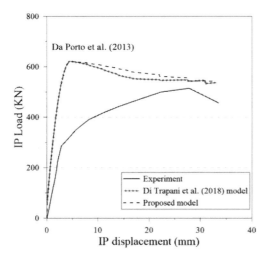

Figure 10. Comparison of IP responses for Da Porto et al. (2013).

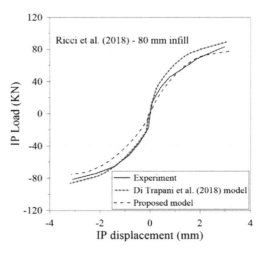

Figure 12. Comparison of IP responses for specimen #1 of Ricci et al. (2018).

Numerical simulations were performed by assigning the empirical strength strain parameters (f_{mo}, f_{mu}, ε_{mo}, ε_{mu}). The ultimate strength 'f_{mu}' was taken as 60% of the peak strength 'f_{mo}' as in Angel (1994)'s. specimens. The simulations were performed assigning these mechanical characteristics in order to obtain responses similar to experimental results. The parameters were fixed in such a way that the minimum offset was observed in both IP and OOP responses for all selected specimens under investigation. To this point a correlation was established between the effective mechanical parameters and the actual mechanical parameters of the masonry wall. Numerical simulations were repeated then by using the parameters calculated on the basis of the found

correlations. The correlation between the infill properties will be discussed in the next section. The geometrical properties of the struts (transformed) are provided in Table 3 while the final effective parameters used for the numerical simulations are given in Table 4. Some of the results obtained from the numerical simulations are shown in Figures 8-13.

It was observed that both IP and OOP responses obtained from the numerical analyses are close to the experimental results in all cases. The numerical simulations performed using the four strut macro-element model of Di Trapani et al. (2018) using the same parameters, are also kept in the same figures (Figs 8-13). IP responses from both models are good and almost similar. However, the proposed model

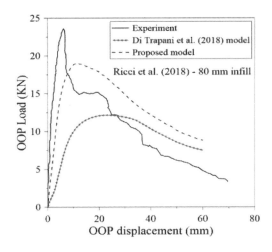

Figure 13. Comparison of OOP responses for specimen #1 of Ricci et al. (2018).

better captures the OOP response of the URM infill wall.

4 CORRELATION BETWEEN EMPIRICAL PARAMETERS AND MECHANICAL PROPERTIES OF MASONRY

A standard approach to select three empirical strength-strain parameters f_{mo}, ε_{mo} and ε_{mu} has been defined. To develop it, these parameters were chosen and numerical analyses were performed with subsequent optimizations of the parameters based on the observed numerical results. These parameters were taken based on the value of the infill wall properties i.e. the product of f_m and E_m (simply $f_m E_m$). As mentioned in the earlier section, the analysis were performed in a way to establish certain degree of correlation. The correlated parameters established for all the later experimental specimens along with those

Table 5. Empirical properties used for correlation.

Experiment	Infill properties $f_m E_m$ N²/mm⁴	Strut properties f_{mo} N/mm²	ε_{mo}	ε_{mu}
Calvi & Bolognini (2001) #6	1505	0.77	0.00046	0.0045
Ricci et al. (2018) #1	2467	0.79	0.00052	0.0056
	2533	0.79	0.00054	0.0058
Da Porta et al. (2013)	7370	1.4	0.00064	0.0074
De Risi et al. (2019)	8456	1.5	0.00068	0.0076
Angel (1994) S2	8730	2.6	0.0038	0.018
Angel (1994) S3	5280	3.6	0.0030	0.015

used for Angel's (1994) specimens were put together and plotted to make best fitting curves. The infill and strut parameters used to establish the relations are given in Table 5. The empirical equations to select the required parameters are shown alongside the curves in Figure 14. The value of f_{mo} increases

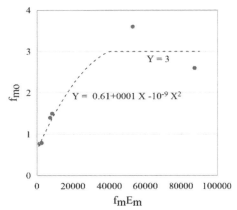

a)

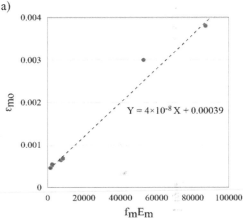

b)

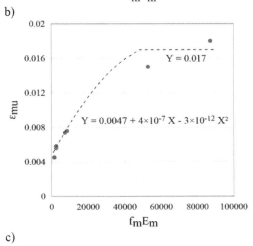

c)

Figure 14. Relationship to obtain the empirical parameters: a) f_{mo}, b) ε_{mo} and c) ε_{mu}.

nonlinearly with the increase in the value of $f_m E_m$ up to 40000 N^2/mm^4 and beyond which it becomes constant. Similar behaviour was also observed for the case of f_{mu} where its value increases with $f_m E_m$ up to a value of 48000 N^2/mm^4 and maintains a constant value after that point. The value of ε_{mo} increases linearly with the increase in the value of $f_m E_m$.

The Angel's (1994) specimens 2 and 3 were resimulated using the parameters calculated according to proposed equations. The values of f_{mo}, ε_{mo} and ε_{mu} used for specimen 2 & 3 are 3, 0.0038 & 0.017 and 3.0, 0.0025 & 0.017 respectively. Figures 15-18 compares the numerical results obtained with parameters used for the validation purpose (before correlation) and the parameters according to the proposed equations (after correlation). The obtained results are also close to the experimental results.

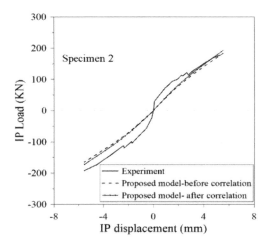

Figure 15. Comparison of IP responses for specimen 2 of Angel (1994) before and after correlation.

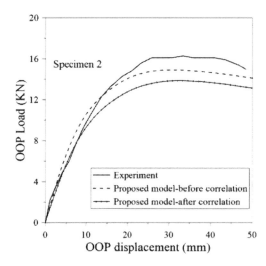

Figure 16. Comparison of OOP responses for specimen 2 of Angel (1994) before and after correlation.

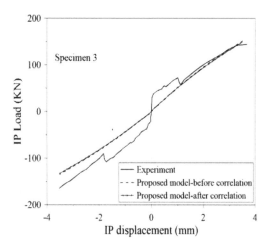

Figure 17. Comparison of IP responses for specimen 3 of Angel (1994) before and after correlation.

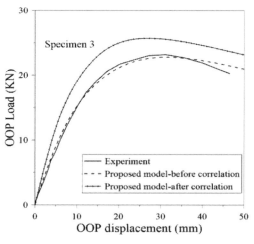

Figure 18. Comparison of OOP responses for specimen 3 of Angel (1994) before and after correlation.

5   CONCLUSION

A macro-element model which is able to simulate the IP and OOP behaviour of URM infill wall is presented in this paper. The proposed model is a modification of the four strut macro-element model developed by Di Trapani et al. (2018). The proposed model has been validated with number of experimental data available in the literature. A standard relation has been developed to determine the empirical strength-strain parameters required for the struts based on the actual mechanical properties of the URM infill wall. The proposed macro-element model is simple and has the potential to be used in practice for the analysis of infilled frame structures in simulating responses of URM infill wall under the action of IP and OOP loads and their interaction.

REFERENCES

Angel, R. 1994. Behavior of reinforced concrete frames with masonry infill walls. Ph.D. thesis, Univ. of Illinois. Urbana-Champaign.

Asteris, P.G., Cavaleri, L., Di Trapani, F., Sarhosis, V. 2015. A macro-modelling approach for the analysis of infilled frame structures considering the effects of openings and vertical loads. *Struct. Infrastruct. Eng.* 12(5): 551–566.

Calvi, G.M. & Bolognini D. 2001. Seismic Response of Reinforced Concrete Frames Infilled with Weakly Reinforced Masonry Panels. *Journal of Earthquake Engineering* 5: 153–185.

Cavaleri, L. & Di Trapani, F. 2014. Cyclic response of masonry infilled RC frames: Experimental results and simplified modeling. *Soil Dynamics and Earthquake Engineering* 65: 224–42.

Cavaleri, L., Fossetti, M., Papia, M. 2005. Infilled frames: Developments in the evaluation of cyclic behavior under lateral loads. *Struct. Eng. Mech.* 21(4): 469–494.

Chrysostomou, C.Z., Gergely, P., Abel, J.F. 2002. A Six-Strut Model for Nonlinear Dynamic Analysis of Steel Infilled Frames. *International Journal of Structural Stability and Dynamics* 2: 335–353.

Crisafulli, F.J. & Carr, A.J. 2007. Proposed Macro-Model For The Analysis Of Infilled Frame Structures. *Bulletin of New Zealand National Society for Earthquake Engineering* 40, 9.

Da Porto, F., Guidi, G., Benetta, M.D., Verlato, N. 2013. Combined In-Plane/Out-of-Plane Experimental Behaviour of Reinforced and Strengthened Infill Masonry Walls. *The Masonry Society, ed. 12th Canadian Masonry Symposium 2013*: 12.

De Risi, M.T., Di Domenico, M., Ricci, P., Verderame, G.M., Manfredi, G. 2019. Experimental investigation on the influence of the aspect ratio on the in-plane/out-of-plane interaction for masonry infills in RC frames. *Eng Struct*. 189: 523–40.

Di Trapani, F., Shing, P.B., Cavaleri, L. 2018. A macro-element model for in-plane and out-of-plane responses of masonry infills in frame structures. *J. Struct. Eng.* 144 (2): 04017198.

Dolšek, M., Fajfar, P. 2005. Simplified non-linear seismic analysis of infilled reinforced concrete frames. *Earthq Eng Struct Dyn* 34: 49–66.

Durrani, A.J. & Luo, Y.H. 1994. Seismic Retrofit of Flat-Slab Buildings with Masonry Infills, *in: NCEER Workshop on Seismic Response of Masonry Infills. National Center for Earthquake Engineering Research (NCEER), 1994*. Buffalo, NY.

El-Dakhakhni, W.W., Elgaaly, M., Hamid, A.A. 2003. Three-Strut Model for Concrete Masonry-Infilled Steel Frames. *Journal of Structural Engineering* 129: 177–185.

Flanagan, R.D., & Bennett, R.M. Bidirectional behavior of structural clay tile infilled frames. *J. Struct. Eng.* 3(236): 236–244.

Furtado, A., Rodrigues, H., Arede, A., Varum, H. 2015. Simplified macro-model for infill masonry walls considering the out-of-plane behavior. *J Earthq Eng Struct Dyn*. 45:507–524.

Hashemi, S.A. & Mosalam, K.M. 2007. Seismic evaluation of reinforced concrete buildings including effects of infill masonry walls. *PEER 2007/100 2007, Univ. of California*. Berkeley, CA.

Henderson. R., Jones, W., Burdette, E., Porter, M. 1993. The effect of prior out-of-plane damage on the in-plane behavior of unreinforced masonry infilled frames. *In: The Fourth DOE Natural Phenomena Hazards Mitigation Conference. 1993*.

Hendry, A.W. 1998. Structural Masonry, II. ed., 1998. Palgrave.

Holmes, M. 1998. Steel Frames with Brickwork and Concrete Infilling. *in: Proceedings of the Institution of Civil Engineers 1961*. 473–478.

Kadysiewski, S. & Mosalam, K.M. 2009. Modeling of unreinforced masonry infill walls considering in-plane and out-of-plane interaction. *PEER 2008/102 2009, Univ. of California*. Berkeley, CA.

Kent, D.C. & Park, R. 1971. Flexural members with confined concrete. *J. Struct Eng.* 97(ST7): 1969–1990.

Liauw, T.C. 1972. An Approximate Method of Analysis for Infilled Frames with or without Opening. *Building Science* 7(4): 233–238.

Madan, A., Reinhorn, A.M., Mander, J.B., Valles, R.E. 1997. Modeling of masonry infill panels for structural analysis. *J. Struct. Eng.* 10(1295): 1295–1302.

Mainstone, R.J. 1974. Supplementary Note on the Stiffness and Strengths of Infilled Frames. P*resented at the Building Research Station, 1974*. Garston, UK.

Mosalam, K.M. & Günay, S. 2015. Progressive collapse analysis of RC frames with URM infill walls considering in-plane/out-of-plane interaction. *Earthquake Spectra* 31(2): 921–943.

Palieraki, V., Zeris, C., Vintzileou, E., Adami, C-E. 2018. In-plane and out-of-plane response of currently constructed masonry infills. *Eng Struct*. 177: 103–16.

Panagiotakos, T.B., Fardis, M.N. 1996. Seismic response of infilled RC frames structures. *Proc., XXI World Conf. of Earthquake Engineering, 1996*. Acapulco, Mexico.

Papia, M., Cavaleri, L., Fossetti, M. 2003. Infilled frames: Developments in the evaluation of the stiffening effect of infills. *Struct. Eng. Mech.* 16(6): 675–693.

Paulay, T., Priestley, M.J.N. 1992. Seismic design of reinforced concrete and masonry buildings. *Wiley, c*, 1992. New York.

Polyakov SV. On the Interaction between Masonry Filler Walls and Enclosing Frame When Loading in the Plane of the Wall. Translation in Earthquake Engineering. *Earthquake Engineering Research Institute* 1960: 36–42.

Ricci, P., Di Domenico, M., Verderame, G.M. 2018. Experimental assessment of the in-plane/out-of-plane interaction in unreinforced masonry infill walls. *Eng Struct*. (173): 960–78.

Smith, B.S. & Carter, C.A. 1959. Method of Analysis for Infilled Frames, *in: Proceedings of the Institution of Civil Engineers 1959*. 31–48.

Thiruvengadam, V. 1985. On the Natural Frequencies of Infilled Frames. *Earthquake Engineering & Structural Dynamics* 13: 401–419.

Žarni´c, R. & Gostič, S. 1997. Masonry infilled frames as an effective structural subassemblage. Seismic design methodologies for the next generation of codes, 1997; P. Fajfar and H. Krawinkler (eds), 335–346. Rotterdam: Balkema.

Brick and Block Masonry - From Historical to Sustainable Masonry –
Kubica, Kwiecień & Bednarz (eds)
© 2020 Taylor & Francis Group, London, ISBN 978-0-367-56586-2

A new macroelement-based strategy for modelling reinforced masonry piers

S. Bracchi
Department of Civil Engineering and Architecture, University of Pavia, Pavia, Italy

M. Mandirola
Department of Emergency Support, EUCENTRE Foundation, Pavia, Italy

M. Rota
Department of Buildings and Infrastructures, EUCENTRE Foundation, Pavia, Italy

A. Penna
Department of Civil Engineering and Architecture, University of Pavia, Pavia, Italy

ABSTRACT: Macroelement models are widely used within an equivalent-frame approach to study the seismic behaviour of URM buildings. Their application to reinforced masonry structures has not yet been verified, although appearing promising for the expected compromise between computation burden and accuracy of the results. In this work, a new macroelement-based strategy to model the in-plane nonlinear behaviour of RM piers is proposed, starting from a macroelement model widely adopted for URM and implemented in the TREMURI software. The strategy consists in discretising a masonry pier into sub-macroelements, representative of masonry and horizontal (shear) reinforcement, with nonlinear beams representing vertical reinforcement. To test the efficiency of this model, experimental tests performed on RM piers were simulated. Afterwards, assemblies of piers were studied. The results of these simulations appear promising and suggest the application of the proposed model for nonlinear analyses of entire RM buildings, as a future development of this work.

1 INTRODUCTION

Macroelement models, in combination with an equivalent-frame idealisation of the buildings' structure, are widely used to study the behaviour of unreinforced masonry buildings subjected to horizontal actions, due to their proven efficiency and simplicity of use (e.g. Penna et al. 2016, Kallioras et al. 2019). The application of this modelling strategy to the case of reinforced masonry elements and structures poses a number of issues, that still need to be explored.

Several attempts were proposed in the literature for modelling the behaviour of reinforced masonry elements (e.g. Magenes & Baietta 1998, Maleki et al. 2005, Peruch et al. 2019, Shakarami et al. 2019) or systems (e.g. Mojiri et al. 2015, Abdellatif et al. 2019), typically based on finite element approaches. These models are usually calibrated on experimental tests on masonry components (e.g. Shing et al. 1989, Voon & Ingham 2006, Mosele 2009, Penna et al. 2015) and can be used to define fragility curves of entire buildings (e.g. Lofty et al. 2019).

The applicability of macroelement approaches has not been deeply explored yet, at least to the authors' knowledge. This approach potentially has several advantages with respect to other proposed strategies,

mainly consisting of a limited computational time, without a proportional reduction in the reliability and quality of the results.

This work proposes a new macroelement-based strategy to model the in-plane nonlinear behaviour of reinforced masonry piers, by considering a number of macroelements arranged in a series configuration. The model was validated through the comparison with the results of experimental tests on piers and then extended to model elements assemblies. The results obtained appear to be promising and encourage the application of this type of modelling approach to the case of larger elements assemblies and even entire buildings.

2 MODELLING STRATEGY

The strategy developed to model reinforced masonry piers is based on the equivalent-frame idealisation of the building, with structural members modelled by means of macroelements. In particular, the starting point is represented by the macroelement model proposed by Penna et al. (2014) and implemented in the TREMURI computer program (Lagomarsino et al. 2013).

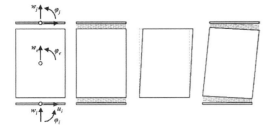

Figure 1. Macroelement proposed by Penna et al. (2014).

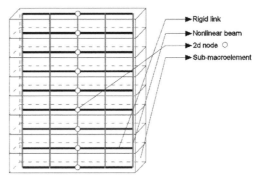

Figure 2. Scheme of the adopted modelling strategy for piers, with indication of the sub-macroelements in which the pier is subdivided and the nonlinear beams representing the longitudinal reinforcement.

As shown in Figure 1, the element can be ideally subdivided into three parts: a central body where only shear deformation can occur and two interfaces, where the external degrees of freedom are placed, which can have relative axial displacements and rotations with respect to those of the extremities of the central body. The two interfaces can be considered as infinitely rigid in shear and with a negligible thickness. Their axial displacements and rotations emulate those of a distributed system of zero-length springs, featuring a no tension law with bilinear degrading behaviour in compression.

These assumptions simplify the macroelement kinematics and allow to reproduce the flexural and shear response of unreinforced masonry buildings through a limited number of degrees of freedom. In particular, the macroelement kinematics can be described by means of eight degrees of freedom, six nodal generalized displacement components (u_i, w_i, φ_i, u_j, w_j, φ_j) and two internal components (w_e, φ_e).

The shear response is modelled through a damage model characterized by a Coulomb strength criterion, depending on equivalent values of cohesion and friction coefficient.

The proposed strategy to model reinforced masonry piers consists in assembling a certain number of sub-macroelements to create a larger macroelement representing a masonry member with shear reinforcement, coupled with nonlinear beam elements, modelling longitudinal (vertical) reinforcement. The contribution of the transversal shear reinforcement is considered through a proper choice of the equivalent shear strength parameters of the sub-macroelements, defined applying the strength criterion prescribed by the Italian building code (NTC18, 2018).

The sub-macroelements modelling masonry have a no-tension behaviour, with possible cracking of the end sections, whereas the nonlinear beams modelling longitudinal reinforcement have a tensile strength limited by the yield strength of the steel.

The division into sub-macroelements is motivated by the possibility of better capturing the flexural deformed shape and, therefore, the contribution of the longitudinal reinforcement. Figure 2 reports a scheme illustrating the proposed modelling strategy for a masonry pier.

In cracked section conditions, axial displacements and rotations are coupled and hence the model is also suitable to model uplift phenomena occurring in case of rocking behaviour.

According to the Italian building code (NTC18), the shear strength of a reinforced masonry pier is given by:

$$V_s = dtf_v + 0.6d \, \frac{A_{sw}}{s} f_y \leq 0.3 f_m t d \qquad (1)$$

where f_v is the contribution to the shear strength due to masonry, i.e.:

$$f_v = f_{v0} + 0.4\sigma_n \leq 0.065 \, f_b \qquad (2)$$

and t is the thickness of the element, d is the distance between the compressed side of the section and the center of mass of the bars in tension (assumed in this work equal to the length l of the section minus $t/2$), A_{sw} and s are the area and step of the transversal reinforcement, respectively, f_y is the yield strength of the steel of the transversal reinforcement, f_m is the masonry compressive strength, f_{v0} is the initial shear strength (cohesion) of masonry, f_b is the compressive strength of the unit and σ_n is the axial stress acting on the considered section over the area dt.

The equivalent cohesion and friction coefficient to be used in the strength criterion implemented in the macroelement proposed by Penna et al. (2014) can be obtained from the linearization of the strength criterion expressed by Equation 1. As it can be noted from Equation 1, the general formulation is intrinsically linear with the value of the axial force acting on the pier. It can be rewritten as follows:

$$V_s = c_{eff} \, dt + 0.4N \leq c_{max} \, dt \qquad (3)$$

where:

$$c_{eff} = f_{v0} + 0.6 \frac{a_{sw}f_y}{st} \qquad (4)$$

$$c_{max} = \min\left(0.065 f_b + 0.6 \frac{A_{sw}f_y}{st}; 0.3 f_m\right) \qquad (5)$$

The calibration of the equivalent strength parameters was performed before running the analysis as no variation in the axial force occurred during the considered tests.

In addition, shear drift limits need to be defined at the sub-macroelement level. Hence, it is necessary to identify an equivalent drift limit, starting from the drift limits for the entire panel. Since the drift tends to concentrate in a single sub-macroelement, a geometrical proportion was used to amplify the sub-macroelement drift with respect to the entire pier's drift.

3 SIMULATION OF EXPERIMENTAL TESTS

To test the efficiency of the developed modelling strategy, a series of experimental tests performed at the University of Pavia (Italy) within the DREMAB project (Magenes 1998) on single reinforced masonry piers was simulated, as discussed in more detail in the following.

3.1 Experimental campaign

The considered experimental campaign included seven walls made of clay bricks, tested in cantilever boundary conditions by applying a cyclic horizontal shear force with constant axial load.

The walls were made of perforated clay bricks, with a void ratio of 45%. Bricks' dimensions consisted of 300 mm width, 230 mm length, 185 mm height. Walls' thickness was equal to 300 mm.

The vertical reinforcement was inserted in the grip holes (80 mm x 85 mm), filled with the same mortar used for bedding. The horizontal reinforcement was laid in the bed-joints.

The mortar was a cement/lime/sand mix with a 1.5/1/4 proportion (in volume). The reinforcement consisted in high bond bars of steel with characteristic yield strength f_{yk} equal to 430 N/mm^2.

Each wall specimen was designed to exhibit a specific failure mode, by varying the aspect ratio (and consequently the shear ratio) and the amount and distribution of vertical and horizontal reinforcement. Table 1 reports the geometry and axial load of each wall, in terms of applied stress at the top of the wall.

Walls 4c and 4s were designed to exhibit pure flexural failure. They had the same amount of total vertical reinforcement, but different distribution along the length of the section. Walls 9, 10 and 11 were designed to exhibit shear failure; these walls

Table 1. Geometry and applied axial load, for each wall.

Wall	l [m]	h [m]	t [m]	$\sigma_{0,top}$ [MPa]
2	2.23	3.015	0.3	0.250
3	2.23	3.015	0.3	0.278
4c	2.23	3.015	0.3	0.285
4s	2.23	3.015	0.3	0.472
9	2.71	1.615	0.3	0.238
10	2.71	1.615	0.3	0.244
11	2.71	1.615	0.3	0.235

Table 2. Reinforcement of each wall.

	Longitudinal reinf.		Transversal reinf.
Wall	Spread	Concentrated	Spread
2	1φ10/48 cm	2+2φ18	2φ6/40 cm
3	1φ14/48 cm	2+2φ18	2φ6/20 cm
4c	1φ6/48 cm	1+1φ18	2φ6/40 cm
4s	1φ12/48 cm	-	2φ6/40 cm
9	1φ10/48 cm	2+2φ20	2φ6/40 cm
10	1φ14/48 cm	2+2φ20	2φ6/20 cm
11	1φ10/48 cm	2+2φ20	1φ6/40 cm

differ only for the spread reinforcement ratios. Walls 2 and 3 were designed to have similar flexural and shear capacities (mixed flexure-shear failure mode).

Horizontal reinforcement consisted of closed hoops, with the exception of wall 11, which had a single 6 mm diameter bar every course. The vertical steel was continuous along the wall height and was anchored into the reinforced concrete foundation at the base and into the r.c. spread beam at the top. Table 2 reports the reinforcement configurations of the walls.

As expected, the tests on Wall 2 and Wall 3 were characterized by flexural and shear cracks: in particular, corner-to-corner diagonal shear cracks formed during the test, followed by toe crushing. Wall 4c had flexural cracks with very moderate shear cracking. Wall 4s was also characterized by a flexural response, with crushing failure of the compressed masonry corners.

Walls 9 and 11, having lower spread reinforcement ratios, had a major diagonal crack developing from the upper corners down to the base of the wall in each direction of loading. On the contrary, Wall 10, which featured the highest spread reinforcement ratio, was characterized by a higher number of inclined shear cracks of lower width.

3.2 Numerical simulation

All tests carried out by Magenes (1998) were simulated using the proposed modelling strategy. Each wall was discretized into eight sub-elements of equal height and the mechanical properties of masonry were assumed equal to the values derived from the characterization tests performed by Magenes et al. (1996). In particular, for all the sub-elements of all the walls, the following values were adopted:

- elastic modulus $E = 11218$ MPa,
- shear modulus $G = 240$ MPa,
- compressive strength $f_m = 10.76$ MPa,
- density $\rho = 900$ kg/m³,
- initial shear strength $f_{v0} = 0.586$ MPa
- compressive strength of bricks $f_b = 17.8$ MPa.

The shear deformability parameters χGc_t and β were assumed equal to 8 and 0.4, respectively.

The equivalent shear strength parameters were calculated for each sub-element, as a function of the geometry, axial stress in the mid-height section of the element and transversal reinforcement.

The strength of the steel was differentiated among the different bar diameters (Table 3), according to the value obtained from the characterization tests.

In this paper, the results of the numerical simulation are reported only for two walls, one characterized by flexural behaviour (Wall 4s) and the other failing in shear (Wall 9).

Figure 3 Shows the comparison between the hysteresis curves, i.e. base shear versus displacement curves, obtained from the experimental tests and the numerical simulations. It can be noted that the numerical model is able of reproducing the experimental response in terms of strength, displacement capacity and hysteretic behaviour. The dissipated energy is slightly underestimated, particularly in the initial phases for the wall having a flexural failure mechanism, i.e. Wall 4s, whose response is shown at the top of Figure 3.

Figure 4 shows the damage mechanism obtained from the simulations of the two experimental tests. It can be noted that the simulated response of Wall 4s is actually characterised by a flexural mechanism, similar to the one observed during the experimental test. No shear cracks are visible in the mechanism. The adopted discretization into sub-elements also allows to reproduce the variation of cracked length of the section along the height, as indicated in the plot by the nearly horizontal lines.

For the case of Wall 9, experimentally failing in shear, the damage mechanism obtained from the

Table 3. Strength of reinforcement, as a function of the bar diameter φ.

φ [mm]	f_y [MPa]
6	592.9
10	589.7
12	597.3
14	487.0
18	608.3
20	510.6

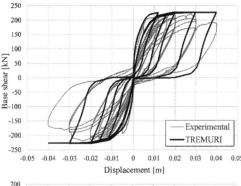

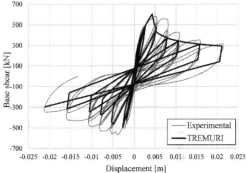

Figure 3. Comparison between the force vs. displacement curves of experimental tests and numerical simulations, for Wall 4s (top) exhibiting a flexural failure mode and Wall 9 (bottom), failing in shear.

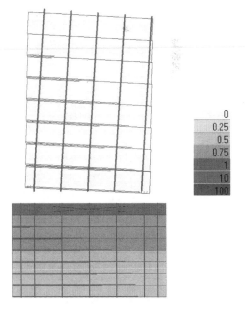

Figure 4. Damage mechanism obtained from the numerical simulation of the tests on Wall 4s (top) and Wall 9 (bottom). The colour legend indicates the level of shear damage in terms of the internal damage parameter α.

903

Table 4. Failure mechanism and comparison among the experimental, numerical and theoretical shear strengths.

Wall	Failure Mechanism	$V_{res,exp}$ [kN]	$V_{res,num}$ [kN]	$V_{res,theor}$ [kN]
4s	Flexure	222	227	221
9	Shear	582	603	623

simulations actually shows a shear failure, as indicated by the "X" marker in the top element. Damage in the sub-elements is increasing moving from bottom to top, as indicated by the darker colour. The legend indeed reports the colour scale with reference to a parameter internally defined in the code (α), which represents the level of shear damage within the element. When this parameter reaches the value of 1, the element fails in shear and a "X" marker appears. In the numerical results, it is evident that the failure is concentrated in a single sub-element (the top one in this case), due to the discretization of the pier in different sub-macroelements, behaving in series. This implicates that the failure of one of the sub-macroelements limits the shear force in the others and hence the failure does not propagate to the other elements.

Table 4 compares the values of strength obtained for the two walls from experimental tests, numerical simulations and theoretical calculations. The theoretical shear strength was calculated with Equation 1, whereas the theoretical flexural strength was obtained assuming a rectangular stress block for masonry in compression (with stress equal to $0.85f_m$), a bilinear constitutive relationship for steel and a linear distribution of normal strains along the section.

Observation of the values in the table shows that, in case of a flexural mechanism, the three values of shear strength provided by numerical, experimental and theoretical evaluation are very similar.

In case of shear failure (Wall 9), the difference is larger, with the theoretical shear strength slightly higher than the numerical value, which is again higher than the experimental value. In any case, the percent difference of the values obtained numerically and theoretically, with respect to the experimental result, is equal to 3.6% and 7% in the two cases, and hence the similarity in the results can be considered acceptable.

4 MODELLING OF MEMBERS' ASSEMBLIES

The strategy developed for single piers was then extended to model structural members' assemblies, by applying it to bidimensional elements (e.g. portals) and tridimensional assemblies (e.g. L-shape or T-shape flanged walls). A parametric study was carried out, considering different geometries of the assemblies.

The study of these assemblies had two main scopes: on one hand, it allowed to account for the flange effect and evaluate its impact on the response of flanged walls, with respect to the case in which the contribution of the perpendicular wall, in terms of out-of-plane stiffness and strength, is neglected. On the other hand, the consideration of multiple coupled elements was required to demonstrate the feasibility of using the proposed modelling approach for modelling entire reinforced masonry buildings, which will be the natural development of this work.

In this paper, the response of a bidimensional portal and of a tridimensional T-shape flanged wall is reported, as an example, to show the applicability of the proposed approach for the case of multiple elements, which is a step forward the application of the methodology to the case of entire reinforced masonry buildings.

The structural members' assemblies were obtained by means of the single piers previously modelled. The portal was obtained by coupling two piers with the same characteristics of Wall 4c, with the addition of an elastic rigid beam connecting the top bidimensional nodes of the piers.

For the tridimensional assembly, the bidimensional nodes at the ends of each sub-element were transformed into tridimensional nodes. The sub-elements of the orthogonal walls are defined on these tridimensional nodes. A bidimensional node was then added at the top of the flanges to allow the application of the masses; this node is connected to the tridimensional node at the top of the intersection by means of an elastic rigid beam.

In a TREMURI model of a building, tridimensional nodes are located at the intersection of walls belonging to different planes, whereas bidimensional nodes are inserted in only a single wall: the two node types are therefore characterized by a different number of degrees of freedom. Figure 5 shows the difference among bidimensional and tridimensional nodes.

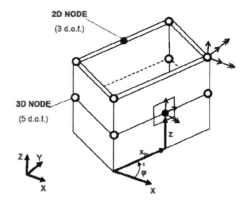

Figure 5. Bidimensional and tridimensional nodes in a TREMURI model.

This modelling approach allows locating the bars in the exact position, avoiding superposition and/or duplication of bars at the corners. Indeed, the bars at the intersection of the walls were modelled only once, attributing them to the wall which is excited in its plane during the numerical analysis.

Figure 6 Shows the adopted modelling strategy for the two considered examples of a portal and a T-shape flanged wall.

To test the proposed modelling strategy, a cyclic displacement was applied at the top of each structure. Figure 7 reports the results for the portal obtained replicating Wall 4c in terms of hysteresis curves and damage pattern, whereas Figure 8 and Figure 9 show the results for the T-shaped flanged wall, again obtained replicating Wall 4c, analysed in the direction of the flange and in the direction parallel to the web, respectively.

In the case of the portal (Figure 7), at the first peak of displacement, a flag-shape hysteresis curve with hybrid damage mechanism was obtained. In particular, due to frame behaviour, the downwind pier increases its axial load showing a shear damage, whereas the windward pier unloads and is characterized by a flexural response. When inverting the load, also the other pier is obviously damaging in shear.

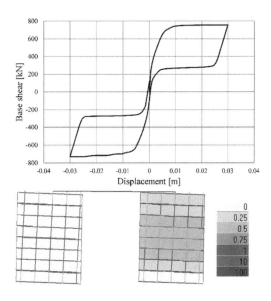

Figure 7. Force vs. displacement curve (top) and damage mechanism (bottom) for the portal defined by an assembly of two piers with the characteristics of Wall 4c. The colour legend indicates the level of shear damage in terms of the internal damage parameter α.

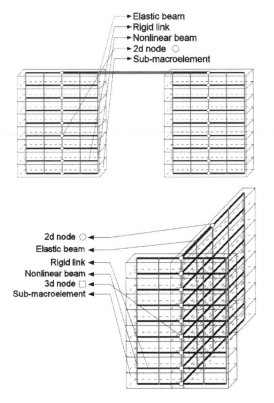

Figure 6. Examples of application of the adopted modelling strategy to the case of a portal (top) and a T-shape flanged wall (bottom).

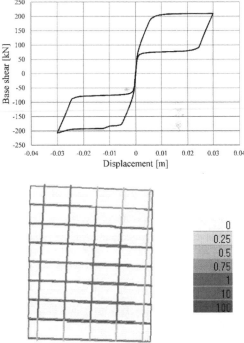

Figure 8. Force vs. displacement curve (top) and damage mechanism of the front wall (bottom) for the T-shaped flanged wall, analysed in the direction of the flange. The colour legend indicates the level of shear damage in terms of the internal damage parameter α.

905

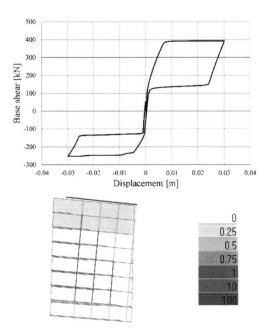

Figure 9. Force vs. displacement curve (top) and damage mechanism of the front wall (bottom) for the T-shaped flanged wall, analysed in the direction of the web. The colour legend indicates the level of shear damage in terms of the internal damage parameter α.

Observation of Figure 8 and Figure 9 show that, as expected, the hysteretic curve is symmetrical when the analysis is performed in the direction of the flange, showing again a flag-shape hysteretic diagram. In the other direction, instead, the cyclic response is affected by the direction of application of the applied displacement, producing a lower strength when the flange wall is compressed (negative displacement in the plot). In the opposite direction, the strength is higher thanks to the contribution of tensioned bars in the flange.

5 CONCLUSIONS AND FUTURE DEVELOPMENTS

This paper proposed a novel strategy to model reinforced masonry piers, developed starting from the macroelement proposed by Penna et al. (2014) and implemented in the software TREMURI, in the framework of an equivalent-frame modelling approach. A new element was developed, by assembling a certain number of sub-elements representing masonry, with nonlinear beams representing the longitudinal reinforcement; the contribution of transversal reinforcement was accounted for by means of properly selected values of the equivalent shear strength parameters of the sub-elements.

The proposed model was then used to simulate the response of a series of experimental tests on reinforced masonry piers, with different geometries and reinforcement configurations. The model was able to satisfactorily reproduce the experimental response, in terms of hysteresis curves and damage patterns, both for walls dominated by a flexural failure mechanism and walls exhibiting a shear failure mechanism.

The proposed strategy was then applied to model structures consisting of assemblies of reinforced masonry piers. Portals, L-shaped and T-shaped flanged walls were modelled and numerical analyses were performed to test the applicability and the numerical efficiency of the new modelling strategy. The results appeared promising and suggest, as possible future developments, the simulation of additional experimental tests and then the application of the proposed strategy for modelling the seismic response of entire reinforced masonry buildings.

In particular, the aim of the continuation of this work will be to simulate more experimental tests on elements or structures (e.g. Shing et al. 1989, El-Azizy et al. 2015), characterized by different masonry typologies (e.g. concrete blocks) and reinforcement configurations (low/high reinforcement ratio). It would also be possible to carry out nonlinear analyses to study the seismic vulnerability of entire reinforced masonry buildings, as already done in several literature works for the case of unreinforced masonry, using a similar equivalent frame approach with macroelements (e.g. Bracchi et al. 2015, Bracchi et al. 2016, Cattari et al. 2018).

This modelling approach could be also used to model masonry structures with strengthening solutions (e.g. with NSM rebars, strongbacks or wood/steel frames).

ACKNOWLEDGMENTS

The work presented in this article was partially carried out within the activities of the DPC-ReLUIS-EUCENTRE 2019-21 research program, funded by the Italian Department of Civil Protection. However, opinions and conclusions do not necessarily reflect those of the funding entity.

REFERENCES

Abdellatif, A., Shedid, M., Okail, H., & Abdelrahman, A. 2019. Numerical modeling of reinforced masonry walls under lateral loading at the component level response as opposed to system level response. *Ain Shams Engineering Journal* 10: 435–451.

Bracchi, S., Rota, M., Penna, A., Magenes, G. 2015. Consideration of modelling uncertainties in the seismic assessment of masonry buildings by equivalent-frame approach. *Bulletin of Earthquake Engineering* 13(11):3423–3448.

Bracchi, S., Rota, M., Magenes, G., Penna, A. 2016. Seismic assessment of masonry buildings accounting for limited knowledge on materials by Bayesian updating. *Bulletin of Earthquake Engineering* 14(8):2273–2297.

Cattari, S., Camilletti, D., Lagomarsino, S., Bracchi, S., Rota, M., Penna, A. 2018. Masonry Italian code-conforming buildings. Part 2: nonlinear modelling and time-history analysis. *Journal of Earthquake Engineering* 22(sup. 2):2010–2040.

El-Azizy, O.A., Shedid, M.T., El-Dakhakhni, W.W., Drysdale, R.G. 2015. Experimental evaluation of the seismic performance of reinforced concrete structural walls with different end configurations. *Engineering Structures* 101:246–263.

Kallioras, S., Graziotti, F., & Penna A., 2019. Numerical assessment of the dynamic response of a URM terraced house exposed to induced seismicity. *Bulletin of Earthquake Engineering* 17:1521–1552.

Lagomarsino, S, Penna, A, Galasco, A & Cattari, S. 2013. TREMURI program: an equivalent frame model for the nonlinear seismic analysis of masonry buildings. *Engineering Structures* 56(11): 1787–1799.

Lofty, I, Mohammadalizadeh, T., Ahmadi, F. & Soroushian, S. 2019. Fragility functions for displacement-based seismic design of reinforced masonry wall structures. *Journal of Earthquake Engineering* https://doi.org/10.1080/13632469.2019. 1659881.

Magenes, G., Calvi, G.M. & Gaia, F. 1996. Shear tests on reinforced masonry walls. Scientific Report RS-03 96. University of Pavia, Department of Structural Mechanics.

Magenes, G. 1998. In-plane cyclic testing of reinforced masonry shear walls. *Proceedings of the 11th European Conference on Earthquake Engineering*, Paris, France.

Magenes, G. & Baietta, S. 1998. Cyclic modelling of reinforced masonry shear walls for dynamic analysis: indications coming from experiments. *Proceedings of the IV STRUMAS*, Pratolino, Italy.

Maleki, M., El-Damatty, A.A., Hamid, A.A., & Drysdale, R.G. 2005. Finite element analysis of reinforced masonry shear walls using smeared crack model. *Proceedings of the 10th Canadian Masonry Symposium*, Banff, Alberta.

Mojiri, S., El-Dakhakhni, W.W. & Tait, M.J. 2015. Seismic fragility evaluation of lightly reinforced concrete-block shear walls for probabilistic risk assessment. *ASCE Journal of Structural Engineering* 141(4).

Mosele, F. 2009. In-plane and out-of-plane cyclic behaviour of reinforced masonry walls. *PhD Thesis*, University of Trento, Trento, Italy.

NTC08. 2008. Ministero delle Infrastrutture. D.M. 17 Gennaio 2018 - "Aggiornamento delle Norme Tecniche per le Costruzioni", S.O. alla G.U. No. 42 del 20.2.2018, (in Italian).

Penna, A., Lagomarsino, S. & Galasco, A. 2014. A nonlinear macro-element model for the seismic analysis of masonry buildings. *Earthquake Engineering and Structural Dynamics* 43(2):159–179.

Penna, A., Mandirola, M., Rota, M. & Magenes, G. 2015. Experimental assessment of the in-plane lateral capacity of autoclaved aerated concrete (AAC) masonry walls with flat-truss bed-joint reinforcement. *Construction and Building Materials* 82:155–166.

Penna, A., Senaldi, I.E., Galasco, A. & Magenes, G. 2016. Numerical simulation of shaking table tests on full-scale stone masonry buildings. *International Journal of Architectural Heritage* 10(2-3):146–163.

Peruch, M., Spacone, E. & Benson Shing, P. 2019. Cyclic analyses of reinforced concrete masonry panels using a force-based frame element. *ASCE Journal of Structural Engineering* 145(7).

Shakarami, B., Kabir, M.Z. & Sistaninejad, R. 2019. Numerical modeling of a new reinforced masonry system subjected to in-plan cyclic loading. *Scientia Iranica, International Journal of Science and Technology*, DOI:10.24200/sci.2019. 5376.1237

Shing, P.B., Noland, J.L., Klamerus, E. & Spaeh, H. 1989. Inelastic behavior of concrete masonry shear walls. *ASCE Journal of Structural Engineering* 115(9):2204–2225.

Voon, K.C. & Ingham, J.M. 2006. Experimental in-plane shear strength investigation of reinforced concrete masonry walls. *ASCE Journal of Structural Engineering* 132(3):400–408.

Brick and Block Masonry - From Historical to Sustainable Masonry –
Kubica, Kwiecień & Bednarz (eds)
© 2020 Taylor & Francis Group, London, ISBN 978-0-367-56586-2

Seismic fragility assessment of masonry aggregates with identical structural units in row

L. Battaglia, N. Buratti & M. Savoia
University of Bologna, Bologna, Italy

ABSTRACT: In civil engineering field, it is a common simplification to consider a building as isolated, even if it belongs to an aggregation of structural units. Thus, the first part of this work is focused on the fragility analysis of masonry structural units, considered as a part of a row aggregate and analyzed at first as isolated. Some variabilities and uncertainties involved in the problem are considered, generating a set of simulations through the Response Surface statistical procedure. The data obtained from the simulations were used to plot the fragility curves, by applying the Monte Carlo method. Afterwards, the masonry structural units have been aggregated in row in order to compare their seismic fragility with that of the isolated ones. The results showed relevant differences if different directions of the seismic action and different positions of the structural unit along the aggregate are considered.

1 INTRODUCTION

Aggregate masonry buildings have been generated over the years, allowing the interaction of different homogeneous or inhomogeneous structural units under seismic action. Therefore, their seismic analysis cannot ignore the inevitable interactions resulting from structural contiguity between adjacent buildings (Formisano et al. 2010, Lagomarsino et al. 2014).

Starting from the idea that buildings located in similar geotechnical conditions and with similar geometrical and structural properties are expected to have similar seismic performances, it is possible to individuate some prototypes having similar characteristics and representative of some classes of buildings, whose seismic fragility is studied.

Thus, since the common simplification in civil engineering field to consider a building belonging to an aggregate structure as isolated, the selected classes of masonry buildings were at first considered as isolated structural units and then belonging to aggregations in row of 5 structural units. As a first step, in order to start from a simplified structure, it was decided to consider the structural units along the masonry aggregate identical each other. Moreover, it is very common to find aggregations of identical or similar structural units in the Italian historic centers, as very often they were built in the same historic period, characterized by the use of similar materials and construction techniques.

This work is focused on the seismic vulnerability and fragility assessment of three stories clay brick masonry buildings, sited in Bologna (Italy), with reference, at first, to single isolated structural units. In order to account for some variabilities and uncertainties involved in the problem, the Response Surface statistical method is used, where the expected value of a response parameter (the peak ground acceleration (PGA) corresponding to the attainment of the Life Safety limit state) is approximated through a polynomial function of a set of chosen variables. The Response Surface model (Franchin et al. 2003, Faravelli 1989) is calibrated through numerical data obtained by non-linear static analyses and used to determine the fragility curves, by applying full Monte Carlo simulations. 384 simulations (for each direction of the seismic action) were performed, considering the thickness of the masonry panels as explicit variable and the uncertainty of the seismic action as implicit variable. The seismic action was defined by means of a group of selected registered accelerograms, in order to analyze the effect of the variability of the earthquakes, also considering two different and orthogonal directions of the seismic action, to analyze the differences on the global seismic behavior due to the different geometrical configurations of the resisting masonry walls in the two directions. The purpose is to show how a masonry structural unit cannot be studied as isolated if it belongs to an aggregation of structures, as the adjacent buildings affect its behavior against the seismic action.

Furthermore, the comparison was made considering the masonry structural units in different positions along the row aggregate, as differences in the seismic response are expected if the structural unit is externally located in the corners of the row or internally between adjacent structural units.

2 SEISMIC FRAGILITY PROBLEM

The assessment of the seismic fragility is a key issue in seismic engineering, as the use of the fragility curves allows to evaluate the probability of exceeding certain levels of damage of a structural system, when exposed to an assigned seismic action. In a seismic reliability framework, the seismic fragility function is defined as the probability of failure of a structure conditional to the ground-motion intensity.

Considering the Life Safety (LS) limit state, the structural failure is attained when the limit state function, defined as the difference between structural capacity (C) and demand (D), both dependent on a set $\mathbf{x}$ of random variables and time t, is less than or equal to zero (Buratti et al. 2010). In the field of seismic engineering, according to Veneziano et al. (1983), Casciati & Faravelli (1991), Shome et al. (1998), in the definition of the limit state function, the explicit dependence on time is eliminated because the minimum value over the entire ground-motion duration is taken. In this contest, the quantities C and D can be expressed in terms on spectral acceleration, corresponding to the first natural period of the structure, and therefore, the limit state function can be written as:

$$g = S_{a,C}(\mathbf{X}) - S_{a,D} \qquad (1)$$

where $S_{a,C}$ is the spectral acceleration corresponding to the attainment of the Life Safety limit state and $S_{a,D}$ is the spectral demand acceleration. The structural capacity depends on the properties of the structure, which are defined in statistical terms. In this way, the Response Surface (RS) statistical model is used to represent the structural capacity $S_{a,C}$ in Equation 1. The data to calibrate the polynomial function of the RS model are estimated by means of non-linear static analyses (push-over) in order to approximate the dependence of $S_{a,C}$ to $\mathbf{x}$.

Finally, a range of values of the spectral demand acceleration is chosen and the generated values of the spectral capacity acceleration are used to solve Equation 1 using full Monte Carlo simulations in order to develop the fragility curves (Pinto 2001).

3 THE RESPONSE SURFACE MODEL

The Response Surface statistical method is based on a probabilistic procedure defining a statistical model to express a response parameter as a function of a set of variables (Searle et al. 1991, Franchin et al. 2003,

Box & Draper 1987). In this work, the response parameter is represented by the peak ground acceleration corresponding to the attainment of the Life Safety limit state (Battaglia et al. 2018). The variables are random and they are defined in prescribed ranges, using defined normal distributions.

In order to reduce the number of the simulations, the unknowns are divided in explicit ($\mathbf{x}_E$) and implicit ($\mathbf{x}_I$) variables (Khuri & Cornell 1996, Searle et al. 1991); the effects of the implicit variables are considered in additive form, thus they do not interact with the explicit variables. A model that also accounts for the implicit variables is called *mixed model*. Considering N observations, i.e. the results of numerical simulations, it is possible to approximate the expected response value $E(Y)$ of a response parameter Y through a polynomial function of a set of $\mathbf{x}$ explicit variables. In matrix form, the definition of the polynomial function is as follows (Buratti et al. 2010):

$$Y = \mathbf{X}\beta + \sum_{j=1}^{r} \mathbf{Z}_j \delta_j + \varepsilon \qquad (2)$$

where Y is the vector collecting the observed response values $Y = [Y_1,\ldots,Y_N]$, β is the vector collecting the regression parameters, $\mathbf{X}$ is the *design matrix* containing the values assumed by the explicit variables and ε is the vector collecting the errors of Y with respect to $E(Y)$.

Each vector δ_j, which represents a different implicit variable, is divided in b_j blocks; a block is a homogenous group of simulations affected by the same value of the variable δ_j. Finally, each $\mathbf{Z}_j$ is a $N \times b_j$ Boolean matrix, with values equal to 1 if the corresponding block is associated to the considered observation or otherwise equal to 0. According to Khuri & Cornell (1996) and Searle et al. (1991), the errors ε are normally distributed with zero mean and constant variance (σ^2_ε).

4 DEFINITION OF THE IDENTICAL STRUCTURAL UNITS ALONG THE MASONRY AGGREGATE

The main purpose of this work is the comparison between the masonry isolated structural units and the masonry aggregates. Thus, starting from the study of the isolated structural unit (ISU), performed by Battaglia et al. (2018), an aggregation in row of 5 structural units (AS) was analyzed. As a first step, in order to start from a simplified structure object of the study, it was decided to consider the structural units along the masonry aggregate identical each other.

Thus as a first step, the isolated structural unit studied in Battaglia et al. (2018) was analyzed with a simplified Response Surface, to compare its seismic behavior with the one of an aggregation of the same identical structural units in row. The purpose is to show how a masonry structural unit cannot be studied as isolated if it belongs to an aggregation of structures, as the adjacent buildings affect its behavior against the seismic action.

The structure is selected as representative of a class of buildings existing in Bologna, in Italy. It is not referred to a real case, but it was selected according to some common geometrical and structural properties belonging to existing masonry buildings in Bologna. It is a three-storey masonry building (plus the roof storey) and it has a rectangular and non-regular shape plan.

The building was modelled using TreMuri software (Lagomarsino et al. (2008)). The external and internal walls are defined with the same properties: masonry clay brick walls, characterized by a medium value of thickness equal to 0.25 m, assumed to be uniform over the entire height of the building. The horizontal elements are selected as hollow-core concrete slabs, modelled considering reinforced concrete joists, alternated with perforated bricks and a continuous layer of concrete above and characterised by an equivalent thickness equal to 0.05 m, defined in the software. Whereas, the pitched roof is made by timber beams with cross section 0.10 m x 0.10 m and spanned in 0.50 m, a timber plank above and it is covered by roof brick tiles. Figure 1 shows the tri-dimensional view and Figure 2 the plan of the structural ground floor of the 5 identical structural unit, defined in Battaglia et al. (2018), aggregated in

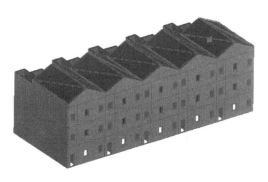

Figure 1. Tri-dimensional view of the aggregate structure.

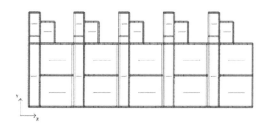

Figure 2. Plan of the structural ground floor of the aggregate structure.

row. The thickness of the common walls between two adjacent buildings is twice as that of the other walls, to ensure that the aggregate structure is a combination of identical structural units (Figure 2).

The structural properties of the masonry walls are chosen according to Table C8A.2.1 of the Italian Code (Commentary to the NTC 2008, 2009).

5 SELECTION OF THE VARIABLES

The variables are chosen according to the Response Surface model defined in Battaglia et al. (2018). Since the purpose of this work is to compare the seismic fragility of the isolated structural unit to that of an aggregate structure, a reduced number of variables is first selected.

5.1 Explicit variables

The thickness (s) of the resisting walls was chosen as explicit variable, as in the RS defined in Battaglia et al. (2018) it was considered as implicit variable: the assumption of s as explicit variable allows to obtain a direct relation with the collapse PGA (PGA_C) referred to the attainment of the LS limit state, representing the response parameter used to calibrate the RS model. In this way, the RS is simplified, but a large number of simulations was considered, compared to that reported in Battaglia et al. (2018).

In order to consider realistic cases of thickness (s) of the walls in civil applications, 8 different values are selected, taking into account the most common typologies of masonry structures in the Italian historic cities. In addition, following the RS rules (Section 3) in order to define the statistical properties, the values of the thickness (s) of the walls belong to a chosen normal distribution. Table 3 gives the 8 wall thicknesses selected for the simulations.

5.2 Implicit variables

In this work, in order to consider the possible variability of the seismic action in a site and the uncertainty related to the definition of the ground motion, a group of registered accelerograms referred to past earthquakes was considered. The accelerograms were chosen based on the data referred to previous earthquakes present in the PEER Ground Motion Database (https://ngawest2.berkeley.edu), created in collaboration with the NGA project (Power et al., 2008).

Thus, the uncertainty of the seismic action (δ_{sis}) was chosen as implicit variable, using the same group of 48 accelerograms defined in Battaglia et al.

Table 3. Values assumed by the thickness of the walls (s), expressed in meters.

s	0.125	0.15	0.20	0.25	0.30	0.375	0.40	0.50

(2018). They were scaled to the same reference peak ground acceleration of the site of Bologna (a_g), imposing some limits to the scaling in such a way as to be compatible with the LS limit state spectrum in that site, in the range period between T = 0.1s and T = 1.0s, but also usable until T = 3.0s. Furthermore, the selection was done avoiding recordings with impulsive characteristics, considering fixed ranges of epicentral distance and of the average shear wave velocity V_{s30} (NTC, 2018), in such a way to make the selections compatible with the considered site.

Figure 3 Shows the group of 48 acceleration scaled spectra, obtained starting from the correspondent accelerograms: the acceleration spectrum defined by the Italian Code (NTC, 2018) is also reported.

In this application, the division in blocks allows to associate each of the 48 accelerograms to each of the simulations defined by the 8 values of the explicit variable s. Thus, the group of the 8 simulations is repeated 48 times, as the number of the selected accelerograms: the total number of the blocks is 48. The partition in blocks, associated to the groups of the explicit variable s, generates 384 simulations in total.

6 PUSH-OVER ANALYSES

In this section the same procedure adopted in Battaglia et al. (2018) was followed, performing non-linear static analyses (push-over), using TreMuri software to obtain the data required to calibrate the Response Surface models.

Two orthogonal directions (x and y) of the seismic action are considered (Figure 2) and the distribution of the forces applied (proportional to the masses) was considered with both signs (+F and -F), generating 384 capacity curves for each studied case. Since the analyses were performed both for the masonry isolated structural units and the row-aggregations of identical structural units, the number of the type of analyses doubles; furthermore, in y-direction the analyses

beyond the attainment of the LS limit state were performed, to evaluate the collapse of the structural units in different positions along the aggregate (further details in the following).

Figure 4 shows the capacity curves obtained from the analyses considering the isolated structural units (ISU) and Figure 5 those considering the aggregate structures (AS), showing the differences between the two cases in terms of capacity and ultimate displacements in the two different directions of the seismic action. The curves are reported in terms of average displacements of the last floor nodes (d) and ratio between the total base shear and the total mass (V/M).

These curves highlight different behaviour of the structures when the seismic forces are considered with different sign (+ or -). Even if the curves "+ F_x," show higher capacity with respect to those "- F_x", the ultimate displacements reached are lower, because of the progressive decrement of the capacity in the case of positive forces (+ F_x). The loss of capacity is due to the configuration of the resisting masonry walls in

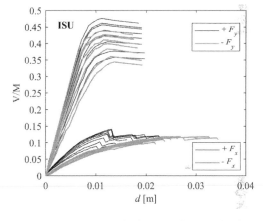

Figure 4. Capacity curves for the isolated structural units.

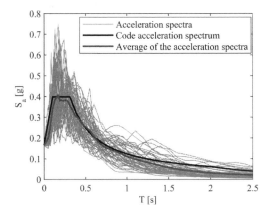

Figure 3. Group of the 48 acceleration spectra for the site of Bologna.

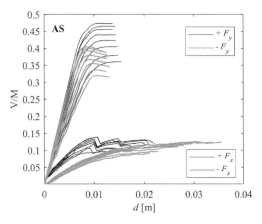

Figure 5. Capacity curves for the aggregate structures.

x-direction, allowing the presence of the flexure as main failure mechanism in this direction.

Looking at the geometrical configuration of the walls in this direction, the left-sides are the weaker due to a greater presence of the openings (Figure 2). Thus, if the forces $+ F_x$ are considered those weaker portions of the walls are the most solicited, causing the progressive decrement of the total capacity of the building.

On the contrary, if the forces $- F_x$ are considered the most solicited portions of the walls are on the right-side, where the reduced presence of openings allows to experience a more gradual loss of capacity with the attainment of higher displacements.

These results highlight how the geometrical configuration of the walls, in particular the presence of the openings, substantially affects the seismic response. In this specific case, the geometrical properties of the walls make the buildings weaker to the positive seismic action in *x*-direction ($+ F_x$). Thus, considering this latter case, lower values of the collapse PGA are expected, with respect to the negative seismic action in *x*-direction ($- F_x$), i.e. higher seismic fragility.

The same buildings referred to the 384 simulations exhibit greater capacity in *y*-direction, due to the arrangement and the geometry of the walls in this direction: they have a greater length and most of them are without openings. Thus, since the masonry walls in *y*-direction are stocky, the main collapse mechanism in this direction is the shear one. In this case, the two behaviours are very similar, due to the presence of the openings just in one panel, in *y*-direction, in each Structural Unit.

Considering the non-regularity in plan, the building results to be weaker to the negative seismic action in *y*-direction ($- F_y$), as the activation of flexural mechanisms in some spandrels and in some ground-floor-piers of the smaller panels; furthermore, the torsional effects are more accentuated, in this case depending on the asymmetry resulted in the upper part (in plan) of the model. However, in correspondence of the same displacement reached considering the forces $+ F_y$, most of the panels are still in the elastic or plastic phase. Thus, considering this latter case, higher values of the collapse PGA are expected, with respect to the negative seismic action in *y*-direction ($- F_y$), i.e. lower seismic fragility.

Moreover, if the *x*-direction is considered, the curves show a little increment of the capacity of the aggregate structures, as well as the attainment of larger ultimate displacements (Figure 5). Thus, since the length of the walls is about 5 times that of the walls of the isolated structural unit and the dominance of the flexure as main global failure mechanism, an increment of the collapse PGA is expected, if structural units in *x*-direction are aggregated.

In *y*-direction, the aggregation of structural units leads to a different geometrical configuration, which causes it to exhibit different levels of vulnerability due to the torsional effects mainly affected the external units: the *y*-direction walls do not increase their length, as for the *x*-direction aggregation, and their

seismic behaviour is affected by the action of the adjacent buildings. Therefore, the attainment of the total displacement of the aggregate structure is smaller than that of the isolated structural unit, if the LS limit state is considered as the limit for the global seismic response of the aggregate structure (Figure 5). Thus, contrary to what happens in *x*-direction, a decrement of the collapse PGA is expected if the global seismic behaviour of the aggregate structures is compared with those of the isolated structural units.

As an example, Figure 6 shows the *y*-direction deformed configuration of the aggregate structure with $s = 0.30$ m at the end of the push-over analysis, considering the seismic forces $+ F_y$. Due to the torsional effects, the external Units 1 and 2 reach larger displacements with respect to units 3, 4 and 5. Nevertheless, only the resisting walls of the external Units 1-2 fail for shear, with all other resisting walls still belonging to the plastic field. Thus, the displacement can still increase until the walls of the other Structural Units experience the failure for shear. For this reason, the analyses beyond the attainment of the LS limit state were performed, to allow the resisting walls in *y*-direction of the other Units (3, 4 and 5) to reach the shear collapse.

Figure 7 highlights the points on the push-over curves (beyond the attainment of the LS limit state) corresponding to the failure of the various Structural Units, considering the $+ F_y$ direction of the seismic action as an example: middle Units 3 and 4 have larger

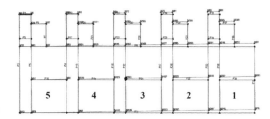

Figure 6. Deformed configuration of the model with $s = 0.30$ m, considering a seismic action in *y*-direction ($+ F_y$).

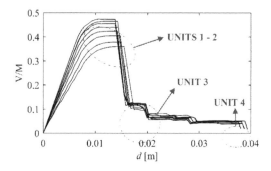

Figure 7. Capacity curves from the analyses in *y*-direction ($+ F_y$) beyond the attainment of the LS limit state.

912

values of displacement capacity, corresponding to the shear collapse of their masonry walls. The results referred to the external Unit 5 were neglected, as its ultimate displacements are so large to make the results not reliable.

7 EVALUATION OF THE COLLAPSE PGA

The 48 different acceleration spectra were applied to each of the models described in order to obtain the collapse PGA (PGA$_C$) associated with each model. As proposed in EN 1998-1 (Eurocode 8, 2004) and in the Italian Code (NTC, 2018), and according to Fajfar (2000), the N2 Method was used to obtain the structural capacity and the structural demand of the models, measured in terms of displacement, starting from the 48 displacement spectra, obtained by dividing the spectral accelerations by the frequency squared (ω^2). Afterwards, the ratio between the displacement capacity and the displacement demand was used to scale the acceleration spectra (Battaglia et al., 2018). According to this procedure, the acceleration corresponding to the structural failure of the model (PGA$_C$) can be obtained from the product between that ratio and the peak ground acceleration of Bologna.

8 RESPONSE SURFACE MODELS

The natural logarithm of the PGA corresponding to the attainment of the Life Safety limit state (PGA$_C$) is the response parameter chosen for the calibration of the Response Surface model. Adopting log(PGA$_C$), instead of PGA$_C$, avoids the prediction of negative values for the response. According to Equation 2, the polynomial function generated is quadratic:

$$\log(PGA_{C,i,j}) = \beta_0 + \beta_1 x_{1,i} + \beta_2 x^2{}_{1,i} + \delta_{sis,j} + \varepsilon_{i,j}$$
(3)

where x_1 is the selected explicit variable (s), i stands for the i-th simulation, j for the j-th δ_{sis} block and ε represents the errors. The regression obtained through the Ordinary Least Squares method.

The following figures show the sections of the RS models (continuous lines) obtained changing the values of the variable s and the sections (dashed lines) obtained adding and subtracting the RS variance $\sigma = \sqrt{\sigma^2{}_\varepsilon + \sigma^2{}_{sis}}$; the points are those corresponding to the various simulations used to calibrate the RS models. The reported results are those referred to the weaker directions ($+ F_x$ and $- F_y$).

Regarding the x-direction, the regression parameters β_1 related to the variable s are positive, for both the ISU and the AS: as expected, the value of the PGA$_C$ increases as the values of the thickness of the walls increase. The comparison between the RS model

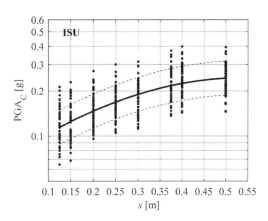

Figure 8. Response Surface section for the Isolated Structural Unit (ISU) in x-direction ($+ F_x$).

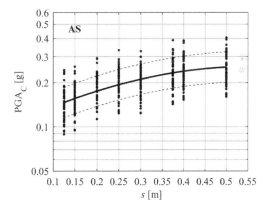

Figure 9. Response Surface section for the Aggregate Structure (AS) in x-direction ($+ F_x$).

referred to the ISU (Figure 8) and that referred to the AS (Figure 9) shows that the aggregation of identical structural units in row leads to a decrease of the vulnerability, due to the consideration on the geometrical properties of the walls discussed in Section 6.

For the seismic action in y-direction, the coefficients of the linear terms β_1 are also positive. Figure 10 shows the RS models obtained for the ISU and Figure 11 shows that referred to the AS, indicating a decrease of the collapse PGA, due to the torsional effects, when the aggregation is considered in y-direction.

As mentioned before, the failure of the aggregate structure is due to the shear failure of the masonry walls of Units 1 and 2. The RS related to the failure of the Units 3 and 4 were obtained continuing the analysis beyond the attainment of the LS limit state, allowing to reach larger values of displacement, corresponding to the failure of the central Unit 3 and Unit 4, which are associated to higher values of the collapse PGA. To summarize, the RS referred to Units 3 and 4 have not been reported.

913

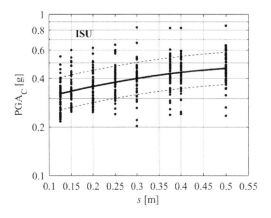

Figure 10. Response Surface section for the Isolated Structural Unit (ISU) in y-direction (- F_y).

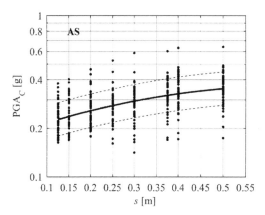

Figure 11. Response Surface section for the Aggregate Structure (AS) in y-direction (- F_y).

The results also confirm that the values of the collapse PGA for the y-direction are larger than those obtained for the x-direction.

9 FRAGILITY CURVES

The obtained RS models were used to estimate the fragility curves of the masonry isolated structural units and the masonry aggregate structures. The fragility analysis was assessed adopting the limit state function in Equation 1, rewritten in the form:

$$\log(\text{PGA}_C) - \log(\text{PGA}_D) = \beta_0 + \beta_1 x_1 + \beta_2 x^2_1 + \delta_{sis} + \varepsilon - \log(PGA_D)$$

(4)

Once obtained PGA$_C$ and fixed PGA$_D$ and being the behavior of the structure non-linear, in order to solve the Equation 4 Monte Carlo method is used.

Thus, four fragility curves were then obtained for the seismic action in x-direction (Figure 12) and eight for the y-direction (Figure 13).

For each direction, the fragility curves are shown distinguishing the positive (+ F_x and + F_y) and negative (- F_x and - F_y) seismic actions, highlighting the same considerations on the geometrical properties of the walls in Section 6. The curves indicate greater fragility for a seismic action in x-direction due to the geometry, the number of openings and the arrangement of the resisting walls in this direction. Moreover, in x-direction aggregating identical structural units in a row decreases the fragility, compared with that of the isolated structural units. On the contrary, the fragility of the aggregate is higher in y-direction, due to the torsional effects affecting the external Units (1-2), decreasing the total PGA$_C$, when

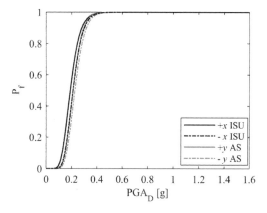

Figure 12. Fragility curves considering the x-direction of the seismic action.

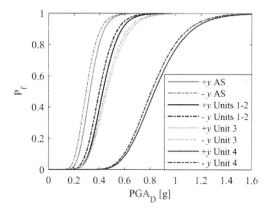

Figure 13. Fragility curves considering the y-direction of the seismic action.

referred to the first attainment of the displacement capacity. If, instead, higher values of the ultimate displacements are allowed, corresponding to the attainment of displacement capacity of the central Units 3 and 4, higher values of collapse PGA and lower fragility are obtained.

10 CONCLUSIONS

This work is focused on the seismic fragility analysis of masonry structural units, considered as part of row aggregate structures, by statistical procedures.

Identical structural units chosen by the RS generated simulations are aggregated, in order to compare the fragility referred to the isolated structural unit and the one referred to the entire aggregate. The results showed an increment on the values of the collapse PGA, aggregating identical isolated structural units and considering the seismic action in x-direction: the length of the walls about 5 times that of the walls of the isolated structural unit and the dominance of the flexure as main global collapse mechanism lead to a decrement of the fragility. On the contrary, in y-direction the different geometrical properties of the resisting walls and the torsional effects deriving from the aggregation of structural units in row cause an increase of the vulnerability, because the external units are affected by torsional effects decreasing the values of the collapse PGA obtained.

Higher values of the collapse PGA are associated to the internal structural units along the aggregate structure, obtained continuing the analyses beyond the attainment of the LS limit state and allowing the walls of the more internal units to reach the shear collapse. These analyses allowed to make a hierarchy of collapse of the various structural units along the aggregate, for the presence of the rigid slabs: the fragility curves show a decrement of the fragility if more internal units are considered.

REFERENCES

Battaglia, L., Buratti, N. & Savoia, M. 2018. Fragility analysis of masonry structural units by Response Surface method. *10th International Masonry Conference, 9-11 July 2018*. Milan, Italy.

Box, G.E.P. & Draper, N.R. 1987. *Empirical Model-Building and Response Surfaces*. New York: John Wiley and Sons.

Buratti, N., Ferracuti, B. & Savoia, M. 2010. Response Surface with random factors for seismic fragility of reinforced concrete frames. *Structural Safety* 32(1): 42–51.

Casciati, F. L. & Faravelli, F. 1991. *Fragility analysis of complex structural system*. New York: John Wiley and Sons.

Commentary to the NTC 2008. 2009. *Circolare 2 febbraio 2009, n. 617 Istruzioni per l' applicazione delle "Norme tecniche per le costruzioni" di cui al D. M. 14 gennaio 2008*. Rome, Italy: Italian Ministry of Infrastructure and Transportation. (In Italian).

Eurocode 8. 2004. EN 1998-1:2004. *Design of structures for earthquake resistance*. European Committee for Standardization.

Fajfar, P. 2000. A Nonlinear Analysis Method for Performance-Based Seismic Design. *Earthquake Spectra* 16(3): 573–592.

Faravelli, L. 1989. Response-Surface Approach for Reliability Analysis. *Journal of Engineering Mechanics* 115 (12): 2763–2781.

Formisano, A., Mazzolani, F., Florio, G. & Landolfo, R. 2010. A quick methodology for seismic vulnerability assessement of historical masonry aggregates. *COST Action C26 Urban Habitat Constructions under Catastrophic Events, 16-18 September 2010*. Naples, Italy.

Franchin, P., Lupoi, A., Pinto, P.E. & Schotanu, M.I. 2003. Seismic Fragility of Reinforced Concrete Structures Using a Response Surface Approach. *Journal of Earthquake Engineering* 7(1): 45–77.

Khuri, A.I. & Cornell, J.A. 1996. *Response surfaces: design and analyses*. New York: Marcel Dekker.

Lagomarsino, S., Penna, A., Galasco, A. & Cattari, S. 2008. *User Guide of TreMuri: Seismic Analysis Program for 3D Masonry Buildings*. User Guide. Genoa, Italy.

Lagomarsino, S., Cattari, S., Abbati, S. D. & Ottonelli, D. 2014. Seismic Assessment of Complex Monumental Buildings in Aggregate: the Case Study of Palazzo Del Podestà in Mantua (Italy). *9th International Conference on Structural Analysis of Historical Constructions, 14-17 October 2014*. Mexico City, Mexico.

NTC 2018. 2018. *Norme Tecniche per le Costruzioni*. Ministerial Decree. Rome, Italy: Italian Ministry of Infrastructure and Transportation. (In Italian).

Pinto, P.E. 2001. Reliability methods in earthquake engineering. *Progress in Structural Engineering and Materials* 3(1): 76–85.

Power, M., Chiou, B., Abrahamson, N., Bozorgnia, Y., Shantz, T. & Roblee, C. 2008. An overview of the NGA project. *Earthquake Spectra* 24(1): 3–21.

Searle, S.R., Casella, G., & McCulloch, C.E. 1991. *Variance Components*. New York: John Wiley and Sons.

Shome, N., Cornell, C.A., Bazzurro, P. & Carballo, J.E. 1998. Earthquakes, records, and nonlinear responses. *Earthquake Spectra* 14(3): 469–500.

Veneziano, D., Casciati, F. & Faravelli, L. 1983. Method of seismic fragility for complicated systems. *2nd CNSI Specialistic meeting on probabilistic method in seismic risk assessment for NPP, 16-18 May 1983*. Livermore, California.

Brick and Block Masonry - From Historical to Sustainable Masonry –
Kubica, Kwiecień & Bednarz (eds)
© 2020 Taylor & Francis Group, London, ISBN 978-0-367-56586-2

Numerical study on the seismic out-of-plane performance of URM walls

H. He, S.J.H. Meijers & F.H. Middelkoop
Advanced Technology & Research, Industry & Buildings, HaskoningDHV Nederland B.V., Rotterdam, The Netherlands

R.A. Vonk
Infrastructure, HaskoningDHV Nederland B.V., Rotterdam, The Netherlands

ABSTRACT: Groningen area of the Netherlands is now under the increasing induced seismic risk. The out-of-plane (OOP) performance of unreinforced masonry (URM) walls is critical for the safety of buildings in that area. A comprehensive masonry material model has been developed in a finite element software DIANA. However, its performance in a dynamic OOP bending test has not yet been fully verified due to the lack of experimental evidences. Fortunately, recent shaking table tests performed in EUCENTRE (in Italy) provide the valuable references of the dynamic one-way OOP bending behaviors of URM walls. Therefore, a numerical comparison study using a dynamic non-linear analysis method was realized in this study. The relative OOP responses in the whole testing process were compared, especially in the near-collapse situation. The influences of various material and model parameters were assessed in the comparison. Results and conclusions drawn in this study have been used for model validation and parameter refinements.

1 INTRODUCTION

The seismic risk in Groningen area of the Netherlands is induced by the gas extraction. The frequency of the shallow earthquakes is generally increased year by year (over 80 times per year in the recent decade), which is a great concern in the Dutch society. Therefore, a framework (called VIIA) for the seismic assessment and retrofitting design of structures in the Groningen region was initiated and developed from 2015. Nonlinear time-history analysis (NLTHA) was one of the approaches utilized for the project (He et al. 2018). It can provide inelastic seismic structural response in a moderate or strong earthquake. The material degradation causing nonlinearity as well as the geometric nonlinearity can be directly considered by this method.

Unreinforced masonry (URM) wall is one of the most vulnerable structural elements for the buildings in the area. A comprehensive masonry material model (EMM) has been developed and applied in the NLTHA of URM buildings in DIANA software (DIANA FEA 2016), which can consider the ortho-tropic mechanical properties (in tension, compression and shear) and hysteresis damping effects (Schreppers et al. 2016). It has been validated by comparison studies with a number of quasi-static in-plane and out-of-plane tests (Schreppers et al. 2016). However, the dynamic performance of EMM has not yet been fully verified due to lack of experimental evidences. Recent tests performed in EUCENTRE (in Italy) provide valuable references of dynamic

OOP response of URM walls (Graziotti et al. 2016a, b). Therefore, a numerical comparison study on the OOP tests of URM walls is realized and reported here. The dynamic relative OOP responses in the whole testing process will be compared, especially in the near-collapse situation. The influences of various model and material parameters are assessed in the study. Conclusions drawn in this study will be useful for model validation and parameter refinements in the NLTHA of the URM buildings in Groningen area in the Netherlands.

1.1 EUCENTRE shaking table tests

OOP performance of URM walls is one of the most important issues of Groningen buildings in the NLTH analysis. There are several relevant dynamic OOP tests performed at EUCENTRE in Italy, both on single-leaf walls and cavity walls (Graziotti et al. 2016a, b, Tomassetti et al. 2016). The main setup details of the tests on the shaking table are illustrated in Figure 1. The dynamic one-way OOP bending tests are done only in one-dimensional (in Y direction) excitation. Meanwhile, one rigid steel frame is used to simultaneously input horizontal motion for the top steel beam, which is lying on the top edge of URM specimens. The overburden stress applied on the top edge of specimen is realized by the two side springs linked between the top beam and the concrete foundation as shown in Figure 1 (Graziotti et al. 2016a).

For clarification and simplification purposes, only the tests on a single-leaf calcium silicate (CS)

Figure 1. Main setup details of dynamic shaking table tests in EUCENTRE, after Graziotti *et al.* (2016a).

masonry wall will be addressed in this study. The tests SIN-03 and SIN-01 are executed on the same specimen SIN (only overburden stress is switched from the 0.3 MPa to 0.1 MPa) (Graziotti *et al.* 2016a). For the test SIN-03, 0.3 MPa overburden stress is applied by two springs with a stiffness of 164.7 N/mm. Meanwhile, 0.1 MPa overburden stress is applied by two springs with stiffness of 53.5 N/mm in the test SIN-01. The dimensions of the wall are: 1.438 (m, length) × 0.102 (m, thickness) × 2.754 (m, height). Thus, the aspect ratio of the specimen is 1.915 (height/length). There are three types of signals (with different PGAs) inputted in the shaking table (Gr-1, Gr-2, RWA), which are plotted in Figure 2.

The sequence of the scaled inputted signals, the corresponding PGAs and the peak mid-height (relative to upper and bottom edges) responses of all tests are presented in Table 1 (Graziotti *et al.* 2016a). The collapse happened in the last step (Test 28) of tests for SIN-01 with the Gr-2 signal (PGA = +0.85g).

The maximum absolute values of relative (MAVR) OOP responses of all 28 subtests can also be illustrated in Figure 3, which is clearer for the comparison.

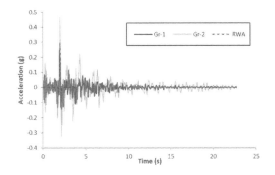

Figure 2. Three basic input base excitation signals for the shaking table tests, after Graziotti *et al.* (2016a), PGA are 0.247 g, 0.469 g and 0.3 g for Gr-1, Gr-2 and RWA, respectively.

Table 1. Test sequence and the OOP responses of the single-leaf wall (SIN-03 and SIN-01), after Graziotti *et al.* (2016a).

Test	Test nr.	Signal	PGA (g)	Peak relative OOP dis. (mm)
SIN – 03 (overburden stress : 0.3MPa)	1	Gr-1	+0.04	+0.06
	2	Gr-1	+0.09	+0.11
	3	Gr-1	+0.16	+0.40
	4	Gr-1	+0.20	+0.57
	5	Gr-1	+0.32	+0.92
	6	Gr-1	+0.42	+1.22
	7	Gr-1	+0.52	+1.40
	8	Gr-1	+0.74	+1.70
	9	Gr-1	+0.96	+4.93
	10	RWA	-1.11	-1.97
	11	RWA	-1.63	-9.63
	12	RWA	-1.05	-2.68
	13	RWA	-1.88	-14.1
SIN – 01 (overburden stress : 0.1MPa)	14	Gr-1	+0.08	-0.35
	15	Gr-1	+0.17	-0.73
	16	Gr-1	+0.21	-0.92
	17	Gr-1	+0.34	-1.28
	18	Gr-1	+0.41	+1.94
	19	Gr-1	+0.51	-7,42
	20	Gr-1	+0.60	-14.42
	21	Gr-1	+0.73	-16.60
	22	RWA	-0.26	-0.38
	23	RWA	-0.48	-1.88
	24	RWA	-0.72	-16.05
	25	RWA	-0.96	-52.98
	26	Gr-2	+0.44	+2.87
	27	Gr-2	+0.64	-9.86
	28	Gr-2	+0.85	Fail

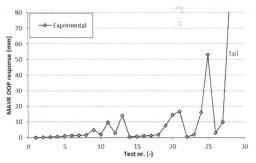

Figure 3. Maximum absolute value of relative (MAVR) OOP responses measured in each test, after Graziotti *et al.* (2016a).

1.2 Masonry material model

Masonry is an important historic construction material, which has been used for buildings worldwide. In the Groningen region of the Netherlands, large portions of walls of buildings are also made of masonry. During seismic loading, the masonry walls are easily damaged and vulnerable to collapse. Many of URM walls require seismic retrofitting measures. The NLTHA is helpful for the assessment of the performance of buildings and specification of the retrofitting of masonry walls. That goal can only be realized by using an accurate and robust masonry material model in the NLTH analysis.

The traditional masonry model in DIANA is based on a total strain cracking model (TSCM), which is isotropic and based on a secant unloading and reloading curve (DIANA FEA 2016). There are four types of masonry materials stated in the Dutch code NPR 9998: 2015 (NEN 2015), which were used in the NLTHA of masonry buildings in Groningen. Compared with the available material parameters, there are no orthotropic properties and shear cracking explicitly described by TSCM. It may largely underestimate the dissipated energy during the seismic analysis (Schreppers *et al.* 2016).

To avoid the limitations of TSCM, a new type of masonry model, called Engineering Masonry Model (EMM), was proposed by DIANA FEA and TU Delft (Schreppers *et al.* 2016). The new model is a total strain-based continuum model, which considers the tensile, shear and compression failure modes. It describes the unloading behaviour more realistically by strong strength decay with original linear stiffness. Shear failure is influenced by a Mohr-Coulomb criterion in the EMM. Its theoretical background and the extensive verification information (mainly on the quasi-static in-plane and OOP tests) can be found in the report (Schreppers *et al.* 2016).

2 MODEL AND PARAMETERS

A basic DIANA model (in the version of 10.1) (DIANA FEA 2016) is established based on the experimental setup details of specimen SIN, shown in Figure 4. The regular curved shell elements with 7 integration points in the thickness direction are used for URM wall and the element size is set to 0.1 m. The overburden stress is realized with pre-deformation of springs, similar to the experiments. The initial stiffness of two springs in experiments is evenly spread out by 15 springs in the DIANA model. The bottom edge of model is translationally constrained and upper side is only translationally constrained in the OOP direction (in Y direction). The rotational freedom along the length direction (along X direction) is constrained both for bottom and upper sides. In the first step of the analysis, the self-weight and pre-deformation of springs are added. Afterwards, the base excitation is applied in the OOP direction.

Based on the experiments, different scaled signals are applied to the specimen with specified sequence. This is also reproduced by the DIANA model. Therefore, one long combined signal is proposed

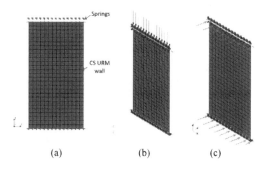

Figure 4. Model of the single-leaf CS wall (a) model, (b) load at the first step: overburden and self-weight, (c) base excitation in the following step.

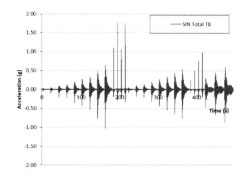

Figure 5. Input base excitation signal for the OOP tests.

based on the measured signal in the experiments. The input signal for all tests can be shown in Figure 5.

EMM material parameters are initially proposed based on the available experimental information (Graziotti 2016, Graziotti *et al.* 2016a, b) and some mechanical relationship assumptions (Schreppers *et al.* 2016). The proposed EMM parameters are presented in Table 2. The parameters will be further discussed during the parameter study in the next section to reveal their influences.

3 RESULTS AND DISCUSSION

3.1 *Basic results*

As indicated in Section 2, a total continued numerical test is proposed based on the experiments. Overburden stress is changed from 0.3 MPa to 0.1 MPa after finishing the test SIN-03. The spring overburden stiffness is also changed following the experimental instruction.

Figure 6 presents the relative OOP responses of mid-height nodes (to the top beam and foundation) through the entire signals. In the numerical tests, the wall survived throughout the entire signal but with much lower response in the large excitation. This is

918

Table 2. Proposed EMM parameters in the tests.

Parameter	Notation	Values	Notes
Density	ρ	1.84E+03	kg/m^3
Shear Modulus	G	1.22E+09	N/m^2
Young's modulus	E_x	1.63E+09	N/m^2, perpendicular to head joint
	E_y	3.26E+09	N/m^2, perpendicular to bed joint
Tensile fracture energy	G_f	1.50E+01	N/m
Tensile strength	f_{ty}	2.16E+05	N/m^2, perpendicular to bed joint
	f_{tx}	3.49E+05	N/m^2, perpendicular to head joint
Compressive strength	f_c	6.20E+06	N/m^2
Compressive strain ratio	n	4.00E+00	-
Compressive fracture energy	G_c	4.07E+03	N/m
Compressive unloading factor	λ	0.00E+00	-
Cohesion	c	2.10E+05	N/m^2
Shear fracture energy	G_{fs}	2.00E+01	N/m
Frictional angle	ϕ	3.98E-01	rad
Rayleigh damping factors	a	2.93E+00	-
	b	5.05E-05	-

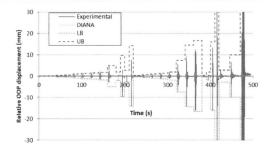

Figure 6. Comparison of relative OOP responses of mid-height of wall (LB: the lower bound, UB: the upper bound).

more explicitly illustrated in the Figure 7 for the comparison of the maximum absolute values of relative (MAVR) OOP responses of all the tests. Therefore, the dynamic OOP responses can be largely underestimated with the proposed model. The influences of some important model and material parameters will be separately assessed.

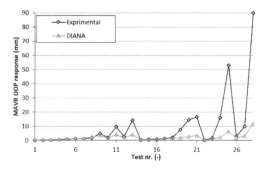

Figure 7. Comparison of MAVR OOP responses of mid-height of wall.

3.2 Rayleigh damping

Rayleigh damping is a viscous damping combining mass and stiffness. It is generally adopted in the EMM of DIANA for the definition of damping coefficients. a and b are the constants of proportionality of mass and stiffness, respectively, in the definition. Their influence will be firstly evaluated. Therefore, a new Rayleigh damping value with a wide range low damping ratio (2% at 1 Hz and 1% at 447 HZ, respectively) is proposed (a = 2.5105E-01, b = 7.0892E-06, called Ray5, respective to original Ray3, see Figure 8). One model without Rayleigh damping (Ray_off) is also proposed for the comparison. The OOP response and maximum OOP responses are illustrated in Figure 9 and Figure 10, respectively.

The results show Rayleigh damping has quite significant influences on the model. Without Rayleigh damping, the wall has significant resonating effect at the beginning of the excitation signals, which is eliminated after enough damage is accumulated (at about the end of GR-1 signal for SIN-03, Test 9). The OOP responses at the large excitations is also significantly improved (to about half of the experimental results).

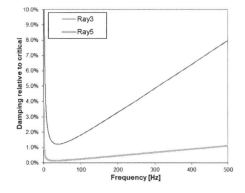

Figure 8. Rayleigh damping curves with different definitions.

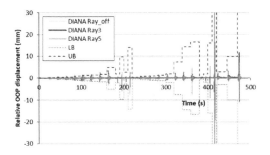

Figure 9. Relative OOP response of the mid-height nodes with different Rayleigh damping.

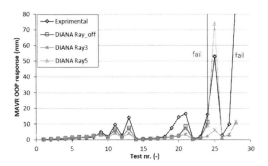

Figure 10. MAVR OOP responses with different Rayleigh damping.

However, early divergence occurs in the largest Ricker wavelet (RWA) in the tests SIN-01. The model with the new proposed Rayleigh damping (Ray5) has closer responses compared the model using original Ray3, while avoiding the early divergence problem. The new Rayleigh damping (Ray5) will be used for the further tests on other parameters. Meanwhile, it indicated that a suitable Rayleigh damping could be proposed for the masonries in the VIIA project, which can partly avoid the early resonance phenomenon, early divergence or the too low responses.

3.3 Tensile fracture energy

In EMM of DIANA (DIANA FEA 2016), tension softening behavior is assumed to be linear, which is partly due to the simplification purpose. Tensile strength of the masonry has been indirectly tested in the experiments. Meanwhile, it indicates the limited influence of tensile strength on the OOP response through a parameter study. Furthermore, tensile fracture energy, G_{ft} (or called GF), is defined as the crack energy in the direction normal to the bed joints. It influences brittleness of tension behavior. This value is also used for crack energy in the direction normal to the head joints. As it has been found that the responses of DIANA model are satisfactory at low excitations, the fracture

parameters are supposed to be more relevant for the responses with large excitations. Tensile fracture energy is not determined by testing in the experiments or indicated in the report (Graziotti 2016, Graziotti et al. 2016b). The initial value proposed in Section 2 is based on a recommendation by NPR 9998: 2015 (NEN 2015) for CS masonry (GF0=15 N/m). Two extra trial GF values are then tested (GF2=5 N/m, GF3=45 N/m). The tensile stress-strain curves with different GFs are illustrated in Figure 11.

The maximum relative OOP values are illustrated in Figure 12. The results show the significant influence of this parameter on the response, especially on the second part of the tests (SIN-01, with overburden 0.1 MPa). Supposedly, the wall is largely damaged in this part and the tension softening behavior plays an important role. Obviously, model with a lower GF reaches a better response compared with the experimental results, which is relevant for the GF values used for the EMM materials in the VIIA project.

3.4 Compressive strain ratio

Similar to the discussion on the tensile fracture energy, the compressive fracture behavior is also

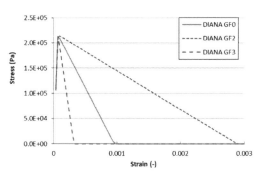

Figure 11. Stress-strain curves of tension in perpendicular to the bed joint with different fracture energies (GF0=15; GF2=45 N/m; GF3=5 N/m).

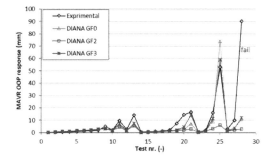

Figure 12. MAVR OOP response with different tensile fracture energies (GF0=15 N/m; GF2=45 N/m; GF3=5 N/m).

tested in the comparison study. As compression stress of a thin wall in the OOP tests is frequently below the compressive capacity (He 2016), the compressive behavior before reaching the strength is assumed to be important for the OOP response. The compressive strain ratio (n, or EPSCFA) is the only value, which could define the curve without changing other properties (DIANA FEA 2016). It defines the ratio of the real strain at compressive strength to the nominal strain at compressive strength based the linear Young's modulus. A model is tested with a larger strain ratio (n=8), based on some compressive curves in the EUCENTRE tests (Graziotti 2016, Graziotti et al. 2016b). Figure 13 shows the stress-strain curves for different values of n. Figure 14 presents the maximum responses of the tests. The results show the compressive strain ratio has a limited influence on the response of most of tests. However, a larger value seems more positive for the response, especially the last test (almost the double response). As indicated by some compressive experiments in the reports (Graziotti 2016, Graziotti et al. 2016b), somewhat larger values could be proposed for the EMM materials.

3.5 The refined models

Besides above-mentioned factors, some other parameters, e.g. tensile strength, mesh sensitivity, frame

stiffness are also studied. However, they will not be presented in this paper due to space limitation. The results indicate the current tensile strength and mesh size can be kept due to limited influence or experimental results. However, frame stiffness (presented by the horizontal OOP springs added at the top edge of the specimen) seems important for the OOP responses. Therefore, the original model is revised to get closer responses compared to the experiments. New model will consider the frame spring (stiffness: 1E12 N/m), lower tensile fracture energy (5 N/m), a larger compressive strain ratio (6) and a lower Rayleigh damping in a wide range of frequencies (Ray5). The results are presented in Figure 15, which reveals better response results of the refined model in almost all the tests compared with the original model. With the influence of the frame spring, the OOP response in the largest Ricker wavelet test (test 25) is getting even larger than the original one. However, no failure is found in the last test (Test 28). which is a major limitation of this model. Considering the influence of compressive strain ratio in Section 3.4, compressive behavior can be assumed to be the most important factor on the collapse. Therefore, a further refined model is proposed for the study, shown in Figure 16.

This model is based on the assumption that the wall is seriously damaged in the last few tests due to

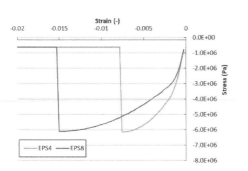

Figure 13. The stress-strain curves with different compressive strain ratios (EPS4: n=4, EPS8: n=8).

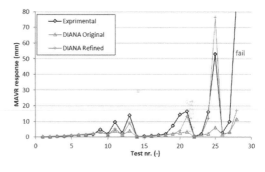

Figure 15. MAVR OOP response of the refined model and original model.

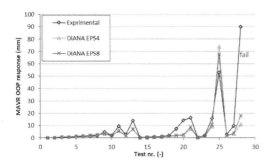

Figure 14. MAVR OOP response with different compressive strain ratios (EPS4: n=4, EPS8: n=8).

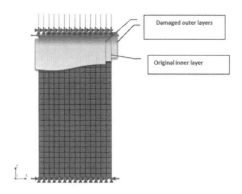

Figure 16. A further refined layered URM model.

921

the tensile fracture, especially on the outer layers of the wall. Meanwhile, the joint material in the cracks could be deteriorated due to the cyclic loading. Therefore, it is assumed that the wall can be separated into several layers, which have different compressive behaviors. The thickness of model wall is separated into three layers, two damaged outer layers and one original inner layer. The thickness of each outer layer is assumed to be 10 mm, similar as the joint thickness. The only difference between the inner layer and outer layer is the compressive behavior. The inner layer is assumed to have initial compressive behavior ($f_{c, inner}$= 6.20 MPa, n =6), shown in Figure 17.

The compressive behavior of the outer layer is proposed based on the OOP performances. Three models are tried in the tests: LayM ($f_{c,outer}$= 5.27 MPa, n =6), LayM2 ($f_{c,outer}$= 5.58 MPa, n =6) and LayM3 ($f_{c,outer}$= 5.58 MPa, n =8), shown in Figure 17. Final maximum OOP responses of different models are presented in Figure 18. It shows three trial models have similar OOP responses in most of the early stages (Test 1 - Test 24). Model of LayM has an early crash (divergence) at Test 25. Therefore, it could be assumed the compressive strength of the outer layer in model LayM is too low. However, model LayM2 (with a bit

higher compressive strength in the outer layer) results a stable OOP response without any collapse at the last step. Model LayM3 is then proposed to have a bit more ductile compressive behavior of the outer layers (n=8). The results of model LayM3 show the most promising responses. The collapse (based on the irreversible OOP displacement larger than 400 mm) happened in the last range of tests corresponding to the experiments.

4 CONCLUSIONS

From the comparison study of dynamic one-way OOP tests on a URM wall, some conclusions can be drawn as follows.

1. OOP behavior is influenced by many factors, not only the materials but also test setups. Therefore, it might be difficult to have a perfect model including every detail of the experiments.
2. Under the low excitation, DIANA EMM model reveals quite satisfactory results compared with the experimental results. However, the response of numerical model with initial proposed parameters is much lower than the experiments under the high excitation.
3. Rayleigh damping has significant influence on the OOP response of the URM wall. It seems that a low damping ratio (about 1or 2 percent) over a wide range of frequencies is reasonable to have more proper responses, avoiding resonance (at the beginning) if no damping is involved or too low response if a high damping ratio is applied.
4. It shows the tensile fracture energy is a crucial parameter for the OOP responses. This parameter was not determined in the experiments by testing or indicated in the references (Graziotti 2016, Graziotti et al. 2016b). With lower tensile fracture energy, the OOP responses of the model show a better fit to the experimental results. This has been adopted by the recent Dutch code NPR 9998:2018 and all the tensile fracture energies of masonry have been largely reduced (NEN 2018).
5. Compared with the tensile properties, compressive fracture behavior has limited influence on the OOP behavior as long as the compressive stresses in the wall are clearly below the compressive strength. The influence of the parameter of the compressive strain ratio is then studied as it is almost the only one parameter in EMM to adjust its behavior in compression. The results show it has a moderate influence on OOP responses in most excitations. However, it does improve the response for the last Test 28 (see Figure 14).
6. With proper frame spring, tensile fracture energy, compressive strain ratio and Rayleigh damping, the refined model can reach better results compared to the experiments. However, collapse is still difficult to be simulated.

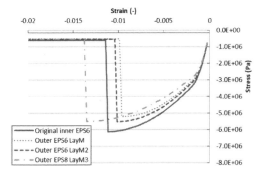

Figure 17. Compressive stress-strain curves of different proposed EMMs for the inner layer and outer layer ($f_{c, inner}$= 6.20 MPa, n =6; LayM: $f_{c,outer}$= 5.27 MPa, n =6; LayM2: $f_{c,outer}$= 5.58 MPa, n =6; LayM3: $f_{c,outer}$= 5.58 MPa, n =8).

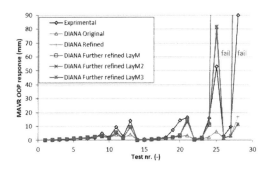

Figure 18. MAVR OOP response of the mid-height node with different model.

7. A further refined model (three layered model) is also tried in the simulation of the tests. The three-layered model is based the assumption that the compressive strength of the outer layers is reduced due to the cyclic crack opening and closing. Model LayM3 (having the outer layers with a lower compressive strength and more ductile compression) reaches collapse at the end of tests corresponding to the experiments.

ACKNOWLEDGEMENTS

Authors are grateful for providing the relevant experimental information and valuable discussion by Dr. F. Graziotti, Mr. U. Tomassetti & Dr. R. Pinho at EUCENTRE in Italy. The discussion and suggestions from Dr. G.-J. Schreppers & Dr. A. Garofano of DIANA FEA are also highly appreciated.

REFERENCES

DIANA FEA 2016. *Finite Element Analysis*, release 10.1, DIANA FEA BV, Delft, the Netherlands.

Graziotti F. 2015. *Experimental campaign on cavity walls systems representative of the Groningen building stock*, Technical report EUC318/2015U, EUCENTRE, Italy.

Graziotti F., Tomassetti U., Penna A. & Magenes G. 2016a. Out-of-plane shaking table tests on URM single leaf and cavity walls, *Engineering Structures*, 125: 455–470.

Graziotti F., Rossi A., Mandirola M., Penna A. & Magenes G. 2016b. Experimental characterisation of calcium-silicate brick masonry for seismic assessment, *Brick and Block Masonry – Trends, Innovations and Challenges*, Modena, da Porto & Valluzzi (Eds), Taylor & Francis Group, London: 1619–1627.

He H. 2016. *The OOP bending capacity of EMM walls*, report VIIA_QE_R376_N015, VIIA, Groningen, the Netherlands.

He H., Meijers S.J.H. & Vonk R.A. 2018. Promoting efficiency of NLTHA using an indirect soil-structure interaction modelling approach, *Preceding of 16th European Conference on Earthquake Engineering (16ECEE)*, June 18-21, 2018. Thessaloniki, Greece.

NEN 2015. *NPR 9998: 2015, Assessment of buildings in case of erection, reconstruction and disapproval – Basic rules for seismic actions: induced earthquakes*, Nederlands Normalisatie-instituut (NEN), Delft, the Netherlands (in Dutch).

NEN 2018. *NPR 9998: 2018, Assessment of structural safety of buildings in case of erection, reconstruction and disapproval - Induced earthquakes – Basis of design, actions and resistances*, Nederlands Normalisatie-instituut (NEN), Delft, the Netherlands (in Dutch).

Schreppers G.M.A., Garofano A. Messali F. & Rots J.G. 2016. *DIANA validation report for engineering masonry model*, TNO DIANA report 2016-DIANA-R1601, TU Delft Structural Mechanics CITG report CM-2016-17, Delft, the Netherlands.

Tomassetti U., Graziotti F., Penna A. & Magenes G. 2016. Out-of-plane shaking table test on URM cavity walls, *Brick and Block Masonry – Trends, Innovations and Challenges*, Modena, da Porto & Valluzzi (Eds), Taylor & Francis Group, London: 1939–1947.

Brick and Block Masonry - From Historical to Sustainable Masonry –
Kubica, Kwiecień & Bednarz (eds)
© 2020 Taylor & Francis Group, London, ISBN 978-0-367-56586-2

Macroelement model for nonlinear static pushover analysis of confined masonry walls with openings

E. Ortega-G, J.C. Jiménez-Pacheco & J.A. Quinde
Red Sísmica del Austro, Department of Civil Engineering, University of Cuenca, Cuenca, Azuay, Ecuador

H.A. García
Department of Architecture and Urbanism, University of Cuenca, Cuenca, Azuay, Ecuador

ABSTRACT: Seismic analysis of confined masonry buildings is complex due to the interaction between the masonry panel and the surrounding reinforced concrete frame. This complexity increases when the confined masonry wall has openings, due to the instability of the failure modes. In this context, the main objective of this work is based in the implementation of a Equivalent Strut Model for nonlinear static pushover analysis of confined masonry walls with openings.

1 INTRODUCTION

Masonry has been used worldwide as an easy solution due to the facilities of its construction, simple layering of bricks and mortar. Confined masonry (CM) is a particular type of masonry construction that consists in masonry panels strengthened with reinforced concrete (RC) confining elements.

CM construction started in Chile in 1930's after 1928 Talca earthquake, Magnitude 8.0, and subsequently this typology was introduced in Mexico, Peru, Argentina and other regions of Latin America for lower rise buildings. As a result, this type of construction has been considered as a typical solution for low- to mid-rise buildings especially in countries with extremely high seismic hazard (Marques & Lourenço 2019).

According to earthquakes experiences and several studies, CM buildings show adequate seismic behavior, however, the presence of openings has a negative influence on the seismic resistance of CM walls (Ghaisas et al. 2017); (Singhal & Rai 2018); (Tripathy & Singhal 2019). The damage is mainly related to the stresses at corners of openings, starting with the formation of diagonal cracks and consequently failure of the masonry walls (Singhal & Rai 2018).

Ecuador lies on the eastern rim of the seismically Pacific Ring of Fire. The map for seismic design included within the Ecuadorian Building Code (NEC 2015) shows that the most vulnerable areas falls within the very high seismic hazard zone (Zone VI), with design ground accelerations greater than 0.50g.

In addition, CM constructions represent a significant percentage of buildings (INEC 2016). Despite this, the Ecuadorian Building Code (NEC 2015) provides guidelines for masonry structures that does not include standards and technical manuals for CM structures, not even assess the influence of openings in this type of construction.

In this context, the seismic response of CM walls with and without openings is investigated. With the aim of understanding the influence of openings on the seismic behavior of CM walls, a parametric study based on pushover nonlinear static analysis was carried out. The macroelement used to represent CM walls, developed by Torrisi (2012), was implemented and validated in the program Ruaumoko-2D (Carr 2007). Ruaumoko is a research-oriented program, which performs nonlinear static and dynamic and analysis. This program has been widely used for the seismic analysis of reinforced concrete and steel buildings. Practically, there are no applications for masonry structures. Accordingly, it is worth noting that Jiménez-Pacheco et al. (2020) developed and validated an Equivalent Frame model for nonlinear static analysis of unreinforced masonry (URM) buildings using Ruaumoko. About CM-buildings, this work would constitute a first application of seismic analysis modeling in Ruaumoko (authors have not found applications with this program in the literature).

2 OVERVIEW OF CONFINED MASONRY

2.1 Types of masonry

Masonry has different configurations as a structural element. These configurations vary from unreinforced masonry (URM), reinforced masonry (RM), confined masonry (CM), and masonry-infilled RC

924

frame. URM has no steel reinforcement and no RC confinement; it is the typical configuration of masonry in countries with low or without seismic demand because this type is very vulnerable to the earthquake shaking. RM includes reinforcement by steel bars embedded in the mortar and/or bricks hollows. The horizontal reinforcement helps to improve the shear resistance and the vertical reinforcement helps to improve the flexural resistance. CM achieves the confinement of the masonry panel through a RC confining elements: columns and beams (Brzev 2007).

Masonry-infilled RC frame construction is a type of structure that in appearance may look similar to a CM construction; however, these systems show some differences. The CM construction is characterized because the masonry walls are constructed first, and RC confining elements are cast later. On the other hand, in masonry-infilled RC frame construction, the RC frame is constructed before the masonry infill wall. Therefore, in the masonry-infilled RC frame, the frame is moment-resisting while in the CM the wall shares a major portion of gravity load as compared to the confining columns.

2.2 *Modes of failure of CM walls*

Modes of failure of CM depend on different aspects such as wall geometry, boundary conditions, vertical load, aspect ratio and quality of materials. Based on past earthquakes experiences and experimental results, Alcocer (1996) deduced different modes of failure for CM walls: compressive failure, shear sliding, diagonal-tension and rocking (Figure 1).

Compressive failure is due to a large vertical load and/or low masonry compressive strength causing crushing of the masonry (Figure 1a). Shear-sliding failure occurs when the wall has low vertical load and the lateral force exceeds the sliding shear resistance (Figure 1b). Diagonal-tension failure occurs when the stress on the wall exceeds the masonry tensile strength and cracks propagate from de center of the masonry panel to the corners of the wall (Figure 1c). Flexural or rocking failure is associated with the yielding of the longitudinal reinforcement in the columns and the crushing of the compressed edge of the wall (Figure 1d).

According to Marques & Lourenço (2019), the four modes of failure are possible theoretically, however, diagonal cracking failure has been mainly observed due to seismic loads, while rocking failure is in general not observed due to the confining effects. In regard to RC confining elements modes of failure different authors such as Yekrangnia et al. (2017), Tomaževič & Klemenc (1997) and Mehrabi et al. (1994) have identified: concrete crushing in the column (Figure 2a), tie´s shear failure (Figure 2b), and flexural/shear failure (Figure 2c) mainly due to an inadequate reinforcement and a high-quality masonry.

2.3 *Modelling strategies*

Since CM construction has emerged as a building technology, many modeling techniques has been developed: simple models, macro models and micro models (Lang et al. 2014).

Simple models represents the behavior of the wall only at the initial stage, within this group are the monolithic wall model and the wide-column method. Macro models represent a solution between simplicity and precision, one of the simplest macro model is the Equivalent Strut Model (ESM), in which a compressive equivalent strut represents the masonry wall, and the RC confining elements are modeled with column-beam elements. In micro models, brick units and mortar can be modeled separately, in this group numerical analysis finite

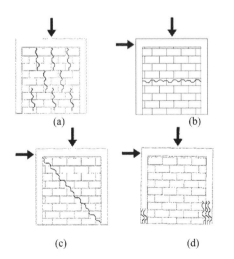

Figure 1. Modes of failure of CM subjected to lateral loads: a) compressive failure, b) shear sliding, d) diagonal-tension and d) flexural or rocking.

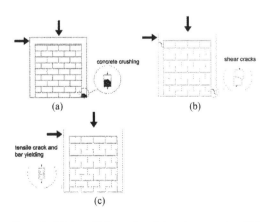

Figure 2. Modes of failure of confining elements when CM is subjected to lateral loads: a) concrete crushing, b) ties's shear cracks, and c) flexural [Modified from Yekrangnia et al. 2017].

elements methods (FEM) can be employed (Camp-bell 2012, Zucchini & Lourenço 2009, Paulo et al. 2011).

3 EQUIVALENT STRUT MODEL (ESM)

The ESM is considered a promising modelling technique for CM. The ESM was initially proposed by Polyakov (1963); later, it was studied and developed for many researchers in order to achieve a better understanding of the seismic behavior of CM walls (Figure 3).

Since one diagonal model (Figure 3a) cannot describe properly the internal forces induced in the confining elements, multi-strut models have been developed (Crisafulli 1997) to obtain realistic values of the bending moments and shear forces in the frame (Figsure 3b,c).

3.1 *Application of equivalent strut model to CM walls*

In principle, the ESM was developed to model masonry-infilled RC frame walls due to the separation observed between the masonry and the surrounding RC frame. As already mentioned, CM differs from infilled RC frame structures; however, Torrisi et al. (2012) state that the response of these structural systems subjected to in-plane seismic loading is similar to a large extent. In both cases, the behavior is mainly controlled by the nonlinear response of the masonry wall panels and the surrounding RC confining elements.

Based on FEM analysis, Torrisi et al. (2012) demonstrated that the structural response of CM un-der lateral loading is initially controlled by the ma-sonry wall, behaving as a monolithic element (in a quasi-elastic way). Then, as the lateral force increases, the masonry-frame interface are not able to resist the tensile stresses and cracking occurs in the interface zones (Figure 4). This effect causing that the masonry panel partially separates from the surrounding RC confining elements, as a result of this, a diagonal compression

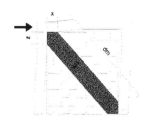

Figure 4. Deformed shape of a single solid panel of CM wall after stress redistribution due to cracking and/or interface separation.

strut is forming across the masonry panel, typically resulting in shear friction failure (Lang et al. 2014).

In some cases, such separation occurs by cracking and opening of the panel-frame interfaces, whereas in other cases, the strength of the interfaces is high and cracking occurs in the masonry adjacent to the interfaces. Considering the two cases as equivalent, it is proper to generalize that structural separation occurs between the masonry panel and the surrounding frame in CM structures. As a result, the masonry is mainly subjected to compressive stresses along the loaded diagonal (Torrisi et al. 2012).

4 MACROELEMENT MODEL

4.1 *General description*

The macroelement implemented in Ruaumoko, based on the ESM, corresponds to that developed by Torrisi (2012). This macroelement includes two components, namely: (i) a refined beam-column component (Figure 7) able to capture the bending moment, shear, and axial forces that are developed in the RC confining elements and (ii) a panel component formed by a multi-strut system to represent the masonry panel assembled as depicted in Figure 5.

Ruaumoko-2D (Carr, 2007b) has in its catalog a multi-spring element, which can function as flexural, shear or axial spring by means of a central

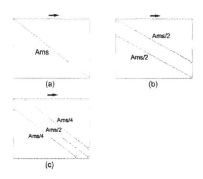

Figure 3. Struts models considered to CM walls: a) one strut model, b) and c) multi-strut models [Modified from Crisafulli, 1997].

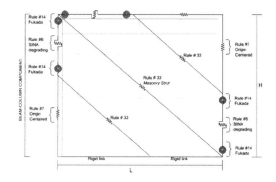

Figure 5. Diagram of the assembly of beam-column and multi-strut components to represent CM wall implemented in Ruaumoko.

package of rotational, transversal and longitudinal springs, respectively (Figure 6). This element is formed by four nodes, comprising two rigid links at the ends and the deformable central part delimited by nodes 3 and 4. With the aiming to simulate a rigid-body behavior a very high stiffness value is assigned to the corresponding spring.

4.2 Description of the beam-column component

The beam-column component, as referred in the present work, was proposed by Torrisi (2012). It was implemented in Ruaumoko as depicted in Figure 7, which consists of an assemblage of 4 non-linear multi-springs integrated into 2 members, M_1 and M_2. In Figure 7, the most highlighted springs are those which their inelastic behavior is controlled, while the less visible springs has their deformability inhibited (by assigning very high stiffness values) to simulate quasi-rigid behavior.

The member M1 is composed by three inelastic springs: 2 flexural and 1 central shear springs, 4 nodes: 2 internal and 2 external and a total of 12 DoFs. The springs are not considered coupled and their stiffnesses are computed directly in Ruaumoko. Equation 1 gives the stiffness matrix of any spring of the beam-column component, Ksi.

$$K_{Si} = \begin{bmatrix} KX_i & 0 & 0 & -KX_i & 0 & 0 \\ 0 & KY_i & 0 & 0 & -KY_i & 0 \\ 0 & 0 & GJ_i & 0 & 0 & -GJ_i \\ -KX_i & 0 & 0 & KX_i & 0 & 0 \\ 0 & -KY_i & 0 & 0 & KY_i & 0 \\ 0 & 0 & -GJ_i & 0 & 0 & GJ_i \end{bmatrix}$$
(1)

whereKX$_i$,KY$_i$ and GJ$_i$ correspond to axial, shear, and flexural stiffness for the i-th spring respectively, i= 2 to 4 refers to the springs S2, S3, and S4 of the member M_1. Equations 2-4 represent the springs stiffness where A= area of cross section; E=elastic modulus; L=length of the member; V_{cr} and γ_{cr} are the cracking shear force and the drift respectively; and M_{cr} and φ_{cr} correspond to the cracking moment and cracking curvature.

$$KX_i = \frac{EA}{L}$$
(2)

$$KY_i = \frac{V_{cr}}{\gamma_{cr}}$$
(3)

$$GJ_i = \frac{Mcr}{\varphi_{cr}}$$
(4)

The stiffness matrix of M_1 is obtained by assembling the stiffness matrix of each spring resulting in a 12×12 matrix. However, for the adequate behavior of the beam-column component, only the external degrees of freedom must be left in relation to the internal ones by means of static condensation. This complete procedure is explained by Torrisi (2012). The final stiffness matrix of member 1, KM$_1$, is a 6×6 matrix

The member M_2 consists of an inelastic axial spring, 2 nodes and 6 Dofs, it does not need static condensation and its stiffness matrix, KM$_2$, is similar to that given by Equation 1. After the definition of members M_1 and M_2, the global stiffness matrix of the beam-column component, K, is assembled as depicted in Figure 8.

4.3 Description of the panel component

The panel component considers six diagonal struts (three in each diagonal direction), each diagonal is represented by an inelastic axial spring which stress-strain relationship (SSR) correspond to the backbone of the hysteresis rule proposed by Crisafulli (1997).

The stiffness matrix of diagonals is similar to the 6×6 matrix of Equation 1, however, the axial stiffness of the diagonals, KX, is given by the Equation 5 where Em = masonry elastic modulus; As = strut

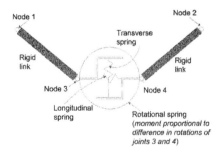

Figure 6. Multi-spring element of Ruaumoko.

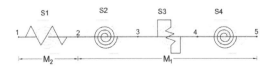

Figure 7. Diagram of the arrangement of springs (S) of the beam-column component: S1) axial, S2) flexural, S3) shear and S4) flexural.

Figure 8. Graphic scheme of the assembly of stiffness sub-matrices KM to obtain the global matrix of the beam-column component, K.

area; and L_s = strut length. To transform from local into global axes Equations 6-7 are used.

$$KX = \frac{E_m A_s}{L_s} \tag{5}$$

$$T = \begin{bmatrix} \cos\theta & \sin\theta & 0 & 0 & 0 & 0 \\ -\sin\theta & \cos\theta & 0 & 0 & 0 & 0 \\ 0 & 0 & 1 & 0 & 0 & 0 \\ 0 & 0 & 0 & \cos\theta & \sin\theta & 0 \\ 0 & 0 & 0 & -\sin\theta & \cos\theta & 0 \\ 0 & 0 & 0 & 0 & 0 & 1 \end{bmatrix} \tag{6}$$

$$K_S = T^T K_{S-L} T \tag{7}$$

where K_{S-L} = diagonal stiffness matrix in local coordinates; K_S = diagonal stiffness matrix in global coordinates; and T = coordinate transformation matrix. Once the stiffness matrix of each diagonal is obtained, they are assembled into the panel matrix and the same into the structure matrix (Torrisi 2012).

4.4 Force-displacement relationships (FDRs)

It is important to note that although Ruaumoko has hysteresis laws for the nonlinear dynamic analysis, for pushover analysis the required parameters are derived from their respective backbone curves, as shown in Figure 9. In Ruaumoko, tri-lineal FDRs employed (Figure 9b) represent the backbone curves corresponding to the rule 14 named Fukada for flexural springs, and the rule 8 named SINA-degrading for shear springs. Bi-lineal FDRs (Figure 9c) for the axial springs correspond to the rule 7 named Origin Centered.

The flexural FDR (moment vs. curvature diagram) was obtained from theory of reinforced concrete developed by (Park & Paulay, 1980). The shear FDR (shear force vs. shear strain) was calculated by the model proposed by Sezen (2008), which uses the Modified Compression Field Theory (MCFT) and experimental formulation.

4.5 Stress-Strain Relationship (SSR) for masonry

The behavior of the masonry struts is defined by the backbone curve of the hysteretic rule proposed by Crisafulli (1997), and implemented in Ruaumoko as hysteresis rule 33 (Figure 9a). The required parameters for the implementation are detailed in Ruaumoko manuals (Carr, 2007).

Many parameters have been taken directly from Torrisis's (2012) proposal, such as the total area of struts, A_{tot}, as percent of the diagonal length, dm, of the panel ($0.15dm$ to $0.25dm$) multiplied by the thickness of the wall. In addition, the percent of area assigned to the central strut is specified usually as $0.35A_{tot}$ to $0.70A_{tot}$.

4.6 Stiffness reduction factor on ESM for openings

In order to investigate the effect of openings on CM walls, a stiffness reduction factor, λ, was implemented. This factor is expressed as a function of the opening percentage (opening area/masonry wall area) of the wall, this factor was developed by (Asteris et al. 2011) using FEM analysis leading the following relationship:

$$\lambda = 1 - aw^{0.54} + aw^{1.14} \tag{8}$$

where αw = opening percentage of the masonry wall. This stiffness reduction factor was employed due to it can be easily applied on the equivalent width of the strut, and it can be used to modify the equations of the Crisafulli model (Asteris et al. 2011).

5 FINITE ELEMENT ANALYSIS FOR CM WALLS

With the aiming to assess the influence of openings on the seismic behavior of CM walls, a FE study (using OpenSees software) is developed in this section.

5.1 General description of the model

In the proposed FE model, masonry was modelled at the macroscopic level; brick and mortar are taken into account in a single basic piece of masonry.

Masonry panel and RC confining elements were modeled as 3D solid elements of type Bbar Brick.

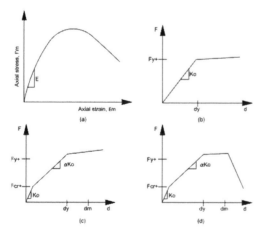

Figure 9. Backbone curves for: a) diagonal, b) flexural and shear, and c) axial.

This is a command used to construct an eight-node mixed volume/pressure brick element object, which uses a trilinear isoparametric formulation. The element has 3 degrees of freedom at each node.

To capture the nonlinear and plastic behavior, the Drucker Prager material was chosen. This is used to construct a multi-dimensional material object that has the Drucker Prager yield criterium. The Drucker Prager parameters, ρ and σy, are related to the Mohr-Coulomb friction angle, ω, and cohesion, c. Cheen and Saleeb (1994), by evaluating the yield surfaces in a deviatoric plane, determined the following expressions.

$$\rho = \frac{2\sqrt{2}\sin\phi}{\sqrt{3}(3-\sin\phi)} \quad (9)$$

$$\sigma y = \frac{6c\cos\phi}{\sqrt{2}(3-\sin\phi)} \quad (10)$$

Other parameters are the Bulk modulus, K, and shear modulus, G, which can be obtained with the following expressions, where E= Elastic modulus and v= Poisson coefficient:

$$K = \frac{E}{3(1-2v)} \quad (11)$$

$$G = \frac{E}{2(1+v)} \quad (12)$$

Table 1. Summary of test specimens validated in Ruaumoko.

Specimen label	Literature reference	Specimen* validated	Wall type
S1	Mehrabi (1994)	S8	solid
S2	Alcocer et. al (2001)	M4	solid
S3	Yañez et. al (2001)	HCBM**	solid
S3W1			window
S3W2			window
S3W3			window
S4	Campbell (2012)	MV1	solid

* Specimen validated correspond to literature reference walls.
** Hollow clay brick masonry

Table 2. Assumed mechanical properties of masonry and concrete for specimens.

| ReSpecimen | Material | Elastic | | Compressive |
		E(MPa)	G (MPa)	fc (MPa)
S1	Masonry	3893	1557	12
	Concrete	21000	8400	21
S2	Masonry	4372.5	480.98	5.83
	Concrete	22589	9036	23.1
S3	Masonry	4849	528	6.89
	Concrete	22977	9190	23.9
S4	Masonry	3750	1500	5
	Concrete	24870	9948	28

6 VALIDATION: MACROELEMENT AND FEM MODELLING

6.1 Macroelement implemented in Ruaumoko

The macroelement, which incorporates the beam-column and panel components previously described, has been implemented in Ruaumoko for the analysis of CM walls. Its effectiveness was validated against experimental results available in the literature. Specimens tested by Mehrabi et al. (1994), Perez Gavilan et al. (2012), Yáñez et al. (2004), and Campbell (2012) were considered for the validation. Table 1 presents a summary of the specimens considered.

The developed models were subjected to a monotonically increasing lateral load, which was applied to the top of the confining column. The lateral loading was simulated to the one applied in the experimental programs. The mechanical properties of CM walls are listed in Table 2.

Comparison of the pushover curves obtained from the aforementioned benchmark with the corresponding backbone curves experimental is shown in Figure 10.

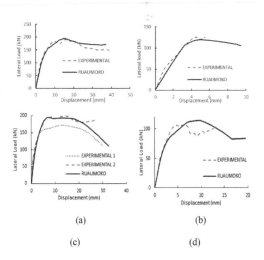

Figure 10. Comparison of the pushover curves obtained with the proposed macro-model with the experimental backbone curves: a) S1, b) S2, c) S3 and d) S4.

As can be seen, exists a good agreement between the results obtained and the experimental ones. Particularly, the elastic stiffness and the strength capacity are well predicted. In more precise terms, Table 3 summarizes the results of the strength capacities.

6.2 FEM model implemented in OpenSees

The pushover curves obtained from the FE OpenSees model were examined against the experimental results for each S3 series walls, see Table 1. These results correspond to CM walls with window opening and are shown in Figure 11.

FEM model results show a good approximation with the experimental results, in terms of strength capacity, elastic stiffness and post- peak decay behavior. However, pushover curves show that after reaching the ultimate displacement, the FEM obtained curves does not complete the decay according to the experimental results.

In regard of Ruaumoko results, these show a slightly difference about the elastic stiffness. This difference could be associated to the configuration and location of the openings within the wall, since in the walls S3 the openings are not centralized.

Table 3. Numerical comparison between experimental and obtained results for the maximum lateral strength of solid CM walls.

Specimen	Experimental Vmax (MPa)	Ruaumoko Vmax (MPa)	Error %
S1	190	189	53
S2	160	145	9.38
S3	172-199	194.2	12.9-2.41
S4	108	114.6	6.11

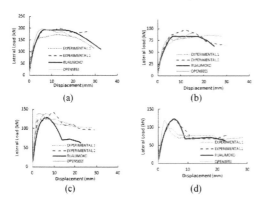

(a) (b)

(c) (d)

Figure 11. Comparison of the proposed macro-model with experimental backbone curves: a) S3, b) S3W1, c) S3W2 and d) S3W3.

Table 4. Comparison between experimental and numerical results for the maximum load of perforated CM walls.

Specimen	Experimental Vmax (MPa)	OpenSees Vmax (MPa)	Error %
S3	172-199	196.38	14.17-1.32
S3W1	85-96	89.54	5.34-6.73
S3W2	135-141	138.59	2.66-1.71
S3W3	100-126	119.67	19.67-5.02

Results from the FE analysis show that the model is capable to capture the elastic stiffness, maximum load and post peak degrading for CM walls. The validation study confirmed that the developed FE model could predict the shear strength of CM wall.

7 PARAMETRIC STUDY: WINDOW OPENING

Since macroelement captures the behavior of CM walls in a satisfactory way, in terms of shear maximum load and elastic stiffness, the macroelement was used in this section to evaluate the influence of openings in CM walls. For the parametric study, only opening percentages of 10%, 20%, 30% and 40% were considered. Tested walls correspond to specimen S4, which were added four opening wall models (window opening) as shown in Figure 12.

Figure 13 shows the pushover curves obtained from Ruaumoko. The plots show that openings significantly influence the in-plane shear capacity of CM walls. In addition, it is evident that the presence of openings reduce the lateral stiffness. It is also observed that as the size of the opening increases, both the elastic stiffness and the maximum load are reduced.

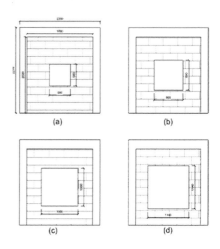

(a) (b)

(c) (d)

Figure 12. Specimens used for parametric analysis a) S4W1 with 10% opening, b) S4W2 with 20% opening, c) S4W3 with 30% opening and d) S4W4 with 40% opening.

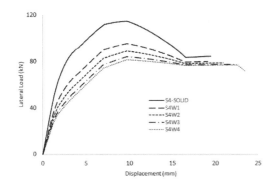

Figure 13. Comparison of pushover curves of solid wall against walls with openings obtained from the proposed macroelement implemented in Ruaumoko.

8 CONCLUSIONS AND FURTHER RESEARCH

A macroelement model for pushover non-linear static analysis of confined masonry walls has been presented. The macroelement was built by assembling the beam-column and masonry panel components.

Four solid and four perforated CM walls taken from literature were analyzed to validate the macroelement in Ruamoko, finding a good agreement with experimental results. The experimental works were selected considering the availability of the mechanical properties of the materials and the adaptability of the test conditions. In addition, a parametric analysis was carried out to study the influence of the openings on the CM walls behavior with four numerical specimens having concentrically located 10%, 20%, 30% and 40% of the opening.

Based on the obtained results in terms of pushover curves, the macroelement implemented is able to represent in a satisfactory way the shear strength capacity and the elastic stiffness of CM walls with and without openings. The average error in strength capacity for solid walls and walls with openings was of 5.63% and 7.10 %, respectively. However, capturing the strength degradation requires a more detailed study in terms of formulation and modeling techniques referred for both macroelement model and FE model.

According to the results of the parametric analysis, presence of openings has an important influence on the shear strength capacity and elastic stiffness of CM walls. It was observed that as the opening size increases the strength capacity and elastic stiffness decreases.

The presented work is considered as a basis for the seismic evaluation of CM buildings. Future lines of work constitute, in the first instance, the analysis of the influence of the openings in the walls. Then, incorporating a floor-diaphragm macroelement, the model will be extended for nonlinear static analysis of buildings.

REFERENCES

Alcocer, S. M. 1996. Comportamiento sísmico de estructuras de mampostería: Una revisión. In *XI Congreso Nacional de Ingeniería Sísmica Veracruz, México* (164–191).

Asteris, P. G., Chrysostomou, C. Z., Giannopoulos, I. P., Frames, I., & Response, S. 2011. In Masonry infilled reinforced concrete frames with openings. *Computational Methods in Structural Dynamics and Earthquake Engineering Corfu, Greece.* (15).

Brzev, S. 2007. Earthquake-resistant confined masonry construction. In *National Information Center of Earthquake Engineering NICEE Kanpur, India* (99).

Campbell, J. A. 2012. *Numerical Model for Nonlinear Analysis of Masonry Walls*. RWTH University.

Carr, A. J. 2007. Ruaumoko Manual Author: Appendices, 5.

Crisafulli, F. J. 1997. *Seismic Behaviour of Reinforced Concrete Structures with Masonry Infills*. University of Canterbury.

Ghaisas, K. V., Basu, D., Brzev, S., & Pérez Gavilán, J. J. (2017. Strut-and-Tie Model for seismic design of confined masonry buildings. *Construction and Building Materials*, 147: 677–700.

INEC. (2016). Encuesta de Edificaciones 2016 (Permisos de Construcción). *Instituto Nacional de Estadísticas y Censos*.

Jiménez-Pacheco, J., González-Drigo, R., Pujades Beneit, L. G., Barbat, A. H., & Calderón-Brito, J. 2020. Traditional High-rise Unreinforced Masonry Buildings: Modeling and Influence of Floor System Stiffening on Their Overall Seismic Response. *International Journal of Architectural Heritage* 00(00): 1–38.

Lang, a F., Crisafulli, F. J., & Torrisi, G. S. 2014. Overview and Assessment of Analysis Techniques for Confined Masonry Buildings. *10th National Conference in Earthquake Engineering*.

Marques, R., & Lourenço, P. B. 2019. Structural behaviour and design rules of confined masonry walls: Review and proposals. *Construction and Building Materials* 217: 137–155.

Mehrabi, A. B., Shing, P. B., Schuller, M. P., & Noland, J. L. 1994. Performance of masonry-infilled R/C frames under in-plane lateral loads. *Structural Engineering and Structural Mechanics Research Series* 259.

NEC-SE-RE. 2015. *Riesgo sísmico, evaluación, rehabilitación de estructuras*.

Paulo, M. F., Neto, M. F., Dias, J. E., & Lourenço, P. B. 2011. Behavior of masonry infill panels in RC frames subjected to in plane and out of plane loads. In *7th International Conference amcm2011 Kraków, Poland* (20).

Perez Gavilan, J. J., FLores, L. E., & Alcocer, S. M. 2012. An experimental study of confined masonry walls with varying aspect ratio. *Earthquake Spectra*.

Sezen, H. 2008. Shear deformation model for reinforced concrete columns 28(1): 39–52.

Singhal, V., & Rai, D. C. 2018. Behavior of confined masonry walls with openings under in-plane and out-of-plane loads. *Earthquake Spectra* 34(2): 817–841.

Tomaževič, M., & Klemenc, I. 1997. Seismic Behavior of Confined Masonry Walls. *Earthquake Engineering and Structural Dynamics* 26: 1059–1071.

Tripathy, D., & Singhal, V. 2019. Estimation of in-plane shear capacity of confined masonry walls with and without openings using strut-and-tie analysis. *Engineering Structures* 188(January): 290–304.

Yánez, F., Astroza, M., Holmberg, A., & Ogaz, O. 2004. Behavior of confined masonry shear walls with large openings. In *13th World Conference on Earthquake Engineering*. Vancouver,B.C., Canada.

Yekrangnia, M., Bakhshi, A., & Ghannad, M. A. 2017. Force-displacement model for solid confined masonry walls with shear-dominated failure mode. *International Association for Earthquake Engineering* 056: 26.

Zucchini, A., & Lourenço, P. B. 2009. A micro-mechanical homogenisation model for masonry: Application to shear walls. *International Journal of Solids and Structures* 46(3–4): 871–886.

Partitions & infill

Brick and Block Masonry - From Historical to Sustainable Masonry –
Kubica, Kwiecień & Bednarz (eds)
© 2020 Taylor & Francis Group, London, ISBN 978-0-367-56586-2

Experimental assessment of an innovative isolation technique for the seismic downgrade of existing masonry infills

V. Bolis, A. Paderno & M. Preti

DICATAM - Department of Civil, Architectural, Environmental, Engineering and Mathematics, University of Brescia, Italy

ABSTRACT: The seismic vulnerability of infilled reinforced concrete (RC) frames built in the Mediterranean earthquake prone regions before the 70's has been assessed by many post-earthquake surveys and several experimental studies. Beyond the lack of seismic-resistant detailing in the frame elements, a relevant source of vulnerability for these structural typology is represented by the in-plane interaction between the frame and the infills. Infills are typically made of masonry, selected for its good thermal and acoustic insulation performance and characterized by high stiffness and strength, but coupled with a brittle post-peak behavior. During an earthquake, the different stiffness and deformation capacity characterizing the infill and the frame can lead to severe damage, including widespread cracking and crushing in the former, and brittle shear failures in the latter. When dealing with the seismic retrofit of an existing infilled RC building, this local interaction cannot be ignored, as it could jeopardize the efficiency of the seismic retrofit intervention by triggering unexpected early collapses in the existing frame. In order to mitigate this issue, in the present paper an infill isolation technique is presented, conceived to reduce the infill-frame in-plane interaction (downgrade). The downgrade is obtained by isolating the infill from the surrounding frame with a cut on a portion of the infill perimeter. A specific innovative wall-to-frame beam connection is implemented, which promotes the masonry arching mechanism against out-of-plane actions, while ensuring in-plane relative sliding. In the paper, the experimental cyclic in-plane and out-of-plane response of a real scale specimen downgraded with the proposed technique are presented and some issues about the conceptual design of the intervention and its invasiveness are discussed.

1 INTRODUCTION

The reinforced concrete (RC) buildings built between 60's and 80's revealed themseves to be particularly vulnerable against seismic actions, suffering severe damages and collapses during the last decades earthquakes (Augenti and Parisi. 2010; EERI, 2000, 2009; Perrone et al. 2018). Such a vulnerability is mainly due to the poor reinforcement in the frame elements against horizontal loads, as these structures were designed for gravity loads only. In addition, their vulnerability has been in many cases amplified by the presence of stiff and strong masonry infills within the frame bays (Basha and Kaushik. 2016; Cavaleri and Di Trapani. 2015; Di Trapani and Malavisi. 2018; Mehrabi Armin B. et al. 1996; Preti and Bolis. 2017a). Although for low imposed displacements these elements could contribute to the lateral strength of the structure, their interaction with the frame could also produce local high stress concentrations in the frame elements. As a result, early brittle collapses can occur in the structure (e.g. columns shear failures, soft story collapses, short column mechanism, etc.), thus limiting the overall building deformation capacity.

In order to reduce the vulnerability of existing buildings, several strengthening interventions were designed and applied with the aim of increasing their strength against seismic actions (Di Ludovico et al. 2008; Gioiella et al. 2017; Metelli et al. 2017; Pampanin. 2012; Passoni. 2016; Riva et al. 2010). These interventions typically consist in strengthening the existing structural elements (by means of FRP, RC or FRC jacketing, etc.) or in introducing additional seismic-resistant elements connected to the existing frame structure (e.g. RC shear walls, steel bracings, external towers, etc.). Beyond a superior performance in terms of overall seismic strength, such interventions are typically designed to provide also a ductile response to the building.

However, as demonstrated in (Preti et al. 2019), even in presence of a strengthening intervention with external RC shear walls, the deformation capacity of an existing building is still limited by the effects of the infill-frame interaction, which causes collapses in the existing structure before reaching the design capacity of the additional seismic-resistant elements. As a consequence, the issues connected to the infill-frame interaction are not mitigated by the sole structural strengthening, but an additional downgrade intervention on the infill walls is helpful in order to exploit the full deformation capacity of the existing frame and optimize the efficiency of the strengthening.

In order to limit the seismic interaction between the infill and the surrounding frame, different construction techniques have been proposed in literature in the last years. According to the proposed solutions, such interaction can be reduced by adopting two alternative approaches: (i) the reduction of the in-plane strength and stiffness of the infill or (ii) the in-plane isolation of the wall from the surrounding frame. The former solution can be achieved by adopting deformable materials for the masonry (Markulak et al. 2013; Pallarés et al. 2017) or by subdividing the infill panel into several sub-panels by means of the introduction of sliding (Mohammadi et al. 2011; Morandi et al. 2018; Preti and Bolis. 2017b; Preti et al. 2012; Tasligedik and Pampanin. 2017) or deformable joints (Verlato et al. 2016). Instead, the second approach consists in providing a structural gap between the frame and the masonry along the infill perimeter, capable of accommodating the frame in-plane deformation during the seismic response, thus limiting the interference with the masonry (Butenweg and Marinković. 2018; Canbay et al. 2018; Tsantilis and Triantafillou. 2018). In both cases, the adopted solutions are meant to ensure also the out-of-plane stability to the wall, by avoiding overturning or brittle failures.

It is worth noting that all the cited techniques are proposed in application to new constructions, when the masonry infills are designed and built by implementing the proposed solutions. When dealing with existing buildings, a possible solution should be that of demolishing all the infills and replace them with one of the technique above recalled. However, this procedure would result in high costs (mainly associated to the restoring of all the technical systems and fixtures and debris disposal) and long building downtime.

A first downgrade technique for the application to existing infills was proposed by (Preti and Bolis. 2017b). According to this solution, the infill-frame interaction is reduced by performing full-height vertical cuts in the masonry where vertical sliding joints (namely wooden boards) are inserted, connected to the frame beams. As a result, the infill wall is subdivided into sub-panels, which activate a rocking mechanism around their base corners. The experimental results reported in (Preti and Bolis. 2017b) showed: (i) an in-plane strength lower than 10kN, thus allowing to almost nullify the interaction with the surrounding frame, and (ii) an out-of-plane resistance in excess to the design demand, ensured by the vertical joints acting as strengthening elements.

In order to limit such invasiveness associated to the infill cutting into subpanels, in the present paper, an alternative downgrade intervention for existing masonry infills is proposed. It consists in isolating the infill from the surrounding frame by means of a few centimetres thick cut, running along a selected portion of the infill perimeter. As a result, the infill-frame contact is limited and the interaction stresses are reduced, thus protecting both the infill and the frame from damage and failure. For the out-of-plane infill stability, a partial contact is restored between the wall and the top beam with a specific detail, in order to activate a compressed arch mechanism in the masonry, spanning between the frame beams, while allowing the frame-to-infill in-plane free sliding. The conceptual design of the technique, which in the following is named as "hybrid infill isolation", is better detailed in par. 2, together with some considerations about the operational procedure for the achievement of the downgrade and its invasiveness on the building functionality.

The performance of the proposed downgrade intervention against seismic loads is validated by means of an experimental campaign on a real scale prototype infill, which was tested both in- and out-of-plane. The in-plane tests showed an infill response characterized by reduced infill-frame interaction (up to 2.2%) and limited damage. In particular, the damage was local and negligible up to 1% drift, which represents a reasonable target deformation demand for retrofitted existing buildings. Moreover, the test demonstrated the out-of-plane stability of the downgraded infill, even in presence of the damage suffered during the in-plane test phases.

2 CONCEPTUAL DESIGN OF THE INFILL "HYBRID ISOLATION" TECHNIQUE

The downgrade technique proposed in the present paper is meant to achieve a partial isolation of the infill from the surrounding frame by means of a cut in the masonry along a selected part of the infill perimeter. In detail, the cut runs along the infill-top beam interface and along part of the lateral interfaces with the columns (Figure 1a), starting from a height of about 1m from the column base ("contact height"). The thickness of the lateral gap is designed to accommodate the maximum inter-story drift expected for the frame, without interfering with the infill top corner.

It is worth noting that the contact maintained at the lower part of the column could induce some damage in the masonry, due to the in-plane interaction with the frame, however it is supposed to be limited, also in relation to the particular deformed shape of the column, acting as a double fixed element (see Figure 1a). The adopted layout of the cut is chosen in order to find a trade-off between the isolation of the infill and the mitigation of the interference with possible technological systems, which, in existing buildings, are typically located in the lower part of the infill. In this choice, maximum care need to be given to the avoidance of the local shear mechanism like short (or captive) column mechanism, which depends on the ratio of columns resistance to infill local trust.

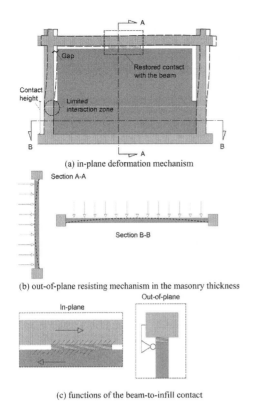

(a) in-plane deformation mechanism

Section A-A

Section B-B

(b) out-of-plane resisting mechanism in the masonry thickness

(c) functions of the beam-to-infill contact

Figure 1. Conceptual scheme of the "hybrid infill isolation" downgrade intervention.

The "hybrid isolation" is obtained by partially restoring the infill-to-beam contact, which is conceived to achieve simultaneously two functions (Figure 1c): (i) provide a restraint against out-of-plane actions and (ii) allow the in-plane relative sliding between the frame beam and the masonry infill, with a limited shear stress transfer. For this reason, a low friction sliding interface is created in at the contact, which is lumped at the central portion of the infill. Such configuration allows to accommodate the drop of the beam in the frame in-plane lateral deformation (Figure 1a) and, consequently, mitigate the contact stress transferred to the infill.

As for the out-of-plane infill stability, the restored contact at the infill-to-beam interface promotes the activation of a vertical arching mechanism within the masonry (Figure 1b). In addition, the conservation of a partial contact of the infill with the columns contributes improving the stability of the infill, as it allows to create an additional horizontal arching mechanism spanning between the two frame columns.

From an operational point of view, the "hybrid isolation" technique allows to limit the invasiveness of the intervention on the existing infills by limiting the number of cuts and by localizing them along the infill perimeter, where it is typically lower the

probability of interfering with the presence of technological facilities. Moreover, the proposed intervention can be operated mainly from outside the building, limiting the required actions from the inside to minor operations. Such retrofit is suitable both for single or double leaf masonry infills.

3 SPECIMEN AND TEST SETUP

The tested specimen (Figure 2) was built within a steel frame already designed and adopted for previous experimental campaigns (Preti and Bolis, 2017b). The frame was made of HEA240 steel (S235) profiles connected by means of comb hinges placed at the columns ends in order to generate a zero-strength in-plane sway mechanism.

The specimen was meant to reproduce the internal leaf of a typical double layer infill adopted in the Italian buildings built between the 60's and the 80's. Therefore, it was built with 8cm thick hollow blocks (65%void ratio), arranged with holes in horizontal direction and M5 cementitious mortar. To be representative of the real configuration, one side of the wall was covered with a 1.5 cm thick plaster, while the opposite one was kept un-plastered. The masonry was built by maintaining a 5 cm gap from the upper frame beam and from the lateral columns, except for the first meter from the base, to reproduce the configuration achievable after the cut. The restoring of the contact between the infill and the upper beam was detailed as reported in Figure 3. A steel IPE140 profile was embedded over the infill by means of mortar, spanning on the entire infill length. Thereafter, a shrinkage-compensated cementitious mortar was casted to fill the gap in the central portion, with a 1m length. In order to limit cohesion and allow a low friction sliding along the restored contact, two polyethylene sheets were placed between the IPE140 and the mortar above. Moreover, a certain vertical deformability was introduced by means of a 1mm thick polyurethane sheet below the

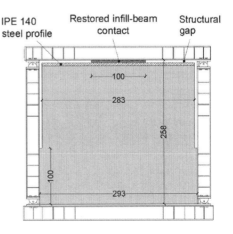

Figure 2. Geometry of the tested specimen.

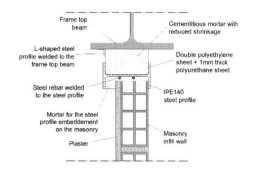

Figure 3. Detail of the restored contact between the infill and the frame top beam.

polyethylene sheets, in order to accommodate small downlifts of the beam during the sway mechanism.

After restoring the contact, the transversal restraint to the contact system was provided, for experimental easiness, by an L-shaped steel profile welded to the upper frame beam instead of a specific shear connection suitable for real application.

It is worth noting that the performed intervention on the specimen was carried out with available on-the-shelves materials in order to satisfy the requirements presented in the conceptual design (par. 2). Of course, in a real application the solution could be improved and refined in terms of materials and detailing.

The experimental campaign consisted in applying a sequence of in-plane and out-of-plane tests, following the loading protocol reported in Figure 4. In detail, three out-of-plane tests were performed, each of them subsequent to an in-plane test at progressively increased maximum drift level. The aim was to assess the out-of-plane stability of the infill also in presence of an increasing pre-existing damage, produced during the in-plane response.

For the tests, loading systems already adopted in previous works (Preti and Bolis. 2017b) were adopted. The cyclic in-plane load was applied by means of a hydraulic jack hinged at the top frame beam. The one-way out-of-plane excitation was applied by means of a statically determined system capable of generating 8 equal point loads distributed on the infill, in order to simulate the effects of a uniformly distributed load.

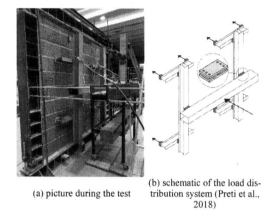

(a) picture during the test

(b) schematic of the load distribution system (Preti et al., 2018)

Figure 5. Details of the out-of-plane loading system.

4 RESULTS

In this paragraph the results of the experimental campaign are reported and commented. As for the in-plane test, the results are presented in terms of lateral force-vs.-drift and damage pattern, while the out-of-plane response is presented in terms of applied load-vs.-displacement measured at the infill centroid.

Figure 6 shows the results of the first set of in-plane cycles, up to a 0.8% drift. The response remained elastic up to 0.5% drift in both the loading directions, reaching a peak strength of about 45kN and 35kN for positive and negative drift, respectively. The obtained lower strength for positive drifts can be justified by the damage cumulated during the previous cycle at the same drift level in the opposite

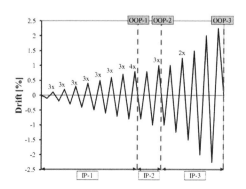

Figure 4. Experimental loading protocol.

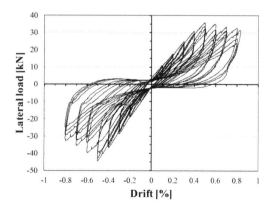

Figure 6. Result of first in-plane test (IP-1).

loading direction. The further increase of imposed drift above 0.5% resulted in a significant strength reduction (of about 25% and 15% for negative and positive loading direction, respectively, measured at 0.8% drift).

As for the suffered damage, Figure 7 shows a limited cracking up to 0.5% drift, when a horizontal crack spanning between the two columns created a sliding surface along the third bed mortar joint. At the further increase of the imposed drift (Figure 7c) the damage evolved, concentrating mainly in the lower part of the infill, near the columns. Such a crack pattern is produced by the local infill-columns contact, that activated compressed struts in the masonry, between the windward column and the base beam. On the contrary, the upper part of the infill remained undamaged, thanks to the surrounding gap and the limited in-plane shear stress transferred by the top contact.

The result of the out-of-plane test (OOP-1) carried out on the damaged is reported in Figure 8, where the resultant of the horizontal loads is plotted as a function of the out-of-plane displacement measured at the infill centroid. By considering the weight of the infill (about 5.5kN), the specimen withstood an out-of-plane load equivalent to a 2g nominal horizontal acceleration, without any additional damage or stiffness degradation.

During the following in-plane test (IP-2), an interstory drift up to 1% was applied to the specimen. The result reported in Figure 9 shows a further reduction of the cycles peak strength with respect to IP-1, associated to a progressively increased damage in the masonry. As shown in Figure 10a, such

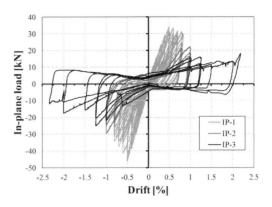

Figure 9. Complete in-plane test results.

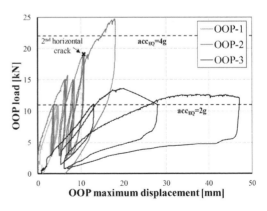

(a) after 1% drift   (b) after 2.2% drift

Figure 10. Damage in the specimen at the end of IP-2 (a) and IP-3 (b) test.

additional damage was represented by masonry crushing and spalling in correspondence of the corners in contact with the frame columns.

In presence of such a damage pattern, the test OOP-2 was carried out and the obtained response is reported in Figure 11. The infill still exhibited a relevant out-of-plane capacity, as it resisted to a transversal load equivalent to a nominal acceleration higher than 4g. In correspondence of an OOP load of about 19kN, a horizontal crack activated along a bed mortar joint in the undamaged infill upper portion (Figure 12).

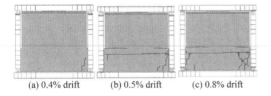

(a) 0.4% drift   (b) 0.5% drift   (c) 0.8% drift

Figure 7. Crack pattern at different steps of the IP-1 test.

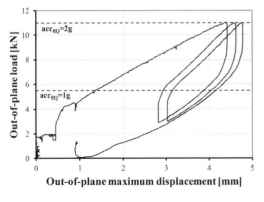

Figure 8. Result of first out-of-plane test (OOP-1). (acc_{EQ} is obtained by dividing the load by the infill mass).

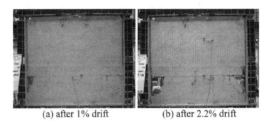

Figure 11. Complete out-of-plane test results.

939

Figure 12. Activation of the horizontal crack in the infill upper portion, during OOP-2 and OOP-3.

As a consequence, in the following cycle, the further increase of applied load is characterized by a slightly reduced stiffness, due to the activation of a different out-of-plane deformation mechanism produced by the additional horizontal crack. However, the stability of the infill was not jeopardized by the observed damage pattern.

Such a crack pattern affected also the results of the following IP-3 test (black line in Figure 9), which was characterized by a further reduction of the infill in-plane strength at 1% drift. Moreover, the following increase of imposed drift (up to more than 2%) produced a further reduction of the specimen resistance down to values of about 10kN. At the end of the in-plane test, the damage pattern was characterized by a severe damage of the lower part of the specimen, concentrated in correspondence of its contact with the columns (Figure 10b). In order to test the maximum capacity of the solution, a cycle with an amplitude larger than 2% was carried out. In correspondence of an imposed drift of about 2.2% a hardening behavior was observed in the load-vs.-drift curve when the top displacement of the column closed the isolation gap. At this drift level the test was stopped, as the maximum isolation capacity was reached and any higher imposed displacement would have produced a significant increase of lateral strength and stiffness and severe damage in the upper part of the infill.

A further out-of-plane test (OOP-3) was carried out after IP-3 and the results are reported in Figure 11 with a black line. The severe damage experienced in the previous test phases produced a reduction of the peak out-of-plane strength to about 13kN, correspondent to an equivalent acceleration higher than 2g. Despite the limited strength, the response is characterized by a significant deformation capacity, exhibiting a maximum deformation of 46mm, without any further strength reduction. Such behavior pointed out the activation of an arch mechanism within the masonry, which is typically characterized by a stable strength coupled with a large deformation capacity, up to the crushing of the masonry. However, the test was stopped before reaching the ultimate capacity to avoid dangerous brittle collapses.

The reduced strength observed in OOP-3 with respect to OOP-2 can be justified by the significant damage at the lower part of the infill that generated an almost complete detachment of the infill from the frame columns. As a consequence, the activation of a horizontal resisting arch in the masonry is prevented and the whole OOP resistance is provided by the sole vertical arching mechanism.

5 RESULT DISCUSSION

The efficiency of the proposed technique is here discussed by comparison with the response of a similar infill in the unretrofitted configuration. Lacking a corresponding experimental benchmark, in this section a first comparison, in terms of in-plane strength and stiffness, is made with the analytical prediction proposed in (Bertoldi et al. 1993) for existing masonry infills. In terms of damage, comparison is made with the results of previous experimental tests on a similar unretrofitted specimen tested by (Calvi and Bolognini. 2001). The same test was used also as reference for the additional mechanical properties required for the analytical prediction (Table 1).

Based on the assumed material properties and the experimental specimen geometry, the equations proposed by (Bertoldi et al. 1993) predict a lateral strength for the ideal solid specimen equal to 98kN and an initial stiffness of about 50000 kN/m, significantly higher than the experimentally obtained results for the downgraded infill. In particular, the infill peak lateral strength reduces by more than 50%, down to 45kN, while the stiffness is reduced to 3500 kN/m, almost 7% of that of the corresponding un-retrofitted infill.

As for the damage, the crack pattern reported in (Calvi and Bolognini. 2001) (Figure 13) shows that the unretrofitted infill suffered a widespread crack already at 0.2% drift and experienced a more severe damage reaching 0.4% drift, characterized by wide cracks and significant failures spread in the whole masonry wall. The comparison with the damage pattern of Figure 7a (registered at the same 0.4% drift) shows the better performance of the downgraded infill, which is only affected by a minor and local

Table 1. Properties of masonry from experimental material characterization tests (Calvi and Bolognini, 2001).

f_{wh} [MPa]	f_{wv} [MPa]	f_{wu} [MPa]	f_{ws} [MPa]	E_{wh} [MPa]	E_{wv} [MPa]	G [MPa]
1.18	2.02	0.44	0.55	991	1873	1089

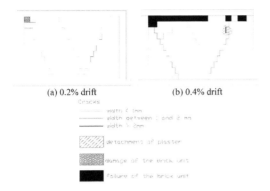

(a) 0.2% drift   (b) 0.4% drift

Cracks

.... width < 1mm

- - - - - - - width between 1 and 2 mm

─────── width > 2mm

▨▨▨ detachment of plaster

▦▦▦ damage of the brick unit

■■■ failure of the brick unit

Figure 13. Experimental crack pattern obtained for the weak solid infill tested by (Calvi and Bolognini, 2001).

crack pattern. The most significant damages in the weakened specimen was observed only after exceeding 1% drift, but they were anyhow limited to the lower part of the infill in contact with the columns.

As for the safety verification of the RC frame columns in presence of the downgraded infill and with regard to the possible captive column mechanism, a first evaluation of the contact force between the column and the infill is here proposed, assuming, as a conservative assumption, the static scheme reported in Figure 14. Based on equilibrium considerations on the windward column, in presence of the peak experimental lateral strength (45kN) the trust acted by the infill on the column at the contact position is about 114kN. As a consequence, the column is subjected to a peak shear action of about 70kN and a peak bending moment of about 70kNm.

As an example, such action could be easily supported by a column with a 30x30cm cross-section, a longitudinal reinforcement with a minimum of 4 Φ16 bars and a Φ6 transversal reinforcement with less than 20cm spacing.

Based on the real configuration of the structure, the verification will drive the choice on the contact height left after the cut, in order to ensure the column resistance.

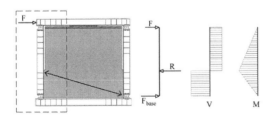

Figure 14. Internal actions in the frame column produced by the interaction with the infill.

6 CONCLUSIONS

The paper describes the experimental response of an existing masonry infill downgrade with an innovative retrofit technique. The intervention, here referred as "hybrid isolation" consisted in isolating the masonry wall from the surrounding frame by creating a gap along the infill perimeter, starting from a height of 1m from the base, where the technical systems are typically located. In order to ensure the necessary out-of-plane stability, a local contact was restored between the masonry and the frame beam, detailed to ensure the activation of a compressed arch in the masonry, against out-of-plane actions, but, at the same time, to limit the transfer of shear stress during the in-plane deformation.

According to the experimental results, the adopted downgrade technique ensured a significant reduction of the infill lateral stiffness and strength, with respect to a traditional solid infill. Moreover, given the same drift level, also the suffered damage is extremely reduced, thus allowing the infill to sustain high frame deformation without experiencing damage. Such a damage limitation is fundamental in order to maintain the infill out-of-plane stability, achievable thanks to the activation of compressed arches in the undamaged masonry thickness.

The obtained results highlight the efficiency of the intervention in limiting the possible seismic infill-frame interaction, which could be the cause of severe post-earthquake damage in both the infill and the structure. As a consequence, the proposed intervention could significantly improve the deformation capacity of the existing building, allowing the retrofit strategy to exploit deformation capacity up to 1% without significant infill frame interaction. In fact, even after minor seismic events, widespread damage to non-structural elements could be observed in typical existing buildings, thus resulting in loss of functionality and high repair costs. Thanks to the proposed intervention for the infills, the damage activation could be delayed to higher inter-story drift levels, only produced by seismic events characterized by higher intensities and longer return periods.

It is worth noting that in the present paper the performance of a prototype application of the technique was investigated, with the only aim of verifying the efficiency of the studied conceptual design reported in par.2. In the future, such a technique can be detailed and optimized for an easier and cost effective application in the seismic retrofit of different typologies existing infills, characterized by different materials and geometries (higher thickness, presence of openings, etc.).

ACKNOWLEDGMENTS

The authors are thankful to the student Maurizio Lorini and to the technicians of the P.Pisa Laboratory of the University of Brescia for the support during the experimental campaign.

The presented study was partly developed in the research program financed by the "Presidenza del Consiglio dei Ministri - Dipartimento della Protezione Civile" within the Reluis research program; the present publication, however, does not necessarily reproduce the Department position and judgments.

REFERENCES

Augenti, N., and Parisi, F. 2010. Learning from construction failures due to the 2009 L'Aquila, Italy, earthquake. *Journal of Performance of Constructed Facilities 24*, 536–555.

Basha, S.H., and Kaushik, H.B. 2016. Behavior and failure mechanisms of masonry-infilled RC frames (in low-rise buildings) subject to lateral loading. *Engineering Structures 111*, 233–245.

Bertoldi, S.H., Decanini, L.D., and Gavarini, C. 1993. Telai tamponati soggetti ad azioni sismiche, un modello semplificato: confronto sperimentale e numerico. *Atti Del 6*, 815–24.

Butenweg, C., and Marinković, M. 2018. Damage reduction system for masonry infill walls under seismic loading. *Ce/Papers 2*, 267–273.

Calvi, G.M., and Bolognini, D. 2001. Seismic response of reinforced concrete frames infilled with weakly reinforced masonry panels. *J. Earth. Eng. 05*, 153–185.

Canbay, E., Binici, B., Demirel, I.O., Aldemir, A., Uzgan, U., Eryurtlu, Z., and Bulbul, K. 2018. DEGAS: An innovative earthquake-proof AAC wall system. *Ce/Papers 2*, 247–252.

Cavaleri, L., and Di Trapani, F. 2015. Prediction of the additional shear action on frame members due to infills. *Bulletin of Earthquake Engineering 13*, 1425–1454.

Di Ludovico, M., Prota, A., Manfredi, G., and Cosenza, E. 2008. Seismic strengthening of an under-designed RC structure with FRP. *Earthquake Engineering & Structural Dynamics 37*, 141–162.

Di Trapani, F., and Malavisi, M. 2018. Seismic fragility assessment of infilled frames subject to mainshock/aftershock sequences using a double incremental dynamic analysis approach. *Bull Earthquake Eng.*

EERI 2000. 1999 Kocaeli, Turkey earthquake reconnaissance report. Earthquake Spectra 16.

EERI(2009. The MW 6.3 Abruzzo, Italy, Earthquake of April 6, 2009.

Gioiella, L., Balducci, A., Carbonari, S., Gara, F., and Dezi, L. 2017. An innovative seismic protection system for existing buildings: external dissipative towers. 11.

Markulak, D., Radić, I., and Sigmund, V. 2013. Cyclic testing of single bay steel frames with various types of masonry infill. *Engineering Structures 51*, 267–277.

Mehrabi Armin B., Benson Shing P., Schuller Michael P., and Noland James L. 1996. Experimental Evaluation of Masonry-Infilled RC Frames. *Journal of Structural Engineering 122*, 228–237.

Metelli, G., Feroldi, F., and Marini, A. 2017. Numerical study on the seismic response of a 1970s reinforced concrete building strengthened with concentric bracings. (Venezia), p.

Mohammadi, M., Akrami, V., and Mohammadi-Ghazi, R. 2011. Methods to Improve Infilled Frame Ductility. *Journal of Structural Engineering 137*, 646–653.

Morandi, P., Milanesi, R.R., and Magenes, G. 2018. Innovative solution for seismic-resistant masonry infills with sliding joints: in-plane experimental performance. *Engineering Structures 176*, 719–733.

Pallarés, F.J., Pallarés, L., Ferrer, I., and Orozco, J. 2017. Isolation device for masonry infills in building frames. *In 16th World Conference on Earthquake Engineering, (Santiago Chile)*.

Pampanin, S. 2012. Reality-check and renewed challenges in earthquake engineering: Implementing low-damage structural systems - from theory to practice.

Passoni, C. 2016. Holistic renovation of existing RC buildings: a framework for possible integrated structural interventions. PhD Thesis. Ph. D. thesis, University of Brescia.

Perrone, D., Calvi, P.M., Nascimbene, R., Fischer, E.C., and Magliulo, G. 2018. Seismic performance of non-structural elements during the 2016 Central Italy earthquake. *Bull Earthquake Eng 1–23*.

Preti, M., and Bolis, V. 2017a. Seismic analysis of a multi-story RC frame with infills partitioned by sliding joints. *Ingegneria Sismica Volume 34*, 175–187.

Preti, M., and Bolis, V. 2017b. Masonry infill construction and retrofit technique for the infill-frame interaction mitigation: Test results. *Engineering Structures 132*, 597–608.

Preti, M., Bettini, N., and Plizzari, G. 2012. Infill Walls with Sliding Joints to Limit Infill-Frame Seismic Interaction: Large-Scale Experimental Test. *Journal of Earthquake Engineering 16*, 125–141.

Preti, M., Neffati, M., and Bolis, V. 2018. Earthen masonry infill walls: Use of wooden boards as sliding joints for seismic resistance. *Construction and Building Materials 184*, 100–110.

Preti, M., Bolis, V., Paderno, A., and Metelli, G. 2019. The masonry infill downgrade in the seismic strengthening of existing reinforced concrete buildings. *In Proceedings of XVII Convegno ANIDIS, L'ingegneria Sismica in Italia, (Ascoli Piceno 15-19 Settembre 2019)*, 227–234.

Riva, P., Perani, E., and Belleri, A. 2010. External r.c. structural walls for the repair of earthquake damaged buildings. (Rome).

Tasligedik, A.S., and Pampanin, S. 2017. Rocking Cantilever Clay Brick Infill Wall Panels: A Novel Low Damage Infill Wall System. *Journal of Earthquake Engineering 21*, 1023–1049.

Tsantilis, A.V., and Triantafillou, T.C. 2018. Innovative seismic isolation of masonry infills using cellular materials at the interface with the surrounding RC frames. *Engineering Structures 155*, 279–297.

Verlato, N., Guidi, G., Da Porto, F., and Modena, C. 2016. Innovative systems for masonry infill walls based on the use of deformable joints: Combined in-plane/out-of-plane tests. 1359–1366.

Brick and Block Masonry - From Historical to Sustainable Masonry –
Kubica, Kwiecień & Bednarz (eds)
© 2020 Taylor & Francis Group, London, ISBN 978-0-367-56586-2

Definition of out-of-plane fragility curves for masonry infills subject to combined in-plane and out-of-plane damage

F. Di Trapani & M. Malavisi
Department of Structural, Building and Geotechnical Engineering, Politecnico di Torino, Turin, Italy

P.B. Shing
Department of Structural and Material Engineering, University of California at San Diego, La Jolla, CA, USA

L. Cavaleri
Dipartimento di Ingegneria, University of Palermo, Palermo, Italy

ABSTRACT: The paper presents the outcomes of a probabilistic assessment framework aimed at defining out-of-plane fragility curves of unreinforced masonry infills walls which have been subjected (or not) prior in-plane damage. A recently developed in-plane (IP)/out-of-plane (OOP) four-strut macro-element model is used to model masonry infills within frames. Out-of-plane incremental dynamic analyses are performed, for a reference infilled frame, based on a suite of 26 ground motion record selection. Peak ground acceleration (PGA) and OOP relative displacement of the midspan node of the infill, are used as intensity measure and damage measure. The outcomes show fragility curves representing the probability of exceeding out-of-plane collapse at a given earthquake intensity as a function of a different combination of geometrical and mechanical parameters, in-plane damage level and supporting conditions. Results are finally summarized by curves relating in-plane interstorey drifts and out-of-plane average collapse PGA.

1 INTRODUCTION

In the last years, the out-of-plane (OOP) earthquake response of masonry infills and its mutual dependence with the in-plane (IP) damage level has received special interest by researchers (e.g. Mazza et al. 2019, Di Trapani et al. 2020a). New experimental and numerical investigations (e.g. Furtado et al. 2016, Ricci et al. 2018, De Risi et al. 2019, Cavaleri et al. 2019) have been carried out and, at the same time, new simplified models able to predict both in-plane and out-of-plane responses have been developed following different mechanical approaches (Mosalam and Günay 2014, Furtado et al. 2015, Di Trapani et al. 2018a). The response of masonry infilled reinforced concrete (RC) frame buildings subject to ground motions inducing in-plane and out-of-plane actions is not of easy generalization as this depends on several aspects as the geometrical configuration of the frame, the position of the infills along the height, the reciprocal earthquake intensity along the two orthogonal directions. Some recent studies referred to the analysis entire buildings (Ricci et al. 2019, Longo et al. 2019) have confirmed this. In fact, infills located at the highest floor are subjected to major accelerations but at the same time lower in-plane damage. Conversely, infills at the lowest floors undergo reduced accelerations demand but suffer large in-plane drift demand that waken their out-of-plane capacity. Considering these premises, a prediction of

the most critical conditions for the infill walls in a frame structure cannot be carried in a simple way in case of combined IP-OOP actions.

In this paper the issue is faced using a probabilistic assessment framework aimed at evaluating out-of-plane fragility curves of infill walls which have been subjected (or not) prior in-plane damage. The fiber-section macro-element model by Di Trapani et al. (2018a) is used to model the infill wall. Fragility curves are obtained performing out-of-plane incremental dynamic analyses (IDA) based on a suite of 26 ground motion records. IDA curves, and the associated fragilities are obtained by varying the slenderness ratio, the in-plane drift level, and the frame stiffness with respect to the out-of-plane stiffness of the infill.

2 DESCRIPTION OF THE MACROELEMENT MODEL AND DETERMINATION OF THE EQUIVALENT OOP MASS

2.1 Definition of the microelement model

The macro-element model by Di Trapani et al. (2018a) provides the replacement of the infill panel with 4 pinned struts, each one divided into two elements. The overall scheme is illustrated in Figure 1a. Each strut is defined using distributed plasticity

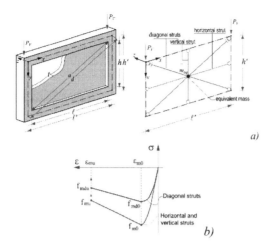

a)

b)

Figure 1. Definition of the 4-strut macro-model: a) Geometric definition; b) Stress-strain models for the struts.

fiber-section beam-column elements available in OpenSees.

In this way, internal cross-section forces of these elements are related to the corresponding deformations (axial deformation and curvature), hence, after the elastic stage axial load is coupled with bending moment and the arching mechanism is then naturally accounted. The diagonal, horizontal and vertical struts have typical concrete-type stress-strain laws, modelled with the OpenSees Concrete02 material (Figure 1b). Moreover, in order to explicitly consider the effects of strength degradation, the Concrete02 material model is combined with the MinMax material so that, once the ultimate strain is achieved at a generic fiber, the corresponding stress drops to zero. The diagonal struts are constrained with pins at the ends. They provide the whole in-plane response of the infill as well as the main OOP contribution. The horizontal and vertical struts provide a complementary OOP contribution to strength, reproducing the 2-way bending effect of the panel. The 4 struts do not share the mid-span node but are constrained to move together along the z direction. In this way each strut can provide its strength contribution to the OOP response. Geometrical and mechanical identification of the struts is performed starting from the diagonals, whose force-displacement behaviour can be assigned by adopting any method. Once that the force-displacement law is determined, this can be easily converted into an equivalent concrete-type stress-strain relationship and assigned to the cross-section fibers. To perform this step, the reference cross-section can be simply obtained by fixing the width (w_d) as 1/3 of the internal diagonal length (a) and the thickness as the actual thickness (t) of the infill. A direct definition of the stress-strain relationship and width of the diagonal cross-sections can also carried out by using the approach by Di Trapani

et al. (2018b). When defining the stress-strain relationship of the diagonal struts it can be easily found that the peak strength f_{md0} is lower than the actual compressive strength of masonry (f_{m0}). This is due to the fact that f_{md0} expresses a fictitious resistance which summarized the complex response of the infill subject to lateral forces. On the other hand, the out-of-plane strength is proportional to the actual compressive strength of the masonry (f_{m0}). In order to compensate the drawback, the cross-section thickness of the strut is increased by the ratio f_{m0}/f_{md0}, so that:

$$\tilde{t} = \frac{f_{m0}}{f_{md0}}t \qquad (1)$$

while, in order to maintain unaltered the cross-section area, the width is reduced as:

$$\tilde{w}_d = \frac{t}{\tilde{t}}w_d \qquad (2)$$

The residual OOP strength is provided by the horizontal and vertical struts. The latter are defined using the actual thickness t of the infill and the actual strength f_{m0}, while the widths are obtained as the difference between the height and the length of the panel and the horizontal and vertical projections of the diagonal initial width w_d on the infill perimeter. The widths of the horizontal strut (w_h) and vertical strut (w_v) are therefore evaluated as follows:

$$w_h = h - \frac{w_d}{\cos\theta}; \; w_v = \lambda - \frac{w_d}{\sin\theta} \qquad (3)$$

where θ is the inclination of the strut with respect to the horizontal direction.

2.2 Definition of the equivalent mass

In order to perform dynamic simulations, the model needs the definition of an equivalent mass to apply at the midspan node of the struts. The mass (m_{eq}) of the so defined single degree of freedom (SDOF) system is of course a percentage of the total mass of the infill. In order to identify this percentage, an experimental/numerical identification procedure has performed using the results of the experimental tests by Angel (1994). The same tests which were also used for the validation of the aforementioned model. The tests regarded reinforced concrete infilled frame specimens subject to in-plane cycles and then pushed out-of-plane using an airbag. The identification procedure consists of the following steps: *a)* determination of the experimental out-of-plane stiffness K_{exp} from the experimental force-displacement diagrams; *b)* identification of the out-of-plane period of the

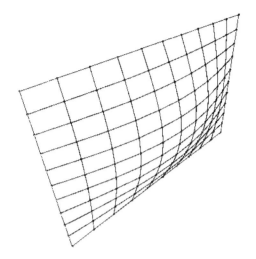

Figure 2. First modal shape of the FE model of the infill.

infill ($T_{i,FEM}$) by defining an elastic finite element model of a plate pinned at the sides (Figure 2), verifying that the stiffness of the plate (K_{FEM}) was the same as the experimental stiffness; c) determination of the equivalent mass of the SDOF system as:

$$m_{eq} = \frac{T_{i,FEM}^2}{4\pi^2} K_{exp} \qquad (4)$$

In the procedure it is assumed that the out-of-plane stiffness provided by the macro-model is the same as that the experimental stiffness. This was proved by the validation tests presented in Di Trapani et al. (2018a) for the same specimens. It should

be also observed that that since the specimen were subjected to moderate cycles before being tests out-of-plane, their experimental stiffness was lower with respect to that estimated by the elastic FE model of the plate. Therefore, the matching between K_{FEM} and K_{exp} was obtained by modifying the elastic modulus of the infill (E_m) into a lower one (E_m^*). This manipulation has no influence on the equivalent mass value found by Eq. (4), with respect to an undamaged case, since the reduction of K_{FEM} due to the reduction of E_m in the FE model, is compensated by the elongation of the period. Geometric dimensions and masses of the infills of specimens by Angel (1994) are reported in Table 1. Mechanical properties of the infills are reported in Table 2 together with the obtained experimental stiffness values, periods and equivalent masses. It is noteworthy observing that independently on the different geometric and mechanical combinations of the specimens the percentage equivalent mass resulted 55% of the total mass on average. This allows concluding that the macro-element model can be adapted to perform dynamic simulations by assigning a 55% equivalent mass at the central node.

3 ANALYSIS FRAMEWORK

3.1 Analysis program

The investigation is first addressed to the derivation of out-of-plane fragility curves of a one-storey infilled frame as a function of different boundary conditions, that is the slenderness ratio and strength of the specimen, the extent of the previous in-plane damage, the out-of-plane vibration period of the infill with respect to that of the supporting frame

Table 1. Geometric dimensions and masses of specimens by Angel (1994).

Spec.	λ (mm)	λ' (mm)	h (mm)	h' (mm)	t (mm)	h/t (-)	γ (kN/m³)	mass (kg)
2	2440	2740	1630	1930	47.6	34.2	19	359.7
3	2440	2740	1630	1930	47.6	34.2	19	359.7
4	2440	2740	1630	1930	92.0	17.7	19	695.2
5	2440	2740	1630	1930	143.0	11.4	19	1080.6
6	2440	2740	1630	1930	98.4	16.6	19	743.6

Table 2. Mechanical properties and identification parameters of specimens tested by Angel (1994).

Spec.	E_m (MPa)	G_m (MPa)	E_m^* (MPa)	f_{m0} (MPa)	K_{exp} (N/mm)	$T_{i,FEM}$ (s)	m_{eq} (kg)	$m_{eq\%}$ (%)
2	8040	8040	1900	10.85	1052.6	0.083	184.5	51.3
3	5208	5208	1736	10.13	1300	0.078	202.3	56.2
4	12429	12429	5000	22.90	20000	0.027	364.2	52.4
5	11616	11616	9000	22.82	148000	0.012	582.6	53.9
6	2136	2136	650	4.60	3500	0.070	429.0	57.7

structure. Two reference specimens among those tested by Angel (1994) have been selected to perform fragility assessment, namely Specimen 2 and Specimen 6. These specimens have different slenderness ratios and masonry strength as it can be observed from Tables 1 and 2. The OOP resistance of a masonry infills (F_r) can be estimated through the EC6 expression as:

$$F_r = f_{m0} \left(\frac{t}{h'}\right)^2 \lambda h \qquad (5)$$

From Eq.(5) one can determine the OOP resisting pressure (f_r) as:

$$f_r = \frac{F_r}{\lambda h} \qquad (6)$$

This parameter combines both strength and slenderness ratio, and so it can be used to identify a class of infilled frames. For specimens 2 and 6 f_r was 6.6 MPa and 11.95 MPa respectively.

In order to simulate the influence of the supporting frame, the reference infilled frame is modelled as in Figure 3, where besides the equivalent mass at the midspan node, the model has 4 mass-spring (m_f, k_f) systems at the corner nodes. The whole system has therefore two degrees of freedom, one related to the frame, the other related to the infill. The vibration period associated with the frame considered alone can be obtained as:

$$T_f = 2\pi \sqrt{\frac{M_f}{K_f}} \qquad (7)$$

in which $M_f = 4m_f$ is the total mass of the frame $K_f = 4k_f$ is the total stiffness of the parallel springs.

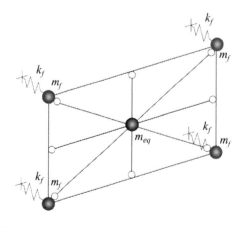

Figure 3. Reference scheme for the infilled frame model.

The vibration periods associated with the infills have been obtained as described in the previous section ($T_i = T_{i,FEM}$). The springs have elastic behaviour and are modelled in OpenSees using zero-length elements. For the two specimens under investigation, the response of the system is analyzed considering five different supporting conditions, namely rigid support ($K_f = \infty$), $T_f = T_i$, $T_f = 3T_i$, $T_f = 5T_i$, $T_f = 7T_i$. Given that the period of the infills is fixed, and attributing a conventional mass (m_f) of 2000 kg to the nodes of the frames, the stiffness of the frame producing the aforementioned period ratios can be obtained as:

$$K_f = 4\pi^2 \frac{M_f}{T_f^2} \qquad (8)$$

and then:

$$K_f = 4\pi^2 \frac{m_f}{T_f^2} \qquad (9)$$

The effect of prior in-plane damage is also investigated considering 4 cases, namely: a) no in-plane damage; b) IDR=0.5%; c) IDR=1.5%; d) IDR=2.5%. The IDR (interstorey drift) is assumed as measure of the in-plane damage. The analyses are carried out in OpenSees through two-steps. First a cyclic static analysis consisting of three cycles having amplitude as the fixed IDR is performed. Subsequently the IDA sequence is started. A summary of the analyses is reported in Table 3. Details about periods, and stiffness of the different systems obtained for Specs. 2 and 6 are listed in Tables 4 and 5.

3.2 Definition of IDA curves

Incremental Dynamic Analysis (IDA) (Vamvatsikos and Cornell, 2002) has been used many times in recent years for the assessment of seismic fragility of structures (Basone et al. 2017, Di Trapani & Malavisi 2019, Di Trapani et al. 2020b). In the current case IDA has been carried out using the peak ground acceleration (PGA) as intensity measure (IM) and the out-of-plane net displacement (Δ_{OOP}) as damage measure (DM).

Table 3. Summary of the test conditions.

Support conditions	Specimen 2 IDR (%)	Specimen 6 IDR (%)
$T_f = 0$ (Rigid frame)	0/0.5/1.5/2.5	0/0.5/1.5/2.5
$T_f = T_i$	0/0.5/1.5/2.5	0/0.5/1.5/2.5
$T_f = 3T_i$	0/0.5/1.5/2.5	0/0.5/1.5/2.5
$T_f = 5T_i$	0/0.5/1.5/2.5	0/0.5/1.5/2.5
$T_f = 7T_i$	0/0.5/1.5/2.5	0/0.5/1.5/2.5

Table 4. Period, mass and stiffness values of the systems obtained by Spec. 2.

T_f/T_i (-)	T_i (s)	T_f (s)	M_f (kg)	K_f (kN/m)	k_f (kN/m)
0	0.083	0	8000	∞	∞
1	0.083	0.083	8000	3801.3	950.3
3	0.083	0.249	8000	1267.1	316.8
5	0.083	0.415	8000	760.3	190.1
7	0.083	0.581	8000	543.0	135.8

26 ground motions records have been considered. The spectra of the selected ground motion are shown in Figure 4. The choice of PGA as IM instead of the usual spectral acceleration at of the first vibration period is due to the fact that, as explained in the previous section, different combinations of periods are considered. The choice of PGA allows using the same ground motion scaling to analyse the different combinations of periods. In detail the ground motions are first scaled so that their respective spectra have the same PGA. The subsequent scaling during IDA uniformly increases/decreases the amplitude. For each ground motion IDA are stopped in correspondence of the achievement of dynamic

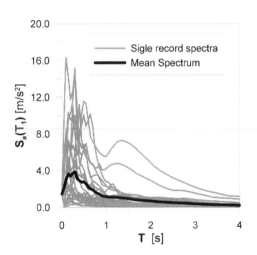

Figure 4. Ground motion selection spectra and average spectrum.

instability (which represents the OOP failure of the infill). After this point, a constant flatline is conventionally represented.

3.3 Definition of fragility curves

Fragility curves express the probability of exceeding a specified limit state as a function of the specified IM (PGA). Fragility curves can be represented using a lognormal cumulative distribution function as:

$$P[C \leq D | IM = x] = \Phi \left(\frac{ln(x) - \mu_{ln_x}}{\sigma_{ln_x}} \right) \quad (10)$$

where $P[C{\leq}D|IM{=}x]$ is the probability that a ground motion with IM=x will cause the achievement of a limit state, Φ is the standard cumulative distribution function, $ln(x)$ is the natural logarithm of the variable x representing the intensity measure (PGA) and μ_{lnX} and σ_{lnX} are the mean and the standard deviation of the natural logarithms of the distribution of x, respectively. Fragility curves are derived considering the collapse limit state, which is attained when the dynamic instability occurs during IDA or when the midspan relative displacement of the infill (Δ_{OOP}) is the same as the thickness of the infill. In this latter case, in fact, it is supposed that the arching action vanishes, and the equilibrium is no longer achievable. Cumulative discrete distribution functions are also overlapped to the analytically obtained fragility curves to verify the adequacy of the distribution model.

4 RESULTS FOR THE ONE-STOREY INFILLED FRAME

4.1 IDA and fragility curves

IDA curves are illustrated in Figure 5-8 for the one-storey infilled frame for different considered combinations of T_f/T_i ratios and in-plane interstorey drifts. For sake of space, IDA curves are only reported for specimen 2. The curves show a reduction of the average collapse PGA when increasing T_f/T_i up to 3. After the collapse PGA tends to increase again, denoting that T_f/T_i ratio has a relevant role, as it influences the accelerations

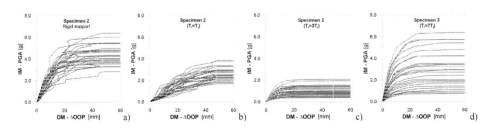

Figure 5. IDA curves of Specimen 2 without in-plane damage for: a) rigid support; b) $T_f{=}T_i$; b) $T_f{=}3T_i$; b) $T_f{=}7T_i$.

947

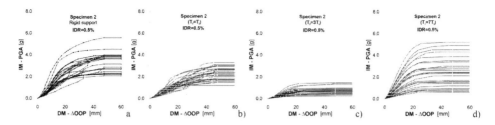

Figure 6. IDA curves of Specimen 2 with IDR=0.5% for: a) rigid support; b) $T_f = T_i$; b) $T_f = 3T_i$; b) $T_f = 7T_i$.

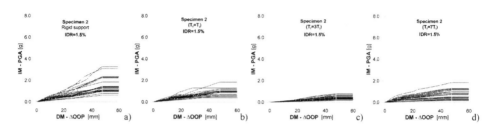

Figure 7. IDA curves of Specimen 2 with IDR=1.5% for: a) rigid support; b) $T_f = T_i$; b) $T_f = 3T_i$; b) $T_f = 7T_i$.

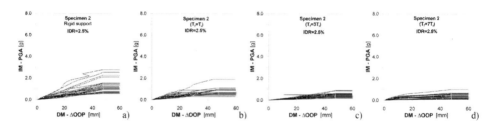

Figure 8. IDA curves of Specimen 2 with IDR=2.5% for: a) rigid support; b) $T_f = T_i$; b) $T_f = 3T_i$; b) $T_f = 7T_i$.

experienced by the infill wall. On the other hand, it should be also observed that the presence of in-plane damage (measured by the in-plane IDR) tends to reduce the influence of T_f/T_i. In fact in presence of severe damage (IDR=1.5%-2.5%) the collapse PGA is dramatically reduced. Under these conditions no

substantial differences can be observed when by varying T_f/T_i ratio. Fragility curves of the one-storey infilled frame are shown in Figure 9.

The latter reflect what already highlighted by IDA curves. For the case of no-IP damage and moderate IP damage (IDR=0.5%) (Figures 9a-9b) fragility tends to

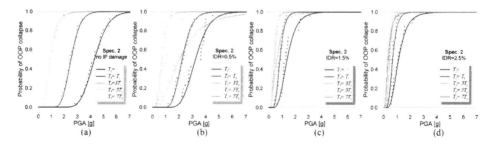

Figure 9. Fragility curves of specimen 2 for different T_f/T_i with: a) IDR=0 %; b) IDR=0.5 %; c) IDR=1.5 %; d) IDR=2.5 %.

increase when increasing T_f/T_i up to 3. After this point fragility tends to be reduced, although significantly larger dispersion of collapse IMs is observed. For the cases of high IP damage (Figures 9b-9c) fragility curves result significantly shifted on the right denoting minor sensitivity to variation of T_f/T_i. In some cases, collapse PGA values may result very high, so that they can exceed a hypothetical collapse PGA for the frame. This is due to the elastic behaviour of the springs used to represent the response of the frame. In this cases OOP fragility curves lose significance since an overall collapse occurs before the OOP collapse of the infills.

In the next section the investigation is extended to the case of a multi-storey frame under the simplified assumption of linear distribution of floor accelerations.

5 EXTENSION TO MULTISTOREY INFILLED FRAMES

Results shown in the previous section are referred to the elementary one-storey infilled frame, which is supposed to be subjected to an acceleration history at the supports. In the case of a multi-storey frame, floor accelerations tend to grow when increasing the height, as also provided by EC6 through the expression:

$$S_a = PGA \cdot S \cdot \beta \tag{11}$$

where S is the soil factor, S_a the pseudo acceleration of an infill wall at the center of mass, positioned at the altitude (Z) from the ground in a building having total height (H) and β is a modulation factor defined as:

$$\beta = \frac{3(1 + Z/H)}{1 + (1 - T_i/T_f)^2} - 0.5 \tag{12}$$

It is easy to demonstrate that according to Eq. (11) S_a increases with increasing the altitude of the infill

with respect to the total height of the building, according to a linear relationship. Eq. (12) can be adapted to estimate floor acceleration demand variation for the analysed infilled frames. In this case, given that the effect of the variation of T_f/T_f ratio has been already taken into account in the previous analyses on the one-storey infilled frame, in Eq. (12) it can be set $T_i/T_f=0$. Finally, for consistency with the assumed scheme for the one-storey infilled frames, the values Z' and H' are used instead of Z and H, where Z' is the quote of the center of mass of the infill measured with respect to the center of mass of the first ground floor infill, and H' is the quote of the center of mass of the infill at the highest floor. Under these assumptions one obtains:

$$\beta^* = \frac{3(1 + Z'/H')}{2} - 0.5 \tag{13}$$

It can be easily observed that if $Z'/H'=0$ (case of infill wall at the ground floor or one storey infilled frame), $\beta^*=1$ (no amplification is provided), while if $Z'/H'=1$ (case of infill wall at the top floor) one obtains the maximum amplification factor ($\beta^*=2.5$).

By defining $\overline{PGA}_{c,0}$ as the 50% probability PGA inducing the collapse of the one-storey infilled frame, it can be reasonably supposed that the average PGA inducing the collapse of an infill wall at the generic Z/H position $(\overline{PGA}_{c,(Z/H)})$ can be obtained by reducing $\overline{PGA}_{c,0}$ by β^*, therefore:

$$\overline{PGA}_{c,(Z/H)} = \frac{\overline{PGA}_{c,0}}{\beta^*} \tag{14}$$

The values of $\overline{PGA}_{c,0}$ extrapolated from the fragility curves of specimens 2 and 6 can be represented as in Figures 10a and 11a as a function of T_f/T_i and IDR. This are coincident with $\overline{PGA}_{c,(Z/H)}$ at $Z'/H'=0$. Diagrams in Figures 10b-c and 11b-c represent $\overline{PGA}_{c,(Z,H)}$ for $Z'/H'=0.5$ and $Z'/H'=1$. The latter are obtained from the first two diagrams by using Eq. (14). Diagrams in Figures 10-11 show that

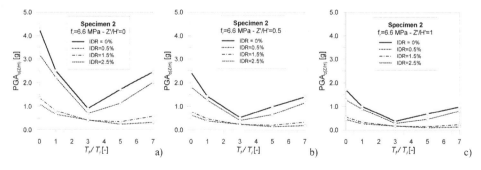

Figure 10. Average collapse PGA for Specimen 2 with: a) Z'/H'=0; b) Z'/H'=0.5; c) Z'/H'=1.0.

949

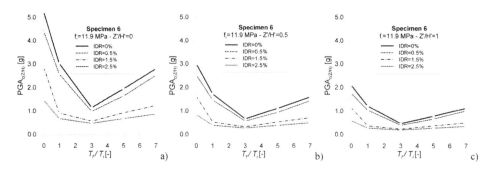

Figure 11. Average collapse PGA for Specimen 6 with: a) Z'/H'=0; b) Z'/H'=0.5; c) Z'/H'=1.0.

infill at the higher stories undergo major spectral accelerations and therefore their collapse may occur with significantly reduced PGA levels. At the same time infills positioned at lower stories suffer major in-plane damage. This means that their collapse may occur with lower PGA values, with respect to those inducing collapse of infills at the upper stories.

Diagrams in Figures 10-11 can be also used as assessment tools by entering with the characteristics of the infilled frames and comparing the resulting average PGA with a design PGA level.

6 CONCLUSIONS

In the paper an existing infilled frame macro-element model (Di Trapani et al. 2018a) has been updated to perform dynamic simulations. The model consist of four fiber-section struts and is able to account for mutual in-plane and out-of-plane damage. Out-of-plane fragility curves for a reference one-storey infilled frames have been derived considering different prior IP damage levels and different ratios between frame and infill periods (T_f/T_i). Incremental dynamic analysis was used to derive fragility curves. Results have shown that for the cases of no-IP damage and moderate IP damage the average collapse PGA tend to increase when increasing T_f/T_i up to 3. After this point collapse PGA tends to be reduced. For the cases of severe IP damage, collapse PGA dramatically decreases denoting minor sensitivity to T_f/T_i. The analysis has been extended to multi storey infilled frames under the simplified assumption of linear distribution for floor accelerations. It has been shown that infills at the higher stories undergo major spectral acceleration and their collapse may be achieved in correspondence of reduced PGA values. However, infills positioned at lower stories undergo major in-plane damage potentially causing their anticipated collapse with respect to the upper stories infills. Therefore, the location of masonry infills subject to major OOP collapse risk in not predictable a priori as this depends on the combination between floor acceleration and in-plane drift at each storey.

Finally, average collapse PGA diagrams have been provided as a function of T_f/T_i, IDR and Z'/H'. The latter can be used as assessment tools by comparing the average PGA associated with a an infilled frame to design PGA level.

REFERENCES

Angel, R., 1994. Behavior of reinforced concrete frames with masonry infill walls. PhD thesis, *University Illinois at Urbana-Champaingn*, Illinois, USA.

Basone, F., Cavaleri, L., Di Trapani, F., Muscolino, G., 2017, Incremental dynamic based fragility assessment of reinforced concrete structures: Stationary vs. non-stationary artificial ground motions, *Bull. Earthquake Eng*, 103, 105–117.

Castaldo, P., Gino, D., Bertagnoli, G. Mancini, G., 2020. Resistance model uncertainty in non-linear finite element analyses of cyclically loaded reinforced concrete systems, *Eng Struct*, 211, 110496.

Castaldo, P., Gino, D., Mancini, G., 2019. Safety formats for non-linear analysis of reinforced concrete structures: discussion, comparison and proposals, *Eng Struct*, 193, 136–153.

Cavaleri, L., Zizzo, M., Asteris, P.G., 2019. Residual out-of-plane capacity of infills damaged by in-plane cyclic loads, *Eng Struct*, 209, 109957.

De Risi, M.T., Di Domenico, M., Ricci, P., Verderame, G. M., Manfredi, G., 2019. Experimental investigation on the influence of the aspect ratio on the in-plane/out-of-plane interaction for masonry infills in RC frames, *Eng Struct*, 189, 523–540.

Di Trapani, F., Malavisi, M., 2019. Seismic fragility assessment of infilled frames subject to mainshock/aftershock sequences using a double incremental dynamic analysis approach, *Bull. Earthquake Eng*, 17(1), 211–235.

Di Trapani, F., Giordano, L., Mancini, G., 2020a, Progressive Collapse Response of Reinforced Concrete Frame Structures with Masonry Infills, *J. Eng. Mech.*, 146(3), 04020002.

Di Trapani, F., Bolis, V., Basone, F., Preti, M., 2020b, Seismic reliability and loss assessment of RC frame structures with traditional and innovative masonry infills, *Eng Struct*, 208, 110–306.

Di Trapani, F, Shing, P.B., Cavaleri, L., 2018a. Macroelement model for in-plane and out-of-plane responses of masonry infills in frame structures, *J Struct Eng*, 144:04017198.

Di Trapani, F., Bertagnoli, G., Ferrotto, M.F., Gino, D., 2018b. Empirical equations for the direct definition of stress-strain laws for fiber-section based macro-modeling of infilled frames, *J Eng Mech*, 144(11), 04018101.

Eurocode 6. Design of Masonry Structures. Part 1-1: General Rules for Reinforced and Unreinforced Masonry Structures. Brussels, 2005.

Furtado, A., Rodrigues, H., Arêde, A., Varum, H., 2015. Simplifed macro-model for infll masonry walls considering the out-of-plane behaviour, *Earthq Eng Struct Dyn* 45(4), 507–524.

Furtado, A., Rodrigues, H., Arêde, A., Varum, H., 2016. Experimental evaluation of out-of-plane capacity of masonry infill walls, *Eng Struct*, 111, 48–63.

Longo, F., Wiebe, L., da Porto, F., Modena, C., 2018. Application of an in plane/out of plane interaction model for URM infill walls to dynamic seismic analysis of RC frame buildings, *Bull Earthquake Eng*, 16, 6163–6190.

Mazza, F., 2019. In-plane–out-of-plane non-linear model of masonry infills in the seismic analysis of r.c.-framed buildings, *Earthq Eng Struct Dyn* 48(4), 432–453.

McKenna, F., Fenves, G.L., Scott, M.H., 2000. Open system for earthquake engineering simulation. University of California, Berkeley.

Mosalam, K.M., Günay, S., 2015. Progressive collapse analysis of RC frames with URM infill walls considering in-plane/out-of-plane interaction, *Earthq Spectra*, 31(2), 921–943.

Ricci, P., Di Domenico, M., Verderame, G.M., 2018. Experimental assessment of the in-plane/out-of-plane interaction in unreinforced masonry infill walls, *Eng Struct*, 173, 960–978.

Ricci, P., Di Domenico, M., Verderame, G.M., 2019. Out-of-plane seismic safety assessment of URM infills accounting for the in-plane/out-of-plane interaction in a nonlinear static framework, *Eng Struct* 195, 96–112.

Vamvatsikos, D., Cornell, A.C., 2002. Incremental dynamic analysis. *Earthq Eng Struct Dyn*, 31(3), 491–514.

Brick and Block Masonry - From Historical to Sustainable Masonry –
Kubica, Kwiecień & Bednarz (eds)
© 2020 Taylor & Francis Group, London, ISBN 978-0-367-56586-2

The in-plane and out-of-plane behaviour of masonry panels with thermo-insulating attachment

G.C. Manos, L. Melidis, V. Soulis & K. Katakalos
Lab. Strength of Materials and Structures, Aristotle University of Thessaloniki, Greece

A. Anastasiadis
Civil Engineer, Thessaloniki, Greece

ABSTRACT: Multi-story buildings are composed of multi-bay steel or R/C frames having unreinforced masonry panels considered as non-structural elements not included in the structural design. Such structures are subjected to strong earthquake motions, leading to potentially damaging conditions for the masonry in the form of in-plane damage or/and its dislocation and partial collapse. Several seismic design code revisions include provisions attempting to take into account such problems of unreinforced masonry– frame structure interaction in an indirect way. It was shown in the past that the contact boundary between the masonry infill and the surrounding frame, is of importance. It was also shown that introducing thermal insulation, without due consideration, can increase the vulnerability of such masonry panels. This work presents results from an on-going investigation on the behaviour of thermo-insulating masonry panels. This is done by studying through testing the influence of thermo-insulation on the in-plane and out-of-plane behaviour masonry panels built with hollow clay bricks having horizontal holes. The relevant measured response is presented and discussed together with the corresponding effort to numerically simulate the observed performance.

1 INTRODUCTION

Unreinforced masonry panels are used in multi-story buildings made of steel or reinforced concrete (R/C) to form the exterior facades or the interior partitions. Thermo-insulating panels are also attached on the exterior facades of these masonry panels in order to improve the energy efficiency of these building as well as to reduce noise and moisture penetration. This type of masonry façades is widely applied in many countries. These unreinforced masonry panels are considered as non-structural elements not included in the structural design. Such un-reinforced masonry panels interact with the surrounding structural members, when such structures are subjected to strong earthquake motions. This is because these masonry panels are forced to follow the displacement response of the supporting surrounding structural members (slabs, columns and beams) leading to potentially damaging conditions. The resulting damaging patterns for these masonry panels bear resemblance, up to a point, to bearing masonry structural elements when they are subjected to earthquake type forces. Thus, it is helpful to distinguish damage due to either in-plane or out-of-plane seismic forces. One of the most serious consequence of either of these forcing scenarios, when the seismic forces are considered as acting either separately (in-plane or out-of-plane) or combined, is the dislocation and partial collapse of such un-reinforced masonry panels.

Figure 1 depicts a typical damage pattern observed in numerous mutli-story buildings in Durres, Albania due to the recent strong earthquake sequence (26th November, 2019). Similar damage patterns have been observed in many past strong earthquake sequences in Greece (Kozani 1995, Aigio 1995, Athens 1999, Kefalonia 2014) as well as in many other countries (Italy, L' Aquila 2009, Emilia Romana 2012) (Manos 2011, Earthquake Reconnaissance 2014).

Thus, the seismic vulnerability of un-reinforced masonry panels has been demonstrated by numerous prototype earthquakes in many countries. At this point, one can distinguish two types of masonry panels. The first type is a masonry panel that is confined within a frame-bay formed by two columns as the one depicted in Figure 2. Such a bay is a "masonry infilled frame unit" and it has been the subject of extensive research for some-time (Manos 2012b, 2014). The second type is depicted in Figure 3 whereby masonry panels are not built within such R/C frame bays as described before. In this case the masonry panels have very limited confinement. Therefore, their progressive damage can easily lead to partial collapse.

The necessity to deal with the impact of strong earthquake on masonry infilled R/C frames by relevant

Figure 1. Collapse of un-reinforced masonry facades of a multi-story R/C building located in Durres, Albania (2019).

Figure 2. Collapse of un-reinforced masonry facades of a multi-story R/C building located in L'Aquila, Italy (2019).

Figure 3. Collapse of un-reinforced masonry facades of a multi-story R/C building located in Durres, Albania (2019).

seismic design provions has been long recognized. This is done by relatively such new updated seismic design provisions like the ones included in Euro-Code 8 (1996), American FEMA-306 (1999) and the newest Greek Structural Intervention Regulation (2009). These provisions are based on several methodologies which have been proposed in the past to take into account the influence of masonry infills on the overall behavior of masonry infilled R/C frames when they are subjected to horizontal seismic loads. However, the performance of the infill walls and the attached components should be further investigated, as the possible economic losses and casualties due to collapsed walls cannot be ignored.

An additional feature that must be considered, when examining the vulnerability of un-reinforced masonry panels subjected to seismic forces, is the incorporation of thermo-insulating materials. This is shown in Figures 4a and b whereby double wythe masonry panels are constructed leaving a cavity in between whereby a layer of thermo-insulating material is placed. As shown by Figures 4a and b, such a construction practice increases the vulnerability of a masonry panel when subjected to seismic forces.

Based on the above observations, it is essential to study the performance of masonry panels incorporating thermo-insulating layers, when these masonry panels are subjected to forces that produce within these panels stress-fields resembling those that are resulting from seismic actions. This is the objective of an ongoing research at Aristotle University which aims to investigate both the in plane and the out of plane behavior of masonry panel specimens together with External Thermal Insulating Systems (ETICs). The thermo-insulating materials are produced by "FIBRAN Anastasiadis Dimitrios S.A." and are applied on the specimens to be tested in the same way that are applied in prototype construction, as will be explained in the following. The specimens are masonry sub-assemblies which are constructed and tested at the Laboratory of Strength of Materials

Figure 4. Collapse of double wythe brick masonry facade with thermo-insulation layer in its cavity a) L'Aquila, Italy, 2009 earthquake, b) Athens, Greece, 1995 Earthquake.

953

and Structures (Aristotle University of Thessaloniki, Greece), as discussed in sections 2 and 3. In section 4, summary results from a number of numerical simulations are presented. These numerical simulations were performed aiming to numerically replicate the observed response of the masonry sub-assembly specimens during testing. All the information of the geometry and material characteristics of mortar, bricks and ETICs material specimens used in building these masonry assemblies were utilized during these numerical simulations.

2 EXPERIMENTAL PROGRAM

All specimens were built with the same 12 hole clay brick unit of nominal dimensions length = 320 mm, height = 180 mm and thickness = 150 mm, having a mean gross compressive strength equal to 2.95 MPa (EC6 2005, EN 772-1 2011); this brick unit, with 12 horizontal holes, is commonly used in prototype construction for this type of un-reinforced masonry panels in multi-story buildings in Greece. Similarly, a relative weak mortar, with an average cubic strength equal to 2.42 MPa was used for all specimens. The thermo-insulating layer was added to one side of all specimens two months following their construction, as shown in Figure 5, following the relevant construction practice. Three different thermo-insulating materials, with code names XPS, EPS and Petro, were investigated, having a panel thickness of either 50 mm or 100 mm, all produced by "FIBRAN Anastasiadis Dimitrios S.A." All specimens were built by builders following the relevant prototype work conditions. Specimens of all materials used for building these specimens were taken during construction and tested for determining the relevant mechanical characteristics. Due to space limitations they are not reported here.

These specimens were built in sizes belonging to two different categories. The first category, named "wallets", is of rather medium size and were used to investigate the behaviour of the thermo-insulating panel attached on the masonry when subjected to specific relatively simple loading conditions. The second category are much larger and are designed to be subjected to combined loading to

subject a specimen, with an almost square shape, to a monotonic compressive force along its main diagonal, as shown in Figures 6a and b, thus resulting in a mainly in-plane state of stress. Figures 6a and b depict such wallet specimens, one without and the other with the thermo-insulating attachment, ready to be tested. A second loading condition was designed to subject a wallet specimen, with a rectangular shape, to an out-of-plane forcing, as depicted in Figures 7a and b. In this case, the wallet was placed on a supporting steel frame having its upper and lower horizontal sides simply-supported, whereas the vertical sides completely free. The out of plane load was introduced in A cyclic manner through an electronically controlled servo-actuator through its back side in a way that it could be considered as spreading along a middle line of the specimen thus subjecting it to flexure (Figure 7c). As can be seen Figure 7a depicts a specimen without thermo-insulation whereas Figure 7b is a specimen of the same geometry that has a thermo-insulating attachment (XPS).

Figure 6. Diagonal compression of a masonry wallets with dimensions 1340 mm × 1340 mm and 150 mm thick.

Figure 7. Out-of-plane flexure of a masonry wallet.

1. Masonry wall

2. Adhesive mortar interface used to attaching the thermo-insulating panel to the masonry façade

3. Plastic anchors also used to attaching the thermo-insulating panel to the masonry facade

4. Thin mortar layer protecting the thermo-insulating panel

5. Thin plastic reinforcing net that is attached with special thin mortar layer to form the final

Figure 5. Construction detail of the masonry specimens with the thermo-insulating attachment.

The applied load in either in-plane or out-of-plane loading conditions is continuously recorded together with the resulting displacement response utilizing a number of displacement transducers, as can be seen in Figures 6 and 7. Due to space limitations this discussion is limited to only essential information.

3 LABORATORY MEASUREMENTS

3.1 *Flexural response*

Initially, simple flexural tests were performed utilizing triplets specimens made of three clay bricks and two bed mortar joints constructed with the same materials as the wallets (Figure 8). The results in terms of flexural stress versus the out of plane displacement are shown in Figure 9.

The measured out-of-plane response of a wallet specimen 2000 mm in length and 940 mm height, constructed in the way described in section 2, is depicted in Figure 10. This figure depicts the variation of the amplitude of the applied out-of-plane load versus the horizontal out-of-plane displacement measured at the center of the specimen. Based on the support conditions of this wallet through the applied out-of-plane cyclic load the flexural stress, at the bottom fiber of a mid-span horizontal cross section, normal to one horizontal central mortar joints could be obtained. The ultimate state is the fracture of this horizontal

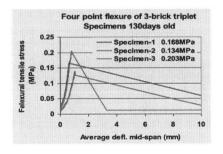

Figure 8. Experimental set up of out of plane triplet bending test.

Figure 9. Flexural stress versus the out of plane displacement obtained by triplet bending tests.

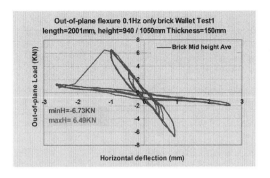

Figure 10. Variation of the applied out-of-plane load versus the horizontal out-o-plane displacement at the center of the specimen.

mortar joint when the tensile flexural stress reaches the value of approximately 0.20 MPa, a value which is in agreement with the corresponding range of ultimate tensile flexural stress values found from the simple triplet tests. As can be seen in Figure 10, after the ultimate state is reached for a maximum value of the out-of-plane load approximately equal to 6.5 KN a sudden drop is observed in the flexural capacity of the specimen, in terms of the applied out-of-plane load, which is accompanied by a substantial increase in the out-of-plane displacement response. This is also evident in the displacement response of this wallet as depicted in Figure 11. The out-of-plane displacement patterns of a vertical cross-section going through the center of the specimen are plotted in this figure. Four different curves are plotted corresponding to four specific time instants of the loading sequence when the cyclic load and displacement response is maximized. The value of the applied out-of-plane load at each time instant is also noted in this Figure.

As can be seen in Figure 11, the flexural out-of-plane response of the un-reinforced clay brick masonry wallet, after maximum load, is dominated by large displacement values that develop for relatively small values of the applied load (approximately 2 KN).

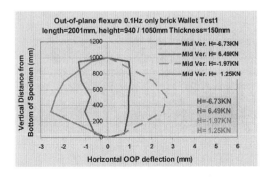

Figure 11. Out-of-plane displacement response for a vertical cross-section passing through the center of the specimen.

955

Next, the corresponding measured out-of-plane response of a wallet with thermo-insulating material is presented. This wallet is of the same geometry and is built with the same materials as the wallet without the thermo-insulating material presented in Figures 10 and 11. This time, a thermo-insulating panel (EPS) with a thickness of 50mm is attached on one of the facades of this specimen, as described in section 2 (Figure 5). The out-of-plane cyclic load was again applied aiming to subject in flexure the thermo-insulating panel and the adhesive mortar interface connecting the thermo-material to the brick façade. The measured displacement response of this EPS-wallet is depicted in Figures 12 and 13. As can be seen in these figure, the maximum value of the applied horizontal out-of-plane load is equal to 33.35 KN, approximately five times the corresponding value measured during the flexure test of the wallet without the thermo-insulating material.

The out-of-plane displacement response is again measured along the panel height at two vertical cross-sections passing through the specimen approximately 150 mm left and right from its central vertical axis of symmetry. Figure 13 depicts the out-of-plane deflection patterns of the specimen measured along the

height of these two vertical cross-sections (East and West). Four pairs of deflection curves are plotted in this figure. One curve of each pair is for the East and the other one for the West vertical cross-section. The first pair of deflection pattern curves corresponds to the time instant when the applied out-of-plane load is maximized as absolute value (Figures 12 and 13, 33.35 KN). The second, third and fourth pair of deflection curves in Figure 13 corresponds to the time instant when a limit state condition develops whereby the significant drop in the flexural capacity of the specimen (5.35 KN), in terms of the amplitude of the applied out-of-plane load, is accompanied with the detachment of the bottom part of the thermo-insulating panel (Figure 14) as well as an increase in the amplitude of the out-of-plane displacement response (25 mm).

A wallet with the same thickness (150 mm) but with relatively smaller dimensions (1000 mm length, 570 mm height) was also tested. This time the thermo-insulating material was XPS 50mm thick (Figure 15). Figure 16 depicts the variation of the amplitude of the axial flexural stress versus the horizontal out-of-plane displacement measured at the center of this specimen. As can be seen in Figure 16, the obtained maximum axial flexural stress value in this case is approximately 0.30 MPa, which is distinctly larger than the value of 0.20 MPa, which was measured during a similar test with a wallet without thermo-insulating material.

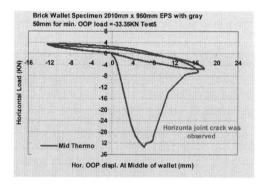

Figure 12. Variation of the applied out-of-plane load versus the horizontal out-o-plane displacement at the middle of the specimen with thermo-insulating material (EPS 50 mm).

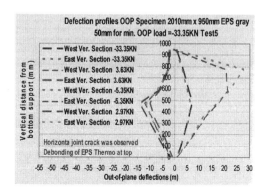

Figure 13. Variation of the applied out-of-plane load versus the horizontal out-o-plane displacement at the center of the specimen with thermo-insulating material (EPS 50 mm).

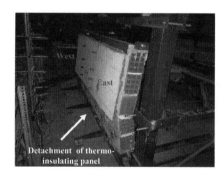

Figure 14. Limit state of the out-of-plane displacement response for the specimen with thermo-insulating material (EPS 50 mm) having the bottom part of the thermo-insulating panel detached from the masonry façade.

Figure 15. Limit state of the out-of-plane displacement response for the specimen thermo-insulating material (XPS 50 mm) developing a detachment of it upper part.

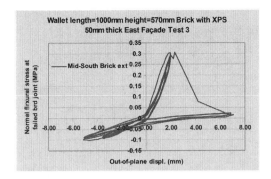

Figure 16. Flexural response, in terms of maximum axial stress perpendicular to the mortar horizontal bed joint versus the corresponding horizontal out-of-plane displacement.

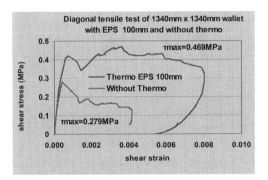

Figure 17. Comparison of diagonal compression response for brick wallets with and without thermo insulation.

3.2 In-plane diagonal compression response

The two specimens depicted in Figure 6, one with 100 mm EPS thermo insulation and the other without any thermo insulation, were subjected to a monotonic diagonal compressive load. The measure response is depicted in Figure 17 in terms of equivalent shear strain versus shear stress for the brick panel geometry. As can be seen in this figure, the presence of the insulating attachment resulted in an increase of the observed bearing capacity.

4 NUMERICAL SIMULATION

In what follows a preliminary numerical investigation is presented aimed to predict the non-linear behavior of clay brick masonry sub-assemblies tested at the laboratory. The 3-D finite element simulation was formed utilizing the capabilities of commercial software (Abaqus). These numerical simulations employed all the geometric and mechanical characteristics of the tested specimen obtained by material testing (see sections 2 and 3.1). Due to space limitations more details will be provided in a future publication.

These numerical simulations adopt a simplified micro-modeling approach, with the clay bricks being simulated as elastic blocks with linear elastic properties, while the non-linear behavior is included as interfaces at the mortar joints between the blocks assigned with non-linear stress-strain constitutive laws. Such micro-modelling numerical simulation of masonry panels has been employed in the past by many researchers (Sandoval et al. 2011).

Initially, such a numerical simulation was tried for the flexural triplet test specimens (see Figure 8). The geometry and the boundary conditions are depicted in Figure 18. The measured response is compared with the numerical predictions in Figure 19. Figure 20 depicts the distribution of the axial stresses normal to

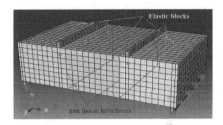

Figure 18. Numerical model simulating the triplet bending.

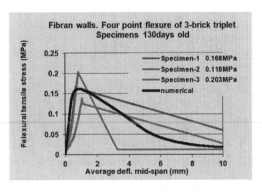

Figure 19. Comparison of measured and predicted behavior in terms of flexural stress versus the vertical mid-span deflection for the triplet tests (see Figures 8 and 9).

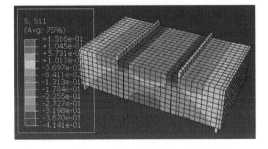

Figure 20. Distribution of the axial stresses (MPa) normal to the vertical mortar joints at the step when the ultimate load is reached (triplet test).

the mortar joints at the step when the maximum load is reached. As can be seen, the measured response is in good agreement with the numerical predictions.

Similar numerical simulations were formed in order to predict the out-of-plane flexural response of the wallet specimens with and without any insulating panel presented in section 3.1. The insulating panel, either EPS or XPS, is represented by a solid part with linear elastic properties attached to the masonry substrate with a 5mm layer of adhesive mortar. The adhesive mortar is numerically simulated with solid elements assigned with non linear properties, using the "Concrete Damaged Plasticity" constitutive law included in the used commercial software (Abaqus), with parameters derived from material testing in compression and four-point flexure. The adhesive mortar and the masonry are connected with a non-linear interface having a tensile strength normal to the wallets plane equal to 0.1MPa. The numerical simulation of the specimen without thermo-insulation is depicted in Figures 21-23. The maximum axial flexural stress is about 0.21 MPa almost equal to the one recorded at the experimental sequence described before (Figure 10). Figure 22 depicts the out-of-plane displacement response whereas Figure 23 the variation of the axial flexural stress with the corresponding horizontal displacement.

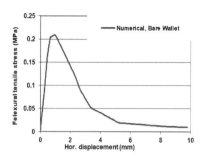

Figure 23. Numerical prediction of the out of plane response of the wallet without thermo-insulating material.

Figures 24-26 depict the corresponding numerical response predictions for the wallet with a thermo-insulating panel (XPS 50 mm). As can be seen in Figure 25, at the ultimate state the numerically predicted pattern is that of detachment of the thermo-insulation panel from the masonry, which is quite similar to the observed behavior (see Figures 15 and 16). The numerically predicted maximum axial flexural stress in this case, whereby the thermo-insulation is included in the numerical simulation, is again approximately equal to 0.3 MPa.

The comparison of the obtained numerical flexural response without and with thermo-insulation (XPS) is

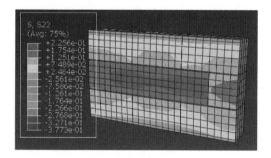

Figure 21. Distribution of normal to the joints stresses (MPa) at the step of the ultimate load, bare wallet.

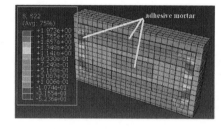

Figure 24. Distribution of normal to the joints stresses (MPa) at the step of the ultimate load, wallet with 50 mm XPS.

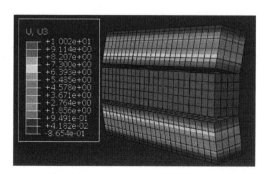

Figure 22. Mode of failure of the bare wallet numerical model. U3 displacement (mm) represents the out of plane displacement.

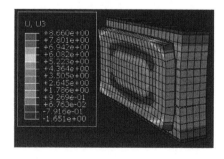

Figure 25. Deformed shape for an horizontal displacement 4 mm. Bed joints failed flexural, while the adhesive mortar develops plastic strains and debonding between wall and the insulating panel occurred. U3 displacement (mm) represents the out of plane displacement.

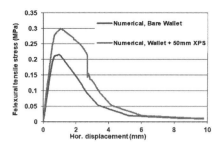

Figure 26. Numerical prediction of the out of plane response, for the wallets without or with thermo-insulation (XPS).

shown in Figure 26. The predicted axial flexural stress increase, due to the presence of this particular thermo-insulating panel, is in agreement with the corresponding observed response (Figures 9 and 16). Due to space limitation, a discussion with reference to relevant work by other researches is not done here.

5 CONCLUSIONS

- The current research effort focuses on the seismic response of un-reinforced masonry panels, which form the exterior facades or interior partitions of multi-story buildings. Such un-reinforced masonry panels interact with the surrounding structural members, when such structures are subjected to strong earthquake motions, leading many times to considerable damage even collapse of such panels. The performance of masonry panels with and without incorporating thermo-insulating layers is studied, when they are subjected to in-plane and out-of-plane forces.

- Summary results from the observed out-of-plane behavior of masonry wallets constructed with prototype materials, with or without thermo-insulation (EPS, XPS), are presented and discussed. It was observed, through the limited testing executed up to now, that the out-of-plane flexural bearing capacity of the specimens including thermo-insulating panels, is larger than the corresponding bearing capacity of similar specimens without the used thermo-insulating attachments. The observed limit state for the specimens with thermo-insulation was partial debonding of the insulating panel. The used plastic anchors prevent, up to a point, the complete debonding of such thermo-insulating panels. The same effect was also observed when these panels were subjected to in-plane diagonal compression.

- 3-D finite element simulations of the out-of-plane flexural behaviour were formed utilizing the capabilities of commercial software. In these numerical simulations all the geometrical, loading and support details of tested specimens were numerically simulated. Such numerical simulations included nonlinear interfaces in an effort to numerically simulate the observed behaviour. In this way, the observed

response was successfully captured by these numerical simulations. The methodology adopted here includes testing samples of the materials used to verify their basic mechanical properties and sub-assemblies of masonry wallets with insulating panels subjected to seismic-type loading, combined with numerical models developed, and it is considered, up to a point, satisfactory. This methodology will be further validated with additional experimental results and parametric investigation in a variety of specimen geometry and materials used.

ACKNOWLEDGEMENTS

All materials for the construction of the specimens were provided by "FIBRAN Anastasiadis Dimitrios S.A.". Part of the aforementioned research has been co-funded by Greece and European Union through the Operational Program "Erevno – Dimiourgo - Kainotomo" which are gratefully acknowledged.

REFERENCES

Manos G. 2011. Consequences on the urban environment in Greece related to the recent intense earthquake activity. *IJCEA* 5(12) (49): 1065–1090.
GEER - EERI - ATC - Cephalonia GREECE Earthquake Reconnaissance January 26th/February 2nd 2014 Version 1: June 6 2014.
Provisions of Greek Seismic Code 2000, EPPO, Earthquake Planning and Protection Organization Athens, Greece, December 1999.
Eurocode 6. 2005. Design of masonry structures - Part 1-1: General rules for reinforced and unreinforced masonry structures.
FEMA-306. 1999. Evaluation of Earthquake damaged concrete and masonry wall buildings – Basic Procedures manual. Federal Emergency Management Agency, Washington, D.C.
EN 772-1. 2011. Methods of test for masonry units - Part 1: Determination of compressive strength.
Eurocode 8. 2003. Design of structures for earthquake resistance - Part 1: General rules, seismic actions and rules for buildings.
Manos G.C. et al. 2012a The behavior of masonry assemblages and masonry-infilled R/C frames subjected to combined vertical and cyclic horizontal seismic-type loading. *I. J. Advances in Engineering Software* 45: 213–231.
Manos G.C. et al. 2012b. A nonlinear numerical model and its utilization in simulating the in-plane behaviour of multi-story R/C frames with masonry infills. *Open Constr. Build. Technol. J.* 6 (Suppl 1-M16): 254–277.
Manos G.C. & Soulis V. 2014. Simulation of the in-plane seismic behaviour of masonry infills within Multistory Reinforced Concrete Framed Structures. *Proc. 9th Int. Masonry Conference, July.*
Manos G.C. et al. 2015. Field experiments for monitoring the dynamic soil-structure-foundation response of model structures at a Test Site. *J. Struct. Eng.* 141(1).
Abaqus Unified FEA - SIMULIA™ by Dassault Systèmes.
Sandoval C. et al. 2011. Testing and numerical modelling of buckling failure of masonry walls. *Constr. Build. Mater.* 25(12): 4394–4402.

Brick and Block Masonry - From Historical to Sustainable Masonry –
Kubica, Kwiecień & Bednarz (eds)
© 2020 Taylor & Francis Group, London, ISBN 978-0-367-56586-2

The interaction between in-plane and out-of-plane seismic response of modern strong masonry infills

R.R. Milanesi & G. Magenes
Department of Civil Engineering and Architecture, University of Pavia, Italy

P. Morandi
Department of Structures and Infrastructures, EUCENTRE Pavia, Italy

S. Hak
Department of Structural Safety, Basler & Hofmann

ABSTRACT: Although during recent post events surveys it has been observed that several masonry infills have been collapsed or severly damaged due the interaction of the in-plane and out-of-plane seismic actions, the standard prescriptions on the in-plane/out-of-plane interaction of traditional rigidly attached masonry infills are often limited or even missing. Many studies have been accomplished on this topic, also recently, being often focused on infills made of bricks, or "weak" masonry infills with high slenderness. However, the researches and the experimental campaigns related to the out-of-plane behaviour of "strong"/"robust" clay masonry infills, for example made by vertically perforated blocks with thickness larger than 20-25 cm, are very limited. The importance of the topic is also related to the wide adoption of this typology of masonry infill in many countries also due to its thermal and acoustic performance. This paper presents a study on the out-of-plane response of a relatively "strong" infill constructed using tongue and groove clay block masonry units with a thickness of 35 cm. The work is based on the results of an experimental campaign conducted at the University of Pavia and Eucentre on real scale one-storey, one-bay reinforced concrete infilled specimens. A series of cyclic static in-plane and out-of-plane tests has been carried out; the out-of-plane experiments have been executed on an undamaged specimen and on infills previously damaged in-plane, reaching levels of maximum drift equal to 1.00, 1.50 or 2.50%, representing the attainment of different performance levels, in order to evaluate the related out-of-plane resistance reduction. The aim of the study is to define a criterion to compute the out-of-plane resistance of strong masonry infills, through the developmenet of an in-plane/out-of-plane interaction curve, where a reduction coefficient of the out-of-plane resistance of the undamaged panel is function of the in-plane drift of the infilled frame and every data have been obtained from experimental results.

1 INTRODUCTION

Although in European design practice masonry infills are commonly deemed as non-structural elements, damage to masonry infills induced by earthquakes can represent threat to human lives and lead to extensive economic losses, as demonstrated by several field observations, once more in recent years, as in L'Aquila 2009 (see e.g. Braga et al. 2011, Ricci et al. 2011), in Emilia 2012 (see e.g. Manzini & Morandi 2012) and in Central Italy (see e.g. Fragomeli et al. 2017), also on recently constructed RC buildings (see Figure 1). As noticed in many post-seismic inspections (see e.g. Figure 2), the more extensive damages on infills are not located at the top part of infilled frames, where larger out-of-plane seismic actions are expected, but at the bottom/intermediate levels, where higher in-plane drift demand occurs, proving that masonry panels decrease their out-of-plane stability when are subjected to in-plane

action. The simultaneous effects of the in-plane and out-of-plane seismic action produces a decrement of the in-plane deformation capacity due to out-of-plane action and a reduction of the out-of-plane capacity due to in-plane action.

In the past, the experimental investigations have been focused mainly on the in-plane seismic response on selected masonry typologies like clay bricks and concrete blocks. Fewer tests have been conducted for studying the out-of-plane response and even less on the mutual in-plane/out-of-plane interaction (e.g. Angel et al. 1994, Calvi & Bolognini 2001, Morandi et al. 2013/Hak et al. 2014, da Porto et al. 2013/Guidi et al. 2013, Furtado et al. 2016, Ricci et al. 2019). Moreover, some tests on simultaneous in-plane/out-of-plane effects have been very recently performed on different innovative infill solutions, by Manzini et al. (2018) through shaking table tests and by Butenweg et al. (2019) through static cyclic tests.

Figure 1. Example of damage on infills from after 2016-2017 earthquake in central Italy.

Figure 2. Collapses of infill in 2019 in Albania due combined in-plane and out-of-plane earthquake loading direction.

However, limited experimental tests have been carried out on real scale infilled RC specimens and, among these, only the research reported by da Porto et al. (2013) and Guidi et al. (2013) has been devoted to study infills with thick/strong masonry representing a modern typology in newly designed buildings, whereas the others were mainly related to study a weak masonry, for example characterized by horizontally highly perforated thin clay block masonry (8 to 12 cm thick infills), typical of old constructions.

Although the present Italian (NTC18, 2018) and European seismic design codes (EC8 – Part 1, 2004) do not consider the in-plane/out-of-plane interaction in the seismic response of masonry infills, the importance to account for the possible out-of-plane strength reduction due to previous in-plane damage was to some extent recognised (e.g., Angel et al. 1994, Morandi et al. 2013, reinterpreting the experimental outcomes by Calvi & Bolognini 2001 on weak unreinforced and lightly reinforced masonry infills, Furtado et al. 2016 and Ricci et al. 2019, dealing with weak masonry infills).

Since the studies and the experimental campaigns related to the in-plane/out-of-plane behaviour of "strong"/"robust" clay masonry infills (vertical holes with a thickness of 30-35 cm) are limited, the following sections present the results of an experimental study on the in-plane/out-of-plane response of "strong" infills made of clay masonry with

a thickness of 35 cm. In particular, in-plane cyclic tests followed by out-of-plane cyclic tests on previously in-plane damaged specimens on full-scale, single-storey, single-bay RC frame infilled with the considered strong infill typology have been carried out at the Eucentre and the University of Pavia. While the results and the interpretation of the in-plane tests have been already discussed in Morandi et al. (2018a), in this paper the attention is focused on the results of the out-of-plane tests and the assessment of the out-of-plane strength in function of previous in-plane damage.

2 EXPERIMENTAL CAMPAIGN

2.1 Description of the specimens and material characterization

Within the scope of the experimental campaign, a series of cyclic static in-plane and out-of-plane tests has been carried out on bare and fully or partially infilled full-scale single-storey, single-bay 4.22 × 2.95 m RC frames (see Figure 3), newly designed according to European (and Italian) code provisions. After a detailed characterization of all material components (i.e. concrete, reinforcing steel, mortar, masonry units and masonry), the experimentation has been accomplished on many frame specimens, as summarised in Table 1.

All the results and the interpretation of the in-plane tests have been already discussed by Morandi et al. (2018a).

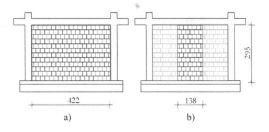

Figure 3. A) Fully infilled specimens (TA1, TA2, TA3, TA6); b) vertical stripe specimen (TA5).

Table 1. Summary of the experimental campaign.

No.	In-plane	Out-of-plane	Configuration
TNT	3.50 %	—	Bare frame
TA1	1.50 %	75 mm	Fully infilled
TA2	2.50 %	75 mm	
TA3	1.00 %	75 mm	
TA5	—	75 mm	Infill stripe
TA6	—	Max force	Fully infilled

The out-of-plane experiments have been carried out on the specimens TA1, TA2 and TA3 previously damaged in-plane, in order to allow the evaluation of the related out-of-plane resistance reduction. Moreover, two specimens were tested in out-of-plane direction without any previous in-plane damage: a 1.38 m wide stripe of the infill (TA5) and a fully infilled one (TA6), with the aim to evaluate the out-of-plane strength of previously undamaged specimens respectively under vertical single-bending and double-bending conditions. The specimen TA6 was tested under monotonic load. All the out-of-plane tests have been pushed up to the attainment of ultimate conditions.

The selected typology represents a commonly adopted, traditional strong single-leaf unreinforced masonry infill of 35 cm thickness, consisting of vertically hollowed lightweight tongue and groove clay units, having nominal dimensions of 235 × 350 × 235 mm and a nominal volumetric percentage of holes of 50%. The application of a general-purpose mortar type "M5" (compression strength of 5.0 MPa) was considered a suitable choice with respect to common construction practise. The out-of-plane slenderness λ (height over thickness) of the infills is equal to about 8.4.

The infills have been constructed after full hardening of the RC frame, adopting traditional bed joints, having a thickness of about 1.0 cm and dry head joints. The bed joint mortar was applied in two stripes in the longitudinal direction of the wall with an intermediate cavity of about 2.0 cm (Figure 4). Full contact between the infill and the surrounding RC members was assumed to be achieved filling the remaining vertical gaps on the two sides of the infill and the horizontal gap at the top of the infill with mortar (Figure 5).

The design of the specimen has been carried out following the European code provisions for new buildings (EC 8-Part 1 2004, EC 1-Part 1-1 2001, EC 2-Part 1-1 2004). More information regarding the design, the dimensions and the reinforcement details of the RC frame are reported in Morandi et al. (2018a). Furthermore, a complete characterisation of the relevant properties for all materials utilised for the construction of the specimens has been carried out, as reported in more detail in Morandi et al. (2018a, b), including the evaluation of unit, mortar and masonry

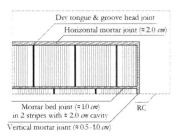

Figure 5. Mortar joints adjacent to the RC members.

properties for the selected infill typology. A summary of the results is reported in Table 2.

2.2 Out-of-plane test setups

The in-plane and out-of-plane cyclic static tests were carried out at the laboratory of the Department of Civil Engineering and Architecture of the University of Pavia. A detailed description of the in-plane test setup, the adopted instrumentation and the testing protocol is given in Morandi et al. (2018a) and in Hak et al. (2014). Furthermore, the construction of a newly designed steel reaction frame was accomplished, aimed to serve as a reaction frame, as an out-of-plane restraint, during the out-of-plane cyclic pseudo-static tests and also during the in-plane cyclic tests. The out-of-plane tests of the specimens previously damaged in-plane have been carried out immediately following the in-plane experimentation, since they were performed on masonry infills that have previously sustained a certain level of in-plane damage. The steel setup was assembled on a 35.0 cm thick RC foundation slab in order to provide a sufficiently rigid support and to ensure a satisfactory transfer of forces to the strong floor. The steel structure was conceived consisting of a central reaction plane serving as a support for the actuator introducing the horizontal load on the masonry infill and two external restraining planes, as illustrated in Figure 6a-b.

The three structural planes were connected in the transversal direction with a steel beam allowing the attachment of the actuator. Additionally, above the actuator, the two restraining planes were connected

Table 2. Mechanical properties of the clay units, mortar and masonry used in the specimens.

Property	Mean	C.o.V.
Vert. normalized compr. strength units	9.81 MPa	9.2%
Later. norm. compr. strength units	3.15 MPa	10.8%
Flexural strength mortar	2.15 MPa	19.4%
Compression strength mortar	7.68 MPa	20.6%
Vertical compr. strength masonry	4.64 MPa	14.1%
Vertical elastic modulus masonry	5299 MPa	8.6%
Lateral compr. strength masonry	1.08 MPa	14.5%
Lateral elastic modulus masonry	494 MPa	32.8%
Initial shear strength of bed-joints	0.359 MPa	26.0%

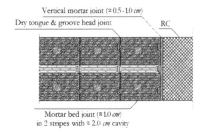

Figure 4. Sketch of the mortar bed joints.

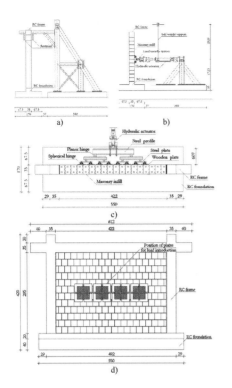

Figure 6. Layout of the out-of-plane test setup: a) restraining plane; b) reaction plane; c) plan view and d) front view of the system for out-of-plane load transfer from actuator to masonry infill.

with a steel profile that provided support for the self-weight of the actuator. In the restraining planes, the steel reaction frame was connected to the beam of the RC frame specimen. The load was applied at the mid-height of the masonry infill by means of a servo-controlled hydraulic actuator with an internal load cell.

In order to transfer the load from the actuator to the masonry infill, a system consisting of a series of hinged steel beams and plates was developed, allowing the introduction of the horizontal force in a relatively large number of discrete points, as illustrated in Figure 6c and d. In particular, for the fully infilled frame specimens previously damaged in-plane (TA1, TA2 and TA3) the forces were applied in two lines and eight points per line, reproducing an approximately linear load distribution at the mid-height of the panel. For the infill stripe (TA5) the load introduction through eight central points was aimed to reproduce vertical single-bending. Further information regarding the experimental setup are reported by Hak et al. (2014).

The undamaged fully infilled specimen TA6 has been subjected to a simple out-of-plane monotonic load test up to the collapse. The specimen was over-turned horizontally to make vertical the direction of

the out-of-plane loading (Figure 7a). The RC frame was supported by a series of concrete blocks to guarantee a continuous vertical support along the perimeter of the frame. The force was applied laying layers of bags containing dry sand (weight of each bag equal to 25 kg) to cover the whole area of the masonry panel up to collapse (Figure 7b). At the end of laying each layer, the damage pattern was monitored through pictures, while the measurements of the displacements were continuous during the whole duration of the test.

2.3 Instrumentation and loading protocols

For the evaluation of the out-of-plane displacements of the infill panel (horizontal for TA1 to TA5, vertical for TA6) and to monitor eventual displacements of the surrounding RC frame during the out-of-plane test, displacement transducers (linear potentiometers) have been adopted. In total, 30 potentiometers have been used for the fully infilled frame specimens TA1, TA2 and TA3, 28 for the infill stripe TA5 and 23 for TA6.

For all infill configurations, with the exception of TA6, the out-of-plane displacements of the RC frame have been monitored as described by Hak et al. (2014) and reported in Figure 8a. For TA6, as shown in Figure 8b, six potentiometers have measured the deformation of the top beam, while four potentiometers have controlled the deformation at mid-height of the column, installed in couples in order to monitor also the rotation of the RC members.

The tests in the out-of-plane direction of TA1 to TA5 have been accomplished using the steel structure designed to act as a reaction frame and to provide adequate restraining of the specimen, preventing out-of-plane displacements of the frame. At the beginning of the tests, the actuator and the out-of-plane restraints have been brought in position and the system for the horizontal load transfer has been placed in contact on the masonry infill. The cycles of horizontal out-of-plane load were imposed on the infill by means of the actuator, loading in one direction (push) and back to zero force. Firstly, two different levels of force-controlled loading were accomplished; subsequently, displacement-controlled loading cycles at increasing

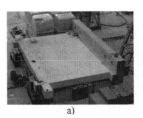

Figure 7. Specimen TA6: a) before the test; b) load application.

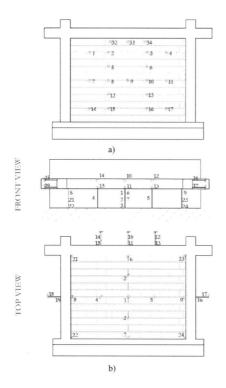

a)

b)

Figure 8. A) TA1, TA2, TA3; b) TA6 instrumentation layout.

3 INTERPRETATION OF THE OUT-OF-PLANE EXPERIMENTAL RESULTS

3.1 *Out-of-plane force-displacement response*

The results of the cyclic out-of-plane tests in terms of hysteretic force-displacement curves for the fully infilled frame configurations (TA1, TA2 and TA3), carried out following the corresponding in-plane tests and for the undamaged vertical infill stripe (TA5), have been evaluated in detail by Hak et al. (2014). The outcome of the monotonic test for specimen TA6 has been instead considered in terms of pressure (total load divided by the area of the infill) or total applied load versus displacement. The cyclic force-displacement responses and the corresponding maximum envelopes obtained for specimens TA1, TA2, TA3 and TA5 are presented in Figure 10a-d, whereas the monotonic pressure-displacement response for specimen TA6 is reported in Figure 10e. Further information is reported in Hak et al. (2014) and Milanesi et al. (2019).

The different loading protocol between TA6 and the other specimens has requested an interpretation of the experimental response in order to define the equivalent force that would have been applied according to the experimental procedure adopted for TA1-TA2-TA3. The equivalent force of TA6 has been computed is resulted equal to 274 kN on the basis of an "energetic" approach. The equivalent force according to the "energetic" approach, which

levels of maximum displacement at the centre of the panel were imposed. For each level of loading (i.e., target force or displacement) three complete loading cycles have been carried out and the duration of the application of the load has been kept approximately constant. During unloading, a minimum force of about 5.0 kN has been maintained in order to ensure complete contact between the loading system and the panel. In Figure 9a, the loading protocol adopted for TA2 is reported.

The loading protocol for TA6, which is shown in Figure 9b, was based to a typical monotonic increasing load test and it was stopped at the collapse of the infill.

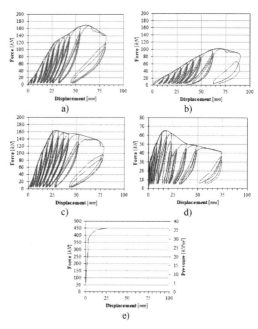

a)

b)

c)

d)

e)

Figure 10. Force-displacement out-of-plane response of: a) TA1; b) TA2; c) TA3; d) TA5; e) TA6.

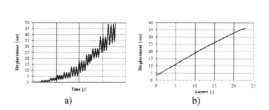

a)

b)

Figure 9. Testing loading protocols: a) for TA2; b) for TA6.

will be object of future studies, is based on the equivalence of the "applied energy" between the two loading protocols, and a "volume of the deformed shape" that is in agreement with the results coming from TA1, TA2 and TA3.

3.2 Sequence of damage propagation and deformed shapes

The damage propagation observed during the test on specimens TA1, TA2, TA3 and TA5 is described in detail by Hak et al. (2014).

Specifically, for TA2, the first residual deformations of the infill occurred at a displacement of 10 mm, when also the creation of a crack at the joint between the masonry panel and the beam was observed. The initial behaviour of the infill indicated a cantilever wall response; however, at the following loading cycles, the formation of an arching mechanism has been observed and horizontal cracks in the mortar bed joints propagated further, in particular in the joint at mid-height of the panel. Subsequently, the arching action continued to develop, residual displacements became more pronounced and local damages within the blocks have been observed, especially in the corners and the top and bottom areas of the infill. At the last target displacements, the damage was spread throughout the infill due to the formation of new cracks and due to previous in-plane damage. The damaged specimen TA2 during the test at the maximum level of imposed out-of-plane displacement (75 mm) is shown in Figure 11.

A comparable response was observed during the out-of-plane tests of the fully infilled frame specimens TA1 and TA3, that were subjected to lower levels of previous in-plane drift, i.e., 1.5% and 1.0%, resulting in correspondingly less extensive damage, as well as higher out-of-plane resistance. In the case of specimen TA1 (Figure 12), the creation of a predominant horizontal crack at the central joint of the infill and the development of a diagonal stepwise crack pattern towards the corners of the infill were identified. Nevertheless, on the bottom left side of the panel, the stepwise cracking was less pronounced and the local failure of blocks in the region along the

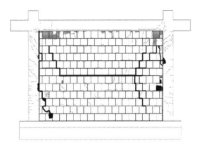

Figure 12. Specimen TA1, crack pattern (in back – out-of-plane damage; in grey – previous in-plane damage).

left RC column and in the left bottom corner was observed instead. On specimen TA3 (Figure 13), in addition to the central horizontal crack and the characteristic diagonal stepwise crack pattern towards the corners of the infill, which was particularly pronounced in the upper half of the panel, extensive damage occurred at the intersection between the upper right stepwise crack and the central horizontal crack. The failure of several masonry blocks was observed and a significant slip of the masonry blocks along the diagonal crack remained after the test.

A damage pattern in line with the other specimens has been observed also for TA6. The creation of horizontal cracks approximately at the centre of the panel due to the vertical arching mechanism and the detachment and sliding of the infill at the beam-panel interface has been noticed. Moreover, diagonal cracks from the corners of the frame have also been detected.

As expected, due to the typology of the specimens, the fully infilled RC specimens (TA1, TA2, TA3, TA6) have shown a double arching resisting mechanism, while specimen TA5 has resisted to the out-of-plane action only through a vertical arching mechanism.

The progressive propagation of damage during the test on specimen TA5 is illustrated in Figure 14a for a cross section view of the specimen, while the damaged specimen during the test at the maximum level of imposed out-of-plane displacement (75 mm) is shown in Figure 14b. The evolution of the damage (Hak et al. 2014) has involved the infill/frame interfaces and the

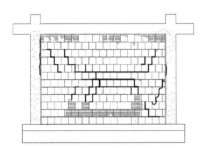

Figure 11. Specimen TA2, crack pattern (in back – out-of-plane damage; in grey – previous in-plane damage).

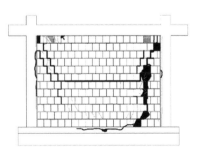

Figure 13. Specimen TA3, crack pattern (in back – out-of-plane damage; in grey – previous in-plane damage).

965

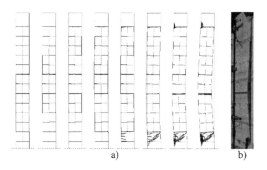

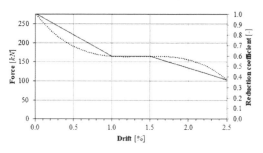

Figure 15. Out-of-plane strength reduction in function of in-plane drift for strong infill.

Figure 14. Specimen TA5: a) progression of damage; b) damage at 75 mm target displacement.

vertical arching mechanisms related cracks (horizontal cracks within the mortar joints at the edges and approximately at mid-height of the infill stripe). Moreover, achieving a displacement demand of 40 mm, an inclined sliding surface formed through the masonry blocks of the bottom course, starting from the joint above the first course at the backside of the panel towards the interface between the infill and the foundation at the front side of the specimen. Furthermore, the horizontal crack above the fifth course resulted to be the most pronounced and started to open evidently.

4 OUT-OF-PLANE STRENGTH REDUCTION IN FUNCTION OF IN-PLANE DAMAGE

The associated experimental values of out-of-plane resistance obtained following the in-plane tests (i.e. at 0.0%, 1.0%, 1.5% and 2.5% drift) are reported in Figure 15. Alternatively, the out-of-plane strength can be expressed as the ratio between the resistance of damaged with the one of the undamaged panel.

Even though only limited data is currently available, resulting in three values of reduced strength at three levels of previous damage, the existing test results indicate that the reduction of the out-of-plane resistance can be assumed from the experimental outcomes. Therefore, the out-of-plane resistance descends for increasing levels of previously imposed in-plane drift. Subsequently, the results of the present study can be adopted for possible design applications to define a simple and effective approach where the out-of-plane strength reduction coefficient may be defined in function of the expected in-plane drift demand δ_w of the infilled frame. The in-plane drift demand δ_w can be evaluated, for example, from analyses on bare frame configurations or considering simplified approaches, as the one proposed by Hak et al. (2018).

For design/assessment purposes, possible approximations of the out-of-plane resistance in function of increasing in-plane drift demands δ_w may be represented by a linear reduction by parts or with a polynomial equation in function of the drift as reported in Figure 15.

Moreover, similarly of what has been observed for "thin/weak" masonry infills (e.g. Furtado et al. 2018 and Ricci et al. 2018), a sharp reduction of the out-of-plane stiffness as the in-plane drift increases has also been observed. According to the preliminary results of the research, that is currently ongoing, the values of stiffness have been computed at the point corresponding to 0.7 times the peak force of the experimental force-displacement envelope.

5 CONCLUSIONS

The present paper is focused on some aspects of an experimental study carried out with the aim to improve the understanding of the cyclic response of rigid masonry infills, commonly adopted for building enclosure systems in seismic regions, with particular interest in the combined action of in-plane and out-of-plane horizontal loads. Within the scope of this work, a strong masonry infill typology constructed using tongue and groove clay masonry blocks has been considered. Following a brief description of the experimental campaign, major results of the new test set-up developed for the needs of this study has been described and the testing protocol has been introduced. The results of fully infilled real-scaled frame configurations, undamaged and previously damaged in-plane, obtained from the accomplished out-of-plane tests, have been summarised, and the related propagation of damage has been discussed. The obtained values of out-of-plane strength, equal to 168.5 kN, 102.7 kN, 163.9 kN and 274 kN for specimens TA1, TA2, TA3 and TA6 respectively, correspond to equivalent accelerations of 4.4g, 2.7g, 4.3g and 7.4g, indicating that for such strong masonry infill typologies without opening, adherent to the RC frame, the out-of-plane stability should not present a critical issue unless in very high seismic regions. Moreover, for the given infill typology, the formation of a resistance mechanism based on two-directional arching action has been found, and a typical out-of-plane failure mechanism has been identified, characterised by the opening of a predominant horizontal crack at mid-height of the panel and the creation of a stepwise crack pattern, starting from the central crack and developing diagonally towards the corners of the infill. Furthermore, the out-of-plane response of a previously undamaged masonry infill has been evaluated according to the results of TA6, being the resistance of the undamaged "strong" masonry infill computed from an experimental interpretation that considers the double arching effect.

Based on the obtained results, the out-of-plane stiffness and strength can be related to previous in-plane damage. Accordingly, the out-of-plane strength (and stiffness) may be expressed in function of expected in-plane drift. Hence, a simplified criterion describing the out-of-plane strength reduction has been proposed on the basis of experimental results only.

ACKNOWLEDGEMENTS

The present research has been conducted by University of Pavia and at EUCENTRE Foundation and it has been sponsored by the former ANDIL and its associated companies and by Executive Project DPC/ReLUIS 2019-2021. The financial support received is gratefully acknowledged.

REFERENCES

Angel, R. et al. 1994. Behaviour of reinforced concrete frames with masonry infills, *University of Illinois, Urbana-Champaign, Illinois*.

Braga, F. et al. 2011. Performance of non-structural elements in RC buildings during the L'Aquila, 2009 earthquake. *Bulletin of Earthquake Engineering* 9(1): 307–324.

Butenweg, C. et al. 2019. Experimental results of reinforced concrete frames with masonry infills under combined quasi-static in-plane and out-of-plane seismic loading, *Bulletin of Earthquake Engineering* 17: 3397–3422.

Calvi, G.M. & Bolognini, D., 2001. Seismic response of RC frames infilled with weakly reinforced masonry panels. *Journal of Earthquake Engineering*, 5 (2):153–185.

CEN, 2002. Eurocode 1 - Actions on structures, Part 1-1: General actions - Densities, self-weight, imposed loads for buildings, EN 1991-1-1, European Committee for Standardisation, Brussels, Belgium.

CEN, 2004. Eurocode 2 - Design of concrete structures, Part 1-1: General rules and rules for buildings, EN 1992-1-1, European Committee for Standardisation, Brussels, Belgium.

CEN, 2004. Eurocode 8 - Design of structures for earthquake resistance, Part 1: General rules, seismic actions and rules for buildings, EN 1998–1, European Committee for Standardisation, Brussels, Belgium.

da Porto, F. et al. 2013. Combined in-plane/out-of-plane experimental behaviour of reinforced and strengthened infill masonry walls. *Proc. 12th Canadian Masonry Symposium, June 2-5, Vancouver, Canada*.

Fragomeli, A. et al. 2017. Comportamento degli edifici in muratura nella sequenza sismica dell'Italia centrale del 2016 – Parte 1: Quadro generale [in Italian]. *Progettazione Sismica*, 8(2): 49–77.

Furtado, A. et al. 2016. Experimental evaluation of out-of-plane capacity of masonry infill walls. *Engineering Structures* 111: 48–63.

Furtado, A. et al. 2018. Effect of the panel width support and columns axial load on the infill masonry walls out-of-plane behaviour. *Journal of Earthquake Engineering*" 1–29.

Hak, S. et al. 2014. Out-of-plane experimental response of strong masonry infills. *Proc 2nd ECEE, August 25-29, Istanbul, Turkey*.

Hak, S. et al. 2018. Prediction of inter-storey drifts for regular RC structures with masonry infills based on bare frame modelling. *Bulletin of Earthquake Engineering* 16(1): 397–425.

Guidi, G. et al. 2013. Comportamento Sperimentale nel Piano e Fuori Piano di Tamponamenti in Muratura Armata e Rinforzata (in Italian). *Proc. XV ANIDIS Conference, Padova, Italy*.

Manzini, C.F. & Morandi, P., 2012. Rapporto preliminare sulle prestazioni ed i danneggiamenti agli edifici in muratura portante moderni a seguito degli eventi sismici emiliani del 2012 (in Italian), *Eucentre press*, http://www.eqclearinghouse.org/2012-05-20-italy/.

Manzini, C.F. et al. 2018. Shaking-table test on a two-storey RC framed structure with innovative infills with sliding joints. *Proc. 16th ECEE, June 18-21, Thessaloniki, Greece*.

Milanesi, R.R. et al. 2019. Risposta fuori piano di tamponature robuste in laterizio (in Italian). *Proc. XVIII ANIDIS Conference, September 15-19, Ascoli Piceno, Italy*.

MIT, D.M. 17/01/2018. NTC18: Norme tecniche per le costruzioni [in Italian], G.U. n.42 20/02/2018.

MIT, Circolare 21/01/2019. Circolare NTC18: Circolare esplicativa alle Norme tecniche per le costruzioni 2018 [in Italian], C.S.LL.PP. Circolare n.7.

Morandi, P. et al. 2013. Simplified Out-of-plane Resistance Verification for Slender Clay Masonry Infills in RC Frames. *Proc. XV ANIDIS Conference, Padova, Italy*.

Morandi, P. et al. 2018a. Performance-based interpretation of in-plane cyclic tests on RC frames with strong masonry infills. *Engineering Structures* 156: 503–521.

Ricci, P. et al.2011. 6th April 2009 L'Aquila earthquake, Italy: reinforced concrete building performance. *Bulletin of Earthquake Engineering* 9: 285–305.

Ricci, P. et al. 2018. Experimental assessment of the in-plane/out-of-plane interaction in unreinforced masonry infill walls. *Engineering Structures* 173" 960–978.

Ricci, P. et al. 2019. Out-of-plane seismic safety assessment of URM infills accounting for the in-plane/out-of-plane interaction in a nonlinear static framework. *Engineering Structures* 195(C): 96–112.

Brick and Block Masonry - From Historical to Sustainable Masonry –
Kubica, Kwiecień & Bednarz (eds)
© 2020 Taylor & Francis Group, London, ISBN 978-0-367-56586-2

Preliminary in-plane shear test of infills protected by PUFJ interfaces

A.T. Akyildiz, A. Kwiecień & B. Zając
Cracow University of Technology, Cracow, Poland

P. Triller & U. Bohinc
ZAG - Slovenian National Building and Civil Engineering Institute, Ljubljana, Slovenia

T. Rousakis
Democritus University of Thrace, Xanthi, Greece

A. Viskovic
University "G. d'Annunzio" of Chieti – Pescara, Chieti, Italy

ABSTRACT: This paper presents results of in-plane shear tests carried out in ZAG laboratory in Ljubljana (Slovenia) on an RC frame with masonry infill made of clay blocks (KEBE OrthoBlock). The frame fixed at the bottom and loaded vertically at columns' tops was excited by horizontal quasi-static cyclic loads at the top beam level. These loads forced various and gradually increasing drift levels (positive and negative) of the frame, with 3 repetitions at each level. Acquired forces and measured displacements allowed drawing hysteresis loops for determination of dissipation energy. Additionally, Digital Image Correlation (DIC) system was used for visualization of the behavior of the frame and the infill. There were tested 3 infill specimens of different configurations at interfaces between RC frame and infills. Type A (reference) was made of the infill bonded classically to the RC frame by mineral mortar. In type B, spaces at sides and top interfaces of 2 cm thickness were cut by saw and filled by polyurethane injection making PolyUrethane Flexible Joints (PUFJ). In type C, all interfaces were constructed from prefabricated PUFJ laminates and then the infill was erected. The specimen A was forced up to 1.6% of the frame horizontal drift when initial crushing at the corners and complete detachment at the frame-infill interface occurred. At this stage, the infill A could collapse even with small out-of-plane forces. The specimen B and C were forced up to 3.5% and 4.4% of the frame horizontal drift, respectively. B and C infills suffered crushing at the corners but the PUFJ interface protected the damaged infills against out-of-plain failure.

1 INTRODUCTION

Utilization of the brick masonry infill walls is very popular across the world. They are preferred due to several reasons for, including but not limited to; creating partition on the floors, providing insulation, decoration purposes and being as either principal or auxiliary load carrying members in structural systems. The construction technique varies depending on different needs, technical regulations, regional and geographical features etc. In most instances, the brick blocks are made by brittle materials and connection between the blocks as well as bonding of them to the surrounding frames, i.e. columns and beams, provided by also brittle materials such as mineral mortars. This practice does not seem as a major concern when the constructions are designed for carrying only vertical loads. In this case, the loads are most likely predictable, hence design phase can cover the damage scenarios precisely.

On the other hand, when the horizontal loads play a significant role in terms of external forces acting on the building, high displacement and ductility demands are emerging on the masonry members. Such incidents are mostly visible in seismic zones, where earthquake loads detrimentally affect masonry elements, particularly the infill walls in RC framed constructions (EERI 1994, Decanini et al. 2004, Li et al. 2008).

Despite the popularity and highly vulnerable nature of this construction type in seismic zones, the contribution and possible effects of the walls on entire construction systems are mostly neglected during the design phase by practitioners (Preti et al. 2015, Razzaghi & Javidnia 2015, Longo et al. 2016). On the other hand, some seismic codes (FEMA273 1997, EC8 2004) already cover this topic by means of making provisions (Tasligedik et al. 2011). Today, it is well-known that infill walls increase the rigidity of structural systems thus they lead buildings to receive higher horizontal forces than the bare framed structures (Fiorato et al. 1970, Murti & Jain 2000, Asteris 2003).

However, robustness and strength correlation cannot be made easily due to the intrinsic fragile features of infill brick walls. Aforementioned brittle characteristics of the traditional infill wall elements, which are brick blocks, mortars and surrounding concrete frames in RC structures, make such systems to get sudden fragile damages under the loads over their elastic capacities. This drawback is due to the lack of ductility in the structural system itself.

Especially the connection between the infill and RC frame has substantial importance in this topic. Some researches already focus on this challenge recently in order to bring new solutions (Dafnis et al. 2002, Preti et al. 2015, Vailati & Monti 2015, Muthu Kumar & Satyanarayanan, 2018) and possible protection methods have already started to find place in technical norms (EC8 2004, TBEC 2018). The most common failures are mainly due to the interaction effects occurring on wall-frame interfaces. This leads to connection loss on the boundaries and eventually causes either partial or total collapse of the walls. Such damages are one of the main reasons of casualties during earthquakes that their destructive effects are seen in different forms, e.g. out-of-plane failures, deterioration of the structural and dynamic characteristics of buildings (Hermanns et al. 2014, Akhoundi et al. 2015, Pasca & Liberatore 2015).

In order to overcome the brittle and sudden failure of infill masonry walls in RC buildings, an innovative approach is presented in this paper which uses a polymer based product "Polyurethane PM" as a joint connection in such systems. The material is developed and tested for different purposes in the past (Kwiecień 2012, Kisiel 2015, Kwiecień et al. 2017a, b) while pilot analytical studies have revealed the high potential in RC frames (Rousakis et al. 2017, Rouka et al. 2017). In this study, in-plane excitation results of the experiments are given. Three different real size specimens are constructed that each represents a different joint connection of the infill-frame systems. Frame A, has cementitious mortar as joint material and shows the traditional construction method. Frame B and Frame C are built to represent the implementation of PolyUrethane Flexible Joints (PUFJ) in old (existing) and in new (to-be-built) constructions, respectively. Figure 1 shows the schematic view of specimens.

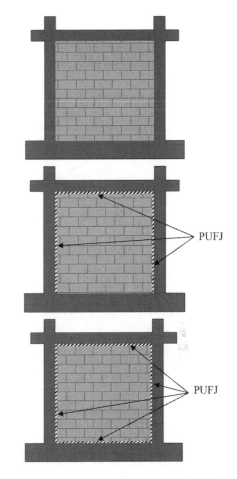

Figure 1. Schematic views of the samples; Frame A (top), Frame B (middle) and Frame C (bottom).

2 TEST SETUP

The experiments were carried out at the Slovenian National Building and Civil Engineering Institute (ZAG) in Ljubljana. In total, three identical RC frames were constructed. Infill walls were built by hollow clay blocks (KEBE OrthoBlock), see Figure 2. PUFJ implementation was done differently for the frame types of B and C. For the Frame B, in-situ casting was preferred while the precast PUFJ laminates were used to

Figure 2. Arrangement of the hollow clay blocks.

construct joint interfaces in the Frame C. Details of the materials are given in Table 1 and Table 2. PUFJ information is taken from the previous researches (Kisiel 2018, Gams et al. 2017), whereas the rest of materials were tested at ZAG.

One-bay and one-story RC frames consist of beam, column and foundation elements. Both beam and column members have section dimensions of 25 × 25 cm². The foundation is designed to be stronger than the upper frame and its dimensions in height and width are 40 cm and 30 cm, respectively. Reinforcement details are as shown in Figure 3.

Infill walls are created by hollow clay blocks with 10 cm thickness. The walls have 220 cm length in both horizontal and vertical directions. Test setup is shown in Figure 4 also with the details of load actuators.

2.1 Preparation of the test samples

For all of the frame types, thin layer mortar (M10) is used as bonding material on the brick interfaces. Cementitious mortar is preferred for providing connection between the masonry and RC frame. Frame A, has across its entire perimeter stiff connection with the cementitious mortar, see Figure 5. Frame B is built in the same way of the Frame A, since it is chosen to represent PUFJ implementation on the existing buildings. Afterwards, the infill wall was cut in three sides, except the bottom one, by a saw for providing the gap thickness of 2 cm. The cut parts then were filled by Polyurethane PM and made ready for the tests. Implementation is shown in Figure 6. Different than the other frame types, precast laminates with 2 cm thickness were first created for the Frame C. The laminates were then implemented on the internal perimeter of the bare-frame as shown in Figure 7. The brick blocks were constructed afterwards in a similar approach as with the other frames.

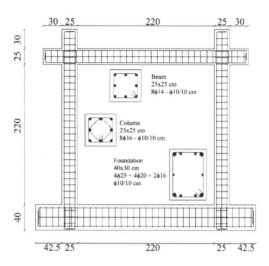

Figure 3. Reinforcement details.

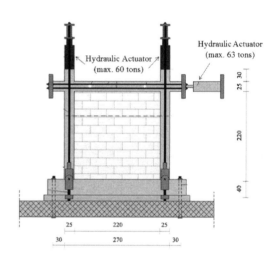

Figure 4. Test setup.

Table 1. Material properties of the concrete and mortar.

Material	Compressive strength [MPa]	Flexural strength [MPa]
Cementitious mortar	30.4	7.2
Thin layer mortar	10.2	4.5
Injection grout	53.1	4.7
Concrete	46.9	-

Table 2. Polyurethane PM mechanical properties.

Material	E_{PM} [MPa]	ν	ε
Polyurethane PM	4.5	0.47	0.5 - 1.5*

* Values dependent on a strain ratio during a tensile test.

Figure 5. Construction process of the Frame A.

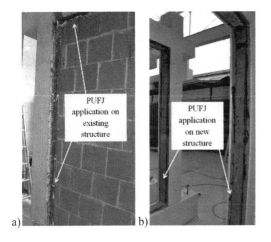

PUFJ application on existing structure

PUFJ application on new structure

a)   b)

Figure 6. Construction process of the Frame B (a) and of the Frame C (b).

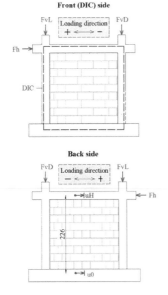

Figure 7. Schematic view of the measurement devices [cm].

2.2 Measuring method

In order to acquire results, horizontal measuring devices (displacement transducers - DT) were placed on the back side of specimens, see Figure 7.

In addition, Digital Image Correlation (DIC) technique was used on the front side of frames as schematically shown in Figure 8, which enables the visualization of the behavior of experiment. In this method, for each test specimen, multiple reference points were identified both on the frames and infill

walls. Localization of these points were determined in a way that entire frame-wall perimeter aimed to be covered. Figures 8 and 9 show the setup and DIC points for the Frame B. DIC technique was applied to measure and visualize deformation of PUFJ and connected by it a RC frame and an infill. Only comparison of measurements accuracy, provided by displacement transducers (DT) and DIC technique, is presented further in this paper.

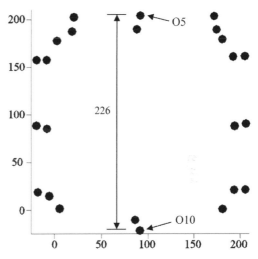

Figure 8. Coordinates of the DIC points on the Frame B [cm].

Figure 9. A view of the DIC localization on the Frame B.

971

2.3 Details of the test loads

All of the frames were exposed to 375 kN vertical load affecting on each column during the experiments. As in this way, a behavior of the lower floor frame in a multiple-story building is represented. In order to examine the in-plane behavior, horizontal hydraulic actuator excited the frames in quasi-static cyclic loads. The steps of cycles were identical for all of the frames. However, due to the load carrying capacity differences between the frames, the loadings were terminated in various cycle steps that are corresponding to different drift ratios. Details of the typical loading scheme is given in Figure 10, in form of a drift ratio, calculated from $(u_{1f}-u_0)/226$ cm (data from DT - see Figure 7). The followed practice was forcing the frame by means of three cycles in each drift ratio.

3 RESULTS OF THE TESTS

At first Frame A, which is the reference frame with stiff mortar, is tested. It is one of the most common construction practices being followed for the unreinforced masonries (URM) in RC buildings. Direct contact of single layer brick blocks to the frames are actualized in this technique. The specimen was forced up to 1.6% of the frame horizontal drift when crushing at the corners and complete detachment at the frame-infill interface occurred (Figures 11 and 14). Since all of the boundaries had connection loss, the experiment was terminated at this step. Otherwise, a slight forcing increment would lead to the total out-of-plane failure. The corresponding base shear force of the Frame A was 165 kN (Figure 11). Frame B, which is the specimen with PUFJ on the three sides of wall (up, right and left as shown in Figure 1) was also performed under the same cyclic loads. The performance of flexible joints was visible and the frame could withstand the lateral drift ratio of 3.5% (Figure 12). At that drift the base shear force of the frame B was 134.5 kN, equal to 0.8 of the maximum 166.6 kN for pull direction. For push direction the maximum base shear was 192.8 kN and at 3.5% drift 179 kN (0.92 of maximum). Therefore, the failure occurred at pull direction, having 20% load drop

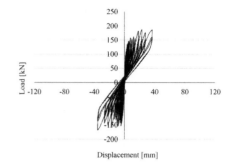

Figure 11. Load-displacement curve of the Frame A.

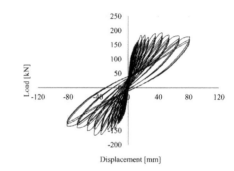

Figure 12. Load-displacement curve of the Frame B.

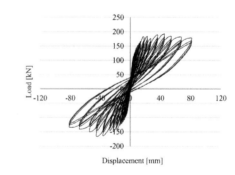

Figure 13. Load-displacement curve of the Frame C.

while the shear force drop on push direction was marginal and still higher than in Frame A.

Despite corner crashes were observed, the wall was still able to carry loads (Figure 15). Thanks to the hyper-elastic features of PUFJ, middle zones across the columns and beam lengths sustained bonding, hence the wall could resist against the out-of-plane collapse. The specimen with laminated PUFJ across the entire wall peripheral, which is referred here as Frame C, exhibited the best performance with 4.4% drift ratio (Figure 13). At that drift the base shear force of the Frame C was 164.6 kN, equal to 0.92 of the maximum 177.6 kN for pull direction. For push direction the maximum base shear

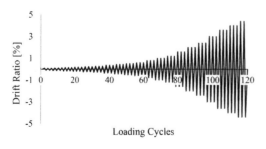

Figure 10. Loading cycles of the horizontal excitation.

was 222.5 kN which is far higher than Frame A or Frame B. The push shear at 4.2% drift was 207.5 kN (0.93 of maximum). Therefore, no frame structural failure occurred at pull or push direction as the shear force reduction was marginal and still higher than in Frame A or Frame B. The observed damages of the Frame C were very similar to that of the Frame B. Namely corner crushes on the infill wall and no substantial damage neither on RC frame nor on the masonry central body. The wall could still carry the loads without the out-of-plane failure (Figure 16).

Comparative backbone curves of the hysteresis loops in Figure 17 illustrate the upgraded behavior of the PUFJ protected frames B and C in terms of base shear and lateral drift.

Throughout the cyclic shear tests both, horizontal measuring devices as well as DIC technique were used to measure imposed horizontal displacements. In order to check these measurements, a study, comparing the results, provided by both measuring techniques, has been performed. The displacement transducers (Novotechnik TEX 200 and TRS 100) have a combined accuracy of approximately ±0.25 mm, while the DIC system (two cameras with 5 Mpx resolution) with a resolution of approximately 0.02 px, considering the field of view of 3 × 3 m, has accuracy of 0.02 mm. In order to compare the results

Figure 14. Frame A at the maximum drift ratio of 1.6%.

Figure 15. Frame B at the maximum drift ratio of 3.6%.

Figure 16. Frame C at the maximum drift ratio of 4.4%.

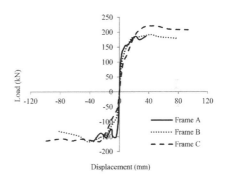

Figure 17. Load-displacement backbone curves of the frames.

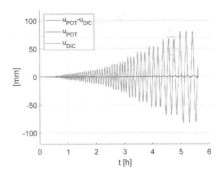

Figure 18. Imposed horizontal displacements, measured using displacement transducer and DIC method and their differences for the case of Frame B.

of the abovementioned measuring techniques, the differences in values of the horizontal displacement imposed to the specimen, have been compared. Values $u_{DT}=u_{1T}-u_0$ (displacement transducers; see Figure 7) and $u_{DIC}=u_{O10}-u_{O5}$ (DIC system; see Figure 8) have been considered. Figure 18 shows the imposed horizontal displacements as well as their differences for the case of Frame B. As can be seen, the measured

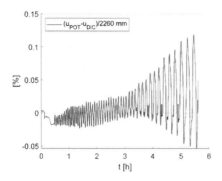

Figure 19. Drift ratio error between DT and DIC measurements, determined for the case of Frame B.

values are quite close and the total difference does not exceed 1 mm in the first part of test sequence.

The difference between the two measuring methods increases significantly in the second part of the loading sequence where it can reach up to 3 mm. Probable cause for the increase can be tracked to the limited accuracy of the a-posteriori synchronization of both measurement systems.

Taking into consideration drift ratios accuracy, presented above differences in horizontal displacements lead to maximum drift error ranging between 0.044% (1 mm/2260 mm) and 0.133% (3 mm/2260 mm) – based on data from Figure 7. Graphical presentation of the error changes, in relation to the drift ratios calculated for DT ($d_{POT} = u_{POT}/226$ cm – Figure 7) and for DIC ($d_{DIC} = u_{DIC}/226$ cm – Figure 7), is presented as $e_{POT-DIC} = d_{POT}-d_{DIC}$ in Figure 19. It can be compared with data presented in Figure 10. Values of the error changes (Figure 19) indicate that differences in drift ratio calculations using DT and DIC are very low (less than 0.18% (peak to peak).

4 CONCLUSION

In this study, seismic performance of the PUFJ usage on masonry infill walls is investigated. Three real-size, one-bay and one-story RC frame specimens were created. Following that, infill walls and PUFJ were implemented on the relevant frames. The frames were tested under quasi-static in-plane cyclic imposed top horizontal displacements while carrying constant axial load through their 2 RC columns.

Frame A experienced the connection detachment along its boundary zones where RC frame and infill wall were fixed with stiff cementitious mortar, following the common application practice. The complete connection loss was observed at the drift ratio of 1.6%. The test was terminated at this point, since the out-of-plane failure was inevitable.

Frame B and Frame C could withstand higher drift ratios that were 3.5% and 4.4%, respectively. Despite the corner crushes, both frames were still

able to keep the infill safely in terms of out-of-plane failure. PUFJ presence led to high strength connections on the frame-wall boundary zones.

In terms of base shear force it is concluded that Frame B reached its structural failure with 20% horizontal load drop from 166.6 kN to 134.5 kN at 3.5% drift. For push direction the horizontal loads were higher from 192.8 kN maximum to 179 kN at 3.5% drift. The marginal horizontal load drop of 8% denotes strength reserve in that direction at levels higher than the maximum of 165 kN for Frame A. Frame C revealed marginal horizontal load drop of 7% for push direction and 8% for pull direction at 4.4% drift suggesting a high reserve in strength and ductility while reaching actuators limits. The levels of base shear forces of Frame C were as high as 222.5 kN at maximum and 207.5 kN at 4.4% drift in push direction, far higher than 165 kN for Frame A.

Ductility is a key element in earthquake resistant building design. The experimental results suggest that RC frames with infills protected with PUFJ reveal a significant ductility upgrade both in terms of overstrength (strength enhancement) but most important in terms of increased lateral drift, having suppressed efficiently out of plane infill failure in addition to RC frame collapse. Out-of-plane infill failure is usually neglected but may reduce significantly the usable ductility of frames as it may cause detrimental life loss or severe injuries, violating the targets of seismic resistant design.

The experimental results validate that lateral drift in PUFJ protected frames is twice or higher than that of frames with traditional mortar at the infill – RC interface. Besides, it is crucial to maintain the stability of infill walls during the ground shakings. Particularly, out-of-plane failures must be prevented, which threaten the human life even under low intensity earthquakes. PUFJ can be an efficient alternative solution to achieve this crucial task since high bonding features of the material decreases the connection loss hazard on boundaries. In conclusion, preliminary outcomes are promising that the innovative joints can be used on orthoblock masonry infill walls-RC frame interfaces in order to provide higher safety against severe earthquakes than common infill construction practice.

The low error value between drift ratios calculated using DT and DIC techniques indicates that further analysis of PUFJ deformations, using only DIC technology, can be reliable.

ACKNOWLEDGMENTS

This research activity was within the framework of the Research Project: "In- and out-of-plane tests of masonry-infilled RC frames with flexible joints" - Cooperation Agreement between ZAG - Slovenian National Building and Civil Engineering Institute (Ljubljana, Slovenia), Cracow University of

Technology (Cracow, Poland) and Democritus University of Thrace (Xanthi, Greece).

Authors acknowledge the materials support provided by Sika Poland company and KEBE Northern Greece Ceramics.

REFERENCES

Akhoundi, F. et al. 2015. Out-of-plane behavior of masonry infill walls. *7th International Conference on Seismology & Earthquake Engineering, Tehran, 18-21 May 2015.*

Asteris P.G. 2003. Lateral stiffness of brick masonry infilled plane frames. *Journal of Structural Engineering* 129(8): 1071–1079.

Dafnis, A. et al. 2002. Arching in masonry walls subjected to earthquake motions. *Journal of Structural Engineering* 128(2): 153–159.

Decanini, L.D. et al. 2004. Performance of reinforced concrete buildings during the 2002 Molise, Italy, Earthquake. *Earthquake Spectra* 20(S1): S221–S255.

EERI. 1994. Northridge earthquake preliminary reconnaissance report. Oakland, CA.

European Committee for Standardization. 2004. Eurocode 8: Design of structures for earthquake resistance – Part 1: General rules, seismic actions and rules for buildings, Brussels.

Federal Emergency Management Agency. 1997. FEMA 273: NEHRP guidelines for the seismic rehabilitation of buildings, Washington DC.

Fiorato, A.E. et al. 1970. An investigation of the interaction of reinforced concrete frames with masonry filler walls. *Urbana, Illinois: University of Illinois.*

Gams, M. et al. 2017. Modelling of deformable polymer to be used for joints between infill masonry walls and R.C. frames. *Procedia Engineering* 193: 455–461.

Hermanns L. et al. 2014. Performance of buildings with masonry infill walls during the 2011 Lorca earthquake. *Bull Earthquake Eng* 12(5): 1977–1997.

Kisiel, P. 2015. The stiffness and bearing capacity of polymer flexible joint under shear load. *Procedia Engineering* 108: 496–503.

Kisiel P. 2018. Model approach for polymer flexible joints in precast elements joints of concrete pavements. *Doctoral Dissertation. Cracow: CUT.*

Kwiecień, A. 2012. *Polymer flexible joints in masonry and concrete structures.* Monograph No. 414. Wyd. Politechniki Krakowskiej, Seria Inżynieria Lądowa, Kraków (in Polish).

Kwiecień, A. et al. 2017a. Validation of a new hyperviscoelastic model for deformable polymers used for joints between rc frames and masonry infills. *Engineering Transactions* 65(1): 113–121.

Kwiecień, A. et al. 2017b. Use of polymer flexible joint between RC frames and masonry infills for improved seismic performance. Zurich: SMAR.

Li, B. et al. 2008. Wenchuan earthquake field reconnaissance on reinforced concrete framed buildings with and without masonry infill walls. *The 14th World Conference on Earthquake Engineering, Beijing, China.*

Longo, F. et al. 2016. Seismic response history analysis including out-of-plane collapse of unreinforced masonry infill walls in RC frame structures. *Proceedings of the 16th International Brick and Block Masonry Conference, Padova, Italy.*

Murty, CVR. & Jain, SK. 2000. Beneficial influence of masonry infill walls on seismic performance of rc frame buildings. *12th World Conference on Earthquake Engineering, Auckland, New Zealand.*

Muthu Kumar, S. & Satyanarayanan, K.S. 2018. Study the effect of elastic materials as interface medium used in infilled frames. *Materials Today: Proceedings* 8986–8995.

Pasca, M. & Liberatore, L. 2015. Predicting models for the evaluation of out-of-plane ultimate load carrying capacity of masonry infill walls. *Earthquake Resistant Engineering Structures X.* Southampton: WIT Press.

Preti, M. et al. 2015. Experimental testing of engineered masonry infill walls for post-earthquake structural damage control. *Bull Earthquake Eng* 13: 2029–2049.

Razzaghi, M.S.& Javidnia, M. 2015. Evaluation of the effect of infill walls on seismic performance of RC dual frames. *Int J Adv Struct Eng* 7: 49–54.

Rouka, D. et al. 2017. Response of RC buildings with Low-strength Infill Walls Retrofitted with FRP sheets with Highly Deformable Polymer – Effects of Infill Wall Strength. *25th International Conference on Composites/Nano Engineering, Rome.*

Rousakis, T. et al. 2017. Fast Retrofitting of Strong Wall Infill of RC buildings with Fiber Sheets Impregnated with Highly Deformable Polymer. *25th International Conference on Composites/Nano Engineering, July 16-22, 2017 Rome.*

Tasligedik, A.S. et al. 2011. Damage Mitigation Strategies of 'Non-Structural' Infill Walls: Concept and Numerical-Experimental Validation Program. *Proceedings of the Ninth Pacific Conference on Earthquake Engineering Building an Earthquake-Resilient Society, Auckland, New Zealand.*

TBEC - Disaster and Emergency Management Presidency of Turkey. 2018. Turkish Building Earthquake Code, Ankara.

Vailati, M. & Monti, G. 2016. Earthquake-Resistant and Thermo-Insulating Infill Panel with Recycled-Plastic Joints. In: D'Amico S. (ed.) *Earthquakes and Their Impact on Society,* Springer Natural Hazards: 417–432.

Reinforced masonry

Brick and Block Masonry - From Historical to Sustainable Masonry –
Kubica, Kwiecień & Bednarz (eds)
© 2020 Taylor & Francis Group, London, ISBN 978-0-367-56586-2

Behaviour of stitching bars in the masonry wall

Ł. Drobiec
Department of Building Structures, Silesian University of Technology, Poland

ABSTRACT: Stitching of cracked masonry elements with the reinforcement is a quite popular method of repairing damaged walls. Scientific publications often describe the effect of repair in the form of crack stitching on specimens tested under compression, bending or shearing. However, it is difficult to find a paper, whose authors made an attempt to estimate the impact of different types of reinforcement on strengthening issues. This paper presents results for tensile tests performed on different types of bars with a diameter of 6, mm, pull-out tests for twisted and ribbed bars in a masonry wall, and tests for flexural strength of mortar prismatic specimens reinforced with a stitching bar. Obtained results were the basis for an attempt to evaluate the effectiveness of the interaction between the stitching reinforcement and the masonry wall. The tested reinforcement was composed of smooth bars made of common or stainless steel, ribbed bars, and twisted bars. The most effective reinforcement proved to be the twisted bar because it showed better deformability and adhesion to the mortar. Test results indicated that the anchorage length of stitching bars should be greater than 250 mm. In practice, anchor lengths of approximately 500 mm are used.

1 INTRODUCTION

Cracks in historical masonry structures are often repaired by the stitching method (near-surface strengthening), which consists in placing the reinforcement in mortar in the grooves (Cook et al. 1995, Modena, 1997, Valluzzi et al. 2005, Bhattacharya et al. 2014, Akcay et al. 2017, Corradi et al. 2018). Grooves are usually made in head bed joints by cutting them out mechanically to the depth of 60 mm. An example of a stitching method is shown in Figure 1. The near-surface strengthening by stitching is often used in seismic areas (Bhattacharya et al. 2014, Corradi et al. 2002, De Lorenzis & Teng, 2007, Corradi et al. 2018). Both steel and non-metallic reinforcing bars are used for that purpose (Modena, 1997, De Lorenzis & Teng, 2007, Petersen, 2009, Willis et al. 2010).

Many papers on repairs of cracked masonry walls recommend the application of proprietary system solutions with twisted bars. Tests on wall models confirm the effectiveness of such a reinforcement under vertical loading (Drobiec et al., 2000, Drobiec et al. 2008), shearing (Ismail et al. 2011, Masia et al. 2010), out-of-plane bending (Ismail & Ingham, 2012) and in brick masonry vaults (Drobiec et al. 2019). Pull-out tests of behaviour of twisted stainless steel bars in mortar were made, demonstrating the high bonding characteristics of the bars in mortar (Moreira et al. 2014). These studies did not take into account the influence of vertical compressive stresses that occur in the masonry bed joints.

The twisted reinforcement is also used in newly built masonry walls to prevent the formation of cracks (Jasiński & Drobiec, 2016, Drobiec et al. 2017, Drobiec & Jasiński, 2019). However, there is no published information explaining the difference between the system solution with twisted bars and the traditional solution (rebars embedded in joints made of cement mortars). This paper presents results for tensile testing of different types of bars with a diameter of 6 mm, pull-out tests for twisted and ribbed bars in masonry structures, and tests on flexural strength of mortar samples reinforced with twisted and ribbed steel stitching bars, which are the base for describing the effect of strengthening.

It also attempts to answer a question about the required anchorage length of the stitching reinforcement. According to the papers (Corradi et al. 2002, Bhattacharya et al. 2014) and recommendations of manufacturers, the required length should be minimum 50 mm beyond the crack plane.

2 TENSILE TEST

Tensile testing of reinforcing steel was performed in accordance with EN ISO 6892-1. Tests were conducted on bars with a diameter of 6 mm (the outer diameter for twisted bars) made of following steel grades: St3SY (smooth), 18G2 (ribbed), stainless steel 2H13 (smooth) and twisted bars (Figure 2). Three samples of each type of rod were tested. During testing, the tensile force was measured and

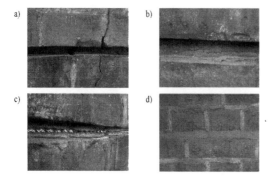

Figure 1. Method of repair by the stitching of the cracks: a) making the groove in the bed joint, b) laying the first layer of mortar, c) laying the reinforcement, d) filling the groove.

Table 1. Results from tensile testing.

Type of bar	maximum tensile force, kN		Yield strength of steel, N/mm^2	Tensile strength, N/mm^2	Modulus of elasticity, N/mm^2
	sample	average			
St3SY (smooth bar)	11,99 12,03 12,10	12,04	362	425	210000
18G2 (ribbed bar)	16,91 16,78 16,89	16,86	445	596	208000
2H13 stainless steel (smooth bar)	23,52 24,01 23,27	23,60	564	834	212000
Twisted bar	9,21 9,23 9,28	9,24	854	1141	168000

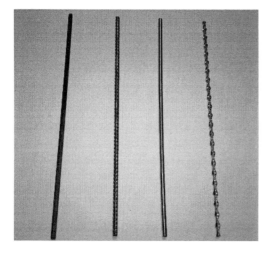

Figure 2. A view of bars used in the research, from the left St3SY (smooth), 18G2 (ribbed), stainless steel 2H13 (smooth) and twisted bar.

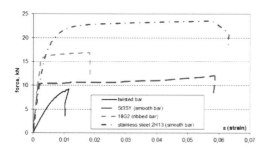

Figure 3. Results from tensile testing of different types of bars.

the displacements of the bars were recorded using the optical extensometer. The average test results illustrated as a force-strain curve are presented in Figure 3. Table 1 shows the values of the maximum tensile forces all tested samples and averaged values of yield strength, tensile strength and elasticity modulus. The yield strength and tensile strength of the twisted bars are determined by assuming the actual field of the bar's intersection, which is equal to 8.1 mm^2.

Results shown in Figure 3 and Table 1 indicate different strain characteristics of twisted bars when compared to ordinary bars used in the construction industry. A lower modulus of elasticity enables greater strains, then twisted bars exhibit better interaction with the masonry wall even at low tensile stresses. In case of smooth bars made of common and stainless steel and ribbed bars, greater strains can be achieved after the plastification of steel.

3   TESTING OF MORTARS

Besides tensile testing of bars, fundamental material tests on the system mortar were conducted in accordance with EN 1015-11. For comparison of results, the same tests were performed for the standard cement mortar (1:3) having the following weight proportion: 450 g of cement, 1350 g of standard sand, and 225 g of water. Three prismatic specimens with dimensions of 40x160x160 mm were made from both mortars. The method of preparing and curing of specimens complied with recommendations provided by EN 1015-11. Figure 4 illustrates specimens prior to tests and during testing.

a)

b)

c)

d)

Figure 4. Testing of mortars in accordance with EN 1015-11 [21]: a) specimens in moulds, b) specimens prior to tests, c) bending test, d) compressive test.

Tests were performed after 28 days of specimen curing. Test results are presented in Table 2. Results presented in Table 2 lead to conclusions that the system mortar has ca. a two-fold greater bending and compressive strength than the cement mortar. The problem can be created by old, historical masonry walls built with lime mortar, and local stiff lining can act as a wedge.

Table 2. Test results for cuboid specimens.

| Mortar type | Bending strength | | Compressive strength | | |
| | | | $f_{m,i}$ N/mm² | | |
	$f_{mt,i}$ N/mm²	$f_{mt,mv}$ N/mm²	½ of beam	beam	$f_{m,mv}$ N/mm²
Cement mortar	5.90	5.89	32.92	32.91	32.02
			32.90		
	5.59		30.86	31.44	
			32.02		
	6.19		31.72	31.71	
			31.71		
System mortar	9.91	10.86	59.50	61.52	58.96
			63.55		
	11.69		55.56	54.84	
			54.13		
	10.97		63.29	60.53	
			57.77		

4 FLEXURAL TESTS OF REINFORCED SPECIMENS

Along with bending (flexural) and compressive testing of mortars, also tests not specified in standards were performed on specimens with rebars. Specimens had a section of 10x40 mm and a length of 160 mm to specimen the geometry of mortar in the joint cut out in the groove. Twisted and ribbed bars were placed in the mid-width and mid-length of specimens. Bars were made of 18G2 steel grade, had a diameter of 6 mm and their ends were bent to facilitate anchoring (Figure 5).

a)

b)

Figure 5. Preparing specimens with reinforcement: a) specimens of system mortar and twisted reinforcement, b) specimens of cement mortar (1:3) and reinforcement of ribbed bar (18G2).

Specimens of the system mortar were reinforced with twisted bars, and specimens of the cement mortar (1:3) were reinforced with a ribbed bar. Specimens were subjected to bending testing like the above specimens with dimensions of 40x160x160 mm. Figure 6 shows specimens before and during tests.

During bending tests, an additional force causing the first crack was recorded. Test results are presented in Table 3. Due to the small dimensions of the samples, the results should be considered only in terms of quality, not quantity.

a)

b)

c)

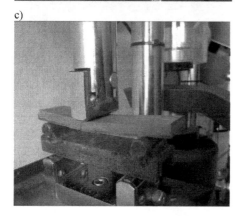

Figure 6. Tests on reinforced prismatic specimens with dimensions of 10x40x160 mm: a) specimens prior to tests, b) testing of specimens of cement mortar reinforced with a ribbed bar, c) testing of specimens of system mortar reinforced with a twisted bar.

Table 3. Test results for specimens with dimensions of 10x40x160 mm.

Mortar type	Stress at cracking		Bending strength	
	$f_{mcr,i}$ N/mm^2	$f_{mcr,mv}$ N/mm^2	$f_{mt,i}$ N/mm^2	$f_{mt,mv}$ N/mm^2
Cement mortar and ribbed bar $\phi6$ mm (18G2)	26.55 30.98 29.21	28.91	42.04 35.84 48.23	42.04
System mortar and twisted bar	30.96 31.86 32.75	31.86	37.61 34.51 35.84	35.99

Figure 7. A view of specimens after tests, left - specimens reinforced with a bar made of 34GS steel grade, right - specimens reinforced with a twisted bar.

Scratching of samples with spiral bars occurred about 30% later than similar samples with ribbed bars. It should be emphasized that the surface area of the twisted bar covered less than 30% of the surface area of the standard reinforcement. However, yield strength of the twisted reinforcement was nearly twice as much. Hence, it can be assumed that the level of cracking stresses and flexural strength are similar in the ordinary and system mortar.

Failure of specimens also requires attention. Specimens reinforced with a bar made of 34GS steel grade crumbled after tests, whereas specimens reinforced with a twisted bar were cracked, but remained whole (Figure 7). It indicates better adhesion of the system mortar to the twisted bar.

5 PULL-OUT TESTS OF BARS IN MASONRY

Tensile testing and the above tests on specimens with reinforced mortar did not provide information on effectiveness of the rebar anchorage. Therefore, pull-out tests of bars in masonry were carried out. For tests, two specimens made of two ceramic bricks with a strength of $f_b = 18.7$ N/mm^2 were prepared in accordance with EN-772-1, and bound with the cement mortar (3 specimens) or the system mortar

(3 specimens). 18G2 steel bars and twisted bars were placed in the mid-width and mid-height of bed joints in the mortar respectively. Anchorage length of the bar in the mortar was 250 mm. In practice, larger anchor lengths, usually 500 mm, are used. Bricks were soaked prior to preparing specimens. Figure 8 illustrates specimens.

Specimens were tested after 28 days of their preparation, and the test consisted in pulling out the bar from the specimen. Compressive stress influenced the anchorage of the rebar, so a tendon prestressing system was made. It consisted of pressure plates, four prestressing screws and a dynamometer (Figure 9b). The dynamometer was used to measured compressive force perpendicular to bed joints. The stresses were compensated by tightening the screws.

Specimens with the system mortar and twisted reinforcement were tested without prestressing (the specimen marked as S-0.0) and under prestressing equal to 0.1 and 0.2 N/mm2 (specimens marked as S.01 and S.02 respectively). Specimens with the cement mortar and the bar made of 18G2 steel grade were tested without prestressing (Z-0.0) and under prestressing of 0.1 N/mm2 (Z-0.1) and of 0.2 N/mm2 (Z-0.2). Figure 9 presents the testing scheme with and without prestressing.

The pulling-out force, displacement of the bar with respect to the specimen face (with elongation of the bar) and the prestressing force (in prestressed specimens) were recorded during tests. The displacements were measured using linear variable differential transformer (LVDT).

Figure 10 presents specimens during tests. Test results as a graph showing the pulling-out force and bar displacement are illustrated in Figure 11. Test results are also presented in Table 4. The pulling-out force was increased by the weight of the specimens and the prestressing system (in prestressed specimens).

Failure of tested specimens also requires attention. No breaking of rebars was observed in any specimens. However, for the specimen with the ribbed bar under

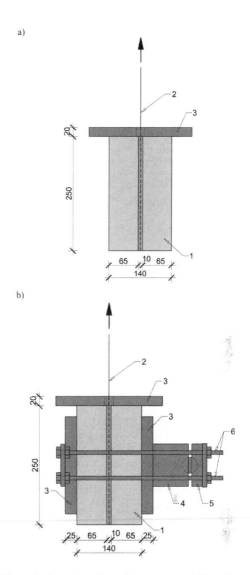

Figure 9. A testing scheme for unprestressed (a) and prestressed (b) specimens, 1- brick, 2- bar, 3- steel plate, 4- dynamometre, 5-ball joint, 6- prestressing screws.

Figure 8. A view of specimens used in tests for bar anchorage.

the prestressing value of 0.2 N/mm^2, the determined pull-out force was smaller only by 25% from the breaking force during tensile testing. Thus, the plastification of the bar was achieved. For the specimen with the twisted bar, under the prestressing value of 0.1 N/mm^2, the determined force constituted 89% of the breaking force, and under the prestressing of 0.2 N/mm^2, it was nearly 99% of the breaking force. Failure of specimens with the cement mortar and the ribbed bar (ϕ6, 18G2) was caused by the loss of adhesion between mortar and the brick (Figure 12 and 13), whereas straightening of the bar (its twisted) and its pulling out from the mortar

a)

b)

c)

d)

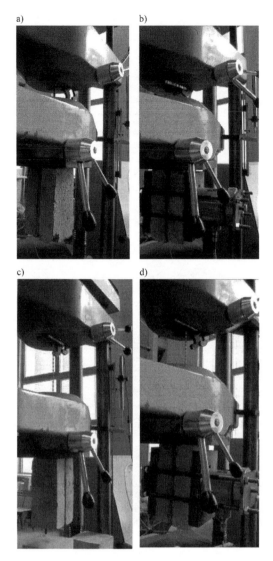

Figure 10. Pull-out tests of unprestressed (a) and pre-stressed (b) ribbed bars in the masonry with cement mortar, and of unprestressed (c) and prestressed (d) twisted bars in the masonry with system mortar.

Figure 11. Results from pull-out tests of bars.

Table 4. Results from pull-out tests of bars.

Specimen No.	Pulling-out force, kN	Maximum displacement, mm
Z-0.0	9.86	2.6
Z-0.1	10.39	2.9
Z-0.2	12.75	3.6
S-0.0	6.2	10.3
S-0.1	8.31	14.8
S-0.2	9.17	18.9

Figure 12. A view of unprestressed specimens of cement mortar and the ribbed bar after tests.

Figure 13. A view of prestressed specimens of cement mortar and the ribbed bar after tests.

led to failure of specimens with the twisted reinforcement.

The pull-out test confirmed that the anchorage length of stitching bars should be greater than 25 cm. In that case, failure would occur as breaking of the reinforcement. Because the force and the strain determined in pull-out tests of twisted bars were similar to parameters observed at breaking of the bar, it can be assumed that slightly greater length of anchorage should result in breaking of the bar. For ribbed bars pulling out from the joint, yield strength of steel was achieved. However, strains constituted ca. 50% of all strains observed during the break of the bar. It can be concluded that ribbed bars require greater length of anchorage than twisted bars.

Maximum displacements in specimens with ribbed bars and the cement mortar were achieved at the maximum pull-out force. When the maximum force was achieved, an increase in displacements with a slow drop in force were observed for specimens with the system mortar and the twisted reinforcement. Twisted bars showed much greater displacements at the time of the specimens failure than specimens reinforced with the ribbed bar.

6 CONCLUSION

Conducted tests indicate that the system solution with twisted bars can transfer greater strains than the equivalent traditional solution. Greater deformability of the system ensures better interaction between the reinforcement and the masonry wall. However, it should be emphasized that stiff lining in historical masonry walls made on lime mortar, can cause damage related to a lack of compatibility of the applied solution with the masonry wall. For this reason, each reinforcement should be analysed individually. The analysis of causes of damage should be a base to undertake measures to remove causes of such damage, and then develop the appropriate method of repairing cracks.

The anchorage length for bars in the masonry wall (beyond a crack) should exceed 25 cm. The bars did not break off in tested specimens, whose failure was caused by breaking adhesion of the mortar to masonry units (specimens reinforced with ribbed bars) or by pulling out the bar from the mortar (specimens with twisted reinforcement). It should be noticed that for the specimen with the twisted bar, under the prestressing value of 0.1 N/mm^2, the determined force constituted 89% of the breaking force, and under the prestressing of 0.2 N/mm^2, it was nearly 99% of the breaking force. The proper anchorage of bars in the masonry wall requires the length slightly greater than 250 mm. For ribbed bars, the anchorage length should be considerably greater than for twisted bars. However, it should be remembered that displacement measurements include slipping of the bars from the mortar. Hence, they cannot be used as a basis for the calculation of strains.

REFERENCES

Akcay, C. & Sayin, B. & Yildizlar, B. 2017. The conservation and repair of historical masonry ruins based on laboratory analyses. *Constr. Build. Mater.* 132: 383–394.

Bhattacharya, S. & Nayak, S. & Dutta, S.Ch. 2014. A critical review of retrofitting methods for unreinforced masonry structures. *International Journal of Disaster Risk Reduction*, 7: 51–67.

Cook, D., Ring, S., Fichtner, W. 1995. The Effective Use of Masonry Reinforcement for Crack Repair. *Proceedings of 4th International Masonry Conference*: 442–450.

Corradi, M. & Borri, A. & Vignoli, A. 2002. Strengthening techniques tested on masonry structures struck by the Umbria-Marche earthquake of 1997-1998. *Constr. Build. Mater.* 16: 229–239.

Corradi, M. & Di Schino, A. & Borri A. & Rufini R. 2018. A review of the use of stainless steel for masonry repair and reinforcement. *Constr. Build. Mater.* 181: 335–346.

De Lorenzis, L. & Teng, J.G. 2007. Near-surface mounted FRP reinforcement: an emerging technique for strengthening structures. *Composites Part B* 38: 119–43.

Drobiec, Ł. & Jasinski, R. & Kubica, J. 2000. Investigation of efficiency stitch boss on capacity of cracked masonry wallets. *12th International Brick/Block Masonry Conference, Madrid, Spain*: 953–970.

Drobiec, Ł. & Jasiński, R. & Kubica, J. 2008. Strengthening of cracked compressed masonry using different types of reinforcement located in the bed joints. *ACEE Architecture, Civil Engineering, Environment* 4: 39–48.

Drobiec, Ł. & Jasiński, R. & Mazur, W. 2017 Precast lintels made of autoclaved aerated concrete – test and theoretical analyses. *Cement Wapno Beton* 5: 339–413.

Drobiec, Ł. & Niemiec, T. & Kawulok, M. & Słowik, L. & Chomacki, L. 2019. The method of strengthening the church building in terms of the planned mining exploitation. *ICSF 2019. MATEC Web of Conferences* 284.

Drobiec, Ł. & Jasiński, R. 2019 Possibilities for optimal use of properties of autoclaved aerated concrete. *Cement Wapno Beton* 5: 383–399.

EN 1015-11:1999 *Methods of test for mortar for masonry. Determination of flexural and compressive strength of hardened mortar.*

EN 772-1:2011 *Methods of test for masonry units. Determination of compressive strength.*

EN ISO 6892-1:2016: *Metallic materials. Tensile testing. Part 1: Method of test at room temperature.*

Ismail, N. & Petersen, R.B. & Masia, M.J. & Ingham, J.M. 2011. Diagonal shear behaviour of unreinforced masonry wallettes strengthened using twisted steel bars. *Constr. Build. Mater.* 25: 4386–4393.

Ismail, N. & Ingham, J. M. 2012. In-situ and laboratory based out-of-plane testing of unreinforced clay brick masonry walls strengthened using near surface mounted twisted steel bars. *Constr. Build. Mater.* 36: 119–128.

Jasiński, R. & Drobiec, Ł. 2016. Study of Autoclaved Aerated Concrete Masonry Walls with Horizontal Reinforcement under Compression and Shear. *Procedia Engineering* 161: 918–924.

Masia, M. & Petersen, R. & Konthesingha, C. & Lahra, J. 2010. In-plane shear behaviour of masonry panels strengthened with stainless steel high stress reinforcement. *8ᵗʰ International Masonry Conference, Dresden, Gremany.*

Modena, C. 1997. Criteria for cautious repair of historic buildings. In: L. Binda, C. Modena, editors. *Evaluation and strengthening of existing masonry structures.* RILEM: 25–42.

Moreira, S. & Ramos, L.F. & Csikai, B. & Bastos, P. 2014. Bond behaviour of twisted stainless steel bars in mortar joints, *9th International Masonry Conference, Guimarães, Portugal.*

Petersen, R.B. 2009. *In-plane shear behaviour of unreinforced masonry panels strengthened with fibre reinforced polymer strips.* PhD-Thesis, The University of Newcastle.

Valluzzi, M. R. & Binda, L. & Modena, C. 2005. Mechanical behaviour of historic masonry structures strengthened by bed joints structural repointing. *Constr. Build. Mater.* 19: 63–73.

Willis, C.R. & Seracino, R. & Griffith, M.C. 2010. Out-of-plane strength of brick masonry retrofitted with horizontal NSM CFRP strips. *Eng. Struct.* 32: 547–55.

Brick and Block Masonry - From Historical to Sustainable Masonry –
Kubica, Kwiecień & Bednarz (eds)
© 2020 Taylor & Francis Group, London, ISBN 978-0-367-56586-2

Strengthened clay brick and lightweight aggregate concrete block walls tested under eccentric axial loading – a feasibility study on different strengthening techniques

J. Niklewski & M. Molnár
Faculty of Engineering, Lund University, Lund, Sweden

ABSTRACT: Strengthening of masonry walls is a means to improve the energy performance of modern external masonry walls without further increasing their total thickness. The present paper evaluates the feasibility - here referring to ease-of-application and structural performance - of several techniques aimed at strengthening masonry walls subjected to eccentric axial loading. Four full scale walls made of solid clay bricks and lightweight aggregate concrete blocks were strengthened using externally bonded reinforcement of conventional steel mesh, high-strength steel wire, masonry reinforcement and mechanically fastened strips of steel sheet. Each of the studied strengthening techniques is feasible from a structural perspective. Mechanically fastened steel strips require the least work, and no render is required since the steel is not exposed to fire. Steel mesh can be applied directly on the wall surface, which lowers the thickness of the render layer. Continued research with more replicates will confirm the presented conclusions.

1 INTRODUCTION

Traditionally, massive load-bearing unreinforced masonry walls (URM) have been commonly used for residential housing as well as for larger buildings. Typical exterior walls may consist of a load-bearing unreinforced clay brick wall, a layer of mineral wool, an air gap and a clay brick façade or alternatively of an unreinforced lightweight aggregate concrete wall, a layer of mineral wool and render (Figure 1a). Due to environmental concerns and increasingly stringent explicit requirements, energy-performance has become decisive in the design of exterior walls and ultimately for the choice of building system. Reducing the thickness of the load-bearing wythe enables the designer to add extra insulation without increasing the thickness of the wall as a whole, thus increasing the energy performance without impinging on interior space (Figure 1b). Such a reduction in wall thickness requires strengthening of URM walls.

According to the present practice in the Nordic countries, unreinforced load-bearing lightweight aggregate concrete walls have a thickness of 150 mm. It is anticipated, that by using adequate strengthening, the thickness of this type of walls can be reduced to around 100-120 mm, as shown in Figure 1. Modern, load-bearing, unreinforced clay brick masonry walls usually have a thickness of 120 mm, which already makes these walls slender. Nevertheless, it might be possible to further decrease the wall thickness or alternatively, increase the load bearing capacity of the wall, by strengthening.

There has been much research on strengthening of URM masonry walls in the most recent decades. A vast majority of the methods employed are based on reinforced cementitious render, either by embedding textile materials such as basalt- (Harajli et al. 2010), glass- (Harajli et al. 2010, Bernat et al. 2013, Prota et al. 2006) or carbon fiber (Babaeidarabad et al. 2014, Bernat et al. 2013), or by conventional steel mesh (Molnar et al. 2018). Other methods include mechanically fastened steel strips (Farooq et al. 2014), slotted in bars (Ismail & Ingham 2012), textile sheets fastened with epoxy (Albert et al. 2001) and shotcrete (Lin et al. 2016, Shakib et al. 2016).

The development of strengthening techniques has been driven largely by a need for increased wall performance in seismic areas, where the in-plane shear resistance is a key feature in preventing overstress during earthquakes (Papanicolaou et al. 2007, Prota et al. 2006, Ismail et al. 2011). Research has also been aimed at the application of similar techniques for increasing the flexural capacity under out-of-plane loading, either in bending (Galati et al. 2006, Babaeidarabad et al. 2014, Willis et al. 2010, Papanicolaou et al. 2008, Valluzzi et al. 2014) or under eccentric axial loading (Bernat et al. 2013, Papanicolaou et al. 2014, Molnár et al. 2018). In general, surface reinforcement is able to provide a considerable increase in flexural capacity, especially in the case of slender walls (Bernat-Maso et al. 2015).

In the present study, eight full-scale URM walls, including four clay brick walls (2400x450x87 mm^3)

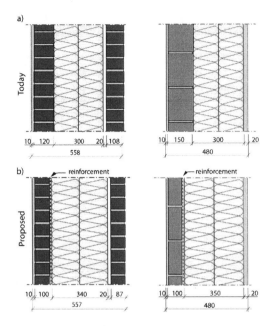

Figure 1. (a) Modern wall designs of lightweight aggregate concrete blocks and clay bricks, respectively, compared to (b) the target design in the present project.

and four lightweight aggregate concrete block walls (2400x590x90 mm³), were strengthened using externally bonded reinforcement on one side of the specimens. The walls were tested under eccentric axial loading and the measured load, deflection and failure mode are here used to describe the behavior of the strengthened URM walls. In considering axial loads, generally one sided-reinforcement is sufficient since slabs and the roof tend to rest on the inner part of the wall and thus give an outward deflection (this is further discussed in section 4.2.2).

The present work is part of a larger effort aiming to develop and verify methods that enable the use of slender load-bearing masonry wythes as part of exterior wall systems. There is no doubt, based on existing evidence, that surface reinforcement can greatly enhance the load-bearing capacity of URM walls. However, the techniques employed in the present paper have mainly been tested for in-plane or flexural loading. In addition to evaluating their structural behavior during eccentric loading, there are several practical issues related to implementation that need to be addressed before the construction industry can adopt the developed technologies for practical use. These include practical aspects for applying the surface reinforcement to the URM, such as ease-of -application and time required for execution. Further, reinforced walls must remain resilient to fire hazard and should not complicate the installation of non-structural elements.

2 METHOD

2.1 General

The tests were carried out during 2019 in the structural engineering laboratory of Lund University. The walls were built, strengthened and tested in an indoor climate.

2.2 Production of URM walls

The URM walls were stacked by a professional brick-layer in order to obtain a consistent degree of workmanship for each wall. Solid clay bricks of size 87x55x228 mm³ and lightweight aggregate concrete blocks of size 90x190x590 mm³ were used. The same brick-layer also constructed smaller specimens for evaluating the compressive strength, Young's modulus and limit strain of the masonry. The material properties are described in Table 1. The walls were strengthened 4 weeks after their construction.

2.3 Application of reinforcement

The wall types and the characteristics of the strengthening materials are presented in Table 2

Table 1. Compressive strength (f_c) of materials.

	f_c
	MPa
Mortar*	2.5
Solid bricks*	>20
Lightweight aggregate concrete blocks*	5
Solid brick masonry	9.8
Lightweight aggregate concrete masonry	3.5
Render* (base)	CS-IV (A)
Render* (top)	CS-III (B)

* Value from manufacturer.

Table 2. Details on the brick (B) and lightweight aggregate concrete walls (L), including total area of reinforcement (A_s), yield strength (f_{sy}) and Young's modulus (E_s).

#	type	A_s	f_{yk}	E_s
		mm²	MPa	MPa
B1	Steel mesh	85	500	200
B2	Masonry reinforcement	50	690	-
B3	Steel strips	200/168*	235	200
B4	High-strength steel wire	15	1525	150
L1	Steel mesh	113	500	200
L2	Masonry reinforcement	75	690	-
L3	Steel strips	100/84*	235	200
L4	High-strength steel wire	15	1525	150

* Gross/net area, the latter accounting for the holes
** f_{yk} and E_s are based on the manufacturer's specification

988

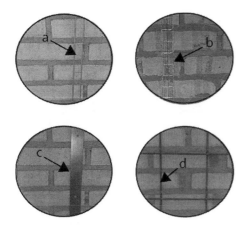

Figure 2. Products used for reinforcement, including (a) masonry reinforcement, (b) high-strength steel wire, (c) steel strips and (d) steel mesh.

Figure 3. Lightweight aggregate concrete block wall after fixing the steel mesh to the wall (left and top right) and after applying the base render (bottom right).

while the actual products employed are shown in Figure 2. The amount of reinforcement was selected so as to obtain approximately the same load-bearing capacity.

The render was applied in two layers consisting of a base layer of type CS-IV and a top layer of type CS-III. The steel mesh was mechanically fastened directly against the dry wall using screws and washers. In practice, due to the wall surface not being perfectly plane, there was a small gap of 0-2 mm between the surface of the masonry and the steel mesh. Base render was then applied onto the walls as shown in Figure 3. The masonry reinforcement and the high-strength steel wires were placed directly in a thin layer of wet base render. The base render was allowed to cure for approximately 24 hours before application of the top render.

The thickness of the base layer varied between 5-8 mm and was adapted to the thickness of the reinforcement. The top render was then applied up to a total thickness of 20 mm (base layer + top layer). The steel strips were mechanically fastened without render using screws with a centre-to-centre distance of 200 mm. The type of fastener used for the respective wall types are shown in Figure 4.

2.4 Testing of the walls

The walls were tested in a custom rig consisting of a welded steel frame and a hydraulic jack (Figure 5). The walls were placed so that the bottom back-edge became aligned with a steel beam and the bottom side was resting on a plane steel section (5x60 mm²) to obtain the eccentricity. A sideways moveable steel beam at the top of the rig was then aligned with the top back-edge in a similar way. Another plane steel section (5x60 mm2) was placed on top of the wall to obtain the same eccentricity. Self-levelling concrete

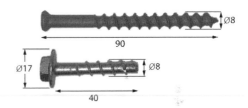

Figure 4. Screws employed for fastening the steel strips to the lightweight aggregate concrete block wall (top) and clay brick wall (bottom). Units in mm.

was used on the top of the clay brick walls to obtain a smooth surface and uniform pressure.

The relative eccentricity of the load, taken as the quota between the eccentricity of the axial load and the section thickness, was 0.22 and 0.23 for the brick and the lightweight aggregate concrete walls, respectively. Having the same eccentricity at the bottom and the top, the walls were subjected to a constant moment along their height.

The axial load was applied in steps using a hand-pumped hydraulic jack. The load and transversal deformation at mid height of the walls were measured. The strain was measured in two locations on the reinforced side and three locations on the unreinforced side, although these results are not presented in this paper.

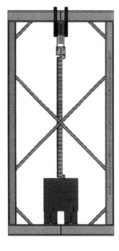

Figure 5. Back- (left) and lateral-view (right) of a wall mounted in the test-rig.

Table 3. Summary of the test results, including maximum axial load (N_{max}), lateral deflection at failure (d_{max}) and failure mode (ductile/brittle = d/b; masonry/reinforcement m/r).

#	type	N_{max}	d_{max}	failure mode
		kN	mm	-
B1	Steel mesh	223	26	b/m
B2	Masonry reinforcement	133	>44	d/r
B3	Steel strips	100	>55	d/r
B4	High-strength steel wire	115	36	b/r
L1	Steel mesh	120	20	b/m
L2	Masonry reinforcement	115	29	b/m
L3	Steel strips	98	19	b/m
L4	High-strength steel wire	163	28	d/m

3 RESULTS

3.1 Experimental results

Axial load and lateral deflection are shown in Figure 6; Table 3 presents a summary of the results, indicating, among others, the mode of failure of the walls.

With the exception of the clay brick wall strengthened with steel strips masonry reinforcement, all other walls were loaded to failure, i.e. when the measured axial load decreased. For all tested walls, a lateral deflection at mid height of the walls larger than 19 mm was observed at failure.

The clay brick wall strengthened with high-strength steel wire (B3) collapsed due to rupture of the reinforcement while the remaining seven walls maintained some level of integrity and were hoisted out of the rig in one piece.

Each strengthening technique was successful in that no bond failure occurred, although it is suspected that the brick wall having masonry reinforcement (B2) suffered from poor bond between the masonry reinforcement and the render.

3.2 Failure modes

The brick wall reinforced with steel mesh (B1) exhibited a brittle failure due to splitting of the masonry along the edge of the loading plate at the upper edge. The failure zone involved 4 courses of bricks, i.e. approximately 270 mm in the axial direction of the wall, as shown in Figure 7.

The failure stress below the loading plate is consistent with the compressive strength of the masonry determined in this project. The failure load of

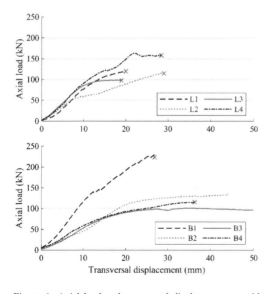

Figure 6. Axial load and transversal displacement at mid height of the walls measured on the strengthened lightweight aggregate concrete walls L1-L4 (top) and clay brick walls B1-B4 (bottom).

Figure 7. Brittle failure in the brick masonry of the B1 wall.

approximately 223 kN corresponds to the capacity of the wall section in compression and is not indicative of the capacity of the reinforcement. The corresponding lightweight aggregate concrete wall (L1) also failed due to crushing of the masonry approximately 250 mm from the top of the wall.

The brick wall strengthened with masonry reinforcement (B2) exhibited a ductile post-peak behaviour, indicating yielding of the steel, and did not reach failure. The corresponding lightweight aggregate concrete wall (L2) failed in a brittle manner by crushing of the masonry and formation of a hinge at mid-height of the wall.

The walls reinforced by mechanically fastened steel strips (B3 and L3) exhibited a fairly ideal linear load-deformation pre-peak behavior with high stiffness. The load reached a plateau when the steel stress reached yielding. The lightweight aggregate concrete wall then maintained a constant load until crushing of the masonry around mid-height of the wall. The clay brick wall did not reach failure.

The clay brick wall reinforced by high-strength steel (B4) wire was the only wall where the reinforcement ruptured, which was followed by the complete collapse of the wall. However, while rupture occurred, the capacity of the wall was limited also by crushing of the masonry. The same behavior was not observed in the corresponding lightweight aggregate concrete wall, where the masonry was crushed in a very ductile and controlled manner with no rupture of the reinforcement.

4 DISCUSSION

4.1 Structural and energy optimization

As aforementioned, the thickness of presently used load-bearing clay brick walls in the Nordic countries is usually 120 mm, while that of lightweight aggregate concrete walls is 150 mm. Comparing the vertical load bearing capacity of these unreinforced walls calculated according to Eurocode 6 with the load bearing capacity of strengthened walls in the present study, an increase in the load bearing capacity of about 15-150 % is obtained. The largest increase in load bearing capacity is observed for the clay brick walls.

It should be kept in mind that the increase in load bearing capacity is obtained in spite of reducing the wall thickness - in the case of the of lightweight aggregate concrete walls with as much as 40 mm, from 150 mm to 110 mm. Thus, the saved space can be used to improve the thermal performance of the building envelope by increasing the thickness of the thermal insulation without increasing the total thickness of the external walls.

In the case of the clay brick walls, the reduction in thickness is only 10-13 mm. Yet, the increase in strength is considerable, extending the potential use of this type of walls to taller buildings.

4.2 Practical aspects

The main aim of the present work is to explore different alternatives for externally bonded reinforcement. The preliminary results presented in this section indicate that all of the included strengthening techniques are able to provide a load-bearing capacity that is considered sufficient for the main application areas of the studied walls. Load-bearing capacity is important, however it is not the only parameter in the design of a wall system. Practical aspects have been the subject of much discussion within the project group, which consists largely of representatives from the Swedish masonry industry. Some of the problems relating to both practical application of the reinforcement and to its effect on other functional requirements of walls as a whole are aforementioned. Possible solutions, albeit, mostly speculative, and subject to future investigation, are also discussed.

4.2.1 Ease-of-application

It is recommended, in general, to apply a base layer of high-strength render before application of the reinforcement. For lightweight products, the reinforcement may be placed directly into the wet base render without further support. However, heavier products, are difficult to be attached in the fresh base layer. This type of products can be fixed to the wall before the application of the base render at the desired distance by screws and washers. In this way, cavities that can be detrimental for the anchorage of the reinforcement can be avoided (Molnár et al. 2018).

Consequently, a base layer was applied before the application of the reinforcement in all cases except for the steel mesh, in which case the mesh was mechanically fastened to the wall without a gap before application of the base render (Figure 3). The mesh was aligned so that the vertical bars lay close to the wall surface, making it so that a distance of at least 6 mm was obtained between the transverse welded bars and the wall. In practice, the wall surface was not perfectly plane, increasing the distance between the mesh and the wall surface by an additional 0-2 mm. Despite the possible presence of gaps, making unlikely that the entire circumference be covered in render, no problems indicating poor bond were observed in the tests. Considering the time that application of bolts and washers require, being able to apply a steel mesh on the wall surface without tedious work is a significant improvement.

The high-strength steel wire delivered as a band has at least two practical advantages over conventional masonry reinforcement. First, the band is lightweight and easy to cut. Second, due to its low flexural stiffness it can be delivered on rolls. The flexural stiffness of the band is not zero, however, and as it is stored and delivered on rolls, its natural shape is slightly curved. Consequently, when applying the bands, they tended to bend out of the fresh mortar. In order to fix them to the wall, another layer of fresh mortar had to be applied while holding the bands in place. Future experiments will show if this problem can be avoided by fastening the high-strength steel wire to the wall before application of the base render.

The steel strips were mechanically connected to the walls using special screws. In the lightweight concrete walls, the steel strips with predrilled holes were attached by direct screwing using special fasteners. In practical applications, it would be rational to use perforated steel strips. In the case of the clay brick walls, holes were predrilled. Yet, fastening of the screws close to the upper, free end of the clay brick wall caused cracking of the joints. This can be explained by low bond strength between the mortar and the bricks.

In the early stages of the project, a system with nailed steel strips was also considered. The idea was to use an electric hand-held nail-gun without pre-drilled holes, i.e. enabling to shoot nails through the steel into the masonry. The system was abandoned due to poor anchorage of the nails in the lightweight aggregate concrete. In addition, nailing in the clay bricks was often unsuccessful due to the bricks' inherent hardness and susceptibility to cracking. Nailing in the mortar joints of clay brick walls was however possible.

The masonry reinforcement was the easiest product to apply. It was light enough so that the bars stuck to the wet base render without additional support. Yet, applying the bars on walls having major irregularities might raise difficulties.

4.2.2 *Building system*

The main function of the reinforcement is to resist bending moments resulting from eccentric axial loading and lateral loads, mainly wind. The main vertical actions include live load, snow load and self-weight. Loading from slabs and roof tend to rest on the inner part of the wall and will thus cause an outward deflection, with tensile stresses on the exterior side of the wall. Application of reinforcement on the exterior side of walls is considered relatively unproblematic.

Wind loads, on the other hand, will often lead to tensile stresses on the inside of the wall. Application of surface reinforcement on the inside is more problematic as the reinforcement is more susceptible to fire hazards. While a sufficient embedment depth in the render could provide adequate protection, this would also increase the thickness of the wall and counteract much of the effort the minimize their thickness.

Another drawback with reinforcement on the inside of the walls is the installation of electrical and water systems. One of the competitive advantages with lightweight aggregate concrete blocks is that slots can easily be cut out of the wall and used for cables and other utilities. Reinforcement placed in the render on the inside would make placement of utilities more difficult, as horizontal slots would intersect the reinforcement. While it is not common practice, a solution could be to place the installments prior to application of the externally bonded reinforcement. Another solution might consist in applying reinforcement only to the central part of the wall height and placing the installations in the upper and the bottom parts of the wall.

While solutions exist that would make double-sided surface reinforcement possible, it would be economically as well as practically convenient to avoid them. In many situations, the weight of the structure alone would be sufficient to counteract tensile stresses caused by wind loads. Anecdotal experience also indicates that surface reinforcement on the side of the wall acting in compression could provide improved, albeit, limited bending capacity.

5 CONCLUSIONS

The work presented in this paper is part of a larger effort aiming to reduce the total thickness of external masonry walls. Four techniques for strengthening masonry walls were tested on clay brick and lightweight aggregate concrete walls. The four techniques contributed to a substantial increase in load bearing capacity of the walls subject to eccentric axial loading. From an ease-of-application point of view, masonry reinforcement and mechanically fastened steel strips exhibited advantages over the other techniques. There are several technical aspects that need to be solved before practical implementation is possible, e.g. reduction of the load-bearing capacity at fire exposure of the externally bonded reinforcement.

ACKNOWLEDGEMENTS

Financial, material and technical support is gratefully acknowledged from The Swedish Energy Agency (grant 37582-3), Brukspecialisten AB, Combimix AB, Fasadgruppen AB, Joma AB, Kåver& Mellin AB, Leca Sverige AB, Nyströms Cement AB, Randers Tegel AB, Tegelmäster AB, Tomas Gustavsson Konstruktioner AB, WSP Sverige AB, Weber Saint Gobain AB.

REFERENCES

Albert, M.L., Elwi, A.E. & Cheng J.R. 2001. Strengthening of unreinforced masonry walls using FRPs. *Journal of Composites for Construction 5 (2):76-84.*

Babaeidarabad S., Caso, F.D. & Nanni, A. 2014. Out-of-plane behavior of URM walls strengthened with fabric-reinforced cementitious matrix composite. *Journal of Composites for Construction* 18(4): 04013057.

Bernat, E., Gil, L., Roca, P. & Escrig, C. 2013. Experimental and analytical study of TRM strengthened brickwork walls under eccentric compressive loading. *Construction and Building Materials* 44: 35–47.

Bernat-Maso, E., Gil, L. & Roca, P. 2015. Numerical analysis of the load-bearing capacity of brick masonry walls strengthened with textile reinforced mortar and subjected to eccentric compressive loading. *Engineering Structures* 91:96–111.

Farooq, S.H., Shadid, I. & Ilyas, M. 2014. Seismic performance of masonry strengthened with steel strips. *KSCE Journal of Civil Engineering* 18(7):2170–2180.

Galati, N., Tumialan, G. & Nanni, A. 2006. Strengthening with FRP bars of URM walls subject to out-of-plane loads. *Construction and Building Materials* 20:101–110.

Harajli, M. ElKhatib, H. & San-Jose, J.T. 2010. Static and cyclic out-of-plane response of masonry walls strengthened using textile-mortar system. *Journal of Materials in Civil Engineering* 22(11): 1171–1180.

Ismail, N. & Ingham J.M. 2012. In-situ and laboratory based out-of-plane testing of unreinforced clay brick masonry walls strengthened using near surface mounted twisted steel bars. *Construction and Building Materials* 36:119–128.

Ismail, N., Petersen, R.B., Masia, M.J. & Ingham, J.M. 2011. Diagonal shear behaviour of unreinforced masonry wallettes strengthened using twisted steel bars. *Construction and Building Materials* 25(12):4386–4393.

Lin, Y., Lawley, D., Wotherspoon, L. & Ingham J.M. 2016. Out-of-plane testing of unreinforced masonry walls strengthened using ECC shotcrete. *Structures* 7:33–42.

Molnár, M., Jönsson, J. & Gottsäter, E. 2018. Experimental study of masonry walls strengthened with externally bonded steel-grid under eccentric axial compressive loading. *In 10th International Masonry Conference*, G. Milani, A. Taliercio & S. Garrity (eds.), Milan, Italy, July 9-11, 2018.

Papanicolaou, C.G., Triantafillou, T.C., Karlos, K. & Papathanasiou, M. 2008. Textile reinforced mortar (TRM) versus FRP as strengthening material of URM walls: out-of-plane cyclic loading. *Materials and Structures* 41(1):143–157.

Papanicolaou, C. G., Triantafillou, T. C., Karlos, K., & Papathanasiou, M. 2007. Textile-reinforced mortar (TRM) versus FRP as strengthening material of URM walls: in-plane cyclic loading. *Materials and structures*, 40(10):1081–1097.

Papanicolaou, C., Triantafillou, T.C. & Roca Fabregat, P.R. 2015. Increase of load-carrying capacity of masonry with textile reinforced rendering. *Mauerwerk* 19(1):40–51.

Prota, A., Marcari, G., Fabbrocino, G., Manfredi, G. & Aldea, C. 2006. Experimental in-plane behavior of tuff masonry strengthened with cementitious matrix-grid composites. *Journal of Composites for Construction* 10(3): 223–233.

Shakib, H., Dardaei, S., Mousavi, M. & Rezaei, M.K. 2016. Experimental and analytical evaluation of confined masonry walls retrofitted with CFRP strips and mesh-reinforced PF shotcrete. *Journal of Performance of Constructed Facilities* 30(6): 04016039.

Valluzzi, M.R., Da Porto, F., Garbin, E. & Panizza, M. 2014. Out-of-plane behaviour of infill masonry panels strengthened with composite materials. Materials and Structures 47(12):2131–2145.

Willis, C.R., Seracino, R. & Griffith, M.C. 2010. Out-of-plane strength of brick masonry retrofitted with horizontal NSM CFRP strips. *Engineering Structures* 32(2):547–555.

Brick and Block Masonry - From Historical to Sustainable Masonry –
Kubica, Kwiecień & Bednarz (eds)
© 2020 Taylor & Francis Group, London, ISBN 978-0-367-56586-2

Influence of horizontal reinforcement type on shear strength of hollow concrete block masonry shear walls

S. Calderón
Pontificia Universidad Católica de Chile, Santiago, Chile & Politecnico di Milano, Milano, Italia

C. Sandoval, E. Inzunza-Araya, G. Araya-Letelier & L. Vargas
Pontificia Universidad Católica de Chile, Santiago, Chile

ABSTRACT: Partially grouted reinforced masonry (PG-RM) shear walls are used as a structural system in several countries. As a result of this, horizontal reinforcement schema varies depending on local construction practices. In some places, horizontal reinforcement bars are placed in the inner center of blocks to form the so-called bond-beams. In other locations, horizontal reinforcement is embedded in bed-joints. Recent studies have proposed to combine both types of reinforcement layouts in order to improve the seismic performance of PG-RM walls, although available experimental evidence is still limited. Considering this, the experimental results of three in-plane cyclic load tests of PG-RM walls with different arrangements of horizontal reinforcement are compared. One wall was only provided with bed-joint reinforcement type, one only with bond-beam reinforcement type, and one with both types of reinforcement layouts. Results indicate that the combination of both reinforcement types is the most suitable option, in terms of maximum lateral resistance and crack distribution.

1 INTRODUCTION

Partially grouted reinforced masonry (PG-RM) shear walls are widely used as a structural system in several countries. As a consequence of this, different construction practices and methodologies can be distinguished among different locations. One of the main differences is the way to place the horizontal reinforcement in the shear walls. For instance, ladder-type steel reinforcement elements are commonly embedded in bed-joints (Figure 1a) in Latin America. In contrast, horizontal reinforcement bars are usually placed in the center of blocks to form the so-called bond-beams (Figure 1b) in the United States of America (USA) and Canada. In general, a higher quantity of shear reinforcement can be used in bond-beam reinforced walls due to the diameter of rebars is not limited by the thickness of mortar joints, as in bed-joint reinforced shear walls. As a result of this, bond-beam reinforced walls might be designed to carry higher lateral loads because of the higher steel contribution. On the other hand, some studies (Baenziger & Porter 2011, Ramírez et al. 2016, Sandoval et al. 2018, Schultz et al. 1998, Stathis et al. 2018) have recognized the ability of bed-joint reinforcement to limit crack widths and to force the distribution of cracks through the masonry panel. Several investigations have studied the structural response of both aforementioned

masonry typologies (El-Dakhakhni & Ashour 2017). However, research findings and the related design recommendations have not been successful enough to avoid severe damages during recent earthquakes (Astroza et al. 2012, Schultz & Johnson 2019).

Yancey & Scribner (1989), Bolhassani et al. (2016), and Schultz & Johnson (2019) have advocated combining bond-beam and bed-joint reinforcement types in order to take advantage of the benefits of both reinforcing systems. Yancey & Scriber (1989) tested 10 PG-RM shear walls (height-to-length ratio of 1.12) under lateral cyclic loads and constant axial load, with a double-bending boundary condition. Three of those specimens had only bed-joint reinforcement, four had only bond-beam reinforcement, and two had a combination of both reinforcement types. However, none of those walls were provided with vertical reinforcement. Bolhassani et al. (2016) tested three square PG-RM shear walls under cyclic lateral loads and constant axial load, employing a cantilever setup. Two of those walls had only bond-beam reinforcement type, and the remaining had a combination of bed-joint and bond-beam reinforcement types. Nonetheless, these walls had different horizontal reinforcement ratio and disposition of vertical reinforcement bars. Schultz & Johnson (2019) tested two C-shaped PG-RM shear walls of aspect ratio 0.6 under cyclic lateral loads and constant axial load, employing a cantilever test setup. Both walls had a centered

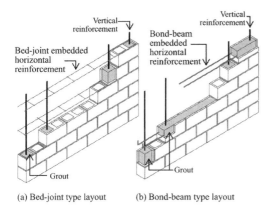

Figure 1. Layouts types of horizontal reinforcement.

window-like opening. One wall had only bond-beam reinforcement type, while the other had a combination of bed-joint and bond-beam reinforcement types. However, these walls also had different distributions of the vertical reinforcement, implementing the detailing proposed in Bolhassani et al. (2016). Despite the differences in the layout of vertical reinforcement, the authors indicated that the wall with a mixed reinforcement layout exhibited a more distributed crack pattern than the wall with only bond-beams. In addition, providing bed-joint reinforcement contributed to increasing the efficiency of bond-beam reinforcement before reaching the peak strength. This improvement in the behavior was attributed to the capacity of bed-joint reinforcement to reduce the growth of cracks, forcing the generation of new cracks. Although the reported results are promising, the available experimental evidence is still limited, particularly in terms of comparing the seismic performance and failure modes of different reinforcing arrangements. In fact, none of these studies had explicitly focused on the effect of combining bed-joint and bond-beam reinforcement types on the

response of walls with similar horizontal reinforcement ratios and the same disposition of vertical reinforcement.

Considering the presented background, this research further studies how the shear reinforcement type affects the in-plane seismic performance of PG-RM shear walls. The experimental results of three full-scale in-plane lateral load tests of PG-RM walls with different arrangements of horizontal reinforcement and the same horizontal reinforcement ratio are compared. The first wall was only provided with bed-joint reinforcement type, the second wall only with the bond-beam reinforcement type, whereas the third wall was a combination of both reinforcement layout types. The three specimens had the same amount and disposition of vertical reinforcement. The results are analyzed in terms of force-displacement curves and maximum resistance.

2 DESCRIPTION OF TESTS

Three hollow concrete block (HCB) PG-RM shear walls were tested under constant axial load and cyclic lateral loads. The walls are schematized in Figure 2, and some of their design characteristics are presented in Table 1. All walls had the same external dimensions, being 2640 mm long, 2270 mm high, and 140 mm thick. Consequently, walls had an aspect ratio (h_w/L_w) of 0.86. In addition, the same layout of vertical reinforcement was provided to the three walls, corresponding to four steel rebars of 22 mm diameter, one was placed at each wall's edge, and two were put in the interior of it. It is worth noting that walls had a high vertical reinforcement ratio ($\rho_v = 0.41\%$) to force them to fail under a shear failure mode. Moreover, reinforced concrete beams were built above and under each wall, whose dimensions are indicated in Figure 2, to ensure a proper anchoring of the vertical reinforcement and a homogeneous distribution of loads. The walls differ in the layout of horizontal

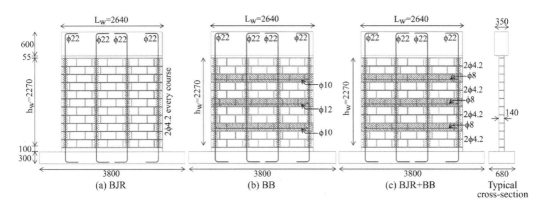

Figure 2. Design of tested walls. (Dashed zones are cells filled with grout).

995

Table 1. Properties of tested walls.

Wall	Gross cross sectional area mm²	Horizontal reinforcement				Total area mm²	Reinforcement ratio, ρ_h %	Vertical reinforcement ratio, ρ_v %	Axial stress, σ_n MPa
		Bed joint		Bond-beam					
		Description	Area mm²	Description	Area mm²				
BJR	369600	10-2 φ 4.2 mm	277.1			277.1	0.087	0.41	0.5
BB	369600	-	-	2 φ 10 mm + 1 φ 12 mm	270.2	270.2	0.085	0.41	0.5
BJR+BB	369600	4-2 φ 4.2 mm	110.8	3 φ 8 mm	150.8	261.6	0.082	0.41	0.5

reinforcement. The first wall, shown in Figure 2a and named BJR, was provided with ladder-type reinforcement embedded in every mortar bed-joint. These reinforcement elements were composed of two longitudinal steel bars of 4.2 mm diameter and transversal steel bars of 4.2 mm diameter every 30 cm. The second wall, presented in Figure 2b and named BB, had three bond-beams equally spaced every three block courses. In that wall, the first and the third bond-beams had one steel bar of 10 mm diameter, and the second bond-beam had one steel bar of 12 mm diameter. The third wall, depicted in Figure 2c and named as BJR+BB, had a combined reinforcing layout. This wall was provided with ladder-type bed-joint reinforcement elements (2φ4.2 mm) every three block courses and three bond-beams, which had one steel bar of 8 mm diameter each one. It is worth noticing that all walls had almost the same horizontal reinforcement ratio, being equal to 0.087%, 0.085%, and 0.082%, respectively. It is also important to remark that only the block cells containing vertical reinforcement and bond-beam reinforcement were grouted, as illustrated in Figure 2.

The employed HCBs (Figure 3) were 390 mm in length, 190 mm in height and 140 mm in thickness, and had a void ratio of 45.2 % (coefficient of variation (CV) of 0.7 %). These standard HCBs were cut to place the horizontal reinforcement in bond-beams, as also depicted in Figure 3. The masonry compound had an average compressive strength of 9.7 MPa (CV = 13.8%) and an average Young's modulus of 6486 MPa (CV = 18.6%). Vertical reinforcement had an average yield strength of 644.9 MPa (CV = 0.4%) and an average tensile strength of 769.2 MPa (CV = 0.7%). While, bed-joint and bond-beam reinforcement bars had an average yield strength of 644.9 MPa (CV = 5.0%) and 395.3 MPa (CV = 1.8%), and an average tensile strength of 667 MPa (CV = 2.7%) and 506.1 MPa (CV = 0.8%), respectively. It is important to notice that the steel of ladder-type reinforcement elements had a small difference between yield and tensile strengths in comparison with the other two employed steel types.

A cantilever boundary condition was provided to walls with the experimental setup that is illustrated in Figure 4. The inferior head beam was fixed to the strong floor of the laboratory. Additionally, out-of-plane displacements were prevented at the level of the upper head beam. A uniform constant axial load of 0.5 MPa was applied at the top of the three tested walls, which is equal to a ratio of 5% of the compressive strength of the masonry composite. Incremental cyclic lateral displacements were applied at the center of the upper head beam. Deformation increments were determined following FEMA 461 guidelines (2007) (See Figure 4), and two cycles

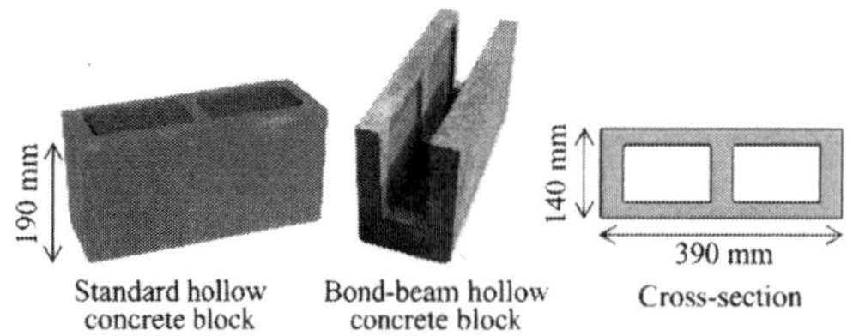

![Standard hollow concrete block, Bond-beam hollow concrete block, Cross-section]

190 mm
140 mm
390 mm

Standard hollow concrete block   Bond-beam hollow concrete block   Cross-section

Figure 3. Employed blocks.

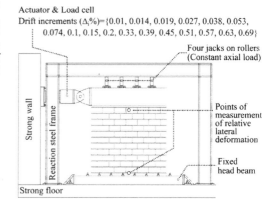

Actuator & Load cell
Drift increments (Δᵢ%)={0.01, 0.014, 0.019, 0.027, 0.038, 0.053, 0.074, 0.1, 0.15, 0.2, 0.33, 0.39, 0.45, 0.51, 0.57, 0.63, 0.69}

Four jacks on rollers (Constant axial load)

Strong wall

Reaction steel frame

Points of measurement of relative lateral deformation

Fixed head beam

Strong floor

Figure 4. Experimental setup.

were performed per deformation increments. The lateral relative deformation of the masonry panels was measured between the first and last block curses with a linear variable differential transformer (LVDT), as also shown in Figure 4.

3 RESULTS AND DISCUSSION

3.1 Hysteretic response

The in-plane hysteresis curves recorded during the tests are presented in Figure 5, where shear stresses were calculated over the gross cross-sectional area of walls. In general, an elastic stage can be identified on all walls in the first loading cycles. Afterward, progressive degradation of the secant lateral stiffness can be observed up to the maximum resistance was achieved. Subsequently, all walls showed a sudden loss of their lateral resistant capacity. It also can be mentioned that the shape of hysteresis curves was related to the damage progression since all walls exhibited narrow cycles at the beginning of tests (incipient damage) and wider cycles at the end of tests (severe damage). Despite the differences in the layout of horizontal reinforcement, the hysteretic curves of the three walls had similar shapes, and none of them showed pinching. However, it can be appreciated that the wall with the hybrid reinforcing system (BB+BJR) exhibited wider cycles, particularly after reaching the maximum lateral load. This suggests that the BJR+BB wall had a higher energy dissipation capacity and a better seismic performance than the other walls. On the other hand, it can be noticed that the walls that had distributed reinforcement (BJR and BJR+BB) exhibited a symmetric response between both load directions, unlike the wall with only bond-beams. The symmetric behavior suggests that bed-joint reinforcement provides integrity to the wall and can control of the propagation of cracks, verifying the observations of previous studies (Baenziger & Porter 2011, Ramírez et al. 2016, Sandoval et al. 2018, Schultz et al. 1998, Stathis et al. 2018).

Figure 6 superimposes the backbone curves of the three walls, and Table 2 presents the information that describes the instant when each wall reached its maximum lateral resistance. As commented previously, the shapes of the curves are quite similar. However, for drifts lower than 0.1% in the push load direction, the walls that had bed-joint reinforcement

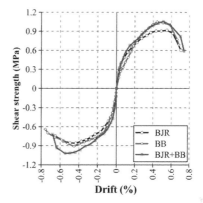

Figure 6. Comparison of backbone curves.

Table 2. Maximum resistance of tested walls.

Wall	Load direction	Displacement mm	Drift %	Force kN	Peak shear stress MPa
			Maximum lateral force		
BJR	Push	12.44	0.55	338.46	0.92
	Pull	11.22	0.49	316.62	0.86
BB	Push	10.05	0.44	387.80	1.05
	Pull	10.25	0.45	337.27	0.91
BB + BJR	Push	11.54	0.51	390.43	1.06
	Pull	12.50	0.55	376.47	1.02

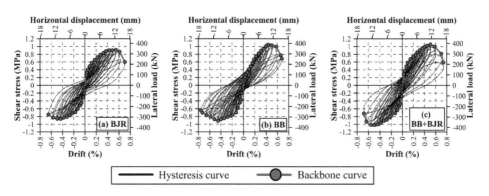

Figure 5. Hysteretic response of tested walls.

997

(BJR and BJR+BB) exhibited higher resistances and lower reductions of the tangential lateral stiffness than the wall with only bond-beams. In fact, wall BB experienced a change in their tangent lateral stiffness in the push load direction at a drift about 0.05%, although the lateral stiffness is relatively constant for higher drifts. It grabs the attention that this behavior did not take place in the opposite load direction of wall BB nor any other wall. This suggests that an early crack formed in the BB wall in the push direction, which remained active until developing the maximum load in that loading direction. This latter observation was also commented on (Araya-Letelier et al. 2019). Conversely, in the pull load direction of wall BB and the other walls, the formation of cracks and the loss of lateral stiffness were progressive. Therefore, the different behaviors of walls indicate that bed-joint rein-for-cing elements avoid the activation of only one deformation mechanism and promote the distribution of the damage in different zones of the masonry panel.

Regarding the maximum resistance, walls with bond-beams (BB and BJR+BB) exhibited value approximately 15% higher than the BJR wall, in the push load direction. This variation can be due to the fact that grouted cells of bond beams incremented the net cross-sectional resistant area in some zones and also subdivided the masonry panels into smaller parts that might damage independently. On the other hand, all walls experienced a lower maximum resistance in the reversal load direction than in the push load direction. Those differences were equal to -6.5%, -13.0%, and -3.6% for walls BJR, BB, and BJR+BB, respectively. As the wall with the combined reinforcement layout (BB+BJR) experienced the smallest reduction, it can be mentioned that the mixed reinforcement layout provided a higher degree of integrity to the panel than standard layouts (BB or BJR), as commented in Bolhassani et al. (2016) and Schultz & Johnson (2019). In addition, the wall BJR exhibited the intermediate reduction, which confirms that the damage control effect was a result of the distributed bed-joint reinforcement. On the other hand, it can be indicated that all walls

reached their maximum resistance at similar lateral deformations (around a drift of 0.5%), in both load directions.

3.2 Progression of damage

In general, damage progressed similarly in the three walls, although considerable differences were appreciated at the end of the tests. In all walls, some horizontal cracks were identified during the first load increments in the lowest mortar bed-joints in the heels of panels. The generation of those cracks could be attributed to the bending moment that is higher at the base of the walls. For this reason, those cracks will be called as bending horizontal cracks hereinafter. Afterward, new cracks appeared in block-mortar interfaces mainly, forming some stair-stepped cracks below the 3rd block course nearby the edges of masonry panels. It was observed that those cracks had a higher extension in the wall BJR because, in the other two walls, the grout in bond-beams was able to impede the propagation of cracks, at least when damage was slight.

When lateral deformation incremented in the walls with bond-beams (BB and BJR+BB), cracks grew along the top and the bottom of bond-beams. This was more evident in the wall BB than in the wall BJR+BB because the distributed bed-joint reinforcement forced the spread of cracks. On the other hand, in the wall only with bed-joint reinforcement, stair-stepped cracks spread through the panel, covering its whole height.

At the end of the tests, cracks spread through the walls forming crack patterns similar to the typically observed in walls that fail under a shear failure mode, as shown in Figure 7. Cracks concentrated mostly at block-mortar interfaces due to the bond strength of block-mortar interfaces is lower than the tensile strength of the concrete block. Therefore, the interfaces made up a preferable path for the growth of cracks. For this reason, blocks remained mostly undamaged until advanced stages in all tests, when a mixed shear-compression stress state produced the

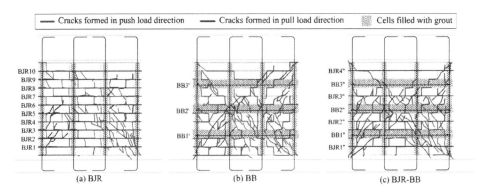

Figure 7. Crack patterns at the end of the tests.

cracking of blocks at walls' heels. This phenomenon can be appreciated in the cracking pattern of the three walls. Despite the abovementioned, blocks at the center of walls with bond-beams (BB and BJR+BB) presented damage. In these zones, the failure of block-mortar interfaces was constrained by grouted zones (vertical cells and bond-beams), and as a result, blocks became the weakest zones in relative terms.

Therefore, the evidence indicates that the presence of grouted cells influences how cracks grow on walls. It also can be mentioned that the presence of bed-joint reinforcement controls the propagation of existent cracks and promotes the generation of new cracks due to its ability to keep masonry tied after cracking (Schultz et al. 1998, Tomaževič & Lutman 1988). Additionally, it is essential to mention that the three tested walls were provided with a high amount of vertical reinforcement, and that the behavior of these walls would have been different if the contribution of a flexural failure mode had been higher.

4 CONCLUSIONS

Three hollow concrete block walls with different horizontal reinforcement types were tested under constant axial and cyclic lateral loads. The first wall was provided with ladder-type bed-joint reinforcement, the second with steel rebars embedded in bond-beams, and the third with a combined layout of the two other walls. It is important to remark that the three walls had approximately the same horizontal reinforcement ratio.

Experimental results suggest that a combination of bond-beams and bed-joint reinforcement arrangements is the most suitable reinforcing strategy. The wall with the combined reinforcement layout exhibited the same maximum resistance than the wall with bond-beams, but the crack pattern was more distributed through the masonry panel. Additionally, the wall with the combined reinforcement layout showed the most symmetrical behavior among the tested walls. These findings corroborate the observations made in previous investigations (Bolhassani et al., 2016; Schultz & Johnson, 2019; Yancey & Scribner, 1989), which demonstrates the potential of employing a combination of bed-joint and bond-beam reinforcement types as shear reinforcement. On the other hand, it can be mentioned that the bed-joint reinforcement is better to limit the width of cracks in comparison with the bond-beam reinforcement, although somewhat lateral resistance is sacrificed.

ACKNOWLEDGMENTS

The authors want to thank the financial support of the *FONDEF Research Project No. ID17I10264*. Sebastián Calderón wants to acknowledge the financial support provided by the scholarship *CONICYT-PFCHA/Doctorado Nacional 2017-21170992*. Laura Vargas thanks the support provided by CONICYT by its program CONICYT-PCHA/Doctorado Nacional/2019-21191181.

REFERENCES

Araya-Letelier, G. et al. 2019. Fragility functions for partially-grouted masonry shear walls with bed-joint reinforcement. *Engineering Structures, 191*.

Astroza, M. et al. 2012. Seismic performance of engineered masonry buildings in the 2010 Maule earthquake. *Earthquake Spectra* 28(SUPPL.1): 385–406.

Baenziger, G. & Porter, L. 2011. Joint reinforcement for masonry shear walls. *Eleventh North American Masonry Conference*.

Bolhassani, M. et al. 2016. Enhancement of lateral in-plane capacity of partially grouted concrete masonry shear walls. *Engineering Structures* 108: 59–76.

El-Dakhakhni, W. & Ashour, A. 2017. Seismic Response of Reinforced-Concrete Masonry Shear-Wall Components and Systems: State of the Art. *Journal of Structural Engineering*, 143(9): 1–25.

FEMA. 2007. *FEMA 461. Interim testing protocols for determining the seismic performance characteristics of structural and nonstructural components*. Federal Emergency Management Agency.

Ramírez, P. et al. 2016. Experimental study on in-plane cyclic response of partially grouted reinforced concrete masonry shear walls. *Engineering Structures* 126: 598–617.

Sandoval, C. et al. 2018. Experimental cyclic response assessment of partially grouted reinforced clay brick masonry walls. *Bulletin of Earthquake Engineering* 16(7).

Schultz, A. E. et al. 1998. Seismic performance of masonry walls with bed joint reinforcement. *Structural Engineers World Congress*.

Schultz, A. E. & Johnson, C. 2019. Seismic resistance mechanisms in partially grouted shear walls with new design details. *13th North American Masonry Conference*.

Stathis, O. et al. 2018. Effects of horizontal reinforcement distribution on in-plane performance and post-peak behaviour of masonry shear walls. *10th Australasian Masonry Conference*.

Tomaževič, M. & Lutman, M. 1988. Seismic resistance of reinforced masonry walls. *Proceedings of Ninth World Conference on Earthquake Engineering*: 6.

Yancey, C. & Scribner, C. 1989. *Influence of horizontal reinforcement of concrete block walls (NISTIR 4202)*.

Brick and Block Masonry - From Historical to Sustainable Masonry –
Kubica, Kwiecień & Bednarz (eds)
© 2020 Taylor & Francis Group, London, ISBN 978-0-367-56586-2

The seismic performance of partially grouted reinforced masonry

G.C. Manos, L. Melidis, L. Kotoulas & K. Katakalos
Lab. Strength of Materials and Structures, Aristotle University of Thessaloniki, Greece

ABSTRACT: The in-plane behavior of partially grouted masonry wall specimens, built using clay bricks with vertical holes, hosting vertical steel reinforcement, is studied. They were rigidly attached at a reaction frame and were subjected at their top to a constant vertical load together with a horizontal seismic-type cyclic load. A numerical simulation was formed, employing micro-modeling, in an effort to reproduce numerically the observed performance. In this numerical simulation the clay brick units, the mortar joints and the longitudinal reinforcement were modeled separately, following all the geometric and construction detailing together with the measured non-linear mechanical characteristics of the mortar joints and the steel reinforcement. A critical factor of the observed performance was the behavior of the employed clay brick unit which developed local compressive-type modes of failure, which influence the overall seismic response. The various steps taken to develop clay brick units capable to resist such local compressive-type loads are also presented and discussed.

1   INTRODUCTION

Extensive past research dealt with the performance of partially grouted reinforced masonry walls, when subjected to simulated seismic loads. Of particular interest is the in-plane seismic behavior. As was shown by Manos and co-workers (Manos 1983, Gulkan 1990) for low-rise well-built masonry structures, the in-plane behavior can be studied separately from the out-of-plane behavior. When subjecting such masonry walls simultaneously to combined in-plane horizontal and vertical loads (Hidalgo et al. 1978, Tomazevic et al. 1993), the most significant parameters that are usually examined are the type and strength of the mortar and masonry units, the geometry of the masonry walls and their reinforcing arrangement (in quantity and structural details), and the level of axial compression (Tasios 1987). Oan & Shrive (2009) examined the shear behavior of concrete masonry walls whereas Sandoval et al. (2018) investigated the in-plane cyclic response of partially grouted reinforced clay brick masonry walls in a way similar to the one employed in the present work. A first objective of the present study is to validate the local materials and construction practices towards building earthquake resistant low-rise partially grouted reinforced clay brick masonry structures in areas of moderate seismicity of Greece (Manos 2000a, b). A second objective is to numerically simulate the observed behavior of such partially grouted reinforced masonry wall specimens. The following summarizes the main features of the experimental program with partially grouted reinforced clay brick masonry wall specimens (Manos 2000a).

2   EXPERIMENTAL PROGRAM

This investigation examined the influence of the variation of the specimens' geometry, the mortar type, the horizontal and vertical reinforcement ratio values, the level of compressive axial load applied together with the horizontal force at the top of each specimen (pseudo-dynamic or dynamic in nature). This variation resulted in a considerable number of partially grouted reinforced masonry wall specimens. Almost all wall specimens were 154 mm thick built with the clay brick unit shown in Figure 1. Specimen Wall-34N, which was 320 mm thick, was also built with the same brick unit of Figure 1 (alternatively either two bricks or one brick forming the width of the wall). Most specimens were built with clay brick type C (see Figure 1). A small number of specimens were built with a slightly different clay brick (type D). The overall dimensions of both brick types were 154 mm wide, 320 mm long with a height of 155 mm. According to Table 3.1. of Eurocode 6 (2005) both these clay bricks are classified as a group 2 masonry units. The measured values of the mechanical properties of the all used materials are given in the next section.

The clay brick unit of Figure 1 was initially developed as a pilot masonry unit by the Filippou Structural Clay Products (Manos 2000a). Moreover, apart from the current study the potential of such a clay brick unit to be used for partially grouted reinforced masonry was also examined by Psylla et al., 1996.

a. Geometry. Most wall specimens (twenty-eight) were 1330 mm long, 1330 mm high and 154 mm thick; these were relatively "wide" specimens with a height over length (*h/l*) ratio equal to 1. In addition,

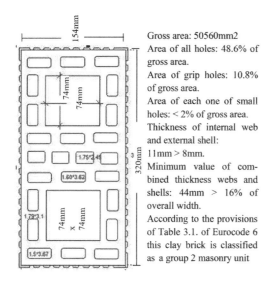

Gross area: 50560mm2
Area of all holes: 48.6% of gross area.
Area of grip holes: 10.8% of gross area.
Area of each one of small holes: < 2% of gross area.
Thickness of internal web and external shell:
11mm > 8mm.
Minimum value of combined thickness webs and shells: 44mm > 16% of overall width.
According to the provisions of Table 3.1. of Eurocode 6 this clay brick is classified as a group 2 masonry unit

Figure 1. Type C clay brick.

eight (8) specimens were 660 mm long, 1330 mm high and 154 mm thick; these were relatively "slender" specimens with a height to length (*h/l*) ratio equal to 2. Apart from the above geometry, two specimens that were 2700 mm long and 2475 mm high were also tested. One of these specimens was 154 mm thick (Wall-27N, Figure 4), representing a one-story high interior wall, whereas the second specimen with the same length and height was 320 mm thick (Wall-34N), representing a one-story high exterior wall. Both these specimens had a height over length (*h/l*) ratio equal to 0.917. All specimens were constructed in running bond with bed joints and head joints 10 mm thick.

b. Mortar type. All wall specimens were selectively built with mortar types characterized by their compressive strength. A small number of specimens (six) used mortar with a target compressive strength

equal to 2.5 MPa, representing a relatively weak mortar (marked with the letter O). Alternatively, mortar with a target compressive strength equal to 5.0 MPa, representing a relatively strong mortar (marked with the letter N) was used for most specimens (thirty-two).

c. Variation of the longitudinal and/or transverse reinforcement. The amount of transverse (horizontal) reinforcement, in terms of ratio of the area of this reinforcement over the corresponding gross cross-sectional area, was varied from 0.05%, to 0.150%. For the largest specimens Wall-27N and Wall-34N the value of this ratio was equal to 0.102% and 0.097%, respectively. The amount of the longitudinal (vertical) reinforcement, in terms of ratio of the area of this reinforcement over the corresponding gross cross-sectional area of each wall specimen, was equal to 0.125% for most specimens. For Wall-27N and Wall-34N the value of this ratio was equal to 0.086% and 0.082%, respectively

d. Variation of the vertical load imposed at the top of each specimen. Twenty-two specimens were subjected to 4% of the compressive strength of the brick masonry (*f*$_k$) whereas eleven specimens were subjected to 8% of that strength. A limited number of specimens (three) were tested with no axial compression. The strength of each masonry type was found by testing masonry wallets constructed with the same clay brick unit using either weak (O) or relatively strong (N) mortar (see Table 1).

e. Variation of the seismic-type load. The horizontal seismic-type load was imposed to each wall specimen together with the vertical compressive load. In all cases, this horizontal load was a series of three sinusoidal cycles of progressively increasing amplitude as depicted in Figure 2, as used by many researchers. In this figure the maximum amplitude at the end (13th step) of this horizontal sequence is 20 mm and it is reached by thirteen (13) similar 3-cycle consecutive loading steps of continuously increasing amplitude.

Table 1. Compression and shear strengths derived from wallet tests.

Specimen Name	Compr. Mortar Strength fm (MPa)	Brick Name	Dimensions of Pier (H/B) (mm) thickness 155mm	Type of Test	Measured Strength (MPa)	Predicted Strength (MPa)
(1)	(2)	(3)	(4)	(5)	(6)	(7)
Wallet 2N	4.03	C	1300/663	Compression	4.56	3.385
Wallet 3N	4.03	C	1300/661	Compression	4.53	3.385
Wallet 5N	4.03	C	1295/655	Compression	5.10	3.385
Wallet 2O	2.09	C	1300/663	Compression	3.48	2.778
Wallet 3O	2.09	C	1295/663	Compression	3.37	2.778
Wallet 2DO	2.09	D	660/660	Diagonal Tension	f_t 0.204	f_{vko} 0.10
Wallet 2DN	4.03	D	975/1000	Diagonal Tension	f_t 0.24	f_{vko} 0.15
Wallet 3DO	2.09	D	975/1000	Diagonal Tension	f_t 0.199	f_{vko} 0.10

*f_b = 9.82 MPa For Brick C, f_b = 7.85 MPa For Brick D

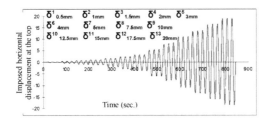

Figure 2. Imposed horizontal seismic-type loading sequence.

The amplitudes of each one of these thirteen 3-cycle loading steps are listed at the top of Figure 2. The frequency of each one of these cycles was kept constant. For thirty specimens, including large dimensions specimens Wall-27N and Wall-34N, the frequency of the horizontal loading cycles was equal to 0.048 Hz, representing a pseudo-dynamic type of loading. For eight wall specimens the frequency of these loading cycles was much higher set equal to 1 Hz thus representing a dynamic type of loading sequence.

3 MATERIAL CHARACTERIZATION

The mechanical properties of the used clay brick units, mortar types O and N, horizontal and vertical reinforcing bars and grout were defined by laboratory testing of samples of the used materials taken during construction of the wall specimens described before. Together with the construction of these wall specimens a number of wallets were also constructed using the same clay brick units and the same mortar types described before. These wallets, with dimensions listed in column (4) of Table 1, were subjected to either axial or diagonal compression, as described in column (5) of the same table. Column 1 of this table lists the tested wallets, indicating with the letter N (strong mortar) or O (weak mortar) the used mortar type in their construction which was the same mortar used in the construction of the corresponding wall specimens described before. The corresponding measured mean compressive mortar strength value of these two mortar types is listed in column (2) of this table. The type of brick used for building these wallets is listed in column (3) of Table 1. At the bottom of this table the values of the normalized mean compressive strength of these clay brick units type C and D, as defined in Eurocode 6, is also listed. These values were found from compressive tests of these clay bricks in a direction normal to their cross-section (Figure 1) performed according to EN 772-1 (2011). The dimensions in millimeters of these wallets are listed in column 4, (height/length/thickness). From the maximum value of the axial or diagonal compressive load, measured during testing these wallets, the corresponding compressive strength (f_k) and tensile strength values (f_t) are

derived (RILEM 1991) and listed in column (6) of Table 1. Column (7) of this table lists corresponding predicted values making use of the relevant provisions of Eurocode 6 (2005) together with the measured mortar compressive strength values (type N or O) and the measured clay brick unit (type C or D) compressive strength values. Listed is also the limit shear strength value for zero normal stress (f_{vko}), for this type of mortar and clay brick units, according to table 3.4 of Eurocode 6. Use is made for predicting the compressive strength value f_k of the following empirical formula (Eq. 1) of Eurocode 6 (2005) as defined after the latest revision of this document, where f_b is the normalized mean compressive strength of the brick unit in N/mm2 and fm is the compressive strength of the mortar, in N/mm2. The value of the constant K ($K = 0.45$) is found according to table 3.3 of Eurocode 6 having previously ascertained that the clay brick unit belongs to group 2, according to table 3 of this document (see also Figure 1). The Young's modulus values, can be approximated from the compressive tests of the wallets listed in Table 1 as equal to 4000 MPa for the wallets built with mortar type N and 3000 MPa for those built with mortar type O.

$$f_k = K f_b^{0.7} f_m^{0.3} \qquad (1)$$

Deformable reinforcing bars with diameter either 8 mm or 10 mm were used as vertical reinforcement with measured mean yield stress 523 MPa and 501 MPa, respectively. The corresponding measured tensile strength values were 635 MPa and 647 MPa for these deformable reinforcing bars exhibiting a tensile strain of 11% at maximum load. Deformable reinforcing rods of 4 mm diameter formed the horizontal reinforcement for most specimens with a measured yield stress and tensile strength equal to 590 MPa and 757 MPa, respectively. The mean measured grout compressive strength was equal to 9.35 MPa and the mean measured bond strength was equal to 0.45 MPa and 0.65 MPa for the horizontal and vertical reinforcing bars, respectively.

4 LOADING ARRANGEMENT

Figure 3 depicts the testing layout whereby all wall specimens were placed, one at a time, within a steel reaction frame. The vertical load was kept almost constant at a predetermined level, whereas the horizontal displacement, applied at the top, was varied in a cyclic manner (Figure 2). When loaded, the specimen has its foundation anchored to the reaction frame. A reinforced concrete beam was built at the top fully connected to each specimen. The horizontal actuator was attached both to this concrete beam and to the reaction frame with a system of hinges in order to allow the in-plane top horizontal displacement and

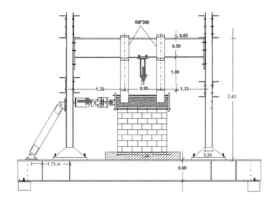

Figure 3. Masonry wall being subjected to the cyclic test.

the in-plane rotation of the specimen during loading. The out-of-plane response of the specimen was prohibited with a system of roller-sliders.

The frequency of this cyclic loading is also one of the studied parameters and was specified at the beginning of the test for each wall (see relevant description in the section named experimental program).

5 CONSTRUCTION DETAILS AND INSTRUMENTATION FEATURES

All specimens were constructed in accordance with standard practices of reinforced masonry construction regarding the laying of the bricks and mortar, the placement of the longitudinal (vertical) and bed-joint (horizontal) reinforcement and the pouring of the grout in the vertical cells. Figure 4 depicts the partial reinforcement of a typical wall-specimen. For all wall specimens with a height 1330 mm the vertical reinforcement was anchored directly to the R/C foundation block, which was already cast before building the masonry.

For walls specimens with a height 2700 mm (Wall-27N and Wall-34N) the vertical reinforcement had a lapse length of 800 mm with 10 mm diameter

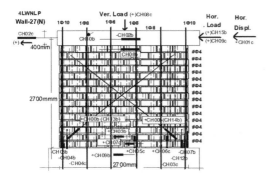

Figure 4. Typical reinforcing details of specimen Wall-27N and instrumentation scheme.

dowels protruding from the R/C foundation block. The cells hosting the vertical reinforcement were the only ones grouted being previously properly cleaned. In all cases the horizontal reinforcement was embedded within the mortar of the bed-joints. For most specimens reinforcing rods of 4.0 mm diameter were used as shown in Figure 4 for Wall-27N. Alternatively, the maximum diameter of the transverse reinforcement was either 4.9 mm or 5.5 mm in some cases. The variation of the diameter of the horizontal reinforcing bars together with their placement either in every bed joint (Figure 4) or every other bed joint resulted in the variation of the transverse reinforcing ratio in the range from 0.05%, to 0.150%. In all cases of wall specimens with a nominal thickness of 154 mm this transverse reinforcement was formed of two parallel reinforcing rods lying within the mortar at a distance of approximately 20 mm from either façade of the wall specimen.

6 INSTRUMENTATION FEATURES

During the loading sequence a number of sensors were employed together with a data acquisition system in order to record continuously the level of the applied horizontal and vertical forces as well as the horizontal, vertical and diagonal displacements at various points of the loaded specimens. Figure 4 depicts the used instrumentation scheme for specimen Wall-27N. During the loading sequence the level of the applied horizontal and vertical forces was monitored in real-time in-order to check at all-times the progress of the loading sequence and prohibit any possibility of accidental over-loading. Moreover, during all the stages of the test the development of damage was recorded at both sides of each wall specimen. Figure 6 showing the damage of Wall-27N represents the state of this specimen at the end of the loading sequence. The main types of damage that were recorded are the following: a) Breaking of a brick unit and disintegration of its outer face. b) Breaking and disintegration of the whole brick unit. c) Compression failure of the brick unit and grout fill at the toe of the wall. d) Rupture or buckling of the longitudinal reinforcements. The measured response during the cyclic tests of the partially grouted reinforced walls has been treated in such a way as to deduce the most significant modes of deformation.

7 OBSERVED RESPONSE

Due to space limitations only the observed response of specimen Wall-27N (Figure 4) is presented here, discussing its most significant features for "wide" wall specimens with shear reinforcement ratio just below 0.1%. The presented response is typical of relatively "wide" specimens with a height over length (h/l) ratio equal to 1 (see section 2). All the results demonstrating the influence of the varied

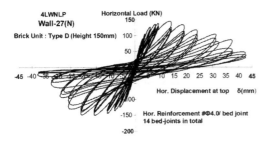

Figure 5. Overall cyclic response of specimen Wall-27N.

Figure 6. Wall-27N damage at end of all cycles.

parameters is included in the full report. Figure 5 portrays the overall measured response in terms of imposed horizontal load at the top of the specimen versus the corresponding horizontal displacement at the same location. As can be seen in this figure an initial approximately elastic behavior is evident with a stiffness value 27 kN/mm in the range up to 4 mm horizontal displacement and 110 kN horizontal load. Next, the non-linear response of the specimen becomes evident till the maximum load value equal to 148 kN is reached for an imposed horizontal displacement equal to 13 mm. From this point any further increase in the imposed horizontal displacement (up to 40 mm) is accompanied with a gradual deterioration of the horizontal load which reaches 25% of the maximum load at the end of test. Throughout all this cyclic loading sequence the response remains almost symmetric absorbing at the same time considerable plastic deformation energy, which is beneficial for the seismic resistance.

The observed damage for Wall-27N (Figure 6) follows similar trends to the damage which was observed for "wide" wall specimens with height over length ratio approximately equal to 1. This is initially of flexural type, but as the imposed level of horizontal displacement increases the shear type of damage also becomes evident. However, for Wall-27N there was no disintegration of any part of the brick masonry within the wall. Moreover, the large plastic rotation at the wall-foundation interface was accompanied with compression brick failures at the toes of the wall and with signs of overstress of the longitudinal reinforcement at these locations during the cycles after the maximum horizontal load was reached. The deterioration of the bearing capacity of Wall-27N must be attributed to the contribution of the shear response together with the disintegration of

the compressive zones at the two toes of this wall during large plastic rotations (Figure 6).

As with Wall-27N response, most of the specimens tested developed local compressive-type modes of failure. The main types of damage that were recorded are breaking of a brick unit and disintegration of its outer face or of the whole brick unit and compression failure of the brick unit and grout fill at the toe of the wall. These modes of failure influence the overall seismic response and they are critical factor of the observed performance. Therefore, an additional objective was to produce an alternative clay brick unit in order to avoid such local compressive-type modes of failure. The various steps employed to achieve this objective are presented and discussed below (section 9).

8 NUMERICAL SIMULATION

An attempt is made here to numerically simulate the observed behavior of Wall-27N presented before. This was done employing a micro-modeling technique whereby the bricks were simulated by elastic plane finite elements separately from the simulation of the mortar bed and head joints and of the longitudinal reinforcement. A brief outline of this numerical simulation is presented together with a selection of the obtained results. The mortar joints in this micro-model are represented with non-linear link elements aimed to simulate the cut-off tensile capacity normal to the mortar joint-brick as well as the non-linear compressive behavior when, as was observed (Figure 6) the axial compression attains relatively large values at the narrow compressive zone along the toes of this wall. At this stage the numerical simulation does not simulate the non-linear shear response of the mortar-joint brick interface and the horizontal reinforcement nor the dowel action of the longitudinal reinforcement. Therefore, this type of the response that developed in Wall-27N, as shown by the relevant measurements, is not numerically simulated at present. The vertical reinforcement was simulated explicitly by including in the numerical model non-linear links with properties based on the measured mechanical tensile characteristics of the actual longitudinal reinforcement as these properties were measured (see section on material characterization). These non-linear links are at the same locations where the vertical reinforcing bars were placed and grouted within Wall-27N specimen (Figures 4 and 7). Similarly, from the

Figure 7. Links placed as reinforcing bars (c).

measured mortar mechanical characteristics the properties of the non-linear links representing the mortar joints were derived. The loading conditions of the experimental sequence were simplified in the numerical simulation through a step-by-step time history algorithm. Initially, the vertical load was applied gradually followed by the gradual application of the horizontal. At this stage the cyclic nature of the horizontal loading sequence (Figure 3) was not simulated. Instead, in the present numerical simulation the horizontal load was applied in a monotonic fashion whereby this horizontal load was gradually increased from zero to a value approximating the maximum value attained during the experimental sequence. An alternative process would be to apply a gradually increasing horizontal top displacement.

Figure 8 depicts the deformation patterns as resulted from the described numerical simulation the vertical and horizontal loading. As can be seen in Figure 8 the observed during testing dominant plastic rotation at the region near the wall-foundation interface was successfully reproduced by this non-linear micro-modeling. In Figure 9 the overall response of Wall-27N specimen is depicted in terms of overturning moment at a horizontal section near the wall-foundation interface versus the horizontal displacement at the top of the specimen. The measured response is depicted as an envelope curve of the cyclic response depicted in Figure 5. As can be seen in Figure 9 the used numerical simulation predicted reasonably well the observed overall response up to the point of maximum overturning moment. From this point onwards the numerical simulation fails to reproduce the observed deterioration in the overall response. This is attributed to the fact that this

numerical simulation does not represent any deterioration arising from the disintegration of the compressive zone, which was discussed in the previous section (Figure 6). Moreover, as already underlined, the used numerical simulation does not include at this stage any non-linearities arising from the shear response of the mortar joints.

Figure 10 depicts the distribution of the normal stresses in the vertical direction (σ_{22}) for Wall-27N specimen at the loading stage where the horizontal forces attain the maximum value (together with the vertical forces described before). As can be seen, this distribution of normal stresses in the vertical direction (σ_{22}) is quite realistic showing, as expected, a narrow compressive zone at the right bottom toe of this masonry wall whereas the tensile forces at the mortar bed-joints at the left side of the wall attain very small values; instead, the reinforcing bars at this part develop, as expected, the required tensile forces, depicted in Figure 11.

9 INVESTIGATION OF THE BEHAVIOR OF THE ALTERNATIVE CLAY BRICK UNIT

The methodology used to investigate the behavior of the new clay brick unit with vertical holes to be used for reinforced masonry low-rise buildings will be briefly presented here. First of all, an extensive experi-

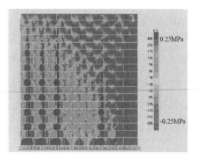

Figure 10. Normal stress distribution of masonry Wall-27N in the vertical direction.

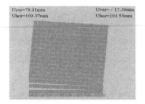

Figure 8. Wall deformation for vertical and horizontal loading Deformations measured at the two ends of top section.

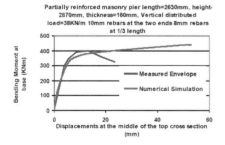

Figure 9. Observed and predicted overall response.

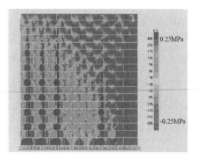

Figure 11. Distribution of axial forces of the links representing the longitudinal reinforcing bars.

mental sequence was carried out focusing on monitoring basic properties of the ceramic material. All clay material specimens are made from the same clay material and burning process that was also employed in the production of clay bricks of Figure 16. Specimens were subjected to either axial compression (Figure 12) or four-point-bending tests (Figures 14 and 15), recording the brittle nature of their behavior together with the corresponding axial compression or flexural tensile strength. A 3-D non-linear finite element simulation of these tests was formed including all the geometrical, loading and support details, in an effort to numerically simulate the observed behavior. The numerical models captured, up to a point, the measured brittle load-deformation response (Figures 13 and 15). Moreover, the observed actual damage at the end of the tests resembles the distribution of the plastic strains after the ultimate load is reached, as predicted by these numerical simulations.

Next, a number of clay bricks with vertical holes commercially produced were subjected to axial

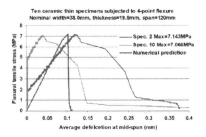

Figure 15. Stress-strain response from the measurements obtained during the four point bending tests of prismatic specimens made with ceramic material.

compression test. The geometrical details of the brick tested are shown in Figure 16. These specimens were subjected to compression according to the relevant standard EN 772-1. The specimens were subjected to two different directions of compressive load. The first is vertical to the direction of the bed joints (direction a-a) whereas the second is parallel to the bed joints (direction b-b).

A three dimensional (3-D) finite element representation of the axial compression test of the clay brick tested at the laboratory was formed. The same methodology is used, as it was used in the previously mentioned numerical simulations of the ceramic specimens. The presented results depicted in Figure 17 are plots of the distribution of plastic strains within the volume of the numerical simulation when the ultimate load is reached. Moreover, a comparison between the observed behavior and the numerical predictions are shown in Figure 17, in terms of the maximum load reached. In both cases there is a good agreement between the measured and the numerical predictions of the maximum load reached. Therefore, this method can also be used with a degree of conformity for the design of a new clay brick unit with the desired geometry and the desired compressive strength to be used to partially grouted reinforced masonry buildings.

Figure 12. Observed damage and numerically predicted patterns of the plastic strain during the axial compression test.

Figure 13. Stress-strain response from the measurements obtained during the axial compression tests of prismatic specimens made with ceramic material.

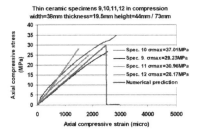

Figure 14. Observed and numerically predicted damage patterns of the four point bending test.

Figure 16. Observed and numerically predicted damage patterns for the two different compression tests.

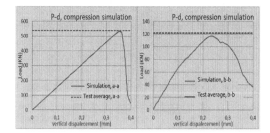

Figure 17. Comparison between experimentally measured response and numerical prediction for the two different compression tests. The dash line denotes the measured maximum.

10 CONCLUSIONS

- A number of partially grouted reinforced brick masonry wall specimens, constructed with a brick unit produced by a Greek manufacturer, were subjected to vertical forces combined with horizontal cyclic seismic-type loads applied simultaneously. The longitudinal reinforcing ratio was 0.12% whereas the horizontal reinforcing ratio was varied in the range from 0.05%, to 0.15%. For walls with a height over length ratio equal to 1 and for horizontal reinforcing ratio values larger than 0.085%, the flexural response together with the rather ductile plastic rotation response at the bottom of the wall, similar to a plastic-hinge mechanism, dominated the observed behavior at maximum horizontal load.
- This observed mainly flexural behavior was successfully reproduced by a micro-modeling numerical simulation featuring all the geometric and construction detailing together with the measured non-linear mechanical characteristics of the mortar joints and the vertical reinforcing bars of the described Wall-27N specimen.
- A critical factor of the observed performance was the behavior of the employed clay brick unit which developed local compressive-type modes of failure, which influence the overall seismic response. Therefore, an additional objective was to produce an alternative clay brick unit in order to avoid such local compressive-type modes of failure. The methodology used to investigate the behavior of the new clay brick unit, to be produced, is briefly presented here.

ACKNOWLEDGEMENTS

All materials for the construction of the specimens were provided by "KEBE S.A. (Northern Greece Ceramics)" a leading partner of this research

Co-financed by Greece and the European Union

Figure 18. The research partnership.

(Figure 18). Part of the aforementioned research has been co-funded by Greece and European Union through the Operational Program "Erevno – Dimiourgo - Kainotomo" which are gratefully acknowledged.

REFERENCES

Computers and Structures Inc. 2010. Structural and Earthquake Engineering Software.

Eurocode 6. 2005. Design of masonry structures - Part 1-1: General rules for reinforced and unreinforced masonry structures.

European Standard EN 772-1. 2011. Methods of test for masonry units - Part 1: Determination of compressive strength.

Gulkan P. et al. 1990 Seismic Testing of Single-story Masonry Houses: Parts 1 and 2. *Journal of Structural Engineering ASCE* 116(1): 235–274.

Hidalgo, P. A. et al. 1978. Cyclic Loading Tests of Masonry Single Piers. *EERC Reports* 78/27, 78/28, 79/12. University of California at Berkeley.

Manos, G.C. 1983. Shaking Table Study of Single-Story Masonry Houses. *EERC Report* 83/11, University of California, Berkeley.

Manos, G.C. 2000a. The earthquake performance of partially reinforced masonry piers subjected to in-plane cyclic loading. *Report to the Greek General Secretariat of Research and Technology* (in Greek).

Manos, G.C. et al. 2000b. The observed performance of partially reinforced masonry piers subjected to combined horizontal cyclic and compressive loads. *12th World Conference Earthquake Engineering.*

Modena, C. et al. 1996. Reinforced Masonry for Buildings in Seismic Zone. *Report of University of Padova in the framework of Brite Euram Project* No 4001.

Oan, A.F. & Shrive, N.G. 2009. Shear of concrete masonry walls. *11th Canadian Masonry Symposium, Toronto, Ontario.*

Psilla, N. 1996. Seismic Behavior of Reinforced Masonry. *Greek Conference on Reinforced Concrete Structures, Cyprus* Vol. II: 284–294 (in Greek).

Sandoval, C. et al. 2018. Experimental cyclic response assessment of partially grouted reinforced clay brick masonry walls. *Bulletin of Earthquake Engineering* 16(7): 3127–3152.

RILEMTC76-LUM: Diagonal tensile strength tests of small wall specimens. Technical 549 Report TC76-LUM, RILEM Publications SARL, Brussels, Belgium, 1991.

Tasios, T. P. 1987. The Mechanics of Masonry, Athens, (in Greek).

Tomazevic, M. et al. 1993. In-Plane Behavior of Reinforced Masonry Walls Subjected to Cyclic Lateral Loads. *Reports ZRMK/PI-92/06 and 08, Institute ZRMK, Ljubljana, Slovenia.*

Brick and Block Masonry - From Historical to Sustainable Masonry –
Kubica, Kwiecień & Bednarz (eds)
© 2020 Taylor & Francis Group, London, ISBN 978-0-367-56586-2

Shear strength prediction of partially-grouted concrete masonry walls with openings

L. Vargas, C. Sandoval, P. Ramírez & G. Araya-Letelier
Pontificia Universidad Católica de Chile, Santiago, Chile

S. Calderón
Pontificia Universidad Católica de Chile, Santiago, Chile & Politecnico di Milano, Milano, Italy

ABSTRACT: As part of an ongoing research, this paper adresses the prediction of the shear strength of partially grouted reinforced masonry (PG-RM) shear walls with openings. For this purpose, one perforated wall was built and tested under cyclic lateral loading, and its lateral resistance has been estimated through different approaches. Three existing shear equations were used to estimate the shear strength of the wall through the sum of the contributions of its piers. Additionally, the overlap of envelopes of hysteresis curves of previously tested isolated walls with similar characteristics of piers was also used. Lastly, the shear strength was predicted by means of fragility curves. Results show that there is considerable dispersion among the applied methods. Horizontal displacement field developed on the wall surface suggests the forming piers should be defined according to their boundary conditions by means of an equivalent aspect ratio.

1   INTRODUCTION

Partially grouted reinforced masonry (PG-RM) walls are used as a lateral load resisting system in many countries with moderate to high seismicity. Despite the publication of many studies on their behaviour during the last years (e.g. Ramirez et al. 2016, Calderón et al. 2017, among others), most of them have focused on solid walls. In fact, actual walls have different configurations due to the presence of openings such as doors and/or windows. These openings divide the wall into connected segments, called piers, that behave as a set and not as individual isolated walls, if they are connected by a rigid diaphragm (Siyam et al. 2016a, b), adding complexity to their seismic response (Voon 2007).

Few studies have been carried out to determine the effects of openings on the seismic behavior of PG-RM walls. Voon & Ingham (2008) tested eight PG-RM shear walls subjected to cyclic lateral loads but without axial compression load. They analyzed different sizes of openings and lengths of trimming reinforcement under the opening. Fortes et al. (2017) tested six half-scale multistory PG-RM walls with openings where two sizes of openings were analyzed separately, that is, three walls with a door opening, and the other three with a window opening, but walls with more than one opening were out of the range of that study. Koutras & Shing (2017) conducted shake table tests on a PG-RM one-story system, and evaluated the accuracy of the US design standards. The authors concluded that the strength of

a perforated wall is a function of the differences in stiffness of its piers, whereby, it should not be calculated as simply the sum of maximum shear capacities of the piers. Calderón et al. (2017, 2019) tested three full-scale walls to study the shear response of PG-RM walls with a window-type central opening, and evaluated their resistance by means of different methods, concluding that damage also extents to the bottom portions of piers and openings.

Regarding the shear strength expressions for PG-RM walls, the majority of the available design expressions has been calibrated with experimental results of solid walls (e.g. Shing et al. 1990, TMS 402/602 2016, CSA S304 2014). This implies that none of the existing expressions explicitly quantifies the effect of openings on the shear response of PG-RM walls. For this reason, an acceptable approach to estimate the shear strength of PG-RM walls with openings is to sum the maximum shear capacity of each one of the piers formed by these openings. Such an approach implicitly assumes that all piers reach their maximum shear capacity at the same time, but this is not accurate for piers with different aspect ratios. An alternative method is to evaluate the behavior from an statistical point of view, employing a specific engineering damage parameter. Recently, Araya-Letelier et al. (2019) developed fragility functions for PG-RM walls with bed joint reinforcement derived from a large experimental database.

Given this background, in this paper the force-displacement response of one PG-RM wall with two openings tested under cyclic lateral loading is presented. In addition, the experimental shear capacity of such wall is compared with the predictions of different methods, including three existing shear equations, the overlap of envelopes for previously tested solid walls, and experimental fragility curves. The tested wall was monitored with a 2D digital image correlation (DIC) measurement system. This technique allowed to observe the displacement field experienced by each pier during the test. This information could be useful in order to determine the real boundary conditions of the piers in any perforated PG-RM wall.

2 EXPERIMENTAL PROGRAM

2.1 Specimen description

One full-scale PG-RM shear wall with two openings was built and tested under cyclic lateral loading. This wall is representative of an actual facade of single-story houses. Figure 1 shows the characteristics of the wall including the piers formed at the adjacent sides of the openings, which are identified

as shaded areas in Figure 1c. The characteristics of the piers are summarized in Table 1.

The wall (named wall HBCO) was built with hollow concrete blocks, with nominal dimensions of 390 mm (length), 190 mm (height), and 140 mm (thickness). As shown in Figure 1d, only cells containing vertical steel bars were grouted. The wall was built with an average thickness of 20 mm mortar joint. Horizontal reinforcement, which consisted of a pre-fabricated ladder-type reinforcement with two 4.2 mm diameter rod and welded cross bar every 300 mm, was embedded in all bed joints. A total of six vertical steel bars of 22 mm diameter were placed as depicted in Figure 1a, providing a vertical reinforcement ratio of 0.53%, as an attempt to generate a shear predominant failure mode during the test.

2.2 Material properties

Three ungrouted masonry prisms were tested as stated in the Chilean reinforced masonry design code NCh1928 (2009), to determine the compressive strength of the masonry as a composite material (f'_m).

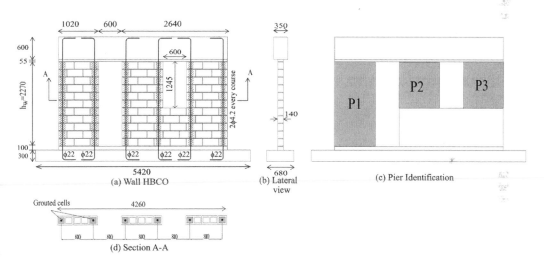

Figure 1. Specimen characteristics (dimensions in mm).

Table 1. Characteristics of piers.

Pier	Dimensions			Aspect ratio	Vertical reinforcement			Horizontal reinforcement				Axial compression stress
	Length 'd' mm	Height 'h_w' mm	Thick 't' mm	h_w/d	Description	A_v mm^2	ρ_v %	Description	s_h mm	A_h mm^2	ρ_h %	σ_n* MPa
P1	1020	2270	140	2.23	2ϕ22	760	0.53	10-2ϕ4.2mm	205	277	0.09	0.48
P2	1020	1245	140	1.22	2ϕ22	760	0.53	6-2ϕ4.2mm	205	166	0.10	0.48
P3	1020	1245	140	1.22	2ϕ22	760	0.53	6-2ϕ4.2mm	205	166	0.10	0.48

* σ_n was calculated considering top RC beam self-weight.

Table 2. Mechanical properties tested for unreinforced and ungrouted masonry.

Property	Average MPa	CV* %
f'_m **	8.1	6.1
f_{yv}	521.3	0.7
f_{yh}	644.9	5.0

* CV = coefficient of variation
** f'_m was calculated based on gross area

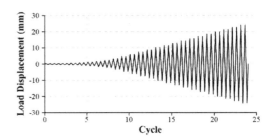

Figure 3. Loading protocol.

The nominal dimensions of the prisms were 390 mm (length), 610 mm (heigth), and 140 mm (thichness). The yield strength of the horizontal (f_{yh}) and vertical reinforcement (f_{yv}) as well as f'_m are presented in Table 2.

2.3 Test setup

The test setup is shown in Figure 2, which consisted of a steel reaction frame with a horizontal actuator that induced cyclic lateral displacements, while four uniformly distributed hydraulic jacks placed on the top of a reinforced concrete (RC) beam applied constant axial compression load. Two steel pipes were connected from the RC beam in the top of the wall to a strong RC wall to avoid out-of-plane displacement. Additionally, the RC beam at the bottom of the wall was braced with two stiff built-up steel angles to prevent any sliding of the base. The specimen was anchored to the laboratory strong floor and free to move at the top, simulating a cantilever system. The response of the wall was monitored by a 2D digital image correlation (DIC) measurement system in order to visualize the displacement field developed on the surface of the wall during the test.

2.4 Instrumentation and loading protocol

The instrumentation used during the test was composed of two load cells, to measure lateral and vertical load, and seven linear variable differential transformers (LVDTs) to capture horizontal displacement at different locations. LVDT number 1 measured the overall lateral displacement at the top of the wall, and LVDT number 2 to 7 captured horizontal displacement at top and bottom of each pier. Figure 2 shows the location of the instrumentation on the wall.

The specimen was subjected to lateral cycling loading under displacement control. The loading displacement history is shown in Figure 3 and was defined according to FEMA 461 guidelines (FEMA 2007). Two complete cycles were performed for each target displacement level, incrementing the displacement until observing a 20% degradation of the maximum recorded resistance, in both loading directions.

3 WALL BEHAVIOR

The hysteresis curve and force-displacement envelope of the wall are shown in Figure 4, depicting the lateral displacement at the top of the wall as a function of the total applied lateral force. The hysteretic behavior is fairly symmetric with a slightly pinched shape. The pinching behavior can be attribute to significant shear deformation and opening and closing of the cracks.

Regarding the damage progression, pier P3 was the first one to crack, and then pier P2. Both of them experienced a diagonal crack pattern as shown in Figure 5, characteristic of a shear failure mode. These cracks first appeared as stair-stepped cracks, running

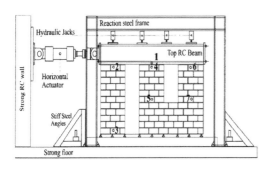

Figure 2. Test setup and instrumentation.

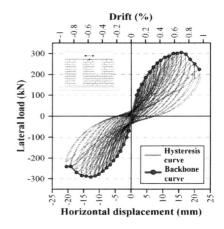

Figure 4. Hysteresis response of HBCO wall.

1010

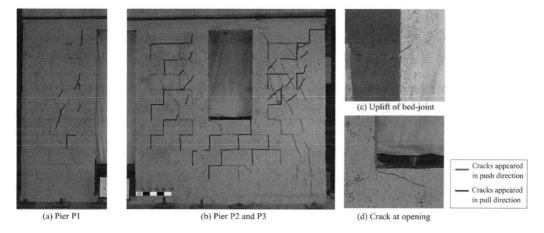

(a) Pier P1   (b) Pier P2 and P3   (c) Uplift of bed-joint

Cracks appeared in push direction
Cracks appeared in pull direction

(d) Crack at opening

Figure 5. A) Flexural and shear cracking in pier P1, b) diagonal cracking pattern of piers P2 and P3, c) Uplift of bed joint at the bottom of pier P1, d) Damage at the lower corner of the opening.

along the mortar bed and head joints, afterward, they grew and went through the units and mortar joints. Sliding of the course located at the bottom of the pier P2 was observed, as a consequence of a crack that appeared in the lower right corner of the opening as shown in Figure 5d. Pier P1 presented flexural cracks, identified by the uplift of bed joints located at the bottom of the pier as shown in Figure 5c, and diagonal cracks, as shown in Figure 5a, which were narrower than those observed in the other two piers.

With respect to shear capacity, the wall had a peak resistance of 305.53 kN and 291.35 kN in the push and pull directions, respectively. Once the wall reached the maximum load, it presented a relatively rapid degradation of the lateral capacity due to the acumulated damage, as shown in Figure 4.

The horizontal displacement field developed on the wall surface was obtained by the DIC method. Figure 6a shows the displacement field in the push direction when the wall reached its maximum resistance, and Figure 6b the displacement field in the pull direction also when the maximum resistance was reached. In the push direction, the wall segment at the right side of the window opening presented displacements along almost the entire height of the wall, in contrast to pier P2, which experienced displacements below the bottom of the opening between only 0 to 2 mm. Similarly, in the pull direction, the wall segment located at the left side of this opening presented considerable displacements nearly along the entire height. Pier P3, however, experienced minimum displacements at the bottom of the opening. Therefore, the boundary conditions, which are mainly a function of the height of the openings, determine the displacement fields that occur in each segment. Considering the aspect ratios of piers 2 and 3, as shown in Figure 1c, the displacements (and consequently the rotations) that occurred at the bottom of the piers are ignored and, therefore, the rigidity of the piers is overestimated.

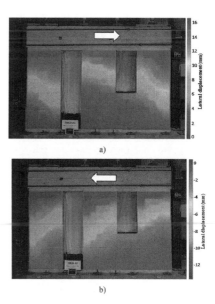

a)

b)

Figure 6. Horizontal displacement field developed in a) push direction and b) pull direction (both at the moment of maximum lateral load).

4 SELECTED METHODS TO ESTIMATE THE SHEAR CAPACITY

The conventional approach to predict the shear strength of a perforated wall considers the contribution of each pier as its maximum shear capacity. Following this method, shear expressions proposed by Shing et al. (1990), CSA-S304 (2014), and TMS 402/602 (2016) were used in this study. In addition, the overlap of envelopes of force-displacement curves of walls with similar characteristics to the piers under study but individually tested have been used (Ramirez et al. 2016). Shear strength was also estimated using experimental fragility curves proposed by Araya-Letelier et al. (2019).

4.1 CSA-S304 (2014)

The Canadian masonry design standard states that the shear strength can be estimated by quantifying the contribution of masonry shear strength, axial compression load, and horizontal reinforcement. The expression is given by the Equation 1.

$$V_{n1} = \left(\left[0.16 \left(1 - \frac{M}{Vd_v} \right) \sqrt{f'_m} \right] td_v + 0.25P \right) \frac{A_n}{A_T}$$
$$+ \left(0.6 A_{sh} f_{yh} \frac{d_v}{S_h} \right)$$

(1)

where t and d_v are the thickness and effective length of the wall or segment of wall, respectively. P represents the axial compression load; γ_g corresponds to a factor equal to A_n/A_T, but not greater than 0.5 for partially grouted walls. A_n represents the effective area of the cross section and A_T is the gross area of cross section. A_{sh}, f_{yh} and s_h are the area, yield strength and vertical spacing of horizontal reinforcement, respectively. f'_m is the compressive strength of masonry. M and V are the maximum moment and shear at the section under analysis; the relationship $M/(Vd_v)$ should not be taken less than 0.25 or more than 1. Lastly, V_n is limited to $0.40\sqrt{f'_m} t_w d_v \gamma_g$ if $h_w/L_w > 1$ and $0.40\sqrt{f'_m} t_w d_v \gamma_g (2-h_w/L_w)$ if $0.5 \leq h_w/L_w \leq 1$.

4.2 TMS 402/602 (2016)

According to the US masonry code, the shear strength can also be calculated by adding the contribution of three components: masonry, axial compression load, and horizontal reinforcement, as follows:

$$V_{n2} = \gamma_g \left(\left[4.0 - 1.75 \left(\frac{M}{Vd_v} \right) \right] A_{nv} \sqrt{f'_m} + 0.25P + 0.5 \left(\frac{A_{sh}}{S} \right) f_{yh} d_v \right)$$

(2)

where f'_m represents the compressive strength of masonry. P is the axial compression load. A_{sh}, f_{yh} and s are the area, the yield strength, and the vertical spacing of horizontal reinforcement, respectively. M and V are the maximum moment and shear at the section considered and d_v is the effective length of the wall or segment of wall. A_{nv} corresponds to the net cross sectional area and γ_g is the reduction factor for PG-RM walls, which is equal to 0.75. Expression for $V_{n,max}$ is a function of $M/(Vd_v)$; if $M/(Vd_v) \leq 0.25$, $V_{n,max}$ is $0.50A_n\sqrt{f'_m}$, if $M/(Vd_v) \geq 1$, $V_{n,max}$ is

$0.33A_n\sqrt{f'_m}$, or else $V_{n,max}$ corresponds to $[0.56-0.23 M/(Vd_v)]A_n\sqrt{f'_m}$.

4.3 Shing Et Al. (1990)

The shear expression proposed by Shing et al. (1990) also defines the shear strength as the sum of independent strength provided by masonry, axial compression load, and horizontal reinforcement. In contrast to the CSA-S304 (2014) and TMS 402/602 (2016) expressions, Shing et al. (1990) expression accounts for the contribution of vertical reinforcement.

$$V_{n3} = (0.166 + 0.0217\rho_v f_{yv})A_n \sqrt{f'_m} + 0.0217\sigma_n A_n$$
$$\sqrt{f'_m} + \left(\frac{L_w - 2d'}{S_h} - 1 \right) A_{sh} f_{yh}$$

(3)

where ρ_v and f_{yv} are the vertical reinforcement ratio and yield strength of vertical reinforcement, respectively. L_w is the wall length, and d' is the cover of reinforcement with respect to the vertical edge.

4.4 Envelopes of force-displacement curves

Taking into account that the piers in the wall under study might not develop their maximum resistance at the same level of lateral deformation, experimental results of two previously tested solid walls were used to predict the maximum shear capacity of the HBCO wall. The isolated specimens were tested in two different experimental programs, one as part of this research, and the other one by Ramirez et al. (2016). Both walls were constructed with hollow concrete blocks, similar to the ones used in the HBCO wall, horizontally reinforced in every mortar bed joint with ladder-type steel elements, and tested under similar conditions.

The first wall, named HBC-1, had the same length and height dimensions of pier P1, with an aspect ratio (h_w/d) of 2.23. The second wall, identified as M3, had an aspect ratio of 0.97 and was used to estimate the shear strength of piers P2 and P3. Their characteristics are summarized in Table 3. In this approach, the lateral capacity of the entire wall HBCO will be controlled by wall M3 (i.e., piers P2 and P3) since it has the lowest aspect ratio and consequently reaches its maximum capacity at a lower lateral displacement.

Figure 7 shows the force-displacement envelopes of walls HBC-1 and M3. Wall M3 reached its capacity at a lateral deformation of 10.3 mm (Δ_m). It was assumed that wall M3 and piers P2 and P3 were subjected to the same shear stress ($\tau_{\Delta m} = 1.02$ MPa). The shear stress was multiplied by the piers' gross area to obtain the lateral capacity of piers P2 and P3.

Table 3. Characteristics of isolated walls.

ID	h_w/d	ρ_h %	σ_n MPa	$f'm$ MPa	$V_{\Delta m}$ MPa	$\tau_{\Delta m}^{*}$ MPa
HBC-1	2.23	0.09	0.48	8.10	68.3	0.48
M3	0.97	0.09	0.56	8.65	283.6	1.02

* $\tau_{\Delta m}$ was calculated based on gross area.

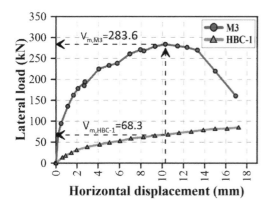

Figure 7. Envelopes of isolated walls.

The lateral capacity of the wall HBCO would be controlled by the lateral deformation of piers P2 and P3 (due to its lowest aspect ratio in relationship to pier P1). Therefore, the deformation Δm = 10.30 mm was used to determine the contribution of pier P1 to the overall capacity of the wall HBCO. Finally, the lateral capacity of wall HBCO was calculated as the sum of the individual resistance of piers P1, P2, and P3. The results obtained are presented in Table 4.

Table 4. Shear strength of piers and entire wall.

	P1 kN	P2 kN	P3 kN	Total shear strength kN	V_n/V_{exp}	Error %
CSA-S304 (2014)	79.8	79.8	79.8	239.5	0.82	-17.8
TMS 402/ 602(2016)	44.8	79.3	79.3	203.4	0.70	-30.2
Shing et al. (1990)	122.3	122.3	122.3	366.8	1.26	25.9
Envelope curves	68.3	145.3	145.3	359.0	1.23	23.2
DS5 fragility function	76.5	135.3	135.3	347.2	1.19	19.2

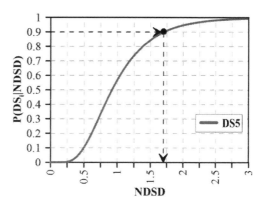

Figure 8. DS5 fragility function for hollow concrete block walls (Adapted from Araya-Letelier et al. 2019).

4.5 Estimation with fragility functions

Fragility function developed by Araya-Letelier et al. (2019) for severe shear damage state (DS5) was used to estimate the maximum shear strength as shown in Figure 8. This function is based on a normalized diagonal shear demand (NDSD), which relates the maximum shear wall resistance (V_{exp}) with the shear resistance estimated by means of TMS 402/602 (2016) (V_{exp}/V_n).

To use this method, the probability of reaching or exceeding DS5 state needs to be defined. According to FEMA P-58-6 (FEMA 2018), there are three confidence levels that can be chosen: the median, the mean, and the 90th percentile value. For the purpose of this study, the 90th percentile was considered. To determine the NDSD parameter, the cumulative probability of 0.90 is intersected with the DS5 fragility curve, corresponding to a NDSD parameter equal to 1.71. This value does not depend of any other characteristic of the wall, and for this reason, the same value has been considered for the three piers. The lateral resistance of wall HBCO was then calculated as the sum of the individual resistance of piers P1, P2, and P3. The results are presented in Table 4.

5 SHEAR STRENGTH PREDICTIONS

The predicted shear strengths of wall HBCO are compared with the experimental shear capacity, i.e., the minimum shear strength between peaks at push and pull directions. Table 4 shows the calculations made for each of the piers and the total shear strength of the wall as a set (as the sum of piers P1, P2, and P3). In addition, each shear strength is normalized by the experimental lateral resistance of the wall. Hence, a V_n/V_{exp} ratio greater than 1.0 indicates the corresponding prediction method overestimates the shear strength. Conversely, a V_n/V_{exp} ratio less than 1.0 indicates the method underestimates the shear strength. The error of each prediction was also calculated as $((V_{ni}-V_{exp})/V_{exp})*100$.

The normalized results are shown in Figure 9. TMS 402/602 (2016) and CSA-S304 (2014) expressions underestimate the shear strength by approximately 18% and 30%, respectively; that is, provide conservative predictions. The Shing et al. (1990) expression, the method by overlapping envelope curves and the method based on experimental fragility curves overestimate shear strength and provide non-conservative predictions.

According to these results, the expression by the Canadian masonry code (CSA-S304 2014) was the most accurate. The method based on experimental fragility curves by Araya-Letelier et al. (2019) provided a prediction with a percentage error lower than 20%, however, it was on the unsafe side. The TMS 402/602 (2016) expression provided the least accurate prediction, although a conservative estimation. CSA-S304 (2014) and the Shing et al. (1990) expressions provided the same shear strength value for all piers and the reason is that the CSA-S304 (2014) expression limits the influence of the aspect ratio to an upper value of 1.0 while the Shing et al. (1990) expression does not depend on the aspect ratio.

The approach based on the overlap of envelope curves provided a non-conservative prediction, even though its accuracy was satisfactory (a percentage error slightly higher than 20%). This overestimation may be due to the influence of other design parameters such as the difference in compressive strength of masonry and the thickness of mortar joints of the wall tested by Ramirez et al. (2016). However, the greatest source of error may be in the definition of an equivalent aspect ratio, which takes into account the rotations observed under the line of the openings. This hypothesis needs to be corroborated with further research.

Regarding the method based on the fragility functions (Araya-Letelier et al. 2019), the over-prediction of the shear strength, and hence the unsafe estimation, could be because the functions do not distinguish between design parameters such as aspect ratio and amount of horizontal reinforcement. Besides, they are also influenced by the error embedded into the TMS 402/602 (2016) expression.

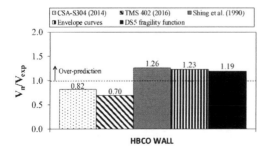

Figure 9. Predicted/experimental shear strength.

6 CONCLUSIONS

One full-scale hollow concrete PG-RM wall with two openings was tested under lateral cyclic loads. The objective was to study its shear capacity and damage progression as well as to assess the accuracy of five different approaches to estimate its capacity. Cracks located in the piers at either side of the openings were observed. However, diagonal cracks were also observed in the zone below the opening.

Regarding the accuracy of the shear expressions, the CSA-S304 (2014) formula provided the most accurate result. The method based on the fragility functions proposed by Araya-Letelier et al. (2019) also provided appropriated results, but non-conservative. The Shing et al. (1990) expression gave the most unsafe estimation, and contrarily, TMS 402/602 (2016) expression provided the most conservative prediction. In general terms, considerable differences were obtained among the predictions with different methods, those differences were influenced by the accuracy of each method and the definition of the piers that contributed to the shear strength of the wall with openings.

The total strength of the wall HBCO was calculated by the sum of the shear strength of the piers, and the height of each pear was determined by the height of the openings. However, the horizontal displacement field obtained by the DIC method suggests that it is necessary to consider different heights for piers P2 and P3, based on their boundary conditions. Due to the results obtained in this research, it is clear that there is a need for further investigation on the study of PG-RM walls with openings, in order to improve the understanding of their seismic behavior, and to develop more accurate methodologies to predict their shear strength.

ACKNOWLEDGMENTS

The authors wants to thank the funding provided by the Fondo Nacional de Desarollo Científico y Tecnológico (FONDECYT-Chile) under *Grant N° 1181598*. Laura Vargas thanks the support provided by CONICYT by its program *CONICYT-PCHA/Doctorado Nacional/2019-21191181*. Sebastián Calderón also wants to thank the support provided by the scholarship *CONICYT-PFCHA /Doctorado Nacional 2017-21170992*.

REFERENCES

Araya-Letelier, G. et al. 2019. Fragility functions for partially-grouted masonry shear walls with bed-joint reinforcement. *Engineering Structures* 191: 206–218.

Calderón, S. et al. 2017. Shear response of partially-grouted reinforced masonry walls with a central opening: Testing and detailed micro-modelling. *Materials and Design 118*: 122–137.

Calderón, S. et al. 2019. Influence of a window-type opening on the shear response of partially-grouted masonry shear walls. *Engineering Structures* 2019: 109783.

CSA Standard S304-14. 2014. Design of masonry structures. Rexdale, ON: Canadian Standards Association.

Federal Emergency Management Agency FEMA. 2007. *FEMA 461: Interim Testing Protocols for Determining the Seismic Performance Characteristics of Structural and Nonstructural components*. Washington: FEMA.

Federal Emergency Management Agency FEMA. Seismic performance assessment of building. FEMA P-58. 2012. Volume 1 – Methodology. Washington D.C., U.S.

Fortes, E. et al. 2017. Influence of openings on quasi-static cyclic behaviour of partially grouted masonry walls. In *13th Canadian Masonry Symposium*.

Koutras, A. & Shing, A.E. 2017. Shake-table testing and performance assessment of a partially grouted reinforced masonry building. *Experimental vibration analysis for civil structures*: 480–493.

INN. Instituto Nacional de Normalización. Norma Chilena Oficial NCh1928.Of93. 2009. Albañilería Armada, disposiciones para el diseño y cálculo. Santiago, Chile.

Ramírez, P. et al. 2016. Experimental study on in-plane cyclic response of partially grouted reinforced concrete masonry shear walls. *Engineering Structures* 126: 598–617.

Shing, P.B. et al. 1990. In-plane resistance of reinforced masonry shear walls. *Journal of Structural Engineering* 116(3): 619–640.

Siyam, M. A. et al. 2016a. Seismic Response Evaluation of Ductile Reinforced Concrete Block Structural Walls. II: Displacement and Performance-Based Design Parameters. *Journal of Performance of Constructed Facilities* 30(4): 1–15.

Siyam, M. A. et al. 2016b. Seismic Response Evaluation of Ductile Reinforced Concrete Block Structural Walls. I: Experimental Results and Force-Based Design Parameters. *Journal of Performance of Constructed Facilities* 30(4): 1–14.

TMS 402/602. 2016. *Building Code Requirements and Specification for Masonry Structures*. The Masonry Society (TMS), Longmont, CO.

Voon, K. C. 2007. In-plane Seismic Design of Concrete Masonry Structures. *PhD Thesis*.

Voon, K., & Ingham, J. 2008. Experimental In-Plane Strength Investigation of Reinforced Concrete Masonry Walls with Openings. *Journal of Structural Engineering* 134(5): 758–768.

Seismic & limit design

Brick and Block Masonry - From Historical to Sustainable Masonry –
Kubica, Kwiecień & Bednarz (eds)
© 2020 Taylor & Francis Group, London, ISBN 978-0-367-56586-2

Local mechanism analysis in unreinforced masonry buildings according to a new procedure based on floor spectra evaluation

L. Sbrogiò, M. Salvalaggio & M.R. Valluzzi

Department of Cultural Heritage (DBC), Padova, Italy

ABSTRACT: Local rather than global seismic behaviour is a well-known feature in both masonry aggregates and monumental buildings with large internal spans, such as churches or palaces. In such cases safety assessment through local mechanisms kinematic analysis is generally considered a viable solution. However, the definition of the seismic forces acting on secondary architectural elements (pinnacles, upper portions of facades, turrets, etc.) or masonry macroblocks, which may interact dynamically with the main structure, is not banal.

Therefore, the recent update of Italian seismic code stresses the role of a building's global dynamic response and floors' stiffness in the evaluation of seismic actions on the macroblocks in which it can be subdivided. This result in a complete new definition of floor spectra which are strongly dependent on the dynamic parameters (damping, frequencies) of both, the building and the local mechanism, which also change at the different limit states.

The paper aims at the implementation of the new procedures in an existing unreinforced masonry building (Palazzo Carraro in Noale – Venice) with flexible horizontal diaphragms. Modal analysis is used to detect the possible local mechanism and its results are compared to the evidence of the visual inspection of vulnerability factors, showing some correspondence. Safety evaluations, in linear and non-linear field, according to the previous and the current Italian seismic codes are carried out and compared. For the case study, the new procedure is much more pejorative, since acceleration and displacement demands are more than twice the ones obtained in the old one.

1 INTRODUCTION

In unreinforced masonry (URM) buildings, safety assessment through the means of the limit analysis (Heyman 1966, Giuffrè 1993) is often a viable option, in both terms of time expense and complexity of calculations, especially in professional practice. This schematization is considered conservative and especially valid when a building's overall behaviour is precluded either by a complex geometrical configuration (aggregated URM buildings), or the lack of rigid diaphragms at floor and roof levels (palaces, churches), or a combination of the two factors.

The method, codified by the *Italian Technical Standards* (MIT 2008) and the attached *Instructions for application* (MIT 2009) as early as 2008-09, has recently (MIT 2018, 2019) underwent quite a radical revision.

In particular, changes appear:

1. in the evaluation of the collapse load of the local mechanism, since:
   a. frictional forces are now explicitly mentioned by the code, which also provides a formulation for their evaluation in case of regular masonry;
   b. the floor loads on an overturning masonry macroblock vary according to the floor's

stiffness: with flexible floors the whole area instead of the regular half of it (provided that floor joists are normal to the wall) may be considered;

2. in the verification of the local mechanism, since:
   a. dynamic parameters of both the main structure and the local mechanism at the limit states of interest are considered;
   b. a general review of the procedure, with important changes in many of the key parameters used in the previous version.

These novelties are mainly attributable to the results obtained by (group 1 of changes) Casapulla & Maione (2011), Casapulla & Argiento (2016), Argiento et al. (2017) and (group 2) Lagomarsino (2015), Degli Abbati et al. (2015a, b, 2017, 2018).

However, apart the general theoretical framework, the practical application of the update requires the definition of:

- The main steps of a correct procedure (modal analysis, definition of floor spectra etc.; see § 2);
- Some details of the procedure, especially in the the seismic demand parameters (e.g. how to decide whether a modal shape is relevant or not; the use of the equivalent damping at different limit states; see § 2.2);

– Some parameters mentioned in the law text do not find a corresponding formulation (e.g. the increased period of the first mode in a building's non-linear field; see § 2.2);

This paper presents the methodology and the results of the safety assessment of a simple out-of-plane overturning local mechanism according to the updated code for an ordinary URM building, palazzo Carraro in Noale (Venice, IT). The palace's global behaviour has been previously analysed by the authors (Salvalaggio et al. 2019, Pavanetto et al. 2019).

The case study is purposely chosen to avoid the usual simplification to the first mode in calculating the seismic demand. A simplification which should be possible only with an important modal participant mass to the first mode, which is not the case (see § 3.2).

2 SAFETY ASSESSMENT PROCEDURE OF A LOCAL MECHANISM ACCORDING TO THE ITALIAN SEISMIC CODES

The following result from a merge between the practical indications given by Degli Abbati et al. (2017, 2018) and the theoretical framework of MIT (2019).

2.1 Calculation of the seismic coefficient

In MIT (2009) the calculation of the static multiplier (seismic coefficient or α) was made according to the hypotheses (Heyman 1966, Giuffrè 1993): a) rigid blocks till their collapse; b) no tensile strength; c) infinite compressive strength; additional hypotheses are d) no frictional interaction among the blocks and among the blocks and other structures (floor, roof); e) masonry blocks and other structural elements simply resting one on top of the other; f) no connections between the blocks. A more 'realistic' value of α may be obtained considering the frictional forces and therefore removing assumptions d) and f), at the cost of more articulated calculations. This possibility is now contemplated by MIT (2019).

2.2 Definition of floor spectra

MIT (2019) implements a formulation of floor response spectra which provides a stricter modelling of the seismic action on a building's parts when they are placed in the upper stories than MIT (2009). This stems from the interaction between the actual ground motion and the dynamic properties of the structure (Degli Abbati et al. 2017, 2018).

Said constructive element may be a secondary architectural element (e.g., parapet, balcony, pinnacle) or even a macroscopic portion of the structure such as masonry macroblocks, as the ones used in kinematic analysis.

It is a common assumption in the verification of a local mechanism that a building's first mode is the most relevant one. On this basis, the formulations used by MIT (2009) defined the floor spectrum attributing it entirely to the first mode (see § 2.3).

However, when floors are flexible or geometry is complex, a free vibration analysis results typically in a participant mass which is dispersed among many modes, each of which excite a specific part of a building. These modes singularly or collectively considered may be 'relevant' for that precise part when they excite it in the out-of-plane direction. Apart from these 'visual' rules, neither Degli Abbati et al. (2017, 2018) nor MIT (2019) provide an objective criterion to choose between relevant or irrelevant modes. However, an eigenvalue analysis on an overall model becomes mandatory before any local analyses take place.

The definition of these more precise floor spectra is much more complex than before since their formulations depend on (Degli Abbati et al. 2018):

1. Overall dynamic features of the main structure (fundamental period, equivalent damping, modal frequencies);
2. Dynamic features of the secondary element or local mechanism (period, equivalent damping);
3. Contribution of superior vibration modes of the main structure (participation factor and modal shape);
4. Amplification factor (constant multiplying factor or a function of the first mode);
5. Nonlinearities of the main structure (equivalent damping, ductility, periods in non-linear field);
6. Nonlinearities of the secondary element/local mechanism (equivalent damping);

Since the main structure filters, amplifying it, the seismic input, for each *k-th* period of the main structure, a peak in the floor spectra at a certain level may be observed. The peak is expressed as:

$$a_{zk} = AMP_k PFA_{zk} \qquad (1)$$

where PFA_{zk} is the Peak Floor Acceleration, obtained as:

$$PFA_{zk} = S_a(T_k, \xi_k) |\gamma_k \psi_k| \left(1 + 0.004\xi_k^2\right)^{1/2} \quad (2)$$

and AMP_k the amplification factor defined as:

$$AMPk = 1.1\, \eta_s / \left(\xi_k\right)^{1/2} \qquad (3)$$

where:

– $S_a(T_k)$ is the acceleration response spectrum at ground level at the period T_k of the *k-th* mode;
– $\eta_s = [10\,(5+\xi)]^{1/2}$ is a factor (MIT 2018) linked to the damping of the main structure;
– $|\gamma_k \psi_k|$ are the dynamic parameters of the *k-th* mode at the secondary element's height resulting from the eigenvalue analysis;

– ξ_k the equivalent damping of the secondary element (i.e. the masonry macroblock).

The complete formulation of the floor spectrum is reported by MIT 2019 (§ C7.2.3) while a deeper discussion of the terms is offered by Degli Abbati et al. (2017, 2018). All the floor spectra $S_{zk}(T_k)$ corresponding to a building's modes considered relevant for the secondary element are then to be combined with a SRSS combination to obtain the design one:

$$Sa(T) = \Sigma_k^{1/2}[S_{zk}(T_k)]^2 \qquad (4)$$

According to the new procedure, the equivalent damping ξ, for both the overall structure (subscript 's') and the secondary element (subscript 'k') (Table 1) and the fundamental period (5) change along with the limit state considered.

To evaluate the increased fundamental period, MIT (2019) itself does not provide any formulation, which may be otherwise found in Degli Abbati et al. (2018):

$$T_{kNL} = T_{ke}\left[1 + (\mu)^{0.5}\right]/2 \qquad (5)$$

Assuming the ductility $\mu=2$ the resulting 'non-linear' period is increased by a 20% than the elastic one. Degli Abbati et al. 2018 maintain that the increase of both the equivalent damping and the period, in order to consider the nonlinearities due to seismic damage at the ultimate limit states (ULS), ought to be applied only to the fundamental mode, while higher ones may be still considered elastic. The law however is not clear if Equation 2, S_a should be reduced once more at ULSs according to a convenient η factor (Table 1).

2.3 Safety assessment

The logical framework of the verification procedure between the old (MIT 2009) and the updated version (MIT 2019) have not changed. If the 'secondary element' recalled by the code is a masonry macroblock, one must calculate the seismic coefficient,

Table 1. Equivalent damping factor at different limit states according to MIT (2019).

	SLD	SLV	SLC
ξ_k	5%	8%	10%
ξ_s	5%	10%	20%

Table 2. DLS acceleration demand.

	a_z	
	MIT 2009	MIT 2019
Expression	$Se(T_1)\,\psi(z)\gamma_1$	$\Sigma_k S_{ek}(T=0)$

corresponding to the activation of the mechanism, and the corresponding capacity curve. Once these values have been converted into spectral terms, the safety assessment passes through the evaluation of:

1. Activation of the local mechanism ('linear', Damage Limitation State or DLS)
2. Collapse of the local mechanism ('non-linear' Ultimate Limit States: no-collapse state or NCSL and collapse state or CLS)

2.3.1 Linear verification (mechanism's activation)
The linear verification refers to the activation of the local mechanism. As a general rule, the mechanism does not activate if $a_0^* \geq a_z$ where a_0^* is the mechanism's spectral acceleration and a_z the expected acceleration (demand) at the height z, where the hinge is situated.

In MIT (2009) a_z (for $z>0$) was obtained from the SLD spectrum according to the following expression:

$$a_0^* \geq \cdot Se(T_1) \cdot \psi(z)\gamma_1 \qquad (6)$$

where $Se(T_1)$ is the spectral ordinate (in acceleration) at the first period, $\psi(z)=z/H$ is the modal shape, $\gamma_1=3N/(2N+1)$ the participation factor and T_1 is calculated by the simplified expression $0,05H^{3/4}$ (MIT 2009, 2019). N is the number of out-of-ground floor and H the total height.

Conversely, in the new procedure, the mechanism's activation must be compared simply to the PFA at the DLS, independently from the number of modes considered.

It is worth noting that the new PFA may be higher than the plateau of the ground spectrum at DLS.

2.3.2 Non-linear verification (kinematic analysis)
As a general rule, the kinematic analysis is satisfied if $d_u^*>d_d$, where d_d is the spectral demand. In both the old and the new procedure (MIT 2009, 2019), d_u^* is a fraction of the ultimate (spectral) displacement d_0^* of the capacity curve of the involved macroblocks, conventionally assumed as $d_u^*= 0,4d_0$ for the NCLS. Being the capacity value untouched, the demand has quite changed. The well-known relationship between period, displacement and acceleration is expressed by:

$$T = 2\pi(d/a)^{0,5} \qquad (7)$$

In MIT (2009) a conventionally reduced period T_s, corresponding to an additional reduction of the displacement to $d_s^*=0,4d_0^*$, is used to calculate the amplification factor of the spectral displacement at ground level at the first period from which d_d is obtained from (Table 3).

Conversely, according to MIT (2019) the period T_u at the ultimate limit state (NCLS in this case) is defined similarly to Equation 7 but where the factor

Table 3. NCLS displacement demands. $S_{de}(T_1)$ is the spectral displacement at the first period, $A=T_s/T_1$, while the other parameters have already been defined in the text.

	d_d	
	MIT 2009	MIT 2019
Expression	$S_{de}(T_1)\gamma_I\psi_I A^2$ $((1-A)^2+0.02A)^{1/2}$	$\Sigma_{kSdek}(T=T_{ULS})$

2π is changed into $1,68\pi$. The obtained period must be directly inserted in the floor displacement spectrum at NCLS to obtain d_d. Assuming a linear capacity curve in the space displacement-acceleration it is possible to express the latter as $a=a_0(1-d/d_0)$. With the limit displacements previously expressed and using expression (7) one can easily obtain that at NCLS $T_u=1,56T_s$, that is almost a 60% increase.

3 CASE STUDY: PALAZZO CARRARO AS AN EXAMPLE OF TRADITIONAL PALACE IN VENETO

3.1 General description and identification of the main vulnerabilities

Palazzo Carraro in Noale (Venice) stands for the traditional palace in the Veneto area, in North-East Italy. It follows the traditional so called 'late Gothic' scheme (Maretto 1960), used from the XV to the XIX cent., which is obtained from the repetition of a modular masonry unit 4.5-5.5 m wide and 10-12 m long, typically with the shorter side on the nearby road. This depth is the maximum for the light coming from the windows on the short side to reach the centre. Generally, one of the cells serves a main entrance hall (Figure 1a, Figure 2a) at the ground floor and as a parlour at the first. The hall also provides access to the surrounding rooms and it can be recognized from the outside from the group of two or three close windows forming the *loza* (loggia, Figure 1b).

The vertical bearing structure consists of the peripheral walls and of the transversal septa both built with common clay bricks embedded in lime mortar according to a Flemish bond (27 cm thickness). Internal walls are often interrupted at the last level, before they reach the roof, to obtain a larger space serving as hayloft. As a result, the roof is supported either by isolated pillars (side of 60 cm, Figure 2b) or slender masonry walls (13 cm thick) on which the ridge beam rests. Internal partitions are in general non-structural, since they are timber-made to reduce the weight on the floors (Figure 2c).

At every level, floor joists are parallel to the main façades and they are very closely spaced due to the heavy load (up to 400 kg/m²) coming from the

traditional paving finishing. This is known as *terrazzo*: a slab of lime mortar, brick and stone fragments polished with flaxseed oil, which can be up to 30 cm thick (Crovato 1999).

Roof rafters (Figure 2c), roughly squared, are generally weakly connected to the ridge beam and they thrust on the peripheral walls; the waterproofing layer is made of thin clay tiles (*pianelle*) in addition to the roof tiles: once again the resulting structure is at the same time flexible (cross-section within 16x16 cm on a 5 m span) and heavy (around 200 kg/m²). Horizontal connections are ensured generally by anchor bolts at the ends of both floor joists and roof girders (Figure 2d).

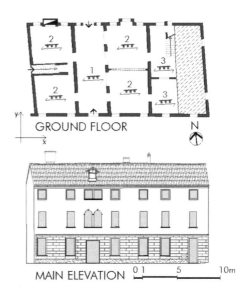

GROUND FLOOR — N

MAIN ELEVATION — 0 1   5   10m

Figure 1. Palazzo Carraro: a) ground plan; b) elevation. Key to numbers in the plan: 1. Hall; 2. Apartment; 3. Rooms; the arrow indicates joists direction.

Figure 2. Palazzo Carraro: a) interior of the main hall; b) pillar supporting the roof girder and wooden tie beams; c) roof rafters and internal wooden partitions; d) anchor bolt (white arrow).

In horizontal diaphragms, beams are usually under-dimensioned than structural needs since the ancient builders put a certain trust on the 'group effect' made possible by the close spacing. This allows to keep joists and rafters close to their flexural resistance (cross-section within 18x18 cm) at the cost of greater out-of-plane flexibility of the diaphragm (Doglioni & Mirabella Roberti 2013).

To sum up, one may state the main vulnerability factors of this class of buildings and of Palazzo Carraro as follows:

– Unidirectional, flexible and heavy floors;
– Flexible, heavy and thrusting roof;
– Different distribution of resistant masonry walls in the two main directions of the building, generally greater along the main façade;
– Limited presence of ties and other horizontal connections;
– Clusters of windows;
– Slender masonry walls;
– Windows adjacent to corners;
– Reduction of masonry walls with the storey.

A building with such features cannot show a clear overall behaviour, since seismic masses may be considered equally distributed among the walls and the floors. Otherwise they may be considered 'regular' in plan and elevation

3.2 Recognition of the main local mechanisms according to empirical vulnerability analysis and comparison with eigenvalue analysis

According to the vulnerabilities described and in relation to the building's features, one may verify the following local mechanisms (Figure 3):

1. Simple out-of-plane overturning of the west gable wall, at each level;
2. Horizontal flexure of the gable in the east façade under the hammering of the ridge beam;
3. Out-of-plane overturning of the last level of the main facade pushed by roof rafters;
4. In-plane shear mechanism in the masonry piers at the ground floor.

Those observations may be confronted with the result of the eigenvalue analysis carried out by Salvalaggio et al. (2019) on a continuum model of the actual state. The model has been analysed in the DIANA FEA 10.2 software (TNO 2018) where a total strain crack model (Figure 4) is used. The geometry is discretized by 4-noded 2-D shell elements.

The modes (Table 4) along the main facade (X direction) with relevant participating mass (Nos 1, 3, 9) excite mainly the gable walls in the out-of-plane direction. In the short dimension of the building (Y), each portion of the facade in between two transversal walls have a clear out-of

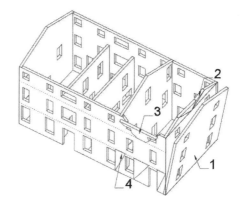

Figure 3. Local mechanisms which may be determined by a traditional vulnerability analysis.

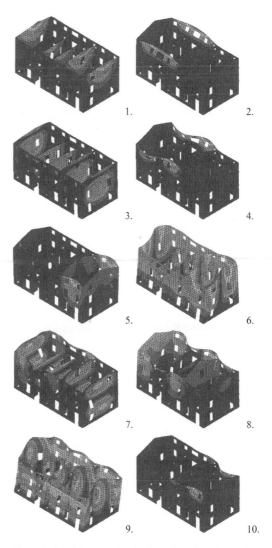

Figure 4. Modal shapes for the first 10 modes (the number next to the sketch indicates the mode).

Table 4. Parameters of the first 10 modes.

Mode	T	Mx	My	γkψk
	s	%	%	-
1	0.358	29.22	0.02	0.23
2	0.284	0.01	9.90	0.13
3	0.242	22.53	003	0.87
4	0.202	0.00	0.01	-
5	0.175	0.06	13.84	-
6	0.168	0.08	3.01	-
7	0.142	1.87	1.07	0.51
8	0.133	0.10	18.22	-
9	0.128	27.44	0.01	0.87
10	0.126	0.34	0.01	-

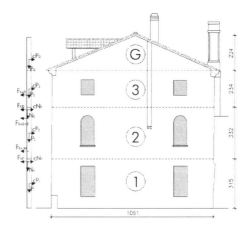

Figure 5. Identification of macroblocks involved in the local mechanism.

-plane shape, independent from other parts. Relevant participating mass corresponds to the out-of-plane of the last storeys for one (No 2) or two levels of the longest facade (Nos 5 and 8). As is generally observed in buildings with flexible diaphragms, superior modes are actually relevant looking at the participant mass.

According to the eigenvalue analysis, out-of-plane local mechanisms may be identified as follows and compared to the previous one:

1. Out-of-plane horizontal flexure of the last two levels of the gable walls;
2. Out-of-plane vertical flexure of the gable walls;
3. Out-of-plane overturning of the last level of the main façade
4. Out-of-plane overturning of the last two level of the main façade between two transversal walls.

4 LOCAL MECHANISM ANALYSIS AND SAFETY ASSESSMENT

The mechanism considered in the following is the overturning of the compound of blocks G, 3 and 2 (Figure 5), with the features presented in Table 5, around a hinge placed at the first floor ($z=3,15$ m).

The resulting static coefficient is $\alpha=0.033$.

4.1.1 *Definition of the seismic input*

The seismic hazard for the site of palazzo Carraro is defined by MIT 2018 for each relevant LS starting from the parameters given in Table 6. Noale is not in an active seismic area but resents from earthquakes coming from the piedmont area of Veneto. The equivalent damping and the other parameters needed have been determined as shown previously in § 2.

The floor spectra as defined by MIT (2009) and MIT (2019) at DLS and NCLS have been respectively evaluated; with reference to the former, they are compared in Figure 4. For the selected mechanism, modes

1, 3 and 9 (cfr. Table 4) are relevant, looking at $|\gamma\psi|$ values. It is worth noting that since $|\gamma\psi|$ is higher for mode 3 than mode 1, the peak in the floor spectrum is shifted toward this value. The simplified formulations, proposed by MIT (2009, 2019) and exposed in § 2.3.1 yield quite different values. In fact, assuming the total height $H=9.67$ m at the level of the centroid of macroblock G, one can found $T_1=0.29$ s, $\gamma_1=1.29$ ($N=3$) and $\psi_1(z=3.15) =0.33$, that is $\psi_1\gamma_1=0.43$.

There is a substantial difference between the two spectra, in terms of: a) peak value; b) anchoring value; c) the peak's horizontal position; d) ordinate values for long periods (i.e. outside the ground spectra plateau).

Table 5. Seismic masses and geometries in the overturning mechanism (see Figure 5).

Storey	Masonry	Floor	hi
	kN	kN	m
1	157	-	1.52
1	-	71	3.15
2	172	-	4.81
2	-	34	6.35
3	133	-	7.64
3	-	0	8.92
Gable	58	22	9.67

Table 6. Seismic hazard parameters for the site of palazzo Carraro according to MIT (2018).

Limit state	Tr	a_g	F_0	T^*_C	ξ
	y	×g	-	s	%
DLS	50	0.042	2.566	0.260	5
NCLS	475	0.111	2.525	0.333	15
CLS	975	0.149	2.516	0.352	20

1024

4.1.2 Linear verification

The linear verification happens trough the means described in § 2.3.1. The spectral acceleration is $a_0{}^*= 0.30$ m/s^2 and in both the procedures (MIT 2009, 2019) the expected acceleration (0.61 m/s^2 and 1.91 m/s^2 respectively) is higher than the mechanism's one and therefore the mechanism activates (Figure 7). It is worth noting that in MIT (2009) the comparison value is the peak in the floor spectra, in MIT (2019) it is the anchorage value. However, the latter is more than twice the former, which by itself is only a little higher than the ground spectrum.

4.1.3 Non-linear verification (kinematic analysis)

The maximum (spectral) displacement of the capacity curve is $d_0{}^*=0.114$ m; therefore, the ultimate spectral displacement, conventionally assumed as the 40% of that value, is $d_u{}^*= 0.058$ m at NCLS.

If one uses the MIT (2009) procedure, the (§ 2.3.2, Table 3) the capacity demand amounts at d_d =0.027 m.

According to MIT (2019) the longer 'ultimate' period defined by the modified Equation 7 yields a spectral displacement d_d=0.069 m.

Therefore, the kinematic analysis results satisfied in the older procedure but not in the new, as it is graphically shown in Figure 8.

The safety assessment in MIT (2019) is substantially aggravated, since the displacement demand is twice as high as the one calculated with MIT (2009) and it surpasses the capacity.

The procedure followed so far, bears an important simplification if compared to the requirements of MIT (2009). In fact, in the code each verification passes firstly through the calculation of an expected acceleration at ground level $a{}^*_{LS}$ whose value must be compared to the actual hazard at the chosen LS as defined in the second column of Table 6.

The increase of the first period at NCLS (Equation 5) has been neglected in previous calculations, since the highest participation corresponds to the third mode.

It is worth noting that, if the non-linear verification is made with MIT (2019) expressions but using the approximated dynamic parameters (T_1=0.29 s, $\psi_1\gamma_1$=0.43) the verification becomes satisfied since d_d stops at 0.037 m. Similarly, the linear verification terms reduce to a value compatible with the one calculated with MIT 2008.

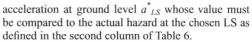
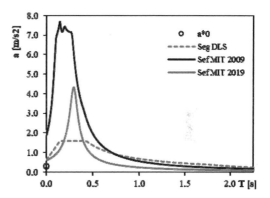

Figure 7. Graphic representation of the safety assessment at DLS. The continuous black line is the floor spectrum according to MIT 2019, the continuous grey line is the MIT (2009) floor spectrum and the dashed grey line the ground response spectrum. Note that in the ordinates.

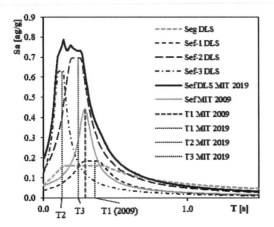

Figure 6. Graph of the floor response spectra at DLS (Spectrum-period plane). Dashed black lines are the spectra according to MIT (2019) for each of the first three modes; continuous black line is the SRSS combination of the three. Dashed grey line is the ground spectrum while the continuous grey line is the floor spectrum in MIT (2009). The dotted lines indicate the periods of the first three modes.

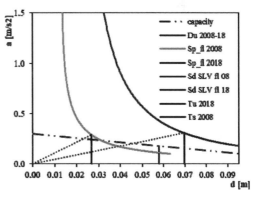

Figure 8. Graphic representation of the safety assessment at NCLS in the ADSR. The continuous black line is the floor spectrum according to MIT (2019), the continuous grey line is the MIT (2009) floor spectrum; the thin grey line is the capacity curve, the dotted lines is the representation of the ultimate periods, the vertical thick lines represent the demand (grey dashed) and the capacity (black continuous).

Eventually, if the first period increase is considered, in the simplified procedure, the shift in the floor spectrum slightly worsens the spectral demand but not enough to alter the positive result of the verification.

5 CONCLUSIONS

The article presents the safety assessment of a local mechanism in an ordinary URM building with flexible masonry floors and clearly lacking an overall behaviour to test the new procedures inserted in the 2018-19 update of the Italian seismic code.

As for the definition of the seismic action (particularly the floor spectra), the main critical aspects are:

- The identification of the base parameters requires to be somewhat familiar with the outputs of an eigenvalue analysis;
- Only sophisticated pieces of software are able to extract those parameters (e.g. DIANA or ANSYS but not TRE Muri);
- The use of different pieces of software may possibly yield to different modal parameters, as has been recently showed by Valluzzi et al. 2019;
- The quick formulations to estimate dynamic parameters provided by the codes under the hypothesis of rigid floors results inaccurate for URM buildings with a relatively simple structure but with flexible floors;
- An objective criterion for the choice of relevant modes is still missing;
- The construction of the floor spectra is laborious.

As far as the safety verification is concerned:

- The linear verification has changed and now it may be considered more straightforward, provided that floor spectra have been calculated;
- The non-linear verification can be also judged simpler than before but there are more parameters to take into account, in particular the equivalent damping at different LSs;
- An overall worsening of the verification requirements both in the linear and non-linear field which result twice and three times as onerous respectively;

It may be argued that the use of such complicated formulations somewhat prevents the 'friendliness' of the local analysis which allowed the designer a greater control over the procedure.

Finally, the code has gone in an opposite direction than the one suggested by Al Shawa et al. (2015), who suggested a further simplification of the verification formulas proposed by MIT (2009). As a result, the verification procedure here presented (§ 2.3.1, § 2.3.2) resembles the old version, instead of the new, which refers to the acceleration even with non-linear assessment.

The worsening of the seismic demand in both linear and non-linear field may be related with the introduction of the frictional forces in the evaluation of the static coefficient. This leads to higher collapse load and piecewise linear capacity curves of local mechanisms as presented by Lagomarsino (2015).

REFERENCES

Al Shawa O., Liberatore D., Sorrentino L. 2015. Influenza dell'effetto filtro su meccanismi locali di collasso. *XVI convegno ANIDIS, L'ingegneria Sismica in Italia*, L'aquila, Italy, 13-17 September 2015.

Argiento, L.U., Casapulla C., Speranza E. 2017. Evaluation of in-plane rocking-sliding failure modes of multi-storey masonry walls for quick analysis at urban scale. *XVII convegno ANIDIS, L'ingegneria Sismica in Italia*, Pistoia, Italy, 17-21 September 2017.

Casapulla C. & Argiento, L.U. 2016. The comparative role of friction in local out-of-plane mechanisms of masonry buildings. Pushover analysis and experimental investigation. *Engineering Structures* 126: 158–173.

Casapulla, C. & Maione, A. (2011). Out-of-plane local mechanisms in masonry buildings. The role of the orientation of horizontal floor diaphragms. *9th Australasian Masonry Conference, Queenstown, New Zealand*.

Crovato A. 1999. *I pavimenti alla veneziana*. Treviso, IT.

Degli Abbati, S., Cattari, S., Lagomarsino, S. 2015a. Proposta di un aggiornamento della formulazione NTC-08 per gli spettri di piano e sua applicabilità nella verifica dei meccanismi locali. *XVI convegno ANIDIS, L'ingegneria Sismica in Italia*, L'Aquila, Italy, 13-17 September 2015.

Degli Abbati, S., Cattari, S., Lagomarsino, S. 2017. Proposta di spettri di piano per la verifica di elementi non strutturali e meccanismi locali negli edifici in muratura. *XVII convegno ANIDIS, L'ingegneria Sismica in Italia*, Pistoia, Italy, 17-21 September 2017.

Degli Abbati, S., Cattari, S., Lagomarsino, S. 2018. Theoretically-based and practice-oriented formulations for the floor spectra evaluation. *Earthquakes and Structures* 15 (5): 565–581.

Degli Abbati, S., Cattari, I. Marassi, S., Lagomarsino, S. 2015b. Seismic out-of-plane assessment of Podestà Palace in Mantua (Italy). *Key Engineering Materials* 624:88–96.

Doglioni F. & Mirabella Roberti G. (eds.) 2013. Venezia. Forme della costruzione, forme del dissesto. Venezia.

Giuffrè A. (ed.) 1993. *Sicurezza e conservazione dei centri storici il caso Ortigia*. Laterza, Roma-Bari.

Heyman J. 1966. The stone skeleton. *International Journal Solids and Structures* 2:249–279.

Maretto P. 1960. *L'edilizia gotica veneziana*. Roma.

MIT 2008 = Ministry of Infrastructures and Transportations, Regulation 14/ 01/2008, *Norme Tecniche per le Costruzioni*.

MIT 2009 = Ministry of Infrastructures and Transportations, Regulation 617/2019, *Istruzioni per l'applicazione delle Norme Tecniche per le Costruzioni*.

MIT 2018 = Ministry of Infrastructures and Transportations, Regulation 17/ 01/2018, *Aggiornamento delle 'Norme Tecniche per le Costruzioni'*.

MIT 2019 = Ministry of Infrastructures and Transportations, Regulation 7/2019, *Istruzioni per l'applicazione dell'aggiornamento delle 'Norme Tecniche per le Costruzioni'.*

Lagomarsino S. 2015. Seismic assessment of rocking masonry structures. *Bulletin earthquake engineering* 13:97–128.

Pavanetto M., Sbrogiò L., Salvalaggio M., Valluzzi M.R. 2019. Equivalent frame modeling of a URM building in finite element environment. *XXIV AIMETA 2019 Conference, Rome (Italy)*, 15-19 September 2019.

Salvalaggio M., Sbrogiò L., Pavanetto M., Valluzzi M.R. 2019. Evaluation of the effect of compatible interventions applied to horizontal components of URM buildings with EFM and FEM models. The case of Palazzo Carraro in Noale (Italy). *COMPDYN 2019 7th International Conference on Computational Methods in Structural Dynamics and Earthquake, Crete (Greece)*, 24-26 June 2019.

TNO 2018 = TNO Building and Construction Research Institute. 2018. DIANA FEA 10.2 [computer software]. Delft, the Netherlands.

Valluzzi M.R., Salvalaggio M., Sbrogiò L. 2019. "Repair and conservation of masonry structures". In: Ghiassi B. and Milani G. (Eds.), *Numerical modeling of masonry and historical structures. From theory to application*, Woodhead Publishing Series in Civil and Structural Engineering, Elsevier.

Brick and Block Masonry - From Historical to Sustainable Masonry –
Kubica, Kwiecień & Bednarz (eds)
© 2020 Taylor & Francis Group, London, ISBN 978-0-367-56586-2

Vertical collapse mechanisms in masonry buildings due to seismic vertical component

F. Comodini
University eCampus, Novedrate(CO), Italy

M. Mezzi
University of Perugia, Perugia, Italy

ABSTRACT: The acceleration histories recorded in the epicentral areas during the last earthquake in Central Italy (2016-17) show very intense vertical components. During the post-earthquake damage assessment operations in the epicentral areas, some damage frameworks different from those described in the literature have been observed, with a macro-element collapse mode attributable to the action of the earthquake vertical component. The paper presents some of these buildings characterized by the presence of horizontal cracks, by the displacement of the highest levels, by vertical cracks in the sub-window walls. The masonry walls are undamaged without the typical shear or flexure failure cracks and the lower levels do not show any damage. The crack pattern and the associated collapse mechanisms cannot be included within the classic in-plane and out-of-plane mechanisms. A collapse mechanism characterized by the loss of vertical connection of entire structural bodies is therefore hypothesized. The activation of this type of mechanism would seem to exclude the formation of the classic collapse mechanisms on which the seismic capacity checks of the masonry structures provided by the technical codes are based. A simplified linear kinematic analysis able to analyze the described behavior is proposed and illustratively applied to a case study building. Criteria for the identification and verification of the proposed mechanisms are described, with the aim to introduce the method in design codes.

1 INTRODUCTION

Currently the Italian technical standard (NTC, 2018) for the linear analysis of the structures, in general does not provide for the combined application of the horizontal and vertical components of the seismic action. The vertical component has to be taken into consideration only in some specific cases. In the case of ordinary buildings and in the absence of particular geometric conditions, indicated by the technical code, the seismic design of the structures is performed only considering the horizontal components of the seismic motion.

The provisions of the code are based on the consolidated principle that ordinary structures, in the absence of local amplification phenomena, are weak with respect to horizontal forces and therefore more vulnerable to these components of seismic motion. However, many research experiences (Broderick et al., 1994; Goltz, 1994; Elnashai et al., 1995; Youssef et al., 1995; Watanabe et al., 1998; Naeim et al., 2000; FEMA355, 2000) showed the influence of the vertical component of the earthquake on the damage of some structural typologies such as masonry buildings, churches, precast buildings. The greatest effects on the structural damage of the seismic vertical component were recorded in the epicentral areas (zone of 5 km radius) where the vertical accelerations reached, in some cases, values much higher than gravity (Collier & Elnashai, 2001; Bouchon, 2000). In areas far away from the epicenter the effects of the vertical component are attenuated with respect to those of the horizontal component and therefore have a minor influence on the seismic response of the structures (Grimaz & Malisan, 2014).

The presence of a relevant seismic vertical component can induce significant changes in the dynamic behavior of the structures (especially considering the dynamic combination with the horizontal component).

The vertical component concentrates its energy in the high frequency range (Collier & Elnashai, 2001; Elnashai & Papazoglou, 1997) commonly higher than 5 Hz, which generally includes the frequency of the first vertical structure mode (both for reinforced concrete and masonry structures). This implies a significant amplification of the response in the vertical direction and consequently a possible change in the type of expected damage.

The action of the vertical component of the earthquake, with a direction opposite to gravity, generates a reduction in static axial loads and therefore generally leads to a reduction in the capacity of the structural elements to resist shear and bending. Also the capacity of deformation in the plastic field suffers a strong reduction (Di Sarno et al., 2012).

In masonry elements, the reduction of axial loads also involves a reduction of the capacity to resist shear sliding and bending and this can lead to particular collapse modes. The vertical earthquake component can also produce an increase of the axial force on the vertical structures and for this reason particular effects are expected (Papazoglou & Elnashai, 1996; Di Sarno et al., 2012) especially on the elements that have larger static axial force.

It is known that the peaks of intensity of the vertical component generally occur for frequencies different from those of the horizontal components (Collier & Elnashai, 2001; Shreshta, 2009), therefore the effect of amplification of the damage occurs only when the component vertical acts simultaneously with the horizontal earthquake component. However, to better understand the effects of the vertical earthquake on the response of buildings, it is necessary to observe the damage detected in buildings located near the epicenter of a seismic event.

In this paper the damage of some buildings in the village of San Pellegrino di Norcia located epicentral area of the earthquake of Central Italy in 2016 are illustrated, analyzed and used to formulate a mechanical interpretation and a check procedure.

Figure 1. Damage state of building A.

Figure 2. Particular of damage state of building A.

2 POST-EARTHQUAKE DAMAGE ANALYSIS

Damage analysis is fundamental to understand the effects of the earthquake on buildings. The scientific and professional community is used to analyze the types of damage according to established procedures and methods recognized by technical regulations. However, sometimes the application of consolidated procedures does not allow a realistic analysis of the actual phenomena that produced "uncommon" effects.

The seismic event of October 30, 2016 of the Central Italy sequence of 2016-17 provoked in some masonry buildings, even of recent construction, damage mechanisms not included among those codified so far. A thorough evaluation of this unconventional damage on some buildings located in San Pellegrino, a village near Norcia, has been performed. Figures 1, 2 and 3 report some examples. It is evident that in these cases it is not possible to identify the classic 1st mode (out-of-plane) and 2nd mode (in-plane) collapse mechanisms (NTC, 2018).

The singularity of the crack pattern can be considered due to a vertical seismic component, well above the conventional values, producing a decrease in the static axial loads of the upper floors, such as to produce a sudden reduction in the masonry shear sliding resistance at the floor level thus favoring the activation of a horizontal slip mechanism subjected to the action of horizontal accelerations.

Analyzing the crack paths, in can be observed that there are no appreciable failure attributable to the "classic" in-plane mechanisms with diagonal cracks, it

Figure 3. Damage state of building B.

is also evident that at the lower story damage is practically absent. This situation is in contrast with the classic damage expected for masonry structures under seismic actions, which involve more severe stresses and damage at the lower stories. The vertical cracks in correspondence of the sub-windows are probably caused by the instant difference of axial load, under dynamic excitation, between the main walls and the adjacent unloaded sub-window elements.

It is clear that, in the specific case, the effects of local amplification of the vertical seismic action have influenced the global seismic response of the structure. However, the study of these effects cannot be pursued considering the shear strength reduction of the masonry assessed through the interaction with the vertical stresses provided by the standards.

The reduction of the building global seismic capacity due to the activation of the identified mechanism cannot be reproduced with the classical finite elements numerical methods even if considering constitutive relationships of the material that take into account the reduction of the masonry shear strength due to the decompression generated by the vertical seismic action. Indeed, by operating a nonlinear dynamic analysis on a numerical model subjected to alternated cyclic actions of the vertical and horizontal component acting simultaneously, it can be assessed that the sudden alternation of the direction of application of the vertical earthquake would produce an oscillation of the shear resistance around the value obtained for static vertical loads. It results that the force-displacement cycles, characterizing the seismic capacity of the structure, show negligible variations with respect to those computed with the only horizontal component of the earthquake (Lagomarsino, 2019).

The most correct approach seems to be that of the kinematic analysis with evaluation of the progressive loss of constraint in correspondence with the solution of continuity between masonry walls and the underlying floor curb. The global slide of the analyzed rigid body must be interpreted as the resultant of the instantaneous slips occurring under specific conditions and combinations of the earthquake components.

3 ACTUAL SEISMIC ACTION

A comparison of the response spectra of the acceleration records of the seismic event of October 30, 2016 recorded at the station of the Italian network called NRC, located in the municipality of Norcia (Italy), with those provided by the Italian standard is presented.

Figure 4 reports the elastic response spectra of the two recorded horizontal components (channels HGE and HGN) and those provided by the Italian code for the life safety (SLV) and collapse prevention (SLC) limit states.

The spectral values of the actual records in the interval of periods ranging from 0.15 s to 0.35 s, typical of masonry structures with two or three floors, exceed those of the conventional elastic spectra (both for SLV and SLC limit state).

Specifically, the spectrum of the HGE channel has a maximum peak equal to 1.93 g at a period of 0.26 s while the HGN channel spectrum has a maximum pseudo-acceleration value of 1.63 g for a period of 0.22 seconds.

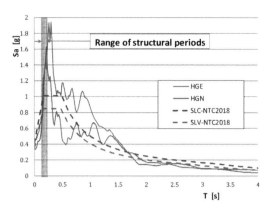

Figure 4. Response spectra of horizontal components.

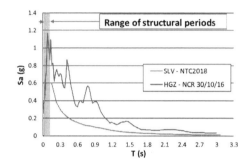

Figure 5. Response spectra of vertical components.

Similarly, Figure 5 reports the elastic response spectrum of the recorded vertical component (channels HGZ) and that provided by the technical standards for the SLV limit state. In the interval of the periods relevant for the considered masonry buildings typology, ranging from 0.02 s to 0.1 s, the response spectrum of the vertical component reaches a maximum peak equal to 1.18 g, almost double than the values of the conventional spectrum.

The comparison of the spectral intensities shown in the graphs of Figure 4 and 5 shows how the seismic response of the acceleration histories recorded in the epicentral areas was considerably higher than the conventional seismic response provided by the standard for the design limit states.

4 SHEAR-SLIDING COLLAPSE MECHANISMS

The damage framework found on some buildings near the epicenter of the Norcia earthquake (October 30, 2016) shows mechanisms characterized by a rigid body sliding of the top floors. It seems that the activation of the global mechanism is related to the formation of horizontal sliding surfaces correspondingly to plane of lower resistance. The planes of minimum resistance can be identified with the masonry-floor slab or masonry-floor curb interfaces.

The shear sliding resistance of the masonry can be represented through the Mohr-Coulomb criterion. The criterion, following a rule similar to that of the friction between two surfaces, is represented by the following relationship between tangential stress τ and normal stress σ

$$|\tau| = \tau_0 - \sigma \cdot \tan\varphi \qquad (1)$$

The shear strenght depends on two parameters: the cohesion τ_0 and the friction angle ϕ. They can be expressed as a function of the compressive strength f_c and of the tensile strength f_t. With simple geometric considerations on the scheme of Figure 6 it results

$$\begin{cases} \tau_0 = \dfrac{\sqrt{f_c f_t}}{2} \\ \tan\varphi = \dfrac{f_c - f_t}{2\sqrt{f_c f_t}} \end{cases} \qquad (2)$$

Assuming that the constitutive low of the potential sliding surfaces follows the Mohr-Coulomb criterion, the damage found in the buildings of San Pellegrino can be interpreted in the following way.

In the specific case, at the site in question vertical earthquake components of high intensity acting simultaneously with high horizontal components have been recorded. When the vertical component of the earthquake acts in direction opposite to the gravity, it leads to a sudden decrease of vertical stress and therefore of the shear sliding resistance, and in some instants it can produce vertical stress higher than the tensile strength of the masonry, completely

cancelling the sliding resistance for a few instants. This effect, combined with the horizontal action of the earthquake, causes a series of instantaneous movements of the entire story of the building.

Sliding is defined as discontinuous as the direction of the vertical seismic action is alternated and only when it is directed upwards it decreases or even cancels the sliding resistance, on the contrary when the vertical stresses are directed as the gravity additional the sliding resistance is increased. An excessive increase of the compressive stresses can also lead to the achievement of the compressive strength of the masonry causing vertical cracks on the seismic-resistant elements.

5 KINEMATIC ANALYSIS

An appropriate approach for the study of the pinpointed mechanism defined "3rd mode mechanism" seems to be the kinematic analysis.

The linear kinematic analysis allows to evaluate the multiplier α_0 of horizontal loads leading to the activation of the failure mechanism, through the application of the Principle of Virtual Works (PVW) to rigid bodies.

The general expression corresponding to the PVW application is

$$\alpha_0 \left(\sum_{i=1}^{n} P_i \delta_{x,i} + \sum_{j=n+1}^{n+m} P_j \delta_{x,i} \right)$$
$$- \sum_{i=1}^{n} P_i \delta_{y,i} - \sum_{h=1}^{o} F_h \delta_h = L_{fi} \qquad (3)$$

where
P_i is the dead load of the masonry block applied at the center of gravity;
P_j is the weight of the superimposed loads, whose mass, due to the seismic action, generates horizontal force on the elements of the kinematic chain;
$\delta_{x,i}$ is the horizontal virtual displacement of the point of application of the i-th force P_i;
$\delta_{x,j}$ is the horizontal virtual displacement of the point of application of the j-th weight P_j;
$\delta_{y,i}$ is the vertical virtual displacement of the point of application of the i-th force P_i, assumed positive if directed upwards;
F_h is the generic external force;
δ_h is the virtual displacement of the point of application of the h-th force F_h;
L_{fi} it is the total work of the internal forces.

The procedure is illustrated in the following with reference to a single masonry wall. In the specific case, the external force opposing the sliding of the rigid body consists of the shear sliding resistance applied at its base. This hypothesis is confirmed by t-
he

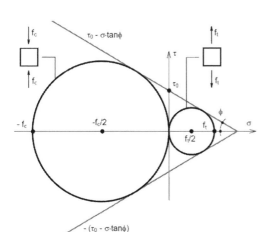

Figure 6. Failure criterion of Mohr-Coulomb.

observation of the damage framework characterized by failure surfaces, with consequent sliding of the masonry block, at the base of the masonry walls correspondently to the masonry-floor slab/curb interface.

It is necessary to specify that at the initial detachment the sliding resistance is associated to the apparent friction coefficient which is a function of the ratio between tangential and normal stress acting on the sliding surface. Once the mechanism is activated, the sliding is governed by a resistance that is function of the dynamic friction angle. In the present discussion, in a simplified way, the sliding-resistant force is evaluated as a product of the shear sliding surface resistance, evaluated according to the Mohr-Coulomb criterion, and of the contact area, the latter assumed equal to the area of the base of the masonry wall. This assumption is valid under the hypothesis that the base section is not partialized by bending.

Equation (3) can be written for the specific case of the introduced 3rd mode mechanism with reference to the scheme of a single wall shown in Figure 7 where the symbols have the following meaning:

W_g is the dead load applied at the center of gravity of the block;

W_s is the weight of the superimposed loads, whose mass, due to the seismic action, generates horizontal force on the elements of the rigid body;

δ_x is the horizontal virtual displacement;

V_R is the sliding resistant horizontal force

$$V_R(\sigma_n) = f_{vm}(\sigma_n) \cdot A_s \tag{4}$$

where

A_s is the area of the sliding surface, assumed equal to the total base contact area between the rigid body and the sliding plane: assumption valid under the hypothesis of whole compressed section;

$f_{vm}(\sigma_n)$ is the average shear sliding resistance calculated according to the Mohr-Coulomb criterion

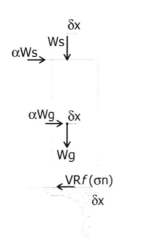

Figure 7. 3rd mode collapse mechanism – single wall.

$$f_{vm} = f_{vm.0} + \sigma_n \cdot tan(\varphi) \tag{5}$$

In equation (5)

$tg(\varphi)$ is the friction coefficient, hereinafter also referred to with the symbol μ and

$$\sigma_n = \frac{(W_g + W_S) \pm E_V}{A_S} \tag{6}$$

is the normal stress acting on the sliding plane;

E_v is the parametric vertical seismic force (with negative sign when it is opposite to gravity).

Therefore the effect of the vertical component of the earthquake on the sliding resistance of the rigid body is considered by introducing alinearly increasing parametric vertical force E_V.

The horizontal sliding mechanism of arigid body presupposes adisplacement δ_x equal for all the application points of forces. Applying the expression of the PVW it results

$$a_0(W_g + W_S)\delta_x - V_R(\sigma_n)\delta_x = 0 \tag{7}$$

from which the multiplier of the horizontal forces that could activate the sliding mechanism is derived

$$a_0 = \frac{V_R(\sigma_n)}{(W_g + W_S)} \tag{8}$$

Once the multiplier a_0 is known, to conduct the checks against the expected demand acceleration it is required to determine the participating mass M* in the mechanism and the spectral seismic acceleration a_0*. It can be assumed that the participating mass is that associated with the totality of the vertical load

$$M^* = \frac{(W_g + W_S)}{g} \tag{9}$$

consequently the participating mass ratio

$$e^* = \frac{gM^*}{(W_g + W_S)} \tag{10}$$

and the capacity spectral acceleration

$$a_0^* = \frac{V_R}{FC \cdot M^*} = \frac{a_0 \cdot g}{e^* \cdot FC} \tag{11}$$

can be computed according to the Italian code. FC ($\geq$1) is the reduction factor to be applied to the material strength accounting for the confidence level in the mechanical characterization of the material.

If the local mechanism affects a portion of the building at a certain elevation, it must be taken into account that the absolute acceleration at the elevation of the portion affected by the mechanism is generally amplified with respect to that at the ground level. An acceptable approximation of the check is given by the following expression

$$a_0^* \frac{S_e(T_1)\psi(Z)\lambda}{q} \qquad (12)$$

where:

$Se(T1)$ is the pseudo-acceleration calculated, for the period T_1, from the elastic spectrum defined by code or obtained from accelerations records, corresponding to the event with the exceeding probability of the design limit state and the design lifetime V_R;

T_1 is the period of the first vibration mode of the entire structure in the considered direction, rigorously determined through a preliminary modal analysis of the entire building or assumed according to simplified expression (i.e. $T_1 = 0{,}05 \cdot H^{3/4}$).

$\psi(Z)$ it is the lateral deformation of the first mode of vibration, normalized to a unit value at the top of the building, in the considered direction: in the absence of more accurate assessments it can be assumed $\psi(Z) = Z/H$, being H the total height of the building with respect to the foundation, and Z the elevation of the connection of the blocks affected by the mechanism to the remaining portion of the structure;

λ is the corresponding modal participation coefficient (in the absence of more accurate assessments it can be assumed $\gamma = 3N/(2N+1)$, with N number of floors of the building);

q is a behavior factor attributable to the mechanism, whose value depends on the design limit state.

6 CASE OF STUDY

In the following a case study concerning the application of the kinematic analysis according to the described methods is analysed.

The case study concerns an external wall located at the last level of a three story building. The building has tuff stone masonry walls, mixed cement and hollow bricks floor slabs, wooden roof. There are r/c curbs at all floors and at the top of the walls.

Figure 8 shows the analysed scheme. Table 1 lists the mean values of the mechanical characteristics, assumed according to the Italian technical standard.

The mean values of the resistances are subsequently divided by the confidence factor FC = 1.35 corresponding to the assumed knowledge level.

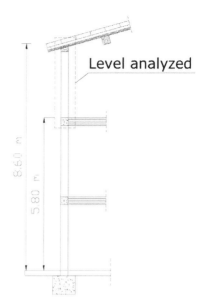

Level analyzed

Figure 8. Case study: section of the wall.

Table 1. Mechanical characteristics of masonry.

f_m N/mm^2	f_{vk0} N/mm^2	E N/mm^2	G N/mm^2	f_{tm} N/mm^2	W k N/m^3
1.40	0.028	4500	1350	0.0017	12

Table 2. Modal analysis: periods of the main modes.

1st Mode X-dirTrans	2nd Mode Y-dir Trans	3rd Mode Z-dir Rot	6th Mode Z-dir Rot
0.15 s	0.12 s	0.10 s	0.035 s

The periods of the main vibration modes resulting from the modal analysis carried out on the entire structure are summarized in Table 2.

Figure 9 shows the outline of the evaluated alignment. The entire wall is subdivided into elementary rigid blocks, each subjected to dead load W_g, supported loads W_s and sliding resistance $V_R(\sigma_n)$ for the blocks at the base.

Applying Equation (7) to the case study the collapse multiplier α_0 of the horizontal loads relative to the sliding mechanism at the base of the third floor is obtained.

The sliding resistance force will vary depending on the normal stresses acting at the interface between walls and floor-curb, that in their turn depend on the gravitational loads and seismic forces. The latter are considered to act in the direction opposite to gravity as this leads to determine the minor collapse multiplier.

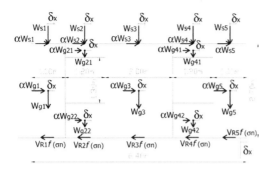

Figure 9. Case study: view of the wall.

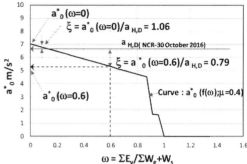

Figure 11. Capacity acceleration a^*_0 vs. vertical seismic action factor ω for $\mu = 0.4$ (reference response spectra of NRC record of October 30, 2016).

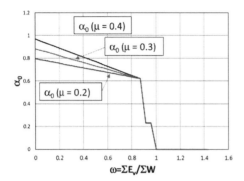

Figure 10. Collapse multiplier α_0 vs. parameter ω for different values of friction coefficient μ.

The vertical seismic action is reproduced by incrementally increasing the vertical acceleration a_{gv}/g. The equivalent vertical seismic force E_v is obtained by multiplying the acceleration by the mass associated to each elementary block.

Figure 10 shows the graphs of the collapse multiplier α_0 as a function of the parameter ω that is the ratio between the overall seismic action ΣE_v and the sum of the gravitational loads W_g and W_s:

$$\omega = \frac{\sum E_{v,i}}{\sum W_g + W_s} \qquad (13)$$

The curves are reported for three different values of the friction coefficient $\mu = 0.4, 0.3, 0.2$.

The values $\mu = 0.3$ and 0.2, lower than the value 0.4 indicated by the standards, have been used to simulate sliding surfaces with low friction. This situation can occur when the first row of bricks does not have an adequate thickness of mortar or the latter is not adequately adherent to the floor curb.

As it can be seen from Figure 11 the collapse multiplier α_0 decreases in a way inversely proportional to the factor ω. This means that as the vertical seismic action increases, the horizontal forces that would produce sliding are lower. Practically, the trend of the collapse multiplier follows the Mohr-

Coulomb law with which the value of the resistance force $V_R(\sigma_n)$ was evaluated.

Indeed, the first section of the graph decreases as a function of friction and normal stress, but when the normal tensile stress reach a value such as to eliminate the effect of the gravitational loads, the sliding resistance is only given by the masonry cohesion f_{vk0}. The latter rapidly proceeds to the null value for small increments of the vertical tensile stress.

For values of ω grater than 1.0 the sliding resistance force is annulled because the normal stress are equal to or greater than the tensile strength of the masonry and the sliding can occur for small horizontal forces.

A particular trend of the graph results for values close to $\alpha_0 = 0.2$. This is due to the fact that the sub-window elements have a lower seismic mass with respect to the masonry walls, therefore with the same vertical acceleration a smaller axial decompression is determined whose effect is reflected on the relative sliding-resistant force.

Figure 11 shows the graph of the capacity acceleration a_0^* calculated according to Equation (11) and for the conventional value of the friction coefficient $\mu = 0.4$. On the same graph it is possible to report the level of the demand acceleration $a_{H,D}$ calculated with the Equation (12) making reference to the spectrum of the acceleration record of the NCR station for the event of October 30, 2016 correspondingly to the period of vibration of the fundamental mode of the structure in the direction of the considered alignment.

To evaluate the outcome of the checks, the safety index

$$\xi = \frac{a_0^*}{\alpha_{H.D}} \qquad (14)$$

has been defined as the ratio between the capacity and demand acceleration. Values $\xi > 1$ correspond to satisfied verifications.

From the graph of Figure 11 it can be observed that in the absence of vertical earthquake component

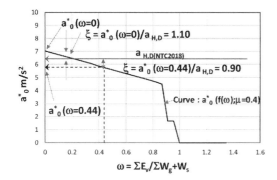

$$\omega = \Sigma E_v / \Sigma W_g + W_s$$

Figure 12. Capacity acceleration a^*_0 vs. vertical seismic action factor ω for $\mu = 0.4$ (reference response spectra of Italian code).

($\omega = 0$) the index ξ results greater than one, therefore the sliding mechanism cannot be activated.

On the same graph the capacity acceleration correspondent to the demand vertical seismic component can be estimated. The demand vertical seismic force is calculated with reference to the spectrum of the acceleration record of the NCR station for the event of October 30, 2016 correspondent to the period of the vertical translation mode of the entire structure. The ratio of the vertical seismic action to the gravitational loads gives a value $\omega = 0.6$. For this value of ω the index ξ results less than 1.0, which means that the sliding mechanism can be activated.

The same type of graph can be plotted (Figure 12) with reference to the conventional seismic action provided by the Italian code with reference to the SLV limit state and a behavior factor $q = 1$.

From the graph of Figure 12 it can be observed that although the vertical component of the spectrum is lower than that resulting from the site records, the check of the mechanism activation is not satisfied, even if slightly.

This result shows that the 3rd mode mechanism, for some building types and for the seismic action values provided by the standard (lower than those recorded near the epicenter), could represent an element of vulnerability that should be accounted for in the design.

7 CONCLUSIONS

In the epicentral areas of a seismic event the ground accelerations and the relative spectral components are much higher than those provided by the technical standards for the reference performance limit states.

The combination of horizontal and vertical seismic components of high intensity can generate collapse mechanisms not attributable to those coded as I-mode (out-of-plane) and II-mode (in-plane) mechanism. So a III-mode (vertical) mechanism can be introduced.

In particular, from the damage analysis of some buildings located in Norcia, near the epicenter of the October 30, 2016 earthquake, it was possible to find rigid body sliding mechanisms of the top floors, whose activation resulted from the effects of the high intensity vertical component of the earthquake. The less is the weight of the structural organism the more is the effect of the vertical earthquake.

A possible kinematic approach for the evaluation of the collapse multiplier of the horizontal forces, as a function of the vertical seismic component, able to activate the hypothesized mechanism has been illustrated and its application to a sample wall has been shown. The method, based on the kinematics of the collapse mechanism, allows to obtain the capacity acceleration, as a function of the normal stress acting on the single walls, that can be compared with the demand acceleration expected at the specific site.

REFERENCES

Bouchon, M., Gaffet, S., Cornou, C., Dietrich, M., Glot, J.P., Courboulex, F., Caserta, A., Cultrera, G., Marra, F., Guiguet R. 2000. Observations of vertical ground accelerations exceeding gravity during the 1997 Umbria Marche (central Italy) earthquakes. *J. of Seismology* 4: 517–523.

Broderick, B.M., Elnashai, A.S., Ambraseys, N.N., Barr, J. M., Goodfellow, R.G., Higazy, E.M. 1994. The Northridge (California) Earthquake of 17 January 1994: Observations, Strong Motion and Correlative Response Analysis. *Engineering Seismology and Earthquake Engineering, Research Report No. ESEE 94/4.* Imperial College, London.

Collier, C.J. & Elnashai, A.S. 2001. A procedure for combining vertical and horizontal seismic action effects. *J. Earthquake Engineering* 5: 521–539.

Di Sarno, L., Elnashai, A.S., Manfredi, G. 2010. *Seismic response of RC members subjected to the 2009 L'Aquila (Italy) near-field earthquake ground motions. ReportNo.01-2010.* Mid-America Earthquake Center, University of Illinois. Urbana-Champaign, U.S.A.

Elnashai, A.S. & Papazoglou, A.J. 1997. Procedure and spectra for analysis of RC structures subjected to strong vertical earthquake loads. *J. Earthquake Engineering* 1: 121–155.

Elnashai, A.S., Bommer, J.J., Baron, I., Salama, A.I., Lee, D. 1995. Selected Engineering Seismology and Structural Engineering Studies of the Hyogo-ken Nanbu (Kobe, Japan) Earthquake of 17 January 1995. *Engineering Seismology and Earthquake Engineering, Report No. ESEE/95-2.* Imperial College, London.

Federal Emergency Management Agency 2000. *State of Art Report on past performance of steel moment frame buildings in earthquakes. Report No. FEMA 355E.* Washington DC, USA.

Goltz, J.D. 1994. *The Northridge, California Earthquake of January 17, 1994: General Reconnaissance Report. Report No. NCEER-94-0005.* National Centre for Earthquake Engineering Research, Buffalo.

Grimaz, S. & Malisan, P. 2014. Near field domain effects and their consideration in the international and Italian seismic codes. *Bollettino di Geofisica Teorica ed Applicata* Vol. 55, n. 4: 717–738.

Lagomarsino, S. 2019. Gli effetti delle accelerazioni verticali su un edificio in muratura in caso di terremoto. *Seminario: Studio e progettazione del miglioramento delle strutture esistenti in fase di quiete sismica e in fase sismica*, La Spezia. https://youtu.be/ye5otu1iznk (site Ingenio, category Scienze e Tecnologie).

Mezzi, M., Comodini, F., Rossi, L. 2011. Base Isolation Option for the Full Seismic Protection of an Existing Masonry School Building, in B.H.V.Topping (editor) *Proc. 13th Int. Conf. on Civil, Structural and Environmental Engineering Computing- Chania, Crete, Greece. CIVIL-COMP PROCEEDINGS*. Stirlingshire, UK: Civil-Comp Press.

Naeim, F., Lew, M., Huang, C.H., Lam, H.K., Carpenter, L.D. 2000. The performance of tall buildings during the 21 September 1999 Chi-Chi earthquake Taiwan. *The Structural Design of Tall Buildings*, 9(2): 137–160.

NTC 2018. *Norme tecniche per le costruzioni*. D.M. 16 Gennaio 2018. Roma.

Papazoglou, A.J. & Elnashai, A.S.1996. Analytical and field evidence of the damaging effect of vertical earthquake ground motion. *Earthq. Eng. Struct. Dyn.*, 25:1109–1137.

Shrestha, S. 2009. Vertical ground motions and its effect on engineering structures: a state-of-the-art review. *Proceeding of International Seminar on Hazard Management for Sustainable Development* - 29-30 November 2009. Kathmandu, Nepal.

Youssef, N.F.G., Bonowitz, D., Gross, J.L. 1995. *A survey of steel moment-resisting frame buildings affected by the 1994 Northridge earthquake. Report No. NISTR 56254*. National Institute for Science and Technology. Gaithersburg, Maryland, USA.

Watanabe, E., Sugiura, K., Nagata, K., Kitane, Y. 1998. Performances and damages to steel structures during 1995 Hyogoken-Nanbu earthquake. *Engineering Structures* 20(4-6): 282–290.

Brick and Block Masonry - From Historical to Sustainable Masonry –
Kubica, Kwiecień & Bednarz (eds)
© 2020 Taylor & Francis Group, London, ISBN 978-0-367-56586-2

Seismic response of an unreinforced masonry building with structural irregularity; Blind prediction by means of pushover analysis

A. Aşıkoğlu, G. Vasconcelos & P.B. Lourenço
Department of Civil Engineering, ISISE, University of Minho, Portugal

A. Del Re
Department of Architecture, Alma mater studiorum University of Bologna, Italy

ABSTRACT: Structural masonry has been widely used in construction since early civilizations and consti-tute a large part of the building stock in both developed and developing countries. Important developments of new structural systems, such as reinforced concrete and steel structures, resulted in a considerable reduction in new unreinforced masonry (URM) constructions. Still, owing to its simplicity and sustainable characteris-tics of masonry, i.e. acoustic and thermal insulation, and fire protection, new URM is a promising structural typology. However, past seismic events confirmed that further developments are required to improve the seis-mic behavior of masonry buildings since there is a lack of seismic design rules on these structural systems. Although masonry structures have a high seismic vulnerability, it is not reasonable to assume that its construc-tion would not be applicable in regions with low to moderate seismic hazard. Proper precautions and improved design rules would contribute to the construction of low- and medium-rise masonry buildings in seismic prone zones. Within this framework, the present study aims to develop experimental and numerical analysis on quasi-static cyclic testing on URM building with geometrical complexity. An experimental build-ing is planned as a half-scale two-story building with structural irregularities in plan and elevation, such as having set back in one corner and irregular opening distributions, respectively. While the seismic codes cover designing rules mainly based on regular structures and impose certain penalties on structural irregularities, regular configurations are not representative ones due to architectural and functional concerns. This study focuses on understanding and assessing the global seismic behavior of a masonry building imposed to tor-sional effects. Thus, the numerical pushover analysis is based on blind prediction by means of finite element method to obtain capacity curves and decide a reasonable loading protocol for the experimental campaign.

1 INTRODUCTION

Masonry construction dates back to early civiliza-tion, being the oldest structural system, which in fact, can be considered as the basis of civil engineer-ing. Naturally available materials inspired the com-position of masonry, and, therefore, have been applied in different forms over centuries as structural systems. In fact, sun-dried mud bricks and stone units were the most common materials used in ancient times and many examples of the particular constructions still stand and can be found all around the world. The use of clay bricks is at least 10,000 years long. However, its use as a construction mater-ial was accelerated after 1858 when the Hoffman Kiln was invented. Subsequently, a significant devel-opment was achieved in the production process of the bricks by using machines which enabled to fire bricks in a continuous process (Drysdale and Hamid, 2008).

Most of the traditional masonry structures were designed based on vertical loads only. This resulted in massive walls in order to prevent the occurrence of the tensile stresses and provide lat-eral stability on the structure under eccentric ver-tical loads and horizontal loads. Thus, in the early time, the masonry building codes covered empirical design rules within this context. How-ever, designing the size of the masonry structure in such a way leads to enormous dimensions that are not feasible neither with aesthetical and archi-tectural present contexts nor with economic restrictions. Therefore, new design rules have been provided in masonry design codes by con-sidering both vertical and lateral loading in order to achieve reliable safety and viable cost. In par-ticular, structural response of unreinforced masonry, which does not include any reinforce-ment, only depends on the strength of the masonry that has high resisting capacity in com-pression but very limited capacity against tensile stresses. Nevertheless, the appearance of other alternative solutions for masonry structures, such as reinforced concrete or steel, resulted in a significant reduction in the number of new masonry constructions as loadbearing structural solutions. Consequently, in several countries, the use of masonry has been narrowed to low-rise residential buildings or being used as infill non-structural panels, cladding or artistic features of the constructions (Drysdale and Hamid, 2008).

Nonlinear dynamic analysis is considered as the most accurate approach for the seismic structural analysis of URM masonry buildings because it pro-vides a seismic response under more realistic cir-cumstances. However, the response of the structure is highly dependent on the dynamic motion and, therefore, a set of analyses with different levels of intensities is needed. The application of such a method is very complex and demanding for prac-tice and requires very high computational effort and expertise. Therefore, nonlinear static (pushover) ana-lysis has been preferred for the seismic design/ assessment of structures (Lourenço, 2009). Pushover analysis provides fundamental information about the seismic performance of buildings by simply pushing the structure with an increasing lateral load until the collapse. In fact, the pushover curve is a persuasive tool to attain the seismic behavior associated with damage limit states.

Recently, the seismic design approach is evolved based on a philosophy of performance rather than strength aiming at achieving struc-tural safety and controlled damage. According to existing knowledge, the application of the per-formance-based design to masonry structures is not straightforward, and it has been mostly applied on frame systems, such as reinforced concrete and steel constructions (Fajfar, 2000; Priestley et. al., 2005; Fajfar et. al., 2011; Kre-slin and Fajfar, 2012). Thus, further investiga-tion is required to adopt a more systematic strategy for masonry structures, particularly for new buildings once they usually impose complex geometry and results in undesired torsional effects under seismic actions (Magenes, 2010; De Stefano and Mariani, 2014). Such consider-ation is crucial because seismic codes for seis-mic design and analysis guidelines for new and existing buildings cover rules mainly based on regular structures and impose certain penalties on structural irregularities (Magenes, 2010). However, regular configurations are not repre-sentative once the design is mostly focused on architectural and functional concerns. Within this context, more representative analysis procedures for irregular structures, mainly in the plan, have been developed and suggested (Bhatt and Bento, 2014; De Stefano and Mariani, 2014; Azizi et. al., 2019). Nevertheless, there are no avail-able regulations/recommendations on designing and assessing irregular masonry buildings.

In this context, the present work proposes a study focusing on understanding and assessing the global seismic behavior of unreinforced masonry buildings having the irregular structural configuration, in which to torsional effects are important.

2 SEISMIC BEHAVIOR OF MASONRY BUILDINGS AND STRUCTURAL IRREGULARITY

Seismic behavior of masonry buildings highly dependent on the direction of the horizontal loading due to the fact that the resistance capacity of the loadbearing walls in the in-plane direction is consid-erably higher than its out-of-plane direction. Further-more, there are several factors affecting the failure mechanisms, such as the geometry, quality of the masonry materials and boundary conditions. The damage on unreinforced masonry buildings can be observed as different forms, such as cracks in struc-tural walls and spandrels, partial collapse, partial dis-integration of the walls, cracks at the corners and intersections. The structural walls suffer vertical cracks at the corners and in the middle portion of the walls when subjected to seismic motion in the orthogonal direction of the walls. Once the seismic action is parallel to the in-plane of the walls, hori-zontal and diagonal cracks due to bending and shear occur, respectively.

Yet, the design of modern masonry buildings is regulated by box-behavior where the in-plane per-formance of the structural walls controls the resist-ance and, therefore, premature out-of-plane mechanisms are prevented. The in-plane behavior of the structural walls is controlled by geometric prop-erties of piers, openings, and spandrels. Tomaževič (1999) identifies the seismic behavior of the pier elements by three modes of failure, such as flexural, diagonal shear and sliding shear (Figure 1). Due to the cyclic feature of the seismic action, structural walls with openings particularly with different open-ing heights impose uncertainties in the lateral load capacity according to the orientation of the dynamic forces. The effective height of the pier changes with respect to the height of the opening which follows the pier in the same direction of the imposed force.

In what concerns the seismic behavior of span-drels, damage occurs in a similar manner of pier elements except having an axis parallel to the bed joints and being independent of the load direction. The mechanism is observed first in the weak

Figure 1. Examples of diagonal shear damage (Augenti and Parisi, 2010; Parisi and Augenti, 2013).

(a)   (b)

Figure 2. Examples of spandrel damage, (a) diagonal, (b) unit cracking/flexural (Augenti and Parisi, 2010).

spandrel elements as joint sliding, unit cracking, and diagonal tension, and then circulates deformations in the piers (Figure 2).

Furthermore, it is widely recognized that geometrical configuration has an important role in the global behavior of the structures. In fact, past seismic events demonstrated that buildings with structural irregularity suffer more damage than their regular counterparts. Irregularities in the force transfer can result in destructive and concentrated inelastic behavior at the sections where the load is not well distributed. In this regard, most seismic code covers designing rules mainly based on regular structures and impose certain penalties on structural irregularities. However, irregularities in a structural system and geometry are inevitable, particularly in masonry buildings. Although seismic codes encourage to design regular structures, complex geometry is usually formed due to limitations, architectural, economic, and/or functional concerns. It is important to notice that loadbearing masonry elements serve as both structural and architectural components which require a multidisciplinary approach and strict collaboration between engineer and architects during the design process.

A good lateral resistance is achieved by symmetrically designed plans once the center of mass, where the resultant force is imposed, coincides with the center of rigidity. Thus, regular structures are less likely to suffer significant torsional effects. Otherwise, the eccentricity in between results undesired behavior which is mainly controlled by a combination of lateral and torsional response (Giordano et. al., 2008). Furthermore, abrupt changes in the plan, such as the presence of setbacks, discontinuities in the in-plan stiffness of the floors due to openings or variable slab thickness, contribute to the torsional damage (Drysdale and Hamid, 2008). For instance, Figure 3(a) demonstrates that an increase in irregularity in plan decreases the torsional resistance considerably. The main reason is due to the fact that the loadbearing walls are relatively closer to the center of rigidity resulting in less effective resistance against the torsion. Similarly, the location of stairways or service shafts in a building, which cause openings in the floor slabs, increase eccentricity and decrease the stiffness of the

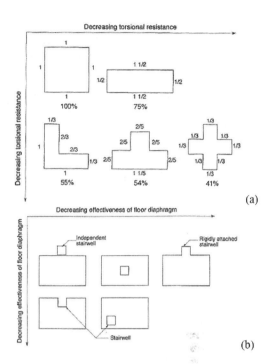

(a)

(b)

Figure 3. Influence of (a) building plan on torsional resistance, (b) stairwell location on the effectiveness of floor diaphragm (Drysdale and Hamid, 2008).

diaphragms. Figure 3(b) shows that the effectiveness of the diaphragm to distribute the lateral loads to the loadbearing walls considerably changes.

As mentioned previously, the openings on the loadbearing walls itself play a crucial role in the lateral load capacity. Regarding the irregular distribution of the openings in elevation, the damage is articulated in a more concentrated way as proven by past seismic events, which in fact increases vulnerability and drift demands (Figure 4). Yet, the response of the structure is not only influenced in terms of lateral loading but also the distribution of the gravity loads uniformly between the masonry panels (Parisi and Augenti, 2013).

3 CASE STUDY

In the present work, a half-scale two-story unreinforced masonry building having a similar geometry properties with the experimental model tested by (Avila et. al., 2018) was investigated aiming at performing a preliminary analysis as a blind prediction in order to have an insight for the experimental campaign of the present project that involves cyclic bidirectional pushover analysis. However, vertical perforated clay masonry brick units have been chosen instead of using concrete blocks (Figure 5). Nevertheless, the geometry has similar features such as plan

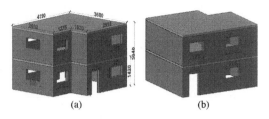

Figure 6. The structural configuration of the experimental model, (a) North-West façade, (b) South-East façade.

Figure 4. Irregularly distributed window openings in a masonry wall subjected to damage after the 2009 L'Aquila earthquake (Parisi and Augenti, 2013).

Figure 5. Vertical perforated clay brick units.

irregularity, a setback in one corner, and asymmetric openings in elevation. The definition of irregular masonry structures is not well defined and general irregularity descriptions available in the codes, which are mainly based on frame structures, are not reliable. To what extent concerns the irregularity of the proposed model, plan irregularity is imposed by a setback in one corner which influences torsional resistance considerably. Considering the asymmetric opening distribution imposes a lack of uniformity, uncertainty in loading paths and variety in bi-directional resistance and stiffness. The dimensions of clay masonry brick units are 24.5 cm x 10.8 cm x 9.8 cm. As per Eurocode 6 (2005), the clay masonry material is classified as Group 3 having a compressive strength of minimum 15 MPa. Class M10 ready-mixed mortar is chosen for bed joints and head joints. The masonry arrangement has been decided as a running bond with the interlocking of the intersecting orthogonal walls. Half-scale two-story masonry building has a plan with dimensions of 419 cm x 368 cm (Figure 6). The inter-story height is 152 cm with a slab thickness of 10 cm. The box-behavior of the building will be ensured by a rigid diaphragm composed of reinforced concrete slabs. In the proposed building model, structural irregularities are achieved both in plan and elevation.

The description of the structural irregularity is limited to mass and stiffness distribution in elevation and plan layout (Eurocode 8, 2004; NTC, 2018; TEC, 2019). Regularities in elevation mainly cover requirements to satisfy axial symmetry, continuity of vertical structural elements from foundation to top, and slight changes in inter-story stiffness. To what concerns the plan irregularity, quantitative provisions implemented in (Eurocode 8, 2004; NTC, 2018; TEC, 2019) are presented in Table 1. A set of indices is calculated in order to assess the severity of the irregularity. It is found that the metrics are slightly than the provisions and, therefore, plan irregularity is confirmed. However, it is significant to point out that irregularities, especially in elevation, appear to be defined based on frame systems. In case of masonry in which architectural components generate the structural elements, i.e. structural walls with openings, so that some of the conditions cannot be satisfied. For instance, the continuity of structural element from foundation to top in case of masonry load bearing wall with window and/or door openings. Therefore, more detailed description of the in-plan and elevation irregularities for masonry buildings are required.

A three-dimensional finite element model was prepared with solid brick elements (CHX60) having a mesh size of 57 mm. The finite element model assumes the masonry as homogeneous continuous material behavior (Lourenço, 2002). The masonry mechanical properties adopted in the numerical models are gathered in Table 2. In the present study, the material properties of the masonry were defined according to Eurocode 6 (2005) and literature review since the material characterization tests have not been started. As per Eurocode 6, the compressive strength of the masonry panel is calculated by Eq. 1 considering M10 class mortar and 25 MPa compressive strength of the brick unit (the brick company says that the compressive strength of the brick is minimum 15 MPa, and based on the preliminary experiments, the capacity is higher than 15 MPa therefore 25 MPa is considered). The K value is selected according to Table 3.3 given in Eurocode 6 for the class of Group 3 clay brick with a general-purpose mortar. Following that, the code suggests using $E=1000*f_k$ relation to calculate the modulus of elasticity of masonry in the absence of the

Table 1. Quantitative plan irregularity metrics.

TEC 2019	• Torsional Irregularity

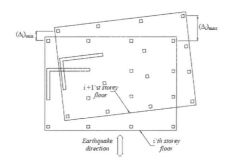

In the case where floors behave as rigid diaphragms in their own planes:

$(\Delta i)_{ort} = 1/2 \, [(\Delta i)_{max} + (\Delta i)_{min}]$

Torsional irregularity factor:

$\eta bi = (\Delta i)_{max}/(\Delta i)_{ort}$

Torsional irregularity: $\eta bi > 1.2$

For +X direction;

$(\Delta 1)_{ort} = \frac{1}{2} \, [1.786+0.959] = 1.372$ mm

$\eta b1 = 1.786/1.372 = 1.3 > \mathbf{1.2}$

$(\Delta 2)_{ort} = \frac{1}{2} \, [3.601+1.994] = 2.797$ mm

$\eta b2 = 3.601/2.797 = 1.3 > \mathbf{1.2}$

• Setback in plan

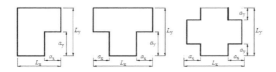

$ax > 0.2 \, Lx$ and $ay > 0.2 \, Ly$

$ax = 1020$ mm, $ay = 1275$ mm

$Lx = 3680$ mm, $Ly = 4190$ mm

1020 mm > (0.2*3680) mm

1275 mm > (0.2*4190) mm

Eurocode 8, NTC 2018	• Slenderness in plan

$\lambda = L_{max}/L_{min} \leq 4$

$L_{max} = 4190$ mm

$L_{min} = 1020$ mm   Therefore, $\lambda = 4190/1020 = \mathbf{4.1 > 4}$

• Setback in plan

$A_{set}/A_t \leq 0.05$

A_t: total area

A_{set}: outline area of setback

$A_t = 17.64$ m^2

$A_{set} = 1.30$ m^2 Therefore, $1.3/17.64 = \mathbf{0.07 > 0.05}$

Table 2. Material properties.

Modulus of Elasticity	E (MPa)	3350
Poisson's ratio	υ	0.25
Shear modulus	G (MPa)	-
Specific weight	γ (kN/m^3)	870
Tensile strength	f_t (MPa)	0.1
Fracture energy	Gf$_t$ (N/mm)	0.012
Compressive strength	f_c (MPa)	6.7
Fracture energy	Gf$_c$ (N/mm)	10.72

Table 3. Modal properties.

Mode	T (sec)	f (Hz)	Mx (%)	My (%)	Mx Sum (%)	My Sum (%)
1	0.030	33.6	0.4	86.4	0.4	86.4
2	0.026	38.9	74.3	1.55	74.7	87.9
3	0.020	50.0	13.9	1.8	88.6	89.7

compressive test. However, half of the calculated modulus of elasticity was considered in order to consider the possible drawbacks of finite element modeling in which very stiff models are constructed due to the modeling approach. Accordingly, the shear modulus (G), which is defined by means of the Poisson ratio in DIANA FEA due to the coupled definition of the E and G parameters, has been taken as 40% of the modulus of elasticity.

$$f_k = K * f_b{}^{0.7} * f_m{}^{0.3} \quad (1)$$

where K=constant according to Table 3.3 (EC6), f_b = compressive strength of the brick and f_m = compressive strength of the mortar.

In the case of the finite element model, the non-linear behavior was defined by the total strain rotating crack model that is available in DIANA FEA (2017), see Figure 7. The constitutive model in tension was

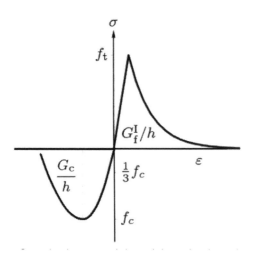

Figure 7. Constitutive material model (DIANA FEA, 2017).

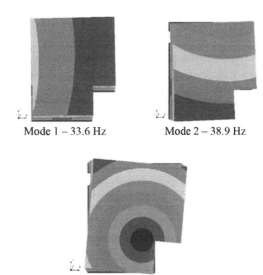

Mode 1 – 33.6 Hz   Mode 2 – 38.9 Hz

Mode 3 – 50.013 Hz

Figure 8. Modes of vibration for the first three global modes.

based on exponential stress-strain relation, while a parabolic relation for both hardening and softening was adopted for compression. Moreover, the value of the tensile strength of masonry was taken as 0.1 MPa based on Tomaževič, (1999). According to Angelillo et. al., (2014) an average ductility index in compression (ratio between fracture energy and strength) equal to 1.6 mm was considered to evaluate the fracture energy in compression whilst, the fracture energy in traction was fixed as 0.012 N/mm.

3.1 Eigenvalue analysis

Eigenvalue analysis was performed in order to obtain the dynamic properties of the representative model. Modal properties and mode shapes are gathered in Table 2 and Figure 8, respectively. It is found that the frequency of the first three global modes ranges from 33.6 Hz to 50.0 Hz. The first mode of vibration has mass participation of 86.4% in the longitudinal (Y) direction having a translational response combined with a torsional effect because of the fact that modal displacements at each corner differ in the slab level. Similarly, the second mode, with 38.9 Hz, shows translational mode in transversal (X) direction having a contribution of 74.3% of the total mass of the building influenced by irregular stiffness distribution through the structural plan resulting in rotational behavior. The third mode shape appears to be highly dominated by torsion which is concentrated where a sudden change in plan configuration is present. Although the torsional mode is the third mode shape, coupling between the translational masses in both X and Y directions is observed for the first two modes.

3.2 Nonlinear static analysis

As aforementioned, a preliminary pushover analysis was carried out in order to have an insight about the seismic behavior of the URM building with torsional effects. It is intended to perform pushover analysis with different loading patterns, such as mass proportional, mode proportional, displacement controlled with uniform and inverse triangle loading, and force-controlled with uniform and inverse triangle loading. Therefore, an extensive comparison is planned to be carried out in order to decide the most reasonable loading protocol based on a representative simulation. In fact, it is stressed that since the material characterization tests are still in progress and the results are not available, the extensive comparison will be performed once the tests are completed. Within this context, the presented paper addresses only the results of the preliminary pushover analyses that were executed by adopting displacement controlled with an inverse triangle loading pattern (1 unit on the second floor, 0.5 unit on the first floor). The finite element analyses were conducted by implementing the secant iterative step-solution method with arc-length control (DIANA FEA, 2017). The energy norm was considered to have a tolerance of 0.001 in order to compute equilibrium at each load step.

The results of pushover analysis are presented in Figure 9 for each direction. It is found that the URM building has a higher capacity in the transversal direction (X) comparing to the longitudinal (Y) one. Such a difference is expected due to the irregularities in plan and opening distributions. In both directions, there is a significant variation in the lateral load capacity of the building among the negative and positive counterparts which can be explained by the different effective height of the piers influenced by the loading direction. For instance, the peak load reaches to 139% of the total structure weight in -X direction while it is 128% in the +X direction. In the case of the longitudinal (Y) direction, the higher capacity is

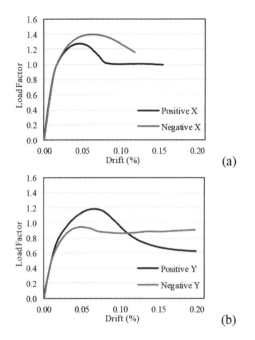

Figure 9. Pushover curves obtained for each direction, (a) transversal (X), (b) longitudinal (Y).

achieved in the +Y direction with a load factor of 1.18. Yet, the representative model has a lateral resistance capacity of 95% of the total weight. Accordingly, it is also noticed that there is a difference in the energy dissipation capacity of the model in different directions.

In the present paper, the focus is given to the responses at the steps where the peak load factor and roughly 0.1% drift is achieved as summarized in Table 4. Accordingly, the post-peak regime is already attained when the drift reaches 0.1% (which the displacement is approximately 3 mm). Although present analysis is a blind prediction, it is possible to say that the displacement values are reasonable since they seem to be in the range of displacements obtained by Avila et. al., (2018) in which similar building configuration was tested on a shaking table. It is stressed

that the final loading step of the presented pushover curves does not present the total collapse. The aim is given to the capacity and evolution of the damage in between the response at peak load and 0.1%.

In this regard, the strain distributions (E1) at the aforementioned load steps are illustrated in Figure 10 and Figure 11. For all directions, global response is

North – West façade   South – East façade

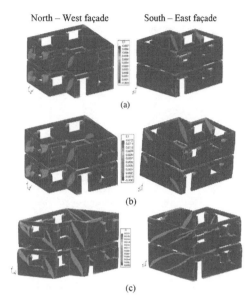

Figure 10. Damage patterns at (a) peak load in -X direction, (b) 0.1% drift in -X direction, (c) 0.1% drift in +X direction.

North – West façade   South – East façade

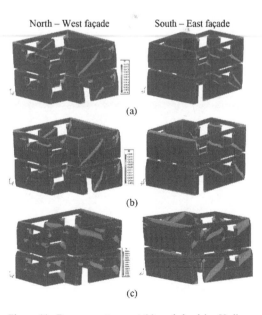

Figure 11. Damage patterns at (a) peak load in -Y direction, (b) 0.1% drift in -Y direction, (c) 0.1% drift in +Y direction.

Table 4. Summary of the results in terms of peak load factor and drift.

		X Direction		Y Direction	
		Pos.	Neg.	Pos.	Neg.
Peak Load	Load Factor	1.28	1.39	1.18	0.95
	Displacement (mm)	1.36	1.95	1.84	1.48
	Drift (%)	0.05	0.07	0.06	0.05
Drift ~0.1%	Load Factor	1.00	1.24	0.89	0.86
	Displacement (mm)	3.14	3.13	3.18	3.11

governed by a diagonal shear mechanism, as expected. Nevertheless, it is noticed that several moderate cracks are observed on the out-of-plane walls (North façade). The damage appears to be intense once the deformation increases (0.1% drift). This aspect can be explained by the influence of the structural irregularity which exposes torsional rotation. For instance, the influence of torsion is observed in transversal (X) direction. Owing to its plan configuration, in fact, the impact of rotation appears to be significant (Figure 12). Especially in the positive direction in which the capacity is relatively lower moderate cracks are observed on the out-of-plane walls (East façade).

Similarly, the deformed shape when 0.1% drift occurs in +Y direction as shown in Figure 13(a), strain concentrations on the out-of-plane walls are relatively less intense than the -Y direction, as seen in Figure 11(b) and (c). Combined behavior of in-

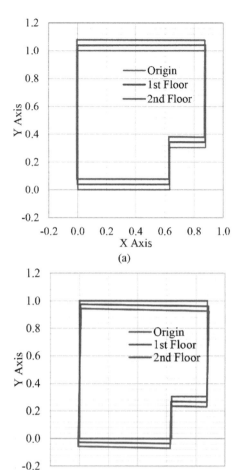

(a)

(b)

Figure 13. Normalized deformed shape in plan view, normalized deformation with a scale factor of 100, loading in (a) +Y direction, (b) -Y direction.

plane (shear diagonal) and out-of-plane (flexural) can be concluded based on the strain distributions and normalized plan deformations (Figure 13). The strain concentration observed on the second floor, where the slab is connected to the slab on the South façade, might be due to the modeling assumption of the slab which is assumed as fully fixed.

4 CONCLUSION

The present work addresses the influence of structural irregularity in URM buildings in both plan and elevation. Even though the design of the URM building is covered by the design codes, further investigation is required in the presence of irregularities. Within the scope of the project, an extensive numerical and experimental pushover analysis is intended to be

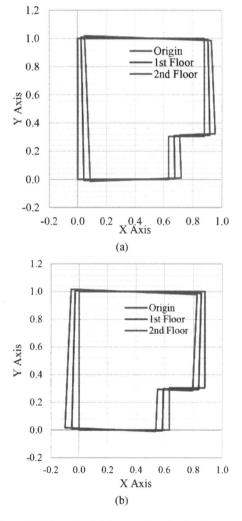

(a)

(b)

Figure 12. Normalized deformed shape in plan view, normalized deformation with a scale factor of 100, loading in (a) +X direction, (b) -X direction.

performed. A blind prediction numerical analysis is conducted by considering code specifications for material properties. The design of the experimental campaign covers an extensive numerical analysis with different loading protocols. In the present paper, an inversely triangular loading pattern is applied as displacement-controlled in order to simulate the experimental test. It appears to be feasible for experimental testing and the influence of irregularities seems to be a considerable impact on the deformation.

ACKNOWLEDGMENTS

The first author acknowledges the financial support from the Portuguese Foundation for Science and Technology (FCT) through the Ph.D. Grant SFRH/BD/143949/2019. This work was financed by national funds through FCT - National Foundation for Science and Technology, I.P., in the scope of the research project "Experimental and Numerical Pushover Analysis of Masonry Buildings (PUMA)" (PTDC/ECI-EGC/29010/2017).

REFERENCES

Angelillo, M, Lourenço, P. and Milani, G. (2014) 'Masonry behavior and modelling', in Angelillo, Maurizio (ed.) *Mechanics of Masonry Structures*. Udine: CISM International Centre for Mechanical Sciences, Springer, pp. 1–24.

Augenti, N. and Parisi, F. (2010) 'Learning from construction failures due to the 2009 L'Aquila, Italy, earthquake', *Journal of Performance of Constructed Facilities*, 24(6), pp. 536–555. doi: 10.1061/(ASCE)CF.1943-5509.0000122.

Avila, L., Vasconcelos, G. and Lourenço, P. B. (2018) 'Experimental seismic performance assessment of asymmetric masonry buildings', *Engineering Structures*. Elsevier, 155(December 2016), pp. 298–314. doi: 10.1016/j.engstruct.2017.10.059.

Azizi-Bondarabadi, H., Mendes, N. and Lourenço, P. B. (2019) 'Higher Mode Effects in Pushover Analysis of Irregular Masonry Buildings', *Journal of Earthquake Engineering*. Taylor & Francis, 0(00), pp. 1–35. doi: 10.1080/13632469.2019.1579770.

Bhatt, C. and Bento, R. (2014) 'The extended adaptive capacity spectrum method for the seismic assessment of plan-asymmetric buildings', *Earthquake Spectra*, 30(2), pp. 683–703. doi: 10.1193/022112EQS048M.

DIANA FEA (2017) 'User's Manual Release 10.2'. The Netherlands.

Drysdale, R. G. and Hamid, A. A. (2008) *Masonry Structures Behavior and Design*. 3rd edn. The Masonry Society.

Eurocode 6 (2005) *Design of masonry structures - Part 1-1: General rules for reinforced and unreinforced masonry structures*.

Eurocode 8 (2004) *EN 1998-1: Design of Structures for Earthquake Resistance - Part 1: General rules, seismic actions and rules for buildings*, European Committee for Standardization. Brussels.

Fajfar, P. (2000) 'A Nonlinear Analysis Method for Performance-Based Seismic Design', *Earthquake Spectra*, 16(3), pp. 573–592. doi: 10.1193/1.1586128.

Fajfar, P., Marusic, D. and Perus, I. (2011) 'Torsional Effects in the Pushover-Based Seismic Analysis of Buildings', *Journal of Earthquake Engineering*, 9(6), pp. 831–854. doi: 10.1080/13632460509350568.

Giordano, A., Guadagnuolo, M. and Faella, G. (2008) 'Pushover Analysis of Plan Irregular Masonry Buildings', *Proceedings of 14th World Conference on Earthquake Engineering*. Available at: ftp://128.46.154.21/spujol/Andres/files/14-0257.PDF.

Kreslin, M. and Fajfar, P. (2012) 'The extended N2 method considering higher mode effects in both plan and elevation', *Bulletin of Earthquake Engineering*, 10(2), pp. 695–715. doi: 10.1007/s10518-011-9319-6.

Lourenço, P. B. (2002) 'Computations on historic masonry structures', *Progress in Structural Engineering and Materials*, 4(3), pp. 301–319. doi: 10.1002/pse.120.

Lourenço, P. B. (2009) 'Recent advances in Masonry modelling: micromodelling and homogenisation', in Galvanetto, U. and Aliabadi, M. H. F. (eds)

Brick and Block Masonry - From Historical to Sustainable Masonry –
Kubica, Kwiecień & Bednarz (eds)
© 2020 Taylor & Francis Group, London, ISBN 978-0-367-56586-2

Effect of mortar-brick cohesive interface on seismic response of masonry walls

B. Jafarzad Eslami
Department of Civil Engineering, University of Ghiasedin Kashani, Abyek, Iran

H. Darban
Department of Engineering, University of Naples Parthenope, Naples, Italy

A. Del Grosso
Department of Civil, Environmental & Architectural Engineering, University of Genova, Italy

ABSTRACT: Masonry, as a heterogeneous structure made by joining bricks with mortar each of them having their own mechanical properties, is one of the oldest and most widespread structural materials with many ap-plications in various constructions such bridges, traditional buildings and historical monument. The overall behavior and load bearing capacity of such structures are highly affected by the mechanical and geometrical properties of bricks and mortar and their cohesive interface.

In this work, cohesive elements are used in finite element model to simulate the damage occurs at the mortar-brick interface in a representative masonry wall element under seismic loadings, which are derived from some real case earthquakes occurred in Iran. Different cohesive laws are assumed and their effects on the failure load and collapse behavior of the masonry wall is investigated. Comparative results are presented together with discussions on different failure scenarios.

1 INTRODUCTION

The developments in numerical methods and advent of computational power, have paved the way to study masonry behavior without expensive and time-consuming series of laboratory experiments. Yet the effect of the presence of mortar joints and propagation of cracking in the masonry is a challenging problem (Thamboo & Dhanasekar 2016).

The geomorphology of Iranian territory is responsible for the occurrence of several earthquakes which induce medium to high seismic hazard levels (such as Manjil, Bam and Kermanshah). In addition, most of the historical buildings lying in this region are made of masonry, and are characterized by significant seismic vulnerability (Eslami & Del Grosso 2019).

Typically, the surface-base cohesive model has been widely used to analyze and describe fracture behavior at the interface level. This constitutive model is implemented in conjunction with the finite element method by introducing zero thickness interface elements along potential crack surfaces (bed and head joints). In this paper the surface-base cohesive model is implemented in a standard finite element framework of ABAQUS 6.13 to simulate the fracture behavior (the crack initiation and its propagation) occurs at the mortar-brick interface in a representative unreinforced masonry wall (URW) under seismic loadings, which are derived from a real case earthquake occurred in Bam city in Iran.

2 STRATEGEY OF MODELLING

Masonry is modelled numerically either as a homogenized material with no distinction between unit and mortar (macro model) or as a composite of constituent materials interacting at their interfaces (micro model), depending on the type of the problem and the level of details desired in a detailed manner (Thamboo & Dhanasekar 2016).

In this work, the simplified micro-modelling method is used by expanding masonry units (brick, block, etc.) with adding the mortar thickness as continuum elements and the partial interaction between units is modelled using cohesive elements available within the software.

Firstly, brick elements are created and assembled through cohesive interfaces, whose constitutive behavior is defined by a cohesive law that relates the tractions to the relative displacements between masonry units. To model damage in bricks, the principles of concrete damage plasticity is implemented. The damage models will be detailed in the next sections.

The constitutive behavior of the interface is defined by surfaced-based cohesive model and cohesive traction-separation relationships, which can be classified as either bilinear, parabolic, sinusoidal and exponential (Salve & Jalwadi 2015).

The traction-separation law used in this work is shown in Figure 1. The law has an initial branch to define the linear elastic behavior of mortar interface until the onset of damage initiation. It is assumed that the cohesive traction increases linearly with separation till it reaches a finite cohesive strength in the linear elastic regime, and then it decreases monotonically till the separation reaches an ultimate value, where the complete damage occurs and the crack propagates. (see Figure 1).

3.1 Elastic behavior of the joint interfaces

In this work a linear uncoupled traction-separation law is assumed for the initial elastic response of interfaces. This law has been extensively used in the literature and is given by Eq. (1):

$$t = \begin{Bmatrix} t_n \\ t_s \\ t_t \end{Bmatrix} = \begin{bmatrix} k_{nn} & 0 & 0 \\ 0 & k_{ss} & 0 \\ 0 & 0 & k_{tt} \end{bmatrix} \begin{Bmatrix} \delta_n \\ \delta_s \\ \delta_t \end{Bmatrix} = K\,\delta \quad (1)$$

The initial stiffness of the interface depends on the elastic properties and thickness of mortar through Eq. (2):

$$K_{nn} = \frac{E_m}{h}, \quad k_{ss}, k_{tt} = \frac{G_m}{h} \quad (2)$$

where $E_m = 2000$ Mpa, $G_m = 870$ Mpa are elastic and shear modules of mortar and $h_m = 10$ mm is thickness of mortar, which result in K_{nn}, K_{ss} and K_{tt} to be equal to, respectively, 200, 87, and 87 N/mm^3.

3.2 Damage and softening of interfaces

3.2.1 Damage initiation

A coupled damage initiation criterion is used as given in Eq. (3). The damage is assumed to initiate, when the quadric of maximum interface stress ratios reaches the value of unity:

$$\left(\frac{t_n}{t_n^{max}}\right)^2 + \left(\frac{t_s}{t_s^{max}}\right)^2 + \left(\frac{t_t}{t_t^{max}}\right)^2 = 1 \quad (3)$$

where, t_n^{max} and t_t^{max}, t_s^{max} are the limiting tensile bond (in normal direction) and shear bond strength (in two tangential directions) of thin-layer mortar. With the combined shear-compression stresses, the mortar exhibit Mohr-Coulomb failure behavior (the shear strength mortar bed joint is linearly depending on normal compression) as shown in Eq. (4):

$$\tau \; (or \; t_s^{max} = t_t^{max}) \; = \; \tau_0 + \mu t_n \quad (4)$$

The post-failure domain of the shear sliding behavior is illustrated in Figure 2, in which, τ_0 is bond cohesion (or initial shear strength in the absence of compression) and μ is the coefficient of friction of the mortar bed joints. The parameter μ ranges from 0.7 to 1.2, for different unit mortar combinations, and value of 0.7 is recommended in the absence of information (Van der Pluijm 1997).

The interface properties of mortar used in the FE analysis results reported in Table 1. The maximum normal and shear stresses for the interface elements are chosen as reported in (Graziotti & Rossi 2016, Van der Pluijm 1997).

3.2.2 Damage evolution (softening)

Once the damage initiation criterion is reached, the propagation of cracks in the masonry joints causes stiffness degradation at a defined rate which leads to

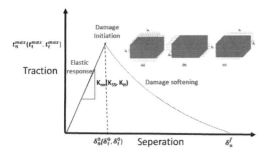

Figure 1. Typical Exponential Traction-Separation curve.

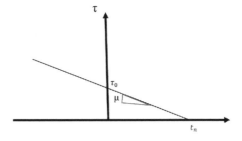

Figure 2. Interface damage activation domains in tangential behavior.

Table 1. Interface properties used for FE model.

Interfacial properties	value
Normal stiffness, k_{nn} (N/mm^3)	200
Shear stiffness, $k_{ss,}$ k_{tt} (N/mm^3)	87
Maximum normal stress, t_n^{max} (MPa)	0.061
Maximum shear stress, t_t^{max} and t_s^{max} (MPa)	0.093
Friction coefficient (μ)	0.7

total strength loss and failure of joints. Post damage initiation response is defined by scalar damage equation as Eq. (5):

$$t = (1 - D)\, K\delta \qquad (5)$$

where D (scalar damage variable) is used to redefine the interfacial traction stress components. D ranges from 0 (for undamaged, there is no degradation of elastic properties) to 1 (for fully damaged, there is no interface stiffness). This parameter is calculated based on an exponential softening (Stuparu & Constantinescu 2012) as Eq. (6):

$$D = 1 - \left\{\frac{\delta_m^0}{\delta_m^{max}}\right\}\left\{1 - \frac{1 - exp\left(-\alpha\left(\frac{\delta_m^{max}-\delta_m^0}{\delta_m'-\delta_m^0}\right)\right)}{1 - exp(-\alpha)}\right\}$$

$$(6)$$

where α is a dimensionless material parameter which defines the rate of damage evolution. This parameter is zero for the linear softening; as softening becomes exponential and α increases, damage propagates more rapidly. Therefore, a linear softening is less conservative, and damage propagates in a longer time. In this study for modeling post damage initiation response of cohesive interfaces an exponential softening function with exponential parameter equal to 10 and a total separation displacement of 1 mm are used (Eslami & Del Grosso 2019).

4 ELEMENT MODELLING PROCEDURE

4.1 Geometry and mesh type

The dimensions of the unreinforced masonry walls are 800 × 900 × 85 mm. The wall is built with 13 rows of wire cut solid clay bricks, with dimensions of 185 × 85 × 60 mm.

In the simplified micro model, the bricks are expanded by adding 10 mm due to the mortar thickness and two concrete beams are created at the top and bottom of the wall for the uniform application of surcharge and lateral loads.

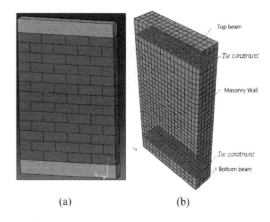

(a)　　　　　(b)

Figure 3. Finite Element Model: (a) geometric model; (b) connection and mesh model.

The load is applied to the upper beam while the bottom beam is fixed to the floor (see Figure 3-a). In the model, the contacts between adjacent masonry bricks are defined through the above-mentioned interface model and for connection between masonry wall and concrete beams, a tie constraint is used.

The masonry then is modelled using 3D hexahedral shaped eight node linear brick elements with reduced integration (type C3D8R).

4.2 Material model

The Quasi-brittle materials, such as brick and mortar undergo several damage states, such as tensile cracking, compressive crushing failure associated with stiffness degradation and post-peak softening.

In order to define the fracture of masonry materials, Concrete Damage plasticity has been chosen as an appropriate failure model. This failure model, which is recommended for brittle materials with appropriate adaptability to laboratory results, includes two cracking and crushing mechanisms, in tension and compression, respectively. We set the values of all the parameters except the dilatancy angle (which depends on the level of the confining stress, so that for high confining pressures, ψ decreases to zero) in the default state (Eslami 2016), these parameters are shown in Table 2.

Table 2. Parameters used in CDP model.

Dilation Angle	Eccentricity	fb0/fc0	K	Viscosity
10	0.1	1.16	0.67	0.001

Table 3. Tensile behavior of masonry.

N	Yield. Stress (Pa)	Cracking strain (εpl)
1	200000	0
2	100000	0.001
3	50000	0.003

Table 4. Compressive behavior of masonry.

N	Comp.stress (Pa)	Plastic strain (ε_{pl})
1	1224666.667	0
2	1304875	0.0001
3	1381722.222	0.0002
4	1455208.333	0.0003
5	1525333.333	0.0004
6	1592097.222	0.0005
7	1655500	0.0006
8	1715541.667	0.0007
9	1772222.222	0.0008
10	1825541.667	0.0009
11	1875500	0.001

The compressive and tensile behavior of the materials are defined according to the strain-stress curve (model of Kent & Park) obtained from the laboratory sample of masonry prism under tension and compression (Eslami 2016).

4.3 Loading

The masonry walls are subjected to an imposed vertical compressive stress (applied as uniformly distributed equal to 0.39 MPa) via the top beam and gravity load are applied to all parts of the model. After applying the compressive load, the vertical movement of the top beam was restrained, and then finally the lateral loads (earthquake load) are defined with "Time history method". In this method, records that relate the longitudinal component of the Bam earthquake with the pick ground of the Earth's acceleration PGA = 796 cm/s^2 have been used. It should be noted that all records are extracted from the data base of Iran's seismograph Site, affiliated to the Housing Research Center [6], and after applying the necessary corrections (Signal Processing) by Sesimosignal software. In this step, the set of displacement signal according to the time history is defined as "Load Amplitude" as a software input (Eslami, & Del Grosso 2019).

5 DESTRUCTION PROCESS

The process of destruction of the wall under time history analysis was predicted when the wall was

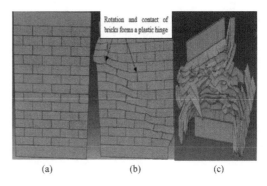

Figure 4. Destruction process of URW: (a) Elastic phase; (b) propagation of crack: (c) collapse.

completely destroyed within less than five seconds. To better understand this process, three phases have been studied. In the elastic phase only invisible cracks formed. Then diagonal cracks were created overtime appeared in the central part of the wall due to opening of the vertical joint and plastic hinges are formed as bricks slide and rotate against each other. ($0 \leq T \leq 1.9$ s). Initiation and propagation of diagonal crack with crushing of the wall happens for $1.9 \leq T \leq 4.29$ s and finally, the masonry wall loses its integrity and tends to break down under the effect of the dynamic loads, which alternating compressive stresses to tensile stresses. This leads the structure to a catastrophic condition and its final collapse (Figure 4).

5.1 Dynamic response of masonry in earthquake load

In the FE model, the masonry wall undergoes a combination of vertical compressive stresses and seismic loads. The failure modes of the walls were due to a combination of diagonal cracks and cracking of units themselves, followed by crushing under compression. Increasing shear base stress makes many changes in the dynamic characteristics; when the cracks propagate much larger displacements happen within the structure compared to those related to the initial cracking before approaching instability. The shear base diagram during a strong motion of Bam earthquake (when a significant earth-moving effect occurred) is illustrated in Figure 5.

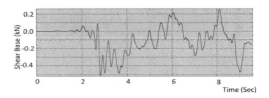

Figure 5. Shear base diagram of unreinforced masonry wall.

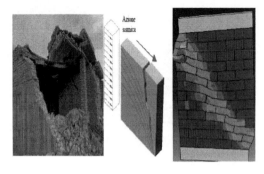

Figure 6. Actual and numerical sample of the demolitions in the Bam earthquake.

5.2 *Validation of the numerical model*

Damage in the numerical model is similar to the real break from the Bam Earthquake. In fact, this process has been occurred during the earthquake for most unreinforced masonry building in the city of Bam. The common modes of failure of these buildings were shear failure of walls, separation of walls from the roof, and separation of roof beams from each other, shown in Figure 6 (Eslami 2016).

6 CONCLUSIONS

Masonry is a heterogeneous, anisotropic composite material whose mechanical response characterized by high non-linearity. Mortar joints act as planes of weakness in masonry and their mechanical proper-ties influence the global behavior of the structure. In this paper, the cohesive interface and finite element models are coupled to study the influence of the brick–mortar interface properties on the mechanical response of masonry walls subjected to seismic loading. This has been done by implementing surfaced-based cohesive model for simulation pre- and post-cracking (fracture) behavior of masonry wall in ABAQUS software. Collapse patterns obtained numerically have good agreement with actual collapse patterns happened in Bam earthquake in Iran.

The processes of destruction in numerical models happened very rapidly, as it was the case also in the real earthquake. Although, the Bam earthquake happened during a noticeably short time, but it is one of the most destructive earthquakes in Iran with many casualties. This is because of the high displacement amplitudes generated in a shallow depth, together with the poor design and construction of the civilian buildings.

REFERENCES

Eslami, B.J & Del Grosso, A. 2019 Retrofit of Masonry Buildings through Seismic Dampers. *Key Engineering Materials* 817: 293–300.

Eslami, B.J. 2016. Impiego di smorzatori sismici nella protezione dei beni architettonici. *PhD thesis*. University of pavia.

Graziotti, A. & Rossi, F. 2016. Experimental characterization of calcium-silicate brick masonry for seismic assessment. *16th international Brick and block masonry conference*, Padua, Italy.

Salve, A.K. & Jalwadi, S.N. 2015. Implementation of Cohesive Zone in ABAQUS to Investigate Fracture Problems. *National Conference for Engineering Post Graduates RIT N ConPG-15*.

Stuparu, A. & Constantinescu, D.M. 2012. Interface damage characterization through cohesive parameters. *Journal of Engineering Studies and Research* 18: 129–139.

Thamboo, J.A & Dhanasekar, M. 2016 Nonlinear finite element modelling of high bond thin layer mortared concrete masonry. *Int. J. Masonry Research and Innovation* 1: 15–26.

Van der Pluijm, R. 1997. Non-linear behaviour of masonry under tension. *Heron* 42(1): 25–54.

Sustainability & innovation of masonry

Brick and Block Masonry - From Historical to Sustainable Masonry –
Kubica, Kwiecień & Bednarz (eds)
© 2020 Taylor & Francis Group, London, ISBN 978-0-367-56586-2

The assessment of confined semi-interlocking masonry buildings using macro-modelling approach

M. Hemmat, Y.Z. Totoev & M.J. Masia
The University of Newcastle, Australia

ABSTRACT: Confined Semi-Interlocking Masonry (CSIM) is an innovative system which is constituted of the semi-interlocking masonry panel together with confining reinforced concrete elements. The semi-interlocking masonry (SIM) has already been studied as infill panels for framed structures and its suitable seismic performance was proven. It is claimed that CSIM building is an alternative to currently used confined masonry buildings. In a CSIM building, not only CSIM walls resist both lateral and gravity loads, but they are also supposed to be energy dissipation devices to the building. In this study, A CSIM building was assessed using a new macro-model suitable to simulate the behaviour of CSIM. In this way, the resettable semi-active damper model was employed to simulate the behaviour of SIM panels in CSIM walls. The responses of a CSIM building to a cyclic displacement excitation were assessed in terms of strength, displacement, and energy dissipation. The hysteresis behaviour of the CSIM building could be appropriately presented by the macro-model. Besides, it was indicated that the CSIM building had considerable energy dissipation thanks to the presence of in-plane sliding joints in the body of SIM panels that led to higher ductility and displacement capacity for the whole building.

1 INTRODUCTION

Confined masonry is an earthquake-resistance system for buildings in many developing countries. A confined masonry building consists of masonry panel enclosed by vertical and horizontal reinforced concrete elements (tie-columns and tie-beams) in both directions of the building. This is the common method for the construction of buildings in many Latin American, Asian, and some European countries (Meli et al., 2011). Figure 1 shows a confined masonry building under construction in Iran. The simplicity and economical construction method of confined masonry buildings increased its popularity for the housing in remote regions with concerning seismic hazards.

Confined masonry has been a retrofitting method for the seismic performance of unreinforced masonry buildings (ElGawady et al., 2004). However, strong earthquakes caused several damages to confined masonry buildings mainly because of the weakness of mortar and the low bond strength of the mortar-brick interface (Yekrangnia et al., 2017). Therefore, developments were required to improve the behaviour of confined masonry buildings against severe earthquakes. One method to improve the behaviour of confined masonry buildings was to replace the traditional masonry panel with a mortar-less masonry panel.

The semi-interlocking masonry (SIM) has been under development at the Centre for Infrastructure Performance and Reliability at the University of Newcastle, Australia. Concrete SIM bricks with special geometry (topological SIM) were able to form a mortar-less masonry panel in which bricks slide on joints in the plane of the wall. Besides, the SIM bricks were designed so that they were interlocked against the out-of-plane relative displacement (Totoev, 2015). Figure 2 shows the topological SIM bricks.

The energy dissipation of the SIM infill panels was found satisfactory in the previous experimental tests (Totoev&Lin, 2012). Literally, the SIM infill panels decreased the stiffness of the system by adding natural dampers to the system. Indeed, the frictional forces on sliding joints in the SIM panel mitigated a significant amount of the energy induced in the masonry wall by the earthquake. In addition, the SIM infill panel was added to a steel frame and noticeable ductility as well as ultimate displacement achieved under the in-plane cyclic displacements (Hossain et al., 2017).

Based on the performance of SIM as an infill panel, it was a motivating reason to employ SIM panel in a confined masonry system. By doing this, a new masonry system called confined semi-interlocking masonry was developed (Hemmat et al., 2018). It is expected that the confined semi-interlocking masonry (CSIM) might discard the weaknesses of confined masonry buildings in structural performance. However, extensive experimental studies were required to justify the capability of the CSIM system as a reliable structural system for

Figure 1. A confined masonry building.

Figure 2. The topological SIM bricks.

masonry buildings. In this paper, the performance of a CSIM building was assessed using numerical modelling.

2 CONFINED SEMI-INTERLOCKING MASONRY BUILDING

A confined semi-interlocking masonry building is provided with a number of CSIM walls in principal directions of the building. CSIM walls are the major structural elements that resist gravity and lateral loads. In this method, SIM panels are assembled first and later thin reinforced concrete frames are attached to the SIM panels. Figure 3 describes how a SIM panel is constructed within a CSIM wall. Another advantageous of CSIM buildings is that the architectural changes such as openings can be performed easily with the minimum costs thanks to the mortar-less masonry panel.

The integrity of the traditional confined masonry walls was guaranteed by applying shear toothing between the masonry panel and tie-column (Meli et al., 2011). Previous numerical simulations of CSIM walls proved that the existence of the shear toothings between the SIM panel and tie-columns limits the freedom of sliding of SIM bricks (Hemmat et al., 2019). Hence, it is recommended to construct CSIM walls without shear toothings. Figure 4 shows a schematic design of a typical CSIM building.

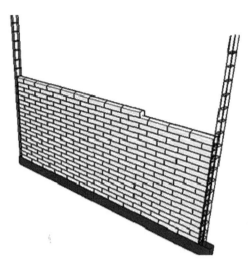

Figure 3. The location of the SIM panel within a CSIM wall.

Figure 4. A typical CSIM building.

3 NUMERICAL MACRO-MODELLING STRATEGY FOR THE CSIM BUILDING

Micro and macro-modelling approaches were the main strategies to predict the performance of masonry. In this regard, macro-models were particularly suitable to capture the behaviour of masonry buildings (Lourenço, 2013). In this paper, both micro and macro–modelling strategies were employed. Although the capacity of CSIM walls was assessed through micro-modelling method (Hemmat et al., 2019), the required simplicity for the simulation of masonry in the scale of a building pointed out the necessity of developing a suitable macro-model.

The numerical simulation of the CSIM building was performed using a new macro-modelling approach available in the RUAUMOKO computer program (Carr, 2006). In the macro-model, the resettable semi-active damper model represented the behaviour of SIM panels. The resettable semi-active device was studied to relieve the responses of framed structures to dynamic excitations (Chey, 2007,

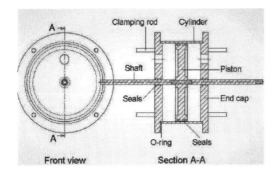

Front view ……… Section A-A

Figure 5. Schematic of the resettable device(Franco-Anaya et al., 2017).

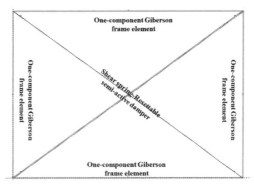

Figure 7. The macro-model to simulate a given CSIM wall.

Franco-Anaya et al., 2017). The resettable device that is shown in Figure 5 was comprised of a cylinder, a piston, and enclosed air. This set dissipated the energy from dynamic loads entering structures (Franco-Anaya et al., 2014). On the other hand, the SIM infill panel dissipated the energy in framed structures through the frictional forces between the sliding joints (Totoev&Al Harthy, 2016). The shear sliding behaviour of the SIM panel was similar to the non-linear model of the resettable semi-active damper.

A single-story box CSIM building without opening on walls was considered for this study. The dimensions of this building are shown in Figure 6. Each CSIM walls of the building were simulated using the macro-model with eight nodes and six elements. SIM panel was represented by two diagonal shear spring elements with the allocated semi-active damper model. The surrounding confining reinforced concrete elements were modelled using the one-component Giberson frame element with the bi-linear hysteresis behaviour (Giberson, 1967). Figure 7 shows how a CSIM wall was modelled in the RUAUMOKO computer program. Finally, the CSIM building was simulated using 24 nodes and 20 elements including eight diagonal shear spring elements and 12 frame elements for the tie-columns and tie-beams.

The in-plane shear capacity together with the ultimate displacement of walls of the building was first

obtained through a detailed micro-modelling approach using DIANA finite element software (DIANA, 2017). It was used to give an approximate range for the parameters of the macro-model for the entire CSIM building. The force capacity of the tie-columns and the tie-beams were reached directly from the pushover analyses on CSIM walls through the micro-model of walls in DIANA software. The force capacity in the confining elements plays a major role in the overall performance of the CSIM building. The failure occurred when the force in confining elements reached their capacity.

The area of the piston was another important parameter in the macro-model. This parameter implied the amount of dissipated energy by the SIM panel. In addition, the displacement of walls depended on the free movement of the piston inside the cylinder. A try and error process was performed for the parameters to match well with the load-displacement curve obtained from the pushover analyses on CSIM walls using the micro-model. Moreover, the friction between the piston and the cylinder pointed to the friction force in the SIM panel due to the self-weight of the panel. The CSIM building in this study had only four walls with similar dimensions. This made the process for simulation convenient. Table 1 summarizes some of the parameters for macro-model simulation of the SIM panels. Finally, Figure 8 shows the CSIM building modelled in the RUAUMOKO.

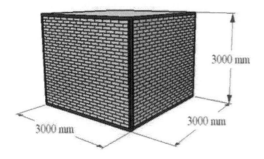

Figure 6. Dimensions of the box CSIM building under study.

Table 1. The summary of parameters for the shear spring element used in the macro-model.

F_s^*	K_s^{**}	A_p^{***}	D^{+-****}	F_r^{*****}
kN	kN/mm	mm2	mm	kN
40	73,000	40,000	50	4

* The spring force capacity
** The spring stiffness
*** The piston area
**** The free movement of the piston from centre
***** The friction force

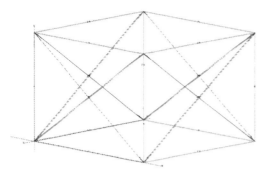

Figure 8. The macro-model of the CSIM building using RUAUMOKO.

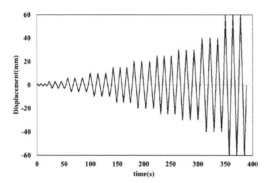

Figure 9. The input cyclic displacement time-history.

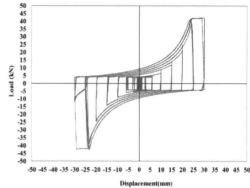

Figure 10. The hysteresis load-displacement curve for the CSIM building.

It was not necessary to simulate the roof of the building to guarantee the connection between walls. Alternatively, the weight of the roof was divided between walls as the uniform gravity load. In addition, the rigidity of the roof was provided in the simulation by slaving the nodes of walls to the top corner node of the building that is directly subjected to the lateral excitations. In this way, the master node acted as the roof diaphragm representative and as a result, the displacement of the building was depended on the displacement of the master node.

The model of the CSIM building was subjected to a cyclic displacement to assess its hysteresis behaviour. The cyclic displacements were applied to the corner node of the roof that was selected as the master node for the building. Figure 9 shows the cyclic displacement-time-history applied to the building. These cyclic displacements pushed and pulled the building in the X-direction. Each cycle was replicated three times to induce the hysteresis load-displacement to the building.

4 RESULTS AND DISCUSSION

The analysis was performed on the CSIM building macro-model. The history of nodal displacements in the X-direction for one of the corner nodes of the

building together with the variations of the axial forces in one of the diagonal shear spring elements for both walls parallel to the direction of loading were recorded. In Figure 10, the resulted hysteresis load-displacement curve is illustrated. The vertical axis shows the average of the axial load induced in the diagonal shear springs in walls parallel to the loading direction. The horizontal axis shows the nodal displacement of the top corner node in the X-direction. The CSIM walls parallel to the direction of the input excitations mainly provided the strength of the CSIM building.

From the resulted cyclic load-displacement curve, it is evident that the building initially performed in the linear stage that showed the small movements of the piston inside the cylinder. The linear performance at this stage implied the small sliding between SIM bricks. Later, increasing the cyclic displacements induced higher displacements of the piston as well as greater axial force in the cylinder. The rate of the axial force variations in the diagonal shear spring (semi-active device) increased gradually with the input displacements. This part represented the greater sliding of SIM bricks and additional friction force in the SIM panel due to the effect of compression from the confining elements. When the direction of input displacements changed, the axial force in diagonal spring dropped sharply to a non-zero constant amount of load. This load is equal to the friction force in the SIM panel due to its self-weight. Finally, when the values of input displacements were higher, the ultimate movement of the piston inside the cylinder achieved and a constant load was obtained when the values of displacements increased. It means that after this stage, the model of the CSIM building failed. It was an indication of the failure of CSIM walls of the building where the confining elements reached its capacity.

No significant contrast was observed in the stiffness degradation between the three cycles at the input displacements. This is because of the shear sliding of the SIM panel at these displacements. When crushing of the concrete and yielding of the reinforcing bars occurs in the critical zones of the tie-columns, the hinges in the frame element model for these elements reached its capacity and degradation in the strength of the building was obtained.

Overall, the numerical macro-model predicted that the CSIM building under the input cyclic displacement showed in Figure 9 was able to reach the strength of almost 41.96 kN. In addition, the ultimate displacement capacity for this building was 30 mm.

5 ENERGY DISSIPATION

The ability of a structural system to dissipate energy is an important factor in the seismic performance of the whole system. According to the experimental studies, the relative slip, and sliding between SIM bricks were the main reason for the good energy dissipation in SIM infill panels compared with the traditional masonry panels (Totoev&Lin, 2012, Lin et al., 2016).

To estimate the energy dissipation, the area inside the hysteresis load-displacement curves was calculated per different values of the displacements. The energy dissipation was calculated for the average of three cycles per displacement. Details of such calculations are summarized in Table 2. Figure 11 shows the rate of variation in the energy dissipation with drift in the CSIM building.

Table 2. The Energy dissipation of the CSIM building.

Displacement	Force	Energy dissipation
mm	kN	Kn.mm
2.5	3.74	46.01
5	4.2	97.29
10	9.2	231.76
15	14.82	423.60
20	25	721.925
25	36	1171.92
30	41.5	1747.88

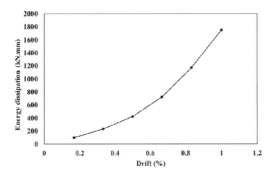

Figure 11. The variation rate of the energy dissipation with drift.

The amount of energy dissipation in the building increased by enhancing the values of input displacements. In the initial steps of the excitations, the constant frictional force in the CSIM walls due to their self-weight was the main cause of the energy dissipation by the CSIM building. When higher values of the input displacements applied to the CSIM building, the surrounding confining elements compressed the SIM panel and added extra frictional force in walls. This led to higher ductility and growth for energy dissipation in the building.

The whole scenario could be justified by interpreting the performance of the semi-active device model for the diagonal shear springs in the CSIM walls of the building. The shear springs performed linearly when small values of displacements were applied to the building. The piston moved slightly and reset to its zero position. By increasing the excitations to the building, the shear springs reached its nonlinear stage of performance. Higher values of piston movements resulted in higher quantities of the energy dissipation in the building.

6 CONCLUSION

Confined semi-interlocking masonry (CSIM) was introduced as a new masonry system with the potential of better seismic performance compared to traditional confined masonry buildings. In this paper, the hysteresis performance of a CSIM building was assessed numerically through a new macro-model. The resettable semi-active device model was employed in shear springs in order to simulate the sliding joints of the SIM panels in the CSIM building. The macro-model could successfully present the behaviour of the CSIM building. Finally, the energy dissipation of the building was assessed. The high values of the energy dissipation due to the presence of sliding joints in the CSIM building was captured by the appropriate performance of the semi-active damper in the numerical macro-model in which the movements of the piston inside the cylinder was able to dissipate the significant amount of energy.

REFERENCES

Carr, A. 2006. RUAUMOKO-Inelastic dynamic analysis program. Computer Program Library. Department of Civil Engineering, University of Canterbury, Christchurch, New Zealand.

Chey, M. H. 2007. *Passive and Semi-Active Tuned Mass Damper Building Systems*. University of Canterbury, Christchurch, New Zealand.

DIANA 2017. DIANA Finite Element Analysis. 10.2 ed. Delft, The Netherlands TNO Building and Construction Research.

Elgawady, M., Lestuzzi, P. & Badoux, M. A review of conventional seismic retrofitting techniques for URM. 13th international brick and block masonry conference, 2004. 1–10.

Franco-Anaya, R., Carr, A. & Chase, J. Shaking table tests of a model structure with semi-active resettable devices. Proceedings of the 10th US National Conference on Earthquake Engineering Frontiers of Earthquake Engineering, 2014.

Franco-Anaya, R., Carr, A. & Chase, J. 2017. Experimentally Validated Analytical Model of a Semi-active Resettable Tendon for Seismic Protection.

Giberson, M. F. 1967. The response of nonlinear multi-story structures subjected to earthquake excitation.

Hemmat, M., Totoev, Y. Z. & Masia, M. J. 2018. Application of semi-interlocking bricks in confined masonry walls. *10th International Masonry Conference*. Milan, Italy.

Hemmat, M., Totoev, Y. Z. & Masia, M. J. 2019. Finite element simulation of confined semi-interlocking masonry walls. *The 13th North American Masonry conference*. Salt Lake City, Utah, USA.

Hossain, M. A., Totoev, Y. Z., Masia, M. J. & Friend, M. 2017. In-Plane Cyclic Behavoir Of Semi Interlocking Masonry Panel Under Large Drift. *13 th CANADIAN MASONRY SYMPOSIUM*. Halifax,Canada.

Lin, K., Totoev, Y. Z., Liu, H. & Guo, T. 2016. In-Plane Behaviour of a Reinforcement Concrete Frame with a Dry Stack Masonry Panel. *Materials*, 9, 108.

Lourenço, P. B. Computational strategies for masonry structures: multi-scale modeling, dynamics, engineering applications and other challenges. Congreso de Métodos Numéricos en Ingeniería, 2013. SEMNI, 1–17.

Meli, R., Brzev, S., Astroza, M., Boen, T., Crisafulli, F., Dai, J., Farsi, M., Hart, T., Mebarki, A., Moghadam, A. S., Quiun, D., Tomazevic, M. & Yamin, L. 2011. Seismic design guide for low-rise confined masonry buildings. *Earthquake Engineering Research Institute (EERI)*. Oakland,CA,USA: Network,Confined masonry.

Totoev, Y. 2015. Design Procedure for Semi Interlocking Masonry. *Journal of Civil Engineering and Architecture*, 9, 517–525.

Totoev, Y. & Al Harthy, A. 2016. Semi Interlocking Masonry as Infill Wall System for Earthquake Resistant Buildings: A Review. *The Journal of Engineering Research [TJER]*, 13, 33–41.

Totoev, Y. Z. 2012. Frictional energy dissipation and damping capacity of framed semi-interlocking masonry infill panel. *Proc. 15th International Brick/Block Masonry Conference, Florianapolis, Brazil*.

Yekrangnia, M., Bakhshi, A. & Ghannad, M. A. 2017. Force-displacement model for solid confined masonry walls with shear-dominated failure mode. *Earthquake Engineering & Structural Dynamics*.

Brick and Block Masonry - From Historical to Sustainable Masonry –
Kubica, Kwiecień & Bednarz (eds)
© 2020 Taylor & Francis Group, London, ISBN 978-0-367-56586-2

Review on CO₂ capture, storage and sequestration in non steel reinforced precast concrete

L.R. Fortunato & G.A. Parsekian
Federal University of São Carlos, Sao Paulo, Brazil

A. Neves
Federal University of Mato Grosso, Mato Grosso, Brazil

ABSTRACT: Climate changes has been a cause of concern to the population, governments and scientists around the world. Such changes are caused by the growing increase in greenhouse gas emissions as a result of human activities. The construction sector stands out as responsible for the emission of one-third of the total greenhouse gases emitted. The process of extracting and manufacturing the raw materials needed to obtain concrete emits large amounts of CO_2 into the environment. However, recent research has shown that Portland cement based materials have the ability to store CO_2 in the form of stable carbonates, especially in the early ages, through the accelerated carbonation process. This technology can be applied to structural elements without reinforcing steel. This paper presents a review on state-of-the-art studies in this emerging technology and provides methods for quantifying CO_2 sequestration in cement matrices, as well as technical information on CO_2 capture from stationary emissions, transport, storage, costs and incentives financial resources.

Keywords: CO_2 capture, cure, accelerated carbonation, concrete

1 INTRODUCTION

Regarding 2016, global CO_2 emissions exceeded 32 billion tons. In this context, an industry has been responsible for the global emission of the atmosphere from 5 to 8% of carbon dioxide (ALI, SAIDUR, HOSSAIN, 2011). The society wants to develop for a fairer world in than all have access to hospitals, schools and homes, and nothing of that becomes viable without cement. Therefore, the consumption of cement materials is related to the economic development of a nation, whereas CO_2 is emitted by the atmosphere (SCRIVENER, JOHNM GARTNER, 2016). Therefore, we have reached a dilemma, which can be solved by developing new technologies to reduce the emission and/or capture of pollutants. Since the 1930s it is explored the carbonation and their harmful consequences to the structures of reinforced armed concrete. In fact, since the academicals banks, the civil engineers learn that the carbonation chemical reaction is harmful, once it carts the destruction of the film that involves the steel, leaving the susceptible material to the corrosion, reducing long term the durability of the structure of reinforced concrete. However, this work wants to change the checked negative focus at the carbonation when analyze her by the sustainable optics which is promoted when it is just used the concrete, being excluded the reinforcing steel. It is observed that to expose for

a short time (some hours) the cement-based material without reinforcing steel recently manufactured, in other words, with early little age to the cure in carbonation camera with high concentrations of CO_2, appropriate conditions of humid-of relative, pressure and temperature, happens the phenomenon of accelerated carbonation occurs, which is able to capture the CO_2 presents in the atmosphere and obtain mechanical gains whose values are comparable to those of conventional curing method at 28 days (SHAO et al. (2006), GALAN et al. (2010), WANG et al. (2012, B. ZHAN et al. (2013)).

For so much, the present article presents the subjects related to the greenhouse effect, generation and it captures CO_2, mechanisms and methods to quantify absorbed CO_2 in cementitious materials, as well as techniques for capture and storage of CO_2 starting from industry of cement.

2 THE CIVIL CONSTRUCTION AND EMISSION OF GREEN HOUSE EFFECT GASES

It is known that the greenhouse effect is responsible for maintaining global warming, once it acts avoiding that the solar rays are reflected to the space. In case him no existed the Earth would possess inferior medium temperatures to -10°C. The great problem is

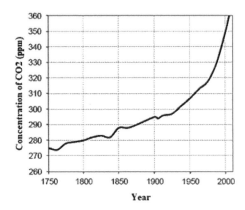

Figure 1. Global concentration of CO_2 in the atmosphere not over the years. Source: Fonte: Yoon, Çopuroğlu e Park (2007).

caused by the human being, that with their intense activities, it emits enormous amounts of CO_2 to the atmosphere causing the increase of the greenhouse effect (BEIROZ, 2011). It projects that up to 2100 the concentration of CO_2 can reach values that vary from 535 to 985 ppm, corresponding her/it an increase from 41 to 158% in relation to the current landings IPCC (2007). Figure 1 demonstrate that after the Industrial Revolution theCO_2 global concentration increased exponentially, corroborating with the fact that the excess of the emissions of GEE in the atmosphere is happened by the human activities.

In agreement with VITO (2001), the procedures used for the production/obtaining of the employed inputs in the building site are responsible for great portion of the emissions causing of the greenhouse gases effect, among them they stand out: cement, whitewash, steel, sand and basalts (retreat and transport), it burns of fossil fuels, among others. The focus will be given to the components of the concrete.

2.1 Cement

According to the Cement Technology Roadmap (2009), between 2000 and 2006 cement production increased by 54% worldwide. It is estimated that between 2006 and 2050, due to the population growth and the consequent need for urbanization which demand the consumption of concrete for construction of buildings, housing, schools, hospitals, among others; cement production will increase from 0.8 to 1.2% per year, reaching between 3.7 and 4.4 billion tons in 2050 (BID, 2010). While global CO_2 emissions from cement in 1990 were 576 million tons (Boden et al., 2011b), it almost tripled in 2006 to 1.88 billion tons. If this trend continues without any action, the amount of CO_2 emissions in the world from the cement industry is expected to reach 2.34 billion tons by 2050. The International Energy Agency (IEA) hopes that through the use of appropriate strategies, such as efficiency energy in kiln,

alternative fuels, clinker replacement and CO_2 capture and storage, these emissions will be reduced and eventually reach 1.55 billion tons by 2050 (CEMENT ROADMAP TARGETS, 2009).

In 2015, at the COP 21 in Paris, a government agreement was established to reduce the global average temperature rise to well below 2 ° C. The WBCSD issued a Global Declaration of Ambition requesting the collaboration of all cement plants to reduce CO_2 emissions by 20-25% by 2030 (SCRIVENER, JOHN, GARTNER, 2016).

The fact is that cement is the most manufactured product of the Earth by mass, and when combined with water and mineral aggregates gives rise to cement-based materials (eg concrete), this is in turn the second used material in the world after water. The major problem is due to the cement manufacturing process, which promotes large release of CO_2 into the environment (ECO- EFFICIENT CEMENTS, 2016). Traditional Portland cement is basically composed of clay and limestone which when calcined at high temperatures give rise to clinker. Clinker grains are grinding with gypsum and give rise to cement. In order to produce 1,000 kg of clinker, 825 to 1150 kg of CO2 are generated, 50 to 60% of which are generated in the chemical calcination reaction of limestone and 30 to 40% are due to the combustion of fossil fuels in the kiln (FREITAS, et al, 2010). Table 1, below, demonstrates the emission of CO_2 due to the production of clinker according to several authors.

The variation in CO_2 emissions is related to factors such as fuel type and type of kiln used for the production of clinker, since dry kilns with preheating and precalcination emit less carbon dioxide than ovens by humid route (WBCSD/GNR, 2011).

Currently the cement industry has employed strategies to reduce CO_2 emissions from clinker, among them, the use of alternative fuels, the search for greater energy efficiency in furnaces during the production process, besides the substitution of clinker for mineral additions such as blast furnace slag and fly ash (WBCSD/CSD, 2009; WBCSD/GNR, 2011).

Table 1. Variations in CO2 emissions related to clinker production (DAMINELI, 2013).

Author	kg of CO_2 per ton of clinker produced
Yamamoto et al (1997)	824,25 a 1151,7
Humphreys; Mahasenan (2002)	870
Gartner(2004)	859
Josa et al (2004)	821,1 a 900
USGS (2005)	960
IEA (2006)	913 a 1125
Damtoft et al (2008)	840 a 1150
John (2011)	855,07
WBCSD/CSI (2009)	842 a 1043

2.2 Aggregates: Sand and stone

According to Stachera (2008), most Brazilian constructions use natural sand, which comes from riverbeds. Dredging powered by diesel oil is used for the extraction. The crushed stone is extracted from quarries and the exploration follows the steps: a) Mechanical crushing of the rock through drill borers or pneumatic hammers; b) Establishment of lines of fire; (c) Fire (explosion): after the placement of explosives; d) Transport in the quarry: through trucks, wagons and/or mats; (e) crushing; f) Transport to the work (ALVES, 1987).

The use of these aggregates causes great impact to the environment, since they are extractive activities and over time promote the exhaustion of the natural resource, since it is not renewable. In addition, the extraction process launches discards into the soil and bodies of water contaminating and degrading them and emits gases into the atmosphere due to the use of combustion engines (STACHERA, 2008).

3 CO_2 CAPTURE

The planet naturally balances the CO_2 present in the Earth's atmosphere through capture mechanisms such as photosynthesis by plants, carbonate-silicate cycle and physical-chemical and biological processes carried out by seas and oceans (KASTING E CASTLING, (2003); PLASYNSKI, 2013); However, these are not enough. In this context, the technology covered in this study deals with the absorption of CO_2 in cement materials without reinforcing steel, which is provided by accelerated carbonation cure, where CO_2 is incorporated into the cement matrix in the form of calcium carbonate ($CaCO_3$) and precipitates in the pores under the mineralogical form of calcite, vaterite and aragonite, promoting the environmental gain related to the CO_2 capture from the atmosphere and cementitious matrix benefits such as the improvement of physical and mechanical properties (YE (2003), SHAO et al. (2006), GALAN et al. (2010), WANG et al. (2012, B. ZHAN et al. (2013), YUAN et al. (2013), NEVES JR (2014), JANG et al (2016), BERTOS et al. (2017), D. ZHANG et al. (2017)).

In developed countries, such as the United States and Canada, carbon sequestration technology in cement-based materials is already being commercially exploited in industries; so much so that CarbonCure Technologies Inc. developed a process for curing their concrete using CO_2 collected from coal mills and cement plants. The company claims that this technology does not only allow the storage of CO_2 in the concrete in the form of chemically stable carbonates, but also relieves efflorescence and shrinkage, as well as improving the resistance to freezing and thawing (MONKMAN e MACDONALD, 2016). Other companies have also used CO_2 beneficially in curing, manufacturing and modifying cements, among which stand out: Solidia Technologies, Calera Corporation, TecEco Pty. Ltd, Calix Ltd. and Kajima Corporation (JANG et al, 2016).

4 ACCELERATED CARBONATION CURE

The cure of cement materials without reinforcing steel through accelerated carbonation comes directly to meet the industrial need related to increased production and time-gain, since dead time (cure time) is greatly reduced. In addition, curing with CO_2 in cement materials without reinforcing steel promotes the absorption of CO_2 from the medium and rapid gains in mechanical strength. Several studies by Ye (2003), Shao et al. (2006), Galan et al. (2010), B. Zhan et al. (2013), Neves Jr (2014), Jang et al (2016), Bertos et al. (2017), D. Zhang et al. (2017), show that this cure, besides capturing CO_2, promotes the densification of the microstructure and changes in the chemical bonds of cement products, leading to increased mechanical strength and durability. In addition, it is possible to verify the good results regarding the resistance gain verified for situations in which the CO_2 cure is replaced totally or partially by the curing and/or thermal cure traditionally employed. Shao e Morshed (2015), found that CO2 curing lasts considerably short duration than traditional curing, as well as important values related to the high resistance and low permeability of the concrete. Boyd et al (2012) compared accelerated carbonation curing with steam curing and found that after curing by carbonation, the concrete showed higher resistance to chloride permeability and sulfate attack and lower damages were observed by freezing and thawing; such facts were attributed to the significantly reduced content of calcium hydroxide on the surface of the carbonated concrete. Some studies have associated CO_2 cures with traditional cures, for example, He et al. (2016) subjected to curing with CO_2 and subsequent hydration, post-hydration by curing with water increased the formation of C-S-H gel, ettringite and monocarboaluminate, which promoted a decrease in the porosity of the cementitious matrix. Pauletti (2004) analyzed the humid curing performed before CO_2 curing and found that submerged or wet curing, while promoting satisfactory conditions for cement hydration, resulted in pore closure, reducing the rate of carbonation reaction, but it is not the most suitable for the purpose discussed in this paper. Lo and Lee (2002), compared an air curing with humid curing prior to CO_2 curing and found that water-cured concretes have only 72% of the carbonation of cured concretes in air; specialists concluded that the air curing leads to a more open pores are more open, what propitiates the increase of the amount of channels intercommunicated and consequent increase of the speed of carbon dioxide diffusion.

5 CARBONATION REACTION

Historically the focus is given to the "weather carbonation" of the reinforced concrete, which happens long term and it is harmful to the steel, once it causes the reduction of the pH of the concrete, promoting the destruction of the passive film layer that involves the steel, leaving the material susceptible to the corrosion, carting the decline of the durability of the structure of reinforced concrete (Neville (1963), Ho e Lewis (1987), Parrot (1987), Saetta e Vitaliani (2004), Pauletti (2004), Villain et al. (2007). JANG et al., 2016).It is well-known that the carbonation and their effects are harmful to the durability of the structures of reinforced concrete; however, recent studies, among them, Ye (2003), Shao et al. (2006), Galan et al. (2010), B. Zhan et al. (2013), Neves Jr (2014), Jang et al (2016), Bertos et al. (2017), D. Zhang et al. (2017), have analyzed and demonstrated that the carbonation reaction promotes benefits when used as a cure process in cement materials without reinforcing steel, among them the potential capacity of the concrete to absorb CO_2 from the environment and gains related to the mechanical properties.

This study deals with curing by "early carbonation", which takes place within a short time (a few hours) and promotes benefits to the cement matrix and the environment.

Through the carbonation reaction, which is exothermic in nature, the cement hydration products, the main one being calcium hydroxide ($Ca(OH)_2$), react with CO_2, generating carbonate and gel silica based products The main carbonation reaction occurs between $Ca(OH)_2$ and CO_2 in the presence of water and is presented in reaction (1.1). The main product is formed by the reactions between carbon and calcium carbonate ($CaCO_3$), which is insoluble and appears in the mineralogical form, as calcite, vaterite and aragonite (SMOCZYK, 1976). However, C-S-H can be carbonated. In fact, a carbonation of the calcium hydrate silicate is reported, it occurs in a second moment, when most of the calcium hydroxide has already been consumed (MORANDEAU et al., 2014), reaction (1.2).

$$Ca(OH)_2 + CO_2 \rightarrow CaCO_3 + H_2O \qquad (1.1)$$

$$3CaO \cdot 2SiO_2 \cdot 3H_2O + 3CO_2 \rightarrow 3CaCO_3 \cdot 2SiO_2 \cdot 3H_2O \qquad (1.2)$$

The mechanism of the carbonation reaction in its sequential sequences occurs as described below Maries (1985):

1) Diffusion of CO_2 into the air;
2) Permeation of CO_2 through the solid;
3) Solubility of CO_2 (gaseous) to CO_2 (aqueous);
4) Hydration of CO_2 (aqueous) to H_2CO_3. This is a slow and determinant step of the rate;
5) Ionization of H_2CO_3 to H +, $_{HCO3-}$, $CO_{3}2-$. Occurs almost instantaneously, causing the pH to drop approximately 3 units, typically 11 to 8;

6) Dissolution of the more soluble phases of the hydrated oil, being one of the easiest solutions to dissolve $Ca(OH)_2$.
7) Nucleation of $CaCO_3$, C-S-H. Nucleation is favored by dirt and the presence of finely divided material;
8) Precipitation of solid phases. At first, vaterite and aragonite may be produce, but reverted to calcite. Amorphous calcium carbonate can be contained in the final product.
9) Secondary Carbonation. The C-S-H gel is formed and progressively decalcified, ultimately converting to S-H and $CaCO_3$.

6 QUANTIFYING CAPTURED CO_2

For the experimental process as newly-fabricated cementitious matrices should be inserted in a carbonation chamber configured for temperature, humidity, pressure and CO_2 concentration, to start the carbonation reaction. Figure 2 shows the insertions in the carbonation chamber.

The system relied on the continuous supply of high-purity carbon dioxide in a sealed chamber in which prefabricated cement products are inserted after production. The pressure and CO_2 concentration remain constant throughout the procedure and the peak temperature varies according to the cement content used in the products (SHAO et al. (2010)). There are several strategies used to measure the amount of CO_2 captured in cement products, the main techniques will be presented below.

6.1 Method of mass gain

It is well documented by Shao et al. (2006), Shao et al. (2010), Shao e Lin (2011), Rostami et al (2012), B. Zhan et al. (2013), Neves Jr. (2014), Fortunato et al (2018) that accelerated carbonation in cement-based materials at early age, following certain experimental conditions of humidity, temperature, CO_2 concentration, time of exposure, among others; increases the mass of the sample by precipitation of $CaCO_3$ in the micro structure, decreasing the porosity, increasing the

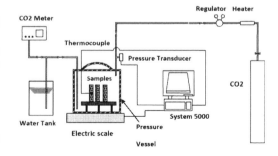

Figure 2. Carbonation chamber - carbonation curing configuration using pre-grounded recovered CO_2. Source: (SHAO et al.,2010).

density and finally raising the resistance, being the method of mass gain fully effective. The purpose of this experimental procedure is to evaluate the influence of accelerated carbonation at early ages of cement based materials from the moment they are inserted in the carbonation chamber and begin to capture CO_2 through the reactions with the cement hydrates ($Ca(OH)_2$, CSH, among others), producing $CaCO_3$ and H_2O, thus providing the mass gain of the sample.

During the curing period in the carbonation chamber, the reactions of vaporization, hydration and carbonation occur simultaneously, so the increase in mass results from a sum of several effects arising from these reactions. The water to be computed in the equation that estimates CO_2 capture comes from both the internal humidity of the sample and the carbonation reaction, so it participates in the mass of the sample and must be computed in the equation. This water is measured through absorbent papers placed inside the chamber (SHAO et al., 2006).

we cannot affirm that samples cured outdoors, for they present smaller initial humidity, they will introduce potential minor of carbonation, because to smallest hydrates formation due to the dry cure can propitiate a smaller alkaline barrier, allowing the front progress of carbonation and consequently larger formation of $CaCO_3$ and H_2O (EL-HASSAN E SHAO, 2013). Therefore, so that the calculation of the CO_2 capture is accomplished in a correct way is not possible to neglect the lost water.

In this experiment, the amount of absorbed CO_2 can be estimated through the mass gain given by the equation 1.3 that it considers the mass of the sample before and after the carbonation, including the mass of lost water during the reaction and the initial mass of cement Portland of the mixture.

CO2 uptake (%)

$$= \frac{((Mass\ after\ CO2\ +\ (Water\ lost)) - (Mass\ before\ CO2)}{Mass\ dry\ binder}$$

(1.3)

In the above formula *(Mass) after CO$_2$* is a mass of the mixture after carbonation; (Mass) *Lost* treats the mass referring to the loss of water from the system; *(Mass) before CO$_2$* refers to the mass of the sample prior to carbonation; and *(Mass) dry Binder*, is initial mass of the reactive binders.

6.2 *Method of the curve of mass*

The second method used to quantify the absorption of CO_2 is the method of the mass curve. The mass curves are obtained by using an electronic scale (as demonstrated in Figure 3) to register the mass gain in situ of the pre-molded samples in function of the time during whole the carbonation process, besides sensor of pressure of injection of gas in the system.

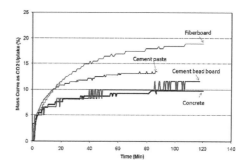

Figure 3. Mass curves of precast products subject to carbonation curing. Source: (SHAO et al.,2010).

The scale is reduced to zero starting from the moment in that the samples are inserted in the camera and the carbonation process begins. The mass gains is registered starting from the moment in that the gas is injected. At the end of the process the gas is liberated to the atmospheric pressure and the residual mass registered by the scale represents the total absorption of CO_2 mass (SHAO et al. (2006), SHAO et al. (2010), SHAO e LIN (2011), ROSTAMI et al (2012), B. ZHAN et al. (2013), NEVES JR. (2014)). Figure 3 regards the study of Shao et al (2010) and it demonstrates the curves of corresponding masses the absorption of CO_2 along the time of the products fiberboard, cement paste, cement bead board and concrete. Through the curves it is possible to observe that the significant mass gain regarding absorption of CO_2 happen in the first 30 minutes for the concrete cement paste and cement bead board, while for the fiberboard the earnings are important even more than a hour of the beginning of the carbonation reaction, resulting in an absorption significantly larger in the plate with fibers. The final values of the mass curve represent the absorption of CO_2 in the studied period. It is important to note that even though longer treatment times promote a greater absorption of CO_2, there is a limit from which the mass gain begins to decrease due to the densification of the layers and the loss of moisture during the reaction (SHAO et al, 2006).

6.3 *Thermogravimetry*

Authors as Papadakis et al. (1992), Fukushima et al. (1998), Huijgen et al. (2005), Chang e Chen (2006), Huntzinger (2006) are used analytical techniques such as the thermogravimetry (TGA) to obtain the amount of absorbed CO_2 during the carbonation reaction. Thermogravimetry is a process for which the sample is heated up and the mass reduction is registered in several temperatures. In the carbonated samples the growth of carbonation is observed through the reduction of the $Ca(OH)2$ peak and concomitant increase of the CaCO3 peak The mass loss can happen due to evaporation of the water or the chemical decomposition of certain compounds releasing gases such as

CO_2 (LIV HASELBACH, 2009).According to Perkinelmer (2018), thermogravimetry is the technique in which by performing the heating of a material will increase or decrease its weight. Papadakis et al. (1991) performed TGA analyzes on samples exposed to carbon dioxide and various relative humidity conditions and obtained decomposition corresponding to dehydroxylation of Ca (OH) 2 ranging from (400-500 °C) and decarbonation of CaCO3 ranging from (600 -800 ° C). Chang and Chen (2006), performed TGA on many samples of hydrated cement at different ages and under varying conditions. They estimated that the bulk of the mass loss at (425-550 ° C) refers to the decomposition of Ca (OH) 2 and that the mass loss between (550 and 950 ° C) is due to the decomposition of CaCO3, and other products.

reinforcing steel they have been accomplished for the development of a cure process that can be implemented in industrial scale and that is capable not just of creating the ideal conditions for the product to reach resistance to compressive strength demanded by norm, but it is also sustainable while it promotes the absorption of CO2 of the atmosphere, reducing the global warming and carbon credits trademarked, which every day represents·an important trade coin among developed countries. The Table 02 below contains several works within the last ten years, from which information related to the factors that influence the carbonation of the concrete were extracted, such as, time of curing, CO2 concentration, relative humidity, temperature, in addition to the results obtained and the advances achieved.

7 STUDIES ON THE CURE PROCESS FOR ACCELERATED CARBONATION IN PRECAST CEMENT MATERIALS WITHOUT REINFORCING STEEL

Several studies involving the use of the cure by accelerated carbonation in prefabricated without

8 CEMENT CAPTURE AND STORAGE IN THE CEMENT INDUSTRY

Carbon capture and storage technology is considered by the International Energy Agency (IEA) as essential for reducing CO_2 emissions in the cement sector

Table 2. References on carbonation cure.

Ref.	Spec.	Initial cure time (h)	Carbonation cure time (h)	Carbonation chamber CO_2 (%)	Pressure (MPa)	T (°C)	RH (%)	CO_2 Absorption (%)	Compressive Strength result
Shao et al (2010)	CMU paver	- - -	2	100	0,5	45-95	- - -	9,8 - 19,0	7,8-10,5MPa
Shao and Lin (2011)	paver	2 - 19	4 - 5	99,5	0,15	56	65	3,4 -7,4	31- 46 MPa
Shao and Lin (2011)	concrete slump = 0	- - -	2	99,5	0,15	40 - 56	65	11,7	increase of 60%
Rostami et al (2012)	Cement paste, w/ c=0,36	18	2	100	0,15	ambient	- - -	8,9	increase of 61%
Boyd et al (2012)	concrete slump zero w/c=0,36	18	2	100	0,15	25	- - -	7,0 - 9,0	post-carbonation water spray ≈ non-carbonated with thermal curing
El - Hassan et al (2013)	CMU	0, 4, 6, 8, 18	2, 4, 96	100	0,01	25	50	9,0 - 35,0	Higher strengths for longer carbonation
Zhan et al (2013)	CMU	- - -	6, 12, 24	100	0,01	- - -	- - -	22,49- 43,96	Resistance gains varying from 108% to 151%
El - Hassan and Shao (2015)	CMU	0, 4, 6, 8, 18	2, 4, 96	100	0,1	25	50	8,3 - 35,1	post-carbonation water spray ≈ non-carbonated with thermal curing
Fortunato et al (2018)	Paver	12	4, 16	20	0,1	23	65	1,5 - 5,1	39,7- 46,4MPa

by up to 56% by the year 2050 (Eco-Efifcient Cements, 2016). It is known that CO_2 emissions of this sector are due to the combustion of fuels and calcination of lime in the furnace; through the cited technology, CO_2 is captured as it is emitted, compressed to a liquid, transported in oil pipelines for later storage. The report also estimate that investment of US $ 321 to 592 billion will be needed to capture CO_2 from the cement industry (IEA, 2009). Table 2 indicate several references on concrete block unit (CMU) and pavers studies.

9 CEMENT CAPTURE AND STORAGE IN THE CEMENT INDUSTRY

Carbon capture and storage technology is considered by the International Energy Agency (IEA) as essential for reducing CO_2 emissions in the cement sector by up to 56% by the year 2050 (Eco-Efifcient Cements, 2016). It is known that CO_2 emissions of this sector are due to the combustion of fuels and calcination of lime in the furnace; through the cited technology, CO_2 is captured as it is emitted, compressed to a liquid, transported in oil pipelines for later storage. The report also estimated that investment of US $ 321 to 592 billion will be needed to capture CO_2 from the cement industry (IEA, 2009).

9.1 *CO_2 capture from stationary emissions*

Three potential methods for capturing CO_2 from stationary emissions are presented below.

- **Post-combustion capture**

Post-combustion capture technology can be implemented in both newly built cement plants and existing plants for modernization. According to IEA (2009b); CSI/ECRA (2009); Bosoaga et al., (2009) et al., this technique that is already commercially mature in other applications, will probably be applied on a large scale in the cement industry. Through said technology the flue gas leaving the plant is collected and sent to the amine solvent, where the carbon dioxide in the gas is captured by the solvent. The CO_2-rich solvent is then pumped to a desorbator where the released carbon dioxide is purified before being compressed and transported to the storage site (IEA (2009b); CSI/ECRA (2009); BOSOAGA et al., (2009)).

- **Oxyfuel**

This technology is promising to capture CO_2 from the cement plant in the long term even if its technical development takes longer to be implemented than that of post-combustion capture (CSI/ECRA, 2009). In the Oxicorte process, the clinker furnace is fed with oxygen instead of ambient air. In order to do this, it is necessary to have a state-of-the-art Air Separation Unit, in this sense, changes in the kiln where limestone calcination occurs and in other facilities of the industry. At present, some plants have increased the oxygen content in the furnace content to increase the rate of reaction during the calcination process. This in itself is a positive indication, as the result of any addition of pure oxygen may increase the CO_2 concentration in the flue gas. According to studies, the reduction of carbon emissions to the cement plant that implements the Oxicorte process should be between 63% and almost 100%, but the application was only carried out in a pilot project (CSI/ECRA, 2009).

- **Carbonate looping**

The carbonate looping for capturing CO_2 for any CO_2 emitting source is theoretically proposed for many years. Similar to other emerging technologies, the looping carbonation process needs to be proven through pilot and large-scale demonstration plants. In general, the substitution of the amine-based solvent by a circulating solid material that absorbs CO_2 occurs, which is then released in the pure form by the application of heat. Theoretically, this system can offer advantages because it involves less energy to remove the CO_2 of the solid material, besides presenting lower level of corrosion and emission of SO2. The gas released by the combustion reacts with the solid calcium-based sorbent (calcium oxide) to form the calcium carbonate. The carbonates are heated to liberate carbon dioxide (IPCC, 2005; BOSOAGA et al., 2009; NARANJO et al., 2011; BLAMEY et al., 2010; LI et al.,2013).

9.2 *CO_2 capture from stationary emissions*

Transporting CO_2 from capture facilities to storage sources is the most important step in the capture and storage process. In order to do so, a system consisting of pipelines and compression stations similar to the pipeline systems developed for the natural gas pipelines has been used, as this is believed to be the safest and most economical way to route CO_2 to future commercial units. The storage of CO_2 captured can be done in geological formations and in pressurized tanks for use in prefabricated industries, such as CarbonCure Technologies Inc. BENHELAL et al., 2013; BARKER et al. (2009).

10 CONCLUSIONS

The desire of the society to move towards a more fair world implies an increase in the built environment. This task maintains or further increases the demand for cement-based materials over the years. Hospitals, schools, dams, bridges, viaducts, towering skyscrapers, single-family homes - none of this would be possible without cement. To do this, the development of sustainable alternatives to reduce the concentration of CO_2 in the atmosphere are necessary. In this sense, a CO_2 capture and storage technology in cement based materials without steel reinforcement is promising, contributing to the mitigation of carbon dioxide present in the atmosphere. Through the state-of-the-art research in the area, has been checked CO_2 capture and resistance gain in all surveys. In most studies, the CO_2 concentration used inside the carbonation chamber was 100%, the maximum concentration of CO_2 inside the chamber allows maximize the accelerated carbonation cure, that is, bigger production of CaCO3 in the cementitious matrix. In the studies by Lin and Shao (2011), Rostami et al (2012), Boyd et al (2012), El - Hassan et al (2013), El - Hassan and Shao (2015) and Fortunato et al (2018) has observed that an initial cure previously the carbonation cure, directly interfered with the hydration of the non-reinforced product, advance of the carbonation front, CO_2 absorption and resistance gain. In all researches, CO_2 capture and resistance gain were observed. Therefore, there is no doubt that accelerated carbonation curing is a promising process for mitigating CO_2 from the atmosphere and can be implemented on an industrial scale, as has already been done by the CarbonCure, Calera and Calix industries, among others.

REFERENCES

Ali M B, R. Saidur, M.S. Hossain. 2011. A review on emission analysis in cement industries. *Renewable and Sustainable Energy Reviews*, 2252–2261.

Barker, D.J., Turner, S.A., Napier-Moore, P.A., Clark, M., Davison, J.E., 2009. CO_2 capture in the cement industry. In: *Proceedings of the Ninth International Conference on Greenhouse Gas Technologies, GHGT-9*, WashingtonDC, USA. Energy Procedia 1, 87–94.

Beiroz, H. 2011. Efeito estufa. Rio de Janeiro: Simonsen – faculdades e colégios, 5. Apostila.

Benhelal A, Gholamreza Zahedi Ezzatollah Shamsaei, Alireza Bahadori. 2013. Global strategies and potentials to curb CO_2 emissions in cement industry. Elsevier Ltda, *Journal of Cleaner Production 51*.

Bertos Fernández M, S.J.R. Simons. 2004. A review of accelerated carbonation technology in the treatment of cement-based materials and sequestration of CO_2. *Journal of Hazardous Materials B112*.

Bosoaga, A., Masek, O., Oakey, J., 2009. CO_2 capture technologies for cement industry. In: *Proceedings of Ninth International Conference on Greenhouse Gas* Technologies, *GHGT-9*, Washington DC, USA, EnergyProcedia 1, pp. 133–140.

Boyd, Andrew J., Rostami, V., Shao, Y. 2012. Carbonation Curing versus Steam Curing for Precast Concrete Production. *Journal of Materials in Civil Engineering, Vol. 24, No. 9*, September 1.

Damineli, Bruno Luis, 2013. Conceitos para formulação de concretos com baixo consumo de ligantes: controle reológico, empacotamento e dispersão de partículas. Tese Doutorado – Escola Politécnica da Universidade de São Paulo. Departamento de Engenharia de Construção Civil. Ed.Rev. São Paulo, 237p.

D. Zhang, Y. Shao. 2016. Early age carbonation curing for precast reinforced concretes, *Constr. Build. Mater. 113*; 134–143.

El-Hassan, H, Shao Y, Ghouleh, Z. 2013. Effect of Initial Curing on carbonation of Lightweight Concrete Masonry Units, *ACI Materials Journal*/July-August.

El-Hassan, H, Shao, Y. 2015. Early carbonation curing of concrete masonry units with ortland limestone cement, *Cement & Concrete Composites 62*; 168–177.

Fortunato, L.R, Parsekian, G.A, Neves Junior, A. 2019. CO_2 capture in concrete pavers through accelerated carbonation cure. Dissertation (Master in Structures and Civil Construction) – Federal University of São Carlos, São Carlos.

Galan, I.; Andrade C.; Mora, P., Sanjuan, M. 2010. Sequestration of CO_2 by Concrete Carbonation. *Environmental Science and Technology. Vol. 44*.

He P., et al., 2016. Effect of further water curing on compressive strength and microstructure of CO_2-cured concrete, *Cem. Concr. Compos. 72*; 80–88.

IEA (International Energy Agency), 2009a. Cement Technology Roadmap 2009, Carbon Emissions Reductions up to 2050. IEA/OECD. Paris <http://www.iea.org/papers/2009/Cement_Roadmap.pdf>. Acesso em 27/ 12/2018.

IEA GHG, 2002. Opportunities for Early Application of CO_2 Sequestration Technology. IEA GHG PH4/10. Cheltenham, UK.

IEA GHG, 2008. CO_2 Capture in the Cement Industry. Cheltenham, UK, Report no. 2008/3.

IEA, WBCSD, Int Energy Agency, OECD/IEA; WBCSD, 2009. Disponível em: http://www.iea.org/publications/freepublications/publication/name,3862,en.html.

IPCC - INTERGOVERNMENTAL PANEL ON CLIMATE CHANGE. Climate change 2007: the physical science basis - Summary for policymakers". Intergovernmental Panel on Climate Change, Geneva, 2007.

INSTITUT WALLON DE DEVELOPPEMENT ECONOMIQUE ET SOCIAL ET D'AMENAGEMENT DU TERRITOIRE ASBL. IDD – Institut Wallon – VITO. Greenhouse gas emissions reduction and material flows. 2001. Disponível em <http://www.belspo.be/belspo/home/publ/pub_ostc/CG2131/rappCG31_en.pdf>. Access on 27/ 11/2018.

Jang, J.G., Kim, H.J., Park, S.M., Lee, H.K., 2015. The influence of sodium hydrogen carbonate on the hydration of cement. *Constr. Build. Mater.* 94.

Lo, Y., Lee, H. 2002. Curing effects on carbonation of concrete using a phenolphthalein indicator and Fourier-transform infrared spectroscopy, *Build. Environ. 37 (5)*; 507–514.

Morandeau A., M. Thiéry, P. Dangla, 2014. Investigation of the carbonation mechanism of CH and C-S-H in terms of kinetics, microstructure changes and moisture properties, *Cem. Concr. Res. 56*.

Morshed A.Z., Y. Shao, 2013. Influence of moisture content on CO_2 uptake in lightweight concrete subject to early carbonation, *J. Sustainable Cement-Based Mater. 2 (2)*;144–160.

Neves, J.A.; 2014. Captura de CO_2 em Materiais Cimentícios através da Carbonatação Acelerada/Alex Neves Junior – Rio de Janeiro: UFRJ/COPPE.

Pauletti, C. Análise comparativa de procedimentos para ensaios acelerados de carbonatação. Dissertação de Mestrado, Universidade Federal do Rio Grande do Sul, 2004.

Rostami V., Shao Y., Boyd Aj. 2012. Microstructure of cement paste subject to early carbonation curingl. Cement Concrete Research, n.42, v.1, 186–93.

Scrivener, K.L., John V. M., Gartner, E. M. 2016. Eco-efficient cements: Potential, economically viable solutions for a low-CO_2, cement-based materials industry. UNEP – United Nation Environment Program.

Shao Y., Zhou X., Monkman S., 2006. A new CO_2 sequestration process via concrete products production. *EIC Climate Change Technology*, Canada, Ottawa, 10-12 May 2006.

Shao Y, Monkman S, Wang S. 2010. Market Analysis of CO_2 Sequestration in Concrete Building Products. *Second International Conference on Sustainable Construction Materials and Technologies. 2010*; Ancona, Italy, 28-30 June.

Shao Y, X Lin. 2011. Early-age carbonation curing of concrete using recovered CO2. Concrete International, 2011 - concrete.org. McGill University, Montreal, QC, Canada.

Smoczyk, H. G. Physical and chemical phenomena of carbonation. In: RILEM COLLOQUIUM ON CARBONATION OF CONCRETE, [s. ed], 1976. Proceedings… 1976. Paper 1.1, 10p.

Zhan, B; Poon, C, Caijun, S. 2013. CO_2 curing for improving the properties of concrete blocks containing recycled aggregates. *Cement & Concrete Composites 42*, 1–8.

Wang, Yc., Chang, Ee., Pan, Sy., et al. 2012. CO_2 capture by using blended hydraulic slag cement via a slurry reactorl, *Aerosol and Air Quality Research, v.12, n.6*, 1433–1443.

Yoon, In-S.; Çopuroğlu, O.; Park, Ki-B. 2007. Effect of global climatic change on carbonation progress of concrete. *Atmospheric Environment, v 41, n 34*, 7274–7285.

Yuan, C., Niu, D., Chen, N., Et Al. 2013. Influence of carbonation on microstructure of concrete, *Guisuanyan Tongbao, v.32, n.4*, 687–691.

Brick and Block Masonry - From Historical to Sustainable Masonry –
Kubica, Kwiecień & Bednarz (eds)
© 2020 Taylor & Francis Group, London, ISBN 978-0-367-56586-2

Freeze-thaw durability of glass textile-reinforced mortar composites

A. Dalalbashi
Department of Civil Engineering, ISISE, University of Minho, Guimarães, Portugal

B. Ghiassi
*Centre for Structural Engineering and Informatics, Faculty of Engineering, University of Nottingham,
Nottingham, UK*

D.V. Oliveira
Department of Civil Engineering, ISISE & IB-S, University of Minho, Guimarães, Portugal

ABSTRACT: Application of textile-reinforced mortars (TRMs) for externally bonded reinforcement of existing masonry structures has received a considerable recent attention. The mechanical behaviour of these composites, which are composed of continuous fibers embedded in an organic matrix, and their effectiveness in improving the performance of strengthened structures are highly dependent on the fiber-to-mortar bond behaviour as well as the bond between TRM system and substrate.

Understanding the long-term performance of these mechanisms is therefore of critical importance for design of durable TRM composites and ensuring the safety of strengthened structures. To address this aspect, the effect of freeze-thaw cycles on the textile-to-mortar bond behaviour is experimentally investigated and discussed in this paper. The results illustrate a significant deterioration of the textile-to-mortar bond performance in the studied composites.

1 INTRODUCTION

The advantages of textile reinforced mortar (TRM) composites with respect to fiber-reinforced polymers (FRPs) have made the former very interesting for externally bonded reinforcement of masonry and reinforced concrete structures (Carozzi & Poggi, 2015; Papanicolaou, Triantafillou, Papathanasiou, & Karlos, 2007; Razavizadeh, Ghiassi, & Oliveira, 2014).

Composed of continuous fibers embedded in a matrix, TRMs show a pseudo-ductile response with distributed cracking when designed appropriately, which makes them interesting for seismic strengthening applications. The large variety of available fibers and mortar types allow development of TRM composites with a large range of mechanical properties (D'Antino & Papanicolaou, 2017; Donnini, Chiappini, Lancioni, & Corinaldesi, 2019; Mazzuca, Hadad, Ombres, & Nanni, 2019; Younis & Ebead, 2018).

The important role of fabric-to-mortar bond behavior on the post cracking response and strength of these composites has been previously discussed in several recent studies by the authors (Dalalbashi, Ghiassi, Oliveira, & Freitas, 2018a, 2018b; Dalalbashi, Ghiassi, & Oliveira, 2019; Ghiassi, Oliveira, Marques, Soares, & Maljaee, 2016).

While several recent studies have addressed the short-term mechanical performance of TRM composites at different levels, their long-term performance and durability under environmental conditions (Ghiassi et al., 2015; Maljaee, Ghiassi, Lourenço, & Oliveira, 2016) is very poorly addressed (Al-jaberi & Myers, 2018; Heshmati, Haghani, & Al-Emrani, 2017; Maljaee et al., 2016). This study is aimed at addressing this gap and presents an experimental investigation on the degradation of the fabric-to-mortar bond behavior in TRM composites under freeze-thaw cycles. The tests include exposing the samples to 360 freeze-thaw cycles and evaluating the changes of the mechanical properties of the material constituents and the bond behavior through performing appropriate mechanical characterization tests.

2 EXPERIMENTAL PROGRAM

The experimental program consisted of a series of glass-based TRM composites exposed to the freeze-thaw conditions. The specimens were taken from the climatic chamber after different periods of exposure in order to examine possible changes in the textile-to-mortar bond behavior as well as in the mechanical properties of the fiber and the mortar. In addition, control specimens cured in environmental lab conditions were used to compare results with specimens exposed to freeze-thaw actions.

2.1 Materials characterization tests

A commercially available hydraulic lime-based mortar (Planitop HDM Restauro) was used as

matrix. A high-ductility hydraulic lime mortar is prepared by mixing the powder with the liquid provided
by the manufacturer in a low-speed mechanical
mixer to form a homogenous paste. For mechanical
characterization of the mortar, compressive and flexural tests were performed at different ages, according
to ASTM C109 and EN 1015-11. Five cubic
(50×50×50 mm^3) and five prismatic (40×40×160
mm^3) specimens were used for each test and at each
age. Specimens were cured for seven days in a damp
environment and then stored in the laboratory (20°C,
67% RH) until the age of 90 days.

The reinforcing material is a woven biaxial fabric
mesh made of an alkali-resistance fiberglass (Mapegride G220). The mesh size and area per unit of
width are equal to 25×25 mm^2 and 35.27 mm^2/m,
respectively. Direct tensile tests were performed on
dry fibers (5 specimens) to obtain their tensile
strength and elastic modulus. A universal testing
machine with a maximum load capacity of 10 kN
and a rate of 0.3 mm/min were used for these tests.

2.2 Bond characterization test

The single-sided pull-out test setup developed by the
authors (Dalalbashi et al., 2018a) is used again in this
study for investigating the fiber-to-mortar bond performance. The specimens consist of fibers embedded
in disk shaped mortar with a thickness of 16 mm (see
Figure 1). The free length of the fiber is embedded in
an epoxy resin block with a rectangular cross-
sectional area of 10×16 mm^2, as shown in Figure 1,
to facilitate gripping during the tests. For detailed
information on the procedure followed for preparation
of the specimens, the reader is referred to (Dalalbashi
et al., 2018a). The specimens are demolded after 72
hours of preparation and are cured in a damp environment for seven days. After that, the specimens are
stored in the lab environmental conditions (20°C,
67% RH) until the age of 90 days. The bonded length

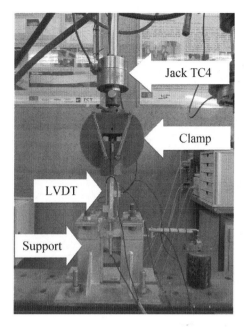

Figure 2. Pull-out test setup details.

of the glass fibers was 50 mm as it was previously
reported to be around the effective bond length of
this TRM system (Dalalbashi et al., 2018b).

For performing the tests, a u-shape steel support
was used for fixing the specimens to a rigid frame.
A mechanical clamp was used to grip the epoxy resin
(and thus the fiber) from the top and performing the
tests (Figure 2). Three LVDTs with 20 mm range and
2-μm sensibility were located at both sides of the
epoxy block to record the slip. The average of these
LVDT measurements is presented as the slip in the
experimental results. All the tests were carried out
using a servo-hydraulic system with a maximum capacity of 25 kN at a displacement rate of 1.0 mm/min.

2.3 Freeze-thaw exposure

After 90 days of curing, the specimens were stored in
a climatic chamber and exposed to freeze-thaw cycles
as shown in Figure 3. The cycles include thawing of the
samples for 2 hours at 30°C and 90% RH and then
freezing at -10°C for 2 hours. Between the thawing and
freezing periods, the temperature was varied at the rate
of 0.111°C/min. This cycle is repeated for 360 times. In
the absence of proper standards, the temperature range
of freeze-thaw cycles was selected from the literature
(Ghiassi, Oliveira, & Lourenc, 2014; Uranjek & Bokan-
bosiljkov, 2015). This allows to compare the durability
of two common strengthening systems (TRM vs. FRP).

At each 40 days (around 60 cycles) and when the
chamber temperature was at 20°C, five specimens
were taken from the climatic chamber to perform
post-exposure tests. Generally, 45 specimens were
tested in which 15 specimens were tested as control
specimens and 30 were exposed to freeze-thaw

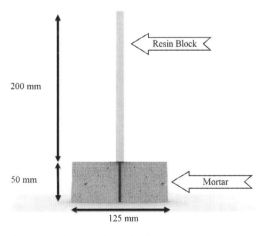

Figure 1. Pull-out specimen details.

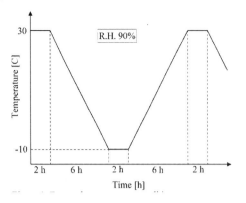

Figure 3. Freeze-thaw exposure condition.

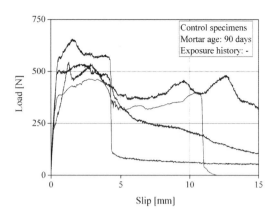

Figure 4. Pull-out response of the control specimens tested at zero cycle.

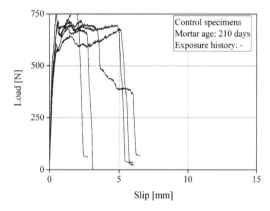

Figure 5. Pull-out response of the control specimens tested at 180 cycles (control).

cycles. The specimens were initially weighed and visually inspected. Then, the specimens were stored in environmental lab conditions for 7 days before testing (Maljaee et al., 2016). The post-exposure tests include pull-out tests and mechanical tests on the constituent materials (mortar and glass yarns).

3 RESULTS AND DISCUSSION

3.1 *Material properties*

The mean compressive and flexural strengths of the mortar as well as coefficient of variation in percentage (provided inside parentheses) after different exposure cycles were presented in Table 1. The average compressive strength of the control specimens (90-day mortar age) was found to be 16.76 MPa. It seemed that with exposure to freeze-thaw conditions, the compressive strength increased until 180 cycles, when it reaches its maximum value (19.48 MPa). At the same time, the compressive strength showed a less conspicuous (18.76 MPa). Comparison between the control and the exposure specimens showed that freeze-thaw cyclic caused the compressive strength to decrease slightly.

Table 1. Mortar mechanical properties under freeze-thaw cycles*.

Number of cycles	Compressive strength [MPa]	Flexural strength [MPa]
0 (control)	16.76 (12)	4.50 (2)
60	17.02 (11)	-
120	18.99 (25)	-
180	19.48 (6)	5.79 (6)
180 (control)	20.02 (14)	4.54 (14)
240	17.46 (4)	-
300	17.32 (2)	-
360	18.76 (3)	4.96 (5)
360 (control)	17.26 (11)	4.65 (7)

Although the flexural strength was only measured at 180 and 240 cycles, the results show a similar trend as what was observed in the compressive tests. The flexural strength increased after 180 cycles (5.79 MPa) and it decreased afterwards. In addition, it was interesting to observe from Table 1 that freeze-thaw exposure led to increasing the flexural strength, especially at the 180-cycle point.

The average tensile strength, Young's modulus, and rupture strain of the control glass roving were 875 MPa (13 %), 65.94 GPa (5 %), and 1.77 % (10 %), respectively, with coefficients of variation provided inside parentheses. These values for the specimens exposed to 360 freeze-thaw cycles were 899 MPa (6 %), 70.72 GPa (3 %), and 1.86 % (8 %), respectively, showing no visible deterioration in the mechanical properties of the glass roving under considered exposure conditions and period.

3.2 Textile-to-mortar bond response

The experimental pull-out response of the control and freeze-thaw exposed samples were shown in Figure 7-6 and Figure 7-12, respectively. It can be observed that the bond strength of the control samples tested at 180 cycles (control) was larger than the samples tested at zero cycle (control). This could be due to the further curing of the mortar during this period. However, the bond strength decreased significantly after 360 cycles (control) of curing. The failure mode of

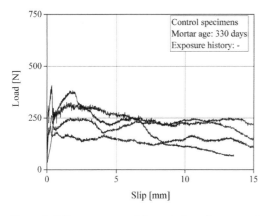

Figure 6. Pull-out response of the control specimens tested at 360 cycles (control).

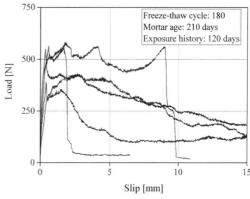

Figure 9. Pull-out response exposed to 180 cycles.

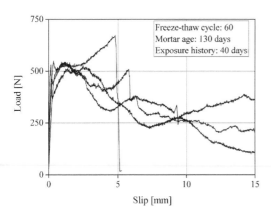

Figure 7. Pull-out response exposed to 60 cycles.

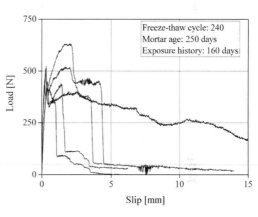

Figure 10. Pull-out response exposed to 240 cycles.

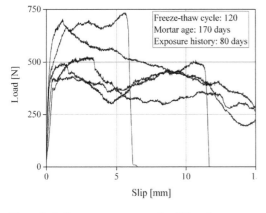

Figure 8. Pull-out response exposed to 120 cycles.

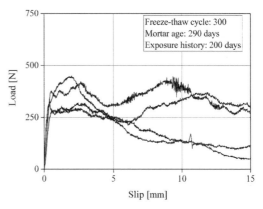

Figure 11. Pull-out response exposed to 300 cycles.

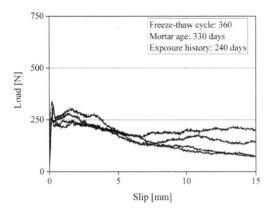

Freeze-thaw cycle: 360
Mortar age: 330 days
Exposure history: 240 days

Figure 12. Pull-out response exposed to 360 cycles.

the samples also changed from the fiber slippage at the beginning (zero cycle) to rupture after 180 cycles (control) and again fiber slippage at 360 cycles (control).

As for the samples exposed to freeze-thaw conditions, a noticeable change of bond behavior could be observed by increasing the number of cycles. Until 240 cycles of exposure, the samples illustrated a negligible change in the bond strength. At this point (240 cycles), the bond strength between the fiber and the mortar showed a sudden decrease (that can be due to incremental damaged induced by previous freeze-thaw cycles) that causes the peak load to drop, in contrast to specimens exposed to fewer cycles. Afterwards, the bond strength decreased until the end of the exposure period. The failure mode of the samples changed from fiber slippage (until 180 cycles) to fiber slippage and rupture (at 240 cycles) then again to fiber slippage (at 300 and 360 cycles).

The main characteristics of the pull-out response of the samples (including peak load, slip corresponding to the peak load, the toughness and the initial stiffness) were summarized in Table 2. In addition, coefficients of variation in percentage were provided inside parentheses. The toughness or the absorbed energy defined as the area under the load-slip curve (Alwan, Naaman, & Hansen, 1991; Naik, Sharma, Chada, Kiran, & Sirotiak, 2019; Zhang, 1998).

It can be observed that the bond behavior was affected by both the freeze-thaw conditions and the mortar age. The bond properties of the exposure specimens showed a downward trend. For example, the peak load at the end of the freeze-thaw cycles reached to the minimum value. At the same time, the initial stiffness as well as the slip corresponding to the peak load was almost constant, while the pull-out energy decreased with time. In addition, the control specimens experience a similar behavior, so that the peak load decreased dramatically from 550.91 N to 308.41 N.

The pull-out behavior was similar to the observed changes in the mechanical properties of materials. For instance, the changes in the peak loads of both the control and the exposure specimens were in-line

Table 2. Effect of freeze-thaw conditions on the pull-out properties of glass-based TRM*.

FT cycle	S [mm]	P [N]	E [N.mm]	IS [N/mm]
0 (control)	2.10	551	4421	2049
	(37)	(15)	(35)	(37)
60	0.73	540	4424	5297
	(76)	(2)	(28)	(79)
120	1.69	567	5341	2815
	(44)	(21)	(24)	(35)
180	0.5	468	3372	3639
	(24)	(22)	(47)	(65)
180 (control)	0.64	692	2700	3006
	(35)	(8)	(36.8)	(40)
240	0.34	470	1962	5476
	(26.6)	(8)	(77)	(61)
300	1.86	384	3839	911
	(55)	(16)	(30)	(19)
360	0.54	314	2605	2862
	(112)	(8)	(15)	(46)
360 (control)	0.69	308	2950	1866
	(114)	(28)	(25)	(69)

* FT: freeze-thaw; S: slip corresponding to peak load; P: peak load; E: energy; IS: initial stiffness.

with the observed changes of compressive strength of the mortar.

4 CONCLUSIONS

The effect of freeze-thaw conditions on the fabric-to-mortar bond behavior in a glass-based TRM composite system was experimentally investigated and discussed in this paper. The following conclusions can be drawn from the analysis of the obtained experimental results: (a) the environmental conditions generally reduced the bond properties of the glass-based TRM. For example, the peak load decreased by approximately 57 %. (b) The changes of the mortar strength seemed to be a good indication of the changes in the bond behaviour and were suggested to be considered in further investigations for estimating the long-term behavior of TRM systems.

More analysis should be performed to determine the contribution of the mortar age and the freeze-thaw environmental conditions to the degradation of the pull-out response as well as the mortar strength.

ACKNOWLEDGEMENTS

This work was partly financed by FEDER funds through the Competitivity Factors Operational Programme (COMPETE) and by national funds through the Foundation for Science and Technology (FCT) within the scope of project POCI-01-0145-FEDER-007633. The support to the first author through grant SFRH/BD/131282/2017 is kindly acknowledged.

REFERENCES

Al-jaberi, Z. K., & Myers, J. J. (2018). Effect of long-term environmental exposure on EB-FRP or FRCM-reinforced masonry system. In G. Milani, A. Taliercio, & S. Garrity (Eds.), *10th International Masonry Conference*. Milan.

Alwan, J. M., Naaman, A., & Hansen, W. (1991). Pull-Out work of steel fibers from cementitious composites : analytical investigation. *Cement & Concrete Composites, 13,* 247–255. https://doi.org/10.1016/0958-9465 (91)90030-L.

Carozzi, F. G., & Poggi, C. (2015). Mechanical properties and debonding strength of Fabric Reinforced Cementitious Matrix (FRCM) systems for masonry strengthening. *Composites Part B: Engineering, 70,* 215–230. https://doi.org/10.1016/j.compositesb.2014.10.056.

D'Antino, T., & Papanicolaou, C. (2017). Mechanical characterization of textile reinforced inorganic-matrix composites. *Composites Part B: Engineering, 127.* https://doi.org/10.1016/j.compositesb.2017.02.034.

Dalalbashi, A., Ghiassi, B., Oliveira, D. V., & Freitas, A. (2018a). Effect of test setup on the fiber-to-mortar pull-out response in TRM composites: experimental and analytical modeling. *Composites Part B: Engineering, 143,* 250–268. https://doi.org/10.1016/j.compositesb.2018.02.010.

Dalalbashi, A., Ghiassi, B., Oliveira, D. V., & Freitas, A. (2018b). Fiber-to-mortar bond behavior in TRM composites: effect of embedded length and fiber configuration. *Composites Part B: Engineering, 152,* 43–57. https://doi.org/10.1016/j.compositesb.2018.06.014.

Dalalbashi, A., Ghiassi, B., & Oliveira, D. V. (2019). Textile-to-mortar bond behaviour in lime-based textile reinforced mortars. *Construction and Building Materials, 227,* 116682. https://doi.org/10.1016/j.conbuildmat.2019.116682.

Donnini, J., Chiappini, G., Lancioni, G., & Corinaldesi, V. (2019). Tensile behaviour of glass FRCM systems with fabrics' overlap: Experimental results and numerical modeling. *Composite Structures, 212* (October2018), 398–411. https://doi.org/10.1016/j.compstruct.2019.01.053.

Ghiassi, B., Oliveira, D. V, & Lourenc, P. B. (2014). *Hygrothermal durability of bond in FRP-strengthened masonry.* 2039–2050. https://doi.org/10.1617/s11527-014-0375-7.

Ghiassi, B., Oliveira, D. V, Marques, V., Soares, E., & Maljaee, H. (2016). Multi-level characterization of steel reinforced mortars for strengthening of masonry structures. *Materials and Design, 110,* 903–913. https://doi.org/10.1016/j.matdes.2016.08.034.

Ghiassi, B., Xavier, J., Oliveira, D. V., Kwiecien, A., Lourenço, P. B., & Zajac, B. (2015). Evaluation of the bond performance in FRP-brick components re-bonded after initial delamination. *Composite Structures, 123,* 271–281. https://doi.org/10.1016/j.compstruct.2014.12.047.

Heshmati, M., Haghani, R., & Al-Emrani, M. (2017). Durability of CFRP/steel joints under cyclic wet-dry and freeze-thaw conditions. *Composites Part B: Engineering, 126,* 211–226. https://doi.org/10.1016/j.compositesb.2017.06.011.

Maljaee, H., Ghiassi, B., Lourenço, P. B., & Oliveira, D. V. (2016). FRP– brick masonry bond degradation under hygrothermal conditions. *Composite Structures, 147,* 143–154. https://doi.org/10.1016/j.compstruct.2016.03.037.

Mazzuca, S., Hadad, H. A., Ombres, L., & Nanni, A. (2019). Mechanical Characterization of Steel-Reinforced Grout for Strengthening of Existing Masonry and Concrete Structures. *Journal of Materials in Civil Engineering, 31*(5), 04019037. https://doi.org/10.1061/(ASCE)MT.1943-5533.0002669.

Naik, D. L., Sharma, A., Chada, R. R., Kiran, R., & Sirotiak, T. (2019). Modified pullout test for indirect characterization of natural fiber and cementitious matrix interface properties. *Construction and Building Materials, 208,* 381–393. https://doi.org/10.1016/j.conbuildmat.2019.03.021.

Papanicolaou, C. G., Triantafillou, T. C., Papathanasiou, M., & Karlos, K. (2007). Textile reinforced mortar (TRM) versus FRP as strengthening material of URM walls: out-of-plane cyclic loading. *Materials and Structures, 41*(1), 143–157. https://doi.org/10.1617/s11527-007-9226-0.

Razavizadeh, A., Ghiassi, B., & Oliveira, D. V. (2014). Bond behavior of SRG-strengthened masonry units: Testing and numerical modeling. *Construction and Building Materials, 64,* 387–397. https://doi.org/10.1016/j.conbuildmat.2014.04.070.

Uranjek, M., & Bokan-bosiljkov, V. (2015). Influence of freeze – thaw cycles on mechanical properties of historical brick masonry. *Construction and Building Materials, 84,* 416–428. https://doi.org/10.1016/j.conbuildmat.2015.03.077.

Younis, A., & Ebead, U. (2018). Bond characteristics of different FRCM systems. *Construction and Building Materials.* https://doi.org/10.1016/j.conbuildmat.2018.04.216.

Zhang, S. Y. (1998). Debonding and cracking energy release rate of the fiber/matrix interface. *Composites Science and Technology, 58*(3–4), 331–335. https://doi.org/10.1016/S0266-3538(97)00073-0.

Brick and Block Masonry - From Historical to Sustainable Masonry –
Kubica, Kwiecień & Bednarz (eds)
© 2020 Taylor & Francis Group, London, ISBN 978-0-367-56586-2

Author Index